ENERGY FROM BIOMASS AND WASTES X

Edited by

Donald L. Klass
Institute of Gas Technology, Chicago, Illinois, U.S.A.

ELSEVIER APPLIED SCIENCE PUBLISHERS
LONDON

INSTITUTE OF GAS TECHNOLOGY
CHICAGO

ELSEVIER APPLIED SCIENCE PUBLISHERS
LONDON

INSTITUTE OF GAS TECHNOLOGY
CHICAGO

Sole Distributor Outside the USA and Canada
ELSEVIER APPLIED SCIENCE PUBLISHERS LTD
Crown House, Linton Road, Barking
Essex IG11 8JU, U.K.

Sole Distributor in the USA and Canada
INSTITUTE OF GAS TECHNOLOGY
3424 South State Street, Chicago,
Illinois 60616, USA

ISBN 1-85166-0720 (Elsevier edition)
ISBN 0-910091-57-9 (IGT edition)

PREFACE

This book is the culmination of the effort to organize and present IGT's 11th conference on fuels and energy from biomass and wastes.* This program was cosponsored for the first time by the Council of Biomass Energy Technology Sponsors (CBETS) and was presented in Washington, D.C., on April 7-10, 1986. Over 300 registrants from 23 countries attended the 3½-day meeting. This particular program in the highly successful series, which began in 1976, was unique compared with IGT's previous programs on biomass because it involved both oral and poster papers as well as a trade exposition, review and original papers on technical and scientific topics, and papers on important business and government issues that affect the contributions of biomass and wastes to primary energy demand.

The conference papers in this book are divided into several subjects. The categories for the oral presentations are Introduction (Papers 1 and 2); Biomass Production — Trees (Papers 3-11), Herbaceous Crops (Papers 12-16), Aquatic and Marine Crops (Papers 17-21); Thermochemical Conversion — Combustion (Papers 22-31), Gasification (Papers 32-40), and Liquefaction (Papers 41-48); Bioconversion — Alcohol Fuels (Papers 49-54) and Methane Fermentation (Papers 55-66); Energetics, Economics, and Markets (Papers 67-82); National and Regional Programs (Papers 83-89); and Government Policies and Issues (Papers 90-94). The categories for the poster presentations are Combustion (Papers 1 and 2); Thermochemical Gasification (Papers 3-10), Methane Fermentation (Papers 11-18), Liquefaction (Papers 19-25), and National Programs (Papers 26-29). The oral papers are produced in full with minimal editorial changes, but all discussions were edited. The poster papers are presented as produced by the author.

Special thanks are in order to several organizations and many people for their participation in the conference.

---

*Prior to this conference, technical symposia on biomass energy have been organized and presented by IGT. They are: "Clean Fuels From Biomass, Sewage, Urban Refuse, Agricultural Wastes," January 27-30, 1976, Orlando, FL; "Clean Fuels From Biomass and Wastes," January 25-28, 1977, Orlando, FL; "Energy From Biomass and Wastes," August 14-18, 1978, Washington, D.C.; "Energy From Biomass and Wastes IV," January 21-25, 1980, Lake Buena Vista, FL; "Energy From Biomass and Wastes V," January 26-30, 1981, Lake Buena Vista, FL; "Energy From Biomass and Wastes VI," January 25-29, 1982, Lake Buena Vista, FL; "Energy From Biomass and Wastes VII," January 24-28, 1983, Lake Buena Vista, FL; "Energy From Biomass and Wastes VIII," January 30- February 3, 1984, Lake Buena Vista, FL; and "Energy From Biomass and Wastes IX, January 28-February 1, 1985, Lake Buena Vista, FL. The symposium "Management Assessment of Energy From Biomass and Wastes" held on January 21-23, 1979 in Orlando, FL, was directed to executive management.

First, the organizations that provided financial support were

U.S. Department of Energy	Tennessee Valley Authority
Gas Research Institute	Battelle Memorial Institute
New York State Energy Research and Development Authority	

Without their help, we would not have been able to develop a comprehensive program and present it at a substantially reduced registration fee to permit people from all sectors of the biomass energy community to attend the conference. The assistance of these organizations was invaluable.

The CBETS Executive Committee was also extremely helpful in the discussion and review of the papers for the program. The committee members were

Peter H. Benson, Gas Research Institute
Beverly J. Berger, U.S. Department of Energy
James A. Fisher, Fiber Fuels Institute
Earle E. Gavett, U.S. Department of Agriculture

Their familiarity with the subjects covered at this conference simplified the difficult task of selecting papers and obtaining commitments from speakers.

Next, the effort of the people who chaired the individual sessions was greatly appreciated. The session chairpersons were

Michael J. Antal, Jr., University of Hawaii
Donald Augenstein, Consultant
Peter H. Benson, Gas Research Institute
Paul F. Bente, Consultant
Beverly J. Berger, U.S. Department of Energy
Earl E. Gavett, U.S. Department of Agriculture
Robert E. Hansen, Legislative Commission on Minnesota Resources
J. Bradford Hollomon, New York State Energy R&D Authority
Larry Mansueti, American Public Power Association
Joseph Roetheli, Tennessee Valley Authority
Wayne H. Smith, University of Florida
Jon Veigel, North Carolina Alternative Energy Corporation
David Zoellner, National Rural Electric Cooperative Association

Their hard work kept the program moving and made the discussions more lively and interesting.

Finally, all the authors and presenters of the oral and poster papers are due a special thanks for the time and effort spent in the preparation of the papers, almost all of which are presented in this book as a permanent record of the conference. Many of these papers will serve as archival reference sources for years to come.

Donald L. Klass
Institute of Gas Technology

TABLE OF CONTENTS

BIOCONVERSION, ALCOHOL FUELS

BIOCONVERSION, METHANE FERMENTATION

ENERGETICS, ECONOMICS, AND MARKETS

NATIONAL AND REGIONAL PROGRAMS

GOVERNMENT POLICIES AND ISSUES

INTRODUCTORY REMARKS

Donald Klass

Ladies and gentlemen, I would like to welcome all of you to our 10th annual meeting on energy from biomass and wastes, and I would like to extend a special welcome to those of you from other countries. I hope you will be able to take a few hours off during the sessions that we have planned for you to visit some of the historic sites here in our National Capital. But if you do decide to take the grand tour, please don't be gone too long because we have a super program ahead of us.

I'm Donald Klass and will serve as your chairman during the meeting. I would also like to mention the IGT staff who are here to help you in the event you need assistance — Jared Smith, Susan Robertson, and Bill Thompson. If any of you have any difficulties, please contact us, and we will try to help.

To help you arrange your schedule so you can attend the sessions of particular interest and still have some time off, most of the papers are included in the notebooks. A few are still missing, and as they arrive, they will be available at the registration desk. Also, we have grouped the papers on the same subjects together to help you develop a schedule.

Before we get on with the program, just a few housekeeping items. We would appreciate it if you would wear your badge at all times while attending the program. This will promote exchange of information and opportunities for developing new friends at the conference and also make it easier to enter the meeting rooms. Also, each of you should know that we are taping the discussions after each presentation, but please do not let this inhibit any comments you would like to make or questions you would like to ask of the speaker. Each speaker will have ample opportunity to review the final draft of the transcribed discussion that occurs after each presentation. After the oral papers are presented, the poster papers, exhibition, and a reception will be presented simultaneously in the exhibition hall each evening — tonight, Tuesday night, and Wednesday night — from 5:00PM to 8:00PM.

This conference marks several firsts for IGT. It is the first time in 8 years that the meeting has been held in Washington. I think many of you know that most of our meetings have been presented in Florida. This meeting is also the first one in which we have combined poster papers, oral papers, and an exhibition. And finally, this is the first time we have had cosponsorship, in particular, cosponsorship by the Council of Biomass Energy Technology Sponsors. We are especially grateful for the financial support provided by several of the CBETS Members. This has made it possible for us to reduce the registration fee to $130 from the $740 top figure last year. This of course opens the program to many more people. We now have over 300 people registered. This was our desired aim and goal.

I would like to point out that almost all of the session chairmen are CBETS Member representatives. I would also like to call your attention to the last document in Volume II of the notebooks. This document describes many of the research programs funded by the CBETS Members. It is quite interesting. I recommend it for your reading when you have an opportunity.

Now, I would like to introduce Dr. Beverly Berger, who is the current president of CBETS. She will make a few comments on behalf of the organization.

Beverly Berger

Thank you Don. The Council of Biomass Energy Technology Sponsors is very pleased to have the opportunity to join with IGT in sponsoring this meeting. CBETS was founded in 1983 to provide a forum for coordination and cooperation among managers of major biomass energy research programs, including federal and state government organizations, industry institutes, and associations. There are 19 CBETS Members — 13 regular Members who sponsor biomass R&D, and 6 affiliate Members, who do not fund R&D, but who have a significant interest in the development of biomass energy. You may wish to seek out CBETS Members to compare your ideas and suggestions. Let me go over quickly who they are — the American Public Power Association, the Electric Power Research Institute, the Gas Research Institute, the National Rural Electric Cooperatives Association, and for the states, the North Carolina Alternative Energy Corporation, the Hawaii Natural Energy Institute, the Legislative Commission on Minnesota Resources, the New Mexico Energy Research and Development Institute, the New York State Energy Research and Development Authority, the Florida Center for Biomass Energy Systems, and 3 federal agencies, the Tennessee Valley Authority, the U.S. Department of Agriculture, and the U.S. Department of Energy. The affiliate Members include the American Gas Association, the Biomass Energy Research Association, the Fiber Fuels Institute, the National Wood Energy Association, the Renewable Fuels Association, and the Wood Heating Alliance. All of the CBETS Members join with Dr. Klass and IGT in welcoming you to this meeting.

Donald Klass

It is a great pleasure to introduce our keynote speaker for the first presentation this morning, Miss Donna R. Fitzpatrick. She was nominated by President Reagan on July 31, 1985, confirmed by the Senate on December 19, 1985, and sworn in as Assistant Secretary for Conservation and Renewable Energy with the U.S. Department of Energy on December 20, 1985. She has served as Acting Assistant Secretary since March 11, 1985. In her present position, her responsibilities include management of development programs in energy efficiency and alternative fuels for industry, buildings, and vehicles. The programs on renewable energy resources involve research in active and passive solar, photovoltaics, wind, ocean thermal, geothermal, and biomass technologies. In addition to this responsibility, she administers federal grant programs for low-income home weatherization and energy efficiency improvements in schools and hospitals. She also coordinates 3 of the 5 federal power marketing administrations. Before becoming involved in these activities, she was involved in the practice of law with the Washington firm, O'Connor and Hannon, and then in her own law practice. In addition to renewable energy and conservation, she is experienced in civil litigation, including administrative law and contracting issues. She received her bachelor's degree in physics from the American University in 1972, and after teaching high school chemistry and mathematics, entered law school and graduated from the National Law Center of George Washington University in 1980. We are very happy to have her with us this morning.

KEYNOTE ADDRESS

Donna R. Fitzpatrick
Assistant Secretary
Conservation and Renewable Energy
United States Department of Energy
Washington, D.C. 20585

Thank you. It's a great pleasure to be here this morning. I'd like to welcome all of you to the 10th annual conference on energy from biomass and wastes, which is being cosponsored this year for the first time by the Institute of Gas Technology and the Council on Biomass Technology Energy Sponsors. I congratulate both of those organizations on their efforts to organize the conference and support the development of biomass and wastes technologies. This is the first opportunity I've had to address the combined communities of biomass and wastes. I'm very pleased to be here and to see such a wide representation from all parts of the industry, and from different levels of government including the federal government, municipalities, our national laboratories, and also academia. I'd like especially to welcome our international visitors, because it is becoming more and more evident that whatever technological progress we're able to achieve is going to be done on the basis of the interplay among researchers from many countries. Conferences such as this help to promote that interplay because they provide an important opportunity for members of the biomass and municipal waste communities to gather in one place and exchange views and information on scientific and technological developments and trends. The potential for a breakthrough is greatly enhanced by having so many knowledgeable people together in one place. So I hope that this conference will strengthen the cooperation among different researchers and also between government and industry. What I would like to do this morning is to make some comments on biomass technologies in the context of all the renewable technologies, and also in the context of the world energy picture, with particular emphasis on the United States.

Where do we stand now in energy? I think it's incontrovertible that we are in a better position than we were in 1973 or 1974, immediately prior to the first oil embargo. We are operating much more efficiently than before the crisis began. We consumed last year in the United States only about 1-1/2% more energy than we consumed in 1973 despite the fact that our gross national product has increased by 31%. We also filled more of our energy needs by domestic production.

Now I don't want you to think that the entire gain in energy efficiency or the disproportional reduction in consumption came from actually becoming more efficient in energy usage. Unfortunately, it did not. We have had a rather significant structural shift in the way that we consume energy in the United States. We have had a shift away, for example, from smoke-stack industries, which are very energy intensive, into some high technology service industries. We also experienced reduced economic growth, along with a good part of the rest of the world, so that when we compare projections for energy consumption in 1985 or 1984 with what we did consume, we see a significant difference. A great deal of it, however, is due to increased energy efficiency. We have also seen a growth in the contribution that renewable technologies are making to our energy supply needs.

Meanwhile in the world scene, OPEC is in disarray, and oil prices are down. It is unknown, of course, whether prices will continue to fall, or whether they have leveled off, or are even beginning to climb in the last few days. However, I think that's far too short a term to make any kind of projection. Some people say we are in an energy glut. The truth is we have always been in an energy glut, at least with respect to the energy that has been available to us.

Our problem is making proper and intelligent use of our resources. The energy is available. The problem is that we still have a bad petroleum habit. So we need to be smarter about how we produce and use energy, and we need to stop wasting so much of the 75 to 77 quads of primary energy that we consume each year. We need to develop our supply resources in an environmentally safe and economically sound way. We need to make choices and public policy based on accurate and unbiased information. And we need to develop our renewable technologies so that they are reliable and cost competitive.

In the United States, we are now getting about 9% of our total energy needs from renewable technologies and that number is growing proportionally faster than any other sector of the supply. About half of that supply comes from biomass, which is also growing rapidly. We also need to understand the way that renewable technologies fit into our entire energy infrastructure. But we are not going to see a major and extremely rapid shift to renewable technologies, simply for economic, technological and infrastructure reasons. Therefore, we must understand that renewable technologies must compete in an energy market that has many facets. Those facets must be allowed to remain healthy, economically viable, and energy efficient for as long as they are going to be needed, and as long as they offer reasonable supply options.

That means that as a Nation, we need to continue to develop coal technologies which are clean. We need to deregulate natural gas to allow it to come to market in an economically efficient way. We need to base all of our decisions, including decisions about nuclear power, on the facts of technology, safety, environment, and economics. All of this presents a continuing challenge to anyone who is working in the energy field.

We know that our economy and our technologies today are still largely dependent on resources which are not replaceable. Oil and natural gas someday will run out, or at least become extremely difficult and expensive to produce. We don't know when, probably in a few decades. But decades are not too long when we are planning a major national or even world energy strategy.

I think the opportunities for improving our energy health even more than we have done in the last few years are still impressive. We can still make substantial increases in efficiency. We are using 14 quads less per year now than we would have if 1973 practices had continued. If we adopt presently available technologies and are reasonably successful in research and development, which is going on both under industry and government sponsorship, we can expect to reduce our consumption by another 16 quads per year by the year 2000.

At the same time the contribution from renewable resources is growing, and as I said, it is now about 9% of total U.S. energy production. Again assuming reasonable success in public and private research and development, renewable energy options and most particularly, biomass options, should continue to become more economically competitive and should be competitive for as much as 75% of our needs within 25 years.

Now please note I am not saying that in 25 years we are going to be producing 75% of our needs from renewable technologies. What we are projecting is that those technologies will be economically competitive. The choice to use them, of course, is based on other factors as well as economic competitiveness. The choice, after all, is what is most important. The freedom to choose, so that we will never again find our national security threatened by interruption of a vital supply source; or our economic growth stymied by inflexibility, inaccessible costs, or short supplies; or world peace shattered in a battle for oil. That is why conferences such as this one provide an important opportunity for members of the biomass and municipal waste community to exchange information and to enhance our energy security so that we will not again face the problems that we have encountered in the past.

This time of lower energy prices should not be a time to be complacent. We must be sure that we do not close off long-term options which can prove to be invaluable in the future. Our national energy policy must continue to emphasize the development of renewable resources, and in particular, energy from biomass and municipal wastes, because these can make very substantial contributions. That is why the Department of Energy is committed to providing the technology base for future industry development of biomass and municipal wastes technology.

In planning our research activities, we look at four key criteria for deciding which projects to support. First, we need to know the potential energy contribution. Next is the potential for technical success and economic viability if the technology is developed. There must be high industry interest, involvement, and cost sharing, and we need to have a time frame for R&D that is too long for industry to sustain, and which justifies government investment.

The key to success in this process is industry, because our research efforts will be successful only to the extent that industry is involved in the planning and execution of the research and in the development of marketing applications. Therefore, we take steps to involve industry in all phases of our decision-making process, from program planning activities to evaluation, with emphasis on cost sharing.

I hope that this conference will serve to strengthen the government and industry cooperation which has been evolving in the biofuels area. For example, in 1978 the Department began supporting research conducted by the Aerospace Research Corporation, which proposed development of an innovative wood-burning gas-turbine system. Today, a 3 megawatt prototype has been constructed and the private sector cost sharing is 80%. There are other cooperative efforts which are continuing in medium-Btu gasification, anaerobic digestion of municipal wastes, short-rotation intensive culture of trees, microalgae assessment, and herbaceous energy crops.

In short, most of the Department's outright successes have had one thing in common, and that is the involvement of the private sector at some stage, preferably in all stages of our research. Government-industry cooperation and greatly increasing private industry involvement in and support of emerging technologies are extremely important. The federal role in the early stages must be to develop a strong industry capability and commitment. Concept feasibility, verification, and technical readiness require increasing levels of industry involvement. The potential contributions of biomass and waste energy are significant and those at this conference are the key to realizing that potential.

Successful marketing of new technologies requires a great deal of patience. We must have the patience and the foresight to maintain our course, even in the face of short-term setbacks, because we realize that we are working for long-term benefits. We know that the time required for new energy technologies to move from invention to significant market penetration has historically been measured in decades. For example, heat pumps took 50 years, diesel engines about 80 years, and low-head hydropower about 70 years. Improved information transfer and more efficient markets should shorten these decades. But it's still up to those who understand both our needs and the opportunities to keep working and planning for the future, not to give up.

I'm very conscious of this because I wear a couple of hats at the Department of Energy. In some respects, we appear to work against ourselves. We work very hard to increase our energy efficiency. We work on conservation, and for the most part, we applaud reduced energy prices, because you know that they are good overall for the economy. But they have an effect on renewable technology which many of us find a little hard to reconcile ourselves to. Obviously, renewable energy technologies, which are just emerging and struggling to become cost competitive, have a tougher time of it if conventional energy resources are becoming cheaper, or if demand is being reduced overall because of increased efficiency.

I think the answer is simply that we have to take the macroeconomic view and understand that whatever we do in renewable technologies has to be premised on the fact that eventually, renewable technologies must be available to us. They must be a source of supply that we can turn to increasingly in the future, and we must have technology development proceed independently to some extent from commercial development. That is why the Federal Government is involved. We understand that we are making an investment that is going to contribute to the future economic well being of the Country and of much of the world. We are not overly concerned that we don't get a payback on that investment within the next year. We don't keep a quarterly balance sheet that says we have to make a profit this quarter, otherwise what do we tell the stockholders. We understand, on the other hand, that private industry to a large extent has to operate under that kind of constraint. That's why I say that technological development, particularly basic research that is supported by the Federal Government, must continue to proceed and be supported, regardless of whether the technology is going to be in demand in the marketplace in the near term.

So I ask those of you who are involved in basic research not to be overly discouraged because of what appears to be short-term setbacks in the economic feasibility of some promising new technologies. Continue to take the long view and to understand that these technologies will have their day. Do not get bogged down in predictions about when this day will occur. You know that it's coming. And we also know that the future is very unpredictable, so it really doesn't pay to try to predict too much.

I heard for example about the visit that the Mayor of Chicago made about 100 years ago to an exhibition of new technologies. He was shown a telephone, which he pronounced to be a marvelous invention. And he said that he was going to get one for the City of Chicago. That story was told to me by the Mayor of Cleveland, by the way.

The point is that very often we can hold the future in our hands and not recognize it. And if you don't think that is true, think about the Mayor of Chicago and his telephone in 1880 or thereabouts. Think even about the level of technological development, and the penetration of new technologies in the marketplace that existed as recently as 15 years ago. Fifteen years ago, many of us would never have predicted some of the things that we now take for granted — everything from desk-top computers to plastic garbage bags. I sometimes wonder what we did before plastic garbage bags.

At any rate, we are seeing growth in all areas of technology, which in the long term is steady and continually more promising. And it's our duty to be patient, to be farsighted, to be resolute in understanding what we are about, and to continue to pursue our goals.

By the way, one of the short-term difficulties that I didn't mention, and I think in all fairness I have to mention, is the level of federal effort. We are in a budget crunch. I don't need to tell you about that. But we are determined that the budget difficulties that the entire federal government is facing should not cause the total elimination of any of our technology programs. In particular, we are determined that all of the technical goals that we have set for ourselves will continue to be pursued, and will continue to be supported at the necessary level. So I want you to know that we are committed to that, however difficult the times may be.

DISCUSSION

C. Setterstrom, Harrington Research Co. In your forecast that renewable resources could be priced cost competitive for 75% of the uses of energy in 25 years, what sort of cost range did you use for a barrel of oil? What are you projecting for the cost of oil?

Fitzpatrick I think most everyone assumes oil will continue to go up in price over the long term. I personally expect that oil in the 1990's will cost about as much as it cost in the early 80's corrected for inflation. The Department does an annual projection of what we expect the price of oil to be for three cases — the low, mid and high cases. But nobody wants to be in the oil price projection business, at least not if you want to have any kind of respectability. So I stay away from it. That is why I said renewables would be cost competitive. But that doesn't mean that people are going to make the investment, or that they will have the need to make the investment in new technologies. It depends on the demands. Are they going to be fairly flat as they have been for some years, or are they going to start to climb again? Also, what will the state of our infrastructure be at that time? Will we have a large proportion of plant capacity that is approaching the end of its lifetime and therefore require investment, or will we be in a position of a rather large capacity-to-demand ratio? All I'm saying is that under a reasonable scenario, we think that renewable energy will be cost competitive and an available choice. Whether that choice will be made depends on many factors.

Klass It's very difficult to deal with numbers and project crude oil spot market or posted prices years from now. But the essence of what you're saying I think is more important than what the price is going to be numerically in dollars per million Btu. The essence is that biofuels will be cost competitive according to Donna Fitzpatrick, and I happen to agree.

Fitzpatrick I also think there's one thing that can distract us from the real facts. The general publicity about oil prices centers on the spot market, and we see prices even under $10 per barrel. The spot market is really a rather small part of all the oil that is delivered. What's much more important is the posted prices. A great deal of the contracting in progress is in the mid-teens to high-teens, and therefore $20 to $25 oil could return in the fairly near future. It could happen, and it would not be as great a spike compared to today's price as we think, simply because not very much of the oil that is marketed today is actually delivered in the $10 per barrel range. That's simply something that makes good headlines.

G. Grassi, Commission of the European Communities, Belgium

Could you describe your position on ethanol?

Fitzpatrick

In the last couple of months, we've seen the price of ethanol in the United States drop by about 30¢ a gallon from around the $1.50 to $1.60 range to about $1.30. This of course has had a significant impact on the market, but feedstock prices have not fallen as fast as product price. This has a great impact on people who are thinking about entering the ethanol market. They have to think not only about today, but also about prices in 2 years when they actually bring a plant on-line, and during the lifetime of that plant. Ethanol is a very attractive fuel. It's obviously an octane enhancer for gasoline, and it can be an attractive fuel for other uses. I think what we have to do is reduce the production cost by getting smarter about how to convert feedstocks, and to develop ways of fermenting cellulosic feedstocks. In some people's minds, the problem of converting feedstuffs such as corn into fuel still exists. But I think that ethanol fuel has a future. There's no doubt that in the very short term, ethanol will have trouble competing against gasoline price and other octane enhancers.

ENERGY FROM BIOMASS AND WASTES: 1985 UPDATE

Donald L. Klass, Ph.D.
Vice President, Education
Institute of Gas Technology
Chicago, Illinois 60616

ABSTRACT

The gradual and then rapid reduction of crude oil prices, the impact of Gramm-Rudman-Hollings legislation on R&D budgets, the partial elimination of renewable energy tax credits, and the possible elimination of all energy tax credits and forgiveness via tax reform, are all taking their toll on energy research and commercialization ventures. But it appears biofuels will survive and continue to exhibit modest increases in contributions to primary energy demand. A comprehensive assessment of renewable technology options has shown that biofuels is the only option capable of making significant contributions to all energy sectors. The total contribution of renewable energy to primary energy demand in the U.S.A. today is about 6.8 quad/yr (3.2 million BOE/d); biofuels provide about 40% of the total. Research on SRIC trees and herbaceous and aquatic biomass has targeted specific species for development. Some of these species appear to be near economic feasibility as energy crops now. Commercialization of biomass and waste combustion systems has shown a spurt in growth because of the business opportunities offered by PURPA for cogeneration. Thermochemical gasification research on advanced technology systems has reached the point where scale-up is being considered. Commercialization of state-of-the-art technologies has slowed, but applications for small-scale producer gasifiers, particularly in developing countries, continue to expand. Thermochemical liquefaction research has concentrated mainly on high-temperature pyrolytic conversion methods that can increase yields and selectivities. Significant research achievements have been made in anaerobic digestion, but whatever its future as a large-scale source of methane, it is clear that digestion capacity will have to increase many-fold before biological methane can attain its potential. The reduction in lead content of gasolines by the EPA presents substantial marketing opportunities to alcohol fuel producers. If projections for ethanol fuel markets materialize, conversion of low-cost lignocellulosics will be necessary. Considerable research is now in progress to develop this technology.

ENERGY FROM BIOMASS AND WASTES: 1985 UPDATE

INTRODUCTION

Last year, the development of energy-from-biomass-and-waste technologies was affected by a totally new combination of political and business conditions, which began to have an adverse impact on biofuels energy systems as well as other renewable energy resources. The situation in 1984 was quite different.[1] In 1985, the worldwide decline in oil price, the introduction of Gramm-Rudman-Hollings legislation in the United States and the accompanying reductions in R&D budgets, and the elimination of some of the U.S. energy tax credits for biomass energy all tended to slow the development of biofuels.[2] Nevertheless, significant advances were made in 1985 that are expected to result in greater contributions of biofuels to primary energy demand. The purpose of this paper is to review the highlights of this activity and to provide critical commentary where appropriate.

U.S. DEPARTMENT OF ENERGY RESEARCH PROGRAMS

Overall Funding of R&D

The U.S. Department of Energy (DOE) has been the lead agency in energy research since its inception. Although budget deficit problems required large reductions in many federal programs, DOE's R&D budget request to Congress for fiscal 1987 is still slightly higher than the estimated expenditures for fiscal 1986 as shown in Table 1. There is a significant decrease of about 17% in the budget for the energy component of this program, while the budgets of the other two components, particularly for defense-related energy research, are increased.

Table 2 illustrates how the overall funding levels of solar energy R&D have been reduced. In this case, the reduction is about 50% for fiscal 1987 as compared to the adjusted appropriations for fiscal 1986. Biofuels and wind energy systems were given the largest reductions of all the solar energy research components. Presuming that DOE's proposed budget for fiscal 1987 for solar energy research is enacted into law, the reduction will correspond to a 58% cut from the fiscal 1985 appropriation (Table 2), and a 74% cut from the fiscal 1982 appropriation.[1] Support of renewable energy research has thus been severely reduced in contrast to other federal R&D funding appropriations requested by the Administration; they have been increased by 17% in fiscal 1987 over fiscal 1986.[5] The large reduction for research on renewable energy also contrasts with the President's belief that the renewable energy industry can supply over 10% of the Nation's energy needs by the Year 2000 and also make an important contribution to world stability.[6]

Table 1. U.S. DEPARTMENT OF ENERGY OVERALL RESEARCH BUDGETS[a]
Fiscal 1985–Fiscal 1987
($$10^6$)

Research Component	Actual FY85	Estimated FY86	Requested FY87
National Defense	2,251	2,246	2,609
General Science	515	515	602
Energy	2,135	2,025	1,674
R&D Facilities	909	813	759
	5,809	5,598	5,645

[a]Ref. 3.

Table 2. U.S. DEPARTMENT OF ENERGY OVERALL SOLAR ENERGY RESEARCH BUDGETS[a]
Fiscal 1985–Fiscal 1987[a]
($$10^6$)

Research Component[b]	Appropriation FY85	Appropriation FY86[d]	Requested FY87[e]
Photovoltaic Energy Systems	54.649	40.678	20.6 (-49)
Solar Thermal Energy Systems	33.895	25.885	15.3 (-41)
Biofuels Energy Systems	30.031	27.326	12.2 (-55)
Wind Energy Systems	28.355	24.825	8.25 (-67)
Solar Buildings Energy Systems	9.185	8.181	4.7 (-43)
Ocean Energy Systems	4.008	4.811	2.5 (-48)
Solar Technology Transfer	5.100	2.983	2.0 (-33)
Program Direction	4.864	4.516	4.1 (-9)
Others[c]	1.500	5.419	2.642 (-51)
	171.587	144.624	72.292

[a]Ref. 4.

[b]Each research component includes construction funds and capital equipment if any.

[c]Includes funds for International Solar Energy Program, SERI, Program Support, and Resource Assessment.

[d]Adjusted for reduction of FY86 appropriation.

[e]Figures in parentheses indicate percent reduction from fiscal 1986.

Biofuels Energy Systems Funding

In the recent past, the biofuels research program conducted by the Biofuels and Municipal Waste Technology Division of DOE[7] was the largest single research program on biomass energy funded by the Government.[1] It was also the largest single program on biomass energy of those conducted by the International Energy Agency countries.[8] The reductions presented in Table 2 suggest that this will not continue to be the case. The available details of the research appropriations by subject for DOE's biofuels program are presented in Table 3 for fiscal 1985 through fiscal 1987. The basic categories of R&D are feedstock production, thermochemical conversion, biochemical conversion, and the activities associated with the regional programs. In fiscal 1985, the respective allocations of the total budget to these categories were about 33%, 36%, 15%, and 13%, but after fiscal 1985, the allocations were not broken down in the final appropriations to individual line items except for the regional programs and capital equipment. DOE staff estimates of the allocation of the fiscal 1987 budget request of $12.2 million are indicated in Table 3. In this projection, feedstock production R&D is increased to about 40% of the total budget and each category of conversion research is allocated about 25%. Testimony has been presented by the Biomass Energy Research Association (BERA) to Congress for a biofuels R&D budget of $23.4 million for fiscal 1987.[13] This corresponds to an 18% reduction from fiscal 1986, but is still about the same reduction proposed by DOE for energy R&D in fiscal 1987. The outcome of these deliberations remains to be established. The allocation of funds recommended by BERA for each category of biofuels research is shown in Table 3.

Biofuels Program Accomplishments

As pointed out last year, federally supported research on biofuels has been one of the major driving forces behind the gradually increasing consumption of energy from biomass and wastes.[1] Federally supported research on biomass energy, urban wastes, and alcohol fuels has been in progress for about the last 12, 10, and 8 years, respectively.[1] DOE consolidated each of these activities into one division about two years ago.[1] Recent accomplishments of the R&D managed by the resulting Biofuels and Municipal Waste Technology Division are listed in Chart 1. This program is comprehensive in scope and involves field laboratories as sub-program managers (Argonne National Laboratory, Oak Ridge National Laboratory, Pacific Northwest Laboratory, and the Solar Energy Research Institute).

Because of the dispersed nature of biomass, its great diversity, and the preference of different biomass species for certain geographic regions, transition from research concept to commercialization is a slow process. To help alleviate some of the problems associated with this transition, DOE established the Regional Program at Congressional direction in 1983 to "carry out activities

Table 3. U.S. DEPARTMENT OF ENERGY RESEARCH BUDGETS IN FISCAL 1985, 1986, AND 1987 FOR BIOFUELS ENERGY SYSTEMS ($$10^6$)

Activity	FY85 Estimated Expenditures[a]	FY86 Administration Request[b]	FY86 House Recommendations[c]	FY86 Senate Recommendations[e]	FY86 Conference Report[g]
Feedstock Production:					
Short Rotation Intensive Culture:	--	--	--	--	--
— Genetic selection, tissue culture, N-fixation	4.2	--	--	--	--
— Experimental plantings, hardwood monoculture studies (20% industry cost-share)	--	3.0	--	--	--
— Genetic selection, tissue culture	--	1.7	--	--	--
Production Research:					
— Generic research for energy applications and proof-of-concept experiments, use of genetic techniques to increase productivity (both herbaceous and aquatics included)	5.97	--	--	--	--
Herbaceous Species:	--	--	--	--	--
— Field trials in Midwest, Lake States, Southeast, Plains States, field tests including hybrids of oilseed crops, studies of hydrocarbon synthesis pathways in plants	--	1.1	--	--	--
— Small-scale plots less than 5 ac of species selected from previous research	--	0.5	--	--	--
Aquatics Species:	--	--	--	--	--
— Screening of microalgae for liquids production, study of growth parameters	--	0.6	--	--	--
— Genetic selection studies on microalgae to improve liquid yields, growth rates, temperature tolerance	--	0.9	--	--	--
— Operate 2 small-scale units, technology transfer	--	0.5	--	--	--
	10.17	8.3	--	--	--
Thermochemical Conversion:	--	--	--	--	--
— Operate process research gasifiers, cold flow modeling; Research on reactor design, product upgrading catalysts, process conditions, Conduct liquid fuels research, Conduct combustion research on biomass and MSW	11.15	--	--	--	--
— Construction of 50-200 ton/d medium-Btu gasification facility (25% industry cost-share)	--	5.0	--	--	--
— Medium-Btu gasification research to optimize process and characterize waste streams	--	1.9	--	--	--
— Laboratory research on MSW to obtain basic understanding of thermal conversion, formation of trace organics (dioxins), gasification-pyrolysis mechanisms; Large-scale MSW combustion tests; Collect base-line data for solid fuels from MSW and small-scale processing equipment	--	2.5	--	--	--
	11.15	9.4	--	--	--

Biochemical Conversion:	--	--	--	(3.1)	
— Acid and enzymatic hydrolysis and integrated experiments for liquid fuels, Anaerobic digestion of cellulosics and MSW for gaseous fuels	--	--	--	--	--
— Develop improved organisms, genetic engineering of hemicellulose-utilizing organisms, Incorporation of acid recovery in TVA test unit, Integrated experiments for ethanol	--	5.5	--	--	--
— Methane fermentation research on physiology and biochemistry of anaerobes, digester designs for high-solids MSW feeds, microbial populations in sanitary landfills, genetic manipulation of MSW anaerobes, waste water treatment; Technology transfer	--	2.6	--	--	--
	4.5	8.1	--	--	--
Regional Programs:					
— Technology transfer of biomass research to industry in 4 regions (Pacific Northwest, Northeast, Great Lakes, Southeast)	4.0	2.0	--	4.0	--
	4.0	2.0	--	4.0	--
Capital Equipment:	--	--	--	--	0.6
— Laboratory apparatus for microalgae and integrated enzymatic hydrolysis	1.0	--	--	--	--
— Laboratory apparatus for biomass production and conversion	--	0.6	--	--	--
	1.0	0.6	--	--	0.6
	30.82	2[illegible].4	28.4[d]	28.4[f]	28.4[h]

[a]Adapted from Ref. 9, p. 61.

[b]Adapted from Ref. 9, pp. 55-61.

[c]Ref. 10.

[d]The House appropriation bill states that DOE is to continue medium-Btu scale-up and should consider locating the facility in a region with a diverse variety of biomass; It makes $500,000 available from within the Biofuels Program for Hawaii to assist them in utilizing idle sugarcane acreage.

[e]Ref. 11.

[f]The Senate appropriation bill states that a minimum of $5 million from within the Biofuels Program should be used for MSW R&D, and that the regional programs are to be continued at $4 million for fiscal 1986.

Table 3, Cont. U.S. DEPARTMENT OF ENERGY RESEARCH BUDGETS IN FISCAL 1985, 1986, AND 1987 FOR BIOFUELS ENERGY SYSTEMS ($$10^6$)

Activity	FY87 Administration Request[u]	FY87 BERA Recommendations[k]
Feedstock Production:		
Short Rotation Intensive Culture:	(2.5)	--
— Genetic selection, tissue culture, N-fixation	--	--
— Experimental plantings, hardwood monoculture studies (20% industry cost-share)	--	--
— Genetic selection, tissue culture	--	--
Production Research:		
— Generic research for energy applications and proof-of-concept experiments, use of genetic techniques to increase productivity (both herbaceous and aquatics included)	--	--
Herbaceous Species:	(1.4)	--
— Field trials in Midwest, Lake States, Southeast, Plains States, field tests including hybrids of oilseed crops, studies of hydrocarbon synthesis pathways in plants	--	--
— Small-scale plots less than 5 ac of species selected from previous research	--	--
Aquatics Species:	(1.3)	--
— Screening of microalgae for liquids production, study of growth parameters	--	--
— Genetic selection studies on microalgae to improve liquid yields, growth rates, temperature tolerance	--	--
— Operate 2 small-scale units, technology transfer	--	--
	(5.2)	5.2
Thermochemical Conversion:	(3.3)	
— Operate process research gasifiers, cold flow modeling; Research on reactor design, product upgrading catalysts, process conditions, Conduct liquid fuels research, Conduct combustion research on biomass and MSW	--	--
— Construction of 50-200 ton/d medium-Btu gasification facility (25% industry cost-share)	--	--
— Medium-Btu gasification research to optimize process and characterize waste streams	--	--
— Laboratory research on MSW to obtain basic understanding of thermal conversion, formation of trace organics (dioxins), gasification-pyrolysis mechanisms; Large-scale MSW combustion tests; Collect base-line data for solid fuels from MSW and small-scale processing equipment	--	--
	(3.3)	8.9

Biochemical Conversion:	(3.1)	
— Acid and enzymatic hydrolysis and integrated experiments for liquid fuels, Anaerobic digestion of cellulosics and MSW for gaseous fuels	--	--
— Develop improved organisms, genetic engineering of hemicellulose-utilizing organisms, Incorporation of acid recovery in TVA test unit, Integrated experiments for ethanol	--	--
— Methane fermentation research on physiology and biochemistry of anaerobes, digester designs for high-solids MSW feeds, microbial populations in sanitary landfills, genetic manipulation of MSW anaerobes, waste water treatment; Technology transfer	--	--
	(3.1)	6.8
Regional Programs:		
— Technology transfer of biomass research to industry in 4 regions (Pacific Northwest, Northeast, Great Lakes, Southeast)	--	--
	--	2.5
Capital Equipment:	0.6	--
— Laboratory apparatus for microalgae and integrated enzymatic hydrolysis	--	--
— Laboratory apparatus for biomass production and conversion	--	--
	0.6	--
	12.2	23.4[k]

[g]Ref. 12.

[h]Became Public Law 99-141 on November 1, 1985; No specific budget items indicated in conference report other than capital equipment and total appropriation.

[i]Ref. 4.

[j]Specific budget items other than capital equipment not indicated in Ref. 5; Items in parenthesis obtained from DOE staff; $3.3 million Thermochemical Conversion request includes $1.3 million on MSW research and $3.1 million on Biochemical Conversion request includes $900,000 on MSW research.

[k]From Biomass Energy Research Association Testimony (Ref. 13).

Chart 1. SOME RECENT ACCOMPLISHMENTS OF U.S. DEPARTMENT OF ENERGY BIOFUELS RESEARCH PROGRAM

BIOMASS PRODUCTION

Trees:	•	Evaluated the potential of intensive management and interplanting of natural stands for increased biomass production in the Great Lakes States.[a]
	•	Identified superior genotypes and clones of at least two short-rotation woody species.[a]
	•	Identified candidate short-rotation woody species capable of genetic improvements to increase average yields as much as 40%.[b]
	•	Compared first growth yields with regrowth yields of short-rotation woody species to document that regrowth yields are 20-100% higher.[b]
Herbaceous:	•	Explored biosynthetic pathways of hydrocarbon production in herbaceous plants to determine whether methods for increasing hydrocarbon production are feasible.[a]
	•	Provided preliminary evaluation of herbaceous species suitable for growth on marginal sites in the Southeast and Midwest/Great Lakes Regions.[a]
Aquatics:	•	Identified and characterized 10 promising candidate microalgal species for further research and evaluation.
	•	Initiated technical and economic analysis of microalgae as an energy source.[a]
	•	Determined properties and composition of oils and lipids for salt-tolerant microalgae.[a]

CONVERSION

Combustion:	•	Conducted full-scale combustion tests of municipal solid waste in an operating test facility to enhance combustion efficiency and reduce emission.[a,b]
	•	Completed successful shakedown operations of a 3-MW, wood-fueled, combustor-gas turbine generator system.[b]
Thermochemical Conversion:	•	Completed assembly of a bench-scale pyrolysis reactor, and recorded data on off-gases and condensed liquids produced from pyrolysis of selected municipal solid waste components.[a]
	•	Analyzed the formation of trace organic substances (dioxins) during thermal conversion of municipal solid waste.[a]
	•	Identified range of heat transfer coefficients in indirectly heated gasifier utilizing a metal fire-tube heat exchanger.[a]
	•	Quantified the effect of methane/wood mass ratio on the synthesis of fuels in fast pryolysis.[a]
	•	Defined temperature range for optimum oil-gas yield and developed a model for ablative fast pyrolysis.[a]

Chart 1, Cont. SOME RECENT ACCOMPLISHMENTS OF U.S. DEPARTMENT OF ENERGY BIOFUELS RESEARCH PROGRAM

Thermochemical Conversion, Cont.:

- Conducted base matrix of experiments in process research unit gasifiers.[a]
- Conducted cold flow model testing to evaluate larger scale problems in thermochemical conversion.[a]
- Characterized selected high-moisture feedstocks for thermochemical conversion.[a]

Anaerobic Digestion:

- Completed the anaerobic upflow bioreactor (Anflow) project which utilizes fixed films of bacteria in a packed-bed column to purify sewage waste water and produce methane gas.[a]
- Improved high solids concentration (75%) of a municipal solid waste in the anaerobic digestion process.[a]
- Continued development of a system for the recovery of short-chain organic acids.[a]

Alcohol Fuels:

- Initiated concentrated sulfuric acid hydrolysis experiments to determine mass balance.[a]
- Initiated evaluation of ethanol and methanol R&D in light of estimated energy supply and world markets.[a]
- Began laboratory testing of enzymatic hydrolysis to verify parametric analysis.[a,b]

[a]Adapted from Ref. 9.

[b]Adapted from Ref. 4.

related to technology transfer, industry support, resource assessment, and matching local resources to conversion technologies."[14] To date, DOE has established four regions — Northeast, Great Lakes, Northwest, Southeast — for participation in the Regional Program as shown in Table 4. This program currently includes 36 States. The goal is to encourage the production of biomass energy feedstocks and their conversion to fuels by the private sector through support of regionally-specific biomass energy projects.[14] During 1984, for example, many projects were sponsored or cosponsored by the Regional Program — 83 in the Northwest, 12 in the Great Lakes, 16 in the Northeast, and 31 in the Southeast.[14] Transfer of the technology is facilitated by conducting a wide variety of program activities, including demonstrations, workshops, publication of case histories and directories, distribution of general information publications, and provision of technical assistance. In addition, assistance is provided in the conduct of resource, technical, and economic assessments, and in the analysis of feasibility, markets, legislative-regulatory affairs, and health and safety matters. These activities are a valuable adjunct to the R&D work on biofuels because the private sector is given the opportunity to make first-hand informed judgments regarding initiation of commercial biomass energy ventures.

OTHER FUNDED R&D PROGRAMS ON ENERGY FROM BIOMASS AND WASTES

United States

Several federal programs on biomass energy, other than the DOE program, are in progress as reported last year.[1] Among these are the departmental programs of the Departments of Agriculture, Commerce, Defense, Housing and Urban Development, and the Treasury; the independent agency programs of the Agency for International Development, Appalachian Regional Commission, Consumer Product Safety Commission, Environmental Protection Agency, National Science Foundation, and Tennessee Valley Authority; and the program of the Office of Technology Assessment.[15] Many states are also continuing to fund energy-from-biomass-and-waste programs.[16]

Table 5 is a summary of the R&D activities in 1984-85 of several independent and state organizations, most of which are not directly involved with federal programs.[17-20] It is apparent that a broad range of activity is in progress, and that there are no geographical limitations to the conduct of research on biomass energy.

Among the programs listed in Table 5 is that of the Gas Research Institute. This program is focused on the production of methane from biomass and wastes and has been in progress for several years. A summary of the R&D budgets since 1982 and future planned obligations up to 1990 are presented in Table 6.[21] It is noteworthy that R&D on processes and systems utilizing waste feedstocks received

Table 4. REGIONAL BIOMASS ENERGY PROGRAMS

Region	States
Northwest (5 states)	Alaska, Idaho, Montana, Oregon, Washington
Great Lakes (7 states)	Illinois, Indiana, Iowa, Michigan, Minnesota, Ohio, Wisconsin
Northeast (11 states)	Connecticut, Delaware, Maine, Maryland, Massachusetts, New Hampshire, New Jersey, New York, Pennsylvania, Rhode Island, Vermont
Southeast (13 states)	Alabama, Arkansas, Florida, Georgia, Kentucky, Louisiana, Mississippi, Missouri, North Carolina, South Carolina, Tennessee, Virginia, West Virginia

[a]Source: Ref. 14.

Table 5. SOME NON-FEDERAL FUNDED BIOMASS RESEARCH PROGRAMS IN THE UNITED STATES

Research Activity[b]	Organization/Location[a]: American Public Power Association Washington, DC[c]	California Energy Commission, Sacramento, CA[d]	Fiber Fuels Institute, St. Paul, MN[c]	Florida Institute of Food and Agricultural Science, Gainesville, FL[c]	Gas Research Institute, Chicago, IL[c]	Hawaii Natural Energy Institute, Honolulu, HI[e]	Illinois Department of Energy and Natural Resources, Springfield, IL[f]
Research	X		X	X	X	X	X
Product/Process Development	X		X	X	X	X	X
Technology Application/Transfer	X	X	X	X	X	X	X
Funding Sources							
Taxes		X		X		X	X
Membership	X		X		X		
Other Revenues						X	X
Feedstocks							
Wood	X	X	X	X	X	X	
MSW and Sewage					X		X
Herbaceous/Farm Residues	X	X	X	X	X	X	X
Grains, Oil Seeds						X	X
Other					X	X	
End Products							
Fuel Gas	X	X		X	X	X	X
Electricity	X	X				X	X
Liquid Fuels		X		X		X	X
Chemicals						X	
Heat	X	X				X	X
Processes							
Biological Conversion	X	X			X	X	X
Thermochemical Conversion	X	X			X	X	X
Direct Combustion	X	X	X	X		X	X

[a]Of managing office.

[b]"X" in table denotes that organization is active in indicated R&D activity with indicated feedstocks, end processes, or processes. "X" also indicates source of funding. Some of the organizations listed fund R&D with outside contractors and/or conduct projects internally.

[c]Adapted from Ref. 17.

[d]Ref. 18.

[e]Adapted from Ref. 19.

[f]Ref. 20.

Table 5, Cont. SOME NON-FEDERAL FUNDED BIOMASS RESEARCH PROGRAMS IN THE UNITED STATES

Research Activity[b]	Organization/Location[a] New Mexico Energy R&D Institute, Santa Fe, NM[c]	New York State R&D Authority, Albany, NY[c]	North Carolina Alternate Energy, Corp., Research Triangle Park, NC[c]	Tennessee Valley Authority, Muscle Shoals, AL[c]
Research	X	X	X	
Product/Process Development	X	X	X	
Technology Application/Transfer	X	X	X	X
Funding Sources				
Taxes	X	X		X
Membership			X	
Other Revenues		X		X
Feedstocks				
Wood	X	X	X	X
MSW and Sewage	X	X		
Herbaceous/Farm Residues	X	X		X
Grains, Oil Seeds	X			
Other		X		X
End Products				
Fuel Gas	X	X		
Electricity		X	X	X
Liquid Fuels	X	X		X
Chemicals		X		
Heat	X	X	X	X
Processes				
Biological Conversion	X	X		X
Thermochemical Conversion		X	X	X
Direct Combustion		X		X

Table 6. GRI's 1985 BUDGET AND FIVE-YEAR PLAN BUDGET ALLOCATION SUMMARY FOR PROJECTS ON METHANE FROM WASTES AND BIOMASS ($$10^6$)

	Approved				Planned Obligations[f]				
	1982[c]	1983[d]	1984[e]	1985[f]	1986	1987	1988	1989	1990
Methane from Wastes (1.2.4)[a]	0.22	1.7	1.8	1.40	2.4[g]	2.1	1.95	1.7	2.6
Methane from Biomass (1.2.5)[a]	8.45	6.7	3.9	0	2.15[h]	2.2	2.2	2.2	2.85
Energy Conversion Economics (2.1.2)[b]	0.4	--	0.1	--	--	--	--	--	--
SNG Supply Environmental and Safety Research (3.1.2)[a]	0.35	0.75	0.5	0	0.39[i]	m	m	m	m
Gasification Chemistry-Organic Sources (5.1..)[a]	--	--	0.2	0.68	0.35[j]	m	m	m	m
Biotechnology Applications (5.5.2)[a]	--	--	--	--	1.645[k]	m	m	m	m
Total Budget	9.42	9.15	6.5	2.08	6.935				

[a]Number designation used in Ref. 21.

[b]Number designation used in previous GRI 5-year plans.

[c]From Table 5, Ref. 22.

[d]From Table 4, Ref. 8.

[e]From Table 8, Ref. 1.

[f]Ref. 21.

[g]Continuing 1986 Projects: Biogasification of Biomass and Sewage Sludge (012), $950,000; Technical/ Economic Evaluation (312), $150,000; Landfill Gas Enhancement (313), $50,000; Biogasification of Municipal Solid Waste/Sewage Sludge (314), $1,150,000; New 1986 Project: Biomass/Waste Gas Cleanup (608), $108,000.

[h]Continuing 1986 Projects: Methane from Biomass System (225), $1,750,000; Biomass Systems Modeling and Analysis (308), $400,000.

[i]Continuing 1986 Projects: Landfill Gas Environmental Analysis (196), $140,000; Biomass and Wastes E&S Research: Digester Residues (236), $150,000; New 1986 Project: SNG Gas Clean-Up Systems (610), $100,000.

[j]Continuing 1986 Projects: Biological Conversion of Biomass (368), $220,000; Plant Physiology and Genetics (370), $130,000.

[k]Continuing 1986 Projects: Advanced Bio-Process Systems (31[illegible]), $605,000; Advanced Plant Systems (311), $1,040,000.

[m]Cannot be listed because project areas 3.1.2, 5.1.1, and 5.5.2 contain non-biomass and waste components not specified in Ref. 21.

relatively little funding compared to biomass in 1982 to 1984, but future planned obligations are about the same for both types of feedstocks. Last year was an exceptional one because no new starts were initiated for biomass R&D (see Footnote i, Table 8, Ref. 1). Chart 2 lists recent reported accomplishments of the GRI program on energy from biomass and wastes, several of which appear to be significant technical and economic advances over the state-of-the-art. The Federal Energy Regulatory Commission has approved GRI's 1986 research program, which will be shifted somewhat from the 1985 program toward mid- and long-term research. In 1986, about 35% of GRI's total budget has been allocated to mid- and long-term efforts and 65% to near-term activities, compared with 20% and 80%, respectively, in 1985.[23]

Non-U.S. Programs

The government-sponsored programs on biofuels in other countries such as Canada, China, India, Japan, and several countries in Europe have been summarized before.[1,8,22] Notable among these is the program of the European countries that are members of the European Commission.

EVALUATION OF BIOFUELS OPTIONS BY DOE

Why Evaluate

The large multiplicity of feedstock, conversion process, and energy product combinations that comprise biomass energy technologies is one of the major problems that must be considered when formulating a biomass research program. The question of where limited research funds can best be invested to meet the objectives of program management is usually a difficult one to answer in terms of selecting specific projects for funding, particularly when mid- to long-term research objectives are targeted. The principal barrier that must be overcome at the outset is the selection of specific feedstock-conversion-product options out of many possible combinations. After the options are selected, the effort must be focused on individual projects and project goals. Again, a selection process is necessary because there are many potential projects, each having specific objectives, that can be initiated for each technology option. This is characteristic of biomass energy R&D more so than most other research areas and is caused by the fact that there is not just one or a few routes to a research goal; there are many. This makes the entire research process for biomass energy, even mission-oriented applied R&D, a complex undertaking. If the R&D is not carefully planned, structured, and performed, the end results may be of little interest and value to the technology users.

Chart 2. RECENT ACCOMPLISHMENTS OF GAS RESEARCH INSTITUTE RESEARCH ON METHANE FROM BIOMASS AND WASTES[a]

- Achieved start-up and stable operation of the RefCoM MSW-sewage sludge plant. During the third quarter of 1984, the system generated 2.6 SCF methane/lb VS added, thereby exceeding the yield predicted from laboratory experiments.
- Observed increases of 60% in methane yield after addition of a buffer slurry to five 5,000-ton landfill test cells when compared to a sixth control cell. The methane content of the gas from the treated cells stabilized at about 50 vol % in half the time needed for stabilization of the control cell.
- Achieved water hyacinth yields of 35 dry ton/ac-yr in small-scale field tests in which hyacinth is grown on sewage effluent.
- Determined from site-specific systems and economic analyses that napier grass and water hyacinth could produce pipeline quality gas at \$5.70 to \$6.75/MBtu[b] (levelized constant 1983 dollars) if system goals can be met. The systems are sized at 2.7 to 8.1 MCF/d.[b]
- Updated the BIOMET (biomass to methane) simulation model to generate more comprehensive estimates of integrated production and harvesting costs and more accurate evaluations of the most cost-effective operating conditions.
- Observed methane yields of 7 SCF/lb VS from one variety of sorghum, or a conversion efficiency of more than 90%.
- Successfully regenerated sterile somatic embryos (napier grass-pearl millet hybrids) using tissue culture techniques and transplated them from the laboratory into the field with a survival rate of up to 100%. Data from the first harvest indicate that biomass yield of the tissue-culture plants is 63% higher than that of vegitatively propagated plants.
- Produced methane yields of 6.5 SCF/lb VS added at low solids concentrations (>90% water) by two-phase digestion as compared to single-stage yields with the same feedstock of 5.5-6.0 SCF/lb VS added.
- Developed a "finger printing" technique to detect early stress conditions in digesters so that corrective action can be tkaen before an upset occurs.
- Completed analysis of raw and processed landfill gas.
- Identified a Hall detector as a promising candidate for measuring trace concentrations of halogens and sulfur in landfill gas.
- Developed an improved method for cultivating acetate-utilizing methanogenic bacteria.

[a]Adapted from Ref. 21.

[b]In this paper, "M" denotes 10^6, so 1.0 MBtu and 1.0 MCF/d represents one million Btu and one million CF/d.

DOE's Assumptions and Methodology

In 1985, DOE conducted a comprehensive evaluation of renewable technology options, including biomass technology options, to prioritize R&D objectives over the next 5 years, and to identify the government's role.[24] The overall assumptions made in this analysis were:

- World oil prices stable through 1985 at about $26/bbl (1982 constant dollars) and will increase to $84/bbl by 2010.
- Gross national product grows at 2.3%/yr compared to 1.3%/yr growth in energy consumption.
- Overall primary energy consumption increases from 73.3 quad in 1982 to 102.7 quad in 2010.
- The sectoral energy consumption pattern in 2010 is 20% Residential, 18% Commercial, 44% Industrial, and 18% Transportation.
- Real energy price increases are about 1%/yr for coal, nuclear, and electric.

The methodology for rating each renewable energy category was complex. Without presenting the details, it consisted of a sequence of steps: (1) Selected 72 resource-conversion-energy product combinations out of hundreds, (2) Estimated the Energy Contribution Potential score (ECP) of each option and selected 43, most of which could contribute more than 0.5% to primary energy demand in 2010, (3) Estimated the Effectiveness score of individual R&D projects within these options in attaining 5-yr objectives, (4) Estimated the Federal Role score in funding each R&D activity in terms of risk, payoff time, and industry funding, (5) Prioritized each R&D project in order of the product of the ECP, Effectiveness, and Federal Role scores, (6) From the 1,353 possible priorities, selected 643 and placed them in three integrated groups at low, medium, and high funding levels relative to the fiscal 1985 program.

Energy Contribution Potential Scores

The ECP was estimated by summing the products of the Payoff, or percent contribution of energy to a given sector that a particular technology option was expected to be able to provide, and the Percent of Energy Consumption for that sector, for all four sectors. Thus, the arithmetic formula for the ECP score is:

$$\text{ECP Score} = \sum_{\text{Sector}} (\text{Payoff})\ (\text{Percent of Energy Consumption})$$

The ECP scores of the 17 biofuels technology options selected by DOE are compared in Table 7. The ECP score for each option and each sector, since the energy consumption estimated for the Year 2010 is near 100 quad (102.7 quad), is approximately equal to the quad contribution of that

Table 7. ENERGY CONTRIBUTION POTENTIAL (ECP) OF VARIOUS ENERGY-FROM-BIOMASS-AND-WASTE OPTIONS IN THE YEAR 2010[a]

	Residential	Commercial	Industrial	Transportation	
Energy Consumption, quad[b]	20.5	18.5	45.2	18.5	
Percent of Energy Consumption	20	18	44	18	
Option (Energy Product)[c]	Payoff[d]				ECP
Direct Combustion Conventional Wood (Heat)	3	3	3	--	246
*Thermochemical Gasification SRIC Trees (Gas)	1	3	3	--	206
*Thermochemical Conversion SRIC Trees (Methanol)	--	--	--	10	180
*Microalgae Extraction (Lipids)	--	--	--	10	180
*Biochemical Conversion Herbaceous Crops (Ethanol)	--	--	--	10	180
*Biochemical Conversion SRIC Trees (Ethanol)	--	--	--	10	180
*Thermochemical Conversion Conventional Wood (Electric)	3	3	1	--	158
*Biochemical Conversion MSW (Fuel)	3	3	1	--	158
*Thermochemical Conversion MSW (Fuel)	--	--	3	--	132
*Thermochemical Conversion Conventional Wood (Methanol)	--	--	--	3	54
*Pyrolysis SRIC Trees (Fuel)	--	--	---	3	54
Biochemical Conversion Herbaceous Crops (Gas)	--	--	1	--	44
Biochemical Conversion Agricultural Residues (Gas)	--	--	1	--	44
Photobiological Conversion Water (Hydrogen)	--	--	1	--	44
Direct Liquefaction Conventional Wood (Fuel)	--	--	--	1	18
Photobiological Conversion Wastes (Hydrogen)	--	--	--	--	0
Electrochemical Conversion Wastes (Feedstock)	--	--	--	--	0

[a]Adapted from Figures B-3 and B-4 in Ref. 24, Vol. 2.

[b]Energy consumption projected to be 102.7 quad in 2010 by DOE personnel performing analysis (Ref. 24).

[c]Those options marked with asterisk were selected for detailed evaluation by DOE.

[d]This nomenclature was used in original reference, and refers to the percentage of the percent energy consumption by sector that the particular option is projected to be able to contribute. Thus, a payoff of 3 for direct wood combustion in the Residential Sector means that this option has the potential of contributing 60 ECP units (3X20). The overall score is the sum of the product of Payoff X Sector Percentage for each sector. The overall score divided by 100 is the approximate quad contribution potential because the energy consumption projected for 2010 is approximately 100 (102.7) quad.

option after division by 100. It is noteworthy that the total ECP score for biofuels of 1,878 corresponds to about 19 quad. This is near the 17-quad upper limit estimated for the potential contribution of biomass energy by the U.S. Office of Technology Assessment for the Year 2000.[25] It is more important to note, however, that the potential contribution of only four biofuels options — thermochemical conversion of short-rotation intensive culture (SRIC) trees to methanol, microalgae production of lipids, and biochemical conversion of SRIC trees and herbaceous crops to ethanol — have the potential of supplying about 40% of the liquid fuels requirement of the transportation sector in 2010.

All of the ECP's for the renewable technology options selected are compared in Table 8. Biofuels and ocean energy are the only two categories that have the potential of supplying energy to all four sectors. In the opinion of the DOE personnel who estimated the ECP's for ocean thermal energy conversion (OTEC) to fuels, they projected OTEC could make a significant contribution to the transportation sector in 2010. Presumably, this contribution would be liquid fuels, in which case it is unlikely to occur since electricity is the primary energy product of OTEC. Conceptually, OTEC power could be utilized for hydrogen generation and subsequent hydrogenation of organic materials or of inorganic carbon resources to produce liquids. But then all options for electric power production should be listed by DOE as contributors to the transportation sector; they are not. The other option is electric or hydrogen-powered vehicles, which again would require the listing of all electric options as transportation contributors. If this rationale is correct, biofuels is the only renewable fuels option that can contribute to the transportation sector. It should be emphasized, then, that the total ECP score for biofuels is the second highest of all the renewable energy options evaluated by DOE, and is the only option that contributes to all sectors.

Prioritization and Project Ranking

Of the 17 biofuels options originally selected for analysis, ten were chosen for further detailed evaluation (indicated by asterisk in Table 7). The highest ECP option, direct combustion of conventional wood production for heat, was excluded because it is considered to be commercial technology,[26] and the six options having scores less than 54 were also excluded. The arithmetic products of the ECP, Effectiveness, and Federal Role scores were then used to prioritize project activities, 119 of which were identified to meet DOE's requirements at three hypothetical funding levels of 60, 80, and 100 investment units. Of the 357 investment opportunities evaluated, 59 scored a high priority, 25 were medium, and 94 were lower. The 25 highest priority projects and the 25 lowest priority projects (out of a total of 178) are listed in Table 9.

Value of Evaluation Results

Numerous observations can be made regarding the results of this exercise; that is beyond the scope of this paper. But a few general comments are in order. First, it is obvious that the ranking methodology is very complex and subject to revision and improvement. It is commendable that the decision was made by DOE to perform the evaluation for all renewable energy systems because as alluded to previously, there are many routes to the same goal, but the R&D plan must still be focused on the proper targets. All cannot be pursued. It is also an easy task to criticize the results of the analysis. For example, it has been pointed out several times in this series of papers that biomass feedstocks are not expected to play a major role in the large-scale production of methanol fuel, at least in the next 20 to 30 years, because of the limitation of biomass transport distance, which in turn makes the economics of biomass-derived methanol plants very site specific.[1] Yet several methanol projects were rated high in DOE's analysis. Such criticism should be limited because evaluation of projects making up a research program should be a continuing activity. It is hoped therefore that the research program underway is flexible enough to incorporate the important results of the analysis. The Biomass Energy Research Association has recommended the formulation of a detailed 5-year plan which is updated on an annual basis.[13]

One of the observations that can be made based on this first attempt to prioritize project activity in biofuels is that it appears almost all (or perhaps all) of the final 178 project activities that "filtered" through the system appear to be on-going projects or projects that were active in the recent past. The methodology used should be capable of evaluating new project concepts also so that the most promising ones can be added to the program, perhaps at the expense of the lower ranked projects or those that are near completion or transfer to commercial use. Program management will always have to utilize a certain amount of subjective judgment to facilitate consideration of new concepts and ideas.

In any event, this first attempt to prioritize biofuels projects is a large step in the right direction.

U.S. TAX FORGIVENESS AND TAX CREDITS

Last year was a crucial one not only for R&D funding appropriations, but also for federal tax subsidies that apply to many segments of biomass energy. Among the business subsidies that received detailed scrutiny and strong lobbying by groups on each side of the issue were the:

Alcohol Fuel Excise Tax Exemption
Renewable Fuel Tax Credits
Production Tax Credit

Table 8. COMPARISON OF ENERGY CONTRIBUTION POTENTIAL (ECP) OF VARIOUS RENEWABLE ENERGY OPTIONS[a]

Percent of Energy Consumption in Year 2010	Residential 20	Commercial 18	Industrial 44	Transportation 18	
Option (Energy Product)	Payoff				ECP[b]
Biofuels:					
Direct Combustion Conventional Wood (Heat)	3	3	3	--	246
Thermochemical Gasification SRIC Trees (Gas)	1	3	3	--	206
Thermochemical Conversion SRIC Trees (Methanol)	--	--	--	10	180
Microalgae Extraction (Lipids)	--	--	--	10	180
Biochemical Conversion Herbaceous Crops (Ethanol)	--	--	--	10	180
Biochemical Conversion SRIC Trees (Ethanol)	--	--	--	10	180
Thermochemical Conversion Conventional Wood (Electric)	3	3	1	--	158
Biochemical Conversion MSW (Fuel)	3	3	1	--	158
Thermochemical Conversion MSW (Fuel)	--	--	3	--	132
Thermochemical Conversion Conventional Wood (Methanol)	--	--	--	3	54
Pyrolysis SRIC Trees (Fuel)	--	--	--	3	54
Biochemical Conversion Herbaceous Crops (Gas)	--	--	1	--	44
Biochemical Conversion Agricultural Residue (Gas)	--	--	1	--	44
Photobiological Conversion Water (Hydrogen)	--	--	1	--	44
Direct Liquefaction Conventional Wood (Fuel)	--	--	--	1	18
	200	216	616	846	1,878
Photovoltaics:					
Flat Plate Thin Film (Electric)	10	10	3	--	512
Flat Plate Silicon Ribbon (Electric)	3	6.5	3	--	309
Concentrators (Electric)	3	6.5	3	--	309
Advanced Conversion (Electric)	1	1	1	--	82
Silicon Ingot (Electric)	1	1	--	--	38
	360	450	440	0	1,250
Wind:					
Vertical Axis Wind Turbine (Electric)	3	3	3	--	246
Horizontal Axis Wind Turbine (Electric)	3	3	3	--	246
Small Wind Systems (Electric)	3	3	1	--	158
Small Wind Systems (Heat)	1	1	1	--	82
Augmenters/Offshore (Electric)	1	1	1	--	82[d]
Augmenters/Offshore (Heat)	1	--	--	--	20[d]
Contained Systems	1	1	--	--	38
	260	216	396	0	872
Solar Thermal:					
Central Receiver (Electric)	10	10	3	--	512
Central Receiver (Heat)	--	--	3	--	132
Central Receiver (Fuel)	--	--	3	--	132
Parabolic Dish (Electric)	3	3	1	--	158
Parabolic Trough (Electric)	--	--	1	--	44
Hemispheric Bowl (Electric)	1	1	--	--	38
	280	252	484	0	1,016

Solar Buildings:					
Flat Plate Collectors (Heat)	10	3	--	--	254
Passive Residential (Heat)	10	--	--	--	200
Passive Non-Residential (Heat)	--	10	--	--	180
Multi-Effect Regenerative Absorption (Heat)	3	1	--	--	78
Hybrid Solid Desiccant (Heat)	3	1	--	--	78
Solid Desiccant (Heat)	3	1	--	--	78
Stationary Concentrating Collectors (Heat)	1	1	--	--	38
Rankine Cycle (Heat)	1	1	--	--	38
Liquid Desiccant (Heat)	1	--	--	--	20
	640	324	0	0	964
Geothermal:					
Hydrothermal —					
Binary Cycle (Electric)	10	10	3	--	512
Direct Heat (Heat)	3	10	1	--	284
Flash Steam (Electric)	3	3	1	--	158
Geopressure —					
Methane (Fuel)	10	3	1	--	298
Hot Water (Heat)	1	1	3	--	170
Hot Water (Electric)	1	1	--	--	38
Hydraulic (Electric)	1	1	--	--	38
Hot Dry Rock (Electric)	3	3	3	--	246
Hot Dry Rock (Heat)	--	1	1	--	62
Magma (Electric)	1	1	3	--	170
	660	612	704	0	1,976
Ocean Energy:					
OTEC Fuel Production (Fuel)	--	--	--	20	360
OTEC Electricity Cabled to Shore (Electric)	3	3	1	--	158
	60	54	44	360	518
Hydroelectric:					
Large Scale (Electric)	10	10	3	--	512
Small Scale (Electric)	1	1	1	--	82
	220	198	176	0	594
	2,680	2,322	2,860	1,206	9,068

[a]Adapted from Ref. 24.

[b]All technology options having an ECP of zero were deleted.

[c]Original text has a total ECP of Augmenters/Offshore at 102, but the sum of the components in the text is 58; a payoff of 1.0 was therefore added in the Industrial Sector.

Table 9. U.S. DEPARTMENT OF ENERGY 1985 RANKING OF BIOFUELS ACTIVITIES
FIFTY OF 278 ACTIVITIES IN ORDER OF HIGHEST TO LOWEST RANKINGS

Research Option	Code[g]	Activity Description[h]	Effectiveness[i]	Federal Role[j]	Funding Level[k]
25 Highest Rank Projects[a]					
BCC of Lignocellulosics[b]	709.06	Integrate Components in PRU	H	P	Base
SRIC Trees	701.02	Screening, Breeding, Species Selection	H	P	Base
BCC of Lignocellulosics	709.08	Optimize Integrated Conversion Process	H	P	Base
SRIC Trees	701.03	Match Species and Regional Sites	H	P	Base
SRIC Trees	701.04	Improve SRIC Management	H	P	Base
BCC of Lignocellulosics	709.01	Investigate Feedstock Pretreatments	H	P	Base
BCC of Lignocellulosics	709.02	Study Enhanced BCC via Strain Improvements	H	P	Base
TCC of Wood to Methanol	705.01	Determine Gasification Process	H	P	Base
SRIC Trees	701.07	Environmental Evaluations	M	P	Base
SRIC Trees	701.01	Physiological Growth Studies	H	P	Base
TCC of Wood to Methanol	705.03	Design-Optimize Partially-Integrated Process	H	P	Base
BCC of Lignocellulosics	709.04	Conduct Fermentation Studies	H	P	Base
Herbaceous Crops	702.03	Breed Improved Energy Characteristics	H	P	Base
SRIC Trees	701.06	Economic Evaluations	H	P	Base
TCC of Wood to Gas	706.01	Determine Gasification Process Design	H	P	Base
SRIC Trees	701.01	Physiological Growth Studies	H	P	Incr
SRIC Trees	701.02	Screening, Breeding, Species Selection	H	P	Incr
BCC of Lignocellulosics	709.03	Develop Applications for Lignin	H	P	Base
SRIC Trees	701.05	Develop Harvesting Systems	M	C	Base
Microalgae to Lipids	703.03	Conduct Screening Experiments	H	P	Base
Herbaceous Crops	702.01	Screen Existing Species for Energy Potential	H	P	Base
Microalgae to Lipids	708.06	Design and Construct Outdoor Production Unit	H	P	Base
Conventional Wood Production	704.01	Advanced Silvicultural Research	H	P	Base
TCC of Wood to Gas	706.05	Optimize Integrated Processes	H	P	Base
Microalgae to Lipids[c]	703.04	Design, Construct, and Evaluate Production Components	H	P	Base
25 Lowest Rank Projects[a]					
Microalgae to Lipids[d]	703.06	Design and Construct Outdoor Production Unit	H	P	Incr
Herbaceous Crops	702.06	Preliminary Economics	L	C	Base
BCC of Municipal Solid Waste	710.08	Theoretical Technology Applications, Economics	M	P	Incr
TCC of Municipal Solid Waste	711.04	Identify Corrosion-Errosion Mechanisms	M	P	Incr
TCC of Municipal Solid Waste	711.05	Identify and Select Concepts	M	P	Incr
Wood Pyrolysis	708.02	Develop Process Design	H	P	Incr
TCC of Wood to Electric	707.02	Perform Economic Analysis of PRU	M	M	Base
BCC of Municipal Solid Waste	710.03	Determine Feedstock Parameters	M	C	Incr
Microalgae to Lipids	703.05	Develop Processing Concepts	M	P	Incr
Wood Pyrolysis	708.05	Test Oils for Combustion and Corrosivity	L	C	Base
TCC of Wood to Gas	706.02	Identify Advanced Tar Removal System	H	P	Incr
TCC of Wood to Gas[e]	706.06	Identify Process Control Needs	M	C	Incr
TCC of Wood to Gas[e]	706.06	Identify Process Control Needs	M	C	Incr

TCC of Wood to Gas	706.04	Evaluate Waste Stream Clean-Up Needs	M	C	Incr
Wood Pyrolysis	708.01	Analyze and Upgrade Oils	L	P	Base
TCC of Municipal Solid Waste	711.07	Evaluate Applications, Economics	L	C	Incr
Microalgae to Lipids	703.08	Analyze System Economics	L	C	Incr
TCC of Wood to Methanol	705.06	Develop Flowsheets	H	C	Incr
TCC of Wood to Gas	706.05	Optimize Integrated Process	H	P	Incr
Herbaceous Crops	702.06	Conduct Preliminary Economics	L	C	Incr
TCC of Wood to Gas	706.03	Economics of Selected Flowsheets	M	C	Incr
TCC of Municipal Solid Waste	711.07	Evaluate Applications, Economics	L	C	Incr
TCC of Wood to Gas	706.04	Evaluate Waste Stream Clean-Up Needs	M	C	Incr
TCC of Wood to Gas	706.07	Identify Industrial Co-Sponsors	M	C	Incr
Herbaceous Crops[f]	702.06	Preliminary Economics	L	C	Incr

[a]Adapted from Table B-4, pp. B-16 to B-23 in Ref. 24; BCC, biochemical conversion; TCC, thermochemical conversion; SRIC, short-rotation intensive culture.

[b]Number 1 priority out of 178.

[c]Number 25 priority out of 178.

[d]Number 154 priority out of 178.

[e]Appears twice in Table; May be error or there is a difference in project activity.

[f]Number 178 priority out of 178.

[g]Code designation of DOE.

[h]PRU, process research unit.

[i]H, high; M, medium; L, low; these are effort-weighted scores and correspond to numerical scores of 300 to 500 (H), 150 to 300 (M), and 0 to 150 (L).

[j]P, primary; C, complementary; M, minimal; these are effort-weighted scores and correspond to 300 to 500 (P), 150 to 300 (C), and 0 to 150 (M).

[k]Each activity was considered in terms of three levels of support of 60, 80, and 100 hypothetical investment units. Base denotes the first level of support for that activity and Incr denotes the next level of support of an additional 20 units. A given activity can therefore appear two or three times if it was allocated additional incremental funding of 20 units from 60 or 80 units. The absolute level of funding was not indicated.

Alcohol Fuel Tax

There is a major difference between a tax exemption and a tax credit. The former results in tax forgiveness, wherein the tax is not collected by anyone in the energy production-to-marketing chain and paid to the government; the energy consumer does not pay the tax as part of the fuel price. In theory, the consumer benefits because the cost of the fuel is less, and the energy supplier benefits because the price of the fuel is more competitive. In many cases, at least a part or all of the exemption ends up as a portion of the profit margin for the energy producer and/or supplier because competitive market prices are high enough to include it.[27] The primary reasons for federal tax forgiveness legislation have been to stimulate entry of new energy technologies into the market place and to generate alternative sources of fuels and thereby reduce dependence on oil imports. Such is the case for alcohol fuel excise tax forgiveness, which was estimated to result in a federal tax loss of $215 million in 1984.[28,29]

The last tax forgiveness rule changes applicable to alcohol fuels were signed into law by the President as the Deficit Reduction Act of 1984 on July 18, 1984 (Public Law 98-369). This law increased the federal excise tax exemption to 6¢ from 5¢ per gallon of ethanol-gasoline blend, increased the blender tax credit to 60¢ from 50¢ per gallon of ethanol used in ethanol-gasoline blends, and reduced the excise tax on neat methanol fuel by 50% to 4.5¢ per gallon. In addition, the tariff on all imported ethanol fuel, except shipments from the Caribbean Basin Initiative nations, was increased to 60¢ from 50¢ per gallon. The law became effective for ethanol fuels on January 1, 1985, while the provision for methanol was made effective on October 1, 1984.

However, these legalities were not the final word on alcohol fuels in 1985. Further actions, some of which could have a large impact on future ethanol fuel markets, were initiated to attempt to plug loopholes that foreign suppliers are apparently able to use to avoid import duties. First, the President's Tax Reform Act proposal of May 1985 (Treasury II) eliminates the excise tax exemption, does not mention the tariff, and makes the blenders' tax credit applicable to plants in existence on December 31, 1985 regardless of their location, thus making foreign alcohol production eligible.[30] This is the first option proposed by the U.S. Treasury Department; two other options were also proposed. One would limit the blenders' tax credit to ethanol produced by plants located in the U.S.A. and the Caribbean Basin Initiative countries and eliminate the import tariff. The other option would eliminate the tariff and replace the blenders' tax credit with a direct subsidy to domestic and Caribbean Basin Initiative ethanol producers.

Near the end of 1985 and the beginning of 1986, there was still much disagreement on this issue. The U.S. House of Representatives did not wish to retain the 60-¢/gal tax credit in the tax reform act for blenders who use less than 10 vol % ethanol in ethanol-gasoline blends, the U.S. Senate did, and the President supported elimination of the excise tax subsidy.[31]

Renewable Fuels Tax Credits

In 1985, the federal tax code provided an energy investment tax credit for business of 10% of the "biomass property," and a regular investment tax credit of 10% for a "qualified investment" (IRS Publication 572, Rev. November 1985). In addition, most equipment could be depreciated over 5 years. In the President's Tax Reform Act proposal of November 1984 (Treasury I), all three benefits are eliminated, and in Treasury II, the tax credits are eliminated, but the 5-yr depreciation schedule for equipment is extended. Interestingly, Treasury II contains tax credits for oil and gas, while Treasury I eliminates all energy tax credits to create a so-called level playing field. Although attempts were made by legislators in 1985 to extend the credits in a new law for renewable energy,[31] the existing credits expired as scheduled on December 31, 1985. The original concept of the new law, the Renewable Energy and Conservation Transition Act of 1985 (introduced in House as Bill 2001 on April 4, 1985 and in Senate as Bill 1220 on May 24, 1985) would, if enacted, provide no residential tax credits for biomass energy, extend business tax credits for biomass energy applications through 1987, reduce them to 5% in 1988, and then phase them out thereafter. The credits for biomass and those of the other renewable energy technologies in this proposed bill are compared in Table 10. The reasons for the large difference and the disproportionately higher proposed credits for non-biomass technologies such as photovoltaics are questionable. As of February 1986, the House of Representatives had developed legislation to extend the tax credits for solar-thermal and photovoltaic energy sources retroactive from January 1, 1986.[33] But even if new legislation were enacted, it would probably only be effective for 6 months and not include biomass energy.[33]

The consensus among those who support the extension of the energy tax credits is that Treasury I is preferred to Treasury II because it makes all fuels compete on an equal basis.[33] Baring enactment into law of the provisions that eliminate all energy tax subsidies, the position of those active in biomass energy is to support passage of energy tax credit extension bills.[34]

One of the important questions regarding the energy tax credit is whether or not biomass is defined to include wastes in the actual language of the law. In the Conference Report to the Crude Oil Windfall Profit Tax Act enacted on April 7, 1980 (Public Law 96-223), which includes the

Table 10. RENEWABLE ENERGY AND CONSERVATION TRANSITION ACT OF 1985[a]

Technology	1985 Law	Proposed Extension 1986	1987	1988	1989	1990
	% Credit ($1,000 Maximum Expenditure)					
Residential:						
Solar Heating & Cooling	40 (10)	35 (6)	30 (6)	25 (6)	20 (6)	15 (6)
Photovoltaics	40 (10)	40(10)	40 (10)	40 (10)	40 (10)	40 (10)
Wind	40 (10)	35 (20)	30 (20)	25 (20)	0	0
Geothermal	40 (10)	40 (10)	30 (10)	20 (10)	0	0
Conservation[b]	15 (2)	25 (0.7)	25 (0.7)	25 (0.7)	0	0
Business:						
Solar (<300°F)	15	15	15	15	15	15
Solar (>300°F)	15	25	25	25	25	25
Photovoltaics	15	25	25	25	25	25
Wind	15	10	10	5	0	0
Geothermal	15	15	15	15	d	d
Ocean Thermal (OTEC)	15	15	15	15	15	15
Hydropower[c]	11	d	d	d	d	d
Biomass	10	10	10	5	0	0

[a]Adapted from Ref. 32.

[b]Could be claimed only by households with income of $30,000 or less; 1985 law had no income ceiling.

[c]Affirmative commitments for hydropower in 1985 law through 1988.

[d]Affirmative commitments at previous year's percent level.

renewable energy tax credits, "biomass" is defined as follows:

> "Biomass is generally any organic substance other than oil, natural gas, or coal, or product of oil or natural gas or coal. For this purpose, biomass includes waste, sewage, sludge, grain, wood, oceanic and terrestrial crops and crop residues and include waste products which have a market value. The conferees also intend that the definition of biomass does not exclude waste materials, such as municipal and industrial waste, which include such processed products of oil, natural gas or coal such as used plastic containers and asphalt shingles."

The definition of biomass cited here is clear; wastes are biomass. But it has been reported that the equipment for biological gasification of nonfossil waste materials is not eligible for the energy tax credit because the statutory language uses the term "qualified fuel" instead of the phrase "synthetic liquid, gaseous, or solid fuel" to define eligible biomass conversion equipment.[35,36] "Qualified fuel" is defined as:

I. Any synthetic solid fuel, and

II. Alcohol for fuel purposes if the primary source of energy for the facility producing the alcohol is not oil or natural gas or a product of oil or natural gas.

This interpretation may not be totally correct because the IRS literature (Publication 572) states that among the types of property that qualify as biomass property are a boiler that uses an alternate substance as the primary fuel, and a burner for a combustor other than a boiler if it uses an alternate substance as the primary fuel, including equipment located on-site at the burner and that is necessary to bring the alternate substance to the burner. The term "alternate substance" is defined as "any organic substance other than oil, natural gas, coal (including lignite), and any product of oil, natural gas, or coal." The intermediate-Btu gas from an anaerobic digester is an alternate substance by this definition, and although the digester may not be eligible for the energy tax credit, the equipment downstream of the digester would seem to be eligible (if new). This argument may be academic, however, since the energy tax credit expired at the end of 1985.

Production Tax Credit

Another tax credit available for biomass energy is the production tax credit. This credit is also included in the Crude Oil Windfall Profit Tax Act of 1980, and is apparently limited to biomass gasification processes.[37] In contrast to the energy tax credit, the production tax credit is available, if not changed or eliminated by tax reform, for projects completed by December 31, 1989, with credits on fuel sales to December 31, 2000.[37] This little known tax credit is $3.00/BOE adjusted for inflation, and was estimated to be $0.75/MBtu for mid-1985.[37] Assuming an overall federal and state income tax rate of 50%, this corresponds to added revenue of double this amount.[37] The subsidy is large and appears to be one of the best kept secrets of the tax code because few have taken advantage of it. The Act applies to "gas" produced from "biomass" sold to an "unrelated person".[37]

Taking the most optimistic view of the language based on the information presented here, wastes are included in the definition of biomass, and the gaseous products from both thermal and biological gasification processes for wastes and/or biomass are eligible for credits. It is still desirable that a position statement be obtained from the Internal Revenue Service by those who plan to use this credit. Since a tax credit of 75¢/MBtu is a significant portion of the cost of the gas from either process, any company involved in biomass gasification should examine the possibility of obtaining the production tax credit early in a project.

Impact of Future Energy Tax Credits on Biomass Energy

Almost without exception, the proponents of energy tax credits believe there will be a substantial decrease in the rate of commercialization of technologies classified as renewable energy systems without these credits.[28] Indeed, some feel the loss of the federal tax credits and the reduction or loss of some of the state tax credits coupled with the decline in oil prices present a very uncertain future for renewable energy.[2] The vagaries of tax law suggest that all entrepreneurs engaged in or contemplating renewable energy projects should follow these precepts:

- Do not initiate a project that depends on tax credits or tax forgiveness to show a profit. They may not exist when one needs them.
- Do initiate a project that has a high probability of standing on its own "two feet" because the IRS giveth and the IRS taketh away.
- Do obtain a ruling from the IRS on the applicability of tax forgiveness or an energy tax credit before key assumptions are cast in stone.

Material Benefits of Energy Tax Subsidies

The question of how the energy consumer benefits from the various energy tax subsidies provided by the government can often be answered quantitatively if it is known how much of the subsidy is passed to the individual energy user. A corollary to this question is to ask how the country benefits as a whole from these subsidies, particularly in view of the fact that new federal energy taxes are now under consideration to help balance the budget.[38] An interesting study, the results of which were reported in 1985, addressed this question and covered the magnitude of the subsidy and the energy supplied per dollar of subsidy.[29,39,40] The results of the study are summarized in Table 11. The energy subsidy in this work consists of the tax losses that result from energy tax credits and tax forgiveness, federal agency payments to each of the energy sectors, including DOE R&D costs, and the costs of federal loans and guarantees on energy projects.

It is surprising that the total subsidy as defined in this way was about $46 billion in 1984, as shown in Table 11, or about one quarter of the federal budget deficit. Also, it is clear that the subsidy is unevenly distributed among the various energy sectors. When energy consumption is calculated for each dollar of federal subsidy, the return is relatively low for nuclear electric power and much higher for fossil fuels and renewables. Thus, there is more "bang for the buck" for these latter fuels according to the analyst who performed the study.[29] It might be concluded from this work that expenditures by the government on energy subsidies are inefficiently used because the benefits are not uniformly distributed. This conclusion, however, is misleading because of several obvious factors — all energy forms are not equivalent on a simple Btu basis, energy availabilities and production costs are not the same per Btu for each energy form, and the reasons for the energy subsidy vary widely for each energy form. It is also not possible to calculate with accuracy how many incremental Btu or BOE were discovered and/or produced as a direct result of an energy tax subsidy after it was made available (see below), especially for established fuels that contribute large amounts energy. It is nevertheless unexpected that the total subsidy and tax loss portion of the subsidy are so large.

A somewhat similar analysis of the government's expenditures, specifically on renewable energy from 1975 through 1984 and renewable energy systems that came on-line during this same period, was published by DOE in 1985.[41] In this study, using information from the Energy Information Agency and the Internal Revenue Service, federal outlays and tax incentives in renewable energy were $4 billion for R&D and $2 billion in tax credits. The incremental renewable energy produced, 1.2 billion BOE, was estimated to be worth $39 billion at the prevailing world oil price. On a "bang per buck" basis, this corresponds to 1.2 million incremental Btu/$ of subsidy, or on a direct dollar basis, $6.50 of incremental BOE/$ of subsidy. DOE concluded that these systems can be expected to produce over 3 billion BOE of additional energy during their operational life-times.

ENERGY IMPACT OF RENEWABLE ENERGY

Since most of the U.S. energy tax credits for renewable energy expired at the end of 1985, it is again apropos to examine the contributions of renewable energy to primary energy demand. Table 12 is an update of the annual contributions of each renewable energy form utilizing the most recent data or conservative assumptions where data were not available. The total contribution of renewable energy is about 6.8

Table 11. U.S. FEDERAL ENERGY SUBSIDIES IN ENERGY SUPPLY IN FISCAL 1984[a] (1984 Dollars)

	Federal Subsidy ($$10^9$)					Energy Supplied			
Energy Resource	Tax Losses	Agency Payments	Cost of Loans and Guarantees	Total Subsidies	% of Total Subsidy	quad	% of Total Supplied	% Supplied to % Subsidy Ratio	MBtu Supplied Per Subsidy $
Nuclear Electric	10.236	2.284	3.320	15.840	34.4	1.110	1.4	0.04	0.07
Crude Oil and Natural Gas Liquids	7.310	1.271	NA	8.581	18.6	20.957	25.9	1.39	2.4
Fossil Electric	5.523	0.211	1.425	7.159	15.5	6.002	7.4	0.48	0.8
Natural Gas	4.292	0.319	NA	4.611	10.0	17.750	22.0	2.20	3.8
Coal	1.275	2.124	0.012	3.411	7.4	19.696	24.4	3.30	5.8
Hydroelectric	0.947	1.076	0.605	2.628	5.7	1.096	1.4	0.25	0.4
Other Renewables[b]	1.406[c]	0.290[d]	NA	1.696	3.7	2.929	3.6	0.97	1.7
Efficient Use	0.280	0.510	0.074	0.864	1.9	11.26	13.9	7.31	13.0
Synthetic Fuels	0.360	0.184	0.100	0.644	1.4	0	0	0	0
Nuclear Fusion	NA	0.606	NA	0.606	1.3	0	0	0	0
	31.629	8.875	5.536	46.04		80.80			

[a]Adapted from Tables 2 and 3 of Ref. 29.

[b]Includes wind, geothermal, photovoltaics, solar thermal, solar thermal electric, ocean thermal, and biomass.

[c]Includes $560 million for investment tax credits and accelerated depreciation, $325 million for residential supply incentives, $220 million for business supply incentives, and $215 million for alcohol fuel tax forgiveness.

[d]Includes DOE research support of $261 million.

Table 12. CONTRIBUTION OF RENEWABLE ENERGY TO PRIMARY ENERGY CONSUMPTION IN THE UNITED STATES

Fuel or Resource	Conversion Technology[g]	Energy Product[h]	Energy Production Units	For Year	BOE/d[o]	Reference
Biomass & Wastes						
Wood, Wood Wastes[a]	C	H,S,E	1.7 quad	1983	803,000	42,43
Wood, Wood Wastes[b]	C	H,S	0.9 quad	1983	425,000	42,43
Corn, Molasses, Cane	F	Ethanol	0.065 quad[i]	1985	30,700	44
MSW	C	E	384 MW[j]	1984	8,230	45
Sewage	BG	Methane	0.007 quad[k]	1984	3,310	46
MSW[c]	BG	Methane	0.005 quad	1984	2,360	1
Farm & Industrial Wastes	BG	Methane	0.002 quad[l]	1984	940	47,48
Biomass & Wastes	TG	Methane	0.001 quad[l]	1983	470	48
					1,274,010	
Falling Water	H	E	3.863 quad	1983	1,825,000	43
Geothermal	T	E	0.1312 quad	1983	62,000	43
Geothermal	HE	H	0.06 quad	1985	28,300	49
Solar[d]	HE	H	0.03 quad	1984	14,200	29
Solar[e]	HE	H	0.01 quad	1984	4,700	29
Solar[f]	HE	H	0.003 quad	1984	1,420	29
Solar	PV	E	50 MW[j,n]	1984	1,070	44,50
Wind	T	E	190 million kWh[m]	1984	930	51
					1,937,620	
					3,211,630 (6.80 quad)	

[a]Industrial sector usage.

[b]Residential sector usage.

[c]In landfills.

[d]For passive space heating.

[e]For domestic hot water.

[f]For active space heating.

[g]C, combustion; F, alcohol fermentation; BG, biological gasification or methane fermentation; TG, thermal gasification; H, hydroelectric; HE, heat exchange; T, turbines; PV, photovoltaics.

[h]H, heat; S, steam; E, electric; Methane, low-, medium-, and high-Btu gas.

[i]Estimated assuming equal volume of gasoline displaced by ethanol in blends with gasoline; If high heating value of ethanol used instead, energy production figures are 0.042 quad and 19,700 BOE/d; ethanol fuel production in 1985 assumed to be 550 million gal. as indicated in Ref. 28.

[j]Electric capacity converted to BOE/d by assuming an efficiency of 33% with fossil fuel and a load factor of 0.5. Estimated from data in Ref. 45.

[k]Estimated from total flow rate of 3.76×10^9 gal/d of wastewater to 209 treatment plants that utilize digester gas as fuel as reported in Ref. 46, assuming 300 mg/L total influent solids, 80% of which is volatile solids, and 60% of which is digested to yield methane at 4 SCF/lb volatile solids added.

[l]Estimated from operating systems listed in indicated references.

[m]Converted to BOE/d for oil-fired plant at 33% efficiency.

[n]Estimated from information in Refs. 44 and 50.

[o]1.0 barrel oil equivalent assumed to be 5.8×10^6 Btu.

quad/yr (1.2 billion BOE/yr), or about 8 to 10% of total energy demand in the United States for each of the last three years. Energy from biomass and wastes provides about 40% of the total renewable energy contribution.

According to recent reports, biomass energy consumption in the United States is approximately equal to the total energy requirements of, for example, Greece, Portugal, and Turkey together.[52] Relative to biofuels utilization in European countries, the United States is comparable or on the low side. The biofuels contributions to total primary energy demand is about 3% in Norway, 4% in Austria, 7 to 8% in Sweden, and 10% in Portugal.[52] In developing countries, examples of the energy contributions of biomass to total energy demand are nearly 100% in Nepal and Ethiopia, 75% in Kenya, 50% in India, 33% in China, 25% in Brazil, and 20% in Egypt and Morocco.[53]

BIOMASS PRODUCTION

Although the potential contribution to primary energy demand of organic wastes — excluding logging and wood-manufacturing residues — is substantial in the United States (about 2.2 quad/yr),[1] land- and water-based biomass represent a much larger energy resource. Research on each category of biomass is continuing to develop the technology needed to plant, grow, and harvest biomass as energy crops.

Silviculture

Numerous projects are underway to assess the woody resource base in the United States such as in the Southeast;[54,55] some of these assessments are being conducted on a county-by-county basis.[56-59] A recent nation-wide assessment indicates that about 600 million dry ton of unused wood are available annually.[60] This is equivalent to about 10.2 quad (Table 13), and is about four times the amount of wood energy now utilized in the United States.

The short-rotation intensive culture (SRIC) program, being sponsored in the United States by DOE, to attain high productivity in the least amount of time at a competitive cost is emerging as an important technology that can assert direct competition between the existing forest products industries and a growing wood energy industry.[61] Test plots of various hardwood species are under evaluation at different spacings, rotations, and management conditions throughout the United States. Table 14 summarizes those species that are receiving the most attention. Recent accomplishments in this program indicate that average productivity rates of several promising species grown on several site types range up to 6.7 dry ton/ac-yr (15 dry t/ha-yr) or up to 13.4 dry ton/ac-yr (30 dry t/ha-yr) for the best species on the best sites under optimum management conditions.[61] Production costs were estimated to range from $2.94 to $4.28/MBtu inclusive of profit and taxes on a range

Table 13. ESTIMATED UNUSED WOOD AVAILABLE FOR ENERGY IN UNITED STATES PER YEAR[a]

	Weight, 10^6 dry ton/yr	Energy,[b] quad/yr
Commercial Forest Land		
Excess Growing Stock	215	3.66
Mortality	95	1.62
Harvested Forest Sites		
Logging Residues	160	2.72
Standing Live and Dead Trees	20	0.34
Forest Products Industrial Waste	20	0.34
Waste Wood from Land Clearing	20	0.34
Urban Tree Removals and Wood Wastes	70	1.19
	600	10.2

[a]Adapted from Table 1 of Ref. 60.

[b]Estimated assuming wood energy content is 17 MBtu/dry ton.

Table 14. HARDWOODS UNDER SERIOUS EVALUATION IN SRIC GROWTH SYSTEMS[a]

Common Name	Species	Group[b]	Regions Where Species Under Evaluation							
			PNW	W/SW	GP	LS	MW	S/SE	NE	SB
Eucalyptus	*Eucalyptus grandis*	1								X
Eucalyptus	*Eucalyptus saligna*	1								X
Eastern Cottonwood	*Populus deltoides*	1			X		X	X		
Hybrid Poplar	*Populus* spp. hybrids	1	X		X	X	X	X	X	
Black Cottonwood	*Populus trichocarpa*	1	X							
Black Locust	*Robina pseudoacacia*	1			X	X	X	X	X	
Silver Maple	*Acer saccharinum*	2			X	X			X	
European Alder	*Alnus glutinosa*	2				X	X	X		
Red Alder	*Alnus rubra*	2	X							
Fourwing Saltbush	*Altriplex canescens*	2		X						
Autumn Olive	*Elaeagnus umbellata*	2					X			
Sweetgum	*Liquidambar styraciflua*	2						X		
Quaking Aspen	*Populus tremuloides*	2				X				
American Sycamore	*Platanus occidentalis*	2					X	X		
Mesquite	*Prosopis alba* and hybrids	2		X						
Chinese Tallow Tree	*Sapium sebiferum*	2						X		
Siberian Elm	*Ulmus pumila*	2			X					
Tree of Heaven	*Ailanthus altissima*	3				X	X			
Paper Birch	*Betula papyrifera*	3				X				
Green Ash	*Fraxinus pennsylvanica*	3				X		X		
Leucaena	*Leucaena leucocephala*	3						X		X
Willow	*Salix* spp.	3				X				

[a]Adapted from Table 25 and Appendix II of Ref. 61.

[b]Group 1 shows considerable promise, based on maximum yields from 10-30 dry Mg/ha-yr (4.46-13.4 ton/ac-yr); Group 2 shows apparent promise, based on maximum dry yields from 4-10 Mg/ha-yr (1.78-4.46 ton/ac-yr); Group 3 shows promise, based on individual tree data, needs to be verified or species has limited geographic range in U.S.A.

[c]PNW, Pacific Northwest; W/SW, West/Southwest; GP, Great Plains; LS, Lake States; MW, Midwest; S/SE, South/Southeast; NE, Northeast; SB, Subtropics.

of experimental plots as shown in Table 15. The average percent of the major cost components from seven research projects was harvesting, 35%; taxes and profit, 20%; land, 16%; planting, 9%; management, 6%; transportation, 6%; administration, 5%; and site preparation, 3%. A more recent report on this program presents the range of incremental yields from first rotations and some initial coppice growth results (Table 16) as well as production costs without profit (Table 17).[62]

Experimental tests of the wood grass concept continued in 1985 to confirm (or deny) the extraordinarily high yields of biomass reported when hybrid poplar is planted at very high densities and is harvested like perennial grasses.[1] Tests were underway or continued in 1985 (and 1986) in Missouri,[63] California,[64] Wisconsin,[65,66] and China.[67] The Missouri test consists of eight 50 ft x 50 ft plots in which 6-in. cuttings of strain D-01, the best clone developed in the Oregon work,[1] was hand-planted at 1-ft spacings on May 15, 1984.[63] This corresponds to a planting density of 43,560 trees/ac. Fertilization and straw mulch were used. The first cutting was made in March 1985, but no weight yields were recorded. At that time, the trees were about 10 to 14 ft high and 1 to 1-1/2 in. in diameter. One tree was left on each plot, and since the first cutting, it grew to a 3-in. diameter tree in one year. Also, since the first cutting, rapid coppice growth occurred. The next cutting will be made in March 1986, at which time dry weight yields will be determined. The new coppice growth may then be cut again in the juvenile stage in 1986 for cattle feeding tests. Fertilization effects were determined in these tests at 50, 100, 150, and 400 lb nitrogen/ac; no significant difference was observed at the two lowest rates, but large differences in growth were found between the 100- and 150-lb dose rates.[67]

A similar test in Wisconsin using the same strain (D-01) of hybrid poplar on a 20-ac test plot at different spacings was initiated in March 1985.[65,66] Dry yields ranged from about 8-10 ton/ac-yr for the first year. One of the key cost factors is to reduce the establishment costs to less than $800/ac,[63-66] so the Wisconsin test included "scatter planting" of the cuttings with a farm manure spreader. Paper mill sludge was also used as a fertilizer with the scatter planting technique. This test is continuing.

The test in China was started in 1985 and consists of the planting of hybrid poplar strains D-01 and D-25 at 6 in. x 6 in. spacings on 4-ac plots in a sub-tropical location, Fujian Province.[67] A survival rate of about 60% was observed. In view of this result, the Chinese Government is now considering expansion of the test program to eight other provinces. The applications of wood grass that the Chinese are interested in include fuels and extrusion processes for the manufacture of building materials.

Table 15. AGGREGATED SRIC TREE PRODUCTIVITY AND COST[a]

Common Name	Hybrid Poplar	Black Cottonwood	Black Locust	Slash Pine	Eucalyptus[c]	Hybrid Poplar	Eucalyptus[d]
Location	Central PA	East KS	Central KS	Northern FL	Southern FL	Northern WI	HI
Planting Density, trees/ac	8,090	570	1,300	2,700	650	2,800	1,210
Rotation, yr	4	7	6	10	3	5	6
Yield, dry ton/ac-yr	4.8	5.2	4.6	5.75	6.3	3.2	8.0
Planning, $/ac	20.23	20.23	20.23	20.23	20.23	20.23	20.23
Establishment, $/ac	409	146	146	456	149	363	625
Management, $/ac	390	140	140	21	21	26	520
Harvesting, $/ton	22	22	22	22	22	22	22
Transportation, $/ton	3.60	3.60	3.60	3.60	3.60	3.60	3.60
Administration, $/ac	12	12	12	12	12	12	12
Land Rent, $/ac	34	39	39	30	50	40	101
Estimated Cost,[b] $/MBtu	4.12	3.07	3.38	3.53	2.94	4.24	4.28
$/GJ	3.91	2.91	3.20	3.35	2.79	4.02	4.06

[a]Adapted from Table 4 and Figure 5 of Ref. 61; All costs in 1983 dollars; No costs included for processing the harvested wood chips.

[b]Includes tax and profit.

[c]*Eucalyptus grandis*.

[d]*Eucalyptus saligna*.

Table 16. SRIC HARDWOOD PRODUCTIVITY IN 1984[a]

Common Name	Age, yr	Mean Increment,[d] ton/yr	Soil Type[e]	Fertilizer[f]	Irrigation[f]
Red Alder	9	5.4-6.7	MA	No	NS
Hybrid Poplar	7	5.1-11.7	AA	NS	NS
Black Locust	6	6.5-9.3	MA	No	No
Silver Maple	6	5.3-7.7	MA	No	No
Eucalyptus[b]	4	6.4-18.4*	G,ST	NS	NS
Hybrid Poplar	4	4.9-8.7	MA	NS	NS
Black Cottonwood	4	4.2-21.2	EX	No	Yes
Black X Eastern Cottonwoods	4	14.6-27.3	EX	No	Yes
Autumn Olive	2-4	2.4-8.3**	MA	No	NS
American Sycamore	2	3.7-5.7	G,IF	NS	NS
Eucalyptus[c]	1.3-3.4	3.6-22.3	MA,T	NS	NS

[a]Adpated from Table III, Ref. 62.

[b]*Eucalyptus grandis*.

[c]*Eucalyptus saligna*.

[d]All are first rotation yields, presumably dry but not stated; Asterisk indicates coppice yield; Double asterisk indicates first rotation and coppice yield.

[e]MA, marginal agricultural soil; AA, abandoned agricultural soil; EX, excellent cottonwood site; G, good soil; IF, industrial forest land; T, tropics; ST, subtropics.

[f]NS, not stated.

Table 17. ESTIMATED SRIC TREE PRODUCTION COSTS[a]

Location	Common Name	\$/MBtu[d]	\$/GJ[d]
Florida	Eucalyptus	1.47	1.39
	Slash Pine	2.00	1.90
	Sand Pine	3.01	2.85
Hawaii	Eucalyptus	2.42	2.29
Kansas, Eastern	(Several Species and sites)	1.56-2.09	1.48-1.98
Central	(Several Species and sites)	1.77-1.81	1.68-1.72
Pennsylvania	Hybrid Poplar[b]	2.46	2.33
	Hybrid Poplar[c]	2.74	2.60
Texas	Mesquite	1.60	1.52
Wisconsin	Hybrid Poplar	2.31	2.19

[a]Adapted from Table IV, Ref. 62.

[b]Control.

[c]Fertilization used.

[d]Assumptions: 30-year planning period; 5% real discount rate; Harvested at cost of \$12/dry tonne; Delivery distance of 15 km; All other data include local conditions, cost of land, realized productivity rates from research plots having minimum size of 1/16 ha; No cost attributed to drying or profit; All costs in 1983 dollars.

Among one of the more interesting larger scale SRIC tests underway is the program for production of eucalyptus in Hawaii.[68] More than one million eucalyptus seedlings were planted (1200-1750 trees/ac) at three major locations representing varying elevations, climatic conditions, and soil types (abandoned cane land, waste land, and forest land). The entire gamut of growth parameters and species is being evaluated in 87 experiments. Planting was initiated in 1979 and at a rotation time of 6 years, coppice growth studies are expected to begin this year after harvesting.

LEBEN, or a large European bioenergy project, was reported last year to be scheduled for initiation in Abruzzo, Italy in June 1985 and be established at the end of 1988.[69-70] This project integrates SRIC tree production, agricultural residues, energy crops, and conversion to fuels and energy, and is one of the largest biomass energy projects in the Western World. About 450,000 ton/yr of biomass consisting of 285,000 ton/yr of woody biomass from 173,000 ac, and 130,000 ton/yr of agricultural residues from 173,000 ac of vineyards and olive and fruit orchards, will be used. Later, 120,000 ton/yr of energy crops from 259,000 ac will be utilized.

It was reported last year that the largest man-controlled SRIC plantation attempted to date — primarily with Gmelina arborea on 127,000 ac (51,400 ha) in Brazil — has had both technical and economic difficulties.[1] Although gmelina failed on some of the sites, it is reported to be doing well on about one-third of the planted sites that have the best soil conditions.[71] On the other two-thirds of the sites with sandy and transition soils, eucalyptus and Caribbean pine are apparently being grown successfully.[71] The project's managers anticipate that ultimately, there will be 62,000 ac (25,000 ha) of gmelina, 104,000 ac (42,000 ha) of pine, and 79,000 ac (32,000 ha) of eucalyptus.[72] Current rotations are 5 to 7 yr for gmelina and 10 to 12 yr for pine, and all species are now planted at 6.6- by 9.8-ft (2- by 3-m) spacings. At short rotations, gmelina can reach a diameter of 19.7 in. (50 cm) and the pine can reach 13 in. (33 cm); average annual growth for all properly managed plantations is about 186 ft^3/ac-yr (13 m^3/ha-yr).

Herbaceous Crops

In the mid-to-late 1970's, the food-vs.-energy controversy relegated herbaceous crops to a secondary position as energy feedstocks. Opponents of biomass energy felt that the use of tillable lands for purposes other than food production was almost, if not in fact, a mortal sin. In 1980 when the Global 2000 Report to the President was published, it was predicted that world demand for food would increase greatly in the next 20 years, that real food prices would double, and that developed countries would have to supply most of the increase.[73,74] These projections were almost as far removed from reality as those that were made at about the same time for U.S. energy consumption and oil and natural gas supplies for 1985.[75] World agricultural production is now at an all-time high and is climbing fast, especially in the developing countries.[74] Modern agriculture is producing more food per capita nearly everywhere in the world despite the most rapid rates of growth in population and food demand in history.[74] Modern agriculture is the reason, for example, why the world record corn yield is now 370 bu/ac,[76] and why stover is the single most abundant agricultural residue in North America.[77] It is the prime cause of the mounting surpluses of farm products and depressed markets; genetic engineering advances are expected to make still more improvements in crop production.[78] All of this suggests that herbaceous energy crops may offer alternative outlets for agricultural products and give the farmer more flexibility in coping with fluctuations in market prices. Certainly, the great diversity of herbaceous species, their broad distribution throughout the world, and their growth characteristics suggest that land-based plants should be examined in detail as energy feedstocks and fuels.

The key to assuming that these opportunities will be utilized by the farmer lies in production costs, and whether or not a reasonable profit can be realized. The current price structure for many herbaceous species indicates that considerable research is needed to develop the technology to the point where these opportunities are real. For example, the average price of common hay in Illinois in the fall of 1985 was $98/ton (about $6/MBtu if dry);[79] the production of "biocrude" liquids by extraction of arid-adapted crops was estimated to range from $12.50 to $18.60/MBtu;[80] and even Brazilian officials reported that Brazil could not afford to manufacture cane-derived ethanol fuel, which is much cheaper to manufacture than ethanol from corn,[1] if world oil prices continue to fall.[81] Yet there are other cases that indicate that herbaceous species may be able to compete with today's fuel prices in the U.S. For example, commercial farming of buffalo gourd in the U.S. high plains may result in yields of 100 and 325 gal/ac-yr of seed oil and ethanol, respectively, with cultivation and harvesting costs, respectively, of $0.28 to $0.44/gal of liquid fuels.[82] Average on-farm processing cost, including amortization of capital, is estimated to be about $0.55/gal. Since much of the farm community in the United States is now operating at a very narrow profit margin or a loss, the barriers to the improvements needed to grow herbaceous energy crops for profit do not appear to be insurmountable, especially if valuable co-products can be produced simultaneously.[82,83]

DOE's program, which was started in 1984 to develop herbaceous energy crops,[1] is now directed to lignocellulosic crops, including forage and hay crops, as the primary target.[84] Oilseed crops are given secondary emphasis.[84] Current non-irrigated hay yields and costs in the United States outside the subtropics are estimated to range from 1.8-4.9 dry ton/ac-yr (4-11 t/ha-yr) and from about $3.70-$9.50/MBtu ($3.50-$9.00/GJ).[84] Species selection and adaptation of agronomic practice to energy crops are expected to raise yields and reduce costs by 1992 to 4.9-

10.3 dry ton/ac-yr (11-23 dry t/ha-yr), and \$2.74-\$3.32/MBtu (\$2.60-\$3.15/GJ).[84] The DOE program initiated five projects in the Southeast and Midwest-Lake states in 1985 on sites that are marginal for conventional row crops. Field trials were established at 21 different sites for 15 annual and perennial grasses. At the end of 1985, the highest productivities were observed with the annual species pearl millet (Pennisetum americana) in Alabama; the yields averaged 5.4 ton/ac (12.2 t/ha). A listing of the lignocellulosic species being screened in this program is shown in Table 18. This program is continuing.

Other recent reports on herbaceous energy crops concerned guayule,[85] buffalo gourd,[83,86] jojoba,[87] and sorghums[88] in the Southwest; and sweet sorghum,[89] rape,[90,91] and napier grass,[92,93] in the East and Southeast. The trend in much of this work seems to be in the direction of no-till methods on marginal lands.[94-96] In the arid to semi-arid regions of the Southwest, current technology for herbaceous crops seems to offer two options: biomass production at low yields without irrigation for high value chemicals, and biomass production at higher yields with irrigation for energy.[96] Interesting advances have been announced on the use of small amounts of dried blue-green algae as a substitute for chemical nitrogen fertilizers[97] and in the development of tissue culture techniques for propagating biomass.[98,99] In the case of the biofertilizer, 1 lb of dried blue green algae on one acre of land is claimed to produce the same amount of nitrogen as 60 lb of chemical fertilizer, thereby offering the possibility of much cheaper fertilization methods.[97] The development of successful embryonic tissue culture techniques for propagating sorghum,[98] napier grass,[98] and sweet potato[99] offers the potential of producing large numbers of seedlings rapidly and mass field production at reduced labor costs.

It should be emphasized that with the limited exceptions of ethanol fuel and bagasse combustion, herbaceous crops currently provide little or no contribution to primary energy demand in the United States. The resource base of herbaceous energy crops and relevant agricultural residues has recently been estimated to be a maximum of about one billion tons per year (about 15-16 quad/yr if dry).[100] Another recent estimate made on the assumptions that grasses can be grown as energy crops on available rangelands in seven southwestern states at yields of 5 dry ton/ac-yr indicate an energy potential of about 17 quad/yr.[96] Dry-matter yields of 20 dry ton/ac-yr have already been achieved for sorghum and napier grass in field test plots, and larger scale experiments are currently underway to demonstrate high sustainable yields.[17] Thus, even at 10% of these idealized estimates, the energy potential of herbaceous energy crops is high; additional research is needed to make some of these systems economically practical.

Table 18. LIGNOCELLULOSIC SPECIES SELECTED FOR SCREENING IN THE SOUTHEAST AND MIDWEST-LAKES STATES[a]

Common Name	Species
Grasses	
Bermudagrass	*Cynodon dactylon*
Johnsongrass	*Sorghum halapense*
Oats	*Avena sativa*
Pearl Millet	*Pennisetum americana*
Reed Canarygrass	*Phalaris arundinacea*
Rye	*Secale cereale*
Smooth Bromegrass	*Bromus inermis*
Sorghum (sweet or forage varieties)	*Sorghum bicolor*
Sorghum X Sudangrass	*S. bicolor* X *S. sudanensis*
Sundangrass	*Sorghum sundanensis*
Switchgrass	*Panicum virgatum*
Tall Fescue	*Festuca arundinacea*
Timothy	*Phleum pratense*
Weeping Lovegrass	*Eragrostis curvula*
Legumes	
Alfalfa	*Medicago sativa*
Arrowleaf Clover	*Trifolium vesiculosum*
Birdsfeet Trefoil	*Lotus corniculatus*
Crimson Clover	*Trifolium incarnatum*
Crownvetch	*Coronilla varia*
Flat Pea	*Lathyrus sylvestris*
Lupine	*Lupinus* spp.
Red Clover	*Trifolium pratense*
Servicea Lespedeza	*Lespedeza cuneata*
Sweet Clover	*Melitotus alba*
Vetch	*Vicia sativa*
Other	
Kale	*Brassica oleracea acephala*

[a]From Table 1 of Ref. 84.

Aquatics

The overall goals of the energy research conducted over the last several years on aquatic biomass have generally been directed to either feedstock production, often with simultaneous waste treatment, for subsequent conversion to fuels by fermentation, or to species that contain valuable products. The aquatics that continue to be studied and their main applications are microalgae for liquid fuels, the macrophyte water hyacinth for wastewater treatment and conversion to methane, and marine macroalgae for specialty chemicals or conversion to methane.

Microalgae. Most of the research in progress on aquatics for energy last year was focused on the development of microalgae growth systems that utilize saline or brackish waters. The DOE program on this subject is the largest of its kind. It consists of several projects and emphasizes the development of oil-yielding microalgae that grow in saline waters of the desert in the Southwest.[101] The work in progress includes projects on siting studies; collection, screening, selection, and characterization of microalgae; the growth of certain species in laboratory and small-scale field units; conceptual design and economic analysis of full-scale production systems; and exploration of innovative approaches to microalgae production.

Microalgae have been investigated as a source of liquid fuels in the form of vegetable-type oils (lipids) for over 40 years. They have not been utilized commercially for this purpose yet because of the high capital cost of the production systems — over 10-fold higher than other biomass production systems but justifiable because of the very high productivities of up to 357 lb/ac-d (40 g/m^2-d) — and the value of the carbon dioxide which represents about \$181/ton (\$200/t) of biomass.[102]

In the siting studies done for the Southwest, the Sunbelt States have been "stratified" into zones of varying suitability for microalgae production systems based on insolation, length of growing season, land and water stewardship, legal and environmental constraints on land and water, sustainable water yields, and development of an infrastructure to supply nutrients.[103] These studies indicate that up to 25 million acres may be highly suitable for microalgae cultivation.[104] In the production technology, key problems concern oil yields and separation of the microalgae from the growth medium. Most of the flocculation, filtration, and centrifugation methods are too expensive for microalgae separation, but some methods are reported to be suitable.[102] The separation involves the concentration of microalgae from suspensions with 150-600 mg/L solids into slurries with 100-200 g/L, or a concentration factor of 160-600.[105] Sand filtration, filtration through strainers, and dissolved air flotation with and without flocculation have been tested, but none of these techniques is totally satisfactory.[101,102,105]

An updated economic analysis was reported in 1985 for the production of microalgae in a 2,500-ac facility (Table 19).[101] It is apparent that the operating cost is a large fraction of the total cost, and that the nutrient cost, most of which is accounted for by carbon dioxide, is over one-half the total cost. If the carbon dioxide were available free, the microalgae costs would be reduced by about 40%. It is also noteworthy that separation of the oil fraction of the microalgae is not included in this analysis; extraction and oil processing costs (if transesterification were employed, for example, for diesel fuel) would have to be added. Overall, even with improvements in production technology, this analysis suggests that microalgal oils cannot compete with petroleum liquids until crude oil costs rise to about $45/bbl (1984 dollars).

Last year, the French reported the results of their continuing work on the green hydrocarbon-producing microalgae Botryococcus braunii.[106-108] Nitrogen limitation was not found to be necessary for high hydrocarbon production, the highest productivities of which were observed during exponential growth.[106] Light intensity did not affect the structure of the hydrocarbons (mainly C_{27}, C_{29}, and C_{31} alkadienes), produced in continuously illuminated "air lift" batch cultures. But adjustment of the light intensity to the proper level gave maximum biomass and hydrocarbon yields.[107] Screening work on various strains of B. braunii has established the existence of two types of organisms; one produces straight-chain alkadienes and trienes as odd-numbered chains from C_{23} to C_{31}, and the other produces triterpene hydrocarbons of the generic formula C_nH_{2n-10} where n varies from 30 to 37.[108] The hydrocarbon content varied from 20-52% of dry weight.[108]

The future of microalgae production for fuel seems to be in the direction of integrated systems. One system announced in 1985 has been "set up" in Belgium (size not specified) and consists of integrated poultry, fish, microalgae, and anaerobic digestion units in which the algal product is converted to methane in a two-phase system; the digester effluent is recycled to the microalgal system or used for "animal breeding"; the wastes from which are recycled to the microalgae ponds.[109] Thus, the outputs are methane, eggs, poultry, fish, and possibly "high-value chemicals." This report states that the system will be further developed in a pilot plant of industrial size. Another system on a 40-ac farm near Barstow, CA integrates an alcohol fuel facility, wind energy, greenhouses, microalgal ponds, fish and prawn production, and photovoltaic modules to produce foods, fuels, and electricity.[110,111] The principles employed in this "Integrated Life System" could be used to meet most of the material needs of communities of 20,000 or more according to the system developers.[110,111]

Macroalgae. Research on marine macroalgae is continuing with the red seaweed Gracilaria tikvahiae and the green seaweed Ulva lactuca in Florida to study their growth

Table 19. COST PROJECTIONS OF MICROALGAE PRODUCTION[a]

	Reference Case[b]		Attainability Case[c]	
Productivity, ton/yr (t/yr)	56,200 (62,000)		--	
Energy Content, MBtu/ton (GJ/t)	21.4 (24.9)		25.9 (30.1)	
Capital Cost	$/ton	$/t	$/ton	$/t
Culture System	32.7	36.0	15.3	16.9
Harvesting System	16.6	18.3	7.3	8.1
Engineering Fees	5.4	5.9	2.4	2.7
Contingency Fees	8.3	9.1	3.7	4.1
Land	1.9	2.1	0.9	1.0
	64.9	71.4	29.6	32.8
Operating Cost				
Labor and OH	50.9	56.1	23.8	26.2
Utility	16.9	18.6	4.7	5.2
Nutrients	187.6	206.8	109.5	120.7
Water	16.1	17.8	8.5	9.4
Operating	24.4	26.9	11.2	12.4
Maintenance	34.4	37.9	15.8	17.4
	330.3	364.1	173.5	191.3
Total Microalgae Cost	395.0	435.0	203.0	224.0
Total Energy Cost, $/MBtu ($/GJ)	18.46 (17.51)		7.85 (7.43)	

[a]Adapted from Ref. 101; All costs are 1984 dollars.

[b]Facility is 2,500 ac (1,000 ha); Contains support facilities and 43 20-ha culture ponds 300 m long x 30 m wide x 0.2 m deep; The harvesting system is a microstrainer followed by a centrifuge to convert solids concentration from 0.05% (500 mg/L) to 10%; Detention time is 4 d; Productivity is 25 g/m^2-d; Growing season is 320 d/yr; Microalgae composition is 30% lipid, 20% carbohydrate, 32% protein, 10% metabolic intermediate, and 10% ash.

[c]Improved case in which nutrient costs reduced and lipid and productivity values increased; Productivity not specified.

characteristics in saline waters of the Southwest (they grow well in seawater), and to evaluate the effects of various environmental stresses and metabolic inhibitors.[101,112,113] G. Tikvahiae is reported to be a good substrate for bioconversion to methane and can effectively scrub nutrients from wastewaters.[113] Productivities for both species are high and range from 20-35 MAF dry g/m^2-d (32.6-60.0 MAF dry ton/ac-yr).[113] G. Tikvahiae also produces an economically valuable polysaccharide, agar, which can be extracted.[113] Currently, floating macroalgae (Sargassum spp.) are being developed as possible biomass energy crops for Florida's near-shore, shallow waters.[113] Dry yields of 20 g/m^2-d have been reported, but they have not yet been sustained for more than a few months.[113]

Research progress has also been described on the work done with the brown kelp Laminaria saccharina, the green seaweed Codium fragile, and G. Tikvahiae to evaluate large-scale cultivation on a buoyed submerged latticework off New York.[114] In the early work, it was found that L. saccharina could be grown for about 10 months of the year in New York waters, while the other two species grew rapidly during the summer months. Since then, the work has been focused on the concept of multiple species farming. A 120 ft x 41 ft seaweed farm was designed and put into place in January 1984, after which the individual plants were attached.[1,114] The L. saccharina was harvested in July 1984, and then the farm was planted with the summer crop species.[114] As reported last year, dry ash-free yields of L. saccharina were about 5 ton/ac-yr.[1]

The experiments to grow the giant brown kelp Macrocystis pyrifera as an energy crop off the California coast on a near-shore, 1.2-ac (0.48-ha) test farm at Elwood, CA are continuing.[115]

Floating Macrophytes. The floating aquatic macrophyte water hyacinth (Eichhornia crassipes), which has considerable potential for cleanup of sewage influent and simultaneous growth, and for conversion to methane, continues to be studied for these applications.[116-121] It was predicted in 1974 that water hyacinth would be a strong candidate as a non-fossil carbon source for methane production if the high yields reported at that time could be sustained on a steady-state basis.[122] At that time, the reported yields were in the range 4.9-14.7 dry ton/ac-yr in Mississippi, and higher yield projections (67 dry ton/ac-yr) were estimated if the plant was grown in good climates, the young plants were dominant, and the water surface was always completely covered. In recent Florida tests, use of an optimum planting density of 7.4 lb/ft^2 has given dry yields of 27-31 dry ton/ac-yr, and it has been estimated that further increases of 30-50% are possible through the discretionary use of aeration.[117] In the temperate climates of France, yields of 13.4-22.3 dry ton/ac-yr have been reported for water hyacinth grown on wastewater.[120] As shown in Chart 2, the pipeline gas price estimated for water hyacinth conversion to methane is $5.50-$6.60/MBtu (in

Table 19. COST PROJECTIONS OF MICROALGAE PRODUCTION[a]

	Reference Case[b]		Attainability Case[c]	
Productivity, ton/yr (t/yr)	56,200 (62,000[illegible]		--	
Energy Content, MBtu/ton (GJ/t)	21.4 (24.9[illegible]		25.9 (30.1)	
Capital Cost	$/ton	[illegible]/t	$/ton	$/t
Culture System	32.7	3[illegible].0	15.3	16.9
Harvesting System	16.6	1[illegible].3	7.3	8.1
Engineering Fees	5.4	[illegible].9	2.4	2.7
Contingency Fees	8.3	[illegible].1	3.7	4.1
Land	1.9	[illegible].1	0.9	1.0
	64.9	7[illegible].4	29.6	32.8
Operating Cost				
Labor and OH	50.9	5[illegible].1	23.8	26.2
Utility	16.9	1[illegible].6	4.7	5.2
Nutrients	187.6	20[illegible].8	109.5	120.7
Water	16.1	1[illegible].8	8.5	9.4
Operating	24.4	2[illegible].9	11.2	12.4
Maintenance	34.4	3[illegible].9	15.8	17.4
	330.3	36[illegible].1	173.5	191.3
Total Microalgae Cost	395.0	43[illegible].0	203.0	224.0
Total Energy Cost, $/MBtu ($/GJ)	18.46 (17.51[illegible]		7.85 (7.43)	

[a]Adapted from Ref. 101; All costs are 1984 dollars.

[b]Facility is 2,500 ac (1,000 ha); Contains support facilities and 43 20-ha culture ponds 300 m long x 30 m wide x 0.2 m deep; The harvesting system is a microstrainer followed by a centrifuge to convert solids concentration from 0.05% (500 mg/L) to 10%; Detention time is 4 d; Productivity is 25 g/m^2-d; Growing season is 320 d/yr; Microalgae composition is 30% lipid, 20% carbohydrate, 32% protein, 10% metabolic intermediate, and 10% ash.

[c]Improved case in which nutrient costs reduced and lipid and productivity values increased; Productivity not specified.

characteristics in saline waters of the Southwest (they grow well in seawater), and to evaluate the effects of various environmental stresses and metabolic inhibitors.[101,112,113] G. Tikvahiae is reported to be a good substrate for bioconversion to methane and can effectively scrub nutrients from wastewaters.[113] Productivities for both species are high and range from 20-35 MAF dry g/m^2-d (32.6-60.0 MAF dry ton/ac-yr).[113] G. Tikvahiae also produces an economically valuable polysaccharide, agar, which can be extracted.[113] Currently, floating macroalgae (Sargassum spp.) are being developed as possible biomass energy crops for Florida's near-shore, shallow waters.[113] Dry yields of 20 g/m^2-d have been reported, but they have not yet been sustained for more than a few months.[113]

Research progress has also been described on the work done with the brown kelp Laminaria saccharina, the green seaweed Codium fragile, and G. Tikvahiae to evaluate large-scale cultivation on a buoyed submerged latticework off New York.[114] In the early work, it was found that L. saccharina could be grown for about 10 months of the year in New York waters, while the other two species grew rapidly during the summer months. Since then, the work has been focused on the concept of multiple species farming. A 120 ft x 41 ft seaweed farm was designed and put into place in January 1984, after which the individual plants were attached.[1,114] The L. saccharina was harvested in July 1984, and then the farm was planted with the summer crop species.[114] As reported last year, dry ash-free yields of L. saccharina were about 5 ton/ac-yr.[1]

The experiments to grow the giant brown kelp Macrocystis pyrifera as an energy crop off the California coast on a near-shore, 1.2-ac (0.48-ha) test farm at Elwood, CA are continuing.[115]

Floating Macrophytes. The floating aquatic macrophyte water hyacinth (Eichhornia crassipes), which has considerable potential for cleanup of sewage influent and simultaneous growth, and for conversion to methane, continues to be studied for these applications.[116-121] It was predicted in 1974 that water hyacinth would be a strong candidate as a non-fossil carbon source for methane production if the high yields reported at that time could be sustained on a steady-state basis.[122] At that time, the reported yields were in the range 4.9-14.7 dry ton/ac-yr in Mississippi, and higher yield projections (67 dry ton/ac-yr) were estimated if the plant was grown in good climates, the young plants were dominant, and the water surface was always completely covered. In recent Florida tests, use of an optimum planting density of 7.4 lb/ft^2 has given dry yields of 27-31 dry ton/ac-yr, and it has been estimated that further increases of 30-50% are possible through the discretionary use of aeration.[117] In the temperate climates of France, yields of 13.4-22.3 dry ton/ac-yr have been reported for water hyacinth grown on wastewater.[120] As shown in Chart 2, the pipeline gas price estimated for water hyacinth conversion to methane is \$5.50-\$6.60/MBtu (in levelized 1982 dollars). This cost is much lower if higher credit can be taken for wastewater treatment.[1]

CONVERSION

Combustion

Cogeneration. Cogeneration is getting intensive study by both small power producers and the electric utility industry.[123-126] Some surveys indicate that as much as 100,000-200,000 MW of power could be cogenerated in the U.S. by 2000; this is about 25% of today's installed capacity.[123] DOE has projected that 5-20% of U.S. electricity consumption could be supplied by cogeneration in 2000.[125] According to federal statistics, cogeneration capacity increased from 13,000 MW in 1979 to 21,694 MW in 1983,[124] and nearly 5,000 MW of additional capacity was announced in 1984.[123] Current investments in cogeneration capacity could reach $50 to $60 billion through the next decade and greatly affect electric utility planning plus raise questions before FERC, utility boards, and PUC's about the need for new central station generation.[123]

The relationship of all this activity to biomass and wastes is that they will have a major role in this technology as fuels. To be eligible for benefits under the Public Utility Regulatory Policies Act of 1978 (PURPA), the combustion system is limited to 80 MW and must receive 75% or more of its total energy input from renewables; cogeneration plants are not limited in size or fuel.[45,127] At the beginning of 1985, 233 non-utility, biomass-fueled electric power-producing facilities having a total capacity of about 3,400 MW were either in operation, under construction, or planned for the United States (Table 20). Only 18 facilities having a total capacity of about 245 MW were utility-owned (Table 20).[45] According to PURPA, the non-utilities can sell any surplus power they may have or do not choose to use to the utilities at the utility's full avoided cost (the amount saved by not having to produce it themselves). Although this approach to power production is a great concern to the electric utilities, it has been well received by organizations in a situation to take advantage of PURPA. For example, currently there are at least 36 operating cogeneration units in Florida having a total capacity of 765 MW ranging in size from a few hundred kilowatts to 100 MW; 410 MW of additional cogeneration capacity is in the advanced planning stage.[126] Of the 45 firms in Florida that were identified as either cogenerators or planned cogenerators and small power producers, the fuels for the facilities are distributed as follows: natural gas, 13; wood and wood wastes, 9; sulfur, 8; municipal solid wastes, 7; bagasse, 6; coal, 1; and landfill methane, 1. In California, in excess of 120 MW of cogeneration capacity fired by biomass and wood wastes exists on Pacific Gas and Electric's system.[123] Cogeneration applications have greatly exceeded estimates in California, so the PUC has

Table 20. ELECTRIC POWER PRODUCTION BY COMBUSTION OF BIOFUELS IN THE UNITED STATES[a]

	Wood Waste	MSW	Ag Residues	Landfill CH_4	Sewage CH_4	Manure CH_4
Small Power Production, Cogeneration, and Municipal Facilities:[b]						
Planned, kW	304,475	541,550	152,600	52,925	164	94,000
No. Facilities	21	25	4	18	2	2
Under Construction, kW	214,820	335,300	56,000	24,850	2,700	10
No. Facilities	18	11	3	16	1	1
Operating, kW	1,092,597	350,410	144,050	40,565	6,720	1,225
No. Facilities	59	18	8	19	4	3
Subtotal Capacity, kW	1,611,892	1,227,260	352,650	118,340	9,584	95,235
Subtotal No. Facilities	98	54	15	53	7	6
Grand Subtotal Capacity, kW			3,414,961			
Grand Subtotal No. Facilities			233			
Utilities						
Planned, kW	50,000	--	--	--	--	--
No. Facilities	1	--	--	--	--	--
Under Construction, kW	--	18,400	--	--	--	--
No Facilities	--	2	--	--	--	
Operating, kW	130,725	15,400	22,500	--	7,970	--
No. Facilities	8	3	2	--	2	--
Subtotal Capacity, kW	180,725	33,800	22,500	--	7,970	--
Subtotal No. Facilities	9	5	2	--	2	--
Grand Subtotal Capacity, kW			244,995			
Grand Subtotal No. Facilities			18			
Grand Total Capacity			3,659,956			
Grand Total No. Facilities			251			

[a]Adapted from Tables 3 and 4 of Ref. 45. All plants included in this survey except one are in the contiguous 48 states. The exception is an MSW combustion plant (56,250 kW) in the planning stage in Honolulu, HI. This means all bagasse combustion plants in Hawaii are not included. Several large industrial power producers who manufacture power from biomass and wastes for captive use are also not included.

[b]Primarily small power producers facilities that do not exceed 80 MW, and must receive 75% or more of fuel from renewables. Cogeneration facilities are not limited in size. These systems have been qualified by FERC for PURPA benefits.

suspended PG&E's standard long-term offer for oil- and gas-fired projects over 50 MW.[123]

Clearly, many firms are seizing the opportunities presented by PURPA, and cogeneration is growing rapidly. The biomass-fueled power plants will generally be 50 MW or less in capacity unless the firm has a captive supply of fuel such as a paper pulp mill. An example of a company that concentrates on small power plants is Ultrapower Inc.[128] This company has built three 11-MW plants in California to convert wood waste to steam and electricity in a traveling-grate, spreader-stoker, watertube boiler capable of producing 105,836 lb/hr of steam at 900 psig and 825°F. They are in the initial stages of constructing their first 25-MW waste wood-fired plant in California. This plant will utilize a fixed fluidized bed watertube design capable of producing 208,640 lb/hr of steam at 950°F and 1,250 psig. Two 26-MW projects are in the advanced stages of development in Maine.

Municipal Solid Wastes. Disposal of MSW and simultaneous electric power production appears to have been resurrected last year after a few years of dormancy, that was probably caused by a few initial failures with some of the RDF plants.[129-132] The Europeans have been generating power by MSW combustion for at least 20 years,[132] while the United States is currently burning about 5% of its MSW in resource-recovery plants.[129] Today, direct MSW combustion is preferred to RDF preparation and combustion because the mass burn technique is simpler, and the development of rotary grates that allow movement of MSW in the combustion unit ensures complete burning.[132] Market projections indicate there is a need for about 100 large plants that process 1,500 ton/d or more of MSW, and 300-400 smaller plants.[130] The capital costs average about $250 million for the large plants and $30 million for the smaller plants, so the total capital requirement is about $35 billion. Construction of the largest MSW-to-energy plant in the United States, a 65-MW, 4,000-ton/d plant in Detroit, MI, has just been started.[133]

Although there is a considerable difference in the survey results of the number of MSW-to-energy plants in operation in the United States,[134,135] one of the most recent reports indicates there are 60 plants in operation or shakedown, 18 under construction, and 31 that are in the advanced planning stages as shown in Table 21. In addition, there are five operating mass-burn plants in Canada, and five plants (or eight, see Footnote d, Table 21) that are shut down in the United States. It is apparent that there are more mass burn plants than RDF plants in the United States. According to another recent survey that concentrated on RDF plants, of the 23 principal RDF plants built in the United States, ten are closed down, six are operating again after being closed down for modification, and the remainder continue to operate.[136] Interestingly, the largest operating plant in the United States (3,000 ton/d, Dade County, FL) and the largest plant under

Table 21. MSW-TO-ENERGY PLANTS IN THE UNITED STATES AND CANADA[a]

	Operating or Shakedown	Under Construction	Advanced Planning	Shut Down
0-300 ton/d				
Mass Burn	38	7	6	1
RDF	3	--	--	--
Other	1	--	--	1
301-800 ton/d				
Mass Burn	3	5	8[c]	--
RDF	3	--	1	--
Other	1	--	--	--
801 ton/d and Above				
Mass Burn	8	4	9	2
RDF	8	2	7	1
Other	--	--	--	--
	65[b]	18	31	5[d]
		114		

[a]Adapted from Ref. 134; All systems listed are surface units; Landfill systems are not included.

[b]Includes five operating Canadian mass burn plants of 108, 220, 500, 1000, and 1200 ton/d capacities.

[c]Re-count of data in Ref. 134 indicates this figure is 8; Ref. 134 indicates 6.

[d]Ref. 134 indicates 8 facilities are shut down; According to the data presented in Ref. 134, it appears to be 5.

construction (4,000 ton/d, Detroit, MI) are both RDF plants. Of 84 MSW-to-energy plants built, operating, under construction, or recently financed in the United States, 22 are mass burning waterwall systems, 27 are starved air (modular) systems, 13 are refractory systems, and 22 are RDF-dedicated boiler systems.[135] The energy products from these same plants are RDF for market, 15; steam, 42; electric power, 11; and cogeneration, 16.[137]

Recent combustion tests with pure cellulose saturated with sodium chloride and with polyvinyl chloride (PVC), which occurs in MSW, have provided experimental data that show no detectable chlorinated organic compounds are formed with cellulose, but concentrations of about 1 ppm of dioxins are detected in the combustion products of PVC at 800°C.[138] The inference from this work might be that biomass combustion should not pose a hazard as far as dioxin toxicity, but MSW combustion may. A comprehensive examination of what is known about dioxins in the environment ended with no consensus on the subject.[138] It has been pointed out before that the hazards of dioxins from MSW and RDF combustion appear to be minimal.[1]

Wood. The world's largest 50-MW wood-fueled generating station in Burlington, VT has been on-line now for over one year.[139-141] The final project cost as of January 1, 1985 has been established at $64,681,779, which represents an underrun of about 19% of the budgeted $80 million.[139] Initially, there was concern about whether sufficient wood could be acquired for sustained operation, but fuel supply turned out to present no problem even when the burn rate was over 2,000 green ton/d. Fuel prices have been slightly less than those projected in 1981. Prices in March 1985 were $19.75/green ton of wood chips delivered to the station by truck, while the total delivered cost by rail including rehandling costs was $22.89/green ton. The requirement to utilize 75% of rail-delivered fuel and the unloading and handling costs plus equipment maintenance price the fuel into the boiler at $24.85/green ton, or 39.64 mills/net kWh. The Btu cost for the fuel is estimated to be $3.11/MBtu.* The plant had start-up problems that were primarily non-technical, such as dust, noise, "vibrations" from the railroad car unloading operations, and an odor emanating from the wood chip piles. Burlington Electric has taken action to eliminate these problems.[141] For those utilities that are planning to build similar facilities, this project is well worth detailed study; it is a valuable source of knowledge that can be used to improve future installations.

Other interesting commercialization projects reported on during the year concerned an atmospheric fluidized bed combustor for green wood to provide nearly 100 MBtu/h of clean process air at 1000°F for a paper dryer;[142] an

---

*Assuming 50% moisture content and 16 million Btu/dry ton; $24.85/(0.5 x 16 x 10^6).

automated 1.5-MBtu/h ag-residue fueled furnace for heating buildings and drying grains;[143] and an innovative combustion unit for high-ash or high-silica biomass residues such as rice, peanut, soybean, pecan, and cotton hulls to recover energy and a high-quality ash.[144]

An approach to potentially competitive-cost power production in small-scale plants is undergoing field testing in Red Boiling Springs, TN. This 3-MW, wood-fueled system combusts pulverized green wood, and after the particulates are removed, the combustion gases are fed to a gas turbine.[145,146] Consideration is being given to water and steam injection to increase power outputs. Water injection lowers the efficiency, but the increase in power output may offset this loss. However, injection of low-pressure steam produced from exhaust gas heat affords both higher power outputs and efficiencies.[146] Preliminary cost estimates of single-turbine modules with and without steam injection indicate substantial technical and economic advantages of a wood-fired, steam-injected gas turbine plant over the corresponding case (same hardware before steam production) without steam injection (Table 22).[147] It is apparent from this analysis that power cost is highly sensitive to wood fuel costs.

Thermochemical Gasification

Research. In North America, research on small-scale gasifiers has been somewhat limited over the last few years; the emphasis is on larger scale systems, particularly for production of medium-Btu gas. Exemplary U.S. research projects include IGT's single-stage, pressurized, fluidized bed process; SERI's pressurized, fixed-bed, downdraft gasifier; the fire tube-heated fluidized bed system of the University of Missouri-Rolla (UM); and Battelle Columbus Laboratories (BCL) indirectly heated, dual, fluidized bed gasification process.[148] The IGT and SERI gasifiers use oxygen to generate medium-Btu gas, while the UM and BCL processes use air and indirect heating to generate medium-Btu gas. Each of these process developments has progressed to the process research unit stage (IGT, 12-in. gasifier, 900 lb/h; SERI, 5.9-in. gasifier, 83 lb/h; UM, 20-in. gasifier, 450 lb/h; BCL, 6-in. and 10-in. gasifiers, 3000 lb/h).[148] Typical data from these gasifiers are presented in Table 23. The next step is a competitive scale-up solicitation by DOE to design, construct, and test a 50- to 200-dry ton/d system capable of producing medium-Btu gas of 300 Btu/SCF exclusive of sensible heat.[149] DOE anticipates the facility will cost about $10 million and will be cost-shared by the private sector at a level of 50% or more.

Although limited research has been carried out on small-scale, low-Btu producer gasifiers for biomass in recent years, significant design advancements have been reported even though they have been used for over 100 years. One of the most interesting developments is the open-top stratified downdraft gasifier in which air is drawn in through successive reaction strata.[153] The unit is

Table 22. COMPARATIVE COSTS OF ONE MODULE OF WOOD-FUELED GAS TURBINE PLANT WITH AND WITHOUT STEAM INJECTION[a] (Mid 1985 Dollars)

	One Gas Turbine		One Gas Turbine + Steam	
	Public[d]	Private[e]	Public	Private
Capital Cost, $$10^3$				
Before Steam Production	1244.9	1244.9	1244.9	1244.9
Waste Heat Boiler/Superheater	--	--	244.0	244.0
Steam Condenser	--	--	173.0	173.0
	1244.9	1244.9	1661.9	1661.9
Other Plant Charges[b]	632.0	632.0	811.0	811.0
	1876.9	1876.9	2472.9	2472.9
Other Capital Costs[c]	282.7	162.5	355.8	205.4
Total Capital Cost	2159.6	2039.4	2828.7	2678.3
Annual Operating Cost, $$10^3$	284.2	234.2	306.8	306.8
Electricity Cost, mills/kWh				
Wood at $1.00/MBtu ($17/dry ton)	33	39	22	27
Wood at $2.00/MBtu ($34/dry ton)	47	52	31	35
Capacity, kW	3519	3519	5780	5780
Overall Thermal Efficiency, %	24.8	24.8	32.8	32.8

[a]Adapted from Tables 8-12 of Ref. 147.

[b]Includes costs for general facility, engineering and fee, contingency fees, taxes, and insurance.

[c]Includes fixed and variable operating and maintenace costs.

[d]Designates ownership by public utility.

[e]Designates ownership by private industry.

Table 23. TYPICAL DATA FROM MEDIUM-Btu GASIFIERS

Source	IGT[a]	SERI[b]	UM[e]	BCL[g]
Run No.	GT-10	19-D	B	3.23
Feedstock	Maple Wood Chips	Wood Chips	Oak Flakes	Sawdust
Pressure, psi	105	162	--	3.41
Temperature, °C	796	--	649	699
Feed Rate, lb/h (wet)	399.5	25.15[c]	210	340
Moisture, wt %	12.1	--	7.0	40.0
Steam, lb/lb Feed (wet)	0.62	--	1.93	1.26
Oxygen, lb/lb Feed (wet)	0.23	1.91	--	--
Dry, Inert-Free Gas Yield, SCF/lb Feed (wet)	44.85	--	0.69[f]	--
Dry Inert-Free Gas Composition, vol %				
H_2	7.52	15.79	19.12	14.16
CO	7.24	24.08	27.97	51.75
CO_2	11.89	22.37	22.69	12.02
CH_4	4.87	7.16	20.19	16.26
C_2H_4	0.45	--	3.74	4.74
C_2H_6	0.37	--	1.14	1.06
C_3H_8	0.01	--	3.15	--
C_6H_6	0.60	--	--	--
H_2/CO Ratio	1.04	0.66	0.68	0.27
Carbon Conversion to Gas, %	95.4	95.0[d]	--	47.95
Product Gas HHV, Btu/SCF	405	228	540	448

[a] Adapted from Table 1 of Ref. 150.

[b] Adapted from Table 3 of Ref. 151 (February 1985); Text states this run is oxygen-blown, but data in Table 3 indicate product gas contains 30.60% nitrogen; Ref. 152 indicates gasifier was in process of being converted to operate on oxygen (October 1985), but conversion has been delayed so Run 19-D is an air-blown run.

[c] Dry.

[d] Labeled Gas Efficiency in Table 3 of Ref. 151.

[e] Adapted from Table 2 in Ref. 151.

[f] Volumetric yield not specified; This figure is the ratio of the dry gas produced (kg/h) to the wood feed rate (kg/h).

[g] Adapted from [illegible]

simple to operate, inexpensive, and can be close-coupled to an engine-generator set without complex gas-cleaning equipment.[154] The gasifier dimensions are sized to deliver gas to the engine based on its fuel-rate requirements, and no controls are needed.[154] In the developing countries, a wide range of producer gasifiers, some of which are open-top units, have been reported to be used for water-pumping, electric power production, vehicular power, and industrial process heat.[155] Fuels for these units include wood chars and woods, rice husks, coconut shells and husks, corn cobs, cotton stalks, peat, and densified biomass. There appear to be no major problems with the use of producer gas in engines other than those caused by ineffective gas-cleaning equipment.[155] However, it appears that many of the systems in use today are labor intensive; low-cost automation is needed. A U.S. work-shop on low-Btu gasification near the end of 1984 recommended that additional research is needed to determine gas cleanup requirements; char and ash effects on engines; the performance of feeding equipment including storage, handling, drying, screening, and feeding; performance of existing cleaning devices; and emissions to the environment.[156] It would appear that much of this information has already been generated in other countries because of the large-scale use of low-Btu gasifiers.

One of the interesting low-Btu gasification projects in progress in the United States is the testing of a down-draft, wood gasifier (130 lb/hr wood consumption rate, sized for 100-kW unit) coupled to a 50-kW internal combustion-engine generator set in North Carolina.[157] The unit is designed for on-farm use and has been operated at full load for over 200 h, which also included unattended operating periods. Approximately 17% of the wood energy is converted to electricity, and 44% is converted to recoverable heat (20% from the exhaust, 18% from the engine radiator, and 6% from the gasifier). Cost estimates indicate the payback on the initial cash investment ($7,195) of a 100-kW system is less than one year if it constructed primarily by the farmer and 40% of the recoverable heat is used.

Considerable basic research has continued on biomass gasification. Noteworthy reports in 1985 include studies on the kinetic parameters for the pyrolytic gasification of cellulose by indirect radiant heating,[158] the development of a thermodynamic equilibrium model of a gasifier to analyze the effects of changes in operating conditions on gasifier performance and product quality,[159] an innovative method of catalytically gasifying high-moisture biomass to a methane-rich gas (exceeding 40% methane and 6 SCF/dry MAF lb) at high pressures but relatively low temperatures (400-450°C) in 95% water slurries,[160] the activation and deactivation of commercial catalysts for biomass gasification in bench-scale fixed-bed and fluid-bed reactors,[161] the detailed analysis of gasification condensates,[162] and the kinetic rate constants for the formation of each gas evolved during the thermal decomposition of cellulose.[163]

Commercialization. The commercialization of thermochemical gasification processes for biomass and wastes in North America slowed in 1985, primarily because of three factors — the reduction in petroleum and natural gas prices, the apparent perception that direct combustion of wood and solid wastes is cheaper, and the fact that few successful gasification plants are perceived to be based on established off-the-shelf technologies. Nevertheless, some projects that have recently been completed or started were reported in 1985. One of these projects was announced near the beginning of 1985 and is based on the pressurized, fluid-bed process developed in Canada by Omnifuel Gasification Systems Ltd. of Toronto and adapted by Biosyn, a subsidiary of Nouveler Inc. of Montreal.[164] The $9 million contract is for a 6.7-MW installation in French Guiana for the conversion of sawmill waste and tropical wood and is expected to save $3 million per year in heavy oil costs. The French firm, Alsthom-Atlantique has developed and will supply a diesel engine-generator set compatible with gas produced from biomass for the project, and will also market this type of power plant jointly with the Canadians. Another project resulted in the start-up of one of the largest wood gasification plants in the United States.[165] Alternate Gas Inc. (AGI) and Energy Products of Idaho (EPI) designed and constructed a 95-MBtu/h gasification plant for wood obtained from forest products operations near the plant (Lowe's Southern Clay, Inc.) in southeastern Missouri (Bloomfield). The green wood is dried to 25% moisture and converted to 150-Btu/SCF gas which replaces fuel oil in a rotary kiln for mineral drying (kitty litter). According to AGI, standard designs and automation are used; EPI is reported to have installed 50 fluid-bed gasifiers for various feedstocks over the last 12 years. The plant operates 3 shifts/d, 5 d/wk, and uses one person per shift. AGI's unique marketing approach[1] employs long-term contracts to sell the low-Btu gas to the user at a specified rate about 10-20% below prevailing fuel costs,[165] but it is not known whether this plan applies to this plant since it is owned by another firm, Time Energy Systems, Inc.[165] Similar projects have been announced for the Florida Panhandle and Newburgh, NY.[165]

As already mentioned, there are many small-scale producer gasifiers for biomass in operation in several developing countries.[155] One of these countries, the Republic of the Philippines, is in the midst of installing a nationwide power production system consisting of 70 small plants (200 MW total capacity) fueled with the ipil-ipil tree (Leucaena spp.).[166,167] Most of the power plants in operation are wood combustion systems, but a few are reported to be 1-MW gasifier-engine systems consisting of dual downdraft gasifiers and dual fuel diesel engines of French design. Presumably, this program will make it possible for performance comparisons to be made from on-site field operations between direct combustion for boilers and gasification for motor-generator sets.

As for the perception alluded to above that direct combustion is cheaper than gasification, the benefits of gasification are sometimes not obvious. In a study reported last year, the estimated economics of a hypothetical 800-ton/d pulp mill whose energy needs are met by the BCL dual-reactor fluid-bed gasifier supplied with 570 ton/d of wood and a package gas boiler, were compared with those of a new wood-fired combination boiler.[168,169] For the same steam output, installed equipment cost for the gasification plant is about 56% of that of the boiler system ($15.2 million vs. $27.0 million). Fuel for a lime kiln can also be supplied by the gasifier at about 63% of the capital investment needed for a new boiler ($17.0 million vs. $27.00 million), and there is an annual savings of $3.7 million in purchased natural gas or oil. Similar results were obtained when a 2,000-ton/d MSW mass-burn power plant was compared with a gasifier-turbine plant fueled with the same amount of MSW. The capital cost of the mass-burn plant was 225% more than that of the gasifier-turbine plant ($254 million vs. $112.5 million); the power production of the mass-burn plant was only 52% as much (473×10^6 kWh vs. 903.7×10^6 kWh); and the power production cost at the same MSW tipping fee ($15/ton) was about four times higher than that of the gasifier-turbine plant ($0.084/kWh vs. $0.023/kWh). Thus, in these applications, the economic advantages of gasification are clear.

Thermochemical Liquefaction

Research. Thermochemical liquefaction of biomass and wastes can be accomplished by direct pyrolytic, low-temperature chemical, or severe hydrogenation methods, and by indirect methods via intermediates such as synthesis gas and methanol. Indirect methods can yield products with high selectivity that are identical to conventional hydrocarbon fuels.

Direct biomass pyrolysis produces pyrolytic oil, char, gas, and water products. In recent work with entrained flow, 6- and 8-in. pyrolysis reactors and hardwood particles, product oil yields on a dry ash-free feed basis ranged up to 51% with both reactors.[170] An economic analysis of this system for plants of 200- and 1,000-ton/d of feedstock was conducted for production of fuel oil substitutes at slightly lower yields as shown in Table 24. It was concluded from this analysis that entrained flow pyrolysis technology seems to be cost competitive with "current fuel oil selling prices" (last quarter of 1985) at a wood cost of $25/dry ton ($1.5/MBtu). Product oil production costs were highly sensitive to wood feedstock costs and product oil yield and quality, but less sensitive to capital costs and by-product char credits.

Since the product oils from direct thermal pyrolysis processes still contain substantial quantities of oxygen, hydropyrolysis offers one approach to making products that are more hydrocarbon-like in composition. The results of one such investigation reported in 1985 are summarized in

Table 24. ECONOMIC ANALYSIS OF ENTRAINED FLOW PYROLYSIS SYSTEM[a]

DESIGN BASIS[b]

Feedstock: 50 x 60 Mesh Mixed Hardwood, 6.26 wt % Moisture.

Pyrolysis: Atm. Pressure, 471°C, 59 lb/h Feed Rate, 223 lb/h Gas Rate.

Products: Oil, 0.46; Char, 6.6; Gas, 2.33 (as dry wt/dry Feedstock wt)

HHV's: Oil, 9693; Char, 11073; Gas, 479 (as Btu/lb MAF)

PLANT ECONOMICS[c]		
Wood Feed, ton/d	200	1000
Pyrolytic Oil, ton/d	92	460
Char, ton/d	13.2	65.8
Total Capital Requirement, $	6.02×10^6	16.63×10^6
Annual O&M Costs, $	2.65×10^6	10.16×10^6
TOTAL ANNUAL PRODUCTION COST		
Utility Financing, $	3.53×10^6	12.71×10^6
Private Financing, $	4.05×10^6	13.89×10^6
PYROLYTIC OIL PRODUCTION COST		
Utility Financing, $/MBtu	6.00	4.33
Private Financing, $/MBtu	6.75	4.74

[a]Adapted from Tables 1, 4, 5, and 6 of Ref 147.

[b]Obtained from 6-in. PDU operating at a temperature range of 400 to 550°C at a nominal feed rate of 50 to 100 lb/h of hardwood particles (Ref. 170).

[c]All dollar figures in mid-1985 dollars; Wood costs assumed to be $25/dry ton.

Table 25. The product oils (methylenedichloride solubles) obtained by catalytic hydropyrolysis of water slurries of biomass at elevated temperature and pressure were still quite high in oxygen, even for the lignin feedstocks. The bulk of the product oils was shown to be weakly acidic (carboxylic acids and phenols) to neutral in character from extraction experiments. These products are far removed from non-oxygenated structures, and it appears that the particular catalytic conversion procedure used had minimal effect on oxygen content of the product oil. In another study using iron powder catalyst and hydrogen-helium mixtures over poplar wood slurries at 340°C and gas pressures of 1 to 39.5 atm, product oil yields were not favored by hydrogen atmospheres.[172] Product oil yields of 45% (acetone solubles) were obtained, however, when helium was used without hydrogen. Another investigation reported that with water alone at pressures that maintained liquid-phase conditions, different poplar wood components depolymerized at different temperatures.[173] Hemicelluloses were solubilized at 200°C, cellulose was converted to oligomeric and monomeric carbohydrates and further degradation products such as hydroxymethylfurfural and furfural at about 280°C, and lignin was degraded at temperatures above 300°C. Overall, it appears that a successive increase in temperature can improve product selectivity because of the differences in conversion rates of individual biomass components. This approach has been suggested as part of a sequential conversion scheme in which proper choice of temperature, solvent, and "mechano-chemical" effects result in fractionation of lignocellulosics.[174] Presumably, the step-wise formation of solubles would improve product selectivities also if the solubles were converted separately.

It should be pointed out that the catalytic conversion (nickel/alkali carbonate catalyst mixture) of biomass-water slurries at 400-450°C and autogenic pressures of 4,000-5,200 psig mentioned in the Gasification Research section of this paper is reported to give high methane yields,[160] so selectivity was improved by use of slurries, but not for liquids. Other work done with slurries of cellulose in aqueous alkali has shown that at 300-350°C, the major components are cyclopentanones and hydroxybenzenes; without alkali, the major components are furfurals.[175] Also, it has been reported that cellulose in aqueous hydriodic acid yields hydrocarbon-like liquids at relatively low temperatures (<130°C).[175] More information was published in 1985 on this rather unique approach to liquefying biomass.[176] Products corresponding to 50% deoxygenation in 1 min, 75% in 30 min, and 92% in 24 hr are obtained, and char is not formed. Since about 90% of the HI reacting is converted to molecular iodine, this may support the intermediate formation of alkyliodides which in turn are known to react with HI to form hydrocarbons and iodine.[1] If this in fact is the mechanism, a cyclic process can be conceptualized in which HI is regenerated.

Table 25. PRODUCTS FROM HYDROPYROLYSIS OF BIOMASS AND BIOMASS COMPONENTS[a]

Feedstock			Product Oil					
Type	Heating Value, Btu/lb	% Oxygen	Yield, %	Heating Value, Btu/lb	% Oxygen	Acetone Solubles Yield, %	Water Solubles Yield, %	Solids Yield, %
Spruce Wood	7,740	43.9	46.0	13,240	20.2	4.5	14.7	1.4
Birch Wood	7,270	45.9	41.0	12,980	21.4	1.5	17.5	0.9
Sugarcane Bagasse	6,490	48.7	41.5	13,460	20.6	4.0	19.4	4.2
Barley Straw	6,840	47.4	40.3	13,370	20.3	2.7	20.1	0.7
Pine Bark	8,600	38.6	20.7	13,110	21.8	17.7	7.4	29.9
Cellulose	6,490	49.9	29.3	13,890	17.4	3.7	13.1	1.5
Spruce Holocellulose	6,960	46.9	31.0	12,250	25.3	10.5	16.2	0.5
Birch Holocellulose	6,580	48.6	31.0	13,980	18.9	1.0	20.3	0.8
Spruce Organosolv Lignin	11,270	27.4	64.0	12,040	24.2	29.3	3.0	6.8
Bagasse Organosolv Lignin	10,700	29.5	60.7	13,310	27.7	31.0	5.1	3.1
Birch Wilstetter Lignin	10,020	32.9	33.0	11,390	27.7	50.7	5.8	4.1

[a]Adapted from Tables 1 and 3 of Ref. 171; All samples ground to particles from 0.1-0.5 mm, mixed with water, treated with palladium catalyst at 1% Pd based on oven-dried wood, placed in 25-ml stainless steel autoclave at an initial hydrogen pressure of 59.2 atm (60 bar), heated to 375°C in 15 min, immediately quenched under running water, and extracted with methylene dichloride (oil), acetone, and water; The indicated yields are percentages of dry feedstock.

Thus:

$$ROH + HI \longrightarrow RI + H_2O$$

$$RI + HI \longrightarrow RH + I_2$$

$$(I_2 + H_2S \longrightarrow 2HI + S)$$

$$(I_2 + H_2 \longrightarrow 2HI)$$

Development of this kind of process for biomass liquids would be much more efficient from a thermal and chemical energy conservation standpoint than high-temperature pyrolysis. The selectivity should also be much higher than a conventional pyrolysis process, and the process should yield liquids that are neutral and distillable.

The obvious approach to improvement of the fuel characteristics of the product oils from direct thermal treatment of biomass is hydrogenation, which results in deoxygenation. Catalytic hydrotreatment of wood-derived oils is being studied using a continuous-feed, fixed-bed reactor.[177] In tests on the product oil made by the PERC Process at the Albany, OR pilot plant (reaction of wood slurried in recycle oil with H_2/CO), a gasoline-like product was obtained on catalytic hydrotreatment over sulfided cobalt/molybdenum catalyst at 340 400°C and 2,000 psig of hydrogen. However, the product oils made in the entrained flow pyrolysis reactor referred to above (mainly acetic acid, small aldehydes, phenolic ethers, aldehydes, and ketones) could not be similarly treated; the partially hydrotreated pyrolyzate had properties more like the Albany high-pressure liquefaction oils. Further research is necessary to develop upgrading procedures for the product oil made by direct pyrolysis.

Direct thermal conversion of biomass at high temperatures and heating rates, short residence times, and rapid quenching tends to yield higher energy non-equilibrium products such as olefins. This is the reason research is continuing to develop advanced flash or ultrapyrolysis methods for direct conversion of biomass. Studies continued in 1985 on the flash pyrolysis of biomass in atmospheres of methane to give high yields of ethylene and BTX (Brookhaven National Laboratory),[178] real-time analysis of fast pyrolysis reactions (SERI),[179] use of an ablative fast pyrolysis system to determine the kinetics of primary product oil cracking to secondary gases and tars (SERI),[180] use of an "ultrapyrolysis" system in a turbulent vertical contactor ("Thermovortactor") to develop pyrolysis kinetics at temperatures between 700-900°C and residence times of 40-700 ms (National Research Council of Canada and U. of Western Ontario, Canada),[181] and flash pyrolysis of wood-particles in a 6.6-lb/h fluid-bed reactor at 500-700°C to obtain liquid yields of 60-65 wt % of the feedstock (U. of Waterloo, Canada).[182] Studies of biomass pyrolysis at reduced pressure to obtain high liquid yields at lower temperatures are also continuing (U. of Sherbrooke, Canada).[182]

Much less research is in progress on the indirect liquefaction of biomass via intermediates even though the end products can be identical to gasoline and diesel fuel components. One of the notable exceptions is the study of biomass gasification to synthesis gas and its conversion to paraffins by Fischer-Tropsch chemistry.[183] Approximately 100 biomass materials comprising industrial wastes, agricultural and forest residues, and energy crops have now been evaluated. Liquid hydrocarbons suitable for use as diesel fuel have been produced; they can be upgraded to high octane gasoline by catalytic reforming if desired. A larger scale system (10 ton/d of feedstock) is now being designed. It has been reported that a plant producing 300 ton/d of liquid products with biomass delivered at $20/dry ton can offer break-even costs.[183]

Commercialization. Up to the 1930's, biomass pyrolysis processes were utilized on a large scale to produce a wide range of chemicals from the pyrolytic oils. These processes have since been displaced by non-pyrolytic processes based on petroleum and natural gas. A few attempts were made several years ago to commercialize biomass pyrolysis processes, including flash pyrolysis, for the manufacture of liquid fuels (See Energy from Biomass and Waste Series). The main objective was to produce pyrolytic oils as substitutes for residual fuel oils. In the developed countries, most of these efforts were terminated in the late 1970's and early 1980's because of operating problems, the characteristics of the non-upgraded pyrolytic oils, and unfavorable economics.

In 1985, the results of two interesting comparative economic studies were reported in which several different biomass liquefaction processes were analyzed.[184,185] Table 26 shows the results of the analysis of various pyrolytic and non-pyrolytic processes for production of pyrolytic oil products and methanol from wood and peat, and of gasoline from wood. The price range for the liquids from a wood-fueled plant (2,203 wet ton/d feed rate) is far in excess of today's oil prices; the most costly liquid fuel is gasoline made from methanol intermediate by the Mobil Process ($128.80/BOE). Table 27 shows the results of a similar analysis. Again, the cost range on a barrel-of-oil-equivalent basis, although somewhat lower than the range in Table 26, is still quite high compared to today's oil prices.

If these cost figures have any validity at all, it is apparent that direct and indirect biomass liquefaction processes must be considerably improved before commercialization can be justified strictly on the basis of economics. The sensitivity of product cost to variations in the economic parameters (capital, discount rates, feedstock, etc.) would have to be examined in great detail to focus on those items that can have the greatest impact on fuel cost. It is interesting to note, however, that the pyrolytic oil costs in Tables 26 and 27 afford the lowest energy costs of the alternatives analyzed. Direct

Table 26. COMPARATIVE ECONOMIC ANALYSIS OF BIOMASS LIQUEFACTION PROCESSES[a]

	PERC[c]	Flash Pyrolysis[d]	Methanol[e]	H-Peat[f]	Flash Pyrolysis[g]	Methanol[h]	PERC[i]	Flash Pyrolysis[j]	MTC[k]
Feedstock	Wood	Wood	Wood	Peat	Peat	Peat	Oil	Oil	Methanol
Feed Rate, ton/h	91.8	91.8	91.8	91.8	91.8	91.8	21.2	32.6	26.1
Product[b]	Oil	Oil	Methanol	Oil	Oil	Methanol	Gasoline	Gasoline	Gasoline
ton/h	21.2	32.6	26.1	12.3	15.2	27.2	--	--	--
By-Products	--	--	--	--	Char	--	--	--	--
ton/h	--	--	--	--	10.4	--	--	--	--
Overall Thermal Efficiency, %	72	60	51	60	43	49	50	46	42
Total Capital Cost, $\$10^6$	120	60	170	220	60	175	--	--	--
Total Operating Cost, $\$10^6$/yr	35.3	30.4	44.4	28.5	18.2	35.4	--	--	--
Product Cost, $/MBtu	10.50	10.00	17.20	14.30	8.50	14.20	18.50	20.30	22.20
$/BOE	60.90	58.00	99.80	82.90	49.30	82.40	107.30	117.70	128.80

[a]Adpated from Tables 1 and 2 of Ref. 184; Wood cost, $33/wet ton; Milled peat cost, $17.60/wet ton; Cost figures are January 1984 dollars.

[b]"Oil" denotes primary product oil produced in process.

[c]High-pressure hydrogenation in recycle oil as originally developed by Pittsburgh Energy Technology Center.

[d]Process under development at U. of Waterloo, Canada.

[e]Pressurized fluid-bed gasification of wood and methanol synthesis.

[f]High-pressure hydrogenation in recycle oil as developed by VTT, Finland.

[g]Similar to U. of Waterloo process, but peat is feed.

[h]Similar to process in footnote e, but peat is feed.

[i]Catalytic hydrogenation of primary product oil from wood-based PERC Process; No additional capital or O&M costs stated.

[j]Catalytic hydrogenation of primary product oil from wood-based flash pyrolysis; No additional capital or O&M costs stated.

[k]Catalytic conversion of methanol from wood-based process; No additional capital or O&M costs stated.

Table 27. COMPARATIVE ECONOMIC ANALYSIS OF WOOD LIQUEFACTION PROCESSES[a]

	PERC[d]	Methanol[e]	MTG[f]	Fischer-Tropsch[g]
Feed Rate, dry ton/d	1,103	1,103	1,103	1,103
10^6 Btu/h	731	731	731	731
Product, 10^6 Btu/h				
Product Oil	411	--	--	--
Methanol	--	415	--	--
Gasoline	--	--	327	180
Diesel	--	--	--	115
LPG	--	--	57.3	18.6
Chemicals	--	--	--	26.1
Efficiency,[b] %	56.2	56.8	52.6	46.5
Total Capital Cost, $\$10^6$	120	125	150	150
Total Product Cost,[e] $/MBtu	10.60	10.90	13.50	15.30
$/BOE	61.50	63.20	78.30	88.70

[a]Adapted from Tables 1, 2, and 3 of Ref. 185; Wood cost, $22.68/dry ton; Cost figures are 1984 dollars.

[b]Product HHV's divided by feedstock HHV.

[c]Includes capital charge, O&M, feedstock, and labor costs.

[d]High-pressure hydrogenation in recycle oil as developed by Pittsburgh Energy Technology Center; In Ref. 185, product oil is listed as diesel fuel substitute.

[e]Wood gasification by Winkler or Westinghouse-type fluid-bed gasifier followed by methanol synthesis by a Lurgi or ICI low-pressure process.

[f]Methanol production followed by Mobil process.

[g]Gasification followed by F-T synthesis in a liquid-phase reactor, and then upgrading to transport fuel.

conversion of biomass to liquids would be expected to have the lowest cost because of process simplicity, but additional costs would be incurred if the primary product oil were upgraded.

It should also be pointed out that it is usually not valid to compare the results of different economic analyses unless the same economic ground rules were used. Therefore, the cost figures in Tables 24, 26, and 27 for similar products should not be considered to be capable of real comparison. But within a series of economic analyses, such as shown in Table 26, comparison should be possible because the economic ground rules are internally consistent.

BIOCONVERSION

The major bioconversion methods for biomass are methane fermentation or anaerobic digestion for methane production, biological liquefaction for liquid fuels production such as alcoholic fermentation for ethanol, and the natural processes that occur in microalgae during the formation of lipids. The latter subject has already been discussed in the Aquatics Section.

Anaerobic Digestion

Research. Anaerobic digestion has by far received the most attention as a potential source of gaseous fuels. Research has continued worldwide to develop improved processes for combined waste disposal-methane production systems and for methane production from biomass. Last year was no exception; in fact, the R&D effort seems to be expanding.[69,186] Much of the current work is aimed at improvement of the digestion process to reduce methane costs through increases in methane yield and production rates. Several projections of the sensitivity of methane costs to various parameters have been made. One projection indicates that increasing operating and kinetic rates by a factor of 3 to 4, increasing feedstock conversion from 50% to 90%, recycling cellular biomass up to 50%, enhancing the culture, and reducing material costs, provide potential cost reductions for the product gas in \$/MBtu of 3.00-5.00, 1.00-2.50, 1.00-2.00, 1.50-5.50, and 1.00-3.00, respectively.[101] Each of these parameters is not independent. For example, pretreatment of the feedstock or post-treatment of digested solids and recycling might improve biodegradability and result in better gas production rates (kinetics), better gas yields (conversion), and smaller digestion equipment (material costs). The additional cost of pre-treating is a trade-off against the value of these benefits.

In another projection made with a specific feedstock, napiergrass, for a specific site, substantial reductions in gas costs were estimated from a base-case grass yield of 26 ton/ac-yr (60 t/ha-yr) and a methane yield of 4.97 ft^3/lb VSA (0.31 m^3/kg VSA).[187] It was estimated that a 35%

reduction in end product gas costs might be achieved if the methane yield could be increased to 7.06 ft^3/lb VSA (0.44 m^3/kg VSA) by improving both plant composition and the conversion process, reducing chemical inputs by 70%, and reducing planting costs by 90% via tissue culture propagation (Table 28). A 30% reduction in end product gas cost was estimated if the methane yield of 7.06 ft^3/lb VSA could be obtained with concurrent SRT reductions from 60 to 30 days, a reduction of 15% in materials cost, and an increase in the methane content of the product gas to over 90%. An advanced two-phase process design was used for this analysis. It was based on a leaching-bed, packed-bed, first-phase, batch reactor supplied with a 20-35 wt% feed slurry to produce acids at 20-60-d SRT's, and a small packed-bed continuous methanogenic reactor.

These types of analysis are very important because they indicate those areas that should be targeted as primary objectives in a mission-oriented research program to obtain the greatest economic benefits.

A different approach to development of economic bioconversion processes for energy was reported last year in which state-of-the-art fermentation of whole crops was evaluated for both ethanol and methane production.[188] In this case, the starchy tubers or roots were converted to ethanol and the residuals were digested (Table 29). The maximum total energy yield was obtained with sugar beet (95.5 MBtu/ac). It is apparent from the split in ethanol and methane yields obtained with sugar beet, Jerusalem artichoke, and potato that the energy yield per unit area for methane can be higher or lower than the corresponding ethanol energy yield, but the economic yield per unit area is significantly higher for ethanol because of its higher energy value. For sugar beet, with ethanol at $1.25/gal and methane at $3.00/MBtu, the respective energy values are $1135/ac and $80/ac. It is necessary to point out, however, that this is state-of-the-art technology. Crops that are especially amenable to methane fermentation such as napier-grass and water hyacinth have been grown at dry yields of about 25 ton/ac-yr (57 t/ha-yr) and 33 ton/ac-yr (75 t/ha-yr), and methane yields of some of the advanced processes can range up to 7 ft^3/lb VSA (0.44 m^3/kg VSA).[187,189] This is equivalent to energy yields of about 300 to 400 MBtu/ac. Nevertheless, the concept of using whole crops for multiple energy purposes (and also non-energy applications) adds flexibility to the process.

Basic and applied research on anaerobic digestion continues to expand our knowledge of the biochemical pathways involved in the anaerobic conversion of organic substrates,[190] the autohydrolysis and pretreatment of organic materials to increase biodegradability (Stanford U. and SERI),[101] the application of genetic techniques to improve methanogenesis (Jet Propulsion Laboratory and others),[101,191] the physiology of acetogenic bacteria (U. of California),[101] the kinetics and rate limiting steps in anaerobic digestion (U. of Missouri, U. of Arkansas),[101]

Table 28. ANTICIPATED COST REDUCTIONS THROUGH R&D[a]

Base Case: Napiergrass Yield, 26 ton/ac-yr (60 t/ha-yr)

Methane Yield, 4.97 ft^3/lb VSA (0.32 m^3/kg VSA)

Production/Harvesting Cost, \$2.43/MBtu (\$2.30/GJ)

Conversion/Gas Cleanup Cost, \$4.75/MBtu (\$4.50/GJ)

	Cost Reduction	
	% of Base	From Base, \$/MBtu
Production/Harvesting Improvements		
Plant Characteristics	9	-0.219
Chemical Input Reductions	3	-0.073
Planting Technology	8	-0.194
Conversion	15	-0.364
	35	-0.85
Goal, Net \$/MBtu		1.58
Conversion/Gas Cleanup Improvements		
Plant Characteristics	5	-0.237
Conversion	10	-0.475
CO_2 Management	14	-0.665
	29	-1.38
Goal, Net \$/MBtu		3.37
Total Gas Cost Goal,[b] \$/MBtu		4.95

[a]Adapted from Figure 2 of Ref. 187.

[b]Presumes successful development of each improvement.

Table 29. EXPERIMENTAL ETHANOL AND METHANE YIELDS FROM SELECTED CROPS[a]

Feedstock	Biomass Yield[b]		Ethanol[c]		
	TS, ton/ac	VS, ton/ac	gal/lb TS	gal/ac	MBtu/ac
Sugar Beet (var. Brita)					
Roots	7.15	--	0.0635	908	68.7
Fresh Tops	3.06	2.33	--	--	--
Jerusalem Artichoke (var. Urodny)					
Tubers	4.90	--	0.0443	434	32.8
Silage Stalks	5.41	4.81	--	--	--
Potato (var. Tylva)					
Tubers	5.94	--	0.0467	555	42.0
Silage Tops	1.14	0.904	--	--	--
White Cabbage (var. Vernida)					
Silage Leaves	3.00	2.34	--	--	--
Maize (var. Pioneer 3995)					
Fresh Stalks	3.26	3.00	--	--	--
Spring Barley (var. Tyra)					
Straw	2.58	2.45	--	--	--

[a] Adapted from Tables 1, 2, and 3 of Ref. 188.

[b] Harvested last days of October 1984.

[c] Heating value of 75,670 Btu/gal used; Highest reported ethanol yield in reference used from fermentation with Saccharomyces cerevissiae at 30°C for 72 h.

[d] Heating value of 1,012 Btu/SCF used; Methane yields are from mesophilic (35°C) experiments in batch digesters.

[e] Error observed in calculated result from experimental data; This figure is 39.3 MBtu/ac (102.469 GJ/ha) in reference, but is 29.2 MBtu/ac based on reported data.

Table 29, Cont. EXPERIMENTAL ETHANOL METHANE YIELDS FROM SELECTED CROPS[a]

Feedstock	Methane[d]		
	ft^3/lb VS	10^3 ft^3/ac	MBtu/ac
Sugar Beet (var. Brita)			
Roots	--	--	--
Fresh Tops	5.69	26.5	26.8
Jerusalem Artichoke (var. Urodny)			
Tubers	--	--	--
Silage Stalks	5.05	48.6	49.2[e]
Potato (var. Tylva)			
Tubers	--	--	--
Silage Tops	3.94	7.12	7.21
White Cabbage (var. Vernida)			
Silage Leaves	7.22	33.8	34.2
Maize (var. Pioneer 3995)			
Fresh Stalks	4.12	24.7	25.0
Spring Barley (var. Tyra)			
Straw	4.39	21.5	21.8

the development of advanced reactor designs and digestion concepts (IGT, U. of Florida, U. of Arkansas, and others,)[192-196] and the evaluation of numerous biomass species and wastes as substrates for anaerobic digestion.[69,186]

One of the highlights of the year was the operation of the Walt Disney World water hyacinth-sewage sludge experimental test unit (ETU).[197,198] This 160-ft^3 unit has been producing methane continuously for more than one year. The first two tests were conducted in the upflow mode with 2:1 and 1:1 water hyacinth:sludge weight ratio feeds at a 0.2 lb VS/ft^3-d loading rate. During a 4-week steady-state test, methane yields averaged 4.6 SCF/lb VSA for the 2:1 blend, and 7.0 SCF/lb VSA for the 1:1 blend. In 1985, the ETU mode was changed to the downflow mode with feed injection at the top. The third steady-state run with the 1:1 blend afforded average methane yields of 7.9 SCF/lb VSA, or over 90% of the maximum biodegradable yield. The product gas contained 64 vol % methane, and was produced at a rate up to 1.5 vol/vol-d. This excellent performance was partially attributed to the larger SRT's in the downflow mode. Since the accompanying hyacinth growth in the sewage influent has already been demonstrated to typically remove 75-95% of BOD_5 and TSS, the system is a strong candidate for combined wastewater treatment, water hyacinth growth, and methane production in scaled-up integrated systems.[199] It appears that a metropolitan sanitary district in Central Florida is now designing a commercial system utilizing water hyacinth for wastewater treatment.[200]

The largest research project of its type for the biological gasification of RDF-sludge blends, the REfCoM (Refuse Conversion to Methane) proof-of-concept plant in Pompano Beach, FL was initiated in 1978 and after several start-up problems, was made fully operational.[201-202] Although the plant has not been operated at its full design capacity of 100 ton/d during this period, valuable data on conversion of RDF-sludge blends in two large CSTR digesters were compiled. The experimental protocol for this project was completed in 1985 for RDF-sludge blends at thermophilic conditions, and the plant was shut down.[203] The final report on this project is expected from Waste Management, Inc., the firm that managed the project. Data reported in 1985 indicate that average steady-state total gas yields were 5.41 and 7.74 SCF/lb VSA at 10- and 18-day retention times.[202]

Commercialization. Of the 6 or 7 million "family" digesters installed in the world, fewer than 3,000 are used to drive a motor, produce electricity, drive an irrigation pump, or turn a grain milling machine.[204] Also, the vast majority of these units are utilized in four countries — Brazil (3,000 to 7,500), China (4.5 million), India (280,000), South Korea (29,000) — as shown in Table 30. The medium-Btu gas produced is almost always used for domestic cooking or lighting, the feedstock is usually human or animal wastes, and the designs are usually the so-called

Chinese-type unit having a fixed-dome gas holder or the Indian-type unit having a floating steel gas holder. The use of other biodegradable materials such as crop residues and urban wastes as feedstocks is increasing, but is still limited to a few countries.

A few farm-scale anaerobic digestion systems were built in the United States as early as 1972. Between 1972 and mid-1985, 96 farm-scale and industrial wastewater cleanup systems were built.[210] Most of these units are much larger and more complex than the family units in the developing countries. Eighty-seven of the 96 U.S. systems are 100 m^3 or larger and of these, 44 are farm-scale systems having a total capacity of 44,000 m^3, and 14 are for industrial or agricultural food wastes with a total capacity of 108,000 m^3. All of these were built after 1981. Excluding sewage treatment digestion systems, of which several thousand exist in the United States, it is clear from Table 30 that Europe has about five times as many systems as the U.S.A.

Recent GRI-funded studies conclude that, with the development of new technologies and a significant reduction in systems costs, conversion of wastes and biomass to pipeline-quality methane could contribute up to 5 quad/yr to the U.S. energy mix after the turn of the century.[211] GRI has projected that up to about 1997, the potential annual contribution of landfill methane (LFM), methane from wastewater-hyacinth and MSW systems, and methane from industrial and agricultural wastes is about 1 quad. Terrestrial and aquatic biomass then begin to contribute to gas demand. In the year 2020, the potential methane contribution from all wastes is about 4 quad; about 6 quad is projected from biomass.

Whatever the future holds for anaerobic digestion as a large-scale source of methane, it is clear that digestion capacity would have to increase many-fold for biological methane to attain its potential. For each incremental quad of new methane from anaerobic digestion, about 500 times the present U.S. digestion capacity of about 5.65×10^6 ft^3 (160,000 m^3) would be required (excluding landfills and wastewater treatment capacity).* According to a recent summary, only a few digesters are currently (November 1985) in various stages of design or development.[210]

As pointed out last year, LFM recovery systems were being installed at a relatively rapid rate in the United States.[1] About 0.005 quad of LFM was recovered from MSW

---

*Calculated assuming 1.5 vol gas/culture vol-d productivity at 65 vol % methane in product gas; 1.0 incremental quad methane/yr or 10^{12} ft^3/yr then requires about $2,740 \times 10^6$ ft^3 of working culture volume. If the gas productivity were doubled and the methane concentration in the product gas were increased to 85 vol %, the required working volume capacity would decrease to about $1,075 \times 10^6$ ft^3 for 1.0 incremental quad/yr of methane.

Table 30. INVENTORY OF COMMERCIAL ANAEROBIC DIGESTERS*

Area and Country	Number of Digesters[h]	Remarks[i]
Latin America and Caribbean		
Argentina[a]	4	--
Barbados[b]	4	--
Bolivia[a,b]	24,[a] 17[b]	--
Brazil[a,b]	7530,[a] 3033[b]	2,000-2,200 Operational; 50% IT; 35% CT; About 40 villages have motor-generator sets for gas; About 500 distilleries use UFSB units for stillage.
Chile[a]	4	--
Costa Rica[b,c]	22,[b] 100[c]	Plastic-bag and PF units.
Cuba[b]	552	--
Columbia[a,b]	4,[a] 31[b]	--
Ecuador[a,b,c]	28	20 IT.
El Salvador[b]	7	--
Guatemala[b,c]	105,[b] 100[c]	38 CT; 34 PF.
Grenada[b]	10	--
Guyana[b]	1	--
Haiti[b]	9	--
Honduras[b]	25	--
Jamaica[b]	16	--
Mexico[b,c]	26	40-m^3 PF unit under test.
Nicaragua[b]	12	--
Panama[b]	7	--
Paraguay[a]	4	--
Peru[a,b]	44	--
Dominican Republic[b]	8	--
Uruguay[a,b]	4,[a] 9[b]	--
Venezuela[a,b]	7,[a] 5[b]	--
	3,948/8,567	81 Batch; 2,569 IT or PF; 996 CT, 307 others; Note sum is 3,953.
Middle East and North Africa[c]		
Egypt	50	Mostly IT; One 200-m^3 unit for poultry manure, One 150-m^3 unit for kitchen refuse.
Lebanon	1	Small 40-L unit.
Morocco	12	10 CT; One 6-m^3 IT; One plastic-bag unit.
Syria	1	400-L unit.
Tunisia	1	10-m^3 unit.
	65	

Subsahara Africa[c]		
Botswana	--	One 75-m^3 IT; Smaller IT and CT units; One 120-m^3 unit under construction.
Burkino Faso	4-10	--
Ivory Coast	1	One 400-m^3 PF unit for electricity.
Kenya	--	Tests in progress.
Mauritus	--	Tests in progress.
Rwanda	--	Tests in progress.
Sudan	--	Tests in progress.
Tanzania	~200	IT units.
Uganda	25	Tests in progress.
	--	
Asia		
China[c,d]	>7million,[c] 4.48 million[d]	Rural domestic CT units; 6 to 8 m^3 or 10^3 capacity; About 1 billion m^3 gas produced annually; More than 10,000 large and medium size units were built with total volume of about 310,000 m^3; 422 Power stations with 5,849 hp total capacity; 822 Electric stations with 7,836 kW total capacity; Many units down because of serious leakage.
India[c]	~280,000	60% IT; 40% modified CT.
Nepal[c]	1,624	1,600 Family-size units; 24 Community-scale units.
Philippines[c]	600	Numerous farm units; 10,000 m^3/d output; Large system for 50,000 hogs.
South Korea[c]	29,000	Modified IT; Built between 1969-1975.
Sri Lanka[c]	>300	Both CT and IT; One 90-m^3 unit for electricity.
Thailand[c]	5,010	5,000 Family-size units; 10 Community-scale units.
	4.8 million/7.3 million	
Europe[e]		
--	420	378 Full-scale and 42 pilot-scale plants (not individual digesters) for ag wastes; Total capacity of 95,000 m^3; 291 for cattle and hog manures; 45 for manure with bedding; 77 for mixed ag wastes mainly manures; 68% are CSTR units without recycle; 16% are PF units without recycle.
--	89	69 Full-scale and 20 pilot-scale plants for industrial wastes or wastewaters; Total capacity of 174,000 m^3; 44 UFSB units; 15 Fixed-film units; 28 Anaerobic contact units; 15 CSTR or PF units; Note sum is 99.

Table 30, Cont. INVENTORY OF COMMERCIAL ANAEROBIC DIGESTERS*

Area and Country	Number of Digesters[h]	Remarks[i]
North America		
United States	209[f]	Municipal wastewater anaerobic digestion treatment plant unit processes (not individual digesters) that use digester gas; Total flow capacity of 3.76 million gal/d.
	(3,873)[f]	Total municipal wastewater anaerobic digestion treatment plant unit processes (not individual digesters); Total flow capacity of 22.09 billion gal/d.
	96[g]	Systems (not individual digesters) built between 1972 and mid-1985; 87 Have capacities of 100 m^3 or more; 60 operational; 7 Temporarily shut down; 44 Farm operating feedlot systems have total capacity of over 44,000 m^3; 14 Other operating systems are for breweries or food processing plants and have total capacity of 108,000 m^3.

[a]This compilation is not a complete inventory. There are also several discrepancies in the numbers of commercial digesters reported in some of the references. Also, for the United States, wastewater treatment by anaerobic digestion is listed.

[b]Excerpted from Ref. 205; pertains to survey in April 1985.

[c]Excerpted from Ref. 206; pertains to survey in 1983-1984.

[c]Excerpted from Ref. 207; pertains to discussions at conference in November 1984.

[d]Excerpted from Ref. 208; pertains to survey in 1985.

[e]Excerpted from Ref. 209; pertains to survey in 1983.

[f]From Table C-10 of Ref. 46; pertains to number of wastewater treatment plant unit processes in use in 1984; One unit process includes complete anaerobic digestion system; Multiple or parallel processes counted only once for any plant.

[g]Excerpted from Ref. 210; pertains to survey in 1984-1985.

[h]Note that figures for Europe and United States are number of plants and do not necessarily correspond to number of digesters.

[i]CT denotes Chinese-type displacement digester with fixed masonry dome gas holder; IT denotes Indian-type digester with floating steel gas holder; PF denotes plug-flow horizontal digester; CSTR denotes complete-mix, stirred-tank digester; UFSB denotes upflow sludge blanket digester.

landfills in 1984, when 37 systems were in operation or under construction. The A.G.A. Gas Supply Committee projected in 1985 that 0.02 to 0.03 quad will be produced in 1990, and 0.04 to 0.10 quad will be produced in 2000.[212] In 2010, LFM production will start to decline according to these estimates to 0.02 to 0.08 quad as landfill sites become less available for methane-recovery operations.

Analysis of the wastes landfilled in Japan and Western Europe range from 47% in Japan to 90% in the United Kingdom.[213] This study indicates that LFM recovery is technically feasible in many large metropolitan areas throughout the world.

Ethanol

Adverse effects on ethanol fuel consumption would be anticipated from elimination of ethanol fuel tax forgiveness (potential) and the continued decrease in the market price of crude oil, but the event that is expected to have a large positive impact on ethanol fuel consumption in ethanol-gasoline blends is the phasedown of lead additive usage in gasolines required by the U.S. Environmental Protection Agency (EPA). The EPA reduced the lead concentration to 0.5 g/gal on July 1, 1985 from 1.1 g/gal, and further reduced the allowable limit to 0.1 g/gal effective January 1, 1986.[214] There are several approaches that can be used to maintain octane levels such as by addition of substitute additives (e.g., methyl-t-butyl ether, MTBE), increasing reforming severity to increase aromatic content, use of hydrocarbon isomerization and alkylation, and addition of alcohols or alcohol mixtures to suboctane gasoline. The blending of suboctane gasolines with ethanol is used by many refiners to maintain octane because it is less expensive than other methods in many areas that have state fuel tax subsidies to add to the federal subsidy.[214c] At mid-1985 prices, the cost per octane point for MTBE, toluene, and ethanol was 1.14¢, 1.12¢, and 0.58¢, respectively.[215] If all of the octane deficiency created by the lead phasedown were met by ethanol, the potential market would be about 7.7 billion gallons in 1986 and then undergo reductions each year to about 4.9 billion gallons in 1994 (Table 31).[214a] About 500 million gallons of ethanol fuel were produced in 1985 in the United States, which has a total production capacity of about 700 million gal/yr,[216] so there is clearly no possibility of sustaining the U.S. octane pool with domestic ethanol. If, however, this were feasible and all of the ethanol were manufactured from corn, 3.0 billion bushels of corn from about 28 million acres of farmland would be required. Obviously, this portends an excellent market forecast for ethanol fuel, all other things being as projected (and they usually are not in the energy business), even if ethanol cannot satisfy all of the octane requirement.

Another possibility is to satisfy part of the requirement with ethanol-methanol mixtures, particularly since the EPA is considering modifying the waiver that it granted to DuPont for use of a blend of 5% methanol (max.)

Table 31. POTENTIAL ETHANOL REPLACEMENT MARKET FOR LEAD ADDITIVE PHASEDOWN[a]

Year	Gasoline Demand, 10^9 gal		Reduction in Lead, 10^9 g	Ethanol Equivalent,[c] 10^9 gal
	Unleaded[b]	Leaded[b]		
1985	68.4	32.2	18.5	3.70
1986	71.5	28.8	38.4	7.68
1987	74.4	25.6	35.8	7.16
1988	77.2	22.4	33.4	6.68
1989	80.1	19.2	30.9	6.18
1990	82.6	16.4	28.8	5.76
1991	84.1	14.9	26.3	5.26
1992	85.6	13.4	25.9	5.18
1993	86.6	12.4	25.2	5.04
1994	87.5	11.5	24.6	4.92

[a]Adapted from Ref. 214a; Table in this reference mislabled Leaded Gasoline Demand; It is Unleaded Gasoline Demand.

[b]Assumes no illegal fuel switching; i.e., leaded for unleaded.

[c]Based on 1 gal ethanol equivalent to 5.0 g lead as octane enhancing additive.

and 2% cosolvent ethanol, propanols, and/or butanols (min.).[217] In 1986, the actual annual U.S. operating capacity to manufacture methanol is about 1.4 billion gallons.[218]

Another approach to production of large volumes of ethanol fuel is to utilize low-cost lignocellulosics as feedstocks for fermentation ethanol. Much of the current research to develop improved ethanol processes is concentrated on this objective.[219] Recent reports on the conversion of lignocellulosics to sugars have been made on enzymatic and acid hydrolysis;[220-237] and on pretreatment by steam explosion, oxidation, steam hydrolysis-extraction, ensiling, and autogenic methods to enhance biodegradability.[238-245] Other areas receiving concentrated research include the utilization of immobilized enzymes and organisms for ethanol fermentation to attempt to increase alcohol production rates and yields,[245-250] and the conversion of the pentoses predominant in the hemi-celluloses.[251-253] Conversion of both C_5 and C_6 sugars in lignocellulosics would improve overall ethanol yields and feedstock utilization.

One approach to the fermentation of lignocellulosics that is reported to solve the sugar and ethanol yield problems experienced in the past is a patented technique in which an aqueous acetone solution and a minor amount of mineral acid is used for hydrolysis.[254] This process, called ACOS (Acid Catalyzed Organosolv Saccharification), was specifically designed to decompose wood into sugars and lignin in a single step. The lignin and sugars are quantitatively recovered without degradation. Total biomass refining is thus claimed for ACOS.[255] In operation, the lignocellulosic feedstock is supplied continuously to the pressurized reactor and extracted with the acid-aqueous acetone mixture at 180-220°C.[256] Solvent removal from the reactor effluent results in quantitative precipitation of lignin and a concentrated sugar solution. Total saccharification can be achieved, depending on temperature, in 25-50 s with both softwoods and hardwoods. The key to this innovative process appears to be the formation of acetonides (cyclic ketals) formed by reaction of acetone and sugars. The acetonides protect the sugars from further reaction and are then hydrolyzed after removal from the reaction zone. Brazil is currently evaluating this process in a 100-L/d pilot plant supplied with 200 kg/d of bagasse.[255] Use of 10-20% aqueous acetone affords 95% yields of the sugars from which ethanol can be produced at a cost of 57¢/gal (15¢/L).[255] As shown in Table 32, one of the unique features of ACOS is that it has the potential of doubling the ethanol yield from corn and tripling the yield from sugarcane by conversion of almost the entire plant. If this type of process is in fact capable of achieving conversion of all the cellulose and hemicellulose in lignocellulosic biomass to sugars as reported, it will have far-reaching implications on the future of grains as an alcohol feedstock, and on our ability to produce large amounts of ethanol.

Table 32. POTENTIAL ETHANOL YIELDS FROM CORN AND SUGARCANE BY WHOLE-PLANT UTILIZATION[a]

Sugar Source	Yield		Sugars, lb/ton		Ethanol Yield, gal/ton			Ethanol Yield		Lignin,
	wet ton/ac	dry ton/ac	C_5	C_6	C_5	C_6	$C_5 + C_6$	gal/ac	$C_5 + C_6$ gal/ac	ton/ac
Corn										
Kernel	7.14	6.25	--	1,480[c]	--	106.4	106.4	491[f]	491	--
Cob	1.34	0.89	840	740	51.0	60.4	111.4	60	99	0.12
Stover	11.60	4.01	350	870	21.6	71.2	90.6	285	372	0.64
Biomass	20.1	11.1	220	1,180[d]	13.7	74.0	87.7	826[g]	978[g]	1.78
Sugarcane										
Cane Juice Only	(35.69)	--	--	--	--	16.8[e]	--	599[f]	--	--
Total Cane Only	35.69	9.46	--	--	--	32.6[e]	--	1,163	--	--
Bagasse Only	8.92	5.00	--	--	--	63.5[e]	--	566	--	--
Bagasse Integral	18.02	9.01	--	--	--	58.2[e]	--	1,050	--	--
Biomass[b]	42.8	13.56	--	--	--	40.3[e]	--	1,725[g]	--	--

[a]Adapted from Tables 1 and 2 of Ref. 254.

[b]Tops and leaves that are normally burned off in field plus cane.

[c]Starch only.

[d]Includes starch.

[e]On wet weight.

[f]Conventional ethanol yield.

[g]Does not equal sum of individual ethanol yields.

REFERENCES CITED

1. D. L. Klass, "Energy from Biomass and Wastes: 1984 Update," Energy from Biomass and Wastes IX Lake Buena Vista, FL, January 28-February 1, 1985 (Institute of Gas Technology, Chicago, IL, December 1985) pp. 1-83.

2. K. Wells, "Clouded Outlook, As a National Goal, Renewable Energy Has An Uncertain Future," Wall Street J. LXVI (86) 1, 16 (1986) February 13.

3. U.S. Department of Energy, The Budget for Fiscal Year 1987, Special Analysis K, p. K-13.

4. U.S. Department of Energy, Congressional Budget Request FY 1987, Vol. 2, DOE/MA-0064/4, February 1986.

5. J. R. Long, D. J. Hanson, and W. Lepkowski, "Funds for R&D Are Up 17% in Administration's Budget Proposal," C&EN 64 (7) 9-14 (1986) February 17.

6. President's Message, May 15, 1985 to Conferees at Renewable Energy Technologies Symposium and International Exposition, Anaheim, CA, June 6, 1985.

7. Report of the Solar Panel of the Energy Research Advisory Board to the United States Department of Energy, DOE/S-0039, December 1985.

8. D. L. Klass, "Energy from Biomass and Wastes: 1983 Update," Energy from Biomass and Wastes VIII Lake Buena Vista, FL, January 30-February 3, 1984 (Institute of Gas Technology, Chicago, IL, September 1984) pp. 1-97.

9. U.S. Department of Energy, Congressional Budget Request FY 1986, Vol. 2, DOE/MA-0064/3, February 1985.

10. U.S. House of Representatives, Energy and Water Development Appropriation Bill, 1986, Report 99-195, July 1985, p. 78.

11. U.S. Senate, Energy and Water Development Appropriation Bill, 1986 Report 99-110, July 25, 1985, pp. 100-101.

12. U.S. House of Representatives, Conference Report to Accompany Energy and Water Development Appropriations for FY86, Report 99-307; Public Law 99-141, November 1, 1985.

13. D. L. Klass, Biomass Energy Research Association, Testimony and Statement presented to Subcommittee on Energy Development and Applications of the Committee on Science and Technology, February 27, 1986, Washington, D.C.

14. Meridian Corporation, "Regional Biomass Energy Program, Project Symposium," Prepared for U.S. Department of Energy, Biomass Energy Technology Division, July 1985.

15. Meridian Corporation and Science Applications International Corporation, A Guide to Federal Programs in Biomass Energy, Contract No. DE-AC01-83CE30784, September 1984.

16. M. Carney, Biomass Energy, A Profile of State Energy Office Programs, Meridian Corporation, July 1983.

17. Council of Biomass Energy Technology Sponsors, A Guide to Non-Federal Programs in Biomass Energy, May 1985; A Guide to R&D Programs in Biomass Energy, Draft, January 1986.

18. Hawaii Natural Energy Institute Annual Report 1984, 10th Anniversary Special Edition, U. of Hawaii at Manoa, Honolulu, Hawaii.

19. R. Tuvell, "Progress in the California Energy Commission Biomass Demonstration Program," Energy Applications of Biomass, M. Z. Lowenstein, Ed. (Elsevier Applied Science Publishers, London, August 1985) pp. 119-134.

20. Private Communication, D. Loos (Illinois Department of Energy and Natural Resources, Springfield, IL) February 19, 1986.

21. Gas Research Institute 1986-1990 Five-Year Research and Development Plan and 1986 Research and Development Program, April 4, 1985.

22. D. L. Klass, "Energy from Biomass and Wastes: 1982 Update," Energy from Biomass and Wastes VII, Lake Buena Vista, FL, January 24-28, 1983 (Institute of Gas Technology, Chicago, IL, June 1983), pp. 1-68.

23. E. O. Webb, "FERC Approves GRI's 1986 Research Program," GRID 8 (4) 18-19 (1985-1986) Winter.

24. U.S. Department of Energy, Office of Conservation and Renewable Energy, Deputy Assistant Secretary for Renewable Energy, "Renewable Energy Research and Development Outlook, DOE/OR/20837-4, Vol. 1, October 1985; DOE/OR/20837-4, Vol. 2, December 1985.

25. Congress of the United States, Office of Technology Assessment, Energy from Biological Processes, July 1980.

26. Private Communication, R. A. Moyle (Science Applications Corp.), February 5, 1986.

27. Cf. Discussion of H. M. Kosstrin and D. A. Himmelblau, "Economic Feasibility of a Small-Scale Wood Gasification to Methanol Plant," in Ref. 1, pp. 1305-1341.

28. Renewable Energy Incentives, Hearing Before the Subcommittee on Energy Conservation and Power of the Committee on Energy and Commerce, House of Representatives, June 20, 1985, Report Serial No. 99-25.

29. H. R. Heede, "A Preliminary Assessment of Federal Energy Subsidies in FY 1984," in Ref. 28, 56-85.

30. "Text of Treasury Ethanol Import Option Paper," Alcohol Week 6 (40) Special Supplement (1985) October 7.

31. "Tax Reform Act Passes House: Senate Being Asked for 60¢ Blenders' Credit," ibid. 6 (51) 1, 9 (1985) December 23; "Reagan Renews Attack on Federal Blend Tax Break; Seeks Cut in Budget," ibid. 7 (6) 1-2 (1986) February 10.

32. R. T. Hayden, Renewable Energy Institute, Draft of Fact Sheet, Renewable Energy and Conservation Act of 1985," March 26,1985.

33. Private Communication, David Moulton (Office, Representative E. J. Markey, Washington, D.C.) February 11, 1986.

34. Cf. "Battle Rages for Tax Credit Equality," and "The Facts on H.R. 2001 and S. 1220," "The Energy Tax Situation — Simplified," Biologue 2 (3) June-July 1985.

35. M. J. Riedy, Testimony of Bacardi Corporation, in Ref. 28, pp. 246-261.

36. M. J. Riedy, Testimony of A. O. Smith, Harvestore Products, Inc., in Ref. 28 pp. 232-245.

37. Private Communication, C. W. Robinson (Energy Transition Corporation, Santa Fe, NM) June 4, 5, 10, 1985.

38. J. H. Birnbaum and D. Rogers, "An Energy Tax Boost is Appealing to Congress as Oil Prices Decline, But the Issue is Devisive," Wall Street J. LXVI (100) 46 (1985) February 20.

39. Rocky Mountain Institute, A Preliminary Assessment of Federal Energy Subsidies in FY 1984, Publication 85-7, 1985.

40. H. R. Heede, "Federal Energy Subsidies, A Look at the Bang Per Buck," Alternate Sources of Energy 76 4 (1986) November/December.

41. U. S. Department of Energy, Office of Renewable Energy, "Renewable Energy Profile 1975-1984," February 1985, 10 pp.

42. Energy Information Administration, "Energy Facts 1984," DOE/EIA-0469, Washington, D.C., May 1985.

43. Energy Information Administration, "State Energy Data Report, Consumption Estimates," DOE/EIA-0214(83), May 1985.

44. H. R. Schwandt, Testimony and Statement, Ref. 28, pp. 174-188.

45. J. Easterly, S. Lees, and B. Detwiler, "Electric Power from Biofuels: Planned and Existing Projects in the U.S.," Revised January 1985, DOE/CE/30784/1, Contract No. DE-AC01-83CE-30784, August 1985.

46. U.S. EPA and Roy F. Weston, Inc., "1984 Needs Survey Report to Congress, Assessment of Needed Publicly Owned Wastewater Treatment Facilities in the United States," EPA 430/9-84-011, February 1985, p. C-27.

47. J. H. Ashworth, Y. M. Bihun, and M. Lazarus, "Universe of U.S. Commercial-Scale Anaerobic Digesters: Results of SERI/ARD Data Collection," SERI Subcontract 4-04096-1, May 30, 1985.

48. D. L. Klass, "Energy from Biomass and Wastes: A Review and 1983 Update," Resources and Conservation 11 157-239 (1985).

49. U.S. Department of Energy, Fiscal Year 1987 Congressional Budget Request, Energy Supply Research and Development, Vol. 1, Solar and Other Renewables, p. 12.

50. Fiscal Year 1986 Department of Energy Authorization, Conservation and Renewable Energy, Hearings Before the Subcommittee on Energy Development and Applications of the Committee on Science and Technology, U.S. House of Representatives, March 5, 7, 1985, No. 31, Vol. III, pp. 81-103, 128-138.

51. A. Duncan, Testimony and Statement, in Ref. 28, pp. 189-193.

52. International Energy Agency, Energy Policies and Programmes of IEA Countries 1984 Review, Paris, France, 1985, p. 48.

53. D. O. Hall and P. J. DeGroot, "Biomass for Food or Fuel: A World Problem," Energy Applications of Biomass, M. Z. Lowenstein, Ed., (Elsievier Applied Science Publishers, London, 1985) pp. 21-35.

54. <u>Energy from Biomass: Building on a Generic Technology Base</u>, Proceedings of the Third Technical Review Meeting, November 27-29, 1984 (Argonne National Laboratory, Argonne, IL, April 1985) 324 pp.

55. E. S. Groton, R. T. Brooks, and T. A. Waldrop, "Summary of Woody Biomass Assessment in 13 Southeastern States," in Ref. 54, pp. 109-116.

56. U.S. Department of Agriculture, Sixth Annual Solar, Biomass and Wind Energy Workshop, February 25-27, 1986, Atlanta, GA, 247 pp.

57. P. C. Badger, "Southeastern Regional Biomass Energy Program" in Ref. 56, pp. 39-44.

58. W. J. Herz, Biomass Energy in Alabama, An Inventory and Assessment, Energy Division, Alabama Department of Economics and Community Affairs, June 1985.

59. H. M. Draper, C. C. NeSmith, and W. H. Smith, "Florida Biomass Resource Assessment: Available Land, Energy Wood, and Crop Residues," in Ref. 56, pp. 45-49.

60. J. I. Zerbe "Forest Service Response to the National Energy Situation," in Ref. 56, pp. 33-38.

61. J. W. Ranney <u>et al</u>., "Short Rotation Woody Crops Program: Annual Progress Report for 1984," ORNL-6160, Oak Ridge National Laboratory, August 1985.

62. J. W. Ranney, R. D. Perlack, J. L. Trimble, and L. L. Wright, "Specialized Hardwood Crops for Energy and Fiber: Status, Impact, and Need," <u>Tappi J</u>. <u>68</u> (12) 36-41 (1985) December.

63. Personal Communication, D. Branson and J. Aycock (Northwest Missouri State University, Maryville, MO) March 4, 1986.

64. Personal Communication, S. P. Bryant (Foster Farms, Livingston, CA) December 12, 1985.

65. Personal Communication, W. Heckrodt (Kimberly Clark Corp., Neenah, WI) December 12, 1985.

66. Personal Communication, J. Tanner (Kimberly Clark Corp., Neenah, WI) September 30, 1985.

67. Personal Communication; S. Shen (Argonne National Laboratory, Argonne, IL) March 3, 1986.

68. T. Crabb, "Eucalyptus Plantations for Energy Production in Hawaii, Six and One Half Years Later – The Bioenergy Development Corporation Story," Paper presented at RETSIE 85, June 3-6, 1985, Anaheim, CA.

69. P. F. Bente, "International Bio-Energy Activities," Paper presented at RETSIE 85, June 3-6, 1985, Anaheim, CA.

70. a. Commission of the European Communities, Energy from Biomass 3rd E. C. Conference, W. Palz, J. Coombs, and D. O. Hall, Eds., Venice, Italy, March 25-29, 1985 (Elsevier Applied Science Publishers, London, 1985) 1211 pp; b. G. Grassi, U. Miranda, C. Baldelli, and F. Gheri, "L.E.B.E.N. – Large European Bioenergy Project," ibid., pp. 164-171.

71. J. I. Zerbe, Private Communication (U.S. Department of Agriculture, Forest Products Laboratory, Madison, WI, February 7, 1985).

72. J. R. Hornick, J. I. Zerbe, and J. L. Whitmore, "Jaris Successes," J. of Forestry 82 (11) 663-667 (1984) November.

73. The Global 2000 Report to the President (Government Printing Office, Washington, D.C., 1980).

74. D. Avery, "U.S. Farm Dilemma: The Global Bad News in Wrong," Science 230 (4724) 408-412 (1985) October 25.

75. K. Maize, "How It Didn't Turn Out: The Forecasters Who Failed (And One Other)" The Energy Daily 14 1, 4 (1986) January 2.

76. R. Orr, "World Record Corn Yield Will Be Tough Act to Follow," Chicago Tribune No. 320, November 16, 1985, p. 18.

77. "Costs Probed for Fuel, Feed from Corn Waste," C&EN 63 (2) 59 (1985) January 14.

78. Cf. "Biotech Research for Agriculture Advances on Several Fronts," C&EN 63 (39) 71, 72 (1985) September 30; "Genetic Engineering: Progress on Herbicide-Resistant Crops," C&EN 63 (45) 4, 5 (1985) November 11; "A Fresh Starting Point for Agricultural Research," National Research Council News Report XXXV (4) 18, 19, 29 (1985) April; Report of the Research Briefing Panel on Biotechnology in Agriculture," National Academy of Science Seminar on Research Briefings, June 12, 1985.

79. U.S. Department of Agriculture, Illinois Cooperative Extension Service, Farming Facts, Winter 1986.

80. J. J. Hoffmann, "Resinous Plants as Economic Alternative to Bioenergy Plantations," Energy 10 (10) 1139-1143 (1985).

81. "Brazil Scuttles Plan for Ethanol Production Stepup, May Drop Financing Bid," Alcohol Week 7 (9) 1 (1986) March 3.

82. L. Icerman, "Arid Land Crops," Science 227 (4686) 466 (1985).

83. M. F. Balandrin et al., "Natural Plant Chemicals: Sources of Industrial and Medicinal Chemicals," Science 228 (4704) 1154-1160 (1985).

84. J. H. Cushman, A. F. Turhallow, and J. W. Johnston, Draft, Herbaceous Energy Crops Program: Annual Progress Report for 1985, ORNL-6263, Oak Ridge National Laboratory, in press.

85. W. Miller, "Extraction of Latex from Guayule," Paper presented at Southwest Industrial Biomass Conference, September 29-October 1, 1985, Phoenix, AR.

86. E. B. Shultz, Jr., J. S. DeVeaux, and L. Icerman, "Buffalo Gourd as an Energy Crop in New Mexico," paper presented at Southwest Industrial Biomass Conference, September 29-October 1, 1985, Phoenix, AR.

87. National Academy Press, Jojoba: New Crop for Arid Lands, New Material for Industry, Washington, D.C., 1985.

88. E. A. Hiler et al., "Sorghums for Methane Production, Annual Report April 1984-March 1985," GRI Report 85-0101, March 1985.

89. J. S. Cundiff, D. H. Vaughan, and D. J. Parrish, "Pith Separation Procedure for Processing Whole-Stalk Sweet Sorghum: in Ref. 56, pp. 66-71.

90. K. A. Mabler and D. L. Auld, "Development and Uses of Winter Rapeseed Adapted to the Southeastern United States, in Ref. 56, pp. 167-170.

91. D. L. Thomas and F. daSilva, "Rape as a Winter Cover Crop in South Georgia," in Ref. 56, pp. 171-176.

92. 1984 International Gas Research Conference, Proceedings, September 10-13, 1984 (Government Institutes, Inc., Rockville, MD, March 1985).

93. S. F. Chandler, K. Rajasekaran, and I. K. Vasil, "Large Scale Propagation of Napier Grass and Giant Napier Grass by Tissue Culture," in Ref. 92, pp. 561-566.

94. D. J. Parrish et al. "Perennial Herbaceous Biomass Species for the Piedmont," in Ref. 56, pp. 60-65.

95. S. L. Ott, "Reducing Energy Inputs in Crop Production by Supplying Nitrogen from Legume Cover Crops: An Economic Case Study," in Ref. 56, pp. 201-206.

96. D. L. Klass, "Bioenergy in the Southwest, Potentials and Unique Aspects," Paper presented at Southwest Industrial Biomass Conference, September 29-October 1, 1985, Phoenix, AR.

97. "Biofertilizer to Debut," Ag. Biotechnology 2 (1) 6 (1985) January/February.

98. M. N. Nabors, Tissue Culture Investigation Into Mechanisms of Biomass Enhancements, Annual Report June 1984-July 1985, GRI Report 85/0175, July 1985.

99. J. R. Liu and D. J. Cantliffe, "Tissue Culture Propagation Development and its Application to Energy Crops," in Ref. 92, pp. 622-629.

100. E. S. Lipinsky et al., "Review of the Potential for Biomass Resources and Conversion Technology Status," in Ref. 92, pp. 594-604.

101. Solar Energy Research Institute, Draft, SERI Biofuels Program, FY 1985 Annual Report, SERI/PR-231-2849, November 1985.

102. J. R. Benemann, D. M. Tillett, and J. C. Weissman, "Microalgae as a Source of Liquid Fuels: Economic Analysis and Experimental Status," in Ref. 56, pp. 237-242.

103. E. L. Maxwell, A. G. Folger, and S. E. Hogg, Resource Evaluation and Site Selection for Microalgae Production Systems, SERI/TR-215-2484, DE85008807, May 1985.

104. R. P. McIntosh, "The Potential for Microalgae and Cultivation in the Southwest United States for the Production of Liquid Fuels," Paper presented at Southwest Industrial Biomass Conference, September 29-October 1, 1985, Phoenix, AR.

105. G. Shelef, A. Sukenik, and E. Sandbank, "Separation and Harvesting of Microalgae from Saline Media," SERI/CP-231-2700, Contractors Meeting Proceedings, March 20, 1985, pp. 244-270.

106. F. Brenckmann et al., "Influence de La Nutrition Azotee Sur La Croissance et La Production D'Hydrocarbures de L'Algue Unicellulaire Botryococcus Braunii," in Ref. 69, pp. 717-721.

107. F. Brenckmann, "Influence of Light Intensity on Hydrocarbon and Total Biomass Production of *Botryococcus Braunii*, Relationships With Photosynthetic Characteristics," in Ref. 69, pp. 722-726.

108. P. Metzger *et al.*, "Screening of Wild Strains of the Hydrocarbon-Rich Alga *Botryococcus Braunii*, Productivity and Hydrocarbon Nature," in Ref. 69, pp. 727-731.

109. A. Legros *et al.*, "An Integrated System: Mass Algal Culture in Polluted Luke-Warm Water for Production of Methane, High-Value Products and Animal Feed," in Ref. 69, pp. 369-373.

110. R. N. Young, Personal Communication (United Energy Corporation, Foster City, CA) July 31, 1985.

111. UEC, "Something New Under the Sun: A Community Lives on Its Energy," *The Solar Revolution* 2 (1) 4, 5 (1984).

112. J. H. Ryther, T. A. DeBusk, and J. E. Peterson, "Studies With Marine Macroalgae: Saline Desert Water Cultivation and Effects of Environmental Stress on Proximate Composition," SERI/CP-231-2700, Contractors Meeting Proceedings, March 20, 1985, pp. 179-227.

113. K. Bird and J. Ryther, "Aquatic Plant Aquaculture: A Multiproduct/Multiuse Approach," in Ref. 54, pp. 277-282.

114. B. H. Brinkhuis, H. G. Levine, and C. G. Shlerk, "Recent Progress in the New York Marine Biomass Program," in Ref. 54, pp. 295-304.

115. M. Neushul and B.W.W. Hunger, "Studies of Biomass Yield from a Near-Shore Macroalgal Test Farm" *J. Solar Energy Eng.* 107 (1) 93-96 (1985).

116. Government Institutes, Inc., *Energy Technology XII, Prices and Uses*, March 25-27, 1985, Washington, D.C., R. F. Hill, Ed. (Government Institutes, Inc., Rockville, MD, August 1985) 1820 pp.

117. T. D. Hayes *et al.*, "Methane from an Integrated Hyacinth Treatment Anaerobic Digestion Facility," in Ref. 116, pp. 1782-1793.

118. K. R. Reddy, W. F. Debusk, and L. O. Bognall, "Water Hyacinth Biomass Production in Eutrophic Lake Water," in Ref. 92, pp. 630-644.

119. D. P. Chynoweth *et al.*, "Biogasification Products of Water Hyacinth Wastewater Reclamation," in Ref. 54, pp. 283-293.

120. F. Sauze, "Epuration Des Eaux Et Products De Haute Valeur Tires De La Jacinthe D'Eau," in Ref. 69, pp. 364-368.

121. A. Larigauderie, J. Roy, and A. Berger, "Photosynthesic Basis of Biomass Production by Water Hyacinth Grown Under High CO_2 Level," in Ref. 69, pp. 403-406.

122. D. L. Klass "A Perpetual Methane Economy — Is It Possible," CHEMTECH 161-168 (1974) March.

123. Special Report, Cogeneration's Future — Will It Keep Booming?," DeedDigest 5 (2) 1, 3 (1985) Spring.

124. "Cogeneration," Compressed Air 90 (10) 9-19 (1985) October.

125. "Cogen Big Factor in PG&E Plans," Cogeneration 2 (2) (1985) March-April; I. Stambler, "California Evaluates Future Cogen Plans," ibid. 2 (3) 15-16 (1985) May-June; "See Cogen Power 5-20% by 2000," ibid 2 (5) 5 (1985) September-October.

126. Governor's Energy Office, State of Florida, Report, Renewable Energy: Cogeneration, February 1985.

127. D. L. Klass, "Renewable Energy," Part 2, CHEMTECH 14 (10) 610-615 (1984) October.

128. S. A. Noll, "Small Wood-Fired Power Plants: The Ultrapower Experience," in Ref. 116, pp. 1580-1586.

129. B. Paul, "Waste-to-Energy Plants are Becoming Popular With U.S. Cities and Towns," Wall Street J. CCVI (118) 21, Eastern Ed., (1985) December 16.

130. J. Cook, "Political Garbage," Forbes 136 (1) 58, 59 (1985) July 1.

131. C. Leinster, "The Sweet Smell of Profits from Trash," Fortune 111 (7) 150-154 (1985) April 1.

132. "Waste to Watts, What's Burning with Municipal Solid Wastes?," DeedDigest 5 (3) 1, 3 (1985) Summer.

133. Detroit Approves Nation's Largest Planned Waste-to-Energy Plant," Waste-to-Energy Report 1, August 14, 1985.

134. Report on Semiannual Survey: Resource Recovery Activities, City Currents (United States Conference of Mayors, Washington, DC, October 1985) 24 pp.

135. H. W. Gershman, "Solid Waste-to-Energy Overview and Status Report," in Ref. 116, pp. 1641-1655.

136. M. L. Smith, "Early and Current Systems Utilizing Refuse Derived Fuels," Paper presented at Energy from Municipal Wastes, Opportunities in An Emerging Market, Sponsored by McGraw-Hill Energy Publications, Washington, DC, October 24-25, 1985.

137. J. F. Bernheisel, "Business Arrangements for Energy Products from Solid Waste-to-Energy Projects," in Ref. 116, pp. 1665-1676.

138. "Doxins In the Environment: No Conversion on Human Hazard," C&EN 63 (21) 41-44 (1985) May 27.

139. T. R. Carr, "The Burlington Electric Experience with 50 MW McNeil Municipal Power Plant," in Ref. 116, pp. 1587-1596.

140. J. Wong, "Vermont Wood-Powered Electric Plant Generates High Costs, Disappointment," Wall Street J. LXV (219) 32 (1985) August 23.

141. "Burlington, Vermont Rectifies Problems at Woodchip Plant," Power Lines, November 15, 1985.

142. M. Murphy, "Hot Gas Heat Exchanger from the Fluidized Bed Wood Combustor," in Ref. 116, pp. 1597-1604.

143. E. G. Sukup, "A Completely Automated System Using Biomass Fuel for Heating and Drying," in Ref. 56, pp. 111-114.

144. D. R. King and C. E. Chastain, "Particulate Waste Product Combustion System," in Ref. 56, pp. 115-120.

145. Draft, Proceedings of the 1985 Biomass Thermochemical Conversion Contractors' Meeting, October 15-16, 1985, Minneapolis, MN (Pacific Northwest Laboratory, Richland, WA, February 1986) 381 pp.

146. J. T. Hamrick, "Development of Wood as an Alternative Fuel for Gas Turbines," in Ref. 145, pp. 263-270.

147. E. I. Wan, "Technical and Economic Evaluation of Biomass-Based Fuel Processes," in Ref. 145, pp. 167-192.

148. G. F. Schiefelbein, "Thermochemical Breakthroughs," Paper presented at RETSIE 85, June 3-6, 1985, Anaheim, CA.

149. U.S. Department of Energy, "Expression of Interest, Medium Btu Gas from Biomass," 4751K/0500G, 1985.

150. R. J. Evans et al., "Development of Biomass Gasification to Produce Substitute Fuels," in Ref. 145, pp. 271-294.

151. G. F. Schiefelbein, "Biomass Gasification Research: Recent Results and Future Trends," PNL-SA-12841, February 1985.

152. T. B. Reed *et al.*, "Moving-Bed Air and Oxygen Biomass Gasification," in Ref. 145, pp. 295-318.

153. H. LaFontaine, "Emergency Use of Simple Inexpensive Gasifiers," in Ref. 56, pp. 161-166.

154. Personal Communication, H. LaFontaine (Biomass Energy Foundation, Inc., Miami, FL), February 25, 1986.

155. D. B. Mahin, June 1985: International Producer Gas Conference, Bioenergy Systems Report, U.S. Agency for International Development, 28 pp.

156. Pacific Northwest Laboratory, "Some Results of the Low-Btu Gasification Workshop," Atlanta, GA, November 1984.

157. J. M. Veigel, "On-Farm Wood Gasification/ Cogeneration," Paper presented at CBETS Meeting, January 31, 1986, Washington, DC.

158. L. D. Clements *et al.*, "Development of an Indirectly Heated Biomass Pyrolysis Reactor," in Ref. 145, pp. 211-232.

159. J. M. Double and A. V. Bridgwater, "Sensitivity of Theoretical Gasifier Performance to System Parameters," in Ref. 69, pp. 915-919.

160. R. S. Butner, D. C. Elliott, and L. J. Sealock, Jr., "Low Temperature Thermochemical Conversion of High-Moisture Biomass Feedstocks," in Ref. 145, pp. 193-209.

161. L. K. Mudge *et al.*, "Bench Scale Studies in Catalytic Gasification of Biomass," in Ref. 145, pp. 235-255.

162. D. C. Elliott, "Comparative Analysis of Gasification/ Pyrolysis Condensates," in Ref. 145, pp. 361-381.

163. G. W. Simmons, Biomass Gasification — Phase II, Final Report (July 1982-November 1984), GRI-85/0038, May 1985, 40 pp.

164. "Biomass Power Plant for Guiana," *Canada Weekly* 13 (8) 2 (1985) February.

165. "EPI Installs Another Fluidized Gasifier System," *Biologue* 2 (2) 1 (1985) April-May; T. Baer, "Former DOE Official Opens Wood Gas Plants, Debates Myths About Their Commercialization," *Renewable Energy News* 8 (4) 20 (1985) May; Nuts and Bolts, *DeedDigest* 5 (4) 11 (1985) Fall.

166. Third Southern Biomass Energy Research Conference Program and Abstracts, Gainesville, FL, March 12-14, 1985.

167. W. S. Bulpitt and J. L. Walsh, Jr., "Utilization of Biomass Fuel for Production of Electric Power," in Ref. 166, p. 51.

168. H. F. Feldmann, M. A. Paisley, and H. R. Appelbaum, "Conversion of Forest Residues to a Methane-Rich Gas in a High Throughput Gasifier," in Ref. 145, pp. 339-360.

169. M. A. Paisley, H. F. Feldmann, and W. D. Watt, "Wood Gasification Combined Cycle Retrofit in a Conventional Pulp Mill Cogeneration System," Paper presented at AIChE 1985 Summer National Meeting, Seattle, WA, August 25-28, 1985.

170. J. A. Knight et al., "Entrained Flow Pyrolysis of Biomass," in Ref. 145, pp. 99-113.

171. D. Meier, D. R. Larimer, and O. Faix, "Direct Thermochemical Liquefaction of Biomass Using Hydrogenating Conditions," in Ref. 69, pp. 929-932.

172. C. Bestue-Labazuy et al., "Study of the Direct Liquefaction of Wood in the Presence of Iron Additives," in Ref. 69, pp. 924-928.

173. G. Bonn et al., "Hydrothermalysis of Short-Rotation Forestry Plants," in Ref. 69, pp. 953-958.

174. R. P. Overend and E. Chornet, "Le Pretaitement, L'Hydrolyse, La Pyrolyse et La Liquefaction de La Biomasse: Vers une Approache Unifiee," in Ref. 69, pp. 933-936.

175. Ref. 1, page 62; Refs. 186 and 190 in Ref. 1.

176. W.J.M. Douglas and S. B. Sabade, "Cellulose Liquefaction Under Mild Conditions," Bioenergy 84 Vol. III, H. Egneus and A. Ellegand, Eds. (Elsevier Applied Science Publishers, London, 1985) pp. 38-46.

177. D. C. Elliott and E. G. Baker, "Upgrading Biomass Liquid Fuels," in Ref. 145, pp. 81-97.

178. M. Steinberg, P. T. Fallon, and M. S. Sundaram, "Flash Methanolysis — The Flash Pyrolysis of Biomass with Methane Gas," in Ref. 145, pp. 15-29.

179. R. J. Evans and T. A. Milne, "Applied Mechanistic Studies of Biomass Pyrolysis," in Ref. 145, pp. 57-79.

180. J. P. Diebold, J. W. Scobill, and R. J. Evans, "The Thermal and Catalytic Upgrading of Oxygenated Primary Biomass Pyrolysis Oil Vapors," in Ref. 145, pp. 31-55.

181. R. G. Graham *et al.*, "Fast Pyrolysis of Cellulose," in Ref. 69, pp. 860-864.

182. "Wood Pyrolysis Oils: Refining Product Recovery to Yield Fuels and Chemicals," *Bioprocessing Technology 7* (6) 3 (1985) June.

183. J. L. Kuester, "Diesel Fuel from Biomass Residues," Paper presented at Southwest Industrial Biomass Conference, Phoenix, AR, September 29-October 1, 1985.

184. Y. Solantavsta and P. J. McKeough, "A Techno-Economic Comparison of Biomass Thermochemical Liquefaction," in Ref. 69, pp. 937-941.

185. L. A. Michaelis, "The Economics of the Thermochemical Routes from Wood to Liquids," in Ref. 69, pp. 1039-1043.

186. *Anaerobic Digestion 1985*, Proceedings of the Fourth International Symposium on Anaerobic Digestion, Guangzhou, China, November 11-15, 1985, 720 pp.

187. J. R. Frank, T. D. Hayes, and P. H. Smith, "The Production of Methane from Biomass in the United States: Economics, Tradeoffs, and Prospects," in Ref. 69, pp. 484-488.

188. J. Zuber, "Photosynthetic Solar Energy Capturing in a Cropping System with External Exploitation of Biomass for Fuel Production," in Ref. 69, pp. 435-440.

189. W. H. Smith, Methane from Biomass and Waste, Annual Report for 1984, GRI-85/0203, January 1985.

190. J. G. Zeikus, R. Kerby, and J. A. Krzycki, "Single-Carbon Chemistry of Acetogenic and Methanogenic Bacteria," *Science 227* (4691) 1167-1173 (1985) March 8.

191. J. Konisky, "Genetic Engineering of Methanogens: Current Status," in Ref. 186, pp. 497-503.

192. S. Ghosh, "Innovative Anaerobic Digester Design," Paper presented at Alternative Energy in the Midwest," Schaumberg, IL, February 21-23, 1985.

193. P. H. Smith, "Multiphase Anaerobic Digestion of Biomass," in Ref. 166, p. 95.

194. D. P. Chynoweth *et al.*, "Biological Gasification of Renewable Sources," Annual Report (March 1984-February 1985), GRI-85/0151, September 1985.

195. G. C. Magruder *et al.*, "High Solids Digetion of MSW to Methane in a Stirred Tank Reactor," in Ref. 56, pp. 144-149.

196. Cf. Ref. 186, pp. 259-383.

197. "Biogas Plant Passes One-Year Mark," GRID 8 (4) 31 (1985/1986) Winter.

198. R. Biljetina et al., Biogasification of Walt Disney World Biomass Waste Blend, Eleventh Quarter Report, July-September 1985, for GRI, November 1985.

199. T. D. Hayes et al., "Methane from an Integrated Hyacinth Treatment Anaerobic Digestion Facility," in Ref. 116, pp. 1782-1793.

200. Personal Communication, W. H. Smith (U. of Florida, Gainesville, FL) January 30, 1986.

201. J. E. Albrecht, "Florida Waste Conversion Project Producing Methane," GRID 8 (1) 14-16 (1985) January/February.

202. H.T.D. Sjoberg and H. P. Mooij, "Anaerobic Digestion of Municipal Solid Waste, Potential Market Implications (Up-Date)," in Ref. 116, pp. 1794-1802.

203. Personal Communication, D. K. Walter (U.S. Department of Energy, Washington, D.C.) October 30, 1985.

204. P. Chartier, G. Donat, D. L. Klass, and E. J. Nyns, "Biogas: Present Situation and Development Perspectives," Paper presented at 13th World Energy Conference, Cannes, France, October 5-11, 1985.

205. N. A. da Silva, "Biogas in South America, Some Remarks on the Situation of Anaerobic Digestion in South America with Special Attention for Brazil," in Ref. 186, pp. 83-94.

206. R. Caceres and B. Chiliquinga, "Experiences With Rural Biodegesters in Latin America," in Ref. 186, pp. 31-39.

207. D. B. Mahin, Bioenergy Systems Report, March 1985: International Conference on Biogas Technology, U.S. Agency for International Development, Washington, D.C.

208. C. Xuan and X. Zhi-heng, "An Outline on the Biogas Development in China," in Ref. 186, pp. 3-14.

209. E. J. Nyns, "Success or Failure of Biogas plants in Europe, The Paradox of a Mature Technology," in Ref. 186, pp. 15-29.

210. J. H. Ashworth, Problems With Installed Commercial Anaerobic Digesters in the United States: Results of Site Visits, Revised Edition, Associates in Rural Development, Inc., November 6, 1985.

211. GRI, Methane from Biomass and Wastes Research, Renewable Resources for Localized Energy Production, (GRI, Chicago, IL, July 1984) 11 pp.

212. Policy Evaluation and Analysis Group, The Gas Energy Supply Outlook Through 2010 (American Gas Association, Arlington, VA, October 1985) p. 44.

213. H. W. Peter, "Recovery of Methane from Landfill," 16th World Gas Conference, Munich, Federal Republic of Germany, June 24-27, 1985.

214. Cf. a. E.V. Anderson, "Lead Cut Gives Alcohols Crack at Gasoline Blend Market," C&EN 63 (14) 7 (1985) April 8; b. "References Grappling Lead Phasedown in Gasoline," ibid., 63 (47) 30-31 (1985) November 25; c. E. V. Anderson, "Use of Gasoline-Ethanol Blends May Rise," ibid., 64 (1) 18 (1986) January 6.

215. Apparent Price/Octane Values, Alcohol Outlook 2 (1985) May.

216. J. R. Norton, III, Remarks of Deputy Secretary of Agriculture, 1985 National Conference on Alcohol Fuels, Washington, D.C., September 18, 1985.

217. E. V. Anderson, "EPA May Modify DuPont Waiver for Methanol Fuel Blends," C&EN 63 (35) 14-15 (1985) September 2.

218. Methanol, C&EN 64 (7) 19 (1986) February 17.

219. Cf. American Chemical Society Symposium, Lignocellulosic Pretreatments for Enzymatic or Acid Hydrolysis, Sponsored by Division of Microbial and Biochemical Technology, American Chemical Society, Annual Meeting, September 8-13, 1985, Chicago, IL.

220. P. K. Chidambareswaran et al., "Enhanced Enzymolysis of Cotton Fibres and Cotton Plant Stalks," in Ref. 219.

221. H. W. Park and D.I.C. Wang, "Dilute Acid and Thermal Modification of Lignocellulose," in Ref. 219.

222. W. Y. Chen and G. C. April, "Ethanol from Southern Hardwood Mixtures," in Ref. 166, p. 53.

223. C. M. Ostrovski and J. Aitken, "Fuel Ethanol from Poplar Wood Following Hydrofluoric Acid Solvolysis," in Ref. 166, p. 55.

224. E. C. Clausen and J. L. Gaddy, "Production of Ethanol from Wood by Acid Hydrolysis and Fermentation," in Ref. 166, p. 56.

225. J. W. Barrier et al., "Experimental Production of Ethanol from Agricultural Cellulosic Materials Using Low-Temperature Acid Hydrolysis," in Ref. 166, p. 57.

226. T. A. McCaskey and M. D. Rice, "Evaluation of Methods for Removal of Yeast Inhibitors from Acid-Hydrolyzed Hardwoods," in Ref. 166, p. 84.

227. A. Ferretti, "Acid Hydrolysis-Wood Hydrolysis Demonstration in New York State," Paper presented at Retsie 85, Anahaim, CA, June 3-6, 1985.

228. E. C. Clausen et al., "Effects of Acid Concentration and Temperature on the Hydrolysis of Wood Using Hydrochloric Acid," in Ref. 219.

229. M. C. Hawley et al., "Observations on the Gas-Phase Hydrogen Fluoride Saccharification of Wood Versus Pure Cellulose," in Ref. 219.

230. M. C. Hawley et al., "Reaction Rates for Gas-Phase Hydrogen Fluoride Saccharification of Wood," in Ref. 219.

231. R. Matharasan, "Acid Hydrolysis of Aspen Wood in Concentrated Slurries," in Ref. 219.

232. J. D. Wright, "The Batch Hydrolysis Reactor," in Ref. 219.

233. R. J. Cox, "Ethanol Production-Two-Stage Acid Hydrolysis of Hardwood, Pilot-Plant Design," in Ref. 56.

234. A Garcia III, E. R. Engler, and E. J. Soltes, "Biochemical Processing of Lignocellulosics," in Ref. 56, pp. 83-87.

235. J. L. Cabler, "Comparison of Lignin in Hardwood Residues from Dilute Acid Hydrolysis," in Ref. 56, pp. 88-93.

236. J. W. Barrier et al., "An Integrated Agri-Refinery System Using Concentrated Acid Hydrolysis of Lignocellulosic Materials," in Ref. 56, pp. 94-98.

237. H. E. Grethlein and A. O. Converse, "Understanding How Pretreatments for Wood Increase the Rate of Enzymatic Hydrolysis," in Ref. 219.

238. H. H. Brownell and J. N. Saddler, "Steam Pretreatment of Lignocellulosic Material for Enhanced Enzymatic Hydrolysis," in Ref. 219.

239. J. C. Linden, V. G. Murphy, and G. E. Maciel, "Ensiling and Sweet Sorghum as a Pretreatment for Lignocellulose Hydrolysis," in Ref. 219.

240. D.D.Y. Ryu and S. H. Kling, "Enhancement of Enzymatic Hydrolysis of Sugar Cane Bagasse by Steam, Explosion," in Ref. 219.

241. T. Ghose and P. Ghosh, "Rice Straw Pretreatment by Aqueous Solvent Process," in Ref. 219.

242. J. M. Gould, "Alkaline Hydrogen Peroxide Pretreatment to Enhance Utilization of Lignocellulose," in Ref. 219.

243. B. E. Dole and L. L. Herk, "Response of Lignocellulosic Materials to Ammonia Freeze Explosion," in Ref. 219.

244. T. P. Schultz and G. D. McGinnis, "Comparison of the Pretreatment of Lignocellulosic Materials by Four Different Processes," in Ref. 219

245. F. F. Seratone, R. W. Tock, and R. S. Chen, "Immobilized Cellulose Activity," in Ref. 219.

246. I. Mucke and W. Hartmeier, "Basic Trials to Co-Immobilize Algae and Yeasts for Ethanol," in Ref. 69, pp. 660-664.

247. W. Hartmeier, V. Forster, and C. Giani, "Ethanol from Unconventional Substrates Using Yeast Co-immobilized with Non-Yeast Glycosidases," in Ref. 69, pp. 665-669.

248. L. Leulliette, M. Henry, and D. Gros, "New Continuous Process for Production of Ethanol Using Immobilized Cells Reactor," in Ref. 69, pp. 689-691.

249. M. K. Handy and K. Kim, "Immobilized Yeast in Bioreactor for Alcohol Fermentation," in Ref. 166, p. 88.

250. R. A. Clyde, "Ethanol from Sugar Cane," in Ref. 166, p. 89.

251. M. J. Beck, "Kinetics of Hemicellulose Conversion to Ethanol by Pachysolen tannophilus," in Ref. 166, p. 85.

252. R. C. Strickland and M. J. Beck, "Evaluation of Several Fermentation Schemes for Ethanol Production from a Cellulosic Hardwood Hydrolyzate," in Ref. 166, p. 86.

253. J. Wiegel and J. Puls, "Ethanol from Pentoses and Pentosans by Thermophilic and Mesophilic Microrganisms," in Ref. 69, pp. 670-674.

254. L. Pasner, A. A. Quinde, and M. Meshgini, "ACOS-Accelerated Hydrolysis of Wood by Acid Catalyzed Organosolv Means," Paper presented at International Symposium on Wood and Pulping Chemistry, Vancouver, BC, August 26-30, 1985.

255. Personal Communication, L. Pazner (The University of British Columbia, Vancouver, BC) December 12, 1985.

256. Ethanol Pilot Plant for Brazil, Bio-Joule 7 (3) 12-13 (1984) September.

FOREST RESOURCE AVAILABILITY: MODELING THE SUPPLY CURVES

Sinyan Shen, Ph.D.
Argonne National Laboratory
Argonne, IL 60439

ABSTRACT

We analyze considerations that allow us to model the supply curves of energy wood as a coproduct of other forest-product wood. Examples in the Midwest are used to illustrate components of the model. It is pointed out that regional supply curves are not useful for most analyses and supply curves for marketing regions should be the target. These supply curves can be obtained by developing a harvesting model and applying it to a dynamic forest growth model.

FOREST RESOURCE AVAILABILITY: MODELING THE SUPPLY CURVES

The cost of wood fuel is often the key uncertainty for industrial or utility decision makers when considering the use of wood fuels on a large scale. The reason is that wood only fuels a small fraction of the industrial and utility facilities today, and most examples in the literature cite sources of wood fuel which are in the wood waste category, having negative value. Users of the wood waste normally just pay for the delivery. When considering using wood fuel at a scale beyond what waste wood can supply, the cost of wood fuel becomes highly uncertain and is site-specific. In establishing cost coals for biofuels derived from advanced fuel cycles under development by the U.S. Department of Energy, we also find the supply curve of wood for each region under consideration a critical missing link.

A general rule for industrial or utility decision maker's considering wood fuel is that if one can afford to pay for wood chips* produced in the region, one can have a steady supply of wood fuel. In this paper we analyze ways to establish the supply curve of energy wood in the form of chips. Regional supply curves are aggregates of supply curves to marketing areas within the region. Establishing the supply curve to a marketing area is thus the first step in the whole game. In fact, except for very global order-of-magnitude analyses, regional supply curves are seldom useful.

In the past, estimates for quantities of energy wood were often derived from applying residue factors to forest product industry's harvesting and production activities. A large fraction of the harvesting residue is not economically available, and many of these estimates are far from realistic. Further, the delivered costs of these residues are very uncertain.

It has now been recognized[1] that the coproduction of energy wood with traditional forest-product wood reduces the cost of managing, harvesting, and transporting wood for the traditional forest products. This recognition is the result of the ever-increasing commitment to achieve energy self-sufficiency by the forest product industry. Many paper and container producers procure wood chips jointly for energy and for pulp. Energy wood will probably always be a coproduct of traditional forest-product wood and to understand the supply of energy wood chips one must understand the close interaction between the production of high-value wood for the traditional forest product industry and the production of lower-value energy wood in the same forest. The overall management practice for the wood coproduction depends on the cover type, species diversity, land use, and location with respect to the marketing area in point.

In the Midwest, land clearing for strip mining has provided a low-cost source for energy wood, and in some cases for both energy wood and pulp wood to be used by paper and container producers[2]. This is one example in which the current land use activity combines

---

*Whole-tree chips today range from $15 to $22 per ton delivered. The average energy value as received is 9 million Btu per ton.

favorably with the cover type and the trees species diversity available on the land.

Dow Corning Corporation, in 1982, replaced their two oil-fired boilers for process steam and heat in Midland, Michigan with a wood-fired plant. Their wood fuel procuring method sets an example for ways to assure a reliable energy wood supply. Midland is situated in the maple-birch, the aspen-birch, and the elm-ash-cottonwood central hardwood region[3]. Most chippers chip to the north of Midland where wood chip production per unit area is higher than areas south of Midland. Dow Corning selected chippers to bid on contracts to harvest and haul Dow Corning stumpage purchased from private landowners. In addition, sawmills in the area were selected to provide waste wood.

The Dow Corning example shows a reasonable way of securing wood supply. In developing the supply curves for a marketing area, we adopt similar approaches. The first step is to characterize the natural resource available. The net annual growth of woody biomass for the 10 U.S. Department of Agriculture regions are shown in Figure 1. Such a map shows the regional distribution of forest growth, but provides no real insights to the cost of wood to a marketing area, such as Midland, Michigan. The supply curves to a marketing area are estimated by developing likely management scenarios for wood resources which have the following as driving inputs: biomass maps, access road location, terrain, landowner goals, and harvest schedules.

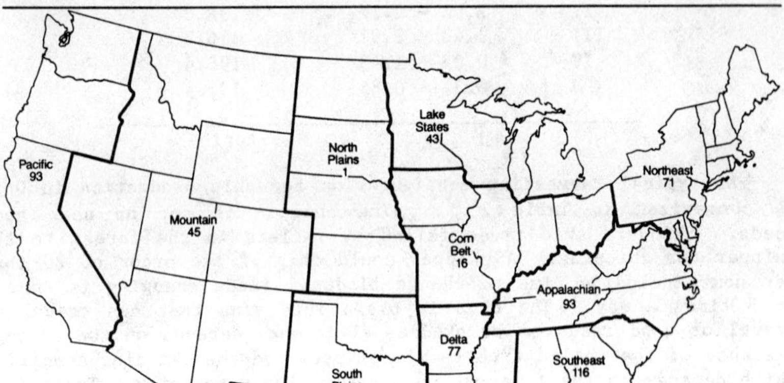

Figure 1 Annual rate of growth of woody biomass, 1980.

The biomass map shows the amount of stumpage, which may or may not be economically available. The access road location provides indication of accessibility as trees are felled and bunched and skidded to road heads for chipping. The longer the skidding distance, the higher is the cost of the wood chip. The terrain through which the skidder has to move determines the skidding speed. The goals of the landowner will determine to whom he is going to sell his stumpage. In our model, we will assume that all contracts are signed with chippers who decide where to get their stumpage. Finally, harvest scheduling is an art, and depends on all the above inputs as well as weather conditions, which can be considered a stochastic input.

Biomass maps in terms of the quantity of biomass per unit area have been established for a number of states in the Midwest[4]. Other regions are working on establishing similar maps or databases. The annual growth in a forest can usually be estimated by determining its site class as classified by the U.S. Department of Agriculture. The range of annual growth for different site classes in the 495 million acres of commercial forest land is presented in Table 1[5].

Table 1 Forest Annual Growth for Different Site Classes

Site Class	Annual Growth (dry ton/acre)	Area (million acre)
I	2.77 +	13.6
II	2.02 - 2.77	38.0
III	1.43 - 2.01	116.3
IV	0.85 - 1.43	195.4
V	0.34 - 0.85	131.4

As typical harvesting configuration for chip production in Ohio is summarized in Table 2[6]. The chipper is set up near good roads. Skidders skid trees felled by fellers in the forest to the chipper for chipping. A chipper could chip of the order of 20 tons per hour including time to change blades. Blade changing is done 3 to 5 times a day. The chip is blown into vans that can travel on gravel or good roads. The skidder efficiency depends on the average distance of the felled tree to the chipper and the skidding condition which depends on the time of the year and the topography. Typically, skidding up hill is slow and expensive. Up to 5% slop is easily handled by skidders usually.

The cost of the wood chips as they get blown into the van is broken down into three parts: the felling cost C_f, the skidding cost C_{sk}, and the chipping cost C_c. To be added to these costs are the storage cost C_{st} and the hauling cost C_h, and this procedure completes the estimate for the delivered cost of the wood chip. The factors affecting C_c include chipper capacity (in tons/hour) and how dirty the trees are, and C_c does not vary a lot for a given chipper. C_f and C_{sk} on the other hand, are very site-specific. C_f depends on the average tree diameter and diameter distribution. Also

Table 2 Typical Configuration of Forest Harvesting/ Chipping Operation in Ohio

Equipment	Number	Unit Cost ($)	Capital Cost ($)
Fellers*	2	300-400 per chain saw which lasts 3-4 months	-
Skidder	3	60,000	180,000
Chipper	1	150,000	150,000
Vans	3	30,000	90,000
Trucks	varies	-	30,000
Total			450,000

*Usually men with chain saws, or mechanical feller-buncher

the tree species may affect branching pattern, thereby restricting the use of mechanical feller-bunchers. For example, oak has many branches, making the use of feller-bunchers almost impossible. Other factors causing C_f to go up include wet and muddy site condition, and small size of cutting area which decreased the cutter's familiarity with the cutting area. C_{sk} is mostly influenced by the intermediate topography, average diameter and diameter distribution and site conditions.

To summarize, the modeling of the supply cost of wood chip can be divided into several steps. For a single chipping point near a road and involving a single type of chipping equipment, C_c is relatively well defined. The average topography and average annual site condition influence C_{sk}. The average species type and average diameter class both influence C_f, and the average diameter class also influences C_{sk}. Further, since a single chipper set-up requires almost a full working day, the production cost of all the wood chip just beyond the chipper for a single chipper set-up point should be uniform.

With the above analysis out of the way, it is now possible to determine the quantity of wood available at a given price to a given marketing area, if forests did not grow and there were no replanting activity. Further, Bradley[7] pointed out that "while it surprises many, trees are perishable; especially in the vast areas of over-mature, short-lived species in the Lake States", and that "current methods to estimate harvest potential or supply have inhibited the recognition of the mortality problem we now face — a problem largely unavoidable in the short run." To accurately describe the dynamic situation in forest growth, mortality, and reforestation, the harvesting model has to be linked to a forest growth model which takes into account of the dynamic situation.

In order to schedule harvests in predesignated time periods, we must also estimate future demands in each of these periods. We have to know the demand for each of the species within the cover type. Certain species of certain sizes are definitely higher-valued than others, but an overall management strategy taking into account joint energy and non-energy wood production decreases the cost of both, and at the same time permits more effective natural resource enhancement, making our forest more productive.

The prerequisites to the determination of accurate supply curves for energy wood chip are thus: (1) accurate assumptions of likely demand scenarios for non-energy and energy wood, (2) a dynamic growth model for the forests likely to supply the given marketing area, and (3) accurate observation of felling and skidding rates for subareas of the forests under consideration for average climatic conditions. The first prerequisite is thus far the most difficult to achieve. Dynamic growth models are available for a number of well-known forest areas, and the development of a harvesting model appears to be under control.

REFERENCES

1. "An Analysis of the Timber Situation in the United States, 1952-2050," U.S. Department of Agriculture, Forest Service, Forest Resource Report No. 23 (1980).

2. Stone Container Corporation, unpublished information (1985).

3. Spencer, Jr., J.S., "Michigan's Fourth Forest Inventory: Area," U.S. Department of Agriculture Resource Bulletin NC-68 (1983).

4. Bancroft, D., Council of Great Lakes Governors, personal communication (1985).

5. Hyde, W.F. and F.J. Wells, "The Potential Energy Productivity of U.S. Forests," in Energy Sources, Vol. 4, Number 3 (1979).

6. Kasile, J. Argonne National Laboratory and Ohio State University, unpublished information (1985).

7. Bradley, D.P., Forest Service, U.S. Department of Agriculture, personal communication (1984).

RESEARCH STRATEGY FOR ATTAINING HIGH YIELDING *POPULUS* ENERGY PLANTATIONS

Edward A. Hansen, M.S.
Project Leader
Forestry Sciences Laboratory
North Central Forest Experiment Station
Rhinelander, WI 54501

ABSTRACT

Large production increases of 3-5x have been attained during the last 20 years through application of agronomic principles to forest plantations. Yields of 15-20 $Mg \cdot ha^{-1} \cdot yr^{-1}$ from test plots are not uncommon. But high yields attained in small plot trials have not been achieved in larger plantations elsewhere because of three factors: 1) poor adaptation of the genotype to site (generally poor moisture stress tolerance), 2) disease susceptibility, and 3) weed competition. Past yield studies have been directed towards achieving high yields by optimizing environmental variables. High yields have been obtained in those instances when the correct genotype (clone) was planted on the proper site, good weed control was practiced and no disease problem occurred. However, the costs of intensive culture of tree plantations are high and the risks of low yields or failure are great. New hybrid clones are needed to reduce the occurrences of failure and bring average yields closer to the potentials observed in test plantations. New efforts in traditional breeding and in biotechnolgy hold promise for developing high yield genotypes that are both herbicide and disease resistant and tolerant of low levels of soil moisture. Identifying physiological traits related to internal moisture control will assist in selecting individuals better adapted to low levels of soil moisture. Developing herbicide resistant poplar clones will allow more effective weed control with reduced costs. Along with continued improvements in cultural management, these efforts will increase average energy yields from intensively cultured plantations by producing clones that attain high yields on a variety of sites while reducing risks of losses from disease and weed competition.

RESEARCH STRATEGY FOR ATTAINING HIGH YIELDING *POPULUS* ENERGY PLANTATIONS

Research and demonstration of growing high yield hardwood plantations under short rotations and intensive culture (SRIC) began in the early 1960's (7,8) The objective of this early work was to produce high fiber yields for a variety of forest products. With the 1973 oil embargo, interest accelerated greatly in growing SRIC plantations for energy production. Although energy was a new use for such biomass, there was little difference in the establishment, culture or harvesting of SRIC plantations, whether the biomass was used for wood products or fuel.

More than 20 years have elapsed since the original introduction of the SRIC concept, and about 10 years since the acceleration of research directed specifically at developing biomass plantations for energy. The development of SRIC technology is to the point that we know that large yields are possible, but they are not obtained consistently in field plantations. In this paper, I will briefly examine reported high yields from small experimental plots and present limited data on field plantation yields. Using these data and the experience of historical yield gains in agricultural crops as an analogy, I will speculate on the possibility of further yield increases from SRIC field plantations. I will then identify factors currently limiting field plantation yields from reaching those achieved in small plots and discuss a research approach to obtain substantial biomass yield increases from SRIC plantations.

AGRICULTURAL YIELD TRENDS

Records show large gains in agricultural productivity in recent decades. For example, in Champaign County, Ill., corn yields tripled (Figure 1) and soybean yields doubled from 1935 to 1975 (2). To measure genetic gain, old and new hybrids were planted at modern plant population densities in a common fertile field environment. Hybrids released in recent years performed better than older hybrids. The adaptation of new genetic material to fertile environments contributed about half of the yield gain with the remaining half due to improved cultural practices.

SRIC YIELD TRENDS

Two years ago we began a study to measure yield gains achieved during the past 15 years of SRIC research. A clone in use 15 years ago was planted under both the original old-culture conditions and present-day culture. Two recently identified faster-growing clones were also planted under present-day culture, all in the same field environment. Although we only have preliminary data from this projected 4-year study, there are already some interesting trends apparent.

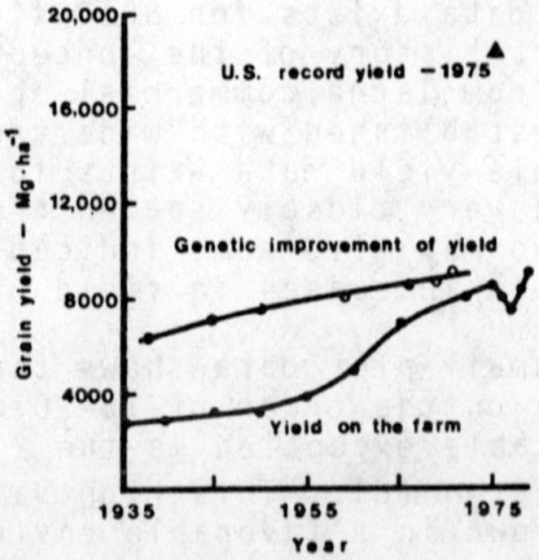

Figure 1. Genetic improvement of yield, grain yield on the farm, and record yield for corn during a 40-year period. Genetic improvement was determined by growing introductions from the years indicated in a common field environment in Iowa; current planting densities and techniques were used. Grain yield on the farm was determined from records for Champaign County, Illinois. (Adapted from Boyer 1982.)

clone	culture	yield $Mg \cdot ha^{-1} \cdot yr^{-1}$
NC-5260 (old)	old	2.6
NC-5260 (old)	new	4.5
NE-386 (new)	new	5.3
NE-41 (new)	new	8.6

First, there have been substantial yield increases with the new clones and culture on the order of 5 $Mg \cdot ha^{-1} \cdot yr^{-1}$ by the end of the second year (yields won't peak until the fourth year). Second, these yield increases result from both cultural gain and genetic gain. These observations parallel that of agricultural experience. Two "new" clones were used because clone NE-41 is believed to be a high-risk clone due to its poor climatic adaptation and disease susceptibility. However, if the short rotation period coincides with favorable weather, NE-41 may achieve very high yields indicative of the potential of future genetically improved material.

THE YIELD GAP

Record field yields of agricultural crops far surpass average field yields. For grain crops (corn, wheat, oats, soybeans, etc.) record yields are 4-7 times average yields (2). For crops with marketable structures (potatoes and sugar beets) record yields are 3 times average yields. These data indicate that there is potential for further yield gains.

No comparable data exists for SRIC field yields because of the short history of the concept. Only limited yield data exists from large commercial scale plantations which are usually established with wide tree spacing. However, considerable yield data exists for small SRIC research plots with very closely spaced trees. A comparison of these two may give some indication of the potential for future yield increases in field plantations.

A summary of small-plot data shows that higher reported yields are on the order of 15-20 $Mg \cdot ha^{-1} \cdot yr^{-1}$ (Table 1). One notable exception is the 27.8 $Mg \cdot ha^{-1} \cdot yr^{-1}$ reported for Washington (6). This high value reflects new improved hybrids grown in a favorable environment.

Table 1. Woody Biomass Production of Populus[1/]

Clone	System[2/]	Age Years	Yield $Mg \cdot ha^{-1} \cdot yr^{-1}$	Location[3/]
Var. hybrids	first	4	11.3	KY
NE: 388, 252, 49	"	3	6.0-8.0	PA
Triochocarpa	"	8	9.6	WA
'Tristis #1'	" SRIC	4	15.7	WI
'Tristis #1'	" SRIC	5	8.4	WI
'Tristis #1'	" SRIC	10	10.4	WI
NE-299 (NC-5331)	" SRIC	6	12.3	WI
Trichocarpa	"	5	10.0	GB
Euramericana I-476	"	10	11.9	I
I-476	"	14	13.7	I
I-214	"	10	7.0-11.0	I
Triploid aspen	sucker	15	7.3	LS
Euramericana	coppice	3	10.0	IA
Euramericana	" SRIC	4	7.6	WI
Euramericana	" SRIC	2	5.0-19.1	O
Euramericana	" SRIC	2	15.0	O
Trichocarpa	"	2	11.7	WA
Trichocarpa	"	2	10.9	WA
Trichocarpa	" SRIC	4	27.8	WA
Var. hybrids	"	5	9.5-18.0	F
Deltoides x trichocarpa	" SRIC	1	16.0-18.0	S
Alba x glandulosa	"	2	18.8	SK

1/ Adapted from Zavitkovski and Isebrands 1983 with additions.

2/ First and coppice (or sucker) refer to plantations established with cuttings (or seedlings) and by resprouting, respectively. SRIC means short rotation intensive culture including fertilization and irrigation.

3/ Abbreviations: KY, Kentucky; PA, Pennsylvania; WA, Washington; WI, Wisconsin; GB, Great Britain; I, Italy; LS, Lake States; IA, Iowa; O, Ontario; F, France; S, Sweden; SK, South Korea.

In contrast to these reported high small-plot yields are the considerably lower SRIC field yields. More than 1000 ha of hybrid poplar plantations planted in the lower peninsula of Michigan in the late 1970's reached less than 2 m height in 3 years (13). For practical purposes, most of those plantations could be considered failures. The most extensive hybrid poplar planting program in North America is that of the Ontario Ministry of Natural

Resources. More that 10,000 ha have been planted to date. Initial harvests of 10-year-old field plantations yielded about 2 $Mg \cdot ha^{-1} \cdot yr^{-1}$ (R. Evers, personal comm.), about the same as the average from natural stands in the North American temperate zone (14). A later harvest in 1977 gave a yield of 2.9 $Mg \cdot ha^{-1} \cdot yr^{-1}$ at 6½ years with a projected yield of 3.8 $Mg \cdot ha^{-1} \cdot yr^{-1}$ at 10 years (Ontario Ministry of Natural Resources 1983). Other plantations presently 6 years old are projected to produce yields of 3-5 $Mg \cdot ha^{-1} \cdot yr^{-1}$ when harvested at about age 10. Still younger plantations recently planted with better adapted clonal stock are expected to produce even higher yields (e. g. 6.4 $Mg \cdot ha^{-1} \cdot yr^{-1}$) (1).

These data show that gains are being made in SRIC field plantation yields and that yields are 2- to 3-times that of average natural stands. Yet SRIC field yields lag considerably behind record yields from small plots. Record small plot yields of 15-20 $Mg \cdot ha^{-1} \cdot yr^{-1}$ are 3- to 4-times that of present field yields. But agricultural data, and our own intuition, tell us there will always be some gap between record and average yields. The question is, can we realistically expect to increase average field yields? To help answer this, let's first examine some of the reasons for the difference between small-plot and field yields.

Small plots are typically characterized by highly intensive culture (usually tended by hand), closely spaced trees and consequently shortened rotations of often only 2-5 years. Shortened rotations increase the chance that at least some plots will have their entire rotation encompassed by a period of better than average growing conditions and hence record yields would tend to increase. As plot size decreases so does soil heterogeneity, which tends to increase record yields from those small plots situated entirely on the best site. With small plots many more clone-site combinations have been tried than with the large field plantations. Consequently we would expect some small-plot yields to be highest because of a greater chance for achieving a good clone-site combination. Although these factors will always result in a yield gap between small plots and field plantations, other factors depressing field yields are controllable. Based on our experience (5) and reports of others (4,9,10), I propose that inadequate weed control, disease, and poor adaptation of clone to site are the three major factors contributing to growth losses in field plantations. Consequently, these areas are where future research may achieve the greatest yield gains in field plantations. Although these factors also reduce yields in small research plots, small plots make excellent weed control possible, and the large number of plot trials results in at least some plots being established with disease resistant and well-adapted clones. Therefore the occasional record small-plot yield indicates what the field yield could be when these factors are optimized. The influence of each of these factors that reduce field yields will be discussed briefly in turn.

Cultural advances have been rapid in SRIC forestry because principles and practices developed in agriculture have for the most part been directly transferable. Many of the techniques in site preparation, planting, mechanical weed control and fertilization were directly transferable to SRIC and required little or no equipment modification (5). However, weed control with herbicides has lagged because herbicides used on agricultural crops were not always suitable for hybrid poplars, new rates and timing of application needed to be developed, and the herbicides were not usually labeled for use on trees. Poor weed control results in increased stress to trees, early growth loss, and occasional failure of the plantation.

Susceptibility of hybrid poplars to a number of important diseases is a major problem in every poplar program (10). Each geographic region must begin with a program of testing and selecting clones that are disease resistant for that area. To be fully effective such a testing and selection program should be supported by tree breeding and/or biotechnology to provide a continuous flow of new genetic material to the program. The majority of hybrid poplar clones presently in use in the United States were not bred specifically for SRIC. They are used simply because of availability.

Poor adaptation of clone to site is the third area where considerable yield losses occur. After eliminating disease susceptible clones and clones unsuited to the local clim-ate, large clonal variations in yield still remain. The largest factor in agriculture contributing to low yields is droughty and shallow soils. Together they are a problem on 45 percent of the agricultural soils in the United States (2). More than 40 percent of the insurance indemnities for crop losses from 1939 to 1978 were for drought. Most past efforts to improve plant productivity have involved selecting for superior yield in fertile environments. But most environments are less than ideal for plant growth. Selecting varieties (clones) that grow best on moderately stressed sites might be an effective way to raise average field yields. This should be particularly true for SRIC which will not utilize irrigation to the extent agriculture does.

It was previously noted that half of the yield gain for corn and soybeans from 1935-1975 was due to genetic improvement. Further examination showed that the yield gain of the newer soybean cultivars correlated with an improvement in their afternoon water potential when growing under adequate soil moisture conditions (2). This correlation was discovered after-the-fact. If it had been known that maintenance of afternoon water potential resulted in higher yields, this trait might have been improved sooner (or to a greater degree) by breeding and selecting specifically for it. Similarly it has been shown that fast growing hybrid poplar clones maintain higher leaf water potential under conditions of soil moisture deficit than do slower growing clones, and some fast growing clones show pronounced stomatal sensitivity to vapor pressure gradient (11). Therefore, the potential exists to breed and select for specific physiological traits to increase yield of hybrid poplars growing under moderate soil-moisture stress. An added advantage of identifying physiological traits that correlate well with yield is that such traits may be relatively stable with tree age. If this proves to be the case, then such traits may allow early selection of clones for specific environments. Present clonal selections are based on growth and are reliable only after tree age approaches the length of the planned rotation (3,12), and then only for the soil moisture environment under which the clone is tested.

RESEARCH STRATEGY

Cultural advances in SRIC have been rapid because of the transfer of technology from agriculture. It is possible that the major yield gains from cultural improvement have already been achieved. In contrast, genetic advances in SRIC have lagged because genetic gains in agriculture are not transferable and the lengthy breeding cycles for most tree species seriously reduces the rate of genetic improvement. The initiation of biotechnology promises to greatly accelerate the rate of genetic improvement, particularly for long-generation plant groups such as trees. Nonetheless, whether genetic gains prove to be rapid or slow, it seems likely that genetic improvement will be the largest contributor to future increases in SRIC yields due to the advanced state of culture and the relatively youthful status of tree breeding and biotechnology. However the relative contributions of culture and new genetic material will vary by the specific problem addressed. The linkage between the major research approaches and the three identified major problems depressing field yields are depicted in Figure 2 and discussed next.

Weed Control

Weeds can be successfully controlled in SRIC plantations but due to the many combinations of weed populations and soil types and the vagaries of weather, good weed control is not consistently attained. Unlike agricultural crops, SRIC plantations do not develop a large canopy during the first growing season to shade the site and suppress weed growth. Typically, "site occupancy" does not occur for 2 or more years. To control weeds during the

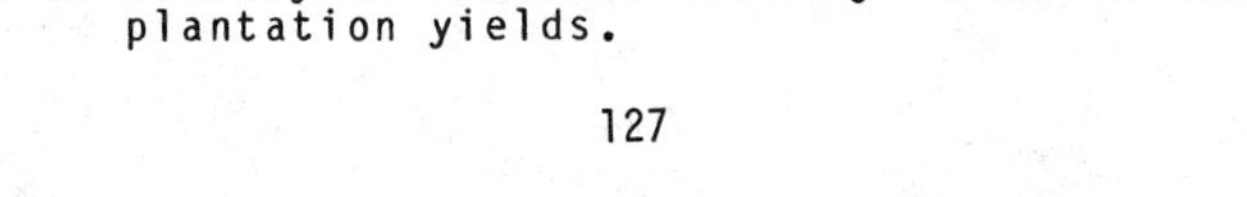

Figure 2. Summary of research strategies to increase SRIC plantation yields.

lengthy period of plantation establishment, longer-lasting pre-emergent herbicides are needed. When weeds do invade during the growing season, an effective contact herbicide not harmful to poplars is needed. One approach to this problem is further testing of rates and timing of herbicide application to develop more effective herbicide prescriptions. This approach seems essential since by law herbicides can only be used on crops that are specified on the herbicide label (4). Poplar crops are seldom included on the labels and to be added requires a substantial body of research information. Another possible approach to controlling weeds is to develop poplar variants that are tolerant to existing approved and effective herbicides. It is through this approach that current biotechnology efforts may make a major contribution by developing herbicide resistant poplars.

Disease

Diseases are "controlled" in forestry (and agriculture) primarily by substituting a disease resistant variety for a susceptible variety. Likewise the strategy with SRIC has been to identify and use disease resistant clones. Although intensive culture tends to reduce stress and hence disease severity, its effect on ameliorating disease is marginal at best. The proven approach has been the continual breeding and selection of new varieties from an ongoing breeding program. It is common knowledge that active breeding programs are the basis for our highly productive agricultural crops. Less well known is the fact that tree breeding programs by Dr's Steenackers and Zsuffa are the basic ingredient for the extensive poplar programs in Europe and Canada respectively. Again, biotechnology holds the promise of quickly producing and identifying variants that are resistant to specific diseases.

Clonal Adaptation

There is a need to identify disease resistant clones that grow rapidly on specific sites. The usual approach is to plant a set of clones on several sites and observe their growth. This approach needs to be continued. Intermediate sites with various degrees of moisture stress need to be included. Identifying clones that have fairly high yields even on moisture-stressed sites will help boost average yields from large plantations that cover several soil types. Identifying physiological traits that correlate with fast growth under moderately stressful growing conditions would be highly beneficial. If such traits can be identified, they may allow early selection of fast growing clones, and also provide a means for selecting parents for a breeding program to produce drought tolerant progeny.

SUMMARY

The major factors preventing attainment of high yields with SRIC hybrid poplar energy plantations are inadequate weed control, disease susceptibility and poor adaptation of clone to site (particularly on areas with moisture stress).

A multifaceted approach is required to provide further major increases in SRIC yield. This approach involves contributions from culture, biotechnology, breeding and physiology. More specifics on potential contributions from breeding and biotechnology will be given in papers that follow in this session. SRIC yields have increased considerably from research conducted during the past 15 years. Based on the substantially higher small-plot yields in SRIC and the historical agricultural record of yield increases, it is plausible to expect that further research will result in additional major yield gains from SRIC field plantations.

REFERENCES CITED

1. Barkley, B. A. A silvicultural guide for hybrid poplar in Ontario. Ontario Ministry of Natural Resources, Brockville, Ontario. 35 p. (1983).

2. Boyer, J. S. Plant productivity and environment. Science. 218: 443-448. (1982).

3. Cooper, D. T. and R. B. Ferguson. Avoid early selection for growth rate in cottonwood. Proc. Southern Forest Tree Improve. Conf. 15: 52-58 (1979).

4. Dickman, D. I. and K. W. Stuart. The culture of Poplars in Eastern North America. McNaughton & Gunn, Inc., Ann Arbor, MI. 168 p. (1983).

5. Hansen, E., L. Moore, D. Netzer, M. Ostry, H. Phipps and J. Zavitkovski. Establishing intensively cultured hybrid poplar plantations for fuel and fiber. USDA Forest Service. Gen. Tech. Rpt. NC-78. 24 p. (1983).

6. Heilman, P. E. and R. F. Stettler. Genetic variation and productivity of *Populus trichocarpa* and its hybrids. II. Biomass production in a 4-year plantation. Can. J. For. Res. 15: 384-388 (1985).

7. Herrick, A. M. and C. L. Brown. A new concept in cellulose production Silage Sycamore. Agr. Sci. Rev. 5(4): 8-13 (1967).

8. McAlpine, R. G., C. L. Brown A. M. Herrick and H. E. Ruark. "Silage" sycamore. Forest Farmer 26(1): 6-7, 16 (1966).

9. Ontario Ministry of Natural Resources. New forests in eastern Ontario, hybrid poplar. Science and Technology Series, Volume 1. 336 p. (1983).

10. Ostry, M. E. and H. S. McNabb. Diseases of intensively cultured hybrid poplars: a summmary of recent research in the north central region. In: Intensive plantation culture: 12 years research. USDA Forest Service Gen. Tech. Rpt. NC-91 p. 102-109 (1983).

11. Pallardy, S. G. and T. T. Kozlowski. Water relations of Populus clones Ecology 62(1): 159-169 (1981).

12. Wilkinson, R.C. Inheritance and correlation of growth characters in hybrid poplar clones. Proc. N.E. Forest Tree Improve. Conf. (1972) 20: 121-130 (1973).

13. Woods, R. F., L. M. Moore, L. F. Wilson, M. E. Ostry and D. I. Dickmann. Performance of 3-year-old hybrid poplar clones in relation to soil texture, natural soil drainage, and pest incidence. Proc. 19th North American Poplar Council Annual Meeting. Rhinelander, WI. July 20-22, 1982. p. 74-83 (1982).

14. Zavitkovski, J. and J. G. Isebrands. Biomass production and energy accumulation in the world's forests. The Seventh International FPRS Industrial Wood Energy Forum '83. Volume I. Nashville, Tennessee, Sept. 19-21, 1983. p. 12-22 (1983).

DISCUSSION

W. Heckrodt, Consultant, Kimberly-Clark: I'm planting hybrid poplar extensively. One of my questions concerns stressing or competition. Does hybrid poplar recover from either a dry season or from weeds or whatever, assuming that you have the site and the growth potential? There is mixed opinion on this as to whether it would recover.

Ranney: We're learning an awful lot about each of these clones and how they perform over their life or rotation, but we've got an awful lot more to learn. Very little work has been done on below ground changes during times of stress, drought, or disease problems. We don't know what the effects are, and how to anticipate the response to some kind of stress. There's no answer yet as to exactly how they will respond. It's going to be a variable by species and by clone. I would ask the same questions with many species. Tolerance to stress, and being able to bounce back should be built into these plantations.

D. Pratt, U. of Minnesota	You mentioned the importance of site variation. In your judgement, is that due more to the range of conditions that you find in these marginal sites that you're dealing with, or is it due to the fact that the kinds of species and varieties that you're working with are less adaptable than let's say a typical agricultural crop?
Ranney	It's both of them; you can't get away from either one. We know some of the poplar clones are quite adaptable across some sites, but their productivity is not quite as high. We're paying some price for that adaptability. We are trying to find the highest productivity clones that are the most adaptable. With hardwoods, we generally have a harder time defining the kinds of sites that are appropriate than with conifers. I have seen a lot of evidence that indicates we can get higher productivity than that from the pines.
J. Keppeler, Ronco Consulting Corp.	You assigned one-half of yield improvement to genetic development, and one-half to culture. Did I understand that correctly, and is that based on your experience and judgement, or do you have some empirical data for that?
Ranney	The data presented were from Ed Hansen's experiments at the age 2, and they are empirical evidence. In general, one would have to add a lot of caveats, because I could say for sycamore, for example, that all our improvement has been pretty much derived from cultural treatments. We have not really even tapped the genetic improvements that are possible. It depends very much on the species. Black locust and silver maple, for example, are excellent short-rotation species. But we don't know anything about what genetic variability there is to improve. So any improvement that has been made has come from cultural treatments to this time. With hybrid poplar, we know more about them. There have been some genetic improvements made for increased productivity. The evidence that I gave this morning from Ed Hansen's work is about as close as we get to solid empirical evidence of that contribution.

THE ROLE OF GENETIC IMPROVEMENT IN THE SHORT ROTATION WOODY CROPS PROGRAM

P. A. Layton and L. L. Wright
Research Associates, Short Rotation Woody Crops Program
Environmental Sciences Division
Oak Ridge National Laboratory
Oak Ridge, Tennessee 37831

ABSTRACT

The U.S. Department of Energy is sponsoring a long-range research program, the Short Rotation Woody Crops Program (SRWCP), to develop short-rotation intensive culture (SRIC) approaches to woody biomass energy production. A major effort in the SRWCP is species screening and genetic improvement of selected species. Of the 125 species initially evaluated for SRIC, only 20 are being seriously considered now, with most of the emphasis in the SRWCP being on 16 hardwood species.

Range-wide seed collections of 12 species were provenance tested; these include Platanus occidentalis (sycamore), Alnus glutinosa (European black alder), and Robinia pseudoacacia (black locust). Based on the results of these tests, highly productive, site-specific seed sources are being chosen for several geographic regions. Three of these species are currently being bred for increased productivity in SRIC systems.

Genetic improvement is viewed as a major tool for increasing productivity, having anticipated gains of 40 to 50%. In addition to breeding, the techniques of somaclonal screening and genetic engineering are being evaluated for their usefulness in the SRIC improvement program. Currently, salt-tolerant Atriplex canescens (four-wing saltbush) and herbicide-resistant Populus spp. are being sought via somaclonal screening.

The SRWCP plans to increase the emphasis on genetic improvement of SRIC species through new initiatives encouraging use of a combination of classical breeding techniques and biotechnology.

THE ROLE OF GENETIC IMPROVEMENT IN THE SHORT ROTATION WOODY CROPS PROGRAM

INTRODUCTION

In response to the energy crisis of the mid-1970s the Department of Energy (DOE) sought significan renewable energy sources within our national borders. On of these renewable sources of energy was biomass, and, o the various kinds of biomass, wood soon emerged as one o the best energy feedstock candidates. With the increase utilization of biomass, however, it became apparent that highly productive sustained source of woody biomass woul eventually be needed to ensure dependable supplies fo energy use. As the demand for wood and wood product increases, short-rotation intensive culture (SRIC) offer the most viable option for ensuring that supply keeps i step with demand.

SRIC is a silvicultural system that utilizes clos spacings (1.5 to 2 m between trees), improved tre species, fertilization, and intensive weed control. Th most economically viable stand reestablishment schem relies on coppicing cycles of 4 to 10 years (19). Because conifers generally do not coppice, hardwoods ar by far the most desirable species for SRIC programs Hardwoods also exhibit a high degree of rapid juvenil growth and are very responsive to optimized cultura regimes.

DOE named Oak Ridge National Laboratory as technica advisor to the national biomass program. The Shor Rotation Woody Crops Program (SRWCP) evolved from thes beginnings at ORNL with the directive to concern itsel with long-term, high-risk research and with as muc industrial involvement in research and development a practical. One of the highest priorities of the progra was the genetic improvement of noncommercial tree specie for high productivity and special energy qualities. Othe research included optimization of management practice such as spacings, nutrients, competition control, an moisture regulation.

Zobel and Talbert (35) stated that all beginning tre improvement programs start with a determination of th species, or geographic sources within a species, tha should be used in a given area. The second stage consist of determining the amount, kind, and cause of variatio within a particular species. Many times this informatio is obtained during provenance or progeny testing. Th third phase involves the breeding and genetic engineerin of the desirable qualities into individual trees. Thes improved trees are then mass produced for planting Breeding is a continuing process that searches fo beneficial plant attributes to combine with current traits. The SRWCP began genetic improvement efforts in a similar manner. Species screening was clearly the first step.

SPECIES SCREENING

More than 800 woody species are indigenous to North America, and many have been candidates for SRIC selection. In 1975, Dickmann (3) summarized the attributes he considered essential in a species chosen for intensive culture. The ideal species is characterized by (a) rapid juvenile growth, (b) continuous growth throughout the growing season, (c) an upright growth pattern with upsweeping branches, (d) easy establishment and regeneration, (e) acceptable wood properties, and (f) no major insect or fungal pests. Within the SWRCP, 142 of 181 species suggested for use in SRIC systems have received preliminary screening. This process eliminated many species and showed the high potential of a few. Currently, 76 species are involved in the screening process, including 40 species recently planted in screening trials in the Southwest and a large number of Eucalyptus spp. being evaluated in the subtropics (Table 1). Many species showing promise for energy use have had little economic potential in conventional forestry.

Species-selection research has proceeded on a regional basis because each region has different soil and climatic characteristics. Species screening will continue as long as there is interest in growing woody crops, especially exotic species, for energy. However, it is not currently the major thrust of the SRWCP, as several species have been identified for most geographic regions of the country. Much of the species-screening research was completed during the last 2 years and now the emphasis has moved to the next step, provenance testing.

Species evaluated for SRIC in the Pacific Northwest (PNW) were Alnus rubra (red alder) and Populus trichocarpa (black cottonwood) and its hybrid with Populus deltoides (eastern cottonwood). Current mean productivity for these species obtained under optimum SRIC conditions in the PNW are 7.4, 12.5, and 20.3 dry Mg ha^{-1} year^{-1} at age four, respectively (12). Rotation length recommendations for A. rubra are the longest of all SRIC species at 10 years (4,11). The Populus hybrids showed the highest productivity of any species evaluated for SRIC in the continental United States. Currently there are no research projects to evaluate other species for use in the PNW, although interest has been expressed in evaluating Acer macrophyllum (bigleaf maple).

Dickmann (3) recommended 14 conifers and 14 hardwoods for SRIC in the Lake States region. Twenty-three of these

Table 1. Species currently being screened for use in short-rotation intensive culture systems.

Southwest	
Acacia abyssinica	*Acacia pennatula*
Casuarina equisetifolia	*Cupressus arizonica*
Leucaena diversifoli x leucocephala	*Leucaena leucocephala* (several varieties)
Pinus eldarica	*Sapium sebiferum*
Sambucus mexicana	*Prosopis juliflora*
Prosopis specifera	*Prosopis tamarugo*
Zizilphus jujuba	*Acacia arabica*
Acacia karoo	*Acacia tortilis*
Acacia victoriae	*Acacia senegal*
Acacia nylotica	*Albizia falcataria*
Albizia julibrissa	*Albizia lebbeck*
Dalbergia sissoo	*Eucalyptus gunnii*
Eucalyptus globulus	*Eucalyptus camaldulensis*
Eucalyptus torquata	*Casuarina cristata*
Casuarina glaca	*Melia azedarach*
Ailanthus altissima	*Chrysothamnus nauseosus*
Elaeagnus angustifolia	*Fraxinus velutina*
Acacia minuta	*Eucalyptus microtheca*
Prosopis chilensis	*Ziziphus spina-cristi*
Quercus virginiana	*Cucurbita foetidissima*
Southeast	
Alnus glutinosa	*Catalpa speciosa*
Fraxinus pennsylvanica	*Paulownia tomentosa*
Quercus nigra x phellos	
Lake States	
Populus trichocarpa	*Fraxinus pennsylvanica*
Salix spp.	*Populus deltoides*
Gleditsia triacanthos	*Populus grandidentata*
Quercus	*Populus balsamifera*
Subtropics (includes Hawaii and South Florida)	
Acacia auriculiformis	*Acacia confusa*
Acacia koa	*Acacia mangium*
Acacia melanoxylon	*Albizia falcataria*
Casuarina equisetifolia	*Copaifera langsdorfi*
Eucalyptus alba	*Eucalyptus botryoides*
Eucalyptus camaldulensis	*Eucalyptus citriodora*
Eucalyptus dunnii	*Eucalyptus globulus*
Eucaluptus grandis	*Eucalyptus amplifolia*
Eucalyptus maidenii	*Eucalyptus microcorys*
Eucalyptus robusta	*Eucalyptus saligna*
Eucalyptus tereticornis	*Eucalyptus urophylla*
Eucalyptus viminalis	*Mimosa scabrella*
Eucalyptus nitens	

species were tested in species trials on old-fields in Michigan. Based on fifth-year results, the hardwood species recommended for Michigan were Robinia pseudoacacia (black locust), Populus tremuloides x Populus alba (hybrid aspen), and Ailanthus altissima (ailanthus) (10). Although a few conifers, such as Larix spp. (larch) and Pinus banksiana (jack pine), were also recommended for biomass production on some of the poorer, more droughty soil types (29), SRWCP-funded research is emphasizing hardwoods due to the economic advantages of coppicing. Other researchers in the Lake States region, the USDA Forest Service Experiment Station in Rhinelander, Wisconsin, and the Agroforestry Research Group at the University of Minnesota are testing additional species for use in SRIC plantations in Minnesota, Wisconsin, and northern Michigan. Salix (willow) hybrids performed well on peat soils (21). But because of the limited acreage of this soil type in the United States, the SRWCP has not continued support of that research. However, the research has continued with funding from other sources. Populus hybrids performed very well in tests in both Wisconsin and

Minnesota. Pure Populus spp. [i.e., P. deltoides, P. trichocarpa, and P. balsamifera (balsam poplar)], are now being screened to evaluate their potential relative to the previously tested hybrids.

Fourteen species considered for testing as SRIC-produced fuel sources for the Northeast were selected by Laing (13) based on their tendency to occupy old-field sites (indicative of rapid juvenile growth). After six years, the recommended species were Robinia pseudoacacia, Acer saccharinum (silver maple), and Populus hybrids (14). There was substantial variation among sites, and productivity of the three recommended species ranged from 3.4 to 15.3, 5.4 to 14.7, and 7.5 to 25.7 dry Mg ha^{-1} $year^{-1}$, respectively. Laing stated that matching species to particular site types was critical in obtaining high productivity rates during the early growth phase. The particular Populus hybrids recommended were of a limited genetic background and, in many cases, were susceptible to disease.

Two species were identified as having high potential for SRIC-produced fuelwood in the midwestern region after being screened for 5 years by researchers at the University of Illinois. The two were Robinia pseudoacacia and Eleagnus umbellata (autumn olive), both nitrogen-fixing species (22). This region, which generally includes Illinois, Indiana, Ohio, and portions of Iowa, Kentucky, Tennessee, Missouri, and Arkansas, is also expected to effectively utilize Platanus occidentalis and Populus spp. and their hybrids in SRIC plantations once appropriate sources are identified.

Species selection in the diverse Southeast has identified five candidate species for SRIC. These are Platanus occidentalis, Liquidambar styraciflua (sweetgum), Populus deltoides, Robinia pseudoacacia, and Sapium sebiferum (Chinese tallow). The selection of any one of these species is site-specific. Species/site relationships for candidate SRIC species were identified by researchers at North Carolina State University. Liquidambar styraciflua is recommended for uplands and bottomlands with clay-textured soils. Platanus occidentalis, which outperforms L. styraciflua in initial height growth, is recommended only for coarse-textured bottomland sites. In general, after 5 years, L. styraciflua height surpasses that of P. occidentalis on most sites. For example, on an upland slope ridge site at 4 years, P. occidentalis averaged 2.6 m in height as compared to 2.3 m for L. styraciflua; at 9 years, P. occidentalis was 4.2 m tall, whereas L. styraciflua was 6.4 m tall (7). Populus deltoides was not intensively tested in the North Carolina State University studies but has been found to be very productive on bottomland sites, especially in the Mississippi Delta area. Robinia pseudoacacia, a nitrogen-fixing species, is also considered a good species for certain sites in the Southeast, averaging 5.4 dry Mg ha^{-1} $year^{-1}$ in a Georgia study (27). Sapium sebiferum has not been extensively tested in the Southeast but has been planted in Texas (22). Species screening is no longer a major project in the Southeast.

Four species have been identified for use in the Great Plains region (8) based on their performance (4 to 7 years) on numerous sites in Kansas. These are Populus deltoides, Acer saccharinum, Ulmus pumila (Siberian elm), and Robinia pseudoacacia. Populus deltoides and A. saccharinum performed best in the eastern portion of the state where rainfall levels are higher (greater than 70 cm), whereas U. pumila is better adapted to the drier (less than 50 cm rainfall), western portion of Kansas. Robinia pseudoacacia is the only species that demonstrated respectable performance in both dry and moist areas. Extensive species screening has not been conducted nor is it planned in this area.

In the Southwest, Reddy (23) conducted a literature review of 35 hardwood species and concluded that 9 of the species considered were likely to perform well. Of those species, five were in the genus Prosopis (mesquite). Prosopis spp. and four other woody shrubs have been evaluated by the SRWCP in the Southwest since 1979; however, a major species trial was not established until the summer of 1985 (33). This trial, established at five diverse sites in the Southwest, includes 47 native and exotic species. From the early evaluations performed by researchers in the region, Prosopis alba and its hybrids (5) and Atriplex canescens (18) have been the most promising.

A number of different species are being tested in the subtropical regions of Hawaii and Florida. Eucalyptus grandis (eucalyptus) is the leading candidate over Eucalyptus saligna (eucalyptus) in both areas on many sites (25,31). Additional species recently tested in southern Florida include Eucalyptus amplifolia, Eucalyptus camaldulensis, and Eucalyptus dunnii. In Florida, cold tolerance is a significant problem. The University of Florida planted a species-screening test containing sources of E. amplifolia, E. dunnii, Eucalyptus nitens, and Eucalyptus viminalis near Gainesville (north-central Florida) in the fall of 1984. After the severely cold winter of 1984-1985, E. amplifolia was the only species that evidenced frost hardiness (33). Fifteen Eucalyptus and five Acacia species have been established in species trials in Hawaii since 1979.

The SRWCP, through the efforts of many researchers, has selected several species (a total of 21) for most of the geographic regions that have been identified as biomass-producing areas (Table 2). Only 16 of those species are currently receiving funding for additional work through the SRWCP, but research is being pursued on the other five through other funding sources. Future SRWCP research efforts are being focused on model species for each area except for the Southwest. The model species, identified on the basis of previous screening efforts, are Populus spp., Robinia pseudoacacia, Platanus occidentalis, Liquidambar styraciflua, and Acer saccharinum. Research will also continue on Eucalyptus spp. The potential ranges for these species that may be used in SRIC plantations are shown in Figure 1. Although the search for additional biomass species will continue,

Table 2. Hardwood species identified for high potential as SRIC species as of 1985 for various regions of the United States.

Species	Regions of the United States[a]							
	PNW	W/SW	GP	LS	MW	S/SE	NE	SB
Acer saccharinum			x	x			x	
Alnus glutinosa				x	x	x		
Alnus rubra	x							
Ailanthus altissima				x	x			
Atriplex canescens		x						
Elaeagnus umbellata					x			
Eucalyptus grandis								x
Eucalyptus saligna								x
Liquidambar styraciflua						x		
Platanus occidentalis					x	x		
Populus deltoides			x	x	x	x		
Populus trichocarpa	x			x				
Populus spp. hybrids	x		x	x	x	x		
Prosopis alba		x						
Prosopis spp. hybrids		x						
Robinia pseudoacacia			x	x	x	x	x	
Salix spp.[b]				x				
Sapium sebiferum[b]						x		
Ulmus pumila[b]			x					

[a]PNW - Pacific Northwest, W/SW - West/Southwest, GP - Great Plains, LS - Lake States, MW - Midwest, S/SE - Southeast, NE - Northeast, SB - Subtropics.
[b]Research on these species is currently receiving little or no funding through the SRWCP but is nevertheless ongoing through alternate funding sources.

the development of SRIC technology may best be aided by more intensive development of the selected species than by continued species screening.

PROVENANCE TESTING

An important aspect of species selection is determining if variation exists within a particular species that can be attributed to geographic origin. Testing may be accomplished by sampling the range of the species and planting all sources in common gardens on sites similar to those where they will eventually be planted. The test plantings are examined for performance, and the best performing seed sources can then be used in further improvement programs. The first stage of development for many of the species was in the area of provenance testing or seed source selection.

Progress has been made with several species that show promise for SRIC plantations. Twelve species have been evaluated, though not all of the species were being evaluated strictly for their performance in SRIC systems (Table 3). The species that have been specifically

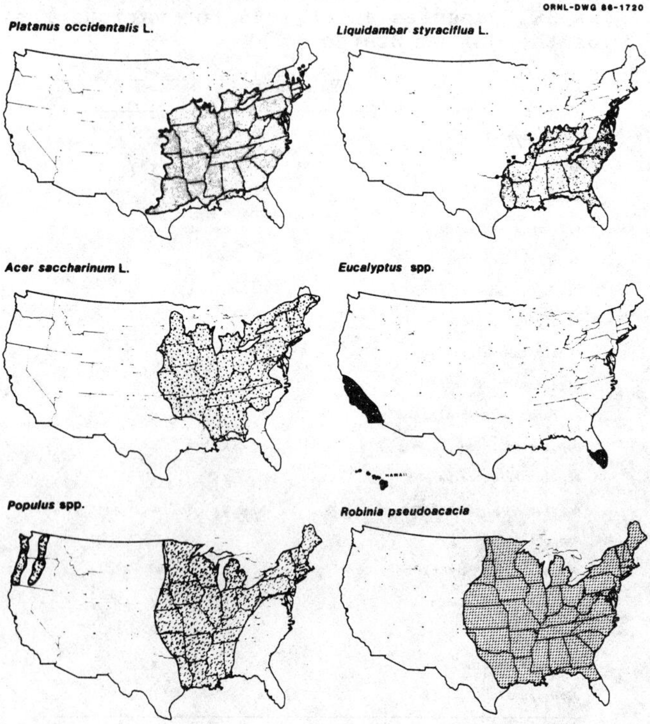

Figure 1. The potential range of five major species and the genus Populus for use in short-rotation intensive culture.

Table 3. Species provenance tested by the Short Rotation Woody Crops Program.

Acer saccharinum	*Platanus occidentalis*
Alnus glutinosa	*Populus deltoides*
Alnus rubra	*Populus tremuloides*
Eucalyptus grandis	*Populus trichocarpa*
Eucalyptus saligna	*Robinia pseudoacacia*
Liquidambar styraciflua	*Sapium sebiferum*

evaluated for use in SRIC systems include Platanus occidentalis, Alnus glutinosa, Alnus rubra, Robinia pseudoacacia, Populus trichocarpa, Eucalyptus grandis, and Eucalyptus saligna.

Michigan State University researchers collected over 400 sources of Robinia pseudoacacia for a provenance study. Based on the results of an analysis of the nursery performance of 264 of these, there were no differences in height of the sources attributable to geographic origin (10). The implication is that superior trees may be selected as seed sources regardless of their location. Four field tests were planted in 1985 to further evaluate provenance performance. The absence of provenance variation may be attributed in part to the movement of this species from its two origins, the Ozark and Appalachian mountains. It is now naturalized in much of the United States and Europe.

Alnus glutinosa is an example of a species with large amounts of provenance variation. Forty-eight provenances from Europe were planted in Iowa, Wisconsin, Pennsylvania, Illinois, and New Brunswick. Hall and Miller (9) reported that Central Euorpean sources performed equally at all test locations, in contrast to other sources. For example, Polish and German sources performed above average in all locations, but the value of Scottish, Danish, and Italian provenances depends on where they are grown. Currently, Hall and Miller (9) recommend not moving seed sources more than two to three degrees in latitude.

Eucalyptus provenance testing is being conducted in both Hawaii and Florida for the SRWCP. Reddy (24) stated that on the basis of coppice growth of E. grandis in Florida, the sources from Queensland, Australia, were the best performers and further introductions should be from that provenance. Provenance testing of this exotic species should continue after appropriate cold-tolerant species have been determined.

Platanus occidentalis provenances were tested at five sites in Mississippi. Provenances varied in their resistance to a lethal complex of canker diseases (15). Trees from southern sources near the Mississippi River were more resistant than trees from northern sources of the mid-South region. Additional studies on geographic variation are being conducted by the North Carolina State University Hardwood Research Cooperative in cooperation with the SRWCP.

Populus trichocarpa sources were collected and tested by the University of Washington. Heilman and Stettler (12) found that provenance variation existed. Southwest provenances had higher production rates at age four than did other sources of clonal material. In general, growth differences were found that correspond to a geographic gradient from southwest to northeast. Southwestern clones develop smaller leaves on more numerous branches. These branches were more erect, and the trees continued to grow later into the fall (30).

Selection of the correct provenance is essential in developing a tree improvement program. For many of the species in SRIC the difference may be the success or failure of a plantation. With other species it may only mean an increase in the profit margin. The next step in tree improvement is breeding.

BREEDING

It has been said that breeding is the genetic adjustment of plants to the service of man. Breeding capitalizes on the natural variability that exists and packages the best traits into a tree. It is a long-term project with few shortcuts. In annual crops the process is more rapid than with trees or farm animals. Most trees take at least 3 years to reach sexual maturity, prolonging the time involved in breeding. Additionally, selecting for specific traits is time consuming if the traits can only be observed after maturity or at rotation age.

As demonstrated in Figure 2, breeding is a cyclic process. As the process continues, new sources of genetically improved material are available to growers. Thus, in the coppice system described for SRIC, rootstocks

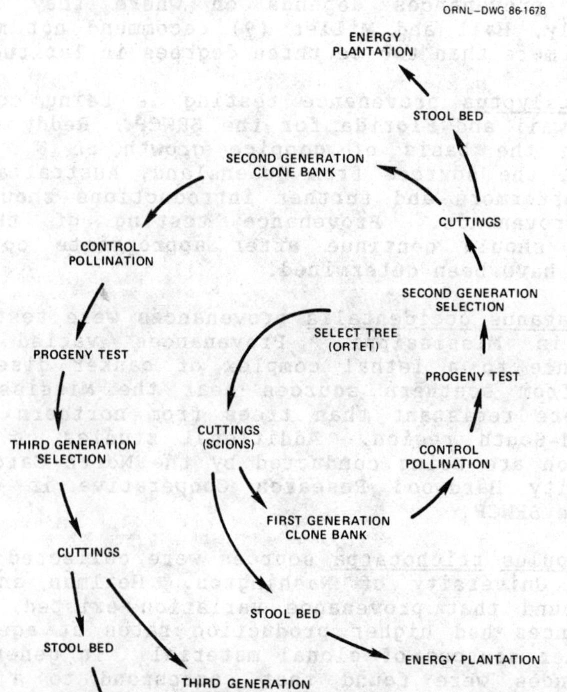

Figure 2. The cycle of breeding used in developing genetically improved hardwoods for biomass.

should be replaced with genetically improved material as it becomes available. Select trees grown in energy plantations will be the best of the trees involved in the breeding cycle (Figure 2). Vegetative propagation via tissue culture will be the best method of clonally producing the trees needed for energy plantations.

With each generation of selection and breeding, specific gains in productivity can be achieved (Figure 2). Preliminary estimates of genetic gain from the first generation for several SRIC hardwood species range from 20 to 100% (averaging about 40 to 50%) (22). The estimates reflect the ability of breeding to capture additive genetic effects into a single tree. Additive genetic effects are those in which each of a large number of genes contributes a small amount to a particular trait. For example, the height of a tree may be controlled by hundreds of genes, each of which might contribute an inch or more in growth, and when all the height genes are combined the tallest tree possible is produced. Therefore, the genetic gain in height may be described as an expression of how well those genes may be combined into a tree. The ability of a tree to mate with many other trees and pass these additive traits to its progeny is termed its general combining ability.

Additional genetic gains in some traits may be achieved through crossing two specific trees. In many cases these traits are controlled by one or a few genes. Clonal propagation of the best progeny from a cross such as this can capitalize on this gain. This is often called the specific combining ability of a tree.

Breeding and selection programs are designed to investigate and utilize genetic gains. Allard (1) described four major breeding methods used with cross-pollinating crops (1) mass selection, (2) backcross breeding, (3) hybridization of inbred lines, and (4) recurrent selection. The following is a discussion of the methods, accompanied by a description of some of the SRWCP research that utilizes the methods.

Mass Selection

In this method desirable trees are chosen, and seed or vegetative propagules are collected for planting. Selection is based on the female parent's performance only. This method has been used in establishing field plots for much of the SRWCP cultural research but is not commonly used in the projects where the major objective is genetic improvement.

Mass selection techniques are effective in modifying traits in plants that can be easily seen or measured (26). Shelbourne (26) stated that mass selection has proven ineffective in modifying yields of well-adapted species. This is probably because genetic variation is less in these species that have been undergoing natural or man-induced selection for many generations.

Mass selection combined with progeny testing is more beneficial. The progeny in this case would be from seed collected directly from select trees and the male parent would be unknown (progeny are open-pollinated). After the family performance (progeny from a select tree constitutes a family) has been evaluated in the test, the worst families are removed (rogued) and only the best individuals of the best families are left. Seed or vegetative propagules from these individuals are then used for establishing plantations. After the individuals are allowed to randomly mate with each other, the process begins again with progeny testing.

Provenance selection may be considered one aspect of this method. Mass selections are obtained from the population range, and once a superior provenance has been determined further mass selections are made for production of seed or vegetative propagules.

In essence, this methodology is being used in a project to identify salt-tolerant individuals of Atriplex canescens. Superior plants of this semiarid shrub were selected and then tested for salt tolerance. Selected individuals are being vegetatively propagated for planting. Another semiarid plant that is being developed with mass selection is Prosopis alba. Dr. Peter Felker, Texas A&I University, evaluated many original selections (5) and has developed vegetative propagation techniques for mass producing the best selections.

Researchers at North Carolina State University recommend that breeding efforts with Platanus occidentalis and Liquidambar styraciflua follow mass selection plus progeny test methods (6). Purnell and Kellison (20) reported that 231 P. occidentalis and 213 L. styraciflua were being tested in open-pollinated progeny tests. Additional families will be collected and tested to expand the number of trees in the breeding population. However, these tests are not exclusively designed to make selections for energy plantations. In later generations, control-pollination and other breeding methods may be necessary to reduce inbreeding depression, a decrease in productivity resulting from the mating of related individuals.

Backcross Breeding

The backcross method is particularly suited to crop plants for transferring specific genes to a good variety that is deficient in one or a few characteristics. The method works best when the trait is controlled by one or a few genes. Its use is somewhat limited in trees because of their long generation time.

The method is not currently used by any of the genetic improvement projects of the SRWCP. However, a great potential exists with genetically engineered traits. For example, if a single gene can be successfully inserted into a chromosome of a single cell of a tree, and the tree propagated from the cell, future crosses with the 'engineered' tree will have a 50% chance of inheriting the trait.

Hybrid Varieties

The superiority of these varieties depends on the hybrid vigor (heterosis) associated with crosses between inbred lines, clones, races, varieties, or even species. This particular method has been used extensively in breeding corn. The basic premise is that the progeny of a cross performs better than either of the parents. Shelbourne (26) stated that hybrid vigor could be used in tree improvement in three ways: hybrids between (1) species, (2) races, or (3) individuals whose specific combination is outstanding.

Hybrid Populus breeding has shown evidence of hybrid vigor in growth and drought tolerance. Figure 3 demonstrates the growth of a hybrid and its parents. The parents, P. deltoides and P. trichocarpa, flanked by the hybrid are much smaller and less vigorous in appearance. Heilman and Stettler (12) found that a hybrid clone after 4 years produced 27.6 dry Mg ha^{-1} $year^{-1}$ while its P. trichocarpa parent population averaged 12.5 dry Mg ha^{-1} $year^{-1}$.

Populus trichocarpa, commonly found growing in river bottoms, grows very slowly on dry sites, limiting its current use. Some of its hybrid progeny with P. deltoides have been found to be more drought tolerant than either parent. Undersurface leaf color appears to be an early indicator of this trait. In a recent study described by Stettler et al. (28) a correlation was found between the thickness of the spongy mesophyll (appearing as a white

Figure 3. Performance of hybrid Populus (extreme right and left) and its parents (two center rows), P. trichocarpa and P. deltoides, planted at Boardman, Oregon.

undersurface) and the capacity to regulate stomata during a severe, short-term drought.

Hybrid Populus breeding is in its infancy in the United States. Euorpeans have been breeding this genus since the 18th century. Current programs, such as those in Belgium and in The Netherlands, develop many new clones for release to the public annually. Two of the main species utilized in their programs are the North American species P. trichocarpa and P. deltoides. The trihybrid produced among these two and P. nigra has shown great potential.

Hybrid Eucalyptus selections have been obtained from Brazilian breeding programs and planted in Hawaii to evaluate their potential (32) in SRIC. These hybrids were selected on the basis of improvements in fiber yield, disease resistance, and ease of vegetative propagation.

Recurrent Selection

Recurrent selection is based on the premise that both parents are superior. Simple recurrent selection is one form of this method. It is similar to mass selection except that both parents have been judged superior. Researchers at North Carolina State University have suggested that after Platanus occidentalis selections have been tested, select trees should be grafted in an orchard and, therefore, all seed obtained from the orchard will have both parents of superior quality (20).

Other forms of recurrent selection, each with its own positive attributes, have been described in detail by Shelbourne (26). However, none of these are currently being used in tree improvement programs for biomass.

BIOTECHNOLOGY

This relatively new avenue of science, often called genetic engineering, is an important tool in the genetic improvement of trees for biomass energy. It must be emphasized that it is only a tool and certainly warrants a separate discussion.

One important part of the biotechnology boom is in the area of tissue culture. Tissue culture can provide the vast amount of vegetative propagules necessary for research or establishing plantations. Tremendous gains in productivity can be achieved by planting superior, vegetatively propagated, clones. For example, 50 Populus trichocarpa clones averaged 12.5 dry Mg ha^{-1} $year^{-1}$ after 4 years in a University of Washington study, but the best clone produced 23.1 dry Mg ha^{-1} $year^{-1}$. The best hybrid, described previously, produced 27.8 dry Mg ha^{-1} $year^{-1}$ (12). Mass propagation of the best clone can greatly increase productivity in this case.

Conkle (2) presented two other reasons for developing tissue culture techniques to grow trees from single cells. The first was that dioxyribonucleic acid (DNA) fragments containing specific genes can be incorporated into a cell's DNA by micro injection or by using a bacterial vector. The second reason was to allow selection of variant cells that either arose as part of the tissue culture process or contained mutations before tissue culture. Cell-to-cell variability is quite common, but large numbers of cells must be screened to find a specific type of variation.

The potential of biotechnology in forest trees is currently less than that in agricultural crops. The two main reasons for this are the lack of specific knowledge of the gene and chromosome structure in trees and the problems involved in reproducing some species from a single cell. The location of a specific trait on a chromosome is unknown in trees. The identity of single-gene traits exists, but generally these traits are of limited value and include things such as isozymes, terpenes, and disease resistance (16). None of these genes have been located on a specific chromosome. The answers to these problems will greatly increase the potential of biotechnology for genetic improvement.

Past SRWCP research was aimed at developing tissue culture systems for several southeastern hardwoods, including Liquidambar styraciflua (Figure 4) and Robinia pseudoacacia, at the University of Georgia (21). Ranney et al. (21) summarized research on tissue culture sponsored by the SRWCP. However, current SRWCP-sponsored

Figure 4. Tissue culture plantlet of L. styraciflua produced at the University of Georgia.

research in biotechnology is more involved with somaclonal screening. Atriplex canescens is being evaluated for salt tolerance using somaclonal screening techniques (34), and several clones have been selected. Somaclonal screening is also being utilized by the U.S. Forest Service North Central Forest Experiment Station Biotechnology Program to screen Populus trichocarpa for tolerance (17) to two broad-spectrum herbicides--glyphosate and sulfometuron methyl (E. A. Hansen, 1985 personal communication).

Biotechnology methods, therefore, are being used to develop mass propagation techniques and to aid in the selection of clones for breeding and planting. The full potential of this tool will not be realized, however, until much more research has been done in contributing areas.

SUMMARY

The SRWCP has determined that under optimum management regimes, additional biomass yields can best be obtained through genetic improvement (22). In some instances, these yield improvements may be substantial enough to warrant replanting rather than reestablishment by coppicing. The SRWCP has made important contributions to hardwood genetic improvement by financing research on species site relationships, provenance testing, and breeding (22). The success of SRIC is heavily dependent on traditional tree improvement strategies to provide high-yielding trees. The SRWCP expects to continue to stimulate advances in hardwood genetic improvement by supporting research in biotechnology. It is expected that traditional breeding and biotechnology will work synergistically to make rapid advances in productivity of hardwood biomass for energy.

ACKNOWLEDGMENTS

Research supported by the Biofuels and Municipal Waste Technology Division, U.S. Department of Energy, under Contract No. DE-AC05-84OR21400 with Martin Marietta Energy Systems, Inc. Publication No. 2709, Environmental Sciences Division, ORNL.

REFERENCES CITED

1. Allard, R. W., Principles of Plant Breeding, New York: John Wiley & Sons, Inc., 1960.

2. Conkle, M. T., "Decoding Tree DNA," J. For. 84, 34-37 (1986) January.

3. Dickmann, D. I., "Plant Materials Appropriate for Intensive Culture of Wood Fiber in the North Central Region," Iowa State J. Res. 49, 281-286 (1975).

4. Dolan, L. S., Schroeder, P., and Gillham, M., "The Cultural Treatment of Selected Species for Woody Biomass Production in the Pacific Northwest," Seattle City Light Department, Seattle, Washington, 1984.

5. Felker, P., Cannell, G. H., Clark, P. R., Osborn, J. F., and Nash, P. "Biomass Production of Prosopis Species (Mesquite), Leucaena, and Other Leguminous Trees Grown Under Heat/Drought Stress," For. Sci. 29, 592-606 (1983).

6. Frederick, D. J., Kellison, R. C., and Lea, R., "Species Selection and Silvicultural Systems for Producing Fuels from Woody Biomass in the Southeastern United States," Unpublished annual report submitted to DOE's Short Rotation Woody Crops Program. North Carolina State University, Raleigh, 1984.

7. Frederick, D. J., Kellison, R. C., and Lea, R., "Species Selection and Silvicultural Systems for Producing Fuels from Woody Biomass in the Southeastern United States," Unpublished annual report submitted to DOE's Short Rotation Woody Crops Program. North Carolina State University, Raleigh, 1985.

8. Geyer, W. A., "Great Plains Energy Forest," Unpublished annual report submitted to DOE's Short Rotation Woody Crops Program. Kansas State University, Manhatten, 1985.

9. Hall, R. B., and Miller, G. A., "Selection and Breeding Strategy for an Exotic Species: Alnus glutinosa in North America," 233-244. In Proceedings of the 3rd North Central Tree Improvement Conference. Dept of Forestry, University of Wisconsin, Madison, 1983.

10. Hanover, J. W., "Tree Species and Management Strategies for Biomass Production in the Lake States," Unpublished annual report submitted to DOE's Short Rotation Woody Crops Program. Michigan State University, East Lansing, 1985.

11. Harrington, C. A. "Factors Influencing Initial Sprouting of Red Alder," Can. J. For. Res. 14, 456-459 (1984).

12. Heilman, P., and Stettler, R. F., "Genetic Variation and Productivity of Populus trichocarpa T. & G. and its Hybrids. II. Biomass Production in a 4-Year Plantation," Can. J. For. Res. 15, 384-388 (1985).

13. Laing, F. M., "The Production of Biomass for Energy on Abandoned Farmlands in the Northeast," Unpublished annual report submitted to DOE's Short Rotation Woody Crops Program. University of Vermont, Burlington, 1979.

14. Laing, F. M., Sendak, P., and Aleong, J., "Species Trials for Biomass Production on Abandoned Farmland," North. J. Appl. For. 2, 43-47 (1985).

15. Land, S. B., Jr., "Genetic Selection of American Sycamore for Biomass Production in the Mid-South. Final report for the period March 9, 1979 to May 31, 1982." ORNL/Sub/81-9051/1. Oak Ridge National Laboratory, Oak Ridge, Tennessee, 1982.

16. Ledig, F. T. and Sederoff, R. R., "Genetic Engineering in Forest Trees," 4-13. In Proceedings of the 18th Southern Forest Tree Improvement Conference (R. C. Schmidtling and M. M. Griggs, eds.). National Technical Information service, Springfield, Virginia, 1985.

17. Nelson, N. D., "North Central Forest Experiment Station Biotechnology Program - Application to Tree Improvement," 14-22. In Proceedings of the 18th Southern Forest Tree Improvement Conference (R. C. Schmidtling and M. M. Griggs, eds.). National Technical Information service, Springfield, Virginia, 1985.

18. Newton, R. J., Goodin, J. R., Magar, D. L., and Puryear, J. D., "Biomass from Unconventional Sources in Semiarid West Texas," 167-217. In Energy from Biomass and Wastes VI. Institute of Gas Technology, Chicago, Illinois, 1982.

19. Perlack, R. D., Ranney, J. W., and Wright, L. L., "An Economic Evaluation of the Competitiveness of Short-Rotation Intensive Culture for Energy," 1245-1263. In Proceedings of Energy from Biomass and Wastes IX. IGT, 1985.

20. Purnell, R. C. and Kellison, R. C., "A Tree Improvement Program for Southern Hardwoods," 90-98. In Proceedings of the 17th Southern Forest Tree Improvement Conference. University of Georgia, Athens, 1984.

21. Ranney, J. W., Trimble, J. L., Wright, L. L., Perlack, R. D., Wenzel, C. R., and Cushman, J. H. "Short Rotation Woody Crops Program: Annual Progress Report for 1983" ORNL-6085. Oak Ridge National Laboratory, Oak Ridge, Tennessee, 1984.

22. Ranney, J. W., Wright, L. L., Trimble, J. L., Perlack, R. D., Dawson, D. H., Wenzel, C. R., and Curtin, D. T., "Short Rotation Woody Crops Program: Annual Progress Report for 1984" ORNL-6160. Oak Ridge National Laboratory, Oak Ridge, Tennessee, 1985.

23. Reddy, S. J., "Potential Fuelwood Species for the Semiarid Southwest," Unpublished final report submitted to DOE's Short Rotation Woody Crops Program. University of Arizona, Tucson. 1982.

24. Reddy, K. V., "Genetic Improvement of Eucalyptus grandis (Hill) Maiden for Coppice Productivity," Ph. D. Dissertation, University of Florida, Gainesville, 1985.

25. Rockwood, D. L., "Genetic Improvement Potential for Biomass Quality and Quantity," Biomass 6, 37-45 (1984).

26. Shelbourne, C. J. A., "Tree Breeding Methods," New Zealand Forest Service Technical Paper No. 55, Wellington, 1969.

27. Steinbeck, K., Brown, C. L., Sommer, H. E., and Bongarten, B., "Increasing the Biomass Production of Short-Rotation Forests," Unpublished final report submitted to DOE's Short Rotation Woody Crops Program. University of Georgia, Athens, 1983.

28. Stettler, R. R., Heilman, P. E., and Hinckley, T. M., "Evaluation and Genetic Improvement of Black Cottonwood for Short-Rotation Coppice Culture," Unpublished annual report submitted to DOE's Short Rotation Woody Crops Program. University of Washington, Seattle, 1985.

29. U.S. Department of Agriculture, Forest Service, "Energy and Wood from Intensively Cultured Plantations: Research and Development Program," U.S. Department of Agriculture Forest Service, General Technical Report NC-57, 1980.

30. Weber, J. C., Stettler, R. F., and Heilman, P. E., "Genetic Variation and Productivity of Populus trichocarpa T. & G. and its Hybrids. I. Morphology and Phenology of 50 Native Clones," Can. J. For. Res. 15, 376-383 (1985).

31. Whitesell, C. D., Crabb, T. B., Schubert, T. H., and Melidosian, M. W., "Eucalyptus Plantations for Energy Production in Hawaii," Unpublished annual report submitted to DOE's Short Rotation Woody Crops Program. BioEnergy Development Corporation, Hilo, Hawaii, 1985.

32. Wright, L. L., Perlack, R. D., Wenzel, C. R., Trimble, J. L., and Ranney, J. W., "Short Rotation Woody Crops Program. Quarterly Progress Report for the Period March 1 to May 31, 1985," ORNL/TM-9698. Oak Ridge National Laboratory, Oak Ridge, Tennessee, 1985.

33. Wright, L. L., Perlack, R. D., Layton, P. A., Wenzel, C. R., Trimble. J. L., and Ranney, J. W., "Short Rotation Woody Crops Program. Quarterly

Progress Report for the Period June 1 to August 31, 1985," ORNL/TM-9832. Oak Ridge National Laboratory, Oak Ridge, Tennessee, 1985.

34. Wurtele, E. S., Balandrin, M., and McKell, C. M., "Tissue Culture of Elite Biotypes of <u>Atriplex canscens</u> as a Short Rotation Woody Biomass Crop," Unpublished annual report submitted to DOE's Short Rotation Woody Crops Program. NPI/Plant Resources Institute, Salt Lake City, Utah, 1985.

35. Zobel, B. and Talbert, J., <u>Applied Forest Tree Improvement</u>, New York: John Wiley and Sons, 1984.

DISCUSSION

J. Ranney, Oak Ridge National Lab. The optimism about biotechnology and improving productivity has caught a lot of people's imagination, perhaps a little bit too much in some cases. Do you have any idea about the probability of passing on genes that we don't want by the biotechnology work or through breeding schemes, but accidentally putting in a gene that's makes it completely unacceptable?

Layton Any biotechnology work is going to have to be rigorously tested. We learned earlier this year in a seminar I went to that some people are going to consider exotic gene pollutants and that there's a possibility that the EPA will even come into the regulation of this. So it will have to be extremely well tested before it's ever put out into the open.

S. Kresovich, Texas A&M U. How easy is it to make inter-specific and inter-generic crosses with a species?

Layton It depends on the species. For example, within populus, there is a certain section that crosses very easily, and there are some sections that don't cross at all. Some alders will cross the between the species, but not very well. The potential is there for some, but not for others.

L. Zsuffa, U. of Toronto (Canada) When you gave a list of attributes for SRIC trees, you didn't mention coppicing. Is that important? You also never mentioned willows.

Layton — Our program has gone away from the willows because they would be used on a limited land base in the United States, so willow is not one of our model species. We are trying to pick more wide-ranging species. Coppicing is very important; we're just beginning our breeding research on that. We're getting some of our breeding programs to the point where coppicing can be evaluated. It is an important attribute, and I probably wouldn't have looked at a species that didn't coppice.

Zsuffa — When mentioning somaclonal variation, you said that it probably couldn't be exploited without having a fairly complete gene map. I'm not so sure about it.

Layton — You can exploit it without having a complete gene map, but I think we need more information on the genes within the chromosomes. Exploitation is possible in the transition from the cell to the plant.

L. Gaines, Argonne National Laboratory — It sounds like the possibilities for improving productivity through genetic improvement are quite large. Do you have any idea what kinds of gains you might expect?

Layton — In our past programs using clonal material, we've gotten 100% or better gains over older material through some selection and use of vegetatively propagated material. Early estimates are in the range of 40-50%. Our gains are higher, for example, in hardwoods than they are going to be in the conifers, for which they average about 10 to 20% per generation because they can't vegetatively propagate. With vegetative propagation, you can usually double yields.

TISSUE CULTURE-BASED BIOTECHNOLOGY FOR <u>POPULUS</u> CLONES

Bruce E. Haissig, Ph.D.
Project Leader
USDA-Forest Service
North Central Forest Experiment Station
Forestry Sciences Laboratory
P.O. Box 898, Rhinelander, Wisconsin 54501

ABSTRACT

Poplars efficiently produce biomass that is a source of fuel, livestock feed, and raw material for manufacturing paper, fiberboard, and other products. Productivity of poplar plantations has been increased through development and use of sophisticated silviculture, e.g. short-rotation intensive culture (SRIC). Further increases in productivity can probably be achieved by tailoring poplar genotypes to SRIC. Breeding has already increased productivity of poplars, albeit not necessarily targeting SRIC. Biotechnology may be a genetic improvement strategy for poplars that will complement breeding. This paper discusses uses of biotechnology to genetically improve SRIC poplars, including: 1) specific traits that might presently be improved, 2) methods of biotechnology that may have immediate application , and 3) research that is testing the hypothesis that biotechnology can achieve herbicide resistance in SRIC poplars.

TISSUE CULTURE-BASED BIOTECHNOLOGY FOR POPULUS CLONES

INTRODUCTION

Most crops have been improved to meet human needs through breeding. The long generation time of many trees slows the advance of genetic improvement via breeding. Nonetheless, increased awareness of forest trees as potential commodity crops has increased interest in their genetic improvement, as evidenced by the forest biology literature.

Biotechnology may contain strategies for rather quickly obtaining specific, genetic improvements in plants. Definitions of biotechnology differ. For example, the USDA (67) has defined biotechnology as "the use of the activities, processes, or products of organisms (including their component parts, analogs, or copies of parts) in genetic engineering (excluding techniques of classical genetics) to develop, make, modify, improve, or change organisms, or to develop products". In comparison, Torrey (65) recently defined biotechnology as "the application of existing techniques of plant organ, tissue, and cell culture, plant molecular biology, and genetic engineering to the improvement of plants and of plant productivity for the benefit of man". To understand the foregoing definitions, one must understand the meanings of "classical genetics" and "genetic engineering". Classical genetics is "genetic manipulation of organisms by procedures which occur in nature or in conventional breeding, including but not restricted to such methods as hand pollination, artificial insemination, and use of mutagens" (67). Genetic engineering is "the process of causing hereditary variation in an organism through the introduction of change in its DNA or RNA structure or function" (67). Definitions of biotechnology differ for several reasons. Some definitions may be flavored by legal or regulatory requirements (67); others may have humanistic inclinations (65). In addition, definitions may attempt to encompass all organisms (67) or only a specific set of organisms, such as higher plants (65). And, new technical methods or new applications of existing methods are added to biotechnology each year. In the present paper, all methods of higher plant biotechnology will be termed "biotechnology" for simplicity.

Several recent studies have identified woody perennial crops as prime targets for specific genetic improvement by biotechnology (13, 59, 66). A major conclusion was that combining biotechnology with classical genetics may produce greater economic returns in forestry than in any other agricultural area (59). The foregoing hypothesis presently lacks supporting evidence but has a strong supporting rationale for the genetic improvement of forest trees (43, see also 12, 50):

- Biotechnology and conventional forest tree breeding can be complementary rather than competing technologies.
- Biotechnology can potentially reduce the long time periods required for forest tree improvement using classical genetic technology.
- Biotechnology may rather quickly introduce rare or foreign, economically important genetic traits into otherwise desirable forest tree germplasm.

Short-rotation intensive culture (SRIC) of poplars has been one important instance where a genus of forest trees has received the silvicultural attention due a commodity crop. The initial successes of SRIC have prompted interest in optimizing productivity through properly matching poplar genotype(s) with silviculture. This consideration has arisen because the poplar genotypes presently used with considerable success in SRIC were not bred specifically for SRIC. Breeding of poplars for special purposes is a logical extension of the important role that breeding has played in the domestication of poplars [(29) and refs. therein].

For poplars, as for other forest trees, the known alternatives for genetic improvement are conventional breeding (proven for some genetic traits but slow) or biotechnology (fast for some genetic traits <u>but totally unproven</u>), or a combination of these approaches. Given today's information, an experimental genetic improvement strategy for poplars should test both conventional breeding and biotechnology, based on the hypothesis that the two methods can be combined into a single coherent technology for expediently attaining desired genomic changes and perpetuating them in breeding populations.

Thus, the present paper deals with the inclusion of specific, desirable genetic traits into SRIC poplar genomes by means of tissue-culture based biotechnology. Where possible, I will use examples of herbicide resistance research in progress at North Central Forest Experiment Station, often in collaboration with others, that was initiated within the last two years (43, 44). The primary scientists involved in this research are: Ms. Joanne Fillatti (Calgene, Inc., Davis, CA); Dr. Brent McCown (University of Wisconsin, Department of Horticulture, Madison), Dr. Karl Wolter (USDA-Forest Service, Forest Products Laboratory, Madison, WI); and Drs. Don Riemenschneider, Thomas Sebastian, and Charles Michler (USDA-Forest Service, Forestry Sciences Laboratory, Rhinelander, WI). Various aspects of our biotechnology research program have been reviewed elsewhere (43, 44, 45, 46).

TISSUE CULTURE-BASED BIOTECHNOLOGY

According to Torrey's (65) previously stated definition, higher plant biotechnology contains all elements of "tissue culture" (sometimes called microculture), which is a term alluding to all types of in vitro culture of protoplasts, cells, cell groups, tissues, and organs. However, the foregoing definition of biotechnology does not adequately portray how critical tissue culture is to many applications of plant molecular biology and genetic engineering in biotechnology. Much current biotechnology cannot proceed without tissue culture to maintain living cells or cell aggregates in vitro (where genetic modification occurs) and to regenerate the genetically modified plants (micropropagation). For example, tissue culture supports somatic cellular hybridization, somaclonal variation, and various recombinant DNA applications, which will be defined below. Probably because of the dependency of biotechnology on tissue culture, the cloning of plants in vitro by micropropagation has often been considered to be biotechnology (44), even though the method does not (by intent) modify the genome. Micropropagation is strongly supportive of biotechnology but is only a modern version of types of clonal propagation that have been practiced for centuries. Thus, it is a misnomer to consider micropropagation as biotechnology.

Somatic cellular hybridization is the union of nonsexual plant cells devoid of cell wall (i.e. protoplasts), which may or may not include nuclear fusion. Such unions can be accomplished by use of an electrical field, termed "electrofusion". Other methods of achieving fusion are also known and workable. If nuclear fusion is not involved, the technique has been termed "cybridization", resulting in "cybrids", joined cytoplasms with a single nucleus. The single nucleus is obtained by destroying the nucleus of one of the donor cells before fusion. Somatic cellular hybridization has limited immediate application to poplars because there are no published reports of poplars having been regenerated directly from individual protoplasts. However, such research is underway (3, 4, 6, 52, 53, 54, 60).

Somaclonal variation is mendelian and non-mendelian genetic variation sometimes found in plants regenerated from in vitro cultures (e.g. 36, 37). Somaclonal variants (variant "somaclones") arise from preexisting cellular genetic variation or mutation during the culture process, or both [(26, 27, 35, 36, 37, 41), additional refs. in (64)]. Variant somaclones may have economically valuable traits that can be directly exploited in sexually or vegetatively propagated progenies, or indirectly exploited through further somaclonal or sexual improvement programs (27, 36). Somaclonal technology offers substantial promise for use in genetic improvement of poplars, but there is only one published observation of somaclonal variation in poplars (39).

The apparent low frequency with which somaclonal variation has been observed in poplars and other forest trees brings forth the question: "Can somaclonal variation be found, given the apparent low frequency of occurrence in forest trees?" This question cannot presently be answered, although we are attempting to do so. Several factors may have conditioned the number of reports of somalconal variation that are found in the literature. For example, somaclonal variants in a single or a few genes may not be morphologically different from "normal" plants. Also, the past history of tree tissue culture has primarily been aimed at plant propagation. Thus, morphologically aberrant forms may have been discarded during propagation, given human tendencies for selection toward perfection. In addition, specific micropropagation methods are chosen based on the probability that they will not change the genotype, as stated above. Thus, most current woody plant micropropagation is based on rooting of microcuttings that are made from shoots that develop in culture from preexisting (as opposed to newly originated or adventitious) buds. The cell lineage and genome of these "non-adventitious" buds can be traced directly to the cells and genome of the apical or axillary buds of the original source plant. Somaclonal variants are thought to arise only when new (adventitious) buds are regenerated in vitro. Therefore, somaclonal selection programs utilize pieces of organs or tissues that do not initially contain buds. Interestingly, the reported somaclonal variation in poplars occurred in plants regenerated from callus tissue (39). In comparison, no morphological (genotypic) variation has been observed in several genotypes of poplars that have been regenerated from shoot cultures in Dr. Brent McCown's or our laboratory. Only substantial further experimentation will determine with what frequency, if any, specific types of poplar somaclonal variants can be obtained.

"Recombinant DNA technology encompasses the isolation, characterization, and expression of foreign DNA in organisms or vectors" (67). "A vector is a cloning vehicle which provides a suitable origin of replication necessary for production of foreign DNA" (67). Bacterial transformation systems (e.g. Agrobacterium and plasmid vectors) are one method of inserting foreign DNA into higher plant cells. As will be discussed below, there are presently examples where Agrobacterium transformation has been used to insert specific genes, including an agronomically important gene, into a poplar. Agrobacterium transformation techniques are potentially one of the most powerful tools for genetic engineering of poplars, but that promise will only be fulfilled to the extent that economically valuable genes or gene sets become available. There is presently a dearth of such genes and gene sets.

Other recombinant DNA techniques such as microinjection, electroporation, and "simple uptake" are methods for rather direct insertion of DNA into plant cells [(34, 49) and refs. therein]. These methods, like somatic cellular hybridization, may depend on the regeneration of poplars from protoplasts, single cells, or small cell groups, which has yet to be completely achieved with high frequency and efficiency. Electroporation, facilitating introduction of nucleic acids into plant cells via pulsing electric current, is a promising technique (34). However, electroporation is among the newest elements of biotechnology and, therefore, poorly understood.

POTENTIAL FOR APPLYING BIOTECHNOLOGY TO POPLARS

Given the potential of biotechnology and the desire to genetically improve SRIC poplars, it must still be demonstrated that poplars are biologically suitable candidates for biotechnological improvement, that genetic attributes exist for improvement, and that biotechnology can be used to attempt the improvements. I will do so in the following discussion.

Biological Characteristics of Poplars

The genus *Populus* has characteristics that make its member species ideally suited for biotechnology:

- Poplars are of worldwide importance for very intensive plantation forestry such as SRIC for production of biomass, wood pulp, wood flakes, animal feed, etc. [e.g. (48)].

- All poplar species are diploid, which facilitates genetic analysis.

- The genus *Populus* is composed of four taxonomic sections, which have different physiological (genetic) characteristics. Such distinct stratification within the genus facilitates genetic analysis and selection of good candidate species for forestry and, therefore, for genetic improvement via biotechnology.

- Poplars have a small genome size of about 1.5 pg DNA per nucleus (22), which facilitates genomic analysis.

- The genus *Populus* contains the most readily *in vitro* propagated tree species (3, 4, 5, 6, 14, 17, 18, 19, 23, 33, 39, 56, 68, 69). Good ability or potential for *in vitro* regeneration at high frequency is probably the single greatest factor favoring immediate use of poplars in biotechnology. Low regeneration frequency *in vitro* is the limiting factor in applying biotechnology to many other commercially important forest tree species.

Because of the foregoing, the USDA-Forest Service National Genetic Engineering Workshop recommended poplars as model species for inclusion in forest tree biotechnology programs (in press).

Characteristics for Genetic Improvement

Biotic and abiotic stresses probably result in the greatest total loss of plantation and forest productivity, compared with any other factors. Example biotic stresses are insects, diseases, and weed competition; example abiotic stresses are sulfur dioxide, ozone, herbicides, and heavy metal ions. Therefore, it can be hypothesized that the genetic improvement of poplars to withstand various stresses would be highly beneficial. It may be possible to breed poplars that will withstand certain stresses, such as drought or salinity. However, other stresses may not be effectively countered with breeding programs. For example, breeding for disease resistance is difficult because evolution favors the pathogen, which always has a much shorter generation time than the tree host. It is doubtful that a tree breeding program for disease resistance could keep pace with an evolving pathogen. However, SRIC poplar plantings, especially plantings of certain very productive clones, can be devastated by some diseases, which indicates a need for disease resistance (42, 47). It would be equally difficult to breed for characteristics that occur with low frequencies (i.e. rare or "foreign" traits), which limits possibilities for selection of parents for breeding. Herbicide resistance falls into this latter category, and would be another desirable target for SRIC poplars (7, 8, 9, 10, 45). Herbicice resistance would allow the most effective use of an abiotic stress (the herbicide) to eliminate a biotic stress (the weeds). Herbicide control of weeds in SRIC plantings is usually cheaper and often more effective than mechanical weed control. Sub-optimal use of herbicides may result in increased costs of weed control, reduced productivity of the poplar crop, or both. However, the most effective contact herbicides readily damage or kill poplars during their seasonal growth period. Thus, without herbicide resistance, eliminating the stress of weed competition may introduce serious or fatal herbicide stress, and offset the value of chemical weed control. There is ample evidence from agronomic crops suggesting that biotechnology can probably be used to obtain disease or herbicide resistance in poplars in a short-term improvement program [refs in (45)].

Herbicide selectivity for weeds versus crop plants can be achieved by avoidance, tolerance, or resistance [reviewed by (45)]. Broad spectrum herbicide selectivity in tree plantations is currently limited to avoidance and species-level tolerance. Herbicide tolerance or resistance can result in several ways (45), such as reduced uptake, reduced translocation, or intracellular sequestration. However, some of the most promising ways of achieving herbicide tolerance or resistance in poplars are through detoxification (metabolism) of the herbicide, overproduction of the target enzyme of the herbicide, or alteration of the target enzyme of the herbicide (Figure 1). The foregoing mechanisms would probably be much more effective in achieving tolerance or resistance if they occurred in combination. Detoxification would change the structure of the herbicide to a non-effective form. Detoxification could result if the poplar genome were changed in a gene(s) that codes for an enzyme(s) that chemically modified the herbicide molecules. Overproduction of the target enzyme could make poplars resistant or tolerant of the herbicide because a normally lethal dosage would not inhibit enough target enzyme to kill the tree. Overproduction of target enzyme would result if extra copies of the target enzyme were produced, either because the poplar genome contained more genes for the target enzyme or because enzyme production per gene were enhanced, or a combination of these. Alteration (mutation) of the target enzyme could reduce or eliminate its inhibition by the herbicide, thereby rendering the herbicide ineffective in causing tree death. Genes for these characteristics might be incorporated into poplar genomes with recombinant DNA technology using Agrobacterium, or by somaclonal selection, as will be discussed below. In the case of somaclonal selection, the requisite genes need not have been isolated or even identified because somaclonal selection is based on spontaneous genetic change (mutation) of unknown origin.

Other desirable traits of SRIC poplars would be fast growth rate, insect resistances (42), ability to withstand drought (30), etc. However, it is presently uncertain whether biotechnology could be used to improve such apparently quantitative traits. Two observations suggest that biotechnology may be useful in improving quantitative traits. As stated above, biotechnology has been used to obtain some disease resistances in agronomic crops. Disease resistances are probably polygenic. In addition, the one published observation of somaclonal variation in poplar describes changes in morphological traits that are seemingly polygenic (39).

A further primary use of biotechnolgoy in genetically improving poplars might be to achieve species hybridizations, particularly between taxonomic sections, that have proven difficult or impossible by conventional methods (29, 62). Such hybridizations could possibly be achieved through somatic cellular hybridization, although present tissue culture technology for poplars does not seem up to the task.

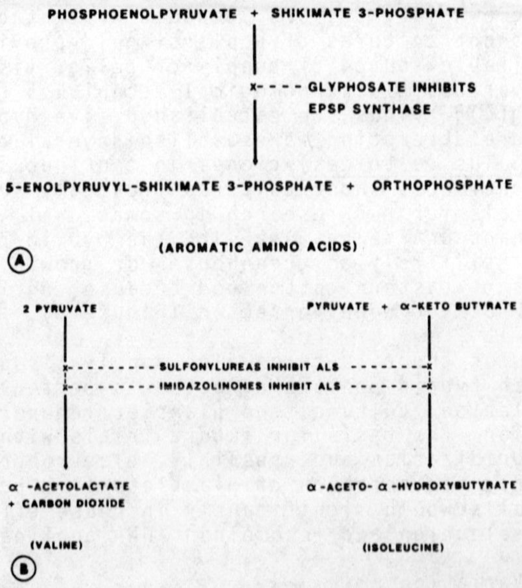

Figure 1. Target enzymes of the herbicides (A) glyphosate and (B) sulfonylureas and imidazolinones. Glyphosate inhibits the activity of 5-enolpyruvyl-shikimate-3-phosphate (EPSP) synthase, the sixth reaction of the Shikimate Pathway, which is essential for biosynthesis of aromatic amino acids. Sulfonylureas and the unrelated imidazolinones inhibit the activity of acetolactate synthase (ALS), which catalyzes the first reactions in the biosynthesis of valine and leucine. In addition, ALS catalyzes the second reaction in biosynthesis of isoleucine.

Types of Biotechnology

Somaclonal selection and recombinant DNA, using Agrobacterium as the transforming agent, may have immediate application to the genetic improvement of poplars, particularly to obtaining herbicide resistance. Thus, our experimentation for the past two years has largely centered around the foregoing techniques, with some preliminary research directed at somatic cellular hybridization.

APPLICATION OF BIOTECHNOLOGY TO POPLARS

Continuous Shoot Culture

There are two major problems in conducting physiological and biochemical experiments with forest trees that involve tissue culture. First, it is often difficult to obtain a physiologically uniform supply of plants or tissue so that large experiments can be conducted and properly replicated. Second, it is difficult to obtain such material in a sterile condition. We addressed this problem early in our experimentation and, based on the research of Dr. Brent McCown (40, 56, 57), decided that stabilized, continuous shoot cultures of poplars would provide the best and most reliable source of supply of cells, tissues, and organs for various biotechnological techniques [Figure 2, reviewed by (4)]. Thus, we established five hybrid poplars clones and are attempting to establish several eastern cottonwood (Populus deltoides) clones in continuous shoot culture. Continuous shoot cultures presently supply the basic materials for our research in somaclonal variation and recombinant DNA techniques with the hybrid poplars. However, we still rely on greenhouse- or growth chamber-grown plants of eastern cottonwood because continuous shoot culture has not yet been perfected (Figure 2).

Continuous shoot cultures also supply tissues for research with hybrid poplars directed at perfecting protoplast isolation, culture, and plantlet regeneration, which will form the basis for future trials with somatic cellular hybridization and, possibly, electroporation. High frequency regeneration of plantlets from protoplasts or single cells would significantly increase efficiency of somaclonal selection and recombinant DNA applications.

Somaclonal Variation

The purpose of our research concerning somaclonal variation is to obtain hybrid poplar and eastern cottonwood clones that are resistant to the herbicides glyphosate (Roundup, Monsanto Chemical Co.), the sulfonylurea, sulfometuron methyl (Oust, Du Pont Co.), or the imidazolinone, 2-(4-iso-propyl-4-methyl-5-oxo-2-imidazolin-2-yl) (Arsenal, American Cyanamid Co.). Glyphosate is presently used for weed control in SRIC poplar plantings; some sulfonylureas and imidazolinones are promising candidates for use in forestry applications. The rationale for our selection of these particular herbicides has been presented in detail elsewhere (45), but is primarily based on chemical efficacy and modes of action.

The phytotoxicity of glyphosate, imidazolinones, and sulfonylureas has been related to their inhibition of amino acid biosynthesis [Figure 1, (11, 31, 55, 58)]. Glyphosate inhibits the enzyme 5-enoylpyruvyl-shikimate-3-phosphate (EPSP) synthase, which is involved in the biosynthesis of aromatic amino acids (31). Imidazolinones (e.g. Arsenal, Scepter) and sulfonylureas [e.g. chlorsulfuron (Gleen), sulfometuron methyl (Oust)] deter branched chain amino acid biosynthesis by inhibiting acetolactate synthase (ALS, also known as acetohydroxyacid synthase). ALS is the initial enzyme in the biosynthesis of valine and leucine, and the second enzyme in the biosynthesis of isoleucine (24, 31, 58). It has been demonstrated with maize (Zea mays Pioneer var. 3547) that three different imidazolinone herbicides markedly inhibited ALS. Similar results have been obtained in tests of sulfonylureas (15, 38, 51).

Herbicides that inhibit ALS and EPSP synthase have low mammalian toxicity and appear to be good candidates for

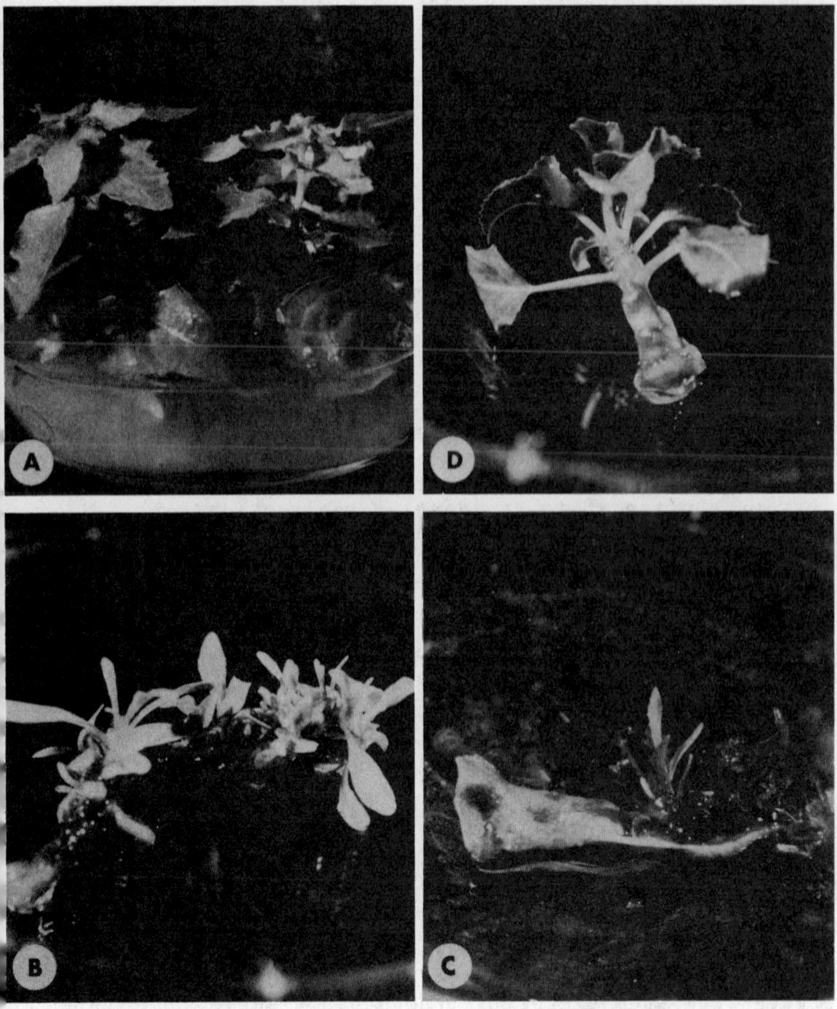

Figure 2. Continuous shoot stock culture (A) of hybrid poplar NC-5339 (Populus alba x P. grandidentata). Shoots regenerating from (B) an internode of hybrid poplar NC-5331 (Populus nigra cv. betulifolia x P. trichocarpa) and from (C) a leaf of hybrid poplar NC-5272 (Populus nigra x laurifolia). Shoot culture (D) removed from the node of eastern cottonwood (Populus deltoides). Shoot-forming tissues were obtained from continuous shoot stock cultures (B and C) or greenhouse plants (D). Shoots regenerated in vitro are readily micropropagated (microcuttings) out of culture, under non-sterile conditions to produce whole plants. (Courtesy of Drs. Thomas Sebastian and Charles Michler)

somaclonal selection programs with forest trees, based on evidence obtained with herbaceous crop plants (45). For example, a 100-fold increase in tolerance of maize to the imidazolinone herbicides was recently obtained in a somaclonal selection program (1). Similarly, sulfonylurea resistant tobaccos were obtained from *in vitro* cultures (16). In the foregoing tests with imidazolines and sulfonylureas, resistance resulted from mutation of a single dominant or semidominant nuclear gene (1, 16). Homozygosity in the tobacco mutation resulted in resistance to 100 times the concentration of chlorsulfuron that produced phytotoxicity in normal tobacco plants (16). It has been demonstrated with *Salmonella* that sulfometuron methyl also inhibits ALS and that a mutation at or near the *ilvG* gene imparts resistance to the herbicide (38). The *ilvG* gene codes for isozyme II of ALS (38). Mutations that confer resistance to imidazolinones may similarly confer resistance to the sulfonylurea herbicides, and the reverse (1, 31). Similar cross-resistances have been found recently for herbicides that inhibit photosynthesis (25).

Wheat, barley, and oats tolerate chlorsulfuron because they metabolize it to an inactive *o*-glycoside (63). Poplars have similarly been shown to metabolize the herbicide simazine (7, 8, 9). Thus, as previously stated, it may be possible to obtain genetic variants that are tolerant of or resistant to glyphosate, imidazolinones, or sulfonylureas based on mutations that confer the ability for herbicide metabolism or conjugation, in addition to those mutations described above that insensitize the target enzymes EPSP synthase or ALS.

Success in obtaining specific somaclonal variants partly depends on the frequency of regeneration of plantlets by the process of adventitious bud formation from the tissue cultures (Table 1). Thus, we initially optimized the culture media and other culture conditions to obtain very efficient regeneration from hybrid poplar tissue (i.e. regeneration by 90%+ of individual tissue pieces, each producing from 20 to 200 adventitious buds). Hybrid poplars are presently being regenerated from leaves (and callus developing therefrom) that originated in continuous shoot cultures; eastern cottonwood is being regenerated from stem pieces of greenhouse plants. Continuous shoot culture and efficient regeneration techniques have been substantially more difficult to develop for eastern cottonwood, compared to most of the the hybrid clones with which we work.

We designed a scheme to develop somaclonal poplar variants that tolerate or resist herbicides, based on initial experimentation and theory (Figure 3). These tests are presently limited to glyphosate and sulfometuron methyl. Cultures are being challenged either by growing tissue directly on culture media containing the herbicide (glyphosate or sulfometuron methyl) or by dipping tissue into herbicide solution before placing it on the culture medium (glyphosate). The different treatment methods were selected because we have found that sulfometuron methyl is quite chemically stable in tissue culture media. Glyphosate may be less chemically stable under equal conditions, which necessitates the need to test different methods. Cultures are being exposed to herbicide before or after they have begun to initiate adventitious shoots.

Table 1. Frequency of regeneration and occurrence of the variant gene influence recovery of a specific genetic variant [After (32)].

Start with one million cells to look for a specific genetic variant	Number of variants recovered if regeneration frequency is	
Condition	50%	0.1%
If 1 of 100,000 cells contain the variant gene	5	none
If 1 of 10,000 ...	50	none
If 1 of 100 ...	5,000	10
If 1 of 1 ...	500,000	1,000

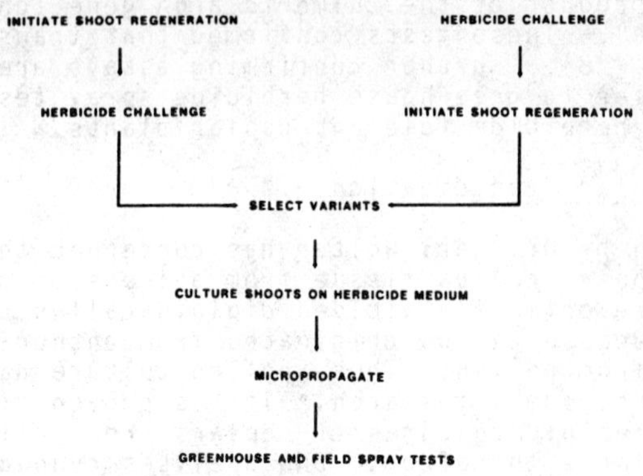

Figure 3. Scheme for challenging shoot-forming poplar cultures with herbicide for the selection of somaclonal variants for herbicide resistance.

Timing of herbicide application may influence selection of cells that will potentially regenerate tolerant or resistant adventitious shoots. However, an equally plausible hypothesis is that herbicide may inhibit adventitious shoot regeneration even from herbicide tolerant or resistant cells. These opposite hypotheses must be tested before a final method of challenge is selected. Our challenge experiments have not been conducted long enough to indicate whether or not they will result in poplar variants that tolerate or resist the herbicides.

Recombinant DNA

Recombinant DNA research with a poplar hybrid has been conducted by Ms. Joanne Fillatti, Dr. Brent McCown and Mr. J. Sellmer, in cooperation with our laboratory at Rhinelander. The purpose of this research has been to produce a hybrid poplar clone that is resistant to the herbicide glyphosate. Scientists at Calgene had previously isolated a bacterial gene, a chimeric *aroA* gene (20, 61), that has been shown to confer tolerance to glyphosate in tobacco (21) and soybean, cotton, and tomato (2). This is the first gene engineered for agronomic crops to be protected by U.S. patent (2).

The transforming agent in these tests was a bacterial strain (*Agrobacterium*) containing two genes for resistance to the antibiotic Kanamycin and the chimeric *aroA* gene (28). Whether or not transformation of poplar plantlets occurred was initially tested by growing them on a Kanamycin-containing medium. Kanamycin deters shoot growth of non-transformed plantlets, i.e. those that do not contain the Kanamycin resistance genes. In addition, enzyme activity was measured. Western blot, a definitive immunochemical assay, was also used to determine presence of the protein product of the chimeric *aroA* gene (chimeric EPSP synthase). These tests confirmed that transformation had occurred (28). Further confirming assays are being conducted prior to greenhouse herbicide spray testing of the putative herbicide tolerant poplar plants.

Somatic Cellular Hybridization

Research by Dr. Karl Wolter has concerned the culture of haploid poplar callus tissue from anthers or pollen. Anther culture primarily yielded diploid callus cultures, apparently because callus originated from anther wall cells rather than from pollen. Thus, pollen culture has been the subject of subsequent research. It has proven difficult to obtain pure haploid cultures of poplars from pollen and to regenerate plants therefrom. DNA analyses conducted on these callus cultures by Dr. S. Dhillon (Department of Botany, North Carolina State University) indicated a mixture of haploid and, probably, dihaploid cells, although the presence of diploid or polyploid cells could not be discounted. Substantial further research will be required to obtain and maintain stable, pure lines of haploid poplar callus that can regenerate plantlets.

ACKNOWLEDGEMENTS

Part of the herbicide resistance research reported in this paper is supported by the U.S. Department of Energy, Biomass Energy Technology Division, Short Rotation Woody Crops Program under Interagency Agreement No. DE-AI05-80OR20763. I gratefully acknowledge the preprints of papers, data, and general information supplied by Dr. S. Dhillon, Ms. Joanne Fillatti, Dr. Brent McCown, Dr. Charles Michler, Dr. Don Riemenschneider, Dr. Thomas Sebastian, Mr. J. Sellmer, and Dr. Karl Wolter. I also thank Mr. E. Bauer and Mr. K. Hansen for preparing figures, and Ms. Lois Moulton and Ms. Sandra Haissig for assistance in preparing the manuscript.

DISCLAIMER

Mention of trade names does not constitute endorsement by the U.S. Department of Agriculture or the Forest Service.

PESTICIDE PRECAUTIONARY STATEMENT

This paper reports research involving pesticides. It does not contain recommendations for their use, nor does it imply that uses discussed here have been registered. All uses of pesticides must be registered by appropriate state and/or federal agencies before they can be recommended. CAUTION: Pesticides can be injurious to humans, domestic animals, desirable plants, and fish or other wildlife -- if they are not handled or applied properly. Use all pesticides selectively and carefully. Follow recommended practices for the disposal of surplus pesticides and pesticide containers.

REFERENCES CITED

(1) Anon. "Selection of herbicide-tolerant corn plants," Agricel Report, p. 5 (1985) January.

(2) Anon. "Calgene secures patent for glyphosate tolerant gene," Agric. Biotech. News (1985) September/October.

(3) Ahuja, M. R. "Isolation, culture, and fusion of protoplasts: Problems and prospects," Silvae Genetica 31, 66-77 (1982).

(4) Ahuja, M. R. "Short Note: Isolation and culture of

mega and normal protoplasts in aspen," Silvae Genetica 32, 5-6 (1983).

(5) Ahuja, M. R. "Somatic cell differentiation and rapid clonal propagation of aspen," Silvae Genetica 32, 3-4 (1983).

(6) Ahuja, M. R. and Muhs, H. J. "Control of growth and differentiation in tissues and protoplast derived callus in different genotypes of aspen," 177-178. Proc. 5th Int. Cong. Plant Tissue and Cell Culture. Plant Tissue Culture, 1982.

(7) Akinyemiju, O. A. and Dickmann, D. I. "Contrasting effects of simazine on the photosynthetic physiology and leaf morphology of two Populus clones," Physiol. Plant. 55, 402-406 (1982).

(8) Akinyemiju, O. A. and Dickmann, D. I. "Variation among 21 Populus clones in tolerance to simazine and diuron," Can. J. For. Res. 12, 708-712 (1982).

(9) Akinyemiju, O. A., Dickmann, D. I. and Leavitt, R. A. "Distribution and metabolism of simazine-tolerant and -intolerant poplar (Populus sp.) clones," Weed Sci. 31, 775-778 (1983).

(10) Akinyemiju, O. A., Isebrands, J. G., Nelson, N. D. and Dickmann, D. I. "Use of glyphosate in the establishment of Populus in short rotation intensive culture," 94-102. Proc. NA Poplar Council Meeting, (J. Zavitkovski and E. A. Hansen, eds), Rhinelander, WI. (1982).

(11) Amrhein, N., Deus, B., Gehrke, P., and Steinruchen, H. C. "The site of the inhibition of the shikimate pathway by glyphosate. II. Interference of glyphosate with chorismate formation in vivo and in vitro," Plant Physiol. 66, 830-834 (1980).

(12) Bliss, F. A. "The application of new plant biotechnology to crop improvement," HortScience, 19, 43-48 (1984).

(13) Burg, A. W., Sacco, G. R. and Wheat, D. Biotechnology: opportunities in the biomedical, chemical, agricultural and food industries. Arthur D. Little Report. New York: Arthur D. Little & Co., 1983.

(14) Bychenkova, E. A. and David, A. "Callus formation and organogenesis in tissue of leaves of Populus balsamifera L. cultivated in vitro," Soviet Plant Physiol. 25, 216-222 (1978).

(15) Chaleff, R. S. and Mauvais, C. J. "Acetolactate synthase is the site of action of two sulfonylurea herbicides in higher plants," Science 224, 1443-1445 (1984).

(16) Chaleff, R. S. and Ray, T. B. "Herbicide-resistant mutants from tobacco cultures," Science 223, 1148-1151 (1984).

(17) Chalupa, V. "Control of root and shoot formation and production of trees from poplar callus," Biol. Plant. 16, 316-320 (1974).

(18) Chardenon, J. et Taris, B. "Recherches sur la reprise de l'activate du cambium du cultivars de Populus cultive in vitro," C.R. Acad. Sci. 251, 120-121 (1960).

(19) Christie, C. B. "Rapid propagation of aspens and silver poplars using tissue culture techniques," Int. Plant Prop. Soc. Proc. 28, 255-260 (1978).

(20) Comai, L., Sen, L. C. and Stalker, D. M. "An altered aroA gene product confers resistance to the herbicide glyphosate," Science 221, 370-371 (1983).

(21) Comai, L., Facciotti, D., Hiatt, W. R., Thompson, G., Rose, R. E. and Stalker, D. M. "Expression in plants of a mutant aroA gene from Salmonella typhimurium confers tolerance to glyphosate," Nature 317, 741-744 (1985).

(22) Dhillon, S. S., Miksche, J. P. and Cecich, R. A. "DNA changes in senescing leaves of Populus deltoides," Plant Physiol. (Suppl.) 75, 120 (1984).

(23) Diaz-Colon, J. D., Bovey, R. W., Davis, F. S. and Baur, J. R. "Comparative effects and concentration of picloram, 2,4,5-T and dicamba in tissue culture," Physiol. Plant. 27, 60-64 (1972).

(24) Dixon, M. and Webb, E. C. Enzymes. Third Edition, New York: Academic Press, 1979.

(25) Erickson, J. M., Rahire, M., Rochaix, J. D. and Mets, L. "Herbicide resistance and cross-resistance: Changes at three distinct sites in the herbicide-binding protein," Science 228, 204-207 (1985).

(26) Evans, D. A. and Sharp, W. R."Single gene mutations in tomato plants regenerated from tissue culture," Science 221, 949-951 (1983).

(27) Evans, D. A., Sharp, W. R. and Medina-Filho, H. P. "Somaclonal and gametoclonal variation," Amer. J. Bot. 71, 759-774 (1984).

(28) Fillatti, J. J., Sellmer, J. C. and McCown, B. H. "Regeneration and transformation of Populus tissue," Proc. XXIInd International Horticultural Congress. Davis, CA: University of California (1986, in press).

(29) Guries, R. P. and Stettler, R. F. "Pre-fertilization

barriers to hybridization in the poplars," Silvae Genetica, 25, 37-44 (1976).

(30) Hansen, E. A. "Irrigating forest plantations," 46-71. Intensive Plantation Culture: 12 Years Research, General Technical Report NC-91. St. Paul, MN: USDA-Forest Service, North Central Forest Experiment Station, 1983.

(31) Hardy, R. W. F. and Giaquinta, R. T. "Molecular biology of herbicides," BioEssays 1, 152-156 (1984).

(32) Kidd, G. H. Strategic plan in biotechnology for forest trees. Final Report for USDA Forest Service, North Central Forest Experiment Station, Forestry Sciences Laboratory, Rhinelander, WI. Milwaukee: L. William Teweles & Co., 1984.

(33) Kouider, M., Skirvin, R. M., Saladin, K. P., Dawson, J. O. and Jokela, J. J. "A method to culture immature embryos of Populus deltoides in vitro," Can. J. For. Res. 14, 956-958 (1984).

(34) Langridge, W. H. R., Li, B. J. and Szalay, A. A. "Electric field mediated stable transformation of carrot protoplasts with naked DNA," Plant Cell Reports, 4, 355-359 (1985).

(35) Larkin, P. J. and Scowcroft, W. R. "Somaclonal variation -- a novel source of variability from cell cultures for plant improvement," Theor. Appl. Genet. 60, 197-214 (1981).

(36) Larkin, P. J. and Scowcroft, W. R. "Somaclonal variation and eyespot toxin tolerance in sugarcane," Plant Cell Tissue Organ Culture 2, 111-121 (1983).

(37) Larkin, P. J., Ryan, S. A., Brettell, R. I. S. and Scowcroft, W. R. "Heritable somaclonal variation in wheat," Theor. Appl. Genet. 67, 443-455 (1984).

(38) LaRossa, R. A. and Schloss, J. V. "The sulfonylurea herbicide sulfometuron methyl is an extremely potent and selective inhibitor of acetolactate synthase in Salmonella typhimurium," J. Biol. Chem. 259, 8753-8757 (1984).

(39) Lester, D. T. and Berbee, J. G. "Within-clone variation among black poplar trees derived from callus culture," For. Sci. 23, 122-131 (1977).

(40) McCown, B. H. "From gene manipulation to forest establishment: shoot cultures of woody plants can be a central tool," TAPPI J. 68, 116-119 (1985).

(41) Meins, F., Jr. "Heritable variation in plant cell culture," Ann. Rev. Plant Physiol. 34, 327-346 (1983).

(42) Moore, L. and L. F. Wilson. "Recent advances in research of some pest problems by hybrid Populus in Michigan and Wisconsin," 94-101. Intensive Plantation Culture: 12 Years Research, General Technical Report NC-91. St. Paul, MN: USDA-Forest Service, North Central Forest Experiment Station, 1983.

(43) Nelson, N. D. "North Central Forest Experiment Station Biotechnology Program - application to tree improvement," 14-22. Proc. 18th Southern Forest Tree Improvement Conference, Long Beach, MS. Gulf Park, MS: Univ. of Southern Mississippi, 1985.

(44) Nelson, N. D. and Haissig, B. E. "Biotechnology in the Forest Service's North Central Forest Experiment Station," 139-154. Proc. International Symposium of Recent Advances in Forest Biotechnology, Traverse City, MI. East Lansing, MI: Michigan Biotechnology Institute, 1984.

(45) Nelson, N. D. and Haissig, B. E. "Herbicide Stress: Use of biotechnology to confer herbicide resistance to selected woody plants," Ch. 10. [Book title not final] (T. Hennessey, ed). The Hague, The Netherlands: Martinus Nijhoff/Dr. W. Junk Publishers, 1986 (in press).

(46) Nelson, N. D., Haissig, B. E. and Riemenschneider, D. E. "Applying the new somaclonal technology to fiber supply in the pulp and paper industry," BioProcess Engineering, 1986 (in press).

(47) Ostry, M. E. and McNabb, H. S. "Diseases of intensively cultured hybrid poplars: A summary of recent research in the North Central Region," 102-109. Intensive Plantation Culture: 12 Years Research, General Technical Report NC-91. St. Paul, MN: USDA-Forest Service, North Central Forest Experiment Station, 1983.

(48) Phelps, J. E. "Biomass from intensively cultured plantations as an energy, chemical, and nutritional feedstock," 131-138. Intensive Plantation Culture: 12 Years Research, General Technical Report NC-91. St. Paul, MN: USDA-Forest Service, North Central Forest Experiment Station, 1983.

(49) Potrykus, I., Shillito, R. D., Saul, M. W. and J. Paszkowski. "Direct gene transfer. State of the art and future potential," Plant Molecular Biol. Reporter, 3, 117-128, (1985).

(50) Qualset, C. O. "Integrating conventional and molecular genetics," Calif. Agric. 36, 29-30 (1982).

(51) Ray, T. B. "Site of action of chlorsulfuron. Inhibition of valine and isoleucine biosynthesis in plants," Plant Physiol. 75, 827-831 (1984).

(52) Russell, J. A. and McCown, B. H. "Thidiazuron-stimulated shoot differentiation from protoplast-derived calli of Populus," Proc. Sixth International Congress of Plant Tissue and Cell Culture. Minneapolis, MN: University of Minnesota, 1986 (in press).

(53) Russell, J. A. and McCown, B. H. "Isolation, culture, and regeneration of Populus mesopyhll protoplasts," Proc. XXII International Horticultural Congress. Davis, CA: University of California, 1986 (in press).

(54) Russell, J. A., Zeldin, E. L. and McCown, B. H. "Leaf grinding improves release of mesophyll protoplasts," Proc. 82nd Annual Meeting of the American Soc. for Horticultural Sci. Blacksburg, VA: Virginia Polytechnic Institute and State University, 1985.

(55) Sauers, R. F. and Levitt, G. "Sulfonylureas: New high potency herbicides," 21-44. Pesticide Synthesis Through Rational Approaches (P.S. Magee, G. K. Kohn, and J. J. Menn, eds), ACS Symposium Series 225. Washington, DC: American Chemical Soc., 1984.

(56) Sellmer, J. C., Russell, J. A., Zeldin, E. L. and McCown, B. H. "Utilization of cytokinin response curves in tissue evaluation of Populus for biotechnology research," Proc. 82nd Annual Meeting of the American Society for Horticultural Science. Blacksburg, VA: Virginia Polytechnic Institute and State University, 1985.

(57) Sha, L., McCown, B. H. and Peterson, L. A. "Occurrence and cause of shoot-tip necrosis in shoot cultures," J. Amer. Soc. Hort. Sci. 110, 631-634 (1985).

(58) Shaner, D. L., Anderson, P. C. and Stidham, M. A. "Imidazolinones. Potent inhibitors of acetohydroxyacid synthase," Plant Physiol. 76, 545-546 (1984).

(59) Skelsey, A. F. "Biotechnology in agriculture -- new tools for the oldest science," 229-266. Reference Document: Needs Assessment For The Food and Agricultural Sciences. Washington, DC: Joint Council on Food and Agricultural Sciences, 1984.

(60) Smith, M. A. L. and McCown, B. H. "A comparison of source tissue for protoplast isolation from three woody plant species," Plant Sci. Lett. 28, 149-156 (1982/1983).

(61) Stalker, D. M., Hiatt, W. R. and Comai, L. "A single amino acid substitution in the enzyme 5-enolpyruvylshikimate-3-phosphate synthase confers resistance to the herbicide glyphosate," J. Biol. Chem. 260, 4724-4728 (1985).

(62) Stettler, R. F., Koster, R. and Steenackers, V.

"Interspecific crossability studies in poplar," Theor. Appl. Genet. 58, 273-282 (1980).

(63) Sweetzer, P. B., Schow, G. S. and Hutchison, J. M. "Metabolism of chlorsulfuron by plants: Biological basis for selectivity of a new herbicide for cereals," Pesticide Biochem. and Physiol. 17, 18-23 (1982).

(64) Thomas, B. R. "Selection of phenotypes in plant cell cultures: A bibliography," Plant Molecular Biol. Reporter 2, 46-61 (1984).

(65) Torrey, J. G. "The development of plant biotechnology," Amer. Sci. 73, 354-363 (1985).

(66) US National Academy of Science. Report on the Briefing Panel on Agricultural Research. Washington, DC: National Academy Press, 1983.

(67) USDA Policy and Procedures for Research and Regulatory Activities in Agricultural Biotechnology. Draft of December 23. Washington, DC: United State Department of Agriculture, 1985.

(68) Venverloo, C. J. "The formation of adventitious organs. I. Cytokinin-induced formation of leaves and shoots in callus cultures of Populus nigra L. 'italica', Acta Bot. Neerl. 22, 390-398 (1973).

(69) Whitehead, H. C. M. and Giles, K. L. "Rapid propagation of poplars by tissue culture methods," Int. Plant Prop. Soc. Proc. 26, 340-343 (1976).

DISCUSSION

S. Schank, U. of Florida	I wonder how difficult it is to obtain a patent on a plant? Was this considered a plant patent, and why did you pursue that route rather than the Plant Variety Protection Act?
Haissig	First of all, the patent work was done by Calgene; I was not involved in it. I'm not fully capable of discussing United States patent law, but I do know that genetically engineered material and genes can now be covered under the U.S. letters patent law. I believe that it gives you better protection.

PRODUCTION OF CHINESE TALLOW THE FUEL OIL SEED CROP

H. W. Scheld, Ph.D.
Director of Research
PhytoResource Research, Inc.
College Station, Texas 77840

ABSTRACT

The Chinese tallow tree is a member of the family Euphorbiaceae; native to subtropical China where it has been cultivated for fourteen centuries as a specialty oil seed crop, medicinal plant and source of vegetable dye. It is now well distributed over the world within the climate zone delineated by the 10-15° C (50-60° F) winter isotherms. The tree resembles a small aspen, seldom exceeds forty feet in height, is typically bushy or multiple-stemmed with a gnarled trunk and rough, gray bark. Among its cultural attributes are rapid early growth, seed production as early as the third season, the capability for growth over a wide range of soil types, resistance to insect or microbial pests, tolerance of infertile, water-logged, or saline soils and a remarkable genetic variability with respect to every measurable characteristic of growth, chemistry and form.

The Chinese tallow tree possesses several present or potential economic values. It is currently valued as an ornamental and is an important bee forage plant. Its susceptibility to direct seeding, rapid early growth and ability to resprout render it potentially useful as a wood-fuel/chemical crop, and because of its unique abrasion resistance, it could be a valuable specialty wood. The major economic potential derives from the unusual variety, unique composition and high yield of its oil products. The seed yields both a hard vegetable tallow and a liquid oil which together comprise 45-50% of the seed weight. An impermeable shell (40%) by weight which separates the tallow and oil, and a seed meal fraction (10%) by weight comprise the remainder of the seed. Tallow may consist of over 95% of a single triglyceride, the palmitic-oleic-palmitic triglyceride. The oil is comprised of glycerides of varying proportions of mono-, di- and tri-unsaturated 18-carbon fatty acids with a small, variable, but potentially valuable fraction of highly reactive 8- and 10-carbon fatty acids. These various components may be converted into a diverse array of valuable chemical products.

PRODUCTION OF CHINESE TALLOW
THE FUEL OIL SEED CROP

INTRODUCTION

In recent years, the threats of diminished petroleum suppl potential instability in foregin supplies, the need for agricultura diversification/improvement of the farm economy, and the necessity f finding ways to utilize land areas which have heretofore been consi ered marginal for agriculture have all combined to generate a stro interest in the discovery and development of new industrial crops fr among the world's inventory of wild, or little-used domestic cr plants. Neglecting food crops, fruits and nuts there are current over two dozen "new crop plants" under evaluation for their potentia in domestic production of industrial raw products, gums, oils, resin fiber products, which are either of strategic importance to the natio al security although obtained from foreign sources and in short suppl or are produced by plants which are suitable for cultivation on marg nal lands. For example the best organized effort is the USDA office Critical Materials effort in new farm and forest products. (19 Anonymous. Growing Industrial Materials. Office of Critial Materia U.S. Department of Agriculture) Chinese tallow is included among t new crops being considered. The discusson below attempts to summari the major available information regarding Chinese tallow and to defi the approach that is being taken and continued development of Chine tallow as a multiple use crop plant.

BACKGROUND

Description - General

The Chinese tallow Sapium sebiferum (L.) Roxb., is a small, dec duous tree seldom growing beyond a height of 40 feet and superficial resembling an aspen with oppositely or spirally arranged, gloss heart-shaped leaves and a rough, gray bark. Form ranges in differe strains from low, spreading, and multi-forked (commonly) to a slende columnar bole with small pendant branches (rarely). Boles of old specimens become twisted and fluted. Large catkins of yellow flowe are formed in spectacular numbers in the spring (May - June in t U.S.); seeds are borne in clusters of green, usually 3-lobed capsul which dry, blacken, and split in the fall (September - November in t U.S.) to expose white, pea-sized seeds. The white color is due to aril-like covering comprising a hard white crystalline tallow embedd in a fibrous matrix. Tallow tree leaves are notable for their abili to produce brilliant fall color (yellow, orange, red, purple-bronze) warm regions where fall color does not ordinarily occur.

Distribution and Useful Range

The Chinese tallow tree is a semitropical plant native to centr and south China. It has been widely introduced and has been report in Japan, Taiwan, Vietnam, Burma, northern India, Java, Pakista Australia, Hawaii, Brazil, Cuba, St. Vincent, Martinique, southe Europe and the Sudan. It is very likely present but simply not reco nized throughout the semi-tropical and milder temperate zones of the world. In the U.S., it has become naturalized along the Gulf and the southern North Atlantic coasts from North Carolina to Texas and in California. From reports currently in hand the useful range appears to be approximately between the 10° and 15° C ($50-60^o$ F) mean January isotherms. When introduced into warmer regions it grows but apparently does not become naturalized. This latter observation does not necessarily preclude its economic use in warmer regions but does indicate the need for caution.

Origins & History of Cultivation

Origins of the Chinese tallow are lost in antiquity. According to Chinese authorities, records of Chinese tallow culture and utilization extend back at least fourteen centuries. There is general agreement that the original wild-type tree no longer exists, although modern techniques have not been applied to comparative studies of domesticated trees with the several existing wild species of the genus. Reports from the 1850's indicate that it was widely cultivated for its seed as well as for use as a source of a leaf tannin dye and as a medicinal plant. Culture at that time was seldom in formal orchards, but rather, as an "odd corner" tree along roads, canals, and in other waste places. Although the tree is currently cultivated in eleven provinces of Southern and Southeast China, the center of cultivation apparently has always been the Zhejiang Province which is at essentially the same latitude as Houston, Texas and Jacksonville, Florida. Within the last thirty years a massive program of crop improvement and the development of a formalized cropping system have been carried out. In the Zhejiang Province alone, over six million individual trees were evaluated, and from these, four clones having superior cultural characteristics of low form, large seed and cluster size and late flowering and ripening were selected for propagation and planting in producing orchards.

The introduction of Chinese tallow into the U.S. began very early in the history of this country. A letter from Benjamin Franklin to Noble Wimberly Jones of Savannah, colony of Georgia, dated 1772, transmits the seeds of Chinese tallow and comments upon the useful qualities of the plant. A few years later Francois Michaux, the French botanist, apparently introduced seeds of Chinese tallow into the vicinity of Charleston, South Carolina. In both areas the tree became naturalized very quickly. United States plant introduction records indicate a large number of separate introductions; very probably there were many more informal, unrecorded introductions. The most important introduction occurred in the early 1900's when the U.S. Bureau of Plant Industry attempted to establish local Chinese tallow industries in Houston, Texas and Jacksonville, Florida. Records of the results of this early work are relatively sparse and the original plantations have long been uprooted, but continued spread and naturalization of Chinese tallow has been mediated by its popularity as an ornamental and the attraction the white seed has for various kinds of birds. It is currently spread throughout the southern North Atlantic coast and coastal plains from Wilmington, North Carolina to lower Texas, and all observations indicate that it is continuing to spread.

UTILIZATION - POTENTIAL

Wood

Because of its rapid early growth, vigorous sprouting, and ability to grow in saline lands which are marginal for conventional agriculture, the Chinese tallow tree has been evaluated for its potential as a short rotation fiber, woodfuel, or chemical crop. The wood is, on the average, moderately dense but varies considerably, with observed densities ranging from .35 to .5 with a mean of about .45. Fuel value averages around 4200 kCal/kg, within roughly the same range as other hardwoods. Burning qualities of adequately dried wood are good with little smoke and no tendency to "spit" when used as a fireplace wood. Use for paper pulp is probably not practical because of short fiber lengths. Fibers measured from trees in Texas averaged 0.8mm while fibers from Taiwan were only slightly longer at 1.3mm.

Freshly cut Chinese tallow tree wood is soft, weak, white, and without obvious heartwood or prominent grain pattern. Moisture content is relatively high (41-45% wet weight) and considerable shrinkage, with associated cupping and warping, can occur upon drying. An extractive component of the wet wood or bark reacts with iron tools to produce rapid rusting of the tools and a black stain in the wood. Drying results in interesting changes in the character of the wood. Quick-dried wood remains white, but wood allowed to dry slowly in-the-round assumes a color and grain pattern very similar to black walnut. The wood is easily worked and capable of taking a high polish. Dried wood is low in structural strength but very resistant to penetration and abrasion, and thus roughly equivalent to oak in its suitability for use in flooring or furniture.

Extractives

Records dating to ancient times indicate that the tree has been used extensively in Chinese herbal medicine, and it may contain compounds of pharmacological value. Relatively few compounds have been extracted and identified from parts other than the seeds. Pods, leaves, and green stems produce small amounts of a milky (latex ?) sap. Several sources describe the sap as being toxic, but supporting data are not presented to indicate manner or degree of toxicity. Direct contact with the latex or the clear sap from the bark causes eye irritation, and there are several reports extant in which children ingesting leaves or the green pods have suffered from severe stomach aches, but apparently no deaths have resulted. Diterpene esters (phorbol) which are used in cancer research as tumor promoters have been reported from bark, roots and kernel oil, but there are conflicting reports regarding their presence in leaves. The current market value of phorbol esters as listed in research biochemical catalogs is approximately $4,000,000/lb. Cases of dermatitis caused by casual contact with the sap are rare.

The leaves contain large quantities of uncharacterized anthocyanin-type pigments which are responsible for the tree's colorful foliage and probably for the black "tannin" dye reported to be prepared from leaf extracts and used in the dying of silk. It is possible that leaves could be useful as a source of industrial tannins.

Bee Forage

The showy yellow catkins of the male flower first appear during the third growth season. The flowering season extends over a period of one to one- and-one-half months yielding a major honey and pollen crop which is notable for extreme dependability in comparison to other common honey plants. Two glands at the junction of the leaf blade and petiole produce nectar which provides additional bee forage in the fall. The honey is moderately dark but has a good flavor. Seeding efforts on the part of professional and amateur beekeepers appear to be an important factor in its recent rapid spread.

Amenity Planting

The Chinese tallow has been used extensively as an ornamental tree. Because of its rapid growth, it quickly produces shade, and some varieties are very colorful in the fall, even in warm climates. It tends to be "messy", producing copious amounts of wind-borne pollen (apparently not responsible for allergic reactions), and the large male flower parts abscise after pollen production. Leachate from the leaves and seed pods is highly destructive of auto finishes and exposed metal. The large number of seeds produced by some varieties attracts birds which eventually deposit them in gardens, flower beds, and agricultural lands.

Seed Oils

The major commercial future of the Chinese tallow tree is as a specialty oil seed. Two types of fatty substances are produced; namely a saturated solid fat, or tallow, on the outside of the hull and an unsaturated liquid kernel oil inside the hull. Proportions of tallow and oil vary widely among individuals, but tallow typically represents 25-30% of the seed weight, oil 15-20%, the very hard hull about 40%, with the remainder taken up in seed protein and fiber. The tallow (known in the oil trade as Chinese vegetable tallow) consists in some strains primarily of a single triglyceride, the palmitic-oleic-palmitic (POP) triglyceride, and thus with proper processing is a source of high-purity oleic and palmitic acids. Oleic acid may be easily converted to azelaic acid and pelargonic acid which have diverse industrial uses, including the production of plasticizers and baking finishes, respectively. The tallow could also be used in the food industry both as a cocoa butter blending component and as a shortening. It could more conventionally be used in the production of surfactants and emulsifiers. The Chinese tallow kernel oil (known in the oil trade as stillingia oil) is highly unsaturated, composed largely of glycerides of oleic, linoleic and linolenic acids, chemically similar to linseed oil and dehydrated castor oil, and therefore is useful as a conventional drying oil in the paint and varnish industry. Further, 8-10% by weight of the oil is composed of two unusual short-chain fatty acids which appear from preliminary examination to have possibilities in a number of chemical syntheses. Some 30% of the seed oil is composed of unsaturated estolide (tetraester triglycerides), which contain in addition to the conventional 18-carbon acids, the short chain fatty acids 8-hydroxy-5, 6-octadienoic acid and trans-2, cis-4-decadienoic acid. Both seed oil and tallow potentially may be utilized as lubricants, and in the production of diesel, turbine, and heating fuel. The seed meal, which is apparently non-toxic when fed to rats at a level equivalent to 10% protein, seems to be potentially useful as a protein concentrate in animal feed.

ENVIRONMENTAL REQUIREMENTS

Temperature

The tree will tolerate frosts and, in the dormant state, will tolerate short-duration drops to -10 degrees C. Once dormancy is broken, exposure to a light freeze can result in widespread killing, although there is a considerable variability in cold-tolerance among individuals. Environmental factors influencing disturbance are not well understood, but there may be a sensitivity to high temperatures and/or a requirement for a moderate period of cool weather and dormancy in order for effective flowering to occur.

Altitude

In the U.S. the tree is naturalized mostly in the coastal lowlands but has been observed as a planted ornamental at elevations of several hundred meters. In China, it is often cultivated on upland sites to elevations of around 1,000 meters. In India and Pakistan, it has been reported at elevations of about 2,000 meters, a probable practical upper altitude limit.

Rainfall & Water

The tree tolerates a wide range of annual rainfall. It is generally considered to be a high-moisture plant and there is probably no upper moisture limit. Trees have survived for over two years in areas flooded by dam construction. In one instance, good growth and seed yield has been observed in a year having a total annual rainfall as low as 710mm. Local soil conditions and the availability of water to the deep root system are probably critical in setting an absolute lower rainfall limit, and where these conditions are optimal, the lower limit is probably somewhat below 500mm of annual rainfall. There is no apparent reason why Chinese tallow cannot grow in very arid areas as long as irrigation is feasible. Because of the tree's tolerance to salinity the low quality water often available is arid regions may be adequate. Transpiration appears to be continuous, and thus water use by the tree is inefficient.

Soil

The Chinese tallow tree possesses a remarkable capacity for utilization of widely varying soil types. In the U.S. it has been found growing on barrier islands, competing effectively in pure sand with oak, pine, and palm, growing in salt marshes and along salt creek borders, in rocky upland soils and in dense clay soils subject to extended periods of flooding. Seeds have been observed germinating, and the seedlings produced have become established, in beds of oyster shell. Chinese workers report tests in which seeds germinated and grew in soil irrigated with sea water. The tree tolerates and competes well with other vegetation on poorly-drained, nitrogen-and phosphorous-poor sites, but response to fertilization is dramatic. Interestingly, in recent greenhouse studies it has exhibited apparent toxicity response when grown on potting media containing a calcined montmorillonite clay product known as Turface.

Establishment of Plantings

Plantations may be established from cuttings, seedlings or seed. Cuttings generally root with difficulty and rooting should be induced prior to planting by means of medium- to high-strength rooting compound. Individuals vary greatly in their capability for rooting. Seedlings grown in a nursery are ready for planting 6-8 weeks after germination. Special care in handling the seedlings is not necessary. A convenient practice is to root-prune (10-15cm) and top-prune seedlings to a uniform length for ease of handling.

The most convenient and economical method of stand establishment is direct planting. The seed is relatively large, the size and shape of a garden pea. It is conveniently hand-planted and it feeds well in standard mechanical row crop planters. Dense plantings (0.60-0.75m spacing) for wood plantations are practical.

Roots and stumps cut at any time of the year sprout prolifically and numbers of sprouts and rate of growth are apparently insensitive to method of cutting. Coppicing is thus a feasible system of management. In areas having previously-established wild stands, it is feasible to clear-cut and establish coppice management directly. Where stocking is sparse, additional stems may be obtained by induction of sprouting from injured roots.

All plantings for seed production utilise selected clonal stock; Chinese tallow is very easily grafted. In the past, Chinese practice was to "top work" or graft several scions into stumps of older trees. Trees used in establishing modern orchards are started in seedling nurseries from wild seed. In the early spring of the second year of growth the seedling is severed at the ground line, a scion is placed in the cleft and the graft is covered with soil to prevent drying and to prevent sprouting from the cut stem. Using this simple method a skilled worker can produce one thousand grafted seedlings per day. After a season of growth the grafted stock is lifted, root-pruned and planted in orchards.

Seed Handling

Seed for planting is harvested after pods begin to turn black and split, and before fungal attack of the tallow coat causes death of the embryo. Germination is greatly reduced if pods are removed before a significant amount of natural opening has occurred. Cold-strage (approximately -4° C) of seeds in sealed glass containers preserves viability for two or more years in most varieties. Pretreatment of the seed to remove the tallow coat, as suggested by some authors, is not necessary. In some varieties, a significant number of the seeds do not germinate until the second season after planting, and seeds planted in the late spring and summer often do not germinate until the following spring. Planting in the fall is a useful procedure that appears not to significantly affect germination and allows the earliest possible start for growing seedlings.

Yields

Wood-yield in unselected stock is adequate but not spectacular. In Texas, Chinese tallow plantations with 2′ x 2′ (0.6 x 0.6 m) spacing produced 17 oven tons of wood per acre (38.1 MT/ha) at the end of four seasons of growth in soils which were considered unsuitable for conventional crops. Well-stocked 15-year-old natural stands placed under coppice management produced in excess of 20 tons/acre (44.8 MT/ha) during four years of regrowth. In general, yield falls off after four to five years because of a shift from vegetative to reproductive growth in most individuals.

The Chinese tallow tree as it is known today is a product of a long-time selection for high seed yield. Reports from early experiments in the U.S. suggest the possibility of yields in excess of 10,000 lbs./acre (11.2 MT/ha) in orchards having spacing of approximately 16 x 16 feet. Such yields place it as a strong contender with tropical oil palm as the highest yielding oil seed known. Recent measurements of yields from selected wild trees confirm the possibility of seventy or more pounds per tree in individuals of the size utilized in the early test plantations. If planted at a similar spacing of 16 x 16 feet, such trees should yield approximately 12,000 lbs/acre (13.4 MT./ha). Producing plantations in China today yield less than 2,000 lbs. of seeds per acre because of the Chinese practice of interplanting food crops in widely spaced orchards. These plantations represent true, practical, agriforestry systems which exploit to the maximum the capacity of relatively poor soils.

Weed Tolerance

Given adequate initial fertilization the large seedling is capable of growing ahead of most annual weeds; in closely spaced plantings (<.75m) crown closure occurs early in the second season of growth, thus eliminating further significant competition from grass and annual weeds. Grass is the more significant competitor. While naturally seeded tallow tree forests typically become established in abandoned crop land or other disturbed areas, grassland is seldom invaded by tallow trees except in areas subject to flooding. The tree appears to compete effectively with grass in poorly drained areas, and flooding, where feasible, may be a useful method of weed control.

Pests and Diseases

Most of the observations on pests have been made either in China or Texas, USA. In the Chinese tallow tree's native range, a number of insect pests appear to have evolved with it. A publication on the practical problems of cultivation in the Zhejiang Province of China singles out for discussion the "poisonous moth", "Ailanthus silkworm", the "water green moth" and the "aphid". Personal communications with Chinese horticulturists indicate that there are at least twenty different insect pests of Chinese tallow including grubs, earthworms (nematodes ?), moths (?), poisonous caterpillars, (Io moth ?), bag moths, and red spiders (mites ?). In India, the larvae of a moth, *Ophiusa melicerte*, have been reported to defoliate the tree. In the U.S. and probably elsewhere outside its native range, there are no significant insect pests. Only two instances of significant leaf damage or defoliation by typical chewing insects have been recorded in the U.S. More typically, the only signs of insect attack are small irregular holes or nibbled edges, never completely consumed leaves. One exception as indicated above, is the defoliation by the leaf cutting ant, *Atta texana*, which in the sandy regions of south Texas appears to preferentially defoliate the Chinese tallow. In this case, the ant does not directly consume the leaf, but rather cultivates a fungus garden on the leaf material, and subsequently grazes upon the fungus mycelium. In another case, the Yellow Striped Army Worm, *Spodoptera ornithagalli*, removed most of the cotyledons from the emerging seedlings in a planting experiment. There was no further attack on the true leaves which sprouted from the apparently undamaged apical meristem.

While evidence of serious damage to Chinese tallow by chewing-insects is rare, sucking-bugs are often found. Large masses of a scale insect (*Pseudaulacaspis pentagona*) are present on the stems of the young trees. A single species of lady beetle, *Adalia bipunctata*, is very commonly found in association with the tallow tree and probably is the main predator on these insects. A large species of leafhopper (*Homalodisca coagulata*) is found constantly on succulent terminals in early or mid-summer. A leafleg bug (*Hyalymenus tarsatus*) that is a remarkable mimic of ants in the nymph stage is present, often in large numbers, in the late summer and fall. There is no evidence that any of these associated insects have a significant effect upon the productivity of the trees, however, quantitative data on growth have not been taken.

The incidence of microbial or fungal diseases is uncertain but appears to be minimal. A canker, apparently of fungal origin, is occasionally found on saplings up to 6 cm dbh. Presence of such a canker results in girdling or weakening of the stem, causing death or breakage of the stem at the point of attack. A number of fungal species attack and degrade the tallow coat of the seed causing discoloration and eventually, total blackening of the seed. In these later stages of degradation, the embryo is killed.

Cultural Limitations

The Chinese tallow tree can be an aggressive weed. The tallow-covered seeds are spread widely by birds, and, once established in agricultural lands, lawns, or gardens, seedlings and sprouts are difficult to eliminate because of the high capacity for sprouting from stumps and roots. In general, there is a high resistance to herbicides; both 2,4-D / 2,4,5-T and Glyphosate are relatively ineffective. On the other hand, difficulty has been experienced in establishment of direct seeded plantings of tallow tree in agricultural lands previously planted to corn, sorghum, or soy beans and treated intensively with "Trifleralin" (Treflan), "Profluralin" (Tolban 4E) and "Alachlor" (Lasso). Tallow tree seedlings are extremely sensitive to some component of a Malathion spray mixture. Pichloram (Tordon 10K) is one herbicide which appears effective if applied early in the spring.

The tree appears to be best adapted to semitropical climates and, although it apparently will grow in the tropics, there is some doubt, until further investigation, as to whether it will compete and reproduce well in the tropics. Most of the varieties currently growing in the U.S. originated from stock selected for seed productivity rather than good qualities as a forest tree. There is thus a need for careful selection from available stock for improved timber qualities.

The reproductive strategy of the Chinese tallow tree is such that breeding and production of seed of known characteristics will be complex. Timing of flower development is such that outcrossing between two major morphological types is assured in approximately 90% of the population. Thus, in development of seed orchards, it will be necessary to include individuals of the two flower types and, in general, crop improvement *via* selective breeding will be a somewhat complex undertaking.

DISCUSSION

L. Wright, Oak Ridge National Laboratory	I believe when our program stopped funding your project 2 or 3 years ago, you were starting to bring new selections into the country and start looking at them. Has that work progressed?
Scheld	There are a few in this country. Ed Glumac is also starting a program. I basically feel that we probably got the most valuable selections of those available. The material in this country must have been brought in by dozens of different people from all over China, and if anything, it's better adapted than the material that we're bringing in now.
Bente	What do the Chinese use the oil for?
D. Rockwood, U. of Florida	They use it for the same sort of things that we would use linseed oil for. They use it in boat finishes and furniture, and things of that sort. It's a drying oil.
Bente	It's not a cooking oil?
Rockwood	It's not a cooking oil. It's very highly unsaturated.

HARVESTING SYSTEM DEVELOPMENTS FOR BIOMASS PLANTATIONS

James A. Mattson, Ph.D.
Principal Mechanical Engineer

Nels S. Christopherson, Ph.D.
Research Mechanical Engineer

Forestry Sciences Laboratory
North Central Forest Experiment Station
USDA Forest Service
Houghton, Michigan 49931

ABSTRACT

Promising research results on the agro-forestry concept of growing trees in short-rotation, intensively cultured biomass plantations have led to continuing interest in this concept as a renewable source of energy supplies. The close spacings and short-rotation ages in biomass plantations produce trees that are small by conventional forestry standards. Economics requires the harvesting of many of these small trees per unit of time; thus equipment designers are faced with a unique challenge for the development of suitable harvesters. Research has begun in the United States, Canada, Sweden, and Ireland to develop appropriate harvesting equipment.

The work to date has produced several new machine concepts which, in prototype stage, have demonstrated several-fold increases in productivity over conventional forestry harvesting equipment. The continued development of these concepts could lead to a new generation of forestry harvesting equipment for biomass plantations. This new technology could also apply to small-tree harvesting problems in conventional forest operations, such as early thinnings in conifers.

HARVESTING SYSTEM DEVELOPMENTS FOR BIOMASS PLANTATIONS

INTRODUCTION

Fiber crops produced under intensive management in short-rotation biomass plantations represent a promising renewable source of energy for the future. The final decision on the commercial applicability of biomass plantations will depend heavily on the availability of suitable equipment and systems to harvest and process the biomass. The unique nature of the small, closely spaced trees grown under these conditions requires a new look at harvesting and processing concepts to develop appropriate machinery.

CHARACTERISTICS OF BIOMASS PLANTATIONS RELEVANT TO HARVESTER DESIGN

In principle, biomass plantations will consist of selected, rapidly growing, hardwood tree species (6). The crop will be harvested at appropriate rotations, and the succeeding crops will arise by coppicing (sprouting from stumps), precluding the need to replant after each harvest.

Suggested row spacings range from 1.2 to 3.0 m (4 to 10 feet) with spacings between trees within the rows ranging from 1.2 to 3.0 m (4 to 10 feet). Rotation age would range from 8 to 15 years, depending upon the spacing. These combinations will produce trees ranging from 5 to 25 cm (2 to 10 inches) in diameter at groundline and from 10 to 20 m (33 to 66 feet) tall. Most of the trees will be 5 to 25 cm (2 to 10 inches) in diameter, but some along the edges of the plantation may reach 30 cm (12 inches). These stands are expected to produce from 10- to 15-dry tonne equivalents of biomass per hectare per year (4.5 to 6.8 tons per acre per year) from trees ranging from less than 50 kg (110 pounds) to more than 200 kg (440 pounds) in the first harvest. Tree size will decrease with successful coppicing of the stands, and the number of trees per hectare will increase. The coppicing of the original stumps to produce new sprouts may produce a wider plant structure in subsequent harvests. The harvesting system may have to be able to handle a sprout cluster up to about 1 m (3 feet) wide.

Because most envisioned systems use coppicing to regenerate the stand after harvesting, it may be necessary to maintain a consistent stump height to ensure a suitable level of regeneration. The stump must have a clean-cut surface, such as would be produced with a saw. Lateral pressure against the stump will have to be minimized to avoid disturbing roots below ground. The harvesting equipment will have to stay within the rows of the plantation to avoid damaging the stump system. The need to obtain adequate coppicing may also limit harvesting to the dormant season when ground conditions may be adverse. To aid mobility of the harvest equipment and avoid excessive damage to the site and root system, vehicle ground pressure will have to be minimized, possibly to as low as 15 to 30 kPa (2 to 4 psi) in extremely low bearing sites. The number of vehicle passes over the stand during harvest should also be minimized to avoid rutting, soil compaction, and other site damage.

The terrain on most sites envisioned for biomass plantations is suitable for vehicle travel. The harvester will probably have to be capable of operating on side slopes up to 10 percent and direct slopes up to 20 percent. The plantations will also probably be free of rocks.

HARVESTER CONCEPT CONSIDERATIONS

The tree severing/accumulating phase of the harvesting operation is where current technology is most deficient for harvesting small trees. The development of equipment that can sever and handle small trees on a high-speed, nonstop basis is the key to the problem. In short, a continuous cutting "swath harvester" is needed.

The requirement of coppicing to produce subsequent crops from the original stumps in biomass plantations further precludes the use of conventional harvesting systems. Preliminary studies have shown that small stumps are susceptible to mechanical damage by equipment working in the stand (2). Conventional skidding, in particular, would be unacceptable in a biomass plantation because the machine's maneuvering and turning would certainly cause excessive damage. Skidding loads of trees across the plantation would also likely cause damage. Thus harvesting will have to be done so that whatever machine is used can stay tracked between rows in the plantation.

The fewer the passes through the stand the better in order to minimize soil compaction and rutting, so a machine needing only one pass would be desirable. Such a machine would have to be capable of simultaneously severing and collecting the stems and forwarding them to the plantation edge.

A machine that could simultaneously fell and chip the trees and transport the chips to the plantation edge is conceivable. Such a machine would be much heavier than one that only fells and collects the severed trees and forwards them to the plantation edge. The lighter machine would cause less soil compaction and cost less. Considering that the harvester will probably be used for only a portion of the year, a smaller, less costly machine would be preferable. If the harvester could be built to utilize a standard prime mover as its motive and functional power source, it could cost even less.

The possibility of the harvest being limited to the dormant season has some definite implications for the design of the harvesting equipment because the fiber must be stored for year-round use. Storing the material on the edge of the plantation would preclude any initial transport because the harvester could offload at the storage site, thus eliminating the need for extensive storage space at the use point. Harvesting would also be more efficient because any delays caused by interaction with the transportation system would be eliminated.

From the utilization standpoint, storing whole trees at the plantation edge appears to have some advantages over chipping the material in the stand and then storing the chips. Chips do not store well over the long term because heating within the chip piles and deterioration of the material can occur. Dirt and grit can get into the material when chips are recovered from the ground for transport. If the trees are stored as whole trees at the plantation edge, conventional chipping equipment can efficiently convert them into chips and load the chips onto trucks for transport. Also, the whole trees could be converted at less cost into alternate product forms such as chunks or blocks, which may be preferred for special uses such as burning (1).

In summary, a suitable biomass harvester would be able to cut one or more rows of a plantation as it moved continuously down the rows. The stumps would be cut cleanly without exerting excessive pressure on them. The severed trees would be collected on the machine and transported to a convenient point, such as the field edge or an access road within the plantation, where they would be offloaded and piled to await subsequent processing and/or transporting. Ground pressure exerted by the harvester would be minimal to avoid damage to the soil and to the roots of the stumps. The harvester would be as simple as possible and would use a separate prime mover to provide the motive and functional power.

HARVESTER DEVELOPMENTS RELEVANT TO BIOMASS PLANTATIONS

The great interest in growing woody biomass for energy production has prompted several research and development organizations around the world to investigate harvester concepts for biomass plantations. This interest has also led to the formulation of an international cooperative project on the development of biomass plantation harvesters. The project is organized under the charter of the International Energy Agency (IEA) forestry energy program; its objective is to facilitate the exchange of ideas and information among the interested groups. Canada, Ireland, Sweden, and the United States are all active participants. The concepts being investigated by the various designers show the diversity of approaches that may be taken to solve the harvesting problems associated with biomass plantations. Curtin and Barnett (3), working under contract to the U.S. Department of Energy, have extensively surveyed the current state of biomass harvesting equipment. The following descriptions of the machine development efforts associated with the IEA project are taken largely from their final report to the Department of Energy. Numerous individuals and manufacturers are continuously working on new concepts for forest harvesting equipment. It is possible, even probable, that these efforts, many of which we are currently unaware, will result in equipment suitable for biomass plantations. Discussing here only those projects associated with the IEA is in no way intended to slight the significance of any other ongoing work.

United States

The United States has two projects in the development of biomass plantation harvesters in cooperation with the Department of Energy, the U.S. representative to the IEA. Virginia Polytechnic Institute and State University (VPI), through a contract with the Short Rotation Woody Crops Program (SRWCP) of the Department of Energy (DOE), developed a harvester for severing and crushing multiple small stems from coppice stools. The U.S. Department of Agriculture, Forest Service, is developing a plantation harvester using an auger cutting device to sever stems between 5 and 18 cm (2 to 7 inches) anywhere across its 61-cm (24-inch) cutting bar.

The VPI/DOE Prototype Short-Rotation Plantation Harvesting Attachment (Figure 1). The VPI/DOE attachment is designed to harvest multiple stems less than 8 cm (3 inches) in diameter cheaply and rapidly from stools in previously harvested plantations. It consists of three major components--severing, collection, and crushing mechanisms (9).

Severing is done by two hydraulically powered, counter-rotating, 61-cm (24-inch) circular saws. The saws are equipped with carbide-tipped, pegged teeth and rotate between 2,500 and 3,000 rpm. This arrangement of teeth and speed leaves a smooth stump, uses minimal power (7 kw or 9 hp), and causes little wear (5).

Three elements gather and convey the stems. Each functions in sequence to guide them into the severing mechanism, which thrusts them into a conveying mechanism and carries them away from the saws to a crushing mechanism. A set of headers separates the stems in the row being cut from those in adjacent rows and guides them into the saws and feed reels. The feed reels are linked to the conveying mechanism and located to provide a positive thrust of stems into the conveying mechanism.

Severed stems are conveyed from the saws to the crusher by two banks of opposing vee belts which are individually tensioned using air-spring idler arms and sheaves. The lower set of belts is approximately 0.6 m (2 feet) above the cutting point. The upper set of belts is variable from 0.6 to 5 m (2 to 16 feet) above the lower belts, depending on the height of stems being harvested. The upper set of opposing belts should be located at or above the center of

Figure 1.--VPI/DOE prototype short-rotation plantation harvesting attachment.

gravity of severed stems. The conveying system elevates the stems vertically while at an angle of approximately 30 degrees from horizontal to a point approximately 1.3 m (4 feet) above the ground. The stems are then dropped into the crushing mechanism.

The crushing mechanism is a double-roll, double-drive crusher. It is hydraulically driven at a speed of approximately 137 lineal meters (450 feet) per minute. The high feed rate permits complete processing of the stems from one stool before stems from the next stool are fed into the crusher, thus minimizing power requirements and improving crushing uniformity. Crushed stems dry faster and are more flexible than uncrushed stems. Tests conducted by Teel (9) indicate a twofold increase in drying rates after crushing.

This prototype proved that high production rates are feasible using minimal power and low-cost components. It was able to harvest and crush 10 to 15 stools (with 4 to 15 stems per stool) per minute while operating at ground speeds of up to 4.8 km (3 miles) per hour. Production rates were between 8 and 19 tonnes per productive machine hour (9 to 21 tons), depending on tree size and spacing.

The USDA Forest Service Prototype Short-Rotation Harvester (Figure 2). The USDA Forest Service short-rotation harvester is being developed by the North Central Forest Experiment Station in Houghton, Michigan. The machine is a front-end attachment that severs trees from 5 to 18 cm (2 to 7 inches) in diameter using a two-flute milling cutter that is 5 cm (2 inches) in diameter and 0.6 m (2 feet) long (7). The cutter is mounted on an assembly that allows it to retract 46 cm (18 inches) if the milling rate is less than the forward speed of the harvester. When cutting is complete, the milling head springs forward to regain its retracting capability before the next stem is encountered. This feature ensures that the carrier vehicle can maintain a constant forward speed.

Severed trees will be pulled into a 0.4 m^2- (4 $feet^2$) accumulating area. When full, the accumulator will be tipped forward to discharge accumulated stems. Sequencing of the accumulating arms will be automatically controlled by an onboard microprocessor. Using the described techniques, the harvester should be able to maintain a continuous forward speed of between 0.8 and 2 km/h (0.5 and 1.2 miles/hour) until the accumulation area is filled. Forward speed will ultimately depend on stem diameter and spacing. The harvester will be capable of severing and accumulating single stems or coppice sprouts.

Canada

Most of the biomass plantation harvester development work in Canada has been funded through the National Research Council of Canada (NRCC). Currently, four harvesters are in various stages of development (FB2, FB7, FB12, and a coppice harvester).

NRCC FB7 (Figure 3). The NRCC FB7 was designed by Hyd-Mech Forestry, Ltd., Woodstock, Ontario, to process 1,000 trees per productive machine hour (pmh) while harvesting single trees from rows. It consists of severing, accumulating, and offloading components assembled into a front-end attachment for a four-wheel drive, articulated farm tractor. Specifically, the prototype is mounted on a 45-kw (60-hp) tractor. All functions except steering and offloading are automatic-sequenced functions controlled by an onboard microprocessor. When initiated by the operator, offloading is also automatically sequenced.

Figure 2.--USDA Forest Service prototype short-rotation harvester.

Sensors located in the cutting opening begin the accumulation sequence simultaneously with severing. Two 60-cm (24-inch) diameter, counter-rotating, inserted tooth saws do the severing. Accumulating arms push severed stems off the saws into one of two holding areas. As the holding area is filled with 8 to 10 trees, the operator activates the offloading sequence with an electric switch. An offloading grapple closes around the trees, swings them to the side and down, placing them on the ground parallel to the direction of travel. An accumulating mechanism, holding area, and offload mechanism on each side of the machine allow the operator to choose from which side of the machine bunched trees will be offloaded. The NRCC FB7 has harvested an average of 850 trees per hour, including turnaround time at the end of each row. Within a row, the average cut was 1,057 trees per pmh.

Power requirements of this attachment appear to be minimal. The saws are powered by hydraulic motors requiring 7 l/min (24 gpm) at a pressure of 140 kg/cm^2 (2,000 psi). This converts to approximately 21 kw (28 hp).

NRCC FB2. The NRCC FB2, being developed by Hyd-Mech Forestry, is a farm tractor attachment designed to harvest coppice sprouts from plantations. It consists of five components which: (1) gather clumps into severing and accumulating mechanisms; (2) sever clumps; (3) accumulate clumps; (4) tie bundles; and (5) discharge bundles.

Gathering is done by two sets of chain-mounted paddles located approximately 2 m (79 inches) above the severing mechanism. These paddles separate sprouts from adjacent rows and guide them into the collection mechanism.

Severing is done by a bandsaw operating at a speed of 3,700 m (12,000 feet) per minute, using about 6 kw (8 hp), and producing 15-cm (6-inch) blocks at a rate of 0.6 m per second (2 feet/second). After severing, pusher arms force the stems into the accumulating pocket. Accumulated clumps are tied using an agricultural tying mechanism and ejected to the side so they are oriented perpendicular to the direction of travel. These tied bundles can later be picked up by a forklift.

NRCC FB12. The NRCC FB12 is likely destined for use in conventional harvesting operations because of its size and

Figure 3.--Hyd-Mech Forestry, Ltd. prototype harvester FB7.

capabilities. The FB12 is being designed to continually harvest trees up to 30 cm (12 inches) in diameter at a rate of 500 trees per pmh. The developers, Hyd-Mech Forestry, see a need for a machine of this capability to harvest conifer plantations.

The severing mechanism will be a large circular saw approximately 1 m (39 inches) in diameter. About 1 tonne of severed trees will be pushed into an intermediate holding area where they will be held vertically. Then a transfer mechanism will lay them flat and deposit them in a second horizontal holding area on the main chassis. As this holding area is filled by several transfer cycles, it will be emptied onto the ground. Bunches left on the ground should be the right size for the inwoods transport machines.

NRCC Coppice Harvester (Figure 4). The NRCC coppice harvester under development by J. D. Reed Machine, Ltd., Brantford, Ontario, is a trailing attachment to a tractor in the 40- to 50-kw (53-to 67-hp) class. A pair of chains mounted with paddles guides the trees into the severing mechanism and conveys severed trees between two sets of rails, across the back of the harvester, and discharges them singly perpendicular to the direction of travel. A reciprocating saw with "M" teeth, driven by a 10-kw (14-hp) hydraulic motor through a crankshaft and connecting rod, does the severing.

Figure 4.--J. D. Reed Machine, Ltd. prototype coppice harvester.

This attachment is intended to continuously harvest multiple stems less than 13 cm (5 inches) in diameter from stools in rows 1 m (3.3 feet) apart. Initial tests were conducted during the winter of 1984 to 1985, and the results were satisfactory. For best results, the forward speed of the harvester must be synchronized with the speed of the gathering chain.

Sweden

Sweden has evaluated many possible machine configurations that could become effective biomass plantation harvesters. Significant thought and effort have been devoted to develop and evaluate concepts long before alternatives were selected and detailed designs prepared. The evaluations routinely included plantation spacings, rotation length, tree size, species, plantation size, the probable type of machine owner (e.g., farmer or corporation), and the final consumer of the biomass. This process has resulted in sound concepts supporting two machines currently undergoing testing.

The ESM Energiskogsmaskiner AB (Figure 5). The ESM is a side-mount attachment for a farm tractor in the 50-kw (75-hp) size class. A pair of counter-rotating circular saws sever stems less than 8 cm (3 inches) in diameter. Stems are then conveyed vertically by two sets of opposing belts along the side of the tractor and to its rear where they are accumulated vertically. When the accumulator is filled, the operator stops the tractor and manually ties the stems into 2.7-tonne (3-ton) "supersheaves" which are discharged backwards when the platform is tilted. A production rate of 16.2 green tonnes/pmh (18 tons/pmh) was established during tests in June 1985, including transport of the sheaves off the site.

The Brunn AIB Harvester (Figure 6). The Brunn AIB is a significantly larger machine than the ESM. Although the severing and initial conveying systems are very similar to the ESM, two sets of counter-rotating augers have been added to convey the severed stems to the bunks of the forwarder. This allows the Brunn AIB to accumulate loads up to 6.7 tonnes (7.5 tons).

The large payload of the Brunn AIB allows the operator to devote most of his time to harvesting before having to start transporting. It also allows a single man/machine to deliver clean stems (no soil contaminants) to roadside. It does have the disadvantage of being a single-purpose, expensive machine.

Ireland

Two biomass plantation harvesters have been developed in Ireland. Planning and development of these high-productivity machines began as early as 1976. A number of alternative configurations were evaluated before construction of either harvester was initiated.

The Bord Na Mona Harvester (Figure 7). The Bord Na Mona harvester is a fully trailed unit requiring a farm tractor in the 90-kw (120-hp) class. It is designed to harvest willow 8.5 cm (3.3 inches) in diameter, growing in double rows with rows spaced 1 m (3.3 feet) apart and double rows 3 m (10 feet) apart.

Severing is done by a pair of 1-m (3.3-feet) diameter circular saws, powered by two 26-kw (35-hp) hydraulic motors. Severed stems are directed to the rear, butt first, by a combination of chain-mounted deflector bars, spring-loaded vertical rollers, conveyor, and horizontal rollers.

This system feeds two horizontal chopping cylinders that cut most of the trees into billets 15 cm (6 inches) long (4). The billets are transferred to a flail unit that lacerates small diameter branches, preventing any material that was not cut by the cutting cylinders from interfering with the flow of billets material into the 7-m^3 (247-$foot^3$) container. When the container is filled, it is hydraulically tipped for emptying.

Figure 5.--ESM Energiskogsmaskiner AB prototype plantation harvester.

The machine was successful in harvesting single and multiple stems from plantations containing up to 6 tonnes per ha (3 tons/acre) of dry material. The problems of drying and relatively low yield led to a recommendation that trees simply be severed, bundled, and stored on or near the growing site until needed. Transportation in tree form and chipping at user facilities were also recommended (4).

Figure 6.--Brunn AIB prototype harvester.

The Loughry Coppice Willow Harvester (Figure 8). Because of drying and feeding problems associated with chips and billets, development of the Loughry harvester began in 1980. It was developed as a feller/bundler, leaving alternate uses open while allowing moisture reduction in storage (8). It mounts on the three-point hitch of a 44-kw (55-hp) farm tractor on which the controls have been reversed. Power is obtained through a hydraulic pump attached to the tractor power take-off. The harvester weighs approximately 1,160 kg (2,560 pounds) and is designed to harvest 2- to 3-year-old willow coppice.

Gathering is done by a pair of spring-loaded, counter-rotating, inclined augers that have a scrolling speed slightly faster than the forward speed of the tractor and harvester. This causes the stems to be "leaned" toward the harvester as they are cut by a single 760-mm (30-inch) diameter saw, rotating at approximately 1,000 rpm. A pair of solid tines with almost parallel motion push the stems off the saw at 90 degrees to the direction of travel into a bundling chamber.

Figure 7.--Bord Na Mona prototype biomass harvester.

Stems are accumulated in the bundling chamber until is filled with approximately 30 kg (66 pounds). They a then separated from new stems entering the bundling chamb and tied with two twines. The tied bundles are then eject to the side of the machine.

The Loughry harvester was tested, evaluated, and mod fied annually from 1981 through 1984. It worked better rows 1 m (3.3 feet) apart than it did when the rows we 0.7 m (2 feet) apart. It harvested at the rate of approx mately 1.13 ha/hour (2.8 acres/hour) and produced betwe 8.6 and 10.8 green tonnes (9.5 to 11.9 tons) per hou Frame and structure strength are marginal. The machine wi be assessed during the 1985 to 1986 harvesting season on 2-ha (5-acre) plot of willow.

SUMMARY

A number of projects are developing equipment appl cable to harvesting biomass plantations. Many solutions a possible, and undoubtedly several different workable desig will evolve. No one machine or system can solve the biomas harvesting problem world-wide, but the technology is vari enough from group to group to provide solutions for a given circumstance. The technology must be developed a ready if we are to commercially harvest biomass plantatio to meet the energy demands of the world.

Figure 8.--Loughry prototype coppice harvester.

REFERENCES CITED

1. Arola, Rodger A., Radcliffe, Robert C., and Winsauer, Sharon A., "Prototype Wood Chunker Used on Populus 'Tristis.'" In: Edward A. Hansen, comp. Proc., Intensive Plantation Culture: 12 Years Research, January 26-27, 1983, Rhinelander, WI. Gen. Tech. Rep. NC-91, 125-130. U.S. Department of Agriculture, Forest Service, St. Paul, MN (1983).

2. Crist, John B., Mattson, James A., and Winsauer, Sharon A., "Effect of Severing Method and Stump Height on Coppice Growth." In: Edward A. Hansen, comp. Proc., Intensive Plantation Culture: 12 Years Research, January 26-27, 1983, Rhinelander, WI. Gen. Tech. Rep. NC-91, 58-62. U.S. Department of Agriculture, Forest Service, St. Paul, MN (1983).

3. Curtin, Dennis T., and Barnett, Paul E., "Development of Forest Harvesting Technology: Application in Short Rotation Intensive Culture Woody Biomass (SRICWB)." September 1, 1985. DOE/OR/21478-1. 89 p. Tech. Rep. Tennessee Valley Authority, Office of Agricultural and Chemical Development, Muscle Shoals, AL (1985).

4. Keville, Bernard J., "Short Rotation Forestry Harvester Chipper." In: Energy from Biomass: Proceedings of the EC Contractors Meeting, 83-89. Brussels (1982) May.

5. Marley, Daniel S., "An Evaluation of Existing and Conceptual Short Rotation Energy Plantation Harvesting Machines and Systems." 213 p. Master's Thesis. Industrial Forestry Operations Program, School of Forestry and Wildlife Resources, Virginia Polytechnic Institute and State University, Blacksburg, VA (1982) May.

6. Mattson, James A., "Harvesting Developments for Short Rotation Intensively Cultured Forests." For. Prod. J., Vol. 33, No. 3, 31-38 (1983) March.

7. Mattson, James A., and Wehr, Michael A., "A Prototype Harvester for Short-Rotation Plantations." In: Edward A. Hansen, comp. Proc., Intensive Plantation Culture: 12 Years Research, January 26-27, 1983, Rhinelander, WI. Gen. Tech. Rep. NC-91, 119-124. U.S. Department of Agriculture, Forest Service, St. Paul, MN (1983).

8. McLain, D. H., "The Loughry Coppice Willow Harvester." Interim Report. Loughry College (Etsu B-1081(9)). 7 p. Department of Agriculture for N. Ireland (1984) March.

9. Teel, James B., "The Design and Development of a Prototype Short Rotation Harvesting System." 175 p. Master's Thesis. Virginia Polytechnic Institute and State University, Blacksburg, VA (1983) June.

DISCUSSION

__________ I was interested in your statement that harvesting in the wintertime would lower the coppicing ability of the trees. I have some doubts about it. Harvesting in summertime would probably have a negative effect on coppicing.

Walgren They have found just the opposite for populus and some of the willow. The number of coppice from each stump was less in the wintertime than it was in summertime.

__________ I think that it deserves further attention.

Walgren Yes. For example, I noticed that for Eucalyptus in Brazil, there are critical times of the year when they cannot harvest if they depend on coppice reproduction.

__________ I believe very strongly that it is so also with populus and willow. Maybe harvesting too early in the fall or sometime in winter may do damage by frost. It's still probably much better than harvesting in the summertime as you have shown. Also, I think another factor could be very significant in countries with a lot of snow in the wintertime. How will you use these machines in 3 ft of snow for harvesting?

Walgren You won't use them.

__________ Then harvest times would be very much restricted?

Walgren That's right.

__________ There are a lot of critical questions about costs.

Walgren In developed countries of the world, you see the big machines such as I've shown. They're cost-prohibited in some of the other countries of the world. I know one of the key goals in the Philippine operation, where they planted vast plantings, is to provide enough wood. The critical operation there was harvesting at about 4-year rotations to supply a continuous feed to a 75-MW plant. Harvesting was the key, and it had to be a simple operation. Everything that we looked at throughout the world was just cost prohibited.

K. Porter, U.S. AID, (Jamaica) — It appears as if the machines you showed were especially built for flat land applications. What about rolling lands?

Walgren — The premise is that most of these plantations will be on fairly level land. Some of these machines could not tolerate 15, 20, or 30% slopes.

P. Mitchell, Aberdeen U. (U.K.) — The machine developed in Northern Ireland the Lockery machine, is actually designed for working on rolling terrain. So it is different from some of the other machines. It's not a flat-land agriculture-type system. It is for adverse conditions if you like. The other thing to point out about that machine is that it has undergone a redesign since the photograph you showed. It's now suitable for bigger bundles, like the ESM machine being developed in Sweden.

Walgren — Conventional harvesting is undergoing a tremendous change even in the United States and a lot of other countries because of the environmental constraints and much steeper land. The machine should be much smaller than they have been in the past.

W. Heckrodt, Kimberly-Clark — On the West Coast, I've seen a beefed-up sickle bar for smaller trees, maybe up to 1-inch in diameter. Is there much work going on in a two-step operation where you go through with a sickle bar arrangement and then gather?

Walgren — Other than on the West Coast, some work is underway in the Southeast, where they first started looking at what we call silage sycamore. We used a conventional farming sickle machine, but nothing more has been developed because they were envisioning something a little larger.

INTEGRATED SHORT ROTATION INTENSIVE CULTURE TREE PRODUCTION AND HARVESTING AS A CASH CROP

Thomas Jorgens
Northwest Regional Development Commission

Wendell Johnson and William Berguson
University of Minnesota

James Fisher
Fiber Fuels Institute

ABSTRACT

Producing wood for fuel on Agroforestry biomass farms is a concept with the potential to increase future energy supplies as well as help stabilize future energy costs. Minnesota is one of several states that import most of their energy sources like coal, fuel oil, natural gas and other hydrocarbon derivatives. About 90 percent of Minnesota's energy needs including electricity have been met by traditional fuels. Agroforestry or short rotation forestry, refers to a production system in which plantations of closely spaced hardwood tree species are managed to achieve maximum wood yields. These plantations are established at densities of 4,000 to 43,500 plants per acre thereby allowing harvest rotations of two to seven years compared to eight to thirty year rotations common in natural stand forestry. Cultural practices used in growing agronomic crops also are very effective in increasing yields in hardwoods.

An agroforestry study, funded by the Legislative Commission on Minnesota Resources (LCMR) during the 1983-84 Biennium, has established that a yield of 8-12 dry tons per acre per year is achievable in Minnesota. This finding was based on test sites utilizing hardwood species of Populus and Salix located in Anoka, St. Louis and Roseau Counties. Using the aforementioned biomass output and assuming a five-year rotation for a sustained yield wood production system, an economic analysis conducted by the alternative energy division of the Minnesota Department of Energy and Economic Development in 1984 indicated that a hybrid poplar biomass production system may be financially competitive with other investments. It is apparent, however, that five years may be too long a time frame to attract most medium sized farm operators. If a bioenergy crop can be harvested in two years and be economically viable, there is a possibility that Minnesota farmers would adopt biomass farming as an alternative use of their marginal productive land.

A five-year biomass cash crop production project was established in the northwestern region of Minnesota to demonstrate the viability and sustainability of woody biomass farming.

The preliminary information appears to justify commercial biomass production, however, more work needs to be completed to attract a significant number of growers. Our project's design was based on previous work and will continue the research and development in the following targeted areas: 1) clonal screening of hybrid poplar; 2) planting techniques; 3) development of simple automated harvesting technologies; 4) requirements, fertility needs, spacing and pest resistance and weed control; 5) economic analysis.

INTEGRATED SHORT ROTATION INTENSIVE CULTURE TREE PRODUCTION AND HARVESTING AS A CASH CROP

I. INTRODUCTION

Minnesota is one of a number of states that are heavily dependent on energy supplies, in the form of coal, oil and natural gas, from other parts of the United States and from foreign sources. Recent data indicates that 95% of Minnesota's energy needs are met by traditional fuels, including electrical power. The heavy reliance on external energy provides a major drag on long range economic development potentials for a state well into the frost belt. Economic multiplier effects of externally derived energy supplies are negligible. In contrast, economic multiplier effects of home grown energy supplies are quite dramatic.

A number of studies have indicated that Minnesota has some interesting potentials for import substitution in the energy realm. The key resources that show considerable promise are forest resources, biomass production, hydropower and wind power. All of these sources are currently being used to a limited degree and efforts are underway to greatly expand use of these resources in the future. As an integral part of this effort, the Legislative Commission on Minnesota's Resources has provided significant funding for clonal screening by the University of Minnesota and more recently for the biomass cash crop development effort. The latter project is a joint effort of the University of Minnesota, Department of Soils at St. Paul, the University of Minnesota Technical College at Crookston, the Northwest Experiment Station at Crookston, the Northwest Regional Development Commission at Crookston, the Minnesota Department of Energy and Economic Development, and the Fiber Fuels Institute. The expertise represented in the project team ranges through all aspects of plant materials, soils, stand management and alternative energy utilization.

The goals of the project are twofold, namely to develop woody biomass production as a viable opportunity for area agriculture, and to promote the expansion of area markets for biomass fuel. Within the context of these central goals, a wide range of supporting objectives come into play. These objectives cluster around the need to select the optimum plant materials, establish optimum management practices, develop or modify equipment to mechanize the operations, to lower the costs of stand establishment and to determine the best methods of linking growers and consumers. The state of knowledge in nearly all aspects is advancing rapidly, but much remains to be learned to ensure the success of Short Rotation Intensive Culture (SRIC) tree production in the field with numerous growers.

II. PROBLEMS ADDRESSED

A. Soils and Climate

The location of field work on the Biomass Cash Crop Project is in Northwestern Minnesota. The area overall is heavily agricultural in character with a base of lacustrine, mineral soils. The portion of the area which contains the project's plantation is in the transition zone between the prairie lands to the west and the forested lands to the east. It is an area where native stands of aspen have been cleared over the past several decades to allow traditional cash crop agriculture to be practiced. A major limitation to successful cultivation of crops arises from the fact that these lands are quite flat and poorly drained. Crop failures due to excess moisture have been common in this area and have accentuated negative economic effects due to general agricultural conditions for farmers in this zone. Consequently, farming operations in the zone are facing severe stress and/or financial disaster. Since successful drainage development in this area is difficult at best, the advent of alternative crops with the ability to tolerate wet soil conditions is important to the future of agricultural operations. The high calcium content of these soils lends suitability to the cultivation of these calcium demanding crops. The inclusion of SRIC as a viable option in this zone will not compete with other agricultural uses, but rather will enhance the area economy.

The climate of the area provides a conducive environment for the growth of hybrid poplars. During the growing season, moisture and temperature conditions promote rapid growth of the plant species in test plots. The winters are relatively harsh, combining temperatures down to -40 F, a low relative humidity. However, extensive snow cover minimizes winter kill. The northerly location also increases the potential for late spring and early fall frosts than can shorten growing seasons.

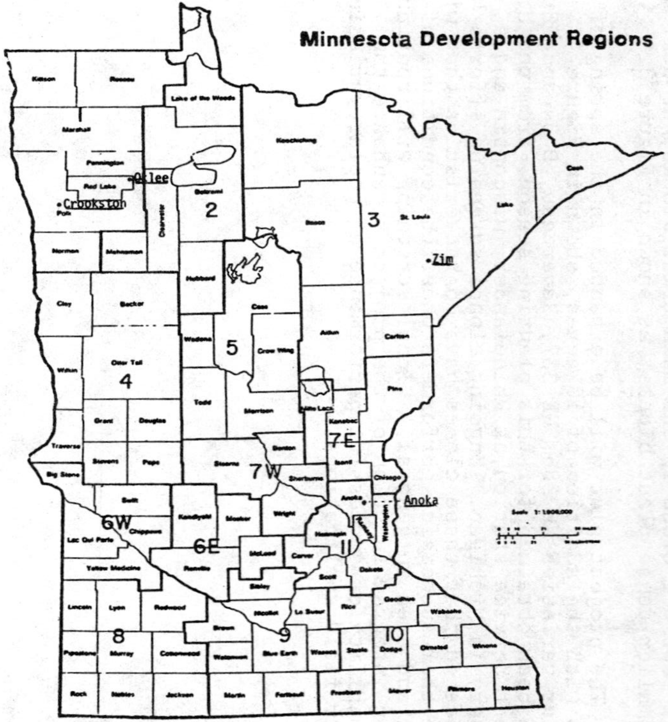

FIGURE 1.

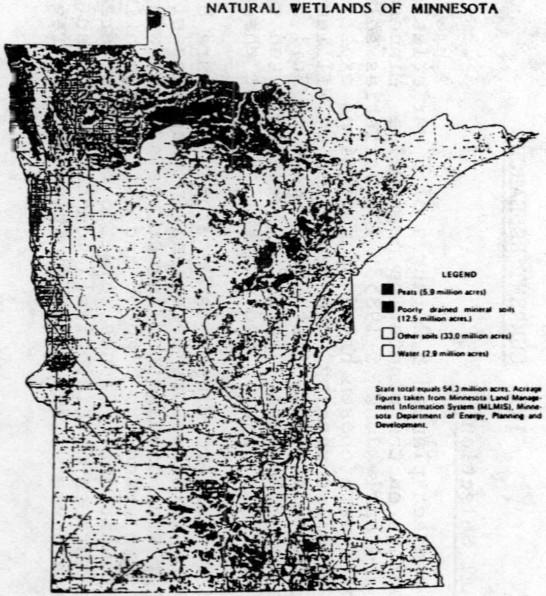

FIGURE 2.

III. CURRENT RESEARCH

A. Clonal Selection

The project planted 9.25 acres of 5 varieties of hybrid poplars on the plantation near Oklee, Minnesota in late May and early June of 1985. The varieties selected for the plot were chosen based upon previous experiences from the clonal screening work performed by William Berguson, Associate Scientist, Department of Soil Science, University of Minnesota. Varieties were selected to meet criteria for the location and mineral soils at the plot.

In Northwestern Minnesota Populus clones are being screened for agroforestry potential in the areas of; rapid juvenile growth, disease and insect resistance, winter hardiness, tolerance to high pH and ability to thrive in wet soil conditions. Clones selected for other areas of Minnesota through physiological, morphological and growth variables are useful but not definitive for our regional needs.

The five poplar clones used at the project plot in 1985 were Raverdeau, NE 386, NE 387, DO 1, and DN 29. Cuttings were obtained from the U.S. Forest Service at Rhinelander, Wisconsin, from University of Minnesota plots and from a source in Oregon. They were planted in strips of 12 rows per clone in the main trial blocks, as well as in smaller dense plant blocks, as shown in Figure 3.

The project plot will be expanded in the spring of 1986 with the addition of 10 acres, shown in Figure 4. Poplar varieties NE 386, NE 387, Raverdeau, D 34 and NE 308 have been obtained for this planting season, from the U.S. Forest Service station at Rhinelander, Wisconsin and a nursery in Michigan. Ongoing clonal screening efforts have indicated that these clones have characteristics that will make them successful performers under the conditions of the plot and area in general. Clonal screening will continue as an integral function of the project to enhance the probability of successful performance of poplar varieties under field applications.

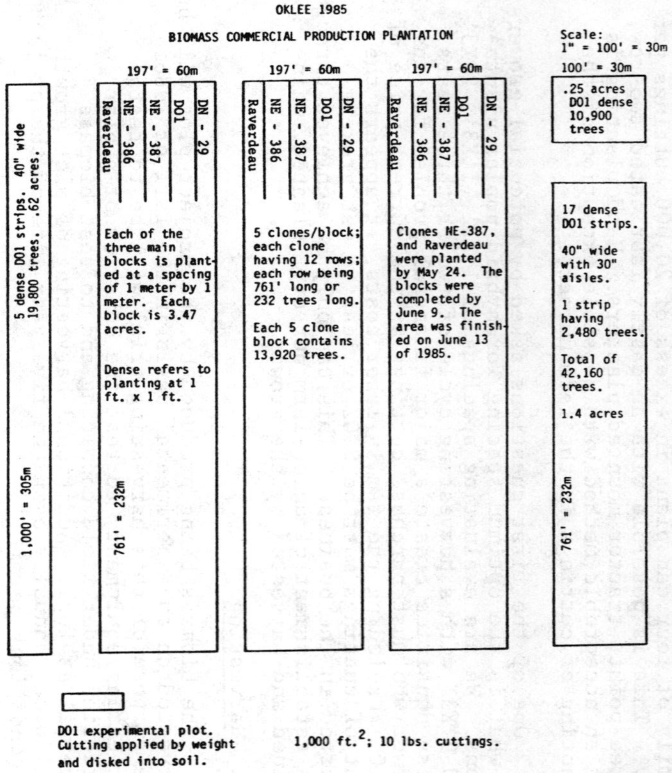

FIGURE 3.

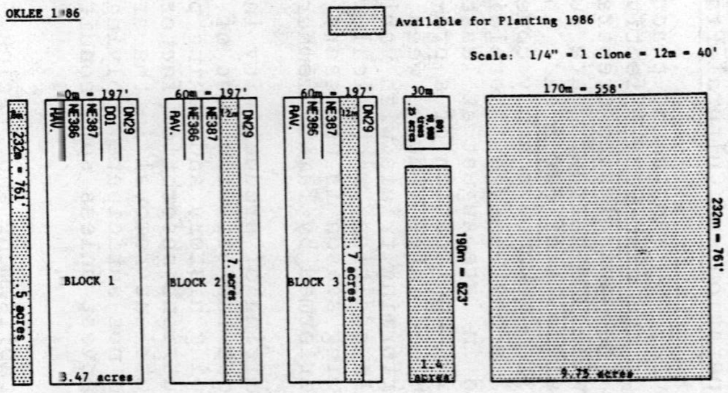

1985: Each one of the clones in a single block contains 2,784 poplar (12 rows x 232 trees/row). Block 1 has five different clones and, therefore, 13,920 trees. Block 2 has four different clones; so, has 11,136 trees. Block 3 likewise has 11,136 trees. Including, the dense 1/4 acre, the grand total of hybrid poplars kept reaches 47,092. Actual acres planted stands at 9.25 acres.

1986: There is available for planting an actual thirteen acres (shaded above). The total plantation (including ditches and alleys) will cover 25 acres.

FIGURE 4.

B. Site Preparation and Cultural Practices

The cultural and management practices suited for agronomic crops are also very effective in growing Populus in Minnesota. Mechanical site preparation and weed control utilizing plowing and harrowing are suitable in the poorly drained mineral soils of our area. In fields, grasses like quackgrass, often make up a heavy cover of vegetation. The application of a post-emergent herbicide, such as Round-up can be applied in late August at least two weeks ahead of plowing. Spring application of a pre-emergent herbicide at locally determined rates reduces weed competition. The project is performing trials with various pre-emergent herbicides. In addition, mechanical weed control during the first growing season is necessary and economically feasible is performed by family members.

The maximizing of productivity in Populus cannot be achieved without the efficient use of soil testing, evaluating of cropping history and making projected yield forcasts. The nutrient absorbtion factor we used was based on forage crop nutrient removal in the area. In a two-year rotation we do not anticipate applying additional fertilizer until after harvest unless soil conditions become exceptionally wet.

C. Stand Establishment and Spacing

Because of the short period of time available for planting in Northwestern Minnesota, it is important that a family of four can plant in excess of 20,000 cuttings per day. This is possible with an easily fabricated two-row, three-point, tractor mounted planters. Hand planting is not an acceptable method with possible exception of the planting of cuttings in the farm nursery.

One of the first questions asked by potential growers is what is the optimum spacing for hybrid poplar plantations. We are evaluating spacings of 1'X1', 1'X3', 2'X3' and 3'X3' with a harvesting cycle of two or three years. These densities create a major financial problem for the grower who must purchase cuttings at the current price of $.10. Excluding the family labor costs, it appears the cost of cuttings must be $.02 or less to get growers started in the business. This can only be achieved through the establishment of an on-farm nursery, planted, maintained and harvested by the grower.

D. Harvesting

The biomass being produced by the project plot will be harvested on an experimental basis in the fall of 1986. The purpose of this harvesting trial will be to test harvesting equipment, to test the behavior of poplar varieties under field coppicing and to test biomass material as fuel. Additional harvesting of plot growth will occur annually over the five year life of the project as conditions permit.

Project plans at this point are to utilize a Hyd-Mech Forestry Limited, FB-3 prototype Feller Buncher, to cut and bunch poplar wands. Bunched wands will be off-loaded to common farm wagons and piled at a collection point for air drying. Materials will then be chipped to appropriate sizes for direct use by burning equipment. Other approaches to harvesting may also be given trials based on experience and indicated need to pursue other systems. Trials involving pelletizing biomass from the plot will also be pursued to test characteristics for use in pellet burning systems.

The project's objective relative to intervals between planting and initial harvest, as well as subsequent coppicing intervals is to minimize those intervals consistent with optimum sustainable yields. A short term return on investment is essential to the small grower/diversified farmer that the project envisions as the primary producer. Since capital is limited for these producers, it is important that the harvest process be developed in a cost effective manner that utilizes existing equipment as much as possible.

Current expectations are that biennial coppicing is the most frequent rotation likely to succeed. The initial harvest trials in fall of 1986 will provide critical information to the project team that will be helpful in determining optimum performance under field conditions. Refinements in harvesting equipment and harvest procedures are also important outcomes expected from the 1986 harvest trials. This is particularly critical given the relatively short window of ideal harvest conditions in the area and the need for diversified growers to fit biomass harvesting into harvest cycles for other crops they may be producing.

IV. ECONOMIC CONSIDERATIONS

A. Small Diversified Growers

The orientation of the Biomass Cash Crop Project to the setting of the diversified family farm operation in a marginal agricultural area poses several interesting problems. First, the typical family farm in the project area has limited capital to invest in the addition of a new crop to the operation. Secondly, the typical farm unit has continuing cash flow requirements that do not allow for long delays in recovering returns from investments. Finally, the small grower needs a broad margin of stability in demand and price for this production, particularly in a new venture such as biomass production, in order to remain in the business over time.

Because of the increasing size of farm equipment, the increasing use of energy and chemicals, and the continuing movement toward specialization on farms, many farms do not provide full-time employment for even one family member.

Yet, one of the major ways for the family farm to increase income is to utilize its labor more efficiently.

In an effort to make grain farming a full-time occupation, many farmers tried to purchase or rent additional land. This practice has been for the most part only partially successful in Northwestern Minnesota.

Many farmers purchased equipment several years ago during an expansion era. Suddenly, the farmer was told if he wished to be eligible for target-price payments or price-support loans, he had to set aside 20-25% of his crop base. The farmer could not easily sell off a comparable percentage of his equipment. The diversification in short rotation trees may not only be labor intensive but also make better use of the equipment owned by the grower.

The project team recognizes that current costs of developing and maintaining a woody biomass plantation are quite high and consequently out of the reach of most prospective growers. The Project faces the continuing challenges of reducing the costs of all aspects of the production cycle for hybrid poplar biomass in order to meet affordability thresholds. The first impediment on the road to commercialization is the cost of cuttings, now typically at \$.10 or more per cutting. This places the cost of material for an acre in the range of \$400 to \$1,200 for densities in one meter width rows of 1 plant per meter and 3 plants per meter respectively. The project team sees the necessity of reducing cutting costs to a level of \$.02 or less to make it feasible for the typical operator to establish stand densities likely to yield the best results. The automation of the cutting production process utilizing material from quality stands of superior varieties will foster cost reduction. The project team also envisions that growers will develop their own nurseries to produce a high degree of self sufficiency in cuttings.

Another element of cost containment being pursued by the project is the automation of the planting, stand management and harvesting processes. The project objective is to adapt to the greatest extent possible to equipment already in use in the area. The minimization of the need to purchase expensive specialized equipment is a critical dimension of the development process. An example of progress made to date is in the planting process, where a standard vegetable planter used in 1985 has undergone considerable transformation for use in 1986. The improvements will reduce labor in the planting process and will significantly improve the speed of planting.

Other aspects of the cost containment effort are trials involving various fertility treatments, weed control trials and pesticide applications. In these, as in the other cases, the objectives of these activities are to attain optimum growth results, while limiting the use of chemical and mechanical practices to necessary levels.

Responding to the evident need for prospective growers to meet cash flow requirements is a critical part of the project's objectives. Essentially, the response requires both the minimization of intervals between harvests on a plot and optimization of the yield of biomass over time. The development of reliable data under area conditions on these parameters is an expected outcome of project activities in 1986 and beyond.

B. Market

The need to establish a predictable band of price and demand stability is quite evident in order to encourage the prospective grower into production. There is currently an established micro-market for biomass fuels within the area of Northwestern Minnesota. The market grew very rapidly until prices for substitutes, such as natural gas and fuel oil, began to decline significantly. The project team is engaged in assessment of market potentials and the feasibility of a grower cooperative to supply biomass to larger users. The project team has held discussions with American Crystal Sugar Company, a very large energy user with 5 plants in the general region, concerning biomass as a supplemental fuel. Other large energy users of the industrial, commercial and institutional are being identified and surveyed as prospective customers for biomass fuels. The project team has concluded that the basic ingredient in a successful market for biomass fuels is the large volume user. Such users provide the critical mass of volume that is essential for continuing development of the industry and stability for producers.

V. Summary

The Biomass Cash Crop Project is directed toward the goal of bridging the gap between the research plots involved in the development of superior plant materials and the commercial application this biotechnology to biomass production. The project is made possible due to the fact that research results have provided data and materials needed for such an effort. The wide range of questions posed by the project team prior to the initiation of the project have proven valid and a number of new questions have arisen from project experience. The process of obtaining answers to the many questions that must be addressed before reliable results can be predicted under on farm conditions is at the core of the project effort.

The project experience reinforces the assumption that the plant materials can be successfully grown in mineral soils in conditions present on the plot. It has also been confirmed that there is considerable opportunity to drive costs sharply downward and that on farm replications of results continue to be promising. Finally, indications are that market evolution to capture the evident economic benefits of locally produced biomass is likely to occur on a compatible timeframe with the onset of commercial scale production. The latter conclusion must be tempered, however, by the assumption that the prices of major competitive fuels will return to long term trend levels by the final years of the decade. An extended continuation of current price levels will not in any way negate the long range prospects for biomass fuel, but will significantly retard economic incentives for expansion of biomass markets.

DISCUSSION

______	It appears that you are aiming your research towards a traditional farming-type operation. Does that explain the short rotations and small material size, so that you can use more conventional farming equipment? Are you going to look at what you might gain at slightly longer rotations?
Jorgens	In the particular projects that I've talked about, and that most of the slides are from, we're trying to gear that to agricultural operations in that area. Therefore, a short rotation that uses traditional equipment as much as possible and that farmers already have available reduces the up-front capital investments. At the same time, we are allowing material to grow for longer periods of time in other research plots. So we will also be able to evaluate whether there is an advantage with longer rotations.
______	Could you tell me a little bit about the densities or the spacings that you're working with?
Jorgens	We're trying densities of 1 ft by 1 ft, and we're doing a small planting to see how it works. But our primary approach is going to be in rows. The densities that we're trying this year are 36-inch rows, at 1-ft, 2-ft, and 3-ft spacings.
______	Are the yields you showed us at age 2 total yields at age 2, or averaged over the 2 years?
Jorgens	The average over the 2 years.
J. Ranney, Oak Ridge National Laboratory	The production that you showed at the site is much more significant than I I think most folks here would realize. This is a peat-land type site where just

2 or 3 years ago, many said you couldn't grow poplar at all. Some of the sites that are set aside as relatively unusable have potential for biomass production. A second point is that we found in many of our experiments that a break-even point for the short-rotation system, assuming a cost of about $16 a ton to harvesting, is between 8 and 10 dry ton/ac-yr, or roughly 20 ton/ha-yr. On the site, you're getting close to that and I think with improved clones and culturing, you're going to be able to do it.

Jorgens

Someone asked a question earlier about willows. We do have willows in a site in Northeastern Minnesota. We're also going to have a very small trial on our plantation in Northwest Minnesota this year because we think that willows will extend the range of soil types and land conditions that we can deal with. We think we can move to slightly wetter conditions with willows. In fact, some grower may have a mix of poplar and willows as his site conditions dictate.

W. Heckrodt, Kimberly-Clark

You have some of your plantings in quite late. Did they recover?

Jorgens

It was noticeable that they were planted late throughout the growing season, but they did do a lot of catching up. We ran into some very wet conditions last spring, so we got to about mid-June by the time we planted. We would like to plant about the latter 2 weeks in May.

L. Zsuffa, U. of Toronto (Canada)

In the table that showed cost factors, you included the land preparation, fertilization, and planting, but you never mentioned the cost of management. Of this first category, I wonder which is the most critical and is planting, cost of planting stock, and the planting itself, a very critical factor in the overall cost? Have you considered also the cost of management?

Jorgens

The costs of establishing the densities that we're talking about in these 36-inch rows range from $400 to $1200 on current prices. Those are significant costs. We left out management costs because we're trying to relate this to the small grower. The small farmer generally doesn't put that into his cost equation.

DISCUSSION, CONT.

Bente

I have a question as one who has never used any cuttings to plant trees. Does it matter which end of the cutting is put into the ground, and also how long or how short can the cuttings be, and does it vary from species to species?

Jorgens

It makes a great deal of difference how you put them in. They have to go right side up. Generally, you can tell simply by looking at the direction the buds are growing. That's one of the problems in trying to move to mechanical planting. How do you get them all right side up? The cuttings work best at lengths of 8-12 inches. Shorter than that, there's not really sufficient energy in the plant to guarantee a high rate of survivability. We generally found that the bigger the cuttings are, and the longer they are, the better the results. It doesn't seem to make a great deal of difference what type of clone you're dealing with.

COMMERCIALIZING HIGH DENSITY ANNUAL ROTATION POPLARS - A STATUS REPORT -

Stephen P. Bryant, MBA
General Manager, Corporate Development
Foster Farms
Livingston, California 95334

Robert G. Bartikoski
Lewis Scott

ABSTRACT

This paper presents a status report on efforts to commercialize high density, short rotation poplars. The three authors all work for different companies, but have supported each other in this endeavor.

The paper is divided into four major sections - planting, cultural, harvesting and marketing. In each section, the authors' experience and results of their research to date is summarized. Economic, cultural and technological issues are addressed. Focus of efforts planned for the coming year are described.

COMMERCIALIZING HIGH DENSITY ANNUAL ROTATION POPLARS - A STATUS REPORT -

The three authors of this paper - all working for separate companies - have been evaluating different aspects of growing annual rotation poplars and supporting one another's efforts. Our common philosophy has been not to search for the optimum solution, but rather for an acceptable solution in terms of both economics and technology. Thus, quite simply, we have been struggling with such questions as:

1. Will poplars compete for ground against alternative crops? That is - could returns be good enough to convince a farmer to grow poplars instead of another crop?

2. What are the economic or technical barriers in dealing with poplars as a crop?

3. What are some alternative uses or markets for the crop that could enhance the stability and/or the attractiveness of the returns to the crop?

We are sure there is much to be done to optimize the question of whether there is a role for tree crops in agriculture - and that ultimately, poplars may not be the right tree. However, we do believe there is merit in establishing whether poplars can at least work, encouraging efforts at some level of commercialization, and then allowing improvements to come out of the participants' efforts to improve their economic returns.

As a way of sharing what we have done, what we know, and what we don't know, we have organized this paper around the traditional steps of planting, growing, harvesting and marketing a crop.

PLANTING

A 10 acre plot was established, divided into 12 separate beds, totaling approximately six acres of actual trees. Six to eight inch cuttings were planted according to the regimen in Table 1.

Table 1

Planting Regimen

Variety	Row Configuration	Planned Density
Cottonwoods	30" rows	1/square foot
	15" rows	3/square foot
DO-1 Poplar	30" rows	1/square foot
	15" rows	3/square foot
	15" rows	4/square foot
	30" rows	.5/square foot
	Cultipacked	.2/square foot
	Cultipacked	3.5/square foot
DO-2 Poplar	30" rows	.5/square foot
Scandinavian Willows	30" rows	.5/square foot

Except where cuttings were broadcast and cultipacked, cuttings were planted with a modified carousel planter that dropped individual cuttings between ground opening and closing mechanisms. The results of this approach were as follows:

1. 90% or better actual planting density versus plan with densities of 1/square foot or <u>less</u>, but actual being 50% or less than plan with higher densities.

2. A simple average of 74% germination.

The key lessons we learned were:

1. Larger cuttings (1/4" or larger) do better.

2. Planting cuttings upright improves germination.

3. At least 1-2" of the cutting should be above ground.

4. Plant cuttings prior to the buds swelling.

5. 30" rows were easiest to work with in terms of off the shelf farm equipment.

6. Early moisture from either rain or irrigation is critical.

In terms of the actual planting process, the approach we used could not be recommended for commercial stand establishment - it was too slow, too labor intensive, and at the cost, not accurate enough. There are probably two alternatives worth pursuing:

1. "Low Cost - Low Tech" - In general, this approach would entail gathering cuttings in a random manner, broadcasting them on beds 4-6 feet wide and cultipacking them. This approach would provide little control on orientation or depth of planting and very approximate control on density. It would probably assume a relatively low value for the cutting, but could use the simplest of equipment.

2. "High Cost - High Tech" - This approach would start with collecting the original trees or whips in a manner that maintains cutting orientation, processing the whips and packaging the cuttings while maintaining orientation, and some form of highly mechanized precision planting.

Our judgement at this point would be to recommend an "on farm" (i.e., low cost, low tech) commercialization approach, since a major objective should be to get the commercial stand establishment cost to $600-$800 per acre. This approach would have the following steps:

Year 1. Purchase 40,000-50,000 cuttings at $.10/$.15 per cutting. Use hand labor to maximize germination and plant one acre.

Year 2. Using a tractor mounted saw similar to the one we have developed, "mow" the trees. Using a cross cut saw, randomly cut the trees (including the branches) into 6" lengths. Broadcast the cuttings and cultipack them in beds 4-6 feet wide (depending on equipment). At a target of one per square foot, this approach should yield an increase of a minimum of 20 acres and a maximum of 35-40 acres.

Year 3. Continue as in Year 2 up to the desired size of stand, harvesting only enough acres for increasing the stand, while converting the balance of the crop to the intended commercial use.

CULTURAL

Weed Control

When the first stand was established, Devrinol was used as a pre-emergent for weed control. If the crop is to be harvested only once a year, it appears that the trees provide enough shade so that additional weed control practices are not required. On plots which were harvested during the summer, mechanical cultivation was required.

Fertilization

200 units of nitrogen were applied each year. Based on plant response, the approach used in the second year of applying nitrogen in the water at each irrigation was preferable to the two 100 unit applications used in the first year. We currently have no information on the optimal amount of fertilizer, so the trees may do as well with less, or it may pay to use more.

Water

Poplars will take large amounts of water, particularly in the Central Valley of California where transpiration rates are very high in the summer. As a guideline, the four acre feet required for corn in California appears to be adequate.

Pest Control

No pesticides have been used to date. We have observed some isolated instances of the American plum borer and will continue to monitor the trees. We have also heard of some grasshopper predation, but have no direct knowledge of the severity or extent.

HARVESTING

Harvesting activities were carried out in both 1983 and 1984. In 1983, all plots but one were harvested in early July and again in late August, using a sickle bar and reel mounted on a corn chopper. Although harvesting during the same growing season as the trees were planted may have improved coppicing, we would not recommend the practice. Yields were very low, and root development was probably retarded. Even so, by cutting at 6-12" and leaving the lowest branches, very little death loss occurred.

In 1984, sample plots were harvested in early June, mid July and late October. The first two harvests were carried out with the corn chopper; however, the trees were too large in the fall and a number of hand methods had to be used. Yield data on a green weight basis for plots consistently harvested are illustrated in Table 2.

Table 2

Harvest Yields

Plot No.	Pounds Per Acre by Date Harvested 7/83	8/83	6/84	7/84	10/84	Total 1984
4	544	2,076	22,133	7,726	35,470	65,329
2A	1,500	4,960	21,679	9,327	27,104	58,110
1	530	2,226	14,743	6,199	21,054	41,996
2C	0	0	0	0	43,831	43,831
5	5	45	0	0	36,079	36,079
11	160	780	0	0	29,040	29,040
2B	1,500	0	0	0	20,212	20,212

Although the data is not conclusive, it does suggest that total tonnage per acre can be increased by cutting at least twice per year - once mid summer and once before leaf drop. The desirability of this approach would depend on the ultimate crop use, but if a corn chopper approach were used, one could expect harvest costs on the order of $10-$15 per acre, or $1.50-$2.00 per green ton.

On the basis that a harvesting machine would have to handle larger stems than a sickle bar would be capable of, we began development of harvesting equipment in 1984. The first step was to develop a chain and bar system that would handle multiple stems of different sizes without binding. Secondly, this bar was tractor mounted, with a very simple PTO driven hydraulic drive. The equipment looked virtually identical to a mowing machine. The machine would cut, but the farmer was still left with the material handling problem of what to do with the trees next. We involved an outside firm to assist in the development. As of early 1986, the companies have built and successfully demonstrated a machine that cuts, orients and collects poplars. The machine will cut up to a five foot swath, with the ability to handle up to 6" diameter stems or large clumps of coppiced material up to 20 feet in height. The significant fact of this development is that now traditional material handling techniques such as chipping or rolling can be used for further processing of the poplars. We expect to see larger scale commercial testing of the machine during the 1986 crop year. A full scale, self propelled machine will probably cost $85,000-$100,000. Chip or roll processing equipment would add $30,000-$50,000 to the cost.

MARKETING

We have been considering four potential commercial markets for annual rotation poplars: fuel, broiler feed, ruminant feed and "Scrimber" feedstock.

Fuel

Grown as a fuel only, poplars are not a particularly attractive crop, at least in California today. Even before the recent drop in natural gas prices, we had estimated that $2.50 per million BTU's would be an appropriate delivered price for biomass fuel. Allowing $.50 per million for transportation, this meant the crop would be worth $2.00 per million roadside. The range of revenue per acre, based on the achievable yields we saw, is illustrated in Table 3.

Table 3

Revenue Per Acre

	Range	
Green Yield	20 tons	32.0 tons
Dry Matter at 55%	11 tons	17.6 tons
MBTU's/Ton	14	14
Gross Revenue	$308	$493

If we use grain corn economics in California as a guide for what it would take to attract land away from growing corn, we see that typically corn revenue per acre would be $420, typical costs (with one harvest) would be $325 per acre, with a return of $95.00. While out of pocket costs for growing poplars are expected to be at or below the cost of growing corn, some factor for the additional investment in the stand establishment and probably one extra harvest should be added to the $325 corn cost. A 10% return on the stand establishment cost and another $15-$20 per acre for the extra harvest would raise the poplar cost to the $400 per acre level - resulting in a return comparable to corn, if 32 ton yields could be consistently achieved.

In terms of specific market opportunities in California today, biomass plants either planned or in operation are trying to source their feedstock at the $1.80 to $2.00 per million BTU level (delivered). At these price levels, only a demonstration of significantly higher yields or lower cultural costs would cause an investment in growing poplars for fuel.

Feed and Fuel

If we look at the concept of farming poplars for feed and fuel, we can somewhat improve the economics. Of the total dry matter harvested from the five largest yielding plots, leaves and petioles were 11%, 17%, 13%, 8%, 16%. The 8% came from a plot that had not been cut for two years. A simple average of the remaining plots was 14%. Using the range of yields from Table 3, an acre would produce a low of 1.5 of leaves and 9.5 tons of stems to a high of 2.5 tons of leaves and 15.1 tons of stems.

The leaves and petioles alone, as well as the whole first year growth plant, have potential value for both chickens and ruminants. We have not carried out any experimental work with feeding ruminants, although a poplar silage evaluation is planned. Dr. Kamstra of South Dakota State University has published information on feeding aspen silage. Due to a relatively high fiber content, and the fact that much of the fiber - at least from first year growth - is available through the ruminant, we would expect a higher value for cattle than for chickens, which are typically fed a more concentrated, low fiber diet.

However, we have developed a fair amount of information about poplar meal - proximate analysis, amino acid analysis, carbohydrate analysis and xanthophyll analysis. In addition, we have carried out battery trials to establish the TME (True Metabolizable Energy) of leaf meal. This information is included in Exhibits I, II and III.

Since xanthophyll is utilized in many parts of the country as a source of natural pigmentation for broiler skin color and table egg yolk color, the xanthophyll in leaf meal adds additional value over and above its energy and protein components. In today's market, xanthophyll is worth about $.30 per gram.

Poplar leaf meal - just like any other feed ingredient - has value only in relation to the price and availability of competing ingredients at a given point in time. Consequently, the value we assign to leaf meal today is totally dependent on the price of corn and soybean meal in Chicago, as well as the price of more minor ingredients (including California by-products) that may be available.

We have not yet completed parametric analysis of leaf meal. However, based on the product characteristics of comparable products, $60-$80 is a reasonable range.

Achieving this price for leaf meal would improve per acre revenue by a range of $70-$100.

Scrimber Feedstock

"Scrimber" is a trademarked name for structural lumber produced by a patented process that orients and glues linear wood fibers. The process was developed in Australia by the Australian government and REPCO, and is being marketed by a company called SIROTECH.

The purported advantage of this system is that almost any type of scrap or cull wood can be used to produce products with equal or greater structural strength than their natural counterparts. The resulting Scrimber is an engineered product that can be used to meet precise building specifications - in place of "over building" to allow for the variation in natural products.

If we accept the premise that the driving economic force behind natural wood products is the inventory carrying costs of 20-50 year cutting rotations, then a manufactured product such as Scrimber can successfully compete if it is based on a very short rotation crop. Our current hypothesis is that short rotation poplars may meet the requirements of such a crop.

In terms of getting closer to commercialization, we believe we've made some progress on two fronts. The harvesting machine previously described does work in terms of cutting, collecting and orienting stems. Although the machine has not been scaled up, we believe it is feasible to approach a cutting rate of two acres per hour. In addition, a small rolling mechanism, consisting of four 5" rollers in tandem driven by an 18 hp motor, has been built and operated. At this scale, the machine debarks 1" stems and produces a linear wood fiber appropriate for use in the Scrimber process. In addition, research work on rolling wood at V.P.I., and work done by REPCO on rolling as part of the Scrimber process should be helpful in moving to full scale commercialization. At this point, we are reasonably optimistic that poplars can be cut and rolled in the field on an economic basis.

In addition to strictly commercial opportunities, annual rotation poplars may have a role in waste water clean up and absorption for municipalities operating methane fired cogeneration systems. In research supported by the Gas Research Institute, it was found that first year growth poplars have a very rapid rate of digestion - generating 80% of the total gas yield in 20 days. So the concept of using the trees to polish waste water, increasing the amount of water absorbed, and harvesting the trees as a methane feedstock to generate additional revenues may have some application. We have had some initial contacts to provide material to a municipality to test this concept.

FUTURE FOCUS

The primary focus of our efforts will fall in two areas:

1. Evaluate two site specific opportunities for a feed and fuel operation. One is a domestic opportunity for producing silage and boiler fuel for a stockyard/ beef processing complex; the second is a foreign operation for producing leaf meal for chickens and fuel for energy production.

2. Refine the harvesting equipment and capital cost, particularly as it relates to integrating a chipping or rolling capability with the machine.

Given the amount of work being done by others in the area of growing the crop, we will not likely spend much additional time or money in this area.

We are looking forward to cooperating with other members of the energy-to-biomass community in pushing this concept toward full commercialization.

EXHIBIT I

Poplar DO-1 - 1985

Sample	Protein	Fat	Moisture	Ash	Fiber	Xanthophyll As Is
Leaves (petioles starting)	6.70	1.56	65.51	3.22	4.07	47.20
Whole Plant (starting)	5.60	1.66		0.71	4.85	54.80
Leaves/ Petioles, Dried	16.4	3.93	11.99	9.84	8.64	68.30
Leaves/ Petioles ($NaHSO_3$)	18.6	4.28	0.37	13.36	10.83	73.70
Whole Plant, Dried	20.7	4.77	2.47	14.25	11.66	22.9

Amino Acid Profile	Whole Plant	Leaves/Petioles
Tryptophan	0.16	0.09
Aspartic Acid	1.19	1.19
Threonine	0.80	0.62
Serine	0.64	0.69
Glutamic Acid	1.65	1.43
Proline	0.55	0.67
Glycine	0.75	0.87
Alanine	0.82	0.84
Cystine	0.08	0.07
Valine	0.70	0.84
Methionine	0.33	0.26
Isoleucine	0.81	0.91
Leucine	1.35	1.42
Tyrosine	0.62	0.83
Phenylalanine	0.76	0.83
Histidine	0.90	1.02
Lysine, total	0.61	0.37
Arginine	0.80	0.54

TME

GE	4.0504 kcal/gram	1,839 kcal/lb.
TME	1.9339 kcal/gram	878 kcal/lb.

EXHIBIT II

Poplar

Carbohydrate Fraction

	1st Year New Growth %	3rd Year New Growth %
Protein	4.49	2.74
Fat/Oil	1.37	1.51
Moisture	78.57	75.86
Ash	1.51	1.89
Fiber, Crude	4.50	5.50
Fiber, Dietary	6.80	8.30
Carbohydrates	14.06	18.00
Cellulose	7.60	8.70
Hemicellulose	5.40	6.90
Liginin	1.63	2.40

EXHIBIT III

Xanthophyll Concentrations

10/26/84 Harvest

Plot Number	Xanthophyll Concentrations on a Dry Basis Leaves (Mg./Lb.)	Stems (Mg./Lb.)
4	197	47
2A	142	56
1	196	52
2C	130	16
5	127	17
11	135	13
2B	154	46

DISCUSSION

______ Different densities had different amounts of germination or sprouting from the root when you start from the cutting. If you had the same clones and you're under the same culture regimes, why do you feel you got differences in sprouting?

Bryant I think it was just bad farming. We were late, and we probably didn't have water consistently across the field in a timely manner.

______ Are you saying that it wasn't density, it was other factors?

Bryant There are two different issues. One is planting success and the other is whether it lived or not. Those are different numbers. Of the numbers planted, how many lived? The other factor is did I plant as many as planned?

R. Houston, Agrecology Inc. In the one-year cuttings, is the material suitable for scrimber-type use? Maybe it's too immature.

Bryant I don't know the answer to that question. Louis has been evaluating the process and you would think not as much lignification has apparently taken place, because there is such a dramatic difference between 1- and 2-year material.

B. Karaim, Alberta Department of Forestry (Canada) How long are you going to use the original crop?

Bryant We don't know the answer to that yet. That's one of the risks that seems to me the next farmer is going to take. Can he cut it for 10 years? On the other hand, I don't know about Canada. In California, a stand of alfalfa lasts 3 years, and then it must be replaced.

Karaim Converting virgin forest to agriculture is a major cost. Have you included that in your overall cost of production over a 5- or 10-year period?

Bryant No. I did destroy about one acre of poplar simply by ripping them. But they were only 2 years old, and they didn't come up from the roots. If you can operate 10 years, that's not an issue.

DISCUSSION, CONT.

If you only operate 3 years, then obviously that's an issue.

L. Zsuffa, U. of Toronto (Canada) — We have run experiments on a similar concept, meaning harvesting more than once a year and harvesting in summertime as well. In most cases, this approach had a detrimental effect in our climatic conditions. Although we know some clones survive much better than others, I have the feeling that unless you do very careful selection work and determine very carefully how to handle them, it won't work for very long.

Bryant — Frankly, I was surprised about our experience because we harvested three times a year for the first 2 years.

Zsuffa — It must make a big difference.

Bryant — No, we could harvest a large crop today.

Zsuffa — With regard to cost of planting stock and planting, we have been experimenting with a different concept for a couple of years now. Instead of making individual cuttings, we cut bundles of sprouts, chop them up, and instead of planting them individually, we use a kind of ditch planting which keep the rows very distinguished in the line. The cuttings are placed in a horizontal way more or less. We are thinking of applying the kind of machine used in planting sugarcane, mix up the cuttings in a sludge, put them in a big tank, and plant them in the ditch. I think that would significantly lower the cost of stock production in planting. And with clones that show good rooting ability, some poplar clones and especially willow clones, we have had good success.

Bryant — If you can get the on-farm cost down low enough, you can probably afford to give up some germination and use just cultipacking. We did some cultipacking, but germination wasn't anywhere near as good. You couldn't afford to throw the money away, but if you did a very cheap collection with an on-farm system, then perhaps you could afford to put up with that.

DISCUSSION, CONT.

Heckrodt — Did you notice any difference in the soil type?

Bryant — No, other than a few isolated spots where we had some hardpan, it was pretty sandy soil. All we really did is supply nitrogen. The second year we gave it a little bit of zinc when we saw some deficiency. We didn't have any specific guidelines available about poplar, and just treated it like corn. I would say we don't have enough variation in our soil in that plot to give anybody any guidance one way or the other.

PROSPECTS FOR COMMERCIALIZATION OF EUCALYPTUS BIOMASS PLANTATIONS IN FLORIDA

D. L. Rockwood, Ph. D., and D. R. Dippon, Ph. D.
Department of Forestry
University of Florida
Gainesville, Florida 32611

ABSTRACT

Eucalyptus biomass plantations will be commercially viable in Florida when productivity is assured, harvesting is economical, and markets are established. To date, substantial increases in yields have been achieved through genetic selection and cultural options, harvested and transported biomass is estimated to have very positive economic and energetic ratios, and utilization of woody biomass for energy is increasing.

Achieved yields have been as high as 23 dry Mg/ha/yr. In southern Florida, *E. grandis* has grown well on drained organic and seasonally wet sandy soils, and coppice yields may be equal or greater if harvests are properly timed and conducted. For central Florida, *E. camaldulensis* can be very productive on sandhills sites and may have suitable frost-hardiness. On agricultural sites in northern Florida, *E. amplifolia* has excellent frost-resiliency and coppicing.

Genetic selection promises to increase productivity and frost resistance, and vegetative propagation may be effective in realizing near-term potential. High quality *E. grandis* seed may be obtained from a 17 ha seed orchard, and four clones combine frost resistance with good growth. Several *E. camaldulensis* sources appear well adapted, and some excellent clones have been selected. Sources of *E. amplifolia* differ in frost resistance and growth.

Associated inputs and costs suggest that short-rotation, close spacing culture can make woody biomass competitive with other feedstocks available for energy generation. Breakeven costs per delivered Gj range from just under $2.00 for *E. grandis* in 2-year rotations on muck soils and for *E. amplifolia* in 3-year cycles on agricultural lands to under $4.00/Gj for *E. grandis* on prairie and *E. camaldulensis* on sandhills.

PROSPECTS FOR COMMERCIALIZATION OF EUCALYPTUS BIOMASS PLANTATIONS IN FLORIDA

INTRODUCTION

Since 1977, research with eucalypts in short-rotation, close-spacing culture in Florida has identified the biological potential of *Eucalyptus grandis* Hill ex Maiden on drained muck and seasonally wet sandy soils in southern Florida (13), *Eucalyptus camaldulensis* Dehnh. on dry sandy sites in southern and central Florida (14), and *Eucalyptus amplifolia* Naudin on agricultural soils in northern Florida (14). Responses of these species to several exceptional freezes have identified the necessity and opportunity for emphasizing frost resistance in genetic improvement. Preliminary analyses suggested that *E. grandis* biomass produced at 10,000 trees/ha in 24 months could cost as little as $11.27/dry Mg with an associated energy output/input ratio of 20.18 (5, 15). This paper utilizes the most recent yield data in conjunction with predicted genetic gains and updated economic factors to evaluate the prospects for commercial use.

PROCEDURES

Yield Estimation

E. grandis. Yields previously obtained from studies initiated in 1979 on seasonally wet sandy soil (18) and 1980 on drained organic soil (16) were compared to five-month results from ORNL-35, described below, to validate early growth trends.

E. camaldulensis. Survival and tree size data from SS-2 and SS-8, also described below, provided the basis for projecting rotation length productivity.

E. amplifolia. The species was established near Gainesville in April 1981 in a 100-tree square plot at a 1 x 1 m spacing (DP-1). The first coppice rotation was induced by a January 1982 freeze that killed the seedlings to the ground. The December 1983 freeze killed the 23-month-old first coppice to the groundline, and a second coppice was initiated when the plot was cleared. The *E. amplifolia* coppice was again killed by the January 1985 freeze, thereby producing a third *E. amplifolia* coppice. Tree height and diameter breast height (DBH) on the interior 36 trees and survival on the whole plot were measured through December 1985. Biomass yields were estimated from dry weights derived from trees harvested in December 1983 and January 1985.

Genetic Improvement

E. grandis. The genetic base population (GPOP77), consisting of 529 progenies planted in July 1977 near LaBelle in a completely randomized design with single tree plots constituting 31,725 trees (approximately 60 trees/progeny) on 17.3 hectares, was developed through four generations of selection by the U.S. Forest Service (Figure 1). Conversion of GPOP77 to a seedling seed orchard (GORCH77) involved subdividing the population into 1575 selection areas of uniform microsites and retaining the best tree in each area, with a second tree occasionally retained in an area due to other considerations. Felling of trees was conducted in February and March 1986.

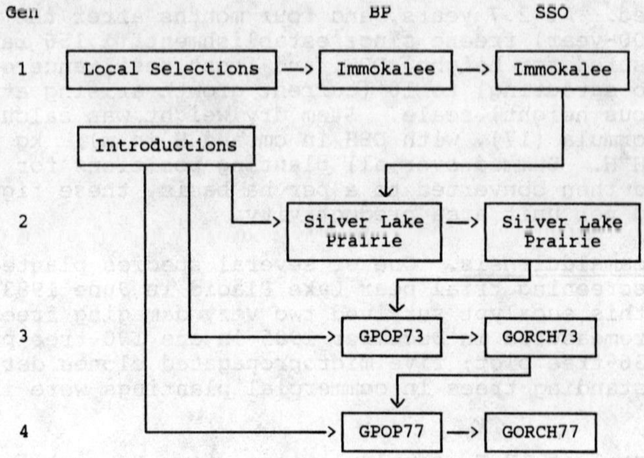

Figure 1. Development of genetically improved *E. grandis* in Florida through four generations (Gen) of genetic base population (BP) establishment and conversion to seedling seed orchard (SSO).

GORCH77 contains 1,677 trees in 405 progenies. Fourth-generation progenies are nearly all represented in GORCH77 with in average of some 8 trees per each of 47 progenies (Table 1). First-generation progenies, on the other hand, average 1.6 trees each for 58 progenies. In addition selected trees in GORCH77 include 13 trees with unknown pedigrees.

Eucalyptus grandis and *E. amplifolia* were planted in July 1985 at Belle Glade (ORNL-35). The study effectively had six replications of 10-tree row plots of 27 *E. grandis* progenies and six *E. amplifolia* sources. Two replications also included 25-tree square plots of five *E. grandis* progenies. Survival and height were measured in December 1985.

In August 1982, 55 clones of *Eucalyptus grandis* and hybrids were planted as rooted cuttings near Palmdale

Table 1. Summary by generation of trees retained in GORCH77.

Generation	Number of Progenies Planted	Number of Progenies Retained	Number of Trees Retained Total	Number of Trees Retained Per Progeny
?			13	
1	144	58	94	1.6
2	211	183	633	3.5
3	126	117	554	4.7
4	48	47	381	8.1
Total	529	405	1677	

(ORNL-37). One clone was represented by only one ramet, and another by two; the other 53 clones had 11 to 42 ramets outplanted. At 2.7 years, and four months after the second major (100-year) freeze since establishment, 1,156 ramets were measured for height, DBH, and frost resilience on a 0 (dead, no sprouting) to 10 (current growth arising at 95% of previous height) scale. Stem dry weight was calculated by the formula (17), with DBH in cm and H in m): $kg = .0196*DBH^2H$. Summed over all planting positions for each clone and then converted to a per ha basis, these figures estimated per unit area productivity.

E. camaldulensis. One of several species planted in a species screening trial near Lake Placid in June 1983 (SS-2), this eucalypt survived two very damaging freezes and was remeasured in December 1985 in one 100-tree plot and one 36-tree plot; five micropropagated clones derived from outstanding trees in commercial plantings were involved.

A total of 18 E. camaldulensis, four E. amplifolia, and 32 E. grandis sources were outplanted near LaBelle (SS-8). Northern sources of E. grandis, 21 from the Atherton area in Queensland not represented in GPOP77, were chosen for concentration. The randomized complete block design with three replications of 8-tree row plots was planted July 1984. Individual sources were assigned to the row plots, and species or source groups were utilized to avoid competitive effects. Spacing was 1.5 x 1 m. Survival and height were measured in December 1985.

E. amplifolia. The species was one of four eucalypts outplanted near Gainesville in September 1984 (SS-9). Twelve sources, each generally a bulked seedlot from several trees in a portion of the species' range, were obtained from CSIRO and other seed suppliers in Australia. The sources were represented in 10-tree row plots with three replications in a randomized complete block design. Spacing was 1 x 1 m. Survival was determined in April 1985.

Ten sources previously in SS-9 were planted near Gainesville in August 1985 in two 25-tree plots at 1 x 1 m spacing and was measured for height and survival in December 1985 (SS-11). Frost damage from a January 1986 freeze with a low of 17°F was evaluated using a scoring system adapted from Wilcox (20). On February 8, foliage damage was scored on a 0 (no damage) to 5 (killed by frost) scale, and stem damage was graded on a 0 (no damage) to 3 (lower 1/3 dead) scale. "Resistant" trees had scores below 4 (90% killed) on the foliage scale and 2 (middle 1/3 killed) for stem damage.

Economic Analyses

Four *Eucalyptus* short-rotation, close-spacing systems were considered: *E. grandis* on muck and sandy soils in southern Florida, *E. camaldulensis* on sandhills in the southern/central sections, and *E. amplifolia* in the northern part of the state. Plantation establishment and management costs are reported on a per hectare basis except for the seedling propagation and planting costs (per seedling) and chipping and transportation costs (per dry Mg).

Most costs were common among the sites since the same types of equipment and methods should be employable. The sites are assumed to be fertilized once each time the plantations are re-initiated by planting since the intensive repetitive removal of biomass is expected to deplete nutrients in the soil. It needs to emphasized that since there currently exist no operational short-rotation intensive-cultured plantation systems, the actual costs for seedling, planting, harvesting, processing, and transportation are based on small sample, experimental data. If these types of systems became more common, it could be expected that improvements in seedling propagation and equipment engineering could substantially reduce these costs.

A key cost that was variable among the four systems was the annual rental price for land. If acreage is to be placed into woody biomass plantations, than it must be sufficiently profitable as to pay the competitive price for the use of the land. The types of land on which the three eucalypts have been tested have different levels of agricultural profitability. The muck soils in southern Florida are valuable vegetable cropping areas with land rentals varying between $190 and $380 per ha, depending upon the depth of the muck soil and therefore the type of vegetables which can be cropped. This analysis used a rental price nearer the lower end ($247 per ha) since *E. grandis* does not need to be on the best muck soils in order to produce efficiently. The southern Florida sandy soil, "palmetto prairie," was less expensive with an average rental price of $128.50 per hectare. The northern Florida flatwoods are typically subdivided between row cropping, grazing, and timber activities. Common rental prices are about $64.25 per hectare. Finally, the central Florida sandhills are in a land-use transition since citrus has been extensively damaged by freezes over the past five years. Hence, rental prices have declined and landowners wish to keep the land in some form of managed agro/forestry system in order to retain Florida's Green Belt Exemption, or a favorably modified property tax assessment scheme. Under these circumstances, a rental price of $25.70 per hectare per year was used. An implicit assumption is that the intensive silvicutural systems are only competing with other agricultural/grazing/forestry systems and not the commercial developmental use of the land.

Computations in determining the break-even costs per Mg and per Gj were based on the assumption that 1000 ha are placed under intensive management for 25 years. The analysis for all four scenarios figured production and costs based on a planting density of 10,000 seedlings per ha. The E. grandis seedlings on the muck soils were grown for two years before the first coppice with a second coppice two years thereafter. Geary et al. (8) suggested that such a system maintained the site's biomass productivity before the original stools coppice vigor diminished and demonstrated that Eucalyptus coppice can grow more rapidly per unit of time as compared to the seedling stand. Therefore, the analysis assumed that coppice growth would be 20 percent greater than seedling growth. Since site preparation and planting costs are a major operational cost, this strategy employed three harvests for each complete planting to re-planting cycle. This schedule of activities resulted in a 2/2/2 growing and harvesting system where E. grandis seedlings were planted only every six years. The other cropping systems employed in this analysis provided that the E. grandis on the "palmetto prairie" and the E. amplifolia were managed under a 3/3/3 year cycle while the E. camaldulensis was managed with a 5/5/5 year scheme.

The only other cost which varied by site was the seedling reestablishment of the plantation after a coppice/harvesting cycle has been completed. In all cases, this operation was considered to be much more expensive than that experienced by the initial establishment of the plantation. This assumption was made since the conversion from existing agricultural or grazing systems provide a site which is fairly easy to plant to the high intensity plantation. It is after the final harvest, when the site must be prepared for mechanical planting, that the remaining stems and root systems need to be removed or modified.

These data were used as input values for the program BIOCUT which is a comprehensive model designed to facilitate the investigation of the economics of short-rotation intensive-silvicultural energy plantation systems (4).

RESULTS AND DISCUSSION

Achieved Yields

Yields of E. grandis in southern Florida followed earlier trends. After five months on drained muck soil, trees in ORNL-35 were similar in size to six-month-old trees in a previous study (15) and suggested that 24 Mg/ha/

yr (Table 2) was an achievable productivity. On "palmetto prairie" after 2.7 years, clones in ORNL-37 that reached sizes reported earlier for seedlings undamaged by freeze (17) supported a growth rate of 10 Mg/ha/yr.

Yields of E. camaldulensis were projected from early performance of sources in SS-8 and clones in SS-2. The species' excellent survival under rigorous freezes suggests that its relatively low projected growth (Table 2) is achievable.

Table 2. Maximum achieved yield (Mg/ha/yr) of *Eucalyptus grandis* (EG), *E. camaldulensis* (EC), and *E. amplifolia* (EA) in short-rotation, close spacing systems in Florida.

Species	Site/Rotation Length/Spacing	Yield
EG	Drained muck soil in southern Florida/ 18 months/1 x 1 m	23.9
	Palmetto prairie in southern Florida/ 33 months/40,000 trees per ha	10.0
EC	Sandhills in south-central Florida/ 2.7 years/1 x 1 m	12.6
EA	Pasture and cropland in northern Florida/ 2 years/ 1 x 1 m	22.8

E. amplifolia coppiced well (Table 3). The vigor of the third E. amplifolia coppice was similar to earlier coppices, and stool survival, which decreased slightly after the first coppice, has remained constant from the second coppice. Each stool typically retained one well-formed stem from an initial proliferation of shoots. Annual productivity in the first year was more than doubled by the second year's growth, indicating that coppice rotations should be at least two years. The 23 dry Mg/ha/yr (Table 2), however, must be confirmed in further studies.

Table 3. Coppice performance of *E. amplifolia* in DP1.

Age	Ht. (m)	DBH (cm)	Surv. (%)
8-mo Seedlings	1.6	-	81
23-mo 1st Coppice	7.9	8.2	76
10-mo 2nd Coppice	3.7	2.0	69
10-mo 3rd Coppice	3.8	2.6	69

Genetic Improvement

E. grandis. Collection of seed from all trees currently in GORCH77 would result in a estimated gain of 58% in coppice stem volume over the GPOP77 average (Table 4). Achieved gain is expected to be higher due to further roguing to be done and to selective seed collection. The impact of additional roguing on gain will be relatively minimal compared to selective seed collection, however.

Maximum gain can be made by utilizing the best individuals in better progenies. Selective collection from the best tree in each selection area would result in about a 81% gain (Table 4). A more selective set of seed trees, e.g., the best 50 trees (presumably from the top 50 progenies), would further increase genetic quality. While the background pollen in GORCH77 would on the average be poorer quality, of course, the potential for commercial seed collection from a limited number of seed trees has been demonstrated by Forest Service commercial collections through 1980 (11).

Table 4. Estimated genetic gains in 64-month coppice volume to be achieved through various seed collection options for GORCH77.

Option	Gain (%)
All current orchard trees	58
1575 orchard trees, 405 top progenies	81
50 orchard trees (50 top progenies)	179

The opportunity for maximizing gain by selective seed collection in GORCH77 will depend on the seed quantities needed. Using as a guide the estimates that a mature seed orchard tree will produce 2 kg of seed annually yielding 75,000 plantable seedings, 1,575 seed trees in GORCH77 could eventually produce over 100 million seedlings per year. Assuming that operational E. grandis biomass plantations would be established at 10,000 trees/ha, GORCH77 could support the planting of more than 10,000 ha/year. Realistically, at least some three years with non-damaging winters are needed before GORCH77 trees develop crowns capable of full seed production. The first crop of improved seed could however be harvested in March 1987.

Considerable variation was observed among E. grandis progenies planted in row plots in ORNL-35 (Table 5). Some of the large variation among 27 E. grandis specifically selected based on performance in other tests was attributable to nursery effects, as some progenies were propagated later and in a second nursery. Five E. grandis progenies in block plots were similar, however.

Table 5. Five-month survival and height of Eucalyptus grandis (EG) and E. amplifolia (EA) in ORNL-35.

Species	No. of Progenies	Survival Mean	Survival Range	Height Mean	Height Range
		(%)		(m)	
EG	27	87.9	71-100	2.25	1.39-2.49
EA	6	75.0	63-85	1.96	1.71-2.20

Within the current E. grandis base population and commercial plantations, exceptional individuals can be selected and evaluated in clonal tests. Four of 55 Eucalyptus clones in ORNL-37 were top choices at 2.7 years, performing at least 2.1 standard deviations above the mean for the majority of traits evaluated (Table 6). With respect to per ha productivity, the top four clones exceeded the average by more than five times. Consequently, clones such as 2814 and 2817, which had particularly excellent performance, represent the potential benefits of clonal propagation.

Table 6. Mean 2.7-year performance of all 55 and top four Eucalyptus clones in ORNL-37.

Trait	All Clones	Clone 2786	Clone 2798	Clone 2814	Clone 2817
Survival (%)	90	100	100	100	100
Height (m)	1.7	5.2	4.6	6.0	6.0
DBH (cm)	1.6	6.3	5.3	7.0	7.1
Weight/tree (kg)	.87	4.79	4.00	6.13	7.00
Frost resilience	2.8	9.0	8.3	8.9	9.0
Dry weight/ha (Mg)	1.17	6.65	5.55	8.51	9.72

E. camaldulensis. Earlier research by the U.S. Forest Service indicated that of four eucalypts suited to southern Florida, E. camaldulensis was the best choice for dry sand ridges (8). Genetic improvement efforts with this species were necessarily limited initially to three small first-generation seed orchards. From these orchards and other introductions, CT-74 was developed by the Forest Service and converted to a seedling seed orchard in 1984. The orchard has not been progeny tested.

SS-8 has a broad representation of E. camaldulensis and permits a comparison with other species for frost-hardiness. The superiority of E. camaldulensis in SS-8 increased during the 1985 growing season (Table 7). After 17 months all but one E. camaldulensis source surpassed an E. grandis genetic check for survival, while 16 E. camaldulensis sources were taller than this check. Individual E. camaldulensis trees were frequently more than three times as tall as the best tree of the E. grandis check.

The strong performance of E. camaldulensis both in terms of frost-hardiness and general vigor was unexpected. While there had been earlier indications of its adaptability to central Florida (8), worldwide experience (e.g., in South Africa (12, 3) and Argentina (10)) with E. camaldulensis provided no basis for expecting high survival after unusually low temperatures. The apparent lack of any freeze damage and the good vigor in comparison to the E. grandis standard further increase its prospects.

Table 7. 17-month performance of Eucalyptus sources in SS-8.

Species	No. of Sources	Survival (%)	Height (m)
E. amplifolia	4	60b[1/]	.31b
E. camaldulensis	18	82a	.97a
E. grandis	32	16c	.58ab

[1/] Species means not sharing the same letter are significantly different at the 5% level.

That none of the E. grandis sources had better survival than the genetic check agrees with comparisons in GPOP77 of first-generation trees to subsequent generations. Of the 31 new sources, 10 exceeded 20% survival after the January 1985 freeze. Such inversion freezes in southern Florida can be especially damaging to short trees, leading to differentials in assumed frost resistance actually being related to tree size (7). Within E. grandis after seven-months however, no meaningful height differences existed among the sources. While only one of these 10 was taller than the check, they like other new introductions currently in GPOP77 have good potential for broadening the E. grandis population and imparting some frost hardiness.

Five E. camaldulensis clones have exhibited good frost tolerance in SS-2, which is representative of some 100,000 ha of sandhill areas that formerly supported citrus. Survival was excellent with good growth (Table 8). Individual trees were up to 9 m in height, and the five clones varied in performance. Clone 172 had particularly exceptional survival and vigor.

Table 8. Twenty-nine month survival, height, and DBH of Eucalyptus camaldulensis in SS-2.

Clone	Survival (%)	Height (m)	DBH (cm)
171	74	3.8	3.4
172	100	6.1	5.2
174	94	4.1	3.5
178	84	1.9	1.6
BH	79	3.8	2.9
Mean	84	3.7	3.1

E. amplifolia. For northern Florida, E. amplifolia consistently outsurvived three other previously promising species following the rigorous January 1985 freeze, a 100-year event. Six E. amplifolia sources exceeded 50% or more survival, with four of these sources incurring 30% or less apparently freeze-related mortality (Table 9).

Table 9. Survival (Sur), height (Ht), and frost resistance (FR) of E. amplifolia sources in SS-8, SS-9, SS-11, and ORNL-35.

Source	SS-9 7-mo. Sur (%)	SS-11 5-mo. Sur (%)	SS-11 5-mo. Ht (m)	SS-11 9-mo. FR (%)	SS-8 17-mo. Sur (%)	SS-8 17-mo. Ht (m)	ORNL-35 5-mo. Sur (%)	ORNL-35 5-mo. Ht (m)
1-92					63	.4		
1-93	33	78	.7	75				
2-1	70	82	.7	36				
2-2	37	60	1.0	0			85	1.6
2-3	33	82	1.3	54			85	1.8
2-4	33	90	.5	73	63	.3		
2-5	60	44	.6	12	63	.3	63	1.5
2-6	57	32	.9	16				
2-7	50	76	.7	21			69	1.5
2-8	40	94	.7	48	52	.2	83	2.0
2-9	30	70	.9	27			74	1.6
2-10	67							
2-11	63							
Mean	48	71	.8	36	60	.3	75	2.0

The 10 E. amplifolia sources reestablished in SS-11 grew well through five months and varied in response to a more typical January 1986 freeze (Table 9). Source 2-3 was exceptionally vigorous with some trees as tall as 1.8 m. No trees were totally undamaged by the freeze; Sources 1-93 and 2-4 had more than 70% "resistant" trees. Sources 2-1 and 2-7 had lower proportions of "resistant" trees but had been very resilient in SS-9. Such differences in response indicate the difficulty in determining frost resistance under variable field conditions.

Continued genetic improvement of these species is dependent on several factors, one of which is the germplasm available (Table 10). By far, E. grandis has the most appreciable genetic base. The considerable area of plantations provides additional opportunity to locate clonal candidates, and the large number of pedigreed trees in GORCH77 is suitable for traditional breeding approaches. The E. camaldulensis orchard, although not tested, is a base for further orchard development in conjunction with expansion of the population by inclusion of source material already or soon to be tested. As represented by the good clones in SS-2, plantations of E. camaldulensis may provide clones for testing. An E. amplifolia genetic base population can be created from sources now in tests and new introductions.

Table 10. Genetic resources for *Eucalyptus grandis* (EG), *E. camaldulensis* (EC), and *E. amplifolia* (EA) in Florida.

Species	Planted Area	Seed Orchard Trees	Introductions		
			Tested	Untested	
			Sources	Sources	Progenies
	(ha)	(no.)			
EG	>6,000	1,677	31	0	0
EC	>100	>200	18	46	0
EA	0	0	12	0	97

Economic Analyses

The relative break-even costs per Gj of woody biomass under the cost assumptions (Table 11) for the four systems in this analysis. The costs are nearly equal for the *E. grandis* on the southern Florida muck and the *E. amplifolia* in northern Florida (Table 12). Both systems could deliver woody biomass at a cost of under $2.00 per Gj, if the discount rate net of inflation is equal to zero. Ranging the real discount rate up to 8 percent causes the estimated costs to increase by approximately 11 and 16 percent, respectively. The net cost for *E. grandis* on "palmetto

Table 11. Costs associated with management of *Eucalyptus grandis* (EG) on muck soil and palmetto prairie, *E. camaldulensis* (EC) on sandhills, and *E. amplifolia* (EA) on crop and pasture land in short-rotation, close-spacing systems in Florida.

Management Activity	EG on Muck	EG on Prairie	EC	EA
Infrastructure Development	50.00	50.00	50.00	50.00
Site Preparation				
Initial	123.50	123.50	123.50	123.50
Replant	300.00	300.00	200.00	300.00
Fertilization				
Once/rotation	93.00	93.00	93.00	93.00
Establishment				
Propagules (per/tree)	.15	.15	.15	.15
Planting (per/tree)	.055	.055	.055	.055
Annual Activities				
Land rent	247.00	128.50	25.70	64.25
Taxes & Adm.	30.00	30.00	30.00	30.00
Removal				
Harvest+forward	150.00	150.00	150.00	150.00
Chip (per Mg)	6.60	6.60	6.60	6.60
Transport+Storage (per Mg)	3.30	3.30	3.30	3.30

Table 12. Projected breakeven prices per Mg and gigajoule for E. grandis (EG), E. camaldulensis (EC), and E. amplifolia (EA) as a function of interest rate.

Species/Site	Basis	Interest Rate 0	.02	.04	.06	.08
		(Dollars)				
EG/Muck	Mg	38.82	39.87	40.94	42.02	43.12
	GJ	1.96	2.02	2.07	2.13	2.18
EG/Prairie	Mg	65.47	67.85	70.48	73.36	76.49
	GJ	3.31	3.43	3.57	3.71	3.87
EC/Sandhills	Mg	65.15	69.42	74.41	80.20	86.84
	GJ	3.30	3.51	3.77	4.06	4.39
EA/Cropland	Mg	38.07	39.36	40.80	42.37	44.10
	GJ	1.93	1.99	2.06	2.14	2.23

prairie" and E. camaldulensis on the sandhill sites are both significantly higher. Whether these systems can produce woody biomass for energy will depend on better matching planting density and productivity to these sites. It is not that the costs of management and implementation are significantly greater than those associated with the other site/species combinations. The factor which contributes to the poor cost per Gj is the relatively low productivity of biomass by weight over time which reduces the attractiveness of the latter two options.

The delivered costs per Gj of the four woody biomass sytems can be compared to the prices reported by Asbury et al. (2) of approximately $2.00 per Gj for delivered coal. Long term prices expected to remain below $3.00 through the end of this Century. Of course, very recent declines in crude oil prices have significantly dropped energy prices below these levels. Whether these prices will remain in the low teens per barrel assumes that the efforts of OPEC will fail and that the world's oil markets will never again be dominated by a cartel.

Market Development

Energy wood is being increasingly utilized in Florida (6). Forest industry used over two million wet Mg of wood for energy production in 1984, and more than 300,000 Mg of wood chips were sold to other wood energy users. While these quantities were almost exclusively derived from commercial forest lands, several applications indicate the potential for short-rotation intensive culture plantations to supply energy wood.

One such application is a wood-fired steam generating facility at Union Correctional Institute in northeast Florida (19). Using some 20,000 wet Mg annually to produce steam at a cost of approximately $2 per Gj, the Institute saves an estimated $360,000 per year compared to alternative fuels. This annual wood consumption corresponds to the approximate yearly production of 400 ha of E. amplifolia at the growth rate observed in experimental conditions.

Another localized application could be a citrus juice plant, of which there may be as many as 100 in operation in Florida. At the rate of 65 Mj per box of fruit processed (9) and a 1984-85 production of 102 million boxes (1), the citrus industry is a major energy consumer. Some 90% of the energy consumed by a processing plant, i.e, by the feed mill, evaporators, and boilers (9), could be derived from wood. Lands in the vicinity of many juice plants are suitable for growing E. camaldulensis and/or E. grandis.

A third potential user of energy wood could be electrical generation plants. Assuming a 10 MW power plant, Draper et al. (6) estimated that cropping of eucalypts on 2,700 ha in southern Florida would be required to meet the energy demand.

The land base required for woody biomass production depends on the energy facility, time aspects of energy wood demand, species productivity, and rotation length. Assuming that 20% of pasture and rangeland would be available for woody biomass production, some 500,000 ha in Florida could be used for energy wood production (6). Non-merchantable commercial forest land provides additional potentially available area.

ACKNOWLEDGEMENTS

Results reported are from research performed under Subcontract No. 19X-09050C with Oak Ridge National Laboratory under Martin Marietta Energy Systems, Inc., contract DE-AC05-840R21400 with the U. S. Department of Energy and from a project that contributes to a cooperative program between the Institute of Food and Agricultural Sciences of the University of Florida and the Gas Research Institute entitled "Methane from Biomass and Waste." The assistance of the following cooperators is gratefully acknowledged: Belle Glade Agricultural Research and Education Center, Belle Glade, FL; Consolidated Tomoka Land Company, Sebring, FL; and Lykes Brothers, Inc., Palmdale, FL.

REFERENCES CITED

1. Anon. Statistical summary 1984-85 season. Fla. Citrus Processors Assn. (1985)

2. Asbury, J., Gaines, L., and Vyas, A. Survey of recent energy price forecasts: cost coal committee report. Argonne National Lab. Argonne, Il. 19 pp. (1983)

3. Darrow, W. K. Provenance studies of frost-resistant eucalypts in South Africa. Proc. IUFRO Frost Resistant Eucalypts Symp.: 448-468. (1983)

4. Das S., Perlack, R. D., Barron, W. F., and Kroll, P. BIOCUT: a microcomputer based econometric model for wood energy plantations. Model description and users guide. Oak Ridge National Laboratory. ORNL/TM-9576. 80pp. (1985)

5. Dippon, D. R., Comer, C.W., and Rockwood, D. L. Cost sensitivity of Eucalyptus woody biomass systems. In: Biomass Energy Development: (in press). (1986)

6. Draper, H. M., Nesmith, C. C., and Smith, W. H. Florida biomass assessment: available land, energy wood, and crop residues. Proc. 6th Solar Biomass and Wind Energy Wkshp.: 45-49. (1986)

7. Franklin, E. C., and Meskimen, G. Choice of species and provenances in cold summer rainfall climates. Proc. IUFRO Frost Resistant Eucalypts Symp.: 341-357. (1983)

8. Geary, T. F., Meskimen, G. F., and Franklin, E. C. Growing eucalypts in Florida for industrial wood production. USDA For. Serv. Gen. Tech. Rpt. SE-23. 43p. (1983)

9. Holladay, K. L. Results of energy audits conducted in various citrus plants during 1979-80. Transact. 1981 Citrus Eng. Conf.: 1-11. (1981)

10. Mendonza, L., and Alliani, R. Preliminary trial of some Eucalyptus in the north of the province of Buenos Aires. Proc. IUFRO Frost Resistant Eucalypts Symp.: 440-447. (1983)

11. Meskimen, G. F. Realized gain from breeding Eucalyptus grandis in Florida. USDA For. Serv. Gen. Tech. Rpt. PSW-69: 121-128. (1983)

12. Nixon, K. M. A test of fifteen Eucalyptus species and provenances for frost resistance in Natal. Proc. IUFRO Frost Resistant Eucalypts Symp.: 367-376. (1983)

13. Rockwood, D. L., Comer, C. W., Dippon, D. R., and Huffman, J. B. Woody biomass production options for Florida. Fla. Agr. Exp. Sta. Tech. Bull. 856. 29p. (1985)

14. Rockwood, D. L., DeValerio, J. T. Promising species for woody biomass production in Florida. Biomass. (accepted for publication.) (1986)

15. Rockwood, D. L., Dippon, D. R., and Comer, C. W. Potential of Eucalyptus grandis for biomass production in Florida. Proc. BioEnergy 84 2:86-93. (1984)

16. Rockwood, D. L., and Geary, T. F. Genetic variation in biomass productivity and coppicing of intensively grown Eucalyptus grandis in southern Florida. Proc. 7th N. Amer. For. Biol. Wkshp.: 400-405. (1982)

17. Rockwood, D. L., Geary, T. F., and Bourgeron, P. S. Planting density and genetic influences on seedling growth and coppicing of eucalypts in southern Florida. Proc. 1982 South. For. Biomass Working Group Workshop: 95-102. (1982)

18. Rockwood, D. L., and Meskimen, G. F. Genetic, spacing, and genotype x spacing influences on growth of Eucalyptus grandis in South Florida. Proc. 16th S. For. Tree Imp. Conf.: 47-54. (1981)

19. Shroeder, R. M. Institutional experience with woody biomass for fuel. Proc. 7th. S. For. Biomass Wkshp.: (in press). (1985)

20. Wilcox, M. D. Genetic variation in frost tolerance, early height growth, and incidence of forking among and within provenances of Eucalyptus fastigata. NZ J. For. Sci. 12(3): 510-524. (1982)

DISCUSSION

K. Porter, U.S. AID (Jamaica)

It has been reported that Eucalyptus has been described as living water pumps. Have you studied the impact on soil moisture, and what the implications are for larger scale developments in the environmental sense?

Rockwood

There were some very early observations based on one planting site at LaBelle, at which the lysimeters that were in place monitored water consumption and the impact on water quality. Unfortunately, due to the freeze that occurred there, they did not get rotation length observations. But for the period of time that the lysimeters were in place, there appeared to be minimal impact on both water quality and quantity. That's not to say that's the way it would be over the full rotation, but for the period observed, there didn't seem to be any difference between leaving the land in Palmetto prairie versus having it in Eucalyptus.

QUALITY OF SORGHUM BIOMASS FOR METHANOGENESIS

G.G. McBee, Ph.D., Professor
F.R. Miller, Ph.D., Professor
R.E. Dominy, M.S., Research Assoc.
Soil and Crop Science Department
Texas A&M University
College Station, Texas 77843

R.L. Monk, Ph.D., Geneticist
Cargill, Inc.
Plainview, Texas 79072

ABSTRACT

A total of 17.8 million acres is planted to sorghum, annually, within the United States. Biomass of the plant, excluding panicle, consists of more than 70% structural (SC) and nonstructural carbohydrates (NSC). Feedstock of this material offers high potential as a candidate for methane production.

Grain, sweet and high energy groups of sorghum were investigated for quality of biomass. Cultivars included within these studies produced a range of 22.2-46.9% NSC and 38.1-64.6% SC. Structural components were primarily composed of cellulose, hemicellulose and lignin with respective percentage ranges of 21.2-37.1, 13.6-25.9 and 5.7-10.5. Two types of hemicellulose were precipitated from the hemicellulose fraction. Results indicate the highest percent of SC is localized in the rind as compared to the pith of the stem. Stems and blades were similar in levels of SC but stems were significantly higher in NSC. Hybrids are being developed that contain high or lower percentages of either carbohydrate fraction as may be desired for different digester systems.

Volatile solids of stems averaged above 92% with the highest value of 97.34% obtained for the cultivar MN1500. Potential methane yields have been obtained for a large number of the cultivars using a bench scale technique. Methane yields exceeded 5.13 standard cubic feet per pound of volatile solids.

QUALITY OF SORGHUM BIOMASS FOR METHANOGENESIS

A total of 17.8 million acres of sorghum was planted in the United States during 1985 (1). The state of Texas accounted for 3.9 million acres of this total during that period and normally leads all states in total acreage. This represents a significant quantity of renewable biomass for potential conversion to methane.

Sorghum Groups

Sorghums [*Sorghum bicolor* (L.) Moench] may be divided into four general groups: grain, sweet, forage, and broomcorn. The grain and sweet groups have been more extensively studied for use in methanogenesis. Additionally, the grain and sweet groups have been crossed by scientists of the Texas Agricultural Experiment Station thus adding a type referred to as high energy (HE). Range of partitioning among blades, stems and panicles summarized from several previous unpublished studies is listed in Table 1. As is typical, the highest percent of dry weight is partitioned to stems for the sweet types. They also contain a slightly higher range for percent of blades which conforms to the objective of fodder production. By contrast, more dry weight is partitioned to panicles in grain sorghum, since the objective is high kernel yields. When the sweet and grain types are crossed, the new type (HE) exhibits characteristics of both types of parents. The typical HE plant contains more dry matter in the panicles than the sweet type and is more vegetative than the grain type parent. Total fresh weight yields can be noted in Table 2. HE cultivars produce a taller plant than the grain parent and thus more biomass tonnage per acre. Additionally, this group also produces a nonstructural (NSC) and structural (SC) carbohydrate profile closer to the sweet group pattern.

MATERIALS AND METHODS

Plant Culture. Sorghum cultivars, as shown in Figures 1-5, were planted on a Norwood silt loam at the Brazos River Field Laboratory near Texas A&M University during April, 1984. Rows were 75 cm apart and plants were spaced approximately 10 cm apart within rows, 7 meters in length. A randomized block design of three replications was used. Nitrogen, as NH_4NO_3, was applied, sidedress, at the rate of 100 kg per hectare. The plots were irrigated as needed to avoid stress. Plants were harvested when kernels were at the 10 day, postblack layer stage.

Analytical Procedures. Special care was exercised during harvest of the plants to preserve the NSC fractions. Immediately upon harvest, the plants were sectioned and the samples for analysis placed in an oven at 100^oC for 90 minutes. They were then transferred to another oven set at 60^oC and allowed to remain there until dried. Subsequently, the tissue was ground to pass a 1-mm screen, homogenized, and stored in sealed, brown bottles at 25^oC until analysis.

Table 1. General Ranges for Partitioning of Sorghum Dry Weight Into Plant Parts

Type	Blades	Stems	Panicles
	----------- % of total dry wt. -----------		
Sweet	16-22	66-73	5-16
Grain	13-17	25-30	55-61
High Energy	11-15	35-41	45-52

Table 2. Yield of Fresh Weight and Partitioning of Carbohydrates Among Three Categories of Sorghum.

Cultivar	Yield tons/Acre	Total NSC†	Hemi-cellulose	Cellulose	Lignin
		Mg/G O.D.W‡	-------- % O.D.W. --------		
-Sweet-					
Rio	23.1	403.6§	19.6	23.6	5.3
Wray	25.9	376.4	18.3	25.1	7.0
MN1500	36.1	311.8	22.1	30.1	9.1
-Grain-					
ATx623 x RTx430	12.3	211.0	19.2	33.6	7.3
ATx378 x RTx430	11.7	151.2	22.4	34.2	10.5
-H.E.-					
ATx623 x Rio	20.9	302.9	21.1	26.8	7.5
ATx623 x Wray	21.7	249.6	19.9	32.2	9.6

† N.S.C. = Glucose + Fructose + Sucrose + Starch; ‡ O.D.W. - oven dry wt.;

§ Mean of 3 reps.

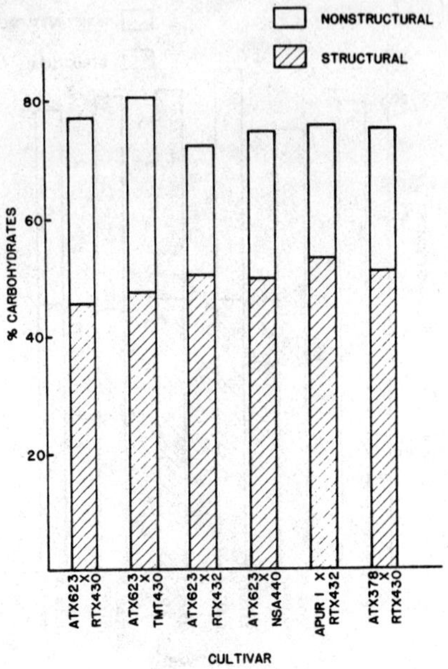

Figure 1. Total Percent Carbohydrates in Grain Sorghum Cultivars, 1984.

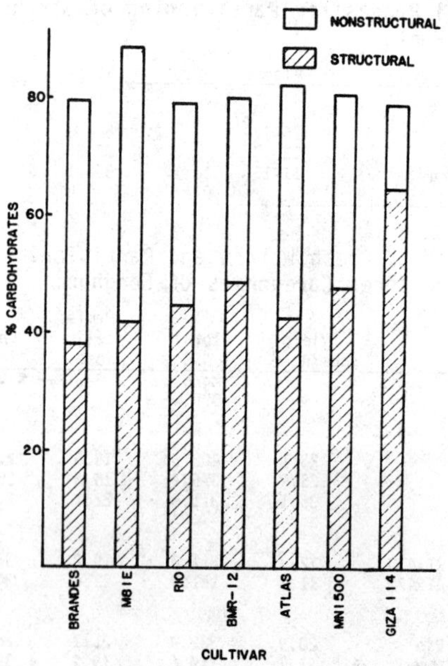

Figure 2. Total Percent Carbohydrates in Sweet Sorghum Cultivars, 1984. (Giza 114, not a sweet type.)

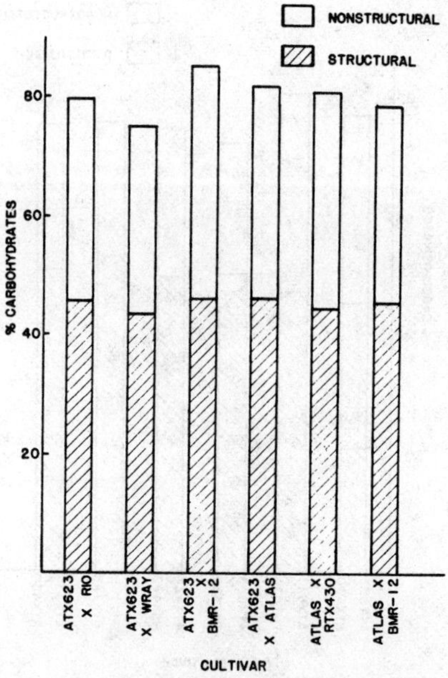

Figure 3. Total Percent Carbohydrates in High Energy Sorghum Cultivars.

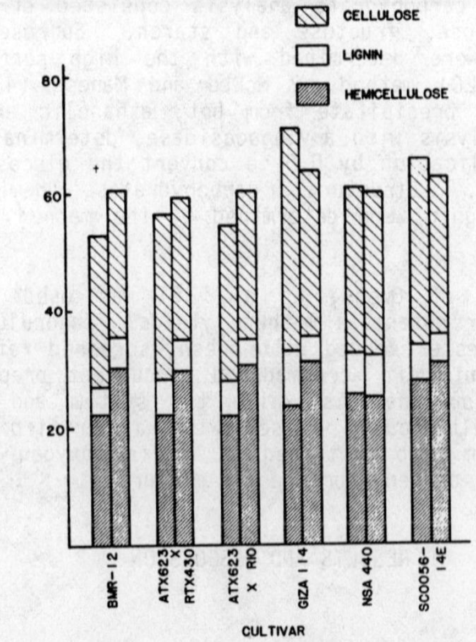

Figure 4. Structural Carbohydrate Content of Stems and Blades From Selected Sorghum Cultivars. (+ paired bars - stems and blades, respectively.)

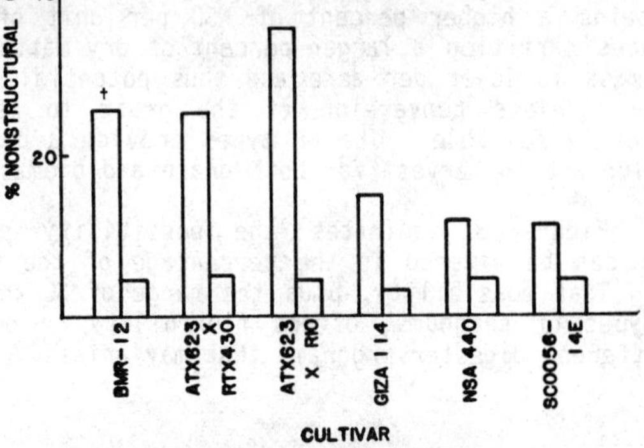

Figure 5. Nonstructural Carbohydrate Content of Stems and Blades From Selected Sorghum Cultivars. (+ paired bars - stems and blades, respectively.)

Nonstructural carbohydrate analysis consisted of determinations for sucrose, glucose, fructose and starch. Sucrose, glucose and fructose content were determined with the high performance liquid chromatography (HPLC) method by McBee and Maness (4). Starch was determined on the precipitate from hot, ethanolic extracts of the tissue after hydrolysis with amylogucosidase, determination of glucose content and multiplication by 0.9 to convert the glucose value to the starch equivalent. Structural carbohydrates (hemicellulose and cellulose) plus lignin were determined by the method of Goering and Van Soest (3).

The procedure of Owens, et al. (5) was used for evaluating sorghum biomass for potential methane yields. Innoculum was obtained from a 4-liter digester seeded with fresh, screened refuse from dairy cows. The treatment unit accommodated concurrent preparation of six sorghum samples for analysis while the system and Wheaton serum bottles were simultaneously gassed with a purified 80:20% N_2-CO_2 mixture. The system also contained an in-line oxygen trap to remove remaining traces of oxygen from the gas mixture.

RESULTS AND DISCUSSION

Carbohydrates

Quantities in Sorghum Groups. Carbohydrate analyses of stem biomass from several cultivars representative of the three groups (grain, sweet, HE) of sorghums are listed in Figures 1-3. Giza 114 is not classified as a sweet type but was included for comparative purposes. Total carbohydrates for all groups approached 80%. The range of NSC was 22.2-33.2, 31.5-46.9, and 31.3-39.3 %, respectively, for the grain, sweet and HE entries (excluding Giza 114). SC ranges for the same sequence of types, was 45.4-52.8, 38.1-48.8 and 43.5-46.1%, respectively. The sweet sorghum cultivars partitioned more carbohydrates into NSC forms, whereas the range for SC levels was greater for the grain types.

From the yield data presented in Table 2, it may be noted that sweet sorghums produced the most biomass per acre, primarily in the form of vegetative matter. As previously discussed, the tissue generally contains a higher percent of NSC per unit of dry weight. Since grain types partition a larger percent of dry matter to kernels, vegetative biomass is lower per acre and thus potential methane yield would be lower, unless conversion of the grain to methane should become economically feasible. The HE types provide a possibility for dual roles, which are to harvest for both grain and biomass.

Study of Figure 3 indicates the possibility of developing cultivars that can be altered in the percentage of the two groups of carbohydrates. That possibility, plus the range of SC and NSC present in parental types of sorghums, offers flexibility in development of biomass for different digester programs that may arise.

Hemicellulose. A limited number of reports are available on the role of cellulose and lignin in sorghum, but little is known about hemicellulose. As shown in Tables 2 and Figures 1-3, the quantity of hemicellulose ranged from 12-25 %. The method of Goering, et al. (3) removes hemicellulose but does not differentiate it into forms. Currently, investigations in our laboratory have included analysis of sorghum stems for the types of hemicellulose present. Methods as outlined by Ericson and Elbein (2) have been employed. Initial results have shown two types of hemicellulose (designated A and B) plus a pectin fraction to be present. Results of one series of extractions is shown in Table 3. The cultivars represent three types of tissues with Giza 114 containing the largest percent of residue after benzene extraction and Brandes, the lowest. When hemicellulose was fractionated, however, the A type was present in a larger amount with a ratio of approximately 3:1 for Giza 114 and 2:1 for the other two cultivars. Studies are being continued on hemicellulose to determine their composition and possible role in stem strength and hardness.

Tissue Fractions

Pith and Rind. The possibility of using pith for more rapid conversion to useable energy sources and incorporation of the rind into products such as pressed board has been investigated. Six of the sorghum cultivars, representing variable stalk composition from a physical aspect, were selected for study of SC localization. At grain maturity, panicles and blades were removed, and the stems were separated into pith and rind and processed for analysis. Results of the SC composition are presented in Tables 4 and 5. Total cell wall content was reduced significantly in the pith as compared to the rind. Lignin values of the pith were also significantly reduced which should be important for efficiency of reduction to energy.

Blades and Stems. Removal of total biomass from the soil would seriously deplete the organic matter content over a period of time. Leaving the blades on the field would partially reduce this problem from both a nutritional and structural aspect. Since a high percent of phenolic compounds within sorghum are located in the blades (6), their omission may also be beneficial due to possible inhibition of the anaerobic fermentation process.

Since blades represent part of the plant composition (Table 1), studies were conducted to determine carbohydrate value of this component. Panicles from the cultivars were removed at maturity and the plants were separated into blades and stems. Both SC and NSC for the blades and stems, from six cultivars representing the array of entries, are shown in Figures 4 and 5. The SC are higher in blades than stems of BMR-12, ATx623 x RTx430 (64) and ATx623 x Rio (6R) whereas the reverse was true for Giza 114, NSA 440 and SC005614. Observation of the cultivars growing in the field helps to explain these results. Both BMR-12 and 6R contain a softer stem than the other types and slightly more than 64. All three contain the highest levels of stem NSC, however. This indicates a depression for partitioning of carbohydrates to SC and results in lower stem levels. Since NSC in the blades are low, this results in a comparatively

Table 3. Structural Carbohydrate Fractionation After Benzene Extraction of Stems From Three Sorghum Cultivars.

Cultivar	After Extract	Lignin	Cellu-lose	Pectin	Hemi-cellulose	
	% O.D.W.†					
Giza 114	82.4	11.2	33.9	8.7	28.6	75A‡ 25B
ATx623 x Rio	64.6	9.0	23.5	10.5	21.6	65A 35B
Brandes	54.8	8.1	21.7	8.1	16.9	68.5A 31.5B

†O.D.W. - oven dry wt.; ‡Represents percent of hemicellulose fraction partitioned between hemicellulose A and B.

Table 4. Structural Carbohydrates in Pith of Selected Sorghum Cultivars.

Cultivar	Cell Wall	Hemicell-ulose	Lignin	Cellu-lose
	% O.D.W.†			
BMR-12	27.94‡	12.98	1.94	13.09
Giza 114	48.35	21.01	5.35	21.96
NSA 440	32.69	14.10	3.43	15.16
ATx378 x RTx7000 (37)	44.93	20.38	3.23	20.88
ATx623 x Rio (6R)	28.38	13.74	2.84	11.90
ATx623 x RTx430 (64)	39.41	17.80	3.29	18.16

†O.D.W. - oven dry wt.; ‡ Values are means of 3 reps.

Table 5. Structural Carbohydrates in Rind of Selected Sorghum Cultivars.

Cultivar	Cell Wall	Hemicell-ulose	Lignin	Cellu-lose
	% O.D.W.†			
BMR 12	56.60‡	22.63	5.65	27.99
Giza 114	76.51	21.66	13.22	41.66
NSA 440	55.96	22.53	7.15	24.64
ATx378 x RTx7000 (37)	70.39	25.75	11.13	33.59
ATx623 x Rio (6R)	62.73	21.57	9.35	31.68
ATx623 x RTx430 (64)	64.11	24.21	8.09	31.50

†O.D.W. - oven dry wt.; ‡ Values are means of 3 reps.

higher percent of SC in the blades than the stems. Giza 114 is taller in height and the outer portion of the stem is extremely hard, whereas NSA 440 and SC005614 are short in height but possess very rigid stems. It may be noted that SC for stems of Giza 114, NSA 440 and SC005614 are higher than in the other three cultivars and also contain lower levels of NSC. When the patterns for NSC accumulation are compared, levels are significantly lower in the blades than stems as might be expected based on the previous discussion.

Volatile Solids and Potential Methane

An important quality of biomass for conversion to energy is the quantity of volatile materials present. Sorghum tissue appears to possess a rather high percent of volatile solids (V.S.). Data has been obtained on a large number of cultivars (not included in this report) representing the numerous types and all have exceeded 90$^+$%. Cultivars listed in Table 6 are selections that are representative for sweet, grain and HE groups. As may be observed, the range of values is 92.39-97.34%. It appears that the sweet types may contain a slightly higher V.S. content, however the grain types represent a very favorable level.

Results of methane yields and volatile solids reduction are included in Table 7. The cultivars represent entries from the three groups of sorghums previously described. As may be noted, higher yields of methane do not appear to be related to higher levels of NSC or SC for the total digestion period. The standard cubic feet of methane per pound of V.S. ranged from 3.68 to 5.50 with V.S. reduction ranging between 69.91-81.61%. Additional and repeated studies are needed for thorough evaluation of sorghum for methane production but these values appear to represent a reasonable potential of sorghum biomass for methanogenesis.

Acknowledgement

Financial support for these studies by the Gas Research Institute, Chicago, Ill. is gratefully acknowledged.

Table 6. Volatile Solids Content of Selected Sorghum Cultivars.

Cultivar	% V.S.
Rio	96.14
MN 1500	97.34
ATx623 x Rio	93.25
Atlas x Rio	94.31
ATx623 x RTx430	93.17
ATx623 x Pickett	92.39

Table 7. Methane Yield, Per Cent Volatile Solids Reduction and Carbohydrate Composition of Sorghum Biomass.

Cultivar	Tot. Sugar Mg/g, O.D.W.	% Cell Wall	CH_4 SCF†/ lb V.S.	% Reduction, V.S.‡
ATx623 x RTx432	197.44	62.36	4.90	69.91
ATx623 x BMR-12	413.27	56.85	5.45	76.89
ATx623 x Atlas	347.38	52.10	5.50	69.97
ATx623 x Orange-1	179.59	64.78	4.59	75.40
Rio	369.26	52.40	4.11	79.63
BMR-12	319.55	50.04	4.07	80.20
M81E	428.29	47.25	3.68	81.61

†Standard Cubic Feet per lb. volatile solids.

‡Digestion period: 07/05-11/01/85

REFERENCES CITED

1. Anonymous, Texas Agricultural Facts., Texas Crop and Livestock Reporting Service., SM 24. 1-4 (1985) Nov 18.

2. Ericson, M.C. and Elbein, A.D., The Biochemistry of Plants, Vol. 3, 589-616. New York: Academic Press, 1980.

3. Goering, H.K. and Van Soest, P.J., "Fiber Forage Analysis"., Agric. Handbook No. 379, 1-20, USDA-ARS, (1970).

4. McBee, G.G. and Maness, N.O., "Determination of Sucrose, Glucose, and Fructose in Plant Tissue by High-Performance Liquid Chromatography," J. Chromatogr. 264, 474-478 (1983).

5. Owen, W.F., Stuckey, D.C., Healy, Jr., J.B., Young, L.Y. and McCarty, P.L. "Bioassay for Monitoring Biochemical Methane Potential and Anaerobic Toxicity," Water Research 13, 485-492 (1979).

6. Ring, A.S. Phenolic Compounds in Sorghums From Boot Stage to Post Harvest Maturity by Folin-Ciocalteau and High Pressure Liquid Chromatography Methods. M.S. Thesis, Texas A&M Univ., Col. Sta. Tx. 1984.

DISCUSSION

Berger — Is it practical to think about separating the rind from the pith?

McBee — The process that we used was from a borrowed Tilby machine. I think others could probably answer the question better than I, but I believe the machine was developed to the point where it could be adapted for the separation. If more use could be made of the rind for pressed wood or fuels, it certainly has practical possibilities for application.

APPLICATION OF CELL AND TISSUE CULTURE TECHNIQUES FOR THE GENETIC IMPROVEMENT OF SORGHUM, Sorghum bicolor (L.) Moench

Stephen Kresovich, Robert E. McGee, Lee W. Panella,
Ann A. Reilley, and Fred R. Miller

Texas Agricultural Experiment Station,
Weslaco, TX 78596 and College Station, TX 77843

ABSTRACT

During the past two decades, an integration of plant biochemistry, plant physiology, and molecular biology has allowed for the development of basic cell and tissue culture techniques, which have the potential of improving our understanding of plant biology. The objective of this paper is to present the "state of the art" with regard to the use of these techniques for the genetic improvement of sorghum. Much of what has been accomplished to date has been based in empiricism and our fundamental understanding of the biology of the sorghum plant is limited. Nonetheless, in the short-term these techniques will provide model systems for the study of developmental biology and stress physiology. As "problem-oriented" basic research becomes focused, cell and tissue culture techniques may be useful to improve our understanding of the genetics of sorghum, to elicit and control desired variability, and to serve as a selection tool in improvement programs.

APPLICATION OF CELL AND TISSUE CULTURE TECHNIQUES FOR THE GENETIC IMPROVEMENT OF SORGHUM, *Sorghum bicolor* (L.) Moench

INTRODUCTION

Having evolved over a great period of time and a broad spectrum of species, plant breeding is a science-based technology directed towards economic objectives (Simmonds, 1982). It is one component of agricultural production and, along with agronomic practices, agricultural chemistry, plant pathology, entomology, and agricultural engineering (mechanization, etc.), has been responsible for the past successes associated with increased yields of food, feed, fiber, and fuel at low costs to the consumer. Plant breeding has long since passed from being an "art", as some of the older plant breeders would lead one to believe, and is now heavily dependent on the disciplines of genetics and biometry.

During the past two decades, an integration of plant biochemistry, plant physiology, and molecular biology, has allowed for the development of basic techniques, i.e. cell and tissue culture, which have the potential of improving our understanding of plant biology. Early proponents of cell and tissue culture techniques expounded their advantages and suggested that these tools might supplant conventional plant breeding. It was suggested that their implementation might cause an agricultural "revolution" which greatly would exceed preceding "revolutions" initiated through the application of fertilizer and mechanization to the agricultural production system.

Fortunately, we all have mellowed a bit from the initial rhetoric associated with cell and tissue culture technologies. Plant scientists now realize that these techniques really cannot supplant plant breeding in all of its many-faceted forms, but rather serve as an adjunct to it. The objective of this paper is to present the "state of the art" with regard to the use of these techniques for the genetic improvement of sorghum, *Sorghum bicolor* (L.) Moench. Sorghum is recognized not only as a source of food and feed in marginal production areas, but as a potential "biomass" crop because of its productivity, efficiency, and adaptability. Within this paper, a framework is developed in which the potential applications of cell and tissue culture techniques to the genetic improvement of sorghum may be viewed and, also their current limitations may be identified.

BACKGROUND

The origin of cultivated sorghums is agreed generally to have occurred on the African continent; however, disagreement exists whether the origin has a monophyletic or a polyphyletic basis. Therefore, the systematics of sorghum are quite complex (De Wet *et al*., 1970; Doggett, 1970). *Sorghum bicolor* (L.) Moench includes a diverse collection including grass types and cultivated sorghums, all having a diploid chromosome complement of 2n=20. Harlan and de Wet (1972) have devised a working classification of cultivated sorghums that encompass five basic races including bicolor, guinea, caudatum, kafir, and durra that can be identified by mature spikelet and panicle type. In addition, ten hybrid races are recognized, of which kafir-caudatums are the source of most hybrid grain sorghums grown in the United States (Harlan, 1972). Other important sorghum types grown include sorghums (sorgos) for syrup and sugar, grass types for forages, and specialty types such as popping sorghum and broomcorn.

Although sorghums encompass a great diversity of types, the basic morphology and anatomy are consistent. This is particularly important when tissues for explantation to cell and tissue culture are considered. The seed germinates with the emergence of a coleoptile and coleorhiza. The shoot apex is forced to the soil surface by the expansion of cells in the mesocotyl region, while one to several primary roots may emerge from the embryonic axis. A node forms at the juncture of the mesocotyl and the shoot apex, from which secondary roots and adventitious buds develop. The primary roots and mesocotyl deteriorate at this early stage, and it is these secondary roots that continue to support the sorghum plant. Secondary shoots may form from the adventitious buds at this basal node. The stem, or culm, is comprised of nodes and internodes. Internode tissue may be thick or pithy, juicy or dry, and either insipid or sweet. Leaves arise from the nodal areas and consist of a sheath region and blade. The sheath base is primarily meristematic tissue which allows the sheath to elongate.

In cultivated sorghums, floral initiation may occur 30-75 days, or greater, following germination, with anthesis (flowering) occurring from 15-30 days later. The sorghum inflorescence is a panicle that matures from the apex towards the base. The rachis branches contain paired spikelets, one sessile and one pedicellate. The sessile spikelet generally contains one fertile and one sterile floret; sometimes both are fertile resulting in twin seededness. The pedicelled spikelet may contain a floret with functional anthers but usually no functional ovary. Anthesis generally occurs during the early morning when paired lodicules at the base of the spikelet swell forcing the glumes apart and exposing the stigmas and anthers (around sunrise). This flowering process takes about 10-30 minutes.

The caryopsis reaches physiological maturity about 30-35 days following pollination. The mature seed coat consists of a pericarp and fused testa. The pericarp color may appear as red, yellow or white. The testa may also contain pigmented compounds. These pigments are composed mainly of phenolic compounds and tannins (Rooney and Miller, 1982). (For greater detail see Artschwager, 1948; Artschwager and McGuire, 1949; Doggett, 1970; Paulson, 1969; Wall and Ross, 1970.)

GOALS IN BREEDING

Regardless of the crop involved, a breeding program must be defined with set goals. In addition, the program must have the means to achieve those goals. Naturally, the selection criteria vary with the crop, location, and goals. A plant breeder may try to address any one or any combination of these goals. The earlier in the selection process the breeder can identify the trait(s) desired, the more quickly progress can be made toward the desired goal. In some instances improvement of one trait results in the loss of another desired characteristic. For example, in sorghum, phenol accumulation gives the developing grain a bitter taste which makes it less susceptible to damage by birds (Brugger and Jaeger, 1982). The same phenol, however, makes the grain less palatable for humans and less digestible for animals (Butler, 1982; Hahn et al., 1984). The best way to prioritize these criteria is to identify the problems in the area where the crop will be grown.

House (1985) suggests a list of general criteria which can be used in selecting phenotypes of sorghum. These criteria include general adaptability, pest resistance, and high yield of quality grain. Sorghums produced specifically for energy production may be selected for total biomass yield, high carbohydrate content of the stalk, and low lignin content. With self-pollinated species such as sorghum, plant breeding methods are based on the principle that selfing, or backcrossing to a homozygous parent leads to homozygosity (Allard, 1960). A new cultivar with better agronomic traits is created and replaces older cultivars, or is used as a parent for hybrid seed production.

The method of plant breeding most often used in sorghum is the pedigree system. Ideally, a pedigree provides a name for each discrete type of plant which clearly shows its breeding history and its relationship to sister lines (House, 1985). In this system (Table 1) a cross is made, the F_1 hybrid grown, and seed for the F_2 generation is bulked. The F_2 generation exhibits maximum diversity and recombination potential. Each succeeding generation of self-pollination brings the plants closer to homozygosity. Individual selections are made from the F_2 population and the seed from selected panicles are grown out as a F_3 family. Selections within families are grown out this way until the F_5 or F_6 generation. The selections then should be homozygous enough to be treated as populations.

These populations are grown in small plots and evaluated for their agronomic performance and their combining ability as a hybrid parent. This testing continues from three to five years in a number of locations. By the F_{11} or F_{12} generations, final selections have been made, the seed is purified, and is ready to be increased and proceed through the certification process prior to release. The duration of some of the steps may vary, but generally a 10- to 16-year period is necessary to develop a new cultivar. (For greater detail see Allard, 1960; Briggs and Knowles, 1967; and Simmonds, 1979.) Because the time needed to develop a new cultivar is so long, any technique which could shorten this time, such as the potential tools provided by plant biotechnology, would be embraced.

The plant breeder is limited in achieving desired goals by the natural variation within the species and its close relatives, i.e.

sexually compatible species. The world collection of sorghum currently contains over 21,000 accessions (Mengusha and Prasada Rao, 1982). Most of these are tall, photoperiod-sensitive, tropical accessions unadapted to the temperate areas of this country. Nonetheless, an increasing amount of this exotic germplasm is finding its way into U.S. breeding programs. This is primarily due to the Sorghum Conversion Project, a joint effort between the USDA and the Texas Agricultural Experiment Station. In the program, knowledge of the maturity and height genes in sorghum is utilized, allowing tall, tropical cultivars to be converted to short, earlier types adapted to the temperate zone. This is done by backcrossing exotic cultivars to shorter, temperately-adapted types in Puerto Rico and selecting plants adapted to temperate areas in Texas (Stephens et al., 1967; Rosenow, 1967; Johnson et al., 1971; Quinby, 1974; Miller, 1979). Material from this project has introduced new sources of disease resistance (Fredericksen and Rosenow, 1979), insect resistance (Johnson and Teetes, 1979), and improved plant characteristics (Miller, 1979).

This infusion of new germplasm is manifest in the rapid gains in sorghum yields and grain quality seen during the last ten years. Clear goals in sorghum production and breeding have been set, and, using conventional techniques, plant breeders are moving to achieve them.

PROGRESS IN CELL AND TISSUE CULTURE

Inspired by the successes of cell and tissue culture work in sugar cane (Heinz et al., 1977) in the early 1960's, much attention has since been given to other forage grasses and cereals. Successes have been arguably limited when considering the variability in response of agronomically-important genotypes within a species. This is especially true with sorghum despite the effort that has been expended.

In sorghum, as in most monocots, only very young tissue temporally close to a meristematic state (seedling, immature inflorescence, and embryo) has proven responsive in culture. Difficulties are also due in part to phenol production by monocots, and sorghum may be an extreme case. Retarded growth in sorghum culture with black and purple pigments have been observed (Gamborg et al., 1977; Brettell et al., 1980; Davis and Kidd, 1980). Despite these problems, progress has been reported in the establishment of sorghum cell and tissue culture techniques.

Callus Culture

Although success with callus induction from aseptically germinated sorghum seedlings has been reported (Strogonov et al., 1968; Masteller and Holden, 1970; Brar et al., 1979a; Davis and Kidd, 1980; Smith et al., 1983), the work of Cure and Mott (1978), Mott and Cure (1978), and King et al. (1978) with other monocots suggests this growth may be aberrant meristematic tissue and not undifferentiated callus cells. This growth generally forms at the basal node of the sorghum seedling in response to 2,4-dichlorophenoxyacetic acid (2,4-D), an auxin analogue. Callus formation from immature (10-18 days post-pollination) sorghum embryos has also been reported (Gamborg et al., 1977; Brar et al., 1979a); however, the cultured tissues described as callus subsequently have been shown to originate from the primary explant by a folding of the scutellar tissue (Dunstan et al., 1978, 1979). Dunstan (1978) did show a recalcitrant callus did arise in some cultures. Callus also has been induced from cultured unemerged inflorescences (Brettell et al., 1980; Boyes and Vasil, 1984) and immature leaves (Wernicke and Brettell, 1980, 1982; Wernicke et al., 1982). Reports of callus induction from mature somatic tissue, however, are lacking; and this deficiency in the current technology precludes selection of inbred lines through callus culture. Without a routine method of callus induction from field-grown mature plants, cell and tissue culture systems are not useful in early generation selection systems.

Regeneration and Somatic Embryogenesis

For the new gene-splicing and genetic manipulation technology to be applicable in a crop improvement program, it is necessary to have an efficient method of plant regeneration from single cells and/or callus. Reports of plant regeneration from sorghum tissue cultures are limited, especially when the question of aberrant meristematic growth is considered. Masteller and Holden (1970) reported limited plant regeneration from primary callus explants derived from the basal nodal tissue of sorghum cultivars 'Norghum' and 'North Dakota 104'. Gamborg et al. (1977) observed morphogenesis and plant regeneration from callus of immature embryos (12-18 days post-pollination) of the sorghum hybrid, Northrup King 'X4004A'. Prolific shoot formation was reported by Thomas et al. (1977) from immature embryos (10-30 days post-pollination) of the sorghum hybrid 'Oro'. Brar et al. (1979a) reported embryogenesis from immature embryos (10-14 days old) of sorghum hybrids 'GPR-168', 'NK-300', and 'X4004A'. It also must be noted that these efforts employed "immature" embryos collected 10-30 days post-pollination from self-pollinated sorghum hybrids as their explant sources. The description of 10-30 day-old embryos as immature (Brar et al., 1979) is inconsistent with the findings of Artschwager and McGuire (1949) and Paulson (1969) which suggest that embryos of sorghum may reach maturity by 12-25 days. Also, a zygotic embryo from a selfed hybrid or heterozygous line will not necessarily have the same genetic constituent as the parent. While documenting the *in vitro* responsiveness of this tissue source, these investigators again have failed to realize the potential breeding problems associated with the use of immature embryos in segregating populations of germplasm, i.e. F_1 hybrids and selections from early generations, F_2-F_6. The responsiveness in culture becomes confounded due to genetic recombination and segregation of the immature embryos. Thomas et al. (1977) also reported a similar regeneration response using mature embryos excised from two-year old seed of the hybrid 'Oro', but this is the only report referencing this important explant source. Bhaskaran et al. (1983) have reported regeneration of 'IS3620C', (a margaritiferum), from long-term (10-month old) callus cultures. Although these reports are encouraging, regeneration from callus of sorghum is generally of low frequency, limited to a few genotypes, and with great variability in culture response.

Plant regeneration via somatic embryogenesis has also been reported in sorghum. The work of Thomas _et al_. (1977) was further studied by Dunstan _et al_. (1978), who examined the development of these embryo-like structures. Although some embryos were of single-cell origin, no evidence was found that they arose from proliferating callus cells. Instead the shoots and embryos formed initially from the primary explant (the zygotic embryo). Brettell _et al_. (1980) described embryogenesis from cultured immature inflorescences of several sorghum cultivars and hybrids including 'CK-60', 'Regular Hegari', 'Plainsman', 'WAC692', 'G522DR', 'FS302', and 'G83F'. All entries except 'Regular Hegari' produced plants, although the hybrid 'G522DR' (Funk Seed International) was superior in culture response. Wernicke and Brettell (1980, 1982) and Wernicke _et al_. (1982) reported somatic embryogenesis from immature leaves of 'G522DR'. This group also noted the further spatially and temporally, i.e. more mature, leaf tissue was from the apical meristem, the less responsive it became. This finding suggests that competence _in vitro_ (the ability to regenerate) is correlated with continued meristematic activity. These reports indicate that somatic embryogenesis from the primary explant (zygotic embryo, inflorescence, or immature leaf) is possible in a number of sorghum genotypes, although the technique is not yet routine.

Protoplast Culture

Relatively little research has been done on sorghum protoplast isolation, culture, and regeneration. This is an important area because many genetic modification techniques (micro-injection, protoplast fusion, etc.) rely on an efficient protoplast system. The earliest report of sorghum protoplast cultures was by Brar _et al_. (1979b). A cell suspension of the sorghum cultivar 'GPR-168' was established from immature embryos, and protoplasts were isolated from three- to four-day old suspension cultures. Fusion was reported between sorghum protoplasts and leaf protoplasts of maize (_Zea mays_ L.). Fusion products underwent several divisions and formed colonies of 8 to 10 cells. Chourey and Sharpe (1985) described cell suspension and protoplast isolation from a sorghum hybrid and two cultivars ('NK-300', 'M35-1', and 'SC420'). Protoplasts underwent sustained cell division and callus formation. In some instances, protoplast-derived calli have been put back into suspension giving rise to "protoclone cell suspension cultures". These results indicate that protoplast isolation and subsequent callus formation is possible in sorghum. No observations of plant regeneration from protoplast-derived cultures have been reported.

STATUS AND POTENTIAL APPLICATIONS

Using plant cell and tissue culture systems as model systems to study and affect plant behavior has been a primary goal of biotechnology. Applications include understanding cellular and whole-plant mechanisms involved in stress physiology, selection and rapid screening techniques at cellular and tissue levels, increasing, identifying and controlling variation, and as adjuncts to genetic studies. A fundamental understanding and control of the in vitro system is necessary to undertake such studies. A basic understanding of sorghum culture techniques is lacking; however, several applications have been attempted and potential exists in these and other areas.

Cell and Tissue Culture as a Model System

Many processes which ultimately affect whole-plant characters occur at the cellular level. By studying these processes using cell and tissue culture systems, free from complicated whole-plant interactions, researchers can gain knowledge into the physiology and biochemistry of the cell. The work of Bhaskaran et al. (1985) is an example of this type of approach in which proline accumulation in callus cultures of sorghum was measured in response to water stress.

Research in our laboratory indicating a cellular basis for salinity tolerance in sugar cane (Kresovich et al., in press; McGee et al., in press) has encouraged us to begin examination of cell suspension cultures of sorghum as a model for studying the effects of salinity stress at the cellular level. By exploiting cell and tissue cultures of sorghum in this manner, it is hoped that a better understanding of stress physiology and biochemistry will evolve. This model may prove useful in understanding mechanisms such as carbon partitioning and utilization, mineral uptake and compartmentation, membrane stability, etc. This approach, when integrated with an improved understanding of the interactions between tissue types, media, pre-conditioning effects, and plant regeneration, offers the advantage of better being able to isolate and propagate variants with these particular characters.

Although there are several well-established model systems used in other studies (tobacco, carrot, tomato, etc.), it is often difficult to extrapolate results across species. Therefore, the use of sorghum cell and tissue culture systems to verify and study these phenomena is important.

Selection Techniques

The phenomenon of callus cultures paralleling whole plant response to salts (Orton, 1980) and heavy metals (Meredith, 1978) in the growth media has been observed in several species. Relatively little work has been done with sorghum regarding this application. Smith et al. (1983) tested callus from four sorghum genotypes ('BTx3197', 'BTx623', 'BTx399', and 'IS3620C') on media containing 0, 100, 200, and 400uM aluminum-EDTA. Differences were reported in growth among the cultivars in response to aluminum. Plants were regenerated from 'IS3620C' after several passages on selection media, but no field test results are yet available.

Bhaskaran et al. (1983) reported plant regeneration from 'IS36-20C'-derived callus grown on medium containing a high concentration of sodium chloride. This callus, obtained by prolonged subculture on a selection medium, was better able to grow on a medium containing sodium chloride than donor cells. Regenerated plants grown to maturity exhibited low seed set hindering any attempts at progeny analysis.

Smith et al. (1985) tested callus from 10 sorghum genotypes on a polyethylene glycol (PEG) selection medium. Indications were that drought tolerance and performance of cultured tissues might be related. Further work suggests extreme variability in callus growth makes discerning differences difficult (Reilley, 1986).

The ontogeny of tissue cultures becomes critical when using such a selection system. Dunstan et al. (1978, 1979) and Wernicke et al. (1982), based on their conclusions that regenerated plantlets arose initially via somatic embryogenesis, albeit from the primary explant only, suggest that variation may be more directed when the selection can be made on single cells. This again points to the need for a culture system containing undifferentiated cells that can be readily induced to differentiate via somatic embryogenesis into new plants. Because relatively little work has been done with sorghum selection for salt, heavy metal, and drought tolerance, it is difficult to predict the value of this potential application of tissue culture to traditional breeding programs. In order for these new techniques to be useful, a strong correlation between in vitro and field responses must be established. In vitro selection must become simpler, more reliable, and more cost-effective to be useful in a breeding program.

Somaclonal Variation

Somaclonal variation, i.e. the variability expressed in plant phenotype derived from cell and tissue culture, does occur in many, if not all, species. This variation, depending upon how it is expressed, is similar to that caused by a chemical mutagen, and is generally detrimental. Smith et al. (1982), studied variability in sorghum phenotypes ('IS3620C') following regeneration. Selfed seeds from sorghum plants regenerated from tissue culture were collected, and the progeny of tissue culture-derived plants were compared to the original sorghum line in terms of drought tolerance characters. In two characters, wax load and heat tolerance, the progeny of tissue culture-derived plants was significantly less desirable than the parent population. Desiccation damage, however, was less for the tissue culture-derived progeny. In a related study, Vaughn (1983) examined the F_1, F_2, and backcross F_1 progeny of a tissue culture-derived line of 'IS3620C'. No significant differences in chlorophyll content or chromosome number, pairing, homology, or configuration were observed between the tissue culture-derived line and the parent. Observed differences in seed weight could not be separated from inherent genetic variation and inbreeding effects. In order to determine whether cell and tissue culture techniques are viable tools for creating variability in sorghum, it will be necessary to use established lines and evaluate them in terms of agronomic characteristics. It is important to point out that variability may be utilized more easily in traditional sorghum breeding programs.

Recognizing the status of application of culture systems to sorghum improvement, the potential exists for genetic improvement in several other areas. Shoot tip or meristem culture offers the potential of isolating and maintaining desirable disease-free germplasm. As an adjunct to these systems, cultures may be maintained at some initial level of development for either long-term storage or to facilitate movement of disease-free germplasm.

Although attempts at using sorghum in wide crosses has been successful in closely related genera such as Saccharum as well as allied sorghum species, other efforts have not been successful. For example, attempts at pollination of maize by sorghum have not produced viable embryos (Helsop-Harrison et al., 1985). Somatic hybridization via protoplast fusion (Brar et al., 1979) may be a means of overcoming such pre-zygotic obstacles, although a regeneration systems is not yet in place for such products. Further, post-zygotic problems, e.g. embryo-endosperm incompatibilities, may be overcome using embryo rescue techniques.

Tissue and cell culture systems potentially offer an adjunct to genetic studies such as the identification and isolation of monosomics or other valuable genetic markers. Additionally, anther culture and regeneration systems should result in the production of haploids. This could prove useful not only in genetic studies, but also in rapid development of homozygous germplasm for sorghum breeders. With a better understanding of protoplast culture systems, novel ideas such as gene transfer and somatic hybridization may facilitate the work of plant breeders.

Competition with current sorghum improvement technologies, i.e. traditional breeding programs, however, may be hard to overcome. This statement is made in view of the fact that, until the recent work of Chourey and Sharp (1985), extended callus and suspension cultures of sorghum have been suspect. Furthermore, plant regeneration has been limited to a restricted number of genotypes, one ('IS3620C') of which has no value in current sorghum improvement programs. Within a given genotype, the frequencies of regeneration are low and, at best, variable.

SUMMARY

Much of what has been accomplished utilizing cell and tissue culture techniques for the genetic improvement of sorghum has been based in empiricism and much of what has been reported may have been misinterpreted or misunderstood. Our fundamental understanding of the biology of the sorghum plant is limited, both at the whole-plant and cellular level. At the present time, only a handful of groups worldwide are actively attempting to integrate cell and tissue culture techniques with the conventional breeding of sorghum. With this limited amount of effort, progress will be slow. Ultimately, the tools that a plant breeder uses are based on (1) the familiarity of the breeder with the tool, (2) the ability of the tool to affect the desired change or recombination of desirable characters, (3) the rate or frequency at which the desired change or recombination is obtained, (4) the cost associ-

TABLE 1: PEDIGREE BREEDING SYSTEM

F_1	Cross between parents with desirable traits all seeds are genetically uniform
F_2	Maximum Variation, Heterosis, and Recombination single plants are selected
F_3-F_6	Increasing Homozygosity and Uniformity single plants within families are selected
F_5-F_{10}	Uniform Lines Effectively Homozygous yield testing and testing for combining ability as a hybrid parent
F_7-F_{15}	Homozygous Cultivars line purification, further testing and seed increase for certification and release

ated with the use of the tool, in terms of money, personnel, and time, and (5) the resources available to the breeder. Applications of cell and tissue culture techniques will become common when they are advantageous for the plant breeder to employ. Therefore, the "revolution" involving plant biotechnology and the genetic improvement of sorghum may be long in coming.

However, the rate of "evolution" in the genetic improvement of sorghum may increase as a result of the application of cell and tissue culture techniques. In the short-term, cell and tissue culture techniques open avenues for an improved understanding of the cellular biology of sorghum. These techniques will prove useful for the study of developmental biology and stress physiology. Plant breeders and plant biotechnologists must work cooperatively to find niches for employment of these tools. This synergism should lead to "problem-oriented" basic research from which tangible results are envisioned. The potential application of these tools and their subsequent pay-offs may be great in the longer-term; however, specific goals must be established and efforts directed at those goals.

ACKNOWLEDGMENTS

The authors gratefully acknowledge support of our research from the U. S. Department of Energy through Contract Number 19X-27452C with Martin Marietta Energy Systems, Inc. and from the Gas Research Institute through Contract Number 5083-223-0766. Funds supplied through the Texas Agricultural Experiment Station Expanded Research Program were integral in the establishment of the Plant Cell Culture Laboratory at Weslaco. The authors also thank Janet N. Reed and Robert E. Reed for their insight and critical suggestions.

REFERENCES

Allard, R. W. 1960. Principles of Plant Breeding. John Wiley and Sons, Inc., New York.

Artschwager, E. 1948. Anatomy and morphology of the vegetative organs of *Sorghum vulgare*. U. S. Dept. Agr. Tech. Bull. 957.

Artschwager, E., and R. C. McGuire. 1949. Cytology of reproduction in *Sorghum vulgare*. J. Ag. Res. 78:659-673.

Bhaskaran, S., R. H. Smith, and K. F. Schertz. 1983. Sodium chloride-tolerant callus of *Sorghum bicolor* (L.) Moench. Z. Pflanzenphysiol. Bd. 112:459-463.

Bhaskaran, S., R. H. Smith, and R. J. Newton. 1985. Physiological changes in cultured sorghum cells in response to induced water stress I. free proline. Plant. Physiol. 79:266-269.

Boyes, C. J., and I. K. Vasil. 1984. Plant regeneration by somatic embryogenesis from cultured young inflorescences of *Sorghum arundinaceum* (Desv.) Stapf. var. sudanense (sudan grass). Plant Sci. Lett. 35:153-157.

Brar, D. S., S. Rambold, O. L. Gamborg, and F. Constabel. 1979a. Tissue culture of corn and *Sorghum*. Z. Pflanzenphysiol. Bd. 95:377-388.

Brar, D. S., S. Rambold, O. L. Gamborg, and F. Constabel. 1979b. Isolation, fusion and culture of *Sorghum* and corn protoplasts. Z. Pflanzenphysiol. Bd. 96:269-275.

Brettell, R. I. S., W. Wernicke, and E. Thomas. 1980. Embryogenesis from cultured immature inflorescenses of *Sorghum bicolor*. Protoplasma 104:141-148.

Briggs, F. N., and P. F. Knowles. 1967. Introduction to Plant Breeding. Reinhold Publishing, New York.

Bruggers, R. L., and M. M. Jaeger. 1982. Bird pests and crop protection strategies for cereals of the semi-arid African tropics. In: Sorghum in the Eighties: Proceedings of the International Symposium on Sorghum. ICRISAT, Patancheru, A. P., India.

Butler, L. G. 1982. Polyphenols and their effects on sorghum quality. In: Proceedings of the International Symposium of Sorghum Grain Quality. ICRISAT, Patancheru, A. P., India.

Chourey, P. S., and D. S. Sharpe. 1985. Callus formation from protoplasts of *Sorghum* cell suspension cultures. Plant Sci. 39:171-175.

Cure, W. W., and R. L. Mott. 1978. A comparative anatomical study of organogenesis in cultured tissues of maize, wheat, and oats. Physiol. Plant. 42:91-96.

Davis, M. E., and G. H. Kidd. 1980. Optimization of sorghum primary callus growth. Z. Pflanzenphysiol. Bd. 98:79-82.

De Wet, J. M. J., J. R. Harlan, and E. G. Price. 1970. Origin of variability in the Spontanea complex of Sorghum bicolor. Am. J. Bot. 57:704.-707.

Doggett, H. 1970. Sorghum. Longmans, London.

Dunstan, D. I., K. C. Short, and E. Thomas. 1978. The anatomy of secondary morphogenesis in cultured scutellum tissues of Sorghum bicolor. Protoplasma 97:251-260.

Dunstan, D. I., K. C. Short, H. Dhaliwal, and E. Thomas. 1979. Further studies on plantlet production from cultured tissues of Sorghum bicolor. Protoplasma 101:355-361.

Frederiksen, R. A., and D. T. Rosenow. 1979. Breeding for disease resistance in sorghum. In: Biology and Breeding for Resistance to Arthropods and Pathogen in Agricultural Plants. Texas A&M University, University of California, and U. S. Agency for International Development. College Station.

Gamborg, O. L., J. P. Shyluk, D. S. Brar, and F. Constabel. 1977. Morphogenesis and plant regeneration from callus of immature embryos of sorghum. Plant. Sci. Lett. 10:67-74.

Hahn, D. H., L. W. Rooney, and C. F. Earp. 1984. Tannins and phenols of sorghum. Cereal Foods World 29:777-779.

Harlan, J. R. 1972. Genetic resources in sorghum. In: Sorghum in the Seventies. Oxford & TBH, New Delhi.

Harlan, J. R., and J. M. J. de Wet. 1972. A simplified classification of cultivated sorghum. Crop Sci. 12:172-176.

Heinz, D. H., M. Krishnamurthi, L. G. Nickell, and A. Maretzki. 1977. Cell, tissue, and organ culture in sugarcane improvement. In: Applied and Fundamental Aspects of Plant Cell, Tissue, and Organ Culture. Springer-Verlag, Berlin.

House, L. 1985. A Guide to Sorghum Breeding. (2nd edition), ICRISAT, Patancheru, A. P., India.

Johnson, J. W., D. T. Rosenow, F. R. Miller, and K. F. Schertz. 1971. Sorghum breeding and improvement. Texas Agric. Expt. Stn. Prog. Rpt. 2942, College Station.

Johnson, J. W., and G. L. Teetes. 1979. In: Biology and Breeding for Resistance to Arthropods and Pathogens in Agricultural Plants. Texas A&M University, University of California, and U. S. Agency for International Development, College Station.

King, P. J., I. Potrykus, and E. Thomas. 1978. In vitro genetics of cereals: problems and perspectives. Physiol. Veg. 16:381-399.

Kresovich, S., S. J. Wadsworth, J. J. Molina, and R. E. McGee. In press. Salinity tolerance of interspecific hybrids of Saccharum spp.: ranges and mechanisms of adaptation exhibited in cellular suspensions of commercial cultivars. Int. Soc. Sugar Cane Tech.

Mascarenhas, A. F., M. Pathak, R. R. Hendre, and V. Jagannathan. 1975. Tissue cultures of maize, wheat, rice, and sorghum. I. Initiation of viable callus and root cultures. Ind. J. Exp. Biol. 13:103-107.

Masteller, V. J., and D. J. Holden. 1970. The growth of and organ formation from callus tissue of sorghum. Plant Physiol. 45:362-364.

McGee, R. E., S. J. Wadsworth, and S. Kresovich. In press. The applications of cell and tissue culture methods as tools in the study of plant-water relations. Proc. Inter-American Sugar Cane Society.

Mengesha, M. H., and K. E. Prasada Rao. 1982. Current situation and future of sorghum germplasm. In: Sorghum in the Eighties: Proceedings of the International Symposium on Sorghum. Patancheru, A. P., India.

Meredith, C. P. 1978. Selection and characterization of aluminum-resistant variants from tomato cell cultures. Plant Sci. Lett. 12:2534.

Miller F. R. 1979. The breeding of sorghum. In: Biology and Breeding for Resistance for Arthropods and Pathogens in Agricultural Plants. Texas A&M University, University of California, and U. S. Agency for International Development, College Station.

Mott, R. L., and W. W. Cure. 1978. Anatomy of maize tissue cultures. Physiol. Plant. 42:139-145.

Orton, T. J. 1980. Comparison of salt tolerance between Hordeum vulgare and Hordeum jubatum in whole plants and callus cultures. Z. Pflanzenphysiol. Bd. 98:105-118.

Paulson, I. W. 1969. Embryogeny and caryopsis development of Sorghum bicolor (L.) Moench. Crop Sci. 9:97-102.

Quinby, J. R. 1974. Sorghum Improvement and the Genetics of Growth. Texas A&M University Press, College Station.

Reilley, A. A. 1985. In vitro determination of nuclear and cytoplasmic components of drought tolerance in Sorghum bicolor (L.) Moench. M.S. Thesis, Texas A&M University, College Station.

Rooney, L. W., and F. R. Miller. 1982. Variation in the structure and kernel characteristics of sorghum. In: Sorghum in the Eighties: Proceedings of the International Symposium on Sorghum. Patancheru, A. P., India.

Rosenow, D. T. 1967. Conversion of alien sorghums. In: 5th Biennial Grain Sorghum Research and Utilization Conference. Grain Sorghum Prod. Assn., Lubbock.

Simmonds, N. W. 1979. Principles of Crop Improvement. Longmans, London.

Smith, R. H., S. Bhaskaran, F. R. Miller. 1985. Screening for drought tolerance in sorghum using cell culture. In Vitro Cell. and Dev. Biol. 21:541-545.

Smith, R. H., S. Bhaskaran, and K. F. Schertz. 1982. Variation in drought tolerance characteristics from tissue culture-derived sorghum. In: Plant Tissue Culture: 1982. Jap. Assn. for Plant Tissue Culture, Tokyo.

Smith, R. H., S. Bhaskaran, and K. F. Schertz. 1983. Sorghum plant regeneration from aluminum selection media. Plant Cell Reports 2:129132.

Stephens, J. C., F. R. Miller, and D. T. Rosenow. 1967. Conversion of alien sorghums to early combine genotypes. Crop Sci. 7:396.

Strogonov, B. P., E. I. Komizerko, and R. G. Butenko. 1968. Culturing of isolated glasswort, sorghum, sweet clover, and cabbage tissues for a comparative study of their salt resistances. Sov. Plant Physiol. 15:173-177.

Thomas, E., P. J. King, and I. Potrykus. 1977. Shoot and embryo-like structure formation from cultured tissues of Sorghum bicolor. Naturwiss. 64:587.

Vaughn, S. F. 1983. Characteristics of a tissue culture-derived line of Sorghum bicolor (L.) Moench: genetic variation during regeneration in vitro. M.S. Thesis, Texas A&M University, College Station.

Wall, J. S., and W. M. Ross. 1970. Sorghum Production and Utilization. Avi Publishing, Westport.

Wernicke, W., and R. I. S. Brettell. 1980. Somatic embryogenesis from Sorghum bicolor leaves. Nature 287:138-139.

Wernicke, W., and R. I. S. Brettell. 1982. Morphogenesis from cultured leaf tissue of Sorghum bicolor - culture initiation. Protoplasma 111:19-27.

Wernicke, W., I. Potrykus, and E. Thomas. 1982. Morphogenesis from cultured leaf tissue of Sorghum bicolor - the morphogenetic pathways. Protoplasma 111:53-62.

DISCUSSION

Berger: How well understood is the sorghum genome?

Kresovich: We know bits and pieces about it. We know chromosome numbers, and we know where certain genes for certain characters are located on the chromosomes. I think there is one program in Florida and one program funded by USDA in Texas, that are doing some work on sorghum genome analysis. It's not very far along.

Berger: So you have plenty of germplasm, plenty of variability, and not too much understanding of the genome?

Kresovich: Exactly. It's a lot of hit and miss.

________: If I understand you, you're saying that any work on tissue culture, in particular embryogenesis, is just sort of a hit or miss work, until you do have a very defined geno-type. Is that correct?

Kresovich: To a certain extent it is. We're basically scratching the surface and a lot of the mechanisms that elicit the embryogenesis process aren't well defined at this time. One of the things we're doing in our lab is to screen cultures for the presence or absence of embryogenic proteins. We don't know what those proteins are, but we're comparing profiles of cultures that regenerate versus cultures that don't regenerate, as a first step. We don't know what the enzymes are, and we don't really know if we're going to find differences in sorghum. They have been reported in carrot and tobacco, but findings in other crops are limited.

THE POTENTIAL FOR GROWING TERRESTRIAL ENERGY CROPS IN THE SOUTHWEST UNITED STATES

Martha S. Salk
Oak Ridge National Laboratory
Oak Ridge, Tennessee 37831

A. Gray Folger
Solar Energy Research Institute
Golden, Colorado 80401
(Current address: 5 South Pearl Street
Denver, Colorado 80209)

ABSTRACT

The southwest United States - an area of diverse climate, topography, terrain, soils, and vegetation - was assessed to determine which parts of the region would be suitable for growing terrestrial energy crops and then to characterize those areas. Minimal conditions for biomass energy crop production were considered to be an average annual precipitation of at least 30 cm (12 in.) and a frost-free period of at least 120 days per year. Approximately 7.7×10^7 ha (19.1×10^7 acres), just over 40% of the study region, met both criteria.

The data compiled for this study can be used both to provide a set of species for screening as biomass energy crops for the Southwest and to identify areas in this region where previously selected energy crops would grow best. Species found to perform best in a given set of environmental conditions can be identified and exploited rather than attempting to alter the environment to suit the needs of preselected plant species.

Species with some or all of the following characteristics would be the most likely biomass energy crops in the Southwest: sprout from roots or stumps, have C_4 or crossulacean acid metabolism photosynthesis, fixed nitrogen, are perennial, withstand moisture limitations and other climatic extremes, have multiple uses, and produce valuable by-products. The most common ecosystems in the area with minimal conditions for growing terrestrial energy crops (i.e., grama-buffalo grass, juniper-pinyon woodland, mesquite-buffalo grass, grama tobosa shrubsteppe, and trans-Pecos shrub savanna) should be surveyed to identify the native species with the highest productivity. Those species would be the best candidates for biomass energy crops.

Since productivity of most ecosystems in the Southwest is low, multipurpose projects in which energy crop production is one aspect should be considered. Site-specific research will be necessary to determine if biomass energy crop production should be encouraged in this region.

THE POTENTIAL FOR GROWING TERRESTRIAL ENERGY CROPS IN THE SOUTHWEST UNITED STATES

INTRODUCTION

Background

The southwestern United States has not been systematically investigated for bioenergy production on a region-wide basis in terms of appropriate plant species, site requirements, management practices, and research and development opportunities. This study attempts to characterize the potential of the Southwest as a region for terrestrial energy crop production.

Use and management of arid and semiarid ecosystems by man must recognize the limitations of the environment. Overuse of vegetation can result in increased susceptibility of the soil to wind and water erosion. Prudent stewardship involves avoiding excessive risks by, for example, managing in accordance with the minimum annual precipitation. Since plant growth depends on the amount of soil moisture that is available, flexible management technologies are required that are capable of both maximizing production during particularly wet years and stabilizing production under drought conditions (18).

Energy crop production is a relatively undeveloped technology with somewhat different objectives and possibilities than conventional agriculture. Innovative, sustainable, and cost-effective production systems must be developed if energy crop production is to be feasible under conditions in which most conventional agricultural crop production is not viable.

The Southwest United States

There is no generally accepted definition of the region called the Southwest United States. For this investigation the Southwest is defined to include all of the states of Colorado, Utah, Nevada, New Mexico, and Arizona, and the portions of California, Texas, and Oklahoma between the 100th meridian, which bounds the east side of the Texas Panhandle, and the 120th meridian, which forms the western border of Nevada (Figures 1-4). The study area encompasses a great variety of natural environmental features and conditions in terms of elevation, physiography, topography, drainage, temperature, precipitation, evaporation, soil type, and vegetation.

Although the Southwest is generally arid, precipitation in the region varies greatly, not only within and among seasons and years but also among locations separated by only short distances (12). Average annual precipitation varies from less than 4 in. (10 cm)/year to over 48 in. (122 cm)/year, with most of the region receiving the lesser amounts. The most arid parts of the area may occasionally experience precipitation in a single event that approaches or exceeds the mean annual value. In contrast, months and even years may pass without significant precipitation (12).

The percent of natural vegetation remaining in each state in this Southwest region varies from a high of 96% for Nevada to a low of 56% for Oklahoma (13), much higher, in general, than in other sections of the country. Plant communities in the region include chaparral, cold desert and semidesert, hot desert, bushland, ecotone woodland, montane forest, alpine, and northern and southern temperate grassland (24). The various plant communities in the region exhibit pronounced patterns of zonation along elevational gradients as a result of different regimes of moisture and temperature related to altitude (20). In addition to altitudinal differences, different assemblages of plant species may be apparent at different times on a particular site, as well as across the landscape, because of differing plant responses to changing conditions (11).

Seasonal variations of precipitation in the Southwest influence the adaptations seen in plants for exploiting the available moisture (21). Shrubs and grasses are the prominent life forms in many parts of the study region. These species are primarily perennials with most of their structure underground, such extensive root systems being essential for vigorous growth (9, 20).

Land uses historically have resulted in drastic changes in the patterns of biotic communities in the Southwest (30). Original plant productivity has been reduced over large areas by grazing abuses, brush invasion, droughts, and attempts to cultivate nonarable land (12). Climatic fluctuations that produce wind and drought cannot be prevented, but short-sighted grazing and agricultural practices are preventable (20). Long-term productive cultivation is possible in some areas of the Southwest if farming practices are carefully matched to soil and climatic conditions. Since most of the severe damage to croplands and rangelands occurs during droughts, farming and grazing systems must be highly flexible to adjust to yearly fluctuations in weather and plant growth. Each site has different characteristics and must be managed accordingly (12). The challenge for the manager is to find the proper balance between the biological realities of the site and the demand for food and other products by the rapidly growing world population.

METHODS

The objectives of this investigation were first to characterize and map regional data on pertinent climate (i.e., freeze-free period, precipitation, and evaporation) and land resource (i.e., soils, vegetation, slope, land use/cover, and land ownership) parameters, and secondly, to overlay and composite the mapped natural resource data to identify those areas in the Southwest possessing the most favorable conditions for biomass production. Such a reduction of the study region into suitable areas provides a preliminary basis for estimating the potential resource and identifies the parts of the Southwest where field experiments and pilot studies could most profitably be conducted.

This study used a cell-based geographic information system (28). At the map scale of 1:2,500,000, each grid cell represents an area of approximately 12,455 acres (5041 ha). The system can be used to create new maps by arithmetic or logical compositing operations. The procedure used for stratification and characterization of the region for terrestrial energy crops involved both compositing and masking (23).

Computer map data were available for all climate parameters and for slope, land use/cover, and ownership of the land parameters (19). A Soil Conservation Service (SCS) map for soils (29) and Küchler's map of vegetation (14, 15) provided additional data needs.

The analysis was based on an ecological perspective. Instead of estimating the potential of a few select species, the naturally occurring vegetation and its relationships with the environment were broadly examined to evaluate the prospects for energy production. Energy costs associated with irrigation will continue to be a major factor in determining the feasibility of biomass energy crops in the Southwest (7). Thus, for this assessment, irrigation was considered an unlikely procedure except during initial planting and establishment of a biomass energy crop.

RESULTS AND DISCUSSION

The product maps illustrate general trends of climate, land use, and vegetation. However, caution must be taken in interpreting the maps because registration error alone may account for precision within only about 5 miles on the land surface. Additional maps are included in Salk and Folger (23).

Area Reduction

Area reduction was accomplished by the logical compositing of precipitation and freeze-free period maps of the Southwest (23). Areas with less than 12 in. (30 cm)/year of precipitation and/or 120 freeze-free days per year were eliminated from consideration on the basis that these criteria identify minimal moisture and growing season conditions for producing energy crops. In the study region, 58.6% of the area has more than 12 in. (30 cm)/year of precipitation, and 75.4% of the area has a freeze-free period greater than 120 days per year. The area that meets both conditions (Figure 1), approximately 40% of the study region or 7.725 x 10^7 ha (1.908 x 10^8 acres), is called the "reduced" or "acceptable" area.

Not all of the land in the reduced area would be available for growing biomass energy crops. Parks, wilderness areas, and wildlife refuges (land ownership category C, Table 1) cover 2.2% of the area. Land use/cover areas (Table 2) that would not be available for growing crops include the "other" category (urban areas, salt flats, oil and gas fields, and snow and ice), irrigated cropland, and a part of the "no data" category. Thus, at least 15% of the land in the reduced area is unavailable for growing biomass energy crops, leaving a maximum of 6.57 x 10^7 ha (1.62 x 10^8 acres). The amount "available" is probably still high since it does not take into consideration land that is too steep for harvesting, has soils that will not support crop production, or is otherwise unsuitable for cultivation.

Because of the variation in the climate in the Southwest, the weather in any year can be colder and/or drier than the minimum used to define the reduced area. Since the minimum values for average precipitation and frost-free period are so low in much of the area, a variation that would be minor in an area with higher precipitation or with a longer frost-free period could be significant in the Southwest. Thus, within the reduced area, those parts that have higher average rainfall and/or longer frost-free periods, or in which the variance from the average is least, would be the best areas for initial investigation of biomass energy crop production.

Relative Climate Suitability of the Reduced Area

Geographic limits of arid regions are commonly based on the occurrence of a climate regime in which potential

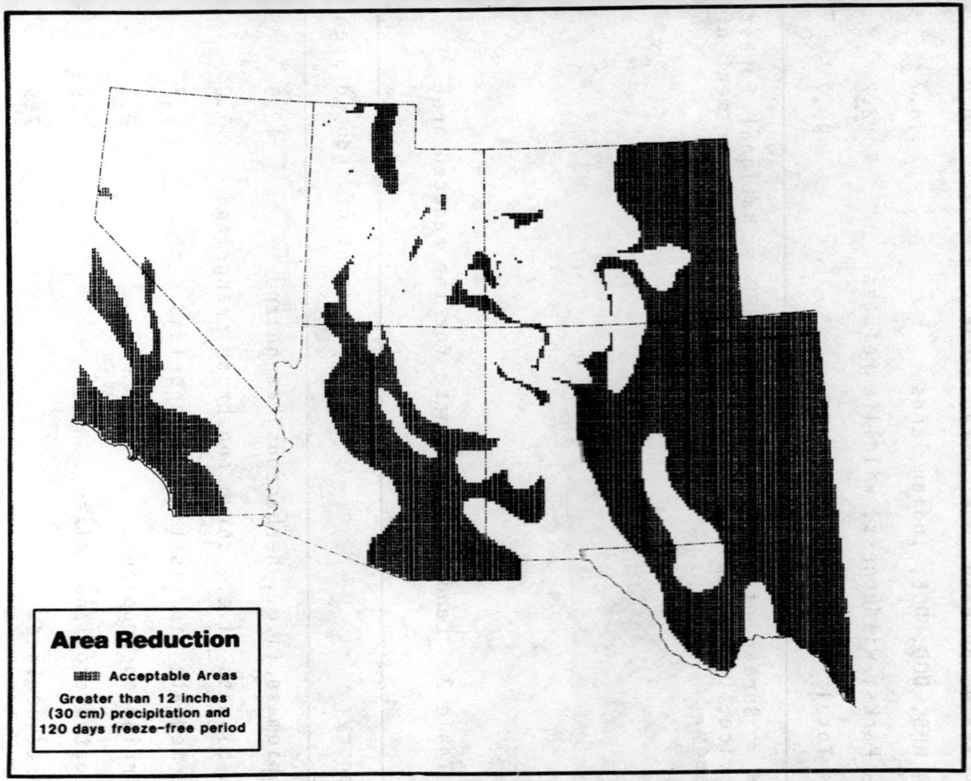

Figure 1. Reduced or acceptable area of the study region.

Table 1. Land ownership of the reduced area

Category group[a]	Percent of reduced area
A -- State, private, BLM, mixed lands	80.0
B -- NFS, DOD, DOE, Indian lands	16.7
C -- Parks, wilderness, wildlife refuges	2.2
Total	99.7

[a]BLM = Bureau of Land Management, NFS = National Forest Service, DOD = Department of Defense, DOE = Department of Energy.

Table 2. Land use/cover data for the reduced area

Category	Percent of reduced area
Low biomass (i.e., herbaceous) rangeland	45.5
High biomass (i.e., shrub and brush) rangeland	13.0
Intermediate biomass (mixed) rangeland	4.7
Nonirrigated cropland	10.3
Irrigated cropland (mixed)	6.9
Forest/woodland	7.9
Other (urban, salt flats, oil and gas fields, and snow and ice)	5.7
No data (Northeast Colorado)	5.9
Total	99.9

evaporation exceeds precipitation (27, 32). 'Effective' precipitation was determined in this study by subtracting the value for evaporation from that for precipitation (23). Data for 'effective' precipitation within the acceptable area are presented in three classes (Figure 2). Less than 12% of the reduced area is in the upper 'effective' precipitation class.

The 'effective' precipitation (Figure 2) and freeze-free period (23) maps were composited (by addition) to derive the map of relative climate suitability (Figure 3). The three classes of 'effective' precipitation (Figure 2) are combined with two classes of freeze-free period (i.e., >180 days and 120-180 days) to give four classes of relative climate suitability. Almost 85% of the reduced area is in the two lower suitability classes.

The environmental suitability composite map (Figure 3) represents a balance between opposite trends in 'effective' precipitation and freeze-free period data. In general, the most favorable moisture conditions (Figure 2) are found in areas of higher latitude and elevation that have shorter frost-free periods. The area in which both conditions are most favorable (i.e., suitability class 1) is just 2.7% of the reduced area and occurs only in California. Suitability class 3 is the largest area, comprising 60% of the reduced area and scattered throughout. This class is divided into northern and southern components, reflecting opposing moisture and temperature trends. Interpreted in terms of crop types, these opposing trends suggest that two different types of plants are desirable for the more continental parts of the reduced area: (1) those that produce rapidly in a short, relatively moist growing season and (2) those that are capable of more or less continuous production under drier conditions. Along the Pacific Coast, a third type of crop, capable of utilizing a long growing season but with relatively less need for drought tolerance, would be suitable.

<u>Vegetation in the Reduced Area</u>

More than 40 of Küchler's (14, 15) potential natural ecosystems are found in the study region (23). Eight of these are not found in the reduced area, while ten occur entirely in the reduced area. Groups of natural vegetation types within the study region and the reduced area are presented in Table 3. Central grasslands and western forests are the most common types in the reduced area.

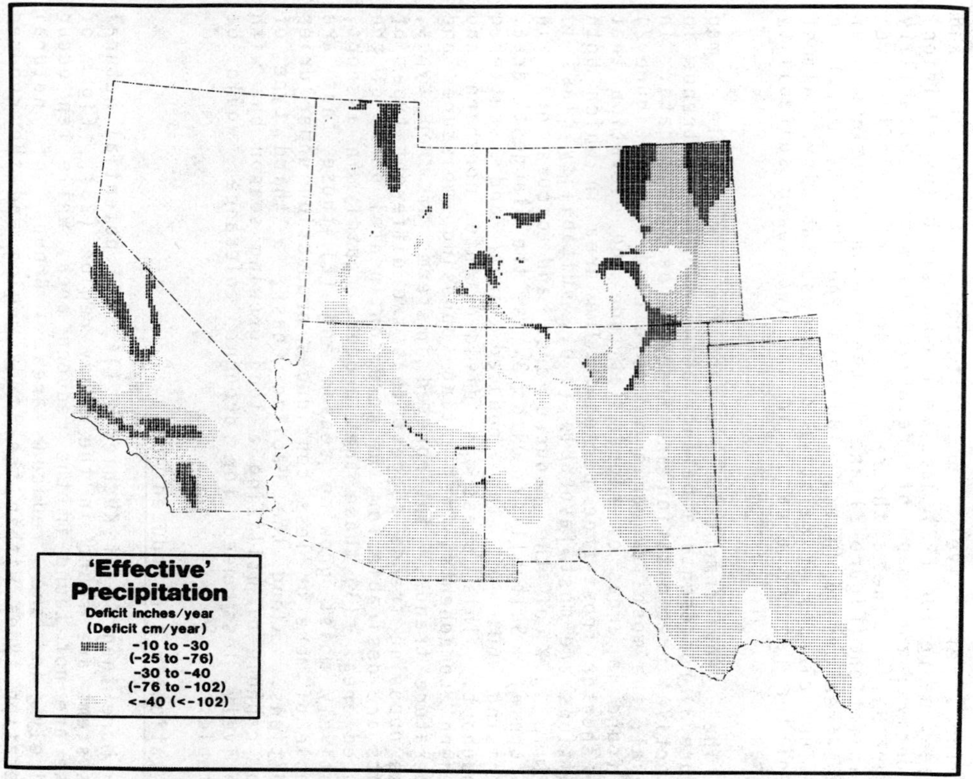

Figure 2. 'Effective' precipitation of the reduced area.

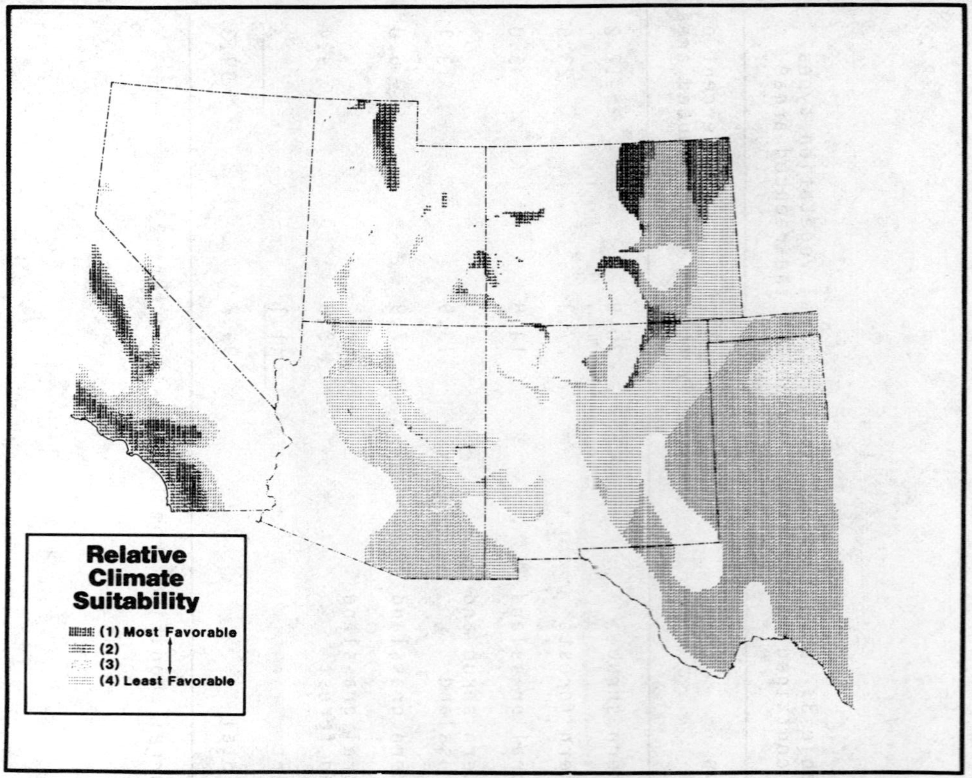

Figure 3. Relative climate suitability of the reduced area.

Table 3. Groups of potential natural vegetation types contained in the study region and the reduced area[a]

Group type	Percent of study region	Percent of reduced area
Western shrub	30.9	12.2
Western forests	27.3	23.6
Central grassland	14.8	35.0
Western shrub and grassland	13.9	13.1
Western grassland	7.9	6.0
Central grassland and forest	3.8	9.4
Desert	1.1	
Total	99.7	99.3

[a]Source: Modified from Table 3-10 in ref. 23.

The five most common ecosystems in the study region cover 48.8% of the land area (juniper-pinyon woodland, 14.4%; grama-buffalo grass, 13%; Great Basin sagebrush, 9.0%; saltbush-greasewood, 7.0%; and grama-tobosa shrubsteppe, 5.4%). Three of these (grama-buffalo grass, juniper-pinyon woodland, and grama-tobosa shrubsteppe) are also among the five most common vegetation types in the reduced area (Table 4). Great Basin sagebrush and saltbush-greasewood, which together cover 16% of the study region, account for only 2.7% of the land in the reduced area.

The grama-tobosa shrubsteppe and the trans-Pecos shrub savanna are characterized by short grasses with dense-to-scattered shrubs and dwarf shrubs (14, 15). These ecosystems are scattered across the southern part of the study region in an area with over 120 frost-free days per year. Thus, 51.9% of the grama-tobosa shrub-steppe and 48.9% of the trans-Pecos shrub savanna, which are not included in the reduced area, are eliminated because the precipitation where they grow is less than 12 in. (30 cm)/year.

The juniper-pinyon woodland type is found scattered throughout the study region. It grows where precipitation is from 10 to more than 20 in. (25.4 to over 50.8/cm)/year, and there are more than 80 frost-free days per year (20, 26). Both of these lower extremes are below the acceptable range for biomass production set for this study, which explains why almost 70% of the juniper-pinyon woodland type does not occur in the reduced area. Differences in this type throughout its range are related to elevation, climate, and soil (20, 26). In general, juniper-pinyon woodlands are found at middle elevations in the mountains as a transition between the arid shrublands and grasslands of lower elevations and the mountain forests at higher elevations. Extreme droughts will result in widespread death of the plants (5).

Slope

The principal limitation of slope for biomass energy production occurs during harvest, since it is difficult for harvesting equipment to operate on steep slopes. A map indicating areas of greater or less than 10% slope (19) was overlayed with an outline of the reduced area to show the location and approximate extent of mountains and other complex terrain (Figure 4). Almost 43% of the study region has a slope greater than 10%. Because minor relief features were smoothed out in the preparation of the digital maps, it is estimated that the area designated as

Table 4. Land area of five most common Küchler vegetation types in the reduced area

Vegetation type	Percent of the reduced area covered	Area[a] ($\times 10^6$)		Percent of each type in study region remaining in reduced area[b]
		Hectares	Acres	
Grama-buffalo grass (Bouteloua-Buchloë)	30.5	23.56	58.19	95.4
Juniper-pinyon woodland (Juniperus-Pinus)	11.3	8.73	21.56	31.9
Mesquite-buffalo grass (Bouteloua-Buchloë-Prosopis)	7.9	6.10	15.07	100.0
Grama-tobosa shrubsteppe (Bouteloua-Hilaria-Larrea)	6.4	4.94	12.21	48.1
Trans-Pecos shrub savanna (Flourensia-Larrea)	5.7	4.40	10.87	51.1
Total	61.8	47.73	117.90	

[a]Based on an area of 7.725×10^7 hectares (1.908×10^8 acres) in the reduced area.
[b]Values based on Table 3-10 in ref. 23.

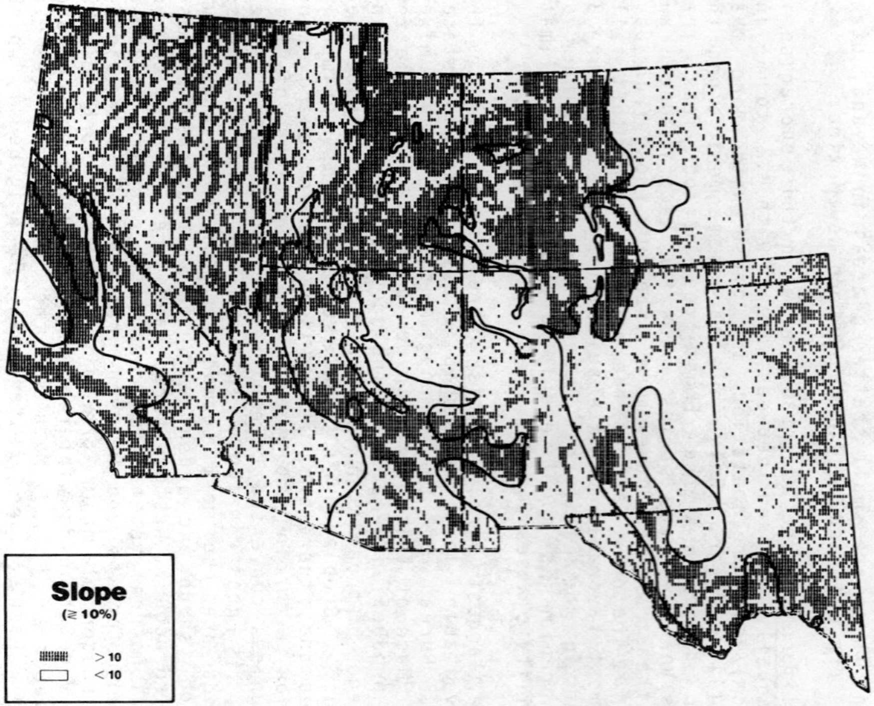

Figure 4. Areas of greater than or less than 10% slope in the study area. Dark line indicates the reduced area.

having a slope less than 10% is probably exaggerated by 30 to 40% (19).

Characteristics of Prime Areas in the Reduced Region

There are three major subregions in the acceptable area: a western area almost exclusively in California; a central area primarily in Arizona; and an eastern area found in Colorado, New Mexico, Oklahoma, and Texas. In addition, there are small scattered areas in Nevada, Utah, southwestern Colorado, and northwestern New Mexico.

California. The acceptable area in this subregion is about equally divided between areas with 12 to 16 in. (30-41 cm)/year precipitation and those with more than 16 in. (41 cm)/year. Only in a narrow segment along the northern border is the frost-free period less than 180 days/year, whereas strips along the Pacific Ocean and the southeast border have more than 300 frost-free days per year. The only part of the reduced region that falls in the most favorable climate suitability class (Figure 3) is found in the California subregion. The remainder of this subregion is about evenly divided between climate favorability classes 2 and 3.

The largest part of this subregion falls into the most available land ownership class, class A [state, private, Bureau of Land Management (BLM), and mixed lands]. Much of this subregion would not be available for biomass production because it includes the Los Angeles and San Diego metropolitan regions and their associated highlands. Cropland in the subregion is primarily irrigated and, therefore, is not suitable for energy crop production according to the basic assumptions of this study (unless converted to nonirrigated production); some, however, is nonirrigated. Intermediate (mixed) biomass rangelands are the most common type of rangelands, but both high (shrub and brush) and low (herbaceous) biomass types also occur. Western shrub lands are the most common potential natural vegetation group in the region, followed by western forests and western grasslands. Since winter is the season of major precipitation (21), much of the natural vegetation is winter-active and summer-dormant and also evergreen (10). Much of the acceptable area in California has a slope greater than 10% (Figure 4), which would make harvesting crops in those areas difficult.

Because of the long freeze-free period combined with the relatively high 'effective' precipitation, this subregion would be appropriate for biomass energy crops that could not be grown in the drier, colder parts of the reduced area. However, only a relatively small portion of the subregion would be suitable for energy crop production.

Arizona. More than half of this subregion falls into the 12 to 16 in. (30-41 cm)/year precipitation category, and only a small part in east central Arizona has more than 24 in. (61 cm)/year. Precipitation occurs primarily in the summer, with a secondary maximum in winter (21). Most of the subregion has 180 to 240 frost-free days per year. This subregion falls almost exclusively in the lower two classes for relative climate suitability because of its very low 'effective' precipitation (Figure 3). However, the third suitability class in this subregion is not uniform. The central and southern blocks in the third class correspond to areas with more than 180 frost-free days per year. The northern tip of the subregion has only 120 to 180 frost-free days per year; however, this area is in the third suitability class because it falls in the middle range for 'effective' precipitation. Therefore, the shorter frost-free period there is balanced by the more favorable 'effective' precipitation value.

A small part of this subregion falls in land ownership class C, whereas the remaining area is about evenly divided between land ownership class A (state, private, BLM, and mixed lands) and class B [National Forest Service (NFS), Department of Defense (DOD), Department of Energy (DOE), and Indian reservations]. This subregion has a higher percentage of category B lands than other subregions of the acceptable area.

Potential natural vegetation in this subregion is about equally divided among western forests, combination western shrub and grasslands, and western shrub lands, with a small part in western grasslands. The most common land use/cover types are rangelands. Nonirrigated cropland is limited in extent, while irrigated, mixed croplands are found primarily in the southern part. Urban areas, including the cities of Tucson and Phoenix, would not be available for growing biomass energy crops. About half of this subregion has a slope greater than 10%, which is unsuitable for biomass crops (Figure 4).

This subregion is drier than other parts of the reduced area. Therefore, plants that can tolerate low levels of rainfall, which vary from year to year, would be the best candidates for biomass energy crops.

Eastern Subregion. The eastern section of the acceptable area is the largest subregion of the acceptable region (Figure 1). The boundaries of the southern half of the subregion are determined by precipitation values, whereas boundaries of the northern half correspond more closely with values of the freeze-free period.

The subregion is about equally divided between the 12 to 16 in. (30-41 cm)/year and greater than 16 in. (41 cm)/year precipitation categories. A very small area in north central New Mexico has more than 24 in. (61 cm)/year. In general, the southern part of this subregion has 180 to 240 frost-free days per year, whereas the northern half has 120 to 180. A portion of southern Texas has more than 240 frost-free days per year, including a small area with more than 300 frost-free days. In terms of relative climate suitability (Figure 3), the third favorability category is the most common.

The subregion falls almost entirely in land ownership category A (state, private, BLM, and mixed land owners). The most common land-use type is low biomass (i.e., herbaceous) rangeland, with nonirrigated and mixed irrigated croplands also common. All potential natural vegetation types except for desert are found in this subregion, the most common being central grasslands. This subregion has less land with a slope greater than 10% than the other subregions (Figure 4).

Since grasslands are the most prevalent ecosystem type here, herbaceous crops would be the best candidates for biomass energy crops in most of the subregion. However, where climatic extremes occur, various other crops might be more suitable. For example, in southern Texas the frost-free period is long enough for subtropical herbaceous and woody species to be grown; whereas at higher elevations, temperate woody species would be more appropriate.

Scattered Areas. Parts of the acceptable region also occur in small scattered areas located in western Nevada, north central and southeastern Utah, western Colorado, and northwestern New Mexico (Figure 1).

The freeze-free period in all of the scattered areas falls in the 120 to 180 days per year category, whereas precipitation is primarily in the 12 to 16 in. (30-41 cm)/year range. The most common climate suitability class in these areas is the second most favorable class followed by the third most favorable class (Figure 3).

The most common land ownership class in the scattered areas is class A (state, private, BLM, and mixed ownership), which is the most favorable for biomass production. The scattered areas have a complex of land use/cover classes, rangelands being the most common. The most common potential vegetation types in the scattered areas are western forests and western shrublands, although western grasslands and combination western shrub and grasslands are also found. The slope in much of the scattered areas is greater than 10% (Figure 4), indicating the mountainous nature of the areas. Although most of these scattered areas are not large, many of them could be suitable for certain biomass energy crops.

Productivity of Vegetation Types in the Reduced Area

Net primary productivity (i.e., the net biomass increase per unit area per year) is an important value for determining the potential of any area for producing terrestrial energy crops. In any landscape, primary productivity can vary over short distances as the result of differences in topography, water availability, soil quality, and the successional stage of the land (16). In addition, productivity can be modified by both natural (i.e., climate) and man-made (i.e., irrigation, fertilization, harvesting frequency, etc.) factors. One of the assumptions of this study was that productivity would not be enhanced by routine irrigation since that could increase the cost of energy crops to a prohibitive level. Therefore, the emphasis is on natural productivity.

Productivity values for terrestrial ecosystems in the world range from a low of almost 0 kg/(ha year) for extreme deserts to over 35,000 kg/(ha year) for tropical rain forests (4, 16). Values for cultivated land vary from 1,000 to 40,000 kg/(ha year), with an approximate mean of 6,500. Productivity rates achieved for short-rotation woody biomass energy crops in test plots range from 4,000 to 15,000 dry kg/(ha year) with an average of about 8,000 to 9,000 dry kg/(ha year) (22). For the most promising species on good sites under optimum management strategies, productivity levels up to 30,000 dry kg/(ha year) are achieved. It is uncertain if such high values can be achieved in larger plantings on a wide range of sites that are marginal for conventional agriculture.

Productivity values have been determined for only some of the ecosystems in the reduced area, including some values from fertilization and/or irrigation studies (23). The highest ecosystem productivity value found for the study region was 11,460 kg/(ha year) for a north-slope montane fir forest in the Santa Catalina Mountains near Tucson, Arizona (33). Several other values for woody ecosystems fall within the ranges reported for short-rotation woody crops in test plots, and, in general, woody ecosystems have a higher natural productivity than either grasslands or shrublands. However, those ecosystems with the highest productivity values tend to be situated in steep areas; consequently, harvesting would be extremely difficult or impossible. Thus, the best sites in the Southwest for biomass crops in terms of natural productivity are probably too steep for such uses.

Selecting Biomass Energy Crop Species

The vegetation of arid regions throughout the world tends to be generally similar in form and function as a result of similar conditions of scanty and irregular moisture supply. The search for biomass energy crops for the Southwest should not be restricted entirely to native species, but should also include plants adapted to similar environmental conditions in other arid and semiarid regions of the world.

Because of the variability of the climate in the Southwest, biomass energy crops should be species that can survive during years when the climate is significantly drier and/or colder than normal. Deep-rooted, perennial shrubs can survive severe moisture limitations more readily than other types of vegetation and would be good biomass energy candidates (9). Ecosystems which occur both in the reduced area and in the nonacceptable area are likely to contain species that would be able to survive when weather conditions are worse than normal. Of the five most common ecosystems in the reduced area (Table 4), three (grama-tobosa shrubsteppe, juniper-pinyon woodland, and trans-Pecos shrub savanna) have significant parts of their range in the nonacceptable areas.

Other ecosystems that have significant parts of their study region ranges both in the reduced area and outside it (23) include three western forest types (mixed conifer forest, Arizona pine forest, and California oakwoods) and two western grasslands (California steppe and grama-tobosa prairie). These ecosystems can survive in areas where climatic conditions are drier and/or cooler than the values used to define the reduced area. Thus, species in those ecosystems are good candidates for biomass energy crops and should be screened to select those with the highest productivity.

Twelve ecosystems have all or almost all of their study region range in the reduced area, including two of the three most common ecosystems in the reduced area, grama-buffalo grass and mesquite-buffalo grass (Table 4). Other ecosystems that are almost exclusively found in the reduced area include two western forests (oak-juniper woodland and a transition community between that and the mountain mahagony-oak scrubland), one central grassland and forest (juniper-oak savanna), one western shrub and grassland (mesquite-acacia savanna), three western shrublands (chaparral, coastal sagebrush, and ceniza shrub), and three central grassland types (bluestem-grama prairie, sandsage-bluestem prairie, and shinnery) (23). The species in these ecosystems should be screened to select those with the highest productivity since they would be likely candidates for biomass energy crops.

Three different photosynthetic pathways are found in plants: the Calvin or C_3 cycle, the C_4-dicarboxylic acid pathway, and the Crassulacean acid metabolism (CAM). Under the same moisture regime, plants that use the C_4 pathway for photosynthesis require about half as much water per unit of biomass produced as C_3 species (8). A number of succulent plants use the CAM pathway. They fix large amounts of CO_2 as organic acids at night and convert them into carbohydrates during the day (3). The lower temperatures that occur at night when the primary assimilation of CO_2 occurs allow CAM plants to limit the amount of water transpired. Species that use the C_4 pathway or the CAM pathway of carbon metabolism could be good candidates for biomass energy crops on marginal lands in the arid or semiarid environments common in the Southwest. Their photosynthesis processes are highly efficient at high temperatures, and they efficiently utilize water during their growth, both characteristics that would increase growth in the arid Southwest (25).

Species native to the Southwest, which sprout from their stumps or roots after cutting or burning are good candidates as biomass energy crops. They would only need to be planted once to produce multiple crops, thus reducing both the costs and the environmental problems associated with crop establishment (22). Half of the species in the California chaparral stump or root sprout as an adaptation to the fires that occur there every 10 to 40 years (10). Spreading rabbitbush (*Chrysothamnus linifolius*), alligator juniper (*Juniper deppeana*), and mesquite (*Prosopis* sp.) are other native species that stump or root sprout (2, 6, 31).

Many woody plants have ectomycorrhizal fungi as symbionts on their roots that substantially increase plant growth (1). Because routine application of fertilizers might not be cost-effective in growing biomass energy crops in the Southwest, those species that have symbiotic relationships with nitrogen-fixing bacteria would be good candidates for such crops (25).

Thus, investigations of native ecosystems should be directed at identifying the most productive species that (1) tolerate climatic conditions that are drier and/or colder than normal, (2) have the C_4 or CAM carbon metabolism, (3) sprout from their stumps or roots when cut or burnt, (4) produce valuable by-products, and (5) have nitrogen-fixing bacterial symbionts.

Matching Site-Specific Conditions and Potential Species

Using the data set prepared for this study, the climate and land characteristics (i.e., precipitation, soils, evaporation, land use, etc.) of each ecosystem can be summarized. For the 24 ecosystems that occur both within and outside the reduced area, 2 lists could be prepared to indicate what conditions, in addition to either the freeze-free period or precipitation limits used to define the reduced area, differ in places where the ecosystem is found in the reduced area and in the nonacceptable part of the Southwest. For the ten ecosystems that are found only in the reduced area, one list would be prepared. Thus, for each ecosystem in the reduced area, the climatic conditions that it grows under and the land characteristics where it is located can be identified.

With these data, species that have the potential to grow on a particular site could be identified. The simplest way to do this is to determine which of Küchler's potential natural vegetation types occurs at that site. However, using the data base summary of ecosystem climatic and land characteristics, the features of the site could be matched with those of other potential natural vegetation types to identify ecosystems that grow under similar conditions. The species in these ecosystems could then be screened to select those that are potential biomass energy crops at a particular site in the Southwest. Thus, the data base could be searched to identify species that could tolerate the natural conditions that occur at a particular site and to identify species to screen as potential biomass energy crops.

The data base could also be used to identify areas where species that have been identified as potential biomass energy crops could be grown. For native species, the natural conditions under which they grow could be identified from the data base, and areas with similar climatic conditions could then be located. For non-native species, this could also be done if the conditions that are required for their growth have been characterized, either under cultivation or under natural conditions. Thus, if the conditions under which these potential energy crops can grow are known, areas that have these conditions could be identified from the data base, and the search for sites to grow energy crops could be narrowed to those areas.

CONCLUSIONS

The southwest United States is an area of diverse climate, vegetation, soils, topography, and terrain. Not surprisingly, the potential for terrestrial energy crops also varies throughout the region. Of the almost 1.90 x 10^8 ha (4.69 x 10^8 acres) in the study region, just over 40% is considered even minimally suitable for further investigation of its potential for growing terrestrial energy crops. This reduced area includes only the land in the Southwest with both an average annual precipitation of at least 30 cm (12 in.) and an average annual frost-free period of at least 120 days. However, because these are average figures, much of this reduced area can have significantly less precipitation and/or a shorter frost-free period in any particular year. Therefore, biomass energy crops to be grown in even the most favorable parts of the Southwest must be adapted to survive in years in which weather conditions are below average.

Natural productivity of vegetation in the Southwest is low. Even in the best years, it may be difficult to grow an economical crop on much of the land in the reduced area. Unless energy prices rise significantly, it may not be possible to justify the production of crops in the Southwest solely for their energy content. However, multipurpose projects in which energy production is one aspect may be feasible (17).

The data compiled for this study can be used both to provide a set of species for screening as biomass energy crops in the Southwest and to identify areas there where previously selected energy crops would grow best. Thus, the species that perform best in a given set of environmental conditions can be identified and exploited rather than attempting to alter the environment to suit the needs of particular plant species.

Species that have some or all of the following characteristics would be the most suitable biomass energy crops in the Southwest: sprout from roots or stumps, have C_4 or CAM photosynthetic metabolism, fix nitrogen, are perennial, withstand moisture limitations and other climatic extremes, have multiple uses, and produce valuable by-products.

A major question remaining is whether terrestrial energy crops can be economically produced in the Southwest. Economic evaluations need to address the trade-offs between productivity; rotation length and number (for woody species); and costs of harvest, transportation, storage, and conversion. Further research, particularly on a site-specific basis, must be undertaken before a definite decision can be made as to whether biomass energy crops are viable in the Southwest.

ACKNOWLEDGMENTS

During the preparation of this document, personnel from the Soil Conservation Service and the Forest Service of the U.S. Department of Agriculture, the Geological Survey and the Bureau of Land Management of the U.S. Department of the Interior, and the U.S. Environmental Protection Agency (EPA) provided helpful information. Safa Shirazi of the Corvallis Environmental Research Laboratory of EPA provided digitized map data for soils and vegetation. Keith Turner of Geometronics International, Inc., generated the computer maps. Janet Cushman, Lynn Wright, and Jack Ranney of the ORNL Environmental Sciences Division reviewed earlier drafts. Publication No. 2712, Environmental Sciences Division, Oak Ridge National Laboratory, Oak Ridge, Tennessee 37831.

Note: Research supported by the Biofuels and Municipal Waste Technology Division of the U.S. Department of Energy, under contract No. DOE-AC05-84OR21400 with Martin Marietta Energy Systems, Inc.

REFERENCES CITED

1. Abelson, P. H., "Plant-fungal Symbiosis," Science 229,617 (1985) August 16.

2. Arnold, J. F., Jameson, D. A., and Reid, E. H., "The Pinyon-Juniper Type of Arizona: Effects of Grazing, Fire, and Tree Control," Production Research Report No. 84, USDA Forest Service, Rocky Mountain Forest and Range Experiment Station (1964).

3. Arnon, I., "Physiological Principles of Dryland Crop Production," pp. 5-145, IN Gupta, U. S. (ed.), Physiological Aspects of Dryland Farming, Montclair, NJ: Land Mark Studies, Allanheld, Osmun & Co. Publishers, 1975.

4. Atjay, G. L., Ketner, P., and Duvigneaud, P., "Terrestrial Primary Production and Phytomass," pp. 129-181, IN Bolin, B., Degens, E. T., Kempe, S., and Ketner, P. (eds.), The Global Carbon Cycle, Scientific Committee on Problems of the Environment

(SCOPE) of the International Council of Scientific Unions (ICSU), New York: John Wiley & Sons, 1979.

5. Darrow, R. A., "Origin and Development of the Vegetational Communities of the Southwest," pp. 30-47, IN Symposium: Bioecology of the Arid and Semi-arid Lands of the Southwest, held at New Mexico Highlands University, April 29, 1958.

6. Felger, R. S., "Mesquite in Indian Cultures of Southwestern North America," pp. 150-176, IN Simpson, B. B. (ed.), Mesquite: Its Biology in Two Desert Ecosystems, US/IBP Synthesis Series 4, Stroudsburg, Pennsylvania: Dowden, Hutchinson, & Ross, Inc., 1977.

7. Foster, K. E., and Brooks, W. H., Assessment of Arid Land Plants as Future Energy Crops for the Electric Utility Industry, EPRI AP-2172-SR, Palo Alto, California: Electric Power Research Institute, 1981.

8. French, N. R., "Principal Subsystem Interactions in Grasslands," pp. 173-189, IN French, N. R. (ed.), Perspectives in Grassland Ecology: Results and Applications of the US/IBP Grassland Biome Study, New York. Springer Verlag, 1979.

9. Goodin, J. R., and McKell, C. M., "Shrub Productivity: A Reappraisal of Arid Lands," pp. 235-246, IN McGinnies, W. G., Goldman, B. J., and Paylore, P. (eds.), Food, Fiber and the Arid Lands, Tucson: The University of Arizona Press, 1971.

10. Hanes, T. L., "California Chaparral," pp. 139-174, IN diCastri, F., Goodall, D. W., and Specht, R. L. (eds.), Mediterranean-Type Shrublands, New York: Elsevier Scientific Publishing Company, 1981.

11. Hastings, J. R., and Turner, R. M., The Changing Mile, Tucson: University of Arizona Press, 1965.

12. Herbel, C. H., "Utilization of Grass- and Shrublands of the Southwestern United States," pp. 161-203, IN Walker, B. H. (ed.), Management of Semi-arid Ecosystems, New York: Elsevier Scientific Publishing Company, 1979.

13. Klopatek, J. M., Olson, R. J., Emerson, C. J., and Jones, J. L., "Land-use Conflicts with Natural Vegetation in the United States," ORNL/TM-6814, Oak Ridge, TN: Oak Ridge National Laboratory, 1979.

14. Küchler, A. W., "Potential Natural Vegetation of the Conterminous United States," 154 pp., 1 map sheet, 1:3,168,000 scale, New York: Special Pub. No. 36, American Geographical Society, 1964.

15. ______________, "Potential Natural Vegetation (Map)," Sheet Number 90, 1:7,500,000 scale, IN USGS 1970, The National Atlas of the United States, Washington, DC: USDI Geological Survey, 1966.

16. Lieth, H., "Primary Production of the Major Vegetation Units of the World," pp. 203-215, IN Lieth, H., and Whittaker, R. H. (eds.), Primary Productivity of the Biosphere, New York: Springer-Verlag, 1975.

17. Lipinsky, E. S., and Kresovich, S., "Fuels from Biomass Systems for Arid Land Environments," pp. 294-306, IN Goodin, J. R., and Northington, D. K. (eds.), Arid Land Plant Resources, International Center for Arid and Semi-arid Land Studies, Lubbock, TX: Texas Tech. University, 1979.

18. Mann, H. S., "Management of Arid-land Resources for Dryland and Irrigated Crops," pp. 479-494, IN Goodall, D. W., and Perry, R. A. (eds.), Arid-land Ecosystems: Structure, Functioning, and Management. Vol. II, International Biological Programme 17, Cambridge: Cambridge University Press, 1981.

19. Maxwell, E. L., Folger, A. G., and Hogg, S. E., "Resource Evaluation and Site Selection for Microalgae Production Systems," SERI/TR-215-2484, Springfield, VA: NTIS, GPO No. DE 85008807, 1985.

20. Mutel, C. F., and Emerick, J. C., From Grassland to Glacier: The Natural History of Colorado, Boulder, CO: Johnson Books, 1984.

21. Pyke, C. B., "Some Meteorological Aspects of the Seasonal Distribution of Precipitation in the Western United States and Baja, California," UCLA-WRC-W-254, Los Angeles: University of California Water Resources Center, 1972.

22. Ranney, J. W., Wright, L. L., Trimble, J. L., Perlack, R. D., Dawson, D. H., Wenzel, C. R., and Curtin, D. T, "Short Rotation Woody Crops Program: Annual Progress Report for 1984," ORNL-6160, Oak Ridge, TN: Oak Ridge National Laboratory, 1985.

23. Salk, M. S., and Folger, A. G., "Characterization of the Southwest United States for the Production of Biomass Energy Crops" ORNL/TM-____, Oak Ridge, TN, Oak Ridge National Laboratory (in press).

24. Shelford, V. E., The Ecology of North America, Seventh ed, Urbana: University of Illinois Press, 1974.

25. Slesser, M., and Lewis, C., Biological Energy Resources, New York: Halsted Press Book, John Wiley & Sons, 1979.

26. Springfield, H. W., "The Characteristics and Management of Southwestern Pinyon-juniper Ranges: The Status of Our Knowledge," USDA Forest Service Research Paper RM-160, Fort Collins, CO: Rocky Mountain Forest and Range Experiment Station, 1976.

27. Trewartha, G. T., and Horn, L. H., An Introduction to Climate, Fifth ed., New York: McGraw-Hill, 1980.

28. Turner, A. K., Weber, J. C., and DeAngelis, M., A Geographic Market Suitability Analysis for Low- and Intermediate-Temperature Solar IPH Systems, 2 vols., SERI/TR-733-1194, Springfield, VA: NTIS, 1981.

29. USDA/SCS, "Distribution of Principal Kinds of Soils: Orders, Suborders, and Great Groups" (Map), Sheet No. 86, 1:7,5000,000 scale, IN USGS, 1970, The National Atlas of the United States, Washington, D.C.: USDI Geological Survey, 1967.

30. ________, Land Resource Regions and Major Land Resource Areas of the United States, Agriculture Handbook 296 rev., 156 pp., 1 map sheet, 1:7,500,000 scale, U.S. Department of Agriculture, Soil Conservation Service, Washington, DC: USGPO, 1981.

31. Van Epps, G. A., Barker, J. R., and McKell, C. M., "Energy Biomass from Large Rangeland Shrubs of the Intermountain United States," J. Range Manage. 35(1), 22-25 (1982) January.

32. Walter, H., Vegetation of the Earth in Relation to Climate and the Ecophysiological Conditions, Heidelberg Science Library, Vol. 15, New York: Springer-Verlag, 1973.

33. Whittaker, R. H., and Niering, W. A., "Vegetation of the Santa Catalina Mountains, Arizona. V. Biomass, Production, and Diversity along the Elevation Gradient," Ecology 56, 771-790 (1975) Summer.

DISCUSSION

Berger	Did you give any thought to how you would go about introducing crops there in terms of the people, the land, and the current use?
Salk	That is part of the economic site-specific analysis and is definitely something that would have to be considered.
E. Shultz, Washington U.	I gather that your study was oriented to the production of woody biomass from the criteria of sprouting from stumps and to perennials rather than annual types of plants.
Salk	You can have herbaceous crops that are perennials and also grassland crops.
Shultz	Do you plan to extend the study to non-woody plants of interest for biomass?
Salk	No, there's no plan right now to continue this study into the annuals. The perennials in the sense of grasses which can be harvested and which then grow back from the roots are the herbaceous crops that we considered.
B. Spaugh, Nebraska Energy Office	I noticed on your graph that a high percentage was in the blue grama-buffalo grass species. They are very low, short-growing grasses, and they're fairly sparse. Did your study look at substitutes for those, or how did you plan to use their wide adaptation in future studies?
Salk	We did not look at specific species, other than in a general way. There are many which have low productivity. The idea was to determine whether there are any particular species within those ecosystems higher in productivity than others and to encourage more research to develop those as herbaceous energy crops. We are planning to do a similar resource assessment of the Great Plains for biomass crops and that will probably look at these species in more detail.

PRODUCTION OF WARM SEASON GRASSES FOR BIOMASS

S. C. Schank, Ph.D.
Department of Agronomy, University of Florida
Gainesville, FL 32611, USA

ABSTRACT

Of all the warm-season grasses, the greatest yield potentials are from sugarcane (Saccharum), sorghum (Sorghum bicolor) and/or napiergrass (Pennisetum purpureum. All three of these grasses have the very efficient carbon cycle in photosynthesis known as the C 4 photosynthetic pathway. Each of the three are sub-tropical or tropical grasses, and are primarily grown between 30° North and 30° South latitudes.

Sugarcane and napiergrass produce only limited amounts of seed, and the seed produced is very small and difficult to handle. Hence the grasses are propagated vegetatively. This requires the planting of stem cuttings, commonly called "bones", which is much more labor intensive than using a seeded crop like sorghum.

Among the grasses, the most promising genera are napiergrass (Pennisetum), sugarcane (Saccharum) and (Sorghum). Biomass dry matter yields of sugarcane and napiergrass approach 50-70 Mg/ha, and yields of 20-30 Mg/ha have been recorded for Sorghum. In this paper a review is given of the selection and breeding work done at the University of Florida on each of these three grasses since the IFAS/GRI program was initiated five years ago. Important characteristics such as resistance to disease, tolerance to cold, heat, and sudden changes in temperature and also the ability of grass to grow under conditions of low soil moisture and low nutrient levels are discussed. Progress in the tissue culture of napiergrass is also presented, as well as the interspecific hybridization of pearl millet and napiergrass. Chemical composition of various plant parts of napiergrass hybrids has also been accomplished and is reported in the Pennisetum section of this paper. Variant plants obtained in the breeding program are being utilized to overcome the lack of seed production and to improve the biomass production potential of hybrid Pennisetums.

PRODUCTION OF WARM SEASON GRASSES FOR BIOMASS

INTRODUCTION

Grasses have a wider range of adaption than any other plant family since they occupy nearly all parts of the earth and exceed any other family in the number of individuals (3). The grass family contains about 620 genera and 10,000 species (11), so it is necessary to carefully select the species that have a potential for biomass. Grasses with the highly efficient C-4 photosynthetic pathway would be more productive than those with the C-3 pathway. Smith and Frank (23) reviewed the comparative biomass yields of energy crops and concluded that among the warm season grasses, the most promising genera were sugarcane (*Saccharum spp.*), napiergrass (*Pennisetum purpureum*) and sorghum (*Sorghum bicolor*). Biomass dry matter yields of the first two grasses approached 50-70 Mg ha^{-1}, and yields of 20-30 Mg/ha were recorded for *Sorghum*. In this paper a review is given of:

1. Sugarcane-- energy type canes
2. Napiergrass and napiergrass hybrids
3. Sorghum-- grain, forage and dual types
4. Miscellaneous grasses-- *Erianthus*, *Cynodon*, *Tripsacum* and *Setaria*.

Breeding objectives for biomass crops can be clearly defined. For example, the plant breeder would select for rapid growth rates, high number of tillers, lodging resistance, disease and pest resistance. The breeder would also select for high cellulose production, low lignin content, and a high dry matter yield. Work to develop superior strains of grasses for specialized uses has been the goal of plant breeders since the beginning of time. Musser et. al. (15) mentioned some of the goals that grass breeders have concentrated on over the years, such as resistance to disease, tolerance to cold, heat, and sudden changes in temperature and also the ability of grass to grow under conditions of low soil moisture and low nutrient levels. Hanson and Carnahan (7) have appraised the problems confronting the grass breeder and have summarized the techniques and methods that are at his disposal to make improvements. They emphasized that the diversity in the grass family affords a wide opportunity to select for the characteristics desired. Special mention was made to have sound objectives in the breeding endeavor since the objectives determine the ultimate success of the program.

Several inherent problems exist with the propagation and multiplication of sugarcane and napiergrass. Both of these warm season grasses do not produce much seed or the seed is small and difficult to handle, and hence the grasses are propagated vegetatively. This requires the planting of stem cuttings, commonly called "bones", which is much more labor intensive than using a seeded crop like sorghum. Sugarcane and napiergrass have been cultivated for centuries using stem cuttings, and will continue to be propagated this way because the perenniality of the grass, and the high biomass production compensate for the labor expended in the establishment phase.

In recent years, great strides have been made in the tissue culture of both sugarcane and napiergrass (19,26,27). With this new biotechnology, it appears possible to mass produce the desired plant genotypes and produce thousands of plantlets for field plantings (19). Interspecific hybridization is often used in sugarcane improvement and also in obtaining hybrids of pearlmillet and napiergrass (5,17). Genetic variation is observed in these F_1 interspecific hybrids because of the high amount of heterozygosity of the parents used in the crosses. The genetic variation may be reflected in the chemical composition of various plant parts as recently reported in sorghum (12). TVA's integrated agri-refinery system proposes to use crop residues and protein-rich cellulosic forage crops to yield protein, lignin for fuel and chemical production, chemicals, fuel ethanol and methane (2). This acid hydrolysis could apply to all of the warm-season grasses mentioned in this paper.

SUGARCANE SELECTION AND BREEDING

Sugarcane (Saccharum officinarum L.) is an effective collector of solar energy. According to Alexander (1), its botanical preference is to store the energy with the structure of new tissues (fiber) rather than to accumulate it as soluble sugars. Sugarcane produces both fiber and fermentable sugars in abundance, and so if sugarcane is grown as a renewable energy crop, it requires changes in agronomic practices to maximize yield. The agronomic changes involve improved land preparation, selection of high-tonnage varieties, increased fertilization, especially nitrogen, adequate irrigation, and inclusion of cane tops and leaf trash in the tonnage calculations. Samuels (20) states that when the best agronomic practices are used, yields of up to 280 Mg ha^{-1} green biomass (112 Mg ha^{-1} dry matter) are possible and have been achieved in Puerto Rico. At present, in most sugarcane growing countries of the Caribbean, the average cane yield is about 75 Mg ha^{-1} hectare green weight (18 Mg ha^{-1} dry weight), when calculated for a 12-month plant crop (21).

Sugarcane breeding work in Florida has been summarized by Grassl (6) where he discusses the problems and potentialities of intergeneric hybridization in the breeding program with Saccharum. It is a fact that one or more species of each of the six genera Erianthus, Miscanthidium, Miscantus, Narenga, Sclerostachya and Sorghum have been successfully hybridized with Saccharum.

More recently, Ozias-Akins et. al. (16) have used protoplast fusion to hybridize Saccharum and Pennisetum.

Each year, a large number of individual seedlings from crosses made at the U. S. Sugarcane Field Station, Canal Point, FL., are grown as single stools at the Sugarcane Field Station, Houma, LA. The literature contains very little information on studies related to the performance of agronomic and chemical characters in sugarcane, especially in regard to their relative performance in single stools and clones. There has been an extensive breeding program in progress in Lousiana since 1919, and many new sugarcane varieties have resulted from this work (8). Tai and Miller (25) have very recently reported their evaluation of crosses of commercial lines of sugarcane with F_1 Saccharum spontaneum lines. They evaluated fifty clones of sugarcane (complex hybrids of Saccharum spp.) for cold tolerance, and found that the estimates of narrow sense heritability of cold tolerance in sugarcane ranged from 12 to 26 in the combined analysis. Although the heritability is low, some genotypes were superior to others in cold tolerance according to the rating on percent green tissue of total leaf area (25).

Sugarcane stocks used in the GRI/IFAS biomass project originated from the USDA sugarcane station. Mislevy et. al. (13) have tested dry matter yields of 15 energy canes and one Erianthus (formerly named Ripidium). They found the dry matter yields of the better selections ranged from 26.2 Mg ha^{-1} at Jay (nortwest Florida to 79.1 Mg ha^{-1} at Ona (south-central Florida). They have presented three years of data on genetic variation found in these lines in Brix %, yield, physiology, lodging, response to N fertilization, and other agronomic traits. For example, the Brix content ranged from 9-19%, which indicates very large genetic differences between lines. Since selection is such a powerful genetic improvement tool, it is believed that not only Brix, but many other of the characteristics of Saccharum may be modified through conventional plant breeding methods. Continued cooperation in the evaluation of sugarcanes is expected with the USDA sugarcane field stations. Screening studies conducted in Florida indicate that certain sugarcane (Saccharum) and (Erianthus) species may be equal or possibly more productive in biomass yield than Napiergrass. In addition, percentage sugar solids or brix were 12.0 for promising sugarcane entries, which was about double that found in Napiergrass. Disease or insect problems did not appear to be a problem with the energy cane selections. Once mature, unlike Napiergrass, senescing leaves cling to the plant, making them available for mechanical harvest. Preliminary observations by Mislevy et. al. (13) indicate that root systems of Saccharum may be stronger than those of Napiergrass thus eliminating the problem of lodging in the sugarcane lines selected.

PENNISETUM SELECTION AND BREEDING

Napiergrass is generally propagated vegetatively since the seeds are very small and lack the vigor for successful establishment in field plantings. Pearl millet, on the other hand, has extremely large seed. The two species hybridize readily and the F_1 progeny have perenniality, vigor, high quality, high yielding, and great immunity to diseases such as rust. Using pearl millet as the female parent has several advantages as outlined by Jauhar (10). Despite the suggestion twenty years ago (17) that the production of hybrid seed could be accomplished in a frost-free environment, the commercialization of the hybrid between _P. americanum_ and _P. purpureum_ has not been achieved. Excellent summary papers describe the hybridization process in _Pennisetum_ and the subsequent genetic variation (5,10,14).

At the University of Florida, we have been involved in the breeding of _Pennisetum_ for the past 12 years. This interest was initiated while I was in Brazil from 1974-76 working at the KM 47 research station. In Brazil, male sterile lines of pearl millet (provided by Dr. Wayne Hanna, USDA, Tifton Ga.) were crossed with various lines of Napiergrass as the male parent in over 50 crosses. To continue the research, see of the following lines or varieties of Napiergrass were sent to Florida: Vrukwona, Merkeron Menlo, Mineirao, Taiwan A-25, Taiwan A-148, Merkeron Pinda, Taiwan A-146, Taiwan A-143, Turrialba, Cameron, Merkeron (common), Cachoeira do Itapermirim, Cubano, Taiwan A-144, Merker- SEA, Napier- SEA, Porto Rico, IPEACO and Taiwan-241. Work also included chromosomal evaluation of 8 _Pennisetum_ hybrids and chromosomal counts on three Napiergrass varieties from Brazil (22).

In 1984, the hybridization scheme included making crosses of pearlmillet to _P. purpureum_, PI 300086 which Dr. Prine had identified as having the highest dry matter biomass potential of 48.3 Mg ha^{-1} (18). A total of 119 crosses were made in the early months of 1984 (in the greenhouse). Seed was germinated in February and March, and the seedlings were tranplanted to the field in late April. We had approximately 12,000 plants of these crosses in the field in 1984. Some of the hybrids showed promising genetic potential. First harvest of the 1984 crosses was made July 1984. Highest dry matter yield was from selection 95, which had 19.5 Mg ha^{-1}. Yields ranged downward to 4.1 Mg ha^{-1} (Table 1). F_1 genetic variation was very apparent in these plots which were replicated five times. Selection for winter survival and persistence will be made from these plots in 1986. This data coupled with biomass production, lodging resistance, disease and insect resistance, and vigor ratings will be combined to give an overall rating of these hybrids of _P. americanum_ x _P._

TABLE 1. Yield Data from 78 different pearlmillet-napiergrass hybrids. First harvest in July 1984.

Selection Number	Dry Matter Yield Mg ha^{-1} 7/25/84	Selection Number	Dry Matter Yield Mg ha^{-1} 7/25/84
95	19.5	101	12.8
35	19.3	28	12.7
38	17.6	60	12.7
1	17.5	90	12.6
55	17.4	109	12.6
40	17.3	41	12.4
25	16.6	78	12.3
6	16.8	72	12.3
54	16.6	61	12.0
85	16.6	12	12.0
13	16.5	11	11.9
17	16.5	98	11.8
82	16.5	50	11.7
103	16.5	45	11.7
15	16.4	93	11.4
14	15.8	68	11.2
18	15.8	7	11.1
56	15.8	108	11.0
111	15.7	74	10.9
32	15.6	71	10.4
88	15.3	59	10.4
84	15.0	31	10.2
27	15.0	43	10.2
53	14.9	24	10.2
106	14.9	47	10.1
76	14.5	112	10.1
2	14.3	9	10.0
91	14.0	44	9.9
48	13.9	94	9.8
75	13.9	52	9.6
4	13.6	8	9.0
96	13.5	23	8.9
62	13.5	63	8.9
36	13.3	46	8.6
118	13.2	3	7.9
97	13.2	SEL #3	6.9
30	13.0	10	5.4
26	13.0	N-75	4.1
102	13.0		
92	13.0		

purpureum for biomass production. The F_1 hybrids of this cross are triploids and therefore sterile, but they can be propagated easily by stem cuttings as has been done on napiergrass for many years.

The problems associated with vegetative propagation can be circumvented by using tissue culture techniques. Rajasekaran et. al. (19) characterized the biomass production and phenotypes in plants regenerated from embryogenic callus cultures of P. americanum x P. purpureum (hybrid Napiergrass). Five hundred and twenty four regenerated plants of a triploid, sexually sterile hybrid Napiergrass were compared to a control population of vegetatively propagated plants in replicated field trials for two consecutive years to compare the biomass yield, phenotype and cytology of the two populations. Vegetatively propagated plants did not show any morphological variability. The tissue culture plants exhibited a high degree of phenotypic stability (96%). There were 23 phenotypic variants in the regenerated population of 524, most of them being more dwarf and later in flowering than the other regenerants. Two of the variants were hexaploid, which are now being tested for seed production, vigor, and other agronomic traits (19).

Initially, emphasis was placed on having vigorous dwarf types for forage, but with the need for higher biomass production, many tall hybrids have now been included in tests. The chemical make-up of tall versus dwarf types is shown in Figure 1, which contrasts protein and soluble carbohydrate content of leaves, sheaths and stems of these differing genotypes. The figure clearly illustrates that the dwarf plants were significantly higher in protein content than the tall genotypes, and that the leaves were highest in protein content (9.5%) compared to the stems (2.6%). Also shown in the figure is the non-structural (soluble carbohydrate) content of the corresponding tissues which ranged from 2.0% in the leaves to 8.3% in the stems. Characterization of the tall versus dwarf plants has been accomplished in more detail as illustrated in Table 2. The dwarf plants have less stem material, higher leaf content, and consequently a much

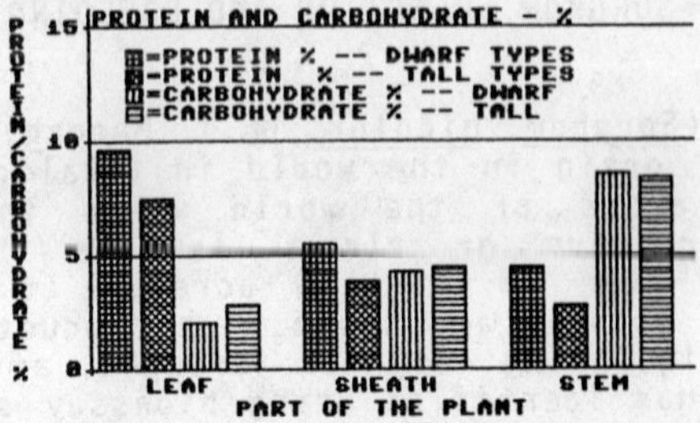

FIGURE 1. A COMPARISION OF DWARF AND TALL TYPES OF PENNISETUM IN PROTEIN AND SOLUBLE CARBOHYDRATE CONTENT.

TABLE 2. VARIATION OF GENOTYPES IN PERCENT OF PLANT IN THE DIFFERENT PLANT PARTS

Genotype	Leaves	Sheath	Rind	Pith	Infl.	PLANT TOTAL DRY WT
						g
N-75 (dwarf)	54.4 A*	14.3 A	25.4 D	5.8 B	0.0 B	79.5
Sel. 3 (dwarf)	37.2 B	6.7 B	41.0 C	10.0 B	5.0 A	150.6
Gahi-3 x 300086	23.8 C	1.8 C	61.4 B	13.1 A	0.0 B	690.5
Gahi-3 x 16	23.2 C	1.6 C	60.4 B	12.1 A	2.7 AB	273.8
23A x 16	18.8 C	2.8 C	61.2 B	12.9 A	4.3 A	250.9
Gahi3 x 10	16.3 CD	1.6 C	63.9 AB	16.7 A	1.7 B	728.2
23A x 300086	16.1 CD	1.3 C	69.2 AB	12.2 A	1.2 B	821.8
Gahi-3 x 14	7.5 D	0.9 C	75.8 A	14.7 A	0.5 B	631.4

Data of S. C. Schank and Karen Bjorndal.
* Means followed by different letters are significantly different at the 5% level according to Duncan's Multiple Range Test.

higher protein content. The stem component was carefully separated into the rind and pith, with clear evidence that in the tall genotypes, the rind portion can make up 60-75% of the plant weight. The rind is highly lignified and so would not contribute much to biomass yield. Barrier et. al. (2) suggest that since lignin is now a waste product of cellulosic materials processing, it might be used as a replacement for wood chips to provide steam for the process or for other industrial purposes. They further state (2), that the leaf portion of the plant could be marketed as animal feed (because of its high protein content), and the stem fraction processed to ethanol (because of its high carbohydrate content). These plant component studies will greatly help the plant breeder design a plant that will be more efficient and productive for specialized purposes in biomass, whether the end product is alcohol or methane.

SORGHUM SELECTION AND BREEDING

Sorghum (_Sorghum bicolor_ (L.) Moench) is the 4th ranking cereal grain in the world in total production and is grown in areas of the world where the climate is unsuitable for corn or rice. In the U. S., it is cultivated on over 15 million acres. It has the C-4 photosynthetic pathway and has a high productivity in most environments including dry or drought affected areas. Chynoweth (4) has identified using bioassay studies in the laboratory some sorghum lines as a high-level producers of methane. There is a great deal of variation among species within the genus, and like corn, plant breeders have worked for decades to improve sorghum. The variability in the genus has been used to tailor genotypes to fit various agronomic situations. Plant breeders have shortened internodes to fit combine heights for grain production, increased stalk juiciness and sugar content for syrup production, and increased leafiness and tillering for use of sorghum for pasture. Work by breeders has only recently been aimed toward production of large quantities of biomass. Cultivars which have been developed with high biomass potential are usually either very tall or they are relatively tall with a high production of grain. In both cases, the potential for lodging is usually large. Grasses with a C-4 photosythetic pathway returning 4-5 calories of energy product for each calories of energy used, having reasonable nitrogen requirements, and possessing drought resistance with relatively low water requirements, characterize prime crops for biomass enterprises.

At Florida, we have not duplicated the plant breeding efforts of Miller and Monk (12), but have simply tried to keep abreast of their work, and have tried many of the newer cultivars, and lines as they have been released by plant breeders in Texas and other stations. Texas data had indicated that, in general, sweet sorghums are the highest for fresh and dry biomass yields and nonstructural carbohydrates (Brix %), but are also tallest, latest in maturity and lowest in percent dry weight. The breeders feel that the principal advantages of grain sorghums include high percent dry weight and early maturity, but reduced biomass production and low Brix %. The hybrids of grain and sweet sorghums, high energy sorghums, attempt to capitalize on some of the advantages of both types. Biomass production, Brix %, and height are increased as compared to grain types while maturity is delayed only slightly. The high energy sorghums have a faster rate of production than either grain or sweet types of sorghum (12).

Production data in Florida (24) indicates that the sorghums are well adapted to the North Florida area. Total biomass production of the forage (or silage) types has been as high as 35.7 Mg ha^{-1} after 2 harvests with grain content from 5 to 20%. When forage and syrup types were compared in the same experiments over a 3-year period, the syrup types produced more than the forage types. Recent entries from the Texas A&M research program have been evaluated for 2 years in North Florida. When harvested twice during a season, highest yield has been 31.6 Mg ha^{-1} for a 2-year average. When evaluated with the older syrup types, none of the Texas entries have produced as much total biomass in Florida as the syrup types. These studies show how genetic selection has been used to identify the best sorghum lines for future testing in the GRI/IFAS Biomass project.

OTHER GRASS VARIETIES

Other forage grasses have been suggested as biomass crops, but none have the production potential of the three mentioned in detail in this paper. For example, _Cynodon dactylon_, cv. Tifton 44 bermudagrass, under heavy fertilization produced an average of 21.3 Mg ha^{-1} dry matter when harvested 6 times at 28 day intervals (9).

Erianthus arundinaceum (Retz.) Jesw. is a tall growing perennial bunch grass which can attain a height of 6 to 7 m. Annual dry matter yields of _Erianthus_ grown over a two year period at Ona averaged 46.4 Mg ha^{-1} (13).

Tripsacum dactyloides L. (eastern gamagrass) is an upright, perennial, warm season, rhizomatous grass which grows 1.5 to 3.0 m in height. Dry matter yields at Jay (northwestern Florida) averaged 23 Mg ha^{-1}, while at Ona, the average yield was 4.3 Mg ha^{-1} (13).

Setaria magna Griseb. (giant foxtail) is a tall annual bunch grass which grows from mid-spring through early fall. Maximum growth rate occurs during warm, moist weather, with dry matter yields as high as 14 Mg ha^{-1} (13).

SUMMARY AND CONCLUSIONS

Of all the tropical or sub-tropical grass species, it is clear that the highest biomass yields are:

1. Saccharum -- energy cane (dry matter yields up to 112 Mg ha^{-1}).
2. Pennisetum -- Napiergrass (3 yr. dry matter yields averaged 48.1 Mg ha^{-1}.
3. Sorghum -- forage sorghums (dry matter yields of 35.7 Mg ha^{-1} (with two harvests)

Additional breeding and selection work is necessary with all three genera to obtain the best genotypes. Sufficient genetic variation is present in each of the three genera to continue to make progress through conventional plant breeding methodology. Interspecific hybridization techniques will allow further genetic variation to be expressed and exploited in the biomass programs.

REFERENCES

1. Alexander, A. G. 1985. The energy cane alternative. Sugar series, 6. Elsevier Science Publishers B.V.

2. Barrier, J. W., M. R. Moore, R. O. Lambert and G. E. Farina. 1986. An integrated agri-refinery system using concentrated acid hydrolysis of lignocellulosic materials. Proc. Sixth Annual Solar, Biomass and Wind Energy Workshop. Atlanta, GA. pp 94-98.

3. Chase, Agnes. 1948. The meek that inherit the earth. In Grass, the Yearbook of Agriculture. pp 8-15. U.S. Government Printing Office, Washington D.C.

4. Chynoweth, D. P. 1983. Gasification of land-based biomass. Final Report No. GRI-82/0058, Gas Research Institute, Chicago, IL.

5. Dujardin, M. and W. W. Hanna. 1984. Micro-sporogenesis, reproductive behavior, and fertility in five Pennisetum species. Theor. Appl. Genet. 67:197-201.

6. Grassl, C. O. 1962. Problems and potentialities of intergeneric hybridization in a sugar can breeding programme. Proc. Congr. ISSCT. Mauritius 11:447-456.

7. Hanson, A. A. and H. L. Carnahan. 1956. Breeding Perennial Forage Grasses. Tech. Bull. 1145. 116p. U.S. Government Printing Office, Washington D.C.

8. Hebert, L. P. and M. T. Henderson. 1959. Breeding behavior of certain agronomic characters in progenies

of sugarcane crosses at the United States Sugarcane Field Station, Houma, LA. Tech. Bull. 1194. 54p. U.S. Government Printing Office, Washington D.C.

9. Hellwig, R. E., J. L. Butler, and J. C. Johnson. 1985. Harvesting and processing forage irrigated with liquid dairy cattle waste. 5th Annual Solar and Biomass Workshop. Atlanta, GA. pp. 95-97.

10. Jauhar, P. P. 1981. Cytogenetics of pearl millet. Adv. Agron. 34:407-479.

11. Lawrence, G. H. M. 1951. Gramineae (Grass Family) In Taxonomy of Vascular Plants. pp 387-391. The Macmillan Company, New York.

12. Miller, F. R. and R. L. Monk. 1984. Breeding and development. In Sorghums for methane production, E. Hiler (Ed.). Annual Report, Gas Research Institute, Chicago, IL.

13. Mislevy, P., J. P. Gilreath, L. S. Dunavin and R. S. Kalmbacher. 1986. Alternate production systems--exotic grasses and unconventional herbaceous species. In IFAS/GRI Five Year Report on Biomass Potential. Elsevier Publishers, Essex, England. (in press).

14. Muldoon, D. K. and C. J. Pearson. 1979. The hybrid between *Pennisetum americanum* and *Pennisetum purpureum*. Herbage Abst. 49:189-199.

15. Musser, H. B., G. W. Burton and H. A. Schoth. 1948. Developing grasses for special uses. In Grass, the Yearbook of Agriculture. pp 367-370. U.S. Government Printing Office, Washington D.C.

16. Ozias-Akins, P. Z. Tabaeizadeh and I. K. Vasil. Somatic hybridization in the gramineae. Abstr. First Intl. Cong. Plant Mol. Biol. 1:44. University of Georgia, Athens, GA.

17. Powell, J. B. and G. W. Burton. 1966. A suggested commercial method of producing an interspecific hybrid forage in *Pennisetum*. Crop Sci. 6:378-379.

18. Prine, G. M. 1985. Leucaena and elephantgrass as energy crops in colder subtropics. 5th Annual Solar and Biomass Workshop. Atlanta, GA. pp. 149-152.

19. Rajasekaran, K. S. C. Schank and I. K. Vasil. 1986. Characterization of biomass production and phenotypes of plants regenerated from embryogenic callus cultures of *Pennisetum americanum* x *P. purpureum* (hybrid napiergrass). Theoret. Appl. Genet. (submitted)

20. Samuels, G. 1984. Potential production of energy cane for fuel in the Caribbean. Energy Progress 4:249-251.

21. Samuels, G. 1986. Growing sugarcane as a renewable energy crop. Proc. Soil Crop Sci. Soc. Fla. 45: (in press).

22. Schank, S. C., R. Mendes de Souza, and R. M. Miranda. 1975. Chromosomal hybrids between Pennisetum americanum and P. purpureum. Abstracts Southern Agric. Workers Conf. New Orleans 2:4.

23. Smith W. H. and J. R. Frank. 1985. Comparative biomass yields of energy crops. In Energy from Biomass, 3rd E.C. Conference, W. Palz, J. Coombs, and D. O. Hall, (Eds.). pp 323-329. Elsevier Publishers, New York.

24. Stanley, R. L. Jr. and L. S. Dunavin. 1985. Potential Sorghum biomass production in north Florida. Biomass Energy Development. Plenum Press, N. Y. (in press).

25. Tai, P. Y. P. and J. D. Miller. 1986. Genotype x environment interaction for cold tolerance in sugarcane. Proc. Congr. ISSCT. Indonesia (in press).

26. Vasil, V. and I. K. Vasil. 1981. Somatic embryogenesis and plant regeneration from tissue cultures of Pennisetum americanum and P. purpureum hybrid. Amer. J. Bot. 68:864-872.

27. Vasil, I. K. 1985. Somatic embryogenesis and its consequences in the gramineae. In: Tissue Culture in Forestry and Agriculture, pp 31-47. Henke, R. R., K. W. Hughes, M. P. Constantin, and A. Hollaender (Eds.). Plenum Press, New York.

DISCUSSION

S. Kresovich, Texas A&M U.

I sit on the USDA Crop Advisory Committee for sugarcane and we just did a study of bio-tech research related to sugarcane. Basically, there are problems with all the data that has been reported. First of all, I think in the early work, people didn't discriminate between escapes and true resistance. Right now there isn't a whole lot of work going on in sugarcane. I think Dr. Irvine's work is probably some of the leading work now. But a lot of programs started that aren't using tissue culture anymore. A lot of things appeared in the literature which stated tissue cultured plants had disease resistance, and things like that. However, the disease resistance turned out to be not as stable as first reported.

Schank — An article that I read in one of the popular magazines stated something like 3 million sugarcane seedlings were going to be planted this year. The article also showed the greenhouses where the seedlings were being grown.

Kresovich — Those are from Crop Genetics in Dorsey, Maryland, and that's really like meri-stem culture. All they are trying to do is clean up genetic material.

Schank — So they have virus problems, is that it?

Kresovich — That's one of the main things they're looking at, and that leads to the other question. When we compare tissue culture to vegetative propagation, we always do an assay first to look for virus diseases. A lot of times when you plant vegetative material, you continue to propagate virus disease. Do you think those differences between tissue cultures and vegetatively propagated material are due to clean material being planted, or do you believe there are other factors?

Schank — No, we do not believe the lower yields are due to virus infection. We believe that it's due to an improved root system in the tissue cultured plants. With a true embryogenic culture, the vascular bundles of the shoot are linked up with the root system. If you use vegetative propagation, the roots come out adventitiously along the stem. When we planted tissue culture plants, we observed that the root system was complete and filled the container, where-as vegetative cultured plants had a less developed root system.

D. Hall, U. of London (U.K.) — Regarding the high yields that you got with the sorghum and the napiergrass, were those under conventional irrigation?

Schank — No, they are not under irrigation at all, but you want to remember that in Florida and in many regions of the Southeast we have rainfall of 75 inches per year and its interspersed nearly every day throughout the summer, so we have some very good growing conditions without irrigation. There is no irrigation provided at the location where Dr. Prine did his work.

Hall	Have you done any trials on energy cane in Florida as has been done in Puerto Rico and Baton Rouge?
Schank	I have not personally, but again other workers in Florida have used energy cane and it is equal to napiergrass in Florida. There's some debate now which will be better, and so they're going to be evaluated a little more stringently. Yields of nearly 50 metric tons per hectare were observed with both napiergrass and also with energy cane.
L. Zsuffa, U. of Toronto (Canada)	Does the increase level of polyploidy in these grasses always result in increased growth? Does it stop at the level of triploid or does it go up to the hexaploid level as you indicated?
Schank	We don't have data on the yields of the hexaploids in comparisons. We just identified them in the Fall of last year, and so we're now in the process of propagating the hexaploid types. However, we think that the ability to seed is a very exciting development, and we do have crosses of the hexaploid to a tall hexaploid type from Tifton, GA. We are very enthusiastic about what is coming out of those crosses.
Zsuffa	You spoke about some dwarfed and tall types, and some additional characteristics being associated with these types. Did you deal there with single clones and additional clonal characteristics of these, or is there a genetic relationship between these genotypes in general?
Schank	The genetic studies are not very complete, particularly with the hybrids of napiergrass and Pearl Millet. We have not gone into some of the specifics as we did on Pearl Millet at Tifton. So we do not have a lot of information about how well these hybrids will perform. But it does look like we could create a plant that is anything from about 3 or 4 ft tall to 20 ft tall by changing the inter-node length, and we have every height of plant in between. It's going to take some selection and some work, but you should remember that we have not been at this very long, so we aren't up with the corn or the sorghum breeders yet.

RAPID IMPLEMENTATION OPTIONS AND OPPORTUNITIES FOR ENERGY CANE BIOMASS PRODUCERS

Alex G. Alexander, Ph.D.
Estación Experimental
University Of Puerto Rico
Río Piedras, P.R. 00928

ABSTRACT

The traditional planting of sugarcane as a monolithic sugar crop is an increasingly uneconomic enterprise. Rising production costs and alternative sweeteners have negatively combined with overproduction and a chronically-depressed world sugar market. Producer diversification to alternative farm commodities is one option in some circumstances. However, for most tropical cane-growing nations, an internal diversification to multiple biomass products from the whole cane plant is a highly-valid alternative.

Cane production as a total biomass commodity, or "energy cane", is a concept of management designed to accommodate cane's latent growth attributes, whole plant harvest, and integration with generically-related tropical grasses. A modular systems approach to the necessary changes in cane agriculture is herein presented. While certain elements of historic cane agriculture are retained, there are new features of personnel training, introduction of new varieties and species, clean harvest of high biomass tonnages, and accommodation of year-round consumer demands for lignocellulose feedstocks. Opportunities are presented for near-term cash flow from fast-growing allied grasses, while long-term maximization of lignocellulose yield is seen in the breeding of second- and third-generation hybrid energy canes.

RAPID IMPLEMENTATION OPTIONS AND OPPORTUNITIES FOR ENERGY CANE BIOMASS PRODUCERS

INTRODUCTION

As previously described (4,2,7) energy cane technology is a concept of growth management for sugarcane and related tropical grasses. It is designed to maximize their total yield of biomass (lignocellulose) rather than sucrose. Fermentable solids remain important byproducts of the cane plant, now managed from the onset as a multiple-products commodity (7,6). Non-sugar bearing grasses, primarily generic-level relatives of <u>Saccharum</u> species, are integrated into the agricultural phase to assure a year-round feedstock supply to biomass-consuming industry.

The largely energy-related issues that prompted energy cane R & D in the 1970's remain valid today. However, newer and added concerns have superimposed themselves over the cane sugar planting world since mid-1984. The outlook for traditional sugar planting has so deteriorated that entire national economies have need to diversify sugarcane in one direction or another. New opportunities abound for this plant (7,6), but they are less clearly perceived than its problems as a monolithic sugar commodity. Nor do the gradual or partial diversification processes visualized in the late 1970's seem appropriate to needs of the late 1980's (7,23,24,16). This paper presents new options and opportunities for cane diversification, together with a modular systems approach designed to minimize expenditures of time and money.

ENERGY CANE ALTERNATIVES

<u>Current Outlook For Cane Sugar (1986)</u>

Owing to limitations of space, the remaining text is confined to sugarcane industry "real" conditions that have emerged since mid-1984, and to conditions in Puerto Rico since January of 1985. Puerto Rico's historic tendency toward leadership in both the good and bad features of sugarcane development (7) provides a suitable model for discussion purposes in this paper.

As an economically-viable farm commodity, cane sucrose has had a remarkably long and successful history. Its recognition as a special feature of <u>Saccharum</u> forms dates back perhaps 22,000 years (7). In modern times, the cane sugar planter has adjusted successfully to market fluctuations, inroads of disease and weather, and social and technological changes in its principal planting regions. However, today's sugar planter has arrived at a new and more sinister crossroads, a more subtle crisis than any of his grandfathers' day (7,24,5,6,8). Its cause is neither visible in the field nor described in periodicals and industry journals. The cause is a composite whose ingredients include an international oil cartel, political stresses in mid-eastern nations, social elevation of formerly low-cost labor, major shifts in consumer tastes and preferences, production of alternative sugar sweeteners, manufacture of artificial sweeteners, and dramatic rises in the costs of modern farming (7).

A persistent decline of world sugar values since June of 1984 is illustrated in Figure 1, together with a comparatively steady-state for the U.S. sugar market during the same interval. At this writing (mid-February), world sugar listings as a "commodity future" for May delivery are in the neighborhood of six cents per pound. This is roughly 6-9 cents lower than production costs in the low-expenditure scenarios of developing tropical countries. In the Philippines, for example, cane sugar production costs 12-13 cents per pound (10,21). The bulk of the crop is sold domestically at around 15 cents per pound, with lesser amounts exported to the higher U.S. market. The outlook for Philippines sugar is regarded as highly unfavorable, with only about 40 percent of milling capacity being realized (10), while various approaches to cane diversification are being studied (21).

The outlook is not much better for suppliers to the U.S. sugar market, which has been stabilized by price supports at 20-21 cents per pound (Figure 1). Sugar planting remains highly uneconomic for Puerto Rico where costs exceed 30 cents per pound (7). Mainland planters are appreciably more cost effective but have barely survived in recent years. Well-publicized efforts to reduce or eliminate sugar price supports have met counter-efforts by sugar interests. Although one group has elequently argued that reduced supports in the neighborhood of 14-15 cents per pound would bring long-term benefits for everybody (15), the outlook for U.S. sugar as the historic, monolithic farm commodity remains cloudy at best.

Cane Outlook as a Multiple-Products Commodity

A vastly improved outlook emerges for cane when evaluated as a multiple-products commodity, that is, with production costs shared among a series of salable products rather than by sugar alone (7,6,8). Potential new products exist even for the sugars of conventional sugarcane, as for example certain "sucro-chemicals" (28,29), or the fuel alcohol of Brazil's cane energy program (11,26). However, the great untapped products potential of sugarcane resides in its lignocellulose components (Figure 2). This fraction comprises over 60 percent of conventional sugarcane dry matter, and nearly 70 percent of energy cane DM (7,30,1). In energy cane technology, clean lignocellulose assumes such importance that the crop's agricultural management is reoriented to maximize its yield. Sugar and molasses remain as important, valuable byproducts of the energy cane dewatering process (7,5).

Lignocellulose "Hyerarchy" Options. In historic cane sugar planting the principal byproduct of sugarcane dewatering processes was bagasse. It was perceived to have only limited value as an in-house fuel. Its day-to-day disposal was of genuine concern to mill engineers who could ill afford to have mill yards and factory operating space cluttered with mounds of debris. About the last thing they needed were efficient boilers that provided daily heat and steam requirements with minimal fuel consumption.

A totally different outlook presides in an energy cane enterprise (7,5). The bagasse itself has become the primary product. There are greater quantities to be delt with per ton of recovered sugar. Its composition bears relatively greater amounts of "whole cane" materials, including leaves, tops, and trash, where formerly the millable stalk predominated. Equally important, the enterprise has a built-in steady

Figure 1

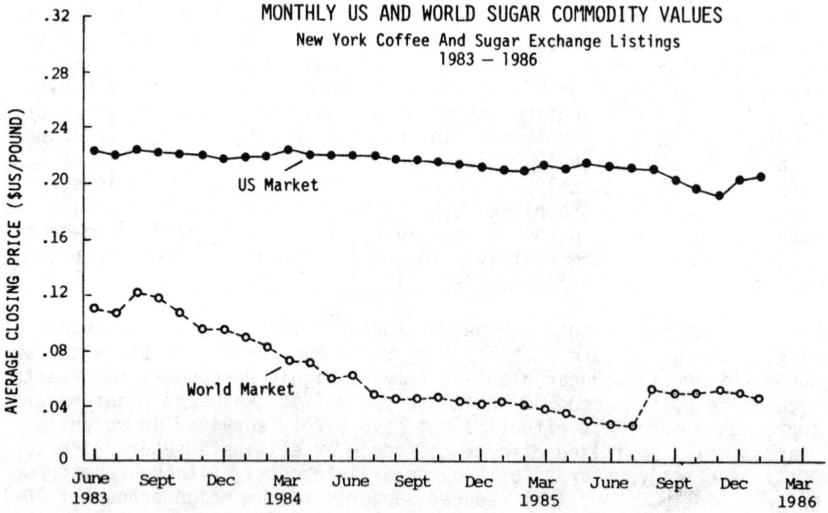

Figure 2

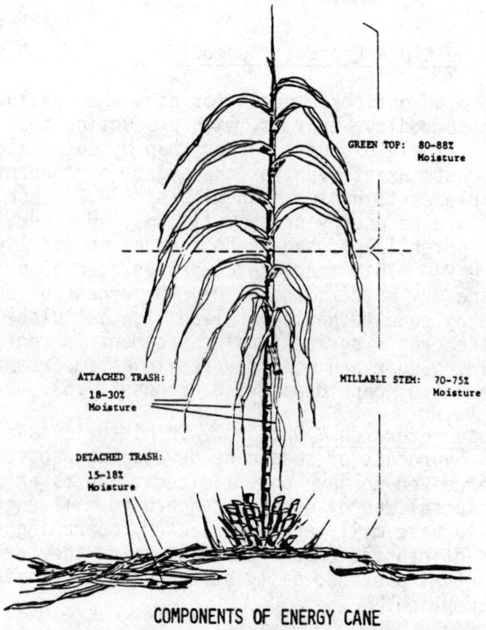

Above-Ground Components of the Energy Cane Plant and Stool Complex. The Illustrated Stem Would Be Approximately Seven Months Old, and the Subtending Crown About 14 Months Old. After Alexander (7). Reproduced by Permission of Elsevier Science Publishers B.V., Amsterdam.

demand for this bagasse as feedstock for a nearby lignocellulose-conversion facility. In this scenario there is no such thing as "excess" bagasse creating disposal problems. Efficient boilers are desired as a means to minimize bagasse diversion from a more useful and economically-attractive feedstock role (7,6). In some circumstances there would be no bagasse consumption at all by the dewatering plant. The latter's needs for heat and electricity would derive from the companion-plant's operations (7,5).

The burning of bagasse to generate electricity is still widely perceived as the "realistic" way to bring sugarcane production to a multiple-purpose basis (7,14,21,13,27). However, an energy cane enterprise designed specifically to maximize lignocellulose in quantity and quality would at the same time strive for maximum value in its conversion-phase products (7,6,5,8). In this context, the product options for lignocellulose-consuming industry can be depicted in a "hyerarchy of values", in which the direct firing of green, unprocessed biomass is a lowest-value opportunity. Major value upscaling is achieved by drying and/or size reduction of the same humid boiler fuel. Progressive improvement opportunities include conversion to proprietary fuels, utilization as feedstock for primary chemicals manufacture, and in feedstock applications for an expansive range of fossil petrochemicals substitutions (7,6,5,32).

Conceptually, the hyerarchy of values for lignocellulose usage is topped off by extremely high-value products (medicinals, perfumes) that have only limited tonnage or volume markets. For an energy cane enterprise, an ideal placement today on the lignocellulose hyerarchy would be in the range of proprietary fuels; however, the physical-plant designs could at the same time accommodate chemical feedstock applications some 15 to 20 years in the future (7,6). The latter point is of more than academic interest. For example, engineers today deal at routine disadvantage with old sugar mill boilers and power plants, designed in the 1940's and 1950's without inkling of present needs for energy and environment conservation. On today's drawing boards, a lignocellulose-conversion facility intended to function some 40 or 50 years can well anticipate product types, needs, and opportunities that are totally different for the plant's final 20 or 30 years. In particular, provision can be made from the onset for adequate physical space, provision of basic utilities, transportation and storage opportunities, and siting relative to integrated production operations for energy cane and food crop commodities (7,5).

Species Integration Options. Cane management as a maximum growth commodity is intended to supply optimal quantities of clean, low-cost lignocellulose to biomass consumers (7). However, whether utilized directly as boiler fuel or processed to higher-value products, the consumer must receive his feedstock supply on a reliable, year-round basis. In most sugar-planting regions it is not possible to harvest and mill cane as a continuous operation, nor is the storage of humid bagasse for off-season applications a realistic alternative (7,8). In energy cane technology a solution is found in concurrent production of high-yielding tropical grasses. As generic-level relatives of *Saccharum*, such species have high-yield capabilities under the same climatic and agricultural circumstances that made cane planting possible, but they do not require transport and milling operations as part of their dewatering process. They are solar-dried and compacted in the field, then stored for utilization as bagasse substitutes at the consumer's convenience (7,6,5).

In past references (7,3) the supplementary grasses have consisted of "short-rotation" species (sorghum, Sudangrass x sorghum hybrids) and "intermediate-rotation" (napier grass, napier hybrids, S. spontaneum spp.). While no sugar is recovered from such plants, their solar dehydration in the open air does in fact offer very distinct advantages over the cane plant. The latter is never dehydrated to a wholly satisfactory state, even after incurring the expense of transportation and milling operations (Figure 3). Its solar-dried relatives are dehydrated cheaply to about 15 percent moisture, as opposed to roughly 50 percent for bagasse, and they are ultimately hauled from the plantation in a compacted, partially size-reduced state offering a more cost-effective BTU payload (7,3).

Since 1984, the concept of allied grasses production as supplements to energy cane has broadened materially with new opportunities for private- and plantation-scale planters of these species. For Puerto Rican growers, these opportunities stem from livestock feed values that always resided in the species but in a largely underrecognized and under utilized state. For example, both short- and intermediate-rotation grasses are excellent cattle feeds when harvested at appropriate intervals in their growth cycles (7,8). There is now evidence that even when harvested in the mature, highly-fibrous periods ideal for maximum lignocellulose, these materials are very satisfactory feeds and feed supplements when blended with small quantities of molasses (7,3), or with high-protein bearing tissues of Leucaena leucocephala (9).

Although recognized as such only recently, the livestock feed sales options for species planted primarily as biomass resources have very real economic interest to Puerto Rico, and practical interest to insular farmers — many of them former sugarcane "colonos" (7,6,8). For example, in 1984 alone, approximately 115 million dollars were spent on imported livestock feed materials (25), virtually all of which could be produced locally or substituted with locally-available plant species. In the single instance of 4 to 6 months old napier grass, a suitable base material for cattle feed blends can be procured locally at less than one-fifth the cost of its imported counter-part (Table 1). The qualified growers are located precisely in those insular areas most needful of new agricultural employment, and many already own cattle in experimental- or commercial-scale operations.

The potential sales outlets have important psychological" as well as economic benefits (7). To a grower newly engaged in biomass planting, and who may never understand completely what lignocellulose-based enterprises are all about, the realization that his products will have alternative sales outlets can be comforting indeed. For the larger energy cane enterprise aiming ultimately for higher-value products, the interim sales of fast-growing grasses would be economically attractive to investors. Cash flow from such sales could begin up to 18 months prior to factory startup (Table 2).

In a broader sense, the expanded opportunities for supplemental grasses producers is consistent with a broadening mainland interest in alternative farm and forestry crops for industrial applications (31). The U.S. Department of Agriculture recently publicized some 11 species believed to have farm commodities potential as industrial feedstock in

Figure 3

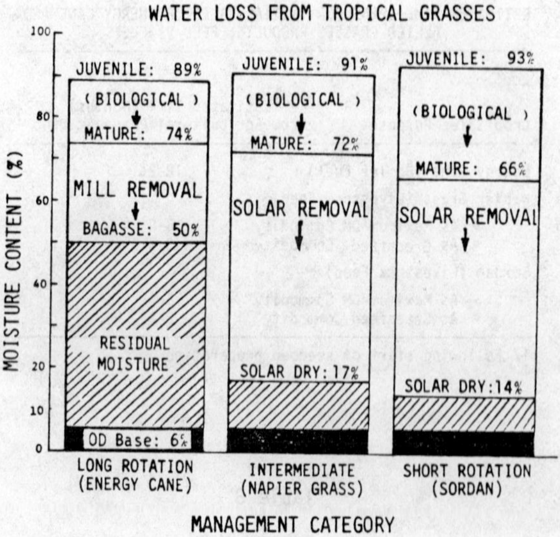

Characteristic Moisture Contents of Energy Cane and Alternative Grasses at Different Stages of Management. Solar Drying in the Field is far More Effective Than Milling for Moisture Removal After Harvest. After Alexander (7). Reproduced by Permission of Elsevier Science Publishers B. V.

Table 1

PR LIVESTOCK FEED IMPORTS AND COSTS; 1984

• Materials: Corn, Grains, Cotton & Soybean Meal	
• Costs: Approximately $115,000,000 Total (19)	$/Ton
○ Materials	126.00
○ Freight	24.00
○ Total Cost/Ton.	150.00
• Cost Of Local Napier Grass, Solar-Dried, Delivered, Size-Reduced	28.50 1/

1/ Estimated from 1982 production data (3)

34 states, Puerto Rico, and the U.S. Virgin Islands (Table 3). Their designation as "industrial materials" crops is perhaps a more practical terminology than the simple "biomass" or "energy" crops designations that often conjure up visions of green boiler fuels or automotive alcohol and little else. Moreover, current U.S. government emphasis on reduced farm subsidies and price supports in favor of free-market agriculture is certain to encourage investigation of alternative plant cropping opportunities.

Table 2

ESTIMATED TIME-SPANS FOR INITIAL SALES OF ENERGY CANE AND ALLIED GRASSES PRODUCTS; FEED Vs FUEL

Crop/Sales Purpose	Estimated Time (Months) From Agricultural Phase Startup 1/
Energy Cane (Boiler Fuel)	18-24
Napier Grass (Livestock Feed)	
• As Maximum DM Commodity	6-8
• As Greenfeed Commodity	3-4
Sordan (Livestock Feed)	
• As Maximum DM Commodity	3-4
• As Greenfeed Commodity	2-3

1/ Following start of seedbed preparation

Table 3

USDA-EVALUATED PLANTS FOR INDUSTRIAL MATERIALS CROPPING

Crop 1/	States Having Potential For Commercial Production
Chinese Tallow	NC, SC, GA, FLA, ALA, MISS, LA, TX, NM, AZ, HAW
Grambe	MKH, OHW, IND, IL, KY, TX, MISS, MD, CO, WY, LA, CT
Cuphea	OR, CAL, IL, IND, IOWA
Guayule	TX, CAL, NM, AZ
Jojoba	CA, AZ, TX, FLA, NM, HAW
Kenaf	NC, ALA, GA, SC, TX, ARK, CAL LA, FLA, NM, MISS, KY, OKLA
Lesquerella	AZ, NM, TX, OKLA, CO, CAL
Lunaria	WASH, OR, ALASKA
Meadowfoam	WASH, OR, CAL, NM, KS, IL, TENN, IND, AZ, TX, ALA, VA, IOWA, MO, PA, OKLA, GA, MD, OHIO, WVA
Winter Rapeseed	WASH, KS, NM, LA, GA, VA, OR, CO, MO, TENN, SC, IDAHO, OKLA, ARK, KY, NC, MONT, TX, MISS, ALA, WVA
Veronia	HAW, PR, VI,
Wood Byproducts	ALL STATES

1/ From "Growing Regions for Crops", ref. no. 31

Integrated Biomass And Food Crops. A long-recognized challenge to energy-planting is a workable integration of biomass and food crops where production resources prohibit their unlimited development as independent commodities (7,16,12,25). In theory, an established energy cane enterprise automatically becomes a purchaser of plant residues, including food-crop residues, from farms suitably close to justify their collection, drying, and delivery (4,7,5). Under Puerto Rico conditions, it was estimated that up to 15 percent of an energy cane consumer's needs for lignocellulose could be supplied in this way (7). However, it was also believed that such residue purchases were as much an act of good "public relations" and a stimulant to local acceptance as they were means of efficient resource utilization. Miscellaneous materials supply was not to be relied upon as a decisive fraction of the energy cane consumer's annual needs.

However, possibly as a result of increased confidence and maturity of thinking, together with progress in biomass conversion technologies, more discrete contributions of food crops to biomass-oriented industry are foreseen (7,6,5). An example of this is found in Puerto Rico's "new agriculture" plans that include major expansion of the insular pineapple industry (20). The pineapple is a CAM pathway representative having very large DM yield potentials from its highly-fibrous leaf residues (22,17). Although less than energy cane, these yields are roughly comparable to those of conventional sugarcane as managed in Puerto Rico today. With expansion of the pineapple industry to the neighborhood of 20,000 acres, a practical interest emerges in the crop's lignocellulose residue. Similarly, there are increased revenue opportunities for pineapple planters from sales of a second product. Moreover, by using the CAM pathway of photosynthesis, this plant would probably utilize more efficiently than sugarcane the available water resources (7).

New Benefits Of Integrated Cropping

The original purpose of integrating the "allied" tropical grasses with cane in energy cane agriculture was to secure a continuous supply of inexpensive, clean lignocellulose to processors and consumers. To the enterprise investors, these are critically-important issues that deserve solutions in a tropical scenario. Energy cane can assure this security of continuous feedstocks supply given the aid of supplemental species (7,6).

However, additional benefits would accrue in terms of improved agricultural employment, and in the technical management and security of the energy cane field operations themselves (7). By "improved" employment it is not meant that the grower's labor costs will be reduced (although they would be lower on a per ton of product basis); rather, the number and quality of farm jobs will be increased in that vicinity (Table 4). This in turn is a form of long-term insurance for the entire energy cane enterprise (7). For some types of cane workers, the integrated production operations will offer year-round employment for the first time in centuries. The "tiempo muerto", or "dead time" as it was known in historic cane sugar planting (18), would virtually disappear. Much greater attention would be given to training and supervision of hired help, from the plantation manager down to the most temporary day labor (7).

Table 4

ESTIMATED COMMODITY ACREAGE PLUS FULL- AND PART-TIME JOBS REQUIRED BY A FULLY INTEGRATED ENERGY CANE INDUSTRY IN PUERTO RICO

Field Operations/Commodity	Projections (8), For — Acres	Jobs
Primary Energy Cane	70,000	21,000
Cane Seed Farms	4,000	1,000
Napier, Napier Hybrids	20,000	3,000
Napier Seed Farms	500	50
Allied Feed Grasses	10,000	1,000
Allied Food Crops 1/	15,000	—
Biomass Storage, Processing	2,000	3,000
Totals:	121,500	29,050

1/ Primarily pineapple and rice.

Opportunities for increased security of energy cane field operations are similarly improved beyond those of conventional sugarcane (7). For example, in addition to the permanency and superior training of employees, the physical integration of non-combustible allied grasses can be done in such manner that fire hazard is reduced to a minimum. Both accidental and "malicious" fires are a significant threat to energy cane stands approaching maturity and bearing many tons per acre of highly-inflammable "trash" (leaf and leaf-sheath tissues that have died and desiccated). However, because the allied grasses are still green and non-combustible at harvest, their carefully-planned dispersal throughout the energy cane plantation offers effective firebreaks (7). As illustrated in Figure 4, even maliciously-started fires would be confined to a small acreage in correctly-integrated plantings of cane and napier grass.

In addition to the more effective utilization of personnel and operations security, the farm machinery base of an energy cane enterprise can be employed more efficiently for multiple species production (7). Costs of farm machinery have risen dramatically in recent years and are seen to be prohibitive for certain "small" landowners otherwise qualified to produce the supplemental allied grasses. As illustrated schematically in Figure 5, many equipment items needed for energy cane can be made available for grasses production as well in a correctly-planned and managed production campaign. As is true of the agricultural phase employees, the "dead time" for costly and otherwise depreciating machinery would be reduced to a minimum.

STARTUP OPTIONS AND OPPORTUNITIES

Whether taken by private enterprise or by government agencies, the decision to diversify sugarcane internally, from a monolithic sugar crop to a multiple-products biomass commodity, would appear to be fundamentally sound for virtually all sugarcane industries in the developing tropical world (7). In fewer instances, particularly in subtropical and "developed" regions, a case can sometimes be made for an external diversification of former cane lands to completely new farm crops, or even to non-agricultural land use alternatives. The balance of this paper considers the changeover of cane to a multiple-products system, by means that will economize expenditures of time, land, and financial resources.

Modular System; Concept And Design

As already noted above, the energy cane "consumer" has a considerable range or "hyerarchy" of product options for the lignocellulose he will receive on a year-round basis. The consumer himself could range from an existing utility seeking only a cheap fuel, to a proprietary

Figure 4

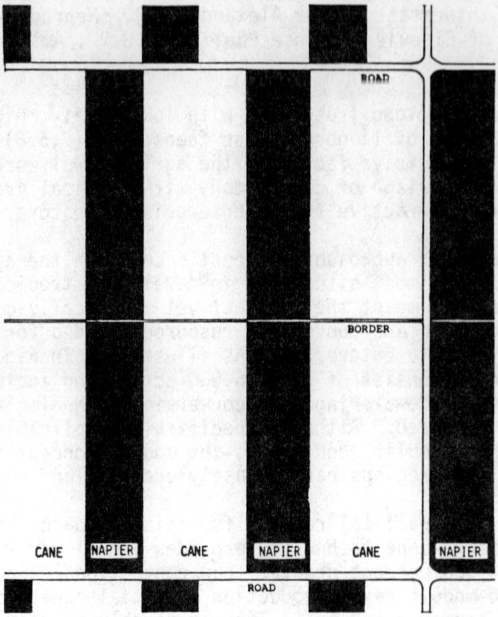

Proposed Field Layout Design for Intercropping of Energy Cane and Napier Grass as a Means of Minimizing Fire Hazard. After Alexander (7). Reproduced by Permission of Elsevier Science Publishers B. V., Amsterdam.

Figure 5

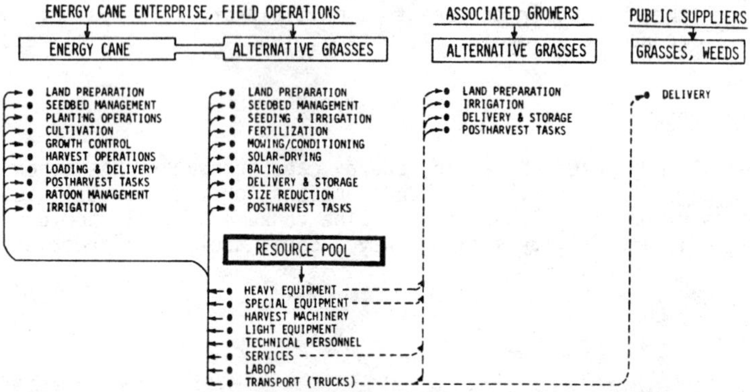

Production Resource Pools Owned by an Energy Cane Enterprise, and Their Availability to Discrete Cane and Allied Grasses Production Interests. After Alexander (7). Reproduced by Permission of Elsevier Science Publishers B.V., Amsterdam.

products manufacturer long frustrated with low quality, high costs, or inconsistent supplies of lignocellulose feedstocks (7,5,8). In any case, a critically-decisive factor is the agricultural workplan that assures a correct marriage of cane botany with tropical growing conditions, in a manner attractive to the enterprise investors.

A modular systems approach will best accomplish the agricultural-phase requirement for most situations in developing tropical countries (7). By "modular" is meant the smallest yet technically-complete package of cane production and conversion resources needed for an efficient and economically-viable enterprise. As illustrated in Figure 6, a single module would consist of 5,000-6,000 acres, and include all necessary agricultural, dewatering, and conversion technologies for the biomass therein produced. Although specifically applicable to Puerto Rico's constrained insular conditions, the module concept also would be appropriate for cane regions having vastly greater land resources.

There are very practical reasons for this approach. Foremost is the fact that energy cane technology provides annual per acre yields in the order of 250-300 percent greater than conventional sugarcane. Hence, the basic module has a production potential equal to 12,000-18,000 acres of former sugarcane. This is roughly the same production capacity that historically evolved around a conventional cane sugar enterprise; however, in the smaller energy cane module, costs are reduced materially for land overhead, land procurement and maintenance (fences, roads, canals, ditches), and transportation of the harvested products. Similarly, historically "fixed" per acre production costs (an acre foot of irrigation water, a hundredweight of commercial fertilizer) have new and favorable impacts in a modular system. A simple economic truth is that for Saccharum and its generic relatives it is vastly cheaper to increase yield per acre than to bring more acres into production (7,8,3).

Also favoring the modular approach is the need for educating management and for on-site training of fieldmen and labor engaged in energy cane production. Energy cane technology is based on the comprehension of cane botany, its untapped growth potential, and management of this potential for maximum lignocellulose production rather than sugar (7). These are features of Saccharum and its botanic relatives that very few people have delt with more than superficially. Conventional sugarcane cannot be reoriented from a sugar crop to a maximum biomass system by decree; rather, a first module will require intensive personnel training throughout much of the first year. Rapid modular expansion occurs from year 2 onward (Figure 7), as the original cadre of trainees become fieldmen and teachers themselves in an acreage expansion requiring 4 to 5 years for completion.

Finally, the modular approach offers not only an economic and technically-systematic means for reorienting sugarcane, it offers a conservation of time as well. Assume, for example, that a large, ailing sugar industry wishes to convert 30,000 acres on a test basis rather than 5,000-6,000 acres. From the first module, sufficient trained personnel, management expertise, and machinery would be available by the end of year 2 to "seed" the startup of 4-5 additional modules in a mutually-supporting "cluster" (Figure 6). By the end of year 6, the 30,000 acres would be largely operational as energy cane, and sufficient expertise would now exist to convert a cane sugar industry of several million acres if so desired (7).

Figure 6

MODULE CONCEPT AND EXPANSION OPTIONS FOR ENERGY CANE

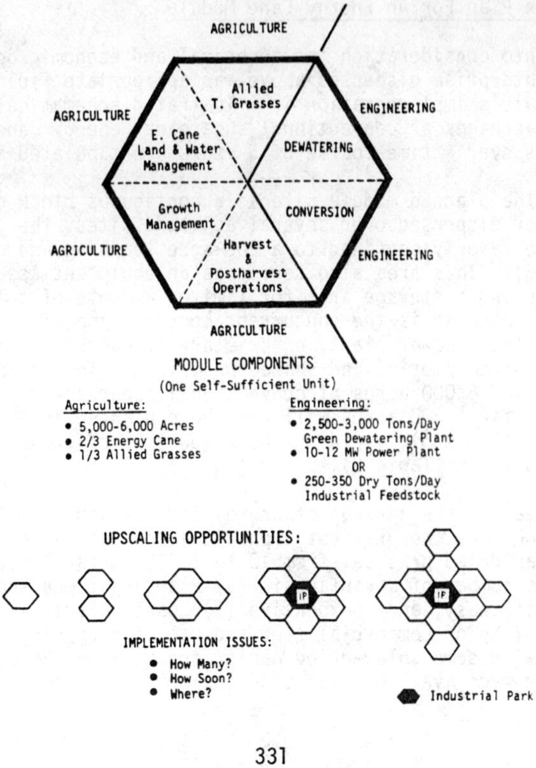

Figure 7

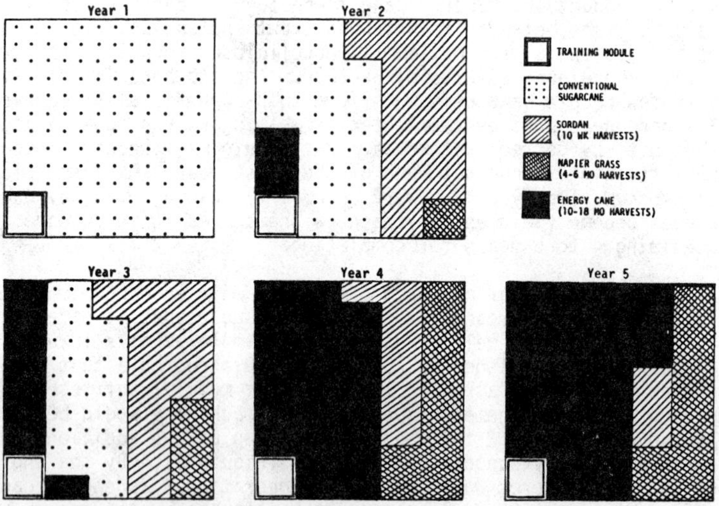

PROPORTIONAL PLANTING DIAGRAM FOR THE INTEGRATION OF SUGARCANE, ENERGY CANE, AND ALTERNATIVE GRASSES OVER A TIME-COURSE OF FIVE YEARS

Implementation Plan For An Energy Cane Module

Taking into consideration the technical and economic options of an energy cane enterprise discussed above, an appropriate field plan for the first-module's implementation is illustrated schematically in Figure 7. Specific acreages of conventional sugarcane, energy cane, and the allied grasses over a time-course of 5 years are tabulated in Table 5.

Whether the planned module site is a contiguous block of land, as illustrated, or dispersed over several adjacent sites, the year 1 operations are largely confined to a 200-acre training and construction area (Figure 6). This area also serves as an equipment-assembly and servicing site and a storage area for limited amounts of solar-dried grasses. Also assumed is the concurrent construction of a biomass-consuming facility (power plant, proprietary products factory) allowing 18-24 months for completion and shakedown trials. The training area is superimposed over 6,000 acres of conventional sugarcane. By the close of year 1, the first 200 acres will have been planted in energy cane and allied grasses (Table 5), yielding about 900 tons of grasses for storage or cattle feed sales (Table 6).

During year 2, the acreage of energy cane is approximately tripled (Figure 7); however, the greatest emphasis is on expansion of the short-rotation, solar-dried grasses, from 10 to 2,400 acres (Table 5). This emphasis has a number of advantages: (a) Planting these species is relatively fast, easy, and inexpensive (7), particularly in relation to ready availability of commercial seed and use of mechanicalgrain drills. (b) Together with some solar-dried napier grass, about 24,000 tons of dry matter are made available for storage, or for cattle feed sales as a

Table 5

MODULE ACREAGE PLANTING OF ENERGY CANE AND ALLIED GRASSES; YEARS 1-5

Year	Approximate Acreage Of — E. Cane	Napier	Sordan	Total Acres[1]
1	120	60	10	190
2	360	180	2,400	2,940
3	1,080	540	1,200	2,820
4	3,240	1,500	800	5,540
5	4,000	1,500	440	5,940

1/ Excluding service and storage space.

Table 6

ESTIMATED DRY MATTER YIELDS FROM AN ENERGY CANE PRODUCTION MODULE; YEARS 1-6

Year	OD Tons/Acre Year For E. Cane[1]	Napier[2]	Sordan[3]	Total Tons
1	0	750	180	930
2	3,600	3,000	21,600	28,200
3	10,800	11,250	32,400	54,450
4	32,400	32,250	32,400	97,050
5	97,200	37,500	11,160	145,860
6	120,000	37,500	7,920	165,420

[1] At 30 OD tons/acre; [2] At 25 tons; [3] At 18 tons (Refs. 2, 3).

means to create significant cash flow. (c) In planting short-rotation grasses, such as Sordan, seedbed preparation for the more difficult napier grass to follow is largely completed (7). In the subsequent two years, napier grass expansion is accomplished on some of this same acreage with little more than discing and rotavation of the Sordan stubble required as preplanting operations. Weeds normally bothersome to the perennial napier grass also are largely eliminated (3).

During the course of years 3, 4 and 5, energy cane itself becomes the vastly predominant species (Figure 7). Also in this time span, the lignocellulose-consuming plant and cane dewatering facility become routinely operational (7). A significant acreage remains permanently in napier grass to provide most of the off-season feedstock needed to supplement energy cane. A smaller acreage (400-500 acres) is still planted in short-rotation grasses. At this late stage the latter can still provide a degree of needed flexibility, a security against interruptions in bagasse or napier feedstocks supply that could still occur under rare circumstances.

Module Sugar And Molasses

Although full treatment lies beyond the scope of this paper, it is worthwhile to note the sugar and molasses recovered from the energy cane module. They are "byproducts" in this scenario, but very important nonetheless, far more valuable to energy cane management than was bagasse to historic cane sugar planters (7,6). Their value is underscored by the fact that lignocellulose has paid virtually all of their production costs (5,8). It is in this context that cane sugar planting has a realistic chance for permanent survival (7,5,8,6)

Noting that energy cane planted in year 5 will not yield its first sugar crop until year 6, the module's year 6 yields for sugar and molasses are conservatively estimated at 12,000 tons and 3.6 million gallons, respectively (Table 7). This amounts to about 3.0 tons sugar and 1200 gallons molasses per acre year. The sugar is "A strike" only, as advocated for an energy cane enterprise (7), while unrecovered sugar (strikes B and C) remain in the molasses. This lends added value to the molasses which is now "high test" (HTM) rather then the lower value "blackstrap" molasses (BSM). On the current world market (6.0 cents/pound for May delivery) the 12,000 tons sugar would be valued at $1.44 million. At 50 cents per gallon the 3.6 million gallons HTM would be valued at about $1.80 million. More favorable prices possibly could be negotiated; however, at over $3.2 million, the fermentable solids contribution to the energy cane module's annual revenue is not insignifican

Table 7

ESTIMATED SUGAR AND HIGH-TEST MOLASSES (HTM) YIELDS FROM AN ENERGY CANE PRODUCTION MODULE; YEARS 1-6

Year	Acres Planted	Tons OD Biomass[1]	Tons Sugar (A-Strike)[2]	HTM (Gallons)
1	120	0	0	0
2	360	3,600	367	110,200
3	1,080	10,800	1,102	330,400
4	3,240	32,400	3,305	991,400
5	4,000	97,200	0,914	2,974,400
6	4,000	120,000	12,240	3,672,000

[1] Based on references (2) and (3). [2] At 85% extraction.

REFERENCES CITED

1. Alexander, A. G., Sugarcane Physiology. Amsterdam: Elsevier Science Publishers. 1973. 753 pp.

2. Alexander, A. G., "Second Generation Energy Cane; Concepts, Costs, and Benefits", Fuels And Feedstocks From Tropical Biomass II, San Juan, P.R. (1982) April 26-28.

3. Alexander, A. G., Allison, W., Chu, T. L., Vélez-Santiago, J., and L. Smith, "Production of Sugarcane and Tropical Grasses as a Renewable Energy Source". Final Report to the U.S. Department of Energy, Contract no. DE-AS05-78ET20071 (1982) August.

4. Alexander, A. G., "The Energy Cane Alternative to Sugar Planting", Energy From Biomass And Wastes VII 186-203 (1983) Jan. 24-28.

5. Alexander, A. G., "Energy Cane as a Multiple-Products Alternative", Pacific Basin Biofuel Conference, Hawaii Natural Energy Institute (1984) Nov. 1-2.

6. Alexander, A. G., "New Alternatives for Sugarcane Utilization", Proc. P.R. Sugar Technol; 1-35 (1984) Dec. 7

7. Alexander, A. G., The Energy Cane Alternative. Amsterdam: Elsevier Science Publishers. 1985. 509 pp.

8. Alexander, A. G., Opportunities for Development of Sugarcane as an Economic, Multiple-Products Farm Commodity". Testimonial to the Agricultural Commission, Puerto Rico House of Representatives (1985) August 7.

9. Anon., Tropical Legumes: Resources For The Future. Washington, D. C.: U.S. National Academy Of Science (1979) 331 pp.

10. Anon., The Wall Street Journal (1986) February 6.

11. Barzelay, M., and S. R. Pearson, "The Efficiency of Producing Alcohol for Energy in Brazil", Economic Development And Cultural Change 31: (1982) Oct.

12. Bungay, H. R., Energy: The Biomass Option. New York, N. Y.: John Wiley And Sons, Inc. (1981).

13. Cabrero, P. M. "Proposal for the Study of Biomass Commercialization at Cambalache, Arecibo, Puerto Rico" (1984) April.

14. Hubbard, H. M., Opening Address, "Pacific Basin Biofuels Workshop Report", Hawaii Natural Energy Institute (1984) Nov. 1-2.

15. Knutson, R. D., Schmitz, A., and T. Earley, "Sugar Policy Alternatives and Impacts: 1985 and Beyond", Sugar Policy Research Task Force Publication, for Sugar Uses Group, Washington, D. C. (1985).

16. Lipinsky, E. S., "Fuels From Bagasse: Integration with Food and Materials Systems", Science 199: 644-645 (1978).

17. Marzola, D. L., and D. P. Bartholomew, "Photosynthesis Pathway and Biomass Energy Production", Science 205: 555-559 (1979) August.

18. Mintz, S., Worker In The Cane. Yale University Press (1980) 266 pp.

19. Personal communication, Federación de Asociaciones Pecuarias de Puerto Rico, Mayaguez, P. R. (1985) July-Dec.

20. Personal communications with the Puerto Rico Department of Agriculture, Autoridad de Tierras, San Juan, P.R. (1985) Nov.— Dec.

21. Personal communications with Eduardo F. Gamboa, Philippine Sugar Commission (1985) Dec. 11.

22. Personal communications with Dr. G. Samuels, Agronomist (retired), Estación Experimental, Universidad de Puerto Rico (1986) February.

23. Samuels, G., "Land Use Analysis of Puerto Rico", CEER-UPR Terrestrial Biomass Division, Internal Publication No. B-15 (1979).

24. Samuels, G., "Puerto Rico Sugar Industry Operations; 1971-1980. A Data Base". CEER-UPR Terrestrial Biomass Division, Internal Publication (1982).

25. Samuels, G., "Sugarcane in Food and Energy Crop Rotations in Puerto Rico", Alternative Uses of Sugarcane for Development, San Juan, P.R (1979) March 26 and 27.

26. Scheller, W.A., "The Use of Ethanol-Gasoline mixtures for Automotive Fuel". Energy From Biomass And Wastes I . Orlando (1977). Jan. 25-28.

27. Torres-Negrón, L. A., "Reforma Energética Depende de Bajos Costos", El Reportero, San Juan, P.R. (1986) Lunes 10 de febrero.

28. Vlitos, A. J., "New Product Developments for Sugarcane Carbohydrates", Alternative Uses Of Sugarcane For Development, San Juan, P.R. (1979) March 26 and 27.

29. Vlitos, A. J.,"Biotechnology in the 1980's and Beyond", Fuels And Feedstocks From Tropical Biomass II. San Juan, P.R. (1982) April 26-28.

30. Van Dillewijn, C., Botany Of Sugarcane. Waltham, Mass: Chronica Botanica Co. 1952. 371 pp.

31. Wheaton, E. R., Tankersley, H. C., Bagby, M. O., Rumburg, C. B., Naegele, J. A., Grinnell, G., Schaub, J. R., Barse, J., Cody, P.H., Johnson, P. B., Bebee, C. N., and D. H. Rahe, "Growing Industrial Materials: Renewable Resources From Agriculture And Forestry". USDA Critical Materials Task Force. Washington, D. C. 1985. 30 pp.

32. Wishart, R. R., "Industrial Energy in Transition", Science 199 (10) 614-618 (1978).

WATER HYACINTH (*EICHHORNIA CRASSIPES*) PRODUCTION IN EUTROPHIC WATERS

K. R. REDDY
University of Florida, IFAS
Central Florida Research and Education Center
Sanford, Florida 32771

L. O. Bagnall
University of Florida, IFAS
Agricultural Engineering Department
Gainesville, Florida 32611

ABSTRACT

Results of a field study on biomass yield, nutrient uptake, harvesting, handling, and processing of water hyacinths (*Eichhornia crassipes* [Mart] Solms) cultured in eutrophic Lake Apopka water are presented. Annual biomass yield of water hyacinths under central Florida's climatic conditions was 62 t (dry wt) ha^{-1} $year^{-1}$. If the biomass is harvested, this system would remove about 930 kg N ha^{-1} $year^{-1}$ and 124 kg P ha^{-1} year from eutrophic lakes.

The harvesting-handling system consisted of a winch driven rake, a cantilevered-flight takeout conveyor, chopper extruder, and pump. Observed system capacity was 32% of its 3220 kg (fresh wt) h^{-1} designed capacity.

Water hyacinths were found to be effective in reducing total N, P, and chlorophyll-*a* concentrations of the Lake Apopka water. Water hyacinth production on 21% of the lake surface and harvesting the biomass from the lake would remove any external and internal nutrient loads to the lake. This would result in the improvement of water quality and a reduction in the eutrophication process.

WATER HYACINTH (*Eichhornia crassipes*) PRODUCTION IN EUTROPHIC WATERS

Aquatic plants are widely dispersed in wetlands, lakes, and streams. Many species effectively use solar energy to fix carbon while absorbing problem nutrients and processing chemical contaminants in the surface waters. Traditionally, aquatic plants were considered as obnoxious weeds because of their rapid growth, which often results in choaking rivers and lakes. More recently, researchers have identified that with proper management these plants can actually improve water quality of polluted lakes and streams. The management of aquatic plants in these systems involves harvesting the biomass periodically to remove nutrients. The harvested biomass can be a potential source of feedstock for anaerobic digesters to produce methane. The characteristics of these plants, the special conditions that exist in aquatic environments, and the potential to use the biomass resource for beneficial purposes encourage design and development of integrated systems.

Among many aquatic plants found in tropical and subtropical freshwater ecosystems, water hyacinth (*Eichhornia crassipes* [Mart] Solms) is the most productive and maximum potential biomass yields of 64 g (dry wt) m^{-2} day^{-1} (104 t dry wt $acre^{-1}$ $year^{-1}$) have been reported under ideal conditons (Reddy and DeBusk, 1984) (Table 1). Private and governmental agencies in many areas of the world spend large sums of money each year to control this weed. In spite of expensive control measures, the water hyacinth problem has not been eradicated, although the populations are significantly decreasing. The characteristics that make this plant grow rapidly in polluted waters make it an ideal candidate for biomass resource and for improving water quality.

About 8% (1.2 million ha) of the total surface area of Florida is occupied by freshwater lakes and streams. As a result of nutrient loading from urban and agricultural activities, these ecosystems have become eutrophic. Nutrients present in these water bodies can be potentially trapped by using aquatic vegetation and the harvested biomass can be used as a feedstock for methane production. In addition, about 25% of the total surface area is under wetlands which can also be used to grow emergent aquatic plants such as cattails, bulrush, and reeds. The economic success of energy production from aquatic plants depends on the availability of low cost technology to manage the plants in natural ecosystems. The Gas Research Institute of Chicago, IL, and the Institute of Food and Agricultural Sciences at the University of Florida at Gainesville, FL, have embarked on a research program on the systematic development of technology for biomass production, harvesting, transportation and conversion to methane. In this program, water hyacinth was selected as a model biomass crop for possible inclusion in "biomass for energy" systems.

Channel III: This channel contained no plants. This treatment was included to demonstrate the effectiveness of water hyacinths in improving water quality.

Channel IV: Plants were stocked at the same density as in channel I and II. During the first 6 months, plants were harvested at constant rate once every month, while during the latter 12 months plants were not harvested. No harvesting treatment was included to demonstrate the effect of harvesting biomass on water quality.

Each channel was divided into 6 cells of equal area to provide replicate growth cultures. A 1 m^{-2} PVC frame with an attached basket of plastic mesh was placed into each cell. Baskets were stocked with water hyacinth plants at the same density as the remaining channel. Once every two weeks baskets were removed, allowed to drain 5 min, and weighed. Net increase in growth was calculated from the difference between initial and final weights. This procedure was repeated for a period of 18 months. Plants in the baskets were harvested at the same time when remaining channel was harvested.

Plant samples were obtained once a month and characterized for shoot/root ratio and for chemical composition. Water samples were obtained twice a week, both at the inflow and outflow of each channel, and analyzed for various water quality parameters. All water and plant samples were analyzed using standard methods.

Studies on Harvesting, Handling, and Processing of Biomass

The concepts of harvesting method development, handling, and processing of water hyacinth biomass are presented in the latter part of the paper. Harvesting was tested in a 0.4 ha pond stocked with water hyacinths. Lake Apopka water was gravity fed to this pond.

RESULTS AND DISCUSSION

Biomass Yields:

Average monthly net biomass yield (dry weight) of water hyacinths cultured in Lake Apopka water is presented in Fig. 1. Maximum growth rates were observed during the month of June, when the solar radiation values were highest associated with warm temperatures. Growth rates during the study period were in the range of 5 to 35 g dry wt m^{-2} day^{-1}. Plant growth was found to be poor during winter months. During the month of January 1985, plant growth was affected by a severe freeze, resulting in damage to photosynthetic tissue, but the rhizomes and the root system were alive. Plants recovered within 4-6 weeks and resumed normal growth. Annual biomass yields during the study period were in the range of 55-65 Mg (dry wt) ha^{-1} $year^{-1}$ (25 to 30 t dry wt $acre^{-1}$ $year^{-1}$).

Table 1. Comparative biomass yields of aquatic plants cultured under nutrient non-limiting conditions.

Aquatic plant	Annual yield	Highest yield during active growing season
	----t (dry wt) acre^{-1}	year^{-1}-----
Water hyacinth	47	105
Water lettuce	32	65
Pennywort	19	39
Salvinia	18	30
Azolla	5	10
Duckweed (L.)	8	20
Giant Duckweed (S.)	5	10
Elodea	6	21
Cattails (shoots)	20	--
Bulrush (shoots)	15	--

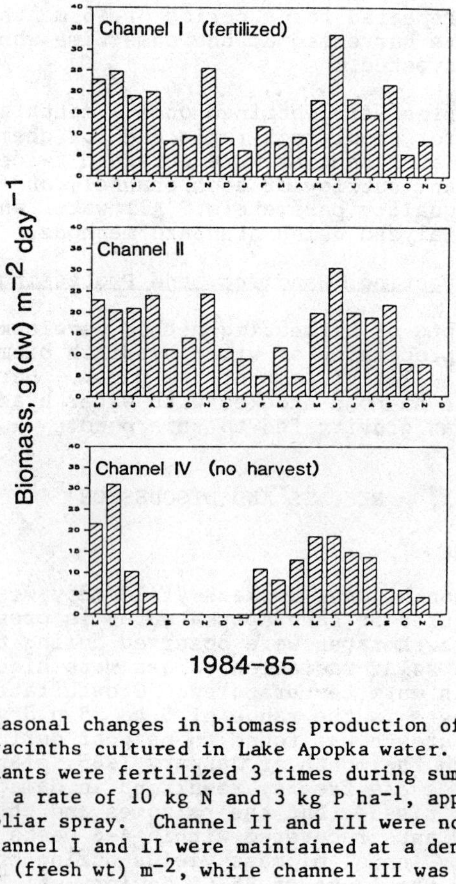

Fig. 1. Seasonal changes in biomass production of water hyacinths cultured in Lake Apopka water. Channel I, plants were fertilized 3 times during summer months at a rate of 10 kg N and 3 kg P ha^{-1}, applied as foliar spray. Channel II and III were not fertilized. Channel I and II were maintained at a density of 15-20 kg (fresh wt) m^{-2}, while channel III was operated with no harvest.

Solar energy conversion efficiency of water hyacinths cultured in Lake Apopka water was about 1.5% as compared to 3.1% by water hyacinths cultured under nutrient non-limiting conditions (Reddy, 1984). Biomass yield distribution showed distinct seasonality. During winter months (January and February) biomass produced represented 5% of the total biomass yield, followed by 7% during November-December, 16% during September-October, 21% during March-April, and the remaining 51% was produced during the summer months (May-August). Anaerobic digesters should be designed based on the average annual biomass yield to have a steady supply of biomass throughout the year. This can be accomplished by harvesting a steady rate of biomass based on average annual biomass yield. Any excess biomass during the summer months can be stored in the lake (Reddy et al. 1984).

Special management strategies such as foliar application of limiting plant nutrients or plant growth regulators were found to increase biomass yields significantly. Biomass yield increase due to foliar applications of nitrogen and phosphorus during June was about 30%. However, the increased yield did not sustain for longer periods. In controlled experiments, an application of plant growth regulators was also found to increase biomass yield by about 36%. Although application of nutrients and plant growth regulators can potentially increase yields, and cost of the chemicals is low, the cost of application over a large acreage of hyacinths can be a limiting factor in using these techniques. Application of these special chemicals may be more appealing when water haycinths are grown in small ponds for wastewater treatment.

Biomass yield responses of water hyacinths under varying culture conditions are presented in Table 2. Data indicate that water hyacinth productivity in polluted lakes is at the base level, and optimization of the system to achieve maximum yields will be controlled by the operating costs and overall economics of the system.

Table 2. Biomass yield responses of water hyacinth under varying conditions.

Culture system	Biomass yield t (dry wt) acre^{-1} year^{-1}
LAKE APOPKA WATER	
1. No management-base line	15 - 20
2. Constant density	20 - 30
3. Foliar fertilization	30 - 35
4. Foliar fertilization + growth regulator (GA_3)	40
BIOLOGICAL YIELD POTENTIAL	
1. Nutrient non-limiting	47
2. Nutrient non-limiting and optimum plant density	60
3. Nutrient non-limiting + summer growing conditions	105
4. Nutrient and environment non-limiting conditions	?
5. Genetic yield potential	?

Quality of Biomass

The biomass quality can be significantly affected by the nutrient composition of the culture medium. Eutrophic waters are usually poor in plant nutrients. This results in elongated root systems and the shoot/root ratios of the plants are usually in the range of 1-2. An increased proportion of roots in the biomass can significantly lower methane yields, since roots usually decompose slowly and produce less methane than shoots.

Plant tissue nitrogen and phosphorus were monitored throughout the year to determine the seasonal variability (Figs. 2 and 3). Nitrogen concentration of the plant tissue was in the range of 14-22 mg N g^{-1}, while the P content was in the range of 1-3 mg P g^{-1}. Foliar application of N and P did not increase plant tissue N and P content significantly. Tissue N content was about 2-3 times lower than the plants cultured in sewage effluent. Tissue P content was about 3-9 times lower than the plants cultured in sewage effluent. High N and P ratios in the plant tissue suggest that Lake Apopka water is limiting in P. Nitrogen and P ratios of the lake water were higher (10-25) than the ratios observed in the plant tissue (6-16), indicating that some P was probably released from the underlying sediments or from the decomposition of detritus tissue. Water hyacinths cultured under nutrient non-limiting conditions have a N and P ratio of 3-5.

Water total N or P showed poor relationship with either biomass yields or plant tissue N and P. In Lake Apopka water, most of the N and P is present in organic form which is not readily available to plant uptake. Any nutrients discharged into the lake are readily assimilated by algae resulting in high chlorophyll-*a* concentrations. Once hyacinth plants are established in the lake, algal populations will be decreased, and external and internal inputs of nutrients will be readily available to hyacinth plants. In other studies, when N and P of the water were present in inorganic form, positive correlations were noted between plant chemistry and water chemistry (Gossett and Norris, 1971).

Nitrogen and P removal by water hyacinths cultured in Lake Apopka water were 0.9-6.5 kg N ha^{-1} day^{-1}, and 0.06-0.74 kg P ha^{-1} day^{-1}, respectively. Both N and P removal rates were found to be lower in winter and highest during summer months. Annual average removal rates were about 850 kg N ha^{-1} $year^{-1}$ and 80 kg P^{-1} $year^{-1}$. Removal of biomass from the lake through harvest can result in export of above shown quantities of N and P from the lake.

Nutrient Availability

Nutrient availability in a sediment-water column can significantly influence the quality and quantity of biomass

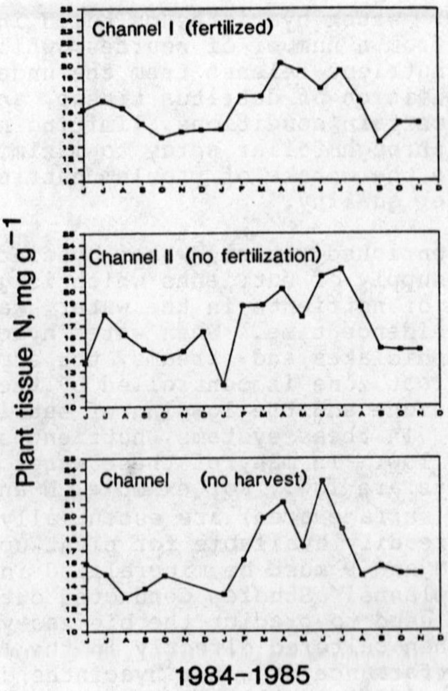

Fig. 2. Seasonal changes in plant tissue N of water hyacinth cultured in Lake Apopka water.

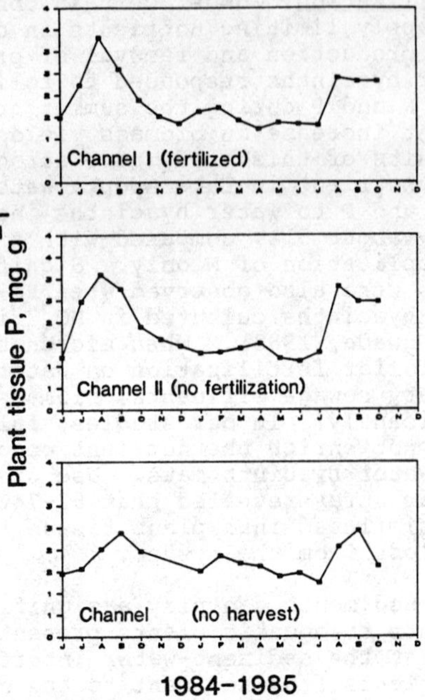

Fig. 3. Seasonal changes in plant tissue P of water hyacinth cultured in Lake Apopka water.

produced in lakes. Water hyacinths cultured in Lake Apopka derive nutrients from a number of sources, which include external inputs, nutrient release from the underlying sediments, decomposition of detritus tissue, and dinitrogen fixation. Under certain conditions, limiting nutrients can also be supplied through foliar spray to maximize plant growth and enhance the uptake of problem nutrients, without deteriorating water quality.

In nutrient-enriched waters, water hyacinth growth is regulated by the supply of nutrients which is governed by the concentration of nutrients in the water, water depth, flow rate, and residence time. When water hyacinths are cultured in eutrophiclakes and streams, the supply of nutrients in the root zone is controlled by the movement of water in the root zone and the loading of nutrients from external sources. In these systems, nutrient supply to the plants is usually low. In many of these water bodies, plant available nutrients are low. For example, N and P in Lake Apopka (12,500 ha surface area) are essentially in organic form and are not readily available for plant uptake. The organic forms of N and P must be mineralized into inorganic forms for use by plants. Studies conducted outside the lake cannot be readily used to predict the biomass yields of water hyacinths when cultured directly in the lake. Field studies on the performance of water hyacinths directly in a lake system are needed to realistically evaluate the biomass production.

Foliar Fertilization. Under certain conditions, it may be necessary to apply limiting nutrients in order to maximize biomass production and removal of problem nutrients. Water hyacinths responded to foliar fertilization of N and P during the summer growing season with a significant increase in biomass yields (DeBusk et al., 1985). Results of this study gave strong evidence that P was a limiting nutrient in Lake Apopka water, since the application of N and P to water hyacinths increased net biomass yields by about 51%, compared with a 6% increase resulting from application of N only. Significant increases in biomass yields were also observed when Fe-EDTA was applied to water hyacinths cultured in NO_3 rich agricultural drainage water (Ogwada, 1983). When micronutrients were applied through foliar fertilization on water hyacinths cultured in primary sewage effluents, biomass yields increased significantly. In our studies, foliar application of nutrients did not enrich the nutrient concentration of the water under water hyacinth mats. Use of ^{15}N-enriched urea in the foliar spray revealed that 62-74% of the foliar applied N was assimilated into plant tissue, while the remaining N was lost from the system.

Sediments. Sediments can play a significant role in supplying nutrients to aquatic plants present in the water columns. Mixing at the sediment-water interface, and diffusion of nutrients from sediment to the overlying water are the primary mechanisms by which nutrients are transported into the root zone of floating aquatic macrophytes. The rate of N transport from Lake Apopka sediment to the overlying water was found to be 42-124 mg N m^{-2} day^{-1}, and these rates are much higher than any other external N sources added to the lake.

Detritus. In a water hyacinth biomass production system, detritus plant tissue can be generated through the aging of plant parts, overcrowding, or freeze damage. Under all these conditions, detritus is generated in the photic zone and eventually deposited at the sediment-water interface. Some of this detritus can also be trapped in the root zone before it is detached and deposited on sediment surface. Previous studies have shown that in a managed system detritus contributes about 10-12% of the total biomass yield (DeBusk et al., 1983; Moorhead, 1986) and upon decomposition, this detritus can release significant amounts of nutrients which can be utilized by the live plants. In experimental channels at Zellwood, about 65% of the plant biomass contained live tissue, while 25% was present as dead plant tissue attached to the plant. About 10% of the total biomass resulted in detritus deposited on the sediment surface. Frequent harvesting of plants for biomass utilization will significantly reduce the production of detritus.

The decomposition rate of detritus was found to be faster in the root zone (k = 0.011-0.014 day^{-1}) than at the sediment-water interface (k = 0.006 day^{-1}). About 21-45% of the detritus N was released when decomposition occurred in the root zone, while 11-17% was released when decomposition occurred at the sediment-water interface (Reddy et al., 1986). Decomposition of detritus was found to be the same when N content of the detritus was in the range of 10-28 mg g^{-1} of tissue. Decomposition rates were higher when detritus P content was >2.2 mg P g^{-1}, as compared to the detritus with low tissue P.

Nitrogen fixation. Although water hyacinth plants do not have the capability of fixing atmospheric N_2, several N_2-fixing algae and bacteria are capable of growing on various plants parts. A few studies (Purchase, 1977; Nayak et al., 1979) have indicated that N_2 fixing bacteria such as *Azotobacter* sp. and *Azospirillum* sp. are capable of growing water hyacinth plant parts. Nitrogen fixation due to this association was found to be 1.7 mg N kg (dw)$^{-1}$ of plant tissue. In this study, no identification was made on the type of algae present on water hyacinth plants.

Cultural strategies

A steady rate of biomass production throughout the year is to some extent controlled by the cultural methods used in the management of the system. The two most important cultural methods, plant density and harveting frequency, and monoculture vs. polyculture systems, will be discussed.

Plant Density and Harvesting Frequency. Three growth phases are usually observed in a biomass production system where surface area available is fixed during the growing period. Phase I of the growth curve represents initial lag phase followed by exponential growth. During this phase, plant density is usually low, and net biomass yield per unit area is also low, but the specific growth rates (percent increase in biomass per day in relation to initial density) are high. Phase II represents linear growth phase, where biomass yields are highest and plant densities are in the intermediate range. Phase III represents exponential growth phase and biomass yields are low and plant densities are highest compared to other phases. The plant densities in the linear phase of the growth curve are considered as optimum plant density to achieve maximum biomass yields. Growth curves of water hyacinths and optimal plant densities were also found to be influenced by seasons (DeBusk et al., 1981; Reddy and DeBusk, 1984). During cooler months, optimal plant density was in the range of 500-800 g (dry wt) m^{-2}, while during warmer months optimal density range was 500-2000 g (dry weight) m^{-2}. However, these results were obtained from the experiments conducted under nutrient non-limiting conditions (Reddy and DeBusk, 1984). A much higher optimal density range was observed for hyacinths grown under nutrient non-limiting conditions while a narrow range was observed for plants grown under a nutrient-limited system (Lake Apopka water). Under both conditions, the growth rate of hyacinths was found to be maximum in a plant density range of 10-20 kg (fresh wt) m^{-2}. In a biomass production system, a harvesting strategy can be developed to control the density in optimal range. Depending on the biomass requirements, plants can be allowed to grow until they reach the higher end of linear phase, and plants can be harvested to the lower end of linear phase.

Monoculture vs. Polyculture Systems. Under central Florida climatic conditions, water hyacinth production is significantly affected by cold temperatures, and temperatures at or below the freezing point for several hours can result in the damage of photosynthetic plant tissue. To obtain sustained biomass yields, it may be necessary to grow cold tolerant aquatic plants with reasonably good growth rates along with water hyacinths in the same system. During the plant selection process, we have identified two cold tolerant aquatic plant species and both, are commonly known as pennywort (*Hydrocotyle umbellata* and *H. ranunculoides*, are native to Florida and predominately seen in winter in freshwater bodies, respectively. In a separate study using microcosm outdoor tanks, we have evaluated the performance of water hyacinth and pennywort for a period of one year under nutrient non-limiting conditions, and the results are presented in Fig. 4. Water hyacinth growth was affected by the freeze during January 1982, and biomass yields were near zero during January and February. In the same system, pennywort growth was not affected and recorded a growth rate of 8 g (dry wt) m^{-2} day^{-1}. During winter months, pennywort occupied the major part of the tank by pushing water hyacinth plants to one side, while during warmer months the situation was reversed, with water hyacinth plants occupying the major portion of the tank. Although pennywort plants offer a great potential as a possible feedstock for winter months in a water hyacinth production system, limited information is available on the growth characteristics and nutrient requirements to optimize the growth. Pennywort plants are intertwined and are hard to separate from one another, and the harvesting methods developed for water hyacinths may not be suitable for harvesting pennywort. The ratio of water hyacinth/pennywort standing crop needs to be determined in order to optimize the biomass production system. Although culturing both plant types in a large-scale system may not be cost effective, polyculture systems can play a significnt role in water hyacinth based sewage treatment systems, where year-round plant growth and water treatment are important.

The purpose of this report is to summarize the results obtained from the field studies conducted to: 1) evaluate the feasibility of growing hyacinths in eutrophic lake water; 2) develop management strategies to maximize the biomass yields; and 3) develop low cost harvesting, handling, and processing of water hyacinth biomass.

MATERIALS AND METHODS

Site description

Field experiments were conducted at the University of Florida's research farm located in Zellwood, Florida. This farm is adjacent to Lake Apopka (12,500 ha surface area) which is highly eutrophic. Four experimental channels (0.045 ha) are located at this site, which can be supplied with the water from adjacent Lake Apopka. These systems were used to evaluate the biomass yields of water hyacinths cultured in eutrophic lake water.

The site chosen for testing the harvesting systems was a 0.4 ha pond also located on the Zellwood research farm. The water depth in the pond was 0.9 m. The pond was located about 30 m north of the dike on the north shore of Lake Apopka. The pond was contained by a 1.1 m high dike. Water was supplied to the pond from the lake through a system of valves and pipes, entering the pond through a grate in the SE corner and overflowing through a drop structure in the NW corner.

Water Hyacinth Production Studies

Four concrete channels each 6.1 m wide and 61 m long and 0.6 m deep were used to evaluate the potential biomass yields of water hyacinths cultured in Lake Apopka water while improving water quaity. Lake Apopka water was allowed to flow into the channels by gravity at about 25 gpm,

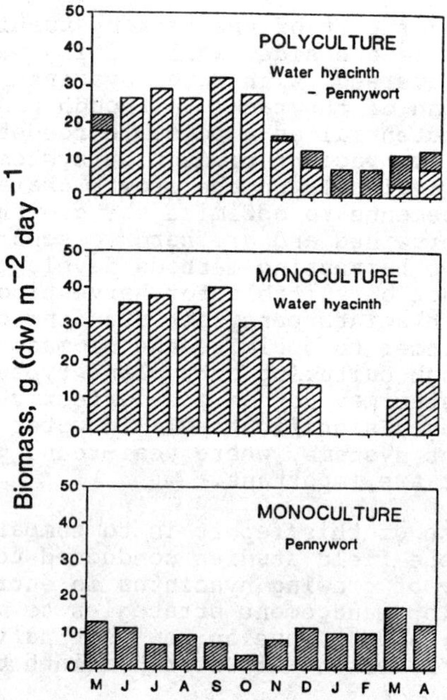

Fig. 4. Growth rates of water hyacinths and pennywort cultured in monoculture and polyculture systems.

resulting in an hydraulic retention time of 1.5 days. Water flow at the inflow was measured manually, while the flow at the outflow was determined using stage recorders. This study was initiated in June 1984 and continued for a period of 18 months. The experiment ended in December 1985.

Operation mode for the channels during the study period was as follows:

Channel I: Stocked with water hyacinth plants at a density of 15 kg (fresh wt) m^{-2}. Plants were sprayed 3 times during summer months, at a rate of 10 kg N and 3 kg P ha^{-1} to provide an additional supply of nutrients. Plants were harvested once a month to maintain initial plant density.

Channel II: Plants were stocked at the same density as in Channel I and received no foliar spray of N and P.

Harvesting, Handling, and Processing

The harvesting system was designed to remove the water hyacinth from the pond in weekly harvests. Based on production in ponds using water from Lake Apopka (Reddy et al., 1984), mean weekly production should have been 6440 kg (14200 lb). During the period of operation, production did not approach the predicted mean, so the harvesting schedule and amount removed per harvest were curtailed to prevent dicimating the population of plants. Production and plant condition improved in September and October, but complete coverage of the pond was never established.

The harvester was designed as a test facility for the concept presented in a study for the Gas Research Institute by Reynolds, Smith, and Hills (Warren et al., 1984). The intent of the concept was a high-capacity system that could be operated by one person from a fixed control station. The plants were to be gathered by a winch-propelled raking boom and moved to an elevating/chopping system where they would be lifted from the water and prepared for transportation and digestion.

Because the site was located 160 km (100 mi) from the operating base of the harvesting crew, the system was designed to harvest 6440 kg in 2 hours of operation once per week. Based on design procedure presented in an earlier paper (Reddy et al., 1984), the width of the rake should have been 5 m (15 ft) and the length of the takeout flights 12 cm (5 in). The rake and takeout were 6 m (20 ft) wide. The takeout flights were nominally 60 cm (24 in) long because this has been observed to be the minimum effective width of components in systems to handle intact water hyacinth.

The triangular system as described by Warren et al. (1984) was modified to fit the small, square pond. The winches were located at the corners of the pond, and powered and controlled through buried wire. The flexible boom was replaced with a rigid rake and the takeout was designed as a cantilevered-flight conveyor, a conept that appeared to have promise, because of the relatively small scale of this system. Because there was no electrical power available on site, power was supplied by a trailer-mounted power unit consisting of an engine-generator and an engine-driven hydraulic power supply.

The rake was 6 m (20 ft) wide and was maneuvered by winches at the four corners of the pond at 0.5 m s^{-1} (100 ft/min). Initially, the rake was a simple frame, supported on pontoons, which pivoted about a horizontal axis 46 cm (18 in) above the water, upward to clear the plants when being towed rearward, downward to engage the plants when being towed forward. Because this design was inappropriate for expansion and inherently weak, it was replaced with a laterally-folding rake shown in Fig. 5, using many of the same components, including frame and flotation. The new rake collapsed to a narrow frontal area for easy passage through the mat when being towed rearward and flared out when being towed forward to gather the widest possible swath of water hyacinth. The winches were driven by reversible gearmotors 0.5 kW (1/2 hp) on the main towing winch and 0.2 kW (1/4 hp) on the other winches. The main pull oline was 3 mm (1/8 in) stainless steel, and the other lines were 6 mm (1/4 in) polypropylene. The rakes were controlled by four double-throw switches on a control station near the takeout conveyor.

Fig. 5. Harvesting system - laterally-folding water hyacinth rake.

The takeout designed for this system, shown in Fig. 6, moved the plants in the water laterally to the approach of the rake, then elevated them up a 45° ramp. The lateral movement effectively formed the narrow chopped feed stream while the plants were still in the water, eliminating the necessity of reforming the stream while it was in the conveying system. The horizontal lower rail of the takeout conveyor was 6 m (20 ft) long, corresponding to the width of the rake. The flights were pulled by 2040 roller chain over 5 cm x 10 cm (2 in x 4 in) tubular steel rails, guided by nylon bearing-line flight shoes. Initially, the flights were 60 cm (24 in) long angles without tines, spaced on 60 cm (24 in) centers. Tines were added to improve aggressiveness of the flight action. The angles were replaced with square tubing when the more aggressive flights twisted the angles. The flight spacing was increased to 120 cm (48 in) and the penetration of the plants into the flight path was improved. Alternate flights were shortened to 56 cm (22 in) and 51 cm (20 in) to improve end clerance at the approach to the ramp where they had been jamming. Design speed of the conveyor was 0.5 m s^{-1} (100 ft min^{-1}). The takeout conveyor was mounted on jackscrews so that its elevation could be changed to match pond water level. Its hydraulic drive motor at the discharge end was controlled by a valve at the control station.

The initially-installed ramp was stationary sheet steel and the entire lifting force was supplied by the takeout. This was soon replaced with a flat-wire-belt conveyor (live ramp) in order to improve traction and delivery and to provide opportunity for greater conveying range and chopper feeding. A flail chopped was soon mounted on the live ramp. The live ramp was later elevated to match the 45° angle of

Fig. 6. Biomass handling system - take out conveyor.

the takeout elevating section to improve delivery to the chopper, with little success. The first live ramp was replaced with a live ramp specifically designed for the system (Fig. 7), having a horizontal chopper feeding section, which was a major improvement. Because the clearance between the ramp and the takeout was fixed, it had to be adjusted everytime the water level changed and jammed frequently. A new coordinated takeout/ramp support structure was built with automatic clearance adjustment based on volume of material passing up the ramp. The ramp was powered by a hydraulic motor in series with the takeout motor. The flail chopped was 60 cm (24 in) wide and had a 40 cm (16 in) diameter rotor which rotated 1800 rpm, powered by a hydraulic motor through a speed-increasing belt drive. The motor was on a separate circuit controlled by a valve on the contol station. Except for minor drive modifications, the chopper has operated unchanged since it was installed.

As is typical of developmental systems, all did not go on the water as it had on paper. The rake did not usually collect as much water hyacInth in a load as predicted. The collected hyacinth tended to spill around or under the rake as it was towed in due to eccentric loading and fragmentation of the collected mat. The lines tended to tangle in the mat, interfering with movement and positioning of the rake relative to the mat. The flights did not strip the plants smootly and cleanly from the rake mat, especially from the deeper laterally-folding rake. As the flights approached the takeout/ramp transition, plants tended to rotate clockwise out of engagement with the flights and

Fig. 7. Biomass processing system - ramp and flail chopper.

spill between the ramp and the rake. The flights tended to ride over the plants, piling them at the transition, creating a ball that could not pass through the clearance between the takeout and the ramp, and leading to jamming and stalling of the takeout. The jamming was exacerbated by the tendency of the cantilever flights to bind on the rail in that corner. The jamming necessitated manual labor to clear and maintain flow in the transition. The accumulation of large clots of biomass also caused irregular feeding, and occasional stalling of the chopper. This litany of problems has been corrected or is being corrected.

System design capacity was 3220 kg h^{-1} (7100 lb h^{-1}). Mean observed capacity was 1027 +/- 354 kg h^{-1} (2264 +/- 780 lb h^{-1}). Design rake capacity was 5100 kg h^{-1} (11260 lb h^{-1}) of active cycle. Observed capacity was 2750 +/- 2210 kg h^{-1} (6060 +/- 4860 lb h^{-1}). Designed takeout/chopper capacity was 33000 kg h^{-1} (72600 lb h^{-1}) of active cycle. Observed capacity was 1640 +/- 724 kg h^{-1} (3600 +/- 1600 lb h^{-1}). The major limitation on system capacity was the areal efficiency and jamming of the takeout. Total rake electrical energy requirement was 0.7 kJ kg^{-1}. Fuel energy consumption by the engine-generator and engine-hydraulic unit was 107 kJ kg^{-1}, of which 84% was by the engine-hydraulic system.

In the context of this development program, handling included the actual material handling operations and preparation of the material for handling and digestion. The intent of the development was to handle the wet plant material as a fluid.

In the initial phases, prior to fine reduction, and in the system in place at Zellwood, the material must be handled as a wet solid, either by continuous conveyors or in batches. The quantity handled in this sytem was well within the capacity of ordinary agricultural chain-and-flight conveyors. Subsequent handling was in batches, for simplicity of weighing for determination of system capacity.

The discharge from the flail chopper was delivered into a 60 cm (24 in) x 7.6 m (25 ft) farm elevator, powered by a hydraulic motor in series with the chopper motor. The product was delivered into a 1.2 m (4 ft) cubic basket for weighing and transportation to disposal, processing, or utilization.

The conveying system operated essentially without problems and had adequate capacity for any slug of biomass that could pass the chopper.

Fine reduction may be accomplished by fine chopping or by extrusion. Extrusion was chosen as the method of fine reduction preparatory to pumping because the equipment and expertise were readily available from prior work.

The extruders used were a 23 cm screw press, a 15 cm screw press, and a 15 cm screw extruder specially designed for pulping and feeding water hyacinth. Restriction on the discharge of the extruders was variable, with greater restriction giving finer reduction, lower throughput, and higher energy requirement. The screw extruders had been tolerant of irregular loading of variable material.

The screw extruders were fed with intact, coarsely chopped or finely chopped water hyacinth. The capacity was higher with the finer material, but the effective size reduction was better with the coarser material, which had a greater natural resistance to flow through the extruder. Some of the product was extruded twice, in order to further reduce the particle size and improve fluidity. The product from the screw presses was reconstituted by adding the juice back to the dry press cake.

The 23 cm screw press extruded 1390 +/- 300 kg h^{-1} (3060 +/- 660 lb h^{-1}) and an energy requirement of 8 +/- 1 kJ kg^{-1} (2.7 +/- 0.3 hp.h/T). The 15 cm screw press extruded 640 +/- 60 kg ha^{-1} (1420 +/- 140 lb h^{-1}). The 15 cm screw extruder extruded 610 +/- 160 kg h^{-1} (1340 +/- 350 lb h^{-1}) at intermediate restriction, 170 kg h^{-1} (370 lb/h) at maximum restriction, and 900 +/- 50 kg h^{-1} (2000 +/- 100 lb h^{-1}) at minimum restriction. The extruder did not perform as well on pre-extruded water hyacinth.

Though water hyacinth is 95% water, the water is trapped in cells and intercellular spaces, so the plant material is clearly a solid until it is very finely divided and an appreciable amount of the water is freed. Water hyacinth may be pumped by suspension of plants or plant fragments in a stream of water or by reduction to the point where the water in the plant structure is freed and the particle size is reduced sufficiently to be suspended in that water. The most appropriate types of pumps for pumping water hyacinth are those designed for pumping thick suspensions, such as the progressive cavity pump.

The progressive cavity pump is a positive displacement, rotary axial-flow pump in which an undulating metal rotor moves a fluid-filled cavity through a flexible rubber stator. They are widely used in waste management facilities. The pumps selected for testing were 10 cm (4 in) single-stage progressive cavity pumps. Most of the testing was done on a pump, rotating at 425 rpm, used to feed a continuously-expanding digester.

Coarsely extruded feedstock was pumped at 770 +/- 450 kg h^{-1} (1700 +/- 990 lb h^{-1}) and required 9 J g^{-1} of pumping energy and 140 +/- 60% supplementary water. Finely extruded feedstock was pumped at 8800 +/- 1400 kg h^{-1} (19300 +/- 3200 lb h^{-1}) and required 4 +/- 2% supplementary water.

Water Quality Improvement

The ultimate establishment of water hyacinth biomass for energy farms in freshwater lakes and streams depends on the impact of plants on water quality improvement.

Results of a field study conducted for a period of 18 months using experimental channels (See Methods section) will be presented in this paper. Although a number of water quality parameters were monitored both at the inflow (Lake Apopka water) and outflow (water leaving hyacinth channel), only changes in N, P, and chlorophyll a concentrations will be presented in this paper. Changes in these parameters are better indicators of the impact of water hyacinths on lake water quality, than any other parameter studied.

Total N concentration of the lake water varied considerably, with values of 2.5-8.1 mg N l^{-1} (Fig.8). Nitrogen in lake water was primarily organic and not readily available to plant uptake. During a 36-hr residence time in the channel containing water hyacinths, total N concentrations decreased significantly. Foliar application of urea during the months of May, July, and September did not alter the quality of water leaving channel I. Both channel I and II were maintained at a constant plant density by harvesting plants once every month. Channel III contained no plants, and in channel IV, plants were not harvested for a period of one year. Total N concentration of the outflow water of the channels containing water hyacinths was decreased by about 50%. Removal rates were significantly lower in the channel with no plants or the channel where water hyacinths were not harvested for a period of one year.

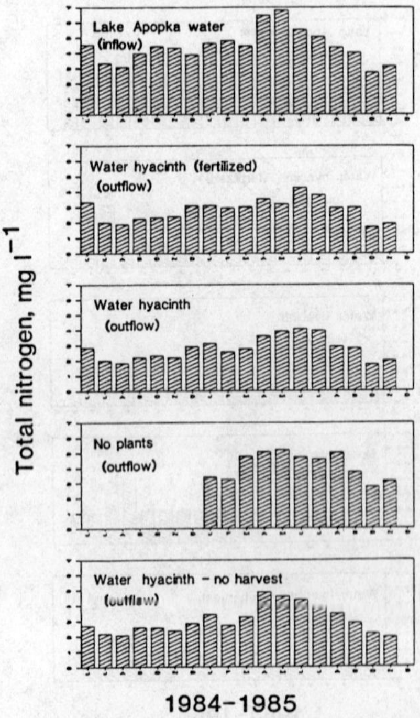

Fig. 8. Seasonal changes in total N of the inflow and outflow of water in Lake Apopka water - hyacinth system.

Maximum P concentration (0.85 mg P l^{-1}) of the lake was found in the month of August, while the minimum values (0.2 mg P l^{-1}) were observed in October (Fig.9). Chlorophyll a concentration of the lake water also showed similar trends as total N and P (Fig. 10). Water hyacinths were found to be very effective in decreasing the chlorophyll a concentrations of the lake water.

Nitrogen removal rates calculated based on reduction in mass loading indicate that about 50-60% of the incoming N was removed by the system, while up to 80% of the incoming P was removed. A significant removal of both N and P in this sytem was due to the settling of the particulate matter, especially algal cells of the lake water. This has resulted in the sediment build up of up to 30 cm over a period of one year. This type of situation will not arise when plants are cultured directly in the lake, because water hyacinth plants shade algae. In the experimental channels, lake water rich in algal cells was allowed to flow through the channel at a relatively rapid rate (36 hr residence time), resulting in a rapid build up of sediment. Frequent harvesting appears beneficial in improving water quality. Although, the results presented are favorable in improving water quality, long-term effectiveness of the system is not known.

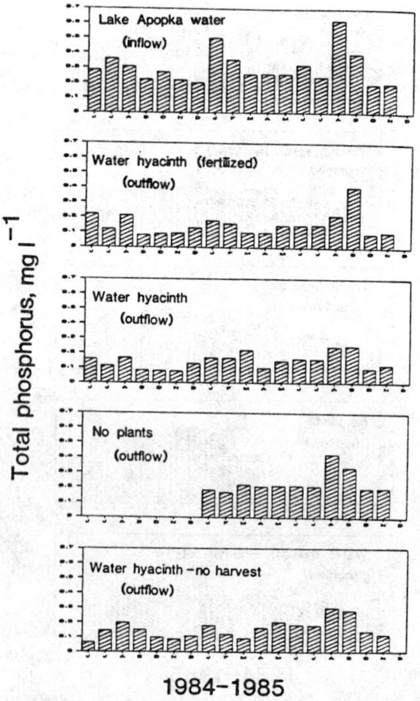

Fig. 9. Seasonal changes in total P of the inflow and outflow of water in Lake Apopka water - hyacinth system.

In Florida, large-scale cultivation of water hyacinths is feasible in freshwater lakes. Nutrients present in these lakes can be trapped by growing water hyacinths and harvesting the biomass on a steady basis. Until recently, water hyacinths were considered as a weed and a nuisance in lakes and streams, and a vast data base was developed for reducing the populations of this plant by using biological or chemical control methods (Gopal and Sharma, 1981). Very little information is available on the beneficial effects of these plants on lake water quality improvement (Yount and Crossman, 1970) and our study reported in this paper is probably the first data set available. Suggestions of the deliberate growing of water hyacinths in lakes will probably create a lot of controversy until a proven data base is developed not only on the biomass production, but also the impact on lake water quality improvement, aquatic biota, and fish populations. All our studies pertaining to the use of lake water were conducted outside the lake to demonstrate the feasibility of growing water hyacinths in lake water and the potential of water hyacinths in improving lake water quality. However, the performance of water hyaciths in the lake itself will be different, thus future studies should be conducted in the lake.

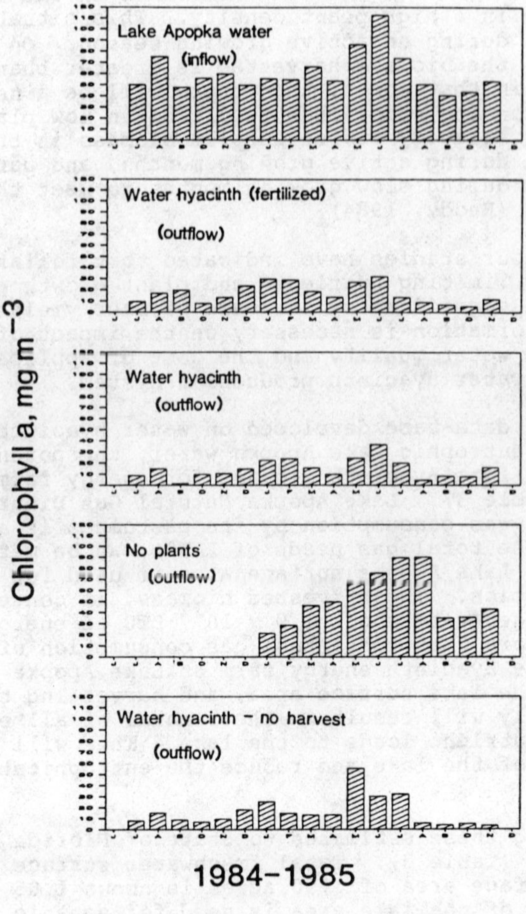

Fig. 10. Seasonal changes in chlorophyll a of the inflow and outflow of water in Lake Apopka water - hyacinth system.

Management strategies for a continuous supply of biomass need to be developed if water hyacinths are grown in the lake as a source of feedstock for methane production. Biomass supply per unit area is varialbe because water hyacinth growth is dependent on solar radiation and temperature. The criticla months with very little or no growth appear to be December, January, and February, and during this period either the methane digester should be supplied with other feedstock such as pennywort or design the digester based on the average annual biomass yield and follow a special harvesting strategy by managing the system at varying plant densities (Reddy, 1984). Harvesting constant amounts of biomass every day will automatically result in varying plant densities, since growth rates are different each month. For example, if the biomass harvested is lower than the biomass yield in a particular month, then there will be a net accumulation of biomass in the system, thus resulting in a high plant density. This situation usually occurs during an active growing season. On the other hand, if the biomass harvested is greater than the biomass yield in that month, then there will be a net decrease in standing crop, thus resulting in low plant density. This strategy stores surplus biomass in the culture system during active growing months, and harvests stored biomass during slow growing months to meet the biomass demand (Reddy, 1984).

Although our studies have indicated that foliar application of limiting nutrients and plant growth regulators can significantly increase biomass yields, additional information is necessary on the impact of foliar application on water quality and the cost of application to a large-scale water hyacinth production system.

Using the data-base developed on water hyacinth production in eutrophic Lake Apopka water, the potential methane yields form hyacinth biomass for energy farm were calculated (Table 3). Lake Apopka Natural Gas District (LANGD) annual gas consumption by its customers is about 1.2 x 10^{12} BTU. The total gas needs of LANGD can be met, if 6,225 acres of Lake Apopka surface area is used for hyacinth biomass production. The harvested biomass, if converted to methane, will generate about 1.9 x 10^{12} BTU of energy. This gas yield is more than the annual gas consumption of LANGD customers. The hyacinth energy farm on Lake Apopka will cover 20% of the lake surface area, and harvesting the biomass annually will result in the removal of all external and internal nutrient loads to the lake. This will improve water quality of the lake and reduce the eutrophication process.

Projecting these estimates to State of Florida is even more appealing (Table 3). Total freshwater surface area for lakes with surface area of >750 acres is about 1.05 x 10^6 acres. If 20% of the lake area is used for aquatic plant biomass production, the harvested biomass will potentially produce an energy yield of about 63 x 10^{12} BTU, while improving water quality. Total natural gas consumption in Florida during 1982 was estimated at 317.3 x 10^{12} BTU and about 20% of this gas can be potentially produced from aquatic plants cultured in eutrophic lakes.

ACKNOWLEDGEMENTS

This paper reports results from a project that contributes to a cooperative program between the Institute of Food and Agricultural Sciences (IFAS) of the University of Florida, and the Gas Research Institute (GRI), entitled "Methane from Biomass and Waste."

Table 3. Projected methane yields from water hyacinth biomass for energy farm in Lake Apopka.

Surface area	12,500 ha
Nutrient loading	
(a) External sources	
Nitrogen	55 kg N ha^{-1} $year^{-1}$
Phosphorus	8 kg P ha^{-1} $year^{-1}$
(b) Internal sources	
Nitrogen	87 kg N ha^{-1} $year^{-1}$*
Phosphorus	19 kg P ha^{-1} $year^{-1}$
Average biomass yield of water hyacinth	62 dry t ha^{-1} $year^{-1}$ (28 dry t $acre^{-1}$ $year^{-1}$
Plant tissue composition	
Nitrogen	15 mg N g^{-1} of tissue
Phosphorus	2 mg P g^{-1} of tissue
Removal of nutrients due to biomass harvesting	
Nitrogen	930 kg N ha^{-1} $year^{-1}$
Phosphorus	124 kg P ha^{-1} $year^{-1}$
Estimated lake surface needed to remove excess nutrients	~2,722 ha (6,750 acres)
Methane yield	5 SCF lb^{-1} dry matter
Methane yield/acre	0.28×10^6 SCF $acre^{-1}$ $year^{-1}$
Surface area of lake needed for water hyacinth culture	6,750 acres (21% lake surface)
Methane yield	1.89×10^{12} BTU
Total natural gas consumption in Lake Apopka Natural Gas District	1.20×10^{12} BTU

*Minimum loading from sediments and N_2 fixation.

Table 4. Projected pipeline quality gas production from aquatic plant biomass under Florida's climatic conditions.

Total freshwater area	2.90×10^6 acres
Area of lakes > 750 acres	1.05×10^6 acres
Area potentially available for biomass production (20% lake surface area)	2.10×10^5 acres
Average biomass yield of water hyacinth	30 t (dry wt) $acre^{-1}$ $year^{-1}$
Methane yield	5 SCF lb^{-1} dry matter
Methane yield/acre	0.3×10^6 SCF $acre^{-1}$
Total methane yield	63×10^{12} BTU
Total natural gas consumption in 1982	317.3×10^2 BTU
Potential gas production from water hyacinth biomass	19.9% of total natural gas consumption

REFERENCES CITED

DeBusk, T. A., Ryther, J. H., Hanisak, M. D., and Williams, L. D., "Effects of seasonality and plant density on the productivity of some freshwater macrophytes," Aquat. Bot. 10, 133-142 (1981).

DeBusk, W. F., Reddy, K. R., and Tucker, J. C., "Management strategies for water hyacinth production in a nutrient-limited system," Proc. 3rd Southern Biomass Conf., Plenum Press (in press), (1985).

Gopal, B. and Sharma, D. P., "Water hyacinth," Hindasia Publishers, New Delhi, India, pp. 128 (1981).

Gossett, D. R. and Norris Jr., W. E., "Relationship between nutrient availability and content of nitrogen and phosphorus in tissues of the aquatic macrophyte, Eichhornia crassipes [Mart] Solms," Hydrobiologia 38, 15-28 (1971).

Moorhead, K. K., "Nitrogen cycling in an integrated 'Biomass for Energy' system," Ph.D. dissertation, Univ. of Florida, Gainesville, Florida (1986).

Nayak, D. N., Swain, A., and Rajaramamohan Rao, V., "Nitrogen fixing Azospirillum lipoferum from common weeds associated with rice and aquatic ecosystems," Current Sci. 48, 866-867 (1979).

Ogwada, R. A., "Growth, nutrient uptake and nutrient regeneration by selected aquatic macrophytes," M.S. Thesis, Univ. of Florida, Gainesville, Florida (1983).

Purchase, B. S., "Nitrogen fixation associated with Eichhornia crassipes," Plant Soil 46, 283-286 (1977).

Reddy, K. R., "Water hyacinth biomass production in Florida," Biomass 6, 167-181 (1984).

Reddy, K. R. and DeBusk, W. F., "Growth characteristics of aquatic macrophytes cultured in nutrient-enriched water. I. Water hyacinth, water lettuce, and pennywort," Econ. Bot. 38, 229-239 (1984).

Reddy, K. R., DeBusk, W. F., and Fowler, R., "Detritus production and decomposition of water hyacinths in eutrophic lake water (in review)," (1986).

Warren, C. S., Bruderly, D. E., Angelieri, M., Bilello, L. J., Bucalo, S., Finger, G. W., Hart, R., Hinton, S. W., Newman, J. R., and Vinzant, J. W., "Evaluation of biomass to methane systems in the Lake Apopka natural gas district," Reynolds Smith and Hills (GRI84/0015.1) (1984).

Yount, J. L. and Crossman, R., "Eutrophication control by plant harvesting," J. Water Poll. Contr. Fed. 42, 173-183 (1970).

DISCUSSION

D. Pratt, U. of Minnesota

There's a prohibition at the present time on growing the plant in the lake. Would you envision doing this in some kind of control system in the lake or in some kind of paddy system alongside the lake?

Reddy

The data I presented was collected outside the lake using lake water. Currently, the State of Florida is in the process of funding a $1 million project to grow hyacinths in the lake itself on a larger scale. This is not for biomass production, but they are convinced now they could use water hyacinth or other aquatic plants to reduce eutrophication in the lake. The proposal is to grow 100 acres of hyacinth in a pilot scale project in the lake. The main reason we didn't do the study in the lake is that we don't have a harvesting technology, and it takes a lot more money to do the studies in the lake. Eventually I think that the study should be done in the lake to relate our results directly to a real system.

D. Burton, Biodex, Inc.

What is the potential for using the system to remove nitrogen and phosphorus farm runoff, say in canals or irrigation ditches?

Reddy

It does a real good job of removing these elements. The design criteria should be set up depending on the concentrations of nitrogen and phosphorus in the wastewater, and how much you want to remove. For example, in secondary sewage, nitrate is a dominant nitrogen species. It can remove up to 80 to 90% of nitrate, not because the plant is capable of taking up all the nitrogen, but because it has the capacity to alter the environment in the water thereby enhancing the denitrification process. In the mass balance, we found the plant only removes about 30 to 40% of the total nitrogen. Other biological processes functioning in the root zone, in the water, and in the sediment contribute more to nutrient removal than plant uptake.

Burton

So you're getting some sort of synergism with bacteria.

Reddy

Yes.

DISCUSSION, CONT.

Burton	Do you think that herbicides and other pesticides that might be in the water would slow down the process?
Reddy	It depends on the concentrations. We did a study in '82-'83 using agricultural drainage water to grow hyacinth. The pesticide concentrations in drainage water were much lower and we never experienced any adverse effect on growth itself. This plant is very resistant to very high concentrations of metals and pesticides.

SUSTAINED PRODUCTION OF CATTAIL FOR ENERGY APPLICATIONS

Dean R. Dubbe, M.S., Assistant Director
Douglas C. Pratt, Ph.D., Director
Bio-Energy Coordinating Office
University of Minnesota
St. Paul, Minnesota 55108

ABSTRACT

Wetland plants, such as cattail (*Typha* spp.), common reed (*Phragmites australis*), and rushes (*Scirpus* spp.), are one type of bio-energy crop under consideration in Minnesota because of their high productivity, chemical composition, and the fact that they are well adapted to a substantial portion of the state's 8.6 million acres of wet marginal lands. Evaluation of the commercial potential of wetland plants as an energy crop depends on an understanding of the tradeoffs between productivity and production costs. Research conducted by the Bio-Energy Coordinating Office at the University of Minnesota has sought to provide the information base that will be required to develop a bio-energy production system that maximizes output while minimizing inputs, resulting in a renewable energy resource that is economically competitive. This paper is organized in a manner that focuses on key steps and issues involved in commercial production, including species selection, establishment methods, and stand management practices.

SUSTAINED PRODUCTION OF CATTAIL FOR ENERGY APPLICATIONS

INTRODUCTION

Minnesota, like many northern states, is a net energy importer with a growing demand for biomass-derived fiber fuels. If fiber fuel use continues to increase, currently exploited sources of these fuels, such as forest product wastes and crop residues, will need to be supplemented with inexpensive biomass grown expressly for fuel use (bio-energy crops). Wetland plants, such as cattail (*Typha* spp.), common reed (*Phragmites australis*), and rushes (*Scirpus* spp.), are one type of bio-energy crop under consideration in Minnesota because of their high productivity, chemical composition, and the fact that they grow naturally on a substantial portion of the state's 8.6 million acres of wet marginal lands.

Evaluation of the commercial potential of wetland plants as an energy source depends on an understanding of the tradeoffs between productivity and production costs. Research at the University of Minnesota, supported by the U.S. Department of Energy, Solar Energy Research Institute, and state of Minnesota, has sought to identify, test, and evaluate potential production practices to better understand these tradeoffs. Ultimately, information gained from this project can be used to develop a bio-energy production system that maximizes output while minimizing inputs, resulting in a renewable energy resource that is economically competitive.

This paper is organized in a manner that focuses on the evaluation of production options for a wetland bio-energy system. It is arranged in sections dealing with the key steps and issues involved in commercial production. These steps include:

- Selection of appropriate species.
- Selection of methods for establishing stands.
- Selection of stand management practices, including fertilization, irrigation, pest control, and harvesting schedule.

SPECIES SELECTION

The first step in establishing a system for bio-energy production on wetlands is identification of appropriate plant species. Early screening was conducted through literature reviews and natural stand surveys to identify productive species which are adapted to wetland habitats and occur naturally in monoculture or in mixed stands with species of similar harvesting requirements (3,15). Eight potential species were identified and subjected to yield trials and nutrient requirement studies.

Cattails (*Typha* spp.) were found to represent the most promising plants for wetland bio-energy production in Minnesota (16). There are, however, differences between the various species of cattails themselves which must be considered in the selection of a species for a particular site. *Typha latifolia* (broadleaf cattail), *Typha angustifolia* (narrow leaf cattail), and *Typha* x *glauca*, a hybrid of the other two species, all occur naturally in Minnesota and are potential bio-energy crops. Traits, which may vary between species, and thus affect species selection, include yield potential, tolerance to various site conditions such as soil pH and water depth, resistance to insect and animal pests, and characteristics as a fuel.

In general, natural stand surveys and field trials have found that *T. angustifolia* and the hybrid have a higher yield potential than *T. latifolia* (16). A study which compared *T.* x *glauca* and *T. angustifolia* found slightly higher yields for *T.* x *glauca* than for *T. angustifolia*. Another trial found that *T. angustifolia* produced nearly twice as much shoot biomass as *T. latifolia*.

In addition to possible variation in yield potential, it appears that cattail species differ in their tolerance to various site conditions as well. Although the optimum pH range for each species is not clearly defined, it appears that *T. angustifolia* is more tolerant of alkaline, saline environments than *T. latifolia*, which can grow under more acid site conditions (1,11). *T. angustifolia* is better suited to deeper water conditions (20-45 inches) than *T. latifolia* because of its taller leaves and larger rhizome storage system (8). Although *T. angustifolia* grows well in shallow water, *T. latifolia* can outcompete *T. angustifolia* in shallow water sites because of its greater leaf area and shade tolerance (8). *T.* x *glauca* is generally found in habitats intermediate between the two parent species. Soil pH and water depth, then, are two factors which might affect site selection, or, alternately, selection of a species for a particular site.

Insect and animal herbivory is another factor which appears to occur differentially among cattail species. *Bellura obliqua* (formerly *Arzama obliqua*) is a stem boring moth whose larvae are found frequently on *T. latifolia*, but rarely on *T. angustifolia* (13). This moth is the most destructive and common insect pest found on cattails in Minnesota, and has been shown to reduce total plant productivity by 45% in affected shoots (12). Depending on the extent of infestation in a cattail stand, this moth could significantly reduce productivity. Muskrats can also significantly reduce the extent of wetland vegetation, including cattails. It has been reported that muskrats prefer *T. latifolia* and *T.* x *glauca* to *T. angustifolia* (9). It is not known how extensively muskrats could reduce productivity in a cattail stand, although, presumably, they could have a major impact on small stands if they are not controlled.

Fuel characteristics of the cattail species under consideration do not appear to differ significantly. The energy content of dry shoots ranges from approximately 7600 to 8130 BTU/lb, with no clear distinction between species. The rhizome energy content is generally lower for all species, ranging from approximately 7190 to 7240 BTU/lb. Ash content, which represents an undesirable by-product in any fuel conversion process, also does not appear to differ between species, with ash contents of shoot samples ranging from 3.7% to 6.8%.

Based on existing information, it appears that *T. angustifolia* or *T.* x *glauca* would be preferred to *T. latifolia* as a bio-energy crop since the yield potential and resistance to insect and animal herbivores is probably lower in the latter. Fuel characteristics are comparable among the three. There may be certain site conditions, however, such as low soil pH on certain peatlands which might still favor the use of *T. latifolia*.

STAND ESTABLISHMENT

Background

Three possible methods have been examined for establishing stands of cattail and reed: seeding, transplanting seedlings, and transplanting portions of the shoot or rhizome system. Criteria for selecting an appropriate method are based on the goals of minimizing establishment costs, and ensuring rapid, uniform stand establishment and development. Costs include those for labor, equipment, materials, and, if needed, storage of plant material and greenhouse space. Also, since establishment only occurs once for these perennial plants, the costs of productivity differences in the establishment season and subsequent years resulting from the choice of establishment method must be considered.

As shown in Table 1, field experiments have demonstrated that cattail stands can be successfully established by transplanting seedlings or rhizome pieces and by planting seed. Since the values shown in the table are aggregates from several field experiments examining various aspects of cattail management, they do not represent maximum productivities that have been achieved within individual experiments. Data for seedling and rhizome transplanted stands have been pooled in Table 1 because no significant differences in establishment success or productivity in the first or subsequent growing seasons has been observed between the two transplantation methods. Table 1 also illustrates that stand establishment by seed under field conditions has not been consistently successful. The seeded stand of *Typha angustifolia* has resulted in the most productive managed stand, whereas the

Table 1. Multiseason Productivity of *Typha* Spp.

SPECIES	EST. METHOD*	ESTABLISHMENT SEASON PRODUCTIVITY (mt/ha)		SECOND SEASON PRODUCTIVITY (mt/ha)		THIRD SEASON PRODUCTIVITY (mt/ha)		FOURTH SEASON PRODUCTIVITY (mt/ha)	
		SHOOT	TOTAL	SHOOT	TOTAL	SHOOT	TOTAL	SHOOT	TOTAL
TYPHA	S	0.5	0.9	13.9	21.5	11.7	25.1	12.6	26.0
ANGUSTIFOLIA	T	1.8	4.5	4.9	11.7	8.7	16.6	--	--
TYPHA	S	0.0	0.0	--	--	--	--	--	--
LATIFOLIA	T	2.2	5.2	6.1	11.0	6.7	12.8	--	--
TYPHA X	S	(NOT ATTEMPTED)							
GLAUCA	T	3.1	7.6	6.5	14.6	8.1	17.5	--	--

* Establishment Method: S = Seeded; T = Transplanted

establishment of the seeded stand of *Typha latifolia* was unsuccessful. Difficulties associated with seeding will be discussed later in this section.

Planting Density

Initial planting density is one of the key variables in determining methods and costs of stand establishment. Planting density also appears to affect eventual stand productivity by determining shoot density in years subsequent to establishment. For example, in Table 1 the *Typha angustifolia* stand established from seed had an initial density of forty plants per square meter compared with five plants per square meter for all transplanted experiments. This resulted in a subsequent equilibrium shoot density of $90/m^2$ for the seeded experiment and about $35/m^2$ for the transplanted experiments. This equilibrium density was fairly constant over three years of sampling, and appears to be closely related to stand productivity.

In addition to affecting overall stand productivity, initial planting density also appears to greatly influence the allocation of biomass between shoots and rhizomes--an effect which is of some importance depending on whether the desired biomass product is shoots or rhizomes or both. The transplanted stands in Table 1 had a first season shoot to rhizome biomass ratio (S/R) of 0.7, whereas the seeded *Typha angustifolia* stand had a S/R of 1.3 in the first season. All stands' S/R increased during the second season, with the seeded stand reaching 1.8 and all other stands reaching 1.0, on the average. Transplanted stands stabilized at a S/R of 1.0 for the third season. The seeded stand's S/R dropped sharply in the third season, stabilizing in the fourth season (not shown in Table 1) at 1.0. It appears, then, that a higher initial planting density which can be achieved by seeding rather than transplanting rhizomes or seedlings results not only in greater overall stand productivity, but also in a greater allocation of biomass to shoots rather than rhizomes, at least during the first few years of the stand.

Planting Method

Transplantation. The principal advantages of establishing cattail stands by transplanting either seedlings or rhizome pieces are similar. Consistently high survival rate of transplants results in uniform stands and rapid growth during the establishment season. Also, transplants are generally tolerant of mechanical handling and a range of moisture conditions from saturated soil to 6" of standing water. This tolerance allows the use of mechanical vegetable transplanters and permits early flooding of the field which minimizes competition from weed species. First season productivities of stands established by either rhizome or seedling transplant methods are generally high enough to warrant aboveground biomass harvest in the first year.

The principal disadvantages of transplanting either seedlings or rhizome pieces are also similar. Costs of materials, labor, and production space are high whether it is for harvesting and storing rhizome planting stock, or for growing seedlings. Given that an initial planting density of at least one seedling or rhizome per square foot appears necessary to achieve high productivities, the number of transplants required for a one hectare field would be 100,000. This number of transplants would be expensive to obtain, since rhizomes must be harvested and processed before using and seedlings must be grown for 90 days in a greenhouse prior to planting. Transplantation is also expensive using current technology because it is a slow and labor intensive process.

With evidence that initial planting density is correlated with eventual stand productivity, it appears that density limitations of transplantation combined with high costs will preclude this method of establishment in most instances.

Seeding. Given the problems of high initial costs and low initial planting density associated with transplanting seedlings or rhizomes, it appears that seeding would be the preferred method of stand establishment. Cattail plants produce up to 200,000 seeds per flowering head which are relatively simple to collect and process. Assuming 100% germination, two flowering heads could provide enough seed to establish a stand approximately one hectare in size with a density of fifty plants per square meter. Seeding also offers the advantage of being able to use inexpensive rapid application methods. These include 1) surface application of dry seed (mixed with an inert carrier) using conventional agricultural equipment such as a fertilizer spreader or grass seeder, 2) hydroseeding using equipment such as roadside grass seeding systems, or, possibly, 3) aerial seeding from planes.

While seeding has several obvious advantages, it also has a major problem associated with a lack of reliability as an establishment method resulting from poor germination and development under a number of field conditions tested to date (2,4,14,17). The field establishment problems appear to result from the sensitivity of cattail seed germination and development to certain environmental factors. It has been demonstrated that optimum germination rates occur under conditions of high temperature (85° F), reduced oxygen concentrations, and exposure to red light (7). This is consistent with observations in nature of dense cattail seedlings occurring primarily in shallow water or mudflats resulting from marsh drawdowns (20). It appears that establishing cattail stands from seed may require mimicking the field conditions seen in natural marsh drawdown in order to meet the physiological requirements for seed germination and development.

In addition to these physiological constraints on field conditions for successful cattail seeding, there are also some technical constraints which have been encountered in past seeding attempts. Dry seeding requires very precise control of soil moisture: too little moisture inhibits cattail seed germination and stimulates weed growth; too much moisture (standing water) can cause a redistribution of the floating dry seeds, resulting in uneven stand establishment. With hydroseeding, germination may be induced prior to field application, giving the seedlings an advantage over weed competitors, but once applied, the seedlings must have a narrowly defined environment for continued growth in the month following application. Also, preplant fertilizer application to the site has been found to be inadvisable, since fertilization frequently results in extensive algal blooms which have been shown to shade and subsequently kill the young seedlings.

Seed preparation for either dry or hydroseeding for past experiments has generally entailed removing the fine downy hairs attached to each seed, but there is now evidence that the hairs may be critical in keeping the tiny seed in the optimum physical environment.

With both the physiological and technical constraints gleaned from earlier seeding trials in mind, an experiment was designed and carried out in an attempt to define a reliable protocol for establishing cattail stands from seed. Figure 1 illustrates the results of this experiment. The bars represent mean percent germination and survival of seed of the two cattail species applied to flooded or saturated paddies with or without the seed hairs removed (H = hairs present, NH = hairs removed).

Overall, it appears that for both species, germination and survival is far better in flooded paddies which are slowly drawn down to mudflat over the course of three weeks than in paddies which had no standing water. This was true whether or not seed hairs were removed. A partial explanation may be provided by the observation that the leaves of very young seedlings, with the seed coat still attached to the leaf tip, floated upright and elongated rapidly in the flooded paddies, but remained curled up on the soil surface in saturated paddies, easily becoming buried in the mud. Whatever other factors may explain the

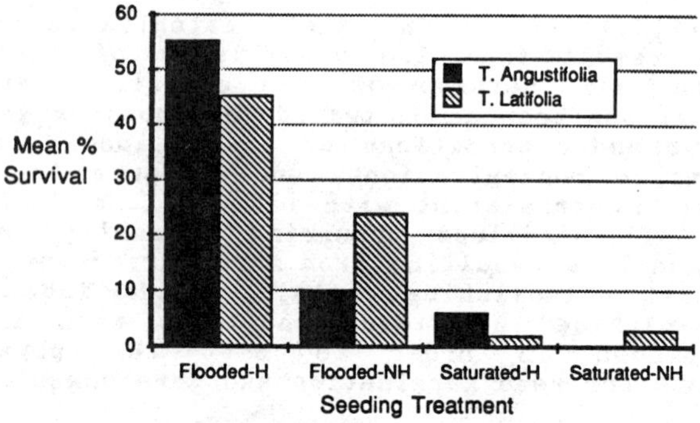

Figure 1. Germination and survival percentages for cattail seeding experiment. Treatments included flooded vs. saturated soil application and seed hairs present (H) vs. seed hairs absent (NH).

differences between saturated and flooded treatments, it appears that water level conditions may be one of the major controlling factors in the success or failure of establishing cattail stands from seed.

Another factor which appears to be of secondary importance but which did make a difference in this experiment was the presence or absence of hairs on the seed. For both cattail species, under flooded or saturated conditions, germination and survival was higher when the hairs were left intact (Figure 1). This difference is possibly explained by the observation that in both flooded and saturated paddies, the hairs aided in keeping the tiny seed from being buried in the mud. To be certain the difference was not caused by seed being damaged during the fluff removal process, a germination check was run under controlled environment conditions in a non-soil medium. Germination of seed with and without fluff was 90% and 86%, respectively.

Based on the results of this experiment, it appears that applying seed with hairs intact onto a flooded site and allowing the water level to slowly drop is the best scenario for establishing a cattail stand from seed. This scenario mimics the establishment of new stands in nature which occur during natural marsh drawdowns, and has the advantage of offering a certain amount of natural weed control (see Stand Management Practices section).

Technical considerations suggest a couple of ways this scenario may be accomplished. Seed may be applied dry in the winter when equipment access is easier and the problem of dry seed redistributing by floating on a flooded site is not present. Or, hydroseeding may be carried out in the winter or spring. The results from outdoor paddies in this experiment demonstrated that even hydroseeded seed remains viable through the winter, as long as the imbibition period is kept short enough prior to application. Other seeding field trials suggest that even spring seeding should occur early enough to precede warm soil and water temperatures, so that the cattail seedlings have a chance to become established before weed species or algal blooms result in serious competition.

STAND MANAGEMENT PRACTICES

Following the initial establishment phase of the production process, management of cattail stands becomes an important factor in achieving high sustained productivity. Management factors are also important for determining the economic feasibility of a cattail bio-energy system. Requirements for fertilizer, irrigation, and pest control can result in significant costs, as in comparable agricultural systems, and must be minimized wherever possible, while maintaining adequate productivity levels. This section discusses fertilizer, irrigation, and pest control requirements for cattail and evaluates stand management options for minimizing these inputs.

Fertilization Requirements

In a natural system, nutrient recycling can greatly reduce the need for additional nutrient inputs (19). In a bio-energy production system, nutrients are removed from the system when the biomass is harvested. High, sustained yields require that these nutrients be replaced. This can be accomplished to varying degrees by natural biological and physical processes (6) and by application of fertilizers. The actual amounts and types of nutrients which need to be replaced through fertilizer application will depend on how much is replaced by natural processes, and the overall nutrient requirements of *Typha*.

In agronomic crops, nitrogen, phosphorus, and potassium are the macronutrients of major concern because of their cost and effect on yield. This will potentially be the case for *Typha* production as well. In addition to these macronutrients, other macro- and micronutrients could prove to be limiting, especially in anaerobic organic soils where cattails may be grown.

In addition to gross levels of nutrients used for *Typha* production over the course of a growing season, seasonal patterns of uptake are of interest since applied nutrients can be lost from the system before being taken up by the plant as is the case with nitrogen losses resulting from denitrification. Several measures can be taken to minimize fertilizer losses, such as selecting the proper form of fertilizer and controlling water levels to minimize leaching and aeration. But timing fertilizer application to coincide with the time of maximum uptake by the plant appears to be one of the most promising control measures, and for this, knowledge of seasonal uptake patterns is needed.

Time of harvest is another factor which can affect nutrient loss from the system. Nutrients are mobilized from the aboveground plant parts at some time late in the growing season and transported to the rhizome system for storage over winter (18). By timing harvest to occur after nutrients have been translocated to rhizomes, unnecessary nutrient removal can be avoided. Again, a knowledge of seasonal patterns of nutrient uptake and movement within the plant is needed.

Nutrient Uptake Patterns. To address the questions concerning seasonal patterns of nutrient uptake, a four year field study was initiated in 1982 (17). Mechanically transplanted seedlings of *Typha latifolia* were used to establish plots on a one-half hectare field, and mechanically transplanted rhizomes of *Typha latifolia*, *T. angustifolia*, and *T.* x *glauca* were used to establish additional plots on a one-quarter hectare adjacent field. The objective of the study was to provide information that could be used to: 1) develop a fertilization schedule that would minimize nutrient losses by timing application to coincide with the period of greatest nutrient uptake, and 2) develop a harvesting schedule that would minimize nutrient removal.

Figure 2 presents results of this study for nitrogen, phosphorus, and potassium during the establishment season, second growing season, and start of the third season. During the establishment season (planting date = June 9), growth was in a lag phase until early August (17), and maximum biomass accrual occurred between the sampling dates of August 4 and September 2. Forty-six percent of the season's total biomass was produced during this period. The greatest amount of each nutrient (g/m^2 basis) was also taken up in this period, with 48%, 46%, and 48% of the season's total N, P, and K accrued, respectively. During the last two months of sampling, changes in partitioning of biomass and nutrients occurred, with total plant biomass and nutrients continuing to increase, but with shoot biomass and nutrients decreasing, and rhizome biomass and nutrients increasing rapidly. The period of maximum biomass production in the second year occurred at the same time as in the first year in terms of shoot age (between 56 and 84 days after plant growth began), but this period was reached one month earlier during the second season. Additionally, in most cases greatest nutrient uptake preceded the period of greatest biomass production in the second season. Biomass and nutrient partitioning was similar to that occurring in the establishment season.

The patterns of nutrient and biomass accrual observed in this experiment can serve as a starting point for testing *Typha* stand management options such as time of fertilization and harvesting. If fertilizer, particularly nitrogen, had been applied to these plots at the beginning of the growing season, much of it may have become unavailable through physical and biological processes before it could be used by *Typha*. Methods of midseason application have been tested and are discussed in the following section.

This study also suggests that harvesting dates and the type of biomass harvested will be important for determining

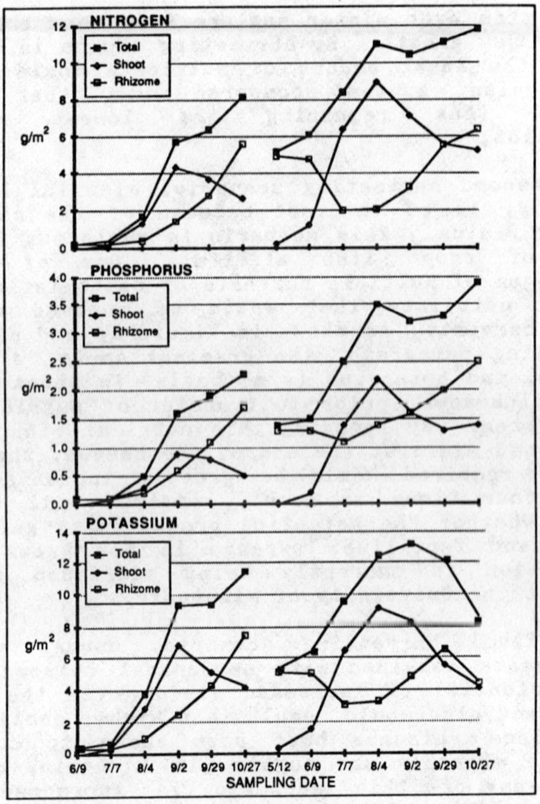

Figure 2. Nitrogen, phosphorus, and potassium uptake patterns in a managed stand of cattail for the establishment, second, and beginning of the third growing season. Values are means of *Typha latifolia*, *T. angustifolia*, and *T. x glauca*.

fertilizer requirements. This is a point worth considering in more detail, since there are several productivity - production cost tradeoff possibilities inherent in the nutrient uptake results. Possible harvesting scenarios for cattail which are discussed later in the report, include: 1) shoot biomass only, harvested annually, 2) shoot biomass only, harvested semiannually, and 3) shoot biomass harvested annually combined with rhizome biomass harvested biennially. Each harvesting scenario will result in different productivity levels and require different levels of nutrient inputs.

The first harvesting scenario, annual shoot biomass harvests, would result in the greatest conservation of nutrients. On average, the Zim study found that 40% of the nitrogen, phosphorus, and potassium (g/m^2 basis) found in the shoots in midseason is transferred to the belowground rhizome system by the end of the growing season. Eighty to one hundred percent of these nutrients are conserved in the rhizome system over winter and are then used the following spring for new growth. By harvesting shoots in October and November, the least amount of nutrients would be removed from the plant system, compared with other harvesting scenarios, thus requiring the lowest amount of fertilization.

The second harvesting scenario, biennial harvests of shoots only, is of interest because of the potential of increasing yields. This scenario is analogous to multiple harvests of crops like alfalfa. One of the great disadvantages of multiple harvests of cattails is the large amount of nutrients that would be removed during the midseason harvest. As shown in Figure 2, and discussed in the preceding paragraph, the greatest amount of nitrogen, phosphorus, and potassium (g/m^2 basis) in shoot biomass is found at midseason, prior to transfer of nutrients to the rhizome system. By removing this nutrient rich biomass at midseason and again at the end of the season, the amount of fertilizer required would be greatly increased (perhaps three to four times as much), compared with the annual harvest. Whether the potential productivity gains warrant the increased fertilizer expense is not known, although this question is currently being addressed in ongoing research at the University of Minnesota.

The final harvesting scenario, annual aboveground shoot harvests combined with semiannual rhizome harvests, has the potential of increased yields over the other two scenarios and also would result in a higher quality biomass product since rhizomes have been shown to contain 40% fermentable starches and sugars (15). Again, one of the disadvantages of this scenario is increased nutrient removal and higher fertilization costs. By removing the rhizomes, the overwintering nutrient storage system of the cattail plant has been lost. Although a study has not yet been done to determine the amount of additional fertilizer that would be required compared with an annual aboveground harvest, it could be estimated from Figure 2 that two to three times the amount of fertilizer would be needed.

Based on these results, it appears that an annual shoot biomass harvest would require the lowest levels of fertilization. The effect of these harvesting scenarios on sustainable total annual productivity will not be known for several years. However, a study for this project looking at aboveground shoot harvest only (see section on Harvesting Effect on Productivity) has demonstrated that shoot yields of 11-14 dry tonnes/ha/yr are possible with minimal nutrient inputs.

Water Requirements

Estimates of water use by cattail species are important for selecting biomass production sites and estimating potential irrigation costs. Although wetlands, with water tables at or near the soil surface, will be used for cattail bio-energy production, it is possible that irrigation will be required to maintain an equilibrium balance of water inputs to water losses. If irrigation is required, sites will need to be located near streams or other water bodies, dikes will need to be constructed, and annual irrigation costs will be incurred. The irrigation requirements and water control methods will also be important determinants of management and harvesting equipment requirements.

Cattails, like terrestrial plants, lose water (transpire) through stomata (microscopic pores) on the leaf surface. These stomata serve as regulatory valves for both CO_2 uptake and water loss. If regulation is poor, as might be the case for wetland plants with surplus water availability, water use could be significantly higher than terrestrial plants. This, in fact, appears to be the case for cattail. Seasonal evapotranspiration (transpiration plus evaporation from the soil or water surface) amounts reported in the literature for cattail (5,10) are approximately twice those for corn and alfalfa grown in Minnesota (University of Minnesota Agronomy Department estimates).

The reported evapotranspiration rates for cattail appear to be quite high, although there is some controversy about the experimental methods which were used. To gain a better understanding of water losses by three cattail species and the effects of water management practices on biomass yield, a study was conducted in outdoor paddies during 1985.

Figure 3 illustrates the mean yield and cumulative seasonal evapotranspiration and transpiration for the different *Typha* species. *Typha latifolia* and *T. x glauca* produced statistically similar amounts of shoot biomass, while *T. angustifolia* produced significantly less than either of the other two. *T. angustifolia* also used significantly less water through evapotranspiration than either *T. latifolia* or *T. x glauca*. Calculation of the amount of water used per unit biomass produced revealed that *T. angustifolia* and *T. x glauca* were less efficient

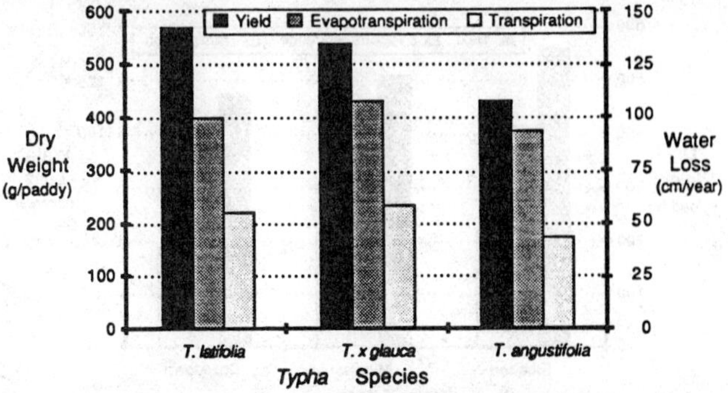

Figure 3. Cumulative establishment season aboveground yield, evapotranspiration, and transpiration for *Typha* as a function of plant species.

than *T. latifolia* in their water usage. Seasonal water use by each species was at the lower end of the range reported in the literature, with *T. angustifolia* using 88.7 cm/yr, *T. x glauca* using 104.3 cm/yr, and *T. latifolia* using 99.7 cm/yr.

Figure 4 illustrates the mean yield and cumulative seasonal evapotranspiration and transpiration for the different water treatments. Plants in the flooded treatment produced significantly more shoot biomass than plants in the drawdown treatment, with biomass production in the mudflat treatment being intermediate to the other two. Flooded treatment paddies also lost significantly more water through evapotranspiration than the other treatment paddies. The efficiency of the plants in using water to produce biomass was statistically similar for all of the water treatments.

Water treatment affected both evaporation and transpiration. Transpiration was similar in flooded and mudflat (saturated) soils, but was significantly lower in drawdown treatments, indicating that the plants may be responding to conditions of limited water availability. Evaporation rates were similar for mudflat and drawdown treatments, with significantly more evaporation occurring in the flooded treatment due to the presence of standing water.

Analysis of the data for this experiment revealed no interaction between the treatment factors of species and water regime. In other words, all three species responded similarly to the various water treatments, with no species showing an advantage under certain water conditions. The fact that *T. latifolia* was more efficient with water usage overall may recommend this species for conditions where irrigation costs are high, but other evidence from past experiments that this species is not highly productive in cultivated stands must also be considered.

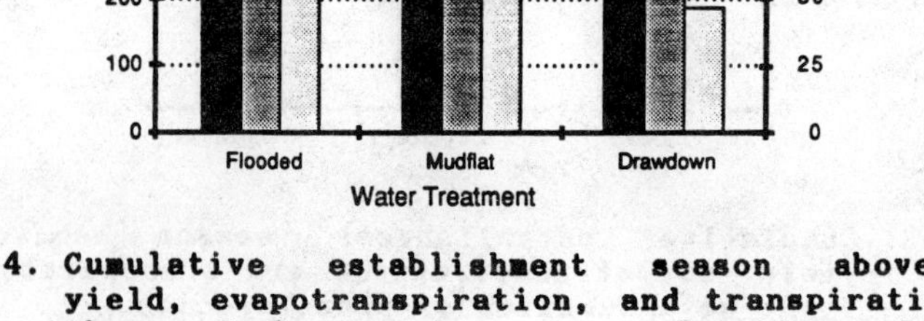

Figure 4. Cumulative establishment season aboveground yield, evapotranspiration, and transpiration for three *Typha* species as a function of water management practices.

In conclusion, the results of this experiment seem to indicate that continuous flooding of a *Typha* stand is optimum for the greatest biomass production, but incurs the greatest expense in water usage. A saturated soil, at least for part of the season, may decrease water usage somewhat without a large sacrifice in biomass production, but it appears that allowing the water table to fall to 45 cm below the soil surface in midseason, as it did in this experiment, may submit *Typha* to water stress and ultimately sacrifice biomass production. This result tends to encourage a harvesting scenario that doesn't require field drying for a fall harvest: either an aquatic harvesting system, or a winter harvest on ice.

Pest Control

Pests of concern in managed cattail (*Typha* species) stands include weeds, insects, disease, and herbivores. Diseases have not been observed in natural or managed stands of cattail in Minnesota and probably do not represent a threat in the future.

Weed Control. Problems with competition from weed species have been encountered in most field trials. The problem is most severe during stand establishment where fluctuating water levels, newly cultivated soils, and the small size of cattail seedlings contribute to the problem. If the stand is successfully established at a uniform density greater than five plants per square meter, weeds do not pose a problem in subsequent years. Additionally, proper site preparation methods and early flooding will preclude the need for chemical controls. A precultivation treatment with a non-selective herbicide like glyphosate is recommended to eliminate perennial weeds which may be present.

Insect Control. Studies have been conducted to identify the common phytophagous insects that feed on *Typha* spp. (cattails) in Minnesota and to quantify the extent of insect herbivory in natural stands (12). Overall, about 15% of shoots showed some evidence of insect herbivory. Eight percent of damaged shoots had damage to upper leaves attributable to lepidopteran miner-borers (chiefly *A. oblonga*). Five percent of shoots had stem damage. Less than one percent of shoots were damaged by leaf-chewing insects. *Typha latifolia* stands were more heavily infested than those of *T. angustifolia* or *T.* x *glauca*. Water depth and shoot nitrogen content did not have a significant effect on infestation rate. The majority of infested shoots suffered minor damage (i.e., 20% reduction in leaf area). Although natural and small cultivated *Typha* stands are occasionally susceptible to heavy infestations of phytophagous insects, it seems unlikely that insect pests will hamper *Typha* bio-energy production in Minnesota.

Harvest Effect on Productivity

Harvesting can affect long term productivity of cattail stands in a number of ways. As discussed in the nutrient section, harvesting can reduce or increase nutrient conservation depending on the time of harvest and the type (shoot, rhizome, or both) of biomass harvested. Harvesting also can alter rhizome carbohydrate storage needed for early season growth and stand vigor. If a portion of the belowground rhizome system is harvested, time may be required for the stand to fill in and become productive again. Finally, harvesting may affect overwinter survival, since removal of shoot biomass eliminates the insulation which normally covers natural cattail stands.

Previous studies have determined cattail biomass yield based on measurements taken in unharvested natural stands and first, second, and third year paddies and managed field stands. However, the effect of repeated annual biomass harvest on sustained productivity has never been evaluated. In 1984, a DOE sponsored multiyear study was initiated to evaluate the effect of various harvesting scenarios on sustained *Typha* productivity. The three harvesting scenarios under consideration are:

1. Harvest of aboveground shoot biomass annually in the fall.
2. Harvest of aboveground shoot biomass semiannually in mid-summer and fall.
3. Harvest of aboveground shoot biomass annually in the fall coupled with harvest of belowground rhizomes biennially in the fall.

Mean *Typha* shoot productivity at the beginning of this experiment was equivalent to 6.2 dry tonnes/ha. Shoots from prior growing seasons and weeds brought the total aboveground biomass yield measurement to 9.7 tonnes/ha. Belowground *Typha* rhizome production contributed an additional 10.1 tonnes/ha to bring total *Typha* biomass production in year 0 to 16.3 tonnes/ha. There were no statistically significant differences in productivity between treatments or blocks, indicating uniform stand characteristics prior to treatment application.

Other stand characteristics of interest at time 0 were biomass moisture content and shoot density. *Typha* shoot moisture in the second week of October was 59% of fresh weight. Including old shoots and weeds reduced the aboveground total biomass moisture content to 54%. Shoot density averaged 26.1 shoots/m² and was not statistically different between treatments or blocks.

During May and June 1985, observations of stand regrowth were recorded. All harvested plots exhibited vigorous new shoot growth during this period, and there were no obvious differences between treatments in which the belowground rhizome material had not been removed. In the treatments where 50% of rhizome biomass had been harvested, the harvested strips showed no signs of stand regeneration from adjacent unharvested strips until late July. By the end of the 1985 growing season, shoot density and aboveground biomass in the rhizome harvest treatments were

7.8 shoots/m² and 3.8 tonnes/ha - considerably lower than the 20.1 shoots/m² and 4.8 tonnes/ha mean values for other treatments. There was, however, no significant difference in rhizome biomass between the rhizome harvest treatment and other treatments by the end of the first growing season.

Plants harvested in July as part of the semiannual shoot harvest treatment produced the highest seasonal biomass yield. Ninety-five percent of this yield was obtained during the July harvest, with only five percent of the yield being obtained from regenerated shoots at the end of the growing season. Although a July harvest made it possible to harvest *Typha* biomass at its seasonal peak (17), the poor shoot regrowth will reduce translocation of carbohydrates and nutrients to the rhizomes. This will likely result in a weakened stand in future years and also increase required nutrient inputs. Somewhat surprisingly, there was no significant decrease in rhizome biomass for the semiannual harvest treatments over the control this year. Increased variability resulting from fewer belowground samples may account for the lack of statistical significance.

Machine travel over plots was expected to either damage stands of *Typha* or have no effect. Unexpectedly, first year results indicate a significant increase in biomass yield and shoot density from machine travel. Excluding the belowground treatment, machine treatments had mean shoot yields of 8.4 tonnes/ha, whereas nonmachine treatments had mean shoot yields of 5.3 tonnes/ha. Similarly, maximum shoot density was 32/m² for machine treatments and 22/m² for nonmachine treatments. Pending further study of this new observation, it appears that machine travel causes physical damage to shoot bases which stimulate additional shoot development resulting in increased productivity. The implications of this discovery, if supported by additional research, will be significant for increasing *Typha* productivity and designing harvesting equipment.

Moisture content of *Typha* biomass was significantly higher for shoots harvested in July (75% moisture) than those harvested in October (58%). The only exception to this observation was semiannual harvest treatments which had moisture contents of 72% in October. This high fall moisture content can be explained by the immaturity of regenerated shoots which remained green until harvest.

In summary, it appears after one year of research that an annual shoot harvest will result in the highest sustained biomass yields of the treatments under consideration. A semiannual harvest in Minnesota results in such low regrowth of shoots following the first harvest that the continued vigor of the stand is doubtful. A 50% belowground harvest results in slow stand regeneration, reducing the annual aboveground productivity in the year following a rhizome harvest. These conclusions must be tempered by the fact that they are based on one year's worth of data obtained in a northern climate. With a longer growing season, shoot regrowth following a midseason harvest or stand regeneration following a rhizome harvest may be sufficient to change these conclusions. The study will be continued for a minimum of one additional season.

CONCLUSIONS

Major requirements for the production of biomass from herbaceous cattail (*Typha* spp.) have undergone preliminary analysis by the University of Minnesota. Results indicate no insurmountable problems for the domestication of this plant. The main challenge will be to balance the tradeoffs between increased yield and increased input costs. Fertilizer and water inputs appear to be the most significant inputs, since establishment costs need only be incurred once for this perennial plant.

Several options are available for minimizing fertilizer requirements. Based on experiments examining seasonal patterns of nutrient uptake, these options include: 1) timing the application of fertilizer to coincide with the period of maximum nutrient uptake by the plants, 2) developing fertilizer application methods which incorporate the nutrients (particularly nitrogen) into the top 6-12" of soil, and 3) selecting harvesting methods which take advantage of the seasonal redistribution of nutrients within the plant. By incorporating fertilizer 30-45 days after the initiation of growth in the spring and harvesting only shoot biomass in October and November, the least amount of nutrients would be lost or removed. Semiannual shoot harvests would result in the greatest nutrient removal, with an annual shoot harvest combined with biennial rhizome harvest resulting in intermediate removal rates.

Water use by cattail species appears significantly higher than comparable water use by common agricultural crops. Since rates for cattail generally exceed seasonal rainfall in Minnesota, it appears likely that at least periodic irrigation will be required to maintain an equilibrium balance of water inputs to water losses - even on wetlands. Further work is being performed to determine water loss under actual field conditions (where relative humidity will be higher) and the effect of various irrigation practices on productivity.

With information presented in this paper and information from ongoing studies, it will soon be possible to characterize the economics of cattail production. A preliminary study in 1985 found internal rates of return of 18% for a large scale cattail stand yielding an attainable 13.5 dry tonnes/ha/yr (16). Although this return is not high enough today to justify the risk associated with a bio-energy plantation, it does show that cattail has the potential to be an economically competitive biomass resource on Minnesota's vast wetlands.

ACKNOWLEDGMENTS

The authors wish to thank the U.S. Department of Energy, the Solar Energy Research Institute, and Legislative Commission on Minnesota Resources for support of the research described in this paper.

REFERENCES CITED

1. Anderson, C. M., "Cattail Decline at Farmington Bay Waterfowl Management Area," Great Basin Nat. 37, 24-34 (1977).

2. Andrews, N. J., Penko, J. M., Mattson, M. and Pratt, D. C., "The Establishment of Cattails on a Northern Minnesota Peatland." Report to the Minnesota Department of Natural Resources, 101 pp. (1981).

3. Andrews, N. J. and Pratt, D. C., "Wetland Energy Crops: The Productive Potential of *Typha* Spp., *Phragmites communis* and *Phalaris arundinacea* in Minnesota." In: W. Palz *et al.*, Eds., Proceedings of "Energy from Biomass": The First European Communities Conference, 75-82. London: Applied Science Publishers, Ltd., 1981.

4. Bedish, J. W., "Cattail Moisture Requirements and Their Significance to Marsh Management," The American Midland Naturalist 78, 288-300 (1967).

5. Bernatowicz, S., Leszczynski, S. and Tyczynska, S., "The Influence of Transpiration by Emergent Plants on the Water Balance in Lakes," Aquatic Bot. 2, 275-288 (1976).

6. Biesboer, D. D., "Nitrogen Fixation Associated With Natural and Cultivated Stands of *Typha latifolia* L. (Typhaceae)," American Journal of Botany 71, 505-511 (1984).

7. Bonnewell, V., Koukkari, W. L. and Pratt, D. C., "Light, Oxygen, and Temperature Requirements for *Typha latifolia* Seed Germination," Canadian Journal of Botany 61, 1330-1336 (1983).

8. Grace, J. B. and Wetzel, R. G., "Habitat Partitioning and Competitive Displacement in Cattails (*Typha*): Experimental Field Studies," American Naturalist 118, 463-474 (1981).

9. Hotchkiss, N. and Dozier, H. L., "Taxonomy and Distribution of North American Cattails," American Midland Naturalist 41, 237-254 (1949).

10. Krolikowska, J., "The Transpiration of Helophytes," Ekol. Pol. 26, 193-212 (1978).

11. Morton, J. F., "Cattails (*Typha* Spp.) - Weed Problem or Potential Crop?," Economic Botany 29, 7-29 (1975).

12. Penko, J. M., "Ecological Studies of *Typha* in Minnesota: *Typha*—Insect Interactions, and the Productivity of Floating Stands." M.S. Thesis, University of Minnesota, 307 pp. (1985).

13. Penko, J. M., Gorham E. and Pratt, D. C., "The Relative Suitability of Two Species of Cattail (*Typha*) as Host Plants for *Bellura obliqua*." Paper presented at the Minnesota Academy of Science 51st Annual Spring Meeting, University of Minnesota, Duluth, p. 13 (abstract) (1983).

14. Pratt, D. C., Andrews, N. J., Dubbe, D. R., Garver, E. G., Penko, J. M., Read, P. and Zimmerman, E. S., "Emergent Aquatics: Stand Establishment, Management, and Species Screening." A subcontract report to the Solar Energy Research Institute. National Technical Information Service No. DE83004697, 56 pp. (1982).

15. Pratt, D. C., Andrews, N. J., Glass, R. L. and Lovrein, R. E., "Production of Wetland Energy Crops in Minnesota." In: Proceedings of Biomass Workshop, Midwest Universities Energy Consortium, 158-175 (1981).

16. Pratt, D. C., Dubbe, D. R. and Garver, E. G., "Energy From Biomass in Minnesota - Part I. Wetland Biomass Production." Final report to Minnesota Department of Energy and Economic Development and the Legislative Commission on Minnesota Resources, 35 pp. (1985).

17. Pratt, D. C., Dubbe, D. R., Garver, E. G. and Linton, P. J., "Wetland Biomass Production: Emergent Aquatic Management Options and Evaluations." Final report to the Solar Energy Research Institute, 74 pp. (1983).

18. van der Linden, M. J. H. A., "Distribution of Nitrogen Among Shoots and Rhizomes During the Growing Season and Loss of Nitrogen Due to Fire Management," Oecologia Plantarum 1, 219-230 (1980).

19. van der Valk, A. G. and Davis, C. B., "Primary Production of Prairie Glacial Marshes." In: Good, R. E., Whigham, D. F. and Simpson, R. L., Eds., Freshwater Wetlands: Ecological Processes and Management Potential, 21-37. New York: Academic Press, 1978.

20. van der Valk, A. G. and Davis, C. B., "The Impact of a Natural Draw-down on the Growth of Four Emergent Species in a Prairie Glacial Marsh," Aquatic Bot. 9, 301-322 (1980).

DISCUSSION

R. Legrand, Reynolds, Smith & Hills	What's the moisture content of cattail when you harvest it?
Dubbe	The moisture content in the middle of summer is about 75%. In the fall, it's about 50 to 55%. In the winter, it drops to about 30%.
P. Benson, GRI	We have digested them in the GRI program. Do you have any data on their bioconversion to methane?
Dubbe	We have not done any of the bioconversion work.

CONTINUOUS MICROALGAE PRODUCTION IN SHALLOW RACEWAYS AT HIGH PRODUCTIVITIES

Edward A. Laws, Ph.D.
Professor
University of Hawaii
Oceanography Department
Honolulu, Hawaii 96822

ABSTRACT

Saline microalgae grown in an experimental flume culture system at the University of Hawaii have achieved daily production rates of 27-37 g ash-free dry weight (AFDW) per square meter and photosynthetic efficiencies of 9.1-10.7% based on visible irradiance over periods of 44-78 days. A key feature of the system is the use of foil arrays to effect systematic vertical mixing of the culture through the production of vortices. An analysis of variance of production results over a 5 1/2 month period showed that the enhancement of production with the foil arrays in place was statistically significant at better than 99% confidence. Photosynthetic efficiencies with the foil arrays in place were a factor of 1.35 times the photosynthetic efficiencies without foil arrays.

CONTINUOUS MICROALGAE PRODUCTION IN SHALLOW RACEWAYS AT HIGH PRODUCTIVITIES

BACKGROUND

Research on this project began in January, 1980, with financial support from the Solar Energy Research Institute (SERI) and the State of Hawaii. The basic hypothesis underlying the project is that production of microalgae in mass culture can be significantly increased through the use of shallow (depth less than 15 cm) outdoor flumes rather than deeper (depth greater than 20 cm) conventional systems. The experimental facility at the University of Hawaii consists of one 48.4 m^2 flume and four smaller 9.2 m^2 flumes (Fig. 1). In all the flumes water is circulated by means of an airlift at a flow rate of approximately 30 cm/sec. A key feature of the flumes is the use of small foils similar in shape to segments of airplane wings to effect systematic vertical mixing in the flumes (Figs. 2-3). It was hypothesized that the systematic vertical mixing created by the vortices which propagate downstream from these foils would cause the cells in a dense algal culture to experience a modulation in irradiance on a time scale sufficiently brief to cause an enhancement of productivity similar to that associated with the flashing light effect. Recently Terry and Hock (1985) have demonstrated experimentally that productivity enhancements of 50-100% can be achieved at flash frequencies as low as 0.5-1.0 Hz in a culture of *Phaeodactylum tricornutum*. This flash frequency is comparable to that experienced by cells swept along by vortices in a dense culture 10-15 cm deep.

The outdoor culture facility at the University of Hawaii has proven very useful for screening various species and strains of algae for outdoor culture, for optimizing species performance, and for developing efficient management strategies. During the early years of this project much time was spent in species screening and optimization of culture conditions. During 1983 it was established that thermophilic strains of microalgae could be grown in the flumes without the aid of expensive cooling, and that marine species could be cultivated successfully in desert saline waters typical of water types found in parts of the southwestern United States.

During 1984 outdoor culture experiments were conducted with a number of species. Excellent results were obtained with the green flagellate *Tetraselmis suecica* and marine diatom *Chaetoceros* sp. Both strains grew well at temperatures above 30°C and produced yields of 30-40 g ash-free dry weight (AFDW) per square meter per day for periods up to one month. A major discovery was that diluting the *T. suecica* culture every third day enhanced production by about 40% over that achieved with a dilution interval of two or four days. In this production mode daily yields during 1984 and

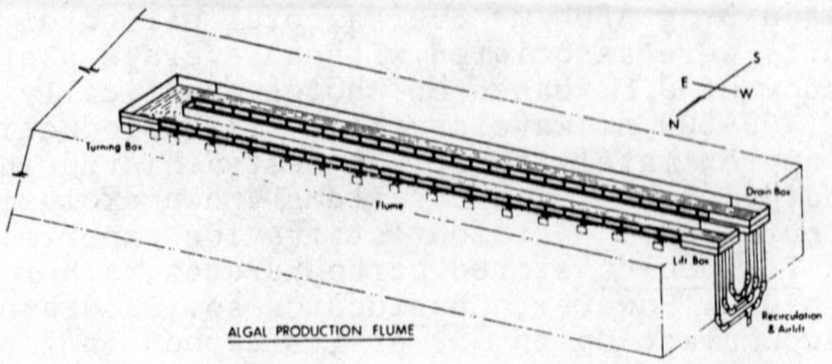

Fig. 1. Perspective of 48.4 m^2 flume.

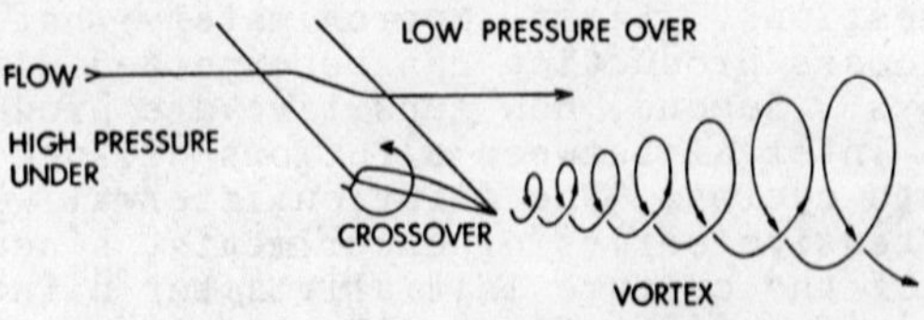

Fig. 2. Design of a single foil indicating mechanism of vortex production.

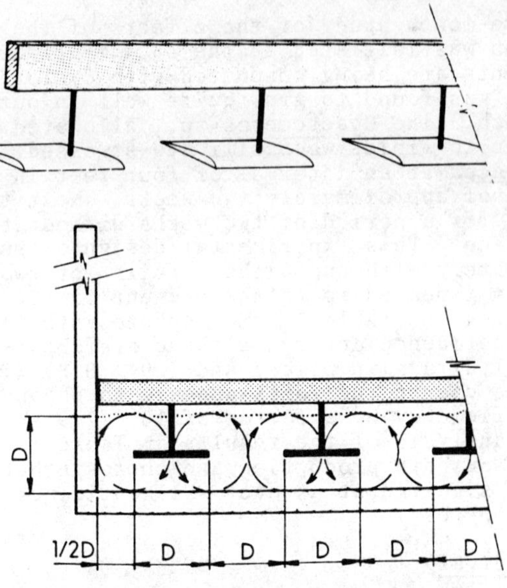

Fig. 3. Positioning of individual foils in foil array. Lower figure indicates positioning of array in the flume. D is the depth of the water. Arrows indicate rotational direction of vortices.

1985 averaged 37 g AFDW/m² over a total period of 78 days. These results were associated with an average photosynthetic efficiency of 9.1% [based on photosynthetically active radiation (400-700 nm wavelength)]. This production efficiency is approximately twice the best achieved in long-term microalgal mass culture systems grown exclusively on inorganic nutrients. Nutrient starvation experiments indicated that T. suecica stored carbohydrates rather than the desired lipids. However, Chaetoceros sp., a diatom, was found to accumulate up to 55% of its carbon in lipids after a 4-5 day period of silicate starvation.

RECENT RESULTS

Research during the last twelve months has focused on two major questions. First, approximately what percent increase in biomass production can be expected with the use of foil arrays. Second, how sensitive are production results to the interval between dilutions of successive batch culture growth cycles. The dilution interval work has required an extensive series of experiments, since the optimum density of the culture initially after dilution is expected to be a function of the dilution interval. An effort was also made to study the compositional and physiological responses of the algae to changing light conditions in the culture over the course of the batch culture growth cycles.

Foil Effects

A twelve-month study of the effects of the foil arrays on production was initiated in the 48.4 m^2 flume on 9/16/85. The experiments are being conducted with Cyclotella sp., a diatom which was found to grow quite well in outdoor culture and which, like Chaetoceros sp., allocated about 50% of its carbon to lipids when silicate stressed. Foil arrays were placed at an interval of four feet in the flume for a period of approximately two weeks. Next the flume was operated for a period of two weeks without the foil arrays in place. This experimental design of running the flume alternately with and without foils for two week periods has been repeated up to the present time. Results are shown in Fig. 4 and Table 1. Mean photosynthetic efficiencies (±95% confidence intervals) have averaged 10.7 ± 1.5% with the foil arrays in place, and 7.9 ± 0.8% without foil arrays. Thus use of the foils appears to enhance photosynthetic efficiency by a factor of 10.7/7.9 = 1.35. A statistical analysis of the results in Table 1 indicates foil effects on both production and photosynthetic efficiency to be significant at over 99% confidence (t-test, Mann-Whitney U-test).

Dilution Interval

Dilution interval experiments have been conducted both in the 48.4 m^2 flume and in the four smaller 9.2 m^2 flumes. During June and July, 1985, a one-month production run was

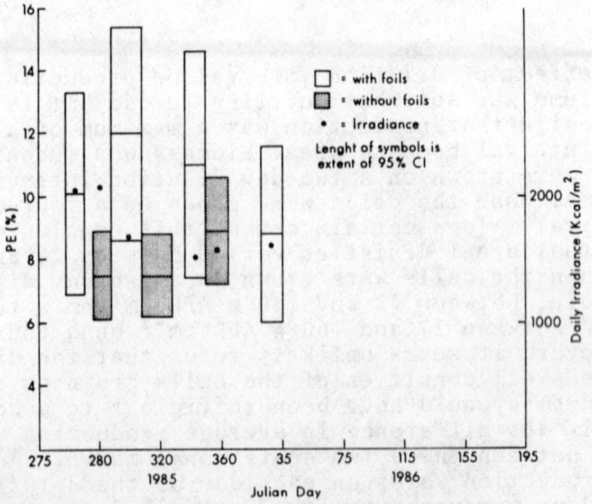

Fig. 4. Experimental results obtained with Cyclotella sp. in 48.4 m^2 flume with and without foil arrays. Width of bars indicates temporal duration of each run. Dots indicate average irradiance (400-700 nm) during the corresponding time period.

Table 1. Results with Cyclotella in 48.4 m^2 flume since 9/16/85.

	With foils	Without foils
g $C/m^2/d$	13.0 ± 2.1	10.3 ± 0.9
E($Kcal/m^2/d$)	1652 ± 145	1790 ± 135
PE (%)	10.7 ± 1.5	7.9 ± 0.8
initial density (gC/m^3)	83 ± 5.0	77 ± 3.5
n	44	40

made with T. suecica in the 48.4 m^2 flume on a three-day dilution interval growth cycle. The principal goal of this research was to determine whether the high production results achieved during 1984 in this production mode could be reproduced. The results are summarized in Table 2. Two aspects of the 1984 results were clearly reproducible. First, the rate of production in 1985 (33.2 ± 4.7 g AFDW/m^2/d) was not significantly different from the rate achieved in 1984 (39.6 ± 8.2 g AFDW/m^2/d) (t-test, $p > 0.05$). Second, as in 1984 production was highest on the third day of the three-day growth cycle (Fig. 5). The difference between production on the third day of the cycle and production on the first two days was highly significant [analysis of variance (ANOVA), $p < 0.05$]. What was not reproducible in 1985 was the high photosynthetic efficiency of 10.2%. The lower photosynthetic efficiency of 7.6% may have been the result of supraoptimal temperatures during the 1985 studies. During July, 1985, the average daily maximum and minimum temperatures in the 48.4 m^2 flume were 31.6 and 21.2°C, respectively. In March-May, 1984, the corresponding temperatures were 28.8 and 18.0°C, respectively. The 3°C higher temperatures in 1985 may well have been supraoptimal for T. suecica during at least part of the day.

The effect of dilution interval on production in the 48.4 m^2 flume was somewhat puzzling to us. It is tempting to hypothesize that production was a maximum on a three-day dilution interval because areal biomass was suboptimal when the cells were grown on a two-day dilution interval and supraoptimal when the cells were grown on a four-day dilution interval. To a certain extent this conclusion may be true. Actual areal denisties varied between 28 and 80 g AFDW/m^2 when the cells were grown on a two-day dilution interval cycle, between 27 and 140 g AFDW/m^2 on a three-day cycle, and between 27 and 160 g AFDW/m^{-2} on a four-day cycle. However, it seems unlikely to us that the difference in average areal densities of the cells grown on three and four-day cycles would have been sufficient to account for the roughly 40% difference in average production rate (Table 2) between these two operational modes. An examination of production rates on each day of the dilution cycles (Fig. 5) does however suggest additional factors which may have contributed to the dilution interval effect. When the cells were grown on a four-day dilution interval, average production declined noticeably on the fourth day of the cycle. One likely explanation for this decline is that the cell density was indeed supraoptimal, and that holding the culture in this high density condition for several days produced an adverse physiological response in the culture. Although the four-day dilution cycle cells displayed production rates comparable to those of the two and three-day cycle cells on days one and two of the growth cycle, production on day three was significantly higher for the three-day cycle cells than for the four-day cycle cells. We postulated that the failure of the four-day cycle cells to display a peak in production on day three was due to a lingering effect of exposure to supraoptimal density conditions.

Compositional and PI Curve Parameters. It was our hope that a study pf photosynthesis-irradiance (PI) curve parameters and compositional characteristics would provide some indication of this phenomenon. Accordingly, experiments were run in the four 9.2 m^2 flumes during July and August, 1985, to study the compositional characteristics and PI curve parameters of *T. suecica* cells grown on two, three, and four-day growth cycles. All PI curve studies were made at approximately noon on each day of the growth cycle. Results are reported as Pm (light-saturated gross photosynthetic rate), α (initial slope of the PI curve) and I_k

Table 2. Production results in 48.4 m^2 flume during 1984 and 1985 with *T. suecica*. Error bars are 95% confidence intervals.

Dates	Dilution Interval	Biomass Immediately After Dilution (g AFDW m^{-3})	Daily Production (g AFDW m^{-2})	Daily Irradiance (Kcal m^{-2}, 400 - 700nm)	Photosyntheti[c] Efficiency (%)
2/23/84 - 3/25/84	two days	219±25	25.9±4.2	2108±87	7.4±1.2
3/26/84 - 4/25/84 5/16/84 - 5/30/84	three days	184±21	39.6±8.2	2325±90	10.2±2.1
4/26/84 - 5/15/84	four days	213±38	25.8±7.0	2442±182	6.2±2.0
6/23/85 - 7/25/85	three days	253±23	33.2±4.7	2643±133	7.6±1.0

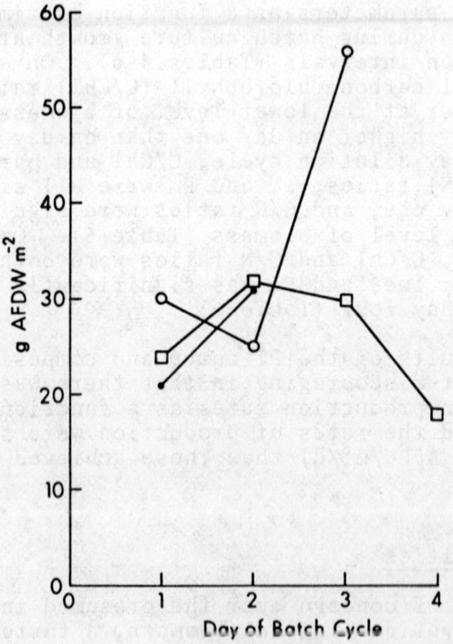

Fig. 5. Production of T. suecica in 48.4 m^2 flume as a function of day in batch culture. Squares = four-day dilution cycle. Open circles = three-day dilution cycle. Solid circles = two-day dilution cycle. 95% CI to data points average ± 7 g AFDW/m^2.

(= Pm/α). Compositional characteristics were measured at approximately 9 AM local time. Experiments were run at two initial densities, 218 ± 25 and 304 ± 44 g AFDW/m^3. Characteristics of the cultures were monitored over three to five growth cycles. The results are summarized in Tables 3-6. A two-way ANOVA on the results in Table 3 revealed no significant treatment effect ($p > 0.05$) on either daily production or photosynthetic efficiency from either the initial concentration of biomass or dilution interval. A two-way ANOVA (initial biomass concentration x time) did however reveal significant changes in compositional characteristics

Table 3. Production results in 9.2 m^2 flumes during 1985 with T. suecica. Error bars are 95% confidence intervals.

Dates	Dilution Interval	Biomass Immediately After Dilution (g AFDW m^{-3})	Daily Production (g AFDW m^{-2})	Daily Irradiance (Kcal m^{-2}, 400 - 700nm)	Photosynthetic Efficiency (%)
8/6 - 8/15	two days	201±26	27.9±6.7	2627±299	6.3±1.3
	"	314±66	23.0±7.4	"	5.2±1.4
7/6 - 7/17	three days	206±49	25.5±4.0	2729±208	5.5±0.8
	"	296±110	24.7±7.2	"	5.4±1.2
7/7 - 7/18	four days	246±122	22.6±5.9	2693±204	5.0±1.2
	"	301±170	21.5±5.7	"	4.7±1.2

and PI curve parameters as a function of time and/or biomass concentration during batch culture growth at each of the three dilution intervals (Tables 4-6). On a two-day dilution interval carbon:chlorophyll (C/Chl) ratios were significantly higher at the lower level of biomass, and Pm was significantly higher on day one than on day two (Table 4). On a three-day dilution cycle, C/Chl and particulate carbon/nitrogen (C/N) ratios, α, and Pm were all significantly higher on day one, and C/N ratios were significantly higher at the lower level of biomass (Table 5). On a four-day dilution cycle, C/Chl and C/N ratios were both significantly higher on day one, and Pm was significantly higher on day one than on day four (Table 6).

The results of the PI curve and compositional studies were somewhat discouraging in that there was no significant difference in production rates as a function of dilution interval, and the rates of production were substantially lower (~24 g AFDW/m^2/d) than those achieved in the 48.4 m^2 flume.

Optimum Densities

Because of concern over the presumed interaction of dilution interval and initial biomass, a factorial experiment was set up in the 9.2 m^2 flumes in September, 1985, to determine the optimum initial density as a function of dilution interval. The effects of foil arrays were also included in the experimental design. Cyclotella sp. was chosen as the test organism. These experiments are still in progress. Fig. 6 summarizes the results to date for cells grown on a dilution interval of one day. A two-way ANOVA revealed both foils and initial density to have a significant effect on photosynthetic efficiency at better than 95%

Table 4. Results of compositional and PI curve studies with T. suecica grown on a dilution interval of two day Compositional units are weight/weight. I_k units are gC/gChla/h. $\alpha = Pm/I_k$.

	Day 1 biomass 1	Day 2 biomass 1	Day 1 biomass 2	Day 2 biomass 2
C/Chl	46 ± 8 n = 4	40 ± 3 n = 4	37 ± 4 n = 4	38 ± 2 n = 4
		biomass (p = 0.04) (1 > 2)		
C/N	6.2 ± 0.5 n = 4	5.7 ± 0.3 n = 4	5.8 ± 0.7 n = 4	5.3 ± 0.6 n = 4
		ns		
α	0.028 ± 0.010 n = 4	0.021 ± 0.001 n = 4	0.025 ± 0.002 n = 3	0.021 ± 0.001 n = 4
		ns		
I_k	282 ± 76 n = 4	278 ± 94 n = 4	282 ± 12 n = 3	324 ± 83 n = 4
		ns		
Pm	7.5 ± 1.2 n = 4	5.6 ± 1.8 n = 4	7.1 ± 0.8 n = 3	5.6 ± 1.1 n = 4
		days (p = 0.02) (1 > 2)		

Table 5. Results similar to Table 4 for *T. suecica* but at a dilution interval of three days.

	Day 1 biomass 1	Day 2 biomass 1	Day 3 biomass 1	Day 1 biomass 2	Day 2 biomass 2	Day 3 biomass 2
C/Chl	40 ± 4	33 ± 4	32 ± 2	44 ± 4	35 ± 6	33 ± 2
	n = 3	n = 4	n = 4	n = 3	n = 4	n = 4
			day (p = 0.0009) (1 > 2 = 3)			
C/N	7.1 ± 0.06	5.7 ± 0.4	5.9 ± 0.2	6.5 ± 0.3	5.6 ± 0.2	5.6 ± 0.3
	n = 3	n = 4	n = 4	n = 3	n = 4	n = 4
			day (p = 0.0001) (1 > 2 = 3) biomass (p = 0.02) (1 > 2)			
α	.022 ± .007	.017 ± .004	.019 ± .001	.029 ± .007	.021 ± .004	.019 ± .003
	n = 3	n = 4	n = 4	n = 3	n = 4	n = 4
			day (p = .015) (1 > 2 = 3)			
I_k	271 ± 32	265 ± 92	191 ± 49	272 ± 59	240 ± 72	233 ± 70
	n = 3	n = 4	n = 4	n = 3	n = 4	n = 4
			(ns)			
Pm	6.0 ± 2	4.4 ± 1.4	3.5 ± 0.8	7.7 ± 1.1	4.9 ± 0.7	4.3 ± 0.9
	n = 3	n = 4	n = 4	n = 3	n = 4	n = 4
			day (p = 0.0004) (1 > 2 = 3)			

Table 6. Results similar to Table 4 but at a dilution interval of four days.

	Day 1 biomass 1	Day 2 biomass 1	Day 3 biomass 1	Day 4 biomass 1	Day 1 biomass 2	Day 2 biomass 2	Day 3 biomass 2	Day 4 biomass 2
C/Chl	[illegible]	[illegible]	[illegible]	[illegible]	[illegible]	[illegible]	[illegible]	[illegible]
	n = 3	n = 3	n = 3	n = 3	n = 3	n = 3	n = 3	n = 3
				day (p = 0.0039) (1 > 2 = 3 = 4)				
C/N	7.6 ± 0.4	5.6 ± 0.2	5.5 ± 0.3	5.7 ± 0.7	7.7 ± 0.6	5.6 ± 0.2	5.6 ± 0.2	5.7 ± 0.3
	n = 3	n = 3	n = 3	n = 3	n = 3	n = 3	n = 3	n = 3
				day (p = 0.0001) (1 > 2 = 3 = 4)				
α	.028 ± .004	.025 ± .006	.024 ± .003	.024 ± .004	.032 ± .016	.024 ± .007	.023 ± .004	.024 ± .002
	n = 3	n = 3	n = 3	n = 3	n = 3	n = 3	n = 3	n = 3
				ns				
I_k	243 ± 37	277 ± 203	208 ± 88	147 ± 51	213 ± 32	208 ± 61	173 ± 20	188 ± 83
	n = 3	n = 3	n = 3	n = 3	n = 3	n = 3	n = 3	n = 2
				ns				
Pm	6.7 ± 0.9	6.1 ± 3.4	4.9 ± 1.6	4.0 ± 1.4	7.0 ± 4.7	4.7 ± 0.4	4.0 ± 0.9	3.6 ± 0.5
	n = 3	n = 3	n = 3	n = 3	n = 3	n = 3	n = 3	n = 3
				day (p = 0.13) (1 = 2 = 3; 2 = 3 = 4; 1 ≠ 4)				

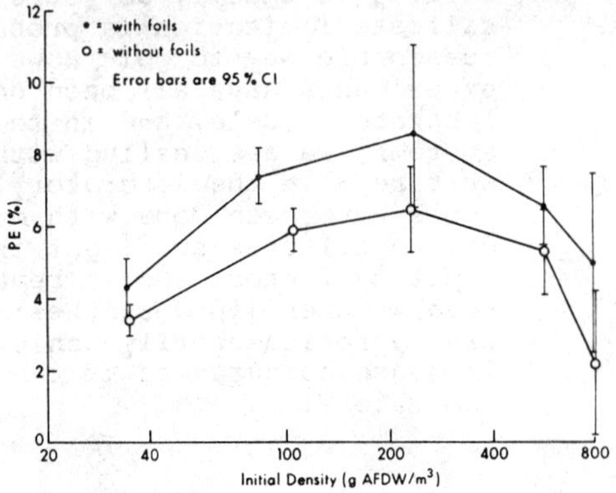

Fig. 6. Production of *Cyclotella* sp. in 9.2 m^2 flumes as a function of initial biomass, with and without foil arrays. Cells were grown on a dilution interval of one day.

confidence. The optimum density for a one-day dilution cycle appears to be about 200 g AFDW/m^3, at which density a PE of slightly more than 8% can be expected. This result compares with the PE of 10.7% achieved with the same species in the 48.4 m^2 flume grown on a two-day dilution cycle and initial density of 171 g AFDW/m^3. This result suggests that for Cyclotella a two-day dilution interval is superior to a one-day dilution interval, but further experimentation is needed to determine whether this tentative conclusion is correct, or whether the result is due to the 48.4 m^2 flume's being inherently superior to the 9.2 m^2 flume.

DISCUSSION

W. Klausmeier, Sylvatex Corp.: It seemed that silicate limitation to enhance lipid production had some potential advantages over nitrogen limitation, so maybe you weren't cutting back on yield as badly as you would with nitrogen limitation. Do you have any data that indicate that might be the case?

Laws: SERI has some data that indicate nitrogen depletion results in conversion of protein and other cellular constituents into lipid without enhancing the actual rate of lipid production very much. In some cases, you get a higher rate of lipid production if you accept 15% lipid then if you try to push the cells to 40 or 50%. So far, one species has been identified which under nitrogen stress increases the rate of lipid production. With silicate limitation, at least in our experiments, we have enhanced the rate of lipid production and intuitively my feeling is exactly as yours, that silicate limitation is probably a more reasonable way to go. However, these experiments have all been done on a laboratory scale, and in these outdoor systems, we are dealing with a very dense culture. In the laboratory, the experiments have been done with much more dilute cultures and light is not a limiting factor. So it remains to be seen whether lipid synthesis stimulation can be satisfactorily achieved outdoors in dense cultures as it can in the laboratory.

BIOMASS PRODUCTION BY MARINE CROPS: GENETIC MANIPULATION OF KELPS

M. Neushul, Ph.D.
Professor of Marine Botany
Department of Biological Sciences
University of California, Santa Barbara, 93106

ABSTRACT

The purpose of this paper is to call attention to the fact that the large scale use of kelps for energy and chemical feedstock production began in California in 1914, and then to consider efforts made since to cultivate these unique plants in the sea, particularly where these have involved genetic selection and hybridization. It has been found that cultures of the microscopic gametophytic phase of the kelp life history can be maintained in liquid culture for many years. As reported here for the first time, cultures isolated from single spores (i.e. monoclonal cultures) ten years ago, have been tested and were found to still be capable of producing gametes that after fertilization give rise to normal sporophytic plants, suggesting that genetic drift is minimal. New methods for collecting kelp spores and for applying mutagens and screening for variants have been developed and are described. Interspecific and intergeneric hybrids have been made, showing that *Macrocystis* species from Alaska and Mexico are interfertile, and that all three species of *Macrocystis* from California can be intergenerically hybridized with *Dictyoneurum*. These results show that the genetic manipulation of these plants via mass- and pedigree selection, mutagenesis and screening of variants is very likely to produce unique, and patentable gametophytic strains that can be crossed to produce high-yielding sporophytes with valuable characteristics.

BIOMASS PRODUCTION BY MARINE CROPS: GENETIC MANIPULATION OF KELPS

INTRODUCTION

The implications of genetically manipulating giant kelps are beginning to be appreciated, and reliable methods for handling these unique plants are being developed. A great deal has been learned in the past few years, as a result of work done both in the U.S. and abroad. The Gas Research Institute has encouraged and funded collaborative research on kelp for several years (15), that has involved visits by U.S. Scientists to Japan and China, as well as work by visitors from China, Japan, and elsewhere (1,2,13,33,39,41) in my laboratory at the University of California Santa Barbara on various aspects of kelp biology. I am also privliged to be working at present with a number of outstanding graduate students and Post-Doctoral visitors who have made and are presently making significant contributions to our knowledge of kelp biology (19,30).

Support from the National Science Foundation and the Chinese Ministry of Education made it possible to initiate collaborative genetic studies at Santa Barbara in September 1984 with the pioneering Chinese geneticist, T. C. Fang, whose death in July 1985, in China, was a great personal and scientific loss. Professor Fang, who worked collaboratively with N. N. Kiang (Mrs. Fang) and others was the first person to produce genetically-distinct kelps, and to show that the mass-selection and breeding of these superior strains has a very signifcant impact on the yield, the temperature tolerance and the iodine content of plants farmed in the sea (8,9,10,11,12). This paper is dedicated to the memory of Dr. Fang (Figure 1).

BACKGROUND

The first large-scale use of float-bearing macroalgae for energy and chemical feedstocks began in California in 1914, when the Hercules Chemical Company signed a contract with the British govenment to produce acetone for the manufacture of cordite, and potash for black gunpowder to be used in the First World War (28). They harvested nearly 400,000 wet tons of kelp from the offshore kelp forests of California and fermented this biomass in a large on-shore plant on aptly-named Gunpowder Point near Chula Vista, California. At this time the method by which kelp reproduced was still unknown, however one year after the Hercules effort began, Savageau (34) discovered that the kelps had a heteromorphic life-history, involving a microscopic sexual stage and a macroscopic spore-producing, sporophytic stage. Gametophyte development, gametogenesis and fertilization has been studied by a number of workers who have considered both the chromosome numbers involved (3,4,5,38,39) and the potential use of tissue culture (40, 32,18) as well as the effects of hormones on gamete release and fertilization (20). Clearly much has been learned since the turn of the century.

The experimental manipulation of the kelp life-history, and the experimental cultivation of the giant kelp, Macrocystis in the sea began in 1956 in the laboratory of F. T. Haxo at the Scripps Institution of Oceanography in La Jolla, California (22). Speculation about possible intergeneric kelp hybrids that were found in the sea by

Figure 1. The late T. C. Fang collecting marine algae at Piedras Blancas Point on the California coast.

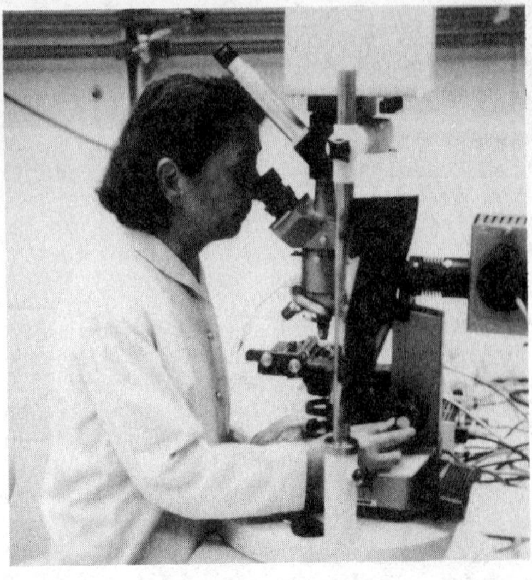

Figure 2. N. N. Kiang (Mrs. Fang) operating a microspectrophotometer used with flurochromes to quantitatively measure DNA levels in kelp gametophytes and sporophytes (6).

the author and James Stewart while diving near the La Jolla submarine canyon at this time, (24) were confirmed after collaborative work with Y. Sanbonsuga when intergeneric hybrids were made in the laboratory from cloned gametophytes (33). It is now known that intergeneric hybrids between the multi-vesiculate genus Macrocystis and the single-vesicle-forming genus Pelagophycus are commonly found in mixed populations of these two very different plants (7). Subsequent hybridization experiments with these and other kelps (25, 19) have shown that it is not difficult to make intergeneric and interspecific kelp hybrids, using cloned gametophytes. It may also be possible to produce, select and hybridize mutant clones.

It is known that ultraviolet light induces mutants in the eucaryotic marine alga, Ulva (14) and that the giant kelp, Macrocystis can be damaged by ultraviolet light (29). Recent work in my laboratory by Chai (1) and unpublished work by the late T.C. Fang, has suggested that kelp spores can be mutagenized by exposure to ultraviolet light. Post-exposure treatments showed that photoreactivation occurs. Chemical mutagenesis of kelps also seems possible. The production of mutants in the red alga Gracilaria has been achieved by van der Meer (36) using ethylmethanesulphonate (EMS). His treatment times ranged from 0 to 100 minutes, with spores being killed after 90 minutes exposure to a 0.2m solution of EMS. Many green and pink color variants were produced and shown, through tetrad analysis, to be true mutants.

The historical record (28) as well as recent advances in macroalgal mariculture in the United States, China and Japan, clearly indicate that marine macroalgae can be used for large-scale production of both chemical feedstocks and energy (21,26,35). Those familiar with the recent spectacular advances made in morphogenetically manipulating land plants (37) will find morphogenically-abnormal kelps (17) and kelp hybrids (33,25) (Figures 3-6) to be provocative. It seems likely that the application of modern genetic techniques to marine macroalgae (14) may allow us to genetically control morphogenesis in these plants, which would have a significant impact on the economics of both near-shore and off-shore farming (26).

As the findings discussed here will show, we have learned how to select and culture specific kelp germ lines, and hold these in laboratory culture for ten years, after which they are still capable of producing seedstock for outplanting in the sea. We have also found that these germ lines can be used to produce interspecific and intergeneric hybrids of plants from as far away as Alaska and Baja California. This suggests that we will not be limited to working with the morphological and physiological characteristics of a single kelp species. Preliminary results discussed below suggest that mutant lines can be produced using physical and chemical methods. These findings lead us toward the conclusion that those who develop, maintain and ultimately license the use of specific macroalgal germlines, are likely to be rewarded for their efforts.

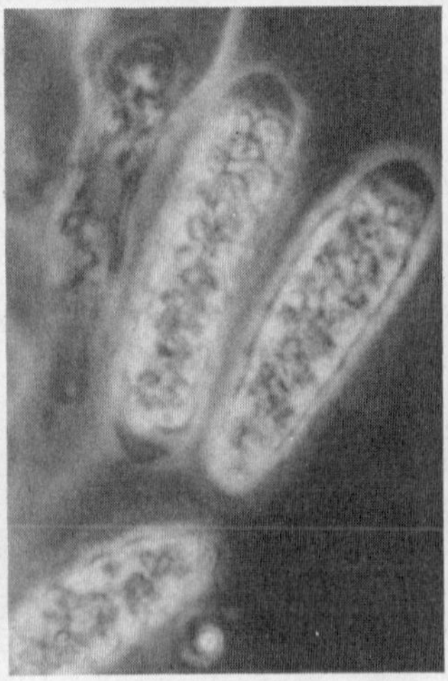

Figure 3.

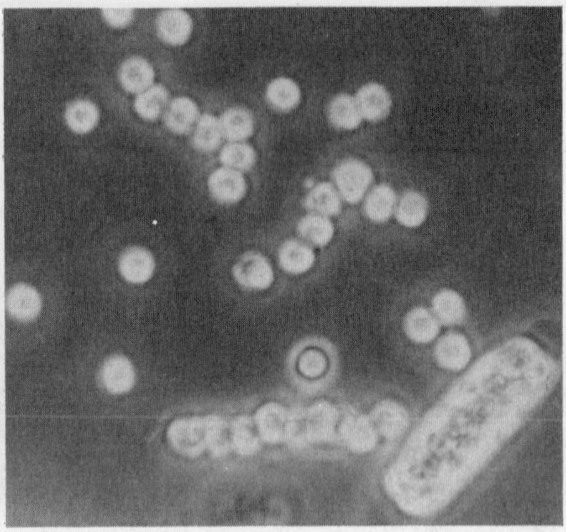

Figure 4.

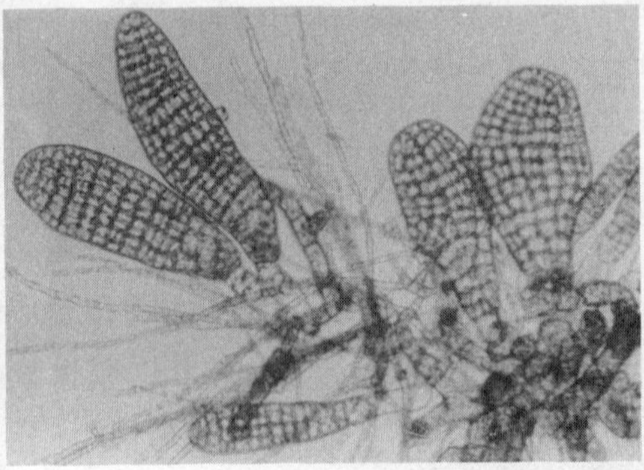

Figure 5.

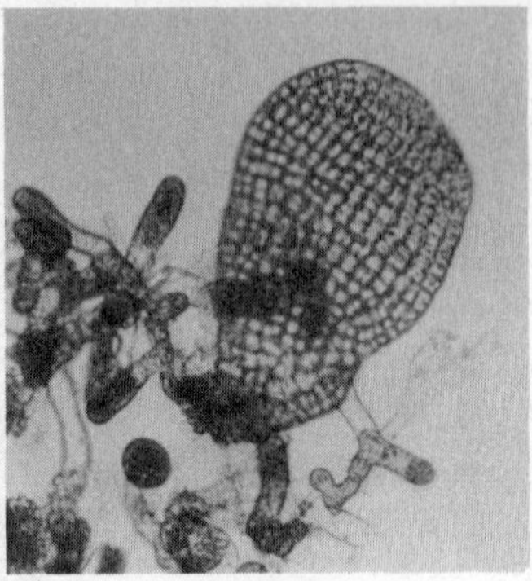

Figure 6.

Figure 3. Macrocystis sporangia containing living spores, can be separated from the sorus, and still show their characteristically thickened end where presumably the wall is softened and release occurs (2).

Figure 4. Separated spores released from a sporangium are seen to adhere to one another in chains or packets a feature that might be useful when genetic assortment is analyzed.

Figure 5. Cloned female (thick) and male (thin) gametophytes of the blade-like kelp, Dictyoneuropsis have produced normal oblong sporophytes, 0.2mm in length.

Figure 6. A female Dictyoneuropsis gametophyte has produced eggs fertilized by a Pelagophycus male, and the resultant sporophyte, nearly 0.3mm in length seems rounded.

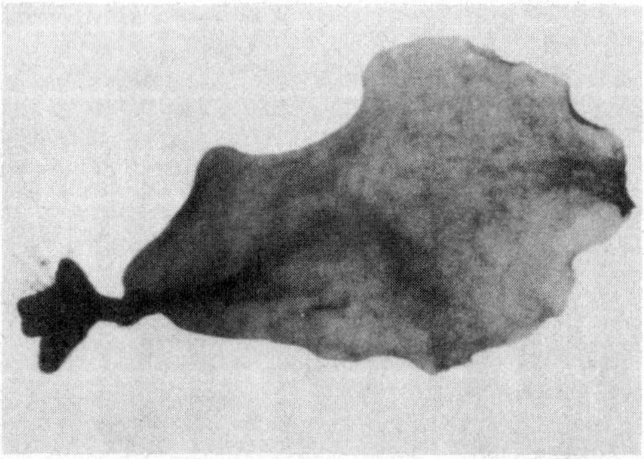

Figure 7. A normal Dictyoneuropsis sporophyte about 7mm in length. Traces of the parental gametophyte can be seen adjacent to the root-lik holdfast.

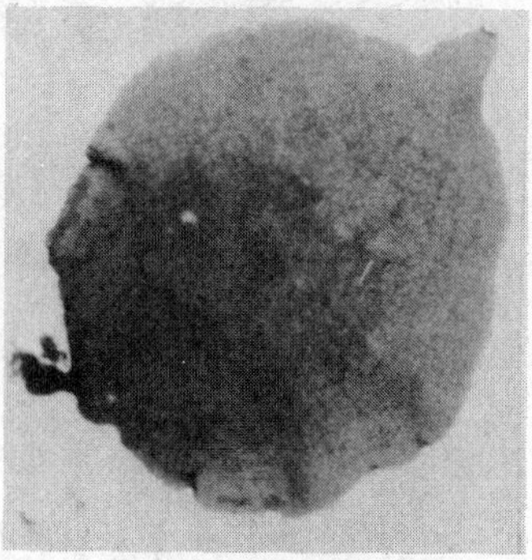

Figure 8. An intergeneric, Dictyoneuropsis x Pelagophycus hybrid, 4mm long, showing the persistance of the rounded phenotype.

MATERIALS AND METHODS

In the present study we have employed methods for the cultivation of gametophytes in the laboratory and their hybridization developed by Sanbonsuga and Neushul (33). Methods employed for the planting, cultivation and harvesting of the large kelp sporophytes in the sea, are those reported by Harger and Neushul (16) and Neushul and Harger (27). New methods have been developed for the collection of kelp spores, and for the application of mutagenic agents to sporangia and spores and determining their effects. These are discussed here.

Kelp Spore Collection

Since spores of all the kelps are produced in sporangia that form extensive sori, special attention has been given to developing methods for exposing these sporangia to mutagens and to collecting spores from them. Small plastic capsules (called Beem-capsules) used for embedding materials for electron microscopy, were held in a plastic framework consisting of centimeter squares, one centimeter thick, so that the open end of each capsule was pressed against the sorus. For *Macrocystis* this apparatus was roughly the same size as a sporophyll, so that a single sporogenous blade could be held against a row of open capsules. Before being pressed against the sporophyll each capsule was filled with filtered seawater. The capsules, capsule holder, and sporophyll were then clamped between two solid plastic sheets, held together with screws. By compressing the sporophyll-capsule-framework sandwich, spores from a given position along the long axis of the sporophyll were released into a specific capsule. After a period of spore release, the collection apparatus was removed from the water, allowed to drain and then carefully disassembled so as not to disturb the contents of the capusles, which were removed and capped. Subsequently the contents of the capsule was stirred with a glass pipette to suspend any settled spores and a hemocytometer was used to count the numbers of spores present in 1 ml samples taken from each capsule.

Ultraviolet-Radiation Mutagenesis

Spores were collected from mature sporophylls that had been abscised from plants in the sea and held in seawater-supplied aquaria. These were wiped clean with paper towls and then placed in an American Scientific Products, ultrasonic cleaner (ME 4.6) and given three five-minute periods of sonication in filtered seawater, to remove surface epiphytes. The sporophylls were then held in darkness at 10 deg. C, for two hours. Pieces of cleaned sporophyll were placed in pyrex culture dishes in Provaosoli's enriched sewater (1,33) and held for 10-20 minutes in the dark, during which time spores were released. The spore-containing medium was then poured into shallow plastic petri dishes containing 18mm square microscope cover glasses, onto which the spores settled and attached. Both radiation- exposed and control spore populations were fixed, stained and made into permanent slides, by fixing first in 5% formalin in seawater, washing and then staining in 1:10 alazarin viridine in distilled water. The cover glasses were mounted on slides in an aqueous mounting medium (Permount). Exposure to ultraviolet light was accomplished at room temperature in a specially-fabricated box where a General Electric G8T5 254 germicidal ultraviolet lamplight source was suspended above the culture dishes, containing the settled spores, at distances of 20 and 26cm. The tube was masked so that only a 4cm length was exposed. Times of exposure varied from 1 to 10 minutes. After irradiation cover slips were moved into new petri dishes containing fresh medium, and held at 17 deg C, under 52 uE/MFsq. per sec. light, after a 24 hour period in the dark to prevent photoreactivtation. After a growth period of two weeks, germinating spores were counted under a Zeiss dissecting microscope. The number of germinating spores growing into gametophytes was counted in a spore-germination-assay, to measure the effect of ultraviolet radiation and the study the process of photoreactivation.

Chemical mutagenesis

The chemical mutagen ethylmethansulfonate (EMS), was disolved in Provasolis enriched seawater plus iodine (PESI) to make a 0.2M solution and placed in the wells of a 24 well culture dish. Disks, 5.5mm in diameter, were cut from the mature portion of a Macrocystis sporophyll with a cork borer and placed in the EMS solution for 10, 20 and 60 minutes and then removed and washed in PESI containing 0.1% sodium thosulfate to neutralize the EMS. The treated disks, and untreated controls, were then transferred to a new 24 well dish with fresh PESI medium and a cover slip in each well and held for 1.5 hours during which time spores were released. Germinating spores on the cover slips were counted after two days of growth, using a spore-germination assay. The disks were then transferred to 0.3% or 1% agar plates for four days to allow further spore release. A simple spore-dispersal assay was used wherein measurements were made of the distances that spores swam out into the semi-solid agar from both the mutagen-exposed, and control disks of sporogenous tissue.

RESULTS AND CONCLUSIONS

The maturation of sporangia along a sporophyll offers a series of developmental phases where mutagens can be applied. Sporangium maturation has been studied ultrastructurally (2) and the meiotically dividing sporangium d the haploid spores produced both seemed a likely places to apply physical and chemical mutagens for maximum effect. Moreover it was felt that a simple, direct spore-killing curve could be established and that lethal damage to the haploid genome of the spore would be immediately expressed in the gametophyte, thereby screening out recessive lethal mutants that might not be evident in a diploid life-history phase.

As might be expected, maximum spore release occured about mid-way between the base of the sporophyll where sporangia are forming to the tip, where spent sporangia are found. Between 250,000 and 400,000 spores were released per ml from the most mature sporangial regions. A total of 32 are released from each mature sporangium (Figures 3 and 4). The application of spore-germination and survival assays show that ultraviolet light was most damaging to spores from the basal and terminal regions of the sporophyll. A one-minute dose from a masked tube, with 4cm exposed, 26 cm above the shallow culture dish containing the attached spores, inactivated 50% of the spores, while 92-98% of the unexposed control spores germinated. Increasing exposures of 3,5,7 and 10 minutes resulted in decreased rates of germination, and although the spores would survive they would not form normal germ tubes. Chai (1) found that spores treated, and presumably damaged by ultraviolet light could be photoreactivated by holding them in the light after UV exposure.

Spore-germination and survival assays, and a spore-dispersal assay were used to measure the effects of the chemical mutagen, ethylmethanesulfonate (EMS). Unfortunately pigmentation mutants, as were produce by EMS treatment of red algae (36) were not found. Swimming spores were observed around sporogenous disks exposed to 10 and 20 minutes in EMS, while no swimming spores were found after 60 minutes in EMS. In the spore-dispersal assay spores were observed nearly 9mm away from the disk, while 10 and 20 minute exposure gave 2-3mm dispersal zones. The disks exposed to 60 minutes of EMS had spores less than 1mm around them, presumably due to physical shaking of the disk in the soft agar. As with ultraviolet exposure increasing mutagenic dose produced decreasing amounts of germination. Ten minutes of EMS resulted in 65% germination, while 20 minutes allowed only 4% germination. EMS treated germlings were stunted and mishapen. Nonetheless by careful searching 16 female and 49 male clones of EMS-exposed, but nonetheless growing, gametophytes were isolated and are now in clonal culture.

Tables 1 and 2 (on the following pages) show the results of intraspecific and intergeneric hybridization of kelps using clonal gametophytic stocks. The intraspecific results (Table 1) show that Macrocystis from Alaska and from Mexico are interfertile, extending the observation that all three of the species found in California will cross (19). It is noteable that even some very old cultures are still capable of being crossed, illustrating the long-term stability of these clones. It is also interesting that while the massive kelp sporophytes can be grown in quantity only in the sea, and are considered to be perennial, they are very susceptible to storm and nutrient-drought damage, while the microscopic gametophytic strains, once thought to be ephemeral are in fact under laboratory culture conditions long-lived and perennial. These clones, not plants in the sea, now serve as a gene bank, much like a seed collection does for land plants.

TABLE 1. HYBRIDIZATION OF CLONED MACROCYSTIS GAMETOPHYTES FROM VARIOUS LOCATIONS (+=sporophyte produced, -=no sporophyte)

FEMALE		MALE Mi-T	Mi-Q	Mi-S	Mp-C	Ma-I	Ma-K	Mp-A	Mp-H	Mp-Ch	Control
Alaska	Mi-T	+	+	+	+	+	+[2]	+	+	+	-
Mont.	Mi-Q	+	+	+	+	+	+[2]	+	+	+	+[1]
Mont.	Mi-S	+	+	+	+	+	+[2]	+	+	+	-
Mont.	Mp-C	+	+	+	+	+	+[2]	+	+	+	-
S.B.	Ma-I	+	+	+	+	+	+[1]	+	+	+	-
S.B.	Ma-K	+	+	+	+	+	+[2]	+	+	+	-
S.C.I.	Mp-A	+	+	+	+	+	+[1]	+	+	+	+[1]
Baja	Mp-H	+	+	+	+	+	+[1]	+	+	+	+[1]
Baja	Mp-Ch	+	+	+	+	+	+[1]	+	+	+	-
Control		-	+[1]	+[1]	+[1]	-	-	+[1]	-	-	

[1]Sporophytes very small, possibly abnormal.

[2]A few small sporophytes observed after 28 days.

note: male and female gametophytes were isolated from Macrocysis angustifolia from Goleta, in October 1975, and clonally maintained for ten years. The male gametophyte strain (Ma-K) did not produce spermatozoids, but the female (Ma-K:Female 6) was still capable of gametogenesis and produced viable sporophytes.

SPECIES	ISOLATES	ORIGIN
M. integrifolia	Mi-T: Female 3, Male 18	Sitka, Alaska
M. integrifolia	Mi-Q: Female 1, Male 9	Monterey, California
M. integrifolia	Mi-S: Female 2, Male 4	Monterey, California
M. pyrifera	Mp-C: Female 1, Male 5	Monterey, California
M. angustifolia	Ma-I: Female 1, Male 1	Santa Barbara, California
M. angustifolia	Ma-K: Female 6, Male 1	Santa Barbara, California
M. pyrifera	Mp-A: Female 1, Male 1	Santa Catalina Island, Calif.
M. pyrifera	Mp-H: Female 10, Male 4	Baja California, Mexico
M. pyrifera	Mp-Ch: Female 1, Male 1	Baja California, Mexico

TABLE 2. HYBRIDIZATION OF CLONED LESSONIACEAN GAMETOPHYTES (+=sporophyte produced, -=no sporophyte, blank=no cross attempted)

FEMALE	MALE Mi-T	Ma-I	Mp-A	Mp-C	Mp-Ch	Dr-A	Nl-A	Pp-B	Pp-C	Control
Mi-T	+					-	-	-	-	-
Ma-I		+				-	-	-	-	-
Mp-A			+			+	-	+	+	+
Mp-C				+		-	-	-	-	-
Mp-Ch					+	-	-	-	-	-
Dr-A	-	-	-	-	-	+				-
Dc-A	+	+	+	+	+					-
Pp-B	-	-	-	-	-			+		-
Pp-C	-	-	-	-	-				-	-
Control	-	-	-	-	-	-	-	-	-	

Macrocystis gametophytes

M. integrifolia:
Mi-T: Female 3, Male 4

M. angustifolia:
Ma-I: Female 2, Male 1

M. pyrifera:
Mp-A: Female 1, Male 1
Mp-C: Female 1, Male 5
Mp-Ch: Female 1, Male 1

Other kelp gametophytes

Dictyoneuropsis reticulata:
Dr-A: Female 9, Male 5

Nereocystis leutkeana:
Nl-A: Male 1

Dictyoneurum californicum:
Dc-A: Female 2

Pelagophycus porra:
Pp-B: Female 3, Male 2
Pp-C: Female 3, Male 2

The intergeneric crosses made in this study (Table 2) show that a female strain of the strap-like, intertidal kelp Dictyoneurum californicum will cross readily with all five species of Macrocystis male plants tried, thus adding a new genus to the other two (Pelagophycus and Nereocystis) that have been crossed with Macrocystis (7,33).

It is particularly interesting to examine the process of intergeneric hybidization in some detail. For example crosses between Dictyoneuropsis and Pelagophycus (Figures 5-8) are distinguishable from selfed Dictyoneuropsisat a very young age. It is also noteworthy that the sporophytes that are produced in these and other kelp breeding work remain attached to the female gametophyte from which the fertilized egg was produced, making a back-selection process possible, where a gametophytic strain giving rise to an abnormal sporophyte could be identified, isolated and cloned, without having to go through the time-consuming task of raising the sporophyte to maturity.

The potential of being able to clone gametophytes from high-yielding strains of Macrocystis have been considered (27) and appear to be considerable, since high-yielding strains have been found to produce two- to three-fold more biomass than average. Of course the realization of this potential will depend on successful isolation and cloning of gametophytes, and on improved field cultivation practices.

Several basic questions about the chromosome numbers and ploidy levels in the nuclei of both gametophytes and sporophytes remain unsolved, however the use of fluorochromes coupled with microspectrophotometery (Figure 2) appears promising. It is particularly interesting to consider the implications of gene-expression and repression in plants where the same genome produces a complex multicellular plant on one hand, and a simple filamentous one on the other.

In conclusion it seems clear that selection, mutagenesis, and the isolation and storage of gametophytic clones that can be hybridized, will make it possible to recombine specific genomes to produce new plants with high-yield, high-temperature-resistance, or perhaps even plants capable of producing specific chemicals, like the oligosaccharines (37) that will be of value as plant growth regulators. The development of large-scale nearshore, and perhaps even open ocean farms, where float-bearing kelps are cultivated, would seem to be well within the realm of possibility at the present time, raising the possibility that once again we will have extensive coastal facilities like those developed early in this century by Hercules Chemical Company, for producing energy and chemical feedstocks from kelps.

ACKNOWLEDGMENTS

The work presented here was supported in part by the U.S.-China Office of the National Science Foundation, the Gas Research Institute, and the Chinese Ministery of Education. Invaluable technical support was provided by D. C. Coon, B. W. W. Harger and R. L. Lewis.

REFERENCES CITED

1. Chai, L. Studies on Ultraviolet Light Sensitivity and Photoreactivation of Macrocystis pyrifera Spores. M.A. Thesis, University of California, Santa Barbara, 59pp. (1985)

2. Chi, E. Y. and M. Neushul. Electron microscopic studies of sporogenesis in Macrocystis. Proc. Intl. Seaweed Symp. (1972) 7: 181-187.

3. Cole, K. Additional Chromosome Numbers in the Phaeophyceae. Can. J. Genet. Cytol. (1968) 10:670-672.

4. Cole, K. Chromosome Numbers in the Phaeophyceae. Can. J. Genet. Cytol. (1967) 9:519-530.

5. Cole, K. Gametophytic development and fertilization in Macrocystis integrifolia. Can. J. Bot. (1968) 46: 777-783.

6. Coleman, A. W., and L. J. Goff. Applications of Fluorochromes to Pollen Biology. I. mitramycin and 4', 6-diamidino-2-phenylindole (DAPI) as vital stains and for quantitation of nuclear DNA. Stain Technology, (1985) 60:145-154.

7. Coyer, J. A. and A. C. Zaugg-Haglund. A demographic study of the elk kelp, Pelagophycus porra (Laminariales, Lessoniaceae), with notes on Pelagophycus x Macrocystis hybrids. Phycologia (1982) 21: 399-407.

8. Fang, T. C. A summary of the genetic studies of Laminaria japonica in China. In: C. K. Tseng, Proceedings of the Joint China-U. S. Phycology Symposium (1983) Science Press: Beijing, China. pp. 123-136.

9. Fang, T. C., C. Y. Wu, B. Y. Jiang, J. J. Li and K. Z. Ren. The breeding of a new breed of haidai (Laminaria japonica Aresch.) and its preliminary genetic analysis. Acta Bot. Scinica (1962) 10: 197-209.

10. Fang, T. C., C. Y. Wu, B. Y. Jiang, J. J. Li and K. Z. Ren. The breeding of a new breed of haidai (Laminaria japonica Aresch.). Scientia Sin. (1963)1011-1018.

11. Fang, T. C., Z. Yan and Z. Wang. Some preliminary observations on Tissue culture in Laminaria japonica and Undaria pinnatifida. Kexue Tongbao (1983) 28: 247-249.

12. Fang, T.C. Some genetic features revealed from culturing haploid cells of kelps. Hydrobiologia (1984) 116/117, 317-318.

13. Fei, X. G. and M. Neushul. The effects of light on the growth and development of giant kelp. Hydrobiologia (1984) 116/117, 456-462.

14. Fjeld and Lovlie. Genetics of multicellular algae. In: R. A. Lewin (ed.), The Genetics of Algae. Univ. of Calif. Press: Berkeley. pp. 219-235. (1976).

15. Flowers, A., and K. Bird. Marine biomass: A long-term methane supply option. Hydrobiologia (1984) 116/117, 272-275.

16. Harger, B. W. W. and M. Neushul. Test-farming of the giant kelp, Macrocystis, as a marine biomass producer. J. World Maricult. Soc. (1983) 14: 392-403.

17. Kuga, V.A. and M. Neushul. The ontogeny of an abnormal specimen of Macrocystis angustfolia Bory (Phaeophyta). Phycologia (1980) 19:334-337.

18. Lee, T. F. Aposporus gametophyte formation in stipe explants from Laminaria saccharina (Phaeophyta). Bot. Mar. (1985) 28: 179-185.

19. Lewis, R. J., M. Neushul and B. W. W. Harger. Interspecific hybridization of the species of Macrocystis in California. Aquaculture. (in press)

20. Muller, D. G., G. Gassmann and K. Luning. Isolation of a spermatozoid-releasing and -attracting substance from female gametophytes of Laminaria digitata. Nature (London) (1979) 279:430-431.

21. Neushul Mariculture Incorporated. Energy from open ocean kelp farms. U. S. Senate Committee on Commerce, Science and Transportation, National Ocean Policy Study. U. S. Government Printing Office 51-285-0. 82 pp. (1980)

22. Neushul, M. Studies on the giant kelp, Macrocystis. 2. Reproduction. Amer. J. Bot. (1963) 50:354-359.

23. Neushul, M. The species of Macrocystis with particular reference to those of North and South America. Nova Hedwigia (1971) 32: 211-222.

24. Neushul, M. Macrocystis and related kelps. Nova Hedwigia (1971) 32:223-228.

25. Neushul, M. Morphology, structure, systematics and evolution of the giant kelp, Macrocystis. In: C. K. Tseng, Proceedings of the Joint China - U. S. Phycology Symposium (1983) Science Press: Beijing, China. pp. 1-27.

26. Neushul, M. Marine Farm Engineering: Genetics and Algal Production Proc. Bio-Expo 85, Boston, Mass. (1985).

27. Neushul, M. and B. W. W. Harger. Studies of biomass yield from a near-shore macroalgal test farm. J. Solar Energy Eng. (1985) 107: 93-96.

28. Neushul, P. and C. Pursell. Energy from Marine Biomass: The Historical Record. in P. Benson and K. Bird Eds, Seaweed Cultivation for Renewable Resources, Elsevier, N.Y. (Msc.)

29. Polne, M. and A. Gibor. The effect of high intensity UV radiation on benthic marine algae, in "The role of solar ultraviolet radiation in marine ecosystems" pp. 573-580. Ed. J. Calkins, Plenum Press, N.Y. (1982).

30. Reed, D. C. Field Studies on the Relationships Between Settlement Density and visible recruitment in two species of kelp. (abstr.) Western Soc. Naturalists, Monterey Calif. December 1985.

31. Saga, N. and Y. Sakai. Axenic tissue culture and callus formation of the marine brown alga Laminaria angustata. Bull. Jap. Soc. Sci. Fish. (1983) 49: 1561-1563.

32. Saga, N., T. Uchida &Y. Sakai. Clone _Laminaria_ from single isolated cell. Bull. Jap. Soc. Sci. Fish. (1978) 44:87.

33. Sanbonsuga, Y. and M. Neushul. Hybridization of _Macrocystis_ with other float-bearing kelps. J. Phycol. (1978) 14: 214-224.

34. Savageau, C. Sur la sexualite heterogameique d'une Laminaire. C. R. Acad. Sci., Paris (1915) 161: 769-799.

35. Tseng, C. K. Phycological research in the development of the Chinese seaweed industry, in, Developments in Hydrobiology, 11th International Seaweed Symposium, Eds. C. J. Bird and M. A. Ragan, Hydrobiologia (1984) 116/117: 7-18, W. Junk, The Netherlands.

36. van der Meer, J. P. Genetics of _Gracilaria_ sp. (Rhodophyceae, Gigartinales). V. Isolation and characterization of mutant strains.

Phycologia (1979) 18: 47-54

37. Van, K. T. T., P. Toubart, A. Cousson, A. G. Darvill, D. J. Gollin, P. Chelf and P. Albersheim. Manipulation of the morphogenetic pathways of tobacco explants by oligosaccharins. Nature (1985) 314: 615-617.

38. Walker, F. T. Chromosome number of _Macrocystis integrifolia_ Bory. Ann. Bot., London (1952) 16: 23-26.

39. Yabu, H. and Y. Sanbonsuga. Mitosis in the gametophytes and young sporophytes of _Macrocystis angustifolia_ Bory. Jap. J. Phycol. (1985) (Sorui) 33: 1-4.

40. Yan Zuo-mei. Studies on tissue culture of _Laminaria japonica_ and _Undaria Pinnatifida_. Hydrobiologia (1984) 116/117, pp. 314-316.

41. Zarmouh, M. M. Bacteria-free cultures of _Macrocystis_. Intl. Phycological Congress Abstracts (1985) 2: 183.

DISCUSSION

P. Benson, GRI
: Temperature is an extremely important parameter with regard to kelp growth. Can you estimate the yields you would have obtained if you hadn't been working in the El Nino season?

Neushul
: The El Nino produced storms followed by a "drought" with nutrient depletion. Where we added nutrients with a fertilizer sprayer during the nutrient drought, we increased the yield by a factor of 4. I think that we could, with effective fertilization, also get very high yields during the winter months. Thus our yields would be much higher without El Nino. The other point that I could raise in answering your question is that one of the biggest success stories of the sea grant program has been the Philippine mariculture program for red algae. Most marine biologists think that macroalgae do not grow well in the tropics, but that is not true. A rapidly growing Eucheuma plant can be the size of a small rose bush, so it is possible to grow macroalgae in warm tropical waters to a fairly large size.

Benson
: Then you'll start crossing or trying to make intergenetic crosses with these temperature-resistant plants?

Neushul
: Yes, I think that would be the thing to do. If for example you could produce and screen gametophyte clones for temperature resistance, and then make the assumption, that the high-temperature-tolerant strains of gametophytic would produce high-temperature-tolerant sporophytes, you could produce high-temperature kelps. It's conceivable that you could do the genetic screening in the laboratory and then make hybrids and crosses and plant those to get a warm-water-tolerant plants. The idea of introducing floats in a specific genetic strain is not completely out of the question. We can now get rid of floats by making certain crosses. It might be possible to make plants that are not floating now into floating strains. Whether this could be done with red algae or not I don't know, but it could probably be done with brown algae.

Benson
: That would certainly be an advantage from a harvesting standpoint.

PRODUCTION OF *LAMINARIA* IN AN EXPERIMENTAL TEST FACILITY

Boudewijn H. Brinkhuis, Ph.D.
Assistant Professor
Marine Sciences Research Center
State University of New York
Stony Brook, New York 11794-5000

ABSTRACT

There are now no commercial kelp farms in North America, nor is *Laminaria* harvested from wild populations. Major interest in *Laminaria* mariculture in North America stems from the selection of this plant as a good biomass crop for conversion to methane in the Northeast. Research supported by the gas industry (through the Gas Research Institute, New York State Energy Research and Development Authority, New York Gas Group, and the New York Sea Grant Institute) developed a plan for *Laminaria* farming in the Northeast. This plan called for preliminary laboratory and field cultivation experiments to obtain basic kelp growth data and responses of kelp to irradiance, nutrients and temperature. Studies were conducted to determine appropriate seeding protocols. Then, a small test farm, modified from both existing Chinese and Japanese designs, was built and deployed in Long Island Sound between 1983 and 1984. This paper describes the results obtained by the New York Marine Biomass Program from that test farm.

PRODUCTION OF LAMINARIA IN AN EXPERIMENTAL TEST FACILITY

INTRODUCTION

Initiation of the Northeast Marine Biomass Program

Since 1980, the New York Marine Biomass Program has conducted applied and basic research aimed at testing the feasibility of a seaweed biomass to energy production syste in New York waters (2,3,4,5,6,10,14). This project was initiated with the coordination of gas industry sponsors an biologists at Stony Brook, and engineers at Stony Brook and Cornell University, by the New York Sea Grant Institute.

Preliminary work required the identification of endemi candidate species suitable for biomass production and conversion to methane in New York waters. A literature search was conducted to determine which species were known to exhibit rapid growth rates, and what growth requirements were known. The selection process identified ten species, including red, brown and green seaweeds with varying modes of reproduction, substratum requirements and growth morphologies (3,14). A greenhouse facility was built at th Flax Pond Marine Laboratory, on the shore of Long Island Sound, to conduct growth studies on each species over a period of about 18 months. Experiments were conducted to determine the effects of different nitrogen additions and irradiance levels on seaweed growth and performance throughout the year (3). Another major component of the selection process involved methanogenesis assays of representative samples from field collected and laboratory grown specimens, which were analyzed by General Electric Co (GE) and the Institute of Gas Technology (IGT).

This program quickly identified the brown kelp *Laminaria saccharina* as the prime candidate for large-scale testing and cultivation in offshore areas. This was principally due to an anticipated growth season of 9 - 10 months and the experience of Oriental cultivation of relate kelp species. It was hoped that several crops could be planted and/or harvested during the growing season that lasts from September to July. The strategy necessitated consideration of impacts on conversion processes, i.e., introduction of feedstock biomass with potentially radical composition differences. Seaweeds exhibit seasonal composition changes that would affect the quality of multiple harvests. Further, the effect of multiple harvesting of the same crop needed to be examined to determine optimal culture density conditions.

Initial Field and Laboratory Studies

Long Island Sound is a unique habitat for *Laminaria* in comparison to those typically reported from higher latitudes. Light is higher during winter and the seasonal maximum of dissolved nitrogen concentrations is also highe Minimal ambient nitrogen concentrations are close to zero i late spring, following a highly productive annual phytoplankton bloom.

Our initial field studies monitored growth rates of several hundred Laminaria saccharina sporophytes grown on artificial substrata (rafts) placed at various depths in a near-shore site located at Crane Neck, Long Island Sound (Figure 1). Growth rate data obtained for plants placed in rafts at depths of 2 and 1 m below mean low water (MLW) in late 1981 indicated that growth of the 25% best growing plants was about 1.5x the raft population average observed, indicating a potential for selection of fast growing plants (12). These same plants were maintained during the summer of 1982, although Schlenk's study terminated in April 1982. Plants at 2 m exhibited greater summer survival than plants cultivated at 1 m; 15 vs. 5%, respectively. Growth rate monitoring continued until mid-summer 1983, after which there were no more survivors. Growth rates of 2 m deep plants during the second year autumn (1982) approached those observed during a similar period in the first year; the shallower plants performed poorly. This pattern is similar to that reported by Boden (2). We concluded that a combination of higher summer irradiance and temperature (1,4) and increased wave exposure in the shallow raft location resulted in poorer performance.

Another group of raft plants were monitored at Crane Neck between September 1982 and August 1983. These plants were initially 15 to 20 cm long and were apparently the result of spore release during the previous May-June period of sporogenesis. Their size range was more homogeneous, and growth rates were consequently less variable. An interesting observation with this second group of plants was that some individuals became reproductive when they were considerably less than 50 cm long and presumably only about 4 months old (from time of spore release in the spring). This is much more rapid than previously observed. Kain (8)

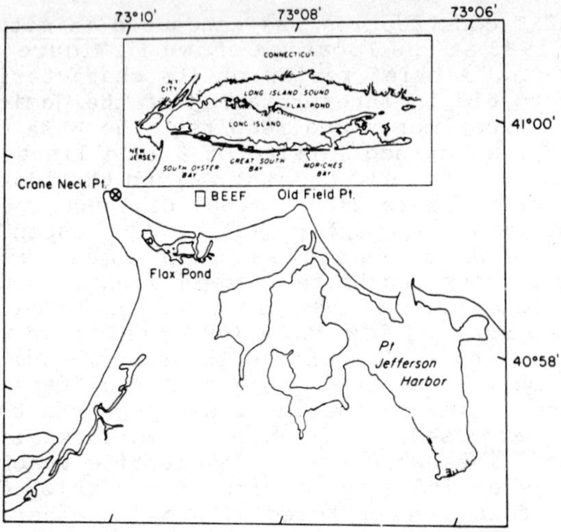

Figure 1. Map showing location of field study sites and BEEF deployment in Long Island Sound, New York.

observed that northeastern Atlantic populations required 6 months in the sporophyte stage before the onset of sporogenesis and plants usually exceed 1 m in length at the time of sporogenesis. Our monitoring has indicated that plants may exhibit sporogenesis between September and July with peaks occurring in October and May (9).

Growth rates at Crane Neck and in the greenhouse decreased rapidly as water temperatures approach 17-18 C and growth virtually ceased in July and August when water temperatures exceeded 20 C. During the summer, distal blade erosion becomes severe and most of the mature sporophytes disappear. Establishment of the autumn sporophyte population is believed to be a result of the April-May sporogenesis peak. Gametophytes and/or small sporophytes apparently survive the summer, although these have not been observed in the field. Gametophytes and juvenile sporophytes have exihibited seasonal differences in maximum survival temperature and growth (Lee, personal communication, 1985). Gametophytes generated from meiospores obtained from fertile sporophytes during winter and early spring exhibited survival only up to 17 C, while those obtained in late spring survived temperatures above 19 C.

By early 1983, we were ready to conduct a larger cultivation experiment in near off-shore waters. The Biological Engineering Experimental Farm (BEEF) was designed in collaboration with engineers from the Department of Materials Science at the State University of New York at Stony Brook and the Engineering Department at Cornell University. The type of structure to be used, determined at a design review meeting held in early 1983, was one modelled after the Chinese and Japanese farms used to cultivate the closely related kelp *Laminaria japonica* (4).

METHODS

The BEEF construction was conducted *in situ* starting in September 1983 at the location shown in Figure 1, in a water depth of 23 m. A brief review of its characteristics is given here to aid in interpretation of the design of plant cultivation experiments and results. The BEEF consisted of a horizontal, submerged lattice of 6 long lines made from wire rope 37 m long held together along the edges by wire rope 15 m long (Figure 2). Nominal distances between long lines were 2.5 m. The entire matrix was suspended below the water surface at depths between 1 and 2 meters by: a) four large corner buoys and b) by 7 small lobster floats attached along each long line. Each corner was anchored by 55 - 75 m of wire rope running down to a 300 kg railroad wheel held in place with a single 40 kg Danforth anchor, thus affording a four-point mooring system. This design differs from the usual Japanese farm in that no cross-supports to separate long lines were used. Similarly, no weights along the outside long lines were used. The lattice assembly, anchoring system and buoy network were completed in January 1984, after delays encountered with bad weather during autumn.

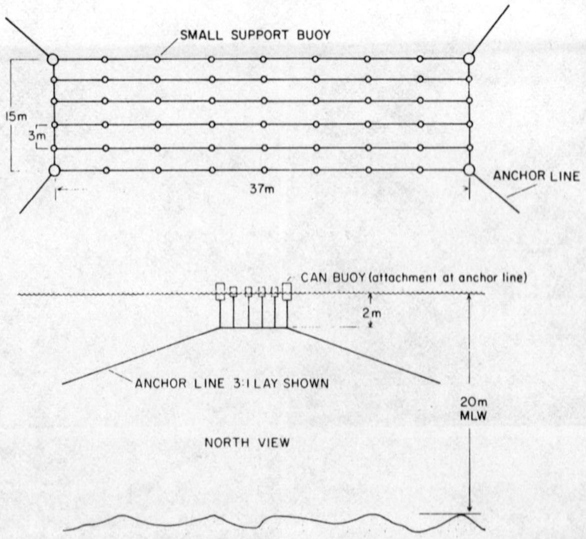

Figure 2. Schematic of BEEF showing construction details.

Based on the findings regarding summer mortality in the monitored field populations, we decided to use the forced cultivation strategy used in warmer waters of the Far East. Plants destined for the BEEF were started from seed (spores) obtained in May 1983. These spores were germinated in PES media (11) under white light irradiances between 6 and 25 $\mu E \cdot m^{-2} \cdot s^{-1}$ and temperatures of 10 - 14 C. Germinating spores go through the gametophyte stage and produce juvenile sporophytes in about 9 days under these conditions (6). These spores were seeded onto seed strings made from polyvinyl alcohol and the resultant sporophytes were cultured in clear plastic containers in an environmental chamber (see Figure 3).

Plants that came off the seed strings were raised in cone cultures consisting of 2 l glass cylinders with a funnel attached to the bottom and a screen cover on top. Seawater was supplied from the bottom by small pumps, suspending the small plants in a moving water column. When these loose and seed string plants were sufficiently large (about 2 cm long), they were transferred to aerated greenhouse culture tanks.

Two methods of deployment were tested: a) the Chinese method of manually attaching individual plants by entwining stipes / holdfasts and b) the Japanese method of inserting segments of seed string bearing juvenile sporophytes, in strands of 16 mm polypropylene culture rope. The seed string material was obtained from Dr. X.G. Fei of the Academica Sinica, Qingdao, Peoples Republic of China.

The unusually warm waters during the fall of 1983 resulted in a delay in plant deployment date. The critical maximum temperature for Laminaria saccharina sporophytes is believed to be 18 C. Normally, water temperatures drop below this limit during the last week of September or the

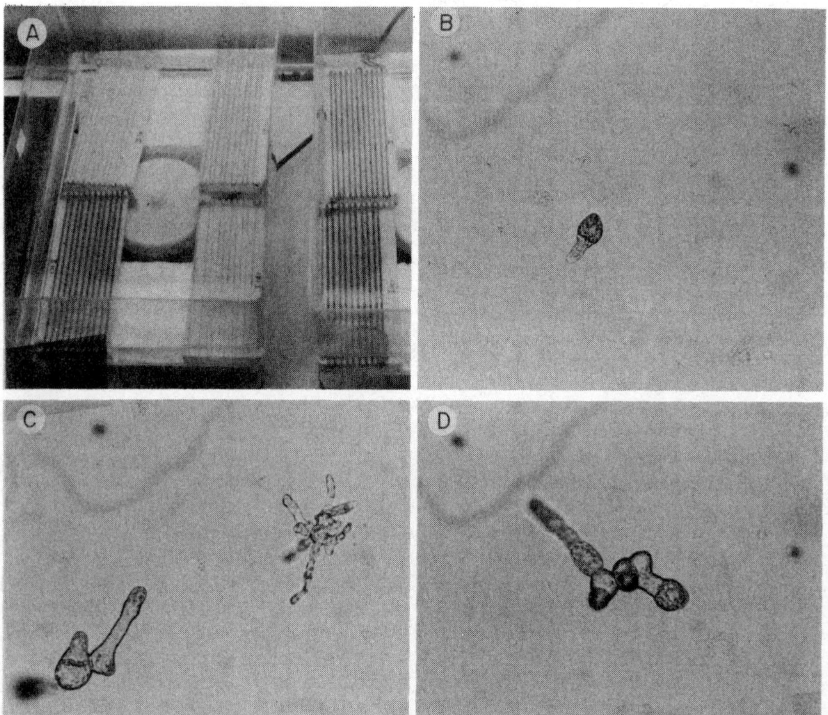

Figure 3. Cultivation of Laminaria involves winding 2 m of seed string around plastic holders that are placed in plastic boxes (A). Spores are allowed to germinate (B) and these develop into male (C) and female (D) gametophytes. After a few weeks of culture in an environmental chamber, many small sporophytes that came off the seed strings (E) were used for the "Chinese" planting method. The "Japanese" planting method used small sporophytes still attached to seed strings (F).

first week of October (unpublished data, 4 years of records maintained in Greenhouse Culture Facility). In fact, ambient water temperatures did not fall below the critical level until mid-November, 1983.

In mid-November, plants still attached to seed strings were prepared for deployment. Seed strings were cut into lengths (5 - 10 cm) and pruned so that approx. 10 plants existed on each seed string piece. These segments were placed on sixteen 2 m long culture ropes with varying spacings (25, 40 & 50 cm). They were held in greenhouse tanks supplied with running seawater until the actual deployment in February 1984. These were among the first plants deployed, and they were attached to long lines "A" and "D". This method of stocking culture ropes is by far the easiest. It took 2 people only a day to prepare these ropes. The second deployment, also in February, consisted of placing Chinese-style ropes bearing individual plants on 2 m ropes using the same spacing as seed string clusters.

The third deployment of Chinese-style plants consisted of 24 culture ropes upon which various plants were designated for growth measurements. Individual plants were attached to short (7 - 10 cm long) segments of 6 mm polypropylene rope by entwining holdfasts between the strands. This was done instead of a more direct attachment of stipes (Chinese method) or holdfasts to 10 mm culture rope because it was felt that the pressure between the strands of the heavier rope would injure the holdfasts and stipes of the plants, possibly resulting in increased losses of plants. The culture rope material we used is much more tightly woven than material used by Chinese workers. These short rope segments bearing individual plants were prepared as soon as plants were judged to be of sufficient size (about 5 cm). Labelled plastic ribbon tags were attached to each rope segment for identification of individual plants. The segments were woven onto strings and held on frames in greenhouse tanks supplied with running seawater.

All of the plants prepared for this third deployment (1382 individuals) were measured for blade length on 27 February, 1984. Blade lengths ranged from 1.2 to 84.2 cm; the majority of the plants were small; the larger plants were ones that generally came off seed strings during handling and had been placed in better conditions in cone cultures and grew to a larger size than plants actively removed from seed strings later on.

One thousand forty-four plants were required for the third deployment onto 24 culture ropes; 6 culture ropes were used for a low density planting, 6 for medium, and 12 for high density. Each 3 m long culture rope held either 21 plants 15 cm apart (low density), 31 plants 10 cm apart (medium density), or 61 plants 5 cm apart (high density). Selection of plants destined for any particular rope was randomly conducted. Generally, plants less than 10 cm long were not used for this deployment, but were held back for the 4th deployment. Some of the larger plants also were not used.

Once plants were designated for a particular culture rope, their position from top to bottom along the rope was randomly selected (13). To accomodate the spacing requirements for medium and high density, segments were placed along twists of the culture rope in an alternating, spiral pattern. The polypropylene segments were attached to culture ropes by using lobster claw rubber bands wrapped around the ends of the segments. Segment ends were "tucked" into the culture rope strands where possible to avoid potential contact with plants (the segment ends were heat-sealed to prevent separation of strands and were often hard and rough). When deployed, the horizontal spacing between adjacent culture ropes on the same long line was 2.7 m for low and medium density areas, and 1.8 m in the high density areas of long lines "B" & "C". As mentioned previously, long lines were designed to be separated by 2.5 m. It took eight people two full days to stock these ropes.

Plants were attached to their assigned ropes and positions on 8 & 9 March, 1984. Total rope weights were recorded at the time of culture rope stocking. Groups of three plants at the top, middle and bottom of ropes were measured again on March 14 and holes were punched 10 cm above the stipe / blade junction for meristematic growth rate determinations. On the day of deployment, 15 March, culture ropes were coiled and placed in about 10 liters of seawater. in opaque plastic garbage can bags. The bags were sealed enclosing a small quantity of air and placed in garbage cans for transport to the field site. Just prior to placement on the BEEF, each rope was surveyed for missing plants. Missing plants were replaced by randomly selected spares. The operation took about 5 hours from packaging to deployment completion using a crew of six people.

Each culture rope was weighted with one half of a "facing" brick weighing approximately 500 g. This type of brick has holes, and 10 mm polypropylene rope assemblies were fabricated so that fireman's safety snap hooks (3" long) could be used to fasten the bricks to a similar setup on the bottom of the culture ropes. This is a mass similar to that of rounded stones used by Japanese mariculturists (Funano, personal communication, 1983). The tops of culture ropes also had a 10 mm polypropylene rope / snap hook setup for attachment to 50 mm diameter galvanized steel rings that were clamped to long lines at appropriate intervals.

Culture ropes were initially hung vertically from a depth starting at 2 m below the water surface (the level of the wire rope lattice). This depth was chosen to minimize winter storm damage to plants, a rationale similar to that used in Japan and China. At the end of February, it was determined that the plants were probably not receiving enough light for good growth and the lattice level was reduced to 1 m across the BEEF. This was done in stages; first the long line lobster buoy support wires were shortened, then the corner buoy support wires. At the time of the third deployment, only the lobster buoy wires had been shortened. About 2 weeks after the deployment, we managed to shorten 2 of the corner buoy wires. Thus, culture ropes at the ends of long lines were at a different depth from those towards the middle (2 vs. 1 m). Due to sagging of long lines between lobster buoys and the fact that connector pieces were used between the long lines and culture ropes, the topmost plants on culture ropes were at a depth of about 2 m. The other two corner buoys were not raised until 24 May, due to loss of one corner buoy in April.

RESULTS

Complications

The BEEF did not assume a rectangular form, partly due to the difficulty in assessing proper lengths of the mooring cables and final positioning of the anchors. The shape of the farm also changed throughout the tidal cycle. The shape was a parallelogram at slack tide, but during ebb and flood, or during heavy winds, the long lines bowed such that downstream (or down-wind) long lines were virtually next to each other. On a few occasions, lobster buoys from adjacent long lines tangled and had to be separated.

Control ropes devoid of plants but with rope segments and lobster claw bans were also weighed during each visit. The degree of biofouling between these control ropes and those bearing plants was different. Plant ropes appeared to be less fouled, especially those with greater numbers of plants. This is probably due to a "sweeping" action of sporophyte blades. However, plants themselves also bore large quantities of hydroids at the distal ends and along upper blade margins, which far outweighed that on culture rope substratum. This biofouling increased over time and was maximal in June.

A violent storm hit the northeast coast on 29 March 1984, shortly after the lattice depth had been decreased. During the storm, wave heights were estimated at 3.5 - 4.0 m from the roof of the Flax Pond Marine Lab. The Coast Guard Station at Eatons Neck, 15 km to the west, estimated wave heights up to 3.2 m. We observed long lines and tops of culture ropes exposed to the atmosphere during the passing of wave troughs. Wind velocities were in excess of 40 kts for 36 hours, with maximum gusts up to 60 kts. The intensity of the storm was more severe than a hurricane due to the duration of heavy winds. The BEEF had been designed to withstand the "hundred year storm", characterized by Long Island Sound wave heights up to 2.7 m (8 feet). Three days after the storm, itit was discovered that only one lobster buoy was missing. Plants at the tops of culture ropes appeared damaged, but this had also been seen before the storm. No culture ropes were lost, but some plants on individual ropes were missing; a survey indicated a loss rate of about 10% was attributable to the storm. Proportionally, more plants were lost from low and medium density ropes than from high density ropes.

Growth Results in the BEEF Experiment

The growth rate data were examined to determine differences with respect to depth in the water column and planting density. All data used in statistical comparisons were first tested for normality (Kolmogorov-Smirnov test) and homogeneity of variances (Bartlett's test). If the data satisfied these requirements, Hochberg's GT2-method for multiple comparisons of means was used to determine significant differences among data sets. If either condition was not fulfilled, a non-parametric Games and Howell method for comparisons of means was used (13).

Mean growth rates (blade elongation rates in $cm \cdot d^{-1}$) of the third deployment plants are shown in Figures 4 and 5. A comparison of rates after combining all depths within each planting density (Figure 4) indicated that medium density plants grew slightly faster than those in other densities. However, the differences were only statistically significant during the last measurement interval when the medium density plants grew faster than high density plants, but not faster than low density plants.

Data were next examined with respect to depth differences (Figure 5) that were expected because of an anticipated light gradient (decreasing light with increasing depth). Here, growth rates were compared among densities at similar depths (top, middle or bottom of culture ropes). There were few consistent patterns. At the middle of culture ropes, low density plants appeared to grow faster than medium density plants, and these, in turn, grew faster than high density plants. Again, the only significant difference was found during the last interval, when medium density plants grew faster than high density plants at the middle depth only.

The data indicate that growth rates on the BEEF were not density limited on the 3 m culture ropes with plant spacings of 7, 10 and 20 plants per meter of rope. We had

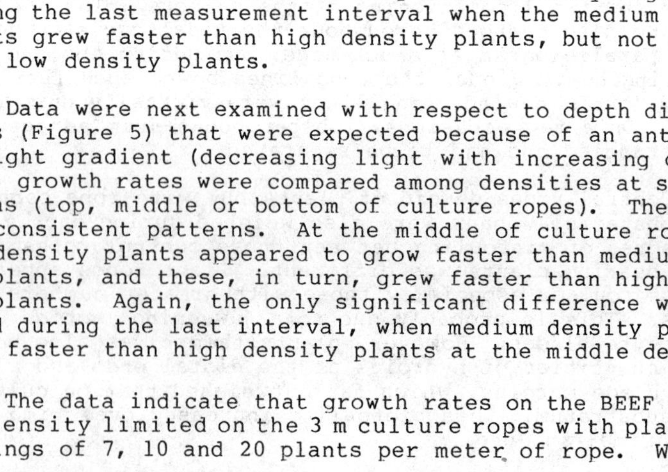

Figure 4. Meristematic growth rates of Laminaria saccharina growing on culture ropes located on the BEEF. The data have been combined by density (low, medium and high). Each bar represents average growth for 7 to 52 plants. The only significant difference noted was that the medium density plants grew faster than the high density plants in the fourth interval (June 1984). The mean standard error for the data shown is 0.05.

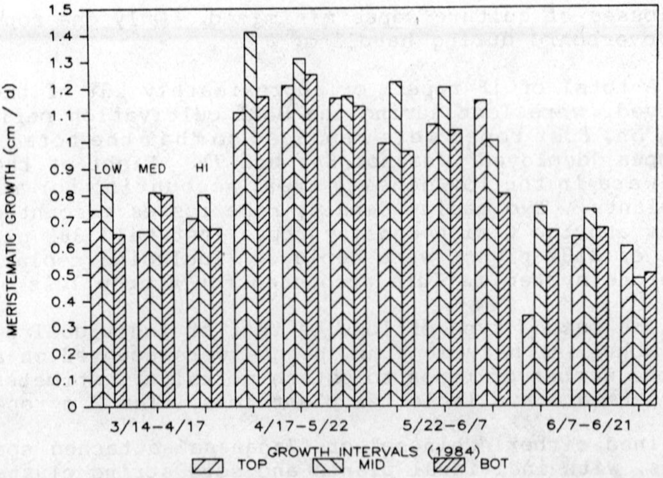

Figure 5. Meristematic growth rates of Laminaria saccharina growing on culture ropes located on the BEEF. The data have been grouped by density and each bar set within a growth interval represents plants at the top (T), middle (M) and bottom (B) of culture ropes. The mean standard error for the data shown is 0.00.

expected to observe an increased growth rate with decreasing depth, however, this was not observed for two reasons. First, since the ropes did not maintain a vertical attitude during most of the tidal cycle (due to high drag), any differences in irradiance received by plants at the tops and bottoms of culture ropes were minimized. Second, the light may have been in the saturation part of the light versus growth response curve. This species shows saturating photosynthetic rates at irradiances above 150 $\mu E \cdot m^{-2} \cdot s^{-1}$. However, a light versus growth curve has not been established for this species.

The plants used on the BEEF varied considerably in size at the time of planting, which was the main reason for the ramdomization of plants during deployment. The growth rates on the BEEF exhibited relatively large, but generally homogeneous, variances. Therefore, the statistical design, incorporating randomization, worked as intended.

Loss of plants occurred throughout the BEEF experiment (February - July) due to individuals breaking off the ropes and ropes breaking free. The loss of entire ropes was due to fraying of the 6 mm polypropylene loops between the culture ropes and the long lines. This problem was largely remedied by wrapping tygon tubing around these loops. Later, metal thimbles were used to protect the loops. Many of the complications that arose were problems created by growth monitoring requirements. Culture ropes were fabricated for easy retrieval by divers or from a boat to facilitate growth measurements. If this had not been a requirement, attachments would have been made more permanent and losses of culture ropes minimized. Only one rope was lost overboard during handling.

A total of 16 ropes, or approximately 23% of the ropes deployed, were lost during the BEEF cultivation period. Early on, lost ropes were replaced so that the total number of ropes "deployed" was not 64, but 71. Eight of the ropes lost were in the low density area, accounting for a loss of 176 plants. Two medium density rope losses accounted for 6 plants and the 6 high density ropes lost held 383 plants. total of 3187 plants were deployed (including replacements) therefore system failure accounted for a crop loss of 19%.

The second type of loss is that of individual plants from ropes. At harvest time, plants were counted on a limited number of ropes to obtain mortality estimates. Two data sets are available for plant loss. The 2 m ropes contained either "Chinese" or "Japanese" attached sporophytes, with individual plants and seed string clusters comparably spaced. These ropes were attached to the BEEF along the two outside (A & D) long lines. Survival estimates for these ropes indicated that the highest survival rate was with the Japanese method. Total numbers can exceed 100% because a few small sporophytes originally not counted (not visible to the naked eye) developed and grew after deployment. The survival of Japanese-style planting is quite impressive since the numbers include losses of entire or partial clusters on some ropes. A tota of 74 clusters were attached to the 2 m ropes. Only one entire cluster was lost; six others lost some plants. In contrast, survival of individual plants attached by the Chinese method was maximum, and only 71%, in the medium density area.

The densities used for the "Chinese" two meter ropes were not the same as that for the three meter ropes used on the two center long lines because spacings were different. Three meter ropes only held Chinese-style plantings. The highest survival (83%) occurred on the high density ropes, while the lowest survival (62%) was found on low density ropes.

One final occurrence in the BEEF experiment is noteworthy. By the middle of July, numerous recently set larvae of blue mussels were observed. These mussels covered rope and seaweed materials alike. By the first week of August, the mussel crop became so massive that most of the farm was submerged, with only the tops of the four corner buoys visible. Subsequently, the farm was cleaned of all mussels and any remaining Laminaria was removed during the third week of August. The farm was demobilized in December 1984.

DISCUSSION

A comparison of the BEEF experiment to Chinese and Japanese kelp cultivation experiences begins with an examination of crop losses. While no hard data on crop loss exists for kelp cultivation in the Far East, conservative calculations based on the number of plants outplanted, expected biomass, and harvested biomass suggests the crop loss during a cultivation season in China (October to July in northern areas) is 60%. The survival of plants on the BEEF ropes from all densities that were maintained throughout the cultivation period was 80%. This number is considerably higher than the conservative estimate for Chinese workers, probably because of the greater care during the stocking process.

The next comparison deals with the biomass yields obtained on the BEEF and those obtained in China and Japan. The spacing of ropes on the farm was arranged to provide low, medium and high density planting areas. The high density planting was determined by calculating the number of plants·m^{-2} required to yield a final biomass of about 20 kg·m^{-2} at harvest time (assuming wet weights of mature individuals of approximately 300 g). This is similar to weights and high densities found in naturally occurring kelp beds and Japanese high quality food kelp is grown at a density of 20 kg·m^{-2} (at harvest). The low and medium density areas were provided to permit determination if the high density area was growth limiting.

A total of 64 culture ropes were attached to the 1/15 acre (0.027 ha.) farm consisting of four long lines, each holding 16 ropes hanging vertically in the water column. For most of the experiment, the tops of culture ropes were at a depth of 2 m and the bottom of the culture ropes were 2 or 3 m below that level. Thus, the cultivation depth was similar to the depth distribution in local waters (2 to 5 m) and that used by Japanese workers.

About half of the seaweed biomass was removed during the first week of July, when water temperatures reached 18 C. The remaining ropes were left in the field to examine summer survival. The harvested wet seaweed weight was 120 kg; the heaviest rope (a 2 m high density seed string cluster) yielded 10 kg, with several other high density seed string ropes approaching that weight (7 - 8 kg). The heaviest Chinese-style high density rope weighed 5 kg, with an average weight of seaweed biomass per rope of about 2 kg. If all of the farm had been harvested in July, the total yield from the structure would have been approximately 215 kg, which translates to about 3.23 metric tons per acre. Assuming that all of the ropes were planted at the same high density and grew to the same extent as the heaviest rope material, yield would have been about 16 metric tons per acre (wet weight). The average individual plant weight was about 80 g wet, considerably less than the anticipated 300 g for mature plants. However, the largest plants (up to 2 m long) did approach that latter figure.

Since growth of plants on the high density ropes showed no indication of being density limited, more seaweed possibly could be stocked on each rope. Chinese workers stock ropes at 3 cm spacings between stipes instead of the minimum 5 cm used on the BEEF. Further, the fact that plants were not deployed early in the growing season resulted in decreased yield. Preliminary calculations involving maximum attained plant size and weight plus number of plants on a rope indicated attainable biomass yields of at least 30 - 40 kg per 3 meter culture rope. That would b similar to the wet biomass per rope achieved by Chinese and Japanese growers mentioned previously (30 - 50 kg).

Japanese estimates for biomass production based on a 400 ha. farm indicated attainable yields of 5.25 x 10^4 dry metric tons, providing 85 dry ash-free metric tons per ha. (36 dry ash-free metric tons per acre at 35 % ash) for a 7 month growing season (Asahi Dow Co., cited in 7). Yields from test farms that were the same size as the BEEF in thre locations along the northern Japan coast were 36 dry ash-free metric tons per ha. (7). The test farms obtained yields of 50 wet kg per meter of culture rope. The Japanes Ministry of Trade estimated attainable yields of 70 wet kg$\cdot$ $m^{-2}\cdot y^{-1}$, which is equal to 7 dry ash-free kg$\cdot m^{-2}\cdot y^{-1}$, or 4(dry ash-free metric tons per ha. (30 dry ash-free metric tons per acre).

Assuming the BEEF had contained 64 x 3 m culture ropes each yielding 30 - 40 wet kg biomass (3 - 5 dry ash-free kg ash content is similar to L. japonica, but L. japonica is a thicker and longer plant) would have yielded a harvest of 7 - 12 dry ash-free metric tons per ha. (3 - 5 dry ash-free tons per acre). The culture rope and long line spacing (each 1.5 m apart) used in Japanese biomass cultivation trials and that proposed for the 400 ha. farm size is less than that used on the BEEF. If the BEEF had similar spacings, it could have held 25 culture ropes on each of 10 long lines in the area provided. Applying a multiplicative factor (250/64 = 3.9), a yield of 28 - 46 dry ash-free metric tons per ha. (11 - 19 dry ash-free tons per acre) would be expected.

This scenario would have to be tested to see if such yields could be attained over a full growing season in Long Island waters. Similarly, the possibility of increasing biomass per meter of culture rope by using horizontally hel culture ropes wound with seed string twisted around it or with greater numbers of inserts warrants further testing. An economics of a Laminaria marine biomass to methane schem based on the yields obtained above and a systems cost analysis would need to be performed. Further, consideratio should be given to alternate farm designs that would simply operations and maintenance.

ACKNOWLEDGEMENTS

This paper is Marine Sciences Research Center Contribution No. 512.

REFERENCES CITED

1. Andrade, S.G., "First Order Analytical Growth Model of Laminaria saccharina with Realistic Dependence on Seasonal Factors." M.S. Thesis, State University of New York, Stony Brook, 86 p. (1984).

2. Boden, G.T., "The effect of depth on summer growth of Laminaria saccharina (Phaeophyta, Laminariales)." Phycologia 18, 405-8 (1979).

3. Brinkhuis, B.H. and Hanisak, M.D., "Development of a marine biomass program in New York." In: Int. Gas Res. Conf. 1981. Govt. Institutes, Inc., Rockville, pp. 682-92 (1982).

4. Brinkhuis, B.H., Breda, V., Tobin, S. and Macler, B.A., "New York Marine Biomass Program- culture of Laminaria saccharina." J. World Maricult. Soc. 14, 360-79 (1983a).

5. Brinkhuis, B.H., Macler, B.A., Hanisak, M.D., Zatorski, R.A., Liu, P. and Tsay, T.K., "Biological studies in the New York Marine Biomass Program." In: McKay, L.B. (Editor), Seaweed Raft and Farm Design in the United States and China. New York Sea Grant Institute, Albany, pp. 6.1-19 (1983b).

6. Brinkhuis, B.H., Mariani, E.C., Breda, V.A. and Brady-Campbell, M.M., "Cultivation of Laminaria saccharina in the New York Marine Biomass Program." Hydrobiologia 116/117, 266-71 (1984).

7. Flowers, A., Bryce, A.J., Harger, B., McVickar, J.T., Neushul, M. and Woessner, J., "Survey of Commercial Seaweed Production for the Marine Biomass Program." Gas Research Institute, Chicago, 29 p. + appen. (1981)

8. Kain, J.M., "A view of the genus Laminaria." Oceanogr. Mar. Biol. Ann. Rev. 17, 101-61.

9. Lee, J.A., "Growth and Reproductive Phenology of Laminaria saccharina (L.) Lamour. (Phaeophyta) at the Southern Limit of its Distribution in the Northwestern Atlantic Ocean." M.S. Thesis, State University of New York, Stony Brook, 72 p. (1984).

10. McKay, L.B., "Marine Biomass: New York State Species and Site Studies, Annual Rept. No. 84-16; December 1, 1982-November 30, 1983." N Y State Energy Res. Dev. Auth., Albany, p. 1.1-6.2. (1984)

11. Provasoli, L., "Growing marine seaweeds." Proc. Int. Seaweed Symp. 4, 9-17 (1963).

12. Schlenk, C.G., "Characteristics of Laminaria saccharina Cultivated in Long Island Sound, New York." M.S. Thesis, State University of New York, Stony Brook, 106 p. (1983).

13. Sokal, R.R. and Rohlf, F.J., Biometry. W. H. Freeman & Co., San Francisco, 859 p. (1981)

14. Squires, D.F. and McKay, L.B., "Marine Biomass: New York State Species and Site Studies, Annual Rept. No. 83-18; December 1, 1981-November 30, 1982." N Y State Energy Res. Dev. Auth., Albany, pp. 1.1-9.5. (1983).

DISCUSSION

K. Bird, Harbor Branch Foundation Inc.

Would you care to reflect on what the biomass yields were in this situation so that we can make some kind of comparison with other types of potential biomass crops?

Brinkhuis

The actual yield that we got in July was approximately 5 dry ash free ton/ac-yr because of the delayed planting season that we experienced. We normally anticipated putting plants in the water in late September when the water temperature drops below 18 to 19°C, the critical temperature for this kelp. Also, the planting density was not a limiting factor. We figure that we can double the amount of biomass production per rope. The heaviest rope we had was about 15 kg wet weight. Minus the fouling weight and the rope weight, we figure we can get somewhere around 30 to 40 kg. This is about what is obtained by the Japanese and Chinese, when they grow related kelp species. We can probably increase the number of ropes in the area that we had. We only had 64 culture ropes on this little grid, and we can probably increase that by a factor of 4. If we do that, we project between 11 and 19 dry ash free ton/ac-yr.

PRETREATMENT OF WOODY BIOMASS FOR LOW-POLLUTION COMBUSTION

K.N.R. Ramohalli, Ph.D., Associate Professor
C.A. Miller, M.S., and J.S. Palmer, M.S.
Aerospace and Mechanical Engineering Department
University of Arizona
Tucson, Arizona 85721

ABSTRACT

The concept of simple, inexpensive pretreatment for low-pollution combustion of woody biomass is explored in a multi-phase program. Based on the belief that a good theoretical understanding of the fundamentals is necessary for an economical program, the micro-details of wood pyrolysis are investigated. The fibrous heterogeneity and the nonlinear Arrhenius degradation term are explicitly included. Analytical solutions are obtained through the method of matched asymptotic expansions; only the final solutions are quoted in this paper.

Guided by the theory, small-scale samples (1" diameter × 2" long) are burned in a furnace. The samples are instrumented with fine bead (52 μ) thermocouples. Morphological structural changes are carefully observed through scanning electron microscopy. Based on earlier successes with graphite/carbon and coal, a 0.5% solution of calcium acetate is used to pretreat the wood, and the experiments are repeated. Time to ignition and the temperature profiles are obtained. The differences in the char, ash, and residue are noted and compared with the pretreatment case. The differences seem to indicate bright promise for the promotion of widespread use of woody biomass as an energy source. Bench scale furnace designs are prepared (for kg/hr of wood burning), and a detailed cost/benefit analysis is performed.

PRETREATMENT OF WOODY BIOMASS FOR LOW-POLLUTION COMBUSTION

INTRODUCTION

The recent dramatic reductions in oil prices are no indication that alternate forms of energy need not be important. Several areas in the United States are geographically suited for the use of biomass energy. In addition, we will need alternate energy sources in the event of any future increases in oil prices (which seem almost inevitable in the long run).

Since the energy "crises" of the 1970's, the use of wood has increased tremendously. Rising fuel costs have been one of the major causes of this increase, which has been encouraged in many cases by local governments and utilities. However, the increase in wood combustion has been occurring without a clear understanding of the environmental consequences of the shift in fuels. Recent studies and air quality trends have shown that the use of wood as a fuel has a significant impact on the environment.[2,7,8,35] Some communities historically enjoying clean air have experienced air pollution alerts due to the amounts of pollutants generated.[2,15] Other cities have taken steps to limit the amount of wood burned or the number of wood stoves to reduce the impact of wood pollution.[1,2]

Not only is the smoky haze unsightly, but the components of the emissions from wood are a mixture of substances which individually have been shown to cause severe health problems and together may be even worse. Some of the more hazardous substances identified in wood smoke are benzo(a)pyrene, a known carcinogen (as well as at least 13 other identified carcinogens),[7,9] respirable particulate matter (<2 μm), and carbon monoxide present in relatively large amounts.[9] In one location, the biomass-generated contribution to the total inhalable particulates in the area has reached 60%,[18] emphasizing the fact that air pollution is not a trivial matter.

Wood is not inherently a "dirty" fuel, as is coal with its high percentage of sulfur. But, the methods of state-of-the-art combustion are such that wood combustion has caused a significant air pollution problem in many parts of the country. To be able to utilize this biomass source of energy to its fullest extent, many researchers have been investigating the methods of wood combustion and the emissions of residential wood-combustion equipment. Many programs are currently in progress on a wide range of subjects dealing with conversion of biomass to energy. Included in these investigations are the study of basic pyrolysis processes at the Solar Energy Research Institute,[11] research into the combustion of pyrolysis products[36,37,21] and the pyrolysis of lignin[4] at the University of Washington, investigation into the pyrolysis of thermally thin wood samples at the Massachusetts Institute of Technology,[12,19,27] work on the use of solar energy as a means of biomass gasification at the University of Hawaii,[10,17] and studies of wood combustion at the National Bureau of Standards.[33] The U.S. Department of Agriculture, through the U.S. Forest Service, has conducted studies on the technical and economic problems encountered in using wood as a primary fuel for large furnaces and small boilers and the feasibilities of installing such systems.[39] The above are only a few of the studies on the subject of the direct combustion of biomass to energy.

The present study (and previously reported work)[29,31,34] has applied the concept of combustion chemistry alteration to wood using certain solutions and investigated its effects on the combustion characteristics of small, thermally thick (1" diameter) pieces of wood. The hope is that the treatment, combined with a more complete understanding of the processes involved in wood combustion, will lead to a method of economically reducing pollution generated by wood combustion in the residential sector and thereby promote the widespread use of this renewable energy resource. The use of wood in an efficient, simple, and environmentally acceptable manner is the ultimate goal of this program.

Looking at such a complex problem from a realistic perspective, the solution must be approached in a step-by-step manner. Attempting to quantify all the variables involved in one step is not a practical method of solving any problem of the complexity of wood/biomass combustion. Therefore, the sequence of steps shown in Figure 1 is the progression taken here. This paper will discuss the findings from Phase I. The following section will present a theory concerning the solid phase details of wood combustion. The third section will present some of the experimental results of a wood treatment process on combustion characteristics. A preliminary cost/benefit analysis of the treatment process will be made in the fourth section, and the design for a realistic bench-scale test furnace is presented in the final section.

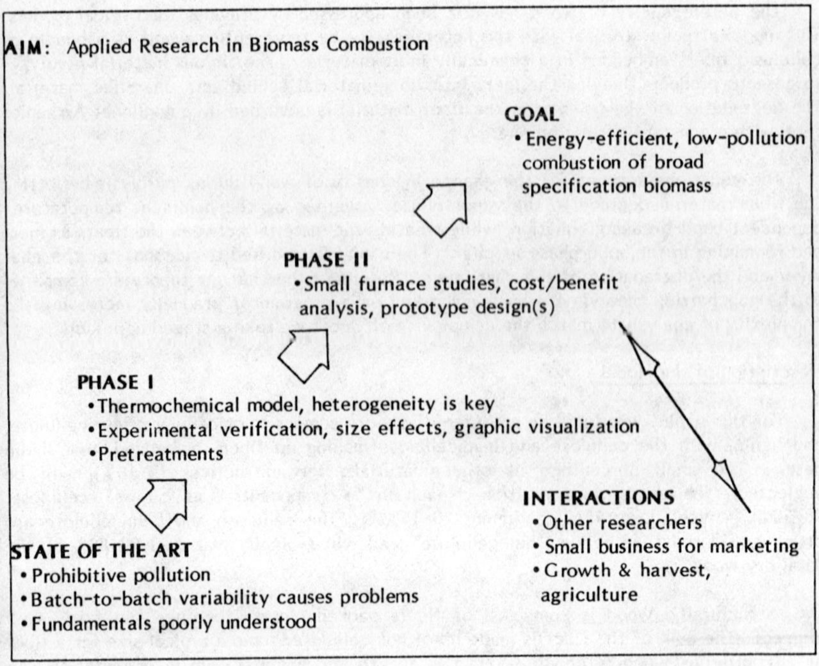

Figure 1. Overall Path of Basic Investigation into Biomass Combustion.

It is believed that several "firsts" were achieved in Phase I. Careful measurements of temperature profiles with fine bead thermocouples, pretreatments for pollution reduction, design and fabrication of a small furnace, and the development of a model explicitly including the inherent heterogeneity of biomass fuels are among the innovations developed in this study.

DEVELOPMENT OF THE THEORETICAL MODEL

A physically relevant analytical model of wood pyrolysis derived from first principles and which accounts for the high degree of heterogeneity of wood would be of great help in shedding light on the pyrolysis processes. In this study, a step toward such an analytical model is developed. The basis for the model is the work done in the aerospace field on the degradation processes of solid polymeric materials.[24,26] The main reasoning behind this approach is that wood is primarily a polymeric composition, as are the compounds upon which the studies above are based. The main problem with using such a theory is that while a polymeric fuel is typically homogeneous and will degrade at the surface directly to the gaseous phase, wood is highly **heterogeneous** and degrades in at least two major steps. The first step (in a greatly simplified representation) is the degradation of the virgin wood to char; the second step is degradation from char to gaseous products. The initial step takes place below the surface and occurs in a fairly well-defined plane sometimes called the charring or pyrolysis zone. This model is compared only with the degradation of wood to char and will not address the vapor phase details.

The heterogeneity of wood has **not** been addressed by previous theoretical models. This model attempts to deal with the heterogeneity by representing wood as a bundle of cellulosic fibers embedded in a chemically inert material. The fibrous material pyrolyzes to gaseous products, leaving the inert interfiber material behind as a char-like material. The degradation of the polymer in the fiber material is governed by a nonlinear Arrhenius term with a constant activation energy.

The equations account for the change in density of wood during pyrolysis by letting the fiber material degrade to the vapor phase, governed by the nonlinear temperature-dependent bond-breaking equation, while treating the material between the fibers as inert and remaining in the solid phase as char. The model is modified to account for the char layer and the degradation of the char layer. Results of the charring theory are compared to the noncharring theory. We have embarked on a program of gradually increasing the complexity of analysis to match the complexity of wood; we take a staged approach.

Description of the Model

For this paper, wood will be considered as a dry composite of cellulose, hemicellulose, and lignin, with the cellulose and hemicellulose making up fibers embedded in a lignin base. The small percentage of other materials, the extractives (3-10%), will be neglected. The ranges of composition of each of the components is as follows: cellulose, 40-50%; hemicellulose, 15-35%; lignin, 20-35%.[20] The cellulose and hemicellulose are often considered together as "holocellulose" and will typically make up 60-80% of the total dry wood.

Structurally, wood is composed of fibers packed closely together and fairly well aligned. The bulk of the fiber is made up of holocellulose,[20] and a typical size for a fiber of a coniferous wood is about 3.5 mm in length and about 35 μm in diameter.[13] An illustration of a woody cell and its components is shown in Figure 2.

The model is constructed by considering a single right circular cylinder of radius r_0 composed of holocellulose and embedded in a square "box" of lignin. If the fiber diameter is 35 μm, then many of these fiber box units are bundled together to make up a thermally thick sample (Figure 3). As the material is exposed to an external heat flux, the fibers begin to degrade endothermically, thereby reducing the fiber radius to zero. The lignin interfiber material will not degrade initially, but remains behind as "char." It is the determination of the time-independent rate of regression of the plane of charring into the solid that is the objective of the analysis.

Note the rough analogy between Figures 2 and 3. The impetus for the physical model shown in Figure 3 is a result of the basic shape of the woody cell structure shown above, although the actual structure of wood is much more complex. However, the physical analogy allows for some similarity between the model and the actual substance.

The model can be analytically handled through the powerful technique of matched asymptotic expansions. The details are available in Miller et al.[30] Space limitations restrict us to giving only the results here:

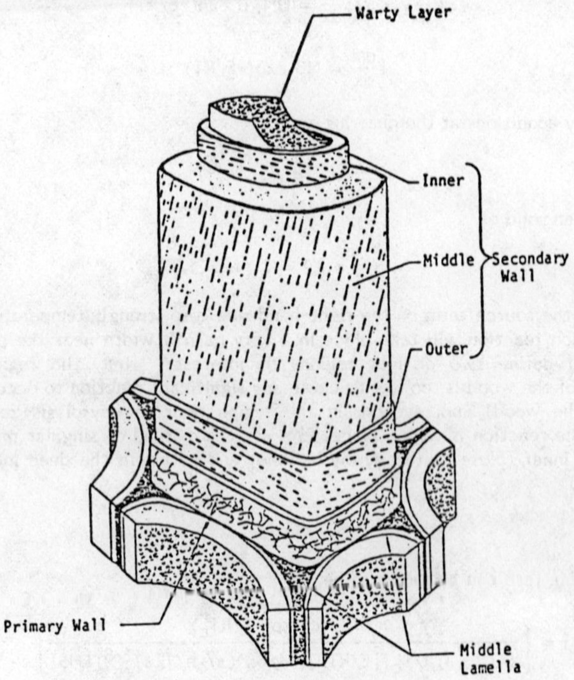

Figure 2. Simplified Representation of a Woody Cell.

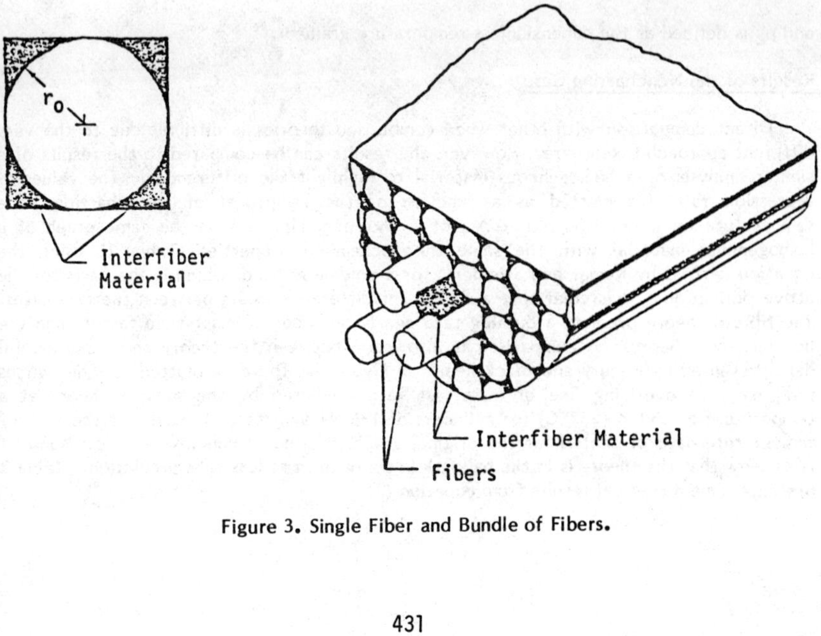

Figure 3. Single Fiber and Bundle of Fibers.

$$k\frac{d^2T}{dx^2} + \rho c\dot{r}\frac{dT}{dx} = DN\rho B\ \exp(-E/RT) \quad (1)$$

$$\dot{r}\frac{dN}{dx} = NB\ \exp(-E/RT) \quad (2)$$

with boundary conditions at the charring zone of

$$x = 0: \ T = T_c;\ N = N_c$$

and in the deep solid of

$$x = \infty: \ T = T_o;\ N = 1.$$

Because the source term on the right-hand side is so strongly temperature dependent, the degradation reaction will take place in a very narrow width near the pyrolysis zone. We can then define two distinct regions of interest: first, the region where the temperature of the wood is too low to cause any significant reaction to occur (in the deep interior of the wood) and, second, in the region near the pyrolysis zone where the majority of the reaction is taking place. We use the method of singular perturbations to evaluate the inner (close to the pyrolysis zone) and outer (in the deep interior) regions separately.

Solution

The burning rate can be shown to be

$$\dot{r} = \left\{ \frac{kB/\rho_f c\ \exp(-E/RT_c)}{[(T_c - T_o)/T_c][E/RT_c](-4/h\pi)(\pi r_o h/12\psi)^{2/3}[LHS]} \right\}^{1/2} \quad (3)$$

where

$$[LHS] \equiv 3(1 + \pi r_o h/12\psi)\{[1 + p_c + \pi r_o h/12\psi]^{1/3} - (\pi r_o h/12\psi)^{1/3}\}$$
$$- 3/4\{[1 + p_c + \pi r_o h/12\psi]^{4/3} - 3(\pi r_o h/12\psi)^{4/3}\}$$

and p_c is defined as the dimensionless temperature gradient.[30]

Results of the Noncharring Case

Direct comparison with other wood combustion theories is difficult due to the very different approach taken here. However, the results can be compared to the results of a similar analysis of a homogeneous material to highlight the differences. The values of regression rate are plotted as a function of the reciprocal of the charring zone temperature on a semi-log plot. A plot is shown in Figure 4 on the same graph of a homogeneous material with the same thermochemical properties (Table 1) using the equation derived by Kumar and Stickler[24] for a polymeric solid, which is the basis for the above derivation. It is clear that a significant difference exists between the two cases. The fibrous theory predicts a burning rate nearly an order of magnitude faster than the homogeneous theory. A comparison is also made between the theory and experimental data presented by Kanury and Blackshear[22] in Figure 5. The data plotted as M/M_i versus t/d^n have an overlying line of mass loss rate predicted by the present theory at a temperature of 630°K (257°C) for cylinders of 1-inch diameter. Because of the lack of temperature data presented for the original graph, it is not a conclusive comparison. It does show that the theory is in the ballpark in terms of mass loss rate prediction. Table 2 presents some numerical results from equation (3).

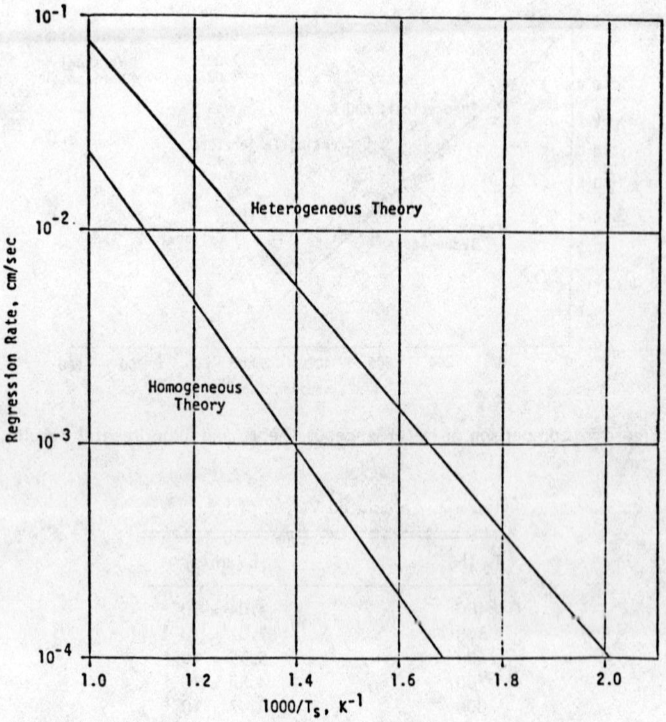

Figure 4. Comparison of Homogeneous and Heterogeneous Theory Results.

Table 1. Property Values Used in Analysis.

Parameter	Value	Source
ρ	0.45 gm/cm^3	Ref. 3
ρ_c	0.12 gm/cm^3	Ref. 3
c	0.33 cal/gm°c	Ref. 3
k	2.6×10^{-4} cal/°C cm s	Ref. 3
r_o	1.75×10^{-3} cm	Ref. 30
D	300 cal/gm	Calculated
E	26050 cal/gm-mole	
B	1.7×10^6 1/s	
ψ	1×10^{-4} cm	

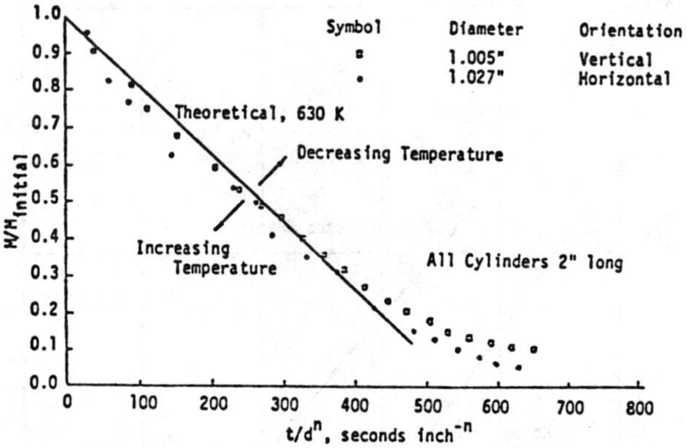

Figure 5. Comparison of Heterogeneous Theory and Experimental Results.

Table 2. Calculated Values of $\dot{r}$.

T_c (K)	$\dot{r}$ (cm/s)
400	4.04×10^{-6}
500	1.07×10^{-5}
600	9.55×10^{-4}
700	4.55×10^{-3}
800	1.47×10^{-2}
900	3.65×10^{-2}
1000	7.57×10^{-2}

Although the values of regression rate are given as a function of a constant pyrolysis zone temperature, the independent variable could as well have been in terms of a heat flux, with minor modifications. In the analysis, the heat flux would have been transformed into an equivalent temperature, and the same results obtained.

Inclusion of Char Layer Degradation

This case was also solved. The results are given in Figure 6.

Comments and Conclusions

The theories developed here are considered a first step in the development of a more complete theory describing the processes occurring during the thermal degradation and combustion of wood. The authors are well aware of the limitations of the work presented in this initial phase of theory development and realize that there is much work remaining for succeeding phases of the program.

The current theory does not account for gas phase details of the combustion of the vaporous degradation products and the resulting heat fluxes back to the wood surface. Neither does the theory take into account the convection component of the energy flow as the degradation material passes from the interior of the sample to the surface. Chemical details such as the methods of char development and the possible changes in kinetic properties due to incubation of the material at high temperatures prior to degradation are not modeled. Also lacking in the theory is the ability to include exothermic reactions within the wood as the char layer is oxidized.

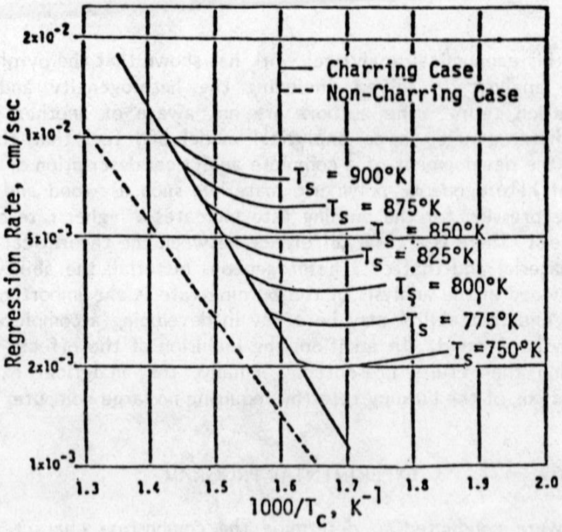

Figure 6. Regression Rate Versus $1000/T_C$ for Different Values of T_S.

The theory does, however, take a step toward painting a more complete picture of the degradation of the wood solid by explicitly considering the heterogeneous nature of wood, along with the changes in the properties of wood as charring occurs. The change in thermal conductivity is assumed here to be a step change at the wood-char interface and is not a continuous change. Density changes are modeled as a continuous function of temperature since density is directly related to the amount of fibrous material remaining at any depth. The theory is able to predict burning rates for small, thermally thick samples within an order of magnitude, as shown in Figure 7.

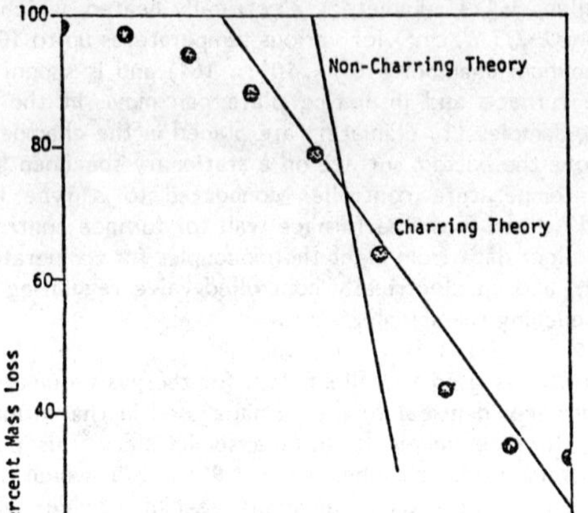

Figure 7. Comparison of Theoretical and Experimental Results.

Summary

In summary, this economical analytical work has shown that the pyrolysis/combustion of wood can be analytically solved, including the heterogeneity and the nonlinear Arrhenius degradation term. The authors are not aware of another example of the inclusion of the heterogeneity in an analytical model and feel that this model is an important step in the development of a complete analytical description of the degradation and combustion of heterogeneous polymeric materials such as wood and other biomass. The closed-form expression for the burning rate indicates a higher rate compared to the homogeneous models. There is a great difference between the theoretical burning rate of a homogeneous material and that of a heterogeneous material; the ability to include the fibrous nature of wood in the analysis of the burning rate is the important point brought out in this study, and that ability may be a key in developing a complete picture of the burning and pyrolysis of wood. In addition, the inclusion of the effects of a char layer were analyzed and the results presented. Finally, the analytical expressions allow economical predictions of the burning rate (by requiring no large computer model).

EXPERIMENTAL PROGRAM

Experiments were conducted to determine the combustion characteristics of wood treated with a calcium acetate catalyst solution compared to the same type of wood left untreated. The objectives of the experiments were to determine ignition times at different temperatures, study the temperature changes over time in small, thermally thick wood samples exposed to sudden heat fluxes, and investigate the composition of products produced during thermal degradation of wood for both treated and untreated wood samples.

Apparatus

Two separate experimental apparatus were used to investigate the characteristics of treated and untreated wood samples. The apparatus shown in Figure 8 consists of a small circular furnace (Mellen, 3-1/4" diameter), electrically heated, which can attain uniform radiative heat fluxes (~1-7.5 W/cm^2) for various temperatures up to 1000°C. The furnace is enclosed in an aluminum chamber (10" × 10" × 16") and is supported by an insulated plate, such that the furnace and insulating plate can move in the vertical direction. Cylindrical sugar pine samples (1" diameter) are placed in the chamber through the small door and mounted above the bottom surface on a stationary specimen holder. External to the chamber are a temperature controller connected to a Type K (chromel-alumel) thermocouple located 7.4 mm from the furnace wall for furnace control, a Commodore 64 computer set up to collect data from eight thermocouples for temperature measurement in the sample (Figure 9), and an electrically controlled valve regulating a flow of nitrogen gas to the jets for quenching the sample.

The second apparatus is used to collect data for the gas chromatograph and consists of a small electric furnace identical to the furnace used in the former setup, but using only one of the two furnace halves to form a semicircle. This particular furnace is enclosed in a refractory brick chamber (9" × 9" × 7") which remains stationary. Semicircular samples are mounted on an aluminum specimen holder which slides into the furnace. Attached to the specimen holder is a hypodermic syringe (Yale #25) which can obtain a gas sample through the wood in the immediate vicinity of the specimen surface. A Varian 3700 gas chromatograph is used for analysis of the gas samples collected.

Sample Preparation

Wood samples (1" diameter, 2" long) were cut from two white pine boards. Calculations confirmed that this diameter would be sufficient to provide thermally thick behavior without using samples so large that the bench-scale apparatus could not be used.

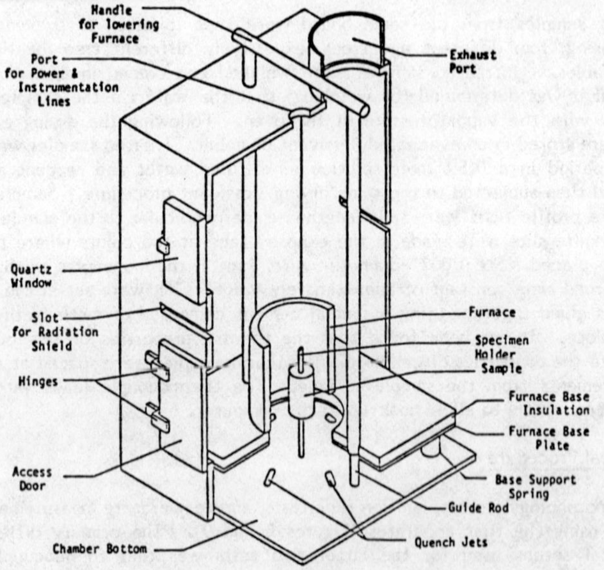

Figure 8. Cut-Away View of Ignition Time/Temperature Profile Test Setup.

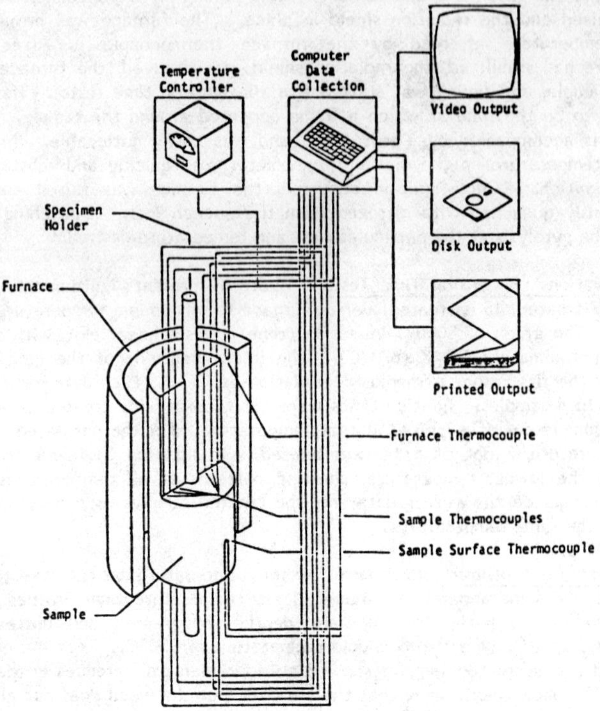

Figure 9. Temperature Profile Measurement Setup.

Batches of samples from the same board were kept together to prevent inclusion of effects arising from different age trees, excessively different tree locations, or other origin variables. Untreated samples were weighed and placed in an oven at 60°C for drying until it was determined (by weighing) that the water in the sample had reached equilibrium with the vapor pressure of the oven. Following the drying procedure, the samples were stored in an evacuated dessicant chamber. Treated samples were soaked for a 30-day period in a 0.5% mole solution of distilled water and reagent grade calcium acetate and then subjected to the same drying/dessicant procedure. Samples used in the temperature profile tests were split lengthwise, perpendicular to the annular rings of the wood, and knife slits were made in the exposed face at the points where thermocouples were to be placed. Six 0.002"-diameter-wire, Type K thermocouples, with 0.055" beads (and a thermal time constant of approximately 0.0067 s^{-1}), were set in the slits, and the sample was glued together using a special cement composed of methyl ethyl ketone and ethyl cellulose. It has been found that the thermal properties of this cement closely approximate the properties of cellulose. The thermocouples were spaced at approximately 1-mm increments from the sample surface. The thermocouple leads were then spot-welded to connectors to allow hook-up to the computer.

Experimental Procedure

Char morphology studies, ignition time tests, and temperature measurement tests were conducted using the first apparatus (Figures 8 and 9). The primary difference in the experimental setups used for the latter two tests was that thermocouples were not embedded in the samples for the ignition time tests.

To begin the tests, a prepared sample was placed on the specimen holder with the furnace raised and the radiation shield in place. The furnace was brought up to the desired temperature as read by the furnace thermocouples. After the furnace temperature had stabilized, the radiation shield was removed, the furnace was lowered over the sample, and timing was started (for the ignition time tests). The ignition time was taken to be the time at which a flame appeared around the sample. The onset of flaming was accompanied by a soft "pop" and was easily noticeable. In addition, the indicated temperature on the digital thermometer rose quickly and substantially at the time of ignition. Once ignition occurred, the furance was raised and the sample simultaneously quenched with nitrogen from the quench jets. Quenching was done to "freeze" the pyrolysis at the particular time and temperature desired.

Observations of Ignition Time Tests. Figure 10 presents ignition time data for 125 samples (77 treated, 48 untreated) versus furnace thermocouple temperatures from 425°C to 800°C. The graph of 1000/T versus log(time) gives a linear plot, with a break in the slope at approximately 900°K (673°C). The linear behavior of the ignition time data emphasizes the first-order Arrhenius degradation behavior. Each data point is an average time for 3 to 8 samples. Ignition times were not recorded for treated samples at 425°C as only 1 sample out of 4 ignited at that temperature. For the untreated samples at the same temperature, 7 out of 8 samples ignited. In addition to fewer treated samples igniting at the lowest temperature, charring patterns at all temperatures were not as distinct and char depths were greater for the treated samples as compared to untreated samples at the same temperatures.

Temperature Profiles. Raw data for the temperature profile investigations were taken using the same apparatus. Figure 11 are temperature-time profiles generated by the Commodore computer for different depths for treated and untreated samples, respectively, at a furnace thermocouple temperature of 350°C. Examination of the data for treated and untreated samples showed no appreciable differences in the temperature profiles. The experiments show that the treatment of the wood does not alter the time-dependent temperature profile at any given point in the sample, indicating the treatment causes no change in the wood's thermal diffusivity.

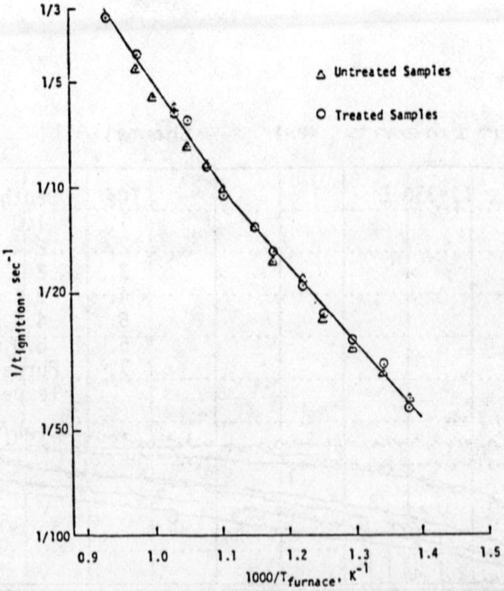

Figure 10. Ignition Times at Different Furnace Temperatures.

Examination of Figure 11 shows that the data are reasonably well behaved; however, some problems were encountered during data collection. The furnace temperature (thermocouple 8 in Figure 11) consistently showed a decrease in temperature of about 100°C as the furnace was dropped over the sample, but the temperature returned to the set temperature within 1 minute. In only one case did poor data require the entire experiment to be discarded; five of eight thermocouples failed during that experiment. Thermocouple failure was a constant and annoying problem and was probably due to the small size of the thermocouples used. In addition, at the conclusion of the experiments, the samples were found to have split at the point of the cut made for thermocouple insertion. However, no significant breaks in the profiles were noted as a result of the splitting. Ignition did not occur during any of the experiments, and the only heat source was the furnace.

Raw data from the temperature profile experiments were transformed using the nondimensionalizations from the theoretical section. The raw data presented in Figure 11 are plotted as temperature, T, versus depth, x, for various times. The resulting curves are shown in Figure 12 for a treated sample. Untreated samples did not show significant differences. Figure 13 shows the theoretical depth-temperature variation for an inert solid, with a constant surface temperature of 350°C for a 1"-diameter cylinder.[3] It can be seen that the calculated temperature of the inert sample is higher for all times and depths than the experimental samples. This can be explained by the fact that the degrading solid requires energy to break the polymer bonds, and the energy into the inert sample is used exclusively in raising the sample temperature.

Gas Chromatograph Observations. Work is currently in progress to determine the composition of the pyrolysis gases, and data will not be presented in this paper. Equipment difficulties and inadequate sample concentrations required alteration of the sampling and testing procedure originally used. One aim of the gas chromatograph experiments is to verify the hypothesis that the degradation of the cellulose is a polymer degradation process to support the theoretical model.

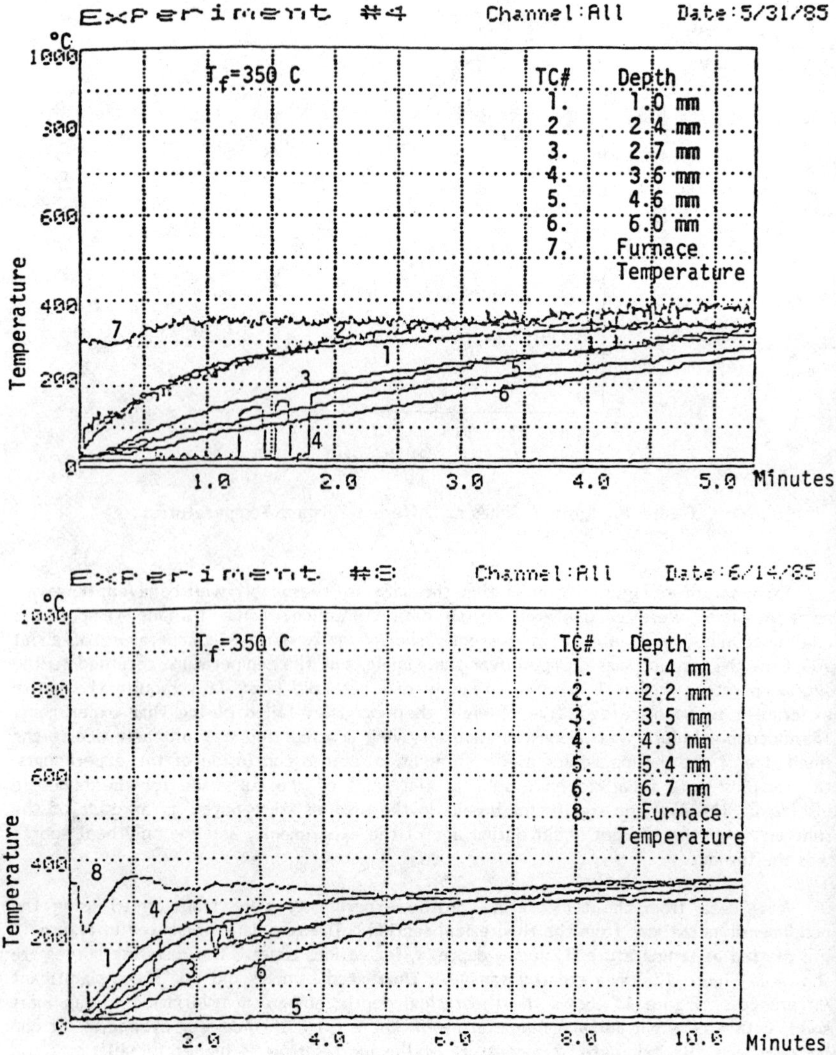

Figure 11. Measured Time-Temperature Data for Treated Sample (Top) and Untreated Sample (Bottom).

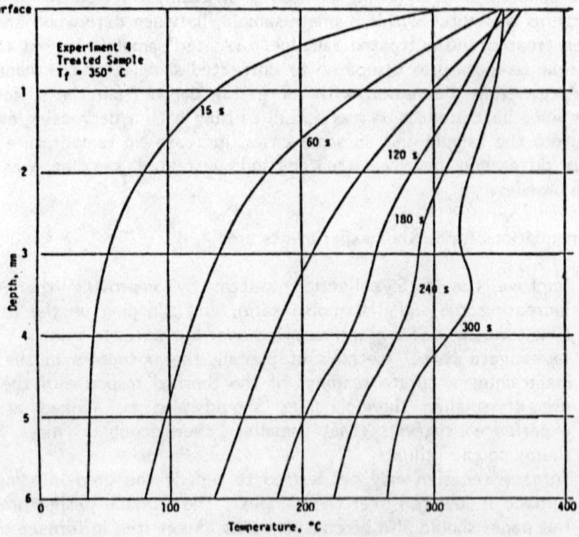

Figure 12. Temperature Profiles, Treated Sample.

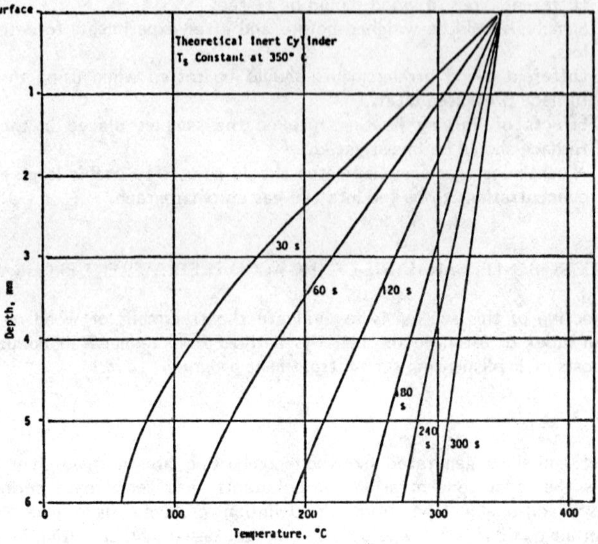

Figure 13. Temperature Profiles, Theoretical Inert Cylinder.

Conclusions and Recommendations. The experiments and microscopic examinations revealed that the basic morphological features of the wood were retained, even after exposure to radiant heat fluxes. In particular, the cell walls remained largely intact in both treated and untreated specimens, although they showed heavy distortion near the surface of the sample. Sphere formation was observed at or near the surface of treated samples, the spheres probably being composed of catalyst residues. A variation in the charring patterns was noted within a single sample, between earlywood and latewood, and also between treated and untreated samples. Treated samples showed a lower tendency to ignite at low temperatures compared to untreated samples at the same temperatures. Temperatures generally decreased with increasing depth from the surface, although at longer times some heat production was noted, causing both a decreasing energy flow from the surface into the sample and, in some cases, increases in temperature with increasing depth. No difference between treated and untreated samples was noted in the temperature profiles.

Recommendations for further experiments are:

1. Improve the data collection system by improving the A/D converter, increasing the signal-to-noise ratio, and speeding up the hard-copy output time (one copy of a graph required over 5 minutes to print).
2. Investigate better methods of placing thermocouples in the samples, while maintaining accurate readings of the thermal response of the samples. The use of smaller thermocouples should also be looked at, although our experience suggests that smaller thermocouples may lead to more thermocouple failures.
3. Furnace redesign may be desired to reduce the drop in temperature as the furnace is lowered over the sample. The furnace design presented later in this paper should also be considered as a next step in furnace design.
4. The amount of catalyst solution should be varied to determine the best level of treatment.
5. Sample size should be varied.
6. Different types of wood should be tested.
7. Samples should be weighed before and after experiments to determine weight loss.
8. Different wood arrangements should be tested when using the bench-scale furnace presented later.
9. Effects of varying heating rate on the samples placed in the bench-scale furnace should be investigated.
10. A continuous gas sampling setup should be used to enable large enough sample concentrations to be fed into the gas chromatograph.

COST-BENEFIT ANALYSIS OF THE WOOD PRETREATMENT PROGRAM

The objective of this section is to evaluate the treatment of wood on the basis of economics in order to estimate the benefits derived from reduced air pollution emissions versus the costs of implementing such a treatment program.

Introduction

Pollutant emissions generated by wood combustion are in themselves unimportant. Rather, it is the local concentrations of pollutants that can create serious problems. Communities traditionally free from air pollution problems have experienced severe reductions in air quality[7,8,32,35] due both to inherent design and operating inefficiencies of wood combustion equipment and circumstances beyond their control, such as meteorology and topography.

One approach to reducing air pollution problems is to treat the wood with a catalyst prior to burning in order to alter the combustion characteristics. If the treatment is successful in reducing emissions, the economic feasiblity of the process must also be demonstrated in order for the approach to be a viable one. The economic analysis must balance the costs of the treatment process with the cost equivalent of benefits derived from a reduction in air pollutant emissions. The costs are relatively simple to calculate, but the benefits are more difficult to quantify. Data on these emissions and resulting concentrations for different locations are scarce, and studies of the effects of the various components of each pollutant are sometimes vague in their ability to quantify the incremental effects of changes in pollutant concentrations.

The approach which will be used in the analysis of the benefits will be to look at the effects of pollution in a few communities and to analyze the benefits in each of these locations, rather than to attempt to derive a value for the entire country, since the key parameter is the local pollutant concentration. Although detailed air quality data are available for almost any large-sized community, the data do not specify all the sources of the pollutants. Therefore, the analysis of specific communities is the most practical and meaningful approach. By far the largest mass of data on wood combustion is available from residential wood combustion studies. While our work is aimed at **industrial** wood combustion, the examination of the vast data on residential wood combustion can form the first step toward an analysis of industrial wood use.

Emissions From Residential Wood Combustion

The characteristics of pollutant emissions from residential wood combustion (RWC) have been determined in the past several years as the result of several studies,[9,14] and the results of these studies have shown that the levels of emissions from RWC are by no means trivial. Table 3 lists the emissions factors from wood stoves and fireplaces in terms of the mass of pollutants emitted per unit of heat released from the wood. Of the emissions listed, carbon monoxide, particulates, and polycyclic organic matter (POM) are of the most concern. Included in the POM inventory is benzo(a)pyrene (BaP), a proven carcinogen in animals and strongly suspected as a potent carcinogen in humans. The emissions of particulate matter are also serious since the particulates emitted are composed of a high percentage of fine particles (<2 μm), which are easily deposited in the human respiratory system.

Nationally, the amount of RWC-generated emissions is substantial. Table 4 illustrates the large amount of materials being released into the atmosphere as a result of wood burning in the residential sector. Although the total amount of POM emitted is not large when compared to the emissions of carbon monoxide, for example, RWC (and particularly wood burned in wood stoves) accounts for almost the entire amount of the nation's combustion-generated POM emissions. Table 5 shows the contributions from the various combustion sources for POM's in 1980.

The ill effects of high levels of air pollution are well known. Table 6 gives a summary of several epidemiological studies relating air pollution to mortality and morbidity rates, with an emphasis on the major pollutants generated by the residential burning of wood.

In order to evaluate the benefits from reduced air pollution levels, some dollar measure of human life must be established. Estimates of this value have ranged from $30,000 to $628 million,[5] a range of over 4 orders of magnitude. A value of $300,000 will be used in this study. This value was chosen because it is cited as the "best estimate," or is within 5% of the "best estimate," of two separate studies on the economic value of a human life.[5]

Costs

The costs of treating the wood will be a function of the amount of wood that is used. There will be some fixed cost associated with the capital cost of the treating facilities (Figure 14), but the majority of the costs will be associated with the chemicals and the process to treat the wood, which will be dependent on the amount of wood treated. Other costs, such as costs associated with initiating and enforcing, if necessary, the treatment program will not be included here.

Table 3. Emissions Factors for Wood-Burning Equipment.[7,23]

Pollutant	Emissions ($lb/10^6$ Btu)	
	Wood Stoves	Fireplaces
CO	22	3.0
NO_x	0.07	0.25
SO_x	0.03	--
Particulates	1.2	1.3
Volatile HC	0.28	2.6
Polycyclic organic matter	0.0314	0.0034
Benzo(a)pyrene*	135,000	45,000

*Measured in $\mu g/10^6$ Btu.

Table 4. Total Emissions From Residential Wood Combustion.[7,14,23,28]

Pollutant	Emissions (tons/year)	
	Stoves	Fireplaces
CO	5,720,000	519,000
NO_x	18,200	43,250
SO_x	7,800	--
HC	72,800	449,800
Particulates	312,000	224,900
POM	7,733	553

Table 5. 1980 National Total Emission Estimates for POM From Combustion Sources.[23]

Source	Emissions (tons/year)	Percentage of Total
Utility Combustion	14.7	0.18
Industrial Combustion	69.0	0.80
Commercial/Institutional Combustion	1.4	negligible
Residential Combustion*	8,495.4	99.0

*Includes oil, gas, and coal.

Table 6. Health Effects of Several Air Pollutants.

Pollutant	Effects
Particulates	Increase in the death rate of 0.3/1000,000 persons/($\mu g/m^3$)/year[25]
SO_2	Increase in the death rate of 4.7/100,000 persons/($\mu g/m^3$)/year[25]
	Additional health costs of \$0.10/($\mu g/m^3$)/person/year[38]
CO	Increase in the death rate of 0.0039/100,000 persons/($\mu g/m^3$)/year[16,25]
B(a)P	Increase in lung cancer rate of 5%/($\mu g/1000\ m^3$)[6]

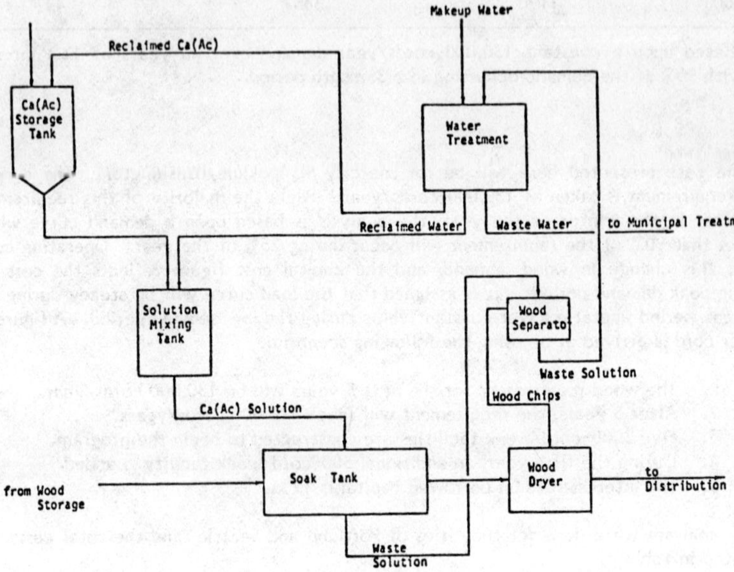

Figure 14. Flow Diagram of Treatment Process.

Table 7 shows the decline in cost per treated cord as plant capacity increases in 100 cord/week increments (each soak tank is designed to work with 100-cord batches). The cost per cord is the lowest for the largest plant size and, hence, the fewest number of plants. On a purely economic basis, the 2400 cord/week facility would make the most sense if a single customer were to be using the entire output of the plant. However, since the treatment program is aimed at reduction of air pollutant emissions by residential and small commercial/industrial wood users, the treatment program should be tailored to meet the needs of these users. So a better plan may be to construct more smaller units to accommodate wood users in different parts of a community, especially in the larger communities.

Table 7. Capital and Operating Costs and Treatment Cost Per Cord for Various Sized Plants.*[30]

Capacity (cords/week)	Plant Capital Cost ($1000)	Plant Operating Cost ($1000)	Cost Per Treated Cord ($)
100	373	208	81
200	403	340	56
300	434	473	48
400	465	606	44
500	497	738	42
600	529	871	41
700	561	1003	38
800	594	1136	40
1200	731	1733	40
2000	1029	2860	36
2400	1190	3457	43

*Based upon a constant 130,000 cords/year demand over 20 years at 12% interest rate with 90% of the demand occurring in a 3-month period.

The case presented here will be for the city of Spokane, Washington. The current wood requirement is taken as 130,000 cords/year. While the majority of this requirement will occur in the winter of the year, the analysis is based upon a demand curve which assumes that 90% of the requirement will occur during 25% of the year. Operating costs reflect this change in wood demand, and the capital cost figure reflects the cost for meeting peak demand periods. It is assumed that the load curve will be steady during the 25% peak period and at a lower constant value during the low demand period. A figure of $38 per cord is arrived at by using the following scenario:

1. The wood requirement for the next 5 years will be 130,000 cords/year.
2. After 5 years, the requirement will reach 140,000 cords/years.
3. Five 2000-cord/week facilities are constructed to begin the program.
4. During the fifth year, an additional 500-cord/week facility is added.
5. The interest rate on borrowed capital is 12%.

Similar analyses were done for the cities of Portland and Seattle, and the total costs are presented in Table 8.

Table 8. Plant Sizes and Treatment Costs Per Cord for Three Cities.[30]

City	Initial Demand[8] (cords/yr)	Peak Monthly Demand	Initial Plant Capacity (cords/week)	Number of Plants	Cost Per Cord
Spokane	130,000	39,000	2,000	5	$38
Seattle	185,000	55,500	2,400	6	$43
Portland	430,000	129,000	3,000	10	
			+2,400	1	$41

Results

With the quantitative values for both benefits and costs per unit of pollution decrease, the total dollar value of both costs and benefits in a given location can be determined if the incremental change in pollutant concentrations are known. Changes in emissions factors are translated into changes in ambient pollutant concentrations on a local scale. Reductions of 15%, 25%, and 50% in the B(a)P emissions factor are used to plot percentage reduction versus benefit value. This allows a determination of the reduction required for benefits to balance costs.

The break-even reductions in B(a)P concentrations (Table 9) are nearly the same for the cities of Seattle and Spokane at approximately 35% reduction for the benefits to equal the costs of implementation. However, the break-even point for Portland indicates that a reduction in B(a)P concentrations alone will not justify the expense of a treatment program since the break-even reduction is over 100%. Nonetheless, it must be recognized that a favorable change in the air quality will lead to benefits other than just a decrease in B(a)P concentrations. The economic effects of these other benefits have not been investigated in this paper, but they will contribute to the economic feasibility of the process since the entire amount of the cost has been included in the analysis but not all the possible benefits.

Conclusions

It appears from this preliminary analysis that the process has a strong possibility of being economically viable. The reductions of one pollutant, benzo(a)pyrene, required to balance the costs and dollar-values benefits are not unrealistic for the cities of Spokane and Seattle. Since only one pollutant was included in the benefit side of the analysis, the dollar value of the benefits presented here are low, and the process becomes even more attractive when the other pollutants are included.

Table 9. Dollar-Valued Benefits Due to Reduction of B(a)P Concentrations for Three Cities.[30]

City	Program Cost ($ million)	Benefits From Reduction of B(a)P Concentrations of: 15% Fewer Deaths	15% Million Dollars	25% Fewer Deaths	25% Million Dollars	50% Fewer Deaths	50% Million Dollars
Spokane	4.98	7.1	2.131	11.8	3.552	23.7	7.103
Seattle	8.20	11.9	3.578	19.9	5.963	39.8	11.925
Portland	19.7	9.6	2.880	16.0	4.794	32.0	9.588

DESIGN OF A BENCH-SCALE WOOD STOVE FOR TESTING KILOGRAM-SIZED SAMPLES

The objective of this section is to develop a design for a bench-scale wood-burning stove which will allow the study of different industrial design features and operating conditions and the effects of such changes on stove efficiency and pollutant emissions. The overall system is shown in Figure 15; instrumentation and wiring has been omitted for clarity. The basic stove is shown in Figure 16 in an isometric cut-away view. The stove is an insulated, cubical structure, with a fuel support situated in the center of the stove floor. The height of the support is adjustable. Below the fuel support is a tray which collects the ash from the combustion of the wood and which is removable to facilitate cleaning of the stove.

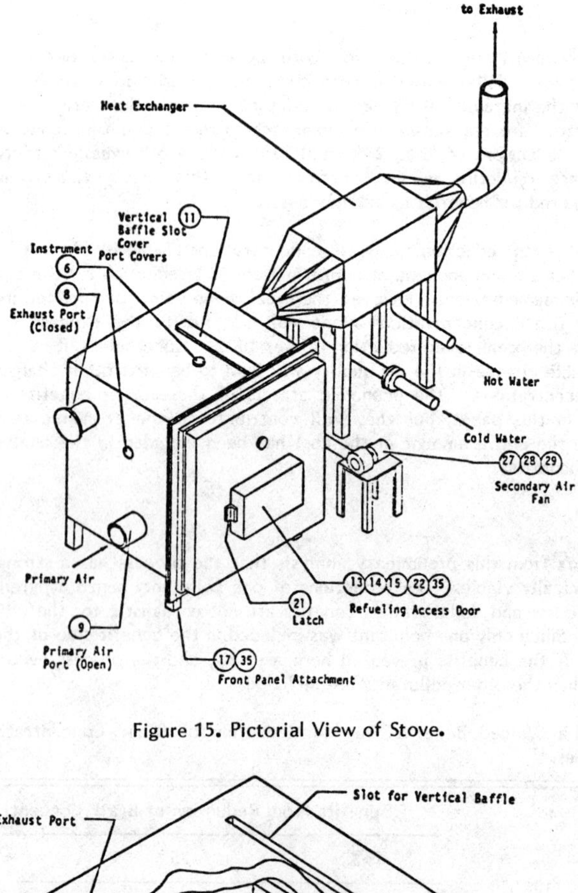

Figure 15. Pictorial View of Stove.

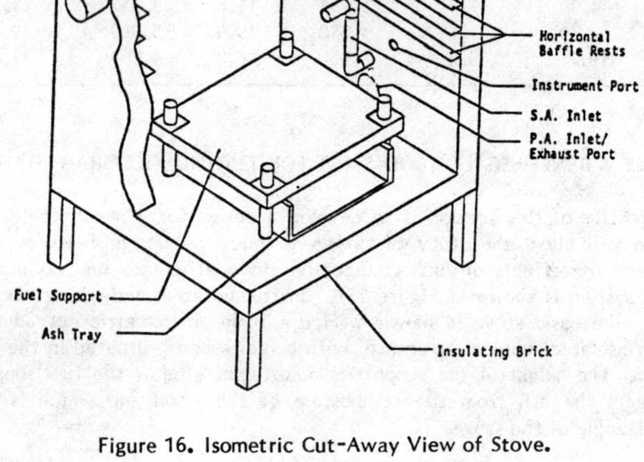

Figure 16. Isometric Cut-Away View of Stove.

The design of the bench-scale wood stove is intended to be as flexible as possible for testing various internal arrangements and the effects of changing those arrangements for increased efficiency and decreased emissions. In addition to the changes in the internal arrangements, the size and layout of the fuel charge may also be changed to determine what effect, if any, changing these parameters may have on stove operation. The air flow rates (both primary and secondary) are also variable, giving an enormous amount of different setups and operating parameters to study. The data acquired from this stove should be of substantial assistance in the investigation of wood combustion as a clean and renewable energy source.

ACKNOWLEDGMENTS

This work was performed at the University of Arizona's Aerospace and Mechanical Engineering Department under a contract from the United States Department of Energy/Battelle Pacific Northwest Laboratories through the Jet Propulsion Laboratory by agreement with the National Aeronautics and Space Administration. The authors would like to acknowledge the work of Kerry Heaton of the University and Bud Ward of Advanced Signaling Technologies for their development of the temperature measurement hardware and software and are highly appreciative of the assitance and information provided by Gary Chandler concerning the Scanning Electron Microscope and Energy Dispersive Spectrometer. The authors thank Drs. Gary Schiefelbein and Don Stevens of Pacific Northwest Laboratories and Si Friedrich of DOE for support and Mr. Karl Ekman of Jet Propulsion Laboratory for technical assistance and contract administration.

REFERENCES CITED

1. Budiansky, S., "Bioenergy: The Lesson of Wood Burning?," Environ. Sci. and Tech. 14, 760-763 (1980).

2. Cannon, J. A., "Air Quality Effects of Residential Wood Combustion," J. Air Poll. Cont. Assoc. 34, 895-897 (1984).

3. Carslaw, H. S. and Jaeger, J. C., Conduction of Heat in Solids. Oxford, England: Oxford University Press, 1959.

4. Chan, R. W. and Krieger, B. B., "Kinetics of Dielectric-Loss Microwave Degradation of Polymers: Lignin," J. Appl. Polymer Sci. 26, 1533-1553 (1981).

5. Cohen, A. S., Fishelson, G., and Gardner, J. L., Residential Fuel Policy and the Environment. Cambridge, Massachusetts: Ballinger, 1974.

6. Committee on Biologic Effects of Atmospheric Pollutants, Particulate Polycyclic Organic Matter. Washington, D.C.: National Academy of Sciences, 1972.

7. Cooper, J. A., "Environmental Impact of Residential Wood Combustion Emissions and Its Implications," J. Air Poll. Cont. Assoc. 30, 855-861 (1980).

8. Core, J. E., Cooper, J. A., and Neulicht, R. M., "Current and Projected Impacts of Residential Wood Combustion on Pacific Northwest Air Quality," J. Air Poll. Cont. Assoc. 31, 138-143 (1984).

9. DeAngelis, D. G., Ruffin, R. S., and Reznik, R. B. Reznik, Preliminary Characterization of Emissions From Wood-Fired Residential Combustion Equipment. Report No. EPA-600/7-80-040. Washington, D.C.: United States Environmental Protection Agency, 1980.

11. Evans, R. J., Soltys, M. N., and Milne, T. A., "Fundamental Pyrolysis Studies." Quarterly Report for Oct. 1-Dec. 31, 1983, Publ. SERI/PR-234-2286, Contract DE-AC-02-83CH10093, Golden, Colorado: Solar Energy Research Institute, 1984.

12. Hajaligol, M. R., Howard, J. B., Longwell, J. P., and Peters, W. A., "Product Compositions and Kinetics for Rapid Pyrolysis of Cellulose," Ind. Eng. Chem. Process Des. Dev. 21, 457-465 (1982).

13. Hale, J. D., "Structural and Physical Properties of Pulpwood." Pages 1-32 in Pulp and Paper Manufacture, ed. R. G. MacDonald. New York: McGraw-Hill, 1982.

14. Hall, R. E. and DeAngelis, D. G., "EPA's Research Program for Controlling Residential Wood Combustion Emissions," J. Air Poll. Cont. Assoc. 30, 862-867 (1980).

15. "Heat Over Wood Burning." Time Magazine, Jan. 16, 1983, pg. 67.

16. Hexter, A. C. and Goldsmith, J. R., "Carbon Monoxide: Association of Community Air Pollution With Mortality," Science 172, 265-266 (1971).

17. Hopkins, M. W., DeJenga, C., and Antal, M. J., "The Flash Pyrolysis of Cellulosic Materials Using Concentrated Visible Light," Solar Energy 32, 547-551 (1984).

18. Hough, M. L. and Kowalozyk, J. F., "A Comprehensive Strategy to Reduce Residential Wood Burning Impacts in Small Urban Communities," J. Air Poll. Cont. Assoc. 33, 1121-1125 (1983).

19. Howard, J. B., Hajaligol, M. R., Longwell, J. P., Nunn, T. R., and Peters, W. A., "Pyrolysis of Biomass." Pages 638-652 in Proceedings of Joint Meeting of Chemical Engineering Chemical Industry and Engineering Society of China and American Institute of Chemical Engineers, Vol. II (1982).

20. Janes, R. L, "The Chemistry of Wood and Fibers." Pages 33-72 in Pulp and Paper Manufacture, ed. R. G. MacDonald. New York: McGraw-Hill, 1969.

21. Kamber, P. D., Thorton, M. M., and Malte, P. C., "High Intensity Woodburner: Emissions as Affected by Operating Conditions." Paper presented at the 3rd Annual Meeting, Pacific Northwest International Section, Air Pollution Control Association, Seattle, Washington, Nov. 16-18, 1983.

22. Kanury, A. M. and Blackshear, P. L., "Some Considerations Pertaining to the Problem of Wood-Burning," Combustion Sci. and Tech. 1, 339-355 (1970).

23. Kelly, M. E., Sources and Emissions of Polycyclic Organic Matter. Document EP.4.2:P76/2, Report on Contract No. 68-02-3818. Washington, D.C.: U.S. Environmental Protection Agency, Office of Air Quality Planning and Standards, 1983.

24. Kumar, R. N. and Stickler, D. B., "Polymer-Degradation Theory of Pressure-Sensitive Hybrid Combustion." Pages 1059-1072 in XIII Symposium (International) on Combustion. Pittsburgh, Pennsylvania: The Combustion Institute, 1971.

25. Lave, L. B. and Seskin, E. P., Air Pollution and Human Health. Baltimore, Maryland: Johns Hopkins University Press, 1977.

26. Lengelle, G., "Thermal Degradation Kinetics and Surface Pyrolysis of Vinyl Polymers," AIAA Journal 8, 1989-1996 (1970).

27. Lewellen, D. C., Peters, W. A., and Howard, J. B., "Cellulose Pyrolysis Kinetics and Char Formation Mechanism." Pages 1471-1480 in XVI Symposium (International) on Combustion. Pittsburgh, Pennsylvania: The Combustion Institute, 1976.

28. Lipfert, F. W. and Dungan, J. L., "Residential Firewood Use in the United States," Science 219, 1425-1427 (1983).

29. Miller, C. A. and Palmer, J. S., "A Model of the Charring and Combustion of Wood." Paper 1-7B presented at the Spring 1985 Combined Meeting of the Central States Section and the Western States Section of the Combustion Institute, San Antonio, Texas, April 22-23, 1985.

30. Miller, C. A., Palmer, J. S., and Ramohalli, K. N. R., "Basic Research in Biomass Combustion: Phase I Final Report." Tucson: Aerospace and Mechanical Engineering Department, University of Arizona, 1985.

31. Miller, C. A. and Ramohalli, K. N. R., "A Theoretical Heterogeneous Model of Wood Pyrolysis," Combustion Science and Technology 46, 249-266 (1986).

32. Murphy, D. J., Buchan, R. M., and Fox, D. G., "Ambient Total Suspended Particulate Matter and Benzo(a)pyrene Concentrations From Residential Wood Combustion In a Mountain Resort Community," Am. Ind. Hygiene J. 45, 431-435 (1984).

33. Ohlemiller, T. J. and Gann, R., Gaseous and Liquid Combustion Products From Wood Burning Devices. Washington, D.C.: Center for Fire Research, National Bureau of Standards, 1982.

34. Palmer, J. S. and Ramohalli, K. N. R., "Pyrolysis and Combustion of Wood With and Without Pretreatment." Paper WSS/CI 84-109 presented at the Fall 1984 Meeting of the Western States Section of the Combustion Institute, Stanford, California, Oct. 22-23, 1984.

35. Romero, A. E., Buchan, R. M., and Fox, D. G., "A Study of Air Pollution From Fireplace Emissions at Vail Ski Resort," J. Environ. Health 41, 117-119 (1978).

36. Thornton, M. M. and Malte, P. C., "Analysis of Condensate and Gases Formed in Combustion of Model Wood Volatile Compounds." Paper presented at the International Conference on Fundamentals of Thermochemical Biomass Conversion, Estes Park, Colorado, Oct. 18-22, 1982.

37. Thornton, M. M., Malte, P. C., Kamber, P. D., and Hertzberg, J. R., "Combustion of Wood Volatiles." Paper WSS/CI 31-81 presented at the Fall 1981 Meeting of the Western States Section of the Combustion Institute, Tempe, Arizona, Oct. 19-20, 1981.

38. Tolley, G. S., Graves, P. E., and Cohen, A. S., Environmental Policy: Air Quality. Cambridge, Massachusetts: Ballinger, 1972.

39. United States Department of Agriculture, Wood Fired Boiler Systems for Space Heating, Vols. 1-2. Publ. EM 7180-2. Washington, D.C.: U.S. Forest Service, 1982.

DISCUSSION

J. Kerstetter, Washington State Energy Office: Have you measured benzopyrene production from catalyst-treated samples?

Ramohalli: No, the pollution measurements, except the particulates that I mentioned, have not been done yet. That is part of our program for this year and that's what we will be doing.

J. Koning, Forest Products Laboratory: What level of treatment are you working with, what percent of pickup, and what do you see as the calcium oxide disposal problem if that's the residue?

Ramohalli: We make a 0.5 mol % solution in distilled water, and we find that experimentally, the weight increase is hardly noticeable. These samples weigh about 100 to 200 grams, and we find that there's hardly any increase in the weight. So we assume that whatever is picked up is probably under 0.1 to 0.5%. We did look at the disposal problem, and we spoke with people who are already using calcium oxide as a sulfur removal agent in coal combustors. But we are not aware of any special problems in the removal of calcium oxide, especially in such low concentrations. If there are any that we have overlooked, we would certainly like to be made aware of them.

M. Lefcort, Heuristic Engineering: Why do you use calcium acetate? Is it a catalyst of some sort?

Ramohalli: We made an extremely extensive search for all possible chemicals that would affect the burning of carbonaceous substances. We did this work in the 1979-1981 time frame. We found that of all the materials, calcium acetate, lithium acetate, and potassium acetate turned out to be some of the very best. That's why we obtained the patent. I may have given the impression that we just pulled the materials out of thin air, but I must assure you that an enormous amount of research went into it. Identification of materials based on the first work function indicated those less than 4.3 electron volts were good. I have a whole set of graphs on other materials also. That's how we chose calcium acetate.

DISCUSSION, CONT.

Lefcort When wood is dried, the moisture comes out of the cells, and you end up with the fibers. There's a point that's called fiber saturation, about 23%, where the rest of the moisture has to be come out of the cell wall. It would appear from the open hexagonal structure that you showed in the electron photographs that the combustion that did take place took place within the cells themselves, and you're left with the walls. Is that related to the fact the calcium acetate affected the fibers in the center of the cells and burned it up?

Ramohalli Maybe one important point that I did not make clear is the way we prepare the sample. The treated and untreated samples are air-dried very thoroughly in an oven for several days to a constant weight. Whether calcium acetate does something inside the cells while leaving the cell walls unaffected, is speculation at this time. It has also been suggested that it may be chemically precipitating some of the chemicals in wood, to ascertain which, we have to look at the solution that's left behind after the treatment. What you said is exactly what we're also speculating on. Calcium acetate treatment is doing something to the cell content while not doing the same thing to the cell walls.

J. Diebold, SERI Your model assumes that the cellulose fibers shrink within the cell walls. Have you seen any evidence that it shrinks to form a smaller radius rather than a larger radius that would open up the center of the fiber? Do you have evidence that what is remaining is lignin rather something that's quite degraded because most chars are quite different than lignin chemically?

Ramohalli We have not seen any evidence of fibers becoming larger in diameter in all of our scanning electron microscopy. I understand that lignin is very hard to characterize and almost impossible to isolate, so we have not identified the remaining substance to be lignin. All I said was that since we know that the cell content appears to be cellulose and since that is gone, we are essentially assuming that what's left behind is non-cellulosic material. If you look at wood composition, we think that it's some lignin or phenyl propane material.

S. Babu, IGT — On the temperature versus the ignition time plot, I realize that wood is not exactly cellulose altogether, but the temperature does not approach the well accepted temperature of 451°F for ignition. Can you comment on this?

Ramohalli — We have found that the "ignition temperature" concept has to be reexamined. We know that it seems to approach a certain temperature, below which ignition is not possible. The actual value will depend I think to a great extent on the physical configuration, the geometry, the way you heat it, and all that. In our configuration, we find that's the limit we are reaching.

FUELWOOD SUBSTITUTES FOR THIRD WORLD DRYLANDS: SEEDS AND ROOTS OF ANNUAL AND PERENNIAL PLANTS

Eugene B. Shultz, Jr. and Patrick M. Carr,
Campus Box 1106,
Washington University,
St. Louis, Missouri 63130,
Harold M. Draper, III,
Governor's Office of Energy, State of Florida,
and Ianto Evans,
Aprovecho Institute, Cottage Grove, Oregon

ABSTRACT

We discuss certain alternatives to wood and charcoal for cooking food in arid parts of the Third World where fuelwood is lacking. Sustainable, high-yield cropping of oil-rich seeds and starchy (non-woody) roots would provide useful fuelwood substitutes, if these unusual fuels can be burned satisfactorily in simple, low-cost cookstoves, and be accepted by the local people. Seeds and roots might be especially helpful in those drylands where trees grow with difficulty and where the firewood crisis is most severe. Further, seeds and roots might be adopted quickly, at low-cost to local people, and without a lengthy and costly research and development process.

Our cookstove tests of combustion of crushed seeds in the form of "fuelcakes" are discussed, and the results compared with the burning of wood fuels. Simple modifications of existing Third World stoves are proposed, to accommodate the new fuels.

Seeds are ground, mixed with ashes and a little water, then squeezed by hand into lumps (fuelcakes) which are allowed to dry in the sun. Combustion characteristics of the fuelcakes are moderated by the presence of the ash.

We have also conducted preliminary tests of combustion of sun-dried starchy roots in several Third World cookstoves, in comparison with hardwood fuels. Of special interest for arid regions is the buffalo gourd (*Cucurbita foetidissima*) which produces both oil-rich seeds and starchy roots. Potential locations are discussed.

FUELWOOD SUBSTITUTES FOR THIRD WORLD DRYLANDS: SEEDS AND ROOTS OF ANNUAL AND PERENNIAL PLANTS

INTRODUCTION

The gap is widening between fuelwood supply and demand in many regions of the Third World. Wood is in short supply for essential cooking and heating needs, so programs are mounted to grow more wood and to burn wood more efficiently. But the crisis generally continues to worsen rapidly. In arid zones firewood plantations establish slowly and trees are slow to grow. Improved stoves take many years to reach everyone.

The lack of fuelwood is urgent in many countries, so a wider range of solutions must be sought vigorously. Animal dung is better used as fertilizer for food crops than as a fuel. Alcohol and biogas production require substantial capital investment. Solar cooking is not desired by those who prefer to avoid cooking in the heat of mid-day. In addition, the fuelwood shortage is of particular concern in dryland regions where native trees grow slowly and tend to be thin, thorny, crooked and tough. Harvesting without good tools is difficult.

It should not be assumed that all potential solutions to the fuelwood problem will require long, expensive, and sophisticated research efforts, and costly implementation. Some innovations, like the ones to be described, might be developed quickly and adopted at low cost without much delay, if reasons for rejection are not found by the local people.

In arid regions where trees grow with difficulty it may be possible to grow non-woody annual and perennial fuel crops which will produce faster than firewood. For example, the buffalo gourd (Cucurbita foetidissima) is native to drylands of the U.S. southwest and northern Mexico. Seasonal unirrigated yields of about 2 tonnes of seeds per hectare are projected conservatively (5). About 11 tonnes of dry roots per hectare can be calculated from fresh root yields reported for Arizona (9). The latter is nearly twice the dry biomass yield reported by McKell and co-workers (8) for unirrigated tree cropping in the arid southwestern United States.

As firewood becomes scarcer Third World cooks are burning a wide range of non-wood fuels - sawdust, dung, agricultural wastes, straw, peat, etc. At times of stress people are often open to trying new techniques. Conditions may then become propitious for introducing new fuels.

REVIEW OF PERTINENT LITERATURE

The use of crushed whole oil-rich seeds as fuel has been under experimental investigation in recent years at Washington University in St. Louis, and at the Aprovecho Institute in Cottage Grove, Oregon (6, 7, 11, 12, 16, 17, 18). Much of this work has been carried out collaboratively. In its latest form, the concept calls for "fuelcakes" to be formed by compacting crushed seeds by hand, together with ashes and a little water, into cakes and pellets of various sizes which are then sun dried. To some extent, the combustion characteristics of the fuelcakes can be controlled by altering the ratio of seeds to ashes. A higher percentage of ash will give a fuelcake that burns more slowly and with less smoke, compared to fuelcakes with a lower percentage of ash. Williams' thesis (17) provides a detailed discussion of this simple technology, applicable in a rural setting at low cost.

The burning of the dried roots of the buffalo gourd (Cucurbita foetidissima) as a novel fuel was first investigated by Shultz and co-workers (13, 14), but tests were performed only in open fires. For example, ugali, a Kenyan staple food, was prepared successfully. Results were encouraging, but indicated the importance of cookstove testing which has now been carried out on a preliminary basis in the present work.

The buffalo gourd is a potential dryland source of oil-rich seeds and starch-containing roots. If adaptable to drylands outside of its native range, it may become an important energy crop for rural development in the arid Third World. In addition to seed fuelcakes and dried roots as fuels, the buffalo gourd might provide a source of fuel ethanol through hydrolysis and fermentation of the starchy roots, and a source of illuminant fuel through extraction and use of the seed oil in appropriately-designed simple lamps (13, 14).

The buffalo gourd is a member of the same genus that includes the cultivated squash and pumpkins. It develops carrot-shaped unpalatable roots containing substantial quantities of starch, varying from 16 to 21 percent of the fresh root (15) or 52 to 68 percent of the dry root. The gourds produce seeds with 30-40 percent oil and the oil has properties similar to those of sunflower and soybean oils. A comprehensive interpretive review of the buffalo gourd literature has been carried out by DeVeaux and Shultz (3), and Gathman and Bemis (4) have reviewed the large body of buffalo gourd literature produced at the University of Arizona under Bemis' direction.

Two modes of cultivation of buffalo gourd have received attention. The "perennial" mode optimizes fruit yield for production of seed, and the annual mode maximizes root yield. In the former, there would be an annual harvest of all the fruit, and in addition, a portion of the roots would be harvested from alternate swaths annually, to prevent overcrowding of the field (1). In the annual mode, root production is optimized with high-density plantings. As many as 550,000 plants per hectare may be appropriate, depending on the elevation, soil, and climate of the location (9). Therefore, the farmer who wishes to produce both roots and seeds as fuels would probably establish separate fields according to the above descriptions. Roots and seeds would be harvested from both, if no photoperiodic limitations were apparent to prevent fruiting from occurring, but the "perennial" field would provide a higher seed yield per hectare than the annual field, and the annual field would yield more roots per hectare than the perennial field.

Buffalo gourd plants can apparently survive with as little as 150 mm (about 6 inches) of water annually. Preliminary tests indicated that about 250 mm (10 inches) is probably necessary to produce seed (15).

TYPES OF STOVES UTILIZED IN TESTS

Our tests of combustion of seed fuelcakes and dried buffalo gourd roots were carried out in a diversity of stoves available to us at the cookstove research facility of the Aprovecho Institute, Cottage Grove, Oregon. Although some tests were carried out over open fires, most of the studies were conducted in the following single-pot charcoal and wood-burning stoves:

a) Z-stove: an insulated cooker with secondary air supply, manufactured, highly efficient.

b) Forneau Malgache: sheet metal, basin-shaped, on cylindrical base, traditional, artisan-built, and used in many parts of Africa, the Caribbean, etc.

c) "Thai-bucket" charcoal stove: artisan-built conical ceramic liner in cylindrical insulated metal container.

Sketches of each are provided below:

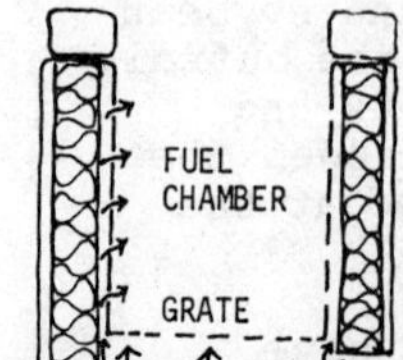

"Z" STOVE

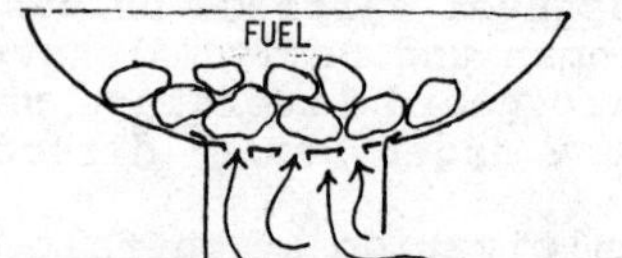

FOURNEAU MALGACHE

THAI BUCKET

In addition to qualitative tests to observe burning characteristics in a diversity of stoves and open fires, standard tests involving the boiling water were conducted. These tests all followed the same format: 500 g of mixed sizes of dried roots were compared with 500g of chopped hardwood in similar sizes. Woods used were Pacific madrone, and Arbutus menziesii. Both were previously air dried. Each pile was divided into four charges, roughly equal in size and in proportion of small to large pieces. One charge was added whenever combustion level dropped significantly. Task: to heat from 20° C to boiling, three liters of water, replacing the pot with a fresh one immediately upon boiling. This is a recognized test for establishing (a) how much work is done by a given fuel in a given stove, and (b) how quickly the stove and fuel respond, by heating water.

COOKSTOVE TEST RESULTS: SEED FUELCAKES

The fuelcakes weighed about 25 grams each, after drying in the sun. They were roughly the size and shape of a chicken's egg, about 6 cm in longest dimension. Fuelcakes of two seed/ash weight ratios were tested, 1.0 and 3.0, termed "lean" and "rich," respectively. Both types of fuelcakes were prepared by mixing dry ground seeds with dry hardwood ash in the desired proportions, adding a little water to form a stiff paste, forming egg-shaped cakes by hand, and allowing the cakes to dry. The dried cakes were typically strong enough to withstand a fall of about 5 ft onto concrete pavement without breaking apart.

In the present series of cookstove tests, sunflower seeds were utilized because buffalo gourd seeds were unavailable. Oil content and hull and oil characteristics were similar to those of buffalo gourd seeds. The lean fuelcakes (seed/ash wt ratio = 1.0) had an estimated heating value on a mass basis similar to that of hardwoods. The rich fuelcakes (seed/ash wt ratio = 3.0) were similar to softwoods in heating value. However, in both cases the organic material in the seed fuelcakes was much more volatile than the organic matter of wood.

It was observed that lean fuelcakes burned more slowly and controllably and with less smoke than rich fuelcakes. Especially in the case of rich fuelcakes, much more air was needed to combust the smoke formed than in the case of burning of wood. Carbon "strings" were often produced from flaming fuelcakes. White smoke and the odor of ammonia were obtained from fuelcakes after the flames had disappeared. Because of the inherent volatility of seed oil, the rate of combustion of fuelcakes was considerably higher than that of wood pieces of similar size burned in the same stove. In short, the fuelcakes were a fuel of much higher power (higher rate of heat release) than wood.

In comparison with wood, much more ash was produced, of course. The cakes tended to retain their shape after combustion was completed, so that ash did not fall through the grate unless the cakes were broken up.

Possible stove modifications to deal with the high ash content, high volatility, and high combustion rate of seed fuelcakes would include means for periodically shaking the grate to break up ash formations, an oversized ash receiver, and the use of a short chimney to induce a draft. The latter would provide sufficient oxygen flow to match the rate of release of combustible volatiles from the fuelcakes. Further, smaller and stick-shaped fuelcakes would allow more surface air circulation, therefore, better mixing of oxygen with fuel.

COOKSTOVE TEST RESULTS: DRIED ROOTS OF BUFFALO GOURD

Our preliminary tests of the burning of buffalo gourd roots indicated that about as much work can be done as by the burning of an equivalent mass of hardwood fuel, probably because the heating value of the roots is similar to that of wood. However, the roots are slower to ignite than wood, apparently due to a lower content of volatile matter. A fire burning pure root fuel is slower to recover than a wood fire after recharging with fuel, but continues to give out heat for a longer time. Therefore, frequent feeding with small charges of roots of varying thickness is needed to maintain a clean burn.

The rate of combustion of the roots is significantly slower than that of wood pieces of comparable size, burned in the same stove. If too many roots are charged at once, more air seems to be needed than in the burning of wood. Therefore, stoves must provide adequate draft, and better air distribution through the grate. An increase in the size of grate openings permits roots to burn at about the same rate as wood on a grate with smaller openings, and appreciably reduces smoke production. In comparison with wood, these roots might be regarded as a fuel of lower power (lower rate of heat release). Therefore, after using wood to quickly reach the boiling point, root fuel might be used for simmering to maintain the cooking temperature for a long time. In most of the Third World the staple food is a root or a grain, brought rapidly to the boil then simmered for a long time, so slow release of heat can be a benefit.

Much more ash was produced from the roots of the buffalo gourd than from hardwood, indicating the need for more frequent clean-out of the ash receiver, or the provision of a larger ash receiver than is usually found in stoves intended for wood. Further, ash from roots of the buffalo gourd tended to retain a rigid form, the shape of the root. It is necessary to vibrate the grate or stir the coals to cause ash to break up and fall through. The ash from the roots may contain volatile constituents; white powdery deposits were observed on the sides of the pots.

After an initial phase of flame production, roots tended to glow like charcoal. After the flames disappeared, the odor of ammonia was evident. Ammonia may be forming from nitrogen-containing components in the roots.

DISCUSSION OF TEST RESULTS

By virtue of their high content of seed oil, seed fuelcakes contain much more volatile matter than wood, and therefore, burn much faster than wood. The challenge is to provide a sufficient rate of flow of air to match the rapid rate of release of oily vapors and other combustible volatiles from the fuelcakes. Stove design modifications, or design of special stoves may be necessary.

By contrast, the dried roots were slower to ignite than wood and burned more slowly than wood, apparently due to a lower content of volatile matter. Even so, smoke production was considerable although it was not the smoke of oily vapors evolved by the seed fuelcakes. Nevertheless, extra air flow was needed to control smoke production; this could be accomplished by a simple modification, the provision of a grate with larger openings than would be required if wood were the fuel.

The explanation of results on the basis of chemical composition of the fuels is more clear in the case of seed fuelcakes, which contain high percentages of easily volatilized and highly combustible vegetable oil. Extra draft, in comparison with wood fuel, was needed to burn the volatile oils. However, the roots also required extra draft to overcome smoke production, compared to wood fuel. This may have been due to bound moisture in the roots. This conjecture will require further research before it can be confirmed or rejected.

POTENTIAL SITES FOR THE BUFFALO GOURD

The wide adaptability of the buffalo gourd to different water and temperature regimes in its native range suggests that it could be grown in many developing nations lying within semi-arid and arid regions. It is found in a variety of climates in North America (Mexico and the United States). It is saline tolerant, it can survive under very dry conditions, and it is well-adapted to the erratic climates of semi-arid lands.

Scheerens and colleagues (10) of the University of Arizona have collected buffalo gourd fruit from a wide range of sites in Arizona, New Mexico, Colorado, and Mexico. There was much variation in foliage shape, fruit shape, and seed yield. This genetic diversity will be especially valuable for ensuring the survival of buffalo gourd under the adverse conditions which prevail in much of the Third World. Dependable yields under poor growing conditions will be more important than very high yields in response to good growing conditions.

An unanswered question concerning the successful introduction of buffalo gourd in certain regions of the Third World concerns its photoperiodic requirements. In the northern part of its range in North America it may have evolved so that it flowers only after the long days of spring have begun. If so, stock from this part of its range might not flower in areas near the equator. It is possible that stock from the southern part of the range, 20 to 25 degrees north latitude, around the Tropic of Cancer, might have evolved differently, but this is speculation at this time. It should also be pointed out that related species such as the garden cucumber do not have photoperiodic requirements, so buffalo gourd may not either.

Given these caveats, Third World countries where buffalo gourd might be introduced successfully as a seed/root crop or just as a root crop are given in Table 1. The soil orders that pertain to these regions are given in Table 2. In essence, no photoperiodic limitations are anticipated outside the equatorial region; both seeds and roots could be obtained. Close to the equator, only roots could be obtained if photoperiodic limitations are found and not overcome. One of us, Carr (2), has given particular attention to Botswana as a potential site for buffalo gourd introduction.

CONCLUSION

The results of preliminary cookstove testing of dried buffalo gourd roots and seed fuelcakes as alternatives to fuelwood was encouraging, but showed that additional studies are needed. A basic understanding of the chemical principles underlying the combustion characteristics of these two fuels, very different from each other and from those of wood, is far from complete.

However, the studies that have been carried out thus far have shown that both of these new fuels can be at least marginally accommodated in existing stoves by providing extra draft and ash removal capability. Work will continue, both on modification of existing stoves and on the design of different but simple and low-cost stoves for these unconventional fuels. As early as is feasible, additional information on the potential acceptability of these alternatives will be sought in Third World kitchens, with the participation of local people.

Table 1

Potential Sites for Introduction of the Buffalo Gourd

Nation	Region	Annual Rainfall, mm
Angola	Lobito, Luanda	279-340
Botswana	Maun	650-700
	Francistown, Gaberone, Kakia	350-500
Brazil	Iguatu	791
Cameroun	Mokolo, Marouna	500-1000
Chad	Abeche, Arada, Ati, Mondo, Ziquey, Bitkine, Bokoro, Bongo, N'Djamena	250-1000
Chile	Concepcion, Punta Arenas, Santiago	368-1000
Ethiopia	Asmara, Dire Dawa	314-910
India	Maharashtra e. of Ghats, ne Karnataka	500-750
	Northern Regions	25-750
Kenya	Gable, Garissa, Magadi, Nairobi, Voi	250-1201
Lebanon	Beirut, Ksara	585-893
Madagascar	Bekily, Toliary	330-800
Mali	Bamako, Gao, Kayes, Menaka Mopti, Nioro du Sahel, Segou, Sokolo, Tombouctou	250-1127
Mauritania	Aioun el Atrouss, Kaedi, Kiffa, Nema, Oulata	184-507
Mozambique	Tete	584
Namibia	Tsumeb	559
	Caprivi Strip, Windhoek	356-660
Niger	Agadez, Niamey, Nguigmi, Tanout, Tahoua, Zinder	38-980
Nigeria	Funtua, Kano, Maiduguri, Nguri	500-1000
Pakistan	Quette, Lahore, Peshawar areas	250-500
Paraguay	Chaco Region	750-1250
Senegal	Dakar, Linguere, Matam, St. Louis	173-901
Somalia	Bardera, El Dere, Erigavo, Hargeisa, Iscia Baidoa, Mogadiscio	57-997
Sudan	El Fula, El Obeid, El Fasher, Gedaref, Kassala, Malakal, Roseires	131-1175
Tanzania	Dar es Salaam, Dodoma	570-1067
Zambia	Livingstone, Lusaka	673-836
Zimbabwe	Harare	810
	Bulawayo	610

Table 2

Main Soils of Potential Sites for Buffalo Gourd Introduction

Nation	Soil Order
Angola	Aridisols
Botswana	Aridisols, Entisols
Brazil	Alfisols
Cameroun	Alfisols
Chad	Alfisols
Chile	Alfisols, Aridisols
Ethiopia	Alfisols, Aridisols
India	Alfisols, Entisols, Inceptisols, Ultisols, Vertisols
Iran	Aridisols
Iraq	Aridisols, Entisols
Jordan	Aridisols
Kenya	Alfisols, Aridisols
Lebanon	Aridisols
Madagascar	Alfisols
Mali	Aridisols, Entisols
Mauritania	Aridisols, Entisols
Morocco	Aridisols
Mozambique	Alfisols
Namibia	Aridisols, Entisols
Niger	Aridisols
Nigeria	Alfisols
Pakistan	Aridisols
Paraguay	Aridisols
Senegal	Ultisols
Somalia	Alfisols, Aridisols
Sudan	Alfisols, Vertisols
Syria	Aridisols
Tanzania	Aridisols
Zambia	Alfisols, Entisols
Zimbabwe	Aridisols

The main objective of these innovations, rapid development and development of inexpensive ways to facilitate cooking in the Third World, has been advanced by this preliminary study. A costly and lengthy research and development effort will not be required. Therefore, it seems likely that implementation in Third World drylands could be pursued without lengthy delay.

ACKNOWLEDGMENTS

The assistance of Larry Winiarski and Jim Tierney in carrying out the cookstove tests is gratefully acknowledged.

REFERENCES CITED

1. Bemis, W. P., J. W. Berry, C. W. Weber, and T. W. Whitaker, "The Buffalo Gourd: A New Potential Horticultural Crop," HortScience **13**(3), 235-240 (1978).

2. Carr, Patrick M., "The Potential of Buffalo Gourd as a Novel Multi-Use Crop in Botswana," presented at the 11th Annual Third World Conference, Chicago, Illinois, USA, April 4-6, 1985, and published in the Conference Proceedings.

3. DeVeaux, Jennie S. and Eugene B. Shultz, Jr., "Development of Buffalo Gourd (Cucurbita foetidissima) as a Semiaridland Starch and Oil Crop," Economic Botany 39(4), 454-472 (1985).

4. Gathman, A. C., and W. P. Bemis, "The History, Biology and Chemistry of the Buffalo Gourd, Cucurbita foetidissima HBK," in The Biology and Chemistry of the Cucurbitaceae, R. Robinson, editor, Cornell University Press, Ithaca, New York, in press.

5. Hogan, L. and W. P. Bemis, "Buffalo Gourd and Jojoba: Potential New Crops for Arid Lands," Adv. Agron. 36, 317-349 (1983).

6. Mathieu, Sandra Lee, Potential Utilization of Oilseeds for Household Energy at the Village Level, Master of Science thesis, Washington University, St. Louis, Missouri, 1982.

7. Mathieu, Sandra Lee, and Eugene B. Shultz, Jr., "Oilseeds as Lighting and Cooking Fuel for Developing Nations," Chap. 12 in Fuels and Chemicals from Oilseeds: Technology and Policy Options, edited by Eugene B. Shultz, Jr. and Robert P. Morgan, AAAS/Westview Press, Boulder, Colorado, 1984.

8. McKell, C. M., G. A. Van Epps and G. R. Barker, "Selection of High-Producing Shrubs of the Western United States for Energy Biomass," Final Report, 1981. Unpublished report to the U.S. Dept. of Energy. Institute for Land Rehabilitation, Utah State University, Logan, UT, cited in J. H. Cushman and J. W. Ranney, "Short Rotation Growth of Hardwoods for Energy Applications Across the United States - Field Results and Economics," Proceedings of IGT Symposium on Energy from Biomass and Wastes VI, Jan. 25-29, 1982.

9. Nelson, J. M., J. C. Scheerens, J. W. Berry, and W. P. Bemis, "Effect of Plant Population and Planting Date on Root and Starch Production of Buffalo Gourd Grown as an Annual," J. Amer. Soc. Hort. Sci. 108(2), 198-201 (1983).

10. Scheerens, J. C., W. P. Bemis, M. L. Dreher, and J. W. Berry, "Phenotypic Variation in Fruit and Seed Characteristics of Buffalo Gourd," J. Amer. Oil Chem. Soc. 55, 523-525 (1978).

11. Schmidt, Edgar W., Sandra L. Mathieu, and Eugene B. Shultz, Jr., "Seeds as Fuels in Simple Cookstoves," in Solutions to the Firewood Crisis, The Aprovecho Institute, Eugene, Oregon, September, 1982.

12. Schmidt, Edgar W., Sandra L. Mathieu, and Eugene B. Shultz, Jr., "Comparison of Oilseed Fuels with

Conventional Fuels in Simple Cookstoves," presented at the 5th National Conference on the Third World, University of Nebraska at Omaha, October 27-30, 1982.

13. Shultz, Eugene B., Jr., Patrick M. Carr, Jennie S. DeVeaux, and Harold M. Draper, III, "Buffalo Gourd as a New Energy Crop for Drylands of the Third World," Presented at the 11th Annual Third World Conference, Chicago, Illinois, April 4-6, 1985, and published in the Conference Proceedings.

14. Shultz, Eugene B., Jr., Patrick M. Carr, Jennie S. DeVeaux, and Harold M. Draper, III, "Buffalo Gourd for Food and Energy in the Third World". Poster presentation, International Arid Lands in the Third World. Poster presentation, International Arid Lands R&D Conference, University of Arizona, Tucson, AZ., October 20-25, 1985.

15. Vasconcellos, J. A., W. P. Bemis, J. W. Berry, and C. W. Weber, "The Buffalo Gourd, _Cucurbita foetidissima_ HBK, as a Source of Edible Oil," in _New Sources of Fats and Oils_, edited by E. H. Pryde, Amer. Oil Chem. Soc., Champaign, Illinois, 1981.

16. Williams, Eric K., "The Planning of Cookstove and Fuelwood Programs: A New Role for Oil-Rich Seeds," presented at the 10th Annual Third World Conference, Chicago, Illinois, March 28-31, 1984.

17. Williams, Eric K., _Seeds as Alternative Fuels for Cookstoves in Developing Nations_, Master of Science thesis, Washington University, St. Louis, Missouri, 1985.

18. Williams, Eric K., and Rob Larson, "New Developments in Seeds as Cookstove Fuels," presented at the 6th National Third World Studies Conference, University of Nebraska at Omaha, 1983.

DISCUSSION

R. Clyde, Clyde Engineering

If it's 2/3 water, why dry it? Why not just use it right away? How much xylose do you get? Do you use the vines?

Shultz

It couldn't be used right away because you would overwhelm the industry. It has to be dried, and it has to be fed to the industry throughout the year, otherwise the market would fall. In fact, the industry really couldn't handle it all. I don't know the xylose yield. On the vines, the report on the distillation studies is still in preparation by those who are carry out that work; the xylose content will I'm sure be in that report. It might at least in principle be possible to use the vines as a source of cellulose and to make even more alcohol from the plant by means of treating the vines. It would add about 20 or 25% more alcohol yield per acre. On the other hand, one has to keep in mind that Eastern New Mexico has high winds in the spring, and a tremendous wind erosion potential. So we feel that it's probably wise to put the vine residue back on the ground, and perhaps even plant something like winter wheat or annual rye as a ground cover. The vines will help avoid the wind erosion problem when spring comes.

B. Hollomon, NYSERDA

You mentioned the water problem as it relates to transportation and storage and the need for what would amount to several months of storage. As I understand it, that's one of the principal problems with root crops as a fuel ethanol feed stock. Can you say a little bit more about how and where the drying and storage will take place?

Shultz

I'd like to, but I don't think that our studies are far enough along on that to be able to say much more. We think that it can be done locally and not require a large centralized facility. It has to be developed as a relatively small local type of activity with only a few miles of transport involved. The sugar beet industry in Curry County, New Mexico, was shut down dead when the railhead for collecting the sugar beets was moved 50 miles away. Transportation is a very important consideration, and can not be neglected.

PELLETIZATION AND COMBUSTION OF STRAW

Carl Wilen, M.Sc.
Pekka Ståhlberg, M.Sc.
Kai Sipilä, Lic.Tech.
Technical Research Centre of Finland
Laboratory of Fuel Processing and Lubrication Technology
SF-02150 Espoo, Finland

Jukka Ahokas, M.Sc.
State Research Institute of Engineering in Agriculture and Forestry
SF-03450 Olkkala, Finland

ABSTRACT

Finland's annual straw harvest is about 4 mill. t. Only 2.2 mill. t of this can be utilized with the present harvesting methods. About 1.8 mill. t, i.e. 0.5 Mtoe, of this could be used as a fuel, primarily in the grain-growing regions of Southern and Western Finland.

A study of straw pelletization technology and of the combustion of pellets has been carried out. Experiments were also made on the drying of straw in storage facilities. Production costs of pellets and costs of produced heat compared to other fuels were evaluated in a techno-economic study. Pelletizing experiments with a portable pelletizing unit indicated that straw can be pelletized without thermal drying over a fairly wide range of moisture content. The moisture content of straw to be pelletized should be less than 20 %, i. e., about the same as that of straw stored in bales. If the straw must be harvested and baled in a moister state, it can be dried in a cold-air drier. The size of the chopped straw has a significant effect on the pelletizability of straw and on the energy consumption during pelletization. Handling and storage properties of straw pellets are also dependent on the moisture content of straw.

Combustion experiments with straw pellets were performed in screw stokers (10 - 30 kW) designed for peat pellets and wood chips. Combustion experiments were also carried out in a fluidized bed (300 kW) combuster and gasification experiments in a fixed bed gasifier with the object of of producing fuel gas. The combustion of straw pellets has been monitored in practice at several farms during the heating season 1984 - 1985. The results were very positive and the annual efficiency of heating high.

The production costs of straw pellets vary over a wide range depending primarily on the straw harvesting method. Straw can be pelletized either in a portable pelletizing unit or in conjunction with some existing commercial operation, for example green fodder drying or sugar milling. The annual consumption of pellets at farms is on average 13 - 26 t. The price of straw pellets should be 66 - 116 USD/t to be competitive with that of light fuel oil. The price of straw pellets made on a farm in a portable unit is about 54 - 84 USD/t. The price of commercially manufactured pellets varies from about 74 to 100 USD/t, depending on the proportion of fixed costs in the price of the product.

PELLETIZATION AND COMBUSTION OF STRAW

1. INTRODUCTION

Considerable attention has recently been paid to the pelletization of biomass, such as wood, peat and straw. The aim of pelletization has been to convert inhomogeneous raw materials of low energy density to homogeneous products with regard to their physical and fuel characteristics.

In Finland, the raw material of most interest for pelletization has naturally been peat. In the United States and in Sweden, the development work mainly involved wood raw materials. In Finland, pellets have been produced both from artificially dried milled peat and from milled peat as produced. The pellets are not only produced for combustion or further processing (gasification, liquefaction), but also for horticultural purposes from weakly decomposed peat. Research on pelletization has been carried out at the Technical Research Centre of Finland (VTT) since the 1960s both on the laboratory-scale and the industrial-scale.

Experience from peat pelletization has later been used in a study of straw pelletization during the period 1981 - 1985. The study was funded by the Finnish Ministry of Trade and Industry. The aim was to study the production and combustion of straw pellets. The work was carried out at the Laboratory of Fuel Processing and Lubrication Technology of VTT and at the State Research Institute of Engineering in Agriculture and Forestry (VAKOLA).

The study on straw pelletization involved the following stages:

- design and construction of a portable pelletization unit; pelletization experiments with different straw types,
- combustion experiments with straw pellets on 10 - 30 kW stoker burners in the laboratory; use of additives to prevent ash from melting,
- monitoring of straw pellet heating at farms,
- preliminary gasification experiments on an updraft gasifier,
- estimates of the production cost of straw pellets and of the cost of produced peat.

2. STRAW AS AN ENERGY RESERVE

Finland's annual straw crop is about 4 mill. t (moisture 25 %). About 2.2 mill. t of this could be utilized with the present harvesting methods. However, only 15 - 20 % is utilized, mainly as cattle fodder and litter or baled as a fuel. The rest, more than 80 %, is either ploughed into the soil or burnt on the field. Thus, a considerable renewable energy resource of about 6200 GWh (0.5 Mtoe) is not utilized. About 70 % of the straw crop is harvested in Southern and Western Finland.

The straw harvest is dependent, for example, on the cereal species, on the length of stubble, and on the geographical position. In Finland, the annual straw harvest is 1600 - 3000 kg/ha with the present harvesting methods (moisture 25 %). The energy content is 5.5 - 10 MWh/ha.

For fuel purposes straw is usually baled: density 70 - 90 kg/m^3, moisture content 15 - 20 % and heat value 3.7 - 4 kWh/kg, i.e. 260 - 360 kWh/m^3. Baled straw can be burnt in some top and bottom combustion boilers designed for solid fuels. However, the operation is labour-intensive and the efficiency of heating is low. The combustion of straw bales or chaff succeeds well in boilers designed for this particular purpose. However, the price of these boiler types is so high, that their use is not feasible, unless the heat consumption exceeds 60 MWh/a. The annual heat consumption of farms is 25 - 60 MWh in Finland.

Compared to other solid fuels, the low energy density of straw results in higher storage and heat costs and also in somewhat higher capital costs for the boiler and the boiler room. These costs can be reduced by processing the straw to fuel briquettes or pellets. The production costs of these products have a decisive effect on the total costs.

3. PRODUCTION OF COMPRESSED STRAW PRODUCTS

For a long time in the U.S.A. and in Europe straw has been compressed into different forms. The products (pellets, briquettes, cobs; in the U.S.A. cubes, wafers) have been used for fodder. Additives to improve the nutritive value, have often been used. The nutritive value is of crucial significance in the production of fodder pellets. When producing fuel pellets, different techniques must be applied, as considerably better strength properties are required for transportation, handling and utilization.

Two different compression techniques can be used: pelletization or briquetting (Figure 1). In pelletization, ring or flat die presses are

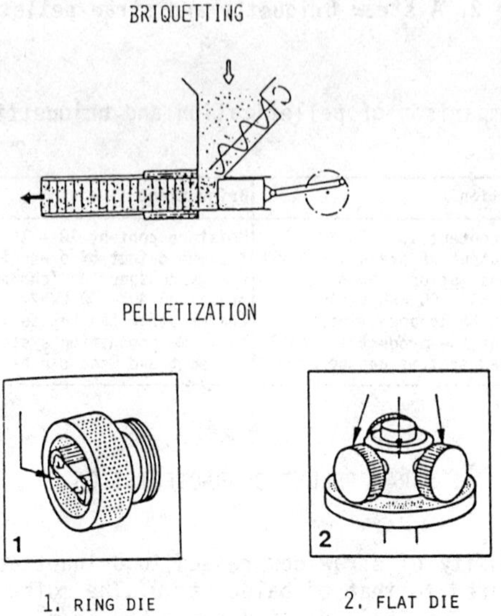

Figure 1. Different compression methods.

used. The material is extruded through a perforated breaker plate, and the products are small cylinders 6 - 20 mm in diameter. The briquettes are larger, 50 - 100 mm in diameter. Screw or piston presses are used in their manufacture. Straw briquettes and pellets are shown in Figure 2. Pelletization is the more modern technique. For wood and peat pelletization technology is already well advanced. Pelletization and briquetting methods are compared with each other in Table 1.

The most significant advantages of pelletization are a higher press output, a wider range of moisture content and a more convenient utilization achieved with the aid of automatized combustion systems. However, the raw material for pelletization must be chopped and possibly ground considerably finer than the chopped straw for briquetting. The mean straw length must be less than 25 mm.

Figure 2. A straw briquette and straw pellets.

Table 1. Comparison of pelletization and briquetting methods.

Pelletization	Briquetting
Moisture content 10 - 20 % Maximum output of press 4 - 6 t/h Energy consumption (chopping, pressing) 60 - 80 kWh/t Straw must be chopped fine. The size of the product is small, automatized systems can be used.	Moisture content 10 - 15 % Maximum output of press 1 - 1.5 t/h Energy consumption (chopping, pressing) 40 - 50 kWh/t Longer straw can be used. The same combustion systems as for sod peat and wood can be used.

4. STRAW PELLET CHARACTERISTICS

The bulk density of straw compressed to briquettes or pellets is 5 - 10-fold compared to that of baled straw. The moisture content of these products is 10 - 15 %, density 450 - 650 kg/m^3, and heat value 4 - 4.3 kWh/kg, i.e. 1.8 - 2.8 MWh/m^3.

Strength and density values of straw pellets compared to those of peat and wood pellets are presented in Table 2. The abrasion strength of the pellets was determined with a square tumbling can used for peat and fodder pellets, and the compression strength with a gauge of the Amandus Kahl type. The bulk density was measured in a 5 l vessel. The strength values of straw pellets are lower than those of peat pellets.

Sprinkling and immersion tests indicated that the pellets absorb water rather rapidly and decompose completely. The equilibrium moisture content of straw pellets was 17 % at 15 ^{0}C and 87 % relative humidity. The straw pellets were more brittle than the peat pellets and also more susceptible to decay by moisture and rain. The straw pellets must be stored protected against the direct action of water. As with the peat and wood pellets, the straw pellets also become moldy if stored at > 20 % humidity without sufficient ventilation.

Table 2. Pellet characteristics.

Pellet type	Moisture content %	Bulk density kg/m^3	Compression strength kp	Abrasion strength %
Straw pellet	10 - 12	560	30	93
Straw pellet	18 - 20	470	10	87
Peat pellet	14 - 16	750	35	95
Peat pellet	30 - 35	650	18	90
Sawdust pellet	10 - 15	650	30	90

5. EXPERIMENTS ON THE PRODUCTION OF STRAW PELLETS

Experiments on straw pelletization were carried out by VTT and VAKOLA with a portable pelletization unit constructed for this particular purpose. The aim was to study pelletization properties of different cereal straws and to produce pellets from different straw types for combustion experiments. The straw was pelletized without thermal drying.

The portable pelletization unit consisted of a straw chopper, to which small bales could be fed, feed equipment for chopped straw, and a flat die press. After the pelletization the pellets were cooled on a belt cooler. The unit is shown in Figure 3. It can also be used for pelletizing other raw materials. The output of the unit is 0.5 - 1 t/h, depending on the raw material. This equipment does not include a power source. Electric power requirement is 120 kW.

There were no significant differences in the quality of rye, barley, wheat and oat straw pellets. The pellets were more brittle than typical drier fuel pellets produced from artificially dried straw, and their bulk density was lower. The appearance of the pellets was also rougher. However, the pellets withstood well the different stages of production, handling and storage in large bags. The quality of pellets compressed from turnip rape straw and from reeds was significantly better than that of pellets produced from other species.

Figure 3. Portable pelletization unit.

For straw the nominal output of the unit was about 800 kg/h. The specific consumption of electricity was about 115 kWh/t, the chopper consuming 25 - 30 kWh/t. The capacity of the unit was restricted by the size of the pellet press. Should larger presses be used, the consumption at the compression stage, and hence the energy consumption of the whole unit, would be lower. The nominal capacity of large presses is 4 - 6 t straw/h. On the basis of experience from peat pelletization the power consumption of the compression stage can be estimated at about 45 kWh/t in this size class. Hence, the specific consumption of electricity would be 80 - 100 kWh/t in a unit producing 3 t/h straw pellets.

6. STRAW ASH

In addition to the low energy density, the combustion of straw pellets is hampered by the high ash content of straw and the melting behaviour of the ash. The ash content of barley, rye and oat is about 5 % and that of wheat 6 - 7 %. The melting temperature of wheat ash deviates from those of other cereals. In addition, the melting temperature of straw ash is affected by the soil composition and by fertilization. The ash melting temperatures of different cereals are presented in Table 3.

The fusion of ash at low temperatures results in difficulties in the operation of grates and ash handling equipment. The compression does not change the melting properties of the ash, and hence, the fusion of straw ash should be considered in the design of boilers and stokers fired with compressed straw products.

Table 3. Melting temperatures of straw ash.

Stage of melting	Temperature range, ^0C				
	Wheat	Rye	Oat	Barley	Turnip rape
Initial deformation	900-1050	800- 850	750- 850	730- 800	1150-1250
Hemisphere	1300-1400	1050-1150	1000-1100	850-1050	1250-1500
Flow temperature	1400-1500	1300-1400	1150-1250	1050-1200	1300-1500

Prerequisites for effective straw burning are that the temperature of embers is 950 - 1100 ^0C and the temperature in the furnace > 800 ^0C. However, the straw ash melts at these temperatures (Table 3), and the operation of grates and ash removal equipment is hampered by the molten ash. As a consequence, straw cannot be combusted in conventional top and bottom furnace boilers designed for wood. Similar difficulties often arise in the combustion of certain peat grades. In burners of small output compressed straw can be burnt, but in these too the operation of the burner is hampered by molten ash.

Compositions of straw ash are presented in Table 4. The greatest differences between different straw types are seen in their SiO_2, K_2O and CaO contents. The fusion temperature is increased by SiO_2 and lowered by K_2O and CaO. This effect is clearly seen in Tables 3 and 4. The fusion temperatures of wheat straw ash that contains an abundance

Table 4. Compositions of straw ashes.

Component	Wheat %	Rye %	Barley %	Oat %
SiO_2	78.2	61.7	44.7	37.3
K_2O	6.6	19.2	37.1	40.3
CaO	5.0	7.4	9.3	12.3
P_2O_5	3.3	3.7	3.8	4.1
MgO	3.6	2.8	2.5	3.0
Al_2O_3	2.0	2.1	0.4	0.8
Fe_2O_3	1.5	1.5	0.5	0.5
SO_3	1.4	1.3	1.4	1.4
Na_2O	0.3	0.3	0.3	0.3

of SiO_2 are clearly higher than those of the other straw ashes. That is why wheat straw pellets can be burnt in burners designed for peat and wood chips, while equipment designed especially for straw should be used in the combustion of the other straw types.

If the fusion temperature of straw ash could be raised by additives, the potential for use of straw would be increased, as straw could be used as briquettes and pellets for the fuel in top and bottom furnace boilers and burners.

7. EXPERIMENTS WITH ADDITIVES

In practice the most convenient method to elevate fusion temperatures would be to mix some, preferably powdery, additive among the straw chuff before compression. Additives of this kind increase the ash content of the fuel and simultaneously, unfortunately, reduce its heat value. The increase due to additives in the energy price should be covered by cost savings due to the use of simpler and cheaper combustion and ash removal equipment.

The effect of additives on the fusibility of straw ash was studied with a test series, where 1, 2, 3, 4, and 6 % of additive were added to chopped oat straw. The fusibility of the straw ash was then determined according to DIN 51730. The additives and their compositions are given in Table 5. The effect of additives on the deformation temperature of ash is shown in Figure 4. The ash melting temperatures of "pure" oat straw were as follows:

deformation temperature	770 °C
hemisphere	860 °C
flow temperature	1050 °C.

Table 5. Additives and their compositions.

Component	Kaolin fine %	Kaolin coarse %	Ceramic felspar %	Talc %	Coal ash %
SiO_2	47.8	48.6	67 - 69	59	44 - 46
Al_2O_3	37.0	36.0	17.8 - 18.3	0.4	22 - 23
Fe_2O_3	0.6	0.82	0.12 - 0.18	0.1	9.5 - 10.5
CaO	0.04	0.82	0.5 - 0.8	0.6	-
MgO	0.16	0.25	-	30	2.8 - 3.2
FeO	-	-	-	2.2	-
SO_3	-	-	-	-	1.0 - 1.5
K_2O	1.10	2.12	7.5 - 8.0	-	-
Na_2O	0.10	0.10	4.5 - 5.0	-	-
TiO_2	0.03	0.05	-	-	-

At least 6 % talc must be added to reach the deformation temperature of 1100 °C, whilst about 4 % coal ash is required to reach this temperature. The effect of kaolin is clearly seen with > 1 % additions and the deformation temperature of 1100 °C is reached with about 2 % addition. With equal additive amounts the hemisphere temperature of oat ash rises to 1200 °C and the flow temperature to 1300 °C. The effect of felspar on the fusibility of ash is so insignificant that its use cannot be justified. Over 6 % talc should be added to reach deformation temperatures > 1100 °C. The large quantity and the high price of talc make its use infeasible.

Coal ash itself has no market price. The price of ash consists mainly of transport costs, about 30 FIM/t (6 USD/t) (transport distance 150 km). The total costs is about 50 FIM/t (10 USD/t). About 4 - 5 % coal ash should be added to oat straw. This would increase the fuel price by only 2 - 2.5 FIM/t (about 0.5 USD/t).

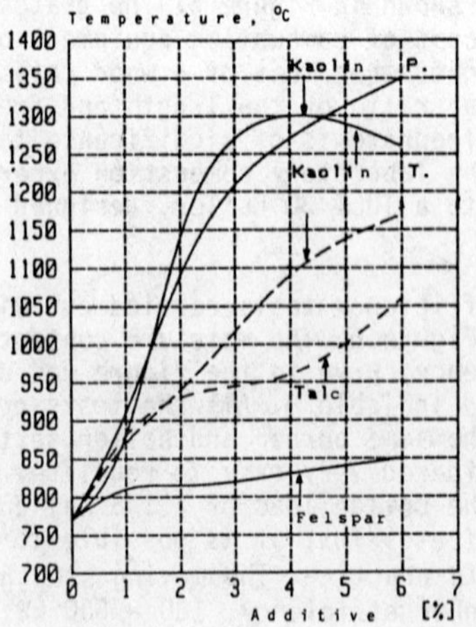

Figure 4. Effect of additives on the deformation temperature of oat straw ash.

Kaolin is used as coating and filling material in paper and ceramic industries. The price of fillers is 360 - 460 FIM/t (72 - 92 USD/t). The price of the fuel increases by 7 - 10 FIM/t (1.5 - 2 USD/t) with a 2 % kaolin addition.

If the price of straw pellets is 400 FIM/t (80 USD/t) and the heat value 4.0 MWh/t, the energy price is 100 FIM/MWh (20 USD/MWh). A 2 % kaolin addition or a 4 % coal ash addition increase the price of the fuel by 5 %. On the other hand, the difficulties due to ash fusion can then be eliminated, and as a consequence, simpler and cheaper combustion and ash removal equipment can be used.

Combustion experiments were carried out with kaolin-straw pellets both in the laboratory and in the longer-term field experiments. The combustion experiments indicated that a 2 % kaolin addition is sufficient to prevent straw ash from melting during combustion.

8. COMBUSTION OF COMPRESSED STRAW PRODUCTS

When straw briquettes are burnt on a stationary grate, the burning is hampered by ash melting soon after ignition. The burning of wheat straw briquettes does not worsen as much as that of the other straw briquettes. Poor burning is due to grate clogging and to embers formed on the grate. Due to the poorer efficiency of combustion, the efficiency of one feed interval (about 6 h) is 10 - 15 % lower than that achieved with wood chips and billets in the same boiler.

Compressed straw products can be burnt in stoker burners, the burner head of which is equipped with ash handling equipment. In the size class of < 50 kW the best results with straw pellets were obtained with the screw stoker shown in Figure 5. The grates are powered by the fuel feed screw. The cost of combustion equipment of this kind is about 500 FIM (100 USD) higher than that of a wood chips burner with the same output. The proper ratio of the length and frequency of grate movement to the fuel feedrate is of significance to the operation of the burner head. In the laboratory combustion experiments, a TP 30 burner was connected to a TULI 30 boiler, designed especially for stoker use.

The results of efficiency tests carried out with barley straw pellets are given in Figure 6. The moisture content of the pellets was 13 - 15 %. The efficiency curve in the figure was drawn on the basis of test results presented in Table 6. All the tests presented in the table were carried out at the same burner and boiler settings. Hence, the equipment can be considered very easy to regulate. An efficiency of > 60 % is achieved at the boiler load of > 1.8 kW, and > 70 % at the load of > 2.8 kW. This indicates that it is possible to achieve an annual efficiency of > 65 % in practice. The moving step grate can also be used in burners of higher efficiency, 100 - 500 kW, but it is difficult to estimate the upper limit of efficiency at this stage of research.

The results presented above were obtained for straw pellets. Briquettes are more suitable for burners with a higher output, approximately > 70 kW. Brittle briquettes can usually also be used as such in small burners.

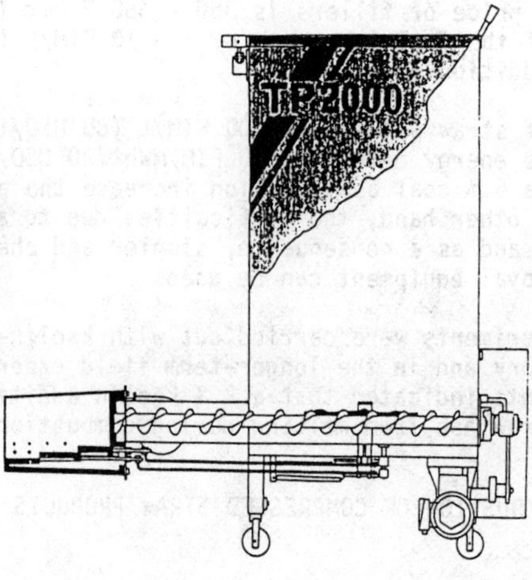

Figure 5. TP 30 burner.

The combustion experiments on the combustion of straw pellets were carried out with a fluidized-bed boiler of 300 kW developed by the Helsinki University of Technology and Neste Oy for research use. The boiler was originally designed for peat pellets. Fluidized-bed boilers are used in the combustion of municipal waste, wood waste, peat, wood chips and coal.

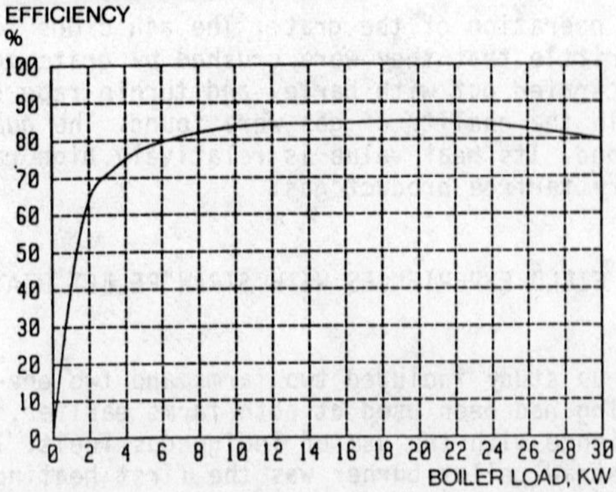

Figure 6. Efficiencies of the TP 30 burner and the TULI 30 burner with straw pellets and at different boiler loads.

Table 6. Efficiency tests with the TP 30 straw pellet burner. The burner was connected to a TULI 30 boiler. The tests were carried out at the same adjustments of the burner and the boiler.

Boiler load, kW	Efficiency, %
29.2	81
20.0	85
10.4	84
4.9	77
2.8	70
1.5	54
0	0

The melting behaviour of straw ash also affects the operation of the fluidized-bed boiler. Therefore, changes in the sand bed, operation of ash removal equipment and fouling of the boiler were, in particular, monitored during the combustion experiments with straw pellets. The experiments indicated that the fluidized-bed can be used for burning straw pellets if the temperature of the sand bed can be kept sufficiently low. Slagging due to ash melting, typical of barley, oat and rye straw pellets, can be most effectively eliminated by using a low temperature, 700 - 730 ^{0}C, in the fluidized sand bed, and additives that elevate fusion temperatures. A temperature under 800 ^{0}C sets a limit of 230 kW to the boiler load. The efficiency, calculated indirectly, ranged from 92 to 98 %. Fouling of heating surfaces was observed in the combustion of barley, oat and rye straw pellets. When kaolin was used as the additive, fouling was clearly less significant and the risk of agglomeration in the sand bed was reduced.

The gasification experiments with straw pellets were carried out on an updraft gasifier of about 200 kW. When the temperature of the gasification zone was 900 - 1000 ^{0}C no significant disturbances were observed in the operation of the grate. The ash clods formed on the grate were so brittle that they were crushed by grate movements. The test runs were carried out with barley and turnip rape straw pellets. No differences in the quality of gas were found. The quality of gas can be considered good. Its heat value is relatively high ranging from 5.3 to 6.8 MJ/m^3n dry tarfree product gas.

9. FIELD EXPERIENCES WITH STRAW PELLET HEATING

The follow-up study included two farms and two one-family houses. Wood chips heating had been used at both farms earlier, and hence, there was experience with the use of indigenous fuels. In the one-family houses, the straw pellet burner was the first heating installation using indigenous fuel.

Three of the houses had a TP burner with a movable step grate and one a Pelikko burner modified for straw pellet combustion. The energy consumptions of the farms were 45 and 55 MWh/a under normal conditions, and those of the one-family houses 24 and 29 MWh/a.

Difficulties appeared in each combustion equipment soon after start-up. However, most difficulties were eliminated during the first months. The operation of the equipment was also hampered by variations in the quality of the fuels. The fuels also contained an abundance of fines, which hampered the operation of equipment.

The boilers often required sweeping and scraping and the amount of ash was large. These were the greatest disadvantages. The need for sweeping was increased by the high content of fines and, in one case, by lack of space in the furnace. The minimum labour requirement was on average 9 min/day and the maximum about 13 min/day. The labour requirement per MWh produced was about 1 h at the farms and about 2.4 h in the one-family houses.

As a whole the operation of the equipment can be considered satisfactory. The annual efficiency of the most inefficient boiler was 52 %. In this case, the Pelikko burner was connected to a Parca Meco 92 oil boiler. The annual efficiency of the best installation, the TP burner and the Tuli burner boiler, was 67 %.

10. PRODUCTION COSTS OF STRAW PELLETS

The production costs of straw pellets vary over a wide range. The variations arise both the production of the raw material and the actual compression. As regards the raw material, it is of significance how the capital costs of the harvesting equipment are accounted for in the price of the compressed product. When the consumer of the compressed straw product delivers the raw material and pays only for the compression, the costs of raw material are 100 - 200 FIM/t (20 - 40 USD/t) if the harvesting equipment is used in a reasonable way. The purchase price of straw is 200 - 280 FIM/t (40 - 56 USD/t).

Compression costs were studied for pelletization only. The compression costs with a movable pelletization unit, capacity about 1.5 t/h and output 1700 - 3000 t/a, are 210 - 250 FIM/t (42 - 50 USD/t). About 10 % of the total operating time (7 months/year) can be used for transportation. If the capacity of the equipment is increased twofold, the compression costs will be about 170 - 220 FIM/t (34 - 44 USD/t). The compression costs with a semimovable unit used by one village and moved once in two weeks are 160 - 180 FIM/t (32 - 36 USD/t), the output being 5000 - 6000 t/a. At a stationary plant, for example at a green fodder drying facility, the compression costs are about 150 FIM/t (30 USD/t) for an output of 10 000 t/a.

The total production costs of straw pellets produced in portable or semiportable units are 270 - 420 FIM/t (54 - 84 USD/t) when the raw material is delivered by the consumer. If the raw material is bought the production costs amount to 370 - 500 FIM/t (74 - 100 USD/t).

11. COSTS OF PRODUCED HEAT

The annual costs of straw pellet heating were compared to those of oil, straw bale, wood chips, and billet heating. In each case, the exact costs were calculated for two heating stations generating heat 25 MWh and 60 MWh. The bases of the calculations are presented in Table 7. An example of annual heating costs is shown in Figure 7. In this case, the heating station is assumed to be installed in an existing building.

In the scale of < 50 kW, heating with straw pellets is competitive with oil heating at the price level of 330 - 580 FIM/t (66-116 USD/t). The use of straw pellets is most unfavourable if new facilities must be built for the heating unit, the repay time is short and the rate of interest high. Straw pellets are competitive with oil if the heating unit is placed in an existing building and the use of indigenous genous fuels is supported by subventions and by loans at a low rate of interest. At farms the use of wood chips is more favourable if the chips are obtained from forests owned by the farmer.

12. CONCLUSIONS

Considerable savings are achievable, if straw is processed to a compressed product. At farms, compressed products can be burnt in solid fuel burners equipped with ash removal equipment. The equipment costs and thecosts of produced heat are of the same magnitude as for the other solid fuels. The storage costs are lower due to high density. It is more favourable to make straw pellets in a portable pelletization unit than in a stationary one. On the basis of calculated production costs the straw pellets are, in certain cases, competitive with fuel oil. At present fuel prices, the production costs of straw fuel pellets should probably be reduced for commercial production to be possible. This would necessitate cheaper harvesting and compression equipment.

Table 7. Bases used in the calculation of fuel costs.

Fuel	Heating system	Annual efficiency %	Heat value	Unit price FIM/MWh
Straw pellet	direct	65	4.0 MWh/t	100
Wood chips	direct	65	0.8 MWh/m³	63
Oil	direct	75	10 MWh/m³	166.6
Billets	charging	55	1.3 MWh/m³	77
Straw bale	charging	40	3.8 MWh/t	40

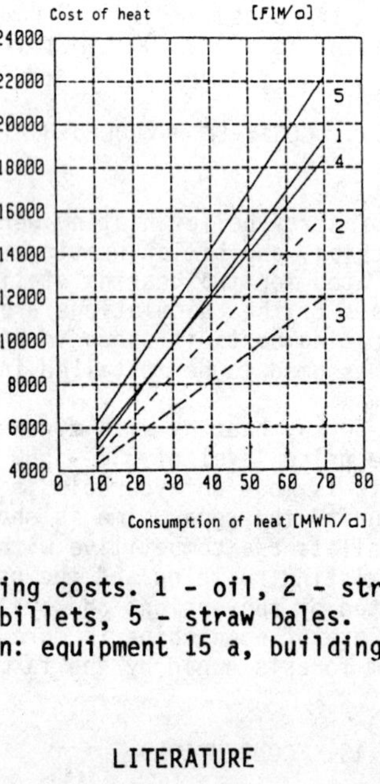

Figure 7. Annual heating costs. 1 - oil, 2 - straw pellet, 3 - wood chips, 4 - billets, 5 - straw bales. Rate of interest 5 %; Depreciation: equipment 15 a, buildings 25 a.

LITERATURE

Wilen, C. and Ståhlberg, P. Oljen pelletointi [Straw pelletizatin]. Technical Research Centre of Finland. Research Note 261. Espoo 1983. 39 p. + app. 12 p.

Ahokas, J., Ståhlberg, P. and Maaskola, I. The State Research Institute of Engineering in Agriculture and Forestry. Research Report 30. 88 p. Vihti 1983.

Peltola, A. Work Efficiency Association, Dept of Agriculture and Building. Duplicate 2/1981. 50 p.

Ståhlberg, P., Wilén, C. and Horvath, A. Oljen pelletointi ja pellettien käyttö polttoaineena [Straw pelletization and use of pellets as fuel]. The State Research Institute of Engineering in Agriculture and Forestry. Research Report 40. 140 p. + app. 12 p.

DISCUSSION

J. Kerstetter, Washington State Energy Office	It appears that straw bales cost more than the straw pellets. Is that correct?
Wilen	Yes, the straw bales are more expensive. The heating efficiency with straw pellets was 65%.
Zoellner	Have you looked at the net energy balance for briquetting and pelletizing?
Wilen	About 2.5% of the energy content of the pellets is needed for the compression.
Zoellner	Is that something that could be done manually?
Wilen	No, I don't think so.
Zoellner	Have you had any expressed interest from other countries outside of Finland for this technology?
Wilen	Yes, from England and West Germany, where straw burning in the fields is restricted already.
Zoellner	Why is the straw burning restricted? Is it for air quality?
Wilen	Yes.
Kerstetter	Did the effect of the additives, kaolin or coal ash, have any effect on your die life?
Wilen	No, perhaps the coal ash has some effect on die life, but not the kaolin; it's very soft.
Zoellner	I noticed that a lot of the mobil and stationary uses were related to farm use. Has there been any interest in commercial use by industry?
Wilen	Not yet in Finland because we have so many other domestic fuels such as wood chips, which are competitive with the straw pellets.
S. Babu, IGT	Can you tell us something more about the gasification experience with these pellets?

DISCUSSION, CONT.

Wilen	We conducted experiments with a small downdraft gasifier in our own laboratory, and as I said the gas was good. We had no problems with ash melting in the gasifier because the gasifier has a moving grate. It breaks up the ash that melts and solidifies under the grate.
Babu	Are the gases maintained under non-condensing conditions between the gasifier and the energy conversion unit?
Wilen	Yes.
Babu	How about the problem with particulates? Being a downdraft unit, you shouldn't have much of a problem with the particulates.
Wilen	No, just after the gasifier.
Babu	Any estimates of efficiency of Btu's in the fuel-to-power output?
Wilen	Not with straw, but for sod peat, the efficiency is very high. I think it's over 85%.
J. Diebold, SERI	Have you tried to determine what it is in kaolin that causes the ash to act properly.
Wilen	Kaolin is aluminum oxide and silicon oxide in the ash. We get a mixture of ash with a higher fusion temperature than only the ash from the straw.

UTILIZATION OF BIOGAS IN DUAL FUEL ENGINES

Franz-Martin Dübel, Dipl.-Ing.
Lehrstuhl für Angewandte Thermodynamik
Universität Aachen, Federal Republik of Germany
Karl Schmillen, Dr.-Ing.
Forschungsgesellschaft für Energietechnik
und Verbrennungsmotoren mbH, Aachen,
Federal Republik of Germany
A.Saez, Dr.-Ing. and G. Reinke, Dipl.-Ing.
Departmento Mechanica Universidad Tecnica
Federico, Valparaiso, Chile

ABSTRACT

In the absence of corrosive agents, biogas is suited to internal combustion engines. The diesel spray ignition process offers many advantages such as high efficiency and alternative operation on straight diesel fuel. Compared to natural gas greater efforts are required to adjust the combustion system while compensating for the detrimental influence of the inert components (CO_2) on the ignition and combustion properties.

In comparison with straight diesel fuel operation, under full load condition the specific fuel consumption with biogas was found to be 5% higher, and the NO_x emissions were reduced by approximately 50%. Higher HC and CO concentrations resulted due to the CO_2 contained in the biogas. Even with higher loads than diesel rating, knocking did not occur.

At high degrees of biogas substitution, smaller spray holes in the injection nozzle are advantageous. However, their use precludes the advantage of alternative diesel fuel operation. At very small pilot fuel injection quantities, the injection nozzle is not sufficiently cooled and carbon deposits form on the spray holes.

The efficiency at part load operation was improved by an advanced start of injection because more optimal injection spray conditions exist and the ignition nuclei are better distributed over the entire chamber. The limit of flamability is beyond RAFR of $\lambda = 2.1$. Specific fuel consumption and HC emissions are reduced considerably, in addition to the advantage of minimized NO_x emissions.

UTILIZATION OF BIOGAS IN DUAL FUEL ENGINES

INTRODUCTION

Alternative energy sources are of growing interest to the world due to limited fossil energy resources. However, decisions on new energy concepts must be based upon sound economic and management principles. In this context, biogas must be considered as it can be cost effectively obtained from waste material by fermentation at commercial facilities or from landfills (dump areas). Biogas can then be transformed into mechanical work and heat at a high level of efficiency by internal combustion engines.

USE OF GASEOUS FUELS IN COMBUSTION ENGINES

Most gaseous fuels can be used in internal combustion engines. Although there are many such fuels, the most important are identified in table 1. One of the unfavorable aspects of using gaseous fuels in engines is the presence of inert gases which cause deterioration of the ignition process and tend to reduce the flame velocity. Biogas contains a large fraction of inert gas in the form of carbon dioxide (CO_2). The detrimental influence of the CO_2 on ignition and combustion must be taken into account when choosing an engine and optimizing the combustion system.

Substances in crude biogas, such as hydrogen sulfides (up to 3 %), halogenated hydro-carbons, ammonia and water vapor, can adversely affect engine operation and may considerably reduce the life of the engine due to their related corrosion mechanisms. Consequently, purification of the gas must be accomplished before it can be used as an engine fuel. The hydrogen sulfides can be removed at low cost by filters containing iron filings. The recommended practice is to reduce the content of H_2S to less than 1.5 g/m^3 in the biogas. Normally, traces of halogenated hydro-carbons are also found in the crude gas. However, it is difficult to establish a quantitative analysis of the impurity fractions because of the large number of chemical compounds which are present. A gas engine is normally affected by direct corrosion or as a consequence of the deterioration of the lubricating oil. Unfortunately, totally effective gas purification methods are still unknown.

For the optimization of a combustion process, exhaust emissions must be considered as well as operational parameters such as specific fuel consumption. Generally, the emissions are CO, NO_x and, of less importance, HC (primarily CH_4). These characteristic substances are formed in the engine in different ways:

Table 1: Fuel Gas /4/

Type of Gas	Main Components	Lower Heat Value $Hu=kcal/m_n^{3x}$	Limits of Ignitability in Air (Vol % Gas)	Density kg/m_n^3
Hydrogen	H_2	2570	4 - 80	0,0899
Carbon monoxid	CO	3020	12 - 80	1,25
Methane	CH_4	8550	5 - 15	0,717
Propane	C_3H_8	22300	2,1- 9,5	1,96
Butane	C_4H_{10}	29500	1,5- 6,5	2,6
Blast Furnace Gas	CO,N_2,CO_2	700- 950	12 -70	1,28
Producer Gas	CO,H_2	1000-1300	20 -70	1,1
City Gas	H_2,CH_4,CO,CO_2	3800-4600	6,5-54	0,58
Biogas	CO, CO_2	5500	7,5-14	1,1
Natural Gas	$CH_4,C_2H_6,C_3H_8,C_4H_{10},N_2$	7000-9500	5 -15	0,76

x 1 Standard meter cube m_n^3 = 1 m³ Gas at 0 °C and 1 bar

NO_x is originated at high temperatures (exceeding 2000 K) behind the flame front when there is excess oxygen during the combustion process.

CO is the product of an incomplete oxidation due to a (local) shortage of O_2. Additionally, it is formed by the dissociation of CO_2.

HC are unburned or partly burned hydro-carbons, primarily formed in quench zones within the cylinder.

Biogas is not suited to all types of gas engines. The gas engines are either spark ignition engines (SI gas engines) or dual fuel engines (spray ignition). In both cases, a homogeneous mixture of air and gaseous fuel is drawn into the cylinder and compressed. In the case of the SI gas engine, the ignition is produced shortly before top dead center (BTDC) by a spark plug (external ignition). Conversely, ignition in the dual fuel engine results from the internal ignition of a small amount of pilot fuel, injected under high pressure into the combustion chamber shortly BTDC.

Successful ignition requires a rich mixture at the ignition source. Generally, SI gas engines have a single ignition source (spark plug). In the case of dual fuel engines, however, a large number of spatially distributed ignition nuclei are produced by the injection of the pilot fuel. This has two consequences on the ignition and the combustion:

1) Due to the large number of nuclei, a rich ignition mixture of gas/air occurs at a number of locations at the moment of ignition even if the gaseous fuel contains a high fraction of inert gas.

2) The flames coming from the ignition nuclei have only short propagation distances to burn the entire fuel charge in the cylinder. These short distances compensate for the low combustion velocities of gaseous fuels containing inert gases and thus assure a quick heat release.

One characteristic of SI gas engines is the potential for knocking. Accordingly, the compression ratio must be kept relatively low (on the order of 12:1). Small compression ratios, on the other hand, produce low efficiencies because of the relatively high thermodynamic losses. Gas Engines with diesel spray ignition allow much higher compression ratio (on the order of 17 : 1). There is no risk of knocking as a result of the small flame propagation distances and short prereaction.

Dual fuel engines, also known as diesel spray ignition engines, are a derivative of the diesel engine. The particular advantage of the diesel spray ignition process is that, when interrupting the biogas delivery, it is possible to switch from the dual fuel operational mode to straight diesel fuel without interruption of engine operation if the injection pump is designed accordingly. These factors may make this type of engine particularly interesting for power plant applications.

At present, there are very little data available for gaseous fuels used in dual-fuel engines such as biogas, what contains a high fraction of inert components. Consequently, a research project was conducted to further the basic knowledge of biogas as an engine fuel. The research work has been conducted by the Universidad Technica Valparaiso, Chile, in conjunction with the Lehrstuhl für Angewandte Thermodyamik at the Technical University in Aachen, West-Germany.

ENGINE MEASUREMENT AND ANALYSIS TECHNIQUES

Engine testing was conducted in Chile where biogas from the municipal dump in Santiago was used. The investigations were run on a commercial 4-cylinder diesel engine (MWM D229-4) with a rated power output of 42 kW at 1800 rpm. For the purpose of comparison, the engine was first tested while operating on diesel fuel. Subsequently, the engine was retested following modifications for biogas operation.

The testing showed that lower smoke values were obtained during operation on biogas than with diesel fuel (Bosch smoke number 1 vs. 3 to 4). The specific fuel consumption using biogas was found to be approximately 25% lower than with diesel fuel and the CO exhaust gas concentration was higher by a factor of 3 (1500 ppm compared to 500 ppm during diesel operation).

The high values obtained for the specific fuel consumption and the CO-concentration indicate an unsatisfactory combustion process. As a result of the testing conducted in Chile, further investigations were undertaken at the Aachen Technical University to define the parameters affecting the biogas combustion process and to improve the combustion process during biogas dual fuel operation.

Tests on the Single-Cylinder Engine

The tests in Aachen were conducted on a single-cylinder AVL test engine. The pertinent engine data are identified below:

Stroke	130 mm
Bore	125 mm
Compression Ratio	17:1
Omega bowl in piston	
4-valve cylinder head	
Shrouded inlet valve for swirl generation	

Fig.1 provides a sectional view of the cylinder head and the piston. It shows the position of the injection nozzle, the piezo-electic pressure transducer and the combustion chamber bowl.

For the engine tests, a synthetic biogas was used with a calorific heat similar to the biogas used for the research in Chile. This synthetic biogas was transformed from natural gas by adding CO_2 (see table 2).

Thermodynamic Analysis of the Combustion Process

Accurate measurement of the cylinder pressure is an essential part of research on reciprocating internal combustion engines. The cylinder pressure versus time (or degrees of crank angle) data can be obtained by a piezo-electric pressure transducer. Trigger marks for TDC and crank angle are recorded concurrently with the cylinder pressure so that cylinder pressure may be determined as a function of piston position.

The time history obtained for the cylinder pressure allows thermodynamic analysis of the combustion process by computer methods. The program used in this project is based on a two-zone model, one zone of completely burned exhaust gas in the combustion chamber and a second zone containing the unburned mixture. These two zones are se-

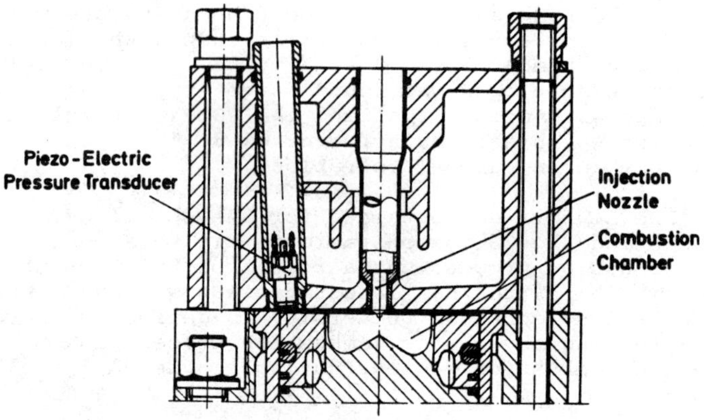

Fig. 1: Cross-Section of Cylinder Head and Piston of the Single Cylinder Engine

Table 2: Gas Composition in Vol %

Type of Gas	CH_4	C_nH_{2n+2}	H_2	O_2	CO_2	N_2	H_u MJ/Nm³
Biogas /8/	50 - 60	-	<1	<1	35 - 46	2	17,9 - 21,5
Biogas /6/	60 - 80	-	<1	-	20 - 40	<1	21,5 - 28,7
Biogas /2/	60	-	3	-	32	5	21,8
Biogas /7/	-	-	-	-	-	-	18,87
Biogas, synth. (nat.Gas + CO_2)	47,3	2,3	-	-	43,5	6,9	18,87

parated by the flame front. The program utilises the first law of thermodynamics for closed systems (equation 1).

$$\text{eq.1} \qquad \int_{IB}^{\alpha} dq - \int_{IB}^{\alpha} p dV = u_{\alpha} - u_{IB}$$

For each degree of crank angle (from the moment the inlet valve closes until the outlet valve opens) the following values are calculated:

Mass fraction burned x_B in terms of the ratio of heat released to the total heat energy of fuel

Burning rate $dx_B/d\alpha$,

Mass weighted gas temperature T_m.

The results of the thermodynamic analysis are shown in fig. 2 The plot of the cylinder pressure versus degree of crank angle and the corresponding mass fraction burned with its first derivative the burning rate are illustrated. Two peaks can be seen in the plot of the burning rate. The first is the result of the sudden combustion of the pilot fuel after the ignition delay, and the second corresponds to the combustion of the biogas fuel.

The mass fraction burned indicates the ignition delay and the duration of combustion. The ignition delay (ID) is defined as the difference of crank angle from start of injection (IB) to the point at which 5% of the fuel has been burned in the combustion process (x_B = 0.05) (equation 2).

$$\text{eq.2} \qquad ID = \alpha(x_B = 0.05) - \alpha(IB)$$

At the completion of combustion, the mass fraction burned converges toward x_B = 1.0. Quite often, however, the end of combustion is delayed or the combustion is incomplete and the value x_B = 1.0 is not obtained. Therefore, the end of combustion has been defined as the crank angle at which the degree of combustion is x_B = 0.90. The duration of the combustion (CD) may be obtained with equation 3.

$$\text{eq.3} \qquad CD = \alpha(x_B = 0.9) - \alpha(x_B = 0.05)$$

<u>Specific Fuel Consumption and Degree of Substitution</u>

The brake specific fuel consumption of internal combustion engines is usually given as the mass of fuel divided by the effective power. For the purposes of this project, diesel fuel was chosen as the basis for comparison. Therefore, in the following paragraphs the brake specific fuel consumption is consistently given expressed in terms of an equivalent quantity of diesel fuel (equation 4).

$$\text{eq.4} \qquad b_e^* = \frac{1}{P_e}\left[\dot{m}_{Diesel} + \dot{m}_{Gas} \cdot \frac{H_{u,Gas}}{H_{u,Diesel}}\right]$$

In the case of the spray-ignition engine, the degree of substitution of the diesel fuel by (synthetic) biogas is called the gas energy fraction. In equation 5, it is defined as the calorific heat of the biogas divided by the total energy of the pilot fuel and the biogas.

$$\text{eq.5} \qquad \psi_{Gas} = \frac{m_{Gas} \cdot H_{u,Gas}}{m_{Diesel} \cdot H_{u,Diesel} + m_{Gas} \cdot H_{u,Gas}}$$

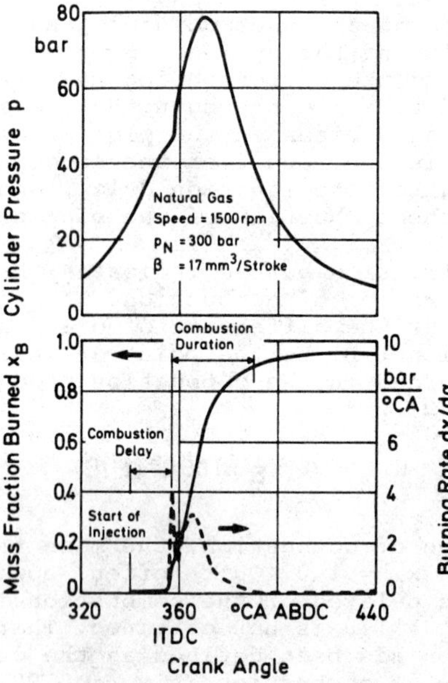

Fig. 2: Cylinder Pressure, Mass Fraction Burned and Burning Rate for Dual Fuel Operation

ENGINE BEHAVIOR AT FULL LOAD

The mean pressure is a characteristic factor for the power of a reciprocating engine. The difference between the indicated mean effective pressure (imep, specific work of the gas at the piston) and the brake mean effective pressure (bmep, specific work obtained at the crankshaft) describes the frictional losses and is, therefore, called friction mean effective pressure (fmep, equation 6).

eq. 6 $\quad$ fmep = imep - bmep

The fmep of the single-cylinder engine used for these investigations was 1.8 bar at a rotational speed of n = 1500 rpm. This value was clearly higher than that of similar multi-cylinder engines exhibiting friction mean effective pressures of approximately 0.8 bar. To assure comparability with the results obtained on multi-cylinder engines, the bmep for the single-cyliner engine at full load was chosen at approximately 6 bar, 1 bar lower than the normal value for multi-cylinder engines.

Impact of the Gas Energy Fraction

The operational behavior at full load (bmep = 6 bar) and the influence of different fractions of diesel fuel and synthetic biogas can be investigated by varying the gas energy fraction. Based on a straight diesel fuel operation at part load biogas is added to the inlet of the engine until full load is obtained. At very high gas fractions, less pilot diesel fuel is required than the quantity corresponding to idling conditions. In this case, the engine is brought to full load as described above and the pilot quantity is then reduced with a corresponding increase in the supplied quantity of biogas.

The results given in fig. 3 provide information about the engine's behavior while the diesel fuel is being replaced with biogas or natural gas. The brake mean effective pressure was constant at 6 bar and the start of the pilot fuel injection was set at 10 degrees BTDC. The Bosch smoke number falls with increasing gas energy fraction. The specific fuel consumption (diesel fuel equivalent) is increasing with a greater gas fraction. At a gas energy fraction of 80%, the SFC with biogas and natural gas is greater by 5% and 2.5%, respectively, than the corresponding SFC for straight diesel fuel operation (upper part of fig. 3).

The emission behavior is also dependent upon the fuel. The NO_x concentration (middle of fig. 3) is approximately the same for both natural gas and diesel fuel. While operating on biogas, however, NO_x concentration decreases by as much as 45% with increasing gas energy fractions. The notable difference in NO_x concentration between natural gas and biogas operation is due to the

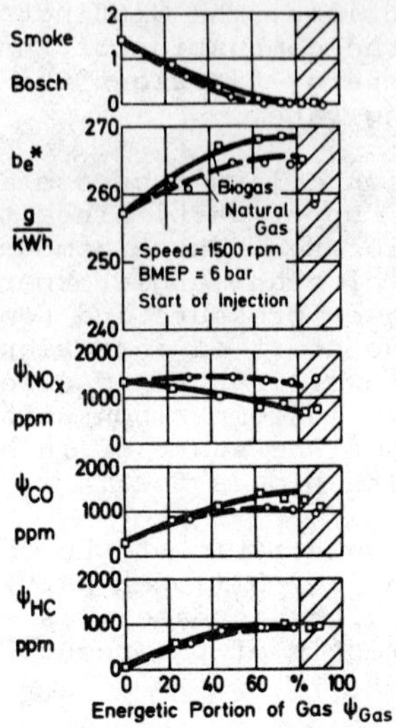

Fig. 3: Fuel Consumption and Pollutants at Dual Fuelled Natural Gas and Biogas Versus Energetic Portion of Gas

presence of the carbon dioxide. The carbon dioxide results in a lower flame velocity and, consequently, lower process temperatures. Moreover, carbon dioxide dissociates endothermically during combustion and the associated temperature reducing effect delays NO_x formation.

The concentration of CO and HC (fig. 3, lower part) increases with increasing gas energy fraction. Unburned fuel in the quench zones, such as the "squish" areas or other "dead spots" results in an increase in HC concentration. Additionally, CO emissions may result from incomplete combustion of the fuel charge in the cylinder. This effect of the carbon dioxide and the lower oxygen concentration of the biogas, in comparison with natural gas, produces HC values which are approximately 10% greater than the corresponding values for diesel fuel. The incomplete combustion occurs because the degree of pilot spray atomization decreases with a decreasing quantity of diesel fuel supplied.

The higher level of HC and CO concentrations while burning biogas as opposed to natural gas, indicates the direct and indirect effects of CO_2:

1) dissociation of carbon dioxide and
2) incomplete combustion due to an decreased oxygen concentration /1/

The combustion calculations provide an interesting insight into the combustion process. Fig. 4 gives the plot of the burning rate versus crank angle for different biogas energy fractions. The distinct first peak is typical and reflects the combustion of the pilot diesel fuel. At very high gas energy fractions the gaseous fuel fixes the maximum burning rate.

The combustion delays increase at higher gas energy fractions (fig. 5, upper part) since the formation of the pilot spray deteriorates. These combustion delays of as much as 2 degrees of crank angle experienced with biogas result from the lower pressure and temperature level in the cylinder at the start of injection. This phenomenon is caused by the lower isentropic exponent of carbon dioxide. Consequently, in comparison with natural gas, it is necessary to advance the start of injection when operating the engine with biogas fuel.

Another characteristic of the combustion duration becomes evident in fig. 5 (lower part). The combustion period is reduced with a greater gas energy fraction. This occurs as a result of the reduction in the quantity of pilot fuel.

With increasing gas fraction, the combustion is increasingly similar to the combustion of the spark ignition engine as can be seen in the oscillograph plots of cylinder pressure. Despite the relatively high compression ratio, the engine did not exhibit knocking even at

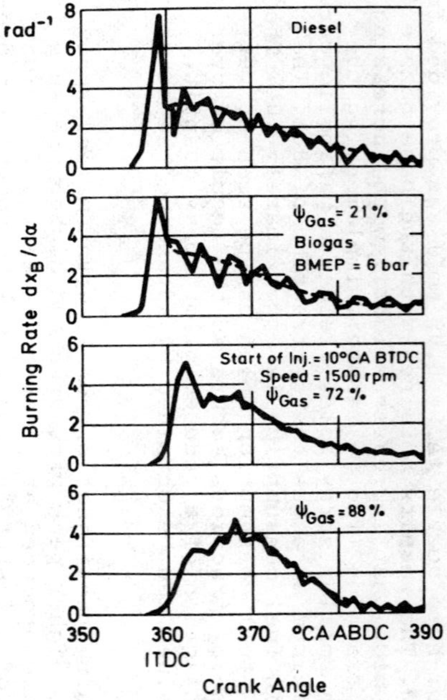

Fig. 4: Burning Rates of Diesel and Dual Fuel Operated Biogas for Different Energetic Portions of Gas

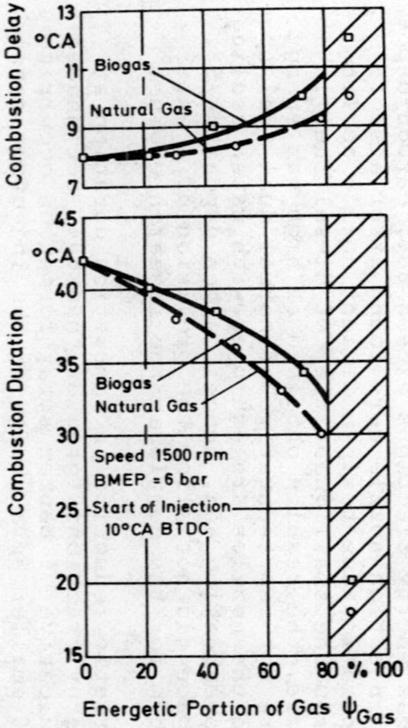

Fig. 5: Combustion Delay and Combustion Duration of Dual Fuelled Natural Gas and Biogas Versus Energetic Portion of Gas

high loads while operating on the biogas fuel. Knocking occurred only during natural gas operation and loads exceeding a bmep of 7.5 bar. This knocking effect can be observed as high frequency vibrations superposed on the normal cylinder pressure curve.

When gas energy fractions exceed 80%, carbon deposits at the injection nozzle were found after a short period of operation due to insufficient nozzle cooling resulting from the small quantities of diesel fuel injected. Therefore, the results obtained at high gas energy fractions are not representative of normal dual fuel operation where the nozzles are equipped with water cooling, preventing overheating and, hence, carbon deposits /5/. Available space did not allow the provision of a water cooling system for the nozzle in the research engine.

The drastic reduction in combustion duration at high gas energy fractions (greater than 80%, see shaded area) is indicative of better spray formation occurring as result of smaller spray holes in the injection nozzle.

Influence of Spray Hole Diameter

To facilitate the use of a spray ignition engine as a diesel engine and, alternatively, as a dual fuel engine, the injection system must be tuned at conditions corresponding to the maximum diesel fuel injection quantity. In particular, the spray holes of the injection nozzle must be sufficiently large. However, this affects the dual fuel operation since the pilot fuel injection spray ia only unsatisfactorily atomized.

In addition to the base nozzle which was suited to alternative operation (spray hole area $A_2 = 0.321\ mm^2$), another four-hole nozzle ($A_1 = 0.09\ mm^2$) was provided for the testing. The results obtained with the base nozzle show a SFC plot that increases slightly with increasing gas energy fraction (fig. 6, above). Implementing the nozzle with smaller spray holes yielded a parabolic curve with its minimum at a gas energy fraction of 80%. The advantage in fuel consumption is approximately 10%. The increase in fuel consumption which can be seen at very high gas energy fractions occurs as the result of overly retarded injection of a reduced pilot spray quantity. Improved values can be obtained by advancing the start of injection.

By using the injection nozzle with smaller spray hole areas, higher peak pressures and higher pressure gradients are produced at greater gas energy fractions, indicative of a faster and more complete combustion process (fig. 6, below).

However, the smaller nozzle can only be used at gas energy fractions exceeding 60%. At lower gas fractions (large pilot fuel injection quantities) the injection

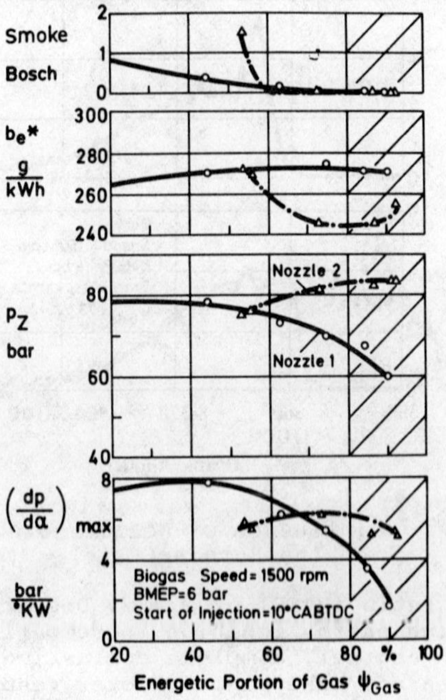

Fig. 6: Influence of Nozzle Hole Size on Fuel Consumption and Operation with Biogas Versus Energetic Portion of Gas

period duration is too long resulting in clearly greater smoke production (fig. 6, above).

The burning rates at a gas energy fraction of 73% are shown in fig. 7 for both injection nozzles. The ignition delay is clearly reduced as a result of the better spray atomization with the small nozzle. The first peak seen in the previously discussed burning rate plot does not occur and the burning rate is globally higher as a result of the spatial distribution of the ignition nuclei.

ENGINE BEHAVIOR AT PART LOAD

Without throttling the dual fuel engine has a higher SFC and produces greater emissions of unburned hydrocarbons under part-load conditions /3/. This is due to the delayed combustion of the lean gas air mixture. While operating the dual fuel engine on natural gas, the operational limits of the single-cylinder engine had already been reached at a bmep of 3 bar corresponding to a rela-

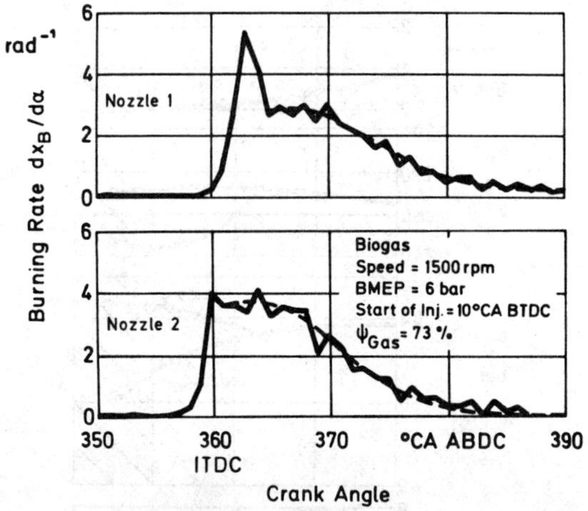

Fig. 7: Influence of Nozzle Hole Size on the Burning Rate

tive air-fuel ratio (RAFR) of λ = 2. Despite the ignition spray (very high energy ignition in comparison with a spark ignition engine), complete combustion of the cylinder charge does not take place. Poorer results were obtained when operating the engine with the biogas fuel.

Throttling avoids overly lean mixtures at partial load and a favorable high pressure process occurs. But there are servere disadvantages with throttling: 1) The pumping losses are increased, 2) the NO_x concentrations are quite high, similar to the case of full load operation, as a result of the high combustion temperatures.

In consequence of the drawbacks of the throttled operation another possibility was examined to improve the part load behavior.

With an injection timing which would yield low fuel consumption at full load conditions, the peak cylinder pressure at part load is between 6 and 9 degrees ATDC. This, in principle, is the optimum crank angle range. Tests conducted with an advanced injection led to improvements in the specific fuel consumption. Fig. 8 indicates the fuel consumption and emissions versus the start of injection at part load (bmep = 3 bar).

The SFC is reduced by approximately 20% by advancing the injection timing to 42.5 degrees BTDC yielding the best fuel consumption values. The unburned hydrocarbons are reduced by 60% and the carbon monoxide emissions are reduced by 50%. If the start of injection is advanced even farther, the CO concentrations in the exhaust gas increase rapidly. Geometric observations reveal that this is caused by pilot spray contact with the cylinder walls which occurs when the injection is started very early.

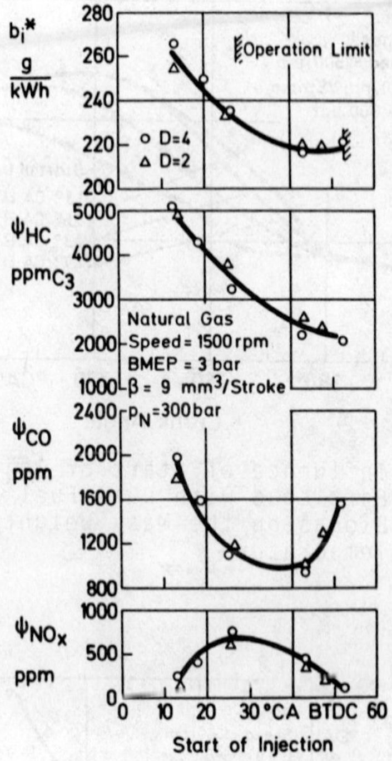

Fig. 8: Indicated Fuel Consumption and Pollutants at Part Load Operation with Biogas Versus Start of Injection

With advancing injection the NO_xx concentrations begin to rise and a maximum is reached at approximately 25 degrees BTDC. Further advancement shows the concentration beginning to decrease. At the optimum start of injection, the NO_x concentrations are reduced by a factor of 2, in comparison with the throttled part load engine condition, and 30% lower than at the maximum NO_x point (start of injection 25 degrees BTDC). Although the maximum mass weighted gas temperatures (fig. 9) are equal, the 30% reduction in NO_x results. The characteristics of the corresponding mass weighted gas temperatures in the crank angle range between 350 and 363 degrees is, however, different. The smoother increase of mass weighted gas temperature at an earlier start of injection produces a shorter combustion duration at a high temperature and, consequently, lower nitrogen formation.

Fig. 10 indicates the results of the thermodynamic analysis. The combustion delays are longer if the start of injection is advanced beyond 25 degrees BTDC. At this point, the pressure and temperature levels in the cylinder are such that, when injection starts, poor conditions for self-ignition are present. The maximum value of mass

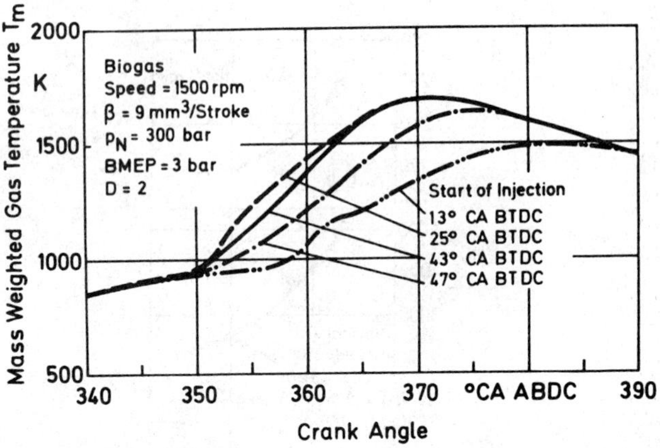

Fig. 9: Influence of Start of Injection at Part Load with Dual Fuel Operated Biogas on the Mass Weighted Gas Temperature

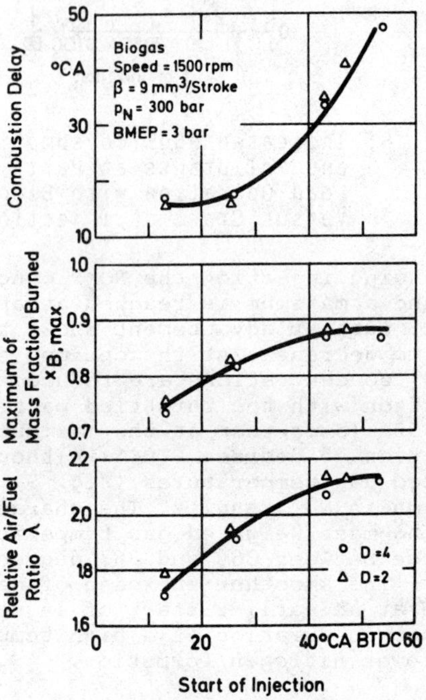

Fig. 10: Combustion Delay, Maximum of Mass Fraction Burned and Relative Air/Fuel Ratio Versus Start of Injection

fraction burned is improved from 75% to 88% despite the higher RAFR of $\lambda = 2.1$ (versus $\lambda = 1.75$). This behavior is present because with a more advanced start of injection there is more time available for spray formation and better distribution of pilot fuel droplets in the combustion chamber. At self-ignition, the more spatially distributed ignition nuclei assure a more complete burning of the lean mixture.

The more homogeneous distribution of the pilot fuel which results also leads to a smoother start of combustion which produces a lower peak in the burning rate shown in the plot of fig. 11. When injection starts early, the plot obtained for the burning rate is similar to the characteristic curve shape for gas

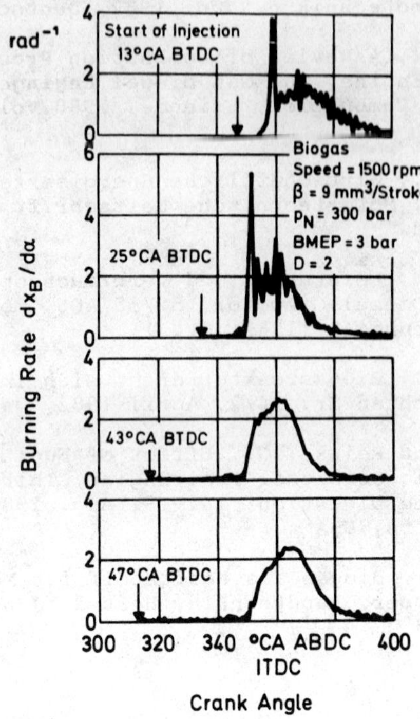

Fig. 11: Burning Rate at Part Load Operation for Different Starts of Injection

ACKNOWLEDGEMENT

Financial support was provided by the Deutsche Forschungsgesellschaft (DFG, German Research Society) and the Comision Nacional de Investigacion Cientifica Y Tecnologica (CONICYT), Republik Chile. The authors would like to express their appreciation for the financial sponsorship.

REFERENCES CITED

/1/ Abthoff, J., Die Messung der Stickoxid-Emission von Dieselmotoren und ihre Beeinflussung durch Maßnahmen am Motor, ATZ 71 (1969), S. 124-130

/2/ Batel, W., Alternative Kraftstoffe für die Landwirtschaft, Landtechnik 6, Juni 1982, Deutschland

/3/ Karim, G.A., A Review of Combustion Processes in the Dual Fuel Engine, The Gas Diesel Enginge Progress in Energy and Combustion, Science, 1980 Vol. 6, pp 227-285

/4/ Klaunig, W., Wirtschaftliche Energieerzeugung durch Gasmotoren, Motortechnische Zeitschrift 38(1977)3, Deutschland

/5/ Klaunig, W., Leistungs- und Verbrauchsoptimierung am M.A.N.-Diesel-Gasmotor, 52/55 AD6, Motortechnische Zeitung 43 (1982) Nr. 1

/6/ Materne, K., Biogasreaktor dreht sich im Wasserbad, VDI Nachrichten Nr. 14/2, April 1982, Deutschland

/7/ Saez, A. and Reinke, G., Effect of Fuel Injection Point on Biogas Diesel Fuel Engine, Third Int. Symp. on Anaerobic Digestion, 14.-19. Aug. 1983, Boston, Massachusetts, USA

/8/ Seifert, A., Biogas als Kraftstoff für Motoren und Ackerschlepper, Landtechnik, Heft 2, Januar 1985, Deutschland

DISCUSSION

H. Kosstrin, Power Recovery Systems, Inc.	I was wondering whether you had done any work with producer gas, which has about one third the calorific value of biogas, and whether you had done that work in a spark-ignited engine as opposed to the dual-fuel engine?
Dubel	We did the work with biogas in the dual fuel engine because biogas has a high content of carbon dioxide. I've done some tests with biogas in spark-ignited engines, but they were not so successful. With a dual-fuel engine, much higher efficiencies can be obtained than with the spark-ignited engine. In the spark-ignited engine, knocking often occurs. This problem is greatly reduced with dual-fuel engines.
D. Jackson, Michigan State U.	I'd like to know if you did any tests on the horsepower output when you were changing the injection timing, and the other modifications that you indicated?
Dubel	We have varied speed and break mean effective pressure too from full load to part load. But you have seen that at part load, lower break mean effective pressures may cause difficulty in burning the biogas with good efficiency.

INTEGRATION OF BIOMASS PRODUCTION AND COMBUSTION INTO A POWER UTILITY

Lars Sjunnesson, Civilingenjör
Chief Engineer, Research & Development
Sydkraft
S-217 01 Malmö, Sweden

ABSTRACT

There is an increasing interest in biomass development in Sweden with focus on the possibility of using waste and various biomass products to secure the future energy supply to the consumer.

The need for technical development is great in several areas in order to ensure a successful commercialisation in the next ten years.

Sydkraft as a power utility has a special interest to develop technical and economical ways of utilising energy sources such as energy forests, energy crops and different kinds of wastes. The background is the need for smaller power plants - often localised close to the consumer. In most cases there is no economical way to transport the fuel over long distances.

The paper deals with the problem to fit biomass production, conversion and utilisation into the system of a power utility. Examples are given to show how problems have been solved in certain cases. Examples of our interest in research and development in several areas are given.

INTEGRATION OF BIOMASS PRODUCTION AND COMBUSTION INTO A POWER UTILITY

Background

Sydkraft is Sweden's second largest electric power utility. The Swedish State Power Board (Vattenfall) is the largest and accounts for about half of all electrical energy generated in Sweden, whereas Sydkraft generates about one-quarter.

The operations of Sydkraft are concentrated to southern Sweden, where a large proportion of Sweden's population is living. But our energy structure is somewhat different than that in the remainder of Sweden, which is primarily due to the smaller proportion of industrial consumers in the southern part of the country.

The Company was formed in 1906, primarily with the objective of supplying the southern part of the country with the electrical energy necessary for maintaining and developing a high standard of industrialisation and industrial production. The first generation plants consisted of water turbines. Other generation units, primarily coal-fired and oil-fired, were subsequently built.

Several nuclear power stations were taken into operation between the early 1970s and the mid-1970s, and these stations have since served as a very strong economic base for the Company. The nuclear stations have performed exceptionally well and have thus made a substantial contribution towards reducing Sweden's oil-dependence.

As in many other countries, the role and orientation of the Swedish power utilities have changed during the past decade. Sydkraft now supplies several different types of products, the main of which are electrical energy, gas and heat, as well as a number of spin-off products.

Several different factors are decisive to future developments for the energy utilities. The most important factor is that the Swedish nuclear power stations will be decommissioned by the year 2010, according to an Act passed by the Riksdag (Swedish Parliament) in the early 1980s. This will obviously be of decisive consequence to our future, particularly to that of the consumers, who will be obliged to pay appreciably more for their electrical energy after nuclear power has been abandoned.

A very important factor in this connection is that the consumption of electrical energy in the future is expected to grow at an appreciably slower pace than it did in the past, even though the growth rate is very high at the present time. As a result of the lower growth rate, it will not be as economical in the future to build generation plant rated up to 1000 - 1300 MW electrical, since it would take far too long for the demand to catch up with the supply, and such large plants would therefore be expensive to operate. As a result, the plant ratings in the future will probably not exceed 100 - 300 MW electrical.

A third factor of importance is that pressure has been exerted for many years to put domestic raw materials to use in the energy supply system. Certain municapilities and consultancy companies were usually at the forefront in propagating for this development. But it was counteracted by the low cost of electrical energy and by the fact that oil prices had not increased as sharply as was expected only about a year ago.

In view of all of the above factors, together with the fact that Sydkraft has expanded its operations to virtually the entire energy field, we are now actively working in several sectors of energy supply, including the biomass sector.

Biomass in southern Sweden

Some energy raw materials are natural for us to use when considering the introduction of a new element into our energy strategy. The raw materials nearest to hand are peat and wood.

The peat and wood resources in southern Sweden are fairly large, and they have long been used to a very great extent, before the large-scale use of coal and oil and also between the two World Wars.

The "new" biomass resources are energy forests, algae and various forms of energy crops. The extent to which these resources will be utilised in the future is largely dependent on the amount of agricultural land, etc. allocated for this purpose.

We are obviously studying traditional methods of peat harvesting, although we are also endeavouring to become involved in new techniques. One of these is a method of extracting gas from peat in situ, without first harvesting the peat. The method involves injecting water into a peat bog and discharging it at a point which is relatively remote from that at which it was injected. Methane is entrained by the water as it flows through the peat bog. The methane can then be separated from the water in a suitable water trap.

This method is already being used today in some projects in central and northern Sweden for extracting gas from peat bogs. What we wish to find out is whether this is more effective in southern Sweden, where the mean temperature over the year is higher than in the parts of the country further north.

Short-term objectives

When a large company such as Sydkraft decides to embark on different areas of operation, one of the difficulties is to find an appropriate role. The path initially chosen often proves to be unfruitful. In smaller projects, it is also fairly difficult to refrain from working with all of the degrees of freedom that are usually available in a larger project.

We attempted to get started on several different areas in different geographical locations at the same time.

Peat firing. The objective in one of the projects was to build a peat-fired heat generation plant for the district heating system of an urban area. It was decided that the boiler rating should be about 10 MW thermal. The project was launched in 1982 and the district heating distribution system has now been virtually completed - but there is still no peat-fired boiler. The electrical energy we generate is still so inexpensive that several electric boilers are used instead for supplying the heat.

We also investigated other alternatives, including refuse incineration and the production and utilisation of peat pellets, but none of these projects has yet materialised.

Wood-waste firing. The purpose of another project, which was appreciably smaller in scope but more interesting from other aspects, was to interlink us with an industrial plant and an urban area. The industrial plant needed steam, the urban area needed hot water, and the distance between the plant and the urban area was no more than about 500 metres. This plant manufactures flooring material, some of which is exported to the U.S.A., and large quantities of sawdust and pulverised wood are a waste product from this manufacturing process. The intention in this project was to build a boiler adjacent to the plant and to fire it with the wood waste.

This fuel has a dry solids content of about 95% and it is thus fairly explosive. Care must therefore be taken when handling the fuel. The dust emission from the plant is also a sensitive matter, since the boiler is located in an area which is densely populated by Swedish standards.

The boiler plant, consisting of one 1.5 MW boiler and one 2.5 MW boiler, was commissioned in the winter of 1985 and has performed very well.

Pellet firing. In a third project, we wished to try to introduce pelletised fuel in a region in south-eastern Sweden. We established collaboration with a local company, which undertook to supply the bark and cutter chips that were to serve as the raw material for the pellets.

The marketing of the pellets was launched by concluding an agreement with pellet manufacturers for the supply of pellets for the first plants fired with this fuel, in order to establish a certain turnover volume. An annual demand for at least about 10 000 tons of pellets was necessary before it would be profitable for us to set up our own production plant.

We worked intensively for about six months to find customers for this particular fuel, but without any marked degree of success. The main reason was that the cost of electrical energy was very low. A normal domestic subscriber now pays about 0.30 SEK/kWh, which is equivalent to about 4 US ¢/kWh. The alternative available today to many prospective customers is to install an electric boiler. The pay-off time for an electric boiler is usually two years, based on the cost difference between continuing to use an oil-fired boiler and changing over to an electric boiler. Another reason for the lack of success in marketing this fuel was that wood chips are a more profitable fuel, particularly for larger boilers.

As a result, we abandoned our efforts and we now intend to wait for a better opportunity for introducing such processed fuels. The company that was set up at that particular time still exists.

In order to make it attractive for potential customers to use pellets and briquettes as a fuel, we entered into a collaboration agreement with a Swedish manufacturer. The purpose was to develop a boiler that was suitable for firing with processed fuels. The new boiler has a stepped grate which ensures fast and efficient combustion of these particular fuels. Different pellets have now been fired in this boiler for 12 months, and no problems have been encountered. The boiler delivers a maximum of about 350 kW of heat. We are planning to launch a new project during 1986 on a somewhat larger boiler, this time using straw pellets as the fuel.

The results of this collaboration were successful, and several such boiler plants have been sold to customers who considered it attractive to use solid fuels, in spite of the low cost of electrical energy. Sydkraft has invested fairly heavily in the development work. This collaboration is now beginning to bear fruit in the form of royalties received from the manufacturer.

Long-term objectives

Our long-term objective is to find a role for ourselves in the biomass area, of a type which is appropriate to a large energy utility. As we see it, this involves trying a variety of approaches that are technically, administratively and economically beneficial to ourselves and our customers.

Energy forests. We are continuing our activities on the energy forest project we presented during the 1985 meeting. Part of the energy forest has been harvested during the past winter. After chipping, numerous boilers have been fired with this fuel, in order to test it in a number of different applications.

The harvesting of the energy forest was the first test of its type. Only a minor part of the forest was harvested during March 1985, and this was done manually. A specially developed harvesting machine is being tested this year. A total of about 25 hectares will be harvested during the winter of 1986. Chipping is carried out in situ, and the wet chips are transported by road directly to a boiler plant. The moisture content of the chips is about 50%.

The rating of the boiler plant is 3.5 MW thermal, and a combustion efficiency of 81% was recorded at a load of about 1 MW. Tests will continue during the winter and spring of 1986, in order to determine various combustion parameters.

We have also studied other opportunities available for using the energy forest. In collaboration with the Lund Institute of Technology, we have studied the opportunities available for gasifying the wood fuel and for producing liquid fuels, primarily ethanol and methanol. These methods are considered to be economically attractive in large-scale utilisation of energy forests.

Municipal refuse. An area of interest today in many countries is the production of gas for direct combustion from municipal refuse. The technique is very old and has been used mainly for generating heat in the immediate vicinity of a refuse disposal plant.

Several different projects with this aim are pursued in Sweden today. The reason for the interest in Sweden is mainly the apprehension caused by the publicised emission of dioxins in the direct incineration of refuse. The construction of new refuse incineration plants is now banned in Sweden. The moratorium will remain in force until the questions related to the emissions of dioxins have been answered.

Sydkraft's interest is to apply our process know-how for utilising municipal refuse in an optimised manner to produce an energy resource. We are therefore trying to find methods that improve the efficiency of gas production. One approach is to contain the refuse in an airtight enclosure and to control the aerobic and the anaerobic decomposition of the refuse, so that the quantity of gas produced will be a maximum.

The enclosure is made from residual products of wet/dry flue gas desulphurisation and is "self-healing" if damage should occur. The method has been tested during the autumn of 1985 and the winter of 1986 and has produced good results. We are now intending to go a stage further and demonstrate the technique on a large scale.

In an experiment in Malmö, we have enclosed about 1000 cubic metres of refuse and we are now intending to control the process by varying the water supply, utilising bacteria, etc., so that a maximum amount of gas will be produced from the refuse. The first reliable test results will be obtained during the spring of 1986, and we intend to test the equipment for a further period of one year.

The gas produced from the refuse consists of about 60% methane and 40% carbon dioxide. For certain applications, we would prefer to have a higher methane content. We have therefore started an experiment aimed at separating the gases in a special membrane. The experiments will start during the spring of 1986.

The technique of using a membrane for gas separation is well known, but we intend to try to develop an inexpensive and effective technique that can be used on a smaller scale. We are pursuing this work in collaboration with several other Swedish companies.

Drying of the gas produced at refuse disposal sites is a technique that has come to our attention in recent years. The gas usually contains water, which may cause difficulties if the gas is to be piped over a longer distance to a user. We have so far not found the best technique for this type of drying, and we were therefore obliged to try to develop a suitable technique ourselves. Freezing is probably the best technique, but we are still looking at other possibilities.

We have found a technique that we believe to be a good solution to the problem, but we have not yet tested it on a larger scale. We are planning to do this during the spring and summer in a gas production project we are pursuing in the town of Värnamo, located about 250 kilometres north of Malmö.

<u>Biomass.</u> Our long-term work also includes a study of the opportunities available for using various energy crops, so that they will play an active role in our energy supply system.

We are collaborating with different parties in this area. Our objective is to obtain electrical energy, gas and heat from various forms of energy crops. It should be possible to employ the residual products obtained in the conversion stages as animal feed and as fertilizers.

The crops that we are mostly interested in are lucerne, Jerusalem artichokes, corn, etc. Algae are also of major interest, since two objectives can be achieved. Algae can first be used for cleaning up lakes or bays. They can then be fermented, and the gas can be used in the usual way.

Certain algae have a high growth rate, and the harvest can be as high as 100 tonnes of dry solids per hectare. We have gained acquaintance with algae in conjunction with the fish farming we pursue at our Barsebäck nuclear power station. We have been officially instructed to clean the water effluent from this plant, and we have found that the best way is to utilise a certain special alga. The fish-farming project is relatively small, and we are now planning to test an alga project on a large scale with other interested parties.

We are now studying two entirely different fermentation processes for energy crops. In one of the projects, the crops will be fermented in a single-stage reactor, in which hydrolysis and methanising take place in the same reactor. In the other project, separate reactors are used for these two processes. We intend to compare the economic and process chemistry results obtained from these two processes, in order to establish which technique is best.

The gas obtained from fermentation contains carbon dioxide and methane. It is usually advisable to maintain a constant quantity of the gas produced. This is also the case when gas is produced at a refuse disposal site. But the consumption of gas may vary within fairly wide limits. This has led us to study the opportunities available for increasing the calorific value of the volume of biogas produced, by running it through a coal bed. We are investigating at present the kinetics of this arrangement, and there are many indications that it may be an interesting development line.

In view of our involvement in the use of gas, our long-term intention is to examine the possibility of using biomass as a fuel for fuel cells. We have no fuel cell in our system yet, but we are planning to install one or more fuel cells for demonstration purposes, if the technique continues to be as promising as it appears to be today. We will then be able to introduce biomass in a natural way into our energy system by utilising an advanced technical process, which is entirely in line with what the power industry can and should do.

Conclusions

The interest of the power industry in the utilisation of biomass and refuse is steadily increasing, in pace with the emergence of new techniques. Most of the techniques will be employed in smaller units than is usual in the power industry. This involves problems but also opens up opportunities.

The area that will probably be of greatest interest to the power industry is the opportunities available for generating gas from different energy raw materials. It will then be possible to use new techniques in the natural manner in the activities of the power industry.

The use of biomass as fuel and conventional gas production from existing refuse disposal sites will be of greatest interest in the short term. In the longer perspective, the production of gas from biomass and refuse will become of interest and importance to the power industry. The greatest interest will then be focused on the high-technology applications, combined with advanced biotechnology.

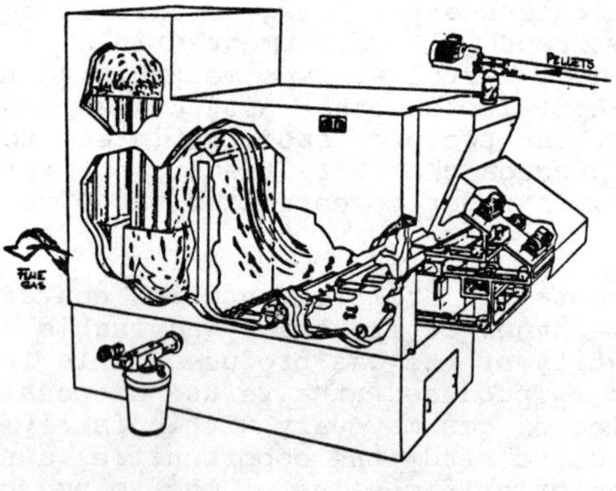

Figure 1. Boiler for pellet firing

Figure 2. Machine for harvesting energy forest

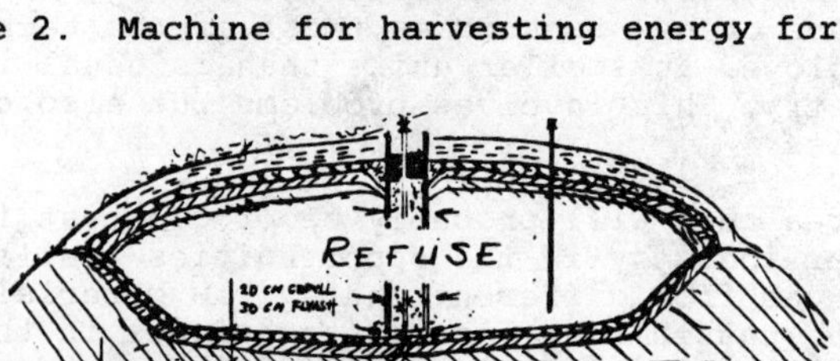

Figure 3. Optimising the production of gas from refuse

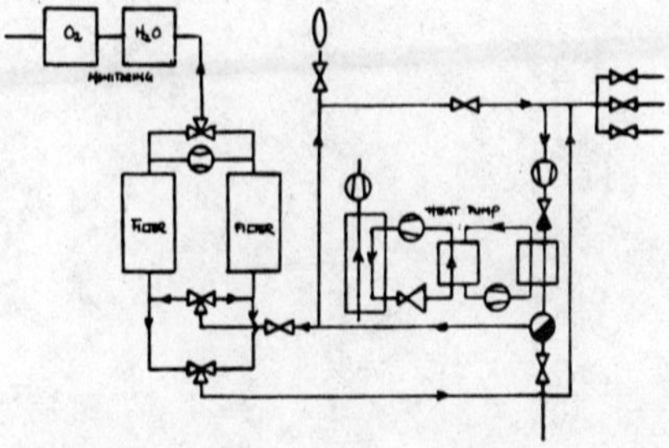

Figure 4. Drying of gas produced at refuse disposal

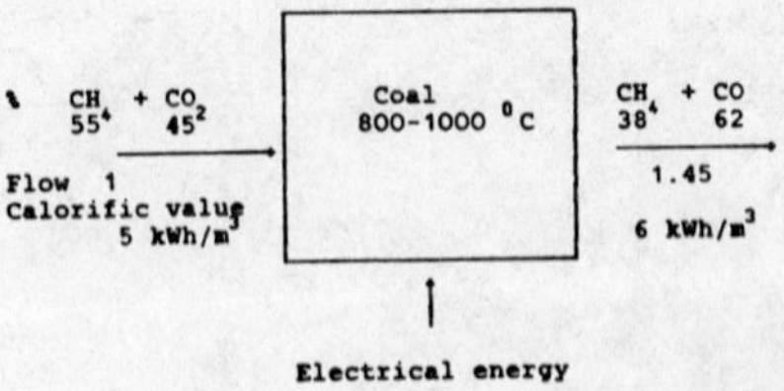

Figure 5. Increasing the calorific value of biogas

DISCUSSION

Zoellner	In one of your diagrams on landfill gas, you showed the disappearance of different elements over time. Nitrogen and oxygen were dropping out in a matter of weeks and months, and carbon dioxide and methane were appearing. In Sweden, are there certain times of the year when production of methane from landfill would be less than adequate?
Sjunnesson	We are investigating this. There should be differences between winter and summer periods. We have three projects at the moment to check what kinds of differences would occur in the gas. But we are not really that much interested in the production of gas. We're more interested in its utilization for producing heat. If there is varying gas production, this will cause problems.
M. Boillot, Electricite de France	How do you use electrical energy in the process to increase the calorific value of biogas?
Sjunnesson	We have done a lot of work to see if that could be a practical way to increase the calorific value. We are now studying the kinetics.

STARTUP AND OPERATION OF A 3-MW WOOD-BURNING GAS TURBINE AT RED BOILING SPRINGS, TENNESSEE

Joseph T. Hamrick, President
Aerospace Research Corporation
Roanoke, Virginia 24014

ABSTRACT

A 3000 kw wood burning gas turbine power generation system for which research was started in 1978 with the support of the Department of Energy is now on line and generating power at Red Boiling Springs, Tennessee. The electrical power which is generated with sawdust produced by sawmills in the surrounding area is purchased by the Tennessee Valley Authority. The system now uses an aircraft derivative gas turbine built by the Allison Division of General Motors. The system was installed and operated at design speed, generating a small amount of electrical power during 1985 with a T-56 engine, the aircraft version of the Allison 501 K engine. The thirty year old T-56 engine was furnished by the U.S. Air Force.

Startup of the system required devising methods for warm up and advancement to design speed without stalling or surging the compressor, protection of the gas turbine from overspeeding due to loss of load and preventing overheating the turbine. Operational problems were dealt with in feeding the combustor with wood against the 130 pound per square inch system pressure, regulation of the wood feed, and sudden depressurization of the combustor system. No erosion of the turbine blades has been detected. Periodic cleaning of the turbine with milled walnut hulls is required. A review of the system operation and the developmental problems are discussed in this report.

STARTUP AND OPERATION OF A 3-MW WOOD-BURNING GAS TURBINE AT RED BOILING SPRINGS, TENNESSEE

REVIEW OF THE SYSTEM

The system which has been described in previous presentations is reviewed herein for those unfamiliar with it. A schematic diagram is presented in figure 1.

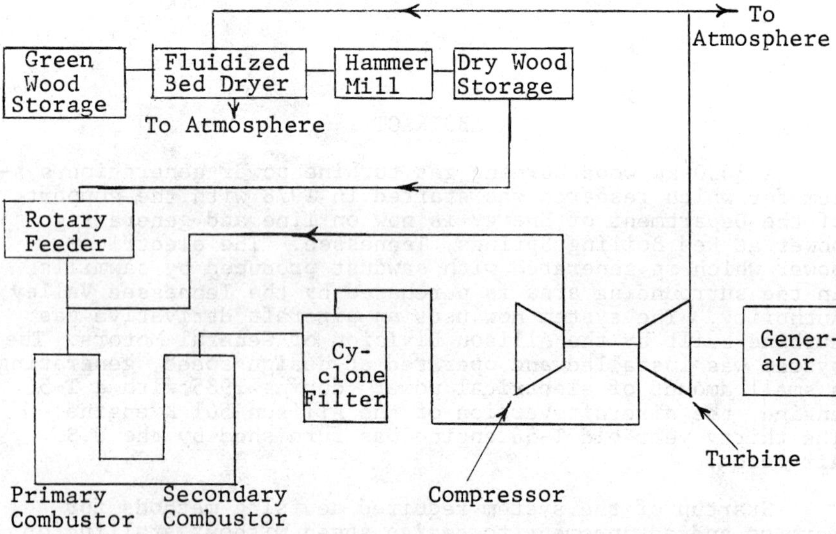

Figure 1 - Schematic diagram of the 3000 kw wood burning gas turbine system.

A view of the installed system at Red Boiling Springs, Tennessee is shown in figure 2.

Figure 2 - View of the 3000 kw Facility

Sawdust is loaded into a green wood storage bin, the outlet end of which is shown in figure 3 for controlled feed into the fluidized bed dryer.

Figure 3 - Outlet end of green wood storage bin (right) and inlet end of fluidized bed dryer (left).

An overhead view of the dryer is shown in Figure 4.

Figure 4 - Overhead view of fluidized bed dryer. Suction fan and cyclone on left, hot gas ducting from gas turbine on right.

Dried sawdust from the dryer is moved by auger to the hammermill at bottom right. The dried sawdust is pulverized and passed through a one quarter inch screen before being moved to storage in the bin shown at left in figure 1. Both the green wood storage bin and the dry storage bin have chain driven bottoms to move the sawdust. They are both tilted to reduce the force on the chains and allow use of small, variable speed drive motors.

The diagonal tubes shown in figure 1 contain flexible augers which convey the wood from the dry wood bin to the live bottom bin at the top. Four augers in the bin meter the wood to the rotary feeder valves by means of variable speed drive motors. The live bottom furnishes the only means of fuel control. Fuel from the live bottom bin and air from the compressor are fed into the top of the primary combustor by means of the rotary valves shown in figure 5.

Figure 5 - View of rotary valves above the primary and secondary combustion chambers.

The combustor is so designed that combustion products and any unburned char swirl into the bottom of the secondary combustor where all unburned particles are centrifuged to the wall and burned. Gases leaving the secondary burner together with the ash are ducted to the cyclone shown in figure 6.

Figure 6 - View of the cyclone and duct leading to the turbine.

A view of the generator set is also shown in figure 6. A view of the gas turbine section of the generator set is shown in figure 7.

Figure 7 - View of the gas turbine installed in the generator set.

The transformer and oil circuit breaker stations are shown in figures 8 and 9.

Figure 8 - View of transformer station.

Figure 9 - View of oil circuit breaker station.

The metering equipment is also installed at the oil circuit breaker station.

OPERATIONAL EXPERIENCE

Starting the system, protection from overload due to line faults, prevention of overspeeding, prevention of turbine overheating, rotary valve performance, and turbine blade cleaning proved to be the prime concerns in the startup program. Each of these is discussed in order.

Starting the Engine

As a gas turbine is brought up to design speed, the compressor usually requires either bleeding of air at intermediate stages or provision for varying the stator pitch angles to avoid rotating stall or surge. In the compressor of the 501K engine, bleed valves at the 5th and 10th stages remain open until the compressor reaches a speed of approximately 13000 rpm. For the conventional configuration of the engine a momentary period of turbine over temperature is used to motor the engine to 13000 rpm while the bleed valves remain open. With a liquid or gaseous fuel the response of the fuel control is so rapid that exposure of the turbine to an overtemperature can be accomplished without undue damage to the blades. With wood fuel the response is much too slow to use that approach. To reach a speed of 13000 rpm with wood fuel it is necessary to use additional bleeds after the 14th stage, use propane gas to warm the system to at least 600°F before admitting sawdust, and to stage the closing of the bleeds as the speed of the engine is increased. A model 455 oldsmobile engine with hydramatic drive is used to motor the system until a turbine inlet duct temperature of 1450°F is reached at a turbine speed of 3600 rpm. At that point the generator with shorted field is energized as a motor to bring the engine up to the design speed of 14,200 rpm. The bleed valve downstream of the 14th stage is closed at 7350 rpm, the 10th stage bleeds at 11,500 rpm and the 5th stage bleeds at 12,500 rpm. If any of these bleeds are closed prematurely, the compressor will experience rotating stall or surge conditions that can cause damage to the compressor. This starting schedule, which took a great deal of experimentation to devise, is satisfactory.

Protection of the Gas Turbine From Overload Due to Line Faults

The possibility of a short circuit in the generator or elsewhere in the electrical system requires that attention be paid to methods of relieving the load on the engine. This is best done by providing a shear section in the drive shaft. There is a shear section in the shaft between the engine and gear box. However the high cost of replacing that shaft and its inaccessibility makes it attractive to put a lower torque shear section in the shaft between the gear box and generator where accessibility is good. The shear section also provides protection for the generator and gear box in the event of a gas turbine failure in which the gas turbine shaft locks up. The effectiveness of the shear section was demonstrated during failure and lock up of the U.S. Air Force T-56 engine turbine.

Protection of the Gas Turbine in the Event of Load Loss

A distance relay in the switch gear system upon detecting a fault in the TVA transmission line will automatically switch the generator off the line. Unless a dummy load is immediately imposed upon the generator or the system pressure is relieved, there is enough residual energy in the pressurized gases in the combustor, cyclone, and ducting to overspeed the gas turbine even though the wood feed is shut off.

An eight second dummy load is in place and is automatically imposed in case of overspeed. In addition, a 12 inch diameter rupture disk is automatically actuated. The rupture disk is located in an outlet to the duct connecting the compressor and combustor. The initial design was such that the rupture disk has actuated unexpectedly twice during operation of the system,once at 100 psi and once at 130 psi. The actuation at 130 psi resulted in lifting the dome in the combustor outlet that partitions off the incoming compressor air from the outgoing combustion products allowing the compressor air to bypass the combustor. To prevent that occurrence in the future it is planned to spring load the dome so that the dome can be lifted up but will return to the initial position. The possibility of unexpected actuation has been reduced by modification of the rupture disk design.

Prevention of Turbine Overheating

It was amply demonstrated early in the program that allowing sawdust to accumulate at the bottom of the primary burner during low speed warm up will result in greatly overheating the system including the turbine as the turbine approaches design speed. One cause of sawdust accumulation results from residual wood in the feed train to the combustor In an emergency shut down it is not possible to clear the wood from the rotary valves and cross auger that feed wood to the valves from the live bottom bin. In an early run uncleared wood in the feed train piled in the bottom of the combustor at start up. The pile which was not consumed in the warmup, burned rapidly as the gas turbine approached full speed and overheated the system severely damaging the cyclone, ducting and turbine. To prevent a recurrence of the problem several measures were taken. They are enumerated as follows:

1. The cross auger between the live bottom bin and rotary valves was fitted so that the auger can be reversed and cleared of sawdust in the event of an emergency shut down.

2. A policy of clearing the rotary valves by turning them on momentarily and off again repeatedly was instituted.

3. The combustor bottom was recontoured to eliminate stagnation pockets where sawdust could collect.

4. Thermocouples were installed in the combustor bottom where a low temperature reading would indicate coverage with sawdust.

5. An air jet was installed at the combustor bottom for stirring up any sawdust that might collect.

6. A temperature controlled automatic water spray was installed in the outlet of the combustor.

The measures that were taken have proven very effective. No further overheating problems have occurred.

Rotary Valve Performance

The system was initially fitted with rotary valves with twelve inch diameter, rotors twelve and one half inches long with eight wiper blades. A schematic cross section of a typical rotary valve is shown in figure 10.

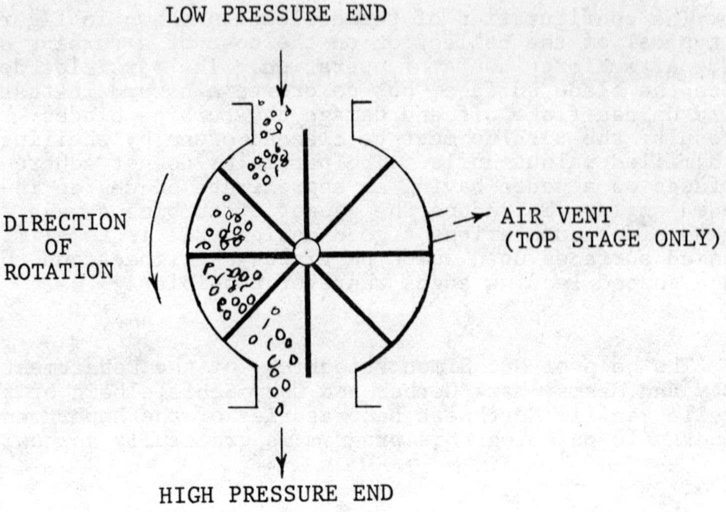

Figure 10 - Schematic of a rotary valve with eight wiper blades.

The twelve inch valves proved to have inadequate capacity. It was necessary to increase the rotor diameter to sixteen inches. The system is currently fitted with four sixteen inch valves operating in series. The two top valves are fitted with twelve blades and the two bottom valves with eight blades. The rotative speeds vary from 22 to 26 rpm. The valves perform well in feeding wood against the 130 psig internal pressure of the combustor.

Turbine Blade Cleaning

The cleaning requirements for a turbine appear to be dependent upon the species of wood in use. The characteris-tics of the ash appear to differ from one species to anothe The oak ash particles generated at Red Boiling Springs adhe to a blunt surface upon impact more readily than the ash particles from Virginia pine. Particles judged to be less than 0.5 microns in diameter tend to agglomerate on rounded surfaces in the manner shown in figure 11.

Figure 11 - Configuration of oak ash particle collection on round rod.

The configuration of the collection shown in figure 11 is typical of the collection on the nose of a turbine stato blade after forty hours of operation. The particles do not erode the blade surface, but do create a hazard in that the build up can flake off and damage the turbine blades. As a result, the turbine must be cleaned often by shelling with milled walnut hulls. The particles do not adhere to surfaces on a wedge having an approximate 50 degree included angle. To reduce the danger of turbine damage from flaked off accumulations a good design rule is to eliminate rounded surfaces upon which particles can impact and to use wedge shaped leading edges wherever possible.

The help of Mr. Simon Friedrich, of the Department of Energy and Mssrs. Mark Gerber and Gary Schiefelbein of the Battelle Pacific Northwest Laboratories of the Department of Energy in pursuing this program is gratefully acknowledge

DISCUSSION

F. Gallagher, Garden State Paper Co.

What is the moisture content of the fuel as it enters the combustor?

Hamrick

It varies anywhere from 12 to 25%. Most of the char is burned by the time it hits the bottom of the primary combustor. Then it swirls into the second combustor and any char that is there is completely burned. You see nothing but clean gases coming out of this system. If there's too much char, say if we're up to 25% moisture, a greater amount is burned in the secondary combustor, and our primary problem then is to keep the temperature on the walls of the secondary combustor below the melting point of the ash. The high pressure generally provides adequate combustion in the primary chamber to keep the temperature down on the walls of the secondary combustor.

D. White, U. of Arizona

There's been a lot of research on the melting, stickiness, and slagging of ashes. Do you think any of this information will be useful in trying to predict a new wood system prior to its testing?

Hamrick

It's something that has to be addressed. PNL is just now looking at that. We've provided them with ash and wood. Our preliminary indications are that manure ash melts at about 1900, and oak ash melts between 2000 and 2100. Work should be done on burners in which the ash doesn't melt. If it's possible by the way to put in small amounts of additives to increase the ash melting temperature, it would certainly help particle accumulation on the blades.

M. Lefcort, Heuristic Engineering

Was there any reason why you didn't try to modify one of the existing pulverized wood burners used in the forest products industry, like the Energex or the McConnel burner?

Hamrick

The problem is one of bringing the air in to reduce temperature to about 1800°. Those burners do not provide that feature. They generally keep the temperatures high enough to get high-radiant heat to the tubes and the boiler. Generally, the burner for a gas turbine is quite different from anything used in a boiler.

M. Boillot, Electricite de France — What is the overall efficiency of such a system?

Hamrick — We have not gotten adequate information on burning to give experimental efficiencies. With 1600° inlet air to the turbine, it's about 26% efficiency. With steam generated by the exhaust gases from the turbine and injection into the engine, efficiency is about 30% overall.

C. Tewksbury, Burlington Electric Dept. — I noticed on the pictures of the facility, the system looked rather makeshift. Have you had any problems with personnel safety and maintenance getting up on the towers?

Hamrick — We use safety belts all the time. Because it's a research facility, we did not provide platforms around the towers as you saw. We do need platforms, but it's secondary right now to other things. Everyone has to have a safety belt on when he's up there. We've had no problems and no one was hurt.

J. Easterly, Meridian Corp. — I wonder what energy penalty might be incurred in grinding the wood up to sawdust size, and whether or not there's a problem with sand coming from bark or when it goes into the turbine stream, or whether you're able to get rid of that beforehand?

Hamrick — We've not had any problem with dirt. Our biggest problem is with big rocks. We don't know where they come from, and as I said, we have a lot of sawteeth in the sawdust. But we have a big magnet that takes the sawteeth out. So the only problem we've really had is with large rocks in the greenwood storage. We have a screen that removes the material at the the greenwood storage trailer. Sand has not been a problem.

ELECTRICITY GENERATED BY A DIESEL ENGINE RUNNING ON LEAN GAS PRODUCED FROM A BIOMASS PRESSURIZED FLUID-BED GASIFIER

Guy Drouin, P. Eng., MBA
General Manager
Louis Béchard, P. Eng.
Process Engineer
BIODEV International Inc.
Montreal, Canada, H2Z 1S8
Philippe Du Chouchet, P.Eng.
Alsthom
Paris, France, 93176

ABSTRACT

The production of electricity from biomass offers most of the oil importing countries a unique opportunity to become energy independent.

BIODEV International Inc. has refined a process which produces a low BTU gas by fluid bed gasification under pressure, and its French partner ALSTHOM offers a Pielstick modified dual fuel diesel engine capable of burning the low BTU gas. A first electrical plant having a capacity of 6.7 MWe is presently under construction in French Guiana for Electricité de France. A pilot plant of 800 kWe has also been built on the site of the BIOSYN Project in 1985. This plant is taking about 900 SCF/min (26,4 Nm^3/min) of low BTU gas coming from the BIOSYN pressurized fluid bed gasifier. The net efficiency of the plant is about 25%, meaning that one (1) ton of wood chips (925 kg) at a moisture content of 20% (wet basis) is needed to produce 1 MWh.

ELECTRICITY GENERATED BY A DIESEL ENGINE
RUNNING ON LEAN GAS PRODUCED FROM A BIOMASS
PRESSURIZED FLUID-BED GASIFIER

INTRODUCTION

BIODEV International Inc. and ALSTHOM are conducting a program for the development of a process which essentially consist to produce from biomass a low BTU gas which is introduced under pressure in an ALSTHOM-PIELSTICK dual fuel diesel engine for generating electricity. A pilot plant of 800 kWe has been installed in 1985 and is currently utilizing about 900 SCF/min (26,4 Nm^3/min) of low BTU gas coming from the BIOSYN pressurized fluidized bed gasifier. A gas cooling and scrubbing system has been developed by BIODEV International Inc. BIODEV International Inc. is also constructing with ALSTHOM a first 6,7 MWe power plant based on this new technology in French Guiana for Electricité de France. This plant is scheduled to be completed by autumn 1986, and will need around one (1) ton of tropical wood chips at 20% moisture content to produce 1 MWh. The net efficiency of the plant is 25%.

BIOMASS POWER PLANT DESCRIPTION

The generation of electricity by the gasification/diesel method comprises four (4) basic sections:

. Biomass preparation.
. Biomass gasification.
. Gas clean-up and cooling.
. Diesel generating set and its auxiliaries.

Figure 1 shows a simplified process description for the 6.7 MWe French Guiana Project.

Biomass Preparation

The gasifier diesel biomass power plant needs around 26 000 bone dry tons per year of tropical wood chips, which means around 10 000 cords (30 000 m^3) of solid hardwood timbers. The main wood species found around Kourou and Degrad des Cannes are wapa (20%), mahot (15%), gaulette (10%) and some other hundred species. Typical wood ultimate analysis is as follow:

Table 1 - French Guiana wood chips - Ultimate analysis, dry WT %.

Carbon	50,0
Oxygen	42,9
Hydrogen	6,0
Nitrogen	0,2
Ash	0,9
	100,0

The silica content of the wood species is very low except for gaulette (1,3%). The average HHV of the wood chips is 8 650 BTU/lb, on a dry basis (5,606 kWh/kg).

The biomass preparation section of the plant consists in the operations that modify the raw biomass properties into a state which is usable by the gasifier. In the specific example of figure 1, the tropical wood logs are fed to a sorting bin and then chipped to appropriate

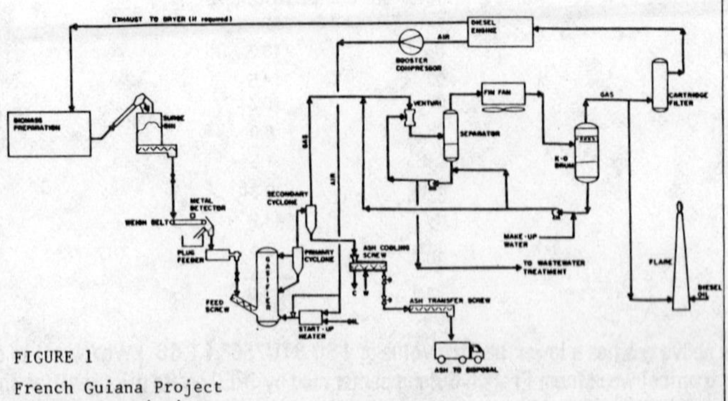

FIGURE 1
French Guiana Project
Process description

dimensions (25 x 25 x 8 mm). A disc screen removes deliterious material from the chips and a metal detector protects downstream equipment. The wood chips are fed to a silo which provides storage for weekend operation. The chips from the silo are conveyed to a rotary drier where exhaust gases from the diesel engine provides the energy to lower the moisture content of the wood chips to 20%. The moisture level is controlled by a recirculation of the drier gases which also keeps the oxygen level in the drum below 7%, thus reducing fire hazards. A cyclone recovers fine particles from the drier exhaust.

Biomass Gasification

From the surge bin, the chips are brought to the gasifier pressure by a plug-type feeder and then introduced into the fluidised bed by a screw conveyor directly into the gasification zone, thus converting almost entirely the biomass. The low BTU gas exiting the gasifier passes through two cyclones, the first one recycles char to the fluid bed and the second one removes ash from the gas.

Compared to fixed bed gasifiers, the pressurized fluid bed gasifier offers the following advantages:

. Because of its heat capacity and its agitation, the fluid bed maintains a uniform temperature permitting a rapid equilibrium of the complex thermo-chemical reactions and therefore a stable gas composition and a low tar content.

. The pressurized operation reduces the equipment size, thus reducing the investment cost, and allows the gas to be fed downstream of the diesel turbo-chargers, avoiding tar deposits on the heather.

The principal thermo-chemical reactions taking place in the gasifier are the following:

$$C + O_2 \rightleftharpoons CO_2 \qquad - \; 394 \text{ kJ/mole}$$
$$C + CO_2 \rightleftharpoons 2\,CO \qquad + \; 172 \text{ kJ/mole}$$
$$H_2O + CO \rightleftharpoons CO_2 + H_2 \qquad + \; 3 \text{ kJ/mole}$$
$$H_2O + C \rightleftharpoons CO + H_2 \qquad + \; 175 \text{ kJ/mole}$$
$$C + 2H_2 \rightleftharpoons CH_4 \qquad - \; 75 \text{ kJ/mole}$$

A gasifier of the above type having a capacity of 10 metric tons per hour and belonging to BIOSYN is in operation since 1985. Gasification of sawdust mixed with bark at a temperature of 1 412°F (767° C) has led to the following dry molar composition for the gas.

Table 2 - Molar gas composition, vol. %.

H_2	10.3
CO	14.5
CO_2	18.3
CH_4	6.0
C_2H_4	1.3
C_2H_6	0.8
N_2	48.2
Other	0,6
Total	100,0

The above gas has a lower heating value of 180 BTU/SCF (1.86 kWh/Nm^3) of dry gas. Tests on tropical wood from French Guiana performed by IREQ on its pilot gasifier indicate no significant deviation from the gas obtained at BIOSYN (1). The characteristics of the biomass gasification system offered by BIODEV International Inc. are the following:

. Operation under pressure (8 bars).
. High carbon conversion (up to 97%).
. Low tar production (up to 2%)
. Turndown ration of 2 to 1.
. Entirely automated operation.

Gas Clean-up and Cooling

The gas clean-up system developed by BIODEV has three basic functions:

. Removes solid and tar particles which are detrimental to the long term operation of the diesel engine.
. Cools the gas to the required temperature.
. Removes water droplets from the gas.

A first clean-up and cooling system has been built by BIODEV and is in operation since 1985 on the site of the BIOSYN plant. The principal characteristics of the system are the followings:

Table 3 - Characteristics of the BIODEV pilot plant.

Inlet gas flowrate	900 SCF/min	(26.4 Nm^3/min)
Inlet temperature	1 450 °F	(790 °C)
Outlet temperature	140 °F	(60 °C)

This system is currently being optimized in order to produce a gas which will meet the properties needed for a long-term operation of the ALSHTOM PIELSTICK diesel engine. Many data gathered on this experimental plant have been utilized for completion of the detailed engineering of the French Guiana Project. In the case of the Guiana Project, the gas leaving the secondary cyclone is cooled to 400°C by injection of water from the K-O Drum. This permits the use of standard piping instead of expensive refractory lined piping. A high velocity venturi scrubber then wets the gas to further reduce its temperature, condense heavy tar, and collect the particulate matter left by the secondary cyclone. The clean gas leaves the cyclonic separator saturated with water vapour and is piped to a fin fan which condenses some of the water and cools the low BTU gas to the desired temperature. The outlet of the fin fan is directed to a K-O Drum to separate the condensate from the gas and to remove droplets entrained by the gas stream. A cartridge filter serves as final protection before the diesel engine.

Diesel Generating Set and its Auxiliaries

The ALSTHOM PIELSTICK engines used is of the Dual fuel type designed and adjusted for burning low BTU gas. Ignition of the air gas mixture is made by injection of a small quantity of diesel oil instead of spark ignition. The advantages of ignition made by pilot injection can be summurized as follow:

. Ignition made by pilot injection leads to a better combustion of the lean gas and permit a good propagation of the flame.
. Pilot injection of diesel oil (about 7%) is small and remains independant of the load i.e. the quantity of burnt gas.
. In the event of a breakdown in the gas supply, the engine switches automatically to 100% diesel oil.

Coming from the gas clean-up and cooling system, the low BTU gas is directly brought under pressure and at a temperature of around 140°F to a carburator where it is mixed with the necessary quantity of air to provide the optimal combustion conditions of the mixture. The carburator is located downstream the turbo-chargers and the air coolers, avoiding tar deposit on these devices. The gas/air mixture is let into the cylinder via inlet valves. On contact with the air/gas mixture, that has been raised to a high temperature by compression, the liquid fuel ignites and causes the gas mixture to burn.

The ALSTHOM PIELSTICK engine utilized in French Guiana is a 18 PC 2 5 V 400 LCV type, meaning that this engine has 18 cylinders developping 400 kWe each on LCV (low calorific value) gas.

PERFORMANCE OF THE PLANT

The rated output of the generating set is 6,700 kWe at the generators terminal. At full load, the diesel engine consumes 17,650 kW, out of which 16,600 kW is due to the low BTU gas and 1,050 kW is due to the diesel oil pilot injection. The overall net efficiency of the plant is 25 % which is roughly twice that of a conventional boiler/steam turbine plant.

For the 6,7 MWe plant in French Guiana, the efficiency is calculated as follows:

Efficiency of the diesel/alternator	38 %
Efficiency of the gasifier	77 %
Consumption by the electrical auxiliary equipments	14 %
Net efficiency	25 %

One (1) ton of wood (925 kg) having a moisture content of 20% on a wet basis is required to generate 1 MWh of electricity. The net efficiency could be more than 80% if cogeneration of steam and electricity is considered. One interesting feature of diesel engine is that it may, in fact, operate an alternator which will produce electricity with the highest efficiency and at the same time it produces heat which could be recovered without affecting in any way the power and the electrical efficiency, unlike what happens with steam turbines.

ADVANTAGES OF GASIFIER/DIESEL PLANTS

There are many advantages of using a gasifier/diesel plant compared to a conventional boiler/steam turbine plant:

. The gasifier/diesel plant is more compact.
. Less maintenance and less supervision is required because of the lower pressures involved.
. The investment cost is lower.
. Operating costs are lower.
. The gasifier/diesel is more flexible in view of its "dual fuel" characteristic.
. The efficiency is greater by a factor of 2.

CONCLUSION

The construction of a biomass gasifier/diesel plant has several positive impacts on the local and regional economy in terms of:

. Job creation for the biomass harvest.
. Improvement in the balance of payment of oil importing countries.
. Increase in national productivity.

We should indeed remind that oil payments from oil importing countries represent 40% of their total export earnings. The total debt of these countries is now approaching 700 billions U.S. $. Many of these countries own large reserve of biomass. A biomass gasifier/diesel plant having a capacity of 1 MWe will save these countries around 1,300 tons/year of imported fuel. BIODEV International has been created to assist these countries in their effort to attain energy independence through the conversion of biomass into electricity.

REFERENCE CITED

(1) CHAMBERLAND A., LABRECQUE R., RICARD D., BENOIT C., Mise au point d'un gazogène à lit fluidifié et résultats de gazéification de bois IREQ, November 1985.

DISCUSSION

R. Gaire, Foster Wheeler	Have you considered the possibility of a gas turbine in place of a diesel powered generating unit downstream? I remember back about 4 years ago talking to people from Omnifuel. They were bidding several plants for gas turbine applications.
Drouin	We chose diesel engines based on the fact that our studies showed that large markets in the oil importing countries range in capacity from 800 kw to 10 MW. We found that gas turbines could be economically attractive for power outputs larger than 10 MW, while reciprocating engines are very good for power outputs ranging from 800 kw to around 10 MW.
J. Kerstetter, Washington State Energy Office	What are your estimated capital costs per kilowatt on that system? How do you treat the water that you use for cleaning?
Drouin	The costs varies a lot depending on the installed capacity. An objective figure would be around 2,500 U.S. dollars per kilowatt installed for a unit ranging from 4 to 8 MW. Below that, it could be higher. We are now designing a smaller plant. There will be significant simplifications for a smaller unit to reduce the investment costs. The wastewater treatment would be by biological devices, which are being designed by us.
B. Miller, Alternate Gas Inc.	Did you say that the cold gas efficiency was 77%?
Drouin	This is the design parameter for the French Guiana project. We chose that figure for design purposes.
Miller	You haven't measured it then? John Black said that the cold gas efficiency on the hot gas side was 80%, and you're only losing 3% in the cleanup?
Drouin	More than that; it's not cold.
Miller	What is the efficiency of the cleanup train?
Drouin	I cannot answer that question.

DISCUSSION, CONT.

C. Tewksbury, Burlington Electric Dept.	I'm operating a wood-fired power plant driving a steam turbine at 28% efficiency, and I was wondering if your claim has to do with the fact that you're running at less than 10 MW? Would the efficiency be higher for a larger size plant such as the one I run?
Drouin	Our claim is based on the fact that the Carnot Cycle efficiency in the diesel engine is higher than the Rankine Cycle. The diesel engine itself has an efficiency of about 38%, and based on that fact, we will end up with a higher efficiency than with a Rankine Cycle.
H. Kosstrin, Power Recovery Systems, Inc.	You mentioned that you have installed an 800-kW engine at Biosyn. Do have you operational experience with that yet?
Drouin	Yes, we began the operation of this experimental power plant in September 1985, and have operated about 100 hours to optimize various process controls which have been integrated into the detailed engineering of the French Guiana project. So we have experience with tar removal, tar formation, and wastewater treatment systems. I should mention that we have a technical advisory committee which includes people from the Hydro Quebec Gasification Research Center, faculty from the University of Sherbrooke who has considerable experience on tars from pyrolysis, and Biosyn and SNC engineers.
Kosstrin	Have you done any emission measurements that would be of interest in the United States, specifically nitric oxides out of the diesel and carbon monoxide?
Drouin	It will be done by Alsthom during future tests at St-Juste.
Kosstrin	Could you tell us what they are?
Drouin	These figures will be available before the end of 1986.

COMMERCIALIZATION OF ENERGY SYSTEMS USING BIOMASS RESIDUES IN ONTARIO: CASE HISTORIES

J.R. Herberman, M.E.S., and P.J. Carrie, B.A.
Technology Development Section
Ontario Ministry of Energy
Queen's Park, Ontario M7A 2B7

ABSTRACT

This paper presents a summary of experience with several waste wood combustion systems in Ontario. The systems are described with respect to technology, economic performance, and environmental emissions.

Ontario has had mixed experience with the technical and economic performance of the systems. Economics are often marginal, even for large commercial systems where economies of scale are present. Most of the systems in place are quite reliable. Where problems have been present, they usually are related to system utilization factors and/or feed handling systems.

Environmental emissions are generally within provincial guidelines. Analysis of emissions for some projects includes trace metals, PAH's, dioxins, furans and particulates.

COMMERCIALIZATION OF ENERGY SYSTEMS USING BIOMASS RESIDUES IN ONTARIO: CASE HISTORIES

1. OVERVIEW

Since the late 1970's the federal and provincial governments have funded a variety of programs to encourage the production of energy from biomass. Wood represents an important energy source in Ontario. Wood fuel is used by industries, institutions, farms and homes. The province has substantial forest resources, with processing mills currently burning or dumping over one million tonnes of wood residue per year. Wood currently displaces over $100 million (Canadian) worth of conventional fuels in the home heating market.

The use of wood waste by industry, particularly the forest products industry, currently displaces over $400 million (Canadian) worth of fuel annually. Wood and wood wastes could provide 4 percent (167 petajoules) of Ontario's total primary end use reqirements (4,174 petajoules) in the year 2000.

2. Programs To Encourage The Use Of Biomass In Ontario

The Canadian Federal Government initiated the Biomass Energy Development Program in late 1985. The objective of the program is to promote the development of technologies which can convert all types of biomass (including forest biomass, peat, municipal wastes and agricultural and food processing wastes) into energy, prepared fuels or chemicals that can significantly reduce traditional fossil fuel consumption patterns. The scope of the 1986/87 $3 million program extends from the preparation of biomass raw material to its transformation into a utilizable form. Cost sharing of above 25% by the proponent is strongly encouraged but is not mandatory. Unproven concepts or ideas will not normally be funded. This program operates in parallel with an accelerated capital cost allowance provision which allows taxable Canadian corporations a 3-year tax write-off (25%, 50%, 25%) on capital equipment designed to produce energy from non-fossil fuels, or recover energy from plant processes.

The Ontario Ministry of Energy supports wood energy development and utilization as part of the provincial government's committment to improve energy efficiency, increase energy self-reliance and recover resources from waste. The following projects are representative of the Ministry's activities in the Biomass area.

3. GRENVILLE - CHRISTIAN COLLEGE

3.1. Background/Rationale For Conversion

The Grenville-Christian College Woodburning Project represents the first modern large-scale (1460 KW; 5.25 GJ/hour) commercial/institutional sector installation of a woodburning system in

Ontario. The system was commissioned in the fall of 1982 and was monitored from that time until December 1984.

This demonstration project was initiated for the following reasons:

- to demonstrate the significant potential for wood energy in the commercial/institutional sector; and,
- to help overcome barriers to the expansion of markets in this sector. The primary barrier was the lack of knowledge of wood energy technology and the consequent perception that it represented a high risk investment.

3.2. System Description

The Anga Varme woodburning system is made up of three basic components; fuel receiving and storage; woodburning and hot water production; and hot water heating.

3.2.1. Fuel Receiving and Storage

Independent contractors gather wood fuel from the surrounding area mills and deliver it to the College in self-dumping trucks. This material is required to conform to specifications that limit particle size and moisture content.

The bunker holds sufficient fuel to run the plant for one week.

A staff member of the College checks the quality of the fuel being delivered. The samples are later analyzed to determine moisture content. Payments to the supplier are based on the weight of dry fuel.

The wood storage bunker is equipped with a "live bottom". From here a system of screw conveyors feeds the furnace.

3.2.2. Wood Burning and Hot Water Generation

The wood burning unit consists of an ignition cell, a combustion cell and a fire-tube hot water boiler.

Wood fuel, conveyed by the stoker screw conveyor, enters the combustion cell where it is exposed to hot gases and radiation from the furnace walls. A considerable amount of moisture is thus vapourized from the fuel before it enters the adjacent ignition cell. Once in the ignition cell the fuel is forced upward through an opening where it spreads over a number of sloping grates arranged in a circle. Underfire air is injected under the grates to assist combustion of the fuel and help cool the grates. Overfire air is introduced above the grates mixing with the gases as they depart to the combustion cell. The narrow passageway between the two cells causes the gases to accelerate to the combustion cell. The burning process is completed in the combustion cell. Here the velocity of the gases slows and causes large unburned particles to drop into the stoker screw conveyor which feeds them back to the ignition cell along with fresh fuel.

The spent combustion gases are induced through the boiler into a fly ash collector. The clean gas is then passed out through the chimney.

A system of screw conveyors automatically removes the ash from the ash pits and deposits it in an ash container outside the boiler house.

3.2.3. Hot Water Heating

The hot water generated by the boiler is pumped to College buildings through a series of pipes located in underground tunnels. In each of the buildings, unit heaters are used to distribute heat.

The boiler is also used to provide domestic hot water via a water to water heat exchanger.

The temperature of the water leaving the boiler is used to control the boiler output. The screw conveyors that feed fuel to the boilers start and stop at intervals which are controlled automatically by timers. When the building requires heat, the screw conveyors run more often.

3.3. Energy Performance

A detailed performance log was maintained from November 1982 to December 1984 and specific tests of efficiency were carried out. The wood burning system had an availability factor of 96% over the two year period. The thermal efficiency averaged 70% which is high in systems that burn wet fuel.

However, the system output was generally not higher than 1.5 GJ/hr, which is only about 28% of the rated capacity (5.25 GJ/hr).

3.4. Emissions Monitoring

Eight separate stack emission tests were conducted during the first year of operation. The tests involved a determination of the following parameters:

- stack gas particulate content and emission rate; and
- sulphur dioxide concentration and emission rate.

Particulate emissions ranged slightly above provincial standards of 100 x 10^{-6} g/m^3. Further study is required to reliably characterize average particulate emissions. SO_2 emissions from the system were measured at levels well below provincial standards at points of impingment near the stack.

3.5. Economics

The final installed cost of the system was $363,000. The measured savings over the two year period, in terms of oil displaced, was $151,000. Figure 1 (Graphics Display of Energy Costs and Savings) graphically illustrates the cost performance of the system.

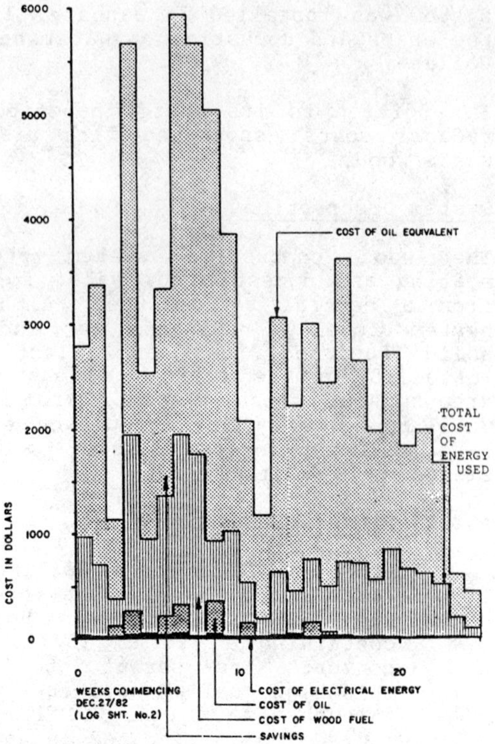

Figure 1 - GRAPHICS DISPLAY OF ENERGY COSTS AND SAVINGS

Savings would be much greater if the system was servicing a larger heat demand. Currently, only 28% of the 5.25 GJ/hour capacity is used under full load.

The system will pay for itself in nominal dollars in under 5 years. This is not attractive to industrial investors but could be attractive to commercial and institutional building operators. If the full capacity could be used the simple payback would drop to 1.4 years.

4. KINGSWAY COLLEGE

4.1. Background/Rationale For Conversion

Kingsway College, an educational institution founded in 1903 and located in Oshawa, Ontario has a large campus with 10 individual college buildings and twenty-five staff residences. The College operates two furniture plants located on campus. Yellow birch is utilized and the wastes have a moisture content in the range of 6-8% by weight. Approximately 2,100 tonnes of this waste material are generated each year. The high calorific value of the waste wood coupled with the low moisture content, make it an ideal fuel for a biomass combustion system. The system was installed in January, 1984 to supply the space and domestic hot water heating for the College.

In addition to its energy benefits, the system reduces costs associated with disposal of the waste wood.

4.2. System Description

The wood combustion system provides space heating and domestic hot water for the campus, from a central boilerhouse. Although the new system totally replaces the individual oil, solid fuelwood, gas and electrical systems installed in the various campus buildings and greenhouses, these systems were left in place until the new system was proven effective, and have remained intact as back-up equipment for emergency situations.

4.2.1. Process Description

From the College woodworking factory the scrap wood, planer shavings and sander dust are put through a hog/blower unit containing a 1.7 cm (3/4") screen which converts the scrap into small chips. This high quality wood fuel is then blown through a 43 m (141 ft) long pipe into a storage silo located adjacent to the factory. A "closed-loop" air system is employed which prevents the escape of dust-laden air into the atmosphere. Also, a spark detection system has been installed to protect the system from fire.

A unique feature of the system is the fuel storage/boilerhouse combination. The structure is a concrete silo 24.3 m (80 ft) high and 11 m (36 ft.) in diameter. Above the boilerhouse is 1,400 m^3 of storage space which can hold up to 350 metric tonnes of wood chip fuel. The fuel is extracted from the silo by a sweep auger unloader. Fuel can also be directed from the silo into trucks during periods of low fuel requirements (i.e. summer months).

A fuel transport fan moves the fuel from the fuel feed auger to the top of a 2.9 megawatt suspension-burning boiler.

4.2.2. <u>The Boiler System</u>

The fully automatic Mattpac boiler has a heating capacity of 10.55 gigajoules/hour (10 million BTU/hr).

The low pressure (100 kPa) 15 psi boiler is a fire tube type capable of 75-80% efficiency.

The conventional stoker has been replaced by a static spreading device situated in the "dropping tube". The fuel enters the boiler through the dropping tube to the top of the combustion chamber. The spreading unit effectively diffuses the transport air and directs the fuel evenly over the live bed area. By diffusing the air, direct impingement on the fire bed is eliminated and problems such as hot spots are prevented. Depending on the particle size of the fuel, much of it can be burned in suspension.

A grit arrester forms part of the boiler package. All heavy particulate from the grit arrester is returned for re-burning, with no ash build-up. This unique system requires no ash removal and heavy metallic grits are collected from the base of the grit arrester every 3-6 months. The boiler is also fitted with a standby natural gas burner controlled to automatically provide ignition should the wood supply be interrupted for any reason.

4.2.3. The Distribution System

The uneven terrain of the campus dictated that energy be distributed to the campus buildings by hot water rather than steam. A condenser transfers the heat from the steam produced in the boiler to the water distribution system. The hot water (110°C) is distributed through 2,000 m of foam-insulated reinforced epoxy pipe arranged in three separate loops. Although the system is equipped with two inter-connected pumps, water can be circulated using one dual speed pump. The pipe is buried to a minimum depth of 1 meter and laid on a bed of packed sand. Depending on the heat demand, the diameter of the pipe varies from 5 to 15 cm. Because the circulating system piping is plastic, minimal treatment of the water is required.

4.3. Energy Performance

During the 1985 operating year 1,889 tonnes of wood waste were consumed in the biomass energy system producing 28,500 gigajoules (27 billion BTU). An additional 413 tonnes were extracted from the silo during periods of excess supply and were sold on the open market. Tests using available dry fuel show that the boiler is operating at 78 to 80% thermal efficiency.

The fuel delivery system which transports fuel from the silo to the boiler was designed and engineered for hogged fuel, sawdust and planer shavings. Late in 1984, College Woodworking installed an abrasive sander which produces a very fine sawdust. When mixed with chips, this fine sawdust created a very heavy fuel that could not be handled properly by the silo extractor. It was therefore necessary to modify the delivery system by lengthening the agitator arms and installing more powerful motors. Repairs and modifications were made during January and February 1985. Consequently, operating costs (natural gas consumption, increased supplies and salaries) are high for 1985.

The monitoring system was set up for detailed computerized monitoring of the combustion and distribution systems. Special emphasis was placed on heat distribution to the respective buildings. After two years of operation, this monitoring system has not been able to supply credible data.

Consequently, the daily boilerhouse logs were the only dependable source of data pertaining to the system's operation.

4.4. Emissions Monitoring

The objective of the stack emission testing program was to measure the following parameters:

- average stack gas velocity and flow rate
- average stack gas pressure and temperature
- average stack gas moisture content
- stack gas particulate concentration and emission rate
- O_2, CO and CO_2 concentration of the stack gas
- sulphur dioxide (SO_2) concentration and emission rate
- total stack gas hydrogen and hydrocarbon content (THC)

All gas sampling and particulate emissions tests indicated levels below those defined as permissible under the Ontario Environmental Protection Act.

4.5. Economics

Capital and operating costs for the wood burning system are detailed below. The system has a payback of 12 years. While this is clearly not attractive to most investors, it was acceptable to the college. The college paid for 62% of the capital cost so the payback on their investment would be 7 years.

Capital Cost (Less Monitoring & Miscellaneous Items)	$1,056,190
Typical Annual Operating Costs:	51,500
Wood Biomass Fuel Cost:	11,346
Total Operating Costs:	62,846
Value of Natural Gas Displacement	150,735
Net Fuel Cost Savings	$ 87,889

Payback Period:

$$\frac{1,056,190}{87,889} = 12.0 \text{ years}$$

5. HILLCREST HIGH SCHOOL

5.1 Background/Rationale for Conversion

During 1982, the Ministry of Energy commissioned a study to establish the economic and technical viability of installing a wood-fired boiler system at Hillcrest High School in Thunder Bay, Ontario. The study concluded that the concept was viable and recommended the replacement of one of the three gas-fired boilers at the school with a wood-fired 585 kW boiler. This provided an opportunity to obtain information on system economics and atmospheric emissions. The emissions work done on this system was quite comprehensive and will be discussed in greater detail later.

The cost of the installation was shared between the Lakehead Board of Education, the Ministry, and the equipment supplier. The emissions study was done by an independent consultant.

5.2 System Description

The system components are fairly typical, consisting of an underground fuel storage bunker, a feed handling system, an Argusfyr boiler, and a stack with a cyclone. Hot water produced by the boiler is used for space heating during the winter.

5.2.1 Fuel storage

The system utilizes underground storage for the hog fuel. This reduces the requirement for heating the storage area in winter (to prevent the fuel from freezing and agglomerating) and provides easier unloading for the dumptrucks which deliver the fuel. The storage bunker has a capacity of 100 m^3, and has loading doors located flush with the ground adjacent to the school parking lot. Fuel is drawn into the boiler room using two large hydraulically-driven rakes.

5.2.2 Feed System

The feed system has proven to be the single most unreliable component of this installation. When using fuel with a high bark content, "bridging" often occurs around feed augers and/or control paddles. At various times, this has resulted in flooding the boiler room with hog fuel, or starving the boiler. As a result, the system requires intermittent operator supervision and it cannot be operated unattended.

Once fuel is drawn into the interior portion of the storage bin by the hydraulic rams, it is "fluffed up" with a rotating breaker bar. A screw conveyer located at the bottom of the storage bin brings the feed to the bottom of a two-stage inclined screw conveyor. The first stage takes the feed to the top of a downcomer shute, where the fuel passes through a star valve. The second stage takes the feed from the bottom of this downcomer directly into the furnace. The system is equipped with photocells and control paddles at various points to control the operation of the augers. Heat sensors in the last feed auger trigger a water quench to prevent fire from backing up through the conveyor system.

Aside from the problem of feed bridging, the feed system has required a lot of maintenance due to wear and tear on the feed augers. This is largely due to the presence of small chunks of wood which bind in the augers and cause premature wear.

5.2.3 Boiler

The boiler used for this system was supplied by Argusfyr. It uses 106 46 mm tubes with a total heat exchange area of 45.6 m^2. The heat exchange area is somewhat oversized for the application, resulting in good thermal efficiency for the system.

Two fans are used to provide combustion air. The primary (underfire) combustion air is preheated by exhaust gases from the stack using an air to air heat exchanger. A second fan draws secondary (overfire) combustion air directly from the boiler room. Water flow is controlled using an automatic valve.

5.2.4 Exhaust Stack and Cyclone Separator

A louvred control valve at the outlet of the boiler maintains a slightly negative pressure in the combustion chamber. After leaving the chamber, exhaust gases pass through an air-to-air heat exchanger to preheat primary combustion air, as mentioned above. Gases are then passed through a cyclone separator equipped with an induced draft (ID) fan. Fly ash collected in the cyclone drops through a star valve onto the ash conveyer from the furnace, and is then transported to an outdoor storage bin.

5.3 Energy Performance

Two types of fuel were tested on this system, sawdust and "hogged" bark fuel. The use of bark fuel was discontinued due to bridging problems in the feed system. A number of "Parr Bomb" tests indicated that the heating value for sawdust was 20,280 KJ/Kg. The value for bark fuel was slightly lower at 19,900 KJ/kg.

5.3.1 Energy Balance

1. 20% of the total heat input was required to evaporate moisture contained in the fuel (about 50% moisture content when received).

2. 16% of the heat input is lost through the stack.

3. The average thermal efficiency for the unit is 64.8%. It ranged from 57% to 71%.

4. The primary combustion air preheater removed only about 3% of the available heat from the stack gases. The use of this heat exchanger is therefore not economically justified.

5.3.2 Utilization factor

Based on the heat load of the building, the wood-fired boiler would be required for 6,300 hours per year. This assumes a 7 day per week, 24 hour per day operation. However, due to the problems mentioned earlier with the feed system and resultant supervision requirements, the boiler operated only 1600 hours during the 1984/85 heating season. Also, the average boiler output during operation was 485 kW, which is somewhat less than the 585 kW rating. This is due to the fact that air and feed flow rates were generally not optimal.

Thus, with respect to its potential utilization, this boiler operated at 21% capacity, as shown below:

$$\frac{\text{Actual utilization}}{\text{Potential utilization}} = \frac{1600 \text{ hrs} \times 485 \text{ kW}}{6300 \text{ hrs} \times 585 \text{ kW}} = 0.21\% \text{ capacity}$$

In absolute terms, it produced about 2793 GJ, out of a potential 13,300 GJ.

5.4 Emissions Monitoring

Extensive emissions monitoring and analysis was done on this system. This can be subdivided into 3 main areas: particulates, metals, and organic compounds.

5.4.1 Particulates

Particulate emissions averaged about 0.88 g/kg of dry wood feed for sawdust, and about 1.07 g/kg of dry wood feed for bark fuel. These numbers are at the low end of the scale for similar projects. The cyclone had an average collection efficiency of 55-70%, with an efficiency of about 95% for particles greater than 12×10^{-6} m in diameter.

5.4.2 Metals

Concentrations of lead, mercury, cadmium, arsenic, chromium and nickel were measured. Of these, only lead and nickel were found in significant quantities. However these levels are well below provincial guidelines.

5.4.3 Organics

A gas chromatograph (GC) and mass spectrometer were used to ascertain levels of poly-aromatic hydrocarbons (PAH's) and other contaminants. Dioxin levels were also analyzed. The results of these test are quite extensive and are not presented here for the sake of brevity. No PAH or dioxin compounds were present at levels which exceed any existing or likely standards or guidelines.

5.5 Economic Performance

The system installed at Hillcrest High School is not an economically attractive proposition for two major reasons:

1. There was a cost overrun on the initial installation due to insolvency on the part of the firm contracted to install the system.

2. The capacity factor is very low due to problems with the fuel feed system.

If the system were able to operate 24 hours per day during the heating season, it would save approximately $27,000 per year worth of natural gas. Since the system cost was $300,000, the simple payback would be about 11 years.

If a 12% discounted rate of return is considered the threshold for economic viability, then the following values would be required for major project variables:

1. wood cost less than $10.60/tonne
2. labour requirement less than one hour per operating day
3. economic lifespan of at least 27 years
4. natural gas cost of at least $0.21/m^3$
5. inflation rate of at least 7.3%

6. Chapleau Cogeneration Project

6.1 Project Background

Although electricity costs in Ontario are generally lower than those in the U.S., the Province recognizes that there are a number of benefits inherent in cogeneration of electricity and steam. Currently, there are 500 MW of cogeneration and parallel generation capacity in Ontario. An additional 500-1000 MW is expected to come on line by the year 2000. The Chapleau cogeneration project (which is being implemente this year) will provide 45 MWh/yr. o electricity and about 100 TJ/yr. of steam an hot water for the Chapleau Forest Products plan in Chapleau.

6.2 System Description

This system is under construction, and i expected to be operational in October, 1986 Three local mills currently produce abou 110,000 tonnes of wood waste per year. I addition to its energy benefits, the plant wil help to reduce costs associated with disposal o this waste.

The 19.41 MW (70,000 lb/hr) boiler will produc steam at about 6200 kPa (900 psi) and 400°C This will drive a 7.2 MW steam turbo-gen set delivering 6.4 MW to the electrical grid. Th generator unit produces 60-cycle power at 13,80 volts.

6.3 Potential Energy Savings

When operational, the system is expected t produce about 500 TJ (500 M lbs.) worth of stea per year, which will be used to produce over 4 million kWh of electricity. Of the residua steam, 20,000 lbs/hr (at 15 psig) will b provided to Chapleau Forest Products - mostl for space heating during the heating season The plant availability factor is expected to b about 85%.

6.4 Economics

The project has a projected capital cost o $10.6 million. After operating expenses, th earnings are $1,155,000 per year. The economic of the project have been made attractive to th private sector by a total contribution of $2. million from the federal and provincia governments. This provides the plant owner with a 15% Return on Investment.

7. Other Ontario Wood/Energy Projects

Wood is processed and used in a wide variety of way in projects throughout Ontario. Following is synopsis of some of these projects:

7.1 Hybrid poplar plantations are a potentially important future energy source. New varietie have short harvest cycles of two to ten years one hectare can produce more than 17 gree tonnes of wood a year. To develop hybrids wit higher energy content, the Ministry of Energ funds research in genetic breeding at th Ministry of Natural Resources Ontario Tre Improvement and Forest Biomass Institute i Maple, Ontario.

7.2 EcoWood Products has established a facility to gather and process waste wood to be sold as fuel. A supply infrastructure for urban wood residues and wastes is now operating successfully and can process up to 10,000 tonnes per year.

7.3 BioShell, a subsidiary of Shell International, turns sawmill residue into compressed wood pellets called Energex fuel pellets. Plants in Iroquois Falls and Hearst can produce 200,000 tonnes of pellets a year. Much of this is used by Abitibi-Price to fuel industrial boilers.

7.4 Kakabeka Timber Limited, a sawmill near Thunder Bay, uses sawdust and wood chips to fuel a drying kiln formerly fired by propane. The system saves about $70,000 a year in fuel costs, giving a simple payback of 2.5 years.

7.5 Foothill Greenhouses Ltd., in Ontario's Holland Marsh area, uses a sawdust-burning system to heat a greenhouse complex producing cucumbers. Since installation in 1982, the $180,000 system has reduced energy costs by about $40,000 to $45,000 a year.

7.6 David Smith, a farmer in Port Ryerse, has a small scale system which burns sawdust, corn cobs, peanut shells, and other agricultural residues to fire a 60-horsepower, restored locomotive boiler. The boiler provides heat to dry grain and heat a small greenhouse for growing tomatoes.

7.7 The Ajax District Heating project is an example of wood waste utilization to cogenerate steam and electricity. Bio-Char Inc. is currently converting the existing district heating system in Ajax to wood burning to dispace natural gas. The first phase of the project consists of installing two wood waste combustors; in the second phase, two more combustors and two steam turbo-generator sets will be installed.

7.8 Near Dorset, Ontario, the Leslie M. Frost Natural Resources Centre demonstrates three different wood-fired central heating systems in homes originally heated by oil. Monitoring has provided valuable data on relative performance and cost-effectiveness.

A summary of the projects reviewed in this paper is presented in Table 1, Representative Ontario Wood Energy Projects.

8.0 CONCLUSIONS

The adoption of biomass energy technologies in Ontario has been accelerating in recent years due to government incentives and increasing interest from the

Table 1 - REPRESENTATIVE ONTARIO WOOD ENERGY PROJECTS

PROJECT/START-UP DATE FEEDSTOCK/ENERGY PRODUCT	CAPACITY TONNES/YEAR	CAPACITY COST ($/tonne.yr)	CAPITAL COST $1,000's
EXISTING PROJECTS			
Shell Woodex/1980 Wood/Solid Fuel	225,000	20	4,500
Grenville College/1982 Wood/Hot Water	1,400	259	363
Kakabeka Timber/1982 Shavings and sawdust/Steam	1,100	163	180
Foothill Greenhouses/1982 Dry Wood/Steam	1,000	180	180
Hillcrest High School/1983 Hog fuel/Hot Water	1,700	176	300
Kingsway College/1984 Dry Wood/Steam	2,100	619	1,300
L.M. Frost Centre/1984 Wood/Hot Air	30	1,666	50
UNDER CONSTRUCTION			
Chapleau/1986 Wood/Steam+Elect.	108,000	92.	10,000
Ajax/estim. 1986 Wood/Steam (Stage 1)	20,000	103	2,075
UNDER STUDY			
Hearst Furnaces Wood Pellets/Hot air	40	1,500	60

private sector. These technologies will play an increasingly important role in satisfying the province's energy requirements, both for thermal energy and electricity.

Currently, system operator experience varies significantly concerning reliability and economic viability. Most reliability problems stem from feed handling equipment, rather than boiler packages or other system components. Economic viability is sensitive to conventional fuel costs, feedstock prices, maintenance costs, and the scale of the system. The effect of scale on system economics is fairly consistent for the projects summarized in this report. Graph 1. summarizes the relative capacity costs of these systems.

GRAPH #1

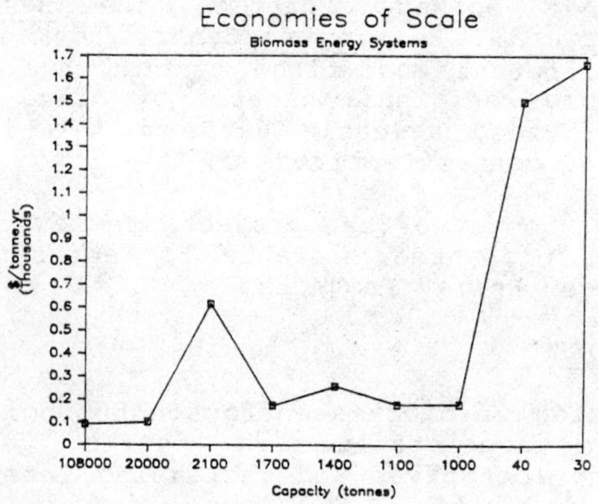

From an environmental emissions standpoint, wood-fired combustion systems are relatively benign with respect to dioxins, PAH's, other organics, and trace metals. The use of cyclones and proper stack heights should ensure the particulate levels are maintained within provincial guidelines.

9.0 Bibliography

Dick Consulting Services Ltd., "Monitoring and Evaluation of Demonstrating Wood Burning System at Grenville Christian College," Ontario Ministry of Energy, 1985.

Fels, N. et al., "Hillcrest High School Wood Burning System Monitoring Study," Ontario Ministry of Energy, 1986.

Jellicoe Resources Associates Ltd., "Kingsway College Monitoring Program," Ontario Ministry of Energy, 1986.

Ontario Ministry on Energy, "Fuelling Ontario's Future," Ontario Ministry of Energy, 1985.

DISCUSSION

J. Koning, Forest Products Laboratory: In light of your problems with the feed system on the school installation, and the knowledge gained from that work, what would you recommend for a feed system that would eliminate the problem?

Herberman: I'm afraid I can't answer that question. The project was managed by my colleague, Peter Carey, the co-author, and I'm afraid I'm just not that familiar with it. [Even with sawdust as a feed, continuous operation of the system without operator intervention cannot be achieved with the present feed arrangement. Extensive modifications to both the feed equipment and their controls are necessary. Some recommendations concerning the modification are as follows: 1) The auger into the furnace as well as the auger feeding it should be replaced with ones of a larger size. It is poor practice to have a small auger following a large one as pieces which can be handled by the large one cannot necessarily be handled by the smaller one. 2) Redundant controls should be used to indicate the state of the wood feed at various points and to control the hydraulic rakes and screw feeders. For example, the rate of wood feed is relatively constant; thus the screw feeders and the hydraulic rakes should be turned off and on at a relatively constant frequency. So if, for instance, the hydraulic rakes have not operated for, say, 15 minutes, some type of alarm would alert the operator that the boiler requires attention. 3) Photoelectric devices to indicate the level of solids should be avoided because of dusting problems; ultrasonic or radioactive sensors would probably be more appropriate. 4) Controls to enable the boiler to run at a reduced feed rate should be installed; this would also help the economics by allowing natural gas replacement even in low-demand situations such as early fall and late spring. 5) The provision of a homogeneous feed would go a long way toward simplifying the feeding and control requirements of the boiler. Towards this end, feed preparation devices (hammer mills, screening) should be investigated for applicability to Hillcrest.]

Zoellner — The main thrust of this program is commercialization?

Herberman — Yes.

Zoellner — Of these demonstration projects, are there some which seem to be more easily commercialized than others? Have you attempted to try to rank these projects that way? Some seem to be very easily accepted into the private sector and others are less acceptable.

Herberman — No, I wouldn't say that. I think that all of the systems that we have installed are equally attractive or unattractive to the private sector. Without grants or some sort of assistance in Ontario at the moment, I don't think any of these would have gone ahead, simply because we have very, very low conventional energy prices. As I mentioned, electricity is about 4¢/kWhr and in fact, the more you use in the industrial sector, the less you pay. So it's a disincentive as far as alternative energy systems are concerned. Technically, there are no problems; the problem is economics.

Zoellner — You're competing with Ontario Hydro.

Herberman — Yes.

DESIGN AND OPERATION OF A 50-MW WOOD-FUELED POWER PLANT

Cleyton Tewksbury, BSME
Production Engineer
Burlington Electric Department
585 Pine Street
Burlington, Vermont 05401

ABSTRACT

The Joseph C. McNeil Generating Station is the largest utility owned wood-fired electric generating plant in the world. This paper reviews the history of the project, discusses in detail the operational strengths and weaknesses as related to firing wood for fuel, outlines considerations for those who are planning a wood-fired power plant, and tabulates costs and operational data.

The concept of a wood-fired generating station was overwhelmingly approved in a 1978 bond vote by the citizens of the City of Burlington. A Certificate of Public Good was received from the Public Service Board of the State of Vermont to construct and operate the power plant in September 1981; construction began immediately. The McNeil Station declared itself available for commercial dispatch on 1 June 1984.

The process and equipment peculiar to the burning of wood are described along with some of the state-of-the-art strengths of the plant design. The bulk of the paper consists of a detailed examination of the major troubles encountered in the first 18 months of commercial operation. Based on these experiences, a short list of general suggestions is offered for the benefit of others who may be in the feasibility, design or even construction stage of a similar project.

DESIGN AND OPERATION OF A 50-MW WOOD-FUELED POWER PLANT

INTRODUCTION AND HISTORY

Burlington Electric Department

The Burlington Electric Department (BED) is a municipal electric utility serving the City of Burlington, Vermont. The City has a population that fluctuates between 35,000 and 45,000 persons, depending mostly on whether or not the University of Vermont is in session. Burlington is the largest city in the State of Vermont. It is located along the eastern shore of Lake Champlain about 45 miles south of the Canadian border.

The Electric Department employees around 180 persons in all aspects of the business including generation, distribution, conservation, finance, power planning, engineering, public relations and management. Along with owning and operating its own generation facilities, the Electric Department has entitlements from New York hydro, Vermont Yankee Nuclear Station and Merrimack Unit #2 (a coal plant in New Hampshire).

Need for Additional Generation

In 1975, in the wake of the Arab oil embargo, BED was faced with a shifting mix of power sources. Vermont Yankee had only been on-line for several years and was the newest addition to the Department's resources. However, contracts with the Power Authority of the State of New York (PASNY) for 150 MW for the State of Vermont (15 MW designated for BED) were scheduled for renewal by 1985 at unknown quantities and costs. BED's Moran Station, with three 10-MW coal fired units that came on-line in 1954, only had about ten useful years left. The City load, along with national averages, had been growing through the late sixties and early seventies at rates ranging from 7 to 14 percent annually with new winter peaks set almost every year. In 1975 the peak load was at 55.7 MW.

As for any electric utility, the problem of long-range power planning involves a blend of trend curve fitting and extrapolation, vision and guess work. Using what seemed to be a conservative 4% growth projection it was clear that additional capacity would be needed by 1985 in order to continue to provide some of the lowest consumer rates in New England with a reliable mix of sources. Dependability was a matter of pride with BED as the Moran Station was one of the few to have been kept on-line during the 1968 blackout.

In the period between 1976 and 1978 the Department wrestled with the choice between building its own new generation or investing in construction of someone else's plant. As a little bit of a hedge, it was decided to do both. The latter option was exercised by original involvement in the Millstone and Pilgrim Nuclear Power Stations. These investments were lost due to well publicised cost overruns and construction delays that led to the cancelling of one plant and the decision to not contribute more funds in order to retain a minimum share in the other.

Fuel Selection

Having decided to build new generation of its own to meet at least some of the projected needs, BED had to make choices concerning size and type. With the anticipated loss of PASNY hydro power as well as decommissioning of the Moran Station, it seemed that there could be a need for as much as 30 MW by 1990.

Because the typical economic size of a nuclear station is at least ten times as big as was needed, as well as the obvious difficulty of licensing and building such a plant by the late seventies, the nuclear option was easily eliminated. It seemed wise to keep as many paths open as possible so both new hydro and new thermal plants would be explored.

Studies and design were done for a 12.5 MW hydro station to be located on the Winooski River between the City of Burlington and the City of Winooski. An application for this facility was submitted to the Federal Energy Regulatory Commission (FERC) in 1980. FERC has still not acted on the application.

For the thermal plant, the primary questions reduced to where and what; that is, where to locate the plant and what to use for fuel. Size would be a function of fuel selection and projected needs. The remaining options were oil, gas, coal and wood. Oil and gas were easily eliminated because of their high prices. Of course, early in the Carter administration it was made illegal to build new power plants using oil or gas. The choice between coal and wood was a difficult one. Wood was the final selection because it could be obtained locally, keeping the large sums of fuel dollars in the local economy, and because it could be burned without having to deal with the sulfur emission problems inherent in coal. Even though a coal plant operates more efficiently and could be expected to dispatch at a lower cost per kilowatt hour, these factors were not considered sufficient to out weigh those stated above in favor of wood.

Although some problems could have been avoided by locating the proposed plant as close to its fuel source as possible, it was preferable to find a place within City limits so that the industrial tax benefits would remain in the City. Sizing the plant to be rated at 50 MW was primarily influenced by studies dealing with both the available fuel and with fuel handling and combustion limitations. BED only needed to retain a 50% ownership of the facility in order to control its operation, the balance would be sold to others.

In addition to feasibility studies, the Department converted Unit #1 of the Moran Station to burn wood in 1977 in order to develop fuel suppliers and gain experience with handling and burning the material. This pilot project was sufficiently successful and cost effective that Unit #3 was also converted to wood in 1980.

Milestones

The Electric Department went to its customers, the citizens and voters of the City of Burlington, with a Bond issue in March 1978 requesting authorization to finance a new wood-fired power plant. The issue was approved with a greater than 70% positive vote.

The architect/engineers Chas. T. Main of Boston were retained to provide design engineering and construction management services. C. T. Main was also used to prepare some of the environmental studies needed to obtain State licenses as well as a financial prospectus for the sale of municipal bonds.

Application was made to the Public Service Board of Vermont to construct the McNeil Generating Station in March 1980. The requisite Certificate of Public Good was issued by that body in September 1981.

Construction began the day after receipt of the Certificate with site preparation which had to be stopped for the winter of 1981-1982. Site work resumed by March 1982. By the end of 1982 and the coming of another winter the building was up and substantially enclosed in its permanent siding; this allowed construction to continue through the winter. In late April 1983 the boiler was hydrostatically tested to 1.5 times design pressure with no leaks on the first try. The generator was phased on-line for the first time on 11 February 1984. The Station was declared commercially available and subject to New England Power Exchange dispatch on 1 June 1984.

By the time the Station was declared commercially available, the 50% not owned by BED had been bought by the Vermont Public Power Supply Authority (20%), Central Vermont Public Service Corp. (19%), and Green Mountain Power Corp. (11%), all Vermont utilities.

DESIGN FEATURES

Conventional

The McNeil Generating Station is really a conventionally designed thermal power plant whose only distinguishing feature is that wood is the primary fuel. The Primary Components (see Table 1) are such as you would find in a similar size coal-fired station, again with the exception that they are sized and shaped for wood.

The plant can be broken down into operational systems such as the Steam Generator and Auxiliaries, Turbine/Generator and Auxiliaries, Feedwater/Condensate, Cooling Water, Draft, Air Pollution Control, Ash Handling, Fuel Handling, Well Water Treatment, Waste Water Treatment, Electrical Transmission and Miscellaneous Support.

The Station was designed as a cycling unit with an anticipated load factor of 70%; of course, it must also be capable of continuous, 24-hour, 100% duty. The water systems are all closed cycle to minimize consumption. Process water is obtained from four wells located about 4,000 feet north of the plant. Waste water is discharged after treatment to the Winooski River about 1,000 feet to the east.

There are three 34.5 KV, 3 phase, transmission lines that take the power away from the plant, both into the City and onto the grid for use by the co-owners or re-sale to NEPOOL.

Peculiar to Wood

Just about everything about the fuel cycle had to be especially designed to provide and handle material such as whole tree chips and saw mill residues including clean chips, saw dust and bark with moisture contents ranging from 35 to 55% by weight.

Forest to Boiler

The process begins with a staff of foresters who identify fuel sources, provide forest management services to land owners who sell stumpage, contract with independent loggers for chipping and delivery, and monitor, along with State regulators, all forest operations. Work conducted in Vermont must adhere to self-styled, State mandated, procedures structured to protect the environment and wildlife, improve forest quality, and be sensitive to visual impact on the public. In its ruling, the PSB stipulated that all locations from which fuel is to be procured in Vermont must receive pre-harvest approval from the Vermont Department of Fish and Wildlife certifying that adequate plans have been made to protect critical wildlife habitats and wetlands. In addition, all harvesting must conform to recognized silvicultural standards of tree stocking, species selection and harvest prescription. Strict standards of erosion control are enforced by BED foresters and clearcut sizes are limited to 25 acres or less except in cases of land use conversion.

TABLE 1

Primary Station Components

Equipment	Manufacturer and Description	Cost
Boiler	Zurn Industries: 2 Drum, water-wall Steam Generator with economizer, tubular air heater, travelling grate. Fire box size 22'X35'X74'; 35,946 square feet of convection heat transfer surface	$8,150,000
Turbine-Generator	Brown Boveri: Turbine: 36 reaction stages with 5 extractions. Generator: 60,000 KVA air cooled, 13.8 KV	$4,310,000
Air Quality	General Electric Environmental Services Cyclones: 24, 56" dia. Electrostatic Precipitator: 9 fields, 45,000 V	$5,500,000
Fuel Handling - Bucket Loaders	Caterpillar: 2, Model 980C, 15 yard buckets	$ 300,000
Conveyors	Fairfield Eng.: 200 TPH, 42" 3-ply rubber belt	$ 410,000
Pile Reclaimer and Boiler Feed Bin	Gauld Equipment: Hydraulic driven moving floors and 8 screw augers	$ 578,000
Railcars	Ortner Freight: 21 Bottom Dumping, 7,000 cubic feet	$1,060,000
Ash Handling - Conveyors	Dravo Wellman: Wet- and dry- drag, synchronized chain for continuous removal from 31 hoppers	$ 790,000
Conditioner	Taunton Eng.: Twin Shaft pug mill	$ 100,000
Cooling Tower	Research-Cottrell: 2-cell, mechanical draft, counter flow, wood-frame	$ 757,000
Stack	Luftek: 257 feet high, 10 feet inside diameter, pre-cast concrete with Corten liner	$ 631,000
Control Systems - Motor Control	Allen-Bradley: Live Backup PLC-3, 160K with I/O hardware	$ 350,470
Boiler and Balance of Plant	Moore Products: On-Line Backup CRT based with individual electronic controllers, data acquisition satellites and generator panel	$ 491,000
Engineering	C. T. Main, Boston: Environmental Studies,licensing testimony, design, construction management and bond issue preparation	$7,509,567

Once the wood has been chipped and loaded into a standard tractor/trailer it will be delivered either to the McNeil Station directly or to a railcar loading satellite yard located 40 miles north of the plant in Swanton, Vermont. The reason for the train deliveries is a condition imposed by the PSB again that 75% of the Station fuel be brought in by rail so as to limit truck traffic through cities and towns on the route to the plant itself.

The Swanton satellite yard is run by an independent contractor for a per ton handling fee. The yard has two parallel truck scales and dumpers along with a computerized system for storing and reporting on truck data. Several bucket loaders are used to move material from the truck dumper to storage and from storage to the train.

The 21-car train is loaded in the early morning and arrives at the plant site around 1130 hours. The PSB has further restricted fuel deliveries of any kind to Monday through Saturday between 0700 hours and 2100 hours. According to the contract with the Central Vermont Railroad, BED has five (5) hours in which to unload the train from the time it is spotted for the first car before a $1,000 "outlaw" penalty is incurred.

The McNeil Station has one truck scale and dumper with the same computer system for handling truck deliveries. Two bucket loaders are also used in the McNeil wood yard to move wood from the truck dumper or rail unloading trestle to storage and from storage to the reclaimer.

The pile reclaimer is constructed in three sections of hydraulically driven cross-bars. The bars are mounted on drive shafts which extend 39 feet into a designated reclaim pile. The bars are wedged-shaped so as to slide under the wood when outbound and pull the wood with them against the flat face when inbound. The bucket loaders maintain a pile of material on the reclaim bed by using a hopper made of the woodchips themselves. Material that is pulled into the reclaim building must pass through a rotating drum with 4 inch spikes welded on it, called a spike roll. The spike roll's function is to break up, if possible, ice chunks and catch foreign material such as branches, concrete blocks and plastic sheets.

The reclaimer/spike roll combination deposits the wood onto a conventional 42 inch rubber conveyor. At an elevation 54 feet higher than the wood yard grade the wood passes under an electromagnet to attract tramp iron. Falling by gravity from the head of the belt, the material passes through a disc screen to sort out proper sized material from the branches, stringy bark, lumber, large stones and other non-ferrous foreign objects. All the material that would not pass through the allowed space between rotating discs falls into one of two hammermill shredders. Eventually, everything is deposited onto a second conveyor for transport into the main boiler building.

Once inside, the fuel is deposited into a metering bin shaped something like an elongated pyramid. The bin extends the width of the boiler and is narrow at the top and largest at the bottom. The floor of the bin has a "live bottom" made from two sections of hydraulically driven cross-bars. In this case, the cross-bars are flat-faced on both sides so as to push wood in both directions and keep it from bridging.

The floor of the bin has eight slots in it to feed the wood into hydraulically driven screw augers. The augers can be controlled in automatic by the plant boiler master or manually biased on an individual basis by the operator. The augers are supposed to have a speed range from 0 to 20 rpm and feed up to 100 tons per hour of fuel to the boiler.

Wood falls from the end of the augers until it is entrained by air from the overfire air fan at 25 inches of water and blown into the fire box. The overfire air for this purpose is fed through rotating dampers so that the wood will continuously feed back and forth across the depth of the fire box similar to a lawn sprinkler.

The fire box itself is oversized, compared to a coal burning boiler, to handle the relatively low density and low Btu fuel. That is, more combustion area and more heat transfer surface is needed. The floor of the boiler is a traveling grate in two sections. This particular floor is one of the largest traveling grates in the world.

<u>Boiler to Atmosphere and Landfill</u>

Material such as sand, rocks, and the heaviest unburned char is deposited off the end of the rotating grate into a submerged drag conveyor. The conveyor de-waters the ash on an incline and drops the bottom ash into a bay accessible by a small bucket loader from the outside.

Combustion efficiency is aided by 100% reinjection from both the boiler and the tubular air heater hoppers. It was intended in the original design that the boiler hoppers, air heater hoppers and cyclone hoppers should all have sand classifiers for partial reinjection. While the former two went one way, the cyclones were modified the other way to 0% reinjection until a successful means for transporting the material closer to the boiler can be designed.

Essentially all the particulate that was not extracted by the cyclones is removed in the electrostatic precipitator (ESP) with a 99.97% efficiency at removing particulate. The four cyclone and 24 ESP hoppers all feed into a metering bin with a vibrating bottom. The bin supplies a constant flow of material to the pug mill conditioner used to mix the ash with water. The pug mill feeds the dust free ash into a 12 yard dump truck which transports the ash to an on-site landfill.

The main flue gas stream exits the ESP, is moved by the clean ID Fan and enters the atmosphere from the top of the 257 feet high chimney. A continuous emission monitoring system is located midway up the stack and is used for environmental compliance and combustion efficiency control. Because wood has negligible concentrations of sulfur, there is no permitted limit on sulfur emissions or control equipment installed, other than limiting the sulfur content of the No. 2 fuel oil used for supplemental firing.

OPERATION

Normal Operation

Staffing. The McNeil Station is staffed by operators, a maintenance crew, foresters, an engineer, and administrative and management personnel. The 25 operators are divided into 5 crews of 5: One shift supervisor responsible for actual operation of the total Station; one Station Operator who does most of the manual valving, floor checks and data recording; one Fireman whose primary task is ash removal although they occasionally have time to assist the Station Operator; and two Yardmen who keep the fuel handling system running from unloading trucks and trains to checking the conveyor belts. Four of the five crews are needed to provide 24-hour coverage; the fifth crew is used to cover vacations, unload the train and as additional maintenance personnel.

The maintenance crew typically consists of a chief mechanic, a mechanic first class, a welder, first and second class electricians, an instrument mechanic, a water treatment specialist, a storekeeper and a janitor.

There is one secretary, a maintenance supervisor, an assistant superintendent and a superintendent to look after the management of the plant. The one engineer provides technical support for the computer systems, environmental compliance and special projects as well as documentation updating.

Fuel procurement is taken care of by a chief forester and three assistant foresters. These people identify the fuel sources, contract with land owners and loggers, provide forest management services and attempt to match the supply to the demand.

The Running Facts. Table 2 lists the quantities of general interest in describing the actual operation of the Station at name plate level. These as-tested figures have turned out very close to design with the only notable exceptions being a slightly lower fuel consumption and a

TABLE 2

Descriptive Statistics for Full-Load Operation
(as measured by plant instrumentation)

Parameter	Quantity	Units
Generator Gross Output	56	MWe
Generator Voltage	13,800	V
Net Station Electric Production	50	MWe
Boiler Pressure	1,300	psig
Steam Temperature	950	degrees F
Steam Flow	500,000	lbs/hour
Fuel Consumption (Wood)	80	tons/hour
Fuel (Wood) Moisture Content (by weight)	50	percent
Ash Production	2	tons/hour
Circulating Water Flow	42,000	gpm
Make-up Water Flow	600	gpm
Waste Water Flow	220	gpm
Condenser Vacuum	2.5	" Hg abs.
Air Flow	714,420	lbs/hour
Air Temperature (Undergrate)	450	degrees F
Boiler Draft	- 0.15	inches H_2O
Gas Flow	270,000	ACFM
Gas Temperature, Boiler Outlet	800	degrees F
Oxygen at Boiler Outlet	3	percent
Cyclone Differential Pressure	3.5	inches H_2O
Stack Gas Temperature	295	degrees F
Carbon Dioxide in Stack	13	percent
Carbon Monoxide in Stack	350	ppm
Particulate (independent test results)	0.0007	gr/DSCF
NOx (independent test results)	0.157	lbs/MBtu

slightly higher circulating water flow rate than originally projected.

The McNeil Station was built for cycling service during its early years with an anticipated load factor of 70% and an annual fuel demand of 500,000 green tons. Table 3 shows the actual results for the first full calendar year of operation, 1985. Table 4 delineates the main costs in building and operating the McNeil Station.

<u>The Control System</u>. Although the control system is not peculiar to the wood-firing feature of the Station, it is one of the earliest state-of-the-art applications of computer technology to power plant operation outside of nuclear stations to be used. It is to be noted, however, that the "state-of-the-art" was in 1982. The available applied technology today has changed dramatically, making the McNeil Station's system almost archaic except by comparison to a traditional 1950's stand-up pneumatic panel.

The distinctive components of the system are the Programmable Logic Controllers (PLC) and the CRT-based (cathode ray tube display) boiler controls. The PLC is used primarily for motor control but also provides a simple path for communicating control data from the field to the control room. The PLC network was and is invaluable in simplifying troubleshooting during startup and day to day maintenance. The PLC has also allowed flexibility in changing the control scheme. Changes that would take almost

TABLE 3

Operating History 1985

Net Electrical Production	280,137,900	KWh
Total Wood-Fuel Consumption	435,060.7	tons
Load Factor	69.8	%
Equivalent Availability Factor	76.5	%
Number of Starts	56	
Turbine Operating Hours	6,542	hours
Total Overtime Man-hours (including safety meetings, training, sick leave coverage and emergency maintenance)	7,539.5	hours

TABLE 4

Costs

Total Project (exclusive of finance charges)	$ 67,000,000	
Average Fuel Cost (by truck), Delivered	$ 20	per ton
Train Delivery (freight only)	$ 4,100	per train
Train Loading Handling	$ 1.75	per ton
Fiscal Year Fuel Budget	$ 12,973,000	
Fiscal Year Operation and Maintenance Budget	$ 1,910,600	
Contribution to the City of Burlington in lieu of taxes	$ 1,474,500	

prohibitive amounts of time and expenses to implement with a hard wired system can be programmed in an hour or so and tested on-line. The McNeil Station is equipped with "live" backup on the central computer modules so that there will be a bumpless transfer of control to a "backup" should the primary processor fault.

The CRT-based boiler controls presently provide the Shift Supervisor with over 350 data points in both analog and discrete displays. With the push of a button or two a single operator can scan almost all the critical pressures, levels, temperatures, flows, motor conditions and conveyor belt status that would take two floor operators an hour to check. Twin CRTs and computer components act as on-line backup as well as providing dual controls. Any of the information available to the CRT system can be trended on chart recorder or floppy disk; alarms can be programmed, changed and enabled/disabled with all the alarm conditions and system faults being logged on a line-printer.

The combination of the two control systems provides a level of automatic operation which eliminates the need for one operator position. The event logging features provide the best way to analyze the cause of operational interruptions by establishing documented sequences and tabulating for later review the multitude of conditions that would be reported by a person in summary with the words: "Everything seemed to happen at once." Of course, the systems also provide critical equipment protection features.

When Things Go Wrong

Fuel Management. Perhaps the most visible problem encountered in operation of the world's largest utility owned wood-fired electric generating station is the procurement and storage of the fuel. There have been some hard lessons learned about basing fuel acquisitions on projected loads and the volatile behavior of large outside storage piles.

The McNeil Station received a great deal of publicity in early Fall of 1985 both locally and nationally because of citizen complaints about noise, dust and bad smells coming from the power plant. The Station has near neighbors located about 120 feet above the ground level at the plant, nearly level with the roof of the main boiler house, on a bluff about 300 yards south. Because of their elevation and prevailing winds out of the north, these neighbors are directly impacted by plant activities. There was a time by late September, though, that just about the entire City of Burlington, most of which, including downtown, is south of the plant, was complaining about at least the odor.

The assigned source of the odor was a 110,000 ton wood chip pile that had material over a year old in it, reached a height of some 80 feet and had been constructed by bucket loader. This pile came into existence because of fuel purchases based on load and dispatch projections for the summer 1985 that never materialized. In other words, sufficient fuel was contracted and delivered to operate the Station at 80% capacity from June through August; the actual load factor for that period was just under 40%.

In early September smoke was observed coming out the west face of the pile. It was decided to cut the pile open, root out all the smoldering material and extinguish it. The top 20 feet of the pile was pushed off and transported to a vacant lot several hundred feet from the main pile. The south end was cleared by placing the Station in "must run", operating at full-load 24 hours a day and consuming the material. This left room to dig out the smoldering volume in the west central region of the pile. Bucket loaders, blade dozers and a back hoe were used to isolate the smoldering section from the remainder of the pile, open it up, spread it out and hose it down.

In the process of controlling the hot wood and exposing large internal areas of the pile, not only smoke was released but other constituents distinguished by their smell. The single largest component was found by gas chromatography and spectroscopy to be furfural, a hydrocarbon with a strong odor. Personnel working on the pile itself wore organic filter masks much of the time; for the City, the odor became a hot political issue.

Once the hot material was fully consumed further smelly emissions were contained by covering the remaining 57,000 ton pile with large sheets of 6 mil plastic. The Station was operated using newer fuel from the rail loading satellite yard until the colder temperatures of winter were present to limit the ability of the furfural to remain evaporated and air born. By early February 1986 all of the "old" wood pile had been consumed in the generation of electricity.

The long term consequence of the events just related was development of a detailed Pile Management Plan. The Plan provides directives for determining the quantities of fuel to be purchased, the schedule for handling the material, explicit limits to future stock pile sizes and ages, and contingency instructions in the event that wood remains in storage beyond the time limits. The Plan spreads responsibility for implementation among the foresters and plant management.

Although there is some research data available on the behavior of woodchips, bark and mill residues in outdoor storage piles, very little of it seemed applicable to the particular circumstances at the McNeil Station. Most study piles were 25 to 30 feet high, conically shaped, conveyor and gravity feed and were monitored for moisture, temperature, fiber and heat value. The McNeil Station's pile was compacted by many bucket loader passes, higher and wider than test piles, and unmonitored as to even how long material had been in storage. The new Plan should also serve to limit material loss due to biological and chemical deterioration.

<u>Fuel Handling</u>. Every aspect of transporting the wood, from railcar unloading to the boiler grate has been a problem at one time or another. Wood has not only been harder to move but much tougher on many equipment components than designed for.

<u>Railcar Unloading</u>. Even under the best of conditions, getting the woodchips to come out of the specially designed "rapid discharge" rail cars involved more than just opening the bay doors and letting it fall out. This lesson became apparent upon arrival of the very first train: it was a spring morning, the sun was shining, it had not precipitated lately, temperatures were well above freezing, the wood had been placed in the railcars in the late afternoon the day before, there was a large crew of yardmen ready to work, managers and TV crews were standing by; after a short delay to determine the manner in which the bottom doors functioned, the six sets of double doors on the first car were opened and nothing happened.

From that day until almost a year and a half later, there was a struggle to unload woodchip trains. After digging the first few cars out by hand over several days, different types of vibrators were experimented with to excite the material into motion. The forced vibration schemes ranged from single car, top mounted monsters to small, multiple point portable units. For just over a year three pneumatically powered eccentric weight vibrators were mounted and removed from each car in turn by hand using two 1 inch square head bolts. In the winter the air lines were constantly freezing up from water. Personnel were experiencing a rash of back, arm, neck and leg injuries. By the end of the year just about every weld on the side of the car where the vibrators were mounted had had to be redone; the steel lip on which the vibrators were bolted was breaking off in large chunks due to stress and fatigue.

The solution in place today consists of one permanently mounted 10 hp eccentric weight electric vibrator on each car at about the mid-point in the center beam. The center beam was doubly cross-braced with welded I-beams to stiffen the section with the vibrator and to better transmit the displacements to the sidewalls of the car. Under good conditions, all the operator has to do is plug in the vibrator and push the start button, after opening the doors, of course.

The outstanding problems with this stage of fuel handling include coping with material that has frozen to the car walls and is layered with snow, and limiting the impact on the neighbors of noise and dust generated during unloading operations. A solution to these problems in the form of a full-length trestle enclosure is being investigated.

<u>Yard Movement</u>. Several different types of loaders and dozers have been tried in an attempt to find the fastest and most reliable means for transporting material around the storage yard. The yard has a floor area of roughly 205,000 square feet or 4.7 acres. The straight line distance from the reclaimer to the farthest corner is 600 feet. Allowing for access roads and clearance between piles the actual distance a conveyance machine has to travel could be over 1,000 feet one way. The material is spongy to drive on yet abrasive. There are wood fines present that plug up cooling systems.

The final solution to moving material around the yard is simply two large bucket loaders with 15 yard buckets. These machines are maintenance intensive because of the heavy service and constant operation, but are generally reliable. One machine is sufficient to keep the reclaimer stoked, but could never keep up with simultaneous train and truck deliveries in addition to reclaim activities.

<u>Reclaimers</u>. The crossbars described earlier in this paper slide over what was originally 1/4 inch A.R. plate.

Inside one year and only 2,500 hours of operation the wear plate was being peeled off by the crossbars and looked like foil. Some of the crossbars had also been broken off. These parts would be found wrapped around the spike roll, plugging the first conveyor feed chute and clinging to the magnet. There were originally three 8 inch hydraulic cylinders on each section fed by a common header from the pump. The middle cylinder was removed during early tests of the system in hopes that the remaining pistons would drive the crossbars faster. There was some gain so the middle piston was left out, this also appeared to help alleviate structural stresses. It was discovered later that the hydraulic pump was undersized to handle the set of three pistons anyway. Running with just two pistons the floor side rails, which are constructed of 20 inch channel iron, were cracking and splitting near the fixed end of the pistons. This causes the whole floor to jump when the piston pushes.

The solution was to install new 1/2 inch A. R. plate for the wearing surface on the floor. Crossbars are still being replaced regularly. The side rails that had cracked were rewelded and reinforced.

The three sections of the reclaim floor were intended to move in sequence and maintain a steady feed onto the first conveyor. The sequencing was controlled by limit switches in communication with the PLC that controlled directional solenoids on the hydraulics. The problem is that the three different hydraulic drives and floor sections all move at different speeds. One section will have to sit and wait for another before it will begin its next cycle. It has also been observed that one section will pull in more material than another. The result is a very uneven feed onto the conveyor and almost sinusoidal loading of the belts. This problem is still being worked on.

Metering Bin and Augers. The floor of the bin showed exactly the same pattern as the reclaim floor. The wear plate wore very thin in a short time and was found wrapped around the augers, stopping the flow of fuel to the boiler. There is also a problem with uneven distribution from the hydraulic header to the drive pistons, causing a twisting of the rods.

The augers, at about 5,000 hours, totally wore through their troughs and stripped the flights from their shafts. The wood exhibited properties that had both erosive and corrosive effects on the wearing surfaces. The pH of the wood from the "old" material in the main pile was lab analyzed as low as 2.8 at one time. The welds that held the flights to the shaft were eaten away.

The floor of the bin was replaced with the thicker plate as was the reclaim floor. The auger troughs were replaced with locally rolled round bottom steel instead of the original equipment manufacturer's half-hex-sided trough. Augers was replaced with various designs including thicker carbon steal, hard facing, stainless steel and flights seam welded on both sides to 10 inch Schedule 80 pipe.

The hydraulic auger drives would not start an auger that had been stopped with wood on it, the gear boxes and couplings were breaking constantly (one or more a week), hydraulic hoses would break and spray 75 gallons of fluid all over the plant before a reservoir low level switch shut off the pump, and the combined drive system was not able to reach design speeds of 20 rpm.

Recent experimentation has shown that the pumps, motors, gear boxes, couplings and pressure relief valves were not properly sized together to match design conditions. The electric motors are adequate but the pumps are in the process of being changed out at the time of this writing. The second set of larger hydraulic motors that was tried is now adequate but all the gear boxes will have to be changed to meet the torque requirements. The new combination of pump, hydraulic motor and larger gear box has been tested on one out of the eight augers and found to run smooth, start from a dead stop after several hours of sitting with a full bin on top of it, and reach 22 rpm at full-speed.

Low speed tachometers which will display individual auger speeds on the Control Room CRT were added in January 1986. These devices will enable the Shift Supervisor to better monitor auger operation, put in zero speed alarms, balance fuel feed across the width of the boiler and possibly even provide a cross-check as to how much fuel has been fed volumetrically.

Boiler Grates. The largest section traveling grate in the world had some installation problems that caused them to hang up periodically since the Station went commercial; in fact, the west grate was the cause of five unscheduled outages in 18 months. It was not until an unscheduled outage was taken just before Christmas 1985 and the west grate section completely dismantled that the extent of the problem was realized.

With the assistance of the manufacturers service engineer, the Station's own personnel worked 20 hours a day for a week to tear down and rebuild the grate. The drive shaft that had been twisted as much as 30 degrees was replaced. New chains were installed. Many new lateral support bars were installed. New drive shaft bearings were installed and properly aligned this time. New sprockets were installed. The sifting ash hopper that is immediately underneath the grate had to be cut free on one side and pulled out from where it was rubbing. The total repair for this in-house rebuild, including lost production, was around $100,000.

Ash Handling and Reinjection. The original design and installation of ash handling and reinjection included sand classifiers for sifting out reinjectable char from ash fines on the four boiler hoppers, the five air heater hoppers and the four cyclone hoppers. One of the early and ongoing problems was air leakage through the sand classifiers. Once fresh air reached the ash, the ash ignited and there was a constant fire in the transport tubes and rotary valves. The heat from the fires would burn through the flexible expansion joints on both sides of the sand classifiers and warp the screens inside.

The worst situation results when a rotary seal valve also leaks air because then the fire burns in the throat of the hopper and forms "clinkers", solid chunks of melted ash. The clinker plugs the throat and the hopper fills with ash. A hopper full of ash is one of the most dangerous conditions to have. There was one major accident in which a man was seriously burned and could not return to work for about six months.

The solutions to these problems were multiple: additional safety precautions were taken including properly mounting safety chains on hopper doors, better instrumentation for level detection, personnel training and publishing written procedures, and improved access around hopper doors.

All of the sand classifiers have been removed. The Station is now running with 100% reinjection from the boiler and air heater hoppers, and no reinjection from the cyclone hoppers. All the rotary seal valves were checked for clearances so as to eliminate air leakage. Since these changes there has only been one incident of "clinkering" in the cyclone hoppers which was cleared on-line by handy operators with poking rods.

On the subject of poking, most of the hoppers did not come with "poke holes"; i.e. a 2 inch angled port with a screw cap allowing access and visibility of critical flow areas, such as the hopper throat and the rotary seal valve inlet. These ports are virtually the only way to confirm the flow of material and clear plugged chutes. Provision for proper operator access and measurement should be made in the original design. The poke holes were added by Station personnel.

After the ash is collected from the cyclone and precipitator hoppers it needs to be contained, transported and disposed of. When the McNeil Station first came on-line the containment and transporting were accomplished with the use of rented concrete trucks. The unfortunate aspect about such machines is that there is big hole in the back which allows all the air born particulate to come blowing out as the drum fills. It did not take long for all the blowing ash to infiltrate the main boiler house and fill everything inside with ash dust, from water filters to electronics.

By 1 January 1985 an ash conditioning system had been installed to perform the function of mixing the ash with water for dust control. The transportation phase saw a single dump truck replace the two $6,000 a month rented cement trucks.

This only leaves the problem of disposal. From the time the plant first fired on wood until the present all the ash has been placed into an on-site landfill. The on-site landfill has a useful capacity sufficient for another two to three years, but this is not long in comparison with the expected life of the Station. Alternatives for disposal include expanding the on-site landfill area, reducing the volume of ash to be disposed of by additional reinjection of combustibles, marketing the ash as an agricultural supplement, and off-site landfilling. The final, long term, solution in this area has yet to be determined but would be likely to consist of a combination of all of the above possibilities.

ENVIRONMENTAL IMPACT

General. A prerequisite to applying to the State Public Service Board is obtaining all applicable permits and licenses from individual State and local regulatory agencies. Of the multitude of such permits that had to be obtained, two are worthy of particular attention because of the time and effort involved not only to apply for them, but to comply with them.

Water Quality. The Waste Water Discharge Permit application required four independent studies projecting the environmental impact of plant discharge into the nearby Winooski River. Along with the consultant studies and time for State Agency of Environmental Conservation review, it took several years to be granted a permit: much longer than was planned or hoped for.

The final provisions of the permit were very stringent as can be seen by the following comparisons: 1) Waste water to be discharged to the River had to meet specifications which were tighter than Federal and State standards for drinking water; the Winooski River is a Class C body, i.e. not even considered safe to swim in (Class A water is for drinking.) 2) Plant waste water had to be an order of magnitude or more cleaner than the ground water from which it was originally taken.

Of course, the control of water quality is always an important function in a power plant in order to protect the miles of pipe and system components. The first stage of treating the well water used for plant make-up and service water is passage through a decarbonator and then through a four-cell, layered gravity filter. The iron content of the ground water increased dramatically above pre-construction test samples, probably because of the transport of the element through the soil surrounding the wells as pump flows ran steadily in the range of 100 to 500 gpm at each well head.

The large iron concentrations resulted in rapid fouling of the decarbonator fill and the requirement for more frequent filter backwashes with longer backwash times, meaning more backwash water, than was ever planned for. The Station Water Treatment specialist devised the means of cleaning the decarbonator fill by removing it from the decarbonator and placing it in the Mixing Basin where all plant waste streams meet before being placed in open air lagoons for cooling and settlement. The mixing basin is equipped with a turbine mixer and the ability to inject chemicals. The combination of agitation and chemical treatment proved effective in scouring the iron build-up off of about half the fill at a time over a 24-hour period. The iron contaminated water was then drained into the lagoons where the ore could settle out.

Perhaps the most difficult area of waste water permit compliance has been dealing with backwash water. Not only have the volumes been excessive relative to permit limits but they also exceed the capacity of the backwash water collection tank, the tank's sump pump and the small underground pipeline installed between the tank and the mixing basin. The problems with moving the backwash water affected the quality of water produced by the filters and, hence, the condition of the cooling and service water systems.

Final solutions to these problems are not fully in place but include development of a means and procedure for air bubbling the filters, a larger flow/larger head backwash tank discharge pump, addition of a turbidity meter for operators to use during backwashing to reduce total water consumption, raising cooling water pH into a range of 7.5 to 8.0, programming a dilute acid feed pump interlock into the computer system, and application for an amended permit allowing increased total discharges.

<u>Air Quality</u>. The terms of the McNeil Station's Air Quality Control Permit are certainly more peculiar to the fact that it is a wood-fired plant than the Water Quality Permit is. The main conditions of that permit were originally 1) "the particulate matter emissions rate shall not exceed 0.007 grains/DSCF", 2) "the heat input . . . by fossil fuel" (oil) "shall at no time be equal to or exceed 250 MBtu/hour", and 3) the Station shall be equipped with in-situ carbon monoxide and opacity monitors. Two years after the permit was issued by the State of Vermont, it was amended by a court order resulting from a suit brought by the City of Winooski, the immediate neighbor of the City of Burlington to the east, and the Sierra Club against the State of Vermont. The court order placed the following

additional constraints: 4) the Station shall be equipped with an in-situ nitric oxide monitor, 5) the nitric oxide emissions attributable to wood shall not exceed 0.2 lbs/MBtu "as computed on an eight-hour average for actual operating hours of wood-burning based on continuously rolling eight-hour averaging periods", and 6) the "Station shall not emit more than 239.4 tons per year of nitrogen oxide from wood."

The requirement for continuous eight-hour rolling averages in determining NO_x concentrations resulted in the need to also install a CO_2 monitor and associated electronics to continuously combine the signals according to the EPA F-factor method to record pounds per million Btu.

Quality assurance requirements for maintenance and operation of the stack gas monitor in addition to reporting tasks takes four to five hours a week, on average, of dedicated engineering time. The monitor has to be tested for relative accuracy annually. Reports must be filed quarterly listing not only excess emissions but all full-scale monitor readings and maintenance activities.

The State is also very concerned about quality assurance procedures and reporting regarding the Station's primary air quality control device, the electrostatic precipitator. Since the first lighting of a wood fire the ESP has performed exceptionally. Any problems that were encountered in the unit were eventually determined to be the result of problems elsewhere in the plant, such as inefficient boiler operation causing excessive carry-over, improperly gauged rotary seal valves and other ash handling problems that cause hopper plugging. There has been some internal damage in the ESP, but only 27 out of 6360 wires have been removed to date.

The effectiveness of the ESP is proven by the compliance test results. Even operating the ESP at half of its design power the particulate emissions were measured at 0.0007 grains/DSCF, ten times less than the permit limit. NO_x concentrations required a full twelve hours of testing to obtain sufficient data for calculating eight-hour rolling averages, but the result was an emission rate of 0.154 lbs/MBtu, comfortably below the 0.2 limit.

Because wood has a nearly immeasurable percentage of sulfur, the McNeil Station does not generate sulfur oxides and is not regulated relative to the same.

Almost the only opacity violations reported to date have been due to firing the boiler on oil for cold starts. Placing an oil gun in service almost invariably causes an opacity spike because of the improper fuel and atomizing media ratio at ignition.

CONCLUSIONS

The world energy scenario in early 1986 is somewhat different than was projected in 1978: the price of oil has just recently taken a dive, although, at the time of this writing, it has yet to be reflected in consumer prices; the nuclear power industry in the United States has not seen any new orders since 1976 and many of the plants ordered between 1970 and 1976 were either abandoned or cancelled, the few that are being completed have suffered tremendous cost overruns; even the coal industry is suffering from the weight of regulation on sulfur oxide emissions and concern about "acid rain"; finally, national, regional and local electric load growths have been on the low side of the range used to predicate the need for the McNeil Station, that is, 2% or less instead of 4%.

Considering this fluctuating milieu it is amazing that actual operating times and costs have come so close to design for the McNeil Station. It was anticipated that the McNeil Station would run at a 70% load factor, the 1985 figure was 64% (Table 3). There was $80,000,000 budgeted to build the facility, it was completed for $67,000,000 (Table 4). It was projected that the plant would dispatch at a price of 39.5 mills/KWh, at the time of this writing the rate is 39.2 mills/KWh.

It was a good time to build, as evidenced by the construction savings. The power is being used by the Station's owners and in sales to the New England Power Exchange at levels very close to those predicted. The environmental compliance exceeds all permit standards. Fuel dollars are staying in the local economy to a large degree. The Station contributes significantly to the tax base of the City of Burlington and the State of Vermont. The forest lands managed to supply fuel for the McNeil Station have been improved for wildlife species diversity and tree health. A market for formerly useless cull and waste material has been created thus improving the cost of land management programs.

As the largest utility owned, wood-fired electric power generating plant in the world, the McNeil Station has been a proving ground for equipment that is needed to operate future facilities with renewable biomass resources. Some of the key points that can be made given the experience detailed in this paper, including ideas that were incorporated with foresight into the McNeil Project and ideas of the "if we had it to do over again" nature, are:

1. Design fuel handling equipment for extremely corrosive and erosive environments. Consider the use of ceramic and/or polymer liners on all wear surfaces.

2. Design on-site fuel processing capabilities so as to allow for quick backup of necessary components and to provide the most uniform material to the boiler.

3. Design the drive and control mechanisms for heavy duty continuous service equipment in the fuel/ash cycle so that they will exceed expected full load capacities and provide the sequencing required to assure consistency of material handling.

4. Listen to operators and maintenance personnel during design, its easier than making field adjustments. Staff early and train well.

5. Remember that solid materials, such as wood and ash do not spontaneously flow. Consider the meteorological conditions which will impact the movement of material. Consider the alternatives for access to all fuel and ash passageways by personnel, avoid inaccessible areas.

ACKNOWLEDGMENTS

This paper was prepared with the help of Mr. Thomas R. Carr, Production Manager; Mr. John M. Irving, Plant Superintendent; and Mr. William Kropelin, Chief Forester.

The author also thanks Mr. James Lauzon, Manager of Engineering and Production, and the Burlington Electric Department for supporting the preparation and presentation of this paper.

DISCUSSION

D. Elliott, Battelle Northwest Laboratory

Have you had any experiences with spontaneous combustion in your wood chip pile?

Tewksbury

That was one of the events we had to deal with in last fall's oversize pile. We didn't get any open flame, but it was apparent that we were getting smoking. We had a lot of charing going on deep within the pile and reached temperatures in the neighborhood of 180°. In some spots, we put in borings and sunk thermocouples to try to spot the hottest locations, and then we'd tear them open and separate them from the main body of the pile. It didn't get out of hand fortunately, but we were close to it, particularly with the older material. With the large size, the great height, and compaction tended to build the heat up inside the pile.

O. P. Vimal, DNES (India)

How much wood is required for this power plant? How do you sustain the wood supply for this plant?

Tewksbury

At full load, we can consume on the order of 85 English ton per hour, wet. Regarding wood supply, the Northwestern part of the State of Vermont alone has been repeatedly shown to have an annual regrowth of material which could fuel 3 plants of this size on a continuous basis. There are some restrictions to protect environmentally sensitive locations such as high elevations, known deer yards, and protected plant and wildlife species. We don't operate in those areas. But there's still a sufficient quantity of material to supply us, and that doesn't even include the resources of Southern Quebec and Northern New York, which also supply fuel to our plant. We're the only major outlet for people who want to improve their wood lots or who have low quality material to dispose of; we give them a market.

Vimal

What is the cost per kWh?

Tewksbury

We dispatch the pool right now at 39 mills or 3.9¢ per kWh. The cost primarily reflects fuel costs. If you include overall capital and financing expenses, it's about 7.2¢/kWh.

H. Shull, Natural Resources Research Institute	How far away can you collect wood economically? Can you tell us how much that wood would cost?
Tewksbury	Of course transportation is one of the major components of the cost of the fuel. If we contract for it directly, we'll generally limit the distance to 50 miles as a maximum. The majority of our fuel has come from within that range. The average delivered cost by truck is about $20 per ton. There are additions to that if the railroads are involved.

COMBUSTION OF RDF - COAL IN A UTILITY BOILER: A CASE HISTORY

Robert H. Herrmann
Associate Program Director
National Ecology, Inc.
Timonium, Maryland 21093

ABSTRACT

Baltimore Gas and Electric Company joined with the Maryland Environmental Service, Baltimore County, and National Ecology, Inc., to help solve Baltimore County's solid waste disposal problem and to reduce the cost of energy at the Baltimore Gas and Electric Company C. P. Crane Station. A ten year contract was signed for co-firing of refuse derived fuel and coal at the C. P. Crane Generating Station.

The RDF is produced at the Baltimore County Resource Recovery Facility in Cockeysville, Maryland, designed and operated by National Ecology, Inc. At this facility, municipal solid waste and selected commercial waste is processed to remove ferrous metal, glass, and other inert materials resulting in refuse derived fuel (RDF), a light combustible "fluffy" material which resembles shredded paper, with a heating value of approximately 6,300 BTU's per pound. The RDF process not only produces a usable fuel, but drastically reduces the volume of waste that is landfilled, thus extending landfill life.

The RDF is transported by trailer to the C. P. Crane Station in eastern Baltimore County where it is co-fired with coal in a cyclone furnace. The C. P. Crane facility consists of two 200 megawatt coal fired units, each powered by four Babcock and Wilcox cyclone furnaces. During the first two years of operation RDF was burned in Unit No. 2 at an average rate of approximately fifteen tons per hour. Up to twenty percent RDF on a BTU basis may be admitted to each cyclone. Inspection of boilers after the first year and second year of operation indicate no deleterious effects as a result of burning RDF although two deposits did form in the super heater section of Unit No. 2 in two instances. Problems with insufficient excess air and problems in controlling oxygen level at that time may have caused the problems.

Plans are currently being implemented to burn RDF in Unit No. 1 as well as Unit No. 2 so that boiler outages will have minimum effect on the ability to burn RDF continuously.

COMBUSTION OF RDF - COAL IN A UTILITY BOILER: A CASE HISTORY

BACKGROUND

In 1980 Baltimore Gas and Electric Company (BG&E) co-fired refuse derived fuel (RDF) and coal to test the combustion characteristics of this fuel mix in the Babcock and Wilcox cyclone furnace installed in their C. P. Crane Unit No. 2 generating station. RDF was prepared at the Baltimore County Resource Recovery Facility (BCRRF), designed and operated by National Ecology, Inc. (NEI). The motivation to burn RDF was primarily financial since it represented potential fuel cost savings for Baltimore Gas and Electric Co., and reduced landfill costs for Baltimore County - a "win-win" situation. In addition, the low sulphur content of the RDF reduced the overall sulphur emissions when burning 2% sulphur coal.

During the test over 2,300 tons of RDF were burned in one cyclone at a rate that was equivalent to 5% of the total furnace heat input, and since there are four cyclones in the furnace, this was equivalent to 20% for one cyclone. As a result of this successful burn test, BG&E entered into negotiations with the Maryland Environmental Service (MES) to purchase RDF for coal-firing at Crane Station.

The contract negotiated with MES called for BG&E to purchase up to 130,000 tons per year of RDF at a maximum value of 20% heat input from RDF or 5% RDF into each of four cyclones. RDF heating value was guaranteed to exceed or equal 5,000 BTU's per pound, and contract duration was ten years from first firing of RDF.

It was felt that three major goals could be accomplished:

1. Displace a non-renewable fuel (coal) with RDF which would otherwise be landfilled.

2. Pass fuel cost savings on to BG&E customers since RDF is bought at a discount.

3. Prolong Baltimore County's existing landfill life, and reduce landfill costs.

The bottom ash from a cyclone furnace is in the form of a molten slag which drops into water and breaks up into a marketable frit, thereby eliminating the need to landfill the bottom ash.

Although twelve projects involving RDF co-firing in large utility boilers had been initiated since 1972, only six are currently operational. An analysis of the problems that occurred in the unsuccessful projects indicated that RDF quality was one of the major problems. RDF must be low in ash and essentially free of glass and grit, and the particle size must be small enough to achieve complete burnout in the boiler. In addition, the boiler itself must have suitable means for handling the additional ash load when burning RDF, and must also be large enough and have sufficient air capacity to accommodate the combustion of RDF. On the basis of this reasoning, it appeared that the low ash, glass free "fluff" RDF produced by National Ecology at the Baltimore County Resource Recovery Facility was ideal for combustion in a utility boiler, and furthermore, the cyclone furnace at the BG&E Crane Station appeared to be ideal for co-firing of RDF and coal.

RDF Production

The Baltimore Resource Recovery Facility in Cockeysville, Maryland, is designed to recover RDF, ferrous metal, and glass from 1,200 tons per day of municipal solid waste. (Fig. 1).

The flow diagram for the facility is shown in Figure 2. Municipal solid waste is shredded in two processing lines each of which has a 1,000 horsepower horizontal shredder capable of shredding fifty to ninety tons of MSW per hour. Ferrous metal is removed from the shredded waste by means of three stage overhead magnets, followed by primary and secondary screening for removal of glass, grit and dirt, as well as non-ferrous metals. A high ash, high moisture residue is also removed and landfilled. The remaining low ash, high BTU combustibles are secondary shredded to form RDF which, in turn, is conveyed (Fig. 3) to sixty-five cubic yard transfer trailers which haul it to the C. P. Crane Station. The utility grade RDF, which BG&E buys, comprises sixty to seventy percent of the total input weight of MSW.

Trailer loads of RDF, approximately fifteen to sixteen tons per trailer, are hauled to the C. P. Crane Station (Figure 4) twenty five miles away, where the RDF is discharged into a specially designed RDF receiving facility designed and operated by National Ecology. This facility can store forty five tons of RDF, sufficient to provide 1.5 hours supply at twenty percent RDF input, or three hours supply at ten percent RDF input with the boiler fully loaded. The RDF is reclaimed from storage by means of an auger and conveyed to a metering bin, which, in turn delivers the RDF to rotary airlock feeders. The airlocks

Fig. 1: Baltimore County Resource Recovery Facility

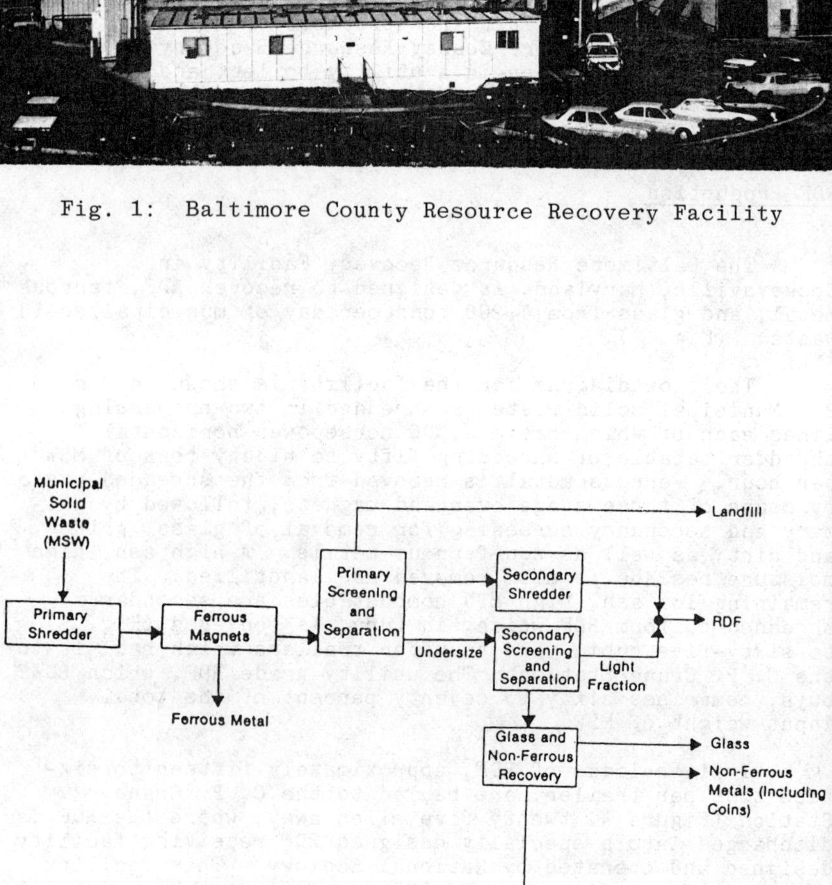

Fig. 2: Process Flow Diagram
Baltimore County Resource Recovery Facility

Fig. 3: RDF Conveyor Baltimore County Resource Recovery Facility

Fig. 4: Baltimore Gas & Electric Company
C. P. Crane Station

deliver the RDF to each of four pneumatic conveying lines which deliver the RDF to the four cyclones in Crane Unit No. 2. A process flow diagram for the RDF Receiving Facility is shown in Figure 5.

C. P. Crane Station Cyclone Boilers

The C.P. Crane Station consists of two Babcock and Wilcox 188 megawatt boilers, each equipped with four cyclones capable of generating the equivalent of forty eight megawatts of electrical power each. RDF is burned twenty four hours per day, seven days per week. Unit No. 2 which, while being a base loaded unit, does reduce load occasionally at night and on weekends, continues to co-fire RDF during these periods. The RDF supply contract stipulates that RDF will not be burned in any one cyclone that is not operating at full load. This means that if the boiler is operating at partial load, at which time the coal feed to one of the cyclones is reduced, RDF cannot be burned in that cyclone. However, the percentage RDF going to the other cyclones can be increased to a maximum of twenty percent per cyclone.

A cyclone furnace (Reference 1, Fig. 6, and Fig. 7) is a water cooled refractory-lined horizontal cylinder in which fuel is fired and heat is released at extremely high rates and at high combustion temperatures. Crushed coal (-¼" particle size) flows through the cyclone along a spiral path and combustion is completed by the time the coal traverses the length of the cyclone. The molten coal ash forms a slag that lines the interior surface of the cyclone permitting particles of coal or RDF fuel to adhere to the molasses-like slag allowing the fuel to burn to completion. About twenty percent of the required combustion air, referred to as primary air, enters the burner tangentially and imparts a spiralling motion to the incoming coal. RDF is blown pneumatically into the high velocity secondary air stream which is also admitted tangentially at the roof of the cyclone to add to the whirling motion. A small amount of air, called tertiary air, is admitted axially at the center of the burner.

The Crane No. 2 Unit generates 1,362,000 lbs. per hour of super-heated steam at 2,475 psig and 1,050°F and sends out approximately 188 megawatts when at full load. Co-firing of RDF has caused no de-rating of the unit to date.

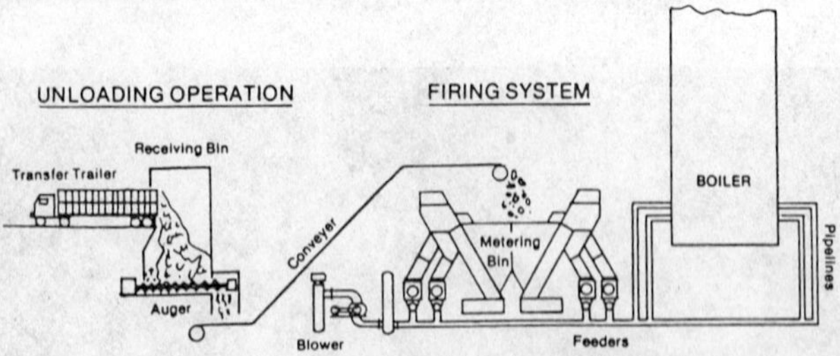

Fig. 5: Process Flow Diagram RDF Receiving Facility, C.P. Crane Station

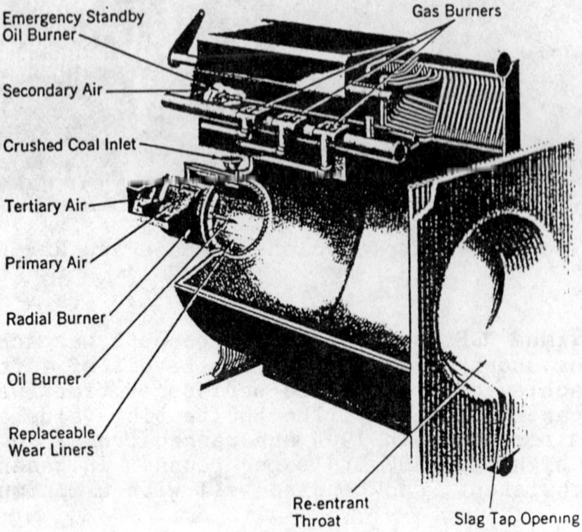

Fig. 6: Cyclone Furnace

First Years Operating Results

Co-firing of coal and RDF commenced in February, 1984, at RDF rates of 330 to 550 day tons per day at feed rates ranging from ten to twenty six tons per hour, the average being approximately fifteen tons per hour. During the first year of operation, approximately 40,000 tons of RDF were burned. This figure could have been much higher if baghouse pressure drop problems, unrelated to RDF firing, had not caused unforeseen load reductions. These reductions limited possible RDF consumption levels during summer months.

Fig. 7: Exterior View of Cyclone Furnace and RDF Feed Pipe.

A maximum RDF input of twenty percent was achieved but only for short periods of time. Levels of fifteen percent were achieved for extended periods of time (weeks) with no apparent problems. The RDF heating value averaged 6,280 BTU's per pound in 1984 and ranged from a low of 5,036 to a high of 8,107 BTU's per pound. In general, the fuel was consistent, and handled well with a minimum of problems.

A typical analysis of coal burned at Crane Station is provided in Table 1. Sulphur is 2.1%, and heating value 13,200 BTU's per pound.

Approximately six months after burning commenced, the boiler experienced excess air problems during RDF co-firing. As is commonly known, RDF, because of its higher moisture content, requires more excess air than coal for combustion. As a consequence, additional excess air must be provided especially when RDF is burned at fifteen to twenty percent of the coal heat value. The Crane boilers have a history of forming deposits if final furnace oxygen is not kept at three percent or higher. During this time the boiler experienced baghouse pressure drop problems, air leakage problems, and swings in oxygen level when firing RDF which served to aggrevate the lack of excess air, resulting in the formation of a deposit. A second deposit formed three months later in November, 1984.

An analysis of deposits that occurred during normal operation in unit No. 1 that burns coal only, and in Unit No. 2 that burns coal plus RDF is given in Table 2. A comparison of the major constituents analyzed in the two units indicates that there is essentially no difference in the constituents making up the deposits found in the two units. The high percentage of iron oxide is significant since it identifies the deposit as primarily coal ash deposit rather than RDF which has a very low iron content.

Since coal that is suitable for combustion in cyclone furnaces must be slagging type coal, it is reasonable to

TABLE 1

TYPICAL ANALYSIS OF COAL
C.P. CRANE STATION

Source:	Pittsburgh Seam Blacksville #1
Moisture	3.5%
Ash	8.5%
Volatile	37.0%
Sulphur	2.1%
High Heat Value	13,200 BTU/Lb.
Fusion:	
o Initial Deformation	2,100°F
o Fluid	2,420°F

TABLE 2

ANALYSES OF DEPOSITS
MAJOR CONSTITUENTS

	Unit 2 Coal + RDF		Unit 1 Coal Only
Compound, %	RH-SH	SSH*	RH/SH*
SiO_2	39.5	41.8	41.2
Al_2O_3	23.8	25.0	22.6
Fe_2O_3	19.1	18.6	18.3
CaO	4.8	4.4	4.9
Na_2O	1.0	0.8	0.7

* RH = Reheater
SH = Superheater
SHH = Secondary Superheater

expect a high percentage of iron since slagging coals normally have a high concentration of iron sulfide. It is significant to note that large deposits have formed in the past with essentially the same composition, when burning coal only.

After the deposit problems occurred, modifications were made to the control system to minimize oxygen swings when firing RDF, and, in addition, changes were made to the baghouse resulting in considerable less pressure drop. A major source of air leakage was identified in the air-heater seals and plans were made to improve these seals in the future.

In late 1984 the unit was removed from service for its annual general inspection. The boiler was inspected from top to bottom and no abnormal corrosion, erosion, or deposit formation was detected.

Second Year's Operating Results

Only 15,000 tons of RDF were burned in 1985, the second year of operation, because of long duration, unexpected outages at Crane Station. Inspection of the turbine in No. 2 boiler revealed major cracks in the turbine blades requiring extensive down-time for repairs. This, coupled with the routine scheduled boiler outage, as well as unscheduled outages, prevented the burning of RDF for almost seven months.

During the second year of operation RDF was fired at essentially the same rate as during the first year, but the moisture content was considerably less as shown in Table 3, in which RDF analyses for 1984 and 1985 are listed both on

TABLE 3

RDF ANALYSIS
(AVERAGE)

Fuel Analysis	1984 As Received	1984 Dry Basis	1985 As Received	1985 Dry Basis
No. of Analyses	51		26	
Moisture, %	21.67	-0-	12.0	-0-
Ash, %	8.99	11.48	9.34	10.61
Higher Heating Value, BTU/lb.	6280	8013	7048	8009
Sulphur, %	0.28	0.36	0.25	0.28

PARTICLE SIZE

Ninety percent (90%) less than 1.5 inches.

an as-received basis and dry-basis. The dry-basis BTU levels remained at approximate 8,000 BTU's per pound for both years, but the ash content dropped slightly from 11.48 to 10.61. The drop in moisture content and dry-basis ash level is probably attributable to the fact that the plant processed greater proportions of clean commercial waste, such as paper products in 1985. It is also of interest to note that the dry basis sulphur content dropped from .36 to .28%.

FUTURE PLANS

Because of the success in co-burning RDF with coal over a two year period, a decision was made to feed RDF to the No. 1 boiler as well as to the No. 2 boiler at Crane Station to minimize the impact of boiler or turbine outages on the use of RDF. Plans call for feeding RDF continuously through the four pneumatic pipe lines so that two of the lines are always feeding two of the cyclones in Unit No. 1, and Unit No. 2. Although the average RDF concentration in each boiler may be ten percent of the coal heat input, the actual concentration in each cyclone will be as much as twenty percent, thus enabling the RDF plant and feeding system to operate up to design capacity. This means that the combined boilers can burn as much as 624 tons per day at an average combustion rate of twenty six tons per hour or 6.5 tons per cyclone per hour.

Modifications to the pipe lines have already been made, and nozzles are scheduled for installation during the spring outage for the No. 1 boiler. Plans call for burning RDF in both boilers upon completion of the outage in late spring.

SUMMARY

RDF can be burned effectively in cyclone boilers at BTU input levels of ten to twenty percent of the total heat input provided the boilers are adequately sized to handle fuel of this type together with the additional excess air requirements. Unusual slagging and fouling problems can be avoided by providing an RDF with minimum ash content - say less than ten percent - and proper oxygen control to assure sufficient excess air at all times. To date, it appears that effects on properly designed pollution control equipment, such as baghouses, is either minimal or non-existent. Handling and conveying problems associated with feeding RDF to boilers can be reduced to a routine maintenance type of operation through proper engineering. Boilers with chronic slagging and fouling problems with coal should be avoided. RDF should be burned in at least two boilers in order to minimize the effect of boiler outages on the RDF processing operation.

BIBLIOGRAPHY

(1) Chon, D. and Herrmann, R. H., "Co-Combustion of Refuse Derived Fuel (RDF) and Coal in a Cyclone Furnace at the Baltimore Gas & Electric Company C.P. Crane Station. For: Maryland Environmental Service. Sponsor: United States Department of Energy Grant No. DE-FG01-80-6524320.

(2) Bourguin, Ralph H. Jr., "Marketing Refuse Derived Fuel (RDF): A Utility Viewpoint", presented at Market Development and Resource Recovery Conference, March 25, 1983, Washington, D. C.

(3) Fiscus et al., "Electric Utility Operating Problems in Co-Firing RDF", ASME/IEEE/ASCE Joint Power Generation Conference, Indianapolis, Indiana, September 26, 1983.

(4) Gunzelman, C. Paul, P.E. "Co-Firing of RDF and Coal in a Cyclone Utility Boiler", 12th Energy Technology Conference and Exposition, Washington, D.C., March 25-27, 1985.

(5) Gunzelman, C. Paul, and Herrmann, Robert H., "Experience with RDF Co-Firing in Utility Boilers", presented at the Energy from Municipal Wastes Conference, Opportunities in an Emerging Market, October 25, 1985, Washington, D. C.

(6) McGowin, C.R., "Municipal Solid Waste: A Utility Fuel?" NCRR Bulletin, The Journal of Resource Recovery, Volume II, Number 4, December, 1981.

REFERENCES CITED

(1) Babcock and Wilcox, STEAM/Its Generation and Use, Page 10-1, New York, New York. Babcock and Wilcox, 1972.

DISCUSSION

J. Kerstetter, Washington State Energy Office	Did you find any disproportionate reduction in sulfur when you were co-firing with RDF? Was there a greater reduction in sulfur emissions than you expected just on a Btu basis?
Herrmann	In other words, was there a synergistic effect?
Kerstetter	Yes.
Herrmann	No, not that I'm aware of. We would like to see it that way. It didn't happen.
E. David, Wisconsin Dept. of Natural Resources	I wondered whether you had any problems persuading your utility operators to take RDF? One of the things we found is that people who run utility boilers tend to think that coal is clean and RDF is not. Is that a problem?
Herrmann	That's a very astute observation, and I think that it's only normal that an operator likes to run things the way they have been running. We found that some operators would throw the switch to 20% RDF, and others would be perfectly happy if you would let them throw it to 0% RDF. That's human nature; you do have to train the operators.
V. Leak, Natural Resource Research Institute	You didn't show the ash fusion temperature on your table as near as I could tell. What was that range and did the RDF additions change it significantly?
Herrmann	The RDF did not affect the quality of the slag or the frit that is used for sand-blasting purposes. It had a fusion temperature that was comparable to that of the coal; it's around 2200°F. I can't tell you whether that's the beginning of the fusion or the end, but it was in that range. It's a low fusion temperature.
J. Herberman, Ontario Ministry of Energy (Canada)	On what basis do the operators decide what percentage RDF to inject?
Herrmann	They were originally given the option ot controlling the RDF in accordance with the oxygen and air demands of the boiler. Now they are being told to burn at whatever level is decided by

management, whether it's 15 or 20%. The decision last week was to burn at 20% RDF.

__________ What is the destiny of chloride in the RDF? Do you have any clean-up system?

Herrmann No we don't. There appears to be no problem from chlorides in this type of operation.

THE KINETICS OF THE FAST PYROLYSIS OF CELLULOSE AT 900°C

R.G. Graham, M. Eng., and B.A. Freel, B.Sc.
Ensyn Engineering Associates
Ottawa, Canada K2P OL9

M.A. Bergougnou, Ph.D. and L.K. Mok, M.Appl.Sci.
Faculty of Engineering Science
The University of Western Ontario
London, Canada N6A 5B9

R.P. Overend, Ph.D.
National Research Council of Canada
Ottawa, Canada K1A OR6

ABSTRACT

After preliminary experiments demonstrated the chemistry of fast pyrolysis in a continuous steady-state reactor, systematic experiments were initiated to determine the kinetics of the production of important gaseous products at temperatures between 650 and 900^{0}C. Residence times were varied between 50 and 900 milliseconds (ms). The novel reactor system utilizes solid and/or gaseous "thermofor" to carry heat, into a turbulent vertical contactor (i.e., a cyclonic mixer termed a thermovortactor) where mixing between the biomass particles and heat carrier occurs within several milliseconds. Rapid heat transfer, which is a consequence of fast efficient mixing, initiates a reaction pathway leading to the production of valuable non-equilibrium chemical intermediates (i.e., olefins and other hydrocarbons). Rapid quenching of the products preserves these intermediates before subsequent decomposition and/or repolymerization can occur. In this paper, experimental results and kinetic parameters are reported for the fast pyrolysis of Avicel cellulose powder at 900^{0}C using a gaseous nitrogen heat carrier. Maximum yields of ethylene, total unsaturates and total non-condensible hydrocarbons are approximately 7, 10 and 18% (by mass) of the dry cellulose feedstock.

THE KINETICS OF THE FAST PYROLYSIS OF CELLULOSE AT 900°C

INTRODUCTION

The fast pyrolysis of wood and cellulose has exhibited promising yields of high quality chemical intermediates (i.e. ethylene, acetylene and light organic liquids). Fundamental research has indicated that olefin yields of 10 to 15% by mass of the dry biomass feedstock can be realized and that little or no char need be formed under conditions of short vapour residence times (less than 500 ms), high reactor temperatures (greater than 650°C), rapid heating rates (thousands of degrees celsius per second) and rapid quenching of the reaction products (1,2, 3,4,5). A complete characterization of fast pyrolysis and an outline of its advantages with respect to product quality and selectivity over the more conventional thermochemical conversion processes, have been presented in earlier publications (1,2).

At the University of Western Ontario, the first phase of a fast pyrolysis program (Ultrapyrolysis) was conceptualized with the intention of engineering a continuous process which would successfully demonstrate the chemistry of fast pyrolysis. In effect, a successful demonstration would require that valuable non-equilibrium thermochemical conversion products (i.e. olefins, BTX and reactive intermediates) and synthesis gas (carbon monoxide and hydrogen) be generated at steady state in equipment that could be used as a conceptual basis for eventual commercial scale-up. The heart of this primary objective was the development of a new ultra-rapid fluidized bed (URF) reactor design in order to exploit and optimize fast pyrolysis product yields. The preliminary results using wood, cellulose and lignin feedstocks are reported, along with details of the original design, in a report to the Canadian government (1). Upon completion of the first phase of the project, the system was modified to enhance operation for kinetic modeling. At the time of writing more than 500 kinetic experiments had been conducted using Avicel cellulose and IEA poplar feedstocks at temperatures between 650 and 900°C and at residence times between 50 and 900 ms. The majority of experiments utilized gaseous nitrogen as the heat carrier although several trials using particulate solids (silica sand) have been conducted to compare the product distributions. This paper outlines the current ultrapyrolysis flow scheme, and the experimental operating procedure and process parameters. Quantitative results are summarized for Avicel cellulose pyrolysis experiments using nitrogen thermofor as the heat carrier at 900°C. Preliminary first-order kinetic parameters are given for the rate of production of total gas, hydrogen, carbon monoxide, carbon dioxide, ethylene and methane.

ULTRAPYROLYSIS FLOW SCHEME

The major components of the current Ultrapyrolysis (UP) process are illustrated in Figure 1. Rapid mixing and heat transfer are carried out in two conical vessels known as vortical contactors or Vortactors. The first has been termed a Thermovortactor and allows heat to be transferred from a hot "thermofor" stream (i..e gaseous nitrogen, suspended particulate solids, or a combination of the two) to the biomass. The second is a Cryovortactor and allows fast quenching of the products by the direct transfer of heat to "cryofor" (i.e. cryogenic nitrogen). The Thermovortactor has two opposing tangential inlets for the thermofor. One tangential stream effectively destroys the momemtum of the other causing severe turbulence. Biomass feedstocks are then injected from the top of the Thermovortactor through an air cooled tube into the turbulent region where mixing occurs within 30 milliseconds. The hot gaseous product is rapidly cooled (i.e. in less than 30 ms) by the injection of a single tangential stream of cryogenic nitrogen.

Mechanical table feeders are used to supply biomass and thermofor solids to the reactor system. Solids pass from sealed hoppers (which have a sufficient inventory of solids for several experiments through a double funnel system and are thereby metered onto a rotating table. Two fixed armatures sit near the surface of the round table and plough the solids off the outer circumference. From the table, the solids then fall into a conical chamber where they are picked up and carried into the transport line by nitrogen carrier gas. The overall range of the feed rate of the solids is controlled by setting the gap between the lower funnel and the table. Fine control is exercised by the rotation speed of the table. Currently the biomass feeder is set in the range of 0 to 20 grams per minute and the thermofor feeder is set to deliver 0 to 200 grams per minute. The biomass feeder is mounted on a 60 kg balance to facilitate the determination of the mass balance. When thermofor solids are required to supply the process heat (i.e. either in place of gaseous nitrogen thermofor or as a supplement), the thermofor feeder sends cold (room temperature particulate solids to a double preheater coil assembly where they are heated to the reaction temperature. These solids are then sent on to the Thermovortactor. Thermofor nitrogen can be sent directly to the preheater coil.

The fast pyrolysis of biomass is initiated in the Thermovortactor and continues in a plug-flow entrained bed downflow reactor. The reactor is simply a one meter length of Inconel pipe in an electrical oven. The mixture of hot gases and biomass passes from the Thermovortactor through the reactor to the Cryovortactor. With the insertion of cylindrical inserts to reduce the reactor volume and by manipulating thermofor/biomass flowrates, the residence time (i.e. the time from the heating of the biomass to the exit from the reactor) can be set between 30 and 900 ms. Reactor temperatures can be set in the range of 650 to 1000°C. The UP flow scheme is described in detail in a previous publication (6).

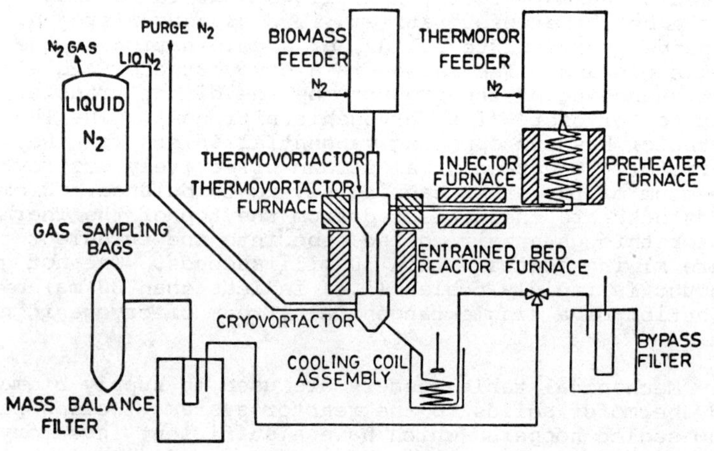

FIGURE 1 ULTRAPYROLYSIS FLOW SCHEME

EXPERIMENTAL

In preparation for an experiment, the system is purged with nitrogen. The furnaces are then turned on and allowed to reach their setpoint temperatures. When the desired temperatures are reached, cryogenic nitrogen is injected into the Cryovortactor. Thermofor (i.e. transported silica sand and/or inert nitrogen gas) is then introduced into the system and the cryogenic nitrogen flowrate is adjusted to ensure that the Cryovortactor temperature is less than 350°C. Gaseous nitrogen was used as the thermofor for all of the experiments reported in this paper. During the startup period, the stream exiting from the reactor system is vented to atmosphere. The biomass feeder is activated and its feedrate is verified by directing the flow to a collection device mounted on an electronic balance. A cellulose feedrate of 4 to 6 grams per minute was typical for the experiments reported in this paper.

Immediately prior to the actual steady state experimental run, the thermofor/cryofor stream (which is exiting from the Cryovortactor) is directed to the mass balance filter and recovery train. The biomass is then directed to the Thermovortactor and the actual fast pyrolysis experiment is thereby initiated. Products which are initially cooled to about 300 to 350°C in the Cryovortactor are cooled to room temperature in a water-cooled coiled tube heat exchanger. Solids and condensibles are deposited in the filter, and the entire non-condensible product gas is collected in the gas collection bags. For extremely short residence time trials (when the yield of "condensibles" increases), an electrostatic precipitator unit can easily be integrated in the downstream gas line. In addition, an efficient cyclonic condenser (which is submerged in an ice bath) has been designed to increase the yield of pure (i.e. non-solvent extracted) liquid sample. The steady state experiment continues for 5 to 20 minutes. At the end of the experiment, the biomass flow is redirected from the reactor system to the bypass line and its feeder motor is turned off. The inert thermofor and cryofor streams, however, are maintained for a short period of time in order to ensure that all products are carried to the recovery system. After purging, the recovery train (including filter, transport lines and cooling coil) is dismantled and washed with methanol to collect any condensibles. When sufficient quantities are produced, pure condensibles are poured from the condensor prior to solvent extraction. Recovery of the methanol-extracted pyrolysis liquids is accomplished using a rotary evaporator. Gas samples are taken from the collection bags and analyzed using a Carlo refinery gas chromatograph (Model 111-H/197A). When gas analysis of the bag contents is completed, the gas is drawn into an evacuated Superior 123 VSWG Proprane tank of known volume. The magnitude of the change in tank pressure and the gas analysis are used to determine the quantity of the product gas.

RESULTS AND DISCUSSION

The results of the fast pyrolysis experiments conducted at 900°C, using Avicel cellulose (a 100 μm microcrystalline cellulose powder from FMC in Philadelphia) as the biomass feedstock and nitrogen gas as the heat carrier, are summarized in Figures 2 to 7 and Table 1. The mass yields of the total gas and its major non-condensible components, expressed as a percentage of the dry biomass input, are plotted as a function of reactor residence time in Figures 2 to 7. These figures are plotted from the experimental data contained in Table 1. The solid lines on the curves represent a first order kinetic model (discussed later) which has been derived from a non-linear regression analysis.

The plot of the mass yield of the total non-condensible gas product vs reactor residence time (ms) is given in Figure 2. The general shape of this curve indicates an initial rapid production rate followed by a tapering off to a constant maximum yield, and is representative of all the individual component yield curves (i.e. Figures 3 to 7). For example, the total gas production rate (i.e. the slope of the curve) is very rapid up to about 100 ms and then tapers off giving a maximum yield of 88% (by mass) after

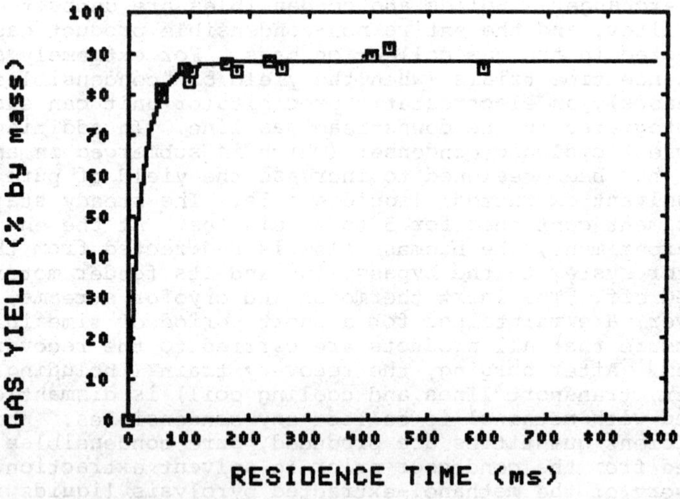

FIGURE 2 CELLULOSE PYROLYSIS KINETICS (900 C): TOTAL GAS YIELD VS RESIDENCE TIME

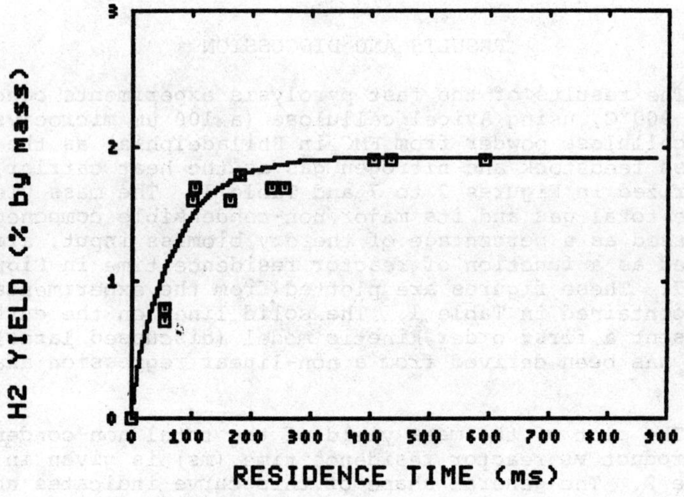

FIGURE 3 CELLULOSE PYROLYSIS KINETICS (900 C): HYDROGEN YIELD VS RESIDENCE TIME

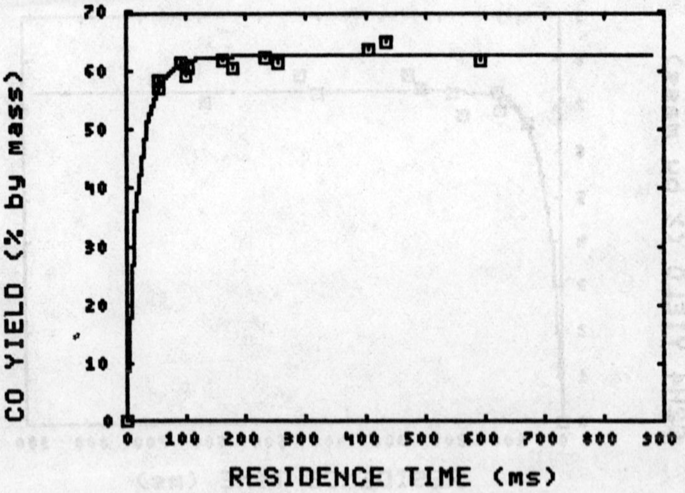

FIGURE 4 CELLULOSE PYROLYSIS KINETICS (900 C): CARBON MONOXIDE YIELD VS RESIDENCE TIME

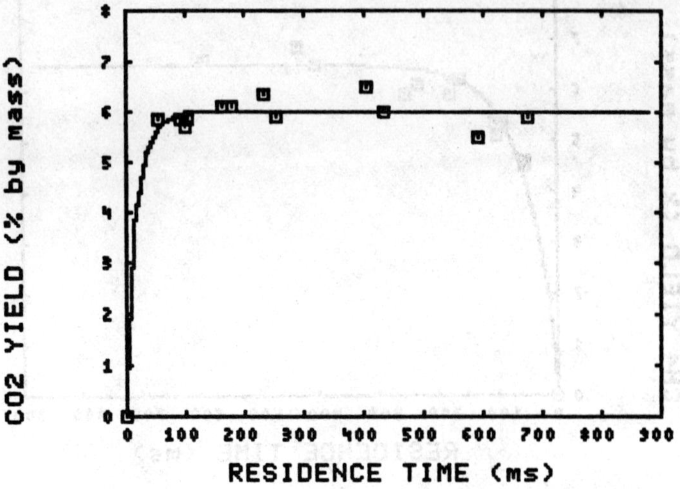

FIGURE 5 CELLULOSE PYROLYSIS KINETICS (900 C): CARBON DIOXIDE YIELD VS RESIDENCE TIME

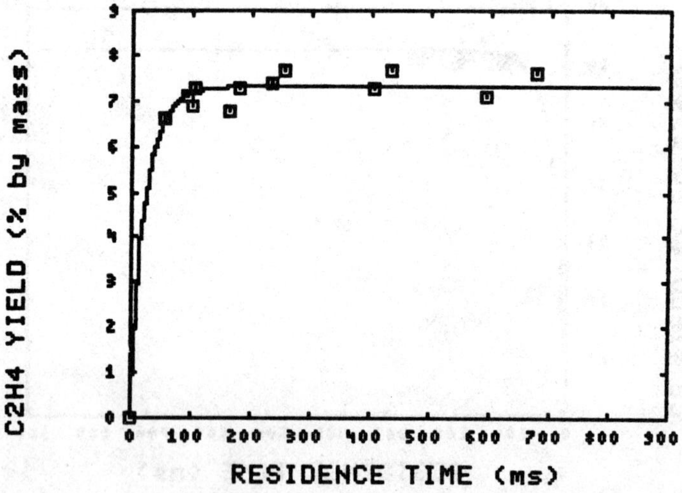

FIGURE 6 CELLULOSE PYROLYSIS KINETICS (900 C): ETHYLENE YIELD VS RESIDENCE TIME

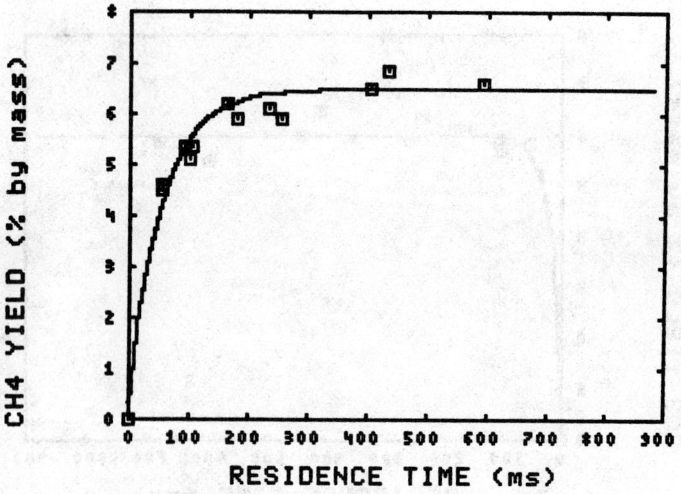

FIGURE 7 CELLULOSE PYROLYSIS KINETICS (900 C): METHANE YIELD VS RESIDENCE TIME

Table 1. CELLULOSE KINETICS EXPERIMENTS (900°C): DATA SUMMARY

Run No.	Residence Time (ms)	Component Mass Yield (%)									
		Gas	H_2	CO_2	CO	C_2H_4	C_3H_6	C_2H_2	CH_4	C_2H_6	B/T*
253	234	87.9	1.7	62.4	6.3	7.4	0.5	1.8	6.1	0.6	1.2
254	164	87.1	1.6	62.0	6.1	6.8	0.4	2.3	6.2	0.5	1.3
280	255	86.2	1.7	61.4	5.9	7.7	0.4	1.9	5.9	0.5	0.8
281	182	85.6	1.8	60.8	6.1	7.3	0.6	1.8	5.9	0.6	0.9
282	105	85.1	1.7	60.7	5.8	7.3	0.8	1.8	5.3	0.7	0.9
381	678	92.8	2.2	66.8	5.9	7.6	0.2	1.6	7.2	0.3	1.0
382	104	82.7	1.6	59.4	5.7	6.9	0.7	1.9	5.1	0.7	0.8
383	435	90.8	1.9	64.8	6.0	7.7	0.3	2.0	6.8	0.4	1.0
384	592	86.1	1.9	61.9	5.5	7.1	0.2	1.7	6.6	0.3	1.0
385	93	86.1	2.0	61.4	5.8	7.1	0.8	1.8	5.3	0.7	1.0
386	408	89.3	1.9	63.6	6.5	7.3	0.3	1.8	6.5	0.4	1.0
396	56	80.9	0.7	58.3	5.8	6.6	1.1	1.7	4.6	0.7	0.9
397	56	79.3	0.8	57.1	5.8	6.6	1.1	1.7	4.5	0.7	1.0

*benzene/toluene

approximately 150 ms (Figure 2). Carbon monoxide (Figure 4) and carbon dioxide (Figure 5) yields also reach maximum values of 62.6 and 6.0 percent by mass, respectively, after 150 ms. The yields of hydrogen (Figure 3) and methane (Figure 7) both seem to exhibit their constant values of 1.9 and 6.5 percent, respectively at a much longer residence time of about 350 ms, indicating perhaps that their appearance may result from a similar reaction mechanism or a common parent compound. Finally, the yield of the major unsaturated component, ethylene (Figure 6) reaches a constant value of 7.3 percent at 150 ms (which is similar to that reported for the total gas, CO and CO_2). The optimum yield of total unsaturates (i.e. ethylene, acetylene and propylene) is about 10 percent by mass at 900°C while the corresponding value for total hydrocarbons is 18 percent.

A consideration of the equilibrium composition and kinetics of the water-gas shift reaction ($H_2O + CO \leftrightarrow H_2 + CO_2$) at 900°C, and the fact that the CO, CO_2 and H_2 yields are relatively constant at longer residence times, leads to the conclusion that the shift reaction does not play an important role in the early pyrolysis reaction mechanism (at least up to 700 ms, at 900°C). Preliminary experimental data using the Ultrapyrolysis system has indicated that the shift reaction is beginning to occur at the longer residence times but is not significant (it certainly does not approach equilibrium). This reaction has therefore been neglected in the overall kinetic modelling of fast pyrolysis.

There does not seem to be any thermal decomposition of the "light" unsaturates, such as ethylene (Figure 6) or acetylene (Table 1), over the residence times studied. On the other hand, the heavier "crackable" hydrocarbons such as propylene and ethane (Table 1) appear rapidly but begin to decompose after 100 ms. Therefore primary pyrolysis appears to consist simply of the unzipping of the biomass polymer followed by rapid thermal fragmentation of the characteristic depolymerized units. There then seems to be a lag time, whose length depends on the nature of the chemical species present, before additional gas phase cracking reactions and the water-gas shift reaction occur.

A simple and convenient approach to kinetic modeling assumes that cellulose decomposes directly to each individual gas component by a single independent pathway and that the kinetics of the decomposition can be represented by a unimolecular first order reaction. This approach has been widely used in representing the kinetics of the fast/flash pyrolysis of biomass and coal (4,7). The following rate equation is thereby derived:

$$V = V^* [1 - \exp(-Kt)]$$

where V represents the mass yield of a component at time 't' and temperature 'T', V* is the maximum attainable yield of the component at temperature 'T' and long residence times, and K is the Arrhenius rate constant [$K = Ko \exp(-E/RT)$]. Best fit values for the kinetic parameters (V* and K) were estimated for all major components at 900°C using a least squares non-linear regression procedures. These parameters are summarized in Table 2 and the corresponding regression curves are shown as the solid lines in Figures 2 to 7.

Table 2. FAST PYROLYSIS OF CELLULOSE AT 900°C: SUMMARY OF KINETIC PARAMETERS V* AND K

Component	Parameter V*	Parameter K
Total Gas	87.7	42.1
Hydrogen	1.9	18.0
Carbon Monoxide	62.6	43.9
Carbon Dioxide	6.0	51.4
Ethylene	7.3	39.9
Methane	6.5	18.6

The fact that the non-gaseous products are liquids rather than solid char or unpyrolyzed cellulose (even at extremely short residence times) would seem to suggest that the solid phase decomposition to primary liquids is practically "instantaneous" in the ultrapyrolysis reactor system. This is likely a result of a combination of the inherent high heat transfer rates and the small cellulose particle size. Therefore the kinetic model is best described as representing the rate of appearance of the gas and its components from the primary pyrolytic liquids. Upon completion of the analysis of the experimental data over the entire temperatures of 650, 700, 750, 800, 825, 850, 875 and 900^{0}C, the activation energies (Ea) and pre-exponential constants (Ko) will then be estimated for the rate of appearance of each gaseous component during the fast pyrolysis of cellulose. Similar experiments have been conducted using the IEA (International Energy Agency) standard poplar feedstock. Kinetic data will soon be published for the trials using wood feedstocks. Refined temperature control, a more accurate reactor system volume determination and enhanced mass balance procedures used to conduct the experiments reported in this paper have resulted in yield differences of a few percent with respect to previously published Ultrapyrolysis data (6,8,9).

ACKNOWLEDGMENTS

This work was generously funded by the National Research Council of Canada, Environment Canada and Energy Mines and Resources Canada. Mr. A. Vogiatzis deserves particular mention for his excellent technical assistance as do Mr. H. Cook and Mr. R. Kager of the University of Western Ontario mechanical shop.

REFERENCES CITED

1. Bergougnou, M.A., de Lasa, H.I., Mok, L.K., Graham, R.G., Freel, B.A. and Hazlett, J.D., "Ultrapyrolysis of Cellulose and Wood Components", Enfor C-147 Final Report. Ottawa: Environment Canada, 1983.

2. Graham, J.P., Bergougnou, M.A. and Overend, R.P., "Fast Pyrolysis of Biomass", Journal of Analytical and Applied Pyrolysis 6, 95-135 (1984).

3. Diebold, J.P., "Gasoline from Solid Wastes by Non-Catalytic Thermal Process", Thermal Conversion of Solid Wastes and Biomass, Jerry L. Jones and Shirley B. Radding (editors). Washington: American Chemical Society, 1980.

4. Hajaligol, M.R., Howard, J.B., Longwell, J.P. and Peters, W.A., "Product Composition and Kinetics for Rapid Pyrolysis of Cellulose", IE&C Process Design and Development 21, 457-465 (1982).

5. Diebold, J.P. (editor), Specialists' Workshop on Fast Pyrolysis of Biomass. Golden, CO: Solar Energy Research Institute (1980).

6. Mok, L.K., Graham, R.G., Freel, B.A., Bergougnou, M.A. and Overend, R.P., "Fast Pyrolysis (Ultrapyrolysis) of Cellulose and Wood Components", Journal of Analytical and Applied Pyrolysis 8, 391-400, 1985.

7. Antal, R.G. Jr., Friedman, H.L. and Rogers, F.E., "Kinetics of Cellulose Pyrolysis in Nitrogen and Steam" Combustion Science and Technology 2, 141-152, 1980.

8. Graham, R.G., Freel, B.A., Bergougnou, M.A., Overend, R.P. and Mok, L.K., "Fast Pyrolysis of Cellulose", In Energy from Biomass: 3rd E.C. Conference. W. Palz, J. Coombs and D.O. Hall (editors), 860-864. London: Elsevier Applied Science Publishers, 1985.

9. Graham, R.G., Overend, R.P., Freel, B.A., Mok, L.K. and Bergougnou, M.A., "Development of the Ultra-Rapid Fluidized (URF) Reactor: Application to the Fast High-Temperature Pyrolysis of Cellulose", Proceedings of the 10th Annual Powder and Bulk Solids Conference, 126-139. Chicago: International Powder Institute, 1985.

DISCUSSION

W. Walawender, Kansas State U. — Could you comment on the particle size of the material that you were running through this, and what your estimate of the heat-up rate of those particles was relative to the residence time?

Graham — The particle size was less than a hundred microns and based on some work done for us by the University of Waterloo and Ralph Overend, heat transfer rates are about 10^4-10^5°C/s. They represent a few percent of the total residence time. I don't have time to mention everything about the program, but we have done kinetic studies on model compounds to compare the published kinetics in more fundamental studies with our own reactor system, and to verify the integrity of the system and the kinetics for ethane cracking for example.

Antal — Eight years ago, I gave a talk on vapor-phase kinetics of pyrolysis at this meeting. I was trying to recall the activation energies that we reported for the evolution of the various species. As I recall they were higher. Ethylene was more like 60 kcal per gram mole. One of the interesting phenomena we noticed that as we went to higher temperatures was that the sensitivity to temperature fell significantly and so we reported the lower temperature values. We attributed that in our case to what we knew were poor heat transfer characteristics of our reactor above 720°C. It would be interesting to compare the two. Do you really feel the numbers you're reporting do not have any heat transfer intrusions?

Graham — Again I can only go on some of the theoretical calculations Dr. Overend and the University of Waterloo group have done. I don't think we have heat transfer limitations. I think we're limited by the chemical kinetics at the higher temperatures.

Walawender — In your system, do you have heating both by convective mechanisms and radiation from the tube wall?

Graham — The calculations show that radiation is relatively minor. Convective heat transfer from the turbulent region is dominant. We assumed that using the solids and larger particles, we would have a greater contribution from radiation in a larger reactor system, and then the ablative mechanisms that Jim and some of the French researchers have talked about will come into play as well.

J. Diebold, SERI — I was wondering if you would comment on how you measured temperatures in the tubular portion of the reactor where cracking occurs?

Graham — Thermocouples are inserted.

Diebold — Inserted axially or radially or both?

Graham — Axially. At the bottom of the reactor, we have an adjustable thermocouple that we can position transverse to the length of the reactor system as well.

INFLUENCE OF OPERATING PARAMETERS ON THE PERFORMANCE OF A WOOD-FED DOWNDRAFT GASIFIER

W. P. Walawender, PhD, Professor,
S. M. Chern, MS, Graduate Research Assistant,
and L. T. Fan, PhD, University Distinguished Professor and Head

Department of Chemical Engineering
Kansas State University, Durland Hall
Manhattan, KS 66506

ABSTRACT

The air gasification of wood chips and wood pellets was investigated in a (0.6m ID) downdraft gasifier to assess the influences of various operating parameters on its performance. The parameters included the type of feed, the type of bed support, and the extent of bed agitation (grate rotation). The gasifier was operated with the fan rpm fixed and the grate rotation variable. Complete material balances were established over an operating range of 200 to 1400 MJ/hr; the assessment was based on the experimental data.

The results of the assessment revealed the following.

1) The type of bed support (ceramic balls, partially obstructed grate, open grate) had little effect on the gasifier performance.
2) The char yield from pellets was higher than that from chips as a consequence of the larger characteristic dimension for the pellets.
3) Increasing the extent of bed agitation (grate rotation) increased the char yield and decreased both the air-to-feed ratio and the cold gas efficiency.

The gasifier performance measures were also compared to those determined for a similar gasifier with a variable fan rpm and the grate rotation proportional to the fan rpm. Most measures were similar, however due to the limited air supply, the gasifier with the fixed fan rpm produced more char and had a lower throughput than the one with variable rpm.

INFLUENCE OF OPERATING PARAMETERS ON THE PERFORMANCE OF A WOOD-FED DOWNDRAFT GASIFIER

INTRODUCTION

Gasification is one means for converting wood or any biomass into a gaseous fuel which is more versatile than the original biomass. Currently, wood provides about 2% of our total energy needs and could contribute up to 8% within the next decade (8). The annual harvest of woody biomass amounts to about 1.4 billion tons in the U.S. alone. Over 700 million tons of this material is not used because it is not of the right species, size, fiber length, fiber morphology etc. (3). Thus the potential for the utilization of wood wastes to ensure a continuous supply of fuels and chemicals is significant.

Some advantages of wood as a gasification feedstock over solid fossil fuels such as coal and oil shale are its renewability, higher reactivity, lower sulphur content, and lower ash content, with the most important being the first. Its disadvantages stem from its low heating value, low bulk density and its high moisture content which makes handling facilities expensive. In this regard, benefication processes (1) are available for improving the properties of biomass, such as drying and densification. Wood pellets are a prime example of 'improved biomass'.

Various types of reactors have been used to gasify wood and other biomass; they include fixed beds, entrained beds, moving beds, rotary kilns, and fluidized beds. Fixed bed and moving bed gasification have been practiced over for a century; some commercial units are currently available. The other technologies are relatively new, and are basically geared toward large-scale operations; only a very few systems have been commercialized. Reed et al. (4) have summarized the state of gasifier research and the status of various gasifier manufacturers.

Downdraft gasifiers for the production of low BTU gas from biomass are generally mechanically simple and relatively easy to operate with suitable feed materials. The product gas has a negligible tar content. This feature makes them superior to updraft gasifiers which produce a high tar yield (about 20%) when biomass is used as the feedstock. The clean gas can be used for different applications, such as heating, and fueling boilers or engines.

Low BTU gas has been produced from moving bed gasifiers since the early nineteenth century. Sweden (5) and other countries used them to power automobiles during World War II. Nevertheless, their performance under various operating conditions has not been well understood. In this regard, research efforts are currently underway to improve our understanding and to develop models to facilitate gasifier design. Despite the limited knowledge of the process, a large number of gasifiers have been constructed for research and/or commercial use.

The Buck Rogers Gasifier is a recently commercialized downdraft gasifier that has undergone some design changes since its invention in 1981. The gasifier used in this study is a variation of a predecessor examined in our previous work (7). The objective of this work was to determine the performance of the Buck Rogers Gasifier under different operating conditions based on material balance data for the gasifier. The operating parameters studied included the grate rotation, the type of feed, and the type of bed support. Two feedstocks, wood chips and wood pellets, were gasified and three different bed supports were investigated at grate rotations ranging from 3 to 14 min/rev.

EXPERIMENTAL FACILITIES AND PROCEDURE

Gasifier Description

The Current Gasifier. Figure 1 presents a schematic diagram of the gasifier used in this study. It has a diameter of 0.6m and is open at the top. There is no throat or choke plate in the unit. A fan located downstream from the gasifier draws air through the bed as well as through the tuyeres. A butterfly valve, inserted in the fan outlet, serves as a throttle for the coarse control of gas throughput (approximately 40% variation). A feed bin and a screw feeder (not shown in Figure 1) supply feed to the top of the gasifier each time the bed height drops to a set level. Feeding is controlled by an electric eye, mounted at the side of the reactor. Bushings on the top and bottom of the reactor hold the 'Airgitator' in place. A chute, seated on the top plate, receives the feed delivered by the screw feeder and admits air. The 'Airgitator' is driven by a hydraulic motor with a hydraulic fluid flow spliter for controlling the grate rotation between 3 and 14 minutes per revolution. The 'Airgitator' has the following functions:

(1) It provides secondary air through the tuyeres.
(2) It levels the feed at the top of the bed.
(3) It slowly mixes material throughout the gasifier.

Gas and char flow out of the gasifier through a grid plate attached to the 'Airgitator' shaft. They are drawn out of the base of the gasifier by the fan which blows them through the cyclone where the char is separated from the gas. The gas stream is then sent to a flare stack (not shown in Figure 1) where it is incinerated.

Comparison with the Previous Gasifier. The current gasifier is basically similar to its predecessor examined in our previous study (7), except for a few design changes.

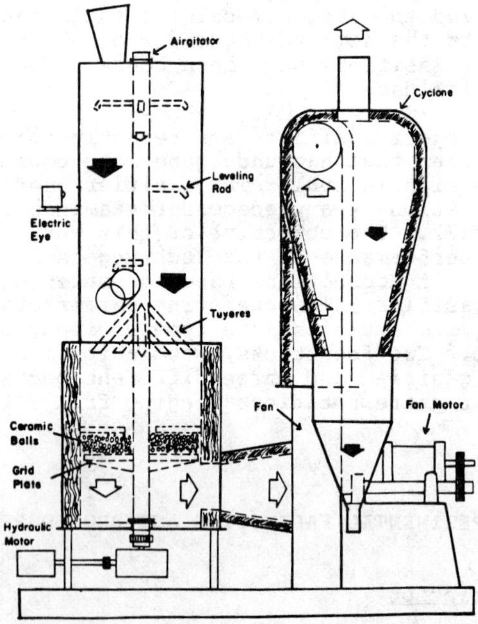

Figure 1. Commercial Gasifier

The current gasifier allows the grate rotation to be controlled independently of the fan rpm, whereas its predecessor had the grate rotation proportional to the fan rpm through a series of pulleys and gears. The fan rpm of the current gasifier is fixed (or more precisely, the applied voltage to the motor driving the fan is fixed), whereas the rpm of the fan in the previous gasifier was electronically controllable. A butterfly valve in the fan outlet serves as a throttle for the coarse control of gas throughput in the current system.

Operating Procedure

The operating procedure for the gasifier is detailed below.

1. Start-up. A portable propane burner was used to preheat the system. The temperature in the empty gasifier was approximately 600 K after 15 minutes. During the preheat period, the fan was started, and its throughput was gradually increased from low to maximum. The fan was then turned down to low throughput at the end of the preheat period, and propane was introduced into the flare stack and ignited. Next, the feeder was manually operated to fill the gasifier so that a shallow bed (about 10 cm thick) was formed. The bed was then set on fire. After fire was uniformly spread over the bed, another layer of feed was introduced. This procedure was repeated until the bed level

reached the tuyere section. The gasifier was then filled to the desired operating level, and automatic feeding was started. The grate was rotated very slowly (or even stopped) when the bed was being developed. The rotation was gradually increased up to the desired operating condition after the bed was established. As the production rate and temperature of the gas increased, the propane injection into the flare stack was decreased and then shut down. The fan throttle was simultaneously brought to the wide open position, and the gasifier was allowed to attain steady state. This procedure normally required two hours to complete.

2. Gas analysis and condensables measurement. After the completion of step 1, a side drawn of the gas was taken for analysis and condensate determination. The specific procedures will be detailed later. This step was conducted for a two hour period.

3. Char measurement. Char was measured by direct collection over a one to two hour interval in the middle of the run.

4. Tracer measurement. Nitrogen tracer gas was introduced into the product gas (at the fan inlet) at a metered rate, and the gas analysis and condensate measurement were repeated. The specific procedures will be detailed later. This step was conducted for a one hour period.

5. Post tracer measurement. Step 2 was again repeated without the tracer gas for 30 minutes to one hour to ascertain that steady state had been maintained by the system.

6. Shut-down. The feeder was stopped, and the bed level was allowed to drop. As the bed was depleted, flame began to spread over the surface of the bed. The whole bed then reduced to char, and the temperature of the bed began to rise rapidly. The fan was shut down when the temperature of the bed reached 1200 K, and the remaining char was allowed to burn slowly. Normally, more than four hours were required to burn the char completely.

<u>Measurements</u>

Because of the configuration of the gasifier and the stream flow rates involved, it was not convenient to measure all of the stream rates directly; consequently, indirect methods were employed to determine these rates where necessary.

<u>Direct Measurements</u>. These measurements consisted of the following.

(1) Char output rate. The char output rate was determined by collecting and weighing the char over a one to two hour interval in the middle of the run, usually covering the tracer gas measurement period.

(2) Gas analysis. The composition of the gas, with and without nitrogen tracer, was determined with an Applied Automation on-line process gas chromatograph (GC) which drew a continuous sample from a side draw downstream from the cyclone. The GC had a cycle time of about 11 minutes and was able to detect the following components: H_2, CO_2, CO, CH_4, O_2 and N_2 along with traces of C_2H_6, C_2H_4 and C_3H_6.

(3) Condensables. Condensables consisted of two parts, tar and aqueous. These were determined by taking a side draw from the main gas stream at a rate of about 0.55 SCM/hr. This stream also provided the GC with sample. The gas was first drawn through a hot filter (packed with glass wool) to remove heavy tar components. Next, the gas was passed through two water cooled condensers in series to remove most of the water. The remaining water removal was accomplished with two receivers in series, placed in an ice bath. The remaining tarry mist was then removed by passing the stream through a tightly packed glass wool filter. Flow through the sample train was maintained with the aid of a 'Gast' compressor which provided suction. The discharge from the compressor was then sent through a wet test meter, followed by a drierite column and then to the GC. The wet test meter readings were corrected for temperature, pressure, water of saturation, and tracer gas when needed. The aqueous condensates were collected and weighed. The tar was not measured because prior experience showed that a negligible amount of tar was present (7). The known volume of dry gas flowing through the sample train allowed the amount of condensate to be converted to a mass per unit volume of dry gas basis.

(4) Temperatures. Temperatures at various locations in the system were monitored over the course of each run. Six thermocouples were placed in the system to monitor the temperatures at the following locations: above the the upper tuyeres (drying zone), 10 cm above the grid plate (one 12 mm from the shaft and one half way between the wall of reactor and shaft), 10 cm below the grid plate, at the fan inlet, and at the outlet of the cyclone. The temperature data were noted manually at 15 minute intervals.

Indirect Determinations. These consisted of the following.

(1) Dry product gas rate. The dry product gas rate was indirectly determined with the aid of a tracer technique. Nitrogen gas was introduced into the product gas stream at a metered volumetric rate at the fan inlet. From the known tracer injection rate and the nitrogen composition of the gas with and without tracer, the dry gas volumetric flow rate could be readily evaluated from a nitrogen balance. The molecular weight of the gas allowed the dry gas volumetric rate to be converted to a mass basis.

(2) Dry air and moisture input rates. The dry gas flow rate, coupled with a nitrogen balance between the inlet and

outlet of the gasifier, allowed the dry air input rate to be evaluated. The temperature and relative humidity of the ambient air over the course of an experiment allowed the moisture input from humid air to be evaluated. Since most experiments were five hours in duration, the air humidity was based on mid-run conditions.

(3) Feed rate. The feed rate for each run was determined from a carbon balance between the gasifier inputs and outputs. Among the inputs, only the feedstock contained carbon (determined by elemental analysis). Carbon existed in two output streams, the dry product gas and char; the carbon content of the former was determined from the gas analysis, and the latter, from the elemental analysis of the char. With the rate of dry gas and char previously determined, the feed rate could then be calculated from the carbon balance.

(4) Condensate rate. The directly measured amount of condensate (mass per unit volume of dry gas) and the calculated dry gas flow rate allowed the total condensate output rate to be evaluated.

Operating Parameters

Three operating parameters were investigated in this study, the type of feedstock ,the type of bed support and the grate rotation.

Feedstocks. Two feedstocks were used; mixed hardwood chips (wood chips) and wood pellets (contaminated with about 10% wood chips). The properties of the two feedstocks will be detailed later.

Bed Supports. The bed was supported by a rotating grate having a diameter of 0.56m and 48 open slots. Each slot had dimensions of 11x52 mm and was rounded at the ends. The open area of the grid plate was estimated to be approximately 12% of the cross section of the gasifier. Three different bed supports were used in this study; the first was a 9 cm deep bed of ceramic balls(19 mm diameter) on top of the grid plate (W/B), the second, without the balls, but with about half of the open area blocked by fractured balls (WO1), and the third, an open grid plate (WO).

Grate Rotation. The grate rotation was independently controllable. The grate rotation ranged from 3 to 14 minutes per revolution.

Chemical and Physical Analyses

These included the moisture analyses, ash analyses, and elemental analyses of the feed and char. Moisture was determined by drying in an oven for three hours at 380 K. Ash was determined in a muffle furnace by the standard ASTM procedure. The elemental analyses of feed and char were conducted with a Perkin-Elmer Model 240b Elemental Analyzer.

Feedstock Properties

Table 1 presents properties of the two feedstocks. The wood chips contained a minor portion of bark. The wood pellets were contaminated with about 10% wood chips. The properties presented in Table 1 include the elemental analysis (C,H,O and N), ash content (dry basis), moisture content (wet basis), bulk density, size distribution and higher heating value.

TREATMENT OF DATA

Calculations

The performance of the gasifier was evaluated in terms of a variety of measures determined from the gas composition, material balance data and chemical and physical analyses. Material balance closure was defined as total mass output divided by total mass input. The inputs included air (with moisture) and the wet feedstock. The outputs included dry product gas, condensate, and char. The dry gas composition provided one measure of the gasifier performance, and from it, another measure, the higher heating value of the dry gas was evaluated from the standard heats of combustion of the dry gas components. The gas and

Table 1. Properties of Feedstocks

		Wood chips	Wood pellets
Elemental and	C	47.07	48.59
ash analysis	H	5.84	5.96
(dry wt. %)	O	46.29	44.23
	N	0.35	0.37
	Ash	0.45	0.85
Moisture (wet wt. %)		9.99	6.50
Bulk density (kg/m^3)		236	585
HHV* (MJ/kg)		16.0	17.1
Size distribution (wt. %)			
	T ~ 7 mm		L ~ 2cm
	W >12.7mm	15.4	D =12mm 33
	12.7> W > 4.7mm	74.1	D =13mm 33
	4.7> W > 0 mm	10.5	D =14mm 33

*Higher Heating Value, evaluated by Dulong's formula.

Notations: T=chip thickness.
L=pellet length.
W=width of screen opening.
D=pellet diameter.

char yields provided other measures of performance. These were calculated on a dry feed basis to eliminate the influence of feed moisture variations. An ash-free basis was not used since the ash contents of the dry chips and pellets were less than 1 %. The char yield was expressed as the mass of char per 100 kg of dry feed (DF), and the gas yield was expressed on a volumetric basis as SCM/kg DF. In this work, SCM was based on the conditions of 289 K and 101.3kPa. Also evaluated were the water yield (the mass of condensate per unit mass DF) and the air-to-feed ratio (the mass of dry air per unit mass DF). The total energy output (cold gas basis) was evaluated in MJ/hr and the energy yield (cold gas basis) in MJ/kg DF. The cold gas efficiency was evaluated from the ratio of the higher heating value of the dry gas produced from a unit mass DF to the higher heating value of a unit mass DF. This efficiency represents the percentage of the energy content of the feed converted to combustible gas. The mass conversion efficiency was evaluated from the ratio of the mass of the dry gas produced to the total mass input.

Statistical Analyses

Means and Standard Deviations. The means and standard deviations were evaluated for the various performance measures, including the dry gas higher heating value, volumetric gas yield, water yield, air-to-feed ratio, energy yield, mass conversion efficiency, cold gas efficiency, the highest observed temperature (either of the two temperatures monitored 10 cm above the grate) and the dry gas composition. For each performance measure, the statistics were evaluated for all runs and for the runs with closures greater than 95%. For the dry gas composition, the statistics were evaluated for all runs, for runs with wood chips and for runs with wood pellets.

Regression Analyses. For the performance measures which appeared to show distinct trends with changes in the operating parameters or other performance measures, regression analyses were conducted using the SAS (Statistical Analysis System) software package. These performance measures included: (1) the char yield, the cold gas efficiency and the air-to-feed ratio as functions of the grate rotation with the type of feed as a parameter, (2) the cold gas efficiency and the air-to-feed ratio as functions of the char yield, and (3) the total energy output as a function of the dry feed rate. The regression models and model parameters are summarized in Table 2, along with the square of the correlation coefficient (R^2) and the probability of falsely rejecting the proposed regression model (prob. F-value > F-statistic in F-test for significance of regression). The parameters within parentheses in the first six models (see Table 2) were determined by a search technique, and the remaining ones were determined by linear regression. The first six models were transformed to logarithmic scales prior to performing linear regression.

Table 2. Statistical Analysis

Feed	y	x	R^2	P>F-value	Significant Regression model
Chips	CY	GR	0.99	0.0027	$y = 30.3(x-1.5)^{-0.911}$
Pellets	CY	GR	0.72	0.0018	$y = 21.2(x-2.5)^{-0.304}$
Chips	CGE	GR	0.98	0.0098	$y = 64.8(x-2.97)^{0.108}$
Pellets	CGE	GR	0.91	0.0001	$y = 62.9(x-2.97)^{0.0679}$
Chips	A/F	GR	0.998	0.0008	$y = 1.46(x-2.97)^{0.0506}$
Pellets	A/F	GR	0.81	0.0004	$y = 1.19(x-2.9)^{0.0887}$
All	CGE	CY	0.93	0.0001	$y = 86.3-1.37x$
All	A/F	CY	0.79	0.0001	$y = 1.72-0.026x$
All*	EOR	FR	0.79	0.0001	$y = 18.6+11.9x$

Notations: CY = char yield (kg of char/100 kg of DF);
GR = grate rotation (min/rev);
CGE = cold gas efficiency (%);
A/F = air-to-feed ratio (kg of air input/kg of DF);
EOR = energy output rate (MJ/hr);
FR = dry feed rate (kg/hr).

*Excluding the two runs with grate rotations of 3.0 min/rev.

RESULTS

A total of 16 runs were conducted, and 14 had material balances within 5% of closure (95-105%); the other 2 runs had closures within 6%. Table 3 presents the operating parameters, including the feed species (with the moisture content of the as-received feed), type of bed support and grate rotation, and various measures of the gasifier performance. The table is arranged according to the feed species and the type of bed support. Included in the table are the dry feed (DF) rate in kg/hr, the gas higher heating value (HHV) in MJ/SCM, the volumetric gas yield in SCM/kg DF, the char yield in kg/100kg DF, the energy yield in MJ/kg DF, the air-to-feed ratio in kg/kg DF, the mass conversion efficiency, the cold gas efficiency and the highest observed temperature. The means and standard deviations for the complete data set and the 14 runs with closures greater than 95% are given at the bottom of the table. The grate rotation varied from 3 to 13.7 min/rev. The dry feed rate ranged from 18.2 to 74.3 kg/hr. The gas HHV varied from 5.64 to 6.45 MJ/SCM with a mean of 6.07 MJ/SCM. The volumetric gas yield varied from 1.35 to 2.20 SCM/kg DF with a mean of 1.95 SCM/kg DF. The char yield ranged from 4.4 to 28.3 kg/100kg DF with a mean of 12.5 kg/100kg DF. The energy yield varied from 8.33 to 13.6 MJ/kg DF with a mean of 11.8 MJ/kg DF. The total energy output ranged from 216 to 824 MJ/hr. The water yield ranged from 0.148 to 0.481 kg/kg DF with a mean of 0.216 kg/kg DF. The air-to-feed ratio varied from 0.97 to 1.60 kg/kg DF with a mean of 1.40 kg/kg DF. The mass conversion efficiency varied between 72.1 and 89.7% with a mean of 85.6%. The cold gas efficiency ranged from 49.1 to 80.3% with a mean of 69.5%. The highest observed temperature ranged from 853 to 1172 K with a mean of 1010 K.

Table 4 summarizes the dry gas compositions (mole %) for all the runs. The principal components of the product gas were H_2, CO, CO_2, CH_4 and N_2 with trace amounts of C_2H_6, C_2H_4 and C_3H_6. The means and standard deviations for all runs, for wood chips, and for wood pellets are presented at the bottom of the table. Although the data show some fluctuations, no distinct trends were observed for most of the gas components as the operating parameters varied. However, the means for the hydrogen composition were significantly different for wood chips, 12.2%, and for wood pellets, 16.11%; this was determined by applying the pooled t-test to determine the significance of difference of the two means, which gave 0.001 for Prob.> |t|.

Figures 2 through 4 present the variations in the char yield, cold gas efficiency, and air-to-feed ratio as functions of the grate rotation. Note that the grate rotation is the reciprocal of the grate rotation speed. The data are presented in 3 groups; (1) for wood chips with the ball bed (4 runs), (2) for wood pellets with the ball bed (2 runs), and (3) for wood pellets without the ball bed (10 runs). For groups 1 and 3, asymptotes were observed as the

Table 3. Performance Summary

Run	Feed	Feed Moisture (wet %)	Bed	Grate Rotation (min/r)	Dry Feed Rate (kg/hr)
08/27	Chips	10.30	W/B	5.25	64.2
08/28	Chips	9.99	W/B	4.66	57.9
08/29	Chips	9.67	W/B	3.00	74.3
08/29	Chips	9.67	W/B	9.90	38.8
09/10	pellets*	6.71	W/B	4.23	39.8
09/11	pellets	6.11	W/B	6.00	18.2
09/24	pellets	6.49	WO1	13.70	44.5
09/28	pellets	6.49	WO1	10.80	49.5
10/01	pellets	6.49	WO1	7.33	56.9
10/03	pellets	6.74	WO	12.50	58.8
10/05	pellets	6.42	WO	9.58	56.6
10/05	pellets	6.42	WO	6.13	70.0
10/08	pellets	6.49	WO	9.65	60.8
10/10	pellets	6.49	WO	3.00	46.3
10/12	pellets	6.49	WO	9.83	41.1
10/15	pellets	6.49	WO	10.00	57.0

*Wood pellets contaminated with about 10% wood chips.
Notations: W/B=With ceramic ball bed.
WO1=Without ceramic ball bed but with the grid plate half blocked.
WO =Without ceramic ball bed.

Table 3. (Continued)

Run	Material Balance Closure(%)	Dry gas HHV (MJ/SCM)	Gas Yield (SCM/ kg DF*)	Char Yield (kg/100 kg DF)	H_2O Yield (kg/ kg DF)
08/27	100.74	5.911	2.00	9.53	0.303
08/28	103.06	6.064	2.00	9.95	0.359
08/29	106.30	5.945	1.55	21.29	0.481
08/29	104.28	6.190	2.20	4.38	0.354
09/10	100.00#	6.261	1.96	10.75	0.214+
09/11	98.38	5.993	1.98	10.64	0.184
09/24	101.72	5.640	2.08	12.22	0.301
09/28	103.94	5.829	2.10	9.90	0.261
10/01	102.01	6.145	1.93	11.71	0.234
10/03	104.12	6.186	1.95	13.12	0.312
10/05	99.56	6.023	2.00	13.14	0.202
10/05	100.00#	6.231	1.89	13.74	0.209+
10/08	101.91	5.982	2.18	8.44	0.251
10/10	94.31	6.171	1.35	28.31	0.171
10/12	98.16	6.450	1.87	12.82	0.148
10/15	100.19	6.142	2.07	10.26	0.187
All data					
Mean	--	6.071	1.95	12.51	0.261
St. Dev.	--	0.193	0.22	5.455	0.0872
>95%					
Mean	--	6.075	2.02	10.75	0.251
St. Dev.	--	0.202	0.10	2.428	0.0658

Run	Air-to-feed Ratio (kg/kg DF)	Energy Yield (MJ/kg DF)	Energy Output (MJ/hr)	Mass Conversion Eff.(%)
08/27	1.53	11.80	757.9	85.81
08/28	1.50	12.11	701.7	85.64
08/29	1.22	9.22	686.0	76.70
08/29	1.60	13.64	529.1	89.71
09/10	1.39	12.27	488.2	87.03
09/11	1.49	11.89	216.5	87.10
09/24	1.60	11.75	523.7	85.99
09/28	1.45	12.23	605.4	89.69
10/01	1.35	11.88	676.3	87.54
10/03	1.35	12.08	710.1	85.95
10/05	1.40	12.05	682.1	86.11
10/05	1.32	11.78	824.4	85.61
10/08	1.52	13.02	791.3	89.07
10/10	0.97	8.33	385.8	72.14
10/12	1.26	12.02	493.3	86.39
10/15	1.40	12.69	723.5	88.53
All data				
mean	1.40	11.80	--	85.56
St. Dev.	0.16	1.29	--	4.65
>95%				
mean	1.44	12.23	--	87.16
St. Dev.	0.10	0.54	--	1.51

Run	Cold Gas Efficiency(%)	Highest Temp. Observed (K)
08/27	69.49	1008
08/28	71.34	986
08/29	54.32	977
08/29	80.32	983
09/10	72.24	852
09/11	70.05	947
09/24	69.22	966
09/28	72.03	1172
10/01	69.97	1005
10/03	71.17	1005
10/05	70.95	994
10/05	69.37	983
10/08	76.69	1093
10/10	49.08	1025
10/12	70.78	1052
10/15	74.71	1166
All data		
mean	69.48	1013
St. Dev.	7.61	78
>95%		
mean	72.02	1015
St. Dev.	3.18	83

*Dry feed.
#Closure was forced to 1 to determine (by difference) the H_2O output, which was not measured.
+By difference.

Table 4. Dry Gas Composition (%)

Run	H_2	C_3H_6	CO_2	C_2H_4	C_2H_6	N_2	CH_4	CO
08/27	11.68	0.31	15.74	1.10	0.20	49.28	3.29	18.39
08/28	12.66	0.25	14.94	1.00	0.19	48.41	3.50	19.05
08/29	10.20	0.34	15.30	1.22	0.21	50.54	3.49	18.70
08/29	14.28	0.23	14.99	1.03	0.17	46.78	3.29	19.23
09/10	14.93	0.21	14.81	0.80	0.17	45.62	3.45	20.01
09/11	12.79	0.24	15.81	0.81	0.20	48.31	3.96	17.88
09/24	14.42	0.18	14.95	0.68	0.18	49.50	3.16	16.94
09/28	17.53	0.08	15.01	0.60	0.10	44.62	2.25	19.81
10/01	15.47	0.16	15.34	0.72	0.17	45.13	3.28	19.73
10/03	17.37	0.14	13.91	0.68	0.11	44.56	2.78	20.45
10/05	17.62	0.09	13.43	0.55	0.09	45.02	2.53	20.66
10/05	16.28	0.18	15.01	0.82	0.15	45.10	3.26	19.20
10/08	17.27	0.09	12.68	0.51	0.08	45.18	2.21	21.98
10/10	15.27	0.19	14.07	0.76	0.15	46.24	3.20	20.11
10/12	16.52	0.21	15.11	0.82	0.18	43.63	3.33	20.20
10/15	17.86	0.08	13.23	0.51	0.08	43.82	2.49	21.93
All data								
mean	15.13	0.19	14.64	0.79	0.15	46.36	3.09	19.64
St. Dev.	2.88	0.08	0.91	0.21	0.05	2.18	0.50	1.33
Chips								
mean	12.20	0.28	15.24	1.09	0.19	48.75	3.39	18.81
St. Dev.	1.71	0.05	0.37	0.10	0.02	1.58	0.11	0.37
pellets								
mean	16.11	0.15	14.45	0.69	0.14	45.56	2.99	19.91
St. Dev.	1.56	0.06	0.96	0.12	0.04	1.73	0.54	1.44

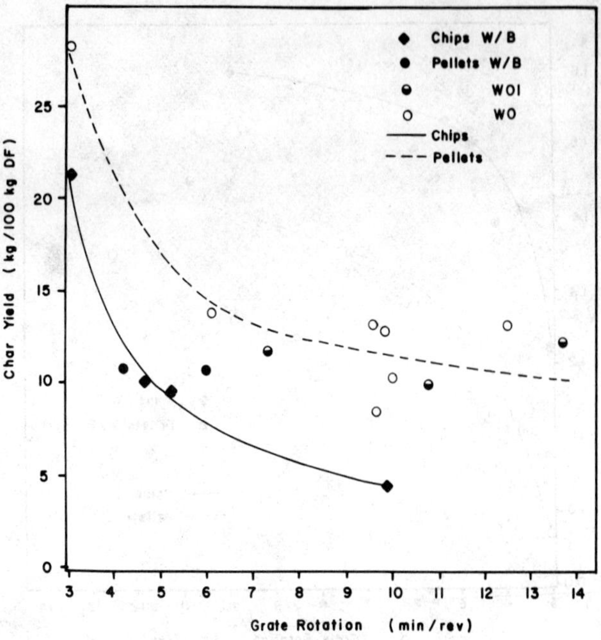

Figure 2. Char Yield versus Grate Rotation

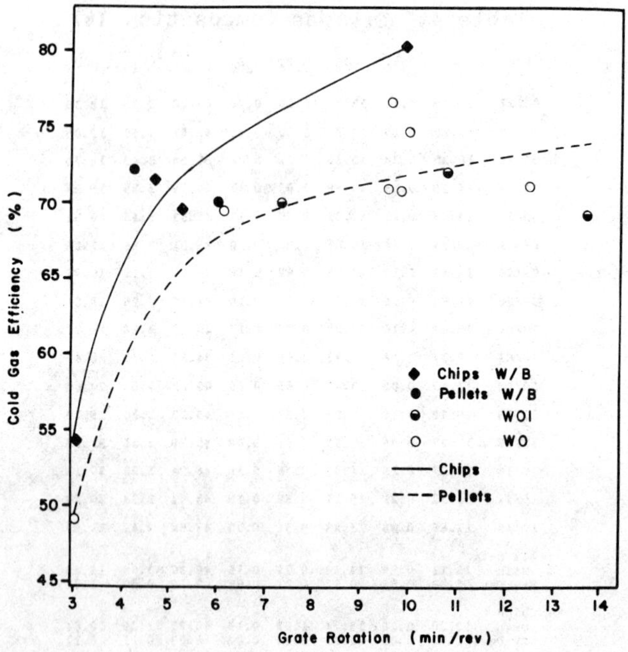

Figure 3. Cold Gas Efficiency versus Grate Rotation

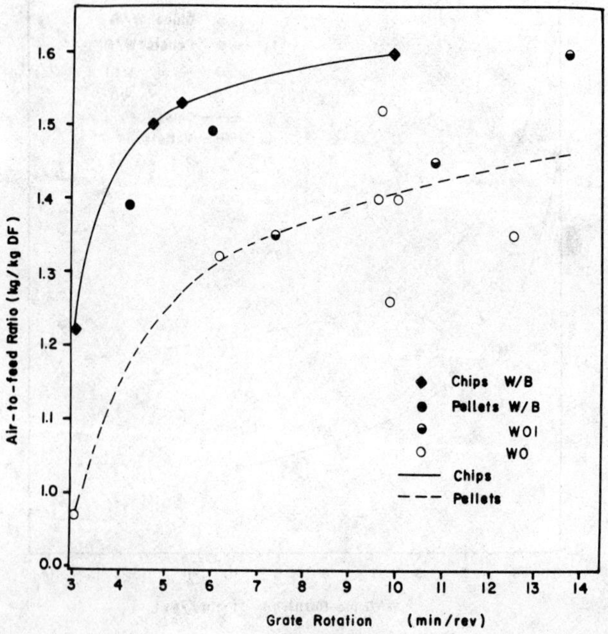

Figure 4. Air-to-Feed versus Grate Rotation

grate rotation approached 3 min/rev. The regression lines (see Table 2) for groups 1 and 3 are included in the figures. Unfortunately, the sample sizes were too small to statistically test for significant differences between the regression models for groups 1 and 3, although a difference appears to exist.

Figures 5 and 6 present the variations in the cold gas efficiency and the air-to-feed ratio as functions of the char yield. Although the data show some fluctuation, a linear correlation was determined by the least-squares method. The parameters and statistics are included in Table 2. The two lowest points in Figures 5 and 6 correspond to the two points at 3 min/rev in Figures 2, 3, and 4.

Figure 7 presents the total energy output as a function of the dry feed rate for the present gasifier, as well as for the previous gasifier (7) for comparison. In fitting the present data by the least-squares method, the two data points at 3 min/rev were excluded. As can be seen from the figure, both data sets were well described by straight lines. Table 2 presents the statistics. It is interesting that these two lines almost coincide with each other, considering the large differences in grate rotation between the two gasifiers. The previous gasifier was operated with maximum grate rotation of the order of 20 min/rev.

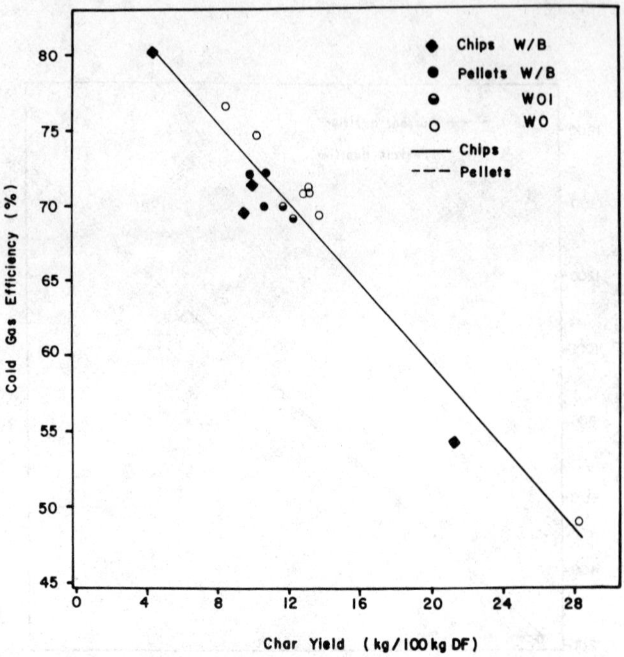

Figure 5. Cold Gas Efficiency versus Char Yield

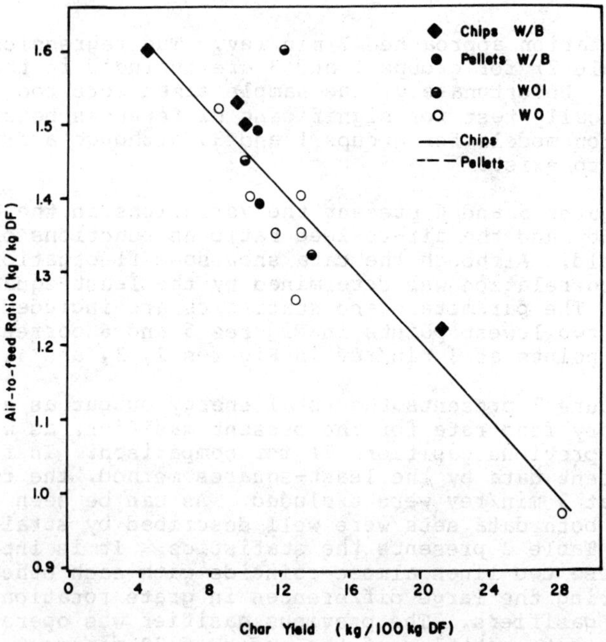

Figure 6. Air-to-Feed Ratio versus Char Yield

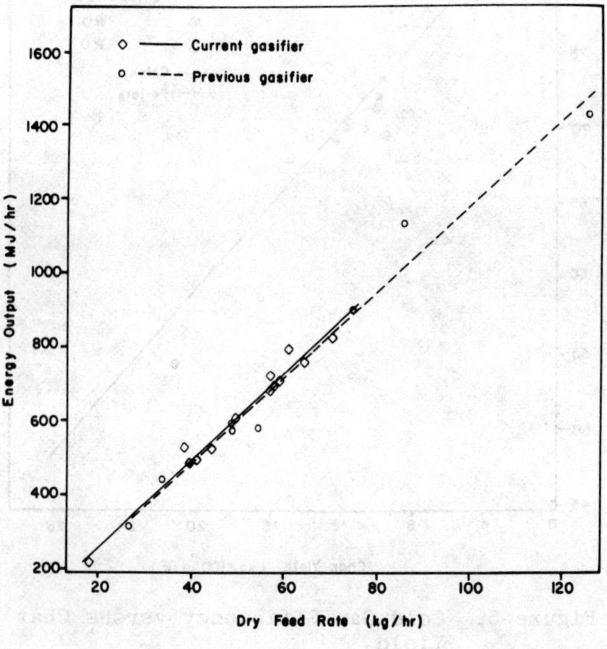

Figure 7. Energy Output versus Dry Feed Rate

DISCUSSION

As shown in Table 3, the 16 experimental runs can be classified, according to feed species and the type of bed support, into three groups, (1) wood chips with the ball bed (4 runs), (2) wood pellets with the ball bed (2 runs), (3) wood pellets without the ball bed, including (a) WO1 (3 runs) and (b) WO (7 runs). Since the sample sizes of the three groups are relatively small, a statistical analysis for determining the effects of the operating parameters on the gasifier performance is not possible. However, a rather primitive qualitative assessment can be made.

Comparison of the results for groups 2 and 3 (a and b) reflects the influence of the bed support on the gasifier performance for wood pellets. Only a few of the performance measures exhibited noticeable trends as the grate rotation increased. These are shown in Figures 2 through 4. As can be deduced from the points in the figures, the data for groups 2 and 3, although scattered, show no definite trend from one group to another as the grate rotation varied (exclusive of the points at 3 min/rev). Therefore, it appears the the three different bed supports have no significant influence on the gasifier performance.

Comparison of the data for groups 1 and 3 illustrates the influence of feed type on the gasifier performance. As can be seen from Figures 2 through 4, different regression lines describe the data for the two groups. The char yield was lower for chips than for pellets, the cold gas efficiency was higher for chips than for pellets, and the air-to-feed ratio was higher for chips than for pellets.

The minimum average residence time of the solid phase in the active zone of the gasifier for the 16 runs was estimated to be of the order of 20 minutes. The solids residence time was evaluated by dividing the length of active zone of the gasifier (approximately 0.3m) by the average solid phase velocity in the active zone. The time required for complete pyrolysis of various types of large particles in a moving bed downdraft gasifier was reported by Reed and Markson (6) to be less than 5 minutes. Therefore, it can be concluded that the residence time of the solid phase in this study was sufficient for complete pyrolysis of the solid phase in all the runs. Consequently, solid phase residence time variations were not responsible for the difference in the char yield between wood chips and wood pellets.

The difference in the char yield between wood chips and wood pellets is probably due to the differences in the dimensions and physical properties (e.g., density and porosity) of the two feedstocks. It is well known that the char yield is higher for large particles than for small particles. Chan et al. (2) reported that the char yields for wood pyrolysis under fire-level radiation were 19% and 24% for pellets with characteristic dimensions of 5 mm and

15 mm, respectively. The smallest dimension (representing the characteristic dimension) of the wood chips ranged from 2 to 5 mm, and that of wood pellets ranged from 12 to 14 mm. The ratio of the characteristic dimensions between pellets and chips ranged from 3 to 7. Therefore, the char yield from wood pellets can be expected to be higher than from wood chips because of the differences in dimensions. In addition, the wood pellets were about twice as dense as the wood chips (see Table 1). Thus, the particle porosity for pellets can be expected to be smaller than that for wood chips (by a factor of about 1/2). The variation in char yield with particle density or porosity has not been determined experimentally; however, it is intuitively expected that it has a role in determining the char yield.

As the grate rotation increased, a sharp increase in the char yield and a sharp decrease in the air-to-feed ratio were observed. The air-to-feed ratio dropped to about 1.2 and 1.0 for wood chips and wood pellets, respectively. Our previous study (7) showed that by maintaining the average air-to-feed ratio at 1.6, we were able to obtain an average char yield of 3.3% with wood chips. Run 08/29 of this study had a char yield of 4.4% with an air-to-feed ratio of 1.6. These observations suggest that the increase in char yield with increasing grate rotation was a result of the decrease in the air-to-feed ratio. This is supported by the decreasing linear relationship between the air-to-feed ratio and the char yield shown in Figure 6. Unfortunately, the air-to-feed ratio for the present gasifier can not be independently controlled. In principle, the air input rate will be determined by the depth and voidage of the bed and the pressure drop across the bed and the fan rpm or suction. The fastest rotations of the grate may have caused a dramatic change in the bed voidage due to the action of the tuyeres, which in turn was responsible for the sharp increase in the char yield. This was suggested by the decrease in the air-to-feed ratio at high grate rotation.

The cold gas efficiency was found to decrease linearly with increasing char yield. This relationship along with the decreasing linear relationship between the air-to-feed ratio and the char yield implies a linear relationship between the air-to-feed ratio and the cold gas efficiency. Note that the cold gas efficiency represents a conservative estimate of the first law efficiency for the gasifier. In other words, by maintaining an adequate air-to-feed ratio a high efficiency will result.

Table 5 summarizes the operating conditions and various performance measures for the present study and the previous study (7). As can be seen from the table, despite the differences in operating conditions used for the two gasifiers, most of the performance measures are similar, including the gas yield, water yield, energy yield, cold gas efficiency, mass conversion efficiency and gas composition. However, the average char yield from the present gasifier was three fold larger than that from the previous gasifier.

Table 5. Comparisons of the Present Study with the Previous Study

Study		Present	Previous (7)
Operating parameters:			
Feedstocks		Chips, pellets	Chips
moisture	(wet %)	9.99, 6.50	10.33
Bed support		W/B, WO1, WO	W/B
Grate rotation	(min/r)	3-14	20-30+
Material balance closure	(%)	94-106	80-115
Input: Dry feed rate	(kg/hr)	18-74	27-126
Air-to-feed Ratio*	(kg/kg DF)	1.44	1.60
Output: Gas yield*	(SCM/kg DF)	2.02	2.19
char yield*	(kg/100kg DF)	10.8	3.34
water yield*	(kg/kg DF)	0.25	0.26
Mass conversion efficiency	(%)	87.2	87.9
Energy aspects:			
Dry gas HHV	(MJ/SCM)	6.08	5.51
Energy yield	(MJ/kg DF)	12.2	12.0
Energy output	(MJ/hr)	220-820	310-1430
Cold gas efficiency	(%)	72.0	71.5
Dry gas compostion:	(%)		
Major components	H_2	15.1	15.1
	CO	19.6	19.1
	CH_4	3.09	2.51
	CO_2	14.6	15.8
	N_2	46.4	46.8

*Mean for the runs within of 5% closure.
+Estimated.

The average air-to-feed ratio for the present gasifier, 1.44 kg/kg DF, was also significantly different from that of the previous gasifier, 1.60 kg/kg DF; this is supported by the pooled t-test, which gives 0.005 for Prob.>|t|. Another difference is in the dry gas HHV, which was 6.08 MJ/SCM for the present gasifier and 5.51 MJ/SCM for the previous one; the Prob.>|t| for the pooled t-test was 0.001. The highest energy output rate of the present gasifier was about 50% lower than that of the previous one; however, the highest throughput was also about 50% lower.

CONCLUSIONS

The results of this work indicate that the performance measures are similar for the present gasifier and the gasifier examined in the previous study. From the viewpoint of total energy output, the gasifier behaves similarly with wood chips and wood pellets as the feedstock.

The char yield was found to sharply increase at high grate rotation and to level out at slow grate rotation. The air-to-feed ratio and cold gas efficiency were found to decrease linearly with increasing char yield, indicating that maintaining an adequate air-to-feed ratio is necessary for high efficiency. The cold gas efficiency was determined to increase linearly with increasing air-to-feed ratio.

The char yield from wood pellets was significantly higher than that from wood chips. The higher yield was a consequence of the larger characteristic dimension of the pellets compared to wood chips.

The type of bed support had no significant influence on the performance of the gasifier when comparable operating conditions were used.

ACKNOWLEDGEMENT

This is contribution no. 86-352-A, Department of Chemical Engineering, Kansas Agricultural Experiment Station, Kansas State University, Manhattan. This work was supported in part by a grant from The Buck Rogers Co., Inc., Industrial Airport, KS.

REFERENCES CITED

1. Bain, R., 'Benefication of Biomass for Gasification and Combustion,' in A Survey of Biomass Gasification, T.B. Reed, Ed., Vol. 3, 1-25, SERI/TP-33-239 (1980).
2. Chan, W. R., Marcia Kelbon and B. B. Krieger, 'Product Formation in the Pyrolysis of Large Wood Particles,' in Fundamentals of Thermochemical Biomass Conversion, Elsevier Science Pub. Co., New York, 219-236 (1985).
3. Goldstein, I. S., 'Wood as a Source of Chemical Feedstocks,' in Energy and Environment Concerns in the Forest Products Industry, AIChE Symposium Series 177, 74 111-114 (1978).
4. Reed, T. B., et al., 'A Survey of Biomass Gasification,' Vol. 2-3, SERI/TP-33-239 (1980).
5. Reed, T. B., and D. E. Jantzen, 'Generator Gas - The Swedish Experience from 1939-1945,' (translated from the work of the Swedish Academy of Engineering) SERI/TP-33-239 (1979).
6. Reed, T. B., and M. Markson, 'A Predictive Model for Stratified Downdraft Gasification of Biomass,' Presented in Proceedings of the 14th Biomass Thermochemical Conversion Contractors Meeting. Atlanta, GA (1982).
7. Walawender, W. P., S. M. Chern, and L. T. Fan, 'Wood Chip Gasification in a Commercial Downdraft Gasifier,' in Fundamentals of Thermochemical Biomass Conversion, Elsevier Science Pub. Co., New York, 911-922 (1985).
8. Zerbe, J. J., 'The Contribution of Wood to the Energy Picture,' presented at Conference on Wood - An Alternate Energy Resource for Application Industry and Institutions, Proceedings published by School of Engineering, North Carolina State University (1981).

DISCUSSION

A. Talib, Mitre Corp. — Within the gasification zone, the temperature should increase because of the increasing air-to-feed ratio. That could have an impact on char yield. As the temperature goes up, the char yield will definitely go down.

Walawender — Unfortunately, in those experiments, we weren't measuring temperatures in that region. We were measuring just above and below the grate.

J. Diebold, SERI — I was wondering if you had data on the char composition and heating values as function of the parameters that you examined?

Walawender — Yes, I have that data. I mentioned that we had to do carbon balances to back up the feed rates, so we have all the elemental compositions for the different conditions.

INTEGRATED PRODUCER GASIFIER-DIESEL ENGINE SYSTEM FOR SOLID WASTE

George J. Trezek
University of California
Berkeley, California

John C. Glaub
EMCON Associates
San Jose, California

ABSTRACT

The results of three separate studies involving the creation of producer gas from densified processed municipal solid waste and agricultural residues are discussed. These include gasification of processed d-RDF, densified rice hulls, and other agricultural residues. The subsequent producer gas was combusted in the following types of diesel engines: a 5 hp single cylinder, a Lombardini two cylinder, and a single cylinder CFR engine. An evaluation of the overall engine operating characteristics is made relative to the type of fuel and the quality of the producer gas. For example, the gas with the highest heating value is not necessarily the most efficient. Studies of additional engine parameters included the effect of injection timing, compression ratio and speed. In some cases, the thermal conversion characteristics of the fuel were evaluated by means of gas chromatography. The performance of the gasifier was shown to be sensitive to the fuel processing methodology and final shape. Results show that solid waste fuels can be used to economically displace up to 80 to 90 percent of diesel fuel in various applications. For some fuels, the engine can be operated completely on producer gas. Preferred conditions correspond to producer gas heating values greater than 3500 kJ/m^3.

INTEGRATED PRODUCER GASIFIER-DIESEL ENGINE SYSTEM FOR SOLID WASTE

INTRODUCTION

The coupling of gasifiers with prime movers had its origin in the late 1800's. During World War II nearly 40 percent of Sweden's motor vehicle fleet was converted to the use of wood and charcoal gasifiers (1). Our studies have shown that producer gas obtained through the gasification of biomass, which includes densified refuse-derived fuel (d-RDF) and carbonaceous agricultural residues, is a suitable supplementary fuel for internal combustion engines. In fact, for a diesel engine, producer gas can effectively be used to meet 70 to 80 percent of the engine's fuel requirements. Using producer gas as the fuel has the added advantage of lowering the exhaust temperature, thereby reducing NO_x air emissions and yielding a smokeless exhaust.

Gasifying agricultural crop residues was pioneered in the Philippines by Professor I. E. Cruz (2). Cruz examined the performance of a gasifier in supplying producer gas to a converted oil-fuel boiler. Performance of the gasifier operating as an updraft reactor was compared to its performance as a downdraft reactor. Higher thermal efficiencies were achieved with the updraft operation , but significant ash and clinkering problems developed. These problems were shown to be greatly reduced or eliminated for operation in the downdraft mode; however, slightly lower thermal efficiencies and reduced gasification rates resulted. The fuel in these studies consisted of coconut waste in the form of shells, husks, coir dust wood, and ipil wood. Later studies using coconut wastes and rice hull fuels demonstrated the feasibility of operating a diesel engine on producer gas (3).

The results of a series of gasifier-diesel engine studies conducted at the University of the Philippines (UP) and at the waste processing laboratory at the University of California, Berkeley (UCB) are given. The studies are divided into three areas: (1) tests with d-RDF made at UCB and evaluated at UP in a downdraft gasifier connected to a 5 BHP Lister single cylinder diesel engine, (2) tests with densified rice hulls made at UCB and evaluated at UP in a downdraft gasifier connected to a Lombardini two cylinder diesel engine and (3) tests with multiple fuels in the UCB gasifier connected to a UCB Waukesha single cylinder CFR engine.

SINGLE CYLINDER d-RDF TESTS

As part of a semi-formal collaboration between UP and UCB in solid waste management, a test program, predicated on the demonstrated feasibility of operating a diesel engine on agricultural residues, was developed for refuse-derived fuel. Since adequate densification equipment was not available at UP, d-RDF fuel pellets were prepared at the solid waste processing laboratory at UCB and shipped to UP. It should be noted that due to significant differences in the waste stream, d-RDF pellets produced from domestic solid waste would probably have different properties than pellets produced from a waste stream in a developing country.

The d-RDF pellets were produced from municipal solid waste that had been subjected to the unit operations of size reduction in a swing

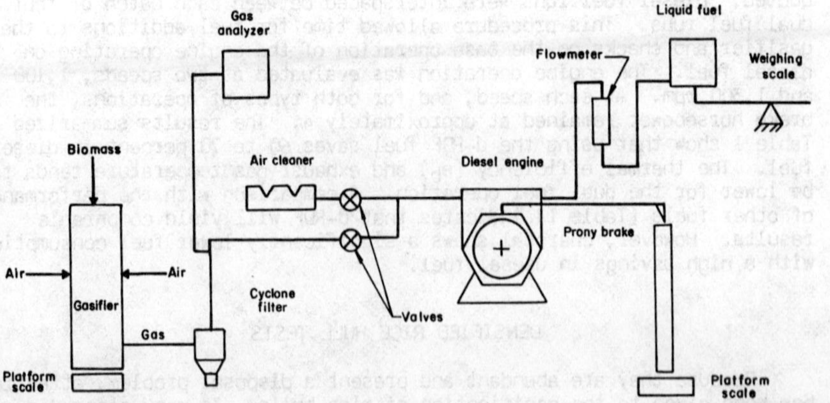

Figure 1. Schematic of UP Gasifier-Diesel Engine System

hammermill, air classification, and trommel screening. The details of these operations in the waste processing facility have been previously discussed (4). The pellets were produced in a densifier manufactured by the California Pellet Mill Co. This device consists of three major components: a mixer/screw feeder, a drive system (speed reducer), and a die. Essentially, loose material is moved by the feeder into the space between the rollers and the rotating die. As the assembly rotates, material is forced through the multiple-hole die. Friction between the material and the wall of the individual die holes supplies the resistance required for compaction. The pellet configuration can be changed by using dies with holes of differing diameters or by changing the die thickness. The pellets used in the tests had diameters of about 1.3 cm and lengths of about 7.6 cm. The analysis of the composition on a dry basis indicated that the results were composed of 50.9% carbon, 6.2% hydrogen, 31.9% oxygen, and 11% ash. The low ash content is due to the final stage of screening which removes a significant portion of the noncombustible fines.

A schematic diagram of the experimental facility used to conduct the gasifier-engine experiments is shown in Figure 1. The gasifier is fabricated from a 55 gal drum and contains an hour glass configuration refractory. Thus, when the unit is operated in the downdraft mode, raw fuel is batch fed at the top. The major zones proceeding downward within the gasifier are dry out or distillation, combustion (contained in the region of smallest cross section, i.e., the mid-section or center of the hour glass), reduction, and ash. Suction of air into the system was provided by the engine. The gasifier contained 12 air holes around the mid-section and a single gas outlet at the bottom. The gas outlet was connected to a cyclone separator that filtered out entrained dust and fines from the producer gas before the gas entered the engine. A Lister single cylinder (4.5 bore by 4.25 stroke) diesel engine with a compression ratio of 14:1 was used. The producer gas together with air, is aspirated into the engine. The gasifier was mounted on a platform scale so that the fuel consumption could be continuously monitored. A similar arrangement was utilized for the liquid fuel. Engine torque and rpm were measured by a prony brake and hand tachometer respectively.

An evaluation of the producer gas based on the procedure given by Cruz (2), which allows for a combined water decomposition of 90 percent, gave the following gas composition on a dry basis: 13.3% CO_2, 12.4% CO, 17.6% H_2, 0.8% O_2 and 55.9% N_a. Twenty 30 minute long runs were conducted. Diesel fuel runs were interspaced between each batch of four dual fuel runs. This procedure allowed time for fuel additions to the gasifier and checks on the base operation of the engine operating on diesel fuel. The engine operation was evaluated at two speeds, 1,100 and 1,300 rpm. At each speed, and for both types of operations, the brake horsepower remained at approximately 4. The results summarized in Table 1 show that using the d-RDF fuel saves 60 to 70 percent in diesel fuel. The thermal efficiency (e_b) and exhaust gas temperature tends to be lower for the dual fuel operation. A comparison with the performance of other fuels (Table 1) indicates that d-RDF will yield comparable results. However, charcoal shows a significantly lower fuel consumption with a high savings in diesel fuel.

DENSIFIED RICE HULL TESTS

Because they are abundant and present a disposal problem, attention has been given to the gasification of rice hulls. In many cases, in addition to the savings arising from the reduction of fossil fuel consumption, other savings become possible through the use of biomass that is indigenous to the location in which the power is to be used. For example, in a developing country, many small rice mills are powered by electricity purchased from a central grid or by an on-site diesel engine-generator system. In both situations, the producer gas resulting from the gasification of rice hulls can be used to fuel internal combustion engine-generator systems and thereby provide a significant portion of the mill's power requirements. The corresponding situation may not be identical in a developed country. Here it may be more advantageous to use the rice hull fuel for drying rather than power generation. Most likely the energy conversion mechanism would take advantage of the ability to use loose instead of densified rice hulls. Thus, a fluidized bed directly combusting loose rice hulls would be more appropriate than a gasifier operating on densified fuel.

Table 1

COMPARATIVE PERFORMANCE OF A 5-BHP SINGLE CYLINDER DIESEL ENGINE WHEN CONVERTED TO DUAL FUEL OPERATION

Fuel	RPM	Brake Horse-power	Brake Specific Fuel Consumption	Solid Fuel Consumption (kg/hr)	Diesel Fuel Displacement (%)	Thermal Efficiency (%)	Exhaust Temperature (°C)
MSW Pellets and Diesel	1,100	4.1	0.370	5.2	62	9.6	444
Diesel	1,074	4.0	0.987	0	-	13.2	>540
MSW Pellets and Diesel	1,333	4.5	0.287	4.5	72	12.0	463
Diesel	1,283	4.0	1.009	0	-	13.0	>540
Coconut Shell and Diesel	1,212	4.7	0.208	5.6	72	10.5	361
Diesel	1,208	4.2	0.730	0	-	17.9	435
Charcoal and Diesel	1,043	4.1	0.148	1.8	83	16.6	396
Diesel	1,000	4.0	0.892	0	-	14.6	>540
Coal and Diesel	1,288	4.8	0.246	2.8	80	12.5	446
Diesel	1,246	4.9	1.261	0	-	10.4	>540
Woodwaste and Diesel	1,221	4.1	0.357	5.3	62	9.5	383
Diesel	1,237	4.1	0.950	0	-	13.7	502

The experience with gasifying loose rice hulls has not been entirely successful. This was true of the experience with rice hull gasifiers in Italy in the 1940's (5) and more recently with the gasification studies in the Philippines and at UCB. The problems have predominantly been the incomplete gasification and, in some cases, the production of a gas with a high tar-content. To overcome these problems and develop a fuel with enhanced storage, transportation, handling, and feeding characteristics, densifying the fuel was considered. Again, rice hulls were densified at UCB and tested in the UP gasifier-diesel engine system.

The densification procedure involved using a lubricant with the fuel and using a die with small diameter holes. The dies, particularly in the 20 to 25 mm diameter range, previously used for RDF had the tendency to jam. The pellets used in the test had a diameter of 6.25 mm and a length of 50 to 70 mm. A small amount of diesel fuel was mixed with the loose rice hulls prior to densification. In effect ,the diesel fuel acted as both a lubricant and binding agent. The pellets were allowed to cool after being discharged from the densifier, thereby allowing some of the diesel fuel to evaporate. It is interesting to note that in comparison with pellets made from sawdust, processed municipal solid waste, and a variety of other agricultural wastes, the rice hull pellets were harder, more difficult to break, and in general retained their length. The heating value of the pellets formed with the diesel fuel lubricant increased by about 10 percent over that of the loose rice hulls.

Both the gasifier configuration and gas cooling and cleaning train were changed during this set of experiments. Further research on the gasifier internal configuration indicated that enhanced operation was possible by designing an air manifold around the throat (Figure 2). The manifold allowed air to enter into the fuel bed through tuyeres positioned immediately below the throat. Thus, the combustion level could be maintained immediately below the throat. A rotary grate ash removal system was also included. The unit was about the size of a 55 gal drum and was operated in the downdraft mode with batch feeding. Typically, these units are started by either (1) starting a small charcoal fire on the grate then the fuel loading or (2) loading fuel to the throat and then igniting it with charcoal.

The gas cleaning train (Figure 3), designed to remove impurities and lower the gas temperature to be consistent with engine operation, consists of a cyclone, a condenser, a mechanical filter, a liquid gas scrubber, and a second mechanical filter. The cyclone serves to remove particulates entrained in the gas stream. The condenser lowers the temperature of the gas causing tars and liquors to condense. The first mechanical filter utilizes rice hull char as the filtering medium. The liquid scrubber uses both water and diesel fuel, with the diesel fuel functioning as a solvent for any tar that may remain in the gas. In the second mechanical filter dried rice hulls are used as the filter medium. The second filter dries the gas prior to its injection into the engine.

The diesel engine is a twin-cylinder, 12 horsepower Lombardini, Model L-20. The engine has a bore of 95 mm, a stroke of 94 mm, and a displacement of 1.332 liters. It is directly coupled to a generator. The generator is an alternating current, single-phase unit rated at 7.5 kW. A rheostat immersed in water is used to apply a load to the generator. The rheostat has a line voltage of 220 volts and a fixed frequency of 60 cycles/second. This frequency corresponds to an engine rpm of 1800.

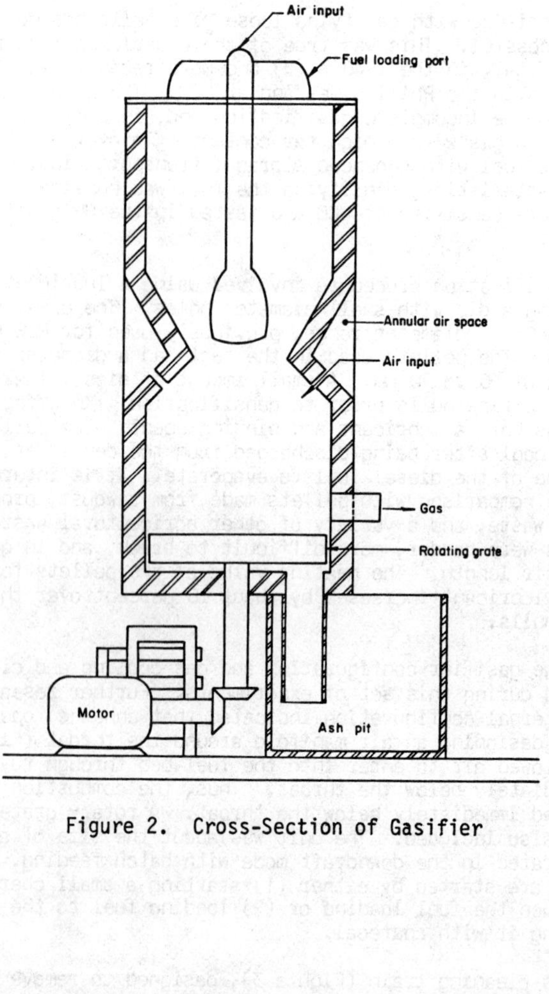

Figure 2. Cross-Section of Gasifier

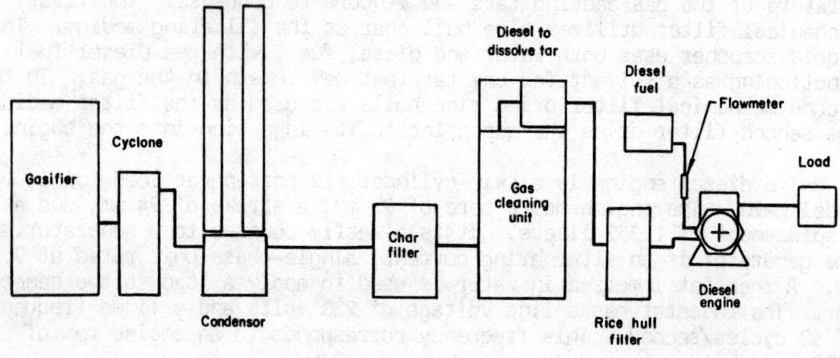

Figure 3. Schematic of UP Gas Cleaning System

Gasification of the densified rice hulls demonstrated several features that were somewhat unique to their use as a fuel. Observations made in the course of the testing include the following: (1) The gasification of pelletized hulls was characterized by a higher efficiency than that attained with loose rice hulls. (2) About 25 percent ash resulted from the gasification process -- a percentage that is comparable to the ash content of rice hulls. (3) Although precise measurements of tar production were not made, the sight glass leading to the engine remained clear. (4) It was possible to operate the engine on producer gas alone. This was the first time during the entire series of UP test runs on various fuels that a twin-cylinder diesel engine could be run solely on producer gas. Typically, the diesel fuel displacement was about 90 percent. (5) The heating value of the producer gas was estimated to be about 4470 kJ/m^3.

The characteristics of the fuel that may have affected the performance of the gasifier include moisture content, size and shape, and volumetric energy density (VED). The densification process typically reduces the moisture content of solid fuels because of the high temperatures generated in the die. A fuel with a low moisture content improves the efficiency of gasification because the sensible heat required to vaporize the water reduces the thermal efficiency. In the present study, the small amount of diesel fuel that was added to the rice hulls would have also had a positive effect on the thermal efficiency. The size and shape of the fuel determines the rate and course of the reactions in the gasifier. Furthermore, size and shape determine the presure drop across the fuel bed. Ideally, since this gas producer operated on the suction from the internal combustion engine, the pressure drop should be as low as possible. Loose rice hulls have the tendency to mat, thus increasing the pressure drop and reducing the gas flow, which results in incomplete combustion. The VED of the fuel has an effect on the size of the fuel hopper and the reactor. Materials with high VED's are considered ideal gasifier fuels (6). Use of these fuels lowers fuel consumption and produces a cleaner gas. In the present study, a larger percentage of diesel fuel could be displaced when pelletized rice hulls were used than when loose rice hulls were used.

The results of the experiments are summarized in Figures 4, 5, and 6. Figure 4 illustrates the percent of diesel fuel displacement measured in the test runs. The curves in Figure 4 indicate that at loads above 3 kW, the diesel fuel displacement was 5 to 10 percent greater when the gasifier was fueled with pelletized rice hulls than when it was fueled with loose rice hulls. At loads of about 5 kW, almost 90 percent of the diesel fuel could be replaced when the pelletized rice hulls were used. As shown in Figure 5, the thermal efficiency of the system operating on pelletized rice hulls was approximately twice that of the system when operated on loose rice hulls. Even though densifying the fuel resulted in improvement in gas quality and hence in the thermal efficiency, it should be pointed out that operation of an internal combustion engine with producer gas is only half as efficient as operation with diesel fuel.

The effect of densification on the consumption of rice hulls and diesel fuel is indicated by the curves in Figure 6. As the figure illustrates, operating the gasifier with pelletized fuel results in an approximately 50 percent reduction in the consumption of rice hulls. The reduction probably was due to the fact that the shape of the fuel improved the gas flow through the reactor, thus making it possible for most of the carbon to be gasified. The more extensive carbon gasification attained with pelletized rice hulls than with loose hulls is evidenced by the lower carbon content of the ash formed in the gasification of the pelletized hulls. Similarly, the consumption of diesel fuel also was reduced, particularly at the higher loadings.

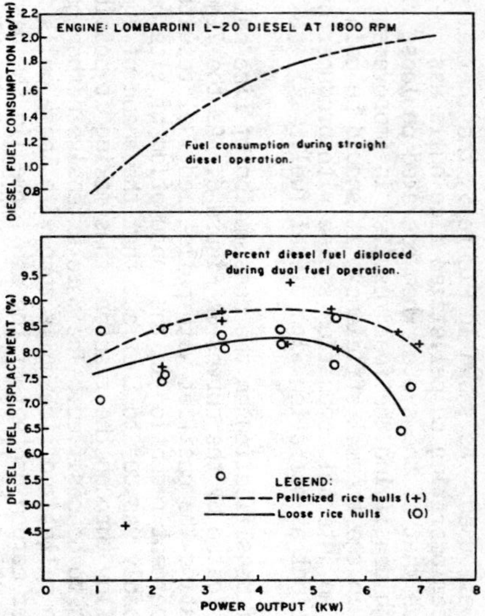

Figure 4. Diesel Fuel Replacement by Producer Gas From Rice Hulls

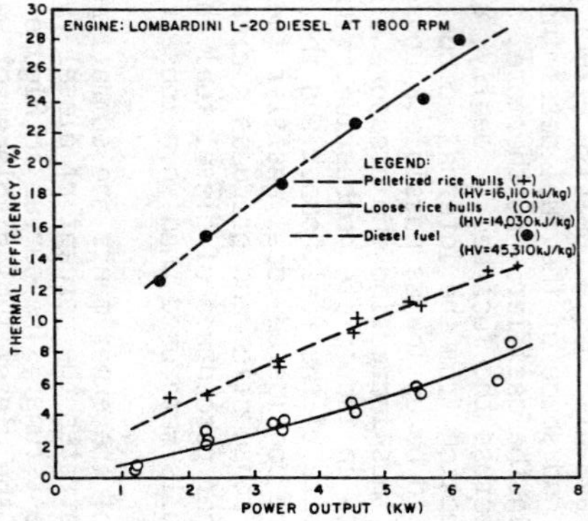

Figure 5. Thermal Efficiency of Rice Hull Gasifier-Diesel Engine System

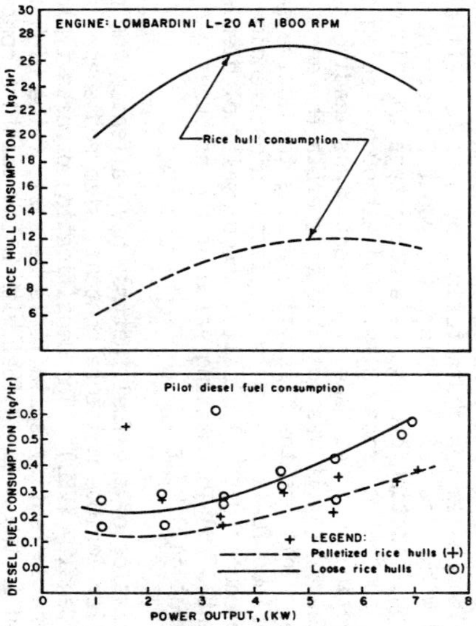

Figure 6. Rice Hull and Diesel Fuel Consumption

MULTIPLE FUEL CFR ENGINE TESTS

Although the gasifier-diesel engine tests at UP demonstrated that producer gas could be a suitable fuel for a diesel engine, the engine parameters governing its performance required refinement. In contrast, the CFR engine is ideally suited for these tests since parameters such as compression ratio and injection timing can varied and the cylinder pressure can be measured. Pressure-volume relationships can be determined from compression ratio and cylinder pressure measurements. Thus, in this set of tests designed to study the behavior of engine parameters, the UCB gasifier was coupled to the CFR engine located in the UCB Mechanical Engineering Laboratory.

The UCB gasifier is similar to the UP unit. It contains 8 cm of Litecrete 70 refractory insulation. The throat is formed by a 16.4 cm cylindrical air distributor suspended in the center of the gasifier by a 5 cm diameter pipe. The pipe is secured to the top of the gasifier and serves as the intake air conduit. A stationary steel grate holds the fuel bed about 15 cm above the reactor bottom but allows ash to fall through. Ash removal and fuel feeding ports are located in the top and bottom of the gasifier, respectively. The details of this unit and its operation have been previously given (7).

The producer gas outlet, located below the ash grate, was connected to a gas cooling and cleaning scrubber column and through by pass valving to the Waukesha CFR engine (Figure 7). The scrubber, designed by Cal Recovery Systems, replaced the cooling and cleaning unit operations of the UP system (Figure 3). Essentially, the unit cools the producer gas and removes tars and heavy hydrocarbons by the interaction of two processes: the flow of the producer gas up through the lower spray, two sieve plates, and the upper spray before exiting at the top of the column; and the downward flow of water from these plates and sprays. In this unit the water level at the bottom of the column must be carefully controlled to avoid water backing up into the gas inlet of the gasifier or producer gas escaping into the water drain. In this regard, the UP system was superior. A moisture trap was included to remove the condensate from the saturated gas leaving the column, and a by pass arrangement was included that allowed unused or excess producer gas to enter into the exhaust system. During operation it is often desirable to run the gasifier at a constant output while varying the gas input into the

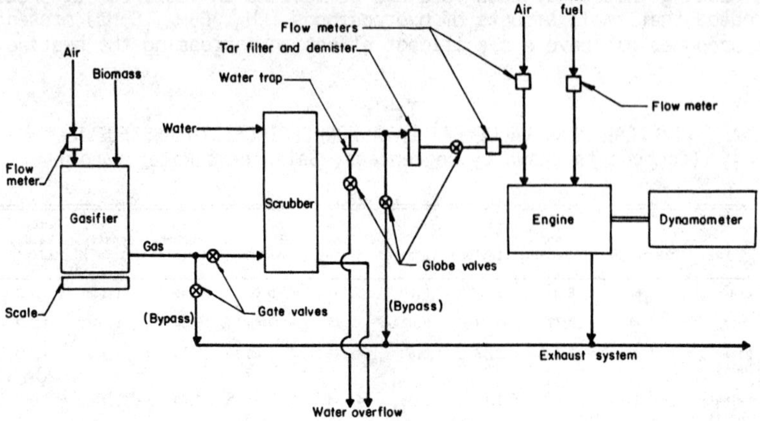

Figure 7. Schematic of UCB Gasifier-Diesel Engine System

engine. The by pass is also used during gasifier warmup when the gas quality is too poor to be effective in the engine.

The single cylinder Waukesha CFR diesel or compression ignition engine has a precombustion chamber. By adjusting a piston in this antechamber the compression ratio can be varied from 8:1 to 25:1. Air to the engine was measured with oriface plates and then with a laminar flow tube. A positive displacement gas meter was used to measure gas flow into the engine because of its low pressure drop. The flow rate of the pilot fuel, commercial diesel oil, was measured by timing the flow of the fuel from a buret. Cylinder pressure was monitored by a quartz pressure transducer on the cylinder wall and displayed on an oscilloscope indicating top dead center (TDC) and twenty degrees before TDC. A pressure transducer was placed in the fuel injector. Injection time is determined by comparing this pressure trace to the TDC tick mark. Injection is assumed to occur at peak pressure. The eddy current dynamometer used to measure engine shaft work operated either in a constant load or constant speed mode. The producer gas was analyzed with a Hewlett Packard Gas Chromatograph (GC) Model 5840A with a 6 foot Molecular Sieve and a 10 foot Porapak Q column balanced and calibrated for the eight gasses tested: hydrogen, carbon dioxide, ethane, ethylene, oxygen, nitrogen, methane, and carbon monoxide. Samples were taken from the scrubber outlet using pressure lock syringes.

Since the gasifier operated in a batch mode, its operation consisted of a series of three phases: warmup, pseudo-steady state or producing phase, and burn out. The initial warmup would take one half to two hours while the warmup phase for reloads was usually less than 10 minutes. The producing phase could last for 20 minutes to more than an hour depending on fuel type, amount of fuel, and air flow rate. The burn out phase occurs when the fuel in the combustion zone becomes spent. In this phase, the hydrocarbon concentration in the gas begins to drop, then the hydrogen concentration drops while the carbon monoxide concentration remains constant or even increases. The overall effect is a reduction in the heating value, which becomes too low for engine operation at 2,500 to 3,500 kJ/m^3.

Five different biomass fuels were used in the gasifier: charcoal, wood, rice hulls, walnut shells, and densified refuse derived fuel (d-RDF). Results varied in gas composition, amount of tar and heavy oils emitted, and the nature of the ash. The producer gas composition and heating values for each fuel are summarized in Table 2. It should be noted that small amounts of hydrocarbons (CH_4, C_2H_4, C_2H_6) present in the producer gas have a significant effect on increasing the heating

Table 2
TYPICAL PEAK VALUE GAS DATA FOR VARIOUS BIOMASS FUELS
(Components Shown by Percentage, Balance is Water Vapor)

Fuel	Hydrogen	Carbon Dioxide	Ethylene	Ethane	Nitrogen	Methane	Carbon Monoxide	Density (kg/m^3)	Heating Value (kJ/m^3)
Charcoal	6.66	8.80	0.00	0.00	61.61	0.00	20.46	1.138	3,211
Wood	12.43	12.79	0.09	0.00	55.35	0.86	16.56	1.102	3,737
Rice Hulls	7.37	17.43	0.76	0.00	53.88	3.00	15.10	1.171	4,202
Walnut Shells	10.36	19.53	1.01	0.38	41.34	5.52	19.40	1.140	6,384
d-RDF	9.95	22.02	0.76	0.40	46.41	3.46	14.52	1.169	4,877

value. On the other hand, lower heating values are attributable to large amounts of N_2 (40 to 60%) and CO_2 (8 to 22%). Even though walnut shells have the highest heating value, their overall fuel efficiency was only 43% as compared to 60% for wood and charcoal. The more efficient fuels produce a cleaner gas, i.e., less tars and oils as evidenced by the color of the scrubber water. Although charcoal gave the cleanest producer gas, slagging problems were encountered with a particular brand. As anticipated from the previous UP experience, loose rice hulls had a high ash content and a quick burn out ; they also had a deleterious effect on the thermocouple probes and refractory. The gas produced from wood was of higher quality than that from charcoal, but not as consistent. As shown in Table 2, the hydrogen concentration was higher in the gas produced from wood than in gases from all the other biomass fuels. This may be due to the higher moisture content, since hydrogen results from the high temperature reaction of water with carbon monoxide or carbon. Methane was usually produced and trace amounts of ethylene were often produced in the gasification of wood. The scrubber water was a dark yellow, indicating that a higher level of condensables were removed from the gas. The ash was similar to that of charcoal.

This particular diesel engine did not perform well on large amounts of producer gas. Unlike the earlier tests at UP, the diesel fuel could never be completely displaced. The limiting factor in the quantity of producer gas that could be utilized was the severe engine knock due to ignition delay. The knock increased directly with the delay between injection and ignition. At low speed, 950 rpm, the delay was 4 ms or 23° of crank rotation, while at higher speeds, in the range of 1450 rpm, the respective values were 1.4 ms and 12°. Although delays could be reduced by advancing the injection timing of the diesel fuel, it ceased to be an effective remedy as more diesel fuel was displaced. As the quantity of producer gas increased there was a reduction in power output of this engine (Figure 8). Power output is sensitive to the gas quality. For qualities greater than 3,500 kJ/m^3 a reasonable level of power could be maintained to nearly a 60 percent displacement. With this engine, diesel fuel operation at the lower speed, 950 rpm, was more efficient (Figure 9). At the higher speed, 1,450 rpm, the dual fuel operation was consistently less efficient than straight diesel operations, while at the lower speed it tended to be a function of power output. The pressure-volume relationships (Figure 10) were obtained from digitized oscilloscope pressure trace photographs. Since the cycle moves more toward a constant volume process with producer gas, the diesel cycle under these conditions tends to resemble an idealized Otto Cycle.

In summary, even when the gas quality was below 3,500 kJ/m^3, this engine could operate with only a minor loss of efficiency. In general, operation with producer gas did not cause more than a 10 percent change in efficiency. Although the diesel engine tended to lose power under dual fuel operation, it could maintain 90 percent power with more than half the diesel fuel displaced by gas if the quality was high enough (4,000 kJ/m^3). As was observed in the UP tests, the exhaust temperature tended to decrease with dual fuel operation.

ACKNOWLEDGEMENT

The authors whish to thank Dr. I. E. Cruz of the University of the Philippines for his valuable contribution. In addition the data collection and organization efforts of John Liebeskind are appreciated.

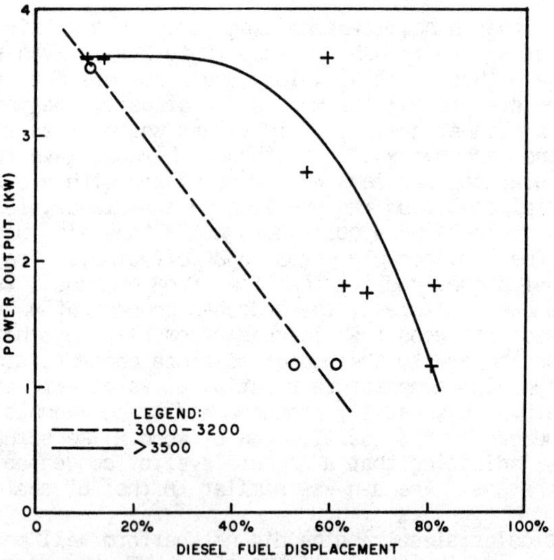

Figure 8. Power Output as a Function of Diesel Fuel Displacement

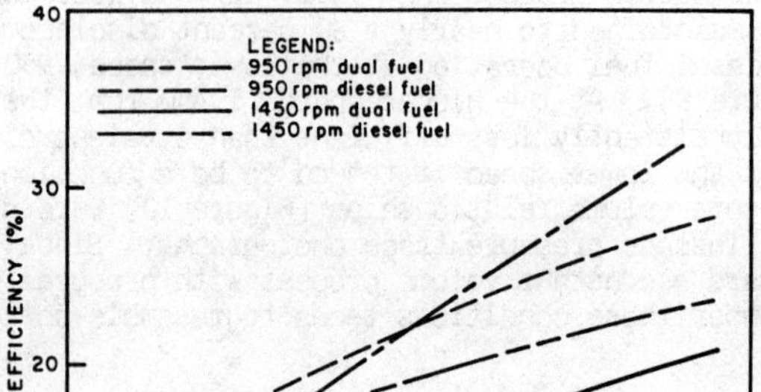

Figure 9. Thermal Efficiency as a Function of Power Output, RPM, and Fueling

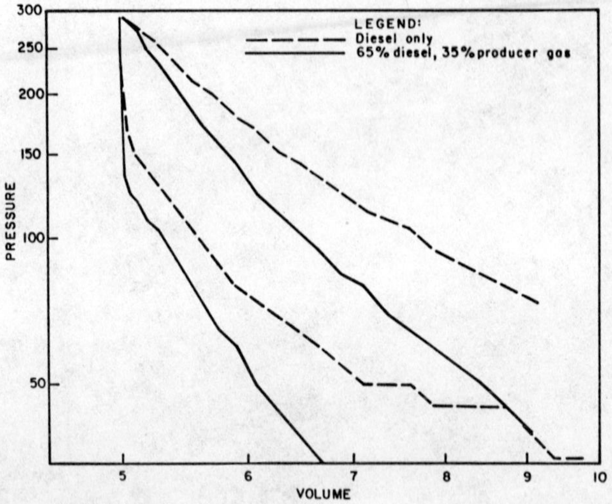

Figure 10. Pressure-Volume Relation to Gasifier-Diesel Engine System

REFERENCES CITED

1. Generator Gas, The Swedish Experience From 1939-1945, Swedish Academy of Engineering, 1950, (translated by Solar Energy Research Institute, 1979).

2. Cruz, I. E., "Producer Gas from Agricultural Wastes-Its Production and Utilization in a Converted Oil-Fired Boiler," Resource Recovery and Conservation, Vol. 2, 241-256 (1976/1977).

3. Cruz, I. E., Producer Gas From Agricultural Residues-Its Production and Utilization in Internal Combustion Engines, College of Engineering, University of the Philippines, 1981.

4. Trezek, G. J. et al., "The Cal Recovery System: A Resource Recovery System for Dealing with the Problem of Solid Waste Management," Compost Science, Vol. 16, No. 5, 18-21 (1975).

5. Beagle, E. C., Rice Husk Conversion to Energy, 213, FAO, Rome, Italy, December 1976.

6. Jones, D., and Jones, J., "Wood Chips vs. Densified Biomass: An Economic Comparison," Energy from Biomass and Wastes, Vol. 4, 223-249, Chicago, Institute of Gas Technology, 1980.

7. Tabatabaie-Raissi, Ali, Studies of Biomass Gasification, Ph.D. dissertation, University of California, Berkeley, 1982.

SPOUTED FLUID-BED GASIFICATION OF BIOMASS AND CHEMICAL WASTES — SOME PILOT PLANT RESULTS

Clarence Zak
Peter B. Nutcher
Process Combustion Corporation
Pittsburgh, Pennsylvania

ABSTRACT

This paper presents a discussion on gasification using a spouted fluid bed combustor with horizontal spouts. The arrangement of the spouts encourages circulation of the bed and feed material. Test results are presented to show that a wide range of biomass materials can be gasified with excellent results. Gas analyses indicate consistent values over a wide range of feeds.

SPOUTED FLUID-BED GASIFICATION OF BIOMASS AND CHEMICAL WASTES — SOME PILOT PLANT RESULTS

BACKGROUND

In the 1970's research and development was started in Australia to develop a small conventional bubbling fluid bed combustor to meet the industrial process market. The initial work found that a conventional fluid bed design was unsatisfactory for several reasons. The spouted fluid bed concept was developed to overcome problems with unsized, high moisture or light weight solids and waste liquids. This spouted technology gave the bed better circulation and more entrainment necessary to overcome the problems. Separately, work was done by an Australian government research group, CSIRO who obtained a patent for the spouted fluid bed combustor. It has been subsequently developed and commercialized by Pyrecon PTY Ltd. who have licensed Process Combustion Corporation in the USA.

For good biomass combustion/gasification, the three "T's" need to be addressed: Time, Temperature and Turbulence. There are many combustion designs that address two of the three "T's", time and temperature, but very few designs that provide proper turbulence.

A rotary combustor for example is a design that tumbles the material, exposing new surfaces and increasing the burning rates. But due to rotating seals, air infiltration is high and therefore the thermal efficiency is low and control of substoichiometric reactions is very difficult. In addition, the capital and operating costs are higher than most other designs.

For consistent gasification of biomass materials, precise air and temperature controls are necessary, but some designs that have these controls do not lend themselves to good mixing of the air and solid material. Fixed bed and chain grate combustor designs provide tight units for good control of air and temperature, but poor mixing.

FLUID BED COMBUSTION

The conventional fluid bed combustor provides the requirement of good temperature control, proper time for the reaction to complete itself and excellent turbulence to obtain high intensity combustion. The fluid bed combustor has proven itself satisfactory for a wide range of solids, sludges and liquids. But there are operating limits that restrict conventional fluid bed applications. Care must be taken as to the size range of the feed material and the uniformity of the bed material. Deep beds are required to insure complete combustion in the bed and low fluidizing velocities are used to minimize bed and feed carry over. The high pressure drop across deep beds increases horsepower requirements and lower velocities slow down the combustion process. Mixing in a conventional fluid bed takes place as the feed material moves vertically through the bed using small bubbles of air for combustion. Even with proper sizing of feed and bed material, and using deep beds with low velocities, problems still persist. Systems have been designed to help overcome the short comings of the conventional fluid bed combustor. Figure 1 shows some of the designs that try to increase either the gas to solid contact (slip) or the solids circulation and mixing rate or both. The problems of properly grading the feed and high carry over still persist with all these designs. The spouted fluid bed combustor pilot plant, Figure 2, was designed to give the high intensity combustion of the conventional fluid bed combustor but eliminate the problems.

The advantages a spouted fluid bed has as shown in Figure 3 are:

1. There is no wind box, air distributor nor grid, which eliminates plugging, variation of air flows, leakage and expansion problems.
2. The horizontal spouts have large openings that will not plug and they can be refluidized without emptying the bed.
3. The lateral bed movement due to the horizontal spouts and recirculation pattern requires less feed preparation, has a wider size tolerance and better temperature uniformity.
4. The bed depth is less than other conventional fluid beds, therefore, requiring less fluidizing air horsepower.
5. A wider turndown is available from the spouted fluid bed as groups of spouts can be shut down three to one and six to one turndowns are standard.

The reason why spouted fluidization is so effective is because the core of an individual spout has extremely high differential velocities between the bed feed material and the spouting combustion air. The core conveys material in a dilute phase upward and spills outward. The annulus, like a continually falling packed bed, also has a higher differential velocity between the gas and solid. As shown in Figure 1, the slip velocity in the core of the spout and the annulus of the spout exceed those of other designs. Material introduced near the top and outer edge of the spout annulus is quickly submerged in the down flow and is distributed by the upward flow of the spout core. Spouted fluid bed combustion has a high gas-solid contact, good bed and feed mixing, low bed attrition/elutriation, low bed pressure drops, and stable operation.

The high circulation pattern in the bed and the high velocity at the base of the spout can entrain very light particles and also lift and circulate large bed and feed particles. Therefore, feed preparation and bed conditioning is minimized which is an advantage because it reduces bed pressure drop and assists stability.

PILOT PLANT

The spouted fluid bed pilot plant is designed with a single spout discharging horizontally into the bed. A patented design of the spout end ensures even distribution of the spouting air in the bed, so that the air quickly turns vertically up. The horizontal spout allows restarts of slumped beds not always possible with vertical spouts. This design also makes it possible to mix liquid or gaseous feeds directly with the spouting air. The spouted fluid bed pilot plant is very flexible and has the capability of taking solid, liquid and gas and in any combination. In order to properly gasify a biomass or chemical waste steam, it should have a Btu value in excess of 6500 Btu/lb depending on oxygen content. Below 6500 Btu/lb, supplimentary fuel may be required to produce a stable gas for burning. In case of upset conditions within the bed, a water spray at the top of the bed can be used to control temperature excursions.

The design of the spouted fluid bed combustor allows the ash and oversized particles to collect below the spout. The bed emptying system removes the build up of ash from this inactive bed area, below the spout. For this reason the spouted fluid bed's operation is more tolerant of large irregular sized particles than the conventional fluid bed design. The removal of the ash is at the coldest part of the bed which facilitates removal and minimizes heat loss.

The selected bed material can be a wide range of low cost minerals, but silica sand is most common because of low cost. Chemically active bed material can also be used for control of gaseous emissions by absorbing harmful chemicals in the bed.

The spouted fluid bed allows fast start-ups, only limited by the rate at which the refractory lining can be heated. Typically, less than twenty minutes are required to bring the combustor to operating temperature. By introducing natural gas or propane into an air spout and igniting the mixture eliminates the need for preheat burners. The mixture quickly burns back into the core of the spout and starts heating the bed. This reduces the capital and operating costs associated with conventional fluid beds. Operator attendance and start-up requirements are also reduced.

Turndown can be as high as three to one under automatic control with a single fluidizing air plenum and higher with multi plenum. Care only needs to be taken that the fuel to air ratio is maintained and that the heat release is greater than the heat loss from the combustor. The spouted fluid bed uses lower pressure centrifugal blowers instead of high pressure positive displacement blowers that are required on conventional fluid bed combustors.

An ultra-high efficiency double-entry cyclone has also been developed which uses two hot inlets, one hot, clean gas outlet and one dust outlet. This design of double-entry cyclone enables it to be installed in either the vertical or horizontal position and minimizes heat losses from the hot gases by refractory lining.

TESTING

The following general procedure is followed for operating the pilot plant (Figure 4). Propane is used for heating up. The fluidizing air blower is started and the air flow adjusted. Bed material is added until 20" w.c. Delta P is reached. The burner is lit and the bed heated to 1500°F. The hopper is filled with feed material, the screw feeder started, and the burner shut off. The feed rate is adjusted until the bed temperature stabilizes at 1200° to 1300°F. The hot gases are piped to a low Btu gas burner where secondary combustion air is added. A continuous pilot is used. The burner fires into a three foot long metal chamber, open to atmosphere. Once the burner becomes stable, the pilot is shut off and the gas samples are taken upstream of the burner.

The different materials that have been tested are:

Rice Hulls	Peat
Wood Chips	Coal Slurry
Saw Dust	TDI Residue
Corn Cobs	Mushroom Mulch
Nut Shells	"Bio Bits" Paper*
Cow Chips	Municipal Sludge

* "Bio Bits" is a manufactured solid fuel, prepared from waste paper, cardboard, and plastics. The waste is shredded, water added and the material extruded into 1/2" to 3/4" diameter pellets, 1" to 3" long. It is available commercially.

Three materials have been selected for presentation, because they differ greatly from each other, wood waste, rice hulls and the commercial solid fuel.

A. WOOD WASTE

Wood wastes (e.g., chips, shavings, saw and sander dust) are an excellent material to gasify. They are usually 6500 to 7000 Btu/lb, low in sulfur, readily available on site, low in ash, and cost very little or nothing. They are generally a waste product that a plant needs to dispose of and landfilling is becoming more and more difficult and expensive.

General Analysis (wt %)

	Dry	15% Moisture	25% Moisture (Maximum)	Green*
Moisture	0	15	25	40
Volatiles	86	73	64.5	51.6
Carbon	13	11	9.75	7.8
Ash	1	1	.75	0.6

Ultimate Analysis (wt %)

	Dry	15% Moisture	25% Moisture (Maximum)	Green*
Water	0.00	15.00	25.00	40.00
Carbon	50.00	42.50	37.50	30.00
Hydrogen	6.00	5.10	4.50	3.60
Nitrogen	0.10	0.08	0.07	0.06
Sulfur	0.05	0.04	0.03	0.02
Oxygen	42.85	36.43	32.15	25.72
Ash	1.00	0.85	0.75	0.60
H.H.V. Btu/lb	8300	7055	6225	4980

Physical Characteristics — Bulk Density — 20 lb/cu ft.

* Material requires drying to gasify.

Test A — Wood Waste Gasification

A.1 Wood Waste Sample — Ultimate Analysis (wt %)

	As Received	Dry
Water	26.27	0.00
Carbon	36.40	49.370
Hydrogen	4.55	6.170
Nitrogen	.04	0.069
Sulfur	.01	0.020
Oxygen	31.75	43.071
Ash	.98	1.300
H.H.V. Btu/lb	6286	8526

A.2 Gas Sample Results (Vol % Dry) Gas Chromatograph

	Test #1	Test #2
Carbon Dioxide	14.00	14.23
Oxygen	2.01	1.55
Nitrogen	56.40	55.69
Methane	4.19	4.35
Hydrogen	4.19	4.12
Carbon Monoxide	16.43	17.25
Ethane, etc.	2.78	2.81
H.H.V. Btu/SCF Dry	158.76	163.88
Sensible Heat Btu/CF	18 (Approx.)	18 (Approx.)

B. RICE HULLS

Unlike wood waste and saw dust, there is very little that rice hulls can be used for. They are low in nutrient value, therefore, a poor feed for animals. They have high ash and lower Btu value than most biomass materials. Because of these disadvantages, there is often an abundant cheap supply of the material for burning.

Test B — Rice Hulls Gasification

B.1 Rice Hull Sample — Ultimate Analysis (wt %)

	As Received	Dry
Water	8.00	0.00
Carbon	35.07	38.12
Hydrogen	4.34	4.72
Nitrogen	0.17	0.19
Sulfur	0.03	0.03
Oxygen	31.38	34.11
Ash	21.01	22.83
H.H.V. Btu/lb	5241	5697

B.2 Gas Samples (Vol % Dry) Gas Chromatograph

	Test #1	Test #2
Carbon Dioxide	13.79	13.70
Nitrogen	63.20	63.04
Methane	2.95	2.86
Hydrogen	2.25	1.69
Carbon Monoxide	15.59	16.29
Ethane, etc.	2.22	2.42
H.H.V. Btu/SCF Dry	127.15	130.26
Sensible Heat Btu/CF	18 (Approx.)	18 (Approx.)

C. COMMERCIAL SOLID FUEL

This fuel is manufactured from waste paper and cardboard obtained from light commercial establishments. It is low in ash and sulfur, dust free, easy to handle and store, high Btu value, 7000 to 8000 Btu/lb, consistent analysis from year to year, easily gasified, inexpensive and abundant in supply.

Test C — Commercial Solid Fuel Gasification

C.1 Solid Fuel Sample

	Test #1 Ultimate Analysis (wt %)		Test #2 Ultimate Analysis (wt %)	
	As Received	Dry	As Received	Dry
Water	9.09	0.00	11.48	0.00
Carbon	41.67	45.84	41.09	46.42
Hydrogen	5.61	6.17	5.68	6.42
Nitrogen	0.06	0.07	0.16	0.18
Sulfur	0.35	0.39	0.17	0.18
Oxygen	41.18	45.28	38.68	43.69
Ash	2.04	2.25	2.74	3.10
H.H.V. Btu/lb	7095	7805	7237	8176

C.2 Gas Sample (Vol % Dry) Gas Chromatograph

	Test #1	Test #2
Carbon Dioxide	14.29	14.17
Nitrogen	58.35	57.30
Methane	3.26	4.26
Hydrogen	2.40	3.74
Carbon Monoxide	18.78	16.58
Ethane, etc.	2.92	3.95
H.H.V. Btu/SCF Dry	159.23	179.44
Sensible Heat Btu/CF	18 (Approx.)	18 (Approx.)

COMMENTS

1. Although particulate in the gas was not measured or analyzed, the carbon soot appeared to burn with the gas in the burner. Some inert particulate was apparent at the burner, particularly firing rice hull gas.

2. Moisture content in the gas was approximately 22% vol for the waste wood and 12% vol for the rice hulls and commercial fuel.

3. At no time was any build up of tars, soot or particulate observed, but the runs were relatively short and continuous operation may be quite different.

CONCLUSIONS

1. All three feed materials produced a good low Btu gas which was easy to burn in a suitably designed burner. The gas consistent in quality, relatively clean and at a temperature of approximately 800°F.

2. The gasifier was easy to start up, operate and control, very tolerant of fluctuations in feed quality and rate, having good turndown capability.

3. From these tests, it would seem feasible that scale up to a commercial sized project should be possible.

ACKNOWLEDGMENT

The authors wish to acknowledge with thanks the use of material obtained from Mr. R. C. Wilkinson, Managing Director, Pyrecon PTY Limited, Sydney, Australia.

MOBILE S.F.B.C. PILOT PLANT DESCRIPTION

General

The PCC/Pyrecon pilot plant is a spouted fluid bed combustor/gasifier/incinerator with a 10-1/2 inch diameter bed or 0.60 square feet of bed area. The pilot plant has been designed to handle a wide range of combustible materials — solid — liquid — sludge. The unit is equipped with a solid feed system consisting of a storage hopper, screw conveyor, rotary air lock, and a screw feeder; a liquid feed system consisting of a liquid metering pump, and patented atomizer; cyclone particulate collector; bed quenching system; fluidizing air blower; start-up burner; control and recording instruments; and electrical control panel. All of the above equipment is mounted on a dual axle trailer for in-plant testing.

Incinerator

Bed Size	10-1/2 inch I.D. of refractory
Bed Area	0.60 square feet
Bed Height	30 inch max
Freeboard Diameter	21 inch
Freeboard Height	60 inch
Heat Release Capacity	Approximately 300,000 Btu/h oxydizing 1,200,000 Btu/h gasifying

Cyclone

Size	10 inch I.D. refractory
Outlet	5 inch O.D.

Start-Up Burner

Manufacturer	Eclipse
Model No.	84-14 TBH
Capacity	250,000 Btu/h 120 SCF/m air max
Fuel	Propane
Ignition	H.T. spark

Fluidizing Air Blower

Manufacturer	Spencer
Model No.	2503
Capacity	120 SCF/m at 40 OSI
Motor	5 HP

Solid/Sludge Feed System

Storage Hopper	15.8 cu ft with plexiglass end plate 6 inch screw conveyor with plexiglass cover. Driven by 1-1/2 HP variable AC drive.
Rotary Air Lock	6" diameter rotor with neoprene sealing strips driven by a 1/3 HP Motor.
Feeder	2" diameter screw feeder. The feeder is driven by a chain and sprocket from a 1/2 HP gear motor at constant speed.

Liquid Feed System

Pump	Metering type with check valve
Atomizer	Special patented type

Instrumentation and Controls

Multi-point recorder with digital indicator
Bed temperature controller
O_2 and combustible analyzer
Burner flame safety system
High and low bed temperature interlocks
Air, water, and feed metering
Miscellaneous pressure indicators

Trailer

Gross Vehicle Weight	9800 lb
Length	20'6"
Width	6'4"
Axle	Dual

DISCUSSION

S. Babu, IGT	Could you tell us anything about the reactivity of rice hulls relative to the wood?
Zak	We're an industrial money-making company, and the proof of the pudding is to put something out in the field that works. When we were satisfied that the burner was self sufficient without a pilot, we said we were making good gas. Rice hulls are much more difficult to burn than wood waste, primarily because of the ash and also the low-Btu value. The system operates by just adding air, getting the bed fluidized, getting the bed hot, and adding the waste material by auger feeding. The bed is stabilized after you shut the burner off.
Augenstein	You said that large pieces of fuel sink down into the ash zone, and they would seem to be relatively cool. Are they completely consumed?
Zak	Large pieces are dropped into the ash in the lower portion of the bed. If they are not consumed or if they're too heavy, or if they become inert material, they can then be augered away. If the ash stays within the bed, it becomes the bed.
K. Sipila, Technical Research Centre of Finland	You mentioned you also tested peat. What form did you test?
Zak	The peat that we tested was the ordinary garden variety. I bought it at the local garden store. It was very difficult to incinerate or to gasify. We had to supplement it with a burner.

EVALUATION OF PROCESSES FOR REMOVAL OF PARTICULATES, TARS, AND OILS FROM BIOMASS GASIFIER PRODUCT GASES

M. D. Brown, E. G. Baker, and L. K. Mudge
Pacific Northwest Laboratory

ABSTRACT

For biomass gasification to make a significant contribution to the energy picture in the next decade, emphasis must be placed on the generation of clean, pollutant-free gas products. This paper attempts to quantify levels of particulates, tars, oils, and various other pollutants generated by biomass gasifiers of all types. Data from a bench-scale fluidized bed is presented which gives pollutant concentrations and compositions. End uses for biomass gases and appropriate gas cleaning technologies are examined. Complete systems analysis is used to predict the performance of various gasifier/gas-cleanup/end-use combinations. Further research needs are identified.

EVALUATION OF PROCESSES FOR REMOVAL OF PARTICULATES, TARS, AND OILS FROM BIOMASS GASIFIER PRODUCT GASES

INTRODUCTION

Biomass gasification has the potential to make a significant contribution to the future energy supply in the United States and is already finding commercial applications primarily in the forest products industry. One area that has not been studied in detail is gas cleaning, primarily for particulate and tar removal. Particulates can cause plugging and erosion of downstream equipment and may also be an environmental problem. Plugging of downstream equipment is the primary problem associated with tars. Once they are separated from the gas, tars may be an environmental hazard and present a disposal problem.

The objective of this paper is to evaluate gas cleaning technology, primarily for particulate and tar removal, as it applies to biomass gasification, and identify potential problem areas and gaps in the technology that could impede development of biomass gasification.

Selection of gas cleanup equipment for biomass gasification depends primarily on two factors: 1) the type of gasifier used and 2) the intended end use of the gas. The type of gasifier, and to some extent, the feedstock, will determine the concentration of particulates and tars in the gas. The end use will define the particulate and tar concentrations which can be tolerated.

Phase I of this study reviewed the state of the art of gas cleaning technology as it applies to biomass gasification. The results of Phase I can be divided into three main areas:

- Biomass Gasification Gas Contaminants
- End Uses for Biomass Gas
- Gas Cleanup Technology

Phase I was based on an extensive literature survey. This survey revealed a profound lack of data on gas cleaning systems for biomass gasification particularly for tar removal. Information on gas cleaning systems for coal gasification and biomass combustion were also studied. This provided a sizeable base of information on particulate removal, but still little on tar removal. Because of this the results focus heavily on particulate removal.

Phase II took specific gas cleaning methods and applied them to various gasifiers. The primary objective was to identify gas cleaning methods which could meet specifications for the end uses of biomass gas. Gaps in the technology are identified and recommendations for additional research were made.

BIOMASS GASIFICATION CONTAMINANTS

When carbonaceous materials are gasified or combusted, char, ash, tar, and permanent gases are produced. The relative fractions of these four products will vary and depend primarily on the type of gasifier, gasifier operating conditions, and the feedstock.

Char/ash particles and higher hydrocarbon vapors or droplets are considered contaminates in a gas stream. Gasification processes either seek to minimize char and tar production or, once they are produced, to maximize their removal. Other contaminants or pollutants which may be present in biomass gas include olefins, hydrogen sulfide, sulfur dioxide, carbonyl sulfide, nitrogen oxides, and trace metals.

Gasifier Types

Over the years many different types of gasifiers have been constructed for experimental and commercial use. However, most all fall into three distinct categories. The three main types are: fixed bed, fluid bed, and entrained flow, referring to the motion of the solids in the gasifier. Each type of gasifier has different characteristics which result in significantly different yields of particulates and tars. The basic types of gasifiers are shown in Figure 1.

"Fixed bed" is a broad term describing a gasifier in which the fuel is fed onto a relatively stationary inventory of biomass already present in the reactor. The fixed bed category encompasses updraft, downdraft, and crossdraft units. Each type has different operating characteristics and emissions. The term "fixed" refers to the condition at the extremes of the bed, which do not vary at steady state.

Updraft units exhibit a countercurrent flow of fuel and gas. Solid fuel is fed from the top by lock hoppers or feeders. The bed of fuel is supported by a grate at the bottom of the reactor. The fuel flows down through the drying zone, pyrolysis zone, reduction zone, and combustion zone. Ash and unreacted fuel (char) exit through the grate at the bottom. Reactant gases (air, oxygen, steam) are introduced into the reactor through the grate. The hot gases from the combustion zone provide energy for the endothermic processes in the upper zones and exit at the top of the gasifier saturated with pyrolysis oils and water. Approximately 20-25 percent of the carbon in wood is recovered as liquid products. The condensed liquids usually are in two phases: an aqueous phase containing highly oxygenated water soluble organics (pyroligneous acids) and a separate tar phase (Mudge, et al., 1980). Because of the low velocity of the gases in the reactor and the filtering effect of the bed, the product gas contains little particulate matter.

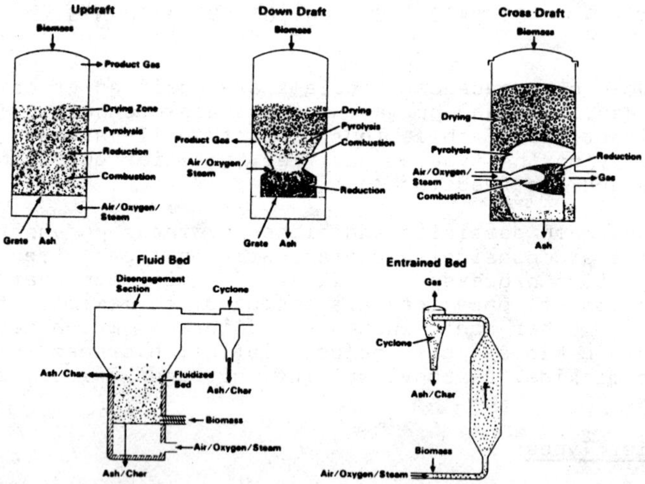

FIGURE 1. Types of Gasifiers

Production of pyrolysis oils is largely eliminated in downdraft gasifiers. As in updraft units solid fuel is fed from the top. However air, which is used in most downdraft units, is introduced concurrently into the combustion zone through a distributor. Pyrolysis oils and moisture from pyrolysis and drying are drawn down through the high temperature reduction and combustion zones where they undergo thermal cracking. The product gases exit near the bottom of the reactor. Ash and char leave through a grate at the bottom of the reactor.

Low gas velocities result in low particulate loadings i the same range as updraft. Downdraft gasifiers can exhibit very low tar yields depending on the combustion zone temperature.

Crossdraft gasifiers exhibit many of the operating characteristics of downdraft units. Air or air/steam mixtures are introduced in the side of the gasifier near the bottom. Producer gases are drawn off the opposite side. Tars and oils are drawn through the reduction zone and some cracking to lighter components occurs. Some particulates (mostly ash) are entrained in the product gases.

In a fluid bed gasifier the incoming and evolved gases maintain the reactor bed in a turbulent fluid-like state muc like a boiling liquid. The result is an expanded reactor be of char particles and, in most biomass gasifiers, an inert solid such as sand. Because biomass is less dense and has less ash and fixed carbon than coal the inert solid is used to maintain proper fluidization (prevent bridging and channeling) and to provide additional heat capacity in the bed.

No distinct zones exist in a fluid bed gasifier as near isothermal operation is maintained with good fluidization. The product gas contains some tars and oils depending on the bed temperature, and does have a fairly large loading of particulates (ash, char, and bed material). These loadings are easily predicted (Zenz 1982). Depending on the design, ash and char may be removed from the top of the reactor with the product gases, from the bottom of the reactor, from the top of the bed, or a combination of the three.

In an entrained flow (or transport) gasifier, finely sized fuel particles are entrained in the feed gas (usually oxygen and steam) prior to entry into the reactor. Gasification takes place with the feed particles suspended in the gas phase. The product gas, ash and char leave the top of the reactor. Limited data indicate particulate and tar loadings similar to or greater than those from fluidized bed units.

Production of Tars and Other Condensible Organics

Tar is a generic term for the higher boiling (>150°C) constituents of biomass gas which are formed during the pyrolysis reactions. Depending on the degree of cracking, the tar may range from light, oxygenated hydrocarbons to heavy, poly-aromatic hydrocarbons (PAH).

Each type of gasifier has different reaction conditions and consequently different tar compositions and production rates. For example an updraft fixed bed gasifier generates a high yield of tar (~20 wt% of the carbon in wood feed is converted to liquids) due to the low temperature at which drying and pyrolysis takes place in the gasifier. The drying and pyrolysis zones in an updraft gasifier will range from 80 to 200°C. The tars will most likely be condensed droplets in the gas phase. Reported tar yields range from 10-100 g/m^3 of gas (Kaupp and Goss 1981; Baker et al. 1984; Fritz 1978; Dravo Corp. 1976). Conversely, downdraft gasifiers pass the pyrolysis gases through the combustion zone which thermally cracks much of the tar at temperatures near 1200-1600°C. Reported yields from downdraft units range from 50-500 mg/m^3 (Kaupp and Goss 1981; Groenevald 1983; Reed 1983; Kumar 1984). Typical tar production rates for each type of gasifier are shown in Figure 2.

Fluidized (or entrained) beds when operating properly are nearly isothermal. Biomass feed is pyrolyzed either submerged in the bubbling bed or entrained in a hot gas stream. High temperature (600-800°C) and good gas-solid contacting results in some cracking of pyrolysis oils. The final concentration of tars will most likely be a function of the reaction temperature and pressure in the bed. The product gases from downdraft, fluidized bed, and entrained bed gasifiers are hot and any tars that are still present will most likely be in the vapor phase.

A recently completed study of biomass gasification/pyrolysis condensates at PNL (Elliott 1985) concludes that there is no typical tar composition which can be adequately used to represent all thermally produced biomass tars. The tar composition, as well as the amount, is dependent on the operating conditions, principally a time/temperature thermal severity-type function. The properties of the tar therefore appear to vary on a continuum from "primary" oxygenated pyrolysis tar collected after a short residence time at low temperatures of around 500°C to highly aromatic, deoxygenated tar which is produced at short residence time at high temperatures of around 900°C.

Table 1 summarizes the properties of tars from different gasifiers as analyzed by Elliott (1985). The properties of tars from fluidized bed or entrained flow gasifiers varied widely depending on the temperature of operation. The properties of downdraft tars were also varying indicating the mode of operation has a significant effect on tars from these units. The analysis of the fixed-bed tar from the Rome, Georgia gasifier is similar to a previous analysis of fixed-bed tar from PNL's fixed bed gasifier (Baker et al, 1984).

TABLE 1. Properties of Biomass Tars (Elliott 1985)

	Fluidized Bed/ Entrained Flow		Updraft		Downdraft	
	480°C	880°C	tar	aqueous	tar	aqueous
Carbon, %	52.7*	84.0*	70.9*	11.3	67.2*-0**	7.0-0.5
Hydrogen, %	6.2	5.7	7.2	--	5.8-0	--
Oxygen, %	40.5	8.7	21.7	--	25.0-0	--
Ash, %	0.6	1.6	0.2	0.1	2.0-0	1.4-0.02
pH	--	--	--	2.1	--	3.0-5.3
viscosity, cps	220-1300 @40°C	9800-26,800 @51°C	410 @40°C	-- --	7500-41,000 @78°C	--
moisture, wt%	16	20-28	8	--	12-13	--
density, g/ml	1.26-1.28	1.14-1.16	1.13	--	1.16	--

* elemental and ash analyses for tars are reported on a dry basis

** downdraft gasifiers may or may not produce a separate tar phase depending on the specific operation

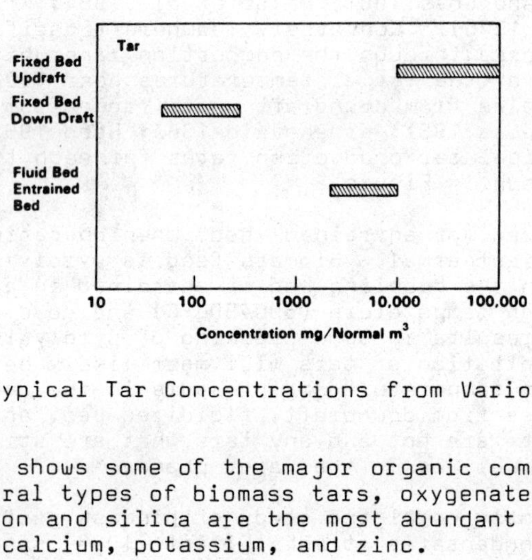

FIGURE 2. Typical Tar Concentrations from Various Gasifiers

Table 2 shows some of the major organic components of the two general types of biomass tars, oxygenated and deoxygenated. Iron and silica are the most abundant trace metals followed by calcium, potassium, and zinc.

TABLE 2. Components of Biomass Tars (Elliott 1985)

Major Organic Components	
Low Temperature Oxygenated Tars	High Temperature Deoxygenated Tars
2-methyl-2-butenal	napthalene
phenol	napthalene
2-methoxyphenol	fluorene
4-methyl-2-methoxyphenol	phenanthrene
4-hydroxy-3-methoxybenzaldehyde	flouranthrene
2,6-dimethoxyphenol	acephenanthylene
3,5-dimethoxy-4-hydroxygenzaldehyde	benz(a)anthracene
levoglucosan	

Particulates

The concentration of particulate matter in the product gas from a biomass gasifier depends on numerous factors including the type of gasifier, feedstock, and gasifier operating conditions. Operating conditions that influence particulate emissions include gas velocity in the bed, temperature, moisture content of the blast, and the rate of gasification. The concentration of particulates from various gasifiers is summarized in Figure 3.

Fixed bed gasifiers, both updraft and downdraft, employ large (2.5-5 cm, 1-2 in) chunks of feedstock that are not entrained by the gas flow; nevertheless, the product gas does contain a significant quantity of particulates, 100-1000 mg/m^3 (Finnie 1979, Hoenig and Cole 1981, Kaupp and Goss 1981, Jacko 1985). The lower values were typically measured after the gas had been combusted in a burner. The higher values are measured directly from the gasifier. The values for downdraft gasifiers are somewhat higher than updraft (Kaupp and Goss 1981) possibly because the higher temperature of the product gas results in higher gas velocities exiting the gasifier.

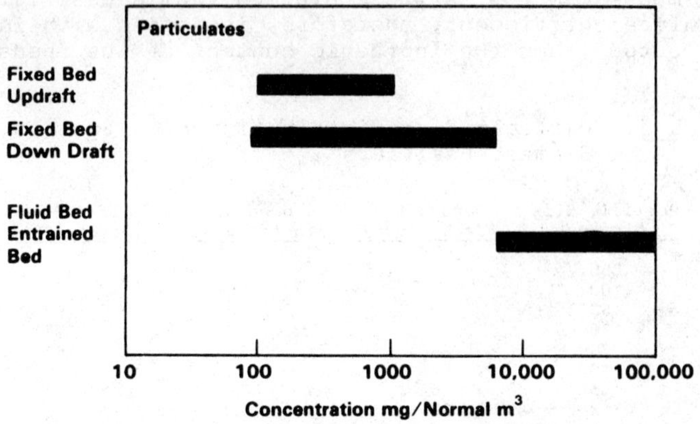

FIGURE 3. Typical Particulate Concentrations from Various Gasifiers

The size of particulates carried out with the product gas will depend primarily on gas velocity exiting the gasifier. Only a limited amount of data on the particle size distribution from biomass gasifiers has been published as shown in Table 3. It is not surprising to find the particulates from an updraft are quite fine (89% smaller than 1 micron) and those from a fluid-bed are somewhat larger, primarily between 5 and 30 microns. It is surprising to see the large particle size distribution reported for Imbert downdraft vehicular gasifiers. This is probably due to the hot gas exiting through a small annular space which results in a high gas velocity. Other downdraft units (Hoenig and Cole, 1981) show more than 99% of the particles being less than 3 μm.

Particulates from the different gasifier types are comprised of the inorganic constituents in the feedstock, any unconverted feedstock, and dirt. In many systems the particulates in the gas stream are collected as a by-product fuel, generally referred to as char because they contain a large percentage of carbon. Char is often a by-product from fluid-bed and entrained-flow gasifiers. Carbon content of solids from fixed-bed gasifiers is very low. Published data on the composition of wood and ag residue ash (Kaupp and Goss 1981) show considerable variations in concentrations of major constituents (Ca, K, Si, Na, and Al). The usual main component in ash is CaO, which accounts for 10 to 60 wt% of the ash and is typically present at about 40 wt%. Concentration of K_2O can be 10 wt% or greater. Ash from agricultural residues consists primarily of silica and alkali. Feedstocks which have a large percentage of dirt or other inorganics can raise the silica content of any ash considerably.

Particulate composition from the different gasifier types is not reported. With fluid and entrained bed gasifiers entrainment of solid particles from the bed is the likely mechanism for particle elution from a gasifier. Particulate entrainment, therefore, increases with increased energy output, and the inorganic content of the feedstock.

TABLE 3. Particle Size Distribution From Various Biomass Gasifiers

Particle Size Microns	Updraft (Jacko 1985)	Downdraft (Seri 1979)	Fluid Bed (Datin 1981)
250		74	
100		50	
50		30	99
30		18	97
20	99+	11	87
10	97		38
5	96		17
2	95		2
1	89		
0.5	81		

In a fixed bed gasifier alkali and possibly other more volatile components of the ash could be volatilized in the combustion zone of the gasifier, condensed or solidified into small particles in the cool zones, and entrained in the gas stream. Analysis of entrained particles would determine if such a mechanism does occur. Corrosion products from the gasifier and outlet piping may also be present in the product gas as particulates.

END USES FOR BIOMASS GAS

The amount of gas cleanup required will depend on the intended end use of the gas. Low-Btu gas (producer gas) from biomass has been used to fire various industrial process burners including direct fired equipment such as dryers and kilns and indirect fired equipment such as boilers and oil heaters. Low-Btu gas has also been used as a fuel for internal combustion reciprocating engines, both gasoline and diesel, and is being considered for use in gas turbine engines. Medium-Btu gas can be substituted for low-Btu gas in these applications. In addition it can be used as a synthesis gas for making fuels as methanol, methane (SNG), and liquid hydrocarbon fuels such as gasoline. This section discusses the gas quality, primarily the tar and particulate levels, that are required for each of these end uses.

The allowable particulate and tar loadings in biomass gas varies significantly depending on the end use. Figure 4 summarizes the requirements for particulates. If the gas is used in a burner to supply energy, particulate removal will be governed primarily by air pollution regulations for particulates. The allowable loadings in the combustion flue gas range from 0.03 to 0.6 lb/10^6 (approximately 70-1500 mg/m^3) Btu depending on the size and location of the facility. The specific allowable limit will be site specific.

For internal combustion engines, particulates must be removed down to 10-50 mg/m^3 to prevent excessive engine wear. Requirements are even more strict to prevent erosion and corrosion of gas turbine blades; particulates must be reduced to 0.001 to 0.01 gr/scf (2-20 mg/m^3) depending on turbine speed and operating temperature.

Less information is available for gases to be used for synthesis of chemicals, SNG, or gasoline. In general the requirements will be similar to specifications for gas turbine applications and will be dictated by erosion of compressor blades in the compressors used to compress the gas to high pressures required for synthesis.

Specifications for tars in the product gas are not well defined. Tars will be consumed in any burner (including gas turbine combustors) and increase the heating value of the gas. However, this requires close-coupling the gasifier and the burner. Separation of the gasifier and the burner by any distance can cause significant problems due to tar

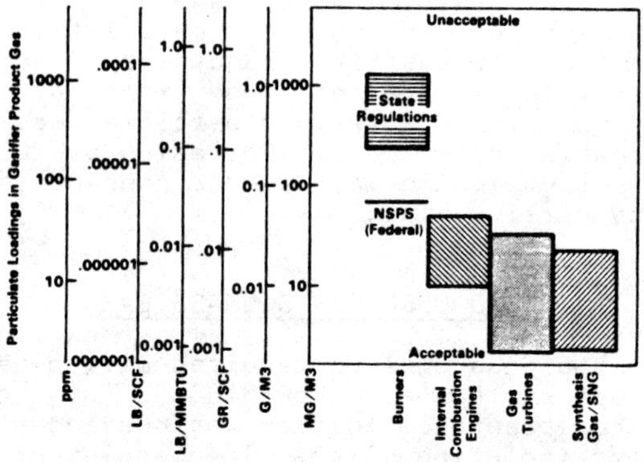

FIGURE 4. Allowable Particulate Loadings for Various End Uses

condensation and precipitation. Tar removal should be considered in this case.

In diesel and spark ignition engines tars must be reduced to 10-50 mg/m^3 to prevent deposition in the gas mixer and inlet valves and gum formation on the valves. The limit on condensible hydrocarbons entering a high temperature gas turbine have been set as low as 0.5 $lb/10^6$ scf (∿8 mg/m^3) although no reasons have been given for the limitation (Robson 1974).

Tars are known to be poisons for various synthesis catalysts, however, their effect is not well understood, nor have limits been set. For synthetic natural gas entering a pipeline a hydrocarbon dew point of <40°F at 1000 psig (∿100 mg/m^3) has been specified. In addition no gum formers are allowed. Tars would probably be classified as gum formers.

GAS CLEANING TECHNOLOGIES

Fuel gases derived from biomass gasification resemble gases derived from coal gasification with respect to major fuel constituents, but differ with respect to major impurities. Because coals are generally much higher in sulfur and ash, these constituents tend to be higher in coal gas than in biomass gas. The spectrum of trace elements derived from coals may also be different from that derived from biomass fuels. Finally, the detailed chemical makeup of tars and oils derived from coals will differ significantly from those derived from biomass gasification. There has been a vast amount of work done on the problem of cleaning coal derived gases and with proper attention to the differences imposed by gasification of biomass fuels much of this technology can be applied.

Various gas cleanup systems exist for removal of both particulates and tars from gases produced by biomass gasification. Often these overlap, particularly when tars are present as liquid droplets.

Figure 5 shows a range of particle and droplet sizes and the type of collection mechanism appropriate for various sizes. Although tars are often present as vapor they are removed from the gas as liquid droplets following condensation. Well-documented data with these systems on biomass gas is limited; however, some actual operating data with biomass is included. Much of the information presented here is based on related experience with coal conversion.

Particulate and Tar Removal

Solid particulate is generally fly ash or char (partially burned feedstock). The class of separators most widely used for removal of particulates are the centrifugal type units (cyclone). For more rigorous removal electrostatic precipitators, baghouses, and other filtering methods are available. These separators are generally operated above the dewpoint of any tars, oils or water that might be present to avoid any sticky combination of tars and char. Depending on the physical properties of the contaminants this may be from 150°C to 500°C.

Tar removal lowers the heating value of the gas produced by a biomass gasifier. It may become necessary, however, because of the effect of tars in plugging lines and instrument ports rendering operation of downstream equipment difficult. Tars may be present in two forms. In a fixed updraft gasifier where the gas exit temperature is low, tars will be present primarily as liquid droplets entrained in the gas. In downdraft and fluid bed gasifiers where the product gas is hot, tars may also be present in the vapor form. Typical tar removal methods include cyclones, centrifugal separators and wet scrubbers of various kinds. Only at high temperatures can particulate and tar removal be separated. At temperatures where tars and oils condense to

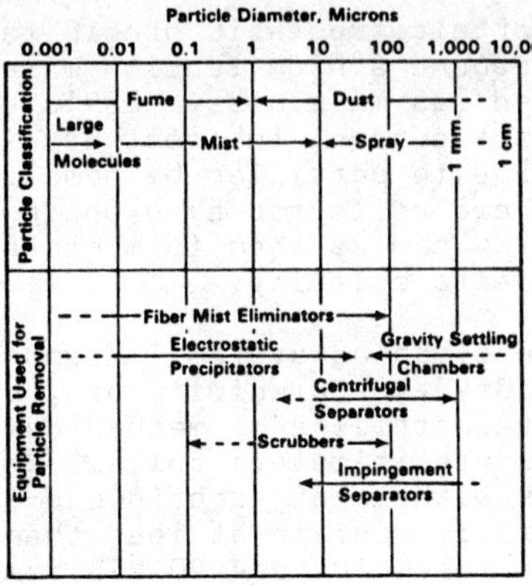

FIGURE 5. Particle Size Classification and Useful Collection Equipment

liquids, removal of tars and oils cannot be divorced from the particulate removal problem (all collect together).

Cyclones

Cyclones, or inertial separators, operate by imparting centrifugal force to the particles to be removed from the gas stream. This is done by directing the gas flow in a circular path, which due to their inertia, the particles cannot follow. Design procedures for cyclones, based on theory and supplemented by empirical techniques accumulated from operating experience, enable design of cyclones with predictable performance (Stone & Webster 1974, Zenz 1982). For this prediction, the nature of the particles and especially the particle size distribution must be well defined. Conventional cyclones designed for a 2 to 3 μm cut off (50% collection efficiency @ this size range) would collect no more than 20% of 1 μm particles (Stone and Webster 1974). Clearly, for collection of sub-micron particles, cyclones will be virtually useless. For collection of particles > 2 μm in diameter they are an excellent choice. Cyclones are manufactured in almost any size, from miniclones at less than 1/2" diameter to large capacity units up to and over 10 feet. Capacity of larger units can be hundreds of thousands of cubic feet per minute. Series cyclones are quite common and as many as three or four units can be coupled.

There are a large number of cases where a cyclone has been employed as a primary particulate remover to be followed by more stringent cleaning devices such as filters, venturi scrubbers or ESP's.

Electrostatic Precipitators

The electrostatic precipitator (ESP) acts only on the particles to be collected and not on the entire gas stream. The gas flow through an ESP occurs with little more pressure drop than through an equivalent length of pipe.

The process of electrostatic precipitation consists of corona formation around a high tension wire with particle charging by ionized gas molecules. ESP's operate by using a high voltage, direct current to creat gas ions that impart an electrical charge to particles by bombardment. The charged particles are collected by exposing them to an electric field which causes them to migrate and deposit on electrodes of opposite polarity.

The electrode cleaning system is dependent on the type of precipitator. Dry type precipitators, which can operate up to 540°C, remove particles by mechanical shaking and rapping. Wet type precipitators collect and remove the particles and tars with a thin continuous flowing film of water. They generally operate at less than 65°C (Kennedy 1979). Typical efficiencies are 90-99% removal for particle diameters 0.1μ and greater.

Baghouses

Baghouse filters are one of the oldest and most widely used techniques for removing particles from gas streams. The primary advantage of bag filters is their very high efficiency in dust removal. Other advantages include low energy use and pressure drop and collection of particles in a dry form and which simplifies disposal. Disadvantages include the large size and high cost. In addition temperatures are limited to about 290°C (Kennedy 1979). For this reason baghouse filters are seldom used in systems where tars are present unless they have been previously removed.

The mechanism of the filtering process is complex. Small particles are initially retained on the fibers of the filter material by interception, impingement, diffusion and settling, and electrostatic attraction. After a filter cake accumulates, further collection is by sieving, which is primarily responsible for the high collection efficiency attainable. Bag filters typically allow collection of <2 μm particles with 99.9% efficiency (Stone and Webster 1974). Maximum pressure differential is approximately 6 inches of water prior to cleaning. Typical capacities are near 1.9-2.8 ACFM/ft^2.

Granular Bed Filters

Granular bed filters employ a stagnant fixed bed of sand or other media to trap particulates by impaction. Fluidizing gas is introduced once a specified pressure differential (1-3 psi) is reached to clean the captured particulate from the much larger filter media. Typical media sizes are around 15-30 mesh and bed depths are 2-4 inches (Zenz 1982).

These devices have been characterized as highly efficient though complex (Razgaitis 1977). In principle a pebble bed filter can be built very large so the filter area is extensive and sufficient to achieve any desired efficiency. The trouble is, sooner or later it has to be regenerated. This can prove quite troublesome, especially with sticky particles.

Other Barrier Filters

Additional barrier filtration methods are available (Henry, Saxena, and Podolski, 1983) but are not widely used due to their cost or complexity. Porous metal filter elements have been used for years in catalyst recovery, nuclear waste calcination, and other situations where particulate removal is desired from hot gas or liquid streams. With proper design and materials (Inconel®) these filters may be operated at 500-700°C (932-1300°F). Porous ceramics may be applicable at even higher temperatures (Cilbreti et al. 1983). Nominal pore sizes range from 0.5 to 100 microns. Typical removal efficiencies can be better than 99.9% for particles greater than 0.05 microns. The inherent high cost, slow sintering at high temperatures, plugging, and possible failure due to thermal shock may make these units uneconomical other than for research scale or special applications.

Wet Scrubbers

Gas scrubbing using various liquids can be highly effective for simultaneous particulate and tar removal. Water is the most common scrubbing liquid although other liquids have been investigated. Wet scrubbers have been used extensively on wood and bark fired boilers in the wood products industry (Scott 1980). Wet scrubbers have also been used in coal-fired plants but on a large scale, economics generally favor ESPs or baghouses (Ford 1980).

There are a wide variety of wet scrubbers, varying primarily in the method of liquid contacting. The most common types are spray towers, packed bed scrubbers, plate scrubbers, impingement scrubbers, baffle scrubbers, and venturi scrubbers. The design and construction of wet scrubbers is covered in detail by Calvert (1972), Lucas (1973), and Kaupp and Goss (1981).

Table 4 shows the pressure drop and the efficiency of various wet scrubbers. Spray towers are the simplest and least expensive and have the lowest efficiency. Venturi scrubbers are the most complex and most efficient.

TABLE 4. Effectiveness of Spray Towers (Lucas 1979; Buonicore 1980)

	Pressure Drop inches water	Particle Size, μm for 80% collection
Spray Tower	0.5-1.5	10
Impingement	2-50	1-5
Packed Bed	2-50	1-10
Venturi	5-100	0.2-0.8

Wet scrubbers all produce a liquid stream which usually requires further treatment for tar recovery prior to disposal. Problems with stable emulsion formation are not uncommon. Oils from biomass gasifiers are quite likely to contain an appreciable fraction of aqueous soluble carboxylic acids, aldehydes, ketones, and substituted phenols such as have been reported in products derived from "flash pyrolysis" and "entrained flow pyrolysis" of wood. (Elliott 1985) In addition, the oils which can be decanted are likely to still contain appreciable amounts of water which lower their heating value and reduce their viscosity.

Operating Conditions and Efficiencies

The main features of commercial and developing particulate and tar removal systems are summarized in Table 5. Cyclones can operate at the highest temperatures followed by dry electrostatic precipitators (ESP's) and granular bed filters. Baghouses are limited to about 290°C by the fabrics they utilize. Liquid scrubbers are limited by the vapor pressure of the scrubbing liquid, usually water.

Filter systems capable of operating at higher temperatures are under development. The effects of operating at pressure are documented for cyclones, electrostatic precipitators, barrier filters, and other hot gas cleaning methods (Henry, Saxen, and Podolski 1982, Gilbert Commonwealth 1978). Generally, pressurized operation reduces the size of gas cleaning equipment. Wet scrubbers, however, are generally used only on atmospheric pressure gasifiers and consequently have not been researched extensively. Research in this area would be beneficial.

TABLE 5. Summary of Gas Cleanup Systems (Stone and Webster 1974)

Commercial Systems	Maximum Operating Conditions Operated °F/PSIG	Projected °F/PSIG	Inlet or Face Velocity CF/min/ft^2	Typical Pressure Drop in H_2O	Collects Condensed Tars
Cyclones	1700/high	>2000/high	4000	4 - 6	Yes
Electrostatic Precipitators	900/15	1700/300	200 to 300	15 - 35	Yes
Bag Houses	550/15	---	5 to 10	15 - 30	No
Granular Bed Filters	>900/high	>2000/300+	100	9 - 12	No
Liquid Scrubbers	175/15	--	9 - 12	Yes	
Aerodyne Tornado	900/500	1700/500	--	13	Yes
Developing Systems					
Fiber Filters	--	1500/high	100	--	No
Porous Metal	800/15	>1500/high	2 to 5	--	No
Ceramic	3450/15	---	2 to 5	--	No
Liquid Scrubbers	1400/15	1400/100	--	--	Yes
Panel Bed	1000/15	>2000/high	30 - 45	--	No
Ducon	600/15	>1500/high	30 - 90	--	No
Pebble Bed	250/15	1500/high	30 - 100	--	No

Typical efficiencies for several of the most common gas cleaning systems as a function of particle size are shown in Figure 6. Cyclone separators are the least efficient. Wet scrubbers are somewhat more efficient and baghouses (fabric filters) and ESP's are the most efficient particularly for submicron particles.

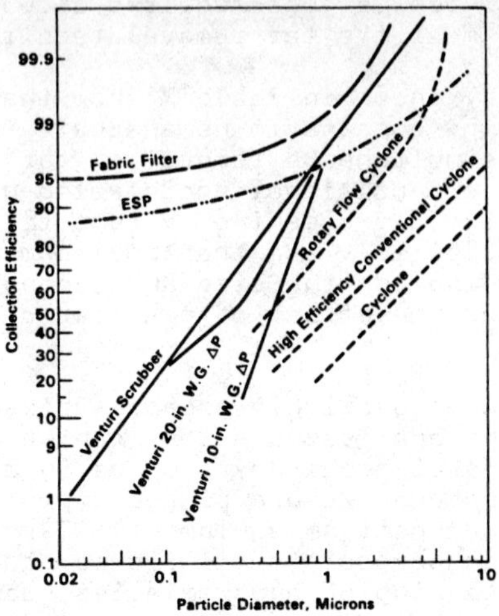

FIGURE 6. Typical Efficiencies for Various Gas Cleaning Methods

SYSTEMS ANALYSIS

There are a multitude of equipment combinations available for cleaning gases produced by biomass gasification. Each type of gasifier has cleanup options which are best suited to that particular gasifier. This chapter combines gasifier pollutant levels, gas cleanup efficiencies, and end use requirements to evaluate total gas cleanup systems which are applicable.

Calculations of system efficiency and end use applicability consider: pollutant concentrations, gas cleanup devices, and end uses. Because no information on tar removal efficiencies was found several assumptions had to be made to complete the calculations: 1) particulates and tar could not be removed in the same step due to the sludge that would result and 2) when tar removal was necessary, wet scrubbing to condense and remove tars would be required (no other removal techniques were considered applicable to tar removal). This method is used for tar removal in commercial fixed-bed coal gasifiers (Kennedy 1979). While these assumptions may not be strictly true, they represent the current state-of-the-art as we see it.

Table 6 tabulates the various particulate emission levels calculated for each type of gasifier and applicable cleaning technology. For fixed-bed gasifiers several technologies were considered not applicable because they would remove particulates and tar together. For fluid bed and entrained bed gasifiers wet scrubbers as a first cleanup step was considered not applicable as a tar and particulates would both be removed. This also results in a significant energy penalty for these high-temperature gasifiers.

Table 7 shows the gas cleanup combinations that achieve particulate performance standards for the various end uses. Where tar removal was required it was assumed the wet scrubbing would meet the tar removal requirements.

The results, shown in Table 7 show that, in general, particle scrubbing appears to be adequate to deal with biomass gas if simultaneous removal of tar is not required. However additional engineering data is necessary to properly design gas cleaning systems to assure that the system operates reliably and that environmental restrictions are met. These include the size and composition of particulates emitted and the effect of gasifier operating variables on emissions.

Tars and aqueous soluble organics present a unique problem. In some applications they can be used as a fuel directly (no removal required) such as in a boiler, but often they condense and cause plugging problems in transfer lines and other downstream equipment. Tars may be present as vapor or as an aerosol (small liquid droplets). There is no basic understanding of how variables such as temperature, tar loading, tar composition, and velocity affect tar deposition.

TABLE 6. Predicted Particulate Emission Levels

Gasifier	Particulate Burden (mg/m^3)	Particle Size Distribution	Single Cyclone	Series Cyclones	Cyclone & Filter	Cyclone & Electrostatic Precipitator	Cyclone & Wet Scrubber	Electrostatic Precipitator	Wet Scrubber	Wet Scrubber & Filter	Wet Scrubber & Electrostatic Precipitator
FIXED BED UPDRAFT											
High	1,000	>10μ 3% 1-10μ 8%	NA	NA	NA	NA	NA	NA	450	18	45
Low	100	<1μ 89%	NA	NA	NA	NA	NA	NA	45	2	5
FIXED BED DOWNDRAFT											
High	6,000	all > 10μ	240	9.6	3.2	3.2	3.2	60	60	0.6	0.06
Low	100		4	0.16	.04	0.4	0.4	1	1	0.01	0.01
FLUID-BED											
High	100,000	>10μ 61% 1-10μ 38%	11,500	2,300	39	77	150	2,060	NA	NA	NA
Low	10,000	<1μ 1%	1,100	230	4	8	15	206	NA	NA	NA
ENTRAINED-BED											
High	100,000	>10μ 61% 1-10μ 38%	11,500	2,300	39	77	150	2,060	NA	NA	NA
Low	10,000	<1μ 1%	1,100	230	4	8	15	206	NA	NA	NA

NA - not applicable

TABLE 7. Applicable Gas Cleaning Options

	CLOSE-COUPLED BOILER	DIESEL OR SPARK-IGNITION ENGINE	GAS TURBINE	SYN GAS OR SNG
Performance Standards	200-1500 mg/m^3	10-50 mg/m^3	1-80 mg/m^3	1-80 mg/m^3
Fixed Bed Updraft (2)	None(1)	WS + F	WS(1) WS + F WS + ESP ESP(1)	WS(1) WS + F WS + ESP ESP(1)
Fixed Bed Downdraft	None(1) C	C(1,4) C + F WS(1) WS + F	C(1) 2C(1) C + F C + ESP	C(1) 2C(1) C + F C + ESP
Fluidized Bed	2C(1,3) C + WS C + F ESP(1,6)	C + F C + WS(1) C + ESP(1,6)	C + WS(1) C + F C + ESP(6) (5)	C + WS(1) C + F C + ESP(6) (5)
Entrained Bed	Similar to Fluidized Bed			

LEGEND

C - Cyclone
2C - Two cyclones in series
F - Fabric Filter (baghouse)
WS - Wet Scrubber
ESP - Electrostatic Precipitator

FOOTNOTES

(1) Lower level of contaminates is acceptable - higher level would exceed limits
(2) Cyclone not effective due to smaller particle size distribution and tar droplets - use wet scrubber to remove tars first if any cleanup is required
(3) Assumes 50% of particulate is char and 90% burns in burner
(4) Cyclones are effective for this application but wet scrubbers are often used instead because gas cooling is also required. Also cyclone efficiency is affected by large turndown ratio required in some engine applications
(5) Pressurized operation may restrict the size or applicability of gas cleanup equipment (particularly baghouses and ESPs)
(6) ESP is not as effective on particulates with high carbon content and may not be applicable

In addition there is no information on tar removal in the literature. We assumed the wet scrubbing is effective for biomass tar removal, but this has not been confirmed in large scale equipment. There are several disadvantages to wet scrubbing. The sensible heat of the gas is lost as is the heating value of water soluble compounds. Wet scrubbing generates a wastewater stream which may present environmental problems. Separation and recovery of biomass tars from wastewater has not yet been proven on a commercial scale.

CONCLUSIONS AND RECOMMENDATIONS

Particulate removal technology appears adequate to deal with biomass gas, especially when simultaneous tar removal is not required. However, more detailed data on particulate loadings, particle size and particle composition from different type gasifiers and more definite limits for various end uses is needed for engineering design of large-scale commercial gas cleanup systems.

Removal of tar from gas streams presents a problem with aspects not previously encountered. Tars from fixed-bed biomass gasifiers are most likely present primarily as very small liquid droplets although no information on particle size was found in the literature.

A study of the mechanism tar formation including the effect of gas velocity, temperature, and wood moisture content on tar loadings and liquid particle diameters would aid significantly not only in gas stream cleanup design but in gasifier design. In conjunction with this, research on the efficiency of various wet scrubbing techniques for tar removal is needed for design of wet scrubbing systems.

Tars from downdraft and fluid bed gasifiers are probably present primarily as vapors although this has not been confirmed. Tar removal will require cooling and water scrubbing. The size of droplets formed by quenching and the effect of the method of quenching on the efficiency of tar removal needs to be studied.

Water scrubbing produces a wastewater stream containing tars and water soluble organics which may present an environmental problem. Research on separation of tars from the aqueous layer and methods of treatment and disposal of the water is needed. Thermal or catalytic cracking of tars to convert them to gas and char should be considered as an alternative to water scrubber.

ACKNOWLEDGMENTS

This report summarizes the technical results of gas cleanup studies on gasification of biomass conducted by the Pacific Northwest Laboratory (PNL) for the Biomass Thermochemical Conversion Program of the U.S. Department of Energy (DOE) under contract number DE-AC06-76RLO-1830. The guidance and support provided by Gary F. Schiefelbein and Mark A. Gerber of the Program Office and Simon Friedrich of DOE are greatly appreciated.

REFERENCES

Baker, E. G., L. K. Mudge, and D. H. Mitchell. 1984. "Oxygen/Steam Gasification of Wood in a Fixed-Bed Gasifier," Ind. Eng. Chem. 23 (4)

Buonicore, A. J. 1980. "Air Pollution Control." Chem. Eng., June 30, pp. 91-101.

Calvert, S., et al. 1972. Wet Scrubber System Study, Volume I: Scrubber Handbook, U.S. Department of Commerce, NTIS PB-213016.

Cilberti, D. F., et al. 1983. "Hot Gas Cleanup Using Porous Ceramic Cross Flow Filters," Proceedings Third Annual Contaminant Control in Hot Coal Derived Gas Streams Contractors Meeting. DOE/METC/84-6 (CONF-8305112). Morgantown Energy Research Center, Morgantown, West Virginia.

Datin, D. L., W. A. LePori, and C. B. Parnell, Jr. 1981. "Cleaning Low Energy Gas Produced in a Fluidized Bed Gasifier" presented at the 1981 Winter Meeting of ASAE, Dec. 15-18, 1981. Chicago, Illinois, Paper No. 81-3592.

Datin, D. L., W. A. LePori, and C. B. Parnell, Jr. 1983. "Characteristics of Particulates Emitted from a Biomass Fluidized Bed Gasifier" presented at the 1983 Winter Meeting of ASAE, Dec. 13-16, 1983. Chicago, Illinois, Paper No. 83-3548.

Dravo Corp. 1976. Handbook of Gasifiers and Gas Treatment Systems FE-1772-11. Dravo Corp. Pittsburgh, Pennsylvania.

Elliott, D. C. 1985. Analysis and Comparison of Biomass Pyrolysis/Gasification Condensates, PNL-5555. Pacific Northwest Laboratories, Richland, Washington.

Finnie, G. 1979. "Haleyon Gasification Systems." Proceedings of 10th Texas Industrial Wood Seminar: Technology and Economics of Wood Residue Gasification. Lufkin, Texas.

Ford, G. C., et al. 1980. "Controlling Particulate Emissions from Utility and Industrial Balers." Power, June 1980, pp. S-1 to S-20.

Fritz, J. J., J. J. Gordon, and V. T. Nguyen. 1978. Status Review of Wood Biomass. ET-78-C-01-2854. Mitre Corporation.

Gilbert Commonwealth. 1978. Particulate Control for Pressurized Fluidized Bed Combustion, HCP/T2220-16, May 1978.

Groeneveld, Michael J., P. E. Gellings, and J. J. Hos. 1983. "Production of a Tar Free Gas in An Annular Co-current Moving Bed Gasifier." Energy From Biomass and Wastes VII, Lake Buena Vista, Florida, Jan. 24-28, 1983.

Henry, R. F., S. C. Saxena, and W. F. Podolski. 1983. Particulate Removal from High-Temperature, High-Pressure Combustion Gases, ANL/FE-82-11. Argonne National Laboratory, Chicago, Illinois.

Hoenig, Stuart A. and Frank L. Cole. 1981. "Cleanup of Producer and Synthesis Gas With a Tube and Wire Electrostatic Precipitator" Solar Energy, Vol. 27, No. 6, pp. 579-580.

Jacko, R. B., M. L. Holcomb, and J. R. Barrett. 1985. "Alternative Energy Source Emissions - PAH, NO_x, Particulates, and Size Distributions From Biomass Gasification," Dept. of Civil Engineering, Purdue University, West Lafayette, Indiana.

Kaupp, A. and J. R. Goss. 1981. State of Art for Small Scale (to 50 kw) Gas Producer - Engine Systems, Final Report to the USDA/USFS on Contract #53-319R-0-141, University of California, Davis, California

Kennedy, D. M., L. Breitstein, and C. Chen. 1979. "Control Technologies for Particulate and Tar Emissions from Coal Converters." Proceedings Environmental Aspects of Fuel Conversion Technology. EPA-600/7-79-217. Research Triangle Institute Research Triangle Park, North Carolina.

Kumar, S., et al. 1984. "Design and Development of a Biomass Based Small Gasifier-Engine System. Suitable for Irrigation Needs in Remote Areas of Developing Countries". Proceedings: Energy from Biomass and Wastes VIII. Institute of Gas Technology.

Lucas, R. L. 1973. "Gas-Solid Systems". Chemical Engineers Handbook, R. H. Perry and C. H. Chilton, ed. McGraw-Hill, New York.

Manhattan College. 1978. Control Technology for Fine Particulate Emissions. ANL/ECT-5.

Moreno, F. E. and J. R. Goss. 1983. "Fluidized Bed Gasification of High Ash Agricultural Wastes to Produce Process Heat and Electrical Power." Symposium Papers Energy from Biomass and Wastes VII, Institute of Gas Technology, Chicago, Illinois.

Mudge, L. K,. E. G. Baker, D. H. Mitchell, R. J. Robertus, and M. D. Brown, 1983. Catalytic Gasification Studies in a Pressurized Fluid Bed, PNL-4594, Pacific Northwest Laboratory, Richland, Washington.

Mudge, L. K., D. G. Ham, S. L. Weber, and D. H. Mitchell. 1980. Oxygen/Steam Gasification of Wood, PNL-3353, Pacific

Northwest Laboratory, Richland, Washington.
Murphy, Michael. 1984. "Case History of the Design, Startup, and Operation of a 54 MM Btu/hr Fluid Bed Gasifier for Steam Production from Wood and Biomass Residue" in Energy from Biomass and Wastes VIII, Jan. 30-Feb. 2, 1984, Lake Buena Vista, Florida.

Oliver, E. D. 1982. Technical Evaluation of Wood Gasification, EPRI-AP-2567. Synthetic Fuel Associates, Palo Alto, California.

Razgaitis, R. 1977. An Analysis of the High-Temperature Particulate Collection Problem, ANL-77-14, Argonne National Laboratory, Chicago, Ilinois.

Reed, T. B. and M. Markson. 1983. "A Predictive Model for Stratified Downdraft Gasification." Proceedings of the 15th Biomass Thermochemical Conversion Contractors Meeting. PNL-SA-11306. Pacific Northwest Laboratory, Richland, Washington.

Robson, F. L. and A. J. Giramonti. 1974. "The Environmental Impact of Coal Based Advanced Power Generating Systems." Symposium Proceedings: Environmental Aspects of Fuel Conversion Technology

SERI. 1979. Generator Gas -- The Swedish Experience from 1939-1945. SERI/SP-33-140. Solar Energy Research Institute, Golden, Colorado.

Scott, J. W. 1980. "A Textile Firm Converts to Wood Fuel." Proceedings - Energy Generation and Cogeneration from Wood. Forest Products Research Society, Madison, Wisconsin.

Stone and Webster Engineering Corp. Purification of Hot Fuel Gases from Coal or Heavy Oil, EPRI-243-1.

Zenz, Fredrick A. 1982. State-of-the-Art Review and Report on Critical Aspects and Scale-Up Considerations in the Design of Fluidized-Bed Reactors - Final Report on Phase 1, DOE/MC/14141-1158.

DISCUSSION

Augenstein — Would you happen to know what fraction of the tars might be carboxylic acids that would be corrosive to mild steel?

Butner — I couldn't give you the answer to that. Doug Elliott may be willing to speak to that.

D. Elliott, Pacific Northwest Laboratory — I guess the answer is that it depends. It's a variable that depends on what kind of gasifier and operating conditions you're working with. Typically, there is not much. The gasifiers that operate at higher temperatures produce essentially none at all. You'd find them typically in updraft gasifiers where there are more of these types of chemicals produced.

S. Babu, IGT — In the tars produced in the process, the total condensate had a pH of about 8. That's because the alkalies also condense and act as a buffer. But I think the phenol content is a good indication. The tars have been characterized and the information is available in the literature.

Augenstein — So if carboxylic acids are formed, they are neutralized?

Babu — Yes, they're neutralized because the potassium oxide combines with the carboxylic acids and forms stable compounds. I don't want people to leave with the wrong impression that fluidized beds are less efficient. If a fluidized bed is operated properly, it will produce thermal efficiencies higher than those of the fixed bed gasifiers that were shown this afternoon. Also, the carbon loss isn't any greater, and char is not produced if you operate the gasifier properly.

PROCESS PERFORMANCE AND ENVIRONMENTAL ASSESSMENT OF THE RENUGAS® PROCESS

Robert J. Evans, Richard A. Knight, Michael Onischak, and Suresh P. Babu
Institute of Gas Technology
3424 South State Street
Chicago, Illinois 60616

ABSTRACT

Adiabatic, pressurized, fluidized-bed steam and oxygen-blown biomass gasification tests were conducted in the 12-tons/day process research unit (PRU) to develop process optimization information for the production of medium-Btu gas. The test program has investigated the effect on gasification performance of temperature, pressure, biomass throughput rate, steam-to-biomass ratio, type of feedstock, feedstock moisture, and fluidized-bed height.

Long-duration gasification tests (up to 3 days of steady-state) were made to establish material and energy balances, to observe the possible breakthrough of low-concentration organic compounds, and to evaluate the mechanical performance of the system.

Test results indicate that the process, which has proven to be simple to operate, can achieve carbon conversions of about 95% with thermal efficiencies of about 75% and with low oil and tar production.

New information on the oil and tar fraction production outline the effect of process operating conditions and feedstock type. Different feedstocks of very similar elemental compositions have produced similar product gas compositions, but have produced different distributions and character of the oil and tar fractions. This information is important environmentally and impacts upon the scope and duty of process cleanup equipment.

PROCESS PERFORMANCE AND ENVIRONMENTAL ASSESSMENT OF THE RENUGAS PROCESS

INTRODUCTION

Program Objective and Scope

The objective of this project is to develop an efficient biomass gasification process to produce medium-Btu gas (suitable as an industrial fuel gas or for upgrading to substitute natural gas, SNG) or synthesis gas.

The project involves the development of a pressurized, single-stage, oxygen-blown fluidized-bed gasifier to produce substitute fuel gases. Fundamental, operational information on the simultaneous physical and chemical processes occurring in biomass fluidized-bed gasification is needed on a large scale for rational design and scale-up of the process.

The overall developmental program included bench-scale experiments leading to the design and construction of a process research unit (PRU). Tests conducted in the 11.5-inch-ID, fluidized-bed gasifier were designed to obtain data on biomass gasification characteristics as a function of significant operating parameters on a large scale (up to 12 tons/day) and at continuous, adiabatic, steady-state operating conditions.

The 11.5-inch-ID PRU is designed for maximum operational flexibility to study the effect of different feeding positions to the fluidized bed relative to the fluidizing-gas distributor and the fluidized-bed height. This gasifier capability is in addition to the program objectives of investigating different feedstocks with different moisture levels and the effect of temperature and pressures up to 1800°F and 500 psig, respectively.

Certain areas where design information was incomplete were evaluated. These include high-pressure feeder operation, the effect of operating parameters on adiabatic gasifier performance, and the characteristics of process effluents.

The experimental PRU test program addressed these areas and quantified the gasification performance. The performance trends found as a function of primary operating variables will identify approaches for system optimization and will indicate the need or extent of various upstream or downstream process operations. This information is available for process assessment and evaluation of the technoeconomics of pressurized biomass gasification.[1,2]

Description of the PRU System

The Institute of Gas Technology (IGT) designed and built the PRU for the Department of Energy (DOE) program at its Energy Development Center in Chicago. The PRU gasifier is an 11.5-inch-ID by 21-foot-high vessel with 12 tons/day gasification capacity. Figure 1 shows an isometric equipment layout of the PRU equipment in a building that extends over 50 feet in total height. The 11.5-inch-ID, Incoloy 800H balanced pressure reactor liner is surrounded by about 12 inches of fiber insulation in a 3-foot-OD by 24-foot carbon-steel pressure vessel. The 12 inches of bulk fiber insulation in the annular region is sufficient to provide adiabatic gasification conditions. The rest of the system has been described previously.(1) Two additions were made to the PRU system early in the test program. These additions have improved the operation of the feed system and the measurement of entrained particulates and the condensable fraction in the product gas.

The first of the changes concerned the mechanical operation of the lockhopper slide-gate valves. The originally purchased slide-gate valves installed in the system developed severe gate misalignment and leak problems after only about 50 cycles of operation. Both the 18 and 24-inch gates were remachined for additional clearance and internal air-operated cylinders were installed to help lift the gate for a seal, however, the problems remained. The manufacturer was unable to repair or rebuild the valves to solve these problems. No other valve manufacturer was known to have suitable valves, so an improved set was designed and fabricated by T. R. Miles Design Engineers in Portland. Many features have been incorporated into the new design based on the early operating and repair experience, which now make inspection, alignment control, and maintenance simple operations. The new valves were installed in January 1985 and have operated very well, now with over 1500 cycles of operation.

The second change to the PRU system improved the closure of the system's material balances. The PRU was equipped with an on-line isokinetic sampling system at the gasifier exit to obtain large raw-gas samples for measuring the solids carry-over and condensable liquids production. The sampling system is operated for 1-hour periods during the steady-state test period and withdraws a raw-gas sample at a rate of approximately 1% of the total flow from the gasifier. The carry-over solids are first trapped in a hot sintered metal filter, then the stream is cooled and the liquids are condensed in a water-cooled condenser followed by an ice-water bath. The condensate-free raw gas is metered and vented. The collection of this large volume of condensate provides a representative sample that is useful to account for the elemental and material balances. The measured steam condensate rate is used to validate the undecomposed steam concentration in the raw gas sample collected from the same location and analyzed by the on-line wet-gas chromatograph. The on-line analytical instrumenta-

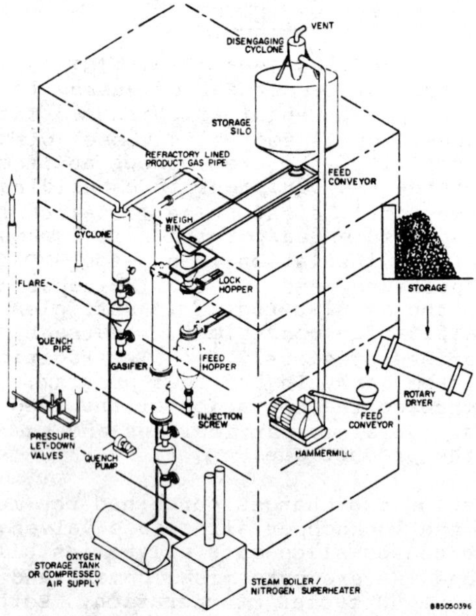

Figure 1. ISOMETRIC VIEW OF PRU EQUIPMENT LAYOUT

tion train consists of a gas chromatograph (GC), CO monitor, and oxygen monitor.

In addition, modifications were also made to the PRU to permit system operation for extended periods. A cyclone was installed inside the existing solids receiver vessel to operate at the system temperature and pressure. The carry-over solids collected in the cyclone are discharged through a small lockhopper. Another lockhopper installed under the gasifier itself can be operated to drain fluidized-bed solids, if required.

PRU GASIFICATION TEST RESULTS

IGT has operated the PRU to conduct a total of 22 gasification tests. The PRU tests were conducted at different gasification conditions to develop process optimization information of pressurized fluidized-bed gasification of biomass to produce medium-Btu gas. Fourteen of these tests were operated to obtain steady-state data under various conditions for approximately 4 hours, while five tests were conducted at longer duration periods of up to 3 days.

The results(2) from these tests were used to determine the effect of temperature, steam feed rate, pressure, biomass throughput rate, type of feedstock, fluidized-bed height, and feed moisture content on pressurized oxygen-blown gasifier performance. The tests were conducted over the range of 1390° to 1800°F, 100 and 300 psig and feed rates up to 1030 lb/h. The steam input rates were varied up to 1.18 lb/lb wood feed, while feed moisture content ranged from 5% to 27% by weight.

The long-duration tests with 12 hours of steady-state PRU gasifier operation and the 3-day steady-state test established the material and energy balances and confirmed the previous measurements of low-concentration constituents in the gasification products. The carry-over solids were effectively removed by the cyclone yielding valuable design information for a scale-up cyclone-train design. After the 3-day operation, no unusual buildup of materials were found in the PRU system components beyond the amounts generally found after a 4-hour steady-state period.

Table 1 summarizes the results of all the gasification tests conducted in the PRU at various conditions. Material and energy balances have been calculated for all tests. Table 2 presents a typical material and energy balance for Test T12-1, with an 11-hour steady-state period. Details for the other tests will be available in the final program report being prepared for DOE.

As shown in Table 1, Tests GT-1 through GT-11, and GT-16 were conducted using a single species, maple wood chip from northern Wisconsin as feed material. Whole tree chips from Pennsylvania were used as feed material for Tests GT-13, GT-14, GT-15 and GT-17, while whole tree chips from Wisconsin were used for the long-duration Tests T12-1, T12-2, T3D-1, T12-3, and T12-4. Representative samples of the prepared wood feed were analyzed for their physical and chemical properties. Table 3 lists the proximate analysis, ultimate analysis, bulk density, and gross heating value of typical samples of maple wood chips and the whole tree chips. The elemental composition of these feeds is remarkably similar, however, gasification performance differences occur, as is described later.

The first six parametric variation tests were conducted to obtain data on the effect of temperature on gasifier performance with maple wood. The gasifier was operated at a pressure of 300 psig and gasification temperatures varied from 1390° to 1800°F. Five of the six tests achieved a steady-state operating period of 4 hours before being voluntarily terminated. Three tests were terminated because of auxiliary equipment malfunctions. A brief discussion on the effect of significant operating conditions on the gasifier performance follows.

Effect of Temperature

The analyzed results from Table 1 are plotted in Figures 2 through 5 to show the effect of temperature on maple wood gasification. The temperature used in these plots is an average of four thermocouples covering a fluidized-bed height of up to 74 inches above the oxygen distributor. In any given test, the variation in temperature between these thermocouple positions was about 45° to

Table 1A. SUMMARY OF ANALYZED RESULTS OF PRU GASIFICATION TESTS WITH MAPLE CHIPS FEEDSTOCK

Test No.	GT-1	GT-2	GT-4	GT-5	GT-6	GT-8	GT-9	GT-10	GT-11	GT-16
Pressure, psia	323.7	323.7	308.7	308.7	311.7	308.7	314.7	104.7	296.7	314.7
Temperature, °F	1500	1650	1800	1390	1500	1510	1500	1465	1500	1450
Feed Rate, lb/h (wet)	813.4	690.6	731.9	724.3	693.5	750.4	765.9	399.5	1029.7	768.6
Moisture, wt %	8.35	9.50	9.70	10.47	10.72	11.15	12.44	12.08	10.02	7.71
Steam Input, lb/lb feed (wet)	0.79	0.71	0.60	0.77	0.86	0.65	0.50	0.62	0.48	0.74
Oxygen, lb/lb feed (wet)	0.21	0.25	0.34	0.17	0.24	0.22	0.21	0.23	0.20	0.22
Nitrogen, lb/lb feed (wet)	0.52	0.45	0.51	0.42	0.50	0.52	0.21	0.76	0.29	0.44
Product Gas Flow Rate, SCF/h	33,446	27,154	31,218	27,662	31,155	30,371	25,103	17,918	34,407	30,164
Superficial Gas Velocity, ft/s	2.20	1.93	2.49	1.80	2.13	2.11	1.70	3.59	2.47	1.99
Gas Composition, vol %										
H_2	10.58	11.82	13.58	9.62	8.08	9.97	9.81	7.52	10.37	10.52
CO	6.09	6.38	9.92	4.07	5.36	6.63	7.05	7.24	9.39	6.09
CO_2	15.63	17.15	15.49	15.92	13.98	15.04	17.64	11.89	15.61	15.89
CH_4	7.39	7.39	5.29	6.20	5.95	6.78	7.96	4.87	7.30	6.17
C_2H_4	0.10	0	0	0.10	0.08	0.19	0.27	0.45	0.27	0.12
C_2H_6	0.27	0.03	0	0.84	0.45	0.31	0.56	0.37	0.54	0.61
C_3H_8	0	0	0	0.02	0	0	0	0.01	0	0
C_6H_6	0.08	0.29	0	0.21	0.12	0.21	0.44	0.60	0.43	0.34
N_2	17.08	15.42	16.23	14.93	15.17	17.45	8.64	22.85	11.71	15.28
H_2O	42.78	41.52	34.96	48.09	45.44	39.00	40.84	36.06	38.11	44.98
Ar	0	0	4.53	0	5.37	4.42	6.79	8.14	6.27	0
Gross Molecular Weight, lb/lb-mol	22.63	22.78	23.36	22.62	23.59	23.74	24.28	25.04	24.09	22.75
Gas Yield, SCF/lb feed (wet)	41.12	39.32	42.65	38.19	44.92	40.47	32.78	44.85	33.41	39.25
Dry, Inert Free Gas, vol %										
H_2	26.36	27.45	30.67	26.01	23.75	25.48	22.43	22.82	23.62	26.47
CO	15.17	14.82	22.40	11.01	15.76	16.94	16.12	21.97	21.38	15.32
CO_2	38.94	39.83	34.98	43.05	41.09	38.44	40.34	36.08	35.55	39.98
CH_4	18.41	17.16	11.95	16.77	17.49	17.33	18.20	14.78	16.62	15.53
C_2H_4	0.25	0	0	0.27	0.24	0.49	0.62	1.37	0.61	0.30
C_2H_6	0.67	0.07	0	2.27	1.32	0.79	1.28	1.12	1.23	1.53
C_3H_8	0	0	0	0.05	0	0	0	0.03	0	0
C_6H_6	0.20	0.67	0	0.57	0.35	0.54	1.01	1.82	0.98	0.86
Gross Heating Value, Btu/SCF	344.20	337.07	292.61	356.93	345.34	354.64	379.35	405.31	382.19	356.62
Dry, Inert Free Gas Yield, SCF/lb feed (wet)	16.51	16.93	18.89	14.12	15.28	15.84	14.33	14.78	14.67	15.60
Cold Gas Thermal Efficiency, %	75.08	76.15	73.85	68.80	71.31	76.42	74.63	82.17	74.94	75.76
Carbon Conversion to Gas, %	88.2	92.2	94.2	82.3	88.2	90.5	88.5	95.4	87.4	90.2
Carbon Conversion to Gas and Liquids, %	92.0	94.4	94.8	87.2	92.0	92.7	92.0	96.0	91.8	94.4

123/renutabs/PAP

Table 1B. SUMMARY OF ANALYZED RESULTS OF PRU GASIFICATION TESTS WITH WHOLE TREE CHIPS FEEDSTOCK FOR PENNSYLVANIA AND WISCONSIN

	From Pennsylvania				From Wisconsin							
Test No.	GT-13	GT-14	GT-15	GT-17	T12-1	T12-2	T3D-1a	T3D-1b	T12-3a	T12-3b	T12-4a	T12-4b
Pressure, psia	323.7	344.7	320.7	320.7	314.7	116.7	323.7	323.7	317.7	317.7	83.7	98.7
Temperature, °F	1520	1510	1520	1530	1530	1500	1538	1562	1672	1413	1509	1516
Feed Rate, lb/h (wet)	742.1	826.9	673.4	684.8	643.9	352.9	427.5	748.0	708.1	708.1	375.4	375.4
Moisture, wt %	9.58	15.02	14.47	26.74	4.94	7.72	10.80	10.80	9.14	9.14	8.98	9.59
Steam Input, lb/lb feed (wet)	0.53	—	0.41	0.46	0.61	0.69	1.18	0.63	0.69	0.68	0.72	0.73
Oxygen, lb/lb feed (wet)	0.23	0.18	0.25	0.33	0.25	0.26	0.26	0.23	0.26	0.18	0.27	0.28
Nitrogen, lb/lb feed (wet)	0.36	0.95	0.87	1.34	0.54	0.91	0.74	0.43	0.41	0.40	0.85	0.93
Product Gas Flow Rate, SCF/h	27,220	27,772	28,271	33,016	24,603	15,693	23,003	27,695	27,472	25,373	16,813	17,439
Superficial Gas Velocity, ft/s	1.81	1.73	1.90	2.23	1.69	2.87	1.55	1.88	2.01	1.63	4.31	3.80
Gas Composition, vol %												
H_2	13.84	5.60	8.14	7.96	12.05	5.90	10.88	12.30	12.69	8.55	7.49	7.38
CO	7.08	7.20	6.89	3.99	8.00	7.72	3.59	5.58	7.22	5.59	7.79	7.59
CO_2	17.28	13.26	12.93	13.42	17.06	11.52	13.86	17.32	17.42	15.72	11.98	11.93
CH_4	6.82	5.83	4.79	3.95	7.37	4.58	4.79	6.56	7.60	6.95	4.32	4.61
C_2H_4	0.01	0.36	0.20	0	0.03	0.65	0.04	0.04	0	0.28	0.65	0.50
C_2H_6	0.51	0.60	0.27	0.09	0.22	0.56	0.17	0.20	0.02	0.90	0.35	0.30
C_3H_8	0	0	0	0	0	0.01	0	0	0	0	0.01	0.01
C_6H_6	0.14	0.84	0.49	0.17	0.27	0.24	0.18	0.81	0.41	0.57	0.47	0.40
N_2	13.16	46.37	28.24	37.53	19.18	27.64	18.63	15.59	14.34	15.13	25.80	27.29
H_2O	35.02	14.53	30.03	32.89	35.82	41.18	47.86	41.60	40.30	46.31	41.14	39.99
Ar	6.14	5.41	8.02	0	0	0	0	0	0	0	0	0
Gross Molecular Weight, lb/lb-mol	23.66	27.59	25.59	24.40	23.27	23.77	22.12	23.04	22.75	23.13	23.58	23.65
Gas Yield, SCF/lb feed (wet)	36.68	33.59	41.98	48.21	38.21	44.47	53.81	37.03	38.80	35.83	44.79	46.45
Dry, Inert Free Gas, vol %												
H_2	30.30	16.62	24.15	26.91	26.78	18.92	32.47	28.73	27.98	22.17	22.66	22.56
CO	15.50	21.37	20.44	13.49	17.78	24.76	10.71	13.03	15.92	14.50	23.56	23.20
CO_2	37.83	39.36	38.36	45.37	37.91	36.95	41.36	40.46	38.40	40.77	36.24	36.46
CH_4	14.93	17.30	14.21	13.35	16.38	14.69	14.29	15.32	16.75	18.02	13.07	14.09
C_2H_4	0.02	1.07	0.59	0	0.07	2.08	0.12	0.09	0	0.73	1.97	1.53
C_2H_6	1.12	1.78	0.80	0.30	0.49	1.80	0.51	0.47	0.04	2.33	1.06	0.92
C_3H_8	0	0	0	0	0	0.03	0	0	0	0	0.03	0.03
C_6H_6	0.31	2.49	1.45	0.57	0.60	0.77	0.54	1.89	0.90	1.48	1.42	1.22
Gross Heating Value, Btu/SCF	331.07	440.15	366.27	292.95	34[illegible].21	384.74	315.72	371.16	346.39	409.61	386.05	377.89
Dry, Inert Free Gas Yield, SCF/lb feed (wet)	16.76	11.32	14.15	14.26	17.19	13.87	18.03	15.85	17.60	13.82	14.81	15.20
Cold Gas Thermal Efficiency, %	75.33	71.05	72.71	69.52	7[illegible].91	70.26	76.99	79.57	80.72	74.95	75.62	76.89
Carbon Conversion to Gas, %	88.3	86.1	90.2	96.6	90.2	87.4	93.0	93.8	96.2	87.1	91.8	93.3
Carbon Conversion to Gas and Liquids, %	89.9	92.1	94.4	96.6	95.6	92.7	96.1	96.8	98.4	93.8	96.9	98.0

123/renutabs/PAP

Table 2. MATERIAL AND ENERGY BALANCE SUMMARY FOR TEST T12-1 (Test Data: June 20, 1985. Data Analysis Period From 1030 h to 2100 h.)

BASIS: 1 hour; all units in pounds.

Input		C	H	O	N	Ash	Total	10^6 Btu
Biomass (Dry)		298.4	38.3	270.7	1.2	3.5	612.1	5.1
Moisture			3.5	28.3			31.8	
Steam to Ring			26.0	207.8			233.8	0.3
Steam to Nozzle			17.4	139.4			156.8	0.2
Oxygen to Nozzle				163.3			163.3	
Nitrogen Purge					347.4		347.4	
Total Input		298.4	85.2	809.5	348.6	3.5	1545.2	5.6
Output								
Entrained Solids		13.0	0.2			3.5	16.7	0.2
Oil/Tars		16.3	1.8				18.1	0.4
Components:								
	H_2		15.6				15.6	
	CO	62.3		83.1			145.4	
	CO_2	132.9		354.4			487.3	
Product	CH_4	57.4	19.1				76.5	
Gas	C_2H_4	0.5	--				0.5	
	C_2H_6	3.4	0.9				4.3	
	C_3H_8	0.0	0.0				0.0	
	C_6H_6	12.6	1.1				13.7	
	N_2				348.6		348.6	
	H_2O		46.5	372.0			418.5	
Total Gas		269.1	83.2	809.5	348.6		1510.4	4.8
Total Output		298.4	85.2	809.5	348.6	3.5	1545.2	5.4
% Balance (Output/Input)		100.0	100.0	100.0	100.0	100.0	100.0	96.4

123/renutabs/PAP

Table 3. TYPICAL WOOD FEED PROPERTIES

	Maple Wood*	Whole Tree Chips** From Pennsylvania	Whole Tree Chips† From Wisconsin
Proximate Analysis, wt %			
Moisture	12.26	10.89	7.72
Volatile Matter	73.94	73.53	77.07
Ash	0.43	0.84	0.75
Fixed Carbon (By Diff)	13.37	14.74	14.46
Total	100.00	100.00	100.00
Ultimate Analysis, wt %			
Ash	0.50	0.94	0.82
Carbon	49.54	48.51	48.40
Hydrogen	6.11	6.17	6.31
Sulfur	0.02	0.04	0.03
Nitrogen	0.10	0.12	0.21
Oxygen (By Diff)	43.73	44.22	44.23
Total	100.00	100.00	100.00
Bulk Density, lb/ft^3	16.6	14.7	10.7
Gross Heating Value, Btu/lb	8306	8210	8228
Ash Fusion Temperature, °F	>2700	>2700	>2700

* Average of two samples (Tests GT-9 and GT-10).

** Average of two samples (Tests GT-12 and GT-13).

† Average of two samples (Test T12-2).

ESTIMATED SPECIE COMPOSITION OF TEST FEEDSTOCKS

Feedstock	Estd Specie Composition, vol %
Maple Wood	100% maple pulpwood chips
Whole Tree Chips From Pennsylvania	90% red oak, balance chestnut, aspen, black birch, some bark, leaves
Whole Tree Chips From Wisconsin	34% maple, 33.5% oak, 19.6% birch, 12.9% pine and brush, some bark, leaves

123/renutabs/PAP

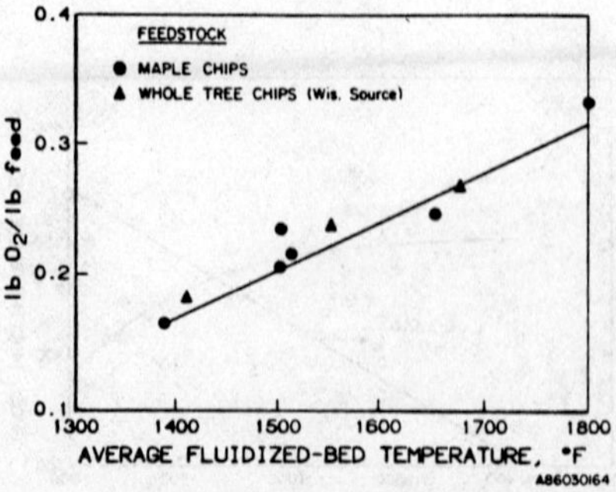

Figure 2. OXYGEN REQUIRED AS A FUNCTION OF GASIFICATION TEMPERATURE AT 300 psig (Maple Feed: Tests GT-1 Through GT-8 and Wisconsin Whole Tree Chips: Tests T12-3a, T12-3b, T3D-1b)

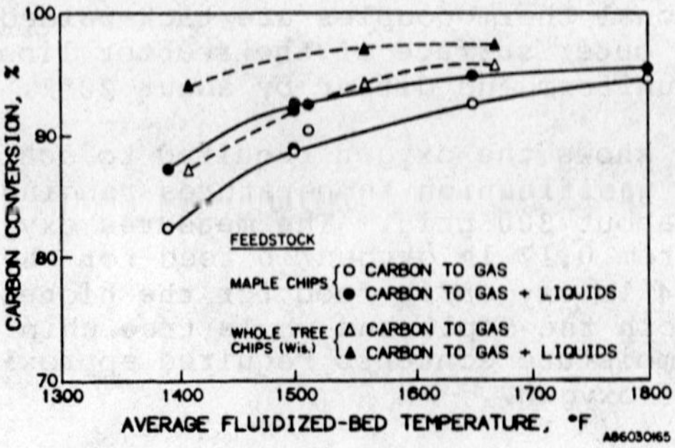

Figure 3. FEED CARBON CONVERSION VERSUS GASIFICATION TEMPERATURE AT 300 psig (Maple Feed: Tests GT-1 Through GT-8 and Wisconsin Whole Tree Chips: Tests T3D-1b, T12-3a, T12-3b)

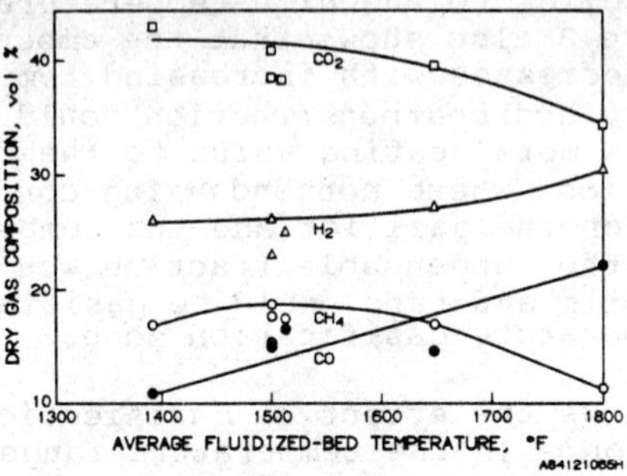

Figure 4. EFFECT OF GASIFICATION TEMPERATURE ON PRODUCT GAS COMPOSITION (Maple Feed: Gasification Tests GT-1 Through GT-8 Conducted at 300 psig, 10% Moisture)

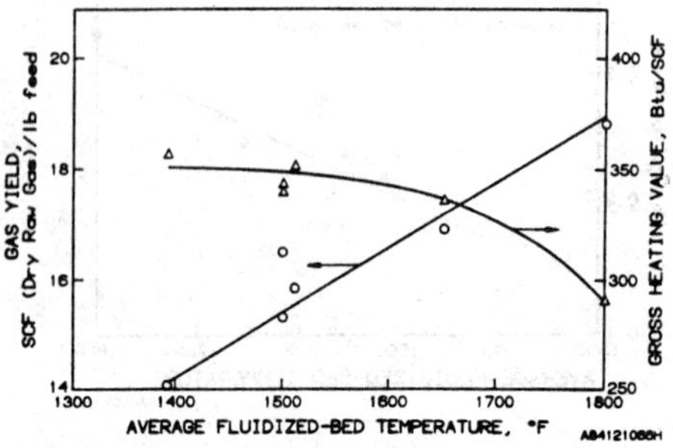

Figure 5. GAS YIELD AND HEATING VALUE AS A FUNCTION OF TEMPERATURE (Maple Feed: Gasification Tests GT-1 Through GT-8 Conducted at 300 psig, 10% Moisture)

90°F. Additional thermocouples are tack-welded in the same region on the outer surface of the reactor liner. These are usually more uniform and differ by about 20°F.

Figure 2 shows the oxygen required to achieve average fluidized-bed gasification temperatures ranging from 1390° to 1800°F at about 300 psig. The measured oxygen requirement varies from 0.17 lb oxygen/lb feed for the test at 1390°F to 0.34 lb oxygen/lb feed for the highest temperature at 1800°F. Both the maple and whole tree chip feedstocks with similar moisture contents required approximately the same amount of oxygen.

As expected, carbon conversion in the gasification process increases when the fluidized-bed temperature increases, as shown in Figure 3. The total carbon conversion for the maple feedstock ranges from about 92% to 95% over the temperature range of 1500° to 1800°F. The conversion is slightly higher for the whole tree chip feedstock. The carbon utilization could be further improved by recycling the elutriated solids to the high-temperature region of the gasifier. Figure 3 also shows that the amount of condensables produced decreases with increasing temperature. The condensable heavy hydrocarbon fraction would contribute approximately 10% more heating value to the fuel gases in on-site applications where noncondensing conditions could be maintained between the gasifier and the combustor. In processes where the condensable fraction would have to be separated, the oils and tars could be gasified by recycling to the high-temperature gasification zone.

Figure 4 shows the effect of gasification temperature on gas composition. In the temperature range of 1390° to 1800°F, the CH_4 content reaches a maximum at about 1500°F and is reduced at higher temperatures by reforming. The hydrogen concentration increases with temperature; the observed hydrogen is produced during biomass devolatilization, steam-char gasification, and the reforming of light and heavy hydrocarbons. Superimposed on these reactions is the water-gas shift reaction, which further alters the overall gas composition. As expected, CO increases steadily with temperature, whereas temperature has an inverse effect on CO_2. The reduction of CO_2 could be attributed to the reverse water-gas shift reaction and to the reaction between CO_2 and biomass char above 1500°F.

The gas yields and the gross heating values are shown as a function of temperature in Figure 5. The results conform with the anticipated trends of increased gas yields coupled with decreasing heating value as the gasification temperature is raised. The reforming of light hydrocarbons, such as methane and ethane, above 1500°F is the reason for the reduction in the fuel gas heating value. The heating value is about 350 Btu/SCF at 1500°F, which is considered a good quality medium-Btu gas. This gas can easily substitute for natural gas in industrial energy conversion systems.

Effects of Other Process Parameters

The effect of the other process parameters such as pressure, wood feed rate, wood feed moisture, input steam-to-wood ratio, and fluidized-bed height were investigated in the tests reported in Table 1, and reported previously.(2) Pressures tested were 60, 100, and 300 psig. Wood feed rates ranged from 350 to 1030 lb/h, with the bulk of the tests conducted at centerline operating conditions of 730 lb/h. Wood feed materials included a single-species maple chip feedstock for the initial tests with the PRU, and then true waste feedstocks composed of whole tree chips from a Pennsylvania source and also a Wisconsin source. The wood feed moisture levels for the tests obtained by dryer control were 5%, 10.5%, 15%, and 27% moisture, with the bulk of the parameter variation tests conducted with about 10% feed moisture.

The amount of steam present in the fluidized bed was varied for the tests with the whole tree chips feedstock. The total steam present relative to the wood feed ranged from 0.18 to 1.44 lb/h dry wood. The total steam present in the gasification zone is considered as that from the feed moisture and from the chemically produced water, in addition to the input steam as the fluidizing gas medium.

The effects of these parameters were reported previously(2) and most show trends that are pertinent to process engineering. The significant effects of the operating conditions and the feed type were observed on the amount and type of the oil and tar fraction.

Operating Conditions Affecting Oil and Tar Production

The quantity and quality of condensables produced during gasification, including oils and tars, is important for the design of effluent treatment processes and for general environmental considerations. Special attention was paid to the production of ammonia, phenols and other organic acids, sulfur gases, benzene, and polyaromatic hydrocarbons

(PAH). Due to the low sulfur content (0.02 to 0.04 weight percent) of the feedstock, the formation of sulfur gases is not expected to be a major concern. Small amounts (0 to 150 ppm) of COS were sometimes found in the product-gas sample bombs, but no H_2S has been detected.

Ammonia, phenol, cresols, acetic acid, methanol, and other water-soluble (that is, heterocyclic) organic compounds were found in the aqueous phase of the gasifier condensate. Due to the presence of ammonia, the pH of the condensate was generally around 8.0, and it was well-buffered, which is favorable for biological wastewater treatment processes. The levels of all water-soluble organic species in the condensate was measured as total organic carbon (TOC). It was found that TOC in the aqueous effluent was dependent on gasification temperature; this is shown in Figure 6.

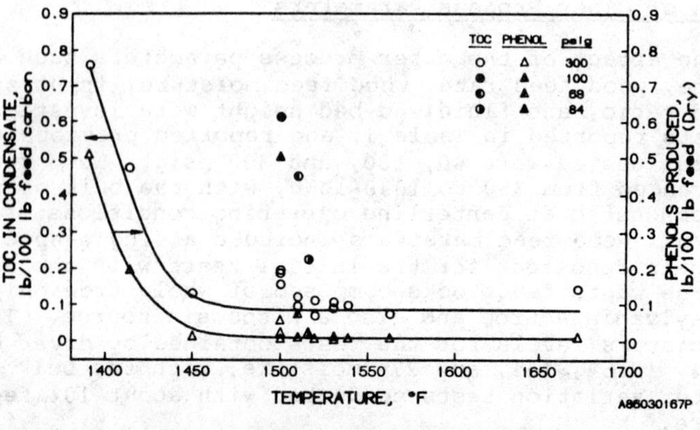

Figure 6. PHENOLS AND CONDENSATE ORGANIC CARBONS VERSUS GASIFICATION TEMPERATURE

An oil phase is also present in the gasifier condensate. This has been analyzed by GC/MS, and the results are tabulated in Table 4. Oil components were predominately aromatic hydrocarbons, with naphthalene and benzene as the most abundant species. In contrast to coal-derived oils, no aliphatic hydrocarbons were identified, although some could be present in the unidentified portions, which comprised 3% to 11% of the total oil. Oxygenated species accounted for 0.4% to 7.5% of the oil collected and included phenols, naphthols, and furans. Nitrogen- and sulfur-containing oil components were very low, which is to be expected due to the low sulfur and nitrogen content of the feed.

PAH are organic compounds containing two or more fused benzene rings. Two- and three-ring hydrocarbons, such as naphthalene and phenanthrene, do not pose any special environmental hazards. Compounds with four or more condensed rings, however, generally include hazardous substances such as fluoranthene, chrysene, and benz(a)pyrene. In this section, PAH will refer to the latter class of compounds, those with more than three fused rings. Although higher temperatures tend to favor the condensation reactions that produce these substances, the data show that the overall yield of oils and tars in the PRU decreases considerably at the higher temperatures. This can be seen graphically in Figure 7, which compares Tests GT-5, GT-8, T12-3a, and T12-3b. Two- and three-ring hydrocarbons, such as naphthalene and phenanthrene, as mentioned, do not pose any serious health hazards; however, they constitute an operating nuisance because they tend to accumulate in the process pipelines.

Table 4. COMPOSITION OF OILS FOUND IN PRODUCT GAS IN PRU GASIFICATION TESTS

Test No.	GT-5	GT-6	GT-8	GT-9	GT-10	GT-11	GT-13	GT-14	GT-15
Temperature, °F	1390	1500	1510	1500	1465	1500	1520	1510	1520
Oil Yield, lb/100 lb biomass*	2.84	2.18	1.33	1.83	0.34	2.33	1.66	3.00	2.26
Products, wt % in Oils									
One-Ring Hydrocarbons	28.6	30.6	24.4	40.8	15.0	39.9	8.4	31.2	26.8
Benzene	8.9	24.9	20.1	36.2	4.1	34.9	6.1	23.7	23.9
Toluene and Xylenes	12.1	2.9	2.3	2.1	2.9	2.3	0.7	4.7	1.2
Others	7.6	2.8	2.0	2.6	8.0	2.7	1.6	2.9	1.7
Two-Ring Hydrocarbons	28.8	39.0	39.5	33.5	40.4	29.7	54.5	32.2	36.5
Naphthalene	17.6	32.2	32.1	28.4	26.5	24.5	48.7	25.7	30.3
Methylnaphthalene	4.5	0.9	0.8	0.6	3.0	0.6	0.6	1.2	0.6
Others	6.7	5.9	6.6	4.5	11.0	4.5	5.3	5.4	5.6
Three-Ring Hydrocarbons	14.0	10.8	13.2	9.1	20.3	10.3	12.3	11.6	11.7
Fluorene	3.5	2.8	2.6	2.2	4.3	2.2	3.2	2.2	2.3
Phenanthrene and Anthracene	5.8	7.0	8.6	5.9	11.8	6.6	7.9	7.5	7.9
Others	4.7	1.8	2.0	1.0	4.3	1.5	1.3	1.9	1.5
Four-Ring Hydrocarbons	9.6	9.7	13.3	8.2	13.2	10.7	10.1	10.8	12.6
Five-Ring Hydrocarbons	2.2	2.2	2.6	2.4	2.5	3.1	4.1	4.4	4.7
Higher Aromatic Hydrocarbons	0.0	0.8	1.4	1.8	1.0	1.9	1.6	2.4	2.6
Total Hydrocarbons	83.2	93.1	94.4	95.8	92.4	95.6	91.1	92.6	94.8
O-Containing Compounds	7.5	1.1	0.9	0.7	3.5	1.0	0.5	1.9	0.4
Phenol and Cresols	3.2	0.1	0.04	0.1	0.4	0.1	0.0	0.9	0.04
Benzofurans and Dibenzofurans	3.7	1.0	0.6	0.5	2.4	0.7	0.004	1.0	0.4
Others	0.6	0.0	0.3	0.1	0.7	0.2	0.1	0.02	0.03
N-Containing Compounds	0.7	0.4	0.3	0.4	0.5	0.5	0.4	0.9	0.7
S-Containing Compounds	0.2	0.1	0.02	0.02	0.04	0.02	0.2	0.0	0.01
Total Heterocyclic Compounds	8.4	1.6	1.2	1.1	4.0	1.5	1.1	2.8	1.1
Unidentified Compounds	8.4	5.3	4.4	3.1	3.6	3.0	7.8	4.6	4.1

* Dry Basis.

123/renutabs/PAP

Table 4, Cont. COMPOSITION OF OILS FOUND IN PRODUCT GAS IN PRU GASIFICATION TESTS

Test No.	GT-16	T12-1	T12-2	T3D-1a	T3D-1b	T12-3a	T12-3b	T12-4a	T12-4b
Temperature, °F	1450	1530	1500	1538	1562	1672	1413	1509	1516
Oil Yield, lb/100 lb biomass*	1.96	3.16	3.32	1.65	1.93	1.12	3.51	2.69	2.48
Products, wt % in Oils									
One-Ring Hydrocarbons	31.5	18.2	23.5	21.8	26.0	17.2	38.8	15.1	11.9
Benzene	26.1	14.9	9.0	19.5	23.2	15.6	27.5	6.6	6.7
Toluene and Xylenes	2.0	1.0	7.1	0.8	1.2	0.6	6.6	2.2	1.1
Others	3.4	2.3	7.4	1.5	1.6	1.0	4.6	6.2	4.1
Two-Ring Hydrocarbons	33.8	38.3	39.0	46.9	39.5	35.9	35.8	41.8	42.3
Naphthalene	28.5	31.9	27.0	41.1	33.4	32.7	27.9	30.2	32.4
Methylnaphthalene	0.6	0.4	3.9	0.6	0.9	0.5	1.9	2.2	1.5
Others	4.7	6.0	8.1	5.2	5.1	2.8	6.0	9.4	8.4
Three-Ring Hydrocarbons	9.2	16.6	12.5	12.1	12.8	11.6	9.8	16.0	14.9
Fluorene	2.3	2.3	2.7	2.6	2.5	1.3	3.0	3.6	2.9
Phenanthrene and Anthracene	5.8	9.7	7.3	8.2	8.4	7.9	4.9	8.9	8.9
Others	1.1	4.6	2.5	1.3	1.9	2.4	1.9	3.6	3.1
Four-Ring Hydrocarbons	8.2	13.4	7.3	10.9	11.3	17.1	5.3	10.6	12.6
Five-Ring Hydrocarbons	3.4	4.9	3.0	3.1	3.3	6.8	1.2	4.1	5.1
Higher Aromatic Hydrocarbons	2.0	2.6	0.8	1.1	1.7	5.7	1.0	2.8	3.0
Total Hydrocarbons	88.0	94.0	86.1	95.9	94.6	94.2	91.9	90.4	89.8
O-Containing Compounds	0.6	0.6	2.6	0.3	0.3	0.1	2.5	2.8	2.0
Phenol and Cresols	0.04	0.02	1.0	0.0	0.0	0.0	1.0	0.3	0.1
Benzofurans and Dibenzofurans	0.5	0.3	1.6	0.3	0.3	0.1	1.5	2.3	1.8
Others	0.01	0.3	0.0	0.0	0.0	0.0	0.0	0.2	0.1
N-Containing Compounds	1.3	0.8	1.4	0.08	0.7	0.3	0.8	0.8	1.1
S-Containing Compounds	0.02	0.3	0.03	0.0	0.1	0.0	0.0	0.02	0.2
Total Heterocyclic Compounds	1.9	1.7	4.0	0.4	1.1	0.4	3.3	3.6	3.3
Unidentified Compounds	10.1	4.3	9.9	3.7	4.3	5.4	4.9	6.0	6.9

* Dry Basis.

123/renutabs/PAP

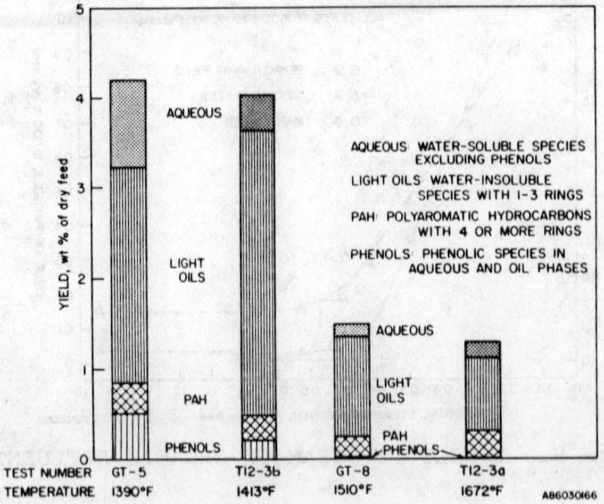

Figure 7. ORGANIC LIQUID YIELD AS A FUNCTION OF TEMPERATURE (Maple Wood Chips: Tests GT-5 and GT-8 and Wisconsin Whole Tree Chips: Tests T12-3a and T12-3b)

It is apparent from the data shown in Table 4 and depicted in Figure 7 that higher operating temperatures, in addition to reducing the oil yield, also cause a sharp decline in the relative concentrations of heterocyclic and alkyl-substituted hydrocarbons in favor of unsubstituted aromatic hydrocarbons. Referring back to Figure 6, it is seen that the overall yield of phenols, which are distributed between the aqueous and organic phases of condensate, declines by 97% to 99%, with an increase in gasification temperature from 1390° to 1510°F and higher.

Also identified in the test results is a relationship between the amount of steam present in the gasifier and the nature of organic liquids produced. Figure 8 shows that the concentration of oxygenated oil components, as well as the overall phenols yield, declines to a low value as the total steam-to-wood ratio is increased (total steam here is the steam from process input plus feed moisture). Although more data points in the range of 0.2 to 0.5 steam-to-wood ratio would be needed to confirm this, it appears that the decomposition of oxygen-containing oils (phenols in particular) reaches a steady minimum at a value of about 0.7 lb total steam/lb wood. The low (0.2 lb/lb) steam input condition in Figure 8 was obtained by replacing steam with hot nitrogen in one test to simulate an air-blown gasification case. In this case, the input steam in the system was derived only from the approximate 15% feed moisture.

The overall oil yield, however, is complex and depends on other factors, such as temperature, pressure, and feed material. The plot in Figure 7 shows the dependence of oil yield on gasifier temperature, and the plots in Figures 9 and 10 show the effects of pressure and feed material type

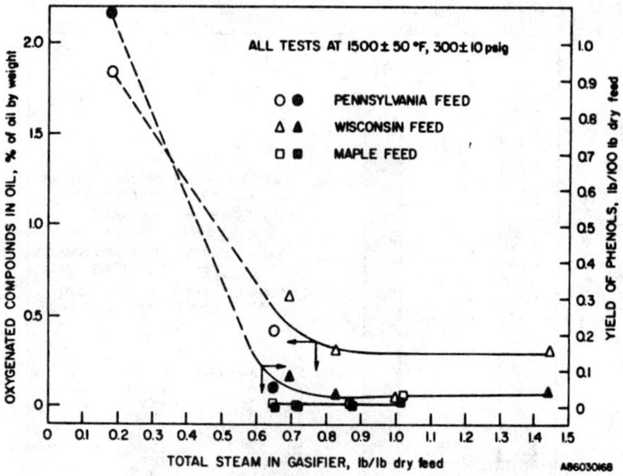

Figure 8. EFFECT OF STEAM ON ORGANIC EFFLUENT

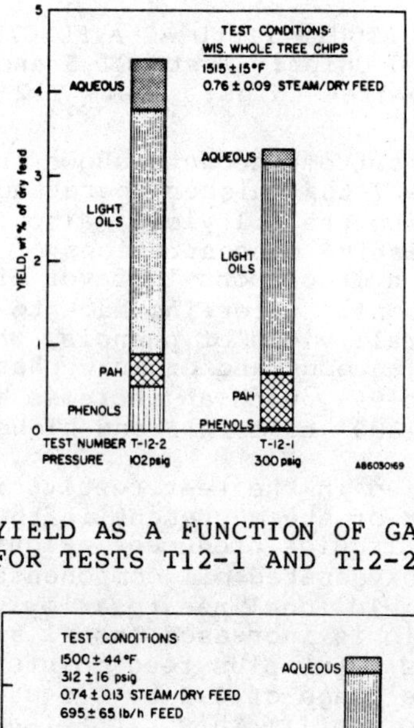

Figure 9. OIL YIELD AS A FUNCTION OF GASIFIER PRESSURE FOR TESTS T12-1 AND T12-2

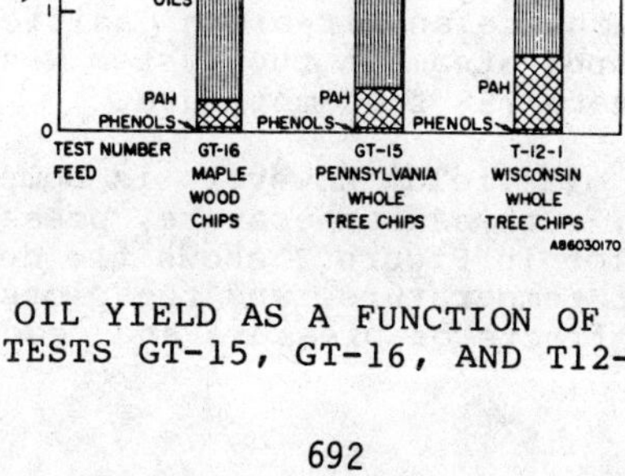

Figure 10. OIL YIELD AS A FUNCTION OF FEEDSTOCK FOR TESTS GT-15, GT-16, AND T12-1

on the yield and composition of organic liquids, respectively. At the lower test pressure, the production of all organic types except pah was much higher than at the higher pressure. The overall yield of oil decreased with increased pressure as well.(2) Most strikingly, the yield of phenols decreased by a factor of 67 with the pressure increase from 102 to 300 psig and the yield of water-soluble organics, excluding phenols, likewise declined by a factor of about 7. The results in Figure 10 shows that the relative proportions of liquid product types did not change as much with varying feed material, although the total organic liquid yield did. The whole tree chips from Wisconsin, with a higher content of bark, leaves, and other non-wood pulp components, did produce more oils than the maple wood chips or the whole tree chips from Pennsylvania.

There was a noteworthy absence of oils in the condensate from Test GT-17. This test was performed with 26.7% feed moisture whole tree chips from Pennsylvania. The reason for the absence of oils is not clear, although it may be related to the high heat demand required to evaporate water from the incoming feed inside the gasifier. A higher oxygen rate was required in that test for combustion to supply this heat, and the combustion of oils, as well as some gaseous hydrocarbons, may have occurred. Additional high feed moisture tests would be useful to quantify this effect on gasification performance.

In summary, the accumulated data on the oil and tar fraction indicate that the production of organic liquid byproducts can be reduced by increased temperature and pressure, but also depends on feed type. From an environmental process cleanup standpoint, conditions that favor lower concentrations of oxygenated, water-soluble components, particularly phenols, are desirable. These conditions include higher temperature and pressure and a total steam-to-wood ratio above a minimum value, which is in the neighborhood of 0.6 to 0.7 lb/lb. Dealkylation and dehydroxylation reactions that favor the decomposition of such undesirable by-products are favored by high partial pressure of hydrogen and high temperature. In this case, hydrogen is supplied by the steam through water-gas shift and steam-carbon reactions, as well as direct reactions of steam with organic liquid species in the gasifier. The relative production of PAH, is not significantly temperature-dependent over the range studied. These are all water-insoluble compounds found in the oil phase, and their elimination from an aqueous process effluent will depend on the efficiency of phase separation in a feedwater recovery scheme. The recovered oils could be burned or recycled to the gasifier to thermally decompose any carcinogens and mutagens contained in them.

<u>Future Work Areas</u>

From the results of the PRU test program, it is seen that the three different feedstocks produced similar product gas compositions and thermal efficiencies, but a different distribution and character of oils and tars. Environmentally, for a commercial process, the characterization of the condensable fraction has to be addressed in detail.

Investigations for process improvement should concentrate on further conversion of the oil and tar fraction and the carry-over char. Recycle of these collectible fractions back into the gasification process or into heat-recovery incineration, for example, would reduce the environmental burden and increase process efficiency. Testing feedstocks with higher ash levels and bark contents, such as mill wastes or crop residues, is needed to expand process applications. In comparison to the lockhoppers employed as the PRU feed system, a successful extrusion-type feeder could save lockhopper compression gases and provide process benefits for high-pressure, combined-cycle applications.

ACKNOWLEDGMENTS

The program guidance provided by the PNL Project Managers, Messrs. G. F. Schiefelbein and M. Gerber, and the support from Mr. Simon Friedrich of the DOE Biomass Energy Technology Division are gratefully acknowledged.

The enthusiastic and conscientious efforts by Messrs. D. Parham, R. Dixon, and K. Hardy in preparation and execution of the tests have been vital to the success of this program.

REFERENCES CITED

1. Evans, R. J., Knight, R. A. and Kosowski, G. M., "Operating Results From a Pressurized Fluidized-Bed Biomass Gasification Process," in Proceedings of Energy From Biomass and Wastes IX, Orlando, Florida. Chicago: Institute of Gas Technology, January 1985.

2. Evans, R. J., Knight, R. A., Onischak, M. and Babu, S., "Development of Biomass Gasification to Produce Substitute Fuels," in Proceedings of 17th DOE Biomass Thermochemical Conversion Contractors' Meeting, Minneapolis, October 14-18, 1985.

123/renugas/PAP

DISCUSSION

R. Gaire, Foster Wheeler — Have you done any tests on the recycling of the carbon char?

Onischak — No, we did not. We've collected the char, characterized it as to particle size and composition, but we had no means in the present test program to reinject or refeed it with the incoming feed in a long-term test, which would be necessary to see that the char does not build up.

Gaire — Do you have any plans to do any recycling in the future?

Onischak — We have plans, but not in the immediate future. We would like to do that. It is important to convert the 3 to 5% carbon carryover.

Gaire — With regard to the moisture in the feed material, most of the data that you showed is for wood at about 10% moisture content?

Onischak — That's correct.

Gaire — Have you tried an excursion of introducing something with a higher amount of water to see if there's any influence on the hydrocarbon liquids?

Onischak — Yes, we have some information in the paper on one test of wood at 27% moisture. It required a higher oxygen demand, which is necessary to dry the wood. Higher oxygen and more steam in the gasifier had an effect, in that we did not measure oils in our system for that one test. It was also more difficult to feed the higher moisture material. It tends to be sluggish and have high friction in the conveying systems. We have only the one test with high moisture wood, but the results indicate lower liquid yields.

F. Gallagher, Garden State Paper Company — Are you planning to do work with industrial waste in this unit?

Onischak — If you mean municipal-type solid wastes, yes, we would like to.

Gallagher — Where is your unit located?

Onischak — At our pilot plant facility in Chicago.

SYNTHESIS GAS PRODUCTION FROM FINNISH PEAT BY FLUID-BED GASIFICATION

Esa Kurkela, MSc(Tech)
Kai Sipilä, Lic.Tech.
Technical Research Centre of Finland (VTT)
Laboratory of Fuel Processing and Lubrication Technology
02150 Espoo, Finland

Jorma Koljonen, MSc(Tech)
Kemira Oy
90101 Oulu, Finland

ABSTRACT

Towards the end of the 1970s, the Laboratory of Fuel Processing and Lubrication Technology, VTT, and Kemira Oy initiated studies on the possibilities of producing synthesis gas from peat. Because of limited resources, it was not considered realistic to initiate development of a new peat gasification process. Furthermore, timetable constraints meant that the studies had to be confined to only the more advanced coal and biomass gasification processes.

As a result of these studies, gasification tests with Finnish peat were carried out in the HTW PDU and pilot plant facilities (West Germany) in 1983 and 1984. In the PDU experiments, peat materials, which had been prepared in different ways, were tested, and, on the basis of these tests, crushed peat pellets were chosen as the feedstock for the pilot plant experiments. The experimental program conducted in the pilot plant proved that the HTW process can be applied to peat, and that there are no significant technical obstacles to the commercialization of synthesis gas production from peat. More recently, Finnish peat has also been used in tests on the U-GAS pilot plant (USA). These experiments confirmed that the U-GAS process is also applicable to peat. The gas composition was, however, a little less favourable than that achieved with the HTW process.

In Kemira's peat-to-ammonia venture, the objective is to convert Kemira's ammonia production line, which now accepts heavy fuel oil as feedstock, to one accepting peat as feed material. The final decision to go ahead with this venture has been made, and the facility is expected to be commissioned in 1988.

SYNTHESIS GAS PRODUCTION FROM FINNISH PEAT BY FLUID-BED GASIFICATION

1 INTRODUCTION

In Finland there are two synthesis gas production facilities of combined capacity about 100 000 Nm^3/h. However, only the smaller of these plants (capacity 28 000 Nm^3/h) is currently in operation. This plant employs heavy fuel oil as feedstock, while the other plant was designed to utilize naphtha. In both cases ammonia is the final product. The plants are part of Kemira Oy's fertilizer complex at Oulu.

In the 1970s, researchers in Finland, as elsewhere in the world, began to examine alternatives for oil-derived feedstocks. The only domestic raw materials, which are suitable for commercial production of synthesis gas, are wood and peat. The utilization of peat as a fuel expanded rapidly in the 1970s, its price being lower than that of wood. As a result, the technology of peat production and utilization quickly developed to a level which made possible the usage of peat in large-scale industrial applications. It was also assumed that the industrial gasification of peat could be realized more readily than the industrial gasification of wood.

In Oulu, synthesis gas had earlier (1952 - 1956) been produced from coal using Koppers-Totzek gasifiers. Around 1960 some limited gasification experiments with peat were also conducted. These experiments, however, were not very successful owing to the physical characteristics of the peat employed in the tests /1/. As well, by the late 1950s, prices of solid feedstocks were no longer competitive with those of oil-derived feedstocks.

In the 1970s, both Kemira Oy and the Technical Research Centre of Finland (VTT) began, again, to examine the possibilities of utilizing peat. (VTT had conducted some laboratory-scale synthesis gas experiments at the end of the 1950s /2/.) Kemira's objective was to find alternative feedstocks for its ammonia production and, at the same time, increase the self-sufficiency of this production. VTT's objective was to determine the potential of producing synthetic liquid fuels from domestic raw materials. Of the different methods for producing synthetic fuels, those based on synthesis gas were nearest commercial realization, so that, for both Kemira and VTT, the central problem became the gasification of peat. However, because of limited resources, it was not considered realistic to initiate development of a new peat synthesis gas process. Experimental gasification research at VTT was instead directed towards development of fuel gas processes which were able to be commercialized rapidly. In the field of synthesis gas production it was decided that the main emphasis be on the application of processes being developed abroad. The Energy Department of the Finnish Ministry for Trade and Industry (KTM) has been the most significant financer of this research.

2 PRELIMINARY STUDIES

At the Laboratory of Fuel Processing and Lubrication Technology, VTT, an extensive literature study, in which various coal and biomass gasification processes were compared, was carried out from 1979 to

1981. The applicability of the different processes to the production of synthesis gas from peat was determined, and methanol production costs were estimated /3/. Although the differences in the estimated production costs were not large, it was concluded that processes employing pressurized gasifiers did offer advantages compared to processes working at atmospheric pressure.

At the same time Kemira Oy had evaluated the possibility of producing synthesis gas from peat at the same capacity as that of the existing naphtha-based plant. The conclusions of Kemira's study were similar to those of VTT. The most promising technology for peat gasification was that based on pressurized fluid-bed technology. An important criterion in these studies was that the process should be capable of commercialization within five years.

The German High Temperature Winkler (HTW) process was already at an advanced stage of development in the early 1980s. Successful pilot plant tests had been conducted with brown coal since 1978. Experiments at a smaller scale had also been performed with wood and peat. Thus the HTW process was deemed to be one of the most promising processes for peat gasification, and, in 1983, VTT, Kemira, and KTM decided to have gasification tests conducted in the HTW test facilities. At the same time, however, it was agreed that the development of other fluid-bed gasification processes should be continually monitored.

3 GASIFICATION EXPERIMENTS IN HTW TEST FACILITIES

Organization of the experiments

During 1983-84 gasification experiments with Finnish peat were carried out in both the PDU and pilot plant facilities. The aim of the experiments was to determine, at large enough scale, the applicability of the HTW process to peat and to produce reliable data for a more detailed techno-economic evaluation of peat-based synthesis gas production.

The experiments were conducted in two phases, the first of which was realized as a Finnish-German co-operature venture. The Finnish partners were VTT and Kemira, while the German partners Rheinbraun GmbH and Uhde GmbH. The main sponsors were the Finnish Ministry for Trade and Industry (KTM) and the FRG Ministry for Research and Technology. In the second phase (pilot plant tests), Kemira Oy had main responsibility for the project, while VTT and Svensk Drivmedelsteknik AB (SDAB) had advisory roles. Kemira Oy and KTM each contributed about one third of the required finance, while the rest came from the Nordic Industry Fund and the National Energy Administration of Sweden. As in the first phase, Rheinbraun GmbH and Uhde GmbH had responsibility for carrying out the actual tests.

PDU experiments at Aachen

The first-phase experiments were performed in the PDU unit located at the Technical University of Aachen. The unit operates at atmospheric pressure. Although brown coal had earlier been the main test material used on the unit, some work had also been done with peat and wood /4/. Initial tests were carried out with peat which had been pretreated in three different ways:

- milled peat,
- 5 mm peat pellets,
- crushed 10 mm pellets.

The objective of the initial tests was to determine which of these materials is best suited to the process. Few differences were observed in the gasification results of the different materials. Crushed pellets were, however, selected as the feed material for the main experiments, mainly because they were relatively easy to feed into the reactor. The continuous feeding of milled peat into the reactor, using the existing equipment, proved to be impossible.

Thus the gasification experiments of longer duration were conducted using crushed peat pellets having a maximum particle size of 5 mm. About 4 000 kg of this material was delivered to Aachen for use in the main experiments. Some characteristics of the various peat materials employed in the tests are presented in Table 1.

In the Aachen experiments crushed peat pellets proved to be a very suitable feedstock for the test gasifier. Beforehand it had been feared that ash might melt and create problems. This was not the case in these tests, when the gasification temperature was kept below 1 000 ^{0}C. The maximum synthesis gas yield was 1 200 Nm^3 ($CO + H_2$)/t dry ash-free peat. Overall, it was concluded that the PDU experiments had revealed no technical problems which would be obstacles to the next phase of the venture, the pilot plant experiments at Wachtberg. On the basis of the PDU tests it was decided to employ crushed peat pellets in the pilot experiments.

Table 1. Characteristics of the peat materials used in PDU tests / VTT analysis.

	Milled peat	Crushed pellets	Pellets 5 mm
1. Bulk density, kg/m^3	240	560	640
2. Higher heating value, dry basis, MJ/kg	22,2	22,3	22,2
3. Moisture, wt-%	11,8	15,0	11,7
4. Ash, dry basis, wt-%	6,1	7,1	6,5
5. Volatile matter, dry basis, wt-%	69,3	67,0	65,5
6. Ash melting, reducing atmosphere, ^{0}C			
- deformation	1095	1115	1070
- hemispherical	1180	1185	1140
- flowing	1340	1340	1310
7. Particle size, %			
- > 10 mm	0,2	4,4	0
- 5 - 10 mm	9,0	44,4	74,1
- 2 - 5 mm	19,0	26,6	17,0
- 1 - 2 mm	13,5	13,1	6,6
- 0,5 - 1 mm	10,7	6,3	1,3
- 0,25 - 0,5 mm	10,6	3,1	0,4
- < 0,25 mm	37,0	2,3	0,6

Description of the HTW pilot plant /5/

The peat test runs were conducted during the period January-August 1984. The HTW pilot facility is situated on the grounds of Rheinbraun's brown coal briquetting factory at Wachtberg near Cologne.

The design parameters of the pilot plant (for brown coal) are:

- throughput (dry brown coal): up to 1 300 kg/h,
- gasification agents: oxygen/steam or air
- pressure: up to 10 bar
- temperature: up to 1 000 ^{0}C.

A flow scheme of the plant is shown in Figure 1. The plant is normally fed with brown coal having a water content of 18 percent (the usual feedstock for the briquetting plant). In a spiral tube pneumatic conveying dryer the water content may be further reduced to 8 to 12 % depending on the experimental objectives. In a lock hopper system the brown coal is then brought to gasification pressure and fed into the fluidized bed by a screw feeder. In the brick-lined gasifier the brown coal first reacts with oxygen/steam or air, charged at different levels, at approximately 700 to 800 ^{0}C in the fluidized bed. Above the fluidized bed additional oxygen and steam are added in order to obtain, at elevated temperatures of about 1 000 ^{0}C, a carbon conversion as complete as possible and the desired gas composition (lowest possible methane content in the case of feeding the gas into a methanol or ammonia synthesis process). All gasification agents are preheated electrically /5/.

Peat as feedstock for the HTW process

Earlier only various coals had been used as feedstocks in the pilot plant. From the viewpoint of gasification the most essential properties of the feedstock are:

- characteristics affecting feeding
  - bulk density,
  - flowing properties
- heating value
- reactivity
- amount of volatiles
- ash melting properties
- fluidization characteristics.

In Table 2 are presented, for various feedstocks, some of the important characteristics affecting gasification.

The peat characteristics which affect feeding are crucially dependent on the selected pretreatment method. A poorly-flowing and light material such as milled peat (bulk density < 300 kg/m^3) could not come into question as a feed material for the pilot plant due to the design of the existing feed equipment. Crushed pellets (~ 500 kg/m^3), on the other hand, had very good feeding characteristics. The reactivity of peat is also very dependent on the pretreatment method. Pelletization breaks up the porous structure of peat, and thus lowers its reactivity. The reactivity of the peat employed in the tests was, in fact, rather low. As a result of this and also of the hardness of the pellets, crushing of the pellets to less than 4 mm particle size was necessary.

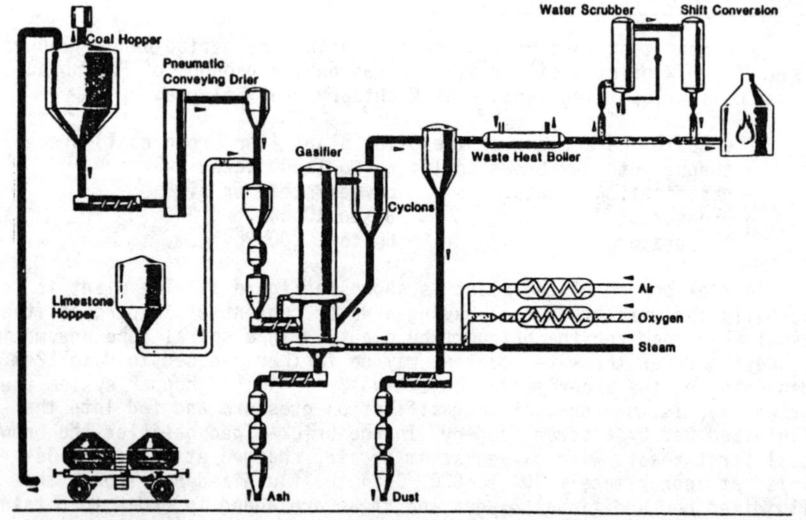

Figure 1. Flow scheme of the Rheinbraun HTW pilot plant at Wachtberg /4/.

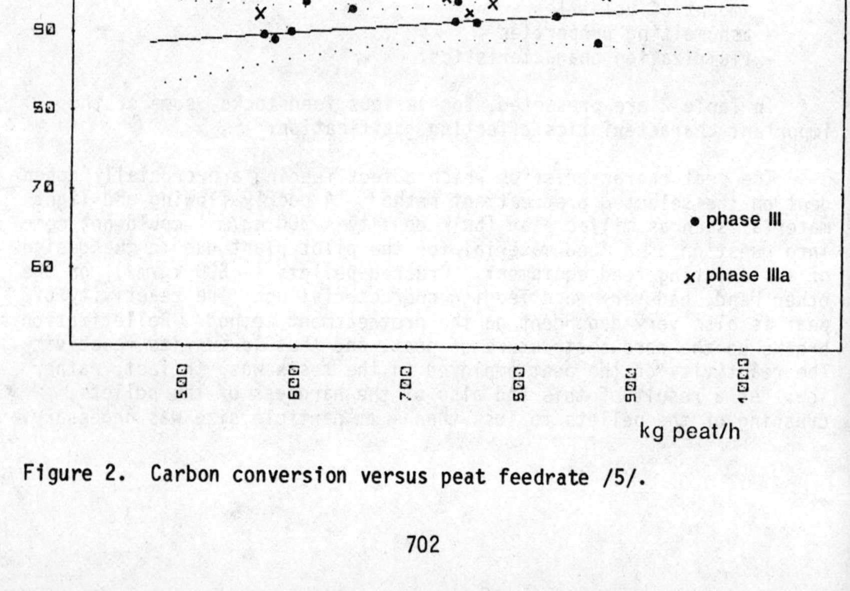

Figure 2. Carbon conversion versus peat feedrate /5/.

Table 2. Feedstock properties affecting gasification /4/.

Material	Coalification index	Calorific value LHV kJ/kg (maf)	Volatile matter maf %	Reaction velocity * k_m cm^3/gs
Wood	0 - 0.36	17 650 - 18 200	81.8 - 86.7	3.4 - 16
Peat	0.18 - 0.36	22 270 - 22 320	61.3 - 73.1	4.13
Soft brown coal	0.45 - 0.48	26 040 - 27 040	48.5 - 55.1	3.6 - 8.0
Hard brown coal	0.52	27 850	45.6	3.8
Bituminous coal, high volatile	0.58 - 0.59	28 350 - 31 480	31.8 - 34.8	0.46 - 0.68
Bituminous coal, medium volatile	0.63	31 200	29.4	0.28

* of char gasified by CO_2 at 900 ^0C and 30 % conversion.

Using peat, primary methane formation in the fluidized bed was higher than when using brown coal. This was a result of the greater amount of volatile matter in peat. In the entrained bed region of the gasifier, the ash melting characteristics of peat (less favourable than those of brown coal) limited the maximum temperature which could be applied.

Results of pilot plant experiments

500 tons of 10 mm peat pellets were delivered to Wachtberg for the experiments. These had been prepared at a pellet factory in eastern Finland. Moisture content of the pellets was 10 - 15 %. The test runs were conducted in six phases between 14.2.1984 and 15.8.1984. Total operating time on peat was about 300 hours. 17 mass and energy balances were determined.

As a whole, reliable operation of the pilot facility was achieved. For example, the feeding equipment operated without problems, as did all the process control systems. Observors from Kemira Oy, VTT, and SDAB were present during nearly all of the tests. The only operating problem was the formation of ash deposits in the product gas train, mainly in the cyclones. This problem was largely solved by altering process parameters and modifying the design of some items of equipment. However, blockages in the internal pipe of the second cyclone could not be entirely prevented during this limited test program.

In all experiments the operating pressure was 10 bar. Temperatures were 750 - 800 ^0C in the fluidized bed and 890 - 930 ^0C in the entrained bed region. The most important variables were the feedrate of peat and the feed ratios of the gasification agents (oxygen, steam, and inert fluidizing gas). Peat feedrate was varied between 550 kg/h and 870 kg/h. Feedrate had a large effect on the gasification results. The specific synthesis gas yield ($m^3(CO + H_2)/100$ t dry ash-free peat) increased significantly with increasing feedrate (refer Figure 3). The increase was largely due to an increased yield of CO, the yield of H_2 remaining nearly constant.

In Figure 2 carbon conversion is plotted against peat feedrate. Only a slight increase in conversion with increasing feedrate was evident.

The dependence of specific synthesis gas output on feedrate is depicted in Figure 4. As was expected, synthesis gas output was directly proportional to the feedrate.

The specific methane yield is plotted against the specific oxygen demand in Figure 5. Increasing the amount of oxygen fed to the process decreased the methane yield, as expected.

Conclusions

The pilot plant experiments proved that the HTW process can be applied to peat and that there are no significant technical obstacles to the commercialization of synthesis gas production from peat. The experimental results are considered to be good in relation to the effort expended. A carbon conversion of 94 % and a ($CO + H_2$) yield of 1100 Nm^3/t of dry ash-free peat were achieved. A typical product gas composition was 40 % CO, 33 % H_2, 22 % CO_2, and 5 % CH_4. Somewhat lower CH_4 and CO_2 contents and higher carbon conversion can be expected in a full-scale demonstration unit designed for peat.

Additional experiments

After the pilot plant program using peat pellets had been completed, some further experiments were made on the PDU unit at Aachen. The feed material in these tests was the more reactive (higher porosity) crushed sod peat. The experiments proved that this material is also a suitable feedstock for the HTW process.

4 GASIFICATION EXPERIMENTS IN THE U-GAS PILOT PLANT

Towards the end of 1985, tests with Finnish peat were also conducted on the U-GAS pilot plant of the Institute of Gas Technology (IGT), Chicago (Figure 6). The main sponsors of this test program were:

- on the USA side, IGT/Gas Development Corporation,
- on the French side, Heurtey Industries and CERCHAR.

A smaller contribution was also provided by Finnish parties, Kemira Oy and KTM. From the Finnish point of view, the objectives of the experiments were to determine the applicability of the U-GAS process to Finnish peat, to gather additional information about the pressurized fluid-bed gasification of peat, and to generate the data required for a process design.

The feed material, which was delivered to Chicago, was crushed sod peat, originating from two different peatlands, Viidansuo and Savaloneva. The Viidansuo peat, in particular, was of low-grade quality (weakly humified), containing a high proportion of fibre material. The two types of peat were mixed to produce two different batches for testing. One batch contained equal parts of each peat, and was dried to 35 wt.% moisture. The other contained two parts of Savaloneva peat and one part of Viidansuo peat, and was dried to 15 % moisture. Despite the poor

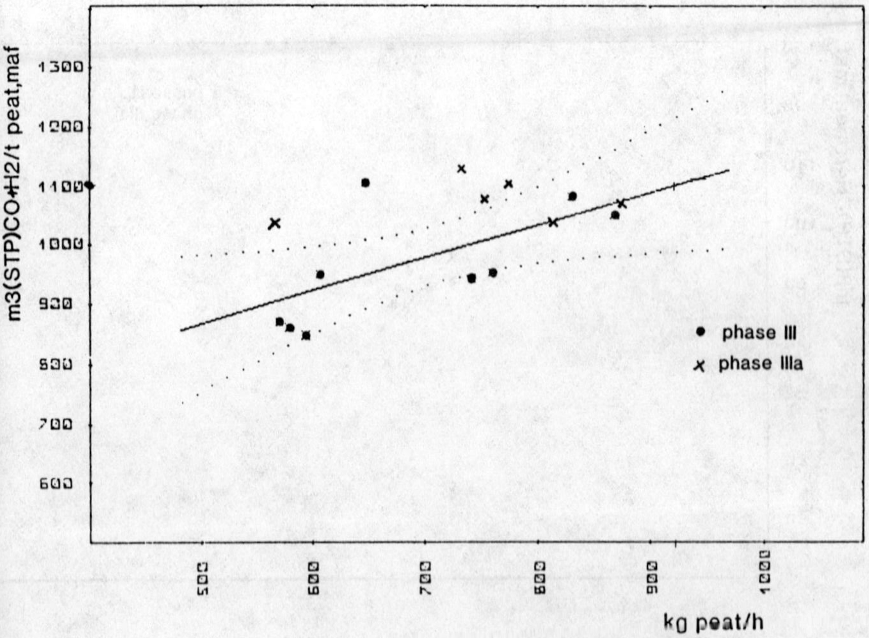

Figure 3. Specific synthesis gas yield versus peat feedrate /5/.

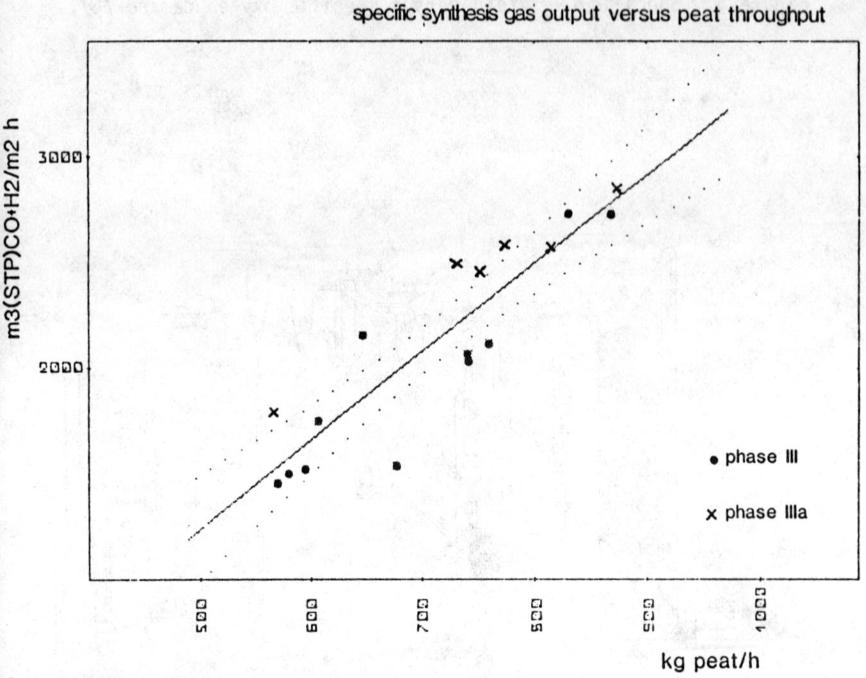

Figure 4. Specific synthesis gas output versus peat feedrate /5/.

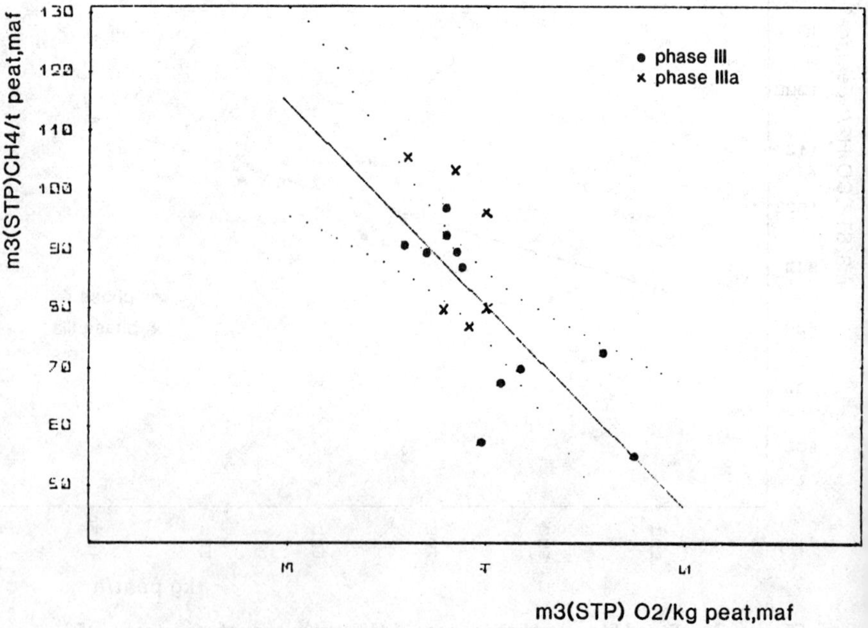

Figure 5. Specific CH_4 yield versus specific oxygen demand /5/.

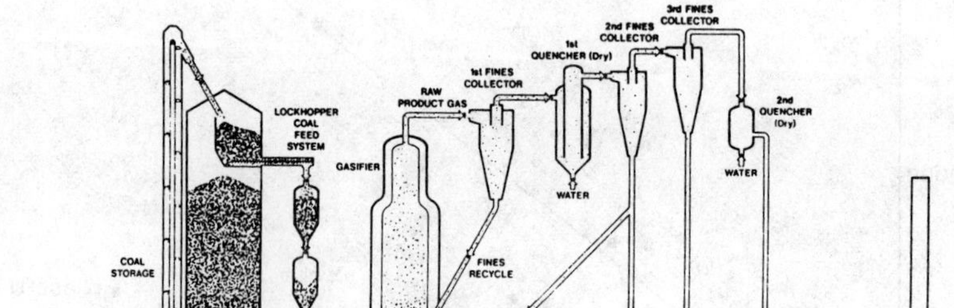

Figure 6. Flow scheme of the U-GAS pilot plant in Chicago.

flowing properties of these materials, the lockhopper system, after some modifications, was able to feed them at controllable rates for the whole duration of the test.

The pilot plant was operated for 5 1/2 days from start-up to shutdown. A total of 42 tons of peat was gasified during 50 hours of continuous operation on peat. Representatives of Kemira Oy and VTT were present as observors. Except for the occasional hold-up of peat in the feeding system, no operational problems were encountered. The peat feeding problems could apparently be solved by designing the lockhopper system specifially for peat.

In the tests the reactor pressure was about 2 bars and the temperature 800 - 980 ^{0}C. The peat feedrate was in the range 650 - 1000 kg/h. 1000 kg/h was the highest rate achievable with the light peat material. Because of the high reactivity of the peat feedstock, a greater feedrate would have been necessary to enable prolonged use of temperatures greater than 900 ^{0}C. For this reason most of the test periods employed temperatures in the range 800 - 870 ^{0}C. Over 90 % carbon conversion was achieved. Both the recycle of fines from the first cyclone and the removal of ash from the base of the reactor proved to be efficient and reliable operations. A typical gas composition (temperature 865 ^{0}C) was, for the main components, 19 % CO, 32 % H_2, 40 % CO_2, and 7 % CH_4. In those test periods, where the temperature was as high as 980 ^{0}C for short times, the methane content decreased to as low as 4 %. In a commercial gasifier designed specifically for peat it should be possible to achieve a more favourable gas composition. In particular, a lower CO_2 content should be achievable through a reduction of the high feedrate of steam, which is peculiar to the pilot plant.

On the basis of this test program it can be concluded that the U-GAS process is also applicable to peat. The operability of the pilot plant was excellent over a wide range of run conditions.

5 RELATED RESEARCH

In the beginning of 1985, a four-year research project, supporting the peat-to-ammonia venture, was initiated by VTT and Kemira Oy. The most important research topics are:

- evaluation of peat pretreatment alternatives,
- classification of the peatlands of the Oulu district with regard to suitability as gasifier feedstock,
- study of the behaviour of peat ash during gasification,
- evaluation of newer gasification processes for possible follow-on ventures.

The main pretreatment alternatives are depicted in Figure 7. Neither milled peat nor sod peat are suitable, as such, as feed material for the synthesis gas facility. The fluid-bed gasifier requires that the material is dried, that the particle size is less than 4 - 10 mm, and that the amount of fines is not excessive. Although a suitable size fraction of milled peat could be produced by screening, the feeding of this very light material into a full-scale pressurized reactor is highly uncertain, if at all possible. Thus the two pretreatment alternatives which remain are:

- pelletization of milled peat and subsequent crushing,
- crushing of sod peat.

Pelletization destroys some of the porous structure of peat and thus lowers the reactivity of the material. Furthermore the pellet alternative has been assessed to be more costly than the sod peat alternative. Thus, at an early stage, it was decided to place emphasis on the optimization of the sod peat alternative. The low strength of sod peat means that the crusher and the dryer must be designed very carefully in order to avoid excessive formation of fines.

During 1985 the Oulu district peatlands, which are available for peat production, were analyzed, and, on the basis of the analyses, were classified according to their suitability as gasifier feedstock. Particular attention was given to the ash melting temperatures of the different peats. The typical Oulu district peat has a relatively high iron content and thus a relatively low ash melting temperature. The reactivities of peat materials of different chemical composition and pretreated in different ways have been determined in a special laboratory-scale fluid-bed reactor at VTT. These experiments have confirmed that the porosity and the pore structure of the particle are the most critical parameters affecting reactivity. For example, the reactivity of crushed pellets (2 mm particle size) at 850 ^{0}C was determined to be about half that of crushed sod peat of the same particle size.

6 THE PEAT-AMMONIA VENTURE OF KEMIRA OY

Kemira Oy is Finland's largest chemical company (outside the field of oil refining). It has operations both in Finland and abroad. Examples of the latter are fertilizer plants in the Netherlands and England and paint factories in England. The company had a turnover of some 750 million USD in 1984. Its employees number about 10 000. The company's main products are fertilizers. Other significant areas are TiO_2-pigments, viscose fibres, pesticides, explosives, paints, and some basic chemicals such as sulphuric acid.

Kemira produces over 1,5 million tons/a of fertilizers in its Finnish plants. Part of this production is exported. However, the domestic production of ammonia, the nitrogen source for fertilizers, covers only about 20 % of the consumption. This situation came about when the naphtha-based ammonia plant was withdrawn from service due to uneconomically high feedstock costs. On the other hand, Kemira does have an ammonia plant in the Netherlands. As mentioned earlier, peat is the only indigenous feedstock alternative. Against this background, Kemira's interest in peat gasification is quite understandable.

Kemira's fertilizer plant in Oulu is the world's most northerly, 200 km south of the Arctic Circle. The plant has been in operation since 1952. (Until 1973 it was owned by an independent company, Typpi Oy.) From 1952 - 1956 a Koppers coal gasification process was also in use. A study of the possibility of applying peat gasification was conducted during the years 1979 - 1980. Most of the commercial coal gasification methods were considered. The plan was to produce synthesis gas at the same capacity as the existing naphtha steam-reforming unit. The plan was not realized because of the low price of ammonia on the world market. However, it was decided to continue reviewing developments in the gasification field, and to promote the utilization of

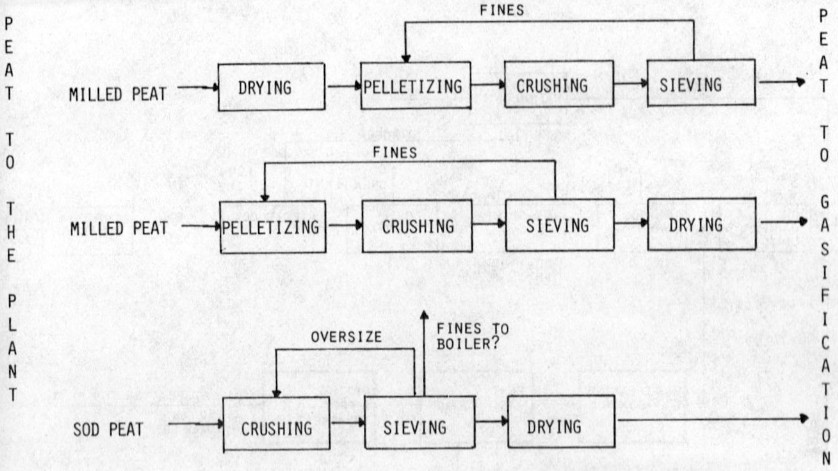

Figure 7. The main alternatives for peat pretreatment.

peat. In fact, a peat-fired boiler, based on Ahlström Pyroflow technology, has been in operation at the Oulu works since 1983.

Oulu is situated near extensive peatlands. Kemira has procured its own peat production areas, sufficient for limited production of ammonia from peat, and is a shareholder in a peat harvesting company. The long-term plan for these peat areas is that, after peat has been raised, the lands will be used for biomass production (hybrid willow).

The Oulu plant offers many advantages as a site for a peat gasification demonstration plant:

- location near peat areas,
- existing personnel and organization,
- beneficial utilization of product gas,
- governmental investment assistance.

Figure 8 reveals the current objective of the project. It is planned to build, with minimum investment, a peat gasification unit of capacity equal to that of the two existing heavy oil gasifiers. The rest of plant will remain unchanged except for some modifications to the gas purification section, required as a result of the altered raw gas composition. The economics of the project necessitate that the gasifier operates reliably, because any lowering of the present production rate (80 000 t/a of ammonia) would mean increased ammonia imports to Oulu and lossed production of side-products (urea, formic acid, carbon dioxide).

A critical factor, affecting the project, is that the existing plant is approaching the end of its service life. Thus, significant postponement of the investment decision has not been possible. Kemira has received quotations for both HTW and U-GAS gasification processes, based on the results of the pilot plant tests discussed in the previous sections. Experiments have also been carried out on the MINO pilot plant in Sweden, and contacts have been maintained with several other

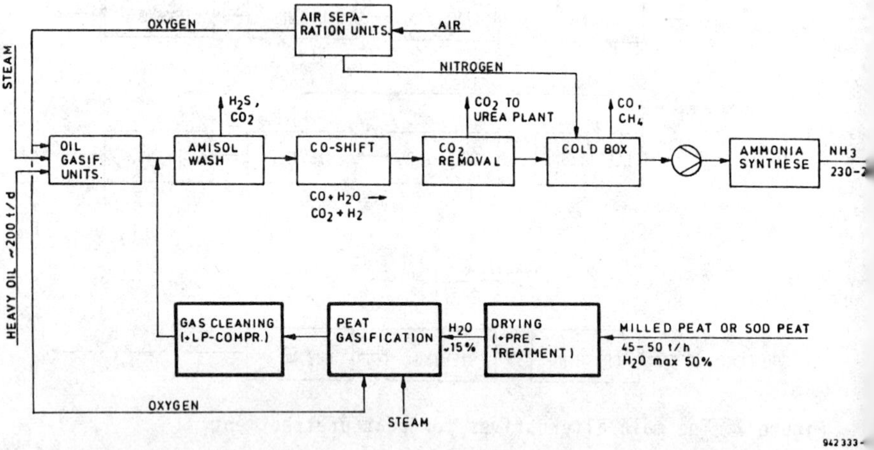

Figure 8. Planned integration of peat gasification into existing ammonia plant of Kemira Oy at Oulu.

companies offering gasification processes. The final decision to go ahead with the venture has been made. The facility is expected to be commissioned in the spring of 1988. Final choice of gasification process will be made during March 1986.

Kemira's objective is to create, during the 1980s, full readiness to produce synthesis gas from peat. The extent and timing of longer-term follow-on projects will depend on feedstock prices in the 1990s and on the experiences gained with peat gasification on the demonstration plant. In the best circumstances, a new and larger peat-to-ammonia plant will be constructed during the next decade.

REFERENCES CITED

1. Kamula, A., Some Observations on the Autothermal Gasification of Peat Dust on an Industrial Scale. Edinburgh: Her Majesty's Stationary Office, 1968. Trans. 2nd Int. Peat Cong., Leningrad, USSR, 1963.

2. Sundgren, A. et al., Manufacture of Water Gas from Milled and Powedered Peat. Edinburgh: Her Majesty's Stationary Office, 1968. Trans. 2nd Int. Peat Congr., Leningrad, USSR, 1963.

3. Solantausta, Y. & Asplund, D., Methanol from Peat. Technical and Economic Aspects in Finland. Chicago: IGT, 1980. Proc. Conf. on Peat as an Energy Alternative, Arlington, USA, 1-3.12.1980. 33 p.

4. Schrader, H. et al., Biomasses as Feed for the Rheinbraun HTW Gasification Process. Chicago: IGT 1984. Proc. Conf. on Energy from Biomass, Buena Vista, Florida, USA, 30.1 - 3.2.1984. 22 p.

5. Bellin et al., Report on the Results of the Gasification Tests on Finnish Peat in the HTW Pilot Plant. Cologne: Rheinbraun-Consulting GmbH, November 1984. 20 p. + app. 18 p.

DISCUSSION

W. Watt, Weyerhaeuser Company	I noticed that when you talked about drying the peat, you did some evaluation of dryers, and you ended up with a steam dryer as your choice. Could you go through your evaluation a little bit on why you used the steam dryer?
Sipila	For peat, there are different types of commercial dryers in operation — flash dryers in Finland for some briquette factories, Peco dryers in Ireland, and drum dryers in the Soviet Union and Germany. The steam dryer is a fluidized bed dryer and the heat is generated from the back-pressure turbine. The advantage is that heat consumption is practically half compared with the conventional flash dryer. It's only about 2 MJ/kg water compared with twice this value for the flash dryer. Another advantage is that it doesn't need any boiler for generating the fluidizing gases. You can use back-pressure steam. For the real case, there is an existing 60-MW power plant with 10-bar steam for the dryers. It proved to be more economical in this particular case.
Watt	What is the fluidizing agent?
Sipila	Steam.
Watt	It's actually a fluid-bed dryer using steam as your contact?
Sipila	Yes.
Watt	Who produces that?
Sipila	In this particular case, it's a Swedish company, M. D. Chemetics.

TESTING AND EVALUATION OF A PRESSURIZED 10-T/H FLUID-BED WOOD GASIFIER

Guy Gravel, Eng.
Director General
Biosyn
Montréal (Québec) H2Z 1S8

André Chamberland,
Leader
Biomass Energy
IREQ,
Varennes, (Québec) J0L 2P0

ABSTRACT

Biosyn, a division of Nouveler Inc. in Montréal, recently put into operation a pilot plant for the production of synthesis gas using an oxygen-blown pressurized gasifier. The gasifier performance proved satisfactory using air at pressures of 100 to 1400 kPa at temperatures ranging from 725 to 850°C.

Preliminary test results do not allow a rigorous model of the performance to be established but are very promising and the performance of the system as a whole surpasses initial expectations. For example, the gross energy conversion rate of wood in hot gas is as high as 78%.

TESTING AND EVALUATION OF A PRESSURIZED 10-T/H FLUID-BED WOOD GASIFIER

Introduction

In the framework of their respective biomass energy development programs, the Canadian and Québec governments signed an agreement in 1981 which led to the creation of Biosyn and the initial of a pilot project aimed at the production of synthesis gas from gasified wood for eventually producing methanol.

For this purpose, Biosyn had to build the plant and install equipment required for the preparation of 10-bd t/h of wood residues to be fed into the oxygen-blown gasifier which is designed to use around 3 t/h of oxygen. The project comprises five phases:

1. Start-up using air at atmospheric pressure
2. Operation with pressurized air
3. Operation with pressurized oxygen
4. Demonstration of the reforming capacity of the product gas
5. Addition of the equipment necessary for the production of fuel-grade methanol from wood residue.

At the time of writing, the construction and commissioning of the gasifier are finished and most of the analytical instrumentation has been installed. Work has almost been completed too on the difficult task of operating the gasifier with pressurized air. The results obtained at temperatures between 725 and 850°C in a pressure range from 100 to 1400 kPa are presented and discussed below.

The Plant

The Biosyn plant is located in St. Juste de la Bretenière on the Québec-Maine border. The entire area is covered with forests, and lumbering activities and sawmills are plentiful. It is the residue from these operations that provides the major feed source for the pilot plant.

The site covers 300 hectares on the banks of the Daquam River, which supplies the water required for operating purposes. The plant itself occupies some 25% of the site while another 25% is used for storing the biomass materials; the remainder is reserved for future expansion.

The equipment used to prepare the wood comprises: a conveyor system, a shredder, a triple-pass rotary dryer heated by means of the wood dust recovered from the operation, storage hoppers and feed screws.

The Gasifier

Figure 1 presents a schematic drawing of the 10-bdt gasifier, which consists of a 2-m-diameter steel cylinder (1) with a 60-cm bottom closure flange, a castable-refractory lining (3) and soft insulation (2). The total height is 13 m. At the reactor outlet, the gases pass through two disengaging (7) cyclones before being evacuated through the vent and flared (9). The granulated wood or sawdust is fed continuously into the sand bed of the reactor via an injector screw through a side port (4). The feeder carrying the wood from the storage hopper to the injector screw is designed to maintain a tight seal between the ambient air and the pressurized reactor. The air or oxygen used for the gasification of fluidization process is supplied through pipes (6) under the grid (5) formed by the various components and their interconnections.

Measuring and Control Equipment

The reactor is equipped with a number of thermocouples, flowmeters and electronic pressure transducers for process evaluation. The product-gas pipe has a number of ports for immediate or later analysis. The four main gas components, namely CO_2, CO, CH_4 and H_2, are analyzed on line by spectroscopy whereas the others are analyzed periodically by gas-phase chromotography. Before they reach the analysers or the expansion valve, the gases pass through a scrubber, which traps all the tar, water and other condensable matter; any solid particles that escape are trapped by filters.

The valves, variable-speed motors and other accessories with which the reactor is equipped can be remote-controlled from a central console. The feed system and the other reactor components are fitted with feedback control loops to ensure maximum operating safety.

Tests

A) Start-up Procedure. The start-up procedure used for this reactor is simple, fast and efficient. It consists in preheating the fluidized bed with the propane burner while the bed is fluidized with air. After a few hours of heating, the temperature exceeds 400°C and wood can start to be fed into the reactor. At this stage, the gasifier is operating with excess air, therefore in combustion mode, and the temperature of the fluidized bed rises rapidly. At around 700°C, the task of establishing the required pressure and temperature conditions begins. Successive adjustments are made to the air pressure, the air flow rate and the wood feed rate until the pre-set conditions are reached.

B) Establishing Performance Levels. Once the initial operating conditions have been attained, all the operating parameters are fixed and it is observed whether the same

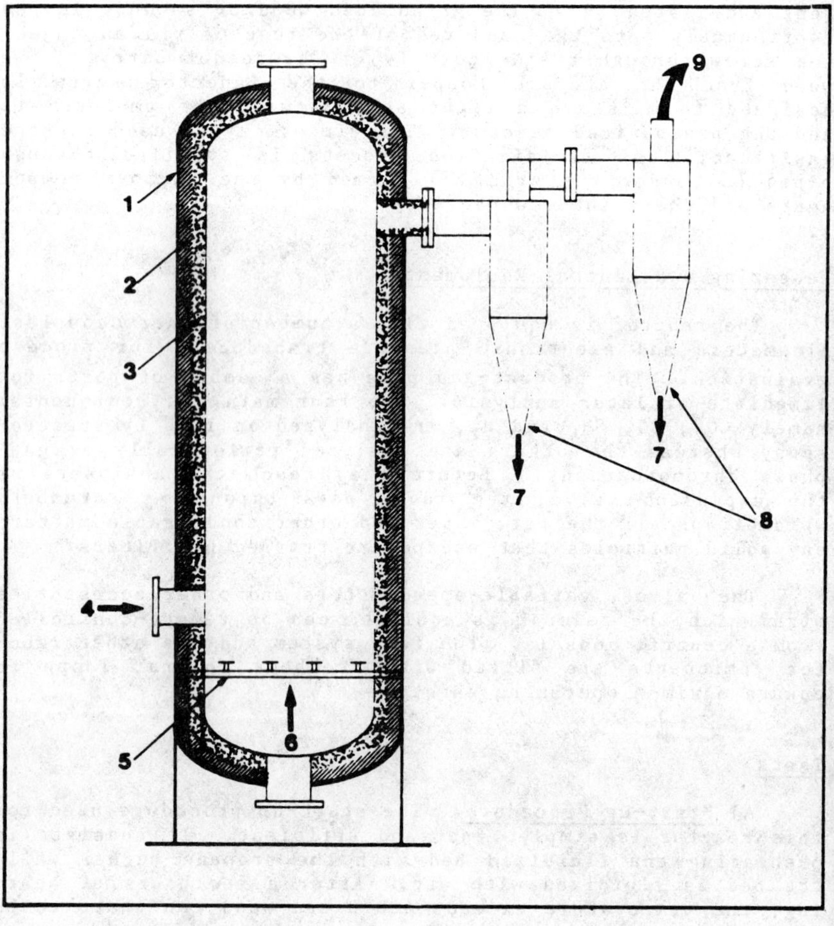

Figure 1 Schematic of Biosyn's wood gasifier 10-bdt

pressure and temperature (to ± 10°C) can be maintained for as long as required for a given biomass feed material. Usually each temperature and pressure level is maintained for over 2 h. Such levels are very easy to detect on the traces of the on-line analysers whose signal becomes constant a few minutes after the set operating conditions have been reached and remains that way so long as these conditions do not change significantly. The results of the analyses reported here are therefore representative of the reactor's behavior in equilibrium under constant operating conditions (maintained for two hours or longer).

Results of the Analyses

A) Wood Composition. The wood used for the tests consists mainly of sawdust and slivers of bark from spruce and similar softwood. This roughly 50:50 mixture is shredded and screened before being fed into the gasifier. The chemical composition of the wood is $CH_{1.4}\ O_{0.68}$ on average. Table 1 presents the results of the thermogravimetric analysis of wood samples from the Blanchette sawmill located some 500 m from the plant.

B) Gases. The results of the chromotographic analysis of the gases as a function of the pressure are presented in Table 2. In all the tests represented here the water content of the feed wood was 20 wt%.

Table 1
Composition of the Wood Samples

H_2O %	Volatile matter	Ash %	Fixed carbon %
18.3	68.5	2.3	10.9

Discussion

Since the results do not represent the mean values obtained for a series of tests but individual tests, it is difficult to accord them a high degree of accuracy. Consequently, the 2% variations observed in the concentrations of H_2, CO and methane could be due simply to differences in the reactor temperature from one test to the other.

The most important result of the 200 hours of gasifier operation is that the system operated satisfactorily at feed rates of up to 6 t/h with air at pressures as high as 1400 kPa.

Table 2
Gas Composition vs Temperature and Pressure

Pressure (kPa)	Temp. (°C)	Gas Composition (%)						
		H_2	CO	CH_4	C_2H_4	C_2H_6	$CO_2 + N_2$	Others
365	825	8.0	13.9	5.5	1.3	0.4	69.5	1.4
630	777	10.3	14.5	6.0	1.3	0.8	66.5	0.6
660	809	9.	12.9	6.7	0.8	0.8	68.5	1.3
700	795	9.7	13.0	6.8	1.1	0.8	68.7	-
1000	800	10.7	13.1	7.3	1.0	1.0	66.1	0.8
1200	812	10.3	12.8	8.6	0.9	1.1	65.4	0.9
1400	800	10.3	11.9	8.2	0.8	1.0	66.8	1.0

Acknowledgments

The authors thank all the members of Biosyn who contributed to the work reported here and are particularly grateful to Mr. John Black and Mr. Maurice Bergougnou for their input to the discussion of the results.

DISCUSSION

B. Miller, Alternate Gas Inc.	Did you say 50% bark and 50% sawdust?
Gravel	Yes, but we intend to try other proportions.
Miller	Have you had problems with the bark content at all?
Gravel	No.
Miller	As long as you shredded it?
Gravel	Yes, it is shredded and it is dry of course before being fed.
Miller	You have two cyclones on this system?
Gravel	Yes.
Miller	Will those cyclones meet U.S. air quality standards? I don't know what the Canadian standards are.
Gravel	I'm looking at John Black because he is more familiar with the design.
J. Black, BBC Engineering & Research Ltd.	No, at the present time, they won't meet U.S. standards.
Miller	What would you do to make them meet U.S. standards, put a multi-clone on? Will that do it?
Black	No, a multi-clone won't work either. Multi-clones are fairly notorious for operating part of the time, but not all of the time. We have porous metal filters, which we plan to test this summer when the weather is a little warmer, and we'll be using that for final cleanup of the particulates.
Miller	Did you say that you were using the fines to fire the dryer?
Black	Yes, that's correct.
Miller	Fines from the gasifier cyclones?
Black	No, not from the gasifiers, from the dryer cyclone.
Miller	How do you measure the hot gas quality?

Black	The hot gas is basically measured from the gas analysis.
Miller	You take a sample out, and measure the tars as well as the particulate content?
Black	We only recently began to do tar analysis. The weather conditions in Quebec during the winter are pretty severe, and we had some trouble trying to get liquid samples from the system during the winter. The first two or three samples were taken within the last two weeks.
Miller	Do you know what the Btu content per SCF is, dry?
Black	We've only been running with air, and it's running a little over 200 now.
Miller	Without tar?
Black	That doesn't include the tar.
Gravel	It's close to 600 MJ/m^3.
F. Gallagher, Garden State Paper Company	What is the ash content of the bark?
Gravel	Typically, it runs a little high on the wood waste and it would be close to 2%.
Gallagher	Is that apt to cause any problems with clinkers in the fluidized bed?
Gravel	No, we've seen no clinker formation at all. The ash all blows over the top.
Gallagher	What temperature is the fluidized bed ash?
Gravel	We have a range of operating conditions. We've been running a little under 1400°F to 1600°F. Of course, we keep the operating temperature below the ash softening point.
G. Schiefelbein, Pacific Northwest Laboratory	I wonder if you could comment briefly on the performance of your compression screw feeders?
Black	They have performed fairly well. Initially during startup, we had problems getting material to feed through the system. We modified the design and now

the system will run under defined conditions at this point up to just a little over 200 psi.

Schiefelbein — What is the longest run duration you've made without any problems with the feeders?

Black — I guess the longest continuous run we've made is about 48 hours. We are limited in the number of personnel we have available at the moment, so that's really the limitation, not the feeding system.

Gravel — On a regular basis, we operate the gasifier a whole day, stop it at night, and the next morning, we go back into production within a very short period of time. The gasifier responds very well in the whole system.

R. Gaire, Foster Wheeler — What overall cold thermal efficiency have you been able to get based on the air gasification system to date?

Black — Typically, we're running a little over 80% with 15-20% moisture fuel.

Gaire — What do you anticipate with oxygen? When do you think you'll be running with oxygen?

Black — We are planning to run with oxygen towards the latter part of this month. The changes are now under way to switch from air to oxygen. The cold gas design efficiencies are closer to 86% with oxygen.

S. Friedrich, U.S. Department of Energy — You mentioned you had over 200 Btu/SCF. That includes sensible heat I assume?

Gravel — No that's strictly the calorific value of the gas. We have a very high hydrocarbon content in the gas.

COMMERCIAL APPLICATION OF WOOD GASIFICATION
- TWO CASE STUDIES -

Bennett Miller
Alternate Gas, Inc.
Washington, D.C. 20007

ABSTRACT

The fundamental problem in commercializing wood gasification involves convincing customers to use wood gas. There are two things that need to be done to solve this problem. The first is to EXPLODE THE MYTHS regarding wood gasification, the second is to DEMONSTRATE THE REALITY OF THE TECHNOLOGY IN THE MARKETPLACE.

Basically there are four myths about wood gasification. They are:

- There is not enough wood available except in rural areas.
- Where adequate supplies are available the price is not stable.
- Wood is not convenient to handle.
- The technology to gasify wood is either unavailable or unreliable.

Let us go through each of these myths in turn and dispell the issues regarding each one of them.

COMMERCIAL APPLICATION OF WOOD GASIFICATION - TWO CASE STUDIES -

THE MYTHS

Myth Number 1: There is Not Enough Wood Available Except In Rural Areas

There are several facts that bear mention regarding the availability of wood supply in the United States. First, if we focus just on non-commercial timber in the U.S., as distinct from commercial timber, i.e., timber that is used by the pulp and paper industry for its needs, then there are 150 quads of standing non-commercial timber today. These figures are substantiated by the U.S. Forest Service and by forestry commissions on a state and local basis. One hundred fifty quads represents about as much energy as all the proven reserves of natural gas in the lower forty-eight.

Second, we are adding 2% each year to that total, or about 3 quads a year is being added to the non-commercial timber in the United States. That alone represents an energy availability equal to that produced by all the nuclear power plants in the United States in 1980.

Third, there is over 1 quad per year of sawdust and other forest residue available from simple sawmill and furniture operations and related operations in the United States. Finally, a fact that people are not generally aware of, is that east of the Mississippi, forest land averages over 50% of the land area.

In sum, focusing just on non-commercial timber and available residue, and limiting our non-commercial timber considerations to only the excess or increased availability each year, there is a total of about 4 quads a year of wood or wood waste available for use in the wood energy industry without competing with the forest product industry.

Myth Number 2: Where Adequate Supplies Are Available, The Price Is Not Stable

There is a general concern that though wood may be available, the price will fluctuate because the competition for the resource is so severe. Let me take an example of our plant in Southeast Missouri. The feed-stock is sawmill residue and sawmill sawdust. The plant needs 40,000 tons per year, on a wet basis, of this material. From county level and state level data, there are 242,000 tons of sawdust and 520,000 of other residue readily available for a total of 762,000 tons within a 50 mile range of the Southeast Missouri Plant. In other words, within 50 miles of that plant we could site 20 such plants. As a result of this kind of availability, it is a straightforward matter to negotiate long term contracts with appropriate escalators for a guaranteed supply of feedstock.

Myth Number 3: Wood Is Not Convenient To Handle

As most people in the forest products industry know, automatic unloading equipment for the handling of wood is available off the shelf from any of a number of vendors. In terms of delivery to the site, wood is routinely shipped over the road in closed and open vans in 30 ton lots without any difficulty. And finally, once on site, handling of wood is completely automatic, and poses no more problems than would any solid fuel.

Myth Number 4: The Technology To Gasify Wood Is Either Unavailable or Unreliable

At latest count there were at least 8 domestic manufacturers of wood gasifiers that can supply a legitimate piece of equipment to a customer. Not all of these are big companies, but several of them are sufficiently large to stand behind their product with credible warranties and guarantees. In addition, there are at least five foreign manufacturers of similar type equipment that are the equivalent of the U.S. manufacturers in terms of financial strength and stability. There are, in addition, more than a dozen operating gasifiers today in the United States with an installed capacity of over 400,000 MMBTU/hour, and there is currently more than 600,000 MMBTU/hour of capacity either being installed or contemplated for installation within the next two years.

In short, the myths that tend to discourage people from looking to wood gasification as a valid technology for use in the industrial marketplace are just that -- myths, and should not be accorded the status of fact.

TYPES OF GASIFIERS

Let me turn now to a brief survey of the types of gasifiers currently available and the advantages and disadvantages of each.

Basically there are three types of gasifiers. Two of them are fixed bed systems in which the material that is to be gasified is maintained on a stationary grate and gasified there, and the third is the fluidized bed technology in which the material to be gasified is injected into a fluidizing medium that provides the heat transfer for the gasification step.

Fixed-bed, updraft systems

There are two variations of the fixed bed systems which are related to the way in which the gas is drawn out of the system. The first of these is the updraft gasifier. The updraft gasifier is characterized by gas percolating up from the grate through the bed and exiting at the top of the gasifier vessel. The advantages and disadvantages of this system are as follows:

The advantages. Fixed bed updraft gasifiers are simple to construct; they are nothing more than carbon steel shells with a well designed grate at the bottom fed by a process air manifold. A lock hopper at the top of the system feeds the material to the unit and a manifold takes the gas out at the top. The operation, as a result, is simple. Fixed-bed, updraft systems are relatively inexpensive, and because the material is not in motion, the particulate loadings in the output gas tend to be low.

The disadvantages. The gas exiting the gasifier tends to be cool because it has percolated up through the bed from the grate. Therefore there is a fair fraction of higher order hydrocarbons that have been formed. These hydrocarbons are condensible and if one is not careful they can condense in the gas distribution system causing problems because they are basically volatile and corrosive. The second disadvantage is that because the gas exits cool there is little sensible heat in the gas stream. The third disadvantage is that the systems tend to be limited to about 50 MMBTU/hour. Not that they can't be built larger, but at this stage in the development of the technology bed stability considerations seem to limit them to about 50 MMBTU/hour. And finally they are limited to chip type fuel for obvious reasons, since such fuel is the only kind that will sit on a fixed grate.

Figure 1 shows a fixed-bed, updraft gasifier installed at the Northwest Georgia Regional Hospital in Rome, Georgia. It has been in operation since 1979 at a nominal level of 25 MMBTU/hour.

Fixed-bed down draft systems

The second kind of fixed bed gasifier is the downdraft gasifier. The downdraft gasifier consists of two concentric shells with the inner shell holding the material on the grate and the outer shell used to transport the gas. The gas is drawn out from under the grate through the outer shell to the outside of the system.

The advantages. Because the gas exits at the combustion zone (at the grate), it is hot. Because it is hot, it tends to have little in the way of long chain hydrocarbons and therefore it is also relatively clean. This kind of system is easily adapted to internal combustion, engine-generator operation.

FIGURE 1.

Like the fixed-bed updraft system and for the same reason, the down draft system has low particulate loadings.

The disadvantages. The system is more complicated to construct because you now have two shells that must be engineered so that you can pull the gas down through the grate and out through the annulus. It is limited to about 10-20 MMBTU/hour, but, again, not for major technical reasons, but simply that is where the technology is at the moment. This technology tends to be more expensive than the fixed-bed updraft, and it, too, is limited to chip type fuel. In Figure 2 you see a

SMALL POWER UNIT OPERATES ON WOOD WASTE

This fixed-bed, self-contained electric power generation unit developed in West Germany by Fritz Werner is fueled by wood waste and develops 22 KVA. Other small power units develop up to 150 KVA. It is distributed by Kleiber & Schulz, Inc., One Huntington Quadrangle, Melville, NY 11747.

FIGURE 2.

small power unit that is fueled by chips, develops 22 kva, manufactured by Fritz Werner and Company in Germany, and distributed by Klieber and Shultz in Melville, New York.

Fluidized bed systems

The advantages. The major advantage of the fluidized bed technology is more efficient gasification than in either of the other two systems, along with the capability of being scaled to much larger sizes. It can be scaled to multi-100 MMBTU/hour systems. Moreover, the fluidized bed system can accommodate a wide variety of fuel types.

Because the gas exits hot, it has a high sensible heat content which when combined with the chemical energy provides a desirable fuel stream. Because the gas exits hot and because of the efficiency of the gasification process, fluidized bed gasifiers tend to produce a gas stream with a low condensible content.

TWO CASE STUDIES

Installation at Lowes Southern Clay, Bloomfield, Missouri

The plant described below provides 100 MMBTU/hour and was designed, engineered and constructed by Alternate Gas, Inc. The application is the direct use of wood gas as a substitute for No. 2 fuel oil in a rotary kiln that is drying clay for use in cat litter. The feedstock in the system is wet sawdust locally available from mills in the area.

Figure 3 shows the elements of the system. The gasifier and the kiln are on the left; the wood handling system on the right. The tower in the middle is the feed unit for the gasifier and the storage silo in the background is the on-site fuel storage system for the unit. The criteria that were used in the design of this system and other systems of similar sort will now be discussed.

Feedstock supply. The basic criteria are to minimize customer involvement so that they (the customer) see nothing more than trucks moving in and out of the plant without any of their own personnel involved. The second criterion is to ensure quality and quantity control of the feedstock through load by load measurements of particle size and moisture content. Third is the assurance of a reliable delivery schedule which is a matter that can only be handled through the careful choice of a local supplier. Once the fuel supply has been contracted for, and the fuel is delivered, it must be handled on-site.

DRAG CHAIN CONVEYOR
100 MMBTU/HOUR GASIFIER
WOOD GAS TO DRYER
DRYER STACK
STORAGE SILO
RECLAIM CYCLONE
KILN
ROTARY DRUM DRYER
GASIFIER SYSTEM

FIGURE 3.

Wood fuel handling. The first step in wood fuel handling is the receiving of the fuel. The design criteria used here are the minimization of labor by completely automating the dumping process and the minimization of the delay in moving the trucks in and out of the system by having an oversized dumping system and an integrated handling bin to receive the fuel from the trucks. Figure 4 shows a typical wood unloading station. The trucks are backed up the entrance ramp onto the truck dump. The trucks are then disconnected from the tractors. The operator goes to the control house where a single operation initiates dumping. The van is tilted to 62° and the material is deposited into the live bottom hopper shown in the lower left. The entire operation is completed in under five minutes. The driver receives a weigh ticket, and removes the truck from the site. The entire wood unloading unit is manufactured by Phelps, Inc.

The next step in the handling of the material is the delivery from the live bottom hopper to storage through a classifier tower that serves three purposes. It serves to remove any tramp metal by the use of permanent magnet installed at the top of the classifier. It serves as a fuel sorter so that only the right size fuel is selected. (This is accomplished through the use of a disk screen and an ancillary wood hog which is used to regrind the "overs" into an acceptable size.) Figure 5 is a schematic of a typical wood classifier station showing the material coming up the conveyor on the left under the metal separator onto the disk screen. The appropriate size material drops through the screen onto the exit conveyor at the bottom; the "overs" go over the screen and into the wood hog which grinds the material and deposits it back on the exit conveyor. Figure 6 shows the classifier tower at the site in Bloomfield.

The material is then transported pnuematically from the output of the classifier tower into the storage silo. It is manufactured by Rader Industries.

The storage component of the system has basically two elements. One is working storage which consists of a 500 ton conventional silo manufactured by Laidig, Inc. The other is contingency storage which is basically built into the system by allowing us to bypass the silo and operate the system directly from the live bottom hopper. This gives us the opportunity to provide full service under any conditions short of major failures of the wood dumping equipment. Figure 7 shows a standard 500 ton silo. This is a bolted metal unit put in place because of seismic considerations. In the next study you will see that it is just as easy to use concrete, and usually cheaper. Figure 8 shows the wood handling and processing components. The classifier tower is at the top of the figure. The pnuematic bypass is to the

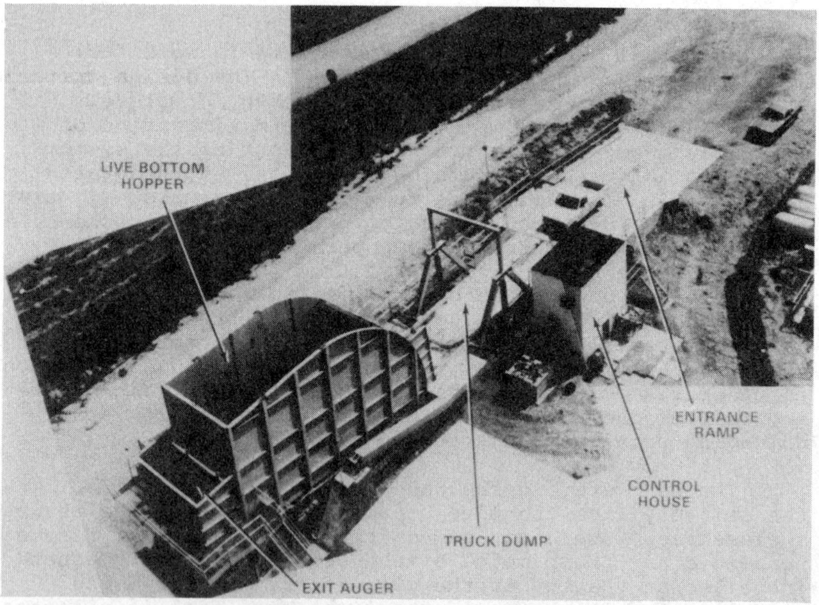

WOOD UNLOADING STATION

FIGURE 4.

FIGURE 5.

TRAMP METAL
MAGNET
INFEED
CONVEYOR
DISC SCREEN
WOOD
HOG
OUTLET
CONVEYOR
CLASSIFIER TOWER

FIGURE 6.

FIGURE 7.

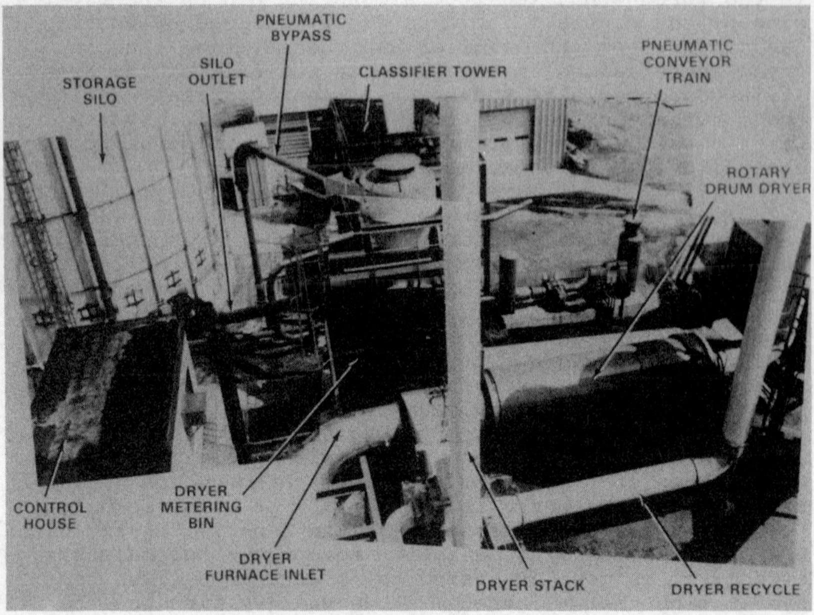

WOOD HANDLING & PROCESSING

FIGURE 8.

left of the classifier tower which allows us, as we mentioned above, to bypass storage if there is a problem. The storage silo is on the left and the silo output at the bottom. The material is delivered from the silo into the metering bin of a wood dryer which is used to dry the feedstock down to a uniform twenty-five percent moisture content.

Wood Gasifier. We now turn to the wood gasification part of the system which consists of a wood dryer followed by the actual gasifier. Figure 9 shows the wood dryer system. The wood dryer is a rotary drum dryer (shown in the foreground) fed from a metering bin on the right side of the figure. The drum dryer is fired by wood gas coming from the gasifier and the material that is dryed is reclaimed in a cyclone at the exit of the dryer. It is deposited onto a drag chain conveyor that carries the material to the gasifier. This particular system was a closed cycle system chosen because it gives good control over environmental emissions.

Figure 10 shows the wood dryer system from another vantage point with a clear view of the dryer burner on the far left and the dryer furnace on the center left. The wood drying step is extremely important in wood gasification because it is only through the availability of a uniform feedstock that one can insure reliable, efficient wood gas quality. The technology that was used in this system for the gasification step itself was the fluidized bed technology. As earlier noted this was for several reasons. One was the efficiency of the gasification that one can achieve with a fluidized bed. Second was for the size of the application which is 100 MMBTU/hour. The third is the ability of this technology to utilize a sawdust feedstock. And finally the fluidized bed gasifier tends to be easier to operate and maintain, even if it is more expensive to do so than a fixed bed system might be. Figure 11 shows the gasifier and the dryer. The gasifier, on the left, was manufactured by Energy Products of Idaho. The wood gas to the dryer train is shown in the middle of the picture and to the right is the dryer burner and dryer furnace for the wood drying system. In the background is the kiln burner house where the primary wood gas is delivered for use in the customer's rotary kiln. Figure 12 shows a closeup of the gasifier, with the gasifier on the left and the metering bin and feed system on the right. The gas from this system is transported hot in order to retain sensible heat and to keep condensibles in the vapor phase. Therefore the gasifier has been situated close to the end use in order to minimize the cost of refractory ducting. Figure 13 shows the last piece of refractory ducting before it was put in place and before the oil burner in the kiln was replaced with a wood gas burner.

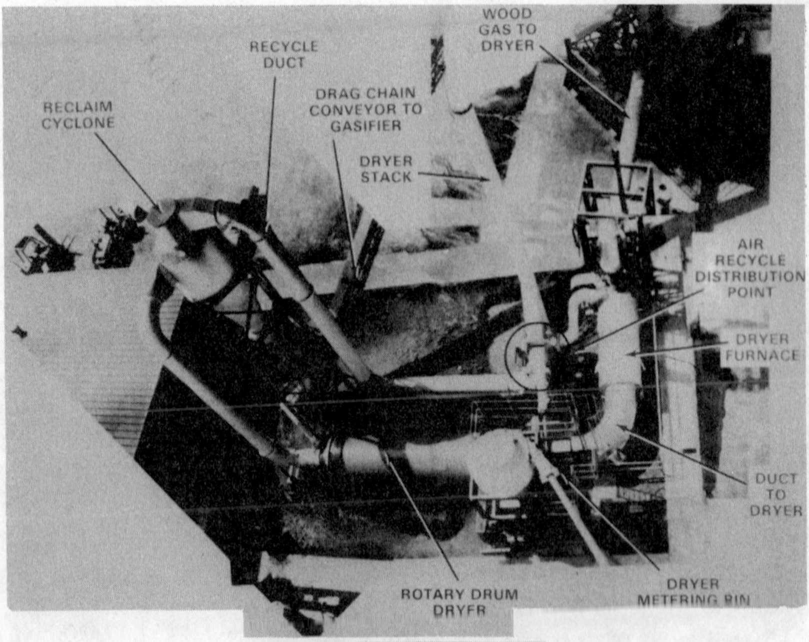

WOOD DRYER SYSTEM

FIGURE 9.

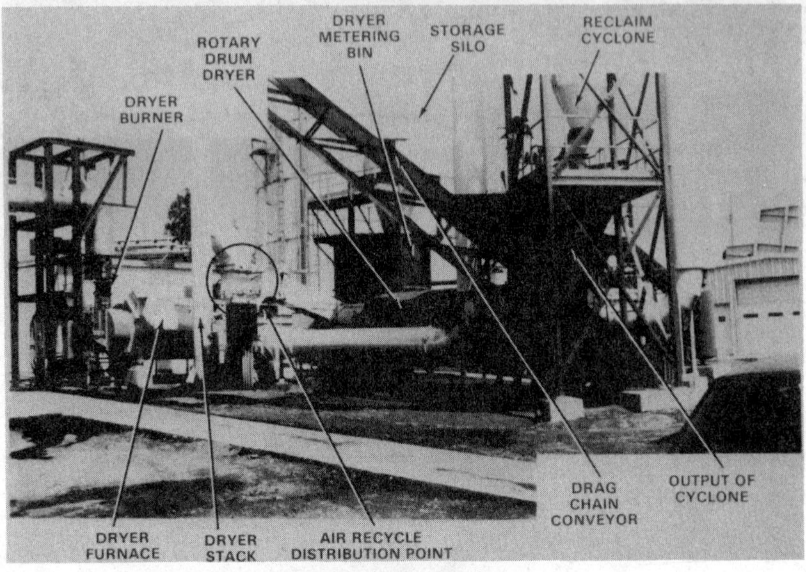

WOOD DRYER SYSTEM

FIGURE 10.

GASIFIER & DRYER

FIGURE 11.

FIGURE 12.

FIGURE 13.

This entire system was designed to be run from a single sixteen-loop process controller (Texas Instrument, PM550) and includes real time measurements of the BTU value of the gas, the flow of the gas and the temperature of the gas. The entire facility can be run by four men.

Installation at Floridin Corporation, Quincy, Florida

We now turn our attention to the second case study, which is a system under construction in the panhandle of Florida. The system is 250 MMBTU/hour and again was designed and engineered by Alternate Gas, Inc. which is also the prime contractor on the job. The application is the substitution of wood gas for natural gas in two fluid bed dryers that are used to dry a locally mined clay. The feedstock here is whole tree chips available from a local supplier. The wood handling and storage is shown in Figure 14. It is virtually identical to that at the Bloomfield site. There is a truck dump of the same manufacturer, along with its integrated live bottom hopper. To the left you see two, 500-ton concrete silos that are used for storage. The classifier tower is similar in design to the one in Missouri and, has virtually all of the identical components. It is interesting to note that the same equipment that handles 100 MMBTU/hour can easily handle 250 MMBTU/hour. The additional requirements on conveyors are minimal. The additional storage is met by using a second silo. In this case, rather than pneumatic transmission, bucket elevators were used to bring the material from the classifier tower to the top of the silo and then distributed into either silo at the direction of the operator. Figure 15 shows a closeup of the two silos and, in the foreground, the output of the live bottom hopper.

This system has twin gasifiers, each rated at 125 MMBTU/hour, manufactured by Power Recovery Systems of Cambridge, Massachussetts. Figure 16 shows the gasifiers in place. The two small cylinders in the foreground are the surge bins, or metering bins, that are used to feed the gasifiers. At the top of the gasifier you see the primary and secondary cyclones used for a more efficient gas cleanup system than at the Bloomfield site. Figure 17 shows the control house in the foreground with the primary and secondary cyclones in the background just before final connection.

In Figure 18 we see the output cleanup unit on the wood dryer, which is needed in Florida to meet the local emissions requirements. Here, in addition to the reclaim cyclone, we have a multi-clone that provides the required level of particulate emission. This particular dryer system is rated at 35 tons per hour maximum throughout and meets all state, local and federal emission standards at that level using wood chips as fuel. This entire gasifier system is scheduled to be put into service in

FIGURE 14.

FIGURE 15.

FIGURE 16.

FIGURE 17.

FIGURE 18.

early July. It is approximately 85% complete at this point and is expected to be fully operational by the early fall.

SUMMARY

Wood gasification is a viable alternative to the use of oil and natural gas in industrial applications ranging from 25 to 250 MMBTU/hour of process heat demand. The gasifiers are capable of providing a hot, clean gas suitable for direct use in any conventional system provided that the burners have been chosen to accept wood gas as a primary fuel. In all designs the burners are fully capable of bumpless transfer between wood gas fuel and natural gas and oil so that the customer retains full capability of operation even when the wood gas system is down.

DISCUSSION

J. Lincoln, Weyerhaeuser Company: How do you clean the bed media in your fluidized bed?

Miller: At the Bloomfield site, it runs 5 days a week, 16 hours a day, so we have both days of the weekend for cleaning. We did not put in an automatic bed clean-out system. It's done by hand. You add a sack of bed sand once a week, and if there is any clinker formation, you go in on Sunday and open up the gasifier. It's a simple matter, a 2-hour job.

Lincoln: Once a week?

Miller: Yes, what you do is monitor unit performance on the controller and when you see bed temperature changes or gas output changes that are characteristic of the clinkers, you know you've got to clean the system. It's not necessarily once a week, it'll go for a couple of weeks. After some operating experience, you know when you're getting the clinkers.

R. Graham, U. of Western Ontario (Canada): If the myths have been exploded, and the economics are there, why are we not seeing more gasification systems implemented in one of the most logical applications, the forest products industry for retrofit?

Miller: The forest products industry has been looking at these systems for a long time. There are two reasons why. You've got to pick the right application. Big plants in the forest products industry are plants with energy demands that are significantly in excess of anything that you see here. The economics of these systems tend to peak at around 250 million Btu's per hour. The place where it fits best in the forest products industry is in the smaller scale sizes. Right now, the technology is such that you can build plants reliably at 250 million Btu/h but not a lot larger, at least I wouldn't. You end up with people in the industry saying it looks like it's a research activity. Several pulp and paper mills that we were talking to very seriously have said they want to wait to see what happens in the oil and gas business.

H. Shull, Natural Resources Research Institute — How much are you paying for wood?

Miller — There isn't one price. It ranges from $3 a ton to $15 a ton. If you're getting sawdust and it really has no other local market, you can get it very cheap. If you're buying whole tree chips, you're going to have to pay in the range of 10 to 12. If there's any competition for the feedstock at all, the price will go up.

Shull — Is there a complete assessment?

Miller — There is not one assessment that you can do. You must do it individually project by project.

J. Kerstetter, Washington State Energy Office — How important are the federal producer tax credits for this type of project, 70 plus cents a million Btu?

Miller — The available tax credits are important in the following sense. They give the investers that little extra added comfort that allows them to put money into what they believe is a somewhat uncertain technology. Anybody in business who has talked to a banker knows that they are very leery of anything that they think has an R&D component. The tax credits provide the little extra added incentive that brings them into projects like this, and so they are very important. But don't misunderstand, these projects make money on their own as compared to almost every other renewable energy alternative to conventional fuels.

ULTRASTRUCTURE CHANGES IN POPLAR CHIPS DURING STEAM LIQUEFACTION

David G. B. Boocock, Ph.D., F. Agblevor, M.A.Sc.,
A. Chowdhury, M.Sc., L. Kosiak, B.A.Sc.,
F. Porretta, M.A.Sc., and E. Vasquez, M.A.Sc.
Department of Chemical Engineering and Applied Chemistry
University of Toronto
Toronto, Canada M5S 1A4

ABSTRACT

A high-pressure cascade system for the steam liquefaction of hybrid poplar chips has been constructed and tested. Wood chips (up to 100 g) are loaded to a preheated nitrogen-purged cylindrical reactor. Steam is injected and the oil, gas and aqueous phase are cooled and collected. Oil yields are in excess of 40 per cent with an oxygen percentage content in the low 20's. The oils are solid at room temperature but soften and flow just above 100°C. Char production is less than 1 per cent. Scanning electron microscopy shows spherical droplets within the still observable cell structure when external temperatures are as low as 280°C. These droplets which are the first indication of oil formation are disrupted by chloroform and soluble in acetone. The droplets combine and fill irregular cavities in the wood as the cell structure disintegrates. These droplets are not seen on the chip surface where the matrix softens and flows due to higher heat transfer rates. The high yield of phenol from some poplar clones, together with spectroscopic evidence of isolated lignin suggests the latter may have significant unmethoxylated aromatic rings.

ULTRASTRUCTURE CHANGES IN POPLAR CHIPS DURING STEAM LIQUEFACTION

INTRODUCTION

At this time last year we reported significant results concerning the liquefaction of poplar chips using only steam and water at temperatures of around 330°C (3). Our research has as its major objective the production of an organic material which can be pumped in liquid form through typical chemical engineering processes. Recognising that biomass inevitably contains water, we opted for a water-based process. We also realised that any successful process must be capable of using wood chips since energy considerations preclude the use of finer material. In addition, we continued our historical involvement with hybrid poplars as studied extensively by the Ontario Ministry of Natural Resources (7). Poplar plantations have shown that significant biomass yields can be achieved for use as pulpwood or construction material.

Our results showed that standard poplar chips (7.6 cm x 0.64 cm x 0.64 cm) were converted to only acetone-soluble oil, water-solubles and gas when heated to 300°C or more in the presence of water (3). Although our heat-up rates were generally high (e.g., 3 minutes to 350°C) lower heating rates (up to 10 minutes to 350°C) gave similar oil yields (≈ 50 per cent) although the chemical composition of the oils were very likely different. The amount of acetone-insoluble material, which was classified as char, was negligible unless sufficient water was present.

Quenching of the reaction at different temperatures showed that the chips initially absorbed water (condensed steam) and swelled. The chips then disintegrated and liquefied, and the absorbed water reappeared. We concluded that the steam, through condensation, was primarily responsible for heating and swelling of the chips. However, once oil was formed sufficient liquid water was required to stabilise it.

The study was made using one particular feedstock, the sapwood of a poplar clone designated DN 70. This results from a cross of populus deltoides and populus euramericana. The latter is sometimes referred to as populus nigra, hence the DN designation. Half the oil from this clone was ether soluble and thus monomeric. Approximately half the monomeric fraction was phenolic, and one quarter was phenol itself at a reaction temperature of 350°C. Thus the phenol was 6 to 7 per cent of the original mass of the wood or close to 25 per cent of the lignin component from which it certainly derives.

During the last year we have focussed on four aspects of the wood liquefaction phenomenon:

(a) the construction of a cascade autoclave system

in which the liquefaction of wood chips could be studied on a larger scale and in which live steam could be injected directly;

(b) the variation in production of phenol among different poplar clones and its relationship to lignin structure;

(c) investigation of the changes in wood ultrastructure during the liquefaction;

(d) the measurement of heat transfer in the wood during the liquefaction process.

In addition, we are also investigating the further deoxygenation of the oil with a view to providing alternate uses (5,6).

CASCADE AUTOCLAVE SYSTEM

A laboratory-scale 'Cascade Autoclave System' has been constructed under funding by Energy, Mines and Resources Canada as part of the Bioenergy Development Program (2). The major objective was to construct a wood liquefaction system in which wood chips could be added to a preheated reactor, followed by immediate treatment with up to 350°C steam and subsequent product discharge. In our single-chip studies, steam was generated from water in the reactor, but the cascade system reflects a more practical version as well as allowing steam to be the primary liquefaction agent. The term 'cascade' was coined to reflect the downward flow in the system, the chips entering at the top of the reactor and the product being discharged through a product cooler at the bottom (8).

The maximum capacity of the reactor was set at 100 grams of dry wood chips which corresponds to a total volume of about 600 mL. It was desirable to load the chips rapidly without the possibility of blocking the inlet valve. To accomplish this, we decided that the feed should be contained in a mesh basket having a diameter slightly less than the internal diameter of the reactor. In the initial design, the plan was to pass the basket completely through the system so it exited with the oil product. This would require three valves having suitable internal diameters. Since the length of the reactor had to be kept within reasonable limits, an internal diameter of 1.5 inches was selected. It then became obvious that the cost of appropriate ball valves would represent almost 75 per cent of the equipment cost. Therefore, to reduce initial costs, it was assumed that only oil gas and water would be collected, and that char (if any) would be retained in the basket which would then be retrieved through the inlet valve. Thus, only the inlet valve required the 1.5 inch diameter.

A schematic diagram of the system is shown in Figure 1, and the unlagged reactor and inlet valve are shown in Plate

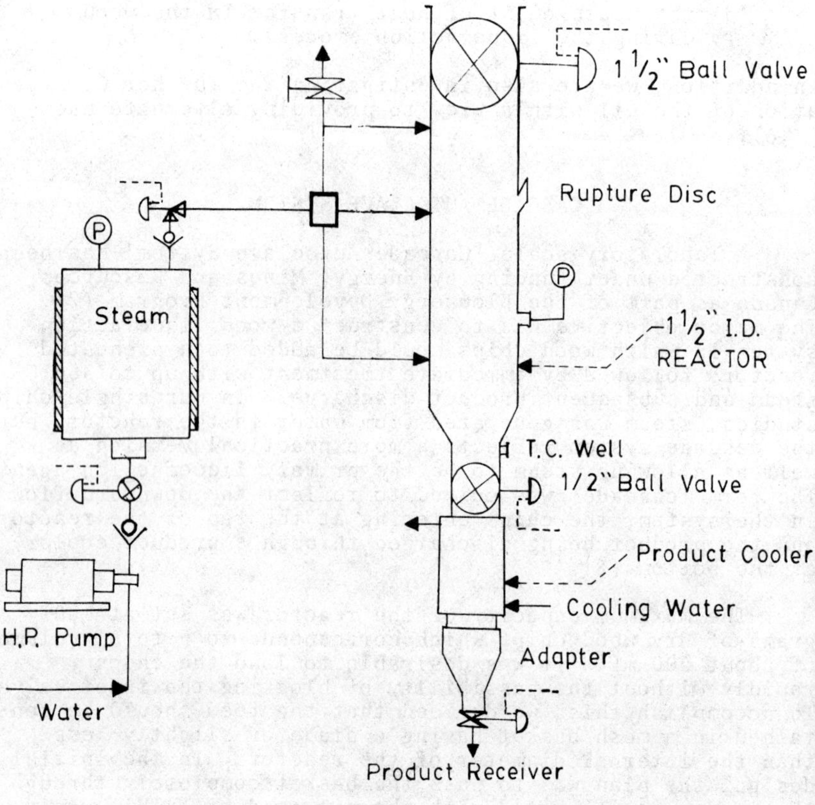

Figure 1: Schematic Diagram of Wood Chip Liquefaction Unit

1. The reactor is machined from a single ingot of 316 stainless steel and has an external diameter of 3 inches, the wall thickness thus being 0.75 inches. It is 18 inches long and at the top end an Oteco hub is threaded and seal welded. The design pressure is 3500 psi at 350°C. The inlet valve (Mogas Industries Ltd.) is a 1.5 inch ball valve (ANSI #2500 series) rated 3600 psi at 370°C. It is mounted on top of the reactor with a matching Oteco hub. The controller is air operated via solenoid valves.

The reactor has nine ports drilled in it. Three of these are for ¼ inch steam injection lines and are located 3 inches, 9 inches and 15 inches from the base of the upper hub. Three of the holes are used for 1/8 inch thermocouples. Of the remaining three ports, one is used for the rupture disc (3000 psi), one for the pressure guage and, at the moment, the third is for optional use.

The current reactor outlet valve has shown some erratic leaks and, from a design standpoint, is the weakest part of the system, being a 0.5 inch ball valve rated 2665 psi at 370°C (Crosby). Since we have already shown that wood is liquefied in the reactor, we have retained the option of replacing the valve with a 0.5 inch non-ball valve rated 18000 psi at 370°C. This is not a straight-through valve, the flow making two right angle turns within 1-2 cm. However, this is not a problem, since only liquids and gases are collected.

Steam is provided from a 2-L autoclave usually maintained at 350°C by upper and lower temperature control. Steam is passed through a two-way valve to the steam injection lines. When the line to the steam line is closed, nitrogen-purge gas can be passed through the valve to the reactor. The latter is heated by two, 6 feet long (1/8 inch OD) heaters joined in paralled and wrapped helically around the outside of the reactor. Four steel strips placed longitudinally and eight circumferential clamps hold the heater against the reactor body. The reactor is insulated with lightweight ceramic brick which is cut and fitted to the external contours of the reactor. The inlet valve is insulated with three layers of 1.5 inch glass wool.

The let-down lock is a 20 inch stainless steel nipple and has an internal diameter of 0.688 inches and a volume of 122 mL. Coolant, which could be water or steam, is passed through an external jacket. Cooled product can be discharged to a glass collection vessel. This vessel has a vent for gas discharge or collection.

The control panel for the system is located in a separate room and the system is equipped with electrical and electromechanical overrides to prevent accidental discharge of either the steam vessel or reactor to the surroundings.

In operation the reactor is preheated and nitrogen-purged with the inlet valve open. The basket containing

PLATE 1: Unlagged Reactor and Inlet Valve

the chips is inserted through the inlet valve which is then closed. Immediately the steam valve is opened (usually for less than 30 seconds). Following valve closure, a further short period (1-2 minutes) elapses before the outlet valve is opened. After a cooling period in the lock, the oil and condensed water are discharged to the product collector. A negligible amount of material remains attached to the basket which is recovered through the inlet after the reactor is cooled.

The oil solidifies on cooling and can easily be separated from the condensed water. The oils typically soften just over 100°C and have oxygen percentages in the low 20's.

ULTRASTRUCTURE CHANGES

Using standard wood chips and the small tube-type reactor described previously (3), we have studied the changes in wood ultrastructure during the early phases of the liquefaction process. Special methods were used for preparing samples for scanning electron microscopy (SEM) both before and after the liquefaction process. In particular, the preparation of samples from inside the chip require special techniques, since the wood structure is weakened as the chip interacts with the steam.

On the surface of the chip at 280°C the most obvious initial change is the disruption of the middle lamella and its separation from the outer cell walls (Plate 2). However, very quickly the matrix softens and flow occurs blurring interfaces and engulfing cell cavities (Plate 3). Inside the wood a different sequence of events occur. Even when the external temperature is 280°C, spherical structures can be seen forming in the matrix (Plate 4), while the open cell structure can still be seen. We believe this is the first signs of oil formation. The spherical structures are unchanged by extraction with ether which usually removes monomeric material. Chloroform, which in our experience dissolves oligomeric material, disrupts the droplets which agglomerate on the surface of the wood. Acetone, however, completely removes the spherical structures showing they are made up of material we would classify as oil, although it is not low molecular weight in nature. At later stages this oil invades irregular cavities forming in the matrix, and gas and/or vapour bubbles can be clearly seen. Beyond this point, the samples have insufficient mechanical strength to allow suitable sample preparations. Studies are underway to characterise the chemical nature of the primary oil. In particular we would like to know the relative contributions to the oil of the carbohydrates and lignin.

LIGNIN STUDIES

In our studies on standard chips using DN 30 poplar, relatively high yields of phenol (up to 7 per cent mass yield on a dry basis) were observed (3). The wood was 4-5 years old and contained very little heartwood, so the feed was limited to the sapwood portion. Other DN clones have shown similar high yields of phenol, but other clones D38, C147, I45/51) yield lesser amounts. The phenol certainly derives from the lignin in the wood, the yield in the best case being close to 25 per cent based on this component.

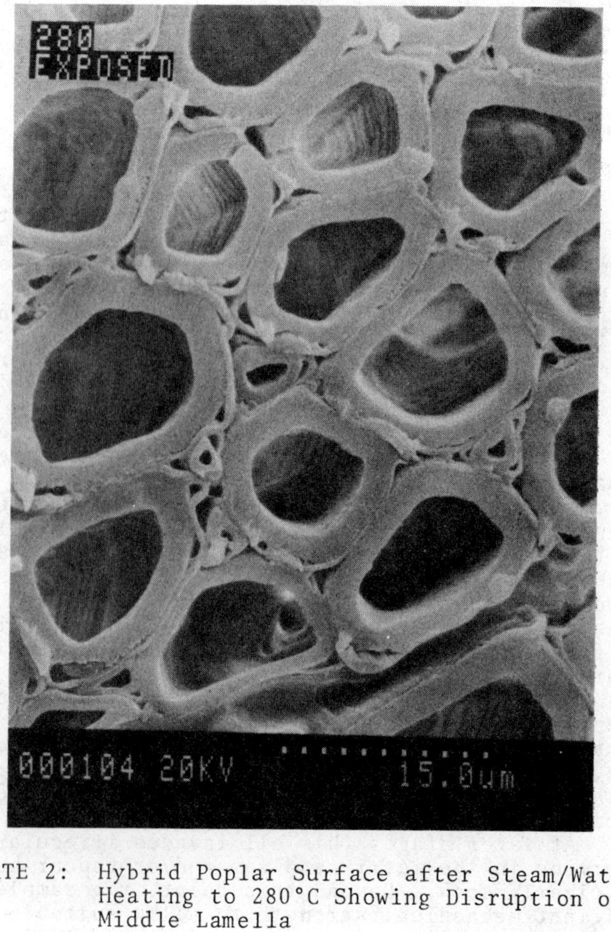

PLATE 2: Hybrid Poplar Surface after Steam/Water Heating to 280°C Showing Disruption of Middle Lamella

This high yield of phenol is unusual given that wood lignins are formed from coniferyl and sinapyl alcohols and thus contain 2-methoxyphenol and 2,6-dimethoxyphenol units. We thus suspect that the structure of the lignin in the DN clones may resemble in part that of grass lignins in which coumaryl alcohol is the precurser and unmethoxylated aromatic rings are present.

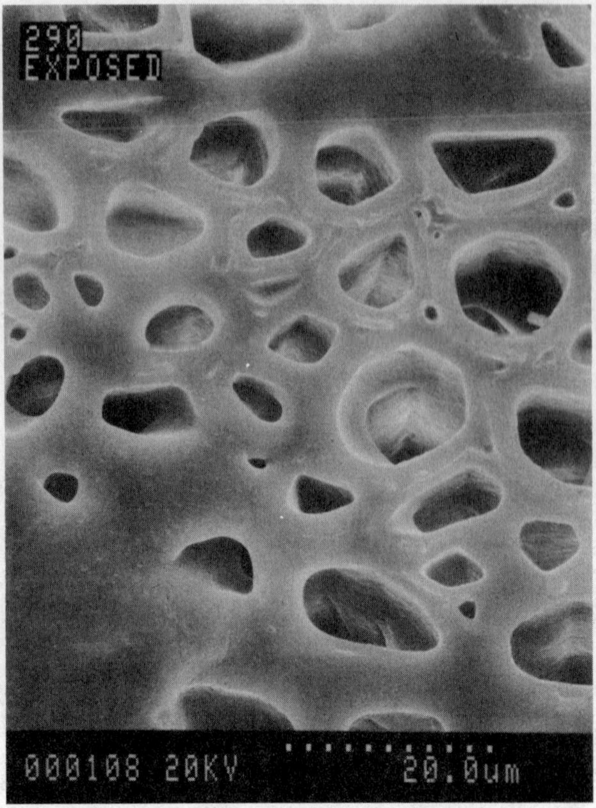

PLATE 3: Hybrid Poplar Surface after Steam/Water Heating to 290°C Showing Flow of Matrix

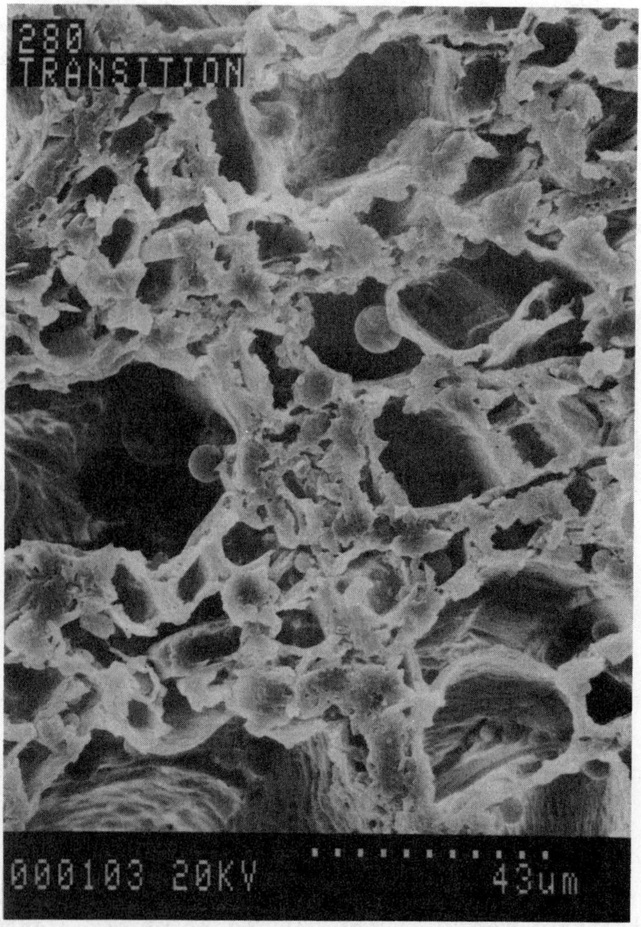

Plate 4: Hybrid Poplar Interior (Surface Heated to 280°C) Showing Spherical Structures. Small spheres can be seen top centre and larger spheres at middle left.

We thus began a study of the lignin structures present in the heartwood, sapwood and bark of different clones. Our aim was to isolate lignins without modifying their structure. We selected a combination of the Bjorkman method (1) and enzymatic removal of cellulose. Techniques such as milling, solvent extraction, dissolution/precipitation, freeze drying and enzymatic treatment were used.

The wood sample is first milled and extracted with EDTA followed by ethanol/benzene (1:2) and ethanol extraction to yield an iron and extractive free wood meal. After ball milling in toluene, the solid material is extracted with aqueous dioxane. The material insoluble in aqueous dioxane is treated with cellulase and the remaining solid (cellulose free) is recycled to the aqueous dioxane extraction step. The material soluble in aqueous dioxane is evaporated, dissolved in 90 per cent acetic acid preciptated with water, centrifuged and freeze-dried. The material which is obtained is dissolved in dichloroethane/ethanol and ether added. This yields, after separation, an ether-soluble lignin and an ether-insoluble lignin, the latter being freeze-dried from dioxane/water.

These lignins give consistent acetone-soluble oil yields of 70 per cent on aqueous thermal liquefaction, thus confirming the major contribution of the lignins to the oil. In addition, combined pyrolysis-mass spectrometry have shown a high abundance ion at m/e 94 (phenol) in the sapwood lignin of clone DN 38. This ion does not appear in the corresponding spectrum of the heartwood lignin or spectra of lignins from other non-DN clones. We are in the process of obtaining similar spectra from other DN clones and populus nigra which we believe is the parent poplar contributing to the special lignin structure.

CONTINUING STUDIES

Research has been extended into all three areas. We are proceeding with the collection of operating data for the cascade autoclave system. The oils will be characterised in this laboratory and also at B.C. Research, British Columbia, Canada. We are also in the process of identifying the contribution of hemicellulose, cellulose, and lignin to the primary oil droplets. The lignin studies are very much entering the analytical phase with the application of pyrolysis-mass spectrometry, C-13 and proton nuclear magnetic resonance spectroscopy, infrared spectroscopy and other spectroscopic techniques to the lignins and the oils formed from them. Chemical reactions coupled with spectroscopic techniques are also being used to elucidate the lignin structures.

ACKNOWLEDGEMENTS

Grants from the Natural Sciences and Engineering Research Council (Operating, Strategic and Equipment), Energy Mines and Resources, and Imperial Oil together with a contract from the Canadian Bioenergy Program (Energy, Mines and Resources) have supported various aspects of this research.

REFERENCES CITED

1. Bjorkman, A., "Studies on Finely Divided Wood, Part I, Extraction of Lignin with Neutral Solvents", Svensk Papperstidn., 59, 477-485 (1956), and "Part 3, Extraction of Lignin-Carbohydrate Complexes with Neutral Solvents", ibid., 60, 243-251 (1957).

2. Boocock, D.G.B., "Construction of a Cascade Autoclave System for the Liquefaction of Wood Chips", Final Report for Canadian Bioenergy Development Program, Contract 05SV.23216-4-6191, Energy Mines and Resources, Ottawa, Canada 27 pp. (1986).

3. Boocock, D.G.B., Agblevor, F., Holysh, M., Porretta, F., Sherman, K., Kallury, R.K.M.R., and Tidwell, T.T., "Liquefaction of Poplar Chips by Aqueous Thermolysis", Symposium Papers, Energy from Biomass and Wastes IX, Lake Buena Vista, Florida, 1107-1125 (1985).

4. Evans, R. J., Solar Energy Research Institute, Personal Communication.

5. Kallury, R.K.M.R., Tidwell, T.T., Boocock, D.G.B., and Chow, D.H.L., "High Temperature Catalytic Hydrogenolysis and Alkylation of Anisole and Phenol", Can. J. Chem., 62, 2540-2545 (1984).

6. Kallury, R.K.M.R., Restivo, W.M., Tidwell, T.T., Boocock, D.G.B., Crimi, A., and Douglas, J., "Hydrodeoxygenation of Hydroxy, Methoxy, and Methyl Phenols with Molybdenum Oxide/Nickel Oxide/Alumina Catalyst", J. Catal., 96, 535-543 (1985).

7. Ontario Ministry of Natural Resources, "New Forests in Eastern Ontario, Hybrid Poplar", Science and Technology Series, Volume 1 (1983).

8. Overend, R.P., National Research Council of Canada, Personal Communication.

DISCUSSION

D. Elliott, Pacific Northwest Laboratory — What is the pressure in the reactor?

Boocock — The pressure is determined strictly by the vapor pressure of the water, which is about 2300 psi.

Elliott — Can you talk anymore about the analysis of the oil product other than just the fluidity?

Boocock — I can tell you it contains oxygen percentages in the low 20's, and the yields are 30 to 50% depending on the reaction conditions. We're currently shut down. I didn't mention that the outlet valve of this reactor leaks and we're currently replacing the ball valve with a non-ball valve which should solve all of our problems. We don't need a ball valve. The initial plan was to pass the basket right through the reactor, going through three ball valves. Our run showed that we didn't need a ball valve, and that all we needed was a valve which would allow us to pass liquid product. I don't have distillation curves or molecular weight distributions, which is what we would really like to know.

D. Meier, Institute of Wood Chemistry (West Germany) — Why did you use pyrolytic methods to describe your wood lignin? I think it's easier to determine the methoxyl group content.

Boocock — We are using a broad range of techniques to characterize the lignin, including methoxy determinations, $C^{13}NMR$, infrared, and pyrolysis mass spectrometry. The only reason I showed the pyrolysis mass spectra is because that's the information we have more of and I thought it was of more interest to this group. We have information on methoxyl, but it's not very clear cut.

W. DeGroot, U. of Montana — I would like to comment on the quality of the micrographs you showed. I thought they were very interesting. I'd like to comment that we have done some work with pyrolysis at atmospheric pressure in flowing nitrogen. In the presence of the ash in wood with poplar species, the cell wall is retained even under gasifying conditions up to more than 95%. The structure is almost indistinguishable from that of unpyrolyzed wood. In the absence of ash though, you get a char that looks as if it were formed from a melt. Have you done any acid washing, or anything to remove the ash to see how that might affect the process?

Boocock — We haven't attempted any pretreatment of the chips. I should say that these chips are dry chips with 2 or 3% water. In the tubular reactor, we did some runs with wet chips, and surprisingly found that the oil yields from wet chips are not a whole lot different than the yields with dry chips. It's possible that the chemical slate in that oil is different, but at least in terms of yields, it's no different. But we've done nothing to remove the mineral content.

DeGroot — In our case, I was attributing the loss of structure to vaporization of the carbohydrate fraction in the absence of ash resulting in just a char derived purely from lignin which could flow. Your micrographs would indicate that flowing also occurs in the carbohydrate fraction of the secondary cell wall. With regard to the mass spec, we found that hard wood is 4 times higher in ash content than sap wood. So that might also point to the importance of ash in terms of the formation of phenol.

Boocock — The formation of phenol is also dependent on conditions in the pyrolysis mass spectra that were run by SERI. The pyrolysis temperatures went up to 500°. In one of the runs where they did a slow pyrolysis, the phenol peak was enhanced when they only went up to 350°, so by playing around with the conditions, you can change the amount of phenol that's produced.

D. White, U. of Arizona — Do you suppose those little nodules on the surface are a little bit of gas being formed that can't break through the viscous material around them?

Boocock — I don't think so. We're getting quite expert at looking at these micrographs. In conjunction with the solvent extraction, where we can actually get these things to agglomerate, we're pretty sure that's solid material.

DISCUSSION, CONT.

J. Piskorz, U. of Waterloo (Canada) — You don't show us total material balances, but you have 50% yields of oil or polymer.

Boocock — We don't have it because we realized that we were going nowhere with the leak on the bottom valve. So we simply terminated the runs, and we're in the process of putting a new valve on the bottom. I can't tell you the overall mass balances, but we should have them very soon.

Piskorz — Have you tried pure cellulose in your system?

Boocock — No, because we're heading in the direction of what we call practical feedstocks. We have studied cellulose in the small tubular reactor.

Piskorz — You have described flowing polymers in your pictures. Is this a polymer of cellulose, hemicellulose, lignin, or what?

Boocock — The only thing I can tell you is that in the solvent extraction work we've been doing on the chips, we first extract with water, then ether, then chloroform, then acetone. The C^{13}NMR shows that the monomeric material in the ether phase is predominantly of lignin origin. The question is where are the carbohydrates? I suspect they're in the water phase, and we haven't seen much of the secondary breakdown products from the cellulose, such as hydroxymethyl furfural or furfural. I suspect that at about 280°, which is where we did the extraction, we probably still have some carbohydrates in the aqueous phase.

HYDROTREATING BIOMASS LIQUIDS TO PRODUCE HYDROCARBON FUELS

Douglas C. Elliott, B.S., M.B.A.
Senior Research Scientist
Eddie G. Baker, B.S.
Senior Research Engineer
Pacific Northwest Laboratory
Richland, Washington 99352

ABSTRACT

Experimentation is underway at Pacific Northwest Laboratory to investigate the thermochemical transformation of biomass to liquid hydrocarbon fuels. Recent emphasis has been on the upgrading of the crude oil from biomass to the finished hydrocarbon. Specifically, catalytic hydrotreating of the crude oil is performed to reduce the molecular weight and remove oxygen. Experimentation has involved the use of a continuous-feed, fixed-bed catalytic reactor to hydrodeoxygenate and hydrocrack the biomass crude oils at conditions of 250°C to 400°C and 2000 psig. Catalysts and conditions similar to those used in petroleum hydrotreating have been used to produce a nearly pure hydrocarbon liquid product. The process optimization has emphasized minimizing hydrogen consumption in order to maintain the aromatic character of the product and minimize gas production. Both wood-derived high-pressure liquefaction oils and flash-pyrolysis oils have been used as feedstocks for hydrodeoxygenation.

Research results are presented pertaining to a two-step hydrotreating process designed for the pyrolysis oils. In the first step, a low-temperature (270°C) catalytic (sulfided cobalt-molybdenum) treatment is used to process the thermally unstable pyrolyzate into a tar similar in characteristics to the high-pressure liquefaction product. This tar can then be hydrodeoxygenated and hydrocracked by conventional hydrotreating techniques to produce a gasoline.

*Operated by Battelle Memorial Institute for the U.S. Department of Energy under Contract DE-AC06-76RLO-1830.

HYDROTREATING BIOMASS LIQUIDS
TO PRODUCE HYDROCARBON FUELS

BACKGROUND

This paper describes the recent efforts at Pacific Northwest Laboratory (PNL) in upgrading the crude liquid products from thermochemical conversion of biomass to hydrocarbon fuels. In order to comprehend fully the work reported, it is necessary to have an understanding of the processing which is used to convert biomass (wood) into the crude liquid products. A recent review of these processes (1) shows the wide range of liquid product composition which can be produced by the different processes currently under development around the world. Two general groups of processes are identified based on their product properties: 1) high-pressure processes with relatively long residence times, and 2) flash pyrolysis processes with relatively higher temperature and short residence times. The properties of these two types of products are summarized in Table 1. These two general classes of liquefaction should be used with the knowledge that the product quality is actually a continuum from one extreme to the other. This continuum can be considered as the result of a process severity function wherein temperature, pressure, time at reaction conditions and added reactants or catalysts effect the transformation from the primary pyrolysis product to the more deoxygenated high-pressure products. Transformations along the continuum can be performed in one or more reaction steps. It is the essence of upgrading biomass liquefaction products to proceed along the product continuum past the point represented by the high-pressure product to the complete deoxygenation and production of hydrocarbon liquids. However, it is important to produce useful yields of liquid products and avoid excessive amounts of solid coke or gas production. This trade-off of liquid versus solid and gaseous products is at the heart of all process optimization for biomass liquefaction.

TABLE 1. Properties of Biomass Liquefaction Products

	High-Pressure Liquefaction	Flash Pyrolysis
carbon content	68-81%	56-66%
sulfur & nitrogen content	0.1%	0.1%
oxygen content (maf)	9-25%	27-38%
water in crude	6-25%	24-52%
viscosity	2900 cP@ 40°C- 55,000 cP@ 60°C	5-59 cP@ 40°C
density, g/ml	1.10-1.14	1.11-1.23

In the PNL hydrotreating research we have studied representatives of both general groups of biomass liquefaction products. Our hydrotreating technique was originally designed for application to the lesser-oxygenated high-pressure liquefaction products as represented by the products from the U.S. Department of Energy Biomass Liquefaction Experimental Facility formerly at Albany, Oregon. This work which has been described previously (8,9,10) successfully tested the catalytic processing of the Albany oils to a gasoline-range hydrocarbon mixture, high in cyclics and aromatics. Our more recent work and the main subject of this paper has been the upgrading of pyrolyzates using similar catalytic methods.

As a review of previous hydrodeoxygenation research both our earlier papers (8,9) and that of Furimsky (12) provide keys to the earlier literature. Hydrodeoxygenation (the removal of oxygen by catalytic hydrogenation) of phenolics or other functional groups has not been studied as extensively as similar sulfur and nitrogen removal methods. The major source of the recent interest in hydrodeoxygenation has been the upgrading of coal derived oils. Our application of the same types of catalytic methods to high-pressure biomass oil tested the applicability of the technology to upgrading biomass oils as well.

Our earlier results with high-pressure biomass oils and phenolic model compounds showed good activity with typical hydrotreating catalytic technology. Temperatures of 350-400°C and pressures of around 2000 psig were used. The catalyst of choice thus far in our research is cobalt-molybdenum which is sulfided prior to use. In our fixed-bed, upflow system we have tested the effect of space velocity on the extent of hydrogenation and have been able to run in an extinction-cracking mode wherein the biomass oil is converted completely to a gasoline-range hydrocarbon mixture. We prefer the cobalt-molybdenum catalyst over other active catalysts such as nickel-molybdenum (sulfided) or nickel metal due to the greater specificity for hydrodeoxygenation and lower consumption of hydrogen in saturating aromatic structures. The product from our processing has been determined to be approximately a 50/50 mixture of aromatics (alkyl benzenes and indenes) and cyclics (cyclohexanes).

UPGRADING PYROLYSIS OILS

A number of attempts have been made in recent years to upgrade biomass pyrolyzates to hydrocarbon fuels. The results have been somewhat mixed. The use of zeolites was first attempted in Canada (2) and was largely unsuccessful due to the large amount of coke production and limited conversion to hydrocarbon liquids. Diebold's paper in this volume (3) describes a more recent effort along the same line of experimentation which involves more sophisticated instrumentation and a broader mandate to understand the chemistry involved. The use of noble metal catalysts has also been attempted in the presence of hydrogen-donor solvents, but only partial deoxygenation was attained.(15) Our own results of catalytic hydrotreating pyrolyzates were also, initially, unsuccessful.(7) Our use of the sulfided cobalt-molybdenum catalyst at 350°C and 2000 psig produced only limited quantities of hydrocarbon product before the catalytic bed plugged with a coke-like material. Based on these results we have realigned our experimentation to test alternate processing conditions. The results are described in the following paper.

Analysis of Pyrolysis Oils

Pyrolysis oil composition can vary greatly depending on the conditions of production.(6) The particular feedstocks we are testing are produced at short-residence time (≤1 second) and atmospheric pressure at temperatures around 500°C. These conditions have been found to produce the maximum amount of liquid product.(13,14) Other studies have suggested that these are also the primary pyrolysis products of the wood feedstock.(11)

Some of the properties of the oils we used as feedstocks are presented in Table 2. The oils are highly oxygenated with large amounts of dissolved water. The components are a mixture of light acids, aldehydes, ketones and furans and a larger fraction of higher molecular weight, more complexly oxygenated compounds. Gas chromatography reveals only a little of the composition of the oil due to the high polarity and/or molecular weight of most of the oil.(6) Poor results were also obtained with nuclear magnetic resonance spectrometry due to the insolubility of some of the oil and its instability in solution suggesting either agglomeration or polymerization of the compounds.(6) What we have seen by gas chromatography/mass spectrometry, other than the light oxygenates mentioned earlier, are primarily lignin structure fragments. In fact, good preservation of the guaiacols (monomethoxyphenolic) and syringols (dimethoxyphenolic) have been reported. These structures can be used as an indicator of the feedstock type: only guaiacols from softwoods, syringols more prevalent than guaiacols for hardwood oils.

TABLE 2. Properties of Pyrolysis Oil Feedstocks

	Georgia Tech		Waterloo	
	As Fed	Dry	As Fed	Dry
Carbon, %	39.5	55.8	49.8	65.9
Hydrogen, %	7.5	6.1	7.3	6.1
Oxygen, %	52.6	37.9	42.8	28.0
Nitrogen, %	<0.1	<0.1	<0.1	<0.1
Ash, %	0.2	0.3	0.03	0.04
Moisture, %	29	0	24.3	0
Density @ 55 °C	1.23	--	1.11	--

When performing thermochemical upgrading on these pyrolysis oils, their thermal instability is a major consideration. This instability is suggested by coking of the oil during batch distillations and decomposition of the oil at elevated temperatures under inert atmosphere. We have reported previously a number of instances of coke formation during distillation of pyrolyzates (4,5,6,7). The batch distillation of these oils under vacuum (10-25 mm Hg) proceeds routinely until a pot temperature of 170-200 °C is attained at which point a strongly exothermic reaction occurs and the bubbling oil solidifies into a sponge-like coke. This phenomenon is totally unlike the distillation of the high-pressure oils which proceeds to temperatures of 300 °C or greater without coking of the distillation residue.

In another type of test we have reported that pyrolyzates will decompose to a solid and a water solution when heated to 270 °C when sealed under inert gas.(4) This effect was found for several whole pyrolyzates as well as an oil distillate, a light-distillate-stripped oil, a water-washed oil, an aqueous sodium carbonate-extracted (acid-free) oil. All these results compare with our earlier report of a lack of change in high-pressure liquefaction products under the same conditions of thermal stress. Apparently, the thermal instability of pyrolyzates cannot be attributed solely to either the low- or high-molecular weight components nor can it be altered by removal of the acid fraction nor by the separation of water soluble components.

Given this lower level of thermal stability for the pyrolyzates as compared to high-pressure oils, processing methods must be designed either to overcome the instability or incorporate it into the processing mechanism. Based on our early results with catalytic hydrotreating, we concluded that we could not overcome the thermal decomposition by a conventional catalytic hydrogenation. We therefore modified our approach to incorporate the decomposition into the process. Whereas, the apparent rate of thermal decomposition to coke is much greater than that of catalytic hydrodeoxygenation at temperatures of approximately 350 °C; we attempted to process the pyrolyzate at a lower temperature where the rate of thermal decomposition would be slower and, potentially, more in balance with the rate of catalytic hydrodeoxygenation.

Reaction System Configuration

The reaction system used for these tests was a continuous-feed, fixed-bed catalyst system as shown in Figure 1. The wood-derived oil, preheated to about 40 °C, is pumped by a high-pressure metering pump. Hydrogen from a high-pressure cylinder is metered through a rotameter and mixed with the oil prior to entering the reaction vessel. The reaction vessel is a thick-walled stainless steel vessel approximately 7.5 cm internal diameter by 25 cm high. It holds about 900 ml of catalyst pellets supported a stainless steel screen. Void volume of the vessel when charged with catalyst is approximately 650 ml.

Our calculations indicate that for the liquid and gas flow rates used in our tests a two-phase flow pattern exists in the reaction vessel. Liquid moves through the reactor very slowly, essentially plug flow. Gas flows through more rapidly, bubbling through the oil. Light products are removed in the gas phase with the hydrogen. There should be essentially no carryover of the liquid with the gas until the liquid level reaches the top of the reactor and overflows into the product line. The pressure within the system is maintained by a back-pressure regulator (Grove valve). Liquid product is recovered in the knock-out pot and the offgas is metered and analyzed before it is vented.

Low-Temperature Hydrotreatment of Pyrolyzates

Since our initial experience with pyrolyzate oil indicated that operating temperature was an important consideration, the first experiments planned were to determine the effect of temperature on the catalytic system. Following the initial determination of allowable operating temperature, catalyst content, space velocity (residence time) and hydrogen partial pressure were varied to determine their effects.

Effect of Temperature. The initial attempt to hydrotreat pyrolyzate at PNL was made with a flash pyrolysis oil from the University of Waterloo (14) at 350 °C over a sulfided cobalt-molybdenum catalyst. As mentioned above, this test was terminated prematurely when the catalyst bed plugged. With this knowledge in hand, a series of tests with pyrolyzate from the Georgia Tech entrained-flow pyrolysis unit were initiated. These tests were to examine the operating temperature range of 250 to 350 °C in the presence of a nickel catalyst which was expected to be a

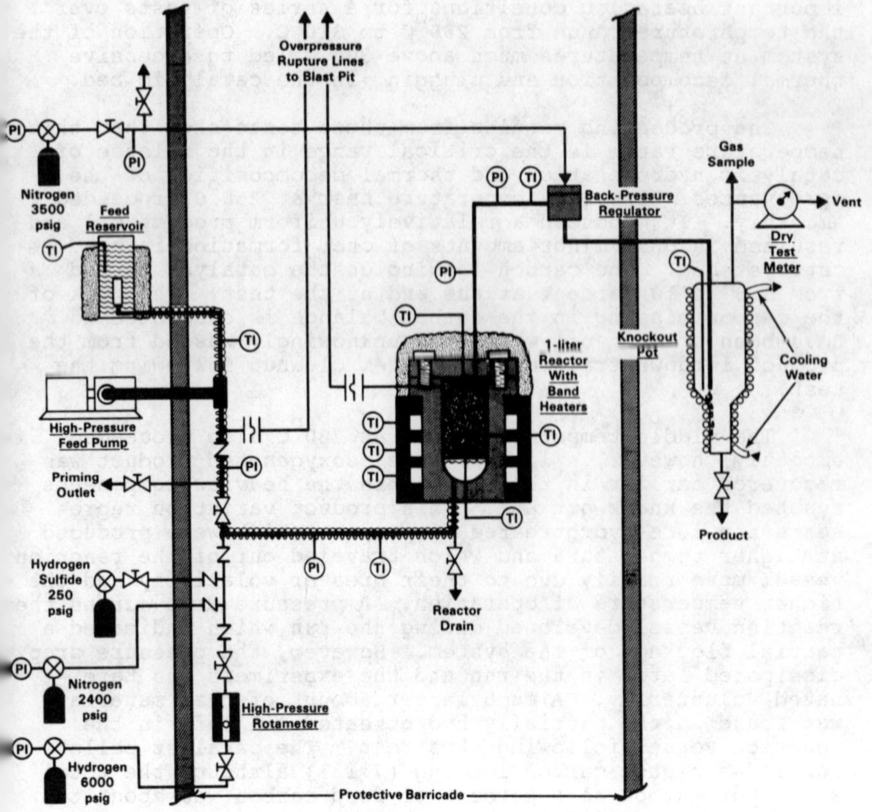

FIGURE 1. Continuous-Feed Reactor System for Catalytic Hydrotreatment

more active hydrogenation catalyst. Given the lower temperatures to lessen thermal decomposition and a more active catalyst it was intended to achieve a balance between decomposition and hydrogenation.

The low temperature operating conditions proved effective in allowing stable operations in the reaction system. However, the product of this type of processing is far from the gasoline-like material produced from the high-pressure oil at higher temperature. Table 3 provides a list of important operating conditions for a series of tests over the temperature range from 250°C to 310°C. Operation of the system at temperatures much above 300°C led to excessive thermal decomposition and plugging in the catalytic bed.

The processing results from these tests show that this temperature range is the critical range in the balance of catalytic hydrogenation and thermal decomposition of the pyrolyzates. The low temperature test at 258°C proceeded smoothly. It produced a relatively uniform product oil and resulted in only minor amounts of char formation in the catalyst bed. The carbon loading on the catalyst varied from 5.5 to 10 percent at the end of the test. The bulk of the carbon missing in the carbon balance is suspected to have been product oil which was unknowingly washed from the product letdown train during system cleanup following the test.

The middle temperature test at 280°C also proceeded smoothly, however, a lighter more deoxygenated product was recovered earlier in the run before the heavier components reached the knock-out pot. This product variation represents the more hydrotreated components which were produced at higher temperature and which traveled out of the reaction vessel more readily due to their greater volatility and the higher temperature of operation. A pressure drop across the reaction vessel developed during the run which indicated a partial blockage of the system. However, the pressure drop dissipated later in the run and the experiment was terminated voluntarily. A much larger amount of char material was found in the partially hydrotreated oil left in the reaction vessel following this test. The catalyst pellets carried a higher carbon loading (7-13%) although the total amount of carbon as a percent of feed carbon was about the same as in the 258°C test.

In the high temperature test at 310°C the reaction vessel eventually plugged and the experiment had to be terminated. The product oil recovered during this test was of a better quality than even the lighter products from the 280°C test. At the end of this test the catalyst bed was completely encased in a brittle, high melting polymer material. The interpretation of this result is that the thermal decomposition leading to coking of the pyrolyzate proceeds at a greater rate than the catalytic hydrogenation process at 310°C. Carbon conversion to gas at 310°C was the highest in any of the three tests.

TABLE 3. Hydrotreating Test Results with Georgia Tech Pyrolyzate

Experimental Operating Conditions			
Catalyst	nickel	nickel	nickel
Temperature, ^{o}C	258	280	310
Pressure, PSIG	2020	2050	2050
Oil feed rate, ml/hr	290	396	405
Hydrogen rate, L/hr	168	216	240
LHSV, vol oil/vol cat/hr	0.32	0.44	0.45
Hydrogen consumption, L/L oil	66	161	252
Experimental Results and Product Analyses			
Carbon conversion, wt%			
to oil/aqueous	36/14	57/10	0/5
to gas (C_1 to C_4	9	11	16
to carbon on catalyst	9	8	--
Oil product yield, ml/ml feed	0.28	0.42	reactor plugged
Carbon balance (based on oil/aqueous/gas	59	78	--
Hydrogen balance	91	100	--
Oxygen balance	100	102	--
Overall mass balance	84	96	--
Total oil feed, ml	1601	1737	2026
Wet product analysis			
H/C, atomic	1.54	1.42	1.49/1.63
oxygen, percent	26.8	25.0	19.4/13.2
density, g/ml	1.1	--	-- /0.96

Effect of Catalyst. The result with the nickel catalyst at 310 C (plugged reaction vessel) was similar to that achieved at higher temperature with the cobalt-molybdenum catalyst. In order to provide a more direct comparison of the two catalysts, tests were undertaken with the sulfided cobalt-molybdenum catalyst at lower temperatures. As shown in Table 4, in nearly all respects the nickel catalyzed tests provided the same results as the cobalt-molybdenum catalyzed tests.

A single test with alumina balls in place of alumina-supported metal demonstrated that the catalytic entity is important to the process. With only alumina in the bed (results in Table 4) the reaction vessel plugged almost immediately. Only limited amounts of water and heavy oil product (24-28% oxygen) were recovered prior to the system plugging. A coke-like material (Hydrogen to carbon atomic ratio 1.04) was recovered from the bottom one-third of the reactor. We conclude that although the net hydrogen consumption in the metal catalyzed tests was relatively small, the metal catalyst must play a key role in interrupting the thermal decomposition/polymerization of the pyrolyzate, probably by hydrogenating active intermediates.

Effect of Residence Time. Since the thermal decomposition reactions are perceived to be relatively fast reactions and since little catalytic hydrogenation appears to be occurring in the reaction vessel (as measured by hydrogen consumption) it was recognized that the length of time in the reaction vessel might play a relatively minor role in this processing. Indeed, as the data in Table 5 shows, there is almost no effect on the product quality as a function of residence time. As the residence time is reduced from 86 minutes to 23 minutes the oxygen content and hydrogen to carbon ratio of the products remains almost unchanged. The differences in the product quality numbers in Table 5 are explained by minor differences ($\pm$ 2%) of water dissolved or emulsified in the product oils. The initial abrupt change in oxygen content of the product tar in going from 86 to 66 minutes is thought to be related to the hydrogen partial pressure change (see next section). Changes which are noticeable as a function of residence time are the decrease in gas production and increase in oil product recovery as the residence time decreases. Coincidentally, the amount of oxygen rejection from the oil phase also decreases. These results suggest that the low temperature hydrotreating to upgrade pyrolyzates can be accomplished with relatively fast throughputs and residence time of 23 minutes or less.

TABLE 4. Additional Hydrotreating Test Results

Experimental Operating Conditions		
Catalyst	CoMo	Alumina
Temperature, ^{o}C	273	254
Pressure, psig	2025	2000
Oil feed rate, ml/hr	392	411
Hydrogen rate, L/hr	168	180
LHSV, vol oil/vol cat/hr	0.44	0.46
Experimental Results and Product Analyses		
Hydrogen consumption, L/L oil	135	-49
Carbon Conversion, wt %		
to oil/aqueous	55/11	--
to gas (C_1 to C_4)	9	10
Oil product yield, ml/ml feed	0.42	reactor plugged
Carbon balance (based on oil/aqueous/gas)	70	--
Hydrogen balance	96	--
Oxygen balance	104	--
Overall mass balance	91	--
Total oil feed, ml	1794	753
Wet Product Analysis		
H/C atomic	1.47	1.14
oxygen, percent	24.6	24.9

TABLE 5. Hydrotreating Results as a Function of Residence Time

Experimental Operating Conditions						
Catalyst	CoMo	CoMo	CoMo	CoMo	CoMo	CoMo
Temperature, °C	273	271	271	274	271	270
Pressure, psig	2025	2020	2020	2010	2030	2040
Oil feed rate, ml/hr	392	515	555	935	1200	1440
Hydrogen rate, L/hr	168	120	120	120	120	120
LHSV, vol oil/vol cat/hr	0.44	0.57	0.62	1.04	1.33	1.60
Residence Time, min	86	66	61	37	28	23
Experimental Results and Product Analyses						
Hydrogen consumption, L/L oil	135	90	60	39	32	28
Carbon Conversion, wt %						
to oil/aqueous	82/11	80/10	87/8	83/5	83/7	87/11
to gas (C_1 to C_4)	9	7	7	5	4	4
Oil product yield, ml/ml feed	0.52	0.56	0.69	0.66	0.65	0.70
Carbon balance	100	97	102	93	94	102
Hydrogen balance	103	104	97	81	81	96
Oxygen balance	100	99	100	76	77	97
Overall mass balance	100	98	101	83	84	99
Total oil feed, ml	1794	3860	3890	2683	1872	2123
Wet Product Analysis						
H/C atomic	1.47	1.47	1.56	1.56	1.48	1.58
oxygen, percent	24.6	30.8	32.7	32.7	31.4	34.2
Oxygen rejection, %	79	70	57	59	62	55

Effect of Hydrogen. A series of experimental conditions were tested in order to evaluate the need for adding hydrogen to the reaction vessel. This series of experiments was performed at the highest oil flow rate tested in the previous section. In the series, the hydrogen flow was reduced in steps to zero then the pressure maintained in the reaction vessel was reduced in steps to only 100 psig. Operations continued successfully throughout the series of conditions without coking and plugging of the reaction vessel until the experiment was terminated and oil flow was stopped. At that point a combination of the thermal inertia in the heaters and exothermic reactions in the vessel led to a temperature increase of over 50 °C and the oil in the vessel coked to a solid.

In addition to the product quality data given in Table 6 there were other indicators of a general product quality loss as the amount of hydrogen in the system was reduced. The viscosity of the product increased from 14,200 centipoise @ 60 °C for the products at the higher hydrogen flows to 18,700 centipoise @ 60 °C for product with low hydrogen flow to 32,700 centipoise @ 60 °C for the product with no hydrogen flow. These viscosities range upward from those measured for the high-pressure oils and are about three orders of magnitude higher than the viscosity measured for the crude pyrolysis oil. An increase in density was also found to correlate with the reduction in hydrogen flow. Densities increased from 1.14 to 1.16 to 1.18 g/ml @ 20 °C for the products of higher flow, lower flow and no flow of hydrogen, respectively. These densities again range upward from that measured for a high-pressure oil while remaining less than the typical density of a pyrolysis oil.

Hydrotreatment to Gasoline

The results given above indicate that the low-temperature catalytic treatment of pyrolyzate transforms the primary pyrolyzate oils into a chemical composition similar to that of the high-pressure oils. Further analysis has confirmed this conclusion.

Product Potential for Catalytic Hydrotreatment. The 258 °C product has been distilled to recover 2% light hydrocarbon, 29% distillate and 59% residual material with 9.5% water dissolved in the oil. The distillation was taken to an endpoint of 205 °C @ 20 mmHg. The presence of small water droplets in the condensate at temperatures approaching the end point indicated that thermal cracking of the oil was occurring and the distillation was terminated. The residual material was still fluid and had not coked at a pot temperature of 280 °C. This behavior contrasts sharply with the thermal decomposition and coke formation at <200 °C experienced with the pyrolyzate feedstock and is more similar to the behavior of the high-pressure oils.

TABLE 6. Hydrotreating Results as a Function of Hydrogen Flow

Experimental Operating Conditions				
Catalyst	CoMo	CoMo	CoMo	CoMo
Temperature, ^{o}C	270	277	276	268
Pressure, psig	2040	2028	2010	200
Oil feed rate, ml/hr	1440	934	959	1062
Hydrogen rate, L/hr	120	40	0	0
LHSV, vol oil/vol cat/hr	1.60	1.04	1.07	1.18
Residence time, min	23	37	37	33
Experimental Results and Product Analyses				
Hydrogen consumption, L/L oil	28	26	0	0
Carbon Conversion, wt %				
to oil/aqueous	87/11	83/9	85/7	85/10
to gas (C_1 to C_4)	4	10	10	5
Oil product yield, ml/ml feed	0.70	0.61	0.61	0.68
Carbon balance	102	102	103	100
Hydrogen balance	96	92	92	101
Oxygen balance	97	97	94	99
Overall mass balance	99	99	98	100
Total oil feed, ml	2123	1900	1618	1300
Wet Product Analysis				
H/C atomic	1.58	1.44	1.33	1.62
oxygen, percent	34.2	30.7	29.5	35.8
Oxygen rejection, %	55	61	62	48

The thermal stability of the low-temperature, hydrotreated product oil as well as its elemental composition (70.7%C, 8.1%H, 0.1%N, 20.9%O, calculated to a dry basis) indicate that it has been significantly upgraded from the original pyrolyzate. Chemical composition analysis by GC/MS was performed on one sample and the identified components are listed in Table 7. The carbonyl side chains which could be a major source of polymerization of the phenolics have been destroyed. Unsaturated alkyl side chains (propenyl) have been saturated. The relative amount of phenolic material appears to have increased at the expense of the phenolic ethers. Saturated cyclic alcohols (cyclohexanols) are also present indicating some hydrogenation of the aromatic rings. Also detected were pure hydrocarbon compounds such as the tetralins (tetrahydronaphthalenes). The acid component is largely removed into the aqueous phase and will no longer interfere in the hydrotreating process. Based on these chemical changes it is reasonable that the partially hydrotreated pyrolyzate can be further processed at higher temperatures by more conventional hydrotreating techniques to produce hydrocarbon fuels.

TABLE 7. Components Identified in Low-Temperature Hydrotreated Pyrolyzate

Major*	Minor*
dimethoxyphenol (syringol)	methylcyclohexanol (2 isomers)
hydroxymethoxybenzoic acid	methylphenols (3 isomers)
propylsyringol	ethylphenols (2 isomers)
ethylsyringol	dimethylphenol
propylguaiacol	phenol
methylguaiacol	cyclohexandiol (2 isomers)
ethylguaiacol	methyltetralins (4 isomers)
methoxyphenol (guaiacol)	ethyl/dimethyl tetralins (2 isomers)
cyclohexanol	3 and 4 carbon substituted phenols (4 isomers)
	indan

* based on relative areas of flame ionization detector peaks, not strictly quantified.

Results of Catalytic Hydrotreatment of Low-Temperature Treated Pyrolyzate

Test results now confirm the production of gasoline-range hydrocarbons from the low-temperature, catalytically-treated pyrolyzate. Catalytic hydrotreating conditions and processing results are nearly the same as those reported earlier for hydrotreating the TR12 high-pressure biomass liquefaction product.(9) In our continuous-feed reactor system, using sulfided cobalt-molybdenum catalyst, at conditions of approximately 350°C and 2000 psig, carbon conversion to gasoline was in excess of 80 weight percent. Product yield was somewhat less for the treated pyrolyzate compared to TR12 (0.77 ml/ml versus 1.0 ml/ml) due primarily to the higher oxygen content of the feedstock. The rate of conversion was slightly higher. Hydrogen consumption was measured at 728 liters per liter of oil feed (4085 SCF/bbl feed) based on gas phase measurements in and out of the reactor. Space velocity in our upflow reactor was relatively low (0.07 volume oil/volume catalyst/hr).

CONCLUSIONS

The main conclusion of this research is that a second route from wood to gasoline via catalytic hydrotreating has been identified. Whereas our earlier reports illustrated the high-pressure route from wood to crude oil to hydrotreated gasoline, this paper suggests a two-step hydrotreating process to convert wood pyrolyzate first to a more thermally-stable tar and then to the hydrotreated gasoline product. Others have demonstrated the high yields of liquid product obtainable by flash pyrolysis of wood flour; our research shows a reasonable processing route to produce gasoline in high yields from the pyrolyzate. The first hydrotreatment step is performed at lower temperature (270°C) and pressures of 2000 psig or less. Hydrogen consumption is relatively low but the combination of thermal reactions and catalytic reactions is sufficient to transform the pyrolyzate in high yields into a useful feedstock for catalytic hydrotreatment to gasoline.

The product yield from this low-temperature hydrotreatment is 85% on a carbon basis with the carbon losses primarily as water-soluble organics and carbon dioxide gas. The second hydrotreatment step to gasoline has a similar yield on a carbon basis with the losses confined almost exclusively to the gas phase, mostly as hydrocarbon gases. The gasoline product from this type of hydrotreatment consists of cyclic and aromatic compounds.

ACKNOWLEDGMENT

The authors acknowledge the technical support of W. F. Riemath and G. G. Neuenschwander who assisted with reactor operations and product analyses. They also acknowledge the financial and managerial support of the Biomass Thermochemical Conversion Program Office at PNL and the Biofuels and Municipal Waste Technology Division of the U.S. Department of Energy.

REFERENCES CITED

1. Beckman, D. and Elliott, D. C., "Comparison of the Yields and Properties of the Oil Products from Direct Thermochemical Biomass Liquefaction Processes," Can. J. Chem. Eng. 63, 99-104 (1985).

2. Chantal, P., Kaliaguine, S., Grandmaison, J. L., and Mahay, A., "Feasibility of the Production of Hydrocarbons from Aspen Poplar Pyrolytic Oils Over H-ZSM5" in Comptes Rendus de l'Atelier de Travail sur la Liquefaction de la Biomasse, pp. 234-248. NRCC 23130. National Research Council of Canada, Ottawa, (1983).

3. Diebold, J. P. et al, "Low-Pressure Upgrading of Primary Pyrolysis Oils from Biomass" in Proceedings of Energy from Biomass and Wastes X pp.801-830, Chicago, Illinois, (1986).

4. Elliott, D. C. "Bench Scale Research in Biomass Direct Liquefaction" in Proceedings of the 16th Biomass Thermochemical Conversion Contractors' Meeting, pp. 399-422, #CONF-8405157, National Technical Information Service, Springfield, Virginia, (1984).

5. Elliott, D. C., "Analysis and Upgrading of Biomass Liquefaction Products" Volume 4 of Final Report, IEA Cooperative Project D-1 Biomass Liquefaction Test Facility Project #PNL-4931, National Technical Information Service, Springfield, VA, 1984.

6. Elliott, D. C., "Analysis and Comparison of Biomass Pyrolysis/Gasification Condensates - An Interim Report." #PNL-5555, National Technical Information Service, Springfield, Virginia, (1985).

7. Elliott, D. C. and Baker, E. G., "Biomass Liquefaction Product Analysis and Upgrading" in Comptes Rendus de l'Atelier de Travail sur la Liquefaction de la Biomasse, pp. 176-183. NRCC 23130, National Research Council of Canada, Ottawa, (1983).

8. Elliott, D. C. and Baker, E. G., "Upgrading Biomass Liquefaction Products Through Hydrodeoxygenation." Biotech. and Bioeng., Symp. No. 14, 159-174, (1984).

9. Elliott, D. C. and Baker, E. G., "Hydrodeoxygenation of Wood-Derived Liquids to Produce Hydrocarbon Fuels" in Proceedings of the 20th Intersociety Energy Conversion Engineering Conference, pp. 1.586-1.592, #SAE P-164, Society of Automotive Engineers, (1985).

10. Elliott, D. C. and Baker, E. G., "Upgrading Biomass Liquid Fuels" in the <u>Proceedings of the 17th Biomass Thermochemical Conversion Contractors' Meeting</u>, #CONF-8510167, National Technical Information Service, Springfield, Virginia, (1986).

11. Evans, R. J., Milne, T. A., and Soltys, M. N., "Fundamental Pyrolysis Studies" in <u>Proceedings of the 16th Biomass Thermochemical Conversion Contractors' Meeting</u>, pp. 349-373. #CONF-8405157, National Technical Information Service, Springfield, Virginia, (1984).

12. Furimsky, E., <u>Catal. Rev.-Sci. Eng.</u> 25, 3, 421, (1983).

13. Knight, J. A., Gorton, C. W., Kovac, R. J., "Entrained Flow Pyrolysis of Biomass" in Proceedings of the 16th Biomass Thermochemical Conversion Contractors' Meeting, pp. 287-297. #CONF-8405157, National Technical Information Service, Springfield, Virginia, (1984).

14. Scott, D. S. and Piskorz, J., "The Continuous Flash Pyrolysis of Biomass," <u>Can. J. Chem. Eng.</u>, <u>62</u>, 404-412 (1984).

15. Soltes, E. J., and Lin, S-C. K. <u>Biotech. and Bioeng., Symp. No. 13</u>, 53-64 (1983).

DISCUSSION

D. Boocock, U. of Toronto (Canada)
When you run the reactor with what you call a high pressure oil, and you sweep out the volatile products, what happens in the reactor to what's left? Is there a deterioration in the material left in the reactor?

Elliott
That's exactly the question we want to answer with longer term tests. First, coke is laid down on the catalyst, in the range of 10 to 20%, which is fairly typical for active hydro-treating catalysts. Secondly, we have seen some coke formation in the reactor bed. This is related to the fact that we have one source of entrance of the fluid into the bottom of the reactor. We tend to get a cone-shaped open area where the gases and the high viscosity material tend to move up and we get coke forming in the side of the reactor where we believe it's stagnant. We don't know if that will continue over time and is eventually closed off, or if that's just a function of the stagnant area in the reactor. The material that we drain out of the reactor isn't set up into a coke and has not become extremely viscous. It appears to be what we see when overflowing the reactor — an intermediate type of product. We don't believe that we have a buildup of high molecular weight material, but we need to study it a little more.

D. Meier, Institute of Wood Chemistry (W. Germany)
Do you think that it is essential to use a sulfided catalyst as your feedstock contains no sulfur? How long can you maintain your catalyst in the reactor, because no sulfur is added from the feedstock?

Elliott
That's a good question and one that we have been concerned with from the beginning. We found that the sulfided catalyst is an order of magnitude more active than just the oxide form. It stays active throughout the runs that we have done so far, up to 10 hours. I can also report that other researchers, at the U. of Utah in particular, have done more extensive studies with model compounds. They conclude the same thing; the sulfided catalyst stays active. They have also reported that if they try to sulfide by putting a carbon disulfide stream in with the feedstock, the

activity of the catalyst is a little bit less, so that the initiallly sulfided form appears to stay active. We tried one test where we injected the sulfide into the system after we shut down, and found very little resulfiding-type activity; the exotherm was not there. We started backup and found only that we had a little more sulfur coming out in our product for an hour or two. Then, sulfur decreased to the low level that we typically see. We don't think it's a problem. It seems to be working very well for us.

Boocock — What is the nature of the support you have for the cobalt-moly catalysts? The reason I ask is that we have been using nickel-moly catalyst with an alumina support for model compounds, and we find alumina is beneficial. The methoxy group on the benzene rings comes off as some kind of active form of methanol, and the methyl group then goes back on the ring. Anisole yields mixtures of benzene and toluene. We think it's catalyzed by the alumina, not the metal content.

Elliott — That's very interesting. We have alumina supported catalysts. We have done some work with silica-alumina catalysts and didn't detect any difference between the two. As to the catalytic effect of the alumina, we think it's not doing much. We saw some of what you're talking about in our early work with model compounds, but we attributed it to nickel activity as opposed to the support.

Boocock — If you don't detect large quantities of simple alcohols such as methanol, it would be a good indication that those are getting back on the aromatic rings, and are alkylating them.

DIRECT HYDROLIQUEFACTION OF SPRUCE WOOD INTO LIGHT- AND MIDDLE-DISTILLATE OILS

Dietrich Meier, Ph.D., Associate Scientist

Karl Fuchs, M.S., Associate Member

Oskar Faix, Ph.D., Chief Department

Federal Research Centre for Forestry and Forest Products
Institute of Wood Chemistry and Chemical Technology of Wood
Leuschnerstr. 91, D-2050 Hamburg 80,
Federal Republic of Germany

ABSTRACT

A slurry oil process has been developed for the direct catalytic hydrogenation of spruce wood. Wood particles are mixed together with the high boiling recycle oil fraction (b.p. > 285°C) from the process. A specially designed autoclave system consisting of a reactor, hot separator and cooling device is used for the experiments. 37.8 % of the wood can be converted into a light and middle distillate oil fraction (net product oil, NPO) which is almost completely soluble in n-hexane. 40 % of the NPO have a boiling range between 60° and 220°C whereas 95 % of the oil can be distilled at 355°C. The average molecular weight ($\overline{M}_w$) as determined by HPGPC is about 122 dalton. Other important oil characteristics are: viscosity (20°C) 8.3 cP, caloric value 40 MJ/kg, density 0.92. Comprehensive analytical work has been done to characterize the NPO and the gases produced. The analytical results obtained by capillary GC/MS, chemical group separation and HPLC separations, the liquefaction procedure, isolation of products as well as results of other different analyses are presented and discuss.

DIRECT HYDROLIQUEFACTION OF SPRUCE WOOD INTO LIGHT- AND MIDDLE-DISTILLATE OILS

INTRODUCTION

Over the last decade many thermochemical methods have been studied as possible routes to convert solid lignocellulosic biomass into liquid products. The aim has been the production of 1) a fuel from renewable feedstocks as substitute for petroleum or 2) valuable liquid intermediates for the chemical industry. After having started with aqueous systems using a carbon monoxide atmosphere and Na_2CO_3 as catalyst (1,2,3) the investigations aimed at maximizing liquid yields focused on high pressure hydrogenation processes (4,5,6,7). An overview about the most promising techniques, developed within the International Energy Agency (IEA) is given by BECKMAN and ELLIOTT (8). Besides three low pressure pyrolytic processes they present eight high pressure processing configurations four of which do not use a catalyst. The other catalytic conversion methods use either Na_2CO_3 or $NiCO_3$ in the presence of CO, N_2, H_2 or CO/H_2. Reaction temperatures range between 330 and 420°C, and pressures between 5 and 28 MPa.

Chemical and physical properties of the crude oils produced differ widely due to feedstock effects and process severity. Important oil characteristics such as asphaltene content, distillation behaviour, oxygen content, heating value, viscosity, and pour point do not meet the standards of the petrochemical industry. Hence, the oils need to be improved by a second catalytic "upgrading" procedure.

Based on experiences with the German direct coal conversion technique a catalytic hydrogenation process for the conversion of spruce wood was developed using a process recycle oil as carrier and a commercial palladium on active charcoal catalyst. Process characteristics and product analyses will be presented in this paper.

EXPERIMENTAL

Liquefaction Procedure and Isolation of Products

Experimental equipment is a versatile reactor system and consists basically of three 1-l capacity autoclaves which are combined by heated tubes. A schematic flowsheet of the system is shown in Figure 1.

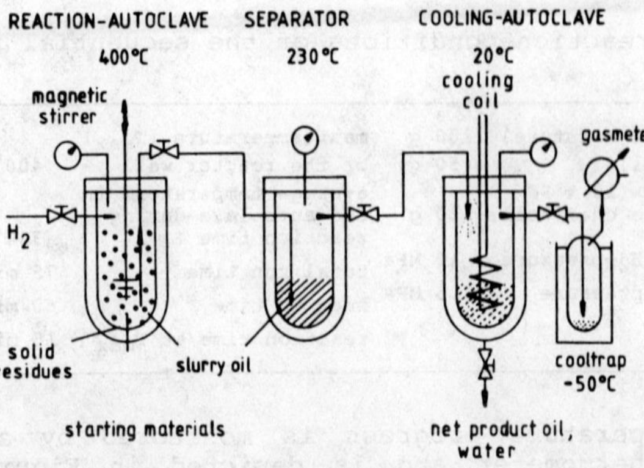

Figure 1 Schematic flowsheet of the reactor system

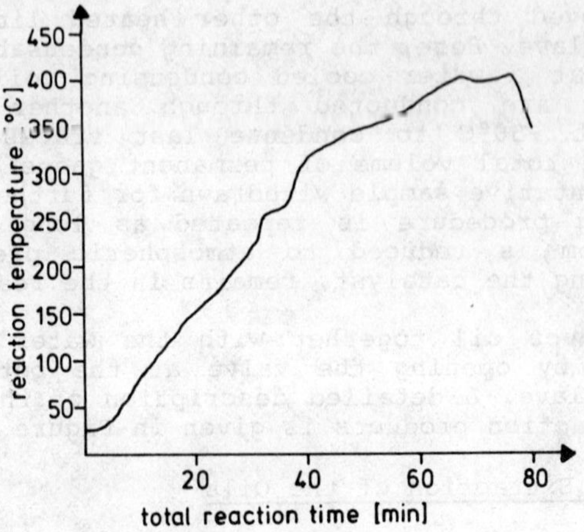

Figure 2 Temperature conditions within the reaction autoclave during the experiments

Liquefaction experiments were conducted in a batch mode. Air-dry spruce wood particles (~ 7% moisture) are mixed together with the catalyst and the carrier oil which comprises the high boiling oil fraction (b.p. > 285°C) of the previous run. After having filled the reactor with the prepared mixture the whole system is flushed twice with hydrogen to remove the air. The reactor hydrogen pressure is adjusted at 10 MPa and the input gas volume is measured. A magnetic mixer provides an intimate contact between solids and the gas. The general reaction conditions are given in Table 1.

Table 1

General reaction conditions of the sequential runs

wood (7 % moisture)	100 g	max.temperature (T_{max}) of the reactor wall	400°C
slurry oil	50 g	average temperature in the autoclave during reaction time	380°C
catalyst (10 % Pd on active char coal)	10 g	total run time	75 min
initial H_2-pressure	10 MPa	heat up time	60 min
maximum pressure	21.5 MPa	reaction time at T_{max}	15 min

The temperature progress is monitored by a Pt 100 resistance thermometer and is depicted in Figure 2. At the end of the run the reaction products are rapidly quenched to approximately 250°C by passing them through the heated line into the hot separator which is preheated to 230°C. Higher boiling products condense in this vessel and after 15 s the uncondensed vapors from the hot separator are removed through the other heated line into the cooling autoclave. Here, the remaining condensable products are trapped at a water cooled condensing coil. The permanent gases are conducted through another cool trap maintained at -50°C to condense last traces of liquid products. The total volume of permanent gases is measured and a representative sample withdrawn for further analyses. The quenching procedure is repeated as long as the gas in the system is reduced to atmospheric pressure. The char, including the catalyst, remains in the reactor.

The product oil together with the water can easily be collected by opening the valve at the bottom of the cooling autoclave. A detailed description of the isolation of the liquefaction products is given in Figure 3.

Liquid-liquid Extraction of the Oils

A liquid-liquid extraction procedure was developed based upon a combination of solubility and acid-base properties of the biomass oils. Figure 4 demonstrates the separation of the product oils into strongly acidic, weakly acidic, and neutral fractions.

Molecular Weight Distribution

Molecular weight (MWt) distribution of the oils were obtained by means of high performance gel permeation chromatography (HPGPC). A stainless steel column (60 cm x 7.7 cm) packed with poly-(styrene / divinylbenzene) spherical particles was used. A recent paper describes the technique and calibration in detail (9).

Capillary GC/MS studies

Separation and identification of individual compounds of the product oils were performed with a KRATOS GC/MS

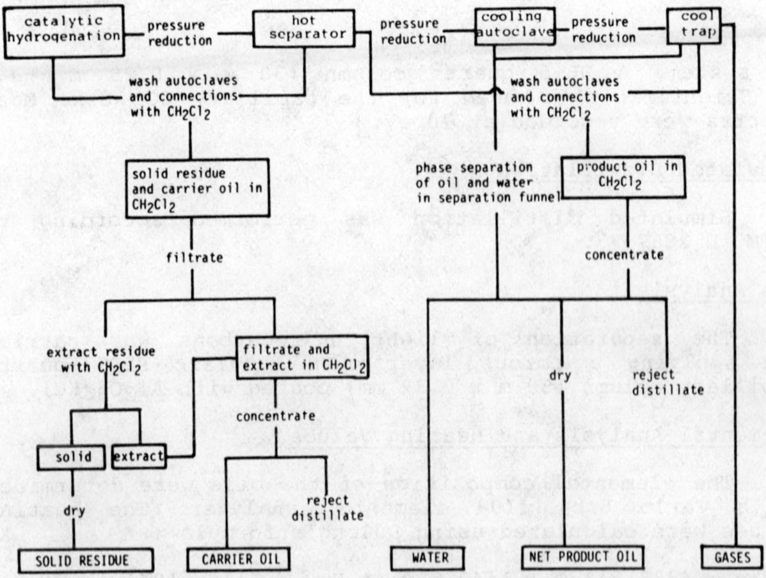

Figure 3 Scheme of the product isolation procedure

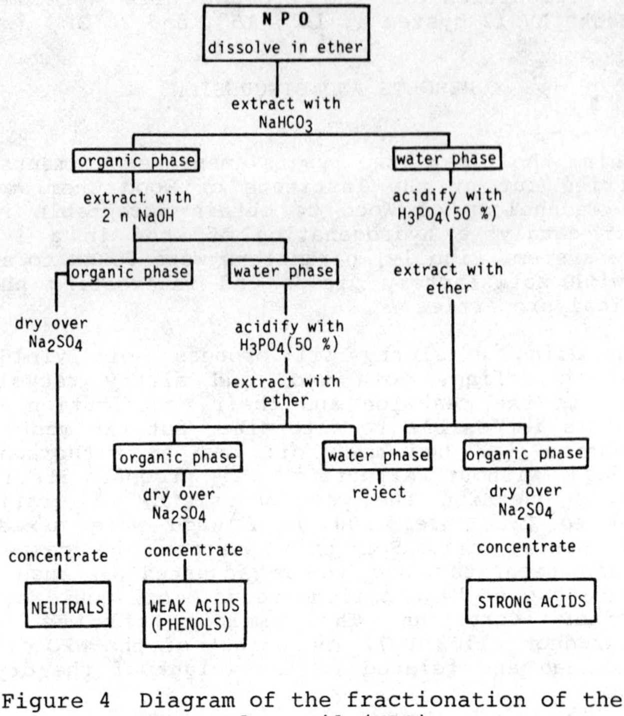

Figure 4 Diagram of the fractionation of the net product oil (NPO)

25 system. A DB-5 quartz column (30 m x 0.25 mm, J.& W. Scientific) was used for the capillary GC work. Mass spectra were recorded at 70 eV.

Simulated Distillation

Simulated distillation was performed according to ASTM D 2887-73.

Gas Analysis

The separation of light hydrocarbons was carried out applying a porous layer open tubular (PLOT)-quartz capillary column (50 m x 0.32 mm) coated with Al_2O_3/KCl.

Elemental Analysis and Heating Values

The elemental composition of the oils were determined on a Carlo Erba 1104 elemental analyzer. The heating values were calculated using Dulong's formula :

$$MJ/kg = (338.2 \% C + 1442.8 \times (\% H - \frac{\% O}{8})) \times 10^{-3}$$

Viscosity

The viscosities of the product oils were measured with a Haake RV 12 system at 10°, 15°, and 20°C.

RESULTS AND DISCUSSION

Within the last two years many experiments have been carried out at the Institute of Wood Chemistry and Chemical Technology of Wood to obtain acceptable results by direct catalytic hydrogenation of wood in a 1 liter-autoclave system. Finally, parameters were found to achieve an oil with satisfactory yields and interesting physical and chemical properties.

When using a slurry oil process oil yields are difficult to define. Both wood and slurry recycle oil take part in the reaction and their contribution to the oil yield is impossible to determine. But the most important question is, how much oil can be withdrawn from the process without affecting the process itself. To answer this question the wood to slurry oil ratio was optimized to 2:1, i.e. 100 g of wood were mixed with 50 g of slurry oil. For this purpose the temperature of the hot separator had to be adjusted in such a way that about 50 g of high boiling recycle oil could condense in it after every run. The remaining oil was removed as net product oil (NPO). The weight of the NPO fraction was determined and related to the weight of the dry biomass.

The NPO yield (%) was calculated as follows:

$$\% \text{ NPO yield} = \frac{\text{oil collected in the cooling autoclave,gx100}}{\text{wood feed, g}}$$

Yields of slurry oil and water were determined in the same manner. The yield of solids was estimated as follows:

$$\% \text{ solids} = \frac{\text{(solids, g - catalyst feed,g) x 100}}{\text{wood feed, g}}$$

Table 2 shows yields of NPO, slurry oil, solids and wastes.

Table 2

Percentage yields based on o.d. wood of NPO, slurry oil, solids, and water of three sequential runs (conditions see Table 1)

1 NPO (%)	2 Slurry (%)	3 1 + 2 (%)	4 Solids (%)	5 H_2O (%)
39.0	53.9	92.9	3.1	28.5
36.5	55.5	92.0	4.2	29.0
37.8	53.7	91.5	6.2	31.1

It is seen from Table 2, that yields are reproducible within certain limits. On an average 37.8 g of oil could be produced by catalytic hydrogenation of 100 g of spruce wood. The NPO yield is in the range of oil yields reported from other processes (8). As the slurry oil yield was kept constant it was possible to simulate a continuous operation. Considering the solid's yield an average conversion rate of 96,5 % can be estimated.

Generally, liquefaction of biomass aims to remove as much oxygen as possible from the raw material in form of gas (CO_2, CO) or water. With regard to the oxygen content of dry spruce wood (wt 43.9 %) and the slurry oil (wt 7.5 %) the input oxygen weight is 47.6 g, i.e. 53.6 g of water can theoretically be produced by the reaction. As Table 2 indicates about 55.1 % of the possible theoretical value are obtained, i.e. 26.2 g oxygen are converted to water. Regarding the wt% O values in Table 4 it can be seen that 4.5 g O remain in the NPO. The residual oxygen is distributed in the high boiling recycle oil, the gas, and the solids.

The NPOs of each run were collected and a sample thereof was analyzed at the petrochemical laboratories of the German SHELL Company in Hamburg. Additional measurements like cP-viscosity and n-hexane solubility were performed at the Institute of Wood Chemistry und Chemical Technology of Wood. Results of the analyses are presented in Table 3.

Table 3

Some properties of the net product oil

density, g/ml	0.9263
refraction index (20°C)	1.4879
viscosity, cSt	3.88 (20°C)
	2.44 (40°C)
viscosity, cP	11.5 (10°C)
	9.0 (15°C)
	8.3 (20°C)
pour point	-24°C
solubility in n-hexane	99 %

In comparison with results of other processes, summarized by BECKMAN and ELLIOTT (8) NPO properties are excellent. Due to the relative low density of the NPO it could be collected as a separate phase from the cooling autoclave. When the oil was exposed to air it changed its colour from light yellow to dark red within a few minutes indicating that oxidation reactions occur with the atmosphere.

In the literature concerning liquefaction of biomass the expression "oil" is often used for every organic liquid product that leaves the reactor regardless of its properties. The so called "oil" is often a sticky brown tar, soluble in acetone, ether or at the utmost in dichloromethane (10,11). However, the NPO is almost completely soluble in n-hexane which emphasizes its unpolar character. Hence, its oil-like properties seem to be well suited for further utilization.

Further important oil criteria are heating value and atom ratios which all can be calculated from the data of elemental analysis. Table 4 shows these results.

Table 4

Elemental composition, heating value and atom ratios of the NPO from three sequential runs (conditions see Table 1)

wt % C	wt % H	wt % O	H/C	O/C	heating value MJ/kg
77.8	11.0	12.2	1.7	0.12	40.1
77.6	10.7	11.7	1.6	0.11	39.7
77.4	10.6	12.0	1.6	0.12	39.4

Elemental analysis of spruce wood showed it to contain 49.9 % C, 6.2 % H and 43.9 % O. Obviously a good hydrogenation effect can be obtained which is reflected by the very high heating value of the NPO in comparison to results of other processes (8,12). The atomic ratio diagram in Figure 5 demonstrates the big step towards desirable analytical properties of liquid hydrocarbons from biomass.

The boiling point distribution is a crucial characteristic for biomass oils. Figure 6 shows the distillation curves of high-pressure liquefaction products from different processes which were comparatively evaluated by BECKMAN and ELLIOTT (8) as well as the distillation curve of the NPO. It should be mentioned that the oils were injected directly into the gas chromatograph without diluting them with a solvent. This was possible because of the low viscosity and polarity of the product oil.

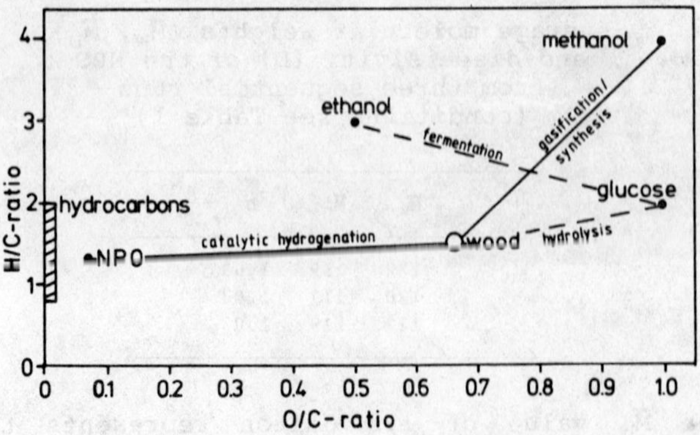

Figure 5 Atomic ratio diagram of different products obtained by wood conversion

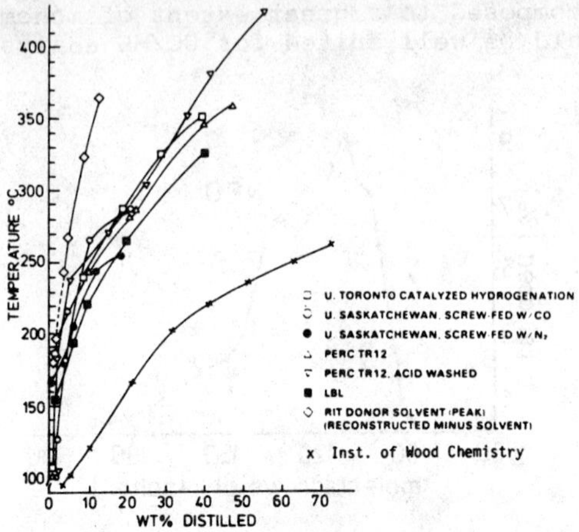

Figure 6 Distillation curves of high-pressure liquefaction products (except x all curves from (8)

According to WOLFRAM et al. (13) oils can be classified as a function of their boiling point ranges. Light-distillate oils have a maximum boiling point at 220°C. Middle-distillate oils range between 220°C and 360°C whereas oils with a boiling point of 360°C are called heavy-distillate oils. If one applies this definition to the NPOs they are composed of 40 % light-distillate oil and 55 % middle-distillate oil.

The amount of low boiling point fractions should also influence the average molecular weight ($\overline{M}$Wt) of the NPO. Therefore the weight average MWt ($\overline{M}_w$) and number average MWt ($\overline{M}_n$) were determined by rapid HPGPC. The results obtained are given in Table 5.

Table 5

Average molecular weights ($\overline{M}_w$, $\overline{M}_n$) and dispersivity (D) of the NPO from three sequential runs (conditions see Table 1)

$\overline{M}_w$	$\overline{M}_n$	D
122	119	1.03
120	117	1.03
122	119	1.03

The $\overline{M}_w$ value of 122 dalton represents the lowest value ever reported from whole biomass oils (14). The dispersivity ($D = \overline{M}_w/\overline{M}_n$) of 1.03 means that there is a very narrow distribution of MWt. This is demonstrated by the distribution curve in Figure 7. The main range includes $\overline{M}$Wts from about 90 to about 200 dalton, i.e., the NPO is composed to a great extent of monomeric substances and should be well suited for GC/MS analysis.

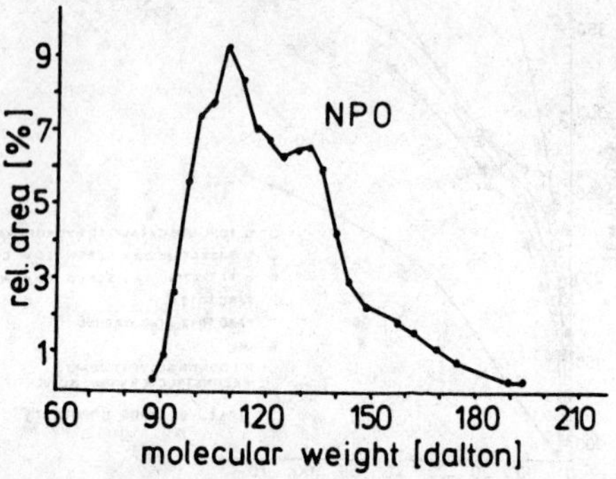

Figure 7 Molecular weight distribution curve of the net product oil

Prior to the GC/MS studies the oils were fractionated into a weakly acidic (phenolic), neutral, and strongly acidic fraction. A diagram of the separation steps is presented in the experimental section. The results of the liquid-liquid extraction procedure are presented in Table 6.

Table 6

Percentage yields based on NPO of liquid-liquid extraction fractions from oils of three sequential runs (conditions see Table 1)

Neutrals	Phenols	Acids	Lossses
63.0	19.05	7.0	10.5
57.0	10.05	8.5	15.0
59.0	20.5	6.0	15.0

The effectivity of the hydrogenation reaction can be deduced from Table 6. Neutral compounds represent the main fraction of the NPO whereas the typical phenolic character of biomass oils is decreased.

Neutral and phenolic fractions were selected for GC/MS analysis. Acids were not analysed. Figure 8 shows the total ion chromatogram (TIC) of the neutral fraction. The main peaks in the TIC belong to n-alkanes of which n-C 17 represents the most important one. Further hydrocarbon group type separation led to the following list of compound classes which is shown in Table 7.

Table 7

Main chemical compound classes of the neutral fraction of the NPO

n-alkanes (C_7-C_{23})	tetrahydro-naphthalenes
C_1-C_7 alkylated benzenes	dihydroxy-benzenes
	naphthalenes
aliphatic ketones	phenanthrenes
cyclic ketones	
dihydro-indenes	

BECKMAN and ELLIOTT (8) reported that straight-chain hydrocarbons (C_8 to C_{26}) could only be detected in peat derived oils and were not present in wood liquefaction products. However, these GC/MS studies revealed that n-alkanes form the main part of the neutral compounds. Further investigations must be done to explain the formation of these unusual wood degradation products. It is supposed that the slurry oil has an effect on the formation of alkanes. The other substance classes mentioned in Table 7 were already reported by other authors (15,16).

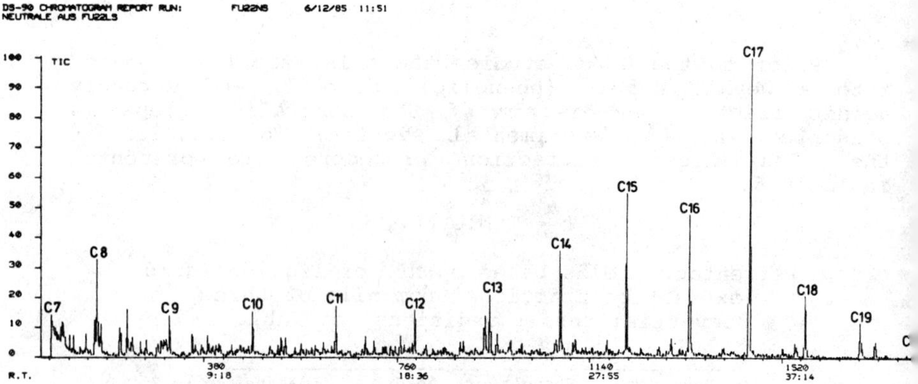

Figure 8 Reconstructed total ion chromatogram of the neutral fraction of the net product oil. (C-numbers refer to straight-chain n-alkanes)

Figure 9 presents the TIC of the phenolic fraction. The main peaks in the chromatogram are identified. The predominant compound is p-cresol which includes small amounts of m-cresol, as the column used does not separate these two isomers. However, single mass chromatograms of mass 107 (greater in p-cresol) and 108 (greater in m-cresol) revealed that peak no. 3 represents only traces of m-cresol. It is remarkable that only guaiacol could be detected in small amounts although guaiacol derivatives are normally the main degradation products of softwood lignins. Hence, the composition of the phenolic NPO fractions is relatively simple compared e.g. with phenolic

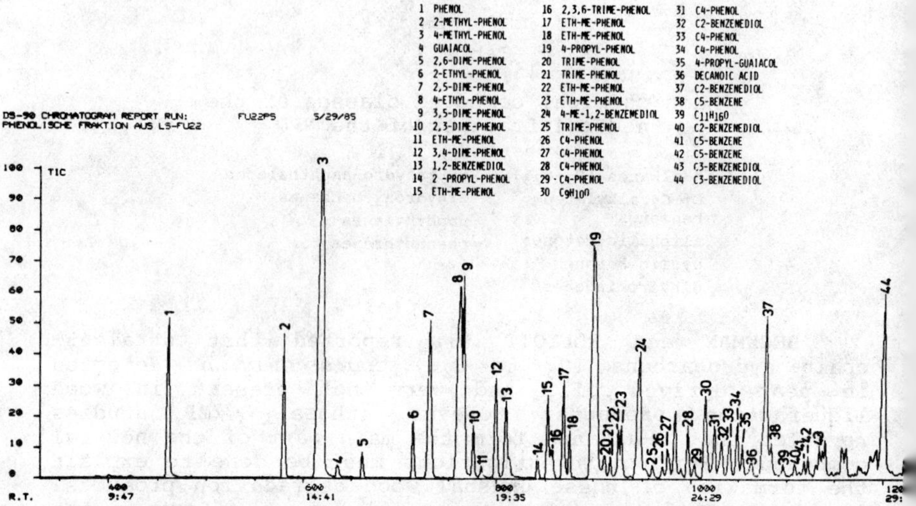

Figure 9 Reconstructed total ion chromatogram of the phenolic fraction of the net product oil (Peak number refer to component list obtained by GC/MS analysis)

degradation products from pyrolytic processes (17). Therefore it might be advantageous to isolate the main phenols by distillation.

After having characterized the NPO the gas composition was analyzed by capillary GC. The corresponding FID-chromatogram is shown in Figure 10. Only hydrocarbons were detected. It is seen that n-C_1-C_4 form the main products. Despite of the hydrogenating conditions some unsaturated gases could be separated. Further gas analysis will focus attention on the quantitation of all single compounds to establish a mass balance and to determine the hydrogen consumption.

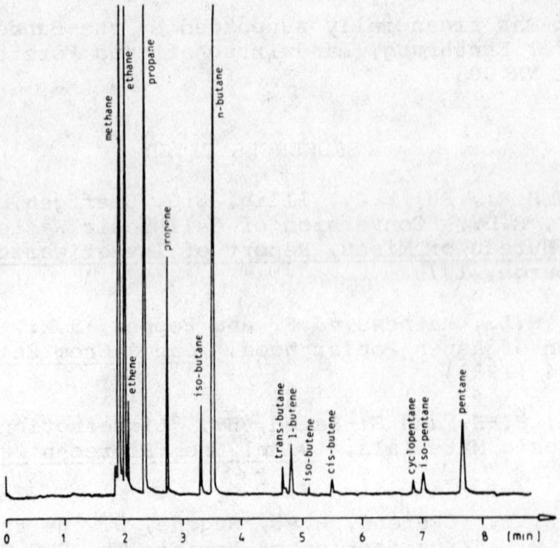

Figure 10 Gas chromatogram of the light hydrocarbons

CONCLUSIONS

Direct catalytic hydrogenation of spruce wood to light- and middle-distillate oils has been successfully demonstrated by use of a slurry oil process and palladium on active charcoal as catalyst. From the results of the experiments it is concluded that the presence of slurry oil, the type of catalyst and the rapid quenching of the products are responsible for the good yields and properties of the oils. Slurry oil helps to dissolve degradation products resulting from the thermal breakdown of wood. Palladium is known to be a good hydrogenating catalyst. On the other hand, it is an expensive material which must be collected and reactivated. The reactor system of three 1 l-capacity autoclaves makes it possible to stop the reaction at a given time and to remove the products immediately from the hot reactor. Thereby further condensation and recombination reactions are avoided.

Chemical composition of the oils indicates that they can be used either as a fuel or as a source for chemical feedstocks.

In the future laboratory scale research will focus attention on selecting and testing other catalysts to improve oil yields and to decrease process costs. Furthermore, lignins should be used as feedstock because they can be expected to reduce undesirable gas and water production, and to show a positive effect on composition and yield of the oils.

ACKNOWLEDGEMENTS

This work was financially supported by the Bundesministerium für Ernährung, Landwirtschaft und Forsten, project number 81 NR 006.

REFERENCES CITED

1. Appel, H.R., Fu, Y.C., Illig, E.G. Steffgen, F.W. and Miller, R.D., "Conversion of Cellulosic Wastes to Oil," Bureau of Mines, Report of Investigation 8013, Pittsburgh, 1975

2. Eager, R.L., Mathews, J.F. and Pepper, J.M., "Liquefaction of Aspen Poplar Wood," Can.J.Chem.Eng. 60, 289-294 (1982)

3. Oelert, H.-H. and El-Saied, H., "Liquefaction of Lignocellulosic Materials," J.Apl.Chem.Biotechn 28, 486-494 (1978)

4. Chornet, E., Overend, R.P., Eugene, D., Heitz, M. and Rubio, M., "Liquefaction of Poplar: The UDS Process Development Unit and Inital Product Characterization," 3rd Meeting of Liquefaction Specialists, Sherbrooke, Quebec, Canada," (1983) September 29-30.

5. Boocock, D.G.B., Mackay, D., Franco, H. and Lee,P., "The Production of Synthetic Organic Liquids from Wood Using a Modified Nickel Catalyst," Can.J.Chem. Eng. 58, 466-469 (1980)

6. Björnbom, P., "Peat and Wood Hydroliquefaction," 33rd Chem. Eng. Conf. of Canad. Soc. for Chem. Eng. (1983) October 2-5.

7. Eager, R.L., Mathews, J.F. and Pepper, J.M., "A Small-Scale Semi-Continuous Reactor for the Conversion of Wood to Fuel Oil," Can.J.Chem.Eng. 61, 189-193 (1983)

8. Beckman, D. and Elliott, D.C., "Comparisons of the Yields and Properties of the Oil Products from Direct Thermochemical Biomass Liquefaction Processes," Can. J.Chem.Eng. 63, 99-104 (1985)

9. Meier, D., Larimer, D.R. and Faix, O., "Direct Liquefaction of Different Lignocellulosics and their Constituents, II. Molecular weight determination, gas chromatography, IR-Spectroscopy," Fuel, in press

10. Russel, J.A., Molton, P.M. and Landsman, S.D., "Chemical Comparisons of Liquid Fuel Produced by Thermochemical Liquefaction of Various Biomass Materials," Proceedings of the 3rd Miami International Conference on Alternative Energy Sources, 307-322 (1983)

11. Wilhelm D.J., Kam, A.Y. and Stallings, J.W., "Transportation Fuel from Biomass by Direct Liquefaction and Hydrotreating," Symp. Paper Energy from Biomass and Wastes V, Lake Buena Vista, Florida, USA, (1981) January 26-30.

12. Moffatt, J.M. and Overend, R.P., "Direct Liquefaction of Wood Through Solvolysis and Catalytic Hydrodeoxygenation: an Engineering Assessment," Biomass 7, 99-123 (1985)

13. Wolfram, E., Wawrzinek, D. and Dockmeyer, W., "Einfluß der Rohstoffeigenschaften auf die Verflüssigung Rheinischer Braunkohlen," Proceedings of the Int. Coal Science Meeting, Düsseldorf, Germany. 434-445 (1981) September 7-9.

14. Davis, H.G., Eames, M.A., Figueroa, C. Gansley, R.R., Schaleger, L.L. and Watt, D.W., "The Products of Direct Liquefaction of Biomass," Investigation Report LBL-15143 from lawrence Berkeley Laboratory, University of California, USA (1982) October

15. Schirmer, R.E., Pahl, T.R. and Elliott, D.C., Analysis of Thermochemically-Derived Wood Oil," Fuel 63, 368-372 (1984)

16. Boocock, D.G.B., Kallury, R.K.M.R. and Tidwell, T.T., "Analysis of Oil Fractions Derived from Hydrogenation of Aspen Wood," Anal. Chem. 55, 1689-1694 (1983)

17. Shen, Y.-H.E., Philip, C.V., Anthony, R.G. and Soltes, E.J., "Separation of Functionalities in Pyrolytic Tar by Gel Permeation Chromatography-Gas Chromatography," J. Chrom. Sci. 22, 497-505 (1984)

DISCUSSION

J. Dhawan, U. of South Alabama: Could you please tell me how much the hydrogen consumption was?

Meier: We estimated the hydrogen consumption and saw that it was 4 to 5% based on dry wood. In the German coal liquefaction process, hydrogen consumption is 6%, and in our process, it's 4 to 5%.

D. Boocock, U. of Toronto (Canada): I think many of us are intrigued by the straight chain alkanes that you reported. Looking at the chromatogram that you showed, the major alkane was C_{17}, and the next most prominent one was C_{15}. Given that palmitic acid and the stearic acids, the two most common acids in lipids, are C_{18} an C_{16}, one could speculate that the alkanes are produced by decarboxylation of lipids that somehow got into your system. Is there any possibility that you picked up some lipids?

Meier: No, we used pre-extracted spruce wood extracted with ethanol and a mixture of benzene-ethanol. We were surprised too about our results, but we have not yet investigated the nature of these products. As yet, we have no explanation for their appearance.

J. Piskorz, U. of Waterloo (Canada): You showed us the carbon-hydrogen content of your product. Have you checked the water content of this product? Traces of water can drastically change the carbon-hydrogen composition.

Meier: We dried the oil over sodium sulfate for several days, and then we analyzed the dried oil.

LOW-PRESSURE UPGRADING OF PRIMARY PYROLYSIS OILS FROM BIOMASS AND ORGANIC WASTES

J. P. Diebold
H. L. Chum
R. J. Evans
T. A. Milne
T. B. Reed
J. W. Scahill
Solar Energy Research Institute
Golden, CO 80401

ABSTRACT

Very rapid heating of oxygenated feedstocks, such as biomass and organic wastes, results in the fragmentation of the macromolecular polymers to produce 60 to 70 wt% primary vapor products composed of oxygenated monomers and polymer fragments. These primary products may be recovered from the pyrolysis zone by minimizing their residence time and exposure to high temperatures. These primary vapor products condense to form an acidic, water-soluble, reactive oil, which has nearly the same elemental composition as the feedstock. Depending upon the desired application, these primary oils can be used without modification or they may be chemically changed to meet the requirements of a specific application, e.g., storable boiler fuels or high-octane gasoline. Our newly initiated approach to the upgrading of these oils is to utilize low-pressure processes, which essentially precludes conventional hydrogenation reactions.

We have tested Mobil's ZSM-5 zeolite catalyst, used in their Methanol-to-Gasoline (MTG) process, with the low-molecular-weight, primary pyrolysis oil vapors as a feedstock. This application appears to be very promising based upon screening tests made possible with SERI's molecular-beam mass spectrometer (MB/MS). These preliminary tests show that primary pyrolysis oil vapors from wood are very reactive with the ZSM-5 catalyst supplied by Mobil Research and Development Corporation and produce a high-octane gasoline product. The lignin-derived portion of the pyrolysis oils is a little refractory to the catalyst under the limited conditions tested, but appears to react to form aromatic hydrocarbons with relatively fresh catalyst. Lignin-derived products may ultimately need to be chemically derivatized to allow their use as a high-octane blending component. Bench-scale testing of these concepts is also in progress. A preliminary estimate of yields and product cost is presented.

LOW-PRESSURE UPGRADING OF PRIMARY PYROLYSIS OILS FROM BIOMASS AND ORGANIC WASTES

INTRODUCTION

The conversion of solid biomass to a more convenient liquid form requires considerable chemical modification. Conventional wisdom in the past has dictated that biomass be catalytically treated at high pressures for several minutes or longer to result in a liquid product suitable for use as a boiler fuel, but which would require additional catalytic, high pressure hydrogenation to result in premium liquid transportation fuels. A more ideal process would have high yields of premium transportation fuels and avoid: the problems associated with feeding solid biomass into a high pressure reactor; the cost of producing high pressure hydrogen; the cost of large, high pressure equipment; and the inefficiencies involved with several pressure and thermal cycles. We are moving toward this ideal by developing a simple, low pressure process which promises to convert biomass to a premium gasoline in a few seconds in one thermal cycle without the addition of hydrogen or carbon monoxide. This process pyrolyzes biomass to a primary, low molecular weight, oxygenated oil vapor at 500°C and then catalytically converts these vapors at 500°C to a high octane gasoline. This paper gives detailed background information on the critical pyrolysis and catalysis steps involved and presents experimental data which support the promise of the proposed process.

FAST PYROLYSIS OF BIOMASS

The heating of biomass in an oxygen-deficient situation is known as pyrolysis and has been used for centuries to produce charcoal, tars, wood alcohol, and other solvents. The traditional slow heating of biomass produces about equal amounts of gases, char, and tarry liquids. These tarry liquids have been promoted as boiler fuels, but they are not suitable for use in internal combustion (IC) engines. Under certain very rapid pyrolysis conditions, valuable products can be recovered, which are not predicted by chemical thermodynamic equilibrium. These nonequilibrium products can be varied considerably depending upon the gas phase temperature and residence time. These products can be mostly oxygenated liquids or almost entirely gases that contain a significant amount of very desirable unsaturated hydrocarbons such as ethylene and propylene. The chemistry of these unsaturated hydrocarbons is very well known and is the basis for a large part of the existing petrochemical industry. In the past, they have been used by the petroleum industry to make gasoline and by the petrochemical industry to make alcohols, plastics, etc. Examples of reactor systems which have demonstrated relatively high yields of primary pyrolysis oils include fluidized beds (49), entrained beds at atmospheric pressure (36)(27) and vacuum pyrolysis (48).

The method we have chosen to achieve very rapid pyrolysis of biomass utilizes an externally heated vortex tube. The biomass particles are entrained tangentially into the vortex tube with a jet of carrier gas at velocities over 100 m/s. This causes the solid particles to be centrifuged to the hot wall of the vortex reactor, where they slide and bump on the surface. Due to the intimate, moving contact of the cold particles on the hot wall, extraordinarily high heat transfer is thought to take place with relatively small temperature differences, i.e., the heat transfer coefficient is very high. (47) This heat transfer mechanism is called solid-convection heat transfer and can be easily shown to result in an ablative or surface pyrolysis that can pass through a biomass particle at rates in excess of 3 cm/s. (2) This type of pyrolysis favors chain-cleavage reactions to form oxygenated, organic vapors rather than char or gases. The ablative nature of this pyrolysis makes the process relatively independent of the particle size of the feedstock. Due to the high heat-transfer coefficients produced with solid-convection heat transfer, small pyrolysis reactors with very high biomass throughputs will be possible.

As shown in Fig. 1, the vortex reactor has a recycle loop, which recycles unreacted feed and large char particles. After attrition of the char into -50 μm particles, the fine char is re-entrained into the vapor stream to be removed in the char cyclone. The vapor stream then enters the vapor cracker section where the vapor may be cracked to gases if desired. Alternatively, a catalyst bed could be inserted at this point. A two-stage gas scrubbing system removes condensates such as water, residual primary oils, and secondary tars. Detailed descriptions of the experimental hardware are given in references (19) and (20) and previous experimental results are given in references (21) through (27). This system design separates the major pyrolysis steps for their individual study. Maximum

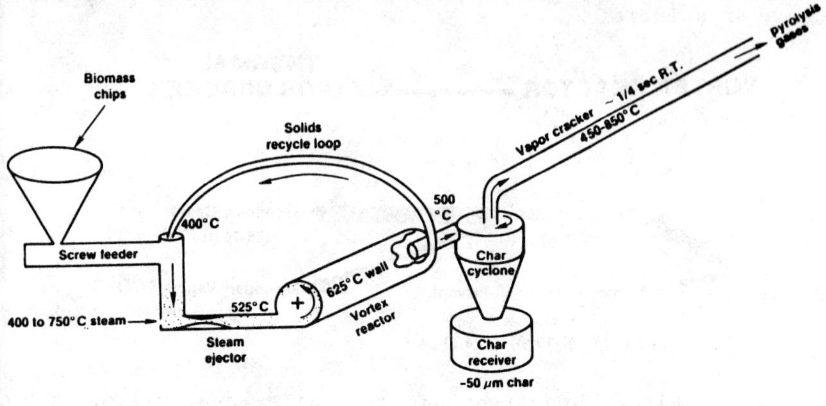

Figure 1. Vortex Reactor Schematic

throughput to date is 32 kg/h sawdust at a carrier-steam-to-biomass ratio of 0.8. The biomass vaporization is thought to take place between 435° and 475°C. By operating the cyclone reactor wall at 625°C, sufficiently rapid biomass vaporization takes place to result in an intermediate product containing 65% to 75% primary pyrolysis oil vapors, 5% to 10% gas, 10% char, and 10% to 15% water formed by the reaction. Although it is not normally done, these organic vapors could be condensed and recovered. The high yields of "primary" biomass vapors suggest that some form of direct catalytic upgrading to a gasoline product would be a significant achievement. The use of zeolite catalysts to achieve this conversion is currently being studied.

THERMAL CRACKING KINETICS OF PRIMARY PYROLYSIS OIL VAPORS (25)(26)(27)

In order to study the interaction of the primary oil vapors with catalysts, one must first know the thermal or non-catalytic reactions taking place. The pyrolysis of biomass is thought to involve a complex set of concurrent and sequential reactions. The exact chemistry of this set of reactions is only now being elucidated (29) and simplistic global explanations are being used to allow the visualization of the effect of controllable parameters on the product slate. Such a set of global reactions for fast pyrolysis is shown in Fig. 2. The key to optimizing a desired product yield out of several possible products is to know what affects the relative rates of the various reactions. A series of experiments was conducted with the ablative pyrolysis system to investigate reaction rates of the decomposition of the primary oil vapors and the products formed during that decomposition. The 5-meter-long vapor cracker with its eight thermocouples and eight sampling ports was used to study the change in gas composition as a function of reactor location during various steady-state conditions. Gaseous residence times were about 1/4 second. Secondary vapors were operationally defined as being water insoluble, whereas primary vapors were defined as water soluble.

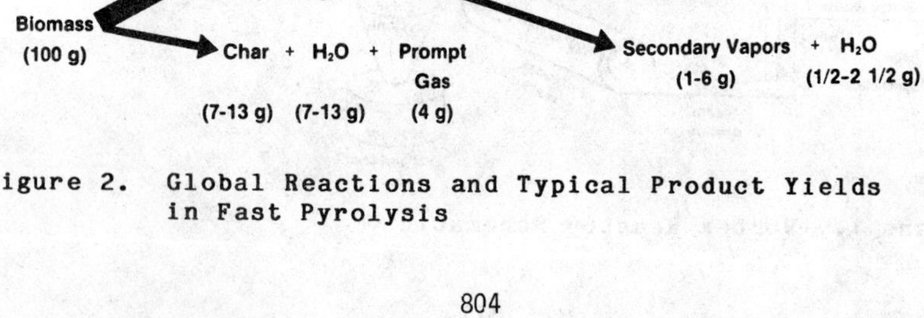

Figure 2. Global Reactions and Typical Product Yields in Fast Pyrolysis

Primary Biomass Oil Vapor

Preliminary analysis indicates that the condensed vapors contain 8% to 18% by weight water of pyrolysis (from dried feed pyrolyzed in nitrogen), and contain a wide range of organic compounds which can be sequentially extracted in a variety of solvents and which contain a very large number of compounds having molecular weights all the way up to 2000. (15)(25) These high-molecular-weight compounds may be original chain fragments from the chain cleavage reactions corresponding to oligomers of the cellulose and lignin polymers, although it is highly probable that many of these compounds were formed from reactive, lower-molecular-weight materials after physical condensation. Condensed biomass vapors from the dry nitrogen runs have a molasses-like consistency at room temperature and a much lower viscosity at higher temperatures. When exposed to air at 100°C for more than a few hours, the condensate will polymerize to form a solid thermoplastic material which is weak and brittle at room temperature. After polymerization, the material is odorless.

Figure 3 shows the compositions of the global products we have recovered from the pyrolysis system after quenching near the outlet of the vortex reactor. The elemental composition of the feed is virtually identical to that of the "primary" oil vapors. These "primary" oil vapors are defined as being those oxygenated, condensible vapors, which we form in the vortex reactor at wall temperatures of 625°C and product stream temperatures of around 500°C. Water is soluble in these primary oil condensates up to about 25% by weight. The "secondary" vapors or tars are water insoluble and have a much lower oxygen content. This strongly supports the hypothesis that a major global pyrolysis mechanism is the depolymerization of the biomass feedstock into molecular fragments without significant gas, water, or char formation. It appears that a separate global reaction may explain the formation of char, water, and prompt gas from the feed. It is interesting to note that the evolution of enough carbon dioxide from the biomass would result in hydrocarbon product, whereas the evolution of water results eventually in char formation.

Thermal Decomposition of Primary Vapors (26)(27)

It was found that the gases produced by the thermal decomposition of the vapors were relatively consistent in their composition, if the reactivity of the hydrocarbon content of the gases was taken into consideration. The overall gas composition initially was very rich in carbon dioxide (about 55 wt%) due to gases formed during the initial char and vapor forming reactions. As more vapor was cracked to gases, the carbon dioxide content of the gases reached on asymptote of about 13 wt%.

The decomposition of primary vapors to gases was found to be very well approximated by a single first-order reaction, having reaction rates of 10, 1, and 0.1 sec^{-1} at 822,

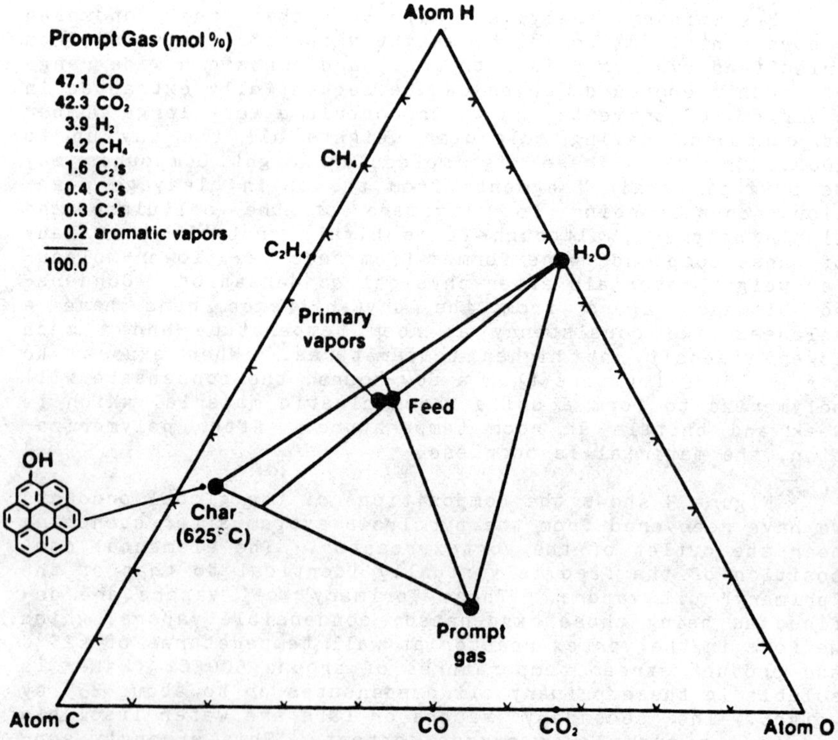

Figure 3. Primary Global Reaction Product Compositions

612, and 469°C (extrapolated) respectively. In all probability, the secondary vapor formation is not a single first order reaction, but rather consists of several consecutive steps involving first and second order reactions. (29) It was concluded that the primary vapors could be pyrolyzed quite readily by a non-catalytic or thermal approach, but that the selectivity to hydrocarbons boiling in the gasoline range was relatively poor. In addition, one would expect relatively slow thermal decomposition of the primary vapors at temperatures of 500°C and below. These kinetic data indicate that pyrolysis processes which utilize high temperature heat sources for pyrolysis, e.g. sand or gases at 1000°C, would tend to thermally crack some of the delicate primary vapors.

ZEOLITE CATALYSIS FOR CONVERSION OF PRIMARY BIOMASS VAPORS TO GASOLINE

Although most catalysis work in the past has been accomplished with noble or transition metal catalysts such as iron, chromium, nickel, platinum, cobalt, molybdenum, etc., a different approach to catalysis has emerged to a prominent position in the last 15 years. The new catalysts are broadly called zeolites, some of which occur in nature as clays. Zeolites can be made synthetically from inexpensive silica/alumina materials, often with small amounts of sodium present. Depending upon the relative percentages of silica and alumina and the precipitation environment, zeolites having differently sized and shaped crystal lattices are formed. These zeolites have channels running through the crystal lattices, the diameter of which determines the maximum molecular size that can enter the catalyst and also determines the maximum molecular size that can leave the channel as a product. The terms "molecular sieve" and "shape selective" are often used to describe this phenomena. In order for the zeolite catalyst to act upon a molecule, the molecule must first enter the channel where catalysis occurs.

Highlights of a Literature Survey of Mobil's ZSM-5 Catalyst

Methanol Feedstock

In the past few years, Mobil's synthetic zeolite, ZSM-5, has received a considerable amount of attention due to its apparently unique abilities. ZSM-5 is the zeolite catalyst used in Mobil's Methanol-to-Gasoline (MTG) process. (40) In the MTG process, methanol vapors are passed through the ZSM-5 catalyst and are dehydrated to form a high octane gasoline (90 to 100 research octane number, unleaded) composed of branched hydrocarbons and methyl-substituted benzenes (e.g., toluene, xylene). The reaction mechanism has several steps which are reported to involve the sequential formation of: the corresponding ether; olefins dominated by isobutene; branched olefinic polymers; and finally branched alkanes and aromatics. These reactions occur quite readily at relatively low pressures, such that most research has been conducted at atmospheric pressure. Overall, the reaction with methanol is exothermic and adequate heat removal from the reactor is an important design consideration. (6)

Higher Alcohols as Feedstocks

Although methanol has been primarily associated with the MTG process, Mobil has found that higher alcohols are also catalytically converted to gasoline. In particular, work has been done extensively with ethanol, and to a lesser degree with t-butanol, and 1-heptanol. When operating a microreactor so as to attain 100% conversion of the alcohol, the hydrocarbon product distribution from the butanol and the heptanol was virtually identical with that obtained from methanol, although ethanol tends to produce a little more ethyl-substituted benzenes (e.g., ethyl benzene) than does methanol. Use of the MTG process by sugar-rich areas, which are already oriented toward ethanol production, is feasible. For example, 40% by volume aqueous alcohol (80 proof rum) was shown to be a good feedstock for the ZSM-5 catalyst. Corn beer (8.7 % by volume alcohol) had a very slow reaction rate, possibly due to competition of the alcohol and water molecules for space in the catalytic zeolite channels. (38)

Other Pure Oxygenated Hydrocarbon Feedstocks

In addition to alcohols, feedstocks which Mobil has successfully evaluated include dimethyl ether, di-n-hexyl ether, tetrahydrofuran, ethylene, acetone, methyl mercaptan, propanal, methylal (dimethoxy methane), acetic acid, t-butyl formate, n-propyl acetate, and tri-n-butyl amine. These organic compounds are representative of many of those thought to be present in primary pyrolysis oils and include sulfur and nitrogen-containing materials. The wide variation in the carbon/oxygen ratio led to varying by-products, but the hydrocarbon product slate is remarkable for its relative independence of the feedstock. (4)(5)(8)

The reactions of various plant extracts over the ZSM-5 catalyst were reported by Mobil for natural rubber, corn oil, castor oil, and jojoba oil. The triglyceride oils have very large molecular weights which would seem to prevent their entry into the molecular sieve catalyst. However, it was rationalized that the triglycerides are essentially long, two-tailed molecules having small effective cross-sectional areas which allow them to enter the catalyst pores. (53) It could also be postulated that the catalyst was not acting upon the feedstocks, but rather upon the products of thermally pyrolyzed feedstocks. More recently, the reaction of canola oil over the catalyst has also been reported. (44)(45) The catalytically modified products from these various plant extracts were very similar in nature to those obtained from methanol.

Aqueous solutions of xylose, glucose, sucrose and acid hydrolyzed starch were sprayed into a fluidized bed of HZM-5 catalyst at 510°C, resulting in 17 to 30 wt% coke and 5 to 10 wt% hydrocarbons. (13) It was not known if the catalyst was converting pyrolysis products or the sugars themselves. The addition of large amounts of methanol to the feed solution significantly reduced the coke and increased the hydrocarbon formation. A comparison of the hydrocarbon product slate of the methanol/glucose with that from methanol, revealed a more methylated aromatic product for the methanol/glucose feedstock at similar space velocities. (14) The conversion of methanol to aromatic hydrocarbons is a series of sequential reactions and the time consuming dimethyl ether formation and conversion may be bypassed with the mixed feed.

There are some compounds that are relatively nonreactive to the ZSM-5 catalyst. In particular, the aromatic ring and phenolic oxygen are very stable as they pass through the catalyst. Experimentation at 350 to 450°C with phenol and cresols showed a distinct tendency to add methyl groups to the aromatic ring rather than to remove the phenolic oxygen. This methylation was especially pronounced when feeding mixtures of phenol with methanol, with the methanol apparently providing the methyl groups. Other lignin-derived phenolic compounds, such as dimethoxy phenol, eugenol, isoeugenol, and guaiacol had lower reactivities than the smaller phenol and cresols, but still had a tendency to alkylate and/or dealkylate and not to cleave the phenolic oxygen. (11) At a temperature of 350°C, methyl phenyl ether (anisole) was found to react to form phenol and cresols. (12) At a lower temperature of 250°C, a variety of zeolite catalysts, including ZSM-5, were found to promote the formation of anisole (methyl phenyl ether) from mixtures of phenol and methanol. (2)

Pyrolysis Vapors as Feedstocks

The use of vapors formed by the pyrolysis of biomass as a feedstock to ZSM-5 catalyst was considered by Frankiewicz at Occidental Research Corporation at the time they were demonstrating the 150 TPD dual-bed, flash pyrolysis plant near San Diego, CA. This plant was designed to convert municipal solid waste to an oxygenated liquid boiler fuel. Their dual-fluidized bed pyrolysis system resembled a fluidized catalytic cracking (FCC) unit commonly seen in the petroleum refining industry. Operational problems were encountered with this system which were never resolved, including the secondary fine grinding of the organic feed to a fluff. (33) The reported zeolite study utilized tetrahydrofurfural alcohol (THFA) as a model compound with a variety of zeolites including a large pore (mordenite) and several intermediate pore zeolites (ZSM-5, TMA offretite, and silicalite). The large pore-sized zeolite coked up very rapidly and was rejected. The concept of shape selectivity of zeolites suggests that the large pore zeolite allowed immobile polycyclic aromatics to form which deactivated the catalyst by plugging the pores. However, encouraging results were reported using both THFA and THFA diluted 1/1 with water with the intermediate pore zeolite catalysts. Approximately 2% to 3% of the carbon in the feed became coke on the catalyst. The carbon converted to gases was only 5% or less of the feedstock carbon. However, 70% of the carbon in the THFA was converted to hydrocarbons of unspecified molecular weight. (30)(31)

This medium pore catalyst was later exposed by Frankiewicz to the pyrolysis vapors formed by the pyrolysis of a 33% aqueous solution of α-methyl glucoside ($C_7H_2O_5$) at 550°C with a rather long pyrolysis vapor residence time of 5 seconds. Thirty percent of the carbon was reported to be in the organic phase of the condensates after a one-second residence time at 375°C in the zeolite reactor. The organic phase was 52% aromatic and was "substantially free" of oxygenated organics. Of the carbon in the feedstock, 15% appeared as pyrolysis char and 11% as catalyst coke. (30)(31) If the α-methyl glucoside vapors are as reactive to form secondary tars as is reported for cellulose vapors, our pyrolysis model for cellulose (24)(25) predicts that 38% of the primary vapors would react during the 5-second residence time in Frankiewicz's pyrolysis reactor to form secondary tars, CO_2, and water vapor. These secondary tars would have a tendency to foul the catalyst. To reduce the secondary vapor formation, a shorter residence time of the primary oxygenated vapors in the pyrolysis reactor, would seem to be indicated to reduce catalyst fouling.

Supercritically Extracted Oils as Feedstocks

Feeding oxygenated oils (extracted from poplar using supercritical methanol) over a ZSM-5 catalyst suggests that small fragments such as formic acid, acetic acid, methoxy-4-furoic acid, and furfural alcohol were eliminated from the process stream, whereas intermediate fragments such as guaiacol, phenol and eugenol were partially eliminated. Larger fragments such as 2,6-dimethoxy-phenol, apparently were unaffected by the catalyst. Liquid hydrocarbon yields were reported to be as high as 15 wt % of the oxygenated feedstock oil. About 21% of the feedstock was converted to coke (which probably reflects to a large extent the difficulty of revaporizing oxygenated oils formed by pyrolysis, as well as inherent coking of the feedstock on the catalyst). This study also suggested that dilution of the ZSM-5 with an inert diluent (silica gel) increased the active life of the catalyst, presumably by decreasing the rate of coke formation. (10)

Oils Formed by CO Reduction of Biomass as Feedstocks

Another material which has been briefly examined as a feedstock for the ZSM-5 catalyst was oil formed by the decomposition of aspen poplar in carbon monoxide at 2000 to 4000 psi pressure at 360°C and a residence time of 2.5 hours. The oil yields reported were in the range of 37 to 42 weight per cent. These oils were fed into the ZSM-5 catalyst reactor held at one atmosphere pressure. The catalytic reactor was held at relatively very high temperatures in the range of 450° to 650°C (most ZSM-5 work has been done between 300° and 450°C). (28) It would appear that perhaps these oils had to be thermally cracked before the oil molecules were small enough to enter the small ZSM-5 pores where the oxygenated oil molecules could be converted to gasoline. Due to the high temperatures employed in the catalytic reactor, much of the oil was apparently thermally cracked, judging by the presence of polycyclic aromatic hydrocarbons in the product stream. Due to the shape selectivity of the ZSM-5, these polycyclic aromatics are not supposed to be formed in the catalyst pores. This interpretation of this work is supported by the stated future plans of investigating the ZSM-5 catalyst in conjunction with an additional cracking catalyst. However, even with the above problems, about one-third of the oxygenated oil was converted to aromatics boiling in the gasoline range. (28)

Coke Formation and Effect of Steam

Use of ZSM-5 with an oxygen rich (hydrogen poor) feedstock has a tendency to convert a few percent of the feedstock to coke on the catalyst. Oxidative regeneration of the catalyst is reported to be employed with quite good results except for the case of acetic acid. Passing pure acetic acid vapors over the catalyst tends to irreversibly poison the catalyst. Mobil found that by feeding acetic acid and methanol simultaneously, a synergism is obtained which increases the reaction rate and decreases the tendency to deactivate the catalyst. (7)(9) The exothermic heat of reaction of methanol may counteract the endothermic heat of reaction of acetic acid to enable maintenance of proper catalyst bed temperatures. In addition, the methanol may be alkylating the acetic-acid derived products without the intermediate formation of dimethyl ether.

An alternate method is claimed by Occidental Petroleum in which the tendency to form coke on the intermediate pore-sized zeolite catalyst (e.g., ZSM-5) was greatly reduced by diluting the catalyst with a low surface area (<25 m^3g) silica material and by the presence of water vapor. The dilution may be as a mechanical mixture or with the catalyst supported on the diluent. (31) A diluent-to-catalyst ratio of 10 to 1 appears to be most preferred. (30) The function of the diluent may be to act as a heat sink to moderate the temperature of the very active zeolite catalyst.

The presence of steam was also reported to be very beneficial to extend the length of time on stream of the catalyst when converting canola oil to hydrocarons with ZSM-5. The coke which was eventually deposited on the catalyst had atomic H/C ratios between 0.6 and 1.0. This coke corresponds to aromatic compounds having up to 7 rings. The extent of the conversion of canola oil was only slightly affected by the gaseous residence time as related by steam-to-oil ratios from 1 to 7, whereas weight hourly space velocities (WHSV) based on the oil flow rate to catalyst weight had large effects on the product slate. (44) This suggests that mass transport of the diluted reactant vapors to the catalyst was relatively rapid, compared to chemical reaction rates of the canola oil to products. However, the nature of the products was a strong function of the WHSV <u>and</u> the gaseous residence time, with aromatic and paraffinic compounds being favored over olefins with an increase in the exposure of the hydrocarbons to the catalyst. Except for the lack of an initial ether forming step, this is consistent with the sequential reaction scheme established for methanol in which low molecular olefins are formed which then disproportionate to produce aromatics and paraffins.

Summary and Conclusions of the Zeolite Literature Survey

Small, linear oxygenated molecules diffuse into the zeolite channels more readily than bulky cyclic ones, which at best must be properly rotated to enter the channels. Although we speculate that linear cellulose oligomers may be able to enter the catalyst pores, it is probable that lignin dimers will not be able to enter unless they are fortuitously linear. For biomass vapors to be catalytically acted upon by the ZSM-5, their molecular size needs to be small enough to allow entry into the catalyst pores, while at the same time not so small as to be permanent gases. This will require a pyrolysis reactor with close temperature control and essentially plug flow characteristics to optimize the yield of suitable biomass vapors.

In a previous section, our high yields of primary vapors were presented. Although sequential solvent extraction followed by gel permeation chromatography has indicated that the condensed oils had molecular weights of 100 to 2000, (15) molecular beam/mass spectrometry MB/MS of the same oils indicates that in the vapor state the primary oils have molecular weights mostly between 28 and 180 with only a little material in the 200 to 300 range. In addition, there is preliminary data from the MB/MS work which indicates that the biomass "monomers" form oxygenated fragments having molecular weights considerably less than the parent "monomer." (29) If a significant fraction of the primary vapors can enter the small catalytic channels of the ZSM-5, then conversion of the oxygenated primary vapors to high octane gasoline would occur. This would result in a significant breakthrough in the conversion of biomass to gasoline, with a relatively simple and low-pressure process having the potential of high efficiencies and low cost.

Experimental

In order to make a quick, qualitative assessment of the effect of the catalyst on the full complement of primary pyrolysis vapors, tests were run using the batch-pyrolysis, direct sampling, molecular-beam mass spectrometer (MB/MS) used in fundamental pyrolysis research at SERI. (29)(41)

Catalyst

Approximately ten grams of pure, ammoniated ZSM-5 were made available to the program by Mobil Research and Development Corporation to be used in our initial testing in exchange for a discussion of our experimental results. The catalyst appeared to be a very fine, crystalline material having a typical diameter estimated to be 5 to 10 μm by microscopic comparison with 4 to 8 μm diameter quartz wool fibers. In order to minimize the rate of coke formation on the catalyst due to overheating, (30) we diluted the catalyst with between 5 and 10 parts by weight of quartz wool per part of catalyst. Microscopic examination of the catalyst-loaded wool indicated that the catalyst appeared to have a mild attraction (possibly electrostatic) to the wool which could survive initial temperature cycling in hot helium. The ammoniated ZSM-5 was converted to HZSM-5 by heating at 550°C for about 18 hours with a low flow of inert gas to sweep out the released ammonia.

Reactors

The catalytic reactor for the initial experiments consisted of quartz tubes 20 to 25 mm in diameter and 25 cm long. One tube was placed on top of another quartz tube, which served to preheat the carrier gas. The reactor tube was loosely packed with either 0.3 g or 0.6 g catalyst on 3 g quartz wool. The sample holder was located near the top of the preheater tube. The preheater tube and the catalytic reactor tube were heated independently by two separate furnaces. The furnace for the preheater was controlled by a thermocouple in the vicinity of the sample holder. The furnace for the catalytic reactor was controlled by a thermocouple located inside the reactor about two inches from the outlet; this furnace was to maintain the temperature attained by the carrier gas in the preheater. Details are given in reference (29).

The effluent from the catalytic reactor was passed directly into SERI's MB/MS for real-time analysis in all tests. Sample sizes used were typically 0.090 g for liquids or 0.030 g for solids. Helium at 2 liters/min (at room temperature) was normally used as the carrier gas, which resulted in gaseous residence times of one to two seconds depending upon the catalytic reactor diameter. The duration of each vapor pulse varied, but was about 30 seconds.

Results

To verify the inertness of the reactor system without the catalyst, blank runs were made with methanol vapors and cellulose (PH-102 Avicel) pyrolysis vapors [The mass spectrum of the cellulose blank is shown in Fig. 4 (bottom)]. The quartz wool was judged to be noncatalytic.

Using a relatively high weight hourly space velocity (WHSV) with the ZSM-5 loaded wool, the conversion of methanol to dimethyl ether (m/z 45^+ 46^+) was very clearly observed at low reaction severities. The appearance of low-molecular-weight hydrocarbons also occurred at the more severe temperature conditions employed. The behavior of methanol in the zeolite reactor appeared to qualitatively conform to that reported in the literature, (8) except that xylene was the first aromatic to appear (perhaps by butadiene homopolymerization reactions followed by aromatization to produce xylene). Major peaks were observed for mass numbers corresponding to ethylene, propylene, butene, pentene, and hexene, as well as for xylene.

Cellulose vapors at 470°C and a relatively high average WHSV of about 20 were very rapidly converted to a new product slate by the ZSM-5 catalyst, as seen in Fig. 4 (top). Most of the peaks in the new product slate were similar to those seen in the more severe methanol runs with the exception of two new peaks at 68 and at 82 m/z. The m/z of 68 could be cyclopentene and/or furan and the m/z of 82 could be cyclohexene and/or methyl furan. However,

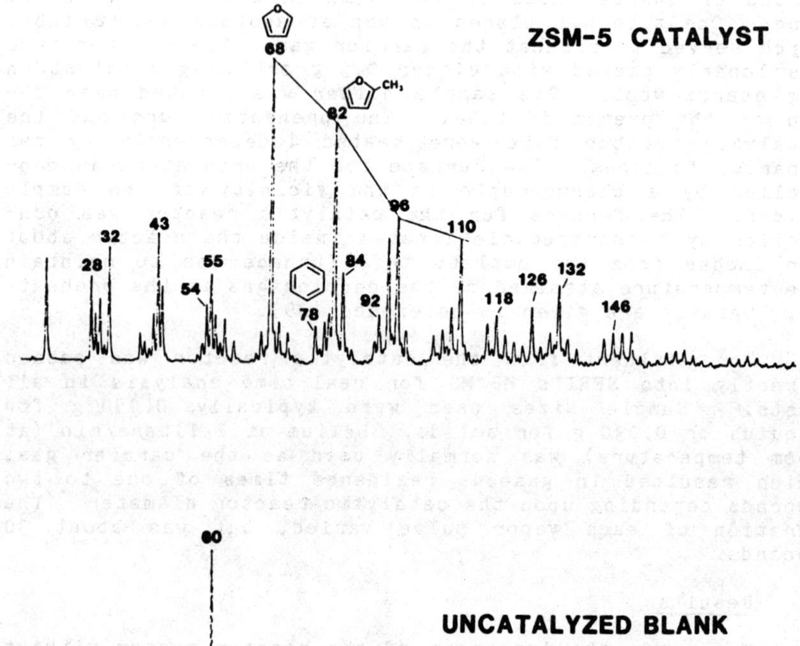

Figure 4. Mass Spectra of the Effect of Zeolite Catalyst on Cellulose-Derived Vapors. Bottom, vapors over quartz wool blank showing preservation of the primary species; top, products slate from zeolite conversion showing nearly complete conversion to deoxygenated and other new products.

using furan vapors as feedstock suggested that furan is refractory to the ZSM-5 and that furan could be a by-product of zeolite catalysis of cellulose vapor decomposition. The aromatic nature of furan may prevent easy removal of the oxygen from this cyclic ether.

To investigate the reactivity of the catalyst on wood vapors, 0.6 g of fresh catalyst on 3 g of quartz wool was placed in the reactor. The approach used was to alternate between pulses of methanol and pulses of primary wood vapor. Helium was used as a carrier gas. Methanol was put into the helium stream by diverting the helium into a bubbler of methanol at 0°C. This resulted in a WHSV calculated to be about 12, assuming the helium to be saturated with methanol. Pine splints were used which had nominal dimensions of 3 X 3 X 10 mm. The pyrolysis of the 33± 3 mg pine splints took place over an elapsed time of approximately 30 seconds at a helium temperature of 500°C, resulting in a pyrolysis wave having a skewed Gaussian shape and generating about 6 mg of charcoal. The arithmetic average WHSV of the wood vapor was between 5 and 6. This WHSV is considerably higher than normally used in ZSM-5 studies and the loss of catalyst activity was very noticeable as it occurred.

The top of Fig. 5 shows the MB/MS spectrum for uncatalyzed primary wood vapors generated in a helium gas stream at 500°C. The spectrum shows pyrolysis products from the three major classes of compounds in the wood: cellulose, hemicelluloses, and lignins. Major cellulose derived pyrolysis peaks are seen at mass numbers 43, 57, 60, 73, 85, 98, 126, and 144. Major hemicellulose derived peaks are at 60, 96, and 114. Major lignin derived peaks are at 110, 124, 138, 150, 164, 180, and 272.

The middle of Fig. 5 shows the product vapor slate observed early in the test series after only two previous wood samples and three methanol pulses. Note that the major wood-derived, primary vapor peaks have all but disappeared. The major peaks correspond to water, oxides of carbon, and a very aromatic hydrocarbon product slate dominated by toluene and xylene, with smaller amounts of benzene, trimethyl benzene, and furan.

The relative catalyst activity at that time with an active catalyst is shown in the top of Fig. 6 by examining the product slate for methanol. With the highly active state of the catalyst, there was no breakthrough of the methanol reactant and the first intermediate product (dimethyl ether) had been consumed to form a mixture of C_2 to C_6 olefins and toluene, xylene, and trimethyl benzene. Note the lack of benzene and alkanes in this early or "immature" product slate from methanol. With lower space velocities, the alkenes are reported to disproportionate to branched alkanes and to more aromatics. The amount of products above 120 m/z formed from methanol is very small in contrast to that observed from catalyzed wood vapors. Species above 115 m/z possibly correspond to homologous

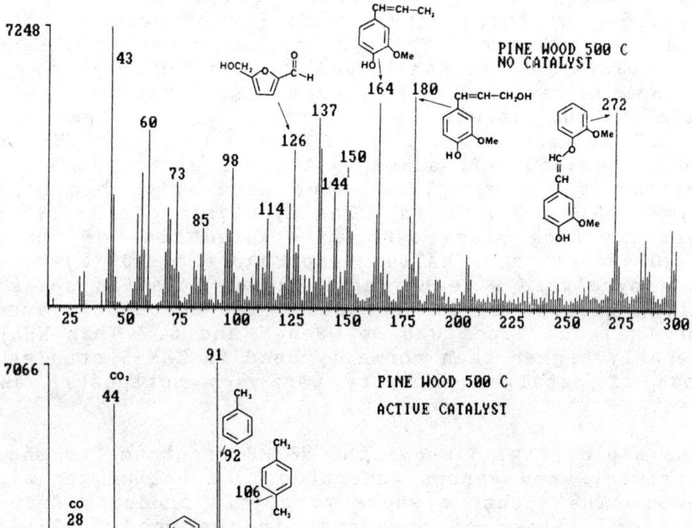

Figure 5. Pine pyrolysis vapors. Top: No ZSM-5 catalyst. Middle: Active ZSM-5 catalyst. Bottom: Partially deactivated ZSM-5 catalyst.

series of benzofurans, indenes, napthalenes, and their partially hydrogenated precursors. These species are not reported in the literature as typical ZSM-5 products and are speculated to be primarily catalytically altered remnants of carbohydrate and lignin derived pyrolysis products rather than products of olefin oligomerization. It is possible, however, that the product slate from the wood vapors is a more "mature" slate due to the absence of the dimethyl ether step. We will use gas chromatography to attain a better understanding of this complex product slate, e.g., to determine if the peaks at M/Z 94, 108 and 124 are due to hydrocarbons or to phenols, cresols, and residual guiacol.

The bottom of Fig. 5 shows the product slate after seventeen wood vapor and 5 methanol pulses. This spectrum shows significant numbers of peaks attributed to lignin and carbohydrate primary pyrolysis products as well as an increase in olefins and other intermediate products with this partially deactivated catalyst. Immediately after the eighteenth wood vapor pulse, a methanol pulse resulted in the product spectrum shown in the bottom of Fig. 6. This spectrum confirmed that the catalyst activity had severely deteriorated, with the breakthrough of methanol (m/z 32), the presence of the early intermediate dimethyl ether (m/z 45 and 46), and the increase in the amount of olefins (m/z 56, 70, 84, and 98).

The catalytic activity was restored by oxidative regeneration with diluted oxygen at 500°C. Using 20% by volume oxygen in helium, after a wood vapor pulse, resulted in a well defined carbon dioxide evolution which was monitored with the MB/MS. The duration of the carbon dioxide pulse was typically about 20 seconds. The carbon dioxide pulse was roughly quantified and found to correspond to between one and two percent by weight of the wood being converted to coke on the regenerated catalyst. Referring back to the series of eighteen wood-pyrolysis-vapor pulses, this yield of coke would correspond to about two percent by weight of coke on the catalyst to achieve deactivation. This corresponds quite well to the reported 2 to 3 wt % coke found on deactivated ZSM-5. (17) The apparently very rapid removal of coke from the catalyst is believed to be both a function of the type of catalyst as well as the relatively short life of the coke itself. The coke precursors in ZSM-5 must migrate to the macrosurface of the crystal lattice before coke may be formed, due to insufficient room within the shape-selective pore. This results in an accumulation of tar at the mouth of the active pores, which eventually can block the entrance of the pore to result in pore deactivation. The formation of coke on the macrosurface of the catalyst makes the coke very accessible to oxidation compared to most large pore zeolites where the coke forms deep inside the pores. In the case of methanol conversion with ZSM-5 where several days of operation between catalyst regeneration is common, this tar accumulation gradually decomposes to form a refractory, graphitic coating on the macrosurface of the catalyst at the temperatures

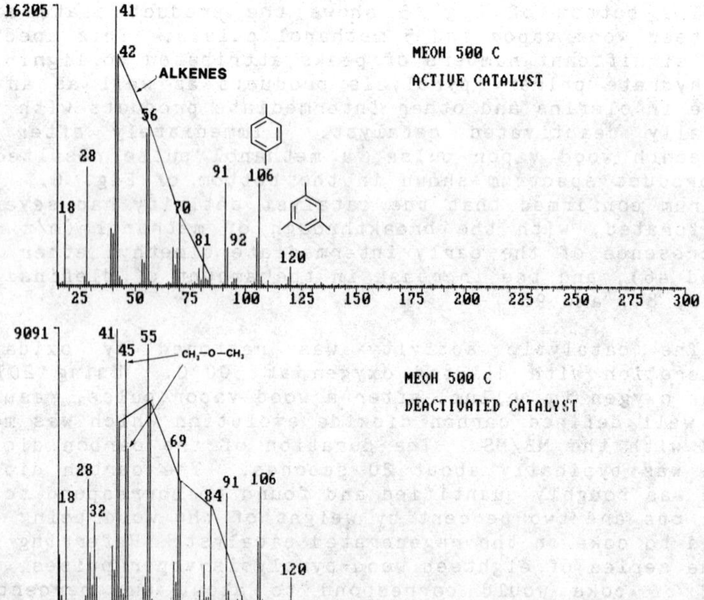

Figure 6: ZSM-5 catalyzed methanol. Top: Active catalyst. Bottom: Deactivated catalyst corresponding to Figure 5 (bottom).

in the reactor. (17) If the catalyst were to be regenerated within minutes of the tar or coke-precursor deposition, the oxidation of that material might be much easier than if the material were allowed to graphitize. This ease of regenerating the ZSM-5 catalyst has been previously reported when using short cycle times. (13)

The tentative conclusions at this early point in the investigation of wood vapors reacting over ZSM-5 catalyst are that: both the carbohydrate and the lignin derived species appear to be converted by fresh ZSM-5 catalyst to volatile hydrocarbon species; the amount of coke deposition is on the order of one to two percent of the softwood feedstock, which is too low to account for the disappearance of the lignin derived species; the catalyst deactivation observed would be expected for the apparent amount of coke deposition; and the regeneration of the catalyst occurs very rapidly at 500°C with 20% oxygen in the carrier-gas stream. Thus, although the rate of coke formation per amount of feed is much higher with wood pyrolysis vapors than with methanol, the amount of carbon in the feedstock converted to coke appears to be small. By proper design of the catalyst reactor/regenerator system, the coking rates should not present an operational problem. It should be noted that in the widely used, petroleum-refining step called Fluidized Catalytic Cracking (FCC), the yield of coke from heavy hydrocarbon feedstocks can be as high as 15% when using zeolite catalysts. (32) In light of these encouraging results, additional experimentation is in progress.

Projected Product Quality Considerations

The presence of some furan and its methylated derivatives in the gasoline product is very probable. Although these furans are oxygenated, they are essentially nonpolar and consequently insoluble in water and soluble in aromatic hydrocarbons. They have good thermal stability and blending motor octane numbers (BMONs). The BMON of 20% furan in an 80 MON gasoline was 122, (42) that of methyl furan in a 103 MON leaded gasoline was 92, (51) and that of 20% dimethyl furan in a 56 motor octane gasoline was 160, (43) and was dependent upon the nature of the blending stock. The high cost of furan at about $7.50 per gallon had apparently precluded serious thought of using it as a gasoline octane improver. The small amount of byproduct furans would most likely be left in the gasoline product.

The residual phenols and other lignin-derived fragments may need to be removed, e.g., by distillation, derivatization, or with a caustic wash to avoid corrosion problems, but they are reported to have excellent blending octane values. For example, p-cresol has a BMON of 158 in a 78 octane base stock. (51) Guaiacol has a reported blending research octane number (BRON) of 129. (3) In addition to having good blending octane characteristics, alkylated phenols have been widely used as antioxidant additives in gasoline, although at a relatively low level of 50 ppm.

(34) Since larger quantities of phenols in gasoline have been reported to be troublesome, (50) there will need to be extensive characterization studies of the gasoline product when it is available in gallon quantities.

Phenol Derivatization to Gasoline Octane Boosters

Methyl aryl ethers (MAE) made from coal liquids have been tested as gasoline extenders and octane improvers by researchers at Gulf Research and Development Co. (50) In a mixture of coal derived liquids, the fraction boiling in the 55°C to 260°C range (about 18 - 20% of the coal fed to the demonstration unit and about 50% of all liquid products obtained) typically contained 45 wt% of phenolic compounds: 5.6 wt% phenol; 23% o-cresol; 3.6% m-cresol; 2.3% p-cresol; 0.8% 2,4-dimethylphenol; 0.4% 2,5-dimethylphenol; 0.2% 2,6-dimethylphenol; 0.1% 3,4-dimethylphenol; 0.8% 3,5-dimethylphenol; 0.3% o-ethylphenol; 0.9% p-ethylphenol; 1.9% propylphenol; 5.5% unidentified phenols. The phenols were caustically extracted from the hydrocarbons and the phenolic mixture was converted into the corresponding methyl ethers by a conventional chemical laboratory method utilizing the reaction of phenolics with dimethylsulfate, which would not be employed in a commercial process. It is worth noting that phenol has a boiling point which is in the upper range of the gasoline boiling range. However, when phenol is derivatized to the methyl phenyl ether, the boiling point is reduced substantially due to the loss of the hydrogen-bonding effects of the phenolic OH. The boiling range of the MAE listed above was from 140° to 226°C.

The following tests were performed: gasoline blending and compatibility, plastics compatibility, microbistatic properties, mammalian toxicity, octane quality, carburetor cleanliness, driveability performance, fuel economy and durability tests. Their conclusion was that the mixture of methyl aryl ethers derived from phenols present in raw coal liquids is an excellent non-metallic gasoline blending agents and octane improver. In a base unleaded gasoline having a 93.1 RON and an 84.1 MON, the blending octane numbers of the MAE were 114 BRON and 93 BMON at the 10 and 15% MAE levels, respectively, or an average blending octane number of 104. The addition of 5% MAE to this gasoline reduced the Reid vapor pressure of the gasoline from 10.7 psi to 9.85 psi. (50)

This approach is being extended to produce a similar or better mixture of phenolic compounds from lignins to be converted into a mixture of methyl aryl ethers, which would be, like those tested from coal liquids, compatible with gasoline and excellent non-metallic gasoline-blending agents and octane improvers. Routes of lignin conversion include conventional high pressure hydrocracking (16) and the promising low pressure pyrolytic approaches. MAE are similar in vapor pressure to xylene and consequently tend to reduce the Reid vapor pressure of the gasoline somewhat. Therefore, their presence would allow refiners to utilize an increased portion of the inexpensive, low molecular weight refining streams (e.g., n-butane), as well as methanol, which raise the Reid vapor pressure of the resulting gasoline. The low vapor pressure in gasoline of the MAE is an important difference from that observed with alcohols in gasoline.

The method of conversion employed at SERI is based on previous research and development to convert lignins into phenol and benzene. While those processes are aimed at the production of a few products, our approach uses only part of that process to produce a mixture of phenolic compounds. These compounds will then be converted to the corresponding methyl ethers by reaction with methanol on a dehydration catalyst. (35) The conversion of phenols to ethers is envisioned as a simple, high efficiency, dehydration process on alumina, although other acidic heterogeneous catalysts will also be tested, such as zeolites which have been reported to be quite effective. (2)

The synthesis of other ethers using these methods is practiced commercially. For instance, ethyl ether is synthesized with greater than 95% yield by a vapor phase dehydration of ethanol on a fixed-bed reactor using an alumina catalyst. (35) Another important ether used as an octane enhancer in gasoline, methyl-tertiary-butyl ether (MTBE) can be easily made by reacting isobutylene with methanol in the presence of an acidic ion-exchange resin catalyst in the liquid phase at temperatures below 100°C. (35)

Zeolite Evaluation Summary

Our conclusion at this point is that the MB/MS clearly shows that the oxygenated vapors derived from the fast-pyrolysis of wood are very reactive with ZSM-5 catalyst. A primarily hydrocarbon slate with some furans and possibly phenols is formed, but which otherwise appears to be more aromatic in nature than that obtained when using methanol as a feedstock. The exciting aspect of this preliminary screening is that the fast pyrolysis of biomass to primary vapors will produce a feedstock for the ZSM-5 catalyst at a fraction of the cost of methanol from biomass, due to the simplicity of the pyrolysis process.

We envision an atmospheric-pressure process involving only one temperature cycle, in which the biomass particles are vaporized at ~500°C, the residual char removed in a hot cyclone, and the oxygenated vapors immediately passed through a bed of ZSM-5 catalyst to form gasoline, furan, water, carbon oxides, and possibly some phenols. The phenols may be a candidate for subsequent etherization or for hydrogenation due to their expected low oxygen content, or they may be used for process energy. We currently have a fast pyrolysis system, which has demonstrated the capability to produce the necessary primary biomass vapors in a continuous fashion in the high yields of interest without the concurrent production of secondary, polycyclic-aromatic tars. (A much larger, 15 TPD biomass pyrolysis system is currently being developed by a private corporation, based on many of our concepts.) (46)

Preliminary Process Evaluation

A process evaluation was initiated, although many of the key operating parameters are not now well established. The study has been useful to identify the sensitivity of the process economics to the various research goals. Four independent economic studies of the production of primary wood oil (oxygenated crude oil or oxycrude) were reviewed. (1)(39)(46)(52) The relative agreement between these four independent studies was quite good after adjustments for feedstock costs and capital costs were made. Oxycrude costs were estimated to be $0.60 ± $0.10 per gallon at $40/dry ton of feedstock and 15% interest, 25 year life (without tax credit considerations). The oxycrude costs were heavily dependent upon the cost of feedstock, with the feedstock costs being between 44 and 65% of the product cost, depending upon capital investment, yields, etc. assumed for each process. At a 70 wt % conversion of feedstock to oxycrude, a yield of 130 gallons of oxycrude per ton of dry feedstock is predicted, having a density of 1.28 and a 0.75 volumetric heating equivalency to petroleum oil (2.3 BOE/ton).

If all of the primary pyrolysis vapors are converted to an aromatic gasoline, the ideal stoichiometric yield would be carbon dioxide, water vapor, and 93 gallons of gasoline per ton of dry feed (formation of byproduct coke, carbon monoxide, and low molecular weight hydrocarbons would reduce this yield of gasoline). If the carbohydrate fraction of the oxycrude is assumed to be 75 wt % and were to be the only source of gasoline boiling range hydrocarbons in the zeolite conversion process, 55 gallons of gasoline per ton of dry wood is predicted as an upper limit. If the lignin derived portion of the oxycrude could be treated to acceptably passivate the chemical reactivity, the gasoline yield would be increased by 67% over that obtained from only the carbohydrates. If 14 gallons of methanol per ton of feedstock were used to derivatize the phenolic content of the lignin derived oils to ethers, the predicted gasoline yield would be increased by 78% over that obtained from only the carbohydrates. The impact of the utilization of the lignin derived portion of the oxycrude oil is an estimated reduction in the gasoline product cost from about $1.63 to $1.05 per gallon of high octane gasoline blending stock. This assumes that the cost of the zeolite conversion of the pyrolysis oil vapors to gasoline can be kept at five times higher than that reported for converting methanol to gasoline with ZSM-5. (40)

A preliminary flow sheet is nearing completion which shows mass and energy balances. A steam-to-biomass ratio of one was selected for this study. Fuel for the system consisted of the byproduct char, coke, pyrolysis gases, and tars. The contaminated waste water generated by pyrolysis and other chemical reactions was incinerated in the pyrolysis furnace. Boiler water was recycled. Typical temperature approaches were over 100°F in the heat exchangers and the analysis was sufficiently in depth to avoid "pinch points", where latent heats were involved. By careful management of the energy in the system, sufficient waste heat was available to dry a feedstock having over 100% moisture on a dry basis (over one pound of moisture per pound of dry feed). In addition, there is some energy available for product distillation purposes.

Although the economic feasibility of this catalytic process will be heavily dependent upon catalyst life and behavior, these initial results are very encouraging with respect to the conversion of primary pyrolysis vapors to hydrocarbons with a simple process operating at atmospheric pressure with modest temperatures and overall residence times of only a few seconds.

FUTURE DIRECTIONS

The catalytic upgrading efforts will begin with the screening of a variety of catalysts using the MB/MS technique in a continuation of the work described here with regard to optimum operating conditions, catalyst life and product characterization. Promising conditions will be tested in the large vortex reactor system using a slipstream catalytic reactor. Liquid products weill be analyzed using GC and GC-MS to determine quality. Phenol derivatization studies will be conducted as described above.

A long, catalytic slipstream reactor is being assembled which will be inserted into the 2-inch diameter vapor cracker reactor. This fixed-bed reactor will use model compounds, or a portion of the pyrolysis vapors generated by the vortex reactor as feed. The catalyst bed is in the annular space between a 3/4-inch diameter tube and 1/4-inch diameter tube and is 4 m in length. The temperature profile in the catalyst bed will be monitored by a long axial thermocouple inside of the 1/4-inch tube. The reactor will be heated with a five zone electric furnace surrounding the vapor cracker. Heat removal from the catalytic reactor will be with a flow of preheated gas or steam in the annular space between the outside of the catalytic reactor tube and the inside of the vapor cracker tube. This reactor will be used to study band-aging in the fixed bed, reaction temperature, coke quantification, catalyst regeneration, and liquid product characterization. Mobil Research and Development has agreed to supply a suitable ZSM-5 catalyst for this reactor system. The use of a simple entrained-flow or fluidized-bed catalytic reactor may be necessary eventually to establish scaling data.

The petroleum industry utilizes a rapid catalyst evaluation technique called the Microactivity Test (MAT). In a related program, this ASTM procedure is being adapted by T. Reed, on leave at the Colorado School of Mines, to allow the feeding of a wooden dowel rod into a pyrolysis reactor to form primary pyrolysis vapors. The heat transfer mechanism depends on the direct application of mechanical pressure in the range of 10-500 psi to produce ablative pyrolysis of the wooden rod. The rate of vapor generation is controlled by the diameter of the rod, the stationary heat sink temperature, and the force exerted on the rod in a manner analogous to that extensively reported by Lede, et al. (37) The vapors will pass over a fixed bed of candidate catalyst to produce 1-10 cc of liquid. The products will be collected and analyzed using GC, GC-MS and MB-MS techniques. This modified MAT system will be used to screen a large number of zeolite and nonzeolite catalysts.

Finally, in a companion program the scalability of the ablative fast pyrolysis concept to the 15 TPD size will be monitored with a pyrolysis unit made available to us and operated by private industry. (46)

ACKNOWLEDGMENTS

We wish to thank the Mobil Research and Development Corporation for their interest in our process and for the ZSM-5 catalyst used in the experimentation. The guidance, direction, and funding by our fast pyrolysis program manager at PNL, Dr. Don J. Stevens, and our DOE sponsor (Biofuels and Municipal Waste Technology Division), Mr. Si Friedrich, is gratefully acknowledged. The lignin upgrading and derivitization work is supported by Mr. Richard Moorer of the same division.

REFERENCES

1. Argonne National Laboratory. 1985. "Biofuels Technology and Fuel Cycle Data", submitted to DOE by the National Committee on Cost Goals for Biofuels Technologies. Private Communication, B. Neenan, SERI.

2. Balsama, S., Beltrame, P., Beltrame, P. L., Carniti, P., Forni, L., and Zuretti, G. 1984. "Alkylation of Phenol with Methanol over Zeolites", Applied Catalysis, 13, pp. 161-170.

3. Burns, L. D. December 1984. "Organic Antiknock Chemicals." Chem. Tech. p. 744.

4. Butter, S. A., Jurewicz, A. T., and Kaeding, W. W. 1975. "Conversion of Alcohols, Mercaptans, Sulfides, Halides, and/or Amines." U.S. Patent 3,894,107.

5. Chang, C. D., Lang, W. H., and Silvestri, A. J. 1975 "Conversion of Ethers." U. S. Patent 3,894,106.

6. Chang, C. D., Silvestri, A. J., and Smith, R. L. 1975. "Production of Gasoline Hydrocarbons." U.S. Patent 3,928,423.

7. Chang, C. D., and Lang, W. H. 1976. "Manufacture of Gasoline," U.S. Patent 3,998,898.

8. Chang, C. D., and Silvestri, A. J. 1977. "The Conversion of Methanol and Other O-Compounds to Hydrocarbons over Zeolite Catalysts." J. of Catalysis. 47:249-259.

9. Chang, C. D., Chen, N. Y., Koenig, L. R., and Walsh, D. E. 1983. "Synergism in Acetic Acid/Methanol Reactions Over ZSM-5 Zeolites." Presented at 185th ACS Meeting, Div. of Fuel Chem., Seattle, WA, 20-25 March 1983.

10. Chantal, P., Kaliaguine, S., Grandmaison, J. L., and Mahay, A. 1983. "Feasibility of the Production of Hydrocarbon from Aspen Poplar Pyrolytic Oils Over H-ZSM-5." Comptes Rendus de L'Atelier de Travail sur La Liquefaction de La Biomasse. Sherbrooke, Quebec, 29-30 September 1983, National Research Council of Canada, NRCC 23130, pp. 234-48.

11. Chantal, P. D., Kaliaguine, S., and Grandmaison, J. L. 1984. "Reactions of Phenolic Compounds on H-ZSM-5." Catalysis on the Energy Scene. Elsevier Science Publishers, Amsterdam, p. 93.

12. Chantal, P. D., Kaliaguine, S., and Grandmaison, J. L. 1985. "Reactions of Phenolic Compounds over HZSM-5", Applied Catalysis, 18, pp. 133-145.

13. Chen, N. Y., and Koenig, L. R. 1985. "Process for Converting Carbohydrates to Hydrocarbons", U.S. Patent 4,503,278.

14. Chen, N. Y., and Koenig, L. R. 1985. "Conversion of Carbohydrates", U.S. Patent 4,549,031.

15. Chum, H. L., Johnson, D. K., Ratcliff, M., Posey, F., and Black, S. 1985. "Preliminary Analyses of Pyrolysis Oils" Appendix B in reference. (25)

16. Chum, H. L., Johnson, D. K., Ratcliff, M., Posey, F., Black, S., Goheen, D. W., and Baldwin R. 1985. "Liquid Fuels from Lignins." In Biochemical Conversion Program, Semi-Annual Review Meeting. SERI/CP-231-2726. Solar Energy Research Institute, Golden, CO., pp. 53-60.

17. Dejaive, P., Auroux, A., Gravelle, P. C., and Védrine, J. C. 1981. "Methanol Conversion on Acidic ZSM-5, Offretite, and Mordenite Zeolites: A Comparative Study of the Formation and Stability of Coke Deposits." J. of Cat, 70, pp. 123-136.

18. Diebold, J. 1980. "Ablative Pyrolysis of Macroparticles of Biomass." In Proceedings of the Specialists Workshop on the Fast Pyrolysis of Biomass, op cit.

19. Diebold, J., and Scahill, J. 1981. "Progress in the Entrained-Flow, Fast Ablative Pyrolysis of Biomass." In Proceedings of the 12th Biomass Thermochemical Conversion Contractors Review Meeting. Washington, DC; 18-19 March 1981. Battelle Pacific Northwest Laboratories, Richland, WA.

20. Diebold, J. P., and Scahill, J. 1982. "Progress in the Entrained-Flow Fast Ablative Pyrolysis of Biomass, Annual Report, October 1980 to September 1981." Proceedings of the 13th Biomass Thermochemical Conversion Contractors Review Meeting; Arlington, VA; 17-19 October 1981. Battelle Pacific Northwest Laboratories, Richland, WA.

21. Diebold, J. P., and Scahill, J. W. 1982. Progress in the Entrained-Flow, Fast Ablative Pyrolysis of Biomass, Annual Report, Oct. 1980 to Sept. 1981. SERI/PR-234-1456. Solar Energy Research Institute, Golden, CO.

22. Diebold, J. P., and Scahill, J. W. 1983. Ablative Pyrolysis of Biomass in the Entrained-Flow Cyclonic Reactor at SERI, Annual Report, 1 October 1981 to 30 November 1982. SERI/PR-234-1883, Solar Energy Research Institute, Golden, CO.

23. Diebold, J. P., and Scahill, J. W. 1984. Entrained-Flow, Fast Ablative Pyrolysis of Biomass, Annual Report, 1 December 1982 to 30 September 1983. SERI/PR-234-2144, Solar Energy Research Institute, Golden, CO.

24. Diebold, J. P., and Scahill, J. W. 1984. "Ablative, Entrained-Flow Fast Pyrolysis of Biomass," Proceedings of the 16th Biomass Thermochemical Conversion Contractor's Meeting. Portland, OR; 8-9 May 1984. Battelle Pacific Northwest Laboratories, Richland, WA.

25. Diebold, J. P., and Scahill, J. W. 1985. Entrained-Flow, Fast Ablative Pyrolysis of Biomass, Annual Report, 1 October 1983 - 30 November 1984. SERI/PR-234-2665. Solar Energy Research Institute, Golden, CO.

26. Diebold, J. P. 1985. "The Cracking Kinetics of Depolymerized Biomass Vapors in a Continuous Tubular Reactor." MS Thesis T-3007, Colorado School of Mines, Golden, CO 80401.

27. Diebold, J. P., Scahill, J. W., and Evans, R. J. 1985. "The Thermal and Catalytic Upgrading of Oxygenated, Primary Biomass Pyrolysis Oil Vapors" Proceedings of 17th Thermochemical Conversion Contractor's Meeting, October 14-18. DOE/CONF

28. Eager, R. L., Mathews, J. F., and Pepper, J. M. 1982. "Conversion of Aspen Poplar into Liquid Fuel," In Spec. Meet. Biomass Liquefaction Proc. Meet. National Research Council of Canada.

29. Evans, R., and Milne, T. 1985. "Applied Mechanistic Studies of Biomass Pyrolysis" Proceedings of 17th Biomass Thermochemical Contractor's Meeting, October 14-18, Minneapolis, MN. DOE/CONF-

30. Frankiewicz, T. C. 1980. "The Conversion of Biomass Derived Pyrolytic Vapors to Hydrocarbons." In Proceedings of Specialists' Workshop on Fast Pyrolysis of Biomass. op cit., pp. 123-36 in reference (1).

31. Frankiewicz, T. C. 1981. "Process for Converting Oxygenated Hydrocarbons into Hydrocarbons." U. S. Patent 4,308,411.

32. Gary, J. H. and Handwerk, G. E. 1975. Petroleum Refining Technology and Economics. Marcel-Dekker, Inc. N.Y., p. 106.

33. Harrison, B., and Vesilind, P. A. 1980. "The San Diego Flash Pyrolysis Project." Design and Management for Resource Recovery, Volume 2, High Technology-A Failure Analysis, ed. P. A. Vesilind, pp. 61-85. Ann Arbor Science, Ann Arbor, MI.

34. Hill, J. B., and Moxey, J. G., Jr. 1960. "Gasoline" in Petroleum Products Handbook, Ed. Guthrie, V.B. McGraw-Hill, N.Y., pp. 4-36.

35. Keeley, D. E. 1980 in Kirk Othmer Encyclopedia of Chemical Technology, Vol. 9, pp. 381-93, 3rd edition, Wiley: New York.

36. Knight, J. 1985. "Entrained Flow Pyrolysis of Biomass for Liquid Fuels" 17th Biomass Thermochemical Conversion Contractor's Meeting, October 15-17, Minneapolis, MN. DOE CONF-

37. Lédé, J., Panagopoulos, J., and Villermaux, J. 1983. "Experimental Measurement of Ablation Rate of Wood Pieces, Undergoing Fast Pyrolysis by Contact with a Heated Wall." Mathematical Modeling of Biomass Pyrolysis Phenomenon Symposium at ACS Annual Meeting in Washington, D.C., September 1. (See also Fuel, 64, pp. 1544-1520. 1985)

38. Lee, W. Y., Chen, N. Y., and Perry, R. H. 1981. "Conversion of Alcohols to High Octane Gasoline." Primero Congresso Brazilero de Petróleo, pp. III. 13-25.

39. McKeough, P., Nissilä, M., Solantausta, Y., Beckman, D., and Ostman, A. (1985). "Techno-economic Assessment of Direct Biomass Liquefaction Processes", Technical Research Centre of Finland, Research Report 337.

40. Meisel, S. L., McCullough, J. P., Lechthaler, C. H., and Weisz, P. B. February 1976. "Gasoline from Methanol in One Step" Chemtech pp. 86-89.

41. Milne, T. A. and Soltys, M. 1983. "Direct Mass-Spectrometric Studies of the Pyrolysis of Carbonaceous Fuels. I. A Flame-Pyrolysis, Molecular-Beam Sampling Technique.", J. Anal. Appl. Pyrol., 5, pp. 93-110.

42. Nisbet, H. B. 1941. "The Blending Octane Numbers of Furan and Furfuryl Alcohol." J. of Inst. Pet.. 27:293.

43. Nisbet, H. B. 1946. "The Blending Octane Numbers of 2:5-Dimethyl Furan." J. of Inst. Pet. 32:162.

44. Prasad, Y. S., Bakhshi, N. N., Eager, R. L., and Mathews, J. F. 1984. "Effect of Steam Addition on Catalytic Upgrading of Canola Oil." Catalysis on the Energy Scene, Elsevier Science Publishers, Amsterdam, pp. 85-92.

45. Prasad, Y. S. and Bakhshi, N. N. 1985. "Effect of Pretreatment of HZM-5 Catalyst on its Performance in Canola Oil Upgrading", Applied Catalysis, 18, pp. 71-85.

46. Pyrotech Corporation. 1985. Private Communication. 8900 Stateline, Suite 357, Leawood, KS 66206.

47. Reed, T. B., Diebold, J., and Desrosiers, R. 1980. "Perspectives in Heat Transfer Requirements and Mechanisms for Fast Pyrolysis." In Proceedings of the Specialists Workshop on the Fast Pyrolysis of Biomass, Copper Mountain, CO; 20-22 October 1980. SERI/CP-622-1096. Golden, CO: Solar Energy Research Institute.

48. Roy, C., DeCaumia, B., Brouillard, D., and Ménard, H. 1985. "The Pyrolysis Under Vacuum of Aspen Poplar" in Fundamentals of Thermochemical Biomass Conversion, edited by R. P. Overend, T. A. Milne, and L. K. Mudge. Elsevier Applied Science Publisher, N.Y. pp. 237-256.

49. Scott, D. S. and Piskorz, J. 1983. "The Continuous Flash Pyrolysis of Biomass", Comptes Rendes de L'Atelier de Travail sur la Liquefaction de la Biomasse, Sept. 29-30, Sherbrooke, Quebec, Canada, MRCC 23130, pp. 116-137.

50. Singerman, G. M. 1980. "Methyl Aryl Ethers from Coal Liquids as Gasoline Extenders and Octane Improvers." Gulf Research and Development Co., Pittsburg, PA. 15230, report prepared under contract No. DE-AC01-79-C50022.

51. Unzelman, G. H., Forster, E. J., and Burns, A. M. 1971. "Are There Substitutes for Lead Antiknocks." API Meeting, San Francisco, May 14. API Preprint No. 47-71.

52. Wan, E. 1985. "Technical/Economic Assessment of Biomass Based Fuel Processes," 17th Thermochemical Conversion Contractor's Meeting, Minneapolis, MN, October 15.

53. Weisz, P. B., Haag, W. O., and Rodewald, P. G. 1979. "Catalytic Production of High-Grade Fuel (Gasoline) from Biomass Compounds by Shape Selective Catalysis", Science, 206, No. 5, pp. 57-58.

DISCUSSION

L. Dao, Inrs. Energie (Canada)	What is the yield from the conversion?
Diebold	Based on the stoichiometry of the primary vapors, we calculated a xylene or toluene product yield, with CO_2 and water as the by-products, of 90 gallons of gasoline per dry ton of biomass. We are not far enough into the research with ZSM-5 to know what percentage of the maximum yield we are acting on both lignin and the cellulosic parts of the vapors with the cellulosic parts of the vapors with the fresh catalyst.
Dao	We did work on the conversion of phenol over ZSM-5 and found it is practically impossible to deoxygenate phenol. I'm not sure that you're right to assume complete deoxygenation.
Diebold	It does appear that we're not making phenols, and that they are not going to be major products. Small amounts of derivatives could form, such as the methyl-phenyl ether, anisole. It has a very high octane blending value, which could be left in the gasoline. We think this could be done by the addition of a second catalytic reactor downstream of the ZSM-5 reactor.
D. Stevens, Pacific Northwest Laboratory	Earlier, you indicated a lot of the vapors might be various furan derivatives. Now, you seem to feel that they are hydrocarbon derivatives. What caused you to change your opinion?
Diebold	It now appears that the furans are an intermediate product, and with fresh catalyst, they are converted to hydrocarbons. As the catalyst becomes deactivated, the furans increase. The weak acid sites on the ZSM-5 apparently are sufficient to create the furans, whereas it takes strong acid sites to convert them to hydrocarbons. We think that the furans will be a very minor proportion of the products rather than a significant fraction.

CONVERSION OF BIOMASS CARBOHYDRATES INTO HYDROCARBON PRODUCTS

Mohammed I. Hanniff, Ph.D., Post doctoral fellow
and Lê H. Dao , D.Sc. Professor
INRS-Energie, Institut National de la Recherche Scientifique
Varennes, Québec, Canada JOL 2PO

ABSTRACT

The catalytic conversion of model compounds found in acid-hydrolysis biomass products have been studied. Furfural, glycerol, glucose, fructose solutions are contacted with crystalline aluminosilicate zeolite such as ZSM-5 to provide hydrocarbon products which are useful as chemicals or fuels. The effects of the addition of methanol and water, of the metal ZSM-5, and of the temperature on the hydrocarbon yields and the coke formation are investigated. A process for the catalytic conversion of biomass carbohydrates into hydrocarbon products is proposed.

CONVERSION OF BIOMASS CARBOHYDRATES INTO HYDROCARBON PRODUCTS

INTRODUCTION

The conversion of biomass materials into hydrocarbon products (chemicals and fuels) could be achieved by thermochemical conversion and fermentation technologies (1). Thermochemical conversion technologies can be divided into two different categories: the integral approach and the fractional approach (Fig. 1). In the integral approach, biomass materials were liquefied at different operational conditions to give oxygenated products ("proto-oils") which contained about 20% of oxygen. Chemical analyses showed five functional types of compounds: derivatives of cyclopentanone and furan, carboxylic acids, carbohydrates (mainly from cellulosic and hemicellulosic materials), and phenolic derivatives from lignins (2). The catalytic up-grading of the proto-oils and model compounds were investigated, but the results obtained were of mixed success because phenol derivatives were difficult to deoxygenate. In the fractional approach, the lignins were separated from the cellulosic and hemicellulosic materials by acid hydrolysis. And depending on reaction conditions further degradation reactions converted cellulosic materials to carbohydrates, carboxylic acids and furfural derivatives (3). The deoxygenation of some carbohydrate model compounds has been reported (4,5).

In this paper, we would like to report:

a- the catalytic conversion of model compounds found in acid-hydrolysis biomass products, particularly furfural, glycerol, glucose and fructose, by ZSM-5 catalysts to hydrocarbon products;
b- the effects of operational conditions on the hydrocarbon yields and the coke formation;
c- the problem of sugar caramelization and the approaches to resolve it;
d- and finally, the processes to convert biomass carbohydrates into hydrocarbon products which are useful as chemicals or fuels.

EXPERIMENTAL

• Materials

Model Compounds and Carbohydrates Feedstocks. Furfural (Aldrich), glycerol (Fisher), glucose (Anachemia), fructose (Anachemia), acetone and methanol (Caledon) were reagent grade. Aqueous carbohydrate solutions were obtained by acid hydrolysis of cellulose (Baker) in a short-residence time reactor (6) (240 to 260°C, 800 psi and 3 minutes residence time).

1,2:5,6-di-O-isopropylidene-D-gluco-furanose and 2,3:4,5-di-O-isopropylidene-D-fructopyranose were prepared from the reaction of glucose and fructose with acetone in the presence of sulfuric acid according to a published procedure (7).

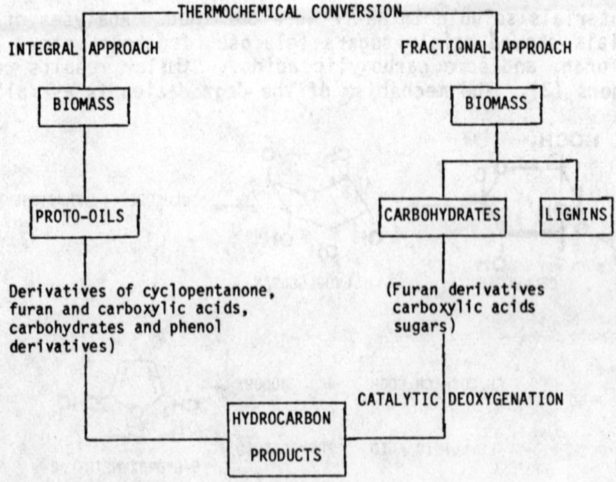

FIGURE 1

PROCESSES TO CONVERT BIOMASS INTO HYDROCARBON PRODUCTS

ZSM5-Catalysts. The catalysts used were ZSM-5 zeolites, synthesized with Si/Al ratio varying from 36 to 400 according to published method (8). The zeolite was converted to the H-form via NH_4NO_3 exchange followed by calcination at 520°C (8). Zn and Mn-ZSM-5 zeolites were obtained by exchanging the NH_4-ZSM-5 form with 10% of $Zn(NO_3)_2$ and $Mn(NO_3)_2$ solutions at 80°C, followed by calcination at 520°C. The properties of ZSM-5 catalysts are reported in Table 1.

• Apparatus and Method

Reactions were carried out in a quartz fixed-bed 10-mm-i.d microreactor (Fig. 2). The reactor contained 10g catalyst (0.35 to 1 mm). The temperature was controlled by an Omega controller and measured by placing a thermocouple in the catalytic bed. Helium was fed through a Matheson flowmeter and carbohydrate solutions were fed with a positive displacement pump. Products were successively collected in 3 traps maintained at 0°C, - 60°C and - 196°C. The outlet gases were collected in a Teflon sample bag. Gaseous and liquid products were analysed by gas chromatography and GC/MS. The reaction conditions and the analytical conditions are reported in Table 1.

RESULTS AND DISCUSSION

• Acid Hydrolysis of Cellulose

Suspensions of 20% cellulose in 1% HCl aqueous solution were hydrolysed in a short-residence time high pressure reactor and in tubing reactor at 200°C to 270°C. After 3 minutes residence time, about 60% of organic materials soluble in water were obtained. Analyses of the organic materials showed mainly sugars (glucose, fructose, etc...), derivatives of furan, and some carboxylic acids. Similar results were observed by others (3). The mechanism of the degradation is as followed.

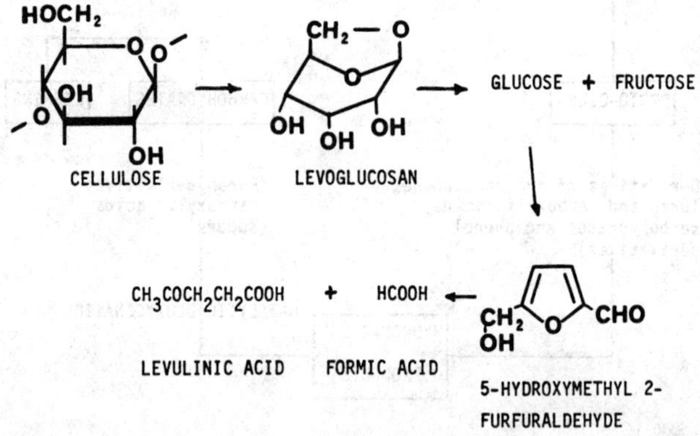

TABLE 1

ZSM-5 PROPERTIES

Catalyst	Si/Al	DC (%) *
H-ZSM-5 (A)	34	100
H-ZSM-5 (B)	333	100
Zn-ZSM-5	333	100
Mn-ZSM-5	333	100

* degree of crystallinity

REACTION CONDITIONS

Catalyst weight	:	10 g (80% ZSM-5 + 20% bentonite)
Temperature	:	350-560°C
Pressure	:	atmospheric pressure
Inert gas	:	Helium (~ 3 mℓ/min)
WHSV **	:	Variable
Reaction time	:	3 hours

** The weight hourly space velocity (WHSV) is defined as:

$$WHSV = \frac{\text{g of injected feed per hour}}{\text{g of catalyst}}$$

ANALYTICAL CONDITIONS

• Gas chromatography	:	HP 5890 GC with DB-5 (SE-54) column (30 m x 0.25 mm, 1.0 μ)
For liquid	:	70°C (4 min), then 4°C/min to 160°C then 20 min at 160°C
Gas	:	33°C (isothermal)
• GC/MS	:	HP 5890 GC and MS detector Pona column and DB-5 column.

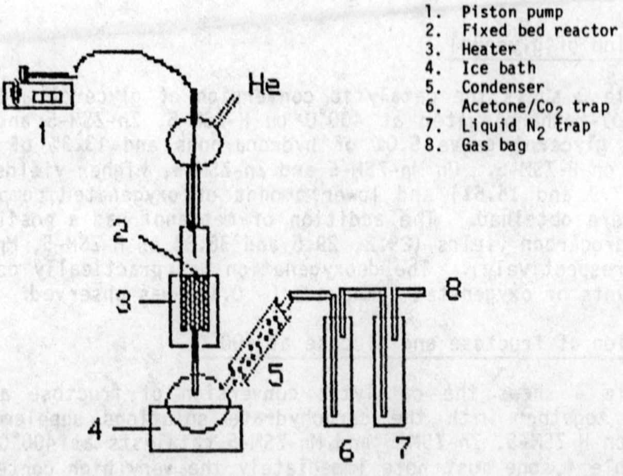

FIGURE 2
EXPERIMENTAL SETUP

In the presence of an hydrogen source, glucose can be converted to sorbitol and glycerol:

$$\underset{\text{glucose}}{C_6H_{12}O_6} \xrightarrow{H_2} \underset{\text{sorbitol}}{HOCH_2\,(CHOH)_4\,CH_2OH} \longrightarrow \underset{\text{glycerol}}{HOCH_2\,CHOH\,CH_2OH}$$

In order to evaluate the catalytical up-grading of the hydrolysed materials, the catalytic conversion of model compounds of representative families were first studied.

• Furfural Conversion

Table 2 shows the catalytic conversion of furfural and a mixture of furfural-methanol-water at 400°C and 500°C on H-ZSM-5, Mn-ZSM-5 and Zn-ZSM-5. At 400°C, in the presence of H-ZSM-5 catalyst, the main reaction was the decarbonylation of furfural to give furan (50% conversion),

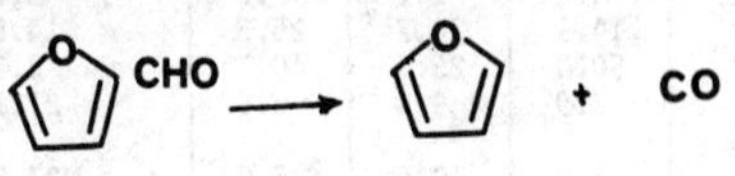

and about 8.9% (carbon basis) of hydrocarbons wich was obtained by catalytic deoxygenation of furfural. The addition of manganese (Mn-ZSM-5) reduced the content of oxygenated compounds (mainly furan) but also gave higher coke yield. Zn-ZSM-5 catalyst reduced the furan content to about half the amount compared with the H-ZSM-5 form. The addition of an in-situ hydrogen donor source such as methanol, reduced dramatically the amount of oxygenated compounds and a higher yield of hydrocarbons was obtained. The deoxygenation of furfural in the presence of methanol on Zn-ZSM-5 was practically complete at 500°C.

• Conversion of glycerol

Table 3 shows the catalytic conversion of glycerol and a mixture of glycerol-methanol-water at 400°C on H-ZSM-5, Zn-ZSM-5 and Mn-ZSM-5. At 400°C, glycerol ga ve 5.0% of hydrocarbons and 13.3% of oxygenated compounds on H-ZSM-5. On Mn-ZSM-5 and Zn-ZSM-5, higher yields of hydrocarbons (7.9 and 15.6%) and lower amount of oxygenated compounds (6.9 and 0%) were obtained. The addition of methanol had a positive effect on the hydrocarbon yields (29.2, 29.5 and 35.8% on H-ZSM-5, Mn-ZSM-5 and Zn-ZSM-5 respectively). The deoxygenation was practically complete and trace amounts of oxygenated compounds (~ 0.1%) was observed.

• Conversion of fructose and glucose at 400°C

Table 4 shows the catalytic conversion of fructose and glucose solutions together with the carbohydrate solutions supplemented with methanol on H-ZSM-5, Zn-ZSM-5 and Mn-ZSM-5 catalysts at 400°C. Looking at the table 4, one must note immediately the very high content of coke for all experiments. These results are not surprising, since it is known that the carbohydrates start to caramelize at about 160°C. The caramelization reaction is in competition with the catalytic conversion, and once the sugars start to caramelize, the catalytic conversion to

TABLE 2

Catalytic Conversion of Furfural With and Without Methanol

• Feeds (w%)*	Furfural 100	Furfural 100	Furfural 100	Furfural/CH_3OH/H_2O 48.3:38.9:12.8	Furfural/CH_3OH/H_2O 48.3: 39 :12.7
• Temperature (°C)	400	400	400	400	500
• WHSV (hr^{-1})	0.31	0.29	0.24	0.23	0.22
• Catalyst	H-ZSM-5	Zn-ZSM-5	Mn-ZSM-5	H-ZSM-5	Zn-ZSM-5
• Total products distribution (% w)					
CO	22.8	36.3	17.9	15.6	24.2
CO_2	0.6	4.8	1.9	1.5	8.2
H_2O	4.4	2.7	7.0	43.7	18.5
Coke	14.1	21.2	25.9	14.8	14.8
Oxygenated compounds*	50.0	25.6	40.7	3.3	0.6
Hydrocarbon	7.9	9.4	6.6	21.1	33.8
• Hydrocarbon yields (% C basis)	8.9	10.0	5.63	27.5	16.8
• Products selectivity (% w)					
Paraffins C_1-C_8	8.8	7.6	10.6	8.6	3.5
Olefins C_2-C_6	2.5	10.6	6.5	5.0	2.2
Aromatics	54.6	48.8	43.9	65.9	79.8
Polyaromatics	34.1	32.9	39.1	20.5	14.5

* mainly furan

TABLE 3

Catalytic Conversion of Glycerol With and Without Methanol at 400°C

• Feeds (w%)	Glycerol 100	Glycerol 100	Glycerol 100	Glycerol/CH_3OH/H_2O 54.2:36.6: 9.2	Glycerol/CH_3OH/H_2O 54.2:36.6: 9.2	Glycerol/CH_3OH/H_2O 54.2:36.6: 9.2
• WHSV (hr^{-1})	0.310	0.249	0.292	0.263	0.193	0.228
• Catalyst	H-ZSM-5	Mn-ZSM-5	Zn-ZSM-5	H-ZSM-5	Mn-ZSM-5	Zn-ZSM-5
• Total products distribution (% w)						
CO	7.8	3.1	6.6	2.5	1.1	2.7
CO_2	0.4	1.4	4.0	1.3	1.0	1.2
H_2O	59.7	59.4	56.7	65.6	58.1	55.9
Coke	14.0	19.3	17.0	14.7	24.9	23.1
Oxygenated compounds	13.3	6.9	0.3	—	0.1	0.1
Hydrocarbon	5.0	7.9	15.6	15.9	14.7	17.1
• Hydrocarbon yields (% C basis)	9.2	9.3	27.9	29.2	29.5	35.8
• Products selectivity (% w)						
Paraffins C_1-C_8	15.8	9.3	5.6	11.2	6.0	2.8
Olefins C_2-C_6	5.9	8.4	9.6	13.0	8.0	7.2
Aromatics	60.1	44.1	63.5	62.9	66.9	74.0
Polyaromatics	18.5	38.2	21.4	16.5	19.0	16.0

TABLE 4

Catalytic Conversion of Fructose and Glucose With and Without Methanol

• Feeds (w%)	Fructose/H_2O 19.8 : 80.2	Fructose/CH_3OH/H_2O 19.8 : 50.4 : 29.8	Glucose/H_2O 20.3 : 79.7	Glucose/H_2O 20.3 : 79.7	Glucose/H_2O 20.3 : 79.7	Glucose/CH_3OH/H_2O 20 : 30 : 50
• WHSV (hr^{-1})	0.056	0.136	0.062	0.055	0.046	0.178
• Catalyst	H-ZSM-5	H-ZSM-5	H-ZSM-5	Zn-ZSM-5	Mn-ZSM-5	H-ZSM-5
• Total products distribution (% w)						
CO	2.6	2.8	2.2	2.3	2.6	1.0
CO_2	2.3	2.1	1.5	4.2	1.9	1.4
H_2O	—	13.0	29.1	39.8	22.3	60.2
Coke	93.3	60.8	65.1	53.1	72.3	28.1
Oxygenated compounds	—	0.1	—	—	—	—
Hydrocarbon	1.8	21.2	2.1	0.7	0.9	9.4
• Hydrocarbon yields (% C basis)	1.55	17.2	2.5	1.14	0.1	6.1
• Products selectivity (% w)						
Paraffins C_1-C_8	17.4	21.5	23.6	35.4	26.2	9.2
Olefins C_2-C_6	18.8	14.2	41.6	2.6	42.8	34.0
Aromatics	48.5	52.1	34.8	62.0	31.0	53.7
Polyaromatics	15.3	12.3	—	—	—	3.5

hydrocarbons stops. Hence the very poor hydrocarbon yields observed (under 2.5%). The addition of methanol reduced the coke formation (60.8 and 28.1% for fructose and glucose respectively, although the degree of caramelization was still too high), and slightly improved the hydrocarbon yields (7.2% for fructose and 6.1% for glucose). The exact contribution of methanol is not yet quantified.

- Conversion of Fructose and Glucose Derivatives at 400°C

Caramelization of sugars is due in part by the large number of hydroxy groups that reduce their volatility. One approach to prevent the caramelization of sugars was to block the hydroxy groups by reacting the sugars with acetone to form isopropylidene derivatives:

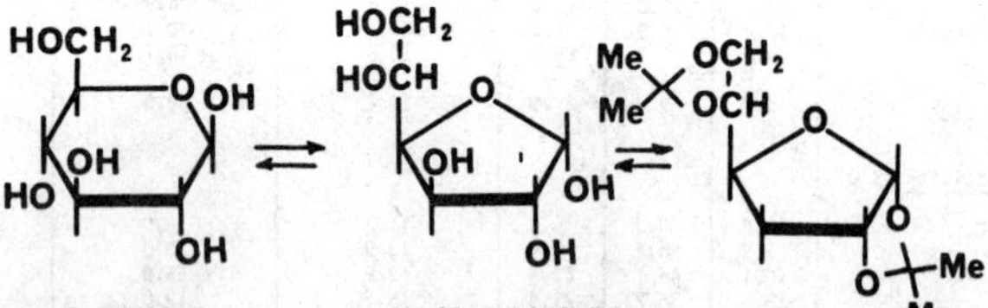

D-GLUCOSE GLUCO-FURANOSE

Table 5 shows the catalytic conversion of aqueous solution of glucose and fructose derivatives in the presence of methanol on H-ZSM-5, Zn-ZSM-5 and Mn-ZSM-5 at 400°C. From the results of table 5, it can be seen that the amount of coke formed by sugar derivatives is significantly reduced compared to pure sugars (table 4) by about a factor of 4. The oxygenated compounds (mostly acetone) are more for the derivatives compared to sugars. This was expected because of the thermal retrocondensation that converted back the derivatives to their starting materials (sugars and acetone). The hydrocarbon yields were also significantly increased. The product selectivity indicates a high percentage of aromatics.

- Effect of Temperature on the Conversion of Glucose and Glucose Derivatives

Tables 6 and 7 show the catalytic conversion of aqueous solution of glucose and glucose derivatives in the presence of methanol on H-ZSM-5 at 5 different temperatures. The results from table 6 indicate that between 350-500°C, there is an increase of the amount of coke as the temperature increase. This increase is due to the increase in caramelization with temperature. At higher temperature, there is some pyrolytic reactions which decrease the coke content. The product selectivity indicates a high percentage of olefins and aromatics. Glucose derivatives g ive less coke (table 7) although the percentage is still high for a catalytic reaction. This is mainly due to the retrocondensation of the derivatives to give acetone and glucose which is known to caramelize in this range of temperature. Higher yields of hydrocarbons were also observed. The product selectivity shows also a high percentage of olefins and aromatics. The maximum yield in hydrocarbons and the minimum formation of coke were observed at 500°C. At lower and higher temperatures, the coke formation obtained increased.

TABLE 5

Catalytic Conversion of Glucose and Fructose Derivatives in the Presence of Methanol

Feeds (w%)*	$GD/CH_3OH/H_2O$ 29.7:60.5: 9.8	$FD/CH_3OH/H_2O$ 27.6:60.2:12.2	$FD/CH_3OH/H_2O$ 27.6:60.2:12.2	$FD/CH_3OH/H_2O$ 27.6:60.2:12.2
WHSV (hr^{-1})	0.243	0.219	0.189	0.157
Catalyst	H-ZSM-5	H-ZSM-5	Zn-ZSM-5	Mn-ZSM-5
Total products distribution (% w)				
CO	0.5	0.8	1.2	1.3
CO_2	1.9	1.2	1.6	1.1
H_2O	63.1	36.5	61.4	58.7
Coke	13.2	24.5	10.7	13.2
Oxygenated compounds	4.3	3.6	13.5	0.3
Hydrocarbon	17.0	33.5	24.9	25.5
Hydrocarbon yields (% C basis)	17.3	21.7	38.0	39.4
Products selectivity (% w)				
Paraffins C_1-C_8	38.7	33.2	9.8	8.7
Olefins C_2-C_6	20.6	5.6	8.1	9.7
Aromatics	38.1	54.0	67.7	71.4
Polyaromatics	2.7	6.3	11.4	10.2

* GD: 1,2:5,6-di-O-isopropylidene-D-gluco-furanose
FD: 2,3:4,5-di-O-isopropylidene-D-fructo-pyranose

TABLE 6

Catalytic Conversion of Glucose in the Presence of Methanol and Water (20:30:50 %W) on H-ZSM-5

Temperature (°C)	350	400	450	500	550
WHSV (hr^{-1})	0.18	0.18	0.19	0.16	0.12
Total products distribution (% w)					
CO	0.8	0.8	0.5	2.3	1.0
CO_2	2.0	1.1	1.4	2.1	1.3
H_2O	66.6	61.9	59.7	77.9	60.2
Coke	26.5	27.5	33.3	50.3	28.1
Oxygenated compounds	0.6	2.0	0.7	—	—
Hydrocarbon	3.6	6.6	4.4	14.2	9.4
Hydrocarbon yields (% C basis)	5.8	6.1	6.5	9.2	13.7
Products selectivity (% w)					
Paraffins C_1-C_8	8.0	24.0	5.2	9.5	9.2
Olefins C_2-C_6	17.3	45.7	33.0	77.3	34.0
Aromatics	70.5	28.7	51.3	51.4	53.7
Polyaromatics	4.2	1.5	7.0	6.6	3.5

TABLE 7

Catalytic Conversion of Glucose Derivative in the Presence of Methanol and Water (41:50: 9 %W) on H-ZSM-5

• Temperature (°C)	350	400	450	500	550
• WHSV (hr^{-1})	0.30	0.33	0.31	0.24	0.29
• Total products distribution (% w)					
CO	0.3	0.5	0.2	0.1	0.7
CO_2	1.0	1.9	0.9	1.0	1.6
H_2O	64.8	63.1	66.1	70.7	75.1
Coke	18.9	13.2	14.1	9.3	10.11
Oxygenated compounds	2.8	4.3	3.9	1.1	0.8
Hydrocarbon	12.3	17.0	14.8	17.8	11.7
• Hydrocarbon yields (% C basis)	11.8	17.3	16.7	21.3	12.1
• Products selectivity (% w)					
Paraffins C_1-C_8	46.8	38.7	17.5	8.1	9.7
Olefins C_2-C_6	8.1	20.8	35.1	57.5	40.9
Aromatics	40.5	38.1	42.6	33.6	46.2
Polyaromatics	4.8	2.6	3.6	3.3	38.0

• Conversion of Hydrolysed-Cellulose Solution at 500°C

Aqueous solutions of cellulose hydrolysed products obtained from the acid hydrolysis of cellulose were contacted on H-ZSM-5 and Zn-ZSM-5 zeolites at 500°C and atmospheric pressure. Table 8 shows that decarboxylation is an important reaction. The coke content and the oxygenated compounds yield (mainly furan) are still too high. About 12% yield of hydrocarbons was obtained. CO and furan were obtained mainly from 5-hydromethyl 2-furfuraldehyde, a thermal degradation product of glucose, and carbon dioxyde from the decarboxylation of organic acids, e.g.:

$$CH_3\ CO(CH_2)_2CO_2H \rightarrow CO_2 + CH_3\ CO\ CH_2\ CH_3$$

With the addition of methanol, furan can be completely deoxygenated to give more hydrocarbons and lowered the coke content to a more acceptable value. Products selectivity indicates a high content of aromatics and polyaromatics.

CONCLUSION

Aqueous sugar solutions can be converted in the presence of ZSM-5 catalysts to hydrocarbons, but in very low yield. The incorporation of metal on ZSM-5 catalysts influences only the product selectivity but not the hydrocarbons yield. The addition of methanol, an in-situ hydrogen donor source, increases slightly the yield of hydrocarbons. The high content of coke formed is due to the caramelization of sugars. In order to prevent caramelization, hydroxy groups of sugars responsable for caramelization are blocked by isopropylidene groups. Higher hydrocarbon yields and lower coke contents were observed. This approach, although

TABLE 8

Catalytic Conversion of Hydrolysed Cellulose Products on ZSM-5 at 500°

	H-ZSM-5	Zn-ZSM-5
• Catalyst	H-ZSM-5	Zn-ZSM-5
• WHSV (hr^{-1})	0.02	0.02
• Total product distribution		
CO	37.3	39.8
CO_2	6.2	6.7
H_2O		
Coke	29.3	24.4
Oxygenated compounds	11.5	12.0
Hydrocarbons	15.7	16.9
• Hydrocarbon yields (% C basis)	11.4	12.4
• Products selectivity (%w)		
Paraffins	17.3	28.3
Olefins	13.1	5.3
Aromatics	42.3	41.4
Polyaromatics	27.3	28.9

it prevents the caramelization and permits higher hydrocarbon yields, is not however cost- efficient. As furfural can be completely deoxygenated to give high hydrocarbon products yield and as cellulose and hemicellulose can be converted in high yield (> 65%) to 5-hydroxymethyl-2-furfuraldehyde and furfural by acid hydrolysis, it is possible to combine these two reactions in a continuous process for converting biomass materials into hydrocarbon products. By first fractionating biomass to lignins and cellulosic materials which can then be degraded into small fragments of oxygenated compounds. Then in a second step the oxygenated compounds can be catalytically up-graded to hydrocarbon products. Presently work is being done in our laboratories to optimise this process.

ACKNOWLEDGEMENTS

This work was supported by grants from the Natural Sciences and Engineering Research Council of Canada and by the Quebec Ministry of Sciences and Technology. The authors express their gratitude to Professor Fernand Rheault for his constant support and Gerald Perron for GC/MS analyses.

REFERENCES CITED

1- Hasnain, S., Editor, Fifth Canadian Bioenergy R&D Seminar, London: Elsevier Applied Science Publishers, 1984.

2- Dao, L.H., Hebert, P., Houle, A., Haniff, M.I., "Two-stage Biomass Liquefaction Process", Montréal: International Solar Energy Society, 1985.

3- Goldstein, I.S., Organic Chemicals from Biomass, Boca Raton: CRC Press, 1981.

4- Frankiewicz, T.C., Process for Converting Oxygenated Hydrocarbons into Hydrocarbons, U.S. Patent 4,308,411.

5- Chen, N.Y., Koenig, L.R., Process for Converting Carbohydrates to Hydrocarbons, U.S. Patent 4,503,278.

6- Hebert, P., "Etudes Catalytiques pour la Liquéfaction Thermochimique Directe de la Biomasse en Carburants Liquides et Produits Chimiques de Base", M.Sc. thesis, INRS-Energie, 1984.

7- DeJongh, D.C., Biemann, K., "Mass Spectra of O-isopropylidene Derivatives of Pentoses and Hexoses", J. Amer. Chem. Soc. 80, 67-69 (1964).

8- Chang, C.D., Hydrocarbons from Methanol, New-York, Marcel Dekker, 1983.

DISCUSSION

J. Dhawan, U. of South Alabama — You mentioned that the zinc catalysts give more favorable results compared to ZSM5 based on the hydrocarbon yield. Also, I observed that the amount of water content in the product is almost double in the case of zinc. Is this a desirable effect?

Dao — Our process is aimed at deoxygenation. We want to remove oxygen as water because there's no other way. The yields are quite low.

Antal — At Copper Mountain, we had a presentation by Frankiewicz, who described a long study of various solid-phase catalysts. He was able to identify one which exhibited little or no coking. I'm curious as to why so much emphasis is being given to ZSM5?

Dao — The catalysts of Dr. Frankiewicz are also ZSM5. He diluted with silica. We used two different types of catalysts, but I cannot describe them here because patents are pending. One of our catalysts is grown on long fibers of silica, and the second one is on asbestos. We leach the magnesium out of the asbestos and create very small holes treated with ZSM5. With this kind of catalyst, coke is not formed either.

Antal — Are you able to offer any explanation as to why, without divulging your patent material, this particular configuration doesn't coke?

Dao — The lack of coke is due to two-phase acidity. With ZSM5, you have medium size pores about 6 A in diameter. If you use pure ZSM5, the acidity forms a lot of coke. If you dilute with a neutral support, as in the case of Dr. Frank Hewitt who used a neutral silica with an area of 0.2 m^2/g, the coke content is reduced.

J. Diebold, SERI — With very reactive feedstocks such as fructose and glucose, why did you choose to use such low-weight, hourly space velocities? This implies a very long residence time within the reactor space. You have competing reactions between the thermal reactions, which make the coke, and the catalytic reactions.

Dao — This study was done about 2 years ago. At that time, we also studied the effect of the space velocity. Last year, we used higher space velocities. The best yield was obtained with a fluidized bed and not a fixed bed. With a fixed bed, the residence time is too long, and permits competition between the catalytic and thermal reactions.

UPGRADING OF LIQUEFIED BIOMASS TO TRANSPORTATION FUELS BY EXTRACTION

Börje S. Gevert and Jan-Erik Otterstedt
Chemical Engineering
Chalmers University of Technology
S-412 96 GÖTEBORG, Sweden

ABSTRACT

Water, inorganic material and high boiling organic compounds of asphaltenic character were removed from directly liquefied biomass oil by extraction. The extraction yield decreased in the same order as the solubilizing power for the solvents acetone, xylene, octane and pentane.

Using decalin about 30% of the biomass oil can be extracted at temperatures higher than 90°C as a fraction which can be further upgraded by hydroprocessing and catalytic cracking to gasoline and diesel. The extraction yield did not depend on the extraction time in the range 8-44 minutes.

UPGRADING OF LIQUEFIED BIOMASS TO TRANSPORTATION FUELS BY EXTRACTION

INTRODUCTION

Processes for upgrading heavy oils such as atmospheric and vacuum residua of petroleum oils, shale oil and biomass oil include steps for removing the heaviest fractions of such oils containing elements, e.g., metals such as vanadium and nickel and heteroatoms such as sulphur, nitrogen and oxygen, which will interfere with downstream process steps such as hydrotreating and catalytic cracking.

Extraction can be used as an alternative to distillation to fractionate heavy oils, e.g., biomass oils, containing components which may undergo polymerization or decomposition at tempertures necessary to achieve efficient separation by distillation. On the other hand, useful solvents for extraction must meet stringent economical and performance criteria and large quantities of them must be circulated and treated in the extraction process (5).

The selection of a proper solvent for extractive separation is facilitated if the content and nature of asphaltenes and resins are known. Speight et al (9) have characterized heavy petroleum oils with regard to their content of asphaltenes and resins. Cavalier and Chornet (1) have shown that oil obtained from direct liquefication of peat contained between 30 and 40 % by wt of bitumen soluble in benzene. Schlager and Davis (8) separated biomass oil into three fractions: oil, asphaltenes and pre-asphaltenes, by using different extractive solvents. The yields of the different fractions could be varied by using solvents of different solubilizing power.

Biomasss in different forms is abundantly available in Sweden and much effort is made to study how it can be used as an energy resource. In the Department of Engineering Chemistry, Chalmers University of Technology, we are developing processes for upgrading liquefied biomass to gasoline and diesel oil. The starting material is a biomass oil, TR12, obtained by the PERC process (2). Preliminary work by us has shown that distillation of TR12 causes coking between 150°C and 200°C. In later work Elliott (3) has managed to distillate at temperatures above 300°C with a vacuum pot according to ASTM D 1160.

In this work we will describe the extractive separating of TR12 into different fractions using solvents of different solubilizing powers. In subsequent work we will study further upgrading by distillation, hydrotreating, catalytic cracking and reforming of extracted TR12.

EXPERIMENTAL

Material

The chemicals used; acetone, decahydronaftalene (decalin), octane and xylene were of analytical grade. The directly liquefied biomass oil used in our programme was produced by the PERC-process. Our sample is referred to as test run 12, TR12, in the Albany Liquefication pilot plant where chips from Douglas fir was liquefied.

Procedures

The extraction experiments were carried out in 250 ml flat-bottomed flasks provided with heater and magnetic stirrer. Biomass oil and solvent in ratios from 0 to 1.5 were added to the flask and the stirrer was started when the temperature reached the desired value, which varied from experiment to experiment but was in the region of from 90 to 200°C for decalin (boils at

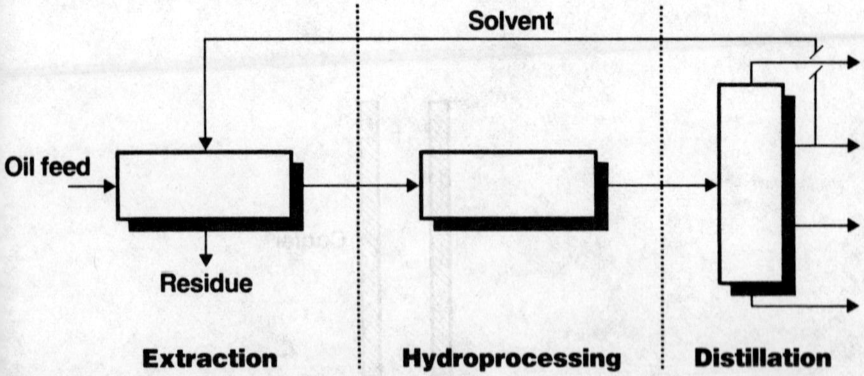

Figure 1. Integrated process for recovery of biomass oil by extraction, hydroprocessing and distillation.

200°C) or the boiling point for the other solvents. The time of extraction was measured from this point and was varied from 8 to 44 minutes. When the extraction experiment was completed the flask was rapidly cooled to room temperature by immersing it in ice. After decantation, the solvent phase was centrifuged at 4000 rpm. The residue in the form of undissolved material in the flask and sediment in the centrifuging tubes were dried and weighed. The yield of extracted material (E) was taken as

$$E = (1 - \frac{R}{B - W})\,100$$

where R = the amount of dried residue
B = the amount of TR12
W = the amount of water in TR12

Elementary analyses of TR12 and extracted fractions of TR12 were carried out using Leco CHN 600 for carbon, hydrogen and nitrogen and Leco SC 132 for sulfur. Oxygen was determined on a Leco RO 116 by a method which had been adopted to liquid samples (4). The molecular size distributions of TR12 and extracted fractions were determined by HPLC using tetrahydrofuran as solvent and by simulated distillation according to ASTM D2887-3. Pour point, water and viscosity at 80°C were determined according to ASTM D97, ASTM D244 and ASTM D445 respectively.

RESULTS AND DISCUSSION

In Table 1, showing the compositions of TR12 biomass oil, the contents of ash and elements were determined on water-free basis. The high contents of ash containing sodium indicate that severe problems would be encountered if TR12 was directly subjected to hydrotreatment in order to remove heteroatoms. The contents of C, H and O correspond to the stochiometric formula $CH_{1.25}O_{.15}$ and H/C ratio of .95 if the oxygen is assumed to be converted to water in a downstream hydroprocessing step. This is a lower ratio than in conventional petroleum feedstocks (7).The nitrogen content is below the detection limit of the Leco R0 116 which is 0.02%. The sulphur content is only 0.009% and sulphur could be present in the organic as well as the inorganic fraction of the oil. The content of water is much higher than in petroleum oil indicating a highly polar nature of TR12. The TR12, used by us, might have polymerized during the storage of about three years. Average molecular weight was determined to 1014.

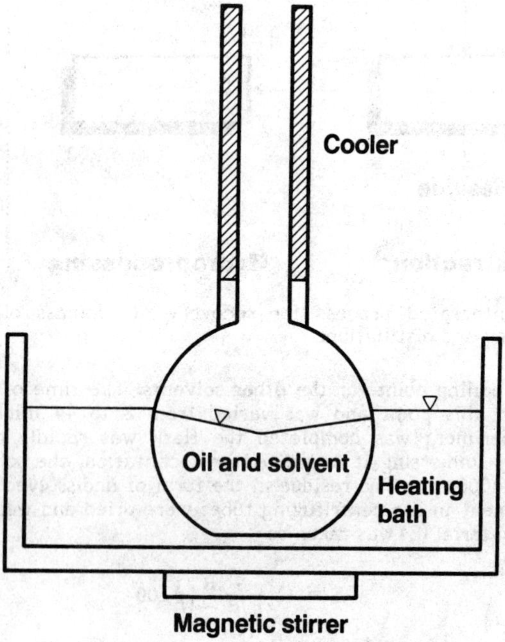

Figure 2. Extraction equipment.

Table 1

Characterization of PERC TR12 Oil

Ash	3.77 %
Carbon	73.8 %
Hydrogen	7.7 %
Nitrogen	<0.02 %
Oxygen	15.1 %
Sulphur	0.009 %
Sodium	2.0 %
Water	7.2%
Pourpoint	35 °C
Weight average molecular weight	1014
Viscosity at 80°C	1066 cst

Table 2 shows that the extraction yield depends on the solvent and increases with increasing polarity as defined by Rohrschneider (6). Acetone, being the most polar solvent used will extract between 47 to 74% of TR12 depending on the extraction time and the amount of solvent, whereas pentane will extract less than 3% . The determination of the molecular weight distribution by HPLC showed that tetrahydrofurane was an even better solvent than acetone, but it was not further studied in this work.

The effect of the amount of solvent used on the extraction yield is shown in Figure 3 for xylene and in Figure 4 for decalin. Curve A in Figure 3 corresponds to a centrifuging time of 5 minutes whereas curve B and Figure 4 correspond to 10 minutes in the centrifuge. In the case of short residence in the centrifuge, Curve A, small particles may not have had time to settle and were included in the yield. At high concentrations of TR12, however, curves A and B are coming together indicating that small particles can agglomerate under these conditions and therefore settle faster. For the same ratio of TR12 to solvent, extraction with xylene gave higher yields than with decalin. When about equal weight amounts of TR12 and xylene were used the solvent dissolved in the biomass oil and only one phase was formed. With decalin the extraction yield drops when using over 30% by weight of TR12 in the extraction mixture as can be seen in Figure 4.

Figure 5 shows that the yield in decalin increases rapidly with temperature up to about 90°C where it reaches a constant value of about 32%. The effect of temperature on the yield is not easy to interpret, since the extracts were cooled to room temperature before they were centrifuged in order to be able to perform this operation at constant temperature. A further complication is that the viscosity of TR12 at temperatures below 90°C is so high that extraction may not be efficient due to poor contact with the solvent. The constant yield at temperatures above 90°C may therefore simply reflect the fact that in this temperature range the viscosity is low enough to ensure efficient extraction and the yield corresponds to the solubility of TR12 in the solvent at room temperature.

Figure 6 shows that the yield of TR12 in decalin is independent of the extraction time in the time interval studied, 8-44 minutes. Some extraction undoubtedly takes place during the heating up period.

The analysis results of the TR12 part of the extracted phase is presented in Table 3. Used solvent is decalin at boiling point, with 33% by weight of TR12. The major fraction, about 72% by weight, is material which correspond to light and heavy gasoil.

SUMMARY AND CONCLUSIONS

The content of ash and sodium in TR12 is so high that severe deactivation of the catalyst would occur if direct deoxygenation of TR12 by hydrotreatment were to be tried.

A fraction corresponding to crude oils can be extracted in yields up to about 30% using solvents with solubilizing powers between those of octane and acetone.

The yields increase with the extraction temperature and reaches a constant level at about 90°C.

The separation of the organic phase containing liquefied biomass dissolved in a solvent from undissolved biomass can be affected by centrifuging for about 10 minutes in order to ensure the removal of small particles of undissolved biomass.

Table 2

Effect of solvent

Solvent	Yield wt %	Solvent Polarity	Normal Boiling Point °C
Acetone	47-74	5.1	56
Xylene	28-56	2.5	140
Octane	11-29	0.1	126
Pentane	0-3	0	36

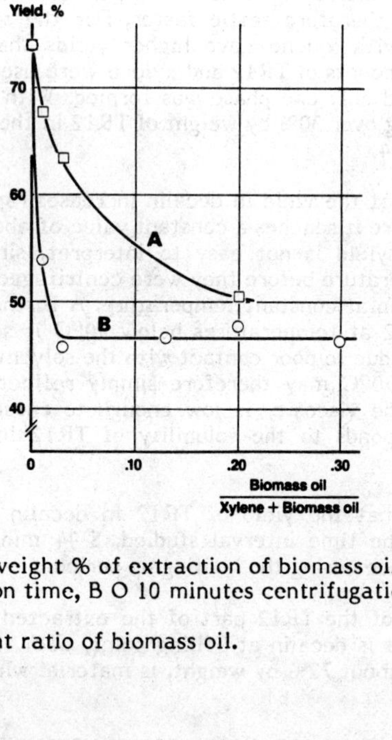

Figure 3. The yield in weight % of extraction of biomass oil with xylene *. A □ 5 minutes centrifugation time, B O 10 minutes centrifugation time.

*as a function of weight ratio of biomass oil.

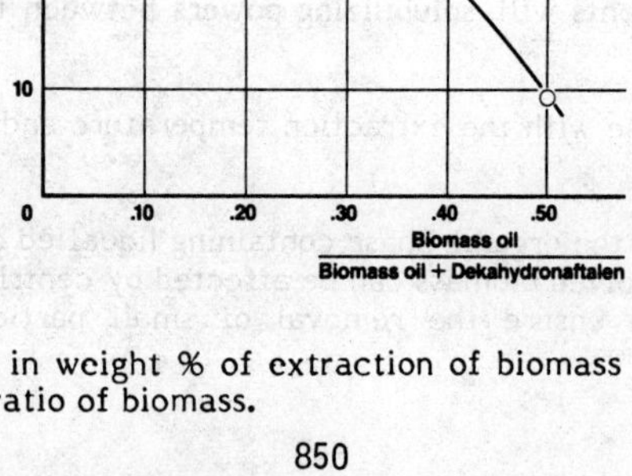

Figure 4. The yield in weight % of extraction of biomass oil with decalin as a function of weight ratio of biomass.

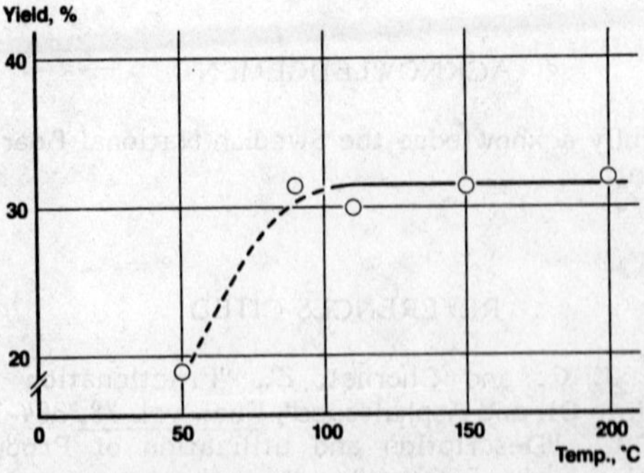

Figure 5. The yield in weight % of extraction with decalin at different temperatures and 20 minutes extraction time.

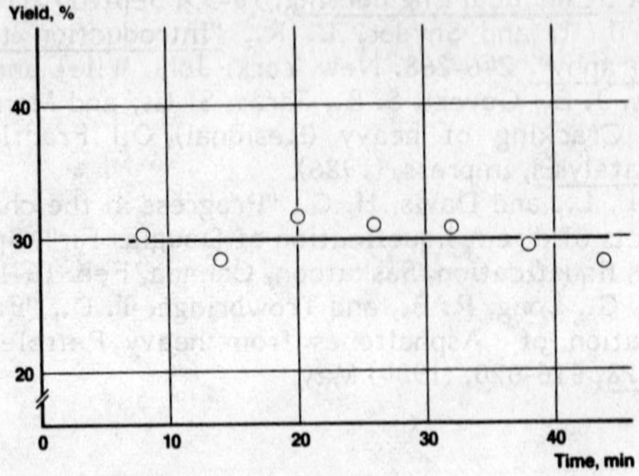

Figure 6. The yield in weight % of extraction of biomass oil with decalin at different extraction times and the amount of solvent constant at 70 wt. %.

Table 3

GC and HPLC analysis results of extracted phase. Extraction with decalin at 33% concentration of biomass oil in extraction.

Type of Fraction	Boiling Range ºC	Relative Amount
Gas	-35	0
Naphtha	36 - 196	9.7
Light gasoil	197 - 350	36.9
Heavy gasoil	351 - 525	35.6
Vacuum residue	525 +	18.9

ACKNOWLEDGEMENT

The authors gratefully acknowledge the Swedish National Board of Energy who sponsored this projet.

REFERENCES CITED

1. Cavalier, J. C. and Chornet, E., "Fractionation of Peat-derived Bitumen into Oil and Asphaltenes", Fuel. vol. 78, 304-308 (1978) May.
2. Elliott, D.C., "Description and utilization of Product from direct liquefication of Biomass", Biotechnology and bioengineering symposium No. 11, 187-198 (1981).
3. Elliott, D. C. "Analysis and comparison of products from wood liquefication". Fundamentals of Thermochemical Biomass Conversion, pp. 1003-1018 (Eds.) Overend, M.T.A. & Mudge, L.K. Elsevier Applied Science Publishers, London (1985).
4. Gevert, S. B., "Determination of Oxygen in Non-volatile Organic Materials", The Pittsburgh conference on analytical chemistry and applied spectroscopy, March 10-14, Atlantic City, (1986).
5. Humphrey, J. L., Rocha, J. A. and Fair, J. R. "The Essentials of Extraction", Chemical Engineering, 76-95, Sept. 17, (1984).
6. Kirkland, J. J. and Snyder, L. R., "Introduction to modern liquid chromatography". 246-268. New York: John Wiley and sons Inc., 1979.
7. Otterstedt, J. E., Gevert, S. B., Järås, S. G., and Menon, P. G., "Fluid Catalytic Cracking of heavy (Residual) Oil Fractions; a Review", Applied Catalysis, in press, (1986).
8. Schlager, L. L., and Davis, H. G., "Progress in the characterization of the Products of direct liquefication of Douglas Fir", Specialist meeting on biomass liquefication, Saskatoon, Canada, Feb. 16-17, (191), 1982.
9. Speight, J. G., Long, R. B., and Trowbridge, T. D., "Factor influencing the separation of Asphaltenes from heavy Petroleum Feedstocks", Fuel. vol. 78, 616-620, (1984) May.

DISCUSSION

D. Elliott, Pacific North-west Laboratory	Was the work done with the oil as received? Did you do an acid wash or any type of extraction of the sodium ?
Gevert	It was done with the oil as received. Pretreatment was not done.
Elliott	I think that it is important to consider then that the high amount of ash and water in the oil, particularly TR12, is somewhat odd in that it has 2 to 3 times more sodium in it than the other products that came out of the Albany plant. The sodium is not just an ash problem, but is combined with some of the phenolic material and converted it into a salt, which may be less amenable to extraction. It might be interesting to try these same kinds of extractions on an oil that contains less sodium to see it more of the free oils might be extractable.
Gevert	One purpose of the extraction is to get rid of the sodium.
Elliott	It might be more effective to remove the sodium by an acid wash rather than by organic solvents separation.
D. Boocock, U. of Toronto (Canada)	It's well known that many of these biomass oils are notoriously unstable and undergo some type of polymerization. Do you know how old this oil was? Has anybody done any work on similar types of extractions as a function of the age of the oil?
Gevert	I don't know of any work on extraction at different ages of the oil. When I did the extraction, the oil was 3 years old.
Elliott	The oil that's produced from TR12 came out of the plant in the Spring 1981, so it's now closing on 5 years.
Gevert	The extraction was done 2 years ago.
Elliott	We have to determine that this type of product is fairly stable when stored in the barrels as it is made. That is based on redistillation and elemental analysis after time. We haven't done molecular weight determinations or solvent fractionation.

DISCUSSION, CONT.

____________	Your last slide showed boiling fractions. Was this done on the decalin extractable fraction? Has it been stabilized in one way or another? Did you do elemental analyses on these fractions, and how do they compare with the original?
Gevert	We have not made elemental analyses on these fractions. We did not separate this oil from the solvent since it was produced by direct hydroprocessing of the material. The figures I presented on constitution of the extracted oil were calculated from analysis of the extract which contained the solvent.

DIESEL FUEL FROM CHEMICALLY MODIFIED VEGETABLE OILS

David J. Hassett
Director, Fuels Analysis Laboratory
Engineering Experiment Station
University of North Dakota
Grand Forks, North Dakota, U.S.A.

ABSTRACT

The University of North Dakota Engineering Experiment Station is currently engaged in research on the use of chemically modified vegetable oils as replacement agricultural diesel fuels. Direct injection diesel engines, which are widely used in agriculture experience a variety of fouling problems when operated on unmodified vegetable oils. These problems such as sticking rings, fouled injectors, and varnish build-up on cylinder walls, appear to be partially related to poor injector performance caused by the relatively high viscosity of vegetable oils. The formation of lower alcohol esters is an effective chemical means to reduce viscosity. Current research has focused on the use of sunflower oil methyl ester. Physical and chemical properties of this fuel at several degrees of refinement were used to determine fuel properties. Engine testing carried out indicated that fuel grade sunflower oil methyl ester, defined as sunflower oil converted to methyl ester of at least 90 percent purity, had about the same tendency to foul a direct injection diesel engine as number 2 diesel fuel.

A pilot plant for the production of fuel grade vegetable oil methyl ester has been designed and constructed. This plant produces 500 gallon batches of fuel grade sunflower oil methyl ester. At the time of this report, 3,000 gallons of this fuel grade methyl ester have been produced. A Case model 2390 tractor with a turbocharged 504 cu. in. displacement engine has used this fuel in a farming operation in North Dakota. There have been no problems either with this tractor or with the operation of the pilot plant.

DIESEL FUEL FROM CHEMICALLY MODIFIED VEGETABLE OILS

INTRODUCTION

Dr. Rudolph Diesel, the inventor of the diesel engine, had determined, as early as 1911, that vegetable oils were suitable for use as diesel fuels. The use of vegetable oils, as well as the use of powdered coal, as diesel fuels, however, was abandoned when inexpensive petroleum based fuels became available. These, besides offering an economic advantage, were free from many of the problems which had been associated with the earlier fuels. Some of the first studies indicated that although vegetable oils could be used as diesel fuels, there were a number of problems. These problems included fouled injectors, sticking rings, and varnish buildup on the cylinder walls (1). Vegetable oils which found their way into the crankcase also had a tendency to polymerize and thicken the lubricating oil. This particular problem has contributed to engine failure in several instances where vegetable oils were used as substitute diesel fuels. Most of the problems associated with the use of vegetable oil diesel fuels can be attributed to both the relatively high viscosity of vegetable oils and to the reactivity of polyunsaturated fatty acid components of the triglycerides making up the oils. Current research has indicated that chemical reduction of viscosity by the formation of lower alcohol esters, such as methyl esters, can reduce or eliminate most of these problems.

Discussion

Methyl ester formation appears to be an answer to most of the problems associated with the use of vegetable oil derived fuels in direct injection diesel engines. Problems associated with the use of vegetable oil diesel fuels are primarily sticking rings, fouled injectors, fouled valve guides, and varnish buildup on cylinder walls. These problems can all be associated with poor injector performance caused by the relatively high viscosity of vegetable oils. The conversion of vegetable oil triglycerides to methyl esters represents a chemical approach to viscosity reduction.

There are drawbacks to the use of vegetable oil methyl esters as diesel fuels. These are:

1. Cost - Vegetable oil diesel fuels are at the present time much more expensive than conventional petroleum based diesel fuels. It costs about $1.00 to convert one gallon of sunflower oil to fuel grade methyl ester. This cost is, of course, independent of the cost of sunflower oil. This makes fuel grade sunflower oil methyl ester about three times as expensive as petroleum based diesel fuel at today's prices.

2. Low Temperature Properties - One of the more serious drawbacks to the use of sunflower oil methyl ester as a diesel fuel is the formation of crystals which occurs at about 4 to 5°C. These crystals are methyl palmitate and methyl stearate. Fuel for use at temperatures below this range would have to be stored in heated tanks and transferred through heated lines. In most climates bulk storage could be underground where the mean temperature would be around 10°C.

Experimental

Sunflower oil was used in all of our research. Sunflower oil was chosen because sunflower oil production is a major local industry and because the chemical makeup of this oil provides desirable chemical and physical properties for its use as a diesel fuel after synthesis into methyl esters. Sunflower oil, a semidrying oil, contains fatty acids with enough unsaturation to insure that its methyl esters will be liquid but at the same time not the conjugated double bonds that are the source of the high reactivity of the drying oils, such as tung or linseed. The term sunflower oil will be used in the rest of this report in place of the generic term vegetable oil. It should be assumed that any vegetable oils from the oleic-linoleic semi-drying category could be substituted.

Research at the University of North Dakota on the synthesis of vegetable oil methyl esters for use as a substitute diesel fuel has been directed by the two following hypothesis:

1. Viscosity reduction is a solution to the engine fouling problems associated with the use of vegetable oils as diesel fuels. This viscosity reduction can be brought about chemically by the conversion of vegetable oil triglycerides to methyl esters. The theory which led to this hypothesis was that poor injector performance, due to the high viscosity of vegetable oils, caused large droplets of oil to be injected into the combustion chamber. These large droplets formed a residue of unburned or partly burned vegetable oil which caused further degradation of injector performance. This cycle eventually led to engine failure.

2. Up to ten percent of unreacted or partly reacted vegetable oils in the methyl ester would not cause fouling problems. This statement concerning allowable fuel impurities is based on two assumptions: First, there have been no reported problems with up to ten percent of vegetable oils as fuel extenders with number 2 diesel fuel; second, if vegetable oil methyl esters perform as well as number 2 diesel fuel, they may also be able to tolerate up to ten percent of unreacted or partly reacted vegetable oils without deterioration of fuel properties. This

hypothesis concerning fuel properties led to the term "fuel grade" methyl ester. This was defined as vegetable oil esters of at least 90 percent purity. The allowable impurities would be unreacted or partly reacted vegetable oil. In all of our syntheses yields have been in excess of 95 percent making this criteria for purity easily attainable, in practice.

Investigation of the use of sunflower oil methyl ester proceeded in three steps. These were:

1. Synthesis of methyl ester and chemical analysis to define reaction conditions and product purity.
2. Synthesis of 9 liter test batches of sunflower oil methyl ester and small scale engine testing.
3. Design and construction of a pilot plant for the production of fuel grade sunflower oil methyl ester. Ester produced in this plant is currently being field tested in a farming operation in North Dakota.

Synthesis of "Fuel Grade" Sunflower Oil Methyl Ester

The conversion of sunflower oil to sunflower oil methyl ester was accomplished by transesterification with methyl alcohol using sodium methoxide as a catalyst. 8.37 kg of sunflower oil (9L) was placed in a 12 L three neck flask fitted with a reflux condenser, stirrer, and thermometer. The flask was heated with a jacketed heating mantle. As the oil was heating but before it reached 40°C, 1456 g of methanol which was reacted with 8.4 g of sodium was added. The reaction mixture was stirred vigorously since sunflower oil and methanol are immiscible until ester formation begins. Methyl ester solubilizes the reactants. The reaction mixture was heated to 60°C and held there with stirring for two hours. The reaction mixture was then allowed to cool overnight without stirring--layers formed. The bottom layer which consisted mostly of glycerol was discarded and the top layer washed with water until the sodium catalyst residue was less than 2 mg/L. This washed layer was dried with anhydrous sodium sulfate and used as fuel grade sunflower oil methyl ester.

In this synthesis, 5 moles of alcohol are used for each mole of vegetable oil instead of the 3 moles required. This excess is used to improve yield and aids in separation of the glycerol layer (2).

In this reaction:

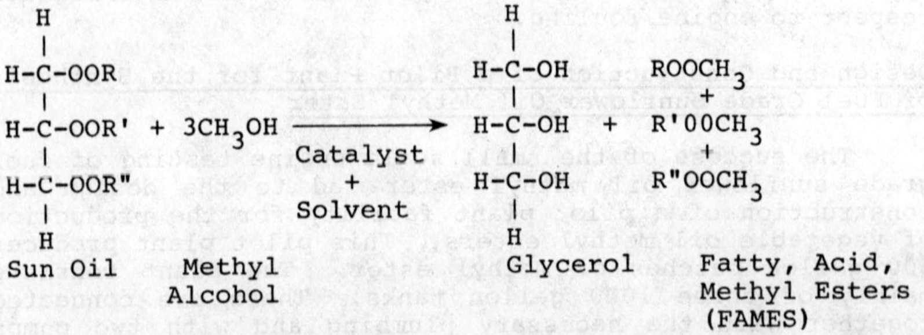

R, R', and R" are primarily 16 and 18 carbon chains. The triglycerides from the sunflower oil in this process are converted into three separate compounds, each a different methyl ester.

All raw oil and ester were characterized chemically. This characterization was done by gas chromatographic analysis and confirmed that the oil and ester were of the composition expected for sunflower oil from northern climates. In the future we feel it may be possible that this characterization, in addition to flash point and viscosity information, could be used as a means of checking vegetable oils and fuel grade esters before use or sale as diesel fuels.

Fuel grade methyl ester was vacuum distilled to a refined sunflower oil methyl ester of about 100 percent purity. This refined methyl ester was used for comparative testing. Engine testing was done using number 2 diesel fuel, "fuel grade" sunflower oil methyl ester, and refined sunflower oil methyl ester. This range of fuels allowed us to compare the relative performance of each vegetable based fuel against that of number 2 diesel fuel.

<u>Engine Testing</u>

Engine testing was done primarily to compare the relative tendency of each of the vegetable oil diesel fuels to foul direct injection diesel engines. The standard for acceptable amounts of deposits was set by running a test with number 2 diesel fuel. Tests were run on a Lister model LT-1 direct injection diesel engine. This 5.25 H.P. engine was coupled to a 10 H.P. electric motor wired so it could be used either as a dynamometer or as a starting motor. All engine testing was done at constant RPM and at constant loading since it has been found in our research and by others (3) that testing in this manner caused direct injected diesel engines to foul in less than 100 hours. This allowed comparative testing in a fairly short time and with a minimum of fuel use. A series of tests were conducted using sunflower oil, fuel grade sunflower oil methyl ester, and number 2 diesel fuel.

These engine tests indicated that fuel grade methyl ester appeared to perform as well as number 2 diesel fuel with respect to engine fouling.

Design and Construction of a Pilot Plant for the Synthesis of Fuel Grade Sunflower Oil Methyl Ester

The success of the small scale engine testing of fuel grade sunflower oil methyl ester led to the design and construction of a pilot plant facility for the production of vegetable oil methyl esters. This pilot plant produces 500 gallon batches of methyl ester. The plant consists mainly of three 1000 gallon tanks. These are connected together with the necessary plumbing and with two pumps and a heater bank are all that is necessary to prepare large (500 gallon) batches of fuel grade sunflower oil methyl ester. The tanks have the following functions:

1. Synthesis - Sunflower oil, methyl alcohol, and catalyst (NaOH) are mixed together in the synthesis tank. The materials are reacted by circulating through an electric heater bank. This synthesis tank has a tapered bottom which leads to a section of glass pipe. The tapered bottom facilitated the removal of glycerol, which is a by-product of the reaction. This glycerol, which has a greater density than the methyl esters, settles to the bottom of the tank. It has a darker color than the top ester layer. The interface between these two layers is easily seen when it reaches the section of pipe at the bottom of the tank. Because of this, it is relatively easy to determine when all of the glycerol has been drained off. Synthesis is the only function in the process that this tank is used for.

2. Washing/drying - After synthesis and after the glycerol layer has been drained off, the reaction mixture is transferred to another 1000 gallon tank. This tank has a narrow vertical glass window mounted near the bottom. The purpose of this window is to be able to see the interface between the oil and the water used for washing. The oil water mixture is agitated gently by means of a stirrer attached to an air motor. Gentle agitation is necessary because the mixture, during the first few washes, will form an emulsion if agitated too vigorously. The window allows the operator to determine if an emulsion is forming and maintain stirring at a rate low enough to prevent this from happening. Five water washes of about 150 gallons each are used. These water washes remove catalyst, glycerol, and unreacted methanol. After the final wash, the water layer is drained and anhydrous sodium sulfate added to remove the remaining traces of water. Fuel grade sunflower oil methyl ester remains, and after filtering, is suitable for use as a diesel fuel. This tank is the only one in which water is ever placed. Water would interfere

with the synthesis, and for this reason, the reaction mixture is transferred to a second tank for washing. The synthesis tank could be used for the washing step but would require drying after each batch before another synthesis could be done. This would not only require extra time but could interfere with the synthesis if not done thoroughly.

3. Storage - The remaining 1000 gallon tank is used for storage of the dried and filtered fuel grade sunflower oil methyl ester. Between batches of ester it can also be used for storage of sunflower oil. As with the synthesis tank, water is never deliberately put into the wash tank.

The tanks are in this manner divided categories by function; one tank for synthesis which never has water added; a second tank for washing with water; and a third tank, just for storage.

The entire pilot plant is constructed so liquid can be pumped from any one tank to another, and so vegetable oil can be pumped into storage, or methyl ester can be loaded into a tank truck using only two pumps. The plant is constructed from mild steel tanks and laid out using no copper based or galvanized materials. Rubber is attacked by vegetable oils and by methyl ester, so where seals or flexible connections must be used, only synthetic materials may be utilized. Reaction conditions used in this plant are the same as described in the experimental section.

Conclusions and Recommendations

1. In a comparison test of Number 2 diesel fuel, fuel grade methyl ester, and sunflower oil, the fuel grade methyl ester (about 90 percent purity) appeared to perform about as well as Number 2 diesel fuel with respect to engine fouling tendency. This comparative testing was done for 180 hours under constant load and constant RPM. Sunflower oil caused engine failure in 87 hours when run under identical conditions.

2. Formation of crystals in fuel grade methyl ester is a problem at temperatures below 4-5°C. It is not a reliable low temperature fuel. Heated fuel lines and tanks would have to be provided for cold weather use. Bulk storage would have to be indoors or underground. The exact temperature of crystal formation is difficult to determine in fuel grade methyl ester because of the tendency for supersaturation to occur with respect to solubility of pure compounds such as methyl stearate and methyl palmitate and because of the polymorphism exhibited by glyceride impurities.

3. Fuel grade sunflower oil methyl ester can be prepared in a 25 gallon/hour plant for $3.25/gallon (4).

Sunflower oil accounts for about 75 percent of this cost and there are possible design improvements which could bring this cost down.

4. There is a great need for the development of fuel standards and standardized test procedures for the characterization and evaluation of vegetable oil based diesel fuels. Many ASTM test procedures are inappropriate when applied to vegetable oils or modified vegetable oils. Some examples are flash point and distillation curves. The primary purpose of determining the flash point in a vegetable oil derived diesel fuel would be for safety reasons to determine if alcohol or solvent contamination was a problem. With respect to a distillation, in a system with only five major chemical components, fuel properties can be defined by a gas chromatographic analysis of fatty acids. Since distillation of glycerides or fatty acid esters at atmospheric pressure causes decomposition, and vacuum distillations are difficult to run and duplicate, a chemical analysis by fatty acid composition could substitute and would provide virtually the same information.

Despite reported drawbacks, sunflower oil methyl ester is an attractive alternative diesel fuel. At the present time a farm tractor is completing field testing of fuel grade sunflower oil methyl ester. Approximately 3,000 gallons of fuel have been consumed during this season's farming operation. At the time of the writing of this report, 500 gallons of additional fuel are being prepared to finish this season's work. No problems have been experienced in field operation, although low temperatures have not yet been encountered.

References

1. Bruwer, J.J.; Van D. Boshoff, B; Hugo, F.J.C.; du Plessis, L.M.; Fuls, J.; Hawkins, C.; van Der Walt, A.N.; Engelbrecht, A.; Sunflower Seed Oil as an Extender for Diesel Fuel in Agricultural Tractors, 1980, Symposium of the South African Institute of Agricultural Engineers, June 11, 1980.

2. Price, D., Sprules, F.J., Transesterification Process, U.S. Patent 2, 543, 421 Feb. 27, 1951.

3. Van Der Walt, A.N., Hugo, F.J.C., Diesel Engine Tests with Sunflower Oil as Alternative Fuel. P. 1927-1933 Proceedings of the Third International Conference on Energy Use Management, West Berlin, October 26-30, 1981.

4. Hassett, D.J., Hasan, A.R., Final Report, Sunflower Oil Methyl Ester as a Diesel Fuel, Bulletin No. 83-05-EES-03, May, 1983.

DISCUSSION

Antal	How much is a gallon of sunflower oil?
Hassett	It's about $2.00 right now. That puts the cost of the fuel at $3.00 a gallon.
W. DeGroot, U. of Montana	You mentioned that you had worked with ethyl esters. Do they have similar problems in terms of crystallization?
Hassett	No, as you would expect from the chemistry, they have better low-temperature properties. But ethanol costs about 15% more than methanol to convert one gallon of sunflower oil to the esters and that's primarily due to the higher cost of the ethanol and because ethanol is a larger molecule. In addition, ethyl esters are a little trickier to make with respect to the amount of water that can be included. Water tends to lower the yields more with the ethyl ester. We like the idea of ethyl esters from the standpoint that there's a larger ethanol industry in North Dakota, and we like to use renewable resources.
P. Bente, Consultant	How large an engine is used in this tractor, and is there a range of sizes that can be operated satisfactorily?
Hassett	The tractor that we're using has a 503 cu. in. engine, and burns about 85 gallons of fuel an hour. Theoretically, you can use any size engine you want to. All of our engine testing at the research level was done with a 5-1/4 horsepower engine. What's different about agricultural engines is that they are direct injection for the most part. Engines that have a precombustion chamber such as cars and some high speed diesels don't have as much trouble with vegetable oil diesel fuels. As a matter of fact, some say that you can probably get by with some vegetable oils without esterification. I don't know if I really subscribe to that theory or not. It hasn't been tested that well, but with direct injected engines, you have to do something. At the end of this season, the diesel engine and tractor will be torn down by diesel mechanics at the Wopton State School of Science in North Dakota. They will try to tell us what we've done to it. Although right now,

DISCUSSION, CONT.

when they hooked it to a dynameter to check it out, the same power output that it produced when new was found.

CATALYZED AND UNCATALYZED CONVERSION OF CELLULOSE BIOPOLYMER MODEL COMPOUNDS TO CHEMICAL FEEDSTOCKS IN SUPERCRITICAL SOLVENTS

Michael Jerry Antal, Jr., Andrew Brittain,
Carlos DeAlmeida, William Mok and Sundaresh Ramayya
Renewable Resources Research Laboratory
Department of Mechanical Engineering
University of Hawaii, Honolulu, Hawaii 96822

ABSTRACT

Products and rates of the catalytic dehydration of ethanol, propanol, ethylene glycol and glycerol in supercritical water are described. These model compounds were selected to mimic the reaction chemistry of the polyalcohol cellulose. In all cases the olefinic dehydration product is rapidly formed with essentially stoichiometric selectivity. These results illustrate the potential of supercritical water as a media for directing the reaction chemistry of biopolymer solutes towards valuable chemical products.

CATALYZED AND UNCATALYZED CONVERSION OF CELLULOSE BIOPOLYMER MODEL COMPOUNDS TO CHEMICAL FEEDSTOCKS IN SUPERCRITICAL SOLVENTS

INTRODUCTION

Efficient thermochemical processes underlie the conversion of crude oil and natural gas liquids to higher value chemicals and fuels. Unfortunately, attempts to develop similar conversion processes for biomass feedstocks (such as wood chips and bagasse) have been frustrated by the non-specificity of high temperature pyrolysis reactions involving biopolymer substrates (1,2). For example, the pyrolysis of bagasse yields a liquid mixture of carbohydrate sirups and phenolic tars, a gas composed primarily of carbon oxides and hydrogen, and a solid charcoal. Variations of the conventional engineering parameters (temperature, heating rate, residence time and pressure) do not provide a good control over the complex set of concurrent and consecutive pyrolysis reactions. Thus it has not been possible to engineer the pyrolysis reactions to produce a few high value products from biomass materials.

The lack of progress towards gaining control of biopolymer pyrolysis chemistry may be explained by the non-specificity of the high temperature free radical reactions which constitute most pyrolytic phenomena. Involving an extraordinary variety of initiation, propagation and termination reactions, the pyrolysis chemistry of biopolymers has defied all attempts to detail reaction mechanisms at the molecular level. In contrast with this situation, the lower temperature heterolytic (ionic) reaction chemistry of many organic compounds in the aqueous phase is well understood, and can be catalyzed to gain desired specificity. For this reason it is of interest to discover conditions which favor heterolytic bond scissions at temperatures which are conducive to the formation of alkenes and related high value chemicals from relatively stable biopolymer materials.

As discussed in the following sections, when the density of supercritical (SC) water exceeds 0.4 g/cm^3 the fluid retains its ionic properties (high dielectric constant and ion product) - even at high temperatures. These properties provide new opportunities to catalyze a variety of heterolytic bond cleavages with a high degree of specificity. In this paper we report progress towards employing the ionic properties of supercritical water as a means for gaining control over the high temperature reaction chemistry of biopolymer materials.

Unfortunately, the reaction chemistry of biopolymers in supercritical fluids is very complex. Consequently, our initial work has emphasized studies of the reaction chemistry of model compounds in supercritical solvents. Cellulose is the major component of most biomass materials, and the hydroxyl group is the principal functionality of the cellulose polymer. Consequently, in this paper we describe studies of the reaction chemistry of alcohols, diols, and polyols in supercritical water. These model compounds mimic the reaction chemistry of the polyalcohol cellulose. Future papers will present research involving four, five, and six carbon sugars, and more complex biopolymer materials in supercritical solvents. In all cases our goal is to define reaction conditions which enable heterolytic reactions to selectively form high value products from a renewable feedstock.

PRIOR WORK

Interest (3-5) in the use of SC fluid solvents as a reaction media is founded upon recent advances in our understanding of their unique thermophysical and chemical properties. Worthy of special note are those thermophysical properties (6) which can be manipulated as parameters to selectively direct the progress of desirable chemical reactions. These properties include the solvent's dielectric constant (7), ion product (8,9), electrolyte solvent power (10,11), transport properties [viscosity (12), diffusion coefficients (13) and ion mobilities (14)], hydrogen bonding characteristics (15), and solute-solvent "enhancement factors" (6). All these properties are strongly influenced by the solvent's density in the supercritical state.

For example, SC water with $\rho = 0.47\ g/cm^3$ at 400° C (P = 35. MPa) enjoys a dielectric constant of about 10 (comparable to a polar organic liquid under normal conditions), an ion product of 7×10^{-14} (vs 10^{-14} at room conditions) and a dynamic viscosity $\rho = 0.57$ millipoise (vs 10 at room conditions). Under these conditions SC water behaves as a water-like fluid with strong electrolytic solvent power, high diffusion coefficients and ion mobilities, and considerable hydrogen bonding. These properties favor chemical reactions involving heterolytic (ionic) bond cleavages which can be catalyzed by the presence of acids or bases.

Dramatic changes occur when the temperature of SC water is raised to 500° C at constant pressure ($\rho = 0.144\ g/cm^3$). Decreases in the dielectric constant to a value of 2 and ion product to 2.1×10^{-20} cause the fluid to lose its water-like characteristics and behave as a high temperature gas. Under these conditions homolytic (free radical) bond cleavages are expected to dominate the reaction chemistry. Thus by using the engineering parameters of temperature and pressure one can dramatically change the chemical properties of the solvent (dielectric constant and ion product) to favor heterolytic or homolytic bond cleavages.

Figure 1 summarizes these phenomena by displaying the dependence of the ion product K_w of water on temperature at constant pressures of 34.5 and 69 MPa. Data taken from reference 11 was used to compile the curves given in Figure 1. At 34.5 MPa water rapidly loses its ionic properties above 400° C; whereas at 69 MPa water retains its ionic properties to about 520° C. The change from an ionic to an aprotic fluid is particularly dramatic at 34.5 MPa: the ion product decreases by six orders of magnitude between 400° and 500° C. Throughout this paper temperatures were maintained below 400 ° C to enhance the role of heterolytic bond scissions within the reactor.

At low temperatures supercritical fluids have found exciting applications as a means for extractions (3,6-18) and chromatographic separations (19,20). These applications exploit the large values of diffusion coefficients and low viscosities enjoyed by many supercritical fluids, as well as their high solvent power. Higher temperature (>200° C) "extractions" of biopolymer materials are recognized to involve decomposition of the biopolymer (21,22), as well as chemical reactions between the biopolymer and the "solvent" (23). Batch (21,29), semi-batch (22,30) and tubular flow reactors (21,31-37) have been used for these studies; however, Metzger et al. (33) note that flow reactors minimize the influence of secondary reactions, and are required at higher temperatures when reaction rates exceed 0.1 $min.^{-1}$ Studies involving methanol (23,24,27,30,31,33), ethanol (31,32), acetone (21-24,26), dioxane (25,26), water (27,28,34-38), and a variety of hydrocarbon (23) solvents have been reported. In general, the underlying chemical reactions are complex, non-specific, and not well understood. Recently reported research involving model compounds (27-29,33) has begun to offer insights into the fundamental reaction chemistry of biopolymer substrates in supercritical solvents. The spirit of our work parallels the pioneering work of Koell, Metzger and their colleagues (22,31-33) in Germany, who recently described fundamental studies (33) of intermolecular C-C bond forming (addition) reactions in SC solvents at 300° C and above.

APPARATUS AND EXPERIMENTAL PROCEDURES

Figure 2 is a schematic of the SC flow reactor. Prior to the initiation of flow, the system is brought up to pressure by an air compressor. Afterwards, an HPLC pump forces a pure solvent into the reactant accumulator at a measured rate of flow. This flow displaces the solvent/solute reactant mixture out of the accumulator, through the reactor and a 10 port valve dual loop sampling system, and into the product accumulator. The flow of products into the second accumulator displaces air through a back-pressure regulator and into a water displacement apparatus, which measures the rate of air flow at ambient conditions. The reactant flow is rapidly heated to reaction temperature by the entry heat guard, and maintained at isothermal

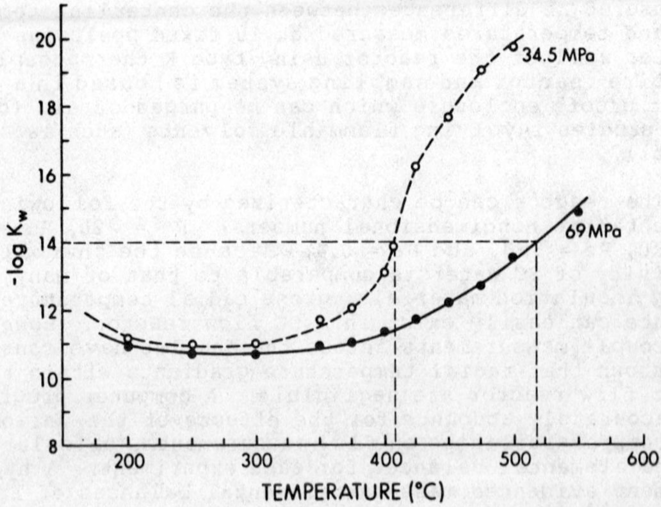

Figure 1. Ion Constant K_W of Water as a Function of Temperature

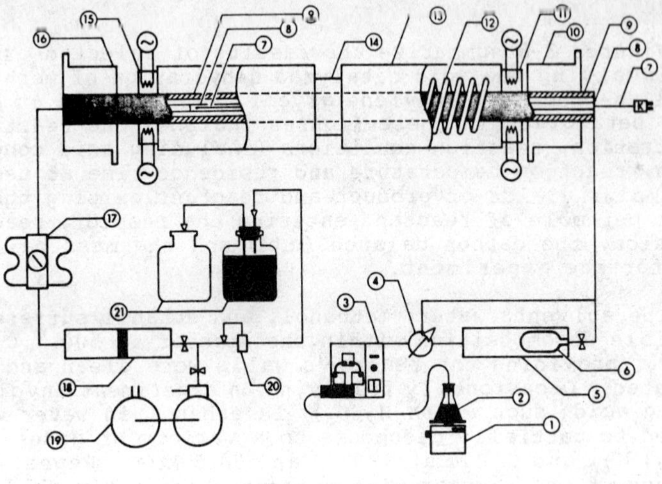

Figure 2. Supercritical Flow Reactor Schematic

conditions by a Transtemp Infra-red furnace and an exit heat guard. Samples captured in 5.4 ml sample loops are released into sealed, evacuated test tubes for quantitative analysis by GC, GC-MS, and HPLC instruments within the laboratory. The outer annulus of the reactor is a 4.572 mm ID Hastelloy C-276 tube, and the inner annulus is a 3.175 mm OD sintered alumina tube, giving the reactor an effective hydraulic diameter of 3.2 mm. The alumina tube accommodates a movable type K thermocouple along the reactor's axis, which provides for the measurement of axial temperature gradients along the reactor's functional length. Radial temperature gradients are measured as differences between the centerline temperature and temperatures measured at 10 fixed positions along the outer wall of the reactor using type K thermocouples. The entire reactor and sampling system is housed in a "bullet proof" enclosure which can be purged of air (oxygen) during studies involving flammable solvents (such as methanol).

The reactor can be characterized by the following representative nondimensional numbers: Re = 420, Pr = 1.9, Sc = 780, Pe = 350, and Da = 0.4. Because the thermal diffusivity of SC water is comparable to that of many high quality insulation materials, gross radial temperature gradients can easily exist in a SC flow reactor. However, thermocouple measurements across the annulus have consistently shown that radial temperature gradients within the annular flow reactor are negligible. A computer program, which accurately accounts for the effects of the various fluid compressibilities on flow measurements, calculates mass and elemental balances for each experiment. A typical experiment evidences mass and elemental balances of 1.00 ± 0.05.

RESULTS AND DISCUSSION

Figures 3-5 summarize the results of selected experiments involving the acid catalyzed dehydration of methanol, ethanol, n-propanol, ethylene glycol, and glycerol in SC water. Data given in these figures includes the reactant concentration, reaction conditions (including acid concentration, reaction temperature and residence time at temperature), molar yields of product and reactant exiting the reactor per mole of reactant entering the reactor, reactant conversion, the carbon balance (c.b.) and the mass balance (m.b.) for the experiment.

The solvents water, methanol, and ethanol suffered negligible decomposition within the reactor to 500 ^{o} C and 34.5 MPa, providing the reactor's walls were clean and passivated. Occasionally following an experiment involving a strong acid (such as 1M H_2SO_4), 1M ethanol in water was observed to partially decompose to a variety of gases (H_2, CO, CO_2, CH_4 and C_2H_4) at 385° C and 34.5 MPa. Repeated cleanings of the reactor with concentrated nitric acid followed by NaOH saturated methanol passiviated the reactor walls and reduced the decomposition of ethanol to a negligible amount. Because ethanol proved to be a sensitive indicator of the activity of the reactor's walls, experiments with ethanol in water were run regularly as blanks, bracketing sensitive reactions to ensure passive walls. Results reported in this paper are believed to reflect solely homogeneous chemistry.

As shown in Figure 3, 0.005M H_2SO_4 catalyzes the dehydration of ethanol to ethene in SC water. The reaction is fast and extremely selective. No polymerization of product ethene is observed under these conditions. Trace

Reactant (Conc.)	Conditions	Unaltered Reactant, Liquid, and Gas Products	Conv.	C.B./M.B.
C_2H_5OH (0.1M)	---------------> 400^oC, 49 s	0.97 C_2H_5OH + 0.005 C_2H_4 + 0.004 H	0.03	0.98/0.99
C_2H_5OH (0.5M)	H_2SO_4(0.01M) ---------------> 373^oC, 23 s	0.82 C_2H_5OH + 0.18 C_2H_4	0.18	1.0/0.99
n-C_3H_7OH (1.0M)	---------------> 400^oC, 8 s	0.95 n-C_3H_7OH + 0.013 C_3H_6 + 0.001 C_3H_8	0.05	0.97/0.98
n-C_3H_7OH (0.5M)	H_4SO_4(0.005M) ---------------> 400^oC, 8 s	0.25 n-C_3H_7OH + 0.16 1-C_3H_7OH + 0.61 C_3H_6	0.75	1.03/0.98

Figure 3

Selected Results of Ethanol and n-Propanol Dehydration Experiments in Supercritical Water at 34.5 MPa

Reactant (Conc.)	Conditions	Unaltered Reactant, Liquid, and Gas Products	Conv.	C.B./M.B.
CH_2OHCH_2OH (0.5M)	---------------> 385^oC, 31 s	0.95 CH_2OHCH_2OH + 0.009 H_2 + 0.004 CO_2+0.002 C_2H_4	0.05	0.96/0.98
CH_2OHCH_2OH (0.5M)	H_2SO_4(0.005M) ---------------> 385^oC, 29 s	0.51 CH_2OHCH_2OH + 0.4CH_3CHO + 0.01 H_2 +0.007 CO_2	0.49	0.92/0.99
CH_2OHCH_2OH (0.5M)	NaOH (0.005M) ---------------> 385^oC, 22 s	0.87 CH_2OHCH_2OH + + 0.005H_2 +0.001CO	0.13	0.87/0.98
$CH_2OHCHOHCH_2OH$ (0.5M)	---------------> 300^oC, 16 s	0.94 $CH_2OHCHOHCH_2OH$ + + 0.03CO_2 +0.004 C_2H_4	0.06	0.95/0.99
$CH_2OHCHOHCH_2OH$ (0.5M)	H_2SO_4(0.005M) ---------------> 325^oC, 39 s	0.77 $CH_2OHCHOHCH_2OH$ + 0.29 CH_2CHCHO + 0.05 CH_3CHO	0.23	1.1/1.00

Figure 4

Selected Results of Ethylene Glycol and Glycerol Dehydration Experiments in Supercritical Water at 34.5 MPa

Reactant (Conc.)	Conditions	Unaltered Reactant, Liquid, and Gas Products	Conv.	C.B./M.B.
HCHO + (0.25M) CH_3CHO (1.0M)	---------------> 385^oC, 21 s	0.04 HCHO + 0.04 CH_2CHCHO + 0.01 $CH_3CHCHCHO$ + 0.05 H_2 + 0.84 CH_3CHO + 0.04 CH_4 + 0.03 CO_2	-	0.88/0.99
CH_3CHO (0.5M)	---------------> 385^oC, 31 s	0.90 CH_3CHO + 0.14 $CH_3CHCHCHO$ + 0.02 CH_3COOH	0.10	1.27/0.99
CH_3CHO (0.5M)	NaOH (0.005M) ---------------> 385^oC, 31 s	0.65 CH_3CHO + 0.23 $CH_3CHCHCHO$ + 0.04 CH_3COOH + 0.08 H_2 + 0.01 CO_2 + 0.01 CH_4	0.35	1.16/0.99
CH_3CHO (0.5M)	H_2SO_4(0.005M) ---------------> 385^oC, 22 s	0.94 CH_3CHO + 0.05 CO + 0.05 CH_4	0.06	0.99/0.99

Figure 5

Selected Results of Formaldehyde and Acetaldehyde Condensation Experiments in Supercritical Water at 34.5 MPa amounts of product diethyl ether were detected in some experiments, suggesting its role as a possible intermediate. Exploring the possibility of the inorganic ester $EtHSO_4$ as an intermediate, we synthesized the ester according to standard procedures and observed its decomposition to ethene, ethanol, and H_2SO_4 at room temperature in excess water. The measured rate of ester decomposition was high enough to ensure its disappearance during the collection and transit time from the reactor to our analytic instrumentation. Consequently, we cannot rule out the ester as a possible intermediate. Finally, the transitory bare 1° carbocation suggested in many textbooks (39) may also have been intermediate in the formation of ethene.

The acid catalyzed dehydration of n-propanol in supercritical water results in the formation of only two products: propene and i-propanol (see Figure 3). Rearrangement of the posited 1° carbocation to the favored 2° carbocation followed by hydration results in the formation of the product alcohol. Apparently, the formation of the double bond is about a factor of four more rapid than the rearrangement/hydration reaction under these conditions.

Ethylene glycol is relatively stable in supercritical water (see Figure 4); however 0.005M H_2SO_4 catalyzes its rapid dehydration to acetaldehyde. In the presence of 0.005M NaOH ethylene glycol undergoes some decomposition. Products of the base catalyzed reaction have not been detected, and simply may be carbon deposited on the reactor's walls.

As shown in Figure 4, glycerol is also relatively stable in water at 34.5 MPa and 300° C. As expected, dilute sulfuric acid (0.005M) selectively catalyzes the dehydration of glycerol to acrolein between 300° and 350° C. Acetaldehyde is formed as a byproduct of the reaction. An earlier paper suggested the role of 3-hydroxypropanal as an intermediate in the formation of both acrolein and acetaldehyde. In that paper it was asserted that acetaldehyde formed from 3-hydroxypropanol via a homolytic C-C bond cleavage. A more satisfying explanation is given by a reverse adol mechanism, which converts 3-hydroxypropanol to formaldehyde and acetaldehyde. Formaldehyde decomposes under these conditions to H_2, CO, and CO_2, leaving acetaldehyde as the observed liquid product.

To test this mechanism an equi-molar mixture of formaldehyde and acetaldehyde was exposed to similar conditions in the reactor. As shown in Figure 5, acrolein was observed to be a major liquid product of the crossed aldol consensation; thereby supporting the role of the reverse aldol mechanism in the formation of acetaldehyde. The other liquid product of the reaction, crotonaldehyde, presumably formed by a conventional aldol condensation of acetaldehyde, followed by dehydration. A confirmation of this presumption is given in Figure 5, which shows crotonaldehyde to be the only significant product of the uncatalyzed and base catalyzed condensation of acetaldehyde in supercritical water. Figure 5 also reveals the role of dilute acid (0.005M H_2SO_4) in slowing the rate of the aldol condensation reaction. Thus the role of the acid catalyst in enhancing the formation of acrolein from glycerol may involve the acid's ability to both increase the rate of the dehydration reaction, and to decrease the rate of the acetaldehyde forming reverse aldol reaction.

CONCLUSIONS

Ethanol, n-propanol, ethylene glycol, and glycerol all undergo facile acid catalyzed dehydration to the corresponding alkene in SC water. Reaction rates and specificity are high. All evidence points to the role of heterolytic (ionic) bond cleavages in product formation. These results illustrate the potential of supercritical water as a media for directing the reaction chemistry of biopolymer solutes towards valuable chemical products.

The extraordinary selectivities reported in this paper are due to the high ion product of water below 400° C at 34.5 MPa. Under these conditions water sustains the dissociation of dilute Arrhenius acids (such as H_2SO_4), thereby catalyzing heterolytic bond cleavages ordinarily observed at much lower temperatures. Thus supercritical water enables acid catalyzed ionic reaction pathways to supplant free radical pathways at high temperatures. Because free radical reactions are notably non-specific, much can be gained by replacing their role in product formation with highly specific, high temperature, heterolytic bond cleavages. Thus the underlying chemistry described in this paper may have implications for the related fields of coal conversion and hydrocarbon processing.

ACKNOWLEDGMENTS

This research was supported by the National Science Foundation under grant CPE 8304381, the Department of Planning and Economic Development of the State of Hawaii, and the Coral Industries Endowment. The authors thank Dr. James Hsu and Dr. Maria Burka (NSF), Kent Keith and Dr. Tak Yoshihara (DPED), and David Chalmers (Coral Industries) for their interest in this work. The authors also thank Dr. Ali Tabatabaie-Raissi, and Professor Maitland Jones (Princeton) for many stimulating discussions.

REFERENCES CITED

1. Antal, M.J. "Biomass Pyrolysis. A Review of the Literature - Part 1: Carbohydrate Pyrolysis" in Advances in Solar Energy, (Eds. K.W. Boer and J.A. Duffie) Vol. 1, American Solar Energy Society, New York, 1984.

2. Antal, M.J. "Biomass Pyrolysis. A Review of the Literature - Part 2: Lignocellulose Pyrolysis" in Advances in Solar Energy, (Eds. K.W. Boer and J.A. Duffie) Vol. 2, Plenum Press, New York, 1985.

3. Paulaitis, M.E., Penninger, J.M.L., Gray, R.D, Jr. and Davidson, P. "Chemical Engineering at Supercritical Fluid Conditions" Ann Arbor Science, Ann Arbor, 1983.

4. Franck, E.U. "Supercritical Water" Endeavor, 1968,**27**, 55-59.

5. Marshall, W.L. "Water at High Temperatures and Pressures" Chemistry, 1975, **48**, 6-12.

6. Franck, E.U. "Thermophysical Properties of Supercritical Fluids with Special Consideration of Aqueous Systems", Fluid Phase Equilib. 1983, 10(2-3), 211-22.

7. Franck, E.U. "Water and Aqueous Solutions at High Pressures and Temperatures".

8. Marshall, W.L and Franck, E.U.,J. Phys. Chem. Ref. Data, 1981, **10**, 295-304.

9. Quist, A.S., Marshall, W.L., and Jolley, H.R. J. Phys. Chem., 1965, **69**, 2726-2735.

10. Marshall, W.L., Rev. Pure. Appl. Chem., 1968, **18**, 167-186.

11. Quist, A.S., J. Phys. Chem., 1970, **74**, 3396-3402.

12. Bruges, E.A. and Gibson, M.R., J. Mech. Eng. Sci., 1969, **11**, 189-205.

13. Lamb, W.J., Hoffman, G.A., Jonas, J., J. Chem. Phys., 1981, **74**(12), 6875-80.

14. Franck, E.U., Hartmann, D., and Hensel, F., J. Phys. Chem., 1968, **72**, 200-206.

15. Gorbatyi, Yu.E., Probl. Fix.-Khim. Petrol., (Eds. Zharikov, V.A., Fonarev, V.I., Korikovskii, S.P.) 1979, 2, 14-24.

16. Williams, D.F. Chem. Eng. Sci., 1981, **36**, 1769-1788.

17. Bott, T.R. Chemistry and Industry, 1980, 228-232.

18. Eggers, R. Angew. Chem. Int. Ed. Engl., 1978, **17**, 751-754.

19. Klesper, E. Angew. Chem. Int. Ed. Engl. 1978, **17**, 738-746.

20. Van Wasen, U., Swaid, I. and Schneider, G.M. Angew. Chem. Int. Ed. Engl. 1980, **19**, 575-587.

21. Koell, P. and Metzger. Angew. Chem. Int. Ed. Engl., 1978, **17**, 754-755.

22. Squires, T.G. et al. ACS FUEL Division Preprint, 1983, **28**(4), 228.

23. McDonald, E.C., Howard, J., Bennett, B. Fluid Phase Equilib., 1983, **10(2-3)**, 337-44.

24. McDonald, E. Howard, J. Proc. - Bioenergy R & D Semin., 1982, 4th, 397-401.

25. Calimli, A. and Olcay, A. Separation Science and Technology, 1982, **17**, 183-197.

26. Olcay, A., Tugrul, T. and Calimli, A. In "Chemical Engineering at Supercritical Fluid Conditions". (Eds. M.E. Paulaitis, J.M.L. Penninger, R.D. Gray, Jr. and P. Davidson) Ann Arbor Science Pub., Ann Arbor, 1983.

27. Abraham, M.A. and Klein, M.T. "Pyrolysis of Benzyl Phenyl Amino Neat and with Tetralin, Methanol, and Water Solvents" submitted to I&EC Prod. Res. Dev.

28. Lawson, J.R. and Klein, M.T. "The Influence of Water on Guaiacol Pyrolysis" submitted to I&EC Fundamentals.

29. Townsend, S.H. and Klein, M.T. "Dibenzyl Ether as a Probe into the Supercritical Fluid Solvent Extraction of Volatiles from Coal with Water", submitted to FUEL.

30. Labrecque, R., Kallaguine, S. and Grandmaison, J.L. Ind. Eng. Chem. Prod. Res. Dev., 1984, **23**, 177-182.

31. Koell, Peter; Broenstrup, Bernd; Metzger, Juergen. Holzforschung, 1979, **33(4)**, 112-16.

32. Koell, P; Broenstrup, B.; Metzger, J.O.. Chem. Eng. Supercrit. Fluid Cond., 1983, 499-514.

33. Metzger, J.O., Hartmanns, J., Malwitz, D. and Koell, P. In "Chemical Engineering at Supercritical Fluid Conditions" (Eds. Paulaitis, M.E.; Penninger, J.M.L.; Gray, R.D., Jr.; Davidson, P.) Ann Arbor Science Pub., Ann Arbor, 1983.

34. Amin, S., Reid and Modell, M. "Reforming and Decomposition of Glucose in an Aqueous Phase". Presented at the Intersociety Conference on Environmental Systems, San Francisco, 21-24 July, 1975.

35. Timberlake, S.H.; Hong, G.T.; Simson, M.; and Modell, M. "Supercritical Water Oxidation for Wastewater Treatment: Preliminary Study of Urea Destruction" SAE Tech. Paper Series 820872, 1982.

36. Whitlock, David R. "Treatment of Organic Material in Supercritical Water", **PCT Int. Appl.** 81 **00,854** 02 Apr 1981, US Appl 79,535, 27 Sep. 1979. 21 pp.

37. Modell, M., Gaudet, G.C., Simson, M. Hong, G.T. and Biemann, K. Solid Wastes Management 1982.

38. Barrows, R.D. and Elliott, D.C. FUEL, 1984, **63**, 4-8.

39. Morrison, R.T. and Boyd, R.N. "Organic Chemistry" (4th Edition) Allyn and Bacon, London, 1978. Also G.M. Louden, "Organic Chemistry" Addison Wesley, N.Y., 1984, and H. Knozinger, "The Dehydration of Alcohols" in "The Chemistry of the Hydroxyl Group, Part 2" (Ed. S. Patai) Interscience, London 1971.

DISCUSSION

B. Boocock
U. of Toronto
(Canada)

What happens to the equilibrium constants up at these temperatures? Can you measure the equilibrium constant and what is it?

Antal

We have calculated the equilibrium constants assuming ideal solutions by the normal textbook approach and ethanol ought to go to ethylene. It's nominally 50 mols of water per mol of ethanol at 1M. We calculated a nominal 70-80% converison of ethanol to ethylene. We are aware that this probably is not an ideal solution. The most recent literature on treating non-ideal solutions appeared in the AIChE Journal a few months ago. This was work out of Mobil on a very complicated correction approach. We applied the correction, and it made absolutely no difference for our calculations; we still got 70 to 80%. There is literature suggesting that hydrogen bonding and other effects produce dramatic changes in the equilibrium constant, especially with ethanol. This is a concern to us. We have had a hard time getting good selectivity and conversion of ethanol above about 30% and I'm not quite sure whether the problem is a selectivity problem or ethylene begins to polymerize or whether it's our inability to really do a good equilibrium calculation. Conversion of ethanol to ethylene in so dilute a solution is probably one of the weaknesses. We are now working with more concentrated solutions and hope to see more promising results. I might mention that methanol with acid and water is converted to dimethylether exactly as you would expect.

____

Have you tried to modify the properties of water with other solvents?

Antal

We have done work with pure methanol, and also water plus methanol, in order to lower the polarity of the solvent. We find less reactivity when at the lower polarity. I mentioned that dioxolane in super critical water goes instantaneously to glycol. If you put dioxolane in methanol, the reaction slows tremendously to yield the expected product. As you move away from solvents that are as polar as water, the ion chemistry slows down.

RECYCLE OF ENZYMES AND SUBSTRATE FOLLOWING THE ENZYMATIC HYDROLYSIS OF STEAM PRETREATED ASPENWOOD

M. Mes-Hartree, C.M. Hogan and J.N. Saddler
Forintek Canada Corp.
800 Montreal Road
Ottawa, Ontario
Canada

ABSTRACT

The commercial production of chemicals and fuels from ligno-cellulosic residues by enzymatic means still requires considerable research on both the technical and economic aspects. Forintek Canada is actively researching the overall process of the conversion of forest residues to chemicals or fuels.

Two technical problems that have been identified as requiring further research are the recycle of the enzymes used in hydrolysis and the reuse of the recalcitrant cellulose remaining after incomplete hydrolysis. Enzyme recycle is required to lower the cost of the enzymes while the reuse of the spent cellulose will lower the feedstock price.

The conversion process studied was a combined enzymatic hydrolysis and fermentation (CHF) procedure which utilized the cellulolytic enzymes derived from the fungus Trichoderma harzianum E58 and the yeast Saccharomyces cerevisiae. The rate and extent of hydrolysis and ethanol production was monitored as was the activity and hydrolytic potential of the enzymes remaining in the filtrate after the hydrolysis period. Since 100% hydrolysis of the cellulose portion is seldom accomplished in an enzymatic hydrolysis process the residual cellulose was used as a substrate for the growth of T. harzianum E58 and production of cellulolytic enzymes.

When a commercial cellulose was used as the substrate for a routine 2 day CHF process, 60% of the original filter paper activity could be detected in the filtrate after this time. When steam treated, water extracted aspenwood was used as the substrate only 13% of the original filter paper activity was recovered after a similar procedure. The combination of 60% spent enzymes with 40% fresh enzymes resulted in the production of 30% less reducing sugars than the original enzyme mixture.

The residue remaining after the CHF process was used as a substrate for the production of cellulolytic enzymes. The production of enzymes from the residue of the Solka Floc hydrolysis was greater than the production of enzymes from the original Solka Floc.

RECYCLE OF ENZYMES AND SUBSTRATE FOLLOWING THE ENZYMATIC HYDROLYSIS OF STEAM PRETREATED ASPENWOOD

INTRODUCTION

An integrated process for the production of higher value products from waste lignocellulosics has been under study at Forintek for the last six years. Both the technical and economic issues which pertain to this topic are being studied by a joint research effort of the economics, biotechnology and chemistry departments. The technical research is now far enough advanced to enable us to input the various process options we are considering into a linear program.

Linear programming is a technique used to determine the optimal solution to a problem where there are a number of competing and usually interrelated choices. The technique requires that each of the restrictions on the problem being modeled be formulated as a linear equation. The model consists of a number of linear equations with more unknowns than equations and thus there are many possible solutions. In order to determine the optimal solution, a criteria such as maximum profit or minimum cost is chosen and equations are set up, giving the amount each variable contributes to this criteria. The linear program then determines which solution will maximize or minimize this criteria.

The linear program was written to determine the best process and set of process conditions for converting steam exploded Aspenwood into butanol, butanediol or ethanol. The linear program is designed to maximize profit based on the sales value of the chemicals produced, the cost of the raw materials and the processing costs incurred. The model is restricted by the raw material availability, the utility and chemical requirements at each process step and the market requirements for each chemical produced.

The process under consideration is shown in Figure 1. The process can be broken down into three major components; steam pretreatment followed by enzymatic hydrolysis of the carbohydrate polymers and subsequent fermentation of the sugars. Two process options exist in the model and these are the direct conversion of the steamed material to butanediol or butanol or the separation of the wood components for different processing of the cellulose and hemicellulose polymers. The processing of the lignin component of the wood was not included in the present model but will be added when the technical research is more advanced.

Each process option considered requires the production of cellulase enzymes and the fungus *Trichoderma harzianum* E58 is used for this purpose. This fungus produces enzymes with equal amounts of filter paper and β-glucosidase activities. Approximately ten percent of the incoming feedstock is diverted to the production of enzymes. After growth of the fungus the extracellular enzymes are filtered by ultrafiltration which not only concentrates the enzymes but also effectively separates the cellulase enzymes from the xylanase enzymes. This separation of enzymes is ideal for our process scheme, in that different enzymes are needed for the CHF2 and CHF3 steps.

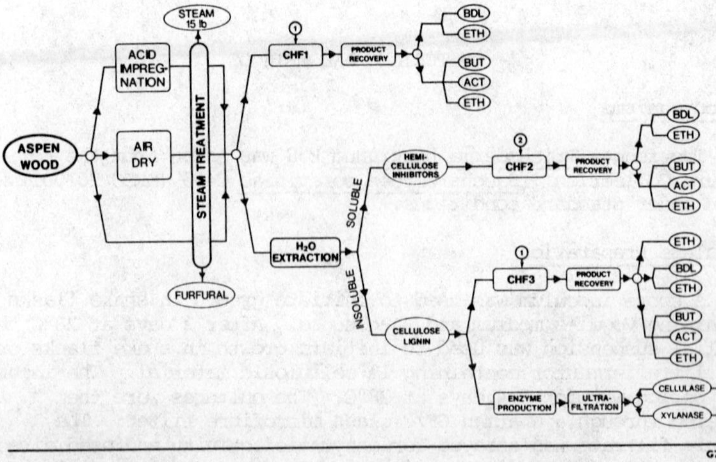

Figure 1. Process options which are being considered in the linear program

The ultimate goal of this approach is to narrow the research effort to the most economical and feasible process options. Since biomass conversion can potentially produce a wide array of products we have restricted the range of products which will be further researched to those that look the most economically favourable. A key factor in this decision is the market assessment of the products. The area that has received the most technical emphasis in our research program has been liquid fuels such as ethanol, butanol and 2,3-butanediol.

The Canadian markets for ethanol and butanol (Table 1) will remain static and an oversupply exists. This market projection does not take into account the demand for ethanol or butanol in gasohol since this market will depend on government policies concerning leaded fuel. If the demand for gasohol follows the same pattern as in the USA the potential market in Canada for these products could be as high as 1.4 million tonnes a year. The butanediol market was not as easy to determine, as the butanediol produced by the fermentation technology is 2,3-butanediol and there is presently no established market for this product as an intermediate organic feedstock. Several derivatives of this diol are traded in volume on chemical markets and the two derivatives which have received the most attention are methyl ethyl ketone and butadiene.

One of the major costs of the enzymatic route of hydrolysis is the high cost of the enzymes. Wilke et al (1) has reported a cost as high as 60% of the total cost of the process and such a high cost component obviously makes this part of the process a prime target for research. One avenue which has been suggested as a means of lowering the cost of the enzyme component of the process is the recovery and recycle of the enzymes. Isaacs (2) has indicated that recycle of 60% of the cellulase enzymes could have a major impact on the cost of the enzymes. Little research has been done on the technical problems and feasibility of these levels of enzyme recycle in a process which uses pretreated wood as its primary feedstock. This was the major objective of the present report.

MATERIALS AND METHODS

Microorganisms

The fungus Trichoderma harzianum E58 was taken from the Forintek Culture Collection. Saccharomyces cerevisiae C495 (NRCC 202001) was grown under standard conditions.

Cellulase Preparation

A spore inoculum was used to initiate growth in shake flasks containing Vogel's medium and 2% glucose. After 2 days at 28°C the mycelial suspension was used to initiate growth in shake flasks or in a 30 litre fermentor containing 1% cellulosic material. The incubation period was for 3-4 days at 28°C. The cultures were then filtered through a Whatman GF/F glass microfibre filter. The culture filtrate was assayed for enzymatic activity and used directly for enzymatic hydrolysis experiments. Some hydrolysis experiments

Table 1. Canadian market information for ethanol, butanol and 1,3 butadiene

Chemical	Bulk Price	Domestic Supply (Kt) 1984	Domestic Supply (Kt) 1988e[1]	Domestic Demand (Kt) 1984	Domestic Demand (Kt) 1988e
Ethanol	$.57/litre	64	65	36	41
n-butanol	$.85/kg	8.2	9	5.2	5.5
1,3 butadiene	$.40/lb	165	172.5	155	172.5

[1]Estimated supply and demand

were performed with culture filtrates which had been concentrated by the Pellicon filtration system using a membrane with a molecular cut off of 10,000 daltons,and freeze dried.

Assays

Soluble protein was determined after TCA precipitation by the method of Lowry et al (3) using bovine serum albumin as the standard. Total reducing sugars were estimated using the dinitrosalicylic acid assay (4). Glucose was determined by the glucostat enzyme assay (5). Ethanol was determined by gas liquid chromatography using a Chromosorb 101 column with flame ionization. Helium which was saturated with formic acid was used as the carrier gas (6).

Hexosan was determined by an adaptation of the anthrone methods of Koehler (7) in which the time was extended to 15 minutes to permit the more labile pentosan color to fade. A correction for the residual pentosan color was made.

Enzymatic activity was determined by the method of Mandels et al (8). One millilitre of a properly diluted culture filtrate was added to one millilitre of 0.05 M citrate buffer, pH 4.8. For the filter paper assay the substrate was a 1 cm x 6 cm strip of Whatman No. 1 filter paper, while for the endoglucanase assay the substrate was 10 mg of carboxymethyl cellulose (CMC medium viscosity, Sigma no. C-4888) and for the β-glucosidase assay the substrate was 10 mg of salicin. All three assays were performed at 50°C. The filter paper assay was incubated for 1 hour while the endoglucanase and β-glucosidase assays were incubated for 30 minutes. For all three assays the reaction was terminated by adding 3 mLs dinitrosalicylic acid reagent to each tube and boiling the tubes for 5 minutes. The tubes were cooled to room temperature and the absorbance read at 575 nm. One unit of enzyme activity was defined as 1 μmol of glucose equivalents released per minute.

Combined Hydrolysis and Fermentation (CHF) Procedure

A five percent substrate concentration (based on dry weight) was added to a 60 mL serum vial or to a 2 litre fermentor. The cellulase enzyme preparation was added at a concentration of 15 IU of filter paper activity per gram of substrate. A three day old suspension of *S. cerevisiae* was added at a 5% inoculum level. The fermentor and vials were incubated at 37°C for 48 hours. The vials were agitated at 200 RPM. The fermentor was first set at an agitation rate of 900 RPM and after two hours the agitation rate was lowered to 200 RPM.

Hyrolysis experiments were performed in a similar manner without the addition of a fermentative microorganism

Adsorption of Cellulase Enzymes

The adsorption of the enzymes onto various substrates was tested by incubating the crude enzyme mixture with the substrate at a pH of 4.8 and a temperature of 28°C for 10 minutes. After incubation the culture filtrate was assayed for endoglucanase, β-glucosidase, filter paper activities, and protein content. These values were compared to the activities of the crude enzyme mixture after incubation without the substrate.

RESULTS

The recycle of cellulase enzymes following a combined hydrolysis and fermentation procedure is dependent on several factors such as the thermal stability of the enzymes, the effect of ethanol on the enzymes and the adsorption of the enzymes onto the cellulose. The amount of recovery of the original enzyme which is considered to be necessary for an economically viable process is 60% (2). In order to test the technical feasibility of this level of recovery, the activity of the cellulase enzymes remaining in the culture filtrate following the combined hydrolysis and fermentation process was studied.

The cellulase enzymes derived from the growth of the fungus *T. harzianum* E58 on cellulose were used for the hydrolysis of a 5% cellulose solution. The yeast *S. cerevisiae* was also added to the hydrolysis system so that the produced glucose would be quickly converted to ethanol and thus optimize the hydrolysis process. The glucose produced by the hydrolysis of cellulose is a known inhibitor of the cellulase enzyme system of *T. harzianum* E58 and its conversion to ethanol enhances the overall hydrolysis (9). The ethanol produced during the CHF process is also inhibitory to the enzymes, however the major effect of the ethanol is not directly on the enzyme activity but rather on the stability of the enzymes under the hydrolysis conditions (10). This fact has direct implications for the recovery and recycle of the enzymes since the presence of ethanol will be detrimental to this procedure.

The combined hydrolysis and fermentation was performed in both small flasks and fermentors (Table 2). Both systems produced similar levels of hydrolysis based on the production of ethanol and reducing sugars. The fermentor residue contained not only cellulose but also yeast cells from the fermentative aspect of the process. Based on the nitrogen content of the residue approximately 19% of the fermentor residue was yeast biomass.

Table 2. The combined hydrolysis and fermentation (CHF) of a 5% concentration of commercial cellulose using the enzyme system derived from the growth of T. harzianum E58

Hydrolysis Conditions	% hydrolysis of cellulose	
	to reducing sugars	to glucose
Hydrolysis alone[1]	52.8	45.5
CHF-Flasks	67.3	58.4
CHF-Fermentor	67.5	59.5

The cellulase enzymes were used at 15 IU of filter paper activity per gram cellulose

[1]The hydrolysis only system means hydrolysis was performed at 45°C for 48 hours and no fermentative organism was added to the system. The flasks and fermentors were both combined hydrolysis and fermentation systems which were performed at 37°C.

The enzymes present in the culture filtrate were assayed in both the CHF systems and in flasks containing enzymes and no cellulose substrate. These flasks allowed the monitoring of the stability of the enzyme system under the CHF conditions. The results presented in Table 3 indicated that the endoglucanase component of the enzyme system was the least stable. Sixty percent of the filter paper activity was recovered from the CHF system but only 50% of its original hydrolytic activity was present. (Table 4) Even the enzymes in the control which had only been exposed to the CHF conditions and no substrate produced 20% less glucose when they were subsequently used to hydrolyze cellulose. This result clearly indicated that even under ideal conditions 100% recycle is unrealistic with this enzyme. A more thermal stable enzyme system would obviously be more useful.

Although the assayed enzyme activities differed for the flasks and the fermentor the hydrolytic activity of the two enzyme preparations were identical. This again raises the question of the reliability of the assays for enzyme activities and points to the need for continual monitoring of the hydrolytic activity of the enzymes as the only means of realistically assessing the potential of a given enzyme preparation.

Table 3. The percent of the original enzyme activities which was measured in the culture filtrate after the CHF of 5% commercial cellulose

Hydrolysis conditions	% activity remaining in filtrate		
	endoglucanase	β-glucosidase	filter paper
Enzyme alone[1]	72.4	96.2	97.0
CHF-Flasks	58.5	116	65.8
CHF-Fermentor	49.0	101	61.3

[1]To assay for the stability of the enzymes, they were treated under identical conditions as the CHF vials in the absence of substrate

Table 4. The hydrolytic activity of the culture filtrates derived from the CHF of 5% cellulose

Hydrolysis[1] conditions	Enzyme Loading IU/gr	Residual CHF Products		% of original hydrolysis	
		Red. Sugars mg/mL	Ethanol mg/mL	Red. Sugars	Glucose
Initial	15	-	-	100	100
Enzyme alone	11.4	-	-	86.3	81.6
CHF-Flasks	7.8	3.7	9.2	51.6	49.2
CHF-Fermentors	6.9	3.0	9.4	50.9	46.4

[1] The culture filtrates from the three systems were added to a 5% cellulose slurry. For this assay one gram of Solka Floc was mixed with 5 mLs of buffer and 15 mLs of culture filtrate. Since the culture filtrate taken from the CHF systems contained both residual reducing sugars and ethanol these products were added to the test hydrolysis.

The results indicated that the cellulose which remained following hydrolysis still exerted a strong adsorptive effect on the cellulase enzymes. The β-glucosidase component of the enzyme preparation was not adsorbed whereas the endoglucanase was strongly adsorbed by the unhydrolysed cellulose. The activity of the β-glucosidase in the culture filtrate after adsorption was always found to be higher than the activity measured in the original filtrate. This apparent increase in activity may have resulted from the removal of the endo and exo glucanase components of the enzyme system which may retard the access of the β-glucosidase to the cellobiose substrate. If we assume that approximately 70% of the cellulose was solubilized then the remaining 30%, or 1.5% cellulose solution would adsorb the enzymes. Thus the total percent hydrolysis was a major contributing factor to the recovery and recycle of enzymes.

Once the base case with a commercial cellulose preparation was established, the feasibility of enzyme recycle was investigated using pretreated lignocellulosic material. The pretreatment procedure currently used as a base case involves the steam treatment of aspenwood at 240°C for 80 seconds and subsequent removal of soluble pentosans and inhibitors with a simple water extraction procedure. This material, which is termed Steam Exploded Aspenwood-Water Insoluble (SEA-WI) was used at a 5% dry weight concentration in a CHF process. The chemical composition of the lignocellulosic material before and after hydrolysis is given in Table 5. The reduction in dry weight was estimated to be about 59% of the original, however due to the presence of the yeast biomass the percent loss of the cellulose component was actually 73%. After the hydrolysis there still remained a 1% cellulose suspension in the fermentor.

The assayed enzyme activity of the culture filtrate before and after the CHF process is given in Table 6. The filter paper activity of the cellulase enzymes recovered from the process was less than 20% of the original filter paper activity. This low enzyme recovery could not be accounted for by the unhydrolyzed cellulose portion. Other components of the steamed material must therefore be affecting the release of the enzyme following hydrolysis.

The hydrolytic activity of the recovered enzymes was monitored and reported in Table 7. This table also illustrates the hydrolytic activity of the recovered enzymes when they were combined with various proportions of fresh enzymes. These results indicated that with the present process conditions the recycle of enzymes does not appear beneficial.

Table 5. Chemical composition of steam treated water extracted aspenwood (SEA-WI) before and after the CHF process[1]

Substrate	Hexosan %	Lignin %	Dry Weight gr
original SEA-WI	66.9	31.2	80
after CHF	36.9	52.7	47.4

1 The dry weight of the residue included the dry weight of the yeast biomass. Based on the nitrogen content of the residue this amounted to 19% of the residue weight. The percent hexosan and lignin were based on the total dry weight of the residue. The percent cellulose hydrolysis was 72.9.

Table 6. Recovery of cellulase enzymes in the culture filtrate following a CHF process of steam treated water extracted aspenwood (SEA-WI)

	Enzyme activity IU/mL			Protein
	endoglucanase	β-glucosidase	filter paper	mg/mL
original activity	15.1	.76	0.75	0.86
recovered activity	1.9	0.39	0.12	0.17

Although the hydrolysis of the commercial cellulose was the same as that obtained with the lignocellulosic residue the recovery of the enzymes was very different. The major difference between these substrates was the amount of lignin present in the (SEA-WI) substrate. Lignin constituted 31% of the original substrate and as much as 51% of the final post-hydrolysis residue. Several researchers have reported (11,12) that lignin strongly adsorbs the cellulase enzymes. This possibility was further investigated.

Table 7. The hydrolytic activity of various proportions of recovered and fresh enzymes

% original/ recovered activity	Enzyme profile IU/gr[1] endo-glucanase	β-glucosidase	filter paper	Hydrolysis Products Red. Sugars mg/mL	Glucose mg/mL
0/100	30	6.0	2.0	5.7	3.1
20/80	94.8	7.9	4.6	16.0	11.3
40/60	212	13.0	9.6	21.0	15.6
60/40	299.2	15.6	13.0	25.2	19.3
80/20	385.6	18.0	16.5	27.9	22.1
100/0	472.0	20.6	20.0	30.8	25.4

The original culture filtrate had an enzyme activity of 23.6 IU endoglucanase, 1.03 IU of β-glucosidase and 1.00 IU of filter paper activity. The hydrolysis was performed at 5% cellulose concentration for 48 hours at 45°C.

[1]Enzyme profile given as IU/gr substrate

The adsorption of the cellulase enzymes onto different lignin fractions were monitored under conditions similar to those found in a CHF process. The results presented in Table 8 indicated that the lignin did not adsorb the cellulase enzymes. However, when the adsorption of cellulase enzymes to various pretreated lignocellulosic residues was compared (Table 8) the results indicated that some sort of adsorption was occurring with the lignin/cellulose complex. Although lignin itself did not appear to adsorb cellulase enzymes, the presence of residual lignin in the lignocellulosic residue appeared to severely affect the recovery of the cellulase enzymes.

In our process design the removal of lignin before or after the CHF process is still one of the decisions that must be finalized. Although the presence of lignin did not appear to affect the cellulose hydrolysis (13), it did influence the recycle of cellulase enzymes. We next monitored the recovery of cellulase enzymes after a CHF process on delignified aspenwood. The substrate that was chosen for this purpose was steamed aspenwood which had been water and alkali extracted which is termed Steam Exploded Aspenwood-Water and Alkali Insolubles (SEA-WAI). The alkali extraction removed 75% of the lignin as can be seen from the data on Table 9. The lignin composition of the material after hydrolysis was 17% which was equivalent to less than half of the total residue remaining after the CHF of the SEA-WI substrate.

The recovery of the cellulase enzymes after the CHF of the water and alkali extracted aspenwood (SEA-WAI) is reported in Table 10 and the results were found to be as poor as with the substrate with a higher lignin content.

The results indicated that for successful recycle of enzymes to occur the percent hydrolysis of cellulose must be at a high enough level so that little cellulose remains in the residual mixture to adsorb the cellulase enzymes. The high conversion of cellulose to sugars will require more research into the mode of action of the cellulase enzymes and into the pretreatment process so that more cellulose is accessible to the cellulase enzymes.

Table 8. The adsorption of cellulolytic enzymes to various lignin samples and to pretreated lignocellulosic substrates

Substrate 5%	% enzyme remaining in filtrate			% protein in filtrate
	endoglucanase	β-glucosidase	filter paper	
Solka Floc	30.5	117.0	32.0	34.8
SEA-WI	11.8	67.7	13.8	26.7
Lignin A	103.6	100.0	97.1	97.6
Lignin B	96.3	103.4	90.0	85.9
Lignin C	100.9	103.4	100.0	96.4
SEA-WI (enz)	66.7	85.0	56.4	67.9

SEA-WI: Steam treated aspenwood which was heated at 240°C for 180 seconds and then water extracted

Lignin A: derived from steamed aspenwood which was treated for 220 seconds at 240°C. The lignin was solubilized with methanol, then filtered and reprecipitated with water. Approximately 50% of the original lignin was recovered by this procedure

Lignin B: derived from aspenwood which was ball-milled for 2 weeks and then partitioned. This material was dissolved in 90% acetic acid to remove carbohydrate contaminants.

Lignin C: derived from steam treated aspenwood which was treated at 240°C for 80 seconds. The lignin was extracted with 0.4% sodium hydroxide.

SEA-WI (enz): The same SEA-WI sample as described above which was enzymatically hydrolyzed by a CHF process. The original cellulose content was 57% and after the CHF the cellulose content was between 3-10%.

Table 9. The chemical composition of steam treated water and alkali extracted aspenwood (SEA-WAI) before and after enzymatic hydrolysis[1]

Substrate	Hexosan %	Lignin %	Dry Weight gr
original (SEA-WAI)	90.0	8.2	80
after hydrolysis	61.6	17.2	38.3

[1]The dry weight of the residue including the dry weight of the yeast biomass. Based on the nitrogen content of the residue this amounted to 19% of the residue weight. The percent hexosan and lignin were based on the total dry weight of the residue. The percent cellulose hydrolysis was 67.2

Enzymatic hydrolysis was accomplished by a CHF process which is described in Materials and Methods.

Table 10. Recovery of cellulase enzymes in the culture filtrate following a CHF process of steam treated water and alkali insoluble aspenwood (SEA-WAI)

Culture Filtrate	Before CHF	After CHF
Enzyme activity IU/mL		
Endoglucanase	14.6	1.6
β-glucosidase	0.83	0.16
Filter paper	0.75	0.09
Protein mg/mL	0.96	0.10

Although the recycle of enzymes is an economically attractive process our results have shown that with the present technology this concept is not worthwhile. Another area of recycle which will also have an impact on the economic feasibility of the process is the recycle of the unhydrolyzed lignocellulosic residue. Since the complete hydrolysis of the cellulose appears to be unrealistic with our present technology we have investigated the possibility of using the unhydrolyzed portion of the cellulose as the growth substrate for the production of cellulase systems.

It was hoped the recalcitrant nature of the cellulose remaining after cellulose hydrolysis might cause the production of a more active enzyme since the physical characteristics of the cellulose substrate is believed to influence the enzyme profile produced (14). The recalcitrant cellulose was produced by performing a CHF process on Solka Floc and using the residual cellulose/yeast mixture as the substrate for the growth of the fungus T. harzianum E58. The results are presented in Table 11. The media for the Solka Floc controls contained either peptone or yeast extract as the production of enzymes may have been influenced by the yeast biomass contained in the cellulose residue from the CHF process. Previous results (12) have indicated that the presence of yeast extract is inhibitory to the action of the enzymes. The addition of yeast extract to the Solka Floc control did not appreciably affect the production of the cellulase enzymes. When spent cellulose was used as the growth substrate higher levels of the cellulase enzymes could be found in the culture filtrate.

The hydrolytic activity of the produced culture filtrates described in the previous table were assayed (Table 12). The culture filtrate originating from the growth of the fungus on the spent cellulose was marginally better than the culture filtrate derived from the control cellulose. These results are very encouraging as they offer another avenue for the reduction of the costs involved in the enzyme production step.

DISCUSSION

The recovery of cellulolytic enzymes derived from the fungus T. harzianum E58 from the hydrolysis mixtures containing lignocellulosic material was less than 20%. This low value has also been reported by Puls (15) who used steamed birchwood as the substrate and cellulase enzymes derived from T. reesei. The removal of the lignin component prior to the hydrolysis step did not improve the poor recovery of the enzymes. Although lignin was shown not to adsorb the cellulolytic enzymes the presence of this material in the starting substrate did influence adsorption. The lack of enzyme adsorption onto lignin is contradictory to results obtained by Klyosov (personal communication). The difference in the results could be explained by a difference in scale of the experimental procedure. The adsorption procedure reported here was designed to test the effect of lignin in a system similar to the CHF process. The ratio of lignin to enzymes was 15 IU of filter paper activity per gram of lignin. At this level adsorption could not be detected. The research protocol used by other workers used enzyme levels 20 times lower than ours which may explain the difference in the results.

Table 11. Production of cellulase enzymes from the growth of *T. harzianum* E58 on spent cellulose

Substrate (1%)	Nitrogen Suppl.[1] (0.1%)	Enzyme activity of filtrate IU/mL Endoglucanase	β-glucosidase	Filter paper	Protein mg/mL
Solka floc	BP	20.6	0.92	1.26	1.28
	YE	19.0	0.78	1.14	1.22
Spent SF[3]	BP	34.3	1.26	1.99	1.95
	-[2]	28.4	0.98	1.47	1.59

[1]BP: Bacto peptone
YE: yeast extract

[2]-:From the nitrogen determination of the cellulose there was 0.23% yeast biomass present

[3]Spent SF was the residual cellulose remaining after a CHF of 5% Solka Floc. The percent hydrolysis was 76%. The cellulose was dried following the CHF process and the nitrogen content was measured.

Table 12. Hydrolytic activity of the culture filtrates described in Table 10

Culture filtrate	percent hydrolysis of Solka floc to Reducing sugars	Glucose
Solka Floc (BP)	55.4	40.8
Solka Floc (YE)	53.8	39.6
Spent SF (BP)	56.9	43.4
Spent SF (-)	53.7	40.6

Hydrolysis was performed on 5% Solka Floc with 15 mLs of culture filtrate per gram of cellulose. The incubation time was 48 hours.

The combination of the lignin and cellulose was found to have a strong adsorptive effect on the cellulase enzymes. In fact this combination adsorbed the β-glucosidase component of the cellulase system whereas commercial cellulose alone did not adsorb this fraction. This result indicates that the phenomenon we observed was not solely due to adsorption and that inhibition of the enzyme activity may have played a role.

From this data it was apparent that the benefit of recycling cellulolytic enzymes will primarily depend on the achievement of high levels of hydrolysis. The complete hydrolysis of cellulose will require more research into the mode of action of the cellulase complex and the pretreatment of the lignocellulosic residue.

An alternative form of recycle which may offer some reductions in the overall cost of the production of the cellulase enzymes is the reuse of the recalcitrant cellulose remaining after the hydrolysis process. This approach was found to be feasible with the commercial cellulose Solka Floc. Future research will now focus on applying this concept to the lignocellulosic residues remaining in the fermentor.

Acknowledgement

A large part of the research presented was funded by the Department of Energy Mines and Resources Canada.

REFERENCES CITED

1. Wilke C.R., Yan R.D. and von Stockar, U. "Preliminary cost analysis for enzymatic hydrolysis of newprint" Biotechnol. Bioeng. Symp. 6, 155-175 (1976)

2. Isaacs S.H. "Ethanol production by enzymatic hydrolysis, parametric analysis of a base case process". SERI/TR-231-2093 Golden Colorado, Solar Energy Research Institute (1984).

3. Lowry, O.H., Rosebrough N.J., Farr A.L., and Randall R.J. "Protein measurement with the Folin phenol reagent". J. Biol. Chem. 193, 265-275 (1951)

4. Miller, G.L. "Use of dinitrosalicylic reagent for the determination of reducing sugars". Anal. Chem. 31, 426-428 (1959)

5. Raabo E. and Terkildsen T.C. "On the determination of blood glucose". Scand. Clin. Lab. Invest. 12, 402-406 (1960)

6. Ackman R.G. "Porous polymer bead packings and formic acid vapour in the GLC of volatile fatty acids". J. Chromatogr. Sci. 10, 560-565 (1972)

7. Koehler L.H. "Differentiation of carbohydrates by anthrone reaction rate and color intensity". Biotechnol. Bioeng. Symp. 6, 21-33 (1976)

8. Mandels M., Andreotti T. and Roche C. "Measurement of saccharifying cellulose". Biotechnol. Bioeng. Symp. 6, 21-33 (1976)

9. Mes-Hartree M., Hogan C.M. and Saddler J.N. "The enzymatic hydrolysis and fermentation of agricultural residues to ethanol". Biotechnol. Bioeng. Symp. 14, 397-405 (1984)

10. Hogan C.M., Mes-Hartree M. and Saddler J.N. "Effect of ethanol on the cellulase and xylanase systems of *Trichoderma harzianum* E58". Proceedings of the Third European Congress on Biotechnology. Munich, West Germany pp II-395 a-d (1984)

11. Deshpande M.V. and Eriksson K.E. "Reutilization of enzymes from saccharification of lignocellulosic materials". Enzyme Microb. Technol. 6, 338-340 (1984)

12. Klyosov A. personal communications

13. Enfor report for project C-181. Forintek Canada Corp. authors. "Conversion of cellulose to ethanol using a two-stage process". DSS contract 0SS83-00027 (1984)

14. Puls J., Poutanen K., Korner H-U and Viikari L. "Biotechnical utilization of wood carbohydrates after the steaming pretreatment". Appl. Microbiol. Biotechnol. 22, 416-423 (1985)

DISCUSSION

R. Clyde, Clyde Engineering

I understand fungus will degrade lignin. Would that be a good way to go, or is the lignin so valuable it should be recovered?

Mes-Hartree

A large part of our program is also on looking at what to do with the lignin. One of the approaches is biomodification which would use a lignin degrader to produce higher value products either for enhanced oil recovery or in adhesives. But the technical research hasn't advanced to a point where I could say what we're going to do with the process. We don't know where we're going to remove the lignin.

H. Esterbrauer, U. of Graz (Austria)

You said that in your cost analysis about 60 to 80% of the enzymes have to be recycled. Is this also true if you use the spent hydrolysis residue for enzyme production? Our cost analyses say that the spent hydrolysis residue is the best substrate for enzyme production. If you go down to 10 filter paper units background substrate, enzyme cost in our analysis is not a major fraction for sugar production.

Mes-Hartree

Our position is based largely on reports in the literature. Since we just looked at recycle of spent residue, we haven't costed it yet, so I'm really not sure. I know that one of our major projects for next year will be to try and enhance the production of the enzymes. We feel the two main areas that need attention are enzyme production and enhanced hydrolysis.

Hollomon

In the United States, there are significant subsidies for ethyl alcohol from biomass, and I understand that in Canada, there are none. If it weren't for the subsidies for ethyl alcohol, you'd be tempted not to produce it at all. Is that also the case with your process?

Mes-Hartree

With the current oil prices, ethyl alcohol doesn't look great. This is why we're using the linear program, as we haven't quite decided on what our final product will be. Once you have glucose, you can go to pretty well anything you want to after that. So we'll be doing a lot of linear programming up to the glucose and xylose fraction, and then depending on market conditions, look at alternatives at that point.

C. Setterstrom, Harrington Research Co.

Would you explain again the relationship between the diol and the 1,3-butadine? Is the thought that Polysar would buy the diol and convert it?

Mes-Hartree

They use 1,4-butane diol for 1,3-butadiene, and we would sell them 2,3-butanediol. Although it would only mean process modifications to them, there's no way you're going to convince the chemical industry to modify a process that's working. But we can convert 2,3-butanediol, and then sell it to Polysar.

ROOTS OF THE BUFFALO GOURD: A NOVEL SOURCE OF FUEL ETHANOL IN THE SOUTHWEST

Barry Goldstein
Solar Energy Institute, New Mexico State University,
Patrick M. Carr, William P. Darby, Jennie S. DeVeaux,
Washington University in St. Louis,
Larry Icerman,
New Mexico Energy Research and Development Institute,
and Eugene B. Shultz, Jr.,
Washington University in St. Louis

ABSTRACT

Buffalo gourd, a native plant of the Southwest and well-adapted to growing conditions there, can produce more starch per acre with less water input than can the conventional southwestern ethanol crops, grain sorghum and corn. The starchy roots of buffalo gourd are an acceptable substrate for fermentation. Root cropping methods are under investigation in eastern New Mexico, where an expanding fuel ethanol industry is presently located. Preliminary economic estimates of cultivating and processing of buffalo gourd roots for ethanol production are encouraging.

The likely characteristics of the organization of production under dryland or limited irrigation conditions, followed by local pre-processing for size reduction, drying, and storage are discussed herein. The soil, nutrient, water, and temperature requirements, and the pest, disease, and harvesting problems are presented along with research and development work that is probably needed to bring this novel ethanol crop to the marketplace.

ROOTS OF THE BUFFALO GOURD:
A NOVEL SOURCE OF FUEL ETHANOL IN THE SOUTHWEST

INTRODUCTION

The buffalo gourd (Cucurbita foetidissima) is native to the southwestern United States and northern Mexico. It is of interest as a novel fuel ethanol crop for semiarid lands because root starch yields are higher and water inputs are lower than those for the conventional grain crops currently used as ethanol feedstocks in the Southwest. Because of these advantages, the economics of buffalo gourd utilization seem likely to be attractive, demand on scarce water resources would be lowered, and the depressed farm economy might be stimulated.

Of special interest is the state of New Mexico. An expanding fuel ethanol industry is centered in the eastern part of New Mexico and is based in part on grains irrigated from the Ogallala aquifer. Grain farming in eastern New Mexico is generally unprofitable, and the high cost of irrigation is a large part of the problem. Even though rainfall in 1984 and 1985 was well above average in Curry, Quay, and Roosevelt counties, the present location of the fuel ethanol industry in New Mexico, estimates of return to land and risk for most dryland and irrigated crops were negative (10). Yields for wheat and sorghum were lower there than in the upper midwest. Production costs including cost of land were also generally lower, but not enough to offset the lower yields, especially if irrigation was utilized.

The Ogallala aquifer is the only source of irrigation water for Curry and Roosevelt counties, and it provides 10% of the needs of Quay County. Water resources in the aquifer are estimated to decrease by 45% by the year 2020 if current consumption patterns continue (8). Already, some irrigated farmland in Roosevelt County is being returned to dryland agriculture because of the increase in pumping costs as the water table drops.

Since 1980, New Mexico has offered a tax incentive on 90% gasoline/10% ethanol fuel blends to encourage the development of the fuel ethanol industry. Initially, the tax deduction was 10 cents/gallon, but this was raised to 11 cents on July 1, 1983, exceeded in the U.S. only by a 16 cents/gallon exemption in Louisiana. With the 6 cents/gallon federal tax exemption on alcohol fuels, the total subsidy is presently 17 cents/gallon for ethanol blended fuel produced in New Mexico. The New Mexico tax deduction program is responsible for stimulating about $90 million in capital investment, and more than a 10-fold increase in ethanol production since 1982. Total installed production capacity is now 31.5 million gallons/year of ethanol, and more than 10% of the gasoline supplies in New Mexico were ethanol blends in 1985.

The state tax incentive program is scheduled to end in 1991, with a phase-out beginning in 1987. Under this provision, the current 11 cents per gallon tax exemption will be reduced to 8 cents between July 1, 1987 and June 30, 1988; 5 cents between July 1, 1988, and June 30, 1989; 3 cents between July 1, 1989, and June 30, 1990; and 2 cents between July 1, 1990, and June 30, 1991. The federal incentive remains unchanged until the scheduled expiration at the end of 1992.

Clearly, switching to a suitable alternative ethanol crop that would give high yields of starch without demanding much water is of interest. In recognition of this need, the New Mexico Energy Research and Development Institute (NMERDI) is supporting the investigation of the roots of buffalo gourd as an ethanol feedstock. The prime contractor is the New Mexico Solar Energy Institute at New Mexico State University (NMSU), Las Cruces. Field work is being carried out at the NMSU Agricultural Science Centers at Las Cruces and Clovis, and ethanol production is under study at the pilot plant operated by the Clovis Center. Development strategy and economics are the responsibilities of the Bioresources Development Group, Washington University, St. Louis, Missouri.

The literature on buffalo gourd now includes about 100 references on morphology, reproduction, culture, diseases, pests, domestication, yields, and uses. Most of these papers have been published by University of Arizona research investigators, headed for many years by William P. Bemis. Gathman and Bemis (6) have reviewed this work. DeVeaux and Shultz (4) published an interpretive review, with emphasis on potential economic uses of the buffalo gourd and development strategies. Icerman and Shultz (9) carried out a preliminary economic analysis of buffalo gourd as a perennial crop for the simultaneous production of starchy roots and high-oil-content seeds. Shultz et al. (16) and DeVeaux et al. (3) reported on the progress of the New Mexico buffalo gourd project.

BUFFALO GOURD AS A FUEL ETHANOL CROP

Root and Ethanol Production

Buffalo gourd roots, as received, are about 17 to 21 wt% starch (17). To maximize root production, Nelson et al. (12) grew buffalo gourd as a summer annual in Arizona at different population densities in three locations of different elevation. The highest root yield was 34,550 kg/ha (as received) at an elevation of 920 m from a density of 177,000 plants/ha. The average root starch content was 63.5 wt% (dry basis), resulting in a starch yield of 6810 kg/ha. Results of New Mexico field trials, using a similar high-density approach, have indicated that about 80% of the Arizona yield was possible. Plant population density must still be optimized.

In a laboratory test conducted by Miles Laboratories, buffalo gourd roots were found to be a suitable substrate that apparently contains no inhibitory substances to impede saccharification and fermentation (5). Fermentation was quick and unproblematic, and a starch requirement of about 1.8 to 1.9 kg/liter of ethanol was estimated. Laboratory tests conducted at the Clovis Science Center resulted in comparable feedstock requirements (14). A somewhat higher value, 2.0 to 2.2 kg/liter, was obtained in a pilot plant test at Clovis, but this result was attributed to certain processing difficulties that are likely to be eliminated in subsequent work.

If the highest Arizona root starch yield of 6810 kg/ha (12) can be confirmed in New Mexico, the ethanol yield would be about 3780 liters/ha (404 gal/acre), representing about 89 million kJ/ha (34 million Btu/acre). By comparison, ethanol yields from irrigated corn and irrigated grain sorghum in Curry County, New Mexico, are about 2800 and 2100 liters/ha (300 and 255 gal/acre), respectively, based on 10-year average yields (13). Dryland grain sorghum yields about half as much starch per acre as irrigated grain sorghum.

The co-product (stillage), although high in protein content, contains a bitter principle termed "cucurbitacins." If the bitter cucurbitacins in the stillage cannot be removed, modified, or bred out of the roots, the dried bitter stillage might be blended successfully with other feed components; however, animal feeding trials will be necessary. If dried to 10% moisture, the stillage would contain about 30% useful protein; however, if the stillage cannot be dried economically, the processing plant would have to dispose of about 10 gallons of wet stillage/gallon of ethanol produced.

Overcoming the bitter flavor may not be the only way to increase stillage value. Alternatively, dried bitter stillage may have potential use in pest management programs of diabroticite beetles which attack cultivated squash. Cucumber beetles are attracted to cucurbitacins, so the dried bitter stillage might be used as bait for beetle poison.

Commercialization Strategy

Consideration must be given to a host of strategic questions on how the farming and processing of buffalo gourd roots might be organized to ensure that farmers have a market and ethanol producers have an assured feedstock supply. For example, fresh roots are nearly two-thirds water, so costs of hauling for long distances will be excessive. The roots dry and harden in a few days making size reduction difficult unless carried out soon after harvest. Cold storage delays drying, but is costly. Therefore, roots will probably be cleaned, chopped or ground within a few days after harvest, dried to prevent fungal and microbial attack, and stored for shipping throughout the year. Although such processing might be done on the farm, a more practical strategy might be to call for the organization of local "pre-processing" facilities - private businesses or farmer-owned cooperatives - that would wash, chop or grind, dry, store, and ship. Also, pre-processors might own special harvesters for rental to individual farmers or provide custom root-harvesting services.

Further, farmers might grow buffalo gourd under contract to pre-processors, or to ethanol producers, to overcome farmer concerns that a market for the roots may not materialize. Pre-processors, both cooperatives and entrepreneurs, may even build or buy ethanol plants, or purchase part interests, to further ensure that markets will exist. Alternatively, ethanol producers may invest in local pre-processing facilities to assure the availability of feedstocks throughout the year.

Economics

The field price that an ethanol producer would be willing to pay for roots was estimated, starting with the assumption that starches from grains and from buffalo gourd roots have the same intrinsic value. Then, adjustments were made for anticipated differences in processing costs and co-product credits. Without these adjustments, the equivalent price for roots would be about $25/ton, based on a starch requirement of 1.85 kg/liter of ethanol. Differing downward adjustments in price from $25/ton were investigated, based on various assumptions of co-product credits and processing costs. The base case (co-product valued at 75% of corn co-product, with an 8% increase over the cost of production from grain) provided a field price estimate of $21.50/ton of roots (16).

A net revenue analysis (2) based on a root yield (as received) of 34,000 kg/ha, an estimated field price of $21.50/ton, and estimates of variable and depreciation costs typical of similar crops in eastern New Mexico, gave a net annual return to land and risk of $26/acre for a limited irrigation system of 5 inches/year (16). By contrast, irrigated corn, irrigated grain sorghum, and dryland grain sorghum had negative returns to land and risk in 1984, with the exception of irrigated corn in one county (10).

The high cost of irrigation is a problem for corn and grain sorghum, and yields are only good with irrigation. By contrast, buffalo gourd appears to have potentially high yields of starch with low irrigation requirements. This characteristic more than anything else boosts the estimated net annual return well above those for the conventional grain crops used for ethanol production.

R&D Strategy

Sensitivity analyses (2) for several major cost parameters indicate that root yield is the dominant variable. Harvesting cost and co-product value also have relatively large impacts on net return, while irrigation levels have a moderate effect. The high sensitivity of net return to root yield suggests strongly that a well-reasoned R&D strategy should continue to emphasize efforts to increase root yields without significantly increasing costs. Attention must also be given to R&D to reduce harvesting costs and to add value to the co-product. Improvements in these three major factors that might be brought about by reasonable expenditures of R&D funds are likely to make major contributions to improving the economics of buffalo gourd farming in New Mexico (16).

Methods of reducing the size of fresh roots, and drying and fermenting the dried fragments need to be studied. Economic feasibility studies should be carried out in parallel with the technical investigations in order to determine the net returns more accurately than can be done at present.

ROOT PRODUCTION

Soils and Nutrients

Within its native range, buffalo gourd is found on all of the dryland soil orders of the world. Further, there are 20 different suborders within the region where buffalo gourd is naturally found. Field tests by Nelson et al. (12) were conducted in Arizona at Mesa and Marana on clay loam, and at Middle Verde on fine sandy loam. The highest starch yield 6810 kg/ha, was obtained without irrigation at Middle Verde, which was also the site of highest elevation (920 m). Sandy land may be better than clay for buffalo gourd root production, but this conjecture remains to be verified. Sands warm up quickly and are easy to till. Soil compaction and development of impermeable subsoils are less likely to occur than in the case of clays. Water percolation is rapid so there is less likelihood of root rot than in soils with a high content of clay; buffalo gourd is very sensitive to anaerobic conditions and will not tolerate water ponding at the surface.

Many root and tuber crops develop best in sandy soil, and Irish potatoes, sweet potatoes, and peanuts grow well in eastern New Mexico in loamy sands and sandy loams. Carrots can be grown without forking in sandy soils; minimization of forking in buffalo gourd roots would also be desirable. However, the nutrient content of sands is generally lower than that of clays, and the high permeability of sands may allow nutrients to be leached out of the root zone before utilization. Also, use of sandy rather than clayey soils would increase the likelihood that some supplemental irrigation would be required. Further, most sandy loams are susceptible to wind erosion, but crop residues could be left on the field after harvest to help mitigate this problem because buffalo gourd produces large amounts of leaves and vines. An alternative would be to plant a winter cover crop after harvesting.

Most soils of New Mexico are phosphorus-deficient (7) so application of phosphorus may be necessary to promote root growth. New Mexico soils generally contain potassium, although this condition may not be true for parts of Quay County. The situation regarding nitrogen is unclear. As yet, buffalo gourd has not shown a response to nitrogen, but, of course, no attempt has been made to breed for nitrogen response.

Water

The consumptive water use of buffalo gourd is intermediate between those of two well-known dryland crops, barley and winter wheat, as measured by Nelson (11) in Mesa, Arizona. Further, consumptive water use tests at Clovis indicate that the average for barley and winter wheat is 26% lower than that of grain sorghum and 42% lower than that of corn (1). Therefore, buffalo gourd should require less water than the two leading feedstocks of the ethanol industry, grain sorghum and corn. The former supplies about 85 percent of ethanol feedstocks in New Mexico.

Coupled with the projection of lower water use is the expected higher starch yield from buffalo gourd roots, probably 35% higher than that of irrigated corn, 58% higher than that of irrigated grain sorghum, and 216% higher than that of dryland grain sorghum, based on 10-year average yields in eastern New Mexico (13). This calculation assumes that New Mexico yields of root starch will equal those obtained at Middle Verde, Arizona, in fine sandy loam without irrigation. This expectation seems to be reasonable in view of the lower average summer temperatures and higher average annual rainfall in eastern New Mexico, compared to Middle Verde, Arizona.

Precipitation in Curry, Quay and Roosevelt counties averages about 13 inches during the growing season. This rainfall is hoped to be sufficient, although a small amount of supplemental irrigation may be needed in some years.

Temperature

Buffalo gourd can tolerate a wide range of temperatures. Vines senesce at the first frost, but roots can survive air temperatures down to -25°C (-13°F) (17) During the growing season, vine and root production can continue in temperatures up to 40°C (104°F), although growth becomes restricted when temperatures reach 35°C (95°F), in regions of low humidity (17).

Summer temperatures in eastern New Mexico would seldom be high enough to stop root production entirely. Maximum temperatures from June through August in Curry, Quay, and Roosevelt counties average about 33° C (91° F). Therefore, root production might carry on throughout the summer at high rates, unlike sites of lower elevation in Arizona.

Pests and Diseases

The resistance and susceptibility of buffalo gourd to diseases and pests has been reviewed and summarized by DeVeaux and Shultz (4). Of particular concern are root rot fungus and the root-knot nematode. The former has been a problem in irrigated test plots in New Mexico, perhaps because rainfall in 1984 and 1985 has been greater than average.

Integration of buffalo gourd into a crop rotation plan in which buffalo gourd is followed by a crop resistant to nematodes, with a fallow period between, may help control nematodes. Shannon et al. (15) suggest that organic matter may encourage growth of certain nematode parasites, so where buffalo gourd is grown, vine and leaf residue should probably remain on the land after root harvest. Further, alluvial and clayey soils often harbor nematodes. Planting after cotton may not be advisable because it tends to be susceptible to nematodes. Finally, resistant cultivars might be developed.

Harvesting

The root of the buffalo gourd is similar to the carrot in shape, but the crown is found 4 to 6 cm below the surface. Two types of mechanical harvesters might be considered for modification for use on buffalo gourd roots, potato harvesters which use a blade to turn up soil and potatoes, and sugar beet harvesters which use digging disks to pull up the beets. A third option is to design a special machine.

Because a modified potato harvester would have to dig significantly deeper than in the case of potatoes, considerable energy would be required and the harvesting rate would be low. However, light sandy soils might be handled in this way. Sugar beet harvesters dig up less soil so fuel and labor costs should be lower than in the potato harvester case. Modifications would be needed, such as special cleaning beds and disk spacing.

The vines will probably need to be removed prior to harvesting roots, unless a special harvester is designed to carry out vine removal and root harvesting simultaneously. Flails are available for removal of sweet potato vines, but buffalo gourd generates heavier vine growth. An oversized flail might be adequate.

CONCLUDING REMARKS

Although the studies reported on here are still in the preliminary stages, buffalo gourd is likely to produce more starch per acre with less water input and at lower cost in eastern New Mexico than the conventional grains used for fuel ethanol. These grains show negative returns to land and risk, while positive returns are likely for buffalo gourd. Corn must be irrigated but grain sorghum can be grown as either a dryland or irrigated crop; however, yields are low without irrigation. Buffalo gourd can probably be grown as a dryland crop in eastern New Mexico, with supplemental irrigation needed in some years, thereby helping to lengthen the life of the underlying Ogallala aquifer.

The profit potential of buffalo gourd will depend upon:

1) the strength of future demand for ethanol as an octane booster;

2) the development of root marketing and pre-processing systems that will ensure that farmers have a market, and that ethanol producers have assured supplies of feedstocks;

3) the ability to harvest roots at reasonable cost;

4) the ability to sell fermentation co-product at a reasonable price, despite the natural bitterness or perhaps because of it;

5) the ability over the long term to increase root yields without significantly increasing root production costs; and

6) the continued availability of government incentive programs.

The development of marketing and pre-processing systems, cost-effective harvesting techniques, improvements in co-product value, increases in root yields and other important objectives of experimental R&D and economic and marketing studies will require a continuing commitment for some years.

ACKNOWLEDGEMENTS

This work was conducted, in large part, with the support of the New Mexico Research and Development Institute. Thanks are due to Koert Lessman of NMSU, Las Cruces, and Ralph Finkner, Director of the NMSU Agricultural Science Center, Clovis, and their staffs. We are also grateful to W. P. Bemis, J. W. Berry, A. C. Gathman, J. M. Nelson and J. C. Scheerens of the University of Arizona for many helpful discussions.

REFERENCES CITED

1. Baker, R. D., Personal communication. Area Agronomist, Coop Extension Service, NMSU Ag. Science Center, Clovis, NM, 1985.

2. DeVeaux, J. S., Development of Buffalo Gourd as a Novel Fuel Ethanol Crop in Eastern New Mexico, Master of Science Report, Dept. of Engineering and Policy, Washington University, St. Louis, MO., 1985.

3. DeVeaux, J. S., William P. Darby and Eugene B. Shultz, Jr., "Buffalo Gourd as a Dryland Fuel Alcohol Crop: Development Strategy and Economics for New Mexico," presented at the International Arid Lands Research and Development Conference, University of Arizona, Tucson, October 20-25, 1985. To be published in Conference Proceedings.

4. DeVeaux, J. S. and E. B. Shultz, Jr., "Development of Buffalo Gourd (*Cucurbita foetidissima*) as a Semiaridland Starch and Oil Crop," *Econ. Botany* **39**(4), 454-472 (1985).

5. Gallagher, Brian, Person communication. Miles Laboratories, Elkhart, IN., 1985.

6. Gathman, A. C. and W. P. Bemis, "The History, Biology and Chemistry of the Buffalo Gourd, *Cucurbita foetidissima* HBK," *In* R. Robinson, ed., *The Biology and Chemistry of the Cucurbitacae*. Cornell Univ. Press, Ithaca, NY, in press.

7. Glover, C. R., R. D. Baker and L. Cihacek, *Fertilizer Guide for New Mexico*, Coop. Extension Service Circular No. 478, 1984.

8. High Plains Associates, *Six-State High Plains Ogallala Aquifer Regional Resources Study*, Report to U.S. Dept. of Commerce and the High Plains Council, Austin, TX., 1982.

9. Icerman, L. and E. B. Shultz, Jr., "A Preliminary Economic Analysis of Buffalo Gourd as a Diesel Fuel and Ethanol Feedstock in the High Plains of New Mexico, Texas and Oklahoma," Energy 10(8), 901-909 (1985).

10. Libbin, J. D., M. J. Schaber, B. E. Mayberry, R. P. Sullivan and B. Dictson, "1984 Crop Cost and Return Estimates, Curry County," New Mexico Cooperative Extension Service, NMSU, Las Cruces, NM, 1985.

11. Nelson, J. M., Personal communication. University of Arizona Ag. Experiment Station, Mesa, AZ., 1985.

12. Nelson, J. M., J. C. Scheerens, J. W. Berry and W. P. Bemis, "Effect of Plant Population and Planting Date on Root and Starch Production of Buffalo Gourd Grown as an Annual," J. Amer. Hort. Sci. 108(2), 198-201 (1983).

13. New Mexico Dept. of Agriculture and New Mexico State University, Agricultural Statistics, 1983. NMSU, Las Cruces, NM, 1984.

14. New Mexico State University, "Laboratory and Pilot-Plant Investigations from Buffalo Gourd Roots Grown Under Different Population Densities and Fertility Levels," NMSU Agricultural Science Center, Clovis, NM, 1985.

15. Shannon, E., S. Thomas and M. Clayshulte, "Control of Nematodes," New Mexico Coop. Extension Service Circular No. 400 W-8, 1983.

16. Shultz, Eugene B. Jr., William P. Darby, Jennie S. DeVeaux and Larry Icerman, "Buffalo Gourd as an Energy Crop in New Mexico," Proceedings of the Southwest Industrial Biomass Conference, Phoenix, AZ, Sept. 29-Oct. 1, 1985.

17. Vasconcellos, J. A., W. P. Bemis, J. W. Berry and C. W. Weber, "The Buffalo Gourd, Cucurbita foetidissima HBK, as a Source of Edible Oil," In E. H. Pryde, et al., eds., New-Sources of Fats and Oils. Amer. Oil Chem. Soc., Champaign, IL., 1981.

DISCUSSION

J. Herberman, Ontario Ministry of Energy (Canada)	Is buffalo gourd native to any developing countries right now, or is it only found in the Southwest?
Shultz	It's only found in the Great Southwest, which does of course include a large portion of Mexico. It's found roughly from Nebraska to the Tropic of Cancer. It is found in a very broad range of climates, soils, and latitude, and therefore we think it's adaptable to many dry lands in the Third World. I might add there's one thing that we're not sure of; nobody knows about the photoperiodic requirements of the plant for sure. We think that there may be photoperiodic requirements for the germ plasm that would be collected in the northern part of its range, but perhaps not down near the Tropic of Cancer. So we hope that would solve the problem, which would be a problem of course of producing seeds if it requires a day-length exchange. Then it isn't going to produce seeds near the equator.
Zoellner	Have you gotten to the point where you have been able to generate some interest in Third World countries? Do you have an expression of interest yet?
Shultz	I'm not really quite sure where we are. We wanted to do enough work here in the States, and also think enough about the social and environmental implications before we go to the Third World and say we have some sort of great thing. We may be at a point now where I would feel comfortable about approaching certain countries and saying we think that we have something that's worth testing. But I certainly would not approach them and say we've got something that is certainly going to work. As you can see, I've talked primarily about technical aspects and probably the social aspects are the ones that are going to control.

GENETIC ENGINEERING OF BACTERIA FOR ALCOHOL FUEL PRODUCTION

Z. P. Shalita, Ph.D., M. D. Yablonsky, B.S.,B.A.,
M. M. Dooley, Ph.D., S. E. Buchholz, B.S.,
S. K. Kahrs, B.A., K. Murphy-Holland, Ph.D., and
D. E. Eveleigh, Ph.D.
Department of Biochemistry and Microbiology
Cook College, Rutgers University
New Brunswick, New Jersey 08903

ABSTRACT

A number of genetic approaches are being taken to allow *Zymomonas mobilis* to utilize biomass for the production of fuel ethanol, such as, the incorporation of cellulase genes into it. Thus, a gene bank of *Microbispora bispora*, which has an active cellulase complex, has been constructed in *Escherichia coli*. Eleven thousand colonies from this bank were screened for the presence of cellulases. Twelve expressed carboxymethylcellulase (CMCase) and twenty-five exhibited β-glycosidase activities. Restriction analysis of plasmid DNA from the clones exhibiting CMCase activity indicated that five different genes were cloned. Fifteen hundred of the above recombinants were screened for the ability to degrade xylan and chitin but no positive recombinants were recovered. Transfer of the cellulase genes to *Z. mobilis* is envisaged. To facilitate transfer of plasmids to *Zymomonas*, a mutant strain with increased sensitivity to kanamycin and ampicillin has been isolated. This mutant can be used to select transfer of recombinant broad host range plasmids which carry these resistances. In order to increase expression of cellulases and other foreign genes in *Zymomonas*, native promoters are being isolated. A new broad host range promoter probe vector, pCMD1, was constructed for this purpose.

GENETIC ENGINEERING OF BACTERIA FOR
ALCOHOL FUEL PRODUCTION

INTRODUCTION

The present production of most fuels and chemicals is bas on crude oil, whose global reserves are not replenishable, and whose current supply is unstable (10,11,19,20). In a search f alternative energy and chemical sources, the option of the utilization of biomass seems attractive. Liquid transportatio ethyl alcohol from biomass can be used as both a gasoline octa booster and as gasohol (10% ethanol in gasoline, 22). In th U.S.A., ethanol derived via fermentation of corn competes at present with the price of that derived synthetically (10). Current estimates of U.S. ethanol fermentative production is 5 10^8 gallons/annum, and represents 75% of total national produc tion (19). Renewable biomass is abundant as agricultural, forestry, and municipal wastes. The current annual use of 32 10^6 tons of feedstock chemicals in the U.S. can potentially be derived from biomass. For the 9 x 10^6 tons of oxychemicals on 10% of total starch (corn) crops or 3% of the currently collectible forest will satisfy the requirements of their production; though it would be preferable to utilize agricultu wastes in fermentation (11). The great majority of biomass is lignocellulose, which is potentially inexpensive, and does not compete directly in other major industrial applications. It i comprised of three major polymers: Cellulose (35-50% of dry weight) in which, β-1,4 glucose units are linked in elongated chains which form crystalline microfibrils, lignin (20-25% of weight), a complex polymer of three phenyl-propane derived alcohols, and hemi-cellulose (20-30% of dry weight) of mainly pentose (mostly xylose) and sometimes hexose polymers (25). T general composition of biomass is dependent on its source. Fo efficient fermentation of biomass, pretreatment with extractio of hemicellulose and disruption of the lignocellulose complex required (40). The latter protocol includes a variety of physical (steaming) and chemical (acidifying) treatments plus particle size reduction (40). Cellulases efficiently degrade cellulose product. Cellulolytic enzymes (16,19,30) occur in a large variety of microbes from diverse habitats (soil, compost hot springs, rumen). The cellulases, based especially on stud of those of the fungus _Trichoderma reesei_ (17,30,34,48), include endoglucanases (EG) which initiate attack on the cellulose, rendering initiating sites for cellobiohydrolases (CBH, exoglucanase) that sequentially cleave cellobiose from t non-reducing ends. The latter disaccharide is readily hydroly by (often cellbound) cellobiase (β-glucosidase) to glucose (1 These enzymes act synergistically to degrade the cellulose polymer. The efficient action of _T. reesei_ cellulase towards crystalline cellulose and its large yields, suggested it as a source of cellulase for industrial applications (19,31). Othe cellulases are now found to be equally effective and to be mor thermostable (14,15,16). We have recently studied such a bacterial cellulase from the thermophilic actinomycete

Microbispora bispora (2,28,46). This bacterial complex contains endo and exoglucanases which act synergistically, are heat stable (60° C), and produced in good yields. The cellobiase of this enzyme complex is remarkably resistant to end product inhibition (46). We have studied this cellulase complex further as it also has the advantage of being of procaryotic origin and thus would be potentially easier to manipulate genetically, it lacks introns, and the cloned bacterial gene(s) may be expressed in other bacterial hosts (promoter or vector promoter recognition). Furthermore, the M. bispora cellulase genes could perhaps be expressed from one operon. It would be advantageous to gain expression of these genes in an effective ethanologen. Several major microbes, include yeasts, such as Saccharomyces cerevisiae, and S. carlbergensis that produce up to 12% ethanol, and S. sake that produces up to 20% ethanol (but in a 30 day slow fermentation), and the bacterium Zymomonas mobilis that grows rapidly and produces up to 12% ethanol (20). Both Saccharomyces and Zymomonas convert glucose to ethanol with similar high efficiencies (1.92 moles ethanol/mole hexose, 96%). As regards ethanol tolerance these candidates are far ahead of others such as Clostridium thermocellum which yields about 4% ethanol plus acetate (51). Ethanol productivities from glucose with Zymomonas optimized in cell recycle system with vacuum fermentation, attain a maximum of 160g/l/hr, while for yeasts it is roughly 80g/l/hr (39,40). In addition, Zymomonas shows uncoupling of growth and ethanol production, is readily handled (as it is not a strict anaerobe), does not require sterols and exhibits some tolerance to low pH (20). Thus Zymomonas is a good candidate for the fermentative production of ethanol.

Presently the use of Zymomonas has some drawbacks: Zymomonas only ferments glucose, fructose and sucrose (43). Genetic systems are adequate but rudimentary, transformation frequencies are low, and there has been poor expression of heterologous genes in it, excluding antibiotic resistance genes (7,8,9,20,26). In addition, its multiple antibiotic resistances, and the presence of 1-5 cryptic plasmids (dependent on the strain) hinder genetic analysis in Zymomonas. However, an antibiotic sensitive strain isolated in our lab has been shown to be a useful recipient for transfer of broad host range plasmids. Additional genetic restraints can be worked around and the exceptionally high ethanol productivity of Z. mobilis make it worthy of further study. Our strategy for gaining an efficient cellulolytic ethanologen include:

a. Construction of a M. bispora gene library in E. coli, and screening it for recombinants containing cellulase genes.

b. Evaluation of broad host range (shuttle) vectors which can replicate efficiently in both E. coli and Zymomonas.

c. Optimizing the transformation of Zymomonas by shuttle vectors.

d. Enhancement of the expression of cloned genes in Zymomonas by application of Zymomonas promoters and strong E. coli promoters.

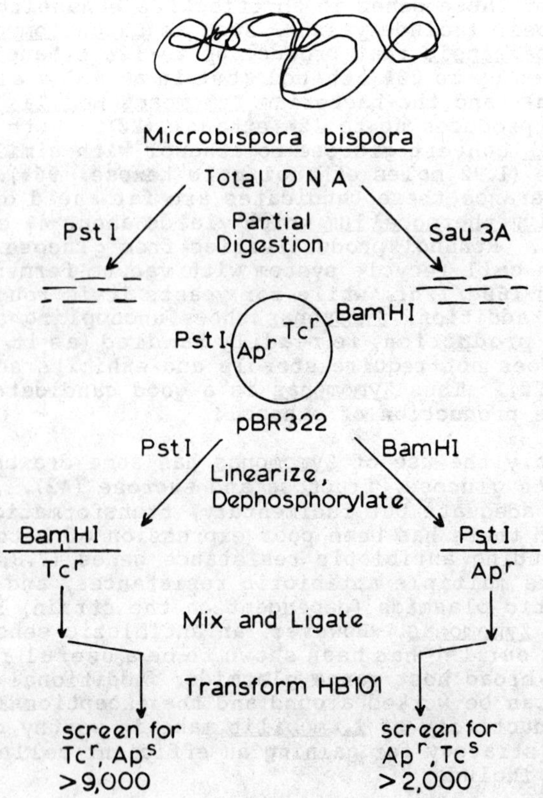

Figure 1. Outline of the construction of the *M.bispora* gene library. Details are in the Results section.

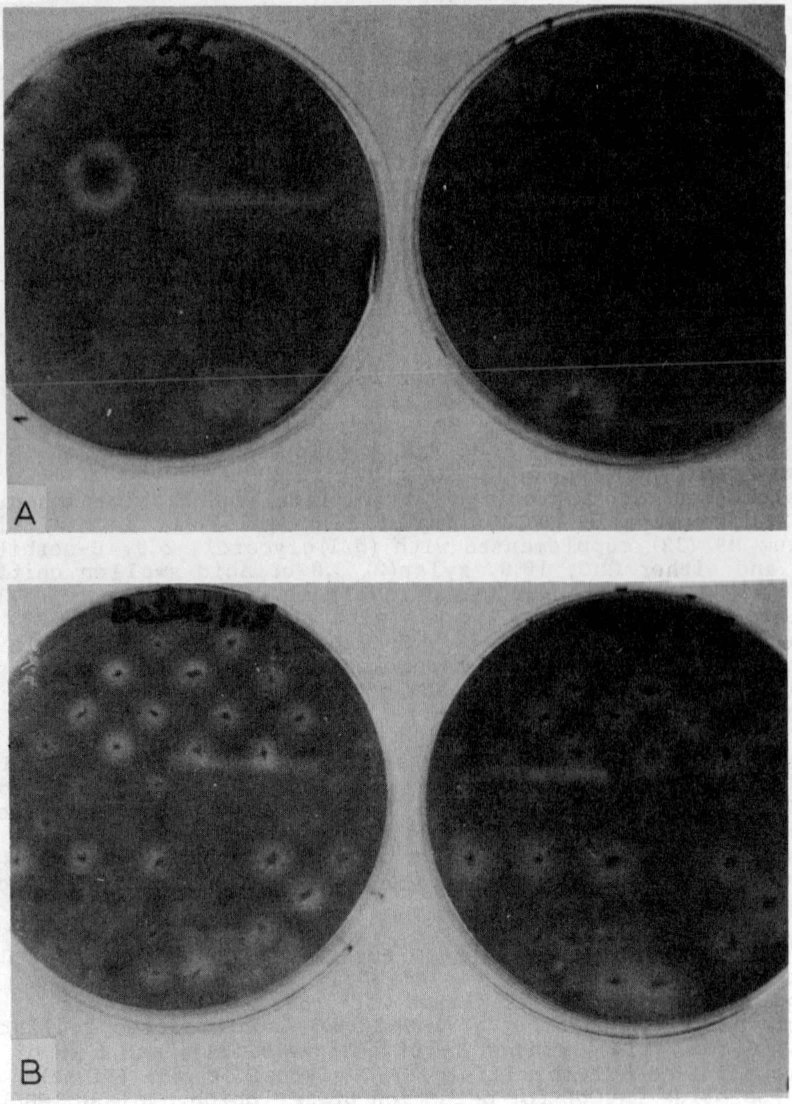

Figure 2. A. Plate detection of CMCase activity in recombinants from the gene library. E.coli transformants containing M.bispora DNA inserts (4-10kb) were grown on M9 with 0.5% CMC. Colonies were washed off, and the plates were flooded with Congo red, followed by 1M NaCl. Clearing zones indicate CMC hydrolysis by CMCase.
B. E.coli colonies transformed by recombinant plasmids from the original CMCase positive strains were checked by the same procedure described in Fig. 2A. E.coli HB101 (pBR322) was streaked as a negative control between the positive clones. The left replica and the right replica plates were incubated for 16 hrs at 37°C, and at 55°C respectively. H.S. stands for heat step.

e. Introduction of cellulase genes into Zymomonas plus analys to evaluate expression, secretion, and synergistic activity.

The initial stages of this project for both characterizing cellulase genes and approaches to transfer them to the ethanologen Zymomonas are described.

MATERIALS AND METHODS

Strains and cloning vehicles. Bacterial strains and plasmids are described in Table 1.

Media. M. bispora was grown in Hägerdal-Pye (HP) salts medium (24), supplemented with 10 g/l crystalline cellulose (Avicel PH101, a gift from FMC corp., Markus Hook, PA.) or carboxymethylcellulose (CMC 7MF, a gift from Hercules, Wilmington, DE.). E. coli was grown in LB medium (33), supplemented with appropriate antibiotics, µg/ml:chloramphenic (Cm)20; tetracycline(Tc)25; ampicillin (Ap)100, or in minimal medium M9 (33) supplemented with (g/l)glycerol, 5.0, D-sorbito 2.5, and either CMC, 10.0, xylan(2) 3.0 or acid swollen chitin 3.0 (2). Z. mobilis was grown in glucose/yeast extract (GYX) medium (43) supplemented (µg/ml) with either Cm 100, Ap 100, kanamycin (Km) 100, or Tc 40. Spheroplasts were prepared in either GYX or in S medium, supplemented with glycine, 2.0% or 4.0%, and regenerated in GYX or in R medium (41,50).

In situ enzymatic screening. To test for endoglucanase activity the procedure of Teather and Wood (44) was essentiall followed. Replicated colonies were grown on M9+CMC plates for 48-72 hrs, incubated at 55°C for 16 hrs, and then the colonies were removed by washing. The plates were flooded with Congo r (Fisher, 0.1%) for 45 min., then with 1 M NaCl for 30 min. Zo of hydrolyzed CMC around the colonies appeared as yellow halos a red background.

Detection of cellobiohydrolase and disaccharases. The CM plate screening procedure was slightly modified as follows. Colonies grown on M9+CMC plates were overlayed with agar (6 g/ 5.0 ml/plate) containing (g/l) CMC, 0.5, sodium dodecylsulfate (SDS, Sigma, 0.1%), Triton X-100, (Sigma 0.02%), and 1 mM of t following 4-methylumbelliferyl-β-D-glycosides (45) (Sigma): -cellobioside (MUCb) for detection of cellobiohydrolase (CBH) activity, -glucoside (MUG) for β-glucosidase (cellobiase), -N-acetylglucosaminide (MUNAG) for chitobiase, and -xyloside (MUX for xylobiase activity. Delta-Gluconolactone (cellobiase inhibitor) was added to MUCb overlay to prevent degradation of the glycoside by cellobiase, and thus focusing on detection of enzymes that cleave off cellobiose from MUCb. Enzymes that cleave these latter aryl-β-glycosides release the UV fluoresce 4-methylumbelliferone product. Cultures were incubated at 55° and screened periodically up to 16 hrs by aid of a UV-360 nm l

Table 1: Bacterial Strains and Plasmids

Strain or Plasmid	Relevant Genotype or Phenotype	Source or Reference
Microbispora bispora	A cellulolytic, chitinolytic, xylanolytic wild type.	2,28,46
Escherichia coli		
HB101	F$^-$Pro, Leu,Thi,lacY,Sm^r, endA, hsdR,hsdM	5
S17-1	RP4-2Tc::Mu-Km::Tn7,Tp^r, Sm^r,Pro,hsdR,Rec	37
MC1060 rec$^-$	(lacIPOZYA)X74, galU, galk, recA, hsdR	13
MXR	(lac-pro), galE, thi, recA	21
Zymomonas mobilis		
CP4	Ap^r(250)Km^r (200)[1]	43
CP.45	Ap^s (80)Km^s (60)	8
Plasmids		
pBR322	Ap^r,Tc^r	4
pSKS105	Ap^r,Lac$^+$	42
pGC91.14 (=RP1::Tn951)	Ap^r,Tc^r,Km^r,Lac$^+$	9
pKT210	Cm^r,Sm^r	1
pKT230	Sm^r, Km^r	1
pKT240	Ap^r,Km^r	1
pRK2013	Km::Tn7	18
pSUP104	Cm^r,Tc^r	37
pSUP106	Cm^r,Tc^r,cos	37
pSUP304	Ap^r,Km^r	37
pCMD1	Gm^r,lacZ, lacY, lacA.	This work

1. Minimal inhibitory concentration (MIC), µg/ml.

Carboxymethyl cellulase (CMCase) assays. Reducing sugar leased by CMCase was measured by the Nelson-Somogyi method 4,38). Units of activity were expressed as µmoles glucose leased per minute per milligram protein (6).

β-Galactosidase assay. Detection of lactose degrading lonies employed the lactose analogue 5-bromo-4-chloro-3-dolyl-β-D-galactopyranoside (X-gal, Boehringer Mannheim) pplemented (40 µg/ml) in the medium (42) to yield positive blue lonies. Further quantitative assays were performed as scribed by Pardee et al., (36).

Chromosomal DNA preparations. The procedure of Collmer and lson (14) was modified: M. bispora cultures were conditioned by o sequential subcultures on HP medium with Avicel at 55°C. The lls were then washed, resuspended to original density, and oculated 1% (V/V) into HP with CMC, in a baffled shake flask 00 ml/500 ml flask). The culture was incubated at 55°C with aking (240 RPM) for 22 hrs. Cells were pelleted and suspended in 10 ml TE (10 mM Tris, 1 mM EDTA, pH 8.0). The spension was homogenized in a tissue homogenizer, and lysozyme (powder, Sigma) was added to yield a final concentration of 10 mg/ml. The mixture was incubated at 37°C for 15 min., then frozen in dry ice. The incubation and freezing cycle was repeated four times. Cells were pelleted and resuspended in fresh TE+lysozyme buffer, and the incubation/freezing cycles repeated four more times. Fresh diethylpyrocarbonate (10% in EtOH), 0.5 ml was added. The cells were lysed by addition of 1 ml of 2% SDS in HTE (50 mM Tris, 25 mM EDTA, pH 8.0) buffer. T lysate was frozen (-20°C), thawed and phenol extracted four times, followed by four ether extractions. Residual ether was evaporated at 60°C for 10 min. The chromosomal DNA was precipitated by addition of K-acetate to 0.3 M and 2 volumes ethanol, at -20°C overnight. The DNA was pelleted and dissolve in 3.0 ml HTE at 4°C overnight. RNAase T1 and RNAase A (Sigma) were added to 5 u/ml and 50 µg/ml respectively, and the mixture was incubated at 37°C for 45 min. Proteinase K (Boerhinger Mannhiem) was added to 50 µg/ml and was incubated at 37°C for 4 min. Two phenol extractions followed by ether extraction were performed. The DNA was ethanol precipitated as above, washed twice with 70% ethanol, and once with 95% ethanol, dried in a vacuum dessicator, and resuspended overnight at 4°C in 200 ul

Isolation of plasmid DNA. Plasmid DNA was isolated by the alkaline lysis method of Birnboim and Doly (3).

Restriction endonucleases. Enzymes were purchased from Boehringer Mannheim, and used as recommended by the manufactur

Electrophoresis. DNA mobilities were analysed in agarose (LE, Seakem) as described (32). DNA fragments were electroelut from agarose gels (GTG, Seakem) into dialysis tubing at 30 mA 7 hrs.

Cloning procedures. *M. bispora* partial chromosomal DNA restriction digests were prepared using either Pst I or Sau 3A using conditions described in Results. Plasmid pBR322 was linearized by Pst I or by Bam HI. *Zymomonas* chromosomal DNA wa digested by Hind III, and plasmid pCMD1 was linearized using th same enzyme. Ligations were performed at concentrations described in Results.

Transformations and conjugations. *E. coli* was transforme as described (32). *Z. mobilis* competent cells were prepared wit 150 mM $MgCl_2$ and transformed as described by Buchholz et al., (9). "Partial" spheroplasts were transformed in transformatio (TF) medium (23) as described by Shalita et al., (41). These "partial" spheroplasts were regenerated (~100%) in GYX medium during the phenotypic expression period (6-16 hrs). Conjugal gene transfers between *E. coli* and *Z. mobilis* (in both directions) were performed as described by Buchholz et al., (8

Mutation. Induction of mutations in *Z. mobilis* was performed by exposure to UV-254 nm illumination so that 95% of the cells were inactivated (36), followed by incubation in GYX broth with mitomycin-C (3 µg/ml) (27) for 16 hrs at 28°C as described (8).

RESULTS AND DISCUSSION

Effects of ethanol concentration and temperature (30°C) on he activity of *M. bispora* cellulase. The effect of ethanol on he cellulase activity of *M. bispora* cellulase was tested by bserving CMC degradation in the presence of 0% ethanol at both 60°C (optimal) and 30°C (optimal for *Z. obilis* CP4) (Table 2). Lowering the assay temperature to 30°C ecreased CMC degradation by 68%. Ethanol at 10% promoted a 16% ecrease in activity at 30°C.

Construction of an *M. bispora* gene library. Chromosomal DNA 135 µg) was partially digested by 26u Pst I, in a total volume f 360 µl, for 20 min. Another DNA aliquot was partially igested by 1.1u Sau 3A, for 20 min. The digests were lectrophoresed in an agarose gel (0.7%), and only fragments of .0-10 kilobases (Kb) were electroeluted from the gels, and recipitated. These fragments were ligated with Pst I linearized nd phosphatased pBR322 at a concentration of 40 µg/ml with a ector to fragment ratio of 2:1. Partial Sau 3A fragments were imilarly ligated with BamHI linearized and phosphatased vector. hese ligation mixtures were used to transform *E. coli* HB101. election was for Tc^r when ligation mixtures from PstI treated NA were used. Of the Tc^r transformants, 95% contained ecombinant plasmids as indicated by sensitivity to Ap. Over 000 Tc^r, Ap^s recombinant clones were recovered. When ligation ixtures from Bam H1 and Sau 3A treated DNAs were used in ransformations, selection was for Ap^r. In this latter case, 82-8% of the transformants carried recombinant plasmids as

Table 2: Effect of Ethanol and Temperature on the Activity[1] of the Cellulase of *Microbispora bispora*.

Ethanol Conc. (%)	60°C	30°C
0.0	4.30	1.370 (68.19)[2]
10.0	2.78 (35.34)[3]	1.138 (16.31)[3]

1. Units expressed activity as µmoles glucose/min/mg protein.
2. Temperature inhibition (%).
3. Ethanol inhibition at same temperature (%).

ndicated by Ap^rTc^s. Over 2000 strains with recombinant plasmids f this latter type were recovered.

Recombinants containing CMCase genes. Screening of the ecombinants on medium M9 supplemented with CMC yielded sixteen olonies that exhibited CMC hydrolysis (49). Plasmid DNA repared from them was used to transform new HB101 cells but ielded only twelve transformants that exhibited CMC hydrolysis. estriction analysis of their plasmids suggested that five ifferent genes were cloned (Figures 3,4). The insert in plasmid CC5188 seemed homologous to that in pCC5434, and the insert in

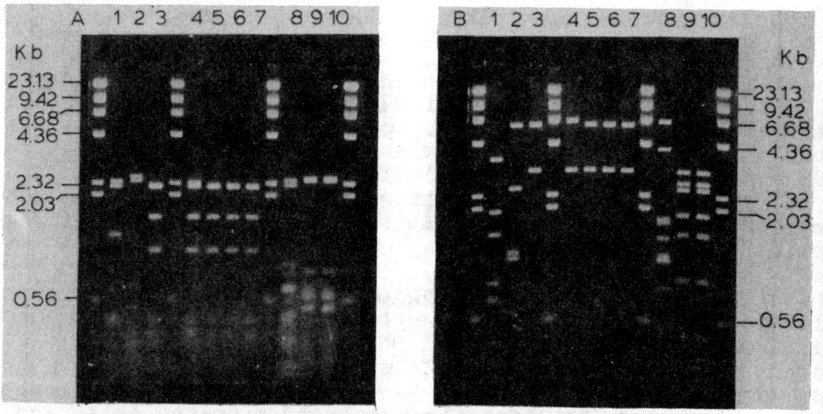

Figure 3. Agarose gel electrophoresis of restriction digests of recombinant plasmids conferring CMCase activity in E.coli. A. All plasmids were cleaved by Ava I. Lanes: 1, pCC1528; 2, pCC6704; 3, pCC3622; 4, pCC3891; 5, pCC3896; 6, pCC6917; 7, pCC7847; 8, pCC1539; 9, pCC5188; 10, pCC5434. Size Marker, phage lambda digested by HindIII, generating fragments of (kb)23.13, 9.42, 6.68, 4.36, 2.32, 2.03, and 0.56. B. All plasmids were cleaved by SalI. Lanes are in same order as in Fig. 2A.

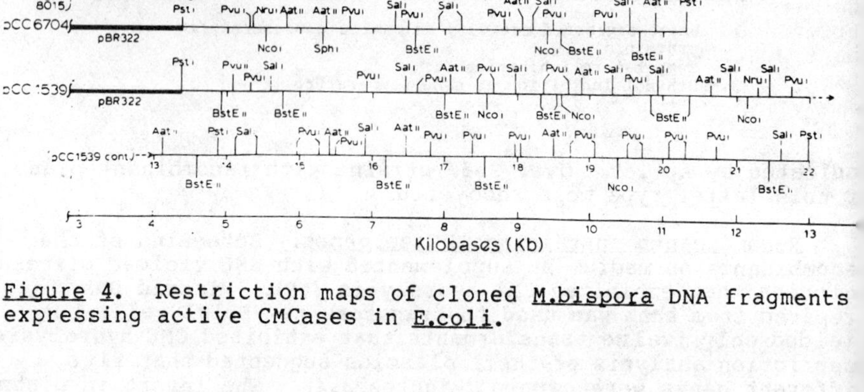

Figure 4. Restriction maps of cloned M.bispora DNA fragments expressing active CMCases in E.coli.

CC3622 appeared homologous to those in plasmids pCC3896, 6917, 847, and to an additional three that harbored a small (0.6 Kb) xtra Pst I fragment, pCC3891, 7947, and 8015. The fact that hese fragments were cloned and expressed in the same orientation uggests that their expression was orientation dependent. That s they were probably transcribed by the pBR322 vector B-actamase promoter, which is close to the Pst I site.

Recombinants containing aryl-β-glycosidases genes. First creening of the PstI cloned (over 9000) recombinants, with a ool of four 4-methylumbelliferyl-β-glycosides (MUG, MUCb, MUNAG, nd MUXy, see Methods) yielded twenty-eight fluorescent colonies 49), indicating that each of them degraded at least one of the ryl-β-glycosides (Figure 5A). When SDS and Triton X-100 were mitted from the top agar, no fluorescent colonies could be etected. This indicated that the β-glycosidases were ntracellular or periplasmic, and only degraded the aryl-β-lycosides upon cell lysis.

Plasmid DNA prepared from each of the twenty-eight aryl-B-lycosides degrading recombinants was used to transform new HB101 ells. This yielded twenty-five strains that exhibited luorescence with the separate aryl-β-glycosides (Figure 5B, able 3). Restriction analysis of these cloned genes is in rogress.

Screening for xylan and chitin degrading recombinants. creening of the first 1500 recombinants of the gene library on edium M9 supplemented with either xylan or acid swollen chitin ailed to reveal any clear zone of hydrolysed polymers. Lysing he cells in situ by SDS and Triton X-100 (see Methods), and xtending incubation at 55°C up to 48 hrs did not result in any ubstrate hydrolysis.

Antibiotic sensitive strain Z. mobilis CP4.45, a new ecipient for cloning vectors. The multiple high antibiotic esistances of Zymomonas remarkably restricts the application of any useful broad host range vectors. The state of Zymomonas ntibiotic resistance genes is unclear, however, it has been hown that some of the antibiotic genes (gentamycin, kanamycin nd streptomycin resistances) are plasmid encoded (47). Several ttempts to 'cure' Z. mobilis strains of plasmids were nsuccessful (not presented). Thus many methods of mutation were ttempted. Mutation of strain CP4 resulted in many mutants xhibiting decreased antibiotic resistances (8). One mutant, P4.45, exhibiting both reduced Ap (80 versus 350 μg/ml original esistance) and Km (60 versus 200 μg/ml) resistance was studied urther. Transfer of plasmid pSUP304 into this strain (by onjugal assistance of pRK2013) conferred resistance to Ap >2000 g/ml, and Km 1500 μg/ml. Transfer of pKT230 conferred esistance to 1750 μg/ml Km. Plasmids pSUP304 and pKT230 were solated from strain CP4.45 and used for transformation of E. oli S17-1 cells. When these plasmids were reisolated from S17-1 nd analyzed by electrophoresis no change in their mobilities ould be observed. Therefore, this antibiotic sensitive erivative may be used as a recipient for cloned genes carried on hese vectors and for their expression in Zymomonas.

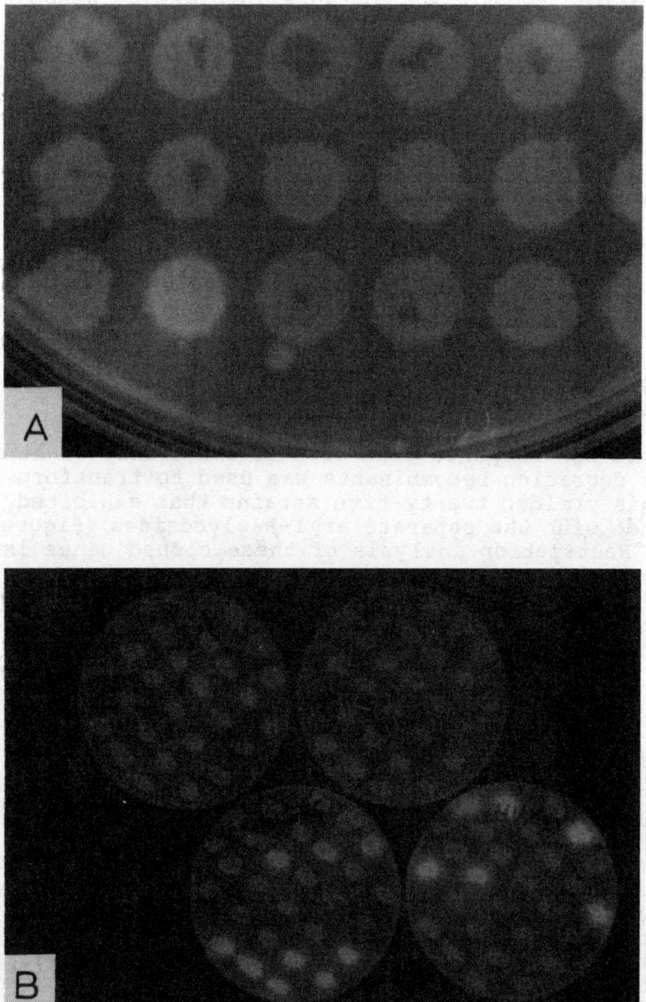

Figure 5. A. Plate detection of aryl-β-glycosides degradation by recombinants from the gene library. Replica plates of recombinants described in Fig. 2A, were covered with top agar containing CMC, SDS, Triton X-100 and four methyumbelliferyl-β-glycosides, MUG, MUCb, MUnAG, and MUXy as described in Methods. Fluorescence under UV lighting indicates glycosidic hydrolysis and release of the fluorescent methyumbelliferyl aglycone. B. E.coli colonies transformed by recombinant plasmids isolated from the fluorescent strains were checked with each of the four methyumbelliferyls separately, in triplicate. There are nine positive group III recombinant clones in the total of twenty eight clones which are illustrated. See Table 4 for results.

Table 3: Evaluation of cloned M.bispora DNA fragments expressing aryl-β-glycosidases degradation.

Plasmid Prep.	Fluorescent Reaction: Degradation of 4-methylumbelliferyl-β-D-*			
#	*glucose MUG[1]	*cello-biose MUCb[1]	*N-acetyl glucosam. MUnAG	*xylose MUXyl
MUG-Type				
1	++[2]	-	-	+
3	+	-	-	+
4	++	(?)[3]	-	-
5	+++	+	-	-
8	+	-	+	+
9	+++	+	-	-
12	+	-	-	-
18	(?)	(?)	(?)	(?)
21	+	-	-	-
23	+	+	+	+
25	++	-	-	-
26	++	-	-	-
MUnAG-Type				
2	-	-	+++	-
6	+	+	+++	-
7	+	-	+++	-
10	+	(?)	+	-
13	-	(?)	++	-
14	-	-	+++	-
15	-	-	+++	-
16	-	(?)	+++	-
17	-	-	(?)	-
22	-	-	++	-
27	-	-	+++	-
28	-	-	+++	-
MUXy-Type				
11	-	(?)	++	+++
19	-	-	-	(?)
20	-	-	-	+++
24	-	-	-	+++

1. D-gluconolactone (β-glucosidase inhibitor) not added.
2. +++ strong reaction, ++ medium reaction, + weak reaction
3. (?): Very weak reaction, possibly positive

β-Galactosidase (lac) expression in Z. mobilis. rior to the search for Z. mobilis promoters, (see below) by heir expression of E. coli lac genes, the expression of the n951 lac genes were compared in E. coli and in Z. mobilis(12). onjugal transfer of plasmid RP1::Tn951 from E. coli to Z.mobilis P4 resulted in X-gal degrading (blue) colonies of exconjugants. owever, none of these could grow on lactose as a sole sugar ource. β-Galactosidase assays of these exconjugants showed bout 15 u/ml and about 75 u/ml for uninduced and induced ultures, respectively. Permease activity was marginally easurable. In contrast, control exconjugants, i.e., X-gal egrading E. coli MXR cells, exhibited in the range of 50 u/ml nd 1000 u/ml, for uninduced and induced cultures, respectively. he reasons for the low lac expression in Zymomonas are unclear. ne suggestion is that the level of lactose premease activity is oo low in Zymomonas. Transfer of galactose utilization operon enes into Zymomonas might enhance the utilization of lactose. uch experiments are now in progress.

Isolation of Zymomonas promoters. Initial attempts to gain trong expression of heterologous genes in Z. mobilis, except for ntibiotic resistance genes, were not successful. Thus, the ikelihood of good expression of M. bispora cellulase genes in Z. obilis using foreign control elements does not appear to be great. In order to enhance expression of foreign genes in Zymomonas, it may be necessary to place these genes under the control of promoters which have good activity in this genus. W have elected to isolate native Z. mobilis promoters for this purpose.

As the broad host range promoter probe vectors which are available are not suitable for use in Z. mobilis, we have constructed a new promoter probe plasmid (Figure 6A) which is designated pCMD1. The replication functions of RSF1010 and a gene coding for chloramphenicol resistance were derived from pSUP104. Digestion of pSUP104 with both Hind III and Sal I removes a small fragment of DNA which contains most of the gene for tetracycline resistance including the Pribnow box of the promoter. Therefore, when this fragment is removed from the plasmid the tetracycline resistance promoter is inactivated. A lactose permease gene and a β-galactosidase gene removed from i native promoter and translation start sequence were derived fr pSKS105. DNA of both parent plasmids was digested with Hind I and Sal I. These digests were mixed, ligated and used to transform MXR, a lactose negative strain of E. coli. Colonies which were Tc^s, Ap^s, Cm^r and white on X-gal, a chromogenic substrate which turns blue in the presence of even low levels o B-galactosidase, were likely to have the desired construction. Restriction analysis of plasmid DNA from colonies with this phenotype confirmed the presence of the new promoter probe vector.

In order to use pCMD1 (Fig 6B), chromosomal DNA digested t completion with Hind III was mixed and ligated with vector treated with the same enzyme and alkaline phosphatase. A 2:1 mass ratio of plasmid to chromosome was utilized at DNA total concentrations of 15 and 60 μg/ml.

Using the above procedure, 42 colonies which contain pCMD1 with Zymomonas promoter containing fragments have been isolated so far. A promoter which will act strongly in Zymomonas should be located and will be used to increase cellulase production in this genus.

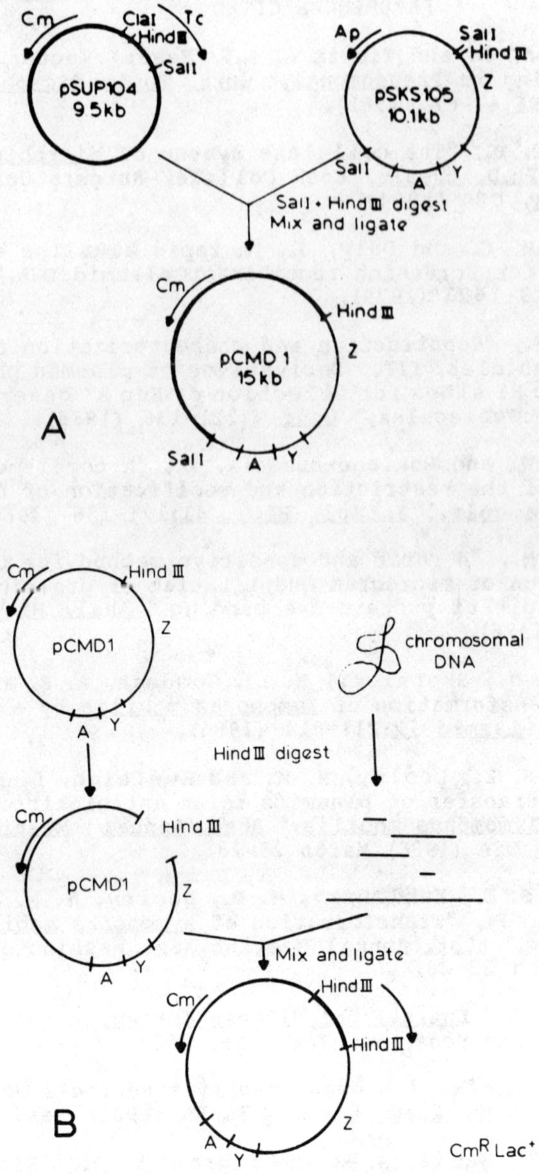

Figure 6. Outline of the isolation of *Z.mobilis* promoters, A. The construction of a broad host range, promoter probe vector, pCMD1. Details are in the Results. B. The procedure for cloning HindIII digested DNA into the probe vector and selection for $Cm^r lac^+$(blue on X-gal) colonies. Details are in the Results.

REFERENCES CITED

1. Bagdasarian, M. and Timmis K. N., "Host: Vector systems for gene cloning in Pseudomonas," Curr. Topic Microbiol. Immunol. 96:47-67 (1982).

2. Bartley, T. D., "The cellulase system of Microbispora bispora," Ph.D. thesis, Cook College, Rutgers University, New Jersey, USA (1984).

3. Birnboim, H. C. and Doly, J., "A rapid alkaline extraction procedure for screening recombinant plasmid DNA," Nuc. Acid Res. 7:1513-1523 (1979).

4. Bolivar, F., "Construction and characterization of new cloning vehicles. III. Derivatives of plasmid pBR322 carryi unique Eco RI sites for selection of Eco RI generated recombinant molecules," Gene 4:121-136 (1978).

5. Boyer, H. W. and Rouland-Dussoix, D., "A complementation analysis of the restriction and modification of DNA in Escherichia coli," J. Mol. Biol. 41:121-136 (1969).

6. Bradford, M., "A rapid and sensitive method for the quantitation of microgram quantitaties of protein utilizing the principle of protein-dye binding," Anal. Biochem. 92: 248-254 (1976).

7. Browne, G. M., Skotnicki, M. L., Goodman, A. E. and Rogers, P. L., "Transformation of Zymomonas mobilis by a hybrid plasmid," Plasmid 12:211-214 (1984).

8. Buchholz, S. E., Dooley, M. M. and Eveleigh, D. E., "Conjugal transfer of plasmids to an antibiotic sensitive mutant of Zymomonas mobilis," Abst. Annual Meeting ASM, Las Vegas, NV, USA (1986) March 23-28.

9. Buchholz, S. E., Yablonsky, M. D., Dooley, M. M. and Shalita, Z. P., "Transformation of Zymomonas mobilis CP4 by plasmid DNA." Abst. Annual Meeting ASM, Washington, DC, USA (1986) March 23-28.

10. Bungay, H. R., Energy, the Biomass Option, p. 347, John Wiley and Sons, New York, NY, 1981.

11. Busche, R. M., "Will biomass supply feedstocks of the future?" Indust. Chem. News, p.34-35 (1983) May.

12. Carey, V. C., Walia, S. K. and Ingram, L. O., "Expression of a lactose transposon (Tn951) in Zymomonas mobilis," Appl. Environ. Microbiol. 45:1163-1168 (1983).

13. Casadaban, M. J. and Cohen, S. N., "Analysis of gene control signals by DNA fusion and cloning in Escherichia coli," J. Mol. Biol. 138:179-207 (1980).

4. Collmer, A. and Wilson, D. B., "Cloning and expression of a Thermomonospora YX endocellulase gene in E. coli," Biotechnol. 1:594-601 (1983).

5. Cornet, P., Tronik, D., Millet, J. and Aubert, J. P., "Cloning and expression in Escherichia coli of Clostridium thermocellum genes coding for amino acids synthesis and cellulose hydrolysis," FEMS Microbiol. Lett. 16:137-141 (1983).

6. Coughlan, M. P., "The properties of fungal and bacterial cellulases, with comment on their production and application," Biotechnol. and Genet. Engin. Rev. 3:39-109 (1985).

7. Cuskey, S. M., Montenecourt, B. S. and Eveleigh, D. E., "Overproduction of cellulase-screening and selection," In Krumphanzl, V., Sikyta, B. and Z. Vanel, (eds) Overproduction of microbial metabolites, 405-416. New York: Publ. Czechoslovak Acad. Sci. and Academic Press, 1981.

8. Ditta, D., Stanfield, S., Corbin, D. and Helinski, D. R., "Broad host range DNA cloning system for gram negative bacteria: Construction of a gene bank of Rhizobium meliloti," Proc. Natl. Acad. Sci. USA 77:7347-7351 (1980).

9. Eveleigh, D. E., "The fermentation of biomass-Current aspects" In Cote, W. A. (ed), Biomass Utilization, 365-391. Syracuse, NY: Plenum Press. 1982.

0. Eveleigh, D. E., Stokes, H. W. and Dally, E. L., "Recombinant DNA approaches for enhancing the ethanol productivity of Zymomonas mobilis," In Wise, D. L. (ed), Organic chemicals from biomass, 69-91. Menlo Park, CA: Benjamin/Cumming: Publ. Co., Inc. 1983.

1. Faelen, M., Resibois, A. and Toussaint, A., "Mini-Mu: An insertion element derived from temperate phage Mu-1," Cold Spring Harbor Symp. Quant. Biol. 43:1169-1177 (1978).

2. Ferchak, J. D. and Pye, E. K., "Utilization of biomass in the U.S. for the production of ethanol fuel as a gasoline replacement. I. Terestrial resource potential." Solar Energy 26:9-16 (1981).

3. Glick, B. R., Brooks, H. E. and Pasternak, J. J., "Transformation of Azotobacter vinelandii with plasmid DNA," J. Bacteriol. 162:276-279 (1985).

4. Hägerdal, B. G. R., Ferchak, J. P. and Pye, E. K., "Cellulolytic enzyme system of Thermoactinomyces sp. grown on microcrystalline cellulose," Appl. Environ. Microbiol. 36:606-612 (1978).

25. Higuchi, T., "The biochemistry of lignification," Wood Research 66:1-16 (1980).

26. Ingram, L.O. and Byun, M. O-K., "Construction of a new vector for expression of foreign genes in Zymomonas mobilis," Annual Meeting SIM, Boston, MA, USA (1985) Augus 4-9.

27. Krug, K. A., and Daugulis, A. J., "Ethanol production usin Zymomonas mobilis immobilized on an ion exchange resin", Biotechnol. Lett. 5:159-184 (1983).

28. Lechevalier, M. P. and Lechevalier, H., "Biology of actinc cetes not belonging to the genus Streptomyces," In Demain, A. L. and Solomon, N. A., (eds) Biology of industrial micr organisms, pp. 315-358. Menlo Park, CA: Benjamin/Cumming Publ. Co., Inc. 1985.

29. Ljungdahl, L. G. and Eriksson, K. E., "Ecology of microbia cellulose degradation" In Marshall, K. C. (ed), Adv. Micrc Ecol. 8:237-299, (1985).

30. Mandels, M., "Applications of cellulases," Biochem. Soc. Transact. 13:414-416, (1985).

31. Mandels, M., Andreotti, R. E., and Roche, C., "Measurement saccharifying cellulase," Biotechnol. Bioeng. Symp. 6:21-3 (1976).

32. Maniatis, T., Fritsch, E. F. and Sambrook, J., Molecular cloning. A laboratory manual. Cold Spring Harbor, NY: Col Spring Harbor Lab. Press. 1982.

33. Miller, J. H., Experiments in molecular genetics. Cold Spring Harbor, NY, Cold Spring Harbor Lab. Press. 1972.

34. Montencourt, B. S., "Trichoderma reesei cellulases," Trend Biotechnol. 1:156-161 (1983).

35. Nelson, N., "A photometric adaptation of the Somogyi metho for the determination of glucose," J. Biol. Chem. 153:375-380 (1944).

36. Pardee, A. B., Jacob, F. and Monod, J., "The genetic contr and cytoplasmic expression of inducibility in the synthesis of B-galactosidase by E. coli," J. Mol. Biol. 1:165-178 (1959).

37. Priefer, U. B., Simon, R. and Pühler, A., "Extention of th host range of Escherichia coli vectors by incorporation of RSF1010 replication and mobilization functions," J. Bacteriol. 163:324-330 (1985).

8. Reese, E. T., and Mandels, M., "Rolling with the times: Production and applications of Trichoderma reesei cellulase", Ann. Report Ferm. Proc. 7:1-20, (1984).

9. Rogers, P. L., Goodman, A. E. and Heyes, R. H., "Zymomonas ethanol fermentations", Microbiol. Sciences 1:133-136 (1985).

0. Rolz, C., de Cabrera, S., Calzada, F., Garcia, R., de Leon, R., Del Carmen de Ariola, M., de Micheo, F. and Morales, E., "Concepts on the biotransformation of carbohydrates into fuel ethanol", Adv. Biotechnol. Proc. 1:97-143 (1983).

1. Shalita, Z. P., Yablonsky, M. D., Buchholz, S. E., Dooley, M. M., and Eveleigh, D. E., "Transformation of partial spheroplasts of Zymomonas mobilis CP4 by broad host range plasmids," Abst. Annual Meeting ASM, Washington, DC, USA (1986) March 23-28.

2. Shapira, S. K., Chou, J., Richard, F. V. and Casadaban, M. J., "New versatile plasmid vectors for expression of hybrid proteins coded by a cloned gene fused to lacZ gene sequences encoding an enzymatically active carboxy-terminal portion of B-galactosidase," Gene 25:71-82 (1983).

3. Swings, J. and DeLey, J., "The biology of Zymomonas," Bacteriol. Rev. 41:1-46 (1977).

4. Teather, R. M. and Wood, P. J., "Use of Congo red-polysaccharide interactions in enumeration and characterization of cellulolytic bacteria from the bovine rumen". Appl. Environ. Microbiol. 43:777-780 (1982).

5. van Tilbeurgh, H., Claeyssens, M. and de Bruyne, C. K., "The use of 4-methylumbelliferyl and other chromatophoric glycosides in the study of cellulolytic enzymes", FEBS Lett. 149:152-156 (1982).

6. Waldron, C. R., Becker-Vallone, C. A. and Eveleigh, D. E., "Isolation and characterization of a cellulolytic actinomycete Microbispora bispora" (submitted for publication).

7. Walia, S. K., Carey, V. C., All III, B. P. and Ingram, L. O., "Self-transmissable plasmid in Zymomonas mobilis carrying antibiotic resistance," Appl. Environ. Microbiol. 47:198-200 (1984).

8. Wood, T., "Properties of cellulolytic enzyme systems," Biochem. Soc. Trans. 13:407-410. (1985)

49. Yablonsky, M. D., Shalita, Z. P., Kahrs, S. K., Murphy-Holland, K. and Eveleigh, D. E., "Cloning and expression of several Microbispora bispora endoglucanase genes in E.coli", Abst. 1st Annual ASM Conference on Biotechnology., Washington, DC, USA (1986) March 20-23.

50. Yanase, H., Yasui, M., Miyasaki, T., and Tonomura, K., "Fusion of spheroplasts and genetic recombination of Zymo mobilis, Agric. Biol. Chem. 49, 133-140 (1985).

51. Zeikus, J. G., "Chemical and fuel production by anaerobic bacteria," Ann. Rev. Microbiol. 34:423-464 (1980).

ACKNOWLEDGEMENTS.

The present work has been supported by the U. S. Department of Energy, Contract No. DE-AC02-82CH10093, the U. S. Department of Agriculture, Contract No. 85-CRCR-1-1701, and the New Jersey Agricultural Experiment Station No. K-01111-03-86, supported by State and Hatch Funds.

DISCUSSION,

R. Clyde, Clyde Engineering: If you can put the gene into Zymononas which will permit the utilization of galactose from lactose, would you think that it would go faster? You mentioned a rate of 160 grams per hour. If anyone would like to find out how to do it quite a bit faster, come to my poster board tonight.

Shalita: Our aim is to degrade cellulose, but we will have to try. We are considering cloning the genes of the galactonate operon.

H. Esterbauer, U. of Graz (Austria): A major problem in cloning is that cellulase enzymes are glycoproteins. Do you have a glycosylation system in any of your bacteria? It's usually believed that you can not express glycoprotein in bacteria.

Shalita: In fungi, it is glycoprotein. In Microbispora, I don't know, but its endoglucanases are active when expressed in E. coli.

INVESTIGATION OF FUNDAMENTALS OF TWO-STAGE, DILUTE SULFURIC ACID HYDROLYSIS OF WOOD

John I. Zerbe, Program Manager
and
Andrew J. Baker, Chemical Engineer
Energy Research, Development, and Application Program
Forest Products Laboratory
Madison, Wisconsin 53705

ABSTRACT

A fundamental analysis of production of ethanol from southern red oak (Quercus falcata) resulted in recommendations for a process that is technically ready for commercial exploitation, but some equipment testing and evaluation are necessary first. Such evaluation is the objective of pilot testing now underway at the National Fertilizer Development Laboratory of the Tennessee Valley Authority. Data for hemicellulose and cellulose hydrolysis were correlated using models for developing and evaluating the process design. Coproducts originating from the hemicelluloses were assumed to be animal-feed molasses, furfural, and acetic acid. The potential for the process could be significantly improved through utilization of the pentose sugars, which might be realized by fermentation to ethanol. It was found that the best existing design for using the hydrolysate of the second stage is one in which all products other than glucose are concentrated and burned. The lignin residue could also be burned as fuel, but the process heat requirements are about 50% greater than could be supplied by the lignin. Advantages of the two-stage process include high concentration of product solutions and somewhat reduced capital requirements compared to the percolation process, but these advantages are offset to some extent by higher yields from the percolation process.

INVESTIGATION OF FUNDAMENTALS OF TWO-STAGE, DILUTE SULFURIC ACID HYDROLYSIS OF WOOD

INTRODUCTION

In 1980 the Forest Products Laboratory undertook selecting, describing, and improving the most promising process for converting wood to alcohol commercially for the 1985-87 time frame. We developed a process which we think is suitable, but some equipment testing and evaluation are necessary before the expense of a commercial plant is justifiable. Such evaluation is the object of pilot testing by the Tennessee Valley Authority at the National Fertilizer Development Laboratory, Muscle Shoals, Alabama.

The process, in its simplest outline, is shown in Figure 1. Wood chips, impregnated with a dilute sulfuric acid solution and drained of all interstitial liquid, are charged to the first stage. Here they are heated with direct steam, resulting in the hydrolysis of most of the hemicelluloses, and then discharged to washers. After being washed free of the material solubilized in the first stage, the lignocellulose is reimpregnated with acid and charged to the second stage. As with the first stage, the liquid content of the second-stage charge is kept to a minimum. Conditions here are sufficient to hydrolyze the resistant cellulose. The resulting mixture is discharged to washers where the glucose solution is separated from the lignin residue.

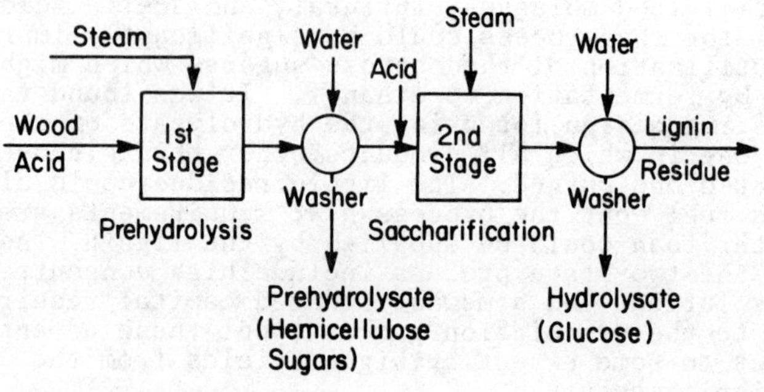

Figure 1. Simple schematic of the two-stage hydrolysis process. (ML84 5484)

The major process elements are the first-stage prehydrolysis and second-stage hydrolysis. The principal emphasis is placed on the prehydrolysis because this process has general application. Prehydrolysis studies were extended to the point of simulating industrial conditions in a small digester, whereas experimental work on the second-stage hydrolysis was restricted to laboratory scale. Since no representative hydrolysis solutions were available from the lignocellulose hydrolysis, no fermentation studies were performed with lignocellulose.

FIRST STAGE (PREHYDROLYSIS)

Hemicellulose Removal

Xylan Removal. The removal of xylan from a hardwood prehydrolyzed under conditions of constant temperature, acidity, and liquid-to-solid ratio (L/S) can best be described by a typical curve on a semilogarithmic plot (Fig. 2). The first portion of the curve, to about the point of 60% removal, is quite linear; the slope is essentially constant. Thereafter, the removal rate continuously decreases until it reaches a minimum value equal to the hydrolysis rate of cellulose at the particular conditions employed. This is deduced from the fact that, on prolonged prehydrolysis, the composition of the carbohydrate remaining in the residue reaches a constant value (4) indicating that a small portion (1-2%) of the xylan is intimately associated with the resistant cellulose.

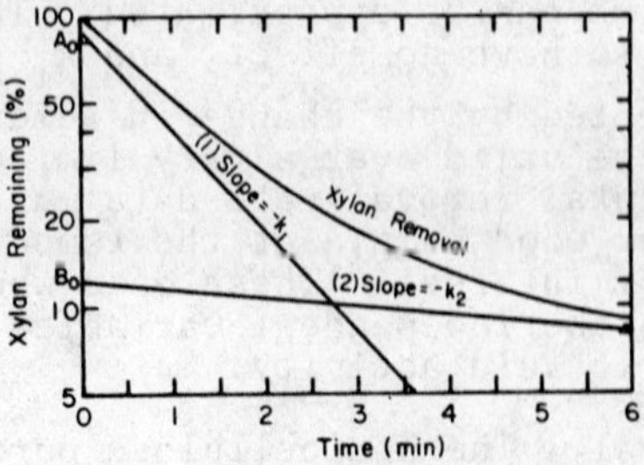

Figure 2. Xylan removal during prehydrolysis. Curves (1) and (2) are linear components of the xylan removal curve. (ML84 5478)

Reasons for the reduction in removal rate after 50-60% of the xylan is removed are unknown. Neither the position nor associated slopes of points on the later portion of the curve in Figure 2 can be predicted. Consequently, it is necessary to collect experimental data in the region of interest. This region for a prehydrolysis aimed at maximum xylose recovery is between 90 and 95% xylan removal. It is convenient to have an analytical expression for the xylan removal curve. The removal curve is assumed to be the sum of two functions, each linear on a semilogarithmic plot:

$$XR = \text{xylan remaining in residue, \% of original}$$

$$= A_o \exp(-k_1 t) + B_o \exp(-k_2 t) \qquad (1)$$

where t = time.

Since the plot is presented as percentage removal, at $t = 0$

$$XR = A_o + B_o = 100 \qquad (2)$$

so that:

$$XR = A_o \exp(-k_1 t) + (100 - A_o)\exp(-k_2 t) \tag{3}$$

An experimental data set can be used with equation (3) to obtain the values of A_o, k_1, and k_2 for the best fit. These values substituted into equation (1) yield the desired analytical expression. One should not be tempted to attach any fundamental significance to this correlation. It is readily apparent from Figure 2 that the values of the parameters will depend on the particular range in which the data were collected.

It has been assumed in the above discussion that the curve of Figure 2 was obtained under constant conditions of temperature, L/S, and catalytic acid activity $[H^+]$, the independent variables that determine the removal rate. The effect of minor changes in the independent variables on the xylan removal curve can be approximated. This is done by assuming the L/S to have no effect, and k_1 and k_2 to be proportionately affected by the change in acidity and temperature. This will be valid over a very limited range. In general, experimental removal rate data must be collected for the particular wood sample at the temperatures, acidities, and times of interest. These data would be expected to correlate with the independent variables, L/S, temperature, and catalytic acid activity.

Glucan Removal. The hemicellulose portion of wood contains, in addition to xylan, other carbohydrate materials such as galactoglucomannan and arbinogalactan. These are hydrolyzed and removed during prehydrolysis in a manner similar to the removal of xylan. The resulting sugars--glucose, mannose, arabinose, and galactose--form a sizable portion of the total carbohydrate content of the prehydrolysate even in the case of southern red oak where xylose is predominant. Rather, more glucose appears in the hydrolysate than could have originated from galactoglucomannan. Although there is probably some amorphous glucan associated with the hemicelluloses (10), most of the additional glucose must originate from the cellulose fraction. Typically, cellulose hydrolysis indicates that cellulose has an easily hydrolyzable component that is removed at a rate comparable to that of hemicellulose (9). It is this easily hydrolyzable component of the cellulose that was solubilized during prehydrolysis.

Xylose and Furfural Yields

From previous work, it was concluded that xylose yields in excess of 80% would be obtained from red oak, and that increasing temperature of xylose and acidity both tend to increase maximum yields.

In this work, data on both residues and hydrolysates were obtained for a variety of hydrolysis conditions with southern red oak using ampoules, a technique described by

Springer (12,13). The calculated xylose yields were all in excess of 80%; they indicated a modest increase with increasing temperature but a slight decrease with increasing acid. Measured values were, with one exception, somewhat greater than the calculated values; this may have resulted from the protection afforded by the presence of oligomers. The experimental yields indicate that the maximum yields increase with increasing acid strength and temperature.

The above data support the conclusion that maximum xylose yields increase with increasing temperature but are inconsistent as to the effect of varying the acid catalyst concentration.

The amount of furfural formed in the prehydrolysis is significant. The xylose degradation can rise to as much as 15% at the point of maximum yield. A model for the complex group of reactions associated with the decomposition of xylose is:

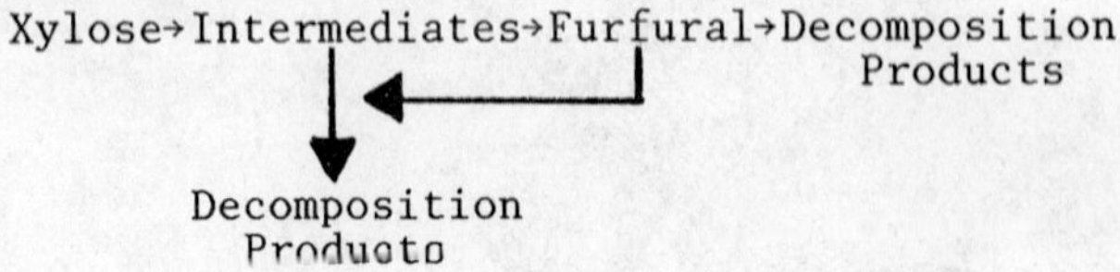

An important point that can be deduced from the above mechanism is that furfural is produced in high yields during prehydrolysis. In the early stages of the reaction, the concentration of furfural is low, and its loss rate is low since this depends on the furfural concentration. The yield of furfural based on xylose reacted is high, beginning at 100%, and decreases as the amount of xylose reacted increases.

Steaming Acid-Impregnated Chips

Studies on the direct steam prehydrolysis of acid-impregnated chips have been carried out in a small jacketed digester.

Conditions in the laboratory digester are intended to simulate, as nearly as possible, the conditions in a continuous digester with direct steam heating. Moist acid-impregnated chips are heated to the desired temperature with saturated steam. Some of the resulting condensate is perhaps absorbed by the chips, and some flows down over the chips below. The movement and location of liquid in the bed is important because it affects the catalyst concentration. The cold inner surface of the digester is also a source of condensate. In this small-scale study, the digester-to-charge mass ratio is much greater than it would be in a commercial unit.

To gather meaningful data, it was necessary to separate the digester condensate from the chip condensate using a chip container (Fig. 3)--a can with a closed bottom, and a wall made primarily of screening (A). The can is equippe with a removable false bottom (B) to drain the chips and hold them above the liquid accumulating in the bottom. The lower portion has sufficient volume to hold the liquor that drains from the chips. The lid (C) for the container diverts any condensate dripping from above. The condensate accompanying the incoming steam and that resulting from digester heating falls to the bottom of the digester and is removed by the sparge line (Fig. 4). With the valving arrangement, it is not possible to remove all condensate from the main steam line without admitting steam to either the inside or the jacket of the digester. A large amount of condensate accumulates in the steam line when the equipment is idle. This is discharged by opening the valve to the jacket before admitting steam to the chips. This results in a preheating (indirectly) of the chips for about 3 minutes, with the temperature in the vapor space rising to 100°C. Following the release of this accumulated condensate, steam is admitted directly into the inner space; this instant is taken as time zero.

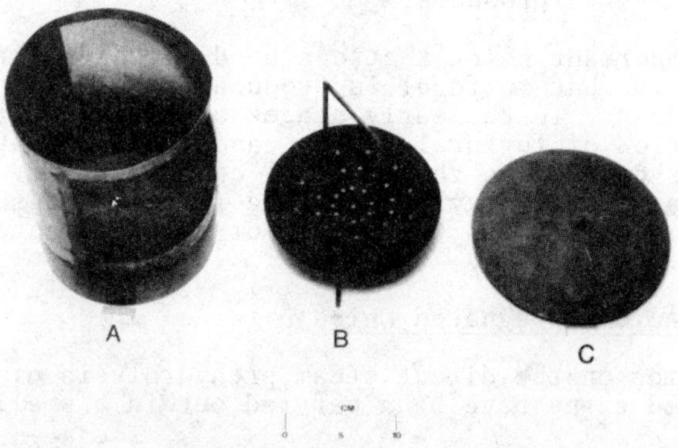

Figure 3. Chip container for prehydrolysis studies.

The xylose yield data pertaining to digester runs 1-18 is compared with calculated yields in Figure 5. The lower curve of xylan removal (Fig. 5 upper), correlated as previously described, is the sum of two exponential terms, specifically:

$$XR = \text{xylan remaining in residue, \%}$$
$$= 88.5 \exp(-\ 0.713t) + 11.5 \exp(-\ 0.0378t) \qquad (4)$$

where t = time (min).

This analytical expression for the liberation of xylose, together with the acid concentration of 0.17% H_2SO_4

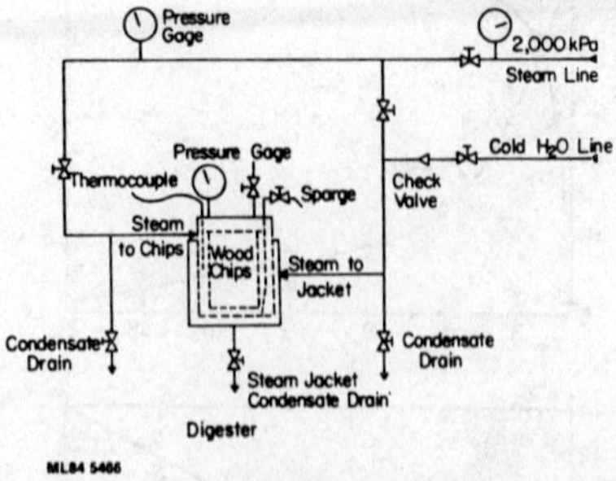

Figure 4. Schematic diagram of prehydrolysis equipment. (ML84 5466)

(pH = 1.77) and the L/S, was sufficient to calculate the yield for xylose in solution (Fig. 5 upper). The calculated maximum yield was 86.2% occurring at 8 minutes; the experimental data indicate a maximum yield of 83.8% at 6 minutes, with the yield at 9 minutes being 83.5%. The logarithmic scale obscures the fit of the data; a much more critical presentation is made in Figure 5 lower. The percentage of xylose reacted is the sum of the xylan in the residue and xylan in solution subtracted from 100. These calculated values are compared to the calculated amount of xylose reacted (the curve in Fig. 5 lower). All the information in Figure 5 lower is contained in Figure 5 upper, but the lower portion is a more stringent test of the data.

The large variations between the experimental points shown in Figure 5 lower are to be expected since they are very sensitive to errors in the xylose determination. The ordinate, "Xylose Reacted," is the difference of two large numbers, one of which is dependent on three separate analyses for xylose. Most of the experimental values fall above the calculated quantity of xylose decomposition--that is, xylose reacted. The data would be best represented by assuming an acid concentration of 0.2% rather than the value of 0.17%, which was used. This difference was judged to be within the limits of error for this group of early runs, and it was concluded that the prehydrolysis is adequately described by the above analysis.

The early runs (1-18) had an acidity considerably lower than the desired pH of 1.45. Runs 19-21 with fully impregnated oak chips were closer to the desired pH. There was a changing acidity during the cook--at 6 minutes the pH was 1.37, corresponding to 0.417% H_2SO_4; by 12 minutes the pH rose to 1.46 or an equivalent acid concentration of 0.338%. The average effective acid concentration over the interval 6 to 12 minutes was 0.38% (pH = 1.42).

The plots for runs 19-21 (Fig. 6) are similar to those for runs 1-18. In this case, the removal curve equation is:

$$XR = 87.0 \exp(-0.914t) + 13.0 \exp(-0.0815t) \quad (5)$$

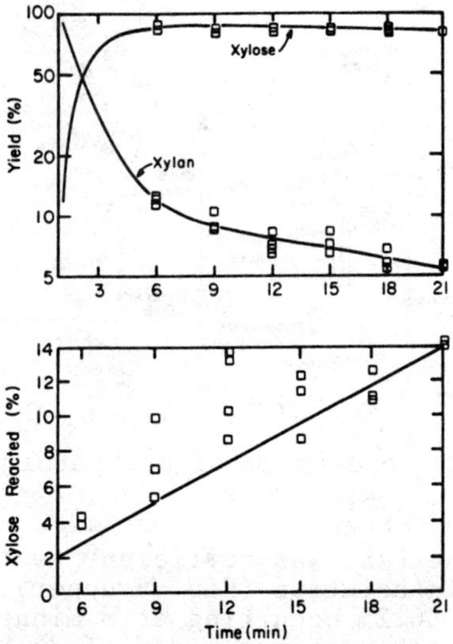

Figure 5. Yields of xylan and xylose over time (upper) and xylose reacted over time (lower) in prehydrolysis of southern red oak, runs 1-18, 170°C, pH - 1.77. Solid lines are calculated. (ML84 5473)

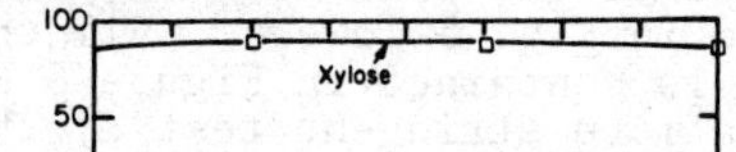

Figure 6. Yields of xylan and xylose over time (upper), and xylose reacted over time (lower) in prehydrolysis of southern red oak, runs 19-21, 170°C, pH = 1.42. Solid lines are calculated. (ML84 5474)

The maximum calculated xylose yield is 84.5% occurring at 6 minutes, whereas the measured maximum is 84.6% at 6 minutes. The experimental values of xylose reacted agree very well with those calculated (Fig. 6 lower).

SECOND STAGE (HYDROLYSIS)

In the past, the usual procedure for estimating glucose yield from cellulose hydrolysis was elementary. It was assumed the system could be modeled simply as two consecutive reactions, each proceeding at the rate predicted by the correlations given by Saeman (11). However, this fails to take into account reversion reactions and other properties of cellulose.

The reversion reactions enter into the considerations in two ways. One is the effect of their presence on the interpretation of the experimental results. The second is that the reversion reactions also enter into the calculation of sugar yields from cellulose since they afford protection for some of the reducing end groups and thus slow degradation.

Other than degradation rate, there are two additional properties of the cellulose that have an effect on glucose yield. One is the amount of readily hydrolyzed material contained in the cellulose; even in native celluloses, this is a significant portion of the total. The second is the basic property of cellulose which lowers the acidity of the reaction mixture; the catalytic hydrogen ion is a function of the concentration and amount of the applied acidic solution and the neutralizing power of the ash.

PROCESS CONCEPTS

Using the prehydrolysis data described, and data gathered at the Forest Products Laboratory over many years to augment the cellulose hydrolysis model, the following concepts for two-stage dilute acid cellulose hydrolysis were proposed.

First Stage

Hydrolyzer Type. The best reactor choice for the first stage seems to be a tubular vessel equipped with a screw auger to move the chips and provided with drainage holes for the escape of liquid during the heatup period. Continuous tubular-type digesters currently used in the pulping industry could perhaps be adapted.

Processing Alternatives. The distribution of the wood components throughout prehydrolysis illustrates an important point about the process (Table 1). The total potential carbohydrates are approximately one-third solubilized and two-thirds in the residue (Table 2). However, yields in the second stage are known to be very low; not more than 50% of the carbohydrates in the prehydrolysate residue will be obtained from the cellulose hydrolysis. Consequently, the yield of carbohydrates from the second-stage hydrolysis will be only about 20 kg/100 kg OD wood, which is less than that obtained in the prehydrolysate. It is also apparent that the total soluble solids content of the prehydrolysate is considerably greater than that of the second-stage hydrolysate. It is obvious that the manner in which the prehydrolysate is utilized has a major effect on the economic viability of the overall process. Unfortunately, few options exist; only four products seemed worthy of consideration: ethanol, single-cell protein, furfural, and feed molasses.

Table 1. Analysis of southern red oak (*Quercus falcata* Michx.) wood samples

Component	Sample number			
	1[1]	2	3[1]	4
	%			
Glucan	40.3	38.8	37.8	41.8
Mannan	2.9	2.7	2.1	2.2
Xylan	19.3	19.1	18.4	18.7
Galactan	(2)	1.4	1.1	.8
Arabanan	(2)	1.4	.71	.82
Uronic anhydride	2.9	2.7	3.3	2.7
Acetyl	3.7	3.9	4.3	3.5
Lignin	21.8	22.2	21.9	19.5
Ash	.24	.50	.72	.29
Extractives	5.5	6.0	6.7	6.6
Total	96.6	98.7	97.0	96.9

[1]Sample 1 was used for ampoule studies; sample 3 for digester studies.
[2]The analytical procedure used does not distinguish galactan from glucan nor arabanan from mannan.

Table 2. Material balance of first-stage prehydrolysis at reaction conditions of 170°C, L/S = 1.5, pH = 1.45, 11 min

Component	Wood	Solution	Residue
	% based on OD process wood		
Total solids	100 (basis)	43.2	62.7
Potential carbohydrates			
Glucose	42.0	3.16	38.6
Mannose	2.37	1.78	.47
Xylose	20.8	16.6	1.38
Uronic	2.84	1.80	.20
Other	2.02	1.57	.28
Total	70.0	24.9	40.9
Acetyl (as HOAc)	6.00	4.50	1.50
Lignin	21.9	3.18	19.2
Extractives	6.67	6.67	--
Ash	.72	.63	.09
Furfural	--	1.51	--
Sulfate (as H_2SO_4)	--	1.73	--
Unaccounted[1]	3.81	--	5.48

[1]Summative analysis can be less or more than 100%.

Ethanol would be preferred since its production is the primary goal. The ethanol production could perhaps be almost doubled by fermentation of the xylose and glucose in the prehydrolysate. Unfortunately, there is no commercial process available to convert xylose to ethanol. Significant progress is being made along this line (3,6,14); however, the lead time required for commercialization of the proposed process makes it untenable at the present.

At the moment, molasses, marketed as an animal feed, appears to be the preferred way of utilizing the prehydrolysate.

First-Stage Process Description. Figure 7 presents a flow diagram for the first stage based on molasses as the hemicellulose product. The quantities associated with each numbered stream are listed in Table 3.

Furfural Recovery. During prehydrolysis, furfural is unavoidably produced. If not recovered, it is a contaminant in the fermentation and a pollutant in the effluent. The yield of furfural during prehydrolysis is a function of time. Under conditions the same as those in Table 2, maximum xylose yield occurs at about 6 minutes, at which time the furfural yield is 0.8% based on OD wood, and it is present in the prehydrolysate at a concentration of 0.4%.

Further consideration of furfural production leads to the conclusion that it is economically beneficial to extend the duration of the prehydrolysis and thus produce more furfural.

Optimizing furfural yield requires more detailed plant design and other data than are presently available. It is assumed that maximum furfural yield would be near the point of maximum xylan utilization--that is, the point at which the total molar yield of xylose and furfural is at a maximum. This is reached at a time of 10.2 minutes when xylan utilization, as described above, is 91.3% and the yield of furfural (on wood) is 1.5%, or about twice that at maximum xylose yield. At this level, furfural production from an 800-tonne/day hydrolysis plant would be 4.2×10^6 kg/year.

Acetic Acid Recovery and Molasses Quality. Recovery of acetic acid is economically viable because of its effect on product (molasses) quality and effluent treatment costs.

Wood molasses must compete with blackstrap molasses. Although there are no market standards, wood molasses must have approximately the same carbohydrate, nitrogen, and ash content as blackstrap molasses. The ash content of wood molasses has three sources:

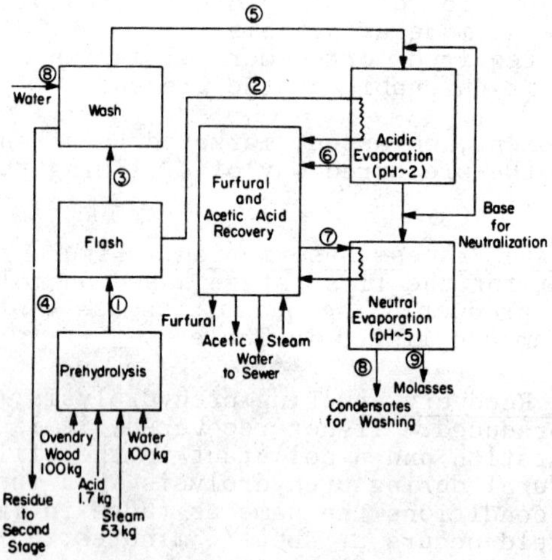

Figure 7. Block diagram of the first-stage processing. (ML84 5475)

Table 3. Components in the streams of Figure 7, first-stage processing

Component	Stream 1	2	3	4	5	6	7	8	9
	kg/100 kg OD process wood								
Water	153	18.3	135	120	210	112	40	89	9.2
Potential carbohydrates									
Glucose	41.8	41.8	38.8	3.0	--	--	--	--	3.0
Mannose	2.25	2.25	.56	1.69	--	--	--	--	1.69
Xylose	18.0	18.0	2.21	15.8	--	--	--	--	15.8
Others	1.85	1.85	.36	1.49	--	--	--	--	1.49
Potential acids									
Uronic	2.0	2.0	.29	1.71	--	--	--	--	1.71
Acetic	6.0	.52	5.48	1.70	3.78	2.32	([1])	--	1.46
Lignin	22.4	22.4	19.4	3.0	--	--	--	--	3.0
Extractives	6.67	6.67	.33	6.34	--	--	--	--	6.34
Ash	.72	.72	.47	.60[2]	--	--	--	--	4.20
Furfural	1.51	.55	.96	.06	.90	.85	([1])	.05	
Sulfate (as H_2SO_4)	1.73	1.73	.09	1.64	--	--	--	--	--
Unaccounted	5.48			5.48	--	--	--	--	--
Total weight	259	19.4	240	184	250	115	42	89	46

[1]Quantities of these components depend on recovery unit design.
[2]Lignocellulose residue has exchange capacity.

· The original ash of the wood.

· The sulfuric acid added for hydrolysis and the base required for its neutralization.

· The base required for neutralization of the acetic and uronic acids liberated on hydrolysis.

Fortunately, with the process scheme shown in Figure 7, the ash content of the molasses can be brought to an acceptably low level without using lime--the process depends on minimizing acid consumption in the prehydrolysis, partially neutralizing with NH_3, and removing some of the acetic acid during evaporation. The ash and nitrogen content of the product would be less than that commonly found in blackstrap molasses.

Second Stage

Hydrolyzer Type. From the prehydrolysis process described the substrate would be a finely divided, fibrous, pulplike material. With such material, the second-stage hydrolyzer would have to be continuous. Commercially available equipment has not been tested at pilot or semi-commercial scale, but significant progress is being made in its development (1,2,5,8). For the subsequent process calculations, it has been assumed that it would be possible to feed a mixture with an L/S of 2. After heating with direct steam, the L/S would rise to 3; and after hydrolysis, at discharge, the L/S would be about 7.

Processing Alternatives. A materials balance for the second-stage hydrolysis was calculated **(Table 4)**.

About one-half of the charge is solubilized (Table 4). Although the residue is primarily lignin, it contains more than a quarter of the original potential carbohydrates; therefore, it has a composition of around 30% carbohydrate. The outstanding property of the soluble solids is the ratio of glucose to other components. Notice that only 61% of the total soluble material is free glucose. The yield of free glucose is only 48.5% of that potentially available. The major nonglucose component is the reversion material, which is also a carbohydrate product and could conceivably be utilized as such. The organic acids are also prominent, and they are of great importance in the economics of processing. The major acidic component is acetic acid, comprising about 60% of the organic acids on a weight basis and about the same on a molar basis. The acetic acid is incompletely removed in the prehydrolysis; 25% is retained by the first-stage solids (Table 2). This figure is probably high; some subsequent experiments indicate that it is 16%. Even at the lower value, the acetic acid is a prominent impurity in the second-stage hydrolysate. HMF, which like formic acid, is toxic in the fermentation, is present in large quantity. Table 4 makes it apparent that the glucose, which is obtained in only moderate yield (48.4%), is accompanied by a large amount of impurities (nonglucose components = 12.0 kg/100 kg OD wood).

Table 4. Material balance of second-stage hydrolysis at reaction conditions of 230° C, 0.8% (added) H_2SO_4 L/S = 3

Component	Charge[1]	Solution	Residue
	- - - kg/100 kg OD wood - - -		
Total solids	62.7	30.7	31.9
Potential carbohydrates			
Glucose	38.6	18.7	10.8
Mannose	.47	.14	
Xylose	1.38	--	--
Reversion material	--	3.1	--
Other	.28	.06	--
Acids			
Acetic	1.50	1.5	--
Levulinic	--	.66	--
Formic	--	.38	--
Sulfuric	--	1.50	--
Lignin	19.2	--	20.2
Ash	.47	.45	.09
Furfural	--	.25	--
Hydroxymethylfurfural	--	1.27	--
Unaccounted	5.48	2.50	4.5

[1]Residue from prehydrolysis (see table 18).

Second-Stage Process Description. The flow plan and stream compositions for the second stage of the process are described by Figure 8 and Table 5. Despite its apparent simplicity, serious questions can be raised about the practicality of almost all the steps. Certain problems require further investigation before plant design can be considered.

MATERIAL AND ENERGY BALANCES

The input and output of materials for the two-stage hydrolysis process are shown in Table 6. Figures 9, 10, and 11 show the distribution of organics, water, and energy. Both the table and figures are based on 100 kg of OD wood entering the first-stage hydrolyzer.

The wood is assumed to be southern red oak of a composition given in Table 1 (sample 3) and a higher heating value of 19.77 MJ/kg (8,500 Btu/lb). The process material delivered from wood preparation is reduced to 9.5-mm chips which are bark free and composed of 50% moisture. The bark and some additional fuelwood are sent to the boiler. It is assumed that after impregnation the acid was uniformly distributed throughout the chip so that the yields on prehydrolysis are those shown in Table 2. The water content of the prehydrolysate is kept to a minimum by recycling acid solution form prehydrolysis for impregnation of chips.

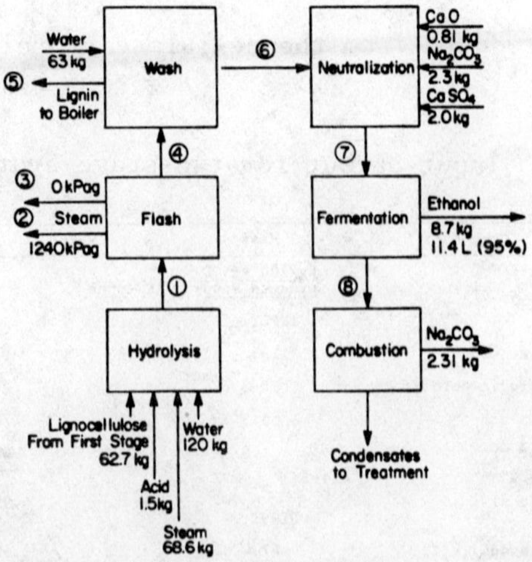

Figure 8. Block diagram of the second-stage processing. (ML84 5469)

Conditions of the prehydrolysis are assumed to be the same as those indicated in Table 2. The acid required to maintain a pH of 1.45 depends on the ash content of the wood. This is assumed to be 0.72%, which is the value found for the particular sample used in this study. Residence time in the prehydrolyzer is 12-14 minutes, resulting in a maximum summative yield of xylose and furfural. In the second stage, the temperature is 230°C, pH = 1.3, and residence time is 35-40 seconds. To decrease the effluent load, the reaction time is set at less than that required to attain maximum glucose yield. Yields for

Table 5. Components in the streams of Figure 8, second-stage processing

Component	Stream							
	1	2	3	4	5	6	7	8
	kg/100 kg OD process wood							
Water	188.6	22.1	27.6	138.9	33	169	169	169
Potential carbohydrates								
Glucose	29.5	--	--	29.5	11.7	17.8	17.8	.9
Mannose	.1	--	--	.1	--	.1	.1	--
Reversion	3.1	--	--	3.1	.2	2.9	2.9	2.9
Other	.1	--	--	.1	--	.1	.1	.1
Organic acids[1]								
Acetic	1.5	.23	.27	1.0	.05	.95	.95	.95
Levulinic	.66	--	--	.66	.03	.63	.63	.63
Formic	.38	.03	.05	.30	.02	.3	.3	.3
Lignin	19.9	--	--	19.9	19.9	--	--	--
Ash	.45	--	--	.45	([2])	.45	--	4.0
Furfural	.25	.11	.08	.06	--	.06	.06	.06
Hydroxymethylfurfural	1.3	--	--	1.3	.06	1.2	1.2	1.2
Sulfate (H_2SO_4)	1.5	--	--	1.5	.08	1.42	--	--
Unaccounted	7	--	--	7	4.5	2.5	2.5	2.5
Total	253	23	28	208	64	198	198	180

[1]May not be present as free acids.
[2]Unknown--probably less than 0.5%.

the various products from the cellulose hydrolyzer are show in Table 4.

Table 6. Input-output for two-stage hydrolysis

Component	Mass	Value	
	kg/100 kg OD process wood	$/tonne	$
	INPUT		
Wood (OD)	119[1]	33	3.93
Sulfuric acid	3.3	88	.29
Soda ash	5.3	132	.70
Lime	.8	38.5	.03
Total			4.95
	OUTPUT		
Molasses	47.4	55	2.61
Furfural	1.15	1,450	1.67
Acetic acid	2	580	1.16
Ethanol (100%)	8.7	585	5.09
Subtotal			10.53
Soda ash (byproduct)	2.3	66	.15
Total			10.68

[1]Includes 19 kg of wood substance used for boiler fuel.

It was assumed that 95% of the soluble solids were recovered in each of the washing units. Boiler efficiency was taken as 60%, and the efficiency of the fermentation as 95%.

Figure 9 shows the flow of organics through the process. Notice that 100 kg of OD wood contains 99.3 kg of organic matter since the wood has a 0.7% ash content. Of the total carbon charged (59.1 kg), 37% is in the marketable products, most of which is contained in the molasses; only 7.7% of the total carbon charged is contained in the ethanol. Most of the remainder leaves as carbon dioxide from combustion and fermentation, but 1.3% is present in effluent streams.

The necessity of the ancillary processing steps to reduce the pollution load is clearly evident (Fig. 9). If the furfural-acetic acid recovery and evaporation combustion units were deleted, the waste streams would contain more than 13% of the carbon charges. Furfural and acetic acid removal decreases the effluent load by 90% to about 1.3%. It is probable that the effluent load could be reduced still further because there is a possibility that some of the effluent from the recovery unit could be recycled into the process. The biological oxygen demand (BOD) from an 800-tonne/day plant would be about 16,000 kg/day (from Fig. 9).

The water content of the effluent streams is shown Fig. 10) as part of the distribution of process water cooling water not being considered). The total volume of oul condensate is moderate: for an 800-tonne/day plant, t would be about 2.3 x 10^3 m^3/day. This results from the ow water input to the process, which amounts to only .2 m^3/tonne of wood.

The process heat requirements are about 50% greater han the heat available from the lignin residue (Fig. 11). 'he additional energy requirement is supplied from bark and

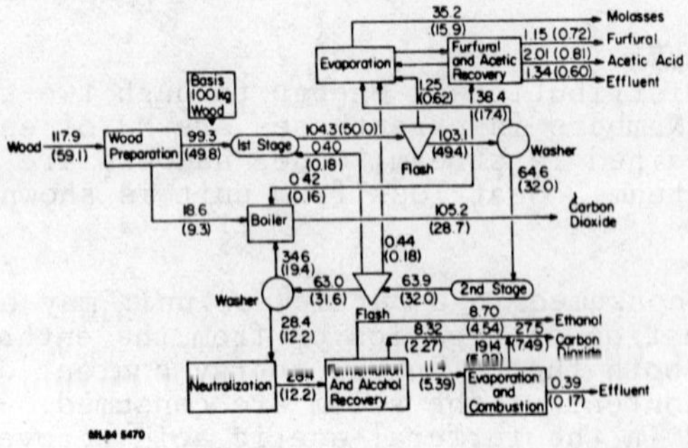

Figure 9. Distribution of organics through two-stage hydrolysis. Numbers in parentheses are kg of carbon; others are kg of organics. (ML84 5470)

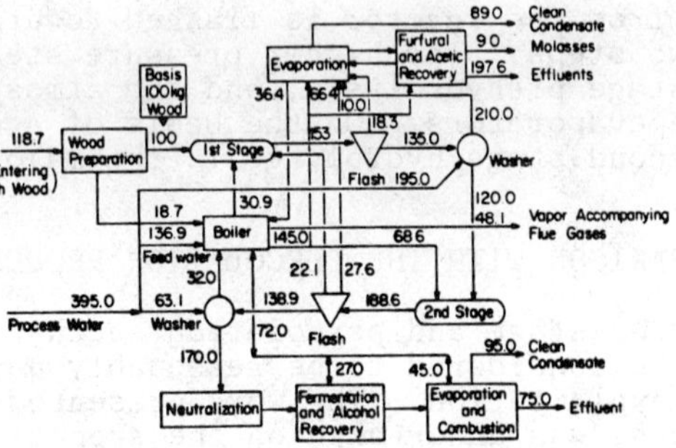

Figure 10. Distribution of process water through two-stage hydrolysis. All numbers are kg of water. (ML84 5471)

additional fuelwood. As indicated, about 16% of the material entering wood preparation is sent to the boiler. No electricity needs were considered.

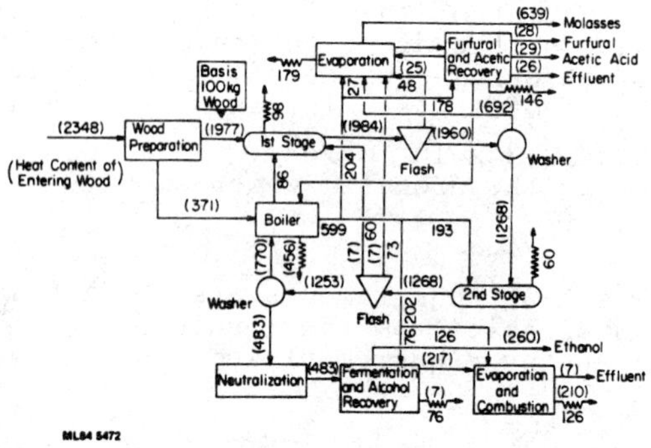

Figure 11. Distribution of energy through two-stage hydrolysis. Numbers in parentheses are MJ of enthalpy of organics contained in stream; other numbers are MJ of enthalpy of steam. Heat loss from unit is shown as ⁓→ (ML84 5472)

The heat consumed in a particular unit may be supplied either by oxidation of organics or from the enthalpy of fermentation--both the potential energy content of organics and the heat content of the steam are consumed. The sum of steam consumed in the furfural-acetic acid recovery and evaporation-combustion units is 279 MJ, which is 40% of the steam generated.

Although more steam enters the second-stage hydrolyzer than any other unit, its energy consumption is the lowest. This is because it is credited with the large amount of low-pressure steam recovered from the second-stage flash unit. The discharge from the reactor is flashed to atmospheric pressure in two steps. The higher pressure steam is used in the first-stage prehydrolysis, and the atmospheric steam is sent to the evaporators. On the basis of available energy, the second-stage hydrolyzer is the major consumer.

COMPARISON WITH THE PERCOLATION PROCESS

Both the two-stage and percolation saccharification processes can be considered to be reasonably near the point of commercial exploitation. The work presented here on the two-stage process was undertaken on the supposition that many of the disadvantages inherent in the percolation process (7) could be overcome. This has proved to be true, but to a lesser extent than it had been hoped. The following discussion compares the two processes. A significant advantage of the two-stage process is the high concentration of the product solutions. As indicated in our analysis, the solution from the first stage, which contains the largest quantity of solubilized material, has a total solids concentration of 18%, with a xylose concentration of 7.5%. Assuming the development of suitable equipment, the second-stage solution contains a total solids concentration of

16.7% and a glucose content of 10%. These concentrations are at least double those obtained from percolation. The energy consumption and equipment size for subsequent processing steps are roughly halved by two-stage processing. However, the capital requirements for the hydrolysis itself are not greatly reduced. In the percolation process, prehydrolysis, saccharification, and washing are all done in a single vessel. In two-stage operation, two reactors and two sets of washing equipment are needed. The net difference for both equipment and energy cost for hydrolysis is only slightly in favor of the two-stage process. For ethanol production the energy requirements of two-stage operation should be about 40% less than the percolation process, and equipment cost, about 25% less.

The purity of the solutions generated by the two-stage process is considerably better than that obtained by percolation. The separation between the hemicellulose carbohydrates and the glucose is much better, and the ratio of degradation products to carbohydrates is lower.

The advantages of the two-stage process are offset to some extent by the higher yields of the percolation process. The yield advantage depends both on the type of wood being considered and the subsequent processing of the solution.

Unfortunately, available data on the percolation process relate primarily to softwoods, and yields are given either as total sugars obtained or as recovered ethanol. No data are available on the changing concentration and composition of the product solution. The carbohydrate yields from each of the component fractions of the wood--hemicellulose and cellulose--are unknown. Because of this, the yield comparisons given below are uncertain estimates, but some points for consideration are brought out.

The southern red oak used in this study contains 37.8% anhydroglucose and 18.4% anhydroxylose (Table 1, sample 3). On hydrolysis, these yield 42.0% glucose and 20.9% xylose, a total of 62.9% potential sugars on the basis of the wood. Complete conversion to ethanol would result in 420 L of 190-proof ethanol/tonne of wood--280 L from glucose and 140 L from xylose.

When only glucose is considered to be fermentable to ethanol, the ethanol produced in the two-stage process (Fig. 9) is 8.70 kg/100 kg of process wood, which is equivalent to 114 L/tonne. The ethanol yield is 40.7% of that theoretically available from the potential glucose of the wood. By the percolation process, the estimated ethanol production is 124 L/tonne, about 9% greater than that obtained by two-stage processing. The yield by percolation is 44% of that theoretically available. The efficiency of saccharification in percolation is actually much greater than this, probably about 55-60%, but much of the glucose produced accompanies the hemicellulose stream and could not be economically processed to ethanol. The percolation method, operating on hardwood and producing ethanol from the glucose-rich fraction only, would have a yield advantage over two-stage processing.

If it is assumed that xylose as well as glucose can be fermented to ethanol with equal efficiency, the percolation process has a much greater yield advantage. This is becaus the xylose-fermenting organism would also utilize glucose, and the poor separation of the sugars is inconsequential. Using this assumption, the percolation-process yield is estimated at 267 L/tonne and two-stage operation 234 L/tonne, a 14% yield advantage for percolation.

Yields for two-stage processing are considerably higher if it is assumed that the reversion material can be converted to ethanol. If this were the case, 133 L/tonne could be obtained solely from the glucose fraction. This increase of 19 L/tonne from the oligomers reverses the ranking and gives two-stage processing a 7.3% advantage under the supposition that xylose is not fermentable to ethanol. If it is assumed that all carbohydrates (xylose, glucose, and reversion products) are fermentable, the percolation process has a 5.5% greater ethanol yield.

ACKNOWLEDGMENTS

This report is a synopsis of work conducted by John F. Harris, Anthony Conner, Thomas W. Jeffries, James L. Minor, Roger C. Pettersen, Ralph W. Scott, Edward L. Springer, and Theodore H. Wegner in cooperation with the authors. A more comprehensive compilation of the work is contained in the publication, Two-Stage, Dilute Sulfuric Acid Hydrolysis of Wood: An Investigation of Fundamentals--Forest Products Laboratory Report FPL-45.

REFERENCES CITED

1. Bender, R. "Method of treating lignocellulose material to produce ruminant feed," U.S. Patent 4,136,207; January 23, 1979.

2. Church, J. A. "Continuous high-solids acid hydrolysis of biomass in a 1-1/2-in. plug flow reactor," Ind. and Eng. Chem. Prod. Res. and Dev. 20(2), 371-378 (1981).

3. Gong, C., Ladisch, M. R., and Tsao, G. T. "Production of ethanol from wood hemicellulose hydrolyzates by a xylose-fermenting yeast mutant, *Candida* sp. XF 217," Biotechnol. Lett. 3(11), 657-662 (1981).

4. Harris, J. F., Saeman, J. F., and Locke, E. G. "Wood as a chemical raw material," Chapter 11. In: Browning, B. L., ed. The chemistry of wood, 535-585. New York: Interscience Publishers, 1963.

5. Iotech Corporation. "Optimization of steam explosion pretreatment," 360 p. Final Rep., DOE Contract No. DE-AC02-79ETZ3050. Ottawa, Ont., Canada: Iotech Corp. 1980.

6. Jeffries, T. W. "A comparison of Candida tropicalis and Pachysolen tannophilus for conversion of xylose to ethanol," In: Scott, C. D., ed. Proceedings of 4th Symposium on Biotechnology in Energy Production and Conservation [1982 May 11-14; Gatlinburg, TN], 103-110. Biotechnology and Bioengineering Symposium No. 12. New York: John Wiley and Sons, 1982.

7. Lloyd, R. A., and Harris, J. F. "Wood hydrolysis for sugar production," Rep. 2029. Madison, WI: U.S. Department of Agriculture, Forest Service, Forest Products Laboratory, 1955.

8. McParland, J. J., Grethlein, H. E., and Converse, A. O. "Kinetics of acid hydrolysis of corn stover," Solar Energy 28(1), 55-63 (1982).

9. Millett, M. S., Moore, W. E., and Saeman, J. F. "Preparation and properties of some hydrocelluloses," Ind. and Eng. Chem. 46(7), 1493-1509 (1954).

10. Rydholm, S. A. Pulping Processes, 1269 p. New York: Interscience Publishers, 1965.

11. Saeman, J. F. "Kinetics of wood saccharification," Ind. and Eng. Chem. 37(1), 43-52 (1945).

12. Springer, E. L. "Rate studies of the hydrotropic delignification of aspenwood," 217 p. Ph. D. thesis. Madison, WI: University of Wisconsin, Department of Chemical Engineering, 1961.

13. Springer, E. L., Harris, J. F., and Neill, W. K. "Rate studies of the hydrotropic delignification of aspenwood," Tappi 46(9), 551-555 (1963).

14. Suihko, M. L., and Enari, T. M. "The production of ethanol from D-glucose and D-xylose by different Fusarium strains," Biotechnol. Lett. 3(12), 723-728 (1981).

COMPARATIVE TECHNICAL EVALUATION OF ACID HYDROLYSIS PROCESSES FOR CONVERSION OF CELLULOSE TO ALCOHOL

John D. Wright
Senior Engineer/Senior Project Coordinator

Arthur J. Power
Principal Engineer

Solar Energy Research Institute
1617 Cole Boulevard, Golden, Colorado 80401

ABSTRACT

This paper evaluates five major acid hydrolysis processes for the conversion of lignocellulosic feedstocks to ethanol. A complete conversion process was defined, and alternative hydrolysis processes integrated into this framework. Differences in ethanol price are directly related to the performance characteristics of the hydrolysis process. The sensitivity of the process economics to technical improvements was also evaluated.

In dilute acid hydrolysis processes, the acid hydrolyzes cellulose to glucose while simultaneously degrading the product sugars. Thus sugar degradation and the consequent low yields are the principal problem. The progressing batch reactor uses several percolation reactors in series to quickly remove the sugars from the reaction zone, minimizing sugar degradation while minimizing product dilution. The benefits of countercurrent operation are obtained, while retaining the mechanical simplicity of the percolation reactor. This process was predicted to produce ethanol for approximately 20% less than the other processes. The high temperature plug flow reactor has low yields, but converts the xylan fraction of the wood to furfural, a valuable chemical intermediate. This may provide a significant near-term advantage, and highlights the importance of making beneficial use of all components of the feedstock.

The concentrated acid processes break down the crystalline structure of the cellulose, rendering it amorphous and readily hydrolyzable at conditions where sugar degradation is minimal and yields approach 100%. However, each process suffers from the expenses associated with handling large quantities of acid. The concentrated sulfuric acid processes consume large quantities of acid, and suffers from high acid replacement costs. The HCL processes recover the acid by distillation, which requires large amounts of energy and extremely expensive corrosion-resistant equipment. The HF-based process has high acid consumption costs because HF is very expensive, and would suffer high capital and operating costs because of the extremely hazardous nature of the catalyst.

COMPARATIVE TECHNICAL EVALUATION OF ACID HYDROLYSIS PROCESSES FOR CONVERSION OF CELLULOSE TO ALCOHOL

INTRODUCTION

Cellulose, the major component of plants, is a polymer of glucose, which can be readily fermented to ethanol. If an inexpensive and efficient method can be developed for breaking down cellulose to glucose, it will be possible to design an economically attractive process to produce fuel ethanol from cellulosic materials.

This paper evaluates the various acid hydrolysis processes for producing sugars for fermentation to ethanol. More detailed documentation is available in Ref. 33. The methodology is that of an engineering study. The process flow sheets are developed, and material and energy balances determined. From these, we calculate the capital and operating costs and the overall economics, and compare the processes on the basis of the selling price of ethanol produced. Evaluations are made of the present status of the technology and the effects of various proposed improvements. Every effort was made to ensure that the processes were compared on a consistent basis.

Because the hydrolysis of cellulose is only a fraction of the overall process, it is important that the design of the remainder of the plant be kept constant. An overall cellulose-to-ethanol plant design, prepared under contract to the Solar Energy Research Institute (SERI) by Badger Engineers, Inc. in 1984, was used as the basis for all process calculations (1). The basis is a plant producing 25,000,000 gallons of ethanol per year from a mixed hardwood feedstock priced at $42/dry ton, located in the midwestern United States, with the design characteristics of the Nth plant. Each hydrolysis process was designed in detail from the best currently available information, and integrated into the overall process defined by Badger. Therefore, changes in the cost of production are directly traceable to the performance of the hydrolysis processes.

To ensure that capital costs were estimated in a consistent manner, we used the ICARUS computer-aided cost estimating program, which produces a detailed time and materials cost estimate. For a completely defined process design the accuracy of the estimating program is ±10%; thus, the limiting accuracy is our knowledge of the process and not the cost estimating technique. Using ICARUS, we estimated the total cost of the base-case Badger design, and then developed estimates for the alternate hydrolysis processes (23).

The hydrolysis processes affect the cost of ethanol through their capital and operating costs, but primarily through their different yields and product concentrations. The yield determines the amounts of feedstock and catalyst required, while sugar concentration affects the size and energy consumption of the downstream process. To rapidly assess the impact of these factors, a mathematical model of the base-case process was developed on a Lotus 1-2-3 spreadsheet. The hydrolysis processes were then integrated into this material balance, allowing easy recalculation of the entire model. The model then calculates the energy balance and capital cost, and from these, the costs of the raw materials, utilities, labor, overhead and maintenance, byproduct credits, annual capital charges, and ethanol selling price. Using the model allows rapid and consistent comparisons between the various processes, and facilitates calculation of the sensitivities to process improvements.

Five acid hydrolysis processes were evaluated. Two of the processes (plug flow and progressing batch hydrolysis) use dilute sulfuric acid as the catalyst, while three use concentrated acid (concentrated sulfuric acid, liquid-phase hydrochloric acid, and liquid-phase hydrofluoric acid).

The base-case designs and costs should not be viewed as real operating plant situations. Their purpose is to define the current state of the technology and serve as a well-defined reference point for investigating the sensitivity of the process to potential improvements. It is quite important to realize that the designs of the various processes are at different stages of development. Some are well understood while the design of others is subject to great uncertainty. Uncertainty in the proper assumptions for determining the revenue required to repay the capital investment further clouds calculation of the absolute selling price of ethanol. Thus, although great effort has been expended to develop accurate designs, cost estimates, and economic evaluations, comparison of the results of this study with the results of studies by different groups or with current market prices is a difficult task. Great care must be taken to be sure that the different studies paid equal attention to the details of design and performance (greater attention to detail usually results in higher estimated costs), and to adjust the selling price to account for different economic assumptions.

Feedstock Characteristics

Crystalline cellulose, amorphous cellulose, and hemicellulose (the carbohydrate portions of lignocellulose) can be hydrolyzed to their component sugars. Cellulose is a linear polymer of D-glucose (a six-carbon sugar) molecules held together by beta-glycosidic bonds. Cellulose fibers are arranged in bundles of parallel chains in which adjacent chains are bound together by hydrogen bonding between hydroxyl groups and hydrogen atoms, forming a crystalline material with great mechanical strength and high chemical stability. Because the bond between the glucose units is the weak link in the chain, the polymer can be broken (hydrolyzed) into its component sugars.

Hemicellulose (in the hardwoods investigated) is a polymer consisting primarily of five-carbon sugars (pentosans or xylans), six-carbon sugars, and organic acids. The six-carbon sugars are readily fermentable to ethanol, but standard industrial yeasts cannot ferment the five-carbon sugars. Unlike cellulose, the structure and composition of hemicellulose may vary widely between species. Hemicellulose is not crystalline and is readily hydrolyzed. Lignin, the noncarbohydrate portion of the cell wall, is chemically bonded to and mixed with the hemicellulose. Lignin is a phenolic polymer, and cannot be converted to fermentable sugars (36).

Theory of the Acid Hydrolysis of Cellulose

By understanding the mechanism of cellulose hydrolysis it is possible to understand the rates and yields obtained during concentrated or dilute acid hydrolysis. To illustrate the hydrolysis mechanism, we look at the hydrolysis of methylglucoside (20,14) (Figure 1). The analogy to cellulose hydrolysis is readily apparent if one envisions a string of glucose units attached by glycosidic bonds. For example, a second glucose unit would be attached where the methyl group is shown. In the proposed hydrolysis mechanism the glycosidic oxygen is quickly protonated. The rate-limiting step is the flexure of the glucose molecule from the chair configuration to the semiplanar configuration. This change in configuration is accompanied by the elimination of the methyl group from the glucose unit. The next step is the rapid addition of water to the reactive end. The final step, also fast, is the regeneration of the proton from the newly formed glucose molecule. It is generally accepted that the rotational energy required in the ring flexure is the rate-controlling factor in hydrolysis. The slow hydrolysis rate of crystalline

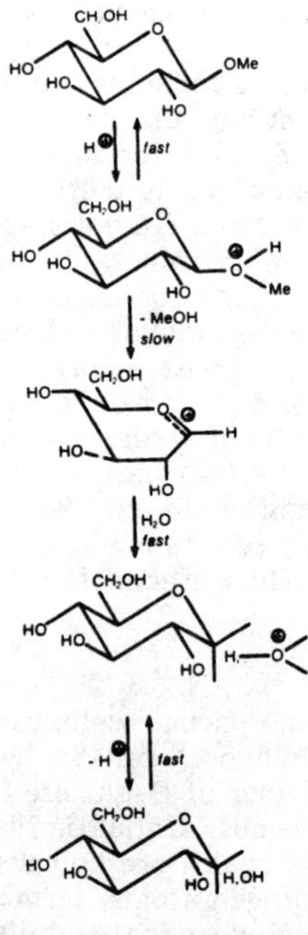

Figure 1. Hydrolysis via the Cyclic Carbonium-Oxide Ion

cellulose is explained by the rigidity of the glucose rings held tightly in the crysta structure. Because cellulose chains are bonded to adjacent chains by hydroge bonding between hydroxyl groups and hydrogen atoms, the close proximity an rigidity of the adjacent chains hinders the flexure of the glucose rings durin hydrolysis. Since the ring flexure is the rate-determining step, the hindrance o adjacent chains decreases the hydrolysis rate. On the other hand, the hydrolysi rates of sugars in aqueous solution and the hydrolysis rate of solid hemicellulos sugars are much faster because their amorphous nature does not hinder the flexur of the ring.

One method of breaking down the chemically resistant crystalline cellulos is with high temperature dilute sulfuric acid. Unfortunately, conditions sufficien to break down the crystalline cellulose are also sufficiently severe to degrade th product sugars. Therefore, sugar degradation is the central problem of dilute aci hydrolysis.

The major reactions of cellulose and hemicellulose degradation are show below (21):

Crystalline cellulose ⟶ glucose ⟶ HMF ⟶ tars
Amorphous cellulase ⟶ glucose ⟶ HMF ⟶ tars
Amorphous xylan ⟶ xylose ⟶ furfural ⟶ tars

Because there is no structure to resist flexure of the glucose ring and slov the hydrolysis reaction, the hydrolysis of the amorphous cellulose and xylan can b carried out at low temperatures and acid concentrations (150°C and 1% H_2SO_4) where the conditions are not severe enough to attack either the sugars in solutio or the crystalline cellulose. This process is called prehydrolysis, and yields o greater than 80% can be achieved on the amorphous components.

Unfortunately, at 180°C, the rates of cellulose hydrolysis and glucose degradation are approximately equal, and yields in a batch reaction are very low. The percolation and progressing batch hydrolysis reactors remove the sugars from the reactor before they have time to degrade. The plug flow reactor operates at higher temperatures (240°-265°C) where the rate of hydrolysis is approximately three times greater than the rate of glucose degradation.

The concentrated acid processes work by breaking the hydrogen bonds between adjacent cellulose chains, rendering the structure amorphous and readily hydrolyzed at low temperatures where sugar degradation is not a problem. The decrystallization and hydrolysis of cellulose may be accomplished at approximately ambient temperature by using concentrated acids such as hydrochloric, sulfuric, and hydrofluoric acids. In all concentrated acid processes, the cellulose feedstock is dried and treated with concentrated acid to decrystallize the cellulose, which is then is hydrolyzed with water at moderate temperatures to form simple sugars. All concentrated acid hydrolysis processes are capable of yields approaching 100%.

PRIMARY BASE CASE

Process Description

This section describes the overall process for converting lignocellulose to ethanol (a 25,000,000 gal/yr plant designed by Badger Engineers) into which the alternative hydrolysis sections are substituted. A simplified flow sheet is shown in Figure 2. The overall design and performance of the plug flow hydrolysis system is also described here but is discussed more completely in a later section.

Fresh mixed hardwood (primarily oak) chips are delivered by rail or truck, washed, and transferred to the wood pile. The chips are reduced to 1 mm particles in the refiner grinders and fed to the prehydrolysis reactor (Figure 3). Dilute acid solution (180°C, 1% H_2SO_4) is added to the ground wood to make a 20% solids mixture, which is held for 10 minutes in the prehydrolysis reactor. This solubilizes about 30% of the dry mass, further reduces the particle size,and forms a pumpable slurry.

Wood slurry mud is pumped into the hydrolysis reactors, where high pressure steam is injected to raise the temperature to 265°C. At the reactor conditions, cellulose is hydrolyzed to glucose and degradation compounds. The pentosan fraction is converted to furfural and other degradation products. The yield of glucose from the cellulose is 55% and the furfural yield from xylose is 67% (24). Outflow from the reactors is flashed in two stages to recover heat and quench the hydrolysis and degradation reactions.

Lignin residue is separated from the hydrolyzate by centrifugation. The lignin solids are dewatered and conveyed to drying and burning. Hydrolyzate from the first stage centrifuges is neutralized to pH 10 with lime slurry to precipitate degradation products. Gypsum and precipitated tars are separated from the liquids and washed using three continuous solid bowl centrifuges in series. Concentrations of residual organics inhibitory to the yeast fermentation are reduced by steam stripping. The treatments required to ferment a dilute acid hydrolyzate solution have been studied at the USDA Forest Products Laboratory (17,31). Fermentation of dilute acid hydrolyzates is more difficult than fermentation of clean sugar streams, and will require additional investigation. Hydrolyzates from concentrated acid processes have few degradation products, require little or no clean up, and can be fermented with high efficiencies.

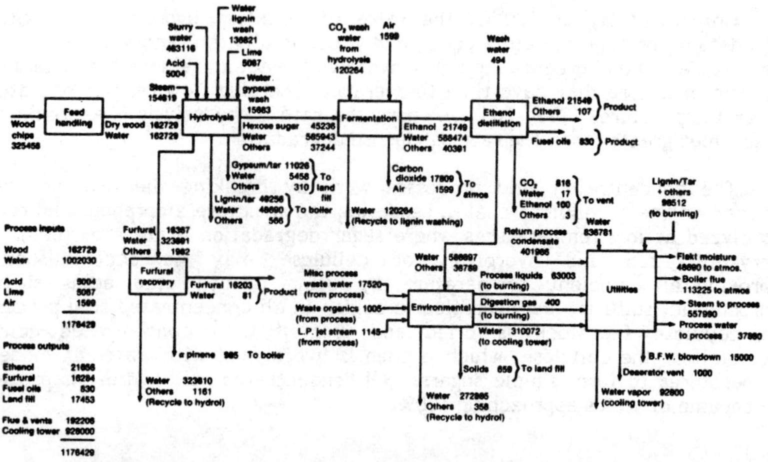

Figure 2. Simplified Process Flow Diagram for the Base Case, High Temperature Dilute Acid Plug Flow Reactor Process for Producing Ethanol from Mixed Hardwoods
Design by Badger Engineers (1)

The batch fermenters are eight chemically sterilized 250,000-gal carbon steel vessels. Yeast (*Saccharomyces cerevisiae*) is added to the hydrolyzate solution to begin the fermentation. A high yeast density in the tank speeds fermentation and helps to overcome any inhibitors remaining in the hydrolyzate. Recycle of yeast provides preacclimatization and minimizes cell growth in the fermentation, improving the yield. The fermentation time is 8 hours, with a vessel turnaround time of 8 hours. Fermentation yield is 95% of the glucose. Xylose and other sugars not fermentable by standard industrial yeasts pass through the fermentation unaffected.

Dilute beer from fermentation is concentrated to 40 wt % in the beer column. Stillage (water, xylose, waste organics) from the bottom of the column is pumped to evaporation. Distillate is fed to the rectification column, which concentrates the ethanol to its azeotropic composition (91.2 wt %). The bottoms stream, containing 3.5 wt % ethanol in water, is recycled to the beer column. Several components, collectively termed fusel oils, tend to accumulate within the rectification column and are removed in a sidedraw purge. Azeotropic ethanol is dehydrated to 199+ proof by molecular sieve absorption.

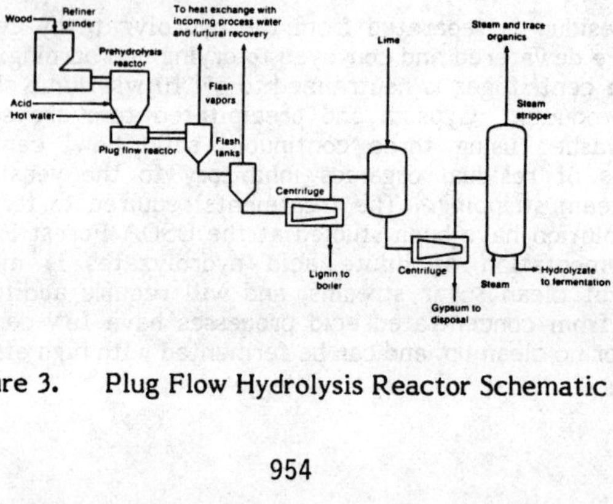

Figure 3. Plug Flow Hydrolysis Reactor Schematic

Furfural recovery is used only in the dilute acid processes, in which it is an unavoidable byproduct. Aqueous furfural is fed to the azeotrope column, which produces an overhead vapor stream of composition close to the azeotropic furfural water mixture. The vapor is condensed and splits into two immiscible phases, a furfural-rich phase and a water phase. The furfural-rich phase from the decanter, containing 48% furfural, is sent to the dehydration column from which a 99.5% furfural stream is taken as the bottoms product. The watery phase from the decanter is recycled as reflux to the azeotrope column.

In the evaporation section, a three-stage evaporator is used to concentrate stillage to produce a solution of 60% mixed organics in water, which is burned as a liquid fuel to produce steam. About half of the recovered condensate is recycled to the acid hydrolysis unit and about half is sent to anaerobic digestion.

Two boilers are designed to burn solid, liquid, and gaseous fuel byproducts from the process units to generate 1100 psia steam for use in the process and for electrical power generation. Gas and liquid fuels are burned directly, but lignin and tar sludge from the acid hydrolysis unit is fed to a Flakt type flash drying system, which dries and fluidizes the solids into the burners using boiler flue gas. A turbogenerator package unit is rated to take 1100 psia 825°F steam and generate 45 MW of electricity.

Discussion of the Process Design

The energy consumption and capital cost by process area are shown in Tables 1 and 2 for the base case and other process designs. This following discussion is based on the base-case process, but the major issues will be seen to be common to all processes. The major energy consumers are the hydrolysis and environmental sections, accounting for some 70% of the total load. Large amounts of electricity are used to grind the wood chips into sawdust-like particles, and large amounts of steam are consumed to bring the feed stream up to the hydrolysis reaction temperature. While all the hydrolysis units evaluated in this report have different configurations, hydrolysis is always a major energy consumer. In the environmental area, virtually all the water in the plant is evaporated and recycled, and the concentrated organic liquid stream produced is sent to the boiler to produce steam and electricity for the process. The overall process of evaporation and combustion is a very large net energy producer. The other sections of the process (distillation, furfural recovery, fermentation, etc.) are much less important. The utility section of the plant is designed to burn all the components of the feedstock that are not converted into useful products. Because only 25% of the feedstock is converted into ethanol, essentially 75% of the feed goes to the boiler. More steam and electricity are produced than are needed to operate the plant, and the excess electricity is sold to the local utility.

The largest capital cost is for the steam boiler and the turbine set, as more than half of the feedstock is burned. In the hydrolysis system the largest single cost is for the refiner grinders, which reduce the wood chips to sawdust. Other expensive items are the centrifuges for solids separation and the multiple-effect evaporators used to concentrate organic liquids for burning and to recycle the water.

Approximately half of the total energy content of the feed leaves in the product streams (21% as ethanol, 15% as furfural, 16% as electricity), and the other half is consumed in the transformation of the feedstock from a solid fuel to a mixture of high-value liquid and energy products.

The cost of production summary for the base case and other production facilities is shown in Table 3. The cost of operating the plant (before byproduct

Table 1. Energy Consumption by Process Area for Lignocellulose to Ethanol Facilities with Various Hydrolysis Processes

	Plug Flow		Progressing Batch		Concentrated H_2SO_4		Liquid HCl		Liquid HF	
	Electrical	Steam	Electrical	Steam	Electrical	Steam	Electrical	Steam	Electrical	Steam
Section	0.5	1.9	0.4	1.4	0.3	1.2	0.3	1.3	0.3	1.1
Feed	32.2	243.6	3.1	77.6	6.0	124.7	4.5	665.1	2.4	388.9
Hydrolysis	2.1	5.3	2.6	6.6	2.5	6.2	1.0	2.6	1.0	2.5
Fermentation	10.0	44.0	12.5	55.4	11.8	52.1	4.7	20.8	4.6	20.2
Distillation	0.2	53.8	0.1	14.2	0.0	0.0	0.0	0.0	0.0	0.0
Furfural Recovery	0.8	0.0	0.6	0.0	0.5	0.0	0.5	0.0	0.5	0.0
Offsite Tank	0.5	164.0	0.7	207.9	0.6	194.9	0.2	74.9	0.2	72.3
Environment	17.0	11.0	13.2	8.5	11.3	7.3	13.9	9.0	11.0	7.1
Utilities	4.0	11.0	4.0	11.0	4.0	11.0	4.0	11.0	4.0	11.0
	67.3	534.6	37.2	382.6	37.0	397.4	29.2	784.7	23.9	503.2

Table 2. Capital Cost by Process Area for Lignocellulose to Ethanol Facilities with Various Hydrolysis Processes

Section	Plug Flow	Plug Flow*	Progressing Batch	Concentrated H_2SO_4	Liquid HCl	Liquid HF
Feed Handling	8.5	8.5	6.7	6.0	6.1	5.6
Hydrolysis	25.7	25.7	15.0	43.8	53.9	34.1
Fermentation	8.1	8.1	9.3	8.9	5.3	5.2
Distillation	13.0	13.0	14.9	14.3	8.3	8.1
Furfural Recovery or Concentration	5.5	5.5	2.5	0.0	0.0	0.0
Offsite Tankage	5.1	5.1	4.4	3.9	4.1	3.9
Environmental	10.9	11.1	12.6	12.1	6.8	6.7
Utilities	39.5	43.3	32.3	33.4	35.8	31.8
	116.2	120.2	97.5	122.5	120.3	95.3

*Furfural concentrated and burned to produce steam and electricity.

credits) is approximately $1.60/gal of ethanol. The major cost is that of the mixed hardwood feedstock ($1.04/gal). The feedstock cost is high because only 45% of the feedstock consists of readily fermentable six-carbon sugars, which are hydrolyzed with an efficiency of 55%. The cost of the catalyst (H_2SO_4) and the lime to neutralize it comes to only 10¢/gal. The remainder of the operating costs (labor, overhead, and maintenance) are roughly proportional to the cost and complexity of the plant (41¢/gal).

Much of the five-carbon sugar fraction of the hemicellulose is converted to furfural. Furfural currently has very high selling price ($0.66/lb) but has limited and shrinking markets. At a much lower price of $0.15-0.25/lb it may be able to compete in much larger markets. Therefore, for purposes of this study we have assigned it a very conservative value of $0.12/lb (essentially equivalent to that of a liquid fuel) (28). The sale of furfural provides 32% of the total plant revenues.

A second major source of income is the sale of surplus electricity, amounting to 16% of the total, which more than offsets much of the cost of the utility section. The major revenue producer is the sale of ethanol, providing 52% of the cash flow.

Table 3. Cost of Production Summary for Lignocellulose to Ethanol Facilities with Various Hydrolysis Processes (25,000,000 gal/yr, mixed hardwood at $42/ton)

	Cost of Production (¢/gal Ethanol)					
	Plug Flow	Plug Flow*	Progressing Batch	Concentrated H_2SO_4	Liquid HCl	Liquid HF
Raw Materials						
Wood	104.6	104.6	77.5	68.2	69.9	62.1
Acid	6.5	6.5	6.5	17.4	11.9	39.9
Lime	3.5	3.5	3.5	9.4	1.6	0.7
Chemicals	3.5	3.5	2.6	2.3	2.3	2.1
Utilities						
Water	0.9	0.9	1.1	1.3	0.9	0.5
Steam	0.0	0.0	0.0	0.0	11.9	2.7
Labor	7.2	7.1	6.0	7.6	7.4	5.9
Overhead & Maintenance	34.0	33.7	28.6	35.9	35.3	27.9
Byproducts						
Furfural	-70.5	0.0	-3.2	0.0	0.0	0.0
Electricity	-35.7	-53.0	-40.7	-37.3	38.2	-30.2
Annual Operating Cost	54.0	106.8	81.9	104.7	102.9	111.6
Capital Charges (FCR=0.13)	60.0	59.4	50.3	63.2	62.1	49.2
Ethanol Selling Price	114.0	166.2	132.2	167.94	165.0	160.8

*Furfural concentrated and burned for energy production.

The charges proportional to capital investment (depreciation, return on investment, income and property taxes, insurance) come to $0.60/gal. This is a fixed charge rate (ratio of the annual capital related charges to total capital investment of 0.13. The required charge rates vary widely across different industries and companies and currently run as high as 0.25. Use of such a rate would add $0.55/gal to the cost of ethanol. The effect would be similar for the other processes. The selling price of ethanol, the sum of the operating cost and capital recovery charges minus the revenues for furfural and electricity, is $1.14/gal in the base case. If the furfural is burned to produce energy instead of sold, the cost of ethanol is $1.67/gal.

Key Issues for Process Economics

From the previous discussion, we can identify key issues common to all hydrolysis processes: yield and sugar concentration, energy efficiency, total capital investment, and cost of catalyst. Increasing the process yield reduces the amount and cost of the feedstock. Equally important, it reduces the amount of the feed that is burned, reducing the size and cost of the utility section. Major improvements in the conversion of cellulose to glucose can be achieved with other hydrolysis processes. Equally important is increasing the fraction of the feedstock converted to saleable products. For example, this hydrolysis process converts much of the xylose to furfural. Most other processes convert the bulk of the hemicellulose to xylose, for which there is currently no high value use. Developing fermentations that can convert the xylose to ethanol or other valuable fuels or coproduct chemicals has the potential to raise the current yield of the process 50% to 100%. A third method of increasing the yield is to recover the lignin in a form where it can be sold as a fuel or chemical product. The value of the lignin will vary depending on the hydrolysis process used. Smaller fractions of the feedstock, such as the acetic acid and extractives (tall oils, turpentine), may also be recoverable. Clearly, the most important factor in determining the competitiveness of these cellulose conversion processes is the fraction of the feedstock converted to saleable products.

The size and energy consumption of virtually every operation downstream of the hydrolysis is inversely proportional to the sugar concentration from hydrolysis. The most important effect is on the energy consumption in the environmental and distillation sections. Also important is the overall energy consumption inherent in the design of the hydrolysis process. The capital investment required by a process strongly affects the cost of ethanol. While total capital investment is most strongly influenced both by the efficiency of the hydrolysis and the concentration of the hydrolyzate product, the cost of the hydrolysis section is also significant. A final important cost is that of the hydrolysis catalyst. In this process the catalyst costs are low, approximately 7¢/gal for the acid, and 3¢/gal for the base to neutralize it.

The following sections describe the operation of the five major acid hydrolysis processes studied, and evaluate the present status of the technology and the process sensitivity to research improvement.

HIGH TEMPERATURE, DILUTE ACID PLUG FLOW REACTOR

At high temperatures (265°C, 1% H_2SO_4) the rate of hydrolysis is three times faster than the rate of glucose degradation (24). At short residence times (4 seconds) the glucose yield is 55%, while xylose is degraded into furfural with a yield of 67%. The proposed mechanical design of the plug flow hydrolysis and furfural recovery units has been previously described.

This process has been operated for approximately 10 years at the bench scale at Dartmouth University in a 1/4-in. diameter reactor, with low solids concentrations (10%) at the pump inlet, primarily for gathering kinetic data (16). Tests at Dartmouth University also confirmed that a 20% solids slurry could be pumped after prehydrolysis. A pilot-scale reactor (200 lb solids/h) of similar design has been operated at SERI for the past year (22,37), and initial results are similar to those obtained at Dartmouth. The reactor is currently being modified to test the pumping of prehydrolyzed slurries at the larger scale. Before this process can be considered demonstrated at the pilot scale, the feasibility of extended operation on prehydrolyzed feedstock must be demonstrated, and the rate and efficiency of the fermentation must be determined. Because of the amount of degradation products produced, the cost and performance of the clean up system is a major uncertainty.

The capital costs, energy consumption, and cost of production were presented earlier. The major improvements in this process would be in the use of the five-carbon sugars, developing a reactor capable of handling a higher percentage of solids, and increasing the glucose yield.

The most important factor in all ethanol processes is the use of the five-carbon sugar fraction. The uncertainty regarding the ultimate price and market size of a coproduct such as furfural could be eliminated if the five-carbon sugars were converted to a fuel such as ethanol. Possible methods include fermentation by other yeasts, fermentation by bacteria, enzymatic rearrangement of the five-carbon sugars to a fermentable sugar, and cloning of the ability to ferment five-carbon sugars into standard industrial yeasts.

Alternative plug flow hydrolysis configurations have been evaluated to produce xylose instead of furfural (1). Such designs increase the capital and energy requirements of the hydrolysis section. The effect of a five-carbon sugar fermentation on the process economics is a dramatic reduction in the selling price of ethanol. In the case where both the five- and six-carbon sugar components are converted to ethanol, and the remainder of the organics are burned to produce steam and electricity, the result is a reduction in selling price from $1.67 to $1.21/gal.

The major process energy uses (steam consumption in the hydrolysis, distillation, and environmental section) and capital costs are all proportional to the reactor product concentration, and therefore, to the solid concentration in the reactor feed. Even after prehydrolysis, effective solids loadings of greater than 20% resemble wet sawdust, have no fluid properties, and cannot be pumped by conventional means. A reactor similar to that proposed by Stake Technologies, in which biomass is moved through a steam-filled reaction chamber by a screw conveyor, will be necessary to process such feeds. Such designs have been designed by Stake Technologies (26), the Tennessee Valley Authority (TVA) (21), and in Germany (29). Development of a high solids reactor would lower the cost of ethanol 25% to $1.27/gal in the case where furfural is sent to the boiler, and by 40% to $0.75/gal in the case where furfural is sold as a co-product. High solids operation may also improve the yield of the hydrolysis reaction by minimizing the amount of free acid available in the liquid phase to catalyze sugar degradation. This may raise the hydrolysis yield from 55% to 65%, further reducing the cost of ethanol by approximately 10¢/gal in the case where furfural is burned or 5¢/gal in the case where furfural is sold as a coproduct. When high solids operation and xylose fermentation are combined, the cost of ethanol is 98¢/gal if furfural is burned in the boiler and 70¢/gal if furfural is sold as a by-product.

PROGRESSING BATCH DILUTE ACID HYDROLYSIS PROCESS

The progressing batch hydrolysis process uses several percolation reactors in series to simulate countercurrent flow of liquids and solids. In this way, the advantages of countercurrent flow are potentially achieved, and the mechanical and operational simplicity of the percolation reactor (all solids handling is done at atmospheric pressure) is retained (39).

Percolation reactor technology was first developed on a commercial sclae in Germany in the late 1920s (36). During the 1940s, an improved percolation process known as the Madison process was developed in the United States (19,13). The percolation reactor technology is currently the basis of an industry in Russia with approximately 40 commercial plants. Yields of 60% and total sugar concentrations of 4% are typically reported. Pilot scale studies have also been conducted in Brazil (6), New Zealand (5,25), and Austria (26).

The countercurrent reactor configuration offers considerable advantages over percolation technology. The rate of sugar production is greatest when the solids are first introduced into the reactor. With countercurrent flow of solids and liquids flow through the reactor, the liquids outlet is adjacent to the solids inlet. Thus, the largest fraction of sugars has only a small distance to travel before it is washed out of the reactor, minimizing the time available for sugar degradation and maximizing the yield and sugar concentration. For similar reaction conditions, the yield is calculated to increase from approximately 60% to 80% and sugar concentration increased by 20% (34).

The continuous countercurrent reactor is attractive on paper, but, the movement of a concentrated stream of biomass and water into and out of a pressurized reactor is a very complicated operation, even when the solids and liquids are moving in the same direction (as in a plug flow reactor).

The progressing batch hydrolysis reactor operates on the principle that a continuous process can be approximated by several batch processes in series. Figure 4 shows several percolation reactors in series. The liquid enters reactor 6, flows from right to and exits from reactor 2. Hydrolyzate is flashed to quench the hydrolysis and degradation reactions. After 15-30 minutes, the reactor stages are cycled by changes in valve position. Fresh liquids enter reactor 5 and exit from

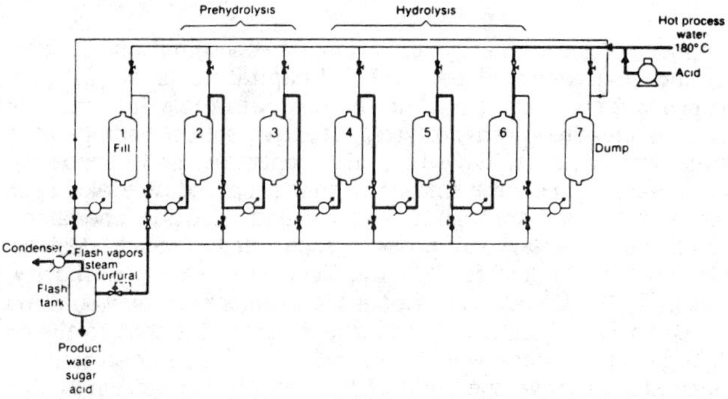

Figure 4. Progressing Batch Reactor Schematic

reactor 1. Reactor 7 is filled with fresh wood and the residual solids in reactor 6 are discharged. The net effect is that fresh solids have entered the reactor train at the left, and spent solids are discharged after leaving the last hydrolysis reactor in the train on the right.

The first reactor in the series is used as a prehydrolysis reactor, while the remaining four reactors are used as hydrolysis reactors. Thus, both amorphous and crystalline sugars can be recovered with high efficiencies. By extensive use of heat recovery a very low energy use is achieved.

The performance of the progressing batch system has been estimated by mathematical modeling. The design conditions for the progressing batch reactor system are a solid residence time of 130 minutes and a liquid residence time of 45 minutes. The solids undergo prehydrolysis at 150°C for 30 minutes and hydrolysis at 189°C for 100 minutes. The acid concentration in the liquid phase is 0.7%. The yield on six-carbon sugars is projected to be 73%, and their concentration after the flash cooling is 4.8%. The five-carbon sugar yield is approximately 95%, and the concentration 3.4%. The hydrolysis reactors are carbon steel pressure vessels lined with acid-resistant brick.

The models used to describe percolation and countercurrent operation are quite similar. Because the models describing percolation reactors reasonably describe the actual operation, we have confidence in the magnitude of the improvement brought about by counter-current operation. However, the mathematical models have not yet been experimentally verified. Work is in progress at SERI to demonstrate actual reactor performance. Additionally, although the progressing batch reactor produces considerably less degradation products than the plug flow reactor, work is necessary to accurately determine the rate and efficiency of the hydrolyzate fermentation.

The cost of ethanol for the assumed base-case design is $1.35/gal with a surplus electrical credit of $0.41/gal. If a credit is taken for the small amount of furfural produced, the selling price of ethanol would be reduced 3¢/gal to $1.32/gal. The major improvements projected for this process over the plug flow reactor are: improved yield, energy efficiency, and reduced capital costs. While the plug flow reactor uses large amounts of electrical energy to reduce the feedstock to millimeter size particles, the progressing batch reactor uses primarily wood chips as a feedstock. Also, because the hydrolysis temperature is much lower than in the plug flow reactor system, less steam is used in hydrolysis. The most important effect is the increase in hydrolysis efficiency from 55% to 73% which reduces the feedstock costs approximately 26¢/gal. Finally, the progressing batch hydrolysis system is considerably simpler and less expensive.

Some benefits may be gained by presoaking the wood chips in acid, creating a more dense packing of the wood into the reactor, or using level control to reduce the effective volume of the reactors as the reaction proceeds. A 10% improvement in outlet sugar concentration would reduce the cost of ethanol by 10¢/gal. Xylose fermentation would reduce the cost of ethanol to approximately 98¢/gal. The progressing batch hydrolysis reactor is an extremely efficient producer of xylose, with a yield of greater than 90%. If we incorporate both xylose fermentation and the potential improvements in the performance of the hydrolysis reactor itself, a final selling price of 92¢/gal is predicted.

LOW TEMPERATURE CONCENTRATED SULFURIC ACID HYDROLYSIS

The low temperature, concentrated sulfuric acid hydrolysis process operates on the principle that the concentrated acid will break down the hydrogen bonds between adjacent crystalline cellulose chains, rendering the cellulose amorphous and easily hydrolyzed at conditions where sugar degradation is slow and yields are high. By using low-cost sulfuric acid, the large capital investment and energy consumption required for acid recovery can be eliminated. Such processes have been studied at the pilot scale in Japan (36), by the USDA Northern Regional Research Laboratory in Peoria, Ill. (10), by Purdue University (35), and by the TVA (2).

The wood is mixed with a recycled solution of sulfuric acid and sugars (primarily glucose) and conveyed to the prehydrolysis tank where it is heated to 100°C and held for 2 hours while the hemicellulose is hydrolyzed with 7% acid and a yield of 90% (Figure 5). After the prehydrolysis, the sugars produced are removed in a countercurrent leaching carried out in progressing batch fashion, and the solids are then centrifuged to remove much of the remaining water. The solids are slurried with water and acid and pumped to a tank, where they soak for 2 hours to allow the acid time to permeate uniformly into the fibers. The loading is 0.25 lb of sulfuric acid per lb of dry solids. The temperature is kept low to prevent significant hydrolysis. The slurry is centrifuged and the liquid stream recycled.

The acid-impregnated solids are dried in rotary dryers to approximately 10% moisture at 85°C. This operation has high capital and operating costs. The drying removes water but not the sulfuric acid, concentrating the acid to 83%. The dry acid-impregnated solids are then mixed with water and heated to 140°C with steam. Now the cellulose can be hydrolyzed to glucose in a few minutes at low temperature, with minimal formation of by-products. The yield of glucose from cellulose is approximately 80% of theoretical. The processing steps following the dryer are relatively fast and are performed continuously.

The product material from the hydrolysis reactor is discharged to a flash drum. The solids and liquids are separated in a centrifuge. The liquid stream, consisting of water, sugars, and acid, is recycled to prehydrolysis, while the solids are burned to provide the energy for the process. The sugars produced in both the hydrolysis and hemicellulose hydrolysis reactors are removed in the leaching process following prehydrolysis. This product stream has a composition of 7.0 wt % glucose and 5.5 wt % xylose (38,2).

The process produces alcohol at $1.68/gal with an electricity credit of 37¢/gal of ethanol. In comparison with the plug flow hydrolysis base case, we see that this process has lower feedstock costs because of higher yield but has a very high investment in the hydrolysis section due primarily to the dryers, larger expenses for acid and base, and lower electrical byproduct credits.

The higher yield is reflected in the feedstock cost which is reduced from \$1.04/gal for the plug flow process to 68¢/gal. This advantage will be shared by all concentrated acid processes. However, the capital investment required for the hydrolysis section is quite high, coming to approximately \$44 million. Another major penalty is the cost of acid and base, which comes to 27¢/gal. The sugar concentration in the concentrated acid process is roughly equivalent to that in the plug flow process. This is not surprising in that both processes use mechanically similar hydrolysis reactors to carry out the cellulose hydrolysis.

The major improvement to the process would be to produce a more concentrated sugar stream by reducing the amount of water needed to extract the product from the prehydrolysis section. A significant reduction in the water usage should be possible by using washing centrifuges to remove the hydrolyzate from the slurry leaving the prehydrolysis reactors. This would lower the cost of ethanol by 10¢/gal to \$1.56/gal of ethanol. The most significant improvement would be the incorporation of the xylose fermentation, which would reduce the price of ethanol to \$1.20/gal. Incorporating both of these improvements would further reduce the cost of ethanol to \$1.13/gal. Recycling the sulfuric acid, sometimes mentioned as a potential improvement, will be difficult to accomplish for less than its purchase price of 3¢/lb.

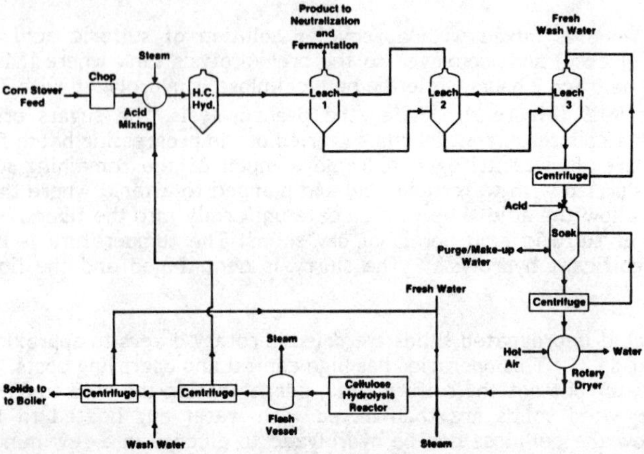

Figure 5. Concentrated-Acid Hydrolysis Process: Block Diagram

LIQUID PHASE CONCENTRATED HYDROCHLORIC ACID PROCESSING

The liquid phase concentrated HCl hydrolysis system uses a solution of 40% HCl to decrystallize the hydrolyzed lignocellulosic materials with a yield of essentially 100% (15). Because HCl is an expensive acid (12¢/lb), an HCl recovery system must be used. This process has been studied extensively. and a plant has been operated in Germany since 1925 (4,30).

The incoming wood is dried to 10% moisture (Figure 6). The hydrolysis system uses six replicate rubber-lined reactor vessels to provide counter-current flow of wood chips and hydrochloric acid. The acid stream is fed to the reactors at a concentration of 41%; it decreases in acid concentration and increases in sugar concentration as it progresses through the four reactors. The reactor residue is washed with water and with 30% acid to recover residual acid in the remaining lignin. The lignin is discharged from the fifth reactor, dried, and burned. The hydrolysis reaction residence time is 1 h at 35°C. Operating pressure is slightly above atmospheric to keep air out. There is no prehydrolysis step because sugar degradation is negligible. The cellulose and hemicellulose are hydrolyzed with essentially 100% efficiency to simple sugars and sugar oligomers.

The solution of acid and sugars is pumped to the evaporators where more than 98% of the HCl fed to the reactors is taken overhead in two stages with water vapor, separating it from the crude water solution of sugars and lignin. The combined stream is approximately 30% HCl in water, which is then fed to the acid reconcentration system. The acid reconcentration system, designed to upgrade 30% hydrochloric acid to 40%-41% HCl, utilizes a two-column system. The columns operate at different pressures to shift the concentration of the azeotrope and take advantage of this shift to remove water. The process energy required is only the net requirement to remove the water by evaporation from the gross feed stream. The system is constructed of glass-lined steel with ceramic packing and includes corrosion-resistant heat exchangers and a vacuum pump. Makeup acid is added in an atmospheric pressure adsorption system.

This posthydrolysis system is designed to hold the crude product stream at 100°F for about 1 h in the presence of a few percent HCl to break down the polysaccharides formed in the main hydrolysis to monomeric glucose and xylose. Approximately 90% of the potential sugars can be fermented (40).

The basic liquid phase HCl process is predicted to produce ethanol for $1.65/gal with electricity as the only byproduct credit. As with all concentrated acid processes, the feedstock costs are low (70¢/gal) because of the high yield. More than offsetting these gains is the extremely high energy consumption required for HCl recovery. The steam usage in HCl recovery is very high, and steam must be imported. Simultaneously, the electrical demand of the plant is relatively low and electricity is sold as a byproduct with a credit of 36¢/gal. The second major drawback to this process is the extremely high capital cost of the acid recovery system. Much of the equipment in the acid recovery system is

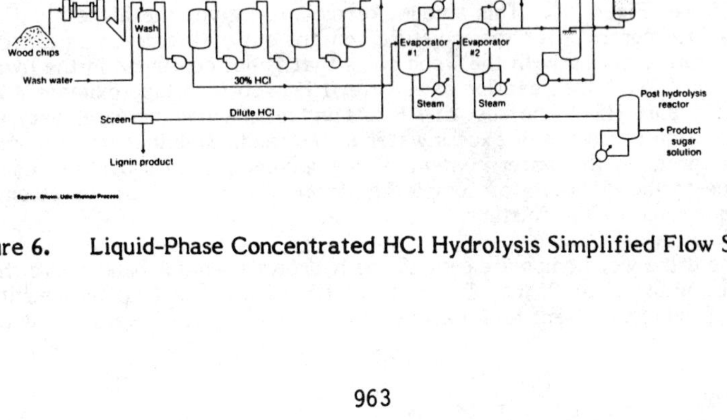

Figure 6. Liquid-Phase Concentrated HCl Hydrolysis Simplified Flow Sheet

constructed of extremely expensive materials such as Hastelloy, Monel, glass lined steel, or graphite. A third expense in this system is the cost of the HCL itself. The base-case design recovers approximately 98.7% of the circulating HCl stream. This comes to approximately 14¢/gal for the acid and the base. This number is not known with great accuracy. Small increases in the amount of HCl lost by the system could greatly increase the cost of ethanol.

The size of the acid recovery system could be reduced by minimizing the wash water requirements, replacing the countercurrent leaching operation employed by the German plants with a system of back-washed centrifuges in series. If the HCl/water circulated could be reduced by 20%, the cost of ethanol production would drop to $1.42/gal.

A second major improvement would be to reduce the cost of recovering the hydrochloric acid. The present distillation system has a long history of development and probably cannot be significantly improved. If major price reductions are to be achieved, an entirely new recovery technology is necessary. An example would be the solvent extraction recovery process patented by Dow Chemical (8) or the electrodialysis recovery technique proposed by Goldstein (14).

Xylose fermentation would reduce the cost of ethanol 20%, from $1.65/gal to $1.32/gal. The improvement is not as dramatic in a process with high yields on crystalline cellulose such as the HCl system, as it is in a low yield process such as the base-case plug flow reactor. Incorporating the improved water balance and xylose fermentation into the process yields an ethanol cost of $1.17/gal. The lignin produced in this process is less condensed than that produced by other acid processes, and could, in principle, be used to generate higher byproduct credits.

LIQUID PHASE HYDROFLUORIC ACID

Like all concentrated acid processes, the liquid phase HF process has yields approaching 100%. Even so, the capital and energy requirements of the hydrolysis are large. Also, losses of HF from the process must be absolutely zero, first because HF is an extremely expensive acid ($0.66/lb), but primarily because of its extraordinarily hazardous nature (40).

Liquid phase HF processes have been studied only at the bench scale. Work has been carried out in Germany (12), in Denmark and France (3,7), and at Michigan State University (32,9). In recent years a gas phase HF process has also been investigated at Hoechst AG in Germany (11) and in Canada (27). This report deals only with liquid phase processes because little performance information has been released on the gas phase processes.

The first step in the hydrolysis section is to dry the incoming wood chips to 5% moisture (Figure 7). This amount of moisture gives the highest yield (near 100%). Close control over the moisture content out of the dryers is a necessity. Any moisture entering with the wood chips that is not consumed in the hydrolysis reaction will leave the reaction as a water/HF azeotrope (approximately 2 parts water to 1 part HF). The cost of HF lost in this manner increases very sharply with even small amounts of excess water in the feed. To eliminate the loss of HF from the process, the water content of the incoming feed should be reduced to 3%. Due to the difficulty of completely drying the wood chips, we assume that the feed is dried to 5% moisture.

The dried wood chips are sent to the hydrolysis reactor bank where they are contacted with HF at 23°C. The ratio of HF to wood is 2.7/1 by weight. The reaction products are sent to the acid recovery section. The sugars are repeatedly

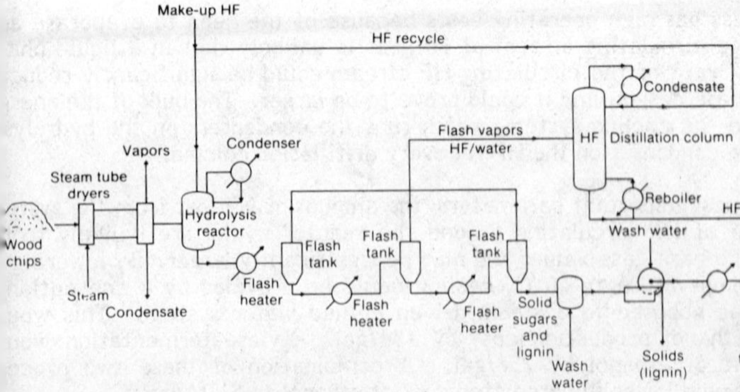

Source: Hawley, Selke, et al. Michigan State University

Figure 7. Overall Flow Sheet for Liquid-Phase HF Hydrolysis

eated and flashed to vaporize the HF and water. The bottom stream from the inal flash tank is an HF-free stream of solid crystalline sugar and lignin. The lash vapors--HF and water--are sent to a distillation column, where water is emoved and the HF recycled. Water is added to the sugars and lignin, and the ignin is removed by filtration. The sugar stream from the filter (containing both nonomeric and oligomeric sugars) is heated and sent to the posthydrolysis reactor where the oligomeric sugars are converted into monomeric sugars. The overall ydrolysis efficiency is 90%. The monomeric sugars are then sent to eutralization and fermentation. A major piece of auxiliary equipment is the efrigeration system, required to remove the large exothermic heat of reaction rom the hydrolysis reactor and to cool the distillation column condenser.

The liquid phase hydrofluoric acid process produces alcohol in the base case or \$1.61/gal with a 30¢/gal credit for surplus electrical energy. As in all concentrated acid processes, the feedstock costs are low (62¢/gal) because of the near stoichiometric yields. This cost is largely offset by the high cost of makeup HF (40¢/gal). HF losses are extremely expensive because of the high price of hydrofluoric acid (66¢/lb versus 12¢/lb for HCl and 3¢/lb for sulfuric acid). There are three major losses of HF. First is the loss to the HF/water azeotrope (20¢/gal). The second major loss is to neutralization of the ash component of the wood. With clean wood from the inside of a tree where the ash component is less than 0.5% of the dry weight, this loss is quite small, amounting to only 6¢/gal of ethanol. Whole tree chips with bark attached have a higher dirt and ash content and would lead to a much greater loss. Third is the HF that remains adsorbed on the carbohydrates, leaving the drying system. In the base-case design, this loss amounts to an additional 8¢/gal of ethanol. All these numbers are uncertain and may easily be much larger.

The capital cost of this system is not significantly greater than that of the plug flow reactor. The major capital items are the reactors, the posthydrolysis system, the cooling system, and the dryers. However, the capital cost could seriously be underestimated due to the leak-tight and minimal maintenance required by the extreme toxicity of HF. Also, the safety measures would significantly increase the amount of labor required to run the plant. These factors have not been taken into account in the economic analysis. Because of the high concentration of sugars leaving the hydrolysis process, there are significant capital savings in the balance of the plant, most noticeably in distillation and the environmental section.

The process has high operating costs because of the need to evaporate a condense a large circulating stream of HF. It is unlikely that in a liquid pha system the flow rate of the circulating HF stream could be significantly reduc from the base-case design, and it could prove to be larger. The bulk of the ener is used to drive the cooling system, which runs the condensers on the hydrolys reactors and the condenser on the HF recovery distillation column.

The two most important parameters, the amount of HF lost from the syste and the volume of HF circulating around the reactor cycle, are unlikely to b lowered from the base-case values and may be significantly larger. By lowering the reaction temperature to 40°C, cooling could be provided by a convention cooling tower as opposed to a steam-driven chilled water system. This wou decrease the ethanol production cost by 18¢/gal. Xylose fermentation wou reduce the cost of ethanol by 23¢/gal. A combination of these two proce improvements would lower the predicted cost of ethanol to $1.14/gal.

The major drawback of this process is its use of a large HF cycle wi intrinsically high capital costs dictated by safety considerations. In general, in process using HF, operators are confined to a central pressurized control roon Crew members required to do outside work must wear complete protective ge and a self-contained air pack. All process equipment is designed to be leakpro and require minimal maintenance, driving the capital costs upward. It questionable whether a company would risk the dangers of a HF process for a lo value product such as a fuel. In reality, halogen acid processing is used in th chemical industry only when there are no alternatives to the process and when th margin of profit on the plant is high. While these considerations cannot b quantified, they suggest that both the capital and operating costs we hav estimated are probably low and that such processes would face severe problems i being adopted at any ethanol price.

SUMMARY

As a broad conclusion, dilute acid hydrolysis processes are preferred t concentrated acid processes. The high yields enjoyed by the concentrated aci processes do not overcome the penalties imposed by the use of large quantities c acid. In the concentrated sulfuric acid process, these liabilities include hig capital cost and high acid consumption, while the halogen acid processes suffe from extreme complexity, high capital and operating costs involved in th recovery and recycling of the expensive halogen acids, the high costs associate with the loss of even small amounts of expensive acid, and the less readil quantifiable disadvantages of working with a highly toxic substance. While all th acid hydrolysis processes have potential for improvement, the potentia improvement appears greater in the dilute acid processes (Figure 8).

The progressing batch hydrolysis concept is projected to be superior to th plug flow and concentrated acid hydrolysis concepts if the objective is to produc ethanol and electricity. The progressing batch hydrolysis system produces sugar with higher yields than the plug flow reactor, requires less capital investment an energy, and is an inherently simple and proven mechanical design. Thes advantages carry over into systems with five-carbon sugar fermentations as th progressing batch reactor appears to be an inherently efficient producer of five carbon sugars. The major need in this process is for operating data on progressin batch operation to determine the actual performance of the process.

However, in the near term the coproduction of furfural by the plug flov reactor is a significant advantage enjoyed by no other hydrolysis process. Larg quantities of a high value byproduct are produced with only a modest additiona requirement for capital or energy. This should allow furfural to be sold as a

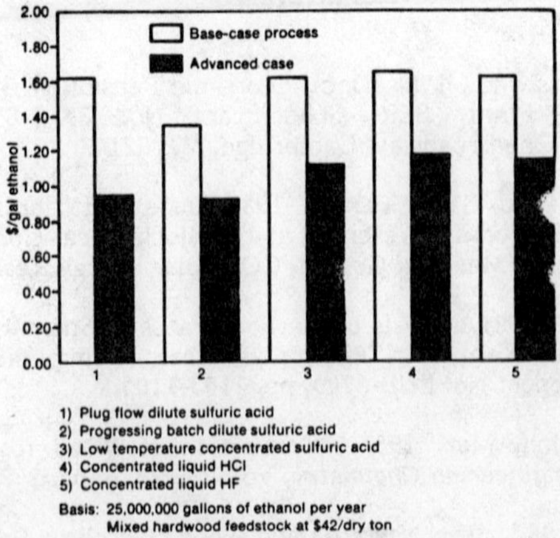

Figure 8. Comparison of Acid Hydrolysis Processes for the Production of Ethanol and Electricity

byproduct at a fraction of its current selling price, expanding its market into readily identifiable and in many cases commercially proven markets and processes.

The concentrated sulfuric acid process has a long history of development. There is little room for improvement in the yield, while the acid consumption costs of the process are inherent to the design. Recovering sulfuric acid from the product stream for less than its extremely low cost (3.25¢/lb) is a difficult task. The high capital cost in the hydrolysis section is also unlikely to be greatly reduced. This is a relatively safe reactor design in that there are no major technical barriers to its working.

The liquid phase HCl process is commercially proven and has a long history of development. As long as a thermal method is used for the recovery of the acid, this process will not be able to compete. The capital and energy costs involved in acid recovery are immense due to the large volume of liquid that is cycled and the exotic materials of construction. Solvent extraction or membrane processes have been proposed to recover the hydrochloric acid, but have not been sufficiently studied to draw conclusions. A useful observation, however, is that in general industry is moving away from the use of concentrated acid processes wherever possible. The maintenance and operation difficulties involved in such processes are large enough that even where they are profitable, alternatives are actively sought.

On paper, the performance of the hydrofluoric acid process is roughly equivalent to the other two concentrated acid processes. It has a high yield, but suffers from the high costs of HF recovery and of HF makeup. The actual losses of HF from this system are unlikely to be smaller than those described, but could easily be much larger. Additionally, the study has not been able to quantify the additional operating costs or hazards inherent in the operation of a cycle using very large quantities of an extremely toxic material. It is unlikely that the profit margins involved in any fuel production business could justify the scale of the risks. Although there are processes which use HF, the chemical industry avoids its use wherever possible.

REFERENCES

1. Badger Engineers, Inc., 1984 (Dec), Economic Feasibility Study of an Acid Based Ethanol Plant, SERI Subcontract No. ZX-3-030-96-2, Badge Engineers, Inc., One Broadway, Cambridge, MA 02142.

2. Barrier, J. W., and G. E. Farina, 1985 (June), "Ethanol from Cellulosi Agricultural Feedstocks," presented at the Biochemical Conversion Prograr Semiannual Review Meeting, Golden, CO: Solar Energy Research Institute.

3. Bentsen, T., 1983, "Hydrolysis of Carbohydrates in Straw Using a Hydroge Fluoride Pretreatment," in *Energy Biomass*, Commission of Europea Communities Report No. EUE-8245, pp. 1103-1105.

4. Bergius, Friedrich, Mar. 1937, "Conversion of Wood to Carbohydrates, *Industrial and Engineering Chemistry*, Vol. 29, No. 3, pp. 247-253.

5. Burton, R. J., 1982, "The New Zealand Wood Hydrolysis Process," *Proceed ings of the Ethanol from Biomass Conference*, Winnipeg, Canada.

6. Carvalho Neto, C. C., and Kling, S. H., Nov. 1983, "Pilot-Plant Aci Hydrolysis Studies Using *Eucalyptus paniculata*," presented at the 10t Interamerican Congress of Santiago-Chile, Fundacao de Technologi Industrial, Lorena-Sao Paulo, Brazil.

7. Defaye, J., A. Gadelle, and C. Pedersen, 1981, "Degradation of Cellulos with Hydrogen Fluoride," *Energy from Biomass*, edited by W. Palz, P Chartier, and D. O. Hall, London: Applied Science Publishers, pp. 292-297.

8. Dow Chemical Co., Dec. 1980, "Process for Separating and Recoverin Concentrated Hydrochloric Acid from the Crude Product Obtained from th Acid Hydrolysis of Cellulose," Patent No. 4,237,110.

9. Downey, K., S. Selke, M. Hawley, and D. T. A. Lamport, Sept. 1983, "HF Solvolysis of Wood for Ethanol Production--Preliminary Design, Economics and Identification of Needed Research," presented at Industrial Wood Energy Forum 83, Nashville, Tenn., 19-21, Sept. 1983.

10. Dunning, J. W. and Lathrop, E. C., Jan. 1945, "The Saccharification o Agricultural Residues--A Continuous Process," *Industrial and Engineering Chemistry*, pp. 24-29.

11. Franz, R., R. Erckel, T. Riehm, R. Woernle, and H. M. Deger, 1983 "Lignocellulose Saccharification by HF," *Second Energy from Biomas Conference*, Berlin, 20-23, Sept. 1983, pp. 873-878.

12. Fredenhagen, K., and G. Cadenbach, 1933, "Der abban der cellulose durch Fluorwasser staff und ein neues Verfahren der Holzerzuckerung durch hoch konzentrierten fluorwasser staff," *Agnew Chem.*, Vol. 46, pp. 113-117.

13. Gilbert, N., I. A. Hobbs, and J. D. Levine, July 1952, "Hydrolysis of Wood Using Dilute Sulfuric Acid," *Industrial and Engineering Chemistry*, Vol. 44 pp. 1712-1720.

14. Goldstein, I. S., 1983, "Acid Processes for Cellulose Hydrolysis and Their Mechanisms," *Wood and Agricultural Residues*, Academic Press, Inc.

.5. Goldstein, I. S., H. J. Pereira, J. L. Pittman, B. A. Strouse, and F. P. Scaringelli, May 1983, "The Hydrolysis of Cellulose with Superconcentrated Hydrochloric Acid," *Proceedings of the Fifth Symposium on Biotechnology for Fuels and Chemicals*, held in Gatlinburg, Tenn., 10 May 1983.

16. Grethlein, H. E., and A. O. Converse, 1982, "Continuous Acid Hydrolysis from Glucose and Xylose Production," presented at the International Symposium on Ethanol from Biomass, sponsored by the Royal Society of Canada, Winnipeg, Manitoba, 12-16 Oct.

17. Hajny G. T., 1981, "Biological Utilization of Wood for Production of Chemicals and Foodstuffs," Research Paper FPL 385, Madison, WI, USDA, Forest Products Laboratory.

18. Harris, E. E., E. Beglinger, G. J. Hajny, and E. C. Sherrard, Jan. 1945, "Hydrolysis of Wood," *Industrial and Engineering Chemistry*, Vol. 37, No. 1, pp. 12-23.

19. Harris, E. E., Beglinger, E., Sept. 1945, "Madison Wood Sugar Process," *Industrial and Engineering Chemistry*, Vol. 38, pp. 890-895.

20. Harris, J. F., 1975, "Acid Hydrolysis and Dehydration Reactions for Utilizing Plant Carbohydrates," *Applied Polymer Symposium No. 28*, pp. 131-144. John Wiley & Sons.

21. Harris, J. F., A. J., Baker, Conner, A. H., T. W. Jeffries, J. B. Miner, R. L., Pettersen, R. W. Scott, E. L. Springer, T. H., Wegner, and J. I., Zerbe, 1985 (Oct.), "Two-Stage, Sulfuric Acid Hydrolysis of Wood: An Investigation of Fundamentals," USDA Forest Products Laboratory, Madison, WI.

22. Hoagland, W., A. H. Brennan, and D. J. Schell, 1986 (May), High-Temperature Acid Hydrolysis of Cellulose Using an Engineering-Scale Plug Flow Reactor, to be presented at the Eighth Symposium on Biotechnology for Fuels and Chemicals, Gatlinburg, TN, May 13-16, 1986.

23. Icarus Corp., 1984, COST System User's Manual.

24. Kwarteng, I. K., July 1983, *Kinetics of Acid Hydrolysis of Hardwood in a Continuous Plug Flow Reactor*, Ph.D. Thesis, Hanover, NH: Portmouth University.

25. Mackie, K., K. Deverell, and I. Callander, 1982, "Aspects of Wood Hydrolysis via the Dilute Sulphuric Acid Process," *Proceedings of the Ethanol from Biomass Conference*, Winnipeg, Canada.

26. National Research Council of Canada, 1983 (Dec.), "A Comparative Assessment of Forest Biomass Conversion to Energy Forms. Phase 1 Proven and Near-Proven Technology, Volume V - Data Book of Unit Processes for Primary Conversion by Thermal, Chemical, and Biological Methods." DSS Contract 43SS, KN 107-1-4416, Energy, Mines and Resources, Canada, Contractors Simons Resource Consultants, Vancouver, B. C., B. H. Levelton & Assoc. Ltd., Vancouver, B.C.

27. Ostrovski, C. M., J. Aitken, and D. Free, 1984, "New Developments in Fuel Ethanol Production by Gaseous Anhydrous Hydrofluoric Acid Hydrolysis of Hardwood, *Populus tremuloides*," presented at Bioenergy '84, Gathenburg, Sweden.

28. Parker, S., M. Calnon, D. Feinberg, A. Power, and L. Weiss, 1983, *The Valu of Furfural/Ethanol Coproduction from Acid Hydrolysis Processes*, SERI/TR 231-2000, Golden, CO: Solar Energy Research Institute.

29. Reitter, F. J., 1984 (Jan.), "Two-Stage Continuous Hydrolysis of Plar Biomass to Sugars," U.S. Patent 4,427,453.

30. Riehm, T., Oct. 1980, "The Production of Fermentable Sugars fro Biomass," presented at the Symposium on the Manufacturing of Chemic Raw Materials from Biomass, Santa Domingo, Dominican Republi 20-25 Oct.

31. Saeman and Andreasen, 1945, "Production of Alcohol from Wood Waste," *Industrial Fermentations*, L. A. Vunderkoflen, and R. J. Hickey, eds., NY Chemical Publishing Co., pp. 136-171.

32. Selke, S. M., M. Hawley, and D. T. A. Lamport, 1983, "Reaction Rates fc Liquid-Phase HF Saccharification of Wood," *Wood and Agricultural Residue* pp. 329-349.

33. Solar Energy Research Institute, February 1986, *Fuel Alcohol Technical an Economic Evaluation*, Draft, Golden, CO.

34. Song, S. K., and Y. Y. Lee, 1982, "Counter-Current Reactor in Aci Catalyzed Cellulose Hydrolysis," *Chemical Engineering Communication* Vol. 17, pp. 23-30.

35. Tsao, G. T., M. R. Ladisch, M. Voloch, and P. Bienkowski, 10 Sept. 198 "Production of Ethanol and Chemicals from Cellulosic Materials," *Proces Biochemistry*, pp. 34-38.

36. Wenzl, H. F. J., 1970, *The Chemical Technology of Wood*, Academic Press New York, pp. 32-149.

37. Wright, J. D., 1983, *High Temperature Acid Hydrolysis of Cellulose fo Alcohol Fuel Production*, SERI/TR-231-1714, Golden, CO: Solar Energ Research Institute.

38. Wright, J. D., and C. G. d'Agincourt, Apr. 1984, *Evaluation of Sulfuric Aci Hydrolysis Processes for Alcohol Fuel Production*, SERI/TR-231-2074 Golden, CO: Solar Energy Research Institute.

39. Wright, J. D., P. W. Bergeron, and P. J., Wendene, 1985 (Sept.), "Th Progressing Batch Hydrolysis Reactor," presented at the American Chemica Society Annual Meeting, Chicago, IL, Sept. 9-13.

40. Wright, J. D., A. J. Power, P. W. Bergeron, 1985 (June), *Evaluation o Concentrated Halogen Acid Hydrolysis Processes for Alcohol Fuel Produc tion*, SERI/TR-232-2386, Golden, CO: Solar Energy Research Institute.

DISCUSSION

G. Grassi, Commission of European Communities (Belgium)

I would like to ask something about the data you presented in your table concerning the plug flow reactor. You referred to experimental research. How did you evaluate it?

Wright

The information on the plug flow reactor is from two places. The kinetic data were originally obtained by Hans Grethlein at Dartmouth. He worked with a 1/4-in. diameter reactor hooked up to a Moyno pump. Since that time, we have built a considerably larger scale reactor at SERI. The reactor is a 1.5 in. in diameter, and processes depending upon solids concentrations up to 400 or 500 pounds of solids per hour. Our results, in terms of yields and sugar concentrations, are essentially the same at that scale as Hans Grethlein reported. Now, that's not surprising because you don't really expect things to change much with scale-up of a plug-flow reactor. We're trying to demonstrate at a reasonable scale the processing of very concentrated slurries and we're modifying our reactor right now to put the hydrolysis unit on the front and show that we can get the same yields as well as the high solids operations to give the concentrated sugar solutions.

J. Roetheli, Tennessee Valley Authority

Regarding the concentrated acid processes, I think you said it operated at about 140°C. We're not operating above atmospheric pressures, so we're under 100°C. This has allowed us to use fiber glass reinforced plastics for all the materials of construction. We could not do that under pressure. The vessels would not stand it.

ETHANOL FUEL FOR SMALL-SCALE MECHANIZATION OF FARMING IN THE THIRD WORLD

David H. Adolph, D.Sc.
Washington University, St. Louis, Missouri
Wayne G. Bragg, Ph.D.
Wheaton College, Wheaton, Illinois
Eugene B. Shultz, Jr., Ph.D.
Washington University, St. Louis, Missouri

ABSTRACT

Rural areas of the Third World often have difficulty obtaining reliable sources of liquid fuels at sufficiently low cost so that selective mechanization of agriculture can be encouraged. Small-scale local production of ethanol from sugarcane, cassava, corn, etc., might be effective in meeting this need (a) if pressure on the land is not great, (b) if small-scale mechanization of farming can expand production without generating undesirable ecological or social side-effects, (c) if adequate supporting infrastructure is available, and (d) if the price and availability of food is not adversely affected by fuel cropping.

The results of a pilot project of this type in a remote part of central Africa have been evaluated by us. A small ethanol plant was employed in the project, utilizing sugarcane as feedstock, to produce 180-proof ethanol. This was used as fuel in small agricultural tractors. The distillery wastes were used to produce biogas for electricity generation. The results of this case study and the general concept of local ethanol for small-scale selective mechanization are analyzed, for potential application in regions with appropriate ecological and social conditions.

ETHANOL FUEL FOR SMALL-SCALE MECHANIZATION OF FARMING IN THE THIRD WORLD

INTRODUCTION

The potential adoption of small-scale fuel ethanol production in the rural Third World raises a number of questions of crop choice and availability of cropland and supporting infrastructure. Local ethanol may be a viable option for locations where land can be used for an ethanol crop without adversely affecting price or availability of food, where a small distillery can be satisfactorily operated and maintained by local people, and where credit, training, and management and other support can be provided.

A somewhat different set of issues surrounds the potential adoption of small-scale selective mechanization of farming in the rural Third World. Less-expensive alternatives such as draft animals should be investigated, and if there are persuasive reasons why they should be disregarded, then small walk-behind or riding tractors might become viable options. There must be unused land available if mechanization is likely to greatly increase farming area per person. Bringing more acreage into production, especially if the land is of marginal quality, might create ecological problems. Social imbalances may occur if only a few members of a community have access to mechanization; equity issues will need to be addressed. Credit, training, and management must also be provided, along with repair, spare-parts supply, and other support functions, as in the case of adoption of small-scale local ethanol production.

The decision, then, to combine small-scale mechanization with small-scale local ethanol production can be seen to be dependent upon a favorable assessment of a number of non-technological variables. Land must not be scarce and the project must not compete with food production, or degrade the ability of the land to feed the people. Social issues of equity, dependence on outside resources, and needs for a variety of social and economic support structures must all be resolved satisfactorily. The decision to introduce such a scheme is clearly not limited to narrow economic and technical considerations.

In the central African case study to be presented in this paper, many of the requirements outlined above were met. It appeared that the others could also be satisfied without much difficulty.

CASE STUDY: CENTRAL AFRICA

Background

The concept was given a preliminary test in Africa in 1984 by one of us (DHA). The location was in the vicinity of Obo in the eastern part of the Central African Republic in an area inhabited by the Zande tribe. This part of Africa is sparsely populated. Land is available, and population is not growing.

The prevailing style of agriculture in Zandeland is "shifting" cultivation, in which the boundaries of a farmer's most important fields will shift from one cultivation cycle to the next. This allows for periods of fallow that restore soil fertility. Adequate fallowing is especially important in the tropics where cleared land rapidly loses its ability to support crops.

Major crops of the Zande include peanuts, cassava, rice and corn. All field work is carried out manually with simple tools. Draft animals are rarely used, apparently for significant cultural reasons, so plot sizes are limited by the use of hand tools to a few acres per adult. The details of Zande agriculture have been discussed by Adolph (1, 2).

The small-scale ethanol production test was part of a selective mechanization project in which small tractors were to be introduced to allow each farmer to expand his farming activities. Sugarcane was planned to be the principal ethanol feedstock, and two years earlier, sugarcane had been planted in preparation for the arrival of the tractors and the ethanol plant. Although cane does grow well in this area, it had been available only in limited quantities as a snack, not as a staple food. The mechanization project and the ethanol fuel project were both sponsored by the Africa Inland Mission, a branch of which has been active for many years in Obo, Central African Republic.

Like many isolated communities, the rural areas of the Central African Republic have difficulty obtaining and maintaining an adequate and reliable source of liquid fuels. Since the country is landlocked and isolated, fuel must be transported from remote coastal areas raising the already high cost of fuel still higher. Therefore, ethanol crops such as sugarcane and cassava have the potential of becoming a more reliable and more economic source of engine fuels than imported gasoline.

Technical Approach

Crops grown by farmers using small ethanol-powered tractors were to supply the primary feedstocks for ethanol production. While sugarcane was to be the principal feedstock, it was also intended that the distillery would occasionally purchase unmarketable surpluses of maize and rice. Surplus maize was being used as a cattle feed, so it was hoped to feed the distiller's grains to the animals. Sugarcane bagasse was the principal heat source. The still bottoms were to be mixed with manure in an anaerobic digester for production of biogas to be used in a diesel system for generation of electricity.

Ethanol Plant. The major pieces of equipment were the sugarcane crusher, fermentation tanks, still, anaerobic digester, and the small tractors. The single-stage crusher was powered by a two horsepower three-phase electric motor. The two fermentation tanks were each of 200-gallon capacity.

The still had been constructed in the United States from the Mother Earth News 6-inch column still plans (3). The still had a 275-gallon beer boiler feeding into a stripping column and then into a packed reflux column. In the latter were water-cooled coils for maintaining optimum temperatures and for condensing the alcohol at the top of the column. The still was rated at 6 to 8 gallons/hr (23 to 31 lit/hr) of 180 to 190 proof ethanol.

The anaerobic digester was fabricated on-site from found materials. It was essentially a metal box about four meters long, one meter wide and one meter high. Gas was stored externally in a water-sealed gas holder, also constructed on the site. Equal weights of stillage and fresh manure were charged to the digester daily. Average retention time was about 14 days.

Small Tractors. Two kinds of small tractors were purchased in the United States by the Mission, and modified to use ethanol as fuel. Each of the 12-horsepower riding tractors and 8-horsepower walk-behind tractors was equipped with accessories for plowing, harrowing, planting, and cultivating. In a controlled test of plowing performance that compared 170-proof ethanol fuel with gasoline, the ethanol-fueled tractor required 6% more energy per unit of plowed area. The details of this and other performance tests of the small tractors carried out in Obo, Central African Republic, have been given by Adolph (1).

Experimental Results with Sugarcane

A synopsis of averaged results of seven tests is given below:

Sugar extraction efficiency, %	56
Fermentation efficiency, %	79
Distillation efficiency, %	49
Proof	181
Ethanol yield, lit/ton	13.4
Biogas production, cubic meters/week	11.6

Discussion of Results

Sugar extraction efficiency could have been higher with a multi-stage crushing system with washing of the bagasse between stages. Alternatively, multi-stage crushing might be simulated labor-intensively with the existing single-stage system.

The fermentation tanks were not equipped for agitation or temperature control, but did have probes for temperature monitoring. They could be sealed air-tight, and were equipped with one-way air vents. Fermentation was carried out for three to five days. Results could have been improved with agitation and better yeast. Regular shipments of fresh yeast and nutrients would have to be arranged.

Distillation efficiency was low and variable, ranging from 32 to 56% in the seven tests, primarily because of difficulty in supplying sufficient cooling water to the condenser. The still was equipped with 1/4-inch copper tubing which should have been adequate if water were clean, and available at pressures typical of municipal water supplies. It had not been anticipated that the small-diameter tubing would pose a problem on-site in central Africa.

Cooling water was obtained from a well, and from rainwater collected from the roof. It was hand-pumped from storage tanks through a filter screen to a holding drum. From there, it was pumped by a small electric centrifugal pump through the cooling coils of the condenser. Flow rates were generally too low, and flow was often interrupted by plugging of the copper tubing by solid matter. Therefore, it was usually necessary to maintain a smaller than normal fire under the boiler to prevent column temperatures from rising excessively. Typically, distillations had to be shut down prior to the separation of a significant fraction of ethanol in the beer. Product ethanol flow rates were erratic, and ranged from one-fourth to one-half of design capacity.

Obviously, the coils should be replaced with tubing of larger diameter. Contamination of the water supply must be prevented so that coil blockages will be unlikely.

In summary, several technical problems were encountered. Some of these were inherent in operating in a remote location, while others were related to inadequacies in the design of the equipment. However, none of the problems was regarded as technically difficult to overcome.

Economic Aspects

Estimates of ethanol plant capital cost in 1983 U.S. dollars, exclusive of shipping, were as follows:

Crushing system	$ 550
Fermentation tanks (2)	210
Still and boiler	4300
Digester	250
Auxiliaries	290
Total	$5600

Total cost was dominated by the cost of the still, which was fabricated in the United States. A significantly lower cost would have been realized if the still parts had been assembled in an African city. Further, because of the aforementioned cooling water problems, yields were low and operating costs were unacceptably high.

However, operation of the anaerobic digester on stillage and manure was acceptable. The weekly production of biogas saved $5.78 in diesel fuel at an estimated weekly labor expense of $3.75, allowing for payback in less than three years. Equipment life of five years was anticipated.

ANALYSIS

General Considerations

Our analysis is based on three general assumptions concerning rural development strategy. These seem to apply reasonably well to the particulars of small-scale ethanol for small-scale agricultural mechanization in the rural Third World:

1. Sustainable and equitable solutions to problems should be sought to minimize ecological and social disruption.

2. Solutions do not necessarily have to call for expensive advanced technologies requiring costly and extended research, development and implementation.

3. Solutions should develop the learning skills and self-confidence of local people, who should be involved as much as possible in decision-making, planning, implementation and operation.

Decentralized, small-scale production of 170-190 proof alcohol for engine use may be socially, environmentally, technically and economically feasible in many rural parts of the Third World. If all four of these elements of the feasibility analysis are eventually deemed satisfactory, and if the mechanization of agriculture survives a similar feasibility analysis, then the combination might be given serious consideration for implementation at a given site.

Specifics of the Central African Case

Issues that must be examined carefully on a site-specific basis are the following:

1. Ethanol crop choice. At present, all practical ethanol crops contain either sugar or starch. Both are food components; therefore, in most cases the ethanol crop will also be a food (or feed) crop and a competition might be induced that would drive up the price of food and/or feed, or limit its availability. In the Central African case, sugarcane was the chosen feedstock but it had never been grown as a significant food crop in the area. Cassava, however, is a staple food of the Zande tribe, so if the project adopted cassava as an ethanol feedstock, a "fuel vs. food" problem might arise. An alternative inedible starch or sugar crop might be sought. A new crop may require special training that may not be within the competence of agricultural extension agents to provide.

2. Land availability. Adoption of ethanol fuel, especially if a new crop is grown for feedstock, will probably place large new acreage under cultivation. Expansion of ethanol use will require an expansion of acreage and this may bring with it adverse ecological consequences, especially if population density is high.

   However, in the central African case, the population is sparse and not growing. The centuries of shifting agriculture have left large areas underutilized. Fallow periods are sufficiently long to allow the soil to recuperate. Even with increased agricultural activity as a result of ethanol-based mechanization, there will still be abundant land to leave fallow for five years on a rotating basis, following traditional Zande practices, which protect fragile tropical soils.

   In other locations where shifting cultivation is practiced, if ethanol were heavily adopted and if population were to increase, pressure might develop to shorten the fallow periods with potentially serious losses in soil fertility. In densely-populated areas, locally-grown ethanol crops might be entirely out of the question.

3. Local infrastructure development. Even though the technology exists to produce ethanol on a small scale, it is likely that it will have to be imported into Third World nations. The numbers of people who understand how to assemble, operate and maintain such plants in rural areas will be limited. These requirements mean that an adequate supporting local infrastructure must be developed to supply the credit, training and organization that are necessary.

   Non-governmental organization (NGOs) dealing regularly with rural development problems and experienced in introduction of small-scale technology may be more effective than governmental agencies. In the central African case, the Africa Inland Mission appeared to be well-positioned to provide the necessary supporting services. However, even though the infrastructure and land resource base seem to be sufficient, care must be taken to work within the cultural and social patterns that have served for centuries. This must be done to avoid creating social power imbalances, and perpetuation of dependencies on outside markets and supplies.

A criticism often leveled at mechanization projects is that they counterproductively substitute for human labor. However, in the central African case a "selective mechanization" approach was adopted by the Mission to increase labor by removing certain production bottlenecks. The goal of the project was to assist the Zande farmers to make an increase in productivity that should lead to more autonomy, not less, and an improved quality of life.

Potential Improvements in the Economics of the Project

It appeared that the economic viability of the distillery was highly dependent on ethanol yield per unit of feedstock. Also, because the operation was new and unfamiliar, more labor than necessary for routine operation was employed, and this had an adverse impact on the economics. Improvement of three aspects of the plant operation will help considerably: multi-stage extraction simulation, arrangements for frequent delivery of fresh yeast, and improved condenser cooling water capacity. Further, about one-third of the labor cost can be avoided by combining certain jobs, and by making a few small capital equipment improvements (1).

The operation of this small ethanol plant brought out the importance of management and coordination skills in cutting costs. The availability of the truck, fuel for the truck, trained labor, water, electricity and other resources were always subject to many constraints, and required considerable attention to detail.

CONCLUSIONS

Small-scale local ethanol production for small-scale agricultural mechanization in the Third World may have merit if certain important ecological and social impact concerns can be satisfied, as well as the need for technical and economic feasibility. We have analyzed a recent test of the concept in the Zande tribal region in the eastern part of the Central African Republic. Although technical problems caused lower than anticipated ethanol yields in preliminary tests, the problems seem to be susceptible to improvement without much additional expense. Therefore, it seems likely that economic feasibility will be demonstrated.

ACKNOWLEDGMENTS

The cooperation of the Africa Inland Mission, and its staff at Obo, Central African Republic, is gratefully acknowledged.

REFERENCES CITED

1. Adolph, David H., Agricultural Mechanization in the Central African Republic Using Renewable Fuels, D.Sc. Dissertation, Dept. of Engineering and Policy, School of Engineering and Applied Science, Washington University, St. Louis, Missouri, December 1985.

2. Adolph, David H., "An African Agricultural System Update: The Zande," Presented at the 12th Third World Conference, Chicago, Illinois, April 3-5, 1986, to be published in the Conference Proceedings.

3. Mother Earth News, "Complete Plans for Building Mother's 6 Inch Column Still," The Mother Earth News, East Flat Rock, NC, 1980.

DISCUSSION

H. Esterbrauer, U. of Graz (Austria)	Why was the decision made for ethanol and not for plant oils? You mentioned the social problems.
Shultz	The use of vegetable oils as diesel fuel extenders is certainly problematic. There are technical difficulties, and furthermore, the processing to solve these technical problems in my estimation is beyond the capabilities of a remote location in Africa. However, the easy way to get out of the question is that the missionary made the decision to go for alcohol. We didn't; we found out about it later on.
W. Bragg, Wheaton College	I don't think he drank it. Actually, they denature it by adding a little bit of gasoline. That's the cheapest way.
Shultz	The fact that it was controlled by the missionary to him meant that it was going to be okay, and he was able to sell the whole program to his hierarchy and get the money. Seriously, the gentleman does have an important question that has to be dealt with. It is a social question, and it arouses suspicion of possible corruption of those in charge of the distillery.
D. Jackson, Michigan State U.	Is there any ongoing work on distillation and the feasibility study, since it seems that it might ultimately be feasible if the missionary or your universities would continue to analyze some of the problems?
Shultz	We just consulted on that. We hope that they do continue. The problem is that the missionary who thought all of this up and who for some many years has run this vocational school and done all of these marvelous things, lost his life in an unfortunate accident recently. We're just not sure what's going to happen.
C. Setterstrom, Harrington Research Co.	How much ethanol was made by the mission station in the course of this study?
Shultz	They ran for about 1-1/2 years, but I don't know exactly how much alcohol was produced.
Setterstrom	Did they run continuously?

DISCUSSION, CONT.

Shultz	There were many ups and downs, largely due to the cooling water problem which could have been solved if they had simply had a larger diameter copper tubing flown in.
Bragg	The missionary had decided that this would be the way to go, and he encouraged people to plant sugarcane. As we said, it was an energy crop on a large-scale basis. Some people were believers, that is they felt that it would happen because he had a good success record.
Shultz	Basically, the mission was guaranteeing a market for the sugarcane, but many people just didn't believe it. So they had a shortage and it continued into the second season.
G. Shahani, Air Products & Chemicals Inc.	Could you comment briefly on the availability of capital as well as technology? For example, the still had to be imported from the States. Do you think this will evolve into technology that is indigenous to the Third World countries?
Shultz	It would have to. You couldn't possibly be importing $4600 stills into Africa from the Chicago area or wherever this was put together. I see this strictly as an experiment, and the eventual development of this if it is going to happen on a fairly large scale would have to be indigenous, at least the East African assembly. The components might be manufactured elsewhere, but the high labor involvement and assembly would have to be indigenous.
Bragg	There are very good capabilities for building these stills in Nairobi, for example. As for capital, the mission gave loans to the farmers to buy the tractors and the tractors were then going to be paid back over a period of 3 or 4 years from profits from the increased production.
Shultz	The capital equipment was donated and the credit was supplied to the farmers so that they could gradually buy the tractors.
J. Wright, SERI	What was the net energy balance on the process? After they had cultivated their sugarcane, did they have enough ethanol

left over to actually expand their other farming operations?

Shultz — Dr. Adolph did do a net energy analysis as best he could under very difficult circumstances, but I can't recall the results.

Hollomon — I was interested in what you thought the prospects were for this kind of approach - improving agriculture as distinct from what might be typical of what would happen in the Third World country where they'd grow a cash crop and simply purchase whatever fuels they would need. Recognizing that this approach is capital intensive, somebody has to provide the up-front investment.

Bragg — If you look at a map of Africa, the Central African Republic, and at Oboe, it's about as far away from any market that you could ever get in Africa. So as far as exporting, I agree with you that it's probably better not to become dependent on the next work crop anyway. There's no major river going into Oboe, the roads are impassible most of the year during the rainy season, and the airplane can't carry large cargoes.

Shultz — I hope you can see that everyone of these sites really has its own characteristics and has to be analyzed separately. Oboe is just one isolated area, and it has its peculiarities. It is very difficult to come up with a generic plan for providing fuel. It's tempting at Oboe because they're paying over $5 a gallon for imported gasoline. That's a very attractive incentive for doing something. Draft animals would be the obvious solution if cultural problems could be overcome. But draft animals have been introduced unsuccessfully in the past.

Bragg — The reasons for that of course go back to their traditional hunting and gathering traditions. The men, rather than taking care of draft animals the year around, would rather be off hunting in the woods.

Clyde — As you know, the big cost in distillation is getting rid of the last 10% of water because of the azeotrope problem. Did I understand you to say that you could use 180 proof alcohol in tractors?

Shultz

The modifications were made by the manufacturer in the United States, and as I understand it they simply put in different gaskets in a few places and changed the carburetion. There was no attempt to go to absolute alcohol; that would have been prohibitively expensive. They found that 170 proof worked quite well in comparisons that they carried out. They compared the two tractors on the same area of land. One burned gasoline, the other burned 170 proof alcohol. Dr. Adolph examined the energy requirements for both duties; the alcohol was 6% more, which was interesting. There were no problems with running the tractor.

REGIONAL AND COMMUNITY APPROACHES TO METHANE FROM BIOMASS AND WASTE: AN INDUSTRY PERSPECTIVE

Peter H. Benson, Ph.D.
President, Renewable Energy System, Inc.
Palos Park, Illinois 60464

Thomas D. Hayes, Ph.D.
Project Manager, Biomass
Ron Isaacson, Ph.D.
Manager, Renewable Resources
Gas Research Institute
Chicago, Illinois 60631

ABSTRACT

Biomass and waste-to-energy systems, although not a panacea, offer the potential to provide secure supplies of cost-competitive energy while mitigating monumental agricultural and waste treatment/disposal problems. The socio/economic and environmental attributes of such systems often outweigh their intrinsic value as energy resources. These positive aspects must be considered when critically comparing energy alternatives. The authors have focused on major benefits while developing potential pathways to commercialization of two renewable energy systems: "Regional Energy Farms" and "Integrated Community Waste/Energy Systems", both based upon the production of methane from biomass and waste feedstocks via anaerobic digestion.

Lack of adequate food like a lack of energy is not necessarily a problem of supply, but is often a problem of distribution and the absence of a positive infrastructure. Surplus crops have been and are becoming a major problem worldwide, particularly in developed countries like the U.S. where a tremendous federal expenditure is used to "shore-up" market prices. Prime cropland has been "set-aside" to no avail. This burgeoning problem can be alleviated by utilizing crop surpluses or by growing crops specifically for the production of cost-competitive methane, thus giving the farmer an alternative market and breaking the vicious surplus crop - depressed prices - overproduction cycle. Prime cropland can be productively utilized through regional energy farms and the misperceived need to grow energy crops on "marginal" land is no longer an issue.

The treatment and disposal of wastes by conventional methods also poses many problems throughout the world and results in the non-productive consumption of energy. In many areas, landfills are approaching capacity and closing and land for the disposal of wastes is no longer available. Costs to dispose of wastes, particularly municipal solid wastes (MSW) and sewage sludge, are escalating. The two alternative technologies discussed reduce the amount of MSW and sludge for disposal by 80%, treat wastestreams at a reduced cost, yield a net production of low-cost methane (less than $2/MMBtu) and result in significant savings in operating costs through energy conservation. By combining both these technologies into an integrated community waste/energy system, 20-25% of a communities residential gas demand can be met. Ultimately, the addition of biomass from a regional energy farm to this centralized facility could supply all of a communities gas demand.

REGIONAL AND COMMUNITY APPROACHES TO METHANE FROM BIOMASS AND WASTE: AN INDUSTRY PERSPECTIVE

INTRODUCTION

In 1984 natural gas supplied 17.9 Quads (18 EJ) or roughly 23% of the primary energy consumed in the U.S. (Figure 1) compared to 17.2 and 31.1 Q (18 & 33 EJ) from coal and petroleum, respectively (Holtberg et al 1985). "Renewables", mainly wood and wood wastes, provided more primary energy, 4.3 Q (5 EJ), than either nuclear, 3.6 Q (3.8 EJ), or hydroelectric, 3.7 Q (3.9 EJ) that same year.

A similar renewable energy use pattern is occurring worldwide. "Biomass energy" supplied 7.5% of Canada's end-use energy demand in 1983, twice as much as coal and three times as much as nuclear, although federal incentives for developing bioenergy were lower than any other energy resource (McCormick 1984). Sweden is considering the possibility of replacing 1/2 of its primary energy, nuclear, with biomass and waste by the year 2010 (Johansson 1986).

Although bioenergy usage continues to increase significantly every year, from 2.7 Q (2.8 EJ) in the U.S. in 1981 (Haggin and Krieger 1983) to 4.3 Q (5.0 EJ) in 1984 (an increase of 59% in 3 years), there is no major biomass-to-energy infrastructure and prevailing attitudes

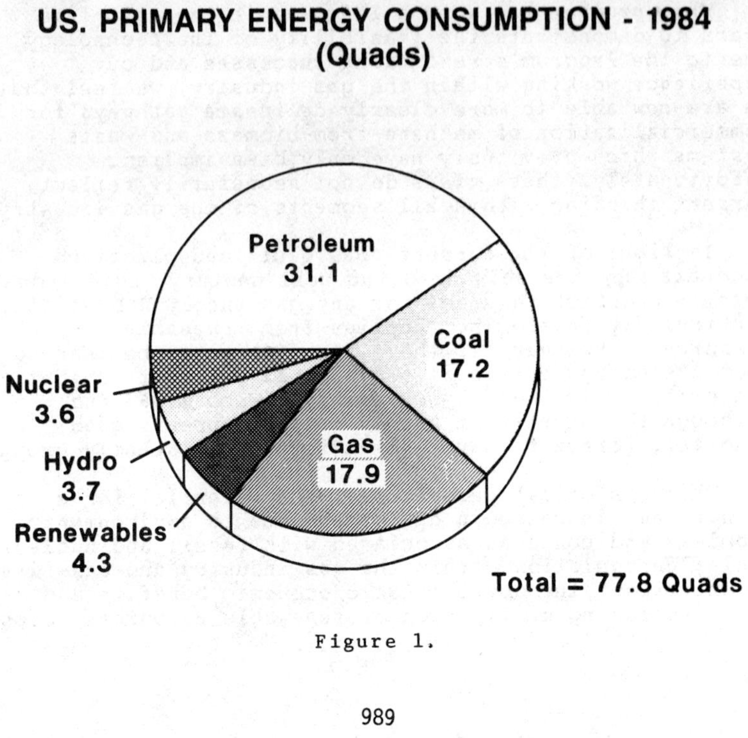

Figure 1.

against developing bioenergy technology still exist (Holden 1985). One possible reason why the technology has been downplayed is that, although the benefits of bioenergy have been established, a clear strategy for large-scale energy production, particularly gaseous fuels from biomass-and-wastes, has not emerged. Much of the present bioenergy growth in the U.S. can be attributed to relatively small, 5-50 megawatt electrical generating facilities which are primarily burning biomass-and-waste feedstocks. It has been estimated that 3,000 megawatts of capacity will be provided by such existing and planned facilities in 1984 (Easterly and Saris 1984). These small facilities provide the flexibility to respond rapidly to unpredicted changes in load growth and reduce technical and financial risk.

The use of renewables for electricity production is a cause of loss in gas demand. The Gas Research Institute (GRI) estimates that the use of "renewables" will increase from the previously stated 4.3 Q (5 EJ) in 1984 to 9.4 Q (9.9 EJ) by 2010. Assuming that supplementary supplies of gas are not provided by GRI's on-going R&D supply program, it has also been estimated that an additional 8-9 Q (8.4-9.5 EJ) of gas, or approximately 40% of the gas demand forecast in 2010, will be required then due to declining natural gas production in the lower 48 states (Ashby et al 1985). As a result of these projected needs, an ambitious program to provide supplementary supplies of pipeline quality methane from biomass-and-waste feedstocks was implemented at GRI in 1981.

This program has gone a long way in the last five years to demonstrate the feasibility of the technology. Due to the Program's resounding successes and our experience working within the gas industry, we feel that we are now able to more clearly delineate pathways for commercialization of methane-from-biomass-and-waste systems which previously have only been implied. Unfortunately, these views do not necessarily reflect current thinking within all segments of the gas industry.

In light of the current "gas glut" and perceived adequate supplies well into the next century, some industry leaders question the need for any gas supply R&D at all, particularly developing supplies from renewable resources. However, if there is anything to be learned from the energy crisis of the past 12 years it is that "uncertainty is the only certainty" (Webb 1986) and although the short-term supply outlook appears good, long-term forecasts have proven to be consistently wrong.

This historical perspective; the potential for a significant increase in gas demand due to environmental problems and concerns associated with fossil and nuclear fuels; deregulation within the gas industry and gas-fired cogeneration; the obvious socio/economic benefits and ever-increasing utilization of renewable resources across the U.S. and in the world, all indicate that now is the time to implement commercialization strategies to establish this technology in the marketplace.

GRI'S METHANE FROM BIOMASS AND WASTES PROGRAM

This program and its results have been discussed extensively (Benson et al 1984a, Benson 1985, GRI 1984). Initially based upon an engineering oriented, marine biomass project, the program was evaluated and restructured in 1981-82 to provide low-cost gas from wastes in the near-term, cost-competitive gas from biomass and waste mixes in the mid-term and biomass in the longer-term, as outlined in Figure 2 (after Benson 1985) which indicates that 10 Q (10.5 EJ) of methane could be derived from these resources by the year 2020.

As indicated, it was assumed that approximately 1.0 Q (1.1 EJ) of supplementary gas could be supplied by the year 1995. This is an ambitious but not totally unrealistic goal in light of the costs of other energy technologies. Multiple facilities to achieve this would cost approximately $5 billion per year which is comparable to, but less than, the Construction Grants Program in the 1960's. From another perspective, the digester capacity to treat the 5.3 million dry tons (4.8 billion Kg) of

POTENTIAL METHANE CONTRIBUTION FROM BIOMASS AND WASTE*

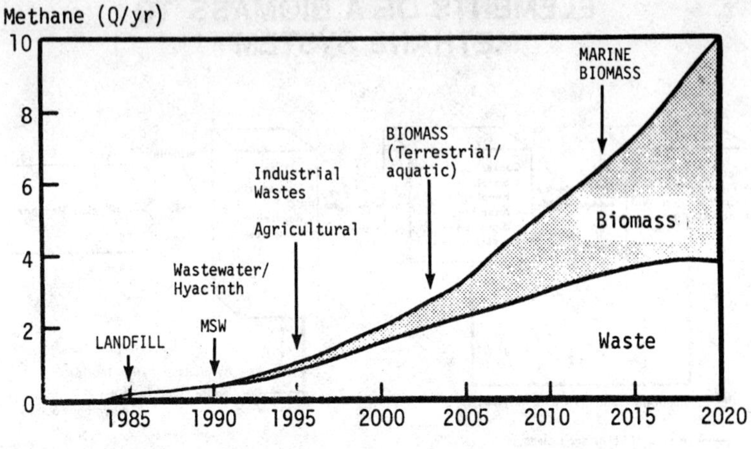

*Year of Commercialization Indicated by Arrow for Each Resource.

Figure 2.

sewage sludge produced annually in the U.S. would need to be tripled, either by the addition of other wastes or biomass feedstocks, to achieve the 0.1Q (0.1 EJ) annual growth curve to 1995.

Based upon our experience in implementing the "Methane from Biomass and Waste Program" and its technical successes, two distinct, but ultimately interrelated, strategies for commercialization evolved; 1) "Regional Energy Farms" and 2) "Integrated Community Waste/Energy Systems". The former concept is the result of a regional approach to developing biomass programs pioneered by GRI (Benson et al 1984b). The latter is the result of recent successes with two waste treatment/energy production projects underway in Florida (Hayes and Isaacson 1986, Mooij 1986).

REGIONAL ENERGY FARMS

The dispersed nature of biomass and waste feedstocks make these renewable energy resources ideally suited to regional utilization with site-specific application dependent upon local resources, energy demand and costs. Accordingly, in late 1981 GRI pioneered a regional approach to the design, development and evaluation of methane from biomass-and-waste systems with the University of Florida's Institute of Food and Agricultural Sciences (IFAS). The research conducted in this comprehensive, multidisciplinary program was focused on developing the critical steps in an integrated biomass-to-energy system (Figure 3) from feedstock selection to gas-cleanup.

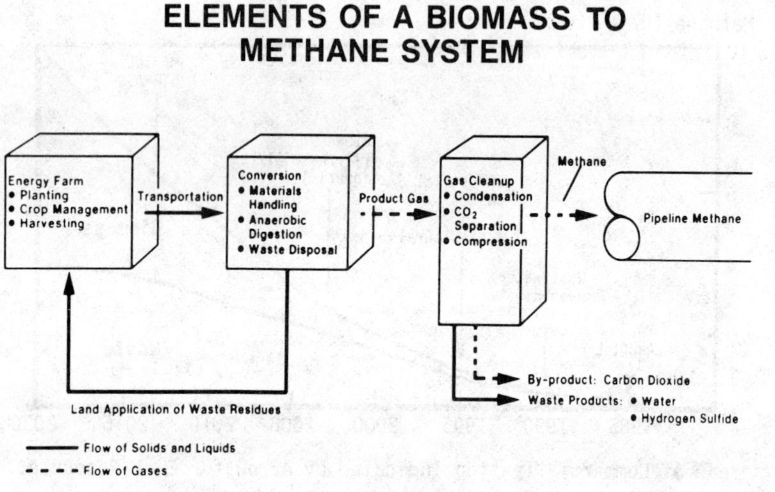

Figure 3.

Each component is interrelated and affects the overall technical feasibility and system economics. Anaerobic digestion (bioconversion) was selected as the energy conversion process since it is applicable to a wide range of feedstocks based upon moisture content and it produces a high methane concentration, typically around 50% which can be manipulated upwards to 90+ % (Isaacson et al 1984). Perhaps of greater benefit is that bioconversion is a non-polluting, environmentally benign process from the standpoint of air quality and waste disposal. The effluent and digester residues are rich in nitrogen and phosphorous which can be recycled back to the soil as fertilizer and to maintain soil tilth.

The establishment of the unique GRI/IFAS Regional Program naturally led us to the concept of "Regional Energy Farms". Obviously, the concept of raising crops exclusively for energy production is not a new idea; however, the many benefits to be derived by this approach have not been fully explored or appreciated. Often the misconception that energy crops will compete with food, fibre and other societal needs and must therefore be grown on marginal land prevails. There is an ironic analogy between food and gas supply. Much of the world food problem is not one of supply but deliverability and other infrastrucure problems (Hall 1984, Avery 1985). Similarly a major problem with gas supply is what can be delivered when demand is at peak. Surplus gas supplies in Texas where virtually non-existent during periods of peak demand recently (Wallace 1984) and when pipelines are operating at peak capacity during the winter load, more gas being produced "downstream" will not improve the situation "upstream" - we need both "upstream" as well as "downstream" supplements (Bonan 1985)

Of greater concern, however, is the growing "World Farm Crisis" (U.S. News & World Report, May 19, 1986) which is "potentially the most serious threat to the world trading system". Big producers such as the U.S. and Europe; as well as many developing nations including China, India, Argentina, Thailand and Brazil have too much food for available markets at home or abroad. Crop surpluses depress prices which motivates the farmer to increase production/acre, further exacerbating the surplus problem. As illustrated by Figure 4, this vicious cycle of agricultural overproduction could be broken by diverting crop surpluses into energy production while alleviating the problem of energy supply uncertainty. The farmer would have an alternative market for his product, an insured return on his investment and both the agricultural and energy sectors of this nation and the world would have increased security and flexibility to respond to unexpected changes.

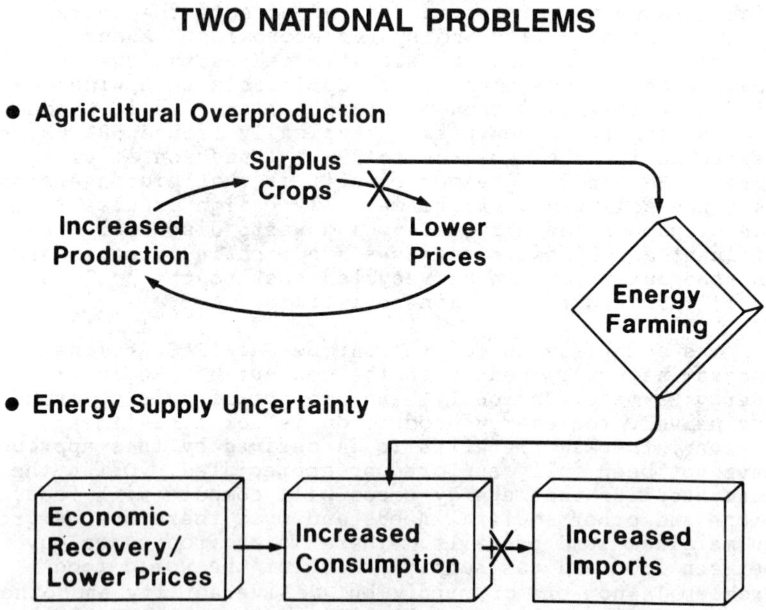

Figure 4.

The potential for methane production from energy crops in the U.S. is illustrated in Figure 5 which compares prime cropland area and biomass yields to energy production potential, assuming a conservative bioconversion efficiency of 60% (efficiencies of over 90% have recently been obtained in the GRI program) and a methane yield of 4.5 SCF/lb volatile solids added (0.27 m^3/kg) to the digester.

At an average biomass yield of 20 dry tons/acre/yr (44.8 tonnes/hectare/yr), 30 million acres (12 million hectares) or less than 10% of the prime cropland in the U.S. would be required to produce nearly 5Q (5.2 EJ) of methane annually, or roughly 25% of the yearly U.S. natural gas demand. Typically 30-40 million acres of prime cropland are set-aside in the U.S. annually and over 80 million acres was set-aside during the PIK (Payment-In-Kind) Program in 1984 in an attempt to relieve the U.S. crop surplus problem. An immense benefit to the nation would be the elimination of the set-aside programs. Energy crops would provide a mechanism for the agricultural industry to compete for a new market and reduce the burgeoning need for the federal government to pay for non-production. The perception that we need to grow energy crops on marginal land is no longer valid other than for erosion control. If bioenergy production is considered on a regional or statewide basis, for example in the State of Florida (Figure 6), there is plenty of land available to satisfy all of that State's natural gas demand, again assuming rather conservative bioconversion efficiencies.

Energy Production Potential from the Anaerobic Digestion of Biomass

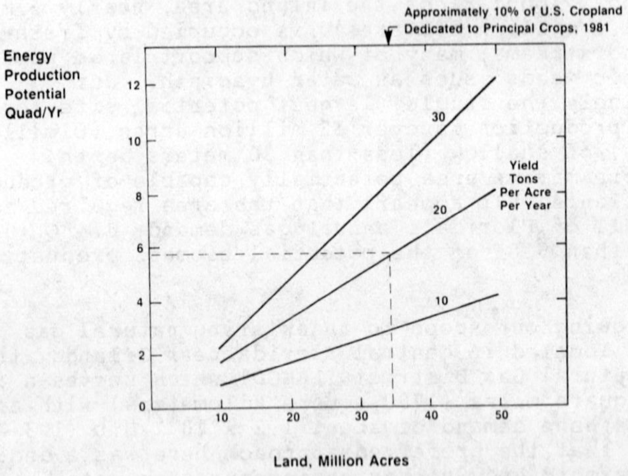

Assumption: Conversion efficiency is 60% with a net energy yield of 4.5 ft³ of methane per pound of biomass added to the anaerobic digestion process.

Figure 5.

Utilization of Florida Biomass Production Areas Required to Satisfy the Total Annual Methane Demand.

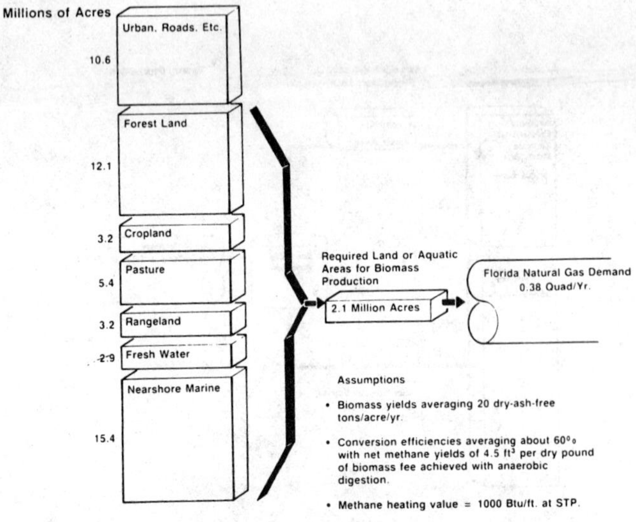

Figure 6.

More than 15 million acres (6 million hectares) in Flor is classified for agricultural crop production and forestry enterprises (Terhume 1983). An additional 8 million acres (3.5 million hectares) are classified as pasture and rangeland. About 8% of the inland area, nearly 3 million acres (1.3 million hectares), is occupied by freshwater lakes and streams, many of which support large populations of aquatic weeds, such as water hyacinth. Not surprisingly the single, largest potential site for biomass production is over 15 million acres (6 million hectares) of shallow (less than 30 meters depth) near-shore marine area potentially capable of producing marine plants. It appears that the area required to supply all of Florida's annual gas demand, 0.4 Q (0.4 EJ), is less than 5.0% of the potential biomass production area available.

Narrowing our scope to an existing natural gas district located in central Florida near Orlando, the Lake Apopka Natural Gas District (LANGD) which serves a region of 680 square miles (1700 square kilometers) with an annual methane demand of about 1.2×10^{12} Btu (1.3 PJ), revealed that the preferred approach here was a dedicated energy farm to grow either napier grass or water hyacinth (Warren et al 1984). Results of that study (Figure 7) indicated that as much as 31 square miles (20,000 acres or 8,100 hectares) of land was available for napier grass production and 51 square miles (32,600 acres or 13,200 hectares) of hypereutrophic Lake Apopka could be used for growing water hyacinth.

Regional Biomass Availability to Satisfy the Lake Apopka Natural' Gas District Methane Demand.

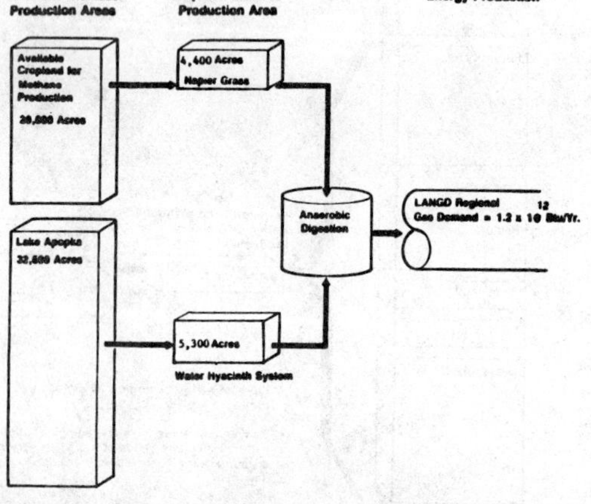

Figure 7.

The potential to supply all LANGD methane demand with 5300 acres (2100 hectares) of water hyacinth or 4400 acres (1800 hectares) of napier grass at gas costs of $5-6/MMBtu ($4.80-5.75/GJ) with advanced conversion systems sized to deliver 1.08 trillion Btu/yr (1.1 PJ/yr), was revealed. Other studies have corroborated this economy of scale at a relatively low methane output (1-3 trillion Btu/yr, 1-3.2 PJ/yr) at conversion facility costs of from $10-30 million (Figure 8). Consequently the technology is available to a wide range of investors. Recent technical improvements within the GRI program, 90+% conversion efficiency and methane yields of 5.5-6.5 SCF/lb VS added (0.34-0.41 m^3/kg) with poplar and sorghum feedstocks (Jewell et al 1986) and high biomass yields of up to 27 dry tons/acre-year (61 metric tons/hectare/year) with napier grass in small test plots (Benson 1985) are reducing gas cost further from the original goal of $5-6 MM Btu, delivered at the city gate, making the technology even more attractive.

The status of energy crop farming for methane production therefore appears most promising although, at current depressed gas costs, these systems are not yet cost-effective. However, they are competitive with and offer distinct advantages over other gas supply options and alternative energy producing technologies. If mid-Btu gas can be considered and waste is mixed with the biomass, as will be discussed later, the concept can be cost-competitive now. Indeed, if mid-Btu gas can be

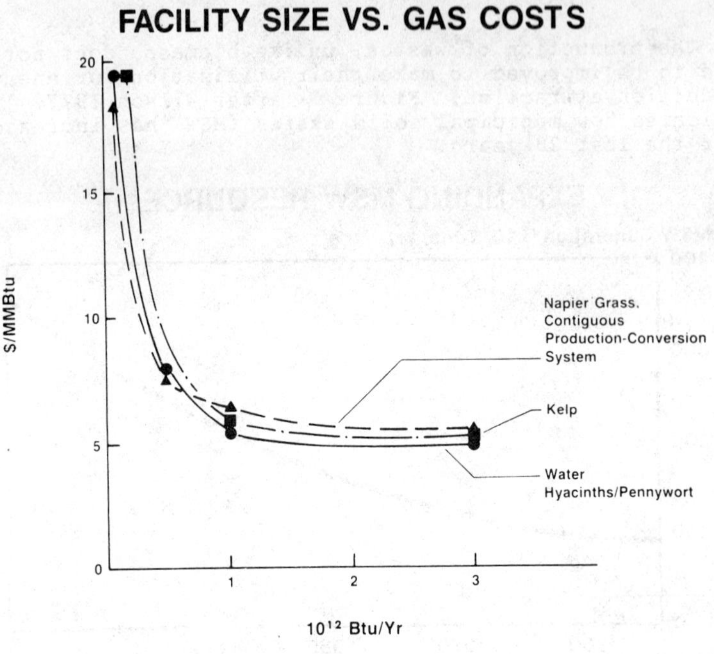

Figure 8.

utilized on-site, these systems offer distinct advantages, particularly when input into fuel cells or gas-fired cogeneration to make electricity under PURPA, or for direct sale to an adjacent industrial user.

When the socio/economic advantages offered by regional energy farms are considered, the concept becomes of overwhelming benefit to the nation, the region and the community. It has been estimated that by doubling the biomass energy utilization in the Great Lakes area between now and 1995 that 42,050 to 54,000 jobs would be created and $43-57 million in state tax revenues would be generated annually, taking into account the number of jobs and income lost due to displaced fuels (Employment Research Associates 1985). This not only stimulates the economy but enhances the area's ability to withstand energy price shocks and supply shortfalls associated with fossil fuels. Therefore, without stressing the importance of resolving the "surplus crop" problem in the nation and the agricultural industry throughout the world, the concept of regional energy farms would be of benefit to many regions and nations now. Perhaps more timely and urgent for some communities to consider, however, are systems which integrate waste treatment/disposal with energy production to alleviate these two problems simultaneously.

INTEGRATED COMMUNITY WASTE/ENERGY SYSTEMS

The production of wastes, unlike biomass, does not need to be improved to make their utilization for energy production attractive. Figure 9 (after Wilson 1977) indicates how municipal solid wastes (MSW) has increased over the last 20 years.

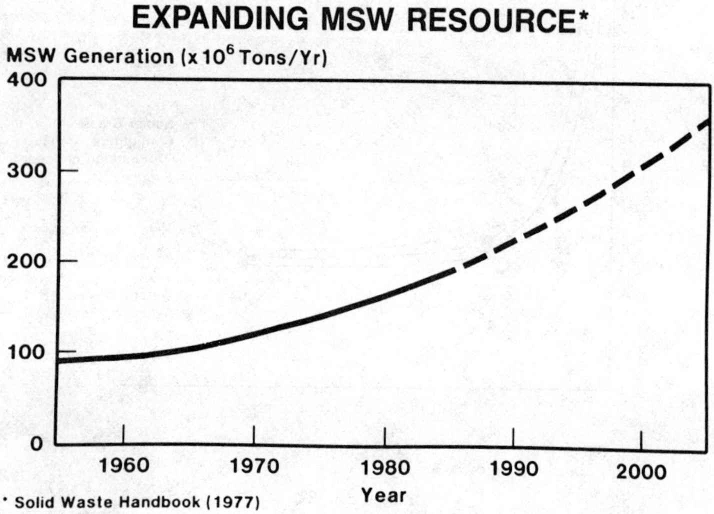

Figure 9.

Indeed, wastes and their disposal are becoming a major problem throughout the world, particularly in urban areas of developed countries. Currently 900 to 1000 million tons of recoverable wastes are produced annually in the U.S. This resource, which doubles approximately every 20 years, could produce 12-15 Q (12.6-15.8 EJ) of methane annually. More realistically, if 50% of this resource were converted to methane; excluding wood waste, the predominate waste effectively utilized now, and landfill gas which is presently under development, it could provide approximately 4 Q (4.2 EJ) of methane annually if efforts are made to capture the resource now (Figure 10).

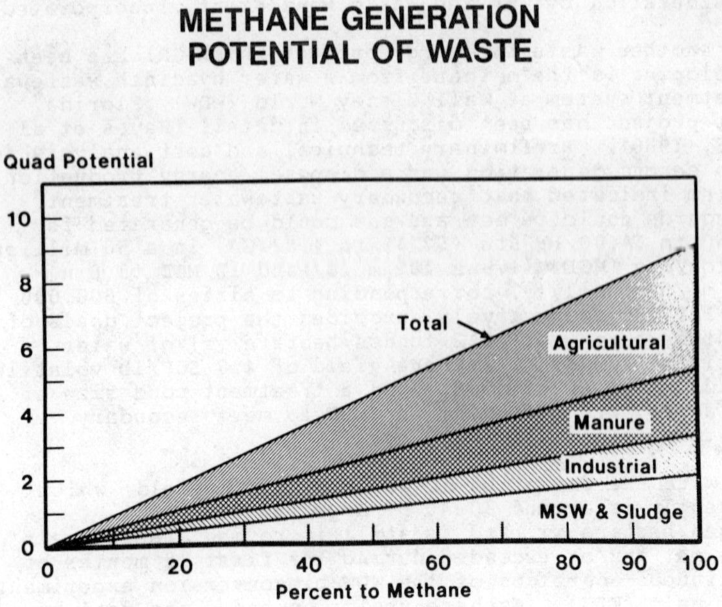

Figure 10.

Pipeline quality gas is presently being produced from landfills for $4-5/MMBtu ($3.80-4.80/GJ) albeit in small but increasing quantities (Rice 1985). As landfills close across the U.S. and tipping fees continue to escalate, the average tipping fee increased by 13% in 1985 (Johnson & Pettit 1986), alternative ways of disposing of MSW will be required. Presently mass burn systems are receiving the most attention; however, air emission problems and economy-of-scale make these systems unsuitable for many locations in the U.S. and they are coming under increasing attack by concerned citizen groups, (Resource Recovery Report 1986).

One alternative to mass burning is "RefCoM", a system where MSW and sewage sludge are fed into a digester. This first generation waste disposal/energy production system has been described (Sjoberg 1984) and the feasibility of the technology has been demonstrated (Pfeffer 1986). Results indicate that a 200 ton per day (TPD) facility; which is equivalent to MSW from a city of 100,000 assuming an average 4 lbs/day of refuse per capita, (5.5 lbs is more realistic for Illinois), would reduce the volume of material requiring landfilling by over 80% (from 500 to 60 yd^3) and produce approximately 800 MCFD (840 GJ/d) of methane for sale at $4-5/MMBtu, ($3.80-4.80/GJ) at a tipping fee of $25/ton (Edwards and Wilkey 1985). The design and construction of such a facility is now under consideration by GRI and Waste Management, Incorporated.

Another waste-to-energy project which GRI has been developing is the methane from a water hyacinth wastewater treatment system at Walt Disney World (WDW), Florida. This project has been discussed in detail (Hayes et al 1985, 1986). Preliminary technical and cost analysis for this second generation waste disposal/energy production system indicated that secondary wastewater treatment standards could be met and gas could be generated for $2.50 to $4.00/MM Btu ($2.41 to 4.82/GJ) in a 50 million gallon/day (MGD) (1.9×10^5 m^3/d) and 10 MGD (3.8×10^4 m^3/d) facility, corresponding to cities of 500,000 and 100,000 respectively, provided the project goals of 50 dry tons/acre/year (112 tonnes/hectare/yr) of water hyacinth biomass, a methane yield of 4.7 SCF/lb volatile solids added (0.29 m^3/Kg) and a treatment pond size of 7.0 acres (2.8 hectares) per MGD to meet secondary treatment standards, were met.

With the exception of water hyacinth yield, which presently is around 30-40 DT/A-yr (67-90 tonnes/hectare-yr) and is still improving, these goals were met and/or exceeded during the first 18 months of continuous operation of the WDW bioconversion experimental test unit (ETU). Methane yields exceeded the goal by 40-70% to 6.6-7.9 SCF/lb VS (0.41-0.49 m^3/Kg) dependent upon whether the ETU was run in an upflow or downflow mode. This reduces gas cost to below $2/MMBtu ($1.90/GJ) for cities of 100,000 or greater and indicated that gas costs of around $3/MMBtu ($2.85/GJ) for cities down to 10,000 inhabitants were possible.

By combining these two systems, RefCoM and the Walt Disney World ETU, we developed an "Integrated Community Waste/Energy System" strategy where 20-25% of the residential gas demand of a city can be generated from its wastes. Assuming a city of 100,000 has a central sewage treatment energy production system which receives its sewage, 200 TPD(182 tonnes/d) of MSW and 100 TPD (91 tonnes/d) of industrial waste which totals 150 dry ash-free tons per day to the digester, at a conservative conversion efficiency of 70%, and a methane yield of 4.5 SCF/lb VS (0.28 m^3Kg), this would generate 1.4 MMSCFD (1.42 TJ/d) or roughly 20% of the residential gas demand. This is a significant portion of any communities energy needs, at low-cost with the added socio/economic and environmental benefits.

The next logical step is to combine biomass from regional energy farms with the community waste concept to supply all of the communities residential gas demand as depicted in Figure 11.

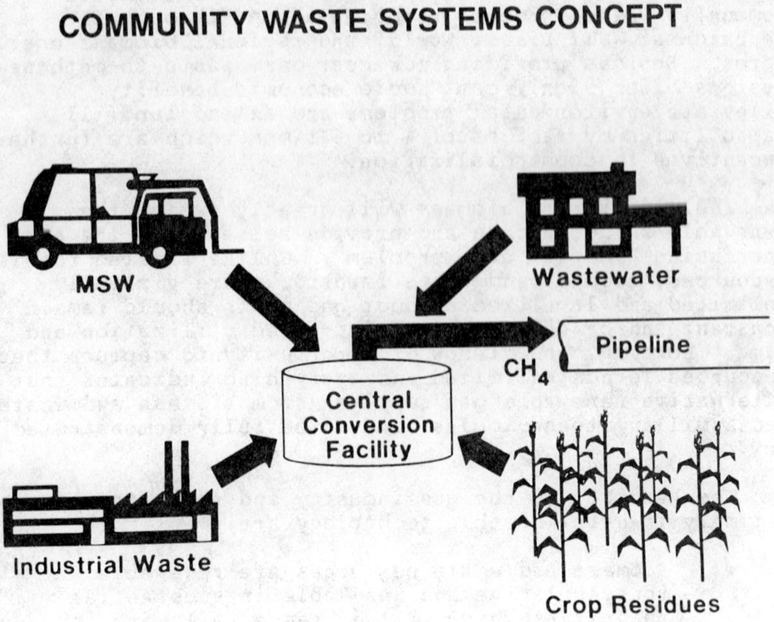

Figure 11.

At a biomass yield of 20 DT/A-yr (45 tonnes/hectare-yr) and a methane yield of 5.5 SCF/lb VS (0.34 m^3/kg) it would require 9,600 acres (3900 hectares) of energy farm, or 16 square miles of energy crops, to provide 5.8 MMSCFD (6.1 TJ/d) of methane or the remaining 80% of the communities residential gas demand. The cost of the biogas produced would be reduced due to a common facility and mixing of low-cost gas from wastes with cost-competitive gas from biomass.

CONCLUSIONS

Hopefully, we have more clearly delineated potential commercialization pathways for methane from biomass and waste systems. Energy derived from gaseous fuels is the cleanest form of hydrocarbon energy usage today and the gas industry should be prepared for a possible increase in gas demand if acid rain and other constraining environmental legislation is passed. Presently most renewable resources are used to produce other forms of energy with a resultant negative impact on gas demand. The RefCoM technology provides a near-term alternative to competing energy forms such as steam and electric power. Further market penetration can be provided by second generation conversion systems such as the integrated community waste concept, being developed through ETU research at Walt Disney World, and regional biomass energy farms. Besides providing low-cost gas, waste-to-methane systems offer significant socio/economic benefit, alleviate environmental problems and extend landfill capabilities by factors of 4 to 6 times which are further incentives to commercialization.

The addition of biomass will greatly expand the renewable resource base and provide relief from the ever increasing "surplus crop problem". Unlike limited fossil resources, biomass and waste feedstocks are virtually unlimited and levelized product gas costs should remain constant and/or decrease with increased utilization and time. However, the window of opportunity to capture these resources is not unlimited and everything indicates that alternative renewable gas supplies from biomass and wastes and ancillary technologies need to be fully demonstrated now.

The benefits to the gas industry and the ratepayer in a timely transfer of this technology are:

- Biomass and waste resources are renewable, nonspeculative and available in substantial quantities which will increase, allowing for continuous expansion of biogas production capabilities. Unlike limited fossil resources, end-product gas costs should hold constant or decrease with time and increased utilization.

- The production of regional, low-cost and cost-competitive gas from these resources should dampen future price increases for natural gas nationwide.

- Both waste and biomass gasification facilities can be commercialized within a relatively short time-frame and without large capital outlays.

- Such facilities offer local utilities an opportunity to control a portion of their gas supply; to increase system reliability, deliverability and user confidence and help regain and maintain industrial load.

- The huge investment in gas pipeline and distribution systems merits expenditure on projects which promise to provide significant quantities of low-cost, supplemental gas in advance of actual need. Performing this R&D now is more productive and cost-effective than during periods of supply shortage.

- Biogasification of wastes provides an innovative alternative to waste treatment/disposal which is environmentally compatible; indeed, as landfills close, it appears to be a most promising, cost-effective means of disposing of municipal solid waste and sludge.

Reference List

1. Ashby, A., P. Holtberg and T. Woods. 1985. "Comparison of the 1984 DOE/EIA Annual Energy Outlook and the 1984 GRI Baseline Projection". Gas Research Institute, Chicago, IL, June.

2. Avery, D. 1985. "U.S. Farm Dilemma: The Global Bad News Is Wrong". Science, 230.

3. Benson, P.H., J.R. Frank and H.R. Isaacson. 1984a. "Gas Research Institute Programs on Methane from Biomass and Wastes". Symposium Papers, Energy from Biomass and Wastes VIII. Institute of Gas Technology, Chicago, IL.

4. Benson, P.H., H.R. Isaacson and J.R. Frank. 1984b. "A Regional Approach to Producing Methane from Biomass and Wastes". 19th Annual Intersociety Energy Conversion Engineering Conference(IECEC) "Advanced Energy Systems - Their Role in Our Future" San Francisco, CA.

5. Benson, P.H. 1985. "GRI Programs on Methane from Biomass and Wastes". (in) Energy Applications of Biomass, M.Z. Lowenstein (ed.). Elsevier Applied Science Publishers, London and New York.

6. Bonan, J. 1985. Brooklyn Union Gas Co., Brooklyn, N.Y. Personal communication.

7. Easterly,J.L. and E.C. Saris. 1984. "A Survey of the Use of Biomass as a Fuel to Produce Electric Energy in the United States". Public Utilities Fortnightly, April 26.

8. Edwards, G.N and M.L. Wilkey, 1985. "RefCoM Economics -- Two Perspectives". Proceedings, Bioconversion: A new Technology for Waste and Sludge Disposal. Argonne National Laboratory, Argonne, IL, ANL/CNSV-TM-174.

9. Employment Research Associates. 1985. "Biomass Resources: Generating Jobs and Energy". Prepared for the Great Lakes Regional Biomass Energy Program, Council of Great Lakes Governors.

10. Gas Research Institute Brochure. 1984. "Methane from Biomass and Wastes Research : Renewable Resources for Localized Energy Production".

11. Haggin, J. and J.H. Krieger, 1983. "Biomass Becoming More Important in U.S. Energy Mix". C & E News, March 1983.

12. Hall, D.O. 1984. "Biomass: Fuel Versus Food, A World Problem?" in Economics of Ecosystem Management. M.S. Margaris (ed.) Publisher Dr. W. Junk, Netherlands.

13. Hayes, T.D., D.P. Chynoweth, K.R. Reddy and B. Schwegler, 1985. "The integration of Biogas Production with Wastewater Treatment" (in) Energy Applications of Biomass, M.Z. Lowenstein (ed.) Elsevier Applied Sciences Publishers, London and New York.

14. Hayes, T.D., H.R. Isaacson, D.P. Chynoweth, R. Biljetina, K.R. Reddy, 1986. "An Integrated Wastewater - Energy Production System", Energy from Biomass and Wastes X, Washington, D.C., April 7-10.

15. Hayes, T.D. and R. Isaacson. 1986. "Community Waste to Methane: Development & Potential". Energy Technology 13 Conference, Wash. D.C., April 7-10.

16. Holden, C. 1985. "Is Bioenergy Stalled?" Science 227.

17. Holtberg, P.D., T.J. Woods and A.B. Ashby. 1985. "1985 GRI Baseline Projection of U.S. Energy Supply and Demand to 2010". Gas Research Insights. Nov. 1985

18. Isaacson, R., T.D. Hayes and D.P. Chynoweth. 1984. "Pipeline Quality Methane from Biomass and Wastes". Proceedings Energy Technology 11 Conf. Wash. D.C.

19. Jewell, W.J., R.J. Cummings, A.M. Foster and B.K. Richards. 1986. "Engineering Considerations for Methane Fermentation of Energy Crops". Annual Report. Gas Research Institute, Chicago, IL, GRI-86/0131.

20. Johnson, C.A. and C.L. Pettit. 1986. "NSWMA's 1985 Tipping Fee Survey". Waste Age, March 1986.

21. Johansson, R. 1986 Swedish State Power Board, Vallingby, Sweden, personal communication.

22. McCormick, R. 1984. "Canadian Energy Supply and Demand 1985-2005". Bio-Joule, Nov., 1984.

23. Mooij, P. 1986. "RefCoM Commercialization Program". Energy Technology 13 Conference, Wash. D.C., March 7-19, 1986.

24. Pfeffer, J.T. 1986. "Evaluation of the RefCoM Proof-of-Concept Experiment". Proceedings, Energy from Biomass and Wastes X. Washington D.C., April 7-10, 1986.

25. Resource Recovery Report. 1986. "NYPIRG Shatters the Incinerator Myth", March 1986.

26. Rice, F.C. 1985. "The Landfill Gas Industry: Overview and Analysis". Waste Age, August 1985.

27. Sjoberg, H.T.D.,1984. "Anaerobic Digestion of Municipal Solid Waste: Potential Market and Implications". International Gas Research Conference, Washington, D.C., Sept 1984.

28. Terhume, F.W. 1983. (ed.) "Florida Statistical Abstracts". The University of Florida Press.

29. US News and World Report, 1986. "The World Farm Crisis". May 19, 1986.

30. Wallace, M. 1984. Chairman, Texas Railroad Commission, Inside FERC, Feb 13, 1984

31. Warren, C.S., et al. 1984. "Evaluation of Biomass to Methane Systems in the Lake Apopka Natural Gas District". Final Report, Gas Research Institute, Chicago, IL, GRI 84/0015.1.

32. Webb, D.O. 1986. Testimony before the Subcommittee on Health and Environment, Energy and Commerce Committee, U.S. House of Representatives, April 29.

33. Wilson, D.G. (ed). 1977. Handbook of Solid Waste Management, Von Nostrand Reinhold, Co., N.Y., N.Y.

DISCUSSION

D. Klass, IGT — When I examined the Black & Veatch numbers and reports, I don't think they took credit for the energy production tax credit. If they didn't, you could reduce the gas price another 79¢ per million Btu based on mid-1985 crude oil prices. And I think this credit would probably be even higher now since the price of crude is much lower. Do you remember if they took credit for the production tax credit?

Benson — I don't believe they did. When we do our economics, we don't take any tax incentive benefits at all. We also don't take benefits for co-products and by-products which can significantly lower product gas costs.

Klass — This is the tax credit I mentioned Monday that hardly anybody knows about with the exception of the people in this room. I hope we all know about it now because it's $3 per barrel of oil equivalent adjusted for inflation. I don't recommend that you base a project on tax credits to show a profit, but if it's there, you might as well use it.

J. Herberman, Ontario Ministry (Canada) — Have you given any thought at this stage to the barriers that you're going to have to overcome to integrate methane production with the existing natural gas industry?

Benson — If you can make supplemental gas for less than the going cost of gas, I would assume that you can sell all that you can make. Every farmer has told us just give me a return on investment, and I'll grow any energy crop for you, as much as you want. So I don't think that's a big problem. We see a lot of agricultural industries right now which are hurting, and they're looking for alternative ways of utilizing their crops.

Klass — About 2 months ago, you mentioned the commercialization of sewage treatment by water hyacinth treatment in Orlando, Florida. Some firm was applying some of this technology and scaling it up to incorporate it into a wastewater treatment plant. Do you have any information on that, what was the name of

	the firm that started this project, and where are they located?
Benson	It's being done at the Iron Bridge plant at Orlando. Essentially, they built a 30-ac water hyacinth pond, and a 250,000-gal vertical flow reactor, based on the ETU reactor for digestion of sludge and water hyacinth at Walt Disney World.

METHANOGENS IN DIGESTERS: IDENTIFICATION AND QUANTIFICATION IN MICROSAMPLES BY DIRECT TESTING WITH ANTIBODY PROBES

Alberto J.L. Macario, M.D. & Everly Conway de Macario, Ph.D.
Wadsworth Center for Laboratories and Research
New York State Department of Health
Albany, New York 12201

ABSTRACT

Elucidation of the methanogenic flora in digestors will help to understand and optimize fermentation. Unfortunately, simple means for identifying methanogens are scarce. Usual enrichment-isolation procedures take time and are likely to alter the original bacterial distribution. We developed immunologic tools for rapid identification of methanogens by direct examination of samples. Bacterial cells, and antigens released by them into the fluid phase of digestor mixtures, can be identified and measured avoiding isolation steps. Panels of poly- and monoclonal antibodies were assembled which detect methanogens and their released antigens in 10 μl samples. Identification is accomplished by complementary micromethods carried out on the reaction areas of a slide immunoenzymatic assay (SIA) designed for immunologic testing, differential staining and microscopic examination. Digestor samples from various sources have been studied; <u>Methanosarcinae</u>, <u>Methanobacteria</u>, <u>Methanobrevibacteria</u>, <u>Methanococcii</u> and <u>Methanothrix</u> were found. Time-course studies to observe population shifts and to define immunotypes of interest are underway. A correlation of the immunologic findings with microbiologic, chemical, and engineering parameters should provide information for assembling antibody kits to monitor digestor function, and for improving process control and energy yields.

METHANOGENS IN DIGESTERS: IDENTIFICATION AND QUANTIFICATIO
IN MICROSAMPLES BY DIRECT TESTING WITH ANTIBODY PROBES

INTRODUCTION

Progress in the understanding of the methanogenic flor of anaerobic digestors will lead to accomplishing better yields in useful molecules than are possible at the present time (10,11,17,18,20-22).

Microbes in digestors have been studied for many years (2,10,11,17-22) but it is still not known with any satisfactory degree of precision which methanogens occur under the variety of conditions inherent to methanogenic biotechnology. The literature abounds in expressions such as "Methanosarcina like", "resembling Methanothrix", or "presumably Methanobacterium" when microbial findings in digestors are described. These expressions reflect the difficulties that state-of-the-art technology encounters when it comes to the point of establishing the microbial composition of digestors and its changes during operation times. Even if determination of species and strain were possible, it would not be enough. A finer distinction within each species or strain found in digestors should be achieved.

Identification of methanogens beyond the species and strain levels is now possible by newly developed immunologi means (14). Distinction of immunotypes within a species ha been accomplished (7). It is, therefore, of great interest to find out which immunotypes occur in digestors and observ population shifts in parallel with chemical parameters.

Learning how individual immunotypes, rather than species which are unknown mixtures of immunotypes, behave i terms of number during digestor operation, will allow for a very precise correlation between a well defined methanogen and any other parameter that one may want to investigate to measure digestor performance for example.

In this report we will review the immunologic tools available at the present time to elucidate the methanogenic flora of digestors, or of any other complex microbial ecosystem, and some of the results obtained thus far.

BACKGROUND

Three major recent developments provided a much needed basis for undertaking studies of methanogens in digestors with a degree of precision never achieved before: a) the systematic characterization of archaebacteria (1); the immunologic classification of methanogens (8); and the advent of the hybridoma technology for monoclonal antibody production (12). In addition, the widespread use of the enzyme-linked immunosorbent assay (ELISA) (9), and particularly the development of the slide immunoenzymatic assay (SIA) (6), helped considerably in bacteriologic determinations, and thus made examination of samples from complex microbial mixtures simpler and more accurate than any previous methodology had allowed. It has also become clear, recently, that direct examination of samples must be done to obtain accurate information on the microbial flora occurring in ecologic niches of scientific or practical interest (3). Methods that rely on measurements of a manifestation of bacterial activity (e.g. gas production) and require manipulations (e.g. culture-isolation) are misleading. These are indirect approaches which will detect only those microorganisms that perform well under the diagnostic laboratory conditions used. For example, bacteria that do not grow fast enough in the culture medium utilized for enrichment-isolation will be missed. This is a serious drawback because one may be overlooking species which are important within the digestor despite their poor "culturability" in diagnostic media.

It is therefore important that the digestors' flora be examined directly, avoiding culture-isolation. For this to be possible, however, certain tools are necessary such as antibody or DNA probes.

Antibody probes for methanogens are now available as described below.

RESULTS AND DISCUSSION

Antisera against Reference Methanogens

A comprehensive antiserum bank has been put together (14), including antisera against the following reference methanogens: *Methanobrevibacter smithii* PS; *Methanobacterium formicicum* MF; *Methanosarcina barkeri* MS; *Methanobacterium bryantii* MoH; *Methanobacterium bryantii* MoHG; *Methanosarcina barkeri* R1M3; *Methanospirillum hungatei* JF1; *Methanobrevibacter ruminantium* M1; *Methanobrevibacter arboriphilus* DH1; *Methanobrevibacter smithii* ALI; *Methanobacterium thermoautotrophicum* GC1; *Methanobacterium thermoautotrophicum* ΔH; *Methanococcus vannielii* SB; *Methanococcus voltae* PSv; *Methanogenium marisnigri* JR1m; *Methanosarcina barkeri* 227; *Methanogenium cariaci* JR1c; *Methanococus mazei* S6; *Methanosarcina barkeri* W; *Methanosarcina thermophila* TM1; *Methanobrevibacter arboriphilus* AZ; *Methanobrevibacter arboriphilus* DC; *Methanomicrobium mobile* BP; *Methanothermus fervidus* V24S; *Methanolobus tindari* T3; *Methanococcus maripaludis* WJJ; *Methanosphaera stadtmaniae* MCB3; *Methanoplanus limicola* M3; *Methanococcus thermolithotrophicus* SN1; and *Methanothrix soehngenii* Opfikon.

Polyclonal Antibody Probes

Calibrated antibody probes were derived from each antiserum by titrating it against the immunizing methanogen, under standardized conditions (14). The S probe, which is used routinely, is the highest dilution in the titration curve's plateau determined by indirect immunofluorescence (IIF). Other antibody probes are used less often and will not be discussed here.

Antigenic Fingerprints

All reference methanogens were assayed with all S probes to obtain their antigenic fingerprints (13). A standard chart of reference fingerprints was constructed and a method was developed, based on this chart, for identifying new isolates (13, 15). Likewise, this method has been standardized to identify methanogens in complex microbial communities, e.g. in digestors (4; see also SIA constellation below).

Hybridoma Collection and Monoclonal Antibody Panels

Table 1. Hybridomas and Monoclonal Antibodies Already Proven Useful to Study Methanogens in Digestors

Immunizing methanogen: Species	Strain	Fusion number	Number of hybridomas studied (antibodies)
Methanobacterium thermoautotrophicum	ΔH	H4	6
Methanobrevibacter arboriphilus	DH1	H3	1
Methanobrevibacter smithii	PS	H6	6
Methanococcus vannielii	SB	H5	6
Methanogenium cariaci	JR1c	H7	2
Methanospirillum hungatei	JF1	H1	1
Methanosarcina barkeri	227	H8	3
Methanothrix soehngenii	Opfikon	H9	6

Table 1 shows the hybridomas we have constructed thus far, including some producing antibodies against species of methanogens often found in digestors (for a reference to the original descriptions of these antibodies see REFERENCES CITED 4,5,7,14,15).

Table 2. Demonstration of Antigens Released by Methanogens in the Fluid Phase of Digestor Mixtures by Inhibition-blocking of Antibody Probes

Digestor[a]	Antibody probe: Made against	SIA reaction after incubation with: Diluent	SIA reaction after incubation with: Sample[b]	Inhibition-blocking (percent)[c]
A	*Methanobacterium formicicum*	1.73[d]	0.40	77
B	*Methanosarcina thermophila*	1.29	0.12	91

[a] Digestor A contained *Mb. formicicum* and digestor B contained *Methanosarcina* as demonstrated by SIA constellation which includes direct visualization of bacterial cells by immunofluorescence with specific antibody probes (see text and reference 16)(SIA; slide immunoenzymatic assay).

[b] Fluid phase of digestor mixture, freed from bacterial cells and particulate materials.

[c] Calculated as follows: 100-[(SIA reaction of antibody probe incubated with sample ÷ SIA reaction of probe incubated with diluent) x 100]. The same type of experiment using antibody probes against methanogens not present in the sample showed no inhibition-blocking. Positive and negative control results were confirmed by direct-binding assays (see REFERENCES CITED 6, 14-16 for technical details).

[d] OD 450nm SIA reaction (at 30 min after addition of enzyme substrate) of antibody probe (incubated with diluent or sample, with immunizing methanogen *Mb. formicicum* (digestor A) or *Methanosarcina thermophila* (digestor B).

Monoclonal antibody probes are currently being used in our laboratory for: a) direct identification and counting of methanogens (whole bacterial cells), and b) measurement of antigens distinctive of immunotypes in the fluid phase of digestors. An example of the latter measurement is given in Table 2.

SIA Constellation

Development of SIA (6), made possible other technologic advances useful for elucidating microbial mixtures. In this regard, SIA constellation (16) has proven very helpful. Identification of methanogens is accomplished by a set of complementary micromethods, requiring only 10 μl sample each.

The circles (reaction areas) of an SIA slide, Figure 1, are coated with the sample, in which the organisms are to be identified, and with control and reference methanogens. SIA, IIF and gram staining are performed to reveal the immunologic reactivity of each morphotype in the sample. Comparing sample with reference circles, methanogens in the former are identified.

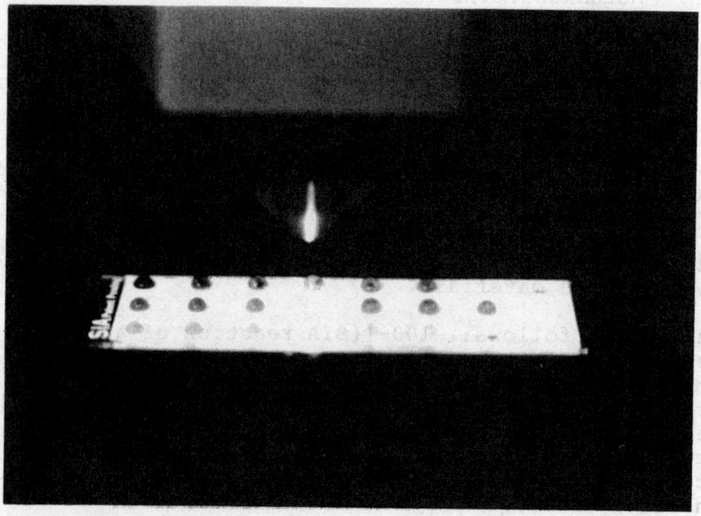

Figure 1. Slide immunoenzymatic assay (SIA). Shown are an SIA slide with 10 μl-drops of enzyme-substrate solution on each circular reaction area, and the head of the vertical-beam spectrophotometer for measuring reaction intensity. Reaction areas were coated with 10 μl samples from a digestor, and incubated with antibody probes for various methanogens. A second incubation followed with the revealing antibody (directed against immunoglobulins and labeled with the enzyme horse-radish peroxidase). Lastly, the enzyme-substrate was added. When a methanogen recog-

nized by one of the antibody probes used was present in the digestor, color appeared in the reaction area incubated with the probe. Results were corroborated by indirect immunofluorescence (IIF) and other methods (SIA constellation) (see text and REFERENCES CITED 4 and 16).

Poly- and monoclonal antibodies are used for SIA and IIF to determine the antigenic fingerprint of each morphotype, and the presence (or absence) of antigens that are known markers of known species of methanogens. All the data are put together for obtaining a complete portrait of every methanogen in the sample.

We have examined several types of microbial mixtures to first test the probes and the procedures, and then to standardize the conditions for routine use of SIA constellation.

Mixtures studied up until now are: a) laboratory made mixtures of known microorganisms mixed in pre-established proportions (4); b) samples from a variety of ecologic niches like dental plaques and the intestinal tract of man and animals (7; unpublished results); and c) digestor and landfill specimens (4,16), Table 3.

Table 3. Illustrative Examples of Digestor and Landfill Specimens Probed for Methanogens in Recent Surveys

Digestor		Landfill	
Characteristics	Number Examined	Depth (m)	Number Examined
thermophilic	6	4	2
whey	4	6	3
sewage sludge	3	8	2
phenol degrading	1	10	1
acetate degrading	1	12	3
expanded bed	3		
fluid bed	2		
UASB[a]	2		
anaerobic filter	2		

[a]Upflow anaerobic sludge blanket

Methanosarcinae, Methanobacteria, Methanobrevibacteria, Methanococci and Methanothrix were found in digestors. Selected cases are being studied further to determine immunotype diversity within species, and time-course variations.

Experimental conditions such as temperature, and addition of toxic materials are manipulated to check how they would affect methanogens. These are determined immunologically, along with other parameters, e.g. gas production.

CONCLUSIONS

We are now witnessing the dawn of a new era in the study of methanogenic communities in digestors and other ecosystems of practical or scientific importance. Molecular biology and immunology are at the center of this new wave of studies aiming at optimizing methanogenic biotechnology.

The distinctive characteristics of the contemporary approach, which makes it far more advanced than older tactics are: a) Small sampling. Important information can now be obtained by examining very small samples (microliter volumes) containing very few bacteria (a few hundred); b) High resolution. Bacterial identification can be done beyond the species and strain levels, i.e. the immunotype level; c) Precision and specificity. Unequivocal identification can be accomplished on account of the precision of quantitative immunologic methods, and the specificity of antibody probes; d) Directness which means distortion-free. Samples straight from the digestor can be examined avoiding culture-isolation and thus eliminating distortions of the microbial composition due to positive or negative bacterial selection. The contemporary, direct approach reveals the methanogenic flora as it actually occurs in the digestor. Culture-isolation instead causes some species to become prominent because they grow well in the diagnostic media used, while others do not show up simply because they poorly grow in these media; e) Multiplicity. The solid (microbial) and fluid phases can be tested with the same antibody probes. In the former phase, whole bacterial cells are identified and counted, while in the fluid phase markers of these cells are quantified. These two sets of data, obtained by at least two types of procedures complement and control each other, as well as data from nonimmunologic tests (e.g. gram staining). This multiplicity yields a series of complementary results for unequivocal bacterial identification; f) Structural connotations. Structural data can be obtained by simple binding assays using monoclonal antibodies of pre-defined fine (molecular) specificities. The morphology and immunologic reactivities of methanogens are determined by assaying whole-bacterial cells with these well characterized monoclonal antibody probes using IIF. In addition, information on the chemistry of the antigens that characterize methanogens and provide a means for their precise identification is obtained testing bacterial cells and fractions thereof, in the fluid phase. For this, inhibition-blocking procedures along with IIF and SIA are employed.

PROSPECTS FOR THE FUTURE

The immediate and long-term futures are promising for both the study and the manipulation-generation-control of methanogens using immunologic means.

Research efforts can now be directed toward answering questions unthinkable only a few years ago. Examples of questions are: Which immunotypes are there in digestors? What are their variations with time? What are the effects of temperature, pH, toxic materials, etc. on these immunotypes? How do immunotypes and numbers of organisms of each immunotype correlate with the type of waste? What is the extent of diversity of the methanogenic flora? Does this diversity parallel the wide variety of wastes? Or, are there a common denominator and a variable fraction spanning all wastes? Which immunotypes predominate when the digestor's performance is optimal (and when it is suboptimal)?

It will soon become possible, after some of the above questions have been answered, to devise a sound strategy for manipulating the methanogenic flora and thereby controlling the fermentation process, using antibodies. This cannot be done yet, because although the antibodies are already available, it is not known which methanogen should be favored and which should be eliminated. Both, enrichment and elimination of methanogens in a digestor can be undertaken using antibodies. The potential therefore exists to treat a waste water system for example with an antibody or with a precisely formulated mixture of antibodies, to promote growth of a methanogenic immunotype and at the same time inhibit others. It should be obvious however, that until "good" and "bad" immunotypes are identified, these process controls cannot be implemented. Likewise, identification of these immunotypes will allow for the preparation of antibody kits for rapid diagnosis of methanogenic flora in digestors. It will then be possible to predict, from the immunotypes present, and from their quantities, the course of functioning of the digestor in the ensuing hours or days. Early detection of malfunction will become a reality.

Lastly, identification of immunotypes will permit choosing those most promising for genetic engineering. Construction of specialized strains will know where to go (i.e. which strain it should make) and will have the antibodies to help in the procedures.

ACKNOWLEDGMENTS

Work done in the authors' laboratories was supported in part by grants No. DE-FG02-84ER 13197 from the U.S. Department of Energy, and No. 261-81/83 from the North Atlantic Treaty Organization (NATO).

REFERENCES CITED

1. Balch, W. E., Fox, G. E., Magrum, L. J., Woese, C. R. and Wolfe, R. S., "Methanogens: Reevaluation of a Unique Biological Group," Microbiol. Rev. 43, 260-296 (1979).
2. Chartrain, M. and Zeikus, J. G., "Microbial Ecophysiology of Whey Biomethanation: Characterization of Bacterial Trophic Populations and Prevalent Species in Continuous Culture," Appl. Environm. Microbiol. 51, 188-196 (1986).
3. Colwell, R. R., Brayton, P. R., Grimes, D. J., Roszak, D. B., Huq, S. A. and Palmer, L. M., "Viable but Non-Culturable *Vibrio cholerae* and Related Pathogens in the Environment: Implications for Release of Genetically Engineered Microorganisms," Biotechnology 3, 817-820 (1985).
4. Conway de Macario, E. and Macario, A. J. L., "Immunologic Probes for Identification of Methanogenic Bacteria in Anaerobic Digestors," Proc. IVth International Symposium on Anaerobic Digestion, Guangzhou (China), pp. 505-514 (1985).
5. Conway de Macario, E. and Macario, A. J. L., "Immunology of Archaebacteria: Identification, Antigenic Relationships and Immunochemistry of Surface Structures," Syst. Appl. Microbiol., (1986) in press.
6. Conway de Macario, E., Macario, A. J. L. and Jovell, R. J., "Quantitative Slide Micro-Immunoenzymatic Assay (Micro-SIA) for Antibodies to Particulate and Non-Particulate Antigens," J. Immunol. Methods 59, 39-47 (1983).
7. Conway de Macario, E., Macario, A. J. L. and Pastini, A., "The Superficial Antigenic Mosaic of *Methanobrevibacter smithii*. Identification of Determinants and Isolates by Monoclonal Antibodies," Arch. Microbiol. 142, 311-316 (1985).
8. Conway de Macario, E., Wolin, M. J., and Macario, A. J. L., "Immunology of Archaebacteria that Produce Methane Gas," Science 214, 74-75, 1981.
9. Engvall, E. and Perlmann, P., "Enzyme-Linked Immunosorbent Assay (ELISA). Quantitative Assay of Immunoglobulin G," Immunochemistry 8, 871-874 (1971).
10. Harvey, M., Forsberg, C. W., Beveridge, T. J., Pos, J. and Ogilvie, J. R., "Methanogenic Activity and Structural Characteristics of the Microbial Biofilm on a Needle-Punched Polyester Support," Appl. Environm. Microbiol. 48, 633-638 (1984).
11. Henson, J. M., Smith, P. H. and White, D. C., "Examination of Thermophilic Methane-Producing Digesters by Analysis of Bacterial Lipids," Appl. Environm. Microbiol. 50, 1428-1433 (1985).
12. Köhler, G. and Milstein, G., "Continuous Cultures of Fused Cells Secreting Antibody of Predefined Specificity," Nature 256, 495-497 (1975).
13. Macario, A. J. L. and Conway de Macario, E., "Antigenic Fingerprinting of Methanogenic Bacteria with Polyclonal Antibody Probes," System. Appl. Microbiol. 4, 451-458 (1983).

14. Macario, A. J. L. and Conway de Macario, E., "A preview of the uses of monoclonal antibodies against methanogens in fermentation biotechnology: Significance for public health. In "Monoclonal Antibodies Against Bacteria," Edited by A.J.L. Macario and E. Conway de Macario. Academic Press, Inc., Orlando, FL, Vol. I, pp. 269-286, 1985.
15. Macario, A. J. L. and Conway de Macario, E., "Monoclonal antibodies of predefined molecular specificity for identification and classification of methanogens and for probing their ecologic niches." In "Monoclonal Antibodies Against Bacteria," Edited by A.J.L. Macario and E. Conway de Macario. Academic Press, Inc., Orlando, FL, Vol. II, pp. 213-247, 1985.
16. Macario, A. J. L. and Conway de Macario, E., "Antibodies for Methanogenic Biotechnology," Trends Biotechnol. 3, 204-208 (1985).
17. Robinson, R. W., Akin, D. E., Nordstedt, R. A., Thomas, M. V. and Aldrich, H. C., "Light and Electron Microscopic Examinations of Methane-Producing Biofilms from Anaerobic Fixed-Bed Reactors," Appl. Environm. Microbiol. 48, 137-136 (1984).
18. Switzenbaum, M. S., "Anaerobic Treatment of Wastewater: Recent Developments," ASM News 49, 532-536 (1983).
19. Toerien, D. F. and Hattingh, W. H. J., "Anaerobic Digestion I. The Microbiology of Anaerobic Digestion," Water Res. 3, 385-416 (1969).
20. van den Berg, L., "Developments in Methanogenesis from Industrial Waste Water," Can. J. Microbiol. 30, 975-990 (1984).
21. Zinder, S. H., Anguish, T. and Cardwell, S. C., "Effects of Temperature on Methanogenesis in a Thermophilic (58°C) Anaerobic Digestor," Appl. Environm. Microbiol. 47, 808-813 (1984).
22. Zinder, S. H., Cardwell, S. C., Anguish, T., Lee, M. and Koch, M., "Methanogenesis in a Thermophilic (58°C) Anaerobic Digestor: *Methanothrix* sp. as an Important Aceticlastic Methanogen," Appl. Environm. Microbiol. 47, 796-807 (1984).

DISCUSSION

W. Smith, U. of Florida	If organisms were isolated from digesters that operate very differently, say a fixed film digester as opposed to a stirred tank, would you expect the organisms to be different?
Macario	The only thing we can say is whether a certain methanogen is there or not. Let's say that in one kind of digester, you would have immunotype 1, and in another, immunotype 2. Then if you found that the second type is better, you probably would like isolate immunotype 2.
M. Neushul, Neushul Mari-culture, Inc.	Could the immuno absorbent techniques be used to develop innocula of specific types?
Macario	I think it's possible, but we haven't used it in a real digester.
J. Deming, Hopkins U.	Are your probes effective with formaldehyde- or glutaraldehyde-fixed samples?
Macario	The probes are developed to recognize antigens on the surface of formalinized bacteria. The samples can be shipped and sent to us from different places. In most cases that we have tested, there is practically no difference between fixed and non-fixed bacteria.
Deming	So bacteria do not actually have to be active or viable per se?
Macario	No.
Deming	Simply present in the sample?
Macario	Simply present.

KINETIC ANALYSIS OF BIOGASIFICATION OF BIOMASS/WASTE BLEND AND ITS ENGINEERING SIGNIFICANCE

Vipul J. Srivastava, M.S.
Manager, Bioengineering Research
Institute of Gas Technology
Chicago, Illinois 60616

David P. Chynoweth, Ph.D
Associate Professor
Agricultural Engineering Department
University of Florida
Gainesville, Florida 32611

ABSTRACT

Process kinetics reflected as methane production rates are important to the design and operation of anaerobic digesters. An important goal of anaerobic digestion processes is to maximize the methane production rate by increasing the loading rate and decreasing the hydraulic retention time. However, kinetic equations show that while methane production rates increase at increasing loading rates, system efficiency reflected as the methane yield decreases. The optimum loading and retention time for an anaerobic digestion system depends on the feed characteristics (including ultimate biodegradability) and the digester design.

This paper describes kinetic models that were developed to predict the methane production rate and methane yield as a function of the organic loading rate in two different digester designs — the continuous stirred tank reactor and the non-mixed vertical flow reactor — and with different biomass-to-waste blend ratios. These models, developed and experimentally checked with laboratory-scale digesters, are currently being evaluated at a pilot-scale experimental test unit located at Walt Disney World, Orlando, Florida. The paper also discusses the engineering significance of such models with respect to digester designs and operation.

KINETIC ANALYSIS OF BIOGASIFICATION OF BIOMASS/WASTE BLEND AND ITS ENGINEERING SIGNIFICANCE

INTRODUCTION

Anaerobic digestion, as a process for the conversion of biomass and wastes to methane and cell mass, has been demonstrated to be technically feasible for a variety of feedstocks and in sizes ranging from small household units to large-scale systems in wastewater treatment plants. Among several advantages of anaerobic digesters, described elsewhere,[1,2] two are very important: production of clean gas fuel with high heating value and reduction of the amount of stabilized effluent solids. In order to rationally design the digester and specify operating conditions for maximum methane production and to be able to predict methane production and effluent quality, an understanding of process kinetics is essential.

Methane production rates can be increased by either increasing the loading rate (or feed concentration) at a constant hydraulic retention time (HRT), decreasing the HRT at constant feed concentrations, or a combination of the two. However, in all three cases, digester efficiency, reflected as the methane yield, decreases. The optimum loading rate and HRT for any anaerobic digestion system depends on the feed characteristics (including biodegradability) and digester design.

This paper describes an advanced digestion process and kinetic models that can be used to predict the methane production rate and the methane yield as a function of retention time. These models were used to describe the digester performance for two different digester designs — a continuous stirred tank reactor (CSTR) and a non-mixed vertical flow reactor (NMVFR) — and for different biomass-to-waste blends. The paper also compares the kinetic constants of different digestion systems and discusses the engineering significance of such models.

BIOGASIFICATION PROCESS DEVELOPMENT

The overall objective of this research program was to develop a data base for the design and operation of a large-scale experimental test unit (ETU) for the gasification of water hyacinth and sewage sludge produced at Walt Disney World (WDW) and its integration with the hyacinth wastewater reclamation facility. The specific objectives were —

- Determination of gasification characteristics of water hyacinth, primary sludge, and hyacinth/sludge blends under baseline conditions of anaerobic digestion

- Evaluation of the effects of blend ratios on anaerobic digestion

- Unconventional digestion studies on different blends of hyacinth and sludge
- Kinetic analyses of anaerobic digestion of hyacinth and sludge blends
- Application of the kinetic model to predict the effects of digester design and operating conditions on the methane formation efficiency (yield) and rate
- Design, construction, and operation of an ETU using different blends of hyacinth and sludge.

Baseline Digestion Studies

Hyacinth, primary sludge, and blends of hyacinth and sludge were studied under conditions of conventional anaerobic digestion. Such experiments provide a data base that can be compared to those obtained in this and other laboratories on various feedstocks and, in addition, provide comparison of performances of unconventional advanced digestion studies with this feedstock. These experiments were conducted in a CSTR operated at 35°C, a loading rate of 0.10 lb VS/ft^3-day, and an HRT of 15 days. The results of these experiments have been described in a previous publication.[3] Methane yields of 3.0, 0.3, and 4.5 SCF/lb VS added were reported for hyacinth, sludge, and a 3:1 blend of hyacinth/sludge which corresponds to volatile solids reductions of 42%, 72%, and 53%, respectively.

Unconventional Advanced Digestion Studies

A non-mixed vertical flow reactor (NMVFR) was designed and operated to promote the retention of solids and solids-associated microorganisms in the reactor and, therefore, results in higher solids retention times (SRT's) than HRT's. This digester design was a modification of the upflow anaerobic sludge blanket digester widely studied and applied to anaerobic digestion of soluble feeds. The NMVFR digester is different in that the feeds contain higher concentrations of particulate matter and the bed is not expanded and contains both unreacted solids and microorganisms. It can be operated either in the upflow mode by entering the feed near the bottom of the reactor and wasting the effluent near the top, or in downflow mode by entering the feed near the top of the reactor and wasting the effluent near the bottom. This reactor concept was first applied to sea kelp in IGT's laboratory.[4] It exhibited the highest performance observed for mesophilic digestion of a particulate feed (a methane yield as high as 6.0 SCF/lb VS added and a methane production rate as high as 3.3 vol/vol-day).

Results obtained from experiments conducted with a 3:1 hyacinth/sludge blend and loading rates in the range of 0.10 to 0.40 lb VS/ft^3-day showed that methane yields were 15% to 30% higher in the NMVFR than in the CSTR and performance was more stable.[5] Table 1 compares methane yields for this digester with those obtained from a CSTR for several different loading rates.

Table 1. COMPARISON OF METHANE YIELDS FROM 3:1 HYACINTH/SLUDGE BLEND (dry weight basis)

Loading Rate, lb VS/ft^3-day	Methane Yield, SCF/lb VS added	
	CSTR	NMVFR
0.10	4.5	5.6
0.12	3.8	5.4
0.17	3.8	4.5
0.20	3.7	4.3
0.30	3.2	3.7
0.40	2.6	4.3

PROCESS KINETICS AND MODELING

Prediction of process performance in terms of methane yield and methane production rate is essential to the effective design and operation of a large-scale anaerobic digestion facility. The objective of this study was to evaluate two well-established kinetic models for their ability to predict performance on bench-scale CSTR and NMVFR digesters receiving two different blends of hyacinth and sludge: the Monod and Standard First Order. Both models were modified to a form that describes the dependence of methane yield and methane production on loading.

In the anaerobic digestion process, the product (methane) formation is a result of bacterial cell growth and substrate (or feed) metabolism. The rate of methane generation can ultimately be related to the fundamental growth rate equation presented below:

$$\frac{dX}{dt} = \mu X \qquad (1)$$

where —

X = microbial cell mass concentration

t = time

μ = specific growth rate

Monod Model

A widely used relationship between the specific growth rate (μ) and an essential compound's concentration was proposed by Monod.[6] Equation 2 gives the Monod model, which describes the kinetics of pure culture utilizing a single substrate:

$$\mu = \frac{\hat{\mu}\, S}{K_s + S} \qquad (2)$$

where —

$\hat{\mu}$ = maximum specific growth rate

S = limiting substrate concentration

K_s = half velocity constant

By substituting Equation 2 in Equation 1, a relationship between the reaction rate ($-r_s$) and substrate concentration can be developed:

$$\frac{dX}{dt} = \frac{X\hat{\mu}S}{K_s + S} = -r_s \quad (3)$$

Application of Equation 3 is impractical for description of anaerobic digestion of particulate feed because measurement of the cell mass concentration, X, is difficult in the presence of non-bacterial particulates.

Many modifications of this equation have been proposed by researchers to accommodate complex organic substrates and mixed cultures. Using a simple equation for solids conversion efficiency, and substituting the values for influent and effluent solids concentrations from Equations 1 and 2 (Figure 1), IGT developed equations (Figure 1) that were able to predict performance of CSTR and NMVFR digesters receiving hyacinth, primary sludge, and a 3:1 blend of these feeds. It was shown that with modification (that is, addition of a methane yield coefficient which is a function of SRT), this model could be used to predict methane yields and rates as a function of loading for the CSTR and NMVFR digesters. The two equations, Equations 3 and 4, developed from the above modifications are shown in Figure 1. Using the data presented in earlier publications,[5,7] kinetic coefficients were developed using Equation 1 (Figure 1) for a 3:1 blend of hyacinth/sludge in both types of digesters. Using Equations 3 and 4 in Figure 1, theoretical plots of methane yield and production rates were developed. Figures 2 and 3 show the comparison of predicted values for methane yield and production rates, respectively, for the NMVFR and CSTR. The actual data presented on these plots show a reasonable adherence to the model and clearly indicate that the performance of the NMVFR in terms of methane yields and methane production rate was better.

Although our data demonstrated that the Monod model is accurate for prediction of performance of both the CSTR and NMVFR when receiving hyacinth/sludge blend, this model has two limitations:

1. It is fundamentally based on a back-mixed reactor principle and therefore difficult to defend conceptually for the NMVFR.

2. Its applications, as described in Equations 3 and 4 of Figure 1, are dependent upon knowledge of the biodegradability of the feed, which varies considerably for this particular feed and is time-consuming to determine.

MONOD KINETIC MODEL

$$\theta = \frac{1}{\hat{\mu}} + \frac{K_s}{\hat{\mu}} \left(\frac{1}{S_1}\right) \qquad (1)$$

where —

θ = hydraulic retention time, days

$\hat{\mu}$ = maximum specific growth rate, days $^{-1}$

K_s = saturation coefficient, g/L

S_1 = effluent biodegradable substrate concentration, g/L

$$S_1 = V - V_O\ (1 - \beta)$$

$$S_b = V_O \beta \qquad (2)$$

where —

V_O = influent volatile solids concentration, g/L

V = effluent volatile solids concentration, g/L

β = ultimate biodegradability coefficient, decimal %

S_b = influent biodegradable substrate concentration, g/L

METHANE PRODUCTION RATE

$$R_m = \alpha\ \frac{(S_b - \hat{\mu}\ \theta) - (S_b + K_s)}{\theta\ \hat{\mu} - \theta} \qquad (3)$$

where —

R_m = methane production rate, vol CH_4/vol culture-day

α = methane yield coefficient, L CH_4/g biodegradable substrate converted

S_b = influent biodegradable substrate concentration, g/L = $V_O(\beta)$

METHANE YIELD

$$Y_m = \frac{R_m}{L} = \alpha\ \frac{(S_b\ \hat{\mu}\ \theta) - (S_b + K_s)}{L(\theta^2\ \hat{\mu} - \theta)} \qquad (4)$$

where —

Y_m = methane yield, Lg VS added

L = loading, g VS/L culture volume-day $^{-1}$

Figure 1. MONOD MODEL FOR ANAEROBIC DIGESTION

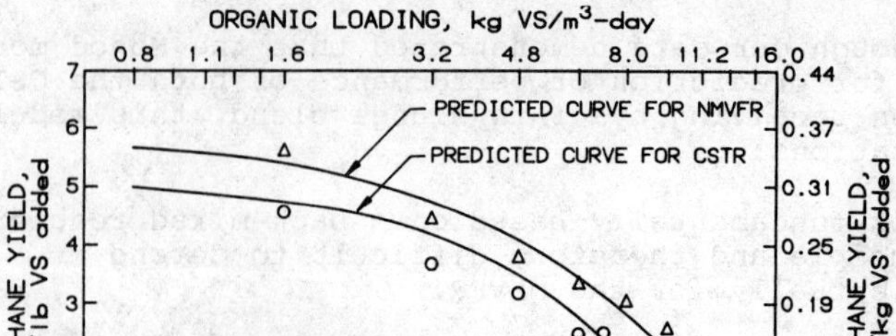

Figure 2. COMPARISON OF ACTUAL AND PREDICTED METHANE YIELDS FO HYACINTH/SLUDGE BLEND (3:1 Dry wt Basis) IN THE NMVFR AND THE CS

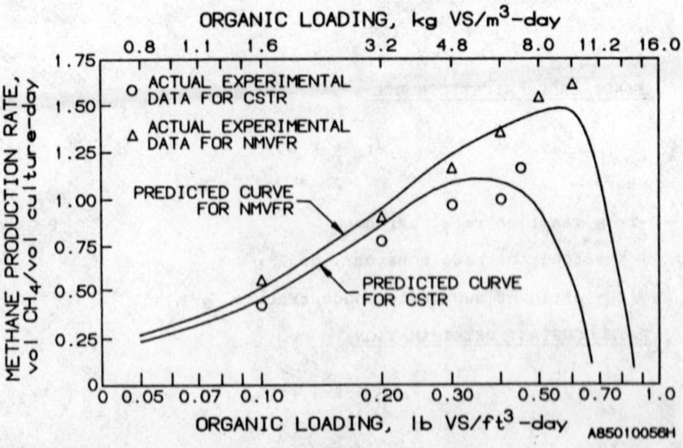

Figure 3. COMPARISON OF ACTUAL AND PREDICTED METHANE PRODUCTION RATES FOR HYACINTH/SLUDGE BLEND (3:1 Dry wt Basis) IN THE NMVFR AND THE CSTR

First Order Model

The more simple first order kinetic equation, applied by others to anaerobic digestion,[8,9,10] was evaluated for hyacinth/sludge blends for digestion in the CSTR and NMVFR. Experiments providing volatile solids reduction and methane yields as a function of loading provide data needed for this mode. The pertinent equations for this model are shown in Figure 4. Using Equation 2 (Figure 4) as a basis for plotting S_o/S versus r, the substrate utilization rate constant (K) can be determined. A sample plot for the determination of K is given in Figure 5. This first order model was rearranged as in the case of the Monod model (that is, addition of methane yield coefficient in the process efficiency equation) to develop Equations 3 and 4 (Figure 4), which describe methane yield and methane production rate as a function of loading and HRT. This model is not dependent upon a known value for feed biodegradability, which is more useful in applications where feed biodegradability is variable.

Using performance data published earlier,[5,7] kinetic constants were determined for digestion of a 3:1 hyacinth/sludge blend in the CSTR and NMVFR and a 2:1 blend in the NMVFR. Using the constants shown in Table 2, theoretical plots of methane yield and methane production rate versus loading were developed, as shown in Figures 6 through 8. Actual data are also shown in these figures.

Table 2. KINETIC PARAMETERS FOR ANAEROBIC DIGESTION OF HYACINTH/SLUDGE BLENDS IN CSTR AND NMVFR USING FIRST ORDER KINETICS

	NMVFR		CSTR	
Feed Type	Methane Yield Coefficient l CH_4/g VS Destroyed	K Rate Constant Day^{-1}	Methane Yield Coefficient l CH_4/g VS Destroyed	K Rate Constant Day^{-1}
3:1 WH/PS	0.532	0.095	0.460	0.088
2:1 WH/PS	0.548	0.121	--	--

FIRST ORDER KINETIC MODEL

$$-r_s = K\,S \qquad (1)$$

where —

$-r_s$ = reaction rate, g/L-day

K = reaction rate constant, day^{-1}

S = effluent substrate concentration, g/L

RATE CONSTANT DETERMINATION

$$\frac{S_o}{S} = K\,\theta + 1 \qquad (2)$$

where —

S_o = influent substrate concentration, g/L

θ = hydraulic retention time, day

METHANE YIELD (Y_m)

$$Y_m = \frac{\alpha\,(K\theta)}{(1 + K\theta)} \qquad (3)$$

where —

α = methane yield coefficient, L CH_4 g substrate destroyed

METHANE PRODUCTION RATE (R_m)

$$R_m = L\,Y_m = \frac{\alpha\,(K\theta)\,L}{(1 + K\theta)} \qquad (4)$$

where —

L = loading rate, g VS/L culture-day

Figure 4. FIRST ORDER KINETIC MODEL

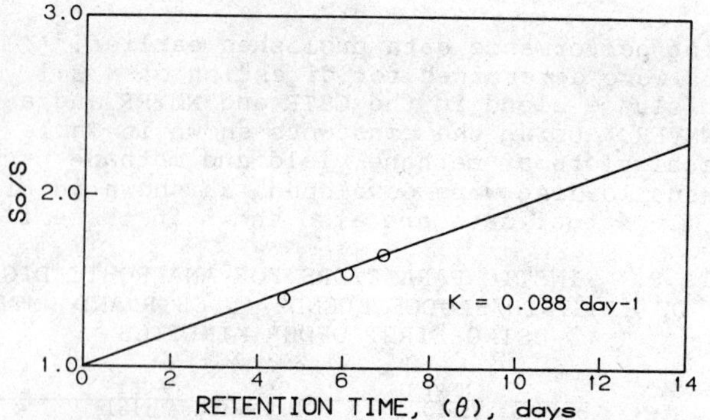

Figure 5. DETERMINATION OF THE RATE CONSTANT K FOR 3:1 BLEND FEED IN THE CSTR USING FIRST ORDER KINETICS

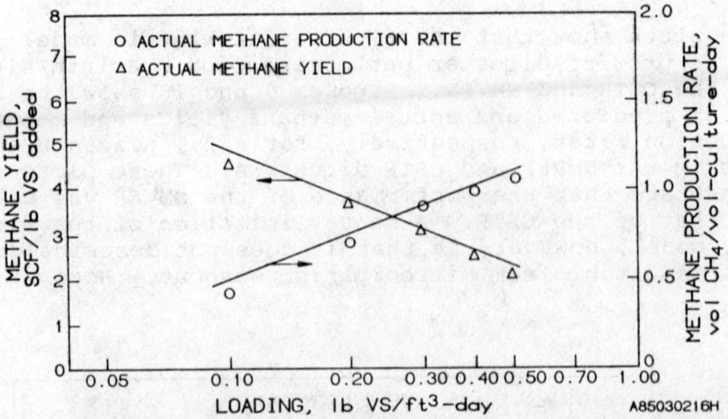

Figure 6. PREDICTED METHANE YIELDS AND PRODUCTION RATES IN TERMS OF LOADING FOR 3:1 BLEND IN THE CSTR USING THE FIRST ORDER KINETIC MODEL

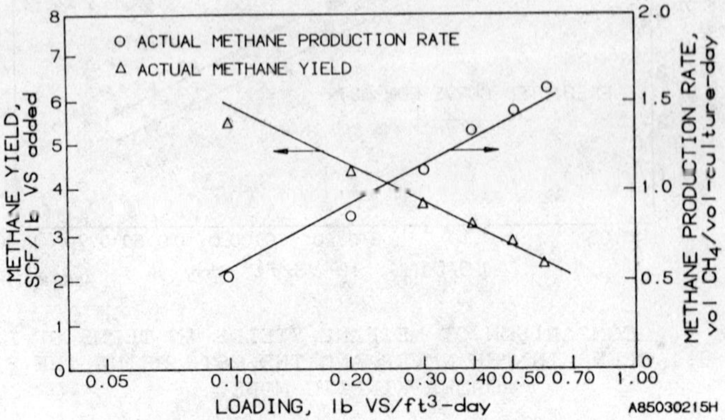

Figure 7. PREDICTED METHANE YIELDS AND PRODUCTION RATES IN TERMS OF LOADING FOR 3:1 BLEND IN THE NMVFR USING THE FIRST ORDER KINETIC MODEL

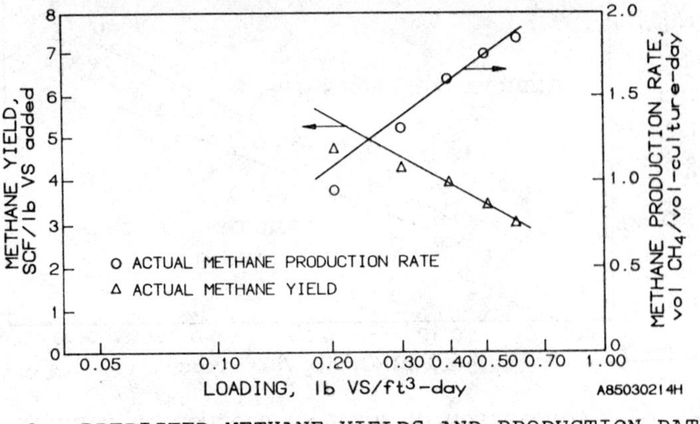

Figure 8. PREDICTED METHANE YIELDS AND PRODUCTION RATES IN TERMS OF LOADING FOR 2:1 BLEND IN THE NMVFR USING THE FIRST ORDER KINETIC MODEL

These plots show that the first order kinetic model can be used to predict digester performance for hyacinth/sludge in both the CSTR and NMVFR. Figures 9 and 10 show the comparison of predicted and actual methane yields and methane production rates, respectively, for a 3:1 hyacinth/sludge blend in the NMVFR and CSTR digesters. These plots also demonstrate that the performance of the NMVFR was better than that of the CSTR. A major limitation of the first order model, however, is that it does not describe conditions resulting from microorganism washout. However, in

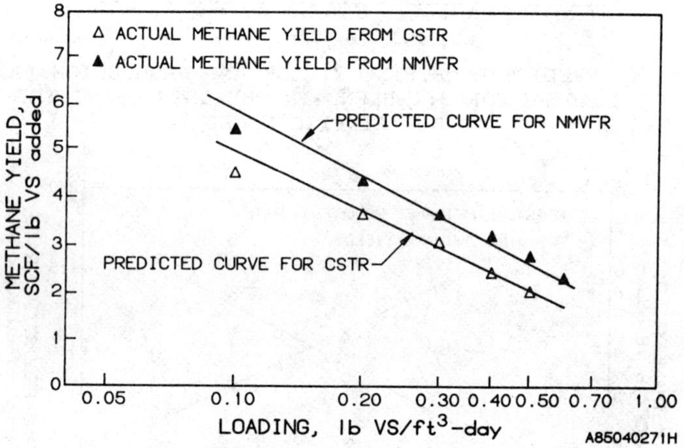

Figure 9. COMPARISON OF METHANE YIELDS IN TERMS OF LOADING FOR 3:1 BLEND IN THE NMVFR AND THE CSTR USING THE FIRST ORDER KINETIC MODEL

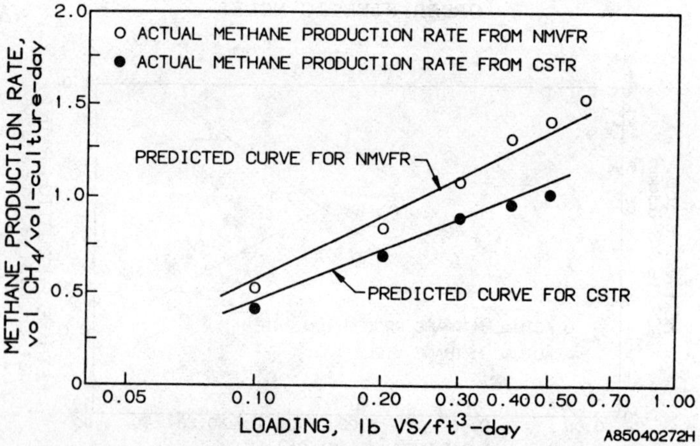

Figure 10. COMPARISON OF METHANE PRODUCTION RATE IN TERMS OF LOADING FOR 3:1 BLEND IN THE NMVFR AND THE CSTR USING THE FIRST ORDER KINETIC MODEL

practice this is not a limitation because digesters would not be operated under conditions where washout occurs.

RESULTS

The predictability of the Monod and first order models was evaluated by two methods: 1) fitness of data to theoretical plots, and 2) average deviation (percent) of actual performance from predicted performance. These results are shown in Tables 3 and 4, respectively. It is clear from this evaluation that the first order model has more validity. As a result, the first order model was selected for future applications for the following reasons:

1. A non-mixed digester was selected for process scale-up, and the first order model does not assume CSTR configurations.

2. Application of the first order model does not require a knowledge of the ultimate feed biodegradability and associated methane yield.

3. The first order model more accurately predicts laboratory results.

Table 3. COMPARISON OF COEFFICIENT OF DETERMINATION (r^2) VALUES OF KINETIC EVALUATION CURVES FOR MONOD MODELS AND FIRST ORDER MODELS

	r^2 Values	
Type of Model	Monod Model	First Order Kinetic Model
CSTR		
3:1 blend	0.65	0.97
NMVFR		
3:1 blend	0.93	0.90
2:1 blend	0.89	0.94

Table 4. COMPARISON OF AVERAGE DEVIATION OF PREDICTED METHANE YIELDS FROM THE ACTUAL METHANE YIELDS FOR THE FIRST ORDER MODELS

	Average Deviations, %	
Type of Model	Monod Model	First Order Kinetic Model
CSTR		
3:1 blend	14	6
NMVFR		
3:1 blend	7	5
2:1 blend	*	<2

*Very erratic data.

ENGINEERING SIGNIFICANCE AND CONCLUSIONS

The kinetic models presented above show that a ration approach to optimize the anaerobic digestion process for methane production is to balance the opposing effects of loading on methane production rates and methane yields. As shown in Figures 2 and 3, while methane production rates increase at increasing loading rates, methane yields decrease. Optimum methane production can be achieved by selecting a digestion design and an operating technique tha increase solids conversion through increased solids and microorganism retention. It should be noted that there is an optimum value of SRT beyond which the digester will star accumulating less reactive material and will experience problems related to mass transfer between incoming feed solids and cell mass. The operating strategy for optimum methane production should also include increasing the feed solids concentration and biodegradability, as indicated by Equations 3 and 4 in Figure 1, and Equations 3 and 4 in Figure 4. The high solids feed will increase the SRT for a given loading rate, which should result in an increased methane yield coefficient and reaction rate constant and therefore a higher methane production. The improvement in feed biodegradability will also increase the methane yield coefficient and thus the total methane production. From th above observations, it can be concluded that —

- Advanced digester design and operating techniques that promote solids (substrate and microorganism) retention time should be employed.
- The highest permissible solids feed slurry should be used to operate the digester.
- A cost-effective and energy-efficient method of feed pretreatment should be investigated to improve the biodegradability of the less biodegradable feedstock because the feed biodegradability is the ultimate determinant of the improvement in methane production.

ACKNOWLEDGMENT

The authors acknowledge the Gas Research Institute (GRI) for the primary support of research conducted at IGT, reported here under GRI Contract numbers 5081-323-0463 and 5083-226-0773. We also acknowledge the technical assistanc of John R. Conrad, Mike P. Henry, and Jana D. Mensinger in running the digestion experiments. Review and comments fro Dharam V. Punwani and Kerby F. Fannin are also appreciated.

REFERENCES CITED

1. Fisher, J. A. et al., "Pilot Demonstration of Basic Designs for Anaerobic Treatment of Petrochemical Wastes," Chemical Engineering Progress, Symp. Ser., Amer. Inst. Chem. Engr., 485, 1970.

2. McCarty, P. L., "Anaerobic Waste Treatment Fundamentals: I. Chemistry and Microbiology," Public Works, 95, 9, 107, 1964.

3. Chynoweth, D. P., Ghosh, S. and Henry, M. P., "Biogasification of Blends of Water Hyacinth and Domestic Sludge," in Proc. of the 1981 International Gas Research Conf., 742-0755, Los Angeles, 1981.

4. Fannin, K. F., Srivastava, V. J., Mensinger, J. D. and Chynoweth, D. P., "Marine Biomass Program: Anaerobic Digestion System Development and Stability Study," Final report for GRI Contract 5082-225-0607 (1982).

5. Biljetina, R., Chynoweth, D. P., Srivastava, V. J., Janulis, J., Razik, A. and Smallwood, S., "Biogasification of Walt Disney World Biomass Waste Blend," Annual report for GRI Contract 5082-223-0567 (1985).

6. Monod, J., "The Technique of Continuous Culture Theory and Applications," Annales de l"Institut Pasteur 79, 390 (1950).

7. Chynoweth, D. P., Jerger, D. E., Srivastava, V. J., Conrad, J. R. and Mensinger, J. D., "Biological Gasification of Renewable Resources," Annual report for GRI Contract 5083-226-0773, 1985.

8. Doyle, O. P., O'Malley, J. D., Clausen, E. C. and Gaddy, J. L., "Kinetic Improvement in the Production of Methane From Cellulosics Residues," Symposium Papers: Energy From Biomass and Wastes VII, Institute of Gas Technology, Chicago, 1983.

9. Pfeffer, J. T., "Domestic Refuse as a Feed for Digesters." Paper presented at The First International Symposium on Anaerobic Digestion, University College, Cardift, Wales, September 1979.

10. McFarlane, P. N. and Pfeffer, J. T., "Biological Conversion of Biomass to Methane," Report for SERI, No. SERI/TR-98357-1, 1985.

DISCUSSION

________	Did you analyze the sludge for heavy metals?
Srivastava	No, we have not analyzed thoroughly for heavy metals, but I don't think that's a problem because the sludge was collected from the Disney World wastewater treatment plant. There aren't that many industries in that area.
J. Garvin, Pitt-Des Moines, Inc.	You emphasized the importance of biodegradability. How does one test for biodegradability under these conditions?
Srivastava	There are several ways. One of the methods we use is a 60-day batch test. We make sure there is no nutrient limitations by supplementing the inoculum with nutrients and vitamins. We use low loading rates, incubate for 60 days, and measure methane production.

THE RELATIONSHIP OF EUBACTERIAL AND METHANOGENIC COMMUNITY STRUCTURE IN ANAEROBIC DIGESTERS

Steven J. Schropp, Ph.D
Tommy J. Phelps, Ph.D.
Al T. Mikell, Ph.D.
David C. White, Ph.D, M.D.
Institute of Applied Microbiology
University of Tennessee
POB X, Bldg. 1503
Oak Ridge, Tennessee 37831

ABSTRACT

Anaerobic digesters contain a complex microbial community that carries out the degradation of organic compounds, converting them to simple end products. Analyses of digester communities using conventional microbiological techniques suffer from several limitations, many of which can be overcome by using a suite of recently developed biochemical techniques. We review some of these limitations and discuss the use of biochemical techniques. Using these techniques, microbial community structure can be defined, nutritional status determined, and the proportion of methanogenic archaebacteria and eubacteria examined in relationship to methane production from a digester.

THE RELATIONSHIP OF EUBACTERIAL AND METHANOGENIC COMMUNITY STRUCTURE IN ANAEROBIC DIGESTERS

INTRODUCTION

Anaerobic digester systems are used to stabilize municipal and animal wastes and to generate methane (9). The degradation of organic material and production of methane in digesters are microbially mediated processes, carried out by a complex microbial community. The community can be divided into several different functional groups based on metabolic activity. The groups include microorganisms that break down organic polymers into simple sugars, organisms that form volatile fatty acids from the sugars, organisms that produce acetate from the simple volatile compounds, and those that form the terminal end products such as methane and hydrogen sulfide (18). These groups include members of the eubacteria and the methanogenic archaebacteria.

In order to understand the carbon conversion processes occurring in an anaerobic digester and to optimize the digester conditions, it is necessary to look at the structure of the microbial community involved. Traditional methods of microbiology have a limited usefulness in the digester environment. Recently developed biochemical techniques, however, offer a means of determining community structure and nutritional status of the microbial consortia in digesters. Further development of these techniques may permit on-line monitoring of digester status.

In this paper we will review the use of biochemical methods to investigate the microbial community in anaerobic digesters. We will discuss the limitations of traditional microbiological techniques and the advantages of biochemical methods. Examples of the use of these methods will be provided to illustrate how they may be used to determine the microbial community structure in digesters and the relationships between eubacterial and methanogenic populations and the production of methane.

BIOCHEMICAL TECHNIQUES USED IN DIGESTER STUDIES

Limitations of Traditional Microbiological Techniques

The microbial community in digesters is a complex consortium of numerous species. For example, Ianotti et al. (10) isolated 130 strains of non-methanogenic bacteria from a swine manure digester. The microorganisms in digesters are attached to substrates and enmeshed in complex extracellular polymers. In such an environment, traditional microbiological techniques tend to give underestimates of the total bacterial populations. Direct microscopic observation does not reveal all of the organisms bound in polymers or buried under other cells or debris. Efforts to quantitatively release microbes from their substrates for subsequent direct observation can suffer from the problem of inconsistent release of the microorganisms.

The culture of microorganisms is of limited usefulness in the study of digester microbial communities because only a percentage of the microorganisms present will grow under artificial culture conditions. It has long been known that culture techniques give underestimates of the total microbial population in aquatic system (11). The microorganisms that grow in culture are not necessarily representative of those present in the natural community but are those that grow best under a particular set of culture conditions. Kirsh (12) and Mah and Sussman (13) reported being able to culture only about 10% of the total non-methanogenic bacteria present in digesters. Cultures, therefore, render an underestimation of total bacterial numbers and do not provide a reliable indication of the metabolic types of organisms present in the consortia.

Determination of metabolic function and state of activity is also a problem. Cellular morphology reveals little or nothing about the metabolic function of a cell. As an example, a well defined functional subset of the digester community, the methanogenic bacteria, exhibit a variety of sizes and shapes (19). Bacterial growth rates may be estimated by direct observation of the frequency of dividing cells (8) but this method suffers from the same limitations as mentioned above for the direct observation of cells. Assays of metabolic activity usually involve the use of artificially high substrate concentrations under artificial conditions. Minimal disturbances of microbial communities can induce unusually high metabolic activity (4). The use of radioactive tracers, an alternative means of measuring metabolic activity, also involves removal of the microorganisms from the digester and incubation in an artificial situation.

Biochemical Methods for Microbiological Studies

Recent efforts in our laboratory have been directed towards developing a suite of biochemical techniques for the determination of microbial community structure and nutritional status. These techniques involve the analysis of biochemical components of microbial cells and, since they involve the extraction of cellular components, can be performed without the necessity of removing the microorganisms from their substrate or growing them in artificial media. Components with a general distribution among cells are utilized as indicators of biomass. Components with restricted distributions among cells are utilized to define community structure.

Biomass Indicators. Phospholipids are components of all cells and under natural conditions occur as a relatively constant proportion of cellular biomass (16). The phospholipids are not storage compounds and have a rapid turnover rate, thus they can be used as an indication of viable microbial biomass. The phosphate from extracted phospholipid is released by an acid digestion and colorimetrically assayed. Biomass estimates based on lipid phosphate correlate well with estimates based on extractable ATP (17). Alternately, the total extractable phospholipid fatty acids may be used as biomass indicators.

Community Structure Indicators. Extracted lipids can be fractionated by silicic acid column chromatography into neutral, glyco, and phospholipid fractions. Ester-linked phospholipid fatty acids can be separated and identified by capillary column gas chromatography after derivatization to fatty acid methyl esters (FAME). The pattern of ester-linked phospholipid fatty acids (PLFA) can be used to define microbial community structure (1, 7). The use of the PLFA fraction is a sensitive indicator of viable microbial community structure because most anthropogenic contaminants and endogenous microbial storage lipids are found in the neutral or glycolipid fractions. Each sample assayed may include up to 150 individual PLFA, so a single assay provides a great deal of information. "Biomarker" PLFA have been recognized for a number of different microbial types (1, 3, 6), consequently the proportions of different PLFA in a sample provide a definition of the community structure.

Nutritional Status Indicators. The endogenous storage lipid, poly beta-hydroxy butyrate (PHB) is formed by certain microorganisms under conditions where the organisms can accumulate carbon but lack all of the nutrients needed for cell division (15). The components of the PHB polymer can be analyzed by a sensitive gas chromatographic technique (5). The ratio of total PLFA to PHB can be calculated to provide information about the nutritional status of the microbial community.

Archaebacterial Indicators. The archaebacteria, a group which includes the methanogens, do not contain the usual suite of ester-linked PLFA found in eubacteria and eucaryotic microorganisms but they do contain a unique assemblage of diphytanyl-glycerol-diether and di-biphytanyl-diglycerol-tetraether lipids (2). These ether-linked lipids can be assayed by high pressure liquid chromatography (HPLC) (14). The ratio of diacyl phospholipid (eubacterial biomass) to ether phospholipid (archaebacterial biomass) can be calculated to estimate the proportion of archaebacteria in the microbial community.

STUDIES OF DIGESTER MICROBIAL COMMUNITIES USING BIOCHEMICAL METHODS

Biomass and Community Structure of Experimental Digesters

Henson et al. (9) used biochemical methods to study the biomass, community structure, and nutritional status of the microbes in experimental methane-producing digesters. Their 4-liter digesters were fed a 3:1 mixture of Bermuda grass and cattle feed. The digesters had a 20 day hydraulic retention time and in addition to control digesters, other digesters were amended with propionate, butyrate, or nitrate at rates of 10 - 15 umole/liter/day.

In the control digesters, microbial biomass, as measured by lipid phosphate, increased three times over that of the initial feedstock (Table 1). The greatest biomass occurred in the nitrate-amended digester while the biomass in the propionate and butyrate amended digesters was intermediate between that of the controls and the initial feedstock. The total PLFA, another indicator of biomass showed the same features as the lipid phosphate measurements.

Analysis of PLFA composition showed that the microbial community structure was different among the different treatments. The feedstock contained a simple pattern of PLFA, predominately composed of saturated, normal fatty acids. The control digesters had high proportions of i15:0, a15:0, i18:0, and 18:1w7 fatty acids. Higher concentrations of 15;0, 17:0, 16:1w7t were found in the propionate-supplemented digester while the butyrate-supplemented digester contained larger amounts of br17:1 and i17:0. The nitrate-supplemented digester had higher percentages of i14:0 and i16:0. The increase in the short chain normal and branched fatty acids are indicative of bacterial colonization of the feedstock material in the digesters and the differing ratios of the PLFA shows that each treatment resulted in the growth of a different bacterial community.

PHB analyses were used to determine the nutritional status of the microbes within the digesters. A comparison of the lipid phosphate to PHB ratios (Table 1) reveals that the microbial community in the nitrate-amended digester was substantially different than the communities in the other digesters. The large amounts of PHB in the nitrate-amended digester indicated unbalanced growth in that system.

Table 1. Biochemical measurements from experimental digesters. (Compiled from ref. 9)

Digester	LPO_4^a (nmole/mg)	Total PLFA (pmole/mg)	PHB (pmole/mg)	LPO_4/PHB
Feedstock	3.5	675.	64.7	53.6
Control 1	12.3	2967.	74.4	165.
Control 2	12.7	2715.	54.1	23.5
Propionate	8.9	1722.	38.6	230.
Butyrate	7.6	1167.	60.0	127.
Nitrate	16.5	2905.	821.	20.1

[a]Lipid phosphate.

Effect of Sulfate on Digester Microbial Communities

The effect of sulfate amendments on a digester was studied by P. H. Smith and P. D. Brooks (unpubl.). After five days of continuous sulfate supplements, methane production decreased by 70%. The total PLFA, indicative of eubacterial biomass, decreased by 66% over the same time period. Increases in the 10:Me16 and the 18:3w3 fatty acids indicated increases in the populations of sulfate-reducing and, perhaps, acetogenic bacteria, respectively.

Community Structure of a Water Hyacinth Digester

Several experiments have been performed by P. D. Brooks, C. A. Mancuso, and D. P. Chynoweth (unpubl.) using biochemical techniques to study a water hyacinth digester. Samples were taken and analyzed from the water hyacinth feedstock and from various levels within the digester. The proportion of unsaturated fatty acids decreased as the feedstock was digested. The decrease was paralleled by an increase in short chain branched and cyclic fatty acids. These changes indicate degradation of the eucaryotic material of the feedstock and the buildup of a eubacterial community. The methanogenic community was determined by assay of the di and tetraphytanylglycerol ether phospholipids. These lipids increased from none in the feedstock to highest concentrations at the top of the digester (Table 2). The ratio of diacyl phospholipid (eubacterial biomass) to ether phospholipid (archaebacterial biomass) decreased from the bottom to the top of the digester, indicating an increase in the proportion of methanogens in the community towards the top of the digester. The maximum methane production, however, was not associated with the increased biomass of methanogenic bacteria but was instead found at the level of greatest acetate concentration just above the bottom of the digester.

Table 2. Biochemical analyses of a water hyacinth digester

Sample type	LPO$_4$			Methane Production (nmole/l/d)
	Diacyl (umole/l)	Ether (umole/l)	Diacyl/Ether	
Hyacinth	29.	<0.1	>16.3	-
Digester-top	8.3	3.4	2.4	-
Digester-middle	9.8	2.0	4.9	0.7
Digester-bottom	5.7	0.8	7.1	9.0

SUMMARY

Biochemical methods have proven to be useful tools for the study of microbial community structure and nutritional status of anaerobic digesters. Using these techniques, it has been possible to define the microbial community structure of functioning anaerobic digesters and to monitor changes in microbial communities preceding a sulfate-induced digester crash. Using the recently developed methods for analysis of signature archaebacterial lipids, it was possible to show changes in the proportion of archaebacteria in a digester and to investigate the relationship between methane production and the archaebacterial and eubacterial populations. The use of these techniques offers the potential of monitoring microbial community structure in order to tune digester conditions to create an optimal microbial community.

REFERENCES CITED

1. Bobbie, R. J., and White, D. C., "Characterization of Benthic Microbial Community Structure by High Resolution Gas Chromatography of Fatty acid Methyl Esters," Appl. Environ. Microbiol. 39, 1212-1222 (1980).

2. De Rosa, M. Gambacorta, A., and Gliozzi, A., "Structure, Biosynthesis, and Physiochemical Properties of Archaebacterial Lipids", Microbiol. Rev. 50, 70-80 (1986).

3. Dowling, N. J. E., Widdel, F., and White, D. C., "Analysis of Phospholipid Ester-linked Fatty Acid Biomarkers of Acetate-oxidizing Sulfate Reducers and other Sulfide Forming Bacteria", J. Gen. Microbiol. 132, (1986, in press).

4. Findlay, R. H., Pollard, P. C., Moriarty, D. J. W., and White, D. C, "Quantitative Determination of Microbial Activity and Community Nutritional Status in Estuarine Sediments: Evidence for a Disturbance Artifact," Can. J. Microbiol. 31, 493-498 (1985).

5. Findlay, R. H., and White, D. C, "Polymeric Beta-Hydroxy-alkanoates from Environmental Samples and Bacillus megaterium," Appl. Environ. Microbiol. 45, 71-78 (1983).

6. Gillan, F. T., and Hogg, R. W., "A Method for the Estimation of Bacterial Biomass and Community Structure in Mangrove-Associated Sediments," J. Microbiol. Met. 2, 275-293 (1984).

7. Guckert, J. B., Antworth, C. P., Nichols, P. D., and White, D. C., "Phospholipid, Ester-Linked Fatty Acid Profiles as Reproducible Assays for Changes in Prokaryotic Community Structure of Estuarine Sediments," F. E. M. S. Microbial Ecol. 31, 147-158 (1985).

8. Hagstrom, A., Larsson, U., Horstedt, P., and Normark, S., "Frequency of Dividing Cells, a New Approach to the Determination of Bacterial Growth Rates in Aquatic Environments," Appl. Environ. Microbiol. 37, 805-812 (1979).

9. Henson, J. M., Smith, P. H., and White, D. C., "Examination of Thermophilic Methane-Producing Digesters by Analysis of Bacterial Lipids," Appl. Environ. Microbiol. 50, 1428-1433 (1985).

10. Iannotti, E. L., Fisher, J. R. and Sievers, D. M., "Characterization of Bacteria from a Swine Manure Digester," Appl. Environ. Microbiol. 43, 136-143 (1982).

11. Jannasch, H. W., and Jones, G. E., "Bacterial Populations in Seawater as Determined by Different Methods of Enumeration," Limnol. Oceanogr. 4, 128-139 (1959).

12. Kirsh, E. J., "Studies on the Enumeration and Isolation of Obligate Anaerobic Bacteria from Digesting Sewage Sludge," Devel. Indus. Microbiol. 10, 170-176 (1969).

13. Mah, R. A., and Sussman, C., "Microbiology of Anaerobic Sludge Fermentation. I. Enumeration of the Nonmethanogenic Anaerobic Bacteria," Appl. Environ. Microbiol 16, 358-361 (1968).

14. Mancuso, C. A., Nichols, P. D. and White, D. C., "A Method for the Separation and Characterization of Archaebacterial Signature Ether Lipids," J. Lipid Res. 27, 49-56 (1986).

15. Nickels, J. S., King, J. D. and White, D. C., "Poly-Beta-Hydroxybutyrate Metabolism as a Measure of Unbalanced Growth of the Estuarine Detrital Microbiota. Appl. Environ. Microbiol. 37, 459-465 (1979).

16. White, D. C., Bobbie, R. J., Herron, J. S., King, J. D., and Morrison, S. J., "Biochemical Measurements of Microbial Mass and Activity from Environmental Samples. In: Native Aquatic Bacteria: Enumeration, Activity and Ecology, Costerton, J. W. and Colwell, R. R. (eds.)., 69-81, ASTM STP 695, American Soc. for Testing and Materials (1979).

17. White, D. C., Davis, W. M., Nickels, J. S., King, J. D., and Bobbie, R. J., "Determination of the Sedimentary Microbial Biomass by Extractible Lipid Phosphate, "Oecologica 40, 51-62 (1979).

18. Wolin, N. R., "The Rumen Fermentation: a Model for Microbial Interactions in Anaerobic Ecosystems," Adv. Microbial. Ecol. 3, 49-77 (1979).

19. Zeikus, J. G., "The Biology of Methanogenic Bacteria," Bacteriol. Rev. 41, 514-541 (1977).

IMPROVING PRODUCT GAS COMPOSITION BY TWO-PHASE DIGESTION OF BIOMASS

Kerby F. Fannin, Ph.D.
Assistant Director, Biotechnology and Environmental Research
Vipul J. Srivastava, M.S.
Manager, Bioengineering Research
Institute of Gas Technology
Chicago, IL 60616

H. Ronald Isaacson, Ph.D.
Program Manager, Renewable Resources
Gas Research Institute
Chicago, IL 60631

ABSTRACT

The economic feasibility of using biomass for the production of methane gas is determined by the costs of production relative to the market price of the gas from other sources. Gas costs from biomass are determined by two primary processes: biomass growth and harvesting;[1] and the conversion of the biomass to gas and the subsequent clean-up of the gas to pipeline quality methane.[2] Growth and harvesting costs are determined by the specific form of biomass used and the availability of economical growth and harvesting methods. Likewise, the costs of conversion are determined by the capital costs and conversion efficiency of the particular process used. Because the product gas contains large amounts of carbon dioxide, the gas clean-up is a major cost component for producing pipeline quality gas. Increasing the methane content by two-phase digestion should improve the overall process economics.

The purpose of this study was to improve the hydrolysis of kelp and to increase the methane content of the product gas. A two-phase digestion system was used which consisted of two upflow solids digesters operated on seawater-diluted kelp in series with each having different retention times. A mesophilic two-phase anaerobic digestion system utilizing USR digesters significantly increased the methane concentration of the product gas. Gas produced in the methane phase had methane concentrations as high as 74.9 mol % at overall system hydraulic retention times of as low as 8.5 days without significantly sacrificing the methane yield. The data suggest that the hydrolysis limitation observed with undiluted kelp was overcome by using diluted kelp in the process.

IMPROVING PRODUCT GAS COMPOSITION BY TWO-PHASE DIGESTION OF BIOMASS

BACKGROUND

Methane fermentation, which occurs widely in nature has been exploited for purposes of waste volume reduction and for the production of methane gas from numerous types o biomass and wastes. However, this process produces gas which, in addition to methane, contains large amounts of carbon dioxide and certain other trace impurities. The proportion of the gas comprised of methane is affected by the conditions of digester operation, which in turn affect the ecology of the major groups of bacteria in the system.

The economic feasibility of using biomass for the production of methane gas is determined by the costs of production relative to the market price of the gas from other sources. Gas costs from biomass are determined by tw primary processes: 1) biomass growth and harvesting; and 2) the conversion of the biomass to gas and the subsequent clean-up of the gas to pipeline quality methane. Growth an harvesting costs are determined by the specific form of biomass used and the availability of economical growth and harvesting methods. Likewise, the costs of conversion are determined the capital costs and conversion efficiency of the particular process used. Because the product gas contains large amounts of carbon dioxide, the gas clean-up is a major cost component for producing pipeline quality gas. Increasing the methane content by two-phase digestion should improve the overall process economics.

By some estimates, the costs of gas clean-up ranges from 49% to 67% of the costs of conversion for a highly degradable form of biomass, such as giant brown kelp.[1] These high costs for gas clean-up can seriously impact the economic competitiveness of producing pipeline quality methane from biomass. The options for reducing these high gas clean-up costs include developing more economical gas cleaning processes or producing gas containing higher concentrations of methane.

Several biological processes exist for converting biomass to methane and carbon dioxide. These processes utilize such conventional digester designs as single-tank continuously stirred tank (CSTR), unconventional upflow solids (USR), attached film (AFR), and plug flow digesters.[2] In addition, certain processes employ two or more of these digesters operating in series as multistage systems. One multistage system exploits the differing growth characteristics of the two major functional types of microbial populations in anaerobic digestion and is referred to as a two-phase digestion process.[3]

Methane production by anaerobic digestion occurs widely in nature in such environments as ocean and lake sediments, marshes, and digestive tracts of animals. This process, illustrated in Figure 1,[4] involves the biological conversion of the organic components of biomass into simple products such as volatile acids (acetate, propionate. etc.), carbon dioxide, and hydrogen by a mixed population of non-methanogenic bacteria.[5,6,7] These products are then utilized by a mixed population of methanogenic bacteria to produce methane and carbon dioxide. The non-methanogenic bacteria are a relatively hearty and fast-growing group of organisms, whereas the methanogens are more fastidious and slow-growing. Because at least two very distinct groups of organisms are involved in anaerobic digestion, the two-phase process is used to promote the predominant activity of each group of microorganisms in each of the two separate digester.

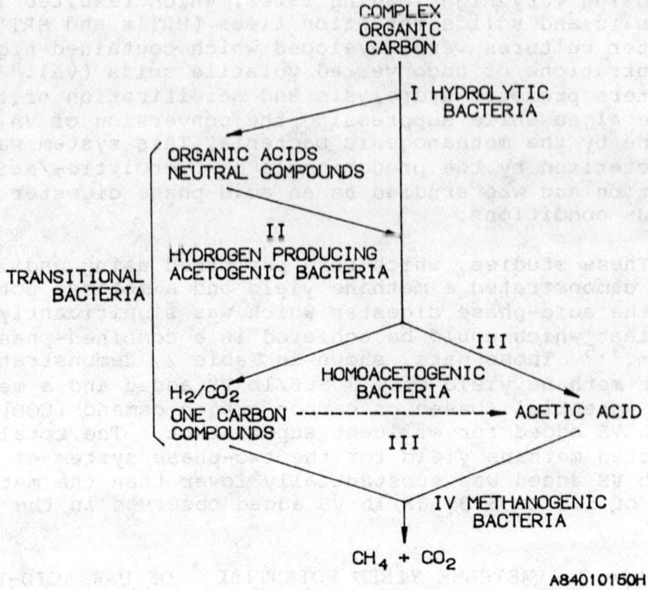

Figure 1. ANAEROBIC DIGESTION PROCESS

Numerous reports suggest major advantages of two-phase over combined-phase digestion.[3,5,8,9] Some of these reported advantages are listed in Table 1.[3]

TABLE 1. ADVANTAGES OF TWO-PHASE DIGESTION[3]

- Maximization of conversion rates through independent control of operating parameters such as pH, temperature, redox potential, retention time, biomass recycle for each phase

- Greater process stability with respect to feedstock loading, pH, and toxic shock

- Reduced gas clean-up and upgrading costs resulting from enrichment of the product gas in the methane phase

Much of the work on two-phase anaerobic digestion has been performed using various types of wastes, such as those from industrial and municipal wastes.[10,11] In addition, techniques have been developed and evaluated that promote high concentrations of volatile acids in the first phase of anaerobic digestion systems using non-waste biomass feeds.[5] Giant brown kelp (Macrocystis pyrifera), a form of marine biomass, was found to be a highly biodegradable feedstock with predictable performance during anaerobic digestion. Because a large data base exists on the anaerobic digestion of this feedstock, it was selected for further studies on the use of two-phase digestion to improve the hydrolysis and to increase the methane concentration of the gas produced by biological gasification in the methane-phase digester.

Using very high loading rates, which resulted in lower hydraulic and solids retention times (HRT's and SRT's), digester cultures were developed which contained high concentrations of unconverted volatile acids (VA).[5] These digesters promoted hydrolysis and acidification of the marine algae while suppressing the conversion of VA to methane by the methanogenic bacteria. This system was characterized by the predominance of hydrolytic-/acid-phase digestion and was studied as an acid-phase digester under various conditions.

These studies, which were performed using undiluted kelp, demonstrated a methane yield and a methane potential from the acid-phase digester which was significantly lower than that which could be achieved in a combined-phase system.[3,5] Those data, shown in Table 2, demonstrated a direct methane yield of 1.3 SCF/lb VS added and a methane yield potential (based on carbon oxygen demand [COD]) of 3. SCF/lb VS added for effluent supernatant. The total projected methane yield for the two-phase system of 4.5 SCF/lb VS added was substantially lower than the methane yield of 5.6 to 6.0 SCF/lb VS added observed in the

Table 2. METHANE YIELD POTENTIAL * OF USR ACID-PHASE DIGESTERS RECEIVING UNDILUTED KELP LOT 54 AT 0.70 LB VS/FT^3-DAY LOADING RATE[3]

Hydraulic Retention Time, days	6.4
Methane Yield (MY) of Acid-Phase Digester, SCF/lb Organic Matter (OM) added	1.4
MY Potential of Acid-Phase Supernatant, SCF/lb OM added	3.1
MY Potential of Two-Phase System Without Solids Recycle, SCF/lb OM added	4.5
MY Potential of Solids, SCF/lb OM added	2.3

*Using COD data. COD analysis accounts for all types of organic matter present in the sample.

combined-phase USR digester. This reduced yield was accounted for in the unreacted solids. The lack of conversion of these solids was attributed to either lack of sufficient retention of solids in the hydrolysis-/acid-phase or inhibition of hydrolysis and fermentation by the accumulation of the fermentation products.

Analyses of VA from those studies at a loading rate of 0.7 lb VS/ft^3-day are shown in Table 3. Total VA concentrations at this loading rate approached 30,000 mg/L. The major acids in decreasing order of concentration were acetic, propionic, and butyric. The significant portion of the COD that could not be accounted for by VA was attributed to other fermentation products such as alcohols and non-VA.

Table 3. DISTRIBUTION OF VA IN THE ACID-PHASE DIGESTER EFFLUENT RECEIVING KELP LOT 54 AT A 0.70 LB VS/FT^3-DAY LOADING RATE AND A 3 DAY HRT[3]

	VA Conc., mg/L	COD Equiv., mg/L
Acetic Acid	20,430	21,700
Propionic Acid	8800	13,300
Isobutyric Acid	130	240
Butyric Acid	1140	2070
Isovaleric Acid	150	330
Valeric Acid	100	200
Caproic Acid	30	70
Totals	29,790	37,990
Total Effluent Supernatant COD, mg/L	45,610	45,610
Percent of COD Accounted For by VA, %	83	83

The purpose of the present study was to develop a method for reducing the effects of hydrolysis limitation on the two-phase digestion of kelp and to improve the methane content of the product gas.

MATERIALS AND METHODS

Feedstock Collection, Processing and Analyses

Giant brown kelp (*Macrocystis pyrifera*) Lot 54, harvested in Southern California by the U.S. Department of Agriculture, Western Regional Research Center, Albany,

California, was used in this study. After harvesting, the kelp was drained of water, chopped, ground and frozen for analyses and use in digestion experiments.

Digester Design and Operation

The USR, illustrated in Figure 2, were used in both the first- and second-phase systems. This digester design, which has been previously described, promoted long microorganism and solids retention times, thereby increasing the loading potential of the conversion system.[4] Digesters were inoculated with a culture originally derived from a mixture of effluents from a pilot-scale digester receiving municipal solid waste/sewage sludge and from a

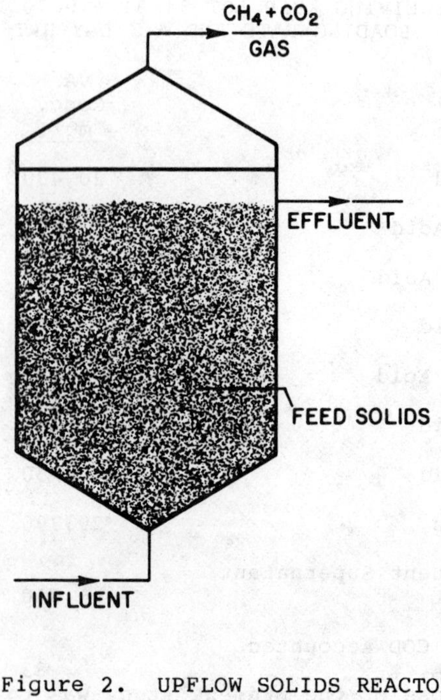

Figure 2. UPFLOW SOLIDS REACTOR

municipal high-rate digester that had been converted since July 1975 to kelp under a variety of operating conditions. Feed was added from the bottom and effluent was removed from the top of the non-mixed vessels. Solids (including microorganisms and feed solids) were passively concentrated by settling, resulting in longer solids than liquid retention times.[4] Previous data showed that performance in terms of methane yield and methane production rate was consistently better using USR than CSTR digesters.[4]

The two-phase digestion system consisted of two USR's operated on simulated seawater-diluted kelp (Lot 54) in series at different retention times. The first USR was operated as a hydrolysis-/acid-phase digester at hydraulic retention times (HRT's) ranging from 4 to 3 days and received simulated seawater-diluted (50 vol %) kelp feed. The second USR was operated as a methane-phase digester at HRTs ranging from 7 to 5.5 days and received the effluent of the first-stage digester. Using the second digester in which the HRT's and SRT's were longer, a culture was developed which promoted the growth of methanogenic bacteria and was referred to as a methane phase digester. Coupling the first-stage hydrolytic-/acid-phase digesters to the methane-phase digester it was possible to develop a two-phase system. Both digesters had separate gas collectors and were maintained in a temperature-controlled chamber at 35°C.

Analytical Methods

The kelp feed was analyzed for total and volatile solids, heating value, elements, and major organic components. Digester gas was analyzed for composition and the digester effluent was analyzed for VA, COD, and pH. All analyses were performed using methods which were previously described.[4]

RESULTS

Feed Characteristics

The characteristics of the kelp lot used in this study are shown in Table 4. The kelp had a total solids content of 11.8%, consisting of 57.9% volatile solids, of which

Table 4. CHARACTERISTICS OF *MACROCYSTIS PYRIFERA* LOT NO. 54

Total Solids (TS), %	11.8
Moisture, %	88.2
Volatile Solids (VS), % of TS	57.9
Ash, % of TS	42.1
Elements, % of TS	
Carbon	25.7
Hydrogen	3.3
Nitrogen	2.2
Phosphorus	0.3
Sulfur	1.2
C/N	11.7
C/P	85.7
Mannitol, % of TS	15.4
Heating Value, Btu/lb VS	7500

mannitol was a major component, and 42.1% ash. The carbon-to-nitrogen ratio was 11.7 and the carbon-to-phosphorus ratio was 85.7. Previous studies with other kelp lots suggest that ratios are not limiting to anaerobic digestion.[2] The heating value of the kelp was 7500 Btu/lb VS.

Two-Phase Digestion Studies

Performance data on the two-phase digestion of diluted kelp in USR's operated at 35°C and at an overall HRT of 11 days are shown in Table 5. At a loading rate of 0.55 lb VS/ft^3-day in the first stage, the system loading rate was 0.2 lb VS/ft^3-day. Ninety-three percent of the methane produced was recovered from the second stage of the system. The 68% methane content observed in the second stage was significantly higher than the 25% methane content of the first stage digester. The VA concentration was about ten times higher in the first compared to the second stage. The methane yield from the combined system was 6.4 SCF/lb VS added, which represents a significant increase over the methane yield of 4.9 SCF/lb VS added which was previously observed with this same kelp lot in a combined-phase USR.[3]

TABLE 5. TWO-STAGE DIGESTION OF DILUTED KELP AT AN 11 DAY HRT

	Stage 1	Stage 2	System
Hydraulic Retention Time, days	4	7	11
Loading Rate, lb VS/ft^3-day	0.55	--	0.2
Methane Yield, SCF/lb VS added	--	--	6.3
Methane Produced, std L	1.0	13.0	14.0
Methane Content, %	25	68	--
VA, mg/L as acetic	7710	730	--
pH	5.5	7.9	--

When the system HRT was reduced to 8.5 days, 92% of the total methane from the system was produced in the second stage. As shown in Table 6, the methane content of the gas in the second stage increased to 74.9% while the methane content of the first-stage gas was only 19.1%. Decreasing the HRT from 11 to 8.5 days decreased the overall methane yield from 6.4 to 6.0 SCF/lb VS-added. However, the methane content of the gas from the methane phase increased by nearly 7%.

The VA composition in the first-stage digester using diluted kelp, shown in Table 7, changed significantly from previous studies performed with undiluted kelp, shown in Table 3.

TABLE 6. TWO-STAGE DIGESTION OF DILUTED KELP AT 8.5 DAYS HRT

	Stage 1	Stage 2	System
Hydraulic Retention Time, days	3	5.5	8.5
Loading Rate, lb VS/ft^3-day	0.7	--	0.25
Methane Yield, SCF/lb VS added	--	--	6.0
Methane Produced, std L	1.4	15.3	--
Methane Content, %	19.1	74.9	--
VA, mg/L as acetic	9590	1076	--
pH	5.4	7.5	--

TABLE 7. DISTRIBUTION OF VA IN THE ACID-PHASE DIGESTER EFFLUENT RECEIVING SIMULATED SEAWATER-DILUTED KELP LOT 54 AT DIFFERENT LOADING RATES

	VA Conc., mg/L	COD Equiv., mg/L	VA Conc., mg/L	COD Equiv., mg/L
Loading, lb VS/ft^3-day	----- 0.55	-----	----- 0.73	-----
Acetic Acid	3220	3430	3480	3700
Propionic Acid	770	1164	1000	1510
Isobutyric Acid	120	220	240	440
Butyric Acid	4500	2170	7380	13,400
Isovaleric Acid	0	0	0	0
Valeric Acid	470	960	260	530
Caproic Acid	1130	2490	470	1030
Totals	10,210	16,430	12,840	20,610
Total Effluent Supernatant COD, mg/L	20,800	20,800	20,430	20,430
COD Accounted for by VA, %	79	79	101	101

Comparing the two digesters operated at loading rates of 0.7 lb VS/ft^3-day (7-day HRT) for the undiluted kelp and 0.73 lb VS/ft^3-day (3-day HRT) for the diluted kelp, the total VA concentration using diluted kelp decreased from near 30,000 to almost 13,000 mg/L. Longer chained acids, such as butyric, valeric, and caproic acids, represented a greater percentage of the total VA in the diluted than in the undiluted kelp digester. It is noteworthy that at the 0.7 lb VS/ft^3-day loading rate, virtually all of the VA was accounted for as COD, suggesting that the other soluble organic compounds were not present in significant quantity in the effluent supernatant.

DISCUSSION

Much of the work on the anaerobic digestion of biomass and wastes is directed toward improving the overall efficiency and rate of digestion. When the objective of the process is to produce methane, especially that which can be considered to be of pipeline quality, consideration should also be given to developing processes that improve the overall quality of the gas. By increasing the methane content of the gas, the costs of gas clean-up may be significantly reduced, which can improve the overall economics of the process and reduce the costs of the product methane.

The methane content of the product gas from the combined-phase anaerobic digestion of undiluted kelp in CSTR and USR digesters has been found to decrease as the loading rate increases and to range from about 48 to 57 mol % at HRT's of 7 to 50 days.[4] Of course, the performance of kelp in anaerobic digestion, like other feedstocks, is also affected by its composition.[7] However, the work performed in this study demonstrates that the combination of USR digesters as part of a two-phase system can increase the methane concentration of the product gas in the methane phase to as high as 74.9 mol % at HRT's as low as 8.5 days without significantly sacrificing the methane yield.

Performance of the two-phase digestion process was significantly improved by the use of diluted rather than undiluted kelp. The data suggest that the hydrolysis limitation observed with undiluted kelp was overcome by using diluted kelp in the process. Two-phase digestion of marine algae is another way of improving the overall economics of the process. Separation of the acid or hydrolytic phases of anaerobic digestion from the methane phase, can help increase the methane concentration of the product gas, thus decreasing gas clean-up costs.

ACKNOWLEDGMENTS

The technical assistance of John Conrad and Jana Mensinger and the technical consultation of David Chynoweth are appreciated. This work was supported in part by the Gas Research Institute through Contract No. 5083-226-0773.

REFERENCES

1. Bird, K. T., "An Historical Perspective on Economic and Systems Costs of Kelp Cultivation for Bioconversion to Methane", in Algal Biomass Workshop of the Solar Energy Research Institute. Bill Barclay, Ed., 1984.

2. Chynoweth, D. P, "Anaerobic Digestion of Marine Biomass," in Proceedings of a Seminar on Biomass Energy for City, Farm, and Industry: Biomass and Alcohol Fuels Production, 185-192. Emmaus, PA: The J. G. Rodale Press, Inc., 1984.

3. Ghosh, S. and Klass, D. L.. "Two-Phase Anaerobic Digestion," in Symposium Papers: Clean Fuels from Biomass and Wastes, 373-415. Chicago: Institute of Gas Technology, 1977.

4. Fannin, K. F., Srivastava, V. S. and Chynoweth, D. P., "Unconventional Anaerobic Digester Designs for Improving Methane Yields from Sea Kelp", in Symposium Papers-Energy from Biomass and Wastes VI, Sponsored by IGT, 375-400, Chicago: Institute of Gas Technology, 1982.

5. Fannin, K. F., Srivastava, V. J., Mensinger, J., Conrad, J. R. and Chynoweth, D. P. "Marine Biomass Program: Anaerobic Digester Process Development." IGT Projects 65044 and 30547 Final Report, Chicago: Institute of Gas Technology, 1982.

6. Fannin, K. F., Srivastava, V. J., Mensinger J. D. and Chynoweth, D. P., "Marine Biomass Program: Anaerobic Digestion Systems Development and Stability Study." IGT Project 30558 Final Report, Chicago: Institute of Gas Technology, 1983.

7. Fannin, K. F., Srivastava, V. J., Chynoweth, D. P. and Bird, K. T. "Effects of the Interaction Between Biomass Composition and Reactor Design on Anaerobic Digester Performance," In Symposium Papers: Energy From Biomass and Wastes VII, 563-584. Chicago: Institute of Gas Technology, 1983.

8. Ghosh, S., Bostian, H. E., Henry, M. P., Sajjad, A., Farrell, J. B., and Salotto, B. V., "Improved Anaerobic Digestion by Staged Fermentation and Advanced Reactor Design," Presented at the American Institute of Chemical Engineers National Meeting, Chicago, 1985.

9. Chynoweth, D. P., Jerger, D. E., Srivastava, V. J., Conrad, J. R., Conrad, J. R. and Mensinger, J. D., "Biological Gasification of Renewable Resources." IGT Project 30571 Annual Report, Chicago: Institute of Gas Technology, 1985.

10. Ghosh, S., Ombregt, J. P., DeProost, V. H. and Pipyn, P. "Methane Production from Industrial Wastes by Two-Phase Anaerobic Digestion," in Symposium Papers: Energy from Biomass and Wastes VI, Sponsored by the IGT, 323-339, Chicago: Institute of Gas Technology, 1982.

11. Ghosh, S., Sajjad, A, Henry, M. P., and Bleakney, R. A., in Symposium Papers: Fifth Symposium on the Biotechnology for Fuels and Chemicals, Gatlinburg, TN, 1983.

DISCUSSION

G. Shahani, Air Products & Chemicals Inc. Can you please describe the configuration of the second-phase reactor and how it was connected to the first phase? Was there a recycle of liquid? What was packed inside phase two?

Fannin These were two upflow solids reactors. In this system, the slurry was passed straight through both phases. Recycle was not done in this work. The major difference between the two tanks was the hydraulic retention times to establish the acid phase and the methane phase in two separate reactors.

Smith Kelp is a rather specific material. How do you think your results might have been had you used some other form of biomass?

Fannin Kelp is a highly biodegradable biomass. When we compare that with some other form of biomass, there are significant differences in biodegradability, but the principles are the same. The hydraulic retention times and the efficiency of conversion may be different. Although kelp isn't necessarily a typical biomass, it is representative of a large amount of biomass that exists on the earth today.

PAP/kff-pap

ENSILAGE STORAGE OF SORGHUM AND HIGH MOISTURE BIOMASS CROPS FOR ANAEROBIC DIGESTION FEEDSTOCKS

Charlie G. Coble, Ph.D., Professor, and
Richard Egg, M.S., Research Agricultural Engineer
Agricultural Engineering Department
Texas A&M University
College Station, Texas 77840

ABSTRACT

A number of seasonal land-based biomass crops are potential methane feedstocks. Sorghum, a leading candidate, normally contains 70% moisture when harvested making drying for storage extremely expensive. Field drying results in excessive dry matter losses. Ensiling is a well developed storage method for preserving digestible nutrients in high moisture forages used as animal feeds. Ensiling can be adapted to preserving fermentable organics in crops grown for methane production.

This paper gives an overview of the ensiling process and how it can be applied to high-moisture biomass storage. Results are reported of a study in which sorghum was ensiled and then digested in a two stage anaerobic digestion system demonstrating the feasibillity of the ensilage storage concept.

ENSILAGE STORAGE OF SORGHUM AND HIGH MOISTURE BIOMASS CROPS FOR ANAEROBIC DIGESTION FEEDSTOCKS

INTRODUCTION

A number of land based biomass species can be raised for energy feedstocks. These include high biomass yielders such as sugarcane, napier grass, and a number of sweet sorghum cultivars. Sugarcane is currently used as an energy crop in Brazil where it is the primary feedstock for ethanol production. Sweet sorghums have received considerable attention in the U. S. over the past several years as an ethanol feedstock. Potential energy production per unit land area for several cultivars is greater than for corn, the most common ethanol feedstock in the U. S.

These species can be used as methane feedstocks as well as for ethanol production. Methane production may prove to be a more efficient conversion process than ethanol fermentation for utilization of the plant biomass. Ethanol fermentation is limited to the conversion of sugars and starches unless additional hydrolysis pretreatments are used. Anaerobic digestion, on the other hand, is capable of converting a large portion of the structural carbohydrates to energy as well as the starches and sugars.

Several factors have prevented the commercial develoment of these land based biomass crops for energy production on a wide scale in the U. S. One important factor is the development of suitable storage systems. With the possible exception of sugarcane raised in the southern extremities of the U. S., energy crop production in temperate climates is a seasonal activity. This necessitates some form of feedstock storage to enable year round production of energy. High biomass yielding crops such as sugarcane and sweet sorghum are generally characterized as containing high levels of sugars and having moisture contents in the range of 65-75% at harvest. These conditions can lead to rapid deterioration or loss of fermentable organic material unless some preservative treatment is employed to arrest the activities of plant enzymes and microorganisms.

The agricultural industry has developed a number of storage and preservation technologies for forage crops. Many of these methods can be adapted with varying degrees of success to the preservation and storage of non-woody, land based biomass energy crops, particularly when the crops are used for methane production. Forage crops are produced primarily for ruminant nutrition. Because a ruminant animal contains an "anaerobic digester," it can be argued that any storage process that conserves plant nutrients from the field to the feedlot would do the same for biomass crops used for anaerobic digestion. In general this is true. However, the storage requirements of biomass used for energy production may not be as stringent as when used for animal nutrition. For example, protein digestibility, palatability, and dry matter intake are of prime importance in animal nutrition. These concerns may be of little consequence in anaerobic digestion systems where the key factor is the preservation of energy during storage.

Energy production from biomass has been the object of substantial research over the past several years. Laboratory studies have explored new conversion processes, adapted existing processes to alternate, and evaluated new plant species. In many cases, the feedstock used in the research has been plant material that was frozen or oven dried immediately after harvest. As a result, biochemical changes in the feedstock that occur during preservation and storage are not considered. These changes can significantly impact the feasibility of scaling up a laboratory system to a commercial system. Numerous systems exist that can be used to store biomass energy crops. Systems that minimize storage losses need to be identified, and the resulting biochemical changes in the feedstocks must be quantified before "energy farms" become a reality.

The objective of this paper is to explore the ensilage process for the storage of high moisture biomass crops used for methane feedstocks. The focus will be on sweet sorghums as the energy crop since the Texas Agricultural Experiment Station is conducting research on the production of methane from sorghums. However, the process is adaptable to, and has been used for, the preservation of a broad range of plant species.

STORAGE SYSTEMS

Existing forage storage systems fall into two broad categories, low moisture and high moisture. Low moisture storage usually takes the form of hay production and storage while ensilage is the basic process for high moisture storage of forage. A number of intermediate moisture storage methods have been developed, such as haylage and high moisture hay with preservatives added, but are not yet in widespread use.

Hay Systems

Hay making depends on the rapid removal of moisture from the harvested crop to preserve the plant nutrients during storage. In producing hay, the crop is first cut, normally with a mower-conditioner that crushes the plant stems to increase the drying rate. It is then allowed to field dry for several days until the moisture content is 20% or less. Field drying is followed by baling and then storage.

The major drawback to hay systems for storage of biomass crops is the high degree of dry matter loss associated with the process. Martin (9) and Rees (14) reviewed a number of studies on dry matter losses associated with hay production. Martin (9) estimated the losses from making and transporting hay to storage could range from 10 to 71 percent, and storage losses were an additional 18-42%. Rees (14) found total dry matter losses from making and storing hay to range from 18 to 30 percent. In a study of storage losses in hay made from three varieties of sorghums, Coble and Egg (4) measured dry matter losses ranging from 8.9 to 35.6 percent.

Losses in hay systems can be minimized by management and proper storage, but some of the losses are mutually unavoidable and others beyond control. Losses due to mechanical operations can be reduced by raking and baling at higher moisture contents. This, however, leads to higher losses during storage. Weather conditions play a significant role in the loss of dry matter during hay making and storage. Leaching of soluble dry matter from cut hay by rainfall has been estimated at 2-30% (14).

Because of the high potential for uncontrollable losses, hay systems do not appear to be feasible for harvesting and storing sorghums and other biomass crops for methane production.

Ensilage Systems

The process normally used for high moisture storage of forages is ensilage. While preservation of biomass as hay is a physical process, ensilage is a biochemical process. Requirements for successfully ensiling a forage crop are exclusion of air from the product during storage and a lactic acid fermentation of WSC (water soluble carbohydrates) contained in the plant. Substantial dry matter and energy losses can occur in ensilage as well as in hay systems, but these losses can be controlled more readily by management practices and storage structures. And because requirements for field drying are greatly reduced, the detrimental affects of weather are minimized.

ENSILAGE

A number of thorough reviews of silage making and the microbiology and biochemistry of ensilage are available (11, 12, 15), and the literature dealing with the subject is quite extensive. Most of this work has been based on grasses and legumes. While the basic process of ensiling sweet sorghums is the same, the end products for individual cultivars will vary according to the physical and chemical properties of the plant biomass.

McDonald defines silage as "the material produced by the controlled fermentation of a crop of high moisture content" (12). Ensilage is the process of silage making, and the structure in which ensilage occurs is the silo.

During ensilage the raw material undergoes an anaerobic fermentation that produces lactic, acetic, and butyric acids from the WSC present in the plant. This causes a pH reduction that prevents the growth of spoilage microorganisms which are intolerant of acid conditions (15).

Silage Making

Silage making today is a completely mechanized operation with two basic harvesting methods employed: direct cut or pre-wilting. The method used depends on a number of factors such as type of crop, type of silo, moisture content of the crop at harvest, and management system.

In direct cut ensilage, the standing crop is harvested with a forage chopper. There are several types of forage choppers, but in essence, the harvester chops the plant into particles that can range in size from less than 1 to greater than 25cm. The chopped forage is blown into a wagon and transported to the silo. With direct cut silage, the dry matter content of the material placed in the silo is the same as the dry matter content of the standing plant in the field. With this method there is no control, other than harvest scheduling, of the dry matter content of the silage which can have a significant affect on the fermentation process.

Pre-wilting is a technique used to control the dry matter content in the silo. When silage is to be pre-wilted, it is first cut with a mower or mower-conditioner and left to partially dry in the field until the desired dry matter content is reached. This wilting period can range from several hours to several days depending on the type of crop, weather, and target dry matter content. The pre-wilted crop is then picked up from the field and chopped with a forage chopper and transported to the silo.

Pre-wilting is used in ensilage for a number reasons. In general, the chance of storage losses is lessened by pre-wilting. Seepage losses, which are often quite high with wet crops, are reduced or eliminated by pre-wilting. Since clostridia bacteria are sensitive to osmotic pressure as well as pH, pre-wilting lessens the probability of a butyric acid fermentation with its associated energy losses. This is especially true for crops that contain low levels of WSC. Pre-wilting can also reduce transportation costs since water removed in the field is not transported to the silo.

There are disadvantages associated with pre-wilting, however. Field losses are normally increased with drying time, and weather conditions become important during the wilting period. With heavily wilted crops, oxygen entrapped in the silo during filling is difficult to remove and can result in high temperatures and silage deterioration.

After harvest, the silo is filled and sealed to prevent the entrance of oxygen and the silage fermentation occurs.

Silos

Almost any type of structure can be used for ensiling as long as it can be effectively sealed and has the required structural strength. Many designs have been employed in the history of ensilage. However, there are three basic types of silos in common use: stack silos, bunker silos, and tower silos. Each has its costs and benefits in terms of economics, storage losses, labor needs, and management requirements.

Stack silos are the simplest and least expensive to construct, but have the highest potential for storage losses. A stack silo consists simply of a pad or floor--concrete, gravel, or packed earth. Chopped forage is stacked on the pad and compacted, usually by driving a tractor over it. The stack is then covered with plastic sheeting which is weighted down with old tires, soil, or some other suitable weight. It can be difficult to achieve an airtight seal around the edges of the plastic, and the plastic cover can be easily torn. As a result, aerobic deterioration can be severe with this type of silo because of oxygen infiltration during storage.

A second basic type of silo is the bunker or trench silo. A bunker silo is a rectangular structure with three walls up to about three meters high. The fourth side contains no solid wall, but is left open for machine operations during loading and unloading. It can be constructed either above or below ground. The filling and compacting procedure is essentially the same as with the stack silo. After it is filled the top and open end are covered with plastic. The same problems of aerobic deterioration described for the stack silo apply to the bunker silo. However, because of the three solid walls, the surface area that might be exposed to air is less, and potential losses are reduced.

Tower silos are the third major category in use. This type is the most expensive to construct, but normally has the least storage losses. Loading and unloading of tower silos can be completely automated. Tower silos can be up to 10 m in diameter and over 27 m tall. They are constructed of concrete staves, reinforced concrete, or glass lined steel. Glass lined steel tower silos are available with breather bags that allow for gas expansion in the silo and, as a result, can be made air tight. This type of silo is the most effective for preventing air infiltration during storage.

One relatively new development in silos is the silo press. This is a machine that forces chopped forage into long plastic tubes or sausages. These filled silos are about 2.5 m in diameter and up to 30 m long.

Silage Fermentation

Silage fermentation begins with an initial aerobic period after filling and sealing the silo. Plant respiration continues, consuming sugars and producing carbon dioxide and water, until all the oxygen is removed. Anaerobic conditions are usually attained in several hours or less. During this period microbial activity begins. Lactic acid production begins within the first few days and continues for several weeks of ensiling.

Four major groups of microorganisms are important in ensilage: coliform bacteria, fungi (yeasts and molds), lactic acid bacteria, and endospore forming bacteria (clostridia and bacilli) (15).

The coliform bacteria predominate the early stages of fermentation but are replaced by lactic acid bacteria after about the first week (15). Fermentation of sugars by coliform bacteria results primarily in acetic acid production with some ethanol, hydrogen, lactate, and formic acid produced as well (12). If a rapid lactic acid fermentation occurs, these bacteria have little effect on the silage fermentation.

Yeasts and molds are present in silage, but their activity is of concern mainly in the aerobic deterioration of silage. This occurs in a poorly sealed silo in which oxygen is allowed to infiltrate or after the silo is opened for feeding.

Lactic acid bacteria are the microorganisms responsible for successful ensilage. They are facultative anaerobes that begin activity during the aerobic phase of ensilage. Lactic acid bacteria are divided into two groups based on their fermentation products. Homofermentative lactic acid bacteria ferment sugars entirely to lactic acid while heterofermentative lactic acid bacteria produce ethanol and acetic acid as well as lactic acid. The primary fermentation substrates for lactic acid bacteria are glucose and fructose, although pentoses resulting from hydrolysis of hemicellulose are also fermented. The major fermentation reactions occuring in ensilage are shown in Table 1.

Lactate silages are produced from crops containing high levels of WSC, a typical characteristic of sweet sorghums and sugarcane, in which the lactic acid bacteria dominate the fermentation. Lactate silages have a low pH, in the range of 3.7 to 4.2, and lactic acid contrations of 8-12% DM. Significant amounts of acetate can also occur, primarily from heterofermentative lactic acid and coliform bacteria. Small amounts of propionic and butyric acids are often found in lactate silages, as well as varying amounts of ethanol produced by yeasts. The amount of butyric acid depends to a great extent on the rate at which lactic acid was produced (12).

If there are insufficient WSC present in the forage or the moisture content is too high, a clostridial fermentation can occur. Clostridia bacteria ferment sugars and lactate to produce butyrate (Table 1). Minor fermentation pathways also produce formate, acetate, propionate, ethanol and butanol (15). Clostridia are sensitive to pH and osmotic pressure. The pH at which clostridia inhibition occurs increases as dry matter increases(12). Dry matter contents above 25-30% generally prevent the growth of clostridia. One of the main reasons for pre-wilting silage is to increase the dry matter content and thereby prevent the growth of clostridia bacteria. Clostridia fermentation is undesirable from an energy preservation standpoint since the butyrate fermentation pathway results in considerable loss of gross energy.

Clostridial silages most often occur in wet crops that do not contain sufficient WSC to allow a rapid lactic acid fermentation. Clostridial silages have pH values ranging from 5.0 to 7.0 and low levels of lactate and WSC (12).

Ensilage Losses

The primary objective in preserving biomass for energy production is to prevent energy losses during storage. Ensilage of these crops can potentially keep these losses to a minimum if a lactic acid fermentation occurs. McDonald (12) summarized the energy and dry matter losses resulting from the major fermentation pathways occurring in ensilage (Table 2). These values clearly show that fermentation losses in lactate silages can be almost negligible.

Losses other than fermentation lossses can occur during ensilage (Table 3). Respiration losses result from the catabolism of sugars during the aerobic stage of ensilage. These losses can be minimized by rapid filling of the silo, dense packing, and effective sealing.

Effluent or seepage losses occur when crops are ensiled too wet. The volume of effluent produced depends on the dry matter content of the crop, the pressure in the silo, and the type of silo. Effluent loss can result in significant dry matter and energy loss since the effluent can contain large amounts of WSC and lactic acid. Pre-wilting is the most common way to prevent effluent losses. McDonald concluded that forage dry matter contents of 26-30% for bunker silos and 35-50% for tower silos would effectively prevent effluent losses (12). While effluent losses are a major concern in ensiling forages for animal

Table 1. Principle silage fermentations.

Homolactic fermentation of hexose

Glucose or Fructose --> 2 Lactate

Heterolactic fermentation of glucose

Glucose --> Lactate + Ethanol + Carbon Dioxide

Heterolactic fermentation of fructose

3 Fructose + Water -->
Lactate + 2 Mannitol + Acetate + Carbon Dioxide

Glucose + 2 Fructose + Water -->
Lactate + 2 Mannitol + Acetate + Carbon Dioxide

Pentose fermentation (Both homo and heterofermentative lactic acid bacteria)

Pentose --> Lactate + Acetate

Clostridia fermentation

2 Lactate --> Butyrate + 2 Carbon Dioxide + 2 Hydrogen

Hexose --> Butyrate + 2 Carbon Dioxide + 2 Hydrogen

Table 2. Dry matter and energy losses for common silage fermentations.

	Losses - %	
	DM	Energy
Lactic acid bacteria		
Hexose (homofermentative)	0.0	0.7
Glucose (heterofermentative)	24.0	1.7
Fructose (heterofermentative)	4.8	1.0
Clostridia		
Lactate	51.1	18.4
Yeasts		
Glucose	48.9	

Source: McDonald (12)

Table 3. Energy losses in ensiling.

Process	Avoidable	Loss%
Respiration	Unavoidable	1-2
Fermentation	Unavoidable	2-4
Effluent or	Mutually	5->7
field loss by wilting	Unavoidable	2->5
Secondary fermentation	Avoidable	0->5
Aerobic deterioration in storage	Avoidable	0->10
Aerobic deterioration after storage	Avoidable	0->15
Total		7->40

Source: Zimmer (16)

consumption, the effluent produced in ensiled biomass could be collected and used as an anaerobic digestion feedstock. This would eliminate the energy loss due to seepage. At the same time, field losses associated with wilting would also be eliminated.

Losses due to secondary fermentation can also occur in ensilage. These losses are caused by the fermentation of lactic acid and WSC by clostridia bacteria to form butyric acid. As previously mentioned, if the crop characteristics are favorable for a rapid lactic acid fermentation, the chance of secondary butyric acid fermentation is unlikely.

Aerobic deterioration, either during storage or after the silo is opened for silage utilization, can result in substantial losses. These losses, however are easily preventable by proper silo construction and sealing and management practices.

Zimmer (16) reviewed the various types and magnitudes of energy losses associated with silage production (Table 3). He concluded that total losses could be as low as 7%. Collection and utilization of silage effluent could reduce these losses even further. Ensilage can be seen as a very efficient method for storing biomass energy feedstocks.

SWEET SORGHUM ENSILING

Type of crop, dry matter and sugar content, and management all have a significant impact on the success of ensiling. Sweet sorghums and high energy sorghums (sweet x grain hybrids) have properties that indicate successful ensilage storage. Recent work at the Texas Agricultural Experiment Station has evaluated numerous sorghum cultivars as methane feedstocks(10,13). Dry matter content and non-structural carbohydrate concentrations for selected sweet sorghum cultivars at harvest are show in Table 4. Dry matter ranged from 25.6% to 33.7%, and total sugar concentrations ranged from 33.0 to 46.9% DM. These characteristics, relatively high dry matter and very high WSC, indicate that sweet sorghums could be successfully ensiled with little or no pre-wilting. A number of laboratory studies have reported success with ensiling sweet sorghums.

Heredia (7) ensiled two varieties of sweet and two varieties of high energy sorghum at three maturity stages in laboratory silos. In all cases he obtained good lactic acid fermentations. Dry matter storage losses ranged from 2.1 to 19.9% (Table 5). Organic acid and ethanol concentrations for these silages are shown in Table 6. Lactic acid concentrations for the four varieties averaged 7.8 to 9.2 %. Ethanol concentrations averaged 1.9 to 6.1% indicating possible yeast activity. Only traces of butyric acid were found. His work demonstrates that these sorghums can be successfully ensiled. It also shows the influence

Table 4. Typical characteristics of selected varieties of sweet sorghums.

Variety	DM yield t/ha	DM content % fresh wt	NSC Content* % DM 1983	1984
M81E	23.4	27.9	40.8	46.9
MN1500	20.7	25.6	26.1	33.0
Wray	16.0	27.6	37.6	39.1
Rio	17.8	34.2	40.4	34.6
Brandes	11.9	27.2	35.5	41.5
Atlas	13.6	33.7	29.2	39.9

*NSC -- non-structural carbohydrates (glucose, fructose, sucrose, starch) as hexose equivalent.

Source: Miller and Monk (13), McBee(10).

Table 5. Dry matter loss resulting from ensiling sweet and high energy sorghums when harvested at various stages of maturity.

Variety	Type	Dry Matter Loss - % A	SD	HD
Rio	Sweet	19.9	12.2	9.9
Wray	Sweet	19.0	15.5	14.9
Atx623xRio	High Energy	6.2	2.1	8.3
Atx623xWray	High Energy	7.1	4.6	2.3

A - harvested at 50% anthesis
SD - harvested at soft dough
HD - harvested at hard dough

Source: Heridia (11)

Table 6. Organic acid and ethanol contents of silages made from sweet and high energy sorghums.

Variety	Rio	Wray	Atx623 x Wray	Atx623 x Rio
Lactic Acid (%)	7.8	9.2	8.3	9.1
Acetic Acid (%)	1.2	1.4	1.2	1.4
Propionic Acid (%)	0.0	0.0	0.01	0.0
Butyric Acid (%)	0.0	0.0	0.0	0.0
Ethanol (%)	4.7	6.1	1.9	2.3

Source: Heridia (11)

of plant stage at harvest and plant variety on ensilage losses.

Garrett and Worker (6) ensiled Brawley sweet sorghum for 90 days. At the end of that period, reported dry matter storage losses were only 3.8 and 4.2%.

Catchpoole (3) harvested and ensiled Saccaline sweet sorghum at seven stages of maturity. After 20 weeks of storage, total weight loss ranged from 63 to 0% (Table 7). Losses decreased with increasing crop maturity. He found that harvest before the boot stage (8 weeks) resulted in high losses and high pH values, while later harvest produced excellent quality silage with little loss.

Unfortunately, none of these authors reported gross energy changes in ensiling. But as McDonald showed (12) energy losses in ensilage are significantly less than dry matter losses. These studies of sweet sorghum ensilage demonstrate that sweet sorghums can be successfully ensiled with minimal losses. However, there is potential for extremely large losses as well. The stage of maturity at harvest appears to play an important role.

ENERGY PRODUCTION FROM ENSILED BIOMASS

Ensilage for preservation of biomass energy feedstocks has gained some attention in the past several years. Bryan (2) used a modified ensiling process to produce ethanol from sweet sorghum. Chopped sorghum was inocculated with distillers dry yeast and ensiled. After 63 hours 80% of the initial sugar was fermented to ethanol.

Linden (8) conducted a 66 day ensiling study of sweet soghum with the objective of cellulose hydrolysis. Sorghum was ensiled with varying levels of cellulase. Enzymatic hydrolysis of lignocellulose produced up to 200 mg sugars/g dry matter after 18 days of ensilage.

Barry and Colleran (1) fed silage effluent to a fixed be anaerobic digester. Reported methane production was 0.49 m^3 per kg COD removed. COD removal in the digester was 75.7%.

Anaerobic Digestion of Ensiled Sorghum

At Texas A&M a system was developed that integrated ensilage storage with a two stage anaerobic digester to produce methane from sorghum (5). The silo/digester system is shown schematically in Figure 1. The system consists of a methane reactor and a number of silos that are used as both storage structures and leaching vessels. With this concept, sorghum is stored in the silos at harvest. A silo is then filled with water and the silo leachate is then fed to the methane phase digester. Effluent from the methane

Table 7. Total ensiling losses and fresh sorghum characteristics for Saccaline sweet sorghum as affected by crop maturity.

Age at harvest weeks	DM content at harvest %	Total sugars at harvest % DM	Silage pH	Ensiling losses % fresh wt
4	9.4	0.0	5.31	63
6	8.4	8.6	5.70	57
8	11.6	9.2	4.09	42
10	18.1	38.1	3.49	14
12	21.3	14.7	3.43	4
14	27.9	7.6	3.74	0
16	31.6	6.1	3.81	4

Source: Catchpoole (3).

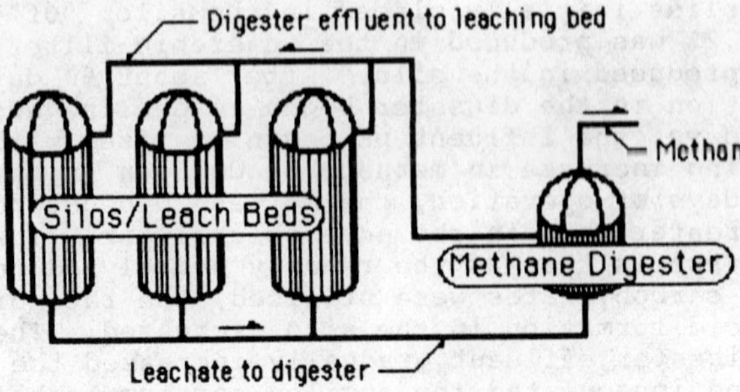

Figure 1. Schematic of two stage silo/digester system.

phase reactor is recycled to the silo to continue to hydrolyze and leach the organic matter in the sorghum.

A silo/digester system was constructed and tested. The system consisted of a 73 liter packed bed anaerobic filter and three 619 liter silos. The silos were filled with chopped sorghum (Atx623 x Rio) and ensiled for approximately 6 weeks. Silo #1 was then filled with water and the leachate was fed to the anaerobic filter, and effluent from the filter was recycled to the silo. Temperature in the anaerobic filter was maintained at 35°C. The temperature in the silo was not regulated. The study was concluded after 203 days of operation. Methane was still being produced but at a very low rate.

Anaerobic Filter Performance. The anaerobic filter performed well throughout the study. COD removal (Figure 2) averaged 71.4% over the 203 day period. COD concentration in the anaerobic filter influent decreased over time (Figure 3), apparently as the more easily digestible organic material in the sorghum was digested.

Methane yield in the anaerobic filter average 0.24 m^3/kg COD added (Figure 4) and 0.35 m^3/kg COD removed (Figure 5). The sharp dip in methane yield on day 91 resulted from a drop in temperature caused by the failure of a heating element in the digester. The temperature dropped to about 10°C for several days. After the heating element was replaced, gas production returned to normal within 24 hours.

The maximum COD loading rate to the filter was 3.03 kg COD/m^3-day (Figure 6). The effluent pH of the anaerobic filter remained stable over the entire digestion period, averaging 7.1 (Figure 7). The influent pH (silo leachate) began low, about 4.2, and gradually rose as the organic acids produced during ensiling were digested.

System Performance. Cumulative methane production from the silo/digester system is shown in Figure 8, and Figure 9 shows the weekly methane produced. Total methane produced from the system was 8.07 m^3 or 0.25 m^3/kg of volatile solids initially placed in the silo. Of this amount, 74.2% was produced in the anaerobic filter, and 25.8% was produced in the silo. After about 90 days, the gas production in the digester began to decline and at about 100 days, the influent pH began to rise, with a corresponding increase in methane production in the silo. After 126 days of operation, the methane production in the silo was greater than in the anaerobic filter. The probable cause is that as the readily hydrolyzed non-structural carbohydrates were utilized, the rate of volatile acid formation in the silo decreased. The recycled digester effluent gradually increased the pH in the silo and innoculated the ensiled sorghum. When the pH reached about 6.0, the methane bacteria were able to survive in the silo, causing the increase in methane production.

Methane concentration of the biogas (Figure 10) averaged 67.0% in the anaerobic filter, remaining relatively constant throughout. In the silo, the methane concentration varied somewhat but averaged only 32.6%. As the methane production began to increase in the silo, the methane concentration also increased.

This study demonstrated the feasibility of anaerobic digestion of ensiled sorghum. Research is continuing with the eventual goal of modeling and optimizing the system.

CONCLUSIONS

Ensilage is a well developed technique for preserving high moisture crops. Though developed for preserving forages in animal nutrition, it can be adapted to preserving biomass crops for energy production as well. Research literature indicates that crops containing high concentrations of sugar can be ensiled with minimal energy losses. Extremely high losses can also occur, but these appear to be avoidable by proper management techniques.

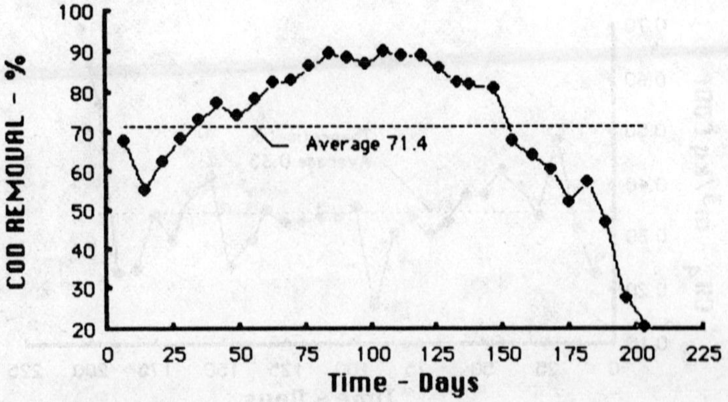

Figure 2. COD removal in anaerobic filter.

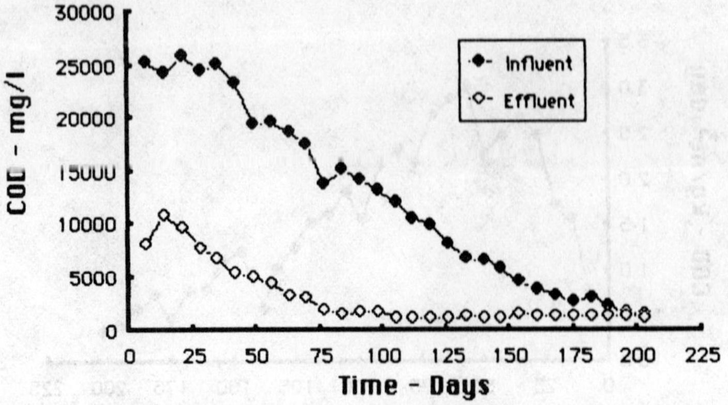

Figure 3. COD concentration in anaerobic filter influent and effluent.

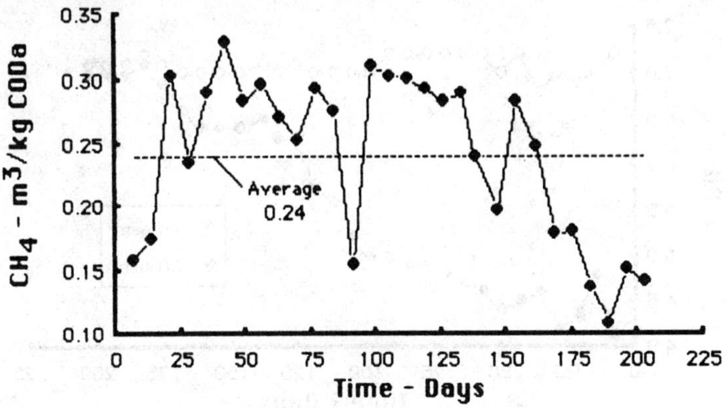

Figure 4. Methane yield in anaerobic filter per kilogram of COD added.

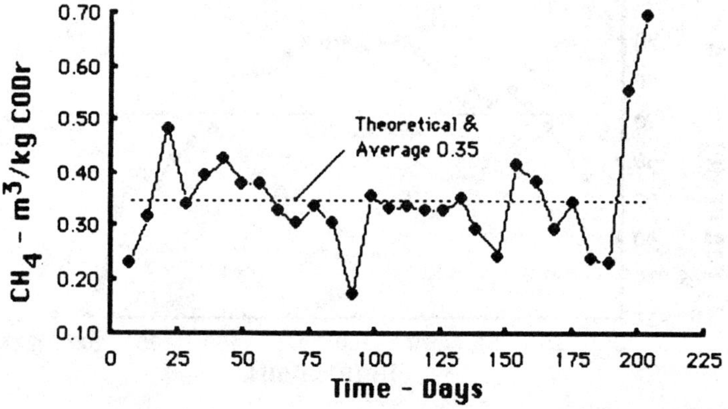

Figure 5. Methane yield in anaerobic filter per kilogram of COD removed.

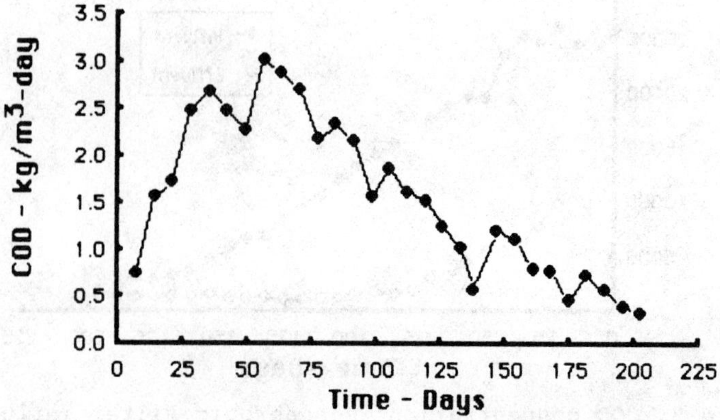

Figure 6. COD loading rate to anaerobic filter.

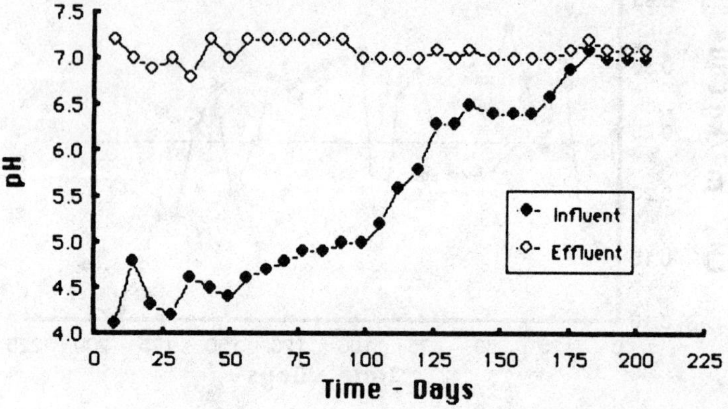

Figure 7. Anaerobic filter influent and effluent pH.

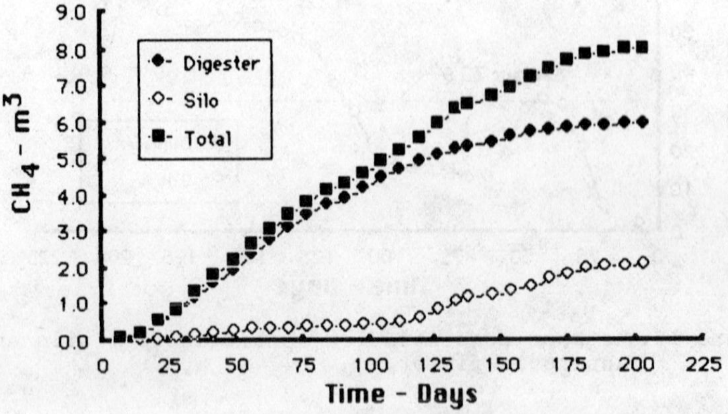

Figure 8. Cumulative methane production from the silo/digester system.

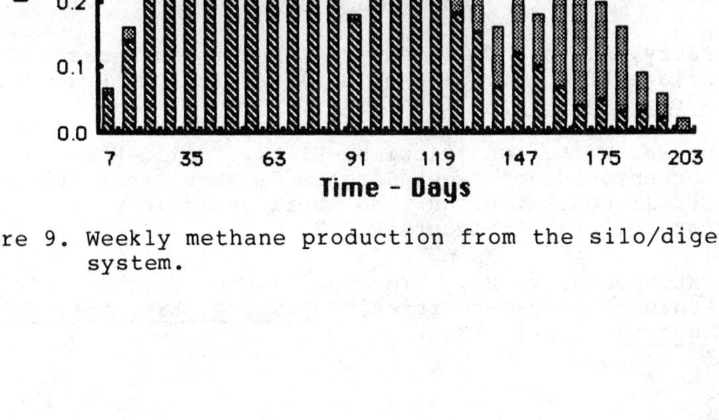

Figure 9. Weekly methane production from the silo/digester system.

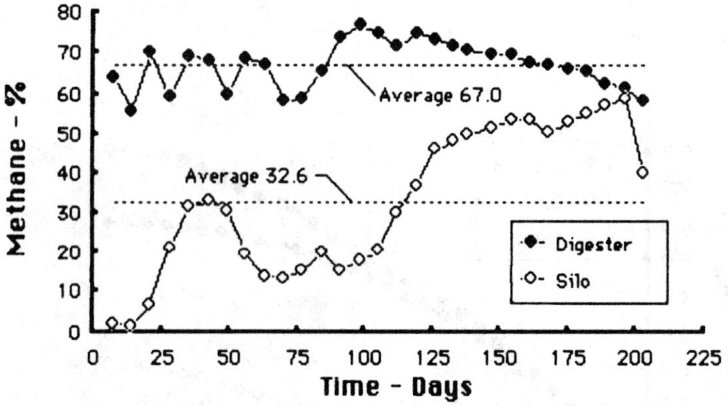

Figure 1Ø. Methane concentration in biogas from silo and anaerobic filter.

It appears that sweet sorghums can be effectively ensiled when used as methane feedstocks without the addition of preservatives or the need for pre-wilting. Ensiled sorghums have been shown to be digestible in methane systems. The rate of digestion and total methane yield may possibly be enhanced by the addition of cellulase during ensilage.

Ensiling can also be used in conjunction with ethanol production. However, additions of enzymes or yeasts are necessary since the lactic acid normally produced in ensilage removes sugars required for ethanol production.

Substantial information on ensilage is available, but additional research is needed to adapt it to storage of biomass energy feedstocks. High biomass producing species need to be tested and accurate economics developed. A viable technology exists that can enhance the feasibility of biomass energy production.

REFERENCES CITED

1. Barry, M., and Colleran, E., "Anaerobic Digestion of Silage Effluent Using an Upflow Fixed Bed Reactor," Agri. Wastes 4, 231-9 (1982).

2. Bryan, W. L., and Parrish, R. L., "Solid-Phase Fermentation of Sweet Sorghum," ASAE Paper #82-36Ø3, St. Joseph, Michigan: The American Society of Agricultural Engineers, 1982.

3. Catchpoole, V. R., "The Ensilage of Sorghum at a Range of Crop Maturities," Aust. J. Exp. Agr. Ani. Husb. 2, 1Ø1-4 (1962).

4. Coble, C. G. and Egg, R., "Low Moisture Storage of Sorghum Used for Methane Production," In: Proceedings of the Sixth Annual Solar and Biomass Energy Workshop. Feb. 25-27, Atlanta, 1986.

5. Egg, R., Coble, C. G., and Hicks, D. D., "A Two Stage Silo/Digester for Methane Production from Sweet Sorghum," ASAE Paper #84-3582, St. Joseph, Michigan: The American Society of Agricultural Engineers, 1984.

6. Garrett, W. N., and Worker, G. F., Jr., "Comparative Feeding Value of Silage Made from Sweet and Dual Purpose Varieties of Sorghum," J. Ani. Sci. 24, 782-5 (1965).

7. Heredia, M. C., High Energy Sorghum Silages Evaluated by In Vitro and In Vivo Digestibility at Parent and Ratoon Harvests, M. S. Thesis, College Station, Texas: Texas A&M University, 1985.

8. Linden, J. C., Henk, L. L., Lange, K. D., and Murphy, V. G., "Ensiled Sweet Sorghum as Fermentation Feedstock," Paper #150, Symposium on State-of-the-Art Review of Fermentation Substrates and Raw Materials, American Chemical Society National Meeting, Chicago, 1985.

9. Martin, N. P., "Harvesting and Storage of Quality Hay," 177-192. In: New Developments in Forages. Lexington, Ky: American Forage and Grassland Council, 1980.

10. McBee, G. G., "Crop Physiology/Biomass Evaluation," In: E. A. Hiler (ed.). Sorghums for Methane Production, Annual Report April 1984 - March 1985 to the Gas Research Institute, College Station, Texas: Texas Agricultural Experiment Station, 1985.

11. McCullough, M. E. (ed.), Fermentation of Silage - A Review, West Des Moines, Iowa: National Feed Ingredients Association, 1978.

12. McDonald, P., The Biochemistry of Silage, New York: John Wiley & Sons, Inc., 1981.

13. Miller, F. R., and Monk, R. L., "Breeding and Development," In: E. A. Hiler (ed.). Sorghums for Methane Production, Annual Report April 1984 - March 1985 to the Gas Research Institute, College Station, Texas: Texas Agricultural Experiment Station, 1985.

14. Rees, D. V. H., "A Discussion of Sources of Dry Matter Loss During the Process of Haymaking," J. Agric. Engng. Res. 27, 469-79 (1982).

15. Woolford, M. K., The Silage Fermentation, New York: Marcel Dekker, Inc., 1984.

16. Zimmer, E., "Efficient Silage Systems," 186-197. In: C. Thomas (ed.), Forage Conservation in the 80's. Occasional Symposium No. 11, British Grassland Society, 1980.

DISCUSSION

W. Smith, U. of Florida: The material that wasn't digested was about 30% or so. Did you look at its chemical nature?

Egg: It was primarily cellulose and lignin. There was very little hemicellulose remaining.

CONTINUOUS ANAEROBIC DIGESTION OF HIGH SOLIDS BIOMASS: MODELING AND EXPERIMENTS

Robert Legrand
Biomass Energy Specialist
Reynolds, Smith and Hills
Post Office Box 4850
Jacksonville, Florida 32201

William J. Jewell
Professor, Department of Agricultural Engineering
Cornell University
Ithaca, New York 14853

ABSTRACT

A steady state mathematical model is presented, based on wet and dry mass balances and assuming a completely mixed reactor. It applies to all possible feeds and incorporates 1st order kinetics. Liquid can be extracted from the reactor effluent and recycled to the reactor. This is shown to allow considerable savings of heating energy, which is crucial to the efficient biogasification of dry substrates. The relationship between feed loading rate, hydraulic retention time and reactor solids concentration is established. With liquid recycle, two of these parameters can be chosen freely (2 degrees of freedom).

This general framework can be used to develop alternative ways of digesting a given substrate, to quantify operational parameters and to optimize operation. Experiments with wheat straw demonstrate the bench-scale feasibility of liquid recycle, illustrate the use of the model and validate it. Four to five volumes of biogas per volume reactor per day were consistently produced, and recycled liquid satisfied most of the water addition needs. No salt or toxics accumulation should occur. In semi-continuous operation, the digesting slurry separates in two layers between batch feeding-mixings and the thickness of the floating solids layer is linearly proportional to the reactor solids concentration.

CONTINUOUS ANAEROBIC DIGESTION OF HIGH SOLIDS BIOMASS: MODELING AND EXPERIMENTS

INTRODUCTION

Anaerobic digestion is the biological process whereby organic matter is partially converted into biogas. Biogas is a mixture of methane and carbon dioxide with usually better than 50% methane. The process requires anaerobic conditions (absence of air), sufficient moisture, and is greatly enhanced by warm temperatures: approximately 35°C for mesophilic digestion, 50-60°C for thermophilic operation. Historically, mostly very wet substrates like wastewater sludge have been subjected to anaerobic digestion. In the last 15 years, increasing attention has been devoted to the biogasification of drier substrates like municipal solid waste (MSW) or crops and crop residues. Water management is often a crucial issue with this type of operation since large amounts of water may have to be added and heated to digester temperature, in the process consuming much of the energy produced by the reactor. At the same time there is a limit to how dry a digester can be operated, so clearly water management is of the essence.

The mathematical model developed here provides a rational basis for the design of anaerobic digestion systems, especially those using high solids feedstocks. It focuses on water management, is based on mass balances and describes continuous operation of a completely mixed reactor (CMR). It is an applied model, no attempt is made to model kinetics and biochemical interactions, it is assumed that an adequate description of these fundamental processes is available. Three water management strategies are envisioned:

1) No intervention regarding water.

2) Continuous addition of water from an external source to the feed. This would be necessary for many drier substrates.

3) Extraction of water from the reactor effluent and recycle of this liquid to the feed. This last technique has great potential for saving water and energy.

MATHEMATICAL MODEL

Model Derivation

The three water management strategies mentioned in the introduction result in four possible water management options:

I) No water addition or recycle whatsoever.

II) Liquid extraction from the reactor effluent and recycle of this liquid to the reactor feed.

III) Same as II, but with supplemental continuous addition of water from an external source.

IV) No recycle of liquid but continuous addition of external water to the feed.

These four options were modeled using a mass balance approach. The following symbols were used:

A = Total Volatile Solids (TVS) fraction of the bulk wet substrate.

B = Ash or Fixed Solids fraction of the bulk wet substrate.

C = Biodegradable fraction of the TVS (= Biodegradable Volatile Solids or BVS).

D = Ratio of BVS degraded to BVS fed.

X = Resulting Total Solids (TS) fraction of the reactor contents.

f = Bulk wet volumetric loading rate in grams of feed added per liter reactor volume per day; does not include any water additions.

X_c = Total Solids fraction of digesting mixture, where water addition or recycle is used.

g = Volumetric liquid recycle rate, grams liquid/liter reactor/day.

w = Volumetric external water addition rate, grams water/liter reactor/day.

P = Total Solids fraction of dewatered effluent.

d = Specific gravity of mixed influent (after addition of water or recycle).

θ = Hydraulic residence time

The following conversion factors may also prove useful:

1 gram/liter reactor volume/day
= 0.0624 lbs/cu ft reactor volume/day

1 Joule/°C/gram feed
= 0.239 Btu/°F/lb feed

A wet mass balance is compared with a dry solids mass balance to obtain TS concentrations at every stage of the process. For example, if no water is added and no liquid is recycled (Option I), the wet mass balance is:

f = amount fed

f ACD = amount converted to biogas

f (1-ACD) = amount of effluent

The dry mass balance looks as follows:

f (A+B) = solids fed (TS)

f ACD = solids converted to biogas

f (A+B-ACD) = solids removed with the effluent

Consequently, the effluent TS concentration is:

$$X = \frac{A + B - ACD}{1 - ACD}, \tag{1}$$

expressed as a fraction.

Since the reactor is completely mixed, this is also the reactor solids concentration.

Example: A feed sludge has 12% TS (A + B = 0.12), 80% of the TS are volatile: TVS/TS = 80% (A = 0.80 x 0.12 = 0.096), 75% of the TVS is biodegradable (C = 0.75) and under existing conditions, 90% of the BVS is expected to be degraded (D = 0.90). The steady state solids concentration in the reactor is given by equation 1 (X) and is 5.90% TS.

Note the wide discrepancy between the solids concentration of the feed and the resulting solids concentration in the reactor. The latter provides a better description of the process than just the feed TS content.

Note also that the solids composition of the substrate has to be known, and that an estimate of biodegradability and actual degradation are used.

The second water management option, extraction of liquid from the effluent and recycle of this liquid to the influent, is schematized in Figure 1. The same procedure is followed as for the first option. This time the reactor solids concentration is given by

$$X_c = \frac{f\,(A + B - ACD)}{f(1 - ACD) + g}$$

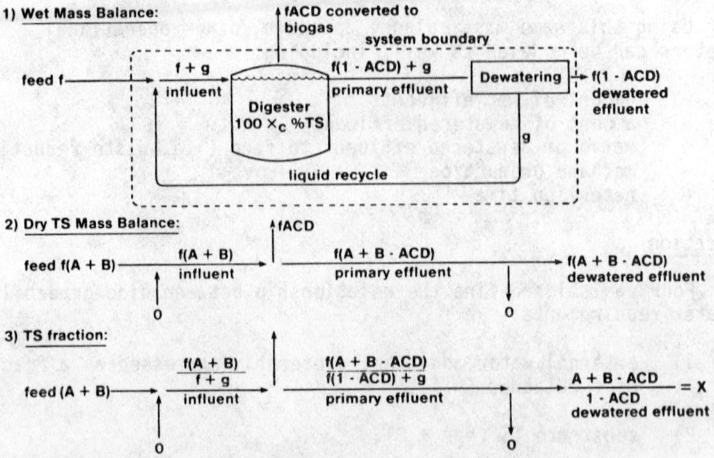

Figure 1: Mass balances and solids concentrations for Option II: Only recycle of liquid.

in which g is the volumetric recycle rate, in grams of liquid per liter reactor per day. From this equation, an expression can be derived for g, namely:

$$g = (f/X_C)\ [A + B - ACD\ (1 - X_C) - X_C] \tag{2}$$

This represents the amount of liquid recycle necessary to achieve a reactor solids concentration X_C. Irrespective of the feed, any reactor solids concentration (between 0 and the solids concentration of the dewatered effluent) can be achieved by manipulating the recycle flow (g). From Figure 1 it can also be seen that the dewatered effluent has a solids content (A + B - ACD) / (1- ACD). As long as the effluent can be dewatered to this TS value, no external water addition is necessary, the recycle will satisfy all water needs.

Some substrates may be so dry that water extraction alone will not suffice to fully conserve water in the system (i.e.,the effluent cannot be dewatered to (A + B - ACD) / (1 - ACD)). In this case, continuous water addition from an external source (w grams of water per liter reactor per day) is necessary. The solids content of the dewatered effluent (P) can be expressed along the same lines as for the second option (see Figure 1) and an expression for w can be derived:

$$w = (f\ /\ P)\ [A + B - ACD\ (1 - P) - P] \tag{3}$$

Note that w is independent from X_C. Here again, any desired (low) reactor solids concentration can be maintained with the same external water addition. For a given substrate, this water addition is determined entirely by the extent of effluent dewatering (P). Finally, where effluent dewatering and liquid recycle are impractical, external water addition will have to satisfy all moisture needs and is given by $(f/X_C)\ [A + B - ACD\ (1 - X_C) - X_C]$ (grams water/liter reactor/day). It is a function of the reactor solids concentration X_C.

Using this same mass balance approach, other operational parameters can be modeled as well, including:

- amount of wet effluent
- amount of dewatered effluent
- ratio of dewatered effluent to feed (i.e., waste reduction)
- methane production
- retention time

Application

Four variables define the relationship between biodegradability and water requirements:

1) external water addition, preferably expressed as a fraction of feed added (w/f),

2) substrate TS (= A + B),

3) total actual feed TS degradation [= ACD / (A + B)],

4) effluent TS after dewatering (P).

All can be quantified using this model. To illustrate this, two of these variables have been kept constant in Figure 2: a) in the digester, 60 and 80% of the feed TS are converted to methane (2 lines on the figure) and b) the digester effluent is dewatered to 25% TS. From the figure, the external water addition required to maintain steady state can be read. For example, using the 60% TS degradation line, and assuming a substrate with 70% TS, then for every kg of substrate fed, 540 grams of water (54% of feed) will have to be added. If the substrate has 45% TS or less, no external water addition is necessary.

Whether liquid is recycled or not, there is a constant flow of water through the system. The evaporative loss of water with the biogas is usually negligible compared to this steady throughflow. Therefore, soluble salts should not accumulate in the system.

Heating Energy Conservation and Extent of Dewatering

The energy required to heat the influent is generally considered to be the major energy input in anaerobic digestion. It equals the heat capacity of the influent multiplied by the required temperature increase. The heat capacity of the influent can be estimated as follows: (disregarding heat losses along the dewatering-recycle circuit)

$f\,[1-(A + B)]\,C_{pw} + f\,(A + B)\,C_{pTS}$ = heat capacity of feed

$+\,w\,C_{pw}$ = heat capacity of added external water

with C_{pw} = specific heat of water (4.17 Joule/gram/°C)

and C_{pTS} = specific heat of the dry solids (0.5 - 2 Joule/gram/°C).

This expression can be simplified to:

$$f\,[4.17\ w/f + 4.17 - (A + B)\,(4.17 - C_{pTS})],$$

or:

$$[4.17\ w/f] + [4.17 - (A + B)\,(4.17 - C_{pTS})]. \qquad (4)$$

per unit of feed loading rate f (gram/liter reactor/day). The value of Equation 4 was estimated for wheat straw (A + B = 0.92, C_{pTS} = 0.8 Joule/gram/°C) and plotted as a function of P on Figure 3. The influent heat capacity (in ordinate) is expressed in Joules per °C of temperature increase, per gram wet feed. It can be seen that increasing P from 20 to 35% TS would cut influent heating needs in half. Similarly, if no liquid recycle is used and the effluent has a concentration of 10% TS (X_c = P = 0.10), dewatering to 27% TS and liquid recycle would reduce influent heating requirements by a factor 3.

As an example let us consider a digester operating at 135° F and 10% TS processing a mixture of wheat straw and water at 55° F. It would require 857,000 Btu per ton feed per day without liquid recycle; assuming 25% heat transfer losses and a gas price of $3/MMBtu this would amount to $3.2 per ton wheat straw fed (or 23% of gross energy recovered in the methane). The implementation of dewatering to 27% TS and liquid recycle would lower theoretical heating cost to $1.04 per ton wheat straw (7.5% of methane gross energy).

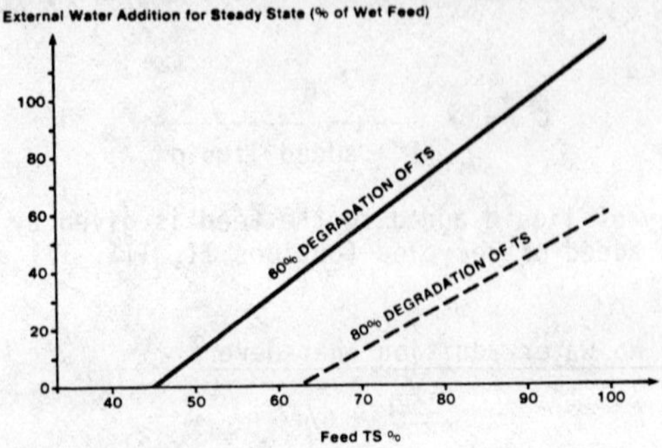

Figure 2: External water addition required for steady state, versus substrate TS %. The effluent is dewatered to 25%.

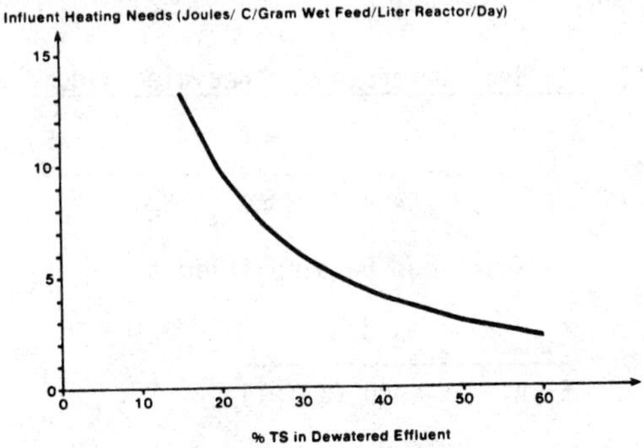

Figure 3: Influent heating needs versus cake dryness for wheat straw (disregarding heat losses over the dewatering-recycle circuit).

Summary of Simplified Model

This simplified model can be used to make decisions about ways to process a given substrate: if it is sufficiently moist and biodegradable it can be digested as such without any water addition or liquid recycle. This can be verified by incorporating substrate characteristics and expected conversion efficiency into equation 1, which yields reactor TS concentration. If the value obtained from equation 1 is too high for a reactor TS concentration, but the effluent can be dewatered to this value, liquid recycle can satisy all water needs. Finally, very dry substrates will require continuous water addition, but liquid recycle will still usually satisfy most of the moisture requirements. All these situations can easily be quantified using this mass balance model. Various kinetic models can be incorporated in this general framework, as we shall see.

Interrelationship of Loading Rate, Residence Time and Reactor Solids Concentration

The average hydraulic residence time (θ) is found by dividing the reactor volume by the flow rate of influent: $\theta = V/Q$ where V = reactor volume and Q = total influent flow. In this model, all flowrates are expressed per unit reactor volume so that $\theta = 1/Q$. Furthermore, mass flowrates were used (grams/liter reactor/day), they have to be converted to volume flow rates by dividing by the specific gravity of the materials. Let d be the specific gravity of the mixed influent. Then,

$$\theta = \frac{d}{f + \text{added liquid}}$$

where the quantity of liquid added to the feed is given by equation 2 wherever water is added or recycled (options II, III, IV) and is zero for option I.

Option I: No water addition whatsoever

$$\theta = d/f \qquad (5)$$

The residence time is inversely proportional to the feed loading rate. For a given substrate and kinetics, once one of these two variables has been selected, the whole system is defined. The system has one degree of freedom.

Options II, III, IV: Water and/or recycled liquid added

$$\theta = \frac{d}{f + f/X_c\,[A + B - ACD\,(1-X_c) - X_c]}$$

which can be simplified to

$$\theta = \frac{X_c\,d}{f\,[A + B - ACD\,(1-X_c)]} \qquad (6)$$

The variables θ (residence time), f (loading rate, wet grams/liter reactor/day) and X_c (reactor TS fraction) are related, and Equation 6 can be rearranged to express f and X_c:

$$f = \frac{X_c\, d}{\theta\, [A + B - ACD\,(1 - X_c)]} \tag{7}$$

$$X_c = \frac{\theta f\,(A + B - ACD)}{d - \theta f\, ACD} \tag{8}$$

Of these three variables, only two are independent and they define the third variable and all other operational parameters, except those for which the input of P (dewatered effluent TS) is needed. Two of these variables can be chosen arbitrarily, the system has 2 degrees of freedom.

Incorporation of Kinetics in Mass Balance Model

The efficiency of BVS destruction (D) has been treated as a constant up to this point. This is fundamentally incorrect since it is a function of reaction kinetics. The approximation is satisfactory as long as residence times are long enough to ensure that the reaction is nearly complete.

In the following approach, it is assumed that the conversion of BVS to biogas can be approximated as a first order reaction (this assumption is made by many authors). The digestor is a completely mixed reactor (CMR or CSTR), with influent substrate concentration So and effluent concentration Se. With k = first order rate coefficient, the following substrate mass balance can be constructed around the reactor (the BVS constitute the substrate):

[Accumulation] = [inflow] - [outflow] + [utilization]

or

$V\, dS/dt = QSo - QSe + V(-kSe)$

At steady state, no accumulation takes place and $dS/dt = 0$.

The previous equation can be rearranged to:

$S_e\,(Q + Vk) = QSo$,

resulting in

$$Se = \frac{So}{1 + k\theta} \tag{9}$$

So = influent substrate/influent volume = f AC/effluent volume

Se = effluent substrate/effluent volume = f AC(1 - D)/effluent volume

With semi-solid substrates, the volumes of influent and effluent can be significantly different. The specific gravity, however, is unlikely to change much and will be assumed constant and equal to d. The volumes and concentrations can now be calculated from the mass flows. (See Appendix.)

Option I: No water addition to the influent

$$D = \frac{k\theta}{1 + k\theta - AC} \tag{10}$$

Options II, III and IV: Water addition to the influent

D is the root of:

$$D^2AC\ (1-X_c) + D\ [AC\ (X_c - \frac{k\theta}{1+k\theta}) - (A+B)] + \frac{k\theta}{(1+k\theta)}(A+B) = 0 \tag{11}$$

Applications

The inclusion of kinetics in this mass balance model allows thorough exploration of the relationship between operational parameters, as illustrated in the following figures (Figures 4 to 8). The substrate modeled here is sorghum, a possible energy crop, and the following characteristics are assumed: 95% of the TS are volatile, 90% of TVS are biodegradable, 350 ml of methane (at 0°C and 1 atm) are evolved from the conversion of 1 gram of Chemical Oxygen Demand (COD) and the COD/VS ratio is 1.20.

1) No water addition. Figures 4, 5 and 6 illustrate the situation when the substrate is digested as such, without any water addition or liquid recycle.

Figure 4: The volatile solids conversion efficiency (Total VS on the left, Biodegradable VS on the right) is depicted versus hydraulic retention time according to equation 10. Four values of the first order decay rate coefficient k are considered. It is assumed that the substrate has 40% TS.

Figure 5: Loading rate versus hydraulic retention time according to $f = d/\theta$. This curve is valid for any feed having specific gravity d of 1000 grams/liter. Feeding 50 grams of wet feed per liter reactor per day results in a hydraulic retention time of 20 days. Doubling the loading rate cuts the retention time in half.

Figure 6: With hydraulic retention time on the horizontal axis, reactor solids concentration is displayed on the right and methane production rate (volumes of methane/unit reactor volume/day) on the left. The latter represents the gas production rate required to maintain solids steady state in the reactor. A 1st order rate coefficient k of 0.1 day $^{-1}$ is assumed and two feed TS concentrations are considered. For example, assuming sorghum is fed at 30% TS and the hydraulic retention time is 30 days, then the digester slurry will stabilize at 11.9% TS and 2.9 v/v/d of methane will be produced.

2) With water addition. For comparison, the same substrate is now modeled in digestion systems using liquid recycle and/or water addition (Figures 7 and 8).

Figure 7: For a digester maintained at 10% TS and assuming a 1st order rate coefficient $k = 0.1 \text{ day}^{-1}$, the loading rate is shown versus hydraulic retention time. Two feed TS contents are considered. The figure would look different for another digester TS concentration (at the same retention time, a higher reactor TS content would yield a higher loading rate).

Figure 8: Under the same assumptions as Figure 7, the methane production rate necessary to maintain steady state is shown versus hydraulic retention time. At a retention time of 30 days (the reactor is maintained at 10% TS), 2.44 v/v/d of methane are required. These results should be compared with those on Figure 6 (no water addition).

EXPERIMENTS

Objectives

One objective of the experiments was to illustrate how this mass balance model can be used to plan an anaerobic digestion system and quantify its operation. Additionally, the experiments would define on a bench scale the feasibility of dewatering effluent and recycling the resulting liquid. Finally, the data thus obtained were used to validate the model.

Materials and Methods

Wheat straw was selected as the substrate, partly because it was available and well characterized. It is also illustrative of a very dry substrate (92% TS), requiring massive additions of water for anaerobic digestion. As such it provided an ideal opportunity to experiment with liquid recycle and other water addition.

The main operational data are summarized on Table 1. Four 9 liter reactors were maintained at four different solids concentrations between 5 and 14% TS. Three ranges of hydraulic residence times were explored: 26-31 days, 28-40 days, 16-19 days; higher loading rates are associated with the shorter residence times. The digesters were operated at thermophilic temperature. Every other day, the reactors were opened, the prescribed amount of effluent was removed, liquid recycle, external water and wheat straw were mixed in according to the model output and the digesters were closed and replaced in the 55° C incubator.

The operational parameters (such as amount of wheat straw added at every feeding, amount of effluent removed, etc.) were calculated using the mathematical model developed earlier. The procedure is summarized on Table 1.

Results and Observations

The results are summarized on Table 2. During the experiments at the shorter residence times, the following stable performance was observed: at a loading rate of 10 grams TVS/liter reactor/day (=11.63 grams wheat straw), 4.2 v/v/d of biogas (50% methane) was produced and 40% of the TVS was converted. Recycling liquid from the effluent satisfied 68% of the water addition needs and the reactor effluent was consistently dewatered to 28-32% TS using 100 psi of pressure.

It was observed that an accurate estimate of the TVS destruction efficiency is essential to maintain constant reactor solids concentration. If the % TVS destruction has been overestimated in calculating operational parameters, these parameters will drive the reactor solids concentration up, and vice-versa. It was also observed that nutrient and buffer additions can contribute a significant amount of solids to the reactor and these should be taken into account in the mass balance.

Stratification. A few hours after each feeding-wasting (carried out every other day), the initially homogeneous digester slurry separated in a floating solids layer (typically 12-18% TS) and an underlying liquid layer (typically 2-5% TS). The float layer tended to trap gas bubbles and therefore to expand, particularly during the period of intensive gas production in the first hours after each feeding. There is a linear relationship between reactor solids concentration on the one hand and the ratio of float volume to initial slurry volume on the other. This relative float volume is given by the following equation:

$$\frac{\text{float volume}}{\text{initial slurry volume}} = (8.56 \times \text{reactor TS \%}) + 2.14 \qquad (r = 0.998)$$

The relative subnatant liquid volume is given by

$$(-\,6.16 \times \text{reactor TS \%}) + 99.67 \qquad (r = 0.999)$$

By extrapolation, it can be calculated that at 16.2% TS, no distinct underlying liquid layer should remain. (See Figure 9)

TABLE 1: EXPERIMENTAL

Substrate Characteristics: Wheat straw; 86% TVS, 6% ash, 8% moisture; chopped to 2 mm size; 33 to 49% of TVS refractory to anaerobic digestion.

Operation: 4 digesters, effective volume = 6 liters, temperature = 55°C, semi-continuous operation, wasting-feeding every other day.

Measurements:	
biogas production and composition:	continuously
pH, TS and TVS of digester effluent and of dewatered effluent:	every other day
Total Organic Acids, Volatile Fatty Acids, Alkalinity:	occasionally

Experimental Design:

INPUT		OUTPUT
Wheat Straw % TS * and % TVS *		Wheat Straw Loading Rate
Dewatered Effluent % TS *		Liquid Recycle Rate
% TVS converted **	MODEL	Rate of External Water Addition
TVS conversion rate ** (grams/liter reactor/day)		Effluent Removal Rate
Reactor slurry % TS ***		Retention Time

* Measured
** Estimated
*** Chosen

TABLE 2

SUMMARY OF RESULTS

	PHASE I				PHASE II				PHASE III			
Reactor number	1	2	3	4	1	2	3	4	1	2	3	4
Retention time (days)	26.4	29.1	30.4	31.0	28.0	28.6	33.8	39.5	19.3	16.8	16.6	16.4
TVS LR (g TVS/l_r/day)	3.32	4.43	5.53	6.63	3.00	3.75	4.44	5.00	3.75	6.00	8.00	10.00
TVS conversion efficiency	47.6	45.2	45.3	39.5	43.1	48.0	37.4	49.0	40.2	39.6	39.4	39.8
Volum. TVS conv. rate, (g TVS destr./l_r/day)	1.52	1.91	2.42	2.68	1.80	2.30	2.28	3.22	1.90	2.86	3.66	4.49
Biogas, v/v/d (48 to 53% methane)	1.34	1.63	2.03	2.51	1.63	1.99	2.13	2.95	1.67	2.63	3.35	4.16
Average reactor TS%	5.4	8.4	10.2	13.1	6.2	8.4	11.2	14.0	6.8	8.8	10.8	13.3

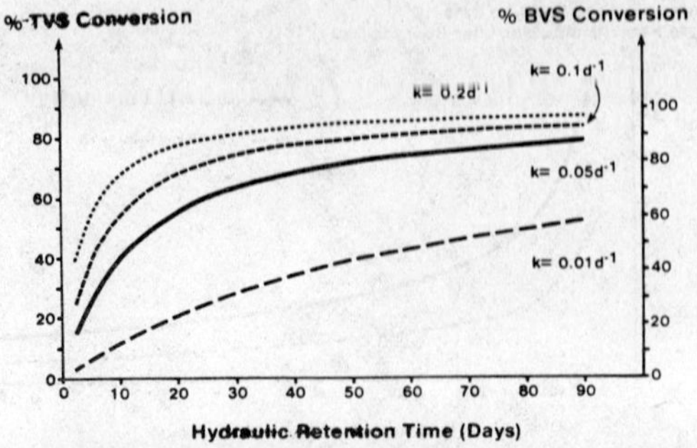

Figure 4: % Solids conversion versus hydraulic retention time, assuming 1st order kinetics, feed = 40% TS, 90% of TVS are biodegradable, NO WATER IS ADDED. 4 rate coefficients k are considered.

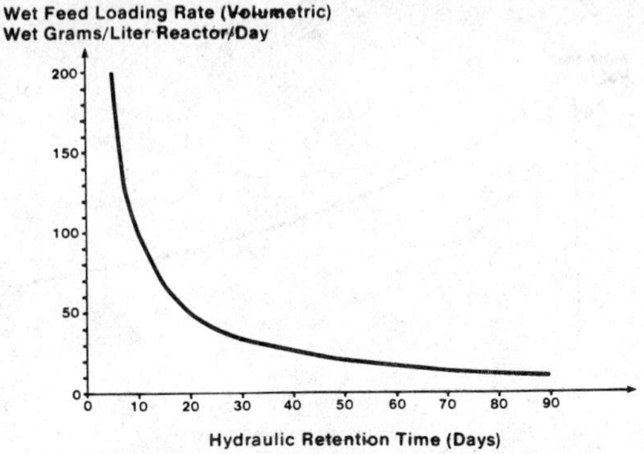

Figure 5: Wet feed loading rate versus hydraulic retention time, NO WATER ADDITION.

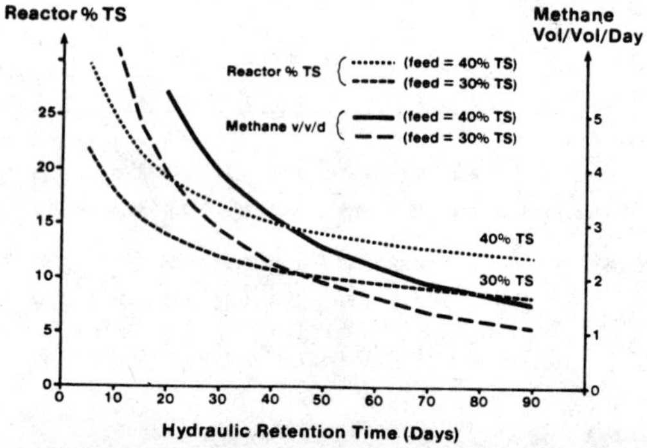

Figure 6: Reactor TS % (left), and methane production rate required for steady state (right), for 2 feed TS %, versus hydraulic retention time. NO WATER ADDED.

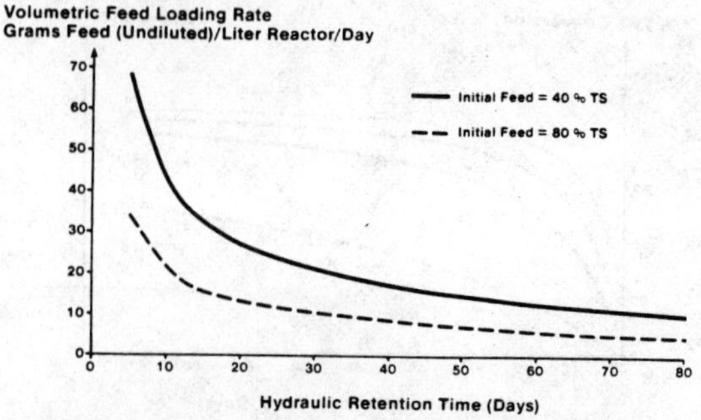

Figure 7: Volumetric feed loading rate versus hydraulic retention time for two feed TS contents. WATER IS ADDED. the digester is kept at 10% TS.

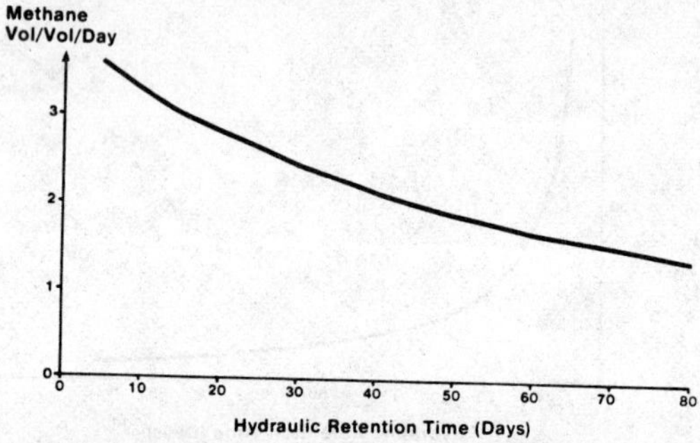

Figure 8: Volumetric methane production rate required to maintain steady state. versus hydraulic retention time. WATER IS ADDED. the reactor is kept at 10% TS.

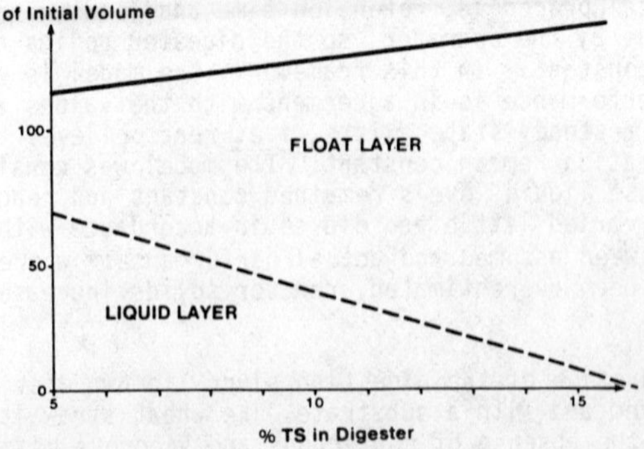

Figure 9: Stratification of reactor contents, relative volumes of liquid lower layer and supernatant float layer.

DISCUSSION

Model

It was shown earlier that net energy production can be optimized by maximizing the extent of effluent dewatering in a system with liquid recycle. In general, whether liquid is recycled or not, the feed should be as dry as possible to minimize the amount of moisture to be heated and so improve net energy production. The inclusion of liquid recycle decouples reactor TS concentration from feed TS content, meaning that conventional (low solids) digestion technology can be used on very dry substrates without necessarily jeopardizing net energy production. At the same time it is desirable to operate at the highest possible reactor solids concentration, because a higher concentration of solids in the digester makes it possible to operate at a higher loading rate with no decrease in retention time; it may also foster a more concentrated active biota. This has to be balanced against increasing handling and mixing problems with high solids slurries. Finally, when describing an anaerobic digestion process, the influent feed solids concentration is generally not a very relevant parameter. The resulting solids concentration in the digester is more closely related to the digestion process.

Experiments

As mentioned in the results section, steady state production of 4.2 v/v/d of biogas was observed, combined with 40% conversion of the TVS. Conversion efficiencies up to 40% were observed at lower loading rates. These figures are not as low as they seem, since the refractory fraction of wheat straw volatile solids is reported to be in the range of 33 to 49% (Jewell et al. 1982, Chandler et al. 1980). Other authors investigating semi-continuous digestion of high solids substrates observed the following biogas production rates: 2.0 v/v/d (with dairy cow manure; Hills, 1980), 1.8 v/v/d (corn stover; Gaddy, 1981) and 1 v/v/d (corn, wheatstraw and grass; Jewell et al., 1983).

Earlier, it was shown that a digestion process with liquid recycle is defined by a set of three dependent variables: reactor solids concentration, loading rate and hydraulic retention time. In this experimental procedure, retention time and loading rate are set at a constant value by the operator, so the digester solids concentration should remain constant. In this framework, the model is validated if the measured performance is in agreement with the values assumed initially, and a steady state exists, i.e. reactor level and reactor solids concentration remain constant. The model was considered validated because liquid levels remained constant and reactor solids concentrations varied little and did so in accordance with the discrepancy between assumed and actual performance: where TVS conversion had been overestimated, reactor solids increased slightly, and vice-versa.

The separation of the digesting slurry in two distinct layers is commonly observed and with a substrate like wheat straw it seems unavoidable in the absence of continuous and vigorous mixing. However, it does not appear to adversely affect reaction rates or conversion efficiency, as long as the slurry is completely mixed at every feeding. Although digesters operated at high TS would require a powerful mixing installation, this observation suggests that mixing energy requirements would be minimal in semicontinuous operation, since actual mixing would only be carried out for a very short time.

CONCLUSIONS

Mass balance modeling indicates that extracting liquid from the digester effluent and recycling this warm liquid to the digester makes it possible to reduce water needs and influent heating requirements drastically, when dealing with high solids feed. To optimize net energy production, a high priority should be given to maximizing the extent of effluent dewatering. If no water or recycled liquid are added to the feed, loading rate is inversely proportional to hydraulic retention time and the system allows 1 degree of freedom. Whenever water is added, the system is defined by 3 dependent variables: loading rate, residence time and reactor solids concentration. Two of these can be chosen freely, the system gains an additional degree of freedom. This allows very dry feedstocks to be digested using conventional low solids technology without necessarily nullifying net energy production. At the same time it is desirable to operate at the highest possible reactor solids concentration, which allows higher loading rates without decreasing hydraulic retention time.

It is possible to anaerobically digest wheat straw in semi-continuous feed mode, producing 4 to 5 v/v/d of biogas containing 50% methane. The residence time was approximately 16 days and 35 to 45% of TVS fed were destroyed. Approximately 68% of the required water addition was supplied by the recycle when the reactor was maintained at 13% total solids.

In the absence of continuous mixing, the wheat straw slurry separates in two layers. Brief thorough mixing every other day is sufficient to maintain optimal performance. There is a linear relationship between reactor solids concentration and relative float thickness; in the case of wheat straw, above 16% TS the float layer occupies the full effective digester volume.

APPENDIX

Inclusion of 1st order kinetics

Option I: No water addition to the influent

Influent volume = influent mass / d = f/d

Effluent volume = effluent mass / d = f(1-ACD) / d

S_o = f AC / influent volume = (f AC) (d/f) = ACd

$$S_e = \frac{f\ AC\ (1-D)}{\text{effluent volume}} = f\ AC\ (1-D)\ \frac{d}{f\ (1-ACD)} = \frac{AC\ (1-D)\ d}{(1-ACD)}$$

Equation 9 now becomes

$$\frac{AC\ (1-D)\ d}{(1-ACD)} = \frac{ACD}{1 + k\theta}$$, which yields

$$D = \frac{k\theta}{1 + k\theta - AC} \qquad (10)$$

Options II, III, IV: Water addition to the influent

Influent volume = (feed + water addition) / d, where the water addition is given by equation 2.

This can be simplified to:

$$\frac{1}{d}\ \frac{f}{X_c}\ [A + B - ACD\ (1 - X_c)].$$

Effluent volume = primary effluent / d

The primary effluent includes what remains of the feed, plus any water that was added:

$$= f\ (1 - ACD) + \frac{f}{X_c}\ [A + B - ACD\ (1 - X_c) - X_c]$$

which simplifies to $(f / X_c)\ (A + B - ACD)$.

Equation 9 now becomes:

$$f\ AC\ (1-D)\ \frac{X_c\ \ d}{f\ (A+B - ACD)} = f\ AC\ \frac{X_c\ \ d}{f\ [A+B - ACD\ (1-X_c)]\ (1+k\theta)}$$

or

$$k = \frac{D}{\theta\ (1-D)}\ \left[1 - \frac{ACX_c}{[A+B - ACD\ (1-X_c)]} \right]$$

Extracting D yields equation 11.

ACKNOWLEDGEMENTS

This work was supported in part by grant No. XB-0-9038-1 from SERI/DOE and by Cornell University departmental funds for which the authors are most grateful.

BIBLIOGRAPHY

Andrews J. F., S.P. Graef. 1971. Dynamic Modeling and Simulation of the Anaerobic Digestion Process, pp 126-162. in F. G. Pohland (Ed.) Anaerobic Biological Treatment Processes, Advances in Chemistry Series, American Chemical Society, Washington D.C.

Chandler J.A., W.J. Jewell, J.M. Gossett, P.J. Van Soest, J.B. Robertson. 1979. Predicting Methane Fermentation Biodegradability. Presented at: Second Symposium in Biotechnology in Energy Production and Conservation, Gatlinburg, TN, Oct. 3, 1979.

Gaddy, J.L. 1981. Economic and Kinetic Studies of the Biological Production of Farm Energy and Chemicals from Biomass. Annual Progress Report 1979-1980. SERI subcontract XJ-9-8020-1

Ghosh S. 1981. Kinetic Basis for Optimization of Methane Production b Anaerobic Digestion. Proceedings of the 1981 International Gas Research Conference. Sept. 28 - Oct. 1, 1981, Los Angeles.

Hawkes D.L. 1979. Factors Affecting Net Energy Production from Mesophilic Anaerobic Digestion. The First International Symposium on Anaerobic Digestion. University College, Cardiff, Wales, Sept. 17-21, 1979, Day 3.

Hayes T.D., W.J. Jewell, S. Dell'Orto, K.J. Fanfoni, A.P. Leuschner, D.F. Sherman. 1979. Anaerobic Digestion of Cattle Manure. The First International Symposium on Anaerobic Digestion. University College, Cardiff, Wales, Sep. 17-21, 1979, Day 4.

Hill D.T. 1983. Simplified Monod Kinetics of Methane Fermentation of Animal Wastes. Agricultural Wastes (1983) 5: 1-16.

Hills D.J. 1980. Methane Gas Production from Dairy Manure at High Solids Concentration. Transactions of the ASAE, 1980, pp 122-126.

Jewell W.J., et al. 1978. Anaerobic Fermentation of Agricultural Residue. Potential for Improvement and Implementation. January 1978. US DOE EY-76-S-02-2981-7. Final Report.

Jewell W.J., et al. 1982. Dry Fermentation of Agricultural Residues, Final Report. SERI Subcontract XB-9038-1 (SERI/TR-9038-10, December 1982).

Shelef G., S. Kimchie, H. Grynberg. 1981. Organic Loadings and Methane Production in Anaerobic Digestion of Agricultural Wastes. Fuel Gas Production from Biomass, 2: 259-271.

DISCUSSION

W. Smith, J. of Florida	At 16.2% solids, you wouldn't have a liquid layer. Did you operate at that level, or would you not want to because of the fact you wouldn't have a recycle volume?
Legrand	No, we didn't operate at that level. We have recycle anyway. You withdraw effluent from the digester that goes to the dewatering installation, and depending on how well you can dewater, you're going to have recycling automatically.
Smith	That's after it leaves the digester?
Legrand	Yes.
T. Nunez-McNally, Tufts U.	Are the percent solids you're quoting on a volume basis or weight basis?
Legrand	On a weight basis.
J. Garvin, Pitt-DesMoines, Inc.	Is operation at thermophilic temperatures necessary for these systems, or could they operate at mesophilic temperatures?
Legrand	You could operate at mesophilic temperature. We elected to operate at thermophilic temperatures because of time constraints and the fact that the reaction is faster at thermophilic temperatures. You can operate at mesophilic temperatures; and the rates will be 2 or 3 times lower.

POTENTIAL FOR BACTERIAL METHANE PRODUCTION AT SUPERHEATED TEMPERATURES

Jody W. Deming, Ph.D.
Associate Research Scientist
Chesapeake Bay Institute
The Johns Hopkins University
Shady Side, Maryland 20764

ABSTRACT

The upper temperature limit of methane production by methanogenic bacteria, isolated and characterized in pure culture, has recently been pushed to a temperature of 97°C. Evidence of the potential for bacterial methanogenesis at superheated temperatures (>100°C) comes from studies of extremely thermophilic, non-methanogenic bacteria, known to grow at temperatures to 110°C, and of mixed cultures of methane-producing "black smoker" bacteria, recovered from superheated (≥350°C), methane-rich, submarine volcanic environments of the East Pacific. Elevated pressure, both as an experimental and evolutionary factor, appears to have played a critical role in the functioning of any microorganism at temperatures above 100°C.

The extreme conditions generally required for the abiotic genesis of methane, either through synthetic or degradative processes, suggest that bacterial activity at previously unimagined temperatures may not be an unreasonable explanation for a recent report of rapid methane production at 300°C and 265 atm pressure. Available data in the literature on fractionation factors of stable isotopes in methane are insufficient to argue against this possibility. Nevertheless, discussions of the potential for bacterial methanogenesis at superheated temperatures will remain highly controversial, until such time as results from new experiments, already forthcoming in microbial ecology, barobiology, biochemistry, petrology, and origin of life processes, become available for consideration.

POTENTIAL FOR BACTERIAL METHANE PRODUCTION AT SUPERHEATED TEMPERATURES

INTRODUCTION

Extremely thermophilic bacteria, by definition, are capable of reproducing at temperatures above 70°C (22). The first extremely thermophilic methanogenic, or methane-producing, bacterium, *Methanobacterium thermoautotrophicum*, was isolated from sewage sludge in 1972 (40). It's maximum temperature for growth and methane production was about 75°C. Subsequent searches also revealed its presence in thermal volcanic environments (41), but some nine years passed before other microorganisms capable of generating methane at temperatures above 75°C were discovered. Therefore, as recently as 1980, the upper temperature limit for the production of methane by living microorganisms was believed to be below 80°C (41). This perceived restriction of biological methanogenesis to relatively moderate temperatures has greatly influenced (and, from some perspectives, restricted) conclusions regarding the history and formation of natural methane reserves present in contemporary geological settings (24,27) or proposed to have existed in Archaean environments (3).

The first half of this decade has seen intensive microbiological explorations of terrestrial and submarine thermal environments, the results of which have caused previous temperature restrictions on bacterial methanogenesis to be revised dramatically. Known strains and mixed cultures of extremely thermophilic methanogens and some of their growth and methanogenic characteristics are listed in Table 1. The maximum growth temperature of 75°C, determined for *Methanobacterium thermoautotrophicum* in 1972, was significantly altered in 1981 with the isolation of new methanogens, especially *Methanothermus fervidus*, from an Icelandic hot spring (32). *M. fervidus* grows and produces methane autotrophically from carbon dioxide and hydrogen at temperatures to 97°C (32). Above 97°C, the organism's usual rod-shaped morphology shifts to "large body" formation, indicative of cell wall synthesis apparently less heat-stable than growth (32). In addition to pushing the upper temperature limit for bacterial methanogenesis upwards by some 22°C, the laboratory characterization of *M. fervidus* raised new questions about the thermostability of DNA in a living microorganism (*M. fervidus* DNA has a low GC-content [33%] and denatures outside of the cell at 83°C). It also led to speculation that extreme thermophiles like *M. fervidus* may be responsible for bacterial methane production in geothermally-heated deep sediments (32), such as those in northern Italy where methane is found in pliocene sediments at a depth of 3000m and temperature of 90°C (27).

Table 1. Extremely Thermophilic Methanogenic Bacteria.

Organism (Ref.)	Growth temp(°C)[a] Range	Optimum	Shortest doubling Time (h)	Methane rate[b]
Methanobacterium thermoautotrophicum (40)	40-75	~65	3	.0013
Methanococcus thermolithotrophicus (16)	37-70	65	0.6	.0054
Methanococcus jannaschii (18)	50-<95	85	0.4	.0032[c]
Methanothermus fervidus (30,32)	65-97	83	2.8	.0048
Black smoker bacteria (5)				
Mixed culture (from smoker rock)	75-100	100	1.1	.41
Mixed culture (from 195°C fluid)	75-100	100	0.9	.32
Mixed culture (from 306°C fluid)	75-100	100	0.6	1.5

[a] At atmospheric pressure.

[b] In umol CH_4/ml/h/ug bacterial protein. For comparative purposes, gas production rates were taken (or estimated) from published data and normalized to weight of bacterial protein in the total volume of the starting culture, converting from O.D. units (ref.30-32,40) or microscopic counts (ref.5) to total bacterial protein, assuming 0.3 O.D. units = 27 ug bacterial protein/ml and 10^7 bacteria/ml = 1 ug bacterial protein/ml. Due to variations and missing details among the studies, these rates should be considered rough estimates only.

[c] Unlike the others, this rate was determined using a cell-free protein extract (18), rather than a bacterial culture.

The purpose of this paper is to explore the potential for bacterial methanogenesis at superheated (>100°C), rather than subcritical (<100°C), temperatures. Since a methanogenic bacterium that can grow or produce methane above 100°C has not yet been isolated and characterized in pure culture, this exercise draws on recent work with non-methanogenic bacteria, known to grow at superheated temperatures, and with mixed cultures of methane-producing bacteria from superheated methane-rich environments. The available information is considered in perspective with the abiogenic production of methane and analyses of bacterial vs. abiotic fractionations of carbon and hydrogen isotopes in methane.

BACTERIAL GROWTH AT SUPERHEATED TEMPERATURES

Non-methanogenic bacteria

The increased pace of discovery of new, extremely thermophilic microorganisms has not been limited to methane-producing bacteria or to the boiling point of water at atmospheric pressure. In 1981, Heinen and Lauwers (15) reported that *Bacillus caldolyticus*, an extremely thermophilic anaerobe isolated from a terrestrial hot spring at Yellowstone Park, could grow at 105°C under slightly elevated gas pressures. A year later, the same laboratory that had characterized *Methanothermus fervidus* reported ultrathin mycelia-forming microorganisms, isolated from shallow submarine volcanic areas and subsequently named *Pyrodictium occultum* (31), that grew optimally at 105°C and very weakly at 110°C (29). Although dramatic, these observations of microbial activity at superheated temperatures (>100°C) were not without precedent, since ZoBell had reported as early as 1958 (42) that a bacterial culture from a deep, hot, oil-well sediment core grew at 104°C when pressurized hydrostatically to 1000 atm.

One of the key factors in all of these observations of bacterial growth at superheated temperatures was the use of elevated pressure: stable microbial culturing conditions at temperatures above 100°C must include liquid water which can be achieved by the application of compensating, stabilizing levels of either hydrostatic or hyperbaric pressure (and, within more restrictive limits, by increased salt content). A second related factor was that the organisms that performed at superheated temperatures were isolated from, and therefore apparently evolved in, pressurized aquatic environments. (Classical spore-forming thermophiles like *Bacillus stearothermophilus*, believed to have evolved in non-pressurized, intermittently-dry soil environments, do not appear able to grow at more extreme temperatures as a result of applying increased hydrostatic pressures [39]).

Thus, pressure acted in the laboratory experiments with *Bacillus acidocaldarius*, *Pyrodictium occultum*, and ZoBell's unidentified bacterium to stabilize culturing conditions above $100^{\circ}C$, compensating for the otherwise destabilizng effects of extreme temperature, but also must have acted over geologic time to drive the evolution of unique microbial adaptations to superheated, pressurized environments.

"Black Smoker" Bacteria

Superheated hydrothermal vents. The most extreme of superheated, pressurized environments yet discovered on this planet occur at remote seafloor depths of about 2650m (and, therefore, hydrostatic pressures of about 265 atm or 3900 psi) along tectonic spreading zones of the East Pacific Rise (28). At specific sites near $21^{\circ}N$, $13^{\circ}N$, the Guaymas Basin of Baja California, and the Juan de Fuca/Gorda Ridge near Oregon and Washington states, molten magma rises sufficiently close to the seafloor to heat and chemically alter seawater that has percolated to subterranean depths and circulated through crustal cracks. After reacting at depth with crustal basalts at temperatures well in excess of $350^{\circ}C$, this water, redefined as hydrothermal fluid, ascends to the seafloor in chemically-altered form. Endmember or pure hydrothermal fluid contains no ambient seawater, is enriched to varying degrees with magmatically-derived elements, gases, and inorganic compounds, and re-enters the ocean, through sulfide deposits known as black smokers, at flow rates of meters per second and temperatures of at least $350^{\circ}C$ (13,20,34-36). In some cases, depending on differences in vent age, mineral composition, and subterranean water/rock ratios, ascending endmember fluids will mix with ambient seawater before venting at the seafloor and, therefore, will emerge from sulfide mounds or "white smokers" at temperatures still superheated but well below $350^{\circ}C$ (e.g., $170-280^{\circ}C$). Hydrostatic pressures at the depths of East Pacific hydrothermal vents theoretically should allow salt water to remain liquid at temperatures to about $460^{\circ}C$ (7); indeed, isolated temperature measurements at Juan de Fuca smokers have registered as high as $420^{\circ}C$ (8).

The first "black smoker" bacteria. While smoker fluids at $21^{\circ}N$ are generally acidic (pH of 3-5) and enriched in He and CH_4, they vary considerably in concentrations of H_2, H_2S, Mn, and Fe (20,21,35,36). Some speculated that such variations in chemical constituents potentially important to microorganisms might be attributable to bacterial activity occurring in the superheated fluids themselves (5,21). Even before the chemical analyses were available, Baross joined the Rise expedition of 1979 and recovered samples of superheated hydrothermal fluids in search of extremely thermophilic microorganisms (5). Sampling gear available at that time was problematic (leaky), and the collection of black smoker, endmember hydrothermal fluids, undiluted by surrounding ambient (2°C) seawater, was virtually impossible (4). To minimize heterotrophic bacterial contamination that might be present as a result of mixing with ambient seawater, the smoker samples were inoculated into media selective for chemolithotrophic bacteria and then boiled for an hour, prior to longterm incubation at a temperature (86°C) that few, if any, non-thermophilic bacteria could survive (5).

The results of these efforts were positive in that mixed cultures of extremely thermophilic bacteria were obtained from seven different smoker samples, including one that, according to its chemically-determined temperature (306°C), represented a relatively successful (minimally-diluted) collection of endmember hydrothermal fluid. Subsequent laboratory studies with these cultures indicated the presence of bacteria capable of rapid growth (doubling times of about 1 h) at the highest temperatures tested without the use of elevated hydrostatic pressure, i.e., at 100±2°C (5). Since the cultures were derived originally from pressurized hydrothermal fluids minimally diluted with seawater, it seemed logical to us to examine their behavior at *in situ* vent pressure (265 atm) and superheated temperatures more characteristic of smoker vents and their subterranean environments.

Growth at 265 atm and 250°C. In 1983, we reported black smoker bacterial doubling times of 8 h at 150°C, 1.5 h at 200°C, and 40 min at 250°C (with equivocal results at 300°C), all under the *in situ* vent pressure of 265 atm (1). Growth was documented microscopically (by epifluorescence and transmission electron microscopy) and biochemically (by colorimetric protein assays and a representative amino acid analysis) and contrasted to negative results obtained with controls (uninoculated media and an inoculated, but poisoned, sample) incubated at 200°C and 265 atm. We followed this report shortly thereafter with a second publication (4), providing details of environmental sampling, preliminary experiments, amino acid analyses of the total particulate protein present in each of the subsamples from experiments at 200, 250, and 300°C, and new transmission electron micrographs of the black smoker bacteria being maintained at temperatures of 95-107°C under slight gas pressures.

Our results and the conclusions we drew from them were attacked by Trent et al. (33) and White (37) on the grounds that pure amino acids in aqueous solution, as well as constituent molecules from non-thermophilic bacteria, were known to be highly unstable at 250°C. [Note that all such experiments at that time had been conducted at 250°C without the application of compensatory levels of hydrostatic pressure]. Therefore, these authors contended that our measurements of the amino acid content of total particulate protein present in superheated cultures of black smoker bacteria (total concentrations increasing with time at 250°C and 265 atm from 1.6 to 145 µg amino acids/ml, or the equivalent of an increase from about 10^7 to 10^9 bacteria per ml) must have been due to laboratory "artefacts" and "contaminants" introduced after the samples were brought to room temperature and pressure (33,37).

Key factors in our experiments, not addressed by these early critics, were the same as those cited above for studies that had led other investigators to detect bacterial growth above 100°C: 1) the culturing medium was stabilized at extreme temperatures by use of elevated pressure; and 2) the bacterial inoculum was derived from a superheated, pressurized environment in the first place. As others have observed with regards to our studies, "it must be irrelevant to examine the stability of mesophilic [non-thermophilic] bacteria and their constituent molecules at 250°C. Thermophilic bacteria capable of growing in superheated hydrothermal environments are expected to have special mechanisms for adaptation to the environments" (38). The major difference (and conceptual roadblock for many readers) between our experiments and those of others showing growth at superheated temperatures to 110°C was our use of temperatures far in excess of any ever before considered by biologists. A supportive biochemical basis for life at 250°C and 265 atm pressure simply did not exist.

In the last three years, that basis has begun to develop. In 1982, Price (25) reported biochemical analyses of sediment cores from deep (6400-7500 m) oil wells, indicating that "high concentrations of C_{15}+ hydrocarbons are thermally stable at high temperatures (at least 300°C) in abnormally-pressured semi-closed systems over geologic time." In 1984, Nickerson (23) developed theoretical bases for interactions between extreme temperature, pressure, and salinity that could lead to stable biochemical precursors and primitive life forms. In the same year, Bernhardt et al. (6) proved that "a number of amino acids exhibit considerable stability [half-lives measured in hours] at temperatures up to 200°C" under a hydrostatic pressure of 260 atm. [The novelty and importance of these findings were largely overlooked by these authors, who chose instead to focus on the lesser stabilities (half-lives of about 50 min) of amino acids at 250°C and 260 atm as evidence that our bacteria could not possibly have reproduced under those conditions. They further attempted to cast doubts on our amino acid analyses by stating that five unidentified chromatographic peaks, which they detected and judged to be degradation products of soluble amino acids at 250°C, "corresponded to those observed by Baross and Deming (1)." In fact, the chromatographic details (never seen by these authors) of the unidentified compounds we extracted from particulate protein, are substantially different, both qualitatively and quantitatively, from the ones they reported]. Finally, in 1985, Yanagawa and Kojima (38) demonstrated that not only amino acids but peptide-like bonds between them are stable at 250°C and 265 atm in the presence of silicates. Silicates are present in excessive amounts in all submarine hydrothermal environments.

New black smoker isolates. In the interim period required to establish an appropriate funding base, personnel, and instrumentation to repeat experiments at *in situ* black smoker conditions (we did the original experiments as short-term guests in another laboratory), we have developed reliable methods for culturing black smoker bacteria at more moderate conditions. Using a new, highly-thermostable, solid culturing medium (12), black smoker bacteria have now been obtained as visible, transferrable colonies at the maximum temperature limit of the medium (120°C) and the highest pressure yet tested (265 atm). In preliminary experiments, some of these bacteria appeared to barophilic at 120°C, requiring elevated hydrostatic pressures for optimal growth (12).

Inocula from which colonies have developed in this new medium at temperatures of 80-120°C include: 1) liquid cultures of the original black smoker bacteria that had been stored anaerobically at room temperature (11); 2) the remaining volume of a sample that had been removed (and similarly stored) from the pressurized culturing chamber at the end of our 250°C experiment in 1982 (2); and 3) freshly-collected samples of superheated smoker fluids from submersible expeditions to Juan de Fuca vents in 1984 (12) and to 21°N in 1985 (unpublished data). The latter samples, which were collected *in situ* by submersible using new and improved (non-leaky) titanium syringe samplers (34), included collections of essentially pure hydrothermal fluid at 345°C (12). We will attempt in the near future to isolate and characterize in pure culture some of the bacteria derived from these superheated hydrothermal fluids and, ultimately, to test their potential for growth and survival at temperatures from >120 to 350°C at *in situ* vent pressure.

METHANE PRODUCTION BY BLACK SMOKER BACTERIA

At temperatures to 100°C

The primary objective of experiments with black smoker bacteria has always been to assess their growth and survival at extreme temperatures and pressures. In the course of early growth studies at 100±2°C, methane production at rates of 22-46 nmol/ml/h (inoculum sizes of 3.0×10^5 to 1.3×10^6

bacteria/30 ml test volume) was discovered. These are the most rapid methane production rates yet recorded for a thermophilic methanogen (Table 1). Methane production at any rate was unexpected, since no precautions had been taken to maintain anaerobic conditions (all known methanogens are anaerobes [17]) with the initial water samples or cultures (5). Apparently, the precautions taken to minimize non-thermophilic contaminants (initial sample boiling) and the incubation temperatures employed subsequently (80-100°C) had excluded enough oxygen (17) to allow survival of at least some of the methanogenic members of the mixed cultures.

At superheated temperatures: bacterial or abiotic?

Gas production was also observed when one of these cultures of black smoker bacteria was incubated at superheated temperatures of 150, 200, 250, and 300°C under vent pressure (1). Unfortunately, the design of the high-temperature, high-pressure apparatus we used did not permit ready collection of this gas. We succeeded in collecting only two samples: one from the uninoculated media control after 16 h at 200°C; the other from media inoculated with the culture and incubated 3 h at 300°C. No methane was detected in the control sample, but 9.6% (4.3 mMol) of the inoculated sample proved by gas chromatography to be methane (1).

Without analysis of a gas sample from an uninoculated (or inoculated but poisoned) control at 300°C and 265 atm, it could be considered premature to conclude that the methane generated by an inoculated sample, incubated under the same conditions, was produced by viable, extremely thermophilic methanogens. However, we entertain that conclusion for the following reasons: 1) the culture medium heated to 300°C was first inoculated with a mixed culture of bacteria, known to contain a methanogenic component (5) and observed to be producing gas actively at 100°C just prior to inoculation (1); and 2) total particulate protein levels in the inoculated sample remained high (did not appear to degrade) during the initial 3-h incubation period at 300°C (4).

The alternative explanation for the unusually high concentration of methane we detected after 3 h at 300°C, i.e, that it was produced abiotically, perhaps as a break-down product of some constituents in the medium or of the starting inoculum itself, seems less tenable. Those who designed and conducted experiments at 250°C and 265 atm (6,33,37), specifically "to look for artefacts arising in the medium itself or in medium supplemented with [non-thermophilic or non-living] cellular material" (33), never reported, nor apparently noticed, gas production of any kind (9). Furthermore, the literature in geochemistry and petrology is replete with observations on the abiotic production of methane, either by synthesis from CO_2 and H_2 or by hydrous or dry pyrolysis of organic matter, that strongly argue against rapid (3 h), abiotic production of millimolar levels of methane at 300°C.

For example, pure carbon combines directly with pure hydrogen to form methane only at temperatures exceeding 1100°C. Recent, novel methods designed to moderate conditions and improve rates of synthesis nevertheless require starting temperatures of 600°C (14). In petrology experiments that examine the degradation rates of organic matter at superheated temperatures, the rate of methane production depends on a variety of factors, including the quality and quantity of the starting material and the pyrolysis conditions. Even so, as noted by Sackett (26), "pyrolysis of 0.5g of n-$C_{18}H_{38}$ at 200°C for 111 days yields methane barely detectable with [sophistocated] gas chromatography, whereas pyrolysis of 60 mg at 500°C for 1 h yields about 2 ml (STP) methane. ...Below 400°C decomposition occurs so slowly that an experiment would take months to complete at temperatures near 400°C and years at temperatures near 300°C, unless extremely large samples (>100 g) were used." (26) The medium we incubated at 300°C contained less than 1 g of dissolved carbon (primarily bicarbonate) and less than 20 mg dry weight of bacteria added in the inoculum.

ISOTOPIC FRACTIONATION OF METHANE

One argument against the production of methane at superheated temperatures by extremely thermophilic, methanogenic bacteria, at least in superheated hydothermal vent fluids, comes from studies of carbon and hydrogen isotopes in environmental samples of methane and in gas samples produced by pure cultures (or cell extracts) of known methanogenic bacteria. Concentrations of methane in endmember (350°C) hydrothermal fluids at 21°N vent fields, which measure about 65 µM (21,22,35,36), have $\delta^{13}C$ values of about -15.0‰ (36). This is much heavier than methane typically considered to be of biogenic origin (-190 to -64‰ [27]). Therefore, geochemists have concluded that hydrothermal methane has an abiotic origin (36).

However, $\delta^{13}C$ values for methane produced by microorganisms have been determined only for pure cultures of methanogenic bacteria growing at temperatures of 65°C or less (20). Following the approach of LaZerte (19), Lilley et al.(20) plotted the available bacterial $^{13}C/^{12}C$ fractionation factors and uncovered a temperature dependence considerably higher than thermodynamic considerations alone would predict. These data indicate that methane produced bacteriologically at 65°C would be 25-34% lighter in ^{13}C than the CO_2 substrate. They also imply that at 86°C, the reduction of CO_2 to methane would not result in any fractionation of the carbon isotopes. Data for the higher temperatures at which new methanogens are now known to function (Table 1) have not yet been obtained, nor can the effects of hydrostatic pressure on the fractionation factor exhibited by these organisms be predicted (20). Until such information is available, it must be recognized that the use of stable isotope data to argue against the biogenic production of methane in superheated vent fluids is not founded in microbiological experimental fact.

CONCLUSIONS

Exploration of the potential for bacterial methanogenesis at superheated temperatures represents a highly controversial area of research (9,24) still in its infancy. However, because it touches a diversity of disciplines, from molecular biology to planetary science, new clarifying pieces of evidence will be forthcoming at a relatively rapid rate. A continuation of the upward revisions in the maximum temperature for bacterial methanogenesis, such as those experienced in recent years (Table 1), seems inevitable, especially as elevated pressures are included more frequently in experimental design (10). According to current trends (Table 1), increased rates of bacterial methane production, which appear to correlate with increased cardinal growth and environmental isolation temperatures, can also be expected. Indeed, recent lines of inquiry in microbial ecology, barobiology, biochemistry, petrology, and origin of life processes, as discussed in this paper, promise a rewarding future for basic and biotechnological (10) research addressing the potential for bacterial methanogenesis at previously unimagined, superheated temperatures.

ACKNOWLEDGMENTS

I thank John Baross and Bob Kelly for stimulating discussions of this topic and the Johns Hopkins University for support during preparation of the manuscript.

REFERENCES CITED

1. Baross, J. A. and Deming, J. W., "Growth of 'Black Smoker' Bacteria at Temperatures of at Least 250°C," Nature 303, 423-426 (1983) June 2.

2. Baross, J. A. and Deming, J. W., "The Role of Bacteria in the Ecology of Black Smoker Environments," Bull. Biol. Soc. of Washington, No. 6, 355-371 (1985).

3. Baross, J. A. and Hoffman, S. E., "Submarine Hydrothermal Vents and Associated Gradient Environments as Sites for the Origin and Evolution of Life," Origins of Life 15, 327-345 (1985).

4. Baross, J. A., Deming, J. W. and Becker, R. R., "Evidence for Microbial Growth in High-Pressure, High-Temperature Environments," 186-195 in Klug, M. J. and Reddy, C. A., eds., Current Perspectives in Microbial Ecology. Washington, D.C.: American Society for Microbiology, 1984.

5. Baross, J. A., Lilley, M. D. and Gordon, L. I., "Is the CH_4, H_2 and CO Venting from Submarine Hydrothermal Systems Produced by Thermophilic Bacteria?" Nature 298, 366-368 (1982) July 22.

6. Bernhardt, H.-D. L., Jaenicke, R., König, H. and Stetter, K. O., "Biomolecules Are Unstable Under `Black Smoker' Conditions," Naturwissenschaften 71, 583-586 (1984).

7. Chen, C.-T. A., "Geothermal System at 21°N," Science 211, 298 (1981).

8. Delaney, J. R., "Hydrothermal Fluid Temperature of 400°C on the Endeavor Seamount, Northern Juan de Fuca," EOS Transactions, Amer. Geophys. Union 65, 973 (1984) November 6.

9. Deming, J. W., "Investigations on Extremely Thermophilic Microorganisms from Submarine Volcanic Vents," 585-591 in The World Biotech Report 1984, Vol. 2: USA 1. Biotechnology. New York: Online Publications, 1984.

10. Deming, J. W., "The Biotechnological Future for Newly-Described, Extremely Thermophilic Bacteria," Microb. Ecol. (1986, in press).

11. Deming, J. W., "Thermophilic Bacteria Associated with Black Smokers Along the East Pacific Rise," Sec. Intl. Coll. Mar. Bact., University of Brest, France (1986, in press).

12. Deming, J. W. and Baross, J. A., "Solid Medium for Culturing Black Smoker Bacteria at Temperatures to 120°C," Appl. Environ. Microbiol. 51, 238-243 (1986).

13. Edmond, J. M., Von Damm, K. L., McDuff, R. E. and Measures, C. I., "Chemistry of Hot Springs on the East Pacific Rise and Their Effluent Dispersal," Nature 297, 187-191 (1982) May 20.

14. Gur, T. M. and Huggins, R. A., "Methane Synthesis on Nickel by a Solid-State Ionic Method," Science 219, 967-969 (1983).

15. Heinen, W. and Lauwers, A. M., "Growth of Bacteria at 100°C and Beyond," Arch. Microbiol. 129, 127-128 (1981).

16. Huber, H., Thomm, M., König, H., Thies, G. and Stetter, K.O., "*Methanococcus thermolithotrophicum*, a Novel Thermophilic Lithotrophic Methanogen, Arch. Microbiol. 132, 47-50 (1982).

17. Jarrell, K. F., "Extreme Oxygen Sensitivity in Methanogenic Archaebacteria," BioScience 35, 298-302 (1985).

18. Jones, W. J., Leigh, J. A., Mayer, F., Woese, C. R. and Wolfe, R. S., "*Methanococcus jannaschii* sp. nov., an Extremely Thermophilic Methanogen from a Submarine Hydrothermal Vent," Arch. Microbiol. 136, 254-261 (1983).

19. LaZerte, B. D., "The Relationship Between Total Dissolved Carbon Dioxide and Its Stable Carbon Isotope Ratio in Aquatic Sediments," Geochim. Cosmochim. Acta 45, 647-656 (1981).

20. Lilley, M. D., Baross, J. A. and Gordon, L. I., "Reduced Gases and Bacteria in Hydrothermal Fluids: The Galapagos Spreading Center and 21°N East Pacific Rise," 411-449 in Bostrom, K., Smith, K., Laubier, L. and Rona P., eds., Hydrothermal Processes at Seafloor Spreading Centers. New York: Plenum Press, 1983.

21. Lilley, M. D., de Angelis, M. A. and Gordon, L. I., "CH_4, H_2, CO and N_2O in Submarine Hydrothermal Vent Waters," Nature 300, 48-49 (1982) November 4.

22. Ljungdahl, L. G., "Physiology of Thermophilic Bacteria," 149-243 in Rose, A. H. and Morris, J. G., eds., Advances in Microbial Physiology, Vol. 19. New York: Academic Press, 1979.

23. Nickerson, K. W., "A Hypothesis on the Role of Pressure in the Origin of Life," J. Theor. Biol. 110, 487-499 (1984).

24. Osborne, D., "The Origin of Petroleum," The Atlantic 257, 39-54 (1986).

25. Price, L. C., "Organic Geochemistry of Core Samples from an Ultra-Deep Hot Well (300°C, 7 Km)," Chem. Geol. 37, 215-228 (1982).

26. Sackett, W. M., "Carbon and Hydrogen Isotope Effects During the Thermocatalytic Production of Hydrocarbons in Laboratory Simulation Experiments," Geochim. Cosmochim. Acta 42, 571-580 (1978).

27. Schoell, M., "The Hydrogen and Carbon Isotopic Composition of Methane from Natural Gases of Various Origins," Geochim. Cosmochim. Acta 44, 649-661 (1980).

28. Spiess, F. N., MacDonald, K. C., Atwater, T., Ballard, R., Carranza, A., Cordoba, D., Cox, C., Diaz Garcia, V. M., Francheteau, J., Guerrero, J., Hawkins, J., Haymon, R., Hessler, R., Juteau, T., Kastner, M., Larson, R., Luyendyk, B., MacDougall, J. D., Miller, S., Normark, W., Orcutt, J. and Rangin, C., "East Pacific Rise: Hot Springs and Geophysical Experiments," Science 207, 1421-1433 (1980) March 28.

29. Stetter, K.O., "Ultrathin Mycelia-Forming Organisms from Submarine Volcanic Areas Having an Optimum Growth Temperature of 105°C," Nature 300, 258-260 (1982) November 18.

30. Stetter, K.O. and Gaag, G., "Reduction of Molecular Sulphur by Methanogenic Bacteria," Nature 305, 309-311 (1983) September 22.

31. Stetter, K.O., König, H. and Stackebrandt, E., "*Pyrodictium* gen. nov., a New Genus of Submarine Disc-Shaped Sulphur Reducing Archaebacteria Growing Optimally at 105°C," System. Appl. Microbiol. 4, 535-551 (1983).

32. Stetter, K.O., Thomm, M., Winter, J., Wildgruber, G., Huber, H., Zillig, W., Janécovic, D., König, H., Palm, P. and Wunderl, S., "*Methanothermus fervidus*, sp. nov., a Novel Extremely Thermophilic Methanogen Isolated from an Icelandic Hot Spring," Zbl. Bakt. Hyg., I. Abt. Orig. C2, 166-178 (1981).

33. Trent, J. D., Chastain, R. A. and Yayanos, A. A., "Possible Artefactual Basis for Apparent Bacterial Growth at 250°C; Reply by Baross, J. A. and Deming, J. W.," Nature 307, 737-740 (1984) February 23.

34. Von Damm, K. L., Edmond, J. M., Grant, B. C., Measures, C. I., Walden, B. and Weiss, R. F., "Chemistry of Submarine Hydrothermal Solutions at 21°N, East Pacific Rise," Geochim. Cosmochim. Acta 49, 2197-2220 (1985).

35. Welhan, J. A. and Craig, H., "Methane and Hydrogen in East Pacific Rise Hydrothermal Fluids," Geophys. Res. Lett. 6, 829-831 (1979).

36. Welhan, J. A. and Craig, H., "Methane, Hydrogen and Helium in Hydrothermal Fluids at 21°N on the East Pacific Rise, 391-409 in Bostrom, K., Smith, K., Laubier, L. and Rona, P., eds., Hydrothermal Processes at Seafloor Spreading Centers. New York: Plenum Press, 1983.

37. White, R. H., "Hydrolytic Stability of Biomolecules at High Temperatures and its Implication for Life at 250°C," Nature 310, 430-432 (1984) August 2.

38. Yanagawa, H. and Kojima, K., "Thermophilic Microspheres of Peptide-Like Polymers and Silicates Formed at 250°C," J. Biochem. 97, 1521-1524 (1985).

39. Yayanos, A. A., Van Boxtel, R. and Dietz, A. S., "Reproduction of *Bacillus stearothermophilus* as a Function of Temperature and Pressure," Appl. Environ. Microbiol. 46, 1357-1363 (1983).

40. Zeikus, J. G. and Wolfe, R. S., "*Methanobacterium thermoautotrophicum* sp. n., an Anaerobic Autotrophic, Extreme Thermophile," J. Bact. 109, 707-713 (1972).

41. Zeikus, J. G., Ben-Bassat, A. and Hegger, P. W., "Microbiology of Methanogenesis in Thermal Volcanic Environments," J. Bact. 143, 432-440 (1980).

42. ZoBell, C. E., "Ecology of Sulfate-Reducing Bacteria," Producers Monthly 22, 12-29 (1958).

DISCUSSION

P. Smith, U. of Florida — Can you take any of the materials that you treated at 250°C, and then take that material and culture your organisms back to 103°C?

Deming — Yes, but we have not yet been in position to reproduce the experiments we did at 250°C because we did the experiments as guests in someone else's lab. We have only one attempt at that. Subsequent to leaving that lab, we took the remaining volume from the 250°C sample with us, and from that, we cultured colonies in the new gel. So it's like having met one of Koch's postulates. We have passed the bacteria through the extreme conditions, whether you believe they grew or not, and after going through those conditions, they grew again at 100°C.

Smith — Is anybody looking at stability in terms of ATP, membrane components, and parameters of that type?

Deming — Dave Karl from the University of Hawaii has been collecting samples from the smokers and extracting ATP directly from those samples. He published a paper a year or two ago and found higher concentrations of ATP in smoker fluids than you normally find in surface-productive waters of the ocean. He draws very cautious conclusions from that work because he feels that the ATP is from contaminating bacteria that were inadvertently collected in his sample. I think the levels of ATP he detected are so high that it would have to be caused by major contamination if it were not the real thing.

R. Legrand, Reynolds, Smith & Hills — What is the critical temperature of seawater? Since it is so hard to operate those reactors at high temperatures and pressures, have you considered the possibility of doing experiments in situ, putting a reactor near a black smoker and see what happens there?

Deming — Seawater boils at 460°C at 4000 psi, so we're working below the critical point of seawater at those pressures. We have considered in situ experiments, and as of May 1st, I am funded to continue this research by the Navy. They are supporting Baross and me to return to the

DISCUSSION, CONT.

vents and do in situ experiments where we will position a multi-chambered titanium apparatus into the flow of a black smoker and have those chambers preloaded with various radio-labeled tracers to attempt to measure uptake of labeled material. We're in the process of engineering designs and laboratory tests of the stability of the precursors under those conditions. There are an incredible host of problems associated with doing those kinds of experiments, everything from the bravado of the pilot operating the submersible to the stability of the reagents we want to use. It's not going to be a trivial experiment, but we're going to try it.

DEMONSTRATION OF LANDFILL GAS ENHANCEMENT TECHNIQUES IN LANDFILL SIMULATORS

James J. Walsh, James W. Stamm, and W. Gregory Vogt
SCS Engineers
Covington, Kentucky

Riley N. Kinman, Ph.D. and Janet I. Rickabaugh
University of Cincinnati
Cincinnati, Ohio

Michael L. Wilkey
Gas Research Institute
Chicago, Illinois

ABSTRACT

Seven techniques to enhance landfill gas production from landfills were tested using 16 landfill simulators. The techniques used were: accelerated moisture infiltration; elevated moisture content; leachate recycle; buffer, nutrient and sludge additions; and elevated refuse temperature. Each test cell contained 380 kg (836 lb)of municipal solid waste. All simulators were constructed in February 1980. Various physical and chemical parameters were measured for five years (through January 1985) to determine gas and leachate quantity and quality.

Data from this study indicated that sewage sludge additions, buffer additions, and nutrient additions each appear to be successful enhancement techniques for gas production from municipal solid waste. The parameters of moisture infiltration and elevated moisture content have shown some beneficial effects.

DEMONSTRATION OF LANDFILL GAS ENHANCEMENT TECHNIQUES IN LANDFILL SIMULATORS

BACKGROUND

The recovery of methane from sanitary landfills can be a partial solution to declining energy resources and the hazards of landfilling solid waste. Failure to control or optimize methane production from a landfill tends to make recovery less attractive economically and yields a longer landfill stabilization period. Due to the sporadic nature of gas production via anaerobic decomposition, the rate and quantity of methane produced from sanitary landfills have been uncertain. The purpose of this research was to test and evaluate environmental factors which should enhance methane production from anaerobic decomposition of municipal solid waste. Enhancement as it is used in this paper refers to the optimization of gas quality (percent methane), and gas quantity (liters of gas per kilogram of solid waste).

In March 1979, the Department of Energy (DOE) announced its intention to fund projects in the areas of research, development, and demonstration of landfill gas utilization. This action came in pursuit of its objective to develop waste to energy processes in the biological area. The thrust of this effort was to investigate optimal methane recovery through improved design, construction, and operational techniques in both the generation and collection of landfill gas from existing or new full-scale landfills. In May 1979, SCS Engineers (SCS) was awarded one of five projects dealing with landfill gas enhancement techniques. In 1982, the Gas Research Institute (GRI) granted a continuation to this ongoing study. The scope of enhancement techniques under investigation was widened at that time.

The objective of the overall program was to evaluate the effectiveness of the following landfill gas enhancement techniques:

1. Accelerated moisture through-put

2. Elevated moisture content

3. Leachate recirculation

4. pH control/buffer addition

5. Nutrient addition

6. Sewage sludge addition

7. Elevated operating temperature

Each of these landfill gas enhancement techniques was included because they attempt to improve environmental conditions within the test cell in order to foster growth of methanogenic bacteria.

INITIAL DESIGN AND CONSTRUCTION

To demonstrate the potential for gas enhancement, four initial techniques were investigated during the first three years of the project, including (1) accelerated moisture through-put, (2) leachate recirculation, (3) nutrient addition, and (4) buffer addition/pH control.

The 16 test cells were located at the EPA Center Hill Laboratory in Cincinnati, Ohio. The cells were identified by the numbers 20 through 35, since they were placed in parallel with a similar EPA-sponsored landfill gas simulator program. Program design details for the GRI/DOE-sponsored project are presented in Exhibit 1.

Exhibit 1. Initial Program Design

Test Cell	Initial Loading Variables	Operational Variables
20, 21	-- --	LI --
22, 23	-- --	HI --
24, 25	-- --	HI LR
26, 27	BA --	HI LR
28, 29	-- NA	HI LR
30, 31	BA NA	HI LR
32, 33	BA --	HI --
34, 35	-- NA	HI --

LI = Low Infiltration
HI = High Infiltration
LR = Leachate Recycle
BA = Buffer Addition
NA = Nutrient Addition

The 16 laboratory scale test cells were constructed to provide durable, gas tight containers of sufficient scale to promote the decomposition processes which occur in an actual landfill. Each test cell was constructed of rolled steel and designed to contain approximately 0.80 cubic meters (1 cubic yard) of shredded municipal refuse. Cylindrical tanks were chosen to prevent "dead" corners of inactive mixing zones. Exhibit 2 shows a cross-section of a typical test cell used in this project. The overall tank dimensions were: height - 1.83 m (6.0 ft); diameter - 0.91 m (3.0 ft); and wall and cover thickness - 4.76 mm (0.19 in.). Bottom and top cover diameters were 1.04 m (3.42 ft). For drainage purposes, concrete block supports were used to elevate the lysimeters. Each lysimeter also contained temperature and gas monitoring systems.

Each lysimeter received 230.34 kilograms (507.90 lbs) of dry-weight refuse and the wet weight moisture content of the refuse was 39.3 percent. Test Cell Nos. 20 and 21 received infiltration rates similar to Midwestern U.S. precipitation/percolation rates. The remaining 14 test cells received accelerated moisture applications at three times this rate. Test Cells 26 through 35 received selected quantities of calcium carbonate (a buffer) and/or ammonium phosphate (a nutrient).

Exhibit 2. Test Cell Cross Section

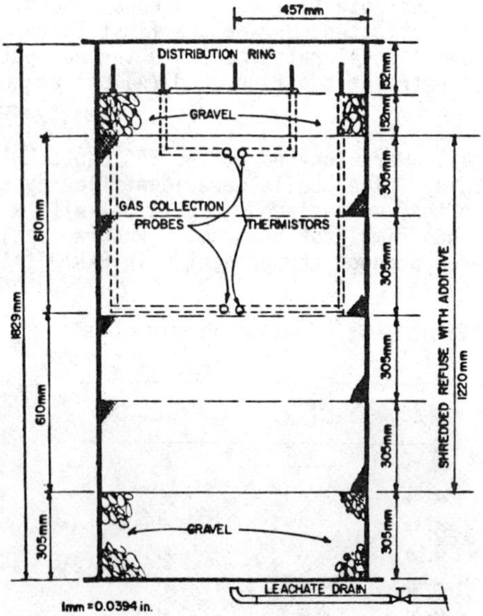

TEST CELL RELOADING

Test cell reloading additives including sewage sludge, sodium bicarbonate, and ammonium phosphate were performed in January 1983 under GRI sponsorship. Other activities which occurred in conjunction with test cell reloading included moisture additions to increase moisture contents, and applied heat to raise internal temperatures of Cells 30 and 31. A summary of the reloading activities is presented in Exhibit 3.

OPERATION AND MONITORING

The test cells were continuously operated from February 1980 to January 1985. Water was added monthly, and leachate and gas samples were collected and analyzed monthly according to a prescribed schedule. Alkalinity, pH, and volatile acids were monitored to evaluate the efficiency of the buffer additions to control the pH of the test cell. The effectiveness of the nutrient additions was measured as a function of the presence of available carbon, total phosphorus, and total Kjeldahl nitrogen. Metal analyses were performed to indicate possible toxicities to the methanogenic organisms and to document the release of metals from a simulated landfill situation. The gas measurements were used to determine gas production and methane enhancement.

Additional monthly operations included leachate recycle (for designated cells). Moisture was added to the test cells each month.

Exhibit 3. Re-Loading Summary

Test Cell	Operational Variables	Initial Loading Variables	Re-Loading Variables
20	LI -- --	-- --	-- -- -- -- --
21	LI -- --	-- --	-- -- -- MC --
22	-- HI --	-- --	-- -- -- -- --
23	-- HI --	-- --	-- -- -- -- --
24	-- HI LR	-- --	-- -- -- -- --
25	-- HI LR	-- --	-- -- SA -- --
26	-- HI LR	BA --	-- -- -- -- --
27	-- HI LR	BA --	-- -- SA -- --
28	-- HI LR	-- NA	-- -- -- -- --
29	-- HI LR	-- NA	-- -- SA -- --
30	-- HI LR	BA NA	-- -- -- -- TI
31	-- HI LR	BA NA	-- -- -- -- TI
32	-- HI --	BA --	-- -- -- -- --
33	-- HI --	BA --	BA -- -- -- --
34	-- HI --	-- NA	-- -- -- -- --
35	-- HI --	-- NA	-- NA -- -- --

Operational Variables:

LI = Low Infiltration
HI = High Infiltration
LR = Leachate Recycle

Initial Loading Variables:

BA = Buffer Addition
NA = Nutrient Addition

Re-Loading Variables:

BA = Buffer Addition
NA = Nutrient Addition
SA = Sludge Addition
MC = Increased Moisture Content
TI = Temperature Increase

RESULTS

The primary purpose of this research project was to study the technical feasibility of enhancing gas production in municipal landfills. Though test cells (or lysimeters) do not precisely duplicate a landfill environment, many key parameters of anaerobic decomposition are encompassed under a controlled set of conditions. The key measurements used to evaluate the success or failure of each technique were gas and methane production.

Gas Production

Exhibits 4 and 5 summarize gas production rates on a year-by-year basis and a cumulative basis for all 16 test cells. Baseline gas production rates and enhanced production rates are presented in these exhibits.

Test Cell 20 was designated the control cell and because of its low rate of infiltration should best represent "typical" landfill conditions. This test cell experienced its greatest level of activity during the first year with a total gas production rate of 5.2 l/kg/year. During the remainder of the project, the control cell had a constant production in the range of 1.3 to 2.0 l/kg/year. The final cumulative gas volume rate for Test Cell 20 as 2.3 l/kg/year.

The first set of enhancement techniques consisted of increased moisture content due to high rates of infiltration (Cells 22 and 23) or a concentrated effort to increase the moisture content of the cell (Cell 21). The high infiltration test cells had gas production rates between 5.0 to 5.5 l/kg/year. These rates fell through Years 2, 3, and 4 of operation. Both Cells 22 and 23 experienced a resurgence of activity during the final year of operation. In general, the high infiltration cells exhibited a high degree of variability with

Exhibit 4. Yearly Total Gas Production Rates

Test Cell	Description	Year 1 (1980) (l/kg/yr)	Year 2 (1981) (l/kg/yr)	Year 3 (1982) (l/kg/yr)	Year 4 (1983) (l/kg/yr)	Year 5 (1984) (l/kg/yr)
20	LI	5.2	1.4	1.3	1.50	2.0
21	LI MC	5.1	1.3	0.70	1.70	6.40
22	HI	5.0	1.9	1.2	1.30	3.70
23	HI	5.5	2.50	0.50	3.50	18.00
24	HI LR	5.97	2.59	0.50	0.37	0.30
25	HI LR SA	3.65	2.18	0.69	49.52	52.16
26	HI BA LR	4.28	5.81	1.53	0.25	0.21
27	HI BA LR SA	5.56	1.77	0.30	42.52	71.14
28	HI NA LR	7.2	1.97	0.61	0.72	0.46
29	HI NA LR SA	7.14	2.27	0.66	11.48	92.26
30	HI BA NA LR TI	7.23	2.81	0.79	0.75	0.12
31	HI BA NA LR TI	7.34	2.26	0.98	1.09	0.17
32	HI BA	6.16	1.35	21.18	28.54	36.66
33	HI BA (BA)	6.41	2.36	1.02	16.86	32.70
34	HI NA	6.54	1.44	0.58	0.82	2.36
35	HI NA (NA)	6.0	2.45	1.50	7.42	32.93

LI = Low Infiltration
HI = High Infiltration
LR = Leachate Recycle
BA = Buffer Addition
NA = Nutrient Addition
TI = Temperature Increase
MC = Increased Moisture Content
(BA) = Reloaded Buffer Addition
(NA) = Reloaded Nutrient Addition

Exhibit 5. Cumulative Total Gas Production Rates

Test Cell	Description	Year 1 (1980) (l/kg/yr)	Year 2 (1981) (l/kg/yr)	Year 3 (1982) (l/kg/yr)	Year 4 (1983) (l/kg/yr)	Year 5 (1984) (l/kg/yr)
20	LI	5.2	2.8	2.6	2.4	2.3
21	LI MC	5.1	3.2	2.4	2.2	3.1
22	HI	5.0	3.5	2.7	2.4	2.6
23	HI	5.5	4.0	2.8	3.0	6.0
24	HI LR	5.97	4.3	3.0	2.4	1.9
25	HI LR SA	3.65	2.9	2.2	12.0	21.64
26	HI BA LR	4.28	5.0	3.9	2.9	2.4
27	HI BA LR SA	5.56	3.6	2.5	12.5	24.3
28	HI NA LR	7.2	4.6	3.3	2.6	2.2
29	HI NA LR SA	7.14	4.7	3.4	5.4	22.7
30	HI BA NA LR TI	7.23	5.0	3.6	2.9	2.4
31	HI BA NA LR TI	7.34	4.8	3.5	2.9	2.4
32	HI BA	6.16	3.8	9.6	14.3	18.8
33	HI BA (BA)	6.41	4.4	3.3	6.7	11.9
34	HI NA	6.54	4.0	2.9	2.4	2.4
35	HI NA (NA)	6.0	4.2	3.4	4.4	10.1

LI = Low Infiltration
HI = High Infiltration
LR = Leachate Recycle
BA = Buffer Addition
NA = Nutrient Addition
TI = Temperature Increase
MC = Increased Moisture Content
(BA) = Reloaded Buffer Addition
(NA) = Reloaded Nutrient Addition

production rates ranging from 15 to 150 percent higher than the control cell. Test Cell 21 underwent increases in infiltration during the final two years of operation. Its gas production rate responded by increasing by a factor of nearly 10 during Years 4 and 5. Due to the increased moisture applications to Test Cell 21, it had a final cumulative production rate 35 percent higher than the control cell (Cell 20).

The second set of enhancement techniques consisted of leachate recycle in conjunction with various chemical loadings of buffer or nutrient slurries. This group encompassed Cells 24, 26, 28, 30, and 31. Test Cells 30 and 31 had their internal temperature raised to approximately 35 degrees C for Years 4 and 5. This group of cells showed distinct signs of accelerated anaerobic decomposition in the first year with a range of production rates from 4.28 to 7.23 l/kg/year. By Year 3 of the project, activity in all of these cells had declined rapidly. During the last two years of operation, gas production almost ceased entirely. A comparison of final cumulative gas generation rates for this group ranged between -17 to +6 percent relative to Test Cell 20. The increased temperature did not affect gas production, but the buffer and nutrient applications to Cells 26 and 28 did lessen the negative gas enhancement effects of leachate recycle.

The third set of enhancement techniques consisted of sludge addition. Cells 25, 27, and 29 were originally duplicate leachate recycle cells and received sludge additions in Year 4 of the project. The first three years of operation saw similar activity as displayed by the other leachate recycle cells. The addition of sludge had an immediate effect on these test cells. The other loading variables of these cells did not seem to radically influence the enhanced gas production caused by the sludge. Test Cells 25, 27, and 29 had respective cumulative gas production rates of 11.4 l/kg/year, 11.3 l/kg/year, and 11.6 l/kg/year. These rates represent an increase of gas production of between 850 and 950 percent as compared to the low infiltration control cell (Cell 20).

The final group of enhancement techniques was loading and/or reloading of various chemical slurries including buffer and nutrient compounds. Test Cells 32 and 33 were the buffer addition cells. By the end of the project, the reloading of buffer had not demonstrated any positive benefits relative to gas generation. Both buffer test cells (Cells 32 and 33) did demonstrate excellent gas production in the range of 420 to 720 percent above the control cell value. The nutrient addition cells (Cells 34 and 35) also enjoyed average to excellent gas production rates. Test Cell 34, which only received an initial loading of nutrient material, had only a 3 percent higher cumulative production rate than the control cell. Test Cell 35, which received both an initial loading and reloading of nutrient slurry, responded in Years 4 and 5 with vigorous anaerobic activity. The final cumulative production rate for Test Cell 35 was 340 percent higher than the control level.

Methane Production

Total gas production is an overall summation of a direct empirical measurement, but methane production was determined by the product of the monthly composition and the monthly cumulative gas generation. These calculated methane yields are the most critical evaluation of the success of the applied enhancement techniques. Exhibits 6 and 7 show methane production rates on an annual and cumulative basis. Both methane production results are presented on a liters/kilogram/year basis.

Test Cell 20 was also considered to be a baseline methane producer because its operating conditions are similar to a normal landfill environment. Methane production determinations for this test cell ranged from 0.05 to 0.60 l/kg/year. The final cumulative methane production rate was 0.22 l/kg/year.

Those cells which were maintained at a high moisture content were considered the first set of enhancement techniques and included Cells 21, 22, and 23. Test Cells 22 and 23 both received high rates of infiltration during the entire length of the project. Both of these

Exhibit 6. Yearly Methane Gas Production Rates

Test Cell	Description	Year 1 (1980) (l/kg/yr)	Year 2 (1981) (l/kg/yr)	Year 3 (1982) (l/kg/yr)	Year 4 (1983) (l/kg/yr)	Year 5 (1984) (l/kg/yr)
20	LI	0.0	0.0	0.05	0.45	0.60
21	LI MC	0.0	0.04	0.06	0.5	2.3
22	HI	0.0	0.0	0.05	0.35	1.65
23	HI	0.0	0.0	0.01	0.99	9.81
24	HI LR	0.22	0.56	0.08	0.07	0.04
25	HI LR SA	0.08	0.23	0.08	28.81	27.74
26	HI BA LR	0.12	0.68	0.19	0.05	0.03
27	HI BA LR SA	0.04	0.19	0.01	24.14	32.05
28	HI NA LR	0.0	0.18	0.04	0.11	0.06
29	HI NA LR SA	0.0	0.14	0.10	6.69	51.03
30	HI BA NA LR TI	0.09	0.60	0.07	0.13	0.01
31	HI BA NA LR TI	0.0	0.13	0.13	0.20	0.01
32	HI BA	0.0	0.06	6.5	16.25	19.27
33	HI BA (BA)	0.0	0.03	0.11	8.49	18.50
34	HI NA	0.0	0.0	0.02	0.15	0.82
35	HI NA (NA)	0.0	0.01	0.33	2.98	18.59

LI = Low Infiltration
HI = High Infiltration
LR = Leachate Recycle
BA = Buffer Addition
NA = Nutrient Addition
TI = Temperature Increase
MC = Increased Moisture Content
(BA) = Reloaded Buffer Addition
(NA) = Reloaded Nutrient Addition

Exhibit 7. Cumulative Methane Gas Production Rates

Test Cell	Description	Year 1 (1980) (l/kg/yr)	Year 2 (1981) (l/kg/yr)	Year 3 (1982) (l/kg/yr)	Year 4 (1983) (l/kg/yr)	Year 5 (1984) (l/kg/yr)
20	LI	0.0	0.0	0.02	0.02	0.22
21	LI MC	0.0	0.0	0.02	0.14	0.57
22	HI	0.0	0.0	0.02	0.10	0.41
23	HI	0.0	0.0	0.01	0.25	2.20
24	HI LR	0.22	0.39	0.29	0.23	0.19
25	HI LR SA	0.08	0.16	0.13	7.30	11.40
26	HI BA LR	0.12	0.40	0.33	0.26	0.21
27	HI BA LR SA	0.04	0.12	0.08	6.10	11.30
28	HI NA LR	0.0	0.09	0.07	0.08	0.08
29	HI NA LR SA	0.0	0.07	0.08	1.73	11.60
30	HI BA NA LR TI	0.09	0.35	0.25	0.22	0.18
31	HI BA NA LR TI	0.0	0.07	0.09	0.12	0.09
32	HI BA	0.0	0.03	2.20	5.70	8.42
33	HI BA (BA)	0.0	0.01	0.07	2.16	5.43
34	HI NA	0.0	0.01	0.01	0.09	0.20
35	HI NA (NA)	0.0	0.01	0.12	0.75	4.32

LI = Low Infiltration
HI = High Infiltration
LR = Leachate Recycle
BA = Buffer Addition
NA = Nutrient Addition
TI = Temperature Increase
MC = Increased Moisture Content
(BA) = Reloaded Buffer Addition
(NA) = Reloaded Nutrient Addition

cells showed little methanogenic activity until the fourth and fifth years of operation. During the last 12 months, Test Cell 23 experienced a large growth in methane production (about 1.95 l/kg/yr).

Alkalinity was greatly reduced during the fourth and fifth years of the project. Recorded alkalinity concentrations were at 3,382 mg/l and 3,921 mg/l for the respective fourth and fifth years of operation for Test Cell 23. These levels were much lower than Test Cell 22 which was under a similar set of operating conditions. The decreased buffering capacity suggests elevated levels of volatile acids, which are the intermediate step in the methane forming process. Apparently, the cell contents of Test Cell 23 provided a strong natural buffering capacity which was able to support vigorous anaerobic activity without allowing a drop in pH. This statement is supported by the fact that Test Cell 23 had a higher pH value (about 5.00) than Test Cell 22 (about 4.79) even though Test Cell 23 was obviously loaded with a larger quantity of acidic material. Both high infiltration test cells had higher methane production rates ranging from 86 to 880 percent above the control cell.

Test Cell 21 received an increased moisture load in the final two years of the project. This cell responded to the increased moisture levels with a dramatic increase in the methane production rate for Year 5 (0.43 l/kg/yr). This rapid loading of moisture allowed Test Cell 21 to have a cumulative methane production rate (0.59 l/kg/year) 160 percent higher than the control cell (Cell 20).

Leachate recycle in conjunction with loading of buffer or nutrient additives encompassed the second set of enhancement techniques. This group of cells included Test Cells 24, 26, 28, 30, and 31. As a group, the leachate recycle cells performed on a subpar level in the area of methane production.

While the recycling of leachate allowed these cells to rapidly undergo anaerobic decomposition, it also produced adverse effects. In general, the leachate recycle cells produced a higher strength leachate with chemical oxygen demand (COD) levels 3 to 5 times higher than control values. The intensification of leachate strength may have caused non-ideal conditions for the growth of methanogenic bacteria within these cells. It is hypothesized that the recirculation of leachate fostered the competitive growth of non-methanogenic bacteria. It is possible that high levels of volatile acids coupled with the high nitrogen levels found in these cells spawned the competitive growth of autotrophic organisms. Many autotrophic organisms are capable of thriving in an anaerobic environment similar to that found in a lysimeter. The other cell parameters, such as buffer or nutrient addition as well as increased operating temperature, could reverse the negative effects of leachate recycle. As a group, the leachate recycle cells had a methane production rate 3 to 65 percent lower than the control cell.

The third group of enhanced cells were those receiving additions of anaerobically digested sludge. This group of cells includes Test Cells 25, 27, and 29. Prior to the addition of sludge, these test cells were duplicate leachate recycle cells. From the project's data base, it was shown that Cells 25, 27, and 29 had average to high levels of volatile acids in the range of 16,546 mg/l to 24,713 mg/l during the third year of operation. The combination of high levels of volatile acids (a "fuel source") with the introduction of methanogenic bacteria present in the sludge (a "consumer"), produced high rates of methane production in these cells. This enhancement technique provided a cumulative methane production rate over 50 times higher than the control test cell. In general, the loading variables (buffer and nutrient addition) did not seem to have a pronounced effect on methane production.

The final set of enhancement techniques utilized the addition of either buffer or nutrient compounds to Test Cells 32 and 33 (buffer addition) and Test Cells 34 and 35 (nutrient addition). The strengthening of the buffering capacity of Cells 32 and 33 provided excellent methane yields. Though the first several years of operation saw minimal methane production, both cells increased methane generation rates in Years 4 and 5. The cumulative methane production rates ranged from 2.4 to 3.5 times the baseline methane producer. By the end of the project, data did not support the reloading of buffering material in Test Cell 33. But due to a lag in methane production shown by Cell 32, it is hypothesized that the buffer reloading would have shown some effect in 12 to 24 months.

Test Cells 34 and 35 represent the nutrient addition test cells. In the case of this parameter, the reloading of nutrient performed in Year 4 produced a dramatic increase in methane generation. While Test Cell 34 (HI NA) had a cumulative methane production rate roughly equivalent to the control cell, Test Cell 35 (HI NA NA) increased its cumulative methane generation rate by a factor of about 20.

SUMMARY

This paper covers the project's entire five years of operation from February 1980 through January 1985. The findings and conclusions presented in this report are based on results obtained under tightly-controlled laboratory conditions. Actual percent increases in gas production or methane generation should likely not be directly applied to "real-world" landfilling operations. Nonetheless, the overall trends seen here may well apply, and thus, the conclusions presented here should be sound for full-scale landfill applications.

The addition of anaerobically digested sludge provided the most enhanced methane production. Using this technique in conjunction with a leachate recycle program should greatly improve the generation of methane from a landfill, without having a significant adverse effect on leachate collection and treatment.

In a landfill situation where economics, location, and geological conditions do not allow use of sludge addition, chemical addition will likely increase methane yields. The choice of buffer versus nutrient addition should be based on the relative cost of the additive and the age of the landfill. A landfill in the first years of operation could use a buffering compound to affect methane generation five to twenty years into its future. A landfill that is nearing its final years of operation may be able to realize increased methane generation in a relatively short period of time by adding nutrients.

While increased moisture content due to high infiltration or leachate recycle appears to accelerate decomposition processes, the small or negative effects on methane yields may make these techniques unattractive. These techniques may also cause the additional problems of high strength or high quantities of generated leachate without commensurate benefits.

ACKNOWLEDGEMENTS

Funding for this project has most recently been through the Gas Research Institute (GRI) Substitute Natural Gas Supply Research Division, under GRI Contract No. 5082-224-0681. Michael L. Wilkey is the GRI Project Manager. Early funding for this program was obtained through the U.S. Department of Energy (DOE), Argonne National Laboratory (ANL) under ANL Contract Nos. 31-109-38-5263 and 31-109-38-6808. Jean Bogner was the ANL Project Officer.

DISCUSSION

J. Papinsick, Air Products and Chemicals, Inc. The sludge addition to the laboratory set up seems to work well. If you can do it to a landfill, do you believe you can get an old landfill to be more productive presuming you can get a permit to use sewage sludge?

Vogt I'm not sure the people in Wisconsin will allow you to do that. They are already adamant about keeping sludge out of landfills. But if the landfill is engineered properly in terms of leachate, you should enhance methane production over and above what the sewage sludge produces itself.

__________ How would you propose to increase the overall temperature of the landfill to shorten the induction period? Your data showed that it took about 3 years. What was the pH?

Vogt Increasing the temperature should lead to enhancement, but our data did not show that occurred in this particular case. I have nothing to propose on how to increase the heat in a landfill. The temperatures of most landfills are higher than room temperature. The interior of the landfill is normally 95°F or more depending on the conditions. You asked about lag time to methane production. This study took quite a long time to produce methane under laboratory conditions. We have no explanation for that because the nutrients and seed were there. The pH was low for about the first 3 years, but incresed as gas was produced. It was about 6.2 to 6.5 in Test Cell 32 as an example, and higher in those with buffer. We're measuring the pH of leachate that sifted through the system into a reservoir which is drained once a month. We may not be measuring the pH within the test cell. I tend to equate the two, but maybe they are not exactly the same.

AN INTEGRATED WASTEWATER-ENERGY PRODUCTION SYSTEM

T.D. Hayes, Project Manager
H.R. Isaacson, Program Manager
Gas Research Institute, Chicago, IL

D.P. Chynoweth, Director of Bioprocess Engineering
University of Florida, Gainesville, FL

K.R. Reddy, Professor of Soil Biochemistry
Central Florida Research and Education Center, Sanford, FL

R. Biljetina, Associate Director
Institute of Gas Technology, Chicago, IL

ABSTRACT

The goal of research conducted at the Community Waste Research Facility at the Walt Disney World Resort Complex near Orlando, Florida is to develop new waste management and conversion concepts capable of producing low-cost, pipeline-quality methane from community wastes at a central location. Waste resources for digestion in a community include municipal solid waste, industrial waste, sewage sludge, and sewage-grown biomass. The first of these concepts to be developed by this project was an integrated hyacinth wastewater treatment system employing anaerobic digestion to convert a blend of sewage-grown biomass and sewage sludge to methane.

This paper describes the results of this research and the economics of a conceptual community system based on the performances of the experimental test unit.

AN INTEGRATED WASTEWATER-ENERGY PRODUCTION SYSTEM

INTRODUCTION

According to U.S. government estimates, nearly a half a trillion pounds of organic municipal wastes (dry ash free basis) enter the U.S. Collection and disposal system each year (1). Urban waste alone could supply up to 3 quad of methane through the successful development and application of anaerobic digestion technology. Residues in the category of urban waste include municipal and industrial wastes, sewage sludge, and sewage-grown biomass.

For more than half a century, anaerobic digestion has been applied to the processing of sewage sludge for the purposes of improved dewaterability and volume reduction. Pollution control, not methane recovery, was the objective of the early digester designs; biogas generated from such facilities was typically flared as a wasted by-product. At present, anaerobic digestion is still used primarily as a pollution control process for the treatment of municipal sludges and high strength industrial wastewaters and is not optimized for the production of low cost methane. If low cost methane production from waste is to be achieved, new community waste conversion concepts need to be developed that maximize methane yields from waste while substantially reducing capital and O&M costs for the overall methane-from-waste system. The development of such concepts is the ultimate goal of the GRI Methane from Community Waste Systems Project at Walt Disney World which is located near Orlando, FL.

The first of these concepts to be developed by this project was an integrated hyacinth wastewater treatment system employing anaerobic digestion to convert a blend of sewage grown biomass and sewage sludge to low cost methane. This community waste system requires the integration of at least three technologies: 1) anaerobic digestion, 2) biomass management for achieving maximum yields, and 3) aquaculture treatment of wastewater to meet federal, state and local standards.

This paper describes the status of research in progress to develop a system that combines methane production with a wastewater treatment concept that produces a high-yielding crop of water hyacinths from sewage. This integrated concept has the potential of providing the pipeline with a stable source of low-cost methane while offering the community a cost-effective process for wastewater treatment and water reuse.

THE LITERATURE

In recent years, the water hyacinth process has gained increased attention as an "alternate and innovative" method for wastewater treatment (7). Hyacinth treatment of sewage is usually conducted in shallow ponds or channels of less than one meter deep. A significant number of pilot plants have been described in the literature (4,5,9,16); most of these have been operated in the warmer regions of the U.S. (such as California, Texas, Florida, Mississippi, Alabama, etc. Reviews of the operation and performance of these systems as well as the state of the art in design are presented in the literature (2,13).

Upon inspection of this literature, it becomes apparent that most of the studies conducted to date have been of a proof-of-concept and demonstration nature. Pilot-scale hyacinth systems have typically achieved removals of 75-95 percent of biochemical oxygen demand (BOD_5) and suspended solids (TSS) under sewage loadings of 20-150 kg/ha-day (18-134 lbs/ac-day) (2,4). Nutrient removal reported for these systems has been moderately successful, with most studies reporting high nitrogen (N_T) but low phosphorus (P_T) removal efficiencies (2). For the most part, these observations of performance relate to non-stressed systems which have not been optimized for pollutant removal and hyacinth productivity. Nevertheless, hyacinth systems continue to be tested and implemented by municipalities as an "innovative and alternate" technology for the cost-effective treatment of sewage, with the majority of these pilot plants being applied to wastewater polishing and nitrogen removal. Information needs for these new systems include: 1) strategies for process improvement based on a sound understanding of plant growth and pollutant removal mechanisms; 2) a data base for rational scale up and design approaches; and, 3) operational procedures that minimize cost and stabilize process performance.

Equally important is the need to identify low cost methods for the disposal and/or utilization of the large amounts of solids that can be generated by the water hyacinth treatment process. In many applications, the production of hyacinths (5% total solids content) could equal or exceed sludge output from a treatment facility, up to 12-15 wet tons per day per 10,000 population served. Anaerobic digestion has the potential of converting this community waste problem into an energy production benefit when effectively integrated into the water hyacinth treatment system.

As a conversion technology, anaerobic digestion has a number of important advantages in the processing of hyacinths, sludge and other community wastes. These include: 1) process compatibility with high and low moisture feedstocks, 2) the potential for the direct production of high-Btu gas (600-950 Btu/scf), and 3) economy of scale at low processing capacities (6).

Several studies have been conducted which show that water hyacinths can be converted to methane with anaerobic digestion (3,8,12,14). Studies conducted in completely mixed reactors operated in a semicontinuous mode, indicated that a methane yield of less than 0.2 m^3/kg of volatile solids added (VS_A) or less than 40% volatile solids conversion could be achieved at retention times of 12 to 16 days (8). Alkaline pretreatment to enhance the performance of two-phase digestion of hyacinths has been reported (14), but the added cost of chemicals could not be justified by the marginal increase in methane production that was achieved. Without pretreatment, however, long solids retention times are needed to achieve hyacinth conversion efficiencies of greater than 50%. Completely mixed conventional reactors are at a particular disadvantage because the solids retention time (SRT) is virtually equal to hydraulic retention time (HRT) and large reactor volumes are required to achieve extended solids retention times necessary for efficient conversion.

The goal of fermentation research conducted at the Community Waste Research Facility (CWRF) of the Walt Disney World Resort Complex is to develop an advanced conversion system that is capable of achieving high methane yields through improved solids management while decreasing mixing requirements to reduce power consumption and maintenance costs. This concept would not be limited to hyacinth/sludge blend, but could be applicable to a wide range of community wastes as well.

INTEGRATED AQUACULTURE/ANAEROBIC DIGESTION CONCEPT

The system concept is shown in Figure 1. As effluent from the primary settler is passed through the water hyacinth channel, pollutants such as organic matter (biochemical oxygen demand or BOD_5), suspended solids (SS), nitrogen (NH_3 and NO_3^-) are removed by the root system through mechanisms of direct plant uptake, microbial conversion, physical entrapment and sedimentation.

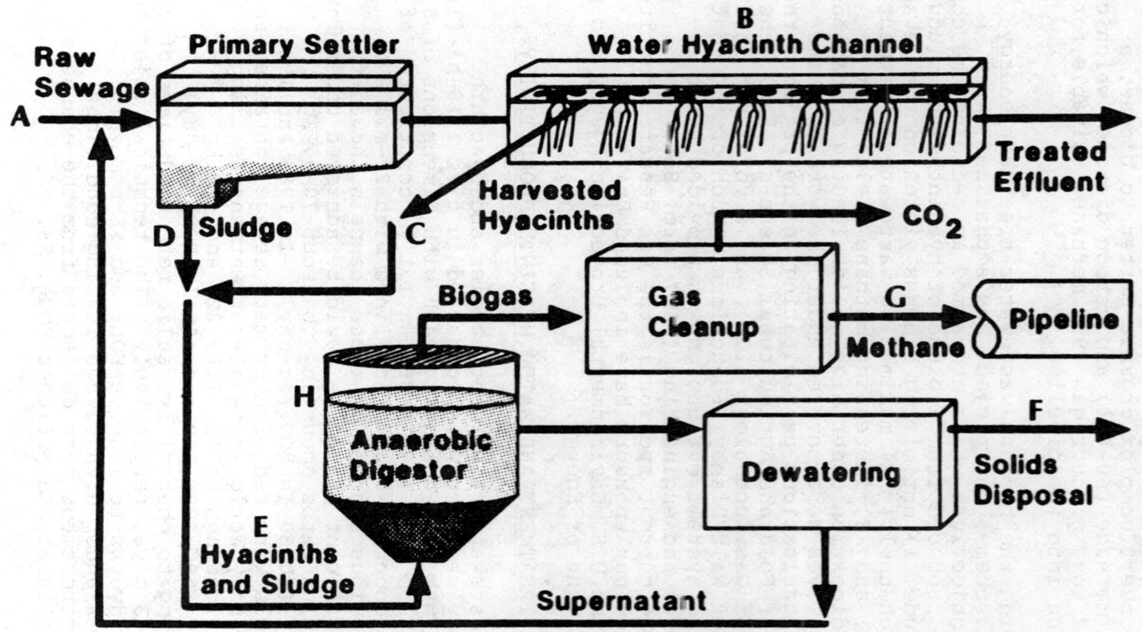

FIGURE 1.
INTEGRATED WATER HYACINTH
WASTEWATER TREATMENT/ANAEROBIC DIGESTION CONCEPT

Hyacinths grow rapidly on the wastewater and are periodically harvested, combined with sewage sludge from the primary settler and introduced to the anaerobic digester. As the feed passes through the digester, bacteria convert complex organic matter to biogas, a mixture of methane (60-65%) and carbon dioxide, which can be upgraded to a product gas (95% methane) suitable for introduction into the pipeline.

In developing the biomass wastewater treatment energy conversion scheme, the project has emphasized three technical objectives which are aimed at reducing the cost of methane produced from a blend of hyacinths and sludge. These include: 1) optimizing biomass yields on the sewage treatment channels; 2) maximizing wastewater treatment efficiency; and, 3) increasing methane yields from anaerobic digestion under high solids loading conditions. These objectives were covered under the work of three research organizations participating in the project, the Institute of Food and Agricultural Sciences (IFAS), the Institute of Gas Technology (IGT), and several subsidiary companies of Walt Disney Productions. Much of the cost analysis and systems evaluation was provided by an architectural and engineering (A&E) firm, Black and Veatch. The current sponsor is the Gas Research Institute (GRI). Previous sponsors have included United Gas Pipeline, the U.S. Environmental Protection Agency and the U.S. Department of Energy.

WASTEWATER TREATMENT AND WATER HYACINTH PRODUCTIVITY

Hyacinth wastewater treatment studies and hyacinth productivity research were conducted in five 1/10-ha (1/4 acre) hyacinth test channels, each with dimensions of 8.8m x 110m x 0.35m deep (29' x 360' x 1.1'), constructed of reinforced concrete blocks and lined with 20-mil PVC sheet. Previous studies on these channels showed that secondary effluent standards could be achieved under low sewage loading rates amounting to about 70-90 kg BOD_5/ha-day (60-80 lb BOD_5/ac-day). In 1983 and 1984, four channels were fed with primary sewage (obtained from the Walt Disney World wastewater treatment settling basins) at loadings of 55, 110, 220 and 440 kg BOD_5/ha-day (50, 100, 200 and 400 lb BOD_5/ac-day) corresponding to channel hydraulic retention times of 24, 12, 6, and 3 days, respectively. The technical objective of this study was to measure BOD_5 and SS removal efficiencies under high sewage loading conditions. Wastewater treatment data collected from the channels included influent and effluent BOD_5, SS, pH, temperature, dissolved oxygen and various forms of nitrogen and phosphorous.

Results from this study over a 12-month test period (November, 1983 through October, 1984) indicate that a single hyacinth channel is capable of removing 72-90 percent of the BOD_5 (81 percent average) and 70-90 percent of the SS (80 percent average) in wastewater at loading rates as high as 440 kg BOD/ha-day (3 days HRT). Average BOD_5 and TSS removal efficiencies for the ponds are summarized in Table 1. Concentrations of BOD_5 and TSS in hyacinth pond effluents satisfied federal standards for BOD_5 ($\leq$ 30 mg/l) and TSS ($\leq$ 30 mg/l) at loadings up to 220 kg BOD_5 (6 days HRT) during all but two of the coldest months of the test period during which time secondary effluent standards were met at the 110 kg BOD_5/ha-day loading rate. The average BOD_5 removal performances of the channels are compared to the performances of pilot scale channels reported in 17 other studies in the plot of BOD_5 removed versus BOD_5 loading, shown in Figure 2 (4). As seen from this graph, the dots that represent literature values fit a linear plot with good correlation (R^2=0.98) and for a wide range of loadings from 110 to 220 kg BOD_5/ha-day (10 to 200 lb BOD_5/acre-day) the BOD_5 removal efficiency approximates 80%. Interestingly, the Walt Disney World points (triangle symbols) fall very close to this same line; even at loadings as high as 440 kg/ha-day (400 lb/acre-day) BOD_5 removal efficiencies were approximating 80%. This strongly suggests that the channels were not stressed with respect to BOD_5 loadings and that the limits of the hyacinth channel for secondary treatment has yet to be determined.

Table 1.
Hyacinth Wastewater Treatment Performance Summary

Sewage Loading kg/ha-day	HRT, days**	Average Percent Removals* BOD_5	Suspended Solids
55	24	89	71
110	12	87	72
220	6	84	83
440	3	81	80

* Based on a 12-month operation from November 1983 through October 1984 and an influent BOD_5 concentration of 200-260 mg/l and an influent TSS concentration of 50-90 mg/l.

** Hydraulic retention time.

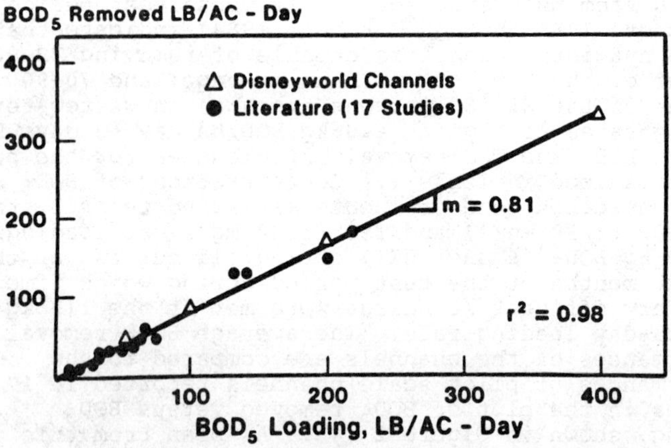

FIGURE 2.
EFFECT OF BOD_5 LOADING ON BOD_5 REMOVAL IN WATER HYACINTH TREATMENT PONDS

A statistical analysis of the influent, effluent, and channel profile data from four years of operation was performed and a wastewater treatment model was developed. Effluent BOD_5 and total suspended solids (TSS) concentrations in mg/l are primarily related to pond hydraulic retention time, t (days), according to the following equations:

i) Effluent BOD_5 = 0.9 x TSS + Influent BOD_5/(1+1.215 t)

ii) Effluent TSS = 10.75 + 0.645 t

The equation for total suspended solids reflects the low concentration of TSS (< 20 mg/l) achieved in all channel effluents throughout the year. The equation for the effects of retention time and influent BOD_5 on effluent quality is graphically shown in Figure 3. This model shows that for a hyacinth pond receiving settled sewage of a BOD_5 typical for municipal wastewaters (120-150 mg/l), an effluent BOD_5 of 30 mg/l could be achieved at a retention time of 6 to 7 days or at a BOD_5 loading of about 220 kg/ha-day. The plot in Figure 3 also suggests that although treatment efficiencies may be improved by extending hyacinth pond retention time, the use of staging unit processes while keeping the total retention time constant may be a cost effective means of improving secondary treatment efficiency while avoiding expansion of pond volumes. Experiments with a high-throughput, two-stage hyacinth system are under way.

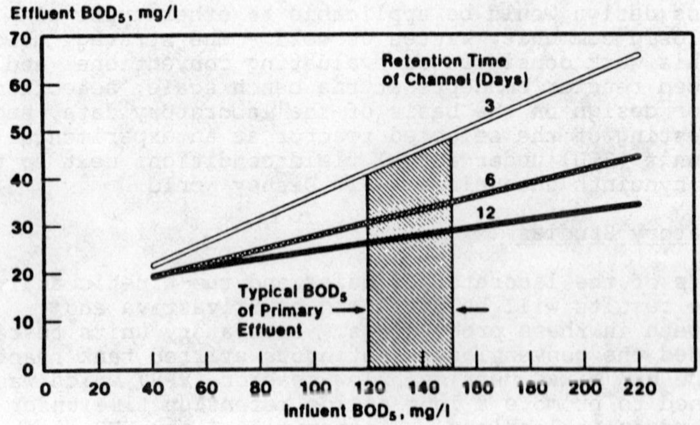

FIGURE 3.
WASTEWATER TREATMENT PERFORMANCE MODEL FOR THE HYACINTH CHANNEL OPERATING AT THE CWRF: EFFLUENT BOD_5 VERSUS INFLUENT BOD_5

Tertiary treatment studies in one of the channels receiving secondary effluent indicated that nitrogen and phosphorus removals of 50-70% and 20-30%, respectively, could be achieved in systems where hyacinths are well managed and harvested on a regular basis. Annual N and P uptake rates of hyacinths grown in tertiary treatment channels were in the range of 1,170-1,200 kg N/ha-yr and 320-390 kg P/ha-yr, respectively.

Hyacinth productivity in the large channels under unoptimized conditions ranged from 45 to 58 kg/ha-yr. Methods used in measuring water hyacinth yields in these channels have been detailed in the literature (10). The use of a harvesting schedule that provides an optimum planting density of 36 kg/m^2 (7.4 lb/ft^2), however, has increased hyacinth yields to 60-70 dry tonnes/ha-yr (27-31 dry tons/ac-yr). Preliminary tests in small field units suggest that additional yield increases of 30-50 percent are possible through the discretionary use of aeration and foliar spray applications of a plant growth hormone called gibberellic acid.

ANAEROBIC DIGESTION PROCESS DEVELOPMENT

The technical objective of this R&D effort was to develop an optimized system for the biogasification of hyacinth and sludge and to integrate this process with the hyacinth wastewater treatment facility. Hyacinth was regarded as a model feedstock and it was expected that the resulting process design would be applicable to other types of cellulosic community wastes as well. The strategy used for this work consisted of evaluating conventional and advanced reactor concepts at the bench scale, selection of reactor design on the basis of the laboratory data, and the testing of the selected reactor at an experimental test unit (ETU) under actual field conditions next to the water hyacinth channels at Walt Disney World.

Laboratory Studies

Details of the laboratory studies and the kinetic analysis of the results will be presented by Srivastava and Chynoweth in these proceedings. Laboratory units tested included the conventional continuous stirred tank reactor and the non-mixed vertical flow reactor (VFR) which was designed to promote a long solids retention time under high hydraulic loadings. A schematic of the VFR is shown in Figure 1 as a part of the integrated system. The VFR employs little or no mixing and can be fed to the reactor at the top or at the bottom (whichever is advantageous from a solids management standpoint).

In the laboratory trials, the VFR was fed from the bottom and liquid effluent was withdrawn at the top. Both the VFR and the CSTR control were fed with a 3:1 blend of hyacinths and primary sludge at loadings between 1.6 and 6.4 kg VS/m^3-day (0.1-0.4 lb VS/ft^3-day), corresponding to HRT's between 31 and 8 days. Steady state data plotted in Figure 4 show that in side-by-side tests, the VFR fed from the bottom or operated in the upflow solids mode (USR) consistently achieved 15-30 percent greater methane yields over a wide range of HRTs. The superior performance of the USR was attributed to the reactor's ability to increase the solids and microorganism residence time significantly above the HRT through sedimentation of particulate solids. Thus, the USR reactor achieved greater methane production with substantially less mixing. These results provided the basis for the selection of the VFR design for testing at the ETU scale.

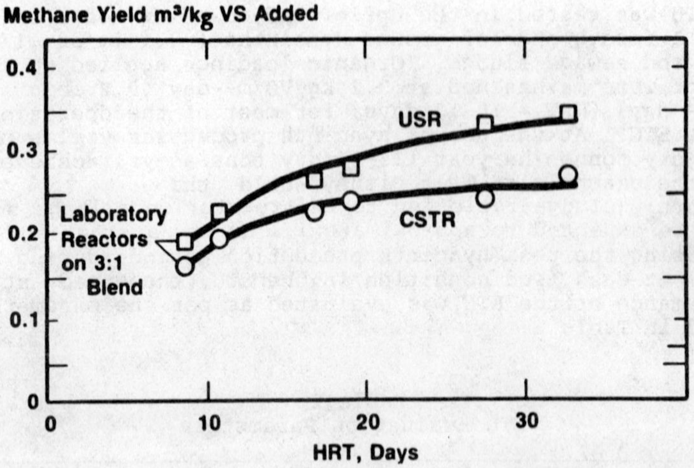

FIGURE 4.
METHANE YIELD VERSUS HYDRAULIC RETENTION TIME
FOR BENCH SCALE USR AND CSTR UNITS
OPERATED ON A 3:1 HYACINTH/SLUDGE FEED BLEND
(USR = VERTICAL FLOW REACTOR FED FROM THE BOTTOM)

Biogasification Experimental Test Unit

In 1983, a 4.5 m^3 (160 ft^3) vertical flow solids ETU was designed and constructed, located beside the five existing hyacinth channels at Walt Disney World. The technical objective of the first phase of the ETU study was to evaluate reactor performance, scale-up, and materials handling parameters at several different mixtures of hyacinth/sludge blends. The ETU facility is capable of processing up to 910 kg or 1.0 wet ton of hyacinth/sludge feed blend (5% total solids) each day. The ETU is sized to ensure that the demand for biomass feedstock does not exceed the availability of hyacinths from the channels during the coldest months of the winter when hyacinth productivity is at its lowest. Major components of the facility include two feed tanks for short-term storage of sludge and chopped hyacinth, a feed blend tank, the vertical flow reactor, an effluent storage tank, and gas compression and storage.

The ETU was tested in the upflow and downflow modes on 1:1 and 2:1 feed blends of ground hyacinths (-3.1 mm or -1/8 inch) and sewage sludge. Organic loadings applied to the reactor were maintained at 3.2 kg VS/m^3-day (0.2 lb VS/ft^3-day) (HRT = 10-12 days) for most of the operation of the ETU. At the annual hyacinth production yields of 56-67 dry tonnes/ha-year (25-30 dry tons/ac-yr) achieved from the channels at Walt Disney World, the hyacinth: sludge ratio fed to the reactor on a large scale would be expected to approximate 1:1 on the average and 2:1 during the peak hyacinth production periods during the year. At each feed condition in the ETU, the steady state performance or the ETU was evaluated as per the parameters listed in Table 2.

Table 2.
ETU Evaluation Parameters

Performance Parameters	Engineering Parameters
Methane Yield	Materials Balance
Methane Production Rate	Solids
Gas Composition	Carbon
Volatile Fatty Acids	Nitrogen
Temperature	Phosphorus
Alkalinity	Materials Handling
Organic Matter Reduction	Scale-up of Laboratory Performance

Results from four steady state conditions, summarized in Table 3, illustrate the effect of feedstock blend ratio and feed mode on methane yield performance. Operated in the upflow mode at a loading of 3.2 kg VS/m^3-day (0.2 lb VS/ft^3-day), the VFR reactor achieved good methane yields on 1:1 and 2:1 blends averaging 0.26 and 0.4 m^3/kg VS added (4.2 and 6.4 ft^3/lb VS added), respectively. This performance exceeded the performance of the completely mixed control operated at the same feed condition by 10 to 15 percent. Solids profile data taken from the reactor revealed that a large float of solids of a thickness of more than 1.2 m (4 ft.) had been maintained during the steady state upflow reactor runs. On the basis of this information, it was decided that improved solids retention might be realized if the reactor feed mode were to be changed to a downflow mode.

Table 3.
Steady State Performance Data
From the Experimental Vertical Flow Reactor
Operated on Mixtures of Hyacinths and Sewage Sludge

PERFORMANCE PERIOD	1	2	3	4	5
DURATION OF STEADY STATE, DAYS	4	3	4	8	4
FEED CHARACTERISTICS					
MEAN WH PARTICLE SIZE, IN.	0.12	0.12	0.12	0.12	0.12
AVE. BLEND RATIO, WH:SLUDGE	2.0:1	0.9:1	0.8:1	2.1:1	2.1:1
SOLIDS CONTENT, WT. %	4.52	3.62	3.76	3.78	4.43
VOLATILE SOLIDS, % OF TS	84.6	87.1	85.5	85.6	83.2
VOLATILE ACIDS, MG/L AS ACETIC	3600	2600	3400	3350	5000
PH	-	5.2	5.2	5.4	5.3
ALKALINITY, MG/L AS $CaCO_3$	-	1900	3400	2950	3860
DIGESTER OPERATING CONDITIONS					
FEED INJECTION	BOTTOM	BOTTOM	TOP	TOP	TOP
DAILY FEED FREQUENCY	1	1	1	1	1
LOADING RATE, KG/M^3-D	3.2	2.9	3.0	3.1	5.0
HYDRAULIC RETENTION, DAYS	12.0	11.0	11.0	10.5	7.0
SOLIDS RETENTION, DAYS	21.0	17.0	30.0	25.0	13.0
TEMPERATURE, ^oC	35.3	34.3	34.7	35.0	35.3
CULTURE VOLUME, M^3	3.9	3.9	3.9	3.9	3.9
RECIRCULATION FLOW, M^3/DAY	4.0	4.0	4.0	4.0	4.0
RECIRCULATION TIME, HOURS/DAY	4.0	4.0	4.0	4.0	4.0
PH	6.7-7.7	7.0-7.9	7.0-7.9	7.1-8.0	7.0-7.9
VOLATILE ACIDS, MG/L AS ACETIC	300-5000	200-1500	50-2300	50-700	200-3600
ALKALI ADDED, MG/L-DAY	0	0	0	0	0
PERFORMANCE DATA					
BIOGAS PRODUCTION, SCFD	208	282	320	269	344
METHANE CONTENT OF BIOGAS, %	60	63	64	63	59
METHANE YIELD, SCF/LB VS ADDED	4.6	7.0	7.9	6.3	4.8
PERCENT OF BIODEGRADIBLE FRACTION CONVERTED TO METHANE	63	84	92	84	62
METHANE PRODUCTION RATE, VOLUME/VOLUME REACTOR-DAY	0.9	1.3	1.5	1.3	1.5
EFFLUENT QUALITY					
EFFLUENT WITHDRAWL	MIDDLE	MIDDLE	BOTTOM	BOTTOM	BOTTOM
TOTAL SOLIDS, WT. %	3.02	1.96	1.91	2.44	2.64
VOLATILE SOLIDS, % OF TS	74.5	78.0	75.1	76.0	73.0
VOLATILE ACIDS, MG/L	2800	600	1100	150	2330
PH	7.3	7.2	7.2	7.3	7.1
ALKALINITY, MG/L AS $CaCO_3$	6000	4700	5200	4500	5240

When the feed mode was shifted from upflow to downflow, methane yields increased to 0.45 and 0.35 m^3/kg VS added (7.2 and 5.7 ft^3/lb VS added) for the 1:1 and 2:1 feed blends, respectively. The methane yield of 0.45 m^3/kg VS added for the 1:1 blend represents a solids destruction efficiency of nearly 70 percent which approximates the anticipated average performance expected from a large production scale facility. A comparison of methane yields achieved with the VFR operated in the down-flow mode are compared with methane yields achieved with the conventional digestion control (CSTR) in Figure 5. Typical solids, pH and volatile acid profiles for the VFR are shown in Figure 6.

Carbon balances conducted on the ETU during each steady state condition showed that at an HRT of 10-12 days, the solids retention time (SRT) of the VFR in the upflow mode was maintained at 17-21 days and in the downflow mode, 24-35 days. Laboratory and ETU data clearly show that higher methane yields are associated with reactors that maintain high SRTs. This ability to maintain SRTs at double or tripple the HRT represents a marked advantage over conventional digestion (CSTR) in which HRT and SRT are virtually equal. This means that high methane yields can potentially be maintained in a vertical flow reactor at less than half the digester volume required for conventional digestion.

Economics

Systems evaluations of the integrated waste conversion concept have been continually performed to assess progress in the research effort and to identify promising new research directions. In 1982, a preliminary A&E economic feasibility analysis was conducted on a conceptual secondary treatment water hyacinth system employing conventional anaerobic digestion and gas upgrading to pipeline quality. Results from this study indicated that a significant amount of methane could be generated with the water hyacinth concept at a cost of $2.40-4.60/GJ at treatment plant sizes of 50 to 10 MGD which correspond to city populations of 500,000to 100,000. These costs were based on some key assumptions of system performance which became the goals of the project. The A&E assumptions included:

- Water hyacinth yields of 110 dry tonnes/ha-yr (50 dry tons/acre-yr) in wastewater treatment channels

- Methane yields from the conversion of the water hyacinth/sludge blend of 0.28 m^3/kg VS added (4.7 scf/lb VS added) achieved with a 28-day reactor retention time

- Water hyacinth pond sizes of about 0.75 ha per 1000 m^3 per day capacity of sewage treatment (7.0 acres/MG)

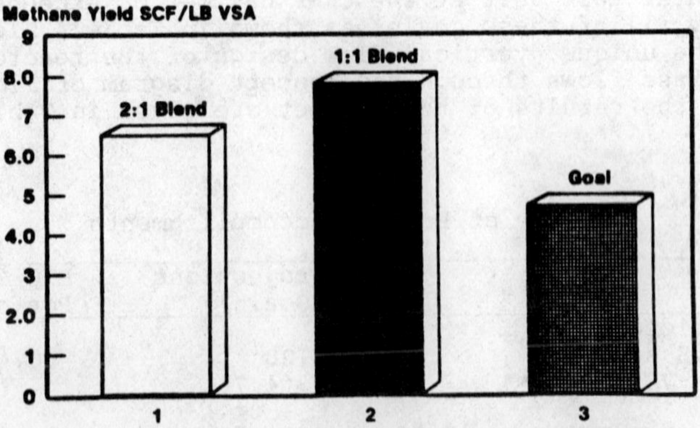

FIGURE 5.
METHANE YIELDS ACHIEVED
WITH THE VERTICAL FLOW REACTOR (ETU) FED WITH
1:1 AND 2:1 BLENDS OF WATER HYACINTHS AND SLUDGE
AT A 3.2 kg VS/m^3-DAY LOADING (35°C)

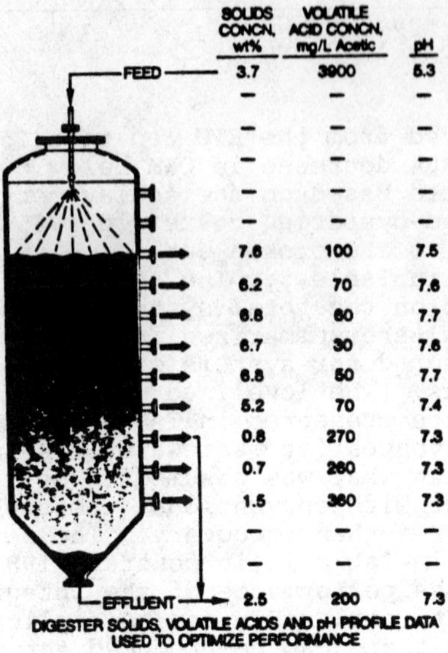

FIGURE 6.
PROFILES OF pH, TOTAL SOLIDS, AND VOLATILE
ACIDS IN SAMPLES TAKEN IN THE VERTICAL
FLOW AT STEADY STATE

Over the last two years, the performance of the integrated experimental test unit at the CRWF has met or exceeded virtually all of these goals, as shown in Table 4, largely due to the unique, vertical flow design of the reactor (VFR). Mass flows through the concept diagram of Figure 1 based on the results of the project are given in Table 5.

Table 4.
Summary of Project Accomplishments

	A&E Projections (Goals)	ETU System Performance
Methane Yield		
m^3/kg VS added*	0.28	0.47
(scf/lb VS added)**	(4.7)	(7.5)
Hydraulic Retention Time, Days	28	11
Water Hyacinth Production		
dry tonnes/ha-yr	114	68
(dry tons/ac-yr)	(50)	(30)
Pond Area Required for Secondary Treatment		
ha per 1000 m^3/day	0.75	0.75
(acres per MGD)	(7.0)	(7.0)

* 1.0 m^3 methane = 0.0367 GJ
** 1.0 scf methane = 1000 Btu

Performance observed from the ETU led to a 15% reduction in capital and a 20% decrease in O&M below costs originally projected based on the achievement of project goals. Capital and operating costs (in 1985 dollars) based on ETU results are broken out for each of the major system components in Table 5. The results of the economic analysis performed on the total system (sized for a 500,000 population) are summarized in Table 6. A similar analysis was conducted for systems sized for 10,000 and 100,000 populations. The levelized costs of all three sizes of systems are presented in Table 7. Reflected in these costs are revenues for wastewater treatment that are 15 percent less than what was assumed to be the cost of treating wastewater with conventional treatment when no credit is taken for methane recovery. The levelized cost results presented in Table 7 (in constant 1985 dollars) indicate that if the performance of the integrated Walt Disney World Resort Complex ETU can be duplicated at a community scale, methane can be produced for less than $2.00/GJ for populations of over 100,000 and for approximately $3.00/GJ for populations of 10,000.

Table 5.
Mass Flows Through an
Integrated Hyacinth Sewage Treatment
Anaerobic Digestion Facility
Sized for a Community of 500,000

Location on the Schematic of Figure 1	Description	Units	Performance
A	Sewage Flow	m^3/day (MGD)	190,000 (50)
B	Hyacinth Pond Area	ha (acres)	142 (350)
C	Harvested Hyacinths	dry tonnes/day (dry tons/day)	25 (27)
D	Primary Sludge	dry tonnes/day (dry tons/day)	23 (25)
E	Digester Feed Blend of hyacinths & sludge	dry tonnes/day (dry tons/day)	47 (43)
F	Digester Effluent to disposal	dry tonnes/day (dry tons/day)	19 (17)
G	Net Methane Produced	GJ/day (mscfd)	550 (530)
H	Reactor Volume	$m^3 \times 10^{-3}$ ($ft^3 \times 10^{-6}$)	11.4 (0.4)

Table 6.
Summary of Economics for the
Integrated Water Hyacinth/Anaerobic Digestion
Facility Sized for 500,000 Population

Assumptions

Return to Debt	9.2%	Construction Period	1 year
Return to Equity	13.7%	Book Life	
Fraction to Debt	0.65	Concrete Structures	50 years
Working Capital Fraction	0.10	All Other	20 years
Inflation	6.0%	Tax Life	10 years
Unit Charge for	53/1000 m^3	Net Methane Output	550 GJ/day

Costs (1985 $)

Total Capital	$24 million
Total Annual O&M	$2.7 million/yr
Annual Wastewater Treatment Revenue	$3.7 million/yr
Levelized Gas Cost	$1.80/mmBtu

Table 7.
Methane Costs from the
Integrated Water Hyacinth/Anaerobic Digestion Concept

Population of Community	Revenue for Wastewater Treatment*	Costs Based on ETU Results**
500,000	53	$1.80/GJ
100,000	74	$1.90/GJ
10,000	200	$3.00/GJ

* $/1000 m^3 of wastewater; revenues are 15% less than the cost of conventional treatment.
** All costs are in 1985 dollars.

FUTURE DIRECTIONS

The successful integration of the vertical flow reactor into aquaculture operations and the favorable economics of the system indicate that low-cost, pipeline-quality methane can be produced from a community waste system and that the VFR technology should be extended to broader concept of community waste management and centralized conversion. The project is now focused on VFR development for handling MSW and industrial waste. If research goals for a multipurpose conversion system are realized, it may be possible for many communities of 100,000 population or more to satisfy up to 15 to 20 percent of their residential gas demand with biogas produced from waste resources.

CONCLUSION

Results to date indicate that the integrated hyacinth wastewater treatment/anaerobic digestion concept can achieve levels of performance that make it possible to produce pipeline quality methane at a cost as low as $2.00/GJ for cities of greater than 100,000 population and $3.00/GJ for cities of 10,000. Because water hyacinth has been used as an example feedstock for the initial development of the VFR, the same reactor design may also be effective in the conversion of other types of cellulosic community wastes such as MSW and industrial wastes. Future research will be directed on these community resources.

It is expected that this research will be of substantial benefit to the community as well as the gas rate payer in terms of providing: 1) a cost-effective technology for wastewater renovation, 2) a method of waste volume reduction that could, if applied to the total organic community waste stream, extend landfill life by a factor of 5 to 7 times, and, 3) a stable, local supply of methane gas to supplement natural gas supplies.

REFERENCES

1. Anderson, L.L.; Energy Potential from Organic Wastes: A Review of the Quantities and Sources, Bureau of Mines Report No. BuMines IC 8549 (1972).

2. Bastian, R.K. and S.C. Reed; Aquaculture Systems for Wastewater Treatment: Seminar Proceedings and Engineering Assessment/MDC-67, Available from General Services Administration, Denver, CO, Document No. EPA 430/9-80-006 (1979).

3. Chynoweth, D.P., et al; "Biogasification of Water Hyacinth and Primary Sludge", Proceedings of the International Gas Research Conference Government Institutes, Inc., Rockville, MD. In Press (1984).

4. Dinges, R.; "Upgrading Stabilization Pond Effluents by Water Hyacinth Culture", Journ. Water Poll. Control Fed. 50:833 (1978).

5. Hauser, J.R.; "Use of Water Hyacinth Aquatic Treatment Systems for Ammonia Control and Effluent Polishing", J. Water Pollut. Control Fed., 56(3):219-225 (1984).

6. Hayes, T.D., D.P. Chynoweth, K.R. Reddy and B. Schwegler; "The Integration of Biogas Production with Wastewater Treatment", In Energy Applications of Biomass, M.Z. Lowenstein Editor, Elsevier Applied Science Publishers, New York, NY, pp. 189-200 (1985).

7. Innovative and Alternative Technology Assessment Manual/ MCD-53, United States Environmental Protection Agency Report No. EPA 430/9-78-009 (1978).

8. Klass, D.L. and S. Ghosh; "Methane Production by Anaerobic Digestion of Water Hyacinth (Eichhornia Crassipes)", in Fuels from Biomass and Wastes, pp. 129-149 (1981).

9. Kruzic, A.P.; "Water Hyacinth Wastewater Treatment System at Walt Disney World", In Aquaculture Systems for Wastewater Treatment/MCD 67, Available from General Services Administration, Denver, CO, Document No. EPA 430/9-80-006, p. 35, (1979).

10. Reddy, K.R.; "Water Hyacinth Biomass Production in Florida", Biomass 6:167 (1984).

11. Reed, S., R. Bastian, and W. Jewell; "Engineering Assessment of Aquaculture Systems for Wastewater Treatment: An Overview", In Aquaculture Systems for Wastewater Treatment/MCD-68, Available from General Services Administration, Denver, CO, Document No. EPA 430/9-80-007, p. 1, (1980).

12. Ryther, J.H. et al.; "Biomass Production by Marine and Freshwater Plants", in the Proceedings of the Third Annual Biomass Energy Systems Conference Proceedings/DOE, Golden CO, pp. 13-23 (1979).

13. Tchobanoglous, G., R. Stowell, R. Ludwig, J. Colt, and A. Knight; "The Use of Aquatic Plants and Animals for the Treatment of Wastewater: An Overview", In Aquaculture Systems for Wastewater Treatment/MCD-68, Available from General Services Administration, Denver, CO, Document No. EPA 430/9-80-006, p. 35, (1979).

14. Vaidyanathan, S., K.M. Kavadia, K.C. Shroff and S.P. Mahajan; "Biogas Production in Batch and Semicontinuous Digesters Using Water Hyacinth", Biotechnology and Bioengineering 27:905-908 (1985).

15. Wastewater Engineering, Metcalf & Eddy, Inc., McGraw-Hill Book Company, New York, NY (1972).

16. Wolverton, B.C. and R. McDonald; "Upgrading Facultative Wastewater Lagoons with Vascular Aquatic Plants", Jour. Water Poll. Control Fed. 51(2):305 (1979).

DISCUSSION

R. Legrand, Reynolds, Smith & Hills	Do you observe the same floating with MSW that you did with water hyacinth?
Hayes	Floating on top?
Legrand	Yes.
Hayes	The best people to answer that question are those who have dealt with MSW in completely mixed systems. I'll defer that question to Dr. Pfeffer.
J. Pfeffer, U. of Illinois	The answer is yes.
A. Ibrahim, Cornell U.	Did you make any measurements on nitrogen uptake in the hyacinth pond? I'm interested to see what level of nitrogen was removed.
Hayes	Again, I'm going to defer. There are three organizations that work on this system. All three organizations are with us today — University of Florida, Institute of Gas Technology, and Walt Disney World. Dr. Reddy from the University of Florida has been doing most of that work with nutrient management in these systems.
K. Reddy, U. of Florida	Are you referring to concentration reduction in the channel?
Ibrahim	Yes.
Reddy	The way the channels are designed now for primary treatment. To achieve secondary goals, nitrogen removal is not the primary goal. The primary goal in that system is BOD and suspended solids removal. With the primary sewage effluent going to those channels, nitrogen removal is very poor, about 10 to 20%. But if you use water hyacinth for tertiary advanced treatment, you can get up to 60 to 70% nitrogen removal.
R. Egg, Texas A&M U.	Is there any work underway on hyacinth treatment efficiencies as a function of climate? This work was done in Florida. How about other locations, say here in Washington with longer winters?

DISCUSSION, CONT.

Hayes — The applicability of hyacinth is probably restricted to about a third of the population of the U.S. living in the southern areas of this country, and in California. For other areas, you would have to go to more cold-tolerant plant systems, and perhaps to nutrient-film technique systems. These are new types of systems that are now being evaluated even as far north as Maine. There are other types of aquatics that would perform better in colder climates.

EVALUATION OF THE RefCoM PROOF-OF-CONCEPT EXPERIMENT

John T. Pfeffer, Ph.D.
Consultant to
Waste Management Energy Systems, Inc.
Tampa, FL

ABSTRACT

RefCoM is an acronym for a process that converts the biodegradable organic material in municipal solid waste (refuse) into a useful fuel gas, methane. The "as received" refuse is processed mechanically to separate the biodegradable organic material from the inert and inorganic components. The processed material is fed into an anaerobic fermentation reactor where the organic material is fermented to methane and carbon dioxide. The gas generated is captured and processed to meet pipeline quality standards. This paper presents the results of an evaluation of the proof-of-concept plant constructed in Pompano Beach, FL. The plant had two fermenters with a capacity of 350,000 gal. each. The system was operated to generate information on the gas yields from refuse and to identify operational problems that must be corrected in the design of a commercial plant. Data are presented on the conversion efficiencies of the biological process and the operational problems encountered in the operation of the fermentation system are identified.

EVALUATION OF THE RefCoM PROOF-OF-CONCEPT EXPERIMENT

INTRODUCTION

The cost for disposal of the massive tonnage of solid waste that is produced annually by municipalities has increased disproportionately during the past two decades. With the increased environmental concerns, the simple inexpensive techniques such as open dumping and uncontrolled incineration are no longer permitted. More sophisticated, and more expensive, processes are required to maintain the desired environmental quality. These higher costs have also stimulated research into new and different techniques for elimination of this material.

Coincidental with the increased disposal cost was the embargo on petroleum that precipitated unanticipated increases in energy costs. Prior claims of unlimited cheap energy supplies were modified. Energy from various fuels was available, but at a substantially higher price. It was recognized that the energy content of urban solid waste was substantial. Reclamation of this energy could provide an income that could reduce the cost of refuse disposal. Consequently, a number of systems for energy recovery from refuse has been studied.

Early research (in the 1930's) demonstrated that the garbage (food waste) component of solid waste was amenable to conversion to methane gas by an anaerobic fermentation process. Later work (in the 1950's and 1960's) demonstrated that the cellulose (paper) was also a good substrate for this biological process. Research in the early 1970's developed a technical base for the conceptual design of a process for the production of a substitute natural gas from urban refuse. The design concept consisted of four basic subsystems; 1) feed preparation, 2) methane fermentation 3) residue disposal and 4) gas clean-up.

The objective of the feed preparation subsystem is the separation of material in the refuse that can not be fermented to methane by the bacteria in the methane fermentation process. It is especially important to remove any refuse constituent that can cause operational problems in the subsequent unit processes. The unit processes contained in this subsystem included the following; 1) size reduction to assist in separation, 2) magnetic separation for recovery of ferrous metals, 3) screening to remove inorganic fines, 4) screening to remove oversize objects and 5) separation of the intermediate size inorganics such as aluminium cans, bottles, stones, etc.

The methane fermentation subsystem is the essence of the process. The biodegradable organic material is fermented to methane by a consortium of bacteria present in the fermentation reactor. The gas produced by this fermentation is released in the reactor and recovered as the product from the process. The unit processes contained in this subsystem include the following; 1) feed slurry preparation tank where the separated refuse is slurried with recycle liquor from the residue disposal subsystem, undigested sewage sludge and any chemicals required to maintain the fermenter pH and provide the nutrients required for microbial growth, 2) heated and gas tight completely-stirred tank reactor (CSTR) in which the fermentation occurs, 3) system for gas collection and transport to gas clean-up subsystem.

The slurry remaining after the fermentation is complete must be processed for final disposal. The slurry is dewatered with the water recycled to the fermenter. The residual cake retains a significant amount of organic material from which additional energy can be obtained by combustion. This leaves only an ash as the residue that must be landfilled. The unit processes contained in this subsystem include; 1) slurry dewatering unit such as a belt-filter press or centrifuge, 2) combustion system to combust the cake and recover steam for inplant use and sale if a market exists, and 3) ash disposal.

The gas produced by the microorganisms contains both methane and carbon dioxide. It is also saturated with water vapor. If this gas is to be marketed to a gas utility, it must be purified to meet the appropriate standards. Removal of all acid gases, including carbon dioxide and hydrogen sulfide, plus drying of the gas will be necessary to meet pipeline standards. Even if used as a low Btu gas, it will generally be necessary to dry the gas to some degree. This subsystem is selected to produce the quality of gas needed to meet the customer specifications.

The RefCoM proof-of-concept-experiment was designed to demonstrate the technical feasibility of the biological processes and to generate operating experience that could be used to improve the design as well as provide cost data for use in evaluating the economic feasibility of the process. The initial installation included only part of the total system; the refuse separation subsystem, the methane fermentation subsystem and the unit process for dewatering the fermenter slurry. This constraint was imposed because of budget limitations. Also, it was felt that the technology for residue combustion and for gas purification was state-of-the-art while the biological process was still at the laboratory stage. A request-for-proposal was issued in 1976 by a federal program that has now become part of the Department of Energy. A two year design and construct period resulted in plant acceptance in May 1978. Since that time the plant has been operating intermittently while a number of modifications were made to the initial installation. The modifications were successful in developing a process that could be operated at will.

BACKGROUND

The RefCoM processing train was selected from among existing unit process technologies. In many cases the selection was based on very limited performance data. This section will highlight the experimental work on the methane fermentation process and the dewatering characteristics of the fermenter slurry. These data provided the basis for the methane fermenter and slurry dewatering process design.

Refuse Separation

The specific purpose of this subsystem is the preparation of a feed material that contains only those materials that can be fermented to methane gas and that are compatible with the mechanical operation of the system. Table 1 shows the composition of a typical municipal refuse.

Table 1. Composition of Municipal Refuse

Constituent	% Wet Wt.	% Dry Wt.
Paper and Paper Products	44	35
Garbage (Food Wastes)	15	7
Yard and Garden Waste	12	5
Metals	9	9
Glass	8	8
Plastic	3	3
Misc. organics	5	4
Misc. inorganics	4	3
Water	-	26

Source: Resource Recovery and Source Reductions, Second Annual Report to Congress, U.S.E.P.A. Pub. No. SW-122, Washington, D.C. (1974)

The constituents that are of particular interest for methane fermentation include paper and paper products, garbage, and yard and garden wastes. These constituents contain primarily carbohydrates (cellulose, starch and sugar) that are the precursors for methane. Therefore, the objective of the refuse separation processes is to maximize the content of the biodegradable organic material and minimize the content of the nonbiodegradable organic constituents and the inorganic solids. Simply because of the increased mass-flow rate, passage of material that cannot be fermented through this system increases the size and operating cost of the fermenter and the residue-disposal system. Properties of certain materials further increase operational problems. Shattered glass, sand, and metal greatly accelerate erosion and abrasion of equipment. Plastic may cause plugging of pipes and pumps as well as problems with the mixing system.

A number of research groups have studied processes for the separation of various components of the solid wastes. This research has generally investigated existing technology to determine what modifications are needed to process urban solid wastes. The EPA demonstration project with St. Louis and Union Electric Co. (1) resulted in the operation of a prototype system for producing a "light" fraction that was primarily organic material. This light fraction was used as a fuel for a coal fired boiler system. Product specifications were poorly defined, but it was thought that this light fraction would be a suitable feed for an anaerobic fermentation system.

A system developed by the U.S. Bureau of Mines (BOM) and the National Center for Resource Recovery (NCRR) (2) concentrated more on material recovery than on the production of a fuel. Also, extensive research by Warren Springs Laboratory in the U.K. (3) and ENADIMSA in Madrid (4) resulted in processes that were very effective in the separation of various components. The processes developed in Europe, which were much less power intensive than the U.S. systems, utilized screening very effectively to separate refuse that had been subjected to a minimum amount of size reduction.

Size reduction is generally the first step in the preparation of solid waste as a feed stock for methane fermentation. The refuse may be shredded to a nominal particle size of 2 inches or less or may simply be passed through a chain (flail) mill that does nothing but open bags and large containers to liberate the refuse. One deficiency in the published literature on size reduction is the lack of specifications relative to the performance of the subsequent separation processes. The separation processes normally employed in the preparation of refuse derived fuel (RDF) have had very loose specifications. Clearly, the particle size distribution and the nominal particle size produced by the shredder must affect the performance of the separation processes. Specific shredder performance specifications should be developed in order to better evaluate the performance of the following separation processes.

Following the initial shredding, the refuse may be subjected to one or more separation steps. Screening, especially rotary drum (trommel) screens, has been found to be a very effective process for separating refuse according to particle size (5). The degree of shredding will have a significant impact on the performance of the trommel. During milling, most of the glass will be shattered, with some of it being imbedded in the organic material. Consequently, sand, dust, dirt, ashes, grass clippings, and some glass might be expected to be passed into the undersize fraction of a screen having openings on the order of 0.5 inch. Depending on the mill used for the primary shredder, the plastics and textiles and some paper products may pass through as large fragments (> 4 inches). This facilitates removal of those materials in the oversize fraction from a screen having 4 inch openings. The particles in the 0.5 to 4 inch range represent the material of prime interest as feed for the methane fermentation system.

The < 0.5 inch stream may be 25 to 50% organic material. This organic fraction can be recovered as feed for the fermenter. Since this material is of a relatively uniform size, it is possible to separate the material based on density. Because of the availability of the recycle stream, it would be possible to use a technology such as a rising current separator. A mini air separator (air knife) may be equally satisfactory. The paper products in the > 4 inch stream would also be a good feed for the fermenter. If desired, this material could be separated from the plastics. However, it may be advantageous to have this dry paper and plastic to add to the process residue that is being incinerated.

Magnetic separation would also be an essential part of this separation system. A magnetic belt station would recover 80 to 90% of the ferrous metal in the output from the shredder (6). Additional magnetic belts and magnetic head pulleys on the conveyor belts could increase this recovery to nearly 100%. Nonferrous metals, especially aluminium, are more difficult to remove. Technology such as the eddy current separator is available for recovery of these metals, which generally account for less than 1% of the refuse stream. However, the cost effectiveness of such recovery must be carefully evaluated.

At the time that the conceptual design was developed, the alternate to screening for separation of some of the undesirable components of the refuse was the air classifier. This process separates material based on the density differences. Under ideal conditions, the dense inorganic material is separated from the less dense organic components (light fraction). Unfortunately, the particle size and shape play a key role in the separation efficiency. Flat particles such as can lids have excellent aerodynamic properties and are likely to be carried over into the light fraction. Also the air velocity must be high enough to transport materials such as orange peels to the light fraction. Consequently, many small particles of shattered glass and related materials are also found in the light fraction.

The efficiency of the air classification is highly dependent on the moisture content of the refuse. High moisture contents result in particle agglomeration. Shattered glass, sand, etc. will stick to larger pieces of wet paper and the wet paper will form larger masses that are difficult to carry over to the light fraction. Continuous adjustment of air flow is necessary to compensate for the changes in particle size, moisture content and other factors. As a result, there has been considerable dissatisfaction with air classification systems applied to municipal refuse (7).

Since the initial conceptual design was developed, studies of the performance of disc screens in separating refuse into size categories have been very successful. The application of this system to the Ames, Iowa RDF plant provided a marked improvement in the quality of the product as well as reduced the operating costs (8). This technology is now available and played a important role in the revisions to the RefCoM facility. The disc screen technology appears to offer many advantages and should be considered as the separation technology for the preparation of the feed for the methane fermentation reactor.

Neither performance nor product specifications existed for those unit processes used in the preparation of a feed stock for the methane fermentation system at RefCoM. Clearly such specifications should be developed. The following discussion presents a set of proposed specifications for the refuse processing and separation processes. The "as received" composition of a"typical" municipal refuse was presented in Table 1. The refuse is first passed through a vertical shaft hammer mill designed to produce a nominal particle size of 4 inches. The quantities of individual components expected in each product stream after the milled refuse is screened and passed under a magnetic belt are shown in Table 2. The 0.5 to 4 inch size fraction from the screens will be the feed for the fermenter.

Most of the solids would be expected to be in the 0.5 to 4 inch size range. This stream will primarily consist of paper products, garbage and yard wastes. It will also contain some metal, mostly aluminium, plus some shattered glass that embeds in the organic fraction. As shown in Table 2, the material in this stream should be about 87% organic and 13% inorganic (ash). Most of the remaining glass and miscellaneous inorganic material will report to the < 0.5 inch stream. This undersize stream from the screens will also contain organic material, especially grass clippings and lint resulting from the shredding of paper and textiles. It may be desirable to recover this material as it is a good substrate for the fermenter. As shown in Table 3, 35% or more of this stream may be organic material.

Table 2. Expected Distribution of Refuse Components after Screening*

Refuse	lb	Screenings			Magnet.
Component	100 lb	< 0.5"	0.5 to 4"	> 4"	Separ.
Paper	44	2	37	4	1
Garbage	15	1	13	1	0
Yard Wastes	12	3	8	1	0
Metal	9	0.25	0.75	0.5	7.5
Glass	8	6	1.5	0.5	0
Plastic	3	0	0.25	2.75	0
Misc. Org.	5	0	2	3	0
Misc. Inorg.	4	3	1	0	0
	100	15.25	63.5	12.75	8.5

* Milled refuse with nominal particle size of 4 in, wet weight

The oversized (> 4 in) from the screens will contain most of the textiles, plastic film and containers, and some paper products. Since plastic and textiles can cause problems with the mixers, it is desirable to keep this material separate from the fermenter feed stream. Recovery of the paper in this stream may be desirable. However, the residue from the fermenter will be incinerated and mixing the dry oversized material, including the paper, with the fermenter residue will substantially improve the combustion properties of the residue.

Table 3. Product Specifications for the Separated Streams

Process	Product Specifications		
	% Total Feed	% Organic Solids	% Inorganic Solids*
Shredder	100	73	27
Screens			
< 0.5"	15.25	35	65
0.5 - 4"	63.5	87	13
> 4"	12.75	83	17
Magnetic Separation	8.5	12	88

* Includes the ash content of the organic fraction at approximately 10% of dry solids

This processing system will prepare an acceptable feed stock for the fermentation system. There will not be any materials that will exacerbate the operation of the system. The embedded glass and inert material will cause wear on the mechanical systems. The aluminium cans will be separated with the proper design of a disc screen system.

Fermentation of Municipal Solid Waste (MSW)

In methane fermentation, organic matter is anaerobically degraded to methane and carbon dioxide with only a small amount of the organic matter being synthesized into bacterial cells. This fermentation consists of three basic steps. The organic solids, primarily polymers of carbohydrates with some lipids and proteins, are hydrolysed by microorganisms that produce the appropriate exoenzymes. The products of this hydrolysis are soluble monomers that are fermented to various intermediate products such as organic acids and alcohols. In the second step, these intermediates are further processed by a group of bacteria referred to as the "acetogenic" bacteria. The resulting products are acetic acid, carbon dioxide and hydrogen. Finally, the "methanogenic" bacteria complete the process by cleaving acetic acid to methane and carbon dioxide and by using the hydrogen produced by the acetogenic bacteria to reduce carbon dioxide to methane.

The methane formed in this last stage, being relatively insoluble in water, is lost to the gas phase. The carbon dioxide produced in the fermentation partially escapes to the gas phase. However, carbon dioxide is relatively soluble in water and it also reacts chemically with any base (OH-). Therefore, a portion of the carbon dioxide will remain in the liquid phase either as a soluble gas or in the form of bicarbonate ion (HCO3). The quality of the gas is highly dependent upon the chemical composition of the organic material used for the fermenter feedstock. Carbohydrates are the primary organic material in refuse. When fermenting this material, the bacteria produce equal volumes of methane and carbon dioxide. The actual gas composition will depend upon how much of the carbon dioxide is retained in the liquid phase.

The amount of methane produced is dependent upon the fraction of organic material that is actually fermented. The biodegradability of organic material is highly dependent upon its chemical structure. Certain materials have a structural form that prevent microbial enzymes from initiating any reaction. Most synthetic materials such as plastics are highly nonbiodegradable. Natural substances such as lignin are very resistant to biodegradation. Therefore, many naturally occurring fibers do not exhibit a high methane production when processed in the methane fermentation system.

Environmental factors are also important in determining the efficiency of the methane fermentation process. By definition, the anaerobic fermentation must occur in a system totally void of oxygen. Methanogens are extremely sensitive to oxygen and oxidized substances such as nitrate (NO3). These bacteria are also sensitive to pH, preferring a neutral pH of 7.0 to 7.2. When the pH drops to below 6.6, severe inhibition of the methanogens results. This in turn can cause process failure since the fermentative bacteria continue to produce acids, driving the pH down to levels as low as 4.5. This results in complete process failure.

The rate of fermentation is directly related to the fermentation temperature. Two temperature ranges have been identified for this process. The most commonly used is the mesophilic range (35 to 40oC). This has generally been found to be the optimum temperature for sewage sludge digestion, primarily because of the dilute nature of the sludge slurry. The heat required to elevate the temperature of the water in the slurry to higher temperatures is greater than the energy content of the additional methane produced. However, if the feed slurry is concentrated or is already at an elevated temperature, conducting the fermentation in the thermophilic temperature range (50 to 60oC) will be more cost effective. This higher temperature results in a much higher rate of methane production. Consequently, the same conversion efficiency can be obtained at this higher temperature with smaller fermentation tank volumes than are required for the mesophilic fermentation temperature.

Methane fermentation has been used extensively since the beginning of the 20th century as a means for stabilizing and reducing the quantity of organic solids found in wastewater. The organic solids are hydrolyzed and fermented to soluble organic compounds that are eventually converted into methane and carbon dioxide. These gases were initially vented to the atmosphere. In the early part of the century, pioneer work on the anaerobic digestion of sewage sludge under controlled conditions was initiated. This then made it possible to capture the gas and use it for its energy. This gas generated by sewage sludge digestion has been used for inplant energy needs for decades.

The application of this process to the treatment of solid waste in the United States was first investigated by Babbitt et al (9) in 1936. The introduction of home garbage grinders posed a question of the compatibility of these solids with the anaerobic fermentation process. This question was answered by conducting a series of laboratory experiments using a mixture of garbage (food wastes) and sewage sludge as a substrate. The results clearly showed that garbage was amenable to anaerobic fermentation.

Ross (10) reported on the dual disposal of garbage (food wastes) and sewage sludge at Richmond, IN. In this process, the garbage was shredded and, after washing for grit removal, pumped into an enlarged sludge digestion system. This digestion system was operated in the mesophilic temperature range. The process worked satisfactorily with the gas used to provide power for operation of the water pollution control plant. The author reported a net profit from the system.

Even though the process was successful, it was not adopted at other locations for disposal of garbage. After a number of years of successful operation at the Richmond Plant, the process was discontinued. There were several reasons for this lack of interest. Extensive natural gas supplies were being developed and this gas was available at very low prices. The economic incentive for this system was eroded. Further more, the characteristic of the garbage was changing. The quantity of food waste was decreasing due to the developments in the food processing industry,and more food was processed prior to its distribution thereby reducing the amount of garbage generated. Also, the quantity of packaging material (paper and plastics) was increasing, and paper, plastic, cans and bottles were assuming a much larger percentage of the feed stream. The lack of a ready technology for separation of these components from the garbage presented additional operational problems.

In 1965 work was initiated at the University of California at Berkeley on the characteristics of solid waste and the effect of solid waste on the methane fermentation process. Golueke and McGauhey (11) estimated that paper accounted for about 67% of the refuse that can be converted biologically to methane. Garbage accounted for only 12% of the biodegradable refuse. The total nitrogen content of MSW was found to be extremely low unless some industrial solid waste was contributing nitrogen. Consequently, supplemental nitrogen from either sewage sludge, animal manure, or other source was necessary in order to have a balanced fermentation.

In 1969, the Bureau of Solid Waste Management of the U.S. Public Health Service (later U.S.E.P.A.) funded a research project at the University of Illinois at Urbana-Champaign. The objective of this research was to investigate the process conditions that would result in the maximum conversion of solid waste to gas. At this time, the concept of energy recovery was not developed. This processing was simply a means of reducing the cost and the weight and/or volume of solid waste for disposal.

Shredded refuse was obtained from a shredding facility in Cincinnati, Ohio, and was used in this investigation. This research was conducted in small laboratory reactors having a working volume of 15 liters. Before the shredded refuse was fed to these fermenters, the obvious pieces of plastic, metal and glass were removed by hand. This study determined the kinetics of methane fermentation as applied to shredded urban refuse. the results of this study are presented in detail by Pfeffer (12, 13). The major finding of this study was the effect of temperature on the rate of methane production at elevated fermentation temperatures. These data are shown in Table 4. Hydraulic retention times of four to five days resulted in high conversion efficiencies. These high rates, coupled with the ability to feed a concentrated slurry to the fermentation tanks made the thermophilic fermentation temperature very attractive.

Table 4. Volatile Solids Destruction (percent)*

Temp. oC	Retention Time - Days					
	4	6	8	10	15	20	30
35	16.5	26.4	28.6	29.8	32.0	32.7	34.0
40	31.0	33.6	35.7	36.8	39.0	39.9	40.7
45	26.7	29.0	31.2	32.3	34.1	35.4	36.4
50	34.9	38.6	41.0	42.2	44.2	45.4	46.1
55	40.4	42.6	44.5	45.2	47.2	48.2	49.1
60	41.9	46.6	48.3	49.6	50.8	51.8	53.4

*Percentage of organic material in the raw refuse that is converted to gas.

The gas production associated with these experiments is shown in Table 5. This table shows the gas production expressed as standard cubic feet (0oC and one atmosphere) dry gas per pound of volatile solids fed to the fermenter. As would be expected, the gas data are comparable to the solids destruction data. The gas composition ranged from 52% to 70% methane, depending upon the temperature and the concentration of solids in the feed slurry. The lower methane contents were associated with the higher fermentation temperatures.

The Illinois studies were continued in 1973 with NSF funding with major emphasis on the process requirements. A labora-

Table 5. Gas Production from Laboratory Tests (SCF/lb V.S. Fed)

Temp. oC	Retention Time - Days						
	4	6	8	10	15	20	30
35	2.23	3.56	3.86	4.02	4.32	4.41	5.59
40	4.19	4.54	4.82	4.97	5.27	5.39	5.49
45	3.60	3.92	4.21	4.36	4.60	4.78	4.91
50	4.71	5.21	5.54	5.70	5.98	6.13	6.22
55	5.45	5.75	6.00	6.10	6.37	6.51	6.63
60	5.66	6.29	6.52	6.70	6.86	6.99	7.21

tory reactor with an operating volume of 100 gallons (400 liters) was employed in this series of experiments. Shredded refuse from a variety of sources was used during the three year study. It was clear that the refuse needed some preparation prior to the fermentation step. Size reduction and inorganic removal significantly improved the operation of the fermenter by eliminating a number of operational problems.

This study provided a variety of information concerning the process for conversion of refuse to methane gas. This reactor was operated for three years without any failure. A major finding of this study was the excellent dewatering properties of the fermented slurry. Because of the fibrous nature of the solids, very high filtration rates were possible. The slurry also dewatered exceptionally well in a solid bowl conveyor centrifuge. Cake solids between 35 and 40% were obtained with a pilot scale centrifuge. The dewatering characteristics of the unconditioned slurry were far superior to chemically conditioned sewage sludge. This greatly reduced the cost of the slurry dewatering step which was necessary for the process to be economically feasible. The details of the dewatering studies were reported by Pfeffer and Liebman (14).

Pfeffer and Liebman (14) also reported on the results of simulation model of the processing system in which they evaluated the sensitivity of the system cost to various operating parameters. They found that the dry cake from the dewatering process was an ideal candidate for incineration. A simple heat balance on an incineration system showed that auxiliary fuel was not required for combustion if the cake solids were greater than about 35%. In fact a significant amount of energy in the form of steam can be recovered from this incineration. This steam can be used to supply the heat needed for the thermophilic fermentation temperature. Incineration of the cake not only increases the energy conversion efficiency, but also greatly reduces the residue disposal cost. Instead of a large tonnage of wet cake for haul to the landfill, a much smaller tonnage of dry ash is produced for ultimate disposal. There is a significant cost savings associated with the incineration.

The liquid produced during the dewatering step must be recycled back to the fermentation tank as slurry water for the dry refuse. The liquid has a high concentration of chemical oxygen demand and suspended solids and cannot be discharged. There are several advantages to this recycle. First, return of this warm liquor to the fermentation tanks greatly reduces the heat required to maintain the thermophilic temperatures. This recycle also conserves the chemicals required for pH control and nutrients for the microbial system. An accumulation of dissolved solids has not been found to be a problem as the water lost with the dewatered cake provides the blow-down stream necessary to keep these solids at an acceptable level.

RefCoM OPERATIONAL DATA

Fermenter Performance

Fermenter Loading Conditions. Although the construction of the facility was completed in May, 1978, it was not possible to achieve a consistent feed for the reactors until late 1981. A substantial number of operational problems were encountered with the refuse separation processes. The installed system was finally modified sufficiently so that refuse could be fed to the fermentation tanks. While the feeding was somewhat inconsistent, the system was essentially operated continuously from November, 1981 through February, 1982. The operation of the facility was terminated at this time due to a lapse of funding.

Funding became available again in late 1982 and the system was restarted with the modified refuse separation processes (air classifier) used to prepare the feed for the fermenter. While it was possible to obtain feed for the fermentation tanks, the existing equipment presented an excessive number of operating problems. Plans were made to replace the existing refuse separation processes with a new system employing disc screens. The facility was shut down to allow for a major rebuilding of the separation facility. Starting in June, 1984, it was possible to operate the plant at will and produce the desired quantity of feed for the fermentation tanks.

Any discontinuity in the operation after June, 1984 represent periods when either new equipment such as the belt-filter press was being installed, operational problems were encountered with the mixers in the fermenters, or the inability to dispose of excess water generated primarily by precipitation caused the plant to be shut down. The facility was operated essentially continuously from March through July, 1985. Operation was stopped on July 17, 1985 when the available funds were exhausted.

The time periods in Table 6 refers to the operational periods when "steady-state" operating conditions were assumed. The data in Table 7 can be used to determine how constant the reactors were loaded. The tons per day of classified lights and the volatile solids, along with the standard deviation, are presented in this table. During certain periods, a very uniform feed of classified lights was maintained. The deviation in the volatile solids fed was always higher. The tons of classified lights was used, especially in the later runs, as the control parameter. Consequently, any variation in moisture or ash content of the refuse would magnify the variation in volatile solids. With the exception for Periods 4,5,8,9 and 10, there was a substantial variation in the feed rates. Much of the variation for the above periods was a result of missing one or two days of operation. These fluctuations do raise the question of whether or not a true "steady state" operation was achieved. This is especially true when the length of the runs are compared to the 20 to 50 days needed to obtain a complete turn over of the fermenter contents when operating at retention times of seven to 15 days.

Table 6. "Steady-State" Operating Periods.

Period	Dates
1	12/14 - 12/31/81
2	2/1 - 2/16/82
3	10/1 - 11/14/82
4	2/27 - 3/12/83
5	3/21 - 4/1/83
6	6/26 - 7/10/84
7	9/5 - 9/20/84
8	3/12 - 4/3/85
9	4/17 - 5/13/85
10	7/3 - 7/17/85

Table 7. Fermenter Loading Conditions.

	Feed Rate - TPD				Feed Slurry Conc.	Fermenter Solids %	
	Clas. Lights		Vol. Solids				
Period	Ave.	S.D.	Ave.	S.D.	%	T.S.	V.S.
1	5.9	2.3	4.3	1.2	(4.5)	2.99	2.19
2	14.7	2.0	7.6	1.1	(6.5)	4.37	2.97
3	9.4	6.7	4.4	3.1	(5.0)	3.09	2.26
4	5.4	0.7	2.9	0.4	(4.5)	2.68	1.93
5	10.0	0.03	5.1	0.5	6.5	4.05	2.95
6	7.1	3.0	3.3	1.4	9.3	2.67	2.00
7	14.3	3.2	6.5	1.6	9.4	3.60	2.76
8	11.8	1.7	6.7	1.1	10.3	6.33	4.65
9	9.5	0.6	5.1	0.5	7.5	4.79	3.64
10	17.6	2.5	7.9	1.4	9.0	4.02	2.91

Table 7 also shows the solids concentration in the feed slurry and in the reactor. The feed solids concentrations shown in the parenthesis are estimates since this parameter was not being measured at this time. It was possible to achieve a concentrated feed but it was not possible to continue to operate at this high level. As the reactor contents exceeded 6% (6.33%) the power draw on the mixer approached the maximum. Because of the age of the mixing system the solids in the reactor were reduced rather than risk failure of the mixers. The results from Period 8 do show that it is possible to feed a slurry with a high solids concentration. If the mixer is designed to accommodate the high concentration of solids in the reactor, the system will operate under these conditions.

Gas Production. Table 8 shows the gas production data for the periods indicated in Table 6. These are the actual meter readings and have not been corrected to any standard condition. The data can be used to illustrate the variations that were experienced in the operation of the fermenters. The variation in gas production as evidenced by the standard deviation was primarily a result of the variation in the feed rate. These data also suggest that "steady state" was not achieved in many of the test runs. The standard deviation of gas production for Periods 4, 5 and 9 was less than 10% of the average. This limited variation would represent steady state. The methane content of the gas as a percentage is not presented in Table 8 since a relatively constant gas composition was obtained, ranging from 51.1 to 54.1% with an average value of 52.8%.

Table 8. Measured Gas Production Rates - CFD

	Total Gas		CH4	
Period	Ave.	S.D.	Ave.	S.D.
1	44,200	15,600	23,600	8,700
2	96,000	19,000	49,400	9,800
3	54,800	25,700	28,000	12,600
4	35,100	2,400	18,700	1,400
5	47,900	4,100	25,100	2,200
6	56,900	12,000	30,800	7,100
7	72,700	9,000	38,800	4,900
8	38,800	6,700	20,600	3,100
9	64,500	6,700	34,700	3,600
10	74,100	16,200	39,300	8,700

The total gas production does not indicate the process efficiency. Conversion to a parameter such as the cubic feet per pound of volatile solids fed provides a measure of the biological efficiency. A performance parameter such as the volatile solids destroyed also measures the biological efficiency. Table 9 presents these data along with a calculated retention time for each period of operation when the feed rate was reasonably constant. The reactor was not necessarily at steady state, but the results are of value.

Table 9. Theoretical and Measured Gas Production Data

Per.	Ret. Time Days	Vol. Sol. Dest. %	Measured Gas CF/LB VS Fed Total	Measured Gas CF/LB VS Fed CH4	Theor. Gas CF/LB VS Fed Total	Theor. Gas CF/LB VS Fed CH4	Theor. Gas SCF/LB VS Fed Total	Theor. Gas SCF/LB VS Fed CH4
1	10.0	43.6	6.8	3.6	7.0	3.7	5.9	3.1
2	6.4	45.6	7.1	3.7	7.2	3.7	6.2	2.7
3	8.6	46.3	6.2	3.1	7.7	3.9	6.2	3.2
4	12.6	52.1	6.0	3.2	8.4	4.5	7.0	3.8
5	9.8	42.6	4.7	2.5	7.1	3.8	5.8	3.1
6	26.6	75.1	8.7	4.7	12.8	7.0	10.1	5.5
7	13.5	66.3	5.6	3.0	11.3	6.1	8.9	4.8
8	10.1	52.4	2.9	1.5	8.7	4.6	7.0	3.7
9	9.0	47.1	6.3	3.4	7.8	4.2	6.3	3.4
10	10.5	63.2	4.7	2.5	10.8	5.7	8.5	4.5

In the column labeled measured gas, Table 9 shows the gas production that was measured with the gas control and measuring system. The other columns labeled theoretical gas were computed from the volatile solids destruction. Since the primary organic material present in the MSW is carbohydrate, a theoretical gas production of 13.5 SCF (0oC and one atmosphere) per pound of volatile solids destroyed was used for the computation. The volume of theoretical gas expressed as CF/LB has been adjusted to the temperature and moisture conditions of the measured gas. Therefore, the computed and measured gas expressed as CF/LB should be comparable.

The purpose of this exercise is to substantiate the measured gas data from another data base. This analysis shows good agreement for the first two periods of operation. However, there is a major deviation between the two data sources for the remaining periods. This deviation is probably due to gas leakage from the fermenters. A number of incidents occurred that caused major damage to the tank covers. Repair or replacement of the concrete tops was necessary. The type of tank used for these fermenters was not conducive to repair, especially when the repairs must be gas tight. In addition, the mixers imparted a significant vibration to the tanks so the any weakness in the tanks were aggravated. This problem was recognized and a daily maintenance routine was employed to minimize the leakage.

The reliability of the measured gas production is questionable, especially in the later periods of operation. The data necessary to compute the volatile solids destruction were incomplete during Periods 1 through 4. A mass balance was possible only if the solids concentration of the feed slurry was assumed. From then on appropriate measurements were made to permit the computation of a mass balance on the fermenter. Consequently, the theoretical gas production is more indicative of the performance of the system than the measured gas production.

A comparison of the gas generation from RefCoM and the laboratory studies is appropriate. The data in Table 5 show that at the thermophilic fermentation temperature an increase in retention time past the about six days resulted in a marginal increase in gas production. These data plus the model studies of Pfeffer and Liebman (14) show that the most cost effective hydraulic retention time will be at the shorted times. This was not the case for the full-scale plant data. Table 9 clearly shows that the longer retention times result in substantially more gas production per unit of volatile solids fed to the reactor. There is enough scatter in the data over the four year period of operation so that it is not possible to identify an optimum retention time. It is clear that there will be advantages to operating at retention times in the 10 to 15 day range. The longer retention times not only produce more gas, but also result in a more stable operating system.

<u>Gas Composition and pH.</u> The gas composition was relatively consistent during the periods of "steady state" operation. The methane content ranged between 50 and 54%. The pH was maintained above 7.0 by the addition of lime and the feed rate was adjusted to prevent a build-up of volatile acids during non-steady state operation. Consequently, the operating conditions were such that fluctuations in gas composition would not be expected. The RefCoM data are essentially the same as the data on gas composition obtained from the thermophilic laboratory studies of Pfeffer (12,13).

<u>Fermenter Operating Conditions.</u> The operating conditions imposed on the fermenters did not encompass a wide range of variables. It was originally planned to conduct a complete parametric analysis of the primary variables of temperature, retention time, pH, feed slurry solids concentration and feed solids particle size. Because of a myriad of problems with the refuse processing system in the initial years of operation, this plan was soon abandoned. The primary effort was placed on developing a refuse separation system that was able to consistently produce an acceptable feed for the fermenters.

During the entire period of operation, the temperature was maintained in the thermophilic range of 135 to 140 oF. The previous laboratory studies and the associated economic analysis strongly support the operation at thermophilic temperature. If RefCoM could not be operated successfully at the elevated temperature, then it was improbable that it would be cost effective.

The data in Table 10 represent the typical operating conditions experienced during the different periods of operation. The retention times ranged from a low of 6.4 days to a high of 26.6 days. This range provided some interesting information on the volatile solids destruction as has been discussed.

The feed slurry solids concentration was not measured during the first four periods of operation. The numbers in the parenthesis are estimates. However, the ability to feed a concentrated slurry was demonstrated. Solids concentrations as high as 10.3% were fed into the fermenters. The high feed solids resulted in a reactor slurry solids concentration of 6.33%. During this period it was necessary to lower the liquid level in the fermenter. The power draw by the mixer drive motor approached the limit. Rather than risk mixer failure, the liquid level in the reactor was lowered inorder to reduce the power draw.

Table 10. Fermenter Operating Conditions

Period	Retention Time, Days	Feed Slurry Conc. %	Fermenter Solids % Total	Volatile
1	10.0	(4.5)	2.99	2.19
2	6.4	(6.5)	4.37	2.97
3	8.6	(5.0)	3.09	2.26
4	12.6	(4.5)	2.68	1.93
5	9.8	6.5	4.05	2.95
6	26.6	9.3	2.67	2.00
7	13.5	9.4	3.60	2.76
8	10.1	10.3	6.33	4.65
9	9.0	7.5	4.79	3.64
10	10.5	9.0	4.02	2.91

<u>Fermenter Operating Problems.</u> A number of operational problems was experienced with the reactor. All of these problems were related to the specific design of the reactor. A major problem was encountered in keeping the reactor gas tight. The tops of both reactors were ruptured more than once. In some cases the top was replaced while in other cases repair was attempted. Because the integrity of the tanks had been disturbed, it was not possible to maintain a gas tight tank. A routine program was initiated on a daily basis to find and seal leaks that could be found. However, it was not possible to prevent significant leakage from occurring. This was shown in the above discussion.

A number of problems were associated with the design of the mixers. The mixer shaft was unsupported at the bottom of the tank. The stress created by a buildup of rope-like material on these shafts resulted in the shafts failing at the bearing support on the bridge that supported the mixer. This accumulation of the material on the mixer shaft severely inhibited the mixing ability of the mixing system. This problem was corrected by changing the feed quality and redesigning the mixers to eliminate the excessive vortex that occurred.

There were no failures of the biological system even though significant stress was placed upon the microorganisms as a result of the physical and mechanical problems encountered. By monitoring the pH and the total volatile acids in the fermenter, it was possible to control the loading of the fermenter such that process failure did not occur. In all cases except one, the buildup of volatile acids occurred during start up, or as a result of process deficiencies. High volatile acids were recorded during part of February and March, 1985. These events were associated with high feed slurry solids in the 12-16% range. The mixing was not adequate to incorporate these solids into the fermenter slurry. This poor mixing resulted in an inhibition of the microbial activity.

From February 14-19, 1983, the total volatile acids were approximately 2000 mg/l as acetic acid. The cause of this high volatile acids was not identified, but a reduction of the loading rate permitted the process to recover in a few days. A slug of phenolic compounds was suspected. At this time the operator of the front end loader complained of a heavy concentration of some type of fumes that caused him considerable discomfort. A sample taken from the reactor several days after this event was analyzed and found to contain several mg/l of phenolics. It is probable that the phenolics inhibited the system to a degree.

Slurry Dewatering System

The initial design employed a rotary drum vacuum filter for the dewatering if the fermented slurry. Laboratory data showed that with the proper loading and feed slurry solids concentration it was possible to obtain cake solids in excess of 30% dry solids (14). The RefCoM installation was never operated under controlled conditions that would permit the evaluation of the process performance. It was operated in a production mode, i.e. process the digested slurry without too much concern about the performance. The dewatered cake was transported directly to the landfill so moisture content was of little significance. The performance of the vacuum filter was good under these conditions. Table 11 presents the results of three periods of plant operation in the fermented slurry was dewatered with the vacuum filter. With a relatively dilute feed solids concentration the cake solids during the first two periods was in the 24 to 25% range. The maximum reported during these periods was 33.6%. The field data support the laboratory data very well.

Table 11. Performance of the Rotary Drum Vacuum Filter.

Operating Period	Fermentor Solids - %	Cake Solids - % Ave.	Cake Solids - % Max.	Filtrate Solids - %
2/1-3/31/83	2.8	24.5	30.6	1.4
5/7-7/15/84	1.7	23.7	33.6	0.8
8/27-9/20/84	2.7	19.9	26.7	1.1

Chemicals were not used to condition the slurry before filtration. It was not necessary to be concerned about a high solids capture efficiency. A coarse weave filter medium was used so the fine solids did not blind the medium. The filtrate was recycled to the fermentor. The presence of these fine solids would not impair the fermentor operation. A steady-state condition is reached in which the fine solids are captured in the fibrous filter mat that is formed on the filter and their concentration does not continue to increase in the slurry.

The vacuum filter capacity limited the rate at which refuse could be processed. A belt filter press with a stainless steel belt was obtained to evaluate the possibility of using this system to more efficiently de-water the fermentor slurry. The process was used in the latter phase of the project and was never operated under optimum conditions. Consequently, the data shown in Table 12 should be evaluated accordingly.

As with the vacuum filter, chemicals were not used to condition the slurry prior to filtration. The concentration of total solids in the filtrate from the belt press was considerably higher than the vacuum filter. The belt was a very porous weave that passed a substantial quantity of small particles. The rate of filtration was much higher than with the vacuum filter. As can be seen from the data, the filtrate solids stabilized at approximately 2.5%. This poor solids capture was not a problem since the filtrate was recycled to the fermentor.

Table 12. Performance of the Belt Filter Press.

Operating Period	Fermentor Solids - %	Cake Solids - %		Filtrate Solids - %
		Ave.	Max.	
March 85	6.0	27.0	31.7	2.6
April 85	5.0	22.4	30.1	2.6
May 85	4.7	22.1	30.8	2.7
June 85	4.8	19.7	26.7	2.2

The data contained in the above tables can be considered tentative because the dewatering units were not operated under optimum condition. It is probable that both dewatering systems would produce a cake with a higher total solids content if the operating conditions were near the optimum. It has been demonstrated that the dewatering characteristics of the digested solid wastes are much superior to the characteristics of digested sewage sludge. Therefore, it is quite incorrect to use sewage sludge dewatering costs as a basis for evaluating the cost of dewatering the slurry from the digestion of solid wastes.

REFERENCES

1. Klimb, D., Resource Recovery and Conservation, 1:225-233 (1976).

2. Sullivan, P.M., N.H. Stanczyk, and M.J. Speedlove, Resource Recovery from Urban Waste, Bureau of Mines Report of Investigation 7760 (1973).

3. Alter, H. Environmental Science and Technology, 11(5):444-448 (1977).

4. Fernandez-Aller, R., ENADIMSA, Serrano 116, Madrid, 6, Spain, personal communication (1981).

5. Campbell, J.A., et al., Solid Waste Management, 24(1):34 (1981).

6. Kinsey, R.D., Solid Waste Management, 23(5):48 (1980).

7. Spencer, D.B., and E. Schomann, Waste Age, 6(10):32 (1975).

8. Hamilton, F., and J. Kelyman, Solid Waste Management, 22(5):40 (1979).

9. Babbitt, H.E., B.J. Leland, and F.H. Whitley, Jr., The Biological Digestion of Garbage with Sewage Sludge, Bulletin No. 287, Engineering Experiment Station, University of Illinois, Urbana, IL (1936).

10. Ross, W.D., Sewage and Industrial Wastes, 26:140 (1954).

11. Golueke, C.G., and P.H. McGhauhey, Comprehensive Studies of Solid Waste Management, Second Annual Report, U.S. Public Health service Report, SW 3rg USPHS, Publ. No. 2039, Washington, D.C. (1970).

12. Pfeffer, J.T., Biotechnology and Bioengineering, 16:771 (1974).

13. Pfeffer, J.T., Reclamation of Energy from Organic Refuse, EPA-670/2-74-016, U.S.E.P.A. National Environmental Research Center, Cincinnati, OH (1974).

14. Pfeffer, J.T. and J.C. Liebman, Resource Recovery and Conservation, 1:295 (1976).

DISCUSSION

R. Legrand, Reynolds, Smith & Hills

When you mentioned parameters such as cubic feet of gas per pound of volatile solids added, do you include the plastics, and do you account for the influence of volatile acids on that balance? Have you explored uses for the dewatered effluent, like composting or combustion?

Pfeffer

We have looked at the heating value of the dewatered effluent and done some calculations as far as what it would take to incinerate it. If you can get about a 35% solids in the cake, it'll be self sustaining in an incinerator, and you'll not need auxiliary fuel. The energy recovery from that is quite substantial in terms of producing steam, and this clearly is of interest for the amount of steam we would use to maintain thermophilic digestion. You will have more than enough steam to maintain digestion temperatures. In terms of calculating the yields of gas, this was based on the feed material, which included everything that went into the digester. The plastics and all solids were ashed. The volatile acids, in general, were usually about 200 mg/ℓ, so we never had significant quantities of volatile acids to deal with.

V. Srivastava, IGT

You showed a table which listed the operating conditions. How long did each period last, and could you tell me how many days there were between each two periods that you listed?

Pfeffer

Table 6 includes the periods and the dates. Some of the runs were very short, maybe 3 weeks; at other times, they may have been 6 weeks or 8 weeks. The period between runs was highly variable, in some cases, a couple of years; in other cases, just a few days.

Srivastava

The reason I asked is because there is a residual effect. In anaerobic digestion you feed today and you'll see the effect several days later. If the interval between two periods is not long enough, the data may be misleading. In one case, you showed that the theoretical methane yield was more than 10 SCF/lb volatile solids added.

Pfeffer — That was total gas.

Srivastava — What is the theoretical yield that you calculated if you can convert the whole?

Pfeffer — To methane?

Srivastava — Yes.

Pfeffer — Assuming cellulose would produce 13.5 cubic feet of total gas at 50% methane, 6.75 would be the maximum using cellulose. We did not make any experimental determinations.

T. Nunez-McNally, Tufts U. — Was the sewage sludge added for moisture benefit, nutrient content, or to alter the carbon-nitrogen ratio?

Pfeffer — The primary purpose of doing that is for the cost benefits associated with getting rid of sewage sludge. We didn't need the water because of our recycling system. The nitrogen was marginal as far as being sufficient in the refuse. Sometimes we had to add it; sometimes we didn't. But the amount added was very small. A small amount of urea was sometimes added to bring it up to the required level. The sewage sludge was added primarily to take advantage of the cost credit when you figure that you might spend $100 to $150 per ton of dry solids for disposal. That's a very attractive cost item for the economics. If we can take that credit from the sewage treatment plant, it gives us a nice cost break.

ENERGY FROM SLUDGE-DERIVED FUELS: THE HYPERION ENERGY RECOVERY SYSTEM

D.F. Bress, M.Ch.E., Staff Consultant
C. Greenfield, B.S.Ch.E., President
Dehydro-Tech Corporation
East Hanover, New Jersey 07936

R.T. Haug, Ph.D.
Consultant and Chief Process Engineer
Hyperion Energy Recovery System
Playa del Rey, California 90293

ABSTRACT

The City of Los Angeles is in the initial stages of startup of the $200 million Hyperion Energy Recovery System (HERS), a project to manage and recover energy from over 400 dry tons per day of municipal sewage sludge. Sometimes referred to as "once used" biomass, the sludge collected at Hyperion has an energy equivalent of over 1000 barrels of oil per day. The HERS Project will convert sludge to gaseous and solid "sludge derived fuels" (SDF). Gaseous fuel is produced by anaerobic digestion. The gas is scrubbed to remove H_2S, compressed, and fired in a gas turbine, cogeneration system. The solid residue remaining after digestion is dewatered and dried in the Carver-Greenfield multiple effect evaporation porcess. Energy will be recovered from the dried solids and air emissions controlled by fluidized bed gasification and staged combustion of the resulting fuel gas. The project will generate about 25 MW of power. Best available air pollution control systems will reduce emissions from present levels at the treatment plant.

ENERGY FROM SLUDGE-DERIVED FUELS: THE HYPERION ENERGY RECOVERY SYSTEM

INTRODUCTION

The City of Los Angeles provides sewerage service to approximately 3.5 million persons living in a 600 square-mile service area. About 370 mgd of wastewater produced in the service area is treated at the Hyperion Treatment Plant. All influent flow receives primary treatment and secondary treatment is provided for 100 mgd. Treated effluent is discharged through an ocean outfall system. Primary sludge currently is anaerobically digested, combined with raw waste activated sludge and discharged through a separate ocean outfall.

In order to completely eliminate ocean dumping of sewage sludge, the City of Los Angeles department of Public Works, in 1981, initated design of the Hyperion Energy Recovery System, which has come to be known by its acronym as the HERS Project. This plant incorporates a series of sludge treatment and energy generation processes selected from all available options after careful evaluation and pilot testing. The result will be the most efficient sludge to energy plant ever built In addition, air emissions from the new HERS Project will be reduced from the levels presently experienced at this location.

As stated, the HERS design represents a highly efficient conversion of sludge fuel values to energy. This is mainly due to the selection of three processes for the project. First, although the present practice of digesting sludge could have been dropped, it was decided to continue with this process by which methane gas is generated while stabilizing and reducing the sludge solids. Second, generating gas permits the use of the 40 percent efficient combined cycle, based on gas turbines, to generate power.

The third process contributing to overall high efficiency is the Carver-Greenfield Process®, (C-G), used to completely dry mechanically dewatered sludge. This process only uses about 340 BTU per pound of water evaporated compared with about 1300 BTU/lb. for indirect steam heated sludge dryers, or about 2000 BTU/lb, for direct fired sludge dryers. Thus the so called Sludge Derived Fuel, (SDF), is dried at a much lower energy input than with the more conventional methods. The SDF is gasified, the gas combusted and power generated via a conventional steam cycle. Because the SDF is dry higher combustion process exit temperatures are possible.

HERS PROJECT DESCRIPTION

Figure 1 is a block flow diagram showing the various steps involved in the HERS Project. The plant is designed to process 405 dry tons per day (dtpd) of Primary and Waste Activated Sludge, (WAS). As can be seen on Figure 1, Thickened WAS is mixed with Primary Sludge and sent to the Digesters. The digested sludge is then mechanically dewatered in a battery of centrifuges raising the solids concentration from about 5 to 18-20 percent. This sludge cake is then dried in the Carver-Greenfield unit to produce dry solid powder and a stream of sewage oil. These Sludge Derived Fuels, (SDF), are converted to energy in a fluidized bed gasification system which includes emission controls for the exhaust gas.

The digester gas is compressed, cleaned and converted to energy in a combined cycle cogeneration system based on gas turbines.

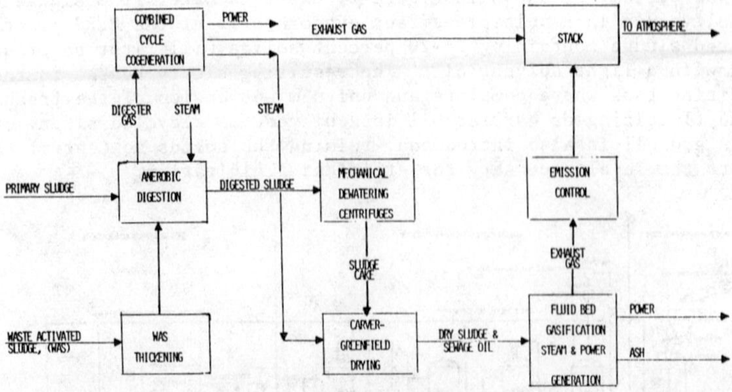

Figure 1. BLOCK FLOW DIAGRAM
HERS PROJECT

A more detailed description of these steps is as follows:

WAS Thickening

The WAS will be thickened from 0.5 percent to 3.5-4.0 percent solids using the Dissolved Air Floation, (DAF), process. The thickening DAF operation is carried out in three 50 foot diameter thickeners operating with a liquid depth of 10.5 feet. Polymer for sludge conditioning may be added if required.

Anaerobic Digestion

The thickened WAS will be mixed with primary sludge and sent to the anaerobic Digestion plant, where the solids are reduced to 265 dtpd by biological fermentation of the organics present.

At Hyperion there are eighteen digesters, each 100 feet in diameter, operating at a 30 foot water depth. These will be refurbished to improve gas mixing, allow for WAS and primary sludge blending and generally upgrade the facility. In addition, a Digester Cleaning Facility is being added which will permit cleaning of the digesters on a regular basis. This consists of screens, hydroclones and belt presses.

The Digesters have sufficient volume to store one day's production of the digested product.

Mechanical Sludge Dewatering-Centrifuges

The digested sludge underflow at about 5 percent solids, is dewatered in a battery of thirteen Ingersoll Rand solid bowl, scroll type centrifuges. The feed sludge is conditioned with polymer fed via a dosing system. The 18-20 percent solids cake made is conveyed directly to the Carver-Greenfield plant or to storage. Storage for 2.2 days production of wet sludge is furnished.

Sludge Drying - Carver-Greenfield Process®

The significant feature of the Carver-Greenfield Process® from an energy standpoint is that the energy recovered from the process is far greater than the energy input as steam and electricity, required to dry the sludge cake produced from the centrifuges.

The system, as shown in Figure 2, extracts water from sludge under partial vacuum in a multiple-effect evaporator. At the HERS plant, digested sludge containing 18-20 percent solids will first be ground and mixed with a light solvent oil. The resulting slurry passes to the fluidizing tank where complete and uniform suspension of the fresh feed in the fluidizing or carrier oil is achieved. A recycled slurry of solids and oil is also introduced, raising the solids content of the feed to the level necessary for efficient fluidization.

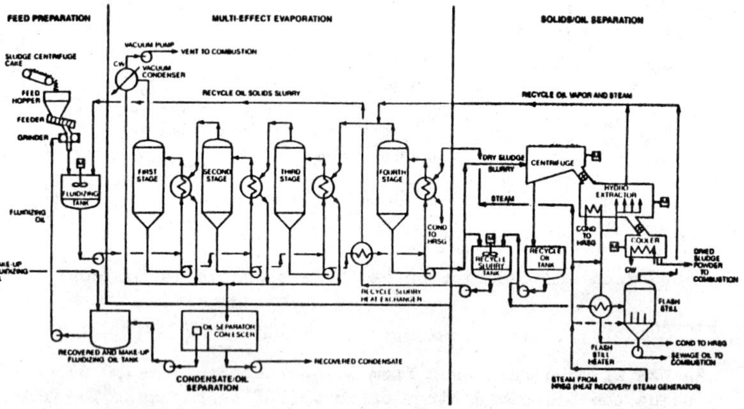

Figure 2. SCHEMATIC PROCESS DIAGRAM FOR A 4-EFFECT CARVER-GREENFIELD PROCESS® AT HERS

From the fluidizing tank the sludge-oil slurry is pumped to a four-stage evaporation system employing the multi-effect principle. The heat required to evaporate water at each stage is obtained by condensing the vapor generated in the following stage at a higher temperature and pressure. The first stage is maintained under vacuum by condensing vapor in a water-cooled surface condenser from which non-condensable gases are extracted with a vacuum pump. Boiler steam is condensed on the shell side of the fourth or final water removal stage. In general, a four-stage evaporation system permits one pound of boiler steam to evaporate four pounds of water from sludge.

After the fourth stage, the slurry of dehydrated solids and oil is divided into two streams, one of which is recycled to the front end of the plant to aid in fluidizing fresh feed. The second stream is centrifuged to recover the bulk of the fluidizing oil. Final recovery of fluidizing oil is achieved using a process known as "hydroextraction" in which the centrifuge cake is heated under vacuum to vaporize the fluidizing oil away from the dried sludge powder.

The equipment is arranged in three parallel trains, one of which is a spare train to care for outages.

The Carver-Greenfield Process® will also extract sewage oil, including metallic soaps, from the sludge feed. To prevent foaming and heat transfer problems caused by an increasing concentration of sewage oil in the fluidizing medium, a portion of the circulating oil is purged to a still. There the lighter oil is vaporized under vacuum with steam employed as an indirect heating and stripping agent. The recovered sewage oil, which has a heating value of about 16,000 BTU's per pound, will be stored and used as fuel.

The dry fuel produced by the Carver-Greenfield plant has a heating value of about 5600 BTU's per pound. This Sludge Derived Fuel (SDF) is comparable in heating potential to other biomass materials.

To make the HERS system as efficient as possible, the oil vapor and stripping steam from the hydroextraction and sewage oil recovery steps are condensed in the multi-stage evaporation train. Additional heat is transferred from the recycled liquid slurry to the fresh feed going through the evaporation stages. For clarity, Figure 2 shows only the heat exchanger between the third and fourth stages.

Because the fluidizing oil must be light enough to be vaporized away from the dry sludge, some oil is vaporized along with the water. The oil-water mixture is separated in a decanter-coalescer. The condensed water is virtually solids free but has a Chemical Oxygen Demand (COD) of about 1000 due to volatile organics in the sludge feed that are water soluble.

Fluidized Bed Gasifier (for SDF)

Figure 3 shows the Fluidized Bed Gasifier System in which the Sludge Derived Fuels from the Carver-Greenfield plant are used to generate steam. Storage is provided for 1.5 days production of powder and one weeks production of sewage oil.

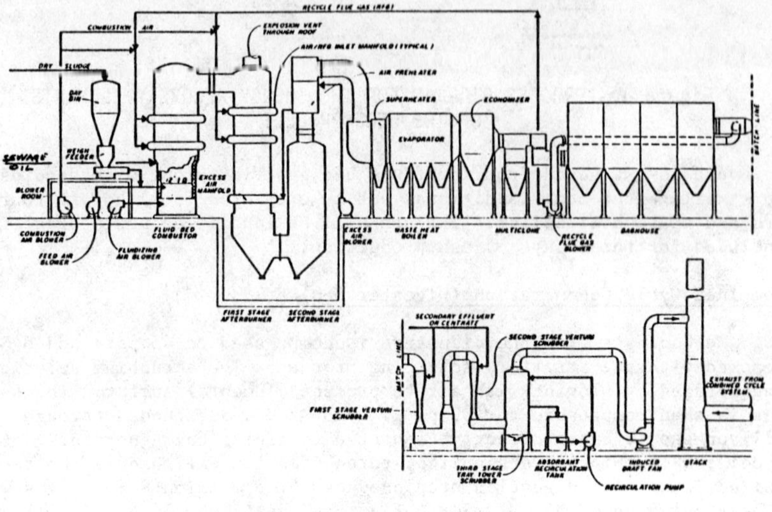

Figure 3. HERS SYSTEM FOR FLUIDIZED-BED GASIFICATION OF SDF, STAGED COMBUSTION AFTERBURNING, WASTE HEAT RECOVERY, AND AIR POLLUTION CONTROL

As for the C-G unit, the SDF combustion system is arranged in three parallel trains, and two are sufficient to handle design requirements.

Both the powdered sludge and sewage oil are fed to a fluidized bed gasifier which employs recycle flue gas followed by two stages of afterburning. This staged combustion of the pyrolysis gas results in minimum NO_x formation.

The hot gases from the second afterburner pass through a waste heat boiler where steam is generated at 620 psig and 750°F. This steam is expanded through a condensing turbine with three levels of extraction. These extraction streams are used for deareation as well as for high and low pressure boiler feed water heating. For a total heat input of about 140 million BTU/hr., power is generated at the rate of 10,120 KW. If necessary, low pressure extraction steam can be diverted to supply digester heating requirements. The steam and power generation systems for the SDF are shown on Figure 4, which also shows the digester gas combined cycle cogeneration system.

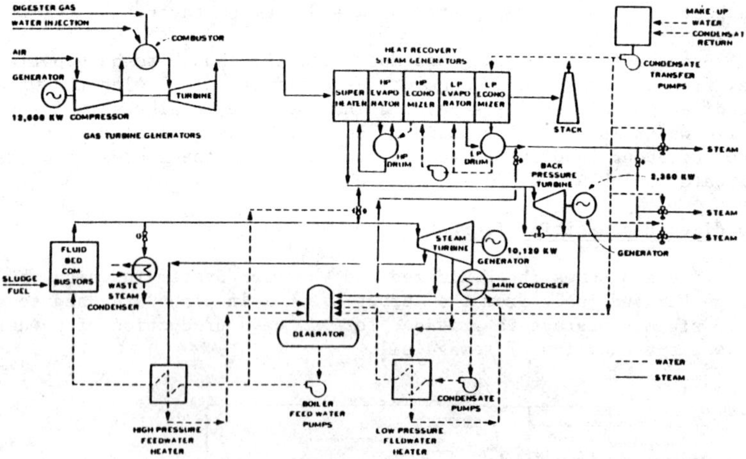

Figure 4. PROCESS DIAGRAM FOR THE ENERGY RECOVERY ELEMENTS OF THE HERS PROJECT

Ash is extracted from the fluid bed gasifier along with sand used to stabilize the bed. Additional ash is extracted from the afterburners and waste heat boiler. The ash will be landfilled and storage is included for three days of ash production.

Combined Cycle Cogeneration (Digester Gas)

Methane gas from the digesters is compressed to 45 psia and H_2S removed with the Stretford scrubbing process. The scrubbing solution i regenerated by blowing with air to produce elemental sulfur. The sweet gas is then compressed to 400 psig, combusted and expanded through Allison gas turbines directly connected to electrical generators. Four model 520K turbints are supplied, three of which will handle the required load. The digestion step produces 6.25 million SCFD of gas with a heating value of 650 BTU/SCF for a total heat input of 4062 million BTU/day, which results in 12,600 KW of power from the gas turbines.

The hot gas exhuast from the turbine is used to generate both high pressure superheated steam (1100 psig, 800°F), and low pressure saturated steam (30 psig) in a combined cycle-cogeneration system. Steam at 165 psig is extracted from a back pressure turbine. This system generates an additional 2350 KW of power and supplies steam requirements of the Carver-Greenfield sludge drying plant at 165 psig and for digester heating at 30 psig.

Overall Fuel and Power Summary

Figure 4 is a combined cycle power generation diagram showing both the digester gas and SDF fuel conversion systems, along with the steam distribution and power generation arrangement.

A heat and power summary for the HERS project is as shown in Table 1.

Emission Control

The HERS project complies with all the applicable air pollution requirements of the Los Angeles Basin Clean Air Act for New Sources.

Table 1. OVERALL FUEL AND POWER SUMMARY
HERS PROJECT

FUEL	HEAT INPUT	BTU/HR.
	DIGESTER GAS	169.3×10^6
	SDF	1400×10^6
	TOTAL	309.3×10^6
	DISTRIBUTION	
	TO CARVER-GREENFIELD	30×10^6
	TO DIGESTION	10×10^6
	TO POWER GENERATION	269.3×10^6
POWER	GENERATION	MW
	DIGESTER GAS COMBINED CYCLE	14.95
	SDF STEAM CYCLE	10.12
	TOTAL	25.07
	DISTRIBUTION	
	CARVER-GREENFIELD	1.50
	ALL OTHER HERS USERS	6.50
	FOR SALE OR USE ELSEWHERE AT HYPERION	17.07

In addition, the City decided that air emissions would not exceed those presently experienced at the Hyperion location. Based on the combustion tests of both sludge and digester gas, processes were selected to minimize polluting emissions, particularly NO_x. As a result, the projected emissions for the year 2000 from HERS will be less than are at present. Table 2, derived from an earlier paper by Haug and Sizemore, shows these reductions.

NO_x, SO_x and CO emissions from the gas turbine system are markedly reduced from those presently experienced at Hyperion using reciprocating gas engines to combust the digester gas. This reduction in pollutants is due to sulfur removal from the gas prior to combustion and the combustion conditions.

The combustion gases from the Sludge Derived Fuels present more of a problem than burning digester gas. As shown in Figure 3, combustion gases from the waste heat boiler pass through a multiclone where larger sized particulate is removed. Final cleanup of particulate takes place in a bag house following the multiclone.

From the bag house, the exhaust gas is scrubbed and cooled with secondary effluent and centrate from the mechanical dewatering, (centrigure), step. The latter having a high degree of alkalinity, will react with the bulk of the acid constituents in the gas. This water scrubbing operation is carried out in a two stage Venturi scrubber. Cooling of the gas will also remove condensable particulates.

Table 2. DATA FOR HERS AIR EMISSION GAS TURBINE COMBINED CYCLE SYSTEM FLUIDIZED-BED COMBUSTION SYSTEM

	NO_x (AS NO_2)	SO_x (AS SO_2)	TSP	NMHC (AS CH_4)	CO
EXISTING RECIPROCATING ENGINES USING DIGESTER GAS (TO BE RETIRED FROM SERVICE)	1800	660	225	650	1740
GAS TURBINE COMBINED CYCLE SYSTEM (YEAR 2000 CONDITIONS)	680	20	55	25	715
OFFSETS AVAILABLE FOR FLUIDIZED BED SYSTEM	1120	640	170	625	1025
ESTIMATED EMISSIONS FROM FLUIDIZED BED SYSTEM	395	145	55	20	65
DIRECT EMISSION REDUCTIONS	725	495	115	605	960
PERCENT REDUCTION	40%	75%	51%	93%	55%

EXCEPT FOR LAST LINE ALL VALUES ARE AS LBS/DAY
TSP = TOTAL SUSPENDED PARTICLES
NMHC = NON-METHANE HYDROCARBONS

From the Venturi scrubbers the exhaust gas enters a tray type scrubber where the gas is contacted with a scrubbing solution working on a sodium sulfite/bisulfite cycle. This solution is made up with aqueous sodium hydroxide to maintain to proper bisulfite level in the fresh solution. Final acid gas removal occurs in this scrubber.

The vent from the final scrubber vents to a stack in which the gas from the combined cycle system is also vented. Temperatures are such that the mixture is heated well above its dew point to prevent a plume.

In addition to controlling pollutants from combustion of Digester Gas and Sludge Derived Fuels, the HERS project includes systems to prevent odor or hydrocarbon emission from the other sludge processing steps. For example, the WAS system includes a scrubbing system to deodorize fumes generated in the thickening operation.

Another possible source of odor or hydrocarbon emission is the Carver-Greenfield Process® in which the mechanically dewatered sludge is dried. To avoid this possibility all vents from the Carver-Greenfield facility will be routed to a collection system into which the vacuum pump also discharges. An exhaust blower will then feed the total vent output to an activated carbon recovery unit and then to the fluidized bed section for combustion of any remaining vapor. This eliminates the possibility of odors and other gaseous pollutants entering the air from the Carver-Greenfield portion of the project.

One other noteworthy fact is that the conditions under which sludge is dried in the C-G process render it sterile and free of pathogens and viruses. This is of interest since cases of illness have been reported in plants converting dewatered sludge to compost.

ECONOMICS

Table 3 lists the capital and operating costs for each of the major processing steps. This Table is derived from one prepared by Haug in mid 1984. No significant differences are expected, although the original 10 percent contingency included for capital may have been eroded somewhat in the nearly two years since the estimate was made.

Table 3. ECONOMIC ANALYSIS HERS PROJECT
JULY 1984
($ X 1000)

	WAS THICKENING	ANAEROBIC DIGESTION SCREENING & CLEANING	CENTRIFUGE DEWATERING & CAKE STORAGE	CAKE DRYING CARVER-GREENFIELD	SDF GASIFICATION COMBUSTION & POWER GEN.	DIGESTER GAS COGENERATION	UTILITIES SUPPORT FACILITIES & MISC. Note 2	TOTAL Note 1
Total Capital Cost	4,339	20,940	15,461	38,752	34,364	57,435	37,647	208,938
Annual Capital Charge	442	2,132	1,574	3,947	3,500	5,850	4,242	21,687
Operation & Maintenance Costs								
Labor	---	1,720	758	909	1,489	1,489	---	6,365
Fuel @ $1.30/gal.	---	---	---	---	---	25	95	120
Electricity @ $0.061/KWH	171	259	566	803	710	992	348	3,849
Polymer @ $2.00/lb.	323	---	2,444	---	---	---	---	2,767
Chemicals	---	---	40	51	530	200	50	871
Fluidizing Oil @ $1.90/gal.	---	---	---	541	---	---	---	541
Maintenance Material	27	148	134	316	450	386	68	1,529
Dump Charge	---	106	---	---	---	---	493	599
Total Annual O & M Cost	521	2,233	3,942	2,620	3,179	3,092	1,054	16,641
Credit for Electricity @ $0.061/KWH	---	---	---	---	(5,408)	(7,989)	---	(13,397)
Net Annual O & M Cost	521	2,233	3,942	2,620	(2,229)	(4,897)	1,054	3,244
Total Cost, O & M and Capital	963	4,365	5,516	6,567	1,271	953	5,296	24,931

Annual Operation & Maintenance Cost per Dry Ton $ 21.94
Total Annual Cost per Dry Ton $168.65

<u>Note 1</u> After allowing for Federal Grants, the Capital Cost to the City of Los Angeles is $32,488,000, which reduces the total annual cost per dry ton to $45.11

<u>Note 2</u> The Utilities, Support Facilities category includes the Distributed Control Instrument System, Air Compression, Cooling Water, Site clearance, Warehouse, Ash hauling & Mobile equipment.

Except for some mobile equipment charged off at 25 percent per year, the annual charge due to capital is 10.19 percent of the total cost.

The total cost, including capital, is $169.65 per dry ton of sludge fed. However, nearly 85 percent of the capital cost for HERS is supplied via Federal Grants so that the actual total cost to the City, including capital, is only $45.11 per dry ton of sludge fed.

Another noteworthy aspect to these costs is that, due to the revenue from electricity, the Operating and Maintenance cost is only $21.94 per dry ton of sludge fed to the complex.

BIBLIOGRAPHY

Anonymous
Power from Sludge
<u>Heat Engineering</u> (Foster Wheeler Corporation House Organ), Vol. L, No. 8, pages 118-122, (1982), Oct.-Dec.

Clark, C. S. et. al.
Biological Health Risks Associated with the Composting of Wastewater Treatment Plant Sludge,
<u>Journal WPCF</u>, Vol. 56, No. 12, page 1269-1276 (1984) Dec.

Haug, R. T.
Introduction to HERS Project
Training Manual for Steam Electricity Generating Plant Operators, Rev. dated Oct. 24, 1984

Haug, R. T. and Sizemore, H. M.
Paper presented at International Conference on Thermal Conversion of Municipal Sludge, Hartford, CT, March 1983

DISCUSSION

L. Mansueti, American Public Power Association

Is the cogeneration facility, owned by the City or owned by Hyperion?

Greenfield

The Hyperion treatment plant is owned by the City of Los Angeles, so the cogeneration plant belongs to the City of Los Angeles.

COMPARATIVE ENERGETICS AND ECONOMICS OF ALTERNATIVE PATHWAYS FOR FUEL PRODUCTION

Stanley R. Bull
Solar Fuels Research Division
Solar Energy Research Institute
Golden, CO 80401

ABSTRACT

An alternative description of conventional fuels, biofuels, and future fuels is provided. The approach taken builds on the commonality of those fuels' origins and the relationship of those origins to their end use. The alternative pathways for fuel production are defined and represented in a logical, visual scheme. All the fuel pathways begin with the basic components of light (hν), water (H_2O), and carbon dioxide (CO_2) and through a variety of processes, result in a utilizable fuel. The conventional fossil fuels, natural gas, petroleum, and coal, were formed by the synthesis of hν, H_2O, and CO_2 into organic matter and the subsequent fossilization over millions of years into underground deposits of these fuels. Biofuels pathways are routes to fuels with the same starting point, but with the time frame for production of a few years or less. Detailed biofuels alternative pathways that integrate the production and conversion steps are defined. Future fuels pathways are direct routes to fuel in the sense that no living organism is required and only a synthetic system, such as a photoelectrochemical system, is used to produce fuels from the three basic components. The utility of the pathway approach for developing a perspective on the opportunities for fuels from biomass as well as future fuels will be illustrated and can be applied with comparative mass and energy flows, economics, energy contributions, and production/conversion efficiencies.

COMPARATIVE ENERGETICS AND ECONOMICS OF ALTERNATIVE PATHWAYS FOR FUEL PRODUCTION

INTRODUCTION

The alternative pathways for fuel production from renewable resources are numerous, and hence many opportunities are available for research and development (R&D) attention. Figure 1 is a typical representation of the many options for biofuels available and the complexity associated with managing an R&D program for biofuels technologies. Given the relationships between the four feedstock categories and the three conversion alternatives (including direct conversion), a total of nine feedstock/conversion options are possible. If the numerous alternatives for each of the feedstock and conversion categories are considered, the number of options multiply very rapidly. An alternative description of pathways is provided here to place conventional fuels, biofuels, and future fuels in a common perspective. Examples are given of the use of this descriptive approach for comparisons and for providing more detailed information.

FUELS ALTERNATIVES

Figure 2 illustrates the relationship between conventional fuels, biofuels, and future fuels. All of the fuel pathways have their origin with the basic components of light (hν), water (H_2O), and carbon dioxide (CO_2) and, through a variety of processes, result in a utilizable fuel. All fuels have their origins in biological processes with the possible exception of nonbiogenic hydrocarbons as advanced by T. Gold.[1] Pathways on the left side of Figure 2 represent mature or existing routes to fuels, and moving from left to right the pathways represent successively more novel, higher potential, but less developed technologies. The pathway on the far left represents conventional or existing fuels from fossilization of organic matter, which provides the fuels we know as natural gas, petroleum, and coal. While in principle this is a renewable fuel route, because of the time required for fossilization this is not a practical renewable route.

The central shaded area of Figure 2 represents the set of fuel production pathways typically designated as biofuels. The goal of R&D efforts in biofuels is to provide alternative pathways to the fossilization route that shorten the required times to practical values. The biofuels pathways are the main subject of this paper, and additional details are provided in succeeding sections.

The pathway on the far right in Figure 2 is labeled "future fuels" and refers to the use of biomimetic systems for fuel production. Biomimetic systems are direct routes to fuel production in that no living organism is required and only a synthetic system, such as a photoelectrochemical system, is used to produce fuels from the three basic components. The advantage is the elimination of the need for organism maintenance and the substitution of an artificial, nonliving system that mimics the photosynthetic process and simultaneously converts the photosynthetic product to a utilizable fuel that provides yields exceeding natural systems. The potential of this pathway is enormous but is clearly in the very early stages of development.

With Figure 2 as a basic framework, many variations are possible for descriptive purposes. Some examples of such applications are given in the following sections.

BIOFUELS ALTERNATIVES

The shaded area of Figure 2 is shown as Figure 3, which shows only the biofuels pathway alternatives. Two major production pathways are shown. One is the use of photosynthetic organisms through "natural and managed production" to produce organic matter such as trees, grain crops, and numerous others. Essen-

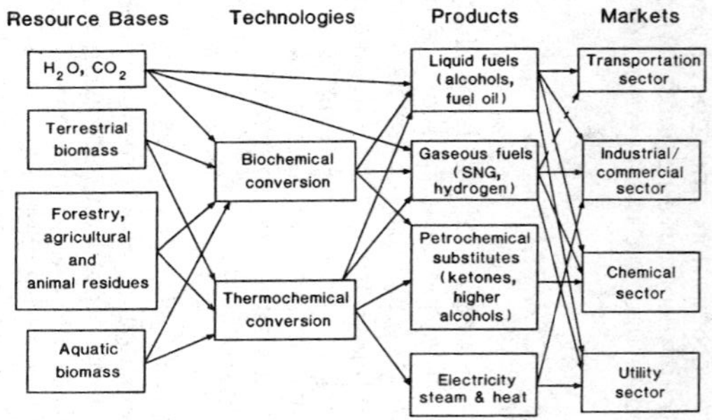

Figure 1. Biofuels Options.

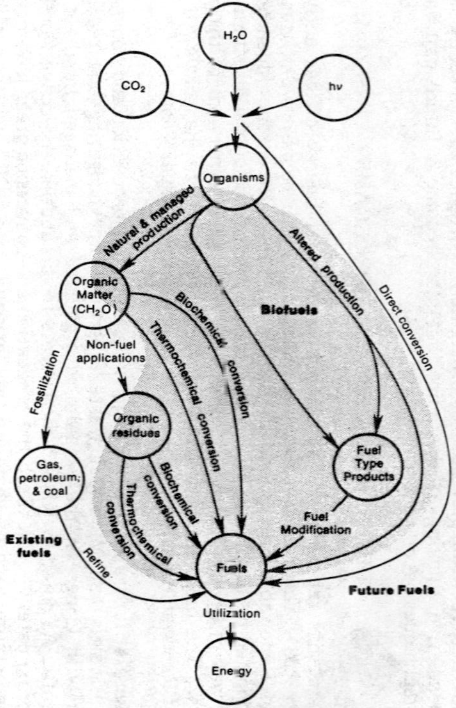

Figure 2. Fuels Alternatives.

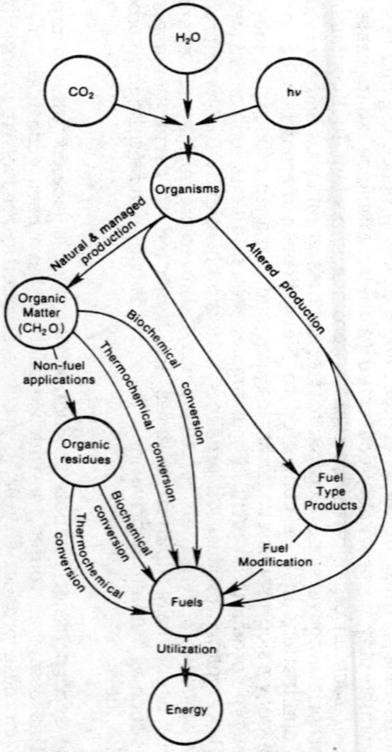

Figure 3. Biofuels Alternatives.

tially all organic matter or biomass produced in the U.S. economy is for nonfuel applications today. Most of these nonfuel applications produce organic residues, some of which may be profitably converted to fuels. Municipal solid waste is an example of such an organic residue that can provide important fuel contributions with appropriate conversion. Organic matter such as grain crops and trees grown for fuel applications may be converted to fuels through thermochemical or biochemical conversion processes. Other natural and managed production may produce fuel-type products directly. For example, seed oil plants such as castor bean and sunflower plants produce oil products that are chemically fuel-type products and may be converted to fuels relatively easy. Lower organisms such as microalgae will accumulate lipids if cultures are subjected to certain management practices such as nutrient deprivation. These lipids can then be refined to produce diesel-like fuels.

The second major biofuels production pathway is labeled "altered production." This describes routes for which organisms have been genetically manipulated to increase productivity or to produce a specific product. It is anticipated that microalgae may be genetically manipulated to increase their lipid accumulation capacities, for example. Altered production to produce fuels directly represents speculative thinking at this point, but may become a future pathway as genetic techniques become more advanced.

DEPARTMENT OF ENERGY PROGRAMS

Klass[2] has reported on the U.S. Department of Energy offices active in bioenergy R&D. The major research expenditures are managed by two offices, the Office of Renewable Energy and the Office of Energy Research. Much of the R&D supported by the Office of Energy Research is focused on the direct conversion pathway of Figure 2, although some R&D is directed to portions of the other pathways. The biofuels portion of Figure 2 corresponds to the purview of the Office of Renewable Energy program in Biofuels and Municipal Waste.

Figure 4 illustrates the two major components of production and conversion for the Biofuels and Municipal Waste Program. For budgeting and reporting purposes with Congress, the program is organized and managed by these two components. As noted by Klass,[2] the budgets for production and conversion have been approximately equal. Both the production and conversion areas have component subprograms, each of which are field managed by a DOE laboratory as shown on Figure 5. Not shown on the chart are the regional programs that have a role encompassing both feedstock production and conversion of biomass to fuels through projects specific to their respective regions. Also not shown is an overall analysis and economics support function managed by Argonne National Laboratory.

COMPARISONS

Figures 6-10 illustrate the kind of comparisons that can be made using the basic framework of Figure 3. The comparisons shown are energetics, economics, and detailed process economics for woody feedstocks and prospects for R&D improvements for microalgae lipid production. It should be noted that while quantitative comparisons are provided, the examples provide only approximate values and should be treated as such. Nevertheless, they are sufficiently accurate to demonstrate the utility of this approach to provide a perspective on numerous aspects of biofuels R&D activities.

Figure 6 shows the relative energetics for production of liquid fuels, specifically alcohols, from woody feedstocks. The cross-hatched areas are proportional to the amount of the energy represented on each path segment. The production path indicates a basis of 1 Btu as the energy content of the woody feedstock. The energy content of the product from biochemical conversion by acid hydrolysis to sugars and subsequent fermentation to ethanol is 0.25 Btu. Alternatively, the energy content of the product from thermochemical conversion by gasification to

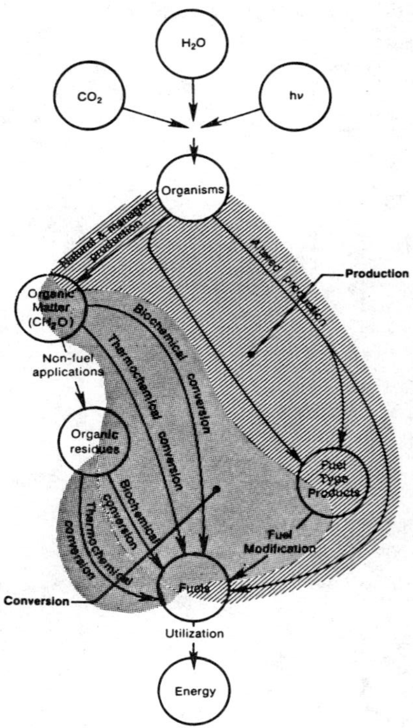

Figure 4. Biofuels Programs.

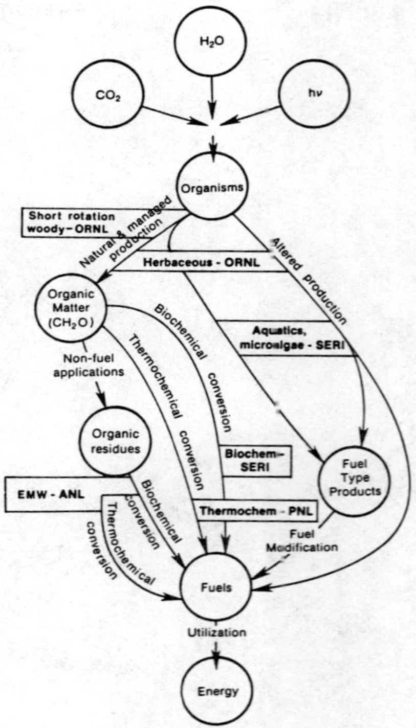

Figure 5. DOE Biofuels Field Management.

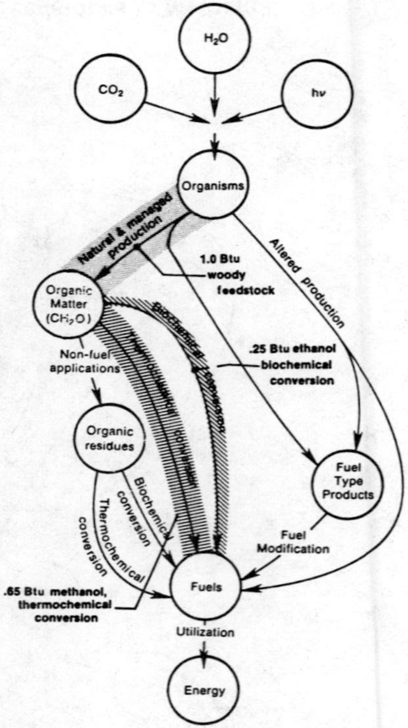

Figure 6. Woody Feedstocks to Liquid Fuels—Energetics.

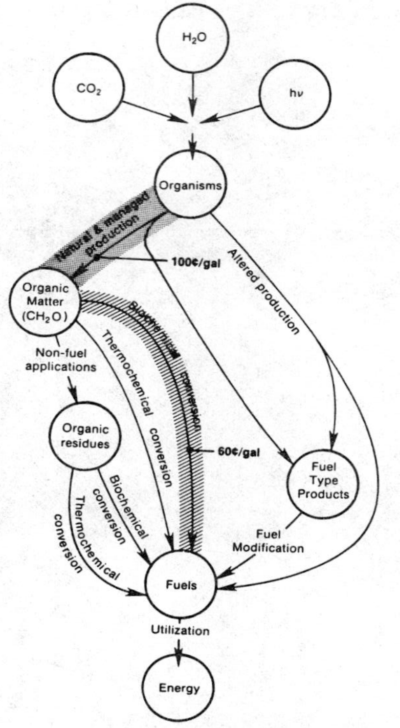

*Includes combustion fuel for electricity

Figure 7. Woody Feedstocks to Ethanol— Biochemical Conversion Economics.

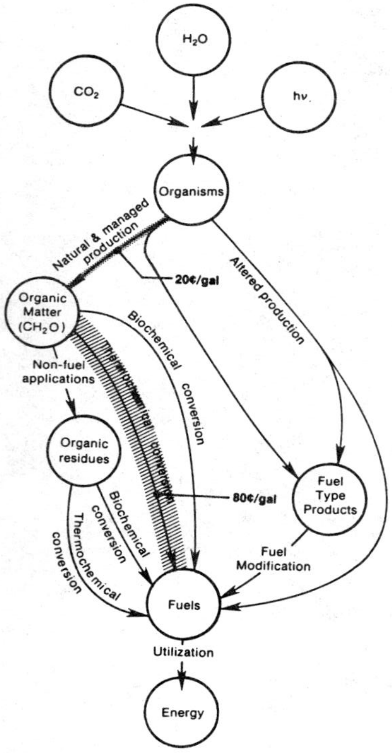

Figure 8. Woody Feedstocks to Methanol— Thermochemical Conversion Economics.

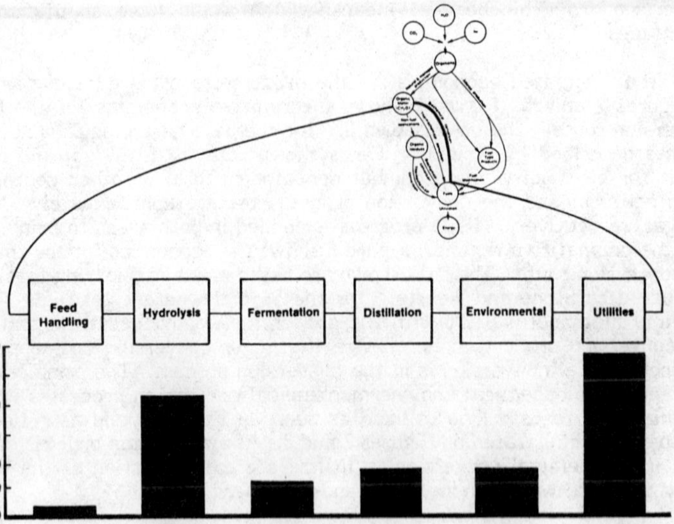

Figure 9. Woody Feedstocks to Liquid Fuels—Ethanol by Acid Hydrolysis Process.

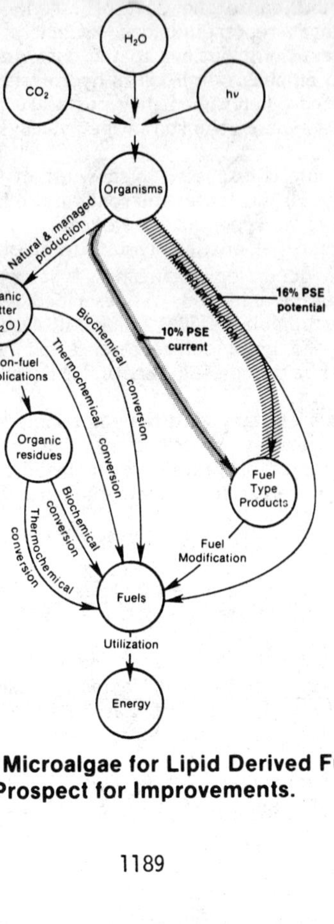

Figure 10. Microalgae for Lipid Derived Fuels—Prospect for Improvements.

syngas and subsequent conversion to methanol is 0.65 Btu. An obvious conclusio from this comparison is that there is an opportunity to improve the efficiency o conversion from biochemical means, which would have significant economi advantages.

The associated economics for the processes considered above are compare in Figures 7 and 8. Figure 7 shows the relative economics for the biochemica conversion route. The costs shown are for the base-case plug flow acid hydrolysi process described by Wright.[5] Costs shown total to $1.60/gal and include onl credit for electricity produced, but none for furfural or other coproducts. Th feedstock cost and the conversion process cost are approximately $1.00/gal an 60¢/gal, respectively. R&D progress is needed in both areas to bring the proces into the competitive regime. Figure 8 shows the economics for the thermochemical conversion route. The data shown are based on an engineering feasibility stud conducted by Stone and Webster[3] for the SERI downdraft gasifier. The cost pe gallon of methanol is $1.00 with 20¢/gal for the woody feedstocks and 80¢/gal fo the conversion process costs. Hence the major conversion cost opportunity fo thermochemical conversion is in the conversion process. The principal differenc between the biochemical and thermochemical conversion processes is the degre to which the feedstock is utilized as seen on Figure 6 and as reflected in th relative costs illustrated on Figures 7 and 8. As a result, major R&D opportunitie exist in biochemical conversion to utilize a greater fraction of the woody feed stock, principally the hemicellulose and lignin fractions.

Further details for comparisons can be illustrated as shown on Figure 9. Th example selected is the acid hydrolysis plug flow reactor base case described b Wright.[5] The major conversion process steps are shown in schematic form, an the associated contribution to the cost of ethanol production is shown by th accompanying bar chart representations. Clearly the hydrolysis and utilitie sections are the largest contributions to the cost, totaling approximately 75% c the process cost. An emphasis on R&D in hydrolysis processes is indicated. Th is further substantiated when the utility costs are allocated to the appropriat process step as about 50% are required in the hydrolysis process.

Finally, an example is discussed to show use of this scheme for prospects fc improvements. Figure 10 illustrates microalgae production for lipid-derived fue and follows technology as reported by Neenan.[4] Current technology, which achieved by culture management, has resulted in a photosynthetic efficiency (PE of 10% with naturally occurring organisms. It is anticipated that with genetic PEs as high as 16% can be achieved. As described by Neenan[4], such an efficienc in an organism that accumulates 50% lipids will result in an economically viabl technology. Hence, the graphical representation here illustrates that for th microalgal program to be successful, genetic R&D appears to be needed.

Many other types of representations not included here are possible such a mass flows, potential energy contributions, and conversion efficiencies. It intended that the approach presented here will aid in describing alternative bi fuels pathways and provide for appropriate comparisons among those pathways.

REFERENCES CITED

1. Gold, T., E. Bilson, and S. Soter, Studies Related to Deep Earth Gas, GR 80/0066, Chicago: Gas Research Institute, 1981.

2. Klass, D. L., "Energy from Biomass and Wastes: 1984 Update," Symposiu Papers Energy from Biomass and Wastes IX, Lake Buena Vista, FL, Januar 30-February 3, 1984 (Institute of Gas Technology, Chicago, IL, Septembe 1984), pp. 1-67.

3. Kosstrin, H. M., "Economic Feasiblity of a Small-Scale Wood Gasification t Methanol Plant," Symposium Papers Energy from Biomass and Wastes I

Lake Buena Vista, FL, January 30 - February 3, 1984 (Institute of Gas Technology, Chicago, IL, September 1984) pp 1305-1336.

4. Neenan, B. F., and D. A. Feinberg, "Fuels from Microalgae: An Assessment of Technology Requirements," Symposium Papers Energy from Biomass and Wastes X, Washington, D.C., April 7-10, 1986 (Institute of Gas Technology, Chicago, IL).

5. Wright, J. D., and A. J. Power, "Comparative Technical Evaluation of Acid Hydrolysis Processes for Conversion of Cellulose to Alcohol," Symposium Papers Energy from Biomass and Wastes X, Washington, D.C., April 7-10, 1986 (Institute of Gas Technology, Chicago, IL).

DISCUSSION

J. Kerstetter, Washington State Energy Office

From your perspective, where do you think the best paybacks are?

Bull

There still needs to be some quantitative study of the various options. I've only given you a few of them. At SERI, we have just recently completed studies on alcohol fuels and microalgae. What we did was to say okay, let's look at what we see as the current technology cost and then try to identify the opportunities. I'm a little hesitant to say too much about what the implications of the current ethanol study are, but generally speaking, we can say that we certainly have to utilize more of the feedstock. That's the first mistake we're making because it costs a lot to deliver the feedstock. If only 50% is used for fermentation alcohol, we need to be concentrating a significant part of our resources on both the xylose fraction and the lignin components. For the microalgae study, we first had to ask ourselves whether both water and CO_2 are available, because those are the two main resources that one needs to be able to culture microalgae in large-scale systems. In addition, can we achieve the kind of photosynthetic efficiencies and concurrent lipid production or fractions that we need? At this time, the program involves organism collection and evaluation. We expect this to be followed by a genetics program. In one of the diagrams, I tried to show where we are now with the current photosynthetic efficiencies, and subsequently where we need to go.

W. Heckrodt, Kimberly Clark

Have you looked at any of the new ester pulping delignification systems for sidestepping the hydrolysis to go directly to the lignin and cellulose?

Bull

We keep a pulse on the forest products industry activities, and look at the various processes, but as far as I know, none of those has offered any alternative that appears more promising than our current areas of emphasis. Some of the pretreatment steps we use for the enzyme work provide a cleaner lignin stream for example. In the acid hydrolysis areas, you don't end up with a by-product that's very useful, and we simply burn it for the most part. There are other options, such as furfural and other by-products.

PERSPECTIVES ON CONVERGENCE IN NET ENERGY ANALYSIS

Sheldon J. Reaven, Ph.D.
Department of Technology and Society
College of Engineering and Applied Sciences
State University of New York at Stony Brook
Stony Brook, New York 11794

ABSTRACT

Net energy analysis requires that there be a point beyond which sufficiently remote (i.e., indirect) energy flows do not significantly affect the ultimate net energy estimate. The need to firmly demonstrate convergence is especially pressing for integrated technologies for using biomass and/or organic wastes. However, the arguments for convergence, drawn from process analysis and input-output analysis, appear to be inconclusive on closer examination, for reasons involving possible structures of the system of energy flows. This conclusion has implications for the role of net energy analysis in deliberations about energy policy.

PERSPECTIVES ON CONVERGENCE IN NET ENERGY ANALYSIS

INTRODUCTION

Many new systems for using biomass and organic wastes are multiple-purpose (providing, e.g., fuel, feed, food, industrial chemicals, and cogenerated electricity) and integrated (e.g., "waste" heat and by-products from one part of the system are used elsewhere in the system, aiming at optimal thermodynamics and minimal adverse environmental impact for the toal system)(1, 18). The net energy analysis (NEA) of such systems is complex: the new system typically replaces several "old" processes, materials, or energy flows at once; system boundaries are harder to draw, owing to multiple direct and indirect energy interactions with the ecosystem and socioeconomic system; the system's relatively tightly coupled internal organization complicates accounting for joint products (17, 20).

There are marked differences of opinion among net energy analysts concerning what we shall call the <u>qualitative</u> problem -- what types of energy flows should be included or excluded in arriving at a net energy estimate? -- and the <u>metrical</u> problem -- what measures, in what energy units, should be used to quantify the energy flows individually and (e.g., as an index of net energy) collectively? (13, 15). Underlying the range of solutions proffered to these problems is a conflict among three research interests that have not always been clearly distinguished: finding out the overall <u>difference</u> the technology or facility makes to society's usable energy resources (so comparing energy flows occurring with the technology with those that would occur without it, or with other technologies); rating the overall engineering performance of the technology as a set of physical processes (so comparing actual, physically essential inputs with actual, physically essential outputs); and understanding the overall performance of the technology as it interacts with its entire surroundings, broadly construed to include everything from sunlight to flora, fauna, and ecosystem services, to other human technologies, and to social and economic structures (so comparing <u>all</u> inputs with <u>all</u> outputs)(14). Differing energy balance estimates for a particular technology or facility often are traceable to tensions among these three research interests.

Each approach assumes, however, that whatever the energy flow selection criteria governing the calculations, as one extends their <u>scope</u> (so as to include, e.g., the energy it takes to make the glass that makes the windows in the factory that makes the tractor that harvests the corn that is to be converted to ethanol, or so as to include, e.g., ever more indirect environmental and social "ripple" effects of operating the ethanol system), there must be a point beyond which such remote energy flows do not significantly affect the final net energy to be calculated. Unless this convergence assumption holds -- unless progressively more remote energy flows have progressively less cumulative effect on the net energy -- NEA would be faced with interminable regresses into the past (to determine embodied energies) and future (to assay environmental impacts or mitigation activities, and other future energy costs), associated with the energy inputs to and outputs from the system under study. The complexity of the computations and the uncertainty or unavailability of the data, as causally more distant flows are assessed, would present nearly insurmountable problems in short order. NEA would be unmanageable unless distant energy flows can be safely ignored, or "truncated".

We shall examine the two main lines of argument for the claim that the sequence of net energies one calculates as one includes causally more distant flows must converge. One involves process analysis, the other derives from input-output analysis. Although the discussion will be general, it needs to be stressed that the question of convergence is significant for many biomass and waste technologies, because of their great "leverage" on past, present, and future energy flows in many domains of activity. Examples of such large actual or potential ripple effects of biomass/waste technologies involve fertilizer production and use; genetic technology; food, diet, and agricultural economics; forest extent and soil quality; substitution of biomass- for petroleum-derived products throughout the industrial and chemical systems; increased and more efficient coal use (e.g., by cofiring a mixture of 50% wood and 50% high-sulfur coal, so greatly reducing undesirable emissions); curbing the "greenhouse effect"; rural and municipal land use; and societal decentralization (6,9,11,1,5,16,8,17). Moreover, the net energy picture of several biomass technologies seems highly sensitive to the boundary of the analysis (e.g., 2,1).

A MODEL OF THE SYSTEM OF ENERGY FLOWS

Let T be the technology under study (here an integrated, multiple-purpose facility based on biomass and/or organic wastes, or a network of such facilities). Let NE(T) = the net energy of T = f[EG(T), EC(T)], where f is a function satisfying the requirements of the metrical problem (13,15), EG(T) = energy generated by T, and EC(T) = energy cost of T. EC(T) "includes" EE(T), the energy embodied in T. Let TS = a total system in which T is embedded. Depending on the research interest, TS can be, variously, the collection of all energy conversion technologies in the nation or world, the collection of all technologies in the nation or world, or the nation or world itself (i.e., all energy flows in nature and civilization).

To avoid extraneous complications, I shall assume that the energy researcher can determine the precise quantity, intensity, and form of each energy flow, past, present, or future, wherever it occurs (as if energy analysts had a system of energy metering devices, like Einstein's rod-and-clock systems, planted in times and places of interest). I also assume that the totality of energy flows can be partitioned into at most denumerably many individual flows such that each is correctly assigned to EG(T), EC(T), or neither. This assignment is assumed to resolve metrical issues (especially the joint product problem) in some reasonable way.

Except where noted, the discussion will revolve around energy flows that (1) take place in the past, so comprising EE(T), and (2) are counted under all three NEA research interests. If convergence cannot be guaranteed even for embodied energies, it cannot be guaranteed for full dress net energies. These provisions allow the criticisms of the rationale for truncation to be formulated somewhat more directly.

As the scope of a NEA expands, the series of intermediate estimates of NE(T) calculated at succeeding stages can be thought of as corresponding to a series of rings around T, as in Fig. 1. Thus $NE_i(T) = f\,[EG_i(T), EC_i(T)]$, where the $EG_i(T)$ and $EC_i(T)$ are, respectively, the total energy yield and total energy cost of T summed over all flows occurring in all rings up to and including the ith ring. Let S(T) be the sequence $NE_1(T)$, $NE_2(T)$,..., $NE_i(T)$,.... NE(T) exists if and only if S(T) converges, in which case $NE_i(T)$ tends to NE(T) as i increases indefinitely.

Let global convergence mean that S(T) converges for every T. Global convergence implies that TS has a well-defined net energy. When $T_a \neq T_b$ implies, in general, that $NE(T_a) \neq NE(T_b)$, when these net energies exist, we shall say that local convergence obtains. The distinction between local and global convergence mirrors the possibility that the structure of TS is such that all or nearly all technologies have identical net energies. In such circumstances, its net energy would not be a distinctive property of a technology, and one technology would not in general have a long-term energy advantage over another. Studies employing truncation that seem to show that two technologies have different net energies could be questioned on the grounds that the differences are merely temporary reflections of differential rates of convergence.

(It is mathematically possible for the $NE_i(T)$ to converge while both the $EG_i(T)$ and $EC_i(T)$ do not. Physical considerations ensure that such degenerate conditions will not arise for reasonable choices of f, and that if the $EE_i(T)$ do not converge, neither do the $EC_i(T)$, nor the $EG_i(T)$, nor the $NE_i(T)$ (15). If the arguments for eventual quantitative insignificance for sufficienly remote flows are invalid when applied to embodied energies, they are invalid for net energies.)

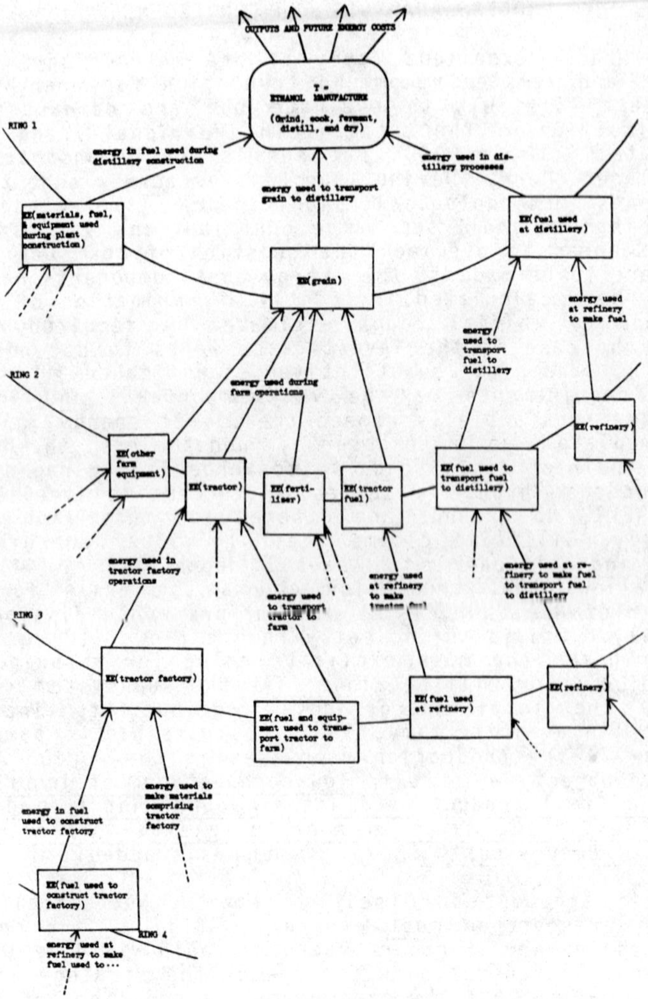

Figure 1. Energy flow model of a (simplified) technology for making ethanol from biomass, Each arrow represents an energy flow. The quantity of energy corresponding to each is added to either EG(T) or EC(T). following some systematic procedure, and is assumed to conform to the requirements of the metrical problem concerning pro-rating, differences of energy form, measures of energy content, etc. Many flows are omitted, including future flows, flows involving farm wastes and distillery by-products, flows involving other components of the integrated, multiple-purpose system of which this is a part, and flows representing the energy "in" human labor, solar influx, knowledge, and the ecosystem (which some energy analysts count). "Fuel" refers to heat and electricity, as well as coal, gasoline, natural gas, etc. as the case may be. EE = embodied energy. The boxes themselves are not included in the calculations of NE(T).

PROCESS ANALYSIS AND CONVERGENCE

In their excellent study of NEA methodology, Perry, Devine, and Reister recommend truncation for energy flows "that are difficult to evaluate but are demonstrably a small fraction of the total (both individually and in the aggregate)" (12, p. 10). But how is the demonstration to take place? Perry, Devine, and Reister acknowledge the question's methodological significance ("references...to such things as small and large contributions as truncation errors appear to overlook the question of how such judgements are to be made if the energy cost components have not already been calculated. Evidently a combination of judgement and of initial rough estimates is required, as is always the case if the investigator wants to use his time wisely"), and its quantitative significance ("indirect energy requirements can be very important, and in some processes substantially exceed the direct energy inputs"), but immediately go on to observe, "nonetheless, the hierarchy of indirect costs tends to converge fairly rapidly, as one proceeds through a series of successive levels" (12, pp. 36-37). No grounds are offered for this last claim. Similarly, Gilliland claims that individual energy flows become insignificant with causal depth (7, p. 44), and Leach allows that truncation errors can arise for some versions of NEA, but claims such errors would disappear if the analyst assays sufficiently remote flows (10, p. 154). Chapman gives the most explicit recipe for ensuring that truncation error will be minor: "As the sub-system is made larger, the total energy cost continues to increase. However, in a finite time, it is not possible to take into account all the production processes in the world. A more feasible objective is to follow each network of inputs back from the final product until it is found that the addition of the next input makes an acceptably small difference to the total energy cost" (3, p. 3, emphasis added).

This argument in itself is not a theoretical proof that the researcher must eventually find a link on each causal chain where further traceback would yield negligible dividends. Rather the significance of the strategy lies in the implication that the energy analyst can find out empirically that S(T) converges for a particular T. I.e., NEA discovers (rather than presupposes) convergence, one case at a time -- in which case global convergence becomes a reasonable generalization to make.

Let us apply Chapman's technique to the model. Chapman does not spell out the exact procedure. Are we to pick a path proceeding outward from an initial branch in ring 1, calculate running net energies en route, stop when a branch is reached that neglgibly affects the net energy calculated up to that point; then carry the running totals for EG(T) and EC(T) to a second branch, again working outward, so iterating until a stopping point is reached for every path leading outward from ring 1? This literal realization of Chapman's method is untenable, as it would lead to absurd truncations for at least the first few paths examined.

Another possibility is to calculate a sequence of net energies ring by ring, beginning with ring 1, proceeding (say) counterclockwise within each ring and taking each branch in the ring in turn, all the while adding the energy flow corresponding to each branch to the running totals for EG(T) and EC(T). When all flows in a given ring are counted, the current totals are carried to the next ring to be assayed. If some branch in ring k does not significantly affect the running net energy calculated up to that point, the branch in question is dropped from further consideration and (say) colored red, along with all branches in all rings beyond ring k connected to that branch by an unbroken path. This process continues until a ring is reached whose branches all have been colored red.

This procedure will not suffice, because it does not preclude the possibility that one or more ignored "outrider" flows lying beyond the innermost completely red ring is large enough to alter the net energy significantly. The difficulty lies in the nature of embodied energy. Consider the energy embodied in the tractor in figure 1. EE(tractor) equals the sum of the energy contents of all flows in the indefinitely extended tree sprouting outward from the box representing the tractor. This total is not necessarily approximately equal to the sum of the energy contents of the flows leading inwrod to the box from the adjacent ring. *Ex hypothesi*, Chapman's method does not evaluate these entire trees. Yet it is constitutive of NEA that *certifiably* approximate embodied energies be ascertained. Chapman's method presupposes convergence instead of revealing it.

Another difficulty is that even though the radially innermost red flows are individually minor, they could have a major cumulative effect on the net energy. This is the point of Perry, Devine, and Reister's stipulation that truncation only follow upon a demonstration of the insignificance of remote flows "in the aggregate." Chapman's method could be modified accordingly: the algorithm now continues until a band of flows is reached which "surrounds" all flows in its interior and whose collective effect on the net energy (calculated for the interior flows) is trivial.

However, the modified procedure may not halt -- where is the proof that the requisite encompassing band exists? Moreover, in principle, there always could be flows *beyond* the encircling belt that singly or jointly significantly affect the net energy calculated to that point. As i increases, the effect of the proliferating number of flows could swamp that of any decrase in their individual magnitudes.

Thus convergence cannot be established merely on the basis of empirical observations of the energy flows in successive rings. However, one might reply that at least embodied energies cannot grow inordinately large with causal depth, because (a) common sense indicates that distant flows eventually must become insignificant, and (b) the total embodied energy in human history provides a physical limit that ensures convergence.

(a) Consider again R = EE(tractor). R simply cannot be too large, because the energy cost of the tractor factory, a component of R, must be divided among the (say) 500,000 tractors produced during its lifetime, and because similar considerations apply to other components of R. Thus the energy embodied in the fuel that runs the tractor can only be a small fraction of the energy embodied in the refinery that produced the fuel. Production systems involve dilution factors of this sort, so the argument would run, throughout society. The intuition being courted is that on the whole TS is organized in a way that diffuses the "draw" (past, present, and future) of a technology on energy resources. Fig. 1 displays this assumed spreading out of the energy costs and benefits.

There are two difficulties with such a picture. (1) It remains possible that the energy cost of the tractor factory really is large enough to outweigh the effect of dilution factors, owing to unexamined "outrider" flows. More importantly, (2) the intuitions suggesting that dilution factors predominate is nullified by equally plausible ones suggesting the opposite. There are items, such as the oil refinery, appearing in several places in Fig. 1. The energy content of the fuels produced by the refinery is divided among many branches (not all shown in Fig. 1), many of which ultimately lead inward to T and are charged to EC(T). Assume for the sake of simplicity that TS contains only one oil refinery and that oil is the only raw material used as an energy source in TS. Could not TS then be so constituted that all products of the oil refinery are connected to T, albeit often by long paths? The energy flows comprising the energy embodied in the refinery (along with the energy content of fuel produced by it) would sooner or later find themselves in the energy accounts of T. A similar possibility applies to the tractor factory. Some of the energy costs of the 499,999 tractors not used directly in T ultimately could redound to T's energy accounts. A tractor shipped abroad, for example, may be used to build a steel plant that makes distillery vats used by T. Such pathways would be very circuitous, often involving comparatively small amounts of energy individually. But there may be large numbers of such pathways.

Hypothetical situations like those outlined will be referred to as examples of reconcentration. For all we know, reconcentration may predominate (over diffusion) within TS. The claim is not that reconcentration does in fact characterize the structure of TS, either globally or locally (e.g., in the "neighborhood" of T), or that it overwhelms the opposing effects of diffusion. Rather the point is that the view that diffusion must typify the structure of TS is based on an intuition that depends on a local perspective for its force: energy flows seem to spread from T in its immediate neighborhood. A global perspective, on the other hand, raises the possibility of extensive reconcentration. The locally-grounded intuition alone does not justify the claim that S(T) converges.

An analogy to the structure of genealogical trees is instructive. The number of one's ancestors seems to grow rapidly (diffusion) as more ancestral (embodied energy) generations (rings) are assayed. So it appears from the local perspective assumed when tracing one's family tree backwards from oneself. Eventually, however, in part because of the finite total number of human beings, one's ancestors tend increasingly to be related to one another. Several centuries back, the family tree begins to shrink (reconcentrate) in a phenomenon known as "pedigree collapse" (19). The entire family tree is diamond-shaped. Similar unexpected large-scale topologies could characterize parts of TS incorporating T.

(b) The second objection to the view that process analysis does not estabish convergence asserts that the total energy embodied in TS is finite, and that this constrains the behavior of the $NE_i(T)$. The crucial insight is that the total annual <u>direct</u> consumption of energy content in human society has mushroomed from the near-zero levels of Promethean times to the Promethean scale of today. The annual consumption is finite, so the annual increase in EE(TS) is finite, so EE(TS) and <u>a fortiori</u> EE(T) are finite. Therefore at some point remote energy flows become unable to significantly change estimates of EE(T).

This argument is beside the point. It imposes convergence on S(T) in a mathematically trivial way, by making the <u>sequence</u> of $NE_i(T)$ finite. Nothing prevents S(T) from being highly erratic until the very end. The argument provides no reason to expect the gradual attenuation (upon which truncation depends) of the fluctuations in the sequence. Another problem is that the argument does not hold for future flows belonging to EC(T) or EG(T). There are many schemes for using small amounts of energy in the present as Archimedean levers triggering vast energy changes in the future (even allowing for dilution factors). Weather control, space colonization, war, agriculture, and various energy technologies provide examples (13,14,15,17).

Convergence, then, cannot be demonstrated empirically by process analysis, due to (a) the possibility of anomalously enormous outrider flows; (b) the possibility that the quantitative effects of proliferation in the sheer number of paths to be analyzed as ever more distant flows are assessed outruns that of any dwindling occurring along each of them; (c) the possibility that reconcentration overwhelms diffusion.

It follows that process analysis cannot provide evidence for global convergence. Local convergence fares no better. Even assuming global convergence, it does not follow that different technologies will have different net energies. The most that can be said of a pair T1 and T2 is that there almost certainly will be many energy flows assigned to the accounts of one but not the other. $NE_r(T1)$ and $NE_r(T2)$ will differ at least for initial values of r, but this does not imply that S(T1) = S(T2). Finally, the prospects for local convergence do not improve if global convergence fails to hold.

INPUT-OUTPUT ANALYSIS AND CONVERGENCE

The appeal to input-output analysis reflects another conception of the structure of the system of energy flows. The radially dendriform image of process analysis is replaced by that of a self-contained hive of circuits connecting system components with one another. A system may turn out to recycle itself superefficiently (yielding a positive net energy), inefficiently (so consuming itself), or perfectly.

Input/output (I/O) analysis is often used in NEA's of large systems, although there are many difficult computational and data requirements. I/O analysis is claimed to provide a theoretical guarantee of convergence. For example, Perry, Devine, and Reister assert that I/O analysis "in principle is free from truncation error" (12, p. 38). What is the guarantee, and what practical import does it have for NEA?

I/O analysis is a modeling technique principally used in economics and developed in its modern form by W. Leontieff and others. The basic idea behind Leontieff models is elegant: an economy is represented as a square matrix A of order n, where the i,j index all sectors of the economy. Each element A_{ij} records the input of money from sector i to sector j per unit output of sector j. The matrix is often called the technology matrix, or matrix of technological coefficients.

The coefficients are considered to be fixed in static Leontieff models (the kind most commonly used in NEA). A is combined with certain vectors and other matrices to form systems of simultaneous equations whose solutions determine the levels of inputs needed to produce given levels of outputs. The accuracy and fruitfulness of static I/O analysis heavily depends on the scope and data base of each application.

Although the term "dynamic Leontieff model" is used in several senses, in the present context it refers to I/O models in which A is allowed to change: in general A becomes a matrix of nontrivial partial differential equations. Many dynamic models used in economics embody simplifications that make them dynamic only in a generous sense of the term. An apropos example is the not infrequent assumption that technological change (and changes in the efficiency and production scale of technologies) is reasonably modeled by varying each A_{ij} linearly with time by the same factor. Yet the more realistic, but more complex, models are comparatively intractable mathematically.

Consider first the static, constant-technology mode of analysis. The transition from economics to energy can be made as follows. Physical quantities are substituted for money in A (this is not always easy since units must be commensurable). A_{ij} now represents the input from sector i to sector j per unit of sector j's output, per unit time; these are converted to energy units. A now models the rates of energy flow within a national, global, or other large-scale technological/economic system. If x is a vector representing a set of n outputs -- say the bill of materials, fuels, and services used to build a biomass and/or wastes system T -- the vector Ax gives the inputs necessary to produce x, A^2x gives the inputs necessary to produce the inputs Ax, etc. The series $K(x) = Ax + A^2x + A^3x + \ldots$ converges (under minor additional conditions) to $[(I - A)^{-1} - I](x)$, where I = the identity matrix. K(x) is the list of all commodities required to produce a specific output within the system covered by the I/O table. EE(T), where T is the technology x represents, is simply the (perhaps weighted) sum of the elements of the vector K(x) that represent energy-consuming sectors of the economy.

The problem with this approach concerns the degree to which the mathematics corresponds to reality. The mathematics shows that (at least) embodied energies converge <u>given a constant, frozen technological/economic system</u>. Yet modern technology changes constantly, often rapidly, often in quantum jumps: the variations in the A_{ij} need not be smooth (15). Embodied energies of real technologies need not behave as nicely as K(x). The loop structure of static I/O analysis sets up convergence by precluding the possibility of encountering surprises as one traces through the causal network, and by precluding the possibility that the proliferating numbers of flows could outrun the effect of their shrinking size individually. But this is an artefact of the modeling: I/O analysis in effect guarantees convergence by imposing global diffusion <u>via</u> the frozen technology matrix.

There are several other difficulties (generallyacknowledged among energy analysts) concerning the appropriateness of static I/O modeling. (a) Much of the economic data is in the form of matrices that omit imports, exports, transportation costs, human labor, externalities, solar inputs, and activities of middlemen. (b) Conversion of economic data into energy units raises questions about the proper relation between money and energy costs for different kinds of goods and services. (c) Sectors are often incorrectly assumed to be independent of one another. (d) The models are linear in the A_{ij}, so that doubling the outputs x automatically doubles EE(x). (e) The effects of aggregation in the model are unknown.

Can dynamic I/O analysis guarantee convergence? Here we encounter problems involving the complexity of the mathematics of complex systems. Assume a dynamic model were to incorporate a set of equations that faithfully model actual energy flows in TS for a significant period of time. The change in the A_{ij} would be mediated by complex partial diffeential equations very likely to be nonlinear in various ways. (And the complexity may well grow explosively with the time interval covered, creating "transcomputable" monsters). It is unlikely that a dynamic model of this kind would reveal much about convergence. Such equations have received a great deal of attention at the frontiers of mathematics and in frontier areas of science where complex systems and intractable systems of differential equations go hand-in-hand. The properties of the mathematical entities involved have been very difficult to fathom (see references in 15).

The relevant questions here are: is there a suitable function (perhaps some form of integration over an analogue of K(x)) that can be used iteratively to compute a net energy? If such a function is found: will the sequence of net energies it computes converge? The properties that ensure convergence with a static A will almost certainly be lacking. If the sequence converges: will it still do so if the system is perturbed by (perhaps arbitrarily small) changes in initial conditions, in the parameters, or even in the form of the equations? If so: will convergence or nonconvergence depend critically upon the T chosen or upon the time interval covered? At the present time, the answer to these questions seems to be "no one can say". Unruly behavior (e.g., punctuated equilibria and chaotic regimes) seems to be highly probable. We cannot be sure that dynamic I/O analysis will establish convergence.

CONCLUSION

In the early stages of a net energy analysis, as one considers in turn direct "process" energy flows, and associated indirect flows from (say) infrastructure repair, facility construction, facility decommissioning or phase-out, environmental impacts and mitigation measures, and so on, the net energy picture is revised significantly and frequently (perhaps swinging from a net yield to a net drain and back again). As noted, this has particularly been the experience with several biomass technologies. Convergence means that the picture will stabilize (and probably do so in reasonably short order). I have argued only that the arguments for convergence are open to question, leaving open the possibility that significant revisions as to the overall energy attractiveness of the technology under study could continue indefinitely (especially if the system of energy flows has an unexpected, even chaotic structure). This conclusion is of some practical relevance to policy makers. That is, more detailed and exhaustive net energy studies, while providing much valuable information about specific flows in the "neighborhood" of the technology, would not necessarily improve the "resolution" of the net energy picture, in that the results need not represent a truer picture of something called the "real" net energy. Such a state of affairs would have the virtue of tempering the confidence greeting the conclusions of earlier net energy "snapshots" in the series. Our overall verdict concerning the technology's energy attractiveness (with respect to any of the three research interests mentioned above) would not in general be stable.

Why, then, have not energy analysts reported such instability? One possible explanation, naturally, is that convergence always does in fact obtain; it is only that an airtight way to demonstrate it has not yet been spotted or invented. But another explanation is that because of the sheer volume of data (and the sheer unavailability of much of it), the sheer number of relevant energy flows, and the sheer complexity of the accounting tasks, involved in systematic and exhaustive assays of increasingly remote energy flows, a full-dress test of the questions raised here has not been forthcoming. Elaborate protocols would have to be devised to ensure the best possible experimental design for such a test. The use of data derived from econometric input-output tables would not settle the issue, since the economic models may themselves embody the assumptions to be tested.

It bears stressing in any event that considerations of net energy are only one element in deliberations about the policy to be adopted with respect to the development of a given technolog (13). Considerations of cost, employment, flexibility, risk, marketability, and simplicity are equally important, and societal and individual values, such as self-sufficiency, regional or national independence, environmental quality, dignity, and fairness are at times overriding. These factors are likely to play a greater role than purely energy characteristics in determining the fate of integrated technologies for using biomass and/or organic wastes.

REFERENCES CITED

(1) Badger, P., Roethli, J., and Stinson, J., "Biomass: America's growing energy resource," U.S. Tennesee Valley Authority: Muscle Shoals, Alabama, 1984.

(2) Chambers, R., Herendeen, R., Joyce, J., and Penner, P., "Gasohol: does it or doesn't it produce positive net energy?" Science 206, 789-795 (1979) November 6.

(3) Chapman, P., "Energy costs: a review of methods," in J. Thomas, ed., Energy Analysis, Boulder Colorado: Westview Press, 1977.

(4) Cleveland, C., Costanza, R., Hall, C., and Kauffmann, R., "Energy and the U.S. economy: a biophysical perspective," Science 225, 890-897 (1984) August 31.

(5) Coughenour, M., Ellis, J., Swift, D., Coppock, D., Galvin, K., McCabe, J., and Hart, T., "Energy extraction and use in a nomadic pastoral ecosystem," Science 230, 619-624 (1985) November 8.

(6) Davis, C., and Blouin, G., "How much energy does

fertilizer consume?" Farm Chemicals 18, 18-22 (1977) June.
(7) Gilliland, M., ed., Energy Analysis: A New Public Policy Tool, Boulder, Colorado: Westview Press, 1978, p. 44.
(8) Goldemberg, J., "Energy problems in Latin America," Science 223, 1357-1362 (1874(March.
(9) Knorr, D., and Sinskey, A., "Biotechnology in food production and processing," Science 229, 1224-1229 (1985) September 20.
(10) Leach, G., "Net energy analysis: is it any use?" in J. Thomas, Energy Analysis, Boulder, Colorado: Westview Press, 1977.
(11) Lipinsky, E., "Chemicals from biomass: petrochemical substitution options," Science 212, 1465-1470 (1981) June 26.
(12) Perry, A., Devine, W., and Reister, D., The Energy Cost of Energy: Guidelines for the Net Energy Analysis of Energy Supply Systems, Oak Ridge, Tennessee: Institute for Energy Analysis, 1977.
(13) Reaven, S., "The concept of net energy I: the problems of net energy analysis," Explorations in Knowledge 1, 191-231 (1983).
(14) Reaven, S., "The concept of net energy II: physical and pragmatic aspects of net energy," Explorations in Knowledge 2(1), 21-28 (1984a).
(15) Reaven, S., "The concept of net energy III: does a technology have a net energy," Explorations in Knowledge 2(2), 25-49 (1985).
(16) Reaven, S., "Value issues in TVA policy formulation," [agro-industry plan], U.S. Tennessee Valley Authority report, 1984b.
(17) Reaven, S., "Methodological and policy dilemmas of net energy analysis: the case of biomass," in J. Bonnet, ed., Energy for the Americas, Oxford: Pergamon Press, 1986.
(18) Roethli, J., Barrier, W., Amin-Arsale, B., and Dale, B., "Forage crop agri-refinery: an alternative for producing food and fuel in developing countries," U.S. Tennessee Valley Authority, Muscle Shoals, Alabama, 1985.
(19) Shoumatoff, A., "The mountain of names," New Yorker 61, 51-101 (1985) May 13.
(20) Smith, L., "Output pricing for biomass energy projects," in J. Bonnet, ed., Energy for the Americas, Oxford: Pergamon Press, 1986.

FINANCING A BIOMASS PROJECT: KEY TAX AND CONTRACT ISSUES

Martin Klepper, Esq.
Lane and Edson, P.C.
Washington, D.C. 20037

ABSTRACT

This paper discusses specific elements of biomass projects rom the perspective of key tax and contracting issues. Part I f the paper begins with a discussion of the definition of iomass facilities under PURPA, and the special criteria these acilities must meet in order to qualify as "small power roducers" and receive the benefits established under PURPA. art II presents a discussion of some of the key tax issues nvolved in connection with biomass projects, including the efinition of "biomass property" under the Tax Code, the elevance of structuring the transaction for the sale of power and steam) as a "service contract" rather than as a "lease", and n up-to-date review of the effect of recent tax reform proposals n the treatment of alternative energy property. Part III of the aper identifies the major contracts connected with the success-ul financing, construction and operation of a biomass project, ncluding the turnkey construction contract, the power purchase greement, the fuel supply agreement, the energy services greement and the site lease.

FINANCING A BIOMASS PROJECT:
KEY TAX AND CONTRACT ISSUES

I. Biomass Facilities Under PURPA

The Public Utility Regulatory Policies Act of 1978 ("PUR and regulations implemented by the Federal Energy Regulatory Commission ("FERC" or "Commission") set forth the general requirements (i.e., size, fuel use, operating and efficiency standards, ownership, etc.) for qualifying cogeneration and s power production facilities. These "qualifying facilities" ("QFs") are entitled to certain benefits (the right to inter-connect with the utility, to obtain back-up, supplementary, maintenance and stand-by power at non-discriminatory rates an sell energy at the utility's "avoided cost") and regulatory exemptions (freedom from state and federal rate, financial an organization regulation) under §210 of PURPA.

PURPA defines a "small power production facility" as a facility which:

1. produces electric energy solely by the use, as a prir energy source, of biomass, waste, renewable resource geothermal resources, or any combination thereof; an

2. has a power production capacity which, together with other facilities located at the same site, is not greater than 80 megawatts.

A "qualifying" small power production facility is a smal power production facility as defined above which meets the Commission-prescribed operating, fuel use and efficiency standards and "which is owned by a person not primarily engag in the generation or sale of electric power (other than elect power solely from cogeneration facilities or small power production facilities)." This latter provision prohibits electric utilities and their subsidiaries from owning greater than 50% of the equity interest in QF projects.

FERC has prescribed specific criteria applicable to qualifying small power production facilities. These include:

1. <u>Maximum Size</u> - The power production capacity of the together with the capacity of any other facilities w (i) use the same energy resource, (ii) are owned by same person, <u>and</u> (iii) are located at the same site, not exceed 80 megawatts.

2. <u>Fuel Use</u> - The primary energy source of the facility must be biomass, waste, renewable resources, geother resources, or any combination of these. At least 75 the total energy input must be from these sources. use of oil, natural gas or coal in a qualifying smal power production facility may not exceed 25% of the total energy input of the facility during any calend year period.

In connection with the fuel use requirements, FERC has efined "biomass" to mean "any organic material not derived from ossil fuels." Included in the definition of biomass are by-roducts of the manufacturing, harvesting and growing of gricultural products, including wood. Peat and plant materials hat are obtained from cultivation or harvested from naturally ccurring vegetation fit within this definition. The regulations urther provide that any primary energy source will be considered s biomass if, on the basis of its energy content, it is 50% or ore biomass. "Waste" is defined under the regulations as "by-roduct materials other than biomass." In most instances, waste s a by-product of fossil fuels. Examples of waste include etroleum coke, refinery gas and plastics.

I. Biomass Property Under the Tax Law

The business energy tax credits officially expired at the nd of 1985. However, the Senate Finance Committee is presently orking on a proposal that would extend the energy tax credits or energy investments, including investments in "biomass roperty." (See below for a discussion of the effects of Tax eform on the energy tax credit.)

A. Biomass Property Defined

Section 48(l)(15) of the Tax Code defines "biomass property" s any "alternate substance" (i.e., any substance other than oil nd natural gas or a product thereof) other than an inorganic ubstance or coal (including lignite) or any product of coal. hus, biomass property includes waste, sewage and sludge, as well s grain, wood and oceanic and terrestrial crops and crop esidues. It also includes waste materials such as municipal and ndustrial wastes, which contain processed products of oil, atural gas or coal.

Biomass property includes boilers and burners which use iomass as their primary fuel as well as equipment used for the nloading, transfer, storage, reclaiming from storage, and reparation at the point of use of the biomass and any necessary n-site equipment to bring the biomass to a burner.

Biomass property also includes equipment for converting iomass into any synthetic solid fuel or an alcohol fuel (so long s the primary source of energy for the alcohol fuel facility is ot oil or natural gas or a product of oil or natural gas). nder Code Section 48(l)(3), biomass property also includes ollution control equipment required by federal, state or local egulations to be installed in connection with otherwise ualifying biomass property.

B. Service Contracts

The tax treatment of property involved in alternative energy ransactions often depends on whether the transaction is lassified as a service contract or as a lease. Although many iomass-fueled small power production facilities involve only the ale of electric power to the local utility, some projects have een constructed adjacent to facilities that have on-site equirements for electric power and/or process steam. For xample, a wood-fueled small power producer could be located next to, or on the site of, a lumber yard, and the small power producer could enter into an "Energy Services Agreement" to me all or part of the lumber yard's power and steam requirements.

If the transaction covering the sale of electric power or thermal energy is treated as a finance lease or installment sa rather than as a service contract, the purchaser of the energy not the service provider -- would receive the benefits of tax credits and depreciation. The use of a service contract provi the means of preserving these tax benefits for the service provider, and if properly structured, can prevent the IRS from successfully recharacterizing the transaction.

Section 7701(e) of the Tax Code outlines the factors used determine whether a specific transaction is a service contract a lease. The Code specifies that the determination of whether service contract should be characterized as a lease is to be m by "taking into account all relevant factors including whether not -

(a) the service recipient is in physical possession of t property,

(b) the service recipient controls the property,

(c) the service recipient has a significant economic or possessory interest in the property,

(d) the service provider does not bear any risk of substantially diminished receipts or substantially increased expenditures if there is nonperformance un the contract,

(e) the service provider does not use the property concurrently to provide significant services to entities unrelated to the service recipient, and

(f) the total contract price does not substantially exce the rental value of the property for the contract period."

The Code includes a special "safe harbor" for determining whether certain energy-related transactions will be treated as service contracts or as leases for tax purposes. In general, transactions involving the sale of electrical or thermal energ produced at a <u>cogeneration</u> or <u>alternative energy facility</u> will qualify as service contracts (rather than as leases) unless th service recipient

(a) bears any significant financial burden if there is nonperformance under the contract (other than for reasons beyond the control of the service provider, not taking into account any temporary shut-down of t facility for repairs, maintenance or capital improvements, nor any financial burden caused by the bankruptcy or similar financial difficulties of the serv provider),

(b) operates the facility,

(c) receives any significant financial benefits if operating costs are less than the standard of performance or operation under the contract (not taking into account any significant financial benefit merely because payments by the service recipient are decreased by reason of increased production or efficiency or the recovery of energy or other products), or

(d) has a fixed or determinable purchase price option (other than at fair market value) or may be required to purchase the facility.

Third party providers of energy generation services must arefully craft their legal and financial arrangements to be ertain they are not violating any of the four "safe harbor" rovisions. For example, a cogeneration facility installed in a tate building may <u>not</u> qualify for the "safe harbor" provision if ne state shares in any reduced operating cost savings of the acility in excess of those projected in the initial contract, ther than cost savings achieved by increases in production, fficiency or recovery of other products.

C. <u>Effect of Tax Reform on the Energy Tax Credit</u>

On March 19, 1986, the Senate Committee on Finance released document entitled "Tax Reform Proposals in Connection with ommittee on Finance Markup," which outlines the proposals that ne Committee has been using as a starting point for its eliberations on tax reform legislation. The document was repared at the request of the Committee's Chairman, Senator obert Packwood (R-OR), and is referred to as the "Packwood roposal." Although the Packwood Proposal is now set forth in reater detail than in the Summary released by the Committee on arch 13, 1986, the actual legislative language and Committee eport will not be drafted until after the Committee finishes its arkup (the Committee drafting process). Thus, many details till cannot be ascertained with assurance. In addition, if the enate approves a bill, differences between that bill and the ouse-passed Tax Reform Act of 1985 (the "House bill") will have o be resolved by a House-Senate Conference Committee.

The Packwood Proposal extends most of the business energy ax credits over longer periods and at higher rates than ontained in the House bill. Biomass property would receive a 0% credit in 1986-1987 and a 5% credit in 1988-1995. In ddition, the Packwood Proposal would increase substantially the eriod over which investments can be recovered through epreciation deductions, and would repeal entirely the regular 0% investment tax credit, effective for property placed in ervice on or after March 1, 1986. The credit would be available or property that qualifies under the transition rules for epreciation. A new provision would refund 70% of the investment redit carryovers existing as of January 1, 1987.

For a comprehensive discussion of the House bill and the arch 13, 1986 Summary, refer to the accompanying Memoranda.

III. Key Contracts for Biomass Projects

The successful financing, construction and operation of a biomass project requires the negotiation of several critical contracts. Depending on the overall structure of the transaction, these contracts could include:

1. Power Purchase Agreement (and Wheeling Agreement, if applicable)
2. Turnkey Construction Contract
3. Fuel Supply Agreement
4. Energy Services Agreement
5. Site Lease

These contracts must reflect consistency and compatibilit and they must all be structured to provide any potential lende with the assurance that the project is financially feasible an operationally sound.

A. The Power Purchase Agreement

The Power Purchase Agreement provides the contractual mechanism through which the small power producer sells electri power to the utility. Even if most electric output is sold to private, non-utility purchaser, the developer would still probably want the ability to sell excess power to the utility an as-available basis, and would also need to contract with th utility for supplementary and back-up power. The Power Purcha Agreement typically sets forth the prices and methodology for computing rates, the effects of interruptions of service in either the utility's or the small power producer's system, performance and continuous service guaranties or warranties, interconnection requirements, and damages and indemnification provisions.

1. Pricing and Rates - The pricing and rate methodology should be structured to take into account escalations and decreases in fuel prices for both the utility and the small power produce Other factors affecting the parties' costs inclu inflation rates, environmental regulation (i.e., acid rain) and regulatory policies. Fixed or levelized payments starting above the utility's avoided cost in the early years of operation can help ensure the availability of suitable financi and provide needed project revenues during the critical initial phases after start-up. If the utility agrees to make these high up-front payments, it may also require adequate security some form of assurance that it will be repaid th high costs in later years. In addition, the sma power producer will likely be required to accept payments that are lower than the utility's avoid cost in later years in return for payments that greater than avoided cost in the early years.

2. Interconnection - In some utility service areas, interconnection obligations are incorporated int the Power Purchase Agreement, whereas in other areas, a separate Interconnection Agreement is

required. The interconnection obligations should outline: the allocation of costs to be borne by each party and the method for payment (i.e., fixed up-front fee vs. periodic payments); specific deadlines that the utility must meet, so that all interconnection components are installed when the small power producer is ready for commencement; and obligations to repair or replace (with all deliberate speed) the interconnection facilities, in order to avoid the problem of having power to sell but no way to get it into the utility's grid. Also, the small power producer should have some degree of control over total interconnection costs; that is, a utility that is allowing its construction subsidiary to build the interconnection facilities, or that is not seeking "low bids" for the construction, should not be allowed "carte blanche" with respect to interconnection costs. Some public utility commissions have specifically outlined the interconnection responsibilities of utilities and qualifying facilities.

3. Damages, Indemnification and Insurance - The utility will most likely require the qualifying facility to indemnify and hold harmless the utility for damages caused by the malfunction or inoperability of the small power production facility. The utility will require the QF to carry casualty, property damage and bodily injury insurance to cover potential losses. These types of contract provisions should be written as a "two-way street"; that is, indemnifications and damage penalties should flow not only from the QF, but also to the QF in the event the utility fails to perform its contract obligations, or the performance is conducted in a negligent manner.

4. Public Utility Commission Approvals - In some states, Power Purchase Agreements must be approved by the PUC. Even where approval is not specifically required, it may be beneficial to receive authorization from the PUC. Such approval could aid in the financing of the project, and in addition, the QF would have on record the fact that the transaction has been "fairly negotiated" and imparts a "benefit to the ratepayers." Project developers should be careful, however, that approval by the PUC, when not specifically required, will not cause the project to become subject to the general jurisdiction of the PUC.

5. QF Status - PURPA only requires utilities to purchase power from qualifying facilities. Utilities will often require that a project remains a QF during the life of the Power Purchase Agreement. In the event of future changes in the law or changes in the QF's operations (i.e., size, fuel use), it is beneficial to provide that the utility's purchase obligations will nonetheless continue.

The small power producer may find it advantageous to contract with a utility that is outside the service territory which the project is located. In this situation, the small pc producer will need to enter into a Wheeling Agreement for the transmission of electric power. The developer should examine state PUC laws and regulations to determine whether wheeling i feasible alternative and also whether transmission rates or policies have been established.

B. The Turnkey Construction Contract

[Key issues involved in the selection, negotiation and compensation of a Turnkey Construction contractor are addresse in detail in an accompanying memorandum.]

C. The Fuel Supply Agreement

One of the most critical elements of a biomass project, f the financing and operational perspectives, is the Fuel Supply Agreement. The fuel supplier must be able to provide a guaranteed minimum quantity and quality of fuel, at a stable price, for the duration of the Power Purchase Agreement and th Energy Services Agreement (if applicable). The Fuel Supply Agreement should set out in great detail the type and quality fuel to be supplied. Key factors for this purpose are Btu content (also known as "calorific value"), moisture content, consistency and size. The Agreement should also specify that fuel will be free of "non-combustibles", such as sand, gravel, dirt or metal (or that the proportion of such materials will n exceed a minimum threshold). In areas where air quality is an important concern, fuel specifications should take into accoun minimum acceptable emission levels.

There are a number of pricing options available to the sm power producer. These should be evaluated based on the particular fuel and financial needs of the project and the willingness of a fuel supplier to share in some of the associa project risks. The pricing options include:

1. Price based on the supplier's actual cost plus a specific percentage above that cost.

2. A fixed price per "unit" of fuel (for example, per volume, weight or Btu content). The Fuel Supply Agreement could provide for a yearly escalation in th fixed price, tied to some readily determinable index (e.g., the Consumer Price Index).

3. The price of fuel could be tied to a percentage of "equivalent" oil or gas prices. (The "tie-in" would b based on Btu content or some similar basis.)

4. The Fuel Supply Agreement could include a performance incentive. In addition to one of the basic pricing mechanisms described above, the supplier could receive negotiated percentage of project revenues. This can help to ensure both the quality and quantity of fuel deliveries during the term of the contract.

The Agreement should provide a schedule for the delivery of fuel at a specific location on the project site. The project developer should retain control over where the fuel will be placed. In addition, the Agreement can provide for a minimum and a maximum amount of fuel that will be available on the project site at any given time. The developer should not be obligated to accept and/or pay for fuel deliveries that exceed this maximum amount. The developer should receive clear title to all fuel delivered to and unloaded on the project site.

The project developer should consider reserving the right to purchase fuel from other suppliers at any time, as long as certain minimum fuel purchase obligations are met with respect to the original supplier. In addition, the developer needs to structure the Fuel Supply Agreement so that any time it reasonably believes the seller will not perform under the contract (i.e., the seller has not provided or will not be able to provide the minimum quantity and quality of fuel), the purchaser can without delay look to other suppliers to fulfill the project's needs.

If the fuel supplier is acting as a "middleman", any contracts between the supplier and the <u>actual</u> providers of fuel should be assignable to the project developer. In order to prevent a disruption in the fuel supply, the developer should reserve the option to cure any default of the supplier under the latter's contracts.

D. <u>The Energy Services Agreement</u>

The small power producer may wish to dedicate all or a portion of the electric power (and/or steam) output of the facility to a private, non-utility customer. This is accomplished through an Energy Services Agreement. Since the sale of power or steam is to a private customer rather than to the utility, the small power producer must first ascertain whether such a "retail sale" is permissible under state law. (PURPA specifically authorizes the sale of power only to utilities.)

The Energy Services Agreement sets forth the obligations of the producer, or service provider, vis-a-vis the customer, or service recipient. The Agreement will outline the responsibilities of each party, specifying such items as the quality of steam or thermal energy, the minimum and maximum energy requirements of the customer, take-or-pay obligations and provisions for indemnification, default and early termination of the Agreement. The pricing structure for electric and thermal energy can be based on cost per unit of energy delivered (i.e., cents per kWh, dollars per million Btus), or payment can be related in some fashion to the performance of the facility.

The project developer should maintain maximum (or exclusive) control over the design, construction, operation and maintenance of the small power production facility. Since the developer is in essence providing a service to the customer, much like the local utility, the customer should be concerned only that it will receive the contracted-for power at the price and in the quantity and quality specified under the Energy Services Agreement.

The small power producer needs to provide itself with adequate remedies in the event the customer does not make payme for energy, fails to accept the minimum quantities of energy or fails to perform any other duties or obligations contemplated b the Energy Services Agreement. Adequate (and necessary) remedi of the service provider may include the suspension of deliverie of electrical or thermal energy until the customer cures all defaults, or removal of the entire facility (or at least the equipment that interconnects the service provider's facility wi the customer's plant). In addition, if the small power production facility is built specifically or primarily to suppl the customer's on-site energy needs, then in the event of a serious or prolonged default, the service provider can consider requiring the customer to purchase the entire facility for a value that reflects the project's cost plus an amount that reflects some element of liquidated damages.

The service provider will generally be required to provide casualty, comprehensive liability and worker's compensation insurance. The customer may require that, in the event of facility damage or destruction, the available insurance proceed must be applied towards the repair or replacement of the damage or destroyed components. However, the service provider should not obligate itself to commit funds in excess of insurance proceeds to repair or replace the project, nor should the servi provider be required to invest additional funds if the operatio of the facility will no longer be economically feasible.

E. The Site Lease

Unless the project developer intends to own the land on which the small power production facility will be located, he will need to enter into a Site Lease covering the project site. The Site Lease must provide for a term (length) that extends beyond the term of both the Energy Services Agreement and the Power Purchase Agreement. Lead time must be given both before the commencement of operations under the two agreements just mentioned (in order to provide necessary time for engineering a design studies and construction) and after the obligations unde both those agreements are terminated (in order to provide for t dismantling and removal of the project).

In some circumstances, the party with whom the developer contracts under the Energy Services Agreement will also be the lessor under the Site Lease. For example, if the developer has entered into an Energy Services Agreement to provide electric power and steam to a lumber yard, it is likely that the lumber yard owns the property on which the project will be built. Although it is possible to encompass the Site Lease obligations within the Energy Services Agreement, it is advisable to create two separate and distinct, yet interrelated, contracts. For example, the small power producer will want to maintain certain vested interests in the project site in the event of early termination of the Energy Services Agreement. These vested interests could include the right to remain on the project site and sell power to the utility or to another third party.

Even though the Site Lease is a separate document, it must nonetheless be tied to the other major contracts in the deal. Specifically, the Site Lease should provide for early terminati in the event the project cannot move forward due to the inability of the developer to obtain a signed Power Purchase Agreement, Fuel Supply Agreement or Energy Services Agreement.

The rental payments under a Site Lease can be fixed throughout the term of the agreement, or they can bear some relationship to the performance and revenues of the small power production facility. Fixed rental payments can be nominal (e.g., $1.00 per month, $10.00 per year), or they can be based upon the fair rental value of the property, as determined by the parties. A Site Lease with payments based on fair rental value could provide that only a nominal rent is paid until such time as the project is operational and earning revenues.

Finally, a Site Lease must cover certain bare essentials to which any lessee is generally entitled. These include rights of free ingress and egress, access to water, rights-of-way or easements for utility interconnections, the right to quiet enjoyment and the right to share in the proceeds of any condemnation award of all or a portion of the project site. The lessor should warrant and/or covenant that it has the necessry ownership interests to convey the leasehold interest in the project site, and that the property is zoned for the intended use. The lessor, in its capacity as owner of the property, should also agree to cooperate with the developer in obtaining all permits, variances and authorizations that may be needed in connection with the construction and operation of the project.

DISCUSSION

Klass How long will PURPA last, or can it be discontinued by the President's tax reform package?

Klepper PURPA is not affected at all by tax reform. PURPA is a separate law that does not have a sunset provision in it, so it does not expire by its own terms on any set date. However, legislation has been proposed over the last 2 or 3 years to change or modify some of the key elements of PURPA. It's also likely that hearings on those proposals will be undertaken this year. It's likely that there will continue to be pressure by the utilities and other interest groups to try to modify or eliminate some of the key parts of PURPA in terms of small power production.

Klass Do you think they will be successful?

Klepper I wouldn't venture to guess on that, but I hope not. It's partly going to be a function of how effective all the renewables and alternative energy industries are in making the case for the importance of PURPA.

J. P. Garvin, Pitt-Des Moines Where there is an alternate use of the power on site, is this generally preferable to PURPA?

Klepper I don't think it's preferable. It's going to depend on the economics of the use on site and the price you can get from the local utility. If you can sell the power for more to the local utility than its value on site, it would be preferable to use PURPA. If the project can work without selling it to the local utility, then you might want to avoid PURPA.

FINANCING A WASTE TO ENERGY PROJECT: KEY CONTRACT AND TAX ISSUES

Robert L. Hoguet, III
Managing Director
Prudential-Bache Securities Inc.
New York, New York

ABSTRACT

Modern waste-to-energy facilities provide an environmentally sound method for municipalities to meet their solid waste disposal needs over a long term. The technologies, while relatively new to this country, have been operating successfully in Europe and the Far East for many years. They are proven and reliable. Many municipalities are turning to waste-to-energy projects as their disposal needs grow, landfill sites become more costly and difficult to obtain and environmental regulations restrict the opening of new landfills or the expansion of existing ones.

Waste-to-energy projects, sometimes called "resource recovery" facilities, dispose of solid waste by burning it. The projects are extremely costly; (a typical 1,000 ton per day plant which can serve a community of 300,000 people costs anywhere from $100-$150 million) They generate revenues first from "tipping fees" that are paid by the users of the facility, and second from the sale of the electric or steam energy which the facility generates in its operation. Putting these anticipated revenues to work to finance a waste-to-energy project requires getting on top of the key contract and tax issues discussed in this paper.

FINANCING A WASTE TO ENERGY PROJECT: KEY CONTRACT AND TAX ISSUES

CONTRACT ISSUES

1. Revenue Bonds. Contract issues are critical in resource recovery because they reflect the business arrangement between the vendor/operator and the municipality, and also because they frequently contain key provisions for providing security bondholders. This situation arises because municipalities generally prefer to finance their waste-to-energy plants by issuing bonds which are backed by project revenues (revenue bonds) as opposed to bonds which are backed by their taxing power (General Obligation Bonds). Revenue bond financing is usually preferred because the capital costs of a resource recovery project are enormous when compared to most other projects which the municipality has undertaken (for example, roads, water and sewer facilities, and public buildings). The use of taking power to support such a costly project would impair the municipality's debt capacity (the amount of G.O. debt that it is permitted raise by statute in most jurisdictions), and in many cases its credit rating as well. Revenue bonds, by contrast, since they are paid from the project's earnings, do not threaten the municipality's tax base or its ability to use that base to borrow for other purposes.

Municipalities prefer revenue bonds for the additional reason that they do want to be responsible for a project's ability (or lack of it) to operates successfully or meet performance standards. Many communities are still skittish of the technological risks waste-to-energy facilities involve. Municipalities don't control the technology, and generally are unable to, or don't want to, operate the plant. By the same token, they don't want to be responsible for repaying the debt in the event the facility fails to perform as expected.

Similar considerations affect the vendor. Again, the capital costs are typically very large relative to the capital base and borrowing capacity of even the largest vendors. Vendors don't want to restrict their borrowing capacity any more than municipalities do. Also, while vendors are typically willing to guarantee project construction and operation, they do not control the siting of disposal facilities or the flow of waste within the community. Accordingly, the vendor doesn't want to assume responsibility for project failure in the event insufficient waste is delivered to the facility to permit it to operate profitably.

In summary, neither the municipality nor the vendor wants to assume responsibility for risks which it can't control, and as a consequence neither wants to guarantee project debt unconditionally. To meet the needs of both parties, most waste-to-energy plants have been financed with revenue bonds. In revenue bond financing, project debt is supported solely by revenues which the facility generates. The principal revenue sources are tipping fees paid for waste disposal and the sale of steam or electricity generated by the facility's operation. The combination of these revenues must be sufficient to repay principal and interest on the debt (debt service) as well as to pay the costs of plant operation and maintenance.

In a properly structured revenue bond financing each party contracts to provide certain components to the project and to pay damages in the event it fails to do so. In a typical structure, the municipality guarantees to provide certain quantities of waste of specified composition to the facility. The vendor guarantees to build the facility for a fixed price, and to operate and maintain it over for a fixed term (typically 20 years to coincide with the maturity of the bonds) for a fixed price, subject to escalation. As part of its operations guarantee, the vendor will commit to meet certain criteria with respect to air emissions, residue content and energy generation. Finally, an energy buyer will commit to buy (take) all of the energy produced by the facility and pay for it.

In order for a project to be financeable these undertakings and guarantees must be tied tightly together so that all of the risks associated with project construction and operation are covered by one of the principal parties to the transaction (that is, the municipality, the vendor, or the energy buyer).

Contract linkages which "hold water" are critical to a successful financing because bond purchasers will take investment risks, but generally will not take operating or technological risks. Investment risks are risks associated with market fluctuations in the value of securities, and the risks associated with the creditworthiness of each of the parties. These are risks an investor can understand and evaluate, difficult as that evaluation may be in a start-up venture that has no operating history the investor can analyze. In any case, however, the bondholder will not take the risk that waste won't be delivered to the plant or that the facility will not operate as designed. These risks must be covered by other parties to the transaction.

In a few cases, insurance has been used to cover some of the technical and operating risks. However, such insurance is frequently not available, and has been obtainable in limited amounts even in periods when it has been available. It is a relatively new product, complex to arrange, and always expensive. While an intractable problem can sometimes be solved using insurance, no one involved with a waste-to-energy project wants to go this route unless it proves absolutely necessary.

I will now briefly describe the principal contracts required for a typical waste-to-energy project and some of the major risks each must cover.

2. Waste Supply Agreement. This agreement between the municipality and the vendor is intended to insure that adequate quantities of waste are delivered to the facility, and that the waste is of an appropriate type for disposal. Often this agreement takes the form of a "put or pay" contract under which the municipality agrees to "put" the waste to the facility or "pay" damages if it fails to do so. The waste supply agreement typically covers risks such as changes in the quantity of waste generated in the service area over the contract term and changes in the composition of the waste (most facilities are designed for waste which falls within certain ranges in terms of BTU content). The waste must be of an acceptable type - for example hazardous waste is generally excluded. The waste supply contract also sets the tipping fee which the municipality pays for the waste disposal service. As I mentioned earlier the tipping fee must be sufficient so that it, along with other revenues (principally from energy sales), provides enough money to cover debt service and operating and maintenance costs.

3. Construction Contract. The construction contract between the vendor and the municipality obligates the vendor to build the facility within a certain time frame for a fixed price. Typically the vendor assumes the risks of most construction delays and construction overruns. The "force majeure" clause, which covers delays caused by events such as strikes or floods, must be carefully negotiated.

Also, the contract must specify exactly what constitutes construction completion (thus triggering the municipality's obligation to deliver waste). Typically an acceptance testing procedure over a period of days or weeks is agreed to. Another provision usually in the construction contract is a so called "buy down" if the facility only partially meets the specified performance criteria (e.g. if it meets only 80% or 60% of performance requirements).

4. Operations Contract. In the operations contract which runs between the municipality and the vendor, the vendor agrees to receive the waste delivered to the facility, to reduce it by combustion to certain weight and volume limits, and to generate a certain quantity of energy. This contract must cover the risk associated with partial performance and unscheduled stoppage, whether due to vendor fault or otherwise.

One troublesome issue which always arises is the possibility of future changes in environmental law which require the installation of new pollution control technology, or even a shut down of the facility. No plant in the U.S. has yet been required to shut down for environmental reasons, but such an eventuality can never be ruled out entirely. Last summer Sweden imposed a moratorium on the construction of new waste-to-energy facilities because of growing concern about air emissions. In this country the EPA is still in the process of determining acceptable standards for dioxin emissions.

5. Energy Sales Agreement. This agreement, which runs between the vendor and an energy buyer, covers the sale of energy output from the facility. If the facility generates electricity, the energy buyer is generally the electric utility which serves the area in which the facility is located. Some municipalities have arranged for the electricity to be sold to other municipal facilities - for example, their wastewater treatment plant. Facilities which produce steam usually sell their output to an industrial user located nearby or use it for a district heating system.

Obviously, the energy sales agreement must fix the price paid by the energy buyer. Under the Public Utilities Regulation and Procedures Act (PURPA) utilities must buy electricity produced by waste-to-energy facilities located within their service area. However, the Act leaves room for negotiation as to price and certain other terms and conditions.

The usual practice is to set price as a percentage of the utility's "avoided cost" (i.e., what it would cost the utility to produce or buy the equivalent amount of electricity from another plant or supplier). The energy sales contract should deal with what happens if the avoided cost declines. It should also address the topic of energy revenue sharing. Typically, municipalities like to give a portion (say 10 to 20 per cent) of the energy revenues to the vendor in order to provide an incentive to operate the facility efficiently and maximize output.

TAX ISSUES

1. Tax-Exempt Financing. The fundamental tax fact about financing resource recovery projects, at least until recently, has been the availability of tax-exempt debt. Currently the Internal Revenue Code permits the use of tax-exempt debt for "solid waste disposal facilities. The use of tax-exempt debt is important because it carries an interest rate which is generally 2-3% lower than the interest rate on taxable debt of comparable quality and

maturity. This lower rate translates into a savings in the disposal costs (tipping fees) of $4-$12 per ton of waste paid by the communities which use the facility. To date most waste-to-energy facilities have been financed with tax-exempt debt, or, as will be discussed later with a combination of tax-exempt debt and equity. Tax-exempt debt has been an important inducement to communities which have sponsored projects because the lower interest rates have helped them to at least partially offset the higher disposal costs associated with a resource recovery facility as compared to the disposal costs at landfills which they had used previously.

In 1984 the Internal Revenue Code was amended to limit the use of tax-exempt financing for certain purposes. Among the restrictions was the imposition of "volume caps" which limit the amount of tax-exempt debt which can be issued in each state for certain purposes, including solid waste facilities. These volume caps put solid waste facilities into competition for available tax-exempt debt with other types of projects such as pollution control equipment for industry and utilities and certain types of airport and other facilities.

In the fall of 1985, the House passed a major tax bill, which is now being considered by the Senate. This new bill, which is referred to by its House number, H.R. 3838, would have a dramatic impact on the use of tax-exempt debt for resource recovery projects. If enacted into law, the bill will drastically reduce the amount of tax-exempt financing available under the volume cap by placing an effective annual limit of $75 per capita on the amount of bonds which can be issued in any state for any of a variety of purposes including solid waste. To put this in perspective, a typical solid waste facility costs abouts $300 per capita, or about four times the annual volume cap. Thus for every community which finances a solid waste facility in a given state, three other communities in that state will have to forego the use of tax-exempt debt for any of the uses covered by the volume in cap that year. Besides solid waste those uses include certain types of airport facilities, docks, sewage, water and mass commuting facilities, as well as student loans.

In its present form H.R. 3838 would also tighten the rules governing the use of bond proceeds in such a way as to prohibit the use of tax-exempt debt to finance the turbine generators used to produce electricity in waste-to-energy projects. This is important because most facilities generate electricity with the heat produced by the combustion process, and revenues from the sale of electricity are a critical ingredient in project economics. In other words, the law would permit the use of low cost financing (tax-exempt debt) for that portion of a facility which disposes of solid waste, but not for that portion which recovers energy.

Currently, the outlook for the passage H.R. 3838 is unclear. However, if it passes in its current form, or anything close to it, the use of tax-exempt debt for solid waste projects will be restricted. The most likely replacement will be taxable debt, which will result in higher costs.

2. The Investment Tax Credit (ITC) and Accelerated Depreciation (ACRS). The second major tax consideration for resource recovery is the use of certain tax incentives, principally the ITC and ACRS depreciation. These tax benefits are available to corporations or investors who own resource recovery plants (but not municipalities which pay no taxes).

It is important to focus on the question of ownership before discussing the ITC and ACRS benefits in detail. Whether the municipality or the vendor will own the facility is an important policy question which should be settled early in the planning process. The availability of ITC and ACRS benefits are only one, and not the most significant, consequence of the ownership decision.

Many communities prefer to own (and thus control) the facility themselves. In such a public ownership structure the municipality contracts with a private company to design and build the facility for a fixed price and to operate it for a set period at an agreed-upon rate. The municipality retains the right to replace the operator if it fails to perform satisfactorily. The municipality retains ownership at the end of the operating contract period. Because the useful life of waste-to-energy facilities can be as much longer than the typical operating contract term of 20-25 years, the municipality can be greatly advantaged by retaining this "residual value."

Other communities prefer to forego ownership of the facility in exchange for lower disposal costs during the 20-25 year operating period. The reason disposal costs can be lower is that a private owner who pays taxes can take advantage of certain tax benefits (principally ACRS and the ITC) which are not available when a municipality, which pays no taxes, owns the facility. In exchange for these tax benefits, private investors are willing to finance a portion of the cost of the facilities with equity. The use of equity reduces the amount of debt which must be issued to pay for the project and thus reduces the amount of payments to be made to repay the debt. The private owner (who can be either the vendor or a third party institutional investor) can pass along at least a portion of these benefits to the municipality in the form of lower disposal costs. How much it passes on and in what form is a matter of negotiation; generally speaking, however, private ownership has historically lowered disposal costs during the operating period by approximately one-quarter to one-third.

It should be pointed out that private ownership results in a much more complex transaction. By necessity it involves more actors - at minimum, an equity participant and its lawyers. Several additional documents are required. There are legal implications and rulings pertaining to tax benefits to be considered. The additional complexity can lead to lengthy contract negotiations with the vendor or equity investor, which in turn can lead to higher project costs. Given the pros and cons of both the public and private approaches, each community must decide for itself which is best for its project.

H.R. 3838 would adversely affect the financing and economics of resource recovery projects first, by eliminating the ITC, and, second, by reducing the depreciation benefits by stretching out the depreciation period. The effect of these changes, if enacted, will be to increase disposal costs paid by municipalities. The impact of these changes will vary from project to project, but a rule of thumb is that the value of the tax benefits will be cut in half, perhaps even more.

Conclusion. For all that waste-to-energy projects are costly and require years to plan and implement, several municipalities over the past ten years have discovered them to offer a reliable, safe and environmentally sound way of meeting waste disposal needs for many years to come. Many communities, confronting the same waste disposal problems, are seeking guidance in this body of experience. While there are variations in the details of how risks have been allocated between the parties, there is today considerable precedent to serve as a guide for projects which are currently in the conceptual or planning stage. There are several investment bankers, financial advisors, lawyers, and consulting engineers who are experienced in the waste-to-energy field and can provide assistance to communities and vendors which are in the process of developing a facility.

DISCUSSION

Klass The Detroit plant is suppose to be the largest MSW-to-energy system under construction today. I think it's 4000 tons per day capacity. How is it being financed?

Hoguet I'm not familiar with the details, but I do know that financing arrangements were made on that project a year or two ago. It was accelerated due to some changes in the tax act and the bond proceeds were put into escrow.

Klass Is that being done with municipal bonds?

Hoguet Yes, the plant is financed with tax exempt municipal bonds. I believe it is a privately owned project. There is some difficulty because of Detroit's credit problems.

Klass I would imagine that 4,000 ton/d of garbage is somewhere in the neighborhood of 25-30% of the total.

Hoguet I don't know, but New York with a population of 8 million and a work-day population of 2 to 2-1/2 times that generates 22,000-26,000 tons a day.

Klass Will the Detroit plant come under PURPA benefits?

Hoguet Yes, as far as I know. Most plants are covered by PURPA in most municipalities, and most utilities have looked at waste-to-energy projects as something that they can't fight. When the mayor of a city like Detroit goes to the chairman of the local utility, you can kind of close your eyes and imagine how the conversation might go and what position the utility might be in for refusing to buy energy from a project that is going to solve a lot of Detroit's municipal solid waste problems. Also, the energy output from a municipal solid waste plant is not large for most utilities. It may be a big deal from the standpoint of whoever is putting the project together, but it's not like a competition from a major power plant as far as the utility is concerned.

DISCUSSION, CONT.

J. Garvin, Pitt-DesMoines, Inc.

Do municipalities find difficulty in awarding construction contracts on a noncompetitive basis for proprietary technology.

Klepper

The general answer is yes. Many of them are constrained by competitive bidding procedures which simply do not fit in these cases because you're not awarding a contract for commodity items. I might point out that typically, most contracts have not been just for construction; some involve construction, plant operations, and in some cases, part of the financing as well. Generally speaking, there are often procedures which municipalities have used which don't allow them to sole source, but which provide a workable competitive process. The usual process involves an RFQ, a short list, an RFP, and a competitive selection process.

THE ECONOMICS OF AVOIDED-COST PRICING FOR WASTE-TO-ENERGY PLANTS

Allen L. White, Ph.D.
Assistant Professor of Geography
University of Connecticut
West Hartford, Connecticut 06117
and
Marie H. Zack
Senior Associate
Charles River Associates Incorporated
200 Clarendon Street
Boston, Massachusetts 02116

ABSTRACT

The Public Utility Regulatory Policies Act (PURPA) of 1978 has emerged as the financial cornerstone of waste-to-energy plants (WEP) in the United States. Mandatory purchase of electric power by private utilities, along with widespread exhaustion of landfills, has accelerated the development of WEPs. PURPA rules and rates vary widely both across and within states according to the allowance (or disallowance) of capacity credits and a utility's fuel mix, excess capacity, and demand growth. Thus, while PURPA has freed WEPs from the uncertainties of identifying and locating near a steam user, project feasibility remains a case-specific determination. In the next decade, local governments pressed by landfill exhaustion will be forced to settle for high tipping fees to offset declining energy and capacity credits occasioned by soft fossil fuel prices and minimal growth in electricity demand. Developers, in turn, will be drawn to areas where PURPA rules and rates will most insulate WEPs from the conventional fuel price swings that have undermined the original cost projections of many recent projects.

THE ECONOMICS OF AVOIDED-COST PRICING FOR WASTE-TO-ENERGY PLANTS

INTRODUCTION

Since the late 1960s, waste-to-energy plants (WEP) have assumed an increasingly prominent role in solid waste management in the United States. Spurred initially by landfill exhaustion and made more urgent in the early 1970s by escalating energy costs, dozens of WEPs are now in the operational, construction, or planning stages. The next decade will witness broad support for WEPs among local governments pressed to find alternatives to landfills in which 90 percent of solid waste is still disposed. Growing national concern about groundwater contamination linked to landfills, disappointing progress in the implementation of recycling programs, and soft markets for recovered materials will reinforce the role of WEPs as a key component in solid waste management.

This paper focuses on one key determinant of the economics of building a WEP -- the rates electric utilities are mandated to pay small electricity producers that generate electric power for sale to the grid. These rates, commonly known as PURPA rates after the federal Public Utility Regulatory Policies Act of 1978, are the cornerstone to insuring WEP developers an adequate rate of return on their investment. Since WEPs are increasingly private ventures -- built, operated, and owned by private firms under contract to host local governments -- the revenues from electric power sales must be adequate to insure long-term viability of a project. Since other federal tax incentives such as the Accelerated Cost Recovery System (ACRS) and the Energy Tax Credit (ETC) face expiration, PURPA rates will become even more critical in determining project feasibility. In cases where tax shelters are not a primary motive, as in the case of publicly-owned facilities selling power to private facilities, PURPA rates become even more decisive in a project's revenue stream. Supreme Court decisions in 1982 and 1983 removed the last cloud of uncertainty over the long-term viability of PURPA rates, thereby bolstering their pivotal role in WEP (and other small power plant) development (1,2).

Our principal objective is to develop some generalizations with respect to how PURPA rates will influence spatial and temporal trends in the implementation of WEPs in the United States. In the next section we present an overview of the geographic distribution of these plants by technologies utilized, size of facilities, and typical capital costs. We then consider in more detail typical project revenue structures and their sensitivities to both state regulations and utility fuel, capacity, and load growth. This is followed by an examination of the revenue structures of five WEP projects, three in operation (in Pinellas County, FL, Westchester County, NY, and North Andover, MA) and two nearing construction (in Bristol, CT, and Bay County, FL). Next we compare PURPA rates in a sample of states and utilities to demonstrate their wide variation. We conclude with a prognosis for the future of the WEP industry in light of near-term trends in fossil fuel prices, tax code revisions, and landfill costs.

NATIONAL TRENDS

WEPs comprise a family of technologies whose primary purpose is to convert municipal refuse into usable energy. In the early stages of U.S. experience, dating to the late 1960s, process steam for industrial use was the most common energy product. Early plants such as those in Hampton, VA, and Hot Springs, AR, exemplify the first generation of relatively small-scale (under 500 tons per day) facilities linked to a nearby process steam customer. These facilities relied on steam sales plus tipping fees from the host municipalities to generate revenues, although one exception, Chicago Northwest (1600 TPD, 1970), sold no energy until 1980. Public ownership and operation was the rule in these early years. The Saugus, MA, facility (1976) was the first major project to be privately developed and financed in part with owner equity, a trend that would soon accelerate.

Landfill pressures were a major impetus to these projects in the pre-PURPA years. Contracts for steam sales were negotiated on a case-by-case basis with industry customers at levels generally pegged to the price of displaced fuel. Not surprisingly, where oil or gas was the boiler fuel (e.g., the Northeast and South), WEPs were a relatively attractive option; where coal dominated (e.g., the Midwest), project economics were considerably less favorable and fewer facilities were built.

Over the course of the 1970s and early 1980s, two trends evolved. First, the spectrum of commercially viable technologies, originally thought to include refuse-derived fuel (RDF), pyrolysis, and mass burn in refractory and water wall incinerators, has gradually narrowed to the mass burn option. Operational setbacks have increasingly caused developers to shun the complex two- or three-step processes in favor of the simpler, untreated, unsegregated mass burn approach. The majority of plants in the planning stage today are designed as mass burners using water wall technology.

Second, early reliance on steam output as the energy product and revenue source gave way to cogeneration and, most recently, electric power production exclusively. This occurred because of the siting constraints imposed by the need to locate near (within approximately one mile) of a steam customer as well as the powerful inducements of guaranteed power sales under PURPA. A 1983 study of operating WEPs in the United States (3) identified only 8 of 49 plants (16.3 percent) that produced electricity or cogenerated steam and electricity. As of 1985, this figure showed virtually no change when all of the estimated 120 facilities operating, under construction, or planned were included. However, the next five years will see a major turn toward electricity-only plants. By 1990, an estimated 60, or half of all WEPs, will be either cogenerators or electricity-only facilities.

The Northeast and South have always been the regions of greatest concentrations of WEPs. Our 1983 survey found that 32.7 percent of all operating WEPs were located in the Northeast and 46.9 percent in the South. Only 1 of 49 was located in a Western state as defined by the U.S. Census. If more current data for the 58 operating or planned WEPs that produce electricity are considered, the Northeast and Western regions increase their fractions to 41.4 and 13.8 percent, whereas the North Central and South decrease to 10.3 percent and 34.5 percent, respectively.

In the Northeast, intense landfill pressures and attractive buyback rates owing to oil-dependent electric utilities have kept WEP activity at a high level. In the Midwest, competition of WEP-generated power with coal-fired utilities, many of which are suffering from surplus capacity, is evidently acting to dampen developer interest. Activity in the South is as notable for its intra-regional shift as for its overall percentage decline. A cluster of projects in Arkansas, all built in the 1970s to generate steam and to lessen industry reliance on natural gas, was a prominent feature of the WEP early profile. These plants, built in Hot Springs (1973), Siloam Springs (1975), Blytheville (1973-1974), and North Little Rock (1977) were followed by two additional plants in Osceola (1980) and Batesville (1981).

This cluster of facilities in a single state was initially spurred by landfill pressures and, in 1976, by an Arkansas Public Service Commission ruling that encouraged natural gas conservation among industrial users. It also contributed to the South's impressive early total of 23 facilities by 1983. As of 1985, however, only 3 of these facilities were still operating (4). Among operating and planned power producing projects, Florida alone accounts for 8 of the South's total of 20. These WEPs have also been prompted by landfill exhaustion in an environmentally fragile state and by a Public Service Commission that has been aggressive in issuing rules and rates for PURPA-qualifying facilities (QFs).

The trend toward cogeneration and electric power WEPs among different-size projects is illustrated in Table 1. Prior to 1986, a total of 19 such plants were identified nationwide: seven small-scale, three intermediate-scale, and nine large-scale facilities. In the next five years, however, this figure will double to 39 (assuming all planned and construction-phase projects materialize), bringing the national total to 58, or nearly half of all WEPs of all types. No size category will dominate; approximately 13 power-producing WEPs in each category will be added during the next five years. The even distribution suggests that electricity production is seen by developers as viable across all size classes, a reflection of both the gradual implementation of PURPA rates and freedom from siting constraints imposed by steam-only projects. This conclusion is corroborated by examining the fraction of each size class and total figures for cogeneration versus electricity-only plants (Table 2). By 1990, electricity-only plants will number approximately twice as many as cogeneration projects in every size class and, nationally,

Table 1

POWER-PRODUCING WASTE-TO-ENERGY PLANTS
BY YEAR OF START-UP AND SCALE

	Design Capacity (Tons per Day)			
Start-up Year	Less Than 500	500-1,499	Greater Than or Equal to 1,500	Total
Before 1986	7	3	9	19
1986 or After	13	13	13	39
Total	20	16	22	58

SOURCE: Calculated from Waste Age, November 1985, and various personal communications.

Table 2

ENERGY PRODUCT BY SCALE FOR POWER-PRODUCING
WASTE-TO-ENERGY PLANTS THROUGH 1990

	Design Capacity (Tons per Day)			
Start-up Year	Less Than 500	500-1,499	Greater Than or Equal to 1,500	Total
Electricity	13	10	14	37
Electricity/Steam (Cogeneration)	7	6	8	21
Total	20	16	22	58

SOURCE: Calculated from Waste Age, November 1985 and various personal communications.

nearly two-thirds of the total of all power-producing projects.

Capital costs are another noteworthy indicator of national trends. Our earlier analysis of the institutional costs of implementing WEPs (3), suggests little relationship between the scale of a facility and its capital costs. Cost equations in which capital cost per TPD design capacity was regressed against size of plant, regional location technology, presence of materials recovery, energy product, and year of start-up yielded no clear evidence of economies of scale. In that study, we concluded that a combination of site-specific variables, coupled with the presence of greater institutional costs for dealing with more complicated multijurisdictional projects, may explain the apparent absence of scale economies. In the case of one technology, RDF, we found evidence of decreasing returns to scale.

Among power-producing WEPs that are either operating or planned, the capital cost estimates show similar trends (Table 3). Average unit costs ($/TPD) for small-scale plants are $69,000, $54,000 for medium-size, and $76,000 for large-scale facilities. When we combine what we learned from the earlier analysis with very recent trends, these numbers seem reasonable. Any type of WEP incurs certain fixed costs distinct from the boiler and turbine

Table 3

AVERAGE CAPITAL COSTS BY SCALE FOR POWER-PRODUCING WASTE-TO-ENERGY PLANTS

	Design Throughput (TPD)		
	Less Than 500	500-1,499	Greater Than or Equal to 1,500
Cost per TPD Design Capacity ($000)	69	54	76
Number in Sample	22	14	22

SOURCE: Calculated from Waste Age, November 1985, and various personal communications.

costs that increase with scale. These fixed costs -- e.g., contract negotiation, engineering fees, site preparation -- vary little with scale. Medium-size projects should enjoy moderate savings per TPD under these assumptions.

In large-scale facilities, however, certain institutional and environmental thresholds are attained. To the extent that more jurisdictions supply waste, power sale contracts experience more delays, public opposition intensifies because of siting controversies, and emissions control equipment becomes more costly, we would expect large-scale plants to show increasing costs per unit throughput. This is not to imply, of course, that such facilities are less profitable, since power sales are greatly expanded with each additional ton of installed capacity. An additional 400-500 KW per hour per ton are generated from each ton of processed waste. Over the course of a year, this might amount to as much as $5 to $10 million in additional revenues from power sales from a 1500- versus a 500-TPD facility, depending on prevailing PURPA rates and other terms of the buyback agreement.

PROJECT REVENUE STRUCTURES

Electricity-producing WEPs depend primarily on tipping fees paid by the disposer of waste and energy revenues to insure a sufficient revenue stream to cover yearly expenses at the facility. While other minor revenue sources routinely appear in financial projections -- e.g., interest earnings on invested bond money, recovered materials sales -- tipping fees and energy sales typically account for at least 80 percent, and often over 90 percent, of revenues.

The view of local governments that participate in WEP projects is that energy revenues should be maximized and tipping fees minimized. The latter, in effect, are seen as a floating revenue source that should be set at a level only high enough to cover shortfalls from energy and recovered materials sales plus investment income. Of course, there are no inconsistencies between this perspective and that of the developers. Maximization of energy revenues is of mutual interest.

The critical role of energy revenues in the economics of WEPs -- typically representing 50 percent or more of annual project revenues -- explains the intensity surrounding power contract negotiations between developers and electric utilities. In these negotiations, host municipalities, though vitally interested in outcomes, are relatively passive observers in the process. The developer seeks to maximize energy revenues, knowing that most, perhaps all, of the differential will have to be made up in tipping fees. If this gap is too great, however, the incentive to municipalities to proceed with project implementation is commensurately less unless landfill exhaustion is imminent. This, of course, is precisely the situation in many areas of the country. Developers are cognizant of this urgency and increasingly able to exercise leverage over local authorities to accept tipping fees that a few years ago would have seemed intolerable.

The control developers exercise over electricity buyback rates is limited by state-specific rules and rates. Once in place, the energy revenue stream from a WEP is calculable at least for the short-term based on the purchasing utility's avoided cost -- that is, the marginal cost of delivering an additional kWh of power above its current capacity. The extremely broad range of PURPA rates derives from: a) differences in state interpretations of what an avoided cost is; and b) the results of applying a given definition to utilities with widely divergent fuel mix, capacity, and load growth profiles.

Consider the determinants of avoided cost and why these calculations vary so widely across states and, more importantly, across utility systems within states. The first component of avoided cost is displaced fuel credit. Prior to the recent precipitous decline in world petroleum prices, a hypothetical New England utility dependent upon oil-fired generating capacity faced relatively high fuel procurement costs. Calculated at the margin, a WEP project could expect an attractive revenues stream even if avoided costs were calculated on a below-100-percent fuel cost basis, which is often the case. In contrast, for a utility system fired primarily by coal, as in the case of most Midwestern systems, avoided fuel costs would be substantially less. Further differences emerge when a WEP contributes to peak versus off-peak power supply, since peaking capacity is left to the most costly power plants. Thus, primary fuel use remains a key determinant of WEP energy revenue forecasts.

The second component of avoided cost is the capacity credit, an allowance for capital costs of new power plants that are rendered unnecessary by virtue of the WEP addition to capacity. Not all states allow for such a capacity credit, and among those that do, actual allowance again varies widely across utilities.

What accounts for these wide divergences? Consider two utility systems, both primarily coal-fired with approximately equal base load and peaking load requirements. System A is located in a mature industrial region of the Northeast with a large industrial customer base, whereas System B is located in a rapidly growing

Sunbelt state with a dominantly residential and commercial customer base. In System A, absolute electricity demand levels are declining and the surplus capacity, now at 30 percent, is expected to increase by 1 percent per year for the next ten years. Furthermore, a nuclear power plan begun in the late 1970s is about to come on line. In contrast, System B electricity demand is growing steadily at 2 percent per year, surplus is now 20 percent and declining, and new capacity will be required within five years.

The consequences for a WEP capacity credit become apparent. In System A, there is no basis for a capacity credit because, at least for the foreseeable future, no ne capacity is either needed or proposed. While the new nuclear plant may create a temporary rate shock, it will only exacerbate an already serious excess capacity situation. Under these conditions, it is reasonable to assume that no capacity credit will be negotiated even if state regulations allow for it.

With respect to System B, however, an entirely different situation exists. A WEP developer can expect a substantial capacity credit because the facility averts th need for at least a small fraction of new generating capacity. This is not to suggest that the purchasing utility will enthusiastically embrace a WEP addition to it capacity (5). Some utilities have been more receptive tha others who tend to view small power producers as administratively burdensome, potentially unreliable, and competitively undesirable, since they may threaten to lure away lucrative industrial customers with independent power supply contracts (1). Nevertheless, capacity credits are standard allowance in many states, and the national trend toward utility deregulation is likely to create continuous pressure to enforce them where they exist and implement them where they are not already in place.

COMPARISON OF REVENUE STRUCTURES

Project feasibility generally rests on the assumption that energy and tipping fee revenues gradually increase over a 20- to 25-year life, while waste quantities, energy produced, and interest earnings remain relatively constant Sale of recovered materials, the fourth component of revenues, either escalates at a modest rate or remains level over the life of the project.

These project revenue patterns are illustrated in the base case assumptions for five recently built or planned WEPs. Financial projections were analyzed for WEPs operating in Pinnellas County, FL, Westchester County, NY, and North Andover, MA, and for the plants nearing construction in Bristol, CT, and Bay County, FL. A survey of these plants reveals fairly consistent revenue structures. Average revenues five years after project start-up are projected to be in the following proportions: 52 percent from electricity, 40 percent from tipping fees, 5 percent from investment interest, and 2 percent from the sale of recovered materials (Table 4). Tipping fee revenues are typically calculated as debt

Table 4

REVENUE STRUCTURES OF SELECTED WASTE-TO-ENERGY PLANTS
IN FIFTH YEAR OF OPERATION
(Thousands of Dollars/Percent)

Facility	Tipping Fee	Energy Use	Interest Earnings	Recovered Materials
Pinellas County, FL	14216/47	11318/37	2319/8	2565/8
North Andover, MA	9665/31	18554/60	2121/7	558/2
Westchester, NY	14287/27	35023/68	1836/4	657/1
Bristol, CT	8951/53	7049/42	840/5	0
Bay County, FL	4322/43	5529/55	152/2	0
Average[1] Percent	40	52	5	2

[1]Rounded to nearest percent.

Calculated from the following sources:

- Pinellas County, FL, (Official Statement for) Solid Waste and Electric Revenue Bonds, July 1, 1980.
- Town of North Andover, MA, (Official Statement for) Resource Recovery Revenue Bonds, April 18, 1983.
- County of Westchester Industrial Development Agency, (Official Statement for) Resource Recovery Revenue Bonds, October 1, 1982.
- Connecticut Development Authority, (Official Statement for) Solid Waste and Electric Revenue Bonds, September 12, 1985.
- Roy F. Weston, Inc., "Consulting Engineering's Feasibility Report for Bay County, Florida, Waste-to-Energy Project." May 28, 1985.

service plus total operating expenses minus electric revenue and investment income.

Three plants predict a substantial increase (10, 18, and 59 percent) in the proportion of energy revenues by the end of the 20th year (Table 5). The same plants show a decrease in contribution of tipping fee revenues of 9, 15, and 53 percent, respectively. The other two facilities show a moderate decrease of 4 or 7 percent of total revenues from energy earnings, both offset primarily by a moderate increase of 6 to 7 percent in the contribution from tipping fee revenues. During the life of the project, the investment earnings that contribute to total revenues decrease between 1 and 4 percent in the sample of five plants. This is offset by varied assumptions concerning the rise and fall of user fees and energy earnings.

Base case assumptions in forecasting the financial operating conditions of a WEP are frequently utilized to provide a benchmark against which more optimistic and pessimistic revenue profiles can be compared. The use of alternative scenarios reflects the multiple uncertainties with respect to waste volumes, inflation and, most critically, energy revenues. The latter is influenced by the volatility in energy prices and, therefore, in the energy credit component of PURPA rates.

Alternative scenarios are cases where the inflation rate is increased and waste tonnage and electric revenues are either increased or decreased. Revenues from tipping fees may be lower because of the need to attract private waste haulers and uncommitted jurisdictions' waste to the

Table 5

THE RELATIVE PROPORTIONS OF REVENUE COMPONENTS FOR SELECTED WASTE-TO-ENERGY PLANTS (Percent of Total Revenue)

Facility	Revenue Component	Year 5	Year 10	Year 20	Change Year 5 to Year 20
Pinnelas County, FL	Tipping fees	47	48	54	+7
	Energy sales	37	36	30	-7
	Interest earnings	8	7	6	-2
	Recovered materials	8	9	10	+1
North Andover, MA	Tipping fees	31	25	-22	-53
	Energy sales	60	68	1.17	+59
	Interest earnings	7	5	3	-4
	Recovered materials	2	2	2	0
Westchester, NY	Tipping fees	27	27	12	-15
	Energy sales	68	70	86	+18
	Interest earnings	4	2	1	-3
	Recovered materials	1	1	1	0
Bristol, CT	Tipping fees	53	56	59	+6
	Energy sales	42	40	38	-4
	Interest earnings	5	4	3	-2
Bay County, FL	Tipping fees	43	44	34	-9
	Energy sales	55	55	65	+10
	Interest earnings	2	1	1	-1

Calculated from the following sources:

- Pinellas County, FL, (Official Statement for) Solid Waste and Electric Revenue Bonds, July 1, 1980.
- Town of North Andover, MA, (Official Statement for) Resource Recovery Revenue Bonds, April 18, 1983.
- County of Westchester Industrial Development Agency, (Official Statement for) Resource Recovery Revenue Bonds, October 1, 1982.
- Connecticut Development Authority, (Official Statement for) Solid Waste and Electric Revenue Bonds, September 12, 1985.
- Roy F. Weston, Inc., "Consulting Engineering's Feasibility Report for Bay County, Florida, Waste-to-Energy Project," May 28, 1985.

WEP. Other fluctuations evaluated in sensitivity analyses are an increase in the actual operating and maintenance costs, a reduction in the availability of the annual effective plant capacity, and a loss of revenues from the sale of recovered materials.

Table 6 illustrates five cases in which different assumptions change projected tipping fees over time for the WEP planned in Bristol, CT. Not surprisingly, the sensitivity analyses indicate lower fees for the cases involving increased electric revenues or tonnage, somewhat higher fees for a case of reduced electric revenues, and significantly higher fees for conditions of increased inflation and reduced tonnage.

COMPARISON OF REVENUES BY STATE AND UTILITY

A survey of states' PURPA cogeneration proceedings and cogeneration buyback rates shows a diversity of positions. Cogenerators larger than 100 kW often must negotiate with the purchasing utility. Standard offerings for smaller

Table 6

SENSITIVITY OF TIPPING FEE IN VARIOUS SCENARIOS
FOR BRISTOL, CONNECTICUT PROJECT
(Dollars per Ton)

Case	Year 5	Year 9	Year 21
	Fiscal Year		
	1992	1996	2008
1. Base Case	48.91	55.63	82.85
2. Increased Inflation[a]	54.22	66.12	126.59
3. Reduced Tonnage[b]	57.03	64.47	94.89
4. Reduced Electric Revenues[c]	49.96	58.43	95.86
5. Increased Electric Revenues[d]	46.77	50.16	60.77
6. Increased Tonnage[e]	43.76	50.42	77.48

[a]The rate of change of the CPI has been increased by 40 percent to an annual rate of 7 percent.

[b]Only current contracting communities supply waste, so that the 17,000 additional tons per year in the base case are not utilized.

[c]Energy payments received from utility never rise above floor price.

[d]Utility's equivalent avoided cost will rise at an average annual rate equal to its projections (8.6 percent).

[e]Waste received is the yearly guaranteed facility capacity.

SOURCE: Adapted from the Connecticut Development Authority, (Official Statement for) Solid Waste and Electric Revenue Bonds, September 10, 1985.

facilities include cost quotes for seasonal and time-of-day variations, and separate estimates for primary, secondary, and transmission levels. Capacity credits vary depending on length of contract as well as near and long-term load forecasts. In virtually all states, the calculation of credits is often a controversial process, the rules are subject to change, and both interstate and intrastate variations are formidable.

Table 7 lists buyback rates for utilities within four states and illustrates these variations. For example, in Pennsylvania seven investor-owned utilities have different rates ranging from 1.98¢/kWh to 5.51¢/kWh, a reflection of differences in fuel use, capacity, utilization, and demand trends.

A sensitivity analysis conducted for the WEP in Westchester County, NY, illustrates how different energy rates can affect a plant's financial projections. Here, the avoided-cost rate is assumed to be 4.2¢/kWh in the first year, compared to 6¢/kWh in the base case. The escalation factor and the amount of electricity generated remains the same (Table 8). After five years of operation, the difference in annual revenues is $15 million for the two scenarios and this more than doubles to $33 million in the 20th year of the project.

Table 7

PURPA COGENERATION BUYBACK RATES
FOR SELECTED UTILITIES BY STATE
(Cents per Kilowatt Hour)

New York

Consolidated Edison		Niagara Mohawk	
Summer		Secondary Voltage Level	
Peak	10.21	Peak	7.172
Off-peak	4.77	Off-peak	3.648
Winter		Primary Voltage Level	
Peak	6.57	Peak	6.924
Off-peak	4.37	Off-peak	3.517
		Transmission Level	
		Peak	5.213
		Off-peak	3.777

California

Pacific Gas & Electric Co.		Southern California Edison	
On-peak	6.423	On-peak	6.1
Mid-peak	6.0	Mid-peak	4.7
Off-peak	4.978	Off-peak	4.0
Weighted Average	5.551	Weighted Average	4.6

Ohio

Cincinnati Gas & Electric Co.		Ohio Electric Co.	
Peak	2.0794	Peak	2.24
Off-peak	1.898	Off-peak	1.70

Pennsylvania (For All Hours)

PA Power & Light	5.51
PA Electric Co.	4.02
Metropolitan Edison Co.	4.89
PA Power Co.	1.98
Duquesne Light Co.	2.33
West Penn Power Co.	2.2
Philadelphia Electric Co.	4.39

SOURCE: "States' Cogeneration Rate-setting Under PURPA, Parts 1-9"; Energy User News: September 9, 16, 23, 30; October 7, 14, 21, 28; and November 4, 1985.

Table 8

ENERGY REVENUES FROM ELECTRICITY FOR TWO SCENARIOS
FOR WASTE-TO-ENERGY PLANT IN WESTCHESTER COUNTY, NY*
(Thousands of Dollars)

	Year 5	Year 10	Year 15	Year 20
Base Case**	35023	51460	75612	111098
Alternative Scenario	24516	36022	52928	77769
Difference in Revenues	10507	15438	22684	33329

*Energy production is 315 million kilowatt hours based on an electric generation rate of 480 kilowatt hours per ton of solid waste.

**The base case of 6¢/kWh from the sale of electricity is reduced to 4.2¢/kWh in the alternative scenario (both expressed in constant 1982 dollars and escalating in accordance with an inflation rate of 8 percent per year).

SOURCE: Adapted from the County of Westchester Industrial Development Agency, (Official Statement for) Resource Recovery Bonds, October 1, 1982.

both operational projects in which energy credits are normally pegged to displaced fuel prices and facilities in either the advanced planning or financing stage.

Ultimately, of course, many local governments will have no choice. Additional landfills are unwelcome virtually everywhere and the tradeoff between a landfill and a WEP simply will not exist, regardless of projected tipping fees. In the near term, however, PURPA rules are likely to direct WEP developers to areas where buyback rates are most insulated from fossil fuel price swings by nuclear and hydroelectric capacity; where growing demand for electricity will yield attractive capacity credits; and where landfill pressures force local governments to be flexible in negotiating long-term tipping fee agreements.

CONCLUSIONS

The increasingly prominent role of WEPs in solid waste anagement has been reinforced by mandatory power buyback equirements imposed on electric utilities through PURPA. he implementation of PURPA rules and rates at the state evel has had the twin advantages of guaranteeing an energy arket and freeing developers from the siting constraints f steam-generation plants. The national trend toward lectricity-only WEPs is a reflection of these conditions.

PURPA rules and rates have been variously interpreted nd implemented across the states. While the fundamental egality of mandatory utility purchases based on avoided osts has been upheld in the courts, there is little niformity or consistency in actual rates within or between tates. The financial implications of this situation are ritical to WEP developers. It is entirely possible (and ronic as well) that a local jurisdiction with an urgent andfill crisis will be a financially infeasible host for a lant if electric utility buyback rates are depressed by ow avoided costs. The steady fossil fuel price escalation f the late 1970s and early 1980s, which provided a strong evenue base for WEPs, has, at least in the short run, been eversed. Host communities will increasingly experience he financial stress of having to offset declining energy evenues with increased tipping fees. This will affect

REFERENCES CITED

. Munson, R., "The Growing Role of Independent Electricity Producers," Public Utilities Fortnightly, 13-16 (1985) May 30.

. Einhorn, M. A., "Avoided Cost Pricing: Who Wins?" Public Utilities Fortnightly, 33-36 (1985) May 30.

. Charles River Associates, "An Institutional Cost Analysis of EMW Projects," prepared for DOE's Energy from Municipal Waste Division, August 1984.

. U.S. Conference of Mayors and Waste Age Staff, "Update: Resource Recovery Activities Report," Waste Age, 99-141 (1985) November.

. Richards, B., "Cogenerated Power Irritates Utilities," The Wall Street Journal, 6 (1985) October 23.

DISCUSSION

. Vimal, IES (India): In Table 3, you mentioned that the average capital cost of the three categories of less than 500, 500 to 1500, and more than 1500 TPD, is $69,000, $54,000, and $76,000. Could you explain that?

White

Under 500, we calculated an average of (the sample here was approximately 50 to 60 plants) $69,000 per ton per day throughput, in the middle category $54,000, and then it went up to $76,000. The data were reasonable and certainly these differences suggested that even if the spread is less than it is or a little different, the relative positions probably would not be changed. There are fixed costs for engineering, contract negotiations, land acquisition, and similar items, which ar predictable and which will decline per ton with plant size increases up to abou 1500 TPD. In the very large facilities, such as those being constructed in New York City, Detroit, and around San Diego we found there is in addition to the costs that are reasonably predictable, a additional level of costs which result from coordination, delay, and the need t accumulate large waste streams, often from multiple jurisdictions. In addition, emission controls for the larger plants require the addition, not of simply more of the same, but additional emission controls for stack emissions. So you reach a threshold at about 1500 TPD, after which you lose the advantage of scale for both environmental and transaction costs, negotiation costs, ar delay costs.

J. Easterly, Meridian Corp.

I think the current tax situation has apparently put energy-from-municipal waste plants on hold because of unfavorable tax considerations like possible elimination of tax exempt bonding, reduc tion in depreciation schedules, and elimination of investment tax credits.

White

These facilities are like other tax-advantaged facilities and are under the gun in Congress. There is certainly a rush now to get as many facilities financed as possible to raise the bond money. So you're seeing a real flurry c activity in anticipation of the loss of these taxes advantages.

Veigel

I've been working in this area myself. There may be such problems, but it also depends on how driven the local communit is by the other choices they have available to them. It may still be the more attractive thing to do.

DESIGN AND ECONOMICS OF A WOOD-FIRED COMBINED CYCLE POWER PLANT: PERSPECTIVES ON TECHNOLOGY DEVELOPMENT

David A. Tillman, Ph.D., Principal Scientist
Amadeo Rossi, Scientist
Envirosphere Co.
Bellevue, Washington 98004

Lawrence Becker, P.E., Senior Engineer
Ebasco Services, Inc.
Bellevue, Washington 98004

ABSTRACT

The indirect fired combined cycle power plant (IFCC) provides the biomass industry with a method for generating electricity that is more efficient than the conventional Rankine cycle power plant. This improved efficiency of the IFCC is most important in a time period when fossil energy prices are declining, improving the competitive economic position of fossil fired electricity generation. While the IFCC potentially has a lower heat rate than the conventional biomass fired power plant, the design has not been commercialized. Further, because it is not yet commercially demonstrated, the IFCC is not economically competitive with the conventional system. The potential exists, however, to commercialize the IFCC. If such commercialization were to be accomplished, the IFCC cycle would be more cost effective than the current system for electricity generation with biomass. This paper examines the design and economics of a 10 megawatt (MW) wood fired IFCC system at the current (pioneer plant) state of technology development. It then examines the potential economic gains from bringing the IFCC technology to maturity. Those gains are examined both at the 10 MW scale and at the 50 MW scale. The paper illustrates that, at maturity, the IFCC can generate electricity at a cost savings of 5 percent relative to the conventional Rankine cycle power plant.

DESIGN AND ECONOMICS OF A WOOD-FIRED COMBINED CYCLE POWER PLANT: PERSPECTIVES ON TECHNOLOGY DEVELOPMENT

INTRODUCTION

Declining world oil prices have increased the costs of biomass fired electricity generation systems relative to fossil fired power plants. The magnitude of that decline is staggering. In 1980, the world price of oil exceeded $43/bbl (in 1985 dollars). Today that price stands at about $15/bbl. As recently as 1984, the prevailing consensus was that world oil prices would experience minor declines and then rebound to reach perhaps $45-$60/bbl prices by the turn of the century (6). That view has largely been abandoned, however. Today there is a more conservative view of world oil prices, with many forecasters believing that the $30-$35/bbl price range is more likely to occur at the turn of the century. Table 1 presents a comparison of 1984 vintage oil price forecasts with the California Energy Commission projections made in 1985. These forecasted differences are presented because they are of enormous significance to the biomass energy industry.

TABLE 1. Comparison of Oil Price Forecasts (in 1985 $/bbl)

Forecast Year	International Energy Workshop Medians (forecasted in 1984)	California Energy Commission (forecasted in 1985)
1980[a]	41.80	
1985	28.70[b]	28.86
1990	37.40	25.56
1995	N/A	27.67
2000	45.50	31.31
2005	N/A	35.42
2010	58.90	N/A

[a] In 1985 $.
[b] 1984 value in 1985 $.

Sources: Manne and Schrattenholzer, Jan. 1985.
California Energy Commission, 1985.

Because world oil prices are declining and making alternative energy systems relatively more expensive, it is necessary for the biomass energy industry to develop the most efficient systems possible. Only gains in thermodynamic efficiency can offset the relative cost penalties associated with solid fuels handling and combustion. One potential system for increasing the efficiency of biomass fuel utilization is the indirect-fired, Brayton cycle-based, combined cycle power plant. Combined cycle power plants provided significant gains in efficiency for fossil fired plants, when heat recovery steam generators (HRSG's) and steam turbines were integrated with natural gas and distillate oil fired combustion turbines. Analogous efficiency improvement potential exists for biomass energy systems.

This paper summarizes extensive investigations concerning the indirect fired combined cycle power plant (IFCC) that have recently been completed (2). It focuses on the levelized cost of power achievable with the IFCC today, and then evaluates what costs of power could be achieved if this technology were to mature. Analysis of the technology at maturity is particularly important, because it provides a meaningful basis for comparative assessment of the IFCC with the conventional biomass fired Rankine cycle system. Analysis of the technology today and at maturity also provides some means for measuring the gains available from technology introduction and development.

In order to evaluate the economic viability of the IFCC, a system was designed in some detail. The system was designed assuming a system capacity of 10 megawatts (MW) nominal, and assuming forest residual chips (wet wood) as the primary fuel. Given these assumptions, the costs associated with that system were estimated both at the current state of technology and at a postulated mature state of technology. Economic factors (e.g., discount rates) were then developed for both states of technology and levelized costs were calculated. Finally the IFCC was compared to the currently employed Rankine Cycle in order to evaluate its economic potential.

IFCC SYSTEM DESIGN

The basic principle behind the IFCC is that biomass or any other fuel is combusted, and the products of combustion are ducted to a flue gas-to-air heat exchanger. The air ducted to that heat exchanger is compressed and, after being heated to some temperature (e.g. 1500-2000°F), is used as the working fluid to drive a combustion turbine. ("Combustion turbine" is a misnomer in this case, as the Brayton cycle turbine is driven with heated air rather than the products of oil or gas combustion.) The exhaust air from the combustion turbine and the exhaust flue gas from the heat exchanger are sent to an HRSG system where steam is produced. The steam so produced is used as the working fluid in a conventional Rankine cycle back end of a combined cycle power plant. The basic principle behind the IFCC is simple, and the concept has been discussed in the literature for at least two decades (see, for example, Fryling 1966). The actual design, however, raises many practical questions which must be addressed. These questions include system configuration; and subsystem selection for the combustor, flue gas-to-air heat exchanger, and gas turbine.

System Selection

Numerous cycle configurations are possible using the IFCC principle. Two potential cycle configurations are shown here in Figures 1 and 2. The system shown in Figure 1 includes the combustor system, a single flue gas-to-air heat exchanger, a single gas turbine, a mixing of air and flue gas streams and a single HRSG, and a consequent conventional steam turbine. The system shown in Figure 2 is different from that shown in Figure 1, in that it ducts the flue gas and the air to separate HRSGs, and employs an induction steam turbine for power generation. A variant on the system shown in Figure 2 uses two steam turbines.

The system shown in Figure 1 has the advantages of using a larger HRSG and a conventional, larger steam turbine. It captures gains from scale both in performance and capital cost in the HRSG and steam turbine as a consequence. The system shown in Figure 1 has the disadvantage of a more complex control system when compared to the system shown in Figure 2. The mixing of flue gas and air to control the inlet temperatures to the heat exchanger, the flows to the HRSG, and to ensure the safety of the system under upset conditions require considerable advanced control systems.

The system shown in Figure 2 has the advantage of simpler controls due to the absence of mixing of the exhaust streams from the flue gas-to-air heat exchanger and the combustion turbine. At the same time it has disadvantages associated with purchasing and operating two HRSGs of small size. The system shown in Figure 2 is generally more costly and less efficient than the system shown in Figure 1. For that reason, the system shown in Figure 1 was chosen for subsequent analysis.

Subsystem Selection

The subsystems of critical importance included the combustor, the heat exchanger, and the gas turbine. Each system was considered separately, and also in the context of its overall system consequences. Because of upstream impacts, the order of consideration here is reversed: the gas turbine is considered first followed by the heat exchanger and the combustor.

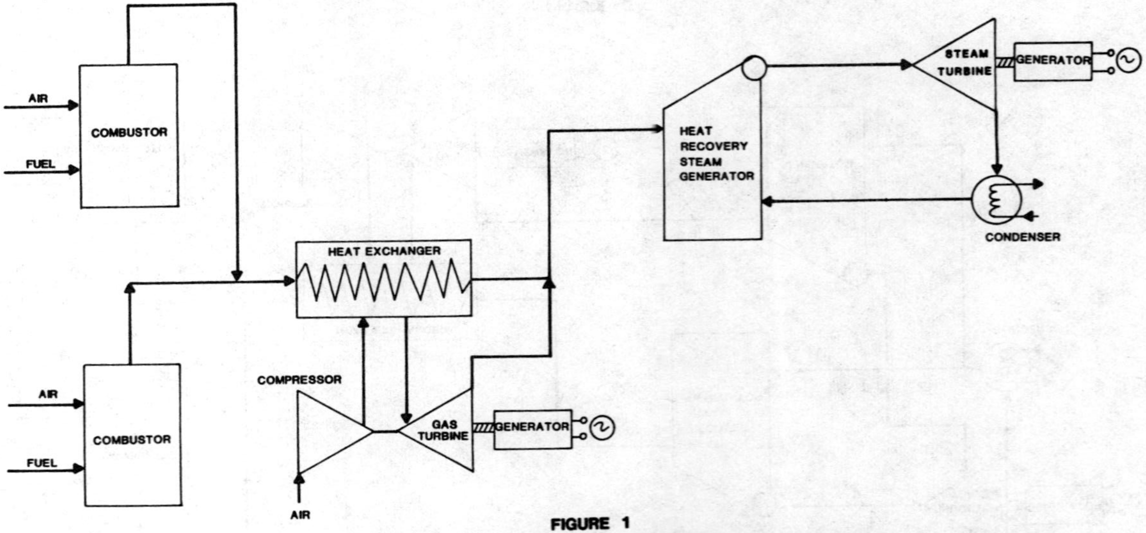

FIGURE 1

CONCEPTUAL SCHEMATIC OF COMBINED CYCLE POWER PLANT

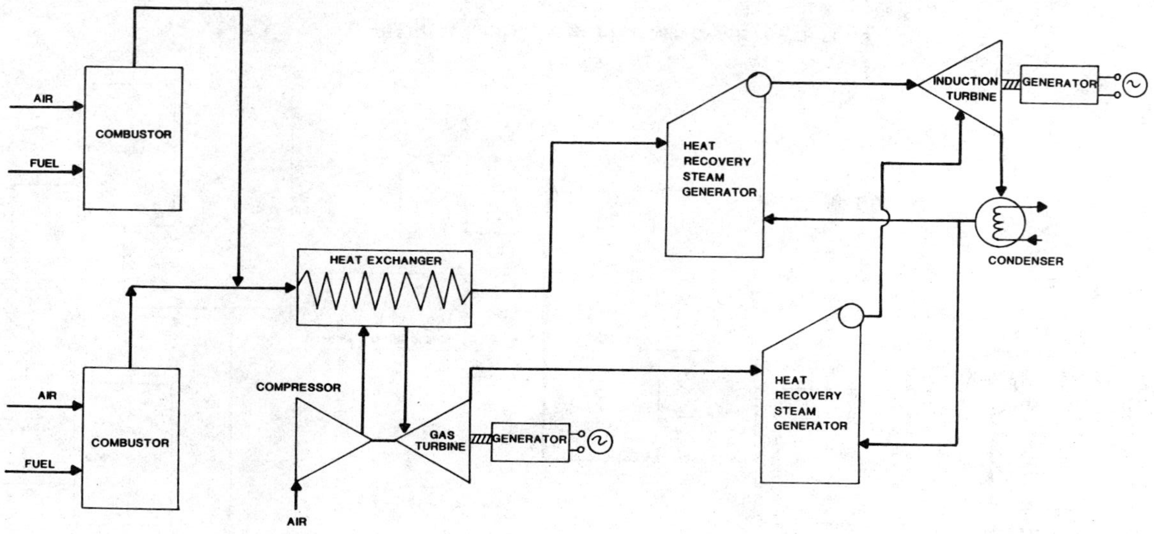

FIGURE 2

CONCEPTUAL SCHEMATIC OF ALTERATE CYCLE CONFIGURATION

Gas Turbine Selection. The gas turbine was selected based upon the maximum temperature differential across the machine, the minimum heat rate (Btu/kWh) or maximum efficiency, and the mechanical compatibility with the entire system (e.g., the ability to duct air from the compressor stage of the machine to a single location). The gas turbine generally had the following specified conditions: (1) inlet air temperature, 1730°F; exhaust air temperature, 920°F; capacity, 3970 kW; and a heat rate of 13,200 Btu/kWh.

Heat Exchanger Selection. The heat exchanger was subject to extensive investigation. Both ceramic and metallic heat exchangers were analyzed, and their consequences on the performance of the system were determined. The ceramic heat exchanger operates at higher temperatures. It accepts flue gases from the combustor in the 2200-2300°F range and produces inlet compressed air for the gas turbine at 1730°F. These performance characteristics can be compared to operating parameters of a high temperature metallic heat exchanger which accepts flue gas at 1675°F and has a safe operating exit temperature for the product, compressed air, of 1450°F. Consequently the ceramic heat exchanger reduces the net station heat rate of the IFCC by 990 Btu/kWh. The ceramic heat exchanger is a less forgiving piece of equipment, however; it can tolerate a temperature swing of only 500°F in a five minute time span, while the metallic heat exchanger can tolerate a temperature swing of 1800°F in a five minute time interval. Both heat exchangers required a clean flue gas in order to prevent serious and rapid deterioration of unit performance.

Because the gas turbine chosen in the evaluation requires inlet air at 1730°F, and because the ceramic heat exchanger could deliver that air directly, it was chosen despite its tight tolerances on temperature excursions. The metallic heat exchanger system requires the addition of a trim burner and the combustion of fossil fuels (e.g., natural gas, propane, or distillate oil) in order to satisfy the working fluid temperature requirements of the gas turbine. While the ceramic heat exchanger is less forgiving and less proven, it permits the design of a simpler and more straightforward unit.

Combustor Selection. While the combustor is at the heart of the system, it is the final system analyzed due to the requirements of downstream components. Initial combustor investigations examined whether to utilize a dryer, pulvarizer, and suspension burner; or whether to combust the wet wood fuel directly. The dry fuel suspension burning system option was not chosen for several reasons including the high capital costs associated with materials handling, the high parasitic power load on the plant caused by the wood pulvarizers, and the significant temperature cycling observed in some suspension burners (e.g., 700°F swings within the space of one minute). Additionally, suspension burning forces all ash and unburned carbon to exit as fly ash; and because of the clean flue gas requirements of the system, a hot cyclone would have been necessary for the system in order to prepare the flue gases for the heat exchanger.

Several wet wood combustors were examined including conventional spreader-stokers, fluidized beds, and refractory lined gasifier-burners such as the Lamb-Cargate wet cell. The spreader-stokers had the advantage of being proven technology, however they typically involved integrating the heat exchanger into the unit design, as is conventional with wood fired boilers. Such integration provided insufficient protection of the heat exchanger against dirty fuels or significant temperature excursions. The fluidized bed option provided for combustion of the wood external to the heat exchanger, and consequently provided for temperature control. Fluidized bed gases could not be produced at temperatures required by the ceramic heat exchanger without slagging up the bed. Further, fluidized bed combustors require hot cyclones if they are to supply a clean gas to the heat exchanger.

Consequently the wet cell type systems has been chosen as the combustor for the IFCC. The wet cell system is a two-stage gasification combustion device where the primary chamber operates as a steam injected countercurrent gasifier, producing hot gases for combustion in the secondary chamber. The secondary chamber combusts those producer gases with minimum quantities of excess air (e.g., 15-25 percent) and generates a clean (0.05 gr/SDCF) hot gas for subsequent use. The gases produced can have temperatures of 2000-2300°F when burning 50 percent moisture wood. A performance profile of the wet cell system specified for this application is shown in Table 2. The wet cell is not without limitations in this application, however. Maximum unit sizes appear to be in the range of 75-100 million Btu/hr due to practical problems of fuel feeding, and the control of gasification reactions on a grate of increasing diameter. This approximate scale limit has been found in other countercurrent gasifiers as well.

System Design

Given the specific configuration selected, and the individual components chosen, a system design was completed. This design includes not only the heat balance shown in Figure 3, and the outline of system characteristics shown in Table 3, but also detailed equipment specifications. Further, the design includes prediction of system performance at part load levels of 60 percent and 80 percent.

It should be noted that the plant was originally designed with provision for a baghouse to achieve airborne emission control. However the wet cell chosen for biomass

TABLE 2. Calculated Performance of the Lamb-Cargate Wet Cells (Basis = 1 hr)

Parameter	Value
Inputs (mass)	
Fuel Feed (lbs)	34,090
Gasifier Air (lbs)	37,720
Gasifier Steam (lbs)	2,390
Combustion Air (lbs)	85,450
Outputs (mass)	
Bottom Ash (lbs)	720
Fly Ash (lbs)	60
Dry Stack Gas (lbs)	128,790
Moisture in Stack Gas (lbs)	30,080
Inputs (energy)	
Fuel (Btu)	150 million
Air (Btu)	8.2 million
Steam (Btu)	2.8 million
Outputs	
Usable Stack Gas Heat (Btu)	124 million
Heat of Vaporization (Btu)	31.2 million
Other Losses (Btu)	5.8 million
Stack Gas Values	
Stack Gas Temperature	2460°K
Stack Gas Flow	44.5 lb/sec
Airborne Emissions	
Particulates: 0.05 gr/DSCF @ 12% CO_2	
NO_x = 61.1 ppmv @ 3% O_2 (DSG) 11.2 lb/h	
SO_2: N/A	

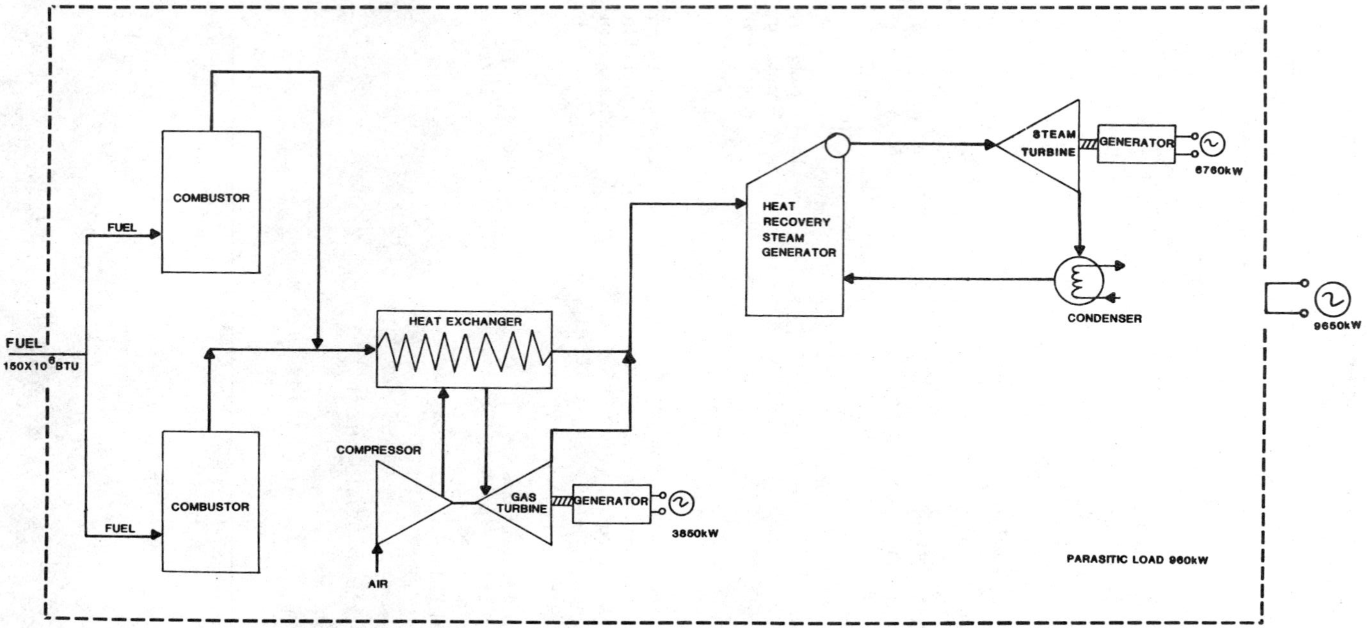

NET STATION HEAT RATE
15,540 BTU/kWH

FIGURE 3

SCHEMATIC OF CONCEPTUAL HEAT BALANCE

TABLE 3. System Characteristics

Parameter	Value
Station Capacity (kW)	
Gross	10,610
Net	9,650[a]
Turbine Capacity (kW)	
Gas Turbine	3,850 (36%)
Steam Turbine	6,760 (64%)
Capacity Factor	0.80
Net Station Heat Rate (Btu/kWh)	15,550

[a]Parasitic power = 960 kW (9%)

combustion produces a stack gas with a measured emission rate of 0.05 gr/DSCF. By the time the flue gas passes through the heat exchanger, the HRSG, and an air preheater (used to raise combustion air to a temperature of 350°F), the grain loading is estimated to be no more than .03 gr/DSCF, however. Control beyond the .03 level is required to meet local regulations or specific local conditions, but not necessarily all Federal regulations. For purposes of this analysis, the fabric filter system has been removed from consideration. Significant evidence existed (see, for example, Junge, 1983) to demonstrate that the baghouse is not needed.

The design described above, and summarized in Table 3, assumes the current state of technology. It also reflects a small scale--10 MW. It is recognized that the efficiency of the IFCC would be improved by the construction and operation of several units. Further the efficiency of an IFCC unit would be improved by increasing the design size of the plant. Wood fired Rankine cycle systems generally improve in efficiency according to an exponential scale factor of -.12, as utilized below:

$$\text{Heat Rate Unit B} = \text{Heat Rate Unit A} \left(\frac{\text{Capacity B}}{\text{Capacity A}}\right)^{-0.12} \quad (1)$$

Underlying equation (1) is the economic attractiveness of increasing the severity of the steam conditions used, increasing the number of feedwater heaters, and similar cycle improvements that can be justified as systems are built with increasing capacity. Such efficiency gains from scale are as available to combustion turbine and combined cycle designs as they are to Rankine cycle systems, although not necessarily at identical rates. Consequently it is believed that equation (1), which has been derived empirically, is as appropriate for the IFCC as for the Rankine cycle system.

ECONOMIC ANALYSIS

The design described above was subjected to detailed economic analysis. That economic analysis was constructed to illustrate the following: (1) the economics of the IFCC system given the current state of technology; (2) the economics of the IFCC system assuming that the technology is brought to maturity, (3) the consequent economic gains available by bringing this technology to maturity; and (4) the economics of the IFCC relative to the conventional Rankine cycle assuming a mature technology.

Economic Assumptions

In order to perform the economic analysis, assumptions were made concerning ownership, the consequent discount rate the projected rate of inflation, tax treatment, and other related concerns. These assumptions are summarized in Table 4.

TABLE 4. Economic Parameters Used to Evaluate the Indirect Fired Combined Cycle Power Plant

Parameters	Value
Ownership	Forest Products Industry
Real (Mature Technology) Discount Rate	10%
Projected Inflation Rate	5%
ACRS Period	5 yrs
Marginal Tax Rate	
Federal	46.0%
State	4.1%
Total	50.1%

Sources: Tillman, 1985
Ebasco, 1985

Of the assumptions made, ownership is most critical, for it determines discount rate and tax treatment. For the bulk of the analysis, it was assumed that the IFCC would be a "Qualifying Facility" under the regulations of the Public Utilities Regulatory Policies Act (PURPA). That implies that private industry, such as the forest products industry (FPI), would own the IFCC. There is ample justification for that assumption. Energy production, and specifically electricity generation, is one of the few growth areas in an otherwise static FPI. Energy today is the third largest market area for that industry; it is ranked behind pulp and lumber, but ahead of plywood and composite products (7).

The assumption of ownership carries with it a capital structure and discount rate. It is considered that the IFCC is sufficiently small that the FPI would not alter its capital structure to finance a given unit. Consequently the conventional industry debt/equity ratio of 0.5 and the conventional real (inflation free) discount rate of 10% was assumed in mature technology cases. That discount rate is based upon an analysis of the cost of capital for such firms as Weyerhaeuser, Crown-Zellerbach, Georgia Pacific, International Paper, Scott Paper, Mead Corp., Fort Howard Paper, Boise Cascade, and other FPI companies. It has been shown that innovations constructed largely with commercial components, where most of those components could be used elsewhere in case of system failure, have discount rates with a risk premium of about 3%. These are referred to as "reversible risk innovations" because of the ability to utilize system parts elsewhere in a given company (7). Consequently the first IFCC units carry a real discount rate of 13 percent.

Because the assumption of ownership is so critical, it has been varied within the analysis in the development of alternative cases. Investor-owned utility ownership, where the real discount rate is about 6 percent, is also assumed in the case of a mature IFCC technology. The real discount rate of 6 percent assumes a capital structure with 43 percent debt, 10 percent preferred stock, and 47 percent common stock as taken from such utilities as Puget Sound Power and Light, Washington Water Power, Pacific Gas and Electric, Portland General Electric, Montana Power, and Idaho Power.

Economic Characteristics of the IFCC

The economic characteristics of the IFCC include the capital cost, the operating and maintenance costs, the project life, and the system load factor or duty cycle. Capital costs are summarized in Table 5. All economic characteristics of the IFCC are summarized in Table 6.

It is important to note that Table 5 constructs overnight capital costs assuming both pioneer plant and mature technology status. These capital costs do not include

TABLE 5. Capital Costs for the 10 MW Indirect Fired Combined Cycle Power Plant (Values in \$ x 10^3)

Cost Category	Plant Status	
	Pioneer	Mature
Direct Costs		
(1) Fuel handling	901	901
(2) Combustor and related	3,547	3,547
(3) Heat exchanger and related	3,662	1,536
(4) Turbine-generators and related	3,465	3,298
(5) Instrumentation	895	895
(6) Electrical and switchyard	943	943
(7) Other Direct Costs	1,751	1,751
(8) Subtotal	15,164	12,871
(9) Engineering @ 10%	1,516	1,287
(10) Contingencies @ 8%	1,244	1,055
(11) Other indirects @ 2.5%	386	322
(12) Total	18,310	15,535
Capacity	9650 kW	9650 kW
Cost/kW	1,900	1,610

interest during construction, finance charges, or other related costs. The difference between the capital cost of the pioneer plant and the capital cost of the mature plant is attributable to the ceramic heat exchanger and related improvements in the gas turbine system. The current costs of ceramic heat exchangers have been estimated based upon vendor quotes. In order to estimate the costs of ceramic heat exchangers at technological maturity, the ratio of ceramic material costs to metallic material costs on a \$/sq ft of surface area was developed based upon vendor quotes. A heat transfer ratio of ceramic to metallic also was developed, and the product of these two ratios was calculated. This product coefficient was then applied to the cost of the Inconel (metallic) heat exchanger as a means for developing a rough estimate of the learning curve gains available to the ceramic heat exchanger industry. These gains from maturity in ceramic heat exchangers carry through the entire cost estimate, from the total materials cost through the indirect costs such as engineering and

TABLE 6. Cost Parameters for the Indirect Fired Combined Cycle Power Plant

Cost Category	Plant Status	
	Pioneer	Mature
'apital cost ($ x 10^6)	18.310	15.535
'ixed O & M cost ($ x 10^6)	0.611	0.611
'lant Output (GWh/y)	67.61	67.61
'ariable O & M costs ($/kWh)		
Fuel	.0310	.0310
Other	.0023	.0023
Total	.0333	.0333
:eal discount rate (%)	13	10

ontingency. This phenomenon is known as the learning curve r experience factor. It holds generally for all manufac-ured components and says for each doubling of production, he product can be made for 70-90 percent of the previous ost.

The estimates presented in Table 6 do not reflect hanges in technology status, except in the area of over-ight capital cost and discount rate. It is recognized that here will be some gains in unit efficiency with experience, owever there is no way to quantify the impact of technolo-ical maturity on net station heat rate. Similarly it is ecognized that maintenance costs will probably decrease as econd and third generation components are manufactured and nstalled. In the absence of any technique for quantifying hese cost reductions, no gains were assumed. The conse-uence is a reasonably conservative estimate of unit econ-mics at maturity.

ase Case Results

The values in Table 6 have been used to calculate the eal (inflation free) levelized cost of power for the IFCC. ue to changes in energy price structure that have occurred ecently, wood fired prices are now estimated to escalate nly at the rate of inflation for purposes of this analysis. nder base case assumptions of private industry (e.g., FPI) wnership, the real levelized cost of power is 85 mills/kWh or the IFCC technology at its point of introduction, and l mills/kWh for the IFCC when the technology becomes ature. These levelized costs do not reflect certain terms uch as land cost, financing cost, and other nontechnology elated charges. The individual components of these levelized costs (e.g.,the contribution of capital cost, fuel cost, etc.) are presented in Table 7. As is shown above, and in Table 7, there is asignificant gain from bringing thi technology from the design stage to maturity.

As mentioned previously, numerous alternative cases have been analyzed. These cases include private industry ownership of a Rankine Cycle system comparable in capacity to the IFCC. The alternative cases also include investor-owned electric utility ownership of the IFCC and the Rankine Cycle system. The real "overnight" levelized cost of power associated with each alternative case is shown in Table 8. A second set of alternative cases has been constructed. These are for IFCC power plants at the 50 MW scale. Such power plants would generate about 20 MW using combustion turbines, and 30 MW with steam turbines. Efficiency gains from scale have been estimated based upon equation (1). Capital cost gains from scale have been estimated using an exponential scale factor of 0.8. Fixed operating and maintenance costs have been estimated assuming a plant staff of 40, comparable to the larger sized Rankine Cycle systems. The results of this analysis, assuming both private industry and investor-owned electric utility ownership, are shown in Table 9. It should be noted that these gains from scale can be considered as additional gains from technology maturity. The history of electricity generation technology demonstrates that new technologies, once proven, are built at increasing capacities until some practical limit is reached.

TABLE 7. Cost Contributions to the Real Levelized Cost of Power from the Indirect Fired Combined Cycle Power Plant

Cost Category	Plant Status	
	Pioneer	Mature
Capital Cost	.043	.029
Fixed O & M	.009	.009
Variable O & M	.033	.033
Fuel	.031	.031
Other	.002	.002
Total	.085	.071

TABLE 8. Levelized Cost of Power for Alternative Cases Used to Evaluate the Indirect Fired Combined Cycle Power Plant (Mature Technology Base)

	Plant Type	
Ownership	IFCC Cycle	Rankine Cycle[a]
Industry (Real discount rate = 10%)	.071	.076
Investor owned utility (Real discount rate = 6%)	.063	.066

[a]Capital cost = $17.23 million (overnight)
Fixed O & M = $429,000/yr
Nonfuel variable O & M = $0.0036/kWh
Heat rate = 16,650 Btu/kWh
Fuel cost = $0.0333/kWh

TABLE 9. Cost Analysis for a 50 MW Indirect-Fired Combined Cycle Power Plant (Mature Technology Basis)

Cost Parameter	Cost
Capital cost (overnight)	$56,300,000[a]
Fixed O & M cost	1,440,000
Plant output (GWH/yr)	350.4
Heat rate	12,830 Btu/kWh[b]
Variable costs ($/kWh)	
Fuel	0.0265
Other	0.0023
Total	0.0278
Tax rate	50.1%
Discount rate (Real)	
Industry	10.0%
Utility	6.0%
Real levelized cost (¢/kWh)	
Industry	.052
Utility	.046

[a] $15{,}560{,}000 \times (\frac{5}{1})^{0.8} = 56{,}300{,}000.$

[b] $15{,}560 \times (\frac{5}{1})^{-0.12} = 12{,}830$

DISCUSSION

The above data demonstrate that the indirect fired combined cycle power plant is significantly more efficient than the conventional Rankine Cycle power plant in stand-alone applications. These data demonstrate that the IFCC can not compete with the Rankine Cycle plant at the present time, however, due to differences in degrees of technological maturity.

Because of declining oil prices and more conservative oil price projections, the biomass industry must move toward the most efficient fuel utilization systems available. As the data above demonstrate, bringing the IFCC technology to maturity is one step that can be made in that direction. If the technology is brought to maturity, the real levelized cost of power can be reduced from the present level of 85 mills/kWh to a level of 46 mills/kWh depending upon plant ownership and capacity. That real levelized cost of power is potentially competitive depending upon regional and utility-specific conditions.

Bringing the IFCC technology to maturity will have additional benefits for the biomass industry. By providing a more efficient means of power generation, the industry will be less sensitive to swings in fuel price should the pulp and paper industry or the wood chemical industry experience renewed growth, hence increased demand for furnish. Further, by providing a more efficient means of power generation, the biomass industry can compete more readily with other alternatives to oil and gas fired generation such as small scale hydroelectric generation.

In conclusion, the detailed analysis of the IFCC at a capacity of 10 MW demonstrates that there is significant potential for developing this electricity generation system. The technologies for the individual components of this system are largely commercially available. The system configuration is readily defined. The technological risks are in the areas of positive pressure ceramic heat exchangers and systems assembly. The detailed analysis demonstrates that while the technology may not be cost competitive at the present time, there are significant gains to be made by obtaining investor experience and by reducing the capital cost of the technology. These gains can be expanded by increasing the capacity of the IFCC from 10 MW to levels approaching 50 MW. When such gains from technological maturity are achieved, the IFCC will be a highly attractive system for electricity generation when using wood or other biomass fuels.

REFERENCES CITED

1. California Energy Commission., "Results of Delphi Analysis of World Oil Prices. Sacramento, 1985.

2. Ebasco, "Oregon State University 10MW Biomass Fired Combined Cycle Cogeneration Plant." Final Design Report, Vol. 1 (Draft). Bellevue, Washington, 1986.

3. Ebasco, "Oregon State University 10MW Biomass Fired Combined Cycle Cogeneration Plant." Feasibility Report. Bellevue, Washington, 1985.

4. Fryling, G. (ed.), Combustion Engineering. New York: Combustion Engineering Co., 1966.

5. Junge, D. C., "Report on the Environmental Envelope for the Operation of the Base Case Biomass Powered Combined Cycle Power Generation Facility. Corvallis, Oregon: Oregon State University, 1983.

6. Manne, A. S., and Schrattenholzer, L., "International Energy Workshop: A Progress Report." June, 1985.

7. Tillman, D. A., Forest Products: Advanced Technologies and Economic Analyses. New York: Academic Press, 1985.

DISCUSSION

R. Price, U.K.A.E.A. (U.K.): Could you tell me what wood prices you assume for your cost calculations and whether, when you're looking forward to a much larger plant of 50 MW, you would assume a higher fuel cost because it would have to be transported further on average?

Tillman: We typically looked at wood fuel cost of $2 per million Btu. We did not vary fuel costs when we got to a larger facility. If you located the power plant in the wrong location, you would experience that. If, however, you put it in say Douglas County, Oregon with a vast resource, a lot of timber harvesting activity, and lots of logging residue, the marginal cost of wood fuel in Douglas County has a very long flat period before it increases.

J. Zerbe, U.S. Forest Products Laboratory: The economics would tend to make this facility look attractive, but would the equipment be ready? Could you find the equipment to build the facility?

Tillman

That I think was the biggest surprise to us; the answer is yes. When it came to the turbine, we looked at an old time model of a GE turbine, Frame 1. The heat exchanger was offered to us by Babcock and Wilcox. The combustor is available. There are 20 of them at varying locations around the country and they've been improved over time such that we believe that's a useful system at this point. Those were the 3 key components. The rest of the equipment — the waste heat boilers, the small steam turbines — are available.

B. Hollomon, New York State Energy R&D Authority

General Electric has been promoting a similar concept, except that one of the differences is that their system involves steam injection. Could you indicate whether or not you would consider steam injection in this sort of system?

Tillman

With steam injection, you're dealing thermodynamically with what is called the Chang Cycle. We've worked with Chang Cycle machines in gas-fired combustion turbines. The combustion turbine is a mass flow machine and the steam simply boosts the mass flow through the turbine. Where we have a steam turbine and a combustion turbine, there's no need to do steam injection into the combustion turbine. We pick up the value of the steam in a turbine designed for it. If you were to build the first system, I would hate to put all of the potential advances and risks on that one system, watch it perhaps not work, and then say, that concept is worthless. I think that's something you would want to do in a third or fourth version. I don't think you pick up overall efficiency if you have a combined cycle. Our calculations on Chang Cycle gas turbines indicate that relative to a combined cycle, you won't increase overall efficiency, but relative to an open-cycle gas turbine, you will increase overall efficiency.

DETERMINING COGENERATION FEASIBILITY AND ECONOMICS OF ELECTRIC POWER FROM SUGAR CANE

John P. Keppeler, B.S. Agr., General Manager
Russel F. Sowers, B.S.I.E., Agriculturist
Energy Cane Project
Ronco Consulting Corporation
Washington, D.C. 20006

ABSTRACT

This paper deals with the important factors to be considered in competently determining the technical, financial and economic feasibility of generating electricity for the national utility's distribution from the combustion of sugar cane waste at the Monymusk Sugar Estate and Factory in Jamaica. This proposed cogeneration scheme is based on the most efficient use of an indigenous and renewable source of biomass fuel -- bagasse (i.e., the fibrous residue of processed sugar cane) and trash (i.e., sugar cane tops and leaves) -- from the intensified production of sugar cane processed at a major Jamaican agroindustrial complex.

DETERMINING COGENERATION FEASIBILITY AND ECONOMICS OF ELECTRIC POWER FROM SUGAR CANE

INTRODUCTION

An important step in the genesis and life of an alternative energy project is the feasibility evaluation. That step can turn a concept into a productive facility. This paper deals with some of the things taken into consideration in determining the technical feasibility of a proposal, its financial and socio-economic attractiveness for investment, and the community and business management conditions that would be required to encourage early placement of such an investment. The magnitude and complexity of certain alternative energy investments in important agroindustrial enterprises necessitates careful preparation of the evaluation in order to assure its conversion to a"bankable document" that presages actual construction and operation.

This paper covers a specific and current proposal to construct a high performance boiler/turboalternator complex of 25 - 30 MW_e in capacity at the Monymusk Sugar Factory in the Clarendon Parish, Jamaica, West Indies. The proposed cogeneration facility, that is planned for construction in 1989 and to be placed into operation in 1990, would far better utilize the sugar cane bagasse (the fibrous residue of the processing of sugar cane) to generate steam and electricity than is now accomplished at the site. The proposed power generation improvements could make the sugar factory and its field operations self-sufficient for its process steam requirements and electricity needs. In addition to the substantial steam and power produced for this important agroindustrial complex, possibly as much as one-third of the power generated during the processing operations could be made available on a "base load" basis for sale into the national utility grid operated by Jamaica Public Service Co., Ltd. (JPS). Priced on a"full-avoided-cost" basis, this level of purchased power could be a significant addition to the company's annual revenues. During the "off-crop" period, when processing operations are suspended, additional power used in the sugar cane crushing and juice processing may be dispatched to JPS during most of the "off-season" to add to that revenue stream.

A proposed source of cost-effective, supplemental biomass that might be developed to significantly extend the indigenous fuel resources is the sugar cane tops and leaf trash currently not utilized. This potential new, but already produced, boiler feedstock is now regularly burned just before harvest to reduce harvest work, but could provide the calorific basis for an important amount of the electricity available for sale to the utility.

The envisioned cane trash recovery system would follow-on a modified practice of reaping (harvesting manually cut green, unburned sugar cane versus the present practice) to solar dry the trash by turning it with a strengthened hay rake and round-baling it for field storage and subsequent transport to the boiler site. The commercial establishment of this trash recovery system from the large acreage harvested annually will require considerable and carefully planned developmental efforts.

It is estimated that this baled and dried material might provide a full one-third of the boiler feedstock (over 100,000 long tons per year) in a highly storable form that has a calorific value similar to bagasse. The trash bales would require shredding in preparation for firing. The storability and portability of baled trash may prove to be financially attractive in allowing the full utilization of the generating complex during the "off-crop" period without having to resort to fossil fuels to meet the feedstock needs during this significant period of the year (34% of the time).

Another possibility that is to be considered in extending the renewable biomass resource from sugar cane to more fully use the generating facility is to increase bagasse's storability by certain densification and drying techniques. This development could allow the bulky bagasse that is prone to spontaneous combustion to be cost-effectively stored for the "off-crop" period. Baled bagasse storage on a large scale and commercial basis enjoys a wider level of experience than does the baled collection of trash.

The "off-crop" period without directly resupplied biomass resources from sugar cane production would be 125 days long and can present a major scheduling hurdle in the adequate use of the proposed facility. Profitable use of this capital intensive facility will require year-round production of electricity. The preferred pricing of the facility's electricity obtained by its "base load" standing with the national utility will require being available to dispatch power on a scheduled basis for all of the year.

The benefits from the proposed facility are expected to flow from increased net operating profits for Monymusk, an increased national energy independence through the displacement of the need for so much imported oil, and decreasing the out-flow of scarce foreign currency required to be spent overseas for oil -- in calendar year 1985, Jamaica brought-in approximately 8.2 million barrels of feedstock at a cost of US$228 million, averaging US$27.80 per barrel (The Hon. Hugh Hart, Minister of Mining and Energy, The Sunday Gleaner, March 2, 1986, pg.1) Part of the retained currency from the displaced foreign purchase of oil might circulate in the Jamaican economy to secure existing sugar factory jobs, increase field labor requirements for the cultivation and harvest of additional sugar cane and be further spread in the economy through the subsequent purchase of other local goods and services.

TECHNICAL CONSIDERATIONS

Facility Size

The proposed complex will be constrained in its sizing by the amount and quality of sugar cane bagasse regularly available to it. Bagasse will be the major biomass resource for this facility. Bagasse is expected to be dewatered to about 47.5% moisture as an output of the sugar cane crushing and juice extraction process with a possibility that a part of that flow may be further dried to about 20% moisture and densified. The daily supply of bagasse will be determined by the sugar cane through-put of the Monymusk sugar cane crushing plant. Sugar cane trash, that is solar dried and baled, is proposed to be the primary supplemental biomass fuel.

Monymusk mill and its cane and sugar production systems are now undergoing an extensive rehabilitation program to regain a previously attained and a higher production norm than has recently been experienced. Currently, the production level at Monymusk has slipped to about forty percent of what could be expected as a normal output. That slide in cane and sugar production has taken place over the last decade since the record production levels were established. Profitability of the sugar cane production and processing operation, as well as the power generating operation, will depend on adequate levels of cost-effective sugar cane production. The profit sensitivity to high and sustainable levels of crop production is due to an inherently high level of fixed operating costs found in this type of capital-intensive agroindustrial enterprise.

Upon rehabilitation, the annual through-put at Monymusk, a function of its effective milling capacity and the efficient use of the scheduled grind time, is anticipated to return to the rate of 3600 long tons per day at an overall efficiency of 75 percent for an annual production of 742,000 long tons of sugar cane and about 70,000 long tons of sugar. The harvest and processing campaign norm after rehabilitation would take 206 days in a 35 week period from mid-December to mid-August each year. This scheduling results in a campaign of 240-245 days (eight months) with an "off-season" or "off-crop" period of 120-125 days (four months).

As part of the planned rehabilitation program, there has been some consideration of the possibility of merging the sugar cane supplies and certain operations of an adjacent sugar estate and factory (New Yarmouth) into

Monymusk. The benefits of a consolidation of the production and processing operations in the Clarendon sugar cane growing area could take advantage of the economies of scale that might be had and further strengthen the Monymusk Factory and Power Station's sugar cane supplies.

In bracketing the range of sugar cane production outputs that might be possible at Monymusk after rehabilitation and merger, a 15 percent plus and minus range (630,700-853,300 long tons per year) was established around the 1989 crop target of 742,000 tons. That range's upper limit of 853,300 tons exceeds the record production of 1975 by only 1.8% (the 1975 record crop at Monymusk processed 837,899.3 tons of sugar cane and 83,664 tons of sugar at a TC/TS of 10.02) and might represent the very best results that could be looked forward to in the near term.

The estimated amounts of biomass fuel that might be derived from the range of sugar cane production and processing possibilities and the sugar cane trash recovered from those fields harvested are tabled below in long tons:

BIOMASS	LOW	MID	HIGH	NOTES
Bagasse	182,000	214,100	246,200	28.85% of net cane
Trash	94,600	111,300	128,000	15% of net cane
Total	276,600	325,400	374,200	
Use Rate/Day	820	964	1109	337.5 power days

In the development of this proposal, very little resource is anticipated from biomass or wastes other than those resources derived from the production of the estate's or growers' sugar cane.

In the long run, the estate's and farmers' efforts at further crop intensification with the growing of new high-performance sugar cane varieties under production optimizing strategies of cultivation (intensive tillage, drainage, irrigation, fertilization, etc.) could result in a significant additional tonnage being available for processing at Monymusk each year. The milling capability would then become the principal constraint on the amount and quality of bagasse being available for power generation. The annual through-put at Monymusk might be increased by materially improving the overall efficiency by up to four hours per day or 1040 hours per year and by increasing the campaign to 43.3 weeks or 260 process days in order to crush 208,000 additional tons of sugar cane for an annual total of 950,000 tons. Production beyond this level would be limited by marked deterioration of cane quality, unfavorable field conditions for harvesting and crop transport and the need for sufficient time to perform the annual maintenance requirements. Production at this level is 28% greater than the base case used with an output of 70,000 long tons of sugar.

During the investment period of thirty years (1989-2019), it is possible that the selection of certain high-performance sugar cane varieties might identify higher fiber clones that would produce no less sugar for the level of production inputs but would have significantly higher fiber and therefore, a higher calorific content than current commercial varieties employed in Jamaica. The degree of increase in the fiber content that would be commercially attractive would be limited by the added cost to harvest, transport and crush the poorer quality sugar canes, and the value of the sucrose entrained and lost in the additional bagasse of the higher fiber canes, as offset by the increased fuel value of the resulting bagasse.

Production expansion beyond both long term developments, significantly increasing the volume of the higher-quality sugar cane crop to be processed (28% greater than the base case or 950,000 long tons per annum) or increasing the fiber content by the same magnitude through a poorer-quality sugar cane crop of the same size as the base case (i.e. 742,000 long tons per annum), will necessitate major crushing plant improvements. Therefore, the near-term sizing of the power generation station should be constrained to the biomass resources derived from a sugar cane crop of 850,000 long tons annually at 4140 tons cane per day at a 86.25% overall efficiency.

Mode of Operations

The processing operations over the eight months of the campaign will dominate the production scheduling at Monymusk. The significant block of electricity available for sale to JPS, after all the factory and field requirements of the sugar company have been met, may be as much as one-third of the total power produced by the cogeneration facility to be installed.

Following processing operations and during the "off-crop" period of 125 days, power normally used by the sugar company (for sugar cane crushing, cane juice processing, raw sugar refining and molasses distillation) will become available in blocks for power sale to the utility. Occasionally, when there is sufficient rainfall to suspend irrigation, the electrical energy used by the irrigation pumps can also be made available for dispatch through JPS for limited periods.

The level of electrical energy to be dispatched to JPS can be scheduled in advance and should be incorporated in the Power Purchase Agreement. It is envisioned, then, that there will be three (3) levels of power exported from the facility: Level C - the highest amount dispatched as the irrigation load is suspended and its energy is re-directed; Level B - the second highest amount dispatched for about 44% of the time, with over 90% of the generation being exported from the facility and Level A - the lowest amount of power dispatched for 56% of the time, with about one-third of the generation being made available for sale to JPS.

The preferred fuel source for this project will be from locally produced biomass over imported fossil fuels. Because the power production is desired throughout the year, considerable biomass storage will be required.

The storage of bagasse, most of which will be at 47.5% moisture and directly from the last mill tandem, should be in an automatic reclaiming house of about 1,000 long tons in capacity, that can be continously operated from the power station's control room. The reclaim of dried and baled bagasse and trash should be processed through the bagasse house after shredding for presentation to the furnace. Trash at 35% moisture is not expected to be a problem, but baled bagasse dried to 20% moisture and then pulverized for ease of feeding to the furnace may present problems of premature combustion, unless adequate safeguards are provided.

The re-supply of the bagasse house from stacked bale storage in the millyard or from field-edge storage sites will be a daylight operation designed to move the daily requirement of 1,000 long tons of baled bagasse or trash to and through the furnace preparation equipment to the bagasse house in an eight hour shift. The storage areas for the baled materials need to be carefully selected to assure good drainage and trafficability with year-round access. These areas need to be reasonably secure and there should be safegaurds set in place to reduce the risk and the consequences of accidental or malicious burning of these stored materials with their comparative oil value.

The collection, storage and retransport of 200,000 bales of solar dried trash and/or the treatment and baling of up to 50,000 MT of bagasse are major, new logistical programs. New cost-effective materials handling systems will have to be developed to keep the cost of these alternative energy resources below or at their economic shadow value of approximately 1.5 times the price of a barrel of oil in Jamaica per ton, times an economic factor reflecting the value to the Jamaican economy of the retained foreign exchange.

Facility Design

The high-performance boiler and the turboalternator design for this facility should be guided by several bagasse-fired generation complexes in successful operation in other sugar cane production and processing areas. The operating experience with these proven units shows a preference for 850 psig at 825 degrees F. At this point the efficiency appears to optimize for the capital expended and the long run operating and maintenance costs in the agroindustrial setting of a sugar estate. These successful units currently in operation employ suspended boiler wall construction and use travelling grates. Varying environmental standards cause differing levels of concern with stack emissions and other waste disposal.

The installation of an efficient steam-electric generation unit should trigger a very careful review of the energy use in the sugar cane crushing and cane juice processing system to insure the greatest conservation of the steam that once was low-valued; a thorough steam balance is to be accomplished as part of the planning discipline in the design of a new facility.

The facility layout should not preclude the easy access of fossil fuels to the furnace. A certain amount of fuel oil storage and feed may be required to properly augment the bagasse flows and to act as short-term back-up to the bagasse. Coal supplies in this region may make an attractive alternate to a full dependence on biomass resources. The design of the coal bin and feeder system should allow for concurrent feed from it as well as the bagasse house.

The generating equipment should be designed so that the annual maintenance can be performed in two or more phases across a 55 day long strech during the "off-crop" period. Additional steps may be designed into the generating capacity to best fit the anticipated loads.

Agricultural Production

The ability of the company's plantation and the adjacent cane farmers to produce an adequate and cost-effective sugar cane crop and deliver it in a regular manner for processing at the Monymusk Factory takes on an added importance when a large, new investment is contemplated. The profitability of the sugar production activity requires that the processing of rather substantial volumes of sugar cane be accomplished each year (i.e. 742,000 long tons from 22,485 acres reaped). This production requirement for a large annual crop is shared by both of the business objectives -- maximizing net revenues from the sales of the traditional products (sugar and molasses and potable alcohol) and the new alternative energy product (displaced electrical purchases and exported electricity sales). The current rehabilitation program, funded by a loan from the World Bank, to restore the production output of the Monymusk Factory and Estate, as well as influence much greater sugar cane production by farmers in the Clarendon Plain and near-by, will warrant very close monitoring as it will greatly affect the feasibility of the cogeneration facility. The possibility of a New Yarmouth/Monymusk merger will bear careful review as it could further enhance the profitability of the successor.

The sugar cane growing area in the Clarendon Plain and certain adjacent areas have been producing cost-effective sugar cane crops for a long period of time. Experimental results and outstanding field production experience support a belief that substantial increases (50% in the near term and 100% in the longer time frame) to the general area's yield could be attained in sugar cane production, provided that production maximizing inputs are made available. There are production problems with poorly drained lands along the sea coast that have developed saline or sodic conditions near this sugar mill, but the primary sugar cane production problem at Monymusk and in the mid-Clarendon area is the lack of good quality water and the timeliness of its applications.

The requisite actions to substantial and sustainable increases in sugar cane production are numerous, and must be treated in detail by each section of cane farm land of a similar soil type, serviced by the same pumped water source. Measurements of the farm's available insolation, maximum-minimum and average temperatures, rainfall, the soil's moisture holding capacity, correlations with evaporation pan data, if available, and so forth are taken into account before the irrigation requirements are determined for each set of cane fields. The irrigation interval and the amount of water required can be determined by time of year and physiological growth stage of the cane crop. Water quality can influence the frequency and sufficiency of application desired. Proper field layout can materially influence the efficiency of the application of valuable water resources. The production limitation of inadequate irrigation water supplies can only be rectified by systematically improving, over time, the pumping capacity and reliability and the conveyance efficiency of that important resource.

The importance of good field layouts cannot be over-emphasized in modern sugar cane plantations. Drainage arrangements for surface run-off as well as the sub-surface requirement, need to be carefully designed with the surface irrigation system so as to facilitate the mechanical operation sequences that may be used in tillage, furrowing, planting, seed covering, ratooning, blank replanting, irrigation ditching, fertilization, weed control, through the loading of harvested sugar cane and its infield transport onward to the mill.

A good job of land levelling, as the first step in a field's renewal, is essential to significantly increasing application efficiency of controlled irrigations. In gaining greater conveyance efficiency, levelling reduces the drainage requirements of the field undergoing irrigation as well as making additional irrigations possible in that plantation section.

An accelerated renewal program of 2,000 to 3,000 acres per year or about a 20% renewal, is being implemented at Monymusk with plans for making field improvements (i.e. new field layouts, land levelling and drainage) as new high performing varieties are being planted.

The renewal program also allows for extensive tillage work to be accomplished to improve soil aeration and drainage as the old stand is destroyed and compacted areas broken-up by the plowing and harrowing operations. A renewal of declined fourth ratoons (a ratoon is the regrowth cropping of sugar cane from the perennial root stock) or a Cropping Cycle of Plant and Four Ratoons is a marginal program to keep production at barely adequate levels. At this rate of renewal, it will take five years or until the early 1990's before the substantial benefits are gained from the improved field layouts, more efficient irrigation, better drainage, and the higher yielding varieties from the Estate's fields. Influencing the cane farmers to make parallel improvements to their areas may take longer.

The assessment of Monymusk Estate's ability to produce croppings large enough, and cost-effectively in the future requires a detailed understanding of its effective production practice. That practice review, beginning with the appropriate selection of cane varieties for an area, fertilization for that selected clone, its irrigation requirements, and every other aspect of the cultural practice and the precision of each operation performed (i.e., seed selection and treatment, seed placement, depth of covering, first irrigation timing, checking adequacy of covering, weed control, fertilization, irrigation, ripening, timely harvest and transport, etc.) needs to be carefully understood to make an accurate production estimate. Because just under half of the sugar cane produced and delivered for processing by Monymusk is grown by cane farmers, their practices and results need to be reviewed as well. Further, the grower relations program for the company should be reviewed in detail for its effectiveness as an extension program to assure a steady and regulated supply of sugar cane for processing at the plant.

There are recommended field services in addition to the out-growers program that need to be in place to assure maximum production. Progressive sugar cane production organizations include a staff department that is charged with the ongoing applied research, including the important varietal selection program, crop scheduling, and crop control functions (i.e. production records, crop sampling, etc.). This staff department provides the intelligence and control monitoring necessary to manage the tens of thousands of acres in cane, and hundreds of pumped water sources, that are farmed by several thousand employees and farmers to produce, harvest and transport a sugar cane crop of adequate commercial size.

A part of the sugar cane production system that will grow in importance in the future is the field machinery -- the tractors and implements for field improvements, tillage, drainage canal cleaning, irrigation ditch maintenance, cultivation equipment, cane loading and transport, etc. -- that is used to hold the cost of production in line as a vital segment of this business. The improvement to field layouts during renewal operations should be made to facilitate the use of farm machinery and transport. During the life of this investment into the 21st Century, serious consideration may be given to mechanical harvesting to allow for cost-effective, one-pass reaping of the sugar cane crop to utilize it entirely for the production of its multiple products including electrical energy. The future progress of cost-effective mechanization of field operations will be very much dependent on the repair and maintenance facilities and programs to be established to support that capital intensive, mobile equipment.

In direct confirmation of the predictive values on annual sugar cane crop size that will be used in this feasibility evaluation, a limited number of crop intensification trials will be run in representative field areas for two croppings -- Plant Crop in 1986-87 and a First Ratoon Crop in 1987-88. These yield trials will demonstrate and confirm the production capability of the sugar cane lands of the Estate and cane farmers with the best yielding varieties and cultural practices that can be recommended by the Jamaican industry's research and crop production experience. A significant part of the sugar cane production area near this factory is affected by poor drainage conditions and, as part of the crop intensification trials, improved drainage techniques are to be evaluated. The general production area's near-term yield target of a cumulative rate of 3 tons of cane per acre-month or 36 tons cane per acre-year under a 20% renewal cropping cycle, and the long-term goal of 4 tons of cane per acre-month or 48 tons per acre-year, can be evaluated against these monitored tests.

The field machinery systems proposed to solar dry and collect the sugar cane trash in bales for field storage and transport to the furnace preparation equipment at the bagasse house will require demonstrations during the next two years to confirm the cost and production estimates made in the feasibility study. The capital requirements for this system of collecting trash may require allocation of the costs of field layout and land levelling that can be attributed to improved productivity of field machinery including the trash recovery operation, if feasible. A certain level of field improvements will be required to assure the productivity and operating cost values used in this evaluation.

FINANCIAL CONSIDERATIONS

The financial soundness of investing in this cogeneration scheme at the Monymusk Factory will be first dependent on the future profitability of the sugar cane production and processing operations. The domestic and European Economic Community (EEC) markets for the raw and refined sugar to be produced at the Monymusk facility are quite secure and at favourable and stablized prices by national and Commonwealth policy. Jamaican raw sugar production beyond the 225,000 long ton annual level during the next few years would not find a ready or lucrative market, but supplies are currently 8% below the available regulated market. Molasses sales through Jamaican rum and livestock feed marketings are also assuredand represent about five percent of the annual revenues of a sugar cane operation. The generation of sufficient steam and electrical energy to make the sugar operation energy self-sufficient except for mobile fuels, especially in that it would be no longer required to purchase power from JPS to run irrigation pumps, will assist in keeping some of the future costs of sugar and molasses production under its own control. The crucial, short-term test of the profitability of the Monymusk Sugar Estate and Factory operation will be how it responds to the planned turn-around, to regain lost ground under new professional management and World Bank financing, in the upcoming few years (1986, 1987 & 1988) before this investment committment finally must be made.

The financial feasibility of the proposed power station in a cogeneration setting, using agricultural wastes as its primary fuel source, will be heavily influenced by the capital costs of putting that machinery and its supporting facilities in place, and the rate that the national utility might be willing to pay for the electricity dispatched to it under a purchase agreement. The facility should be operated to qualify for base-load consideration, and as a result could receive recognition in its pricing on a "full-avoided-cost" basis in the displacement of future requirements for electrical generating capacity on the island by the national utility and other electrical cogenerators. That part of the electricity that is of fluctuating availability may not qualify, and might be priced at a discount reflecting the loss of part or all of the capacity charge.

The financial planning that will go into this feasibility evaluation is focused on the fact that substantial investment capital will be coming into the country by private placement in a joint venture arrangement involving the pooling of resources by Jamaican and international investors. The investment promotion policies of the Government of Jamaica as they pertain to this particular type of investment (i.e. alternative energy development in an important agroindustrial business) must be thoroughly understood as part of this evaluation. Some of the critical policies for the overseas investor will include the customs duty to be charged on the imported big ticket items (i.e. boilers, turboalternators and controls), the speed of customs clearances on these items, all replacement parts and operating supplies, the rate of taxation, the access to the earned foreign exchange to pay for operating and maintenance items required from overseas, and the conditions imposed on the repatriation of earnings.

In the preparation of this project's financial statements for final evaluation, the estimated capital costs, including the expenditures to be capitalized under the rehabilitation program of the World Bank, will need to be carefully detailed. The annual forecasts of operating and maintenance costs for the sugar, molasses and power production, as paired with the yearly estimates of their anticipated revenues, will be set up for an evaluation of its profitability. An analysis of project risk by each component (i.e. cane production, cane processing, power generation and commercial market) will be required to provide the basis for a comprehensive evaluation. The forecasted statements should then be subjected to a sensitivity analysis and the project worth calculations need to be expressed for review.

BUSINESS CONSIDERATIONS

The social and economic setting for this investment will require evaluation as an important part of the feasibility considerations of this project. The government's and the community's attitude towards this project and its fit with the overall social development objectives will need to be clearly determined. In addition to the determination of the economic rate of return and the regional and national benefits that may be derived from this project, an understanding of the specific socio-economic impacts on the effected Clarendon Parish and its agricultural sector will have to be developed to allow a concise determination of the possible business settings for this project.

A wide range of considerations will require attention under this heading (i.e., the facility's environmental impact, the organization and management of the sugar and power corporation, unskilled and skilled labor availability, wage and benefit levels under existing union contracts, sugar and molasses marketing arrangements, JPS purchase power agreements, the condition and management of the community infrastructure, and more) in order to fully describe and characterize the setting of this investment. The possibility of the establishment of positive relationships with elements of the commerce of the surrounding community (i.e. cane growers' association, organized labor, production input suppliers, governmental offices, JPS, etc.) can be as influential on a successful outcome of the project as the technical and financial factors.

The careful development of a comprehensive document that competently describes the Jamaican Cane Energy Project is a serious undertaking. That feasibility document can then act as a vehicle for meaningful business and financial discussions that may lead to enabling the project.

ACKNOWLEDGEMENTS

The authors wish to extend grateful thanks to the following individuals for thier helpful support.

Earl S. P. Smith, P.E. for technical analysis of sugar cane factory and power generation principles on this project, Vanica Cranston and Dainty Rose Burnett for their assistance in preparation of this paper.

BIBLIOGRAPHY

Alexander, A. G., The Energy Cane Alternative, 509. Amsterdam: Elsevier, 1985.

Alexander, A. G., Sugarcane Physiology,752. Amsterdam: Elsevier, 1973.

Austin, J. E., Agro-industrial Project Analysis,214. Baltimore: Johns Hopkins Univ. Press, 1981.

Barnes, A. C., The Sugar Cane,572. New York: John Wiley & Sons, 1974.

Barry, R. D., Sugar & Sweetener Outlook and Situation Report, Washington: U.S,D.A. Economics Research Service, 1985.

Batson, H. F., Ferguson, T. U., Archibald, K. A. E., "Variability in Leucaena and Its Potential in the Caribbean", Ciencia InterAmericana, 24, 12-23 (1984).

Black, C. V., The Story of Jamaica, London: Collins, 1965.

Block, F. E., "Economic Development of Jamaica", Kingston Commemorative Album (1969).

Brewster, H., The Sugar Industry: Our Life or Death, Kingston: World Group, 1967.

Chuck, M., "Sample Survey of Registered Cane Farmers Supplying Cane Factories in Jamaica, 1962 Crop, "Min. Agr. & Lands (1963)

Clements, H. F., Sugarcane Crop Logging and Crop Control: Principles and Practices, 520. Honolulu: Univ. of Hawaii Press, 1980

Cumper, G. E., "Labor Demand and Supply in the Jamaican Sugar Industry 1830-1950, "Soc. & Econ. Studies, I.S.E.R., U.C. W. I., 2 (1954) March.

Davidson, R. B., Evolution of the Jamaican Economy, Kingston: U.W.I. Dept. of Economics, 1971.

Davidson, R. B., "The Labour Force in the Jamaican Sugar Industry: Labour Shortage and Productivity in the Jamaican Sugar Industry, "U.W.I. Mona Inst. of Soc. & Econ.Res. (1966)

Dillewijn, C. Van, The Botany of Sugar Cane, Waltham, Mass: Chronica Botanica, 1952.

Eaton, G., "Trade Union Development in Jamaica", Caribbean Quarterly, 8,1 (1962)

Evans, C. J. and Vickers, D. O., The Climate of Jamaica, 68. Kingston: National Meteorological Service, 1973.

Evans, W. et al, "Sugar Industry Studies - Irrigation, Phase II Review", Jam. Min. Agr. (1985) May.

Giamalva, M. J. and Clarke, S. J., Biomass Production from Saccharum Species, Baton Rouge: Louisiana State Univ., 1983.

Girvan, N. and Jefferson, O., Readings in the Political Economy of the Caribbean, Kingston: New World Group, 1971

Gittinger, J. P., Economic Analysis of Agricultural Projects, Baltimore: John Hopkins Univ. Press, 1982.

Horton, R. W., "Developments in Boiler Design and Installation", Int. Sugar Journal, 82, 308-315 (1980) Oct.

Humbert, R. P., The Growing of Sugar Cane, 799. Amsterdam: Elsevier, 1968.

Jefferson, O., The Post-War Economic Development of Jamaica, Kingston: U.W.I. Inst. of Soc. & Econ. Res., 1972.

Kampf, H., "Energy from Biomass", Int. Sugar J., 82, 39-40 (1980) Feb.

Keenliside, W., "Economics of Bagasse Utilization", Jam. Assn. of Sugar Tech., 39, 26-27 (1979).

King, N. J., Mungomery, R. W., & Huges, C. G., The Manual of Cane Growing, 375. Sydney: Angus & Robertson, 1965.

Kirkaldy, Introduction to Industrial Relations & Labour Laws in Jamaica, Kingston: Univ. of W. Indies, 1979

Lehrer, M., Jamaica - Annual Sugar Report, Washington: U.S.D.A. Foreign Agricultural Service, 1984.

Lipinsky, E. S., and Kresovich, S., Sugar Stalk Crops for Fuels & Chemicals, Columbus: Battelle, 1980.

Lipinsky, Jackson, Kresovich, Arthur & Lawhon, Carbohydrate Crops as a Renewable Resource for Fuels Production, vol. 1: Agricultural Research, Columbus: Batelle, 1979.

Mahin, D. B., "BioEnergy for Agriculture", BioEnergy Systems Report, (1983) Mar.

Mahin, D. B., "BioEnergy from Crop Residues", BioEnergy Systems Report, (1983) Dec.

Mahin, D. B., "Biomass Fuels for Vehicles", BioEnergy Systems Report, (1982) Sept.

Mahin, D. B., "Growing Trees for Fuel and Wood Fuels for Industry", BioEnergy Systems Report, (1982) Dec.

Mahin, D. B., "Innovations in Biogas Systems & Technology", BioEnergy Systems Report, (1984) Apr.

Mahin, D. B., "International Conference on Biogas Technology", BioEnergy Systems Report, (1985) Mar.

Mahin, D. B., "Thermochemical Conversion of Biomass for Energy", BioEnergy Systems Report, (1982) Jun.

Minott, D. A., Biomass: an Overview of Fuels from Biological Sources, Kingston: International Renewables, 1985.

Nettleford, R., "Trade Unionism: Impact on Jamaica", Sunday Gleaner Supplement, (1984) May 6.

Paturau, J. M., By-Products of the Cane Sugar Industry, Amsterdam: Elsevier, 1969.

Ransford, O., The Slave Trade, London: John Murray, 1971.

Rao, S. M. & Shaw, M. E. A., Review of Research on Sugarcane Soils of Jamaica, Mandeville: Sugar Industry Research Institute (Agr. Div.) 1983.

Rao, S. M., "Report on Quality of Water at Monymusk", Jam. Assn. of Sugar Tech., 42, 23 (1981)

Reckord, C., "Jamaica's Sugar Industry", Jamaica Journal, 3, 27-33 (1969).

Runge, P., "The West Indian Sugar Industry", J. Royal Soc.

of Arts, 109, 91-104 (1961) Jan.

Salmon, M. G., "The Social Economy of Sugar", Kingston: Sunday Gleaner Supplement (1984) June 24.

Samuels, G., Foliar Diagnosis for Sugar Cane, 362. Rio Pedras, Puerto Rico: Agricultural Research Publication, 1969.

Sangster, I., Sugar and Jamaica, London: Nelson, 1973.

Shaw, M. E. A., Agronomic & Harvesting Operations Management - Variables Affecting the Production of Excess Quantities of Bagasse in Sugar Factories, Mandeville: Sugar Industry Research Institute, 1984.

Shirley, A. G., The National Energy Outlook 1985-1989, Kingston: Petroleum Corp. of Jamaica, 1985.

Stanford, D. C., "Agro-Industrial Crops with Production Potential on Sugar Estates", Jam. Assn. of Sugar Tech., 42, 14, (1981).

Summers, D., "Mechanical Harvesting", Sugar Cane, 15, 2 (1985) Jun.

Suzor, N. L. C. & Bouvet, E., Identifying the Basic Conditions for Economic Generation of Public Electricity from Surplus Bagasse in Sugar Mills, Washington: World Bank Energy Paper #13, 1983

Suzor, N. L. C., "Pelletization and the Rational Utilization of Biomass Energy in the Cane Sugar Industry", Sugar y Azucar Yearbook, (1984).

DISCUSSION

J. Veigel, North Carolina Alternative Energy Corp.

Would it ease the economics at all and is it institutionallly possible to talk about the utility and your company doing this as a joint venture?

Keppeler

Yes, in this case the utility and the sugar company have the same president, who is the Prime Minister of Jamaica, so it's a consolidated operation already. Our charter is to try to place this as a bankable document for private joint venture investment to carry out the objectives of the government. Privatization would require some overseas investment, so it's structured as a Jamaican overseas joint venture. If we are able to make a case for feasibility, the probability of the project happening is much higher as opposed to packaging it for study only.

J. Earle, U. of Florida

Where do you propose to get the feedstock for your boilers during the off-season?

Keppeler

This is a critical issue. Two strategies are available to us. One strategy is the recovery of cane leaves and tops from green, hand-reaped fields where manual cutters segregate the tops from the stalks and make a physical segregation. The follow-on operation would use a technique that we've titled "barbojo" because we've heard about it from the Dominican Republic where the trash is turned with a modified hay rake for solar drying and baling. We're fully aware that this is an extremely marginal operation economically. Costs and the net energy balance make it very difficult if the fields are not properly laid out. Regenerative growth from root stock begins in Jamaica normally within 5 days of reaping and doesn't stop for Sunday. So if you don't get in and turn it twice for solar drying and get the bailers in, you start to have ratoon damage on regrowth. The second strategy has to do with saving some bagasse and bailing it for future use. But we have a large period of off crop in Jamaica — 125 days or so, and so the storage of 100 metric tons of material is quite an undertaking. This issue is crucial in the determination of whether or not the project is feasible. There is a fall back, and the fall back of course is to use fossil fuels. We may have to use fossil fuels to buy time to perfect biomass recovery systems. But we won't make that determination until we see how far off the perfection of the system is.

__________ The total cost of this cogeneration facility is about $45 million. What is the payback period?

Keppeler We have not yet made the calculations of the return on investment, but the facility will have several revenue streams. One is the exportable sale at three levels from the Jamaica Public Service, a public national utility. It will also sell back on an internal transaction to the sugar factory, both power to run its pumps and other electrical requirements in the factory as well as on the estate, and low grade steam on an interdepartmental transfer basis for the crushing and boiler operations. If you were a stand-alone profit center, it would have several streams of revenue. Our business right now is to quantify these streams to see if we can justify the investments.

TROPICAL MACROALGAL CULTIVATION FOR BIOCONVERSION TO METHANE

Kimon T. Bird, Ph.D.
Associate Scientist
Division of Applied Biology
Harbor Branch Foundation
Ft Pierce, FL 33450

ABSTRACT

Several concepts have been developed for tropical marine biomass cultivation for bioconversion to methane. These concepts take advantage of Florida's large areas of relatively shallow water. One concept, tidal flat seaweed farms, uses currently available macroalgal candidates (*Gracilaria*, *Ulva*) and at biomass yields of 12-25 dry ash free tons/hectare-year can provide delivered low feedstock costs of \$40-25/DAFT, or on an energy basis, \$3.60-2.30/G joule, respectively. These biomass yields are close to those achieved in commercial *Gracilaria* culture in Taiwan. Such systems would be constrained to nearshore waters of 0.5-1.5 m in depth, of which there are 190,000 hectares in northwestern Florida.

Concepts which would work in deeper waters (from 1.5-20 m depths) use floating seaweeds. Such biomass species would need to be produced by genetic breeding and hybridization, as there is not an adequate natural species available which also has high bioconversion rates. Such hybrids may be intrageneric ones of *Sargassum*, or *Sargassum* hybrids with other algae such as *Macrocystis*. A biotechnology approach could provide competitive feedstock costs with a large potential gas production, as there is approximately 1,900,000 hectares between 1.5-20 m depths in northwestern Florida.

TROPICAL MACROALGAL CULTIVATION FOR BIOCONVERSION TO METHANE

INTRODUCTION

Marine biomass cultivation for renewable energy has centered primarily on cultivation of cold water kelps on offshore structures, nearshore farms, and Chinese long line systems. The Harbor Branch Foundation, located in Florida, has been developing cultivation practices for tropical macroalgae as biomass resources. The initial efforts focused on determining maximum productivities of seaweeds, and later, cultivation practices for the red alga, *Gracilaria tikvahiae*, and the green alga, *Ulva* sp. These experiments were largely carried out in land based ponds or high intensity seaweed raceways (8,9,10). Later, the focus changed to "in the sea" cultivation of the brown algae, *Sargassum* spp. *Sargassum*, which can float on the sea surface like water hyacinths on lakes, has great appeal as engineering studies with various off-bottom kelp concepts have revealed that very high biomass yields are required to overcome the high capital costs. A floating seaweed could be contained in floating enclosures which are being developed at University of Florida for water hyacinth. Floating containment systems are already being used for commercial marine fish culture in Japan and Norway. The results of the research on *Sargassum* have been recently summarized (2,7). Economic analyses of water hyacinth cultivation (3,14) have been extrapolated to estimate cultivation costs of floating *Sargassum* (1).

Little attention has been paid to developing a viable concept for using *Gracilaria* or *Ulva* as a biomass resource. *Gracilaria* production can be sustained year round, and seasonal yields are closely correlated to seasonal light availability (9). However, the greatest limitation to high *Gracilaria* yields is water turnover rate in the culture system (8,11). Low water turnover rates lead to high increases in culture pH due to photosynthesis, which in turn reduces the availability of CO_2 (12). In commercial *Gracilaria* cultivation in Taiwan, the water turnover rate in ponds is on the order of once every 20 days, and results in biomass yields of 14 dry tons/hectare-year (ca. 9 dry ash free tons/hectare-year, Ref. 13). In high water turnover rate systems (10 or more exchanges/d), biomass yields can be as high as 150 dry tons/hectare-year (DT/HA-Y), and in small ponds receiving one to two exchanges per day, in the range of 25-50 DT/HA-Y (ca. 15-30 dry ash free tons/hectare-year, Ref. 10). It is therefore important that any concept for *Gracilaria* or *Ulva* biomass cultivation must include a means of providing sufficient water turnover to maintain reasonable biomass yields.

THE TIDAL FLAT FARM CONCEPT

Two concepts were initially explored, one of diking off ponds in shallow water and pumping water in and out, and the other of netting in shallow water areas and letting the tide and currents provide the water exchange. The first concept was quickly rejected as it requires energy consuming pumps and expensive construction of dikes. The other concept, tidal flat farming, has great appeal as tides would provide one to two water exchanges per day, depending on the tidal cycle. In addition, construction methods and materials are fairly inexpensive. The concept involves enclosing areas of 1.5 m or less in depth, using netting supported by pilings. The seaweed would grow in the enclosure, and would be harvested daily by harvesting boats entering through boom gates. The seaweed would be shredded during harvesting and upon return to the dock facility, would be pumped out of the barges to the digester facility. In order to estimate potential economic feasibility of such a concept, a preliminary idealized farm and harvesting system has been costed out, using contractors' quotes or cost data from the engineering analysis of the Macrocystis (kelp) nearshore farm concept (4). Feedstock production costs have estimated at two different biomass yields of 12 and 25 dry ash free tons/hectare-year (DAFT/HA-Y). The lower end of these yields are close to those obtained in commercial scale cultivation (12 vs ca. 9 DAFT/HA-Y). The upper end was estimated based on biomass yields obtained in Florida in ponds receiving 1-2 exchanges/day (ca. 30 DAFT/HA-Y), and allowing for decreases due to loss, herbivory, and scale up effects on production.

The conceptual farm is circular, of 19 kms circumference, and 5344 hectares in area (based on best case of kelp economics). The enclosure is constructed of creosote pilings driven into the sediment every 5 meters, with a stainless steel cable joining the pilings at the top. One inch square nylon fish netting is secured to the cable and the pilings, with the weighted bottom buried in the sediment. Six floating boom gates allow access to the farm at various locations, depending on harvesting schedule, wind conditions, etc. There are 10 kms of drift seaweed fences within the farm, arranged to prevent all the biomass from concentrating in one location, and breaking through the net fence. These drift fences are constructed in a like manner as the farm perimeter. Capital costs are shown in the Table 1. In the event a circular farm is not feasible, a long rectanular farm, 1 km wide, would cost 15% more to construct.

Annual O&M for the farm consists of two components: farm maintenance and seaweed cultivation. The major cost estimated for farm maintenance is the replacement of all the netting every two years in a year round replacement operation. Replacement would be necessary due to biofouling of the netting and material degredation. Seaweed cultivation costs are primarily labor costs of farmers who must ensure that the seaweed does not pile up too heavily in select locations. The farmers are equipped with small boats to move seaweed mats around. Farmers will also be needed for weed and herbivore control. In addition, marine biologists must ensure the seaweed is adequately fertilized, and help plan harvesting operations. A nursery/laboratory has been included in the capital costs in the event that several species or clones are used over the course of a year, to maintain important clones and inocula, and for water chemistry. Possibly, germlings grown on vermiculite could be used for the large scale farm inoculations which would be involved (5).

The harvesting system for a tidal flat farm is fundamentally the same as the kelp system, specialized harvesters pumping kelp into barges. As the water depth is shallow (less then 1.5 m depth) all the equipment will be shallow draft (30 cm) with extra floatation via pontoons. The harvester will have a collection treadmill similar to a

TABLE 1. CAPITAL AND ANNUAL COSTS OF TIDAL FLAT FARM

CAPITAL COSTS

ITEM	NUMBER	COST
FARM		
Pilings, 5" diameter, 2.5 lb/ft^3 CCA treated. Installed cost.	8,448	$278,287
Piling - Cable connections	8,448	845
Cable, 1/2" diameter	130,000 ft.	76,700
Netting	18 bales	14,900
Net anchors		316,800
Net ties	90,133	2,700
Boom gates	6	30,000
Permits, etc.		50,000
Labor		104,000
Boat, equipment rentals, $300/d		37,500
		$ 911,732
15% Contingency		136,760
		$1,048,492
10% profit		104,849
		$1,153,341
4% engineering fee		46,134
Subtotal		$1,199,475
LAND BASED NURSERY/LABORATORY		
Laboratory, 1000 sq. ft.		$ 120,000
Greenhouses, 10,890 sq. ft.		217,800
Subtotal		$ 337,800
TOTAL (Farm and Lab-Nursery)	ca.	$1,538,000

ANNUAL COSTS

Farm Maintenance		
50% of netting		7,450
50% of ties		1,352
3 people, maintenance $8/hr(burdened)		49,920
		$ 58,722
Operations		
6 farmers, $8/hr(burdened)		$ 117,444
Lease		
$12.50/hectare, 5344 hectares		66,000
Subtotal, Farm O&M		$ 242,166
Nursery/Laboratory		
O & M, 3 people		$ 140,000
Fuel		$ 84,000
TOTAL O & M		$ 382,166
TOTAL FUEL		$ 84,000

kelp cutter, but the treadmill will not be fixed, rather be hinged. The seaweed will be piped to a towed barge after it is shredded by the harvester. For the 12 DAFT/HA-Y system, only one non-motorized barge is required. In the 25 DAFT/HA-Y system, motorized barges will be used, and the harvesting speed will have to increase to 1.5 kts. When the barge is full, it will return to the docks as an empty one is rotated in place. Crews from the empty barge will join the harvesting crew. The harvester will also spray digester effluent on the farm as a source of fertilizer. Digester effluent has been demonstrated as an effective fertilizer for *Gracilaria* (6), and recycling such effluents saves both disposal costs and fertilizer purchase. Harvesting will occur over 300 d, with 65 d allowed for inclement weather. Due to

TABLE 2. HARVESTING COSTS AT DIFFERENT BIOMASS YIELDS

FOR 12 DAFT/HA-Y		
Harvesting speed 1 kt		
Capital Costs		
1 Harvester at 1100 K		$1,100,000
1 Barge at 1000 K		1,000,000
2 Small boats at 10 K ea.		20,000
Dock		50,000
Pumps		200,000
Seaweed & effluent pipes to and from digester		2,000,000
		$4,370,000
	15% cont.	656,000
		$5,026,000
	10% profit	503,000
		$5,529,000
	4% eng fee	221,000
		$5,750,000
O & M		
Labor		
Harvesting	$ 1,408/d direct	
	535 38% fringe	
	582,912 for 300 days	
Maintenance	50,000 boat haul-outs & repairs	
2 Dock laborers	57,408	
	690,320	
[illegible]	103,548	
	$793,868	
TOTAL O & M		$794,000
FUEL		$190,000
FOR 25 DAFT/HA-Y		
Harvesting speed 1.5 kts (8.3 h total harvesting time)		
Capital Costs		
1 harvester at 1100 K		$1,100,000
2 motorized barges at 1500 K		3,000,000
2 small boats at 10 K		20,000
Dock		100,000
Pumps		350,000
Pipes		3,000,000
		$7,570,000
	15% cont.	1,135,500
		$8,705,500
	10% profit	870,550
		$9,576,050
	+ 4% eng. fee	383,042
		$9,959,092
Annual Costs	ca.	10,000,000
O & M	$1,408/d harvester	
	285/d motorized barge	
	$1,693/d	
	643	
	$2,336/d	
	$700,800 300 d	
	75,000 boat haul outs	
	57,408 2 dock laborers	
	$833,208	
	114,981	
	$958,189	
TOTAL O & M	ca.	$960,000
FUEL		$230,000

insuffient data, it has not been possible to build calculations for seasonality into this study; therefore, yield has assumed to be constant throughout the year. Seasonal peaks in growth, which probably occur in spring and summer, will have to be accomodated by increasing harvest speed and incorporation of larger capacity barges. Several small boats are also provided for in this fleet for farm maintenance. At the dock, the seaweed is pumped from the barges through pipes to the bioconversion facility. Pipes also return digester effluent to the barges for recycling on the farm. Table 2 details these costs.

Feedstock Costs and Cost Sensitivities

The total capital, O & M, and fuel costs, as well as delivered feedstock costs are shown in Table 3. Feedstock costs on an energy basis for the 12 DAFT/HA-Y yields are \$3.60/G joule and for 25 DAFT/HA-Y, \$2.30/G joule, assuming methane yields of 5.5 SCF/lb V.S. added and 85% net methane production after digester heating requirements are met. Recent bioassays of Harbor Branch Foundation *Gracilaria* clones by the Institute of Gas Technology have indicated a methane yields of 6-7.5 SCF/lb V.S. added. Feedstock costs on a weight basis for this system range from \$40-25/dry ash free ton, respectively. By comparison, feedstock costs for nearshore *Macrocystis* are \$80/DAFT at 25/DAFT/HA-Y; for *Sargassum* in floating farms, \$71/DAFT at 25 DAFT/HA-Y; and for *Laminaria* raised on long line farms, \$132/DAFT at 38 DAFT/HA-Y (1). Typically, these other macroalgal systems require biomass yields in the range of 38-50 DAFT/HA-Y in order to be price competitive, with the exception of *Laminaria*, which must be cultured on cost-prohibitive long line farms. When costs sensitivities for both capital and operating costs were performed at the two different biomass yields, the effect was only significant at low biomass yields of 12 DAFT/HA-Y when O&M costs increased by a million dollars. The greatest cost unknowns in the system are the final farm configuration (site specific), and total lengths of drift seaweed fences required in the farm. In the 25 DAFT/HA-Y system, if the farm is located more than 6 kms from the dock or is rectangular in shape, an additional harvester may be required.

Biological Constraints

The tidal flat farm concept is an untested approach to seaweed farming and as such, will encounter a number of constraints with regards to potential biomass yields. Key problems will be weed species which foul the biomass crop itself, cutting down substantially on production, and the impact of marine herbivores such as amphipods. Current, chemical control technologies for these pests are not well developed, and it may be necessary to develop selective algicides and herbivore control agents. Alternatively, co-culture of important carnivorous fish may provide herbivore control (and a significant economic co-product). Current speeds will also affect the choice of the biomass crop. In confined bays and estuaries, with low water movement, *Ulva* sp. may be better suited as it is well adapted to such water movement, while *Gracilaria* would be best suited suited for areas with stonger currents and greater water exchange. Perhaps most important, however, will be the seaweeds' interactions with the substrate. As the seaweed tumbles and moves across the bottom, a fine particulate substrate can cover the thalli. With wave and wind action,

Table 3. SUMMARY OF COSTS BY BIOMASS YIELD

Feedstock Costs		
DAFT/HA-Y	12	25
Capital	\$7,288,000	\$11,538,000
O & M	1,176,000	1,342,000
Fuel	274,000	314,000
G joule/year*	6.17×10^5	1.23×10^6
\$/G joule**	3.60	2.30
\$/DAFT	40	25
Sensitivities (\$/G joule)		
+\$1million Capital	4.10	2.46
+\$100,000 O & M	4.14	2.48
+\$1million O & M	5.63	3.22

*based on 85% net methane production from the conversion facility.

**based on 5.5 SCF/LB V.S. ADDED

the seaweed can actually be buried in the sediments, greatly reducing farm productivity. The Taiwanese Gracilaria farmers prefer ponds lined with coarse sand (13). Coarse material is less likely to get stirred up by wave and wind action, or to cover the thalli. Should seaweed-sediment actions prove to be deleterious, it may prove useful to engineer several small boats to constantly stir up the sediment and release buried seaweed.

RESOURCE BASE

Florida is rapidly growing state, which will inevitably lead to problems of recreational water use versus seaweed farming. In analyzing locations best suited for seaweed to energy farms, it was felt that the area between Tarpon Springs and Pensacola would incur the least conflicting use. This area is north of the "cold front line" and is less attractive from a recreational and living perspective to immigrants. The area is primarily used by fishermen. In this area, subtidal bottom lands between 0.5-1.5 m depths encompass 190,000 hectares, after elimination of shipping lands, state and federal parks and preserves, oyster fishing and rearing areas, and areas where seaweed does not naturally occur. This analysis has not yet taken into consideration substrate types, annual changes in water chemistry, physical properties and flow, or socio-political restrictions, hence the available area may be further restricted. It should be realized that this large area is only one tenth that which is available in the depth ranges of 1.5-20 m (1,900,000 hectares), hence use of this area would result in far greater gas supplies. Nonetheless, the tidal flat farm concept could make a significant contribution to regional gas needs.

RECOMMENDATIONS

While sustained culture of pelagic, floating *Sargassum* has been recently accomplished, and resulted in biomass yields of 9-12 g dry wt/m^2-d from small (1-2 m^2) enclosures (2), the technology for *Sargassum* culture is still not as well developed as for *Gracilaria* or *Macrocystis*. In addition, all the *Sargassum* species tested to date have shown poor bioconversion performance compared to other seaweeds. A floating crop approach may not be the only way to effectively use deep water areas. More recent economic analyses of bottom anchored, canopy forming species such as *Macrocystis*, in depths of 9-15 meters, have indicated that cost effective gas may be produced at reasonable yields of 38 DAFT/HA-Y. While *Sargassum* can form a canopy in shallow water, it does not achieve the depth range of *Macrocystis* (possibly due to a less efficient translocation system?). Plant breeding could potentially improve *Sargassum* yields, composition for conversion, and develop the plant for either pelagic cultivation or as a canopy species, or perhaps somatic hybrids of *Sargassum-Macrocystis* could provide valuable biomass crops. It is obvious that this approach involves long term research in the areas of genetics, biotechnology, and algal physiology, including studies of translocation processes, biochemical composition, carbon fixation, and plant growth regulators. Understanding the adaptive role of variable morphologies under different environments can also help guide the breeding and biotechnology research. Such long term research, rather than cultivation trails and scale up, is appropriate given the recent decline in worldwide oil prices.

The tidal flat farm concept needs to be tested in field trials and scale up studies to learn more about large scale cultivation technology. While there is little interest in scaling up biomass systems at the moment, Florida's regional energy planners should realize there is greater certainty that this approach can be commercialized than offshore seaweed cultivation. Should regional energy demands ever necessitate an engineering and scale up approach to solve immediate problems, this technology could be developed most quickly. The tidal flat farm also has great appeal for use in the Caribbean islands or Central America, as it could be part of a polyculture system with fish, leading to a high value export product and local energy. A long term, more basic research effort would also benefit this concept, possibly through development of new marine biomass clones and cultivars. Biotechnological research such as somatic hybridization could lead to herbicide resistant strains as a byproduct of the researchs' experimental design which incorporates such resistance in plant lines to aid in selection of new hybrids. These herbicide resistant *Gracilaria* or *Ulva* clones would facilitate weed control on tidal flat farms.

Marine biomass represents a real energy option for Florida. The tidal flat farm concept provides a fairly low risk approach which can help ensure marine biomass technology is implemented when regional biomass based energy becomes price competitive. Should it be possible to develop the higher risk offshore marine biomass technology, Florida may become an energy exporting state. A well planned, long term investment in the genetics, biotechnology, and physiology of Florida's tropical seaweeds could lead to major benefits for the energy business of this state.

ACKNOWLEDGMENTS

Funding for this research came from the U.S. Department of Energy, the Gas Research Institute, and the Harbor Branch Foundation. This is Harbor Branch Foundation Contribution No. 497.

REFERENCES CITED

1. Bird, K.T. "Economics of Macroalgal Cultivation for Methane" In: Seaweed Cultivation for Renewable Resources. (Benson, P.H., and Bird, K.T., eds). Elsevier Press. (In Press).

2. Bird, K., Dawes, C., LaPointe, B., Hanisak, D., and Ryther, J. "Marine Biomass Cultivation in Florida" In: Methane from Biomass- A Systematic Approach. (Smith, W. and Frank, J., eds). Elsevier Press. (In Press).

3. Bird, K.T. and Ashby, A.B. 1984. "Recent Economic Results of Converting Biomass to Methane" Energy from Biomass and Waste VIII: 975-994. (1984).

4. Brehany, J.J. "An Economic and Systems Assessment of the Concept of Nearshore Kelp Farming for Methane Production" Gas Research Institute Final Report, GRI 82/0066. Chicago, IL. (1983).

5. Gibor, A. "A Method for Cultivation of Benthic Algae by Artificial Floatation" Botanica Marina 19, 397-399. (1976).

6. Hanisak, M.D. "Recycling the Residues from Anaerobic Digesters as a Nutrient Source for Seaweed Growth" Botanica Marina 24, 57-61. (1981).

7. Hanisak, M.D., LaPointe, B., and Ryther, J. "Tropical Seaweed Cultivation for Methane" In: Seaweed Cultivation for Renewable Resources. (Benson, P.H. and Bird, K.T., eds). Elsevier Press. (In Press).

8. LaPointe, B.E., and Ryther, J.H. "Some Aspects of the Growth and Yield of *Gracilaria tikvahiae* in Culture" Aquaculture 15, 185-193. (1978).

9. Ryther, J.H. "Cultivation of Macroscopic Marine Algae and Fresh Water Aquatic Weeds" Solar Energy Research Institute, SERI/TR 98133-1A, 1617 Cole Blvd, Golden, CO. (1982).

10. Ryther, J.H. "Biomass Production, Anaerobic Digestion, and Nutrient Recycling of Small Benthic or Floating Seaweeds" Solar Energy Research Institute, SERI/TR-98133-1B. 1617 Cole Blvd.,Golden, CO. (1982).

11. Ryther, J.H. "Marine Biomass Research in Florida" Soil and Crop Science Society of Florida Proceedings. 42, 40-48. (1983).

12. Ryther, J.H. and DeBusk, T. "Significance of Carbon Dioxide and Bicarobonate-Carbon Uptake in Marine Biomass Production" Energy From Biomass and Wastes 6, 221-236. (1982).

13. Shang, Y.C. "Economic aspects of Gracilaria Culture in Taiwan" Aquaculture 8, 1-7. (1976).

14. Warren, C.S., Bruderly, D.E., Angeliers, M., Bilello, L.J., Bucalo, S., Finger, W., Hart R., Hinton, S., Newman, J., and Vinzant, B., "The Methane from Biomass and Waste Program Task 1: Evaluation of the Lake Apopka Natural Gas District", Reynolds, Smill and Hills. Gas Research Institute Topical Report 84/0015.1, Chicago, IL. (1984).

DISCUSSION

M. Neushul, U. of California: Have you considered the possibility of other crops on that vast area of submerged land off the coast of Florida?

Bird: We looked at the sea grasses at one time in the early part of the Florida Marine Biomass program, but we only looked at the sea grasses that are native to Florida. The three native species have low biomass yields and productivities, so we eliminated them early. They're very slow growers compared to Gracilaria. There's a possibility that these could also be bred, and their yields improved quite dramatically through classical breeding and used as a biomass crop for the very shallow areas.

FUELS FROM MICROALGAE: AN ASSESSMENT OF TECHNOLOGY REQUIREMENTS

Bernard F. Neenan*
Daniel A. Feinberg
Solar Energy Research Institute
Golden, CO 80401

ABSTRACT

The Aquatic Species Program (ASP), field managed for the U. S. Department of Energy (DOE) by the Solar Energy Research Institute (SERI), seeks to develop the technology for production of liquid fuels from aquatic plants, primarily microscopic algae. As part of this program, an extensive technology evaluation was performed to determine the extent to which microalgae might replace conventional petroleum-based liquid fuels by the early 21st century, as well as to delineate the critical research areas and issues that must be resolved before such replacements could occur. Several aspects of the overall technology evaluation are discussed in this paper. Two critical methodological decisions were made in the performance of this work: (1) that the multivariate nature of technology evaluation precludes the reduction of alternative configurations to a single-valued measure of worth; and, (2) that the technology evaluation team must reconcile the perspective of the scientists with that of the decision makers. Therefore, the approach chosen includes the development of a series of detailed technology configurations, and attempts to present the results and requirements of these options as dispassionately as possible. The result of such an approach is to turn the data over to the decision makers, allowing them to seek their own expert consensus on the likelihood of success of each technology configuration.

* Current Address: Niagara Mohawk Power Co., Syracuse, NY 13202

FUELS FROM MICROALGAE: AN ASSESSMENT OF TECHNOLOGY REQUIREMENTS

INTRODUCTION

The efficient allocation of federal funding for research projects requires the establishment of a basis for comparing the potential benefits of investments in technology improvements. The standard of comparison is typically multivariate and includes attributes such as the timing and level of benefits expected, the cost and time frame for achieving these benefits, and a variety of additional factors such as the distribution of benefits, the appropriate role of government and the private sector, and the prevailing political and economic environment. Combining all these diverse factors into a single measure of relative value or worthiness is generally not possible since the measures of success or of value do not lend themselves to quantification on a common scale and are not equally important in terms of resource allocation.

This paper reports the work of researchers and analysts at the Solar Energy Research Institute (SERI) to develop a method for evaluating energy production technologies. Specifically, we concentrate on measures of the relative benefits expected from the development of technologies that utilize alternative, nonhydrocarbon feedstocks and produce fuels compatible with the existing transportation fleet. Moreover, the technologies that establish the basis for this method are in a nascent stage of development and therefore subject to considerable uncertainty with regard to their ultimate configuration and cost. This uncertainty poses problems in analyzing costs and benefits but also is indicative of an inherent potential for vast improvement--the range of possible configurations and physical performances generally is distributed normally about an expected cost that suggests economic viability in the early 21st century.

The philosophy behind the method of analysis demonstrated here is that the multivariate nature of technology development evaluation precludes reducing alternative configurations to a single-valued measure of worth. Too many assumptions are required to do so, and many aspects of the technologies that are vital to resource allocation are compromised when they are forced to a common scale for convenience of analysis. Instead, we propose constructing a frame of reference that allows program planners and funding agents the opportunity to compare the requirements for success with the degree of technology improvement necessary for achievement of a given set of benefits. The discussion that follows is directed toward illustrating the framework of analysis in the context of the requirements for an economic technology for the production and conversion of microalgae to liquid transportation fuels.

We emphasize at the outset of this discussion that as analysts we do not approach technology comparisons solely from the perspective of the decision maker. We cannot hope to embody all aspects of the technology nor can we represent the many factors and interests that ultimately must be taken into account in making funding decisions. As analysts, we see our role as developing the data and a framework of analysis that allows such judgments to be made by those vested with the responsibility. Where uncertainties exist, we attempt to reduce them to their lowest common denominator, which often poses the question in very technical terms. This result we consider inevitable and desirable as it allows decision makers to seek their own judgments on the likelihood of success through discussions with experts who can be charged with assessing issues specific to their knowledge without encumbrances such as the imposition of unfamiliar or wide-ranging assumptions. Once decision makers have solicited the required technical expertise and formed their own judgments, the method allows combining all aspects of the technology into a representative cost configuration consistent with conventional benefit/cost analyses. In addition, the method allows for the identification of the specific performance requirements of the technology, and these are readily transformed into research plans that allow planning and measurement of progress toward the established goals.

THE TECHNOLOGY

The concept of fuels from microalgae includes provision for the production and conversion of biological feedstocks, rich in oleaginous polymers (lipids) that will serve as near-perfect substitutes for the hydrocarbon fuels that now constitute the basis for our transportation fleet. The unique chemical composition of lipids lends itself to a number of conversions to products such as gasoline, diesel fuel, or methane. However, the technical basis for these conversions in most cases is not available, so in this analysis we must consider not only the renewable production process, the cultivation and harvest of microalgae, but also the technical requirements of a conversion technology. Since both aspects are at a rudimentary stage of development, we also address the joint requirements for technical and economic viability and trade-offs between the two in achieving a specified goal.

The technology can be visualized as aquatic agriculture. Microalgae, small aquatic plants, are raised in a closely husbanded environment where water serves as the suspension medium for the plants and for most of the nutrient inputs. Algal crops, unlike conventional crops (except rice and cranberries), are not dependent on the indigenous soils for support or nutrients, so site-selecting criteria are dominated by the slope of the land, rather than its tilth, nutrient, and erosion content. Support structures, ponds or raceways constructed or compacted earth, or rigid materials, are constructed to provide a slight grade to allow for gravity flow, but depend on level land, so the slope of the area is an important factor in site selection. Major nutrients are added in a controlled manner, so while local availability is desirable from a cost perspective, sites need not be constrained by the indigenous supply of carbon, water, and organic nutrients.

The product, algal biomass, is concentrated at harvest to about 10% solids content and in this form can be transported to another site for processing. Thus, there are no special or temporal constraints to the siting of production and processing operations as are required with many agricultural products such as fruit and vegetables. One important factor that needs obviation is that the infrastructure does not currently exist for either the production or the conversion of vast quantities of microalgae, in the desert Southwest or anywhere else in the United States. The advent of the technology as a contributor to the nation's fuel needs will require vast expenditures on the accoutrements of production and an organized system of transportation and distribution. By not factoring these elements into our analysis, we do not intend to diminish their importance. Our objective is to derive the technical conditions necessary to make such a technology feasible, since these conditions are vital to the evaluation of technology feasibility and funding decisions. Factors such as the evolution of infrastructure are elements of the rate at which the technology is adopted, once it has been shown to the investment community to be competitive, and should be viewed as institutional constraints that can be addressed as additional information used in the decision process.

MICROALGAE PRODUCTION AND CONVERSION

A conceptual layout of a microalgae facility is illustrated in Figure 1. Modular production units share common water, nutrient, and harvesting facilities to form an extended algal farm, possibly exceeding 2000 acres and located in the desert Southwest to take advantage of abundant and cheap land and saline water resources (2). The algae are grown in a circulating water medium consisting of shallow containment walls, pumps for circulation, and harvesters. Extensive carbon sources must be provided to support growth in the sunlight of the desert environment. The primary factors in algae growth are the incident sunlight that energizes reproduction and growth and leads to the production of storage lipids, the culture depth and density that determine yields, and the maintenance of optimum conditions of culture temperature and salinity. The modular nature of the ponds allows efficient usage of fixed capital assets such as nutrient and water supply systems and can take advantage of economies of scale in harvesting, transportation, and processing. The only limitations to the size of the facility are the availability and cost of water, the amount of contiguous land of acceptable slope, and economies or diseconomies of managerial scale and of markets.

The harvested algal biomass must then be converted into a slate of fuel products usable in the mass market. First it is piped or trucked to the conversion facility, which may or may not be contiguous to the biomass production facility. Some pretreatment, or extraction of lipids, may be required at this point; this requirement depends in particular on the details of the conversion process. The

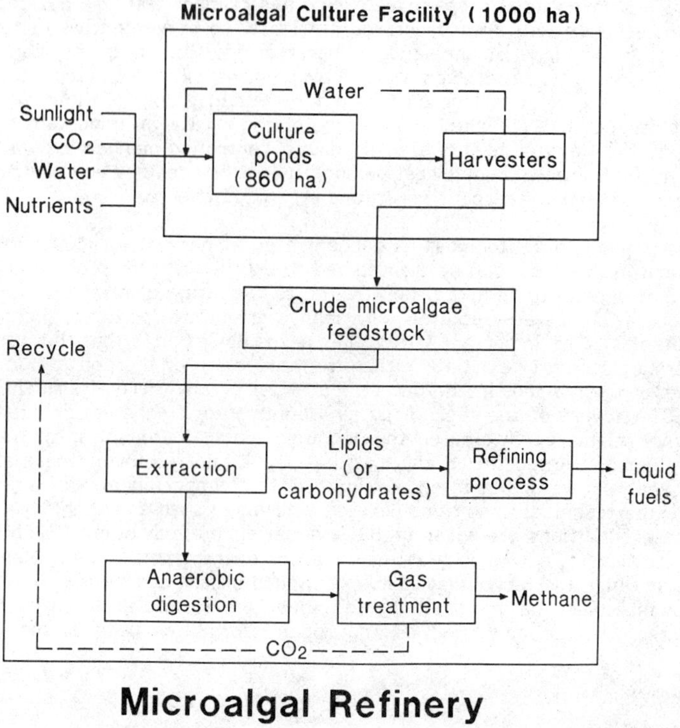

Figure 1. Conceptual Layout of Microalgae Production and Conversion Systems.

conversion facility can be captive to one production facility, i.e., sized to accept the biomass from only one facility, or it can take advantage of the substantial conversion scale economies by being sized to process feedstock from a number of production facilities.

Lipids produced by microalgae have physical and chemical properties similar to those of hydrocarbons and with proper processing could be converted to direct hydrocarbon substitutes (3,1). Thus we examine process configurations that could produce direct, or nearly direct, substitutes for conventional fuels--a lipid-derived ester fuel and a pseudo-vegetable oil (PVO) fuel that could replace or extend diesel fuel, a catalytic process that produces lipid-derived gasoline and other light fuels similar to those produced by distillation of petroleum, and for comparison, two processes that do not specifically utilize lipids--conventional carbohydrate fermentation to ethanol, and anaerobic digestion of the entire biomass to produce methane (2). The options utilizing nonlipid fuels as the primary products (e.g., ethanol and methane) were subsequently discarded, since too many less-expensive sources of carbohydrate are readily available; further, the strategy of maximizing carbohydrate content in microalgae fails to exploit their most interesting (and potentially valuable) ability to accumulate high-energy lipids. Although many unknowns remain concerning exact configurations of these conversion processes, a synthesis of bench-scale process research (on similar though not identical feedstocks) with algal composition data was required to provide program researchers with some preliminary direction concerning potential fuel production routes.

THE ECONOMIC SETTING

The fuel products discussed above are assumed to compete with conventional fuels in the liquid transportation market, and therefore to penetrate these markets they must achieve an equivalent price. We grant the possibility that to compete for, or more correctly, to win, market share, algal fuels might have to sell below conventional fuels, at least initially. However, our goal is to establish equilibrium conditions under which these substitute fuels can compete in the market, and we avoid description of how the market might evolve.

For the purposes of this evaluation, we use fuel price projections based on conditions expected in 2010, a period we believe that will be characterized by rising conventional fuel prices and that allows the time needed to develop the intricate biological and engineering methods required for economic success. The cost goals used as standards of evaluation in this analysis are $1.75/gal of both diesel fuel and gasoline, and all costs and prices are in real 1984 dollars, consistent with standard technology evaluation practices. Figure 2 shows the current (late 1985) fuel prices and the increases projected through 2010.

THE METHOD OF ASSESSMENT

The objective of this analysis is to construct production and conversion configurations that meet the program goal--production of liquid fuels that will compete with hydrocarbon-based fuels. Above we have constructed a goal or target that embodies this concept of success. Next we construct a technology configuration that allows measurement of current technology status as well as determination of the requirements for the production and conversion configurations to meet the goal. These specific requirements, measures of technical and economic performance parameters, can then be evaluated in terms of their likelihood of achievement and subsequently serve as the basis for constructing a technology development plan.

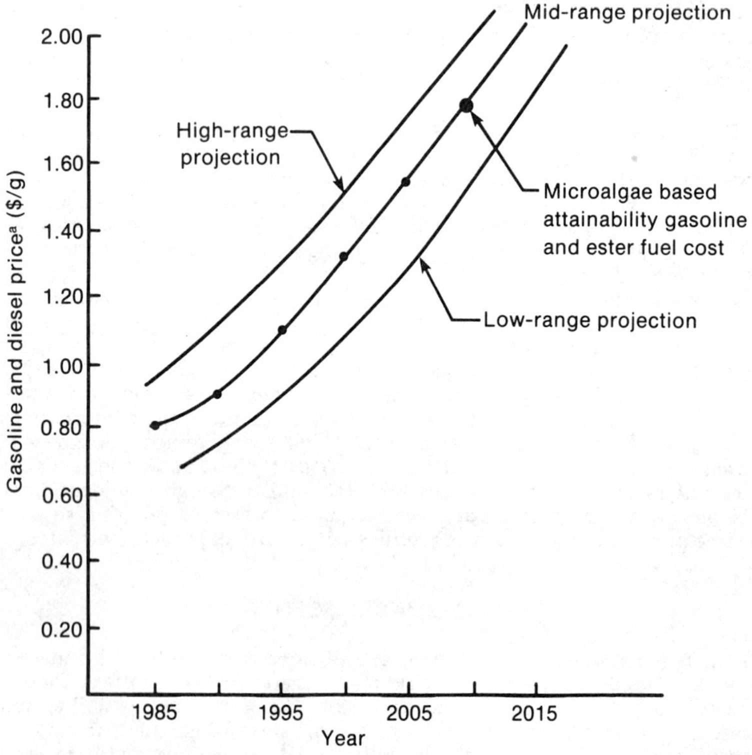

Source: EIA, NEPPV energy projections
[a]Refinery gate price

Figure 2. Gasoline and Diesel Fuel Price Projections.

First a reference case is developed from which the required improvements in production and conversion technologies can be defined. As a basis for comparison, we derived costs for all components based on technology available today. All components are designed according to what would be used today should a facility be designed and operated. Thus, the reference case illustrates the current state of the technology without the benefit of proposed or projected improvements. The major determinants of the reference case are summarized in Table 1. Sensitivity cases are then evaluated to determine areas where proposed improvements have a large impact on costs and serve as the basis for the formulation of alternative performance targets to compare against requirements.

The reference-case production configuration is a 1000-ha facility, located in the desert Southwest, that uses local saline water resources and carbon dioxide of unspecified origin (most likely electric power plants or extensive natural deposits in the area). The cost of water is based on pumping from a proximate saline water well of sufficient size to meet the facility's needs. Carbon dioxide cost is subject to considerable uncertainty because it depends on the cost and availability of pipelines to transport it from regions of abundant supply. All other minerals are priced at existing levels as are labor and other factors of production. The refer-

Table 1. **Major Parameter Ranges Identified for Microalgae Systems**

	Range		
	Low	Reference	High
Category A: Resource Parameters			
Carbon in source water (g/m^3)	10	200	420
Energy cost (\$/kWh)	0.040	0.065	0.10
Water cost (\$/m^3)	0.05	0.067	0.20
Distance to CO_2 source (km)	0	0	100
CO_2 cost (\$/MSCF)	3.67	3.67	4.50
Category B: Facility Design Parameters			
Effective culture area (% of total size)	70	86	90
Effective culture downtime (% of total surface)	30	10	5
Channel width (m)	6	30	60
Depth of culture (m)	0.10	0.15	0.20
Carbon losses (g m^{-2} d^{-1})	0	0.1	3
Nitrogen losses (g m^{-2} d^{-1})	0	0.1	3
Mixing velocity (m/s)	0.05	0.2	0.3
Mixing system efficiency (%)	30	65	75
Harvester solids removal (%)	50	90	98
Category C: Biology Parameters[a]			
Lipid content (% dry wt)	20	30	50
Carbohydrate content (% dry wt)	49	20	12
Protein content (% dry wt)	13	32	20
Salinity tolerance (g TDS/L)	10	35	120
Phosphorus cell content (g/g dry wt)	0.0012	0.007	0.023
Growing season (d)	200	250	365
Photosynthetic efficiency on PAR (%)	4	5	16
Depth (m)	0.10	0.15	0.30
Retention time (d)	7	7	3
Density (mg/L)	500	600	1000
Category D: Financial Parameters			
Return on debt (%)	3.7	4.6	5.6

[a]The chemical composition sensitivity analysis examined three cases where lipid content, carbohydrate content, and protein content were varied as a consistent group or set. Parameter ranges reflect a high lipid producer (high case) or a high carbohydrate producer (low case).

Source: Ref. 2

ce photosynthetic efficiency (PE) is about 5%, and the algae are assumed to ntain 30% lipids. Annual production of microalgae of the reference system is out 44,000 metric tons, and amortized capital and annual operation costs and bt service are \$393/metric ton (dry basis).

The reference algae processing system was configured for the production of hanol, methane, PVO, gasoline, and ester fuel. Representative parameter levels r the conversion processes using primarily lipids are shown in Table 2. For each, e calculate the allowable feedstock cost under the reference processing config- ation cost. The allowable processing cost is the maximum a processing facility uld pay for the feedstock and convert it to the specified fuel for sale at the rget price. The allowable cost includes amortized capital cost, operating costs, d allowance for profits. The allowable costs range from a low of \$33/t for

Table 2. **Parameter Levels for Reference and Attainability Conversion Process Cases (Lipid Processes Only)**

	Reference	Attainability
% Lipid	30	50
% Carbon	20	12
% Protein	32	20
PVO Process		
% of lipid available	50	75
% of available lipid extracted	100	100
Transesterification		
% of lipid available	50	75
% of available lipid converted	98	98
Catalytic Conversion		
% of lipid available	75	90
% of available lipid converted	90[a]	90
Product slate, % of conversion products		
Gasoline	59	59
Diesel	11	11
LPG	24	24
Fuel gas	6	6

[a]GJ basis

alcohol production to a high of \$150/t for gasoline, all well below the calculat \$393/t cost of production of the required feedstock (Table 3). We now have esta lished a basis for considering design and performance improvements that bring t algal production costs and allowable processing costs to equality. Once such configuration is constructed, we can examine how the design requirements chang becoming more or less stringent, as we vary other parameters, especially tho outside of the control of the technology development effort such as carbon dioxi costs. In this way we begin to "map" the solution space for this technology (i. establish design alternatives that meet the implied goals) and construct the ba from which to direct the further development of this technology.

The sensitivity results are too extensive to report here and are availat elsewhere (2) so we will summarize the salient findings. We find that the ma determinants of the cost of production are yield, as represented as PE [measur as a percentage of photosynthetic active radiation (PAR)] and the costs of carb dioxide and water. These factors are then considered as primary targets technology improvements--areas where substantial gains from scientific investig tion will yield large reductions in cost that could lead to a competitive system the production of liquid fuels. Evaluation of conversion technology indicates th the usual economies of scale found in chemical processing can exercise a lar influence on processing costs and therefore final product cost.

Based on the target fuels costs and the performance requirements, construct an attainability case that contains our judgments about likely impro ments and that combines to describe a system that produces fuels at the targ cost. This attainability case, which contains assumptions subject to uncertain is then subjected to further sensitivity analyses to illustrate how the requireme

Table 3. **Comparison of Feedstock Production Cost and Allowable Processing Cost for the Reference Case**

Fuel Processing Option	Reference Case	
	Feedstock Production Cost ($/t)	Allowable Processing Feedstock Cost ($/t)
Ethanol	393	48
Methane	393	83
Pseudo vegetable oil (PVO)	393	85
Ester fuel	393	72
Gasoline	393	97

Note: All parameters for feedstock production and processing are set at the reference value, and all products and coproducts are valued at their cost goal value.

change under alternative (to those of the attainability case) parameter assumptions; the case thus gives a perspective on the set of conditions under which investing in technology development satisfies program goals.

TECHNOLOGY REQUIREMENTS ANALYSIS

Figure 3 illustrates how the production and allowable costs change as the PE is increased from 5% to nearly 20% at lipid contents ranging from the reference level of 30% up to 50%. With all factors of production except PE and lipid content held constant at the reference level, equality of production and allowable cost is achieved only at PE in excess of 20%, over three times the PE currently achievable in large outdoor systems. Clearly, large technical gains are needed before the technology becomes competitive. To establish conditions under which competitive technology could be achieved, we examine various levels of improvement in the system parameters to ascertain feasible configurations--sets of engineering and biological performance parameters that produce fuels at the cost goal levels. Our aim is to develop configurations that are both economically and technically viable and use these as the basis for a technology development plan.

One such configuration constructed to represent improvements over the reference case is detailed in Table 4. This set of parameters, which meets the criteria of competitiveness (algal production cost equal to calculated allowable production cost), is characterized by high PE (roughly 16.2%), a long growing season (300 days), and production of algae with a lipid content of 50%, along with the other improvements listed. This so-called attainability case can be used as the basis for planning a technology development program since it sets out the levels of performance of the key biological parameters consistent with meeting the cost goals for 2010. However, numerous other combinations and permutations of system parameters can be constructed that also pass this test of competitiveness. Rather than derive endless sets of performance parameter configurations, all of which meet the measure of success but imply different development paths, we shall illustrate the sensitivity of the feasible solution space to changes in the main performance parameters to give a perspective on the technology and its implications for alternative development paths.

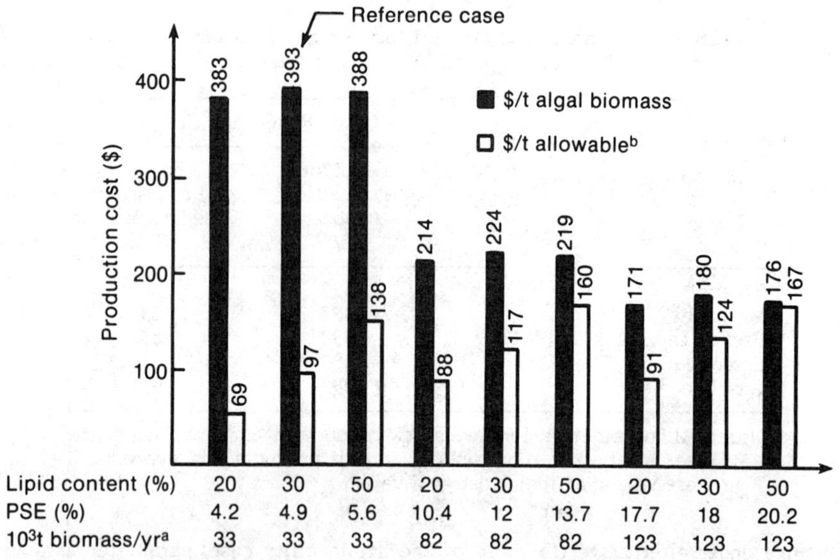

[a]1000 ha microalgae facility
[b]based on catalytic conversion process

Figure 3. Biomass Production Cost Sensitivity to Photosynthetic Efficiency and Lipid Content.

Table 4. Parameter Levels for Attainability Microalgae Production Case: Changes from Reference Case

Parameter	Units	Reference Case	Attainability Case
Resources			
Growing season	days	250	300
CO_2 cost	\$/m^3	0.13	0.10
Facility Design			
Detention time	days	7	3
Depth	in.	870.15	0.15
Density	mg/L	800	1000
Biological parameters			
Lipid content	%	30	50
Carbohydrate content	%	20	12
Nitrogen content	%	32	20
Salinity tolerance	g TDS/L	35	80
Financial parameters			
Return of debt	%	4.5	6.5

A series of static-comparative changes in the main parameters are illustrated in Figure 4a, along with the level of PE required to maintain a cost-competitive fuel price under each condition. In this figure we force improvements in PE alone to compensate for increases in other variables to illustrate trade-offs in development paths and to give a perspective on technology feasibility under conditions more constraining than those used in constructing the attainability case. In other words, we ask such questions as: What level of photosynthetic efficiency must be achieved to mitigate an increase in CO_2 costs from \$0.10/m^3 to \$0.12/m^3? Referring to point A41 on Figure 4a, the answer is roughly 18%. A further increase in CO_2 costs to \$0.14/m^3 would require a PE of very nearly 20%, along with reduced outgassing losses, to mitigate its effects. Since CO_2 cost is not itself a research parameter, but depends entirely on external factors (primarily related to competing demand for CO_2 for enhanced oil recovery: refer Ref. 2), we need mainly to be aware of the high sensitivity of biomass cost to CO_2 costs and the requirement to conserve CO_2 in the culture ponds. Further examination of Figure 4a reinforces the argument that the success of this technology is heavily dependent on factors external to the realm of scientific research, e.g., maintenance cost and return on investment, although there are other important factors upon which research can have an influence (e.g., length of growing season).

The converse situation is shown in Figure 4b, where decreases in the required PE result from more favorable assumptions about the main system parameters; this figure gives a perspective on how early opportunities may arise for

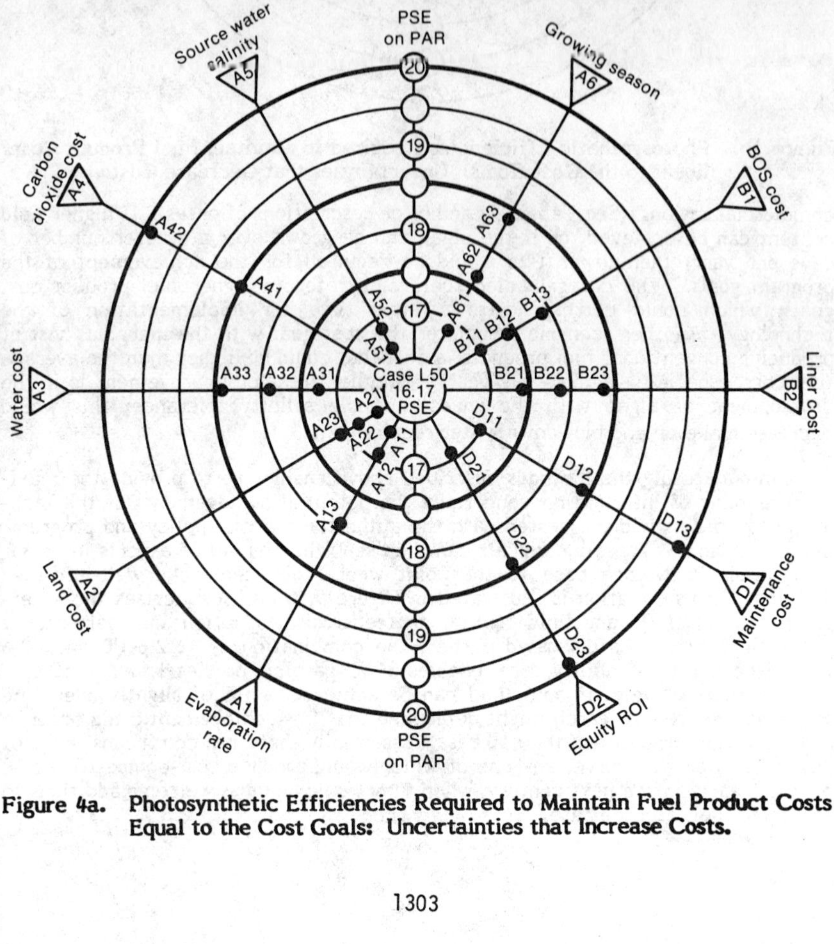

Figure 4a. Photosynthetic Efficiencies Required to Maintain Fuel Product Costs Equal to the Cost Goals: Uncertainties that Increase Costs.

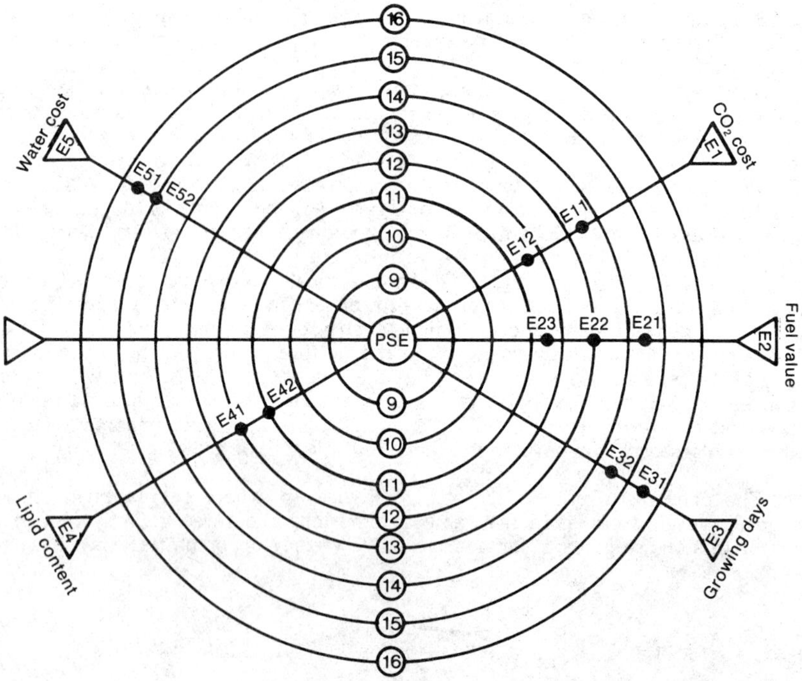

Figure 4b. Photosynthetic Efficiencies Required to Maintain Fuel Product Costs Equal to the Cost Goals: Uncertainties that Decrease Costs.

commercialization. (See Tables 5a and b for descriptions of cases.) If higher lipid content can be achieved, or if the algae can be grown over a greater number of days per year, then lower PEs would be required for the achievement of the program goals. The cases labeled fuel value refer to higher fuel product cost goals, which could correspond to a later time for implementation of the technology. Another scenario associates the cost goal with the marginal cost of producing conventional fuel products, a cost that could be higher than the average fuel cost, especially in the early years. Another area for improvement is in the development of algae with the widest possible salinity tolerance; this would decrease make-up and blowdown water requirements.

In sum, using the methods described above enables us to provide the maximum amount of information concerning the potential achievements of the technology to the individuals vested with the authority to make policy and programmatic decisions. These individuals can then seek the counsel of experts in developing their knowledge base to the point where they can make well-informed decisions. This situation is illustrated in Figure 5, which summarizes the synergistic effects of PE and lipid content on feedstock production and fuel product conversion costs. As discussed earlier, the combination of 16.2% PE and 50% lipids (the L50 attainability case) enables $1.75 gasoline or diesel fuel to be produced. Alternatively, if 55% lipid can be achieved, a PE of slightly over 14% would be required; research might determine that this achievement could occur in a shorter time frame than the L50 case. Essentially there is a continuous range of possibilities on this curve, any one of which would produce cost-competitive fuel products early in the next century, when they would be most welcome additions to the liquid fuel resource base.

Table 5a. Sensitivity Cases for Attainable Microalgae Production Costs: Uncertainties that Increase Costs[a]

Case	Parameter Type	Parameter	Units	Attain-ability	Levels of Change[b]		
					1st	2nd	3rd
A1	Resource	Evaporation rate	$m\ d^{-1}$	0.0035	0.007	0.01	0.02
A2	Resource	Land cost	$\$\ ha^{-1}$	1245	2500	5000	10000
A3	Resource	Water cost	$\$\ m^{-3}$	0.067	0.13	0.17	0.20
A4	Resource	Carbon dioxide cost	$\$\ m^{-3}$	0.10	0.12	0.14	--
A5	Resource	Source water salinity	$g\ TDS\ L^{-1}$	25	35	50	--
A6	Resource	Growing season	$d\ yr^{-1}$	300	285	270	265
B1	Facility Design	BOS cost[c]	%	0	10	15	20
B2	Facility Design	Liner cost	$\$\ ha^{-1}$	5000	7500	10000	12500
D1	Financial	Maintenance cost	% CI[d]	7	10	15	20
D2	Financial	Equity capital	% ROI[e]	30/6.5	100/7	100/8	100/9

[a]Cases represent imposition of additional constraints, through higher levels of indicated variables, on the attainability microalgae production concept.

[b]Levels correspond to a second level of indexing of cases, e.g.; A11 (1st level), A12 (2nd level), A13 (3rd level).

[c]A balance-of-system contingency defined as a percentage off capital cost.

[d]CI = capital investment.

[e]Both the percent debt capital and the return on investment (ROI) on debt capital are varied.

Table 5b. Sensitivity Cases for Attainable Microalgae Production Costs: Uncertainties that Decrease Costs[a]

Case	Parameter Type	Parameter	Units	Attain-ability	Levels of Change[b]		
					1st	2nd	3rd
E1	Resource	Carbon dioxide cost	$\$\ m^{-3}$	0.10	0.08	0.06	--
E2	Conversion	Fuel value[c]	$\$\ gal^{-1}$	1.75	1.93	2.10	2.28
E3	Resource	Growing season	$d\ yr^{-1}$	300	320	340	--
E4	Biological	Lipid content	%	50	55	60	--
E5	Resource	Water cost	$\$\ m^{-3}$	0.067	0.045	0.034	--

[a]Cases represent imposition of additional constraints, through higher levels of indicated variables, on the attainability microalgae production concept.

[b]Levels correspond to a second level of indexing of cases, e.g.; E11 (1st level), E12 (2nd level, E13 (3rd level).

[c]Gasoline or diesel fuel product cost goal for 2005-2010.

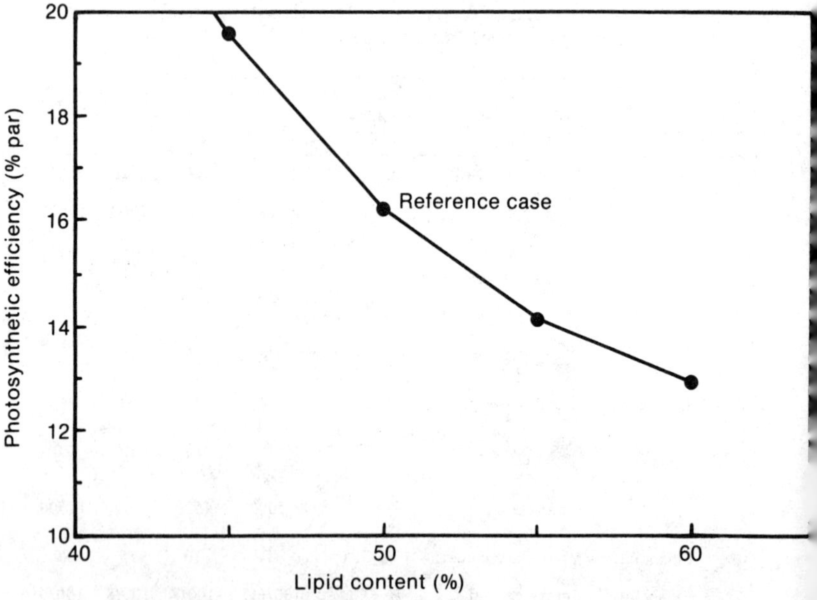

Figure 5. Combinations of Lipid Content and Photosynthetic Efficiency whi will lead to the Production of Cost Competitive Liquid Fuels.

REFERENCES

1. Ben-Amotz, A., 1984, "Development of Outdoor Raceway Capable of Yieldi Oil-rich Halotolerant Microalgae. Identification of Oil-rich Strains," Aquatic Species Program Review, Proceedings of the April 1984 Princip Investigators Meeting, SERI/CP-231-2341, Golden, CO: Solar Energy R search Institute, p. 186.

2. Hill, A., D. Feinberg, R. McIntosh, B. Neenan, and K. Terry, 1986, Fuels fr Microalgae: Technology Status, Potential, and Research Issues, SERI/SP-23 2550, Golden, CO: Solar Energy Research Institute (in press).

3. Tornabene, T. G., A. Ben-Amotz, S. Raziuddin, and J. Hubbard, 1983, "Che ical Profiles of Microalgae with Emphasis on Lipids," Final Report for SE Subcontract No. XK-2-02149-1 (portions also published in Screening for Lip Yielding Microalgae, Final Subcontract Report SERI/STR-231-2207), Golde CO: Solar Energy Research Institute.

DISCUSSION

D. Hall, U. of London (U.K.)	In trying to get higher photosynthetic efficiencies, did you take into account the increased cost which would be required because of different sorts of depths and stirring rates, which increase the cost tremendously?
Feinberg	Yes, the technical evaluation and planning section developed a model through a long period of consulting with experts. The model attempts to take the three primary parameters of depth, detention time, and culture density, and then model those effects through the utilization of solar energy into specific productivity rates. By specifying a chemical composition, the photosynthetic efficiency that you've achieved is then back-calculated.
O. Vimal, DMES (India)	What is the highest level of salinity that can be tolerated? Have you have made efforts to combine the production with a conversion system such as biogas?
Feinberg	Yes, we did. We found that microalgae is not a cheap feedstock, and if you're going to turn it into a gaseous fuel, which generally has a much lower value than a liquid fuel, it's not going to pay unless the same type of cost reductions that can be achieved would also help anaerobic digestion. On a Btu basis, liquid hydrocarbon has a much higher value than methane. Regarding salinity, I'm going to say it's 50 parts per thousand or higher, but I'm not sure that's accurate. I would to have to go back to our researchers to provide you with that. Obviously, the wider the range of salinity that the organisms can tolerate, the less water you'll have to remove in the makeup blowdown cycle.
J. Earle, U. of Florida	A major consideration in the utilization of microalgae is the cost of harvesting. What harvesting techniques did you consider in your evaluations?
Feinberg	Generally, we envision a two-stage system. The first stage would either be a sand filter, belt filter, or microstrainer, which would give you about 1 to 5% solids, and a centrifuge would be used as the second stage, which would increase the solids to 10%.

DISCUSSION, CONT.

Veigel	It has been refreshing to have a speaker who, as a researcher, is willing to say that there may be circumstances under which it would not be worthwhile persisting in the research, and I imagine that shocks the federal government to hear that periodically.
Feinberg	Since this work was sponsored by the U.S. Department of Energy, the models on which this work was based are available, and we certainly solicit people's interest. I can send anyone copies that run on FORTRAN or the IBM PC.

USE OF WOOD FOR ENERGY IN THE UNITED STATES--
A THREAT OR A CHALLENGE?

John W. Koning, Jr.
and
Kenneth E. Skog, Research Forester
Forest Products Laboratory
Madison, Wisconsin 53705-2398

ABSTRACT

The use of wood for energy--including the burning of olid wood and black liquor from pulping--has been growing t a rate significantly greater than that for all other ses such as lumber, pulp, or particleboard. This could hreaten to increase the price of wood for those other ses, or it can stimulate us to seek more creative ways of sing untapped wood resources for fuel.

On the basis of estimates of heavy wood energy use elative to other uses for wood, and estimates of continuing igh costs for fossil fuels, we suggest here the feasibility f meeting the demand for fuelwood through small-scale coop- ratives. Such an approach can improve forestry practices nd can avoid unduly increasing the cost of wood for other nd uses.

USE OF WOOD FOR ENERGY IN THE UNITED STATES--A THREAT OR A CHALLENGE?

INTRODUCTION

The primary use of wood in the world today is for energy, with most of it being used as a cooking fuel. Use for cooking fuel is greatest in less developed countries and levels of supplies have reached crisis conditions in some places. However, it might come as a surprise that, in the United States, the end use of most timber is not lumber or pulp and paper, but fuel for energy--and this energy use has been growing at a substantial rate (Fig. 1). Wood energy use estimates for 1972 to 1983 are given in Table 1 and shown in more detail for 1983 in Table 2.

Though these wood energy use estimates are surprisingly high, they are conservative because they may underestimate the wood burned by industrial firms outside the forest industry; relatively little information about that component is available (11).

Wood is burned both in solid form and in the form of black liquor from the pulping process. Even if the substantial black liquor tonnage used for energy is subtracted, the amount of wood used for energy is still more than the use of wood for all other wood products combined.

Such continued increase in the use of wood for energy could increase the cost of certain wood materials, and challenges foresters and woodland managers to look more carefully at available sources of fuelwood.

THE THREAT

Currently we have ample unused hardwood and waste wood for fuel in the United States (13); their production exceeds use. But, if the rate of use of wood for energy shown in Figure 1 continues in a linear fashion, this Nation could face fiber cost increases depending on local abundance of wood by the year 2000--14 years from now--because of increased competition for wood. Higher wood prices would support better forestry but the payoff in more timber growth would take many years. If the linear trend continues we will use over 280 million tons per year (15% MC) (238 million ovendry tons) for fuelwood by the year 2000. This represents less than 1.5 percent of the total weight of aboveground tree biomass presently on commercial forest land (280 million tons divided by 20.7 billion tons (15% MC)) (8).

Although the increase may not be linear, increases in nonwood fuel prices will cause people to use more wood in place of natural gas and oil in residential, commercial, and industrial applications. Mid-level projections from

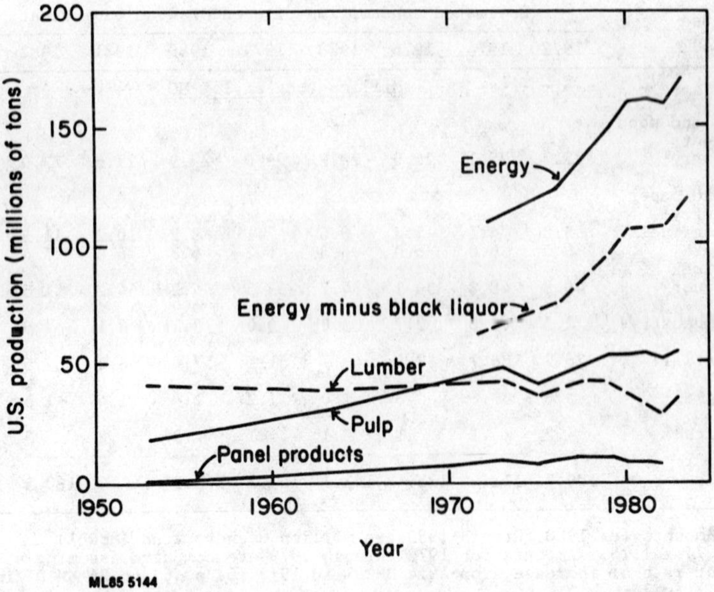

Figure 1. Major end uses of timber and bark in the United States. ("Energy minus black liquor" is the total energy with the black liquor subtracted from it.) Values for lumber and panel products are from "U.S. Timber Production, Trade, Consumption and Price Statistics, 1950-83" (9). Values for pulp are from the American Paper Institute as reported in Paperboard Packaging (1). (ML85 5144)

Table 1. Wood energy use, 1972-1983.

Use	Estimated consumption of wood or wood equivalent								
	1972	1974	1976	1978	1979	1980	1981	1982	198[illegible]
	- - - - - - - - - Million tons at 15% MC - - - - - - - - -								
Lumber and wood products industry[1]	23.7	25.1	26.6	28.2	29.0	29.9	27.6	23.0	27.[illegible]
Pulp and paper industry[2]									
As hog fuel	3.0	4.1	5.3	6.6	7.1	8.2	10.2	11.3	14.[illegible]
As bark	6.0	6.0	6.0	6.3	6.2	6.8	6.9	7.4	7.[illegible]
As black liquor	46.8	48.8	50.1	54.2	53.5	54.2	54.2	51.5	54.[illegible]
Other industry[3]	2.5	2.6	2.7	3.0	3.0	3.1	3.1	2.9	3.[illegible]
Residential[4]	26.7	26.1	33.8	43.7	51.1	57.4	58.0	62.5	61.[illegible]
Commercial[5]	.6	.6	.8	1.0	1.2	1.4	1.4	1.5	1.[illegible]
Utilities[6]	.2	.1	.1	.1	.1	.3	.2	.2	.[illegible]
Totals	109.5	113.4	125.4	143.1	151.2	161.3	161.6	160.3	171.[illegible]

[1]Amounts for 1980 through 1983 are million ovendry tons (MMODT) (11, p.7) x 1.15. Amounts for 1972 through 1979 are computed assuming a constant rate of increase from 19.4 MMODT in 1970 (5, p.46) to 26 MMODT in 1980 (11, p.7).

[2]Amounts of hog fuel and bark are MM tons at 50% moisture content (2) x (1.15 ÷ 2.00). Black liquor is (quads (2) ÷ (17.2 MM Btu/ODT)) x 1.15.

[3]Other industry consumption is computed as 3% of total industry consumpti[illegible] (11, p.17).

[4]Amounts for 1980 through 1983 are MM cords (11, p.27) x 1.16 ODT/cord (this is ((29.1 OD lb/ft³) x (80 ft³/cord)) ÷ 2000 lb/ton) x 1.15 (See fn 2 of Table 2). To compute amounts for 1972 through 1979: MMODT (10, p.95) x .8[illegible] cords/ODT (11, p.27) converts the DOE estimate to a correct cord estimate. Second, x 1.16 ODT/cord to obtain a correct ODT estimate. Next, x 1.15. Finally, ÷ .95 to account for an underestimate (11, p.59).

[5]Amounts for 1980 through 1983 are MMODT (11, p.43) x 1.15. Amounts for 1972 through 1979 are MMODT (10, p.111) ÷ 0.8 (this is a correction factor (11, p.61)) x 1.15.

[6]Amounts for 1980 through 1983 are MMODT (11, p.43) x 1.15. Amounts for 1972 through 1979 are MMODT (10, p.122) x 1.15.

Table 2. Wood energy use in 1983 by four measures.

Use	Estimated wood energy use in 1983				Wood use from primary data source	
	MM tons (15% MC)[1]	MMODT	MM cords[2]	Quads	Amount	Source
Lumber and wood products industry	27.6	24.0	20.7	[3].40	24.0 MMODT	(11)
Pulp and paper industry						
As hog fuel	14.6	12.7	10.9	.21	12.7 MMODT	(2)
As bark	7.8	6.8	5.9	.12	6.8 MMODT	(2)
As black liquor	54.9	[4]47.7	41.1	.82	.82 Quads	(2)
Other industry	3.2	2.8	2.4	.05	2.8 MMODT	(11)
Residential	61.6	[5]53.6	46.2	[6].92	46.2 MMcords	(11)
Commercial	1.5	1.3	1.1	[7].02	1.3 MMODT	(11)
Utilities	.2	.2	.1	[8]--	.15 MMODT	(11)
Total	171.4	149.1	128.4	2.54		

[1]1983 ODT estimates x 1.15.

[2]1983 ODT estimates ÷ (1.16 ODT/cord), except for residential use. 1.16 ODT/cord = (29.1 OD lb/ft^3 x 80 ft^3/cord) ÷ 2000 lb/ton. Hardwood is 32.8 OD lb/ft^3, softwood is 27.4 OD lb/ft^3. Estimating 31% of wood harvested is hardwood and 69% is softwood, then (32.8 x .31) + (27.4 x .69) = 29.1 OD lb/ft^3. Wood at 15% MC contains 1.15 x 1.16 ODT/cord = 1.33 tons (15% MC)/cord.

[3]1983 ODT estimate x 17.2 MM Btu/ODT (11).

[4]1983 quad estimate ÷ 17.2 MM Btu/ODT (11).

[5]1983 cord estimate x 1.16 ODT/cord.

[6]1983 cord estimate x 20 MM Btu/cord.

[7]1983 ODT estimate x 17.2 MM Btu/ODT (10).

[8]Less than .005 quad.

the Department of Energy show that prices for natural gas and oil (without inflation) will be 35 percent to 65 percent higher in 1995 than in 1984 (12). Commercial and industrial users of natural gas and oil, assuming mid-level projections, could afford $60 to $87 per cord of wood (1984 dollars) by 1995 versus $43 to $63 in 1983 (Table 3). These cordwood costs must, of course, cover any higher costs to store wood and to build and operate woodburning plants. Wood is much less likely to replace low-priced coal used by large plants to make steam.

Of the 171.4 million tons (15% MC) used for fuel in 1983, only that for residential use (61.6 million tons (15% MC)) came largely from roundwood from forests. Less than a quarter of residential use comes from timber usable for other products (6). The remaining 109.8 million tons is wood waste from making other wood products. Continued growth in wood energy use will require use of more roundwood from forests and/or untapped sources of wood waste. Data in "A National Energy Program for Forestry" (7) suggest "waste" sources could provide 600 million ovendry tons (690 million tons at 15% MC) per year without diminishing timber for other products. Amounts physically available include (among others):

	Million Tons (15% MC)
Forest	
Harvest sites	
Logging residues from growing stock and nongrowing stock	184
Standing live and dead trees	23
Timberland	
Excess growing stock	247
Mortality	109
Total, forest	564
Urban	
Tree removals and wood wastes	81
Other	
Forest products industrial waste	23
Waste wood from land clearing	23
Total, other	46
Total, all sources	690

However, much of the wood physically available is only available at prohibitively high costs. As use of wood for fuel grows, market forces may not lead us directly to use these dispersed waste materials, particularly where wood fuel markets are not well organized.

Table 3. Wood costs equivalent to projected cost for selected nonwood fuels.[1]

ctor and fuel	Equivalent costs				
	1973	1984	1995[2]		
			Mid price + mid growth	Mid price + high growth	High price + mid growth
	- - - - - - - 1984 dollars per green cord - - - - - - -				
mmercial					
Natural gas	19	56	83	95	82
Heating oil	28	63	87	87	116
Coal for steam	9	21	24	24	24
dustrial					
Natural gas	10	43	71	82	70
Residual fuel oil	16	44	60	59	78
Coal for steam	9	19	24	24	24

[1]Prices for selected nonwood fuels from DOE (12) are converted to rdwood cost assuming a cord of green wood contains 20 MM Btu; 10 MM must be ed to convert water in wood to steam and the remaining 10 MM Btu are nverted to energy at the same efficiency as the fuel that wood replaces. ese assumptions imply a cord of green wood (containing 10 MM usable Btu) uld have a value 10 times the price charged for 1 MM Btu of nonwood fuel. r example, if natural gas costs $10/MM Btu we could afford to pay up to 00/green cord to replace natural gas.

[2]For three assumptions regarding world oil price and U.S. economic owth rate.

If fuelwood needs were satisfied by conventional logging practices, we could face several ecologically or economically unsound situations:

(1) High grading (removal of only high-quality trees) of present timber stands with loss of future high-value large timber;

(2) depletion of the forest nutrient supply;

(3) loss or change of some wildlife;

(4) negative impact on recreation; and

(5) increased imports of wood fiber.

A threat--yes, but not a crisis if we act now.

THE CHALLENGE

If this potential problem is properly addressed within the next 10 years, through research and adoption of appropriate technology, we can turn this threat to opportunity and reap considerable benefits from wood energy.

Already wood energy use has reduced U.S. dependence on imported oil. Replacement of oil with wood and energy conservation efforts in the pulp and paper industry alone have reduced fuel oil use from 500 trillion Btu's in 1972 to 230 trillion Btu's in 1984, a drop of more than 50 percent. At the same time, production of paper has increased.

Over 61 million tons (15% MC) of wood were used for residential fuel in 1983, greatly relieving economic hardships caused by higher oil prices. In the process, this use of wood has created thousands of jobs and helped stem the hemorrhage of U.S. dollars in our balance of payments. (Unfortunately, the home use of wood is not without its problems--such as air pollution; thus, industrial or residential use with proper air pollution controls should be preferred.)

As fuelwood use grows, harvest of large quantities of wood can be used to improve forest management. A substantial portion of this wood can come from less intensively managed nonindustrial private forest land. Fifty-eight percent of U.S. timberland is in nonindustrial private ownership. Twenty percent of this land is in parcels of less than 100 acres (3). Given these factors, the problem becomes one of how to fill the need for wood fuel from many private ownerships.

The challenge is to supply wood fuel from known, existing, untapped sources, as demand increases.

A SOLUTION

One solution is removal of low-quality wood using small equipment--called "low-grading"--from private woodland through farm or farm-like cooperatives.

An independent farmer/landowner could chip undesirable species and multiple-stemmed, improperly spaced, or poorly formed trees up to 6-8 inches in diameter using a tractor-mounted or independently powered chipper. The whole-tree chips would be trucked to a concentration yard run by a local farm co-op, or wood co-op patterned after the farm co-op, for consolidation, storage, and shipment, or they would be trucked directly to the customer. If trucked directly to the customer, the co-op would act as the broker.

The co-op would have at least two marketing options: a) screen and clean the chips to upgrade them for pulp or b) sell the mixed material as fuel. Companies could buy chips from the co-op, and ship them in bulk by rail or truck.

The idea helps meet the challenge to both supply fuelwood and improve forest productivity. This approach satisfies the need for small regional fuelwood markets to meet the demands of small businesses that could not support large-scale chipping contractors. A conscious decision to encourage low-grading by owners of small woodlands recognizes that ownership of the majority of commercial timberland is private, and that stands are typically unmanaged and include large quantities of small diameter hardwood trees.

The cooperative provides a new means for collecting, concentrating, and transporting scattered wood material. The use of existing co-ops reduces overhead costs. Because they are member-owned, the operation and management are accepted by the farmer. Because existing farm co-op facilities can be used, the need for investment in new facilities is reduced. The addition of chips strengthens the co-op by broadening its product line.

The use of a tractor-mounted or independently powered chipper--rather than a large whole-tree chipping system--minimizes landowner capital investment. Co-ops can buy small-scale chipping equipment and rent it to members, or (4) hire a crew and do chipping for landowners. The use of small-size chipping equipment encourages the removal of the small low-quality trees rather than the well-formed larger trees. It results in less damage to the remaining trees and ecosystem, and frees desirable trees for maximum growth. Poor-quality large trees can be cut for firewood or reduced in size and chipped.

The accumulation of chips from a number of small landowners at a co-op helps reduce the high cost per ton for a single small landowner to arrange sale and transport of a small quantity of chips to a buyer. Consolidation of small quantities of chips would allow bulk shipment by rail or truck, and many co-ops are located at railheads. The co-op can upgrade the whole-tree chips to pulp chips or particleboard chips, or can sell unprocessed chips as fuel, depending on market conditions.

Centralized marketing facilitates the acquisition of wood chips by a pulp and paper company, fuel supplier, or fuel buyer by allowing them to deal with a single organization rather than a multitude of landowners. Centralized marketing will also stabilize the supply of wood chips for buyers.

An additional benefit of the cooperative idea is that the rights of the landowner are given protection from interference in land management because buyers are dealing with the landowner's co-op rather than the owner directly.

The increased demand for wood for energy is a reality. That we can meet the demand with an as-yet-untapped resource is a possibility. Current means of timber harvesting have economic and ecological problems when used to supply small regional markets. The alternative of low-grading with the aid of cooperatives can serve as a silvicultural treatment that will yield multiple forestry benefits and return a profit to the landowner. Committing ourselves to innovative ideas can assure tomorrow's bounty--in this case, turning the threat of fuelwood shortages into an exciting challenge.

REFERENCES CITED

1. American Paper Institute, "How Long Will Good Times Last," Paperboard Packaging 69(8), 18-46 (1984).

2. American Paper Institute, "U.S. Pulp, Paper and Paperboard Industry Estimated Fuel and Energy Use," 24 p., New York, NY (1984) March.

3. Birch, T. W., Lewis, D. G., and Kaiser, H. F., "The Private Forest-land Owner in the United States," USDA Forest Serv. Resour. Bull. WO-1, 61 p. (1982).

4. Brusila, B., "Forest Management and Marketing Cooperatives: Are They a Viable Alternative?" p. 110-114. IN: Proceedings, Fuelwood Management Utilization Seminar, 152 p., November 9-11, 1982. East Lansing, MI: Michigan State University.

5. Goetzl, A., and Tatum, S., "Wood Energy Use in the Lumber and Wood Products Industry," Forest Prod. J. 33(3), 44-48 (1983).

6. Skog, K. E., and Watterson, I. A., "Residential Fuelwood Use in the United States: 1980-81," Nat. Tech. Inf. Serv. ADA 131724, 120 p., Springfield, VA (1983).

7. USDA Forest Service, "A National Energy Program for Forestry," USDA Forest Serv. Misc. Publ. 1394, 16 p., Washington, DC (1980).

8. USDA Forest Service, "Tree Biomass--a State-of-the-Art Compilation," Gen. Tech. Rep. WO-33, 34 p., Washington, DC (1981).

9. USDA Forest Service, "U.S. Timber Production Trade, Consumption, and Price Statistics: 1950-83," USDA Forest Serv. Misc. Publ. 1442, 83 p., Washington, DC (1984).

10. U.S. Department of Energy, Energy Information Administration, "Estimates of U.S. Wood Energy Consumption from 1949 to 1981," DOE/EIA-0341, 149 p., Washington, DC (1982).

11. U.S. Department of Energy, Energy Information Administration, "Estimates of U.S. Wood Energy Consumption 1980-1983," DOE/EIA-0341 (83), 61 p., Washington, DC (1984).

12. U.S. Department of Energy, Energy Information Administration, "Annual Energy Outlook, 1984," DOE/EIA-0383(84), 354 p., Washington, DC (1985).

13. Wahlgren, H. G., and Ellis, T. H., "Potential Resources Availability with Whole Tree Utilization," Tappi 61(11), 37-39 (1978).

DISCUSSION

E. Gavett, USDA

How does this take care of the competition for the wood? Why do you see this as being able to solve the problem of competition in localized areas?

Koning

Simply by broadening the base of the supply, that's the primary advantage that I can see. I think if you've got a co-op that is handling wood, it can be built up or decreased very readily by telephone. It's like money in the bank to the farmer. You just let it grow. If the co-op calls and says we've got a customer that needs 100 or 1000 tons, he can go out and harvest it.

W. Smith, U. of Florida

How are the co-ops capitalized? I've been very interested in this situation because the problem with using wood as fuel has been in developing an infrastructure for harvesting and deliveries. The capital that's required for the harvesting and delivery is now controlled by the people who don't want competition. So the co-op idea sounds very attractive with regard to capitalizing a system. Do they borrow from the co-op banks?

Koning

There are a number of ways this can be done. One of course is for the farmer to invest in the chipper, and he's already got the vehicle to get the chips to the concentration yard — that's his truck. He can bring in a load of chips when he goes to get his feed. Another way is the co-op could buy the chipper and rent it to the farmer. Now that's done in other operations, like fertilizing. In some co-ops, they don't even rent the equipment out, they just loan it out. It's possible that the co-op could be making enough on this that they would loan out the chipper. So there would be no capital investment for the farmer that he doesn't already have. My guess is that some farmers are going to find that using a chipper is so helpful that they are going to invest in it just for their own use. A third way is for the co-op to borrow the money, buy the equipment, and then hire a crew, so if you have absentee land owners they could contract with the cooperative. If you really want to talk to an unhappy land owner, talk to somebody who has hired a big chipping operation where they didn't care. They can, with the wrong approach, ruin your wood lot, and then everybody in the next county won't have anything to do with wood chips. The co-op isn't going to let that happen because it's owned by the farmers.

P. Nilsson, Swedish U. (Sweden)

We have 5 generations of planted forest in Europe and we really know what lack of forest means. Here, I think you have the tradition if you want to have a tree you just go out and get one. Fuel wood, when added to the timber and pulp wood markets, will help the whole economy and in fact produce more wood for industrial purposes out of the system. I would like to point out that it is not a threat to the forest industry; it's complimentary and it's badly needed for the future, perhaps not for the energy, but for the benefit of the forest.

J. Ranney, Oak Ridge National Laboratory

I have a question on dealing with farmers and small land owners. Continuation of the harvesting of small diameter material without regeneration or working with uneven age-type approaches to hardwood or conifer management could cause ecological problems. Do you have any comments on that particular concern?

Koning

I do have a problem with the nutrient depletion question. I wouldn't want someone to just wipe out on a continuing basis all the small material. I don't think that's reasonable or going to happen first of all. But we don't really know the carrying capacity and the nutrient impact of any of our forestry practice at this point. There's been some good work started, but it's a long-term study. I was bothered for a long time on the wildlife question, but it turns out that there's a lot of positive things that can happen with this approach to enhancing wildlife. I would go so far as to say as a forester, this approach can also let you expand your mind and you can dream a little bit. What we're taught in forestry is that the ideal forest is one which is like an arboretum; it has a variety of species and ages. We're a long ways from that, but we're much closer if we can develop wood energy to provide a means of getting rid of the junk. You noticed I didn't say the industry should automatically diversify the total forest. We've got to crop som of it.

Gavett — What is the sentiment of the traditional wood users towards something like this?

Koning — The first time I proposed this, I walked into the buyer's office of a pulp mill. I introduced myself and I was welcomed because I have long history with the industry, but when I said cooperative, I was almost thrown out of the office. When I explained the total picture and what my goals were, he not only accepted them, but he went on the program with me right after my talk to give the industry viewpoint which is very positive. As long as there is an honest business deal helping forestry and the landowner, you've got a chance. Where this has failed is when you get a greedy manager, either in the wood end of it, or in the supply end of it. I know of one state that made an attempt very similar to wha I'm advocating, and they abandoned it — they ended up with a manager who squandered the profits. But I think if it's done right and on a professional level, I think the industry will support it.

Gavett — There is one aspect of this that you failed to mention at all, and I think it's absolutely critical that people understand it. The largest use of biomass energy is one for which there is no subsidy. You don't mention that at all.

Koning — That's true.

Gavett — This is operating on its own and competing with OPEC, and here is a way o expanding its utilization by getting int the small farm wood lot operation, and i really needs to have that element added in further presentations.

INDUSTRIAL WOOD FUEL MARKET ASSESSMENT IN WASHINGTON STATE

James D. Kerstetter, Ph.D.
Bioenergy Program Manager
Washington State Energy Office
Olympia, WA 98504

ABSTRACT

Manufacturing industries in Washington State have the potential for using wood fuels to help meet their energy needs. We reviewed the use of purchased fuels in Washington's industrial sector and established criteria to identify potential wood fuel users. We targeted large fuel purchasers with steam and hot water as their primary energy uses. We reduced the population of 6,000 facilities down to a target group of 300.

We surveyed the 300 facilities to determine what criteria they use for making capital investment decisions and what their knowledge and interest was in using wood fuels. We determined that the principal barrier that must be surmounted is economic. We also learned that 45 percent of the firms that have not investigated the use of wood fuels were not even aware of its potential.

INDUSTRIAL WOOD FUEL MARKET ASSESSMENT IN WASHINGTON STATE

INTRODUCTION

Opportunities

Washington's logging industry harvests more than 5 billion board feet of timber from more than 300,000 acres every year. This logging activity generates in excess of 3 million dry tons of forest residues and 4 million tons of mill residues. While 98 percent of the mill residues are used, almost none of the forest residues are used. In fact, up to 2 million tons of forest residues are disposed of through prescribed burns each year. These materials are not only wasted, they create an adverse environmental effect. Approximately 34 trillion Btu of heat energy was wasted through their burning. This compares to 123.3 trillion Btu of fossil fuels purchased by Washington's manufacturing industries at a cost of $505 million dollars.

While these forest residues could be used by the residential sector, this generally just moves the pollution problem from forest lands to urban areas. The industrial sector offers two potential market advantages for forest residues. First, the emissions from their facilities are regulated and thus will not create the pollution problem. Second, the industrial sector is the second largest energy consuming sector in the state, transportation being first. As a large energy consumer, it offers a large potential market. The purpose of this paper is to assess that market potential.

Purpose

The Washington State Energy Office (WSEO) bioenergy program has assumed responsibility for helping to move underutilized forest residues and mill residues to an energy market. The benefits to the state include increased economic activity, reduced emissions from prescribed burns, and use of an indigenous resource. WSEO will not make this transition happen, that is the private sector's responsibility. We can, however, help to identify the potential industries who may become users of wood fuels in Washington. A study by Huff, et al. (9) made the first attempt at market identification for the entire United States in 1982. We have built upon that study while focusing on Washington.

Our objectives are to:

1. Identify facilities that have a "theoretical" potential for use of wood fuels;
2. Determine the basis for capital investments;
3. Determine the real or perceived barriers to wood fuel use; and
4. Determine what industry feels is needed for them to use wood fuels.

Overview

This paper is divided into four sections. The first section examines the industrial manufacturing sector of Washington's economy. The purpose is to identify large fuel purchasers.

Section two presents our method for identifying candidate facilities. We present the data bases used, the identification of target industries based on how they use energy, and our selection of target facilities. The limitations and effects of these selection criteria are shown.

Section three discusses the preparation and results of our survey of the potential candidates. In this section, we outline questions and procedures. The results, as presented, represent the heart of this paper.

The final section presents the conclusions of this study. Many industries were unaware of the potential for wood fuels, but their interest in learning more about economic and environmental issues is high.

ENERGY USE BY WASHINGTON'S MANUFACTURING INDUSTRIES

Our search for the "best" market for wood fuels begins with a look at existing energy use within the state. Figure 1 shows the overall consumption of fuels and electricity by four major sectors: residential, commercial, industrial, and transportation. We have chosen the industrial sector as the prime target because the conversion technologies are available, the market is large, and it should be beneficial to the environment.

Washington Industries

We use the Standard Industrial Classification (SIC) system to identify industries. Data is readily available using this classification method. Each major industry is represented by a two-digit number. For example, the food industry is represented by the number 20, while the paper industry is 26. Each major industry group is further divided into industry subgroups represented by three-digit codes (e.g., meat products by 201 and miscellaneous converted paper products by 264). These subgroups are in turn divided into industries represented by four-digit codes (e.g., sausage and other prepared meats by 2013 and envelopes by 2642). Nationally there are 20 major industry groups, 142 industry subgroups, and 449 industries. Washington has companies represented by 347 different categories.

Fuel Purchases

It is difficult to identify large fuel purchasing facilities as there are no books containing that information. In fact, the use of fuel by specific industries is not available on the state level. The Annual Survey of Manufacturers (3) stopped collecting energy purchase information in 1981, and that survey only reported fuel purchases aggregated to the major industry or two-digit level. We have, however, taken national industry fuel purchases and combined them with known two or three digit state SIC totals for 1980 to derive consistent estimates of fuels purchased by each industry type in the state. We found that a small number of industries -- the top 25 out of 347 total industries -- represents 85 percent of the total purchased fuels. These 25 industries (listed in Figure 2) represent an initial list of good target areas.

SELECTION OF CANDIDATE FACILITIES

There are 6,000 manufacturing facilities in the state. We limited our study to 300 which appeared to be good candidates for conversion to wood fuels. The selection process involved two steps. First, we selected industries that appeared technically able to use wood fuels based on their end use requirements. Second, we identified larger facilities within each of the fuel using industries. The process is illustrated in Figure 3.

FIGURE 1
ENERGY USE BY SECTOR
1982 TOTAL, 940.2 T BTU

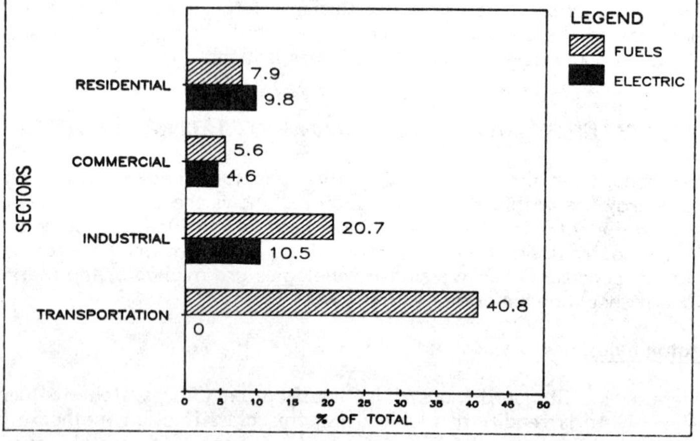

FIGURE 2
MAJOR FUEL PURCHASING INDUSTRIES

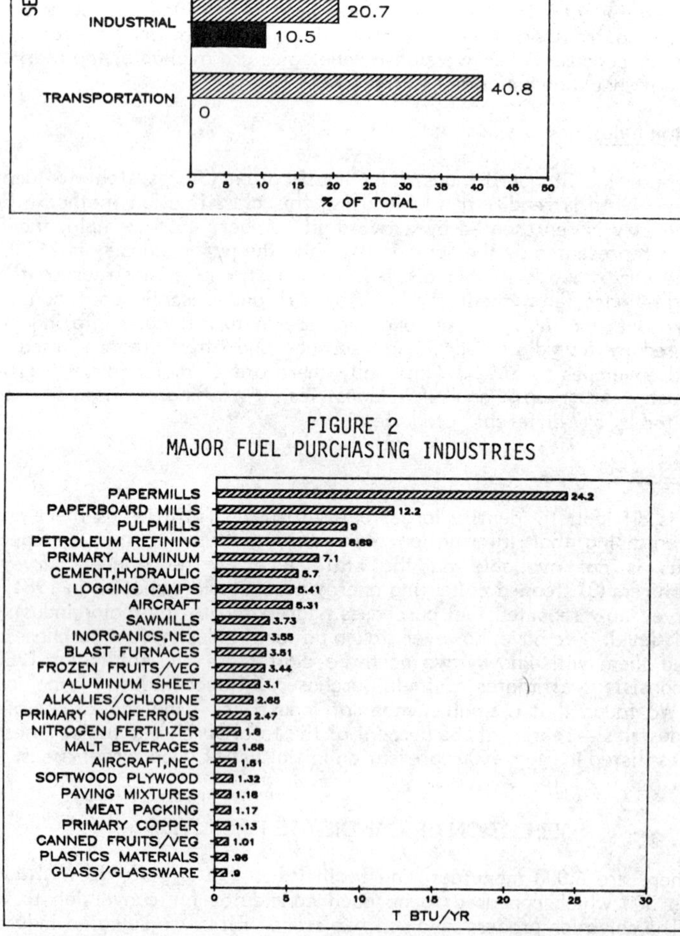

FIGURE 3
IDENTIFICATION OF POTENTIAL WOOD FUEL
USERS IN WASHINGTON INDUSTRIES

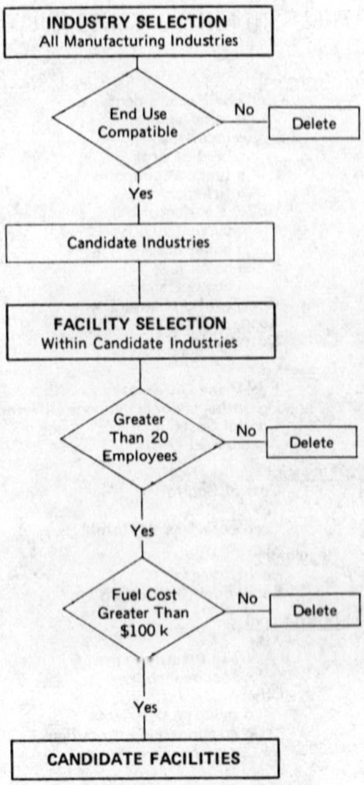

Data Bases

Any facility manager considering the use of wood fuels must consider many different factors. His set of criteria includes many of those listed in Table 1. We do not have access to this information for each facility. Our selection criteria had to be based on the availability of data for industries and facilities.

Several information sources were used to develop the list of candidate facilities. Six general classes of information were required:

1. Energy end use by industry.
2. Energy purchase by state industries.
3. Employment data for facilities.
4. Boiler information for facilities.
5. Wood resources by county.
6. Facility names and addresses.

The information sources are given in Tables 2-7. From these sources, three data bases were developed. The first contains end use information for specific industries. The second contains an estimate of the fuels purchased by industry types in Washington for 1980 and the third has names, addresses, employee numbers, and boiler capacities for individual facilities. These three data bases are stored on a micro computer using Lotus 1-2-3 software.

TABLE 1
FUEL SWITCHING CONSIDERATIONS

Technical
- End Use Requirements
    - o temperature
    - o heat flux
    - o heat control
    - o fuel contaminates
    - o firing rate
- Boiler Considerations
    - o current fuel type
    - o boiler age
    - o boiler size
    - o combuster type
- Cogeneration Feasibility
- Land availability
- Duty Cycle

Economic
- Fuel Price Differential
- Operation and Maintenance Differential
- Capital Costs
- Annual Fuel Use

Financial
- Cost of Capital
- Tax Credits
- Corporate "Hurdle Rates"

Regulatory
- Environmental
- Fossil Fuel Restrictions

Institutional
- Risk
    - o regulatory changes
    - o fuel availability
- Other
    - o industry experience
    - o company/industry outlook

TABLE 2
ENERGY END USE DATA SOURCES

Source	Content
Energy Analysis at 108 Industrial Processes, 1979 (8) Department of Energy	Drexel University analysis of 108 major manufacturing sectors. Data contents are energy requirements (Btu and temperature) for each unit operation per pound of product.
End-Use Matching for Solar Industrial Process Heat, 1980 (1) Solar Energy Research Institute	Industrial process heat requirements for 280 manufacturing sectors. Data contents are fluid used for heat transfer, temperature, annual energy use, operation schedule, and process heat rate.
Cogeneration Technology Alternatives Study, Vol. 2 Industrial Process Characteristics, 1980 (12) Department of Energy	Process heat requirements (Btu and temperatures) for unit operations in 15 major industries.
Analysis of the Economic Potential of Solar Thermal Energy to Provide Industrial Process Heat, 1977 (10) Inter Technology Corp.	Process heat requirements for 78 four digit SIC sectors by unit operation.

TABLE 3
ENERGY PURCHASES DATA SOURCES

Source	Content
1979 Annual Survey of Manufacturers' Fuels and Electric Energy Consumed (3) (States and SMSAs by Major Industry Groups) Bureau of the Census	State data for 1979 by two digit and selected three digit SIC codes. Data on quantity and cost of purchased fuels and electric energy.
1980 Annual Survey of Manufacturers' Fuels and Electric Energy Consumed (3) (States and SMSAs by Major Industry Groups) Bureau of the Census	State data for 1980. Content type same as above. Note: This report is the last in the annual series.
1982 Census of Manufacturers' Fuels and Electric Energy Energy Consumed (5) (States and SMSAs by Major Industry Groups) Bureau of the Census	State data for 1981 by two digit SIC only. Content type same as above.
1980 Annual Survey of Manufacturers' Fuels and Electric Energy Consumed (3) (Industry Groups and Industries) Bureau of the Census	National data for 1980 by four digit SIC codes. Content type same as above.

TABLE 4
EMPLOYMENT DATA SOURCES

Source	Content
1980 Annual Survey of Manufacturers' Statistics for Industry Groups and Industries (4) Bureau of the Census	National data for 1980 by four digit SIC codes. Data on employees, hours, payroll, value added, capital expenditure, cost of materials, and value of shipments.
1982 Census of Manufacturers Geographic Area Series, Washington (6) Bureau of the Census	State and county data for 1982 by four digit SIC codes. Content type same as above plus number of establishments.
County Business Patterns, 1982 Washington (2) Bureau of the Census	State and county data for 1982 by four digit SIC codes. Data on number of employees, payroll, number of establishments, and number of establishments by employment size class.
Employment and Payrolls in Washington State by County and by Industry (14) Quarterly reports Employment Security Department	State and county data for each calendar quarter by four digits for state and two digits for county. Data on number of employer units and employees and wages.
Workers' Compensation Data Base Unpublished Department of Labor and Industries	Individual company data including name, address, classification number, and hours worked.

TABLE 5
BOILER DATA SOURCES

Source	Content
Regional Air Pollution Control Agency Puget Sound and Southwest	Company name and address, boiler, capacity, fuels, and SIC code.
Department of Ecology Unpublished	Company name, address, boiler capacity, fuels, and SIC code. Computer data base.
Labor & Industries Boiler Inspection Office Not open to public	Company name, address, boiler pressure inspection. Sometimes boiler capacity information, manual files.

TABLE 6
FACILITIES IDENTIFICATION DATA SOURCES

Source	Content
Washington Manufacturers Register 1984/85 (11) Times Mirror Press	Name of company, SIC code, address, number employees, type business, sales range, year established, etc. Indexed by geographical area, SIC code, and alphabetical. Approximately 3,000 entries.
Workers' Compensation Data Base Unpublished Department of Labor & Industries	Individual company data including name, address, classification number, and hours worked.

TABLE 7
WOOD SUPPLY DATA SOURCES

Source	Content
Timber Harvest Reports Department of Natural Resources Annual Report	Volume harvested by county, species, ownership, and acres logged.

Industry Selection Criteria

There are 347 different manufacturing industries in Washington purchasing more than 120 trillion Btu of fuel energy each year. We wanted to know which industries are the "best" candidates for conversion to wood fuels.

Wood fuel can be used to meet the energy end use needs of many, but not all, manufacturing industries. These needs include process heating, boiler fuels, and engine fuels. Process heaters include furnaces, ovens, dryers, and kilns. Boilers are used to generate hot water and steam while engines are used for moving products and raw materials. The Energy Information Agency's (7) estimates of fuel uses for the major industries in the U.S. are shown below:

		Fuel Use, Percent	
SIC Code	Industry Group	Boilers	Process Heaters
20	Food	84	16
26	Paper	87	13
28	Chemicals	74	26
29	Petroleum	26	74
32	Stone, Clay, Glass	1	99

Using the sources given in Table 2, we developed a data base to estimate the energy end uses for particular industries. This data base gives an estimate of the fuel use in five categories for 155 different industries. The categories are process heat greater than 600°F, steam, hot air, hot water, and engines.

Current technology imposes limits on the application of wood fuels to some of these categories. Solid wood cannot be used directly in engines and its conversion to a liquid fuel is still under development. Therefore, all industries with greater than 90 percent of their fuel being used for engines are deleted. An example of such an industry is logging camps and logging contractors (SIC 2411). High temperature clean gas applications is another area where the technology for using wood fuels is limited. Thus in general, industries using greater than 90 percent of their fuel use in applications requiring greater than 600°F clean gases were eliminated. Some high temperature applications, such as cement kilns, must be evaluated on an facility by facility basis.

Our final selection of candidate industries was based on the quantity of fuel used by that particular industry. We therefore deleted those industries that did not meet our end use criteria and then ranked the remaining industries in order of their fuel consumption. Our focus is on the large users.

Facilities Selection

The selection of specific facilities as wood fuel conversion candidates was based on three basic criteria. First, the facility had to pass the end use criteria discussed above. The second criteria was that the facility have more than 20 employees. This criteria was chosen for two reasons. First, it reduces the number of facilities that had to be considered and second, a small firm may not want to hire an additional person to deal with a solid fuel combustor.

The third criteria was that a firm must have an estimated fuel bill greater than $100,000 per year. This criteria assumed that facilities with large fuel bills would be more interested in converting and it also recognized the economies of scale involved in the capital costs for a solid fuel combustion system. We applied this criteria indirectly because we do not know how much each facility spends on fuel each year. We assumed fuel costs $5.00/MBtu and the fuel intensity for industries in Washington is the same as similar industries throughout the country. Fuel intensity is the quantity of fuel purchased per employee and it was calculated from the annual survey of manufacturers. Fuel intensities vary from 22,100 MBtu/employee for nitrogenous fertilizers (SIC 2873) to 9 MBtu/employee for women's handbags (SIC 3171). Our $100,000 selection criteria is therefore equivalent to a minimum number of employees for a particular industry or facility. The number of employees is much easier to determine than the fuel expenses. Fuel intensities are very rough estimators, but they do permit a selection criteria to be applied to a facility related to the cost of fuels.

Results

Our selection criteria reduced the number of candidate industries from 347 to 257 and the number of candidate facilities from 6,000 to 1,276. The effect of each criteria on the number of facilities is shown in Figure 4. The deletion of facilities with less than 20 employees had the most dramatic effect. However, in terms of energy use the effect is relatively small. It is still the large facilities in the energy intensive industries that purchase most of the fuel.

We identified 300 facilities that account for 90 percent of the fuels purchased by the state's manufacturing industry. This group was chosen as the list of candidates for conversion to wood fuels. In selecting the candidates we emphasized the large facilities in the large energy using industries. We did not include facilities known to currently use wood fuels exclusively, or those that had known boiler capacities less than 1 MBtu/hr. We did, however, survey some industries, such as concrete products, just to learn more about their end use of fuels.

SURVEY OF CANDIDATE FACILITIES

This section contains the heart of this paper. We move from the theoretical candidate list to the real world. The choice of a mail survey was based on the time resource available to us. The choice of questions was based on our desire to learn as much as we could. This desire had to be tempered by the amount and type of information the facility managers would share with us.

FIGURE 4
CRITERIA SELECTION

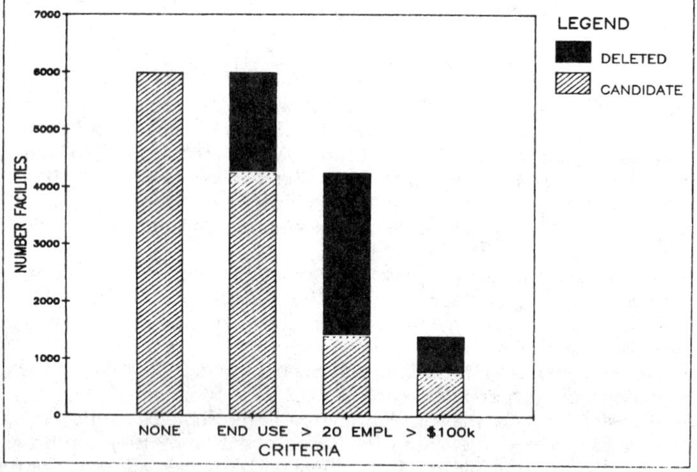

Our survey sought information in three areas: 1) the nature of the company, how much and what type of fuel they used, and what criteria they used for making capital decisions; 2) for wood fuel users the why, how long, how much, and why not using more; and 3) for non-wood users what constraints and obstacles they perceived, what fuel savings they would want to see, and their interest in additional information.

Preparation and Response

The survey contains 34 questions. Each facility was asked to answer the first 17 questions and then only those that applied to their particular situation on remaining questions. The questions included yes and no types, open response, ranking, and multiple choice ranging from one response to selecting all that apply. We solicited comments on the draft questionnaire from individuals familiar with surveys and industry. We incorporated their suggestions and then mailed a pretest to 30 (10%) of the candidates. We received a 30 percent response after three weeks and then mailed a follow-up letter which brought in an additional 30 percent. After few minor changes, it was mailed to the remaining 270 candidates. After receiving a 30 percent response we mailed a follow-up letter that resulted in an additional 25 percent response. Our overall survey response was 162 (54%).

The survey responses were coded and entered into a microcomputer data base. We used a commercial survey analysis program to prepare the tables and distributions.

Who Responded?

The survey was primarily completed by the plant manager or plant engineer (64%). In some cases the chief executive officer or president responded (12%). Table 8 shows the response by industry type. The large energy using industries had a high response rate.

Paper and primary metals are the two largest energy purchasing industries in Washington and 70 percent of the facilities solicited responded. The sectors with the lowest response were the machinery and equipment

TABLE 8
RESPONSE BY INDUSTRY TYPE

Standard Industrial Classification	Industry	Number	% Response
20	Food	46	42
22-23	Textiles	3	75
24-25	Lumber/Furniture	31	49
26	Paper	25	76
28	Chemicals	17	71
29	Petroleum	8	54
32	Stone, Clay	11	40
33	Primary Metals	7	70
34-39	Machinery, Equipment	10	33
		162	54

companies. These sectors, interestingly enough, indicated that their main use of fuels was other than boilers or furnaces, i.e., less easily converted to wood fuels.

Fuel Costs and Number of Employees

We asked for the approximate amount spent on fossil fuels at each facility. We divided costs into five categories ranging from less than $100 thousand per year to greater than $1 million per year. We also asked for the number of employees.

The percent of respondents in each category is shown in Table 9.

TABLE 9
FUEL COSTS AND NUMBER OF EMPLOYEES

Fuel Costs	Percent Respondents	Number Employees	Percent Respondents
<$100k	32	50	26
$ 100 - 250k	16	51 - 100	21
$ 250 - 500k	12	101 - 250	25
$500k - 1 M	10	250	26
>$1 M	22		

The number of employees per facility are evenly distributed in the various classes. The cost of fuels are tilted on each end of the distribution. The large number of firms indicating less than $100 thousand for fuel bills points out the limitations of using fuel intensity data to estimate fuel costs. The relatively large number of firms with fuel bills greater than $1 million is due to our focus on large energy users.

Basis for Capital Investment Decisions

The method used by facilities in deciding upon capital investments was surveyed. We also inquired about current constraints to investments and asked if energy investments were treated differently than other capital investments.

The predominant method chosen for analysis of capital investments was rate of return (47 percent). The second most used method was simple payback (33 percent). Discounted payback and net present value were reported by 8 percent each. The food processing industry was the only sector that tended to slightly favor simple payback, 46 percent versus 39 percent for rate of return. The petroleum industry was predominately (63%) rate of return.

Most firms (86%) reported that they made energy-related investment decisions on the same basis as other capital investments. The responses to our question about which constraint provides the greatest obstacle to energy-related capital investments was not surprising. The requirement for a particular rate of return was listed by 47 percent as the major obstacle. The condition of their particular industrial sector was listed as the major obstacle by 35 percent. The lumber industry, however, indicated that industry conditions were the major obstacles to capital investments (55%).

Current Wood Fuel Users

Twenty-six of the respondents (16%) are currently using wood fuels. We chose to survey known users if we also knew that they also purchased fossil fuels. Our intent was to find out how much of their fuels needs were met by wood, where they got the wood and why they are not using more. Sixty-two percent of the wood users reported meeting more than 90 percent of their fuel needs with wood fuel. Fifteen percent met under 25 percent of their fuel needs and 31 percent met between 26 and 90 percent of their need. The source of wood fuels was predominantely from in-house residues. Forty-six percent obtained greater than 90 percent of their wood fuel from in-house sources. Only 12 percent had to purchase greater than 75 percent of their wood fuel. The reasons given for not increasing their use of wood fuels included economic feasibility (47%), technical feasibility (35%), uncertain fuel supply (29%), and capital availability (24%).

Investigation of Wood Fuel Uses

Twenty-four (18%) of the non-wood users have investigated the use of wood fuels. Most of the investigators -- 15 -- were in the lumber and furniture industries. However, four were in the food industry, two in paper, one in chemicals, and two in other manufacturing. These investigations generally involved internal discussions (70%) and a feasibility study (58%). Less than half of the facilities (42%) actually contacted equipment vendors for information. The larger group of respondents (73%) have not investigated the use of wood fuels.

We asked those who had not investigated wood use what prevented them from such an investigation. Forty-three percent said they were not aware of the potential for using wood fuels; forty-eight percent said they did not have the resources (staff, time, money) to investigate; and 25 percent said that the conditions within their industry precluded investigation.

Constraints to Use of Wood Fuels

We asked both groups to identify what they consider to be the key constraint to using wood; other obstacles they consider important; a ranking of the influence of various alternatives on their decisions; and finally, what level of fuel-related savings they would require before converting to wood fuels.

Table 10 shows the responses to our questions on constraints to wood fuel use for different industry sectors. The rows labeled "important" are a tally of all respondents that indicated that a particular constraint was important. They were allowed to indicate as many items as they wanted. The "combined" rows give the response to the sum of the key constraints and the important. However, if a respondent had a category as both a key constraint and an important constraint they were only counted once. Thus, the "combined" row does not necessarily equal the sum of the "key constraint" and the "important" rows.

In terms of ranking the key constraints, we found that economic feasibility was the major hurdle, 29 percent. The second biggest barrier (14%) was environmental issues and third was uncertain fuel cost savings (10%). This same order was observed with the combined results.

TABLE 10
CONSTRAINTS ON BIOFUEL USE

		TWO-DIGIT SIC CODE								
	TOTAL	20 (FOOD)	22-23 (TEXT)	24-25 (LUM/FUR)	26 (PULP)	28 (CHEM)	29 (PETRO)	32 (STONE)	33 (PRM MET)	34-37 (MANUF)
Number Responding	131	44	3	17	12	16	8	11	7	10
Lack of Reliable Information										
Key Constraint	11	3	1	0	3	0	1	1	0	2
	8%	7%	33%	0%	25%	0%	13%	9%	0%	20%
Important	42	19	2	1	3	5	3	4	1	3
	32%	43%	67%	6%	25%	31%	38%	36%	14%	30%
Combined	52	22	3	1	5	5	4	5	1	5
	40%	50%	100%	6%	42%	31%	50%	45%	14%	50%
Uncertain Fuel Cost Savings										
Key Constraint	12	5	0	1	1	0	1	2	0	2
	9%	11%	0%	6%	8%	0%	13%	18%	0%	20%
Important	50	20	2	1	5	8	4	2	1	6
	38%	45%	67%	6%	42%	50%	50%	18%	14%	60%
Combined	57	22	2	2	6	8	4	4	1	7
		50%	67%	12%	50%	50%	50%	36%	14%	70%
Economic Feasibility										
Key Constraint	38	10	2	6	3	4	4	4	3	2
	29%	23%	67%	35%	25%	25%	50%	36%	43%	20%
Important	60	22	2	6	4	6	2	4	4	7
	46%	50%	67%	35%	33%	38%	25%	36%	57%	70%
Combined	85	28	2	9	7	9	6	7	6	8
	65%	64%	67%	53%	58%	56%	75%	64%	86%	80%

TABLE 10 (Cont'd)

		TWO-DIGIT SIC CODE								
	TOTAL	20 (FOOD)	22-23 (TEXT)	24-25 (LUM/FUR)	26 (PULP)	28 (CHEM)	29 (PETRO)	32 (STONE)	33 (PRM MET)	34-37 (MANUF)
Number Responding	131	44	3	17	12	16	8	11	7	10
Capital Availibility										
Key Constraint	9	1	0	3	2	1	1	0	0	0
	7%	2%	0%	18%	17%	6%	13%	0%	0%	0%
Important	49	13	1	9	3	7	2	4	5	4
	37%	30%	33%	53%	25%	44%	25%	36%	71%	40%
Combined	54	14	1	10	4	7	3	4	5	4
	41%	32%	33%	59%	33%	44%	38%	36%	71%	40%
Management Acceptance										
Key Constraint	5	2	0	0	0	2	0	0	0	1
	4%	5%	0%	0%	0%	13%	0%	0%	0%	10%
Important	23	8	2	0	1	4	1	1	2	3
	18%	18%	67%	0%	8%	25%	13%	9%	29%	30%
Combined	26	9	2	0	1	6	1	1	2	3
	20%	20%	67%	0%	8%	38%	13%	9%	29%	30%
Unacceptable Risk										
Key Constraint	5	4	0	0	0	0	0	0	0	1
	4%	9%	0%	0%	0%	0%	0%	0%	0%	10%
Important	22	6	1	1	3	3	3	1	1	3
	17%	14%	33%	6%	25%	19%	38%	9%	14%	30%
Combined	25	9	1	1	3	3	3	1	1	3
	19%	20%	33%	6%	25%	19%	38%	9%	14%	30%

TABLE 10 (Cont'd)

		TWO-DIGIT SIC CODE								
	TOTAL	20 (FOOD)	22-23 (TEXT)	24-25 (LUM/FUR)	26 (PULP)	28 (CHEM)	29 (PETRO)	32 (STONE)	33 (PRM MET)	34-37 (MANUF)
Number Responding	131	44	3	17	12	16	8	11	7	10
Uncertain Fuel Supply										
Key Constraint	3	1	0	0	0	0	0	1	0	0
	2%	2%	0%	0%	0%	0%	0%	9%	0%	0%
Important	42	16	1	3	5	5	1	4	4	3
	32%	36%	33%	18%	42%	31%	13%	36%	57%	30%
Combined	44	17	1	3	5	5	1	4	4	3
	34%	39%	33%	18%	42%	31%	13%	36%	57%	30%
Fuel Handling/Storage Capability										
Key Constraint	13	7	0	0	2	1	1	1	0	1
	10%	16%	0%	0%	17%	6%	13%	9%	0%	10%
Important	67	24	2	4	8	9	5	3	5	6
	51%	55%	67%	24%	67%	56%	63%	27%	71%	60%
Combined	73	29	2	4	8	9	6	3	5	6
	56%	66%	67%	24%	67%	56%	75%	27%	71%	60%
Environmental Considerations										
Key Constraint	18	8	0	5	0	2	0	1	1	0
	14%	18%	0%	29%	0%	13%	0%	9%	14%	0%
Important	71	23	3	6	8	4	7	3	7	9
	54%	52%	100%	35%	67%	25%	88%	27%	100%	90%
Combined	81	27	3	10	8	6	7	3	7	9
	62%	61%	100%	59%	67%	38%	88%	27%	100%	90%

Our question asking them to rank alternatives in terms of their potential influence on converting to wood fuel is shown in Table 11. Once again, the economic issues are most important, followed by capital availability, assured fuel supply, and technical assistance availability.

TABLE 11
RANK OF FACTORS INFLUENCING WOOD FUEL DECISIONS

		TWO-DIGIT SIC CODE								
	TOTAL	20 (FOOD)	22-23 (TEXT)	24-25 (LUM/FUR)	26 (PAPER)	28 (CHEM)	29 (PETRO)	32 (STONE)	33 (PRM MET)	34-37 (MANUF)
Responded Mean Rank	125	44	3	16	12	16	7	10	7	8
Assured Fuel Supply	3.11	2.84	3.00	3.63	3.43	3.45	2.80	2.88	3.00	3.25
Demonstrate Cost Savings	1.74	1.68	1.33	2.44	1.63	2.09	1.14	1.89	1.80	1.44
Availability of Capital	2.88	3.20	5.00	1.56	3.00	2.90	2.20	2.38	2.50	3.25
Technical Assistance	3.60	3.45	3.00	3.57	3.75	3.80	3.50	3.43	3.50	4.29
Tax Incentives	4.09	4.13	4.00	3.29	4.00	4.00	4.20	4.67	4.20	4.50
Similar Industries Using Wood	5.14	4.97	4.67	5.50	4.38	5.80	5.40	5.00	6.25	5.00

The level of fuel-related savings required for a facility to convert to wood fuels was investigated. Table 12 shows the required savings for each major industry sector. The range of 26-50 percent fuel savings was selected by 40 percent of the respondents. Interestingly, 14 percent of the facilities indicated they would not switch fuels irregardless of fuel savings. The facilities indicating a required savings of 11-25 percent were fairly uniform across the industries with an average response of 26 percent.

Interest in Wood Fuel Information

We asked non-wood users if they were interested in receiving information on the use of wood fuels. We also asked what types of information they would find useful. Forty percent of the respondents expressed an interest in receiving more information. They showed a preference for information on economic feasibility (87%). However, they also wanted information on vendors and technologies (75%), fuel supply (72%), and environmental issues (72%).

CONCLUSIONS

Washington's manufacturing industries are interested in the use of wood fuels. They are open to receiving information and would like to know about

TABLE 12
FUEL RELATED SAVINGS REQUIRED FOR CONVERSION TO WOOD FUELS

		TWO-DIGIT SIC CODE								
	TOTAL	20 (FOOD)	22-23 (TEXT)	24-25 (LUM/FUR)	26 (PAPER)	28 (CHEM)	29 (PETRO)	32 (STONE)	33 (PRM MET)	34-37 (MANUF)
Responded	102	37	3	10	9	13	5	8	6	8
Required Savings										
1-10%	6	2	0	0	1	0	1	1	0	0
	6%	5%	0%	0%	11%	0%	20%	13%	0%	0%
11-25%	26	8	1	3	3	4	2	3	1	0
	25%	22%	33%	30%	33%	31%	40%	38%	17%	0%
26-50%	42	16	2	5	2	4	0	4	3	6
	41%	43%	67%	50%	22%	31%	0%	50%	50%	75%
50%	14	5	0	1	2	2	2	0	0	2
	14%	14%	0%	10%	22%	15%	40%	0%	0%	25%
Will Not Switch Regardless	14	6	0	1	1	3	0	0	2	0
	14%	16%	0%	10%	11%	23%	0%	0%	33%	0%

the economics of using wood fuels. Their concerns about environmental issues are higher than expected. Their use of fuels is similar to data available nationally.

ACKNOWLEDGEMENTS

This work was partially sponsored by funds from the U.S. Department of Energy Regional Bioenergy Program. The Pacific Northwest and Alaska program is managed by the Bonneville Power Administration. The collaboration of Peter Spinney in the preparation and analysis of the survey results is sincerely appreciated. This study could not have been completed without the cooperation of the manufacturing industries taking time to respond to the survey.

REFERENCES CITED

1. Brown, K. C., Hooker, D.W., Rabel, A., Stadjuhar, S.A., and West, R.E., End-Use Matching for Solar Industrial Process Heat, Golden, Solar Energy Research Institute, 1980.

2. Bureau of the Census, County Business Patterns, 1982 Washington, Washington, U.S. Government Printing Office, 1983.

3. Bureau of the Census, 1980 Annual Survey of Manufacturers, Fuels and Electric Energy Consumed, Washington, U.S. Government Printing Office, 1981.

4. Bureau of the Census, 1980 Annual Survey of Manufacturers, Statistics for Industry Groups and Industries, Washington, U.S. Government Printing Office, 1981.

5. Bureau of the Census, 1982 Census of Manufacturers, Fuels and Electric Energy Consumed, Washington, U.S. Government Printing Office, 1983.

6. Bureau of the Census, 1982 Census of Manufacturers, Geographic Area Series, Washington, Washington, U.S. Government Printing Office, 1983.

7. Cohen, B.N., The Substitution of Coal for Oil and Natural Gas in the Industrial Sector, Washington, Energy Information Agency, 1980.

8. Drexel University, Energy Analysis of 108 Industrial Processes, Washington Department of Energy, 1979.

9. Huff, W. J., Kennel, R.P., Schertz-Willett, L. and Steadman, P.E., Industrial Biomass Market Assessment, Golden Solar Energy Research Institute, 1982.

10. InterTechnology Corp., Analysis of the Economic Potential of Solar Thermal Energy to Provide Industrial Process Heat, Department of Energy, 1977.

11. 1984/85 Washington Manufacturers Register, Los Angeles Times, Mirror Press, 1984.

12. United Technologies Corp., Cogeneration Technology Alternatives Study, Volume 2, Industrial Process Characteristics, Department of Energy, 1980.

13. Washington State Department of Natural Resources, Timber Harvest Reports, Olympia, Department of Natural Resources, Annually.

14. Washington State Employment Security Department, Employment and Payrolls in Washington State by County and Industry, Olympia, Employment Security Department, Quarterly.

DISCUSSION

E. Wan, Science Applications International Corp.	Were you able to identify those that are both wood producers and consumers?
Kerstetter	Of those that responded that were using wood, it varied with the individual facilities. But as a mean average, 70% use self-generated materials, and 30% purchase on the open market.

DIRECT COMBUSTION OF HERBACEOUS ENERGY CROPS: MARKET ASSESSMENT

Kathleen A. Dripchak
Senior Analyst
JAYCOR
Vienna, Virginia 22180

ABSTRACT

Assuming petroleum prices again increase in real terms in the mid-term as anticipated, herbaceous crops may prove to be a valuable energy resource. The objective of this study was to elucicate potential opportunities for forage energy crop development. Both economic and technical issues will impact its development. Technical issues include choice of optimal combustion technology, ash slagging, and the potential for co-firing herbaceous crops with other solid fuels. Economic issues include cost tradeoffs available among densification, transportation distance, and combustor size.

Facilities which currently combust solid fuels such as wood and coal are potential areas of opportunity for herbaceous energy crop (HEC) combustion. Facilities fired with agricultural wastes and residues, wood, and coal, were characterized, and several SIC codes emerged as potentially suited to HEC combustion based on facility and boiler size, and geographic location, as well as fuel type. These are SIC 20 - Food and Kindred Products, and SIC 22 - Textile Mill Products in the manufacturing industries; and the institutions in non-manufacturing industries: SIC 80 - Health Services; SIC 82 - Educational Services, and SIC 92 - Justice, Public Order and Safety. Those geographic regions with high forage crop productivity and lower than the national average land values were identified to be the Southeastern region, Appalachia, and the Corn Belt. In total, 274 facilities in these regions use wood or coal as their primary energy source in SIC codes 20, 22, 80, 82, and 92. In addition to the potential for replacing the current solid fuel use with herbaceous energy crops, a further examination of these industries may highlight the potential for HEC use in non-solid fuel combusting facilities within these industries, and in other industries with similar characteristics.

DIRECT COMBUSTION OF HERBACEOUS ENERGY CROPS: MARKET ASSESSMENT

INTRODUCTION

The overall goal of this study was to assist Oak Ridge Laboratory (ORNL), the field manager for the Department of Energy Biomass Energy Technology Division's Herbaceous Energy Crop Program in assessing the market potential for the direct combustion of herbaceous crops. This market assessment is intended to assist ORNL program managers in evaluating their program strategy by providing information on the regions and industrial sectors where the direct combustion of forage energy crops can generate the greatest impact. The study will also help in the development of criteria for selecting R&D strategies that will enhance the economics of this technology.

CURRENT USE OF SOLID FUELS

As a first step towards defining a market for combusting herbaceous energy crops (HEC), it is useful to characterize those facilities that combust solid fuels for process heat, steam, and electricity generation. The direct combustion of herbaceous energy crops may have application in two areas. Firstly, as the fuel of choice for new installations and, secondly, as a replacement fuel in conventional fuel systems. In either case, decisions will be based on the comparative cost advantage of converting various fuels to useful energy. As a solid fuel, herbaceous crops can be burned most readily in systems designed to handle other solid fuels, if not specifically designed for herbaceous energy crops. As a result, it is likely that the economics will favor HEC combustion in industries which currently combust other solid fuels (i.e., wood and coal) before it is economic to combust HEC in industries presently using gaseous and liquid fuels.

Industries which combust solid biomass fuels share some basic characteristics. Because of the low energy content to volume ratio of biomass fuels, facilities which combust these fuels tend to be small-scale. Facility size is determined more often by economic than technical constraints; transportation costs can quickly increase the delivered cost of biomass fuels to make any large-scale potential uses uneconomic. In many cases these facilities are located nearby the source of the solid fuel in order to minimize transportation costs. This location factor should be especially important when considering HEC's, which typically tend to have a much lower energy content to volume rates (in an undensified form) than other solid fuels.

As a result of these factors, a characterization of facilities which currently combust solid fuels could provide insight towards identifying industries which potentially may switch to combusting HEC, and other industries which because of their size, process type, geographic location, and other factors, may demonstrate potential for switching from conventional fuels (e.g., electricity, petroleum, and natural gas) to the direct combustion of herbaceous crops. Of those establishments currently burning solid fuels, commercial and institutional facilities, small scale industrial facilities, and electric utilities powered by wood or herbaceous crops are of the greatest interest. Coal-fired utilities are typically large and thus were excluded from the analysis. The residential sector was excluded for several reasons. Firstly, herbaceous energy crops used in the residential sector would require a highly developed infrastructure, since residential users are dispersed. Secondly, combusting HEC could create pollution problems as are currently being experienced with residential wood combustion. In some cities such as Aspen, Colorado, the pollution from wood-fired stoves has been so severe that the city has limited their use. The problem arises because the stoves lack adequate pollution control devices, and also because the stoves are often not properly operated, resulting in inefficient combustion and higher pollution output.

The characterization of solid fuel users is divided into three groups by fuel type: (1) herbaceous crops or agricultural wastes and residues; (2) wood and wood wastes and residues; and (3) coal. Facilities in Hawaii and Alaska were excluded since our primary area of interest was in forage crops typically produced in the continental U.S.

Currently, direct combustion of herbaceous crops on a commercial basis is limited to agricultural wastes and residues, including rice hulls, bagasse, corn cobs and cotton gin waste. Forage crops are combusted for energy purposes elsewhere, most notably in Denmark and Germany. In the U.S., about 50 facilities which combust agricultural wastes and residues were identified, and a further 20 are in the planning or permit stages. Most of the establishments are food growing or processing facilities, which use the process wastes to produce heat, steam, or electricity.

The forest products industries (SIC codes 24, 25, 26) account for over 95 percent of the 1.6 quads of wood energy currently being used in the industrial, commercial and municipal sectors. Because of the unique relationship between the wood resource and the forest products industries, these were excluded from the characterization. The remaining 5 percent of wood energy use in these sectors was characterized using the EPA's National Emissions Data System (NEDS). Over 50 percent of industrial and institutional wood-fired boilers are used by manufacturing, primarily (75%) within five industries: Miscellaneous Manufacturing (SIC 39); Food and Kindred Products (SIC 20); Chemicals and Allied Products (SIC 28); Textile Mill Products (SIC 22); and Rubber and Miscellaneous Plastic Prodcucts (SIC 30). Many of the industries in SICs 39 and 28 are linked to the wood resource in their manufacturing process, and so are less likely to switch to herbaceous feedstocks. Of the wood-fired boilers in non-mnaufacturing industries, about three-fourths fall within SICs 49 - Electric, Gas and Sanitation Services, 82 - Educational Services, and 80 - Health Services. In the institutional sector, hospitals and schools are the predominate users of wood energy.

The distribution of wood-fired facilities reflects regional and state interest in wood energy. Of the 120 facilities reported by NEDS, 43 are located in the Southeast, 25 in New England and the Mid Atlantic states, and 27 in the Great Lakes Region. Most (55%) wood-fired boilers are small, with boiler design capacities less than 25 million Btu per hour. Virtually all of the larger (100 million Btu per hour) wood-fired boilers are found in SIC 28.

Coal use is the largest category of solid fuel consumption in the U.S., and totaled 737 million tons in 1984. Information on industrial and commercial coal consumption was derived from a database currently being marketed by Resource Data International (RDI). The database is comprehensive, and indicates that about 1500 facilities in the industrial and commercial sectors (excluding electric utilities, coke plants and residences), combusted coal in 1983. The pattern of coal consumption revealed by the RDI database is clearly linked to the coal supply regions. Greatest coal use is in fairly industrialized regions located near coal mining areas such as Ohio, Pennsylvania, Indiana, and Michigan. Manufacturing industries accounted for 76 percent of coal-fired facilities and 86% of coal consumption in 1983. Indeed, three industries (SICs 26 - Paper and Allied Products, 28 - Chemicals and Allied Products, and 32 - Stone, Clay, Glass, and Concrete Products) accounted for about 43 percent of all coal-fired facilities and 59 percent of coal consumption. Three-fourths of all coal buring facilities in the non-manufacturing industries are found in SICs 80; 82; and 92 (Justice, Public Order, and Safety). Small scale coal users (less than 20,000 ton/yr) should have the best potential for combusting herbaceous crops. Non-manufacturing industries tend to be smaller-sized than manufacturing industries, with 74 percent of these facilities (as opposed to 47 percent of the manufacturing facilities) consuming less than 20,000 tons of coal per year. Of the manufacturing industries, Food and Kindred Products (SIC 20) and Textile Mill Products (SIC 22) facilities tend to be the smallest sized.

The most promising markets for expanding the use of herbaceous feedstocks, based on the current use of solid fuels, appear to be the food and textile industries in the manufacturing sector, and educational and health service industries, and public institutions.

TECHNICAL ISSUES

Direct combusion of forage herbaceous crops appears to

be the most promising conversion method in the near term, since the technology is well established for some types of feedstocks and the current economics generally appear more favorable than for the more advanced technologies.

As an energy feedstock, forage crops differ considerably from conventional fuels, in terms of physical and chemical characteristics (e.g. calorific value, ash content, bulk density, moisture content, and volatility). The thermal stability of the ash is perhaps the most significant factor affecting combustion of forage crops. Materials with high ash contents (greater than a few percent) may cause combustion system failures due to ash slagging and fouling. Many herbaceous feedstocks have high ash contents and so require combustors than can either burn the fuel at low temperatures or are equipped with ash removal devices.

Several types of combustors are available for burning solid fuels, including pile-burning, cyclone or suspension burners, and fluidized-bed combustors (FBCs). It is generally agreed that FBCs are the most technically efficient method for combusting agricultural residues and wastes such as straw, grass residues, and rice hulls. However, some researchers suggest that straw and grasses should be gasified and not combusted, particularly if steam is required. Fluidized-bed systems are particularly appropriate for herbaceous fuels, because the combustion temperature can be maintained at lower temperatures in order to minimize ash slagging and fouling. Fluidized-bed systems are more expensive than conventional pile burning or suspension and cyclone combustors, and these cost factors may limit their application.

Co-firing of forage crops with coal appears to have good potential, but there is very little experimental or commercial experience in combusting a mixture of these feedstocks. Forage crops with higher moisture contents (such as the grasses) may require drying before combustion. Several technologies are available for compacting straw and other forage crops into pellets, briquettes, cubes, and other densified forms. However, densification may add as much as 50 percent to the cost of prepared feedstock, and so should be evaluated on a site specific basis. The form of the densified or as-harvested feedstock will impact the system for transporting, storing, and handling the fuel.

Environmental impacts of combusting forage crops may result from the delivery and combustion systems. Emission of particulates is the primary impact associated with combusting these types of crops. Emissions vary depending upon the feedstock type and the combustor type. Fluidized-bed and cyclone and suspension burners tend to emit more particulates than pile-type burners in which the fuel is piled on a grate. High ash fuels such as rice hulls yield more particulates than lower ash biomass fuels such as wood or straw. Equipment available for controlling pollutant emissions includes simple collectors, cyclones, electrostatic precipitators, scrubbers, and baghouses. The most appropriate technology will depend on the combustor type and allowable emission limits.

ECONOMIC ISSUES AND MARKET ASSESSMENT

Economics is an essential component in assessing the market potential of any alternative energy system such as the direct combustion of herbaceous crops. Indeed, in the long run and precluding public subsidies, the development of this energy resource rests solely on its ability to compete in price with conventional fossil fuel systems. From the discussion on technical issues, it is apparent that even though all of the technical issues have not yet been solved, combusting a variety of forage crops and agricultural residues is technically practical. The primary issues impacting the development of this energy source are largely economic.

The economics of forage combustion systems are complex. Component activities are inter-related, and costs are site- and situation-specific, the latter reflecting political and institutional factors. Also, rigorous estimates of the costs associated with producing, harvesting, processing, storing, transporting, and combusting forage crops have not yet been developed, since such systems are not yet commercially used. As with the technical issues, economic studies have focused mainly on burning agricultural residues and not forage crops per se. As a result, the economic data available is more anecdotal than complete. Nonetheless, sufficient information exists to enable a rough estimation of ranges in the costs of producing energy from forage crops.

In lieu of firm cost estimates for herbaceous energy system components, approximations based on forage crops used for agricultural purposes were employed in the analysis. The estimates produced from this data should be conservative. It is anticipated that research on producing forage crops for maximum biomass yield and changes in management practices to suit energy crops should reduce the cost per ton of producing these crops. Transportation costs could prove to be particularly significant. Because of the low bulk density of most biomass fuels in general, and forage crops in particular, the cost of transporting these materials may quickly become a major portion of the delivered cost of the fuels. Tradeoffs are available among the costs of transportation, densification, and combustor size. Typically, economies of scale exist in combustion systems, and the savings achieved by using a larger system may offset the increased transportation costs, up to a point. However, the size of the combustor may be fixed, as in cases where a system previously fired by a conventional fuel is converted to herbaceous crops. Also, the size of a new facility is usually set, and the objective is to meet its energy requirements in the least cost manner. Regional and site-specific factors such as land rental costs, land productivity, and transportation costs will impact the choice of the optimal system. All of this underscores the need for a systems analysis approach to the question of the least-cost forage crop combustion system.

Another major factor which will impact the development of forage energy systems is the presence of an infrastructure of industrial, commercial and institutional facilities to combust the energy crops. For example, costs of producing forage crops may be particularly low in Wyoming, but with the state's low population and thus limited energy requirements, substantial use of energy crop systems may be unlikely.

Currently, the economics of forage energy systems are not favorable. Howerver, certain sectors of the economy and regions of the country are likely to become areas of opportunity when the price of competing fuel rises. Many energy analysts predict an increase (in real terms) in oil prices in the mid-term. This will impact the prices and relative advantage of all types of energy sources including herbaceous-fired systems. Based on the characterization of current solid fuel users, eight SIC codes have emerged as sectors of potential opportunity for herbaceous energy systems:

SIC 20-Food and Kindred Products
22-Textile Mill Products
28-Chemicals and Allied Products
30-Rubber and Miscellaneous Plastics
32-Stone, Clay, Glass, and Concrete Products
80-Health Services
82-Educational Services
92-Justice, Public Order, and Safety

Facilities in SIC's 28, 30, and 32 may be less suited to this energy source, since many of these can be relatively energy-intensive. SIC 28 may also be inappropriate if the product is based on cellulosic materials such as wood. All three SIC codes which include institutions indicate potential for forage energy crop systems.

Opportunities for energy crop systems will be conditioned in large part, on the availability of a low-cost resource. It appears that the Lake States and the Southeastern regions of the United States have such resources. Other regions that appear favorable based on productivity, production costs, and land values are the Mountain States, Corn Belt States, Appalachian States, and the Mid-Atlantic region. The Southeastern region may be especially appropriate due to the area's longer than average growing season. A sub-set of facilities with the greatest potential for HEC use includes SIC codes 20, 22, 80, 82 and 92, in the Southeast, Appalachia, and the Corn Belt states. In total, there are 274 facilities which combust wood and coal in these SIC codes and regions, as displayed in Table 1. Most solid fuel facilities in the Southeast and Appalachia are textile mills (SIC 22). The Corn Belt states use very little wood in the industries identified, but burn coal in at least 25 facilities in each of SIC codes 20, 80, and 82.

In the mid-term, the institutions and industries mentioned above, in the regions with a low-cost forage resource, have the greatest potential for development of forage energy crops systems.

TABLE 1
TARGET FACILITIES -- NUMBER OF ESTABLISHMENTS IN THE REGIONS AND SIC CODES PROJECTED TO HAVE THE GREATEST POTENTIAL FOR HEC USE

REGION AND	SIC 20		SIC 22		SIC 80		SIC 82		SIC 92		
STATES	WOOD	COAL	WOOD	COAL	WOOD	COAL	WOOD	COAL	WOOD	COAL	TOTAL
Southeast											
South Carolina	1	--	--	23	--	--	--	--	--	--	24
Georgia	1	3	1	18	1	--	--	--	--	--	24
Alabama	--	1	1	4	--	--	--	--	--	--	6
Florida	--	--	--	--	--	--	--	--	--	--	0
Appalachia											
Virginia	--	--	--	11	--	6	1	6	--	9	33
West Virginia	--	--	--	--	--	--	--	--	--	1	1
Tennessee	1	--	2	3	--	--	1	7	--	--	14
Kentucky	1	12	--	1	--	2	--	8	--	1	25
North Carolina	--	--	1	30	1	--	--	4	--	--	36
Corn Belt											
Ohio	--	3	--	--	--	25	--	12	--	9	49
Indiana	--	11	--	--	--	4	--	6	--	6	27
Illinois	--	6	--	2	--	4	--	3	--	4	19
Iowa	--	10	--	--	--	--	--	4	--	--	14
Missouri	--	--	--	--	--	--	--	2	--	--	2
Sub Total	4	46	5	92	2	41	2	52	0	30	274

CONCLUSIONS

The objective of this study was to identify areas of opportunity for the direct combustion of herbaceous crops, and to provide ORNL Herbaceous Energy Crop Program managers with information to assist them in planning and conducting their programs. A major conclusion from this study is the need for additional research to lower the costs of growing, harvesting, and transporting herbaceous crops for direct combustion applications.

Thus far, the Herbaceous Energy Crop Program (HECP) has focused primarily on solving production issues, and has also been concerned with the harvesting and transportation components of herbaceous energy systems. The reasons are apparent. Biomass productivity can have a tremendous impact on the overall system economics, and also, this type of R&D requires more time than combustion system R&D. Production and harvesting costs are anticipated to decline, due to increased yields (from species selection and breeding), and modifications to management practices for crops with an energy rather than agricultural end use. Opportunities for decreasing transportation costs are limited to densification options. Densification currently is expensive, and this becomes an issue to be considered as a tradeoff with transportation costs. Several combustion systems are available which should require only minor modification to the storage, handling, and possibly the slag removal and environmental control equipment, to accomodate forage crop fuels. Once developed, combustion sytems modified for herbaceous crops should be tested thoroughly. Development of this technology is not likely to follow the development of equipment for combusting wood. In this latter case, the consuming industry was extremely familiar with the feedstock's chemical and physical properties. As a result of these factors, Federal testing of new herbaceous-fired systems may be required.

The costs of all of the activities mentioned above - production, harvesting, transportation, combustion, and others (e.g. storage)--are interdependent. Thus, the optimal (least-cost) route to producing energy from forage crops should be ascertained through a systems analysis of these activities. Studies of this sort should be conducted for individual potential applications of forage energy systems, since component costs may vary considerably from site to site.

In addition to these recommendations about forage energy systems in general, the study indicates other areas for further in-depth study. Certain regions of the country are endowed with lands potentially well suited to producing forage energy crops. The industries and institutions identified as being well suited to combusting solid fuels within each of these regions should be examined in order to better evaluate their potential. Each facility currently combusting solid fuels should be characterized in more detail, including the following factors:

- SIC code
- Location
- Size
- Boiler type
- Boiler age
- Boiler design capacity
- Seasonality of use
- Current fuel
- Cost per unit energy output
- Importance of energy in operation (percentage of product's final cost attributable to energy)
- Storage

Such an assessment would provide a more detailed description of facilities that currently combust solid fuels. As R&D on herbaceous energy crops becomes available, rigorous economic analyses can be conducted for each site in order to determine the potential for replacing the current fuel with herbaceous feedstocks.

ABBREVIATED BIBLIOGRAPHY

1. Argonne National Laboratory. 1985. Uniform Cost-Estimation Method for Biomass Production and Conversion Technologies.

2. Bente, P.F. Jr. 1985. Bio-Energy Progress: 1984-85 Industry Update and Future Trends. The Bio-Energy Council.

3. Bulpitt, W., G. Curtis, and S. Drucker. 1984. The Industrial Wood Energy Handbook. Georgia Institute of Technology.

4. Dobie, J., G. Miller, and R. Mosley III. 1984. Ground Level Harvest of Rice Straw. TRANSACTIONS of the ASAE 27(5): 1263-1269.

5. Flaim, S., D. Urban. 1980. The Costs of Using Crop Residues in Direct Combustion Applications. Solar Energy Research Institute.

6. Great Lakes Regional Biomass Program, Council of Great Lakes Governors. 1985. Directory, Great Lakes Regional Biomass Energy Facilities.

7. Hellwig, R., H. Sumner, and G. Monroe. 1984. Densification of Biomass for Energy Utilization. Proceedings, 4th Annual Solar and Biomass Energy Workshop. USDA, SAEC, Tifton, GA.

8. Hitzhusen, F., M. Abdallah. 1980. Economics of Electrical Energy from Crop Residue Combustion with High Sulfur Coal. American Agricultural Economics Association.

9. Iverson, G. 1984. Utilization of Straw for District Heating. Proceedings, BIOENERGY '84 Conference, Volume I - Bioenergy State of the Art. June 1984, Sweden.

10. Jenkins, B.M., and H.R. Sumner. 1985. Research Needs and Priorities for Harvest and Handling of Agricultural Residues. ASAE Paper No. 85-3066.

11. Jenkins, B.M., J.F. Arthur, G.E. Miller, and P.S. Parsons. 1984. Logistics and Economics of Biomass Utilization. TRANSACTIONS of the ASAE 27(6):1898-1904, 1910.

12. Jenkins, B.M., and J.F. Arthur. 1983. Assessing Biomass Utilization Options Through Network Analysis. TRANSACTIONS of the ASAE 26(5): 1557-1560.

13. Jenkins, B.M., B.C. Horsfield, J.B. Dobie, and G. Miller. 1981. Agricultural Residues: Renewable Energy for Utility Power Companies. TRANSACTIONS of the ASAE 24(1):197-201, 207.

14. LePori, W. 1985. Thermochemical Conversion of Biomass Using Fluidized-Bed Technology. ASAE Paper No. 85-3071.

15. Maartmann, S. 1984. Control of Particulate Emissions Using Biomass as the Main Fuel. Proceedings, BIOENERGY '84 Conference, Volume IV--Bioenergy Utilization. June 1984, Sweden.

16. Meridian Corporation and Science Applications International Corporation. 1985. Electric Power from Biofuels: Planned and Existing Projects in the U.S.

17. Miles, T. 1983. Biomass Preparation for Thermochemical Conversion. Proceedings, 1st European Workshop on Thermochemical Processing of Biomass. April 1983, University of Aston, Birmingham, U.K.

18. Payne, F.A. 1985. Combustion and Stationary Bed Gasification - Status and Research Priorities. ASAE Paper No. 85-3070.

19. Science Applications, Incorporated. 1984. Estimates of U.S. Wood Energy Consumption -- 1980-1983.

20. Strehler, A. 1984. Commercial and Practical Aspects of Small Scale Combustion of Biomass. Proceedings, BIOENERGY '84 Conference, Volume 1 - Bioenergy State of the Art. June 1984, Sweden.

21. Strehler, A. 1980. Energy from Straw and Woodwaste. Proceedings, Biomass Energy Conference. November 1980, Brighton, England.

22. Waelti, H., and J.B. Dobie. 1973. Cubability of Rice Straw as Affected by Various Binders. TRANSACTIONS of the ASAE 16(2):380-383.

23. Wilen, C., P. Stahlberg, K. Sipila, and J. Ahokas. 1984. Pellitization of Straw. Proceedings, BIOENERGY '84 Conference - Volume III -- Biomass Conversion, Densification, Storage and Drying. June 1984, Sweden.

DISCUSSION

Veigel	You can find brokers to buy wood chips from if you want to. Do you see the same kind of brokering role in the herbaceous area?
Dripchak	Yes, to an even greater extent. In the wood energy industry, the user and producer are often one and the same. But forage crops are not in all likelihood going to be produced by the same people who are using them, so I would indeed see a role for brokers.
D. Pratt, U. of Minnesota	I realize that you haven't looked at natural vegetation. But there are some places where natural vegetation could possibly be harvested. Since production costs are the most significant factor that you've identified here, I wonder it that isn't something that ought to be examined?
Dripchak	Absolutely. Oak Ridge National Laboratory is responsible for the herbaceous energy crop program for the DOE's Biomass Energy Technology Division. They are looking at a whole series of grasses and legumes for all the conversion processes, so a wide variety of herbaceous species is under consideration.

WORLD-WIDE MARKETS FOR BIOMASS GASIFICATION SYSTEMS

A V Bridgwater, PhD DSc
Aston University
Birmingham B4 7ET
UK

ABSTRACT

The potential offered by biomass and related materials to solving some of the world's energy problems appears to widely recognised, and the level of interest requires no further justification. Ultimately the success of any potential opportunity is judged on its performance in the market place. This paper therefore attempts to put into perspective the potential of biomass and relate it to the feedstocks, technology, products, economics, and applications.

Feedstocks include wood, energy crops, agricultural waste, and municipal solid waste or refuse. The essential features of available feedstocks are described in order to relate them to gasification technology and economics. An appreciation of the status of thermochemical conversion technology and it's limitations is necessary to be able to specify a suitable system and match this to market requirements. The primary products from gasification and pyrolysis are described together with derived secondary and tertiary products, and typical plant size ranges for economical exploitation in the short term. A survey of known applications of biomass and waste gasification is included.

Both capital and variable costs are discussed with results from analyses of the literature and other sources of information. The data is used to derive a range of production costs in order to help identify market opportunities, since implementation is more dependant on costs than any other single factor. In order to put biomass and waste gasification into a market context, a superficial examination is included of the main competitors to this technology - combustion and coal gasification

There are two approaches to establishment of a market opportunity: supply and demand. Supply is considered in terms of biomass and waste production and availability, while demand is considered in terms of conventional fuel replacement or substitution. The opportunities and constraints for implementation are described, concluding that the potential market opportunity is considerable, and that the limiting factor is the match between technology push and market pull. This is still imperfect, and it is this matching that requires most attention if implementation is to be successful.

WORLD-WIDE MARKETS FOR BIOMASS GASIFICATION SYSTEMS

INTRODUCTION

The potential offered by biomass and related materials to solving some of the world's energy problems appears to widely recognised, and the level of interest requires no further justification. Ultimately the success of any potential opportunity is judged on its performance in the market place, and this aspect seems to have been neglected in many RD&D activities. This paper therefore attempts to put into perspective the potential of biomass and relate it to the feedstocks, technology, products, applications and economics.

FEEDSTOCKS

Feedstocks considered are wood including energy crops, wood and agricultural waste, and municipal solid waste or refuse. This section summarises the essential features of available feedstocks in order to relate them to gasification technology and economics.

Wood and Wood-waste

Wood wastes are available in a variety of forms: forest residues which have to be removed from existing plantations for silvicultural reasons; wood processing residues which arise in the timber production industry, usually adjacent to the forest resource; and whole trees from thinning. Wood as an energy crop can arise from short rotation forestry as single stem or coppice. Although a considerable proportion of waste is often utilised, ranging from combustion on-site to production of high value speciality products such as cat litter, there is considerable potential for greater and more effective utilisation. Wood from energy crops offers the greatest potential for the most efficient and integrated opportunity for energy production from wood in developing countries.

Wood costs range from \$50/t (all costs are quoted on a dry and ash free basis) in industrialised countries with high labour costs, to as little as \$10/t in a developing country - this last figure includes all replanting and delivery costs.

Agricultural waste

Considerable quantities of agricultural waste, such as straw, bagasse, corn cobs, and rice hulls, arise each year for which no ready market is available. The size of the arisings has caused considerable attention to be paid to the possibility of utilising this material for a variety of different processes. In many cases this material is free and may even have a negative cost due the cost of disposal if it cannot be used. These wastes also have the advantage of arising at a single site without the problems and cost of collection.

Agricultural wastes are difficult to price since they can either be left to accumulate, burned or be disposed of at a cost. Such wastes can thus be charged from \$-10/t to \$+30/t according to local requirements and alternative opportunities.

Municipal Solid Waste (Refuse)

Considerable quantities of municipal solid waste (domestic and commercial refuse) are produced from any society which require disposal at what is often a significant cost. Processes that utilise this waste, therefore, can often justify a credit from the disposal authority, who will save the disposal cost, which can range from \$-40/t to \$+40/t depending on local circumstances and the form of the waste. Raw waste attracts the highest credit but requires the most preparation. There is again the advantage of free collection and often adjacency to a potential market for the energy product.

Industrial wastes

Combustible waste arisings from the industrial sector can also be advantageously used. In the UK about a third of the combustible waste arisings is from industry with about 20% from the timber and furniture sector, 17.6% from the chemical and allied sector, 16.1% from the paper and printing sector, and 12.1% from the food, drink and tobacco sector. On-site utilisation is often the most economic way of realising the energy potential. Costs are thus often negligible, although there may be a utilisation/conversion cost to attribute to the waste.

Characteristics

The basic characteristics of biomass are well known. They have relatively low ash levels, except for particular agricultural wastes such as rice hulls; moderate to high moisture levels according to particular waste, locality and storage; and very low levels of sulphur and similar pollutants. Table 1 shows a representative selection of the range of materials that have been successfully processed.

Table 1 Potential Feeds for Gasification

Alfalfa straw	Almond shells	Bagasse
Bamboo	Barley straw	Bark
Brazil nut husks	Cocoa husks	Coconut husks
Coconut shells	Coffee grounds	Cotton gin trash
Cotton gin stalks	Flax	Ground nut shells
Hardwood	Maize cobs	MSW
Olive pits	Paper	Peach pits
Prune pits	RDF pellets	Rice hulls
Safflower straw	Sawdust	Sewage sludge
Straw	Sunflower husks	Softwood
Tobacco dust	Tyres	Vegetable fibre
Walnut shells	Wood pellets	

TECHNOLOGY

An appreciation of status of the technology and it's limitations is necessary to be able to specify a suitable system and match this to market requirements. The feed as produced or delivered is thermochemically converted to a useful and marketable product in a sequence of operations:

Pretreatment: Reception, storage, handling; *size reduction; screening; drying*
Conversion: Gasification; gas cleaning
Processing: *Wastewater treatment; gas compression; pipeline to user; equipment modification; conversion to higher value products;*

Items in italics may or may not be required, depending on the feedstock, gasification process, and fuel gas application.

Pretreatment

The biomass resource may have to be subjected to a variety of operations between where it arises and its conversion, which may include any of the following:

Reception, storage and handling. Some provision has to be made to receive feedstock, including possibly weighing; storing; retrieving; and handling. Specification of suitable systems will depend on type of feed, quantities handled, and sophistication required. Larger systems justify the provision of enclosed storage such as silos and live bottomed bins; particularly if these are automatically linked to the gasifier feeding system. Costs are not particularly high and technology is readily available.

Size reduction and control. Provision of screening and/or densification of feedstock is entirely dependent on the characteristics of the feed supplied, and the requirements of the gasification system. Generally the most economical conversion system will not require any size control operation. An exception is refuse which, if delivered as raw refuse, will at least require shredding. Straw and similar materials would be expected as bales which require chopping or milling. Wood would be expected to be supplied as ready chipped material since this considerably reduces transport and handling costs. Locally arising wastes may require size reduction.

Drying. This is an expensive operation to provide and operate. In larger systems, the trade-off in improved gasifier performance from use of a drier feed needs to be offset against the cost of the drying operation. Generally drying should be avoided unless essential and cost effective, although it is claimed that drying wet biomass before combustion is economically attractive. Use of waste heat in this respect is probably not as efficient as use of part of the feedstock as fuel.

Classification. This operation is only usually required for shredded refuse which may require air classification prior to gasification to concentrate the combustibles and remove inerts, glass, metal etc. Provision of this operation must be set against reactor requirements, and the trade-offs should be examined both technically and economically.

Conveyance. An essential feature of any solids handling sysem is movement of the solids. Small systems will tend to use batch handling systems such as front end loaders, rather than automatic conveyor systems which are more suitable for large solids handling processes. Where labour is cheap and plentiful, manual or semi-manual feeding may be considered. Each case should be examined on its merits and related to resources and facilities available.

Economics. There are so many variations of front end handling and pretreatment according to the feedstock characteristics and gasifier characteristics, that generalisations are not possible. Pretreatment costs have, therefore, been omitted from the economic assessment carried out later. A basic storage and handling step prior to gasification would cost about 10% of the total gasification system cost.

Gasification Technology

Thermochemical conversion. There are four basic methods of thermochemically converting biomass:

Gasification: The bulk of the carbon is converted into the gaseous form, leaving an inert residue. This is achieved by partial combustion of part of the fuel in the reactor with oxygen in air, or as pure oxygen, or by concurrent reaction with steam. Relatively high temperatures are employed of 900 - 1 100 C with air, and 1 000 - 1 400 C with oxygen. Air gasification is the most widely used technology since a single product is formed at high efficiency and without heat transfer problems as in pyrolysis. A low heating value gas is produced of up to 60% nitrogen with a heating value of 4 - 6 MJ/nm3 (100 - 150 Btu/scf). Oxygen gasification gives a better quality gas of 10 - 18 MJ/nm^3 (250 - 500 Btu/scf) but requires an oxygen supply with concommitant problems of cost and safety. Such a medium heating value gas can also be produced by pyrolysis (qv) and the technologies below, none of which require oxygen: steam gasification encourages steam reforming reactions by adding water/steam and relies on gasification and/or pyrolysis with external heating to drive the process; and the now abandoned Oxygen Donor Gasifier which relies on chemical oxygen gasification in a twin reactor.

Pyrolysis: Heat is added indirectly in a variety of forms, and products in the gas, liquid and solid form are usually formed, the relative proportions of which depend very much on the method and parameters of reaction. Relatively low temperatures are employed of 600 - 900 C. Fast or flash pyrolysis is used to maximise either gas or liquid products. The gas is of medium heating value, 13 - 21 MJ/nm3 (350 - 600 Btu/scf); and the liquids, although often referred to as "oil", are physically and chemically unstable and require upgrading to produce a marketable product. Pyrolysis is a less popular technology because of the multiplicity of products that are difficult to handle and market; lower efficiency in gas production; and problems of heat transfer. The latter has been overcome in twin reactor systems which incorporate a char combustor to provide heat for the pyrolysis and/or steam gasification process. A better quality gas is produced, but at a penalty of lower efficiencies and possibly higher cost.

Carbonisation: This is pyrolysis operated at low reaction rates to maximise yields of charcoal. The gas and liquid products are byproducts which may be used in the process for energy or exported.

Liquefaction: This is an essentially low temperature (250 - 400 C) but high pressure (up to 100 bar) process. The liquid fraction is maximised in yield terms and also in quality terms. The liquid product is of higher quality than pyrolysis liquids, but still require chemical upgrading.

Types of reactor. Reactor technology is classified simply by mode of contact:

Fixed bed and moving bed: Dense phase, ie high solid to reactor volume ratio. These tend to be simple technologies with fewest moving parts.

Fluid bed: Lean phase, ie low solid to reactor volume ratio. These tend to be more complex mechanically, with more moving parts thus requiring a higher level of technical support.

Other .

These can be further subdivided by relative movement of solid feed and gasifying agent:

Fixed bed

- Downdraft : solid moves down, gas moves down, ie co-current.
- Updraft : solid moves down, gas moves up, ie counter-current.
- Co-current : solid and gas move in same direction, eg downdraft, but both can move up.
- Counter-current : solid and gas move in opposite directions, eg updraft, but flows can be reversed.
- Cross current : solid moves down, gas moves at right angles, ie horizontally.
- Variations : stirred bed, two stage gasifier.

Moving bed This implies mechanical movement or transport of solid, usually horizontally. High temperatures cause materials of construction problems, so this is usually used for lower temperature processes such as pyrolysis:

Multiple hearth
Horizontal moving bed
Sloping hearth
Screw/auger

Stirred bed Mechanical agitation of any dense phase reactor to improve gas solid contact.

Fluid bed These are classified by relative speed of fluidising medium and number of reactors:

a) Single reactor

- fluid bed : relatively low gas velocity, inert solid stays in reactor.
- fast fluid bed : inert solid is elutriated with product and may be separated for recycle.
- circulating fluid bed : inert solid is elutriated, separated and recirculated back to the fluid bed; also refers to twin reactor system (see below).
- entrained bed : usually no inert solid, highest gas velocity, can be run as a cyclonic reactor (see below).

b) Twin or double reactor:

Steam gasification and/or pyrolysis occurs in the first reactor, which can be any fluid bed variety. Char produced is transferred to the second reactor, where it is combusted again, in principle, in any fluid bed variety . This system is also referred to as a circulating fluid bed.

c) Variations:

Single circulating horizontal fluid bed or cyclonic reactor; and twin horizontal circulating fluid bed

Other Rotary kiln and cyclonic reactor

Gasifier characteristics. Gasifier characteristics have been derived for all common types of gasifier in Table 2. The level of activity and interest in gasifiers is summarised in Table 3, which is based on gasifier systems that have been claimed to be available commercially in the last few years.

Table 2 Gasifier Characteristics

	Capacity range daf t/h	Feed (a)	Gas HHV MJ/m3	Gas qual. (b)	Gas temp. C	Effic. %	Number available (c)
Pyrolysis (all types)	1-8	5	11-15	1	500-700	40-55	13
Gasification							
Downdraft, air	0.1-0.7	2	4.5-6	4	400-1000	70-90	20
Downdraft, oxygen	0.1-0.7	2	9-11	4	700-1100	80-90	2
Updraft, air	0.2-10	2	4.5-6	3	100-400	60-80	17
Updraft, oxygen	0.2-10	2	11-13	3	100-400	70-80	1
Single fluid bed, air	0.3-15	4	4-6	3	500-900	70-80	16
Single fluid bed, oxygen	2-10	4	11-13	3	700-1100	75-90	5
Fast fluid bed, air	2-3-	3	5-6.5	2	700-1100	80	2
Fast fluid bed, oxygen	0.3-10	3	10-13	3	800-1200	75-90	1
Twin fluid bed	1-10	4	13-22	3	750-1150	65-80	6
Cross flow, air	0.5-2	2	4.5-6	1	600-900	60-80	1
Horizontal moving bed, air	0.5-5	4	4.5-6	2	300-800	60-75	1
Rotary kiln, air	1-20	3	4-6	2	700-100	65-80	1
Multiple hearth	2-20	3	4.5-6	2	300-700	65-75	2
Single fluid bed, steam	1-10	4	12-18	3	700-900	70-80	1
Secondary processing	-	-	-	5	1000-1200	-	5

NOTES
(a) specificity : most specific - 1; least specific - 5.
(b) relative assessment in terms of tars and particulates in raw gas : worst - 1; best - 5.
(c) commercial or demonstration

Table 3 Activities in Gasification by Reactor Type

REACTOR TYPE	NUMBER OF INSTALLATIONS Company Commercial	Company Demonstration	Institution Demonstration	TOTAL
Pyrolysis (all types)	8	3	2	13
Gasification				
Downdraft, air	13	5	2	20
Downdraft, oxygen	-	1	1	2
Updraft, air	13	3	1	17
Updraft, oxygen	1	-	-	1
Single fluid bed, air	8	6	2	16
Single fluid bed, oxygen	1	2	2	5
Fast fluid bed, air	1	1	-	2
Fast fluid bed, oxygen	1	-	-	1
Twin fluid bed	1	3	2	6
Cross flow, air	1	-	-	1
Horizontal moving bed, air	1	-	-	1
Rotary kiln	1	-	-	1
Multiple hearth	2	-	-	2
Single fluid bed steam	-	1	-	1
Secondary processing	1	3	1	5
TOTAL	53	28	13	94

Gas cleaning

The raw gas from a gasifier can have a number of contaminants over a wide range of concentration including condensible, water miscible organics (eg acetic acid); immiscible organics (eg tars); non condensible pollutants (eg H_2S, HCl); ash; char; sand (eg from fluid beds). Cleaning requirements depend primarily on the application and also on the technology chosen.

Technology. The simplest and most efficient approach is to burn the hot raw gas, which uses the sensible heat of the gas and minimises pollution control requirements. This is limited to applications where particulate carry-over is not critical, for examples lime kilns and cement kilns: boiler retrofitting is also acceptable in many instances. The minimum cleaning otherwise is a hot gas cyclone. This provides a very coarse separation, which is inefficient for smaller particle sizes including tar mist. Condensation can also be a problem. A waste heat boiler may be added to recover sensible heat from hot gas. Piping of a fuel gas to a nearby user or combustion in an engine to produce power requires that the gas be cold and clean, and will not deposit water or condensibles in the associated equipment. Water washing or scrubbing is the preferred approach, as physical filtration is generally inefficient and unreliable as a primary cleaning system. Chemical processing to high value fuel and chemical products requires an even higher gas quality specification necessitating costly tertiary treatment. An alternative to treatment is prevention, for examples the use of limestone or dolomite to crack tars in the reactor; or secondary, possibly catalytic, processing to also crack tars; or judicious selection of reactor design to avoid particle carry-over.

Gas quality. The conventional view is that biomass or waste derived fuel gas has to compete with natural gas, and the reaction invariably encountered when asking for quality specifications is "the same as natural gas". There does not appear to be any appreciation of the quality tolerances and requirements of fuel gas in any industrial sector.

Waste disposal. The main products from water washing are dirty water and tars. The latter are generally readily separable from water by simple flotation and should be recycled to the reactor, for example by absorption into feedstock or ash. Waste water treatment is essentially conventional, although oxygenated organics such as phenols may create problems. This aspect of gasification application is often omitted from technology and economic assessments.

PRODUCTS

The primary products from gasification and pyrolysis are summarised in Table 4, together with derived secondary and tertiary products, and typical plant size ranges for economical exploitation in the short term.

Table 4 Products from Thermochemical Processing of Biomass

Process	Primary product	Process	Secondary product	Process	Market	Size range t/h input
Gasification	LHV gas	-	-	-	Fuel gas	0.1-5
		Engine	Power	-	Electricity	0.1-5
		Turbine	Power	-	CHP	1 -10
		Conversion	Ammonia	Conversion	Fertilisers	2 -20+
Gasification	MHV gas	-	-	-	Fuel gas	1-5
		Engine	Power	-	Electricity	1-5
		Turbine	Power	-	CHP	1-10
		Conversion	Methane	-	SNG	10+
		Conversion	Methanol	-	Methanol	10+
		Conversion	Methanol	Conversion	Gasoline	10+
		Conversion	Gasoline	-	Gasoline	10+
		Conversion	Diesel	-	Diesel	10+
		Conversion	Syncrude	-	Gas&liquidfuels	20+
Pyrolysis	MHV gas	-	-	-	Fuel gas	1-5
		Engine	Power	-	Electricity	1-5
		Turbine	Power	-	CHP	1-10
		Conversion	Methane	-	SNG	10+
		Conversion	Gasoline	-	Gasoline	10+
		Conversion	Diesel	-	Diesel	10+
		Conversion	Syncrude	-	Gas&liquidfuels	20+
	"Oil"	Upgrading		Upgrading	Fuel oil	5-20+
	Char	-	-	-	Solid fuel	0.2-5
Liquefaction	"Oil"	Upgrading	Fuel oil	-	Fuel oil	2-20+

Notes: + upper size limited by feed supply

ECONOMICS

Both capital and variable costs are described with results from analyses of the literature and other sources of information. The data is used later to derive production costs, profitabilities, cost analyses and sensitivity analyses for a range of opportunities in order to help identify market opportunities, since implementation is more dependant on costs than any other single factor.

Capital Cost

Biomass conversion. Practical implementation of a gasification system requires provision of all necessary operations to convert feed to the required final product. The total plant cost of a gasification system is the equipment costs for all required steps, together with costs of design, installation and commissioning. The reception storage and handling steps will be about 10% of the system capital cost; and the supply of gas to the consumer, up to and including the burner but excluding compression or long pipelines, will be approximately a further 10%.

Capital costs of gasifiers and pyrolysers for fuel gas and power production, built in industrialised countries, have been collated, covering the process from the gasifier feeding system (feed on the ground) to clean cold product gas, as shown in Figure 1 (1). The capital cost is a total plant cost including equipment, installation, design, project management and commissioning. Equipment cost data was converted to total plant cost by suitable factors: for field erected plant, typically of capacities above about 2 t/h, the factor is 2.25 times the delivered equipment cost; and for skid mounted units which only require a concrete pad and connection to services, the factor is 1.25 times the delivered equipment cost. An intermediate factor of 1.75 may be used appropriately. Known organisations are listed in the appendix and corrections are welcome.

All data lies between two lines: the lower line represents the lower limit of gasifier capital costs, typically achieved by relatively simple gasifier systems and/or supplied by small companies. The type of technology does not appear to affect capital cost. The upper line represents the cost of high technology systems eg twin fluid bed systems and stirred bed systems and/or systems supplied by major companies. The equations of the lines are shown in Figure 2 and are as follows:

Industrialised countries

Higher capital cost:

$\$_{1985} = 970\,000\ (\text{capacity})^{0.65}$ $\quad$ $£_{1985} = 775\,000\ (\text{capacity})^{0.65}$

Intermediate capital cost:

$\$_{1985} = 325\,000\ (\text{capacity})^{0.65}$ $\quad$ $£_{1985} = 260\,000\ (\text{capacity})^{0.65}$

In each case, the capacity is in daf t/h. These correlations apply to systems manufactured in industrialised (developed) countries. Systems fabricated in developing countries will generally cost less than the intermediate figure quoted, and the following correlation is proposed, but with less substantial evidence:

Developing countries

Lower capital cost:

$\$_{1985} = 163\,000\ (\text{capacity})^{0.65}$ $\quad$ $£_{1985} = 130\,000\ (\text{capacity})^{0.65}$

Power generation. A similar economy of scale is assumed (although power costs are conventionally expressed as a cost per installed kW which gives no economy of scale), resulting in the following equation for installed power generation equipment based on an engine:

$\$_{1985} = 400\,000\ (\text{ daf t/h biomass })^{0.65}$ $\quad$ $£_{1985} = 320\,000\ (\text{ daf t/h biomass })^{0.65}$

$\$_{1985} = 4500\ (\text{ kW })^{0.65}$ $\quad$ $£_{1985} = 3600\ (\text{ kW })^{0.65}$

Turbine system costs would be higher, and additional gas cleaning plant may be required.

Methanol production. Similar simple expressions can be derived for add-on costs of methanol production:

$\$_{1985} = 1\,800\,000\ (\text{daf t/h biomass})^{0.65}$ $\quad$ $£_{1985} = 1\,440\,000\ (\text{daf t/h biomass })^{0.65}$

$\$_{1985} = 800\,000\ (\text{t/h methanol})^{0.65}$ $\quad$ $£_{1985} = 640\,000\ (\text{t/h methanol})^{0.65}$

Analogous relationships are derivable for methane, ammonia, gasoline, etc.

Comparison with combustion and coal gasification. Data on equivalent biomass and waste combustion systems, and coal gasification systems, all adjusted to the same basis as the cost data for gasifiers, is also shown in Figure 2. Combustion shows an apparently lower economy of scale, and coal gasification appears to be relatively expensive. It can be concluded that the capital costs of combustion and gasification are similar for wastes and refuse.

Variable costs

Variable costs are feedstock, utilities, maintenance, overheads and labour.

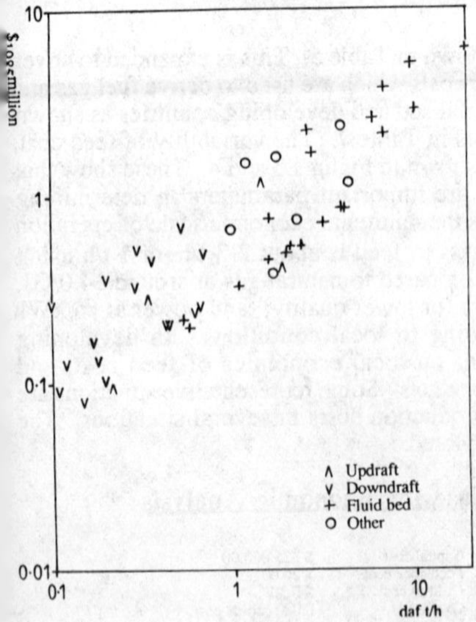

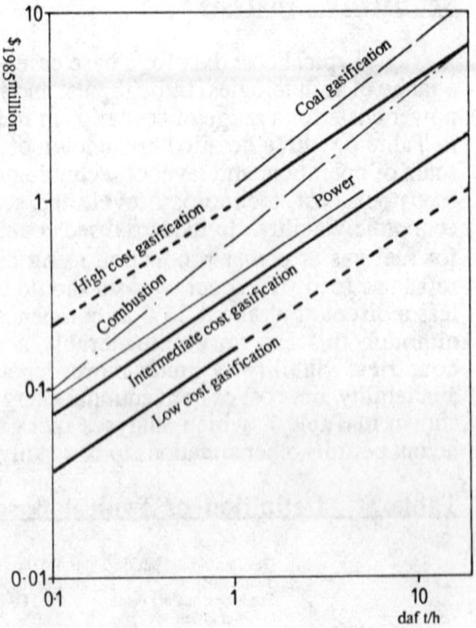

Figure 1 Installed Capital Cost **Figure 2 Capital Cost Correlations**

Utilities. These include electricity, for driving pumps, blowers and feeding systems; wash water for product gas scrubbers; and boiler feedwater for steam raising. Steam, if required, would probably be raised on site within plant battery limits, and probably from a waste heat boiler. Oxygen, reagents and catalysts would be charged to feedstock. It is assumed that all other energy requirements are met in-house without requiring conventional fuels, except for start-up. The average value has been found to be \$0.14/GJ (£0.114/GJ) output for electricity, and \$0.30/GJ (£0.24 /GJ) overall (1).

Maintenance. Annual maintenance cost is usually estimated as a proportion of the gasifier capital cost, and the average cost for gasifiers has been found to be 2.5% of installed plant cost (1).

Overheads. Annual overheads cost is usually expressed as a proportion of the capital cost, similar to maintenance cost. Although little data has been found, 8% on capital cost is an appropriate figure to cover local tax, insurance and company expenses. Payroll overheads are included in labour costs.

Labour. Continuous operation would be expected, but only for a load factor of 70%, or 6000 h/y. Labour requirements and cost will vary considerably with scale, technology, and location; and also with the general level of costs. A high technology plant of say 5 t/h capacity will probably require 5 men for continuous operation: four shifts plus supervisor. Labour costs have been estimated on a shift cost basis to allow for different skill levels and local productivity variations.

Production cost

There are two ways to assess a proposed project : calculation of product cost and comparison with alternative energy prices; or profitability assessment by Payback time, ROI, NPV and DCFrr, and comparison with alternative investments. Product cost is calculated by totalling the feedstock cost, variable cost, and capital amortisation charge which covers recovery of the capital and interest payments on the capital employed. This approach has been adopted here to provide a ready comparative measure of the viability of biomass derived energy products. Profitability assessment also includes all income and expenditure, but does not include charges for interest or capital recovery because it is intended as a comparison with investment of a similar sum in a conventional deposit, such as a bank.

Sensitivity Analysis

Typical cost data for a base case is shown in Table 5. This is expanded to cover a range of technologies, throughputs, and feed costs which are used to derive fuel gas and power costs for a range of scenarios in industrialised and developing countries as shown in Table 6, with a detailed breakdown of costs in Table 7. The variability of feed cost, scale of operation, and level of technology is shown in Figures 3 and 4. These show that feedstock cost, technology level, and scale are important parameters in determining economic viability. In industrialised countries the minimum economic scale of operation for fuel gas or power production using biomass as feed is about 2-3 t/h, or 1 t/h using refuse as feed. The energy costs should be compared to natural gas at around $4.0/GJ, less a discount of about 25% as compensation for lower quality; and power at ¢5/kWh although this can vary considerably according to local conditions. In developing countries, viability is much more dependant on local economics of feed cost, and availability and cost of conventional energy products. Some representative situations are shown in Table 7, which analyses fuel gas production costs in several situations. The layout permits other situations to be readily explored.

Table 5 Definition of Typical Base Case for Economic Analysis

Gasifier throughput	1.0 t/h	Capital cost	$ 325 000.00
Technology	Intermediate	Feedstock cost	$ 20.00
Feedstock heating value	20.0 GJ/t	Utilities cost	$ 0.30/GJ*
Gasifier efficiency	75.0 %	Overheads	0.080 capital cost
Project life	10 y	Maintenance cost	0.025 capital cost
Operation (70% load factor)	6000 h/y	Total cost of labour	$ 25 000/shift/y

(NB All tonnes are a daf basis)
* utilities assumed constant for any feed cost and any location

Table 6 Some Results of Sensitivity Analysis for Fuel Gas and Power Production (See Table 7 for Cost Analysis)

No	Scenario	Size t/h	Technology	Capital cost $ *	Feed $/t	Labour $/shift	Product cost $/GJ	Product cost ¢/kWh¶
1	Industrial country	1	Intermediate	325 000	0	25 000	2.28	5.04
2	Industrial country	10	Intermediate	1 452 000	0	25 000	0.80	1.92
3	Industrial country	10	High	4 333 000	0	25 000	1.58	3.08
4	Industrial country	1	Intermediate	325 000	20	25 000	3.62	7.04
5	Industrial country	10	Intermediate	1 452 000	20	25 000	2.13	3.92
6	Industrial country	10	High	4 333 000	20	25 000	2.91	5.08
7	Industrial country	1	Intermediate	325 000	40	25 000	4.95	9.04
8	Industrial country	10	Intermediate	1 452 000	40	25 000	3.47	5.92
9	Developing country	0.1	Low	36 000	0	2 500	2.39	7.92
10	Developing country	1	Low	163 000	0	2 500	0.85	3.21
11	Developing country	0.1	Intermediate	87 000†	0	2 500	3.75	9.97
12	Developing country	1	Intermediate	390 000†	0	2 500	1.46	4.12
13	Developing country	10	Intermediate	1 742 000†	0	2 500	0.78	2.03
14	Developing country	0.1	Low	36 000	20	2 500	3.73	9.92
15	Developing country	1	Low	163 000	20	2 500	2.18	5.21
16	Developing country	0.1	Intermediate	87 000†	20	2 500	5.08	11.97
17	Developing country	1	Intermediate	390 000†	20	2 500	2.79	6.12
18	Developing country	10	Intermediate	1 742 000†	20	2 500	2.11	4.03

Notes
* Gasification capital cost only - excludes power generation equipment cost.
¶ Cost includes gasifier and power generation equipment.
† Capital costs are 20% higher for imported intermediate technology plants for transport and commissioning; labour costs are lower as shown.

APPLICATIONS

Existing applications

A survey of known commercial/demonstration applications of biomass and waste gasification is summarised in Tables 8 and 9 covering systems manufactured in

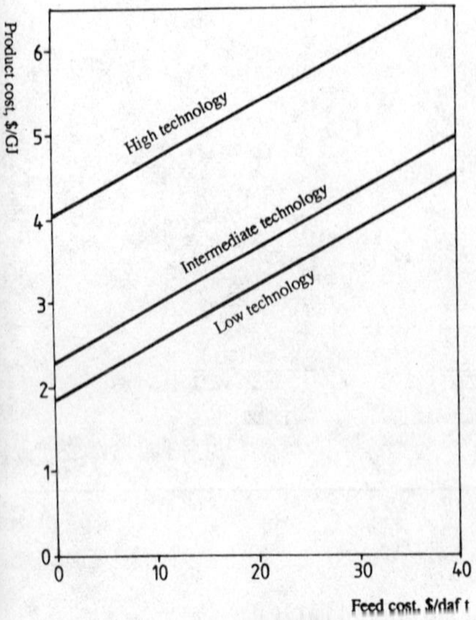

Figure 3 Fuel Gas Production Cost and Feed Cost

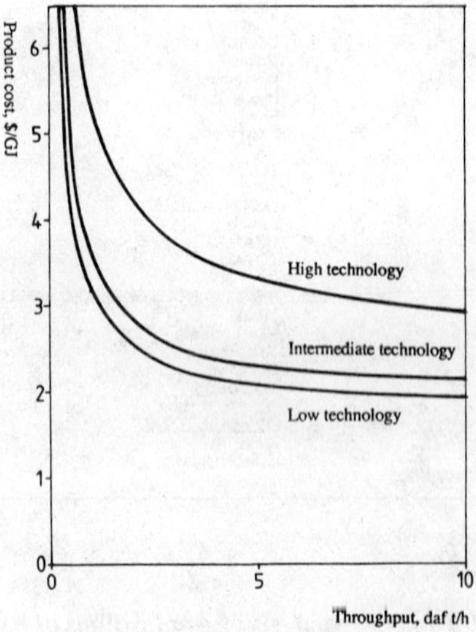

Figure 4 Fuel gas Cost and Scale of Operation

Table 7 Product Cost Analysis for Fuel Gas Production

Number (Table 6)	Total cost $/GJ	Capital cost $/GJ	Variable cost Subtotal $/GJ	Feed $/GJ	Labour $/GJ	Utilities $/GJ	O/H&Maint. $/GJ
Industrialised countries							
1	**2.28**	0.49	1.79	0	1.11	0.30	0.38
2	**0.80**	0.22	0.58	0	0.11	0.30	0.17
3	**1.58**	0.66	0.92	0	0.11	0.30	0.51
4	**3.62**	0.49	3.12	1.33	1.11	0.30	0.38
5	**2.13**	0.22	1.91	1.33	0.11	0.30	0.17
6	**2.91**	0.66	2.25	1.33	0.11	0.30	0.51
7	**4.95**	0.49	4.46	2.67	1.11	0.30	0.38
8	**3.47**	0.22	3.25	2.67	0.11	0.30	0.17
Developing countries							
9	**2.39**	0.55	1.84	0	1.11	0.30	0.43
10	**0.85**	0.25	0.60	0	0.11	0.30	0.19
11	**3.75**	1.32	2.43	0	1.11	0.30	1.02
12	**1.46**	0.59	0.87	0	0.11	0.30	0.46
13	**0.78**	0.27	0.51	0	0.01	0.30	0.20
14	**3.73**	0.55	3.17	1.33	1.11	0.30	0.43
15	**2.18**	0.25	1.94	1.33	0.11	0.30	0.19
16	**5.08**	1.32	3.76	1.33	1.11	0.30	1.02
17	**2.79**	0.59	2.20	1.33	0.11	0.30	0.46
18	**2.11**	0.27	1.85	1.33	0.011	0.30	0.20

Table 8 Locations and Capacities of Known Commercial/Demonstration Gasifiers

Country	Known installations Number	% of total	Output GJ/h	% of total
EUROPE				
Austria	1	-	nk	-
Finland	2	-	360	-
France	7	-	70+	-
Germany	10+	-	70+	-
Holland	1	-	nk	-
Luxemburg	1	-	nk	-
Portugal	1	-	120	-
Sweden	2+	-	120+	-
Switzerland	5	-	7+	-
UK	2	-	nk	
TOTAL	32+	12.8	747	7.3
NORTH AMERICA				
Canada	18	-	608	-
USA	130	-	6438+	
TOTAL	148	59.2	7016	68.5
REST OF INDUSTRIALISED WORLD				
Japan	4	-	655+	-
New Zealand	2+	-	nk	-
South Africa	1	-	10	-
Unknown Location	63	-	1810	-
TOTAL	70+	28	2475	24.2
OVERALL TOTAL	250+	100.0	10 238	100.0

nk = Not known

Table 9 Analysis of Applications of Known Gasifiers (Table 8)

Application	Number installed	%	Output GJ/h	%	¶
a) INDUSTRIALISED COUNTRIES					
Kiln	9	(3.7)	1032.0	(10.8)	
Boiler	39	(15.8)	680.5	(7.1)	
Drying	16	(6.5)	508.0	(5.3)	
Metal processing	3	(1.2)	250.7	(2.6)	
Fuel gas	73+	(29.7)	4845.0	(50.6)	
Power	44+	(17.9)	117.0	(1.2)	
Chemical	1	(0.4)	200.0	(2.1)	
Other / unspecified	61+	(24.8)	1938.8	(20.3)	
TOTAL	246+	(100.0)	9569.0	(100.0)	
b) DEVELOPING COUNTRIES					
Kiln	1	(1.3)	-	-	
Fuel gas	7	(9.3)	nk	-	
Power	56	(74.7)	68.3	(88.0)	
Other / unspecified	11	(14.7)	9.3	(12.0)	
TOTAL	75	(100.0)	77.6	(100.0)	
c) PLANNED					
Kiln	1	(4.5)	1.3	(0.3)	
Boiler	1	(4.5)	50.0	(10.5)	
Drying	1	(4.5)	6.0	(1.3)	
Fuel gas	1	(4.5)	50.0	(10.5)	
Power	10	(45.6)	11.2+	(2.4)	
Chemical	3	(13.7)	250.0+	(52.9)	
Other / unspecified	5+	(22.7)	104.5+	(22.1)	
TOTAL	22+	(100.0)	473.0	(100.0)	
d) TOTAL					
Kiln	11	(3.2)	1033.3	(10.2)	(32.6)
Boiler	40	(11.7)	730.5	(7.2)	(23.0)
Drying	17	(4.9)	511.0	(5.1)	(16.3)
Metal processing	3	(0.9)	250.7	(2.5)	(8.0)
Fuel gas	81	(23.6)	4895.0	(48.4)	
Power	110	(32.1)	196.5	(1.9)	(6.1)
Chemicals	4	(1.2)	450.0	(4.4)	(14.0)
Other / unspecified	77	(22.4)	2052.0	(20.3)	
TOTAL	343	(100.0)	10 119.6	(100.0)	(100.0)

Notes
- nk : Not known
- Power : no differentiation is made between mechanical and electrical power. Includes vehicle power.
- ¶ : analysis excluding fuel gas and unspecified applications

industrialized (developed) countries (1). A complete list of potential applications is given in Table 10.

The area of greatest activity in the industrialised world is North America with about 60% of installations and 70% of the energy production of industrialised countries. (Developing countries are excluded because of the lack of robust data). The average size of an installation there is 50 GJ/h or 2.5 t biomass/h. No information is available on analogous development of combustion technology and application, except that it is thriving. The applications by sector in Table 9 confirms interest in high temperature process heat for kilns (eg lime, cement, brick etc), drying, and boiler applications including retrofitting.

Product Quality

Extensive discussions and contact with researchers, manufacturers, potential users, and advisers failed to produce any guidelines on fuel gas quality constraints for any application: the criterion is natural gas quality. An attempt has been made to characterise quality requirements and constraints by contaminants in Table 11, and by application in Table 12. Both tables include the properties that have been identified as potentially important.

Table 10 Known Applications of Gasification Technology

Heat (general)	Power (general)
Drying (see also Agriculture)	Boat power
Greenhouses (see also Agriculture)	Electricity generation via engine, and CHP
Process heating (see also Chemicals)	Electricity generation via turbine, and CHP
Retrofitting gas and oil fired boilers	Irrigation
Space heating	Shaft power for industry
Steam raising in boilers	Vehicle power
Wood Processing	Agriculture
Lumber drying	Crop drying eg tobacco, grain
Sawmill (see also Power)	Food processing
Timber drying	Greenhouse heating
Veneer drying	
Wood products	
Chemicals and Process Industries (direct and indirect heating)	
Bakeries	Food processing
Brick	Glass
Cement	Lime
Ceramics	Paper production and drying
Chemicals	Pottery
Fertiliser production	Rubber
Foundries	Textile drying

Table 11 Product Quality and Applications

Particulates	Tars	Technology	Applications
High	High	Updraft, crossflow	Direct firing when product contamination acceptable
High	Low	Fast fluid bed,	Direct firing when product contamination acceptable
Low	High	Pyrolysis	Direct firing, retrofitting
Moderate	Moderate	Fluid bed	Direct firing when product contamination acceptable
Low	Low	Downdraft, or *Cleaning needed*	Engine, CHP, retrofitting, boilers, drying
V. low	V. low	*Cleaning needed*	Direct firing when product contamination not acceptable
None	None	*Cleaning needed*	Turbine, CHP, syngas

Table 12 Product Quality Requirements

QUALITY	Fuel gas temperature permissible	Maximum particulates permissible	Maximum tars permissible	Pollutants eg HCl, H2S	Combustion temperature required
APPLICATION					
Drying	Low (less than 250 C)	None (textiles and food) - low	None (textiles and food) - low	Low	Low
Kilns	High	Mod: cement,lime	High	Low - high	High
		Low: bricks, biscuit		Mod. - high	Low - high
		None: ceramics, glass	Low	None - low	High
Metal processing	High	Low	High	Low	High
Boilers and space heat	High to use sensible heat	Low - mod.	Mod.	Low - mod.	Low
Power, engine	As low as possible	Very low	Very low	Low	Low
Power, turbine	As high as possible	None	None	None - low	High
Syngas	As low as possible	None	None	None - low	-

COMPARISONS

In order to put biomass and waste gasification into a market context, a superficial examination is included of the main competitors to this technology - combustion and coal gasification

Combustion

There are two implications on capital cost - equipment size, and instrumentation and control. As the volume of a gasifier system is substantially less than the equivalent combustion unit, capital cost would be lower. However additional control and safety features would counteract this cost advantage. The capital cost analysis demonstrated close similarity in capital costs between combustion of solid wastes and gasification of solid wastes at the intermediate technology level. This is in agreement with analogous comparisons in the USA. A summary of all the the relative advantages and disadvantages is given in Table 13. Biomass and solid waste can supplement or replace conventional solid fuel fired systems, and it is in the steam and Rankine cycle power generation sector that combustion opportunities lie.

Table 13 Summary of Gasification and Combustion Applications

Gasification	Factor	Combustion
Similar	Capital cost	Similar
Similar	Operating costs	Similar
Higher	Product value	Lower
Better	Versatility in application	Worse
Similar	Efficiency	Similar
Worse currently	Reliability	Better
Higher	Hazards and safety	Lower
Similar	Extent of automation	Similar
Similar	Controllability of process	Similar
Better	Controllability of application	Worse
Worse	Technology status	Better
Higher	Expertise requirement	Lower
Better	Atmospheric pollution potential	Worse
Worse	Water pollution potential	Better
Worse	Equipment availability	Better

Coal Gasification

In terms of gasification technology, the technology related to coal gasification is very similar. There are a number of significant differences however:

Advantages Coal is easier and less costly to handle and process through size reduction.
Coal is easier to feed into closed systems, particularly if pressurised, as the fibrous nature of biomass can give rise to substantial sealing problems in feeders and valves.
Coal is available at a single site in substantial quantities and does not require consideration of a collection infrastructure compared to biomass. Alternatively the cost of coal can be considered as including the underground size reduction and collection cost.

Disadvantages Coal has a much lower level of volatiles than biomass, (typically 30% compared to 70% for biomass), and coal char is significantly less reactive than biomass char. This adversely affects reactor sizes with coal requiring longer residence times.
Coal has a higher ash content than biomass, and with different characteristics. Ash handling in coal processing is thus more complex, more significant, and more costly.
Coal often has a significant sulphur content which, if deleterious to downstream application, requires more complex and more costly clean up.

The sum effect of these differences has not been explored. The limited comparison of coal and biomass fed methanol production processes concluded that there was no more than a 10% capital cost advantage to biomass - mostly associated with acid gas treatment required of coal derived gas.

While little overall difference in capital cost would be expected for the same system, analysis of data on small scale coal gasification suggests that coal gasification capital costs are about double those of high cost biomass gasification systems, on a mass throughput basis, but similar on an energy output basis. To further explore the economics of biomass versus coal, the sensitivity analyses have been extended to include coal gasification in terms of capital cost and feed cost representative of coal, as shown in Table 14.

Table 14 Comparison of Coal and Biomass Gasification

	Mean Product Cost 15GJ/h output	1 t/h input		75GJ/h output	5 t/h input		150GJ/h output	10 t/h input	
	$/GJ	$/GJ	¢/kWh	$/GJ	$/GJ	¢/kWh	$/GJ	$/GJ	¢/kWh
Biomass, $0/t	2.28	2.28	5.04	1.30	1.30	2.45	0.80	0.80	1.92
Biomass, $10/t	2.95	2.95	6.04	1.69	1.69	3.45	1.47	1.47	2.92
Biomass, $20/t	3.62	3.62	7.04	2.35	2.35	4.45	2.13	2.13	3.92
Biomass, $30/t	4.28	4.28	8.04	3.02	3.02	5.45	2.80	2.80	4.92
Coal, $30/t	4.45	4.24	7.29	3.03	2.80	4.73	2.75	2.47	4.12
Coal, $40/t	4.83	4.63	7.86	3.41	3.18	5.30	3.13	2.86	4.69
Coal, $50/t	5.21	5.01	8.43	3.80	3.57	5.87	3.51	3.24	5.26
Coal, $60/t	5.59	5.40	8.90	4.18	3.95	6.24	3.89	3.63	5.83

All costs are on the same basis except feed cost, and plant size and cost, as shown; and feed heating value which was taken as 20GJ/daf t for biomass, and 35GJ/daf t for coal.

MARKET

There are two approaches to establishment of a market opportunity: supply and demand. Supply is considered in terms of biomass and waste production and availability; while demand is considered in terms of conventional fuel replacement or substitution. A major problem is the reliability and variability of the data. Representative figures have been used, therefore, to provide trends rather than absolute figures.

Biomass Resource

The annual world production of biomass is many times more than the energy consumed in the world. It is, however, dispersed over large areas; sometimes in accessible locations; and it may be in forms that are difficult to harvest. This production of biomass can be utilised in a number of forms: existing resources that may be used once or replaced by energy crops; energy crops that optimise production for utilisation or conversion; a wastes and residues from existing biomass utilisation operations.

Existing resources . There are many locations in world where biomass as wood and oth forms has accumulated. These resources cannot be regarded as renewable unless they a routinely replaced by energy crops or some other product such as timber stands agriculture. The cost of the resource lies mostly in the harvesting. Replanting for energy then considered an energy crop . Locations of these resources tend to be in developi countries and/or less populated areas.

Energy crops . These may begin as an existing resource which is replaced by energy crop or may be a new development in which single stem woody biomass or coppice is grown short rotation forestry. The cost lies in the complete cycle of planting, cultivatio harvesting and replanting. Efficient production depends on fixation of solar energy whic is highest in equatorial regions, (Table 15) although other aspects of climate and topograph are important.

Wastes and residues . Most forestry and agricultural industries generate waste which ca often be several times more than the product itself. The timber industry, for examp produces about a tonne of waste per tonne of timber; and the sugar cane industry produc about 4 tonnes of bagasse per tonne of sugar (Table 16). While some of these residues a utilised (particularly bagasse) considerable proportions are not; and those that are used a often used inefficiently. In this category of resource the cost of harvesting and collection subsumed into the product cost. The waste is, therefore, often at least free and there m even be a disposal credit if it is utilised. Much of this waste arises in agriculturally base economies, which tend to be in the temperate and equatorial regions of the world.

The distribution of biomass production by latitude is shown in Table 1 demonstrating the influence of solar energy. The availability of wood is indicated b fuelwood production in Table 16 for the seven major regions of the world and the mo productive countries. The significance of fuelwood to the total wood production of th largest fuel users is also included. The significance of wastes is indicated by bagass arisings in the same Table 16. Total fuelwood and wastes and residues in the five maj developing areas of the world is shown in Table 17. It is the developing areas of the wor where most of the biomass is produced and where need will be shown to be greates Fuelwood and wastes can currently contribute about 43% of the total energy demand of th developing world; but this contribution is predicted to fall to 24% by 2000 due to th anticipated considerable growth in total energy demand of about 6.5%/y while fuelwoo plus wastes and residues will only grow by 2.3%.

Table 15 Biomass Production by Latitude (3)

Latitude	% Production
NORTH	
90 - 75	0.1
75 - 60	2.9
60 - 45	8.3
45 - 30	10.8
30 - 15	14.2
15 - 0	23.7
0 - 15	23.6
15 - 30	11.8
30 - 45	3.5
45 - 60	0.1
60 - 75	0.3
75 - 90	0.3
SOUTH	100.0

Some comparative figures for industrialised countries are shown in Table 18 for available and recoverable energy from wastes, residues and energy crops. "Available" refers to technical availability, while "recoverable" usually includes consideration of costs and thus refers only to biomass that is economically recoverable at the time of the analysis. There is no implicit relationship between the two definitions. In the UK, for example, only about 20% of the total available resource is currently economically recoverable.

A further factor is the extent of agricultural surplus/deficit and energy surplus/deficit; the ideal combination for biomass utilisation being agricultural surplus and energy deficit. A limited survey is shown in Table 19. Bagasse contributes a substantial proportion to the solid waste figure, with cereal and grain wastes, rice wastes, and fruit, nut and vegetable wastes contributing to the remainder.

Demand

Statistics on energy demand are prolific, and only one set of data is included as Table 20. The current and future need for energy is unquestionable, and the areas of the world with greatest potential are Africa, South America, and Asia, ie the developing areas or countries (see also Table 17). Demand is for all forms of energy for all applications, although evidence is given above that power production is the greatest demand (Table 19).

Biomass can satisfy all these energy forms; gas as LHV or MHV fuel gas; liquid as gasoline, diesel or methanol; solid as biomass for direct combustion; and power from fuel gas or combustion. The extent to which biomass can contribute to these energy forms in either gasification or combustion technology is indicated in Table 21.

Table 16 Biomass Production: Fuelwood and Bagasse, 1983 (9)

	Fuelwood million m^3/y	% total wood production		Bagasse million t/y
WORLD	1680			213
Production from major world areas				
Africa	421			20
North America	150			50
South America	220			60
Asia	724			65
Europe	70			-
Oceania	12			9
USSR	83			-
Production from individual countries				
India	213	90	Brazil	53.6
Brazil	173	88	India	27.6
China	155	76	Cuba	18.5
Indonesia	114	95	China	11.7
USA	102	4	Mexico	10.0
USSR	83	25	USA	9.4
Nigeria	78	96	Australia	7.5
Tanzania	36		Philippines	6.9
Sudan	36		Thailand	6.9
Thailand	36		South Africa	6.5
Bangladesh	31		Argentine	5.3
Turkey	29		Indonesia	4.9
Zaire	29		Columbia	4.4
Philippines	29		Dominica	3.9
Kenya	28		Pakistan	3.8
Ethiopia	28		Mauritius	2.4

Table 17 Fuelwood and Wastes Production and Energy Demand in Developing Areas (5)

	NOW			2000		
	Fuel Wood PJ/y	Solid Waste PJ/y	Total Energy PJ/y	Fuel Wood PJ/y	Solid Waste PJ/y	Total Energy PJ/y
Middle East and North Africa	710	550	6 910	1 080	760	22 140
Sub-Saharan Africa	4 500	510	6 230	7 110	810	12 310
South Asia	2 430	4 050	10 760	3 510	6 300	21 370
South East Asia	2 810	1 710	9 840	3 730	3 060	25 240
Latin America	2 970	1 510	16 650	2 200	1 840	47 700
TOTAL	13 420	8 330	50 390	17 630	12 770	128 960

Table 18 Contribution of Wastes, Residues and Energy Crops to some Industrialised Countries (7,8,10)

	Total Energy Potential: Technically available: Forestry and timber residues, mtbe/y		Economically recoverable: Agricultural residues and wastes mtbe/y	Total wastes and residues mtbe/y	Energy crop potential mtbe/y by 2000
	Current	2000	Current	Current	
Australia	6.5			5.9	
Canada	66.8		4.8	13.0	
Finland	18.0		0.8	1.0	2.0
France	12.8		10.9	11.9	4.0
Germany	13.1		4.6	5.5	
Italy	3.2		5.3	5.5	
Japan	14.1		0.8	1.8	
Sweden	21.5		1.7	1.7	3.3
Switzerland	1.8		0.2	0.4	
Turkey	12.4		-	-	
UK	2.3	4.0	1.2	1.4	4.0
USA	142.7	500-950		42.4	260.0
EEC				32.1	60.0
OECD				119.7	

(N B excludes MSW)
mtbe - million tonnes biomass equivalent @ 20 GJ/t daf

Table 19 Agricultural and Energy Surplus/Deficit Ranking in 1980(11)
(NB positions may have changed significantly, eg UK, since 1980)

Energy Deficit	Energy Surplus	
Brazil	Australia	Agricultural surplus
Argentine		
Thailand	Columbia	
South Africa		
USA		
Philippines		
Sudan	Canada	
Turkey	Burma	
France	Mexico	
Peru		
India		
Pakistan	Nigeria	Agricultural deficit
Bangladesh	Zaire	
Spain	USSR	
Korea	Poland	
Japan	Egypt	
Italy	Venezuela	
Germany		
(UK)		

Table 20 Energy Consumption per Capita

	Natural gas GJ/capita	Liquid fuel kg crude oil/cap.	Solid fuel kg coal eq/cap.	Electricity kWh/capita
WORLD average	11.7	579	588	1 876
Africa	1.5	167	173	405
North America	52.4	1 992	1 529	7 453
South America	5.9	585	71	1 197
Asia	1.3	216	293	558
Europe	23.8	1 191	1 467	4 731
Oceania	22.8	1 312	1 730	5 646
USSR	60.3	1 806	1 775	5 072

Table 21 Potential Gasification and Combustion Contribution

	Gasification	Combustion
Retrofitting and Boilers		
Gas fuel	100%	0%
Liquid fuel	20%	0%
Solid fuel	0%	100%
Process Heat		
Low temperature eg drying	100%	0%
High temperature eg cement, bricks	60%	0%
Gas fuel	80%	0%
Liquid fuel	20%	0%
Solid fuel	0%	(Charcoal 100%) 50%
Power		
Up to 5 MW	80%	20%
Above 5 MW	20%	80%
Heating		
Gas fuel	100%	0%
Liquid fuel	20%	0%
Solid fuel	0%	100%
Liquid fuels and chemicals	100%	0%

The main limitations to satisfying this demand in the shorter term is the availability of a sufficiently developed and reliable technology to satisfy the purchaser and finance provider.

Products

The range of products realisable by thermochemical conversion of biomass is summarised in Table 22 with suggested economic size ranges and probable time for implementation. Applications in industrialised and developing countries were analysed in Table 9 above and are summarised in Table 23, as support of the time scales and size ranges envisaged in Table 22.

Short term applications are for direct firing and retrofitting in industrialised countries, and power in developing countries. This technology is available now.

Longer term applications are for more sophisticated fuel gas production and power generation with less distinction between industrialised and developing countries. With successful long term commercialisation of the basic technology, attention will turn to larger scale applications for production of commodity chemicals and fuels, of which the more obvious are listed in Table 22. The technology is generally readily available in constituent parts, but its assembly into a working integrated process is unlikely to be realised for a decade.

Size of Gasification System

The minimum (economic) size of a gasification system is set generally by the economies of scale as discussed above, although the technology and demand of the

Table 22 Products, Scale and Implementation Time Scale

Product	Size Range daf t/h biomass	product	Years to commercial use	Application
Fuel gas, LHV	0.2 - 5	3 - 100 GJ/h	0	Retrofitting, process heat, drying
Power, engine	0.1 - 5	0.1 - 5 MW	0	Electricity
Fuel gas, MHV	1 - 10	15 - 200	2	Retrofitting, process heat
Power, turbine	1 - 10	1 - 10 MW	4	Electricity
Ammonia	5+	2 - 20+ t/h	6	Fertilisers
Methanol	15+	5 - 20+ t/h	6	Fuel additive and chemical.
SNG	15+	5 - 20+ t/h	6 - 8	Gaseous fuel
Gasoline	20+	3 - 15+ t/h	6 - 8	Liquid fuel
Diesel	20+	3 - 15+ t/h	8	Liquid fuel
Syncrude	20+	3 - 15+ t/h	8	Synthetic crude oil

+ Upper limit set by logistics and cost of biomass harvesting particular applications have some influence. The maximum size is set by the feasibility and cost of harvesting and collection of the biomass resource. This will depend on the biomass productivity per unit land area, local labour costs, and other factors that are peculiar to a given situation. In high productivity areas, for example, yields of up to 20 dry t/ha/y biomass are possible in equatorial regions. A 20 t/h plant, for example, producing methanol or ammonia, will require 16 000 ha at 50% land utilisation, or 160 km^2 - a circle of radius 7.1 km. Supply chains for larger applications thus assume considerable importance. Large plants will benefit from economies of scale, although multiple gasification reactors would be specified for larger systems for design and availability reasons.

Table 23 Summary of Applications (Table 9)

Application	Industrialised Countries		Developing Countries	
	% energy	Average size GJ/h	% energy	Average size GJ/h
Process heat	37.0	115.0	0	
Boiler and retrofitting	24.4	17.5	0	
Drying	18.2	32.0	0	
Metal processing (process heat)	9.0	84.0	0	
Chemical	7.2	200.0	0	
Power	4.2	2.7	100	-
	100.0			
Average		38.9		-
Overall				
Fuel gas		60.4		
Other/unspecified		26.6		
All gasifiers		**29.5**		

Average sizes of installations to date are reported in Table 23. Process heating (direct firing) applications have tended to be relatively large at around 80 - 120 GJ/h (5 - 7 t/h) although this is not in any way a minimum economic size. Conversely drying, retrofitting and power applications have much smaller unit sizes of 3 - 30 GJ/h (0.2 - 2 t/h). Data on power generation in developing countries is not reliable.

The minimum economic size of a system is indicated in Table 22, although this is subject to feedstock cost which can have a significant effect on minimum size as shown in the cost analysis above. There are not believed to be any size restrictions on LHV gas production, but MHV gas can only be economically produced at capacities above around 1 t/h.

Fuels and chemicals need to take advantage of economies of scale to achieve a low product price in industrialised countries. An analysis of an African developing country for example, suggested that the minimum economic size for nitrogenous fertiliser production was 12 000 t/y ammonia equivalent, requiring gasification of about 5 t/h wood (dry basis) which was available as energy crop at about $10/dry t (2). For gasoline production (gasoline process via methanol) the minimum economic size was 30 000 t/y requiring conversion of about 25 t/h wood (2).

For power production, engine generated electricity offers significant advantages for small scale systems of up to 5 MW compared to Rankine cycle systems from combustion. This is due to the considerably reduced efficiencies of small steam based systems compared to engines, with typical efficiency figures of 15 - 20% for steam and 25 - 30% for engine delivered power at around 1 MW (6). Power production from biomass has been successfully achieved at down to 50 kg/h biomass, although the conversion efficiency is much lower than for larger systems. Turbines are a future development in the 1 - 5 MW range, which require high standards of gas clean-up and preferably a pressurised reactor for optimum performance. These are, therefore, an industrialised country application. Combined heat and power systems from 0.5 MW upwards have much potential if markets are available for both products.

Constraints

Restrictions on developments of a new market can be caused by a number of factors, each of which is briefly discussed.

Product price. The cost at which the fuel, chemical or power products can be produced has to be estimated for the location, technology, feedstock, process size and product required. This cost has to be compared to the same product from conventional energy sources including fiscal allowances and including adequate margins for investment. Product prices is perhaps the most important factor influencing implementation of biomass conversion technology. It is a market aspect over which virtually no control can be exercised, since energy prices are controlled globally. The short term prognosis for industrialised countries is not encouraging for alternative energy technologies. Medium to longer term energy price trends will inevitably show real price increases so the question is really "when" rather than "if" gasification will be economic compared to conventional energy sources. The opportunities for developing countries are, however, significantly different since they often have abundant biomass resources; import conventional fuels at considerable cost; and could benefit greatly from a reduction in this import bill .

Feedstock availability. Utilisation of all theoretically realisable wastes is unrealistic. There will be competition for these wastes, notably from combustion for all wastes; from higher added value products for forestry wastes and from lower costs disposal options for MSW and other wastes. There is unlikely to be any local shortages of biomass until at least 25% utilisation of available resources is achieved.

Capital cost. The intermediate level of capital cost is likely to be the target figure for gasification/pyrolysis technology in industrialised countries. The lower figure will apply to small scale applications in developing countries for charcoal and fuel gas production. There will be cost advantages from learning, replication, technology development and competition; but cost disadvantages from lessons of implementation and environmental and safety problems. An overall reduction of more than 25% in the long term is unlikely.

Variable and feedstock costs. Variable costs are unlikely to change significantly with time but will vary significantly with location. Developing countries in particular will have a significantly lower unit labour cost but higher manning levels; and maintenance costs may be higher. Feedstock costs, particularly energy crops in industrialised countries will tend to follow energy prices. Wastes will fluctuate between disposal costs and energy prices according to market forces. In developing countries where substantial national benefits are to be gained, feed costs will be governed mostly by labour costs of planting, cultivating, harvesting and handling the solid material.

Competition There will always be competition from alternative technologies such as combustion and biological routes to fuels. Gasification technology must, therefore, break through the implementation barrier if it is to succeed. This can be achieved on grounds of cost, different product, performance, environment and safety, and status. The cost comparison with combustion above concluded that neither technology is likely to have a capital cost, operating cost, or performance advantage. Gasification is potentially more versatile since the product fuel gas can be piped to the user, is more readily integrated with natural gas, gives cleaner and more efficient combustion, can be used for power and electricity production, and is more acceptable environmentally.

Technical development Of the 100 or so companies around the world claiming to have a "commercial" system in 1983 (see Appendix), only a relatively small proportion can substantiate that claim with installations in the last three years. The simple fact is that the market opportunity is not perceived to be sufficiently encouraging to justify capital expenditure on developing suitable technology. The total costs of bringing an existing technology to the demonstration stage is $1-5 million according to the status of the technology plus the cost of the demonstration plant of $ 1-5 million according to size. The minimum market size necessary to justify this level of activity is $ 1-10 million/year of business, according to the location of the market and the level of technical sophistication.

Technology. A major technical factor that influences choice between combustion and gasification is the levels of technology. Combustion is seen to be simple, proven, with easy to build plant, which is easy to operate. Small companies can readily enter the combustion equipment market and compete on price, with relatively low risk of complete technical failure. Users also perceive combustion as being essentially proven and trouble free.

Gasification and pyrolysis process plant is viewed in the market place as a much more sophisticated technology, requiring a level of expertise and knowledge that is not widely available. Small companies that have sprung up have relied on both enthusiasm and knowledge to pursue implementation. Larger companies seem to appreciate the potential of the technology but are inhibited from committing the necessary resources to a suitable RD & D program due to the many competing demands for capital, with shorter paybacks and less perceived risks.

Time scale. The time required to bring any existing developed, but unproven, technology to a point where it has been demonstrated successfully and can be offered in the market place is not less than 3 years and could be 5 years or more depending on the development needed, and problems encountered on demonstration. Such time scales are not encouraging to industry, who look for a very much shorter time horizon in their planning.

Market size. The overall ultimate market potential is a sizeable fraction of the world's energy demand. The realisable opportunity within conventional industrial time horizons is only a small, but still substantial fraction.

Growth prospects. The growth in adoption of waste conversion to energy will considerably exceed the growth in overall waste production, although this latter aspect is tempered by availability, accessibility and supply chains. Growth in energy crop production for energy and fuels will probably grow at about the same rate as utilisation. Growth prospects are dependant on several factors, notably energy prices, degree of success of demonstration and initial plants, attitudes, confidence and incentives (in descending order of importance).

Profitability. There are two aspects of profitability to be considered: the equipment manufacturer; and the customer ie the user of the equipment. The process plant manufacturer is concerned with profitable production of process plant, and this conforms to orthodox product life cycle concepts:

a) Entering an established market with a conventional product, implies low prices and consequently low profit in early years in anticipation of a significant market share and hence profits subsequently.

b) Entering the same market with a better and/or more economic technology will give higher profits initially which will reduce as competitors catch up.

c) Entering a new market has the most uncertain outcome since pricing policy is difficult and the market unknown.

Thermochemical conversion of biomass can fit any or all of these categories according to how it is perceived by the equipment manufacturer and the user. It may be directly compared to combustion in (a) or (b) above, or it may be considered as sufficiently distinctive from combustion (and other renewable energies) to be classified as (c).

The waste industry is seen as high risk requiring very high rates of return to justify entry. This applies both to sellers and buyers. Payback times of 2 - 3y have been suggested as an expected return which could be tolerated at the operating or user end, but for manufacturing process plant would be both unusual and difficult to achieve and still successfully enter the market.

Turning points. It is possible to identify significant turning points that will influence the acceptability and implementation of gasification technology:

Increase in real energy prices,
Demonstration plant not unsuccessful,
Increased concern over atmospheric pollution (gasification is cleaner),
Greater awareness of technology (gasification is poorly understood and is therefore currently avoided),
Real increases in waste disposal costs,
Increased concern over current waste disposal management,
Political unrest in sensitive areas.

None of these can be predicted with any accuracy, yet all could have a major impact on the market for gasification technology. There is a high probability that all will happen, but the timing and rate of change are uncertain.

Growth

Predictive simple model. Prediction of future markets is conventionally achieved with empirical S-curves, which relate penetration to time and ultimate market size. In order to derive predictions of saturation, market size in PJ/y, gasifier sales and capital market size, the following model has been derived:

$$St = Ss[1 - \exp(-(t/A)^{2.0})]$$

where St = energy demand from biomass gasification in year t
Ss = saturation level, or maximum potential biomass availability
t = time in years from now
A = growth rate characteristic
= 24.02 to give 50% saturation after 20 years
= 24.02(1+v)
where v is the fractional annual change in the (local) economic environment. This is empirical, and can range from -0.5 for rapid growth due to substantial changes in the economic environment, to +2.5 for minimal growth.

and $Ss = So(1+a)^t(1+g)^t$

where So = current biomass and waste available arisings
a = fractional annual change in availability of resource (feed)
g = fractional annual change in resource size

Thus: $$St = So\,(1+a)^t(1+g)^t\,[1 - \exp(-\{t/24.02(1+v)\}^{2.0})]$$

The time periods involved mean that compound escalation of arisings and availability changes can produce serious distortions. These effects are best handling with their own S-curves, although they have not been included here. There are too many inconsistencies in data on biomass resources and in the range of countries and scenarios possible to give results, although the model has been successfully used in several situations. The principle is applied below to overall prediction of market size.

Detailed model. A thorough market appraisal would assemble the data listed in Table 24 for each country to identify the potential that biomass and waste can offer the energy requirements of each country. This should examine the current, short term, and long term future requirements. It can be readily refined with adequate data, to match technology and demand, and also to match supply of feed and demand of energy by location and size. When the technology of biomass and waste conversion is more assured, such an exercise will be of value in identifying the more rewarding opportunities.

Market Structure

This is relatively complicated with combinations of technology push, market pull, and government intervention. The technology push comes from speculative

Table 24 Country by Country Market Assessment Methodology

Consumption of energy by type : gas, liquid, solid, power,
Conventional fossil fuel resources : indigenous, imported,
Cost of conventional resources,
Shortfall or net requirement,
Biomass and waste resource size : current, short term, long term,
Cost of biomass and waste,
Growth: resource, energy demand,
Infrastructure status and requirement,
Biomass conversion technology and product,
Technical skill availability,
Finance availability,
Economic cost-benefit analysis.

---

industrial and institutional R&D arisings from fears of further oil crises, extensive publicity given to biomass as a renewable resource, growing awareness of the finiteness of energy resources and the need for conservation, fiscal incentives, and the glamour of the "bio-" image. The pull comes from industry to reduce its energy costs and from developing countries to reduce their dependency on imported oil. The latter provide intervention and incentives to speed up the rate of implementation. The match between technology push and market pull is still imperfect, and it is this aspect that requires most attention if implementation is to be successful.

Technology. Low technology, small scale (<1t/h) installations will continue to be built in developing countries, but their relative unreliability, poor performance, and pollution potential will mitigate against substantial market penetration. There is no potential for this technology in industrialised countries.

Intermediate technology processes will have a slower initial rate of implementation in developing countries, but have the greatest medium to long term potential, with particular applications in power generation. This lower initial rate of growth will be due to higher costs and need for importation of equipment. In industrialised countries, this technology will be the cost target for high technology processes as implementation and competition grow. Direct heating, retrofitting and power applications are the more likely applications.

High technology processes will only be able to compete at larger scales of operation where there are identifiable benefits such as a higher quality product or lower pollution potential. This technology is more likely to be developed for CHP and liquid fuels and chemicals applications, when an optimally integrated process can be designed.

Forecasts. Table 25 provides a view of the possible gasifier population and energy supplied as an indexed value for the three levels of technology and their life cycle. The index base is 100 gasifiers of intermediate technology in 20 years time. Index figures are used in order to be able to describe different scenarios, for example multiply by 0.5 for indefinitely constant energy prices and a relatively untroubled world; the figures as quoted would apply to a more real world slowly increasing energy prices in real terms and no major conflict - these are then broadly in agreement with other views (eg 5); and substantial increases in energy prices and/or major conflict could cause these figures to be multiplied by 2-10 according to circumstances. For the process plant and contracting industry, the capital value of the industry in the real world scenario as quoted is given in $\$_{1985}$ million. The likely distribution by main geographical areas is indicated in Table 26.

Table 25 Indexed Gasifier Population and Energy Demand Futures, and Capital Value of the Market

Year	INDEX VALUES						CAPITAL VALUE
	Low technology		Intermediate technology		High technology		$1985 million/y
	No.	Energy	No.	Energy	No.	Energy	
1990	25	8	10	10	2	10	17
1995	40	10	30	30	6	30	59
2000	50	13	55	55	10	50	152
2005	60	15	100	100	25	125	290
2010	70	18	160	160	60	300	690
2015	80	20	230	230	100	500	1025
2020	80	20	305	305	160	800	1970
2025	70	18	380	380	220	1100	2700
2030	60	15	450	450	300	1500	3730
2035	50	13	520	520	400	2000	5260

Numbers of gasifiers and energy are not directly related - both are indexed to base 100 in 2005. Equipment replacement is included.

Table 26 Distribution by Main Geographical Areas

Year	Africa	N. America	S.America	Asia	Europe	Oceania
1990	5	35	15	20	20	5
1995	10	30	15	20	20	5
2000	10	25	20	25	15	5
2005	15	20	20	25	15	5
2010	15	15	25	25	15	5
2015	20	15	25	25	10	5
2020	20	15	25	25	10	5
2025	20	15	25	25	10	5
2030	20	15	25	25	10	5

The growth rates proposed are not restricted by availability of biomass. The entire projected demand in 2000, for example, could be met by 5% of the wastes of the least prolific producer - Middle East and North Africa (Table 17). Applying the geographical distribution to the same region (Table 26), then less than 0.5% of the waste would meet projected demand. The proposals are, therefore, conservative, and could be much higher if demanded by the market. Finally distribution by resource is indicated in Table 27 suggesting a substantial growth of the contribution made by energy crops.

Table 27 Contribution of Different Feedstocks

Year	Wastes and Residues (%)	Energy Crop (%)
1990	95	5
1995	90	10
2000	85	15
2005	75	25
2010	65	35
2015	50	50
2020	40	60
2025	30	70
2030	25	75
2035	20	80

CONCLUSIONS

The status of technology for realising the renewable energy of biomass has been summarised with the conclusion that it is now well established and versatile, but there are technical gaps between biomass production and conversion, and between conversion and application.

Current and short term economic applications lie in utilising wastes and residues to produce power, and fuel gas where quality requirements are less demanding. There is longer term potential for transport fuels and chemicals production, particularly in less developed countries and those with few indigenous conventional energy resources.

The rate and extent of implementation is essentially an economic problem of matching or beating energy market prices. Small variations in price will have a substantial effect on the market opportunity for gasification technology.

There is no shortage of suitable wastes and residues for conversion in any region of the world, at least until the medium term, and by then energy crops will have been developed if justified by demand and cost. There is no lack of demand for any energy product derived from thermochemical conversion of biomass, except that markets for heat, particularly waste heat, are likely to be limited.

A perceived major problem is the imperfect match between market requirement and technological capability, which is currently bridged by promises or hopes. This requires resolution.

Tentative market forecasts for technology development, level of implementation, regional allocation, and capital requirements are proposed, which confirm that a substantial worldwide market opportunity exists. The limitation on implementation lies not in supply or demand for products, or costs, but on technological matching and finance.

ACKNOWLEDGEMENTS

This paper is based in part on work carried out for the UK Department of Energy's Biofuels R&D programme through the Energy Technology Support Unit (1). Work carried out for Aston Science Park is also included. The views expressed in this paper are the opinions of the author and do not necessarily reflect the views of either sponsoring organisation. The assistance of J M Double and D M Earp is gratefully acknowledged.

REFERENCES CITED

1 Bridgwater, A V, Double, J M, and Earp, D M, "Market and Technical Assessment of Biomass Gasification in the UK", Report to UK Department of Energy, (1986)
2 Comben, A, Bridgwater, A V, Greenhalgh, P, and Wolstenholme, J O, "The Viphya Resources Utilisation Study", UKODA, (1980)
3 Federal Institute of Geoscience and Natural Resources, Survey of Energy Resources 1980, (Hanover, Germany, 1981)
4 Grayson, A J, "Wood Resources and Demands: A Statistical Review", HMSO UK (1970)
5 Harlow, I, in "Package Power Stations for Export", Institution of Mechanical Engineers, London, (1985)
6 Heaton, J, and Bridgwater, A V, Ibid.
7 OECD, "Biomass for Energy", (1984)
8 Siren, G, "The Many Aspects of Intensive Energy Forestry", BioEnergy 84, II (Elsevier Applied Science, 1985)
9 UN, "Energy Statistics Yearbook", (1983)
10 US DoE, Biomass Energy, (1984)
11 World Bank, "Alcohol Production from Biomass in the Developing Countries", (1980)

ADDENDUM

Copies of the charts based on the tabulated data in the paper that were projected in the oral presentation are available from the author on request.

APPENDIX

Companies Believed to be at Commercial or Demonstration Scale Operation (since 1983)

Alternate Energy Industries	C,B	KHD Humboldt Wedag AG	G
Alternate Gas Inc	U	KMW AB	S
Agip SpA	I	Keiner Pyrolyse GmbH	G
Ahlstrom Osakeyhtio	Fi	Kim	G
Alberta Industrial Developments	C	Machinenfabrik A Lambion	Be
Alsthom Atlantique	F	Lurgi GmbH	G
American Carbons Inc	U	Lurgi Corporation	U
Andco Torrax Corp	U	E R Mellenger	C
Applied Engineering Company	U	Mannesmann Veba Umwelttechnik GmbH	G
Applied Power Technology	U	Myrtleford Electric Pty	A
Battelle Columbus Laboratories	U	North American Gasifier Co	U
Biomass Energy Consultants &Eng	H	National Synfuels Inc	U
Biomass Corporation	U	Nedalo BV	H
BKMI	G	NEI Fluidyne (NZ) Ltd	NZ
BTU Bio-Technik-Union	G	Nichols Engineering and Research Corp.	U
Buck Rogers Co Inc	U	Nouveler	C
Century Research Inc	U	Novelerg	F
CHH Technology Inc	U	Omnifuel Gasificaiton Systems	C
Clayton and Lambert	U	Petrogas	H
Combustion Power Company Inc	U	Pillard	F,U
Coppee SA	Be	Power Recovery Systems Inc	U
Framatome Divn. Creusot Energie	F	Pyrenco Inc	U
DM International Inc	U	Pyrotech	U
DRP GmbH	G	Pyro-Sol Inc	U
Duvant Moteurs	F	Rheinische Braunkohlenwerke AG	G
East Moors Fabrications Ltd	B	Rutsgerswerke AG	G
Ebara Corporation	J	Saarberg Fernwarme GmbH	G
The Energy Equipment Co Ltd	B	Specialist Engines Ltd	B
Energy Equipment Engineering BV	H	Solar Energy Research Institute	U
EE Howden BV	B	Studsvik Energiteknik AB	S
ENEA	I	Sur-Lite Corporation	U
Enerco Associates	U	Syngas Inc	U
Energy Products of Idaho	U	TAG	G
Energy Resources Co Inc	U	Thalapat AG	Sw
Entropie	F	Tropical Development and Research Inst.	B
EZ Manufacturing Co	U	Tunzini Nessi Entreprises d'Equipements	F
Forest Fuels Manufacturing Inc	U	Tsukishima Kikai Co Ltd	J
Fritz Werner Industrie	G	Union Carbide Inc	U
Foster Wheeler Power Products	B	University of Missouri-Rolla	U
Gasification Associates (UK) Ltd	B	VTT Technical Research Centre of Finland	Fi
Gasification Specialities Inc	U	Vyncke Warmtetschiek	Be
Gotlas Genke	S	Woodhall Duckham	B
Halycon Associates Ltd	U	Wellman Mechanical Engineering Ltd	B
James Howden Ltd	B	Westwood Polygas Ltd	C
Institute of Gas Technology	U		
Imbert Energietechnik	G		
Italenergie SpA	I		
John Brown	B		

COUNTRY CODES

A	Australia	F	France	I	Italy	Sw	Switzerland
B	Britain	Fi	Finland	J	Japan	U	USA
Be	Belgium	G	Germany	NZ	New Zealand		
C	Canada	H	Holland	S	Sweden		

ECONOMICS OF EMERGING BIOMASS THERMOCHEMICAL TECHNOLOGIES

Edward I. Wan, Ph.D.
Manager, Energy Technology and Information Division
Science Applications International Corporation
8200 Greensboro Drive
McLean, Virginia 22110

ABSTRACT

While some conventional fuels produced from biomass are continually needed in certain energy sectors, the recent decline in energy prices may have the effect of impeding adoption of biomass-derived fuels in the marketplace. Emerging and innovative thermochemical technology may present new avenues for converting biomass into high-value fuels and chemicals. This paper presents a technical and economic assessment of an innovative ethanol production process that can be integrated with biomass thermochemical gasification technology. The paper describes an integrated process design and economic analysis of two wood-to-ethanol process plants: (1) a single feedstock (wood) plant, and (2) a combined wood and natural gas feedstock plant. The combined feedstock plant is shown to have impressive economic advantages over traditional fermentation technology.

ECONOMICS OF EMERGING BIOMASS THERMOCHEMICAL TECHNOLOGIES

INTRODUCTION

In principle, any reasonably dry form of biomass can be thermochemically processed to produce high-value fuels and chemicals, but for practical consideration thermochemical processing techniques have focused on replacing low- and medium-Btu gas, heat and electrical energy, and boiler fuels derived from conventional fuel sources.

With the continuing decline in the price of conventional fuels, the long-term economic viability of biomass-derived fuels appears to be increasingly uncertain. Perhaps it is now necessary to re-orient thermochemical research to focus on re-evaluating opportunities for more diverse, high-value fuel forms. It is certainly an opportune time to explore a much broader spectrum of fuel forms and look for expanded options that may enable the thermochemical conversion of biomass to find a niche in the energy marketplace.

One opportunity for fuels from biomass that has long been ignored is the production of ethanol via thermochemical conversion. Ethanol is a logical choice because of its high-value use in gasoline as an octane enhancer. The Federal Government's decision to phase-out lead additives has boosted the demand for ethanol. Nonetheless, economic considerations will ultimately determine whether such fuel products will penetrate the marketplace. A technology to thermochemically convert biomass to ethanol is a viable option only if it has economic advantages.

To examine a potentially promising role for biomass thermochemical technologies for producing high-value fuel products, this paper presents a technical and economic assessment of an innovative process technology to convert biomass to ethanol via the thermochemical route. This technology is commercially available and the overall economics appear to be promising.

Overview of Thermochemical Processing Route

Biomass can be converted to various fuels and chemicals by thermochemical pyrolysis, liquefaction, and gasification technologies. Table 1 summarizes various potential thermochemical processing routes and alternative fuel products.[1]

TABLE 1. Potential Thermochemical Processing Routes to Produce Alternative Fuels From Biomass.

ocessing Route	Raw Fuel	Fuel Conversion/Upgrading	Fluid Fuel Product
eam Gasification ygen Gasification	Medium Btu gas/synthesis gas	Fischer Tropsch Synthesis Gasoline Synthesis Methanol Synthesis Ammonia Synthesis Ethanol Synthesis Methanation	Gasoline Methanol Ammonia Ethanol Methane
ld Pyrolysis st Pyrolysis	Ethylene, Fuel Oil, Aromatics, Olefins	Hydrotreating/Hydrocracking	Fuel Oil Diesel Fuel Jet Fuel Gasoline
quefaction lvolysis traction	Synthetic Liquids	Upgrading/Refining	Distillate Fuels Jet Fuel Diesel Fuel Gasoline

The pyrolysis process, depending on process severity, produces a spectrum of oxygenated liquid fuels and chemicals that can be further upgraded to a variety of fuel forms such as distilled oil, diesel, and jet fuels. More recent efforts have focused on controlling flash pyrolysis processes to maximize yields of ethylene and aromatics. Other efforts have been directed toward producing mixtures of heavy oil and char under mild pyrolysis conditions.

The direct liquefaction/solvolysis processes have received less attention because of the process complexity and the difficulty of controlling process yields. However, renewed interest exists in searching innovative liquefaction concepts operating under mild conditions. Similar to mild pyrolysis, the mild liquefaction process could become feasible for producing a variety of liquid transportation fuels.[2]

Thermal gasification and catalytic synthesis processes seem to offer the greatest potential for the production of diverse, high-value fuels. These processes can be combined in a variety of process configurations to produce hydrocarbon and alcohol fuels (i.e., gasoline, methane, methanol, ethanol). The advantage of this route is associated with high conversion efficiency and product selectivity. Table 2 summarizes the commercial synthesis process technology and its product forms.[3]

TABLE 2. Summary of Liquid Fuel Synthesis Technology.

Synthesis Technology	Chemical Reaction	Catalytic Condition	Development Status	Major Produ
Methanol Synthesis	$CO + 2H_2 = CH_3OH$	Cu/ZnO Catalyst 50-100 atm, 220-260°C	Commercial process by ICI, Lurgi	99% MeOH
	$CO_2 + 3H = CH_3OH + H_2O$	Topsoe Type MK-101 Catalyst 47-77 atm, 220-270°C	Commercial process by Haldor Topsoe	96% MeOH
Higher Alcohol Synthesis				
• Methanol Homologation	$CH_3OH + CO + 2H = CH_3CH_2OH + H_2O$	Co Complex Catalyst 200-300 atm, 175-225°C	Bench plant by PETC	50% EtOH, 20% HAc, 15% MeOAC, 15% Me_2O
		Co Rh Complex Catalyst 200 atm, 200°C	Comm. proc. by Halcon, Bench plant by Gulf	95% EtOH, 4% MeOH
• Ethanol Synthesis	$2CO + 4H = CH_3CH_2OH + H_2O$	Rh Complex Catalyst 60-120 atm, 250-270°C	Bench plant by IPF	35% EtOH, 20% MeOH, 20% PrOH, 15% BuOH
Gasoline Synthesis				
• F-T Synthesis	$nCO + (2n - 1)H_2 =$ paraffins, alcohols, oxygenated HC	Fe-Cu Catalyst, 10-30 atm, 230°C	Commercial process by Sasol	SNG, LPG, gasoline, d fuel, alco
• Methanol	$2\ CH_3OH \xrightarrow{-2H_2O}$ paraffins, aromatics	ZSM-5 Zeolite Catalyst	Commercial process by Mobil	90% gasoli LPG and fu
• Olefins ($C^=_3/C^=_4$)	$C_nH_{2n} \longrightarrow C_5 +$ gasoline/diesel	ZSM-5 Zeolite Catalyst	Pilot plant by Mobil	90% gasoli diesel & f

Ethanol Production via Thermochemical Route

Ethanol is traditionally produced by fermentation or ethylene dehydration. Because of the U.S. government incentive for its use in gasoline as an octane enhancer, efforts have intensified to explore new ways to produce ethanol (see Table 2).

One new process technology developed by the Halcon S. D. Group[4] involves a sequential catalytic process of converting methanol into ethanol via carbonylation, esterfication, and hydrogenolysis reactions, as follows:

$$CH_3OH + CO \longrightarrow CH_3COOH \text{ (carbonylation)}$$

$$CH_3COOH + CH_3OH \longrightarrow CH_3COOCH_3 + H_2O \text{ (esterfication)}$$

$$CH_3COOCH_3 + 2H_2 \longrightarrow C_2H_5OH + CH_3OH \text{ (hydrogenolysis)}$$

net reaction: $CH_3OH + CO + 2H_2 \longrightarrow C_2H_5OH + H_2O$

The carbonylation reaction is carried out in a liquid phase, catalytic reaction; the esterfication reaction is carried out in a novel tower reactor/distillation system; and the hydrogenolysis reaction is carried out in vapor phase to produce an equimolar mixture of methanol and ethanol. Methanol and ethanol are separated and the methanol is recycled to the esterfication reaction system.

The overall ethanol yield from methanol is approximately 95% of theoretical, and approximately 0.73 lbs. of methanol are consumed per lb. of ethanol product.

Wood-to-Ethanol Process Conceptual Design

The wood-to-ethanol production process involves four essential thermochemical processing steps:

1. Generation of synthesis gas via biomass gasification
2. Synthesis gas modification
3. Methanol synthesis
4. Ethanol synthesis

Each processing step considered for development of an integrated ethanol process plant is described below.

Generation of Synthesis Gas. The commercially available Winkler wood gasification system, employing an oxygen-blown fluidized-bed reactor, is used to produce a synthesis gas. Steam and oxygen are used as the gasification and fluidized media at rates of 0.46 and 0.18g/g wood, respectively. With the gasifier reactor operated at 980°F and 10 atmospheres, the product gas has a H_2/CO ratio of 1.12, and contains negligible amounts of methane and tar.

Synthesis Gas Modification. The raw synthesis gas needs to be modified for methanol and ethanol synthesis. These modifications include the following:

- Adjustment of the H_2/CO ratio via water shift conversion of CO to H_2
- Removal of excess acid gas, CO_2 and H_2S, using absorption in hot carbonate solutions and reaction with Fe_2O_3 and ZnO
- Removal of CH_4 and higher hydrocarbons and adjustment of the synthesis gas H_2/CO ratio using steam reforming
- Compression of synthesis gas to pressure levels favorable for methanol and ethanol synthesis

Methanol Synthesis. Methanol synthesis is carried out in a heterogeneous catalytic reactor operated at low pressure (50 atmospheres) under approximately isothermal conditions (260 to 270°C). The feed synthesis gas is characterized by a H_2/CO ratio greater than 2.0, a moderate level of CO_2 (2 to 10%), low levels of inerts (CH_4, N2), and extremely low levels of catalyst poisons (H_2S, O_2). The major pieces of equipment in the methanol synthesis area include a zinc oxide guard bed, a recycle gas compressor, and the methanol synthesis converter.

Ethanol Synthesis. Ethanol is produced from methanol and synthesis gas with an H_2/CO ratio of 2.0 in a sequential process involving carbonylation, esterfication, and hydrogenolysis, under moderate operating conditions (200 atm, 200°C). In the first step, methanol and CO react in the liquid phase in the presence of a homogeneous, non-noble metal catalyst system to form acetic acid. The acetic acid is then recovered and esterfied with recycle methanol to form methyl acetate. This reaction is carried out in a tower reactor system. Dry methyl acetate is recovered and sent directly to the hydrogenolysis step, where methyl acetate is reacted with hydrogen to form equimolar of methanol and ethanol. Methanol is recycled to esterfication and ethanol is separated as product.

Integrated Ethanol Process Configuration

Two alternative integrated ethanol process configurations were developed based on the four processing steps described above.

Process Configuration A—Winkler Wood Gasification. This ethanol process configuration uses Winkler wood gasification for synthesis gas production. Figure 1 is an integrated process block diagram of the sequence of major unit processes and operations as explained below.

Green wood chips containing 50% moisture are transported by truck to the wood receiving and storage area. The wood chips (1) are conveyed to the drying section, where the moisture is reduced to less than 10%. The dried wood (2) is ground to approximately 1/8" particles before entering the fluidized-bed gasifier. An air separation plant provides oxygen (12) to the gasifier. Two streams leave the gasifier: (3) combined purge gas (12) to provide energy to the boiler, and (4) the cleaned product gas. The cleaned gas is compressed and sent to a shift convertor. The H_2/CO ratio is increased to 2.0 after being processed in a shift converter (6). The excess CO_2 (7) is removed in an acid gas removal system. Approximately half of the adjusted synthesis gas is compressed and sent to the methanol synthesis loop (9). The other half of the synthesis gas enters a gas

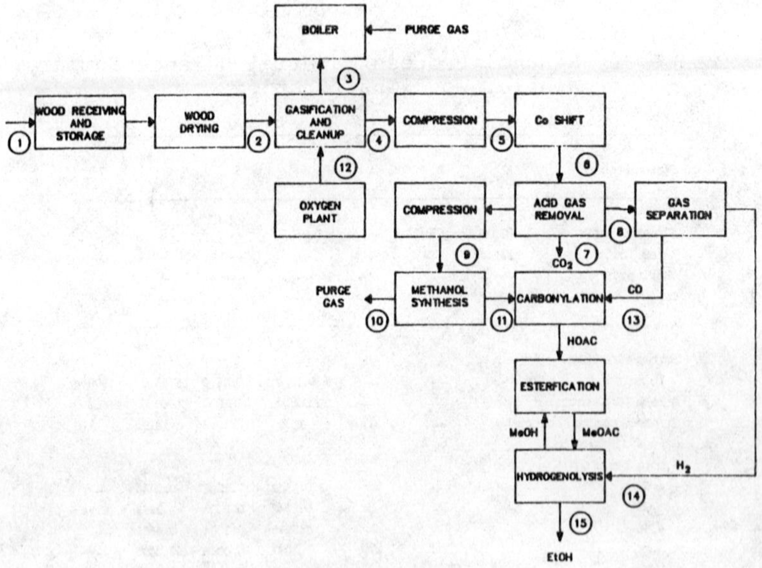

FIGURE 1. Process Block Diagram of Wood-to-Ethanol Process – Winkler Wood Gasification.

separation unit to separate H_2 from CO. The methanol fuel product (11) and CO stream (13) react in a carbonylation reactor to form acetic acid. The acetic acid is esterfied with recycle methanol to form methyl acetate. The methyl acetate is reacted with H_2 (14) to form methanol and ethanol. The methanol is recycled and ethanol is separated out as product (15).

Table 3 summarizes material balance data for the major flow streams based on 1000 TPD wood input. As indicated in the table, 280 TPD of ethanol is produced.

Process Configuration B—Winkler Wood Gasification and Natural Gas Steam Reforming. This ethanol process configuration employs co-feedstock, wood, and natural gas for the production of synthesis gas. The technical advantage of this combined feedstock is that steam reforming of natural gas produces a hydrogen-rich synthesis gas which can supplement a hydrogen-deficient synthesis gas produced from biomass gasification. The proper balance of these combined feedstocks not only can eliminate the shift conversion process step, but can reduce the acid gas removal requirement. Figure 2 is a process block diagram of the combined feedstock processing steps. As the figure shows, the various process steps are similar to that in Figure 1, except that a steam reforming process is added to replace shift conversion step. Table 4 summarizes the conceptual process plant mate-

TABLE 3. Ethanol Process Material Balance Summary: Winkler Wood Gasification.

Stream Number	1	2	3	4	5	6	7
Biomass, lb/hr	83333	67555	—	—	—	—	—
Water, lb/hr	83333	7506	—	—	—	—	—
Char, lb/hr	—	—	504	—	—	—	—
Oxygen, lb/hr	—	—	—	—	—	—	—
Stream Flow Rate, lb mol/hr	—	—	—	4624.0	5145.0	5145.0	1998.6
Stream Temperature, R	—	—	—	560.0	1010.0	660.0	610.0
Stream Pressure, atm	—	—	—	10.9	19.8	18.8	1.0
Stream Composition, mole fraction							
CO	—	—	—	0.3562	0.3202	0.2143	—
CO_2	—	—	—	0.2342	0.2105	0.3163	0.7471
H_2	—	—	—	0.3976	0.3574	0.4632	—
CH_4	—	—	—	0.0061	0.0055	0.0055	—
H_2O	—	—	—	0.0059	0.1064	0.0007	0.2529
N_2	—	—	—	—	—	—	—
CH_3OH	—	—	—	—	—	—	—
C_2H_5OH	—	—	—	—	—	—	—

TABLE 3. Ethanol Process Material Balance Summary: Winkler Wood Gasification (Continued).

Stream Number	8	9	10	11	12	13	14	15
Biomass, lb/hr	—	—	—	—	—	—	—	—
Water, lb/hr	—	—	—	—	—	—	—	—
Char, lb/hr	—	—	—	—	—	—	—	—
Oxygen, lb/hr	—	—	—	—	30911.0	—	—	—
Stream Flow Rate, lb mol/hr	1126.7	1830.0	76.4	534.6	—	375.6	751.1	534.6
Stream Temperature, R	560.0	560.0	560.0	540.0	—	850.0	850.0	850.0
Stream Pressure, atm	1.0	50.0	48.6	—	—	200.0	200.0	200.0
Stream Composition, mole fraction								
CO	0.3015	0.3015	0.2349	—	—	1.0	—	—
CO_2	0.0384	0.0384	0.0962	—	—	—	—	—
H_2	0.6511	0.6511	0.4905	—	—	—	—	—
CH_4	0.0077	0.0077	0.1732	—	—	—	—	—
H_2O	0.0013	0.0013	—	0.015	—	—	—	—
N_2	—	—	—	—	—	—	—	—
CH_3OH	—	—	0.0052	0.985	—	—	—	0.05
C_2H_5OH	—	—	—	—	—	—	—	0.95

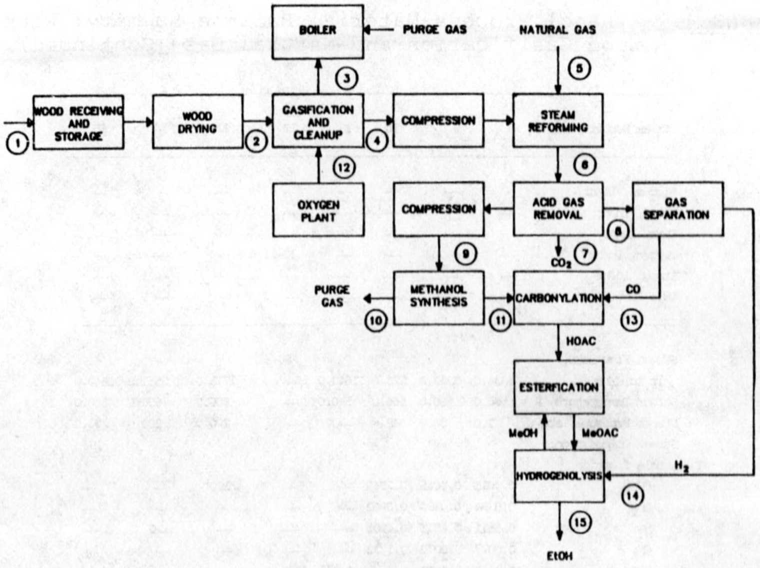

FIGURE 2. Process Block Diagram of Wood-to-Ethanol Process – Winkler Wood Gasification and Natural Gas Reforming.

TABLE 4. Ethanol Process Material Balance Summary: Winkler Wood Gasification and Natural Gas.

Stream Number	1	2	3	4	5	6	7
Biomass, lb/hr	83333.0	67555.0	—	—	—	—	—
Water, lb/hr	83333.0	7506.0	—	—	—	—	—
Char, lb/hr	—	—	504.0	—	—	—	—
Oxygen, lb/hr	—	—	—	—	—	—	—
Steam, lb/hr	—	—	—	—	—	—	—
CH_4, lb/hr	—	—	—	—	29562	—	—
Stream Flow Rate, lb mole/hr	—	—	—	4624.0	—	12014.5	1388.6
Stream Temperature, R	—	—	—	560.0	—	660.0	610.0
Stream Pressure, atm	—	—	—	10.9	—	18.8	1.0
Stream Composition, mole fraction							
CO	—	—	—	0.3562	—	0.2898	—
CO_2	—	—	—	0.2342	—	0.0922	0.7471
H_2	—	—	—	0.3976	—	0.6053	—
CH_4	—	—	—	0.0061	1.00	0.010	—
H_2O	—	—	—	0.0059	—	0.0023	0.2529
N_2	—	—	—	—	—	—	—
CH_3OH	—	—	—	—	—	—	—
C_2H_5OH	—	—	—	—	—	—	—

TABLE 4. Ethanol Process Material Balance Summary: Winkler Wood Gasification and Natural Gas (Continued).

Stream Number	8	9	10	11	12	13	14	15
Biomass, lb/hr	—	—	—	—	—	—	—	—
Water, lb/hr	—	—	—	—	—	—	—	—
Char, lb/hr	—	—	—	—	—	—	—	—
Oxygen, lb/hr	—	—	—	—	30911.0	—	—	—
Steam, lb/hr	—	—	—	—	—	—	—	—
CH_4, lb/hr	—	—	—	—	—	—	—	—
Stream Flow Rate, lb mole/hr	5384.0	5242.0	223.0	1640.0	—	1775.0	3551.0	1640.0
Stream Temperature, R	560.0	560.0	560.0	540.0	—	850.0	850.0	850.0
Stream Pressure, atm	1.0	50.0	48.6	—	—	200.0	200.0	200.0
Stream Composition, mole fraction								
CO	0.3015	0.3015	0.2349	—	—	1.0	—	—
CO_2	0.0384	0.0384	0.0962	—	—	—	—	—
H_2	0.6511	0.6511	0.4905	—	—	—	1.0	—
CH_4	0.0077	0.0077	0.1732	—	—	—	—	—
H_2O	0.0013	0.0013	—	0.015	—	—	—	—
N_2	—	—	—	—	—	—	—	—
CH_3OH	—	—	0.0052	0.985	—	—	—	0.05
C_2H_5OH	—	—	—	—	—	—	—	0.95

rial balance for the ethanol process based on 1000 TPD of wood input and 355 TPD of natural gas input. The ethanol product yield is approximately 860 TPD.

Ethanol Plant Capital and Operating Cost Estimates

The plant investment costs for the two conceptual process configurations described in the previous section are summarized in Table 5. These costs are in 1986 $ and are based on wood input capacity of 1000 TPD. The total plant capital costs are broken down into battery limit capital costs (BLCC), offsite costs, and contingency and engineering costs. The offsite costs are those associated with utilities, site development, and waste water treatment.

The total plant capital cost is estimated to be $146MM and $275MM for the two wood-to-ethanol process configurations, respectively. The total capital investment requirements consist of the plant capital, interest during construction, start up costs, and working capital. The total capital requirement adds up to $175MM and $329MM, respectively. The corresponding capital investment per unit ethanol capacity varies from $5.95 to $3.64 per gallon per year.

TABLE 5. Comparison of Capital Investment Costs of Wood-to-Ethanol Process.

Process Type	Winkler Gasification	Winkler Gasification & Natural Gas Reforming
Total wood feed, TPD	1000	1000
Ethanol production, TPD	280	860
Natural gas feed, TPD	—	355
Battery Limit Capital Cost (BLCC), MM$		
Feed handling and storage	4.49	4.49
Oxygen plant	9.61	9.61
Feed drying and combustion	6.51	6.51
Gasification and cleanup	17.65	17.65
Shift conversion	0.47	—
Acid gas removal	6.12	8.83
Steam reforming	—	29.10
Methanol synthesis	12.57	25.54
Gas compression	7.43	14.26
Carbonylation	13.39	26.31
Esterfication	21.43	42.09
Hydrogenolysis	18.75	36.83
BLCC total	118.42	220.22
Offsite costs	10.21	18.98
Contingency and engineering	17.76	35.90
Total plant capital	146.39	275.10
Interest during construction	17.91	33.66
Startup costs	3.47	6.52
Working capital	7.32	13.78
Total capital requirements	175.09	329.06
$/gal/yr	5.95	3.64

Table 6 summarizes the various operating and maintenance (O&M) costs for the two wood-to-ethanol process plants. These costs are broken down into raw material, utility, and labor-and capital-related costs. The annual O&M costs for the two conceptual wood-to-ethanol plants range from $22.17 to $34.47 MM, respectively. As the table shows, raw material and capital-related costs account for 80% of the total O&M costs.

Ethanol Product Cost Analysis

Table 7 summarizes the total ethanol product cost for two 1000 TPD wood input capacity plants based on the estimated capital and annual O&M costs. The ethanol production costs are calculated for both public utility financing and private investor financing. The economic factors and assumptions used in these calculations are summarized in Table 8. Total ethanol production costs are broken down into three categories: annual O&M excluding feedstock costs, capital charge (including return on investment), and feedstock costs.

TABLE 6. Operating and Maintenance Cost Summary for Wood-to-Ethanol Process.

Process Type	Winkler Gasification	Winkler Gasification & Natural Gas Reforming
Ethanol production, TPD	280	860
Raw material		
Wood, @ 25 $/ton	8.21	8.21
Natural gas, @ 0.5 $/MMBtu	—	2.61
Catalyst and chemical	0.35	0.57
Purchase utility (MM$)		
Power	1.39	2.61
Process and cooling water	0.02	0.03
Labor-related (MM$)		
Operator	1.13	1.13
Supervisor	0.29	0.29
Labor overhead	1.42	1.42
Capital-related (MM$)		
Maintenance	5.85	11.00
General plant overhead	3.51	6.60
Total annual O&M costs (MM$)	22.17	34.47

TABLE 7. Summary of Wood-to-Ethanol Production Costs.

Process Technology	Winkler Gasification	Winkler Gasification & Natural Gas Reforming
Wood input, TPD	1000	1000
Natural gas input, TPD	—	355
Cap. charge, utility financing, MM$	21.01	39.48
Cap. charge, private financing, MM$	26.96	50.67
Annual O&M costs, exc. feedstock, MM$	13.96	23.65
Annual feedstock cost, MM$		
Wood @ $25/ton	8.21	8.21
Natural gas @ $0.5/MMBtu	—	2.61
Total annual production cost, MM$		
Utility financing	43.18	73.95
Private financing	49.13	85.14
Ethanol production cost, $/gal		
Utility financing	1.47	0.82
Private financing	1.67	0.94

TABLE 8. Ownership Economic Factors and Assumptions.

Factor	Public Utility	Private Industry
State Taxes	2% of Gross Revenue	Combined 50%
Federal Taxes	N.A.	Net
Investment Tax Credit (on Applicable Items)	0	10
Depreciation Type	N.A.	ACRS
Depreciation Period	N.A.	15 years
Long-Term Debt Interest	10.0	10.0
Amortization Period	15 years	15 years
Debt/Equity Ratio	100/0	30/70
After-Tax Return on Equity	0	15%

For utility financing, total ethanol production costs, at a wood cost of $25/ton, range from $0.82 to $1.47/gallon. For private investor financing, the comparable production costs are $0.94 to $1.67/gallon. These results show the significant economic advantages of using combined feedstocks of wood and natural gas for ethanol production. Figure 3 shows the distribution of the total ethanol production costs for both utility financing and private financing.

Since ethanol production costs vary with respect to feedstock cost, capital cost and other economic parameters, several product sensitivity analyses were made. The findings are as follows:

- Ethanol product costs decline 12% when the cost of wood feedstock decreases from $25 to $10 per ton for the wood gasification process configuration.
- Ethanol product costs increase 9% when the cost of natural gas increases from $0.50 to $1.00 per MMBtu for the combined wood gasification/natural gas process configuration.
- For the wood gasification process configuration, ethanol product costs rise 25% when the plant capital investment costs increase by 150%.

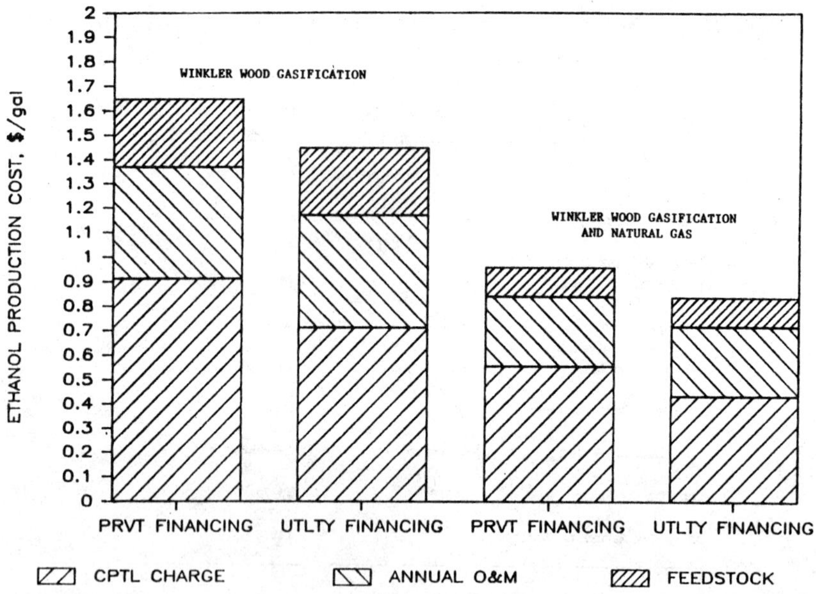

FIGURE 3. Wood-to-Ethanol Production Costs.

The above results indicate that ethanol production costs are highly sensitive to capital investment, but are less sensitive to feedstock costs. This is especially true for the combined wood/natural gas process plant configuration.

CONCLUSIONS AND RECOMMENDATIONS

Technical and economic evaluation of thermochemical wood-to-ethanol technology reveals the following:

1. The production of ethanol via wood gasification and thermochemical ethanol synthesis, using commercially available technology, is a viable alternative to the production of ethanol through fermentation technology.

2. Ethanol production technology using combined feedstocks has a significant economic advantage over the single biomass feedstock process. This is due primarily to the synergistic effect of combining hydrogen-rich gas from natural gas with hydrogen-deficient gas from biomass gasification. For utility financing, ethanol production costs are as low as $0.82/gallon for a 1000 TPD wood capacity plant.

3. Future efforts should focus on optimizing gasification and catalytic processes suitable for processing multiple feedstocks to produce ethanol or hydrocarbon fuels.

4. Research and development activities in thermochemical conversion should be re-oriented to focus on production of high-value fuel products.

REFERENCES CITED

1. Wan, E.I. et. al., "Economic Evaluation of Indirect Biomass Liquefaction Processes for Production of Methanol and Gasoline," Proceedings, IGT Energy from Biomass and Wastes VI (1982) February.

2. Tabak, S.A. and Krambeck, F.J., "Shaping Process Makes Fuels," Hydrocarbon Processing (1985) September.

3. Wan, E.I., Price, J.D., and Simmons, J.A., "Methanol Production from Indigenous Resources in New York State," Volume III: Conversion Process, New York, NYS ERDA, 1983.

4. Juran, B. and Porcelli, R.V., "Convert Methanol to Ethanol," Hydrocarbon Processing (1985) October.

DISCUSSION

W. Klausmeier, Sylvatex Corp.	Have you looked at the case of just buying the methanol?
Wan	Yes, there's a reference in the paper. Halcon Scientific Development has done an economic analysis using methanol at 35¢/gal as a starting feedstock. Although their plant is about twice as large as the ethanol capacity I'm quoting, their cost is about 80¢/gal based on 35¢/gal of methanol feedstock cost.
Klausmeier	I looked at a number of related processes in making acetates, and it turns out that acetate esters are reasonably good octane enhancers and fuels. If you avoid the hydrogenation step, and you buy your methanol, then

you can make an acetate ester using only the carbon monoxide from the biomass. If you only need carbon monoxide from biomass, you can use an adsorption technology called COSORB to pull the carbon monoxide out. You can just use air gasification to avoid the oxygen plant. Have you considered that?

Wan

I think they do produce the acetic acid first and then produce methyl acetate, so if your product slate is correct, I think that could be a similar kind of technology. But rather than producing ethanol, you'll produce another fuel product. I think the point is that there may be other kinds of fuel products, especially high-value fuel products that we have ignored in the past. We're talking most of the time about dollars per million Btu, regardless of what kind of fuel you produce. I think it's now perhaps time to focus on certain specific types of fuel, especially high-value fuels.

J. Kerstetter, Washington State Energy Office

Do you know what the cost would be if you converted 35¢ methanol to gasoline using the Mobil Process?

Wan

Not off hand. We looked at the methanol-to-gasoline production cost, but that was based on something like $1/gal of methanol. At that price, our estimate at the time was about $1.80. I don't know whether there is a direct relation if we started with 35¢ methanol. But again, many of the new processes are still in the final development stage. I think New Zealand has been pioneering in terms of using the Mobil technology. I think there are other technologies that perhaps we ought to look at in the future too.

FERMENTATION OF LIGNOCELLULOSIC FEEDSTOCKS: PRODUCT MARKETS AND VALUES

Linda L. Gaines
Energy and Environmental Systems Division
Argonne National Laboratory
9700 South Cass Avenue
Argonne, Illinois 60439

Michael Karpuk
Solar Energy Research Institute
1617 Cole Boulevard
Golden, Colorado 80401

ABSTRACT

Ethanol is the primary product of biomass hydrolysis and fermentation processes. The quantity and quality of coproducts depend on the feedstock and on the process steps employed. When the feedstock is lignocellulosic -- wood or herbaceous energy crops, such as napier grass or kenaf -- substantial quantities of lignin and five-carbon sugars and derivatives are produced, and the sale of these and other coproducts has a potentially large impact on the economics of fuel alcohol production. For each product, potential markets and materials competing for these markets (or competing routes to the same product) are identified. The value of ethanol as a chemical, octane enhancer, and fuel is estimated. Coproduct credits are calculated, and the potential benefits resulting from coproduct recovery are estimated. The quantity of each material that could be produced by fermentation is compared with the size of potential markets. Where coproducts of substantial fuel ethanol manufacture could saturate their markets, the question of whether early sales could serve as a wedge to ethanol market penetration is addressed.

FERMENTATION OF LIGNOCELLULOSIC FEEDSTOCKS: PRODUCT MARKETS AND VALUES

INTRODUCTION

Wood and other lignocellulosic feedstocks are composed primarily of cellulose, hemicellulose, and lignin. Cellulose, the major constituent, comprises approximately 43-50% of dry wood by weight. A high-molecular-weight linear polymer, when hydrolyzed it yields the simple sugar glucose as the major product. It is from this six-carbon sugar that ethanol is produced by fermentation.

Hemicellulose is intimately associated with cellulose and is also a sugar polymer. However, hemicellulose yields both five- and six-carbon sugars on hydrolysis, and the relative amounts of these sugars vary with species. Hardwoods contain an average of 20-30% hemicellulose, with xylose (a pentose or five-carbon sugar) as the major sugar. Softwoods contain an average of 15-20% hemicellulose, with mannose (a six-carbon sugar) as the main sugar (1).

Lignin makes up 23-33% of softwoods but only 16-25% of hardwoods. Like cellulose, lignin has a macromolecular chemical structure, but its three-dimensional network is far more complex. It can be broken down into a complex mixture of high-molecular-weight phenolic compounds.

Other biomass feedstocks will have the same three polymers as major components but their proportions will be different. Herbaceous crops may have cellulosic contents in the 25-42% range while the hemicellulosic content is 20-40%. Lower lignin yields but high protein contents are associated with these feedstocks. Agricultural residues have about 40% cellulose, 30% hemicellulose, and 15-20% lignin.

The hemicellulose and lignin must be profitably utilized to achieve satisfactory production economics. Ethanol production from lignocellulosic feedstocks will utilize only the cellulose fraction of the feedstock. The production of ethanol from lignocellulosics is similar to petroleum refining in that both use a feedstock with many components. A typical gasoline yield from crude oil is 50%. The production and sale of coproducts such as jet fuel, lubricants, and residual oil are crucial to refining economics. Figure 1 shows the types of products that can be derived from lignocellulosic feedstocks and that are discussed in this paper.

ETHANOL MARKETS AND VALUES

Ethanol now supplies large markets as an industrial solvent and as an automotive fuel extender and octane enhancer. The potentially largest market for ethanol is as a neat automotive fuel; this use could easily consume all the ethanol likely ever to be produced from biomass. The value of ethanol in each market is estimated as the cost to the user of the conventional substitute derived from oil or natural gas.

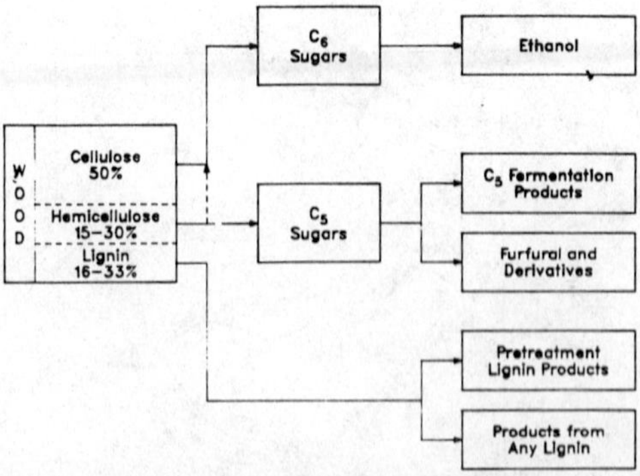

Figure 1. Wood Components and Products

Chemical Market

Industrial alcohol was originally produced from natural sources of carbohydrates by fermentation of molasses, sulfite liquors, and grain. When petroleum and natural gas became available as cheap energy sources around 1950, synthetic ethanol (produced by hydration of ethylene) rapidly replaced fermentation ethanol. The industrial market is now supplied almost entirely by synthetic ethanol, but it could return to fermentation ethanol, if that were the cheaper source, with no user equipment modifications.

Industrial consumption of ethanol in the U.S. in 1984 was 1.1×10^9 lb, or 170×10^6 gal (2). Industrial production peaked at about twice that level around 1970 and has been falling slowly since then (see Fig. 2). Solvents account for almost 60% of industrial ethanol consumption, with the largest solvent use in cosmetics, followed by coatings, cleaning preparations, and pharmaceuticals. Principally because of the steep decline in use of ethanol to manufacture acetaldehyde, a product used to produce acetic acid, peracetic acid, pyridine, and other chemicals, use of ethanol as a raw material for chemical manufacture has been declining. Ethanol use is increasing in the manufacture of ethanol-based vinegar (chiefly in the convenience food industry) and in a miscellaneous category of ethanol-based chemicals that includes ethyl acetate, ethyl acrylate, and glycol ethers. Current chemical uses lack the potential to utilize the product of a large fermentation alcohol industry. However, the price is high, currently about \$1.80/gal (3), and will rise with oil and gas prices. Chemical sales could serve as an early commercial foothold for an emerging industry before its product can compete without subsidies as a fuel.

The most significant chemical potential for biomass ethanol lies in its dehydration to ethylene. This process for ethylene production was common in the U.S. until the early 1950s, when large-volume hydrocarbon cracking became more economical. United States ethylene consumption has risen from 5×10^9 lb/yr in 1960 to 31×10^9 lb/yr in 1984 (4). Most of this growth comes from the ethylene-based major plastics--low- and high-density polyethylene, polyvinyl chloride (PVC), and styrenics--which together account for over 60% of total ethylene demand. Production from ethanol of all ethylene consumed in the U.S. in 1984 would have required over 8×10^9 gal of ethanol, almost 50 times the total 1984 industrial ethanol consumption.

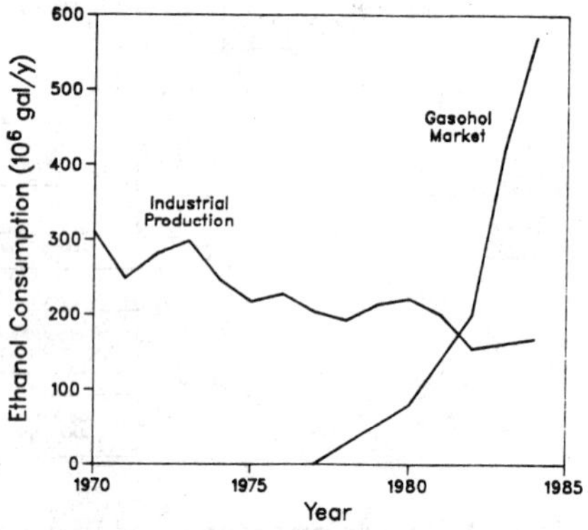

Figure 2. U.S. Ethanol Consumption

Gasoline Octane Enhancer Market

Ethanol-gasoline blends accounted for 5.4% of U.S. gasoline sales in 198 (6). The ethanol content, in a 10%-by-volume mixture, was 570 x 10^6 gal. Thi ethanol served mainly as a fuel volume extender in states where subsidies provid especially high incentives to use ethanol produced from corn. However, ethanc also increases the octane number of the gasoline blend stock into which it i mixed, so that buyers of gasoline extended with ethanol received the bonus o higher octane. Octane enhancement has become increasingly important since th decision by the U.S. Environmental Protection Agency (EPA) to reduce the allow able levels of lead in gasoline from the maximum of 1.1 gram per leaded gallo (gplg) to 0.5 gplg on July 1, 1985, and then to 0.1 gplg by January 1, 1986. The us of lead may be banned entirely by 1988. The 1.1-gplg limit raised the octane o raw gasoline by about 6 octane numbers, and refiners must now supply this adde octane in other ways. Options include more intensive processing of gasoline i refineries, addition of octane enhancers by refiners or blenders, imports of high octane gasoline components or finished gasoline, and changes in refinery produc slates.

The upper bound for ethanol use as an octane enhancer is 10% of the tota gasoline volume (about 100 x 10^9 gal in 1984; Ref. 7). This value is expected t drop slightly and level off at about 90 x 10^9 gal/yr (or 6 x 10^6 bbl/d) by the yea 2000. If 10% of total gasoline demand were satisfied with ethanol, 9 x 10^9 gal/y would be required, about 16 times current consumption. However, this bound i far above the quantity refiners and blenders could be expected to purchase Average reformer utilization in 1984 was about 75% (8), implying the capability t produce another 20 x 10^6 octane-number barrels per day (ON-bbl/d) withou increasing severity. Incremental octane demand caused by lead phase-down an premium unleaded demand growth is projected to be only about 8 x 10^6 ON-bbl/ by 2000. Small and low-complexity refiners will, however, serve as a market fo about 0.4 x 10^6 ON-bbl/d or 2 x 10^9 gal/yr (9). The energy content (higher heatin value) of this quantity of ethanol is about 0.15 Q (1 Q = 10^{15} Btu), and the mate rial input required annually is over 40 x 10^6 tons, equivalent to 15% of all unco lected crop residues.

Additives. Octane enhancers for gasoline can be classified into two categories: organometallic compounds (antiknocks) that are added in small quantities to raw gasoline to raise its octane rating; and organic compounds, themselves high-octane fuels, that are mixed in volume percentages up to about 15 with low-octane blend stocks. The first category includes tetraethyl lead (TEL) and manganese-based compounds. The second category includes ethanol and other oxygenates, along with aromatic hydrocarbons (e.g., toluene).

Organometallic Compounds. Organometallic octane boosters are generally less expensive to use than organic compounds, because small quantities produce substantial response. In addition, antiknocks can be added easily to adjust the octane rating of a batch of gasoline. The only commercially available organic manganese compound is methylcyclopentadienyl manganese tricarbonyl (MMT), currently allowable only in leaded fuels. Its use in unleaded gasoline was banned (except with a waiver from the EPA) by the 1977 Clean Air Act Amendments. Even if MMT becomes available, the EPA is likely to allow so little that supplemental sources of octane will be needed.

Organic Blending Agents. Several physical properties influence a material's utility as an octane enhancer. The most important properties, from the refiner's viewpoint, are octane number [usually expressed as the average of motor and research, (R+M)/2] and Reid vapor pressure (RVP), which indicates the vapor lock tendencies of motor gasoline, as well as explosion and evaporation hazards. From the user's point of view, heat of combustion (as a surrogate for miles per gallon) is also important. Table 1 lists these properties for several commercially available organic octane enhancers, together with EPA limits on their use in gasoline, current production volumes, and prices (when available) (2-5,7,10). Each octane enhancer also has advantages and disadvantages arising from physical properties, availability, and institutional factors.

Table 1. Blending-Agent Characteristics

Agent	EPA Limit (vol. %)	Blend (R+M)/2[a]	RVP Increase (psi)[a]	Heating Value (10^3 Btu/gal)	Price (¢/gal)	1984 Production (10^9 gal)
Methanol	5[b]	116	3	56	37-39[c]	1.25[d]
Ethanol	10	113	0.7	76	180[e,f]	0.17[g]
					121-145[c,h]	0.4
MTBE	11	106	<0.1	94	72-73[c]	0.24[d]
TBA	16	101	0.3	99	115-137[f] 86-88[f] (spot)	—
Toluene		106	<0	132	65-70[c]	0.7[d]-1.6
Unleaded regular gasoline	N.A.	87	0	115	52[c]	102.6[j]

[a]Ref. 10.
[b]With at least 2.5% cosolvent.
[c]Ref. 5.
[d]Ref. 4.
[e]Synthetic, 190 proof.
[f]Ref. 3.
[g]Ref. 2.
[h]Fermentation.
[i]No limit.
[j]Ref. 7.

Alcohols. Several characteristics are shared by alcohols because of their chemical similarity. First, they are polar molecules and soluble in water. If water is present in an alcohol-gasoline blend, a separate alcohol-water phase may settle out, causing problems in distribution and fuel delivery. Therefore, a clean, dry system for handling all alcohol-gasoline blends is necessary.

Current gasoline storage and distribution facilities are not kept sufficient dry to handle alcohol blends, and pipeline systems generally will not transpo them. It is possible to modify the existing systems to tolerate alcohols, b the alternative often used is to blend alcohols with suboctane gasoline local terminals. Second, again because of their polar nature, alcohols c attack metals sometimes used in fuel tanks and can degrade rubber a plastic components. The problems can be overcome by additives and by usi materials not affected by alcohols. The lightest alcohols are very volati (volatility increases as molecular weight decreases) and produce evaporati emissions. If volatile materials are to be blended into gasoline, the refin must back out butanes, which have high octane and vapor pressure coupl with low price, from the blend stock. These disadvantages of alcohols a most serious for methanol, becoming less serious with increasing molecul weight. However, octane number decreases as molecular weight increas (faster for the straight chain compounds), and tertiary butyl alcohol (TBA) the highest alcohol useful as an octane enhancer.

Methyl Tertiary Butyl Ether (MTBE). The maximum MTBE producible fro domestic feedstocks is about 10^9 gal/yr, insufficient to fill the entire octan enhancer market. Most domestic and European MTBE production capacity owned by refiners for their own use, so little MTBE is or will be sold in th merchant market for purchase by octane-deficient small refiners, eve though MTBE has several advantages over alcohols. It has a higher heatin value and a lower vapor pressure, and it is insoluble in water. Because of th insolubility, blends containing MTBE are produced in refineries and are gene rally accepted for pipeline transport.

Toluene. Toluene is an aromatic organic compound produced in refineries b catalytic reforming of naphtha. Production can only be increased at th expense of other high-octane gasoline components. It has a high heat conten and octane number, and low vapor pressure, so over 90% of toluene produce in the U.S. is already used as a gasoline blending component. Toluene shoul not be regarded as an additional source of octane, but some toluene is avai able for sale to octane-deficient refiners.

Octane Enhancement by Refinery Processing. Petroleum refiners can adjus to lead phaseout by changing the crude oil input and the output product slate c each refinery. On the average, the U.S. refining industry has the processin capacity and flexibility to supply its own octane needs economically, at least unt well into the 1990s. It is those refiners who are not in average situations that wi provide the largest market for octane enhancers. Small, low-complexity refi neries depend on purchased octane and generally do not have the equipment t raise their product octane sufficiently through processing. Many such refinerie have already closed or will do so as lead is phased out. Those that survive wi blend gasoline with octane enhancers, unless they can acquire sufficient capital t purchase reformers. Some high-complexity refineries facing unusually high gasc line demands might also elect to purchase octane enhancers, because enhancer could be cheaper than the marginal cost of running equipment at high severit (higher temperature and pressure).

Differences in crude oil input also influence a refiner's need to purchas octane enhancers. A refiner constrained to use crude oil low in aromatic com pounds might need to purchase, while another could shift to a highly aromati crude feedstock and produce enough high octane gasoline with no enhancer purchases. Finally, not all refineries must produce the average product slate; those without the capability to produce sufficient high-octane gasoline can sell a lower-valued product slate. That decision, of course, depends on whether the cost of octane enhancers is greater or less than the increased product slate value they allow. Trade among refineries and interregional shipments could smooth some of the differences discussed here and lower the demand for octane enhancers.

U.S. petroleum refineries have been running well below capacity for several years. Most equipment used to produce high-octane gasoline is not being fully utilized. This situation gives refiners more flexibility to deal with lead phaseout than they would have in periods of strong product demand. Process modifications installed to supply octane needs will continue to be adequate, except for additional requirements brought about by a shift in demand from (unleaded) regular to premium gasoline.

Catalytic reforming is the dominant process used to raise the octane number of refinery streams that are already in the appropriate boiling range for gasoline. More high-octane gasoline can be obtained from an existing unit by running at higher throughput. Alternatively, the reformer can be run at higher severity to produce a higher octane product. However, product yield decreases with severity, and costs rise; at some point, it becomes more economical to expand the capacity of existing units or construct new units. Isomerization can also be employed to increase the octane number of C_4-C_7 material.

Several refinery processes produce more material in the gasoline range, either by breaking down larger molecules (catalytic cracking, hydrocracking coking) or by combining smaller molecules (alkylation, polymerization). Products from these processes are used directly in the gasoline pool, if their octane ratings are high enough, or else as reformer feed. Improvements in refinery processes enable refiners to remain independent and should not be overlooked when examining the options ethanol must compete against in the octane enhancer market. Generally, oil refiners prefer selling oil to buying and selling additives.

Value of Ethanol as an Octane Enhancer. The benchmark for estimating ethanol's value as an octane enhancer is the cost to refiners of supplying the required high-octane gasoline components by refinery processing. Care must be taken that equivalent products are compared. The conventional product is one gallon of unleaded gasoline that meets the appropriate octane-rating and vapor-pressure specifications for a given product grade, location, and season. For this analysis, we assume that an equivalent ethanol-gasoline blend must meet the same specifications (this meeting of specifications is not now required by law) and delivers the same mileage.

The blend stock to be mixed with ethanol for production of an equivalent final product differs from finished unleaded gasoline. It is "suboctane" gasoline, which means that it has lower octane than the final product. Less costly octane-increasing refining processes have been performed on the blend stock. In addition, its vapor pressure must be somewhat below that of the final product in order to compensate for the increase in vapor pressure caused by ethanol addition. Vapor pressure is lowered by the backout of butanes (i.e., less of these inexpensive high-vapor-pressure, high-octane components can be added for blending with ethanol than can be used in conventional unleaded gasoline). The cost of this butane backout partially offsets the credit due to less costly processing.

We have used refinery octane cost estimates from Refs. 8 and 9 to calculate a range of values for ethanol. High-complexity refiners with sufficient flexibility to supply their own octane have the lowest costs, while refiners who must run equipment at high severity or install additional reforming capacity have the highest costs. The cost of reforming is dominated by the fuel consumption; therefore, the cost of additional octane is roughly proportional to the price of crude oil. The addition of 10% ethanol raises the octane number of 84-octane blend stock to 87 and saves about $0.02 in processing cost per gallon of blend or $0.20/gal of ethanol. To compensate for a rise in vapor pressure of about 0.7 psi, 0.014 gal butane must be removed from each gallon of blend stock and replaced by heavier material, at a cost of about $0.02/gal ethanol.

The equivalent product to one gallon of unleaded gasoline is ethanol-gasoline blend in sufficient volume to supply equivalent mileage. Equal conversion efficiency (mi/Btu) is assumed as a base case for gasoline and blend, and the equivalent volume is 1.034 gal [1/(0.9 + 0.1 x 0.67)]. Because of the importance of this Btu adjustment factor, a sensitivity analysis was performed for a range of efficiencies. As an upper bound, the blend was assumed to deliver the same mileage per gallon as gasoline (Btu factor = 1); this would mean that the ethanol was effectively delivering 50% more miles per Btu than the gasoline. Current industry practice of pricing fuels on a per-gallon basis is consistent with this upper-bound case. However, although the Btu factor makes a considerable difference in the value of the ethanol (about $0.25/gal now), the blend's value is only affected by 10% as much. Consumers generally cannot detect any small reduction in mileage that blends deliver, so they do not demand that blends be sold at lower prices than straight gasoline.

Ethanol values calculated using the outlined procedure are shown in Fig. 3 for unleaded gasoline prices ranging from $0.30 to $1.80 at the refinery gate (crude oil priced at $10-$60/bbl) and for a range of efficiencies. Ethanol's Btu value (its value as a fuel) is shown, along with its value as an octane enhancer, as a function of avoided processing cost. A refiner who avoids costly processing can afford to pay about 25% more for ethanol than one facing average costs. About a 25% variation from the base case is observed with conversion efficiency. Conclusive mileage tests are required if the value of ethanol is to be known better. At any conversion efficiency, ethanol's value as an octane enhancer exceeds its effective Btu value by 10%-40% of the price of gasoline.

Neat Fuel Market

Pure ethanol could be used directly as an automotive fuel to displace gasoline. This market could easily consume all the ethanol produced from biomass. Approximately 150 x 10^9 gal (over 200 times current capacity) would be required to satisfy total demand. The energy consumed annually in the U.S. in the form of gasoline is about 11 Q. Approximately 3 x 10^9 dry tons of lignocellulosic feedstock would be required annually to replace this gasoline; this is 20 times the material processed by the pulp and paper industry and 10 times the unused crop residues. Neat ethanol use in existing automobiles requires only minor modifications (11). Emissions are generally similar to or less than gasoline's; no serious objections to ethanol use are expected on the basis of air pollution. Distribution of neat ethanol would require a reasonably dry system to prevent dilution, but phase separation (which could occur with ethanol-gasoline blends) is not possible. With little modification, ethanol could be transported and stored in existing, or in substantially similar, systems.

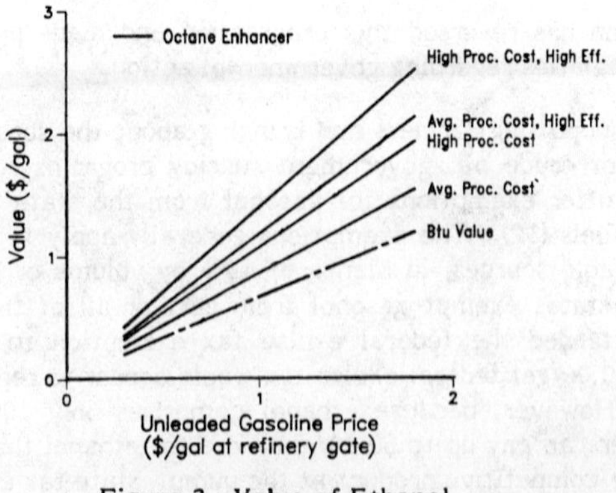

Figure 3. Value of Ethanol

Several substitutes for gasoline exist besides ethanol. An additional complicating factor is that gasoline prices are falling and may continue to fall in the near future. In the long term, however, gasoline prices are expected to rise, opening the market for substitute fuels. Alternative liquid fuels can use existing vehicles and infrastructure and will experience the fewest barriers to market entry. These liquids include methanol from biomass, natural gas, or coal; and synthetic oils from coal or oil shale. The synthetic oils could be processed, transported, and stored in existing facilities, either blended with petroleum products or alone, but development is proceeding very slowly because of low oil prices and the high costs of the large-scale plants required. Methanol, which can now be made cheaply in large quantities from natural gas, may be derived economically from biomass in the future. However, evaporative emissions from methanol are toxic, it is more corrosive than ethanol, and it has only half the heating value of gasoline. On the other hand, dissociated methanol engines could increase the efficiency to give the same fuel economy as ethanol.

Gases can also be used as automotive fuels. Methane (either natural gas or synthetic gas made from coal) can be stored and used as compressed natural gas (CNG) or liquefied natural gas (LNG). Proponents of hydrogen fuel foresee its manufacture in the future by water electrolysis using cheap nuclear or hydroelectric power. Another long-term option is the electric vehicle (EV), either with batteries on board or (further in the future) running on a rail.

In the automotive fuel market, as in the octane enhancer market, the service delivered is mileage, which is roughly proportional to energy content. To a first approximation, then, ethanol should be valued in this market for energy content, on an equal basis with gasoline (in $/10^6 Btu). Ethanol's heating value is only two-thirds that of gasoline, so its per-gallon value should be proportionate. In the event of higher efficiency with ethanol, as expected because of its relatively high thermodynamic availability, ethanol's value would be about 10% higher. However, the effects of increased efficiency here are less dramatic than in the blend case, where they are magnified by a factor of 10.

Outlook for Ethanol

In competitive markets, ethanol from biomass would first compete for its highest-value uses in the chemical market and proceed to the lower-value octane enhancer and fuel uses as the chemical market became saturated. However, government action has reversed this progression and made projections difficult because of uncertainties regarding governmental action.

Aimed at supporting farmers and bringing about the substitution of renewable resources for crude oil, government subsidy programs for alcohol-gasoline blends (gasohol) offer exemptions for gasohol from the state and federal excise taxes on motor fuels (12). The exemptions generally apply to ethanol and methanol from renewable sources, in blends of 10% by volume or greater with gasoline. About 30 states exempt gasohol from part or all of their taxes, and the Congress has extended the federal excise tax exemption to 1992. Exempting gasohol from a \$0.06/gal federal excise tax would appear to represent a relatively small subsidy. However, because ethanol comprises only 10% of the blend, a refiner or blender can pay up to \$0.60/gal more for ethanol than for gasoline and still offer a price-competitive product at the pump. State tax exemptions average \$0.04/gal ethanol, ranging up to \$0.16/gal. When the state exemption is added to the federal exemption, the effective subsidy is often over \$1.00/gal. The combination of state and federal subsidies is sufficient to get ethanol into the fuel market in many states. In Europe, where there are no subsidies, little ethanol is sold in fuel markets.

If fermentation alcohol were not being subsidized into the motor-fuel market, it could gradually displace synthetic ethanol in the higher-value industrial market, at an overall savings and a net gain to the economy. The ultimate size of the industrial market for fermentation alcohol depends on the relative costs of biomass and oil and gas feedstocks. The industrial market could be expected to grow if ethanol recaptures markets previously lost to petroleum-derived organics. A return to the ethanol-based manufacture of acetaldehyde (itself a precursor to several other organics) could triple the industrial market, while exports and conceivably, the use of ethanol as a source of ethylene could also lead to dramatic increases in demand (12).

With continuing subsidies, ethanol will remain an attractive octane enhancer and cosolvent. Its use is expected to double, reaching about 1.2×10^9 gal/yr, by 1992 (13), when the federal subsidy is scheduled to expire. However, the objective of current research is to produce ethanol (from lignocellulosic feedstocks) that is competitive in fuel markets without subsidies. Some market penetration of neat ethanol can be expected thereafter in local or fleet applications, with other applications developing slowly along with a distribution infrastructure. Ethanol is not likely ever to capture the entire gasoline market, because of substitutes and because of the enormous quantity of feedstock required. However, ethanol fuel blends in any percentage are feasible. The 10% blend now sold is an arbitrary choice determined by government subsidy and regulation, rather than by optimization of either performance or economics. The fuel market has the potential for utilizing all biomass-derived ethanol in the long term.

COPRODUCTS

This section examines the options available to utilize the lignocellulosic feedstock components not used in ethanol production. Coproduced fuels and chemicals were identified (using published data) for further analysis and development. The coproducts analyzed range from commercial products with known production processes to products with speculative uses and laboratory-scale production data.

The flow of material from feedstock to intermediate products (five-carbon sugars, furfural, acid hydrolysis and pretreatment lignin) to final products is shown in Fig. 4. Five-carbon sugars result from hydrolysis of the hemicellulose xylan component and are recovered from ethanol distillation or from pretreatment. Furfural can be produced from the pentosan component in the plug flow acid-hydrolysis reactor or from five-carbon (C_5) sugars by acid treatment at high temperature. Lignin obtained by acid-hydrolysis processes is distinguished from lignin obtained by pretreatment associated with enzyme-hydrolysis processes, because pretreatment lignin -- being more reactive -- is suitable for higher-value uses.

Coproduct Values

The analysis of coproduced fuels and chemicals has three primary objectives: to determine the current coproduct market, to determine the credit to ethanol production for each coproduct, and to determine the potential revenues from coproducts.

To perform the coproduct analysis, several parameters were obtained from the open literature or were calculated. These include current market price, U.S. market size, yields, and recovery costs. These data are shown in Table 2. Much of this information is based on conceptual studies and is subject to considerable uncertainty. Results of this analysis are summarized in Table 3 for coproducts derived from C_5 sugars and in Fig. 5 for those derived from lignin. The method employed is described below. The feedstock analyzed was a mixture of hardwoods.

Ethanol Production to Saturate Coproduct Market. The current U.S. market size for each potential coproduct was taken from the open literature. This size represents an upper bound on the actual market that could be captured, unless a price drop stimulates demand growth. The market size (in lb/yr) is divided by the coproduct yield (in lb/gal ethanol) to give the ethanol volume (in gal/yr) for which the coproduct would saturate its current market.

Coproduct Credit. The credit calculated for each coproduct represents an upper bound on the marginal net revenues that an ethanol producer would accrue by producing it. Actual revenues could fall considerably lower as the additional supplies lower the market price. The value of a coproduct ($/gal ethanol) is defined as its current market price ($/lb) times its yield (lb/gal ethanol). The coproduct credit is the market price of the coproduct minus its unit cost of production, where the cost of production is the feedstock cost plus the operating and capital costs of the recovery facility (estimated from published data). The cost per pound of feedstock is assumed to equal the cost of wood; the cost of separating wood into its components is formally charged to ethanol production.

Potential Revenues. The maximum benefit ($/yr) that could accrue to producers from recovery of a coproduct is defined as the coproduct credit ($/gal ethanol) times the ethanol volume (gal/yr) for which the coproduct saturates that market. This potential is important to consider, because high-value coproducts with small markets may yield lower benefits than lower-value coproducts with larger markets. Allocation of research funding should take potential benefits into account, as well as the likelihood of success.

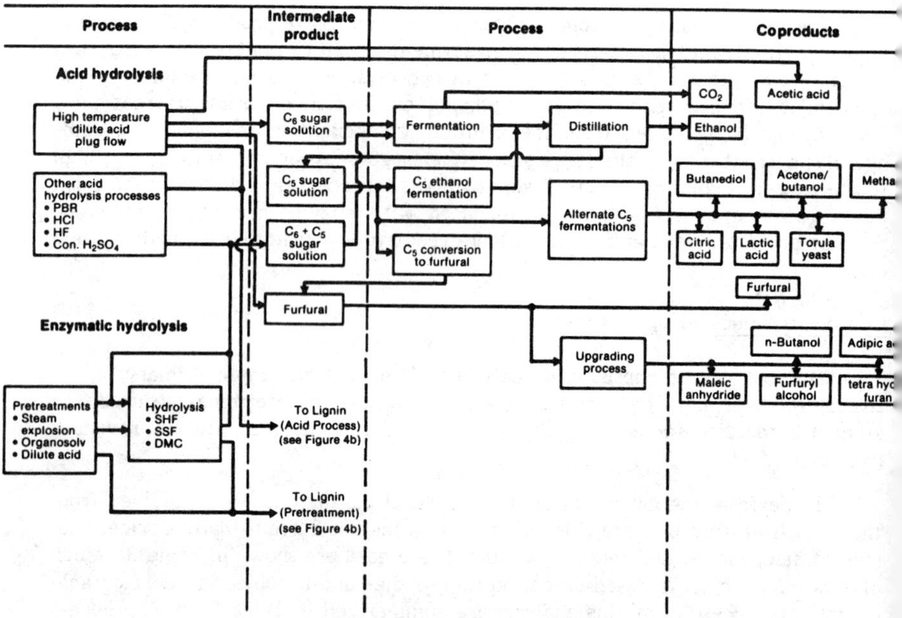

Figure 4a. Coproducts Analysis

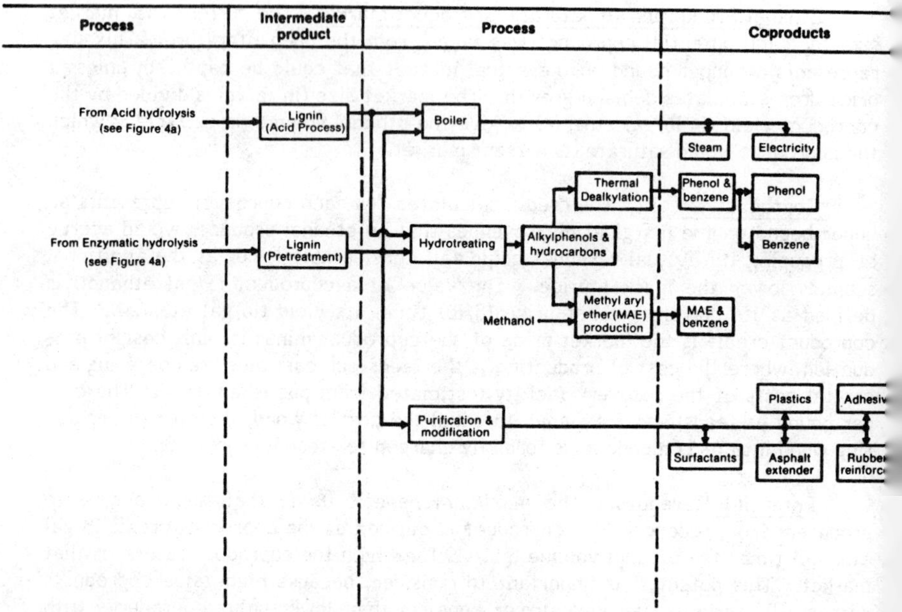

Figure 4b. Coproducts Analysis

Table 2. Coproduct Values and Markets

Fixed Charge Rate =	0.1288	Lign. Frac.=	0.24			
Ethanol Yield (Gal/Ton) =	50.00	Xylan Frac.=	0.24			
Wood Cost, ($/Dry Ton) =	$42.00					

	Current Market Value, $/lb	U.S. Market Million lb/ Year	Theoretical Yield, lb Product/ lb Wood	Theoretical Coproduct Value, $/gallon of Ethanol	Ethanol Production to Support Current Coproduct Market, Million Gal./Year
CO-PRODUCED FUELS					
C5 Fermentation Fuels					
Ethanol	$0.150	60000.00	0.138	$0.83	10870
2,3 Butanediol	$0.150	60000.00	0.137	$0.82	10965
Methane	$0.067	800000.00	0.072	$0.19	276396
Lignin Derived Fuels					
MAE +	$0.150		0.099	$0.59	
Benzene Mixture	$0.150		0.026	$0.16	
Total	$0.150	75000.00	0.125	$0.75	15036
CO-PRODUCED CHEMICALS					
C5 Fermentation Chemicals					
Acetone +	$0.250	1890.00	0.053	$0.53	895
n-Butanol	$0.360	918.00	0.066	$0.95	348
Total				$1.48	
Torula Yeast	$0.077	48600.00	0.142	$0.44	8581
Citric Acid	$0.860	154.00	0.174	$5.99	22
Lactic Acid	$1.060	17.60	0.149	$6.31	3
MEK from 2,3 Butanediol	$0.360	468	0.109	$1.57	107
Furfural Derived Chemicals					
Furfural	$0.750	140.00	0.174	$5.22	20
n-Butanol	$0.360	918.00	0.132	$1.90	174
Adipic Acid	$0.570	1800.00	0.264	$6.02	170
Maleic Anhydride	$0.550	361.00	0.170	$3.75	53
Furfuryl Alcohol	$0.720	66.00	0.178	$5.13	9
THF	$1.020	107.00	0.148	$6.04	18
Chemicals Derived from Pretreatment Lignin					
Plastic Extenders	$0.450	1470.00	0.240	$4.32	153
Adhesives	$0.135	740.00	0.240	$1.30	77
Surfactants	$0.165	2050.00	0.240	$1.58	214
Asphalt Extenders	$0.075	4510.00	0.240	$0.72	470
Rubber Reinforcers	$0.214	1060.00	0.240	$2.05	110
Chemicals Derived from Any Lignin					
Phenols +	$0.360	2175.00	0.083	$1.20	655
Benzene	$0.180	7700.00	0.034	$0.24	5662
Total				$1.44	
Misc. Co-products					
CO2	$0.025	8800.00	0.228	$0.23	965
Acetic Acid	$0.250	2640.00	0.039	$0.39	1692

Table 2. Coproduct Values and Markets (Concluded)

	Estimated Coproduct Yield, %	CoByproduct Feedstock Cost, $/lb Byproduct	Coproduct Recovery		Total Coproduct Recovery Cost, $/lb	Net Coproduct Credit, $/Gal
			Capital Investment $/Annual lb	Operating Cost $/lb		
CO-PRODUCED FUELS						
C5 Fermentation Fuels						
Ethanol	90	$0.041	$0.183	$0.010	$0.074	$0.38
2,3 Butanediol	100	$0.037				
Methane	90	$0.077	$0.870	$0.079	$0.268	($0.52)
Lignin Derived Fuels						
MAE +	100	$0.032	$0.104	$0.049	$0.094	$0.22
Benzene Mixture	100	$0.008	$0.027	$0.013	$0.025	$0.13
Total			$0.131	$0.062	$0.119	$0.35
CO-PRODUCED CHEMICALS						
C5 Fermentation Chemicals						
Acetone +	70	$0.027	$0.242	$0.056	$0.114	$0.20
n-Butanol	70	$0.034	$0.302	$0.069	$0.128	$0.43
Total			$0.544	$0.125	$0.242	$0.63
Torula Yeast	100	$0.036				
Citric Acid	90	$0.032	$0.930	$0.350	$0.502	$2.24
Lactic Acid	85	$0.040	$1.050	$0.493	$0.668	$1.99
MEK from 2,3 Butanediol	70	$0.066				
Furfural Derived Chemicals						
Furfural	65	$0.045	$0.035	$0.006	$0.055	$3.14
n-Butanol	59	$0.065	$0.373	$0.196	$0.309	$0.16
Adipic Acid	49	$0.039	$0.365	$0.108	$0.194	$1.95
Maleic Anhydride	42	$0.070	$1.245	$0.144	$0.375	$0.50
Furfuryl Alcohol	62	$0.046	$0.058	$0.076	$0.129	$2.61
THF	59	$0.058	$0.489	$0.133	$0.254	$2.68
Chemicals Derived from Pretreatment Lignin						
Plastic Extenders	50	$0.042		$0.200	$0.242	$1.00
Adhesives	80	$0.026		$0.020	$0.046	$0.68
Surfactants	80	$0.026		$0.020	$0.046	$0.91
Asphalt Extenders	80	$0.026		$0.020	$0.046	$0.22
Rubber Reinforcers	70	$0.030		$0.020	$0.050	$1.10
Chemicals Derived from Any Lignin						
Phenols +	100	$0.031	$0.201	$0.042	$0.098	$0.87
Benzene	100	$0.013	$0.082	$0.017	$0.040	$0.19
Total			$0.283	$0.059		$1.06
Misc. Co-products						
CO2	100	$0.000				
Acetic Acid	100	$0.021				

Table 3. Value and Potential of Coproducts from Five-Carbon Sugars

Coproduct	Type[a]	Ethanol Volume to Saturate Market (10^6 gal/y)	Credit ($/gal ethanol)	Potential Net Revenues ($\$10^6$/y)	Marginal Benefit over Fuels ($\$10^6$/y)
Furfural	F	20	3.14	63	55
THF	F	18	2.68	48	41
Furfuryl alcohol	F	9	2.61	23	20
Citric acid	C_5	22	2.24	49	41
Lactic acid	C_5	3	1.99	6	5
Adipic acid	F	170	1.95	332	267
Acetone and n-butanol	C_5	348	0.63	219	87
Maleic anhydride	F	53	0.50	27	6
Ethanol (fuel)	C_5	10,870	0.38	4131	--

[a]F = furfural and derivatives, C_5 = C_5 (five-carbon-sugar) fermentation products.

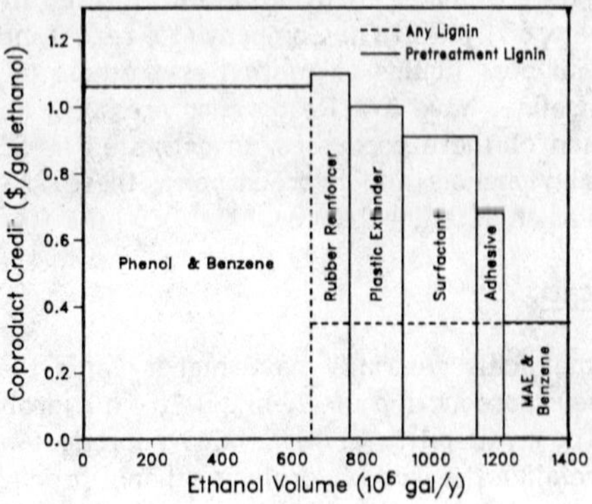

Figure 5. Coproduct Credits: Lignin Only

Coproduced Fuels

From a DOE programmatic viewpoint, the production of fuels is the primary objective. Described below are several potential coproduced fuels:

Fuels from Five-Carbon Sugars. Ethanol, 2,3-butanediol, and methane are three fuels that can be produced by five-carbon-sugar fermentation. Additional ethanol is perhaps the most desirable coproduct from five-carbon sugars. Microorganisms such as ***Candida shehatae*** and ***Pachysolen tannophilus*** can ferment xylose to ethanol, but this fermentation (the subject of current research) is slow compared with glucose fermentation. The production cost was estimated based on a C_5 sugar fermentation rate 25% that of glucose.

Fermentation of five-carbon sugars to 2,3-butanediol has been demonstrated in the laboratory by ***Klebsiela pneunomiae*** and ***Bacillus polymyxa***. While 2,3-butanediol is not currently a commodity chemical, it appears to be a suitable gasoline extender and cosolvent for other alcohols in gasoline. The Solar Energy Research Institute (SERI) has conducted preliminary tests, but additional testing is necessary. Problems may exist with water incorporation and materials compatibility. Other, higher value uses may exist for 2,3-butanediol as a petrochemical feedstock. Due to lack of data, no analysis of recovery costs was made, but costs are expected to be similar to those for ethanol fermentation.

Methane can be produced from five-carbon sugars by anaerobic digestion, but the yield is low. Consequently, the cost of the methane is well above its value, and methane will not be considered further.

Fuel Production from Lignin. A process developed by Hydrocarbon Research, Inc. (Lignol (c)), produces phenol and benzene from lignin (14). Slurried solid-phase lignin in lignin-pasting oil is hydrotreated in the presence of a catalyst to a mixture of phenolic compounds and hydrocarbons. The mixture of alkylphenols is thermally dealkylated to phenol and benzene, which could be separated and sold in the chemical market or further processed to produce fuels. Sales of phenol and benzene in the chemical market have the highest potential benefits of all of the lignin-derived chemicals. The phenolics are also good octane enhancers, but their corrosive nature makes them unsuitable for addition to gasoline. However, a process developed by Gulf Oil Company (15) reacts methanol with the alkyl phenols derived from coal liquids to methyl aryl ethers (MAE), which are fully compatible with gasoline, have low Reid vapor pressure, and are octane enhancers. A combination of these processes, investigated at SERI, then produces a mixed stream of alkylphenols and hydrocarbons; these are converted to MAE, which could be sold as an octane enhancer.

Coproduced Chemicals

Commodity chemicals generally have higher value than fuels, but smaller markets. Therefore, coproduction of chemicals from hydrolysis would be limited to early plants. The availability of high-value coproduct markets could induce industry to commercialize the wood-hydrolysis ethanol-production technologies.

Chemicals from Fermentation of Five-Carbon Sugars. A variety of chemicals could be produced by fermentation of five-carbon sugars. Some of these chemicals are currently commercially produced by fermentation. Several species of the genus ***Clostridium*** will ferment xylose into a mixture of acetone and n-butanol. The production process uses a batch anaerobic fermentor with a residence time of 40-45 h. Process conversion efficiency is estimated to be 70% (16). Although the markets for these products are large, their value is only slightly greater than that of fuels (see Table 3).

Torula utilis is a non-spore-forming yeast that will grow on five-carbon sugars. The body of the yeast has a protein content over 50% and could be used as a food supplement for humans or animals (17). The current torula yeast production in the U.S. is small. Because the product is similar to soybean meal, the market for soybean meal was used to estimate the potential market and price for torula yeast. No process data were found in the literature, so no coproduct credit was calculated. However, the value of the product and its yield are relatively low.

Citric acid, a widely used food and pharmaceutical additive, has many industrial uses. It is produced commercially today by fermentation of molasses with the mold ***Aspergillus niger.*** Production from five-carbon sugars would use a similar process. Fermentation is carried out in agitated, aerated fermentors with a residence time of 72 h (16).

Lactic acid, used primarily as a food additive, is produced from molasses or other sugar solutions by anaerobic fermentation by ***Lactobacillus delbruckii.*** The fermentation is carried out at 49°C for about 5 days (16). There are also synthetic chemical routes to lactic acid.

Methyl ethyl ketone (MEK), a major industrial solvent, can be produced from 2,3-butanediol upon heating with sulfuric acid or by passing 2,3-butanediol vapor over a dehydration catalyst, such as alumina, bentonite, or phosphorus pentoxide. Costs are unknown, but the value ($1.57/gal ethanol) is high, and significant capital and operating costs could be tolerated.

Furfural-Derived Chemicals. Furfural is made by acid degradation of five-carbon sugars. In the commercial process, biomass high in xylan (bagasse, corn cobs, oat hulls) is loaded into a reactor and treated with acid at high temperature and pressure. Furfural is used in petroleum refining as a selective solvent. It is also reacted with phenol to produce thermosetting resins. Furfural is sold as a chemical feedstock to produce furfuryl alcohol and tetrahydrofuran (THF). However, the present market for furfural is only slightly larger than the amount produced in one ethanol plant with a capacity of 25×10^6 gal/yr. A number of furfural derivatives are technically possible but not economic if furfural costs $0.75/lb. The incremental cost for furfural coproduction with ethanol is only $0.055/lb, so some of these processes could be used to market the large volumes of coproduct furfural (18).

Sold as an industrial solvent, n-butanol is also used to make lacquer solvents, such as butyl acetate. It can be produced from furfural by reacting it with steam to produce furan, which is hydrogenated in the presence of a catalyst. Process yield is approximately 60% (mol) n-butanol and 35% (mol) THF. The analysis assumed that THF and n-butanol could be sold at the same price and that process yield was 95%.

Nearly all adipic acid is used in the manufacture of nylon-6,6, and most is used at the location of production. There are two commercial production routes in use today for petrochemicals. DuPont (E.I. DuPont de Nemours, Inc., Wilmington, Del.) operated a process in the early 1950s to produce adipic acid from furfural but shut down the plant due to technical and economic problems. Another process, developed on the laboratory scale, uses an oxo-type reaction in which THF is reacted with carbon monoxide in the presence of nickel carbonyl catalyst and nickel iodide promoter. Process yield is estimated to be 75% (18,19). Adipic acid's relatively large market and very high coproduct credit give it the highest potential benefits of all the chemicals that could be produced from hemicelluloses (see Table 3).

Maleic anhydride is used in various unsaturated polyester resins used for boat construction, fiberglass-reinforced tanks, piping, and electrical insulation. Commercial processes make maleic anhydride from benzene, butane, and butenes. A speculative route would produce maleic anhydride from furfural by oxidizing furan over a vanadium pentoxide catalyst at high excess air levels. Process yield is 65%.

Most of the furfuryl alcohol produced is used to make furan resins for binders in foundry sand molds. Furfuryl alcohol is made commercially from furfural by hydrogenation in the presence of copper chromite catalyst at elevated pressure and temperature (16). The yield is 95%.

Tetrahydrofuran (THF) is used as a solvent for polyvinyl chloride (PVC), polyurethane, and other polymers. It was first made from furfural, but recent capacity uses petrochemical feedstocks. The process for making THF from furfural involves furan production by steam decarbonylation, followed by hydrogenation over a conventional nickel-based catalyst.

Chemicals Derived from Pretreatment Lignin. A recent SERI study (20) analyzed a variety of products that could be made from lignin. Lignin from biomass pretreatment for enzyme hydrolysis is more reactive than heavily condensed lignin derived from most acid hydrolysis processes. Because of its higher reactivity, pretreatment lignin may have several high-value uses.

Lignin, the binder in wood that holds the cellulose fibers together, could play a significant role as an extender in PVC, polyurethane, epoxy resins, and other plastics. Although lignin is not now so used, research efforts in the area are promising. Lignin could be used as an extender or coreactant for phenolic- and urea-based wood adhesives. Although this is not commercial today, research has shown near-commercial results. Lignosulfonate-based anionic surfactants are a major use of lignin today. These surfactants are used in oil-field drilling mud, pelletized animal feed, and concrete mixtures. Potential uses include enhanced oil recovery, coal slurries, and substitution for nonionic surfactants. Several efforts have been made to incorporate lignin into asphalt for street paving. Although these efforts have not been entirely successful, this market could be large (but with a very low value). Competitors for this market include discarded tires, sand, and sulfur. Nearly all of the 3.3×10^9 lb/yr of carbon black produced in the U.S. is used to reinforce rubber. Beginning during World War II, research was conducted to replace carbon black with lignin. A number of other applications are mentioned (20) including large volume agricultural chemicals.

Miscellaneous Coproducts. Two additional coproducts were considered -- CO_2 and acetic acid. Carbon dioxide has a wide variety of industrial and consumer uses, including carbonation of beverages, flash freezing of food, propellants in aerosol spray cans, welding, and enhanced oil recovery. Much of the CO_2 produced is used on site. In the merchant market, CO_2 is sold either as a liquid or a solid. Carbon dioxide is available from the fermentation in a relatively concentrated stream. On a mass basis, CO_2 production is nearly equal to ethanol production, but liquefaction costs are expected to exceed the product value. Acetic acid is a widely used industrial chemical. However, it constitutes a small component and results in a relatively low coproduct credit.

Coproduct Summary

There is a wide variety of fuels and chemicals that can be produced from the hemicellulose and lignin fractions of lignocellulosic feedstocks. Figure 6 illustrates the effect of coproducts on the ethanol production cost. Early capacity will take advantage of high-value coproducts. As ethanol production increases above 1.2×10^9 gal/yr, only coproduced fuel markets will be available. Coproduction of fuels could produce an additional annual benefit of about $1.5 billion over the $2 billion revenues resulting from ethanol production of 2×10^9 gal. Recovery of chemicals could increase this potential to about $2.8 billion. Production of adipic acid from furfural and phenol and benzene from lignin could account for $1 billion of these benefits. These processes require further research.

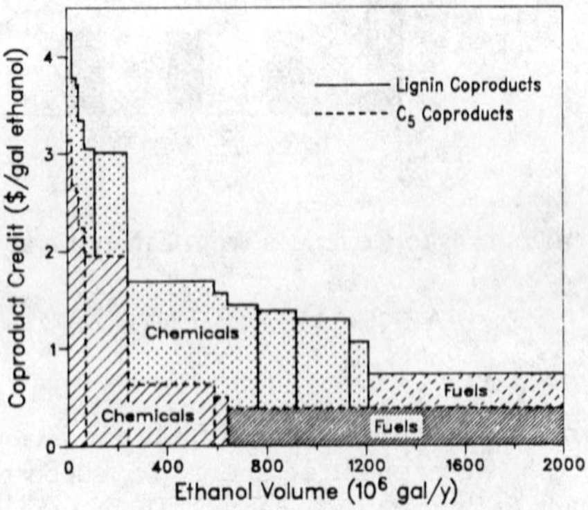

Figure 6. Coproduct Credits

CONCLUSIONS

For both ethanol and coproducts, sales in chemical markets yield higher revenues than sales for fuel use, but chemical markets are smaller and have limited potential. Early plants can maximize their revenues by selling products as chemicals before the markets are saturated. These plants can serve to demonstrate and establish the ethanol production technology, which can then be used to produce fuels, with potentially large benefits to the economy.

The current chemical market for ethanol is 170×10^6 gal/yr at a price of $1.80/gal. Individual chemical markets could consume coproducts from ethanol production in the range of $3\text{-}655 \times 10^6$ gal/yr each, with coproduct credits of up to $4.24/gal ethanol. In the larger fuel markets, ethanol is worth about $0.86/gal as an octane enhancer (based on $1.00/gal gasoline and average avoided processing costs), and coproduct fuel credits are about $0.64/gal ethanol. Figure 7 shows how revenues decline as coproducts saturate higher-value markets and are sold in lower-value markets. Assuming large ethanol production facilities (25×10^6 gal/yr), the first 25 facilities or so can sell products at above their fuel value. Coproduct credits (from which costs have been subtracted) exceed the value of ethanol until all chemical markets are saturated. This demonstrates the importance to process economics of utilizing all of the high-value wood components.

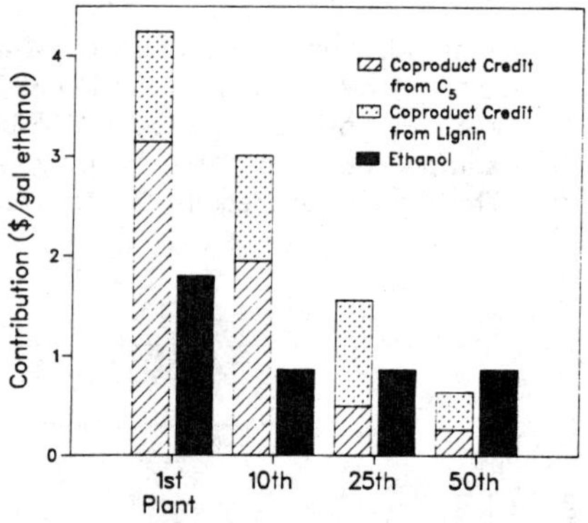

Figure 7. Contribution to Revenues from Ethanol and Coproducts

ACKNOWLEDGMENTS

The authors would like to thank the Biofuels and Municipal Waste Technology Division of the U.S. Department of Energy for supporting this work. They would also like to thank F. Bennett (ANL) and C. Thomas (SERI) for editing, and L Haley (ANL) and C. Hawthorne (SERI) for graphics.

Note: This work was supported by the U.S. Department of Energy, Assistant Secretary for Conservation and Renewable Energy, under Contract W-31-109-Eng-38.

REFERENCES

1. Wood as an Engineering Material, Forest Products Laboratory, U.S. Department of Agriculture, Agriculture Handbook No. 72 (1974).

2. Synthetic Organic Chemicals, U.S. Production and Sales, 1984, U.S. International Trade Commission, Pub. No. 1745, U.S. Government Printing Offi (1985).

3. "Current Prices of Chemicals and Related Materials", Chemical Marketin Reporter, 229, (6) (Feb. 10, 1986).

4. "Chemical Industry Facts and Figures", Chemical and Engineering News, 63 (23):22 (June 10, 1985).

5. "Alcohol Week Price Watch: Key Fuel, Grain, and By-Product Prices," Alcohol Week 7, (10) (March 10, 1986).

6. Lead Phasedown to Spur Alcohol Marketing Changes, Information Resources Inc., Washington, D.C. (March 4, 1985).

7. Monthly Energy Review, U.S. Department of Energy, Office of Energy Markets and End Use, DOE/EIA-0035 (85/10) (Jan. 1986).

8. Alcohols: Economics and Future in U.S. Gasoline Markets, The Pace Company, Houston, Texas, et al. (1984).

9. Van Tassel, B., Refinery Economics of Alcohols and Aromatics, Pace Consultants, Houston, Texas (Sept. 1985).

10. Unzelman, G.H., and G.W. Michalski, "Fuel Supply for the 1980s--at the Crossroads," Proc. 21st Automotive/Petroleum Industry Forum, Southfield, Mich. (1980).

11. Monaco, L.C., "The Brazilian Alcohol Program," in Energy Applications of Biomass, M.Z. Lowenstein, Ed., Elsevier, London (1985).

12. Alston, T.G., and J.G. Asbury, Grain Alcohol: The Right Product in the Wrong Markets?, Argonne National Laboratory Report ANL/EES-TM-79 (March 16, 1980).

13. Unzelman, G.H., unpublished information (1985).

14. Personal Communication from Chem Systems, Inc., to A. Power (SERI). Hydrocarbon Research, Inc., data furnished to Chem Systems, Inc., for preparation of the report: Technical and Economic Assessment of Processes for the Production of Butanol and Acetone--Phase Two: Analysis of Research Advances. Final report for Contact No. AI 01-81CS66001.DOE/CS/6601-5. Additional information: Parkhurst Jr., H. J., D. T. A. Huibers, and M. W. Jones, "Production of Phenol from Lignin," ACS Div. Petroleum Chem. Reprints, Vol. 25, No. 3, pp. 657-667 (1980).
Gendler, J. L., S. T. A. Huibers, and H. J. Parkhurst, Jr., "Hydroxyaromatics from Lignin Hydrogenolysis," in Wood and Agricultural Residues, E. J. Solts, Ed.,Academic Press, N.Y., pp. 391-400 (1983).

15. Singerman, G. M., Methyl Aryl Ethers from Coal Liquids as Gasoline Extenders and Octane Improvers, Gulf Research and Development Co., Pittsburgh, Penn., Contract No. DE/AC01-79 C50022 (1980).

16. Biomass Process Handbook, Technical Insights, Inc., Fort Lee, N.J. (1980).

17. Syncorp, Inc., "Commercial Production of Ethanol from Wood, Market Analysis (May 1981).

18. Parker, S., M. Calnon, D. Feinberg, A. Power, and L. Weiss, The Value of Furfural/Ethanol Coproduction from Acid Hydrolysis Processes, Solar Energy Research Institute Report SERI/TR-231-2000, Golden, Colo., (Aug. 1983).

19. Reppe, W., et al., Ann. der Chemie, 582, p. 87 (1953).

20. Chum, H., S. K. Parker, D. A. Feinberg, J. D. Wright, P. A. Rice, S. A. Sinclair, and W. G. Glasser, The Economic Contribution of Lignin to Ethanol Production from Biomass, Solar Energy Research Institute Report SERI/TR-231-2488, Golden, Colo., (May 1985)

DISCUSSION

J. Kerstetter, Washington State Energy Office	Did you look at methanol at all as an option from wood? Your yields are certainly better, and this is an energy product. You're not really being a chemical producer and the Department of Energy is supposedly dealing just in energy.
Gaines	The scope of this study was limited to ethanol. The entire Department of Energy biofuels program does consider methanol, but the idea of this study was to look at the best ways for producing ethanol and the best coproducts to recover. Once that's determined, ethanol will be compared to all the competition.
Kerstetter	Basically ethanol is a coproduct isn't it for real world economics, if there's no subsidy for ethanol fuel?
Gaines	If you were to build an ethanol-from-wood plant today, I think that would be true.
J. Zerbe, U.S. Forest Products Laboratory	You indicated the coproducts are phenol, benzene, and furfural. Did you consider any others?
Gaines	There's a whole list in the paper. That analysis was done at SERI, and there's a fairly extensive list that they considered, something like 30 materials.

WORLDWIDE REVIEW OF ALCOHOL FUEL MARKETS

William H. Klausmeier, Ph.D.
President, Sylvatex Corporation
10708 Burr Oak Way
Burke, Virginia 22015

ABSTRACT

This paper is based upon a survey of alcohol fuel activities worldwide that was conducted to determine the status of the ethanol industry and to identify successful strategies that have evolved from experience to date. Worldwide fermentation ethanol capacity has increased nearly eight-fold since 1977 and has still more than doubled even if one excludes the major programs of Brazil and the United States. New distilleries and alcohol fuel programs continue to be introduced rapidly. Although oil prices are declining, ethanol's utility as an octane enhancer and depressed world market sugar prices continue to provide impetus to the alcohol fuels industry. Many contrasts can be noted when comparing alcohol fuels programs. In the United States, corn and other grains are the principal feedstocks. While sugar cane predominates elsewhere, U.S. distilleries are, on the average, three or four times larger than elsewhere yet the mass production of small distilleries in countries like Brazil have greatly reduced investment costs. While the production of hydrous alcohol in autonomous distilleries dominates the Brazilian industry, the use of annexed distilleries to produce anhydrous alcohol for blending with gasoline predominates elsewhere. Prospects for the continued growth of alcohol fuel markets are quite good.

WORLDWIDE REVIEW OF ALCOHOL FUEL MARKETS

This paper will review information on the introduction of fuel alcohol by the principal regions of the world. Developments will be summarized for each region in which fuel ethanol fermentation plants have been installed and will address the feedstocks employed and the rationales for instituting alcohol fuel initiatives.

NORTH AMERICA

United States

The fuel ethanol industry was created in the late 1970's as a reaction to America's gasoline shortage and gasohol, a blend of 10% ethanol in gasoline, was first retailed in June of 1978 (1). Since that time, ethanol has gained recognition as an effective octane enhancer rather than merely an extender for gasoline and this has become the principal market for fermentation ethanol. Production of fuel alcohol has risen steadily since 1978 as is evident by the data in Table 1. As noted in Table 1, there has been a growing supply shortfall for fuel alcohol in the United States which has been met through imports from Brazil.

Table 1

Production and Consumption of Fuel Ethanol in the United States

	Alcohol (10^6 Liter)	
Year	Production	Consumption
1978	38	
1979	76	
1980	152	420 estimated
1981	284	416
1982	795	852
1983	1,419	1,628
1984	1,643	1,942
1985	-	2,300 projected
1986	-	3,000 projected
1990	-	3,800 projected

Source: References 1, 2.

The exemption of 10% ethanol blends from Federal excise tax has been increased from 1.3 to 1.6¢/liter and combined with the tax exemptions allowed in most states has provided significant incentives for alcohol fuel production. EPA's ruling that lead in gasoline must be reduced from 0.29 gm/1 to 0.03 gm/1 by 1986 has stimulated demand for alternative octane enhancers. A number of organic liquids, including ethanol, are known to be effective octane enhancers. Figure 1 presents the octane ratings and prices for the principal organic additives.

Ethanol is a somewhat better octane enhancer than many of the other well-established enhancers. Although it would normally be much more expensive than the others, the Federal and State tax exemptions for ethanol make it cost competitive with many of the others. Many refiners can also increase the octane rating of their gasoline by increasing the severity of reforming. However, ethanol blending with sub-octane gasoline is less expensive in many states (retail savings of 2.6¢/gal. in North Dakota and 3.9¢/gal. in Minnesota) (4). Ethanol consumption in the United States is expected to grow rapidly as long as the subsidies remain in place.

Although the growth in fermentation ethanol production has been rapid, the industry is operating at only 52% of capacity (2). Three out of five ethanol plants are currently shut down and only 58 out of 145 plants with annual capacities over 1.9 million l/yr were operating by the end of 1984. There are two explanations for the low rates of capacity utilization reported.

The most fundamental reason for the many shutdowns is the early emphasis on small-scale ethanol production to encourage the development of a broadly based, disaggregated industry. The principal feedstock, corn, is relatively expensive compared to other feedstocks such as sugarcane in Brazil. Large-scale corn processing operations can refine a number of valuable by-products (corn oil, gluten meal) from the corn that reduce the cost of the starch component by as much as 40% (5) but small-scale plants must use simpler dry-milling technologies that afford much smaller by-product credits. Wet millers have a 6.9¢/liter production cost advantage over dry millers (5). In addition, small-scale distilleries have greater capital costs per unit output because of engineering scale effects. A 38 million l/yr distillery using dry milling has roughly twice the capital and operating costs per liter as a 189 million l/yr distillery

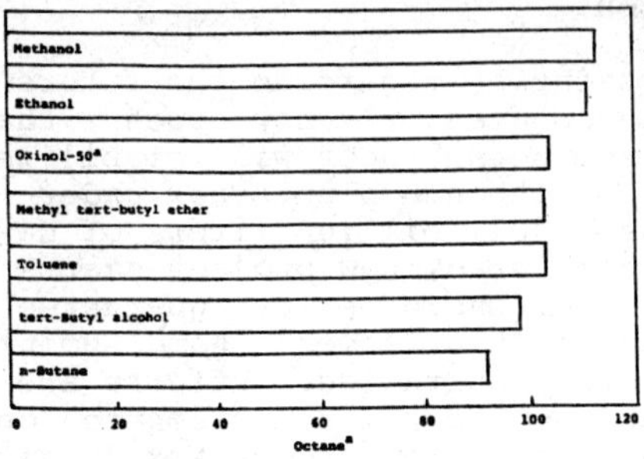

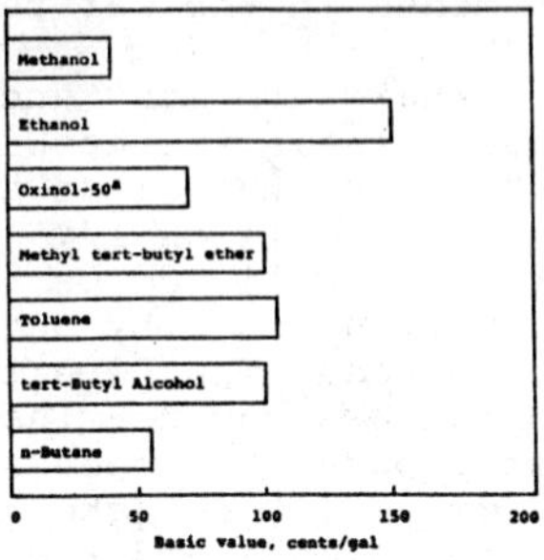

Source: Reference 3.

Figure 1. Octane Ratings of Selected Octane Enhancers

Prices of Selected Octane Enhancers

using wet milling (6). Favorable economics for small-scale ethanol production depended inordinately upon tax breaks and escalating gasoline prices. Consequently, small plants could not be sustained when conditions moderated. Of the 200 companies that initially ventured into the field, 13 companies now control 69% of production. The largest company, Archer Daniels Midland, controls 50% of the domestic market (2). Recent alcohol plants show diversification to a broader range of feedstocks and several small-scale distilleries using processing waste streams demonstrate much more favorable economics than small-scale distilleries not processing waste streams.

A second explanation for poor capacity utilization has been competition from low-cost imports. Although the U.S. Government increased the import duty on fuel ethanol from 13.2¢ to 15.0¢/l in 1984 and added a 3% ad valorem tax, imports of Brazilian alcohol steadily grew to more than 500 million liters in 1984 (16.2% of the U.S. market). This has prompted charges by a consortium of U.S. alcohol producers that Brazil is dumping ethanol in the United States at prices below the cost of production. The International Trade Commission recently ruled that some Brazilian companies were, in fact, dumping ethanol. Also, under the Caribbean Basin Initiative (CBI), Caribbean countries can export anhydrous alcohol to the United States duty-free provided the alcohol is produced in the Caribbean or further purified to increase the product value by 35%. In the past year, perhaps a hundred million liters of upgraded alcohol has been imported under the CBI. Although it is likely that the loophole allowing duty-free status for upgraded alcohol will be blocked, the CBI remains a powerful incentive for fuel alcohol production for export to the United States by Caribbean nations. Even though the large U.S. alcohol fuel producers are vehemently opposed to Brazilian ethanol imports, many others view low-cost imports as a means of meeting the rising demand for ethanol as an octane enhancer and securing that market from competing substitutes. Also, some in the U.S. ethanol industry appear to be softening their views concerning CBI imports and are even considering investments in the CBI region.

There are many uncertainties regarding the future of alcohol fuels in the United States. It now appears that the market for ethanol as an octane enhancer will stabilize at about 3-4 billion l/y (7). Although plants now under construction will add perhaps a half billion liters of new production capacity once completed, new project starts may be limited. The Department of Energy's loan guarantee program has now terminated and the Farmer's Home Administration program has been severely curtailed. Efforts are also underway to terminate both the Federal tax exemption on ethanol blends and the 10% energy investment tax credit. Despite these actions, the production of fuel ethanol in the United States appears at least solidly established if not encouraged to expand further.

The current sales price of fermentation ethanol ranges from 39.1-44.9¢/liter (f.o.b. works) depending upon the level of state tax incentives (8). Corn remains the predominant feedstock but the industry is now diversifying into other feed grains (barley, milo, wheat) and sugar-containing waste streams (molasses, fruit processing rejects, whey). Large-scale corn-based distilleries have changed little in the past ten years. Fermentation and heat recovery efficiencies are just about optimal. Many of the large plants are fueled with low-cost coal. Stillage is usually concentrated to dryness to provide distiller's dried grains with solubles (DDGS) for livestock feeds and a large market for DDGS has been cultivated in Europe. Although this is an energy-intensive solution to stillage handling, energy costs are made up by the revenues of DDGS sales. Another factor limiting further expansion of the U.S. ethanol industry is the saturation of markets for distillery by-products.

Canada

Canada is a relatively recent entrant in the alcohol fuels arena. The country has abundant solid fossil resources and national policy had been directed more towards exploiting tar sands and other hydrocarbon sources rather than by developing a national alcohol fuels program (9). Nevertheless, Canada has an interest in alcohol fuels and endorsed an international cooperation agreement to study alcohol fuels under the International Energy Agency. Canada has also issued regulations calling for the reduction of the lead content in gasoline from 0.77 to 0.29 gm/l as of January 1, 1987, which will result in higher alcohol fuel demand. It also appears that a modest ethanol fermentation industry will develop in Canada to meet the demand for ethanol as a stabilizing co-solvent in methanol/gasoline blends. Mohawk Oil is now operating a distillery in Minnedosa, Manitoba, producing 6.8 million l/y from grains for blending with methanol in gasoline (10). A 7.6 million l/y ethanol plant, planned by Bruce Grain Products, Ltd., is now nearing completion at Kincardine, Ontario (11). The Canadian Renewable Fuels Association, a trade association concerned primarily with ethanol, has been formed with funding from over 43 companies and organizations (12). Although only the beginnings of an alcohol fuel industry now exist, Canada has been quite visible in terms of alcohol fuels research.

CENTRAL AMERICA AND THE CARIBBEAN

This region now seems poised for vigorous alcohol fuels development and substantial new ethanol fermentation capacity has already been introduced. There are several reasons for the high level of interest in alcohol fuels. Most of the countries in the region rely heavily upon oil imports to meet their energy demand. Foreign exchange needed to pay for imported oil greatly exceeds exchange earnings for cane sugar, which is the third largest earner of foreign exchange (foreign debt is $15.1 billion, oil imports cost $1.9 billion yr and sugar exports earn $0.53 billion/yr) (13). Due to the unfavorable terms of trade, most of the countries are finding it exceedingly difficult to service their foreign debt.

The only country of any size in the Caribbean and Central American region that is not dependent on oil imports is Mexico. The second major factor for developing a biomass based fuel alcohol program in these countries is the very depressed state of the sugar industry which directly employs 12.5 million people in the region (13, 14).

The sugar industry expanded rapidly in the early 1970's during a brief period of high sugar prices. Most of the exports went to the United States which imported from 3.5 to 5.3 million metric tons (MT) prior to 1982. In 1982, the United States imposed a quota system to protect its growing high fructose corn syrup sweetener industry and imports immediately dropped to 2.4 million MT/yr and are expected to remain below 3 million MT/yr for the foreseeable future. At the same time, in response to the high world sugar price of the early 1970's, European (EEC) countries' production of sugar beets rapidly expanded. Most of the increased production was put on the world market, resulting in a sharp increase in EEC sugar exports which grew dramatically from 1.9 million MT in 1975 to 5.6 million MT in 1982. Concurrently, the use of artificial sweeteners gained wide acceptance, reducing the growth of demand for sugar. This resulted in a glut of sugar on the world market and sugar prices plunged to a current record low of 3¢/lb. This has created a crisis within the sugar industry that can only be effectively solved in the long term by diversification of the industry into other products. Of the 97 operational sugar mills in the region, 33.9% are currently idle. This represents an excess production capacity of 10.7 million MT/yr.

The third consideration affecting the prospects for alcohol fuels in the Caribbean is the fact that Brazil has already perfected the technology for fuel alcohol production from sugarcane. Turnkey distilleries are now available from the Brazilian firms Dedini and Zanini at a capital cost less than half, on a per unit production basis, of the cost of the most economical U.S. distillery ($4.1-$5.1 million for a 120,000 liter/day distillery) (15). Brazil has also demonstrated that a national fuel alcohol program can be successfully established. The Brazilian firm, Copersucar, recently hosted an International Symposium on Sugar and Alcohol as part of an aggressive effort to market their alcohol technology to developing countries. An evaluation of the potential for Central American and Caribbean countries to produce ethanol using idle cane crushing capacity was recently performed by the Organization of American States (OAS) and the results are shown in Table 2.

Table 2

Ethanol Production Potential in Central American and Caribbean Countries

Country	No. of Distill-eries	Capacity (1,000 liter/day)	Days of Operation	Ethanol Production (1,000 liter/yr)[1]
Costa Rica	3	120	130	47,240
El Salvador	2	60	160	19,380
Guatemala	6	120	150	109,010
Honduras	3	120	180	65,400
Nicaragua	3	120	130	47,240
Panama	3	120	130	47,240
Subtotal	20	660		335,510
Jamaica	2	120	200	48,450
Dominican Rep.	4	120	200	96,990
Subtotal	6	240		145,440
Total	26	900		480,950

[1](4) = (1)(2)(3)
Source: Reference 14.

Nearly half a billion liters/yr of alcohol could be produced in this region using idle capacity for a total additional distillery investment of US$ 227-284 million, which is about the same as the capital investment for a single large corn-based distillery in the United States (16). The two principal factors inhibiting further introduction of alcohol fuel technology in Central America and the Caribbean are a lack of coherent national policies in most countries and a lack of investment capital (15).

The final factor affecting the prospects for alcohol fuels production in the Caribbean is the existence of the Caribbean Basin Initiative which allows for duty-free import of alcohol to the United States that has been produced in Central America and the Caribbean. Some estimates indicate that ethanol could be landed in the United States for 31¢/ liter (based on a feedstock cost of $12/MT) (16), but a recent study of alcohol fuels potentials in Honduras reported higher cane costs of $14.63/MT or more which yielded an alcohol production cost of 37.0-45.4¢/liter (17). At a current ethanol sales price of 35.1¢/liter CIF the gulf coast, alcohol production by CBI countries for export may well not be universally profitable. Nevertheless, the export of alcohol would generate more revenue than would the sale of sugar on the world market at the current price of 6¢/lb (35.1¢/l vs 21.9¢/3.64 lb sugar; 3.64 lb sugar yields 1 l ethanol). In the case of Honduras, the introduction of a distillery is best regarded as an insurance policy against low world market sugar prices. If the CBI countries were to convert all their world market sugar to ethanol, production would exceed the potential market for domestic blending and some alcohol would have to be exported (see Table 3).

Table 3

Alcohol Production and Consumption Potential in Caribbean Basin Countries (1984)

Country	Actual Gasoline Consumption (10^6 liter/yr)	Potential Alcohol Consumption (10^6 liter/yr in 20% blend)	Alcohol Available Using Idle Capacity (10^6 liter/yr)
Costa Rica	159	32	47
El Salvador	174	35	19
Guatemala	303	60	109
Honduras	129	26	65
Panama	257	51	47
Jamaica	246	49	48
Dominican Republic	360	72	97
Total	1,628	325	432

Source: Reference 15.

Despite the appeal of a U.S. export market for CBI alcohol, most potential distillers fear that this opportunity could be eliminated by future legislation and protectionist sentiment is currently high in the United States. Consequently, many feel that ethanol markets are best assured by the creation of domestic alcohol fuel markets through the implementation of national fuel alcohol programs. The ethanol demand so created would be able to support at least two distilleries per country and this would suffice to get the industry off to a good start. Although in most areas ethanol/gasoline blends will require modest tax exemptions to be competitive with gasoline, the potential loss in fuel tax revenues would probably not exceed the savings in foreign exchange from gasoline import reduction. However, the economic benefits to the CBI countries would be much greater for exporting alcohol to the United States to, instead, take advantage of its subsidy. There is, therefore, a trade off between national benefits and market risks. To date, Guatemala, El Salvador and Costa Rica have initiated national fuel alcohol programs. Most of the other countries in the region are also actively considering this option at present. Information of specific developments is provided by country and is summarized in Table 6.

SOUTH AMERICA

Alcohol fuels development in South America has been dominated by Brazil's Proalcohol program. However, many other countries in South America are net importers of petroleum and produce substantial quantities of sugarcane and other potential feedstocks. In addition to Brazil, both Argentina and Paraguay have announced national fuel alcohol programs.

The trend of alcohol fuels development in South America differs in several respects from the strategies that are emerging for producing and introducing alcohol fuels in the Caribbean and Africa. The most practical concept currently envisioned for introducing alcohol production into the sugarcane industry is to annex distilleries onto existing cane processing plants to use either molasses by-product or cane juice obtained by taking advantage of idle cane crushing capacity. This saves at least 30% of the capital investment required for a stand alone autonomous distillery. The existing utility system for the processing plant is often sufficient to also provide for the distillery's process energy requirements. Use of wastes and idle production capability also avoids the need to bring more land into cultivation.

However, at the time that Brazil's Proalcohol program was initiated (1975), the ethanol productivity needed to significantly reduce oil imports greatly exceeded the molasses and idle capacity available from the refined sugar industry. Oil price projections were also extremely pessimistic at the time and sugar prices were fluctuating wildly. It was, therefore, decided that ethanol production should be dissociated from the sugar industry and autonomous distilleries were built to produce ethanol directly from cane juice. It was also necessary to greatly increase the acreage devoted to sugarcane. Most of Brazil's alcohol is now produced in autonomous distilleries.

Another difference in alcohol fuels development in South America and elsewhere is the widespread use of hydrous ethanol (95% ethanol, 5% water) in cars specially manufactured for neat ethanol fuel. This was possible because of the existing automobile industry in Brazil and allowed Brazil to greatly exceed the substitution possible with 20% ethanol blends in conventional automobiles. The quantity of alcohol now used as fuel in Brazil is equal to the consumption of gasoline.

It is, therefore, important to realize that the Brazilian experience with alcohol fuels should not necessarily be emulated in all respects elsewhere. The extremely favorable combination of low capital and feedstock costs in Brazil is not readily apparent in other countries. Although numerous other countries are low-cost sugar producers, the investment costs for distilleries are generally much greater because of the need to import equipment and technical services. Brazil's initiative to make turnkey distilleries available to other countries at low cost improves prospects elsewhere but it is still well advised for others to concentrate on annexed distillery construction initially (18). In order for other countries to convert to hydrous alcohol fuel, it would be necessary to import new cars from Brazil and to create an entirely new infrastructure for the distribution of fuel. This would require an enormous commitment of resources by countries adopting this approach. International financial institutions are not inclined to support the costs for such a transition considering the many other pressing needs for investment.

Brazil has the greatest biomass energy potential of any country in the world and in a recent study was identified as the only country whose potentially usable biomass exceeds its primary energy demand (192% in 1990) (9). Despite the caveat just expressed, Brazil deserves much credit for the outstanding success of its national fuel ethanol program, Proalcohol. At the start of the 1973 oil crisis, Brazilian crude oil production supplied only 23% of a growing oil demand of 753,000 barrels/day. The increasing price of oil had a debilitating effect on economic development and studies by the private sector called for the substitution of alcohol for gasoline. In November of 1975, former President Ernesto Geisel launched Brazil's National Alcohol Program. The initial objective was to blend 20% ethanol into gasoline using the idle milling capacity of existing sugar mills but for the reasons already cited, the focus shifted to the construction of autonomous distilleries completely dissociated from the refined sugar industry. There was strong resistance to alcohol fuels initially but this subsided as transitional problems were resolved and the benefits to the national economy became apparent. The World Bank endorsed the alcohol program and provided much of the financing for distillery construction and technology development. Despite the allegations of critics in developed countries that the program was viable only because it was heavily subsidized, alcohol producers, particularly in Sao Paulo State have achieved major reductions in production costs by improving distillery efficiencies and by enhancing by-product applications.

The alcohol fuel price at the retail level is fixed by the Brazilian Instituto do Acucar e do Alcool and is based on the sugar prices (5, 19). Although the program was launched initially with investment credit incentives, financing is currently available at market interest rates only. Ethanol production data for Brazil is shown in Table 4.

Table 4

Ethanol Production in Brazil

Crop Year	Anhydrous (200 proof) (10^6 liters/year)	Hydrous (185 proof) (10^6 liters/year)	Total
1974-75	216	409	625
1975-76	231	322	553
1976-77	303	367	670
1977-78	1,177	295	1,472
1978079	2,097	394	2,491
1979-80	2,714	693	3,402
1980-81	2,104	1,601	3,705
1981-82	2,453	2,786	4,239
1982-83	3,550	2,279	5,829
1983-84	2,445	5,325	7,770
1984-85	2,650	6,442	9,092
1985-86 (forecast)	-	-	11,600

Source: Reference 3.

Current production is equivalent to 240,000 barrels of oil equivalent/day. Roughly 400 autonomous and 200 annexed distilleries are now operating. The program has created

600,000 new direct jobs and 1,800,000 indirect jobs (20). Sugarcane production increased from 74.5 million MT in 1974-75 to 189.1 million MT in 1983-84 with 77% of production coming from the South-Central region (21); 89.2 million MT (47%) of this sugarcane is devoted to ethanol production.

Ethanol-powered trucks are used to haul sugarcane and products and even the tractors plowing the fields run on ethanol. The large quantity of bagasse available is also able to provide for all distillery energy requirements and an efficient distillery can produce a 30% surplus of bagasse for sale as fuel to others. These factors allow for a very positive net energy balance for alcohol production from sugarcane which is compared to that for corn in Figure 2.

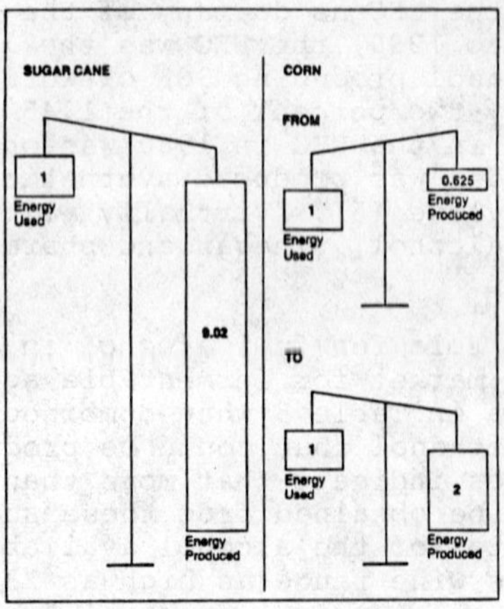

Source: Reference 21.

Figure 2. Energy Balance from Sugarcane and Corn

Although the initial focus was on anhydrous alcohol production for blending with gasoline, the demand for alcohol for blending began to level off by 1980. Attention then shifted to the production of hydrous alcohol (185 proof) for use as a neat fuel (no gasoline whatsoever). The Brazilian Astronautics Technological Center undertook the development of an Otto cycle engine to use hydrous alcohol. The Brazilian automobile industry, which was facing the threat of possible gasoline rationing, also began work on an ethanol engine. Cars for hydrous ethanol fuels were first sold in 1979 and by December of 1984, 1,800,000 automobiles (17% of the total fleet) were running on neat hydrous ethanol. Some 91% of all new passenger car sales are for hydrous alcohol fuel. Another advantage of hydrous alcohol as a fuel is that the capital-intensive dehydration of alcohol can be eliminated.

In addition to the distilleries already in existence, 97 new units (47 expansions and 50 new units), with a combined capacity of 1.9 billion l/yr, have been approved and 187 projects (21 expansions, 166 new units) totalling 2.6 billion l/yr are under evaluation. Brazil hopes to greatly increase alcohol exports and has targeted production expansion for export at 4.9 billion l/yr by 1990 (21).

EUROPE

The prospects for alcohol fuels among countries of the European Economic Community (EEC) are difficult to assess on a collective basis. Countries vary markedly in terms of their petroleum resources, agricultural products and production costs. Some of the strongest proponents of alcohol fuels come from the EEC as do many of the most ardent critics. Prior to 1980, the EEC was the largest producer of fermentation ethanol producing 36% of worldwide output in 1977 (22). Sixty-two percent of the 1.45 billion l/yr of ethanol produced in the EEC in 1980 was obtained from crops while the remainder was produced synthetically by the hydration of ethylene (5). Virtually all ethanol produced was for potable alcohol, vinegar and pharmaceutical applications.

A potential role for fuel alcohol in the EEC would be as an additional market for fermentable agricultural surpluses. The data in Table 5 when combined with the 1.3 billion l/yr of ethanol that could be produced from waste fruit and potatoes indicate that more than 3 billion l/yr of ethanol could be obtained from these surpluses. In addition, estimates of the alcohol available by the fermentation of surplus wine range as high as 231 million l (23).

Gasoline consumption by the EEC was 106 billion l/yr in 1981 (5). With gasoline readily available at present. The phase down of lead in gasoline from 0.4 gm/l to 0.15 gm/l by January 1986 will create a significant demand for new octane enhancers, including ethanol. The potential demand for ethanol as an octane enhancer could be as high as 4.6 billion l/yr and this would be a good match with the productivity currently possible using agricultural surpluses (23).

The principal objection to alcohol fuels in the EEC is the cost of fermentation ethanol. Cost analyses indicate that synthetic ethanol costs 33.8-41.2¢/l versus a cost of 35.1-55.9¢/l for fermentation ethanol (24). However, the data does not take into account the fact that synthetic ethanol is not anhydrous as currently produced and that producers would have to purchase equipment to produce anhydrous ethanol suitable for blending with gasoline. The redistillation of surplus wine is estimated to cost at least 11.5¢/l (23).

Proponents argue that if more of an effort was made to recover marketable by-products in the process of ethanol fermentation and distillation, then alcohol could probably be produced for 30¢/l. Also, government funds might be better used to subsidize alcohol fuel production rather than their continued use to support the prices of crops now produced in surplus. Although a substantive argument for ethanol has been made for the EEC, the petroleum industry is strongly opposed to alcohol fuels and no concerted policy

Table 5

Estimated Energy Potential of Sugar Surpluses[1] in the EEC Crop Year 1978-1979, 1979-1980 and 1980-1981 Average

	Sugar Production (-) (1,000 t) (a)	Sugar Consumption[3] (1,000 t) (b)	Surplus (a)-(b) (1,000 t) (c)	Ethanol Production $10^6 l$[4]	Ethanol Production % of Total Petrol Consumption
Germany	2,785	2,216	569	391	1.2
France	4,218	1,991	2,227	1,388	6.2
Italy	1,614	1,674	-60	-	-
Netherlands	893	574	319	210	4.0
Belgium	847	355	492	317	8.1
United Kingdom	1,094	2,338	-1,244	-	-
Ireland	170	144	26	23	1.6
Denmark	429	202	227	182	9.1
EUR-9	12,050	9,494	2,556	1,798	1.7

1. Sugar surplus = production-consumption.
2. EEC, "The Agricultural Situation in the Community."
3. Donnees statistiques sur l'industrie sucriere, Syndicat National des Fabricants de Sucre en France, 1981.
4. $10^4 l$ = (6)x4658.4 l/ha of sugarbeet.
5. Blended with petrol, 1,000 liters of ethanol represent 0.85 toe.

Source: Reference 5.

has yet been established to use agricultural surpluses for this purpose. Some exceptions to this static situation have, however, been identified and distilleries recently have been introduced in Sweden and Ireland and a 50 million l/yr distillery is contemplated for Abruzzo, Italy.

AFRICA

Alcohol fuels have been aggressively pursued in many African countries currently producing sugar. A number of these countries have modernized their sugar industries and are among the lowest cost sugar producers worldwide. In addition, many of these countries are land locked which means that it is not feasible to sell by-product molasses on the world market and that gasoline is quite costly if imported. The situation in Africa differs from that in the Caribbean in that there is no ready U.S. export market available to African producers that is exempt from import duties. This necessitates a careful determination of alcohol production targets so that quantities do not exceed the 20% blending limit. The gasoline demand in many of the countries is quite small and in several instances would not support more than one or two distilleries. Africa and the Caribbean are similar in that the best approach to alcohol production is through the construction of annexed distilleries using by-product molasses.

Mali, Malawi and Zimbabwe have successful operating fuel alcohol programs while efforts in Kenya have been les promising. The experiences in these countries offer excellent examples of the effects of different strategies for introducing alcohol fuels technology. Approaches that employed the simplest technology and that maximized the use of existing infrastructure were far more successful than those attempting to be more sophisticated. The insistence by many distillery engineering firms that their equipment be purchased for projects they engineer forces developing country buyers to pay a substantial penalty by increasing distillery capital investments. Other countries that are currently planning for alcohol fuel distilleries are Ethiopia, Sudan and Zambia. Although the ultimate demand for alcohol for blending may not be great in Africa, a place f alcohol fuels has been clearly established and is growing.

Zimbabwe

Fuel alcohol now meets about 15% of the gasoline dema in Zimbabwe. The alcohol production cost of 55¢/l compare favorably with a gasoline import price of 58¢/l. Because t government of Zimbabwe had limited capital resources (economic sanctions were in effect), the government elected to use a batch design offered by Gebr. Herman/Buckau Walther of West Germany, a firm which did not stipulate equipment suppliers. Instead of importing distillery components, th locally available fabrication infrastructure was exploited to fabricate 80% of the plant in Zimbabwe. The distillery was annexed to use the surplus molasses available at the Triangle Ltd. sugar mill. By using this approach, Zimbabw was able to build a 42 million l/yr distillery at a capita cost of only $6.4 million (3). This strategy not only sav money, it also helped further the infrastructure for distillery construction in Zimbabwe and thus advanced nationa technical capabilities. There are reports of a second distillery using cane juice but no details are available (16). Although the government has considered a hydrous ethanol fuel program, the World Bank recommended against such a program because of the cost and complexities of vehicle conversion (25).

Kenya

Kenya's experience with fuel alcohol contrasts sharpl with that of Zimbabwe. One of the driving forces was the large availability and poor market for molasses which stimulated processors to consider downstream investment to convert the molasses to a more marketable commodity, ethanol. The gasoline import price of 26¢/l was not particularly high. Nevertheless, two distilleries were constructed.

The Madhvani project was undertaken by a sugar compan recently expelled from Uganda and that, therefore, did not have access to its own sugar mill. The project site was at Kisumu on Lake Victoria and was 25-80 Km away from the source of feedstock molasses. In addition to producing 9.2 million l/yr of fuel alcohol, the complex was to produce 1,000 MT/yr of baker's yeast, 2,200 MT/yr of citric acid and 1.7 million l/yr of vinegar. A turnkey distillery design was purchased from the Brazilian firm Conger. The Swiss firm, Process Engineering Company (PEC) and the Swedish firm AB Sorigona provided engineering and technology to the project along with a portion of the equity investment. Delays, inflation and growing design complexities escalated capital costs to $125 million, more than twice the original estimate. The project ran out of capital, went into receivership and has never produced any alcohol. This unmitigated disaster is the supreme example of the consequences of poor siting (remote from feedstock), overly elegant design and wastefulness of capital (31). Even if one treats the capital investment for this facility as a sunk cost, the estimated production cost of 36¢/l is still 10¢/l greater than the price of gasoline in Kenya (25).

The second project, undertaken by the Mehta sugar group, followed a more conventional strategy. The plant was annexed to the Mehta group's East African Industries mill from which it drew molasses and steam. The 18 million l/yr distillery was fabricated by the Austrian firm, Vogelbusch and was constructed by an Indian firm for a total capital investment of $33.8 million (31). The plant has been producing alcohol since 1982 at an estimated production cost of 24.5¢/l excluding depreciation (25). A third, autonomous distillery was planned for Riana to produce 45 million l/yr of alcohol from cane juice at an estimated capital investment of $76 million. This project was shelved, largely because of the negative atmosphere created by the problems that had been encountered during the Madhvani project. All three projects employed continuous processes.

In addition to the activities highlighted for Zimbabwe and Kenya, several other fuel ethanol distilleries have been built. In Malawi, IFC financed a 6 million l/yr annexed distillery to produce enough alcohol to substitute for 13% of the gasoline demand (25). In Mali, two small industrial alcohol distilleries are being upgraded to produce 1.8 million l/yr of fuel alcohol which will displace roughly 3% of gasoline demand (26).

ASIA

Alcohol fuels activity in Asia has not, to date, been noteworthy. Sugarcane does not dominate agriculture on this continent and most existing sugar producers either produce for local consumption or have favorable trade agreements with developed countries. Strategies for producing alcohol fuels probably will be more diversified than has been the case elsewhere and other feedstocks, such as cassava, may predominate.

Thailand

Thailand is a promising candidate for alcohol fuels ar an alcohol fuels program is under active consideration. Th country depends heavily on the free market for the sale of about the 1.5-2 million MT/yr of sugar exports and is a large exporter of molasses (approximately 750,000 MT/yr). It is also the only country that produces large quantities of cassava as a commercial crop. Dry cassava chips had bee exported to Europe as an animal feed but legislation and competition from U.S. DDGS have greatly reduced this market The opportunity, therefore, exists to use cassava as an alcohol feedstock using the know how that has already been developed by Brazil and an economic analysis indicates that cassava alcohol will cost less than alcohol produced from sugarcane. Thailand already produces 203 million l/yr of industrial alcohol and another 162 million l/yr of capacity is now under construction (27).

OCEANIA

Many of the countries in this region are reasonably good candidates for alcohol fuels. They produce large quantities of sugar and molasses and rely heavily upon exports into the world market. Although none of the countries is land locked, some are remote from commercial shipping lanes and many are oil importers. The Philippines Papau New Guinea and Fiji either have or are in the process of initiating alcohol fuels programs. Again, the most appropriate technology for fuel alcohol production is by the use of annexed distilleries using molasses.

Philippines

The Philippines exports substantial quantities of both sugar (0.6-1.0 million MT/yr) and molasses (500,000 MT/yr) and sugar production costs are relatively low. Access to many of its islands is limited and the country imports nearly all of its petroleum requirements. A national alcohol fuels program was recently announced (Decree 2201) and fuel alcohol is already being produced on a trial basis (28). On the island of Luzon, a distillery using molasses feedstock is producing fuel alcohol for blending up to 10% with gasoline (25). Large-scale annexed distilleries are planned for Luzon and in Negros (5). Research is also in progress investigating alternative feedstocks for ethanol production. On Bohol Island, a mini-distillery has been constructed using Nipa Palm sap as feedstock. This indigenous species produces a sap containing 15% sugar at a rate of 40 l of sap/tree/season (29). Government planners hope to convert all the excess sugar capacity to ethanol production and to export ethanol to Japan and elsewhere when thei own requirements for blending are fulfilled. As mentioned before, transplanting the aggressive Brazilian program to other countries needs to be done with caution and careful planning.

AUSTRALIA AND NEW ZEALAND

With the substantial fossil resources available in this region, there is currently little emphasis on alcohol fuels (30). There is a substantial sugar industry in Australia which has 33 sugar mills and 6,600 cane growers (7). There are currently 4 molasses-based distilleries producing 63 million l/yr of ethanol of which 42 million l/yr is consumed domestically by the beverage and chemical industry and 21 million l/yr is exported. There is some potential for the use of ethanol as an octane enhancer in the future.

SUMMARY

Fermentation ethanol capacity identified in the course of this preliminary survey is summarized in Table 6. The information so far collected should not be considered comprehensive but it is sufficient to demonstrate the trends in alcohol fuel production. Since Brazil and U.S. production far exceed alcohol production in the rest of the world, their contributions to total productivity are subtracted out in Table 7.

In the past 7 years, worldwide fermentation alcohol production has increased 7.5-fold. Even when the two major centers of increased production are excluded, alcohol production has nearly doubled during this period. Virtually all the additional capacity installed is being devoted to fuel alcohol production. Furthermore, the momentum for continued production increases remains strong worldwide with planned projects contributing 23.4% more capacity to current capabilities (excluding Brazil and the United States). Trends clearly indicate that alcohol fuels are gaining in popularity despite moderating oil prices. The principal reason for this appears to be because large surpluses of sugar and other fermentable agricultural commodities have depressed world market prices. This is driving feedstock producers to seek alternative outlets for their products.

It is also possible to make some meaningful observations about the nature of regional alcohol fuel activity. Although capacity growth has been rapid in North America, growth has moderated and is expected to level off by 1988. Growth in South America has slowed somewhat but is expected to continue to increase steadily, at least through 1990. Central America and the Caribbean are poised for explosive growth in alcohol production because of the 34% excess in cane crushing capacity available and because of the foreign exchange drain caused by oil imports. The export market to the United States made possible through the Caribbean Basin

Table 6

Worldwide Alcohol Fuels Production Capacity

Location	Fermentation Alcohol Capacity (10^6 liters/year, 1984) Installed	Under Construction	Planned
North America			
United States	1,628[11]	360	367
Canada	6.8	7.6	
Central America (+ Caribbean)			
Jamaica	91[1]		
Costa Rica	31		0.6
Guatemala	0.2	18	
El Salvador	15.1		159
Honduras			21.6
Dominican Republic			56.8
South America			
Brazil	11,600[2]	508	686
Argentina	380	n.a.[3]	n.a.[3]
Paraguay	26.1		
Equador		5.4	
Colombia	38		
Europe			
United Kingdom	45[4]		
France	260[4]		
Fed. Rep. of Germany	94[4]		
Sweden	6		
Italy			46-53
Other EEC countries	246[4]		
Surplus wine	182		
Africa			
Malawi	6		
South Africa	31[6]		
Zimbabwe	42		
Kenya	38[7]		
Mali	2		45[8]
Sudan			39
Ethiopia			34
Zambia			1.5[8]
Asia			
India	391[4]		
Thailand	203	162	
Oceania			
Papua New Guinea		5.4	
Philippines	10[9]		
Japan	88[6]		
Australia	63[6]		
New Zealand	15		

Alcohol fuel projects have also been reported in Bangladesh, Botswana, Fiji, Indonesia, Ivory Coast, Nicaragua, Tanzania, Thailand, Upper Volta (Bourkina Fasso) and Panama, although no detailed information on these projects has, as yet, been obtained.

[1] Includes 15 million l/yr of rum production and a 76 million l/yr hydrous alcohol upgrading facility. Similar facilities are also planned for Antigua, Aruba and the Virgin Islands.

[2] Total production is changing constantly.

[3] n.a. - data not available but capacity is known to exist.

[4] 1977 data - alcohol is for non-fuel uses, total fermentation alcohol production in the EEC increased to 900 million l/yr by 1980.

[5] Alcohol available by the distillation of all surplus wine. Estimates range as high as 231 million l.

[6] Alcohol produced for non-fuel purposes.

[7] 20 million liter/year of capacity is currently idle.

[8] Plans have been suspended for the moment.

[9] Exact production difficult to determine.

[10] A national alcohol fuels program is in preparation.

[11] Data represents actual production. Estimates of available capacity range as high as 3.2 billion liters/year.

Table 7

Estimated Worldwide Fermentation Ethanol Production

(10^6 liters/year)

	Production		Under Construction	Planned
Sample	1977	1984		
Worldwide Total	1,814	13,938	1,066	1,480
Total less Brazil	1,210	3,938	558	794
Total less Brazil and the U.S.	1,011	2,310	198	426

Initiative provides additional impetus for rapidly increasing capacity in this region. Although Europe produces a substantial quantity of fermentation ethanol, it has made only a few tentative attempts to introduce fuel alcohol. Africa is another region where fuel alcohol has achieved a significant standing and further opportunities remain to be exploited if the necessary capital and management expertise can be found. Enthusiasm for alcohol fuels in Oceania (Southeast Asia) has been strong but development has been inhibited by shortcomings in program planning. Australia and New Zealand do not appear to be very good prospects for alcohol fuels because of the abundance of fossil resources.

REFERENCES CITED

1. Ethanol, Performing for America, Renewable Fuels Association, Washington, D.C., 1983.

2. Anderson, E. V., "Battle Rages over Imports of Fuel Ethanol," C&EN, 63 9, April 22, 1985.

3. Anderson, E. V., "Large Volume Fuel Market Still Eludes Methanol," C&EN, 62, July 16, 1984.

4. Anderson, E. V., "Use of Gasoline Ethanol Blends may Rise," 64, January 6, 1986.

5. Biomass for Energy, Organization for Economic Cooperation and Development, Paris, France, 1984.

6. Paul, J. K., Large and Small-Scale Ethyl Alcohol Manufacturing Processes from Agricultural Raw Materials, Noyes Data Corp., Park Ridge, NJ, 1980.

7. Alcohol Week, 6, April 8, 1985, p. 5.

8. Chemical Marketing Reporter, Schnell Publishing Co., 228, August 19, 1985.

9. Biomass Energy: A Business Opportunities Program, SRI International, Menlo Park, CA, October 1981.

10. Alcohol Week, 6, December 24, 1984, p. 8.

11. Finneran, K., "Characterization of the Renewable Energy Industry," Solar Energy Industries Association, Washington, D.C., July 1985.

12. Synfuels Week, June 25, 1984, p. 4.

13. Cerro, J. A., Prospects of National Programs in Latin America and Caribbean Area, International Symposium on Sugar and Alcohol, Sao Paulo, Brazil, June 23-27, 1985.

14. Rivero, N., Resources and Potential of the Industry for the Production of Energy-Producing Products, presented at the Technical Meeting on the Identification of Opportunities for Developing of Energy-Producing Products Based on Sugarcane and Study of Ways for Financing Projects, Santo Domingo, Dominican Republic, 18-20 July, 1984.

15. Rivero, N., personal communication, Organization of American States, Washington, D.C., August 25, 1985.

16. Hallberg, D., personal communication, WETAG, Washington, D.C., August 8, 1985.

17. Arechaga, F., Klausmeier, W., Richmond, R., Alvarez, M., Steingass, H., "The Feasibility of Alcohol Fuels Production in Honduras," draft report to U.S. AID by RONCO Corporation, Washington, D.C., March 1986.

18. Navarro, L., personal communication Destilaria Alcidia, Sao Paulo, Brazil, June 23, 1985.

19. "Estudo Sectorial Sobre o Alcool," IBRASA, Nucleode Planejamento, Brasilia, 1981.

20. Gusmao, R., Introductory Remarks, International Symposium on Sugar and Alcohol, Sao Paulo, Brazil, June 23-27, 1985.

21. Brazil Ethanol Export Potential, prepared by the Brazilian Ethanol Producers' Special Committee, Sao Paulo, Brazil, 1984.

22. Alcohol Production from Biomass in the Developing Countries, The World Bank, Washington, D.C., August 1983.

23. Coombs, J., Outlook for Oxygenates in the European Market - The Role of Ethanol, International Symposium on Sugar and Alcohol, Sao Paulo, Brazil, June 23-27, 1985.

24. Marchal, E., "Mercuriales Alcool," Biomasse Actualities, Paris, October 1982.

25. Prospects for Ethanol in Developing Countries, World Bank Memorandum, Washington, D.C., February 21, 1985.

26. Mali Biomass Alcohol and Efficiency Project, Industry Department, World Bank, Washington, D.C., May 25, 1983.

27. Trindade, S., "Fuels Ethanol Issues in Thailand," report to UNIDTCD, Ref. INT-120-4-R30-03-X, July 1985.

28. The Energy Transition in Developing Countries, The World Bank, Washington, D.C., August 1983.

29. Alcohol Fuels: Options for Developing Countries, report of an Ad Hoc Panel of the Advisory Committee on Technology Innovation, National Research Council, National Academy Press, Washington, D.C., 1983.

30. Noble, J., Australian Sugar Policy, International Symposium on Sugar and Alcohol, Sao Paulo, Brazil, June 23-27, 1985.

31. Juma C., "Market Restructuring and Technology Acquisition: Power Alcohol in Kenya and Zimbabwe," Development and Change, 16, 39 (1985).

DISCUSSION

Veigel — What price would oil have to be to justify importation in Brazil?

Klausmeier — I'd say that they should be importing oil now. Besides a strict economic analysis, it's very important to define the social component. I would say that the social factors are probably worth $5 a barrel to a government like Brazil or a government in Central America.

D. Hall, U. of London (U.K.) — The employment component in Brazil is 800,000 permanent jobs, and 1.4 million part-time jobs, and that is a very important component at a very low cost. Secondly, I must say that the EEC is not the only group that subsidizes agriculture. The calculation that we've done shows that within Europe, we subsidize agriculture to the tune of $30 billion per annum. The U.S.A. is approximately the same.

Klausmeier — In addition, almost every sugar-producing country subsidizes the domestic market for their sugar. Almost every country has a very high price for sugar if somebody wants to buy it in the country. EEC was just very aggressive in increasing the exports. That is perhaps unique to EEC.

B. Spaugh, Nebraska Energy Office

Your introduction included looking at chemicals and commodities from renewal resources, and one of your slides had small-scale versus large-scale corn to ethanol plants with more by-products. We're interested in industrial chemical uses of crops. Could you tell what you think the future is for some of these areas?

Klausmeier

I think that over the long run you have the potential for a very diversified industry based upon fermentables, such as acetone, butanol, acetic acid, propanol, citric acid, gluconic acid, all kinds of vitamins, and many specialty chemicals. As oil became popular, a lot of the commodity chemicals that had low value per pound shifted to production from oil, but there's no intrinsic merit for producing oxygenates other than the fact that oil is located in one place. I'd say over the long run, the biotechnology base that we have in this country could offer real opportunities. A good analysis could identify some selective near-term opportunities in chemicals.

A. White, U. of Connecticut

Is Brazil the only country which is manufacturing alcohol fuel vehicles on a large scale? Is that not a limiting factor in the Third World countries where you need the vehicles to really create major viable markets for alcohol fuels?

Klausmeier

It is true. Brazil is the only country producing all-alcohol-fuel vehicles. You can use alcohol blended with gasoline up to 20%, and that is a significant market in most countries. It's not going to displace 50% of the gasoline the way it is in Brazil, but it can displace 10 to 20%. In some small countries, you have to do a good market analysis because you might not be able to support more than one distillery. Usually, you can support a couple of distilleries, but nothing like the 600 they have in Brazil.

REGIONAL EFFORTS TO PROMOTE BIOMASS ENERGY TECHNOLOGIES

Phillip C. Badger
Manager, Southeastern Regional
Biomass Energy Program
Tennessee Valley Authority
Biomass Branch
Muscle Shoals, Alabama 35660

ABSTRACT

The U.S. Department of Energy sponsors the Regional Biomass Energy Program to promote the development and application of biomass energy production and conversion technologies. Four geographical regions have been identified to participate in the program: the Northwest, Great Lakes, Northeast, and Southeast. Projects conducted under the program are specifically oriented toward resources, technologies, applications, and issues most significant to the particular region. Projects are usually conducted through the State governments or through private sector solicitations. Although relatively new, the program has been very successful and has gained strong private sector and State government support.

REGIONAL EFFORTS TO PROMOTE BIOMASS ENERGY TECHNOLOGIES

BACKGROUND

Of all the renewable fuel options, biomass offers by far the greatest potential for both liquid and gaseous transportation fuels. Biomass also can provide renewable fuels for stationary applications. Utilization of biomass can provide jobs and markets from resources that are presently unused and enhance agricultural and forestry productivity and quality.

Biomass contributes more to our Nation's energy supply than either nuclear or hydro resources. About 8 percent of the U.S. industrial energy and over 5 percent of our national energy requirements are supplied by wood. The Office of Technology Assessment estimates that by 2000, biomass could provide more than 20 percent of our national energy needs.

Biomass is not only renewable but also contains negligible sulfur, is cost-competitive, provides local jobs, and reduces the export of energy dollars from communities. The ripple effects of the latter advantages can be significant, especially for smaller, poorer communities in rural and more heavily forested regions. For these reasons, community leaders interested in economic development are strong biomass supporters and promoters. These factors have led to continued industrial biomass energy growth despite declining oil prices. Obviously, biomass energy should play a critical role in the national energy supply. Such use would provide positive effects on the national economy, security, and environment.

While biomass energy resources are highly diverse, their distribution in terms of both form and quantity tends to be highly variable and area-specific. Consequently, there are great differences in biomass energy potentials among geographical areas. These differences are related to: regional biomass types and availability; availability and costs of other energy sources; local requirements for energy; State and local laws; climate and terrain; transportation; population distribution; and the needs of energy network distributors and users.

To effectively handle problems associated with these differences, a national regional biomass energy program was established by Congress in 1983. Congress directed that this national program consist of regionally administered programs that were to "...carry out activities related to technology transfer, industry support, resource assessment, and matching local resources to conversion technologies." Since then, the DOE's Biofuels and Municipal Waste Technology Division has established four regional biomass energy programs. Each has a primary goal of promoting development of effective uses of biomass for energy purposes. Until now, total DOE funding for the national regional program has been about $4 million per year.

Regional programs established for the Northeast, Great Lakes area, Pacific Northwest and Alaska, and Southeast involve 36 States. These programs are managed by the Coalition of Northeastern Governors, Council of Great Lakes Governors, Bonneville Power Administration, and the Tennessee Valley Authority, respectively. Regional administration provides flexibility in meeting regional needs and matching local resources and enhances management by being local.

Goal and Objectives

The goal of the Regional Biomass Energy Program is to encourage the production of biomass energy feedstocks and their conversion to fuels by the private sector through support of regionally-specific biomass energy projects. The Regional Program, recognizing the differing biomass energy potentials and needs among regions, encourages private industry to make its own investments in biomass technologies which match regional or local opportunities. Key activities include projects which favor establishing of biomass availability within the various regions through resource assessment studies; transferring to the private sector the results of biomass energy research and development sponsored by DOE; and encouraging cooperation between industry and DOE through cost-shared projects that will build private sector confidence in biomass energy technologies.

General Concepts

The Regional Program carries out its mission based on the following concepts:

- The Regional Program emphasizes transfer of current and reliable information to potential biomass energy users. This includes information and technologies best suited to near-term use and economic and technical information based on case studies, where possible.

- Regional activities are coordinated with organizations, in Federal and State agencies, universities, and private industry, plus groups which further the use of biomass for energy.

- Regional information and education programs are provided and structured to reach potential users who may not be aware of biomass energy opportunities. State agencies identify potential industrial users.

- The Regional Program emphasizes modest expansion of existing knowledge rather than development of new technologies or information. For example, USDA programs in the Forest Service and Soil Conservation Service provide extensive data on biomass resources. Regional biomass resource assessments build on existing programs. Similarly, regional activities dealing with wood harvesting or direct combustion build on existing knowledge in the wood products industry.

- Current and projected nonenergy users of the region's biomass resources, both forestry and agricultural, are considered when assessing regional biomass energy potential.

- In conversion technology, pilot-scale, demonstration, or commercial facilities are used where possible to avoid duplication of effort and to make best use of public funds.

- Cost sharing is used as much as possible for most Regional Program activities.

- Projects are conducted with the States to strengthen individual State biomass programs.

- Projects are conducted through competitive solicitations relating to:

  --Resource Assessments
  --Environmental Concerns
  --Institutional Constraints
  --Fuel Standards
  --Combustion Technologies
  --Harvesting and Fuel Handling
  --Technology Transfer

Each of the four Regional Programs is structured with two basic funding features. The first provides for competitive solicitations for technical projects and cooperative funding of programs within each individual State. The State cooperative program provides money for individual States to pursue biomass energy projects of their choice, subject to Regional Program agreement and to a matching funds requirement. State efforts to promote biomass energy often are fragmented among many agencies involved in biomass energy production and use. A central objective of the State cooperative program is to improve coordination among those agencies by providing the States with an opportunity and an incentive to strengthen and integrate the work of energy, forestry, environmental, and other appropriate agencies in promoting biomass energy.

The second funding element involves contracting, via competitive solicitations, for conducting technical studies and producing information. These activities are directed to the most important biomass energy problems in each region. Planning, advisory, and steering committees composed of State and Federal representatives, university researchers and industry experts provide guidance to the Regional Programs in identifying those problem areas and selecting contract awards. Representatives on these committees also represent such national organizations and institutions as the Biomass Energy Research Association (BERA), Council of Biomass Energy Technology Sponsers (CBETS), National Wood Energy Association (NWEA), Gas Research Institute (GRI), American Boiler Manufacturers Association (ABMA), the Fiber Fuels Institute, and others.

In most cases, Regional Program funds are leveraged through cost-sharing provided by contract awardees. Technology transfer activities are coordinated with DOE field labs where the Regional Programs help obtain cost-share funds; locate industrial vendors, manufacturers, or research and development partners for biomass technologies; and emphasize promotion of DOE-developed technology. Program activities are coordinated between regions to take advantage of inherent regional strengths and to more efficiently utilize funding. For example, most work on agricultural residues is left to the Great Lakes Region due to that region's abundance of such resources.

Program Activities and Accomplishments

Regional Program activities have helped establish and strengthen State biomass programs. When the program began, many States did not have biomass programs. They were able to use Regional Program funds and leverage money from State legislators to begin programs. States with existing programs were able to strengthen their programs. In some cases, others have been able to maintain programs that would have been phased out.

State projects funded by the Regional Program vary from resource assessments and feasibility studies, through technology transfer activities and large-scale development projects. Regional Program approval and cost sharing is required for each project. Cost sharing is frequently two to three times the Regional Program input. One example of a State biomass program is a project by the State of Florida to assess the feasibility of a State biomass trade association. If warranted, this project would establish such an association.

The Regional Programs use different methods to transfer technology to potential producers, users, and developers. A national publication on biomass energy, Biologue, has been supported by the Regional Program by supplying articles, names of subscribers, and funding. The newspaper publication is printed quarterly and is sent free to about 15,000 recipients.

Workshops, conferences, expositions, and short courses are sponsored by the Regional Programs. Most of these are held in conjunction with or cosponsored with complementary institutions and organizations such as the Solar Energy Research Institute, Fiber Fuels Institute, National Wood Energy Association, or DOE contractors review meetings. Workshops on utilizing wood for industrial applications and municipal solid waste to energy have been conducted in most regions.

The Southeastern Region has established a computerized information exchange network consisting of information bases and an electronic bulletin board. The information bases complement existing information bases and contain files on biomass facilities, consultants, equipment vendors, and fuel brokers. The bulletin board provides instant, up-to-date information on conferences and new additions to the information bases as well as providing teleconferencing capabilities.

Factsheets, guidebooks, and brochures have been prepared by each Region. For example, the Great Lakes Region has published a "1985 Directory of Great Lakes Regional Biomass Energy Facilities"; other regions are preparing similar directories. The Great Lakes Region has also published a "Directory of Biomass Combustion Systems 250,000-1,000,000 Btu/hr Output."

Technology development activities are usually funded through solicitations which require cost sharing. Such solicitations are usually conducted annually by each Region. Areas of solicitation depend upon needs established by the appropriate Region's planning or steering committee involving heavy private sector participation. Technologies related to production, harvesting, handling, conversion, and utilization are included. However, because of the abundance of biomass wastes and underutilized resources, production technologies are being given lower priority. These projects stress private sector involvement, impact on biomass energy development, and development of technologies closest to commercialization.

Solicited projects have been very successful in meeting these criteria. Solicitations received are 60-70 percent from the private sector with cost sharing of up to four times the Regional seed monies. This also has promoted cooperation between university and private sector enterprises. Some funded projects include development of standards for biomass fuels and a feasibility assessment of cofiring wood and coal. The latter technology appears to have tremendous environmental benefits in terms of reducing sulfur emissions. Analyses of economic impacts of biomass energy on the economy have been completed for the Great Lakes and Northeast Regions. Results indicate vast potential with the possibility of adding 60,000 jobs in the Great Lakes Region within the next ten years. Numerous biomass utilizing facilities have been installed as a direct result of Regional Biomass Programs.

Other examples of projects include:

- Development of methodology for assessing forest residue volume, leading to an accepted industry-wide application.
- Increased wood fiber utilization standards, resulting in improved land management programs.
- Enhanced woodstove standards, leading to standards development and emissions reductions.
- Demonstration of enhanced residue recovery and collection, i.e., double-entry methods for residue collection, and steep slope mechanical harvesting applications.
- Quantification of air, water, and soil related environmental impacts, leading to emissions reductions from slash burning.
- Focus on institutional factors perceived as barriers to use and acceptance of wood residue by industry, leading to publication of permit guides and project development guidebook.
- A handbook of environmental factors and regulations related to biomass energy.
- A handbook on utilizing biomass for cogeneration.
- Assessment of wood, wood residue, and wood waste resources.
- Development of factsheets/case studies on actual facilities.
- Development of simplified economic assessment procedures for biomass energy facilities and a guidebook on financing options.
- A study on residential catalytic wood stove.
- A handbook on wood heating.
- A residential emissions and safety guidebook for policymakers.
- A national survey of wood energy users.

Program support, especially by the private sector, has increased steadily as the program has become better known and its capabilities have been recognized. Widespread private sector participation has further enhanced success by assuring that program plans and priorities correspond with development needs. Most companies, especially smaller companies, usually have limited research and development budgets. For these companies, regional program seed funds have become an important source of support for finishing the development of new technologies that can significantly impact biomass energy. The regional programs are now looked to for "local" biomass energy leadership and as information brokers for technology, sources of biomass feedstocks, potential business partners, and seed funds.

SUMMARY

The Regional Biomass Energy Program provides a way to focus on regionally-specific needs and opportunities in biomass energy development and application. This is accomplished through cooperative activities funded and conducted by Federal, State, and local governments working with private industry. This type of broad-based interagency, intergovernmental, and industry support enhances the program's effectiveness by adding to the diversity and regional acceptance of the program and ensuring wide dissemination of research results. In fact, an important aspect of the program is the transfer of current and reliable information to potential users of biomass energy. This emphasizes information and technologies best suited to near-term use. In this respect, the Regional Program differs from the long-term, high-risk orientation of other biomass energy programs sponsored by the Department of Energy.

The Regional Biomass Energy program, although less than two years old, has already made a significant impact on the use of biomass energy in the Nation. State governments are either implementing biomass programs or strengthening existing ones. Commercial and industrial companies, especially among the nonwood industry, are increasing their use of biomass resources as they become more aware of resource potential and utilization technologies. Future program growth will depend on many factors including conventional fuel costs. However, the large amount of unused biomass resources is expected to play a significant role in meeting the energy requirements of the Nation.

BIBLIOGRAPHY

Badger, P. C., Annual Report - 1985, Southeastern Regional Biomass Energy Program, Tennessee Valley Authority, Muscle Shoals, Alabama 35630.

Bancroft, David, Annual Report - 1985, Great Lakes Regional Biomass Energy Program, Council of Great Lakes Governors, Madison, Wisconsin 53703.

Department of Energy, Regional Biomass Energy Program, (brochure) 1984.

Energy Information Administration, Estimates of U.S. Wood Energy Consumption 1980-1983, DOE/EIA-0341(83), 1984.

Flavin, C., Renewable Energy at the Crossroads, Center for Renewable Resources, Washington, D.C., 1985.

Fox, Pat, Annual Report - 1985, Pacific Northwest and Alaska Bioenergy Program, Bonneville Power Administration, Portland, Oregon 97208.

Office of Technology Assessment, Energy from Biological Processes, Volume I - Biomass Resource Base, Congress of the United States, Washington, D.C. 20510, 1982.

O'Connell, R. A., DOE's Regional Energy Programs, Proceedings of the 5th USDA Solar and Biomass Workshop, Atlanta, Georgia, 1985.

Rinebolt, David C., Annual Report - 1985, Northeast Regional Biomass Energy Program, Coalition of Northeast Governors, Washington, D.C. 20001.

Schiefelbein, G. F., Thermal Conversion of Biomass: Progress and Prospects, Pacific Northwest Laboratory PNL-SA-12486. 1985.

DISCUSSION

L. von Engelbrechten, West German Embassy	I have a question regarding pollution. You only mentioned it for wood stoves. Can you say anything about pollution problems with the other devices you showed, and if there are EPA rules and regulations to control them?
Badger	In the Southeast, we have not worked directly on residential wood stoves. Regulations are in place for wood stoves in some areas of the country. When a pollution alert is declared because of the amount of wood smoke in the air, everyone in town has to shut off their wood stove until the pollution level decreases below a certain level.
D. Rinebolt, Coalition of Northeastern Governors	I might just say in response to the question that EPA regulates major sources of a million Btu and above, and then individual states regulate in increments. Wood stoves are actually the only small source for which regulations are contemplated at this time.
Klass	Some states have rather severe regulations, like the State of Colorado, and the State of Oregon. Oregon has probably one of the most severe regulations for controlling residential wood-burning fireplaces and furnaces in the entire nation in terms of emissions.
Rinebolt	I might also say that the Northeast Regional Biomass Program publications detail state-by-state air emission regulations for wood-burning devices. You can contact me and I can get those for you.

CANADIAN FEDERAL BIOENERGY PROGRAMS

Allan J. Dolenko, Ph.D.
Renewable Energy Branch
Energy, Mines and Resources Canada
Ottawa, Ontario, Canada K1A 0E4

ABSTRACT

Canadian government involvement in the development of bioenergy as an alternate energy source began in earnest in 1978. In that year the government introduced several program initiatives in conservation and renewable energy. This was a period of great uncertainty and instability with regard to oil price and supply. Canada was a net importer of approximately 300,000 barrels of oil per day. Conventional oil reserves were being rapidly depleted and predictions were that oil prices would rise to $60 a barrel or more by 1990.

It was generally perceived at the time that the potential for increased bioenergy development in Canada was significant. The contribution of bioenergy was already estimated at between 300 to 350 PJ, about 3% of the primary energy demand, and was largely attributed to the combustion of liquors and mill residues in the forest products industries. It was recognized that biomass could be economically attractive as an alternate energy source and could contribute to the national goals of energy self-reliance, job creation and regional development as well.

It was also recognized however, that a number of barriers existed which impeded the widespread introduction of bioenergy as a viable energy alternative. These barriers included the following.

1) A lack of precise knowledge of bioenergy supply, availability and costs. While it was apparent that there was a huge quantity of biomass in Canada, there was a lack of reliable statistical data on its availability and costs.

2) A fragmented and weak supply infrastructure. This of course was part of the chicken and egg situation where the establishment of a viable biomass supply infrastructure was entirely dependent upon the development of an assured market.

3) The high cost associated with the introduction of proven bioenergy technologies. It was recognized that the capital cost of bioenergy equipment can be up to several times greater than the cost of conversion equipment for conventional energy sources such as oil, natural gas or electricity. Even with lower fuel costs, the payback period was often too long to encourage industry to make the necessary capital investment.

4) A lack of awareness of opportunities. There was very little information available for potential bioenergy users as to the technologies and economic opportunities associated with the use of biomass.

5) The high risk and cost of introducing new technology. Canadians are generally conservative in their approach to the introduction of new technologies. We tend to be followers rather than innovators and therefore quite reluctant to take on the risks involved in the introduction of new technologies.

6) Lack of available expertise. There were few individuals who had much knowledge of or first-hand experience with biomass as an energy source.

The initiatives introduced in 1978 and in subsequent years were designed to address the barriers described above and take advantage of the energy and economic opportunities of bioenergy. They were comprised of programs which offered direct incentives to the bioenergy user as well as programs which support research, development and demonstration of bioenergy technologies and applications. The overall objective was to increase the total bioenergy contribution from 3% to 6% of primary energy demand by 1990.

Before giving a description of the programs and their status, it is desirable to first address the biomass supply question.

BIOMASS AVAILABILITY AND PRICE

Current estimates show that the availability of forest biomass alone is in the order of 100 million dry tonnes annually. This is equivalent to approximately 25% of Canada's total primary energy demand. Figures 1 and 2 show the source and supply/price estimates respectively for forest biomass in Canada.

Forest biomass sources are broadly divided into five categories.

1) Mill residues are comprised of bark, sawdust, shavings, pulping liquors, etc. which are the waste byproducts of forest products processing mills. They represent some of the lowest cost biomass at 0 to $15/dry tonne and represent 14% of the forest biomass available. Much of this residue is used for energy by the mills themselves.

2) Logging residues are composed of the residual material left in the forest after normal harvesting operations i.e., slash (tops, branches, etc.), small trees, non-commercial species as well as deformed and defective trees. These residues represent a large potential source of forest biomass at 23.6% of the total available, but tend to be somewhat difficult and costly to collect ranging from $25 to $90/dry tonne.

3) The salvaging of wood killed by fire, insects, disease, wind or flooding, and

4) the conversion of stands of non-commercial trees and scrub represent another large potential source of biomass (24.9% and 9.6% respectively) which are recoverable at costs ranging from $30 to $60/dry tonne.

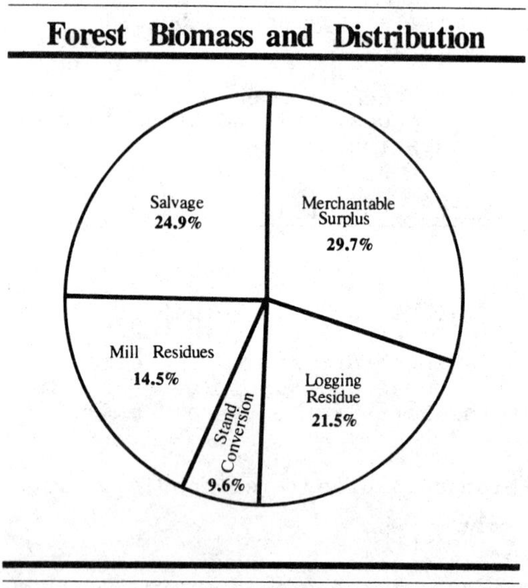

Figure 1 Forest Biomass and Distribution

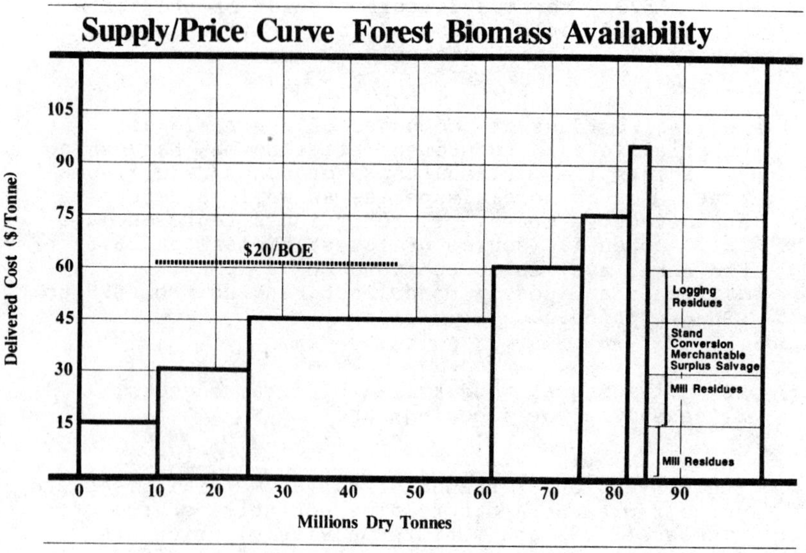

Figure 2 Supply/Price Curve Forest Biomass Availability

5) Finally, there is the surplus material which is over and above the current industry demand for fiber used in the manufacture of conventional forest products. In Canada this is almost 30% of the total forest biomass available at costs between $30 to $60/dry tonne.

The current demand for forest biomass is roughly equivalent to 26 million dry tonnes per annum. This suggests that the utilization of forest biomass could readily be doubled at supply costs not exceeding $45/dry tonne provided suitable markets and applications were available. The utilization of biomass obtained from logging residues, stand conversion and salvaging operations for energy and chemicals can provide the necessary economic return to encourage more intensive forest management and regeneration. This concept is being applied with considerable success in a number of parts of the country.

In addition to the forest biomass, substantial quantities of agriculture biomass are available particularly in eastern Canada as well as municipal solid wastes (MSW) in the densely populated urban areas.

BIOENERGY INCENTIVE PROGRAMS

Beginning in 1978, the Canadian government introduced a number of direct incentive programs to address the barriers of high capital costs of bioenergy conversion equipment, the underdeveloped supply infrastructure and the general lack of awareness of the potential of bioenergy. These programs were aimed at the industrial, commercial and residential sectors.

The most ambitious bioenergy program initiated in 1978 was the Forest Industry Renewable Energy (FIRE) Program. Its purpose was to promote energy security and self-sufficiency by substituting readily available renewable biomass fuels for fossil fuels in the forest products industry. The original program budget was $103M over five years ending in 1983-84. As part of the National Energy Program, announced in 1980, the program budget was increased to $288M and its duration extended to March 31, 1986. In addition, the program was expanded to include non-forest industrial, commercial and institutional organizations as well as other forms of biomass such as municipal and agricultural wastes and peat. The program budget was later reduced to $96M.

FIRE is a capital cost sharing program which contributes between 10 and 20% towards the capital costs of modifying existing equipment or establishing new facilities to convert biomass into primary energy. Contributions are determined by a cash flow analysis which takes into account all eligible capital and operating costs as well as revenues and avoided costs. Almost all costs associated with purchasing and installing the necessary equipment are eligible capital costs.

The program officially ended March 31, 1986 and is now in a two year wind-down period during which time projects still underway will be completed, a number of selected projects will be carefully monitored to obtain actual benefit/cost data. In addition, a concerted effort will be made to transfer this information to the commercial/industrial and institutional sectors in order to promote as many replications as possible.

Table 1 summarizes the results of the program to date. Approximately $96 million have been contributed to 180 projects having a total cost of about $725 million. This represents exceptionally good leverage in private sector investment in Canadian goods and services for the number of federal dollars spent. Once completed, the projects will displace approximately 9.3M BOE per year. As anticipated, the greatest number of projects were undertaken by the forest products industries. Furthermore, projects in the wood industries, i.e. sawmills, furniture factories, particle-board mills, etc. gave the greatest return in terms of program dollars spent per BOE displaced per year. This result is not unexpected since such projects are characterized by the lowest relative capital and feedstock costs. In contrast, energy from waste projects carried out primarily by municipalities are characterized by high capital costs associated with the safe and efficient handling as well as combustion of solid waste.

In response to the 1979 oil price shocks, the Canadian government introduced the National Energy Program in 1980. A major initiative under this program was the establishment of the Canada Oil Substitution Program (COSP). This program was aimed at the residential sector and provided a one-time taxable grant of up to $800 for homeowners to convert from oil to natural gas, electricity, solar or wood. The program was originally scheduled to run for 10 years to 1990. It was terminated in 1985 as part of a general government effort to cut cost and reduce the deficit.

TABLE 1

FIRE PROJECTS

	No. of Projects	Capital Cost ($ x 10^6)	Fire Contribution ($ x 10^6)	Calculated BOE Savings/yr. (x 10^6)	$/BOE/yr
Pulp and Paper	46	426.4	51.5	5.7	$ 9.03
Wood Industries	89	104.5	14.4	2.7	$ 5.33
Municipalities	7	127.2	18.3	0.4	$45.75
Institutions	10	43.5	7.4	0.2	$37.00
Other	28	25.6	4.6	0.3	$15.33
TOTALS	180	725.5	96.2	9.3	$10.34

Residential wood heating in Canada is quite significant with over 2 million (25%) of Canadian homes using wood as a primary (50%) or secondary (10 to 50%) heating source. This does not include open fireplaces. Under COSP, approximately 225,000 grants having a value of $230 million were issued to homeowners to convert off oil to wood representing 23% of all off-oil conversions. The number of installations of wood burning appliances has levelled-off with the market being largely saturated. Installations which have occurred over the past five to ten years are characterized by low technology systems with inadequate attention paid to safety and performance. This has resulted in an increased number of house fires as well as increased levels of pollution in areas having high concentrations of wood-burning appliances. The bioenergy program in the residential sector is focussing on the development of safety standards as well as efficiency and emission standards in cooperation with the Canadian Standards Association and the industry itself; the promotion of public awareness on these issues and on the development of improved technology through R,D&D.

In 1984, a sort of mini-FIRE program known as the Alternate Energy Development Program (AEDP) was initiated in Canada's smallest province of Prince Edward Island. PEI has a small population of less than one hundred and fifty thousand which is largely rural. Energy costs on the Island are considerably higher than the Canadian average as it has no fossil reserves and little hydro. The objective of AEDP is to displace imported fossil fuels with indigenous biomass fuels. A large forest biomass resource exists on the Island which is under-utilized and has little traditional commercial value. Under the program, the commercial/industrial and institutional sectors are eligible for contributions of up to 62.5% of the total cost of the installation of biomass conversion systems, thus guaranteeing a three year pay-back In conjunction with AEDP, a massive forest management and improvement program has been undertaken which provides the biomass fuel in the form of woodchips. The program is scheduled to run through to 1989 and is expected to displace some 50,000 BOE per year.

BIOENERGY RESEARCH, DEVELOPMENT AND DEMONSTRATION PROGRAMS

In order to address several of the barriers impeding the progress of bioenergy such as the high risk and cost of developing and implementing new technologies, the lack of awareness of bioenergy technologies and the lack of technical expertise, the federal government has initiated a number of programs to support research, development and demonstration of new methods and technologies to substitute biomass for non-renewable fuels and chemicals. Because of the diversity of interest in bioenergy technologies, the Canadian R,D&D program has involved a number of federal departments and agencies.

The first major bioenergy R&D program was the Energy From the Forest Program (ENFOR) initiated in 1978. This was a five year, $30M program to support research and development of new methods and technologies aimed at substituting forest biomass for non-renewable fuels and chemicals.

ENFOR was comprised of two sub-programs: Biomass Production, dealing with forest related topics involving resource assessment and raw material supply; and Biomass Conversion, dealing with the transformation of biomass into energy and chemicals. In 1983/84 the final year of its authorized term, responsibility for the conversion sub-program was transferred from the Canadian Forestry Service (CFS), Environment Canada to the Renewable Energy Branch of Energy, Mines and Resources. This transfer was carried out to permit better integration of biomass conversion R&D with other related R&D activities within EMR and to broaden the program to include all forms of biomass. This integration was carried out in 1984-85 resulting in a new, restructured R&D program called the Bioenergy Development Program (BDP). The CFS is still responsible for the Biomass Production sub-program of ENFOR.

Under NEP announced in 1980, energy R&D funding was greatly expanded. Particular emphasis was given to the development of alternative transportation fuels to displace gasoline and diesel. This included the use of biomass as a feedstock for the production of fuels. Federal departments and agencies involved with R&D included the CFS, Energy, Mines and Resources, Agriculture Canada, Environment Canada and the National Research Council (NRC). The programs managed by these departments were largely contracted-out to the private sector, universities, provincial and other non-profit research organizations and were supported in most instances by a small in-house R&D effort. The various departmental programs were co-ordinated by the Bioenergy Program Office of the NRC. Then in late 1984 the government instituted a number of measures to reduce expenditures in an effort to come to grips with the growing federal deficit. Among those cost-cutting measures was a drastic reduction in the energy R&D budget involving the virtual elimination of all bioenergy R&D activities at the NRC. The Bioenergy Development Program managed by Doug Hayes is attempting to fill the gap left by the NRC and the reduction in bioenergy activities by other departments.

Figure 3 traces the estimated bioenergy R&D expenditures beginning with under $1 million in 1977, reaching a peak of about $20 million in 1984, then dropping back sharply to about $8 million where hopefully it will stabilize. Government reductions in spending and lower oil prices have taken their toll on federal bioenergy R&D spending in Canada as they have in many other countries. Of even greater significance is the fact that these extreme swings in budget allocations make it very difficult to plan or implement an effective R&D program which, of necessity, must be long-term in nature.

In parallel with its efforts in R&D the federal government also initiated the Conservation and Renewable Energy Demonstration Program (CREDA) in 1978 managed by Energy, Mines and Resources.

The objective of this program was to demonstrate promising new technologies and applications which exploit renewable energy sources, conserve energy or make its use more efficient. In effect, the program was designed to defer the technical and commercial risk involved in the deployment of new technologies and, in so doing, gain consumer acceptance. A program budget of $114M was allocated over five years ending in 1983-84. CREDA was a bilateral program which involved cost shared agreements with the provinces. The program has been extended an additional year to wrap-up on-going projects. A new federally delivered demonstration program, ENERDEMO-Canada was announced in 1984. With the change in government that year however, the program never became operational. The current goverment's policy is to return to the joint federal/provincial program mode. As a result EMR has negotiated a number of Memoranda of Understanding (MOU) with the provinces under which bioenergy demonstration activities are being undertaken. These demonstration activities have been incorporated into the BDP.

Well over one hundred bioenergy demonstration projects have been carried out under the programs across the country having a total cost of just under $50 million and a federal contribution of approximately $12 million.

Examples of typical bioenergy demonstration projects supported under the programs include: co-generation from energy-from-waste, wood-fueled district heating, energy from solid-waste, wood-chip furnaces and boilers, anaerobic digestion of hog manure, medium-scale wood chippers, whole-tree chipping and large-scale, fully automated wood heating systems.

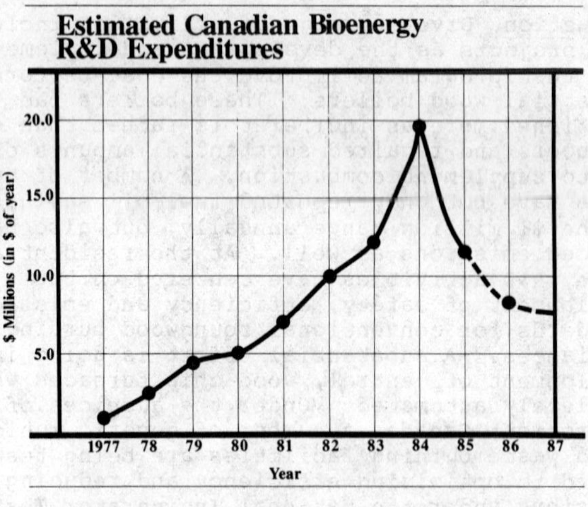

Figure 3 Estimated Canadian Bioenergy R&D Expenditures

As currently structured, the BDP is comprised of R&D as well as demonstration activities and is managed by the Bioenergy Division of EMR. Theoretically, the delivery of an R,D,&D activity by a single group should improve operational efficiency and should also improve the rate at which new technologies achieve commercialization. A wide spectrum of activities are being supported under this program as outlined below:

1) Biofuel Production. Activities range from the development of improved harvesting and forest management techniques, to the development of equipment for the harvesting of logging residues, to the development of a sophisticated computer program (FORCYTE) which assess the impact of various forest management systems on the forest ecosystem.

2) Combustion. Diverse range of activities including such projects as the development and implementation of a test program to improve the cost-performance of industrial wood boilers. These boilers had functioned more as incinerators rather than energy producers and required substantial amounts of fuel oil to supplement combustion. A number of tested units have not only reported fuel oil savings, some in the $1 million range annually, but also have reduced emissions as well. At the residential scale, R&D activities have centered on the development of safety, efficiency and emission standards for conventional roundwood burning appliances. A substantial effort is going into the development of central, wood-chip furnaces which are completely automated. Under the auspices of Environment Canada, a number of generic municipal solid waste burning facilities are being tested with regard to optimizing efficiency and reducing emissions under the National Incinerator Test Program.

3) Thermochemical Conversion. In the area of biomass gasification, work is well underway on the operational phase of the world's largest fluidized-bed gasifier (Biosyn) in Quebec. To date this gasifier installation has been used to demonstrate the production of electricity in engine generators using a low-joule value gas (LJV). Biodev, a sister company of Biosyn, is installing a 6.7MWe, $9M unit, in French Guyana on behalf of Eléctricité de France. Work is continuing on the utilization of oxygen as the combustive gas under pressure for the production of a medium joule value (MJV) gas for use in the production of chemicals or fuels such as methanol. Work is also underway on the development of the two-stage Heuristic gasifier which can operate using wet hog fuel as feedstock.

Research on the direct liquefaction of biomass for the production of fuels and chemicals is proceeding at a fairly high level among several world-reknown Canadian researchers in the field including Dr. Scott's fluidized-bed process at the University of Waterloo and Dr. Roy's Vacuum Pyrolysis process at Laval University. The federal strategy is to wrap-up most of the basic studies in this area within the next couple of years and select one or two processes which have potential commercial interest for scale-up to the pilot level.

4) Biological Conversion. Research is focussed on the enzymatic cellulose hydrolysis process being developed at the Iotech/Monenco 1 tonne wood/day pilot plant using steam exploded wood and the hydrogen fluoride (HF) process originally begun by Canertech. Mohawk Oil Company, Canada's only distributer of gasohol, has announced plans for the 9 million l/day expansion of its grain alcohol plant in Minnedosa, Manitoba based on the Iotech process. The recent drop in world oil prices, however, makes the economics for the process more tenuous.

Regarding anaerobic digestion, work is limited to the monitoring of several installations using various digestion techniques. Applications for the advanced concepts for anaerobic digestion (e.g., fixed film) will be promoted in various industrial installations (as opposed to farms).

5) Materials Handling and Pretreatment. The efficient handling and reliable feeding of biomass is crucial to any biomass conversion process. Therefore, this area continues to receive a high priority. Past work has resulted in rapid commercialization of the technique developed. For example, a new bin and silo design developed by B.C. Research for the efficient storing and transfer of biomass eliminates flow stoppages. A moisture metering device developed by Forintek is undergoing field trials to allow better process control for wood-fired boilers. Steam pretreatment processes developed by Iotech and Stake Technology hold considerable promise in cellulose hydrolysis applications.

6) IEA. Canada plans to continue its active involvement in the IEA through participation in the new Bioenergy Implementing Agreememt.

THE FUTURE OF BIOENERGY PROGRAMS IN CANADA

According to current estimates, bioenergy now contributes approximately 5.6% of Canada's primary energy demand or about 230,000 BOE per day (Table 2). This contribution is mainly in the residential and industrial sectors of the economy where bioenergy contributes approximately 14% and 17% of the primary energy demand respectively (Figure 4). The overall contribution is well on its way to meeting the target of 6% of primary by 1990.

Under the National Conservation and Alternative Energy Initiative (NCAEI) announced by the government last year, bioenergy will continue to receive significant funding in keeping with its role as a significant contribution to the national energy budget. In keeping with fiscal realities however, the trend will be away from large incentive type programs such as the FIRE program toward more selective R,D&D and information transfer. The budget for the various bioenergy activities under the NCAEI is summarized in Table 3. It is anticipated that the budgets for R, D&D will remain fairly constant into the foreseeable future. We shall attempt to gain the maximum leverage on the limited funds available through cost sharing arrangements with the private sector and other levels of government as well as the international community. Our objective is to carry on a viable bioenergy program with particular emphasis on shorter-term applications during the present oil glut while continuing work on promising longer-term opportunities for possible implementation when oil prices rise once again.

TABLE 2

CONTRIBUTION BIOENERGY

(BOE/DAY, PRIMARY ENERGY)

	CURRENT	NEAR TE[illegible] (1995[illegible]
Residential	76,000	85,0[illegible]
Commercial/Industrial	153,000	200,0[illegible]
MSW	1,800	4,5[illegible]
Liquid Fuels	50	2,2[illegible]
TOTAL	230,850	291,7[illegible]
% Primary Energy Demand	5.6%	

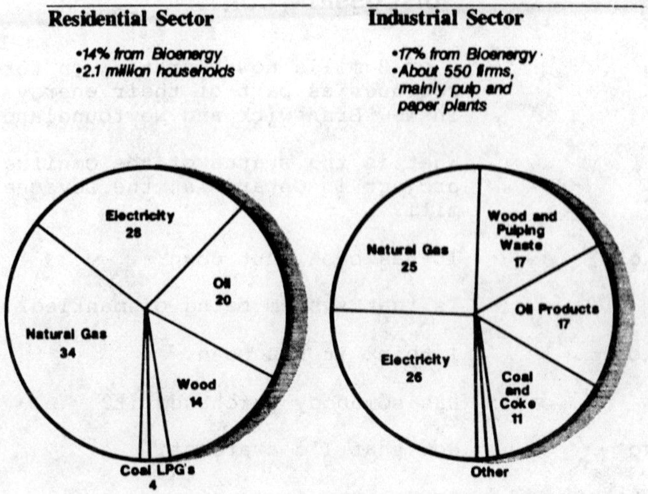

Figure 4 Bioenergy Contribution

TABLE 3

SUMMARY OF BIOENERGY ACTIVITIES AND BUDGETS

	1985/86	1986/87 $ millions	1987/88
1. Forest Industry Renewable Energy (FIRE)	9.8	12.8	4.2
2. Bioenergy Development Program (BDP)			
Research & Development	4.5	5.9	6.0
Demonstrations	.6	3.0	2.5
3. Canada/PEI Alternate Energy Development Program	2.0	2.0	1.7
4. Biosyn	1.0	1.0	0.2
5. Liquid Fuels from Biomass (E from C)	0.1	0.7	0.1
TOTAL	18.0	25.4	14.7

DISCUSSION

G. Marsh, Oregon Department of Energy — You gave an example of forest residue going to a mill. Was that a demonstration and do you have any mills that regularly use forest residue?

Dolenko — Yes, that was a demonstration project, and it is continuing. I think there are

	2 or 3 mills now operating on forest residues as part of their energy supply in New Brunswick and Newfoundland.
Klass	What is the status of the Omnifuel project in Ontario at the Leveque plywoo mill?
Dolenko	It has been shut down.
Klass	Is that system being dismantled?
Dolenko	I think it has been.
Klass	Has somebody purchased it?
Dolenko	Not that I'm aware of.
C. Setterstrom, Harrington Research Co.	In Canada, how serious are the concerns about solid waste disposal and the othe environmental problems mentioned by New York State?
Dolenko	I think we have very much the same concerns that have been referred to alread today. About 3 weeks ago, we had a special conference in Toronto focusing energy from wastes and some of the majo technical, economic, and environmental problems. The program I mentioned is being operated by Environment Canada an is a 3 or 4 year program that will spen just over one million dollars a year, looking at emissions from our existing energy-from-waste plants.
G. Grassi, Commission of the European Communities (Belgium)	Could you say something about future funding for bioenergy over the next few years?
Dolenko	We have received a 3-year mandate to fu bioenergy R&D. Within my department, we'll be funding at a level of about $6 million per year for R&D. In addition that, there's another $2 million of bioenergy R&D being carried out in the Canadian Forestry Service, Agriculture Canada, and Environment Canada. In addition, we have about another $3 million specifically for demonstration projects per year.

NEW YORK STATE RD&D IN ENERGY FROM BIOMASS AND WASTES

Parker D. Mathusa
Program Director
Energy Resources & Environmental Research
New York State Energy Research
and Development Authority
Two Rockefeller Plaza
Albany, New York 12223

ABSTRACT

The New York State Energy Research and Development Authority programs in biomass and wastes have expanded from approximately $2.5 million in new project commitments in FY '85 to approximately $4 million in FY '86. The increase reflects a major emphasis on municipal solid wastes and sewage as energy resources. Existing biomass programs emphasizing wood resources are continuing at historic levels.

This progress report includes highlights of completed projects as well as new program directions. In the areas of municipal waste and sewage, the following topics are described: 1) performance of MSW pretreatment systems at existing installations; 2) progress in managing dioxin emission at MSW incinerators; 3) success of energy conservation and recovery measures at municipal wastewater treatment plants; and 4) recent contributions to anaerobic digestion technology. The biomass topics discussed include: 1) results of a series of fast-growing hardwoods projects, including the commercial plantation in Massena, NY; 2) progress in membrane separation of biogas streams; 3) prospects for refining of wood to produce fuels and chemicals, including the status of a proposed facility near Tupper Lake, NY; and 4) accomplishments of the Authority's wood technology transfer and industrial risk-sharing programs.

NEW YORK STATE RD&D IN ENERGY FROM BIOMASS AND WASTES

INTRODUCTION

The New York State Energy Research and Development Authority was established as a public benefit corporation in 1975. The New York State Legislature, in describing the need for an energy research, development and demonstration agency, declared:

> ...that the need for obtaining and maintaining an adequate and continuous supply of safe, dependable and economical power and energy is a matter of concern to the people of the state; that accelerated development and use within the state of new energy technologies to supplement energy derived from existing sources will promote the state's economic growth, protect its environmental values and be in the best interests of the health and welfare of the state's population; and that such development and use requires special efforts to foster research, development, and demonstration in the methods of production and use of new energy technologies.

Based on its enabling legislation, the Energy Authority's goals are to obtain and maintain an adequate and continuous supply of safe, dependable and economic power and energy for the State, thereby promoting the State's economic growth and protecting its environmental values. The Energy Authority's specific objectives are "the development and utilization of safe, dependable, renewable and economic energy sources and the conservation of energy and energy resources."

In summary, the Energy Authority's mission is to help ensure that the State is provided with adequate and economical future supplies of energy. To accomplish this mission, the Energy Authority's research, development and demonstration (RD&D) program focuses on technologies and resources specifically applicable to New York State and its needs. The RD&D program is designed to avoid duplication of research being conducted at the national level and to supplement, rather than preempt, private sector research efforts.

New York State is heavily dependent upon petroleum products and upon imported oil in particular. This disproportionate reliance upon an unstable energy source has had a major effect in shaping State energy policy. In the area of petroleum use, 1984 statistics show that:

- The share of the State's total energy needs provided by petroleum remains higher than the national average (50 percent vs. 40 percent);

- Most of the petroleum consumed in the State is still imported (60 percent);

- Of the petroleum imported, New York imports a higher total volume of OPEC oil (31 percent) than any other state.

In terms of end use, utilities account in large part for New York's heavy dependence on foreign oil. Many New York State utilities, and to a lesser extent other large-volume oil consumers, rely heavily on residual oil, 90 percent of which is refined from foreign oil. In addition, a large percentage of the refined petroleum products vital to the State's economy is supplied from nearby refineries in the mid-Atlantic region, which themselves are supplied almost entirely with imported oil.

Natural gas is New York State's second major source of primary energy, as it is nationally, with New York currently consuming approximately eight percent less than the national average. Nearly all of New York's current gas supplies originate from domestic sources, primarily outside the State.

New York's own renewable and indigenous resources, including hydro, solar, wood, biomass and wastes, currently supply ten percent of the State's energy needs.

The New York State Energy Master Plan forecasts that the contribution of biomass and waste energy resources to the State's energy supply mix could triple by 1996. With further technological development and application, this contribution could be even greater. Consequently, biomass and waste resources have been designated a major area for RD&D.

In 1985, New York State's biomass and waste energy resources provided 1½ percent of its primary energy consumption, or about 60 trillion Btus. The increased use of these resources is a key State strategy for diversifying its energy mix and reducing its dependence on imported energy supplies.

This presentation has been divided into three sections: Program Organization; Program Strategy; and Findings. Each section focuses on the Energy Authority's program to accelerate the use of biomass and waste resources for the production of primary energy in New York State.

PROGRAM ORGANIZATION

The biomass and waste resources of New York State represent renewable feedstocks for the production of energy and chemicals. The opportunities available to maximize the utilization of biomass in economically attractive applications are as diverse as the various resource materials themselves. To best realize these opportunities, a method of coordinating RD&D and user groups must be established. Recognition of this fact resulted in the Energy Authority taking the lead in 1982 in extending its RD&D programs on a regional basis in collaboration with the utility industry, other northeastern states, and Canada. Currently in its fourth year of operation, the Northeast Biomass Research Center is intended to coordinate and accelerate the development of biomass and waste resources and technologies in order to ensure that the Northeast obtains the energy and economic benefits available from these resources.

Biomass Resources

A coordinated program has been developed to provide a mechanism for combining available resources of participating organizations interested in the development of products derived from biomass resources to achieve mutually defined goals. The overall objective is to formulate, fund and execute cooperative RD&D programs for converting biomass to energy products throughout the Northeast. Research projects evolve through program development and planning, which provide for participation by managerial and technical staff from each cosponsoring organization.

The Energy Authority's efforts emphasize a balanced program to use effectively and efficiently the various biomass and waste resources in specific applications.

The Energy Authority's overall goals for northeast biomass research are to:

- Provide an integrating mechanism for bringing together resources to facilitate the utilization of biomass and waste materials for energy purposes; and
- Provide a mechanism for structuring relationships among policymakers, resource controllers, and user organizations.

One of the program's essential concepts is integration. In the area of biomass and waste energy development, the primary goal of the program is to provide: (1) the needed linkages between various industrial and governmental research organizations, and (2) an effective mechanism for the joint identification of problem areas and opportunities, as well as joint formulation and funding of RD&D programs for dealing with them.

The Energy Authority's specific research goals for the program include:

- Developing methods for converting biomass materials into pipeline-quality substitute natural gas;

- Accelerating the commercial deployment of technologies for converting biomass to energy products, emphasizing combinations of resources and technologies which produce solid, liquid, and gaseous fuels, along with electricity and process heat; and

- Assuring that adequate supplies of biomass resources will be provided in an environmentally sound manner.

Municipal Wastes

The Energy Authority's RD&D program in municipal wastes addresses both the energy and environmental aspects of converting these wastes into supplemental energy sources. Energy consumers and municipalities will benefit from the utilization of the energy potential contained in municipal wastes, thereby helping to decrease energy imports into the State.

The Energy Authority goals for the Municipal Wastes area are to:

- Ensure that required RD&D is performed to meet common needs of the State's municipalities;

- Bring together resources to facilitate the utilization of waste materials for energy purposes; and

- Structure constructive relationships among State policymakers, municipalities and the environmental community.

The Energy Authority's program has been expanded to include direct RD&D support for the continued development of standards and criteria for toxic air contaminants related to resource recovery facilities.

The Energy Authority's specific program goals in Municipal Wastes are to bring about:

- Efficient recovery and utilization of wastes as energy and energy-related resources;

- Effective environmental control systems that will permit the utilization of wastes for energy; and

- Improved methods for converting waste materials into natural gas for the benefit of the utility ratepayers and end-users.

To achieve these goals, the Energy Authority has divided its efforts into two major components: (1) municipal solid wastes and (2) municipal wastewater and sludges. The Energy Authority's Five-Year Funding plan for Biomass and Wastes is attached in Table 1.

Table 1

BIOMASS AND WASTES
FIVE-YEAR FUNDING PLAN

Program Element/Project Area	FY 86 ($000)	FY 87 ($000)	FY 88 ($000)	FY 89 ($000)	FY 90 ($000)
Northeast Biomass Research Center					
• Northeast Comprehensive Regional Biomass Program					
Feedstock Supply	150	150	250	200	100
Engineering Technologies	150	-	-	100	100
Gasification Technologies	200	300	200	100	200
Environmental and Economic Support Analysis	50	150	100	150	150
Subtotal	550	600	550	550	550
• Biomass Energy Applications					
Fuel Production	800	650	700	600	600
Fuel Utilization	750	750	700	750	750
Subtotal	1,550	1,400	1,400	1,350	1,350
Municipal Wastes					
• Municipal Solid Wastes					
Existing Facilities	150	200	250	700	700
Combustion Quality	950	900	700	500	500
MSW Separation & Coproducts	-	450	200	400	200
Ash Utilization & Disposal	200	150	-	250	300
Alternative Conversion Systems	-	-	500	550	800
Landfill Gas Production	-	100	250	100	375
Biogas Separation	-	150	200	100	150
Leachate Treatment	350	-	250	100	75
Subtotal	1,650	1,950	2,350	2,700	3,100
• Municipal Wastewater Treatment	150	350	200	300	-
• Sewage Sludge Utilization	150	200	200	100	400
Subtotal	300	550	400	400	400
TOTAL	4,050	4,500	4,700	5,000	5,400

PROGRAM STRATEGY

The economy of the State of New York and the Northeast has engaged the focus of government for nearly ten years. Energy has always been an essential part of the Northeast's industrial and commercial health. New York State industry and that of the Northeast cannot function in a competitive world without an assured supply of reasonably priced energy. Because the State of New York and the northeastern states rely primarily on imported oil as a major component of their energy supply mix, fuel diversification is critical to the stability of the region's energy economy. The importance of biomass and wastes as energy feedstocks has been realized, since these materials are capable of making a valuable contribution toward stabilized energy requirements, as well as providing needed employment opportunities.

Biomass

Forest resources are expected to provide the base supply of biomass material in New York State in the near term. Primary research at the Energy Authority will include projects that address efficient utilization and sustained yields. Basic to both concepts is the development of a thorough understanding of the forest ecosystem. Of fundamental importance in this effort is detailed information concerning the physical limits of the forest.

Improved forest management techniques can increase output significantly. The small landowner must be encouraged to participate in any forest supply program, since over 90 percent of the commercial forestland in New York is privately owned. Of this total, over 20 percent is owned by farmers, and approximately 10 percent is owned by the forest industry. It is apparent that any major new demand on the resource base must involve multiple land and product use.

Investigations into the long-term effects of intensive forest management and whole tree chipping will be considered by the Energy Authority, with emphasis on the potential for soil depletion and erosion.

To provide the amount of woody biomass that might be required by New York State, silvicultural biomass plantations need to be established. A decade of research has gone into the intense cultivation of trees for biomass production, with efforts in the past focused on providing feedstock for the pulp and paper industry. The Energy Authority is currently demonstrating the use of hybrid poplar plantations for commercial harvesting of fuelwood.

Research in intensive cultivation can be broken down into two major subgroups: (1) management techniques for maximum growth/survival and (2) propagation techniques for optimum clonal development. With output per acre of primary importance, upcoming research must be focused on investigating the effect of alternative culturing techniques on growth. Number of stems per acre, irrigation, fertilization, pest and disease control, and pure-versus-mixed stands are some of the issues to be covered in such research.

Fertilization is an area which merits fundamental research. The potential for developing symbiotic relationships between trees and nitrogen-fixing organisms must be explored, along with the phosphorus-absorption ability of mycorrhizae. Fertilization is an expensive procedure which effectively increases yield and decreases the rate of soil depletion.

The success of intensive silvicultural management is directly correlated to the growing stock. In the case of hybrid poplars, extensive research has been done on the variance in growth and survivability between clone types. Scientists can now optimize a clone of poplar for a given climatic, edaphic and hydrographic regime. Further research is required on the hybridization of desired characteristics such as disease/pest resistance and frost hardiness into high-growth clone. Genetic manipulation of desired genotypes and clone-handling techniques are among the issues that need to be addressed.

Crop residues are another large potential source of biomass feedstocks. These residues can play an important role in supplementing materials grown for energy production. Herbaceous biomass plantations specifically designed for energy production will use marginal lands to avoid the "land-for-food" issue, and developmental efforts must concentrate on species selection and harvesting techniques for wet soils. Management techniques should minimize tillage culturing, soil erosion and depletion.

Fertilization requirements and alternatives, including nitrogen fixation, also must be addressed in the near term. Soil nutrient management is of vital importance, as is the required level of irrigation.

Criteria for the selection of new species should include: growth requirements; genetic variability; strain selection alternatives; response to fertilizers; susceptibility to weeds, insects, disease, pests; yield potential; and a set of cultural practices. Crops must have a high level of stress tolerance to cold, short growing season, wet conditions and low soil fertility. A number of crops with desirable traits have already been identified and classified as suitable for New York State.

Genetic manipulation, hybridization and reproductive techniques will be target of long-term fundamental research. Agronomic techniques developed for conventional crops should facilitate research in this area.

With a short growing season, consideration must be given to storage of biomass resources since some crops have a greater tolerance for long-term storage than others.

The goals for this program are to develop integrated processes, including production, harvesting and conversion, necessary for the generation of energy from biomass and wastes on a commercial basis. Milestones critical to the achievement of these objectives are listed below:

- Select plant species from herbaceous and woody plants and develop energy plantations on the basis of reported biomass yield data;

- Determine maximum biomass yields under controlled laboratory conditions for selected biomass species on marginal lands using specific growth criteria;

- Develop bench-scale processes that optimize energy yields at high conversion rates on selected biomass species;

- Confirm growth and yield data in test farm experiments for selected biomass species;

- Demonstrate pilot plant growth and conversion subsystems capable of scale-up to commercial systems with projected energy costs competitive with other energy sources;

- Integrate high-yield biomass sources and their respective end fuels such as ethanol, methanol and direct combustion;

- Demonstrate that a biomass yield of about 25-50 dry, ash-free tons/acre-year can be sustained from the offshore cultivation of seaweed in Atlantic waters; and

- Complete the peat resource assessment; if the technical and economic evaluations of peat are promising, develop peat as a gasification feedstock through the pilot plant stage.

Biomass RD&D Project Areas

1. Fuel Production

Surplus wood is sufficient, at least in New York and throughout most of the Northeast, to support foreseeable biomass energy applications for the next several years. However, the resource may eventually become limited either through expansion of the traditional forest products industry or through new uses of trees. For this reason, our resource development research addresses potentially large sources of supply for the next decade and beyond. Recent accomplishments include the expansion of the State University of New York College of Environmental Science and Forestry's research into the suitability of different hybrid poplar varieties for intensive energy farming under soil, climate and pest conditions prevalent in New York. The data and institutionalized expertise resulting from Energy Authority-supported work will provide a unique resource for organizations considering raising fast-growing trees for energy. The Reynolds Metals Co. in Massena, which now has approximately 200 acres in cultivation, is an example of the audience toward which research results are being disseminated through technical bulletins and workshops, as well as formal reports.

In addition to working to develop fast-growing trees, we are cooperating with several other institutions in evaluating the mechanisms of possible forest decline and the effects of atmospheric deposition on forest nutrient cycles and tree metabolic processes. The knowledge generated from this activity may help to preserve the existing forest resource for all uses, including energy.

The Energy Authority also has completed a survey of the prospects for energy from herbaceous (non-woody) plants in each of several regions in New York. The study evaluated the advantages and disadvantages of growing 49 different species, and although this resource is generally not cost-effective compared to surplus wood today, the report will be useful for future developers of energy farming concepts.

One of the barriers to implementation of cost-effective industrial and commercial wood energy technologies is uncertainty about the performance of newly developed equipment. The Energy Authority has instituted a risk-sharing program through which it will underwrite the cost of certain equipment to be used in innovative wood energy systems performing according to specifications agreed upon in advance.

2. Fuel Utilization

The Energy Authority is pursuing production of biomass substitutes for energy-intensive chemicals, not only to burn as fuel, but also to use for other purposes. The value added in the conversion process is greater in the case of many non-fuel chemicals, and the fossil energy required to make commodity petrochemicals can exceed a given product's heat of combustion by a factor of two or three.

Ethyl alcohol (ethanol) has the advantage that it can be derived from biomass and blended readily in gasoline for motor vehicles. Although ethanol for fuel blending is now generally made from grain, it also can be produced from wood, which is abundant in the Northeast. The Energy Authority completed an investigation last year into the most promising near-term technology for conversion of wood to ethanol and identified the Brink acid hydrolysis approach as most nearly ready for commercial application. As a result, the Authority is sharing the cost of a testing and evaluation project designed to lead to construction of a $25 million installation to produce 2.75 million gallons of ethanol per year and a variety of co-products at a site in Tupper Lake (Franklin County), New York.

Another milestone in our petrochemical substitutes program was the completion of the first phase of a laboratory effort at Rensselaer Polytechnic Institute, cost-shared by two industrial concerns, to perfect a process for converting sugars derived from wood to useful chemicals that could substitute for petroleum or natural gas-derived materials. The initial results indicate that fermenting 2, 3 butanediol from glucose is feasible and economically promising. Planned subsequent work will examine xylose as a raw material and will attempt to produce glycerol and succinic acid as additional products.

Municipal Wastes

Municipal wastes is a program area that has two components: municipal solid wastes and wastewater treatment and sewage sludge, each of which has its own development path.

Municipal solid waste is the most abundant energy resource of our cities. New York City currently landfills 20,000 tons of residential trash per day. Problems inherent to other wastes, such as collection and transportation to a central location, are addressed through conventional means with municipal solid wastes. Due to the varying composition of the waste, material separation and handling techniques must be addressed in near-term research. A trade-off may develop between the need for material separation versus the development of conversion techniques for contaminated feedstocks. Environmental concerns must be addressed in any research carried out in this area.

New York State has existing landfills that could produce over 1,000 billion cubic feet of methane during the next 10 years using conventional landfill gas technology. About 85% of this energy potential emanates from the most promising of 60 landfills. Landfill gas is composed mainly of methane and carbon dioxide in a roughly 2-to-1 ratio. These gases are formed during the anaerobic decomposition of refuse over time. It has been estimated that it takes about 50 years to fully degrade a given quantity of refuse using conventional landfilling techniques.

The major issues associated with landfill gas recovery and utilization include:

- the slow speed of normal decomposition which requires an extremely large amount of refuse to be in place to produce sufficient quantities of gas to make development economically attractive;

- the leachate produced during anaerobic decomposition which is a nutrient for methanogenic bacteria, and its loss through migration into the ground below the refuse which both creates a pollution problem and lessens the amount of recoverable methane;

- the separation of carbon dioxide and methane for making a pipeline-quality gas which is currently a costly procedure, requiring large volumes of gas for favorable economic development; and

- the definition of an acceptable level of treatment for the noxious pollutants contained in the condensate produced from the landfill gas extraction process.

Models for landfill gas production rates have been developed for the largest landfill in New York State. Generic models are available which correlate the landfill gas composition with potential gas production rates over time. Further work is required to statistically validate current models for a number of different landfills. Landfill gas production enhancement techniques have received a moderate level of experimental attention, with large-scale demonstrations of enhancement techniques being conducted in New York and on the West

Coast. Inherent to this research area is the development of alternative landfill designs adapted to gas recovery and enhancement techniques.

Sewage treatment plant sludges are another waste resource which can be made available for alternative gas production in the near term. Required research for this will concentrate on resource development and optimization of anaerobic digesters for the feedstock.

Solid Wastes. With the barriers to realizing the energy potential of refuse clearly recognized, the following RD&D strategy was developed by the Energy Authority:

1. Conduct feasibility studies and demonstration projects for the purpose of retrofitting existing municipal solid waste incinerators and other refuse-processing facilities for energy recovery -- A 1979 U.S. Department of Energy report noted that New York State had 23 existing municipal solid waste incinerators which were either not operating or operating but not recovering heat. The report estimated the total design refuse capacity of these plants at over 10,000 tons per day, and that 35 trillion Btu's of heat per year might be recoverable with proper plant rehabilitation. With high costs and siting difficulties of new plants, the potential to recover a large amount of heat, and the potential to lower the cost of incineration by utilizing existing equipment and structure, a study of existing facilities was warranted.

   Other solid waste processing facilities in New York State also offer opportunities for increased energy recovery. These include incinerators that recover energy but don't make use of their maximum potential. They also include resource recovery facilities such as those in Monroe County and Hempstead where equipment failures and environmental issues have prevented the realization of their full energy potential. Retrofitting these facilities may be warranted so that energy can be recovered without expending the capital to build an entirely new facility.

2. Conduct RD&D to improve the quality of refuse combustion and air pollution control. The control logic for current incinerators embodies the premise that plant operations should be governed by steam demand, while air emissions quality resulting from changing plant operating characteristics is not directly taken into account. The results can produce undesirable combustion gas effluents including compounds like dioxin. Laboratory tests have indicated substances like dioxin can be eliminated or minimized by proper and complete combustion. A new control logic needs to be developed which relates plant operating characteristics to emissions quality first, and steam demand second. Even with complete combustion of refuse, particulates and acid gas compounds or precursors will be emitted. With the current public concern regarding acid rain and potential health problems due to inhalable particulates, more efficient pollution control devices need to be studied and demonstrated.

3. Conduct RD&D to improve the quality of refuse as a fuel. Refuse inherently is a very difficult material to burn properly because of its highly variable nature in terms of composition and size, and because it contains a large amount of inert materials like metal and glass which cause slagging and erosion difficulties in incinerators. Refuse can be processed to remove inert materials (which in turn can be recycled) and make it more homogeneous, thus improving its combustibility. The processing can be done mechanically, although reliability, efficiency and costs remain as potential problems. The processing also can be done manually, although separation at the source methods are needed. Efforts to date to produce a densified refuse-derived fuel (d-RDF) from processed MSW have failed because of the highly variable moisture content of refuse, and technological problems in mass producing d-RDF pellets and in the handling/storage/transportation of these pellets to markets. The highly variable moisture content of refuse also can create difficulties in maintaining complete combustion conditions at all times during incinerator operating periods. To compensate for wet refuse conditions, a high-Btu density fuel such as waste oil can be stored at a facility and used when needed.

4. Conduct RD&D to utilize the co-products of refuse incineration. Ash, a major co-product of incineration, must be disposed of properly. In most cases, the ash will contain small quantities of compounds like dioxin which are adsorbed during the combustion process. In certain areas of the State, including Long Island and New York City, ash disposal by landfill may be extremely costly and difficult. An (unintentional) co-product of incineration is scrap tires, which sometimes are not burned because of a high sulfur content and resulting potential impact on local air quality. The utilization of such co-products in an environmentally sound manner can improve overall system economics.

5. Conduct RD&D to evaluate the long-term potential of alternative refuse disposal systems. Bioconversion of refuse to such fuels as ethanol and methane offers only a long-term potential benefit because of the heterogeneous nature of refuse, which contains large amounts of non-biodegradable materials which can degrade into toxic substances, as well as a wide variety of materials which can be toxic to biological organisms. Combustion-based systems other than incineration have been tested, but all pose questions which limit their potential and immediate use. Because New York State municipalities will be investing heavily in capital expenditures for waste processing facilities over the next 10 years, RD&D priorities are directed at technologies which can be retrofitted into existing facilities.

6. Conduct RD&D to enhance landfill gas production rates -- The recovery and use of landfill gas from the very largest of landfills has been demonstrated and is a commercialized technology. However, landfill gas production enhancement

techniques are needed to economically develop smaller landfills. For new landfills or new sections of existing landfills which can be lined with impermeable membranes, or for existing landfills with naturally impermeable bottom surfaces, leachate recycling within the landfill is practical and should be environmentally acceptable. With proper pH control and leachate recycle, a doubling of the amount of methane produced in a given time frame can be expected, thus making gas recovery economics more favorable. Important ancillary benefits of this approach are pollutants mitigation and a prolonging of landfill life as the decomposition rate increases.

7. Conduct RD&D to separate methane from landfill gas efficiently and economically. Current technologies for gas separation are efficient and economical for very large gas volumes. To enable more options to be available for smaller landfills (and sewage treatment plants) and to achieve higher efficiencies and reduce costs, alternative gas clean-up systems need to be developed and tested.

8. Conduct RD&D to produce biogas from landfill leachate. For existing landfills having bottom surfaces which are semi-permeable or of unknown quality, leachate must be collected and treated. Traditional aerobic biological or physical/chemical treatments produce sludges which must be disposed of in one of several conventional methods. Anaerobic treatment produces very little sludge and yields a biogas which could be merged with landfill gas collected using existing technology.

Solid Wastes RD&D Project Areas.

1. Existing Municipal Incinerators Program

(a) Heat Recovery from Existing Municipal Incinerators

In a recent study, 19 municipal incinerators were found to exist in New York State which are either operational and do not practice heat recovery, or are in reasonable structural condition but are not operating. These plants, owned by 13 different municipalities, have the potential to produce about 770 million Kwh/year of electricity plus additional thermal energy, if they are rehabilitated and retrofitted for efficient energy production in an environmentally sound manner. Because the structures and some of the existing equipment at these plants can be utilized, capital costs for rehabilitation/retrofit are expected to range from $20,000 to $64,000 per ton per day of capacity, with the lower cost applicable to larger plants. This contrasts to typical costs of $75,000 to $100,000 per ton per day of capacity for completely new facilities. Life cycle costs and revenues, engineering layouts, and a final report remain to be completed.

The planned follow-up to this study includes presentation of results to each municipality along with recommendations regarding next steps to implement the feasibility study results.

For several non-operating facilities, population changes over the years have led to situations where a sufficient amount of refuse is not available to meet the design capacity of the facility. These communities will either have to import refuse from other areas or supplement the refuse with wood chips to provide sufficient fuel to properly operate the facility. A feasibility study followed by a demonstration project would be considered to evaluate the co-firing potential of wood chips and MSW, if a municipality decides to attempt this type of a development program.

(b) Demonstration and Evaluation of the Full Utilization of Steam from a Municipal Refuse Incinerator

A project has been initiated that is intended to demonstrate the operating characteristics and economic potential of adding an electric generator to a small incinerator that was initially planned only to produce thermal energy. Oneida County is completing construction of a steam generating incinerator which will produce more steam than its market can utilize. The County and the Energy Authority are participating in a joint project which has added an electric generator to the existing facility to obtain and analyze long-term data on the operations, reliability and maintenance of the entire facility. This demonstration will provide other municipalities with more realistic data to evaluate the full energy potential of small-scale incineration.

2. Improve Combustion Quality and Air Pollution Control Efficiency

(a) Combustion/Emissions/Dioxin Testing at the Albany Steam Station

A series of tests were performed at the Albany incinerator to develop the relationships between combustion and refuse quality and stack emissions of dioxin and furan. The Energy Authority's contractor worked closely with the State Departments of Environmental Conservation and Health, and the Office of General Services to accomplish the testing. Analyses of the data have assisted in developing measures to ensure that future incinerator operations will minimize dioxin emissions, and what operating/combustion variables to continuously monitor (as a surrogate to dioxin testing) for control purposes.

(b) Other Facility Testing and Demonstrations

Planned future work concerns similar testing at other generic types of incinerators to build up a data base for the four basically different incineration technologies. An extensive testing project at the Pittsfield, Massachusetts, incinerator has been finalized and the first phase of work is complete. This project, cofunded by a number of other States and federal agencies, will determine the emission quality response of the incinerator to a wide variety of operating conditions and refuse quality. With these data bases, the development of automatic real-time feedback control systems can be accomplished to optimize facility control and minimize emission levels of potentially toxic compounds such as dioxin. After development of such systems, demonstrations of automatic feedback control systems are planned at New York State facilities.

Dioxin testing at all incineration facilities is based upon sampling and analytical methodologies finalized at a workshop sponsored by the Energy Authority, New York Power Authority, United States Environmental Protection Agency, and United States Department of Energy. Staff from the New York State Department of Environmental Conservation and New York State Department of Health participated in the workshop. The workshop results have been accepted for processing by the American Society of Testing and Materials as a standard method.

(c) Evaluation of Pollution Control Devices

The Energy Authority and the New York Power Authority are currently co-funding an evaluation of air pollution control technologies which may be more effective than the electrostatic precipitators now being used at most new incinerators in New York State. Fabric filter systems and combination systems using dry scrubbers and fabric filters are both being evaluated and compared to the control efficiency and costs of electrostatic precipitators.

If the ongoing project finds these technologies have good potential, a small-scale system at an existing incinerator may be installed and monitored for performance. It would incorporate dry scrubber, electrostatic precipitator and fabric filter subsystems and would allow actual on-line comparative testing of their effectiveness throughout the range of normal operating conditions.

3. Refuse Fuel Improvement Projects

(a) Air Classification of Refuse

The effectiveness of a rotary drum air classifier (RDAC^{tm}) is being evaluated to remove undesirable materials from the refuse-derived fuel being burned at the Office of General Services steam plant in downtown Albany. Almost 3,000 tons of waste material was processed through the RDAC^{tm}. The RDAC^{tm} improved the quality of the fuel by removing heavy metal items and large pieces of glass, stones and other problem materials. Personnel at the steam plant were very satisfied with the handling and burning characteristics of the improved fuel and have indicated their interest in having a permanent RDAC^{tm} installed at the Albany waste processing plant. The report summarizing the results of this project will help municipalities throughout

New York State to better evaluate whether an RDAC^{tm} would be appropriate for their existing or future resource recovery facilities.

(b) Refuse Shredding

The effectiveness of a different type of shredder is currently being evaluated in a project at Chemung County. Shredding the waste makes it more uniform, should allow for more reliable operation of waste burning boilers, and may allow more waste to be burned in the same furnace volume. Increased reliability and less down time means more waste can be burned for energy recovery and less needs to be landfilled. Smaller, less-expensive boilers also may be used when burning shredded fuel instead of burning the waste as received, i.e., unprocessed. Compared to the currently used shredders, shear shredders should require less maintenance and operate with fewer explosions and fires. To evaluate the effectiveness of the shear shredder, over 25,000 tons of waste has been shredded at the Chemung County shredding plant. Samples have been taken of the shredded material to see how it compares with the shredded product of standard hammermill shredders. The report being prepared will assist New York State communities in deciding whether to utilize shear shredders as part of their waste to energy systems. Chemung County's interest in purchasing a shear shredder to permanently replace its hammermill shredding operation shows that the shear shredder has been found to have significant advantages for that community.

(c) Future Projects for Improving the Quality of Fuel for Waste-to-Energy Facilities

One potential retrofit project would involve utilization of a shear shredder at an existing incinerator that is currently burning unprocessed waste. This project would determine if the reliability and waste throughput increased as a result of shredding the waste.

Another planned project would combine the data obtained in the Chemung County shear shredder project and in the Albany air classification project to develop recommended combinations of such equipment that would maximize energy production considering costs, air emissions, residue disposal and reliability.

A third project being considered would demonstrate methods of improving the quality of waste fuel by removal of glass, metal, household chemicals and other undesirable materials through separate collection programs. This project would be carried out by issuing a Program Opportunity Notice to communities in New York State.

Another planned project would test the effect on combustion and emissions quality of adding waste oil during periods of wet refuse to enhance the Btu value of MSW. This technique could be adopted at existing facilities to help maintain operating conditions which allow complete combustion to take place.

4. Co-Product Utilization

(a) Ash Utilization

Ash from the Southwest Brooklyn incinerator is being tested to determine its suitability as an aggregate for asphalt paving mixes. If this application is found to be technically and economically feasible, its utilization can reduce the overall costs of incineration and reduce the need for residue disposal capacities. The project will include the definition of subsequent steps toward implementation of a full-scale operation, and a description of a developmental methodology which other municipalities can follow. This project is co-funded by the Energy Authority and New York City.

Future work in this area includes determining the leaching characteristics of various materials, including dioxin and various heavy metals, from ash-asphalt pavement, and implementing successive development steps towards a full-scale demonstration.

(b) Scrap Tire Recycling

The Energy Authority and Teledyne are funding an assessment of the technical and economic feasibility of establishing a scrap tire pyrolysis plant in Western New York State. Initial efforts have focused on the availability and cost of scrap tires, and on the markets for system products including oil, metal, carbon black and steam. If the results are positive, an in-depth assessment of the technical aspects of the process and a rigorous cost and system reliability study would be the next steps in using the pyrolysis approach to recover energy from scrap tires. Scrap tire utilization would resolve the disposal problem at landfills and reduce sulfur emissions at MSW incinerators.

5. Alternative Refuse Disposal Systems

(a) Biological Conversion Systems

Biological conversion of solid wastes to biogas or alcohol requires the biodegradable fraction of MSW to be solubilized in water. In turn, this can require refuse to be classified, shredded and wet-pulped. Classified and shredded refuse from the rotary drum air classifier (RDAC^{tm}) project at the Albany Answers Plant is being wet-pulped and elutriated to determine the characteristics of the materials which are separated. The overall degree of separation of metals, glass, plastics and other inert materials from the refuse will be estimated. Preliminary tests will be performed on the remaining material to determine moisture content, amount of chlorine present (a potential toxin to biological organisms), and an ultimate analysis to determine carbon, oxygen, and ash content, and specific metals concentrations. The results of this project will be used in the current planning of a laboratory-scale project to produce ethanol from this type of feedstock.

(b) Gasification

A project is currently in planning to test a laboratory-scale steam gasifier which is capable of producing a medium-Btu gas. The project includes the preparation of a conceptual design and cost estimate for a full-scale system. Medium-Btu gas can be burned directly to produce steam or hot gas, or upgraded to pipeline-quality gas. The New York Power Authority and the Energy Authority are co-sponsors of this project.

6. Enhancement of Landfill Gas Production Rates

(a) Leachate, Condensate, Sewage Sludge Recycling to Accelerate Landfill Gas Production

Nine individually isolated test cells, holding 1000 tons of refuse each were constructed at the Broome County (Nanticoke) landfill. Different amounts and combinations of sewage sludge and buffer compounds were added to each cell (except a control cell). Water, to simulate rainfall, was added to each cell to induce leachate formation. Leachate and condensate (from landfill gas extraction) are recycled within each cell. The enhanced treatment cells have been producing up to 50 percent methane in extracted landfill gas, while the control cell has produced no methane to date. Gas and leachate quality tests will continue for another year. An economic analysis will be conducted to determine if the enhancement treatments are feasible on a large scale.

A demonstration project at a New York State landfill is planned if the economics of enhanced treatment are favorable.

(b) Demonstration and Evaluation of Small- and Medium-Sized Landfills

A future project would utilize a Program Opportunity Notice (PON) to demonstrate and evaluate the potential for recovering energy from small- and medium-sized landfills. The PON would stimulate greater interest in the development of smaller landfills and encourage use of leachate recirculation and other methods found to enhance methane production and treat leachate anaerobically.

7. Biogas Separation

(a) Semi-Permeable Membrane Separation

A demonstration of a plastic-polymer membrane with a high degree of selective permeability for carbon dioxide/methane gas mixtures is ongoing at the Pelham Bay Landfill in the Bronx. This unit is a variation of a commercially available membrane gas separation system used in industrial applications. Operation and maintenance experience from this project will provide information useful to small and large landfill developments as well as sewage treatment plants with anaerobic digesters.

(b) Liquid Membrane Separation

Laboratory tests are being conducted in a development project which utilizes two permeable membranes with a liquid immobilized in between that has selective gas transport characteristics. The carrier gas on the sweep side of the unit is steam. This concept is expected to be more efficient and cost-effective than other types of membrane separation systems because liquids have far more selective permeability than solids, and thinner polymer-plastic membranes can be used because of lower total pressure differentials. This results in the need for a lower input energy requirement and in faster overall permeation rates.

A successful development program would lead to a demonstration project at a New York State landfill.

8. Landfill Leachate Treatment

A project being planned with Chemung County would test the capability of various anaerobic digester systems to treat landfill leachate, yielding biogas and a small quantity of sludge residue.

Wastewater Treatment and Sewage Sludge. New York State has over 500 municipal sewage treatment plants which process more than 4 billion gallons of sewage per day. In general, these plants use aerobic biological treatment processes which remove organic materials from the wastewater and create large quantities of sludge (about 40 million gallons per day) that need disposal. Partial sludge disposal is accomplished at about 60 percent of the plants through anaerobic digestion, which transforms sludge into biogas. Ultimate sludge disposal is accomplished primarily by landfilling or ocean dumping, and by sludge incineration which is used in a few instances.

The major problems with these plants include: the large energy requirements of aerobic treatment plants; the difficulty of sludge disposal in environmentally sound ways; the inefficiency of existing anaerobic digesters; and the vulnerability of anaerobic digesters to shock loadings, changing environmental conditions and toxic materials which enter the wastewater stream. Impediments to addressing these problems include: the large amount of capital already invested in physical plants; the firmly established philosophy and tradition of aerobic sewage treatment and sludge treatment/disposal methods; and a lack of fundamental knowledge about the biology of anaerobic organisms.

With the problems and impediments to energy-efficient sewage treatment and sludge disposal noted above, the following RD&D strategy was developed:

1. Conduct RD&D to produce/save energy at existing sewage treatment plants with minimal additional capital investment. Aerobic sewage treatment plants have large energy requirements for process needs (about 600 kWh of electricity per millon gallons per day - primarily for aeration requirements), heating needs (for space heating and anaerobic digester heating) and sludge disposal requirements. A great deal of capital has been invested in these plants and they will continue to be used in the near future. Energy can be produced and saved at these plants in a variety of ways which require small additional capital expenditures and retrofits which take advantage of recent technical findings.

2. Conduct RD&D to change the sewage treatment process of choice from aerobic to anaerobic. Relative to aerobic sewage treatment methods, anaerobic processes produce up to 90 percent less sludge, are capable or producing twice as much energy as the process consumes, require far less land area for construction, and less capital expenditure. Implementation targets include existing plants requiring an increase in treatment capacity, new plants, and plants requiring re-furbishment.

3. Conduct RD&D to improve the reliability and efficiency of the anaerobic process. Principle reliability problems with the anaerobic process are due to slow reproduction and growth times for anaerobic bacteria. This causes the process to be vulnerable to shock loadings, environmental changes and toxic materials. In addition, very long process start-up times are extremely common, and recovery times due to shocks and toxics can be very long and difficult. Principal efficiency problems with the anaerobic process are due to difficulties in optimizing the process with respect to: environmental conditions for the two primary bacterial groups involved in anaerobic digestion, trace nutrient addition, and bacterial compatibility with different organic wastes. To resolve these problems, a greater understanding of the basic biological mechanisms at work is needed as are improvements in process engineering.

4. Conduct RD&D to produce energy from partially digested sludge. due to process inefficiencies and refractory compounds in sludge, anaerobic digestion reduces the organic content of sludge by only about 40 percent. Remaining materials are primarily landfilled or ocean-dumped. Significant improvements in organic conversion (to biogas) efficiency requires special pretreatment of the sludge prior to digestion. To produce energy economically from a process requiring pretreatment, a large volume of sludge in one location may be necessary. A primary target of opportunity in this case is the New York City/Long Island area, where the existing ocean-dumping barging infrastructure can be used to bring sludge from many sewage treatment plants to a central location for processing to biogas.

Wastewater Treatment and Sewage Sludge RD&D Project Areas

1. Existing Plants Program

   (a) Demonstration Projects at Municipal Sewage Treatment Plants

   Nine municipalities have contracted with the Energy Authority to implement and evaluate the performance of a variety of measures to produce and save energy on-site without reducing treatment efficiency. These demonstration projects involve: improving aeration and air handling systems, utilization of biogas from anaerobic digesters to cogenerate electricity and heat, utilization of heat pumps to extract heat from sewage treatment plant effluent for plant space heating, and improving sludge-dewatering systems. Implementation of these measures at these plants is expected to save and produce the equivalent of 13,500 barrels of oil per year. Simple payback on investment for these projects is estimated at less than four years.

   A project report will document the savings and payback periods associated with each energy-related improvement. These findings will be presented at meetings of local New York State municipalities to induce other municipalities to make similar changes and additions to their plants. It has been estimated that if measures similar to those being demonstrated in this project are applied to the State's 500+ municipal sewage treatment plants, over 650,000 barrels of oil per year would be saved.

   (b) Sludge Dewatering

   A wood chip dewatering process developed by the Revere Copper and Brass Company of Rome, New York, is believed to be applicable to the dewatering of municipal sludge as well. Preliminary calculations have been made for capital and energy costs using a 5-ton-per-hour dryer processing wood chips. In this instance, the capital cost of the Revere process is one-fourth the cost of conventional dewatering equipment per ton of capacity, and the energy cost is less than one twentieth. The process is easily integrated within existing treatment schemes and can be used at several different stages of the sludge handling/treatment process. Assuming cost savings similar to that for wood can be carried over to processing sludge, use of the Revere dewatering process will save municipalities $10 million annually in sludge dewatering costs.

   The results of this laboratory-scale project will form the basis of a planned Project Opportunity Notice (PON) for New York State municipalities. This PON will demonstrate the process at a pilot scale and test the combustibility and emissions quality of dewatered sludge in a sludge incinerator.

(c) Real-Time Process Feedback Control to Minimize Energy Consumption

This planned series of projects would develop and demonstrate an aeration feedback control system which could be retrofitted to existing plants. As indicated previously, the largest single energy consumer at existing plants is the aeration phase of treatment. Experiments in Canada have shown that as much as 50 percent less air by volume is sufficient in many instances to achieve effective treatment -- if the air is introduced at proper times.

Planned projects in this area include: laboratory tests to develop an optimal feedback control system; parallel tests of automatic feedback controlled and manually controlled aeration in adjacent tanks at a municipal plant; installation and demonstration of the system at a full-scale municipal plant; seminars to familiarize other treatment plant operators with the process and its advantages/disadvantages.

2. Change to Anaerobic Process Program

(a) Anaerobic Fluidized Bed Sewage Treatment -- Bay Park Plant, Nassau County

One 10,000 gallon-per-day pilot-scale reactor and two 600 gallon-per-day support test reactors have been treating primary settled sewage anaerobically for over a year. Results to date have been encouraging with chemical oxygen demand removals as high as 80 percent and a biogas quality of 67 percent methane. Economic analyses will be conducted to determine the feasibility of converting one of the Bay Park Plant's five aerobic fluidized beds to an anaerobic mode.

With lab and pilot-scale tests complete, the next step will be for the Energy Authority and Nassau County to fund a full-scale (2 millon gallon per day) reactor conversion, and operate it for at least one year. The 2 million gallon per day reactor would produce 10,000 cubic feet of methane per day, and save over 250,000 kWh/year of electricity relative to comparable aerobic treatment of the sewage. Site visits by other New York State treatment plant operators, papers and reports will be used for technology transfer efforts.

3. Improve Anaerobic Digestion Reliability and Efficiency Program

(a) Development and Testing of an Early Warning Process Control System for Anaerobic Digesters

This project involves the development and testing of an early warning process control system, based on real-time monitoring of hydrogen gas which is generated during the digestion process.

If successful, a pilot-scale test at a municipal plant is expected. It is the Energy Authority's intention to develop a commercially viable product to improve the reliability of anaerobic digestion systems.

(b) Complete Anaerobic Bioconversion with Mixtures of Pure Starter Cultures

This is a laboratory-scale project which is investigating the potential of using combinations of pure cultured anaerobic bacteria to digest cellulose (an important component of sewage and other wastes). It is anticipated that use of pure culture bacteria will minimize the retention time for anaerobic bioconversion of wastes to methane, and minimize the start-up time to achieve a stabilized system.

Favorable project results will lead to pilot and full-scale tests at municipal and industrial plants.

(c) Two-Stage Anaerobic Digestion

This Two-Stage Anaerobic Digestion provides for optimization of the major steps of sludge digestion (acidogenation and methanogenation) in separate reactors (digesters), as opposed to conventional, single-reactor design. Several technical and economic advantages accrue from application of the two-stage process, including:

- Reduced overall digester volume and capital and operating costs;
- Higher methane content in digester gas;
- Higher net energy and surplus methane production due to an expected improvement in volatile solids reduction from 40 percent to greater than 65 percent.

As part of the Northeast Regional Comprehensive Biomass Program, a road-transportable pilot plant will be constructed and tested at a host municipal wastewater treatment plant in Salem, Massachusetts. The pilot plant will then be brought to New York State for a series of demonstrations, leading to a full-scale conversion at an existing plant.

4. Energy From Partially Digested Sludge Program

(a) Methane from Partially Digested Sewage Sludge (New York City/Long Island Area)

Laboratory tests of sludge pretreatment methods to enhance digestibility are being conducted. Pretreatment methods include acid, alkaline, auto, and thermal hydrolysis techniques along with partial oxidation of the material under high pressure. The resulting materials are subjected to anaerobic digestion using three different reactor types.

Tests will continue for another year. Economic analyses of scaled-up systems will be conducted to determine the feasibility of constructing a pilot plant from among the most efficient systems (as determined from the lab scale test results).

Future work would consist of pilot plant construction with the assistance of New York City or a Long Island County. The energy potential of sludge in the New York City/Long Island area using these processes is on the order of 1-2 million cubic feet of methane per day, resulting from treatment of the estimated 500,000 cubic feet of sludge generated daily.

FINDINGS

The following general findings have been established based on the Energy Authority's biomass and waste energy projects conducted over the past several years.

- Sufficient biomass and waste energy resources are available in New York State to economically provide up to 10 percent of the primary energy needs of the State upon completion of the action programs recently initiated.

- The economics of biomass and waste resources hinge on the creation of markets for the feedstocks: establishing long-term contractual obligations on the part of customers and suppliers is crucial.

- The initial business plan for biomass and wastes energy resource development must highlight other positive attributes beyond economics, since higher costs are associated with the beginning of any new energy program involving small- and less-than-optimum-scale operations. But, once in production, biomass and waste-derived fuels can become competitive with fossil fuels.

- Commercially proven technologies are available for the generation of steam from municipal solid wastes and could provide New York State the energy equivalent of approximately five million barrels of oil each year.

- The currently available wood resources of New York State, derivable primarily from the unharvested surplus forest growth and secondarily from wood industry residues, amount to approximately eight million tons of material a year. This low-grade resource has attracted considerable interest for use in energy production. It is postulated, however, that regional shortages of wood may result from increased demand. Also there may be economic pressure on conventional forest product industries. However, by using New York State's extensive abandonded marginal farmlands, these resources may be expanded through such long-range options as the energy plantations. This could contribute significantly to the State's energy needs. For this to be possible, however, a stable, reliable market for wood fuel must be developed, beginning with supplying wood for direct burning and leading to more ambitious uses of wood as an input to liquefaction and gasification methods of producing biomass fuel.

- If biomass and wastes are to become significant components of the State's energy supply system, research is needed on the production as well as conversion phases of biomass and wastes to energy. Production research must address increasing biomass productivity and energy crop management, harvesting and storage. Conversion research is required in the areas of waste gasification, enzyme hydrolysis, and wood combustion with coal.

- Biomass and waste energy industries would have greater employment impact in New York State and recycle more dollars within New York State per Btu than any other energy supply programs. Overall, government and private economic forecasts for biomass and waste energy's role in the national energy budget for the year 2000 do not project a large role for these energy forms, typically one to 1.5 percent. However, based on its unique supply of these resources and its internal energy needs, New York State has designated a much more significant role for biomass and wastes.

DISCUSSION

Klass	I like your thinking regarding the public trough. It isn't a disaster that the public trough is starting to go down a little bit, perhaps run dry in certain areas. I think your approach is refreshing, and possibly a better approach in many cases, particularly when dealing with wastes.
T. Nunez-McNally, Tufts U.	I have heard that NYSERDA is interested in using acetate produced from landfills to make the CMA. Is that true? Is anyone doing that?
Mathusa	Yes, we just signed a contract for the first phase of that work. That work will be starting in the next couple of months. It's a 2-year program. At this point, we're working with a small firm in the Massachusetts area. The program is equally co-funded by that organization and ourselves. The Department of Transportation of New York State and the New York State Thruway Authority are interested in using the product. I expect successful implementation of the results could provide quite an industry for the Northeast, if acetate can be obtained from waste material and used in an environmentally attractive way. We're very much interested about how it would conflict with the salt industries, as a business, and we have found that there are some salt companies who were willing to pick this up as a separate line and sell it as a product to the counties and municipalities who buy salt for spreading on the highways.
D. O. Hall, U. Of London	I'm sure you are aware of the MSW plants in Europe. Does New York State have any agreements or collaboration with the Germans and British, for example, because we've gone through the same problems as you seem to be having here, and have solved them about 10 years ago.
Mathusa	We do not have any formal agreements with either one of those countries, but we'll be pleased to discuss it. The time may be right for our staff to visit you to develop cooperative agreements. We would be pleased to share with you with what we have, and would look forward to working with you.

L. Erickson, Waste-Energy Corp.: Would you elaborate on your program for cooperating with private enterprise firms, say in the conversion of sewage sludge to energy?

Mathusa: Yes, there are a couple of ways this can be done. We're always receptive to what I call unsolicited proposals. We also from time to time issue what I call program opportunity notices, but we haven't issued one on sludge in a couple of years. Some of that work should be done in the near term. We've done a lot of work on anaerobic digestion in terms of pure starter cultures and early warning systems because we feel that those types of systems are required to ensure the reliability necessary to run anaerobic digestion in sewage treatment facilities. There should be early warning systems to predict or indicate when upsets might occur so that the system doesn't go down. If you have something you would like send us, I woul be pleased to receive it.

ENERGY PRODUCT PLANTATIONS: TOWARD COMMERCIALIZATION

Sidney H. Browne, Ph.D.
Assistant to the Director
D. Richard Neill, B.S., M.Div.
Program Coordinator
Patrick K. Takahashi, Ph.D.
Director
Hawaii Natural Energy Institute
University of Hawaii at Manoa
2540 Dole Street, Holmes 246
Honolulu, Hawaii 96822

ABSTRACT

Isolated as it is in the mid-Pacific ocean, and with no indigenous fossil fuels, almost 90 percent of the energy consumed in Hawaii is derived from imported petroleum. The greatest demand is for liquid fuels, over 50 percent of the total energy used. Hawaii has an abundance of renewable energy resources, but biomass is the only one that can readily provide liquid and gaseous as well as solid fuels.

The Hawaiian sugar plantations are, in fact, energy plantations. The products are sugar, molasses (currently sold mostly for cattle feed), and bagasse. Bagasse, and sugar cane leaves and tops are burned in sugar factory boilers to produce almost 10 percent of the total energy consumed in the state, which includes about 4 percent of the electricity supplied through the public utility.

Hawaii's agricultural industry is faced with serious difficulties. Over 100,000 acres of sugar and pineapple lands have been abandoned and the remaining 200,000 acres in sugar has to compete with subsidized foreign sugar and sugar substitutes. HNEI, in cooperation with the State of Hawaii and the sugar industry, is developing a program that seeks first to put back into production the abandoned prime agricultural lands. The energy plantation program will assess the available lands for energy crops; run field trials of the most promising fast growing tropical trees and grasses; support R&D on harvesting systems for these crops; and develop the most promising technologies to convert these crops into useful forms of energy and other products.

ENERGY PRODUCT PLANTATIONS: TOWARD COMMERCIALIZATION

INTRODUCTION

Hawaii lies in the mid-Pacific ocean, 2500 miles from the Continental United States, the nearest major land area, and has no indigenous fossil fuels. There is an abundance of renewable energy resources, albeit largely undeveloped. From 1984 statistics, about 88 percent of the energy consumed in Hawaii is derived from imported oil; biomass supplies about 9.5 percent; and other sources only 2.5 percent. About 33 percent of total energy* used goes to generate electricity. More than 50 percent is consumed for liquid fuels (gasoline, diesel fuel, and jet fuel), mostly for transporation. Jet aircraft are the major transportation fuel users, about 55 percent of the demand for liquid fuels, or roughly 30 percent of total energy demand. Hawaii's economy could be disrupted by any extended interruption of its oil supply.

In 1974 the Hawaii State Legislature created a number of agencies to support development of alternative energy resources, including the Hawaii Natural Energy Institute (HNEI), the Natural Energy Laboratory of Hawaii (NELH), the State Program for Energy Planning and Conservation, now the Energy Division of the State Department of Planning and Economic Development (DPED), and the State Energy Coordinator. HNEI was established as a research unit at the University of Hawaii at Manoa (UHM) to provide direction and assistance to ongoing energy-related research throughout the UH system, and encourage new programs and research on alternatives to seaborne petroleum. One of its important roles is to coordinate R&D interactions among government, private industry, the utilities, community organizations, and individuals in order to maximize cooperation and funding. HNEI works closely with the State Department of Planning and Economic Development (DPED), Energy Division, which has the responsibility for overall energy policy, energy conservation, and assistance to new technologies nearing the marketplace. Thus, from the very outset of its alternative energy development efforts Hawaii has encouraged cooperative development of energy technologies by research organizations and private industry.

Because most of Hawaii's energy supply is imported, energy prices are high so local business and industry have strong incentives to find economical alternatives. However, in spite of the added cost of transportation,

*Total energy consumed in 1984 was 276 trillion Btu (81.0 billion kWh, or 47.9 million bbl oil equivalent). Source: State Department of Planning and Economic Development.

products derived from petroleum are still less expensive than locally produced replacements, if production costs are charged entirely against energy products. The key to successful marketing of energy is in co-product support and, of course, tax incentives or other legislative actions. As will be discussed later, commercial success with alternative fuels has depended upon both factors.

Due to the decline of the sugar and pineapple industries in Hawaii there are more than 100,000 acres of abandoned sugarcane and pineapple lands, not fully utilized, and much more underused grazing and forestland. The "Energy Plantation" concept is to develop these lands to produce fast growing trees and grasses and to harvest and convert the biomass produced to solid, liquid, or gaseous fuels and high-value co-products.

The remaining sections of this paper first will discuss current production and uses of biomass in Hawaii, followed by a description of past and present biofuels research. Finally, plans and proposals for future biofuels research and development will be briefly described.

CURRENT BIOMASS PRODUCTION AND USE

For many years the Hawaiian sugar industry has played a significant role as a producer of energy. Sugar plantations are, in fact, energy product plantations. The products are sugar, molasses (currently sold mostly for cattle feed), and bagasse (the fibrous by-product of milled sugar). Almost 3 million tons of bagasse and sugarcane leaves and tops, roughly equivalent to 3 million barrels of fuel oil, are burned in Hawaiian sugar factory boilers annually, to provide process steam and electricity, used for irrigation pumping, and excess electricity to sell to the public utilities. About 10 percent of the electricity generated in the state is produced by biomass power plants in sugar factories and about 4 percent of the electricity distributed by the public utilities originates in sugar plantations.*

In spite of its increasing reliance on renewable energy sources, the sugar industry still consumes almost 200,000 barrels of fuel oil annually in plantation boilers, largely during periods when sugar cane is not being harvested, making bagasse unavailable. A source of additional biomass fuel for use during off-season periods would decrease substantially the amount of fuel oil used by the industry, and thereby decrease the amount of fuel oil imported to the state.

*Source: Unpublished data from the Hawaiian Sugar Planters Association.

State and federal laws passed in 1977 and 1978 provided incentives for the sugar companies to sell their excess electricity and to expand their power plant facilities. Lihue Plantation Company on Kauai has installed a boiler capable of burning bagasse, sugar cane trash, wood chips, municipal refuse, coal, and fuel oil. The Hawaiian Commercial & Sugar Company on Maui recently added a new turbo-generator to increase its generating capacity. The processing facility at Davies-Hamakua Sugar Company on the Big Island incorporates a pelletizer, which converts bagasse into energy pellets, to increase output capacity and to provide storage capabilities that will make firm power available for 12 months of the year.

Molokai Electric Company's biomass generating facility began operating in May 1982. This biomass boiler, capable of burning a variety of biomass fuels--including hay, wood chips, pineapple chop, and waste paper--and with the addition of a gas turbine and waste heat boiler will be able to provide 60 to 70 percent of the island's electrical energy.

Besides bagasse-burning, wood chips from trees also can be used as a replacement for fuel oil. The State Division of Forestry (SDF), which has been conducting a tree planting program for many years, recently increased the propagation of seedlings at the Waimea State Nursery on the Big Island and is expanding the planting of eucalyptus trees throughout the state. SDF has calculated that the potential demand for eucalyptus as a wood fuel would necessitate 20,000 productive acres. More than 100,000 acres of state forest land have been identified as potential sites for energy tree farms.

Mauna Loa Macadamia Nut Corporation utilizes the nut shells as a high-Btu source of fuel to produce steam and electrical energy for its processing plant. With the rapid expansion of macadamia nut orchards in the state, this energy source will be further exploited.

BIOFUELS PRODUCTION RESEARCH

Other than the continuing research on sugar cane conducted over many years by the Hawaiian Sugar Planters' Association (HSPA), major ongoing biomass production research in Hawaii is in tree farming and algal culture. Studies and research on liquid and gaseous fuels that can be derived from these feedstocks, as well as other available biomass (e.g., agricultural waste and MSW), lead to the conclusion that for at least the next 10 to 20 years the most promising approach for Hawaii is production of selected trees and herbaceous crops for gasification, using the gases themselves as fuels or converting them to liquid fuels.

Tree Farms

Bioenergy Development Corporation, a subsidiary of C. Brewer & Company, Ltd., under a cooperative agreement with U.S. Department of Energy (USDOE), is now in the seventh year of a project to determine the feasibility of developing eucalyptus plantations for energy production in Hawaii. Thomas Crabb of Bioenergy Development is the project manager. The U.S. Forest Service, U.S. Department of Agriculture (USDA) Pacific Southwest Forest & Range Experiment Station, USDA Soil Conservation Service, and the University of Hawaii College of Tropical Agriculture and Human Resources (CTAHR) are providing technical assistance. The total project area covers 714 acres, of which 148 acres are at Ka'u, a dry area, and 566 acres are along the Hilo Coast, a wet area. The objective is to answer eight general questions:

- How do different sites, cutting cycles, spacings, and intensive cultural practices affect the growth, yield, rotation, length, and profitability of short-rotation eucalyptus investments?
- What are the relative cost/benefits of growing pure stands of eucalyptus with various fertilizers vs. admixtures of eucalyptus with nitrogen-fixing species on different sites?
- Will crop-logging--that is monitoring growth periodically with chemical and physical tests--provide guidelines to produce maximum yields?
- Can genetically superior eucalyptus planting stock be identified and mass-produced within a short time-frame through phenotypic selection in seed orchards?
- Which species of eucalyptus will prove best adapted on planting sites differing greatly in elevation, rainfall, and soil properties?
- Can existing equipment be modified at minimal cost to plant and cultivate eucalyptus seedlings, and haul wood fiber to generating plants?
- To what extent do soils and terrain conditions limit acreage in large-scale commercial forest operations on company lands?
- What is the optimum mixture of eucalyptus wood chips and other biomass fuel sources?

Harvesting studies were conducted in 1981 and followed up in early 1985, with encouraging results. Economic studies are underway, and preliminary results suggest that eucalyptus production for fuel can have returns competitive with other land use.

An energy tree farm study, this one on the giant leucaena, commonly known as the giant koa haole, was conducted for the island of Molokai in 1980 by James Brewbaker of CTAHR, funded by HNEI. The objectives were to determine the economic feasibility of this species for the production of wood chip fuel, primarily for electrical power generation, and to evaluate the concept of energy tree farms in general for Hawaii. The final report suggested that this type of tree farm could be a profitable enterprise on Molokai and could provide an economical source of biomass on marginal lands throughout the state.

Arid and semi-arid lowlands in Hawaii (less than 30 inches annual rainfall) suffer protracted drought and variable soil salinity that eliminate many energy fuelwood crops from consideration. Lands marginal to sugar and pineapple are often too rocky or irregular for cultivation, but can be managed for tree crops and have high solar insolation values. Nitrogen-fixing trees like leucaena and prosopis historically have occupied these lands, now in excess of 100,000 acres. Primary considerations in commercialization of these as biofuel trees are: (1) extent of irrigation and type of equipment, and (2) size of tree and type of harvest equipment.

Statewide Energy Tree (SET) trials of leucaena and other N-fixing trees have been conducted collaboratively with landowners on the islands of Maui and Molokai since 1980. Large landowners collaborating with CTAHR in these trials are the Alexander & Baldwin Company (Hawaiian Commercial & Sugar Company plantation on Maui), the Office of Hawaiian Affairs, and the Molokai Ranch. Biomass burning facilities exist on both islands, with Molokai Electric on the Molokai Ranch and at Hawaiian Commercial & Sugar Company on Maui. Attractive rates exist for marketing biofuel power on these island grids (notably on Molokai, where diesel is currently in use).

<u>Conversion Research</u>

The production of ethanol from sugar cane molasses offers the best potential for the near-term development of a liquid fuel from an indigenous resource. Fermentation of carbohydrates to ethanol is a well-known process with numerous distilleries in operation in the U.S. and throughout the world.

A number of studies of the economic feasibility of producing ethanol by fermenting sugar cane molasses have been conducted by HNEI, HSPA, and others.

Pacific Resources Inc. (PRI), in cooperation with Aloha Petroleum, began a marketing test in 1979 to determine the public's acceptance of gasohol produced from imported mainland ethanol. The fuel did not prove to be economically feasible due to the high ethanol cost and was discontinued in 1982. In 1981 PRI suspended plans to build an ethanol-from-molasses plant.

A comprehensive study funded by USDOE and conducted in 1981 by C. Brewer & Co. assessed the feasibility of an 11.4 million-gallons-per-year ethanol plant at their Hilo Coast Processing Company facility on the Island of Hawaii, which was found to be economically feasible with federal and state tax credits. The Governor's approval of House Bill No. 1262 authorized the issuance of $10 million of special purpose revenue bonds for the construction of an ethanol plant. Subsequently, C. Brewer decided to utilize $15 million of private financing instead of state money. Negotiations were progressing for long-term purchase commitments of the ethanol by the oil industry for blending with gasoline; however, after many months of discussion, the board of directors finally tabled plans for a 5-million-gallon ethanol plant because of failure to secure a satisfactory contract with an oil company and concern about the possible termination of the federal tax credit as of December 31, 1985.

The phase out of leaded gasoline mandated by the federal government has resulted in a demand for other compounds as octane enhancers. Ethanol is one candidate, which has led to a revival of interest in Hawaii in ethanol production from molasses.

A recently formed company, Pacific Ethanol Products, Inc., has built a two-million gallons-per-year ethanol plant at Campbell Industrial Park on Oahu. Molasses from Oahu Sugar Company will be purchased for feedstock, and the stillage will be sold to a nearby feedlot for animal feed. The ethanol-gasoline mixtures (10 percent ethanol) will be distributed initially by Aloha Petroleum Ltd., who is already marketing the product with ethanol imported from California. With the sale of stillage, the federal gasoline tax exemption (6 cents per gallon), and the Hawaii state general excise tax exemption of 4 percent through 1992, the venture has an excellent chance to succeed.

Stone & Webster Engineering Corporation of Boston, with the cooperation of HNEI, received two ethanol plant design awards from the Solar Energy Research Institute in 1983: one for acid hydrolysis and the other for enzymatic hydrolysis of lignocellulose. The site selected for these projects was the island of Hawaii, with eucalyptus as the feedstock. Stone & Webster also received a third award to study a modular biomass-to-methanol plant at a Mainland site, but in partnership with HNEI, discussed the prospects of such a pilot plant for Hawaii.

Michael J. Antal, Jr., the UHM/HNEI Coral Industries Professor of Renewable Energy Resources, has been conducting both basic and applied research in thermochemical conversion of biomass since 1982 at the UHM Renewable Resources Research Laboratory (R^3L). Technologies employing thermochemical processes for the conversion of biopolymers (bagasse, woodchips) and oligomers/dimers/monomers (molasses, sugars) into high value chemicals, feedstocks, and fuels offer the greatest promise for gaining independence from imported fossil fuels. To date, these technologies have been hamstrung by the lack of specificity of the thermochemical reactions realized in most biomass gasifiers and pyrolysis reactors. Research in the R^3L is aimed at overcoming problems with specificity through the development of a fundamental understanding of the complex chemistry which occurs during pyrolysis and gasification of biopolymers, dimers, and monomers.

Two projects in the laboratory focus on the solid phase chemistry of biopolymer pyrolysis. One of these projects, in collaboration with the Hungarian Academy of Sciences National Laboratories, involves studies of the chemistry underlying the pyrolytic depolymerization of biopolymers to monomeric sirups and tars. The second project emphasizes doubling the yields of charcoal from biopolymers by an improved understanding of the charcoal-forming condensation reactions that occur during pyrolysis. This work is just beginning, in collaboration with the Asian Institute of Technology.

A third project focuses on the fluid phase pyrolysis chemistry of biomass-related oligomers/dimers/monomers (molasses and sugar) in supercritical solvents. We already have learned how to selectively catalyze the transformation of ethanol and other alcohols to ethylene and other olefins in supercritical water. Work now is beginning on the conversion of glucose to high value olefins and other petrochemical feedstocks. This research has important ramifications for sugar-producing economies such as Hawaii and Brazil.

FUTURE BIOFUELS RESEARCH

In November 1984, HNEI held a workshop to review Hawaii's biofuels research and development activities. Scientists, engineers, and managers from universities, government, and the private sector participated. Workshop objectives were to identify the most promising crops for further development, to examine available and emerging conversion technologies, and to formulate plans for future research.

Biomass resources considered included trees, grasses, algae, and agricultural and urban waste. From review of workshop results it was concluded that large-scale production and conversion of fast-growing trees and grasses offer the earliest possibility of satisfying Hawaii's renewable energy needs. It also was concluded that research on algal culture should be continued. Conversion of waste to energy products already is being developed by the City & County of Honolulu and by two other counties. Anyone who might be interested in further details can obtain copies of the workshop report from HNEI (Volume I: Summary, Volume II: Presentations).

Based upon workshop guidance, a Hawaii biofuels research and development program plan for the next few years was developed. The program comprises three elements: (1) an assessment of potential biomass energy resources, initially for Hawaii with later extension to the American territories of the Pacific, (2) experimental and pilot-scale production and harvesting of selected fast-growing trees and grasses as the first phase of energy plantation development, and (3) biomass conversion research, emphasizing gasification to provide medium-Btu gas to be used as a fuel to generate electricity or to be further converted to methanol or hydrocarbons. The processes and developed technology will have a direct transfer to the entire Pacific Basin.

Biomass Resource Assessment

The purpose of the biomass resource assessment is to develop and make available to planners and decision-makers much of the information needed to plan for large-scale biofuel production. This element includes development of a land resources data bank, computer models to provide expected biomass yields for selected crops and yields of gaseous or liquid fuels for each crop, and an economic model combining all three for energy plantation operations.

Development of the land resources data base is already well along. Over the past 10 years, the CTAHR has established a resource assessment data base for all agricultural lands in the State of Hawaii. It contains most of the land data needed for biomass production planning.

A biomass yield model is being developed for selected species of trees and grasses for various types of land. Integration of this model with the land data bank will enable the user to locate the best land for a crop of interest or, conversely, to determine the best crops for his land. Models for conversion of biomass feedstocks to liquid and gaseous fuels have been developed under USDOE sponsorship. Appropriate ones will be selected and adapted to the requirements of the program. The economic model will be adapted from one of the many already available agricultural economic models, depending on the complexity appropriate to the outputs of the data base and yield models. This model will receive inputs from the yield models to provide product unit costs and return on investment, or other economic variables of interest.

Biomass Production and Harvesting Research

The wealth of data available from the Bioenergy Development Corporation's eucalyptus project on the Big Island of Hawaii will provide information for model development as well as for comparisons with results of field tests of other crops.

Two biomass production projects are planned. One, "Bioenergy Pilot Plantations of Leucaena in Arid Lowlands in Hawaii" to be performed by CTAHR, is a continuation of the leucaena research described in the previous section of this paper. In past years this project has been supported by a combination of HNEI, DPED, and some federal funding.

HSPA, with first-year state funds of $150,000, has initiated a biomass production project, "Research to Compare Various Crops in Hawaii for Energy Yield and Economic Feasibility." Its purpose is to compare different tree and grass crops grown in the same environments for total energy and dry biomass yield and economic feasibility of their use as feedstocks for conversion processes. Test sites will be located on each of the five major Hawaiian Islands: Kauai, Oahu, Maui, Molokai, and Hawaii. For each site chosen, three to four crops will be selected from among leucaena, eucalyptus, sugar cane, napier grass, and sweet sorghum. These are all crops known to be good biomass producers in separate locations, but no side-by-side comparisons have been made previously in any one location. Land will be provided by private institutions. In addition to HNEI, HSPA, CTAHR, and DPED, other participating organizations are A&B (Hawaiian Commercial & Sugar Company, Maui), AMFAC (The Lihue Plantation Company, Kauai, and the former Puna Sugar land, Hawaii), Castle & Cooke, Inc. (Waialua Sugar Company, Oahu), and Molokai Ranch (Molokai).

Biomass Conversion Research

The objective of biomass conversion research is to determine the technical and economic feasibility of producing and using liquid and gaseous fuels by thermal conversion of biomass in Hawaii. It is planned to develop and test a medium-Btu gasifier at a suitable location in Hawaii. The gas produced will be used to fire an existing biomass boiler or to fuel an existing diesel-electric generator. A preliminary study of the suitability of currently available gasifiers is underway. A number of such gasifiers are at or near pilot plant stage of development, or commercially available. A prime candidate is the oxygen-blown, down-draft gasifier developed at SERI and now being developed for commercial use by Syngas Systems, Inc. An important part of this preliminary study is the selection of the best site in Hawaii for a test bed plant.

Concurrently with gasifier development, requirements to convert the syngas to a liquid fuel--methanol or a hydrocarbon--will be investigated, and development of equipment for this purpose will be initiated. Finally, the basic pyrolysis research by the UHM Renewable Resources Research Laboratory will be continued.

SUMMARY

Hawaii's critical renewable energy needs have been discussed, and some of the combined efforts of government at various levels and the private sector to find ways to satisfy these needs have been described. A research and development program to progressively replace fossil fuels has been outlined, first to generate electricity and later to produce transportation fuels, the greatest long-term problem.

Hawaii offers an ideal environment for biomass research. The climate permits essentially year-round biomass growth. The prices of all energy products are high, so local industry has a strong interest in replacing imported oil with indigenous energy sources. Developed technology for production of biofuels and other products not only would help Hawaii's economy, but would be directly transferable to other tropical and subtropical islands as well as to much of the United States and the world.

BIO-ENERGY PROGRESS IN CHINA

Paul F. Bente, Jr., Ph.D.
Consultant on Bio-Energy
5500 Lambeth Road
Bethesda, Maryland 20814

ABSTRACT

Nearly 800 million rural Chinese do not have enough fuel for cooking. Straw and wood together provide 85 percent of rural household energy, virtually all of which is used for cooking. Government authorities have encouraged development of new more efficient cookstoves to burn straw and/or wood, since traditional cookstoves are generally less than 20 percent efficient. A national goal has been set to have cookstoves be 40 percent efficient. Some models are described which have recently been developed to achieve this level.

China has embarked on a comprehensive reforestation plan to increase the amount of land covered by forests from the present level of 12 percent to 20 percent by the turn of the century. Details of reforestation progress are cited. China is also encouraging plantings of trees around farms and ponds, along roadsides and ditches, etc. A recent law requires every able bodied person over 11 years of age to plant 3 to 5 trees per year or do an equivalent service tending trees.

China has modified anaerobic digesters to make them leak proof and has begun to switch operation of digesters over to a lower pressure, collecting the biogas outside the digester in a flexible bladder or a floating "bell" type collector. Dry digesters are also being developed to complement the use of hydraulic digesters. Some 4.5 million improved digesters are now operating; more are being installed at the rate of 500 thousand per year. About 3 percent of national rural cooking needs are satisfied by biogas. Wherever biogas is used, the need to burn straw and/or wood is greatly reduced; this solves the fuel shortage at such locations.

BIO-ENERGY PROGRESS IN CHINA

INTRODUCTION

This paper is based on information received and observations made by a Biomass Utilization Delegation which visited the People's Republic of China during late October and early November, 1985. The writer served as leader of the delegation which consisted of two Canadian and 17 U.S. professionals (and eight spouses). The trip was organized by the People to People Citizens Ambassador Program, Spokane, Washington and was hosted by the China Association of Science and Technology, which is an affiliation of well over 100 national and local technical and professional societies and organizations.

Technical exchanges and assessments took place in formal meetings held in Beijing, Chengdu, Shanghai and Guangzhou and during informal visits made to laboratories and nearby rural sites to observe bio-energy installations in operation. A detailed journal (1) of the visit has been prepared and will be published by the People to People Citizens Ambassador Program.

Unless otherwise noted, this review draws upon that journal to present the current status of bio-energy in China. The information is based upon 1) the unpublished papers and verbal reports presented to the Delegation by Chinese technicians, 2) the collective observations made by the delegates visiting laboratories and rural operations and 3) the various responses of the Chinese to the delegates' queries when inspecting facilities.

THE ENERGY SITUATION IN RURAL CHINA

In rural China meals have traditionally been cooked over small stoves which burn straw and other available vegetation and/or firewood. However, there is a general shortage of these fuels. As a result some 800 million rural Chinese living in 170 million households experience hardships. Also, the extent to which these fuels have been used has deprived farmlands of much needed mulch and has caused serious local deforestation.

Because of the serious nature of the fuel shortage which afflicts nearly all of the rural population, because of the pervasiveness of environmental damage occurring from lack of mulch on farmlands and because of local deforestation, both of the latter two being caused by intensive harvesting of biomass for fuel, the authorities in China have launched broad programs to alleviate these problems. Emphasis is being given to the better use of all available, locally exploitable natural resources. These are straw, firewood, manure, hydropower, local coal, solar energy, wind and geothermal energy. In this paper only the first three, the biomass resources, are considered.

Total annual energy consumption in the rural areas is estimated to be equivalent to burning 330-350 million tons of coal. One estimate breaks this down into 265 million tons of straw and crop stalks and 180 million tons of wood burned each year.

Only 20 percent of this rural energy is used for farming operations; the other 80 percent is used by farmers' households, primarily for cooking. The fuels used in operations are coal, oil and electricity. But most of the farm work is done by human and animal labor, and this is not counted. Farming is done by 300 million farm workers and 50 million head of draught animals, providing 70 million horsepower equal to 1/3 of present farm machinery power in China.

Of four-fifths of rural energy used by households, 85 percent comes from biomass, principally straw (including crop stalks) and firewood. Thus crop residues and firewood combined supply 2/3 of total rural energy. Straw and crop stalks provide somewhat more than half this bio-energy on the average, although in some rural areas wood is the chief source of fuel. Nationally, from 60 to 80 percent of all straw and vegetative agricultural wastes are used as fuel by rural households. Consumption of firewood is about double the amount that local forests can supply without deforestation taking place.

The cooking fuel crisis has been resolved in about 3 percent (national average) of rural households which use biogas made by anaerobic digestion of pig manure, night soil and crop wastes. Sichuan Province has the highest level of usage, namely 7 percent. This process encountered many difficulties after it was introduced and widely promoted in the mid- to late-1970s, but these setbacks have now been overcome, as will be explained later. In warmer climates the process can be operated 9 or 10 months a year, reducing the need to burn crop residues and wood fuels to only 2 or 3 months a year. Where biogas is used, much more straw is now used for mulch, and local deforestation is avoided.

IMPROVEMENTS IN TRADITIONAL COOKSTOVES

The traditional Chinese cookstove in a rural household burns straw and crop stalks or wood. Before 1960 it had an efficiency of only 10 percent. By installing a sealed floor as part of the stove and by connecting it to a chimney, the efficiency was increased to 15 or 20 percent. A large number of stoves featuring such improvements have been introduced. However, much greater improvements are needed and are indeed possible. In the 1980s more work was done on stoves and the efficiency increased to 25-30 percent.

The authorities next introduced competition to develop more efficient stoves and in 1984 adopted standard procedures (8) for measuring efficiency. Some results of this drive were recently published (4). This report showed that six wood burning stoves had achieved efficiencies of 36-44 percent and four cropwaste stoves efficiencies of 27.5-31 percent. It also reported that in 1983 some 7 million approved "fuel saving stoves" were disseminated in 90 selected counties. A national goal of having cookstoves reach an efficiency of 40 percent was finally adopted.

Improvements which brought many stoves up to the 30 percent efficiency level were generally achieved by 1) using a smaller fire chamber in order to bring the hottest part of the flame up close to the underside of the cooking pot and 2) passing the hot flue gases through narrow passageways, one side of which is the wall of the pot, before the gases enter the duct work leading to the chimney. The hot gases thus contact the pot wall in a thin layer at a high velocity, affording greater heat exchange. These principles are shown schematically in Figure 1. One such model has achieved an efficiency rating of 30 percent and has been installed in 3¼ million rural homes.

The Chinese often prefer to have the cookstove built into a rather massive masonry counter made of various combinations of bricks, concrete and often decorative ceramic tile on the surface. Reducing the total mass encasing such stoves, or effectively insulating the stoves that are built into such cooking counters are essential steps that must be taken to improve efficiency, because the energy absorbed in heating up the massive installation is wasted, so far as cooking is concerned. The stored thermal energy is released slowly long after the cooking is done. Such a loss can use up half of the heat that could have been captured for cooking. Many cooking stoves in China, though otherwise rather well designed and even attractive in appearance suffer from such inefficiency. Brick and concrete stoves, even if well insulated, can only reach an efficiency of 25-30 percent.

The most efficient stoves are of relatively compact metal construction. One such stove which has met China's new 40 percent efficiency rating is the "Assembled Model NC-1 Firewood Saving Stove". It is available as a knocked-down, ready-to-be-assembled, do-it-yourself unit, parts of which are cast-iron. The stove is called a "combination" stove, because it uses what would otherwise be wasted heat to boil water for tea or for drinking. (In China, all water is boiled before it is safe to drink.) Because of that feature, this stove readily achieves the 40 percent efficiency rating.

Another very efficient stove observed in operation was an experimental, small-scale, lightweight, sheet metal, updraft gasification-combustion stove, shown schematically in Figure 2. This stove burns small pieces of wood or wood chips in a starved-air mode on a grate at the bottom. The gases and liquids which are expelled are then burned with preheated secondary air in a fire chamber directly above,

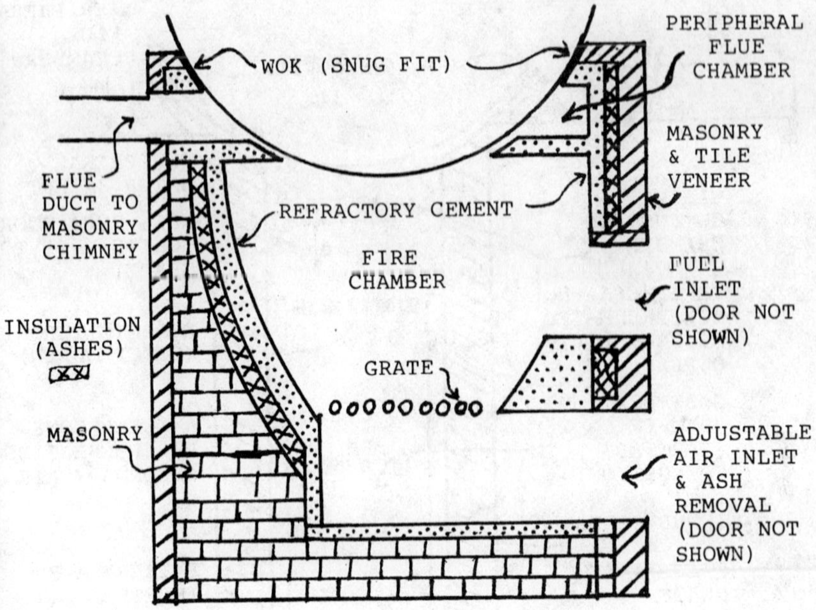

Figure 1. Solid Fuel Saving Stove

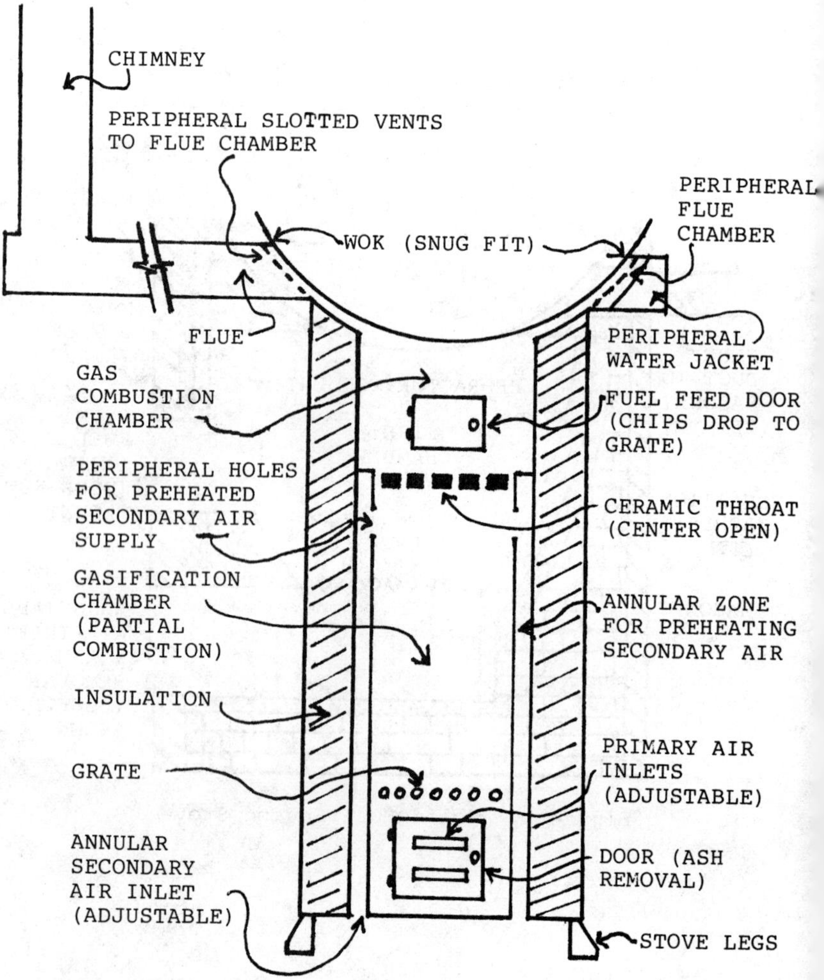

Figure 2. Updraft Gasification-Combustion Stove

producing a smokeless flame around an upper ceramic throat. The cooking pot or wok fits above this flame and close to slotted flue vents which are in the upper periphery area around the side of the cooking vessel. A water jacket also surrounds the top of the unit, extracting heat from the flue gases to produce hot water. The efficiency of this stove was rated at 34 percent without recovering any waste heat to make hot water and at 40 percent with such heat recovery. This experimental stove has very little mass, compared to most models. It was said to cost only 20 Yuan (U.S. $7.35); as is customary this includes little, if anything, for labor.

ADVANCES IN REFORESTATION

According to pamphlets of the Chinese Academy of Forestry (included in Reference 1), when the People's Republic of China was formed, only 8.6 percent of China's land was covered by forests. This has since been increased to 12 percent. China presently has 1.5 million hectares of forests which contain 9 billion cubic meters of standing timber. The annual harvest of timber is 50 million cubic meters. Great as these figures may seem to be, they are considered below world average. China recognizes this and has taken steps to correct the situation.

China's legislation to continue reforestation of the country was completed in September 1984 when the Forest Law was passed by the National People's Congress. This legislation set up the goal of having 20 percent of the land covered by forests by the turn of the century. This should increase the volume of timber from 9 to 10.2 billion cubic meters and the annual harvest from 50 to 100 million cubic meters. Of course, this program will also protect more land and provide more fuelwood.

Establishing a shelter belt of trees, known as the Great Green Wall, across the northern zones of China is a multiphase project that has been underway for 7 years. Ultimately 59.3 million hectares of such windbreaks are planned. Since 1978, 4.4 million hectares of forest shelter belts have been established. About 12 million hectares of cultivated land are now protected by these shelter belts.

In the South, 5.3 million hectares of fir and 1.33 million hectares of bamboo have been planted. Throughout the country 4.74 million hectares of trees of economic value have also been planted and 3.2 million hectares of sparsely populated and unaccessible lands have been seeded by air. Also forest cutting enterprises have replanted 2.43 million hectares of new trees.

A dozen new clones of fast growing poplars are being tested on 100,000 hectares of land. In addition, 7.2 billion trees have been planted around homes and villages as well as along roads and ditches. The area of intercropping of agricultural crops with trees has reached one million hectares. Overall, the improvements during the 1978-1983 period of reforestation amounted to nearly 41 percent.

A voluntary mass afforestation campaign is also forging ahead under a resolution adopted by the Fourth Session of the Fifth National People's Congress. This "Resolution on Nation-Wide Voluntary Tree-Planting Campaign" stipulates that all Chinese above age 11 (except the old, weak, sick and disabled) have an obligation to plant three to five trees every year or do an equivalent amount of work cultivating saplings, tending trees, etc. After adoption, a big drive was launched with good results. Further, in 1985, each person in the armed forces was requested to plant 20 young trees (4).

IMPROVEMENTS IN ANAEROBIC DIGESTERS

The Problem of Leakage

Wide publicity has been given to the fact that in the mid-1970s China embarked on a crash program to build a large number of anaerobic digesters as cheaply as possible using relatively unskilled local labor and materials. About 7 million hydraulic type digesters were built. In these, gas accumulated under the dome of the digester, pushing the digesting liquid slurry up the digester's feed and discharge side arms, the latter of which had built into it a substantial reservoir to handle overflow until the displaced liquor could be laddled out and carried away for fertilizing crops. Due to the gas pressure, which generally rose to 50-80 cm of water column, most of the digesters leaked after operating for a time. Efforts to patch the leaks generally failed, so that many installations had to be abandoned and the national program to install more digesters degenerated into a crisis situation around the end of the 1970s.

Undaunted by this failure, Chinese technicians analyzed the problems and overcame the difficulties with several redesigned digesters which do not leak. The national program (7) was redirected to installation of improved digesters with emphasis on technical adequacy and de-emphasis on cheap construction with inadequate local materials and untrained labor.

Lower Pressure Digesters

These new digesters have essentially the same configuration as the ones which leaked, but the concrete and/or masonry construction is very much better. Moreover, in the best models, biogas which collects under the dome is not allowed to reach pressures as high as before. Pressure is kept at 20-30 cm of water column and often much lower. Instead of being confined to the digester, the gas is led off by a tube for storage in a separate gas holder, such as a flexible, red-mud-plastic bladder located inside or near the house. Another version employs a "bell" type gas holder floating over a water seal, also located some distance from the digester, often below grade. These three types of digesters are shown schematically in Figures 3, 4 and 5.

Such collectors outside the digester are essentially expandable and make it possible to maintain a nearly constant, low gas pressure in the system whether gas is being collected or dispensed. This keeps the digesters from leaking. Keeping pressure constant may be accomplished by placing a weight, such as a brick or two, on top of the flexible plastic gas holder or by placing weights on the floating collector. Since with these collectors much less gas is stored inside the digester, a larger charge can be kept therein. This increases overall production. It was also found that inoperative high pressure digesters could sometimes be repaired for successful low pressure operation.

Using biogas at a lower pressure required a redesign of the gas burners used for cooking. Fortunately, a redesign operating at 3 cm water column produced a hotter flame. This increased burner efficiency to 60 percent. Improved biogas lanterns operating at 4 cm water column produce light equivalent to that of a 40 watt electric light bulb.

The improved digesters were said to cost from 110 to 150 Yuan (U.S. $35 to $50) and to have a payback period of 2 to 3 years. Installation in one case was said to require 17 person days of labor. Labor is so cheap in China that it is an insignificant part of total cost.

Training and Logistical Details

China has established training centers for technicians who oversee installation of the improved digesters. One such center is the Asia-Pacific Regional Biogas Research and Training Center at the Biogas Research Institute in Chengdu, Sichuan Province. The Institute has 6,000 m^2 of laboratory building, 4,000 m^2 of pilot plant, and 6,000 m^2 of service facilities, built at a capital expenditure of 5.1 million Yuan (U.S. $1.7 million). This Institute, staffed by 150 scientific technicians as of 1985, has trained 600 Chinese technicians and 75 foreign technicians from 45 developing countries. New classes are enrolled several times a year. A special manual (6) has been published covering the basics of this training program.

Presently, 4.5 million improved rural digesters have been installed, about 98 percent of which are claimed to be in operation. At U.S. $50 per digester, this equals $225 million of cash outlay and involves about 77 million person-days of labor. If done in the U.S., however, this large a program would cost $1-2 billion. This probably represents only half of the national effort expended thus far, since it appears that at least as many digesters have been abandoned for leakage as are now working well. Some 25 million people now have their food cooked by using biogas as fuel. In addition to rural installations, some housing centers in large cities are being served by digesters. Shanghai has

Figure 3. Fixed Dome, High Pressure Anaerobic Digester with Internal Biogas Collection

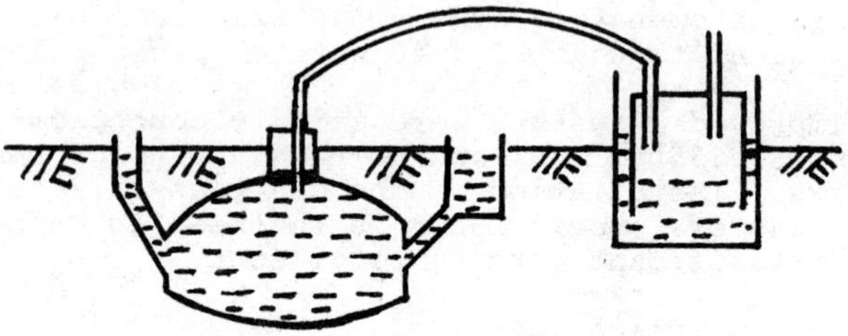

Figure 4. Fixed Dome, Low Pressure Anaerobic Digester with Biogas Collected in Floating Gas Holder

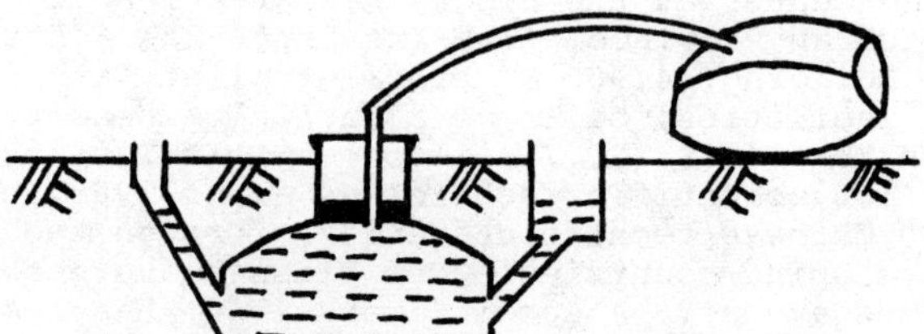

Figure 5. Fixed Dome, Low Pressure Anaerobic Digester with Biogas Collected in Red Mud Plastic Bladder

22 such centers. Overall, the program continues to advance annually with installation of 1/2 million new digesters costing about U.S. $25 million, very little of which is paid for labor which amounts to 8.5 million person-days of work.

Most digesters have a volume of 6 cubic meters and produce enough biogas for a family to cook 3 meals per day for 9 or 10 months and 1 or 2 meals per day for the remaining winter months. However, in northern areas sufficient gas for 3 meals per day can only be generated for the May through November period, and gas for one meal per day can be generated for an additional month. Wood or straw burning stoves must be used when the supply of gas is insufficient.

As a rule of thumb, manure from one 58 kilogram pig will be digested to give 0.2 cubic meters of biogas per day. The average extended family living in one household needs about 5 pigs to have an adequate amount of biogas for cooking. The initial charge for a digester is 50 kilograms of pig manure, 50 kilograms of straw, preferably composted, and 15 kilograms of "starter" residue from another digester to speed up initiation of the microbial digestion of the new charge. Enough water is added to make up 90-92 percent of the charge. If solids content gets too high, the system tends to clog. Manure and night soil are added daily, and composted vegetation wastes from time to time. Retention time of the materials added to the digester is estimated to be about one month. As feed is added, the slurry is displaced into the discharge side-arm. From there it is removed for use as a fertilizer. About twice a year the digester is shut down for clean-out and recharging. At this time the solid residue not needed to start up the next charge is removed and spread as fertilizer on crop lands.

Dry Digestion

"Dry" digestion of crop wastes is also under serious development. This is an extension of work done a few years ago at Cornell University. The dry digester is smaller than the 6 cubic meter conventional Chinese hydraulic digester; moreover it is a single chamber without feed and discharge arms. Therefore, a dry digester costs only 60 Yuan (U.S. $20), compared to 150 Yuan (U.S. $50) for a hydraulic digester. The digester is charged with manure and chopped up pieces of straw (5-7 cm long) and starter residue.

The straw and manure are generally charged in the ratio of 2 to 1 and mixed with an inoculum of 30 percent of total charge. Composting may be done for 3-5 days to raise temperature before the digester is sealed for generation of biogas. These materials are charged without addition of any water except for that present in the materials used. The dry digester therefore has about 25 percent solids or 2½ to 3 times more biomass per unit volume than the hydraulic digester. There is no place for the gas to be collected in the dry digester, and therefore the gas is led off to a separate gas holder which may be located some distance away at a convenient place.

Dry digestion occurs more slowly than hydraulic digestion and lasts from 2 to 5 months, depending on the material charged. Retention time for the hydraulic system is about one month. However, because of the greater amount of solids per unit volume of digester, the amount of gas produced is at least comparable and may sometimes be greater than that made by the hydraulic digester, even though the dry digester is smaller in size. The dry digester produces 0.3-0.4 cubic meters of biogas per cubic meter of digester volume per day (sometimes even up to 1.0 m^3/m^3/day), whereas the hydraulic digester produces 0.2 m^3/m^3/day. Without pig manure a dry digester charge lasts 4-5 months; with pig manure it lasts about 2 months. One modification of dry digestion has been applied to nearly 42,000 digesters in Sichuan Province.

Anaerobic digesters, both hydraulic and dry types, can provide most of the fuel for cooking and thus make it no longer necessary to burn so much straw. More straw thus remains and can be spread on, or left on the fields. Also some of the straw thus spared from burning can be charged into dry digesters.

The dry digester generally finds use as a supplement to, and is run in tandem with the hydraulic digester. The latter handles the discharge from the pigsties and latrine, as well as garbage and inedible daily agricultural wastes. The former is suitable for charging the straw and plant stalks gathered at harvest times (two or three times per year), as well as for farming operations where no pigs are being raised. A family of six generally needs 600 cubic meters of biogas per year. This can be generated by one 5 cubic meter hydraulic digester and one 2 cubic meter dry digester.

The residue from dry digestion (known to be rich in protein) is being evaluated for several special uses. It is being tested as the medium for vermiculture; worms so raised are fed to chickens, whose droppings are then fed to a digester. Mushrooms are also being grown on residue. Eels, a delicacy in China, are being raised in water in which a thick layer of residue is submerged. Other tests are evaluating residue as a feed for livestock. Residue has traditionally been used to fertilize crops and to promote aquaculture.

ADDITIONAL BENEFITS OF ANAEROBIC DIGESTION

In addition to providing fuel for cooking, operation of a digester improves sanitation. Pigsties and latrines are connected directly to the feed arm of the digesters. Many operating digesters were observed. None were malodrous. Not even the pigsties smelled bad, since the manure is washed into the digester daily. It is claimed that the population of insects dwindled with use of digesters, so that use of DDT and other pesticides was reduced. It is also claimed that the health of the rural villagers was significantly better after anaerobic digesters were installed. Consequently, there was less absenteeism from work and higher worker productivity.

The liquor discharged daily from digesters and the residue removed at clean out times are used principally to fertilize crops. This has substantially increased crop yields.

Biogas is also used to provide heat for growing rice seedlings, drying agricultural crops, hatching chickens, heating structures housing silkworms and running irrigation pumps.

In monetary terms, if agricultural wastes used on farms are given a value of 47 Yuan (U.S. $13.33) per ton, the value of a comparable amount of wastes subjected to anaerobic digestion would be increased to 60 Yuan (U.S. $20.00). Thus small biogas installations are claimed to be the key to better rural health and prosperity.

COMMERCIAL USES OF ANAEROBIC DIGESTION

A few standard commercial size digesters have been operated for a long time, and others are under development. For example, two 2,000 cubic meter digesters have been run at an alcohol plant to dispose of distillery wastes from sweet potato fermentation. Also, new 3,600 and 5,000 cubic meter digesters are under construction.

Research and development tests are also being run on more sophisticated anaerobic digester systems to process urban and industrial wastes. The Delegation was provided with details which were subsequently presented at the 4th International Symposium on Anaerobic Digestion (2,3,5) and therefore are not reviewed here. In these efforts, China is adopting Western technology, such as two step digestion, fixed film and anaerobic filter digestion, upflow anaerobic sludge blanket digestion and combinations of these adapted to specific needs. In one of the two step digestion developments, the first step is conducted in a 2 cubic meter uncovered acid-producing pool. The second step is done in a sealed tank in which an anaerobic sludge blanket operation produces biogas. The liquid effluent from this step is returned to the pool. Such an operation is intended for use on a livestock farm.

Two experimental 30 cubic meter, above-grade, insulated digesters were observed in operation. These were being run to compare use of internal and external heat exchangers to get faster digestion by maintaining higher temperatures than can be achieved in those installed below grade.

One distillery that recovers biogas runs one of its trucks on compressed biogas contained in cylinders slung under the truck body. It also uses biogas to run engines driving electrical generators. Many diesel engines are being operated fueled with 80 percent biogas, while engines converted to spark ignition are run on pure biogas. It was reported that there are now over 430 electrical generators in China that are being run on biogas. These have an output of 9,000 Kw.

The Fourth International Symposium on Anaerobic Digestion held in Guangzhou November 11 to 15, 1985 drew an attendance of about 200 foreigners from about 35 countries and an equal number of Chinese, despite the fact that all sessions and papers were in English.

The Symposium included a number of interesting Chinese commercial exhibits featuring a variety of biogas equipment: red-mud-plastic bladders and tubing; biogas burners, including some featuring electronic ignition; biogas-fueled motor generator systems; biogas lanterns; biogas valves, gas meters and manometers; and implements for removing residue from the bottom of digesters.

CONCLUSION

China's claim that anaerobic digesters, wherever used, have significantly improved the standard of living is no doubt true, since its application occurs in situations where there is a need for better sanitation as well as a shortage of fuels, fertilizers and electricity.

REFERENCES CITED

1. Bente, P.F. Jr., "People to People Biomass Utilization Delegation Journal," Citizens Ambassador Program, People to People International, Dwight D. Eisenhower Building, Spokane, Washington 99202, available July 1986.

2. Chen Zepeng, Wu Yongfen, Xu Xueying, Wang Zuxuan, Li Nianguo, "The Application of Two-Phase Anaerobic Digestion on the Disposal of Organic Wastewater," Anaerobic Digestion 1985, Guangzhou, China, pp 375-383.

3. Wang Zuxuan, Chen Zepeng, Qian Zeshu, Li Nianguo, "Status Quo and Prospects on the Study of Anaerobic Disposal for Industrial Wastewater in China," Anaerobic Digestion 1985, Guangzhou, China, pp 259-277.

4. Wu, W., "Biomass Energy Utilization and Its Technologies in China Rural Areas," Energy From Biomass, 3rd E.C. Conference, pp 228-233, Elsevier Applied Science Publishers, London and New York, 1985.

5. Xu Jie-quan, Liao Duo-qun, Shao Qi-yu, Yang Chun-fu, "Pilot Studies on Two-Phase Methane Fermentation with Pig Manure at Ambient Temperature," Anaerobic Digestion 1985, Guangzhou, China, pp 125-134.

6. "Biogas Technology UNDP-ESCAP-FAO-CHINA Biogas Training Course," The Asian-Pacific Regional Biogas Research Training Centre, Chengdu, China, 223 pages, 1983 (in English).

7. "The National Standard of the People's Republic of China (English Text) GB 4750-4752-84
   "The collection of designs for household hydraulic biogas digesters in rural areas
   "The standard for check and acceptance of the quality for household hydraulic biogas digesters in rural areas
   "The operation rules for construction of household hydraulic biogas digesters in rural areas"
   1984-11-10 Date of publication 1985-08-01 Date of implementation, Approved by the State Bureau of Standardization, 103 pages with 59 drawings.

8. "State Standard of the People's Republic of China GB4363-84 Testing Method for the Heat Properties of Civil Firewood Stoves" Ratified by Bureau of State Standard of China, 1984, 12 pages.

DISCUSSION

Klass	How long were you on your trip?
Bente	We were in China for 3 weeks on this trip, and I was there for the anaerobic digestion conference for 1 week, and then another week of touring.
D. Hall, U. of London	The fact that the Chinese authorities now recognize that about 350 million tons of coal equivalent is used in the form of biomass is actually quite an advance. Until recently, this was not recognized by the Central Planners in Beijing. About 80% of the population get about 80% of their fuel from biomass. That is equivalent to about 1 ton of wood per person per year in the countryside, and the fact that this is now acknowledged as being a serious problem is quite a change.
C. Setterstrom, Harrington Research Co.	How do they handle fire and explosives problems?
Bente	With respect to biogas?
Setterstrom	Yes.
Bente	It is not a problem. We saw them smoking all around these units. I understand the explosive limits are rather narrow and not likely to be encountered the way they're operating.
Klass	Pure methane has an explosive limit of about 5 to 15 volume % in air, and when diluted with CO_2, I believe it's quite a bit narrower. But I still feel it wouldn't be a good policy, especially with large digesters, to be smoking around the systems.
Bente	Lots of the Chinese smoke.
K. Porter, U.S. AID (JAMAICA)	From your comments on biogas, what sort of applications do you think there are on a wider scale for developing countries?
Bente	I think that a lot of the progress that the Chinese have made could be used in many developing countries. I think the Chinese are successful because of the fact that there is a shortage and they have the technical knowledge and all the incentives from the very top echelon

right down through every level of the Chinese government and the citizenry. The Chinese concentrate on one village after another, and get these all lined up and operating. I think the Chinese are willing and anxious to train technicians from any other country. They have an institute set up for that. It's equipped with the best of equipment, as good as any we have in the U.S. When you're trained there, you're taught how to build the units under supervision.

V. Bakthavatsalam, NED CAP Ltd (India) — What other types of raw materials are being used?

Bente — The raw materials used are crop residues, which often are composted before they're put into the digestion system and mixed with the pig manure, night soil, and about 30% of the residue from a digester which has essentially finished its operation. The digesters are run for 4 or 5 months, and then they're shut down, cleaned out, and recharged. The dry digesters run for shorter periods of time and are phased in with the crop harvest.

Bakthavatsalam — You mentioned forest area increased from 8.6% to 12%. Could you comment on the seriousness of the fuel wood shortage in China, and its availability for the poor people? Because most of the wood is used for commercial production in the pulp and paper industry, what is the availability of fuel wood for the poor people in China? The forest area you mentioned is very, very small.

Bente — The plantations are often located where the people need the fuel, and wood from local planting along the roadsides, around the ponds, and in the border areas of the farms are available to the people for fuel. The national programs on forestry are generally located too far away from the people. But knowing the Chinese as being careful not to waste any materials, I would assume that what they have as residues and wastes from harvesting of trees for lumber and commercial uses would be transported and made available as fuel.

THERMOCHEMICAL CONVERSION OF PEAT INTO GASES AND LIQUIDS: RECENT DEVELOPMENTS IN FINLAND

Paterson McKeough, Ph.D.
Esa Kurkela, M.Sc.(Tech)
Kai Sipilä, Lic.Tech.
Technical Research Centre of Finland
Laboratory of Fuel Processing and Lubrication Technology
02150 Espoo, Finland

ABSTRACT

Finland is currently utilizing about one million toe of peat per annum, mainly for generating heat and power. The method of peat production and pretreatment has a significant effect, not only on the physical properties of the feedstock, but also on its reactivity in the peat conversion process. For this reason, various pretreatments (drying, mechanical treatments) have been the objects of extensive research in recent years.

The production of fuel gas from peat is now at the commercial level. In particular, the demand for gasifiers of capacities under 15 MW is growing rapidly. Research in this field is currently directed towards increasing the forms of utilization of the fuel gas. Synthesis gas production from peat is also under intensive development. Tests with Finnish peat have been carried out on pilot plants employing pressurized fluidized-bed technology: HTW (West Germany), U-GAS (USA), and MINO (Sweden). The first application of synthesis gas production from peat will be an ammonia-from-peat plant to be commissioned by the Kemira Oy company in 1988. The direct liquefaction of peat is being investigated at the bench-scale level. Peat has emerged as a very suitable feedstock for such a process. In addition, estimates of the economics of the process have yielded promising results.

Overall, it can be stated that the conversion of peat into gases and liquids is being extensively researched in Finland. The research is aimed at the development of processes for both the short and the long term. Prospects for commercialization of these processes are good.

THERMOCHEMICAL CONVERSION OF PEAT INTO GASES AND LIQUIDS:
RECENT DEVELOPMENTS IN FINLAND

1 INTRODUCTION

In 1985 the production of peat in Finland was about 3.4 x 10^6 metric tons, corresponding to 0.9 x 10^6 toe of energy. During the last ten years, the utilization of peat has grown dramatically, as is evident in Figure 1. The strong dependency of peat production on the annual weather pattern is also evident in the figure. The effects of this dependency have, however, been minimized by maintaining a reserve stockpile equal to at least half the annual demand.

Finland has a higher proportion of peatland than any other country. 5 million hectares of its surface are covered by peatlands; however only about 20 % of this area is attributable to peatlands of depth more than 2 m and area more than 50 hectares. 763 000 hectares of peatland have so far been set aside for preservation in their natural state, while about 500 000 hectares (420 x 10^6 toe of energy) are considered suitable for fuel peat production by present harvesting methods. At the present time about 30 000 hectares are under production.

It is expected that peat consumption in Finland will continue to increase significantly and that it could well exceed 7 x 10^6 metric tons/a (about 1.9 x 10^6 toe/a) by the early 1990s. Long-term predictions have indicated that, technically, peat consumption could grow to 18 x 10^6 tons/a by the year 2000. One such evaluation of the technical potential for peat utilization is depicted in Figure 2. However, economic factors will probably limit peat consumption to about 9 - 11 x 10^6 tons/a (2.3 - 2.8 x 10^6 toe/a) by the end of the century.

The commercialization of processes converting peat to gases and liquids is expected to expand, particularly in the long term. At the present moment, the production of fuel gas from peat is already commercial. The demand is growing rapidly, particularly for plants under 15 MW. The fixed bed gasifiers generally employ sod peat as feedstock, while the larger fluidized-bed reactors use waste wood or milled peat. Synthesis gas production from peat is being intensively researched. A

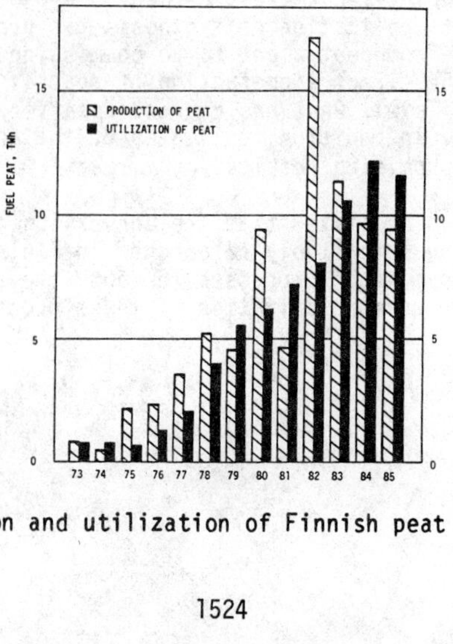

Figure 1. Production and utilization of Finnish peat 1973 - 1985 /1/.

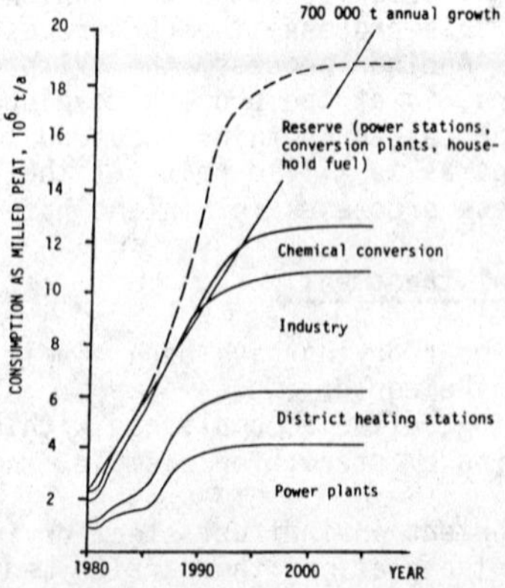

Figure 2. The potential for utilization of peat /2/.

demonstration plant, producing ammonia, will be commissioned within the next few years. The realization of this venture will significantly increase the potential of converting Finnish peat to liquid fuels and chemicals.

2 PRODUCTION AND PRETREATMENT OF PEAT FOR CONVERSION PROCESSES

General

The feedstock requirements for peat conversion processes may well differ from those for peat combustion processes. Should peat be fed, for example, into a pressurized reactor, the use of milled peat, with its low bulk density and poor flowing properties, will be extremely difficult. A better alternative in that case might be to prepare pellets from peat. The bulk density of pellets is twice that of milled peat and the flowing properties of pellets are significantly better. It is also difficult to prepare, from peat, the concentrated feed slurries required for certain other types of reactor. However, the slurrying characteristics of peat can also be improved by certain pretreatments.

In Finland in recent years, various drying and mechanical pretreatment operations, required for conversion processes, have been the object of extensive research. It is possible to prepare peat materials with widely differing physical properties. This technology is, as well, readily applicable to other biomass and waste materials.

About 90 % of peat harvesting in Finland is based on the milled peat method, the rest on the sod peat method. Owing to improvements in harvesting equipment and to a better understanding of weather influences, the annual yield/hectare of milled peat has been increased, and production costs have been reduced. In recent years, methods have been developed to produce peat sods of novel sizes: diameters of 30 mm and 200 mm, in addition to the traditional 80 mm. The aim has been both to increase the yield/hectare and to reduce dependence on weather factors. The use of these methods has, in fact, resulted in technical and economic improvements. Research and development work in this area is continuing.

Methods of artificial peat dewatering have also been the objects of intensive process development work in recent years. A wet carbonization process, which is particularly suited to Finnish climatic conditions, has been developed. This process- the PDF process - reached the pilot plant stage in 1984. Another process, the VVI process, based on suction dryers and heat pumps, is at the process development stage. The primary product from these processes contains about 50 % moisture; that is, about the same amount as in milled peat. At the present time, decisions to commercialize these processes in Finland have not been made.

Drying and mechanical treatment

Two methods of peat drying have been widely applied in the peat industry throughout the world:

- flue gas drying, often accomplished within grinding mills,
- indirect drying by steam; for example, the well-known Peco dryer.

A recent development in indirect steam drying is the use of product steam, evolved from the peat, as the carrier medium. This allows recovery of heat from the product steam. In Finland these steam dryers appear to be the most economic alternative for many applications. One significant potential advantage is the capability to accept milled peat with higher moisture contents (up to 60 %). Such moist peat can be produced significantly more cheaply than conventional milled peat, and, what is more, harvesting of peat is then far less sensitive to weather variations.

Mechanical treatments such as briquetting or pelletizing can be used to increase the bulk density of peat and to improve its flowing properties. In Finland briquettes and pellets are being commercially produced. Pellets can be prepared, like briquettes, from artificially dried peat (15 % moisture), but also directly from milled peat of 30 - 40 % moisture on the peat field. The latter technology is particularly interesting because of the resultant lower investment costs. In Finland the ideal capacity for a transportable field pelletizer is about 10 000 t/a. When pellets are manufactured from moister peat, extra ventilation measures are required during storage to prevent mold formation.

Resultant Characteristics

On the basis of the foregoing, it is clear that a wide variety of peat materials can be prepared as feedstocks for peat conversion processes. The prime effects of feed preparation are on the physical properties:

- bulk density,
- flowing characteristics,
- moisture content.

In addition to these, however, are the effects of pretreatment on the reactivity of peat in the conversion processes. In gasification, for example, reactivity is affected by the porosity of the peat particle. Typical properties of different peat materials prepared from Finnish peats are presented in Table 1.

The important interrelationship between the pretreatment process and the conversion process will be returned to at several points in the following sections.

Table 1. Characteristics of Finnish peat products /3/.

	Moisture content %	Bulk density kg/m³	Hg density g/cm³	Angle of slide ⁰ (steel	Friction coefficient plate)
Milled peat	40 - 45	250 - 450	0.7	35.0	0.70
Sod peat	35 - 40	300 - 400	0.9	35.0	0.70
Pellets Pellets	15 35 - 40	550 - 750	1.3	23.6	0.44

3 PRODUCTION OF FUEL GAS

Status in Finland

Since about 1975, fuel gas production and utilization have been the prime subjects of Finnish research and development on gasification technology. The aim has been to substitute imported fuels by low calorific fuel gas (4 - 7 MJ/Nm^3) derived from the indigenous fuels, wood and peat. The Ministry of Trade and Industry (KTM) has strongly supported the development work, and has, in fact, been the main sponsor.

In Finland at the present time, there are five operative district heating stations which are based on the updraft gasification of sod peat. The plant outputs are in the range 1.5 - 6 MW. Perusyhtymä Oy is the company which has engineered the plants. The other industrial manufacturer of gasifiers is A. Ahlström Oy. The first commercial Pyroflow gasifier began operating in 1983, supplying fuel to the lime kiln of a pulp mill of Wilh. Schauman in Pietarsaari. The maximum output of this gasifier is 35 MW and its feedstock is waste wood. Peat is also a very suitable feedstock for the Pyroflow gasifier. Both gasifier manufacturers have, as well, made sales abroad.

A summary of the present status of development of Finnish fuel gas technology is presented in Table 2.

Development of Commercial Gasifiers

Perusyhtymä's Bioneer gasifier was developed in co-operation with the Technical Research Centre of Finland (VTT) during the years 1979 - 1982. The gasifier is of the updraft fixed-bed type, producing low calorific tarry gas. As such, the gas is suitable only for combustion in the vicinity of the gasifier. Examples of applications are hot water and steam boilers, chip dryers, grain dryers, and process ovens. An extensive experimental program, using a wide variety of feed materials, has been undertaken on VTT's 1.5 MW test gasifier. The experiments ascertained that the gasification method can be applied to most of the solid feedstocks available in Finland. To ensure reliable operation of the gasifier and good combustion of the product gas, the feedstock should meet the following requirements /4/:

- mainly piece-like, with a maximum fines content of about 30 wt-%,
- moisture content less than 55 %,
- ash melting point above 1000 ⁰C,
- no caking tendencies.

Table 2. The status of Finnish fuel gas technology.

Gas utilization method	Stage of development and number of operating plants	Commercial scale	Manufacturer / product name or research establishment
Direct combustion in boilers, process ovens, etc.			
- updraft gasification	Commercial, 5 plants, 1.5 - 6 MW_{th}	1 - 15 MW_{th}	Perusyhtymä Oy/ Bioneer gasifier
- circulating fluidized bed	Commercial, 2 plants, 25 & 35 MW_{th}	5 - 100 MW_{th}	Ahlström Oy/ Pyroflow gasifier
Use as motor fuel			
- circ. fluid. bed/ diesel engine	Demonstration stage (2 MW_e)	{2.3 - 28 MW_e 2.9 - 35 MW_{th}	Ahlström Oy and Wärtsilä Oy
- downdraft gasifier/diesel/ otto engine	Process development stage (50 kW_e)	{50 - 500 kW_e 60 - 600 kW_{th}	VTT *, small local applications VAKOLA**, farm tractors
Clean fuel gas distributed to several users			
- updraft+cracking	Process development stage	1 - 15 MW_{th}	Perusyhtymä Oy and VTT
- downdraft	"	0.5 - 5 "	VTT
- fluidized bed	"	5 - 50 "	Ahlström Oy

* Technical Research Centre of Finland
** Finnish Research Institute of Engineering in Agriculture and Forestry

Table 3 presents product gas compositions obtained from different feedstocks.

A schematic diagram of the Bioneer gasifier-heating station combination is presented in Figure 3. Using a gas-tight feeding system, the fuel is charged into the upper part of the gasifier, from where it moves downward through the drying, pyrolysis, gasification, and combustion zones, respectively. The ash, remaining after combustion, is removed from the base of the reactor. The gasifying air is injected from below

Table 3. Compositions of dry gas produced from different peat and biomass materials.

Fuel	Moisture content %	CO	H_2	CH_4	CO_2	N_2	H_u MJ/m_n^3
[1]Sod peat	33	24.2	19.0	3.0	10.3	43.4	6.2
[1]Sod peat	42	21.6	16.8	2.4	12.5	46.7	5.4
[1]Sod peat	52	16.1	16.7	2.0	14.3	50.9	4.5
[1]Small sod peat	28	23.2	16.7	2.0	14.3	42.9	6.0
[1]Peat pellets	15	23.0	16.7	2.5	10.8	47.0	5.6
[1]Peat pellets	17	28.8	16.9	2.5	9.0	42.5	6.4
[1]Small wood chips	43	29.0	15.4	1.6	6.8	47.2	5.9
[1]Large wood chips	41	30.2	10.8	2.5	7.1	49.2	5.9
[2]Forest waste chips	55	20	17	1.5	13	48.5	4.9
[2]Straw pellets	12	26	15	2	10	47	5.8

[1] Experiments on the 1.5 MW pilot gasifier
[2] Experiments on the 0.3 MW test gasifier

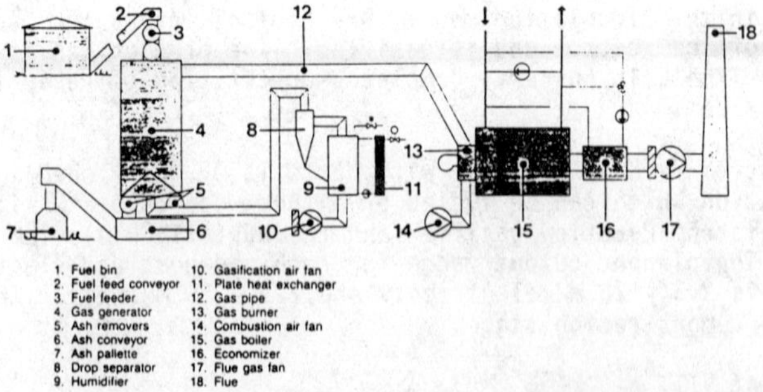

Figure 3. Schematic diagram of the Bioneer boiler plant /6/.

through a rotating cone-shaped grate. The product gas is taken from the top of the gasifier and fed to the burner via a short pipe. The boiler can be either an existing oil-fired boiler or a new one. The first commercial installation was commissioned in 1982 in the town of Kankaanpää. The competitiveness of the Bioneer boiler plant compared to conventional solid fuel boilers is based, in the main, on the following factors:

- low investment costs,
- high reliability and automatic operation,
- high efficiency even at part-load,
- possibility to employ different feedstocks in the same plant.

The first commercial gasifier using Ahlström Pyroflow fluidized bed technology was commissioned in 1983 at the Wilh. Schauman pulp mill in Pietarsaari. A schematic diagram of the facility is shown in Figure 4. The gas is used as fuel in the lime kiln of the mill. In order to provide a sufficiently high temperature in the kiln, the feedstock must be dried to about 15 % moisture. In the facility depicted in Figure 4, drying is accomplished in a flash dryer employing flue gases from the recovery boiler of the pulp mill. Another alternative is to employ the flue gases from the lime kiln itself.

The pyroflow gasifier consists of a refractory-lined cylindrical vessel and a refractory-lined cyclone. A fluidized-bed of hot sand is

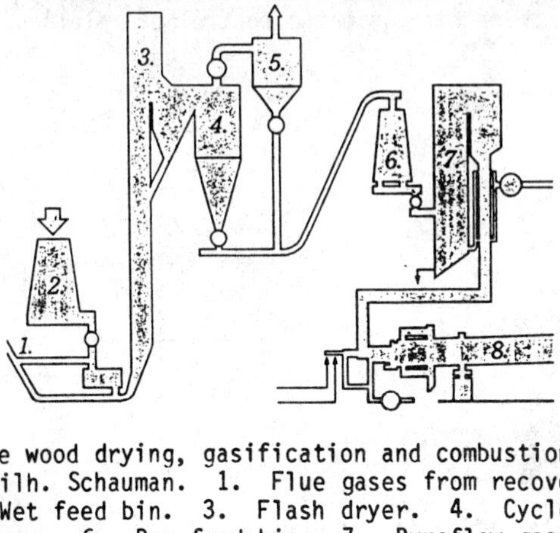

Figure 4. Waste wood drying, gasification and combustion in a lime kiln at Wilh. Schauman. 1. Flue gases from recovery boiler. 2. Wet feed bin. 3. Flash dryer. 4. Cyclone. 5. Multicyclone. 6. Dry feed bin. 7. Pyroflow gasifier. 8. Lime kiln. /5/.

operated in the circulating mode. Dry biofuel enters the bed through the side of the reactor and is gasified. The product gas is led directly to the lime kiln burner. Complete substitution of heavy fuel oil is possible /7/.

Ahlström Oy has, together with Wärtsilä Oy, developed a diesel power station which can be fueled by either wood or peat. It is based on the Ahlström Pyroflow gasifier and the Wärtsilä Pielstick diesel engine. The planned output range for combined heat and electricity stations is 2.3 - 28 MW electricity and 2.9 - 35 MW heat. The project is at the demonstration stage.

Research at VTT

The research at the Laboratory of Fuel Processing and Lubrication Technology, VTT, has, in recent years, been directed towards increasing the utilization possibilities of fuel gas. Figure 5 depicts the classification of low calorific fuel gases according to their level of impurities, which, in turn, defines their possible forms of utilization (also given in the figure). Raw product gas, produced by various gasifiers, is usable only in applications where combustion in the immediate vicinity of the gasifier is possible. Using simple purifying equipment, the raw gas can be partly cleaned. Its potential applications are then much more numerous. For example, the gas can be distributed by pipeline to a number of users in a restricted locality, and the gas can be fired with conventional combustion equipment. The long-distance distribution of gas requires, however, its complete purification. The utilization potential of such a gas is the most extensive but the purification costs are also the greatest.

Since 1983 the experimental fuel gas research at VTT has concentrated on the purification, distribution, and utilization of gas produced by the fixed-bed methods. The layout of the gasification equipment of the Laboratory of Fuel Processing and Lubrication Technology is shown in Figure 6. The experiments made on the downdraft gasifier (500 kW) have been aimed at the development of a gasifier (0.5 - 3 MW) for producing low tar gas from sod peat and wood chips. The gas from this unit has also been used as a testing medium for various purification systems and as a motor fuel. The motor in question is a Valmet 6 cylinder (originally) diesel engine, which was converted to a spark ignition engine and equipped with systems for automatic control of rotation speed and fuel mixture.

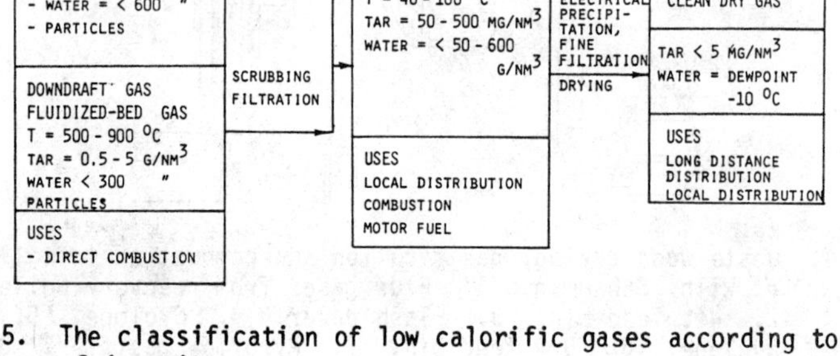

Figure 5. The classification of low calorific gases according to degree of impurity.

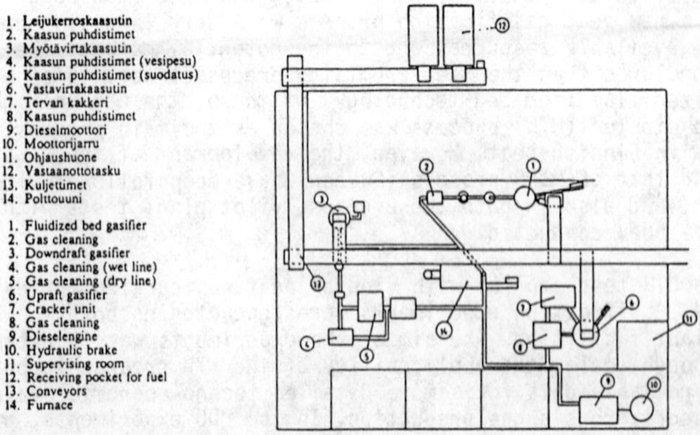

Figure 6. Layout of the gasification equipment at the Technical Research Centre of Finland /8/.

In other research, the product gas from the updraft gasifier has been purified by thermal cracking. The resultant gas had a purity similar to that of a typical product from a downdraft gasifier. The updraft test gasifier has an output of 300 kW. This research has been carried out in co-operation with Perusyhtymä Oy and is aimed at the development of a gasifier producing low-tar gas at capacities 3 - 15 MW.

4 PRODUCTION OF SYNTHESIS GAS

In the 1970s, researchers in Finland, as elsewhere in the world, began to examine alternatives for oil-derived synthesis gas feedstocks. The only indigenous raw materials, which are suitable for commercial production of synthesis gas, are wood and peat. As stated in section 1, the utilization of peat as a fuel expanded rapidly in the 1970s, its price being lower than that of wood. As a result, the technology of peat production and utilization quickly developed to a level which made possible the usage of peat in large-scale industrial applications.

In Finland there is only one synthesis gas production facility currently in operation. The feedstock is heavy fuel oil and the final product ammonia. However, the capacity of this plant is sufficient to satisfy only about 20 % of the Finnish demand. The plant is part of Kemira Oy's fertilizer complex at Oulu.

The research on synthesis gas production from peat has been undertaken by Kemira Oy and the Laboratory of Fuel Processing and Lubrication Technology, VTT. Kemira's objective was to find alternative feedstocks for its ammonia production and, at the same time, increase the self-sufficiency of this production. VTT's objective was to determine the potential of producing synthetic liquid fuels from domestic raw materials. Of the different methods for producing synthetic fuels, those based on synthesis gas were nearest commercial realization, so that, for both Kemira and VTT, the central problem became the gasification of peat. The Energy Department of the Finnish Ministry for Trade and Industry has been the most significant financer of this research.

Between 1979 and 1981 an extensive literature study was carried out. The applicability of the more advanced coal and biomass gasification processes was evaluated with the view to commercialization within five years. Initial studies had indicated that the resources required to develop a new peat gasification process were very large in comparison to both the available resources and to the potential market. The literature study concluded that the most promising processes were those based on pressurized fluidized-bed technology. Of these, the German High Temperature Winkler (HTW) process was chosen as the main process to be tested with Finnish peat. However, the development of the U-GAS process (USA) and that of MINO process (Sweden) have been followed with great interest, and also in both these cases, pilot plant tests with Finnish peat have been conducted.

The HTW test program with Finnish peat was carried out during 1983 and 1984. Gasification experiments were conducted on both the PDU and pilot plant facilities. The aim of the experiments was to determine, at large enough scale, the applicability of the HTW process to peat and to produce reliable data for a more detailed techno-economic evaluation of peat-based synthesis gas production. In the PDU experiments, peat materials, which had been prepared in different ways, were tested, and, on the basis of these tests, crushed peat pellets were chosen as the feedstock for the pilot plant experiments. The experimental program conducted in the pilot plant proved that the HTW process can be applied to peat, and that there are no significant technical obstacles to the commercialization of synthesis gas production from peat. The experiments also indicated that the method of peat pretreatment is a very critical parameter in this process. At the present time VTT and Kemira are studying the effects of various pre-treatments on gasification behaviour and on the synthesis gas production costs. Results of the HTW experiments are considered in more detail in a separate paper /10/.

On the basis of the results of the pilot plant tests, quotations for gasification processes, based on HTW technology, have been drawn up and delivered to Kemira Oy. The objective is to build, with minimum investment, a peat gasifier to replace the existing heavy fuel oil gasifiers at the Oulu ammonia plant. The rest of the existing ammonia production line will remain in use. The final decision to construct the facility has been made and the plant will be commissioned in the spring of 1988. It will produce 80 000 t/a of ammonia from 500 000 t/a of sod peat. More details of this venture are also given in the separate paper /10/.

5 DIRECT LIQUEFACTION

Processes for the direct liquefaction of peat are being developed at several places throughout the world, but, in every case, the development work is still at the bench-scale level. In Finland the bulk of the work is being carried out at the Laboratory of Fuel Processing and Lubrication Technology, VTT. Some catalytic studies have also been made at the University of Oulu.

This paper will concentrate on the research being undertaken at the Laboratory of Fuel Processing and Lubrication Technology. This research is being sponsored by the Energy Department of the Finnish Ministry for Trade and Industry (KTM). Two types of direct liquefaction process are being studied:

- high pressure liquid-phase type,
- flash pyrolysis type.

The research is being conducted on two fronts:

1. Development of process concepts for the direct liquefaction of peat. This work is based on a combination of experimental investigation and techno-economic evaluation.
2. Participation in an international co-operative project (BLTF project) organized by the International Energy Agency (IEA). The ultimate aim of this project is to design, construct, and operate a pilot scale test facility for the direct liquefaction of wood and peat. The main task in the first phase of the project, now completed, was a techno-economic comparison of the various process alternatives /11/.

Research on High-pressure Liquefaction

Objective. The aim of the research to date has been to identify the most promising process concept for the liquefaction of peat by high-pressure liquid-phase treatment. The ultimate products of interest are transport fuels. The important process steps are:
- pretreatment, e.g. drying, grinding,
- slurrying and pumping,
- primary liquefaction,
- separation of primary product from unreacted raw material,
- upgrading of primary product into transport fuels.

These individual steps are interrelated, for example:
- pretreatment determines slurrying properties,
- slurry concentration affects liquefaction capacity,
- pretreatment affects reactivity,
- liquefaction severity determines possible separation methods and the upgrading requirement.

As a consequence of this high degree of interrelation, the entire production chain, from raw material to transport fuel, should be optimized as a whole.

Initial Studies. The following prestudies were carried out:
- an extensive literature review aimed at narrowing down the type of process to be investigated experimentally /12, 13/,
- a study of the stoichiometry of deoxygenation routes which are pertinent to liquefaction /14/,
- a preliminary economic evaluation of a peat liquefaction process /15/.

On the basis of the literature study a primary liquefaction process with the following basic features was chosen for further studies:
- partly-dried peat (10 - 50 % moisture) as feed material,
- recycle oil as carrier (slurrying) medium,
- non-catalytic liquefaction step (but recycle oil may be catalytically treated),
- reducing gas present (H_2 and/or CO).

This type of process differs considerably from those concepts employing undried peat (90 % moisture) as feed material, e.g. /16/.

The stoichiometric study /14/ was aimed at determining the relative effects on process economics of the different deoxygenation mechanisms and the conditions which enable low-rank, highly oxygenated materials, such as peat, to compete with higher-rank materials, such as coal, in these processes. The two main deoxygenation routes lead to the formation of water and carbon dioxide respectively. The main conclusion of the analysis was that the competitiveness of lower-rank materials relative to higher-rank materials need not be limited by the higher deoxygenation requirement. In one particular example, where feed materials were assumed to have the same energy price, competitiveness of lower- rank material required that at least 30 % of the deoxygenation took place via the CO_2-forming route /14/.

Experimental Investigations. To this date the experimental research has been directed towards the following process steps:
- pretreatment,
- slurry and pumping,
- primary liquefaction.

The upgrading of the primary liquid product into transport fuels will be the object of future experimental studies.

Only one type of peat has been employed in the experimental program: a typical fuel peat from the Oulu district. It is a Carex variety of medium degree of humification (H_5 on the von Post scale). An analysis of the peat is shown in Table 4.

Table 4. Analysis of peat.

Proximate analysis (wt-%, dry basis):	Ash	5.7
	Volatile matter	68.7
	Fixed carbon	25.6
Ultimate analysis (wt-%, dry ash-free basis):	C	58.4
	H	6.6
	N	2.4
	S	0.2
	O (by difference)	32.4
Higher heating value (dry basis):	22.3 MJ/kg	

The peat had been excavated by the milling method. Its as-received moisture content was about 45 %. Table 5 lists the pretreated peat materials that have, so far, been selected for study. A maximum particle size of 0.5 mm was imposed because of the small scale of the unit used in the slurrying and pumping tests (see below). However, even in large-scale plants, larger particle sizes have rarely been used.

The unit employed in slurrying and pumping studies is depicted in Figure 7. It was designed to simulate the feeding section of conventional high-pressure direct liquefaction processes. The scale of the unit was, however, small, for example, pipe internal diameters of 6 to 10 mm, check valves with balls of 9.5 mm diameter. The high-pressure loop was operated semicontinuously between 5 MPa and 25 MPa.

The objective of the slurrying and pumping tests was to determine, for different slurrying mediums, the maximum solids concentrations employable in the unit. Two slurrying mediums were chosen: anthracene oil and water. The former simulates a recycle oil of an actual process. The tests with water were performed mainly for reference purposes. Results of the tests are presented in Table 6. This data has been used in determining the optimal pretreatment - liquefaction combination, as will be discussed later. The results in Table 6 indicate that, in general, relatively high solids concentrations are achievable in processes employing recycle oil as carrier (compared, for example, to processes in which raw peat is slurried with water).

The experimental bench-scale unit which has been used for the primary liquefaction research is depicted in Figure 8. It is a batch

Table 5. Materials prepared for experimental studies.

Material	Preparation	Moisture content (approximate), %	Particle size
A	As-received milled peat, ground in a disc mill	45	100 % < 0.5 mm 90 % < 0.25 mm
B	Milled peat, air-dried, ground in a ball mill (15 min)	10	100 % < 0.5 mm 90 % < 0.2 mm
C	Milled peat, air-dried, ground in a ball mill (60 min)	10	100 % < 0.5 mm 90 % < 0.09 mm

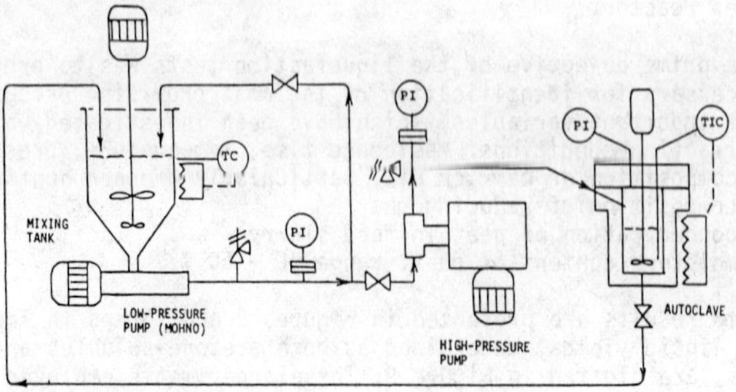

Figure 7. Slurry test unit.

Table 6. Results of slurry tests.

Test	Slurry composition	Temperature °C	Maximum solids concentration (dry basis), %
1	Material A - water	65	25
2	Material A - anthracene oil	110	33
3	Material A - anthracene oil	75	24
4	Material B - anthracene oil	110	40
5	Material C - water	65	24
6	Material C - anthracene oil	110	43

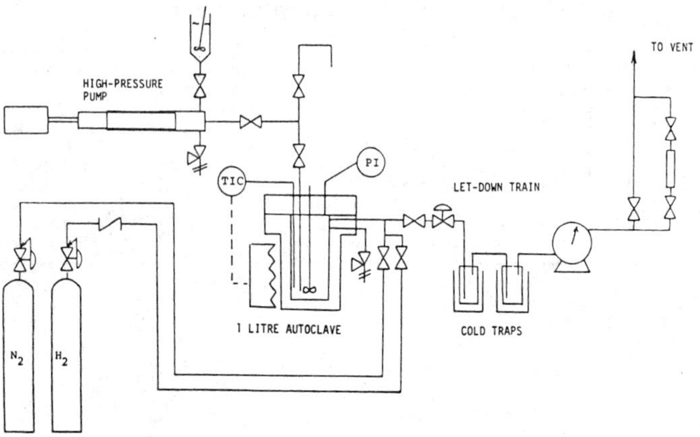

Figure 8. Hot-charge batch liquefaction unit.

reactor system with a facility for charging the feed slurry into the preheated reactor.

The prime objective of the liquefaction tests was to provide the data necessary for identification of the most promising process concept. The most important variables, which have been investigated, are:

- reaction conditions: residence time, temperature, pressure
- composition of carrier oil: particularly H-donor content
- composition of reducing gas
- concentration of peat in feed slurry
- moisture content of peat: range 10 - 50 %.

Some results are presented in Figures 9 and 10 and in Table 7 /18/. Primary liquid yields, determined as both acetone-solubles and pentane-solubles, are plotted in Figure 9. These experiments employed "dry" peat (10 % moisture) as feed material and three different carrier oils:

- anthracene oil,
- tetralin, as a model of a good H-donor carrier /17/,
- 1-methylnaphthalene, not an H-donor but apparently able to transfer hydrogen between peat fragments /17/.

From Figure 9, it can be seen that the liquid yields are fairly independent of residence time for times greater than 10 min. In other words the major liquefaction reactions proceed at a fast rate. Yields are at a maximum in the temperature range 380 to 420 ^{0}C. A significant effect of the H-donor content of the carrier oil is evident. Particularly dramatic is the positive effect of increased H-donor content on the yield of pentane-solubles. In other words, the quality of the primary product is significantly improved by the presence of H-donor.

Some reducing gas consumption results are presented in Figure 10 /18/. For some of the data points the amounts of hydrogen donated by the carrier oil are also shown. The amounts are significant, which is consistent with the higher product quality produced by carriers of greater H-donor ability. Overall, these results indicate that, in the primary liquefaction of peat, low levels of hydrogen consumption (less than 1 % of dry ash-free peat (maf)) are attainable. This is consistent with the fact that, in most experiments, almost 50 % of the peat oxygen was removed as carbon dioxide /18/. This augers well for the competitiveness

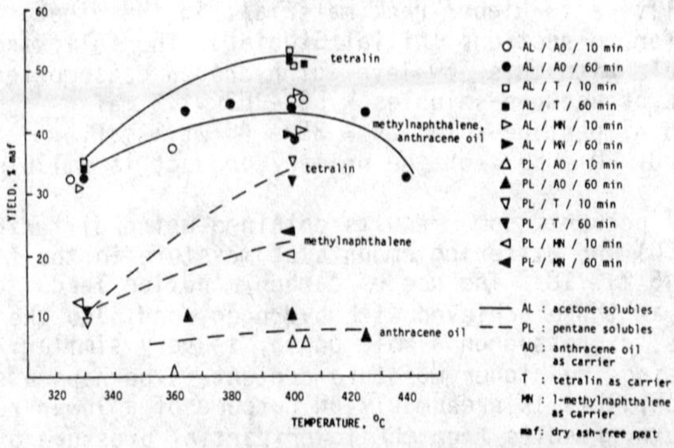

Figure 9. Primary product yields from peat liquefaction. Pressure 20 MPa. Reducing gas H_2. Peat moisture 10 %. /18/.

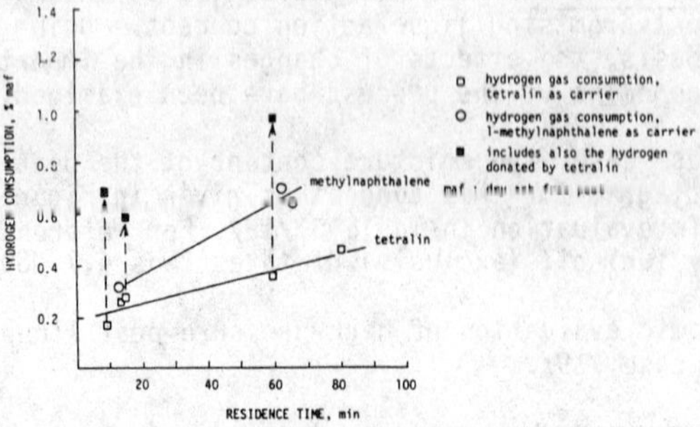

Figure 10. Hydrogen consumption in peat liquefaction. Temperature 400 °C. Pressure 20 MPa. Peat moisture 10 %. /18/.

Table 7. Results of experiments with differing gases and differing feed moisture contents. Temperature 400 °C, residence time 10 min, pressure ~ 20 MPa. AO: anthracene oil, MN: 1-methylnaphthalene, CL: soluble in carrier oil, AL: acetone solubles, maf: dry ash-free peat.

Run	Moisture content wt.%	Carrier oil	Reducing gas	Conversion* %	Liquid yields wt.% maf CL	AL	Reducing gas consumption wt.% maf
152	10	AO	H_2	93	55	35	0.5
72	9	AO	CO	93	52	37	8.9
146	45	AO	H_2	89	49	37	0.5
143	46	AO	CO	89	49	34	8.9
154	10	MN	H_2	83	43	39	0.3
165	10	MN	CO	81	44	32	5.7
166	48	MN	H_2	79	40	39	0.6
168	48	MN	CO	85	47	36	4.6

* Conversion = $100 \cdot [1 - \frac{\text{residue(insoluble in carrier)}}{\text{maf}}]$ %

of peat, relative to higher-rank material, in the direct liquefaction process (refer to section, Initial Studies). The following liquid yields are achievable with this low level of hydrogen consumption:

- yield of acetone-solubles ≈ 50 - 53 wt.% maf
- yield of pentane-solubles ≈ 35 - 40 wt.% maf.

In other words 70 - 80 % of the primary product is soluble in pentane.

Table 7 presents some results obtained using differing reducing gases (H_2, CO) and differing amounts of moisture in the feed material (~ 10 %, ~ 45 %) /18/. The use of carbon monoxide leads to similar product yields as those achieved with hydrogen, and also the consumption of reducing gas, expressed on a mole basis, is very similar. The greatest effect of using the higher moisture content is perhaps a small decrease in conversion. This is presumably an outcome of a lower reaction rate, which, in turn, results from the lower partial pressure of the reducing gas. Drying peat from 45 % moisture to 10 % does not evidently cause any significant reduction in peat reactivity in this type of process. Synthesis gas (H_2/CO mixture), being cheaper than either H_2 or CO alone, should evidently be employed as the reducing gas in the liquefaction plant.

Economic Evaluation. Economic evaluation is the main tool for identifying the most promising liquefaction concept. Using the experimental results as a basis, the effects of changes in the important variables on the overall economics of the process have been examined.

In the base case, the moisture content of the peat feedstock was 10 %. A flow diagram for this concept is given in Figure 11 and results of the economic evaluation in Table 8 /19/. For reference the Finnish price of heavy fuel oil (exclusive of taxes) was 4.7 USD/GJ in mid-1985.

Table 8. Economic evaluation of high-pressure peat liquefaction process. Base case /19/.

Bases	
Costs as of mid-1985, Gulf coast	
Plant capacity:	1000 dry t/d feedstock
Service life:	20 a
Interest rate:	8 % (annuity factor 0.1019)
Feedstock:	milled peat (50 % moisture)
Feedstock cost:	1.5 USD/GJ (at the bog)
Electricity cost:	3 c/kWh
Results	
Production rate of primary product:	395 t/d
Electricity requirement:	5.9 MW
Total capital requirement:	86 x 10^6 USD
Variable operating costs:	11.1 x 10^6 USD/a
Fixed operating costs:	8.4 x 10^6 USD/a
Capital charges:	8.8 x 10^6 USD/a
Production costs:	6.1 USD/GJ primary product (LHV)

In addition to the base case, several sensitivity cases have been evaluated. The results of two such evaluations are compared to the base case results in Table 9 /19/. The specific aim of these two evaluations was to assist in the determination of the optimum pretreatment-liquefaction combination. In the "Moist Peat" sensitivity case, peat with about 45 % moisture was fed directly to the liquefaction reactor. This led to a significant improvement in the thermal efficiency of the process (refer Table 9). However, production costs are somewhat higher than those for the base case, mainly due to increased mass flows (more water and recycle oil, refer Table 6). In the "40 % solids" case, the solids content of the feed slurry was increased from 30 % to 40 %. The resultant reduction in product cost was quite small (Table 9), the improvement resulting from higher feed concentration being largely offset by increased pretreatment costs. Overall these results suggest that "dry" feed material (10 % moisture) should be employed, and that the particle size should be sufficiently small (90 % < 0.2 mm) to enable a solids concentration of about 30 % in the feed slurry. Efforts at improving the thermal efficiency of the process should be directed at the thermal drying operation itself; for example, application of multiple-effect steam dryers.

Table 9. Results of evaluations of different cases. Bases as in Table 8.

	Base case	Moist Peat case	40 % Solids case
Product cost USD/GJ (LHV)	6.1	6.6	6.0
Thermal efficiency* %	61.5	67	62

$$\text{* Thermal efficiency} = \frac{\text{energy out as liquid products (HHV)}}{\text{total energy in (HHV)}} \cdot 100\ \%$$

Using published data for coal and wood liquefaction products, a rough estimation of the costs of upgrading the primary product into high-octane gasoline has also been performed /19/. The resultant gasoline production cost was determined to be in the range 10.5 - 11.5 USD/GJ (LHV) (same bases as in Table 8). This compares favourably with the Finnish gasoline price (exclusive of taxes) of about 10 USD/GJ (mid-1985). However, it should be noted that the Finnish price is quite a lot higher than the world market price. For example, at the same point in time, Rotterdam spot-price was short of 7 USD/GJ:

Overall, the results of both experiments and economic evaluations indicate that Finnish peat is a very suitable feedstock for a high-pressure liquefaction process.

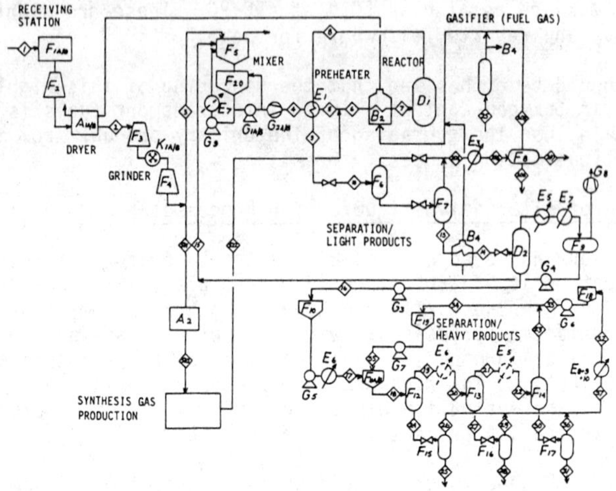

Figure 11. Flow diagram for peat liquefaction concept. Base case /19/.

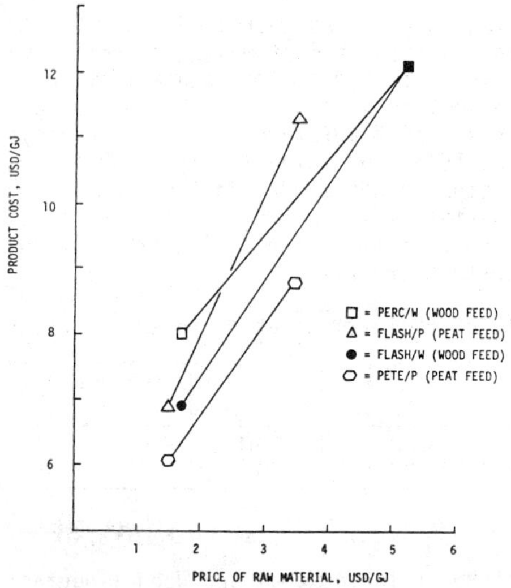

Figure 12. Comparison of primary product costs for various liquefaction processes. Bases as in Table 7. Process descriptions in text. /19/.

Research on Liquefaction by Flash Pyrolysis

Research in this area is at a very early stage. Some experiments have been conducted in co-operation with the University of Waterloo, Canada. The University of Waterloo research group has developed a promising flash pyrolysis process using wood as feedstock. Exceptionally high yields of organic liquids (60 - 70 %) have been observed /20/. Results obtained with Finnish peat (same type as in Table 4) in the University of Waterloo bench-scale unit are presented in Figure 13. Yields of tar are as high as 50 % at 450 ^{0}C. These are significantly higher than any reported elsewhere for peat.

It should be emphasized that the upgrading of this highly- oxygenated peat tar (oxygen content ≈ 15 %) into transport fuels is technically less certain than the upgrading of the primary product from the high-pressure liquefaction process.

Comparison of Alternative Liquefaction Processes

This work has been carried out at the Laboratory of Fuel Processing and Lubrication Technology since 1977, and also, more recently, as part of the IEA BLTF project /11/. In Figure 12, the economics of several direct liquefaction processes, using peat or wood as feedstock, are compared (primary products only). The bases of the assessments are the same as in Table 8. The high-pressure peat process (abbreviation, PETE) is the base case evaluation discussed earlier. The other processes, evaluated in the IEA project /11/ are:

1. PERC/W. High-pressure liquid-phase liquefaction of wood in recycle oil, based on the process originally developed at the Pittsburgh Energy Technology Centre (PERC), USA.

2. FLASH/W. Flash pyrolysis of wood based on the process being developed at the University of Waterloo, Canada.

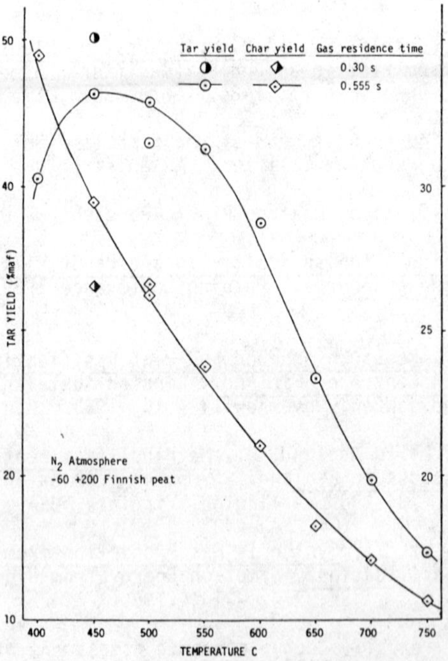

Figure 13. Yields of tar and char from the flash pyrolysis of peat.

3. FLASH/P. Similar to process 2, but employing peat as feedstock.

In mid-1985 prices of peat and wood in Finland were:
- peat: milled peat (50 % moisture)
  at the bog: 1.5 USD/GJ
  35 km from bog: 1.7 USD/GJ
- wood: chips (50 % moisture) 3.0 - 4.5 USD/GJ.

At these feedstock prices, the peat processes have a distinct economic advantage over the wood processes (refer Figure 12). Furthermore, over the entire range of raw material price, the high-pressure peat process yields the lowest primary product cost.

6 CONCLUDING REMARKS

The conversion of peat into gases and liquids is being extensively researched in Finland. Particular attention is being paid to the interrelationship of the pretreatment process and the conversion process. The research is aimed at the development of processes for both the short and the long term. Prospects for commercialization of these processes are good.

REFERENCES

1. Turveteollisuus 1985:1, p. 7.

2. Turvekomitean mietintö (Report of Peat Committee). Komiteanmietintö 1873:4. Helsinki 1983. P. 79.

3. Rautalin, A. et al., Flow Properties of Peat Products and Wood Chips in Silos. VTT Publications. Espoo 1985 (in print).

4. Salo, K. & Kurkela, E., A District Heating Plant Fueled by Gas Generated from Peat in an Updraft Gasifier. Bioenergy 84 World Conference, June 18 - 21, 1984. Gothenburg, Sweden. 4 p.

5. How to Cut Energy Costs and at the Same Time Solve the Waste Wood Problem with Pyroflow Gasifier. A. Ahlström Oy. Brochure. 11 p.

6. Bioneer Gasification Heating Plant. Perusyhtymä Oy. Brochure. 4 p.

7. Siro, M., First Finnish Kraft Pulp and Paper Mill Independent of Oil as an Energy Source. Pulping Conference 1984, Proc., p. 591 - 595.

8. Kurkela, E., Research on Wood and Peat Gasification at the Technical Research Centre of Finland. Finnish-Austrian Workshop on Energy from Wood. Wien, November 14 - 16, 1983. 8 p.

9. Solantausta, Y. & Asplund, D., Methanol from peat. Technical and economic aspects in Finland. Symp. on Peat as an Energy Alternative, Dec 1 - 3, 1980, Arlington, Virginia, USA.

10. Kurkela, E. et al., Synthesis Gas Production from Finnish Peat by Fluid-Bed Gasification. Symp. on Energy from Biomass and Wastes X, Washington, D.C., USA, 7.4. - 10.4.1986.

11. McKeough, P. et al., Techno-economic Assessment of Selected Biomass Liquefaction Processes. Stockholm 1983. IEA Co-operative Project D 1 (BLTF). Final Report, Vol. 5. Publ. National Energy Administration, Sweden.

12. McKeough, P. & Solantausta, Y., The Direct Liquefaction of Peat - the Finnish Program. Specialists' Meeting on Biomass Liquefaction, Saskatoon, Canada, 16.2.-17.2.1982. University of Saskatchewan Publication.

13. McKeough, P. & Solantausta, Y., An Updade on VTT's Direct Liquefaction Program. Comptes rendus de l'atelier de travail sur la liquéfaction de la biomasse, Sherbrooke, Qué, Canada, 29.9. - 30.9.1983, NRCC 23130.

14. McKeough, P., Oxygen Removal during the Direct Liquefaction of Low-rank Materials. Fuel 62, (1983) 4.

15. McKeough, P., Turpeen suora nesteytys. Taloudellinen arviointi. (The Direct Liquefaction of Peat. Economic Evaluation). Espoo 1982. Technical Research Centre of Finland, Laboratory of Fuel Processing and Lubrication Technology. Internal Report (in Finnish).

16. Björnbom, P. et al., Liquefaction of Swedish Peats. Fuel 60 (1981)7

17. Whitehurst, D. et al., Coal liquefaction. Academic Press, New York 1980.

18. McKeough, P. et al., Suoraan nesteytykseen liittyvä koetoiminta, 1983 - 1985 (Experiments on Direct Liquefaction, 1983 - 1985). Espoo 1986. Technical Research Centre of Finland, Laboratory of Fuel Processing and Lubrication Technology. Internal Report (in Finnish).

19. Tulenheimo, V. & McKeough, P., Turpeen korkeapainehydrausprosessin teknis-taloudellinen arviointi (A Techno-economic Evaluation of Direct Peat Liquefaction). Espoo 1986. Technical Research Centre of Finland, Laboratory of Fuel Processing and Lubrication Technology. Internal Report (in Finnish). To be published in English as a VTT Research Report.

20. Scott, D. S. & Piscorz, J., The Continuous Flash Pyrolysis of Biomass. Can. J. Chem. Eng. 62, 404-412 (1984)3.

L.E.B.E.N. — LARGE EUROPEAN BIOENERGY PROJECT

G. Grassi
Commission of the European Communities
Belgium

P. Ciammaichella
Regional Minister for Agriculture and Forestry

N. Pace
Regional Minister for Industry and Energy

F. Gheri
Coordinator L.E.B.E.N. Project at the Abruzzo Region
Abruzzo, Italy

ABSTRACT

The Abruzzo Region is a part of the "Mezzogiorno" of Italy. In this region, there is a large imbalance between energy (electricity) production and consumption. Therefore, if a large amount of thermal and electrical power generation can be produced at mode-rate cost from local resources, a strong contribution to regional development will arise. Vast job opportunities will also present themselves to a wide range of specialists.

The L.E.B.E.N.-Project is based on:

- The exploitation of about 400,000 t/year of biomass to produce liquid fuel (bio-oil/charcoal emulsion) and electricity in a power station (200.10^6 kWh/year);
- The generation of hydro-electricity: 124.10^6 kWh/year.

This energy will be distributed into regional agro-forestry and agro-industrial sectors.

L.E.B.E.N. — LARGE EUROPEAN BIOENERGY PROJECT

1. SUMMARY

The Abruzzo Region (41° parallel east of Rome) is a part of the "Mezzogiorno" of Italy. In this region exists a big imbalance between energy (electricity) production and consumption (40% in the year 1990) and the specific energy consumption in agriculture (0.5 T.O.E.* /Ha) is lower than the Italian average value. Therefore, if a large amount of thermal and electrical power generation produced at moderate cost from local and so far unexploited resources, (i.e. Electricity: 0,03-0,04 ECU** / kwh) can be offered permanently to the regional agro-forestry and agro-industry sectors, a strong contribution to the intensity and quality of the regional development will arise through increase in the technological content of all activities (energy production, more sophisticated energy-intensive agricultural and agro-industrial activities, services). Furthermore, vast job opportunities will present themselves to a wide range of specialists.

The L.E.B.E.N.-Project is based on:

a) the exploitation of about 400,000 t/year of biomass to produce liquid fuel (bio-oil/charcoal emulsion) and electricity in a power station (200.10^6 kWh/year);

b) the generation of hydro-electricity: 107.10^6 kwh/year.

The total amount of energy (electricity, fuels and waste heat) will then be injected into regional agro-forestry and agro-industrial activities.

2. SCOPE AND OBJECTIVES OF THE L.E.B.E.N.-PROJECT

The L.E.B.E.N.-Project, which is already in an early phase of implementation in the Abruzzo Region of Italy, will be the first large European bio-energy project. It is technically and financially supported by the Commission of the European Communities (CEC) together with Italian national and regional authorities. This project embodies new concepts and integrated activities not tried before. It will, it is hoped, serve as a model, not only for the CEC, but also for developing countries. It is already attracting worldwide interest.

---

* T.O.E. = Tonne Oil Equivalent

** ECU 1 = Lire 1472 = 0,904$ USA (Feb.1986)

The main objectives of the L.E.B.E.N.-Project are the following:

1. to show the possibility for the E.C. agro-forestry activities to contribute, in the medium-long term, to the energy market by the "Agro-T.O.E.*" production. For many industrialized and developing countries faced with agricultural surpluses (products and soils) this is attractive since it is a non-saturable market**. Furthermore an important and new characteristic of agro-forestry activities oriented towards the energy market, is that the level of productivity is more important than the quality of the product. For the CEC two different strategies look possible and promising for the exploitation of the large amount of biomass available (85 mio t.o.e./year of net energy in the year 2000):

   a) ethanol production, from good agricultural soils, as an ingredient for motor fuels, derived from agricultural surpluses and (later) from energy crops (medium- and long-term implementation);

   b) Pyrolytic oil/charcoal emulsion production from marginal land for general thermal application and local electricity generation through thermochemical conversion of dry biomass (short- and medium-term implementation);

2. to demonstrate, in particular, at a regional level, the significant contribution that such agro- energy projects can offer to the general economic development of the less advantaged areas (south of Europe, developing countries, etc.) as well as their social interest, by:

   - making available supplementary and useful regional new resources (fuels from residues);
   - creating a large number of new jobs meeting the needs of a wide spectrum of activities resulting from such multi- disciplinary projects (biomass production, harvesting, transportation, conversion, upgrading, utilisation, services etc...). 60% of the energy products cost is due to the cost of personnel.

3. to verify with regard to the conversion technologies the practical validity of the modular concept. Thus the risks associated with large-scale projects will be reduced and the injection of improved and eventually new technologies will become possible.

3. DESCRIPTION OF THE L.E.B.E.N.-PROJECT

The L.E.B.E.N.-Project is an integrated complex project subdivided into four main sectors of activity:

---

* The concept of agro-Tonne-Oil-Equivalent

** At Community level, in the year 2000, only 6-7% of the total energy needs (1,350 million T.O.E.) could be satisfied by the full exploitation of the biomass resources.

3.1. SECTOR OF BIO-ENERGY ACTIVITIES

subdivided as follows :

a) Production and collection of regional biomass resources.

The amount of ligno-cellulosic resources are the following :

1st - 12th year:

250,000 ton/y (residues* from 95,000ha of "Bosco Ceduo" available)
70,000 ton/y (agricultural residues**)

320,000 ton/y : TOTAL

with a humidity content of 30%

12th - 20th year

120,000 t/y (forest res.)
70,000 t/y (agricultural res.)
130,000 t/y (energy crops***)

320,000 t/y : TOTAL

After 20th year

350,000 t/y (forest res.)
70,000 t/y (agricultural res.)
130,000 t/y (energy crops)

550,000 t/y : TOTAL

Furthermore, in future, an area of 10,000 - 20,000 ha could be available for the production of a particular species of sweet Sorghum to be entirely converted into ethanol. Thus, a maximum of 2,000,000 ton of fresh sorghum (50% bagasse/10-15% of sugar content) will be available. In summary :

---

* The total minimum amount of "Bosco-Ceduo" forest Biomass evaluated in detail (1985) by the Forestry Dept. of the Regione Abruzzo is 2,500,000 t ; the actual productivity is about 2.34 t/ha-y.

** Vineyards, olive and fruits trees (about 70,000 ha)

*** S.R.F., broom, acacia, from 15,000/20,000 ha on a total marginal and eventually available area of 115,000 ha.

THE LAND AND BIOMASS RESOURCES AVAILABLE FOR THE L.E.B.E.N.-PROJECT ARE :

Total surface of Abruzzo Region : 1,079 million ha
Forestry area for L.E.B.E.N.-P : 78,000 ha (35% of total forest area)
Total area for energy crops : 20,000 ha (17% of total marginal land) - Biomass production: 130.000 T/y (min.).
Ligno-cellulosic biomass available for L.E.B.E.N.-Project: 320,000 - 550,000 t/y. Max which could be at present extracted: $3,5.10^6$ t.
Sweet-sorghum (fresh) potential : 2 million t/y from 20.000 ha.

b) Harvesting - transportation - chipping of different types of biomass.

Harvesting and transportation of the forestry biomass present great difficulty due to the particular morphologic characteristics of the Abruzzo Region. On the other side, it constitutes a critical subject for the economical viability of the project. Preliminary tests by a medium size harvesting machine have been carried out in 1985. Large experimental harvesting programme will be carried out in September-November 86 (see chapter 5).

In the long term, for very difficult and extended areas, a likely method perhaps of biomass transportation to the conversion site could be by "air ships". Of course, the low cost of biomass raw materials requires low-cost technology (might small remotely-piloted aereostats be a solution?)

c) Network (30 - 40) of dispersed pyrolytic conversion and emulsification units,

for the thermo-chemical conversion of biomass into fuel. Such units showing enough flexibility to match their conversion capacity to the biomass accumulation in different sites. Their energy-conversion efficiency is about 82%. The average fuel production (from 1 ton of dry biomass) is the following:

- gases : 300 kg
- charcoal : 280 kg
- bio-oil : 200 kg

A particular post-treatment (emulsification) will be carried out on these solid/liquid products to facilitate their transportation, handling and utilisation, as mentioned in later chapters.

d) A 27 MWe power station.

The existing (3 sections) oil-operated Avezzano power station will be modified for an optimum utilisation of the emulsion. Its main characteristics are :

- Electricity generation = 200 x 10^6 Kwh/y
- Fuel consumption = 640 x 10^9 Kcal/y
- 3 steam generators (3 x 60 ton of steam at 32 Bar, 425°C
- 1 turbine Brown Boveri with water coaled steam condenser;
- 1 B.B. generator (33,750 KVA - 5,000v - 50 Hz);
- 1 pumping station for the cooling water (5,600 m^3/h)
- 1 pumping station for water feeding to the boilers;

The present efficiency of the power station (operated by ATZ-oil or nature gas) is very low :

= 27%

Perhaps, it will be possible to increase the efficiency at least of few %.

The waste-heat from the power station will be recovered :
- to supply heat to the "green-house" complex;
- to supply process heat to the "agro-industry" complex for thermal conditioning, drying, deshydratation, liophilisation etc... of products. Properly selected industrial "Heat-pumps" will be employed to adjust the processing temperature to the wanted value (60°C in most of cases)
- to supply heat to the ethanol destillation plant.

e) Industrial centre for plants production

having a capacity of about 19 million plants/year and constituted by :

1. A micro propagation centre (capacity 4 million plants/y-area : 15.000 m^2) subdivided into :

   A laboratory for "in-vitro" fast reproduction of plants, growth-rooms (500m^2), sterilized rooms (200m^2), preparation (500m^2), control room (50 m^2), green-houses (tunnel type) 6.000m^2 for plants climatisation....

2. A production centre of plants from seeds (by hot-type seed-beds) having a capacity of 15 million plants/year.

3. A "green-houses" complex (total covered area of about10 ha) for "early" fruit and flower production, commercial trees etc.. (value of commercial production : 350,000 ECU/ha).

   This plant production will be destined to the regional, national and export markets.
   The waste-heat available is 43 x 10^6 Kcal/h. The heating needs are : 10 Kcal/h x m^2 x 1°C (inside-outside temperature difference), the maximum heating period is about 3.500-4.000 hr/year.

Some green-houses will utilise heat-pumps, which need an electricity supply of 4,2-4,8 x 10^6 Kwh/year (each).

f) Ethanol factory

This will have a total capacity of about 30.000 t/year. The ethanol will eventually be utilized as "octane-booster" in the national programme for unleaded-gasoline for motors.
The raw-material will be sweet-sorghum. The amount of about 200.000 ton (fresh)/year needed will be supplied by the surrounding cultivated area (Avezzano) of about 2.000 - 2.500 ha.
The "sorghum-saccharatum" present very interesting characteristics, which are :

- number of existing species : 4,600
- world production (China, India, Africa, Soviet Union, Canada etc..) : 7.10^6 t/y
- European production : 0.6 million t/y
- productivity (max by 2 or 3 cuts) : 70-100 t/y fresh
- vegetation period : 90 - 120 days according to the climate
- composition : cellulose 15 . 25% - Hemicellulose 35-50% - lignin 20-30%
- sugar content (average) : 12-13% (18% max. before grains growth)
- plants needed : 6,000 - 20,000/ha
- man-power : 205 hr/hay (three harvestings/year)
- sorghum does not need good quality soil, needs light irrigation.

The total alcohol production cost (agricultural costs, industrial costs, transportation) from sorghum should not exceed 9,000 ECU/ha. The total income for the agricultural activity could be nearly twice that of sugar-beet production (but it depends on the policy adopted).

3.2. SECTOR OF HYDRO-ELECTRIC ACTIVITIES.

The construction of 12 small hydro-electric power stations is foreseen.
The total installed electric power will be 22,13 MWe and the electricity production will be 107 x 10^6 KWh/y at an estimated cost of 0,02-0,028 ECU/kwh.

3.3. Sector of Agro-Industrial Activities

It consists of a group of agro-industries for processing several types of agricultural products grown in the Abruzzo Region. Requires large amounts of thermal power and electricity (for processing, drying, liophilisation, cold storage, freezing...)

Among them, are the following :
- plant for potatoes processing;
- plant for carrots processing ;
- plant for potatoes-chips production;
- plant for agricultural products drying
- plant for products liophilisation
- plant for animal feed stocks production;
- cold storage and freezing installation.

3.4. SECTOR OF ASSOCIATED ACTIVITIES

Due to the variety and number of the interconnected activities, two types of associated activities have been considered essential and critical for a correct management of the L.E.B.E.N. Project. These activities will be carried out in:
an educational center (8.000 m^2) for training a maximum of 200 people/year in the field of biomass production, harvesting, pre-treatment, conversion and in the field of post-treatment, storage and utilization of conversion products;
a maintenance center to guarantee the correct operation of all dispersed sub-systems of this very complex and diversified agro-industrial Project.

4. CONVERSION TECHNOLOGIES SELECTED FOR THE L.E.B.E.N. PROJECT

These are:

4.1. Pyrolytic Conversion

As already mentioned in Chapter 3, pyrolysis will be one of the basic technologies adopted for the conversion of agricultural and forestry residues. Its energy conversion efficiency is 82%, the biomass fuel conversion efficiency about 78%. From 1 ton of biomass (20% of humidity) the following products can be obtained :

- gases : 300kg (heating power 700Kcal/kg); part of the gas will be supplied to a 70 KWe motor generator and to a biomass drying unit.
- charcoal : 250 kg (heating power 7600 Kcal/kg)
- bio-oil : 200 kg (heating power 5000 Kcal/kg)

Here below a picture of a pyrolysis unit (capacity 1t/h) is given:

Modular type pyrolysis installation (Italenergie SPA - Sulmona/I)

In table n°1 and n°2 some results of the chemical analysis carried out on charcoal and pyrolytic-oil from agricultural and forest residues, are presented. But, a particular post-treatment (emulsification) of these solid-liquid pyrolytic products is foreseen to facilitate their transportation and utilisation. Thus, the final product will be :

A TERNARY OR QUATERNARY EMULSION (h.p. 5000-7000 Kcal/kg),made by:

charcoal powder + pyrolytic oil + water + conventional fuel (oil : methanol) + additives.

Such liquid fuel which can be transported by tank-lorry and can be utilised in existing boilers (only replacement of burner is needed) will greatly increase the general interest for the low temperature, low sophisticated pyrolytic - technology.

The experimental activity is already in an advanced stage of implementation at several Community Laboratories. Preliminary characteristics of such emulsion

- Heating power: 6,000-7,000 Kcal/kg
- viscosity (60°C): 250 C.P.
- composition: 50-55 % charcoal powder (< 200 µm) - 40% bio-oil - 10% oil
- ash-content: 1-1,5% (weight)

Results obtained in U.S.A. (on charcoal-oil slurry) are presented in Table n°3. Table n°4 shows results of combustion of pyrolytic products slurry.

TABLE 1 : COMPARATIVE ANALYSIS OF CONVENTIONAL FUELS AND CHARCOAL*

	COAL	LIGNITE	CHARCOAL
Humidity %	3.2	11.5	4.0 to 10.8
Volatiles %	28.8	42.7	8.0 to 16.4
Carbon fix. %	55.9	23.6	62.5 to 79.6
Ashes %	12.1	22.2	2.9 to 14.7
Volat./Carb.	0.51	1.81	0.11 to 0.24
Volat./Carb.Volat.	0.34	0.64	0.10 to 0.19
C %	69.7	43.4	71.1 to 83.7
H %	4.7	4.8	0.8 to 1.7
N %	1.2	0.9	0.5 to 1.3
S %	0.7	0.9	trace to 0.2
C %	69.7	43.3	75.7 to 87.7
H %	4.5	4.0	0.1 to 1.3
N %	1.2	1.0	0.5 to 1.4
S %	0.7	1.0	trace to 0.2
C %	82.3	65.5	88.3 to 97.8
H %	5.1	5.3	0.1 to 1.4
N %	1.4	1.4	0.5 to 1.5
S %	0.8	1.4	trace to 0.3
O %	10.3	26.5	6.3 to 9.5
C/H	16.1	12.4	50.6 to 163.0
C H/O	8.5	2.7	9.4 to 14.8
N_2 m^2/g	3.9	4.2	28.3 to 149.5
CO_2 m^2/g	46.5	37.1	123.2 to 233.3
HHP Kcal/kg	6740	4000	6200 to7400

* from different types of forestry and agricultural residues.

TABLE 2 : PYROLYTIC OIL CHARACTERISTICS

	Tar Phase	Liquid Phase
Composition		
C %	65.8	5.9
H %	6.6	10.0
N %	1.5	0.1
S %	0.2	traces
Ignition Temp. (°C)	94	-
H.Heating Power (Kcal/kg)	6500	-
Chlorides (ppm)	215	-
Specific gravity (g/cm^3 at 15°C)	1.19	1.00
Ash content (% weight)	1.6	0.76

TABLE 3 : CHARACTERISTICS OF PYROLYTIC EMULSIONS

	Fuel Oil	30% coal/oil slurry	30% char-coal/oil slurry *
Hydrogen % weight	12.2	10.1	9.3
Carbon % weight	86.7	83.2	84.5
Nitrogen % weight	0.2	0.6	0.3
Oxygen % weight	0.1	2.9	4.1
Sulphur % weight	0.8	1.1	0.5
Ashes % weight	-	2.1	1.3
P.C.S. kcal/kg	10470	9550	9480
Viscosity at 60°C cP	82	220	225
kg $SO_2/10^6$ kcal	1.52	2.30	1.05

* 50% oil n° 6 - 20% pyrolytic oil - 30% charcoal

TABLE 4 : COMBUSTION TEST RESULTS WITH PYROLYTIC EMULSIONS

Combustible :	Oil n° 6	Coal/Oil slurry	Charcoal/ Oil slurry*
Solid concentration % weight	-	30	30
Particle dimension mesh	-	93	92
Combustible prewarming temp. °C	100	110	105
Smoke % O_2 Volume	4.8	5.7	4.4
% CO_2 Volume	9.8	9.9	11.2
% CO ppm	50	235	-
% NOx ppm	223	357	235
% SO_2 ppm	340	490	260
Steam production kg/h	1525	1350	1575
Combustible consumption kg/h	100	105	110
Air excess %	27.1	27.0	23.0
Fly ashes: carbon %	46.4	49.8	28.3
Smoke: Ashes %	1.2	6.0	3.28
Stack: carbon %	0.2	1.1	0.3

* 50% oil n° 6 - 20% pyrolytic oil - 30% charcoal

4.2. Ethanol Technology

The technology for the ethanol production proposed for the L.E.B.E.N. project is an innovative high speed continuous (two-stage) acid-hydrolysis process utilizing sorghum, as raw material, for its global conversion (sugars, hemicellulose and cellulose) into alcohols.

In Fig. n°5-6 a schematic diagram of the process is presented.

After washing, dripping, enzymatic treatment, cellular openings and counter current jet washing, ethanol will be produced (after fermentation and distillation). From the residual bagasse, it will be possible to produce ethanol and furfural through the transformation of cellulose in glucose and xylose and through the fermentation of pentosanes, and esanes .

The bagasse hydrolysis procedure is composed of a compression stage, a fast acid biolysis (some tens of seconds) with H_2SO_4 (diluted), a decompression and neutralisation stage. In particular:

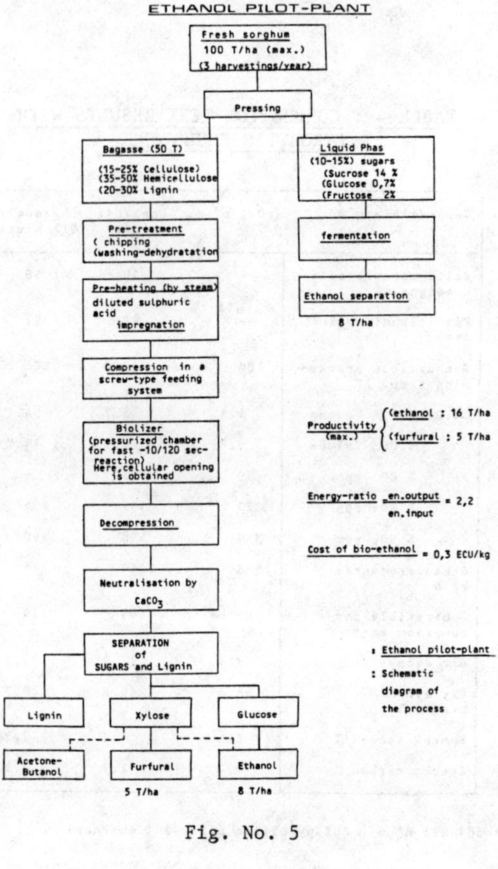

Fig. No. 5

ETHANOL PILOT PLANT

(Bio-Etanolo - Italenergie S.P.A.)
Sulmona - I

Flow sheet

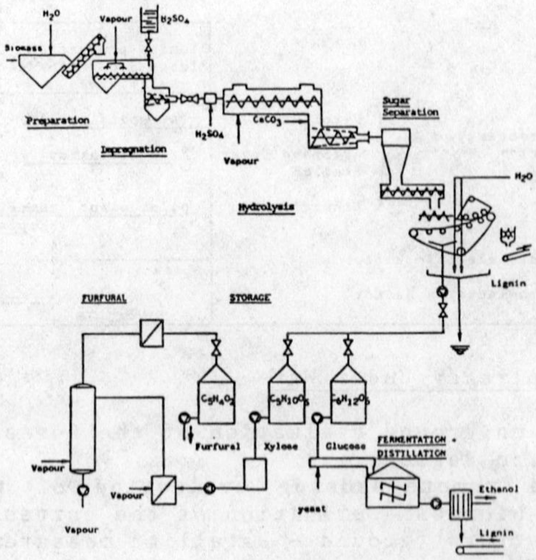

Fig. No. 6

after chipping and mechanical trituration, the biomass will be washed and dehydrated. After, the biomass particles will be pre-heated (by steam), impregnated by the catalyst (diluted sulphuric acid) and mixed outside the reaction chamber.

The biomass will then be introduced in a pressurized reaction chamber (biolizer) through a hermetic screw-type feeding system which is able to eliminate most of the air and liquid. Here, depending on the steam-catalyst conditions, mixing, quality and sizes of particles, a fast reaction (hydrolysis or cellular opening) will be obtained.

The extraction of the biomass (cellulose, lignin, sugar) from the biolizer takes place through a screw-type press; then the sugar separation is carried out.

Lignin, xylose and glucose are obtained after separation (press-filter). Xylose gives furfural and glucose gives ethanol.
About 13% of the sorghum gives a sugar syrup. It is possible to produce 30% of glucose and 23% of xylose. About 27% of the bagasse is composed of cello-lignin, making the process energetically self-sufficient.

5. IMPLEMENTATION AND TIME-SCHEDULE OF THE L.E.B.E.N.-PROJECT ABRUZZO

Completion is scheduled within 5 years, starting January 1986 as shown below :

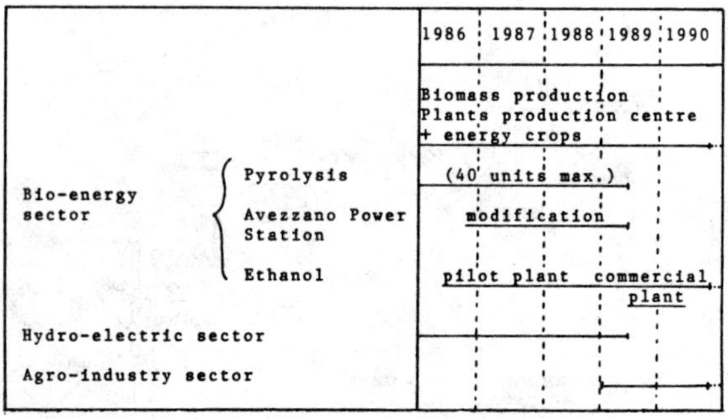

Activities Already Under Way

- Detailed on ground evaluation of the forestry biomass of the Abruzzo Region;
- Satellite remote-sensing evaluation of the amount of forestry biomass. Definition of the corresponding correlation factors (ground - satellite measurement);
- Elaboration of a socio-economic model to be tested by the implementation of the L.E.B.E.N. project to allow new job and economical benefit evaluation for future and similar agro-energy projects;
- Experimental programme on forest coppice harvesting on difficult land. This activity will offer the possibility :
  - to assess different harvesting machines and technology available,
  - to define the best methodology for harvesting in the mediterranean area of the Community
- Intensive testing of the pyrolytic process on a pilot plant in operation since 1983;
- Group of pyrolysis units under construction;
- Large experimental programme for the post-conversion treatment (emulsification) of pyrolytic products. Five organisations and universities are involved;
- Design of the modifications to be carried out on the Avezzano power station in order to operate by pyrolytic emulsion. Modifications of the power station will start in 1987;
- Experimental programme on energy crops, in particular :
  - SRF, broom cultures (on marginal land)
  - sorghum (on agricultural good soil)
- Pilot project for the production of ethanol from sorghum (0.5 t/h);
- Educational programme for training people in these new agro-energy activities
- Two hydro-electric power stations.

6. ECONOMIC EVALUATION

INVESTMENT

			million ECU
	(forest infrastructures (road)	:	9,0
	(Pyrolysis Conversion Units	:	20,0
	(Avezzano Power Station (modification)	:	6,0
	(Ethanol factory	:	33,0
Bio-energy sector	(Green-houses + infrastructure)	:	38,5
	(Laboratory for "in-vitro" reproduction of plants)	:	6,5
	(Centres for training + maintenance + services)	:	6,2
	partial total	:	119,2
Hydro-electric sector		:	24,0
Agro-industries sector		:	70,5
Miscellaneous		:	13,3
	Total	:	227,0
			=====

The budget already authorized by the C.E.C. and Regional Authorities amounts to:

Year 1985: 4,3 million ECU

Year 1986: 29,7 million ECU

Year 1987: under evaluation

PRODUCTION COST

- Fuel emulsion of pyrolytic products: 130 ECU/TOE
- Ethanol (global conversion of sorghum): 0,28-0,3 ECU/litre
- Electricity cost
  from power station (emulsion fired): 0,04 ECU/kWh
  from local hydro-electric power station: 0,02-0,034 ECU/kWh

7. CEC SUPPORT FOR THE L.E.B.E.N. PROJECT

The Commission of the European Community is providing a strong support to the L.E.B.E.N. project. This will be technically supervised jointly by the Commission (DG XII for Science, R & D and DG XVII for Energy), the Italian

authorities for the development of southern Italy and th
Abruzzo regional authorities.

Most of the financial contribution will be provided by:

- regional authorities
- Italian authorities (Organismo Intervento per il Mezzo giorno)
- Commission of the E.C. - DG XVI (Regional Developmen Fund)
- Commission of the E.C. - DG V (Social Fund)
- Commission of the E.C. - DG VI (Agriculture and Fores try)
- other Organisations and DG's of the Commission.

This programme is considered of great strategic impor tance for demonstrating, for some regions of the Communit and other countries the technical and socio-economic via bility of similar large agro-energy/agro-industrial integ rated projects. In this way the Abruzzo region, by th implementation of the L.E.B.E.N. project, could become th first test bench for this kind of project and a trainin centre. It is believed that, in a more or less distan future, some ten similar projects could be implemented a the Italian level, some hundred at European level and seve al thousands in developing countries.

8. INTERNATIONAL COOPERATION

The results of the L.E.B.E.N. project and the experi ence drawn from it are of high interest for cooperatio with third countries, namely developing countries.

As the L.E.B.E.N. project tries to optimise the re sults by an integrated approach between agro-energy an agro-industrial activities on a regional basis, it seem well suited for many developing countries whose economy i based on agriculture.
It is also important to note that this project has an im portant social impact on the population of such countrie (new jobs). The development of local energy resources in a integrated scheme facilitates the social life of the coun try and reduces the need for urbanisation.

Cooperation with developing countries can be foresee at different levels, starting with a transfer of expertis to feasibility studies, demonstration and operation pro jects.

Training programmes can also be set up for personne coming from third countries as well as the participation o specialists in the programme preparation and/or in it implementation. Authorities of Abruzzo region have declare to be ready to accept proposals for a training and collabo ration programme.

9. CONCLUSIONS

To conclude the implementation of this large agro-energy scheme will demonstrate in general:

1. the possibility for agricultural and forestry activities to contribute to the very attractive energy market which for many countries, is a non-saturable market.

In particular, the L.E.B.E.N. project, by the exploitation of a significant amount of the biomass resources of the Abruzzo region (up to 500.000 t/year of agricultural and forestry residues), will demonstrate:

- the possibility of a local massive production of thermal and electrical energy (at moderate cost);
- the consequent specific and high socio-economic benefit. The total expected number of new jobs is about 2.000;
- the practical validity of adopting, for the conversion technology, a modular concept;

2. increasing of the general interest for the pyrolytic process by adoption of the emulsification post-treatment of pyrolytic products;

3. the importance of the penetration of the agro-energy concept into the general development strategy and form of cooperation for emerging and developing countries.

DISCUSSION

Klass — You told me earlier this morning that about 40 billion lire had been committed on this project. Could you tell us what the total commitment is in U.S. dollars, and also what the rate of expenditure would be starting with this year?

Grassi — Last year, the total budget was about $4 million ECU, one of which is about 0.94 U.S. dollars.

Grassi — This year, the budget is about 36 billion lire (24 million US $). Some projects are under negotiation now.

Klass — Is this all distributed equally among the European Community members?

Grassi — No, the budget from the Commission is used in different areas. Agriculture requires a large expenditure by the Commission. Other budgets are used for other activities and programmes for less developed areas like South Italy, Greece, Spain, and Portugal.

Klass — Which country is contributing the most?

Grassi — One can not say because the budget comes, through taxes, so the general budget of the Commission. Each year a program for social funds, for research, for development, for agriculture etc. is elaborated. Each country in general has its own priorities. The budget is proposed by the Commission but must be approved by the National Authorities. To give you an idea of the regional development fund for Italy only, you have 1.5 billion ECU each year. This is distributed in different regions, and if the region prefers to spend money for bioenergy, you cant get it if you have a good program.

H. Esterbrauer, U. of Graz (Austria): A meeting was held in February in Brussels by the European Communities on bioethanol. They talked about production costs of 13.5 ECU/hectoliter, and a raw material cost between 60-80 ECU/hectoliter, independent of whether the feedstock was wheat, sweet sorghum or potato. They also said that the petrochemical industry would buy ethanol at a price of 35 ECU/hectoliter. Do you think that the integrated process you have described could produce bioethanol at a much cheaper price than the figure state in Brussels?

Grassi: Integrated processes can give you a benefit. For LEBEN, we plan to utilize a particular species of sorghum which in general is not cultivated and to get complete conversion.

Klass: Does this mean that the answer to his question is yes, you feel that the integrated process would be able to produce the ethanol at a lower cost?

Grassi: We have three things in the pilot project. The first one is the selection of the particular species of sorghum. The second one is the technology, which we believe will be very interesting from an efficiency point of view. Increased productivity is another point. We believe the data we presented here are realistic.

______________: Are you associated with the agriculture biotechnology group in Italy relative to your use of sorghum?

Grassi: We have had consultation with the member states, and get information from each member and research organization, and also from Italy.

TECHNOLOGY TRANSFER FROM GOVERNMENT TO THE PRIVATE SECTOR: A CASE HISTORY

A. Harvey Abernethy
Partner
RBS Electric
Knoxville, Tennessee 37919

ABSTRACT

The successful transfer of new emerging technologies from the overnment sector to the private sector can only be accomplished by he continued financial support by the government sector. This is articularly true with small research and development companies who o not have alternate sources of revenue in which to fund the ommercialization of these new technologies. Many millions of dollars ave been committed by both the private sector and the government ector in cooperation together for the research and development of new echnologies. But, when it comes to the commercialization of this echnology, many times the private sector has been unable to obtain he continued support of the government, and the project is dropped for ack of funding. The successful transfer of technology from the govern-ent sector to the private sector is greatly enhanced by the continued upport of the government.

TECHNOLOGY TRANSFER FROM GOVERNMENT
TO THE PRIVATE SECTOR: A CASE HISTORY

BACKGROUND

In 1978, with financial assistance from the Department of Energy, research was begun by Aerospace Research Corporation to test the burning characteristics of green wood chips and sawdust at atmospheric pressure. Research expanded to determine the feasibility of using wood to fuel a gas turbine that would have an external combustion chamber. The primary objective was to learn if the combustion gases from the burning of the wood could be adequately cleaned through the use of cyclones and that the gases would not unduely errode the turbine blades of a gas turbine.

Construction was completed on a 375 KW wood burning gas turbine in which the analysis indicated that the gases could be adequately cleaned.

In 1981, Aerospace Research through the assistance of the Tennessee Valley Authority chose a site in Red Boiling Springs, Tennessee, to locate an upscaled version of the 375 KW wood burning gas turbine. The new system would be rated at 3000 KW and incorporate an Allison 501 K gas turbine. The initial assembly and testing of the 3000 KW system at 60% of speed was performed by Aerospace at their Roanoke, Virginia facility. In late 1984, the system was sent to Red Boiling Springs and final assembly was completed in the Spring of 1985.

The Red Boiling Springs site was chosen because of its abundant supply of sawdust from area sawmills and the favorable buy-back terms offered by TVA for the electric power produced.

COMMERCIALIZATION

The Process

The decision to commercialize a technology can be defined as the point in which the development process has reached the stage that the technology has a high probability of functional and economic success. At this point, final design must be completed and the economic commitment be obtained to move ahead.

Even with this high probability, success is still full of many unknowns and surprises that require continual revisions and changes to the system until reliable economic success can be claimed. It is very seldom that the initial costs of the first of anything built can stand on its own economic feet by itself. Changes and revisions are expensive but once these have been incorporated and the new technology proves successful, the cost of a production unit should meet the economic criteria originally set.

Financial Commitments

The benefits from the continued financial support of the government sector are many fold:

1. First, and most obvious is the sharing of the economic burden until the technology reaches the income producing stage.

2. Second, with the continued economic support by the government sector, private enterprise has a much greater chance of securing the financial commitments from other interested industries and private financial institutions. Without this continued support, many organizations will lose interest or will immediately see that unless they are willing to up their financial stake substantially, funds will not be adequate to see the project through. Others may simply not have the financial resources to increase their contribution.

3. Third, the continued sharing of information with the government sector is a two-way street. Beyond the financial support, the sector brings with it an abundant resource of professional assistance, learning experiences and contacts that are many times lost without that continued partnership.

4. It is extremely difficult for small research and development companies to find larger companies who have the R&D funds available willing to take on the funding of others work. The process the new company must take in order to fully understand the scope of the work along with the willingness by their engineers and researchers to pick up on others ideas is very hard to accomplish. Many times the process is so slow and unworkable that the delays alone will ultimately doom a project.

A CASE HISTORY

The first commercialization of a 3000 KW wood burning gas turbine is now being undertaken at Red Boiling Springs, Tennessee. The system was built and final assembly completed in the Spring of 1985.

The short history below gives the reader an understanding of what can happen and why it is so important to keep the working partnership together until success is achieved.

While it was believed that the major concern at start-up would be the burning characteristics of the wood, the ability of the cyclones to adequately clean the gases and the resulting amount of erosion of the turbine blades, history to date has proven these items, while still are essential to success, have not been the major cause of revisions and delays to date.

Feeding wood into the pressurized combustor through rotor valves has now been accomplished, but only after a major redesign of the valves. This caused a shut down of the project and a three month delay.

The ability to produce an even and accurate flow of wood from the metering bin to the rotor valves has now been accomplished, but only after the inaccuracy of the original unit created an excess of wood fuel to air flow ratio causing the piling of wood in the combustor.

The undetected error in wood feed in May, 1985, resulted in a piling of sawdust in the combustor and the sawdust igniting all at once. This large burn of fuel at one time resulted in temperatures in excess of 3000°. Most of the liners in the hot exhaust piping and cyclones were melted down and the Allison 501 K blades in the turbine were destroyed.

In late 1985, start-up was begun again. New rotor valves were in place and the system was fitted with the T-56 aircraft version of the Allison 501 K, which was on loan from the U.S. Air Force. Full power was achieved for a brief time until other non-research functions caused delays.

Upon a subsequent run, the T-56 engine turbine was destroyed by what was believed to be the loss of a blade in late 1985.

At the present time of this writing, the Allison 501 K has been rebuilt by Allison and has been installed in the system. Current expectations are that the system is in good working order and that power will be produced again at any time.

CONCLUSION

As can be seen the commercialization of a technology is still not without major risk. Although the fundamental functions of the system had been proven satisfactorily, prior to the scaled up design of the 3000 KW system for commercial purposes, it is not a production unit.

The additional commitment required to fund and support this project has been substantially above those projected as early as one year ago.

Continued funding has come from the Department of Energy which has been absolutely essential in keeping funding coming from other resources. The private sector and the semi-private sector has supported this project due to the DOE's commitment. Continued support is coming from General Motors' Allison Turbine Division, Aerospace Research Corporation, RBS Electric, a partnership formed to own the plant facilities, and the Electric Power Research Institute. Reports to the DOE will continue through 1987 on all operating and performance data.

Tremendous resources can be spent on research and development. What must be realized that the same commitment must be applied to the commercialization stage if these technologies are to be proven economic successes.

DISCUSSION

Klass — You indicated that private sector support is about 80% of the total cost of the project.

Abernethy — That is true for commercialization of the 3.0-MW facility. There was a large amount of R&D funding before this.

Klass — Do you have any idea as to the total R&D support before commercialization started?

Abernethy — In the neighborhood of $750,000.

M. Lefcort, Heuristic Engineering — In hindsight, would it have been better to use a heat exchanger between the compressor and the turbine? Will you do so in the future?

Abernethy — We have no present plans to do that. We feel that the efficiency with the heat exchangers would not be near as good as direct combustion. We have the alternative problem though of making sure the combustion gases are cyclonically cleaned in an adequate manner so as not to damage the turbine.

J. Waller, Energy, Mines and Resources Canada — What percentage of federal funding do you feel is necessary to leverage the funds from other sources?

Abernethy — It depends on each individual basis. I'll be quite honest with you; I don't know if we would have obtained financing based on the technology itself. We pledged other assets to obtain this financing for the project because we feel it has a long-range tremendous potential. When you're looking at capital costs of $2000 per KW for other plants and our costs of $600-$700 per KW, that's a large incentive for the wood-fueled gas turbine system.

Klass — Isn't it true that for technology that is developed to the point where it's ready for commercialization and is so good that it almost sells itself, government support is not required to provide the little kick necessary to get it over the hump?

Abernethy — Have you got any of those projects?

C. Setterstrom, Harrington Research Co.	That leads to the question, suppose there are inventions or proprietary technology developed in the course of the development work. Do the patents and the proprietary technology belong to you, as the entrepreneur?
Abernethy	The Government passed a bill back in the early 80's that allowed private enterprise to own the patents, and obtain the benefits from research and development where government funds have been used. So we're proceeding on that basis.

OVERCOMING BARRIERS TO COMMERCIALIZATION OF BIOMASS ENERGY PROJECTS

John D.C. Waller
Renewable Energy Branch
Energy, Mines and Resources Canada
Ottawa, Canada K1A 0E4

ABSTRACT

Before biomass energy technologies can be commercialized there are many barriers, both real and perceived, to be overcome. One means by which this can be achieved is to demonstrate the project in such a way as to reduce the risk to the user and at the same time convince others that the technology is viable.

This paper discusses how Canadian programs have demonstrated that barriers can be overcome and gives examples of some of the projects that were implemented.

OVERCOMING BARRIERS TO COMMERCIALIZATION OF BIOMASS ENERGY PROJECTS

Before the oil crisis of the 1970's, Canada supplied about 75% of its own oil, the balance being imported. Canada also was one of the world's highest per capita users of energy. The government of the day was convinced that we could become energy self sufficient within a few years. To achieve this we implemented a number of programs designed to both conserve energy and to enable us to utilize our own vast resources of renewable energy.[(1)] By the end of 1984 while we still import oil, we actually export more than we import. Approximately 4% of our total energy consumption was derived from biomass fuels.[(2)] It is the objective of the Canadian government to double it by the turn of the century and to raise it to 10% by 2010.

Of the government sponsored energy programs introduced in the 1970's several were designed not only to develop new technologies (research and development) but also to hasten the introduction of biomass, as a source of energy, into the marketplace. It is, however, a known fact that a development just out of the laboratory is seldom an instant commercial success. There are several steps to be taken and barriers to be overcome. Some of these steps are shown in Figure 1. One must appreciate that not all steps are applicable to the conversion of an idea to a marketable product. Nor is it necessary in every case to enter the process at the starting point. Like everyone else today we are prepared to transfer technology from other countries to our use. In doing so, however, there may be new barriers to overcome. A significant portion of the solution to the question "How can I get this idea into the marketplace?" comes in recognizing these barriers.

In the field of energy from biomass, there is a full gamut of technologies in various stages of development ranging from laboratory scale models to old conversion technologies nearly forgotten by this modern day high technology society. I have identified in Table 1 for various biomass feedstocks the barriers to be overcome in moving from raw feedstock to the final energy user. The terms used are explained in Annex A to this paper. Some of these barriers are obvious and simple to overcome; others may require years of work. As noted at the outset of this paper, demonstrations are one means by which some of these barriers may be overcome. While our programs have gone through several changes in the past ten years the basic objectives have always included:

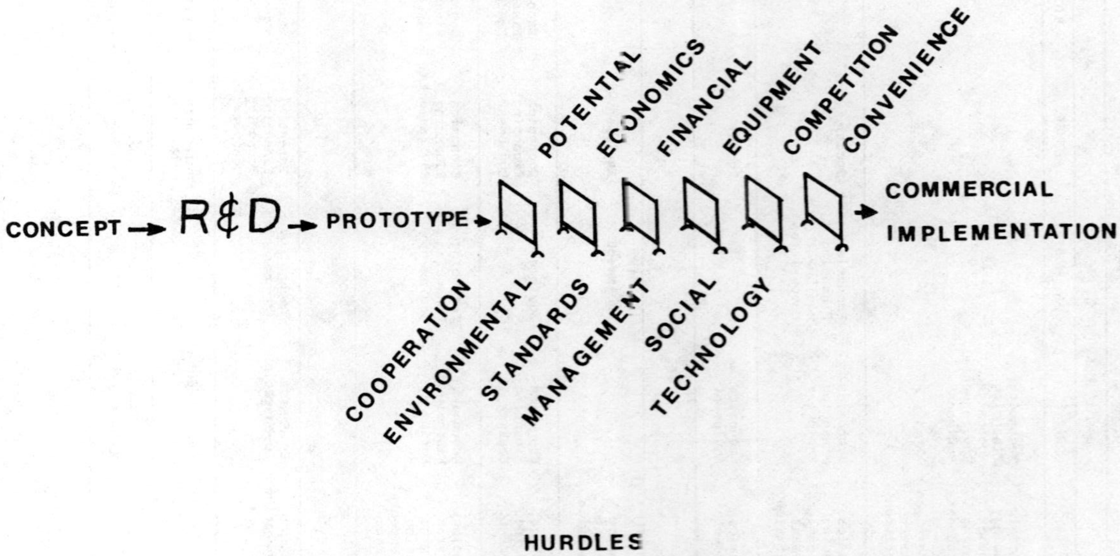

Figure 1 Overcoming Barriers to Commercialization of Biomass Energy Projects

TABLE 1

BARRIERS TO TECHNOLOGY IMPLEMENTATION FOR VARIOUS FEEDSTOCKS

	FEEDSTOCK				
PROCESS	WOOD	PEAT	AG. WASTES	REFUSE	BIO-WASTES
Harvesting	Potential Technology Economics Financing Environmental Social Co-operation Equipment Competition	Technology Financing Economics Environmental Equipment			
Collection	Economics Financing Environmental Equipment Competition	Economics Financing Equipment	Economics Financing Equipment Competition	Social	
Transportation	Economics Equipment	Economics Equipment		Social	
Feedstock Preparation	Economics Equipment Competition		Equipment	Equipment	
Conversion	Economics Financing Equipment	Economics Financing Equipment	Equipment	Economics Financing Equipment	Economics Financing Equipment
Combustion	Technology Standards Economics Financing Environmental Equipment	Economics Financing Equipment	Technology Financing Equipment	Economics Financing Environmental	Economics Financing
Energy Use	Economics Co-operation	Economics Co-operation	Technology Co-operation	Potential Financing Economics Co-operation	Economics Co-operation
Waste Disposal	Environmental	Environmental		Environmental Social	Environmental
Other	Convenience Management	Management		Management	Management

1. to accelerate the adoption and commercialization of conservation and renewable energy technologies;

2. to create business employment in new and/or existing industries.

While the first objective includes conservation, this paper refers principally to biomass fuel or projects in the renewable energy sector.

To satisfy these criteria we saw the need for two basic types of demonstrations:

1. test case of demonstrations of new technologies or new methods by which biomass to commercialization can be overcome;

2. show case or demonstrations of underutilized technologies which are not adequately known by potential users.

What are the fundamentals of a demonstration?

First it must show that the technology works, is economic, safe non-polluting etc. and secondly it must be perceived by others as a success and ready for use without government subsidy.

We therefore break demonstration projects into three phases.

I Construction on implementation of the technology;

II Monitoring of the construction and operation;

III Information transfer.

Phase I is considered in relation to the national and regional strategies and priorities. Contributions are based on the perceived risk and its potential return to the end user. Phases II and III are funded up to 100% of this cost depending on the benefits to the end user. For example a manufacturer of wood fueled funaces would be expected to contribute a significant portion of the cost of information transfer because it is in effect advertsing his product. On the other hand a grocery store installing a waste cardboard combustion system to heat the store would not pay anything for information transfer as his primary interest is selling food.

Demonstrations, however, are not necessarily the solution to all barriers. As noted in the Charpie task force report(3) "in the absence of price controls and rationing, supply always equals demand". The opposite, however, does not hold true, and no demonstration will make it otherwise. Setting up a fuel wood supply business will not create a demand for the wood unless there is also someone selling wood fired furnaces. However, a demand for both may be created by a training program in the schools or colleges. This in fact is a form of a Phase III information transfer only project. In order for this aspect to be effective one must first look at the target audience and the costs associated with getting the information to that audience. For a large project such as an energy from waste plant project, Phase III might include a national seminar and tour of the facility with people coming from across the country to attend. At the other end of the scale a wood chip fueled heating system for a greenhouse might well be demonstrated in each province as the cost of phase I may well be less for several projects than phase II for a single project and bringing all greenhouse operators to it.

How do we start the process of demonstrations? There are two primary methods of creating (from the government point of view) deomonstrations:

1. passive. We put out an announcement that we are in business and we sit back waiting for the proposals to come in;

2. proactive. We take our strategy studies as a starting point, beat the bushes to find suitable demonstrations and assist the potential user to prepare the application and put the technology to work to our mutual benefit.

It must be appreciated that not every demonstration supported by the government has been a complete success. That is if success is measured by the number of copies or replications of the technology. However, these failures have taught us lessons which can be applied to other demonstrations or commercial applications of the technology. The most frequent cause of failure is the lack of recognition of one or more barriers to the use of the technology in that specific application. As an example we, and a proponent, missed the necessary fact that a project to treat hog manure anaerobically and use the biogas to drive an engine generator would not be economic in Canada because we have no PURPA law. The farmer, while selling his excess electric power to the utility, could not get a good price for it and he ended up with a payback of 19 years instead of the estimated seven to nine years.

The following examples will describe some of the projects that we have supported, the barriers that were overcome, and some of the lessons learned from them. In fact some will show that a barrier may become part of the driving force for the project.

INDEX OF BARRIERS OVERCOME BY DEMONSTRATIONS

Barrier	Example
Potential	E1
Competition	E1, E10
Technology	E2
Economics	E1, E3, E4
Financing	E5, E12
Infrastructure	E4, E5, E7, E11, E12
Social	E6, E7
Environmental	E2, E3, E6, E7
Co-operation	E6, E7, E8, E9
Standards	E9
Management	E8
Convenience	E10
Equipment	E2, E4

E1

Barriers	Potential Competition Economics
Feedstock	Fish oil
Title	Fish oil as an alternative fuel
Proponent	Connors Brothers, Blacks Harbour, N.B.
Cost:	$51,000 Fed. support: $30,600

Connors Brothers can sardines and herring. The waste stream is further processed into fish meal and oil. Fish oil is used in margarines, paint and tannery chemicals. The demand and price for this oil is volatile. A demonstration of the burning of fish oil in the plant furnaces showed that it is economical to burn fish oil as a replacement for Bunker C only when the price of fish oil is less then 85 percent that of Bunker C. Capital costs for the conversion were minor.

E2

Barriers	Technology Equipment Environmental
Feedstock	Chicken litter
Title	Chicken litter as fuel in a stoker fed furnace
Proponent	Paul Emile Michaud, St Francois de Madawaska, N.B.
Cost:	$130,065 Fed. support: $66,705 Payback: 4.8 years.

Boiler barns are traditionally heated with oil fired equipment. Chicken litter in the form of wood shavings and chips are spread under the cages to absorb the droppings. Periodically the waste is trucked to landfill sites. This waste consists of 50% wood chips with a moisture content of 20-35%. Ash content is of the order of 15%. Experiments have shown that a stoker fed furnace can be modified to accept this high ash fuel. The farmer was initially unaware of what could be done; but having approached our office he obtained information on a small scale test of similar equipment in another province allowing him to build on that experience. As part of the information transfer activities, other chicken farmers will be invited to visit the site and to bring samples of their own chicken litter to be burned. While this project is just getting underway we are already anticipating twenty replications in the province.

E3

Barriers	Economics Environmental
Feedstock	Swine manure
Title	Anaerobic digestion of swine manure and electrical generation
Proponent	Linden Lee Farms, Cornwall, PEI
Cost:	$201,300 Fed. support: $146,100 Payback: 18.9 years

Having seen reports on the success of anaerobic digestion of animal manure and the use of the biogas as fuel in an engine-generator, this farmer obtained a contribution to assist in installing this process. From an environmental clean-up point of view, the project has been a complete success. However, as there is no PURPA law in Canada, the price he gets for electricty delivered to the grid does not provide the return necessary to reduce the payback to the eight years initially estimated for the project.

E4

Barriers	Infrastructure Economics Equipment
Feedstock	unmerchantable wood
Title	Harvesting forest residue for fuels
Proponent	Repap Miramichi Forest Products, Newcastle, N.B.
Cost:	$233,700 Payback: 1.2 years

When cutting pulp wood in mixed wood forests, the hardwoods and slash are usually left in the woods. This has a significant cost applicable to future silvicultural work. By harvesting the hardwood and chipping the non-merchantable wood, savings of over $100/hectare in site preparation costs can be effected. At the same time the wood chips will displace 3700 litres of oil per hectare harvested. By integrating the harvesting work, the company was better able not only to reduce costs but also to allocate them equitably to the various aspects of the work. The demonstration, however, showed that there was a need for better chip storage and transportation equipment.

E5

Barriers	Infrastructure Financing
Feedstock	wood chips
Proponent	La Cooperative Forestière, Abrams, PEI
Cost:	$193,755 Fed. support: $70,294 Payback: 13 years

A wood supplier realized that he had to create a market for his fuel if he was to remain in business. He also found that schools, community buildings and churches had little or no money to buy wood fueled heating equipment. The company succeeded in negotiating a heating contract with a school. This contract offered heat at 90 percent of oil costs. The firm installed a packaged stoker fed wood boiler on a pad outside the school and connected it to the existing school heating system. This permits the financier or the owner to readily move the equipment to another site if necessary. While this demonstration showed the economics to be marginal, the lessons learned will enable the co-op to expand their operations to other sites at a reasonable cost.

E6

Barriers	Social Environmental Co-operation
Feedstock	Municipal Solid Waste
Proponent	PEI Energy Corporation, Charlottetown, PEI
Cost:	$9,900,000 Fed. support: $4,000,000 Payback: 11 years

The city of Charlottetown considered two potential sites and customers for an energy from waste plant. The first was close to an industrial area and the steam could be piped to two new hospitals. The second was on the edge of a residential area and the steam could be piped to a thermal electric plant. Local residents objected to having garbage trucks passing their homes at all hours and to the possibility of rodents being attracted to the garbage i.e., the NIMBY or never in by backyard syndrome. The corporation finally settled on the first site which involved a 2km pipeline to the hospital. Three levels of government were involved in the approval and financing of this project. Steam heat is sold to the hospitals at oil cost. The hospitals savings are in the reduction in operating and maintenance costs of their own oil fired equipment. Environmental tests show that, even without add-on antipollution equipment, the plant meets all environmental and public health standards. Landfill requirements have been reduced by 90 percent.

E7

Barriers	Co-operation Infrastructure Social Environmental
Feedstock	Wood waste
Title	Mini district heating with wood fueled equipment.
Proponent	Klondike Central Heating, Whitehorse, Yukon
Cost:	$322,500 Fed. support: $141,025

A lumber mill with large amounts of waste wood formed a new company to provide heat to several commercial buildings in downtown Whitehorse. By fully informing the public before putting the project in place the social barrier did not appear. Co-operation between the various approval authorities at all levels of government allowed the environmental concerns to be overcome. The company will provide heat at 80% of oil costs to the building owners on a long term contract. Fuel supplies will come from the various local sawmills thus reducing their disposal problem and increasing their revenues. This will assist in keeping marginally economic mills from going bankrupt.

E8

Barriers	Management Co-operation
Feedstock	Wood chips
Proponent	James Paton Memorial Hospital, Gander, Nfld.
Cost:	$1,008,000 Fed. support: $642,800 Payback: 10 years

This hospital installed a wood fueled wet cell burner to replace three older oil fired units. Initial management and financing problems were not helped by having a management committee running the project. However once a project manager was appointed, the project moved ahead to successful completion. The unit has since been operated on such fuels as sawdust and milled peat as well as on wood chips. The plant is providing all the hospital's heat and consideration is being given to piping some heat to a nearby school. This plant has also reduced the plant maintenance costs as the wood fired equipment requires less maintenance than the oil fired equipment.

E9

Barriers	Standards Co-operation
Feedstock	Wood chips
Proponent	Homeowners in PEI, NB and Nfld.

Gravity fed wood chip furance add-on units were demonstrated in various parts of the Atlantic provinces to show that wood chips were an acceptable heating option to oil. There were no manufacturing or installation safety standards written for this equipment. These were being developed during the period of the demonstration. In one demonstration the consultant met with the fire marshall's office and the insurance companies prior to the demonstration. By insisting on careful installation and monitoring of the units, approval was obtained for the demonstration. In a second demonstration in another province, these prior negotiations did not take place. In these cases the homeowners found it difficult if not impossible to obtain insurance. Some demonstrations were cancelled as a result. This has shown that all authorities must be convinced, before the fact, that no dangerous activities will take place.

E10

Barriers	Competition Convenience
Feedstock	"Junk" wood
Proponent	Maritime Forest Ranger School, Fredericton, N.B.
Cost:	$52,600 Fed support: $31,560 Payback: 1.4 years

The Forest Ranger School was initially heated with a furnace burning four foot roundwood logs. As time went on the school found it difficult to get students to stoke the furnace at night. The furnace was converted to a stoker fed wood chip unit. This still used high quality cordwood as the basis for the chips. By modifying the system to screen the chips the school was able to utilize the low grade wood available from the school's forest stands. This allowed for sale of the quality wood to lumber and pulp operations and at the same time provided for disposal of the waste woods.

E11

Barriers	Infrastructure
Feedstock	Wood chips
Proponent	Newfoundland Hardwoods, Cloverville, Nfld.
Cost:	\$587,500 Fed. support: \$130,000 Payback: 4.9 years

The proponent pressure treats wood for use in Newfoundland. Process steam was produced by an oil fired furnace. At the same time that a wood fueled system was installed, a woods contractor was engaged to provide chips on a long term contract. The forest lot to be used contained a little merchantable wood which the contractor was allowed to keep. This was a situation where the demand created the supply. The presence of a stable supply of fuel enabled other establishments in the town to convert to wood fueled heating systems. After one year of operation, the payback for the project was determined to be two years and not the 4.9 initially estimated.

E12

Barriers	Financing Infrastructure
Feedstock	Wood residues
Proponent	University of New Brunswick, Fredericton, N.B.
Cost:	\$2,420,500 Fed. support: \$428,000 Payback: 3.98 years

The university has a mini district heating system which services all university buildings, a hospital and two apartment buildings. Northeast Energy Services installed an 18,000 Kg/hr steam boiler, fueled with wood waste, designed to displace 69% of the oil previously used. The university is paying for the system over nine years out of its savings in oil consumption. At the end of the first year of operation the system was shown to have saved 72%. The university could not have financed this project on its own. The project has created a large stable wood fuel supply in the area thus allowing other conversions to take place.

ANNEX

Definitions for Table 1

Feedstocks

Wood	- forest wastes mill wastes unmerchantable wood
Peat	- fuel grade peat
Ag Waste	- agricultural field crop wastes not - animal wastes
Refuse	- Municipal solid wastes non toxic industrial wastes
Biological Waste	- animal wastes sewage sludge food processing waste not pathological wastes

Barriers

Potential	- A lack of knowledge, by the proponent, of the potential for energy from biomass on his premises.
Technology	- A lack of knowledge of existing or developing technologies to perform the task required.
Economics	- includes cost benefits
Financing	- source of funds to put a project in place
Environmental	- pollution problems solved or caused by the energy project

ANNEX (continued)

Definitions for Table 1 (continued)

Social	- problems perceived by the public to mitigate against the project. NIMBY - never in my backyard syndrome.
Co-operation	- the lack of, or need for interdepartmental or inter agency agreements
Equipment	- the lack of suitable equipment to perform the tasks required
Competition	- competing uses for the biomass either now or in the future
Standards	- the lack of suitable safety and environmental standards for the equipment and/or its installation and operation
Convenience	- is the extra effort required by the process worth the cost savings?
Infrastructure	- is a supply available to meet the demand or can it be made available?
Training	- are suitably trained personnel available?

BIBLIOGRAPHY

1. William E. Lowe and Allan J. Dolenko "Programs on energy from Biomass and Wastes funded by the Canadian Government". Energy from Biomass and Wastes VI, 1982

2. Anon. "Canadian Energy Statistics" - Enerfacts - Energy, Mines and Resources Canada, 1984.

3. Anon. The Demonstration project as a precedure for Accelerating the Application of New Technology. (Charpie Task Force Report) US DOE 1978.

4. Anon. "CREDA Project Summaries" EMR, Ottawa, 1985.

DISCUSSION

J. Koning, Forest Products Laboratory	You alluded to another barrier and left it up to our imagination what it was. Do you want say what it was, or did I miss it?
Waller	Communication.
J. Roetheli, TVA	From your experience in working with these projects what have you basically learned about how to get started and whom to involve when you lay your cards face up on the table in the initial stages?
Waller	You involve everybody who is likely to have anything to say about it. I know that in this part of the world you have many more public hearings than we do in Canada on various subjects, but if you tell people beforehand what you are hoping to do, not tell them after you've got it so far along that you're going to be in real trouble, I think that you will find that you will overcome many of the barriers.

PUBLIC POLICY ISSUES RELATING TO THE ECONOMICS OF FUEL ETHANOL PRODUCTION AND USE

Gerald E. Grinell
Policy Analyst, Office of Energy
U.S. Department of Agriculture
Washington, D.C.

ABSTRACT

It is a pleasure to be here today to discuss such a timely issue. Economic issues have been a major source of concern in the fuel ethanol industry since its inception. I have two broad topics to discuss: Economics of ethanol production and use and public policy issues. I will discuss the economics of ethanol production and use first.

Economics of Ethanol Production and Use

Relatively little public information is available on the cost of producing fuel ethanol. In addition, many ethanol plants have avoided significant costs by using existing facilities (such as an old fertilizer plant), waste products (such as potato wastes) for feedstock, sharing costs with another product (such as high fructose corn syrup), and so forth. Others have relatively low capital costs because of government loans or loan guarantees. For example, some loans made by the Department of Energy (DOE) call for no interest or principal payments until 10 years after the loans were closed. All existing plants probably took advantage of the special 10-percent energy investment tax credit that expired at the end of 1985.

For these reasons, costs for a given plant may be quite different than the ones I will present today. However, some error in the estimates probably will not have a major impact on subsequent conclusions regarding public policy. Let me further preface my estimates by pointing out that they are for a plant built to produce ethanol from corn, using private sources of capital. The estimates include the benefits of the energy investment tax credit and the regular investment tax credit. The costs also include an allowance for a normal profit since people will not commit funds or leave them in a plant that does not earn at least a normal rate of return. I estimate that with corn selling for $2.30 per bushel, it costs approximately $1.45 per gallon to produce ethanol in a wet milling plant and $1.60 per gallon in a large dry milling plant (at least 40 million gallons per year). Costs would be higher in smaller plants — some may have costs that exceed $2 per gallon including a normal profit. Cost differences between corn wet milling and dry milling plants are due mostly to differences in the value of byproducts.

Now let's turn to the value of ethanol to its users. Its value is largely determined by the cost of other octane enhancing alternatives. Methanol appears to have been the lowest-cost octane booster in recent years but its use is strictly controlled by EPA. The next best alternative appears to be increased reforming. EPA has estimated that the cost of making up for the octane lost when lead levels in gasoline drop from 1.1 grams to 0.1 grams per gallon is less than 2 cents per gallon of leaded gasoline when octane is boosted through the reforming process. Without a subsidy, ethanol costing $1.45 per gallon would add about 13 to 17 cents per gallon of gasoline according to information in a 1985 report by the DOE.[2] These estimates were made prior to the recent sharp decline in oil prices. However, lower oil prices also lower the cost of boosting octane through more severe reforming. Lower oil prices also lower the cost of petroleum-based octane-boosting chemicals such as toluene and MTBE but not the cost of producing ethanol. This places ethanol in a very tight price-cost squeeze.

Volatility issues are a concern for alcohol blenders. EPA approved a Clean Air Act waiver requested by DuPont permitting the use of methanol in a 5 percent (by volume) blend with gasoline provided the blend also includes 2.5 percent of a cosolvent (such as ethanol) and 2.5 percent of a corrosion inhibitor that is proprietary to DuPont. The waiver was expected to increase demand for ethanol since low-priced methanol (selling for about $0.40 per gallon) could help make ethanol more price competitive in the octane

---

[2] Office of Environmental Analysis, Gasoline Octane Enhancement: Technology, Economics, and Environmental, Healt and Safety Considerations, DOE/PE/72013-1, July 1985.

market. However, the DuPont waiver is not being widely used because the resulting gasoline blend does not meet EPA's volatility standards. EPA has been asked to modify its volatility requirements for the DuPont blend. Even if the DuPont blend receives full approval at EPA, it now faces economic hurdles because methanol prices have fallen little while the prices of petroleum-based octane boosting chemicals have fallen sharply. EPA also is considering tightening volatility requirements in general, which could prove troublesome even for regular ethanol-gasoline blends during summer months.

Fuel ethanol is selling for about $1.15 per gallon and reportedly has been below $1 in some places. After netting out the value of direct Federal and State subsidies, this suggests that blenders are valuing ethanol at or below the wholesale price of gasoline. The hope that some have of selling ethanol for more than the price of gasoline does not appear likely to be realized any time soon. Lead phasedown will increase the demand for chemical octane boosters including ethanol. However, ethanol will continue to be used primarily by gasoline distributors rather than by refiners who have other readily available alternative means of boosting octane ratings without exceeding volatility standards.

The U.S. fuel ethanol industry faces strong competitive pressures from Brazil. Brazilian ethanol is produced from sugar cane. Costs of production are significantly lower than in the United States. Various sources put the cost of production under $0.90 per gallon at the distilleries. When transportation and insurance costs are added, the cost of ethanol at U.S. ports is estimated to be about $1 per gallon. These costs may not include the full opportunity cost of capital and probably do not include an estimate of normal profits. Part of the reason Brazil has low ethanol costs is because of the low world price of sugar. If and when sugar prices rise, the opportunity cost of sugar cane feedstock will also rise. The International Trade Commission (ITC) recently ruled that Brazilian ethanol has not materially injured the U.S. ethanol industry. Brazilian producers are expected to continue to be significant competitors in the U.S. market if they choose to do so.

Caribbean countries also are producing or considering producing ethanol. Most lack sufficient quantities of low-cost feedstocks necessary for large-scale efficient production. Instead they are considering building distillation plants to dry wet ethanol produced in other countries. The United States, through the Caribbean Basin Initiative program (CBI) exempts Caribbean ethanol from our import tariff and, thereby, offers to Caribbean producers most of the same subsidies that are available to U.S. producers. Congress is considering legislation that would limit tariff-free status of ethanol which is merely upgraded in Caribbean distillation plants. If such a restriction is imposed, the Caribbean will not become a significant sources of ethanol. If not, large quantities potentially could pass through the Caribbean on their way to the United States.

Several European countries have expressed an interest in producing ethanol, but actually produce little at this time. Informal sources suggest that production costs in Europe are not sharply different from costs in the United States. Much of the interest in fuel ethanol seems to stem from a desire to find ways to use surplus agricultural products. European ethanol will not compete with U.S. ethanol but European ethanol production could reduce demand for U.S.-produced ethanol byproducts corn gluten feed and meal by increasing Europe's own supply of byproduct feeds.

Fuel Ethanol's Competitive Outlook

Ethanol producers are in a crisis situation at the present time due to the low gasoline prices. Costs of ethanol production are very high compared to today's cost of gasoline and other octane-boosting alternatives. Assuming the industry survives the present crisis, lead phasedown could increase ethanol demand in the short run vis a vis what it otherwise would have been. Still, ethanol will not become a readily accepted additive until volatility issues are resolved and supplies become large enough to warrant widescale use by large oil companies. In the long run, research breakthroughs could reduce operating expenses and increase ethanol yields. In the final analysis, ethanol is not likely to be viewed as a long term viable option until it can compete without government subsidies. This will not happen soon.

Public Policy Issues

Several major public policy issues are directly related to ethanol fuel. These include the level of the Federal gasoline excise tax exemption for gasohol and the related blender tax credit; limitations in imports of Brazilian ethanol; production and importation of ethanol from countries participating in the Caribbean Basin Initiative program; the status of loan guarantees for ethanol plants; extension of the energy investment tax credit; creation of a strategic ethanol reserve; fuel volatility limits; and, the latest issue, sale of surplus Commodity Credit Corporation (CCC) commodities to ethanol producers at reduced prices.

With the exception of fuel volatility limits discussed earlier, all of these issues are economic in nature; that is, they involve various means of providing subsidies to ethanol producers or limiting competition from foreign producers. These issues are quite complex and, in many cases, intertwined. The U.S. fuel ethanol industry has received large amounts of financial support and continues to receive very favorable treatment in Congress.

Federal Gasoline Excise Tax Exemption and Related Blender Tax Credit. The Energy Tax Act of 1978 exempted gasohol from the 4-cents-per-gallon Federal gasoline excise tax through October 1, 1984. The Crude Oil Windfall Profit

Tax Act of 1980 extended the exemption to December 31, 1992. Then the Surface Transportation Act of 1982 raised the excise tax to 9 cents and the exemption to 5 cents. Finally, the Tax Reform Act of 1984 increased the gasohol exemption to its present level of 6 cents per gallon. Since gasohol consists of one part ethanol and nine parts gasoline, the exemption is actually worth 10 times as much per gallon of ethanol. The Acts also made provision for ethanol users to claim an equivalent income tax credit if they did not qualify for the excise tax exemption. This provision, known as the blender tax credit, has been used little. The gasoline excise tax exemption reduces the receipts of the Highway Trust Fund — some $450 million in 1985 — whereas the blender tax credit affects the general fund.

Some 30 states also offer gasoline excise tax exemptions or comparable direct tax subsidies which vary widely. They average somewhere around 30 cents to 40 cents per gallon of ethanol but go as high as $1.60 (in Louisiana). Many have restrictions that limit their value to some ethanol producers. Because of frequent efforts to reduce or eliminate the State ethanol subsidies and because little ethanol is sold in States that do not provide subsidies, attempts have been made in recent years to federalize the State subsidies. This would be accomplished by raising the Federal gasoline excise tax exemption for gasohol to 8 or 9 cents and eliminating the State programs.

The Administration's 1985 tax reform proposal, also known as Treasury I and Treasury II (Tax Reform for Fairness, Simplicity, and Economic Growth) proposed to eliminate the Federal gasoline excise tax entirely and limit the blender tax credit to ethanol produced by plants built before 1986. The House of Representatives subsequently reversed this. It dropped the blender tax credit and retained the excise tax exemption in its present form. The Senate has not acted on the proposal.

Limitations on Imports of Brazilian Ethanol. The Omnibus Reconciliation Act of 1980 imposed a per-gallon tariff on fuel ethanol imported into the United States after January 1, 1981. The tariff was 10 cents per gallon in 1981, 20 cents in 1982, and was to be 40 cents per gallon from 1983 through 1992. But, when the Federal gasoline excise tax exemption was raised in 1982 and 1984, the tariff was raised to 50 cents and then 60 cents per gallon. The per-gallon tariff was imposed to prevent foreign-ethanol producers from benefitting from subsidies intended for the U.S. ethanol industry. Ethanol imports also must pay a 3-percent ad valorem duty. The per-gallon fuel ethanol tariff has caused numerous problems. Brazil has been accused of finding various ways to ship ethanol into the United States without paying the full tariff. These efforts include blending ethanol with gasoline offshore to qualify for a duty rate of 1.25 cents per gallon; mixing ethanol with toluene, benzene or other chemicals to qualify for a lower tariff; shipping hydrous ethanol to a Caribbean country for final distillation to qualify for exemption from the U.S. tariff under the Caribbean Basin Initiative program; and attempting to ship ethanol to a duty-free Foreign Trade Zone in the United States where it would be blended with gasoline before being declared for tariff purposes. It has been alleged, in addition, that imported ethanol has been declared to be for industrial purposes to avoid paying the fuel ethanol tariff.

Several bills have been introduced into the Congress to plug the tariff loopholes, and some form of loophole-plugging measure seems to have better than a 50-50 chance of passage if a major tax bill is passed this year. This would have the support of U.S. corn growers as well as ethanol producers. The Administration has been generally supportive of the need to resolve the loophole premium but has had reservations about the specific measures that have been proposed. The Administration would like to be able to eliminate the tariff entirely because it is in technical violation of the General Agreement on Tariffs and Trade (GATT) and is the source of unwanted trade frictions. The U.S. has already granted favorable tariff treatment to Brazilian exports of corned beef to compensate for GATT violations caused by 40 cents of our 60 cents-per-gallon fuel ethanol import tariff. However, the Administration is concerned about how the bills would affect our commitments to the Caribbean and believes that attempts to plug loopholes may not end until the basic need for the tariff itself is eliminated. An interagency staff working group is looking at alternatives at this time.

The Department of Agriculture (USDA) has strenuously opposed allowing fuel ethanol into the United States without paying the 60-cents-per-gallon tariff. Last summer, when large quantities of ethanol-toluene blends appeared headed toward the U.S. subject only to a 5 to 7 percent ad valorem duty based on a letter ruling by U.S. Customs (subsequently overturned), Secretary Block demanded that Treasury Secretary Baker not allow the imports to enter without paying the 60-cents tariff.

The Secretary of Agriculture is required by the Food Security Act of 1985 (Section 1155) to determine the impact that imports of Brazilian ethanol have on the domestic price of grain and the domestic ethanol industry. The Secretary, in consultation with the International Trade Commission and the United States Trade Representative, is to determine what relief should be granted. We are working on that study at the present time and expect to report to Congress by May 22.

This study is being conducted against the backdrop of recent investigations by the International Trade Administration (ITA) and the International Trade Commission (ITC) prompted by petitions filed by 14 U.S. ethanol producers in 1985, requesting that antidumping and countervailing duties by imposed against Brazil. The ITA found that several Brazilian companies were dumping ethanol in the United States at less than fair value and that Brazil did provide small subsidies. However, the ITC ruled that the U.S. ethanol industry was not materially injured by these actions. As a result, no antidumping or countervailing duties will be imposed. The ITC and ITA studies did not examine impacts of Brazilian ethanol on U.S. agriculture.

Importation of Ethanol From Countries Participating in the Caribbean Basin Initiative Program. The Administration is committed to stimulating economic growth in the Caribbean. CBI program participants are fully exempt from U.S. import tariffs on fuel ethanol. The U.S. Customs Service has ruled that ethanol that is merely distilled in a CBI country qualifies for the exemption. Distillation capacity has been built in Jamaica and is planned in other countries. There is a policy concern, not yet resolved, that attempts to end the tariff exemption for ethanol merely distilled in a CBI country will be perceived as reneging on the U.S. commitment to stimulate growth in the CBI. Distillation alone requires relatively few resources and, therefore, may offer only marginal benefit to CBI nations. On the other hand, the BCI does not have large quantities of low-cost sugar cane and is unlikely to build many complete ethanol production plants that would use local feedstocks.

Loan Guarantees for Ethanol Plants. The DOE and USDA both have loan guarantee programs. The DOE has issued some $300 million in loans and loan guarantees. The USDA has issued about $118 million worth of loan guarantees and about $42 million in additional loan guarantee commitments are pending. Five of the plants that have USDA loan guarantees are in liquidation and one has already been liquidated.

The DOE no longer has authority to approve loan guarantees except for three applications already under consideration. USDA still receives loan guarantee applications but it is unlikely that any new ones will be able to qualify unless gasoline prices increase sharply. USDA will not finalize a loan guarantee until a plant is built and operating at rated capacity. USDA scrutinizes applications very carefully in order to reduce the risk of being required to pay on a loan guarantee.

Strategic Ethanol Reserve (SER). The Energy Emergency Preparedness Act of 1982 required DOE, with USDA assistance, to assess the feasibility of an SER. The Congressional Budget Office (CBO) also made an assessment in 1983 and the General Accounting Office (GAO) studied the issue in a 1984 report.[3]

---

[3]U.S. Department of Energy, A Report to Congress: Strategic Alcohol Fuel Reserve, DOE/EP-0079, December 31, 1982; Congressional Budget Office, Possibility of a Strategic Alcohol Reserve as a Supplement to the Strategic Petroleum Reserve, Staff Memorandum, U.S. Congress, March 1983; and General Accounting Office, Importance and Impact of Federal Alcohol Fuel Tax Incentives, GAO/RCED-84-1, U.S. Congress, June 6, 1984.

The DOE study found that a strategic petroleum reserve is more cost effective than an SER. The CBO study reported that an SER would cost more than corn price support programs. GAO found that an SER is more expensive than the combined cost of a strategic petroleum reserve and corn price support program unless surplus corn owned by the Commodity Credit Corporation (CCC) is worth less than $1.60 per bushel. However, GAO cautioned that further analysis was needed. All studies found a strategic petroleum reserve to be preferable to an SER.

Some farm groups sought an SER in the 1985 farm bill as a way to increase demand for corn and reduce agricultural price support program costs. The bill that was finally signed into law (the Food Security Act of 1985) did not require the creation of an SER but did require, in Section 1774, that USDA "conduct a study of the cost effectiveness, the economic benefits, and the feasibility of establishing, maintaining, and utilizing a Strategic Ethanol Reserve." If the study shows that an SER is "cost effective, beneficial to the U.S. economy, and feasible in comparison with the Strategic Petroleum Reserve, the Secretary of Agriculture may establish, maintain, and utilize" an SER. That study, which is due to Congress next December, is now underway.

Sale of Surplus CCC Grain at Reduced Prices. Section 1024 of the Food Security Act of 1985 states that the CCC "may, under terms and conditions established by the Secretary (of Agriculture), make its accumulated stocks of agricultural commodities available, at no cost or reduced cost, to encourage the purchase of such commodities for the production of liquid fuels and agricultural commodity byproducts. In carrying out the program established by this section, the Secretary shall ensure, insofar as possible, that any use of agricultural commodities made available be made in such a manner as to encourage increased use and avoid displacing usual marketings of agricultural commodities."

This authority has not been exercised. Last December, 121,895 short tons of raw cane sugar owned by the CCC was sold to Shepherd Oil, Inc., Jennings, Louisiana, at an average price of $60.75 per ton (just over 3 cents per pound). However, that sale was announced for bidding before the farm bill was passed and the sale was made for normal inventory management purposes rather than to promote ethanol production as specified in the farm bill.

Very recently, however, several ethanol producers have requested that USDA sell CCC grain to them at reduced prices to help them survive during this critical period of low gasoline prices. The merits of the requests are under active consideration in USDA at this time. The Secretary has made no decision and we have no indication which way the decision will go. This may be a life or death decision for several ethanol firms and the Secretary is giving the issue appropriate consideration.

CONCLUSIONS

These are not good times for fuel ethanol producers. The entire industry faces the threat of extinction due to falling oil prices. The industry has benefitted from massive Federal and state subsidies since its inception, and needs even larger ones now. This need has, in turn, led to a number of economic policy issues concerning the industry. The specific issues discussed above are not significantly different from economic issues facing biomass and solar energy in general.

All of these industries, which are relatively high cost sources of energy, were encouraged from the beginning by a concern about depletion of fossil fuels. The same economic forces that created the crises in the petroleum industry and, in turn, the economic incentive for development of alternative energy sources (namely scarcity and higher fossil fuel prices) also encouraged greater production of fossil fuels. It was inevitable that oil prices would fall when supplies increased in Mexico, Great Britain, Norway, the Soviet Union and elsewhere. The Organization of Petroleum Exporting Countries (OPEC) lost its ability to restrict supplies and, therefore, lost its ability to maintain high prices. We all know, also, that the present low oil prices will reduce oil supplies, curb development of alternative energy sources and increase energy demand, eventually causing prices to rise again. It seems likely that ethanol production will become a viable energy source without subsidies at some time in the future. Unfortunately for the industry, but probably good for consumers, that time may not arrive until well into the 1990s or later. That presents the country with a basic economic policy dilemma. Should subsidies be provided to keep the industry alive until fuel ethanol is really needed or should we cut our losses and let the industry disappear now with the expectation that it can be rebuilt again when needed? Neither option has a low cost. The choice clearly calls for a political decision and I will not speculate on the outcome.

I will be happy to address any questions you may have.

Note: The views expressed here are those of the author and do not necessarily represent those of the Department of Agriculture.

DISCUSSION

D. Hall, U. of London (United Kingdom)	What is the cost of MTBE? You gave the cost of all sorts of alternatives, but not the cost of some of the octane boosters.
Grinnell	Prices have been changing rapidly lately and I can't tell you exactly what they are. I'd have to check.
Hall	Are they between 2¢ and 17¢ a gallon?
Grinnell	I don't know the cost of increasing octane using MTBE. I haven't recalculated it since oil and MTBE prices have fallen.
Hall	Can you tell me whether all those octane boosters have been cleared by the EPA for health standards?
Grinnell	I assume MTBE has because it's used all the time. The others that we've been looking at have been cleared. There may be some that have not, but there are several that have been.
Hall	We have these same questions in Europe, and they haven't been cleared.
B. Spaugh, Nebraska Energy Office	Did you say at the beginning of your paper that most of the ethanol plants that are economic now are using waste or by-products?
Grinnell	No.
Spaugh	Do you see perhaps the future of ethanol is not in raising crops specifically for ethanol where we have to take into account direct and indirect energy inputs into raising a crop itself, but as a waste treatment method? In our local area, ADM produces ethanol very economically. They operate I am told primarily to get rid of a pollution problem. They produced two other major products first, and used the waste products for ethanol.
Grinnell	I hope I didn't give the impression that most ethanol producers are using wastes as feedstocks. Most ethanol producers have several ways of cutting costs such as by using wastes, using a plant originally built for some other purpose, sharing costs with other production

processes, or simply getting government loan guarantees to reduce capital costs. But I doubt that waste will become a major feedstock because there isn't enough to sustain a large-scale ethanol industry.

Spaugh — By waste, I don't mean municipal waste, I mean industrial by-products.

Grinnell — Thus far, it hasn't happened. Cheese whey and other waste products have been considered, and fruit and vegetable wastes have been used, but they generally aren't available in large enough quantities or they're not economical in the conversion.

R. Clyde, Clyde Engineering — With the prices of gasoline coming down, I think people are going to use more. I understand campers for example are selling like crazy. Wouldn't that help the situation?

Grinnell — Yes, the increase in use of gasoline will increase its price, and that will definitely benefit ethanol production. Whether or not the price of gasoline will go up in the near term to bail out the ethanol industry to make it competitive is a very big question at this time.

L. Engelbrechten, West German Embassy — Does the USDA or the DOE support research programs to increase the efficiency of ethanol production? You mentioned that the Europeans tried to find other uses for their agricultural production for instance through ethanol. Do you think that could become an issue for the United States as well?

Grinnell — On the latter question, it already is an issue in the United States. We have received considerable interest from the agricultural community to encourage additional ethanol production as a way to solve the farm surplus problem. And yes, we do have research programs that address ethanol production, but they are not as large as they were in the past. Research budgets have been cut, but there is continuing research that benefits the ethanol industry.

THE INFLUENCE OF MARKETING AND GOVERNMENT POLICY ON BIOMASS WASTE CONVERSION TECHNOLOGIES IN LDC'S

John W. Tatom, President
John W. Tatom & Associates
4074 Ridge Road, Smyrna, Georgia 30080

ABSTRACT

Agricultural and forestry process wastes represent a major renewable energy source in the Developing World. Generally, the quantity of wastes produced at a typical processing plant contains energy far in excess of that required to power the plant. From a national perspective, full, efficient utilization of these materials is mandatory. Therefore conversion technologies such as pyrolysis and briquetting, which can efficiently convert all the wastes while providing fuels for plant needs and an array of easily transportable by-products probably offer the most attractive means for utilization. This is in contrast to technologies such as direct burning and gasification which intrinsically require only a fraction of the wastes available, thus frequently leaving the remainder practically unavailable as fuels while simultaneously producing a serious disposal problem.

However, the difficulty with the technologies for full utilization of the wastes is the lack of a proven market demand for the by-products. This has seriously inhibited development of these technologies and unless something is done soon, these wastes will remain unavailable economically due to growing use of direct combustion and gasification.

Thus, there appears to be no practical alternative but for government intervention. For several reasons it appears that demand-side as opposed to supply-side support would be better with assistance initially provided in the form of guaranteed prices, tax incentives, government purchase of by-products, etc., that could be phased out gradually as the industry prospered.

Likewise, experience teaches that medium production scale conversion systems utilizing mid-level technology and an Intermediate Capital Intensive design approach is perhaps most appropriate to LDCs. This is because practical marketing of the by-products demands a phased production program which itself requires a technology applicable at small and medium production scales that does not produce too wide a variety of by-products and is environmentally acceptable.

Because the seriousness of the Energy Problem in LDCs can only increase with time, there is a pressing need for recognition of these factors and for the implementation of government programs to overcome the marketing problems described.

THE INFLUENCE OF MARKETING AND GOVERNMENT POLICY ON BIOMASS WASTE CONVERSION TECHNOLOGIES IN LDC'S

DISCUSSION

In contrast with the Developed World, the major energy demand in Lesser Developed Countries (LDC's) is for domestic fuels. Depending on the country, the domestic energy needs may represent from 70 to 90 per cent of the total national energy requirements. While the industrial energy demand is growing, it will not become dominant in most LDC's for many years. Hence, for the foreseeable future the primary problem in meeting the energy needs of the Developing World is to provide an adequate domestic energy supply. However, while the industrial demand is small, it could play an important role in the establishment of a viable waste biomass fuels industry. This role will be discussed later.

Traditionally wood and charcoal--because of their availability--have been the predominant choices as domestic fuels in LDC's. However, the increasing Developing World population and wasteful farming and logging practices, in combination with other factors, have caused a rapid depletion of forest land. A direct result of this is a severe wood shortage and a corresponding rapid increase in traditional fuel costs, with families frequently forced to spend from 20 to 40 percent of their income just to provide for their domestic energy needs.

In more affluent families and in LDC's where oil deposits are present, kerosene is used domestically to a limited degree, especially for lighting. And clearly there is an ongoing need for diesel fuel and gasoline for transportation needs, in addition to an industrial demand for fuel oil. However, the debt crisis in the Developing World and the continuing shortage of capital make it clear that the use of oil--imported or domestic, and even at the drastically reduced prices recently seen--must be curtailed strongly. Thus from a national perspective the use of any and all other available indigenous energy sources in LDC's is absolutely essential.

<u>Process Wastes as an Energy Source in LDC's</u>

One area with significant potential is in the use of available agricultural and forestry process wastes to supply a portion of LDC energy needs. The word "available" is important because there is already considerable use made of these materials, and thus only a portion of the total is practically useable. Since LDC's typically have mainly agricultural economics, it might be expected that these wastes would be in greater relative abundance than in the Developed World, and so a broad evaluation of their extent would be a useful first step. However, so far as is known, there has been no credible worldwide survey of the quantities of available wastes, and so to get some idea of the potential energy from them we must rely on anecdotal data. Because of the difficulty in estimating the fractions of locally available wastes, and because of the strong site specific[1] nature of waste production and utilization, it is easy to understand why there is a lack of a broad data base. Further, it is important to appreciate that any estimates are in truth rough approximations which can be used only to establish the relative importance of these wastes as energy sources.

To illustrate the approximate potential of energy from these materials, surveys of available wastes in two widely differing locales give the following results:

(1) In Senegal, a semi-arid African nation with a low population density, it is estimated that available process wastes could provide all the domestic energy needs (assuming $5x10^6$ Btu per capita per annum) of five to ten percent of the population (7)[2]

(2) On Java, a tropical Southwest Pacific island with near the world's highest population density, it is estimated that available process wastes could provide all the domestic energy needs for four to eight percent of the population (8).

It is interesting that the results are so similar in spite of the widely differing climates and population densities, and suggests that perhaps five percent might be a conservative, roughly representative fraction of an LDC population whose total domestic energy needs can be met by available process wastes.

In some respects five percent seems disappointing and insignificant in comparison with total national energy needs. But viewed from a local perspective it could be critically important because it might represent all of the energy needs of a city or a region populated by many thousands of people and a livelihood for a large portion of them. Moreover, the unbalance between annual natural forest wood regeneration and firewood cutting in many countries is comparable to the five percent mentioned. Thus the use of process wastes as fuels could play a significant role in reducing the rates of deforestation. These ideas should

---

[1] The site specific nature of the overall energy problem in LDC's also makes it difficult and dangerous to speak in generalities regarding its solution. However, there is a broad, urgent need to better understand how to deal with this problem, and so with this caveat, it is important to recognize that all the conclusions of this paper do not necessarily apply to all LDC's.

[2] Numbers in parentheses refer to citations listed in the REFERENCES CITED section.

remind us that, just as in the Developed World, the total answer to the energy problem in LDC's is composed of many small, interconnected pieces, all of which are vital to the overall needs, and that few if any sweeping solutions will probably be found.

In passing it should be noted that many LDC's[3] have undeveloped or underdeveloped peat, liquite, and/or sub-bituminous deposits in addition to unutilized biomass sources such as papyrus reeds, rice straw and other field residues that could be used in combination with and to supplement process wastes as a domestic energy supply. These materials could also utilize similar conversion technologies and be introduced using a similar marketing program as for process wastes. Depending on the local availability of all these materials, the national significance of a combined program to utilize them could be very substantial indeed. But no matter how significant these energy sources may be, it is important to reiterate that for the foreseeable future, wood will continue to be the dominant fuel in the Developing World, and in recognition of this fact, that massive reforestation and afforestation programs are necessary. In many countries these programs are already underway. Thus non-conventional biomass fuels mainly represent a means to bridge the energy gap in the Developing World--while we wait for the trees to grow.

Biomass Waste Conversion Technologies

In the last few years a number of technologies have been developed in LDC's to utilize solid biomass residues for operating process plants and/or to provide fuels for social needs. These include:

(1) direct burning
(2) gasification
(3) densification
(4) pyrolysis

Others have been developed in LDC's to utilize mainly liquid residues; i.e.:

(5) digestion
(6) fermentation.

Without doubt the most successful have been direct burning and gasification. The progress of others such as briquetting, pyrolysis etc. has not been so promising. A characteristic of gasification and direct burning is that they are generally designed to meet the local needs of a

---

[3] For example, peat, lignite and/or sub-bituminous coal are found in a number of developing nations such as Haiti, India, Thailand, Pakistan, Indonesia, the Philippines, Ethiopia, Botswana, Senegal, Nigeria, Peru, Costa Rica and others.

processing plant and are not focused on providing social needs for energy. A characteristic of the less successful technologies is that they necessarily produce a mix of fuel and other products that can be used both to meet local energy needs and social needs. Thus for the latter to be economically viable requires that external markets for these by-products be found, while the former are not constrained by this need. And therein lies a fundamental problem which has hindered their development and commercialization regardless of the level of performance they may achieve.

From a national perspective the use of all energy sources is mandatory. This need, however, raises some difficult questions as to the choice of process waste conversion technologies. To put this situation in perspective it is important to recognize that typical processing plants produce a waste stream having an energy content many times the energy demand of the facility. To illustrate:

(1) The energy value of wood residues from a sawmill may be 10 to 50 times the mill energy needs--even after the efficiencies of thermal and mechanical conversion are included.

(2) A study of one hundred randomly selected rice mills out of the 12,000 in the Philippines (9) revealed that 84 percent produced more waste than required to run the plant, with an average producing four times the breakeven energy.

(3) Peanut hulls in piles six feet deep line the roads for miles around Dakar, Senegal; a graphic testimony of the quantity of residue available beyond that needed to fire the ancient, inefficient boilers that power the peanut oil mills.

This unused excess waste represents a major pollution source and a large potential loss of energy which cannot be tolerated under the severe fuel shortages faced by most LDC's.

Hence, from a national perspective there is a basic need to encourage use of energy conversion technologies that not only provide for local processing plant needs but also generate a fuel supply for wider applications, especially as domestic energy sources.

It has already been observed that wood and charcoal are the traditional fuels in almost all LDC's. This fact strongly suggests that in the near term solid fuels, which are interchangeable with wood and charcoal and can be reasonably well used in the same stoves as they can, will have significant advantages over liquid or gaseous fuels. In any case the solid nature of the great majority of the available process wastes and the technologies available for their conversion require that primarily solid fuels be produced from process wastes.

This suggests that, of the technologies applicable to solid residues which can produce domestic fuels, briquetting and pyrolysis are the two conversion methods whose use should be encouraged at a national level in LDC's. Before discussing how this might be accomplished, it may be useful to briefly review recent technical developments and experience in LDC's using these two technologies.

Review of Process Waste Briquetting in LDC's

Briquetting was originally developed for use in the Developed World primarily as a means for reducing transportation and handling costs associated with process wastes and other materials. Thus typically the briquettes formed are very hard and dense. The use of briquetters utilizing extrusion or roll techniques usually involves the need for significant capital, mechanical power, maintenance, spare parts and a technical infrastructure; characteristics not typically representative of LDC's. Depending on the form of the fuel and the briquetting technique chosen, binders such as starch may also be required.

Recent programs to produce and market domestic fuels in the Gambia(4), using briquetted peanut hulls, and in Rwanda(3) using papyrus reeds have been conducted. However, there has been only limited success in these projects. Besides the obvious wear and logistical problems arising in their production, the briquettes themselves have been found:

(1) to be slow to light
(2) to burn with a hot smokey, foul smelling flame that tends to burn out the metal used in the local stoves, and
(3) to not be reusable once lit, since water quenching results in disintegration of the unburned portion.

Therefore, an improved stove design has had to be developed. A new program utilizing peanut hulls in the Sudan, funded by the UNDP(1) and which hopefully can take advantage of the experience gained so far, is currently underway with technical support provided by the author. A report on this work will be prepared in a later paper. Most recently a project in Ethiopia is reported to have been initiated by the Danes.

In spite of their marketing and operational problems, these programs have had the advantage that the briquetting equipment itself could be purchased from established manufacturers and has not had to be developed in country. While the equipment used so far has been relatively expensive, there are quality sources of inexpensive briquetters, perhaps better suited to LDC applications, in India and other Asian countries that give hope to the ultimate economic success of these efforts. In addition,

Fuel Program was conducted under funding by USAID, who not surprisingly was interested in the application of this experience to the energy problems of LDC's having underdeveloped coal deposits. The Indian program itself arose in response to an early recognition by that government of the domestic energy and environmental needs of its people and of the necessity for direct government involvement in meeting these needs. The program involved a number of Indian coal research institutes, together with numerous government and private corporations formed to commercialize the processes coming out of the laboratories. In addition to (6), a two-volume, very comprehensive, mainly technical description of this program at a relatively early point in its development can be found in (2) which unfortunately may also be difficult to locate.

Since the overall program involved every aspect of the production of smokeless, solid domestic fuels, it is not practical to describe it completely. Thus only that part which directly relates to the issues drawn previously are discussed here. And further, only the relevant lessons learned from this program, many of which interestingly arose from failure, are described, and these are presented in the context of fuels from biomass. Further, in the following section other ideas drawn from more recent experiences in LDC's are also woven into the discussion. Thus while the Indian experience comprises a major part of this material, it is by no means the sole source of all the ideas presented.

CONCLUSIONS

Marketing of Domestic Fuels

Of basic importance is the fact that the Indian program was intended to primarily provide domestic fuels for that country's urban poor, who in spite of their low economic status, live in a cash economy. The focus on this market influenced the entire program to a profound degree. For that reason, it is extremely convenient that today in most LDC's the urban poor appear to be the logical market for non-conventional domestic fuels--if for no reasons other than their number[5] and that they are now buying[6] wood and

[5] It is estimated from various sources that by century's end a near majority of people in the Developing World will live in cities.

[6] In the country, women and children may spend many hours each day gathering firewood, which though scarce is still free. Hence, until that time when a fair value is placed on these efforts, there is believed to be little chance to sell these fuels in rural areas.

recent work to develop a simpler, less expensive briquetting machine using elevated temperatures to reduce the compression required have also shown some promise(5).

Pyrolysis of Process Wastes in LDC's

Pyrolysis of process wastes in the Developing World has so far been the subject of a number of non-commercial, applied research investigations in universities and government laboratories in six LDC's(10, 11, 12, 13). In contrast to the briquetting experience, development of the pyrolysis equipment itself, rather than commercial field studies, has been the prime focus of these efforts. This largely technical work has resulted in the adaptation of basic partial-oxidation, vertical-bed pyrolysis technology to produce an inexpensive, reliable, Intermediate Capital Intensive design using a minimum of imported components and capable of manufacture in country. On paper, at least, it is profitable and can provide for a wide range of process plant energy needs, while producing a supply of fuels for domestic use. Some limited but successful studies of the burning characteristics and social acceptability of one- to two-inch diameter spherical charcoal "Fireballs" formed by a simple, basic agglomeration process have also been conducted. However, there has been little or no effort or interest in commercializing this overall technology largely because of the problems of marketing the char and oil.

Recently it has been demonstrated in a program in Hawaii that the oil itself can be pyrolyzed to increase char yields by 30 to 40 percent using bone dry feed and/or that the oil can be used as a briquette binder. While this utilization of the oil is perhaps not the most efficient, in the short term it does avoid the associated marketing problems and thus offers a promising avenue for future work in LDC's.

What has Gone Wrong?

The lack of commercial success so far in these two renewable energy areas is certainly of concern to everyone working to improve the domestic fuel supply in LDC's. But the stakes are too high to become disenchanted or to give up just because the problem grudgingly yields to solution. What is needed is a careful study of what has gone wrong:

(1) in the conduct of the previous alternative fuel programs
(2) in the non-technical approach of the programs

and then a solid commitment made to overcome the problems encountered. A recent paper(14) identifies some of the problems in the first category. They include:

(1) a basic lack of understanding by project managers of the capabilities and advantages of the technology used,

(2) a lack of long term support and poor followup by aid agencies who often underfund alternative energy projects,

(3) unrealistic expectations of the ease of success in introduction of alternative energy technologies in LDC's by program managers and funding agencies who tend to become too easily frustrated with slow program progress, and

(4) poor project management, since good managers are often difficult to find in LDC's and those who are talented tend to be abusively overworked and are thus inevitably reduced to mediocrity in spite of their efforts.

There is clearly no question that technical problems also have significantly contributed to the program difficulties encountered. But these problems, while often perplexing, can be and have been solved by persistent applications of basic engineering and economic principles. Therefore, while no doubt there are improvements that should be made in the conversion technologies, at present it is believed that the primary difficulty lies in the latter category; i.e., in the non-technical program approach which too frequently has ignored the people-oriented aspects of the problem of developing non-conventional domestic fuels. Thus it is in improving the delivery system and the marketing and ultimately in encouraging government participation where the greatest needs and opportunities now exist.

The Indian Domestic Fuel Program

How can we be confident that solution of these latter problems will yield the results so urgently desired? In reality we cannot be sure of anything; but besides the recent experience in LDC's there is good evidence to the validity of this belief that comes from study of a huge, fundamentally significant, but largely unpublicized, national effort by India started 25 years ago to produce smokeless domestic fuels for the urban poor through the densification and coking of low quality coal. While the focus in this monumental work was on coal, not biomass, the results were solid domestic fuels produced through use of a wide variety of different technologies for briquetting and carbonization. This national experience therefore is as close as we are likely to come to an ideal to learn from, and we are truly fortunate to have it to guide us, particularly in light of the mentioned recent LDC renewable energy programs.

A recent survey(6)[4] of the Indian Domestic Smokeless

[4] The report on this study, for reasons known to the Indian Government, was never approved for public distribution but is available from the author and/or perhaps United Engineers and the AID Energy Office.

charcoal--sometimes brought hundreds of miles to the cities from increasingly remote forested areas.

The urban character of the market is very convenient since it facilitates efforts to publicize the new fuels. With an urban market, the Indians not surprisingly used every mass media technique at their disposal. They ran radio and newspaper advertisements. They distributed flyers to the public and papered every vertical surface with announcements regarding their product. They demonstrated it at bazaars, technical fairs, in private homes, etc. In short, they did everything we might expect to be done in the West to advertise their product.

Additionally, in their initial efforts, they focused on large commercial fuel users such as hotels, restaurants, bakeries and street food vendors who could readily see the advantages of the new fuels. Not only did this assist in establishing a baseline demand, it provided free advertising and generally raised the public consciousness that a quality domestic fuel could be produced from coal. This strategy has obvious parallels during the startup of any LDC non-conventional domestic fuels program and suggests that both the commercial and the industrial sector--because of their flexibility in using various fuels--could play a vital initial role in the utilization of the fuels, especially before sufficient production reliability is achieved to safely offer them to the domestic market.

The overall process of introducing a new product--such as domestic fuel briquettes--into any market is an expensive, chancy operation. This is true in the Developed World, where only a small fraction of all new products favorably test marketed are actual commercial successes. And it is doubtless even more so in the Developing World. So in spite of all the modern marketing techniques (such as opinion surveys and focused group research) at our disposal, the economic viability of a new product is difficult to achieve or predict--witness the recent embarrassing, but profitable, experience of an Atlanta-based soft drink company. Hence, any efforts to introduce non-conventional domestic fuels must be done with absolutely the greatest care, because the margin between success and failure is likely to be so small.

But some failure is inevitable in introducing a new product anywhere--as noted previously. And so any realistic program to introduce non-conventional fuels into the Developing World must be designed to control and minimize the effects of failure. One method would be a phased introduction of these fuels starting with pilot marketing efforts and small to medium production facilities, as opposed to the Indian emphasis on massive coking plants. But in any case the approach must recognize the need for careful research prior to product introduction. This demands intelligent, sensitive understanding of the local people, their culture, their cooking practices and needs.

While the hearth, as typically the center of family life, tradition and culture, might be expected to represent an area where change would be hardest to achieve, there is considerable evidence from India, Indonesia and other LDC's that cooking practices can be adapted to meet present circumstances. Thus even people in traditional societies can modify their habits if they clearly see it to be in their own best interests. Keeping the consumer's, not the producer's, best interests in mind is therefore a key to successful marketing.

But the marketing problem in the Developing World is even more complicated than in the West. One reason is because of the typical unpredictability of the infrastructure within which the non-conventional fuels must be manufactured, transported, distributed and sold. Thus the on-again-off-again electric power, the chronic fuel shortages, and the breakdowns in the rail and road transportation system so characteristic of many LDC's, can play havoc with even the best plans to meet a basic marketing need; i.e., to provide a reliable product supply to the customer. For example in India, where because special stoves are needed to burn the briquetted coke--just as may be the case with non-conventional fuels from biomass--and these stoves do not operate effectively on other fuels, it is very important for the distributor to maintain a constant supply of briquettes. In fact it has been found that next to price, the maintenance of a reliable supply of briquettes is more important to the consumer than anything else. The concern that a reliable supply might not be maintained has been a primary reason for resisting purchase of the stove and the briquettes. This concern has fundamental implications throughout an entire system for producing and marketing biomass fuels in LDC's.

Production Scale and Technology

The basic need for a reliable product supply, plus the chronic shortages of trained personnel, capital and foreign exchange in LDC's together directly impinge on the level of technology, the production scale and the capital intensity of the plants used in the manufacture of the non-conventional domestic fuels. High technology approaches most likely will require dependable electric power sources, large production scales and corresponding amounts of money. While they may be efficient, they may not be economic except at full production. While they may also produce a range of by-products, if there is no market for these by-products, they are in deep trouble. Since they typically need a relatively long time for construction and startup in order to get into full production, this exacerbates their sensitivity to the success of the marketing program.

The Indians, as mentioned previously, chose this high technology route in several of their most important coking plants because of its expected efficiency and to help reduce the foreign exchange drain due to import of coal tar by-products which they planned to produce in these facilities. The results have been disappointing. Though construction of the plants started 15 to 25 years ago, they have still not reached full-scale production and are only marginally profitable, with poor sale of the coal tar by-products strongly influencing the situation in at least one case. These plants have been financed through the formation of large state-owned corporations. It is easy to understand why large sources of capital were necessary to start up these plants, and in a society with centralized government economic planning, the choice of state-owned corporations to provide this support is not surprising.

On the other hand some of the smallest Indian coke production facilities have utilized basic technology such as primitive beehive coke ovens or incomplete, open burning of coal. These heavily polluting, inefficient plants have only one thing to their credit. They seem to operate at a marginal profit, and because of their nature are largely free of the technical, financial and even marketing problems faced by the largest plants. While there are lessons to be learned from their relative success, their squandering of limited resources and the damage they do to the environment make them undesirable from a national perspective.

A few privately owned Indian plants, in their efforts to economically produce smokeless fuel briquettes, have chosen a mid-level of coking technology and a simple agglomeration process using locally available components and with almost no coal tar by-product recovery. By using a modular construction approach, plants built using this technique can be gradually expanded with demand so that little or no capacity is ever unused--as occurred with the largest Indian plants. And by limiting the production of by-products they can avoid the marketing problems faced by the larger plants. While these mid-level technology plants too have had their problems--for example they are very dependent on the electric power supply, although they are capable of economically generating their own--they appear to be closest to being in a profitable mode, and in the long term and from a national perspective seem to offer the most promising means for supplying briquetted coke to the Indian people. This experience has obvious applications in non-conventional biomass fuel production in other LDC's and is entirely consistent with the results from recent programs in renewable energy development.

The Role of Government

The importance of the government role in establishing a non-conventional biomass fuels industry in an LDC cannot be overemphasized. The problems of providing the startup financing and of establishing a viable market for the fuels and any associated by-products is so broad that without government involvement there is little hope in the short term that a successful industry can develop. There are a number of ways this support can come, but most important is the organizational relationship between the government and the new industry. This relationship could come in at least two forms; i.e.,

(1) in the formation of government-owned companies, as in India, which could exercise a monopoly on non-conventional fuel development. These perpetual organizations would basically be extensions of government operating without the necessity of making a profit. Governments and bureaucrats everywhere are not widely known for their efficiency. So there is ample reason to wonder if this arrangement is the best. Certainly the Indian experience does not convincingly argue its wisdom.

(2) in the formation of privately-owned, privately-financed corporations, initially enjoying government support through subsidies such as low interest loans, tax incentives, and government policies encouraging purchase of the fuels and the by-products and/or direct government procurement of these products. The latter actions would help establish an initial base demand for these fuels, and in so doing would materially assist in overcoming a fundamental problem facing the industry in startinq production of a new product. However, when the public demand for the fuels has reached a sufficient, stable level, the government subsidies could then be gradually removed. The result would be a positive social benefit at a minimum social cost. While this course of action might result in individuals and companies profiting hugely, it is important to recognize that they would be also taking all the risks and in the case of failure would be the losers.

These two options represent basically important modes of operation which can directly influence the production scale and the level of technology chosen for a biomass fuels program.

Regardless of whether or not it is desirable to have direct government involvement in smokeless fuel production and sale, there are other important roles that government must play. One would be in establishing fuel standards, which would be an important regulatory activity needed to guarantee the quality of the biomass fuels and to see that realistic specifications regarding product safety are adopted. More important is the need for internal support for a non-conventional fuels industry not just within one department or ministry, but throughout the entire government. One hidden, powerful opponent to this effort inside or outside a government can ruin the entire program, if not at its beginning, then at some later point when it still is vulnerable. Therefore, in the formation of a non-conventional fuels program leaders must see that the interests of all parties are looked after to the maximum possible extent. It is also important that internal government priorities are consistent with the formation of a fuels from biomass industry. For example, two different agencies may have designs on a waste such as peanut hulls. One may want to utilize them as a fodder for livestock, the other may want to briquette them. In such cases priorities must be defined and hard decisions made to insure that embarrassing confrontations are later avoided. Finally, in the case of private industry financing, governments should inform and encourage their investment and banking communities as to the advantages of a biomass fuels industry. This might come in the form of seminars and workshops in which the investment opportunities together with the technical, marketing and distribution aspects of the industry would be explored.

General

The complexity and uniqueness of every LDC makes it difficult to generalize regarding application of the ideas presented here. And so in using this paper the old saying, "If the shoes fits, wear it" is probably most appropriate. Certainly for a particular country, there are likely important considerations either overlooked or under-emphasized in the work. But if this paper accomplishes only an increasing of awareness of the non-technical aspects of introducing biomass fuels in LDC's, it will have achieved the rudimentary objective in writing it.

REFERENCES CITED

1. Anon., "Fuel Briquette Production and the Promotion of Improved Cooking Stoves for Household Use," UNSO Project Document UNSO/SUD/83/X04, 1984.
2. Council of Scientific and Industrial Research, Low Temperature Carbonization of Non-Caking Coals and Lignites and Briquetting of Coal Fines: Symposium Vols 1 & 2, Sree Saraswaty Press Ltd, Calcutta, 1963
3. Energy Initiatives for Africa, "Fuel from Papyrus," Report to USAID under AFR-0424-0-00-2079-00, August 1984.
4. Joseph, S. and Loose, J., "Gambia Briquette Stove Test Report," Intermediate Technology Development Group, Internal Report, London, August 1983.
5. Joseph, S. and Hislop, D., "Residue Briquetting in Developing Countries," Proceedings 3rd E.C. Conference: Energy from Biomass, Elsevier Applied Science Publishers, London, 1985.
6. Schwartz, M. and Tatom, J., "Study of Coal Carbonization Processes in India for Domestic Fuel," United Engineers Report under AID/SOD/PDC-C-0305, May 1982.

7. Tatom, J., "Demonstration of Alternative Fuel Production Through Pyrolysis of Agricultural Wastes at the UNEP Rural Energy Center in Senegal," Mission Project Report, November 1977.

8. Tatom, J., et al., "Pyrolytic Conversion of Agricultural and Forestry Wastes to Alternate Energy Sources in Indonesia," Georgia Tech EES Report to AID under AID/ASIA-C-1203, February 1977.

9. Tatom, J., "Assistance to Energy Production From Biomass Waste Materials," Mission Report to UNIDO under DP/PHI/78/Ø22/A/01/37, November 1982.

1Ø. Tatom, J., et al., "Third World Applications of Pyrolysis of Agricultural and Forestry Wastes," ACS Symposium Series, 1979 AICHE Conference Proceedings, Washington.

11. Tatom, J., "Pyrolysis Experience in Developing Countries," Proceedings Bio-Energy 80 Conference, Atlanta.

12. Tatom, J., "Pyrolysis of Waste Biomass in Developing Countries," Proceedings Third International Conference on Energy Use Management, Pergamon Press, Oxford 1981.

13. Tatom, J., "Pyrolysis of Waste Biomass in Developing Countries; Costa Rica," Proceedings of 2nd E.C. Energy from Biomass Conference, Berlin 1982.

14. Tatom, J. and Bota, K., "An Intermediate Capital Intensive Pyrolysis System Applicable to Developing Countries" Proceedings 3rd E.C. Conference: Energy from Biomass, Elsevier Applied Science Publishers, London, 1985.

DISCUSSION

W. Klausmeier, Sylvatex Corp. You, I gather, are implying that the government's role pertains to the 80 to 90% of the energy needed by the domestic households?

Tatom Yes.

Klausmeier How about the 10 to 20% of the energy that's tied up in the few industries that do exist, for instance, the sugarcane industry? Do you feel that government participation in resolving those problems is also part of your suggestion or should there be another mechanism for addressing them?

Tatom

I'm not as familiar with that particular area as I would like to be. However, I feel that if there is demand-side involvement by government which can be phased out by tax incentives or low interest loans and so forth, this assistance can be withdrawn as the problem is solved. This is a much more superior way of dealing with these types of problems than to form large corporations which are government controlled and are basically perpetual. As far as the particular problem of industry is concerned, I can't speak to that specifically, unless it's industry that's using conventional fuel, like wood and charcoal. In that case, what I'm saying I think would be perfectly applicable, and certainly a great deal of the industry does rely on traditional fuels.

J. Roetheli, TVA

I've heard that there are quite a number of places around the world where it costs people more for their fuel to cook food than it does for the food that they cook. Can you comment on that from your experience?

Tatom

It's getting to be truer than anyone would like it to be. In the rural areas people pick up twigs and branches and so forth, and it doesn't cost them anything, even though they spend their day gathering it. But in the urban areas, which are growing alarming rapidly, there's no question that that's true. Fuel costs are ridiculous. Families can't afford to cook. It's not true everywhere, but it certainly is increasingly true, particularly in countries in the tropical band across Africa.

P. Bente, Consultant

When you raised the point about an inadequate continuous supply of electric power, it reminds me of a situation I learned about recently where in the U.S., there is a free standing plant that cuts and processes lumber, kiln-dries it, pressure treats it, and so forth, but it runs entirely without electricity by using hydraulic motors. This presumably would also be something that could be done in other countries.

Tatom — Many plants can be made energy self-sufficient. I think some of the plants are too small to be able to justify the cost of some of the equipment that's involved, but for a medium-scale plant, there's no reason why it shouldn't be made energy self-sufficient.

K. Porter, U.S. AID (Jamaica) — I have identified pretty much with your presentation, and in fact I have my own ideas that governments in developing countries need to have more vital role in shifting policies to bring about certain developments. Do you personally know of any firms, that are actually doing work in trying to adopt or modify these technologies that we've been hearing about to fit the needs and conditions in developing countries? I'm interested in obtaining information on companies that are making wood chippers in the range of 10 to 20 ton-per-hour outputs?

Tatom — I'm sure there may be people who manufacture those in the audience. Wood chippers are quite readily available. The problem is getting one that is economical in a developing country. I have some private opinions about the need for transferring what I've called intermediate capital-intensive technology into developing countries to avoid the foreign exchange problems that so many of them face. But certainly, the type of equipment you're talking about is available and it's for sale today.

Klass — There are a number of wood chipper manufacturers, and I'm sure there are a few people here who could provide the names of several wood chipper manufacturers for you. But I would like to suggest that you consider what is called BUN, the Bioenergy Users Network, which is being organized now. This consists of a number of developing country organizations and people who will share this type of information just as they would in a computerized network of information exchange. Wil Klausmeier is involved in this, and will be heading the group here in Washington, D.C. that is managing the program. I think if you talk to him after the meeting, he will be able to help you.

FUTURE FUNDING PROSPECTS FOR RENEWABLE ENERGY
AND THE EFFECT OF GRAMM-RUDMAN-HOLLINGS LEGISLATION

William T. Harvey, Jr.
Harlan Watson
Committee on Science and Technology
U.S. House of Representatives

ABSTRACT

In view of the purpose of this conference, it would appear to be more appropriate to discuss the future funding prospects for biofuel energy than the broader renewable energy concept. I must preface all my remarks with the point that these comments are my own. They do not necessarily reflect the opinions and outlooks of Members of Congress or of any staff people other than myself. My comments address three areas. The first will pertain to energy itself, the second will pertain to biofuels energy, and the third will consider the Gramm-Rudman-Hollings legislation.

FUTURE FUNDING PROSPECTS FOR RENEWABLE ENERGY AND THE EFFECT OF GRAMM-RUDMAN-HOLLINGS LEGISLATION

There was a time, and not too long ago, when energy was a very high priority item on the list of any person, whether private, public, elected, or appointed. Oil prices had escalated from a relatively stable base to an order of magnitude higher, and there was no sign of a change in that trend. Then we saw those energy prices drop, practically with the same speed with which they had earlier risen. During those days of escalating energy prices, Members of Congress were motivated by every means possible to "do something about it". That is not the case now. There are many other issues that crowd out energy as an item of any importance since oil prices have dropped to present levels.

Fortunately, there are a number of Members who do recollect those chaotic days of gas station lines and rapidly increasing fuel oil prices. They continue to provide a high priority to energy supply sources, and to maintain a viable R&D program. It is not easy for them because their ranks are thinning. It is also difficult because so often they get little appreciation for their support and much demands for more money. All this in an environment intent on swinging development funds out of civilian programs into defense-based technologies.

However, it might be stated that energy from biofuels stands a little better chance than the broad energy picture. One thing you have going for you is the use of waste materials as a source of biofuels. This utilization is something that people can understand. Certainly our metropolitan areas are beginning to swim in waste generated by the metropolitan area. Effective waste utilization, reduction in total volume, and economic usefulness make that type of process very attractive. But I don't see how you can expect to get a lot of support from Congress, especially if it costs federal dollars. There are just too few dollars to go around. Your best bet is to leverage your own resources, through cost-sharing ventures. Although not adequately defined, the FY 1987 budget request concept of cooperative R&D ventures could be an attractive vehicle for biofuels projects.

By the time the Conference is held in April, we will know more about the implementation of the Gramm-Rudman-Hollings act. The process for getting from here to October is not certain, since there are so many possible routes. The one thing of which we can be sure is that under the present legislation, the deficit for the fiscal year 1987 budget cannot exceed $144 billion. If we take a look at the FY 1986 spending rate, the revenues which are presently expected, and the programs which are not exempt from reduction, it looks like we would have to reduce those FY 1986 spending levels by 15 percent to get down to this deficit ceiling. Such an equal distribution of reductions might be satisfactory to a number of people but possibly the Administration would object to that size of a defense cut. Then, of course, several things could happen to change such a scenario. For example, additional revenues could be generated through some tax. Another possibility is the repeal or amendment to the Gramm-Rudman-Hollings act.

The message is that forecasting anything is always difficult. Forecasting the budget process for Fiscal Year 1987 at this moment is impossible. For you people who are participating in the FY 1986 budget through various projects and programs, there is the possibility that you could be curtailed to some extent in 1987. Changes to your projects and programs probably would be left to the Department of Energy unless you have a strong support base and could have specific support for your activity written into the legislation or its accompanying report language.

Another interesting feature of the entire process is that FY 1987 is the first of five years in which the Gramm-Rudman-Hollings act is being fully implemented. You can simply regard FY 1986 as a dress rehearsal. What will happen in FY 1988 and subsequent years is anyone's guess at the present time.

DISCUSSION

D. Baxter, Maryland Department of Resources

What is the three judge panel's stand now in the District of Columbia on the constitutional question?

Watson

A provision was ruled unconstitutional that allowed the comptroller general to essentially dictate budget cuts. That was ruled unconstitutional because they said the comptroller general is a legislative official and he would be carrying out the act in a way that would permit him to carry out executive functions. There has been a series of arguments in the Supreme Court, and more are scheduled this month. A decision is not expected before early summer, maybe July.

D. Klass, IGT

So the law is still in effect then until a final decision is made?

Watson

Yes, the act provided that cuts would stay in place.

Klass

Many people who are at this conference are personally involved in the utilization of federal funds for conducting their own research. Now, I do not believe that any agreement is going to be reached by April 15th. Most of us don't even have our income tax done yet, including me. I can't imagine how a complete budget system could be developed by that time in just a few days. Assuming that does not happen, does this mean that there will be something equivalent to a continuing resolution where the budgets for last year would be Gramm-Rudmaned by 15% in the non defense and other program areas? Could you explain that?

Watson

Constitutionally, the House starts the appropriation process. The House is supposed to start considering the appropriation bills by May 15th, and have action completion by 15th of June, and be finished and ready for the President to sign by June 30th. They put in a number of points of order on each side of the Hill to prevent delays in the process. They've also said we can't go home for 4th of July recess until we get this done. But of course all they have to do is to vote to change the rules. Clearly, one possible scenario is that neither side can come to an agreement on a budget resolution, it drags on and on, and we go to some sort of a continuing resolution. This could either be done across the board, but it can't be across the board because the President will not sign a defense appropriation that's reduced 15%. So if the President refuses to increase taxes, everything else has got to go down more than 15%.

Klass — What's your personal opinion about the probability of there being an automatic cut versus a budget that meets the requirement?

Watson — I believe that defense will be protected, and there will be an automatic cut because it's the easiest way to get around it. An alternative of course which some people are talking about is to pass a continuing resolution and come back and have a lame duck session after the election is over.

P. Bente, Consultant — I don't hold this against you personally but I think you should recognize there is a great feeling of frustration among many of us here because of these particular facts. There are three procedures for getting about 3 quads of energy — nuclear power, hydroelectric power, and biomass. We can't get much more nuclear power. The government has been trying since the late '70s without success, although a huge amount of money has been set aside to try to promote nuclear power. There's not many places left for large hydro dams. But when we look at forest wastes, municipal wastes, and that type of fuel all over the country, we could probably expand bioenergy several fold. Yet, there is no real financial support for this. We're behaving as if fossil fuels are going to last indefinitely, but they are limited. Somewhere along the line, we have to wake up and really get busy and provide support for the resources that we have, and those are biomass resources.

Watson — I agree, you won't get any argument from me on that.

Roetheli — You mentioned venture spools in the DOE budget. I think that's a very interesting concept. I guess what perplexes me about it is in terms of the regional biomass programs. I know very intimately from the Southeast Regional Program that we funded projects in a way very similar to the venture pool concept. We have gotten approximately $2 of private funding per dollar of government funding and that seems to have been growing, yet when we look at the budget for '87, the regional programs don't even appear as a line item. They're zeroed out in effect. Could you give us the rationale for what appears to me to be somewhat of a perplexing area where there's basically a system already set up to do what's called for in a joint venture, yet it is being terminated at the same time?

Watson	It was recommended for termination by the Subcommittee. It was a straight money problem; it just feel off the bottom essentially. The idea was to keep the level of spending down below the current level. Although we proposed to double the department's request, it was simply a call to throw as much into the more basic and applied research areas rather than the commercialization activities.
J. Koning, Forest Products Laboratory	I believe Keyworth has left, is that correct?
Watson	Yes, he has.
Koning	I believe he was a very strong proponent of nuclear research. In the past, that budget has gotten a lot of support. Do you see with his departure the possibility that there might be a shift of funds away from basic research in nuclear activity, which was large, to more of an equitable distribution into biofuels and other areas of research?
Koetheli	Yes, the nuclear program is coming down anywhere from $350 million to nearly $500 million. They're either one half or a third of what they were when the Administration took office. Obviously, they started from a higher base, but they're also getting chopped considerably. I think it's a philosophy that has been applied pretty much across the board.
Glass	If you study DOE's budget components in the R&D area, you will find that defense related R&D increased substantially, and

I presume most of that is nuclear. The energy component decreased substantially, and general science went up a little bit. The DOE overall budget as a matter of fact increased slightly in FY87 as compared to FY86. So I would guess that emphasis on nuclear research is included in the FY87 budget recommendation by DOE.

Watson — Yes, that is true. I was referring to the civilian program. What they are doing is now starting to find a lot of things over on the defense side. In fact, there's a defense nuclear program now, which is looking at things for Strategic Defense Initiative power sources and so on. So I agree, they are doing a lot of nuclear work on the defense side. But the civilian side is ramping down. The civilian breeder program I believe is essentially out. I don't see that reviving very quickly.

Roetheli — Could you give us your opinion on what will happen with tax reform? Specifically, how will this affect the biofuels industry. It appeared that the Treasury I proposal basically eliminated subsidies for all of the energy fields, but when Treasury II came out, the subsidies for renewables are basically removed.

Watson — This is an area in which I have to profess nearly total ignorance because I really have not been following tax legislation. I only know what I read in the paper. I'm not familiar enough with the details to know what's in and what's out, but I think there's going to be some form of tax bill before the end of the year.

Klass — I think I mentioned Monday that if Treasury II is passed, all renewable energy tax credits, including the production tax credit, will be eliminated, and that they will be reduced but not totally eliminated for petroleum and other fossils. If something like Treasury I passes, then all of the tax credits for energy disappear, as I understand it, including the 6¢ per gallon alcohol fuel tax forgiveness.

POSTER PAPERS

STUDY OF A COCURRENT BIOMASS COMBUSTOR

Nazrul Islam, M.S., Eng., Research Fellow, Resource System Institute, East-West Center, Honolulu, Hawaii

J.S. Fox, Ph.D., Prfessor, Mech. Eng., Univ. of Hawaii
M.J. Antal, Ph.D., Professor, Mech. Eng., Univ. of Hawaii

K.R. Smith, Ph.D., Research Associate, Resource System Institute, East-West Center, Honolulu, Hawaii

ABSTRACT

This paper presents the results of an experimental study on the behavior of biomass combustion under cocurrent conditions. Cocurrent conditions are obtained when fuelflow and combustion airflow are unidirectional. The study ascertained whether complete combustion is possible in a cocurrent burner using primary air (PA) only or using both primary air and secondary air (SA).

The study involved design and fabrication of a suitable combustor test bed. Two 75 mm dia. tubular reactors were used; one made of quartz, the other of steel. Birch dowels 8mm to 11 mm dia., cut to 37 mm lengths were used as the test fuel. Independent parameters were combustion air velocity (superficial), fuel moisture, and fuel diameter. The reactor's outward heat transfer was varied by changing reactor insulation. Their effect on combustion efficiency, air pollution emissions viz. carbon monoxide (CO) and total suspended particulates (TSP), equivalence ratio, fuel reaction rate and flue gas composition were studied.

One of the most important burner characteristics observed was stability at each combustion air velocity, denoting steady state operation of a continuous flow reactor. With PA combustion, values of combustion efficiencies were found to be as high as 99.7%. Correspondingly, CO emissions were 3 gm/kg, and TSP less than 1 gm/kg. Higher moisture contents or more reactor heat losses required less velocity for complete combustion. With PA and SA (two-stage combustion device) complete combustion was again obtained. In this case, the primary air was controlled to operate the system as a gasifier, so that the fuel gas generated had maximum combustibility and temperature approaching the slowdown of the reduction reactions. This resulted in easy ignition with introduction of secondary combustion air.

The good combustion obtained by PA only or by PA+SA offers options to a designer of actual biomass combustion systems. In systems required to burn moist fuel or transfer heat from the combustion zone (eg. domestic heating stoves), a PA system may be suitable. On the other hand, use of a dry fuel and insulated combustion chamber (like boiler furnace) would tilt the favor towards systems with PA+SA. The PA+SA system would also handle occasional wet fuels by shutting down the SA and operating the burner as only a PA system.

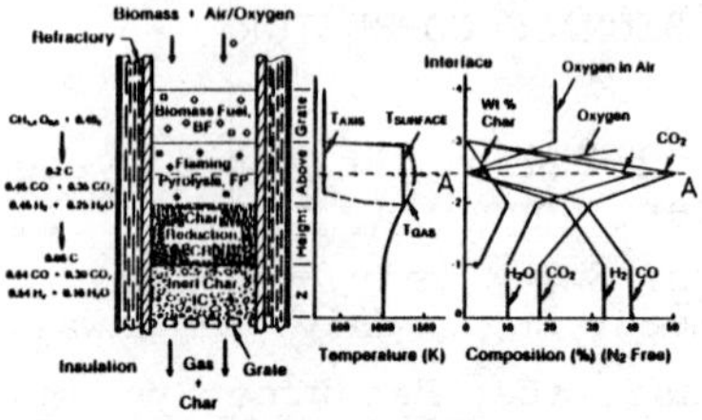

Source: Dr. T.B. Reed, Solar Energy Research Institute

Figure 1: A stratified downdraft gasifier. Oxidant and fuel flow in horizontal layers down through the flaming pyrolysis, reduction and inert char zones. The temperatures and gas compositions at different cross sections are shown alongside the gasifier.

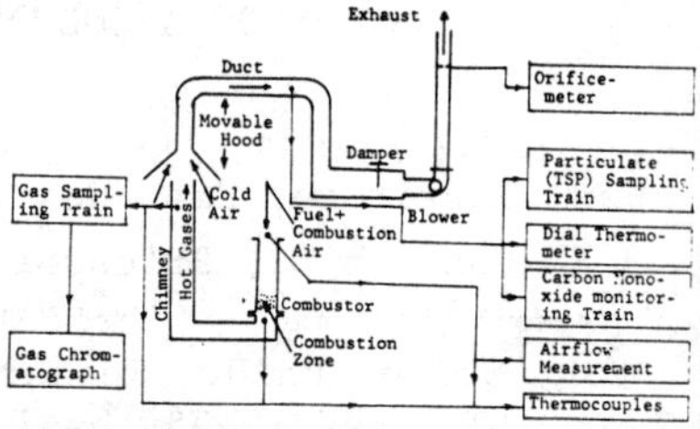

Figure 2: Experimental setup showing location of variou instrumentation. Fresh fuel and combustion air are fed fro the top and react in the combustion zone at the bottom of th combustor. The reactants pass up. and are diluted with col air turning vaporized tar to aerosol.

A cocurrent combustor has similar reaction zones as that of the gasifier shown in Fig. 1. To convert from gasifier to combustor, sec. A-A has to be driven down below the grate by increased air velocity or suppression of pyrolysis.

Fig. 2 shows test bed. System behavior was evaluated by flue gas analysis.

The profiles of fuel reaction rate (R_{fu}) vs superficial air velocity (U_s) obtained after many combination of runs are shown in Fig. 3. The apex of each curve denoted by U_s^* and $R_{fu,max}$ reflect inability of R_{fu} to increase any further; being limited by boundary layer diffusion. Complete combustion occurs very soon afterwards at $U_{s,100}$.

Fig.4 shows CO profiles obtained after many runs. The velocities for maximum CO ($U_{s,g}$), and minimum CO ($U_{s,100}$) are shown. $U_{s,g}$ is the point of operation for a gasifier, or a PA+SA system, while $U_{s,100}$ for a combustor operating on PA only.

Fig. 5 shows that combustion efficiency is very high and TSP emission very low at $U_{s,100}$. The shift of the curves with insulation can be seen clearly.

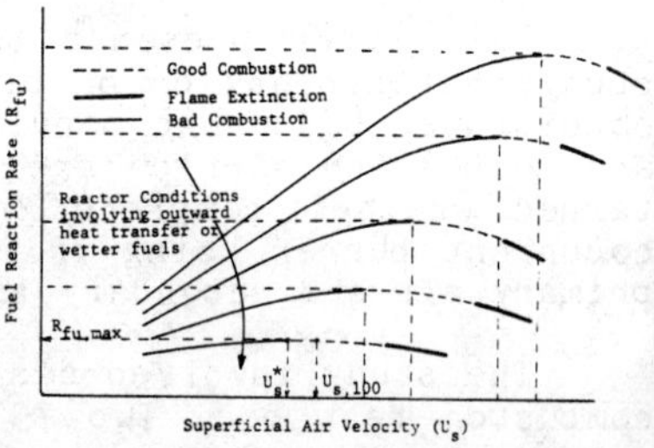

Figure 3: Generalized behavior of fuel reaction rate (R_{fu}) as a function of superficial air velocity (U_s) under differ ent reactor conditions. At low velocities. R_{fu} starts of together. At higher velocities the curves separate. Th maximum fuel reaction rate ($R_{fu,max}$) is higher for drie fuels or heat insulated reactors. The velocity at whic $R_{fu,max}$ occurs (U_s^*) becomes progressively higher for highe $R_{fu,max}$. For each condition. immediately after this velocity complete combustion occurs at a velocity termed as $U_{s,100}$.

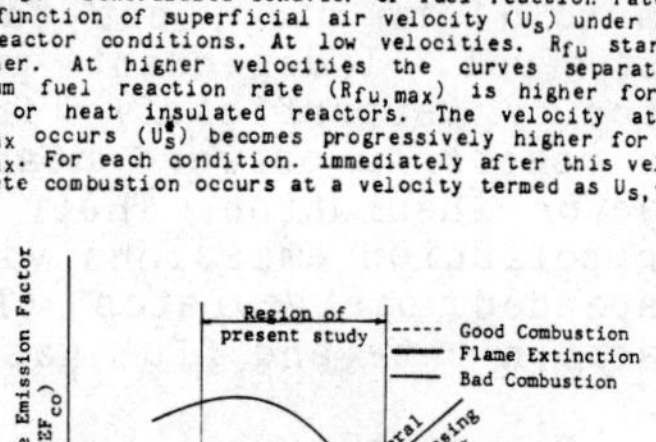

Figure 4: Generalized behavior of carbon monoxide (CO) emis sion factor (EF_{CO}) as a function of superficial air velocit under different reactor conditions. The maximum CO emissic factor ($EF_{co,max}$) increases for drier fuels and heat insula ted reactors. The velocity at which $EF_{co,max}$ occurs ($U_{s,}$ becomes progressively higher with $EF_{co,max}$. For wetter fuel or less heat insulation. the velocity $U_{s,100}$ decreases.

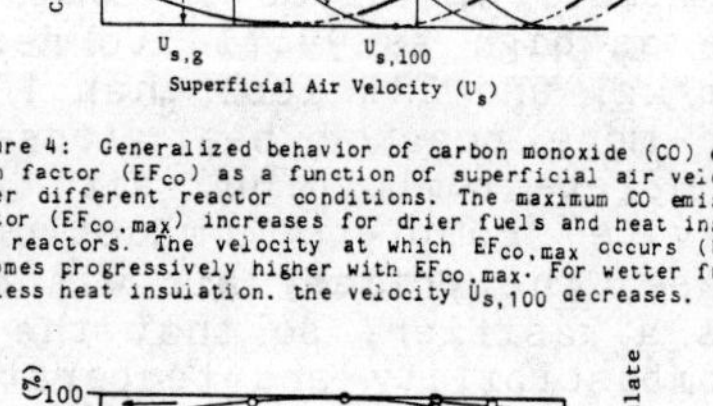

Figure 5: Combustion efficiency and total suspended par iculate matter emission as a function of superficial a velocity for different reactor conditions. Complete combu tion occurs at $U_{s,100}$ for a set of specified reactor con itions. Particulate emission at this point is almost zero.

AN EXPERIMENTAL EVALUATION OF BIOMASS BRIQUETTES AS COOKING FUEL FOR THIRD WORLD COUNTRIES

M. R. Brambley, Ph.D., Assistant Professor; T. Medynski, MS
Washington University, St. Louis, Missouri 63130

ABSTRACT

Firewood is the principal source of energy for more than 1.5 billion people in the less developed regions of the world. In many regions (e.g., many parts of India and semi-arid central Africa), severe firewood shortages are developing or already exist. In many rural areas people have resorted to collecting and burning agricultural wastes and dung to supplement or replace firewood. These wastes would probably provide greater benefits if partially or completely returned to the soil. One method that could contribute to alleviating or minimizing the detrimental impacts of cooking fuel shortages is the use of alternative fuels such as biogas, oilseed fuels, and charcoal produced from other biomass. A technology by which agricultural and some food processing wastes may be used efficiently and conveniently as cooking fuel is briquetting. The utilization of biomass briquettes as cookstove fuel may require little modification of present cookstoves and cooking practices, yet provide a convenient way to utilize wastes. This study addresses the suitability of machine-made briquettes for use as cookstove fuel.

The objectives of this project were to determine the suitability of biomass briquettes for use as fuel in one-pot chimneyless cookstoves. The study was restricted to machine-made briquettes that were available from manufacturers of briquetting equipment and from firms presently producing briquetted fuels. Most of the briquettes that were obtained consisted of wood shavings and scrap. Other materials (including oat hulls and sunflower seed hulls) could only be obtained as much smaller pellets. Preliminary tests indicated that these pellets were difficult to use in cookstoves without substantial modifications to the stoves. Therefore, most of the stove tests involved use of the larger briquettes.

Fuel samples were characterized by shape, size, density, and proximate analysis. With only one exception, the samples had compositions very similar to wood with approximately 70-79% volatile matter, 5.3 to 7.8% moisture, 0.4 to 6.2% ash, and 14 to 20% fixed carbon. All fuel samples were tested in small-scale (approximately 6-inch diameter) fuel beds. Some were slightly more difficult to ignite than wood, but once lit, burned much like wood. Only one sample, which had an extraordinarily high ash content of 13% and a low volatile content of 63%, would not sustain combustion under natural-draft conditions.

Wood blocks and two densified fuels were tested extensively in a full-scale one-pot chimneyless stove constructed from sheet metal. Both densified fuels burned very much like wood after ignition. Stove efficiencies were essentially equivalent for the two briquetted fuels and wood blocks. Maximum efficiencies for all fuels were about 30%.

Two stove characteristics, grate-pot distance and pot-stove gap were found to significantly affect efficiency. This was expected based on the reports of others. Much greater efficiencies (by about 36%) were obtained with a gap around the pot of about 6 mm compared to 25 mm. The distance between the grate and pot also was found to significantly influence stove efficiency. The highest efficiencies for all three fuels were found for the smallest grate-pot separation that was tested (about 10 cm).

The ventilation area below the grate did not affect efficiency substantially over the range tested. Tests were completed with ventilation areas of 7.7 cm^2 (12 in^2) and 155 cm^2 (24 in^2) using two and four holes, respectively, equally spaced around the circumference. Smaller total ventilation areas may have an influence on stove efficiency.

The mass (and total heat capacity) of the grate also had a considerable effect on efficiency. A heavy cast iron grate about 2.3-cm thick reduced efficiency by about 20% (compared to a thin galvanized steel grate) and caused difficulty with ignition and maintenance of the fire. This is expected because the heavy grate acts as a heat sink until it reaches a steady operating temperature.

The presence of a stoking hole that was not closed between fuel additions had a negligible effect on stove performance. Efficiencies with the hole sealed were essentially indistinguishable from efficiencies with the hole open. Interestingly, secondary air holes at approximately the same height as the stoking hole reduced stove efficiency, possibly by introducing cold ambient air that decreased the temperature of the gases passing the pot.

These results suggest that briquettes can be used as efficiently as firewood in one-pot chimneyless woodburning cookstoves. Some fuels apparently are difficult to burn. Consequently, briquettes formed from feedstocks actually available in Third World countries will require further characterization and testing.

Acknowledgement

This study was supported by a grant from Volunteers in Technical Assistance, Inc. (VITA).

EVALUATION OF THERMODYNAMIC EFFICIENCIES FOR A WOOD-FED DOWNDRAFT GASIFIER

S. M. Chern, MS, Graduate Research Assistant,
W. P. Walawender, PhD, Professor,
and L. T. Fan, PhD, University Distinguished Professor and Head

Department of Chemical Engineering
Kansas State University, Durland Hall
Manhattan, Kansas 66506

ABSTRACT

The air gasification of wood chips in a commercial downdraft gasifier (Buck Rogers Gasifier) was thermodynamically analyzed. Material balances, established over an operating range of 0.3 to 1.4 MJ/hr, were used to formulate an empirical stoichiometry for the gasification of mixed hardwood chips. Energy and available energy balances conducted on the gasifier, were used to determine both first and second-law efficiencies by considering various products as being useful. For example, the first-law efficiency, based on dry cool gas, was determined to be 72%; it agrees well with the experimental observations. The second-law efficiency, based on dry cool gas, was 53%. The methodology is based on a general approach for the thermodynamic analysis of systems containing structurally complex materials; it is suitable for the analysis of a variety of biomass gasification processes.

EVALUATION OF THERMODYNAMIC EFFICIENCIES FOR A WOOD-FED DOWNDRAFT GASIFIER

Empirical Stoichiometry For Wood Gasification

$$1.0 \text{ Dry chips} + 1.23\ O_2\ (5.87 \text{ Air}) \longrightarrow 1.44\ H_2 + 1.55\ CO_2 + 1.88\ CO + 0.25\ CH_4 + 0.051\ C_2H_4 + 0.008\ C_2H_6 + 0.011\ C_3H_6 + 0.308\ H_2O + 0.004\ N_2\ (4.64\ N_2) + 0.036 \text{ Char} + 0.003 \text{ Tar}$$

Dry chips, Tar: $C_{4.04}H_{6.05}O_{2.83}N_{0.0088}$; Char: $C_{5.45}H_{34.6}$

The above stoichiometry (ash free basis) was developed using a two-stage adjustment procedure. In the first stage, incremental adjustments were made in the experimental stream rates (Walawender et al. (4)) for chips, air, gas and water to obtain complete overall balance closure and the best possible elemental balance closures (weighted in the order C, H, O, N). The results from the first stage were normalized and averaged for use in the second stage adjustment. Closures of the elemental balances were obtained by assuming that the char contained no O or N and forcing the elemental closures through the C and H in the char. The above stoichiometry was in agreement with the means of the experimentally observed stream rates to within one standard deviation.

First and Second-Law Gasifier Efficiencies

Mode	Discarded Output	E_1(%)		E_2(%)	
		Output 298K	Temp. 700K	Output 298K	Temp. 700K
1	None	81.2	89.5	59.4	62.4
2	Char	74.9	83.1	53.2	56.1
3	Char, Tar	74.5	82.7	52.9	55.7
4	Char, Tar, H_2O	72.0	79.3	52.9	55.4

E_1 = (Enthalpy in useable products)/(Total enthalpy input)

E_2 = (Available energy in useable products)/(Total available energy input)

The above efficiencies were evaluated following the works of Ishimi et al. (2) and Shieh and Fan (3). Enthalpy was the only significant form of energy for the energy balances. Key assumptions included negligible work, ideal gas, a constant pressure of 1 atm, and input streams entering at 298K. Enthalpies and available energies for the structurally complicated materials (chips, char and tar) were evaluated from formulas presented by Shieh and Fan (3), and their elemental compositions. The experimentally determined char composition was used in the computations because of its importance in mode 1. The heat loss from the gasifier was determined to be 10.5% of the input enthalpy; the associated available energy loss was about 7%. The available energy dissipation was determined to be 31% of the input available energy; this loss is due to various irreversibilities in the gasification process. Utilization of the char improves the process efficiencies significantly. Details of the procedural aspects of the analysis are given by Chern (1).

<u>References</u>

(1) Chern, S. M., The Gasification of Biomass in Commercial Downdraft Gasifiers, MS Thesis, Kansas State University, Manhattan (1985).

(2) Ishimi, T., J. H. Shieh and L. T. Fan, Thermodynamic Analysis of A Biomass Pyrolysis Process, presented at 184th ACS National Meeting, Kansas City, MO, (1982).

(3) Shieh, J. H., and L. T. Fan, Energy Sources, 6, 1-45 (1982).

(4) Walawender, W. P., S. M. Chern, and L. T. Fan, Ch. 50 in Fundamentals of Thermochemical Biomass Conversion, Elsevier Science, New York, 911-922 (1985).

<u>Acknowlegment</u>

This is contribution no. 86-357-A, Department of Chemical Engineering, Kansas Agricultural Experiment Station, Kansas State University, Manhattan.

TAR IN PRODUCER GAS WHY AND HOW MUCH?

Parikh P.P., Ph.D., Professor; Paul A., M.Tech., Res.Engr; Bhave A.G., Ph.D., Res.Engr; Uma R., M.Sc., Res.Asst.
Department of Mechanical Engineering,
Indian Institute of Technology, Bombay, India.

ABSTRACT

Thermal gasification of biomass is the primary step towards its utilisation as a source of energy for shaft power as well as thermal applications. In the former case, Producer gas is used for fuelling internal combustion engines used for applications like irrigational water pumping, electricity generation, etc. The gas formed during gasification process contains tar, and condensibles. The tar content of the gas has been of great concern especially from point of view of its use in I.C.Engines. Presence of tar in Producer gas not only adds to the maintenance requirements of the engine but also adversely affects life of the engine due to the corrosive wear it promotes. It is therefore necessary to ensure that the tar content of gas is at the minimum before it enters the engine. Minimisation of tar has two distinct aspects i.e. minimisation of the tar produced in the gasifier and efficient cleaning and scrubbing of gas for tar removal. Quantification of tar content in the gas therefore becomes an important part of research, development and testing of gasifier and its sub-systems. Several methods have been used and suggested for quantifying tar content of the gas. This paper is an attempt to present the various aspects of tar problem of gasification technology including parametric influence on formation of tar in the gasifier, efficient scrubbing and cleaning in the processing system and measurement of tar at various points of the gasifier system. Based on a critical review of existing published information, an attempt is also made to suggest maximum acceptable tar values from point of view of engine applications.

TAR IN PRODUCER GAS WHY AND HOW MUCH?

Introduction :

Tar & Particulate contents of the hot Producer gas generated by gas producers have been a matter of serious concern in engine applications. Rapid deterioration of lubricating oil and consequent abrasive and corrosive wear of engine has been mentioned in numerous publications (1,2,3,4,5). In addition, deposit formation on combustion chamber walls, inlet passages, nozzle tip, valves, piston rings and grooves are not new to the users of Producer gas for fuelling I.C.Engines. Piston ring sticking caused by deposits increases the blowby i.e. flow of tarry gases past piston to lubricating oil sump, which deteriorates lube oil and result is corrosive and abrasive wear. Deposit formation on valves can result in abrupt stoppage of engine. Piston sticking has also been reported due to tar deposition. The problems caused by tar content of Producer gas are thus wide ranging; from increased maintenance of routine nature to serious reduction in engine life, not to mention the drop of performance caused due to these problems. Quantitative data, about the problems caused by tar is not presently available in literature. This paper is an attempt, to analyse important facets of "Tar in Producer gas" problem which could be listed as:

1) Causes of presence of tar in the Producer gas. Effect of various design and operating parameters on tar quantity.

2) Assessment of achievable minimum level of tar in the gas.

3) Assessment of tar tolerance of engines.

4) Methods of tar removal from Producer gas to match the engine requirements i.e. gas cleaning processes and systems.

5) Effective and accurate method of measurement of tar content of gas.

Why Tar ?

In the process of gasification, biomass undergoes drying, pyrolysis, oxidation and reduction-in that sequence. Volatile contents of biomass are released during pyrolysis, the temperature range being from about 150 to 600 C. Depending upon the biomass material used(6), the quantity and nature of volatiles change. Pyrolysis products including tar have to pass through the high temperature oxidation zone where they should get burnt. Uniformity of the oxidation zone in terms of temperature distribution is the primary requirement for complete elimination of tar, in addition to the temperature levels which exist in the oxidation zone. Temperatures lower than 1000 - 1100 C prove inadequate for tar cracking and elimination. Design of oxidation zone of the gasifier and the operating load conditions are the two main parameters in this reference. Gumz's figure (fig.1) (7) very clearly illustrates the importance of oxidation zone design. The dark zones (the lower temperature zones) will permit tar passage without cracking. Such dark zones could exist either due to impro--per design or because of low gas generation rate, which in turn might be either due to lower load requirements (fig.2) (8) or constrictions like clinker. Clinker forma--tion therefore can result in increase in tar content. Higher moisture content of biomass also affects the oxidation zone temperature and therefore the extent of tar in Producer gas (8).

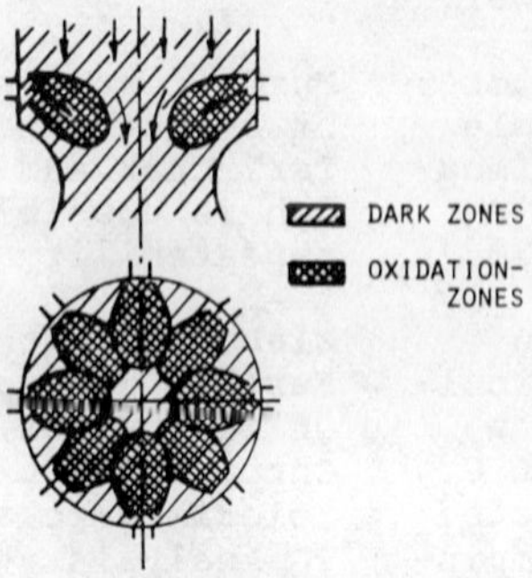

FIGURE 1. HEARTH OF A DOWNDRAFT GASIFIER (7)

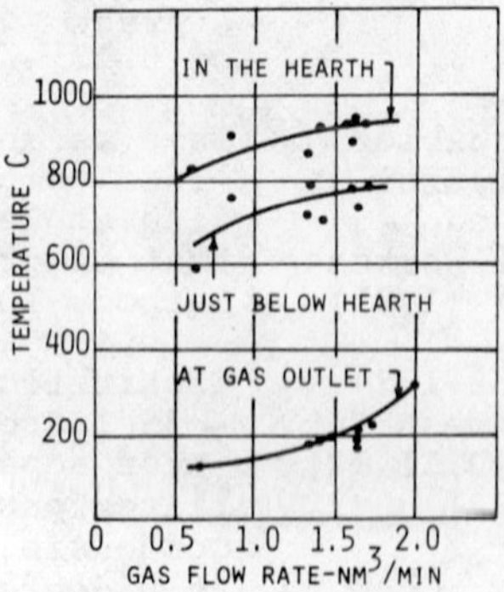

FIGURE 2. TEMPERATURE VARIATION AS A FUNCTION OF OUTPUT (8)

How Much Tar ?

Precise estimation of tar in Producer gas is important from point of view of determining the achievable minimum level, so also the tar tolerance of engines. Investigations regarding parametric influence on tar levels also warrent its accurate determination. Measurement of tar therefore has become an important facet of testing of gasifiers. The first step towards designing a reliable yet simple but acceptable method of estimation of tar is to define what is tar ? In most of the methods described in literature (9, 10,11,12,13,14,15,16) all the condensible matter collected either at room temperature or at 0 degree C has been consi--dered as tar, irrespective of its composition. There is another class of Scientists who have analysed the conden--sates and identified as many as 300 compounds as the tar constituents (17). From point of view of I.C.Engine applica--tions both these are important but isolated and therefore inadequate approaches. The condensates which can be consum--ed by engines need not be named as tar. Linking of tar definition with engine requirements is the lacuna of the published literature. It is only after this linkage is established that attempts can be made in standardizing upon tar estimation method. Here again it will be necessary to decide upon two methods; one for field locations and another for research purposes. The following table gives the summary of the available tar measurements methods and thier relative merits and demerits.

COMPARISON OF AVAILABLE TAR MEASUREMENT METHODS

Grey scale method by SERI(16) and Tipi Workshop (15) (1985)	Contaminant collection on a cool filter paper by definite No.of strokes of hand powered pump using static sampling probe. $3m^3$ of gas is passed through the filter or till pressure drop across the filter doubles i.e.50% grey. Average of many samples of 0.12 - 0.15 mg.weight.	Rapid, simple and easy. Accuracy ± 7.5% suitable for measurements under field conditions.
Weighable/ Analytical Method Tipi Workshop (15) (1985)	Same as above but motorised pump and sample size at least 10 times that of grey scale method, 10-30 mg.	Further analysis of sample into Tar, Char and Ash is possible, repeatability 3%.
Enfor-172 Techwest Ent. (9,10,11,12) (1983)	Isokinetic sampling @ 4-45 litres/m, particulates separated on a filter paper at 200 C. Condensibles collected at 0 degree C in impinger train. Solvent extracted condensate is analysed.	Elaborate method. Further analysis of Tar by CHN, Chromatographic Colorimetric and TG Analysis. Accuracy not giver
UNDP/World Bank Guidelines (14) (1985)	Isokinetic sampling, particulate collection on filter cartridge @ 90-110 C. Tar condensed at 0 degree C.	Gravimetric determination. No further analysis. Suitable for field testing.
Purdue University(J.N.Da Silva's Ph.D Thesis) (13) (1984)	Static sampling, complete chilling at 0 degree C. Condensate dried in vacuum oven for 25-35 hours. Dry residue is considered as Tar.	No further analysis of condensates or particulates. Accuracy not given.

Acknowledgement

Authors are thankful to DNES,(Ministry of Energy, Govt.of India) and Indian Institute of Technology, Bombay for supporting this work.

Selected Bibliography

1. Kaupp A.and Goss J.R.,SMALL SCALE GAS PRODUCER ENGINE SYSTEMS, GATE Publication (GTZ),GmbH, 1984, p 278.

2. Blomquist U.and Bohr E.,ADAPTATION OF ENGINES TO GENER--ATOR GAS OPERATION,GENGAS Edited by T.B.Reed,Published by Tipi Workshop Books,Allenspark, Colo 80510, 1982, pp 199-229.

3. Bohr E.,PRACTICAL OPERATING EXPERIENCE,GENGAS Edited by T.B.Reed, 1982, pp 262-271.

4. Barrett J.R.,Jacko R.B.,Sumner H.R.,CORN RESIDUE FURNACE EMISSIONS, Transactions ASAE,v 26,n2,1983, pp 363-366.

5. Kutz L.J.,Barrett J.R.,Richey G.B.,Jacko R.B.,DOWNDRAFT CHANNEL GASIFIER OPERATION AND PARTICULATE EMISSIONS, Transactions ASAE,v 26,n6,1983, pp 1614-1618.

6. Silva J.N.,Peart R.M.,Ladisch M.R.,Gibson H.G.,Anderson V.,TAR FORMATION IN CORNCOB GASIFICATION, Presented at the 1984 Summer Meeting,ASAE, June 1984, p 8.

7. Gumz Wilhelm ,GAS PRODUCERS AND BLAST FURNACES-THEORY AND METHODS OF CALCULATION,John Wiley&Sons Inc, New York,Chapman&Hall Ltd. London, (1950), p 316.

8. Widell T.,THE THEORY OF GAS GENERATION,GENGAS, Edited by T.B.Reed, 1982, pp 14-60.

9. Aiken M.,McDonald E.C.,TESTING THE PROTOTYPE SAMPLER ON A DOWNDRAFT BIOMASS GASIFIER,DSS Contract 4155 KL299-1-4114,ENFOR Project C-172, Oct.1983, p 47.

10. Aiken M.,McDonald E.,Esplin G.,DEVELOPMENT OF SAMPLING AND ANALYTICAL PROCEDURES FOR BIOMASS GASIFIERS,ENFOR Project No.C-172, By Techwest Enterprises Ltd.CANMET, 1983, p 220.

11. Techwest Enterprises Ltd.,A WORKBOOK FOR BIOMASS GASIFIER SAMPLING AND ANALYSIS,DSS Contract 4155 KL229-1-4114,CANMET, Canada,ENFOR C-172, 1983, p 91.

12. Aiken M.,McDonald E.,Esplin G.,A MANUAL FOR EVALUATING BIOMASS GASIFIERS,ENFOR Project No.C-172, 1984, p 65.

13. Da Silva Jadir Nogueira ,TAR FORMATION IN CORNCOB GASIFICATION,Ph.D.Thesis,Purdue University, May 1984, pp 59-78.

14. World Bank,Energy Department,UNDP/WORLD BANK GUIDELINES FOR FIELD MONITORING OF SMALL SCALE BIOMASS GASIFIERS, Twente University of Technology, 1985, p 84.

15. Das A.,CONTAMINANT TESTING METHOD FOR BIOMASS GASIFIER ENGINE SYSTEMS, Tipi Workshop,Colorado,80510,1985, p 30.

16. Reed T.B.,Das A.,HANDBOOK ON TESTING OPERATION AND MANUFACTURE OF SMALL SCALE GASIFIER SYSTEMS,SERI Report 1984, p 25.

17. Goldstein Irving S.,ORGANIC CHEMICALS FROM BIOMASS, CRC Press,Boca Raton,Florida, 1983, p 310.

CATALYSTS FOR GASIFICATION OF BIOMASS

L. K. Mudge, E. G. Baker, M. D. Brown and W. A. Wilcox
Pacific Northwest Laboratory, Richland, Washington

The U.S. Department of Energy, Thermochemical Conversion Program, sponsored studies at the Pacific Northwest Laboratory under contract DE-AC06-76RLO 1830 on development of catalysts for conversion of biomass to high-valued gas products. Bench scale studies using dual fluid bed reactors in series developed catalysts with a long active lifetime. Conditions for catalyst testing were based on results of previous studies in a fluid bed process research unit.

Supported metal catalysts were developed that remained active indefinitely at 750°C for destroying tars in wood-derived gases. Active catalysts included NiCuMo on silica-alumina and Ni on a α-alumina. These catalysts remained active when contacting wood-derived gases in a secondary fluid bed reactor.

Potential applications for catalysts in biomass gasification include their use to improve synthesis gas yields and to simplify wastewater cleanup problems in commercial gasification processes. The catalysts are effective for reducing the organic carbon content in condensate streams.

SUMMARY

Nickle based catalysts that were found to be active for conversion of wood to synthesis gas in previous studies were evaluated on a bench scale. These catalysts remained active indefinitely in laboratory studies but lost activity rapidly when evaluated in a process research unit (PRU). Bench scale equipment was designed and installed to allow resolution of differences in laboratory and PRU results.

Three different process configurations for testing secondary catalysts were used in the study: 1) catalyst as the fluid-bed material in the gasifier (primary bed), 2) catalyst in a separate fixed-bed reactor where it contacted products from the gasifier, and 3) a configuration similar to (2) where the separate reactor was operated as a fluid bed. In the separate fluid-bed reactor NiCuMo on silica-alumina and Ni on α -alumina remained active during tests lasting 30-40 hours for generation of synthesis gas from wood at 750°C. No sign of catalyst deactivation was noted at the end of these tests. Based on similar previous results obtained in the laboratory where a NiCuMo catalyst was tested for over 1200 hours and remained active, we believe that the activity of these catalysts can be maintained indefinitely in configuration 3. Deactivation of these same catalysts occurred within 10 hours when used in configuration 1 similar to results obtained previously in the PRU. Using configuration 2 these catalysts were deactivated after about 20 hours. Yields of condensable hydrocarbons (tars) are shown for the three configurations in the following figure.

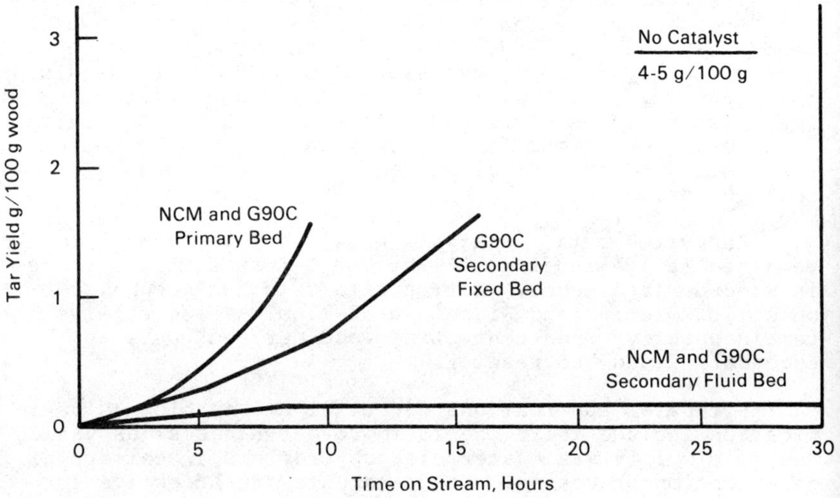

Carbon deposition on the catalyst was identified as the major cause of deactivation. Whereas the carbon content on the catalyst surface leveled off at 2 to 5 weight percent after about 10 hours in configuration 3, it continued to increase with exposure in the other two configurations. Apparently the rate of carbon forming reactions continues to exceed the rate of carbon removing reactions in configurations 1 and 2, and an equilibrium in rates is achieved in configuration 3. Differences in heat transfer to the catalyst pellets in the fixed-bed and fluid bed catalytic reactors are postulated to influence the net rate of carbon deposition on the catalysts surface.

CATALYSIS OF GASIFICATION OF WOOD CHARS IN THE PRESENCE OF CARBON DIOXIDE

William F. DeGroot, Research Specialist
G.N. Richards, Director
Wood Chemistry Laboratory, Department of Chemistry
University of Montana, Missoula, Montana 59812

ABSTRACT

Wood or wood chars provide a low-nitrogen, low-sulfur, and low-ash carbonaceous substrate for the production of gaseous fuels. In the direct gasification of wood the reaction of the char represents the slow, rate-determining step. Effective catalysis is required to achieve complete gasification of the char, which contains 30-50% of the heat content of the original fuel, in a reasonable time. Wood chars can also be considered as a gasification substrate, since they can be prepared in a separate pyrolysis step with recovery of valuable pyrolysis products such as methanol, acetic acid and levoglucosan (1,6-anhydro-β-D-glucopyranose, which can be hydrolyzed and fermented to give ethanol). The yield and reactivity of chars can be controlled by varying the pyrolysis conditions and by addition of inorganic species to the wood prior to pyrolysis. Wood chars have been prepared from untreated wood of a well-characterized hardwood species (*Populus trichocarpa*), and from the same wood treated with low concentrations of selected catalytic species by ion exchange. Ion exchange in this species was found to involve formation of salts of uronic acids in the hemicellulose fraction. Calcium in the untreated wood was also found to be bound in this way. Gasification of chars derived from untreated wood indicates that the natural inorganic species (ash) have significant catalytic activity. This is attributed to natural calcium content. In general the reactivities of wood chars were found to be less dependent on time and temperature of pyrolysis than are coal chars. The following order of effectiveness of ion-exchanged catalytic species has been found: Co > Ca > Ni > Mg, K, Na > Cu. Calcium and cobalt were found to be exceptionally effective catalysts of wood char gasification. Cobalt was especially effective in chars formed at temperatures lower than 800°C. Chars prepared from cobalt-treated wood could be gasified by CO_2 at temperatures as low as 500°C.

Elements of the alkali metal, alkaline earth and transition metal groups are known to catalyze gasification of carbon by steam, carbon dioxide and air. In this study we have measured reactivities of chars prepared from untreated wood and wood teated by ion exchange with selected catalysts. Carbon dioxide (1 atm) was used as the reactant gas. Reactivities of chars were measured under well-controlled conditions using a reactor/detector system developed as part of this study. The rate of combustible gas production is determined during gasification by measuring the depletion of oxygen in a combustion air stream in the detector. A typical gasification rate profile for a char prepared from untreated cottonwood at a heat treatment temperature (HTT) of 800°C is shown in Figure 1. The detector output can be integrated over the entire run to determine the total extent of gasification or above the baseline indicated by the dotted line in Figure 1 to determine the extent of gasification due to reaction with carbon dioxide.

Samples were prepared from the sapwood and heartwood of cottonwood (*Populus trichocarpa*), a representative hardwood species. Chars were also prepared from a lignite sample to allow comparison of the reactivities of wood and coal. Treated wood samples were prepared by acid-washing the sapwood sample to remove the inorganic constituents and then adding cations of selected catalytic species to the wood by ion exchange. The composition of the inorganic fraction of all samples was determined by inductively coupled plasma emission spectroscopy (ICP) in order to relate the reactivity of chars prepared from wood to the type and quantity of catalytic species in the wood. The composition of the inorganic fraction of untreated wood and lignite samples is shown in Table 1. The ash content of lignite is much higher than that of wood. The ash content of wood is primarily composed of calcium, potassium and magnesium. All of these elements are catalysts of gasification reactions. Acid washing (0.01 M HCl) effectively removed the inorganic constituents, and ion exchange treatments consistently resulted in catalyst loadings of 8-10 meq/100 g.

TABLE 1

Composition of inorganic fraction of cottonwood and lignite samples.

SAMPLE	% ASH	COMPOSITION OF ASH (%, dry weight of wood) K	Ca	Mg	Fe	Al
COTTONWOOD						
Untreated						
Heartwood	1.76	0.32	0.50	0.08	--	--
Sapwood	0.51	0.10	0.15	0.02	--	--
LIGNITE[a]	11.88	0.04	1.81	0.64	0.39	1.08

[a]Sample and analysis provided by Penn State Coal Data Base.

Table 2 shows the extents of gasification of chars (HTT 800°C) prepared from untreated cottonwood, representative ion-exchanged cottonwood samples, and lignite. Cottonwood heartwood was one of the most reactive samples, apparently because of its high ash content. Lignite char was intermediate in reactivity between the untreated cottonwood and sapwood chars. The catalytic activity of the indigenous inorganic species in wood is evident from the lack of reactivity of the char prepared from acid-washed wood. Addition of calcium and cobalt to the acid-washed wood resulted in chars of higher reactivity than those prepared from untreated wood. Potassium and magnesium were less effective catalysts.

TABLE 2

Char yields and extents of conversion due to pyrolysis (P) and gasification (G) of chars (HTT 800°C) gasified for 30 min at 800°C in 1 atm of CO_2.

SAMPLE	CHAR YIELD (%, d.a.f.)	PERCENT CONVERSION P	G	Total
Cottonwood Heartwood	15.0	(100% in <20 min)		
Cottonwood Sapwood				
Untreated	11.9	2	32	34
Acid-washed	7.2	2	<1	2
K-exchanged	13.4	2	13	15
Ca-exchanged	9.3	(100% in 30 min)		
Mg-exchanged	8.8	3	15	18
Co-exchanged	9.2	9	73	82
Lignite[a]	55.4	4	66	70

[a]Sample and analysis provided by Penn State Coal Data Base.

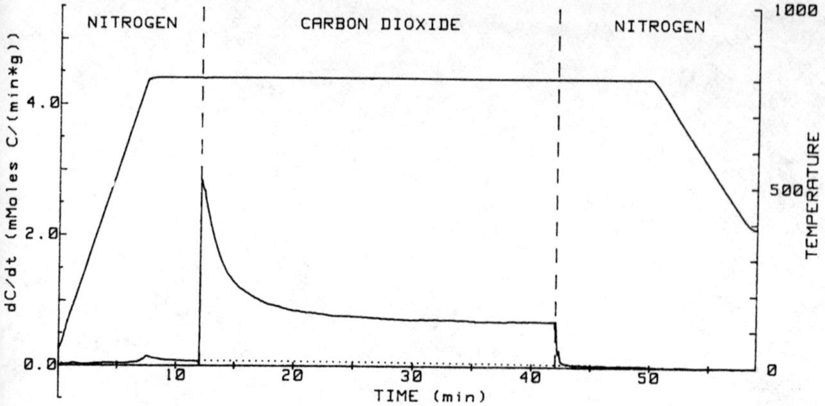

Figure 1. Gasification rate profile for char (HTT 800°C) prepared from untreated cottonwood.

THERMOCHEMICAL CONVERSION OF HIGH-MOISTURE BIOMASS FEEDSTOCKS TO METHANE-RICH GAS PRODUCTS

R. S. Butner, B.S., Research Engineer
L. J. Sealock, Jr., B.S., Project Manager
D. C. Elliott, M.B.A., Senior Research Scientist
Pacific Northwest Laboratory,* Richland, Washington

ABSTRACT

A low temperature biomass gasification concept for the gasification of high-moisture biomass feedstocks (>50% moisture) is under development at Pacific Northwest Laboratory (PNL). A one-liter batch reactor is being employed to determine the effectiveness of the concept, which uses a combination of pressure and catalysts to gasify biomass and produce a product gas which has a raw methane content (on a dry gas basis) of up to 53%. Commercially manufactured, alumina-supported nickel hydrogenation catalysts are used with and without alkali carbonate co-catalysts at reaction conditions of 400-450°C and 4000-5000 psig. Carbon conversion to gas is typically 80-90% after 15 minutes at temperature.

The primary goal of the project is to demonstrate the technical viability of a process for the utilization of high moisture biomass. The following high-moisture feedstocks are being employed in the reactor: agricultural residues (sorghum and corn stover, sunflower stalks and stems), aquatic plants (water hyacinths and macrocystis kelp), food processing by-products (spent brewer's grain, potato steam peeling sludge) and other miscellaneous materials (napier grass and unconverted effluent from an anaerobic digestor). These feed-stocks represent a wide range of moisture contents and elemental compositions.

A second goal of the project is to demonstrate the advantages of a low temperature pressurized reactor for the production of methane from biomass. To this end, several experiments have been performed which demonstrate a strong catalyst influence on methane yield. The best results to date show a yield of 8.4 scf methane/dry, ash-free pound of biomass (using spent brewer's grain) after 15 minutes at 400°C. Further experiments are being conducted to determine whether similar yields can be expected from other feedstocks, and whether these yields can be further improved by the use of different catalysts.

The project is funded by the United States Department of Energy, Division of Biofuels and Municipal Waste Technology through the Biomass Thermochemical Conversion Program, which is operated by PNL under contract to DOE.

---

*PNL is operated by Battelle Memorial Institute for the U.S. Department of Energy.

EFFECTS OF CATALYST ON GASIFICATION OF HIGH MOISTURE BIOMASS

Effect of Nickel Loading

The amount of nickel present in the reactor has a strong influence on both gas quality and methane yield. The nickel catalyst allows the operation of the biomass gasification reactor at the low temperatures required for high methane content in the raw gas.

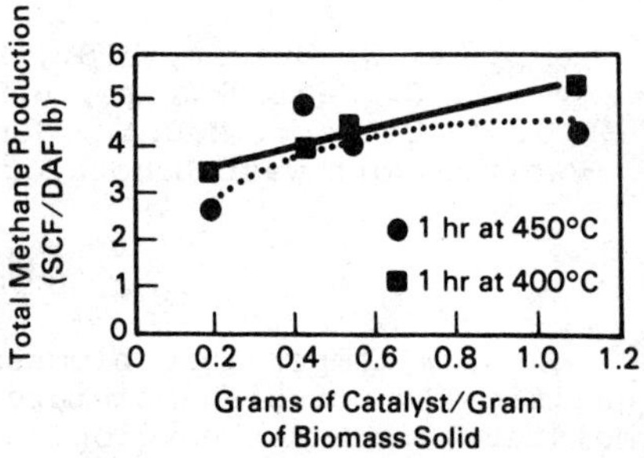

FIG 1 Effect of Ni on CH_4 Yield

Effect of Na_2CO_3 Loading

Various basic salts, including Na_2CO_3, have been used to help promote basic gasification reactions. While increasing gasification rates, these materials may have an adverse impact on methane production by simultaneously promoting competitive reactions such as the water-gas shift.

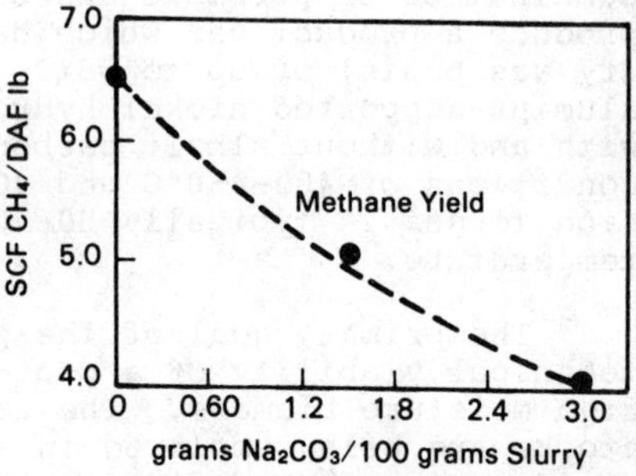

FIG 2 Effect of Alkali on CH_4 Yield

Implications

Work to date has shown a dramatic increase in the maximum methane yields obtained. By carefully optimizing the reactor temperature and catalyst concentrations, methane yields have been obtained which rival or surpass those obtained via anaerobic digestion or other thermochemical processes. Further optimization of these parameters is planned.

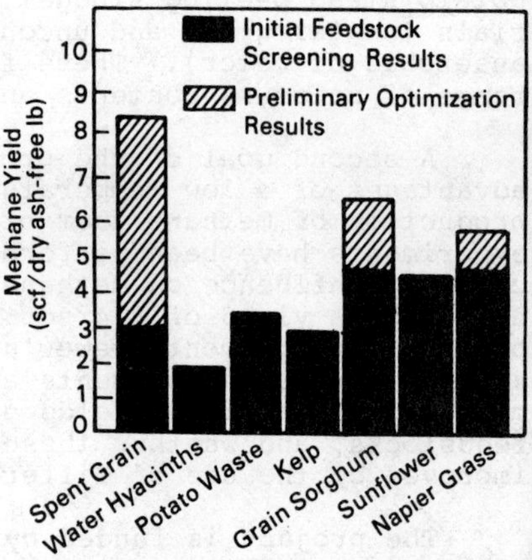

FIG 3 Preliminary Yield Data

Future Work. Continuation of this research is expected to include additional catalyst studies and the development of a continuous reactor system.

ATMOSPHERIC FLUID-BED BIOMASS GASIFICATION

André Chamberland, Ph.D.
Institut de recherche d'Hydro-Québec (IREQ)
Varennes, Québec, Canada JOL 2PO

ABSTRACT

Under a research and development project at IREQ on biomass energy and a demonstration project at Biosyn related to the production of methanol from gasified wood, a fludized-bed gasifier and a cold model were built. In addition, a mathematical model and a factorial design for gasifier performance evaluation were also developed.

The gasifier, which has a 50-kg/h capacity, consists of a steel cylinder with a castable-refractory lining and soft insulation. The inside diameter at the level of the grid is 30 cm and the total height of the reactor is 400 cm. This reactor has already been used to gasify various species of wood, including tropical wood from French Guiana.

Description of the Gasifier

Figure 1 presents a schematic diagram of the 50-kg/h gasifier. The reactor consists of steel cylinders (2) 400 cm high, 80 to 60 cm in inside diameter (1), which have a 6-cm castable-refractory lining (3) and 9 cm of soft insulation (4). The inside diameter is therefore 50 cm at the top of the reactor, in the disengagement zone, and 30 cm at the level of the grid. At the reactor outlet, the gases pass through two disengaging cyclones (8) and (9) before being evacuated through the vent and flared. The particles trapped in the first cyclone (8) are returned to the sand bed of the reactor. During fluidization, the sand bed reaches between 75 cm and 100 cm in height. The wood, granulated or in the form of sawdust, is fed into the reactor continuously by the injector screw passing through one of the side ports (5). The air used for gasification and fluidization is supplied through a pipe (6) under the grid formed by the various components and interconnections.

The operating temperature and gasification conditions are fixed by adjusting the airflow above the minimum fluidization velocity (minimum air velocity) and in terms of the wood feed rate, which is around 1.7 kg of air per kilogram of bone-dry wood.

Measuring Instruments

The reactor is equipped with a number of thermocouples, flowmeters and electronic pressure transducers for process evaluation. All the valves and variable-speed motors can be remote-controlled from a central computer. The product-gas pipe has a number of ports for immediate or later analysis. The main gas components are analyzed on line by spectroscopy whereas the others are analyzed by chromatography.

Major R&D Activities in the Field of Biomass

. Air-blown wood gasification	- André Chamberland Ph.D. Research Scientist IREQ
. Oxygen-blown wood gasification	- Raynald Labrecque M.Sc. Research Scientist IREQ
. Mechanics of fluidization as applied to wood (doctoral thesis)	- Michel Meunier in collaboration with Prof. Maurice Bergougnou University of Western Ontario
. Development of a mathematical model of wood gasification using pressurized oxygen (master's thesis)	- Pierre Marchand under the supervision of Prof. Normand Therien University of Sherbrooke
. Process automation and measuring instrumentation	- John Black BBC Engineering Part-time Research Scientist IREQ
. Statistical analysis of the results and factorial design for gasifier performance tests	- Raynald Labrecque M.Sc. Research Scientist IREQ
. Gasification of homogeneous waste	- Hugues Hamelin Research Scientist IREQ
. Process analysis and control	- Claude Benoît and Denis Ricard Senior Technicians IREQ

Main Results

Air-blown gasification of wood produces a gas whose chemical composition varies according to the temperature, as indicated on table 1.

Cold-model studies were performed to establish the relationship between the product gas and the parameters of fluid mechanics. The design of the reactor and the distributor is among the most important of these parameters as far as process improvement is concerned.

A program has been developed for processing the gasification test results in terms of using them as input for a mathematical regression model based on a statistical approach.

Table 1

Temp. (°C)	Composition								Calculated amount of H_2O (%)	Calorific value (kJ/Nm^3)
	H_2	CO	CO_2	CH_4	C_2H_4	C_2H_6	O_2	N_2		
771	8.6	15.4	14.3	6.3					11	7100
824	7.2	13.9	16.4	4.5	1.92	0.45	0.9	51.6	13	5980

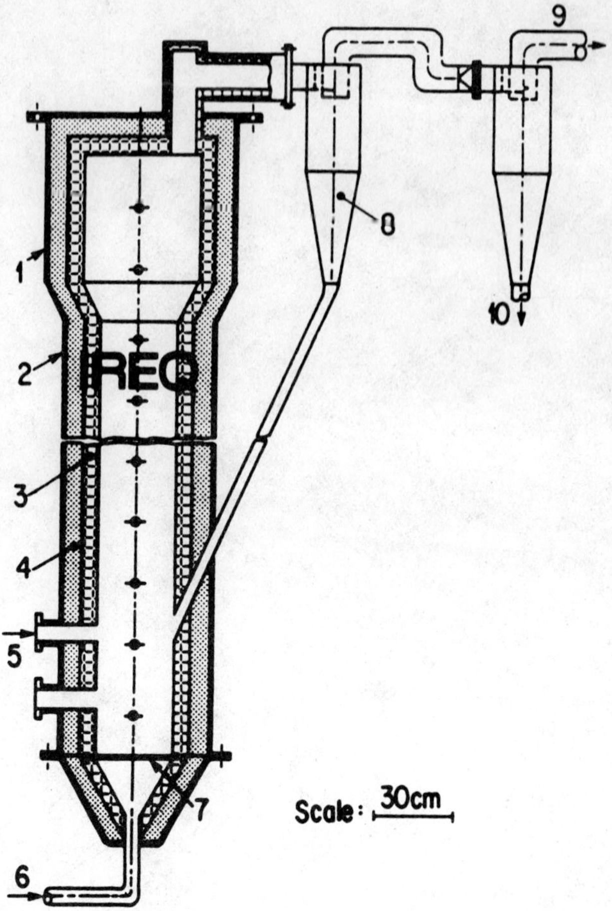

FLUID-BED GASIFICATION OF WHOLE RICE HULLS

VIRGIL J. FLANIGAN, Professor, Mechanical Engineering
W. ERIC HUANG, PhD. Candidate, Chemical Engineering
OLIVER C. SITTON, Professor, Chemical Engineering
University of Missouri-Rolla
Rolla, Missouri

XU BING YAN, Visiting Professor
Guangzhou Institute of Energy Conversion, China

ABSTRACT

The use of rice hulls for the production of energy has been drawing more attention, especially in the rice growing areas of the world, lately. The principal problems, in the existing systems, are the low carbon conversion and the high operating temperature, which leaves high carbon content in the char. This is a result of high ash content in rice hulls (20% by weight) and high silica content (95%) with small amounts of alkali in ash. Classical processes usually resulted in sticky and conglomerating ash residue, which brought about a shut-down of the entire conversion process.

A previous study, at the University of Missouri-Rolla facility, has provided insight into ground rice hull gasification. The initial work was performed in a six inch fluidized bed reactor constructed at a minimum cost. The study indicated that the gasification temperature needed to be carefully controlled at less than 1800°F to prevent forming a glass-like barrier. This barrier inhibits further reaction of the remaining biomass material to usable energy. The system capability, of excellent mixing and control of the temperature, permitted good operation and resulted in excellent carbon conversion of the feed material.

The extension of this work, using the whole hull, presented several initial problems. A very high sand to rice hull ratio was required to achieve normal fluidization of a bed containing the hulls. The reason for this difficulty was the natural cylindrical shape of the rice hull. It was discovered that when the hulls fed on the top of the bed they would settle into the fluidized sand bed and the hot sand would pyrolyze the sinking hull, resulting in the release of the volatile matter while producing rice hull char. Char was much easier to fluidize and the sand to rice hull ratio was reduced from 20 to 5 with normal bed fluidization.

The gasification performance tests were conducted in the 6" fluidized bed, the reactor was 12 ft. high with a 10" disengager at the top of the reactor. Fig.1 shows the dependence of the gas quality and carbon conversion on bed temperature. Theoretically, increasing bed temperature would increase the product gas heating value but in the data presented, gasification temperatures above 1300°F resulted in a decrease in heating value and thermal efficiency. 1300°F, under present conditions, appeared to be the upper limit for the gasification process without lowering the heating value. It required good control of the air to fuel ratio to control the temperature. Fig. 2 clearly demonstrates the required air to fuel ratio to maintain the proper bed temperature and produce the best gas quality. A superficial velocity of air in ranges of 2.2 to 3.2 ft/sec. provides a good fluidization rate and results in an unit easy to operate. Considerable data has been collected in the top feed unit running under steady state conditions. The data provides the experience necessary for operating a small scale fluidized bed gasifier and for any desired scale-up.

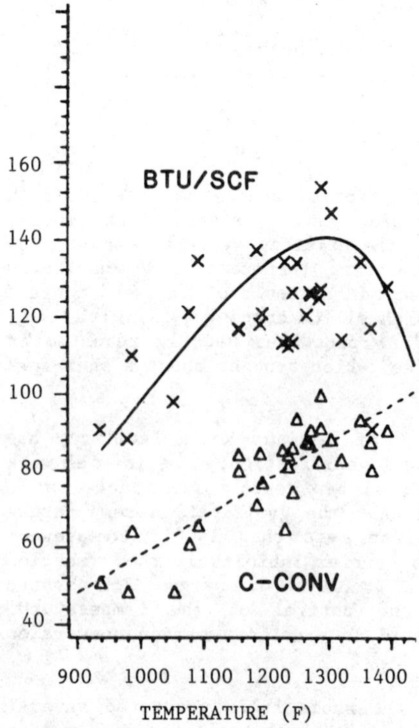

FIG. 1. BTU OF GAS & CARBON CONVERSION VS TEMPERATURE

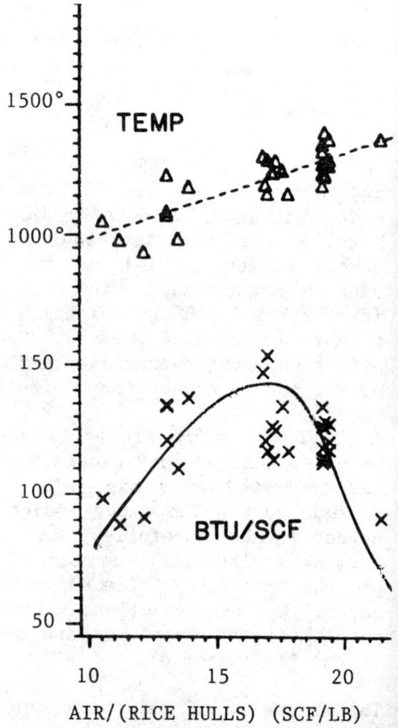

FIG. 2. TEMPERATURE & Btu VS AIR TO RICE HULL RATIO

In conclusion, feeding from the top of the fluidized sand bed is a good solution for the gasification of whole rice hulls. During the gasification operation, it is recommended that the air to rice hull ratio should be approximately 17 and the superficial velocity of air be between 2.2-3.2 ft/sec. Under the above conditions, the heating value of the product gas reached 140 Btu/Scf, the carbon conversion rate ran up to 85 to 90%, heat efficiency ranged from 50% to 60% and the productivity rate was 14 X 10^5 Btu/hr.ft^2.

OPERATING PARAMETERS FOR A SMALL, STRATIFIED BED, GASIFIER-ENGINE SYSTEM

K. M. Eoff, Ph.D., Assistant Professor of Physics and Adjunct Assistant Professor of Forestry, University of Florida, Gainesville, Florida

ABSTRACT

Typical down draft gasifiers such as were widely used during World War II to power engines required an air tight lid for the fuel hopper. Feedworks which would introduce fuel across this air tight barrier and into the corrosive and vapor rich atmosphere of the fuel hopper were difficult and expensive to make and maintain.

Recently, open top gasifier designs have been developed for which feedworks can be very much simplier. However, high temperature alloys are required in their construction. It would be an advantage, particularly for construction and use in developing nations, if the design could be modified so that mild steel could be used.

Such a design is being tested at the University of Florida. The poster presentation will include results obtained from these tests such as:

1. Gasifier dimensions suitable for a particular power requirement

2. Range of power output for given dimensions

3. Fuel size requirements and other pertinent information.

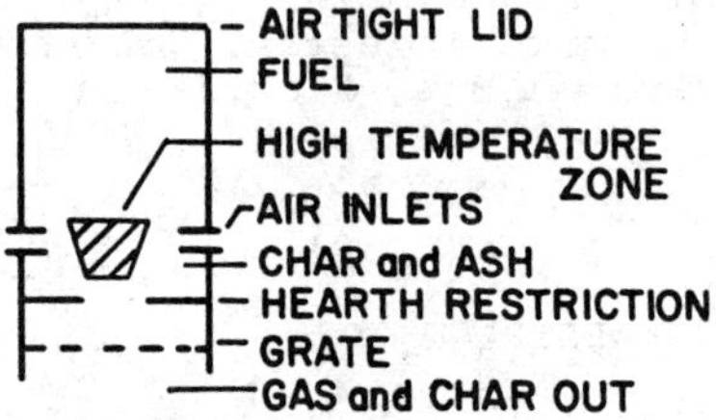

FIG. 1

SCHEMATIC of DOWN DRAFT GASIFIER

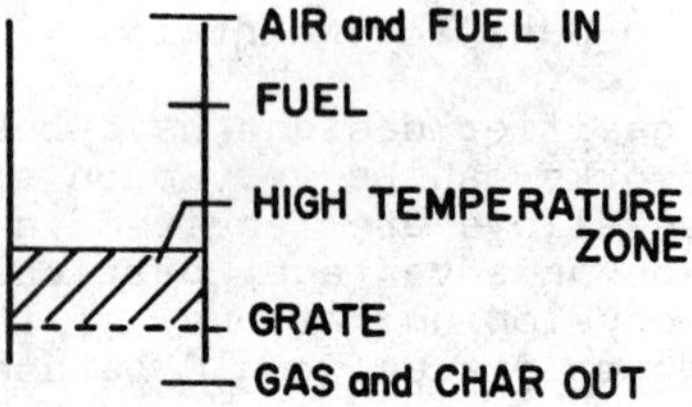

FIG. 2

SCHEMATIC of OPEN TOP GASIFIER

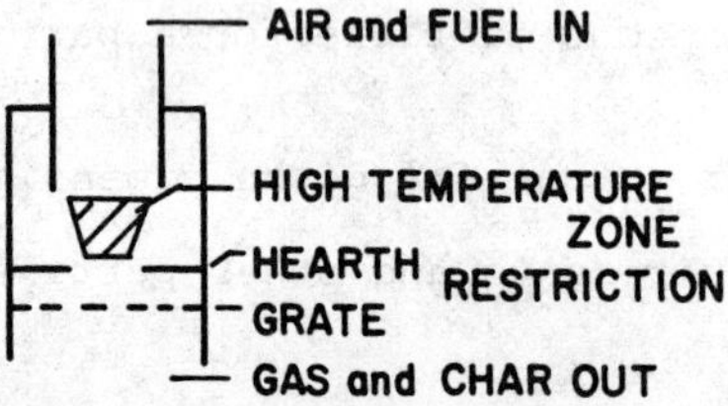

FIG. 3

MODIFIED OPEN TOP GASIFIER

Comparison of the figures will elucidate the rationale for the modified, open top, gasifier design which is being tested at the University of Florida. In the gasifier of figure two the high temperature zone is in contact with the metal wall and with the grate requiring use of high temperature alloys. Notice that in the other two designs the high temperature zone is isolated from the metal parts by a zone of insulating ash and char so that mild steel can be used. Slow erosion of the hearth restriction and the bottom of the fuel entrance pipe will occur but easily replaceable inserts can be utilized in these areas.

A REACTION MODEL DESCRIBING EARLY EVENTS IN ANAEROBIC DISSIMILATION OF NAPIER GRASS

Masa Goto, Ph.D., Postdoctoral Associate;
Ann C. Wilkie, Ph.D., Postdoctoral Associate;
Francis M. Bordeaux, MSE., Associate in Microbiology;
Paul H. Smith, Ph.D., Professor;
Univ. of Florida, Dept. of MCS, Gainesville, FL
John F. Andrews, Ph.D., PE., Professor,
Rice University, Dept. of ESE, Houston, TX

ABSTRACT

Napiergrass was anaerobically digested in leaching bed (LB) reactors connected in series with a packed bed (PB) reactor. Volatile solids (VS) destruction was approximately equal to the soluble cell contents in the feedstock (18 % of total VS).

A mathematical model for describing early events of anaerobic dissimilation of Napiergrass is proposed. The model, which includes gas phase dynamics and internal generation of cations, was evaluated by comparing computer simulation results with experimental data. There was good agreement between model predictions and the experimental results.

EXPERIMENTS AND RESULTS

Effluent from a 600 L PB was used as the inoculum for the 300 L LB's, which were operated at 5 % VS loading (w/v). The LB's were operated either as batch or continuous systems with respect to the liquid phase. Increases in the soluble COD content (CODs) of the liquid phase were recorded *ab initio*. Within 12 hours the CODs had increased to 70 % of its ultimate concentration for the 7 day experimental period. VFA's comprised 0 and 80 % of CODs content at days 0 and 7, respectively. Apparently a substantial proportion of the CODs in the early stages exists as intermediates which are subsequently converted to VFA's. Analysis of biomass after the 7 day experimental period showed that 18.8 and 26.0 % of the VS content was removed, during batch and recycle modes, respectively. Methane production was observed in both modes of operation. However, the cumulative methane produced by the LB operated in batch mode was substantially lower than that from counterparts operated in continual recycle with the PB.

MATHEMATICAL MODEL

The proposed mathematical model was primarily intended to describe the early events which take place in the first stage reactor (LB), of a multi-phase anaerobic biomass to methane conversion system currently under investigation in a GRI/IFAS project at the University of Florida.

The model assumes the dissimilation reaction pathways are affected by product (total VFA) inhibition of conversion of soluble intermediates (SI) to VFA, substrate (un-ionized VFA) inhibition of methane production (Fig. 1), and an internal generation of net cations which are defined as cations in equal molar concentration to the sum of molar concentrations of bicarbonate and ionized fatty acids. The latter assumption was found to be particularly important in achieving more realistic simulation of the process behavior and to express the effects of feedstock on buffering capacity. Although the identities of the cations and the mechanism of their generation are as yet unknown, a comparison of the net cation concentration curve with total soluble and VFA COD curves implies that the generation of net cations occurs as SI are converted to VFA. The proposed model was developed by including this hypothesis.

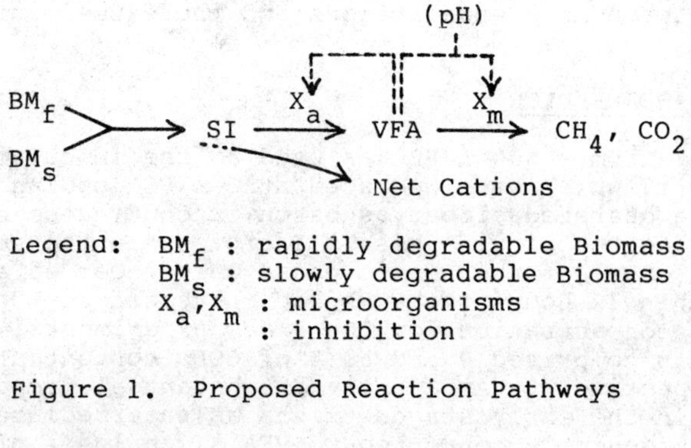

Legend: BM_f : rapidly degradable Biomass
BM_s : slowly degradable Biomass
X_a, X_m : microorganisms
: inhibition

Figure 1. Proposed Reaction Pathways

ACKNOWLEDGEMENT

This paper reports results from a project that contributes to a cooperative program between the Institute of Food and Agricultural Sciences of the University of Florida and the Gas Research Institute, entitled "Methane from Biomass and Waste".

The authors wish to extend special thanks to Mr. Shinji Ide of Rice University for his contribution in development of the mathematical model investigated in this study.

CONVERSION OF BIOMASS AND WASTES IN THE SOLCON DIGESTER

Richard Biljetina, MBA, Director
Vipul J. Srivastava, MS ChE, Manager
Institute of Gas Technology, Chicago, Illinois
David P. Chynoweth, Ph.D., Associate Professor
University of Florida, Gainesville, Florida
Thomas D. Hayes, Ph.D., Project Manager
Gas Research Institute, Chicago, Illinois

ABSTRACT

A novel concept has been developed and scaled-up by the Institute of Gas Technology (IGT) to demonstrate the conversion and stabilization of biomass, sludge and community wastes in a solids concentrating (SOLCON*) digester. This digester is designed and operated to obtain long solids and microorganism retention times. Results from two years of operation on water hyacinth/sludge blends in a 1200 gallon digester located at the RCUC wastewater treatment facility, Lake Buena Vista, Florida have confirmed earlier laboratory observations that this reactor achieves higher methane yields and solids conversion than those observed in continuous stirred tank reactors (CSTR's). Methane yields of 7.9 SCF/lb volatile solids added have been obtained during steady-state operation on a feed of water hyacinth and sludge (1 to 1 solids weight ratio) during which hydraulic retention time was controlled at 11 days.

This work is part of a larger effort sponsored by the Gas Research Institute and is designed to evaluate the use of water hyacinth and other aquatic plants for wastewater treatment and subsequent conversion of the biomass and sludge to methane. In addition, future development efforts include processing municipal, industrial and agricultural waste feedstocks in the SOLCON digester to demonstrate an integrated concept for community waste management. Commercialization of this technology is expected to provide communities with effective wastewater treatment, economic municipal and industrial waste management, and lower energy prices.

* IGT has filed a patent application on the SOLCON digester

SOLCON DIGESTER PERFORMANCE Using Water Hyacinth/Sludge Blend*

Loading Rate, lb VS/ft^3-d	0.2
Feed Solids Concentration, wt%	3.8
Detention Time	
Liquid, days	11
Solid, days	30
Total Gas Yield, SCF/lb VS added	12.4
Methane Yield, SCF/lb VS added	7.9
Methane Concentration, vol. %	64
Carbon Conversion, wt %	62

* 1:1 Solid Weight Blend

DIGESTER PROFILE FOR WATER HYACINTH/SLUDGE BLEND

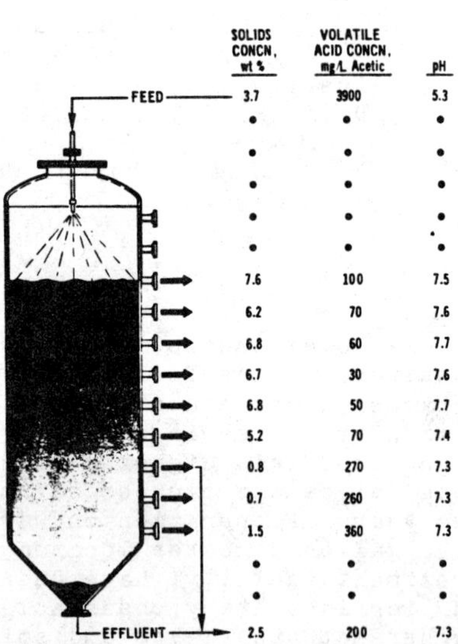

DIGESTER PRODUCTION DATA (1:1 Water Hyacinth/Sludge Blend)

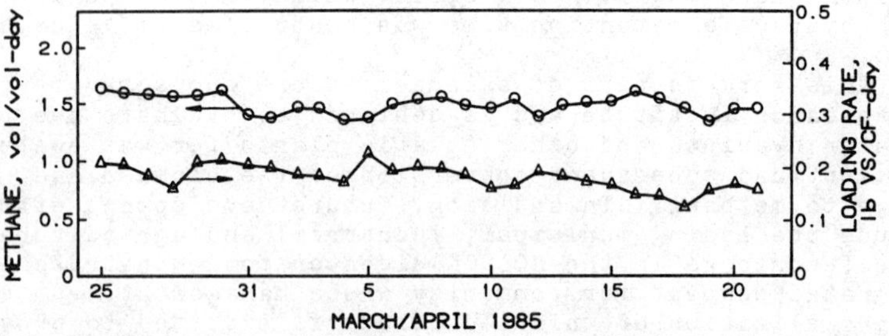

PERFORMANCE COMPARISON

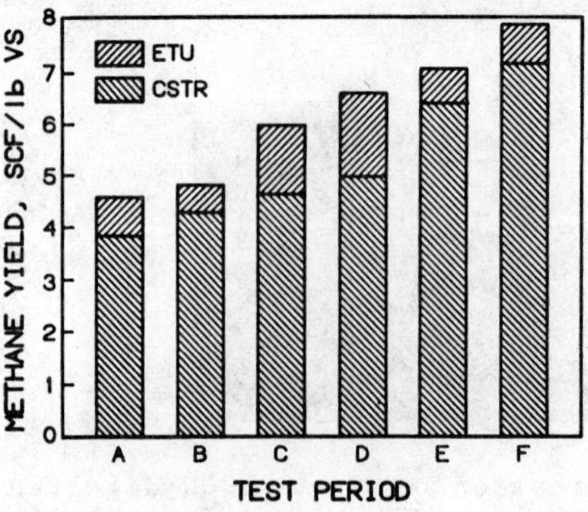

ETU SOLCON DIGESTER OUTPERFORMS CONVENTIONAL CSTR

BENEFITS

- Smaller Digesters
- Higher Conversion
- Improved Stabilization
- Higher Gas Production
- Better Stability
- Higher Net-Energy Recovery

ANAEROBIC DIGESTION TECHNOLOGY : A PROCESS FOR POLLUTION CONTROL OR FOR ENERGY PRODUCTION?

Claude Camilleri, Technical Manager
Michel Arnoux, Process Engineer

SGN,
Société Générale pour les Techniques Nouvelles
France

ABSTRACT

The recent improvements brought into the field of waste water control by the new anaerobic digestion technologies considerably changed the economic aspect of pollution abatement of waste waters. These improvements, in certain cases, make it possible to envisage an investment cost return within a reasonable period of time. This was never the case with the conventional aerobic waste water treatment processes.

The table following this summary gives a partial list of SGN anaerobic treatment plants, all full-scale industrial units built between 1983 and now. Water treatment was the first aim for some of then, mainly in sugar plants or fruit and vegetable processing factories. In these cases energy production is only a bonus, helping anaerobic treatment to be compared favourably with aerobic schemes.

However, other SGN plants have been completed mainly or even only for the purpose of energy production. The SGN anaerobic plant built in 1984 in Cognac (France) replaced the existing concentration/evaporation facilities and produces during the present distillation campaign a daily energy equivalent of 7 to 9 t.o.e. Another full-scale plant is under start-up in Guadeloupe (Franch West Indies) for anaerobic treatment fo sugar cane molasses stillage in a 26,000 lts/day rum distillery : in that case, the scheduled biogas production will reach around 1.8 kg steam per liter of alcohol.

Is it therefore possible to look for anaerobic digestion as an energy production process which would be applicable to substrates other than agro-food effluents. The industrial experience gained by SGN in operating several units in different fields particularly in the food industry enables us to answer this question according to each application.

(1) Also President of the French Trade Association of Waste Waters Treatment Industries.

90 % BOD REDUCTION AND MOREOVER EXTRA-ENERGY BONUS

SGN anaerobic technology has been developed since 1979 to offer an alternate way to conventional energy-consuming aerobic processes. Many industrial waste water treatment units were commissioned because of pollution control needs, biogas economical value being a bonus compared to aerobic treatment :

- Beet sugar plants :
  Two units in Northern France (Beghin Say 1983 ; Générale Sucrière 1985) produce biogas during the two-months sugar campaign and reduce COD outlet by 80 to 95 %.
- Jam Factory :
  A complete anaerobic/aerobic unit in Central France will reach a total 98 % efficiency on BOD5.

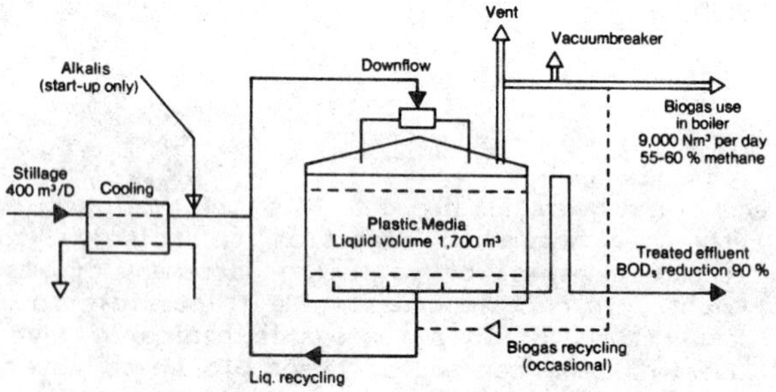

DIAGRAM OF THE SGN FIXED FILM PROCESS

COGNAC WINE STILLAGE PRODUCES 9 T.O.E. PER DAY

SGN anaerobic process has been applied since November 84 in Cognac (France) for treatment of wine and lees stillage in a two-reactor unit :

- An average daily production of 15,000 - 18,000 Nm³ biogas has been achieved during the 1985-86 distillation campaign on the fixed-film reactor (mean feed 25 to 30 tonnes COD per day).
- The completely-mixed reactor designed for treating lees stillage and working since early 1985 at its design capacity (8 to 10 tonnes COD per day) has produced about one third of this value.

Biogas is used :
- in a steam boiler used for wine distillation
- in thermal engines generating electricity and hot water for effluent heating.

A similar unit is under construction for anaerobic treatment of stillage from Armagnac in Condom (France).

BIOGAS PRODUCTION SAVES UP TO 2 KG STEAM PER ALCOHOL LITER IN A TYPICAL MOLASSES DISTILLERY

A full-scale anaerobic digester has been under start-up since April 1986 in Guadeloupe (FRENCH WEST INDIES) in a 28,000 lts per day rum distillery using SGN fixed-film process.

Biogas production (about 4.5 tonnes of oil equivalent per day) will replace fuel in the existing distillery boiler and produce around 1.8 to 2.0 kg of steam per liter pure alcohol covering a large part of the present energy needs. In case of additional improving of distillation technology, near zero-energy input distilleries can be designed.

IS ANAEROBIC DIGESTION TECHNOLOGY FOR ENERGY RECOVERY ONLY ?

Interest in energy recovery from liquid wastes is increasing with the improvement of anaerobic technology, mostly in industries working all the year like wheat starch manufacturing and chemical industries.

Developing countries willing to decrease their dependence toward foreign energy resources favor such technology : SGN fixed-film process is now licenced in INDIA for treatment of sugar cane molasses stillage and commissioning of the first two plants is scheduled in 1987.

TABLE 2 : KEY REFERENCE DATA FOR WINE DISTILLERIES

- BEGHIN SAY COMPANY - Thumeries - France -

- Capacity : 5,000 to 8,000 t of beets per day

- Effluents :
  . Waste water from beet washing
  . Total COD : 16 t/d
  . Flowrate : 125 m^3/h.

- Campaign : 90 days/year.

  Before 1983 the waste water treatment included a 30,000 m^3 lagoon with an aeration basin equipped with six 15 kW aerators and a finishing lagoon.
  The new treatment plant treats the effluents from the lagoon (aeration was definitively discontinued) and is composed of an heat exchanger, a packed bed 1,260 m^3 digester, a flare and a biogas boiler.

- Commissioning : October 1983.

- Efficiency :
  . COD reduction 90 %
  . methane capacity : 5,000 Nm^3/day
  . actual methane production : 220,000 Nm^3/year used for steam generation

- Net energy savings : 400 Toe/year including saving from the shut down of aerators.

- Capital investment : 6,000,000 FF 1983.

- GENERALE SUCRIERE - Aulnois sous Laon - France -

- Capacity : 7,000 t/day of beets

- Effluents :
  . Beet washing, herb pressings and condensates
  . Flowrate : 135 m^3/h - 3,240 m^3/d
  . Soluble COD : over 28 tons/d

- Campaign : 90 days/year

- The unit includes an heat exchanger, a packed bed digester (2,000 m^3 liquid volume) and a flare. Biogas is burned in the existing plant boilers.

- Commissioning : October 1985.

- Load : 35 to 40 tonnes per day treated end of december 85 (100 % of the sugar plant waste waters).

- Capital investment : 3,900,000 FF 1985.

TABLE 2 : KEY REFERENCE DATA FOR WINE DISTILLERIES

- REVICO - Cognac - France -

REVICO is a group of well-known cognac distillers (MARTELL, HENNESY, COURVOISIER...) which was created to treat the effluents coming from more than 100 distilleries.

The first plant built by SGN in 1972 before the energy crisis included a multi-stage evaporation unit with tartric acid recovery. Impressed by the results of the demonstration unit of Condom and faced with increasing energy cost, the REVICO group decided to replace the existing unit by a SGN anaerobic process.

- Effluent :
  . Vinasses from wine and yeast distilleries (flowrate : 100 m³/h, soluble COD : 80 t/d).

- Campaign : 6 months/year

- The unit includes :
  . Existing settling tank.
  . One packed bed 5,660 m³ digester
  . One completely mixed 4,160 m³ digester
  . Final clarifier
  . Existing activated sludges and lagoons for finishing treatment
  . Two electrical generator 500 kWA each feeded with biogas
  . PLC controller for one line monitoring.

- Commissioning : December 1984

- Efficiency :
  . Total COD reduction : 99 %
  . Methane production capacity : 25,000 Nm³/day during campaign 1,700,000 Nm³/ie ; 1,450 Toe/year.

- Energy savings : over 4,300 Toe/year including savings for shut down of evaporation plant.

- Capital investment : 18,500,000 FF 1984.

- S.I.A. CONDOM - France -

S.I.A. is a group of several distilleries of Armagnac.

As for REVICO, S.I.A.'s previous unit was based on multi-stage evaporation.

- Effluent :
  . Flowrate : 400 m³/d ; soluble COD : 11.5 t/d.

- The new treatment plant includes :
  . Two heat exchangers
  . One packed bed 570 m³ digester
  . One completely mixde 1,570 m³ digester
  . Flare and boiler
  . Existing storage ponds.

- Commissioning foreseen in fall 1986.

- Anticipated efficiency :
  . Wine vinasses 90 % soluble COD reduction
  . Yeast vinasses 85 % soluble COD reduction
  . Methane production : 3,500 Nm³/d
  . Net annual production : 720,000 Nm³ ie 600 toe/year

- Energy savings : over 1,600 Toe/year including savings by shut down of evaporation plant.

- Capital investment : 6,500,000 FF 1985.

- TOMELLOSO - Spain -

The TOMELLOSO plant built under SGN licence by PROSER S.A. treats 1,400 m³/day of distillery vinasses and several by-products of the wine industry.

- Effluent :
  . Flowrate : 1,400 m³/day ; COD : 45 tons/day.

- The unit includes :
  . Two packed bed : 1,400 m³ digesters

- Efficiency :
  . COD reduction 90 %
  . Methane production : 12,000 Nm³/day ie 10 toe/day

- Capital investment : approximately 10,000,000 FF 1984.

TABLE 3 : KEY REFERENCE DATA FOR MOLASSES DISTILLERIES

- S.I.S. - La Guadeloupe - French West Indies -

- Rum distillery : 26,000 lts alcohol/day.

- Campaign : 150 day/years

- Effluent : Sugar cane molasses spent wash
  . Flowrate : 420 m^3/day
  . COD : 24 t/day.
  . BOD5 : 18 t/day.

- The unit includes :
  . Heat exchanger
  . Packed bed : 1,700 m^3 digester

- Commissioning : spring 1986

- Anticipated efficiencies :
  . COD reduction 65 %
  . BOD5 reduction 85 - 90 %
  . Methane capacity : 5,400 Nm^3/day
  . Annual methane production : 800,000 Nm^3

- Energy savings : 680 Toe/year

- Total capital investment : approximately 10,000,000 FF 1984.

TABLE 4 : KEY REFERENCE DATA FOR CANNING INDUSTRY

- CANNING INDUSTRY JAM FACTORY -

ANDROS (Bretenoux - France)

- Capacity : 40,000 T/day of jam

- Effluent :
  . Flowrate : 600 m^3/day
  . Max : 100 m^3/h
  . Soluble COD : 4,500 kg/day

- The unit includes :
  . Packed bed : 470 m^3 digester
  . Heat exchanger inlet - outlet
  . Gas storage : 40 m^3
  . Flare
  . Secondary activated sludge system
  . Final clarifier
  . Sludge thickening

- Commissioning expected : Spring 1986

- Anticipated efficiency :
  . COD reduction 97 %
  . Biodegradable COD reduction 99 %
  . Methane production 1,400 m^3/day, 310,000 Nm^3/year

- Energy savings :
  . 260 toe/years

- Capital investment :
  . 4,000,000 FF 1985

- CANNING INDUSTRY VEGETABLES -

LA SEMEUSE (Bierne - France)

- Campaign : 2 months/year

- Effluent :
  . Flowrate : 1,200 m^3/day
  . COD : 6,000 kg/day
  . BOD5 : 3,000 kg/day

- The unit includes :
  . Packed bed : 470 m^3 digester
  . Commissioning : 1984

- Anticipated efficiency :
  . COD reduction over 70 %
  . Methane production : 1,500 Nm^3/day

- Capital investment : 1,600,000 FF 1984

PARTIAL LIST OF S G N FIXED FILM ANAEROBIC DIGESTION REFERENCES

FULL SCALE INDUSTRIAL PLANTS		BEGHIN-SAY Thumeries (France)	GENERALE SUCRIERE Aulnois/Laon (France)	REVICO Cognac (France)	S.I.A Condom (France)	S.I.S Guadeloupe (F.W.I.)	TOLEDO (Espagne)	LA SEMEUSE Bierne (France)	ANDROS Bretenoux (France)
Design Data	INFLUENT	Waste water Beet sugar plant		Spent wash of wine distillery		Spent wash cane distillery	Manure from piggery	Waste water from vegetable cannery	Waste water from jam factory
	Packed bed digestor liquid volume (m^3)	1,260	2,000	5,600	570	1,700	100	430	430
	Loading rate (kg COD / m^3 digestor day)	12.7	14	11.8	11	14.1	12	14	10.5
	Daily COD load (tons COD / day)	16	28	66	6.25	24 (after screening)	1.2	6	4.5
	COD reduction (%)	90	80	90	90	65 (total COD)	55-60	70	90
	Methane specific productivity (Nm^3 / m^3 / reactor day	4.6	4.1	3.4	3	2.8	3 - 4	3.1	3
Biogas utilization		Natural gas / biogas boiler	Boiler	Biogas boiler Thermal Engine (electricity)	Gas / biogas Boiler	Fuel / biogas Boiler	Fuel / biogas Boiler		Boiler
Actual Operating Performance	Flow rate (m^3 / hour)	95 - 125	150	60 - 80	Under construction	Under Start - up	1.4	Under Start - up	Under Start - up
	Daily COD load (tons COD / day)	7 - 13	38 - 40	50 - 80			1.2 (after screening)		
	COD reduction (%) of the digestion unit	90 - 95	75 - 80	90			55 - 60 (total COD)		

SEWAGE SLUDGE RESTRUCTURING AND CONVERSION INTO METHANE

Lennart G. Erickson, President and Co-founder, Waste=Energy Corporation, San Mateo, California

Howard E. Worne, Ph.D., Sc.D., Chairman of the Board and Scientific Director, Waste=Energy Corporation, Medford, New Jersey

ABSTRACT

Waste=Energy Corporation is conducting a process research and development program for restructuring sewage sludge to improve anaerobic digestion and conversion of organic matter into methane. The sludges are processed to reclaim water, to remove non-biodegradable matter, and to convert biologically degradable matter into digester gas with an increased volume of methane.

Mixed primary and secondary treated sludge is restructured by mechanical homogenization to reduce particulate size, followed by enzyme hydrolysis and chemical hydrolysis. Metals are chemically converted into insoluble chelated particulates and removed with other inorganic solids down to levels which will not inhibit microbial activity in the digester. Biologically recalcitrant cellulosic fibers are removed as a separate sidestream.

The BioFeedstock thus produced is characterized by biologically available nutrients in soluble and microparticulate form, practically free of viable microorganisms, inert or non-biodegradable solids, and heavy metals at toxic levels. Such BioFeedstock is metabolized in the digester at a rate approximately eight times faster than for conventional digestion of raw mixed primary and secondary sludge. Approximately 90% of the BioFeedstock biodegradable nutrients are converted into digester biogas having a methane content of about 75%. The remainder is metabolized into anaerobic bacterial biomass. Surplus digester biomass sludge is thickened to remove effluent and soluble inorganics. The digester solids are combined with the inflow raw sludge for processing of the biodegradable fraction into fresh BioFeedstock. This is recycled to the digester for conversion into additional biogas.

Typical digester hydraulic retention time is about two days. A proportion of the separated fibrous solids may be recycled back to the digester to provide substrate surfaces for the anaerobic bacteria and for multiple-pass hydrolysis and eventual conversion of the biodegradable fraction into additional biogas.

WASTE=ENERGY PROCESS
for
RESTRUCTURING SEWAGE SLUDGE and CONVERSION INTO METHANE

* RESTRUCTURE Biodegradable Matter into Soluble Form
* REMOVE Metals and other Non-Biodegradable Matter
* W=E BIOFEEDSTOCK is the the intermediate product
* CONVERT the BioFeedstock in a high performance DIGESTER
* BIOGAS contains 75% Methane
* RECYCLE/RESTRUCTURE DIGESTER SOLIDS into more BIOFEEDSTOCK
* THE BOTTOM LINE:
  MORE METHANE
  LESS RESIDUAL SOLIDS

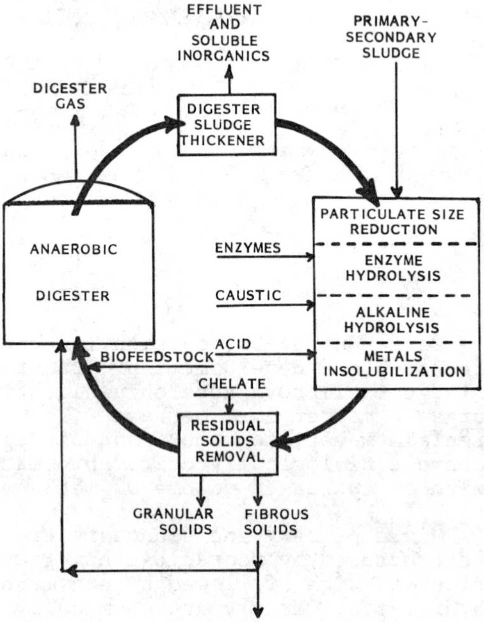

WASTE=ENERGY PROCESS DEMONSTRATION PROJECT
REDWOOD CITY, CALIFORNIA

THE EFFECTS OF REACTOR CONFIGURATION ON METHANE PRODUCTION FROM HIGH-STRENGTH WASTEWATERS

John N. Veenstra, Ph.D., P.E., Associate Professor;
John B. Barber, M.S., Research Assistant,
Oklahoma State University, Stillwater, Oklahoma

ABSTRACT

A research program was conducted to investigate the effects of reactor orientation on the treatment of efficiencies of anaerobic digestion. Three fixed-film reactors were constructed and fed a high-strength wastewater (50 g COD/L) at hydraulic retention times of 10, 5, and 3 days. One reactor was oriented vertically and had a gas-to-liquid interface of 325 cm^2. The other two reactors lay horizontally and had interfaces of 700 cm^2 and 1400 cm^2, respectively. All three reactors were of equal liquid volume (9.5 L) and contained equivalent amounts of media.

At the lowest loading (5 kg COD/m^3/d), the vertical reactor outperformed the two horizontals, removing a greater proportion of the influent COD and, in turn, producing a greater quantity of methane. Though gas composition was approximately 69% methane for all three systems, the vertical unit produced an average of 20 L/d of biogas (at STP) as compared to about 18 L/d for the horizontals.

At the highest loading (16.7 kg COD/m^3d), the results were reversed. The vertical system produced the least amount of biogas (32 L/d) and had the lowest COD removal efficiency (35%). Once again, however, the gas composition was relatively consistent among the three units, averaging at about 67%.

In light of these findings, it is proposed that the gas-to-liquid interface does not play as important a role in gas composition as has been suggested. Under all loading conditions, the difference in percent methane between the reactors was insignificant. Reactor orientation, however, did seem to have an effect on COD removal efficiencies and total biogas production. It is suggested that this is a result of changes in degree of biological solids retention.

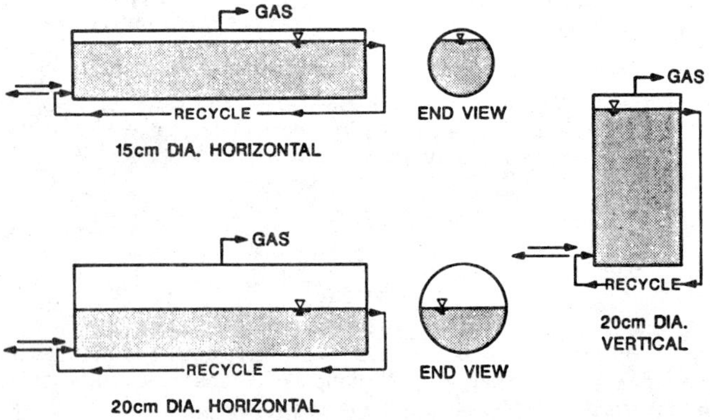

Figure 1. Reactor Schematic

The horizontal units had the following dimensions, 15 cm diameter by 68.5 cm in length and 20 cm diameter by 68.5 cm. The dimensions of the vertical units are 20 cm diameter by 38 cm in height. All three units are illustrated in Figure 1. All three units were constructed out of PVC pipe and contained pall ring type plastic media. The units were operated at 35°C in a temperature controlled hot room. For the purpose of this work, the units were operated in a batch mode, using swine waste with a COD of 50 gm/L.

Results indicating the efficiency of the various steps in the anaerobic process are illustrated in Figure 2.

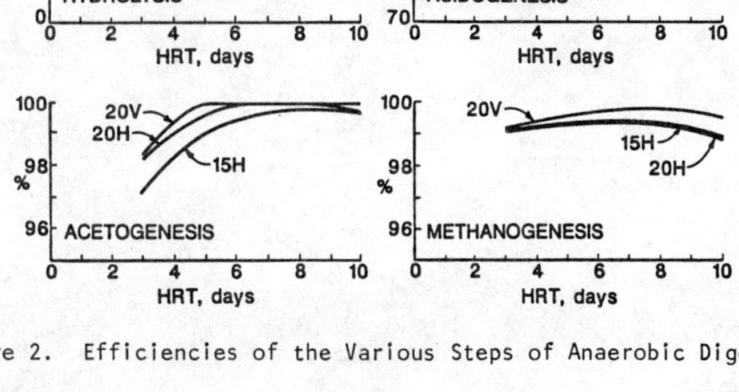

Figure 2. Efficiencies of the Various Steps of Anaerobic Digestion.

PURIFICATION OF EFFLUENTS FROM A PEAT WET-CARBONIZATION PROCESS BY AEROBIC AND ANAEROBIC FERMENTATION

Hilkka Kyllönen, M.Sc., Research Scientist; Tapani Peltola, M.Sc.; Technical Research Centre of Finland (VTT), Laboratory of Fuel Processing and Lubrication Technology, 02150 Espoo, Finland

ABSTRACT

Effluents are a great problem in the peat wet-carbonization process. A wet-carbonization plant, capacity 100 000 t/a of PDF briquettes (moisture 10 % w/w), produces about 3600 m^3/d of waste water, i.e. 12 m^3/t PDF product, when the solid content of raw peat is 9 % w/w. The BOD_7 value of the effluent is 7000 - 8000 ppm.

The fermentation of wet-carbonization effluents to fodder protein was studied by VTT and JP-Energy Oy in early 1984. A similar process, the so-called Pekilo process, has been applied to the purification of effluents from a sulphite cellulose mill in Finland. A BOD_7 reduction of as high as 84 % was achieved in the Pekilo fermentation of wet-carbonization effluents. About 3 kg of Pekilo product with about 50 % w/w protein content could be produced from one m^3 of effluent.

In late 1984 anaerobic purification of wet-carbonization effluents was studied by the Biogas Group of Neste Oy and by VTT both in a laboratory reactor and in pilot-scale reactors at the Haukineva wet-carbonization demonstration plant. A BOD reduction of > 95 % was achieved by the laboratory reactor and about 75 % by the pilot-scale reactors. Gas production was at its best > 3 l of biogas (79 % CH_4, 21 % CO_2) from one litre of effluent.

The costs of the Pekilo process with subsequent anaerobic treatment were compared to those of the ANAMET process, which gives equivalent purification results (BOD_7 < 400 ppm). The estimated purification costs of waste water for one ton of PDF fuel were 7.4 USD for the Pekilo process and 5.1 USD for the ANAMET plant.

Both the Pekilo process and the anaerobic treatment were found to be suitable for treating waste waters from the wet-carbonisation process. The profitability of the purification process is dependent on the price of product and on the capacity of the wet-carbonization plant.

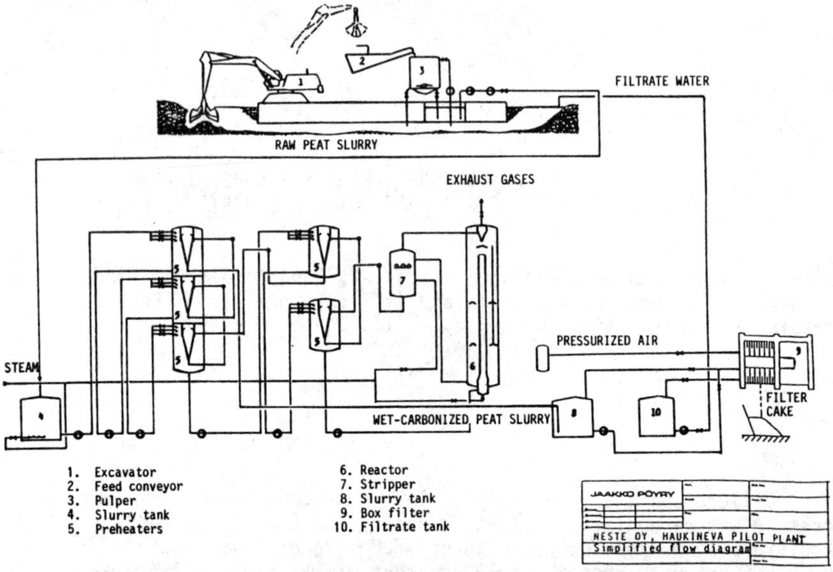

Figure 1. Simplified process flow diagram of the PDF pilot plant.

Table 1. The results of the batch fermentation test with waste water from the wet-carbonization process.

Maximum pekilo concentration	3.1 g/l
Yield for dry matter	27 %
BOD_7 value before fermentation	7.6 g O_2/l
BOD_7 value after fermentation	1.2 g O_2/l
BOD_7 reduction	84 %
Protein content of the product	49 wt-%

Table 2. Some results of the laboratory-scale anaerobic fermentor tests.

COD influent (average)	12 000 ppm
COD load	1.5 - 6.0 kg/m^3 d
COD reduction	45 - 65 %
BOD_7 load	0.7 - 2.9 kg/m^3 d
BOD_7 reduction, max	95 %
Methane generation (average)	0.34 cm^3 CH_4/kg COD_{red}

Table 3. Cost comparison of the Pekilo + anaerobic process and the ANAMET process.

* 10^5 USD	Pekilo + anaerobic	ANAMET
Capital investment	80.5	40.2
Annual capital charges (10 %, 20 a)	10.6	5.3
Annual operating costs	16.1	3.7
Total annual costs	26.7	9.0
Annual income/savings**	19.3	4.0
Annual charges	7.4	5.0

* Costs in 1984, 1 USD = 5.9 FIM ** Pekilo protein valued at 500 USD/t

RECOVERY OF METHANE FROM LANDFILL GAS BY THE SEPAREX® MEMBRANE PROCESS

C. Douglas Houston, B.S., Director of Engineering
Separex Corporation, Odessa, Texas
William J. Schell, Ph.D., Senior Vice President
Separex Corporation, Anaheim, California

ABSTRACT

The first commercial Separex® membrane system for landfill gas recovery, and the world's largest to date, has been operating since August 1984 at Rossman's Landfill in Oregon City, Oregon. The membrane process selectively removes carbon dioxide from the landfill gas to produce approximately 800 MSCFD of high BTU methane product at 500 psig containing less than 3.5% CO_2. Membrane technology in this application offers the advantages over conventional CO_2 treatment methods of: low capital and operating costs, simplicity of operation and maintenance, compact size, and modular construction for flexibility of operation. The membrane system also simultaneously dehydrates the gas to pipeline specification while removing the carbon dioxide.

Process design, operational parameters, economic factors, operating history and photos of the facility are presented, in addition to a description of the spiral wound membrane element and the principles of membrane gas separation that affect the efficiency of this new unit operation. Other biogas applications for this process, such as CO_2, H_2S and water vapor removal, will be discussed in terms of membrane selectivity, recovery and costs.

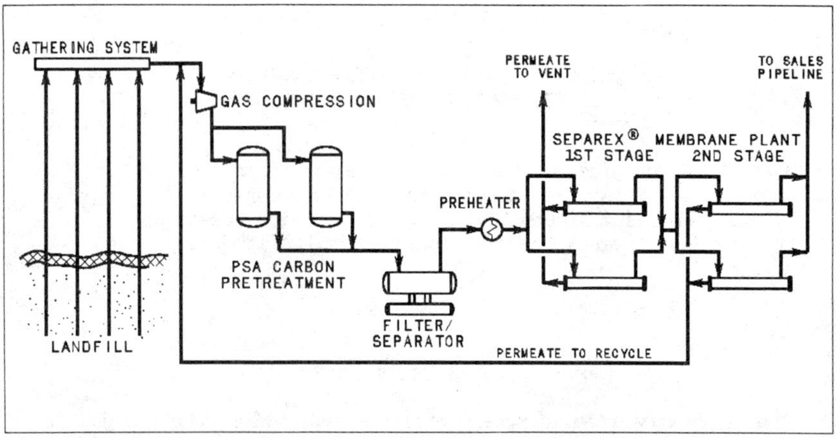

Figure 1. Methane Recovery Schematic For Rossman's Landfill

In the process schematic shown in Figure 1, the gas coming off the gathering system is compressed to 525 psig, passed through dual-bed carbon adsorption towers to remove organics, fed to a filter/separator to collect condensed liquids, preheated to 130 F to prevent water condensation and delivered to the membrane system. The membrane system preferentially permeates carbon dioxide, thereby increasing the BTU value of the product gas while maintaining feed pressure. The CO_2-rich permeate from the first stage of the 2 stages in series is vented, while the second stage permeate is recycled to the front of the compressor to achieve a higher methane recovery. The membrane system, shown in Figure 2, consists of (7) 8-in. dia by 20-ft long tubes, each containing (6) spiral wound membrane elements. This system is capable of processing from 1.5 to 2.0 MMSCFD of landfill gas depending on specified recovery and product gas BTU value.

Figure 2. Separex® Membrane System At Rossman's Landfill

ADSORPTIVE PURIFICATION OF BIOGAS BY THE PSA PROCESS

Erwin Pilarczyk, Dr.-Ing.; Klaus D. Henning, Dipl.-Ing.; Karl Knoblauch, Dr.-Ing.
Bergbau-Forschung GmbH, Dept. Coal Chemistry, Essen
Federal Republic of Germany
Joseph S. D'Amico, P.E.; Herbert E. Reinhold, MS-Eng.
Nitrotec Engineering Co., Div. of Nitrotec Corporation
Glen Burnie, Maryland

ABSTRACT

Using activated carbons and a special carbon molecular sieve of Bergwerksverband GmbH, Essen, FRG biogases from agricultural residues, municipal and industrial wastes, and biological degradation in waste water cleaning can be purified and thereafter processed to SNG which can be fed to a pipeline system.

The preliminary purification includes the removal of hydrogen sulfide and halogenated hydrocarbons. According to

$$2\ H_2S + O_2 \longrightarrow 1/4\ S_8 + 2\ H_2O$$

hydrogen sulfide is oxidized in presence of oxygen to elemental sulfur on a special type of activated carbon. Sulfur remains adsorbed on the activated carbon, which can be regenerated thermally. During regeneration, sulfur is recovered by condensation.

Halogenated hydrocarbons are removed by adsorption on another type of activated carbon. The loaded activated carbon can be regenerated thermally by steaming or purging with a hot inert gas.

After purification, the clean biogas is separated by a PSA-process in three or four adsorbers. Carbon dioxide, oxygen and main part of nitrogen (infiltrated air) are adsorbed on a special carbon molecular sieve at 2 - 9 bar whilst methane flows through. Depending on the nitrogen content in the biogas, methane is enriched up to 99% by vol. at adsorption pressure. After adsorption, desorption takes place by depressurization and evacuation. Then adsorption is started again after previous pressure buildup. A methane recovery up to 98.5% can be achieved.

For commercial plants and technical evaluation, contact Nitrotec Engineering Co., Glen Burnie, Maryland, (301) 766-6595.

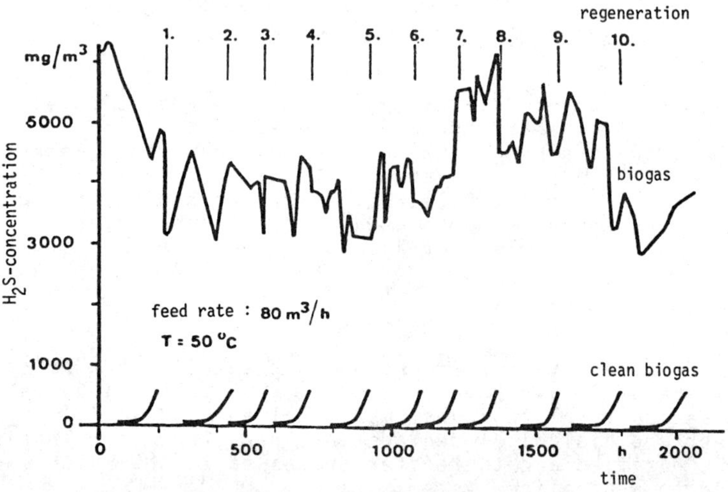

Fig.1: Field tests for H_2S-removal from sewer gas containing 3 - 5 g H_2S/m^3

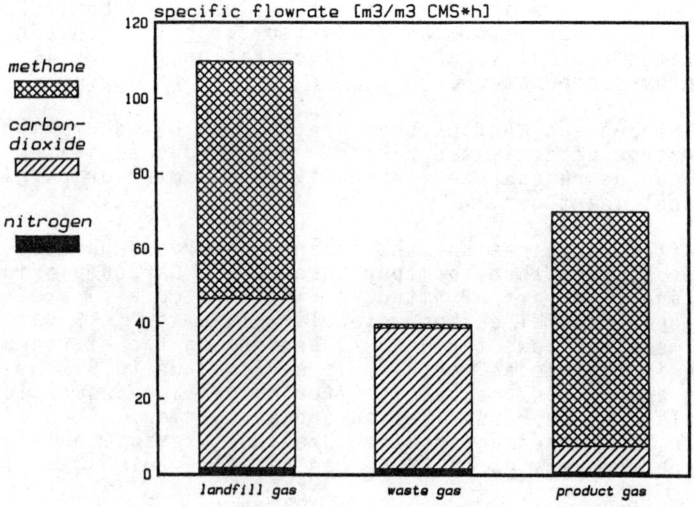

Fig.2: Methane enrichment from landfill gas by PSA (product gas: Dutch SNG-quality)

CELLS ON DIATOMS ENTRAPPED IN FIBERS

Robert A. Clyde, B.S., Pres., Clyde Engineering,
New Orleans, Louisiana, 70174

ABSTRACT

Diatoms are inexpensive and have high surface area for immobilization of cells. When pelletized, diffusion into the interior is poor, and when powder is used, pressure drop is high. By entrapping in fibers, pressure drop is reduced, and high area powder can be used. Fast fermentations are possible. Xanthan gum and electricity can be produced. By running an RBC (rotary biological contactor) half full and shining a light in the top, hydrogen can be produced. Methane can be produced from wastes. Uranium and algae can be removed from the fibers with brushes between the discs.

Without diatoms, ethanol can be produced from glucose (80% yield) in 10 minutes, from sucrose in 2 hrs., and from lactose in 5 hours. Several runs have been made with diatoms and the rate is five times faster. *Zymomonas mobilis* is used with glucose and sucrose and *Kluyveromyces fragilis* with lactose. If cane fibers clog the system, they can be removed as shown in Fig. 2. Brushes between the discs go around when operating. Then a hook goes in through a double seal (sterile between) and pulls the horizontal rod into a slot, so the brush is stationary and scrapes material into the screw conveyor. Sterility is not compromised, so the run can continue. Uranium can be removed from wastewater with *Pseudomonas* and scraped in a similar manner. Algae can be harvested also. The fungus *Phanerochaete chrysosporium* can be grown on polyester discs to degrade lignin. A recent Japanese patent, 4,562,150, describes cellulase production with *Acremonium*. Clyde's patent 4,407,954 describes rotation, 4,446,236 photo reactor, and 4,530,763 metal removal. G. Strandberg from Oak Ridge found that free flowing *Pseudomonas aeruginosa* reduced uranium from 100 ppm to 2 in only 10 seconds, but then they plugged the filter. By immobilizing on fiber, filtration is avoided. Chromium, silver, palladium and vanadium can be removed also.

Xanthan gum has many uses - enhanced oil recovery, slurry for coal in a pipe, and preservative for meat, avoiding refrigeration. Production is described in Enz. Micro. Tech. vol. 7 pg. 235-238. *Xanthomonas campestris* can be immobilized on diatoms but the solution is so viscous it is hard to stir and cool. By putting a slow moving agitator between the discs, more efficient stirring is realized with less power and cooling water can be put inside the agitator. Considerable product stays on the diatoms. The fibers can be pulled out and squeezed.

Electricity production is described in Biotech. Letters Oct. '85 pg. 699-704. We believe that *Proteus vulgaris* could be immobilized in graphite fibers. Bennetto said (personal comm.) that it works well with a low concentration

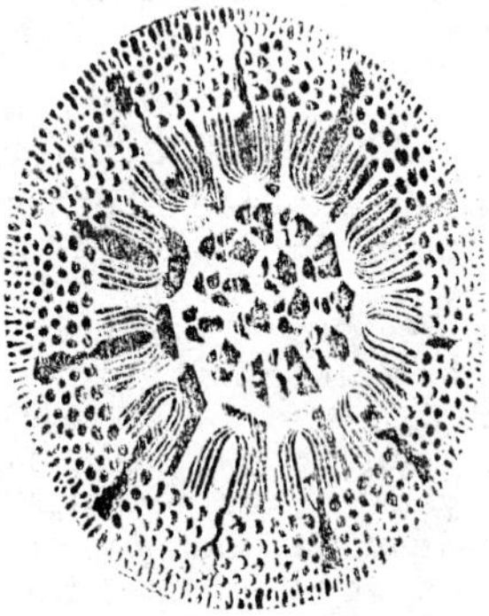

Fig. 1 Diatom 200X

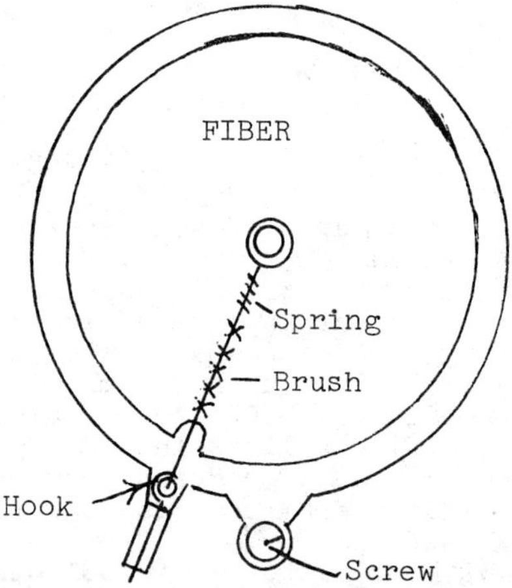

Fig. 2

of sugar. When fermenting sugar to alcohol, it is difficult to ferment the last 1 or 2%. Electric power production is also described in Engelhard's patent 4,522,894 using a monolithic catalyst support (but no cells). By entrapping diatoms in ceramic fiber and coating with Pt or Pd, an inexpensive catalytic reactor can be made.

Hydrogen can be produced in a photo reactor with bacteria such as *Rhodospirillum*, *Rhodopseudomonas*, and *Chromatium* and algae such as *Oscillatoria*, *Chlamydomonas*, and *Phaeodactylum*. In most photo reactors, the colored solution blocks the light, but by running an RBC half full, the light hits more organisms and a thin moving film. Hydrogen is now trucked from New Orleans to Cape Canaveral (rather dangerous). *Lactobacillus* makes acrylonitrile, a component of the shuttle solid fuel. Prof. J. Pos from the Univ. of Guelph has made methane with an RBC of polyester discs, as described at a convention in Miami last December.

In another application, Interleukin 2 (for cancer) can be made by putting its plasmid into yeast, as described by A. Shaw in Cellular Immunology 90 (2) pg. 549. Yeast can be grown faster by Clyde's method. Kolot states in Process Biochemistry Oct./Nov. 1981 pg. 33 that the doubling time of *E. coli* attached to a membrane with medium flowing past was shorter, and gives references where cell concentration was 5 times higher compared with free cells. Dr. B.A. Burdick from Kodak gave a Poster at a Bio/Tech. conf. in New Orleans showing how cells on nylon could produce aspartic acid and alanine, intermediates for the sweetener, aspartame. In the Jan. Genetic Eng. News pg.7, Pall says monoclonal antibodies can be immobilized on nylon and on pg. 53 Nucleopore shows a photo of polyester fiber. Nylon and polyester are covered in Clyde's patent 4,407,954. A new method for growing animal cells is being patented.

ALCOPLEX®, A Fuel Alcohol Production Complex for Rural America

R. T. LeFrois, BS, President, THOMAS ENTERPRISES, INC.,
Minneapolis, Minnesota

ABSTRACT

A concept is proposed to revive the grassroots ethanol fuel industry through the use of centralized dehydration plants and small scale, highly automated satellite plants located in the rural regions of America. The approach, called ALCOPLEX®, would enable farming regions to market grains and cellulose materials as feedstock to these vertically integrated chemical processing production complexes. An application study for the State of Minnesota indicates that to convert twenty percent of the annual corn crop to ethanol would require thirty production complexes (ALCOPLEX®es) of 8.5 million-gallons each located in selected counties of the State. Each ALCOPLEX® would consist of a Community Dehydration Plant surrounded by twenty-five 380,000-gallon-per year Satellite Plants dedicated to the production of hydrous ethanol and de-watered stillage. These Plants would have the state of the art operations and be highly automated and serviced from the Community Plant. Features such as regional development planning, parts standardization, process modularity, equipment mass production, dual feedstock processing, computerized automation, wet stillage distribution, CO_2 recovery with UREA co-product upgrading, energy recovery, local business development and training will enable ALCOPLEX® to compete with large stand-alone Plants of equal capacity in both capital and operating costs. Introduction and widespread use of these Alcohol Production Complexes or ALCOPLEX®es will not only give rural America a viable renewable liquid fuel production capability but will provide the farmer and his community with a much needed business enterprise, especially at this time of crisis in Agriculture.

COMMUNITY/FARM SATELLITE PRODUCTION COMPLEX - ECONOMIC POTENTIAL

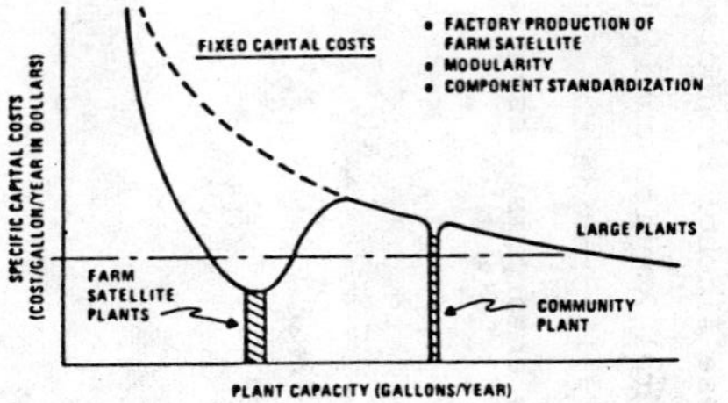

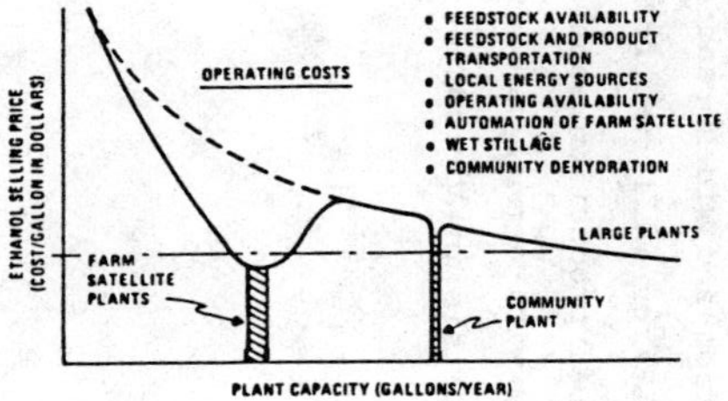

The result of implementing the ALCOPLEX[R] concept with all of its features will give the econonomic benefits depicted on the left. The upper curve shows that average specific capital costs significantly below the trend curve can be achieved. Likewise, the lower curve shows that the ethanol selling price for implementing ALCOPLEX[R] approximates that of large scale centralized plants.

ETHANOL PRODUCTION PLANT CAPITAL COST TRENDS

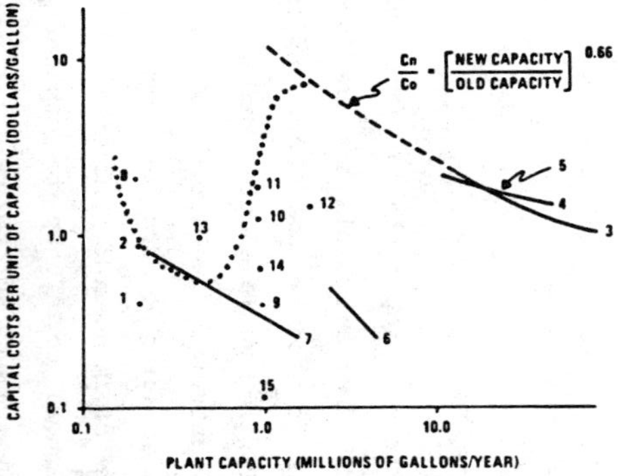

1. EASY ENGR.	182 PROOF	9. USDA CENTRAL DEHYDRATION	190–200 PROOF
2. A.M.F. PADOVAN	184–190 PROOF	10. USDA SMALL COMMUNITY	200 PROOF–WET
3. KATZEN ASSOC.	200 PROOF–DDGS	11. USDA SMALL COMMUNITY	200 PROOF–DRY
4. VOGELBUSH	200 PROOF	12. USDA LARGE COMMUNITY	200 PROOF–DRY
5. PEDCO INT.	200 PROOF–DDGS	13. USDA LARGE ON FARM	190 PROOF–WET
6. A.M.F. PADOVAN	200 PROOF	14. AGRI-FUELS ENERGY INC.	200 PROOF–DDGS
7. ACR	200 PROOF	15. AGRI-FUELS ENERGY INC. CENTRAL DEHYDRATION	190–200 PROOF
8. DOE/INEL	190 PROOF–WET		

SOURCE: OPEN LITERATURE

The plot above shows specific capital costs versus capacity of over fifteen plants obtained from the open literature. The dashed line extension of item (3), the Katzan Study, is the trend curve based on a 0.66 scaling law typical for expressing economies of scale. The dotted line reflects the departure from the expected trend curve and supports the ALCOPLEX[R] approach.

PRODUCTION OF UREA FROM CO_2 FROM GRAIN ALCOHOL PLANT

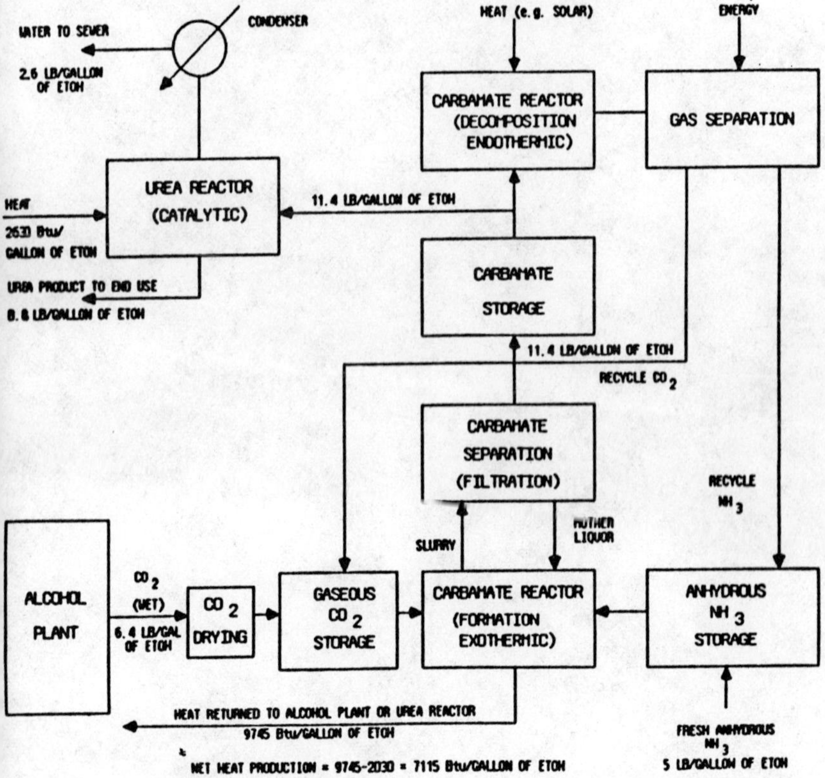

The carbon dioxide from the Alcohol Plant is dried, compressed and stored. Anhydrous ammonia and carbon dioxide are fed to the carbamate reactor. The ammonylsis reaction is highly exothermic and easily carried to completion without a catalyst. The resulting ammonium carbamate is filtered and stored. The heat generated by the reaction is supplied to the Alcohol Plant for process operations. To recharge this chemical storage system, heat from solar, waste heat sources or off-peak electric is used to decompose the carbamate into ammonia and carbon dioxide which are separated and stored. To use the carbon dioxide from the Alcohol Plant on a continuous basis a reactor is provided to hydrolize a portion of the carbamate to urea - a value added high nitrogen (46% wt.) content material used as a solid fertilizer and as a chemical feedstock.

A Concept for Utilizing the Carbon Dioxide Off-Gas From an Ethanol Production Plant

R. T. LeFrois, BS, President, THOMAS ENTERPRISES, INC., Minneapolis, Minnesota

ABSTRACT

The process for the production of ethanol from grains fermentation and distillation generates two potential co-products, i.e., stillage and carbon dioxide. Large alcohol plants are able to recover, finish and sell both of these co-products due to economy of scale of large plants. However, in medium and small scale plants, this is typically not the case. The viability of the smaller plants may be dependent on the ability to dispose of the unprocessed or wet stillage from the process and recover the carbon dioxide off-gas which is generally vented to the atmosphere. Recovery and utilization of this waste gas could significantly improve plant economics and process efficiency.

A concept has been proposed by the author for a chemical heat storage device based on the ammonylsis of carbon dioxide, i.e.,

$$CO_2 + 2NH_3 \rightleftharpoons NH_4CO_2NH2, \quad \Delta H = -67,000 \ \frac{BTU}{LB\text{-}MOLE} \qquad (1)$$

producing ammonium carbamate. This reversible reaction (1) is highly exothermic, and easily carried to completion without a catalyst.

Two basic process applications are highlighted. The first is a reversible chemical storage unit sized to supply all process heat needs to the alcohol plant and is recharged from waste, solar or off peak electric power heat sources. The second process application adds the production of a new co-product, urea. This system is sized to handle the carbon dioxide off-gas and supplies only part of the process heat needs. The carbon dioxide and ammonia are combined in a reactor to produce ammonium carbamate. The recharge heat is at a temperature greater than the decomposition temperature which produces ammonia and carbon dioxide which are compressed, separated and stored under pressure as liquids.

For every 13 pounds of sugar fermented, one gallon of ethanol and 6.4 pounds of carbon dioxide are produced. This amount of carbon dioxide could react with five pounds of annhydrous ammonia to produce up to 8.8 pounds of urea and 7115 BTU/gal. ETOH. Using good design practice an 80% yield of urea is assumed. Based on: 1) the cost of the ammonia being equal to the sale price of the co-product urea, 2) the market value of the carbon dioxide being one cent per pound, and 3) the value of the heat supplied being equal to the annualized capital investment cost of the CO_2 recovery and urea production equipment: Then the net value gained as a result of carbon dioxide ammonylsis and conversion to urea is estimated to be \$0.12/gal ETOH.

THE UTILITY OF PRISTINE LIGNIN FOR HAZARDOUS WASTE TREATMENT

Daniel J. O'Neil, Christopher J. Newman and Paul M. Hawley
Energy and Materials Sciences Laboratory
Georgia Tech Research Institute

E.S.K. Chian, H. Gao and J.M. Kim
Environmental Engineering Laboratory
Georgia Institute of Technology
Atlanta, Georgia 30332

ABSTRACT

Lignin, isolated from steam-exploded Tulip poplar, during the GIT wood-to-ethanol process was evaluated as a potential adsorbent for a series of eight organic and two inorganic model toxic substances.

Several lignocellulose-to-ethanol processes result in the isolation of the lignin component of woody biomass. In the GIT Process(1) preseparation of lignin, followed by hydrolysis of cellulose in a plug flow reactor with recycling, allows conversion rates to sugar of 80%. A pristine lignin is isolated at low energy levels and for low cost. The use of pristine lignin as a low cost adsorbent for hazardous wastes is being investigated.

Hardwood chips (Tulip poplar) were subjected to steam treatment at 350 psig and a residence time of four (4) minutes followed by decompression. The steam-exploded wood was warm water-washed to remove soluble hemicellulose hydrolyzate compounds or derivatives (e.g., acetic acid, methanol, furfural, xylose). Pristine lignin isolates were produced by multiple extraction with 95% aqueous ethanol or 0.1N NaOH. Four-fold increases in yields (to 36%) of pristine lignin were achieved by the alkaline hydrolysis process. Calculated lignin extraction efficiency is 93% based on the initial hardwood. Gel permeation chromatography results show low values of number-average molecular weight (Mn = 800-1100) for both ethanolic and alkaline lignin isolates but a narrower distribution (Mw/Mn=1.4-1.6) for ethanolic isolates when compared to alkaline isolates (Mw/Mn = 2.4-4.2) indicating greater branching in the macromolecular structure of the alkaline lignin isolates. The results suggest a high retention of potential chemical reactivity for hazardous waste treatment.

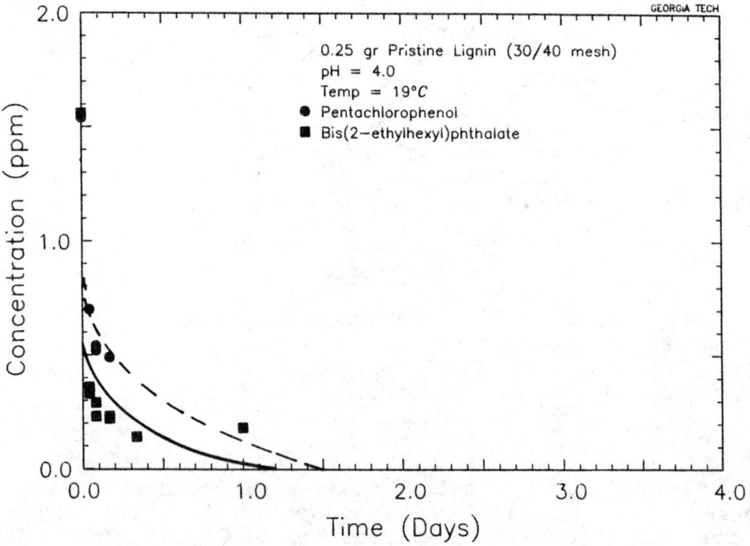

Exemplary Adsorption Rate Curves

Preliminary kinetic studies were conducted on eight organic and two inorganic model toxic compounds by equilibration of the compounds over several days with an ethanolic pristine lignin isolate, prepared as a 30/40-mesh range powder. Equilibration of all compounds was completed within forty-eight hours with maximum rates achieved generally within two to four hours. Phenols, polyaromatic hydrocarbons, chlorinated hydrocarbons and heavy metals were represented in the study. The accompanying adsorption rate curves for pentachlorophenol and bis(2-ethylhexyl)phthalate are indicative of the initial rate studies.

Pristine lignin has potential utility as a low-cost, high-efficiency adsorbent for certain classes of toxic wastes(2). Further physical and/or chemical modification is being undertaken to optimize adsorption capacity.

(1) O'Neil, D.J. et al, Seventy-second Annual Meeting, American Institute of Chemical Engineers, San Francisco, 25-29 November 1979.

(2) Although research described in this article has been funded wholly or in part by the U.S. Environmental Protection Agency through Cooperative Agreement No CR-812223-01-0 with the Georgia Institute of Technology it has not been subjected to agency review and, therefore, does not necessarily reflect the views of the agency and no official endorsement should be inferred.

A CONTINUOUS EXTRUDER-FEEDER FOR REACTOR SYSTEMS FOR BIOMASS FUELS PROCESSING

Don H. White and D. Wolf
Department of Chemical Engineering
University of Arizona
Tucson, Arizona 85721

ABSTRACT

Biomass including wood could provide liquid fuels on a regional basis if an economic liquefaction process can be developed. Through research sponsored by the U.S. Department of Energy (DOE), under its Thermochemical Conversion Program managed by Battelle PNL, the University of Arizona has developed a unique method of pumping concentrated, viscous biomass slurries, characteristic in biomass direct liquefaction systems. A modified single-screw extruder has now been shown to be capable of pumping solid slurries as high as 60 wt. % wood flour in wood oil derived vacuum bottoms, as compared to only 10-20 wt. % wood flour in wood oil in conventional systems.

A Phase II program has integrated the extruder-feeder pumping system with a unique reactor which offers potential for further improving direct biomass liquefaction technology. An experimental facility was constructed during 1983-84, and shakedown operations were successfully completed in early 1985. A three-year experimental program utilizing this advanced feeder-reactor system was started in June, 1985.

A CONTINUOUS EXTRUDER-FEEDER FOR REACTOR SYSTEMS FOR BIOMASS FUELS PROCESSING

INTRODUCTION

During the course of this research, the University of Arizona has developed a unique method of pumping concentrated, viscous biomass slurries, characteristic in biomass direct liquefaction systems. This method utilizes a modified extruder similar to those used for very viscous materials in the polymer industry. Various preliminary process design and economic studies indicate that the utilization of this newly-developed extruder-feeder in biomass direct liquefaction processes could lead to one or more of the process improvements listed below.

1. Elimination of recycle wood oil.

2. Attainment of reactor temperature almost instantaneously by the mode of adding heat.

3. Attainment of near-plug flow in the reactor when using static mixers.

The extruder-feeder acts simultaneously as both a feed preheater and a pumping device into a 3,000 psi pressure reactor in an experimental biomass liquefaction process. It could prove to be of importance in other processes where high concentration of solids in liquids are to be handled, like in the pulp and paper industry, hydrolysis of wood or other cellulosics, other biomass liquefaction processes, coal liquefaction processes and some biomass gasification processes.

BASIC EXTRUDER-FEEDER CONCEPTS

The heart of an extruder-feeder consists of a helical screw, conveying solids, liquids or a slurry through a barrel. It has been proven experimentally and theoretically that an extruder is a good conveyor of solids, as long as the friction of the solids on the screw surface is not excessively greater than the friction of the solids on the barrel surface. The helical geometry results in a "plug of solids" moving down the channel, due both to an axial translation and a rigid rotation of the solid plug. The axial components can be shown as a force balance and the tangential components can be expressed as a torque balance.

The pressure control on an experimental Prodex-type extruder-feeder was accomplished by having a valve on the outlet die, which could be opened or closed to adjust for the output rates and pressures. For example, when the die valve is open, the output rate is at a maximum while output pressure is at a minimum. This is the case of pure drag flow. However, when the die valve is partially open the output rate is lower while output pressure is higher. This is the usual case which then involves both drag and pressure flow. It is to be noted that the output rates in both cases (open and partially closed die valve) are about the same at low RPM but they differ slightly at high RPM. We conclude that the extruder output rate does not drop significantly when the pressure increases significantly.

The rheology of the materials handled is a very important parameter in the performance of the extruder-feeder. Polymers, heavy oils, Albany vacuum bottoms (VB) and Albany wood oil (TR12) are non-Newtonian materials. A comparison of pumping by the experimental extruder-feeder of such materials has been studied experimentally.

SUMMARY PROCESS DESCRIPTION OF EXPERIMENTAL UNIT

An advanced extruder-feeder biomass liquefaction reactor system was designed to experimentally convert wood flour to liquid wood oil fuels by direct liquefaction. The extruder-feeder system developed at the University of Arizona was incorporated into the system to develop 3,000 psi pressure at its discharge and preheat the slurry prior to entering the vertical reactor, as shown in Figure 1. As the wood flour slurry enters the reactor it is mixed with superheated steam and carbon monoxide. Gas dispersion as well as heat and mass transfer will be attempted by the use of static mixer elements inside the tubular reactor. The superheated steam is used to heat the wood flour slurry to the reaction temperature of 350°C almost instantaneously, which is then maintained by the use of electrical band heaters surrounding the reactor. Steam and carbon monoxide also react via the water-gas shift reaction to provide hydrogen for the liquefaction. The liquefaction products are split into the vapor and liquid fractions, condensed, cooled down and brought to atmospheric pressure by means of the let-down system. Computer control and real-time data acquisition have been implemented for the unit.

ACKNOWLEDGEMENT

This research was supported by Department of Energy Contracts DE-AC06-76RLO-1830 (Subcontract B-96249-A-Q from Battelle Pacific Northwest Laboratories) and DOE Contract EW-78-S-05-5679.

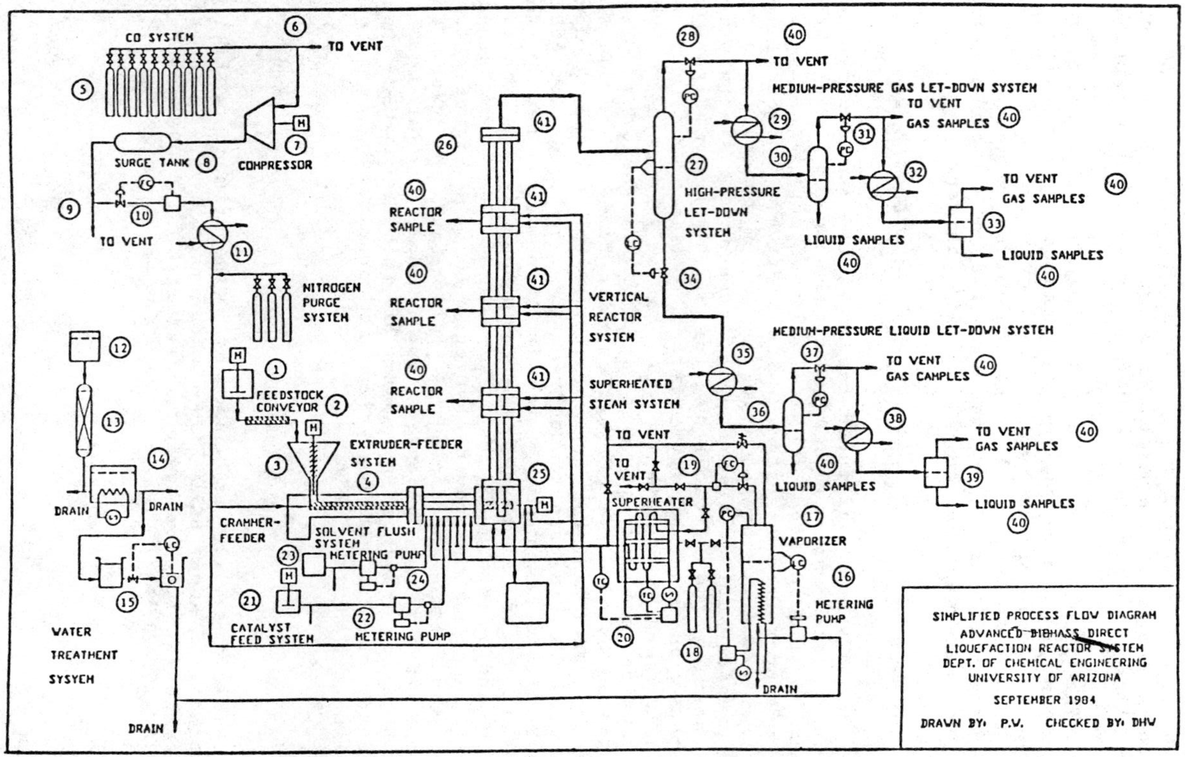

FIGURE 1. Simplified Process Flow Diagram

INFLUENCE OF NITROGEN AND TEMPERATURE ON LIPID PRODUCTION IN MICROALGAE (HANTZSCHIA DI-160)

Shobha Sriharan, Ph.D., Associate Professor
Science Division, Selma Univeristy, Selma, Ala. 36701

Davinderjit Bagga, Ph.D., Associate Professor, Biology Department, Miles College, Birmingham, Alabama 35208

ABSTRACT

Chrysophytes (microalgae-diatoms) are known to accumulate large quantities of various lipid types. The objectives of the studies reported here are to evaluate the effect of induction strategies (varying nitrogen concentrations and temperature) on the lipid production in diatom, *Hantzschia* DI-160.

Hantzschia DI-160 were batch cultured in N-sufficient (NS) and N-deficient (ND) media at two different temperatures under continuous illumination as shown below:

NITROGEN

SUFFICIENT (NS)		DEFICIENT (ND)	
600 uM		300 uM	
20°C	30°C	20°C	30°C

The cell density was measured during logarithmic and stationary phases. The biomass was harvested; dry cell mass and ash-free dry mass (AFD) were determined. From the harveated cells, the total lipids were extracted and these were separated into neutral lipids and polar lipids by column chromatography on silicic acid column.

The growth curve, Fig. 1, shows that the growth of *Hantzschia* DI-160 was rapid in NS medium at 30°C and the growth continued during the logarithmic phase as well as stationary phase. The growth of the cells was not inhibited in suboptional concentrations of nitrogen (ND) at 20°C indicating that the diatom has the ability to tolerate low temperature.

The analytical results, Table 1, show that the AFD at 30°C was 1.420 g/L and 1.120 g/L in NS and ND treatments respectively. At 20°C, the AFD was 0.998 g/L and 0.885 g/L in NS and ND respectively. The total lipids were 0.070 g/L and 0.092 g/L at 30°C; 0.054 g/L and 0.071 g/L at 20°C in NS and ND conditions respectively. The neutral lipids were 0.044 g/L and 0.058 g/L at 30°C; 0.032 g/L and 0.043 g/L at 20°C in NS and ND experimental groups. The polar lipids were 0.026 g/L and 0.032 g/L at 30°C; whereas at 20°C these values were 0.022 g/L and 0.027 g/L in NS and ND treatments. The above results suggest that by growing *Hantzschia* DI-160 in nitrogen limited medium, the yield of lipid content may be increased.

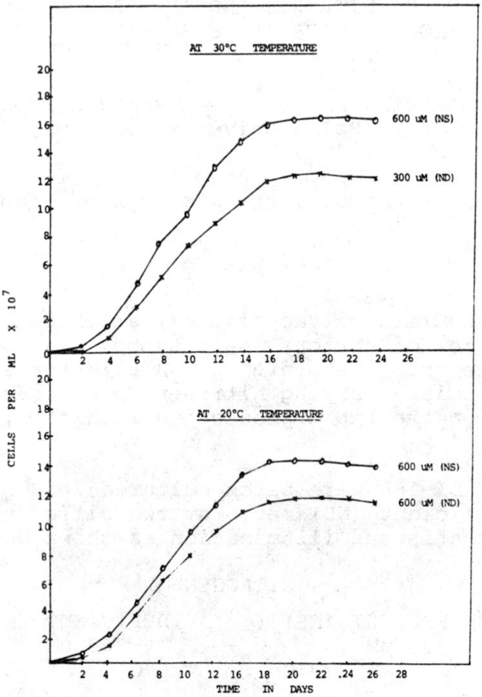

Fig. 1. Growth curve of Hantzschia DI-160 at 30°C and 20°C temperature in nitrogen sufficient (NS) and nitrogen deficient (ND) media.

Table 1. Cellular Mass and Lipid Yields of <u>Hantzschia</u> DI-160 Batch Culture in Varying Nitrogen Supply and Temperature.

Growth Conditions	Ash-free dry mass AFD g/L	Total Lipids g/L	Neutral Lipids g/L	Polar Lipids g/L
		30°C TEMPERATURE		
Nitrogen Sufficient (NS)	1.420	0.070	0.044	0.026
Nitrogen Deficient (ND)	1.120	0.032	0.058	0.032
		20°C TEMPERATURE		
Nitrogen Sufficient (NS)	0.998	0.054	0.032	0.022
Nitrogen Deficient (ND)	0.885	0.071	0.043	0.027

PRODUCTION OF OXYGENATED FUELS FROM MUNICIPAL SOLID WASTE COMPONENTS

Ravindra K. Agrawal; James E. Helt
Chemical Technology Division
Argonne National Laboratory
Argonne, IL 60439

ABSTRACT

There is a worldwide concern over both future energy supplies and the disposal of municipal solid wastes (MSW). Most MSW is cellulosic in nature, comprising paper, newsprint, packaging materials, wood wastes, and yard clippings. Investigation of the basic mechanisms of pyrolysis of these materials is being undertaken in both a TGA and a bench-scale, fixed-bed reactor which generates adequate quantities of gaseous and liquid products to allow them to be characterized analytically.

The end products of MSW pyrolysis are: a highly carbonaceous solid char, a highly oxygenated liquid tar, and a mixture of gases with low to medium Btu content. Depending on the type of fuel preferred, the pyrolysis conditions can be varied to maximize the desired product yield.

TGA data, Fig. 1, were analyzed for kinetic information. Activation energies calculated for the decomposition of several MSW components are given in Table 1. It was found from the TGA studies that the apparent value of E for various cellulosics and synthetic plastics ranges from about 25-50 kcal/mol.

Experimental results on the thermal decomposition of MSW components (Whatman #1, newsprint, kraft paper, and cardboard) over a temperature range of 275-475°C have been gathered. From product distributions during exploratory runs indications are that heating rate (for the range studies, 5-35°C/min) has no significant effect on the yields. However, the final temperature of the pyrolysis reaction does have a significant effect on the yields of the various products. Examples of the analytical results obtained on these runs are given in Tables 2 and 3.

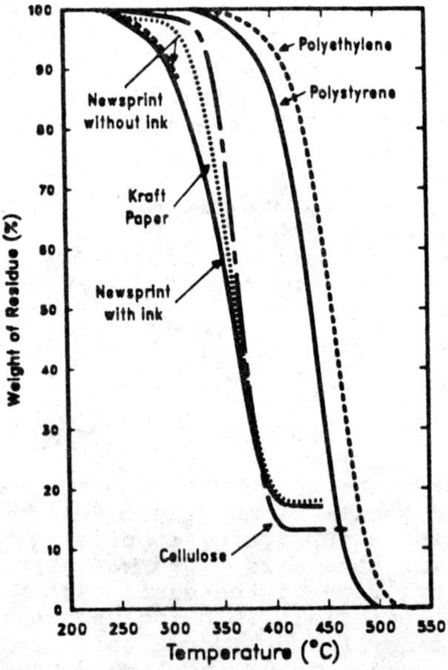

Figure 1. THERMOGRAVIMETRIC STUDIES OF MSW COMPONENTS (Nitrogen Atmosphere, 5°C/min Heating Rate)

Table 1. ACTIVATION ENERGIES FROM TGA STUDIES

Sample	Heating Rate (°C/min)	Activation Energy (kcal/mole)
Whatman #1 paper	4	49.0
Whatman #1 paper	14	42.8
Whatman #1 paper	25	39.6
Newsprint with ink	5	26.3
Newsprint without ink	5	27.1
Newsprint	20	24.2
Kraft paper	5	40.4
Polyethylene	5	41.7
Polyethylene	20	34.9
Polystyrene	5	38.0

Acknowledgment

This work was funded by the U.S. Department of Energy, Bio-Fuels and Municipal Waste Technology Division.

Table 2. ANALYTICAL RESULTS OF CELLULOSE, TARS AND CHARS (Heating Rate 5°C/min, Final Temperature 475°C)

Cellulose (Whatman #1)	Elemental Analysis[a]
Moisture - 4.1%	C - 44.7%
Ash - 0.074%	H - 5.9%
Heating value[a] - 4170 cal/g	O[b] - 49.4%
Cellulose Tars (Yield ~46%)	
Ash ~0.075%	C - 42.3% (43.6)[c]
Heating value[a] - 4330 cal/g (47.8)	H - 6.1% (47.5)
	O[b] - 51.6% (48.0)
Cellulose Chars (Yield ~15%)	
Ash - 0.63%	C - 81.0% (27.3)
Heating value[b] - 7566 cal/g (27.2)	H - 3.6% (0.1)
	O[b] - 15.4% (0.1)

[a] - Dry basis.
[b] - By difference.
[c] - % value of original material (based on the reported yields).

Table 3. COMPARISON OF ATOMIC RATIOS AND HEATING VALUES OF SOME FUELS

Substrate	Atomic Ratio H/C	Atomic Ratio O/C	Heating value Cal/g
Cellulose (Whatman #1)*	1.58	0.83	4170
Wood	1.44	0.56	5000
MSW	1.51	0.52	4565
ECO I	1.61	0.61	3679
ECO II	1.55	0.49	5303
Cellulose chars*	0.53	0.14	7566
Utah bituminous coal	0.92	0.09	7872
Lignite	0.79	0.23	5710
Cellulose tars*	1.73	0.91	4330
No. 2 fuel	1.84	0.01	10,917
No. 6 fuel	1.47	0.02	10,105

*Values obtained in this work.

BIOMASS FLOWS IN THE U. S. ENERGY ECONOMY (*)

Thomas B. Cobb, Ph.D., Department of Physics
Bowling Green State University, Bowling Green, OH 43403

Marc H. Ross, Ph.D., Department of Physics
University of Michigan, Ann Arbor, MI 48109

ABSTRACT

A flow chart, depicting the role and flow of terrestrial biomass in the U.S. economy has been developed in the form of a large, colorful wall chart suitable for office or classroom display. The chart provides an instructive and convenient way to follow biomass resources both quantitatively and qualitatively from source to end product. Energy is discernable from material flows by comparing entries on this chart with those on related diagrams. An accompanying technical memo describes development of the chart and discusses details of its contents.

The chart shows, for example, that the U.S. annually harvests nearly 1700 million tons of biomass - a quantity larger than the annual production of coal and equal to about 40 lbs per day for every person in the United States. About two-thirds of this biomass is accounted for by agricultural crops and by animal forage on pasture and rangeland; the remainder reflects material harvested by the forest products industries.

Uses of biomass are classified by the authors into three main categories: food, fuel, and materials. Materials are further subdivided into durable and non-durable goods. The total dry weight equivalent of all domestic biomass products is about 350 million tons, or 25% of input. Roughly half of the biomass harvest is recycled - either in the form of crop residues or as animal manure, and the remaining 25% is ultimately lost - primarily in animal respiration.

The use of biomass for food and feed clearly dominates the agricultural sector, yet less than 5% of all biomass harvested actually ends up as food for humans. The agricultural system efficiency for the conversion of biomass to food, generally taken to be in the range of 10:1 in terms of energy content, is dramatically illustrated by a ratio of nearly 50:1 of feed input to meat output in terms of the dry mass units used in this chart.

Although fuel uses for biomass have increased significantly in recent years, especially within certain industrial sectors, less than 10% of the total biomass harvest actually is used for fuels, and most of this is accounted for by the pulp, paper, and lumber industries. Household uses of firewood have increased substantially in recent years, yet this segment still only constitutes about half of the industrial use. The total contribution of biomass fuels recently has been estimated at about 3 quads or 4 percent of total U.S fuel use. The chart shows this energy to be supplied from 150 million tons of matter - roughly 10% of the domestically used biomass production.

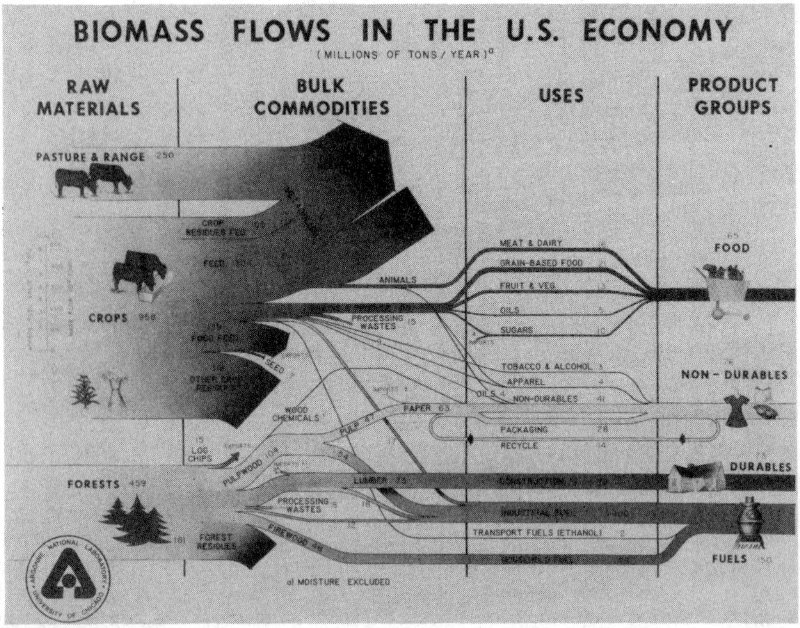

Figure A.

In terms of dry mass units, materials produced from biomass are shown to be comparable to fuel use, thereby leading one to conclude that non-food uses account for over 85% of system output. Inherent losses and alternate pathways within a particular product stream are easily identifiable with the chart, as are recycling paths and other possible ways to increase efficiency through the conversion of wastes and residues. The chart should stimulate theoretical and applied energy analysts alike to think more carefully and realistically about the potential uses for biomass.

Single copies of the chart, either in the form of a 27" x 36" full color wall poster, or in reduced version as a black-and-white print or color slide, are available at minimal charge. Requests for the chart and the related technical memo may be sent to: Special Projects Group, Energy and Environmental Systems Division, Building 362, Argonne National Laoratory, Argonne, IL 60439. A fee of $4.00 is requested to cover the costs of postage and handling for the large wall chart.

(*) Work supported by Argonne National Laboratory and the U. S. Department of Energy through its Bio-Fuels and Municipal Waste Technology Division.

WOOD ENERGY DEVELOPMENT IN INDIA
Programme & Policies

O.P. Vimal and N.P. Singh
Department of Non-Conventional Energy Sources
Ministry of Energy
New Delhi

ABSTRACT

In rural India, wood is the principal source of fuel (68.5%), followed by oil products (16.9%), animal dung (8.3%) coal (2.3%) and others (3.4%). About 70% of this wood is collected from the forest village plantations and other nearby sites. An important consequence of inadequate supply of fuelwood relates to the burning of agricultural residues and animal dung which otherwise would have been used for restoration of soil fertility and increasing food production. The rapid depletion of forest wood and the loss of forest area has led to serious problems of erosion, floods, siltation and desertification. Considering the seriousness of fuelwood and power problems in India, Department of Non-Conventional Energy Sources has initiated a number of programmes for production, conversion, utilization and conservation of wood which involve popularisation of different models of improved wood stoves, biogas plants, energy plantation and power generation systems. Establishment of a chain of Biomass Research Centres under different agro-climatic conditions of the country is one of the main priority areas under this programme. The present communication is aimed at to discuss the multi-dimensional nature of this problem and the approach adopted to meet this challenge.

Energy Plantation Programme: The main objective of energy plantation is to produce energy rather than industrial wood; therefore selection of species assumes greater significance. The specific objectives should be to select species with High biomass yield, High calorfic value,Fast growth,Good coopicing (regeneration after harvest), Ease of establishment, High adaptability, Little after care, Identification of plants for problematic regions, species with nitrogen fixing ability as well as fuelwood sources. In order to carryout investigations on screening, evaluation and agronomic trials of fast growing fuelwood species under such problematic situations, it has been planned to set up a chain of Biomass Research Centres under different agro-climatic conditions of the country. So far, four Biomass Research Centres have been set up by the Department of Non-Conventional Energy Sources (DNES) at National Botanical Research Institute (NBRI),Lucknow for saline sodic soils; Madurai Kamaraj University for Arid and semi-arid regions at Srinagar, Garhwal University (UP) for mountainous areas and Sukhadia University, Udaipur for arid land.

Power Generation: Power generation from energy plantations is a very promising option for substantial power needs. It can make important contributions to the rural energy scene as they can be developed, implemented, maintained and managed after appropriate training of the local people, in other words, decentralised energy supply will help to raise incomes and improve the well-being and quality of life of the people. Existing schemes and funds for IRDP,NREP,TRYSEM,harijan and tribal welfare, adult education and literacy, primary health care, minor irrigation afforestation etc. can be dovetailed with these projects so that they become composite programmes for economic, social and cultural development.

Biomass Gasification Demonstration Programme: Amongst the variety of technologies based on biomass, gasification route offers the most promising scope because it helps in conservation of diesel and saving of foreign exchange. Moreover, this technology can be employed reliably and conveniently to energise even a small engine driven irrigation pumpset in the capacity range of 5-10 H.P. which are common due to small land holdings of average farmer. Gasification technology is more suitable for remote areas. These areas can not be provided with power through the usual transmission lines or transporting fuels to these areas. For such locations, this technology offers a cheap and easily available source of energy.

The Department of Non-Conventional Energy Sources (DNES) has initiated a number of projects relating to the development of a village level gasifier for irrigation pumps based on agricultural residues, utilisation of producer gas in small H.P. utility engines and R&D in reliable wood-based gasifier coupled with I.C. engine for field implomentation. The gasification programme is planned to be substantially expanded during the Seventh Plan.

Charcoal Production: Although wood burns readily to produce heat, the heat value from charcoal is approximately twice as great as that from the same weight of wood. Charcoal burns with very little smoke and can be used in smaller or more efficient stoves than those generally used for burning wood. It is used for domestic purposes and industrial heating. Studies have been initiated on charcoal production using pit kiln, fixed kiln and portable kilns including drum type but the following problems need to be considered in order to ensure the success of this technology: suitability of the design for different regions; temperature control through recycling of fuel gases; availability of raw material throughout the year; loading the kilns; size of the raw materials; duration of carbonisation and cooling; control of air circulation; quality control of charcoal for various applications; briquetting in small scale; production; cost factor in terms of transportable steel kiln and fixed brick kilns; transportation of raw material and charcoal; storage facilities and extension and training facilities. A portable kilns that gives optimum charcoal yields and producer gas utilizing locally available biomass and agricultural residues is the need of the hour.

National Project on Demonstration of Improved Chulhas: 'Chulha" is the central point of household activities. Any programme aimed at development of the village in general, women and children in particular has to start from the 'fireplace'. Traditionally, the chulhas used in the village are very inefficient releasing a lot of smoke, making the thatched huts fully sceared with block soot, causing irritation to the eyes and a host of respiratory troubles. Improving the efficiency of these 'fireboxes' alone would lead not only to the conservation of firewood 5 to 10 times but also make houses clean. Department of Non-Conventional Energy Sources (DNES) launched a Nati-

-onal Project on Demonstration of Improved Chulhas (NPDIC) during 1983. During 1984-85, 7.82 lakh chulhas were constructed or installed against a target of 4 lakhs only, thereby resulting in the savings of more than 6 lakh terms of fuelwood valued at Rs. 24 cores.

Massive programme of extension on a large scale will be taken up by DNES to replace the existing traditional stoves (about 113 million in number) in household all over the country. It is proposed to instal 50 lakhs improved stoves during the seventh plan period. The net saving in fuelwood by installing 50 lakhs improved stoves would be of the order of 42 lakhs tonnes valued at Rs. 294 crores.

<u>National Project on Biogas Development</u>: Keeping in view the multiple benefits of this 'anaerobic digestion technology', the Department of Non-Conventional Energy Sources (DNES), accelerated the activities of the National Project on Biogas development (NPBD) during 1984-85. Against a target of 1,50,000 plants which was double the target for 1983-84 1,80,000 plants were installed. The total number of plants installed sofar under NPBD exceeds 4 lakhs. The people who do not have the required number of animals and who do not posses the required open space to build the plants, miss the benefit of biogas plants under the normal programme. It is in this context that the concept of community biogas plant is receiving attention during recent years.

DNES has commissioned 134 CBPs/IBPs sofar (till December 1985). The plants are hoped to help significantlly in fuelwood conservation and addition of organic manure. These CBPs/IBPs act as focal points for socio-economic upliftment in the village providing some of the minimum needs like sanitation and toilet facilities, improved health, employment drinking water and power (only at some selected places).

<u>Policy Action Programme</u>: Fuelwood occupies a dominant place as an energy source particularly in the rural sector. In order to meet the minimum energy needs, fuelwood programme needs to be given topmost priority, the success of which, however depends upto the following policy measure

- Integration of production, conversion, utilisation and conservation of fuelwood aspects

- Integration of R&D, demonstration and extension activities

The Department of non-conventional Energy Sources (DNES) has initiated a number of projects on the above lines to develop a package which can be delivered to the end-user. In order to accelerate the development activities relating to wood energy, stres needs to be laid on the following aspects;

- Conservation of wood through popularisation of imprved chulhas biogas plants, briquetting of organic residues and technologies based on solar energy
- Selection of fast growing, nitrogen fixing, species, having good coppicing capacity and high calorific value under sub-standard soil conditions

- Development of management practices e.g. spacing, fertilizers irrigation, protection measures and harvesting techniques etc.
- Improvement of fuelwood yielding plants through cytogenetical physiological and cultural practices.
- Development of producer gas technology based on wood and agricultural residues for electrification, lift irrigation, and in automobiles.
- Integration of energy plantation for power generation (EPP concept)
- Reducing transportation costs and increasing efficiency of fuel through briquetting and pyrolysis of agricutural and forestry residues.

The success of this programme, however, depends on the following pre-requisites:

- Involvement of all categories of people-school children, teachers, women, ex-servicemen, volunatary agencies etc.
- Development of integrated training course for wood energy programmes.
- Exchange of visits to selected centres of production, processing and utilization of bio-energy
- Organisation of regional workshops/symposia/seminars on priority areas

In conclusion, it may be stated that wood energy programme is a commondenominator for all the developing countires, varying only by certain degrees. It is hoped that exchange of experience through various forums at the regional level may help to develop many new horizons, many new vistas in this vital area.

PRODUCTION AND UTILIZATION OF AQUATIC BIOMASS IN ASIA

Md. Shahjahan Ali, Ph.D., Associate Professor, Department of Agricultural Chemistry, Bangladesh Agricultural University, Mymensingh, Bangladesh

ABSTRACT

Lately, under the stress of the oil crisis and its effect on the price and availability of Nitrogenous fertilizers considerable emphasis has been laid on the use of Rhizobium or blue green algae as a substitute or supplemental to the usual fertilizer treatment and an approach is attempted to valuation of potential and chances of development of aquatic biomass in the world, based on forseen data of available superficies and biomass yields.

Algae, water hyacinth, water chestnut, azolla are selected for their good productivity and wealth in high value components and a low cost technology is possible and desirable for production of this biomass; earth built basin for cultures, waste waters for fertilization, very simple floating machine for harvest may be included. Another biomass or bio-fertilizer Azolla which is receiving much attention in South and South-East Asia offers a good prospect for its introduction in our crop management in Bangladesh. The recent energy crisis has turned the attention of the developed and developing countries alike towards alternative means for maintaining soil fertility and augmenting agricultural productivity. All kinds of agricultural production wastes can be utilized in the production of Energy and they can be regained. Biogas derived from biomass or organic wastes can be used straight way for cooking, lighting and other purposes in villages in Bangladesh and several gas plants are constructed in different parts of the country to meet the challenge of the shortage of fuels and a natural gas from the organic deposits has been meeting the requirement of fuel of various parts of our country.

The major source of Energy in Bangladesh is bio-mass. The availability of Energy could be divided into Commercial and non-commercial forms of energy giving a total requirement of the country. The bio-mass utilized in this country include crop residues, animal excreta, municipal wastes and various discarded materials from processing natural and industrial products. These are Farm yard manure (cowdung), poultry manure, composts, bone meal etc. and the cowdung is used as a source of bio-energy relating to the production of bio-gas solving to some extent the crisis of fuel. The bio-gas plants are constructed in various villages. Azolla being bio-mass receiving attention in Asia for its introduction in crop management in Bangladesh and blue green algae as a supplemental to the usual nitrogenous fertilizer treatments.

Table-1. Total Energy Consumption

Consumption (toe 000)		
Commercial -	2182	44.3%
Non-commercial	2748	55.7%

Asian Development Bank Report, 1980.

Table-2. Commercial Energy Consumption (toe 000)

Coal	265	12.1%
Petroleum	1333	61.1%
Natural gas	540	24.7%
Hydro-electricity	44	2.0%
		99.9%

Asian Development Bank Report, 1980.

Table-3. Total bio-mass consumption per person in wet and dry season (GJ/ Six months).

Wet season	2.50 (100%)
Dry season	2.40 (100%)

M.N. Islam and Morse
Rural Energy to meet development needs.
(1984)

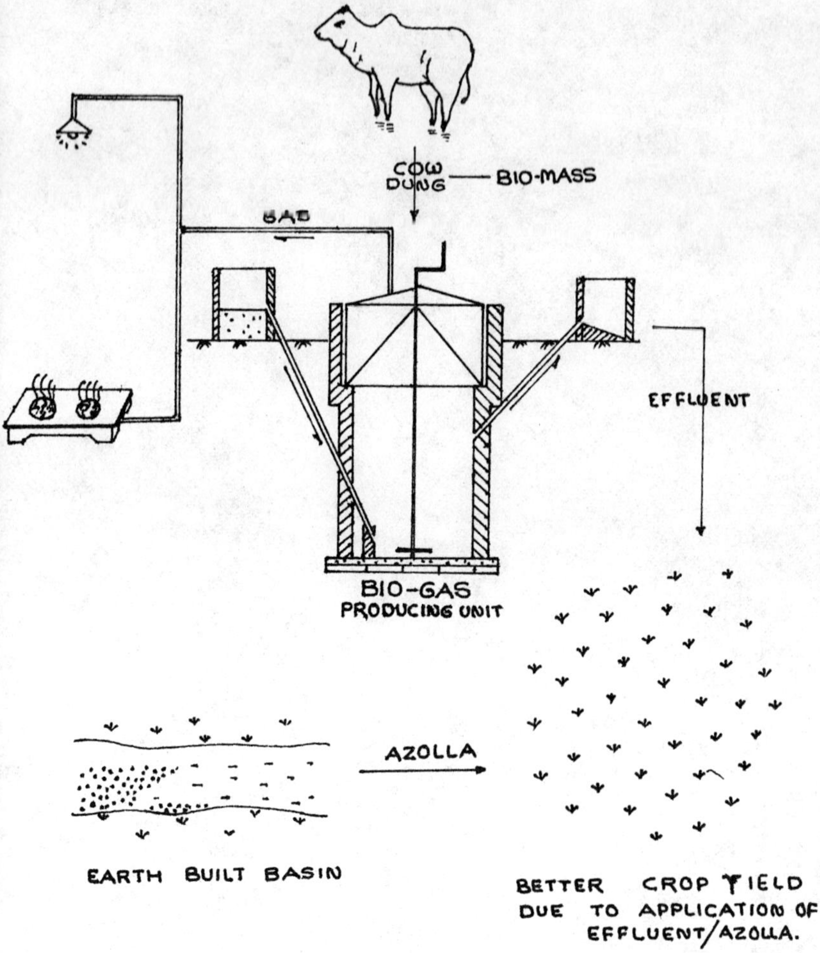

PRODUCT DEVELOPMENT NEEDS OF WASTE MANAGEMENT

Pekka Vilppunen, M.Sc. (Engineering), Laboratory Manager,
Energy Laboratory, University of Oulu, Finland

ABSTRACT

1. Introduction

This project consists mainly research on the energy possibilities from municipal solid waste in Finland. The experiences in other countries have been taken into account. In addition, the most suitable method for the quantity and quality analysis of the municipal refuse has been studied.

The preliminary results show that the separation of combustible and other fractions in the municipal refuse significantly influence the economical utilization of the waste. The sorted municipal refuse enables the preparation of more homogeneous refuse derived fuel (RDF) increasing the possibilities of using the waste in combination with other combustibles in an ordinary burning plant.

2. Purpose

The aim of this research is to find out the R&D requirements to increase the recovery of energy from municipal refuse. A partial objective is to develop a separation-handling system to produce refuse derived fuel. This is a two phase separation system: **at source segregation** and mechanical sorting, followed by a refining stage of the combustible fraction.

The method can be utilized when the recovery of energy and/or materials is pursued by a community.

3. Research realization

During the research the differents stages in the recovery of energy from waste will be studied:

- the preparation of refuse derived fuel (waste composition, possibilities of at source and mechanical sorting,refining methods...),
- the production of energy (combustion, gasification, anaerobic digestion...) and
- the environmental impacts, pollutants and their origin, exhaust gas cleaning systems and equipment.

The realization of the research includes

- a literature review of the present state of knowledge and foreign experiences in energy recovery methods,
- the technical and economical aspects of the most important handling and energy production methods and
- the defination of R&D requirements.

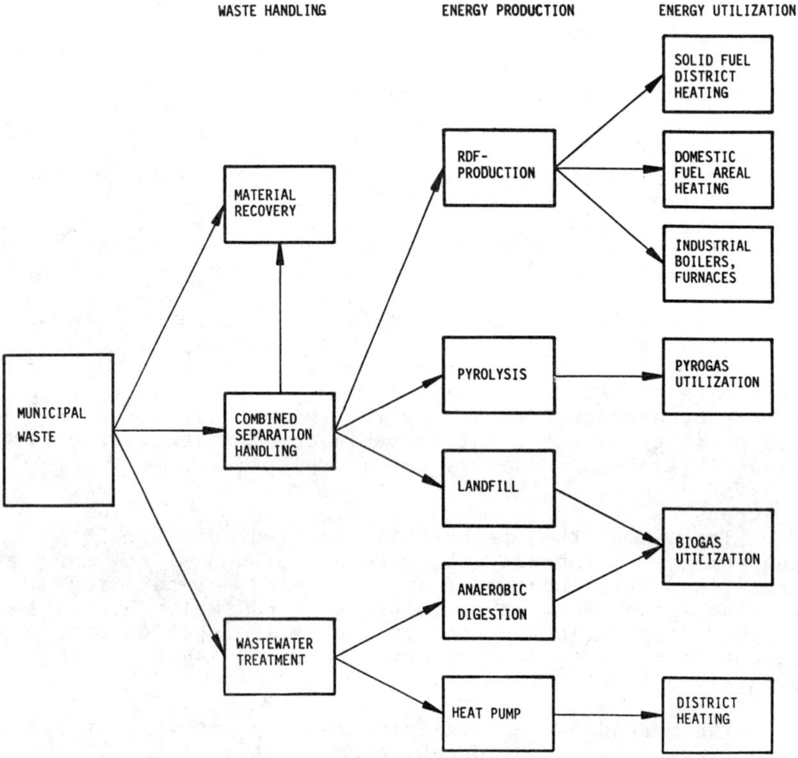

Picture 1. Economical utilization of energy from the solid municipal waste.

4. Result utilization

The result can be utilized to achieve the economical recovery of energy from municipal refuse.

Benefits:

- The system increases:
  * directly, the possibilities to utilize domestic solid fuels (wood and peat) in combination with municipal refuse,
  * indirectly, the organic fraction in the collected municipal solid waste by separation at source of other components which is then more suitable for energy recovery through anaerobic digestion and
  * the degree of recovery of recuperable materials (two phase separation).
- The system decreases the environmental impacts caused by incineration (more homogeneous fuel - less emissions).
- The system facilitates the long term planning of the waste management (e.g. decreases the need of landfills).

CLOSING REMARKS

D. L. Klass

When we started this program 4 days ago, the first slide I projected on the screen was titled "Can Biofuels Survive Declining Crude Oil Prices, Declining Energy Tax Credits, and Declining R&D Budgets?" I commented that we were going to try to convince you during the week that the answer to this question is yes, biofuels can survive all of these problems because our current situation is clearly a temporary one. In the long term, I don't think there is any doubt about the answer to that particular question. As we have seen during the week, there are many exciting new R&D developments in progress and many new commercialization efforts. I believe biofuels are destined for growth and advancement in both the technical and business arenas. And we're doing our best to spread the gospel in the political arena too by informing those who are concerned with energy R&D budgets that biofuels deserve a better shake. They contribute more than the other alternative energy sources to primary energy demand now, but their full potential is still unrealized.

I would like to thank the speakers who participated in this program and made it all possible. This was the largest number of speakers for the poster and oral papers that we ever had at Energy from Biomass and Wastes. The total registration of over 300 is also the largest we ever had. I hope it's mainly because of the subject matter, and not because the registration fee was reduced to $130 from $740 last year. We may have to return to the higher fee next year, but that determination will have to wait until all of the bills are paid. I especially appreciate the financial support that some of the CBETS Members provided.

Next, I would like to extend a special thanks to the session chairmen who did their work very diligently during the program. Almost all of them were members of the CBETS organization.

And finally, I want to thank the IGT staff — Sue Robertson, Bill Thompson, and Jared Smith. They worked very hard to try to make the program a success. If any of you have comments about improving the program that you would like to give us personally, let me hear from you. We'll try to incorporate your ideas and thoughts into the program next year. Meanwhile, thank you all for staying with us to the end. I hope the benefits justified the time you spent with us, and I hope to see you all again next year.